COMPREHENSIVE HETEROCYCLIC CHEMISTRY

IN 8 VOLUMES

JPC	J. Phys. Chem.	PIA	Proc. Indian Acad. Sci.
JPR	J. Prakt. Chem.	PIA(A)	Proc. Indian Acad. Sci., Sect. A
JPS	J. Pharm. Sci.	PMH	Phys. Methods Heterocycl. Chem.
JSP	J. Mol. Spectrosc.	PNA	Proc. Natl. Acad. Sci. USA
JST	J. Mol. Struct.	PS	Phosphorus Sulfur
K	Kristallografiya	QR	Q. Rev., Chem. Soc.
KGS	Khim. Geterotsikl. Soedin.	RCR	Russ. Chem. Rev. (Engl. Transl.)
LA	Liebigs Ann. Chem.	RRC	Rev. Roum. Chim.
M	Monatsh. Chem.	RTC	Recl. Trav. Chim. Pays-Bas
MI	Miscellaneous [book or journal]	S	Synthesis
MIP	Miscellaneous Pat.	SA	Spectrochim. Acta
MS	Q. N. Porter and J. Baldas, 'Mass Spectrometry of Heterocyclic Compounds', Wiley, New York, 1971	SA(A)	Spectrochim. Acta, Part A
		SAP	S. Afr. Pat.
		SC	Synth. Commun.
		SH	W. L. F. Armarego, 'Stereochemistry of Heterocyclic Compounds', Wiley, New York, 1977, parts 1 and 2
N	Naturwissenschaften		
NEP	Neth. Pat.		
NJC	Nouv. J. Chim.		
NKK	Nippon Kagaku Kaishi	SST	Org. Compd. Sulphur, Selenium, Tellurium [R. Soc. Chem. series]
NMR	T. J. Batterham, 'NMR Spectra of Simple Heterocycles', Wiley, New York, 1973		
		T	Tetrahedron
		TH	Thesis
OMR	Org. Magn. Reson.	TL	Tetrahedron Lett.
OMS	Org. Mass Spectrom.	UKZ	Ukr. Khim. Zh. (Russ. Ed.)
OPP	Org. Prep. Proced. Int.	UP	Unpublished Results
OR	Org. React.	USP	U.S. Pat.
OS	Org. Synth.	YZ	Yakugaku Zasshi
OSC	Org. Synth., Coll. Vol.	ZC	Z. Chem.
P	Phytochemistry	ZN	Z. Naturforsch.
PAC	Pure Appl. Chem.	ZN(B)	Z. Naturforsch., Teil B
PC	Personal Communication	ZOB	Zh. Obshch. Khim.
PH	'Photochemistry of Heterocyclic Compounds', ed. O. Buchardt, Wiley, New York, 1976	ZOR	Zh. Org. Khim.
		ZPC	Hoppe-Seyler's Z. Physiol. Chem.

COMPREHENSIVE HETEROCYCLIC CHEMISTRY

The Structure, Reactions, Synthesis and Uses of Heterocyclic Compounds

Volume 6

Chairman of the Editorial Board
ALAN R. KATRITZKY, FRS
University of Florida

Co-Chairman of the Editorial Board
CHARLES W. REES, FRS
*Imperial College of Science and Technology
University of London*

Part 4B

Five-membered Rings with Two or More Oxygen, Sulfur or Nitrogen Atoms

EDITOR
KEVIN T. POTTS
Rensselaer Polytechnic Institute, New York

PERGAMON PRESS
OXFORD · NEW YORK · TORONTO · SYDNEY · PARIS · FRANKFURT

U.K.	Pergamon Press Ltd., Headington Hill Hall, Oxford OX3 0BW, England
U.S.A.	Pergamon Press Inc., Maxwell House, Fairview Park, Elmsford, New York 10523, U.S.A.
CANADA	Pergamon Press Canada Ltd., Suite 104, 150 Consumers Road, Willowdale, Ontario M2J 1P9, Canada
AUSTRALIA	Pergamon Press (Aust.) Pty. Ltd., P.O. Box 544, Potts Point, N.S.W. 2011, Australia
FRANCE	Pergamon Press SARL, 24 rue des Ecoles, 75240 Paris, Cedex 05, France
FEDERAL REPUBLIC OF GERMANY	Pergamon Press GmbH, Hammerweg 6, D-6242 Kronberg-Taunus, Federal Republic of Germany

Copyright © 1984 Pergamon Press Ltd.

All rights reserved. No part of this publication may be reproduced, stored in a retrieval system or transmitted in any form or by any means: electronic, electrostatic, magnetic tape, mechanical, photocopying, recording or otherwise, without permission in writing from the publishers

First edition 1984

Library of Congress Cataloging in Publication Data

Main entry under title:

Comprehensive heterocyclic chemistry.

Includes indexes.
Contents: v. 1. Introduction, nomenclature, literature, biological aspects, industrial uses, less-common heteroatoms –
v. 2. Six-membered rings with one nitrogen atom – [etc.] –
v. 8. Indexes.
1. Heterocyclic compounds.   I. Katritzky, Alan R. (Alan Roy)
II. Rees, Charles W. (Charles Wayne)
QD400.C65   1984   547'.59   83-4264

British Library Cataloguing in Publication Data

Comprehensive heterocyclic chemistry
1. Heterocyclic compounds.
I. Katritzky, Alan R.   II. Rees, Charles W.
547'.59   QD400

ISBN 0-08-030706-X (vol. 6)
ISBN 0-08-026200-7 (set)

Typeset by J. W. Arrowsmith Ltd., Winterstoke Road, Bristol
Printed in Great Britain by A. Wheaton & Co. Ltd., Exeter

Contents

	Foreword	vii
	Contributors to Volume 6	ix
	Contents of All Volumes	xi
4.16	Isoxazoles and their Benzo Derivatives S. A. Lang, Jr. and Y.-i Lin, *American Cyanamid Company, Pearl River*	1
4.17	Isothiazoles and their Benzo Derivatives D. L. Pain, B. J. Peart and K. R. H. Wooldridge, *May & Baker Ltd., Dagenham*	131
4.18	Oxazoles and their Benzo Derivatives G. V. Boyd, *Chelsea College, University of London*	177
4.19	Thiazoles and their Benzo Derivatives J. Metzger, *Université d'Aix-Marseille*	235
4.20	Five-membered Selenium–Nitrogen Heterocycles I. Lalezari, *Montefiore Medical Center and Albert Einstein College of Medicine, New York*	333
4.21	1,2,3- and 1,2,4-Oxadiazoles L. B. Clapp, *Brown University, Providence*	365
4.22	1,2,5-Oxadiazoles and their Benzo Derivatives R. M. Paton, *University of Edinburgh*	393
4.23	1,3,4-Oxadiazoles J. Hill, *University of Salford*	427
4.24	1,2,3-Thiadiazoles and their Benzo Derivatives E. W. Thomas, *The Upjohn Company, Kalamazoo*	447
4.25	1,2,4-Thiadiazoles J. E. Franz and O. P. Dhingra, *Monsanto Agricultural Products Co., St Louis*	463
4.26	1,2,5-Thiadiazoles and their Benzo Derivatives L. M. Weinstock and I. Shinkai, *Merck Sharp & Dohme Research Laboratories, Rahway*	513
4.27	1,3,4-Thiadiazoles G. Kornis, *The Upjohn Company, Kalamazoo*	545
4.28	Oxatriazoles and Thiatriazoles A. Holm, *University of Copenhagen*	579

4.29 Five-membered Rings (One Oxygen or Sulfur and At Least One Nitrogen
Atom) Fused with Six-membered Rings (At Least One Nitrogen Atom)   613
K. UNDHEIM, *University of Oslo*

4.30 Dioxoles and Oxathioles   749
A. J. ELLIOTT, *Schering-Plough Corporation, Bloomfield*

4.31 1,2-Dithioles   783
D. M. MCKINNON, *University of Manitoba*

4.32 1,3-Dithioles   813
H. GOTTHARDT, *Universität Wuppertal*

4.33 Five-membered Rings containing Three Oxygen or Sulfur Atoms   851
G. W. FISCHER and T. ZIMMERMANN, *Academy of Sciences of the G.D.R., Leipzig*

4.34 Dioxazoles, Oxathiazoles and Dithiazoles   897
M. P. SAMMES, *University of Hong Kong*

4.35 Five-membered Rings containing One Selenium or Tellurium Atom and One
Other Group VI Atom and their Benzo Derivatives   947
M. R. DETTY, *Eastman Kodak Company, Rochester*

4.36 Two Fused Five-membered Heterocyclic Rings: (i) Classical Systems   973
K. H. PILGRAM, *Shell Development Co., Modesto*

4.37 Two Fused Five-membered Heterocyclic Rings: (ii) Non-classical Systems   1027
C. A. RAMSDEN, *May & Baker Ltd., Dagenham*

4.38 Two Fused Five-membered Heterocyclic Rings: (iii) $1,6,6a\lambda^4$-Trithiapentalenes
and Related Systems   1049
N. LOZAC'H, *Université de Caen*

References   1071

Foreword

Scope

Heterocyclic compounds are those which have a cyclic structure with two, or more, different kinds of atom in the ring. This work is devoted to organic heterocyclic compounds in which at least one of the ring atoms is carbon, the others being considered the heteroatoms; carbon is still by far the most common ring atom in heterocyclic compounds. As the number and variety of heteroatoms in the ring increase there is a steady transition to the expanding domain of inorganic heterocyclic systems. Since the ring can be of any size, from three-membered upwards, and since the heteroatoms can be drawn in almost any combination from a large number of the elements (though nitrogen, oxygen and sulfur are the most common), the number of possible heterocyclic systems is almost limitless. An enormous number of heterocyclic compounds is known and this number is increasing very rapidly. The literature of the subject is correspondingly vast and of the three major divisions of organic chemistry, aliphatic, carbocyclic and heterocyclic, the last is much the biggest. Over six million compounds are recorded in *Chemical Abstracts* and approximately half of these are heterocyclic.

Significance

Heterocyclic compounds are very widely distributed in Nature and are essential to life; they play a vital role in the metabolism of all living cells. Thus, for example, the following are heterocyclic compounds: the pyrimidine and purine bases of the genetic material DNA; the essential amino acids proline, histidine and tryptophan; the vitamins and coenzyme precursors thiamine, riboflavine, pyridoxine, folic acid and biotin; the B_{12} and E families of vitamin; the photosynthesizing pigment chlorophyll; the oxygen transporting pigment hemoglobin, and its breakdown products the bile pigments; the hormones kinetin, heteroauxin, serotonin and histamine; together with most of the sugars. There are a vast number of pharmacologically active heterocyclic compounds, many of which are in regular clinical use. Some of these are natural products, for example antibiotics such as penicillin and cephalosporin, alkaloids such as vinblastine, ellipticine, morphine and reserpine, and cardiac glycosides such as those of digitalis. However, the large majority are synthetic heterocyclics which have found widespread use, for example as anticancer agents, analeptics, analgesics, hypnotics and vasopressor modifiers, and as pesticides, insecticides, weedkillers and rodenticides.

There is also a large number of synthetic heterocyclic compounds with other important practical applications, as dyestuffs, copolymers, solvents, photographic sensitizers and developers, as antioxidants and vulcanization accelerators in the rubber industry, and many are valuable intermediates in synthesis.

The successful application of heterocyclic compounds in these and many other ways, and their appeal as materials in applied chemistry and in more fundamental and theoretical studies, stems from their very complexity; this ensures a virtually limitless series of structurally novel compounds with a wide range of physical, chemical and biological properties, spanning a broad spectrum of reactivity and stability. Another consequence of their varied chemical reactivity, including the possible destruction of the heterocyclic ring, is their increasing use in the synthesis of specifically functionalized non-heterocyclic structures.

Aims of the Present Work

All of the above aspects of heterocyclic chemistry are mirrored in the contents of the present work. The scale, scope and complexity of the subject, already referred to, with its correspondingly complex system of nomenclature, can make it somewhat daunting initially. One of the main aims of the present work is to minimize this problem by presenting a comprehensive account of fundamental heterocyclic chemistry, with the emphasis on basic principles and, as far as possible, on unifying correlations in the properties, chemistry and synthesis of different heterocyclic systems and the analogous carbocyclic structures. The motivation for this effort was the outstanding biological, practical and theoretical importance of heterocyclic chemistry, and the absence of an appropriate major modern treatise.

At the introductory level there are several good textbooks on heterocyclic chemistry, though the subject is scantily treated in most general textbooks of organic chemistry. At the specialist, research level there are two established ongoing series, 'Advances in Heterocyclic Chemistry' edited by Katritzky and 'The Chemistry of Heterocyclic Compounds' edited by Weissberger and Taylor, devoted to a very detailed consideration of all aspects of heterocyclic compounds, which together comprise some 100 volumes. The present work is designed to fill the gap between these two levels, *i.e.* to give an up-to-date overview of the subject as a whole (particularly in the General Chapters) appropriate to the needs of teachers and students and others with a general interest in the subject and its applications, and to provide enough detailed information (particularly in the Monograph Chapters) to answer specific questions, to demonstrate exactly what is known or not known on a given topic, and to direct attention to more detailed reviews and to the original literature. Mainly because of the extensive practical uses of heterocyclic compounds, a large and valuable review literature on all aspects of the subject has grown up over the last few decades. References to all of these reviews are now immediately available: reviews dealing with a specific ring system are reported in the appropriate monograph chapters; reviews dealing with any aspect of heterocyclic chemistry which spans more than one ring system are collected together in a logical, readily accessible manner in Chapter 1.03.

The approach and treatment throughout this work is as ordered and uniform as possible, based on a carefully prearranged plan. This plan, which contains several novel features, is described in detail in the Introduction (Chapter 1.01).

ALAN R. KATRITZKY
Florida

CHARLES W. REES
London

Contributors to Volume 6

Professor G. V. Boyd
Department of Chemistry, Chelsea College, Manresa Road, London SW3 6LZ, UK

Professor L. B. Clapp
Department of Chemistry, Brown University, Providence, RI 02912, USA

Dr M. R. Detty
Research Laboratories, Eastman Kodak Company, Rochester, NY 14650, USA

Dr O. P. Dhingra
Monsanto Agricultural Products Co., 800 N Lindberg Boulevard, St Louis, MO 63167, USA

Dr A. J. Elliott
62 Seven Lakes Drive, Sloatsburg, NY 10974, USA

Professor G. W. Fischer
Forschungsstelle fuer Chemische Toxikologie der AdW der DDR, DDR-7050 Leipzig, Permoserstrasse 15, German Democratic Republic

Dr J. E. Franz
Monsanto Agricultural Products Co., 800 N Lindberg Boulevard, St Louis, MO 63167, USA

Professor H. Gotthardt
FB-9 Organische Chemie, Universität Gesamthochschule Wuppertal, Wuppertal-Elberfeld, Gaussstrasse 20, Postfach 100127, 5600 Wuppertal 1, Federal Republic of Germany

Dr J. Hill
Department of Chemistry & Applied Chemistry, University of Salford, Salford M5 4WT, UK

Dr A. Holm
Department of General and Organic Chemistry, University of Copenhagen, The H.C. Orsted Institute, DK 2100 Copenhagen, Denmark

Dr G. Kornis
Parasitology Research, The Upjohn Company, Kalamazoo, MI 49001, USA

Dr I. Lalezari
240 Garth Road, 4F-2, Scarsdale, NY 10583, USA

Dr S. A. Lang, Jr.
Medical Research Division, American Cyanamid Company, Lederle Laboratories, Pearl River, NY 10965, USA

Dr Y.-i Lin
Medical Research Division, American Cyanamid Company, Lederle Laboratories, Pearl River, NY 10965, USA

Professor N. Lozac'h
17 rue du Docteur Georges Maugeais, 14000 Caen, France

Professor D. M. McKinnon
Department of Chemistry, University of Manitoba, Winnipeg, Manitoba, Canada R3T 2N2

Professor J. Metzger
Université de Droit, d'Economie et des Sciences d'Aix-Marseille, Laboratoire de Chimie Organique A, Faculté des Sciences et Techniques de Saint Jerome, rue Henri Poincare, 13397 Marseille Cedex 4, France

Mr D. L. Pain
May & Baker Ltd., Dagenham, Essex RM10 7XS, UK

Dr R. M. Paton
Department of Chemistry, University of Edinburgh, West Mains Road, Edinburgh EH9 3JJ, UK

Dr B. J. Peart
May & Baker Ltd., Dagenham, Essex RM10 7XS, UK

Dr K. H. Pilgram
Shell Development Company, Biological Sciences Research Center, PO Box 4248, Modesto, CA 95352, USA

Dr C. A. Ramsden
May & Baker Ltd., Dagenham, Essex RM10 7XS, UK

Dr M. P. Sammes
Department of Chemistry, University of Hong Kong, Pokfulam Road, Hong Kong

Dr I. Shinkai
Process Research, Merck Sharp & Dohme Research Laboratories, PO Box 2000, Rahway, NJ 07065, USA

Dr E. W. Thomas
Diabetes and Atherosclerosis Research, The Upjohn Company, Kalamazoo, MI 49001, USA

Dr K. Undheim
Department of Chemistry, University of Oslo, Blindern, Oslo 3, PO Box 1033, Norway

Dr L. M. Weinstock
Director, Process Research, Merck Sharp & Dohme Research Laboratories, PO Box 2000, Rahway, NJ 07065, USA

Dr K. R. H. Wooldridge
May & Baker Ltd., Dagenham, Essex RM10 7XS, UK

Dr T. Zimmermann
Forschungsstelle fuer Chemische Toxikologie der AdW der DDR, DDR-7050 Leipzig, Permoserstrasse 15, German Democratic Republic

Contents of All Volumes

Volume 1 (Part 1: Introduction, Nomenclature, Review Literature, Biological Aspects, Industrial Uses, Less-common Heteroatoms)

1.01 Introduction
1.02 Nomenclature of Heterocyclic Compounds
1.03 Review Literature of Heterocycles
1.04 Biosynthesis of Some Heterocyclic Natural Products
1.05 Toxicity of Heterocycles
1.06 Application as Pharmaceuticals
1.07 Use as Agrochemicals
1.08 Use as Veterinary Products
1.09 Metabolism of Heterocycles
1.10 Importance of Heterocycles in Biochemical Pathways
1.11 Heterocyclic Polymers
1.12 Heterocyclic Dyes and Pigments
1.13 Organic Conductors
1.14 Uses in Photographic and Reprographic Techniques
1.15 Heterocyclic Compounds as Additives
1.16 Use in the Synthesis of Non-heterocycles
1.17 Heterocyclic Rings containing Phosphorus
1.18 Heterocyclic Rings containing Arsenic, Antimony or Bismuth
1.19 Heterocyclic Rings containing Halogens
1.20 Heterocyclic Rings containing Silicon, Germanium, Tin or Lead
1.21 Heterocyclic Rings containing Boron
1.22 Heterocyclic Rings containing a Transition Metal

Volume 2 (Part 2A: Six-membered Rings with One Nitrogen Atom)

2.01 Structure of Six-membered Rings
2.02 Reactivity of Six-membered Rings
2.03 Synthesis of Six-membered Rings
2.04 Pyridines and their Benzo Derivatives: (i) Structure
2.05 Pyridines and their Benzo Derivatives: (ii) Reactivity at Ring Atoms
2.06 Pyridines and their Benzo Derivatives: (iii) Reactivity of Substituents
2.07 Pyridines and their Benzo Derivatives: (iv) Reactivity of Non-aromatics
2.08 Pyridines and their Benzo Derivatives: (v) Synthesis
2.09 Pyridines and their Benzo Derivatives: (vi) Applications
2.10 The Quinolizinium Ion and Aza Analogs
2.11 Naphthyridines, Pyridoquinolines, Anthyridines and Similar Compounds

Volume 3 (Part 2B: Six-membered Rings with Oxygen, Sulfur or Two or More Nitrogen Atoms)

2.12 Pyridazines and their Benzo Derivatives
2.13 Pyrimidines and their Benzo Derivatives

2.14 Pyrazines and their Benzo Derivatives
2.15 Pyridodiazines and their Benzo Derivatives
2.16 Pteridines
2.17 Other Diazinodiazines
2.18 1,2,3-Triazines and their Benzo Derivatives
2.19 1,2,4-Triazines and their Benzo Derivatives
2.20 1,3,5-Triazines
2.21 Tetrazines and Pentazines
2.22 Pyrans and Fused Pyrans: (i) Structure
2.23 Pyrans and Fused Pyrans: (ii) Reactivity
2.24 Pyrans and Fused Pyrans: (iii) Synthesis and Applications
2.25 Thiopyrans and Fused Thiopyrans
2.26 Six-membered Rings with More than One Oxygen or Sulfur Atom
2.27 Oxazines, Thiazines and their Benzo Derivatives
2.28 Polyoxa, Polythia and Polyaza Six-membered Ring Systems

Volume 4 (Part 3: Five-membered Rings with One Oxygen, Sulfur or Nitrogen Atom)

3.01 Structure of Five-membered Rings with One Heteroatom
3.02 Reactivity of Five-membered Rings with One Heteroatom
3.03 Synthesis of Five-membered Rings with One Heteroatom
3.04 Pyrroles and their Benzo Derivatives: (i) Structure
3.05 Pyrroles and their Benzo Derivatives: (ii) Reactivity
3.06 Pyrroles and their Benzo Derivatives: (iii) Synthesis and Applications
3.07 Porphyrins, Corrins and Phthalocyanines
3.08 Pyrroles with Fused Six-membered Heterocyclic Rings: (i) *a*-Fused
3.09 Pyrroles with Fused Six-membered Heterocyclic Rings: (ii) *b*- and *c*-Fused
3.10 Furans and their Benzo Derivatives: (i) Structure
3.11 Furans and their Benzo Derivatives: (ii) Reactivity
3.12 Furans and their Benzo Derivatives: (iii) Synthesis and Applications
3.13 Thiophenes and their Benzo Derivatives: (i) Structure
3.14 Thiophenes and their Benzo Derivatives: (ii) Reactivity
3.15 Thiophenes and their Benzo Derivatives: (iii) Synthesis and Applications
3.16 Selenophenes, Tellurophenes and their Benzo Derivatives
3.17 Furans, Thiophenes and Selenophenes with Fused Six-membered Heterocyclic Rings
3.18 Two Fused Five-membered Rings each containing One Heteroatom

Volume 5 (Part 4A: Five-membered Rings with Two or More Nitrogen Atoms)

4.01 Structure of Five-membered Rings with Two or More Heteroatoms
4.02 Reactivity of Five-membered Rings with Two or More Heteroatoms
4.03 Synthesis of Five-membered Rings with Two or More Heteroatoms
4.04 Pyrazoles and their Benzo Derivatives
4.05 Pyrazoles with Fused Six-membered Heterocyclic Rings
4.06 Imidazoles and their Benzo Derivatives: (i) Structure
4.07 Imidazoles and their Benzo Derivatives: (ii) Reactivity
4.08 Imidazoles and their Benzo Derivatives: (iii) Synthesis and Applications
4.09 Purines
4.10 Other Imidazoles with Fused Six-membered Rings
4.11 1,2,3-Triazoles and their Benzo Derivatives
4.12 1,2,4-Triazoles
4.13 Tetrazoles
4.14 Pentazoles
4.15 Triazoles and Tetrazoles with Fused Six-membered Rings

Volume 6 (Part 4B: Five-membered Rings with Two or More Oxygen, Sulfur or Nitrogen Atoms)

4.16 Isoxazoles and their Benzo Derivatives
4.17 Isothiazoles and their Benzo Derivatives
4.18 Oxazoles and their Benzo Derivatives
4.19 Thiazoles and their Benzo Derivatives
4.20 Five-membered Selenium–Nitrogen Heterocycles
4.21 1,2,3- and 1,2,4-Oxadiazoles
4.22 1,2,5-Oxadiazoles and their Benzo Derivatives
4.23 1,3,4-Oxadiazoles
4.24 1,2,3-Thiadiazoles and their Benzo Derivatives
4.25 1,2,4-Thiadiazoles
4.26 1,2,5-Thiadiazoles and their Benzo Derivatives
4.27 1,3,4-Thiadiazoles
4.28 Oxatriazoles and Thiatriazoles
4.29 Five-membered Rings (One Oxygen or Sulfur and at least One Nitrogen Atom) Fused with Six-membered Rings (at least One Nitrogen Atom)
4.30 Dioxoles and Oxathioles
4.31 1,2-Dithioles
4.32 1,3-Dithioles
4.33 Five-membered Rings containing Three Oxygen or Sulfur Atoms
4.34 Dioxazoles, Oxathiazoles and Dithiazoles
4.35 Five-membered Rings containing One Selenium or Tellurium Atom and One Other Group VI Atom and their Benzo Derivatives
4.36 Two Fused Five-membered Heterocyclic Rings: (i) Classical Systems
4.37 Two Fused Five-membered Heterocyclic Rings: (ii) Non-classical Systems
4.38 Two Fused Five-membered Heterocyclic Rings: (iii) 1,6,6aλ^4-Trithiapentalenes and Related Systems

Volume 7 (Part 5: Small and Large Rings)

5.01 Structure of Small and Large Rings
5.02 Reactivity of Small and Large Rings
5.03 Synthesis of Small and Large Rings
5.04 Aziridines, Azirines and Fused-ring Derivatives
5.05 Oxiranes and Oxirenes
5.06 Thiiranes and Thiirenes
5.07 Fused-ring Oxiranes, Oxirenes, Thiiranes and Thiirenes
5.08 Three-membered Rings with Two Heteroatoms and Fused-ring Derivatives
5.09 Azetidines, Azetines and Azetes
5.10 Cephalosporins
5.11 Penicillins
5.12 Other Fused-ring Azetidines, Azetines and Azetes
5.13 Oxetanes and Oxetenes
5.14 Thietanes, Thietes and Fused-ring Derivatives
5.15 Four-membered Rings with Two or More Heteroatoms and Fused-ring Derivatives
5.16 Azepines
5.17 Oxepanes, Oxepins, Thiepanes and Thiepins
5.18 Seven-membered Rings with Two or More Heteroatoms
5.19 Eight-membered Rings
5.20 Larger Rings except Crown Ethers and Heterophanes
5.21 Crown Ethers and Cryptands
5.22 Heterophanes

Volume 8 (Part 6: Indexes)

Subject Index
Author Index
Ring Index
Data Index

4.16
Isoxazoles and their Benzo Derivatives

S. A. LANG, JR. and Y.-i LIN
American Cyanamid Company, Pearl River

4.16.1 INTRODUCTION	2
4.16.2 STRUCTURE	3
4.16.2.1 Theoretical Methods	3
4.16.2.1.1 Molecular orbital calculations	3
4.16.2.2 Molecular Dimensions	4
4.16.2.2.1 X-ray diffraction	4
4.16.2.3 Molecular Spectra	4
4.16.2.3.1 UV spectroscopy	4
4.16.2.3.2 Photoelectron spectroscopy	5
4.16.2.3.3 IR and Raman spectroscopy	5
4.16.2.3.4 Nuclear magnetic resonance spectroscopy	5
4.16.2.3.5 Mass spectrometry	6
4.16.2.3.6 Electron paramagnetic resonance spectroscopy	7
4.16.2.3.7 Microwave spectroscopy	8
4.16.2.4 Thermodynamic Effects	8
4.16.2.4.1 Intermolecular forces	8
4.16.2.4.2 Stability	10
4.16.2.4.3 Conformation	10
4.16.2.4.4 Potentiometric properties	10
4.16.2.5 Tautomerism	11
4.16.3 REACTIVITY OF ISOXAZOLES	12
4.16.3.1 Reactions at Aromatic Rings	12
4.16.3.1.1 General survey of reactivity	12
4.16.3.1.2 Thermal and photochemical reactions	12
4.16.3.1.3 Electrophilic attack at nitrogen	20
4.16.3.1.4 Electrophilic attack at carbon	21
4.16.3.1.5 Nucleophilic attack at carbon	28
4.16.3.1.6 Nucleophilic attack at hydrogen	29
4.16.3.1.7 Reactions with cyclic transition states	35
4.16.3.2 Reactions of Non-aromatic Compounds	36
4.16.3.2.1 2-Isoxazolines	36
4.16.3.2.2 3-Isoxazolines	44
4.16.3.2.3 4-Isoxazolines	44
4.16.3.2.4 Isoxazolidines	45
4.16.3.3 Reaction of Substituents	48
4.16.3.3.1 Fused benzene rings	48
4.16.3.3.2 Other C-linked substituents	48
4.16.3.3.3 N-Linked substituents	54
4.16.3.3.4 O-Linked substituents	56
4.16.3.3.5 S-Linked substituents	57
4.16.3.3.6 Halogen atoms	58
4.16.3.3.7 Metal- and metalloid-linked substituents	58
4.16.3.3.8 Substituents attached to ring nitrogen atoms	59
4.16.3.3.9 Special reactions of side-chain substituents: rearrangements involving three-atom side chains	60
4.16.4 SYNTHESIS	61
4.16.4.1 Isoxazoles	61
4.16.4.1.1 Introduction	61
4.16.4.1.2 [3+2] Routes	61
4.16.4.1.3 [4+1] Routes	71
4.16.4.1.4 [5+0] Routes	73
4.16.4.1.5 [3+1+1] Routes	76

4.16.4.1.6 Dehydrogenation of isoxazolines to isoxazoles	78
4.16.4.1.7 Ring transformations leading to isoxazoles	78
4.16.4.2 Isoxazole Derivatives	82
4.16.4.2.1 Isoxazole	82
4.16.4.2.2 Alkylisoxazoles	83
4.16.4.2.3 Phenylisoxazoles	83
4.16.4.2.4 Alkylarylisoxazoles	84
4.16.4.2.5 Isoxazole aldehydes	84
4.16.4.2.6 Isoxazole ketones	84
4.16.4.2.7 Isoxazolecarboxylic acids	85
4.16.4.2.8 Halogen derivatives	86
4.16.4.2.9 N-Linked derivatives	86
4.16.4.2.10 O-Linked derivatives	87
4.16.4.2.11 S-Linked derivatives	88
4.16.4.3 2-Isoxazolines	88
4.16.4.3.1 By cycloaddition of nitrile oxides	89
4.16.4.3.2 From α,β-unsaturated ketones and hydroxylamine	93
4.16.4.3.3 From ylides and nitrile oxides	94
4.16.4.3.4 By the activation of alkyl nitro compounds	95
4.16.4.3.5 From β-halo (tosyl) ketone oximes	96
4.16.4.3.6 From 2-isoxazoline N-oxides	97
4.16.4.3.7 Miscellaneous methods	97
4.16.4.4 3-Isoxazolines	98
4.16.4.5 4-Isoxazolines	99
4.16.4.6 Isoxazolinols	100
4.16.4.7 Isoxazoline N-oxides	102
4.16.4.8 Isoxazolin-5-ones	103
4.16.4.8.1 From hydroxylamine	104
4.16.4.8.2 By the reaction of nitrile oxides	104
4.16.4.8.3 Miscellaneous syntheses	104
4.16.4.9 Isoxazolin-5-imines	105
4.16.4.10 Isoxazoline-5-thiones	105
4.16.4.11 Isoxazolin-3-ones	106
4.16.4.12 Isoxazolin-4-ones	106
4.16.4.13 Bis(isoxazolines)	107
4.16.4.14 Spiro Fused Isoxazolines	107
4.16.4.15 Bis(spiroisoxazolines)	108
4.16.4.16 Other Spiroisoxazolines	108
4.16.4.17 Isoxazolidines	108
4.16.4.17.1 By cycloaddition of nitrones and alkenes	108
4.16.4.17.2 Cycloaddition of nitronic esters	110
4.16.4.17.3 By the reaction of tetranitromethane and halotrinitromethanes with alkenes	111
4.16.4.17.4 Other methods	111
4.16.4.18 Isoxazolidinones	112
4.16.4.18.1 3-Isoxazolidinones	112
4.16.4.18.2 4-Isoxazolidinones	113
4.16.4.18.3 5-Isoxazolidinones	113
4.16.4.18.4 Isoxazolidine-3,5-diones	113
4.16.4.19 Benzisoxazoles	114
4.16.4.19.1 1,2-Benzisoxazoles	114
4.16.4.19.2 2,1-Benzisoxazoles	120
4.16.5 APPLICATIONS	127
4.16.5.1 Pharmaceutical Applications	127
4.16.5.2 Agricultural Applications	129
4.16.5.3 Other Applications	130

4.16.1 INTRODUCTION

Isoxazole (**1**), 2-isoxazoline (**2**), isoxazolidine (**3**), 1,2-benzisoxazole (**4**) (sometimes termed indoxazene) and 2,1-benzisoxazole (**5**) (often referred to as anthranil) are numbered as shown. The chemistry of isoxazoles ⟨79AHC(25)147, 63AHC(2)365, 62HC(17)1⟩, isoxazolines ⟨62HC(17)1, p. 95⟩, isoxazolidines ⟨77AHC(21)207, 62HC(17)1, p. 229⟩ and benzisoxazoles

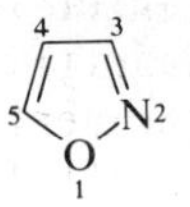

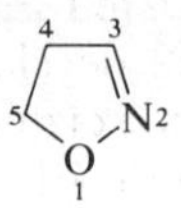

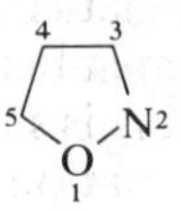

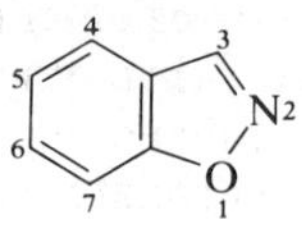

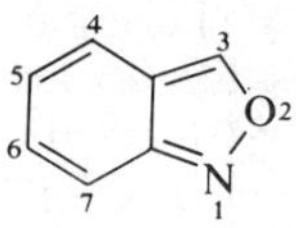

⟨81AHC(29)1, 67AHC(8)277⟩ has been reviewed and these reviews have been supplemented by recent articles ⟨B-73MI41607⟩. In this chapter the literature through 1981 is surveyed with emphasis on developments of the past 20 years.

In 1888, Claisen first suggested an isoxazole structure (**1**) for a product from the reaction of a 1,3-diketone with hydroxylamine ⟨1888CB1149⟩. Subsequently, Claisen and his students laid a solid foundation for the chemistry of isoxazoles ⟨63AHC(2)365, 62HC(17)1⟩. For example, they discovered that isoxazole possesses the typical properties of an aromatic system but under certain reaction conditions, particularly in reducing or basic media, it becomes very highly labile. The next important contribution to the chemistry of isoxazoles was made by Quilico in 1946, when he began to study the formation of isoxazoles from nitrile N-oxides and unsaturated compounds ⟨70S344⟩. At present, these two major routes to isoxazole derivatives embrace perhaps more than 90% of the practical preparations of isoxazoles. In the last three decades the potential of unique properties of isoxazoles first discovered by Claisen has been finally realized. The isoxazole ring has become a major tool as a masked enaminone or 1,3-diketone, particularly for the synthesis of heterocycles and natural products. In addition, 3-unsubstituted isoxazolium salts have been used as peptide coupling agents.

In theory, three isoxazolines are capable of existence: 2-isoxazoline (**2**), 3-isoxazoline and 4-isoxazoline. The position of the double bond may also be designated by the use of the prefix Δ with an appropriate numerical superscript. Of these only the 2-isoxazolines have been investigated in any detail. The preparation of the first isoxazoline, 3,5-diphenyl-2-isoxazoline, from the reaction of β-chloro-β-phenylpropiophenone with hydroxylamine was reported in 1895 ⟨1895CB957⟩. Two major syntheses of 2-isoxazolines are the cycloaddition of nitrile N-oxides to alkenes and the reaction of α,β-unsaturated ketones with hydroxylamine. Since 2-isoxazolines are readily oxidized to isoxazoles and possess some of the unique properties of isoxazoles, they also serve as key intermediates for the synthesis of other heterocycles and natural products.

The preparation of isoxazolidine derivatives was first reported by Bodforss in 1918 ⟨18CB192⟩. The major synthesis of isoxazolidines involves the cycloaddition of nitrones with alkenes, and isoxazolidines have also enjoyed an increasing use as key intermediates in the synthesis of natural products and other heterocycles ⟨79ACR396, 1892CB1498, 1892CB3291, 1882CB2105⟩.

The first 1,2-benzisoxazole, 3-phenyl-1,2-benzisoxazole, was obtained from the treatment of o-bromobenzophenone oxime with alkali in 1892 ⟨1892CB1498, 1892CB3291⟩. 2,1-Benzisoxazole has been known since 1882, being obtained as a reduction product of o-nitrobenzaldehyde with tin and hydrochloric acid ⟨1882CB2105⟩. In general, benzisoxazoles behave much like substituted isoxazoles. Numerous structural ambiguities occur in the early literature of these two systems, and these have been discussed in the above reviews.

4.16.2 STRUCTURE

4.16.2.1 Theoretical Methods

Theoretical and structural studies have been briefly reviewed as late as 1979 ⟨79AHC(25)147⟩ (discussed were 'the aromaticity, basicity, thermodynamic properties, molecular dimensions and tautomeric properties') and also in the early 1960s ⟨63AHC(2)365, 62HC(17)1, p. 117⟩. Significant new data have not been added but refinements in the data have been recorded. Tables on electron density, density, refractive indexes, molar refractivity, surface data and dissociation constants of isoxazole and its derivatives have been compiled ⟨62HC(17)1, p. 177⟩. Short reviews on all aspects of the physical properties as applied to isoxazoles have appeared in the series 'Physical Methods in Heterocyclic Chemistry' (1963–1976, vols. 1–6).

4.16.2.1.1 Molecular orbital calculations

A Hückel model used for calculating 'aromaticity' indicated that the isoxazole nucleus is considerably less aromatic than other five-membered heterocycles, including oxazole and furan. SCF calculations predicted that isoxazole is similar to oxazole. Experimental findings are somewhat difficult to correlate with calculations ⟨79AHC(25)147⟩. PRDDO calculations were compared with *ab initio* values and good agreement for the MO values was reported

⟨79JST(51)87⟩. Measurements of the aromatic character by magnetic properties indicated that both isoxazole and oxazole have non-local out-of-plane contributions that are similar to those observed for furan ⟨74JA7394⟩. Dehydrogenation studies on 4,5-dihydroisoxazoles and other dihydro five-membered heterocycles support the low resonance energy calculations ⟨79AHC(25)147, p. 199, 78T1571⟩.

The spectrochemical data on 1,2- and 2,1-benzisoxazoles were reviewed in 1962 and reported on in greater detail in 1967 ⟨62HC(17)1, p. 159, 62HC(17)1, p. 213, 67AHC(8)277⟩. These reviews discuss π-electron densities, bond order calculations and dipole moment studies for these ring systems.

Polarization and dipole moment studies for alkyl-, aryl-, carbonyl- hydroxy- (keto-) and amino-isoxazoles have been compiled and likewise support the low electron nature of the ring ⟨63AHC(2)365, 62HC(17)1, p. 177⟩. More recent studies predict the order of electrophilic substitution to be 5>4>3 on frontier electron density values of 0.7831, 0.3721 and 0.0659, respectively ⟨71PMH(4)237, pp. 245, 247⟩. This contrasts with earlier reports of 4>5>3 on density values of −0.09, +0.14 and +0.18 in that order ⟨63AHC(2)365⟩.

Dipole moments on carbonyl derivatives of isoxazole in solution demonstrate that the ring and the CO group are coplanar ⟨79KGS1189⟩. Tautomerization of hydroxy and amino substituents has been examined and the enol form has been shown to be predominant for simple 3-hydroxy- and 4-hydroxy-isoxazoles ⟨71PMH(4)265, pp. 312, 314, 63PMH(2)161, p. 319⟩, while in the 5-hydroxy series the tautomers are the Δ^2-5-one in non-polar media and the Δ^3-5-one in polar media ⟨71PMH(4)265, p. 312, 62HC(17)1, p. 177⟩. In a series of GABA analogs where an aminoalkyl side chain is present, the compound is in a zwitterion hydroxide form ⟨74ACS(B)308, 74ACS(B)625⟩. The 3-, 4- and 5-aminoisoxazoles exist solely in that tautomer ⟨71PMH(4)265, p. 312⟩ although there is a significant difference in the site of protonation. The 3- and 5-aminoisoxazoles are protonated on the ring nitrogen, while for 4-aminoisoxazoles the exocyclic nitrogen is protonated ⟨63PMH(1)1, p. 50⟩.

4.16.2.2 Molecular Dimensions

4.16.2.2.1 X-ray diffraction

Many physical methods have been used to study the structural constraints that exist in isoxazoles. X-ray crystal determinations of the parent heterocycle as well as hydroxyl, fused and bis derivatives have been carried out ⟨72PMH(5)1, 79AHC(25)147⟩. Additional X-ray structural determinations have been reported ⟨74ACS(B)308, 74ACS(B)625, 74JCS(P2)1409, 79AX(B)471, 79CSC415⟩, including certain metal complexes ⟨77MI41616, 79MI41617⟩, as well as for isoxazolidines ⟨78IZV850, 74CSC397⟩.

4.16.2.3 Molecular Spectra

4.16.2.3.1 UV spectroscopy

UV spectra of isoxazoles have been recorded and characterized with regards to substitution patterns. These have proved useful in monitoring the Δ^2–Δ^3 tautomerism and the hydroxy–keto tautomerism in isoxazoles ⟨62HC(17)1, p. 177, 63AHC(2)365⟩. Methyl substitution at the C-4 position gave rise to a large shift while similar substitution at the C-3 position had little effect. Similar observations were reported for phenyl substituents ⟨63AHC(2)365⟩. Conjugative effects in 13 arylisoxazoles were studied by UV spectroscopy and the results were in agreement with data previously reported ⟨78KGS327⟩.

A study of the effect of substitution patterns in oxadiazoles and isoxazoles and their effect on the UV spectra in the 10^{-4}–10^{-6} M concentration range was performed. Hypsochromic effects and deviations from Beer's law were observed and were believed to be associated with antiparallel, sandwich-type self-association via dipole–dipole interactions. Beer's law is followed when the molecular dipole moments are small or when self-association is sterically hindered.

Extensive UV absorption data (as well as NMR and IR data) have been reported for a large number of 2,1-benzisoxazolium salts. The salts show three absorption maxima at 270–273, 280 and 335–338 nm ⟨73DIS(B)1434, 71JOC1543⟩.

4.16.2.3.2 Photoelectron spectroscopy

PE spectral data for a number of heterocycles including isoxazole have been compiled ⟨76PMH(6)1, p. 38, 77JST(40)191⟩. The binding energies of electrons have considerable potential in analytical chemistry as this spectrum is reflective of the orbital energy. The spectra contain distinct vibrational fine structure and the ionization potentials have been measured and those of isoxazole were compare with 15 other heterocycles. Classically, five-membered heterocyclic compounds are structurally related to benzene in containing six π-electrons. Differences between the parent and substituted derivatives are useful as fingerprints to an identity aid. The observed ionization potentials for isoxazole are 10.17 and 11.29 eV. The technique has also been applied to isoxazolidine and derivatives ⟨78TL841⟩.

4.16.2.3.3 IR and Raman spectroscopy

Extensive IR studies have been recorded for isoxazole and its derivatives and the frequencies and strengths have been cataloged ⟨63AHC(2)365, 62HC(17)1, p. 177⟩. The ring stretching modes occur in the 1650–1300 cm^{-1} region. The mixing of the ring mode near 1600 cm^{-1} occurs by a scissor deformation in aminoisoxazoles. β-CH modes occur in the 1085–1212 cm^{-1} region and the γ-CH modes occur ~900 cm^{-1}. Ring breathing modes are at ~1028 cm^{-1} ⟨63PMH(2)161, pp. 231, 235, 71PMH(4)265, p. 328, 62HC(17)1, p. 177⟩. Strong electron acceptors at the 4-position or strong electron donors at the 3- or 5-positions give rise to a high intensity band at 355 ± 155 cm^{-1} ⟨63PMH(2)161, pp. 231, 235⟩.

Characteristic bands occur in the 1300–1000 cm^{-1} region for 3,4- and 3,5-disubstituted isoxazoles ⟨71PMH(4)265, p. 330⟩, while bands below 1000 cm^{-1} contain modes for most substitution patterns ⟨71PMH(4)265, p. 332⟩. Total assignments for isoxazole and isoxazole-d have been made ⟨63SA1145, 71PMH(4)265, p. 325⟩ and some of the thermodynamic functions calculated ⟨68SA(A)361, 71PMH(4)265, p. 330⟩.

The carbonyl IR stretching frequencies of isoxazole carboxylates agree essentially with the observed and calculated electron densities at the respective positions, *i.e.* $4 > 3 > 5$. When compared with alkyl benzoates the corresponding higher shifts are observed for positions 4, 3 and 5, respectively (2–8, 9–12 and 17–18 cm^{-1}) ⟨63PMH(2)161, p. 312, 63AHC(2)365⟩. In an experimental and theoretical study of the molecular vibrations of isoxazole and methylisoxazoles the IR and Raman spectra were obtained between 200–4000 cm^{-1}. The observed frequencies were assigned and analysis disclosed two fundamental vibrational modes not previously observed ⟨74MI41612⟩.

The IR (and Raman) spectra of 1,2-benzisoxazole and 2,1-benzisoxazole have been recorded and their fundamental and combined vibrations assigned ⟨80MI41604⟩. Similar studies have been carried out with 2,1-benzisoxazolium salts ⟨74DIS(B)147, 71JOC1543⟩.

The keto–enol tautomerism of 1,2-benzisoxazoles has been examined and the existence of either form can be postulated on the basis of reactivity. IR analysis on the solid indicates the exclusive existence of the enol form, while in CHCl$_3$ solution both appear to be present ⟨71DIS(B)4483⟩.

4.16.2.3.4 Nuclear magnetic resonance spectroscopy

(i) ^{1}H NMR spectroscopy

A number of studies on the NMR spectra of isoxazole has been compiled and this list includes the coupling constants in various solvents as well as the neat liquid. The ^{15}N signal for isoxazole appears at 339.6 p.p.m. relative to TTAI and is at much lower field than in other azoles. Reports of spectra of substituted isoxazoles also abound ⟨79AHC(25)147, p. 201⟩.

The proton spin–spin coupling constants over three to six bonds of isoxazole and 10 alkylated derivatives have been reported and the chemical shifts of ring protons and methyl protons arise from a common mechanism which originates in the ring but is not simply related to the electron densities calculated by CNDO/2 level 2 MO ⟨74CJC833⟩. The ring proton shifts are linked to the total charges on the adjacent carbons with the highest electron density paired with the most shielded hydrogen. The methyl proton shifts are in the same direction but only fractional in intensity; the methyls shield the remaining ring hydrogens and this may be due to inductive effects.

A ^{1}H NMR study of 3-, 4- and 5-phenylisoxazoles partially oriented in a liquid crystal media showed a preferred comformation existed in which an angle of 18–20° appeared between the planes of the two rings ⟨79JCS(P2)572⟩. NMR studies ⟨72G223, 74AC(R)131, 67CB1802⟩ on the *cis* and *trans* orientation of various groups at the 4,5-positions of isoxazolines have appeared ⟨68G42, 67MI41601, 66G1046⟩.

The review by Takeuchi and Ferusaki is quite encompassing and, in addition to synthesis and reactivity, the physical and spectroscopic properties of isoxazolidines are discussed in detail. Additional spectral studies on the parent and derivatives include ^{1}H NMR ⟨68MI41600, 77H(7)201, 78IZV850⟩.

Palmer *et al.* ⟨77H(7)201⟩ have also discussed the use of ^{13}C and ^{1}H NMR with long-range C–H decoupling to assign the stereochemistry of substituents. ^{1}H NMR parameters on 1,2-benzisoxazole, 3-methyl-1,2-benzisoxazole, 2,1-benzisoxazole and 3-methyl-2,1-benzisoxazole were generated by Rondeau *et al.*, and they included an extended discussion on the spin interactions ⟨72JHC427⟩. Extensive NMR data have been reported for a large number of 2,1-benzisoxazolium salts and a characteristic resonance between δ 8.96–10.6 was observed for the C-3 proton ⟨74DIS(B)147, 71JOC1543⟩.

(ii) ^{13}C NMR spectroscopy

^{13}C NMR data have been used to determine the substitution pattern in complex isoxazoles by comparison with simply substituted molecules ⟨75JCS(P1)2115, 76OMR226, 80JOM(195)275⟩. Additional data on the parent system and derivatives are available ⟨77H(7)201⟩.

The position of metalation of 4-substituted 3,5-dimethylisoxazole was determined to occur solely at the C-5 methyl group by ^{13}C examination of the products ⟨76OMR226⟩. The ^{13}C NMR for C-3 in 1,2-benzisoxazole *N*-oxide occurred at 118 p.p.m. ⟨80CC421⟩; the equivalent ^{13}C signal in furoxans occurred at 114 p.p.m. ⟨76OMR158⟩.

(iii) Nuclear quadrupole studies

The weakness of the resonance and the low quadrupole moment make nitrogen containing heterocycles among the most difficult to study and the coupling constants must be obtained from simple molecules. The nuclear quadrupole coupling constants, orbital populations and asymmetry parameters have been determined for a number of nitrogen containing heterocycles. The experimental value for isoxazole is <1 MHz, while that calculated by *ab initio* methods is −6.021 ⟨71PMH(4)21, p. 44⟩. The principal values and axes of the nuclear quadrupole coupling tensor from the microwave spectrum were measured. The values with regard to axes in the nuclear plane are essentially numerically equal, −5.42 and +5.39 MHz. The axis of the most negative coupling constant has an angle of 26° with respect to the ring-angle at the nitrogen atom ⟨78MI41614⟩.

The spectra are useful to predict the axes and coupling magnitude for analogous and related members ⟨79ZN(B)235⟩. The assignments of transition coupling up to $J = 42$ and hyperfine splitting $J = 0 \to 1$ have been partially assigned. The vibrational excited states for species containing ^{13}C, ^{15}N and ^{18}O have been compiled and the structural parameters which optimally reproduce the observed change of all three moments of inertia constructed ⟨75MI41615, 75JCP(63)2560⟩.

4.16.2.3.5 Mass spectrometry

The mass spectra of isoxazoles can be best understood by a rearrangement to azirines with subsequent loss of methyl radicals, methyl radicals and carbon monoxide or an acetoxy ion. Often the rearrangements occur with accompanying loss of ketene, but this is significantly dependent on the kind and pattern of substitution. The overall spectra of diphenylisoxazole and diphenyloxazole are quite similar and a thermal rearrangement of the azirine intermediate into an oxazole ion has been proposed (Scheme 1) ⟨71PMH(3)223, p. 249⟩. Thermal isomerization of isoxazoles and the rearrangement process of isoxazoles into oxazoles have also been examined and correlated with substitution patterns ⟨74KGS457, 74KGS453, 74KGS755⟩.

Electron impact mass spectrometry has been employed to study the fragmentation patterns of isoxazolylmethyl- and bis(isoxazolylmethyl)-isoxazoles and the results are in agreement with proposed pathways ⟨79AC(R)81⟩. Electron impact studies of nitrostyryl isoxazole (**6**) show fragmentation in a variety of ways. The standard loss of NO_2 from the molecular ion

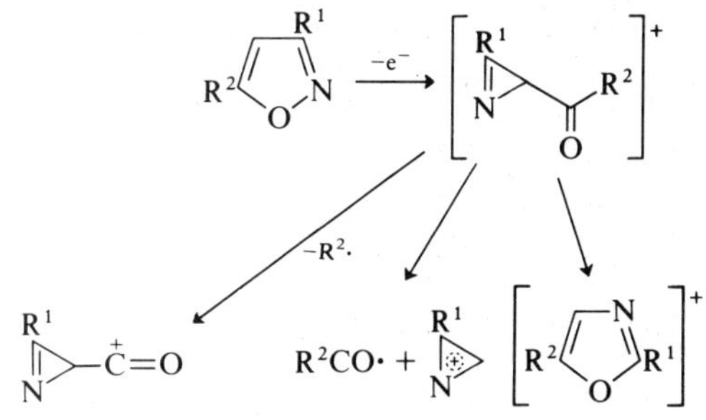

Scheme 1

occurs but is not the predominant pathway. An ion at $[M-17]$ occurs by the loss of HO·
to form cation (**7**). Some rearrangement of $M^{\ddot{+}}$ occurs by formation of the nitrosoketone
(**8**). Rupture of the isoxazole ring also takes place ⟨80JHC585⟩.

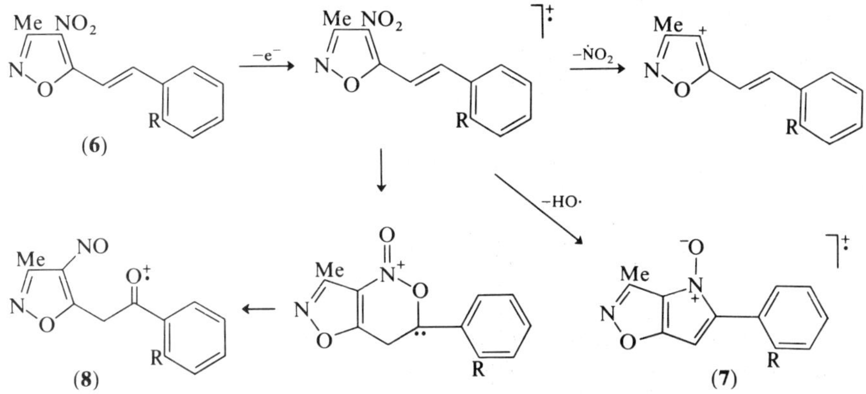

Quasiequilibrium statistical theory was applied to the negative ion mass spectra of
diphenylisoxazoles. Electron capture by the isoxazole leads to molecular ions having excited
vibrations of the ring and of bonds attached to it. The dissociation rate constants were also
calculated ⟨77MI41615, 75MI41614⟩.

The chemical potentials and free energies of the 2-isoxazolines have also been studied
and the electron impact and chemical ionization mass spectra determined ⟨77MI41614⟩.
Fragmentation pathways and retrocycloadditions of various derivatives were discussed in
these reports.

Electron impact fragmentation studies on 1,2-benzisoxazoles and benzoxazole indicate
that isomerization takes place before degradation. Shape analysis and metastable ion
abundances in the mass spectra indicate that isomerization to *o*-cyanophenols occurred
prior to degradation by loss of CO or NCH ⟨75BSB207⟩.

Extensive mass spectral and electron impact studies have been reported for 3-hydroxy-
1,2-benzisoxazole and its ethers. Similar work was also carried out with the isomeric
N-alkyl-1,2-benzisoxazol-3-one ⟨71DIS(B)4483⟩. 1,2-Benzisoxazole *N*-oxide showed a mass
spectral pattern than more closely resembled furoxans. The loss of NO predominated over
the loss of O ($M^{\ddot{+}}$ intense, $[M-16]^+$ weak, $[M-30]^+$ strong).

4.16.2.3.6 Electron paramagnetic resonance spectroscopy

EPR studies on copper complexes of 3-methyl-5-phenylisoxazole and 3,5-dimethyl-
isoxazole have been reported. The isoxazoles were complexed with $CuLiBr_2$ and the EPR
spectra in pyridine were similar to those of powders due to the anisotropic molecular
reorientation in solution. The spin Hamiltonian constants in DMF indicate the presence of
a CuO_6 chromophore ⟨80MI41603⟩. In DMF the spectra consisted of a single line and the
spectra in pyridine had four hyperfine lines with large broadening effects. The $g_{\parallel}$ and $A_{\parallel}$
values of observable species in various solvents were discussed ⟨79JST(55)143⟩.
Hexakis(isoxazole)iron(II) bisperchlorate has been identified as a compound which exhibits
a high spin–low spin transition. The magnetic moment at room temperature is 5.28 μB
⟨74MI41611⟩.

4.16.2.3.7 Microwave spectroscopy

Microwave irradiation has been used to probe aromatic character in isoxazoles ⟨74JA7394⟩ and in the examination of nuclear coupling tensors of ^{14}N, ^{15}N, ^{13}C and ^{18}O isoxazoles ⟨76PMH(6)53, p. 76, 75MI41615, 75JCP(63)2560⟩. Microwave spectra have also been used to study barriers to internal rotation in these molecules. The ground state rotational spectrum of 5-methylisoxazole in the 18 000–36 000 MHz region was observed and the state transitions assigned ($A = 9230.831$, $B = 3559.334$ and $C = 2610.255$ MHz). The three-fold barrier to internal rotation of the methyl group was determined to be 322.7 ± 23.0 J mol^{-1} ⟨75ZN(A)1279⟩. The three-fold barrier value is in agreement with that obtained by the internal axis method ⟨77MI41613⟩.

4.16.2.4 Thermodynamic Effects

4.16.2.4.1 Intermolecular forces

(i) Melting points and boiling points

Isoxazole itself is a colorless liquid with a strong, pyridine-like odor and has the following physical properties: b.p. 94.8 °C/769 mmHg; m.p. < -80 °C; D_4^{20}, 1.0763; critical temperature, 552.04 K; K_b at 25 °C, 2×10^{-12}; dipole moment at 25 °C in benzene 2.75 ± 0.01 D, in dioxane 3.01 ± 0.03 D.

All the theoretically possible methyl derivatives are known. They are also liquids with a pyridine-like odor and possess slight solubility in water. Detailed compilations of physical properties such as boiling point, density, basicity, dipole moment, surface tension and solubility in water are available ⟨62HC(17)1, p. 54⟩. Several generalizations are possible. The majority of alkyl and alkenyl substituted isoxazoles are liquids, as would be anticipated for an organic system with such a molecular weight and no opportunity for intra- or intermolecular association as occurs in five-membered heterocycles containing both a trigonal nitrogen atom and an NH group.

The b.p. of isoxazole, while below that of pyrazole, imidazole, *etc.* is still higher than that of furan (31 °C) and oxazole (60–70 °C). Simple isoxazoles all have boiling points higher than the corresponding oxazoles, indicating a possibility of greater association for isoxazoles than for oxazoles but less than that occurring in pyrazoles and imidazoles (see, however, Section 4.16.2.4.1(ii)). Introduction of aromatic substituents results in crystalline products whose melting points gradually increase with increase in molecular weight. These, and the effects of introducing halogen, nitro, formyl, keto, cyano and carboxy substituents, are illustrated by the data compiled in Table 1.

Table 1 Melting (Boiling) Points of Various Isoxazole Derivatives ⟨62HC(17)1, p. 159⟩

Substituent	M.p. (b.p.) (°C)	Substituent	M.p. (b.p.) (°C)
3-Methyl	(118)	3-Methyl-5-styryl	88
4-Methyl	(127)	3,5-Distyryl	170
5-Methyl	(121)	3-(2-Furyl)-5-phenyl	81
3,4-Dimethyl	(144)	4-Bromo	28 (130)
3,4,5-Trimethyl	(171)	4-Chloro-3-methyl	(135)
3-Ethyl	(139)	4-Bromo-3,5-dimethyl	(169)
3-Isopropyl	(168)	3-Bromo-5-phenyl	68
3-*t*-Butyl-5-methyl	107	3-Methyl-4-nitro	(191)
3-Phenyl	1–2 (252–253)	3,5-Dimethyl-4-nitro	62
4-Phenyl	46	4-Amino-3-methyl	43 (118/25 mmHg)
5-Phenyl	22–23 (256)	4-Amino-5-methyl-3-phenyl	56
3,4-Diphenyl	91	5-Formyl-3-methyl	47 (70/30 mmHg)
3,5-Diphenyl	141	3-Acetyl	16
3,4,5-Triphenyl	212–214	5-Acetyl	52
3-Methyl-5-phenyl	68	5-Benzoyl-3-methyl	50
5-Methyl-3-phenyl	42	3-Carboxy	149
3,5-Dibenzyl	89	3,5-Dicarboxy	213
5-(2-Naphthyl)-3-phenyl	161	3-Cyano	(168)
3-(2-Propenyl)	(131)	5-Cyano-3-methyl	(174)

2-Isoxazolines with alkyl substituents are also all liquids (or low melting solids) and incorporation of aryl substituents results in crystallinity. Introduction of carboxy substituents and endocyclic carbonyl or imino groups also has the anticipated effect, with crystalline products being isolated. These trends are illustrated by the data compiled in Table 2.

Table 2 Melting Points (Boiling Points) of Various Δ^2-Isoxazoline Derivatives ⟨62HC(17)1, p. 95⟩

Substituent	M.p. (b.p.) (°C)	Substituent	M.p. (b.p.) (°C)
3-Methyl	15 (60)	5-Carboxy-5-methyl-3-phenyl	109
3-Phenyl	66	3,4-Dimethyl-5-oxo	123
5-Methyl-3-phenyl	49	5-Oxo-3-phenyl	42
3,5-Diphenyl	75	4-Methyl-5-oxo-3-phenyl	123
3-*p*-Bromophenyl-5-phenyl	138	4-Chloro-3-methyl-5-oxo	86
cis-3,4,5-Triphenyl	167	5-Imino-3-methyl	84
trans-3,4,5-Triphenyl	140	3,4-Dimethyl-5-imino	125
5-Carboxy-3-phenyl	143	3-Phenyl-4-ethoxycarbonyl-5-imino	124

1,2-Benzisoxazole and its simple alkyl derivatives are liquids with b.p.'s of 84 °C/11 mmHg for the unsubstituted system, 92.5 °C/11 mmHg for the 3-methyl compound, and 117 °C/11 mmHg for the 4,6-dimethylbenzisoxazole. 2,1-Benzisoxazole is also a liquid, b.p. 94.4–94.5 °C/11 mmHg, and its 3-methyl derivative has a b.p. of 115.5–116 °C/11 mmHg. Introduction of a 3-phenyl substituent in both systems results in crystallinity, with m.p.'s of 83–84 °C and 52–53 °C, respectively. Polar substituents, as anticipated, also impart crystallinity to these systems.

(ii) Solubility

Isoxazole dissolves in approximately six volumes of water at ordinary temperature and gives an azeotropic mixture, b.p. 88.5 °C. From surface tension and density measurements of isoxazole and its methyl derivatives, isoxazoles with an unsubstituted 3-position behave differently from their isomers. The solubility curves in water for the same compounds also show characteristic differences in connection with the presence of a substituent in the 3-position ⟨62HC(17)1, p. 178⟩. These results have been interpreted in terms of an enhanced capacity for intermolecular association with 3-unsubstituted isoxazoles as represented by (9). Cryoscopic measurements in benzene support this hypothesis and establish the following order for the associative capacity of isoxazoles: isoxazole, 5-Me, 4-Me, 4,5-(Me)$_2$ ≫ 3-Me > 3,4-(Me)$_2$; 3,5-(Me)$_2$ and 3,4,5-(Me)$_3$ isoxazole are practically devoid of associative capacity.

(9)

No detailed study of the solubility characteristics of more complexly substituted isoxazoles has been made. However, qualitative indications of solubility characteristics may be found associated with their synthesis.

(iii) Chromatography

Various types of chromatographic procedures have been utilized in the purification and detection of isoxazoles, isoxazolines and isoxazolidines. No systematic study of their behavior, however, has been undertaken but generalizations are possible. The thermal stability of the isoxazole derivatives, apart from those containing substituents likely to undergo a thermal reaction such as decarboxylation, indicate that they should be amenable to gas chromatographic procedures. However, it must be kept in mind that 3-unsubstituted isoxazoles undergo extremely facile ring-opening under alkaline conditions. Their solubility in most organic solvents makes HPLC, thin-layer and preparative-layer chromatographic procedures readily applicable to isoxazole derivatives. Individual papers dealing with their synthesis provide a useful source of general information on this topic.

4.16.2.4.2 Stability

(i) Thermochemistry

Isoxazoles, isoxazolines, isoxazolidines and benzisoxazoles are all thermally stable, distilling without decomposition, but the stability of the system depends on the substitution pattern. For example, aminoisoxazoles distill unchanged but the isoxazole carboxylic acids usually decompose at or above their melting points without giving the corresponding isoxazole.

The enthalpy of combustion of isoxazole was only determined several years ago ⟨78MI41615⟩. For isoxazole, $\Delta H°_c$ (298.15 K) = $-(1649.85 \pm 0.50)$ kJ mol^{-1}, from which the entropy of formation in the gas phase was derived as $\Delta H°_f(g) = 78.50 \pm 0.54$ kJ mol^{-1}. The enthalpies of combustion of 3-amino-5-methylisoxazole and 5-amino-3,4-dimethylisoxazole have also been determined ⟨73MI41606⟩.

(ii) Aromaticity

Aromatic character in isoxazoles has been studied from a number of viewpoints, and these studies indicate that although isoxazole may be formally considered an aromatic system, the disposition of the ring heteroatoms modifies this 'character' to an appreciable extent. From a qualitative viewpoint, thermal stability and electrophilic attack at the 4-position may be considered consistent with an aromatic character. Furthermore, NMR chemical shifts of the ring protons are consistent with those of an aromatic compound. References related to these studies may be found in Section 4.16.2.3.4.

4.16.2.4.3 Conformation

The isoxazolidine ring exists primarily as an envelope ⟨77AHC(21)207⟩ and the nitrogen lone pair can occupy an axial or equatorial position. Photoelectronic spectroscopy is a useful tool to determine conformational analysis of molecules possessing vicinal electron lone-pairs. Rademiacher and Frickmann ⟨78TL841⟩ studied isoxazolidine and 2-methyl- and 2-t-butyl-isoxazolidine and found mixtures of equatorial and axial (e/a) compounds. The ratios of H, Me and Bu^t in the e/a position were 1:3, 4:1 and 10:1, respectively.

The inversion of the substituent and the lone-pair on nitrogen takes place slowly at room temperature and has been extensively studied ⟨77AHC(21)207, 77IZV2266, 77IZV1687⟩. The inversion takes place by a buckling of the ring and the mixtures obtained are those with the *trans* orientation predominant ⟨77IZV2266⟩.

4.16.2.4.4 Potentiometric properties

The acid–base properties of isoxazole and methylisoxazoles were studied in proton donor solvents, basic solvents or DMSO by IR procedures and the weakly basic properties examined ⟨78CR(C)(268)613⟩. The basicity and conjugation properties of arylisoxazoles were also studied by UV and basicity measurements, and it was found that 3-substituted isoxazoles were always less basic than the 5-derivatives. Protonation increased the conjugation in these systems ⟨78KGS327⟩.

The ligand pK_a values and transition metal chelate stability constants of arylisoxazoles were detected photometrically and the stability of the complexes studied ⟨79JIC1251⟩.

The pK_a values of a number of isoxazoles have been reported and again the weakly basic nature of the ring, being less than oxazole, is demonstrated (see Table 3) ⟨71PMH(3)1, p. 23⟩.

Table 3 pK_a Values of Some Isoxazole Derivatives ⟨71PMH(3)1, p. 23⟩

Compound	pK_a	Compound	pK_a
Oxazole	0.8	4-Chloro-3,5-dimethylisoxazole	−2.48
Isoxazole	−2.03	4-Nitro-3,5-dimethylisoxazole	−3.72
3,5-Dimethylisoxazole	−1.26	3-Phenylisoxazole	−3.18
3,4,5-Trimethylisoxazole	−0.76	5-Phenylisoxazole	−3.22

In a study of a variety of substituents of isoxazole (Me, Ph, H, Cl, Br, NO_2), the $-pK_a$ values ranged from 2.08 to 7.90 ⟨76KGS1029⟩.

Aminoisoxazoles can be determined photometrically by reaction with sodium 1,2-naphthoquinone-4-sulfonate and selective extraction of the resulting dye into CCl_4 for absorbance measurements. This class of compound can be determined in the presence of sulfonamides, sulfanilamides, hydroxylamines and other select amines ⟨74MI41610⟩.

Polarographic studies on haloisoxazoles in anhydrous DMF containing R_4N^+ were performed and the magnitude of the half-wave potentials were recorded. Cleavage of the C—X bond was faster in phenylhaloisoxazoles than in halobenzenes. Substitution patterns affected the reduction ⟨79ZOB1322⟩.

The isoxazolecarboxylic acids are the most intensively investigated of the isoxazole derivatives, and a detailed compilation of their physiochemical properties has been made ⟨62HC(17)1, p. 177⟩. Similar studies have also been carried out with the 5-isoxazolones ⟨62HC(17)1, p. 124⟩, and pK_a values of 1,2-benzisoxazole derivatives are listed in Table 4.

Table 4 pK_a values of 1,2-Benzisoxazole and Methyl Derivatives

Compound	pK_a	Ref.
1,2-Benzisoxazoles	−4.71	B-71MI41605, 77JCS(P2)47
3-Methyl-1,2-benzisoxazole	−2.69	B-71MI41605, 77JCS(P2)47
5-Methyl-1,2-benzisoxazole	−4.03	B-71MI41605, 77JCS(P2)47
3,5-Dimethyl-1,2-benzisoxazole	1.5	79HCA314

4.16.2.5 Tautomerism

Prototropic tautomerism of isoxazole derivatives has been well studied over a number of years and has recently been reviewed in context with similar behavior in other five-membered heterocycles ⟨70C134, 76AHC(S1)1, 79AHC(25)147, p. 202⟩. Several generalizations are summarized below.

Three tautomeric forms, (**10**), (**11**) and (**12**), are possible for 5-substituted isoxazol-5-ones. For the isoxazol-5-ones (**10**; X = O) the OH form (**12**; X = OH) does not contribute to the equilibrium mixture unless some stabilizing influence is present. For example, a 4-carbonyl substituent would stabilize this form by hydrogen bonding. Medium and substituent effects influence the proportion of the CH form (**10**; X = O) and the NH form (**11**; X = O), MS studies showing evidence for the existence of the CH form (**10**; X = O) in the vapor phase ⟨77OMS(12)65⟩. Substituents in the 4-position capable of undergoing a 1,3-prototropic shift also have an influence on the overall tautomerism of the system, often contributing significantly to it. For example, a 4-arylazo substituent has been shown to exist largely in the hydrazono form ⟨76AHC(S1)1⟩. In contrast, the corresponding isoxazole-5-thiones (**11**; X = S) show a tendency to exist in the thiol form (**12**; X = S) and the nitrogen derivatives almost always exist in the amino form (**12**; X = NH or NR).

With 3- and 4-substituted isoxazoles the tautomeric form normally present is the XH tautomer, (**13**; X = O) and (**14**; X = O, N) respectively. However, other influences need to be considered as in cycloserine (**15**), which exists as a zwitterion, as does 5-amino-3-hydroxyisoxazole (**16**).

4.16.3 REACTIVITY OF ISOXAZOLES

4.16.3.1 Reactions at Aromatic Rings

4.16.3.1.1 General survey of reactivity

In the isoxazole nucleus the nitrogen atom is electron withdrawing, and the oxygen is electron donating. Therefore isoxazole undergoes electrophilic substitution more readily than pyridine but less readily than five-membered heterocycles with one heteroatom. Isoxazoles when reacting as neutral molecules are also more reactive than benzene. The presently known electrophilic substitution reactions all occur at the C-4 position of the isoxazole nucleus. Just as in benzene, substituents at C-3 and/or C-5 have a strong activating effect (e.g. NH$_2$, NMe$_2$, OMe, Ph), a strong deactivating effect (e.g. NO$_2$, SO$_3$H, CO$_2$Et), or have relatively little effect (e.g. Me, Cl) on the isoxazole nucleus (or rather the C-4 position). A substituent at C-5 activates or deactivates the C-4 position more strongly than does a substituent at C-3.

Isoxazoles are presently known to undergo hydrogen exchange, nitration, sulfonation, halogenation, chloroalkylation, hydroxymethylation, Vilsmeier–Haack formylation, and mercuration. The Friedel–Crafts reaction on the isoxazole nucleus has not yet been reported.

Isoxazoles are susceptible to attack by nucleophiles, the reactions involving displacement of a substituent, addition to the ring, or proton abstraction with subsequent ring-opening. Isoxazolium salts are even more susceptible to attack by a variety of nucleophiles, providing useful applications of the isoxazole nucleus in organic synthesis. Especially useful is the reductive cleavage of isoxazoles, which may be considered as masked 1,3-dicarbonyl compounds or enaminoketones.

1,2-Benzisoxazoles undergo electrophilic substitution in the benzo ring, but with nucleophiles the reaction occurs in the isoxazole moiety, often leading to salicylonitriles with 3-unsubstituted systems. The isomeric 2,1-benzisoxazoles are characterized by the ease with which they may be converted into other heterocyclic systems.

4.16.3.1.2 Thermal and photochemical reactions

The reactivity of isoxazole in the presence of light, heat or electron impact has been well studied and the various transformations analyzed in terms of reaction pathways and of the potential intermediates. These studies have also been extended to a large variety of substituted derivatives ⟨79AHC(25)147⟩.

The 1,2-bond is homolytically cleaved by both thermolytic and photolytic means to generate a biradical (**17**) which in the absence of reactive groups generally forms a 2H-azirine ⟨79AHC(25)147⟩. No direct evidence for the biradical has been presented, but indirect evidence points to such a species. Acylpyrazines have in some instances been isolated, and these would arise by dimerization of the biradical ⟨70JCS(C)1825, 71JCS(C)2644⟩.

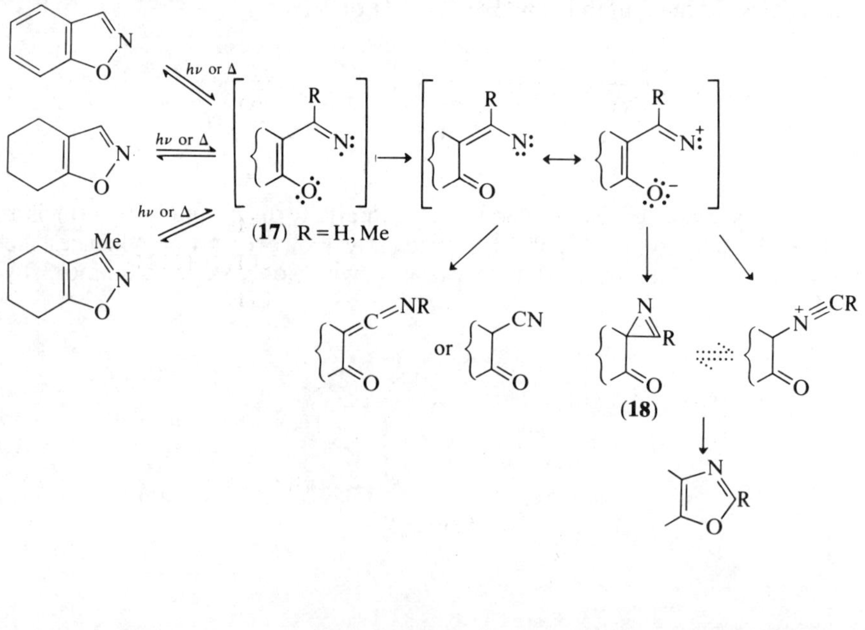

From the type and nature of products generated by photolytic or thermolytic means on suitably substituted isoxazoles, the biradical can again be inferred. The interchange of an acyl moiety between the 4- and 5-position ⟨76HCA2074, 75JA6484⟩ and the product obtained in the presence of a 5-hydrazino moiety ⟨77JCS(P1)971⟩ further support a biradical intermediate. The interconversion of the acyl functionality was studied by deuterium labelling. The biradical can reclose to form a 2H-azirine (**18**), which in many cases can be isolated, or the azirine can undergo intramolecular attack by suitable substituents or react with a nucleophile in the medium to produce a variety of products such as ureas, acetals, ketoketenimines and other heterocycles ⟨79AHC(25)147⟩.

In the absence of suitable reactive moieties, further heat- or light-induced rearrangements of the azirine can occur and are generally believed to proceed *via* nitrene or ylide intermediates ⟨79AHC(25)147⟩. The photochemical rearrangement of the above isoxazoles gave primarily oxazoles by the mechanism outlined ⟨76JOC13, 74JA2014⟩. In a like manner the photolysis of 3-hydroxy-5-phenyl(or 5-methyl)isoxazole produced an oxazolone plus small amounts of benzoic acid and benzoylacetamide (Scheme 2) ⟨74ABC2205⟩.

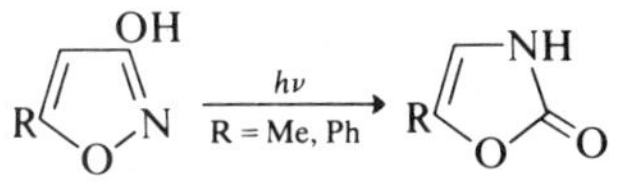

Scheme 2

The isoxazole (**19a**) when photolyzed produced two oxazoles (**20**) and (**21**) which were not interconvertable under the reaction conditions. The corresponding esters (**19b**) and (**19c**) gave (**20b**) and (**20c**), respectively. These studies were instrumental in establishing the photointerconversion of 4- and 5-acylisoxazoles ⟨76HCA2074⟩.

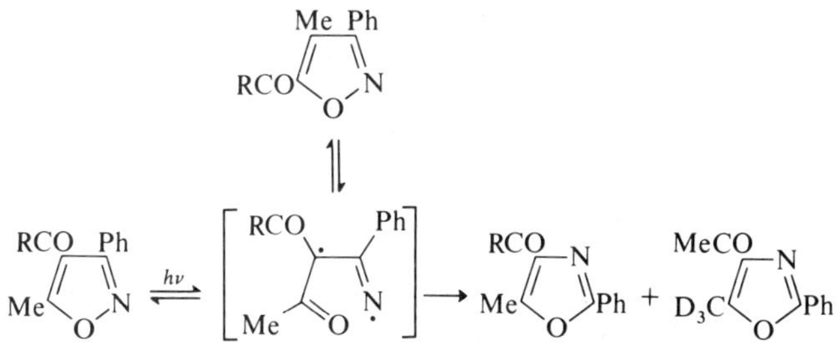

(a) $R = CD_3$; (b) $R = OEt$; (c) $R = OCH_2CF_3$

Irradiation of 3,5-disubstituted isoxazoles in alcoholic solvents gave reaction products such as acetals incorporating the reaction solvent. The use of triethylamine in acetonitrile media produced ketene-aminals by reductive ring cleavage. The reductive ring cleavage product was also obtained by irradiation of the isoxazole in alcohol in the presence of copper(II) salts (Scheme 3) ⟨76JCS(P1)783⟩.

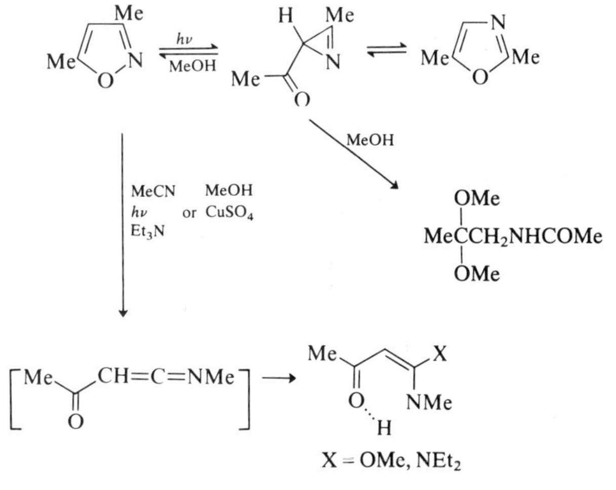

$X = OMe, NEt_2$

Scheme 3

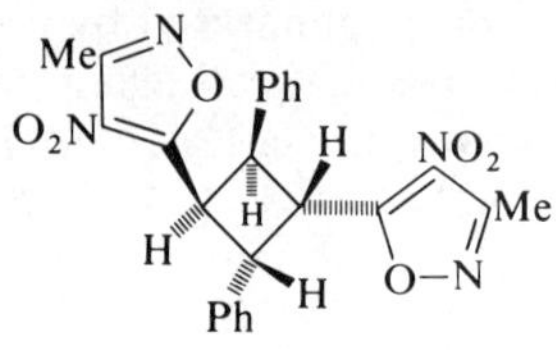

(22)

The photolysis of 3-methyl-4-nitro-5-styrylisoxazole in the solid state produced the *trans* dimer (**22**) ⟨77JHC951⟩. Irradiation of bisisoxazoline (**23a**) in benzene gave three products in equal yield (20%), only one of which retained an isoxazoline ring (**24**). Similar treatment of (**23b**) but using an aqueous nickel sulfate filter gave 28% of (**24b**) (Scheme 4). Photolysis of (**25**) gave (**26**) in 70% yield. ESR spin trapping with 2,3,5,6-$(CD_3)_4C_6HNO$ and 2,4,6-$(Me_3C)_3CH_2NO$ indicated that these reactions proceeded *via* imino and 2-isoxazolinyl radicals ⟨77TL4619⟩.

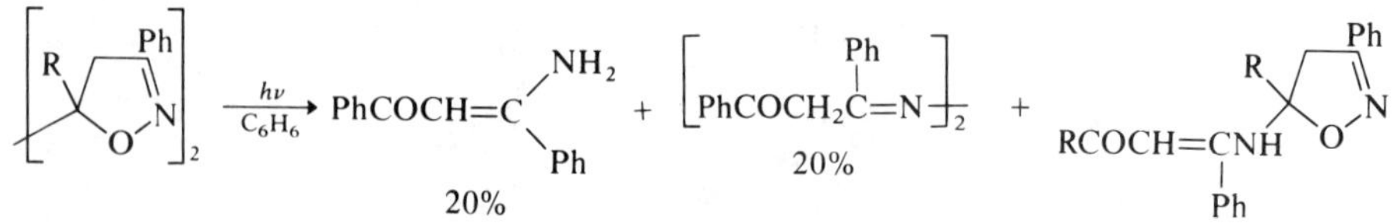

(**23**) **a**; R = Ph
b; R = H

(**24**) **a**; R = Ph; 20%
b; R = H; 28%

Scheme 4

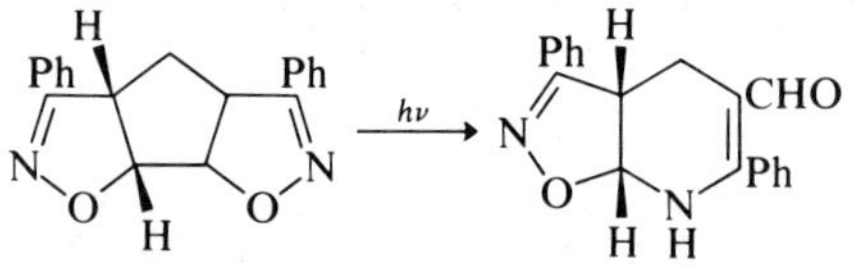

(**25**) (**26**)

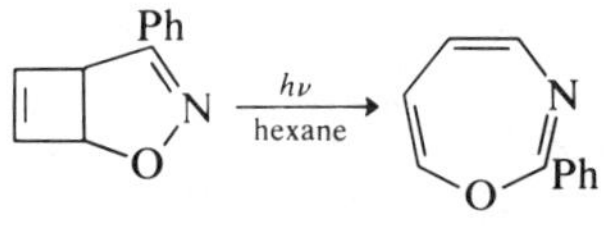

(**27**) (**28**) 80%

Irradiation of the fused ring isoxazoline (**27**) gave the ring expanded product, oxazepine (**28**), in 80% yield.

The photolysis of 4-substituted 2,3-dimethyl-3-isoxazolin-5-ones has been studied. Irradiation in methanol or ethanol with a 100 W high-pressure mercury lamp through a Pyrex filter of a 4-phenylthio compound produced a semithioacetal (Scheme 5). In contrast, an H, Cl or OPh moiety gave no reaction. The use of alkylthio substitution gave similar products. Cyclic compounds yielded cyclic products (Scheme 5), and the photolysis of (**29**) in benzene

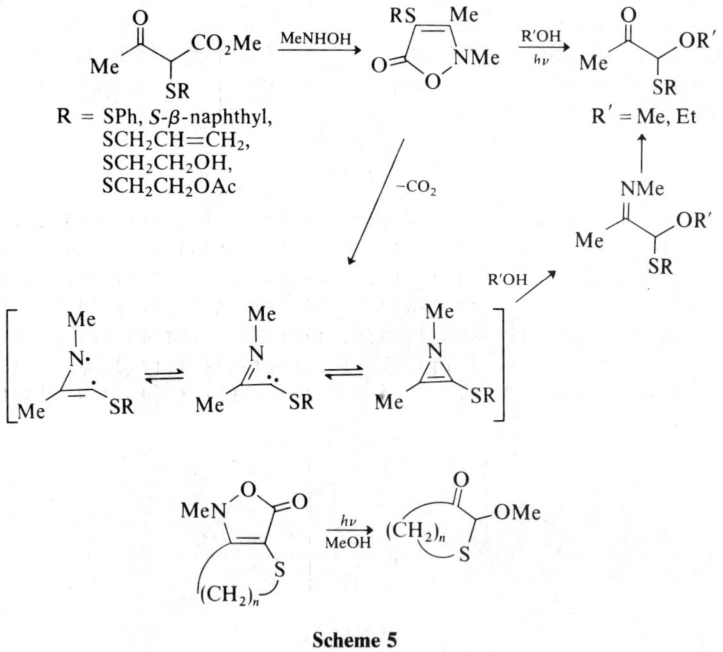

Scheme 5

afforded the cyclic thioacetal (**30**) ⟨80CC1054⟩. A 2*H*-azirine intermediate has been postulated for these transformations.

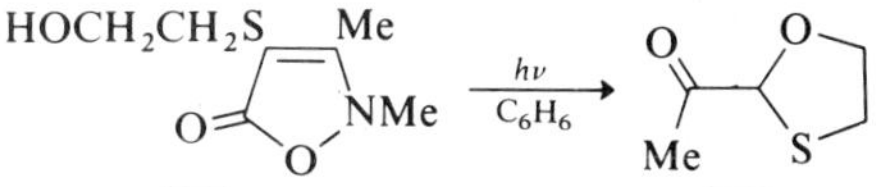

Pyrolysis of isoxazoles produces intermediate azirines which generally rearrange to oxazoles, although decarbonylation and subsequent formation of phenylindoles has been also observed (Scheme 6) ⟨79AHC(25)147⟩. 3,5-Dimethylisoxazole isomerized to 3-acetyl-2-methyl-2*H*-azirine at 500 °C in an equilibrium reaction, whereas at higher temperatures both materials gave 2,4-dimethyloxazole. An indole was also observed and was postulated to arise by decarbonylation of the azirine with subsequent rearrangement (Scheme 6) ⟨76MI41612⟩.

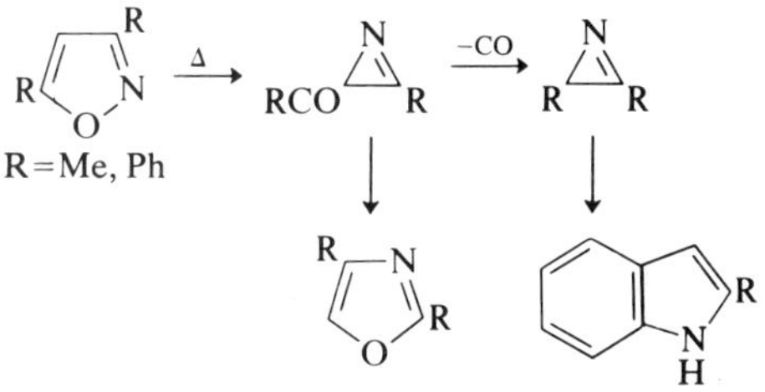

Scheme 6

In the attempted thermolytic preparation of pyrroloisoxazole (**32**) from azidoisoxazole (**31a**), only cinnamoyl cyanide was isolated. The assumed intermediate nitrene (**33**) did not insert into the styryl bond, but rather ring rupture and loss of acetonitrile produced the product. Similar products were obtained from the homolog (**31b**) (Scheme 7) ⟨79TL4685⟩. The stabilized nitrene intermediate is similar to that postulated for diazofuryl- and diazoisoxazolyl-methanes ⟨78JA7927, 79TL2961⟩.

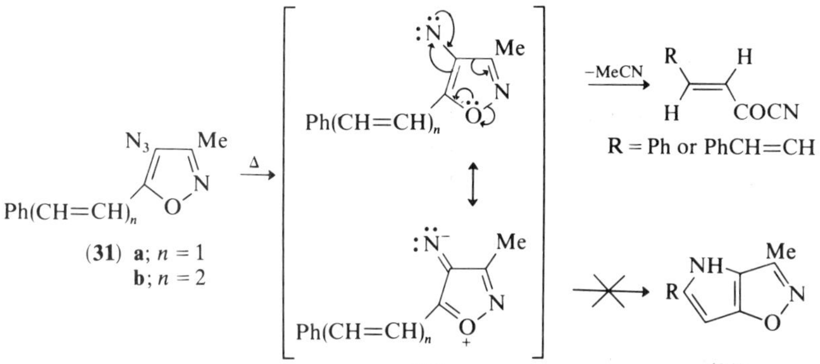

Scheme 7

4-Acylisoxazoles are converted into 4-acyloxazoles on heating and the intermediacy of a 2*H*-azirine is postulated (Scheme 8) ⟨74JOC1976⟩. The thermal isomerization of 5-methoxy-3-arylisoxazoles (**34**) to 3-aryl-2*H*-azirine-2-carboxylates (**35**) has been extensively studied and is a convenient synthetic method ⟨78KGS1053, B-77MI41612, 75KGS1292, 74JCS(P1)1867⟩. The heats of isomerization and kinetics of isomerization have been studied for a large number of derivatives ⟨78KGS1053⟩. Solvent and substituent effects were detected and a biradical intermediate was proposed ⟨B-77MI41612⟩. 5-Aminoisoxazoles (**36**) when heated in HMPA produced indoles (**37**) ⟨74JCS(P1)1867⟩.

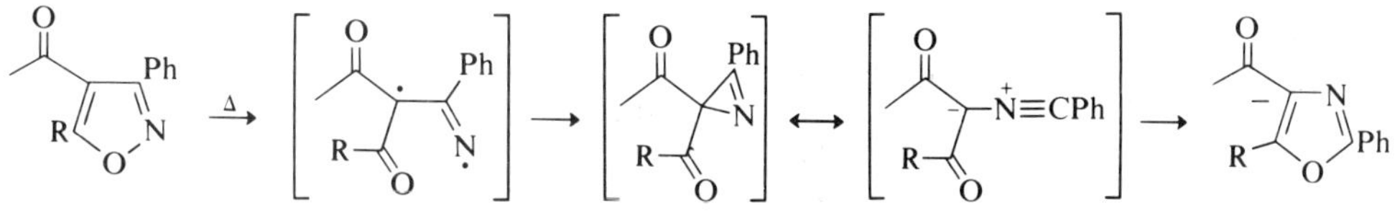

Scheme 8

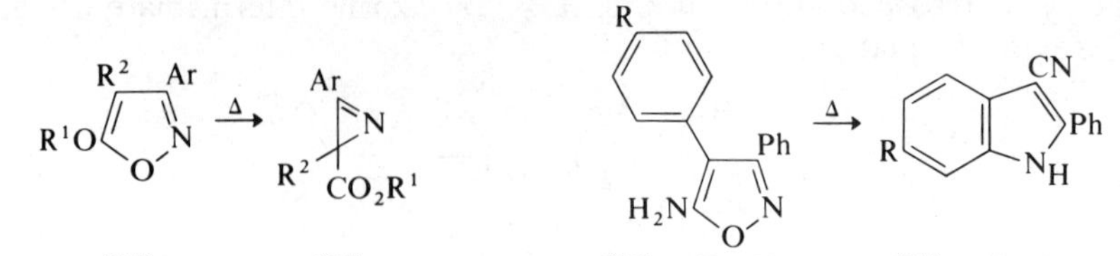

(34) (35) (36) a; R = H (37) a; R = H
             b; R = Cl   b; R = Cl

The photolysis of 1,2-benzisoxazole in the absence of air in acetonitrile gave salicylonitrile and benzoxazole ⟨67AHC(8)277⟩. When air-saturated acetonitrile was employed, 2,2′-dimerization to (38) occurred, accompanied by benzoxazole. Photolysis of the 2,2′-dimer (38) and benzoxazole did not alter the ratio, thus indicating that neither one arose from the other. Selective excitation also ruled out dimer formation from benzoxazole under the reaction conditions (Scheme 9). This dimerization is similar to that observed for benzimidazole, except that in that series no 2,2′-dimerization was observed ⟨74TL375⟩.

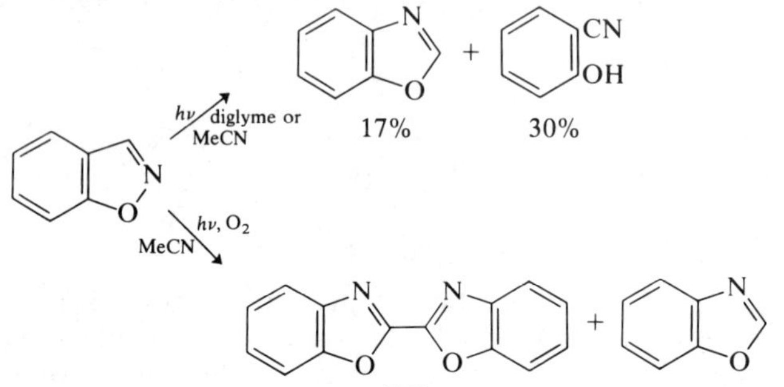

(38)

Scheme 9

Salicylonitrile is believed to arise by direct cleavage with subsequent hydrogen transfer, while the benzoxazoles were produced by an isocyanide intermediate ⟨73JA919, 74HCA376⟩. Photolysis in D_2O tends to confirm this possibility and rule out an aziridine intermediate (39), due to deuterium corporation into the molecule (Scheme 10) ⟨74HCA376⟩.

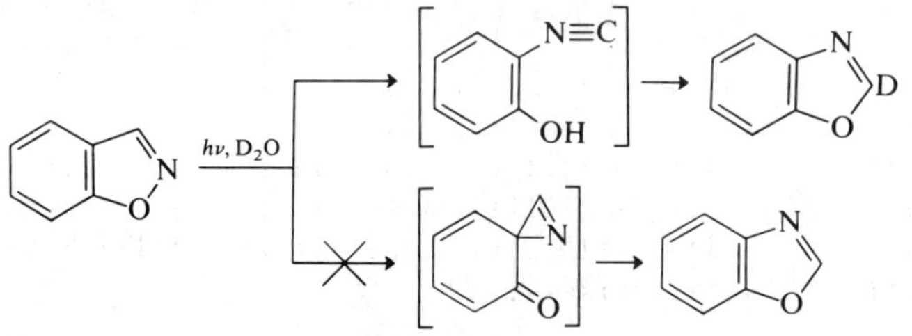

(39)

Scheme 10

Photolysis of 3-methyl-1,2-benzisoxazole in n-hexane/acetonitrile produced a salicylamide. In contrast, photolysis in acetonitrile/methanol (95:5) gave an iminoester which subsequently hydrolyzed to methyl salicylate (Scheme 11) ⟨74HCA376⟩.

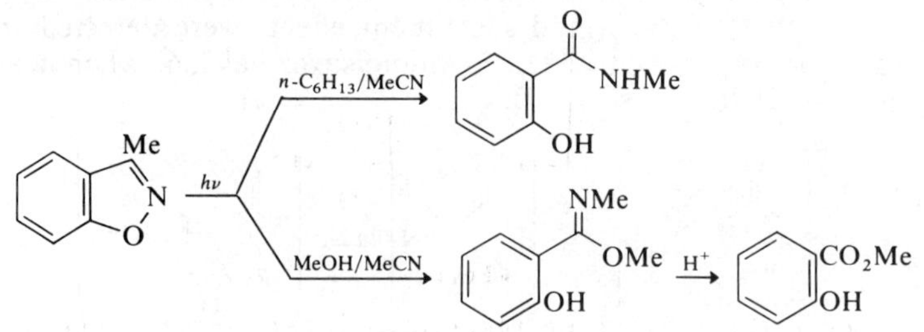

Scheme 11

Photolysis in strongly acidic medium gave rise to cleavage with substitution. 1,2-Benzisoxazole when irradiated in 96% H_2SO_4 gave two phenones which had hydroxyl substituents in the 3- or 5-position. Photolysis of 3-methyl-1,2-benzisoxazole in 96% H_2SO_4 gave similar products. The utilization of 2M H_2SO_4 produced 2-aminophenol, which presumably arose *via* benzoxazole by hydrolysis (Scheme 12) ⟨79HCA314⟩. The inclusion of a suitable nucleophilic anion in the photolysis medium led to replacement of the hydroxy group by the nucleophile. A mechanistic pathway was proposed ⟨79HCA314, 71HCA2916⟩ and is shown in Scheme 13.

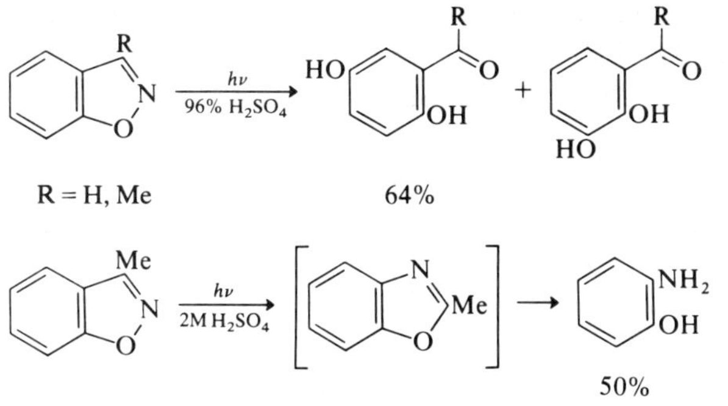

Scheme 12

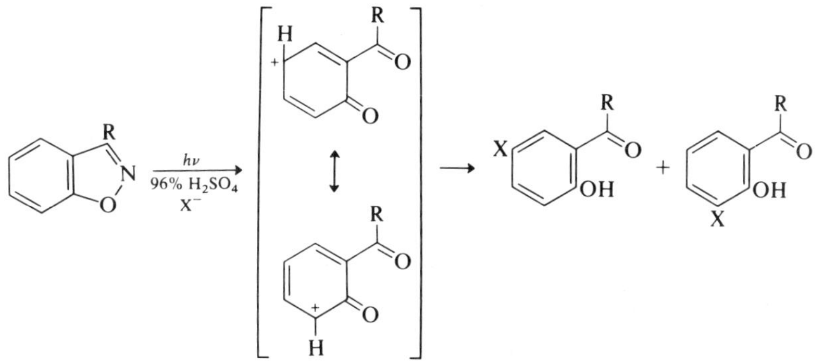

Scheme 13

Photochemical studies on the ring degradation of 3-hydroxy-1,2-benzisoxazole also yielded benzoxazolone, and (**40**), (**41**) and (**42**) (Scheme 14) were believed to be potential intermediates. Low temperature IR measurements indicated the presence of (**42**) during the photochemical reaction ⟨73JA919, 71DIS(B)4483, 71JOC1088⟩. Sensitization studies indicate that the rearrangement is predominantly a triplet reaction, and the keto tautomer is believed to be the key orientation for the photolysis.

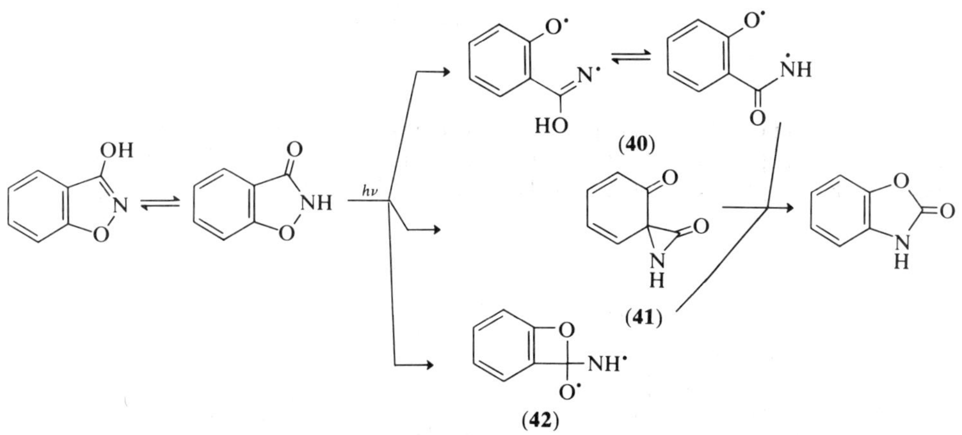

Scheme 14

2-Methyl-1,2-benzisoxazol-3-one on photolysis yielded *N*-methylbenzoxazole and *N*-methylsalicylamide (Scheme 15) ⟨71DIS(B)4483, 71JOC1088, 77MI41611⟩.

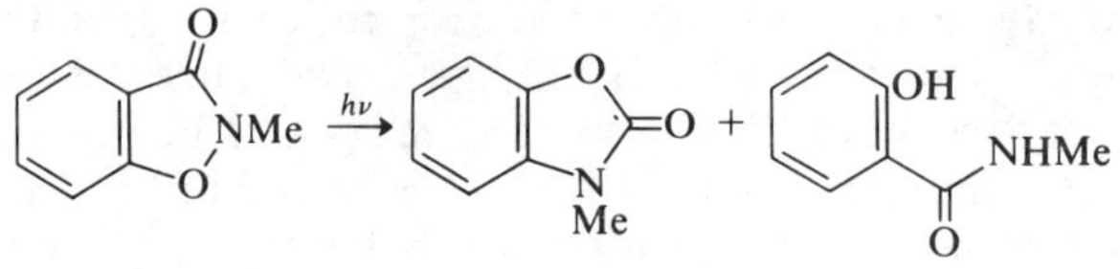

Scheme 15

The mechanism of thermolysis and photolysis of ethers of 3-hydroxy-1,2-benzisoxazole has also been studied. Heating of the allyl ether (**43**) gave minor amounts of (**44**) and two benzoxazoles. Photolysis of (**45**) in methanol gave a benzisoxazole and an iminoester, *via* intermediate (**46**). Thermolysis at 600 °C gave a benzoxazole, a benzoxazolone and cyanophenol (Scheme 16) ⟨71DIS(D)4483⟩.

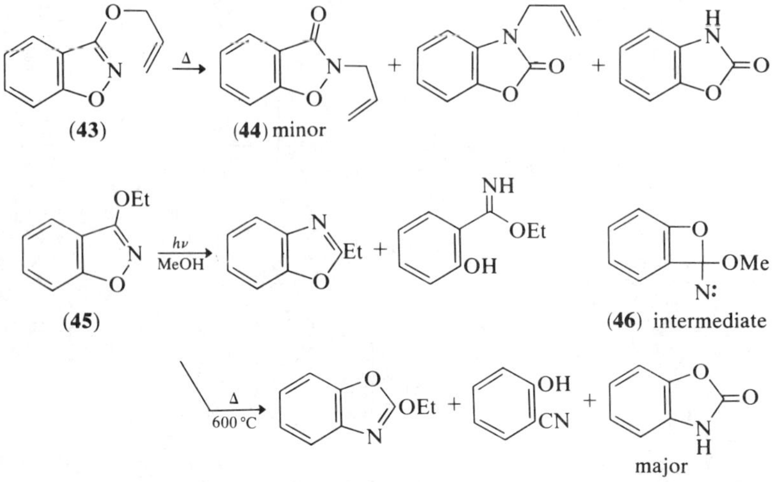

Scheme 16

Photolysis of 2,1-benzisoxazoles cleaved the ring and led to the introduction of a substituent in the 3- or 5-position of the 2-aminophenone ⟨71HCA2916, 72HCA1730, 71HCA2111, 79HCA304, 79HCA198, 79HCA185, 79HCA271, 77MI41611⟩. Photolysis of 3-methyl-2,1-benzisoxazole in concentrated HCl gave 2-amino-5-chloroacetophenone as the major product and in 66% H_2SO_4 2-amino-5-hydroxyacetophenone was produced as the major product (Scheme 17) ⟨71HCA2111⟩. The photolysis of 3-methyl-2,1-benzisoxazole gave a more complex mixture of products ⟨79HCA271⟩.

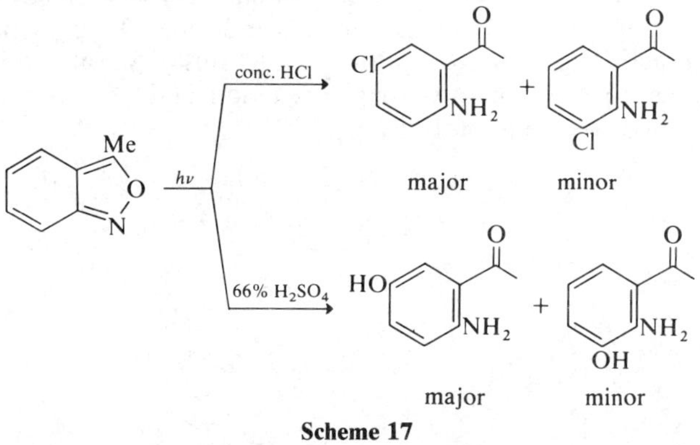

Scheme 17

Photolysis of 3-phenyl-2,1-benzisoxazole in 48% HBr produced reduction and substitution products *via* a proposed triplet state nitrenium ion intermediate ⟨71HCA2111⟩. Photolytic decomposition of 5-bromo-3-phenyl-2,1-benzisoxazole in 48% HBr gave 2-amino-5-bromoacetophenone and 2-amino-3,5-dibromoacetophenone (Scheme 18). A nitrenium ion intermediate was also proposed for the photolytic decomposition of 3-phenyl-2,1-benzisoxazole in concentrated HCl (Scheme 19) ⟨71HCA2111⟩.

Photolysis in concentrated HCl of 3,5-diphenyl-2,1-benzisoxazole gave 2-amino-3-chloro-5-phenylbenzophenone and 2-amino-5-(*p*-chlorophenyl)benzophenone *via* similar intermediates (Scheme 20) ⟨71HCA2111⟩.

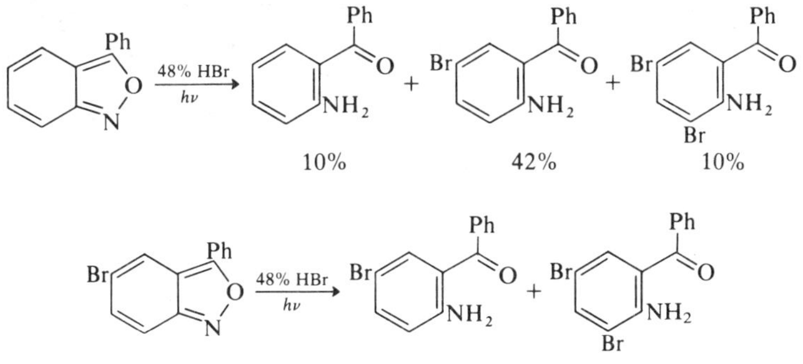

Scheme 18

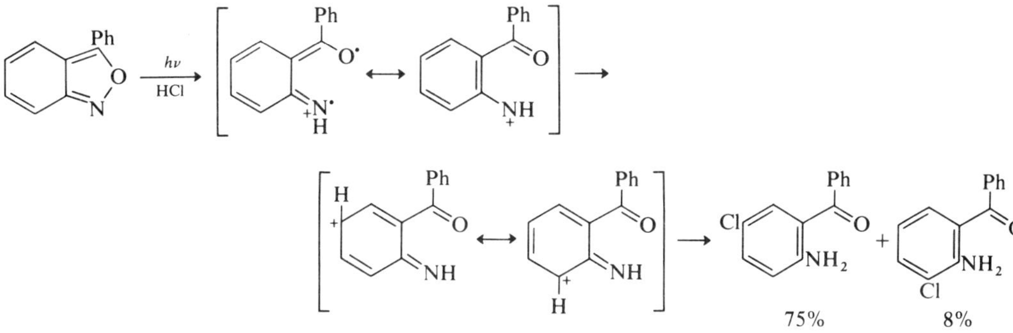

Scheme 19

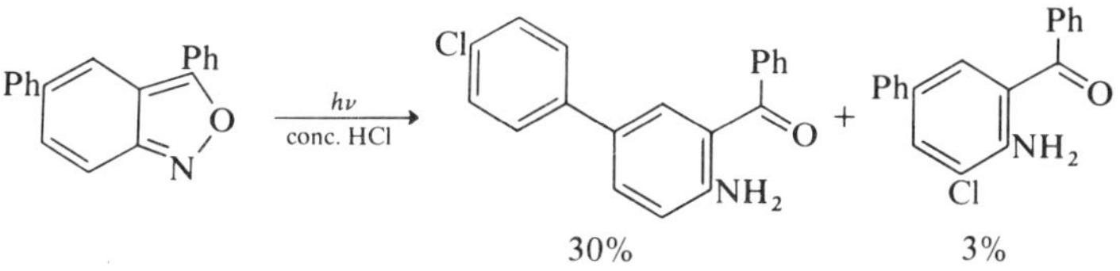

Scheme 20

The photolysis of 3-methyl-2,1-benzisoxazole in H_2SO_4/MeCN in the presence of anisole gave a variety of products, also *via* a proposed nitrenium ion intermediate (Scheme 21). Decomposition of azide (**47**) under the same conditions gave a similar ratio of products (Scheme 21) ⟨79HCA304⟩. Pyrolysis of (**48**) generated an acridone (**49**) *via* an analogous proposed nitrene intermediate (**50**) ⟨68JOC2880, 71DIS(B)4883⟩.

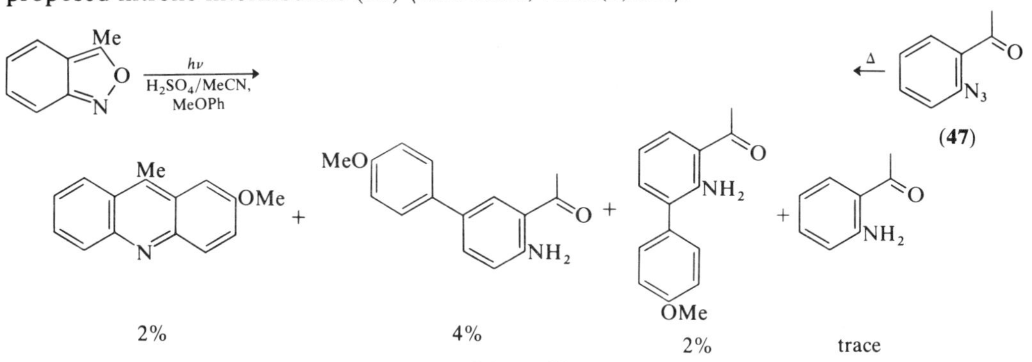

Scheme 21

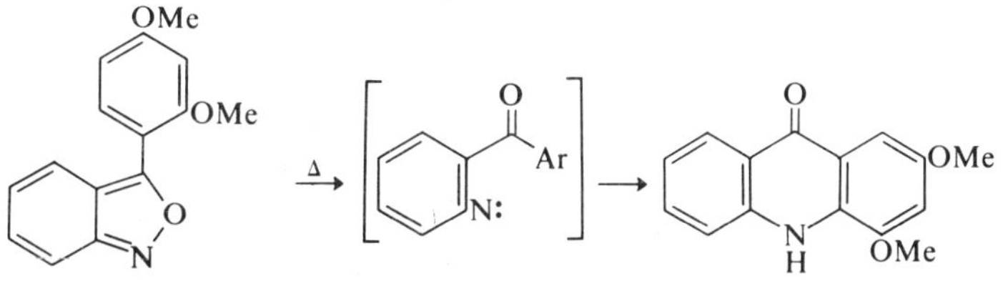

(48)   (50)   (49)

The photodecomposition of 2,1-benzisoxazolium salts gave N-substituted phenones (Scheme 22). In one case the 1-(adamantyl)-3-phenyl-2,1-benzisoxazolium cation (**51**) did not generate a substituted phenone with reductive ring substitution. Rather, adamantyl ring rupture occurred to produce (**52**) (Scheme 22) ⟨78JOC1233, 77JOC3929⟩.

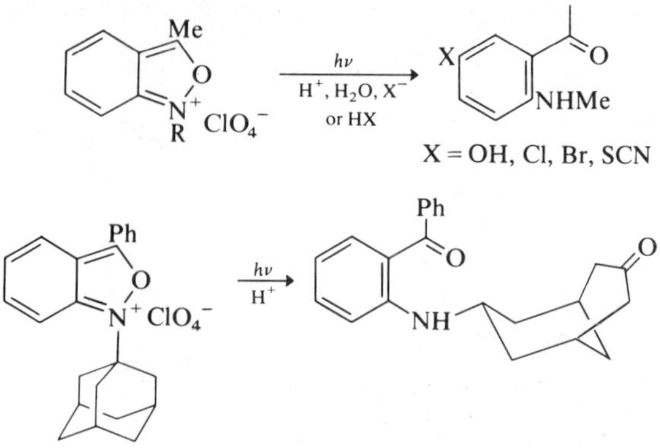

Scheme 22

4.16.3.1.3 *Electrophilic attack at nitrogen*

(i) Basicity of isoxazoles

Isoxazoles are weakly basic. Isoxazole has a pK_a of -2.97 at 25 °C, and the values for its 3-methyl, 5-methyl and 3,5-dimethyl derivatives are -1.62, -2.01 and -1.61, respectively; a 4-nitro group reduces the pK_a of 3,5-dimethylisoxazole to -6.43. For 4-substituted 3,5-diphenylisoxazoles the pK_a values showed a good correlation with the Hammett σ_m constants for the substituents ⟨79AHC(25)147⟩ (see also Section 4.16.2.4.4, Tables 3 and 4).

On the basis of UV studies, 1,2-benzisoxazoles are considered to be weaker bases than the above isoxazoles ⟨81AHC(29)1, p. 9⟩ and the pK_a values for 1,2-benzisoxazole-3-carboxylic acids are comparable with those of *o*-nitrobenzoic acids.

Although isolated salts of 2,1-benzisoxazoles have not been reported, spectral data in acidic media show evidence for protonation.

(ii) Metal ions

Although isoxazoles are comparatively weak electron donors, complexes with numerous metal ions, notable metal(II) ions, have been reported. The ligands include isoxazole and its methyl, phenyl, amino and hydroxy derivatives. They are listed with references in Table 5.

Table 5 Metal Complexex of Isoxazoles

Ligand	Metal ion[a]	Ref.
Isoxazole	Cu, Ag(I), Zn, Cd, Hg, Cr(III), Mn, Fe, Co, Ni, Pd, Pt	77MI41607, 74MI41600, 76MI41606, 76SA(A)1779, 78MI41600
5-Methylisoxazole	Cu, Ag(I), Zn, Co, Ni	77MI41609
3,5-Dimethylisoxazole	Cu, Cu(I), Zn, Cd, Hg, Cr, Cr(III), Co, Ni, Pd, Pt, Mn	76MI41606, 76SA(A)1779, 75MI41601, 77MI41616, 78ZOB418, 80SA(A)143
3-Methyl-5-phenylisoxazole	Cu, Cu(I), Zn, Cd, Hg, Cr, Cr(III), Co, Ni, Pd, Pt, Mn	76SA(A)1779, 78ZOB418, 76MI41603, 75MI41600, 76MI41605, 80SA(A)143
3,5-Diphenylisoxazole	Cu, Zn, Cd, Hg, Cr(III), Co, Ni, Pd, Pt, Mn	76SA(A)1779, 77MI41616, 76MI41603, 76MI41605, 77MI41605
3-Amino-5-methylisoxazole	Cu, Zn, Cd, Hg, Co, Ni, Pd, Pt, Mn	76MI41602, 76MI41604, 77MI41604, 77MI41602
4-Amino-3,5-dimethylisoxazole	Zn, Cd, Hg, Pd, Pt	76MI41603, 77MI41606
5-Amino-3,4-dimethylisoxazole	Cu, Zn, Cd, Hg, Co, Pd, Pt	76MI41604, 77MI41604, 77MI41603
3-Phenylisoxazole-5(4H)-one	Cu, UO_2	77JIC536
Dimethyldiisoxazolone	Cu, Co, UO_2	76JIC439, 76JIC779

[a] M(II) unless otherwise stated.

(iii) Alkyl halides and related compounds

Isoxazolium salts can be prepared by reaction with alkyl iodides or sulfates, although the low basicity of isoxazoles and their sensitivity to nucleophilic attack may necessitate special care. Isoxazolium salts containing bulky *N*-substituents can be prepared by the reaction of isoxazoles with alcohols in the presence of perchloric acid. For example, the reaction of 3,5-dimethylisoxazole (**53**) with some alcohols in the presence of 70% perchloric acid gave isoxazolium salts, (**54a**) in 29%, (**54b**) in 57% and (**54c**) in 82% yield ⟨79AHC(25)147, 68JOC2397⟩. Attempts to quaternize 3,5-dimethyl-4-nitroisoxazole failed ⟨71JCS(B)2365⟩.

$$\text{Me} \underset{\text{Me}}{\overset{}{\bigcirc}}_{\text{O}}^{\text{N}} + \text{ROH} \xrightarrow{70\% \text{ HClO}_4} \text{Me} \underset{\text{Me}}{\overset{}{\bigcirc}}_{\text{O}}^{\text{+NR}}$$

(**53**) (**54**) a; R = CH$_2$Ph
b; R = Bu^t
c; R = CHMePh

The reactivity of isoxazole toward quaternization is compared with those of pyridine-2-carbonitrile, pyridine and five other azoles in Table 6 ⟨73AJC1949⟩. Isoxazole is least reactive among the six azoles and 10^4 times less reactive than pyridine. There is also a good correlation between the rate of quaternization and basicity of the azole.

Table 6 Rate Data for the *N*-Methylation of Some Azoles and Related Compounds with Me$_2$SO$_4$ at 38 °C ⟨73AJC1949⟩

Compound	log k_{rel}	Compound	log k_{rel}
Pyridine-2-carbonitrile	0	1-Methylpyrazole	0.75
1-Methylimidazole	(3.11)^a	Oxazole	0.50
Pyridine	(2.66)^a	Isothiazole	−0.49
Thiazole	(1.34)^a	Isoxazole	−1.33

^{a} The reactions were run with methyl iodide in DMSO at 25 °C. Log k_{rel} for pyridine-2-carbonitrile is also equal to 0. In order to combine two sets of results, the author assumes that the ρ value for each reaction is the same.

In 1,2-benzisoxazoles, quaternization at nitrogen occurred with a variety of alkylating agents ⟨67AHC(8)277⟩. Similarly, 2,1-benzisoxazoles gave quaternary salts by the action of acetic anhydride–perchloric acid ⟨78JOC1233⟩, methyl fluorosulfonate ⟨66JOC2039⟩, an alcohol with tetrafluoroboric acid ⟨66JOC2039, 70JOC2440, 71JOC1543, 77DIS(B)147, 77JOC3929, 75USP3911132⟩ or perchloric acid ⟨77TH41600⟩, and treatment with orthoacids ⟨72CPB2209⟩.

Sheads studied the reactivity of over 100 alcohols with 2,1-benzisoxazoles with 70% aqueous perchloric acid in nitromethane ⟨74DIS(B)147⟩. The rates of quaternization by reaction with MeI in DMSO were also studied and the dynamics of the reaction investigated ⟨74AJC1221⟩.

4.16.3.1.4 Electrophilic attack at carbon

(i) Acid-catalyzed hydrogen exchange

Quantitative information on the reactivity of isoxazole was obtained by studies on acid-catalyzed deuteriodeprotonation reactions ⟨74JCS(P2)399, 71JCS(B)2365⟩. Isoxazole and its 3-methyl-, 5-methyl and 3,5-dimethyl derivatives all underwent exchange in D$_2$SO$_4$ at the 4-position of the isoxazole nucleus. Further heating of these compounds with D$_2$SO$_4$ of various concentrations showed no further change of the NMR spectrum under moderate conditions; under forcing conditions the isoxazoles decomposed. The rate profiles for the isoxazoles indicate that in every case the reaction is proceeding on the free-base form over the entire acidity range. No exchange of the 4-proton occurred when 2,3,5-trimethylisoxazolium bisulfate was heated under similar conditions, which confirms the above conclusion.

The reactivities of the isoxazoles are compared with those of benzene and some five-membered ring heterocycles in Table 7. Isoxazole is more reactive than benzene (by 4.3 log units) and isothiazole (0.8) and is less reactive than 1-methylpyrazole, furan, thiophene and 1-methylpyrrole. A 5-methyl substituent activates the nucleus more than does a

3-methyl substituent; the activation provided by 3,5-dimethyl substituents is nearly the sum of the individual effects of the 3-methyl and 5-methyl groups considered separately.

Table 7 Reactivities of Some Azoles and Other Five-membered Heterocycles in Acid-catalyzed Deuteriodeprotonation ⟨74JCS(P2)399⟩

Compound	log(partial rate factor)[a]			
	C-2	C-3	C-4	C-5
Isoxazole	—	—	4.3	—
3-Methylisoxazole	—	—	5.4	—
5-Methylisoxazole	—	—	6.8	—
3,5-Dimethylisoxazole	—	—	7.6	—
Isothiazole	—	—	3.5	—
3-Methylisothiazole	—	—	4.5	—
5-Methylisothiazole	—	—	4.7	—
3,5-Dimethylisothiazole	—	—	6.7	—
1-Methylpyrazole	—	5.6	9.8	5.6
1,3-Dimethylpyrazole	—	—	10.7	—
1,5-Dimethylpyrazole	—	—	11.4	—
1,3,5-Trimethylpyrazole	—	—	13.4	—
Furan	8.2	ca. 4.5	ca. 4.5	8.2
Thiophene	8.6	5.0	5.0	8.6
1-Methylpyrrole	10.6	10.3	10.3	10.6

[a] Relative to one position of benzene and its log (partial rate factor) = 0.

Table 7 also indicates that the rate enhancements for a 3- and 5-methyl group vary significantly among 1,2-azoles. The difference between the increments in log units for a 3- and 5-methyl group, which should vary directly with bond fixation in the ground state, is larger for isoxazole (1.4) than for pyrazole (0.7) and for isothiazole (0.2). This indicates that the aromaticity increases in the same order and contributes the first quantitative evidence that the 1,2-azoles follow the same aromaticity order as furan < pyrrole < thiophene.

(ii) Nitration

Nitration of alkylisoxazoles and phenylisoxazoles has received considerable attention ⟨71JCS(B)2365, 75JCS(P2)1627⟩. Alkylisoxazoles underwent nitration as the free base at the 4-position of the isoxazole nucleus; the non-reactivity under similar conditions of the 2,3,5-trimethylisoxazolium cation strongly supported this finding. 3-Methyl-5-phenyl-isoxazole underwent nitration in H_2SO_4/HNO_3 as the conjugate acid, substitution occurring at the *para* position of the phenyl group. However, nitration in Ac_2O/HNO_3 gave 4-nitro-5-*p*-nitrophenylisoxazole in poor yield as the only product isolated. 5-Methyl-3-phenyl-isoxazole underwent nitration in mixed acid as the conjugate acid, substitution occurring at the *meta* position, and also as the free base with substitution at the *para* position of the phenyl group. The proportion of the *meta* and *para* isomers depended on the acidity of the reaction medium. The nitration of 5-methyl-3-phenylisoxazole in Ac_2O/HNO_3 gave 4-nitro-3-*p*-nitrophenylisoxazole.

Under carefully controlled conditions, nitration of 3,5-diphenylisoxazole in Ac_2O/HNO_3 at ~20 °C gave 4-nitro-3,5-diphenylisoxazole. However, 3,5-diphenylisoxazole in HNO_3 underwent nitration first at the *para* position of the 5-phenyl substituent and then at the *meta* position of the 3-phenyl group ⟨74KGS597⟩.

The parent isoxazole was nitrated under strictly controlled conditions (35–40 °C) in only 3.5% yield whereas 3,5-dimethylisoxazole was nitrated in mixed acid at 100 °C to give 3,5-dimethyl-4-nitroisoxazole in 86% yield. The 3-anilinoisoxazole (**55**) underwent ready nitration to give the 4-nitroisoxazole derivative (**56**) ⟨63AHC(2)365⟩.

Nitration of 3-ethyl-1,2-benzisoxazole 2-oxide (**57**) and reduction under forcing conditions with triethyl phosphite gave 6-nitro-1,2-benzisoxazole. In contrast, nitration of 2-ethyl-1,2-benzisoxazole gave 5-nitro substitution (Scheme 23) ⟨80CC421⟩.

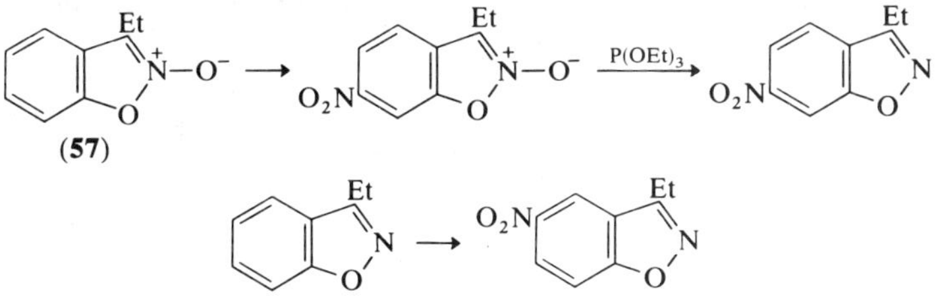

Scheme 23

Bianchi *et al.* speculated on the nature of the species undergoing reaction when 3-methyl-1,2-benzisoxazole was nitrated and concluded that the free base is the species involved in 80–90% H_2SO_4 while at higher percentages the protonated benzisoxazole is the moiety undergoing nitration (Scheme 24). It was speculated that nitration is not charge controlled, but is controlled by the orbital coefficients which are largest at C-5 and next at C-7 ⟨77JCS(P2)47⟩.

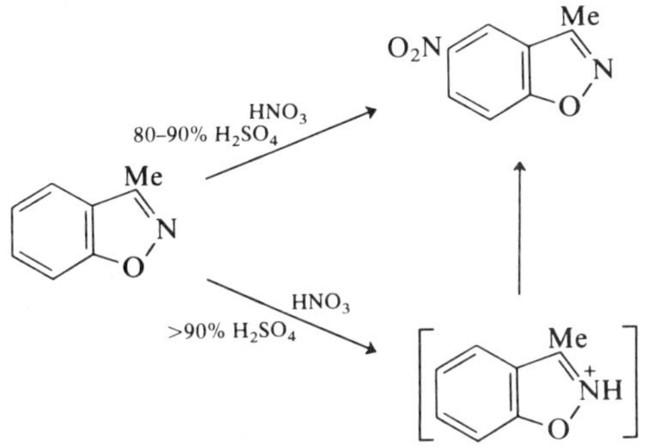

Scheme 24

Nitration of 1,2-benzisoxazoles under a variety of conditions has been reported to give the 5-nitro compound with or without mixtures of an isomeric mononitro compound and/or dinitration ⟨80CC421, 77JCS(P2)47, B-71MI41605, 77IJC(B)1061, 74JCS(P2)389, 75JCS(P2)1627, 67AHC(8)277⟩. Nitration of the unsubstituted nucleus gave predominantly the 5-nitro derivative ⟨26LA(449)63⟩.

Nitration of a series of methyl-1,2-benzisoxazoles was studied by Tahkar and Bhawal using fuming nitric acid and sulfuric acid in acetic acid at 100 °C. 3-Methyl-1,2-benzisoxazole gave a mixture of 5-nitro- and 5,7-dinitro-3-methyl-1,2-benzisoxazole, with the 5-nitro isomer predominant. The product obtained from 3,5-dimethyl-1,2-benzisoxazole was the 4-nitro derivative and not the 7-nitro compound as proposed by Lindemann ⟨26LA(449)63⟩. The synthesis of the 7-nitro compound by an alternative method was used as structural proof. Two products were obtained from 3,6-dimethyl-1,2-benzisoxazole and these were the 5-nitro and 5,7-dinitro derivatives. 3,7-Dimethyl-1,2-benzisoxazole was converted into the 5-nitro derivative (Scheme 25) ⟨77IJC(B)1061⟩.

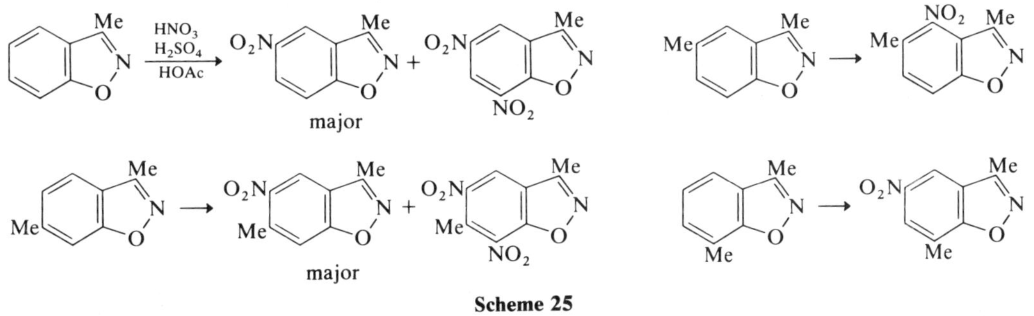

Scheme 25

Nitration of 3-chloro-1,2-benzisoxazole gave the 5-nitro derivative ⟨80ZC18, 79ZC452⟩, as did nitration of 1,2-benzisoxazole-3-acetic acid with fuming nitric acid. Under more forcing conditions, nitration of 1,2-benzisoxazole-3-acetic acid resulted in the four products (**58**), (**59**), (**60**) and (**61**). In contrast, under the same forcing conditions, methyl 1,2-benzisoxazole-3-acetate and 3-(1,2-benzisoxazolyl)acetonitrile gave the 5-nitro products (**62**) and (**63**). Activation of the acid to side-chain degradation was linked to intramolecular hydrogen bonding ⟨78CPB3498⟩.

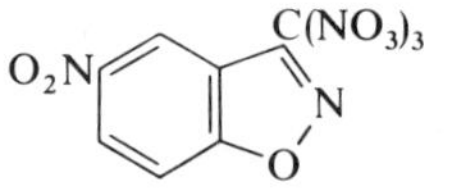

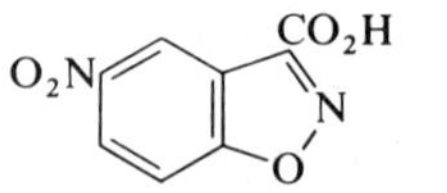

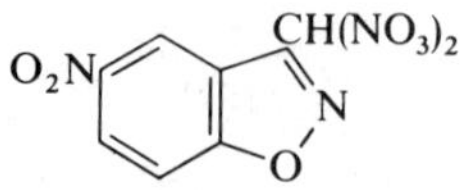

2,1-Benzisoxazole or 3-methyl-2,1-benzisoxazole when treated with KNO$_3$ in sulfuric acid produced the 5-nitro compound as the main product, with some of the 7-nitro as a side product ⟨63MI41601, 65CC408⟩.

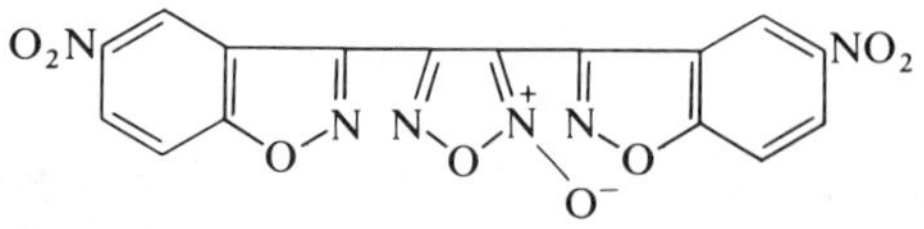

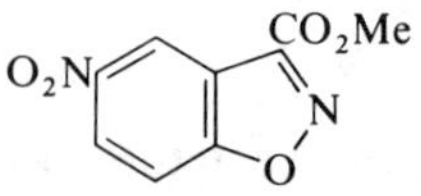

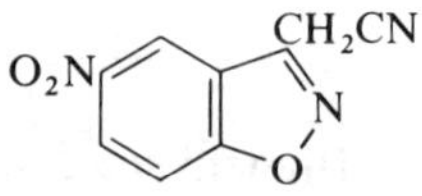

(iii) Sulfonation

The isoxazole ring is rather resistant to sulfonation. However, on prolonged heating with chlorosulfonic acid, 5-methyl-, 3-methyl- and 3,5-dimethyl-isoxazoles are converted into a mixture of the sulfonic acid and the corresponding sulfonyl chloride. The sulfonic acid group enters the 4-position even when other positions are available for substitution. The sulfonation of the parent isoxazole occurs only under more drastic conditions (20% oleum) than that of alkylisoxazoles; isoxazole-4-sulfonic acid is obtained in 17% yield. In the case of 5-phenylisoxazole (**64**), only the phenyl nucleus is sulfonated to yield a mixture of *m*- and *p*-arenesulfonic acid chlorides (**65**) and (**66**) in a 2:1 ratio ⟨63AHC(2)365⟩.

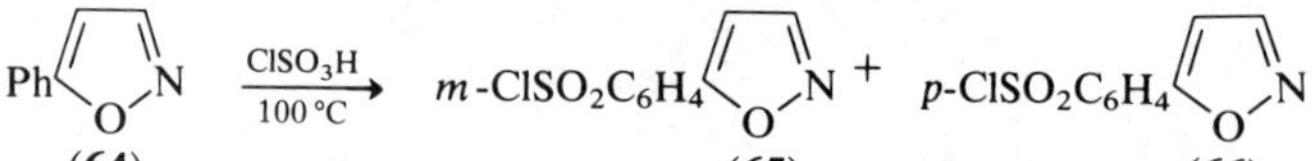

3-Phenyl-1,2-benzisoxazole has been reported to give a disulfonic acid of unknown structure on treatment with 40% oleum ⟨67AHC(8)277⟩. The chlorosulfonation of 1,2-benzisoxazole-3-acetic acid has been reported to give a mixture of the two products shown in Scheme 26.

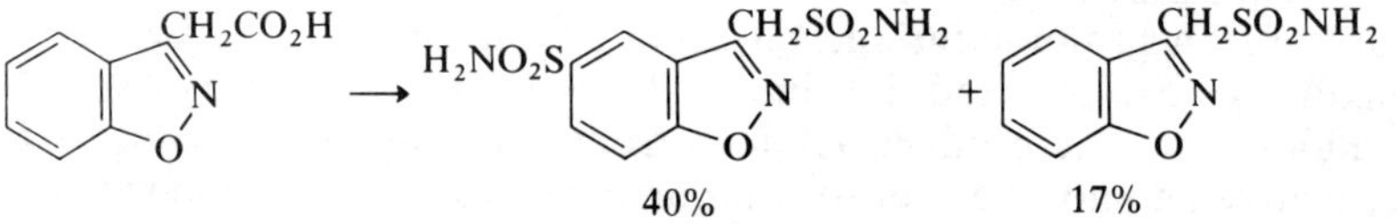

Scheme 26

(iv) Halogenation

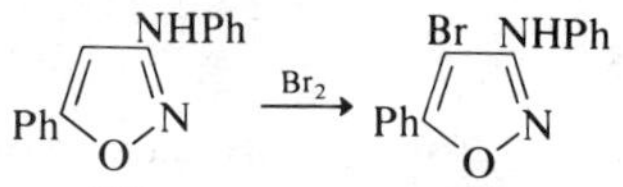

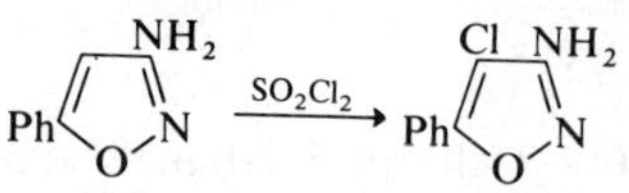

Treatment of isoxazoles with chlorine or bromine may give rise to coordination compounds which on heating or exposure to light give 4-chloro- or 4-bromo-isoxazoles. Thus, the chlorination and bromination of 3-methyl-, 5-methyl- and 3,5-dimethyl-isoxazoles gave the corresponding 4-chloro- and 4-bromo-isoxazoles. The bromination of unsubstituted isoxazole gave 4-bromoisoxazole. In the case of 3- and 5-phenylisoxazoles, the reaction with chlorine and bromine occurred at the 4-position of the isoxazole ring in preference to the phenyl ring. In agreement with these observations, 3-anilino-5-phenylisoxazole (**67**) was brominated extremely rapidly to 3-anilino-4-bromo-5-phenylisoxazole (**68**) ⟨63AHC(2)365⟩, and 3-amino-5-phenylisoxazole (**69**) was chlorinated with sulfuryl chloride to 3-amino-4-chloro-5-phenylisoxazole (**70**) ⟨77JMC934⟩. Recently, the second order bromination kinetics of alkyl- and phenyl-isoxazoles in 85% acetic acid at 150 °C have been reported. The rates of the bromination were 10^2–10^4 times that for bromination of benzene ⟨79KGS746⟩.

Neighboring group participations in electrophilic bromination of isoxazoles have recently been reported. Bromination of the isoxazolylpropanol (**71**) with bromine in pyridine gave 95% of spiro[furan-2,5-isoxazole] (**72**) ⟨74IZV2651⟩. 5-Phenacyl-3-phenylisoxazole (*E*)-oxime (**73**) with *N*-bromosuccinimide gave 59% of the spirobromide (**74**) ⟨76JCS(P1)619⟩. *N*-(3-Phenylisoxazol-5-ylmethyl)benzamide (**75**) with NBS gave the spirobromide (**76**), and 3-(3-phenylisoxazol-5-yl)propionic acid (**77**) with bromine gave the spirobromide (**78**) ⟨76JCS(P1)1694⟩.

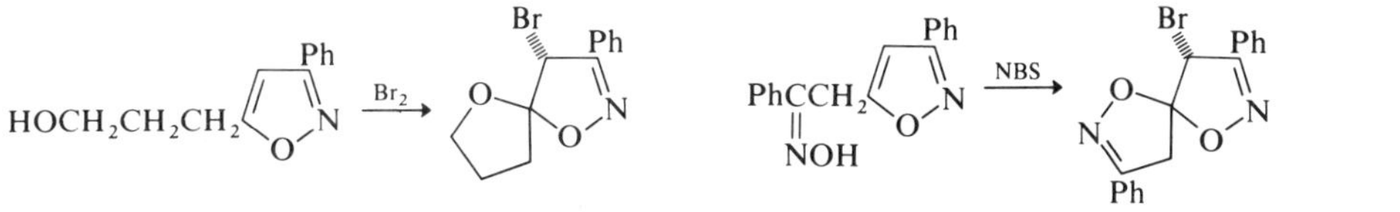

A good general method for iodination or bromination of isoxazoles is to use iodine or bromine in concentrated nitric acid. For example, the iodination of 3,5-dimethylisoxazole (**53**) with iodine in nitric acid gave 4-iodo-3,5-dimethylisoxazole (**79**; X = I) in 85% yield ⟨60DOK598⟩.

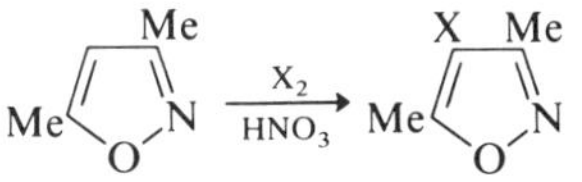

Bromination of 1,2-benzisoxazoles gave primarily the 5-bromo derivative, although other isomers have also been reported ⟨67AHC(8)277⟩. Bromination of 5-methoxy-3-methyl-1,2-benzisoxazole at room temperature gave a mixture of the 4- and 6-bromo compounds, while at elevated temperatures a 4,6-dibromo compound was produced (Scheme 27) ⟨79IJC(B)371⟩.

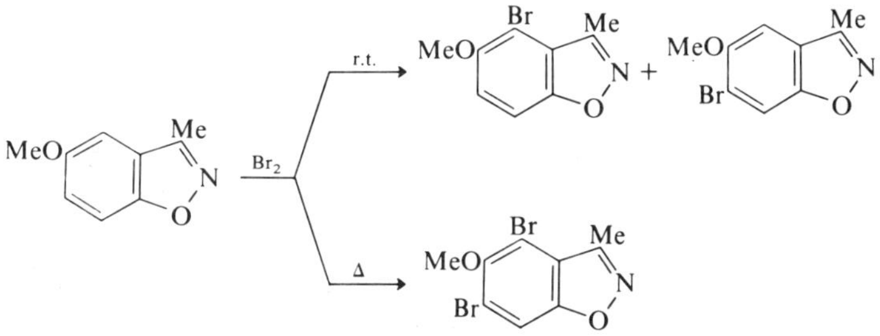

Scheme 27

Reduction of 3,5-dimethyl-7-nitro-1,2-benzisoxazole with SnCl$_2$ and HCl produced both the 7-amino compound and a 4-chloro-7-amino derivative (Scheme 28) ⟨67AHC(8)277⟩.

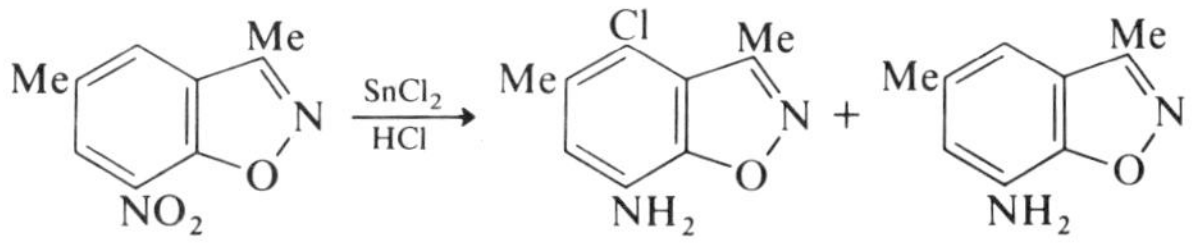

Scheme 28

In 2,1-benzisoxazoles, halogenation with Cl$_2$ or Br$_2$ leads to substitution in the 5-position, although the reaction appears to behave as a simple addition across a double bond. A

4,5-dichloro-4,5-dihydro analogue has been isolated and shown to undergo dehydrohalogenation to the expected product ⟨67AHC(8)277, 66T(S7)49⟩.

A similar reaction occurred when (**80**) was treated with sodium hypochlorite in methanol to give a derivative (**81**) containing the elements of ClOMe with loss of NO_2 ⟨66DIS(B)102⟩.

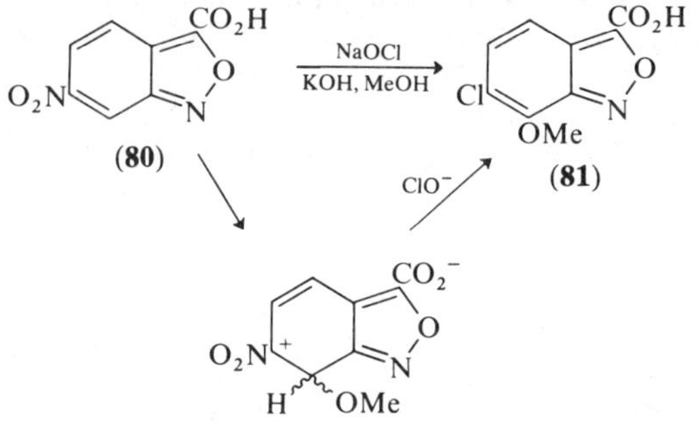

(v) Chloroalkylation and hydroxymethylation

The chloromethylation of isoxazoles (**82**) results in the formation of 4-chloromethyl derivatives (**83**), and their yield decreases in the following order: 5-phenyl > 3,5-dimethyl > 5-methyl > 3-methyl > H (the parent isoxazole). 3,5-Dimethylisoxazole (**53**) also reacts with benzaldehyde in the presence of hydrogen chloride to give 4-(chlorobenzyl)isoxazole (**84**), and with formaldehyde in the presence of sulfuric acid it yields 4-hydroxymethylisoxazole (**85**). It is noteworthy that pyridine and its homologs, often compared with isoxazoles, do not undergo chloromethylation and hydroxymethylation ⟨63AHC(2)365⟩.

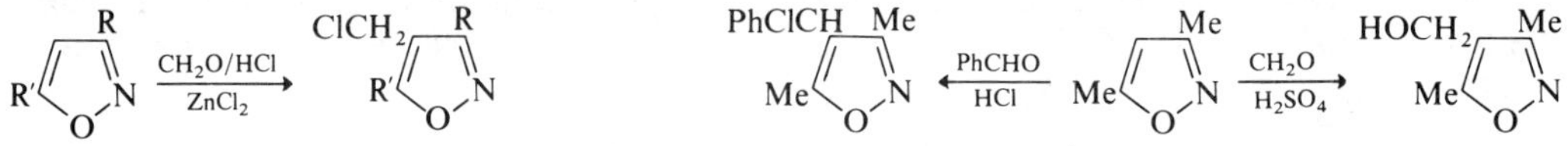

(vi) Vilsmeier–Haak formylation

Vilsmeier–Haak formylation of 5-amino-3-phenylisoxazole (**86**) gave the aldehyde (**87**), which is a useful intermediate for the synthesis of isoxazolopyridines and isoxazolopyrimidines ⟨77H(7)51⟩.

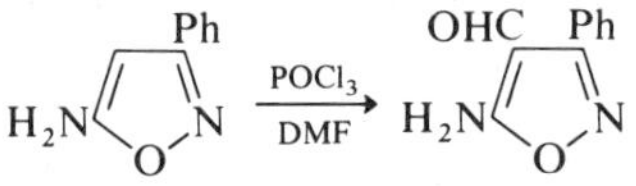

(vii) Mercuration

Electrophilic mercuration of isoxazoles parallels that of pyridine and other azole derivatives. The reaction of 3,5-disubstituted isoxazoles with mercury(II) acetate results in a very high yield of 4-acetoxymercury derivatives which can be converted into 4-bromoisoxazoles. Thus, the reaction of 5-phenylisoxazole (**64**) with mercury(II) acetate gave mercuriacetate (**88**) (in 90% yield), which after treatment with potassium bromide and bromine gave 4-bromo-5-phenylisoxazole (**89**) in 65% yield. The unsubstituted isoxazole, however, is oxidized under the same reaction conditions, giving mercury(I) salts.

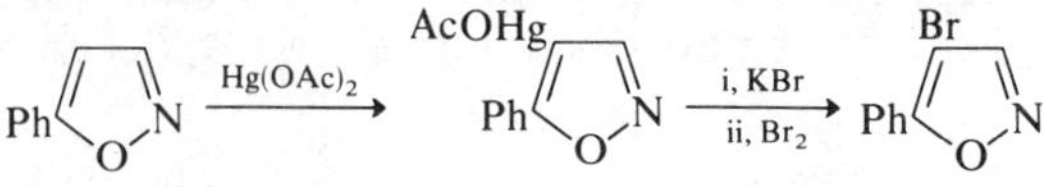

(viii) Oxidation

Isoxazoles are stable toward peracids and are fairly stable to other acidic oxidizing agents such as chromic and nitric acids, and acidic permanganate. 3-Substituted isoxazoles are rather stable to alkaline oxidizing agents, whereas 3-unsubstituted ones are easily attacked and rearrange to cyanoketones. Substituted isoxazoles can be ozonolyzed with ring cleavage to give α-ketooxime esters ⟨63AHC(2)365, 79AHC(25)147, p. 149⟩.

No example of the preparation of isoxazolecarboxylic acids by the oxidation of methyl- or alkyl-isoxazoles has been reported in the literature. However, isoxazoles containing unsaturated side chains or oxygenated functions are easily transformed into the corresponding carboxylic acids on treatment with different oxidizing agents. Peracetic acid has been used, for example, in the oxidation of 3-methoxymethyl-5-methylisoxazole (90) to the corresponding isoxazole-3-carboxylic acid (91). Chromic acid has been used in the oxidation of 3-acetyl-4-chloro-5-methylisoxazole (92) to the corresponding isoxazole-3-carboxylic acid (93), and alkaline permanganate results in the oxidation of 3,5-dimethyl-4-vinyl-isoxazole (94) to the corresponding isoxazole-4-carboxylic acid (95) ⟨76JCS(P1)570⟩, the oxidation of 5-methyl-3-phenylisoxazole-4-carbaldehyde (96) to the isoxazole-4-carboxylic acid (97), and the oxidation of 3-chloro-5-hydroxymethylisoxazole (98) and 3-chloro-5-acetylisoxazole (99) to the 3-chloro-isoxazole-5-carboxylic acid (100). Nitric acid has been used for the oxidation of 3-acetyl-4,5-dimethylisoxazole to the isoxazole-3-carboxylic acid, and acidic permanganate has been used in the oxidation of 3-isopropenylisoxazole to 3-acetylisoxazole ⟨62HC(17)1, p. 43⟩.

The stability of various heterocycles can be also compared using oxidation procedures. Thus, the oxidation of the heterocycles in Scheme 29 with potassium permanganate showed that under these reaction conditions the isoxazole ring is more stable than the furan ring but less stable than the pyrazole and furazan rings.

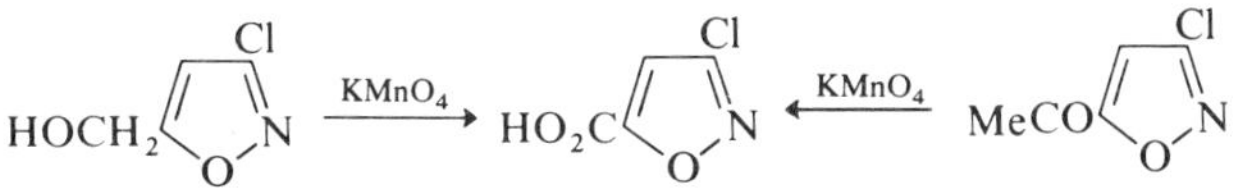

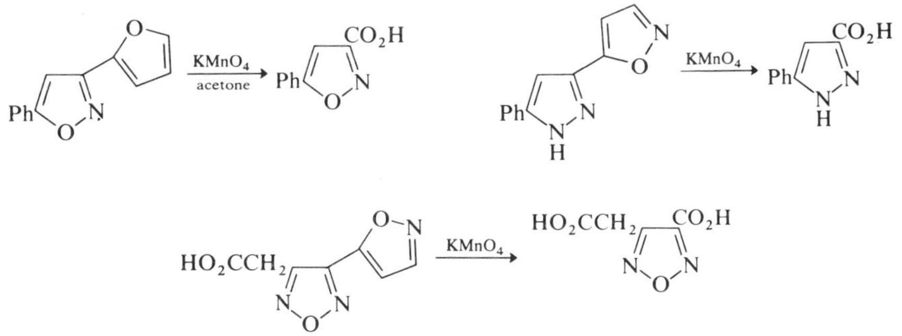

Scheme 29

Treatment of benzo derivatives with oxidizing agents leads to less predictable results. Thus, substituted 2,1-benzisoxazoles with nitrous acid or with $CrO_3/AcOH$ generated a variety of ring-opened products of higher oxidation state, the ratio of which depended on the amount of oxidant. These reactions are illustrated in Scheme 30.

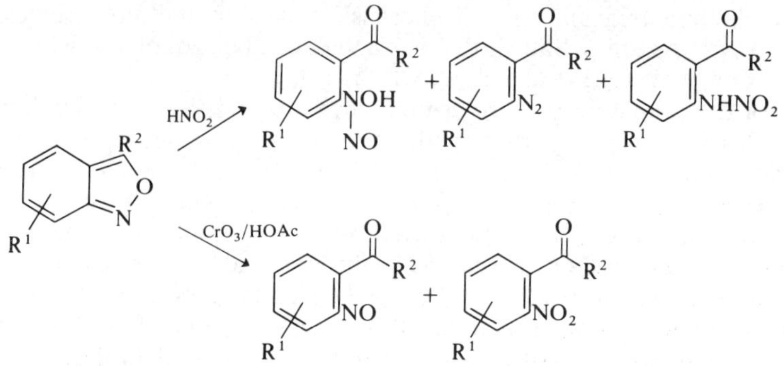

4.16.3.1.5 Nucleophilic attack at carbon

(i) Neutral isoxazoles

The action of nucleophilic reagents with isoxazoles can take a number of courses involving (i) nucleophilic addition to the ring; (ii) nucleophilic replacement of a substituent; and (iii) deprotonation. Other processes such as thermal or photochemical reactions may precede reaction with a nucleophile (see Section 4.16.3.1.2).

Nucleophilic replacement of hydrogen on an isoxazole is unknown and replacement of substituents is discussed in Section 4.16.3.3. In this series it is difficult to identify reactions involving addition to the ring as, in many instances, they are rapidly followed by elimination or ring cleavage sequences.

1,2-Benzisoxazoles are unstable in the presence of base, especially when unsubstituted in the 3-position ⟨67AHC(8)277⟩, and ring-opening reactions predominate (see Section 4.16.3.1.6). Sodium borohydride and lithium aluminum hydride both result in ring-opening reactions occurring with 1,2-benzisoxazoles ⟨81AHC(29)1, p. 15⟩. However, it may be inferred from their method of formation that many 3-substituted 1,2-benzisoxazoles are stable to a variety of nucleophiles as they are prepared by nucleophilic displacement of a 3-chloro substituent under relatively vigorous conditions. Of interest in this context is the LAH reduction of 1,2-benzisoxazolyl-3-acetic acid to the corresponding alcohol. No ring-opened products were observed (Scheme 31).

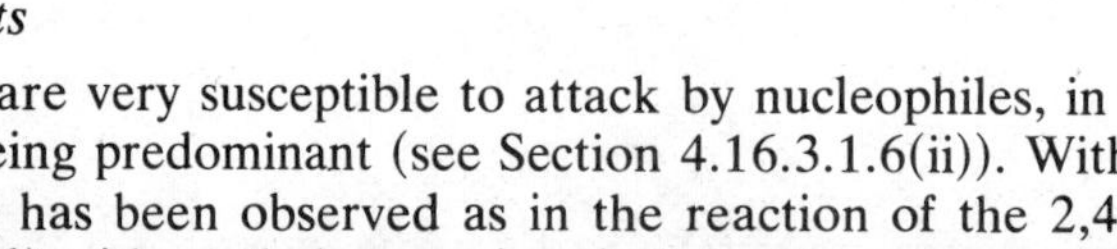

Scheme 31

Similar instability is found for 3-unsubstituted 2,1-benzisoxazole in the presence of base, ring opening to anthranilic acid derivatives occurring readily (see Section 4.16.3.1.6).

(ii) Isoxazolium salts

Isoxazolium salts are very susceptible to attack by nucleophiles, in many instances ring opening reactions being predominant (see Section 4.16.3.1.6(ii)). With some nucleophiles, addition to the ring has been observed as in the reaction of the 2,4-dimethyl-5-phenyl-isoxazolium salt (**101**) with methylmagnesium iodide. The adduct (**102**) together with the ring-opened product (**103**) were obtained in 41.4% and 21.3% yield, respectively ⟨69TL4875⟩. With benzyl- and allyl-magnesium halides the adducts corresponding to (**102**) were obtained in high yields. With a blocked 3-position and a free 5-position, nucleophilic attack occurred at the latter leading to the Δ^3-isoxazoline (**104**) ⟨69CPB2201, 74CPB61⟩.

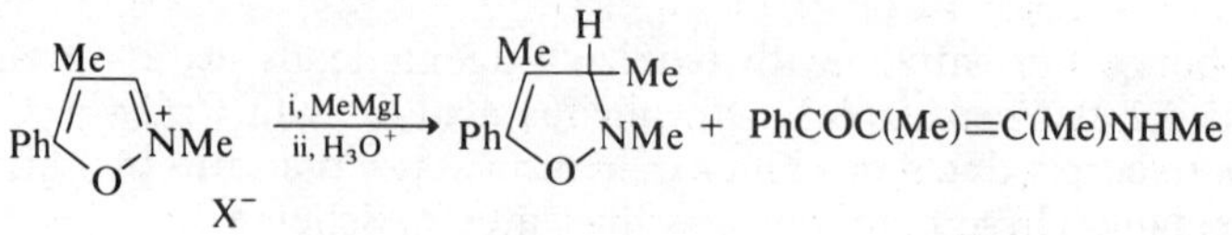

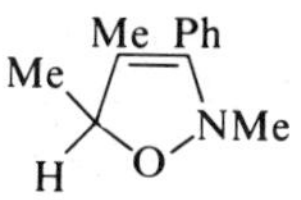

(104) 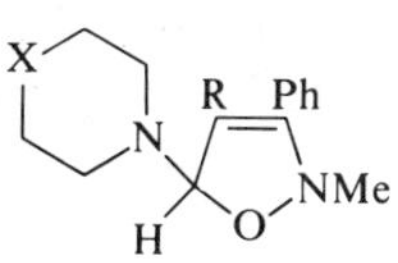   **(105)** X = CH$_2$, O

Sodium borohydride and 3-isoxazolium salts with a 3-unsubstituted position also give isoxazolines, as do the 3-substituted 5-unsubstituted derivatives. With the latter group, further reduction occurs to the isoxazolidines ⟨74CPB70⟩.

Although most nucleophiles (*e.g.* ethoxide, hydroxide and amines) result in ring-cleavage products, in the case of piperidine and morpholine the intermediate adducts (**105**) have been isolated at low temperatures. It is an interesting speculation whether ring cleavage occurs *via* these intermediates or whether an initial proton abstraction is involved ⟨69CPB2201, 74CPB61⟩.

1,2-Benzisoxazolium salts with nucleophiles undergo a facile ring opening to salicylnitrile derivatives (see Section 4.16.3.1.7(ii)). However, with appropriate nucleophiles the structure of the products obtained indicate that ring opening could have been preceded by nucleophilic addition to the ring. Thus, the reaction of the 2-methyl-1,2-benzisoxazolium salt with NaOMe in methanol gave the imino ether which on hydrolysis formed the salicylate shown in Scheme 32 ⟨74HCA376⟩. However, proton abstraction followed by methoxide ion addition to the nitrilium salt cannot be ignored as a possible reaction pathway (see Section 4.16.3.1.7(ii)).

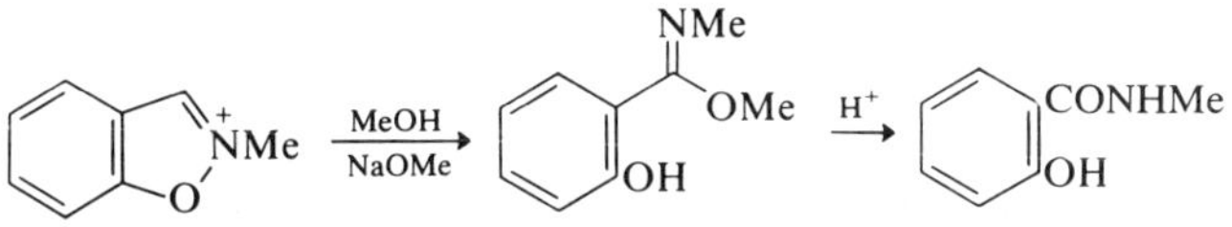

Scheme 32

2,1-Benzisoxazolium salts are especially reactive towards nucleophilic attack at the 3-position and, in many instances, a variety of stable adducts have been obtained with nucleophiles such as cyanide ion, malonate ion, amines and methoxide ion ⟨78JOC1233, 71JA1543, 72CPB2209⟩. Adducts obtained with isocyanate ion and the azide ion are unstable and undergo interesting rearrangements ⟨71JA1543⟩.

4.16.3.1.6 Nucleophilic attack at hydrogen

(i) Cleavage of neutral isoxazoles with nucleophiles

The importance of this group of reactions to the chemistry of isoxazoles is shown by the considerable amount of effort expended on this topic ⟨63AHC(2)365, 79AHC(25)147⟩. The lability of the isoxazole nucleus towards nucleophiles and bases distinguishes this heterocycle from other azoles. The conditions which lead to ring cleavage are quite varied and depend on the position and the nature of the substituents.

The reaction of 3-unsubstituted isoxazoles with bases, leading to the ring-opened products, has been known for more than 90 years and is widely used in organic synthesis. The mechanism of the reactions with hydroxide ion has been investigated ⟨67G173, 67G185, 77JCS(P2)1121⟩. The reactions showed second-order kinetics (first order in base and in substrate). A primary deuterium isotope effect indicated that fission of the C(3)—H bond was rate determining. Furthermore, substituents in the 4- and 5-positions have a substantial effect on the rate of cleavage. In general, the stability of the isoxazole ring decreases as the electron-withdrawing properties of the substituents become more pronounced and the effect of a 4-phenyl group is greater than that of a 5-phenyl group. These observations provide a basis for the assumption that the described reaction proceeds *via* a one-step concerted *E*2 type mechanism with a transition state (**106**) which resembles the cyanoenol anion (**107**) with localization of the partial negative charge on the oxygen atom of the isoxazole ring. In most of the reported reactions the cyanoenol anion (**107**) was protonated and underwent rearrangement to the β-ketonitrile (**108**), which may or may not be followed by further reactions. At low temperatures (−40 °C in the case of isoxazole itself) the *cis* geometry of the enolate is maintained, enabling valuable *cis*-cyanoenol derivatives such as (**109**) to be prepared in high yield. The reaction is also effected by strong bases such as alcoholates, sodium amide, lithium diisopropylamide and *n*-butyllithium or by weaker bases such as ammonia and phenylhydrazine at a higher temperature.

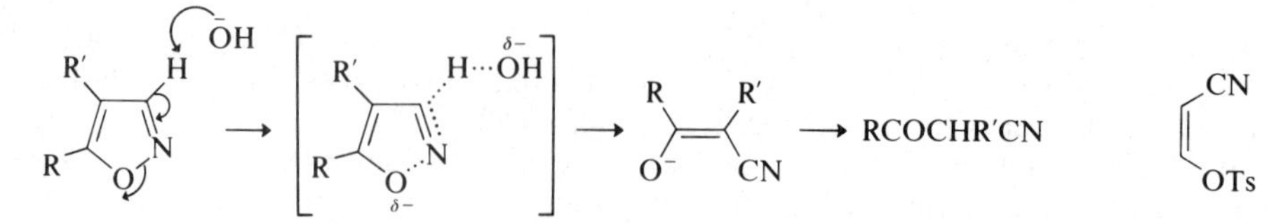

R = R' = alkyl, aryl   (**106**)   (**107**)   (**108**)   (**109**)

The importance of this ring-opening reaction is illustrated by the following. 5-Methyl-isoxazole (**110**) on treatment with ammonia was partly converted into cyanoacetoneimine (**111**), and when refluxed with phenylhydrazine it yielded the 5-aminopyrazole (**113**) through the intermediate cyanoacetone (**112**) ⟨63AHC(2)365⟩.

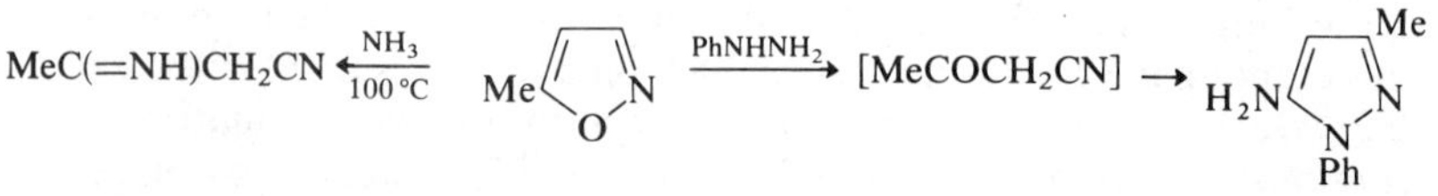

$$MeC(=NH)CH_2CN \xleftarrow{NH_3}_{100\,°C} \text{Me-isoxazole} \xrightarrow{PhNHNH_2} [MeCOCH_2CN] \rightarrow \text{5-aminopyrazole}$$

(**111**)   (**110**)   (**112**)   (**113**)

The ring opening of 3-substituted isoxazoles proceeds differently, and the reaction can take various courses depending on the nature of the substituent. The reaction has been effected by sodium hydroxide and sodium ethoxide in alcoholic or aqueous media and by sodium amide and also *n*-butyllithium in inert solvents.

5-Unsubstituted 3-alkyl- or 3-aryl-isoxazoles undergo ring cleavage reactions under more vigorous conditions. In these substrates the deprotonation of the H-5 proton is concurrent with fission of the N—O and C(3)—C(4) bonds, giving a nitrile and an ethynolate anion. The latter is usually hydrolyzed on work-up to a carboxylic acid, but can be trapped at low temperature. As shown by Scheme 33, such reactions could provide useful syntheses of ketenes and β-lactones ⟨79LA219⟩.

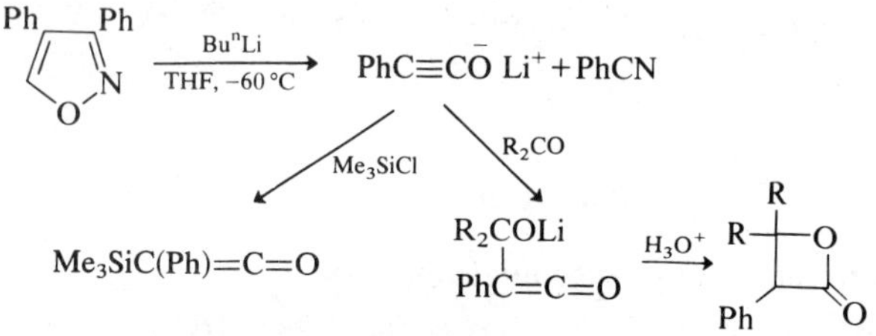

$$\text{3,5-diphenylisoxazole} \xrightarrow[THF, -60\,°C]{Bu^nLi} PhC\equiv\bar{C}O\ Li^+ + PhCN$$

$$Me_3SiC(Ph)=C=O \xleftarrow{Me_3SiCl} \quad \xrightarrow{R_2CO} \begin{array}{c} R_2COLi \\ | \\ PhC=C=O \end{array} \xrightarrow{H_3O^+} \text{β-lactone}$$

Scheme 33

If the substituent at the 3-position is a group that can be eliminated as an anion (such as Cl⁻, CN⁻ and N₃⁻), the reaction proceeds without the cleavage of the C(3)—C(4) bond in the isoxazole ring and involves the ejection of the 3-substituent as an anion. For example, 3-cyanoisoxazole (**114**) reacted with sodium ethoxide at room temperature to give ethyl cyanoacetate (**115**) *via* an intermediate cyanoketene ⟨32G436⟩.

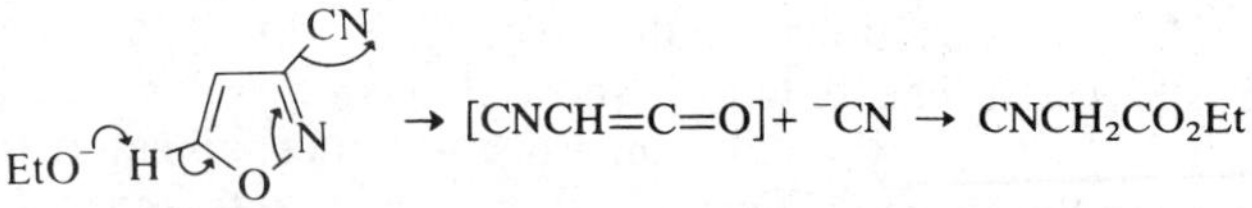

EtO⁻ H   (**114**) → [CNCH=C=O] + ⁻CN → CNCH₂CO₂Et   (**115**)

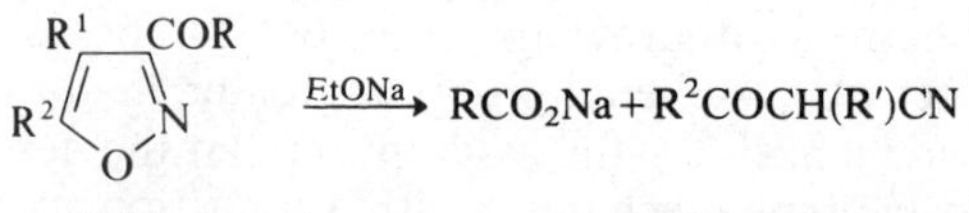

(**116**) $\xrightarrow{EtONa}$ RCO₂Na + R²COCH(R')CN   (**117**)

It should be noted that when 3-acylisoxazoles (**116**) are heated with a base such as sodium ethoxide, the acyl group is eliminated to give a carboxylate salt and a β-ketonitrile (**117**). The reaction probably occurs *via* initial attack at the carbonyl group ⟨46G206⟩.

Equivalent mechanistically to the above ring opening of 3-unsubstituted isoxazoles is that occurring with isoxazole-3-carboxylic acids. Decarboxylation of 5-arylisoxazole-3-carboxylic acids (**118**) occurs at temperatures above 200 °C and is accompanied by the cleavage of the N—O bond of the isoxazole ring to yield aroylacetonitriles (**119**). The reaction is facilitated in the presence of arylhydrazines, yielding 5-aminopyrazoles (**120**) in 80–100% yield ⟨58G879⟩. It has been postulated that on heating 5-aryl-3-carboxylic acids (**118**) in the presence of arylhydrazines, it is the anion (**118a**) which is being decarboxylated, and the degradation then follows the path described above for the nucleophilic cleavage of 3-unsubstituted isoxazoles.

1,2-Benzisoxazoles unsubstituted in the 3-position were initially thought to undergo ring opening by an initial abstraction of the 3-proton. However, mechanistic studies of the ring opening of 5-, 6- and 7-substituted 1,2-benzisoxazoles by hydroxide ion and amines show that ring cleavage takes place by a concerted $E2$ elimination, (**121**)→(**122**) ⟨81AHC(29)1, p. 13⟩.

Compound (**122**) is also obtained by decarboxylative ring-opening of 1,2-benzisoxazole-3-carboxylic acid. It has also been concluded that the reaction involves an intermediateless, concerted loss of carbon dioxide *via* a transition state in which the negative charge is spread over the carboxyl group and the isoxazole ring.

3-Unsubstituted 2,1-benzisoxazoles undergo C(3)-proton abstraction with base to give an intermediate iminoketene which can undergo further reaction with nucleophiles. However, alternative Michael addition pathways are possible and these have been discussed ⟨81AHC(29)1, p. 56⟩.

(ii) Cleavage of isoxazolium salts with nucleophiles

Quaternization of the nitrogen atom increases the reactivity of the isoxazole ring significantly with respect to nucleophiles and bases. It was discovered early this century that 3-unsubstituted isoxazolium salts were cleaved by a variety of mild nucleophilic reagents, including even carboxylate ions in aqueous solution ⟨63AHC(2)365⟩. However, it was not until much later that Woodward and Olofson were able to recognize the significance of this observation and employed 3-unsubstituted isoxazolium salts as coupling reagents for a 'one-pot' synthesis of peptides ⟨66T(S7)415, 79AHC(25)147⟩.

The reaction of 3-unsubstituted isoxazolium salts (**123**) in the peptide synthesis is generalized in Scheme 34. Deprotonation at the 3-position is rapidly followed by ring opening to give the ketoketenimine (**124**). Ketoketenimines have been detected spectroscopically and are sufficiently stable for isolation in some cases. Reaction of carboxylic acid (or carboxylate) with the ketoketenimine (**124**) is followed by transacylation, giving an enol ester (**125**). The enol ester (**125**) is then attacked by a nucleophile. If the nucleophile is an amino acid, the final product is a peptide (**126**) and β-ketoamide by-product (**127**). Some examples of the isoxazolium ions which have been tried as peptide coupling reagents are compounds (**128**)–(**131**). Of these the sulfonium inner salt (**128**) (Woodward's reagent K) is widely used because the by-product (**127**) is water soluble, and it is superior to diimides as far as racemization is concerned. The isoxazolium ions (**129**)–(**131**) are unsatisfactory because of a number of competing side reactions.

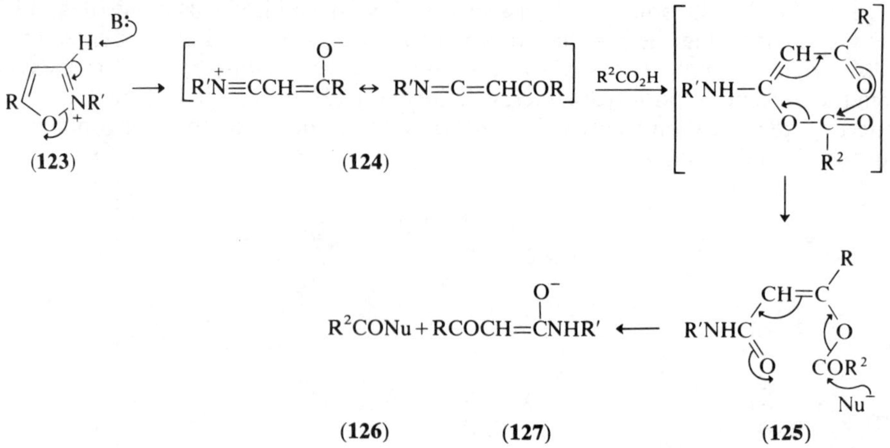

Scheme 34

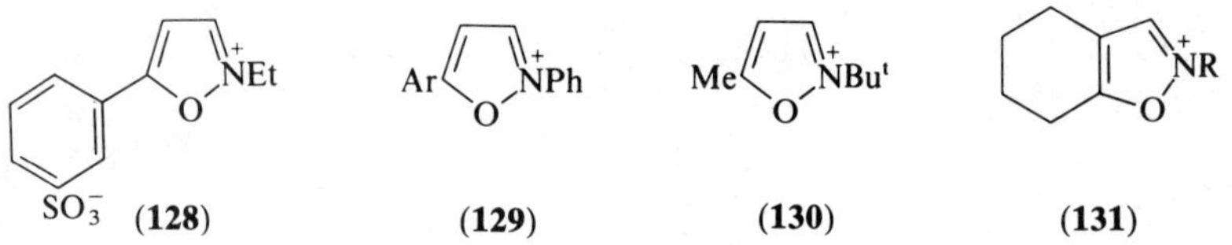

The reactions of 3-unsubstituted isoxazolium salts (**123**) with hydroxide, alkoxide, cyanide and azide ions have also been studied, and they can in general be rationalized in terms of the ketoketenimine intermediate (**124**). The results of these reactions are summarized below. The application of such reactions to 3-unsubstituted isoxazolium salts bearing substituents other than alkyl and aryl groups has received little attention, but 5-aminoisoxazolium salts have been studied ⟨74CB13⟩.

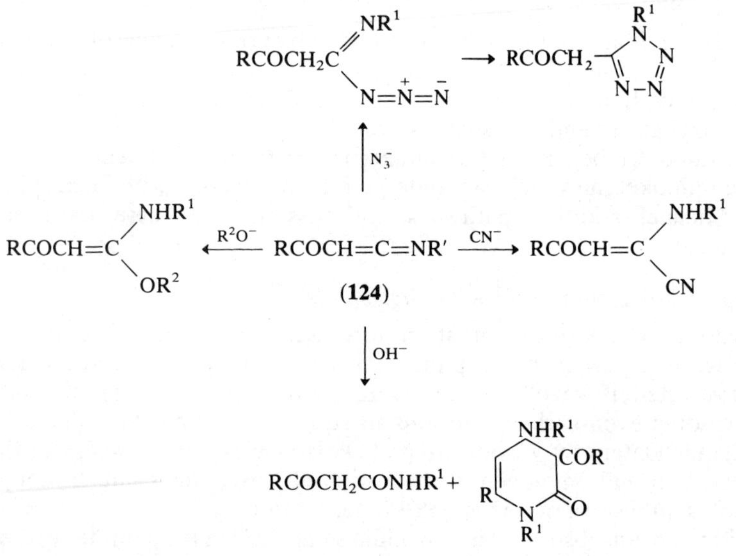

N-Ethyl-1,2-benzisoxazolium salts (**132**) also undergo a facile, general base-catalyzed ring opening to form the resonance stabilized, highly electrophilic N-ethylbenzoketoketenimine intermediates (**133**) ⟨74T3677, 67T2001⟩. Reaction of the salts (**132**) with carboxylate ions to give stable phenolic esters has resulted in a new peptide coupling process, especially with (**132**; R = OH) ⟨74T3677⟩. Under optimum conditions (pH 4.5–5.0 with a >0.3M solution of carboxylate ion in aqueous pyridine) peptide esters are obtained cleanly and efficiently in almost quantitative yields ⟨67JA2743, 74T3955, 74T3955⟩.

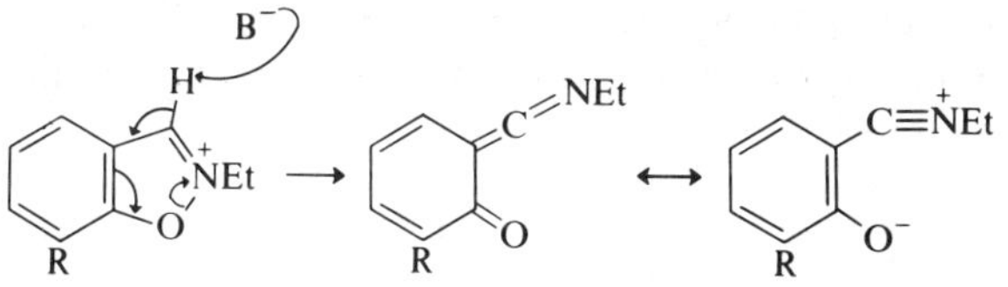

The 2,1-benzisoxazolium salts unsubstituted in the 3-position behave in an analogous manner to their 1,2-isomers above. Particularly interesting is the reaction of the salt (134) with Et₃N. Abstraction of the C(3)-proton is followed by ring opening to the iminoketene (135) which undergoes electrocyclization to its stable valence tautomer (136) in 84% yield ⟨71JA1543⟩.

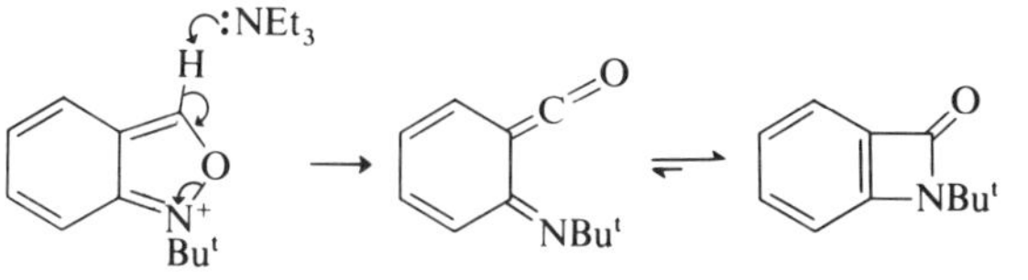

(iii) Reductive ring cleavage

Isoxazoles are readily cleaved under reducing conditions, and many examples have been reported ⟨79AHC(25)147, 63AHC(2)365⟩. In the last three decades the potential of these reactions in synthesis has finally been realized, and the isoxazole ring has become a major tool as a masked enaminone (137) or 1,3-diketone, particularly for the synthesis of heterocycles.

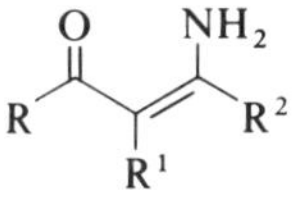

In the earlier work, various reducing agents were used ⟨63AHC(2)365⟩. Catalytic hydrogenation is now almost always employed, and some examples illustrating the high yields attainable and the compatibility with other functional groups are shown in Scheme 35.

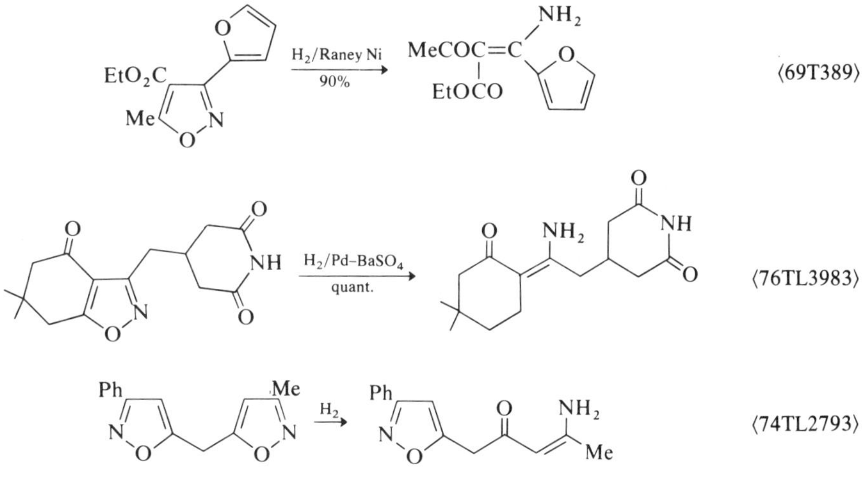

Scheme 35

The enaminones (137) thus obtained may then be transformed in a number of useful ways. Under mild conditions the hydrolysis of the enaminones (137) can give the corresponding 1,3-diketones. They may also be reduced to aminoketones which under the conditions used (sodium and *t*-butanol in liquid ammonia) lose ammonia to give enones. The same reducing system can, in fact, be used to convert an isoxazole into an enone in one step. The sequence is exemplified in Scheme 36, which illustrates an elegant method for transposing the substituents of a conjugated enone ⟨72JA9128⟩. If, alternatively, it is desired to protect and then regenerate a conjugated enone, the enaminoketone can first be benzoylated or acylated and then reduced with borohydride and hydrolyzed, as illustrated in Scheme 37 ⟨75JOC526⟩.

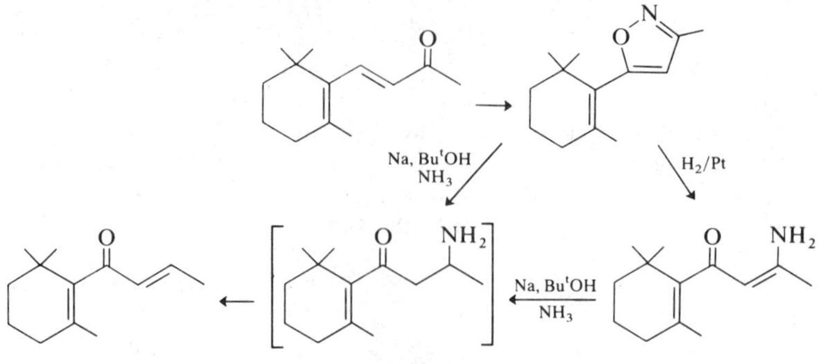

Scheme 36

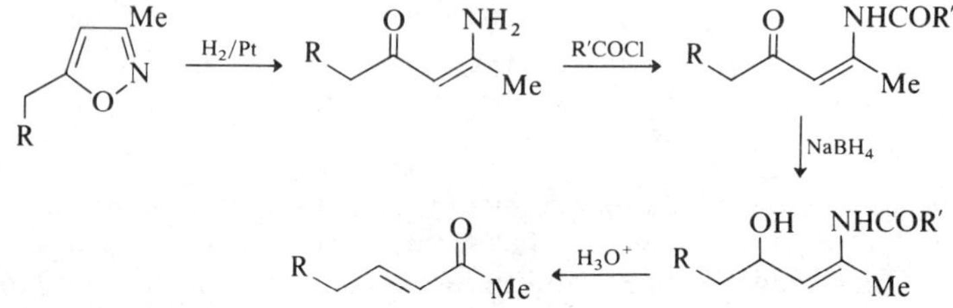

Scheme 37

Applications of the isoxazole ring as a masked enaminone for the synthesis of various heterocycles are shown in Scheme 38.

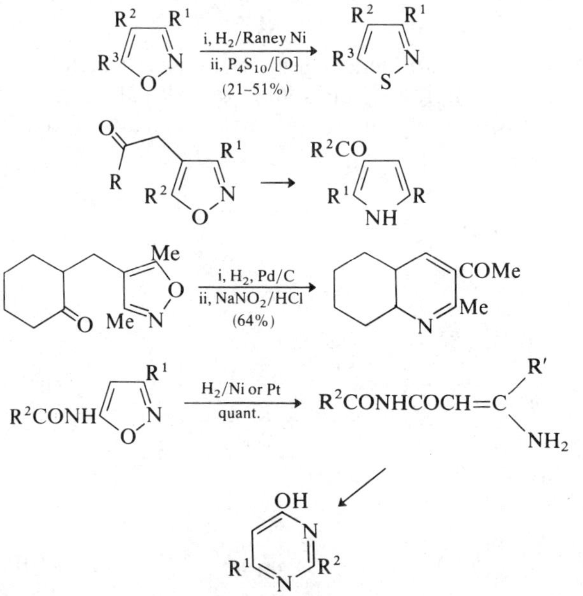

⟨69T389⟩

⟨79AHC(25)147⟩

⟨71JOC2784⟩

⟨54JCS665⟩

Scheme 38

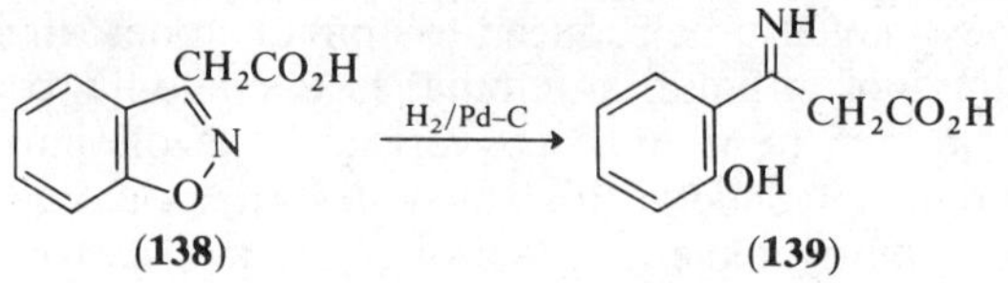

Reduction of 1,2-benzisoxazole derivatives under analogous conditions to those above results generally in ring cleavage taking place. Thus 1,2-benzoxazolyl-3-acetic acid (**138**) with Pd/C and hydrogen gives the imine (**139**). The corresponding nitrile under similar conditions gives the imino nitrile corresponding to (**139**). 3-Hydroxy-1,2-benzisoxazoles are converted by NaBH$_4$ at 80 °C into *o*-hydroxybenzamides ⟨81AHC(29)1, p. 15⟩.

Catalytic reductions using PdC, PtO$_2$ or Raney Ni as catalysts of 2,1-benzisoxazole derivatives results in ring fission and the formation of 2-aminophenones. Reductions are successful with a variety of chemical reducing agents including iron(II) sulfate, zinc or iron in HCl, iron in acetic acid, *etc.* ⟨81AHC(29)1, p. 49⟩. Exceptions to this easy hydrogenolysis are known and substituents are reduced rather than the isoxazole ring ⟨81AHC(29)1, p. 49⟩.

4.16.3.1.7 Reactions with cyclic transition states

Although cycloaddition reactions provide a convenient entry into a variety of isoxazole derivatives, the isoxazole nucleus itself does not behave as a diene in cycloadditions. This is in marked contrast to the behavior of oxazole derivatives which in several instances are active dienes (see Chapter 4.18). This difference in behavior is emphasized in the reaction of diphenylcyclopropenone with derivatives of the two ring systems. Isoxazoles apparently behave as nucleophiles, with nitrogen attacking the cyclopropenone ring to give, after several rearrangements, a pyridone ⟨73CC349⟩.

2,1-Benzisoxazole, however, behaves as a diene with dienophiles. The Diels–Alder addition of *N*-phenylmaleimide was reported to give a mixture of *endo* and *exo* cycloadducts, structural proof resting partially on the conversion of these compounds into acridinic acid by base ⟨58MI41602⟩. These products were later found to be the *exo* isomer and a ring-opened product (Scheme 39) ⟨67JOC1899⟩. Further heating of the *exo* adduct produced the ring-opened aldehyde. Heating of 2,1-benzisoxazole and *N*-phenylmaleimide in dry xylene for 20 hours produced (**140**) ⟨66DIS(B)102⟩. The cycloaddition of dimethyl acetylenedicarboxylate to 2,1-benzisoxazole in refluxing benzene gave the quinoline 1-oxide (Scheme 40).

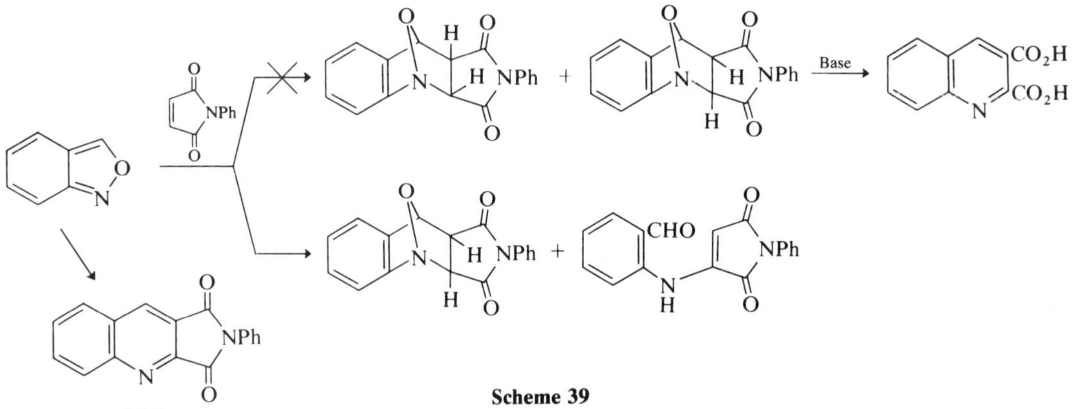

Scheme 39

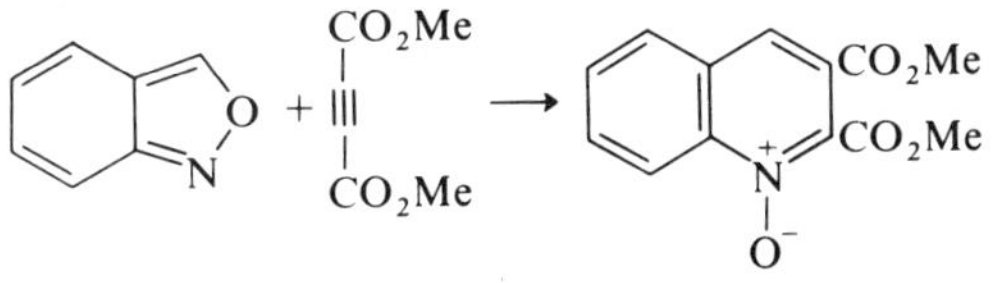

Scheme 40

The Diels–Alder addition of maleic anhydride gave a compound for which two structures were proposed (Scheme 41) ⟨66DIS(B)102, 43JCS654⟩.

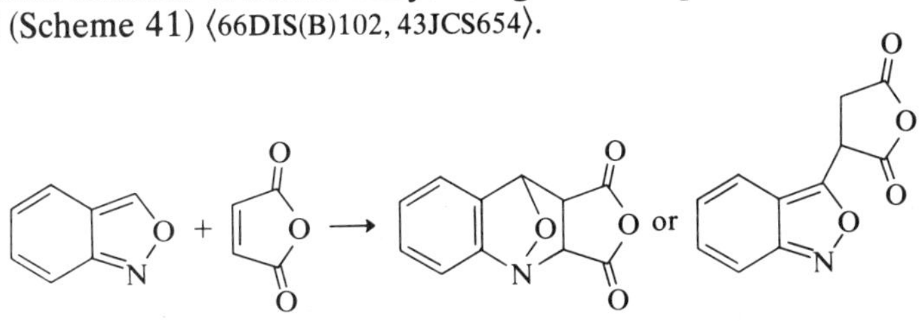

Scheme 41

1,2-Benzisoxazole and alicyclic ketones undergo a mercury sulfate-catalyzed cycloaddition in boiling xylene. The enol tautomers (**141**) are apparently involved and the yields of the 2,3-cycloalkanoquinolines (**142**) range from 11% ($n = 4$) to 39% ($n = 5$). Phenylacetylene and diethyl maleate yield similar products ⟨66LA(698)149⟩.

A particularly interesting cycloaddition of a 2,1-benzisoxazole is that where the mode of addition is reversed, *i.e.* the 2,1-benzisoxazole acts as a 2π-component with a 4π-electron 1,3-dipole. When 3-alkoxycarbonyl-6-nitro-2,1-benzisoxazoles (**143**) were treated with an excess of diazoacetic esters at 120 °C, the 6*H*-pyrazolo[3,4-*g*][2,1]benzisoxazoles (**144**) were obtained in good yields. When (**144**) was methylated with an excess of diazomethane, a mixture of the two isomeric *N*-methyl products was obtained. The 3-alkoxycarbonyl group was found to be essential for this cycloaddition to occur and replacement of the nitro substituent by a halo or alkyl group also resulted in no reaction occurring ⟨82S677⟩. This cycloaddition should be contrasted with the reaction of the corresponding 6-nitro compound (**143a**) and diazomethane. A similar tricyclic intermediate was postulated but in this instance loss of N_2 occurred to give the 7-methyl-6-nitro-2,1-benzisoxazole (**144a**) ⟨79JHC1557⟩.

4.16.3.2 Reactions of Non-aromatic Compounds

4.16.3.2.1 2-Isoxazolines

Since the last review of the field in 1962 ⟨62HC(17)1⟩, the majority of studies still deal with 2-isoxazolines (most of these possessing a 3-substituent) with a minor number for 4-isoxazolines and one study of 3-isoxazolines. The 2-isoxazoline ring system undergoes a variety of reactions including reductive ring opening, oxidative ring rupture and oxidation to isoxazoles. Photolytic and acid and base reactivities have also been documented.

(i) Reduction

Reductions with LiAlH$_4$ are the most common and provide a convenient route to amino alcohols, aziridines or both, depending on reaction conditions ⟨68G74, 80LA122, 80LA101, 70T539, 78TL3133, 68TL5789, 67NEP6606579⟩. The LiAlH$_4$ reduction of (**145**) produced two amino alcohols (Scheme 42) and the isoxazoline (**146**) also gave two products (Scheme 43) ⟨80LA101⟩. Modification in stoichiometry produced aziridines plus amino alcohols. The azidirines produced in this manner have a *cis* orientation, and these results point to an aziridine intermediate in the reduction of oximes in general. The use of LiAlD$_4$ provided a means of obtaining specific, deuterium labelled compounds (Scheme 44) ⟨70T539, 68TL5759⟩. Reduction with sodium amalgam produced a cyano alcohol ⟨79JA1319⟩, Zn/EtOH yielded ketones *via* intermediate oximes ⟨79AG91⟩, Zn/HOAc produced (**147**) ⟨72JCS(P1)437⟩, Pd/C or NaBH$_4$ reduction of 2-isoxazoline-3-carboxylic acid gave amino alcohol acids ⟨64CB159⟩ while, in contrast, NaBH$_3$CN produced isoxazolidines (Scheme 45) ⟨79JA1319⟩. The reduction of (**148**) gave steroidal intermediates (**149**) and (**150**) ⟨79DOK615⟩ and hydrogenation produced amino alcohols. These transformations are shown in Scheme 45 ⟨76KGS891⟩.

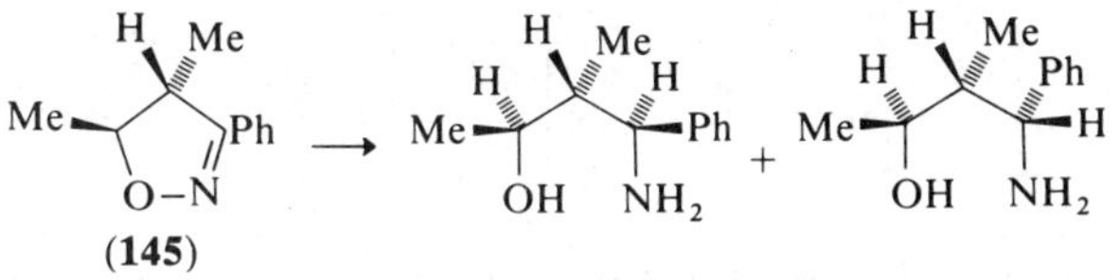

Scheme 42

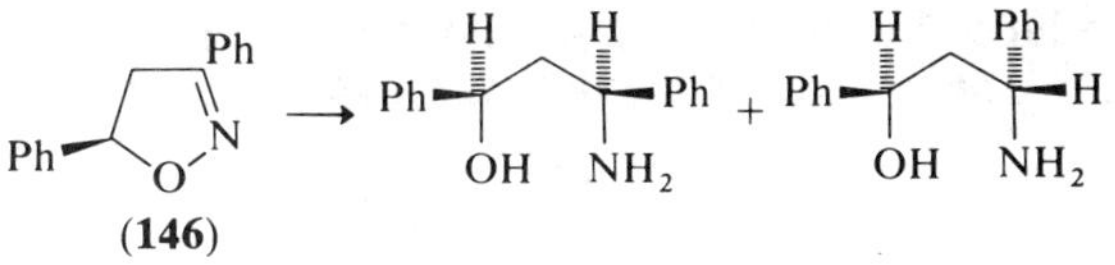

Scheme 43

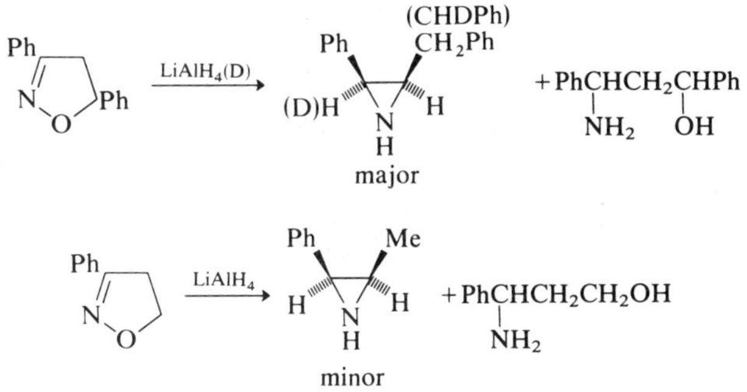

Scheme 44

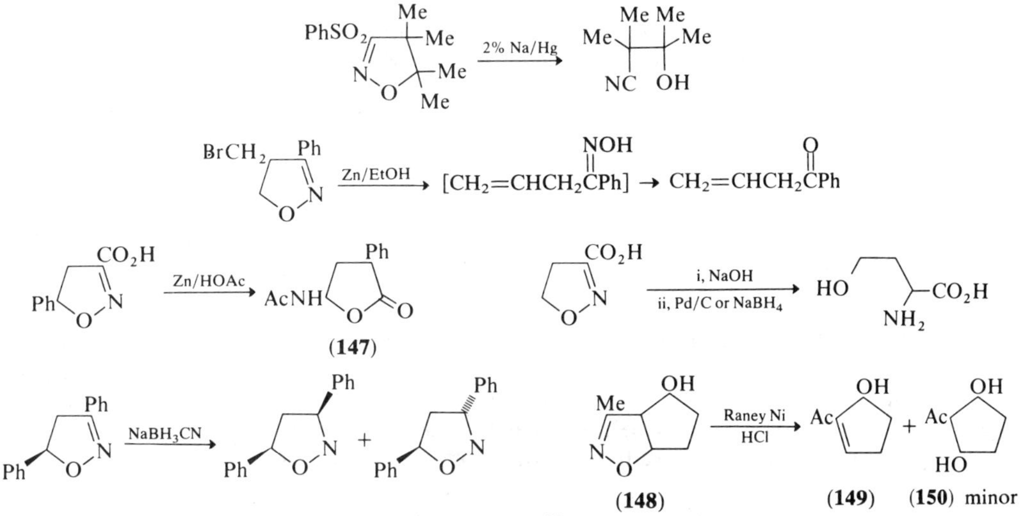

Scheme 45

Although 1,2-benzisoxazol-2-ines are known, their chemistry has received little attention to date ⟨81AHC(29)1, p. 29⟩.

(ii) Oxidation

Oxidation under mild conditions with a variety of agents such as NBS ⟨68G331, 80IJC(B)571⟩, SeO_2, $KMnO_4$ in acetone or CrO_3 in acetone ⟨60G347⟩, DDQ ⟨79JCR(S)311⟩ or $KMnO_4$ on the potassium salt of 4-nitro-3,5-diphenyl-2-isoxazoline produced isoxazoles ⟨79ZOR2436⟩. Oxidation under more forcing conditions produced isoxazole as well as ring rupture to cyano alcohols ⟨72JCS(P1)437⟩.

(iii) Photolysis

Photolysis in general produced oxazoles and a variety of other products including aminochalcones, nitriles, aldehydes and chalcone oximes. A number of photolytic intermediates have been postulated, represented by (**151**), (**152**), (**153**) and (**154**) ⟨77CL1195, 75T1373, 73HCA2588, 73TL2283⟩.

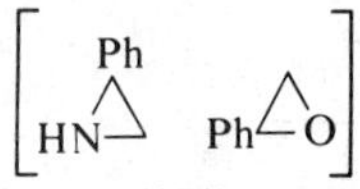

(151) 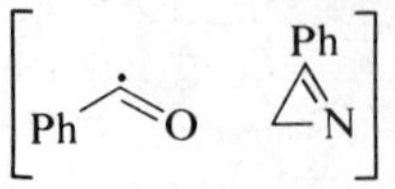 (152) 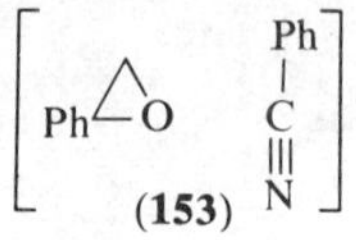 (153) 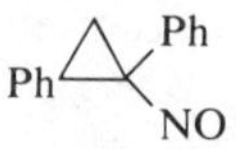 (154)

The photolysis of 3-(*p*-cyanophenyl)-2-isoxazoline in benzene produced a tricyclic product along with six other materials (Scheme 46) ⟨B-79MI41616⟩. Irradiation of the bicyclic 2-isoxazoline (155) produced benzonitrile, β-cyanonaphthalene and polymer *via* a proposed biradical intermediate (156) (Scheme 47) ⟨B-79MI41615⟩.

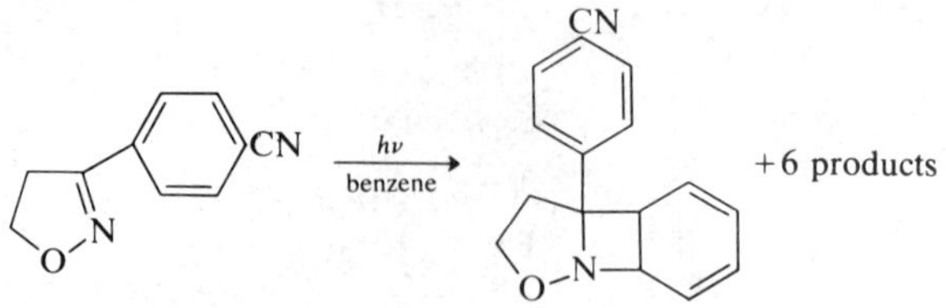

Scheme 46

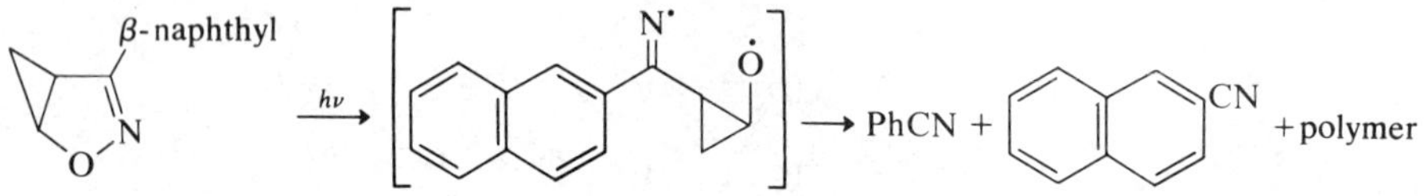

Scheme 47

(iv) Other reactions

The reaction of 3,5-diphenyl-2-isoxazoline with lithium diisopropylamide produced with 2 equivalents of base a chalcone oxime, while in the presence of 1 equivalent and an alkyl iodide, ring alkylation occurred at the 4-position of the nucleus (Scheme 48) ⟨80LA80, 78TL3129⟩.

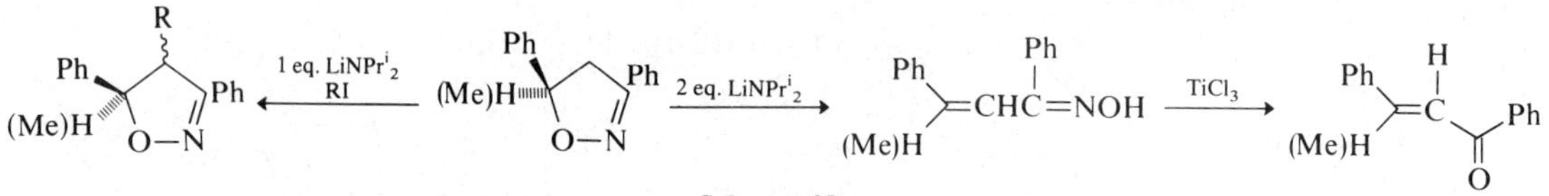

Scheme 48

The reaction of aqueous base and 5-acyl-3-phenyl-2-isoxazoline yielded benzonitrile and biacetyl (Scheme 49) ⟨68JHC49⟩. The base treatment of selected 2-isoxazolines containing a halomethyl group in the 4- or 5-positions resulted in the formation of bicyclo[3.1.0]systems (Scheme 50) ⟨79JCR(S)54, 77AJC1855⟩, whereas the base catalyzed reaction of (157) resulted in a major rearrangement to give the triazole (158) ⟨80GEP2815956⟩.

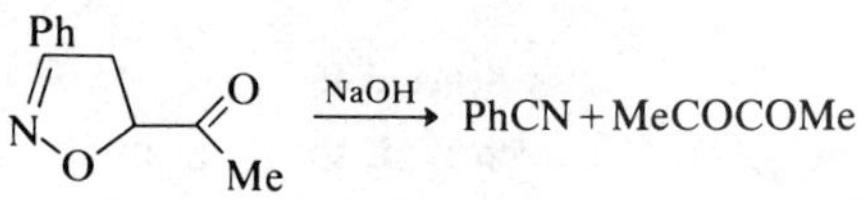

Scheme 49

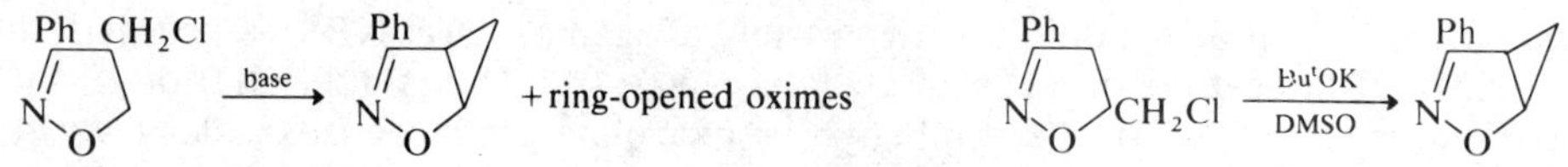

Scheme 50

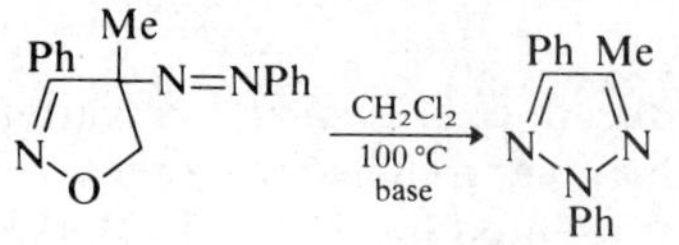

(157) (158)

Treatment of 2-isoxazolines with acid usually leads to ring rupture and formation of chalcone products ⟨62HC(17)1⟩, although 5-methyl-3-phenyl-2-isoxazoline forms a quaternary salt with dimethyl sulfate in the presence of perchloric acid (Scheme 51) ⟨73BSF1390⟩.

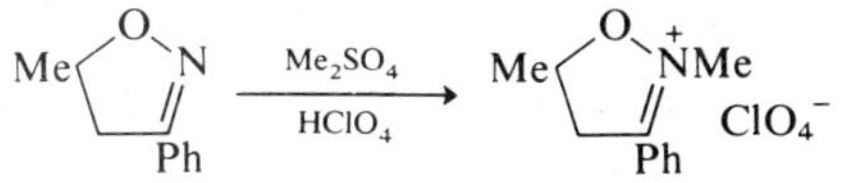

Scheme 51

The attempted nitration of 3-phenyl-2-isoxazoline yielded only nitration in the *para* position of the phenyl ring ⟨67ZOR1532⟩. The pyrolysis of a 2-isoxazolinepyruvate oxime (**157a**) produced three products (Scheme 52) ⟨77JHC523⟩, while heating of a steroidal isoxazoline-3-carboxylic acid (**158a**) also produced three products (Scheme 52) ⟨67JOC1387⟩. Heating of 3-phenyl-2-isoxazoline-5-isocyanate produced 3-phenylisoxazole and cyanuric acid (Scheme 53) ⟨72MI41616⟩. Elimination of other 5-substituents by thermolysis was also examined ⟨72MI41615⟩.

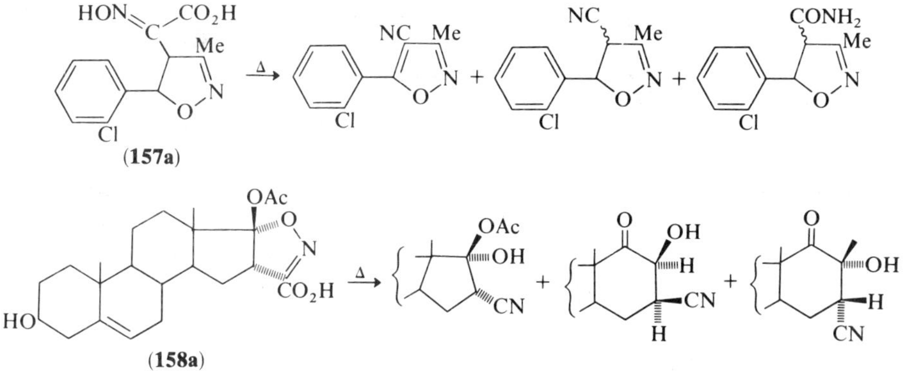

Scheme 52

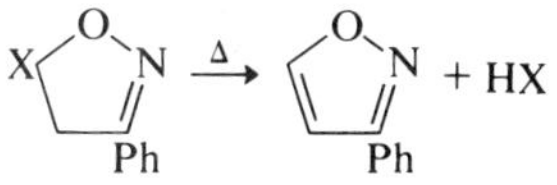

Scheme 53

The 3-substituents in 3-nitro- and 3-phenylsulfonyl-2-isoxazolines were displaced by a variety of nucleophiles including thiolate, cyanide and azide ions, ammonia, hydride ions and alkoxides. The reaction is pictured as an addition–elimination sequence (Scheme 54) ⟨72MI41605, 79JA1319, 78JOC2020⟩.

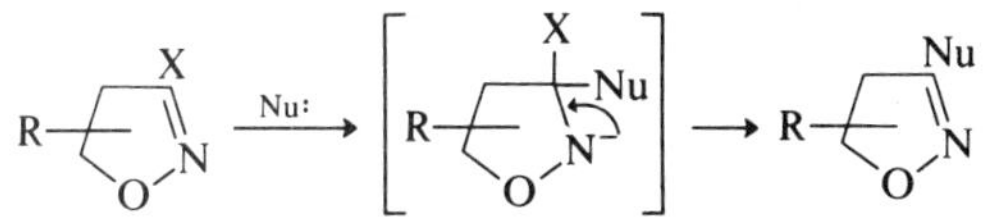

$X = NO_2, PhSO_2$

Scheme 54

A number of manipulations on isoxazolinone can be performed with the nucleus remaining intact, and a variety of reagents cleave the ring. Isoxazolin-5-ones have also been used as synthons for the production of other heterocycles ⟨76ZC270⟩.

Alkylation of, or Michael addition, by salts of isoxazolin-5-one produced *N*-, *O*- and/or *C*-alkylation ⟨62JOC2160, 73JCS(P1)2503, 69TL543, 74MI41609⟩. The reaction of the sodium salt of isoxazolin-5-one with dialkyl sulfate esters or tosyl esters gave *O*-alkylisoxazoles ⟨74MI41609⟩. Methyl iodide yielded an *N*-methylisoxazolin-5-one ⟨62JOC2160⟩. The Michael addition to acrylonitrile produced an *N*-alkyl product. Allyl bromide ⟨73JCS(P1)2503⟩ and the parent produced an *N*-alkylated product. The reaction of allyl bromides and the sodium salt of 3-methyl-4-phenylisoxazolin-5-one gave *N*- and *C*(4)-alkylation in a 2:1 ratio. Heating the mixture in an amino-Claisen reaction changed the ratio to 1:99 (Scheme 55) ⟨69TL543⟩.

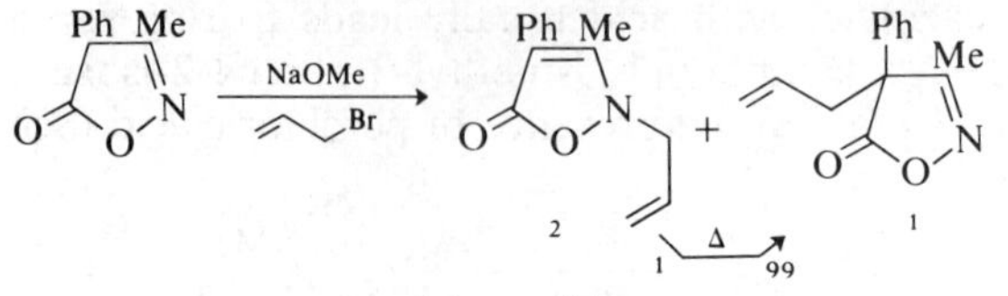

Scheme 55

The reaction of 4,4-dialkyl substituted isoxazolin-5-ones with Grignard reagents or NaBH$_4$ produced 5-ols. Without disubstitution in the 4-position, ring cleavage occurred ⟨73BSF3079⟩. The 4-position in unsubstituted isoxazolin-5-ones is highly reactive and undergoes a variety of transformations. The Mannich reactions produced aminomethyl derivatives ⟨79MI41619, 75JIC231, 72MI41614, 76MI41611⟩. The employment of POCl$_3$ and amides produced aldehydes or aminomethylene compounds ⟨73JCS(P1)465⟩, aldehydes and base produced 4-methylene analogs, coupling with diazonium salts produced hydrazones ⟨75MI41613, 76JPR658, 66MI41602⟩, and reaction with nitrite yielded a 4-oxime ⟨68AC(R)189, 59AC(R)2083⟩. Alkyl groups at the 4-position are reactive. Base treatment and reaction with aldehydes produce styrene derivatives ⟨76MI41610⟩ and reaction with phenylselenyl chloride produced a selenide ⟨79TL4903⟩. Other 4-methylenes are Michael acceptors and undergo addition of phosphorus compounds ⟨72ZOB750, 72JCS(P1)90, 67DOK1321⟩ or malonate derivatives to give formally a 2H- or a 4H-isoxazolin-5-one or a hydroxy-isoxazole. The addition of malononitrile to 3-methyl-4-benzylidineisoxazolin-5-one produced a 4H-compound (Scheme 56). The reaction of the same compound with dimethyl phosphite produced a compound in which the isoxazole was favored. The same reaction using 3-phenyl-4-benzylideneisoxazoline-5-one produced a Michael addition product which was in the Δ^3-5-keto form.

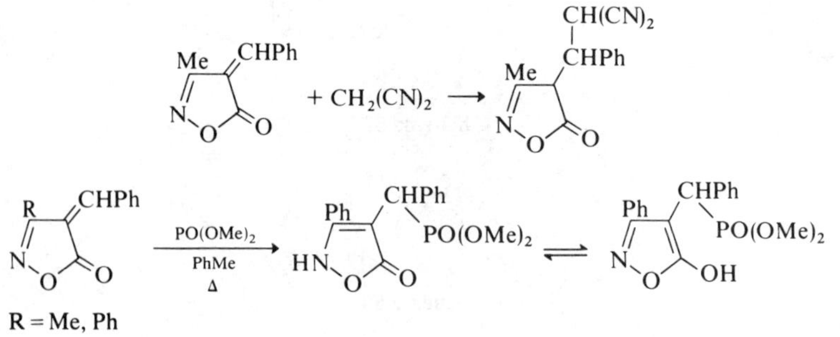

Scheme 56

4-Acyl (or 4-hydroxymethylene) isoxazolin-5-ones react with Grignard reagents to give 4-methyleneisoxazolin-5-ones (Scheme 57) ⟨72MI41613, 73IJC1⟩. In contrast, 4-acylisoxazolium salts under the same conditions produced tertiary alcohols (Scheme 57) ⟨75MI41617⟩.

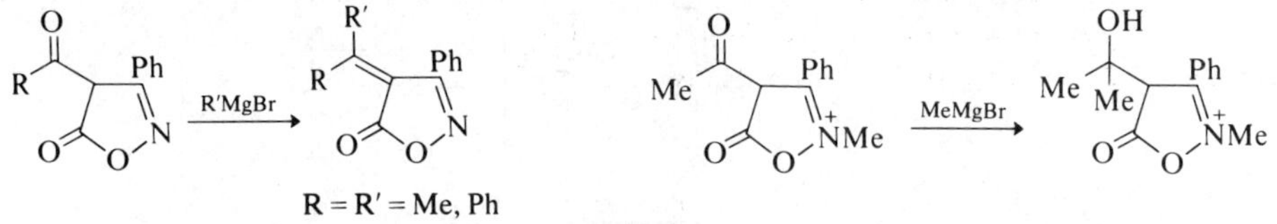

Scheme 57

4-Hydroxymethylene-isoxazolin-5-ones undergo reaction with SOCl$_2$ to form 4-chloromethylenes, with alcohols they form ethers ⟨62HC(17)1, p. 7⟩ and with amines aminomethylenes are obtained (Scheme 58) ⟨73T4291, 60ZOB600⟩.

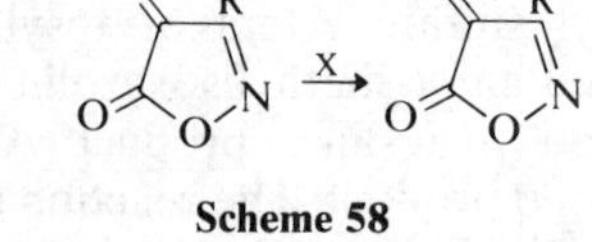

Scheme 58

The treatment of ethyl N-methyl-5-oxo isoxazoline-4-carboxylate with NaOH generated ethyl N-methylmonomalonamide. The reaction in the case of 3-unsubstituted derivatives proceeds by base attack on the 3-position (Scheme 59). In contrast N-alkyl-Δ^3-5-ones substituted in the 3-position are saponified to the free acid (Scheme 60) ⟨62JOC2160⟩.

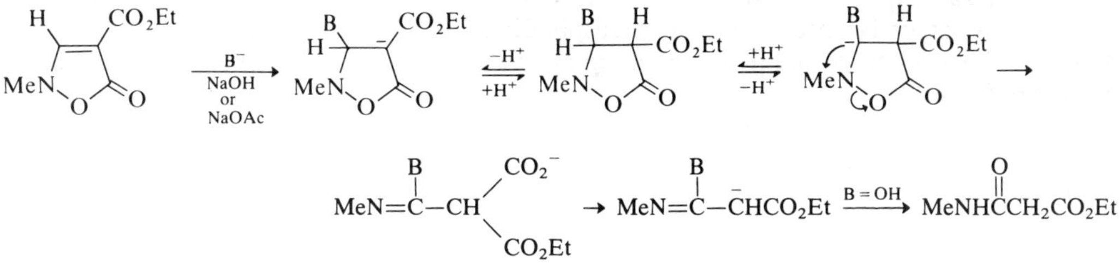

Scheme 59

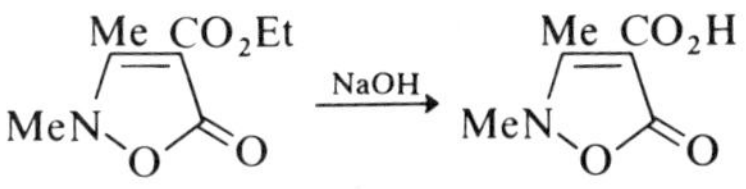

Scheme 60

Base treatment of isoxazolin-5-ones produced the cyano acids ⟨70JOC3130⟩, while ethoxide treatment of N-alkyl-Δ^3-5-ones gave, after H_2O treatment, a malonyl monoamide monoester ⟨69JOC2981⟩ and not an oximoacrylate as was reported earlier (Scheme 61) ⟨23HCA102⟩. The base treatment of a 4-bromoisoxazolin-5-one generated a ring-opened oxime (Scheme 62) ⟨79JOC873⟩.

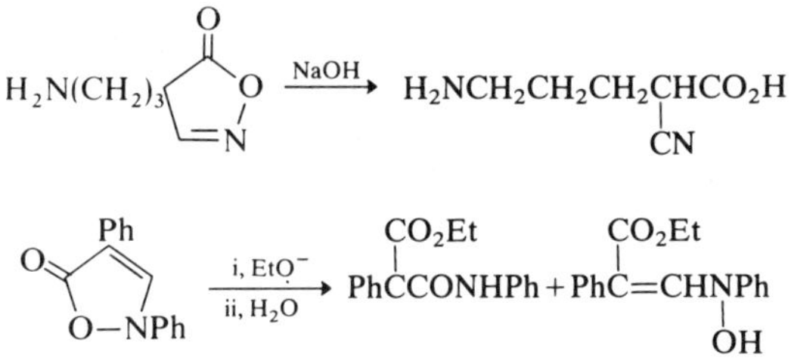

Scheme 61

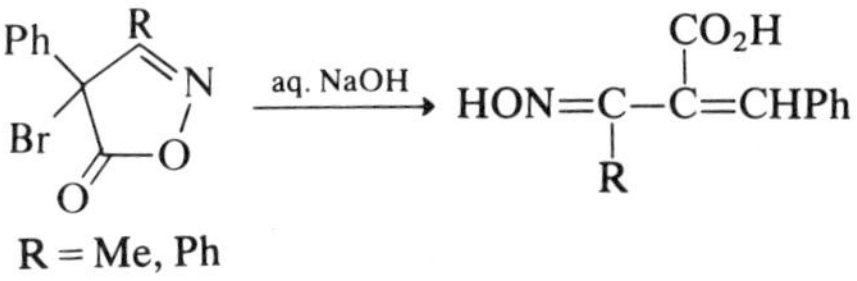

Scheme 62

Reaction of quaternized isoxazolin-5-ones with phenylmagnesium bromide produced a chalcone and dibenzoylethane. Those 5-ones with a 4,4-disubstituent undergo addition of the Grignard reagent to give a 5-ol (Scheme 63) ⟨73BSF3079⟩.

Scheme 63

The hydrogenation of N-substituted isoxazolin-5-one ethyl esters produced aminomethylenemalonates ⟨74G715⟩, while hydrogenation of 4-benzoyl-3-phenylisoxazolin-5-one generated an α-aminomethylene-β-keto acid (Scheme 64).

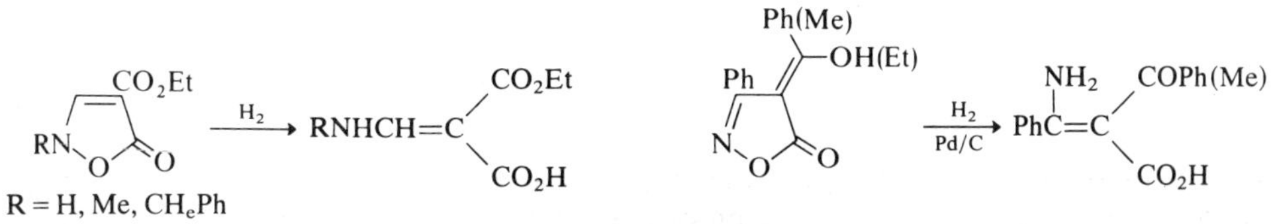

Scheme 64

Thermolysis of 4-methyl(4-phenyl)isoxazolin-5-one produced α-cyanophenylacetic acid ⟨67JHC533⟩. The pyrolysis of 3-methylisoxazoline-4,5-dione 4-oxime generated fulminic acid, which was trapped in a liquid N_2 cooled condenser for further study. Pyrolysis of metal salts such as Ag or Na produced the corresponding highly explosive salts of fulminic acid ⟨79AG503⟩. Treatment of the oxime with amines generated bis-α,β-oximinopropionamides (Scheme 65) ⟨68AC(R)189⟩.

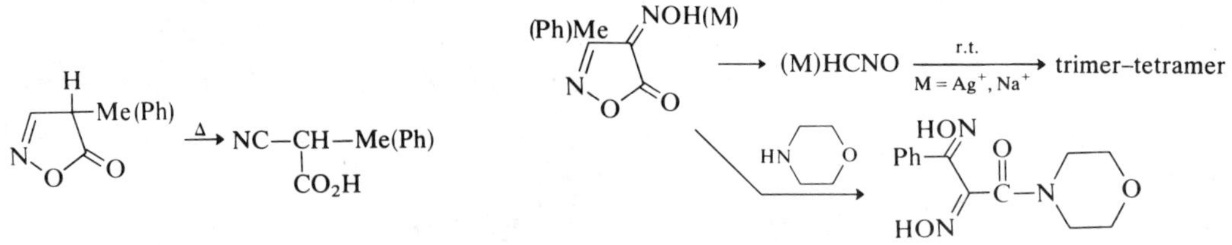

Scheme 65

Flash photolysis of a 4-enamino-5-one at 650 °C generated ethylnylamine, which was stable at −196 °C and its IR spectrum was examined. Warming caused tautomerization to N-phenylketenimine (Scheme 66) ⟨80AG743⟩.

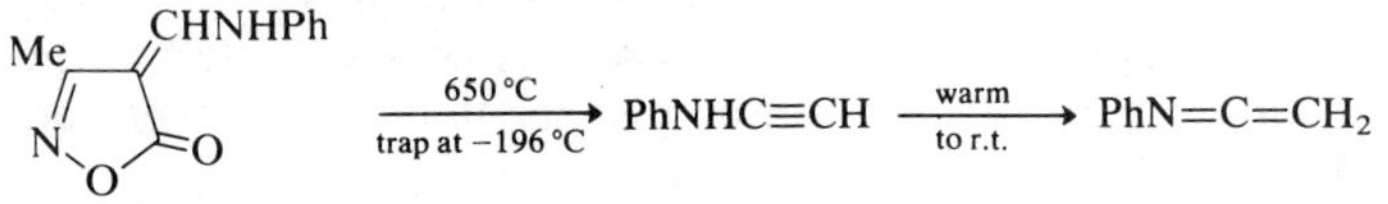

Scheme 66

The reaction of 2,4-diphenylisoxazolin-5-one with amines produced malondiamide and a ketenamine intermediate was suggested ⟨70TL4473, 73DIS(B)1434⟩. Stronger bases gave similar products and amine salts and malonimide intermediates were postulated to be involved in the reaction ⟨76JA6036⟩. The use of N-t-butyl-4-phenylisoxazolin-5-one produced an insoluble ketenamine carboxylate salt which on protonation produced a malonimide (Scheme 67) ⟨77H(7)247⟩.

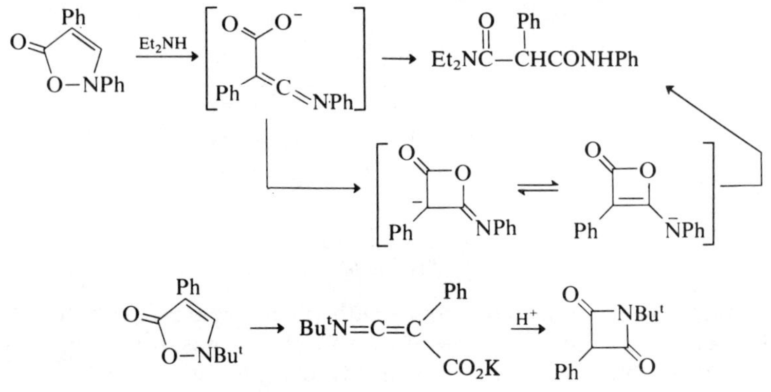

Scheme 67

Lithium aluminum hydride reduction of 2,3,4-triphenylisoxazolin-5-one yielded 1,2,3-triphenylaziridine and dibenzylaniline. The reaction was considered to proceed by a concerted [1,3]-sigmatropic migration of the N to a C atom. HOMO–LUMO calculations show this type of concerted reaction is possible (Scheme 68) ⟨80JA1372⟩.

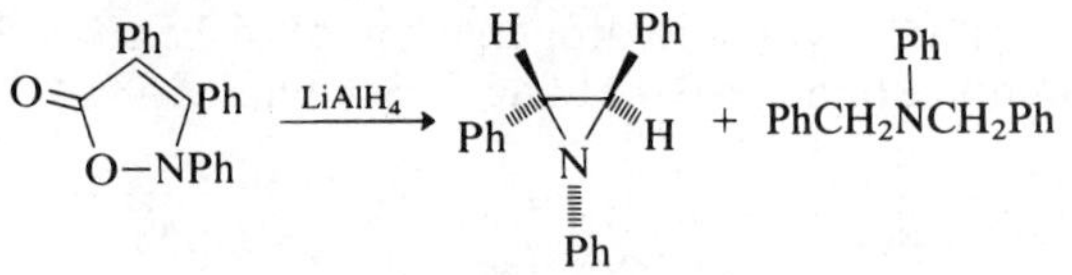

Scheme 68

The treatment of 4-benzoyl-3-phenylisoxazolin-5-one with KOH generated 3,5-diphenylisoxazole-3-carboxylic acid via a ring-opened intermediate as shown in Scheme 69 ⟨61CB1956⟩.

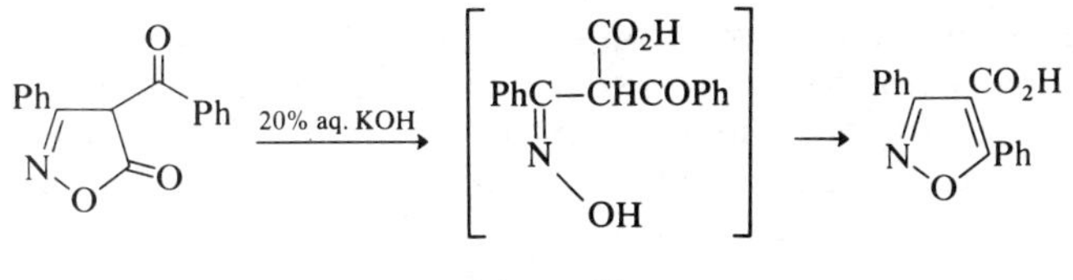

Scheme 69

The photolysis of 2,3,4-triphenylisoxazolin-5-imine generated an imidazolone and an indolone *via* an aziridinone ⟨70LA(732)195, 74CB13⟩. Acid treatment of the imine transformed it into an indole (Scheme 70) ⟨74CB13⟩.

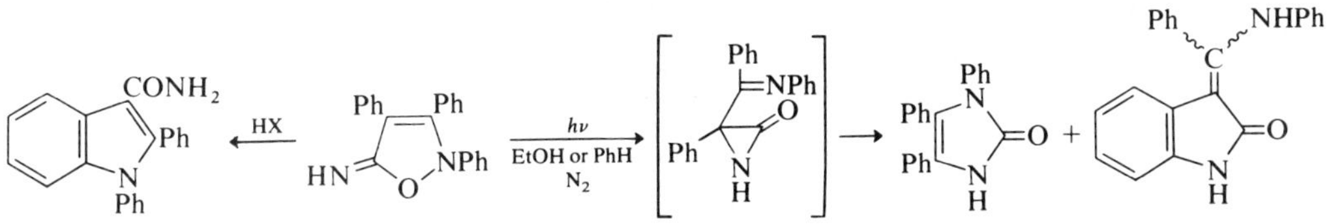

Scheme 70

4-(Arylenamino)-5-ones were converted into a benzodiazepine by Pd/C and H_2 while the alkyl counterpart produced a pyrazole and pyrazolone as shown in Scheme 71 ⟨73CB332⟩.

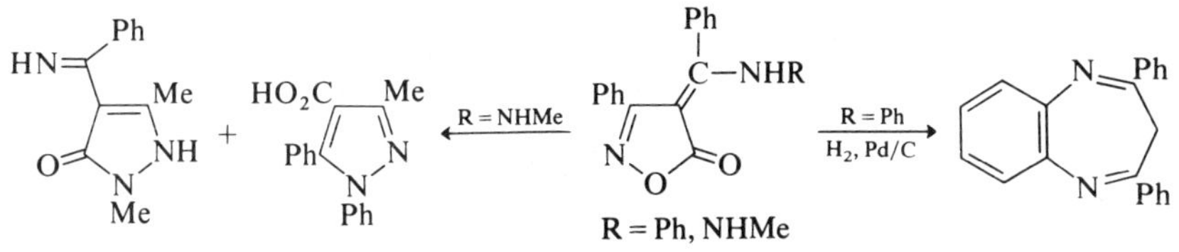

Scheme 71

3-Phenylisoxazolin-5-one condensed with anthranal to give a tricyclic isoxazolylquinoline (Scheme 72) ⟨78CZ264⟩. 3-Methyl-4-phenylazoisoxazoline-5-thione reacted with ethyl chloroacetate for form an intermediate isoxazolethiol, which on heating generated a 1,2,3-triazole (Scheme 73).

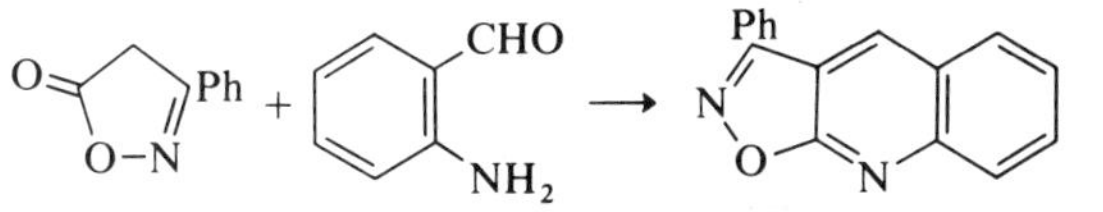

Scheme 72

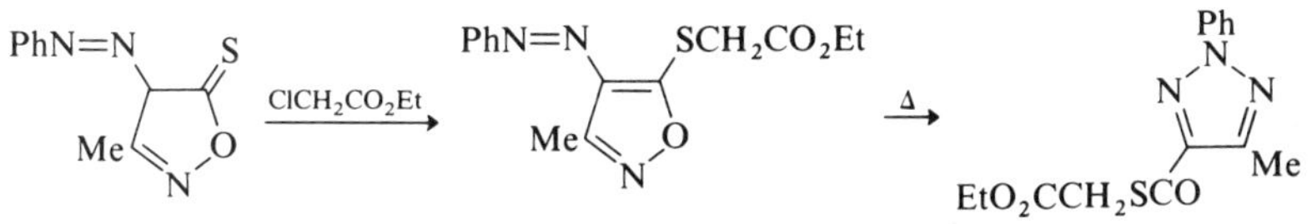

Scheme 73

3-Thiones were produced by the treatment of 3-chloro-*N*-substituted-isoxazolium salts with NaSH ⟨79CPB2398, 68ZC170⟩, which in turn were prepared from isoxazolin-3-one, phosgene or $SbCl_3$ ⟨78JAP(K)7863376⟩. Alkylation of isoxazolin-3-ones with acid halides ⟨74BRP1334882⟩, carbamoyl halides ⟨74JAP(K)7472252, 76JAP(K)7644637, 78MI41613⟩ and the condensation of aldehydes and $SOCl_2$ ⟨80JAP(K)8083766, 78JAP(K)78135971⟩ produced *N*-acyl-, *N*-carbamoyl-, *N*-hydroxymethyl- and *N*-alkyl-isoxazolin-3-ones. 5-Methyl-2-phenylisoxazolin-3-one on heating at 150 °C gave *via* a first-order rearrangement, an oxazolone (Scheme 74) ⟨66TL5451⟩.

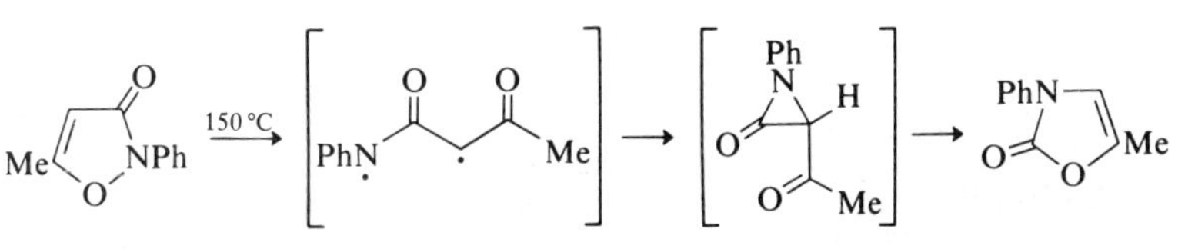

Scheme 74

The reaction of 2,5-dimethylisoxazoline-3-thione with 48% HBr produced, among other products, an isothiazole-3-thione, diisoxazole disulfide and 2,5-dimethylisoxazolin-3-one. In contrast, the reaction of 2-methyl-5-phenylisoxazoline-3-thione produced primarily disulfide ⟨80CPB487⟩. The reaction with H₂S in 48% HBr produced an isothiazole-3-thione and a cyclic thiazole (Scheme 75) ⟨80CPB487⟩.

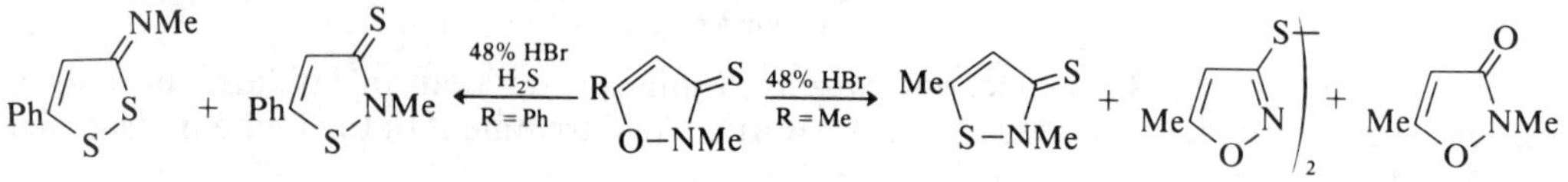

Scheme 75

Reaction of 2,5-dimethylisoxazoline-3-thione with thiophenol produced a ring-opened thioamide, while pyrolysis gave two products *via* a thiocyclopropene intermediate (Scheme 76) ⟨80CPB103⟩.

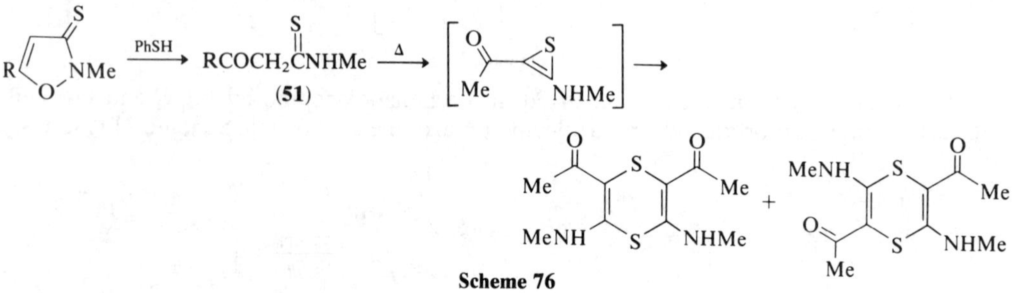

Scheme 76

4.16.3.2.2 3-Isoxazolines

The catalytic reduction of 2-methyl-3-phenyl-3-isoxazoline (**159**) produced β-hydroxypropiophenone (**160**) ⟨74CPB70⟩. Ring fission also occurred on base treatment of the 3,5-diaryl-3-isoxazoline (**161**) to give the α,β-unsaturated oxime (**162**) ⟨70CI(L)624⟩.

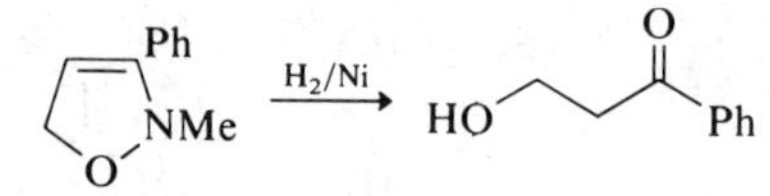

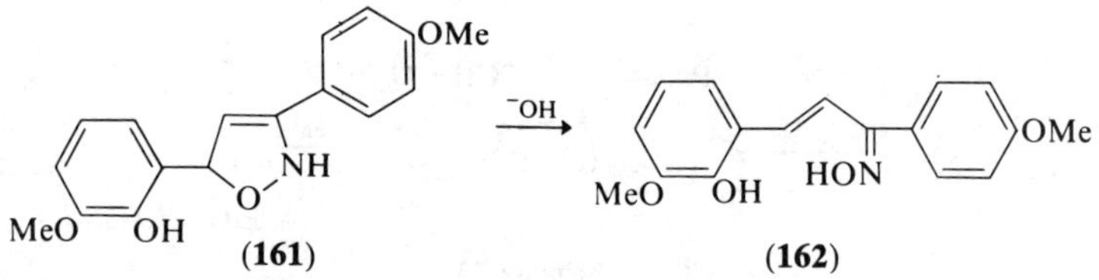

4.16.3.2.3 4-Isoxazolines

4-Isoxazolines, with the substitution pattern shown in (**163**), on warming to room temperature gave (**164**). Similarly, heating (**165**) resulted in formation of the pyrrole (**166**) ⟨70CB3196⟩, while heating (**167**) produced oxazolines, presumably *via* the aziridine intermediate (**168**) ⟨68JA5325, 77H(8)387⟩. 4-Isoxazolines such as (**165a**) on heating in benzene or xylene are also converted into pyrroles (**166a**) in 60–90% yields ⟨76S281, 74CPB61, 70CB3196, 69TL4875⟩. The nature of the rearrangement is dependent on the substituents and aziridine intermediate or a hetero-Cope reaction can be used to rationalize these transformations.

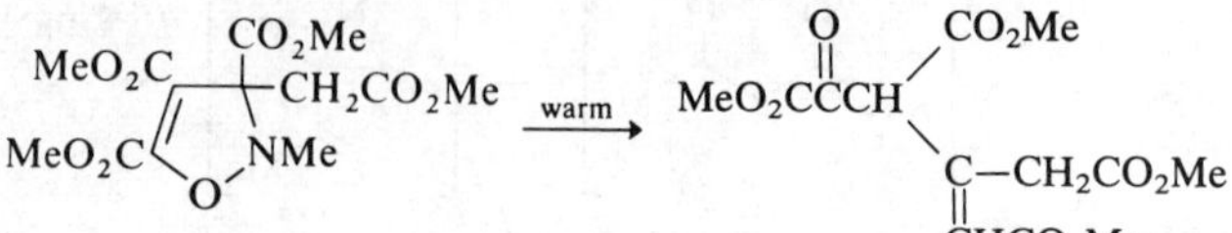

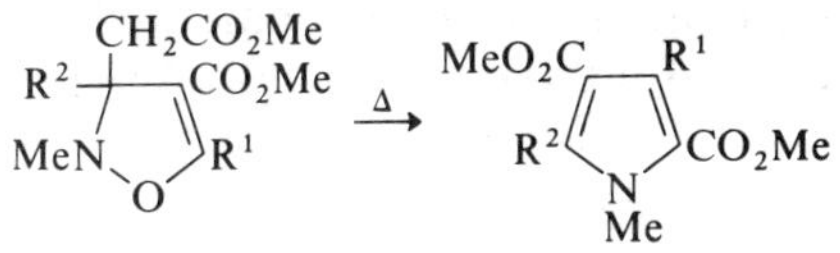

(165) R¹ = R² = H or CO₂Me   (166)   (165a) R¹ = Ar, R² = H,   (166a)
                                               R³ = Ar; R¹ = H,
                                               R² = R³ = CO₂Me

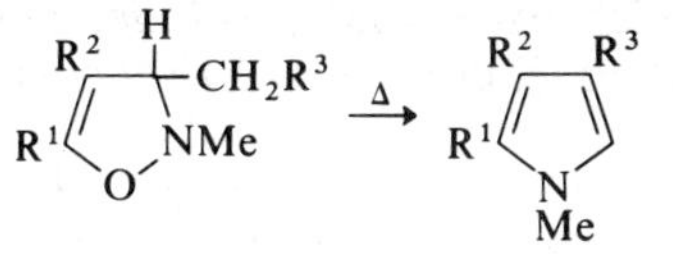

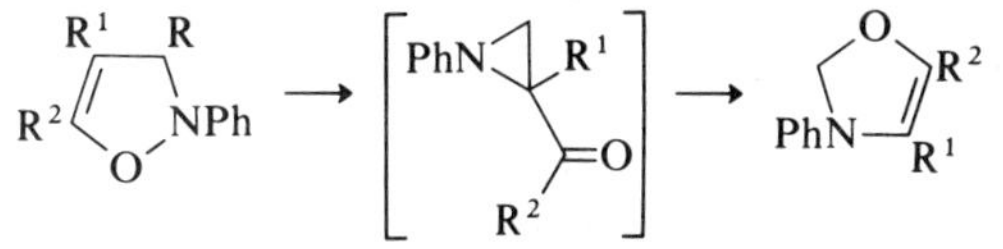

(167) R = H or Ph;   (168)
R¹ = R² = CF₃ or Ph

Hydrogenation of 2-methyl-3-phenyl-4-isoxazoline surprisingly yielded cinnamaldehyde ⟨74CPB70⟩.

4.16.3.2.4 Isoxazolidines

Various reactivities of isoxazolidines have been discussed in detail by Takeuchi and Furusaki ⟨77AHC(21)207⟩. Attention is drawn to the principal reactions below.

(i) Reduction

The hydrogenation of diastereomeric isoxazolidines to give *erythro-* and *threo-* amino alcohols has been well studied ⟨77AHC(21)207, 74CPB70⟩ and a similar cleavage of a fused isoxazolidine ring was an important step in the synthesis of ergot alkoloids (Scheme 77) ⟨80JA4265⟩.

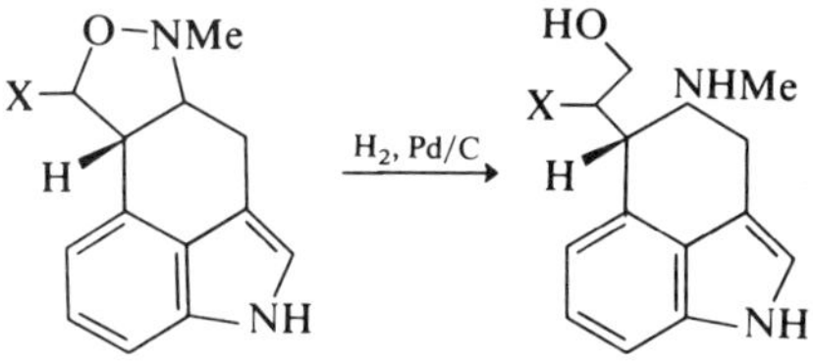

Scheme 77

The hydrogenation of (**169**) with Pd/C produced (**170**), with Raney Ni (**171**) was obtained, and with PtO₂ compound (**172**) was formed.

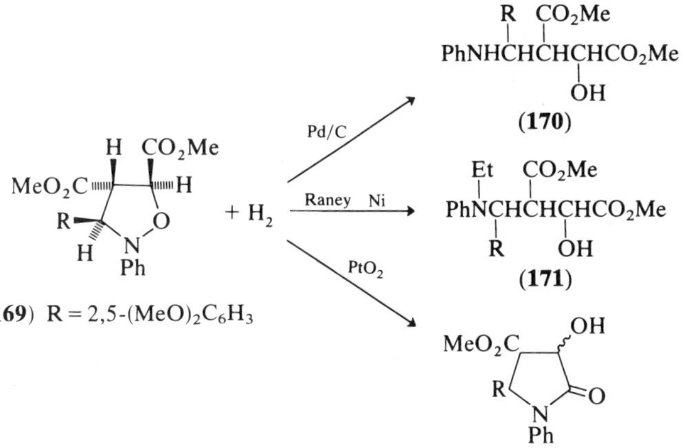

(169) R = 2,5-(MeO)₂C₆H₃

(ii) Oxidation

Oxidation of fused isoxazolidine (**173**) with one equivalent of *m*-chloroperbenzoic acid produced an oxazine. The mechanism of transformation involves formation of an *N*-oxide which undergoes cleavage, rearrangement and cyclization. An additional equivalent of MCPBA produced (**174**) ⟨79JOC1819⟩. These transformations are illustrated in Scheme 78.

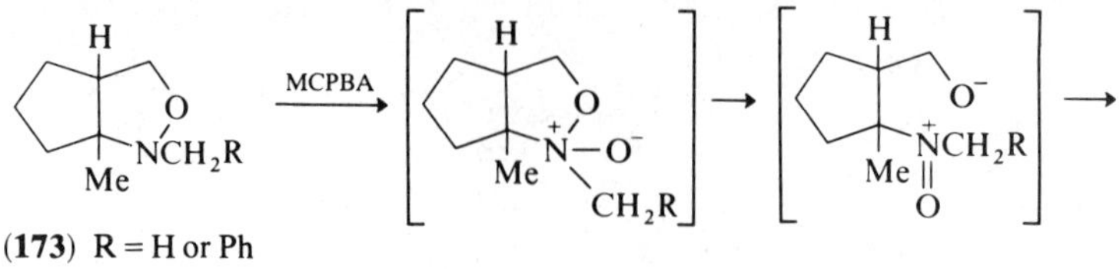

(**173**) R = H or Ph

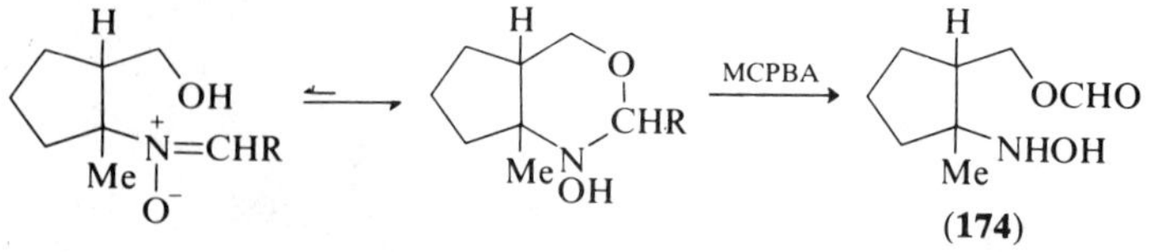

Scheme 78

(iii) Photolysis

The photooxygenation of (**175**) produced a ring ruptured compound (**176**) (Scheme 57) ⟨B-78MI41612⟩ while similar treatment of steroidal isoxazolidine (**177**) resulted in the formation of two major nitro compounds also shown in Scheme 57 ⟨77CC749⟩.

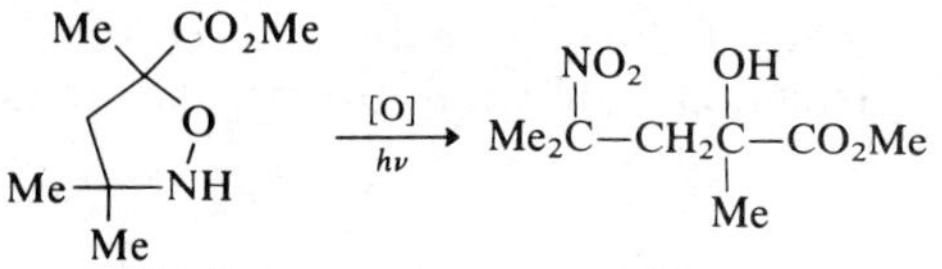

Scheme 79

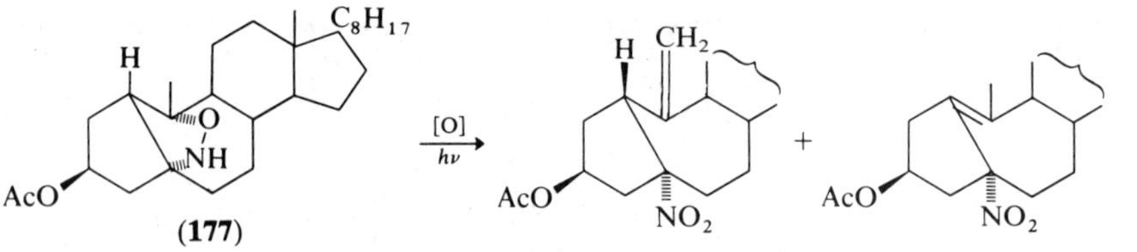

(iv) Thermal reactions

Pyrolysis of isoxazolidines usually results in retro-1,3-addition reactions and/or isomerization of ring substituents ⟨77AHC(21)207⟩. Heating of (**178**) produced the anilide (**179**) ⟨79KGS746⟩, while at 100 °C the isoxazolidine (**180**) underwent ring opening to the oxime (**181**) which upon hydrogenation gave 4-amino-2-hydroxybutanoic acid ⟨75CL965, 76BCJ2815⟩

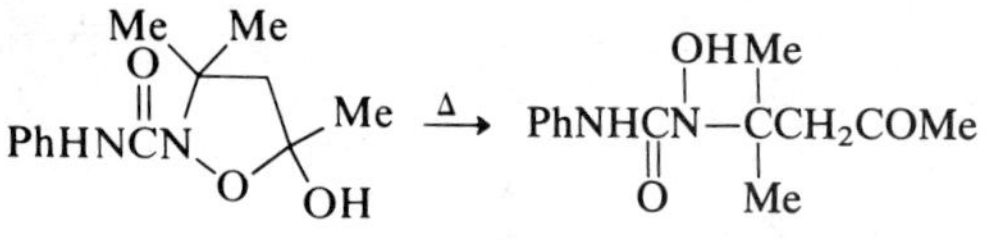

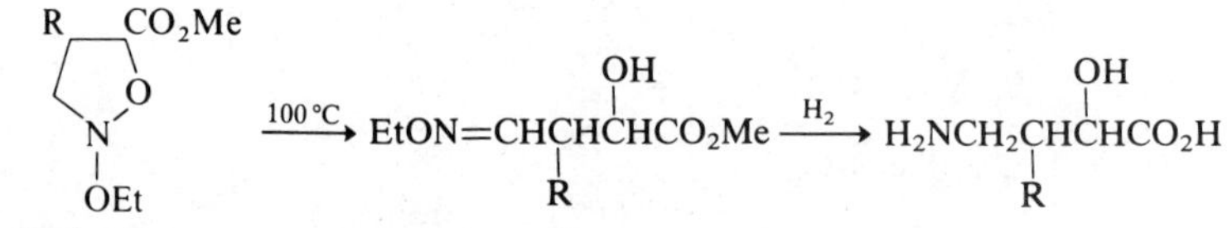

(**180**) R = H, Me

(v) Acid reactions

In general, alkoxy- or silyloxy-isoxazolidines when treated with acid produce 2-isoxazolines ⟨77AHC(21)207, 74MIP41601⟩. Other isoxazolidines are cleaved at the N—O bond with further degradation then following ⟨77AHC(21)207⟩. The treatment of (**182**) with HCl generated cinnamic acid and a small amount of benzoic acid, whereas treatment of (**182**) with acetic anhydride produced a 2-isoxazoline *N*-oxide which gave identical products when treated with HCl (Scheme 80) ⟨76ZOR2028⟩.

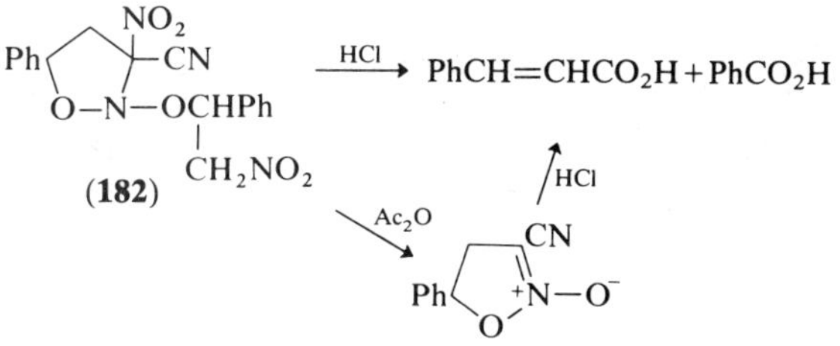

Scheme 80

Strong acid treatment of 2-(trimethylsilyloxy)-5-methoxycarbonylisoxazolidine (**183**) produced (**184**) ⟨78ACS(B)118⟩.

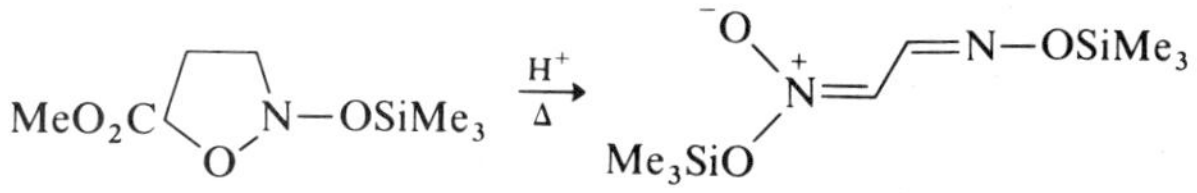

(vi) Base reactions

Mild base does not effect isoxazolidine ring cleavage by fission of the N—O bond; rather, C—O bond cleavage takes place ⟨77AHC(21)207⟩. The reaction of *N*-trimethylsilylisoxazolidine (**185**) with KOH produced a β-hydroxyketone oxime (Scheme 59) ⟨74DOK109⟩.

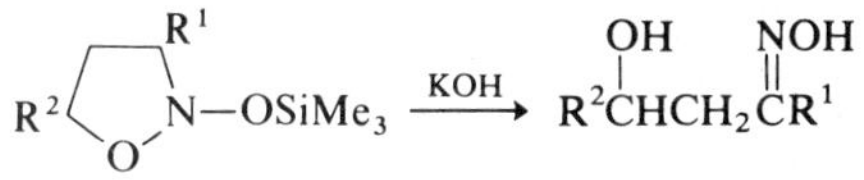

Scheme 81

The treatment of 3-benzoyl-2-phenylisoxazolidine with strong base generated an aldehyde and a ketimine ⟨74T1121⟩. Under these conditions dimethyl 2-α-methoxyisoxazolidine-3,3-dicarboxylic acid (**186**) produced isoxazoline-2-carboxylic acid. Reaction of the monomethyl amide (**187**) gave the corresponding isoxazoline-2-carboxamide (Scheme 60). CD was used in the conformational studies ⟨79T213⟩.

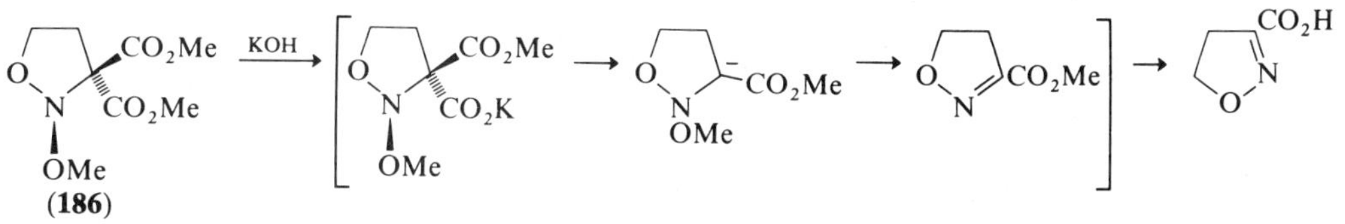

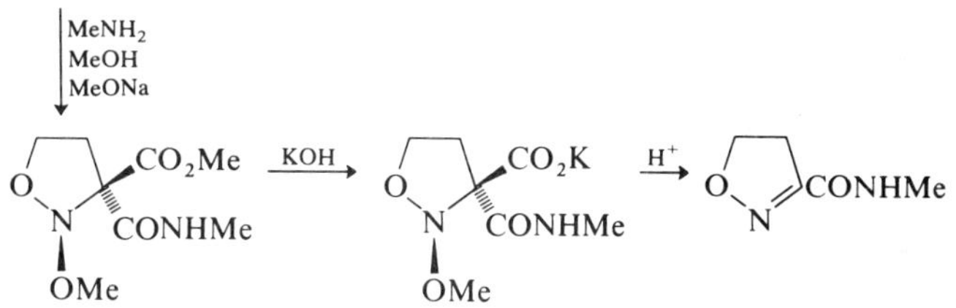

Scheme 82

(vii) Miscellaneous reactions

Isoxazolidine reacted with acetone and perchloric acid to produce the iminium salt (**189**) ⟨80JA1649⟩.

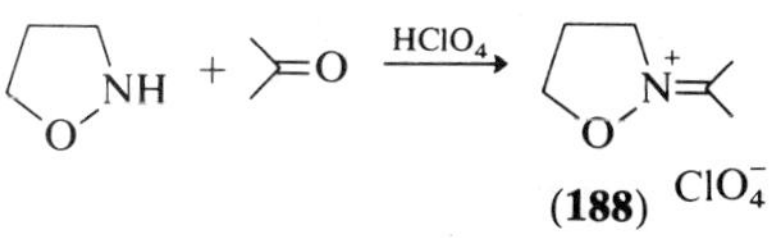

4.16.3.3 Reaction of Substituents

4.16.3.3.1 Fused benzene rings

It would be anticipated that 1,2-benzisoxazoles undergo electrophilic substitution in the homocyclic ring, and this has been borne out by experiment. The absence or presence of substituents in the 3-position has no influence on the course of the reaction. In qualitative terms it would also be expected that substitution would occur in the 5- or 7-position (see Section 4.16.2.1.1) but most electrophilic substitutions take place at the 5-position.

Nitration of 3-phenyl-1,2-benzisoxazole with fuming nitric acid has been shown to give dinitro products of undetermined substitution pattern ⟨67AHC(8)277, p. 290⟩. However, more satisfactory studies have now been described, especially on the kinetics and mechanism of nitration of 3-methyl-1,2-benzisoxazole ⟨77JCS(P2)47⟩. Nitration in cold, concentrated mixed acids yields the 5-nitro derivative exclusively, nitration in 80–90% sulfuric acid occurring on the free base whereas at higher acidities the conjugate acid is the species involved in the nitration.

However, other studies on the nitration of a series of 3-methyl- and 3-ethyl-1,2-benzisoxazoles have shown that a mixture of the 5-nitro and 5,7-dinitro derivatives is formed ⟨77IJC(B)1058, 77IJC(B)1061⟩. The effect of substituents in the benzene ring is also of interest. If the 5-position is blocked, *e.g.* by a chloro group or by alkyl groups, nitration then occurs at the 4-position. 3-Alkyl-7-chloro and 3,7-dialkyl derivatives result in the formation of the appropriate 5-nitro derivative. The isomeric 3-alkyl-6-chloro- and 3,6-dialkyl-1,2-benzisoxazoles yield a mixture of the 5-nitro and 5,7-dinitro compounds. Both ^{1}H NMR measurements and alternate syntheses were used in establishing the structures of these substitution products.

The same substitution pattern is maintained on nitration of the isomeric 2,1-benzisoxazole system. Thus 3-methyl-2,1-benzisoxazole gives the 5-nitro compound on reaction with potassium nitrate in concentrated sulfuric acid. 2,1-Benzisoxazole itself gives predominantly the 5-isomer, together with a little 7-nitro product under the same conditions ⟨67AHC(8)277, p. 322⟩. Introduction of a directing group into the benzene ring as in 6-chloro- and 6-methoxy-2,1-benzisoxazole results in the 7-nitro product being obtained ⟨70JOC1662, 77JOC897⟩.

Other electrophilic substitutions give mixed results. Bromination and chlorination follow the above substitution patterns but are also dependent on the directing influence of substituents already in the ring ⟨67AHC(8)277, p. 290⟩. Thus bromination of 6-hydroxy-3-alkyl-1,2-benzisoxazoles in acetic acid at room temperature yields the 7-bromo derivatives; however, in the 6-methoxy analogs the entering bromo group is oriented to the 5-position ⟨79IJC(B)371⟩. At 110–120 °C the 5,7-dibromo derivatives are the sole products, irrespective of whether the 6-hydroxy compound or its ether is used as substrate. Electrophilic substitution of 1,2-benzisoxazol-3-ylacetic acid always results in modification of the 3-substituent, *e.g.* chlorination, bromination or iodination leads to partially or fully halogenated derivatives of the acid ⟨81AHC(29)1, p. 11⟩.

In the isomeric 2,1-benzisoxazole series halogenation also occurs in the 5-position, but it appears that an initial addition product is formed at the 4,5-positions ⟨67AHC(8)277, p. 321⟩.

Acylation of 3-alkyl-6-hydroxy-1,2-benzisoxazole has also been reported ⟨77JIC875⟩ under Friedel–Crafts conditions to give the 7-acyl product. Fries rearrangement of 6-acetoxy-3-methyl-1,2-benzisoxazole in the presence of AlCl$_3$ at 140 °C also provides a route to the 7-acetyl-6-hydroxy derivatives ⟨73IJC541⟩. Reactions of these kind are rare in this series.

Using Reimer–Tiemann reaction conditions on 3-alkyl-6-hydroxy-1,2-benzisoxazoles results in formylation occurring at the 7-position ⟨77IJC(B)1056⟩.

4.16.3.3.2 Other C-linked substituents

(i) *Alkyl groups*

(a) *3,5-Dimethylisoxazole and its 4-substituted derivatives*. The methyl group of toluene, 2-methylpyridine ⟨65JOC3229⟩ and 2-methylquinoline can be deprotonated by a strong base to give a carbanion which reacts with various electrophiles to give the corresponding derivatives. Likewise, the same reactions can be expected with the methyl groups of 3,5-dimethylisoxazole and its 4-substituted derivatives. In this section the reactions of the methyl groups of 3- and 5-unsubstituted isoxazoles will not be discussed since 3- and 5-unsubstituted isoxazoles under basic conditions lead to ring cleavage products (Sections 4.16.3.1.5 and 4.16.3.1.6).

In theory two carbanions, (**189**) and (**190**), can be formed by deprotonation of 3,5-dimethylisoxazole with a strong base. On the basis of MINDO/2 calculations for these two carbanions, the heat of formation of (**189**) is calculated to be about 33 kJ mol^{-1} smaller than that of (**190**), and the carbanion (**189**) is thermodynamically more stable than the carbanion (**190**). The calculation is supported by the deuterium exchange reaction of 3,5-dimethylisoxazole with sodium methoxide in deuterated methanol. The rate of deuterium exchange of the 5-methyl protons is about 280 times faster than that of the 3-methyl protons ($\Delta\Delta F^{\neq} = 13.0$ kJ mol^{-1} at room temperature) and its activation energy is about 121 kJ mol^{-1}. These results indicate that the methyl groups of 3,5-dimethylisoxazole are much less reactive than the methyl group of 2-methylpyridine and 2-methylquinoline, whose activation energies under the same reaction conditions were reported to be 105 and 88 kJ mol^{-1}, respectively ⟨79H(12)1343⟩.

3,5-Dimethylisoxazole can be metallated by a variety of bases (Bu^{n}Li, LiNPr$^i_2$ and NaNH$_2$/NH$_3$) at -78 °C in THF. Quenching with deuterated methanol at low temperature yields at least 98% deuterium incorporation at the 5-methyl group. The carbanion (**189**) is stable for several hours at -78 °C but decomposes upon warming to 0 °C over 30 minutes. It is interesting to note that Bu^{n}Li does not metallate 3-methyl-5-pentylisoxazole; reaction in THF at -78 °C followed by quenching with D$_2$O/THF gave no deuterium incorporation. A similar reaction using Bu^{t}Li (or Bu^{s}Li) yielded the starting material with 97% deuterium incorporation at the 3-methyl group ⟨81TL3699⟩. Therefore, 3,5-dimethylisoxazole can be metallated and alkylated regiospecifically at the 5-methyl group, or first at the 5-methyl group and then at the 3-methyl group in a sequential manner. This method permits a regiospecific synthesis of various disubstituted isoxazoles. 3-Ethyl-5-*n*-pentylisoxazole (**192**) has been synthesized from 3,5-dimethylisoxazole (**53**) in 81% yield *via* the sequence (**53**) → (**191**) → (**192**). Reactions of the 3,5-dimethylisoxazole carbanion (**189**) with various electrophiles are summarized in Table 8, and reactions of 5-alkyl-3-methylisoxazoles with

Table 8 Reactions of 3,5-Dimethylisoxazole Carbanion (**189**) with Various Electrophiles

Electrophile	Base	Yield (%)	Ref.
MeI	Bu^{n}Li	75	81TL3699
MeI	NaNH$_2$	70	73BCJ310
Bu^{n}I	Bu^{n}Li	90	81TL3699
CH$_2$=CHCH$_2$Br	Bu^{n}Li	97	81TL3699
CH$_2$=CHCH$_2$Br	NaNH$_2$	44	73BCJ310
PhCH$_2$Cl	Bu^{n}Li	98	81TL3699
PhCHO	Bu^{n}Li	96	81TL3699
Cyclohexanone	Bu^{n}Li	95	81TL3699
MeCOMe	Bu^{n}Li	97	81TL3699
(epoxide)	Bu^{n}Li	87	81TL3699
Me$_3$SiCl	Bu^{n}Li	80	81TL3699
I(CH$_2$)$_{10}$I	Bu^{n}Li	78	81TL3699
CO$_2$	Bu^{n}Li	73	70CJC2006
PhCO$_2$Me	NaNH$_2$	—	76BCJ2254
MeCO$_2$Et	LiNH$_2$	40	76BCJ2254
COCl$_2$	Bu^{n}Li	25	76G823
PhCH=NPh	NaNH$_2$	31	73BCJ3533
PhC≡N	NaNH$_2$	49	73BCJ3533
C$_5$H$_{11}$ON=O	NaNH$_2$	—[a]	76OPP87
PhN=O	NaNH$_2$	17	76OPP87
MeSCl	Bu^{n}Li	36	77JHC37

[a] A mixture of a nitrosation product and an amide.

various electrophiles in the presence of a strong base are summarized in Table 9. It should be mentioned that sodium amide in liquid ammonia only deprotonates the α-protons from the 5-alkyl group of 5-alkyl-3-methylisoxazole. For example, 3,5-dimethylisoxazole (**53**), when treated with four equivalents of sodium amide and methyl iodide, gave only the further methylated product, 5-*t*-butyl-3-methylisoxazole (**193**) ⟨79H(12)1343⟩.

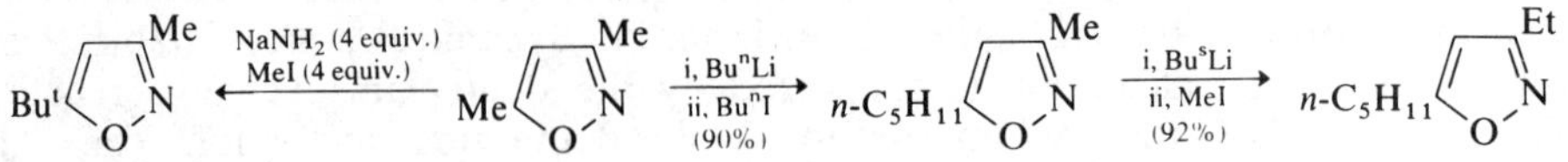

Table 9 Reactions of 5-Alkyl-3-methylisoxazoles with Various Electrophiles
⟨81TL3699⟩

5-Substituent	Electrophile	Base	Yield (%)
n-C$_5$H$_{11}$	MeI	Bu^{t}Li	50
n-C$_5$H$_{11}$	MeI	Bu^{s}Li	92
n-C$_5$H$_{11}$	CH$_2$=CHCH$_2$Br	Bu^{s}Li	80
n-C$_5$H$_{11}$	cyclohexanone	Bu^{s}Li	80
n-C$_5$H$_{11}$	Me$_3$SiCl	Bu^{s}Li	83
n-C$_5$H$_{11}$	PhCHO	Bu^{t}Li	93
PhCH$_2$CH$_2$	PhCH$_2$Cl	Bu^{s}Li	85
PhCH$_2$CH$_2$	Bu^{n}I	Bu^{s}Li	85
PhCH$_2$CH$_2$	Me$_3$SiCl	Bu^{s}Li	90

The deuterium exchange reaction of 4-substituted 3,5-dimethylisoxazoles has also been investigated. It was found that the 5-methyl groups are consistently more reactive than the 3-methyl groups, and an electron-donating group at C-4 retards the deuterium exchange reaction of the 5-methyl group while an electron-withdrawing group accelerates it. Therefore, 4-substituted 3,5-dimethylisoxazoles (**194**) having an electron-withdrawing group such as nitro, cyano, acyl and carboxyl at the 4-position condense regiospecifically with aromatic aldehydes and their anils in the presence of a secondary amine such as diethylamine and pyrrolidine to give the condensation products (**195**). 3,5-Dimethyl-4-nitroisoxazole (**194**; X = NO$_2$) also reacts with *p*-nitrosodimethylaniline and benzalacetophenone to give the condensation products (**195**) and (**196**), respectively ⟨63AHC(2)365, 80JHC621⟩.

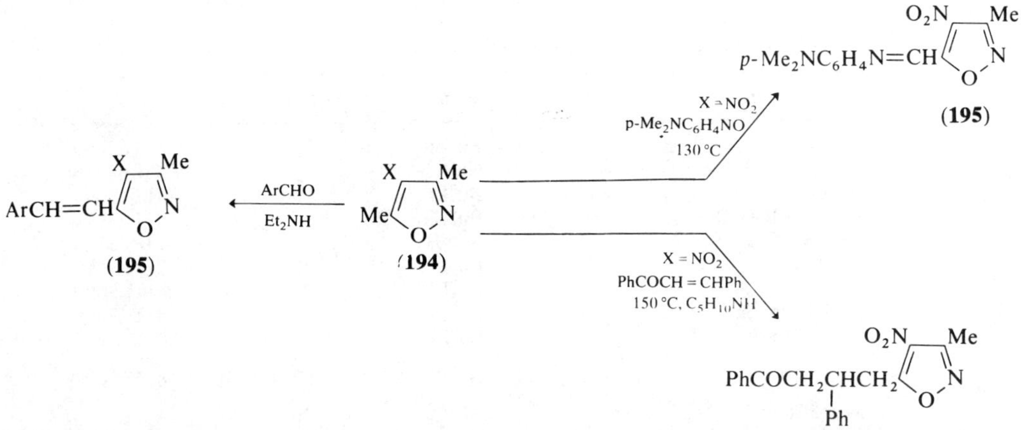

(*b*) *2-Alkylisoxazolium salts.* 2,3,5-Trimethylisoxazolium salt (**197**) can undergo condensation reactions in the presence of comparatively weak bases such as secondary amines, hydroxides and alkoxides. In this compound, however, the 3-methyl group seems to be slightly more reactive than the 5-methyl group. For example, the reaction of 2,3,5-trimethylisoxazolium salt (**197**) with an aromatic aldehyde in the presence of pyrrolidine gave a mixture of 5-methyl-3-styrylisoxazolium salt (**198**) and 3,5-distyrylisoxazolium salt (**199**) ⟨58MI41601⟩, and under the influence of potassium methoxide, 3,5-distyrylisoxazolium salt (**199**) was the only product ⟨77H(7)241⟩.

The 2-alkyl group of 2-alkylisoxazolium salts is also susceptible to α-deprotonation. For example, treatment of 3,5-distyrylisoxazolium salt (**199**) with potassium methoxide followed by acid hydrolysis gave the natural β-diketone curcumine ⟨77H(7)241⟩.

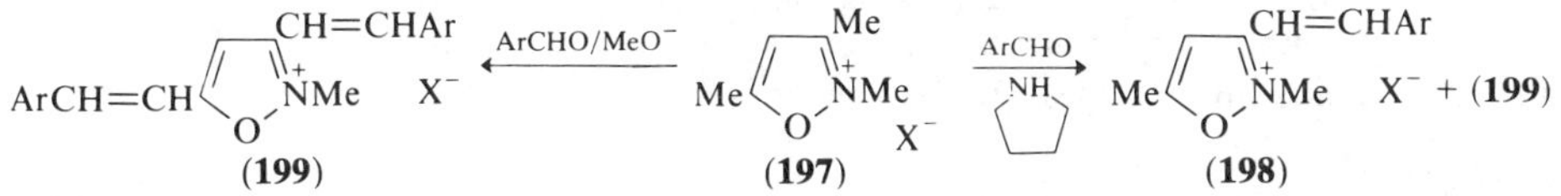

In general, the product of deprotonation of the 2-alkyl group in the above salts is an ylide (**200**) which can rearrange to give the ring-opened intermediate (**201**) or 2H-1,3-oxazine (**202**) ⟨69CPB2201, 74CPB61, 62CJC882⟩.

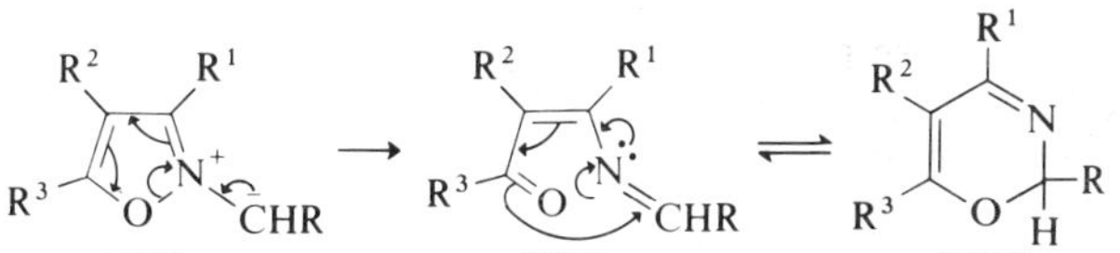

Although there is very little scope for reactions of the above types with 1,2-benzisoxazole derivatives, the quaternary salts such as 2-methyl-3-phenyl-1,2-benzisoxazolium salt underwent base-catalyzed isomerization to the 1,3-benzoxazine shown in Scheme 83. This reaction is analogous to the formation of (**202**) above ⟨67AHC(8)277⟩.

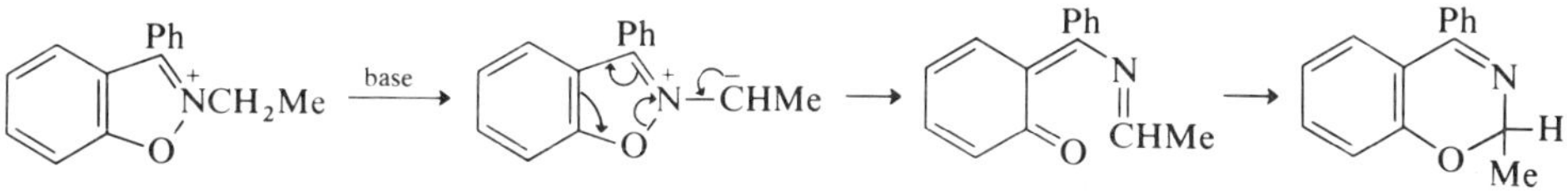

Scheme 83

With 1,3-dimethyl-2,1-benzisoxazolium salts, however, considerable reactivity has been reported. Condensation occurs readily with aldehydes, ketones, orthoesters and diazonium salts to yield styryl, cyanine and azo compounds, respectively ⟨78JOC1233⟩. In the presence of triethylamine, dimerization was observed, and the reactions of the cation were considered to involve the intermediacy of the *anhydro* base ⟨77JOC3929⟩.

(*c*) *Homolytic halogenation of methylisoxazoles.* The halogenation of 4-unsubstituted isoxazoles gave 4-haloisoxazoles. With methylisoxazoles having substituents at the 4-position, homolytic bromination of the methyl group depends on its position in the ring and on the 4-substituent. Thus, 4-chloro-3-methyl-5-phenyl-, 4-benzoyl-3,5-dimethyl- and 3,5-dimethyl-4-nitro-isoxazoles do not react with *N*-bromosuccinimide. Under these conditions 4-bromo-, 4-chloro- and 4-(*p*-nitrophenyl)-3,5-dimethylisoxazoles (**203**) are brominated at the 5-methyl group, giving (**204**). Bromination of trimethylisoxazole (**205**) takes place on the 4- (mostly) (**206**) and 5-methyl (**207**) groups. 3,5-Diphenyl-4-methylisoxazole (**208**) also reacts readily with bromine and with sulfuryl chloride in the presence of benzoyl peroxide to yield 4-halomethyl derivatives (**209**) ⟨63AHC(2)365, 74KGS602⟩. In general the reactivity of the methyl groups in these reactions decreases in the following order: 4-methyl > 5-methyl > 3-methyl.

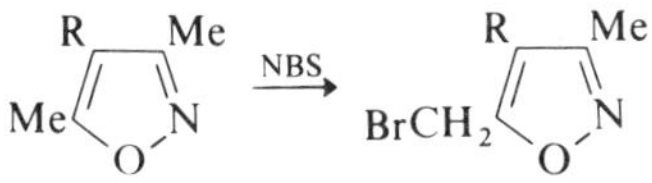

(**203**) R = Cl, Br, *p*-NO$_2$C$_6$H$_4$ (**204**) 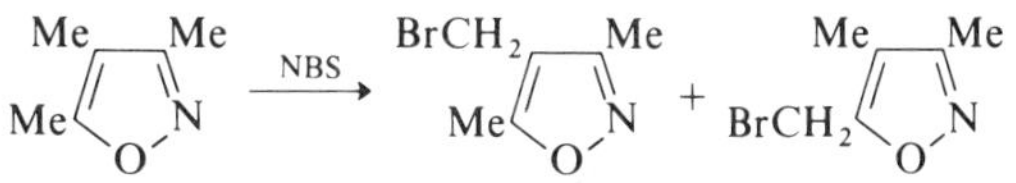 (**205**) (**206**) (**207**)

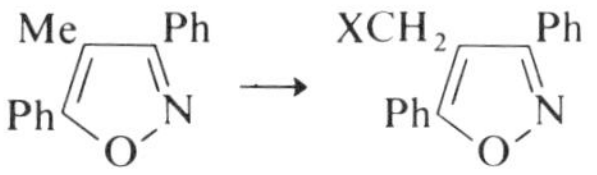

(**208**) (**209**) X = Br, Cl

3-Methyl-2,1-benzisoxazole and its 6-methoxy derivative are both readily brominated at the 3-methyl group with NBS ⟨79TL4687, 74RTC139⟩. The 4-methoxy isomer gives only the 7-bromo compound unless excess of the reagent is used, when 7-bromo-3-(bromomethyl)-2,1-benzisoxazole is obtained in 52% yield ⟨74RTC139⟩. However, there is a report of 3-methyl-2,1-benzisoxazole with NBS affording the corresponding 3-formyl compound ⟨78TL2309⟩.

(d) *Oxidation of alkylisoxazoles.* No preparation of isoxazolecarboxylic acids by the oxidation of methyl- or alkyl-isoxazoles has been reported. However, isoxazoles containing unsaturated side chains or oxygenated functions are easily transformed into the corresponding carboxylic acids on treatment with different oxidizing agents.

(ii) *Aryl groups*

Electrophilic substitution occurs readily in C-aryl groups. 3-Methyl-5-phenylisoxazole underwent nitration in H_2SO_4/HNO_3 as the conjugate acid at the *para* position of the phenyl group; however, nitration in Ac_2O/HNO_3 gave a poor yield of 4-nitro-5-*p*-nitrophenylisoxazole as the only product isolated. 5-Methyl-3-phenylisoxazole underwent nitration in mixed acid as the conjugate acid at the *meta* position and also as the free base at the *para* position of the phenyl group; the proportion of the *meta* and *para* isomers depended on the acidity of the reaction medium. The nitration of 5-methyl-3-phenylisoxazole in Ac_2O/HNO_3 gave 4-nitro-3-*p*-nitrophenylisoxazole. 3,5-Diphenylisoxazole in HNO_3 underwent nitration first at the *para* position of the 5-phenyl group and then at the *meta* position of the 3-phenyl substituent ⟨74KGS597, 75JCS(P2)1627⟩. The sulfonation of 5-phenylisoxazole gave a mixture of the *m*- and *p*-phenylsulfonic acid chlorides in a 2:1 ratio ⟨63AHC(2)365⟩.

The condensation of 5-phenyl-3-(*p*-tolyl)isoxazole (**210**) with benzylideneaniline gave the stilbene derivative (**211**) ⟨78AHC(23)171⟩.

(iii) *Other C-linked functionalities*

(a) *Carboxylic acids.* Isoxazolecarboxylic acids are crystalline solids, and their comparatively low melting points indicate that they do not possess a zwitterionic structure. They are usually decomposed on heating above their melting points without giving the corresponding isoxazoles. However, on pyrolysis, 5-substituted isoxazole-3-carboxylic acids give the corresponding cyanoketones, often in good yield. The reactions are much facilitated by the presence of arylhydrazines, yielding 5-aminopyrazoles *via* the intermediate cyanoketones (see Section 4.16.3.1.6).

Properly substituted isoxazolecarboxylic acids can be converted into esters, acid halides, amides and hydrazides, and reduced by lithium aluminum hydride to alcohols. For example, 3-methoxyisoxazole-5-carboxylic acid (**212**) reacted with thionyl chloride in DMF to give the acid chloride (**213**) ⟨74ACS(B)636⟩. Ethyl 3-ethyl-5-methylisoxazole-4-carboxylate (**214**) was reduced with LAH to give 3-ethyl-4-hydroxymethyl-5-methylisoxazole (**215**) ⟨73OS(53)70⟩.

Treatment of silver isoxazole-3-carboxylate (**216**) with bromine in refluxing carbon tetrachloride gave the 3-bromoisoxazole (**217**) in 42–47% yield. Similarly, silver 5-phenylisoxazole-3-carboxylate gave 3-bromo-5-phenylisoxazole ⟨74KGS1697⟩. Dehydration of isoxazole-3-carboxylic acid amide with P_2O_5 gave 3-cyanoisoxazole. 3-Cyano-5-methylisoxazole, 5-cyano-3-methylisoxazole and 4-cyano-5-methylisoxazole were also obtained from the corresponding carboxamides ⟨38G109⟩.

Similar reactions have been observed with 1,2-benzisoxazole-3-carboxylic acids ⟨67AHC(8)277, p. 294⟩ and the isomeric 2,1-benzisoxazole-3-carboxylic acids ⟨67AHC(8)277, p. 330⟩.

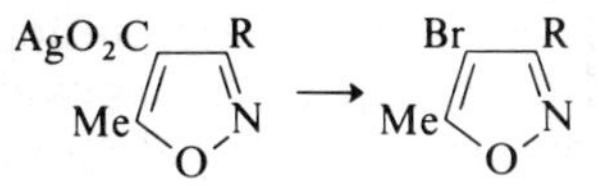

(**216**) R = Me, Ph (**217**)

(*b*) *Aldehydes and ketones.* In general, the properties of 3,5-disubstituted isoxazole carbonyl compounds and those of their benzenoid analogs are similar. Thus, 5-methyl-3-phenylisoxazole-4-carbaldehyde (**218**) and 5-acetyl-3-chloroisoxazole (**219**) were oxidized with alkaline permanganate to the corresponding isoxazole-4-carboxylic acid (**220**) and isoxazole-5-carboxylic acid (**221**), respectively. 3-Acetyl-4-chloro-5-methylisoxazole (**222**) was oxidized with chromic acid to the corresponding isoxazole-3-carboxylic acid (**223**), and 3-acetyl-4,5-dimethylisoxazole (**224**) was oxidized with nitric acid to the corresponding isoxazole-3-carboxylic acid (**225**).

(**218**) (**220**) (**219**) (**221**)

(**222**) (**223**) (**224**)

Other standard carbonyl reactions also occur readily. Thus 5-acetyl-3-methoxyisoxazole (**226**) was converted into 5-(1-aminoethyl)-3-methylisoxazole (**228**) *via* the oxime derivative (**227**) ⟨74ACS(B)636⟩.

The phosphonium salts (**230**) can be prepared in 80–90% yield from the 4-acetylisoxazoles (**229**) ⟨75CI(L)1018⟩.

(**226**) (**227**) (**228**)

(**229**) R = Me, Ph; R′ = H, Me (**230**)

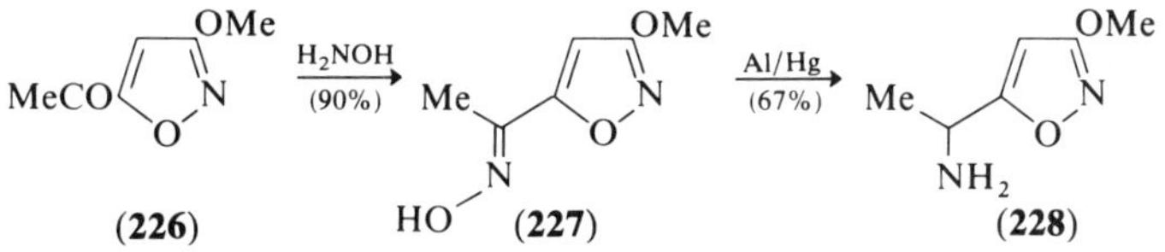

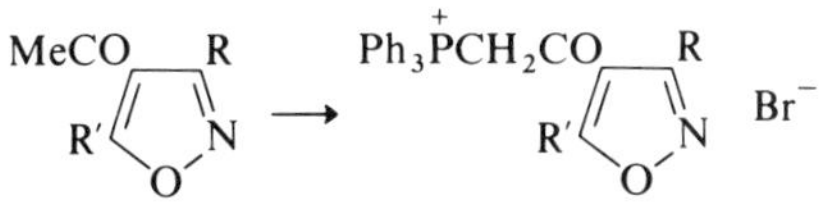

Both 1,2- and 2,1-benzisoxazoles with carbonyl groups in the 3-position show reactions such as those above, characteristic of the carbonyl function ⟨67AHC(8)277, p. 296, 329⟩.

(*c*) *Chloroalkyl derivatives.* The properties of 3,5-disubstituted isoxazoles are in general similar to those of their corresponding benzenoid analogs. Thus, the 4-chloromethylisoxazole (**231**) was oxidized with chromic acid to the corresponding isoxazole-4-carboxylic acid (**232**). 3-(Bromoacetyl)isoxazole (**233**) when treated with ammonia gave the pyrazine derivative (**234**) ⟨63AHC(2)365⟩, behaving as a typical α-aminoketone. The 4-chloromethylisoxazole (**235**) was treated with potassium phthalimide in DMF to give the phthalimide derivative (**236**) in 78% yield ⟨73JMC512⟩.

(**231**) (**232**) (**233**) (**234**)

(**235**) (**236**)

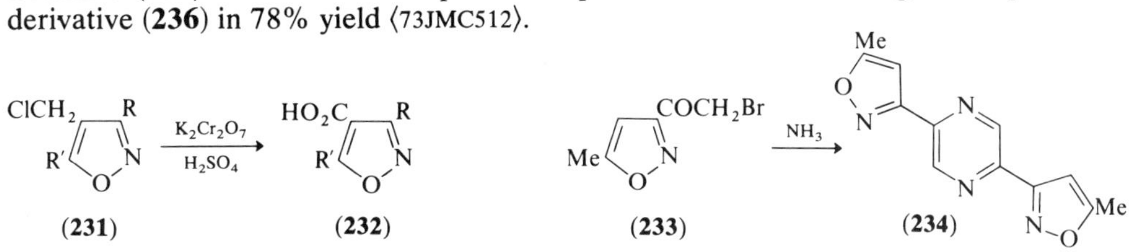

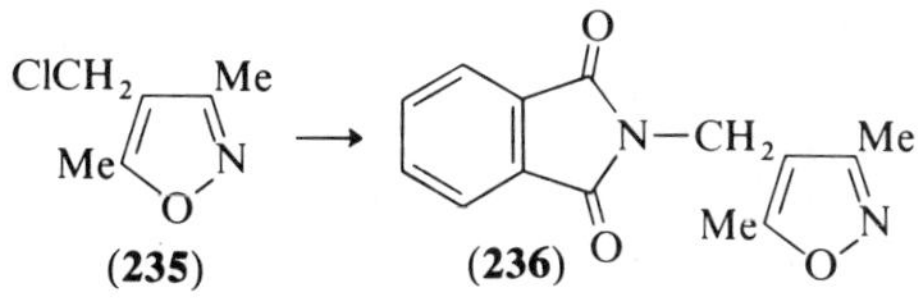

(d) *Vinyl derivatives.* The usual alkenic reactions from appropriately substituted isoxazoles are observed. Thus, 3,5-dimethyl-4-vinylisoxazole was oxidized with alkaline permanganate to 3,5-dimethyl-4-isoxazolecarboxylic acid ⟨76JCS(P1)570⟩ and 3-isopropenylisoxazole with acidic permanganate was converted into 3-acetylisoxazole ⟨62HC(17)1, p. 43⟩. Polymerization of 3-isopropenylisoxazoles has also been reported ⟨72MI41612⟩.

4.16.3.3.3 *N-Linked substituents*

(i) *Aminoisoxazoles*

(a) *Reaction with saturated electrophilic reagents.* The rates of N-methylation of isoxazole and the 1,2- and 2,1-benzo analogs were studied and compared with those of pyrazole and isothiazole (and their benzo derivatives). The rates of quaternization in DMSO with MeI or $(MeO)_2SO_2$ were generally faster for the nonbenzenoid nucleus, with the isoxazoles being the slowest; however, the 2,1-benzisoxazole was faster than isoxazole ⟨74AJC1221⟩. 5-Amino-3-phenylisoxazole (**237**) when reacted with DMF/POCl₃, in addition to the formation of an amidine at the amino group, also underwent formylation at the 4-position to give (**238**). This is an attractive intermediate as a building block to form isoxazolo[5,4-*d*]pyridines (**239**) ⟨80CPB1832⟩.

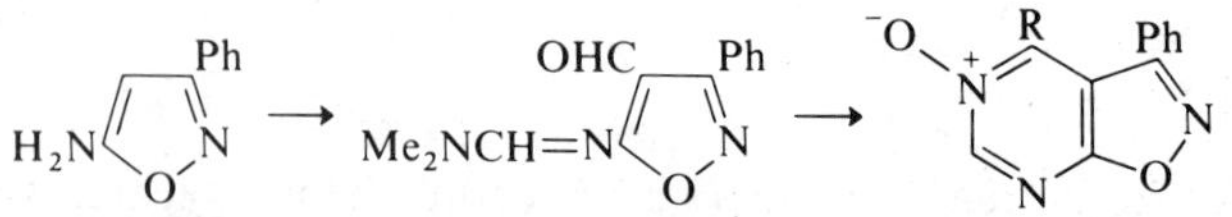

(b) *Reaction with unsaturated electrophiles.* 3-Aminoisoxazoles react with phenyl isothiocyanate to form thioureas (**240**) which subsequently rearrange on heating to thiadiazoles (**241**) ⟨77JCS(P1)1616⟩. 5-Aminoisoxazoles react with ethoxyacrolein to produce isoxazolo[5,4-*b*]pyridines *via* an intermediate isoxazolylacrolein which in some cases could be isolated (Scheme 84) ⟨79S449⟩. Structures of the general formula (**242**) could also be obtained by the reaction of 5-aminoisoxazoles with β-diketones (Scheme 84) ⟨77H(7)51⟩.

Scheme 84

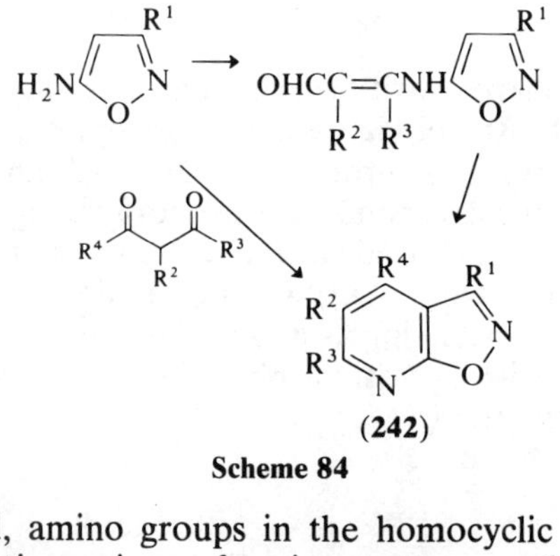

As would be anticipated, amino groups in the homocyclic ring of 1,2-benzisoxazoles behave as normal aromatic amines, forming mono- and bis-acyl derivatives, *etc.* ⟨67AHC(8)277, p. 296⟩. In the isomeric 2,1-benzisoxazoles the 3-amino compound exists as such and not in the tautomeric 3-imino form ⟨65CB1562⟩. Amino groups in 3-phenyl substituents behave as normal aromatic amines ⟨67AHC(8)277, p. 331⟩.

(c) *Diazotization.* 5-Aminoisoxazoles can be diazotized by isopentyl nitrite to generate the diazonium ion which can be decomposed to form an isoxazol-5-yl radical; the latter will dimerize or react with I_2, EtOH/Cu or ArH to give the 5-iodo, 5-hydro or 5-aryl derivative, respectively ⟨79AHC(25)147⟩. Diazotization of 4-amino-3-methyl-5-phenylisoxazole by $NaNO_2/H_2SO_4$ in acetic acid, containing nitrosylsulfuric acid, followed by

SnCl$_2$ reduction produced the 4-hydrazinoisoxazole (**243**). In ethanol the diazonium salt reacted with the 4-aminoisoxazole to produce the linear triazine (**244**) (Scheme 85). Diazoisoxazoles can also be treated with KI or H$_2$O/urea to produce the 4-iodo or 4-hydroxy derivatives ⟨63AHC(2)365⟩. These Sandmeyer reactions have been extended to a variety of isoxazole systems ⟨77JMC934, 63AHC(2)365⟩.

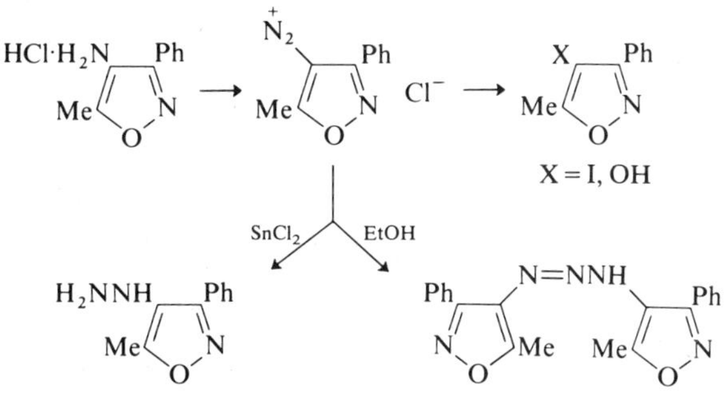

Scheme 85

3-Amino-1,2-benzisoxazoles which are stable to acids and bases can be diazotized in acetic acid, usually giving 3-hydroxy compounds.

(*d*) *Nucleophiles.* The reaction of 5-aminoisoxazoles (**245**) with phenylhydrazine produced the pyrazolinones (**246**) in good yields ⟨75S20⟩. In like manner the 5-hydrazinoisoxazoles (**247**) yield a mixture of pyrazolinones as shown in Scheme 86 with (**248**) being predominant ⟨75S20⟩. These reactions involve nucleophilic ring opening processes which are discussed in Section 4.16.3.1.5.

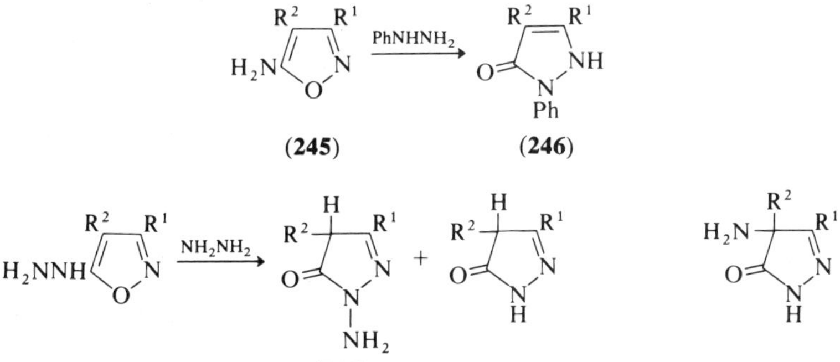

Scheme 86

Other amino substituted isoxazoles undergo ring-opening reactions on treatment with base. Thus the amidine derivative (**249**) gave the triazole (**250**) ⟨64TL149⟩, while the triazene (**251**) on reaction with ammonia gave the tetrazole (**252**) ⟨64T461⟩.

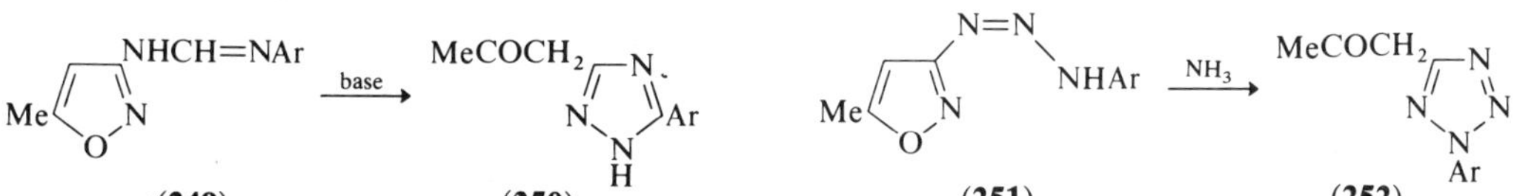

(*ii*) *Nitro groups*

Isoxazoles with nitro groups attached are generally relatively inert and although this group may influence the relative rates of reaction, the fragmentations and rearrangements are the same as those reported elsewhere in this chapter and will not be discussed here. 3-Nitroisoxazoles undergo alkaline hydrolysis and result in either displacement of the nitro group by HO or ring fragmentation, depending on the substituents. HCl/SnCl$_2$ reduction of 3-nitroisoxazoles produced the 3-amino compound. 4-Nitroisoxazoles can be reduced chemically to the 4-amino derivatives ⟨62AHC(17)1, p. 33⟩ and this has been extended to 3-nitroisoxazoles ⟨63MI41600⟩.

Nitro groups in the homocyclic ring of 1,2-benzisoxazoles have been reduced to the corresponding amino groups without opening of the isoxazole nucleus. $SnCl_2/HCl$ and Adam's catalyst/H_2 have both proved effective in this respect. Similar behavior was observed for nitro groups in the isomeric 2,1-benzisoxazoles ⟨67AHC(8)277, pp. 295, 331⟩.

4.16.3.3.4 O-Linked substituents

(i) 3-Hydroxyisoxazoles

(a) Electrophilic attack on oxygen. Acetylation of 3-hydroxyisoxazole gave mixtures of *O*- and *N*-acetyl derivatives, the ratio of which depends upon the nature of the substituents and the acetylating moiety ⟨79AHC(25)147, 77MI141600⟩. The reaction of *N*-protected amino acids with 3-hydroxy-5-methylisoxazole in the presence of dicyclohexylcarbodiimide gave the *O*-ester which slowly changed into the *N*-acyl derivative; the latter was converted into a 1,3,5-triazine by reaction with biguanidines ⟨73CL185⟩.

3-Hydroxy-1,2-benzisoxazoles show a variety of reactions. With $POCl_3/Et_3N$ the 3-chloro compound is formed (see Section 4.16.3.3 for further discussion).

The acetylation of 3-hydroxy-1,2-benzisoxazole generally produced a mixture of *O*- and *N*-acetylation, depending upon the conditions ⟨78MI41611, 69CB3775⟩. Controlled heating of the *O*-acetyl compound gave the *N*-acetyl compound. More extensive heating or irradiation produce benzoxazolones (Scheme 87) ⟨78MI41611, 69CB3775⟩.

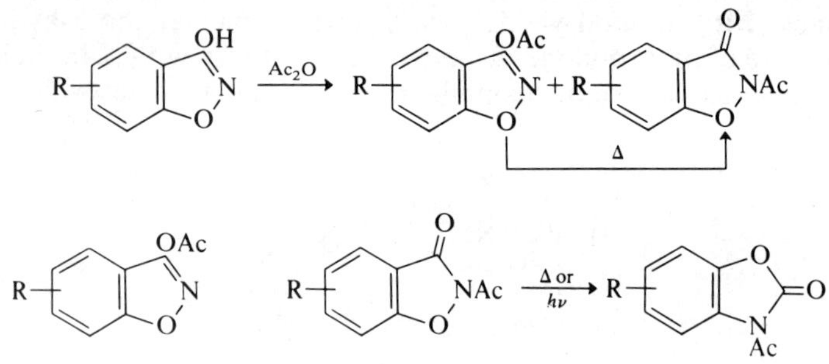

Scheme 87

Additional *O*- or *N*-acyl derivatives undergo similar rearrangements. The carbonate (**253**) on heating at 120 °C for 5 minutes gave the benzisoxazole (**255**) *via* the proposed intermediate (**254**) ⟨79MI41614, 70CB123, 69CB3775⟩. Heating (**253**) at 225 °C gave benzoxazolone (**256**), while heating of (**255**) at 225 °C did not give (**256**) but instead decarboxylation occurred to give *N*-methylbenzisoxazolone (**257**).

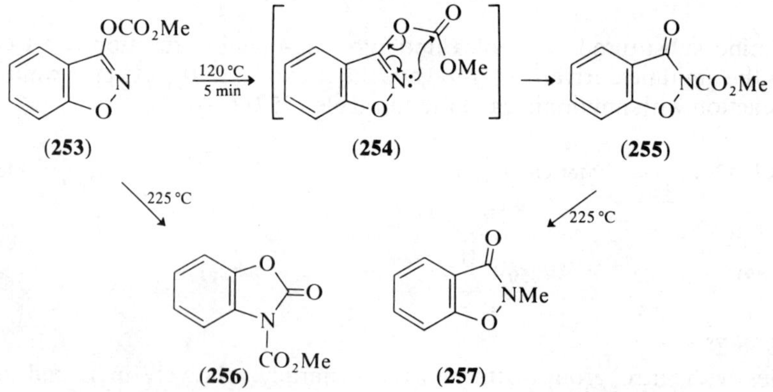

Treatment of 3-hydroxy-1,2-benzisoxazole with benzyl bromide gave a mixture of *O*- and *N*-benzyl compounds. The *N*-benzyl compound gave a benzoxazin-4-one on reaction with base, *via* the intermediates shown in Scheme 88 ⟨78CPB549⟩.

In contrast to the 3-amino derivatives, 3-hydroxy-2,1-benzisoxazoles are relatively labile. With nitrous acid they undergo ring fission to anthranilic acid. Its 3-acetoxy derivative (**258**) reacts with primary amines to form the quinazolone (**259**) ⟨67AHC(8)277, p. 297⟩.

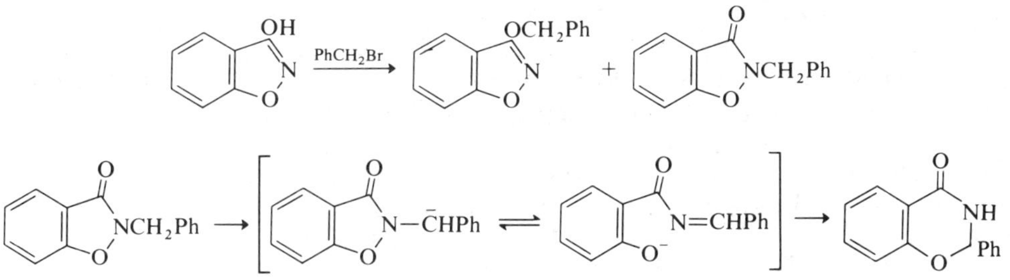

(ii) 4- and 5-hydroxyisoxazoles

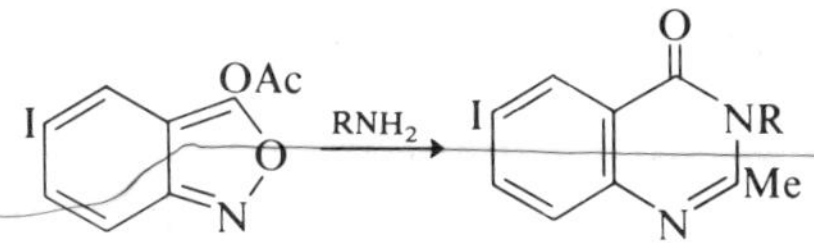

Generally, the 4-hydroxyisoxazole will react with hydrazines and hydroxylamine to form hydrazones and oximes which then undergo further reaction. The 4- and 5-hydroxyisoxazoles react with alcohols to form the alkoxyisoxazoles, and react with $SOCl_2$ and the like to form the halogen derivatives ⟨79AHC(25)147, 63AHC(2)365⟩. The fragmentation reactions of the nucleus are generally affected by the presence of these groups and thus will not be discussed here.

(iii) Alkoxyl groups

The hydroxyisoxazoles can be alkylated by alkyl halides or by tosylates. The alkoxy groups are quite stable and, in general, are unreactive. The alkoxy groups can be removed by the standard ether fragmentors except for the 3-alkoxy derivatives which can be removed by hydrolysis with NaOH ⟨63MI41600⟩. The group is not readily displaced but some reactions have been reported ⟨79AHC(25)147, 63AHC(2)365⟩. The presence of a 4-cyano group can cause activation of the 5-position and the methoxy group of (260) can be displaced by amines to form (261) or by Ph (in PhMgBr) to give (262) ⟨73AC(R)613⟩. Ester groups are also successful as activating groups.

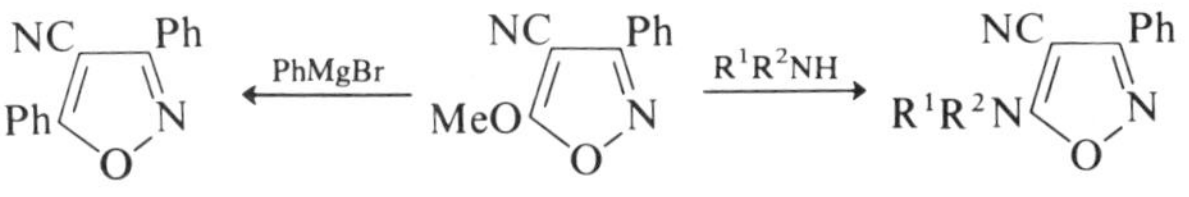

3-Alkoxy-2,1-benzisoxazole-4,7-diones undergo ready nucleophilic displacement of the 3-alkoxy substituent, yielding 3-alkylamino and 3-dialkylamino derivatives with primary and secondary amines, respectively ⟨67TL4313⟩. In this instance the 4-carbonyl group apparently provides an activating effect.

4.16.3.3.5 S-Linked substituents

Among the isoxazolethiols, all are reported as the SH tautomer, although some reports for the 3-thione exist ⟨79CPB2398⟩. Most have been prepared from haloisoxazoles by displacement with sulfides and thiolates (see below). The thione form has been postulated for 4-arylazo-3-methylisoxazole-5-thione ⟨78ZN(B)1056⟩. A series of isoxazole-3-thiols was prepared starting from allylthioisoxazoles, but these were unstable in air and oxidized to disulfides. In alkaline solution 5-phenylisoxazole-3-thiol was fragmented into benzoylacetonitrile and sulfur ⟨80CPB552⟩. The thio group can be directly alkylated or the sulfur can be removed by treatment with select metals and by hydrolysis ⟨79AHC(25)147, 63AHC(2)365⟩.

4.16.3.3.6 Halogen atoms

(i) Nucleophilic substitution reactions

4-Substituted 3-chloroisoxazoles undergo displacement reactions with alkoxide ion ⟨63AHC(2)365, 66CPB89⟩, but 3-alkylthioisoxazoles cannot be synthesized by displacement with alkylthiolates ⟨79CPB2415⟩. Generally, halogen atoms in the 5-position can be displaced when a suitable activating group is present in the 4-position; these groups including ester, cyano and benzoyl moieties ⟨63AHC(2)365⟩. 5-Aminoisoxazoles can be prepared by the reaction of 5-chloroisoxazoles with amines ⟨77JMC934⟩. 5-Chloro-3-phenylisoxazole reacted with allyl alcohol at 90 °C for 30 minutes in the presence of KOH to give 80% of 4-allyl-3-phenyl-3-isoxazolin-5-one. An intermediate 5-allyloxy-3-phenylisoxazole has been postulated for this conversion.

Halogen atoms in the 3- and 5-positions of isoxazoles can also be activated if the ring is quaternized. For example, the reaction of 3-chloroisoxazolium chlorides (**263**) with sodium thiophenolate gave the 3-phenylthioisoxazolium chlorides (**264**), which were then converted into 3-phenylthioisoxazoles (**265**) by thermolysis ⟨79CPB2415⟩. These nucleophilic displacement reactions have been extended to a variety of 3-blocked 5-chloro- and 5-blocked 3-chloro-isoxazoles ⟨77JMC934, 63MI41600, 61G47, 63AHC(2)365⟩.

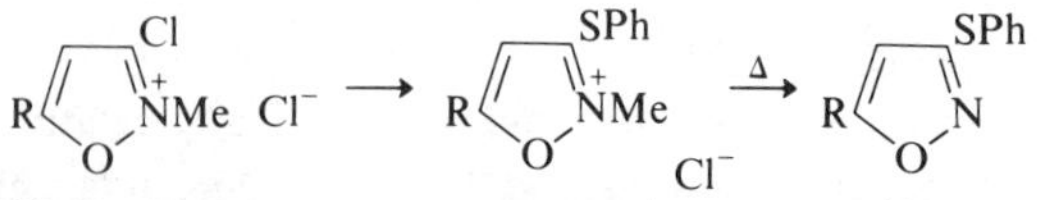

(**263**) R = Ph, Me   (**264**)   (**265**)

3-Chloro-1,2-benzisoxazole, which was prepared by the action of $POCl_3/Et_3N$ on the 3-hydroxy compound, undergoes nucleophilic displacement of the chloro group to form a variety of products. Azide ion reacted to form the 3-azido derivative which was reduced to the 3-amino compound ⟨80ZC18, 79ZC452⟩. Other nucleophiles include hydrazine, ammonia, alkoxide, substituted amino and OH (KOH) (Scheme 89) ⟨80ZC18, 79ZC452, 76CB3326⟩.

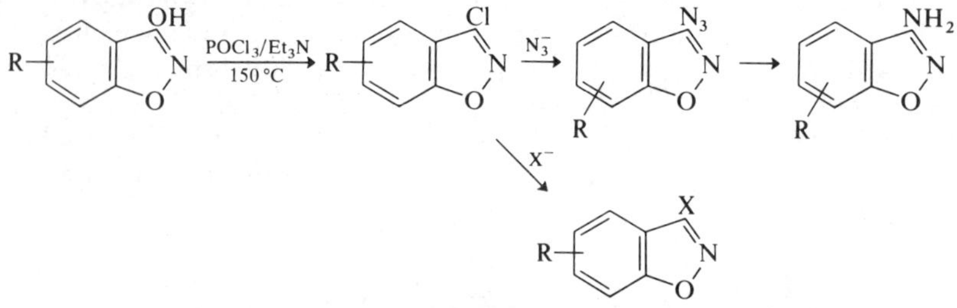

X = $NHNH_2$, NH_2, OEt, NHR_2, OH

Scheme 89

(ii) Other reactions

Grignard reagents can be successfully prepared from 4-iodoisoxazoles and these react with CO_2 and ketones to give carboxylic acids and methanols, respectively. 3-Bromo- or 3-chloro-isoxazoles can be reduced by sodium amalgam to give as intermediates the 3H-compound, or a ring-opened product which then proceeds further ⟨63AHC(2)365⟩.

The reaction of 4-bromo- or 4-iodo-isoxazoles with ethylmagnesium bromide gave the 4-MgBr reagent by an exchange reaction, while the 4-chloro derivative undergoes reductive cleavage (Scheme 90) ⟨63AHC(2)365⟩.

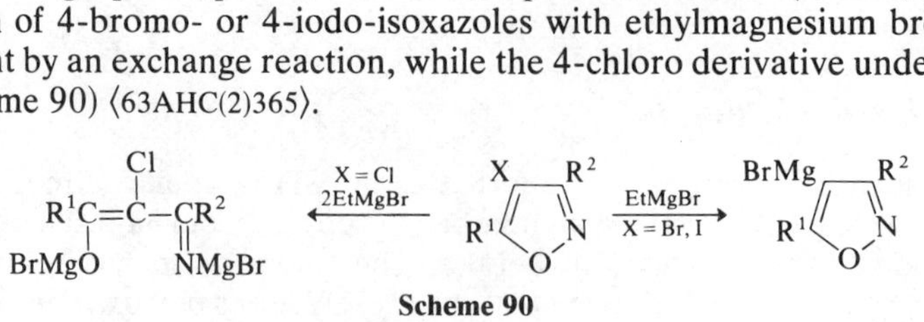

Scheme 90

4.16.3.3.7 Metal- and metalloid-linked substituents

4-Isoxazolyl Grignard reagents react normally in that a 4-carboxylic acid or 4-methanol can be obtained by reaction with CO_2 or ketones ⟨63AHC(2)365⟩. Lithiation of 3,5-disubstituted isoxazoles occurred in the 4-position and subsequent reaction of (**266**) with trimethyltin chloride or trimethylsilyl chloride gave (**267**) ⟨79AHC(25)147⟩. The isoxazolylmercury compound (**268**) reacted with KBr and Br_2 to give the 4-bromo compound (**269**) ⟨63AHC(2)365⟩. The reaction of 3,5-dimethyl-4-iodoisoxazole (**270**) with ethylmagnesium bromide gave large amounts of diisoxazole (**271**) ⟨80JOM(195)275⟩.

4.16.3.3.8 Substituents attached to ring nitrogen atoms

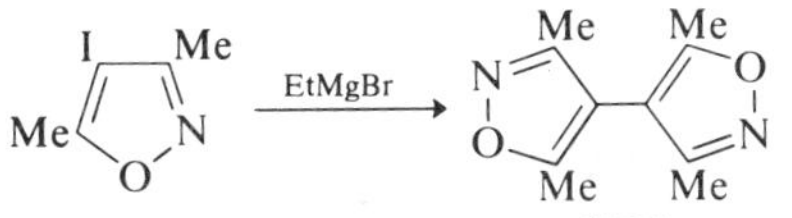

(i) Alkyl groups

Treatment of *N*-benzyl-1,2-benzisoxazolin-3-one with base produced a benzoxazine-4-one (see Scheme 88). The base catalyzed rearrangement of the 2-methyl-3-phenyl-1,2-benzoisoxolium salt to an oxazine is believed to proceed *via* a similar intermediate ⟨67AHC(8)277⟩. A number of other decompositions could possibly proced *via* this proposed route ⟨74HCA376, 67AHC(8)277⟩, which has also been postulated for the rearrangement of a variety of isoxazolium salts, *e.g.* the conversion of (**200**) into (**202**) (Section 4.16.3.3.2(i)(b)).

Alkylation of 3-methyl-4-phenylisoxazolin-5-one with allyl bromide gave a mixture of *N*- and *C*(4)-alkylation in a 2:1 ratio. Heating the mixture changed the ratio to 1:99 and this conversion is believed to take place by an amino-Claisen rearrangement (Scheme 91) ⟨69TL543⟩.

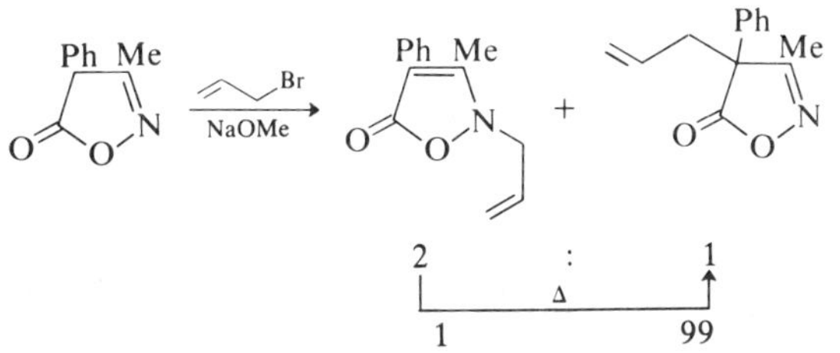

Scheme 91

(ii) Other groups

The oxygen of 1,2-benzisoxazole *N*-oxides can be removed by treatment with triethyl phosphite ⟨80CC421⟩, and the oxygen of 3,4,5-triphenylisoxazoline *N*-oxide with H_2 catalysis and PCl_5 ⟨69JOC984⟩. Isoxazoline *N*-oxide (**272**) reacts with styrene to give cycloadduct (**273**) ⟨66ZOR2225⟩.

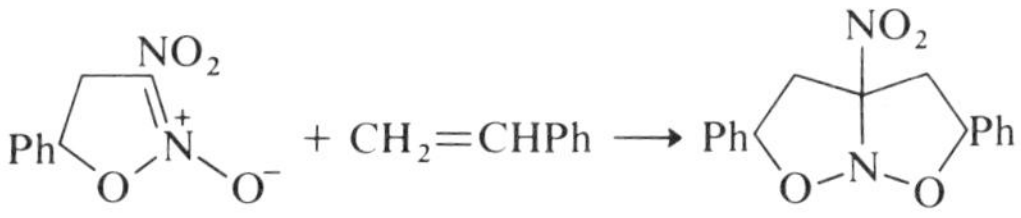

Generation of an anion at the 3-position of certain 2-methoxyisoxazolidines leads to loss of the *N*-methoxyl group (Scheme 92) ⟨79T213⟩.

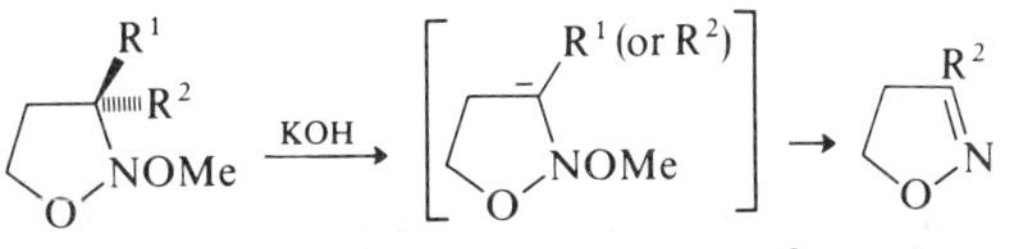

$R^1 = R^2 = CO_2Me$;
$R^1 = CO_2Me, R^2 = CONHMe$

$R^2 = CO_2Me$ or $CONHMe$

Scheme 92

4.16.3.3.9 Special reactions of side-chain substituents: rearrangements involving three-atom side chains

The rearrangement represented by the conversion of (274) into (275) involves derivatives of a considerable number of azoles, including the isoxazoles. These are of general interest and isoxazole derivatives which undergo this rearrangement are shown in Table 10.

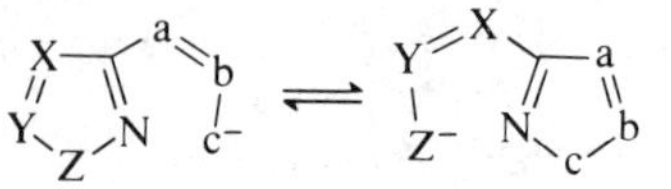

(274) ⇌ (275)

Table 10 Rearrangement of Isoxazoles with Suitable Three-atom Side Chains

Starting isoxazole	Product	Ref.
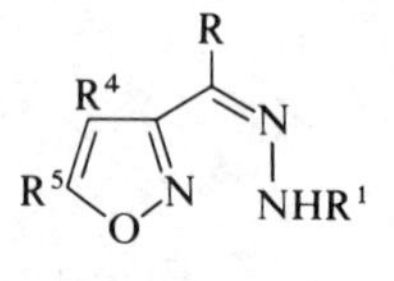	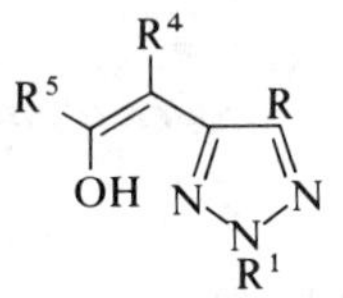	40G770, 47G332, 67JCS(C)2005
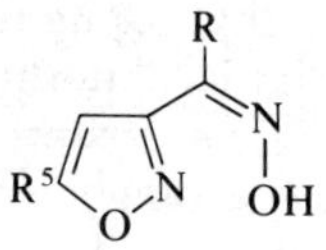	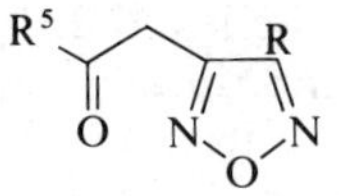 (isolated as oxime)	67JCS(C)2005, 37G779, 38G792, 39G391, 46G1
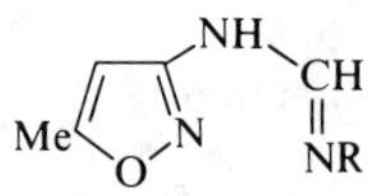	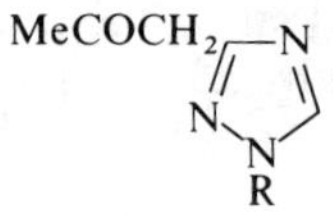	64T159
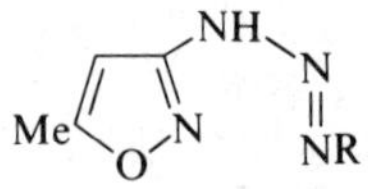	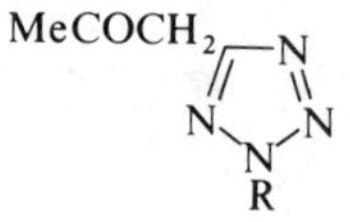	64T461
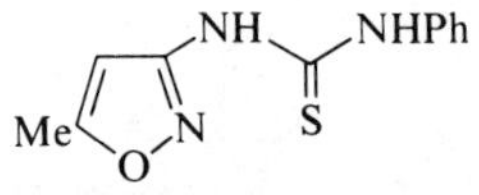	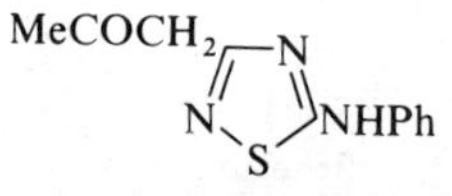	77JCS(P1)1616

Although evidence is not conclusive, indications are that the rearrangements are concerted. Heteroatom compositions required for the rearrangement are at least one N—O bond in the nucleus of the starting material and the formation of a C—N, N—N or C—S bond in the product ⟨79AHC(25)147, p. 193, 81AHC(29)141⟩.

Related 1,2-benzisoxazole derivatives such as 3-(o-aminoaryl)benzisoxazoles (276) undergo rearrangement on refluxing in THF solution in the presence of NaH or LiAlH$_4$, giving the indazoles (277) ⟨74JHC885⟩.

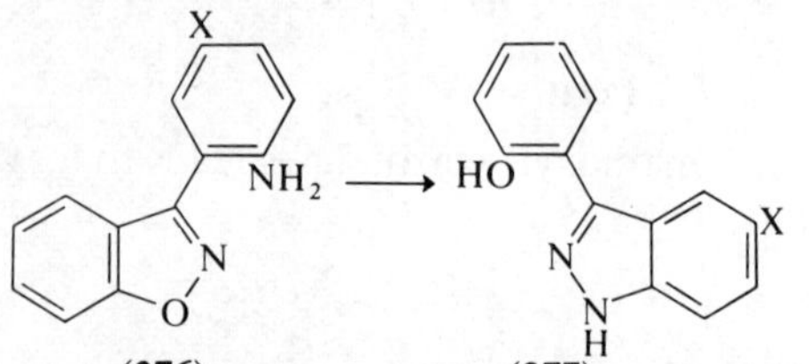

(276) → (277) 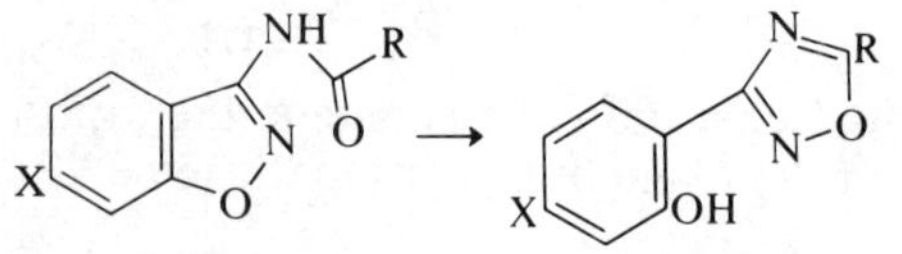   (278) → (279)

Several 3-acylamino-1,2-benzisoxazoles such as (278) underwent rearrangement to 3-(o-hydroxyphenyl)-1,2,4-oxadiazoles (279) on heating with base. It has been questioned whether the same mechanism is operative in these last rearrangements ⟨81AHC(29)141, p. 150⟩.

4.16.4 SYNTHESIS

4.16.4.1 Isoxazoles

4.16.4.1.1 Introduction

Synthetic routes to isoxazoles may be classified according to the number of isoxazole ring atoms in each component synthon (*e.g.* 3+2), which is then subdivided by the type and arrangement of ring atoms in each component (*e.g.* CCC+NO). The [3+2] routes containing (CCC+NO) and (CNO+CC) reactions embrace perhaps more than 90% of the practical preparations of isoxazoles. The [4+1] and [5+0] routes provide new regiospecific syntheses of isoxazoles and the [5+0] routes involve cyclization of synthons containing all five atoms, usually in the sequence (CCCNO). Some [3+1+1] routes have also been reported, while to our knowledge [2+2+1] routes have not yet been described. Isoxazoles have also been prepared by dehydrogenation of isoxazolines and ring-transformation reactions.

4.16.4.1.2 [3+2] Routes

(i) (CCC + NO) reactions

In 1888 Claisen ⟨1888CB1149⟩ first recognized a general synthesis of isoxazoles (**283**) by the condensation–cyclization of 1,3-diketones (**280**) with hydroxylamine. It is now generally accepted that the monoxime (**281**) of the 1,3-diketone and the subsequent 5-hydroxyisoxazoline (**282**) are the intermediate products of the reaction. The isolation of the monoxime (**281**) and 5-hydroxyisoxazoline (**282**), which were both readily converted into the isoxazole (**283**) by treatment with acid or base, has been reported ⟨62HC(17)1⟩.

RCOCH$_2$COR $\xrightarrow{NH_2OH}$ [RCCH$_2$COR ‖ NOH] ⟶ [isoxazoline with R, OH, R] ⟶ [isoxazole with R, R]

(**280**) (**281**) (**282**) (**283**)

The method provides a convenient preparation of 4-substituted isoxazoles having the same substituent at the 3- and 5-positions. For example, the reaction of tetraalkoxypropanes (**284**) with hydroxylamine readily gave unsubstituted isoxazole (**285a**) in 72% yield and 4-substituted isoxazoles (**285b**) in 43–67% yields ⟨63AHC(2)365⟩. The 1,3-dicarbonyl compound (**286**) gave isoxazole (**287**) ⟨63AHC(2)365⟩, and the 1,3-dicarboxylic acid ester (**288**) gave disic acid (**289**) ⟨75T1861⟩.

(R'O)$_2$CHCHRCH(OR')$_2$ $\xrightarrow{NH_2OH}$ [isoxazole with R]

(**284**)

(**285a**) R = H (72%)
(**285b**) R = Me, Et or Ph (43–67%)

2,4-(NO$_2$)$_2$C$_6$H$_3$CH(COMe)$_2$ $\xrightarrow{NH_2OH}$ [isoxazole with Me, C$_6$H$_3$(NO$_2$)$_2$-2,4, Me]

(**286**) (**287**)

ArCH(CO$_2$Et)$_2$ $\xrightarrow{NH_2OH}$ [isoxazole with HO, Ar, OH]

(**288**) (**289**)

It was soon found that the reaction of unsymmetrical 1,3-diketones (**290**) or their derivatives with hydroxylamine results in both possible isomeric isoxazoles (**291**) and (**292**), a complication which not only reduces the yield of desired product but also often leads to separation problems, particularly when R and R' are similar. However, the reaction does give one isomer, or predominantly one isomer, if the right combination of the CCC component synthon and reaction conditions are used. The regiospecificity of the reaction can be influenced by making one of the terminal carbon atoms in the CCC component synthon more or less electrophilic than the other; by protecting one of the terminal carbon atoms with a dioxolane group; or by controlling the pH of the reaction medium. These aspects have been the subject of extensive study ⟨79AHC(25)147⟩.

$$\text{RCOCH}_2\text{COR}' \xrightarrow{\text{NH}_2\text{OH}} \underset{(291)}{\text{N}\overset{R}{\underset{O}{\diagdown}}\text{R}'} + \underset{(292)}{\text{N}\overset{R'}{\underset{O}{\diagdown}}\text{R}}$$

(290)

The CCC component synthons are frequently derived from 1,3-dicarbonyl compounds (**290**) which can exist in any of the tautomeric forms (**290a**), (**290b**) or (**290c**), or as some equilibrium mixture of these tautomers. In general, (**290a**), (**290c**) or a mixture of both is the main form present in the reaction mixture; in acidic media, (**290a**) and (**290c**) are in dynamic equilibrium with (**290b**) ⟨66JOC3193⟩. The structure of the principal product of the reaction can be predicted on the basis of the more or less positive character of the two carbon atoms bearing R and R'. Thus, the reactivity of carbonyl functions in (**290**) is in the order of aldehyde > ketone > ester. The terminal carbon atom can also be derived from the functional groups: —C(=O)NRR', —C≡N, =CR—NR'$_2$, =CR—OR', =CR—SR', =CR—X, —COCO$_2$R, ≡CR, —CHX—CHXR, —CX=CHR, —C(OR)$_2$— and —CR—CR'R''.

$$\underset{(290a)}{\overset{\text{OH}\cdots\text{O}}{\text{RC}=\text{CHCR}'}} \rightleftarrows \underset{(290b)}{\overset{\text{O}\quad\text{O}}{\text{RCCH}_2\text{CR}'}} \rightleftarrows \underset{(290c)}{\overset{\text{O}\cdots\text{HO}}{\text{RCCH}=\text{CR}'}}$$

Hydroxylamine, the major source of the NO component, is commercially available as the hydrochloride or as the sulfate. The pK_a of hydroxylamine has been reported to be 6.0 and 13.0, as shown in equations (1) and (2) ⟨61JPC1279⟩. Here the oxygen atom is more nucleophilic than the nitrogen atom in NH$_3$OH and NH$_2$O$^-$, and the reverse is true for NH$_2$OH. Thus, the pH of the reaction medium can be very crucial to the regiospecificity of the reaction, and the use of hydroxylamine hydrochloride without a base — particularly in acetic acid — is a poor choice of reaction conditions because a high concentration of protonated hydroxylamine in the reaction mixture may lead to the undesirable isomer ⟨63AHC(2)365, 45JA1745⟩.

$$\overset{+}{\text{NH}_3\text{OH}} \underset{}{\overset{\text{p}K_a = 6.0}{\rightleftarrows}} \text{NH}_2\text{OH} + \text{H}^+ \qquad (1)$$

$$\text{NH}_2\text{OH} \underset{}{\overset{\text{p}K_a = 13}{\rightleftarrows}} \text{NH}_2\text{O}^- + \text{H}^+ \qquad (2)$$

To date, a great deal of knowledge about the selection of the appropriate CCC component and reaction conditions has been accumulated. Some regiospecific syntheses of isoxazoles having different 3- and 5-substituents are illustrated by the following examples.

The reaction of the steroidal β-ketoaldehyde (**293**) with hydroxylamine hydrochloride in acetic acid gave a mixture of the 3- and 5-substituted isoxazoles (**294**) and (**295a**). In sodium acetate buffer the reaction provided exclusively the 5-substituted isomer (**295b**) ⟨66JOC3193⟩.

(**293**) R = H   (**294**) R = Ac   (**295a**) R = Ac
                                              (**295b**) R = H

The reaction of β-substituted vinyl ketones RCOCH=CHY (Y = halogen, OR' or NR'$_2$) with hydroxylamine hydrochloride has been extensively investigated ⟨63AHC(2)365, 62HC(17)1⟩. One would anticipate that replacement of hydroxylamine hydrochloride with hydroxylamine in these reactions would result in enhanced regiospecificity and increased yields of products. However, the reaction of alkyl and aryl dialkylaminovinyl ketones (**296**) with hydroxylamine hydrochloride gave exclusively 5-substituted isoxazoles (**297**) in excellent yields, giving this method considerable preparative value ⟨54IZV47, 77JHC345⟩.

$$\text{RCOCH=CHNR}'_2 \xrightarrow{\text{H}_2\text{NOH·HCl}} \underset{(\mathbf{297})}{\text{[isoxazole-R]}} \qquad \text{RCOC(CO}_2\text{Et)=CHOEt} \xrightarrow{\text{H}_2\text{NOH·HCl}} \underset{(\mathbf{299})}{\text{[isoxazole-CO}_2\text{Et, R]}}$$

(**296**)   (**297**)   (**298**)   (**299**)

Since an electron-withdrawing group such as ethoxycarbonyl at the α-carbon atom enhanced the electrophilicity of the β-carbon atom, the reaction of α-ethoxycarbonyl-β-ethoxyvinyl ketones (**298**) with hydroxylamine hydrochloride gave solely 5-substituted isoxazole-4-carboxylates (**299**) ⟨55JOC1342, 59YZ836⟩.

To synthesize isomeric 3-substituted isoxazoles (**301**) the reaction of ethylene acetals of β-ketoaldehydes (**300**) (readily available from β-chlorovinyl ketones ⟨57IZV949⟩) with hydroxylamine was employed. Owing to the comparative stability of the dioxolane group, this reaction gave exclusively 3-substituted isoxazoles (**301**) ⟨60ZOB954⟩. The use of noncyclic, alkyl β-ketoacetals in this reaction resulted in a mixture of 3- and 5-substituted isoxazoles ⟨55AG395⟩.

$$\text{RCOCH}_2\text{CH(OCH}_2\text{CH}_2\text{O)} \xrightarrow[\Delta]{\text{H}_2\text{NOH}} \underset{(\mathbf{301})\ R=\text{alkyl, phenyl (65-77\%)}}{\text{[3-R-isoxazole]}} \qquad \text{ArCOCH}_2\text{COR} \xrightarrow[\substack{\text{pyridine, }\Delta\\56-70\%}]{\text{H}_2\text{NOH·HCl}} \underset{(\mathbf{303})}{\text{[3-R-5-Ar-isoxazole]}}$$

(**300**)   (**301**) R = alkyl, phenyl (65–77%)   (**302**)   (**303**)

3-Alkyl-5-arylisoxazoles (**303**) were prepared by the regiospecific reaction of appropriate 1,3-diketones (**302**) (R = alkyl or perfluoroalkyl) with hydroxylamine hydrochloride in pyridine ⟨79MI41601⟩.

3-Alkyl(or 3-aryl)-5-methylisoxazoles (**306**) were prepared by the regiospecific reaction of phosphonium salts (**304**) with hydroxylamine, followed by the treatment of the resulting isoxazole-containing phosphonium salts (**305**) with aqueous sodium hydroxide ⟨80CB2852⟩.

$$\text{Ph}_3\overset{+}{\text{P}}\text{CH}_2\text{COCH}_2\text{COR}\ \text{Cl}^- \xrightarrow[60-77\%]{\text{NH}_2\text{OH}} \underset{(\mathbf{305})}{\text{[3-R-isoxazole-CH}_2\overset{+}{\text{PPh}}_3\text{]}\ \text{Cl}^-} \xrightarrow{\text{NaOH}} \underset{(\mathbf{306})\ R=\text{Me, Pr}^i\text{ or Ph}}{\text{[3-R-5-Me-isoxazole]}}$$

(**304**)   (**305**)   (**306**) R = Me, Pr^i or Ph

The reaction of α-bromoenone (**307**) with hydroxylamine hydrochloride in ethanol in the presence of potassium carbonate resulted in the regiospecific formation of 3-alkyl-5-phenylisoxazoles (**303**). On the other hand, when sodium ethoxide was used as the base under similar conditions, 5-alkyl-3-phenylisoxazoles (**308**) were obtained exclusively ⟨80CC826, 81H(16)145⟩.

$$\underset{(\mathbf{303})}{\text{[3-R-5-Ph-isoxazole]}} \xleftarrow[\text{K}_2\text{CO}_3]{\text{NH}_2\text{OH·HCl}} \underset{(\mathbf{307})}{\text{PhCO-C(Br)=CHR}} \xrightarrow[\text{NaOEt}]{\text{NH}_2\text{OH·HCl}} \underset{(\mathbf{308})\ R=\text{Me, Et, Pr}^n\text{ or Pr}^i}{\text{[3-Ph-5-R-isoxazole]}}$$

(**303**)   (**307**)   (**308**) R = Me, Et, Pr^n or Pr^i

Only in a few instances, where both phenyl groups were sufficiently different in their substitution patterns, were 3,5-diarylisoxazoles prepared regiospecifically by the reaction of 1,3-diketones (**302**) (R = substituted phenyl) with hydroxylamine ⟨45JA134⟩. Accordingly, other CCC component synthons have been employed for the regiospecific synthesis of 3,5-diarylisoxazoles.

The reaction of the α,β-alkynic ketone (**309a**) with hydroxylamine hydrochloride and sodium carbonate gave 5-p-anisyl-3-phenylisoxazole (**310a**), whereas the α,β-alkynic ketone (**309b**) under the same reaction conditions gave the oxime of β-hydroxychalcone which was then converted into 5-p-bromophenyl-3-phenylisoxazole (**310b**) by hydrochloric acid. However, the reaction of the α,β-alkynic ketones (**309a**) and (**309b**) with hydroxylamine hydrochloride and pyridine gave 3-(p-anisyl)-5-phenylisoxazole (**311a**) and 3-p-bromophenyl-5-phenylisoxazole (**311b**), respectively ⟨68JCS(C)1774⟩. The reaction of chalcone dibromides and chalcone oxides with hydroxylamine hydrochloride and sodium hydroxide to give 3,5-diarylisoxazoles regiospecifically has also been reported ⟨68JCS(C)1774, 61AP769⟩.

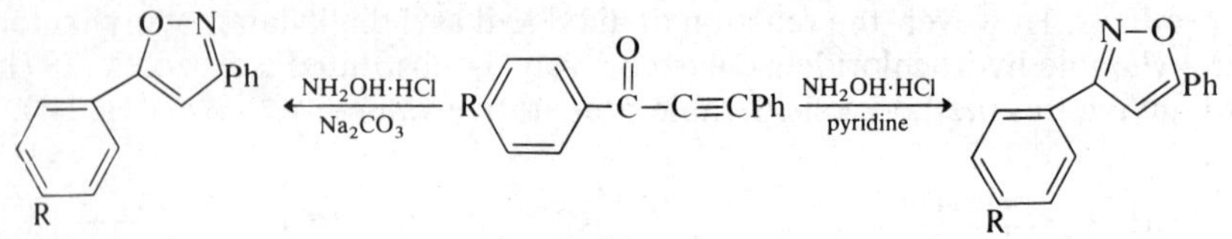

O–N
Ph
NH₂OH·HCl
Na₂CO₃
R
O
C–C≡CPh
NH₂OH·HCl
pyridine
N–O
Ph
R
R

(**310a**) R = OMe (**309a**) R = OMe (**311a**) R = OMe
(**310b**) R = Br (**309b**) R = Br (**311b**) R = Br

The reaction of methyl acetylpyruvate (**312**) with hydroxylamine hydrochloride gave the 3-carboxylate (**313**) in 76% yield together with traces of the isomeric 5-carboxylate (**314**) ⟨78MIP41600⟩. However, the sodium salt (**315**) of acetylpyruvic acid resulted in 3-methylisoxazole-5-carboxylic acid (**316**) as the major product.

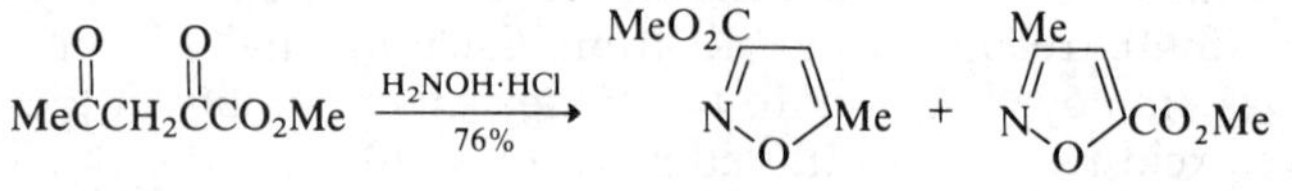

O O
MeCCH₂CCO₂Me
H₂NOH·HCl
76%
MeO₂C
N O Me
+
Me
N O CO₂Me

(**312**) (**313**) (**314**) trace

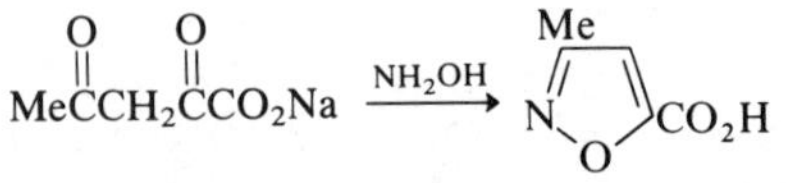

O O
MeCCH₂CCO₂Na
NH₂OH
Me
N O CO₂H

(**315**) (**316**)

The reaction of α,β-alkynic nitriles (**317**) with hydroxylamine gave the 5-aminoisoxazoles (**318**) regiospecifically, whereas in the presence of sodium hydroxide the 3-aminoisoxazoles (**319**) were obtained exclusively ⟨66CPB1277⟩. Similarly, the course of the cyclization of arylhydrazones (**320**) was influenced by a change in the base employed ⟨75JOC2604⟩.

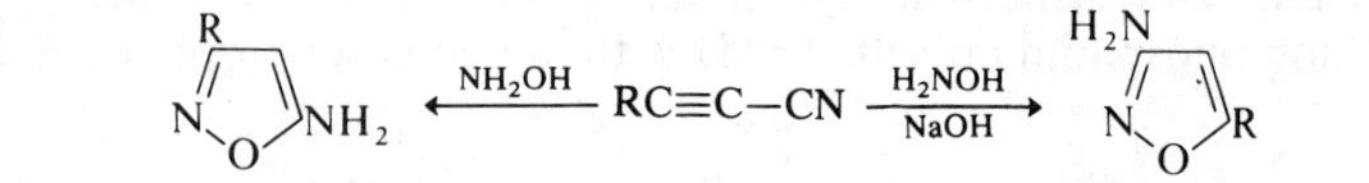

R
N O NH₂
NH₂OH
RC≡C–CN
H₂NOH
NaOH
H₂N
N O R

(**318**) R = H, Me or Ph (**317**) (**319**) R = H or Ph

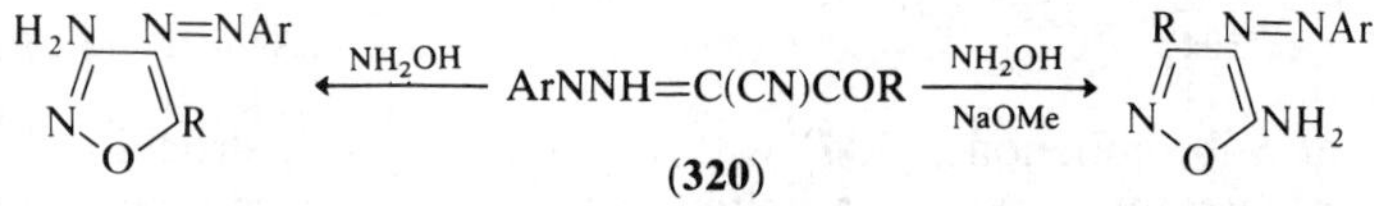

H₂N N=NAr
N O R
NH₂OH
ArNNH=C(CN)COR
NH₂OH
NaOMe
R N=NAr
N O NH₂

(**320**)

The reaction of α,β-alkynic esters (**321**) with hydroxylamine resulted in the regiospecific formation of 5-isoxazolones (**322**), whereas in the presence of sodium hydroxide only the 3-hydroxyisoxazoles (**324**) were obtained. In this latter reaction the alkynic hydroxamic acids (**323**) were thought to be intermediates ⟨66CPB1277⟩. 3-Hydroxyisoxazoles (**324**) were also obtained by the reaction of the glycol acetal of β-ketoesters with hydroxylamine in the presence of potassium hydroxide. Cyclization of the resultant intermediate hydroxamic acids was effected with hydrochloric acid ⟨70BSF1978, 74ACS(B)533⟩. 5-Isoxazolones (**322**) were also prepared by the reaction of β-ketoesters ⟨79MI41601, 62HC(17-1)1, p. 48⟩ or dioxane derivatives ⟨79JAP(K)79106466⟩ with hydroxylamine.

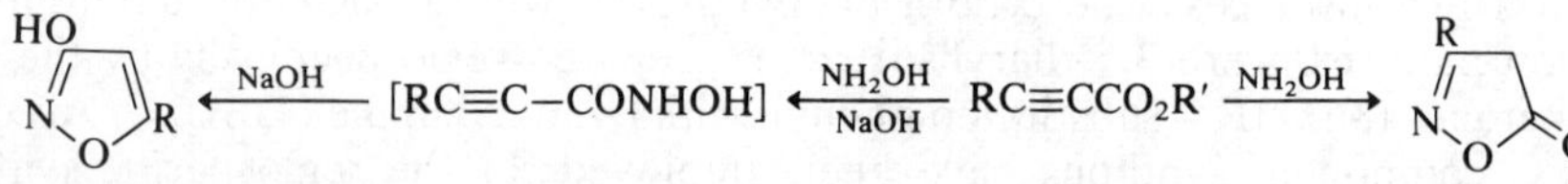

HO
N O R
NaOH
[RC≡C–CONHOH]
NH₂OH
NaOH
RC≡CCO₂R′
NH₂OH
R
N O O

(**324**) (**323**) (**321**) (**322**)

The following reaction sequence provides a regiospecific route to 3,4-disubstituted 5-isoxazolones (**328**) ⟨73ACS2802⟩. The β-ketoesters (**325**) were heated under reflux with benzylamine in benzene in the presence of molecular sieves (3 Å) to give the β-benzylamino-α,β-unsaturated esters (**326**). The latter reacted first with hydroxylamine hydrochloride in

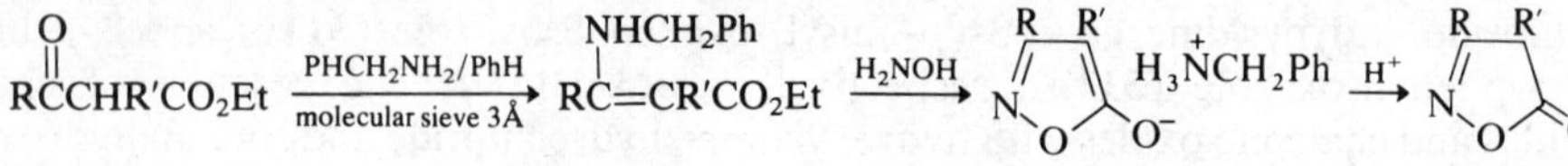

O
RCCHR′CO₂Et
PHCH₂NH₂/PhH
molecular sieve 3Å
NHCH₂Ph
RC=CR′CO₂Et
H₂NOH
R R′
N O O⁻
H₃N⁺CH₂Ph
H⁺
R R′
N O O

(**325**) (**326**) (**327**) (**328**)

the presence of sodium methoxide forming the salt (**327**), then with hydrochloric acid to give the 5-isoxazolones (**328**) in satisfactory yields.

The reactions (3)–(16) show additional regiospecific syntheses of substituted isoxazoles using a variety of starting components. They illustrate the introduction of diverse substituents into the isoxazole nucleus by choice of an appropriate starting material.

$$\text{PhCC(CN)=CNHPh} \xrightarrow[63\%]{\text{H}_2\text{NOH}} \underset{\text{N}\diagdown\text{O}\diagup\text{Ph}}{\overset{\text{PhNH}\quad\text{CN}}{\text{[isoxazole]}}} \quad \langle\text{78JPR585}\rangle \quad (15)$$

$$\text{PhCC(CN)=C(SMe)}_2 \xrightarrow[70\%]{\text{NH}_2\text{OH}} \underset{\text{N}\diagdown\text{O}\diagup\text{Ph}}{\overset{\text{MeS}\quad\text{CN}}{\text{[isoxazole]}}} \quad \langle\text{78JPR585}\rangle \quad (16)$$

Other variations of the (CCC + NO) reactions are worthy of mention. Propargylic halides (**329**) reacted with sodium nitrite to give 3-nitroisoxazoles (**330**) ⟨73TL485, 80ZOR1328⟩. α,β-Unsaturated aldehydes (**331**) or ketones reacted with dinitrogen trioxide or nitric acid to give 4-nitroisoxazoles (**332**) ⟨62HC(17)1, p. 28⟩. Acetonedicarboxylic acid esters (**333**) reacted with nitrosyl chloride to give 4-hydroxyisoxazole-3,5-dicarboxylate (**334**) in quantitative yield ⟨78JHC1519⟩.

$$2\text{RC}{\equiv}\text{CCH}_2\text{Br} \xrightarrow[20\text{-}63\%]{\text{NaNO}_2} \text{(330)} \quad R = H, \text{alkyl}, \text{Ph or CO}_2\text{Me}$$

$$\text{PhCH=CHCHO} \xrightarrow{\text{N}_2\text{O}_3} \text{(332)}$$

(**329**) (**330**) (**331**) (**332**)

$$\overset{\text{O}}{\overset{\|}{\text{C}}}(\text{CH}_2\text{CO}_2\text{Et})_2 \xrightarrow[\text{CaCO}_3]{2\text{NOCl}} \text{(334)}$$

(**333**) (**334**)

(ii) (CNO + CC) reactions

The chemistry of nitrile N-oxides has been well reviewed ⟨79AHC(25)147, 70S344, 70E1169⟩. Although benzonitrile N-oxide has been known since 1894 and the parent compound, fulminic acid, since the beginning of the last century, very little synthetic use of these reactive intermediates was made until 1946, when Quilico began to study the formation of isoxazoles from nitrile N-oxides and unsaturated compounds. This type of reaction was later classified by Huisgen as a typical 1,3-dipolar cycloaddition ⟨63AG742⟩ and was shown to occur not only with alkenes and alkynes but also with many other types of unsaturated systems.

The electronic structure of nitrile N-oxides may be represented as a resonance hybrid of the canonical structures (**335a–e**). The structure (**335a**) is commonly used to represent this reactive species.

$$\text{R}-\overset{+}{\text{C}}{\equiv}\text{N}-\bar{\text{O}} \leftrightarrow \text{R}-\bar{\text{C}}=\overset{+}{\text{N}}=\text{O} \leftrightarrow \text{R}-\bar{\text{C}}=\text{N}-\overset{+}{\text{O}} \leftrightarrow \text{R}-\ddot{\text{C}}-\overset{+}{\text{N}}=\text{O} \leftrightarrow \text{R}-\overset{+}{\text{C}}=\text{N}-\bar{\text{O}}$$

(**335a**)   (**335b**)   (**335c**)   (**335d**)   (**335e**)

Nitrile N-oxides, under reaction conditions used for the synthesis of isoxazoles, display four types of reactivity: 1,3-cycloaddition; 1,3-addition; nucleophilic addition; and dimerization. The first can give isoxazolines and isoxazoles directly. The second involves the nucleophilic addition of substrates to nitrile N-oxides and can give isoxazolines and isoxazoles indirectly. The third is the nucleophilic addition of undesirable nucleophiles to nitrile N-oxides and can be minimized or even eliminated by the proper selection of substrates and reaction conditions. The fourth is an undesirable side reaction which can often be avoided by generating the nitrile N-oxide *in situ* and by keeping its concentration low and by using a reactive acceptor ⟨70E1169⟩.

The rate of dimerization of nitrile N-oxides is strongly influenced by the nature of R. When R = Cl, Br, CO₂ alkyl or COR', the nitrile N-oxide cannot be isolated nor obtained in solution for any appreciable time. Table 11 gives the approximate time required for complete dimerization of some nitrile N-oxides (**335**) to furoxans (**336**) in benzene solution at 18 °C ⟨70E1169⟩. Evidently, steric and electronic effects dramatically increase the stability

$$2\text{R}-\overset{+}{\text{C}}{\equiv}\text{N}-\text{O}^- \longrightarrow \text{(336)}$$

(**335**)   (**336**)

of nitrile N-oxides. The development of several methods for the *in situ* generation of a large number of nitrile N-oxides in the presence of the desired acceptor has made this synthetic approach increasingly important.

Table 11 Stability of Some Nitrile N-Oxides, $R-C\equiv\overset{+}{N}-\overset{-}{O}$, towards Dimerization to Furoxans ⟨70E1169⟩

Substituent R	Approximate time for complete dimerization at 18 °C	Substituent R	Approximate time for complete dimerization at 18 °C
Methyl	<1 min	p-Chlorophenyl	10 days
t-Butyl	2-3 days	p-Tolyl	5-7 days
Phenyl	30-60 min	p-Nitrophenyl	very slow
o-Chlorophenyl	3-6 days	2,4,6-Trimethylphenyl	very stable
m-Chlorophenyl	50-60 min		

The mechanism of 1,3-cycloaddition of nitrile N-oxides has been a subject of extended debate. While these reactions displayed the regioselectivity and stereospecificity appropriate to concerted pericyclic mechanisms ⟨76JOC403⟩, it was suggested that the experimental facts as a whole favored a mechanism involving a spin paired biradical intermediate ⟨77T3009⟩. However, no direct evidence for a radical intermediate has been produced, and the balance of the theoretical argument favors the concerted mechanism. A number of theoretical studies based on a concerted mechanism have been reported ⟨80JA1763, 79AHC(25)147⟩.

The treatment of chloroximes (**337**) with bases, particularly with triethylamine, is the most commonly utilized method for the preparation of nitrile N-oxides (**335**). The chloroximes (**337**) are conveniently prepared from aldoximes and chlorine or nitrosyl chloride. In cases where the nitrile N-oxide (**335**) dimerizes rapidly and the substrate is base sensitive, the dehydrohalogenation can be carried out *in situ* at temperatures as low as −20 °C. Recently, a particularly convenient preparation of benzohydroximinoyl chlorides by the reaction of benzaldehyde oximes with N-chlorosuccinimide in DMF has been reported ⟨80JOC3916⟩. In the absence of a base, most chloroximes (**337**) are stable at room temperature and even at higher temperatures the equilibrium between chloroxime (**337**), nitrile N-oxide (**335**) and hydrogen chloride favors chloroxime (**337**). However, thermal dissociation in an inert solvent in which hydrogen chloride has low solubility can be used to generate the nitrile N-oxide (**335**) in the presence of substrate ⟨79AHC(25)147⟩.

$$R-\underset{\underset{(337)}{}}{\overset{\overset{Cl}{|}}{C}=NOH} \xrightarrow[-HCl]{base} R-C\equiv\overset{+}{N}-\overset{-}{O}$$
$$(335)$$

Nitrile N-oxides can also be prepared by: (1) oxidation of aldoximes with N-bromosuccinimide in the presence of triethylamine, with lead(IV) tetraacetate or with alkaline hypobromite ⟨79AHC(25)147⟩; (2) oxidation of araldoximes with nitrogen dioxide, followed by thermal decomposition of the resulting dehydro dimers ⟨73S156⟩; (3) dehydration of primary nitroalkanes with phenyl isocyanate ⟨79AHC(25)147⟩, acetyl chloride ⟨80CPB3296⟩, acetic anhydride and sodium acetate ⟨74T1365⟩, or with isopropenyl acetate and p-toluenesulfonic acid ⟨72JOC2686⟩; and (4) photolytic or thermal cleavage of a number of heterocycles including furoxanes ⟨81JOC316⟩, furazans ⟨79AHC(25)147⟩, and 1,3,2,4-dioxathiazole 2-oxides ⟨76JOC1296⟩. Table 12 shows some of the nitrile N-oxides which have been reported.

Table 12 Some Known Nitrile N-Oxides, $R-C\equiv\overset{+}{N}-\overset{-}{O}$

R	Ref.	R	Ref.
Ph	70E1169, 75S664, 80JOC3916	CF$_3$	71IZV362
PhCO	75S664	H	70S344
Me	70E1169, 74CB3717	CN	70S344, 66JOC4235
MeCO	72JOC2686	Cl, Br	80TL229, 69G1107, 61G47
CO$_2$Et	75S664	NO$_2$	70E1169, 63MI41600

(iii) Preparation of isoxazoles from nitrile N-oxides

The reaction between a nitrile N-oxide and an alkyne is so facile that it is usually sufficient to leave an ether solution of the reactants at room temperature to obtain the desired isoxazole in good yield. The reaction is in general sensitive to the size of the substituent on the alkyne but not on the nitrile N-oxide. In the case of poorly reactive alkynes, the difficulty may be overcome by generating the nitrile N-oxide *in situ* and keeping its concentration low.

The nitrile N-oxide reacts with a monosubstituted (alkyl/aryl) alkyne regioselectively *via* 1,3-cycloaddition and also by 1,3-addition to give the 5-substituted isoxazole (**338**) and the alkynic oxime (**339**), respectively. The alkynic oxime (**339**) cyclized readily to the 5-substituted isoxazole (**338**) and has been isolated in some cases ⟨79AHC(25)147⟩. For example, the reaction of phenylacetylene with benzonitrile N-oxide in carbon tetrachloride at 25 °C gave 3,5-diphenylisoxazole in ~90% yield. The product was formed in 78% yield through a 1,3-cycloaddition and in only 22% yield through a 1,3-addition process ⟨74JCS(P2)1591⟩. In the reactions with perfluoroalkylalkynes, ethyl propiolate ⟨75JOC2143⟩ and cyanoethyne ⟨68BCJ2212⟩, significant amounts of the 4-substituted isomers have been obtained.

$$R-C\equiv\overset{+}{N}-\overset{-}{O} + HC\equiv C-R'$$
(**335**)

(**339**) → (**338**)

In an intramolecular 1,3-cycloaddition reaction, 3-butynyl carbonocyanidate N-oxide (**340**) was constrained to give exclusively the 4-substituted isoxazole (**341**) ⟨80JHC609⟩.

(**340**) → (**341**)

The nitrile N-oxide reacts with a disubstituted alkyne *via* 1,3-cycloaddition only when at least one of the substituents on the alkyne is electron-withdrawing such as RCO, CO_2H or CO_2R. The nitrile N-oxide does not undergo reaction with methylphenylacetylene, butynediol or diphenylacetylene ⟨70E1169⟩. However, benzonitrile N-oxide reacted with bis(trimethylsilyl)acetylene ⟨74CB3717⟩ and much more readily with benzyne ⟨70E1169⟩. Table 13 shows a selection of the syntheses which illustrate applications of this method and some of the generalizations discussed.

Table 13 Isoxazoles from Nitrile N-Oxides and Alkynes

Alkyne used	Isoxazoles isolated	Ref.
Trimethylsilylalkynes	5-Substituted	74CB3717
Alkynic phosphinate	5-Substituted	80JOC529
Magnesioalkynes	5-Substituted	69G1107
Alk-1-ynes	5-Substituted	78JCR(M)2038
1-Alkoxy- and 1-dialkylamino-alkynes	5-Substituted	71IZV362
1-Acylalkynes	4-Acyl-5-substituted	71BCJ803
Perfluoroalkylalk-1-ynes	4- and 5-perfluoroalkyl	78JCR(M)2529
Cyanoethyne	4- and 5-cyano	68BCJ2212
Propiolates	4- and 5-carboxylates	75JOC2143

On the whole, the cycloaddition of alkynes to nitrile N-oxides is one of the most important routes to isoxazoles, but in spite of its potentially wide application, its synthetic utility is less than that of the corresponding reaction with alkenes for the following reasons. (1)

Alkynes are often less accessible and less stable than the corresponding alkenes. (2) The generally slower reaction rate with alkynes usually leads to lower yields. (3) Alkenes provide better regiospecificity than alkynes because alkenes have more variation in substitution and some substituents have a pronounced directional effect.

For these reasons, isoxazoles are often prepared via the reaction of nitrile N-oxides with alkenes rather than with alkynes and subsequent conversion of the resulting isoxazolines into the desired isoxazoles. This goal can be achieved in two different ways: (1) isoxazolines obtained from nitrile N-oxides and mono- or 1,2-di-substituted alkenes are dehydrogenated by oxidizing agents, or (2) suitably substituted alkenes, R'CH=CR'X, where HX can later be eliminated from the resulting isoxazolines, are used. The X function can be OAc, OCOPh, OEt, OSiMe$_3$, SEt, Cl, NO$_2$, N$_3$, NMe$_2$, NEt$_2$, N(CH$_2$)$_4$ or $\bar{C}$H—$\overset{+}{P}$Ph$_3$ (see Table 14). Orientation in these 1,3-cycloadditions results in the group attached to a monosubstituted alkene entering the 5-position of the isoxazole nucleus. With di- and tri-substituted alkenes, the X functions generally enters the 5-position of the isoxazole ring, particularly those which are electron-releasing, i.e. OAc, OCOPh, OEt, OSiMe$_3$, SEt, N$_3$, NMe$_2$, NEt$_2$ and $\bar{C}$H—$\overset{+}{P}$Ph$_3$. However, in the cases of ClCH=CHCOPh ⟨70S344⟩, ClCH=CHNO$_2$ ⟨75C350⟩ and NO$_2$CH=CHMe ⟨76S612⟩ the X functions (with an underline) enter the 4-position. The advantage of this technique is best demonstrated by the following examples. The parent isoxazole (**1**) was only obtained in 5% yield from acetylene and fulminic acid, whereas the present procedure involving the generation of fulminic acid from iodoformoxime in the presence of vinyl acetate and subsequent elimination of acetic acid from the isoxazoline (**342**) raised the yield of (**1**) to 90% ⟨70S344⟩.

HCNO + CH$_2$=CHOAc $\xrightarrow{90\%}$ [(**342**)] ⟶ (**1**) $\xleftarrow{5\%}$ HCNO + HC≡CH

Reaction of benzonitrile N-oxide with alkynic phosphinate (**343**) gave exclusively 5-phosphinylisoxazole (**344**), whereas reaction with enamine phosphinate (**345**) gave the otherwise inaccessible 5-unsubstituted 4-phosphinylisoxazole (**346**) ⟨80JOC529⟩.

PhCNO + HC≡C—P(=O)(Me)(OMe) $\xrightarrow{81\%}$ (**344**)

(**343**)

PhCNO + Et$_2$NCH=CH—P(=O)(Me)(OMe) $\xrightarrow{\sim 50\%}$ (**346**)

(**345**)

Reaction of benzonitrile N-oxide with benzoylacetylene gave a mixture of 4-benzoylisoxazole (**347**) and 5-benzoylisoxazole (**348**), whereas that with β-enaminone (**349**) gave exclusively 4-benzoylisoxazole (**347**) ⟨71S433⟩ and that with chloroenone (**350**) gave exclusively 5-benzoylisoxazole (**348**) ⟨70S344⟩.

PhCOC≡CH

PhCOCH=CHNMe$_2$ $\longrightarrow$ PhC≡N—O $\longrightarrow$ (**347**) 12% + (**348**) 70%
(**349**)

PhCOCH=CHCl
(**350**)

The reaction of benzonitrile N-oxide with α- and β-azidostyrenes (**351**) and (**352**) gave 3,5- and 3,4-diphenylisoxazoles (**353**) and (**354**), respectively, in good yields ⟨74JOC1221⟩. Clearly, the X function has a pronounced directional effect on this type of 1,3-cycloaddition reaction. Table 14 shows a selection of syntheses illustrating applications of this method.

$$\text{PhCNO} + \text{CH}_2=\text{C}\begin{smallmatrix}\text{N}_3\\ \text{Ph}\end{smallmatrix} \xrightarrow{64\%} \underset{\text{(353)}}{\text{N}\underset{\text{O}}{\diagup}\text{Ph}}^{\text{Ph}} \qquad \text{PhCNO} + \text{PhCH}=\text{CHN}_3 \xrightarrow{82\%} \underset{\text{(354)}}{\text{N}\underset{\text{O}}{\diagup}}^{\text{Ph Ph}}$$

(351)　　　(353)　　　(352)　　　(354)

Table 14 Isoxazoles from Nitrile N-Oxides and C=C Dipolarophiles

$$\text{RC}\equiv\overset{+}{\text{N}}-\bar{\text{O}} + \text{R'CH}=\text{CR''X} \rightarrow \left[\begin{smallmatrix}R' & R\\ R'' & \\ X & O-N\end{smallmatrix}\right] \rightarrow \begin{smallmatrix}R' & R\\ R'' & \\ O-N\end{smallmatrix}$$

		Substituents			
R	R'	R''	X	Yield (%)	Ref.
H	H	H	OAc	90	70S344
Me	H	H	OEt	—	62BSF2215
Et	H	H	OEt	—	62BSF2215
Ph	H	H	OAc	69	62BSF2215
Ph	H	H	Cl	—	70S344
Ph	H	H	NO_2	70	76S612
CO_2Et	H	H	OEt	—	70S344
Ph	Ph	H	N_3	82	74JOC1221
Ph	H	Ph	N_3	64	74JOC1221
m-$NO_2C_6H_4$	Me	COPh	N_3	30	74JOC1221
Ph	COPh	Me	N_3	60	74JOC1221
Ph	H	Me	NO_2	56	76S612
Ph	Me	Me	NMe_2	56	71S433
Ph	CHO	H	NMe_2	49	71S433
Ph	COPh	H	NMe_2	65	71S433
Ph	COPh	H	NMe_2	58	71S433
Ph	COMe	Me	$OSiMe_2$	29	78MI41602
Ph	COMe	Me	OCOPh	8	78MI41602
Ph	CN	p-ClC_6H_4	$N(CH_2)_4$	56	77JCS(P1)2154
Ph	CN	H	N-morpholino	—	70S344
Ph	H	SEt	SEt	—	70S344
COPh	H	Me	OAc	62	72JOC2686
Ph	H	Ph	$OSiMe_3$	45	79MI41603
Ph	H	Et	$OSiMe_3$	35	79MI41603
p-FC_6H_4	PO(OMe)Me	H	NEt_2	51	80JOC529
p-$NO_2C_6H_4$	CO_2Me	H	$CH-\overset{+}{P}Ph_3$	64	76JCS(P1)619
p-$NO_2C_6H_4$	CN	H	$CH-\overset{+}{P}Ph_3$	70	76JCS(P1)619

A variation of this procedure involving the reaction of nitrile N-oxides with doubly activated methylene groups containing at least one carbonyl or nitrile function has made this synthetic method even more versatile. As is evident from Table 15, the acyl group (under X) ends up in the 5-position of the isoxazole ring as alkyl or aryl substituents; the 2-oxoacyl group as an acyl function; the ethoxyoxoacetyl group as an ethoxycarbonyl function; the nitrile group as an amino function. The other electron-withdrawing function (under Y) found in the 4-position of the isoxazole ring can be CO_2Et, COMe, COPh, CN, NO_2, $PO(OEt)_2$, CONHPh, SOMe, $C(NH_2)=C(CN)_2$, $C(NH_2)=C(CN)(CO_2Me)$ and $\overset{+}{P}Ph_3$. The $\overset{+}{P}Ph_3$ function is converted into H during work-up ⟨78ZOR2003⟩.

This variation provides a regiospecific synthesis of isoxazoles with a great variety of substituents. The nitrile N-oxide does not react with the doubly activated methylene group in neutral or acidic medium, but under alkaline conditions the reaction proceeds exother-

$$\begin{smallmatrix}\text{RCOCHSOMe}\\ \updownarrow\\ \text{RC}=\text{CHSOMe}\\ |\\ \overset{+}{O}^-\end{smallmatrix} + \text{PhCNO} \rightarrow \left[\begin{smallmatrix}\text{Ph} & \text{CHSOMe}\\ N & |\\ \bar{O} & CR\\ & \|\\ & O\end{smallmatrix}\right] \xrightarrow[-H_2O]{H^+} \underset{(356)}{\text{N}\underset{O}{\diagup}}^{\text{Ph SOMe}}_R$$

(355)　　　　(356)

mally. As illustrated by the example shown, the isoxazole (**356**) from this variation could be formed by the cyclodehydration of an open-chain intermediate such as (**355**).

Table 15 Isoxazoles from Nitrile N-Oxides and Doubly Activated Methylene Compounds

$$RCNO + Y-CH_2-X \longrightarrow \text{isoxazole}$$

		Substituent			
R	Y	X	R'	Yield (%)	Ref.
Ph	CO$_2$Et	COMe	Me	77	46G148
Ph	CO$_2$Et	CN	NH$_2$	41	46G148
Ph	COMe	COMe	Me	76	46G148
Ph	COPh	COMe	Me	—	46G148
CO$_2$Et	CO$_2$Et	COMe	Me	46	39G322
CH=CHPh	CO$_2$Et	COCO$_2$Et	CO$_2$Et	—	40G89
Ph	CO$_2$Et	COCO$_2$Et	CO$_2$Et	—	37G589
Ph	CN	COPh	Ph	—	62HC(17)1, p. 18
Ph	NO$_2$	CN	NH$_2$	38	75S664
Ph	NO$_2$	COMe	Me	72	75S664
Ph	NO$_2$	COPh	Ph	65	75S664
COPh	NO$_2$	COPh	Ph	53	75S664
CO$_2$Et	NO$_2$	COPh	Ph	54	75S664
Ph	PO(OEt)$_2$	CN	NH$_2$	28	73LA578
CHPh$_2$	CONHPh	COMe	Me	80	78MI41600
CHPh$_2$	CO$_2$Et	COMe	Me	90	78MI41600
COMe	$\overset{+}{\text{P}}$Ph$_3$ [a]	COCOMe	COMe	68	78ZOR2003
COMe	$\overset{+}{\text{P}}$Ph$_3$ [a]	COCOPh	COPh	75	78ZOR2003
COPh	$\overset{+}{\text{P}}$Ph$_3$ [a]	COCOPh	COPh	75	78ZOR2003
Ph	SOMe	COPh	Ph	71	73JHC669
Ph	SOMe	COC$_6$H$_4$Me-p	C$_6$H$_4$Me-p	54	73JHC669
Ph	C(NH$_2$)=C(CN)$_2$	CN	NH$_2$	80	80CB1195
Ph	C(NH$_2$)=C(CN)-(CO$_2$Me)	CN	NH$_2$	55	80CB1195

[a] The $\overset{+}{\text{P}}$Ph$_3$ function is converted into H during work-up.

4.16.4.1.3 [4+1] Routes

These reactions can provide convenient routes to isoxazoles in which the substituents are unequivocally located in the nucleus. They are subdivided into (i) (CCNO+C) and (ii) (CCCO+N) combinations.

(i) (CCNO + C) reactions

The reaction of 1,4-dilithio oximes (**358**) with aromatic esters provided an unequivocal synthesis of the unsymmetrically substituted isoxazoles (**359**) ⟨70JOC1806⟩. However, the yields were only 10–40% based on the more expensive oxime component (**357**). The deficiency was finally overcome by condensing the 1,4-dilithio oximes (**358**) with amides (DMF, ArCONMe$_2$) ⟨78JOC3015⟩. The condensation of the 1,3-dilithio oximes (**358**) with carbon dioxide ⟨71JCS(C)974⟩, benzonitriles ⟨72JHC183⟩, benzaldehydes, benzophenones ⟨76JHC449⟩ and aroyl chlorides ⟨76JHC607⟩ has also been reported. It may be mentioned that the condensation with aroyl chlorides provides a convenient synthesis of 4-aroyl-3,5-diarylisoxazoles not readily available by other procedures. More recently, the related condensation of 2-nitroacetophenone oximes with aldehydes and ketones in the presence of sodium methoxide was shown to give 4-nitroisoxazolines ⟨79ZOR735, 79ZOR2408⟩. This reaction shows promise of being a general and useful method of making isoxazoles since 4-nitroisoxazolines give isoxazoles in good yields at the boiling temperature of xylene

$$R^1C(=NOH)CH_2R^2 \xrightarrow[0\,°C]{2Bu^nLi} \left[R^1C(=NO^-)\overset{-}{C}HR^2 \right] 2Li^+ \xrightarrow[ii,\,H^+,\,heat]{i,\,R^3CONMe_2} \text{isoxazole}$$

(**357**) (**358**) (**359**) 59–91%

Table 16 Condensation of 1,4-Dilithio Oximes and 2-Nitroacetophenone Oximes with Various Electrophiles

Oxime	Base	Electrophile	Product	Yield (%)	Ref.
PhC(=NOH)Me	n-C_4H_9Li	$PhCO_2$Me	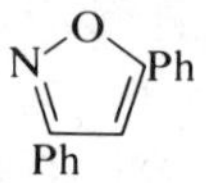 3,5-diphenylisoxazole	30	70JOC1806
		PhCN	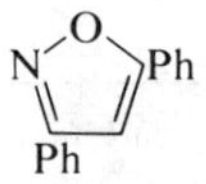 3,5-diphenylisoxazole	73	72JHC183
		CO_2	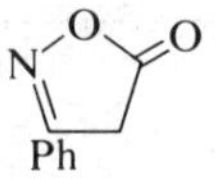 3-phenylisoxazol-5(4H)-one	45	71JCS(C)974
		PhCHO	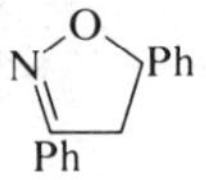 5-phenyl-3-phenyl-4,5-dihydroisoxazole	60	76JHC449
		PhCOPh	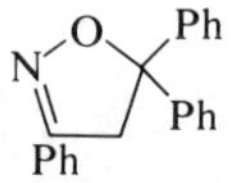 5,5-diphenyl-3-phenyl-4,5-dihydroisoxazole	69	76JHC449
		p-MeC_6H_4COCl	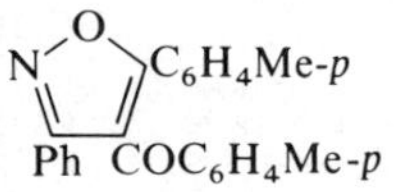 5-(p-tolyl)-3-phenyl-4-(p-toluoyl)isoxazole	54	76JHC607
PhC(=NOH)CH_2Ph		CO_2	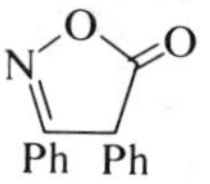 3,4-diphenylisoxazol-5(4H)-one	60	71JCS(C)974
		$PhCO_2$Me	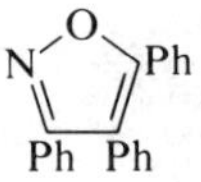 3,4,5-triphenylisoxazole	36	70JOC1806
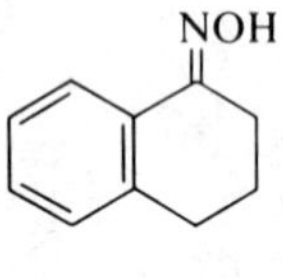 (tetralone oxime)		$PhCO_2$Me	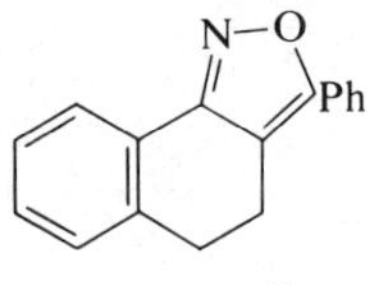 3-phenyl-4,5-dihydronaphth[1,2-d]isoxazole	10	70JOC1806
		DMF	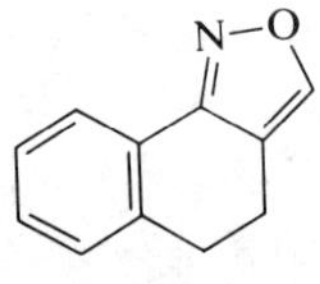 4,5-dihydronaphth[1,2-d]isoxazole	87	78JOC3015
$PhCH_2CH_2$C(=NOH)Me		DMF	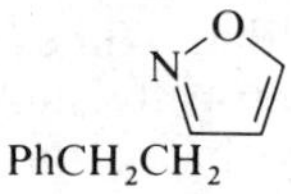 3-(2-phenylethyl)isoxazole	91	78JOC3015
MeC(=NOH)CH_2Me		DMF	 3-methyl-4-ethylisoxazole	77	78JOC3015
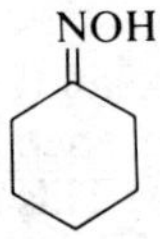 cyclohexanone oxime		$PhCONMe_2$	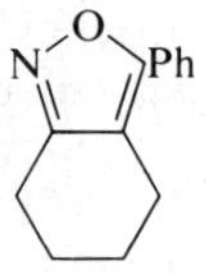 3-phenyl-4,5,6,7-tetrahydrobenzo[d]isoxazole	67	78JOC3015
C_6F_5C(=NOH)CH_2NO_2	MeONa	PhCHO	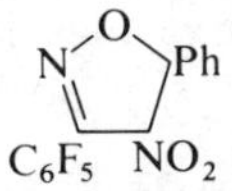 5-phenyl-3-(C_6F_5)-4-nitro-4,5-dihydroisoxazole	67–85	79ZOR2408
PhC(=NOH)CH_2NO_2		PhCHO	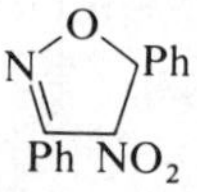 5-phenyl-3-phenyl-4-nitro-4,5-dihydroisoxazole	—	79ZOR735

⟨76S612⟩. However, more work is needed to determine its full utility. Table 16 lists the types of isoxazoles synthesized from the condensation of 1,4-dilithio oximes (**358**) and 2-nitroacetophenone oximes with electrophiles.

The reactions of several other (CCNO + C) synthons are shown in equations (17)–(20).

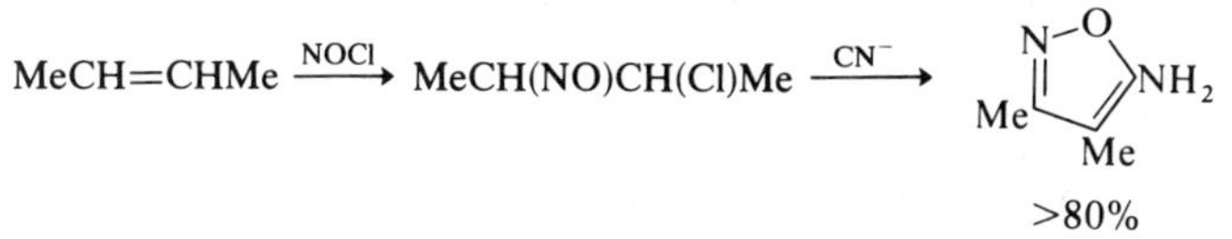

MeCH=CHMe $\xrightarrow{NOCl}$ MeCH(NO)CH(Cl)Me $\xrightarrow{CN^-}$ [3,5-dimethylisoxazol-4-amine] ⟨69TL4817⟩ (17)

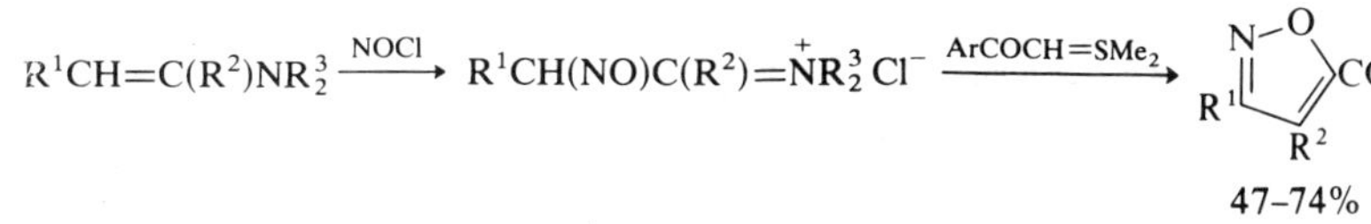

>80%

R^1CH=C(R^2)NR^3_2 $\xrightarrow{NOCl}$ R^1CH(NO)C(R^2)=$\overset{+}{N}R^3_2$ Cl$^-$ $\xrightarrow{ArCOCH=SMe_2}$ [isoxazole with R1, R2, COAr] ⟨75G91⟩ (18)

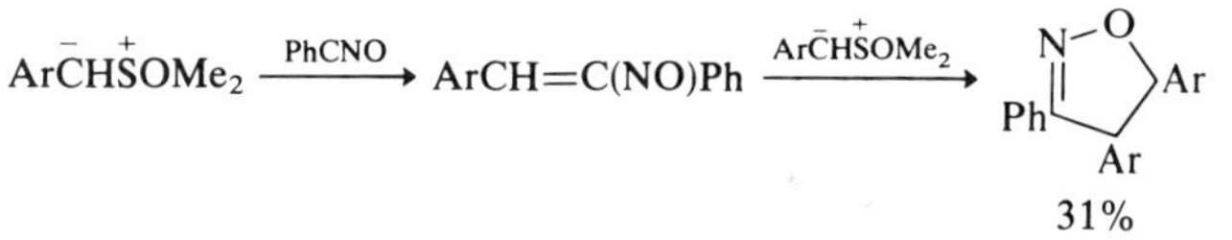

47–74%

Ar$\overset{-}{C}$H$\overset{+}{S}$OMe$_2$ $\xrightarrow{PhCNO}$ ArCH=C(NO)Ph $\xrightarrow{Ar\overset{-}{C}H\overset{+}{S}OMe_2}$ [isoxazoline Ph, Ar, Ar] ⟨77JCS(P1)1196⟩ (19)

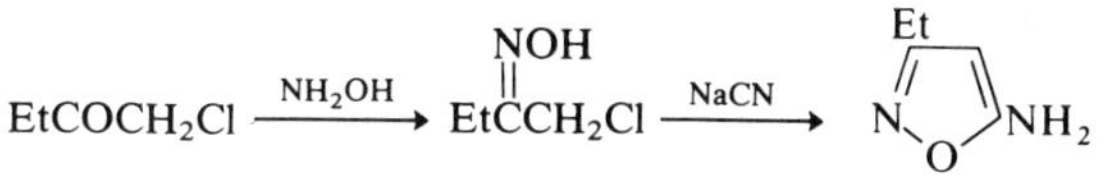

31%

EtCOCH$_2$Cl $\xrightarrow{NH_2OH}$ EtC(=NOH)CH$_2$Cl $\xrightarrow{NaCN}$ [5-amino-3-ethylisoxazole] ⟨80MI41606⟩ (20)

(ii) (CCCO + N) reactions

Two examples are shown in equations (21) and (22).

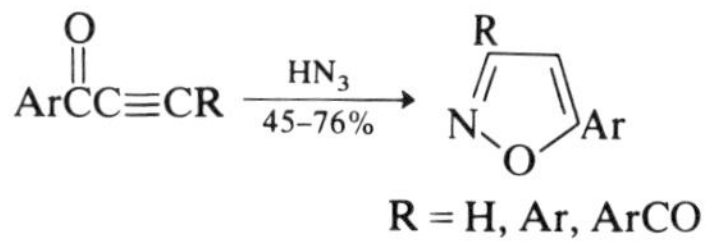

ArCOC≡CR $\xrightarrow[45-76\%]{HN_3}$ [isoxazole R, Ar] ⟨65CB3020⟩ (21)

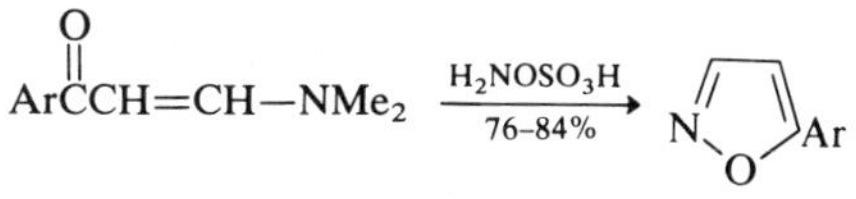

R = H, Ar, ArCO

ArCOCH=CH—NMe$_2$ $\xrightarrow[76-84\%]{H_2NOSO_3H}$ [isoxazole Ar] ⟨80JOC4857⟩ (22)

4.16.4.1.4 [5+0] Routes

Most reactions leading to isoxazoles must involve at some stage cyclization of an intermediate which contains all five atoms of the isoxazole ring. In some cases the acyclic intermediates are short-lived and unisolable, in others they are stable and able to be isolated. In this section we discuss reactions which involve an isolable acyclic precursor. These reactions mostly utilize (CCCNO) synthons although a few examples of (OCCCN), (CCCON) and (CONCC) synthons are encountered. We are unaware of examples involving (CNOCC) synthons.

As with [4+1] routes, the substituents in the resultant isoxazole system are unequivocally located.

(i) (CCCNO) reactions

When an ethanol solution of the oxime mixture of (**360**) and (**361**) was heated with a catalytic amount of hydrochloric acid, only one isoxazole (**362**) was formed. This regiospecific

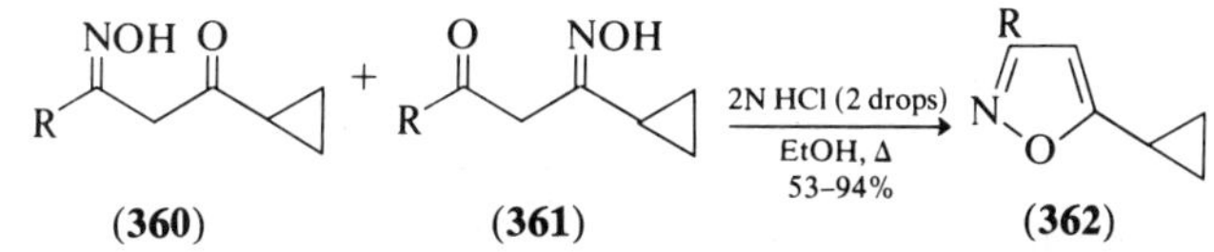

[structures of (360), (361), (362)] $\xrightarrow[EtOH, \Delta]{2N\ HCl\ (2\ drops)}$ 53–94%

R = [pyridyl], [pyrimidinyl], [pyrazinyl], [pyridazinyl]

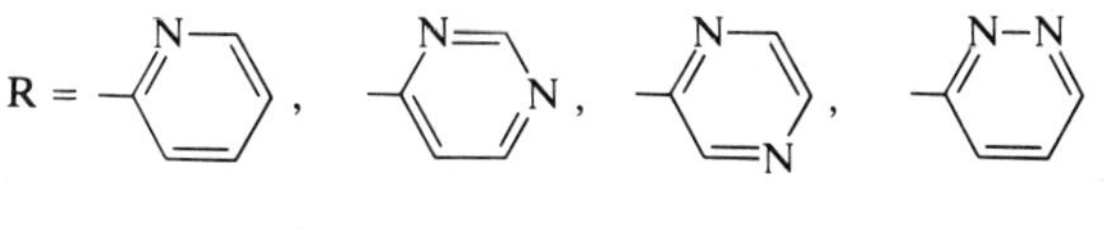

isoxazole synthesis was explained in terms of neighboring group participation of the *ortho* nitrogen atom of the heterocyclic ring in the cyclodehydration step. Cyclodehydration of a 50/50 mixture of the oximes with concentrated sulfuric acid gave approximately equal amounts of the isomeric isoxazoles ⟨77JHC531⟩.

Additional examples of acid- and base-catalyzed reactions of this type are given in equations (23)–(32).

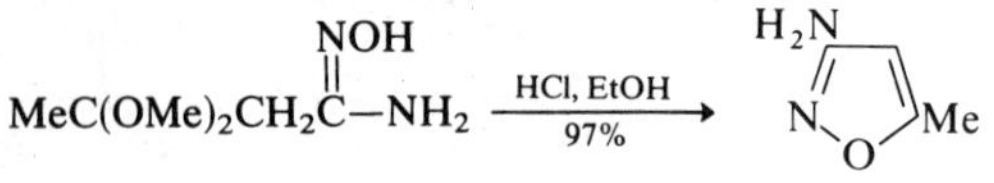

$$MeC(OMe)_2CH_2C(=NOH)-NH_2 \xrightarrow[97\%]{HCl,\ EtOH} \underset{N-O}{\text{H}_2N\text{-isoxazole-Me}} \quad \langle 77USP4225721\rangle \quad (23)$$

$$PhCOCH=CHN(OH)Ph \xrightarrow[88\%]{i,\ H_2SO_4;\ ii,\ NaClO_4} [PhN^+\text{-isoxazole-Ph}]\ ClO_4^- \quad \langle 66JOC2039\rangle \quad (24)$$

$$ArCOCH_2CH_2NO_2 \xrightarrow[\leq 60\%]{HX} \underset{N-O}{X\text{-isoxazole-Ar}} \quad \langle 77JMC934\rangle \quad (25)$$

X = Cl, Br

$$PhCOCH_2COCF_3 \xrightarrow[K_2CO_3,\ 88\%]{NH_2OH\cdot HCl} PhCOCH_2C(=NOH)CF_3 \xrightarrow[\Delta,\ 67\%]{AcCl} \underset{N-O}{F_3C\text{-isoxazole-Ph}} \quad \langle 77JMC934\rangle \quad (26)$$

$$Br\text{-C}_6H_4\text{-COC}\equiv C\text{-Ph} \xrightarrow[Na_2CO_3,\Delta,\ 77\%]{H_2NOH\cdot HCl} Br\text{-C}_6H_4\text{-COCH}_2C(=NOH)Ph \xrightarrow[\Delta,\ 85\%]{HCl} \underset{N-O}{Ph\text{-isoxazole-C}_6H_4Br} \quad \langle 68JCS(C)1774\rangle \quad (27)$$

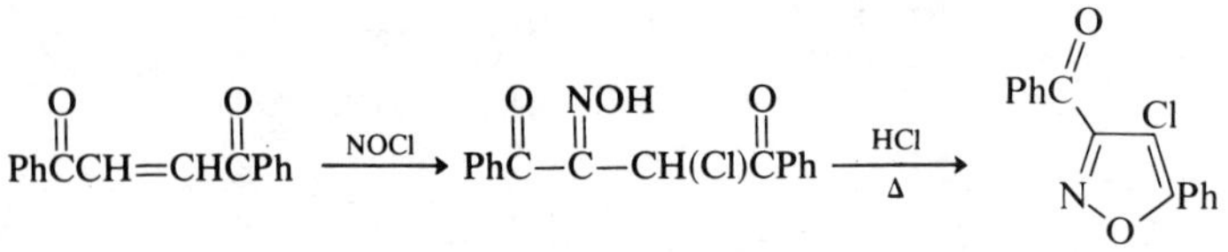

$$PhCOCH=CHCOPh \xrightarrow{NOCl} PhC(=O)C(=NOH)CH(Cl)COPh \xrightarrow[\Delta]{HCl} \underset{N-O}{PhCO,\ Cl\text{-isoxazole-Ph}} \quad \langle 54CJC288\rangle \quad (28)$$

$$\underset{R^1\ CH(R^2)CO_2Ph}{\text{dioxolane}} \xrightarrow[10-65\%]{H_2NOH/KOH} \underset{R^1\ CH(R^2)CNHOH}{\text{dioxolane ketone}} \xrightarrow[80-95\%]{HCl} \underset{N-O}{HO,\ R^2\text{-isoxazole-}R^1} \quad \langle 70BSF1978\rangle \quad (29)$$

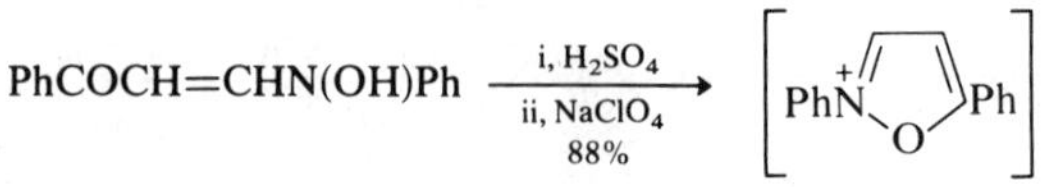

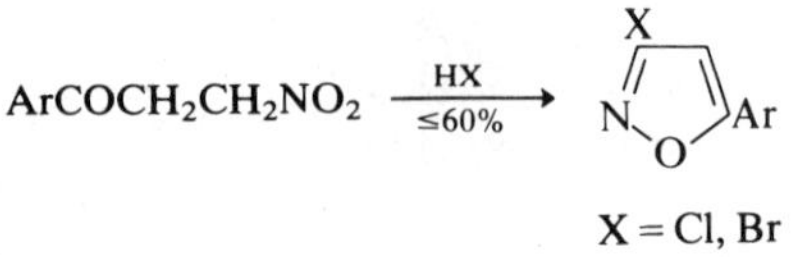

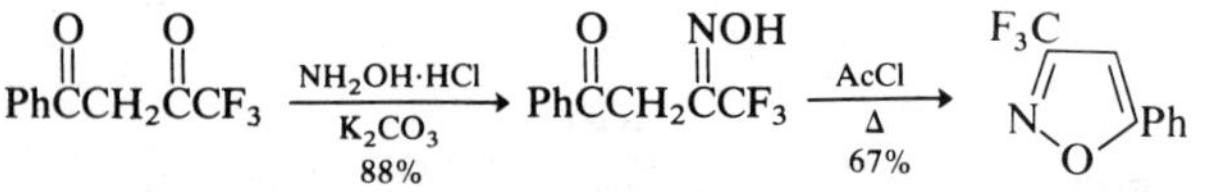

$$EtO_2CCH_2C(=O)C(=NOH)CO_2Et \xrightarrow[CaCO_3,\ 98\%]{NOCl} \underset{N-O}{EtO_2C,\ OH\text{-isoxazole-CO}_2Et} \quad \langle 78JHC1519\rangle \quad (30)$$

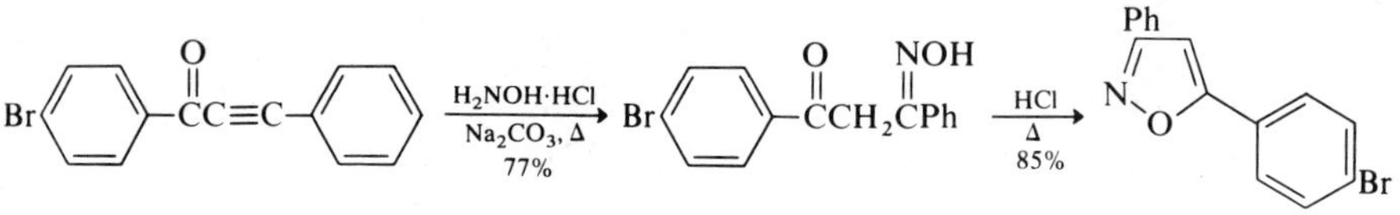

$$MeO_2CCH(NO_2)C(R)=CHCO_2Me \longrightarrow \underset{-O-N^+-O}{MeO_2C,\ R\text{-isoxazoline-CO}_2Me} \xrightarrow{Bu^nNH_2} \underset{N-O}{Bu^nNHOC,\ R\text{-isoxazole-CONHBu}^n} \quad \langle 74CPB477\rangle \quad (31)$$

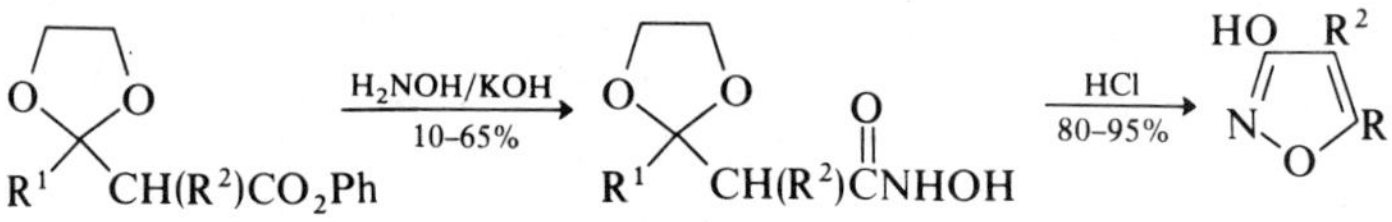

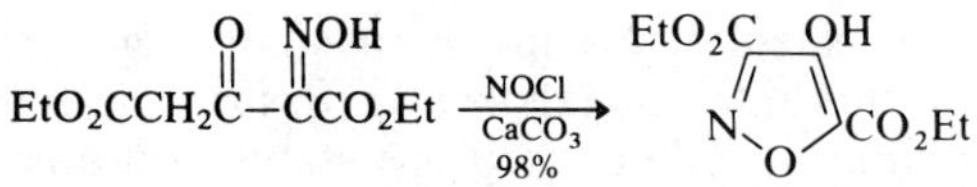

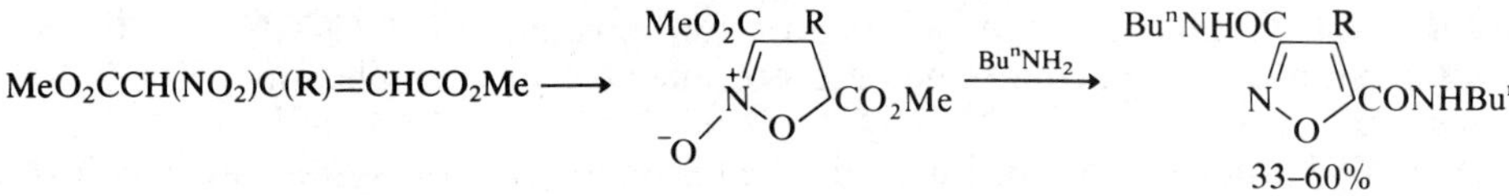

33–60%

$$HONHCOC(Cl)=CHCONHOH \xrightarrow[45\%]{NaOH} \underset{N-O}{HO\text{-isoxazole-CONHOH}} \quad \langle 77JMC965\rangle \quad (32)$$

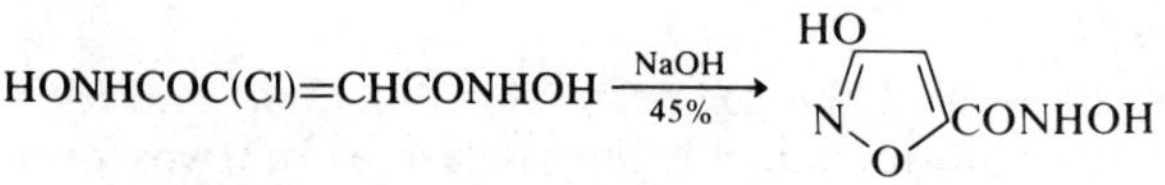

Illustrative examples of oxidative and reductive cyclizations of (CCCNO) synthons are shown in equations (33)–(39).

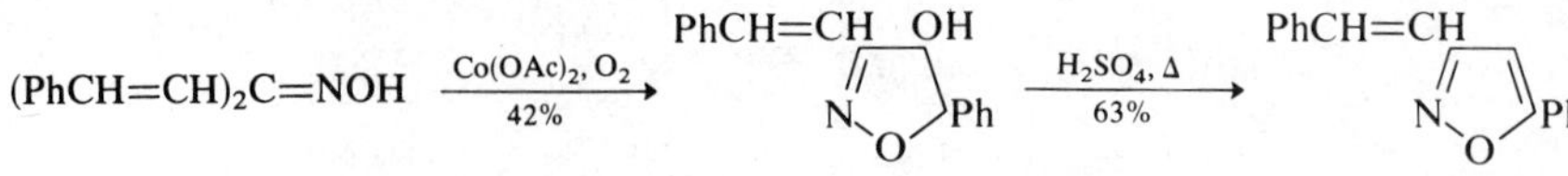

$$(PhCH=CH)_2C=NOH \xrightarrow[42\%]{Co(OAc)_2,\ O_2} \underset{N-O}{PhCH=CH,\ OH\text{-isoxazoline-Ph}} \xrightarrow[63\%]{H_2SO_4,\ \Delta} \underset{N-O}{PhCH=CH\text{-isoxazole-Ph}} \quad \langle 78H(11)187\rangle \quad (33)$$

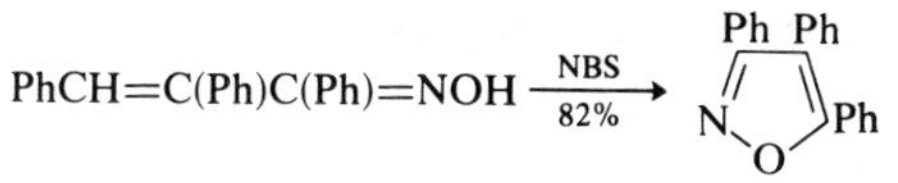

⟨77JHC1289⟩ (34)

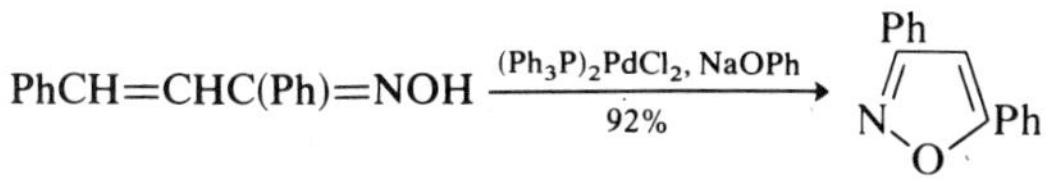

⟨73TL5075⟩ (35)

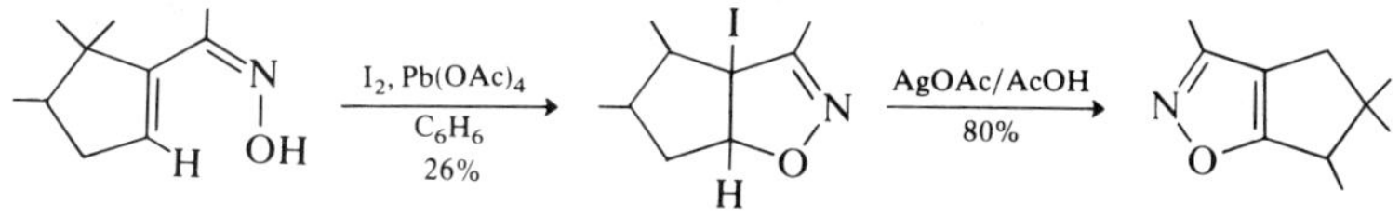

⟨69JOC1618⟩ (36)

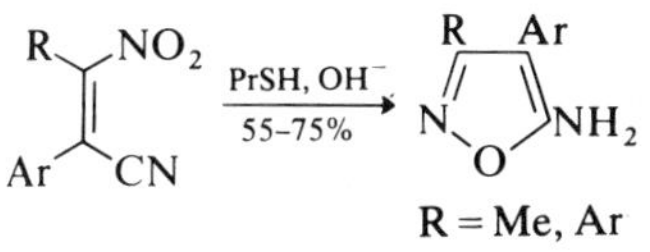

⟨80BSF163⟩ (37)

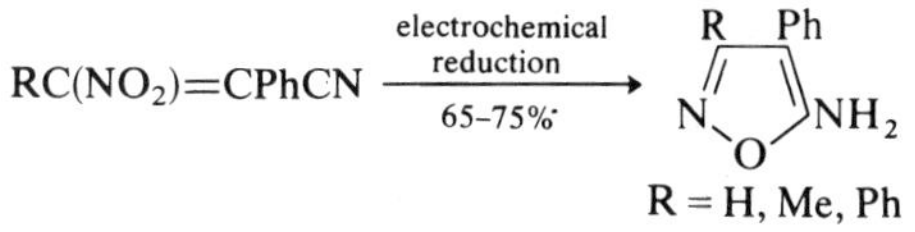

⟨79JHC1611⟩ (38)

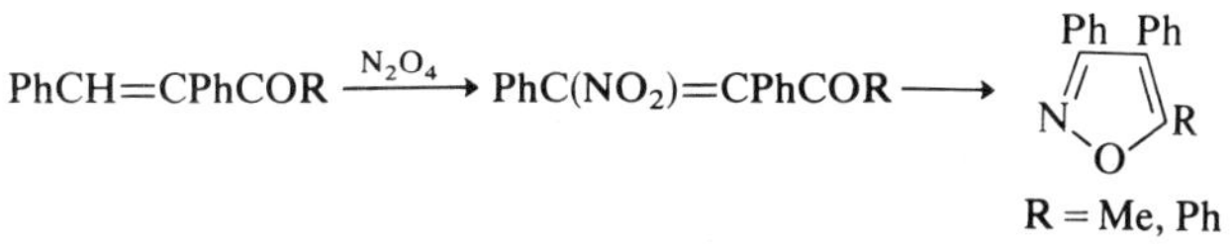

⟨79JHC1657⟩ (39)

(ii) (OCCCN) reactions

A 1:1 adduct from diphenylsulfilimine and a benzoylacetylene underwent an intramolecular cyclization reaction to give an isoxazole in good yield (equation 40). Similarly, the 1:1 adduct from iodoazide and chalcone gave 3,5-diphenylisoxazole (equation 41). These two approaches to regiospecific isoxazole synthesis are of little practical significance. Additional examples of the (OCCCN) reaction are given in equations (42) and (43).

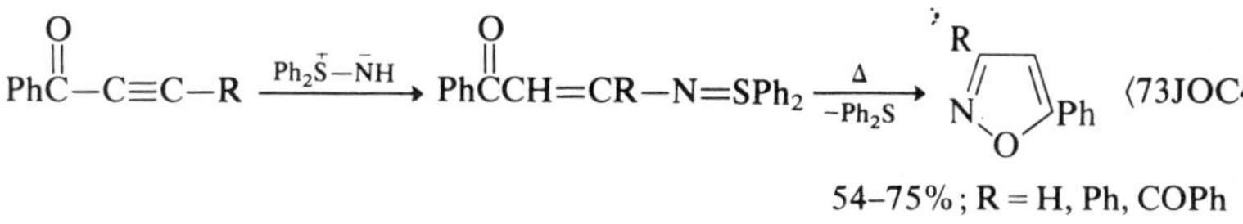

⟨73JOC4324⟩ (40)

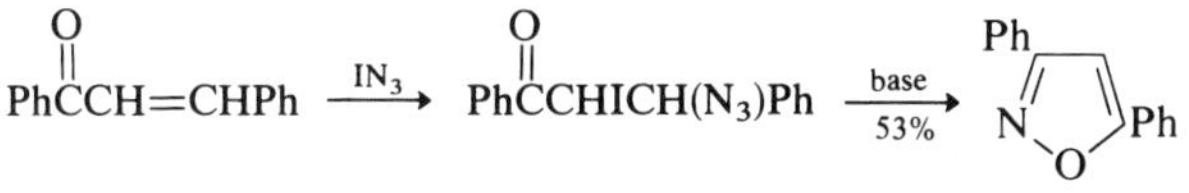

⟨67JA2077⟩ (41)

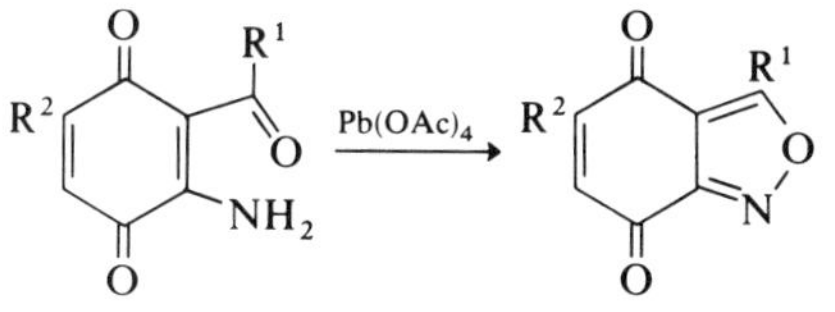

⟨74S30⟩ (42)

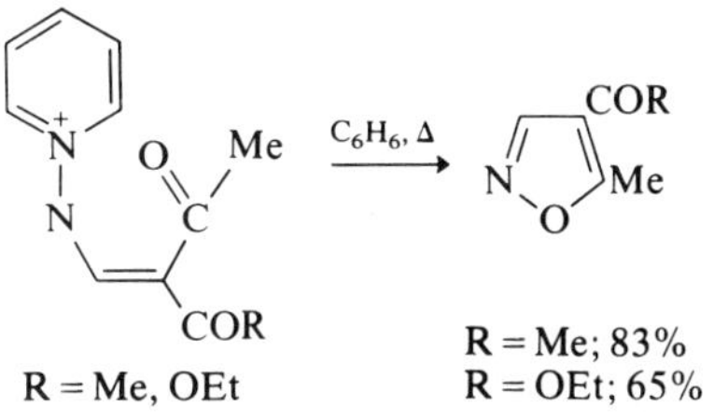

⟨73JCS(P1)2580⟩ (43)

(iii) (CONCC) reactions

Only one example of this type has been reported. Cyclization of α-(acylmethoxyimino)nitriles in the presence of lithium hydroxide provides a convenient synthesis of 5-acyl-4-aminoisoxazoles not readily available by other procedures (equation 44) ⟨80LA1623⟩.

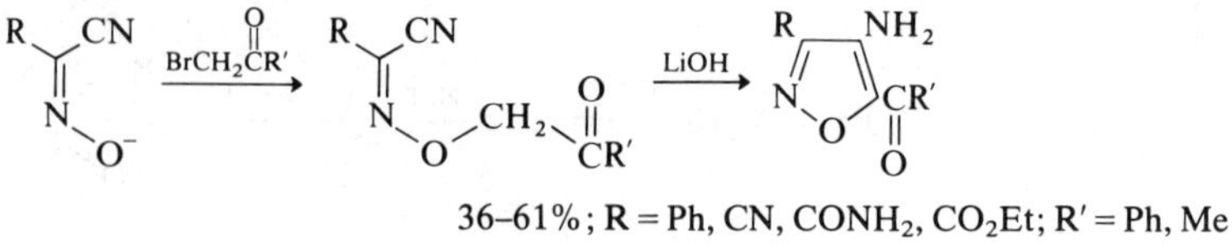

$$36-61\%; R = Ph, CN, CONH_2, CO_2Et; R' = Ph, Me \quad (44)$$

(iv) (NOCCC) reaction

3-Aminoisoxazole has been synthesized in high yield by the hydrolysis and ring closure reaction of cyclohexylideneaminoacrylonitrile under acidic conditions (equation 45) ⟨69GEP1814116⟩.

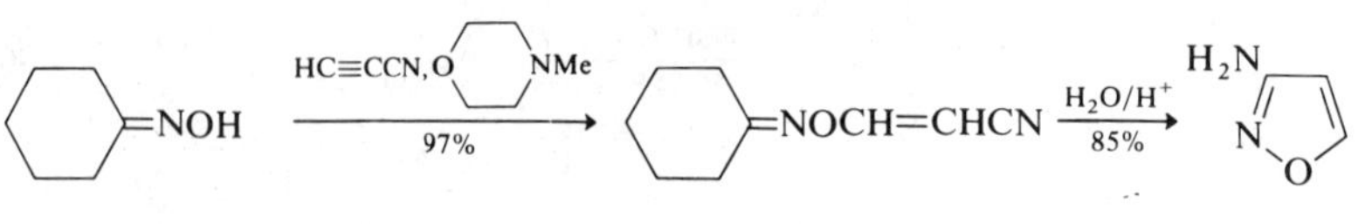

$$(45)$$

4.16.4.1.5 [3+1+1] routes

In 1888 Dunstan and Dymond observed that primary nitroalkanes of more than one carbon atom in the presence of inorganic bases formed trialkylisoxazoles and a significant amount of the corresponding nitriles ⟨1891JCS410⟩. About 50 years later a study of the formation of trialkylisoxazoles in the condensation of primary nitroalkanes under the influence of basic reagents was carried out ⟨40JA2604⟩. Primary nitroalkanes when heated with organic bases (such as diethylamine, *n*-propylamine and *n*-butylamine) gave dioximes (**365**) of the 1,3-diketone RCOCHRCOR in good yields (37–57%) except for one from nitroethane (15%). On boiling with dilute acids, each of these dioximes (**365**) yielded trialkylisoxazole (**366**) and hydroxylamine in quantitative yield. The dioxime (**363**) could be isolated when the reaction was carried out at a sufficiently low temperature (between −10 and +30 °C). It was postulated that the reaction involved two transient intermediates (**363**) and (**364**). These reactions are shown in Scheme 93.

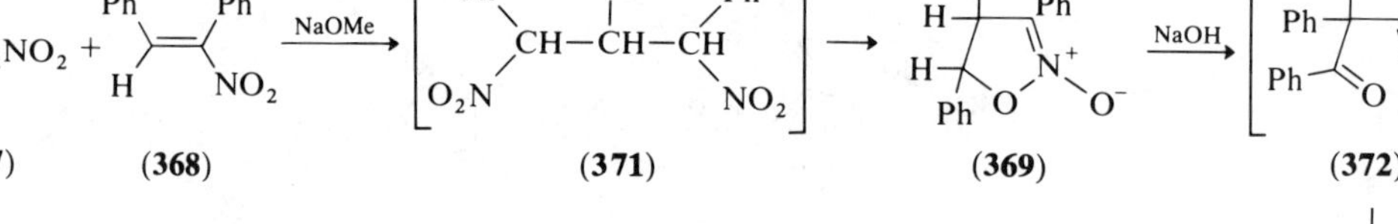

Scheme 93

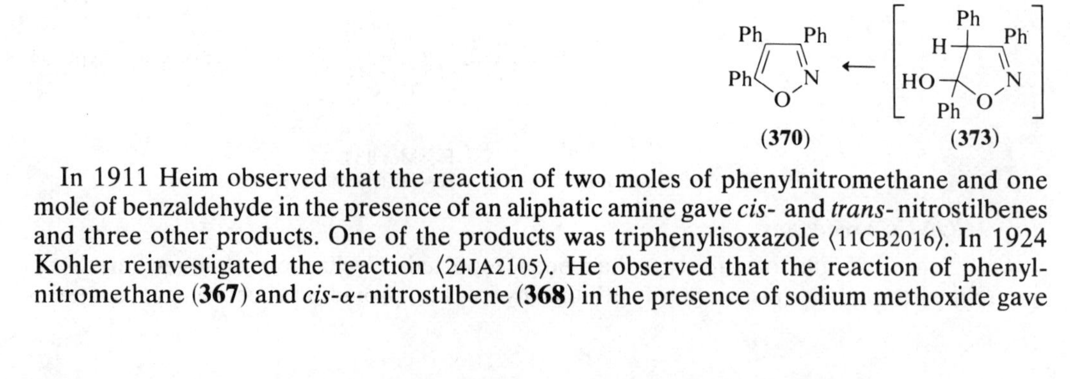

In 1911 Heim observed that the reaction of two moles of phenylnitromethane and one mole of benzaldehyde in the presence of an aliphatic amine gave *cis*- and *trans*-nitrostilbenes and three other products. One of the products was triphenylisoxazole ⟨11CB2016⟩. In 1924 Kohler reinvestigated the reaction ⟨24JA2105⟩. He observed that the reaction of phenylnitromethane (**367**) and *cis*-α-nitrostilbene (**368**) in the presence of sodium methoxide gave isoxazoline N-oxide (**369**). The isoxazoline N-oxide (**369**) was readily converted into triphenylisoxazole (**370**) in aqueous alcoholic sodium hydroxide. It was suggested the dinitropropane (**371**), the open-chain derivative (**372**) and the isoxazoline (**373**) were probably the transient intermediates. Recently, a convenient synthesis of the isoxazoline N-oxide (**369**) has been reported ⟨78S55⟩. A mixture of sodium or potassium arylmethanenitronate (**374**) and silver nitrate in DMSO was stirred at room temperature to give the isoxazoline N-oxide (**369**) in 40–53% yield.

$$3\,ArCH=NO_2^- \, M^+ \xrightarrow[\text{DMSO}]{AgNO_3} \underset{(\mathbf{369})}{\text{Ph, Ph isoxazoline N-oxide}} \xleftarrow[\text{r.t.}]{DMSO} \underset{(\mathbf{375})}{PhCHBr + 2(\mathbf{374})}$$

It was also found that bromonitrophenylmethane (**375**) reacted with sodium arylmethanenitronate (**374**) in DMSO to give the isoxazoline N-oxide (**369**) in 60% yield. Both reactions probably involved the formation of a vicinal dinitroethane derivative (**376**), which lost nitrous acid to give cis-α-nitrostilbene (**368**). As mentioned, the reaction of (**368**) with (**374**) gave the isoxazoline N-oxide (**369**).

$$\underset{(\mathbf{376})}{Ar-C(H)(O_2N)-C(NO_2)(H)-Ar} \xrightarrow{-HNO_2} (\mathbf{368}) \xrightarrow{(\mathbf{374})} \underset{(\mathbf{369})}{\text{Ph, Ph isoxazoline N-oxide}}$$

The method has been extended to the synthesis of 4-substituted isoxazole-3,5-dicarboxylic acids (**379**) ⟨63BCJ1150⟩. The reaction of two mole equivalents of ethyl nitroacetate (**377**) with one mole equivalent of an aldehyde in the presence of six mole equivalents of n-butylamine in refluxing ethanol gave isoxazoles (**378**) as crude products in 21–68% yields. The bis(butylcarbamoyl)isoxazoles (**378**) were then hydrolyzed with aqueous sodium hydroxide to give isoxazoledicarboxylic acids (**379**) in good yields. Recently, the reaction has been reinvestigated ⟨80CPB479⟩. The reaction of methyl nitroacetate (**377**) with the aldehydes in DMA in the presence of diethylamine at room temperature gave isoxazoline N-oxides (**380**) in 67–95% yields without any by-product except in the case when R = Ph (57%). The isoxazoline N-oxides (**380**) were then converted into the bis(butylcarbamoyl)isoxazoles (**378**) in 46–65% yields on boiling with butylamine in methanol. It is probable that with a hindered non-nucleophilic base the isoxazoline N-oxide (**380**) may be dehydrated to the ester (**381**).

$$2NO_2CH_2CO_2R' + RCHO \xrightarrow{BuNH_2} \underset{(\mathbf{378})}{\text{BuNHCO-isoxazole-CONHBu}} \xrightarrow{OH^-} \underset{(\mathbf{379})}{HO_2C\text{-isoxazole-}CO_2H}$$

$$\underset{(\mathbf{380})}{\text{isoxazoline N-oxide}} \rightarrow \underset{(\mathbf{381})}{MeO_2C\text{-isoxazole-}CO_2Me}$$

Another type of (CNO + C + C) reaction has been reported ⟨77JCS(P1)1196⟩. The reaction of dimethyloxosulphonium 4-nitrobenzylide (**382**) with benzonitrile N-oxide gave 4,5-dihydro-4,5-bis(4-nitrophenyl)-3-phenylisoxazole (**383**) in 31% yield. So far, no synthetic utility for this reaction process has been reported.

$$4\text{-}NO_2C_6H_4\overset{-}{C}H\overset{+}{S}OMe_2 \xrightarrow{PhC\equiv\overset{+}{N}\overset{-}{O}} [4\text{-}NO_2C_6H_4CH=CPh\,|\,NO] \xrightarrow{4\text{-}NO_2C_6H_4\overset{-}{C}H\overset{+}{S}OMe_2 \;(\mathbf{382})} \underset{(\mathbf{383})}{4\text{-}NO_2C_6H_4, 4\text{-}NO_2C_6H_4, Ph\text{-isoxazoline}}$$

(**382**)

4.16.4.1.6 Dehydrogenation of isoxazolines to isoxazoles

Δ^2-Isoxazolines are readily available from the 1,3-dipolar cycloaddition of nitrile N-oxides with alkenes and from the condensation reaction of enones with hydroxylamine. Therefore, methods of conversion of Δ^2-isoxazolines into isoxazoles are of particular interest and of synthetic importance.

The lability of the isoxazole ring precludes the use of drastic, conventional reagents for aromatizing Δ^2-isoxazolines. However, three satisfactory methods are now available. The first involves bromination of Δ^2-isoxazolines with N-bromosuccinimide to give an intermediate which is readily dehydrobrominated, either spontaneously or on treatment with weak base ⟨65T817⟩. The second involves dehydrogenation of Δ^2-isoxazolines with active γ-manganese dioxide to give isoxazoles in essentially quantitative yields ⟨77S837⟩. Very recently, it has been reported that dehydrogenation of Δ^2-isoxazoline (**384**) with 2,3-dichloro-5,6-dicyanobenzoquinone gives isoxazole (**385**) in 90% yield ⟨79JCR(S)311⟩. However, more applications of DDQ to dehydrogenation of other functionalized Δ^2-isoxazolines need to be demonstrated for this process to have general applications.

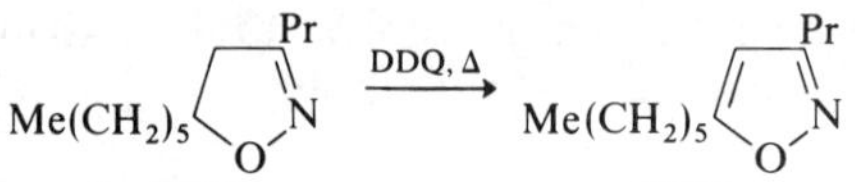

4.16.4.1.7 Ring transformations leading to isoxazoles

Ring transformations of heterocycles leading to isoxazoles have been briefly reviewed ⟨79AHC(25)147⟩. The heterocycles undergoing such transformations may be divided into three classes.

(i) Heterocycles which provide the CCC component synthon

Many of the reported transformations are more of theoretical than practical interest. However, the ring contraction of condensed 4-pyrones continues to play a useful role, particularly in the synthesis of steroidal isoxazoles ⟨73AJC1763, 71T711⟩. The reaction of chromone (**386**) with hydroxylamine can give five products (**387–391**), depending on the reaction conditions used (pH, ratio of substrates). The reaction of chromone (**386**) with excess hydroxylamine hydrochloride in refluxing ethanol gave a mixture of isoxazoles (**387**; 20%) and (**388**; 50%). The isoxazole (**388**) was also obtained by the cyclization of monoxime (**391**) in an acidic medium. The monoxime (**391**) was prepared in 81% yield by dissolving chromone (**386**) in one equivalent of a cold sodium hydroxide solution followed by addition of one equivalent of hydroxylamine hydrochloride. The reaction of chromone (**386**) with hydroxylamine in the presence of potassium hydroxide at <20 °C gave isoxazoline (**389**, 80%), which was thermally converted into derivative (**390**) in good yield ⟨79MI41600⟩. These inter-relationships are shown in Scheme 94.

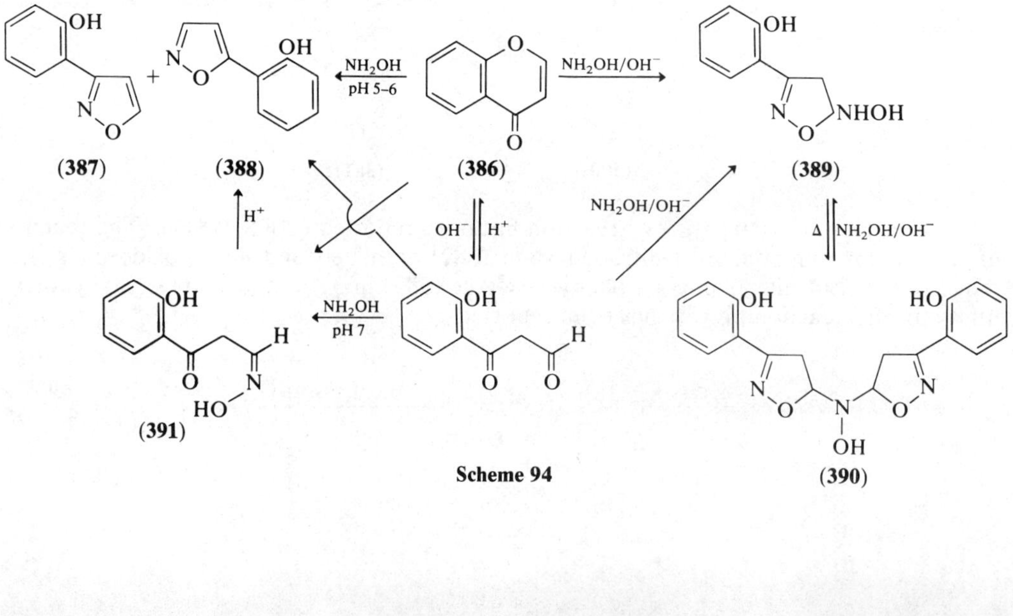

Scheme 94

The fragmentation pattern of isoxazoles on electron impact has been well studied. It has been used as an important tool for the structural assignment of isoxazoles obtained from the reaction of chromones with hydroxylamine ⟨79MI41600, 77JOC1356⟩. For example, the structures of the isoxazoles (**387**) and (**388**) were assigned on the basis of their fragmentation patterns. Ions at m/e 121 (100%) and m/e 93 (19.8%) were expected, and indeed observed, for the isoxazole (**388**), and an ion at m/e 132 (39.5%) was similarly predicted and observed for the isoxazole (**387**).

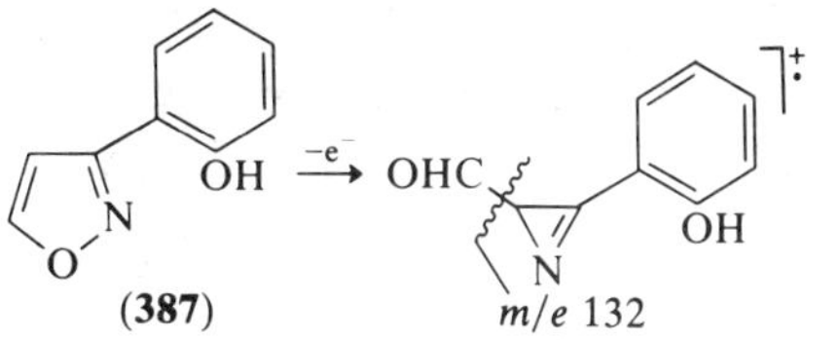

(**387**)

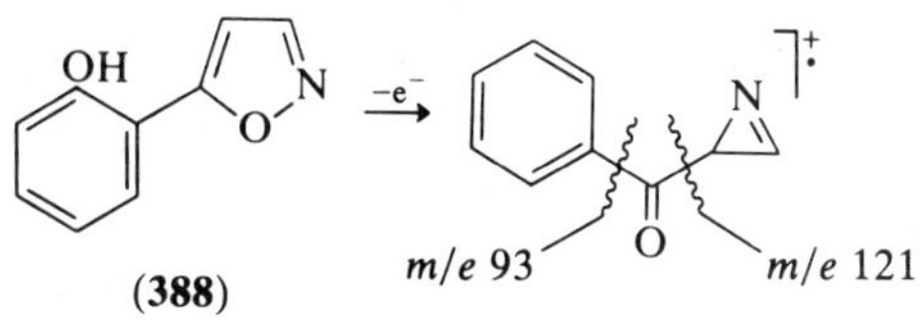

(**388**)

The reaction of hydroxylamine with 2-substituted chromones (**392**) where R = Me, Ph ⟨76MI41601⟩ or CO_2H ⟨79MI41600⟩ gave exclusively 5-(*o*-hydroxyphenyl)isoxazoles (**393**).

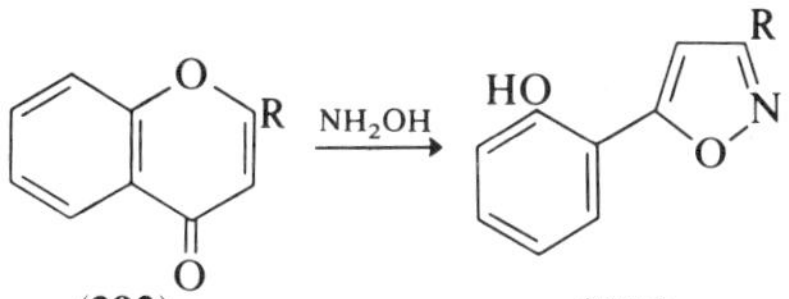

(**392**) (**393**)

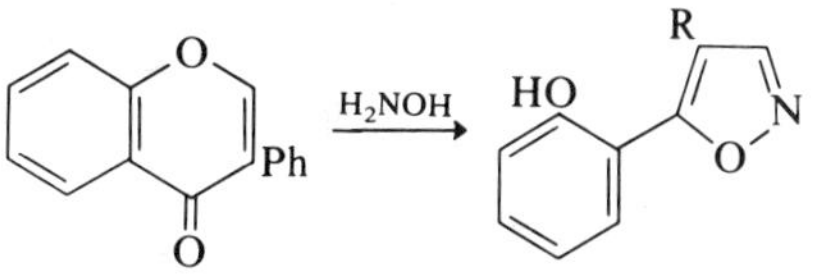

(**394**) (**395**)

The structures of the isoxazoles (**393**) were all consistent with their mass spectral fragmentation patterns. The reaction of hydroxylamine with 3-phenylchromone (**394**) gave exclusively 5-(*o*-hydroxyphenyl)isoxazole (**395**) ⟨78ACH(97)69⟩.

The reaction of the 2,4,6-triphenylpyrylium salt (**396**) with hydroxylamine in aqueous ethanol gave monoxime (**397**) which, on heating, isomerized readily to isoxazoline (**398**). In hot 70% perchloric acid the isoxazoline was converted into 3,5-diphenylisoxazole (**399**) and acetophenone ⟨68T5059⟩. 2,6-Dimethylpyrylium salts (**400**) also gave isoxazoles (**401**) with the elimination of acetone ⟨73KGS1016⟩. The reaction of (**396**) with hydroxylamine in acid medium gave a mixture of pyridine *N*-oxide and isoxazolines, and the corresponding 2,4,6-trimethylpyrylium salt gave a good yield of the corresponding pyridine *N*-oxide, either in acidic or basic medium ⟨70ACS3435⟩. These reactions are shown in Scheme 95.

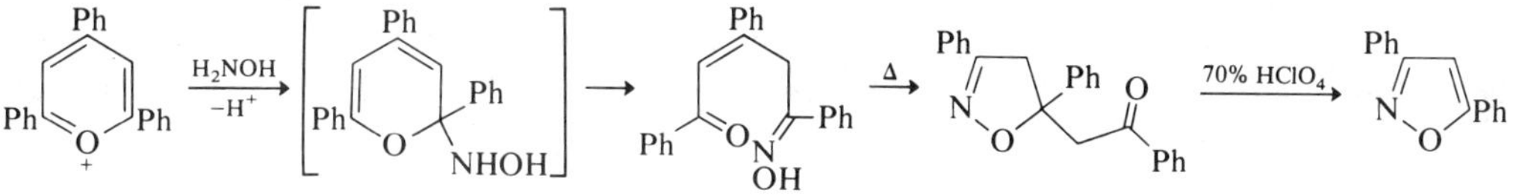

(**396**) (**397**) (**398**) (**399**)

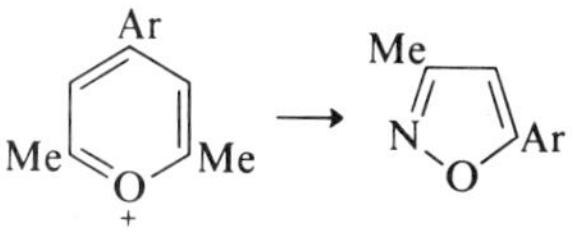

(**400**) (**401**) Ar = benzothiazol-2-yl

Scheme 95

Another interesting reaction of the pyrylium salt (**396**) has been reported ⟨73TL2195⟩. With nitrous acid in alcohol, (**396**) gave an intermediate (**402**) which on heating in acetic acid gave the diacylisoxazole (**403**). The structure of (**402**) was determined by X-ray crystallography. These ring interconversions are shown in Scheme 96.

The reaction of flavylium salts (**403a**) with hydroxylamine in pyridine gave 2,5-dihydroisoxazoles (**404**) in an analogous manner ⟨75T2884⟩. Pyrimidines have also been converted into isoxazoles, and the reaction of the pyrimidines (**405**) with hydroxylamine hydrochloride gave the isoxazoles (**338**).

Highly activated pyrimidine derivatives such as the diquaternary salt (**406**) and the monoquaternary salt (**407**) underwent ring contraction at a lower temperature and gave isoxazole in much better yields. Pyrimidine *N*-oxides (**408**) also underwent ring contraction in a similar fashion to give isoxazoles (**338**). It was found, using 1,3-^{15}N-labelled (**408b**) as

(396) (402) (403)

Scheme 96

(403a) (404)

(405) (338a) R = R′ = H (338b) R = R′ = Me (338c) R = Me, R′ = Ph (406) (407) (408a) R = R′ = H (408b) R = R′ = Me (408c) R = Me, R′ = Ph

substrate, that the isoxazole (**338b**) obtained in this reaction was not ^{15}N-labelled, showing that the N—O moiety of the isoxazole ring originated from $HONH_2$ and not from the *N*-oxide ⟨74RTC225⟩. Another interesting reaction of pyrimidine *N*-oxide has been reported ⟨74RTC58, 74RTC225⟩: 3-chloropyrimidine *N*-oxides (**409**) with potassium amide in liquid ammonia at −75 °C or with liquid ammonia at −33 °C gave a mixture of 5-aminoisoxazoles (**410**) and 4-aminopyrimidine *N*-oxides (**411**). 1,3-Oxazin-4-ones (**412**) also underwent ring contraction on treatment with hydroxylamine hydrochloride to give isoxazoles (**413**) ⟨78H(9)185⟩.

(409) (410) (411) (412) (413)

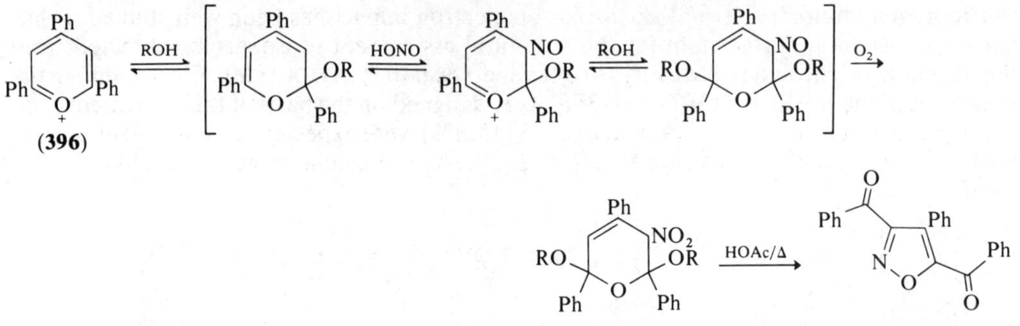

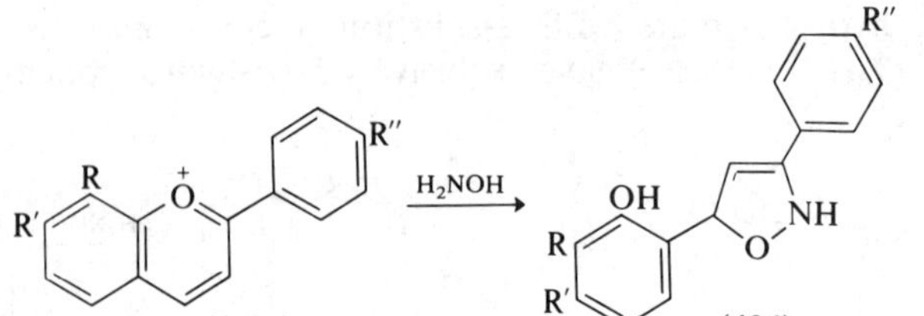

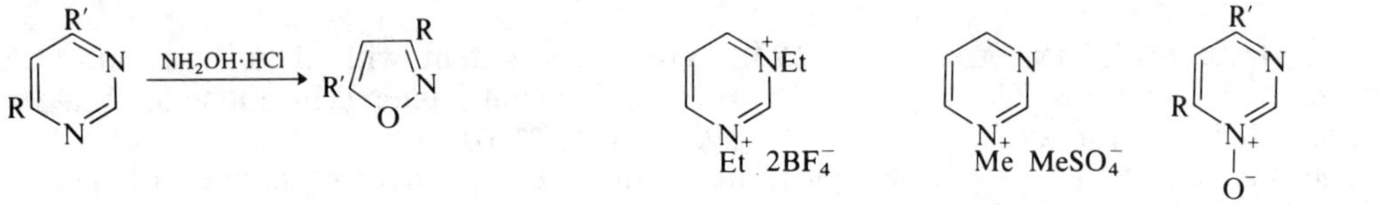

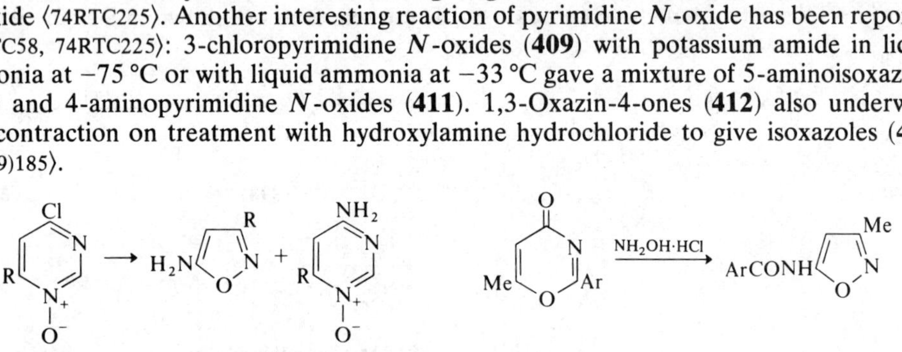

The nucleophilic attack of hydroxylamine at the 5-position of 4-trifluoroacetyloxazoles (**414**) led to a mixture of stereoisomeric isoxazolines (**415**). Dehydration of these isomeric isoxazolines (**415**) in the presence of trifluoroacetic anhydride gave isoxazoles (**416**) in good yields ⟨76JHC825⟩.

(414) (415) (416)

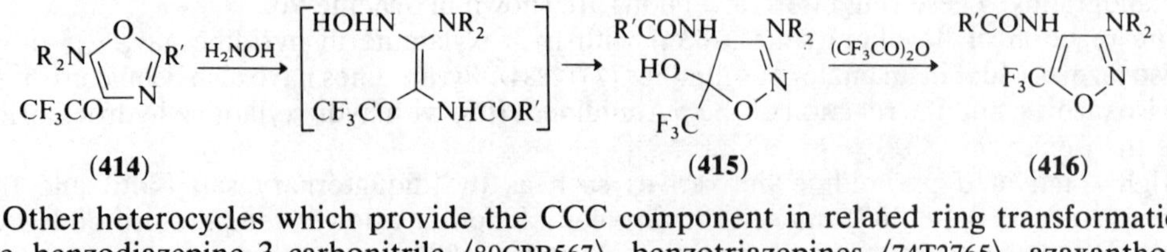

Other heterocycles which provide the CCC component in related ring transformations are benzodiazepine-3-carbonitrile ⟨80CPB567⟩, benzotriazepines ⟨74T2765⟩, azaxanthones ⟨75JOC1734⟩, acyloxiranes ⟨61AP769⟩ and oxetanones ⟨65CPB248⟩.

(ii) Heterocycles which provide the NOC or CNO component synthon

Isoxazoles can be prepared by the thermal or photolytic cleavage of a number of heterocycles, such as 1,3,5-dioxazolidone, furazans, furoxans and 1,3,2,4-dioxathiazole 2-oxides, in the presence of a reactive alkene or alkyne.

Isoxazole (**418**) was obtained from a stepwise polar addition of diethylaminopropyne to the 1,4,2-dioxazolone (**417**) which provided the (NOC) synthon ⟨73TL233⟩.

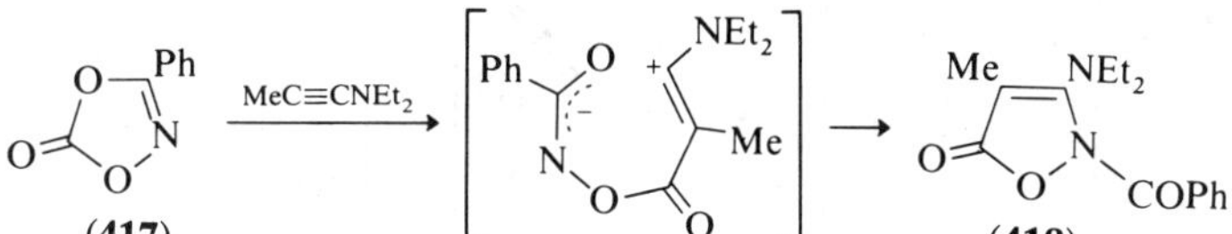

The thermal or photolytic fragmentation of furazans to nitriles and nitrile N-oxides has been reported ⟨73JOC1054, 75JOC2880⟩. The irradiation of dimethylfurazan (**419**) in the presence of cyclopentene, and benzofurazan (**420**) in the presence of dimethyl acetylenedicarboxylate, gave isoxazoline (**421**) and isoxazole (**422**), respectively, in good yields. The thermolysis of acenaphtho[1,2-c]furazan (**423**) in the presence of phenylacetylene gave isoxazole (**424**) in 55% yield.

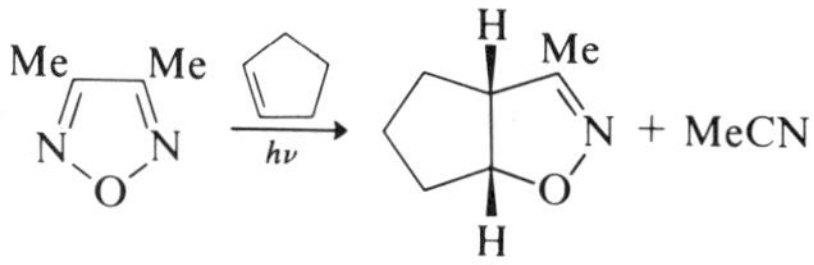

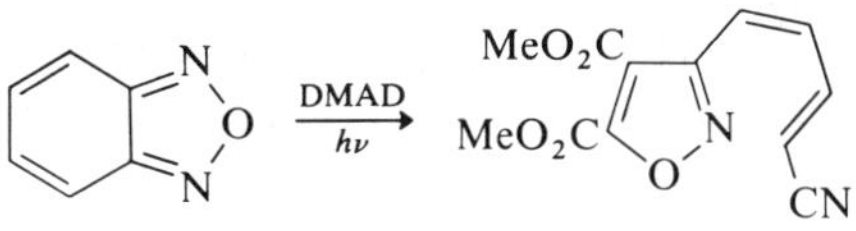

The thermal regeneration of nitrile N-oxides from furoxans has been extensively investigated ⟨81JOC316, 76CC240, 72JCS(P1)1587⟩. Thermolysis of alkyl- and aryl-furoxans resulted in ring cleavage to nitrile N-oxides which could be trapped with alkenes to give isoxazolines in good yield. For example, isoxazolines (**426**) and (**428**) were obtained by thermolysis of furoxans (**425**) and (**427**), respectively, in tetradec-1-ene.

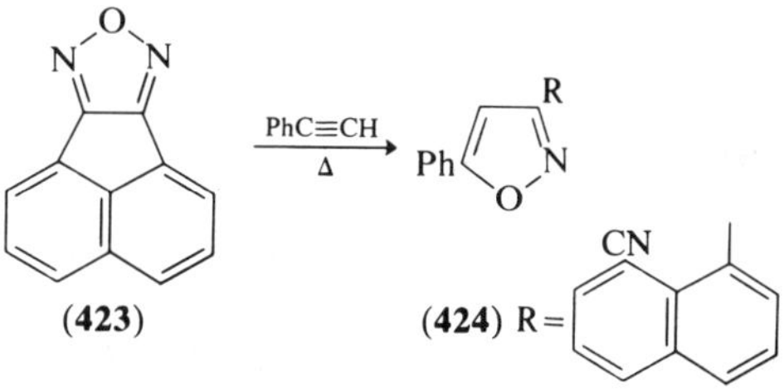

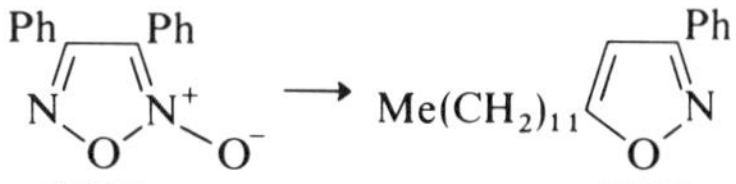

However, the thermolysis of diacylfuroxans (**429**) yielded two types of nitrile N-oxides. An uncrowded diacylfuroxan such as (**429a**) rearranged to the α-acyloximino nitrile N-oxide (**430**); the diacylfuroxan with bulky substituents such as in (**429b**) gave rise to the 'half molecule' acyl nitrile N-oxide (**431**). Both types of nitrile N-oxides (**431**) and (**430**) have been trapped with DMAD and hexafluoro-2-butyne to give isoxazoles in good yield. These reactions are shown in Scheme 97.

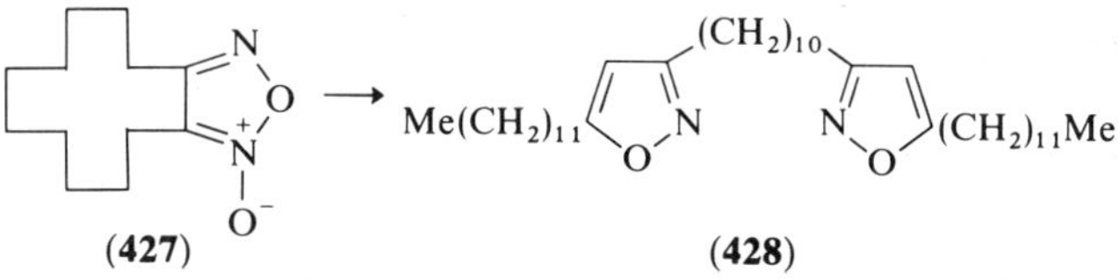

The thermolysis of 5-substituted 1,3,2,4-dioxathiazole 2-oxides (**432**) also gave nitrile N-oxides which were trapped with DMAD to give isoxazoles (**433**) ⟨76JOC1296⟩.

(429a) R = Ph

(429b) R =

(431)

(430)

Scheme 97

(432) (433)

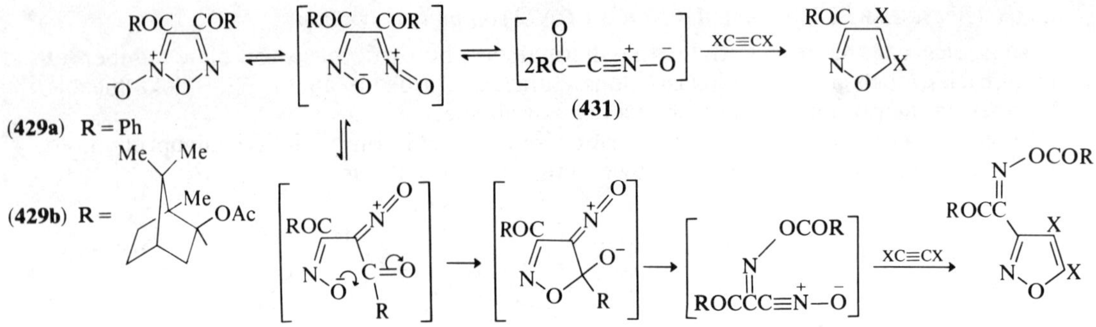

(*iii*) *Heterocycles which provide all the ring atoms of isoxazoles*

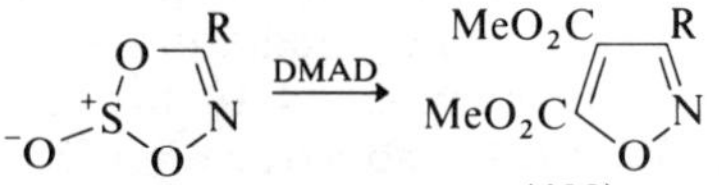

Diaroylfuroxans (**434**) when heated with aniline in ether solution undergo a complicated reaction which leads to 3-anilino-4-nitroso-5-arylisoxazole (**435**). One of the acyl groups is eliminated in the form of an acylanilide. Aromatic hydrazines and benzylamine can be used instead of aniline ⟨62HC(17)1, p. 35⟩.

(434) (435)

(436) (437)

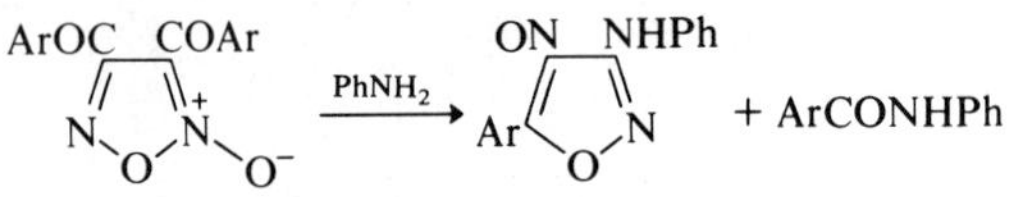

Disubstituted and trisubstituted 3-isonitrosopyrroles rearrange to 3-acylisoxazoles under the influence of hot, dilute mineral acids. For example, isonitrosotriphenylpyrrole (**436**), when treated with boiling alcoholic mineral acid, is converted into 3-benzoyl-4,5-diphenylisoxazole (**437**) ⟨62HC(17)1, p. 34⟩.

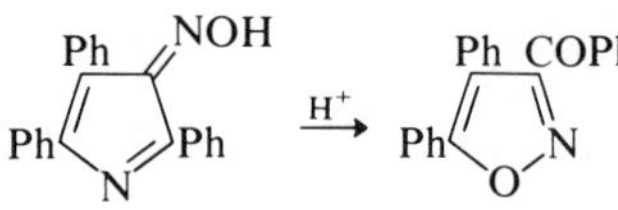

The treatment of 3-acylisoxazoles (**438**) with hydroxylamine hydrochloride gives furazan ketones (**439**). On the other hand, furazan ketones (**439**) rearrange to 3-acylisoxazoles (**438**) with a loss of hydroxylamine under the influence of a mineral acid. Thus, by refluxing phenacylphenylfurazan with concentrated alcoholic hydrogen chloride, 3-benzoyl-5-phenylisoxazole is formed; similarly, phenyl(phenacylphenyl)furazan gives 3-benzoyl-3,5-diphenylisoxazole ⟨62HC(17)1, p. 35⟩.

(438) (439)

(440) (301)

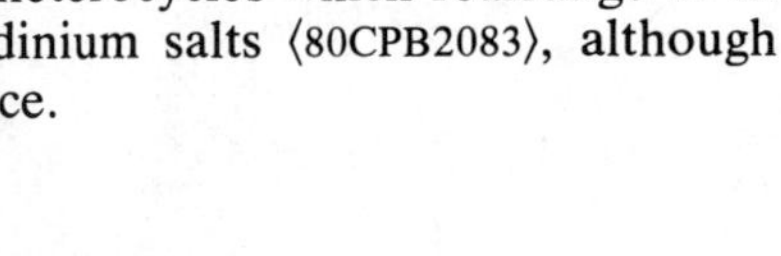

Nitrophenylcyclopropane (**440**) in acetic acid at 40 °C rearranges to 3-phenylisoxazole (**301**) in 87% yield ⟨81ZOR1435⟩.

Other heterocycles which rearrange to isoxazoles are pyridazine 1,2-dioxides ⟨77CC856⟩ and pyridinium salts ⟨80CPB2083⟩, although these transformations are of little synthetic importance.

4.16.4.2 Isoxazole Derivatives

4.16.4.2.1 Isoxazole

Isoxazole was first synthesized by Claisen in 1903 from propargylaldehyde diethyl acetal and hydroxylamine ⟨03CB3664⟩. It has also been obtained by addition of fulminic acid to acetylene in methanol–dilute sulfuric acid solution, by acidic hydrolysis of 5-acetoxyisoxazoline, by reaction of β-chloroacrolein or β-alkoxyacrolein with hydroxylamine hydrochloride

⟨62HC(17)1, p. 53⟩, and by retro-Diels–Alder reaction of the adduct from norbornadiene and fulminic acid ⟨67AG(E)456⟩.

For preparative purposes a mixture of propargyl alcohol and aqueous ammonium dichromate is added dropwise to boiling dilute sulfuric acid. The propargylaldehyde thus formed is passed into a cooled aqueous solution of hydroxylamine hydrochloride. Distillation of the reaction mixture gives an aqueous solution of isoxazole, from which the pure compound is isolated through the cadmium chloride salt. The yield is 25–30% of the theoretical amount based on propargyl alcohol ⟨62HC(17)1, p. 54⟩. Other preparative methods for isoxazole involve the reaction of malondialdehyde bis(diethyl acetal) (**284**; R = Et) with hydroxylamine hydrochloride (72% yield) ⟨63AHC(2)365⟩, the reaction of β-ethoxyacrolein diethyl acetal with hydroxylamine hydrochloride (67% yield) ⟨49CB257⟩, and the reaction of vinyl acetate with fulminic acid (90% yield) ⟨70S344⟩.

4.16.4.2.2 Alkylisoxazoles

4-Methylisoxazole and its homologs have been readily prepared by reaction of hydroxylamine hydrochloride with tetraalkoxypropanes (**284**) ⟨63AHC(2)365⟩, α-alkyl-β-alkoxyacroleins ⟨62ZOB2961⟩ and with α-alkyl-β-dimethylaminoacroleins ⟨60CB1208⟩.

5-Methylisoxazole (**297**; R = Me) and its homologs can be synthesized by reaction of hydroxylamine hydrochloride with 1-alkyl-3-dimethylamino-2-propen-1-one (**296**) ⟨54IZV47⟩, the anilino derivatives of acetoacetaldehyde ⟨47G556⟩, 3-dimethylaminomethylene-1-propyne (equation 7) ⟨69ZOR1179⟩ and the β-ketoaldehyde (**293**) ⟨66JOC3193⟩.

3-Methylisoxazole and its homologs are also prepared by reaction of hydroxylamine hydrochloride with the ethylene acetals of β-ketoaldehyde (**300**) ⟨60ZOB954⟩ and by reaction of alkylnitrile N-oxides with vinyl ethers ⟨62BSF2215⟩. The corresponding 3,5-dimethylisoxazole has been prepared by reaction of hydroxylamine hydrochloride with acetylacetone ⟨62HC(17)1, p. 54⟩. 3,5-Dialkylisoxazoles have been synthesized by a regiospecific metallation and alkylation of 3,5-dimethylisoxazole ⟨81TL3699, 79H(12)1343⟩. The alkylation occurs first on the 5-methyl group and a second alkylation then occurs on the 3-methyl group. For example, 3-ethyl-5-n-pentylisoxazole has been prepared in 81% yield from 3,5-dimethylisoxazole (see Section 4.16.3.3.2(i)(a)). 3-Alkyl-5-methylisoxazoles are also obtained from the reaction of appropriately substituted phosphonium salts (**304**) and hydroxylamine ⟨80CB2852⟩.

Both 4,5-dimethylisoxazole and 3,4-dimethylisoxazole are formed on treatment of the sodium derivative of α-methylacetaldehyde with hydroxylamine hydrochloride. The two isomers can be separated by fractional distillation ⟨62HC(17)1, p. 54⟩. 4,5-Dialkylisoxazole or 3,4-dialkylisoxazole can be obtained as the sole reaction product from an appropriate nitrile N-oxide and an appropriate vinyl acetate.

3,4,5-Trialkylisoxazoles have been prepared by the condensation of primary nitroalkanes under the influence of basic reagents ⟨40JA2604⟩. They can also be obtained from the reaction of a 1,3-diketone RCOCHRCOR with hydroxylamine hydrochloride ⟨62HC(17)1, p. 54⟩.

4.16.4.2.3 Phenylisoxazoles

All the isomeric phenyl substituted isoxazoles are known.

5-Phenylisoxazole was initially prepared in 1891 by Claisen by treatment of benzoylacetaldehyde monoxime with acetyl chloride ⟨1891CB130⟩. It has also been prepared from 3-dimethylamino-1-phenyl-2-propen-1-one (**296**) and hydroxylamine hydrochloride ⟨77JHC345⟩ or hydroxylamine-O-sulfonic acid (equation 22) ⟨80JOC4857⟩; by reaction of benzoylacetylene with hydroxylamine hydrochloride ⟨46JCS953⟩, hydrazoic acid (equation 21) ⟨65CB3020⟩, or diphenylsulfilimine (equation 40) ⟨73JOC4324⟩; by reaction of benzoylacetaldehyde anil with hydroxylamine hydrochloride ⟨47G556⟩; and by addition of fulminic acid to phenylacetylene ⟨74JCS(P2)1591⟩.

3-Phenylisoxazole has been obtained from the ethylene acetal of β-benzoylacetaldehyde (**300**; R = Ph) and hydroxylamine ⟨60ZOB954⟩, and also from benzonitrile N-oxide and acetylene ⟨49G703⟩, vinyl chloride ⟨70S344⟩, vinyl acetate ⟨62BSF2215⟩ or nitroethylene ⟨76S612⟩.

4-Phenylisoxazole was obtained by the reaction of 2-phenyltetraalkoxypropane with hydroxylamine hydrochloride ⟨63AHC(2)365⟩.

3,4-Diphenylisoxazole has been synthesized by the reaction of benzonitrile N-oxide with β-azidostyrene (**352**) ⟨74JOC1221⟩.

4,5-Diphenylisoxazole can be synthesized by the reaction of hydroxylamine chloride with 3-anilino-1,2-diphenyl-2-propen-1-one or formyldeoxybenzoin, PhC(CHO)=C(OH)Ph ⟨62HC(17)1, p. 57⟩.

3,5-Diphenylisoxazole is the most readily available arylisoxazole. Two preparative methods may be used: the reaction of hydroxylamine hydrochloride with dibenzoylmethane ⟨01CB3973⟩, and the addition of benzonitrile N-oxide to phenylacetylene in the presence of a trace of alkali ⟨46G148⟩. Regiospecific syntheses of 3,5-diarylisoxazoles can be carried out in several ways: (1) the reaction of hydroxylamine with α,β-alkynic ketones (**309**), chalcone dibromides ⟨68JCS(C)1774⟩ or chalcone oxides ⟨61AP769⟩ gives the 3,5-diaryl derivatives; (2) the cycloaddition of arylnitrile N-oxides with arylalkynes ⟨46G148⟩ or α-azidostyrene (**351**) ⟨74JOC1221⟩ is also a successful method; (3) the condensation of an appropriate 1,4-dilithio oxime (**358**) with benzonitrile or N,N-dimethylbenzamide ⟨72JHC183, 78JOC3015⟩ also results in these derivatives; (4) the reaction of α,β-alkynic ketones with hydrazoic acid (equation 21) ⟨65CB3020⟩ or with diphenylsulfilimine (equation 40) ⟨73JOC4324⟩; and (5) the reaction of chalcones and iodo azide (equation 41) ⟨67JA2077⟩ may be used. Many other syntheses of 3,5-diphenylisoxazole have been reported ⟨62HC(17)1, p. 57⟩ but offer little preparative advantage over those discussed above.

3,4,5-Triphenylisoxazole can be synthesized by the condensation of 1,4-dilithiodeoxybenzoin oxime with N,N-dimethylbenzamide ⟨78JOC3015⟩ and from the rearrangement of triphenylisoxazoline N-oxide (**369**) in aqueous alcoholic sodium hydroxide ⟨24JA2105⟩. A number of other synthetic routes to triphenylisoxazole have been reported ⟨62HC(17)1, p. 57⟩ but these are of relatively minor importance.

4.16.4.2.4 Alkylarylisoxazoles

Alkylarylisoxazoles can be obtained from the cycloaddition of nitrile N-oxides to substituted alkynes or alkenes (Section 4.16.4.1.2(ii)), and from the condensation of the 1,4-dilithio oximes (**358**) with benzonitriles ⟨72JHC183⟩ or amides ⟨78JOC3015⟩.

The reaction of appropriate 1,3-diketones (**302**) with hydroxylamine hydrochloride in pyridine ⟨79MI41601⟩ has been reported to result in a regiospecific synthesis of 3-alkyl-5-arylisoxazoles, as has the reaction of an α-bromoenone (**307**) with hydroxylamine hydrochloride in ethanol in the presence of potassium carbonate ⟨81H(16)145⟩. Regiospecific syntheses of 5-alkyl-3-phenylisoxazoles also result from the reaction of an α-bromoenone (**307**) with hydroxylamine in the presence of sodium ethoxide ⟨81H(16)145⟩. 3-Aryl-5-methylisoxazoles were prepared from phosphonium salts (**304**) and hydroxylamine ⟨80CB2852⟩.

4.16.4.2.5 Isoxazole aldehydes

Isoxazole-3-carbaldehyde has been obtained as a minor product from the reaction of acetylene with a mixture of nitric oxide and nitrogen dioxide ⟨61JOC2976⟩. Although 3-aryl-4-formylisoxazoles have been synthesized in good yields from the reaction of benzonitrile N-oxides with 3-(dimethylamino)-2-propen-1-one ⟨71S433⟩, the parent member of the series, isoxazole-4-carbaldehyde, has never been reported. It may possibly be obtained by the addition of fulminic acid to 3-(dimethylamino)-2-propen-1-one.

Isoxazole-5-carbaldehyde was prepared by the manganese dioxide oxidation of 5-hydroxymethylisoxazole ⟨67T4697⟩, the latter being formed from sodium fulminate and propargyl alcohol in greater than 90% yield.

A variety of other isoxazole aldehydes has also been reported ⟨62HC(17)1, p. 77⟩.

4.16.4.2.6 Isoxazole ketones

Numerous isoxazole ketones have been described ⟨62H(17)1, p. 79⟩. They have been synthesized by the many different methods available for the preparation of the isoxazole ring and by modification of suitably substituted isoxazoles. Ring syntheses include cycloaddition of a nitrile N-oxide to an appropriately substituted alkene or a doubly activated methylene (Tables 14 and 15), condensation of 1,4-dilithio oximes with aroyl chlorides ⟨76JHC607⟩, reaction of hydroxylamine with tricarbonyl compounds ⟨62HC(17)1, p. 79⟩, reaction of γ-diketones with nitric acid ⟨48G630, 59USP2908688⟩, cyclization of α-(acylmethoxyimino)nitriles (equation 44) ⟨80LA1623⟩, intramolecular cyclization of a 1:1 adduct from diphenylsulfilimine and a benzoylalkyne (equation 40) ⟨73JOC4324⟩, reaction of a 2,4,6-triphenylpyrylium salt (**396**) with nitrous acid in alcohol ⟨73TL2195⟩ and rearrangement of the 3-isonitrosopyrrole (**436**) ⟨62HC(17)1, p. 34⟩. Syntheses from isoxazole derivatives include oxidation of isoxazoles with unsaturated or carbinol substituents ⟨62JCS4234, 49G703, 58G149⟩, reaction of isoxazolealdehydes with diazomethane ⟨43G99⟩ and reaction of isoxazole nitriles with Grignard reagents ⟨46G87⟩.

4.16.4.2.7 Isoxazolecarboxylic acids

The synthesis of isoxazolecarboxylic acids has been well investigated. They can be prepared either from compounds which already contain an isoxazole nucleus or by isoxazole ring-closure methods using appropriate starting materials containing carboxy or alkoxycarbonyl groups.

Ring syntheses leading to isoxazolecarboxylic acids have been widely employed. The cycloaddition of nitrile N-oxides (CNO+CC) with unsaturated esters is probably the most versatile method for preparing various isoxazolecarboxylic acids. Thus, acetoacetic ester gives 4-ethoxycarbonyl and 3,4-diethoxycarbonyl derivatives ⟨46G148, 39G332⟩ and diethyl oxalacetate results in the formation of 4,5-diethoxycarbonyl derivatives ⟨40G89, 37G589⟩. Phenylpropiolic acid ⟨46G148⟩ and 3-methoxycarbonylallylidenetriphenylphosphorane ⟨76JCS(P1)619⟩ give 4-carboxylic acid derivatives. DMAD gives 3,4,5-trimethoxycarbonylisoxazole ⟨78CPB3254⟩. If ethoxycarbonylhydroxamyl chloride, $EtO_2CC(=NOH)Cl$, or nitrooximinoacetic ester, $EtO_2CC(=NOH)NO_2$, is used in these reactions a 3-ethoxycarbonylisoxazole is obtained ⟨40G119, 47G586⟩.

Ring closures with hydroxylamine (CCC+NO) are also particularly useful. The reaction of acyl- and aroyl-pyruvic acid esters with hydroxylamine gives mainly isoxazole-3-carboxylic acid esters ⟨09CB59, 78MIP41600⟩, whereas the sodium salts of acylpyruvic acids result in mainly isoxazole-5-carboxylic acids. Isoxazole-4-carboxylic acids have been obtained from 2-alkoxymethylene derivatives of β-keto esters ⟨55JOC1342, 59YZ836⟩, diacylacetic esters ⟨1893LA(277)162⟩, ethoxycarbonylmalondialdehyde ⟨47G206⟩, diethyl 2-acylmalonate (equation 4) ⟨78JHC1145⟩ and 2-(1-iminoalkyl) derivatives of β-keto esters (equation 3) ⟨78S829⟩.

4-Substituted isoxazole-3,5-dicarboxylic acids have been prepared from ethyl nitroacetate and an aldehyde ⟨63BCJ1150⟩. A related reaction leads to diethyl 4-hydroxyisoxazole-3,5-dicarboxylate (**334**) and involves the reaction of acetonedicarboxylic acid ester (**333**) with nitrosyl chloride ⟨78JHC1519⟩.

Nitric acid reactions often lead to isoxazolecarboxylic acids. Thus, isoxazole-3-carboxylic acid is most conveniently prepared by passing acetylene through fuming nitric acid ⟨29G930⟩. Diacetyl- and acetyl-succinic esters give 5-methylisoxazole-3,4-dicarboxylic acid when oxidized with nitric acid ⟨09CB1886⟩. Under these conditions, phenacylacetic ester gives 5-phenylisoxazole-3-carboxylic acid ⟨38G566⟩ and acetonylacetone is converted into 5-methylisoxazole-3-carboxylic acid ⟨59USP2908688⟩.

A variety of procedures is available for the conversion of isoxazole derivatives into carboxylic acids. Side chains containing alkenic, hydroxyl or carbonyl groups can be oxidized to carboxyl groups using alkaline and acid potassium permanganate, chromic acid, nitric acid or peracetic acid as oxidizing agents. Several specific examples illustrate the usefulness of these procedures: 3-acetylisoxazoles with dilute nitric acid form isoxazole-3-carboxylic acids ⟨48G764⟩ and, on treatment of 5-isobutenylisoxazole with chromic acid, isoxazole-5-carboxylic acid ⟨42G458⟩ is obtained. 3-Styrylisoxazole-4,5-dicarboxylic acid with basic potassium permanganate forms isoxazoletricarboxylic acid ⟨40G89⟩ and 3-ethoxymethyl-5-phenylisoxazole with peracetic acid gives 5-phenylisoxazole-3-carboxylic acid ⟨41G553⟩. Oxidation of 5-phenyl-3-(2-furyl)isoxazole with neutral potassium permanganate leads to 5-phenylisoxazole-3-carboxylic acid ⟨49G683⟩. A few examples are also known of the formation of isoxazolecarboxamides by Beckmann rearrangement of isoxazole ketone

oximes ⟨24JA1733⟩. Thus, 3,4-diphenyl-5-benzoylisoxazole oxime forms 3,4-diphenylisoxazole-5-carboxylic acid anilide when treated with phosphorus pentachloride in ether.
Numerous other isoxazolecarboxylic acids have been reported ⟨62HC(17)1, p. 82⟩.

4.16.4.2.8 Halogen derivatives

3-Chloro- and 3-bromo-isoxazoles have been prepared by hydrohalogen acid cyclization of β-nitroketones (equation 25) ⟨77JMC934, 57CI(L)1650⟩, and by reaction of 1-alkynylmagnesium bromide with carbonyl chloride oxime ⟨61G47⟩ and carbonyl bromide oxime (equation 46) ⟨69G1107⟩. 3-Chloroisoxazoles can also be obtained from the pyrolysis of 2-chloro-2-methylisoxazolium chlorides ⟨79CPB2398⟩, ready demethylation occurring under these thermal reaction conditions. The introduction of an iodo substituent as in 3-iodo-5-phenylisoxazole can be conveniently effected by nucleophilic substitution of 3-diazonium-5-phenylisoxazole with potassium iodide ⟨77JMC934⟩, diazotization of the amino group being carried out with nitrous acid. 5-Phenyl-3-trifluoromethylisoxazole has been synthesized in a two-step sequence from α-trifluoroacetylbenzophenone and hydroxylamine ⟨77JMC934⟩.

$$R-C\equiv CMgBr + HON=C\genfrac{}{}{0pt}{}{X}{X} \xrightarrow{THF} \underset{\underset{O}{N}}{\overset{X}{\diagdown}}R \tag{46}$$

X = Cl; R = H, Ph, OEt
X = Br; R = Me, Ph

In contrast to the 3-substituted products above, 4-chloro-, 4-bromo- and 4-iodoisoxazoles are readily prepared by direct halogenation of the corresponding isoxazoles, from 4-isoxazolediazonium salts by the Sandmeyer reaction, or by reaction of hydroxylamine with α-halo-β-dicarbonyl compounds ⟨62HC(17)1, p. 66, 63AHC(2)365⟩. 3,5-Bis(dimethylamino)-4-fluoroisoxazole has been synthesized by reaction of $(Me_2NCO)_2CHF$ with hydroxylamine ⟨78BSB391⟩.

5-Chloro- and 5-bromo-isoxazoles have been prepared by reaction of 5-isoxazolones with the appropriate phosphoryl halide ⟨77JMC934⟩. 3-Phenyl-5-trifluoromethylisoxazole has been synthesized by reaction of benzonitrile N-oxide with 3,3,3-trifluoropropyne ⟨77JMC934⟩.

4.16.4.2.9 N-Linked derivatives

(i) Aminoisoxazoles

4-Aminoisoxazoles are obtained by reduction of 4-nitroisoxazoles with amalgamated aluminum, tin(II) chloride and hydrochloric acid, or zinc dust and acetic acid ⟨62HC(17)1, p. 73⟩.

5-Acyl-4-aminoisoxazoles have been prepared by cyclization of α-(acylmethoxyimino)nitriles in the presence of lithium hydroxide (equation 44) ⟨80LA1623⟩.

Several methods are also available for the synthesis of 3-aminoisoxazoles. They have been prepared: (1) by the reaction of α,β-alkynic nitriles (**317**) with hydroxylamine in the presence of sodium hydroxide ⟨66CPB1277⟩, (2) by the reaction of 2,3-dibromoalkanoic acid nitriles with N-hydroxyurea in methanol in the presence of sodium hydroxide (equation 12) ⟨70M1109⟩, (3) by the reaction of β-keto iminoethers with hydroxylamine (equation 9) ⟨78GEP2825194, 63CB1088⟩, and (4) by the reduction of 3-nitroisoxazoles with tin(II) chloride and hydrochloric acid ⟨63MI41600⟩. A number of N-alkyl- and N-aryl-3-aminoisoxazoles have also been prepared by condensation of hydroxylamine with β-diketothioanilides (equation 47) ⟨20JA1055⟩ and with alkynic thioamides (equation 48) ⟨38JA1198⟩, 3-alkylthio- (or phenylthio)-1-aryl-3-arylamino-2-propene-1-thiones (equation 11) ⟨73LA256⟩, 1-aryl-3-arylamino-3-ethoxy-2-propen-1-ones (equation 49) ⟨63CB1088⟩, or with 3-anilino-2-cyano-3-methylthio-1-phenyl-2-propen-1-ones (equation 15) ⟨78JPR585⟩. 3-Acylaminoisoxazoles can be conveniently obtained by reaction of N-acylbenzoylacetamides with hydroxylamine hydrochloride (equation 13) ⟨81T1415⟩. 3-Aminoisoxazole has been prepared by reduction of 3-azidoisoxazole with tin(II) chloride ⟨31G759⟩, by reaction of propiolonitrile and hydroxylamine in the presence of sodium hydroxide ⟨66CPB1277⟩, and by the hydrolysis and subsequent ring closure of cyclohexylideneaminoacrylonitrile under acidic conditions (equation 45) ⟨69GEP1814116⟩.

$$\underset{\text{MeCO}}{\overset{\text{MeCO}}{>}}\text{CHC(S)NHPh} \xrightarrow{\text{H}_2\text{NOH}} \underset{\text{N}\diagdown\text{O}}{\overset{\text{PhNH}}{\diagup\diagdown}}\text{Me} \qquad (47)$$

$$\text{Br}\underset{}{\bigcirc}-\text{C}\equiv\text{C}-\overset{\text{S}}{\overset{\|}{\text{C}}}-\text{NHPh} \xrightarrow{\text{H}_2\text{NOH}} \underset{\text{N}\diagdown\text{O}}{\overset{\text{PhNH}}{\diagup\diagdown}}\underset{}{\bigcirc}\text{Br} \qquad (48)$$

$$\text{PhCCH}=\text{C}\overset{\text{OEt}}{\underset{\text{NHPh}}{<}} \longrightarrow \underset{\text{N}\diagdown\text{O}}{\overset{\text{PhNH}}{\diagup\diagdown}}\text{Ph} \qquad (49)$$

5-Aminoisoxazoles have been synthesized by reaction of hydroxylamine with α-alkynic nitriles (**317**) ⟨66CPB1277⟩, with 3-ethoxyacrylonitriles (equation 10) ⟨78KGS969⟩ or with 3-bromoacrylonitriles (equation 14) ⟨80TL3755⟩. They have also been prepared by reaction of nitrile *N*-oxides with ethyl cyanoacetate, nitroacetonitrile or cyanomethanephosphonic acid diethyl ester (Table 15), by reaction of α-chloroketoximes with sodium cyanide (equation 17) ⟨69TL4817⟩, by reduction of nitroacrylonitriles (equations 37 and 38) ⟨80BSF(2)163, 79JHC1611⟩ and by reaction of 5-chloroisoxazoles with amines ⟨77JMC934⟩. 5-Aminoisoxazole has been obtained by reaction of propiolonitrile with hydroxylamine ⟨66CPB1277⟩.

(ii) Nitroisoxazoles

3-Nitroisoxazoles result from the reaction of propargylic halides (**329**) with sodium nitrite ⟨80ZOR1328, 73TL485⟩ and may also be prepared by the reaction of chloronitroformoxime with PhC≡CMgBr ⟨63MI41600⟩. The parent member of the series, 3-nitroisoxazole, has been obtained by the reaction of propargyl bromide with sodium nitrite in DMF ⟨72MI41606⟩.

4-Nitroisoxazoles can be prepared by a variety of methods, including (1) the nitration of the corresponding isoxazoles with mixed acid or Ac$_2$O/HNO$_3$ ⟨75JCS(P2)1627⟩; (2) the reaction of a nitrile *N*-oxide with α-nitroketones, nitroacetonitrile ⟨75S664⟩ or 1-dimethylamino-2-nitroethylene ⟨81JOC316⟩; and (3) by the reaction of α,β-unsaturated aldehydes (**331**) or ketones with dinitrogen trioxide ⟨62HC(17)1, p. 28⟩. The parent member of the series, 4-nitroisoxazole, has been synthesized by the reaction of nitromalondialdehyde with hydroxylamine and by nitration of isoxazole in mixed acid ⟨63AHC(2)365⟩. 3-Methyl-4-nitro-5-phenylisoxazole was obtained by reaction of 3-bromo-1-phenyl-1-butyne with sodium nitrite ⟨80ZOR1328⟩.

5-Nitroisoxazoles have been synthesized by reaction of 1-chloro-2-nitroethylene with nitrile *N*-oxides ⟨75C350⟩.

4.16.4.2.10 *O-Linked derivatives*

There are five convenient routes to 3-hydroxyisoxazoles. They may be prepared by (1) the reaction of the α,β-alkynic esters (**321**) with hydroxylamine in the presence of sodium hydroxide ⟨66CPB1277, 67CPB1025⟩; (2) by the reaction of α-methyl (and ethyl) acetoacetic esters with hydroxylamine ⟨61MI41600⟩; (3) by the reaction of the glycol acetal of β-ketoesters with hydroxylamine followed by the cyclization of the resultant hydroxamic acid; (4) by the nucleophilic displacement of the halogen atom in 3-haloisoxazoles with alkoxides followed by hydrolysis of the resulting 3-alkoxyisoxazoles with hydrogen bromide ⟨61G47, 66CPB89⟩; and (5) by hydrolysis of 3-nitroisoxazoles with sodium hydroxide solution ⟨63MI41600⟩.

Diethyl 4-hydroxyisoxazole-3,5-dicarboxylate (**334**) was prepared by the reaction of acetonedicarboxylic acid ester with nitrosyl chloride ⟨78JHC1519⟩. Other 4-hydroxyisoxazoles have been prepared by the reaction of 2-hydroxy(or acetoxy)-1,3-diketones with hydroxylamine ⟨34JA2190, 62HC(17)1, p. 149⟩, and by hydrolysis of 4-isoxazolediazonium salts ⟨62HC(17)1, p. 149⟩. The parent 4-hydroxyisoxazole has not yet been reported.

The 5-hydroxyisoxazoles which exist predominantly as the 5-isoxazolones have been prepared (1) by reaction of the α,β-alkynic esters (**321**) ⟨66CPB1277⟩, β-ketoesters ⟨62HC(17)1, p. 118⟩ or the dioxane derivatives ⟨79JAP(K)79106466⟩ with hydroxylamine; (2) by cycloaddition of nitrile N-oxides with alkoxyalkynes ⟨59G2466⟩ or ketene acetals ⟨62HC(17)1, p. 118⟩ followed by acid hydrolysis of the resulting isoxazole derivatives; and (3) by condensation of the 1,4-dilithio oximes (**358**) with carbon dioxide ⟨71JCS(C)974⟩. 3,4-Disubstituted 5-isoxazolones can also be obtained by a reaction sequence starting from α-substituted β-ketoesters (**50**) ⟨73ACS2802⟩.

4.16.4.2.11 S-Linked derivatives

Among the three isomeric parent isoxazolethiols (C_3H_3NOS) only isoxazole-3-thiol has been reported recently ⟨80CPB552⟩.

Isoxazole-5-thiols and 5-alkylthioisoxazoles have been obtained by reaction of 5-chloroisoxazoles with potassium hydrogen sulfide and potassium alkylthiolates ⟨67JHC54, 77JMC934⟩. These thiols and thioesters can be oxidized to the corresponding disulfides, sulfoxides and sulfones with 30% hydrogen peroxide. 5-Alkylthioisoxazoles have also been synthesized by the cycloaddition of nitrile N-oxides to ketene dithioacetals ⟨70S344⟩, and 4-methylsulfinylisoxazoles result from the cycloaddition of nitrile N-oxides to β-ketosulfoxides ⟨73JHC669⟩. 4-Phenylsulfonylisoxazole has been prepared by the reaction of the ethoxymethyleneacetonitrile with hydroxylamine (equation 10) ⟨78KGS969⟩. 4-Ethylthioisoxazole has been prepared by the reaction of the bromoethylidenacetonitrile with hydroxylamine (equation 14) ⟨80TL3755⟩.

3-Alkylthioisoxazoles cannot be synthesized by the displacement of the chlorine atom of 3-chloroisoxazoles with sodium alkylthiolates. However, 3-phenylthioisoxazoles have been synthesized from 3-chloro-2-methylisoxazolium chlorides (equation 50) and 3-alkylthioisoxazoles from 2-benzyl-4-isoxazoline-3-thiones (equation 51) ⟨79CPB2415⟩. 3-Alkylthioisoxazoles can also be prepared by the reaction of α-ketoketene dithioacetals with hydroxylamine (equation 16) ⟨78JPR585, 59BSF1398⟩. Isoxazole-3-thiols have been prepared by a series of reactions starting from allylthioisoxazoles (equation 52) ⟨80CPB552⟩. On standing in ethanol, isoxazole-3-thiols were oxidized by air to give bis(3-isoxazolyl) disulfides.

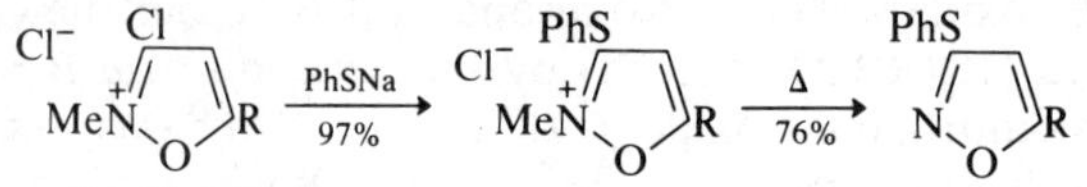

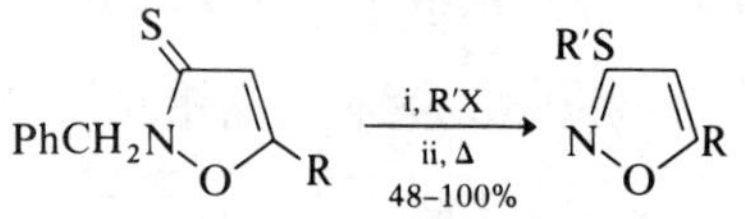

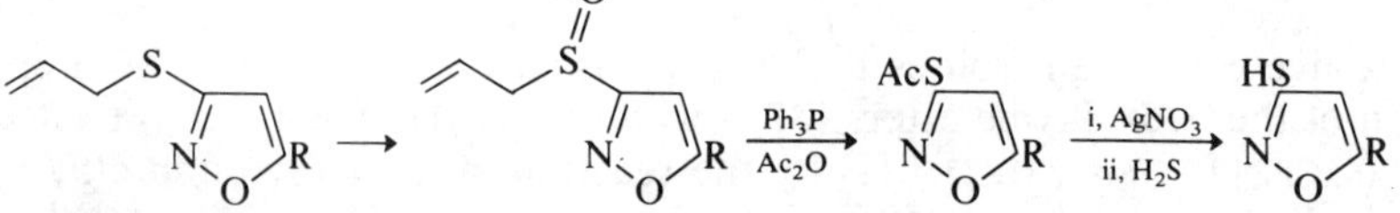

4.16.4.3 2-Isoxazolines

The two major methods of preparation are the cycloaddition of nitrile oxides to alkenes and the reaction of α,β-unsaturated ketones with hydroxylamines. Additional methods include reaction of β-haloketones and hydroxylamine, the reaction of ylides with nitrile oxides by activation of alkyl nitro compounds from isoxazoline N-oxides (methoxides, *etc.*) and miscellaneous syntheses ⟨62HC(17)1⟩.

4.16.4.3.1 By cycloaddition of nitrile oxides

Nitrile oxides react with a wide variety of alkenic compounds and this reaction may be complicated by dimerization of the nitrile oxide to furoxan in the presence of unreactive double bonds (Scheme 98).

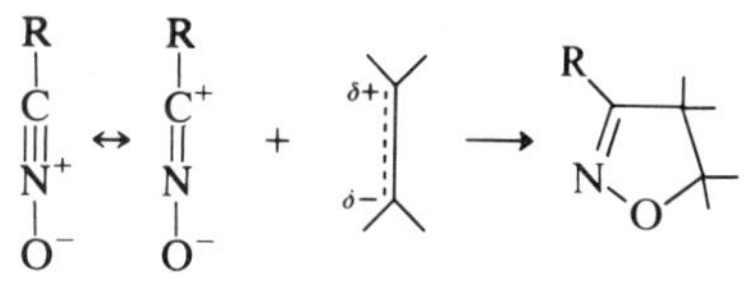

Scheme 98

The effect of steric and electronic factors, orbital interactions and transition state structures in the addition have been reported ⟨73JCS(P1)1148, 80JCR(S)348, 76JCS(P1)1518⟩ and the kinetics of the addition of benzonitrile oxide to styrene have been studied. The use of nitrile oxides in the synthesis of heterocyclic compounds has also been reviewed ⟨70S344, 68MI41604⟩. Shvekhgeimer has studied the cycloaddition and concluded that the products are formed by a radical pathway (Scheme 99). The addition of dibenzoyl peroxide shortened the reaction time by 30–50% while radical inhibitors lengthened the reaction time ⟨72MI41602, B-74MI41608⟩.

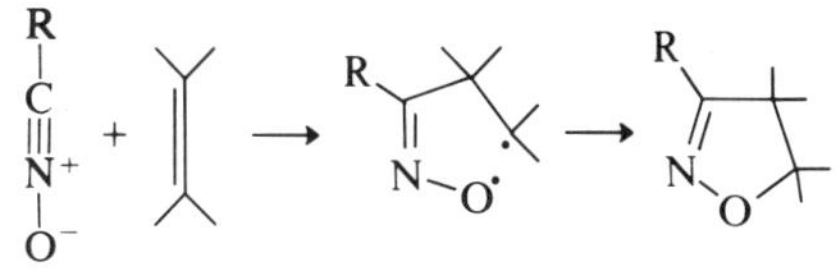

Scheme 99

With benzonitrile oxide, alkenes that do not exceed a bond length of 1.35 Å (80% double bond character) react in the absence of overriding steric factors ⟨62HC(17)1⟩. The largest number of reported articles has been concerned with the addition of benzonitrile oxide and substituted benzonitrile oxides to form 2-isoxazolines, fused analogs and heterocyclic fused analogs ⟨62HC(17)1, 67BCJ2608, 66MI41601, 77JCS(P1)2222, B-74MI41607, 76MI41613, 75KGS180, 76KGS625, 75ZOB2746, 74T3765, 74BSF1479, 71MI41604, 70T5113, 70CJC3753, 70G1144, 70ZOB2612, 71JPR745, 68ZOB1248, 68TL5209, 68ZOB1820, 70CJC467, 70JCS(C)1165, 80MI41601, 67ZOR821, 66MI41600, 78JCR(S)164, 78JCR(S)192, 72JCS(P2)1914, 74JCS(P2)1301, 76CSC75, 79JOC2796⟩.

The addition of benzonitrile oxide to acrylic acid gave only the 4-carboxylic acid (**441**) ⟨59MI41601⟩, while addition to *cis*- and *trans*-cinnamic esters gave *cis* and *trans* diastereomeric pairs of 4-carboxylic acids (**442**) (Scheme 100) ⟨59MI41600⟩. Arbisono repeated the experiment and, when methyl *cis*-cinnamate was used, in addition to the 4-carboxylic acid some 5-carboxylic acid (**442**) was isolated ⟨66MI41600⟩. The reaction of vinyl bromides with benzonitrile oxide yielded only an isoxazole and not a bromoisoxazoline (Scheme 101) ⟨78JCR(S)192⟩.

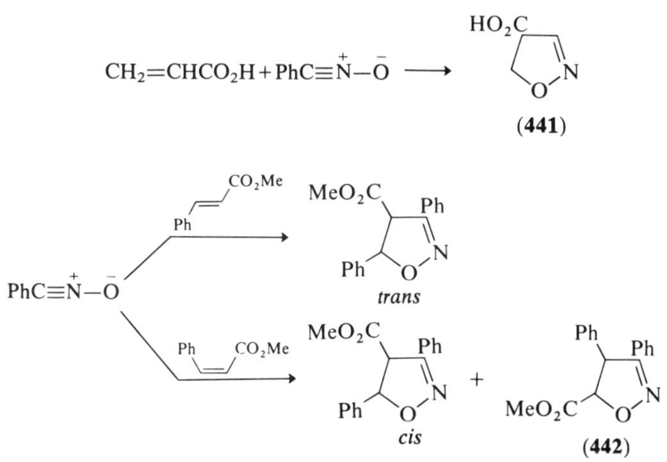

Scheme 100

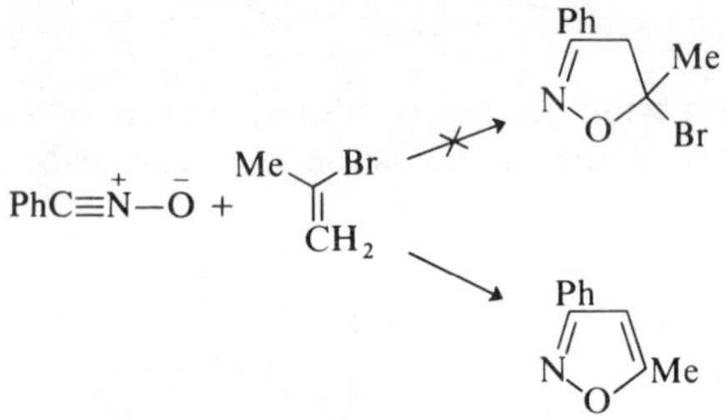

Scheme 101

When compound (**443**), which contains alkene and alkyne moieties, was reacted with benzonitrile oxide, both an isoxazoline (**444**) and isoxazole (**445**) were produced, with the former predominating. Oxidation of (**444**) with permanganate produced 3-phenyl-2-isoxazoline-5-carboxylic acid (**446**) ⟨67ZOR821⟩. The reaction of 1-trimethylsilylbut-1-yne-3-ene produced only a compound which reacted at the alkenic unit. Oxidation of the adduct also produced (**446**) ⟨68ZOB1820⟩. These reactions are shown in Scheme 102.

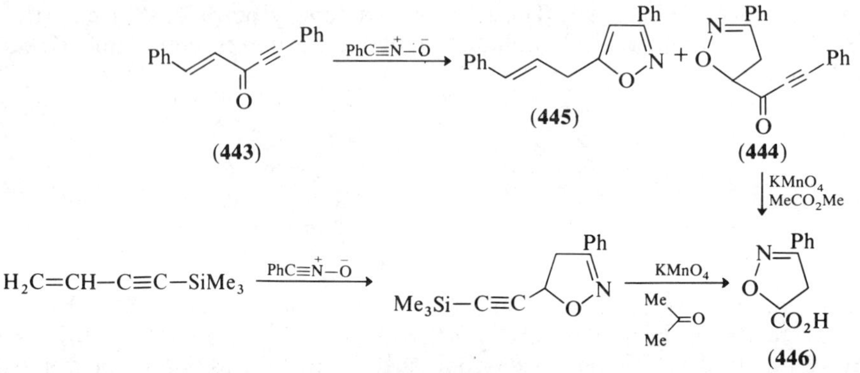

Scheme 102

The reaction of benzonitrile oxide with (**447**) or (**448**) produced only (**449**). No isoxazoline (**450**) was observed in the reaction with (**447**) ⟨80MI41601⟩.

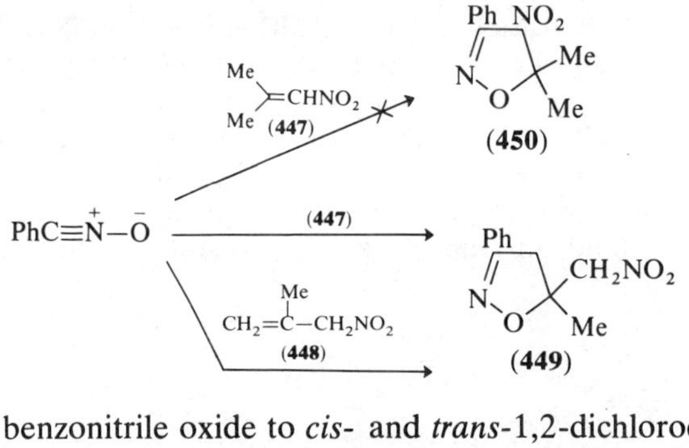

The cycloaddition of benzonitrile oxide to *cis*- and *trans*-1,2-dichloroethylene produced the appropriate *cis*- and *trans*-4,5-dichloro-3-phenyl-2-isoxazoline diastereomers. Base elimination produced only one compound, 4-chloro-3-phenyloxazole (Scheme 103) ⟨70CJC3753⟩.

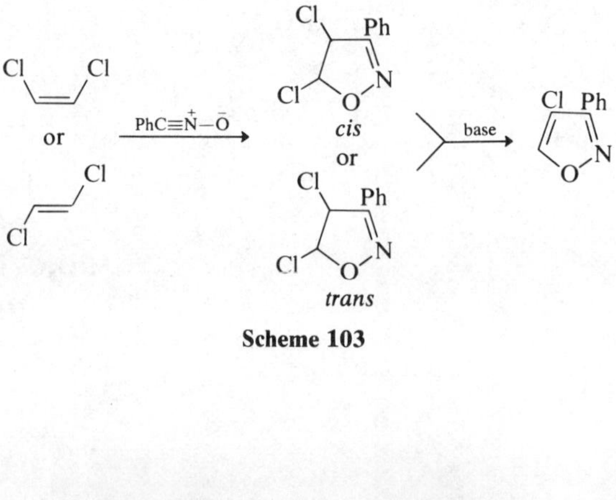

Scheme 103

The addition of benzonitrile oxide to cyclooctatetraene produced a monoadduct which was induced to undergo valence tautomerism to produce a tricycloisoxazoline (Scheme 104). A similar reaction with tropone gave a minimum of eight adducts from which two monoadducts were isolated (Scheme 104) ⟨70T5113⟩.

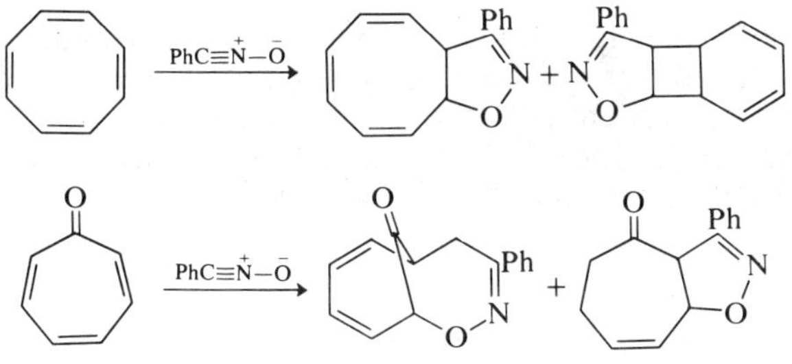

Scheme 104

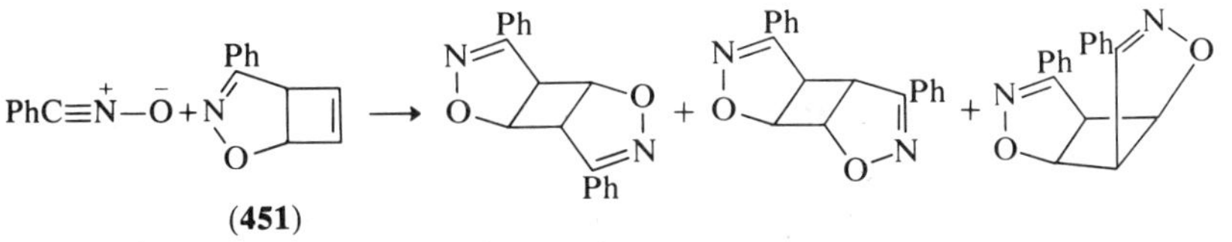

The reaction of benzonitrile oxide with the bicyclic isoxazoline (**451**) produced the three fused diisoxazoles shown in Scheme 105 ⟨77JCS(P1)2222⟩.

Scheme 105

The reaction of arylnitrile oxides with 1,1-diphenylallenes gave a mixture of 4-methylene-2-isoxazolines (Scheme 106) with major attack at the C(2)—C(3) double bond ⟨74JCS(P2)1301, 76CSC67, 76CSC71, 72JCS(P2)1914⟩ and not a mixture of the 4- and 5-methylene compounds. 1-Phenoxyallene and benzonitrile oxide produced a mixture of positional isomers and a spiro compound (Scheme 107) ⟨79JOC2796⟩.

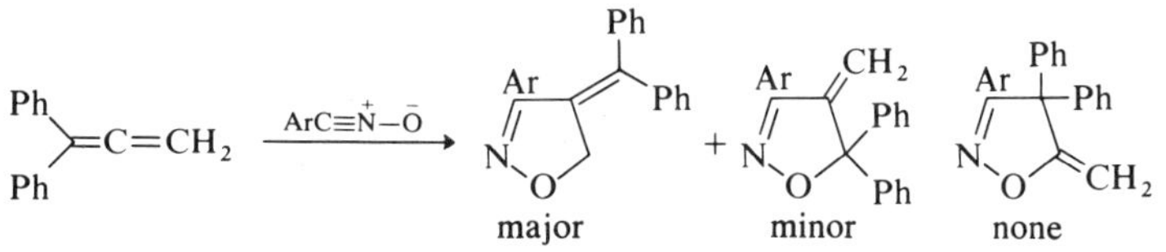

Scheme 106

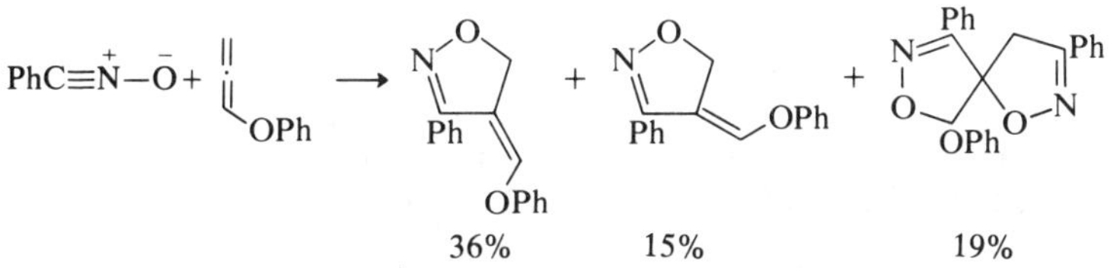

Scheme 107

Other less commonly used nitrile oxides include acetonitrile oxide ⟨77JCS(P1)2222, 71JOC3470, 66ZOR1766, 79TL2443, 80H(14)185⟩, *t*-butylnitrile oxide ⟨59G1525⟩, ethoxycarbonylfulminic acid ⟨62HC(17)1, 78BCJ1261, 75USP3852299, 65EGP37461, 60JOC1160, 64CB159, 62BSF2215, 79JA1319⟩, phenylsulfonylfulminic acid ⟨79JA1319⟩, fulmic acid ⟨79JA1319⟩, fulminic acid halides and their dimers ⟨62HC(17)1, 73CB3291, 76CC795, 80TL229, 74JHC63⟩, heterocyclic nitrile oxides ⟨73MI41601, 71JAP7100026, 67CPB366, 62BSF2215, 76USP3959343⟩ and nitrile oxides generated from an alkyl nitro compound by acetyl chloride ⟨B-79MI41613⟩.

Acetonitrile oxide was generated from 3,4-dimethylfuroxan oxide by flash vacuum pyrolysis and trapped at −40 °C where its ^{1}H and ^{13}C NMR spectra were examined. Warming to room temperature in the presence of propene produced 3,5-dimethyl-2-isoxazoline (Scheme 108) ⟨79TL2443⟩. The oxide could also be generated by photolysis of furoxan ⟨68CC977⟩.

t-Butyl nitrile oxide reacted with diphenylketene to produce a 2-isoxazoline, but benzonitrile oxide did not react with diphenylketene ⟨59G1525⟩. Vaughan and Spencer studied

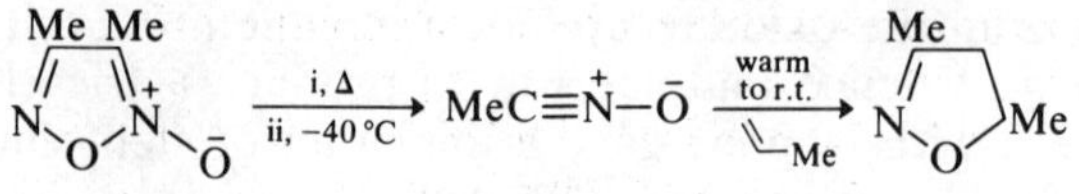

Scheme 108

the reaction of ethoxycarbonylfulminic acid with styrene to produce 3-ethoxycarbonyl-2-isoxazolines and concluded that a carbonium ion intermediate was important (Scheme 109) ⟨60JOC1160⟩. Hutchinson generated ethoxycarbonylfulminic acid *in situ* from ethyl chloroacetate and $NaNO_2$ in DMF at 100 °C which reacted with ethyl acrylate to produce 3,4-bis(ethoxycarbonyl)-2-isoxazoline (**452**) ⟨66USP3852999⟩.

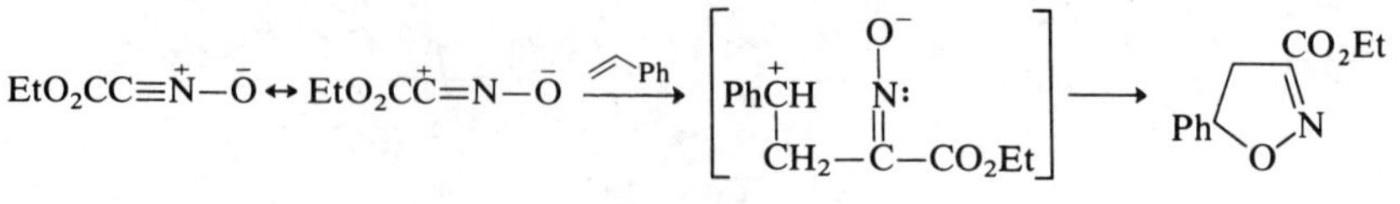

Scheme 109

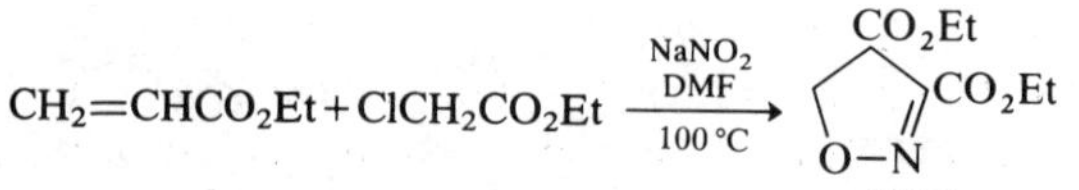

Phenylsulfonylfulminic acid, prepared as outlined in Scheme 110, reacted even with highly hindered alkenes such as tetramethylethylene. The phenylsulfonyl group was easily displaced by a variety of nucleophiles such as methoxide, cyanide and hydride ⟨79JA1319⟩.

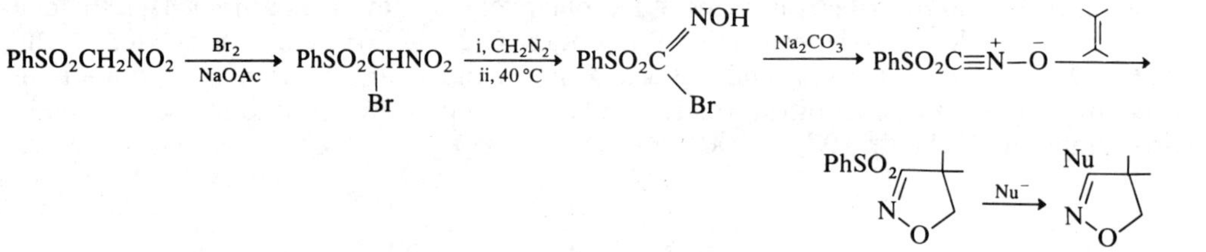

Scheme 110

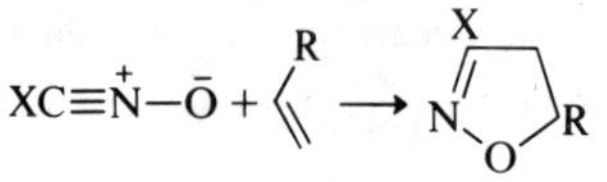

Chloro- or bromo-nitrile oxides, generated from dihalocarbonyl oxime, reacted with alkenic compounds to produce 3-halo-2-isoxazolines (**453**) ⟨80TL229, 76CC795⟩. Fulminic acid, generated from haloformaldehyde oxime or its dimer, reacted with ethylene to give the oxime of 3-formyl-2-isoxazoline (Scheme 111). The reaction of these agents with other alkenic compounds corresponds closely to that for benzonitrile oxide ⟨73CB3291, 74JHC63⟩.

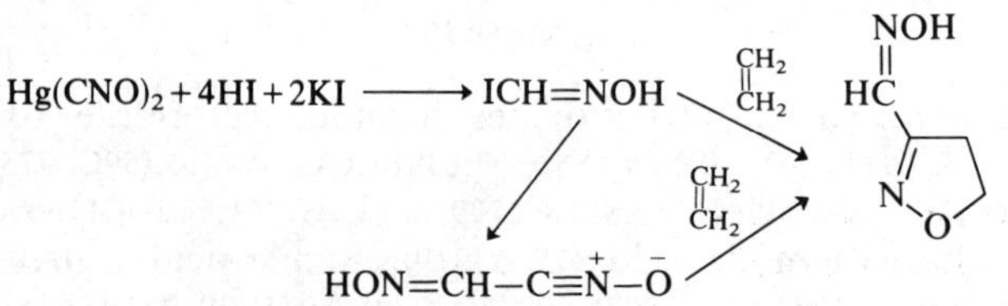

Scheme 111

The reaction of alkyl nitro compounds with acetyl chloride in the presence of an alkenic compound produced a 2-isoxazoline. The mechanism is believed to proceed *via* a nitrile oxide and is illustrated in Scheme 112 ⟨B-79MI41613⟩.

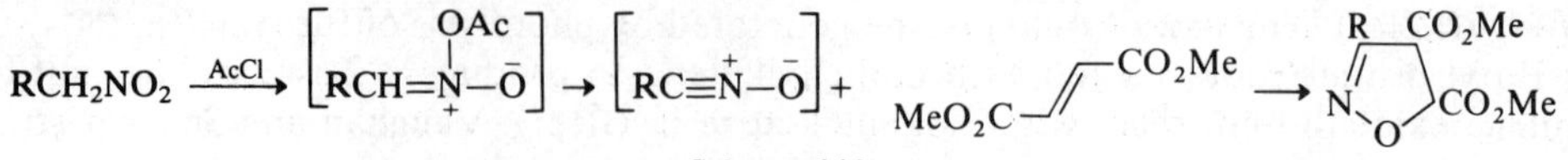

Scheme 112

4.16.4.3.2 From α,β-unsaturated ketones and hydroxylamine

The first isolated isoxazoline was obtained by this method in 1895, and it has been found to be a highly complicated reaction which yields a variety of products depending on, among other conditions, the pH, concentration, temperature and solvent ⟨62HC(17)1⟩.

A number of reports show 2-isoxazolines as the primary, predominate or sole product from the reaction of α,β-unsaturated ketones by attack at the carbonyl carbon atom ⟨78AP817, 78IJC(B)57, 77JHC523, 77ZN(B)443, 77ACS(B)184, 72MI41611, 73MI41600, 65JIC733, 78RRC1541⟩ or by the action of hydroxylamine on α,β-unsaturated ketone precursors ⟨70IJC796, 74CPB1990⟩. The reaction of chalcone with hydroxylamine in pyridine–water gave 3,5-diphenyl-2-isoxazoline (**454**) ⟨65JIC733⟩.

$$PhCOCH=CHPh \xrightarrow{NH_2OH \cdot HCl}_{H_2O/Py} \text{(454)}$$

The reaction of a dibromochalcone with hydroxylamine hydrochloride in pyridine gave three products with the expected 2-isoxazoline product as the predominate compound. A ring bromination product and an isoxazole were also isolated ⟨70IJC796⟩. The reaction of hydroxylamine with β-thiosulfates of propiophenone at reflux produced 3-phenyl-2-isoxazoline (**455**). At room temperature a 'bis-Michael product' (**456**) was produced. The reaction with N-phenylhydroxylamine yielded a 'mono-Michael' type product (**457**) ⟨74CPB1990⟩.

$$PhCOCH_2CH_2N(Ph)OH \;(\mathbf{457}) \;\longleftarrow\; PhCOCH_2CH_2SSO_3Na \;\xrightarrow{NH_2OH, \Delta}\; \mathbf{455}$$
$$\downarrow NH_2OH, r.t.$$
$$PhC(=NOH)CH_2CH_2N(OH)CH_2CH_2C(=NOH)Ph \;(\mathbf{456})$$

Other references list a number of products including simple oximes, oximes–hydroxylamines and Michael type products ⟨75KGS162, 75MI41616, 72BSF330, 68MI41603, 70M704, 80ZC19, 77MI41610⟩. The reaction of (**458**) gave, under various conditions, a Michael type product or an oxime–hydroxylamine. The treatment of the Michael type product (**459**) with hydroxylamine gave both the oxime–hydroxylamine (**460**) and a 2-isoxazoline (**461**).

The treatment of 3-buten-2-one containing various alkyl substituents with hydroxylamine was carried out to study the effect of the substituents and three general products were obtained: a Michael product, a cyclized Michael product (a 4-isoxazoline) and a 2-isoxazoline. With increasing bulk of the substituents the simple oxime was the predominant product to 80% yield ⟨75KGS162⟩. The reaction of 4-methyl-3-penten-2-one with hydroxylamine hydrochloride in methanol with 2 equivalents of sodium methoxide gave five products but with the employment of only 1 equivalent of sodium methoxide a 50% yield of 3,5,5-trimethyl-2-isoxazoline was obtained ⟨72BSF330⟩. The reaction of chalcone with hydroxylamine hydrochloride under acid conditions gave the simple oxime while similar aminoximation with base yielded 3,5-diphenyl-2-isoxazoline (**454**) ⟨68MI41603, 70M704⟩. The treatment of 4-phenyl-3-buten-2-one with hydroxylamine gave three products: 3-methyl-4-phenyl-2-isoxazoline, a Michael product and a simple oxime (Scheme 113) ⟨77MI41610⟩.

Reacting hydroxylamine with the corresponding trichloro derivative (**462**) under acidic conditions gave an oxazinone (**463**) while under basic conditions the 2-isoxazoline (**464**) was produced ⟨80ZC19⟩.

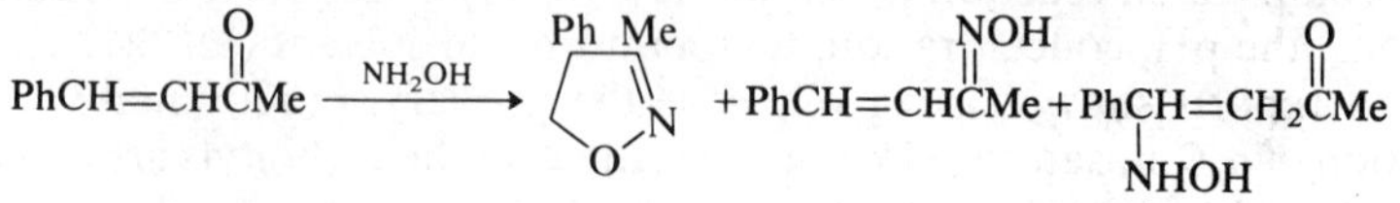

Scheme 113

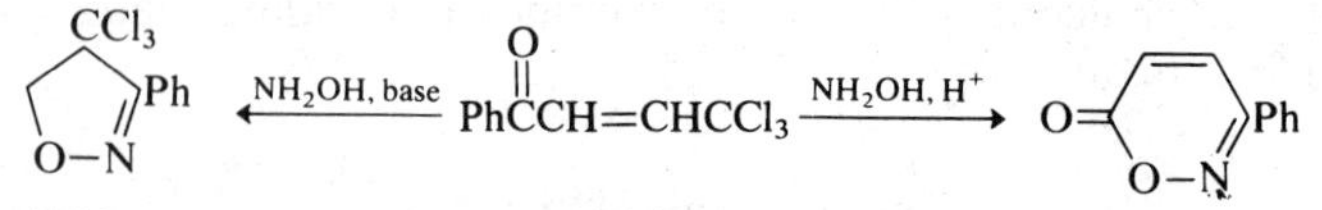

(**464**) ← NH₂OH, base — PhCCH=CHCCl₃ (**462**) — NH₂OH, H⁺ → (**463**)

The reaction of α,β-unsaturated ketone oximes or precursors usually gave, depending on the amount of the corresponding reactant such as NBS, I_2, acid or heat, 2-isoxazolines with some isoxazoles ⟨77JHC1289, 80JHC475, 76JHC449, 75T3069, 71MI41606, 71JCS(C)584, 70TL2993⟩. The reaction of α,β-unsaturated ketone oximes with NBS or I_2 produced either 4-bromo-2-isoxazolines or isoxazoles (Scheme 114).

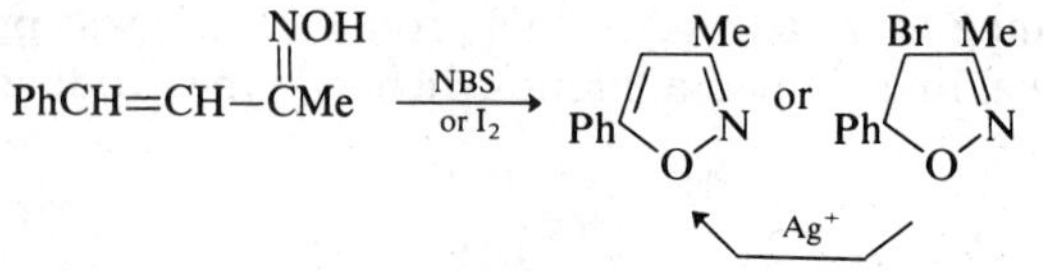

Scheme 114

When the chalcone oxime (**465**) was heated at 190 °C the 2-isoxazoline (**466**) was produced. Further heating at 280 °C led to ring rupture, giving benzonitrile and acetophenone. Heating of (**467**) at 130 °C in H_2SO_4 produced a 2-isoxazoline and an oxazinone ⟨75T3069, 71MI41606⟩. These reactions are shown in Scheme 115. β-Mannich bases of ketone oximes (**468**) yielded 2-isoxazolines (**469**) on base treatment ⟨71JCS(C)584, 70JCS(C)2993⟩.

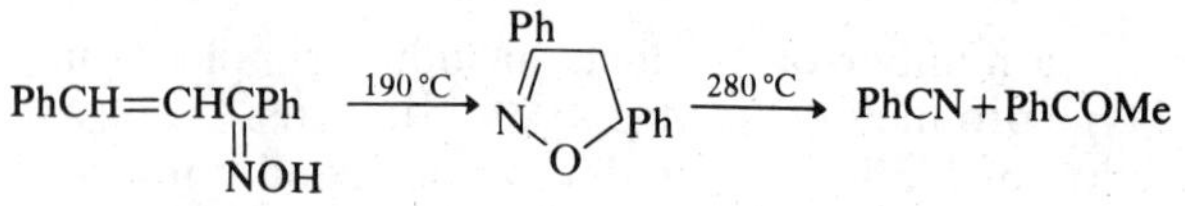

Scheme 115

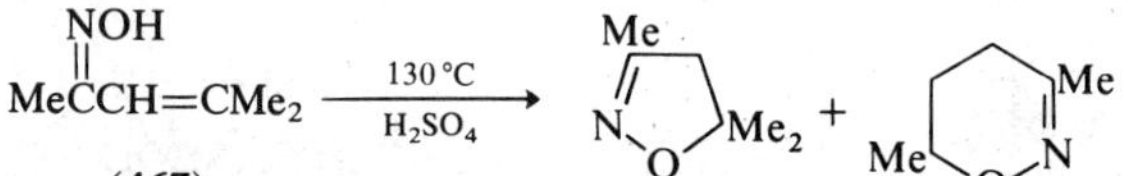

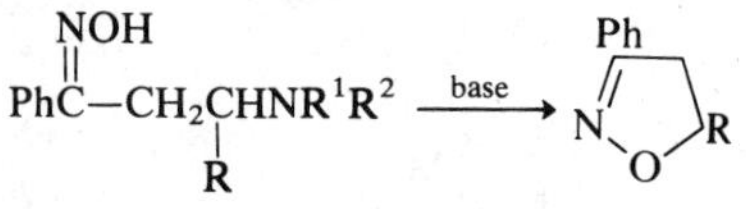

4.16.4.3.3 From ylides and nitrile oxides

The reaction of benzonitrile oxide with dimethylsulfonium methylylide or triphenylarsonium methylylide produced 3-phenyl-2isoxazoline (Scheme 116) ⟨67MI41602, 68G48⟩. A similar reaction of benzonitrile oxide with diazomethane likewise gave 3-phenyl-2-isoxazoline via intermediate (**470**) (Scheme 117).

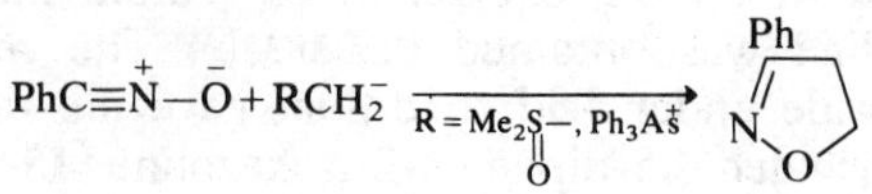

$$PhC\overset{+}{\equiv}N-\overset{-}{O} + RCH_2^- \xrightarrow{R = Me_2\overset{+}{S}-, Ph_3\overset{+}{As}}$$

Scheme 116

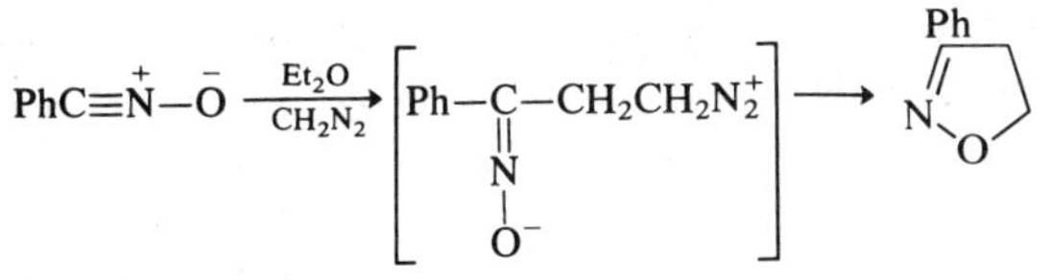

Scheme 117

The action of benzoylmethylides (**471**) and (**472**) with nitrosyl chloride produced the 2-*trans* products (**473**) and (**474**) ⟨72T3845⟩.

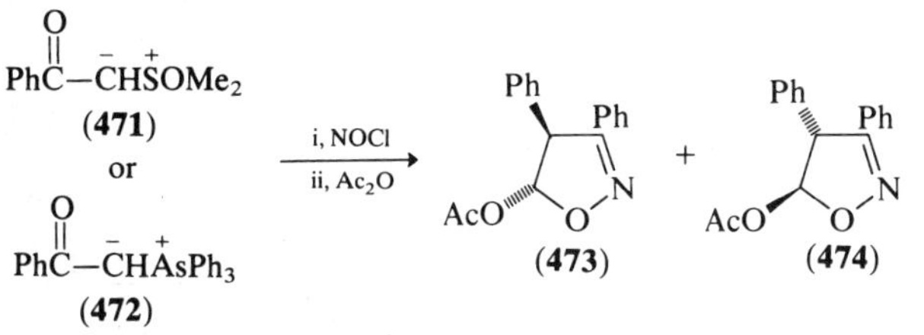

4.16.4.3.4 *By the activation of alkyl nitro compounds*

Many of these reactions are believed to generate an *in situ* nitrile oxide ⟨76JOC122, 75TL2131, 75AJC207⟩ by cycloaddition of an *aci*-nitro moiety to an isoxazolidine *N*-oxide with subsequent loss of H_2O to produce the 2-isoxazolines ⟨75MI41612, 68ZOR236, 78ACS(B)118, 67ZOR980, 72MI41605, 73ZOB1715, 74DOK109, 78MI41610, 65FRP84686, 70JOC2065, 78USP4092327, 80MI41600⟩. The treatment of (**475**) with HCl gave not 3,4,4-triphenyl-2-isoxazoline (**476**) but 3,5,5-triphenyl-2-isoxazoline (**477**). The product arises by fragmentation similar to an abnormal Beckmann rearrangement of oximes to 1,1-diphenylethylene and benzonitrile oxide, which then recombine ⟨75AJC207⟩. The decomposition of (**478**) in the presence of cyclohexene produced (**479**) ⟨75TL2131⟩.

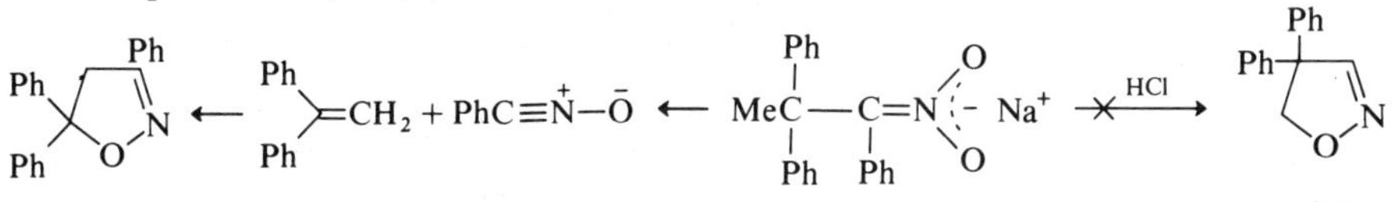

Rahman and Clapp decomposed dinitromethane derivatives in DMF in the presence of alkenes to obtain 2-isoxazolines. Without any alkene present, an acid and KNO_2 were obtained. They proposed a mechanism which proceeded *via* a three-membered ring or a nitrocarbene which rearranged to a nitrile oxide ⟨76JOC122, 75MI41612⟩.

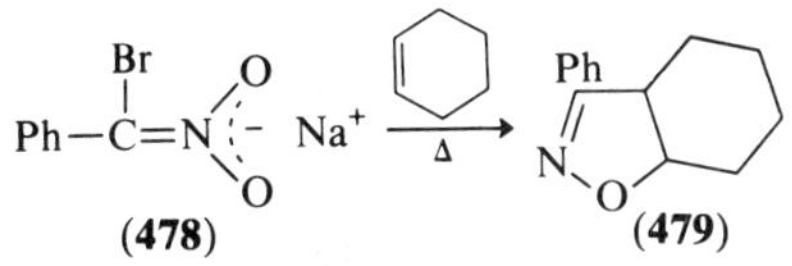

Nitrones or *aci*-nitro esters react with alkenes to give in some cases *N*-substituted isoxazolidines and in others 2-isoxazolines. When the intermediate isoxazolidines were observed, a number of procedures transformed them into the 2-isoxazolines. Acrylonitrile and phenyl *aci*-nitrone esters produced an *N*-methoxyisoxazolidine. Treatment with acid generated a 2-isoxazole while treatment with base generated an oxazine (Scheme 118) ⟨68ZOR236⟩. When an ethoxycarbonyl nitrone ester was reacted with alkenes, no intermediate isoxazolidine was observed, only Δ^2-isoxazolines. Other *aci*-nitro methyl esters used are shown in Scheme 118 and these generate *N*-methoxyisoxazolidines or Δ^2-isoxazolines which can be further transformed ⟨72MI41605⟩.

The trimethylsilyl group has been used to prepare stable *aci*-nitro esters and these react with alkenes to produce intermediate isoxazolidines which were readily converted into 2-isoxazolines (Scheme 119) ⟨73ZOB1715, 74DOK109, 78ACS(B)118⟩.

Alkylnitro compounds when treated with acetic anhydride and triethylamine produced furoxans. With an alkene present, 2-isoxazolines are isolated (Scheme 120) ⟨78MI41610⟩.

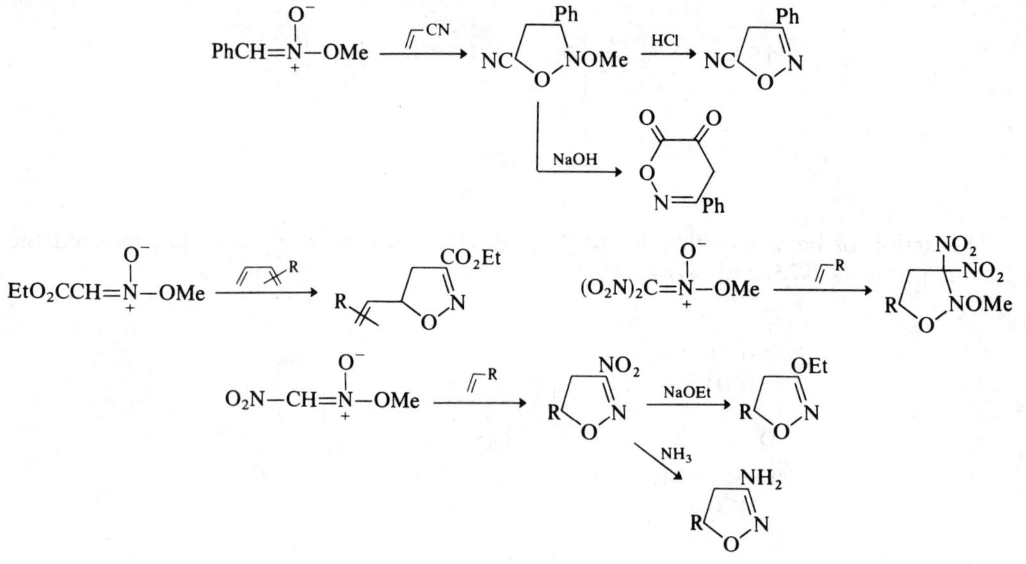

Scheme 118

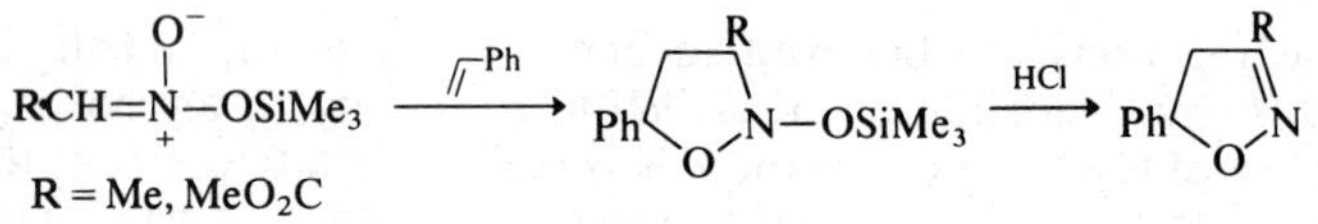

Scheme 119

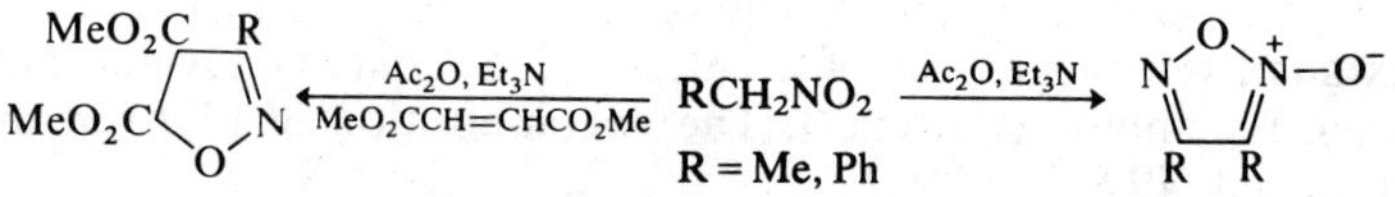

Scheme 120

Aldehydes react with α-nitroacetophenone in refluxing toluene to generate a 3-acetyl-2-isoxazoline ⟨78USP4092327⟩, while α-nitroacetophenone oxides react with alkenes to provide 2-isoxazolines (Scheme 121) ⟨79ZOR735, 79ZOR2408, 80CPB479⟩.

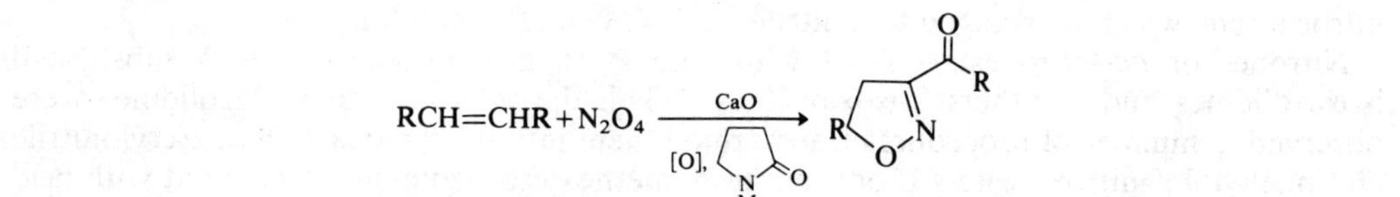

Scheme 121

Alkenes treated with N_2O_4 and CaO in oxygenated N-methylpyrrolidone are converted into 3-acyl-2-isoxazolines as in Scheme 122 ⟨78USP4069226⟩.

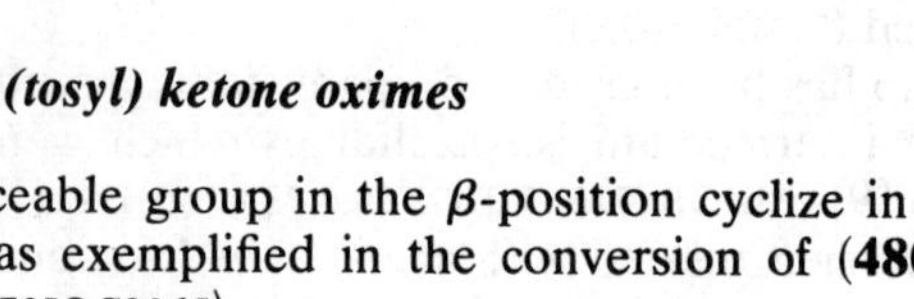

Scheme 122

4.16.4.3.5 From β-halo (tosyl) ketone oximes

Oximes with a displaceable group in the β-position cyclize in the presence of base to produce 2-isoxazolines as exemplified in the conversion of (**480**) into (**481**) ⟨62HC(17)1, 78JOC2020, 66JAP6616384, 70JOC2065⟩.

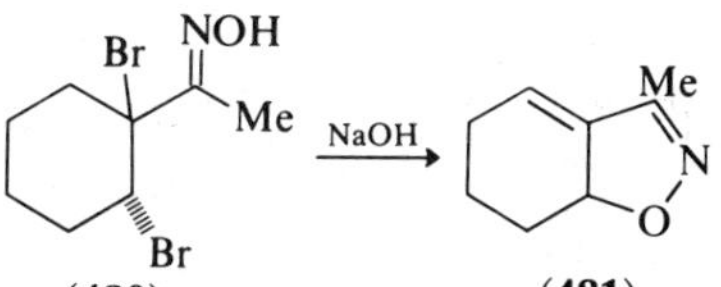

4.16.4.3.6 From 2-isoxazoline N-oxides

N-Oxides are commonly reduced by a variety of reagents including PCl$_5$ and H$_2$, as shown by the reaction of (482) with H$_2$ to give (483) ⟨69JOC984⟩.

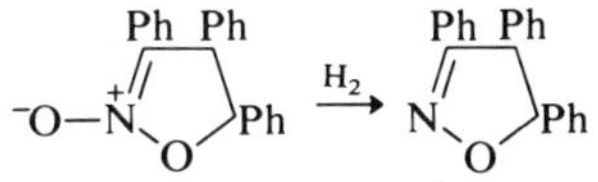

4.16.4.3.7 Miscellaneous methods

Ketones react with formaldehyde and perchloric acid to produce 2-isoxazolines and also with urea in α-methylnaphthalene at 200 °C ⟨75MIP41600, 75ZOB2090, 79MI41612⟩; with N-hydroxyurea they produce 3-hydroxy-2-carbamoylisoxazolidines or, after acid treatment, 2-isoxazolines are formed (Scheme 123) ⟨75TL2337⟩.

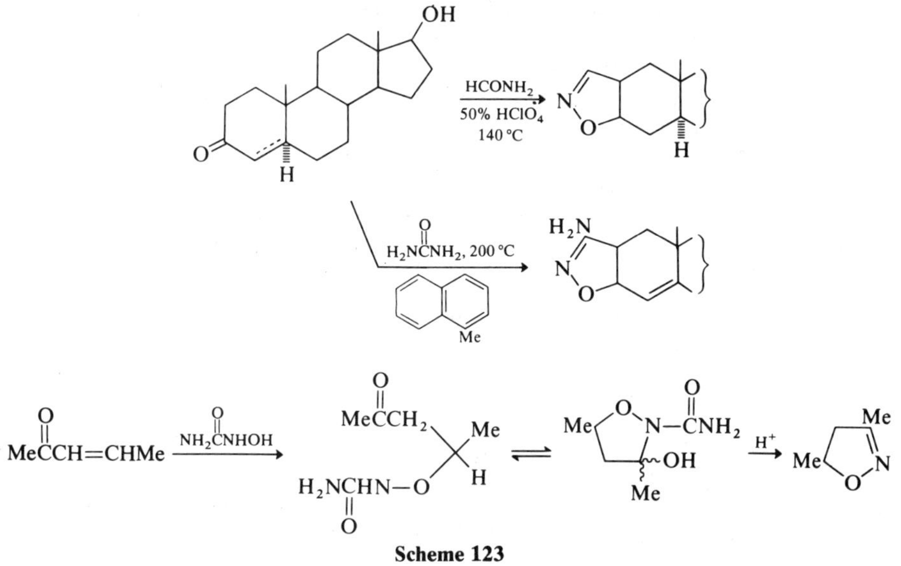

Scheme 123

Phenyldiazonium salts react with malonaldioxime to produce a 2-isoxazoline ⟨71GEP1920245⟩, and the diazo ketone (484) when photolyzed gave a mixture of 2-isoxazoline and an isoxazole by a 1,5 carbon–hydrogen insertion. A phenyl migration was apparently not involved (Scheme 124) ⟨66CC689⟩.

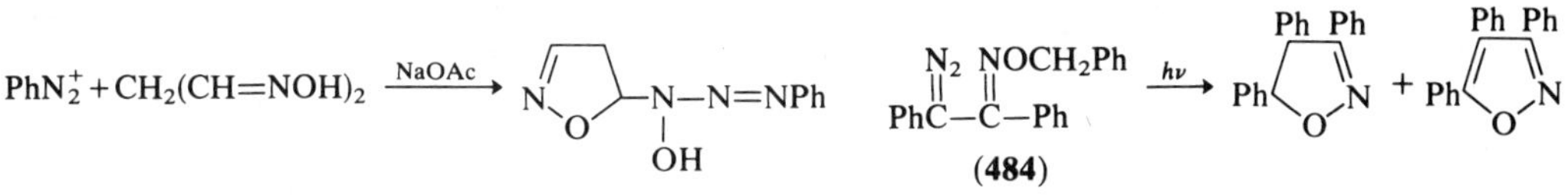

Scheme 124

The reaction of (485) with hydroxylamine was reported to generate an azetoisoxazole and an azetinoisoxazoline as shown in Scheme 125 ⟨76TL1825⟩, but this was demonstrated to be a 4-cyano-2-isoxazoline ⟨76TL3931⟩.

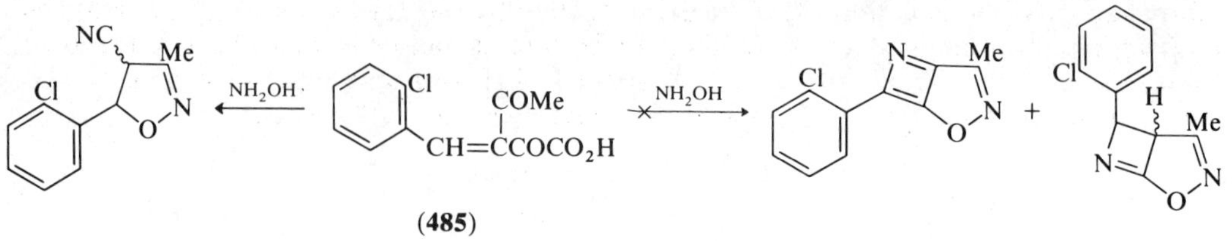

Scheme 125

Acid treatment or thermolysis of aziridinyl phenyl ketone oxime generated a 2-isoxazoline ⟨74JAP(K)74117462⟩, while treatment of a diphenylcyclopropene (**486**) with NOCl generated a 2-isoxazoline in contrast to dialkylcyclopropene (**487**) which produced addition across the double bond (Scheme 126) ⟨73MI41605⟩.

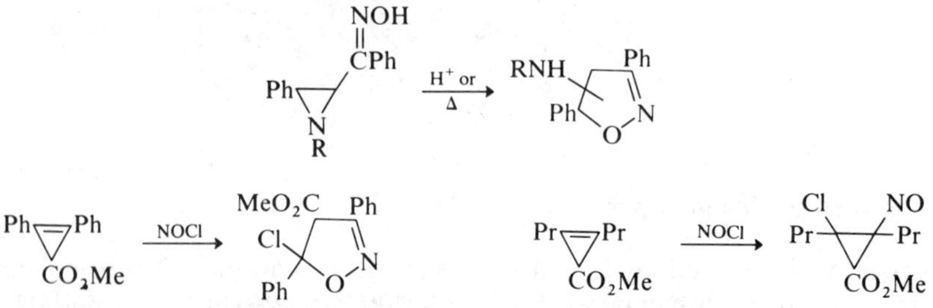

Scheme 126

A number of novel 2-isoxazolines and fused 2-isoxazolines have been generated from other alicyclic and heterocyclic systems either as sole products or as a component of a mixture ⟨74T63, 79T1267, 75JCS(P1)1342, 72JHC1189, 72TL3469, 78IZV1149, 80H(14)1319, 79MIP41601, 68ACS2719, 71MI41603⟩. Bromonitrocamphane when treated with sulfuric acid rearranged to *anhydro*-bromonitrocamphane by a fragmentation–rearrangement recombination path ⟨74T63⟩, and a novel synthesis of AT-125, an antitumor antibiotic, starts from cyclopentadiene monoepoxide ⟨79JA1054⟩.

3-Amino-2-isoxazolines were prepared by the condensation of acrylonitriles with *N*-hydroxyurea ⟨73BSF1138⟩ or of urea and acetophenones, as shown in Scheme 127 ⟨78ZOR2000⟩.

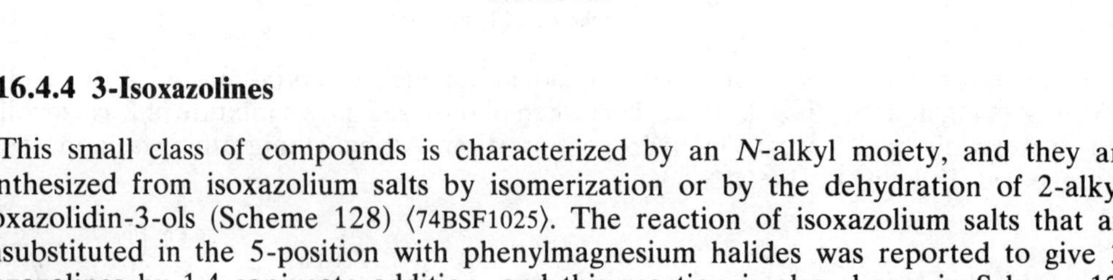

Scheme 127

4.16.4.4 3-Isoxazolines

This small class of compounds is characterized by an *N*-alkyl moiety, and they are synthesized from isoxazolium salts by isomerization or by the dehydration of 2-alkylisoxazolidin-3-ols (Scheme 128) ⟨74BSF1025⟩. The reaction of isoxazolium salts that are unsubstituted in the 5-position with phenylmagnesium halides was reported to give 3-isoxazolines by 1,4-conjugate addition, and this reaction is also shown in Scheme 128 ⟨74CPB70⟩.

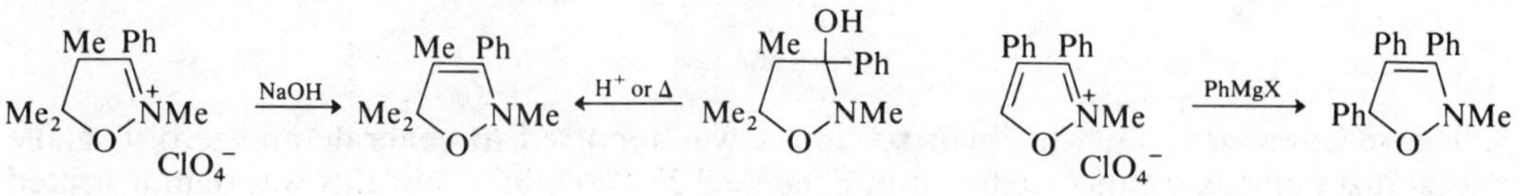

Scheme 128

Jurd reported the isolation of a 2*H*-3-isoxazoline by the reaction of the flavylium salt (**488**) with hydroxylamine. Gentle heating of the material caused isomerization to the more stable 2-isoxazoline. Treatment with base generated an α,β-unsaturated oxime which on photolysis regenerated the starting flavylium salt (Scheme 129) ⟨70CI(L)624⟩.

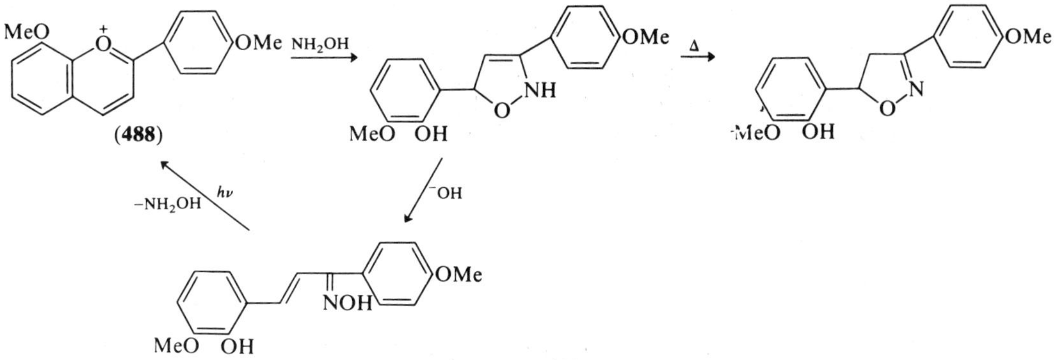

Scheme 129

4.16.4.5 4-Isoxazolines

This is also a relatively small class of compounds and is represented by 2*H*- and 2-alkyl members, but it is also indicated as an intermediate in a number of reactions. The NaBH$_4$ reduction of isoxazolium salts and the addition of Grignard reagents to isoxazolium salts containing a 5-substituent produced 4-isoxazolines plus some ring-opened material (Scheme 130) ⟨69TL4875, 74CPB54, 74CPB61⟩.

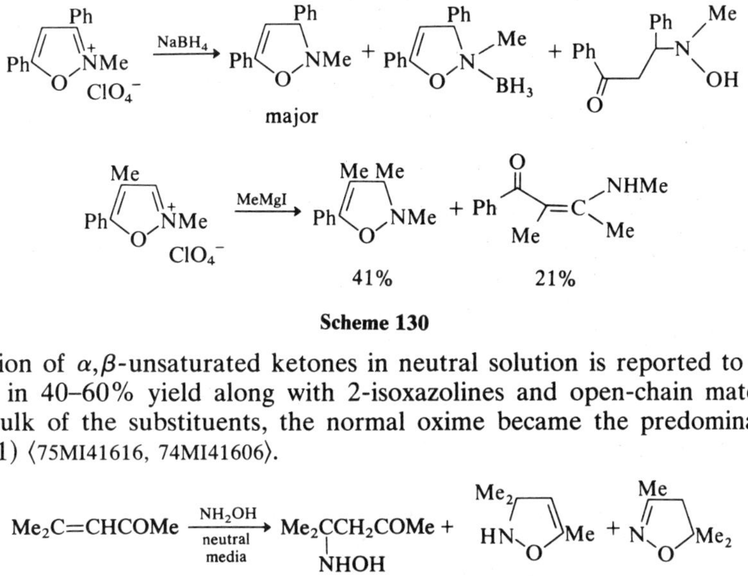

Scheme 130

The reaction of α,β-unsaturated ketones in neutral solution is reported to give 2*H*-4-isoxazolines in 40–60% yield along with 2-isoxazolines and open-chain materials. With increasing bulk of the substituents, the normal oxime became the predominant product (Scheme 131) ⟨75MI41616, 74MI41606⟩.

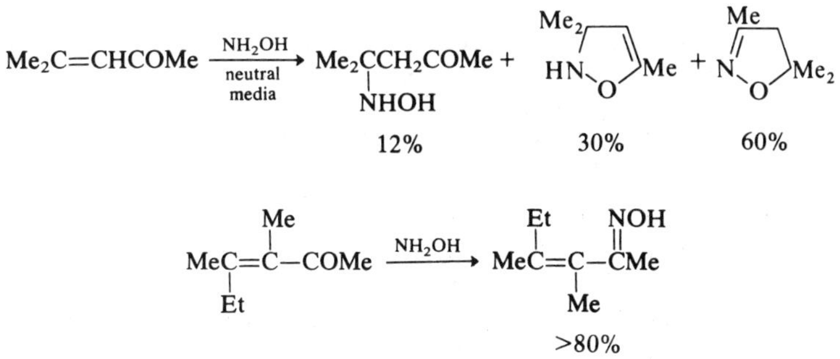

Scheme 131

The chlorination of 3,5-dimethylisoxazole gave the 3,4-dichloro-4-isoxazoline (**489**) ⟨77MIP41602⟩. Additional 2-substituted 4-isoxazolines were prepared by the addition of nitrones to triple bonds ⟨76AP1014, 77H(8)387, 70CB3196, 67AG(E)709, 69CB2346⟩, as shown by the conversion of (**490**) into (**491**) ⟨76AP1014⟩.

The addition of nitronic esters to alkynes to produce aziridines was postulated to proceed through a 4-isoxazoline as one of the intermediates (Scheme 132). A biradical intermediate (**492**) was also included in the mechanistic pathway for the reaction ⟨77JA6667⟩.

The rearrangement of fused triazole (**493**) produced a fused aziridine and a fused 4-isoxazoline (**494**) ⟨75T831⟩.

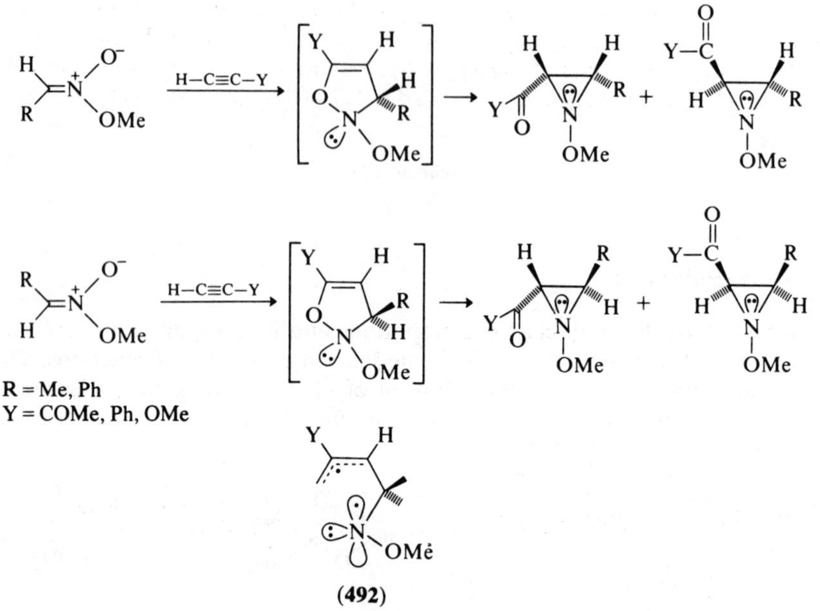

Scheme 132

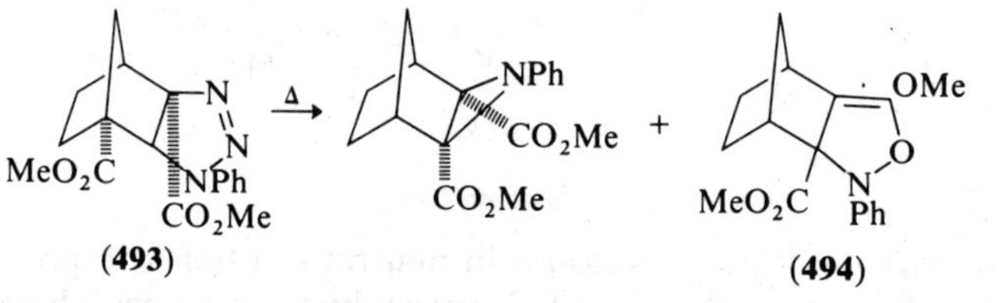

4.16.4.6 Isoxazolinols

The 2-isoxazolin-5-ol system is of interest in that many reports indicate that in solution an equilibrium exists between the ring and open-chain forms. The direction of the equilibrium depends on the bulk and/or number of substituents, with increasing either of the above favoring the cyclic form (Scheme 133) ⟨76T1369, 74BSF725⟩.

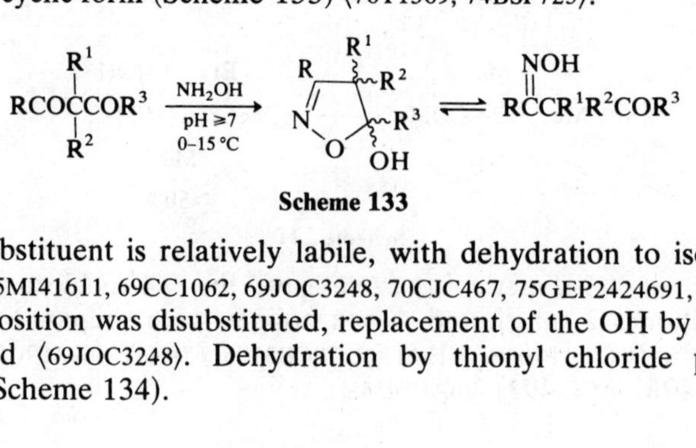

Scheme 133

The hydroxyl substituent is relatively labile, with dehydration to isoxazoles occurring readily ⟨74BSF725, 75MI41611, 69CC1062, 69JOC3248, 70CJC467, 75GEP2424691, 72NKK1452⟩. In one case where the 4-position was disubstituted, replacement of the OH by MeOH under acid conditions occurred ⟨69JOC3248⟩. Dehydration by thionyl chloride produced an *exo*-methylene group (Scheme 134).

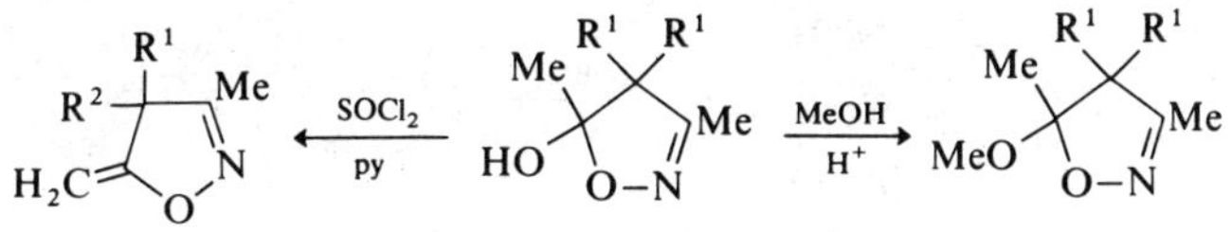

Scheme 134

The reaction of β-diketones with hydroxylamine can generate a variety of products besides the 2-isoxazolin-5-ol and includes dioximes and monooximes ⟨69JOC3248, 69CC1062, 75GEP2424691, 80CB1507, 72USP3629245⟩.

Bravo *et al.* studied the reaction of various ylides with monooximes of biacetyl and benzil. Dimethylsulfonium methylide and triphenylarsonium methylide gave 2-isoxazolin-5-ol and isoxazoles, with the former being the major product. Triphenylphosphonium methylide and dimethyloxosulfonium methylide gave open-chain products (Scheme 135) ⟨70TL3223, 72G395⟩. The cycloaddition of benzonitrile oxide to enolic compounds produced 5-ethers which could be cleaved or dehydrated (Scheme 136) ⟨70CJC467, 72NKK1452⟩.

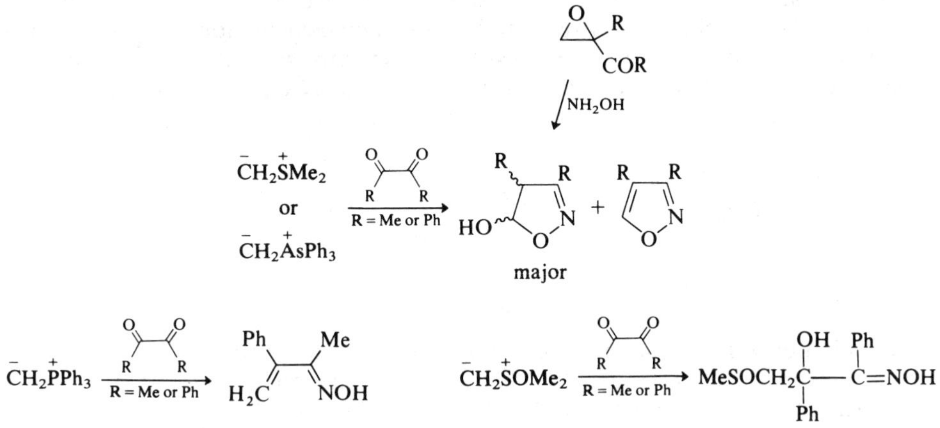

Scheme 135

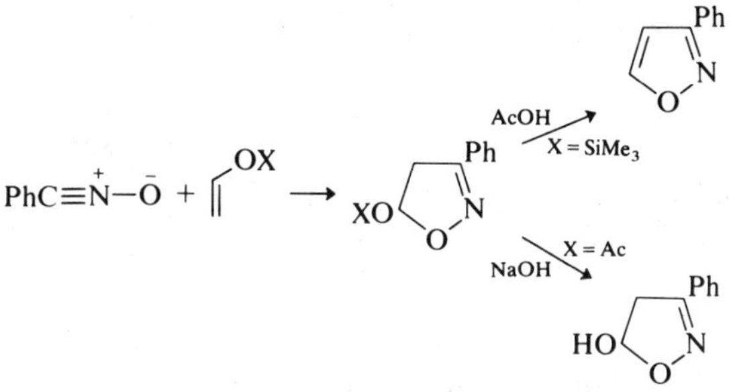

Scheme 136

A series of 2-isoxazolin-5-ols was produced by the reaction of Grignard reagents with 2-isoxazolin-5-one and the interconversion of the groups at the 5-position studied. This synthetic route is shown in Scheme 137 ⟨74TL387⟩. Hydroxylamine also added in a Michael fashion to propargyl ketones to provide a series of 2-isoxazolin-5-ols (Scheme 137) ⟨78MI41609⟩. 2-Isoxazolin-4-ols were prepared by the NaBH₄ reduction of (**495**) ⟨70MI41601⟩ or by a metal catalyzed cyclization of (**496**) ⟨78H(11)187⟩.

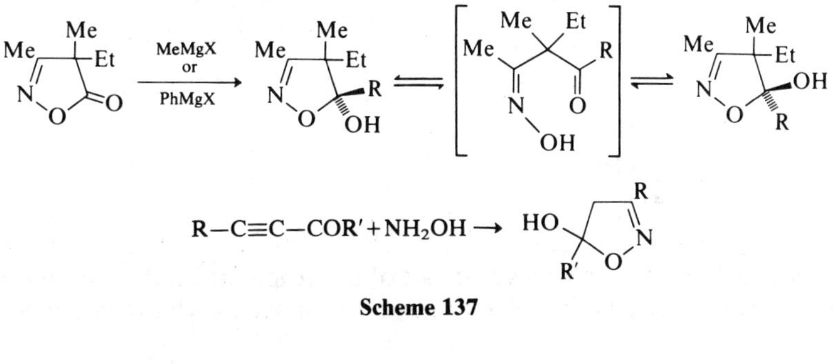

Scheme 137

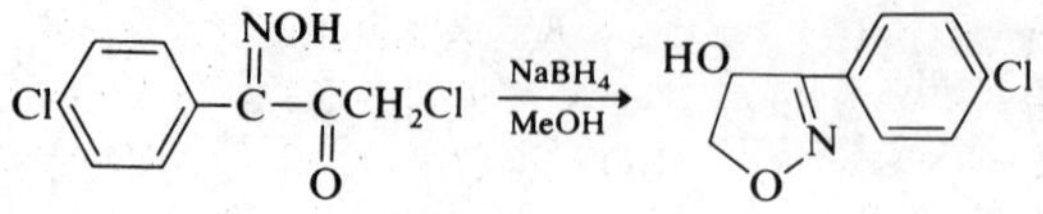

4.16.4.7 Isoxazoline *N*-Oxides

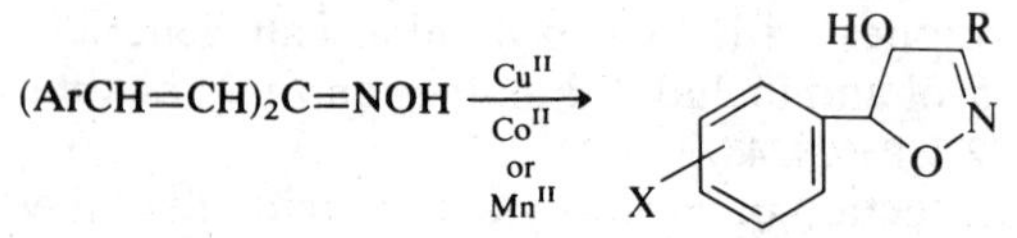

This class was first reported in 1924 and was formed ⟨62HC(17)1⟩ by cyclization of α-bromo-β-aryl-γ-nitroketones. The direct synthesis by oxygenation of 2-isoxazolines has not been reported. To date only 3-substituted derivatives have been prepared. Aryl-nitromethanes react with nitrostilbene to form isoxazoline *N*-oxide by a nitrile ion displacement (Scheme 138) ⟨62HC(17)1, 68TL3375⟩.

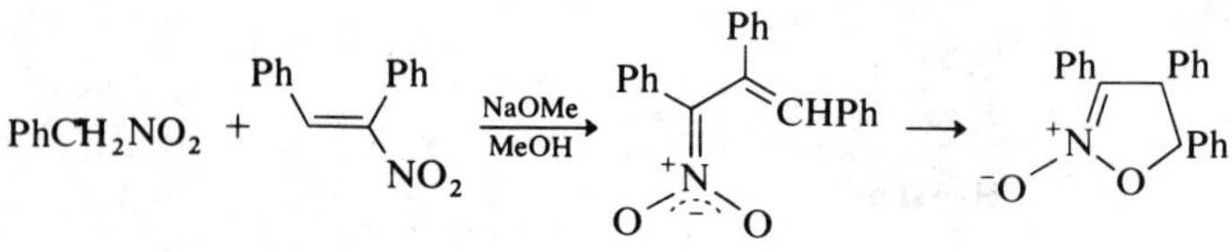

Scheme 138

Bromination of γ-dinitrobutanoic acids and base treatment produced 3-nitroisoxazoline *N*-oxides (Scheme 139) ⟨75MIP41601⟩. Alkylation of the potassium salt of γ-dinitro-2-butenoic acid also gave a similar compound (Scheme 139) ⟨74KGS571⟩.

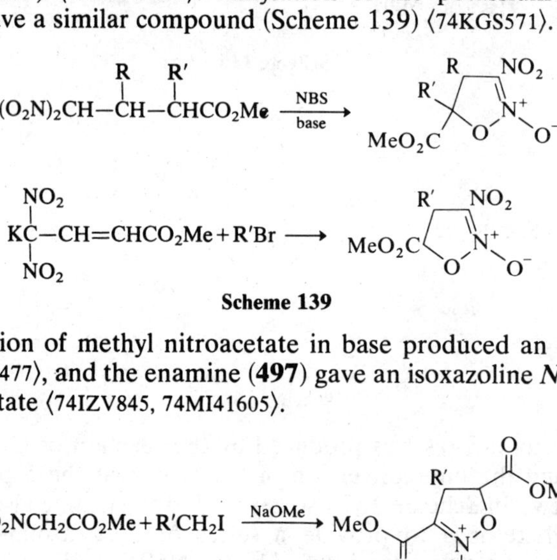

Scheme 139

Attempted alkylation of methyl nitroacetate in base produced an isoxazoline *N*-oxide (Scheme 140) ⟨74CPB477⟩, and the enamine (**497**) gave an isoxazoline *N*-oxide when reacted with methyl nitroacetate ⟨74IZV845, 74MI41605⟩.

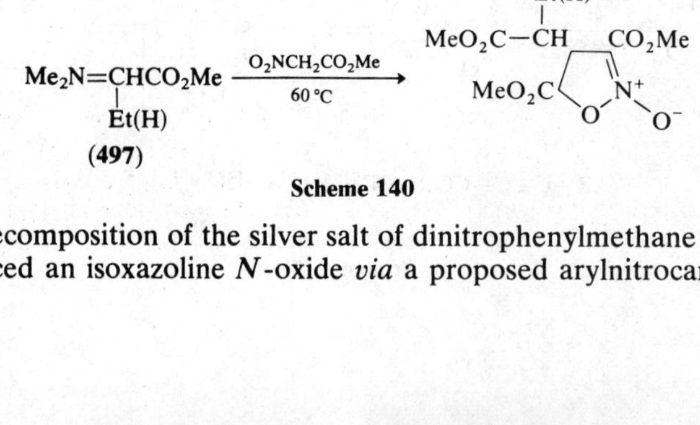

$R' = Me, Et, Pr$

Scheme 140

The thermal decomposition of the silver salt of dinitrophenylmethane in the presence of an alkene produced an isoxazoline *N*-oxide *via* a proposed arylnitrocarbene ⟨80JOC4158⟩, and dimethyloxosulfonium methylide has been reported to react with vinyl nitro compounds to generate isoxazoline N-oxides ⟨77CC7, 76JOC4033⟩. Diazo compounds react with halotrinitromethane [FC(NO$_2$)$_3$ does not react] ⟨68MI41602, 68ZOR2259⟩ or with vinylnitro compounds ⟨72MI41610, 71ZOR1309, 78KGS324, 73MI41605⟩ to produce isoxazoline N-oxides (Scheme 141).

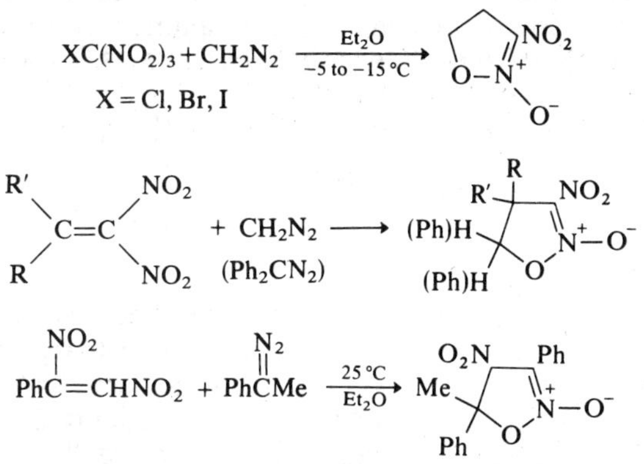

Scheme 141

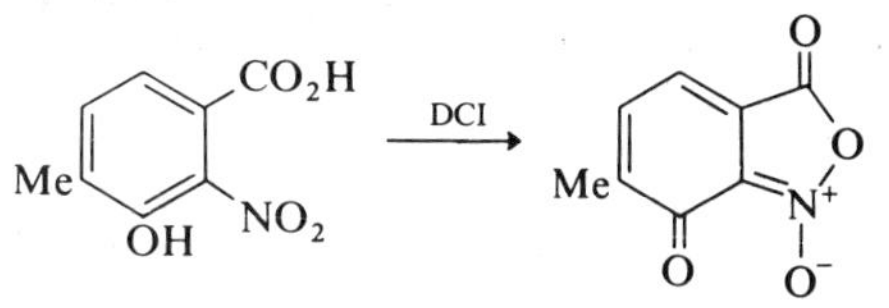

Dicarbonylimidazole reacted with the anthranilic acid derivative (**498**) to produce the fused isoxazolone N-oxide (**499**) ⟨77ZOR462⟩. Methyl nitroacetate reacted with indole-3-carbaldehyde to produce (**500**) ⟨70KGS1505⟩. Treatment of (**501**) with base gave 3,4,5-triphenyl-2-isoxazoline N-oxide (Scheme 142) ⟨69JOC984⟩. The reaction was reported to be a direct displacement as (**502**) did not give a product and no incorporation of deuterium was found using DOMe.

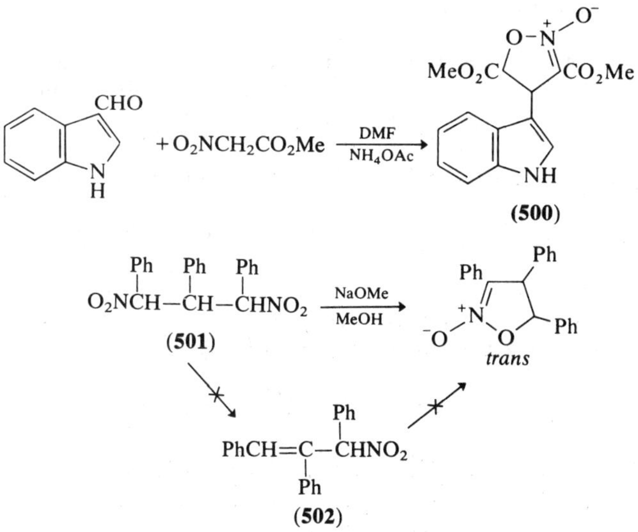

Scheme 142

4.16.4.8 Isoxazolin-5-ones

Generalized methods of preparation include the reaction of β-keto esters (or amides) with hydroxylamine, α-alkynic and α,β-unsaturated esters (or amides) with hydroxylamine (real or generated *in situ*), hydroxylamine and nitrile oxides, and β-keto and α-alkynic nitriles with hydroxylamine ⟨62HC(17)1, pp. 3, 7⟩.

4.16.4.8.1 From hydroxylamine

Quilico reported on a wide range of reactions involving hydroxylamine that led to isoxazolin-5-ones and a kinetic study demonstrated that the formation of an oxime was the rate determining step which was followed by a rapid cyclization. The pseudo-first-order rates were a function of basicity and concentration of the reactants ⟨62HC(17)1, pp. 3, 7, 76MI41609⟩. Malonate derivatives react with hydroxylamine to form isoxazolin-5-ones containing a carboxy function (Scheme 143) ⟨75MI41610, 74G715, 73ACS2820, 68MI41601, 79MI41611, 62JOC2160, 77JHC181⟩.

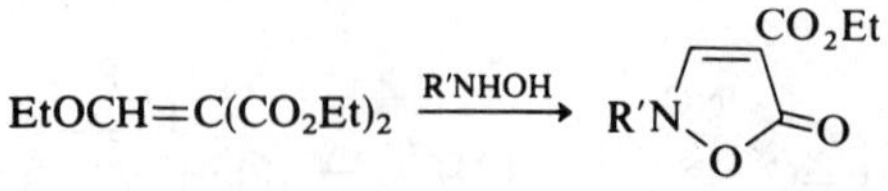

$$EtOCH=C(CO_2Et)_2 \xrightarrow{R'NHOH} \text{[isoxazolinone with } CO_2Et\text{]}$$

Scheme 143

In a number of cases the intermediate oxime has been isolated in the reaction of hydroxylamine and β-keto esters. The reaction of ethyl acetoacetate with hydroxylamine generated an oxime which cyclized on base treatment (Scheme 144) ⟨70MI41600⟩. Likewise, treatment of an analogous amide with hydroxylamine generated a ring opened material which cyclized on treatment with HCl (Scheme 144) ⟨67T831⟩. The presence of a minor contaminant in the standard reaction of ethyl acetoacetate with hydroxylamine was discovered and identified as an isomeric isoxazolin-3-one. The mechanism of product formation has been discussed ⟨70BSF2685⟩.

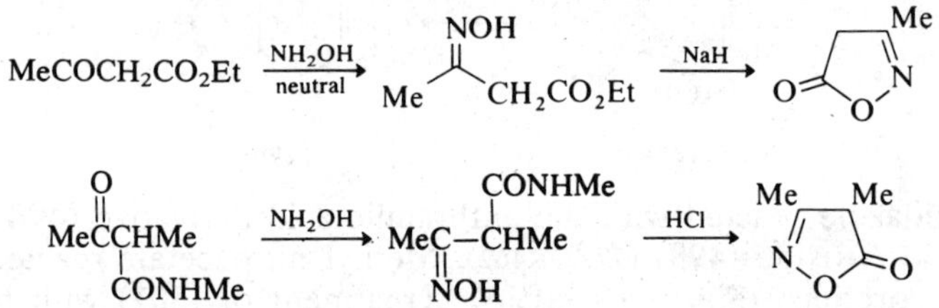

Scheme 144

4.16.4.8.2 By the reaction of nitrile oxides

Alkynic esters react with nitrile oxides in a pH dependent reaction to product isoxazolin-5-ones (Scheme 145) ⟨71JCS(C)86⟩. Alkynic ethers also react with benzonitrile oxide to produce an isoxazole–ether which on treatment with HCl or HBr gave an isoxazolinone (Scheme 145) ⟨63CB1088, 58MI41600⟩. The reaction of benzonitrile oxide with dimethoxyketene yielded a dimethyl acetal which was split with acid into the isoxazolinone (Scheme 145) ⟨59G1511⟩.

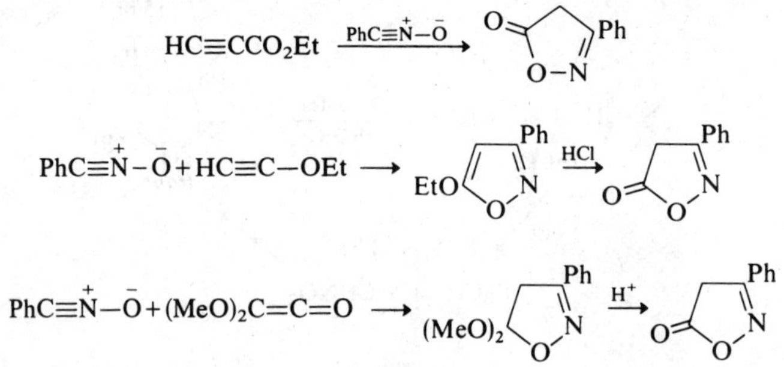

Scheme 145

4.16.4.8.3 Miscellaneous syntheses

Cyclopropenones react with nitrosobenzene by an O-initiated attack at C-1 to produce isoxazolin-5-ones ⟨75TL3283, 78USP4053481⟩, and an isoxazolin-5-one was produced as a by-product in the photolysis of nitroethylene ⟨78AJC113⟩. Substituted oxazolin-5-ones have been formed by the base treatment of (**503**) to give (**504**) ⟨74T2765⟩, and dehydrobromination of (**505**) produced (**506**) ⟨72AP359⟩. The reaction of tetrahydronicotinamide with hydroxylamine gave an isoxazolin-5-one (**507**) (Scheme 146) ⟨70JOC3130⟩.

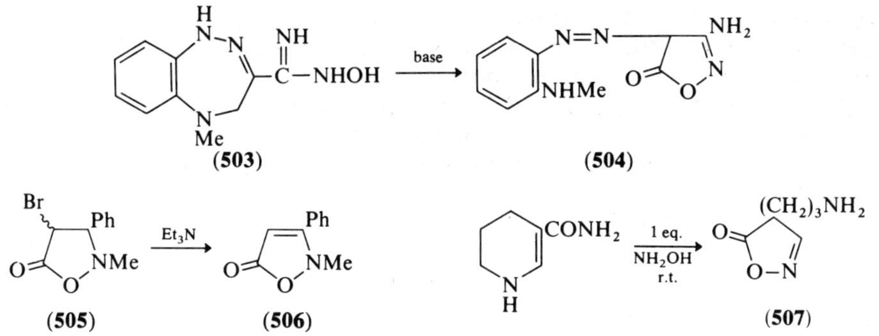

Scheme 146

A series of 4-methyleneisoxazolin-5-ones was produced by the reaction of hydroxylamine with iminoether (**508**) ⟨77ACH(94)403⟩ or with a β-ketoester and triethyl orthoformate (Scheme 147) ⟨65CI(L)36⟩.

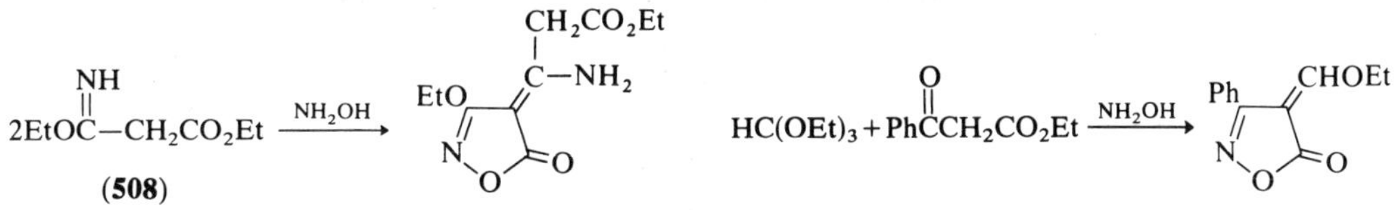

Scheme 147

The reaction of α-hydrazino-β-ketoesters (**509**) with hydroxylamine produced 4-hydrazinoisoxazolin-5-ones (**510**) ⟨71GEP2024393, 70JMC1250⟩.

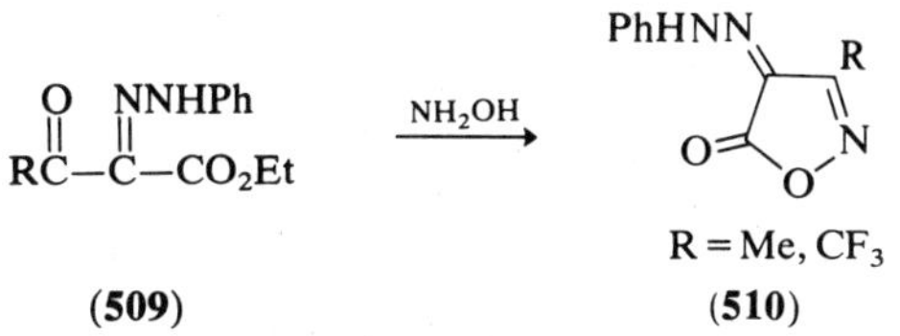

R = Me, CF$_3$

4.16.4.9 Isoxazolin-5-imines

Isoxazolin-5-imines were produced by nitrile oxide addition to cyanoacetates ⟨62HC(17)1, p. 7⟩, by the reaction of nitrones with phenylacetonitrile ⟨74CB13⟩, and by base addition of nitrosobenzene to nitriles (Scheme 148) ⟨72LA(762)154⟩.

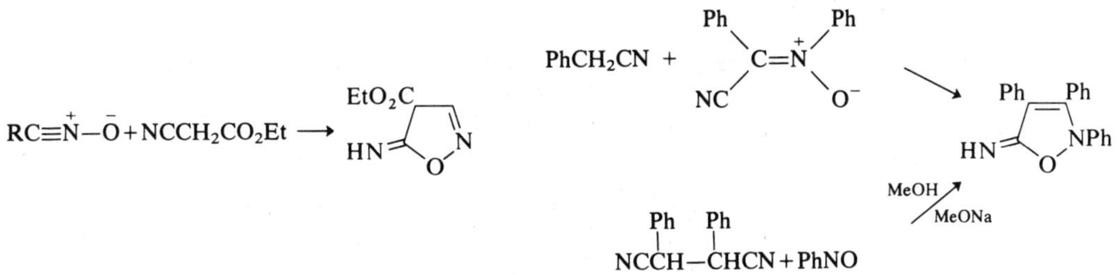

Scheme 148

4.16.4.10 Isoxazoline-5-thiones

5-Haloisoxazolium salts were treated with Na$_2$S to produce the above titled compounds ⟨76ZC270⟩. For example, 5-chloro-2-methyl-3-phenylisoxazolium tetrafluoroborate and Na$_2$S gave 2-methyl-3-phenyl-3-isoxazoline-5-thione.

4.16.4.11 Isoxazolin-3-ones

Relatively few *N*-unsubstituted isoxazolin-3-ones are known and a brief review was published by Quilico in 1962; the ring system was obtained by the hydrolysis of 3-methoxy-5-phenylisoxazole in 1961 ⟨62HC(17)1, p. 3⟩.

Earlier reported syntheses have been shown to give isoxazolin-5-ones. Other isoxazolin-3-ones have been prepared by the reaction of methylacetoacetic esters and hydroxylamine. An additional synthesis was reported by the action at 0 °C of hydroxylamine on ethyl α-benzoylpropionate to produce an insoluble hydroxamic acid which cyclized on acid treatment. The hydroxamic acid acetal was similarly transformed into the isoxazolin-3-one (Scheme 149) ⟨71BSF3664, 70BSF1978⟩.

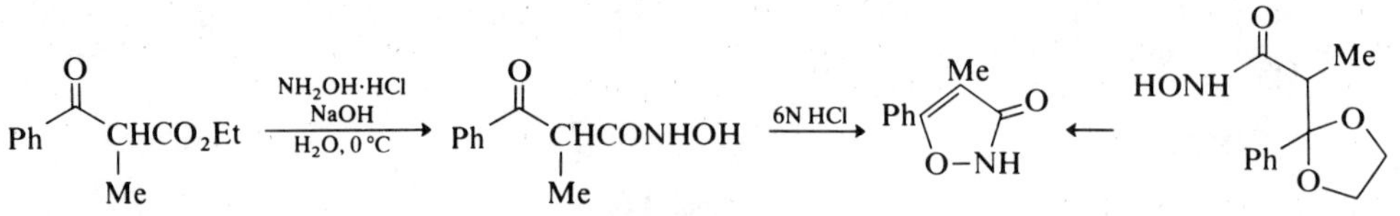

Scheme 149

N-Substituted derivatives have been prepared by the reaction of *N*-methylhydroxylamine with phenylpropiolic esters or acid chlorides ⟨71CPB1389⟩, the cyclization of *N*-substituted β-ketohydroxamic acids or the reaction of phenylhydroxylamine with diketene (Scheme 150) ⟨63GEP1146494⟩.

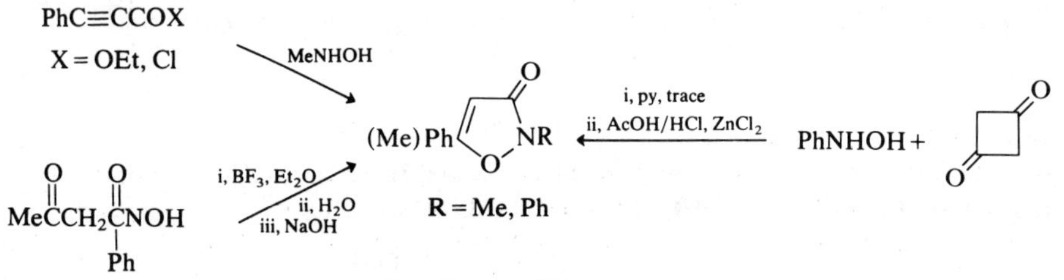

Scheme 150

N-Methyl-5-phenylisoxazolin-3-one (**512**) was prepared by the thermolysis of (**511**) ⟨71MI41602⟩.

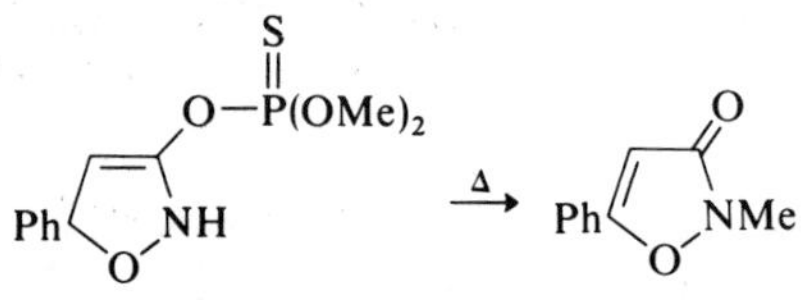

(**511**)   (**512**)

4.16.4.12 Isoxazolin-4-ones

The first in this series to be reported was 4-oxoisoxazoline-3,5-dicarboxylic acid diethyl ester, which was formed by the reaction of nitrous acid on diethyl acetonedicarboxylate in 1891. Quilico described a number of syntheses in his 1962 review and the most general include the reaction of hydroxylamine and α-hydroxy-(or acetoxy)-β-diketones and the conversion of 4-isoxazolediazonium salts to the hydroxy moiety ⟨62HC(17)1, p. 3⟩. Additional syntheses reported were the oxygenation of a 4-boric acid derivative ⟨67JOM(9)19⟩ and peroxide oxidation of a 4-nitro-2-isoxazoline (Scheme 151) ⟨79ZOR2436⟩.

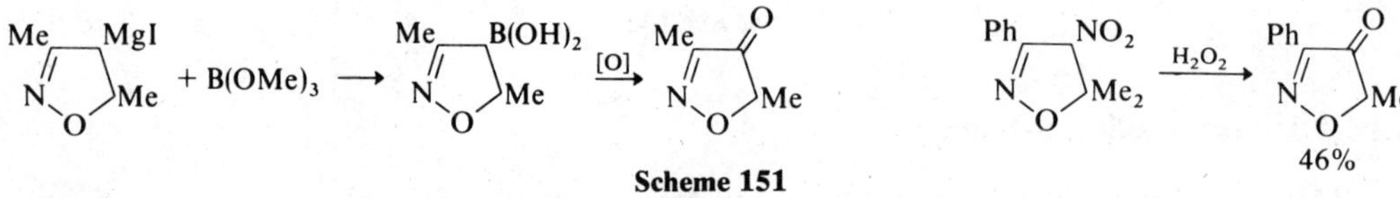

Scheme 151

A 4-hydrazone (**514**) was synthesized by the condensation of hydroxylamine with (**513**) ⟨72JPR815⟩.

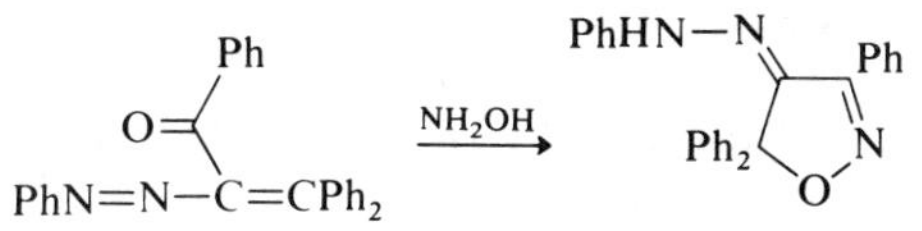

Isoxazoles and their Benzo Derivatives

4.16.4.13 Bis(isoxazolines)

This series of compounds was also discussed briefly by Quilico in 1962 and only a limited number of new representatives have been reported ⟨62HC(17)1, p. 3⟩. The pressure reaction of ethylene and nitric acid in the presence of Ni, Zn or Cu produced 3,3′-bis(isoxazoline) ⟨70FRP94493⟩, and the isoxazoline N-oxide (515) was prepared by the reaction of β-dimethylaminoacrylaldehyde and methyl nitroacetate ⟨74IZV845⟩.

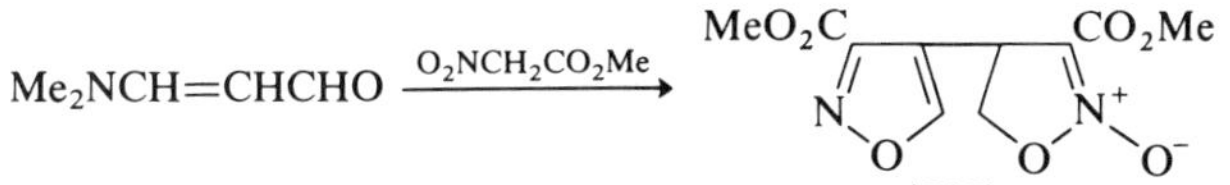

Ethyl acetoacetate and hydroxylamine with a large excess of alkali produced (516) which on heating generated 4-methylene-2-isoxazoline (517), while limited base generated the dimer (518) ⟨80JHC763⟩.

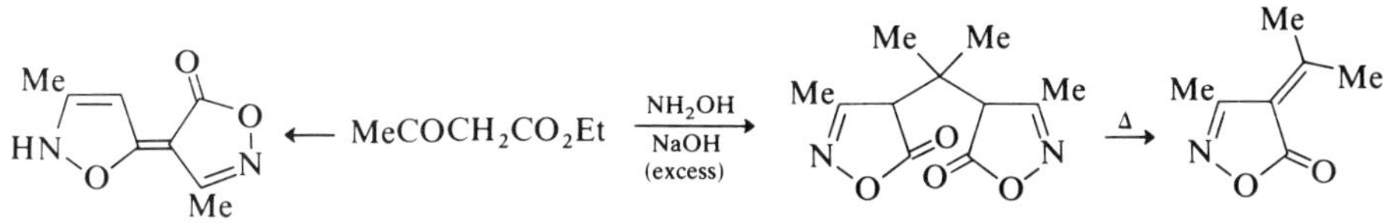

The oxidative coupling of 3,4-dimethyl-or 3,4-diphenyl-isoxazolin-5-one by activated MnO_2 produced a 4,4′-bis(isoxazolinone) (519) and 2,4′-bis(isoxazolinone) (520). Hydrogenation of (519) over PtO_2/HOAc produced a pyrrole derivative while similar reaction of (520) produced an isomeric pyrrole ⟨80JHC763⟩. These reactions are shown in Scheme 152.

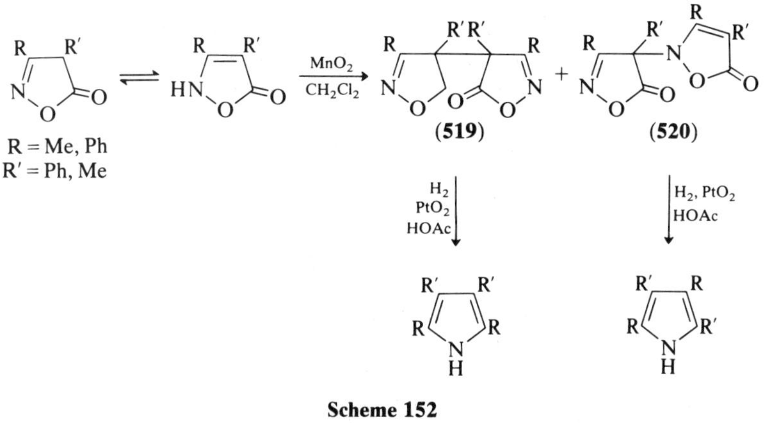

Scheme 152

5,5′-Bis(isoxazoline) decomposes to a number of products depending on the substitution ⟨77H(6)1599⟩; on photolysis using $NiSO_4$, two major products were formed. ESR spin trapping demonstrated the intermediacy of imino and 2-isoxazolinyl radicals ⟨77TL4619⟩.

4.16.4.14 Spiro Fused Isoxazolines

A number of bis(spiroisoxazolines) and spiroisoxazolines fused to another ring have been reported. A number of spiroisoxazolines containing polymers have been reported to have potential commercial utility as photosensitive agents ⟨78MI41608, 75JAP(K)7515840⟩ and as cross-linking agents in butadiene rubber ⟨71MI41601⟩.

4.16.4.15 Bis(spiroisoxazolines)

Benzonitrile oxide reacted with 3-phenyl-4-benzylideneisoxazolinone to produce two isomeric spiro compounds (Scheme 153) ⟨72MI41609, 72MI41608⟩. The reaction of benzonitrile oxide with ketene produced a spiro derivative ⟨67MI41600⟩; with allenes, bis(spiroisoxazolines) along with monoaddition products were formed (Scheme 154) ⟨79JOC2796, 70CR(C)-(271)1468⟩.

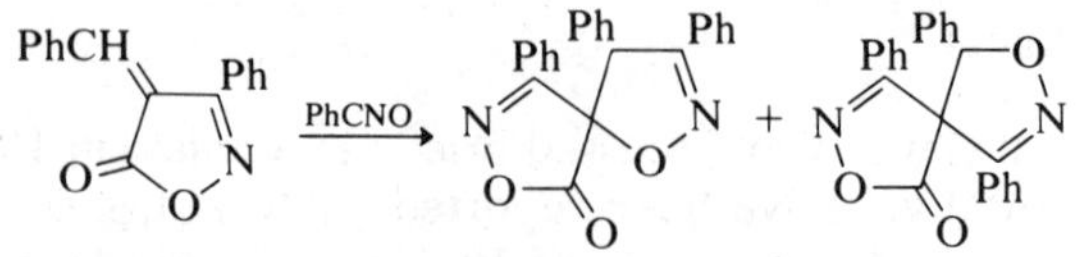

Scheme 153

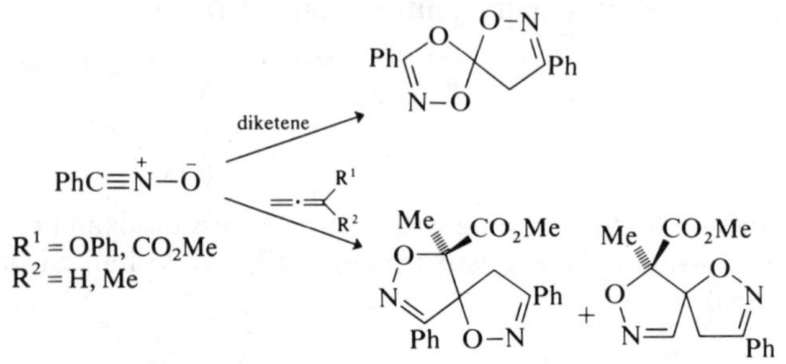

Scheme 154

Bis(spiroisoxazoline) was cleaved by PPh_3 in PhH/MeOH to provide an oxime, and in benzene to produce an amide ⟨75MI41609⟩.

4.16.4.16 Other Spiroisoxazolines

Nitrile oxide addition to *exo*-methylenes was the most common method of formation ⟨76CC210, 76CC209, 79RRC111, 69JPR118, 80JAP(K)8019209⟩. The transformation of (**521**) into (**522**) was accomplished by $BF_3 \cdot OEt_2$ ⟨78IZV2551⟩ and the reaction of (**523**) with *n*-butyllithium and diazomethane yielded (**524**) ⟨75BCJ1675⟩. Other routes to these spiro systems involve the oxidative cyclization of appropriately substituted phenols ⟨75JCS(P1)2340, 75JCS(P1)2348⟩, and the reaction of a sulfonium ylide with an isoxazolinone ⟨76JHC1109⟩.

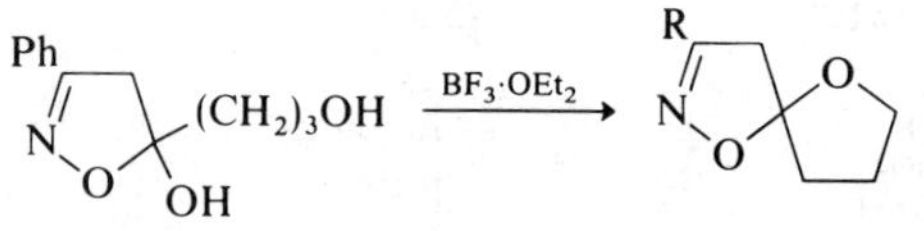

(**521**)   (**522**) 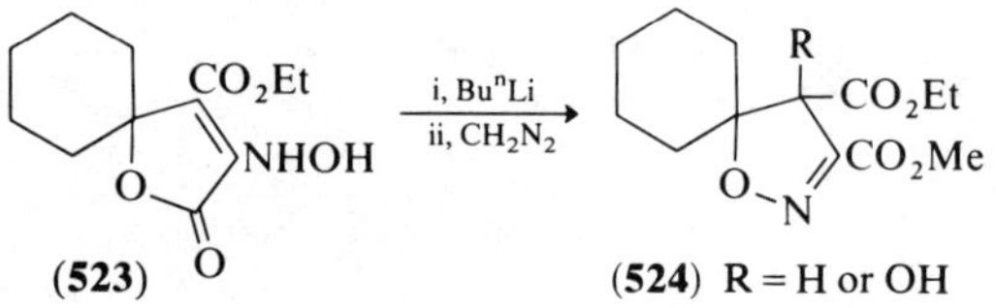   (**523**)   (**524**) R = H or OH

4.16.4.17 Isoxazolidines

Isoxazolidines and isoxazolidinones are a small class of compounds which has received only limited attention. Reviews have appeared dealing with isoxazolidines in 1962 ⟨62HC(17)1, p. 7⟩ and 1977 ⟨77AHC(21)207⟩.

4.16.4.17.1 By cycloaddition of nitrones and alkenes

The most predominant reaction for the preparation of isoxazolidines involves the cycloaddition of nitrones and alkenes ⟨79KGS599, 80IZV207, 80MI41602, 80BPR2024218, 78MI41607, 78IZV1881, 76BCJ1138, 74IZV908, 75MI41608, 76IZV2779, 77USP4010176, 77HCA426, 77USP4018774, 77TL3759, 79JOC835, 79IZV131, 80HCA1706, 78OS(58)106⟩. The regio- and stereo-selectivity of the reaction were discussed at length by Takeuchi and Furusaki ⟨77AHC(21)207⟩ and others ⟨77CC303, 78IZV2588, 78JCR(S)240, 76H(5)109, 79BCJ3763, 78AJC2239, 78ZOR1693, 79MI41610, 80IZV715, 76T675, 78T2459, 79JOC1212⟩, and the product ratio of *cis* and *trans* pairs examined.

(*E*)-Alkenes react with nitrones or with (*E*)-nitrones to produce isoxazolidines of the conformations shown in (**525**) as the predominant product ⟨76H(5)109, 78IZV2588, 80IZV715, 78JCR(S)240, 79BCJ3763⟩.

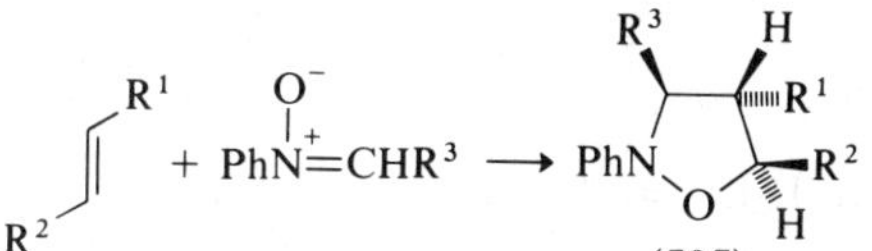

(**525**)

Chiral nitrones react with alkenes to produce 3,5-disubstituted isoxazolidines that are nonracemic diastereomeric mixtures and are oriented predominantly *cis* (equation 53) ⟨77CC303, 79JOC1212⟩.

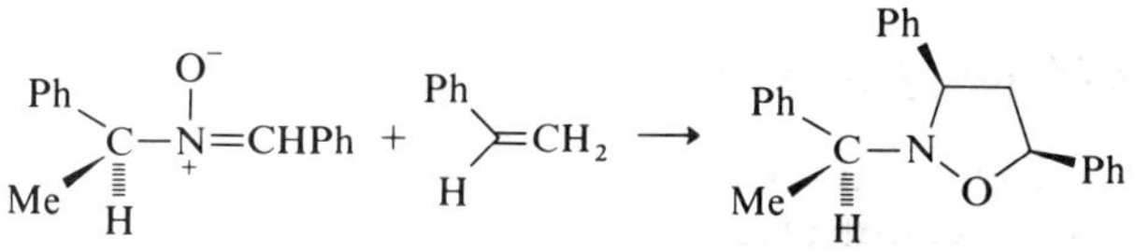

(53)

A theoretical study by LCAO-SCF-MO methods of the reaction of nitrone and ethylene which examined potential energy hypersurface structures, transition state characteristics and geometric parameters within the interaction has been reported ⟨78T2459⟩. Two kinetic studies which were undertaken ⟨78ZOR1693, 78AJC2239⟩ and the second-order rate constants and linear free energy relationships for this cycloaddition indicated that: (1) the reaction proceeded *via* a nonsynchronous addition of a dipolar intermediate or possibly a two-step addition by a discrete zwitterionic intermediate; and (2) the reactivity was governed more by localization energies than by orbital donor–acceptor interactions.

A synthesis of (±)-cocaine proceeded through an initial cycloaddition of (**526**) to (**527**) to produce the bicyclic structure (**528**) ⟨78JA3638⟩.

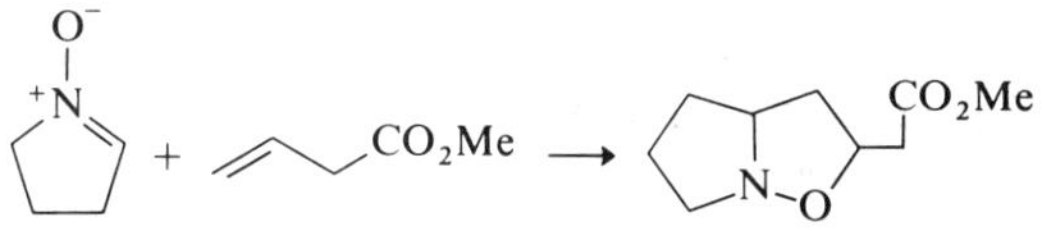

(**526**)   (**527**)   (**528**)

The reaction of nitrones with allenes produced three main products: an azepine, a pyrrolidinone and an isoxazolidine (Scheme 155) ⟨79JOC4213⟩. The intramolecular cycloaddition of nitrones (**529**) produced different products depending on the length of *n* (Scheme 156) ⟨78H(10)257⟩.

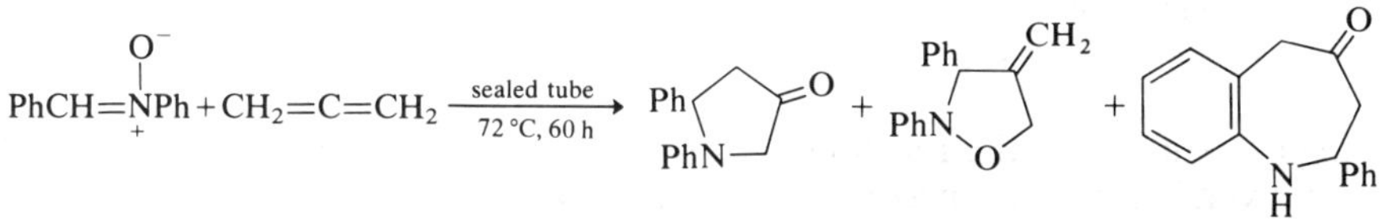

23%   22%   31%

Scheme 155

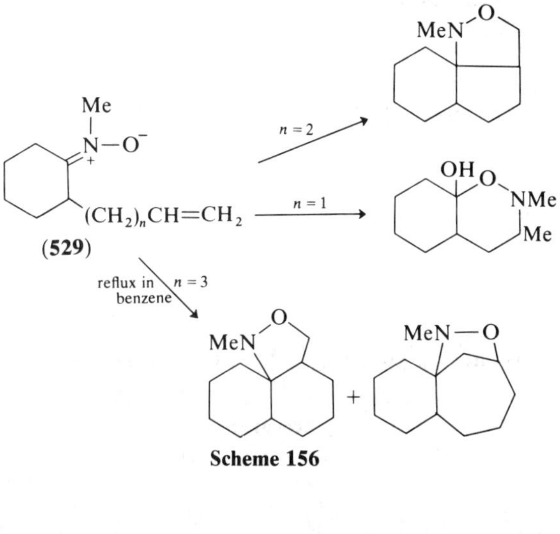

Scheme 156

Nitrone (**530**) exists in thermal equilibrium with vinylamine (**531**) and isoxazolidine (**532**), with (**532**) (a dimer of **530** and **531**) being predominant. The equilibrium in DMSO was studied by ^{1}H and ^{13}C NMR spectra ⟨80TL3447⟩.

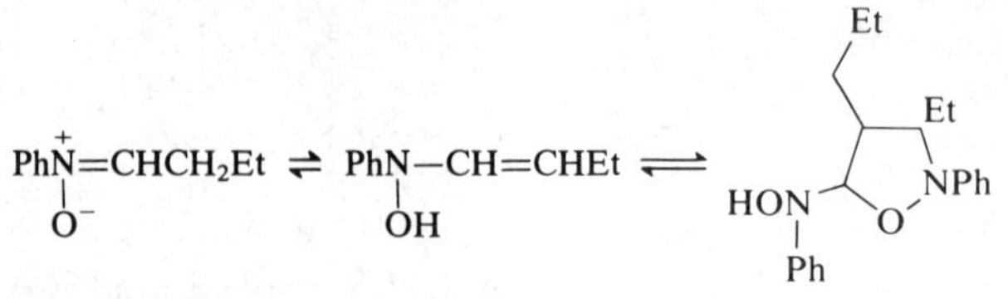

$$\text{PhN}^+ =\text{CHCH}_2\text{Et} \rightleftharpoons \text{PhN}-\text{CH}=\text{CHEt} \rightleftharpoons \text{(532)}$$
$$\quad\;\; \text{O}^- \qquad\qquad\qquad \text{OH}$$

(**530**)   (**531**)   (**532**)

4.16.4.17.2 Cycloaddition of nitronic esters

Nitronic esters are commonly produced by the reaction of alkylnitro compounds with diazoalkanes and these react with alkenes to produce isoxazolidines ⟨77AHC(21)207, 80MI41602, 76BCJ2815, 76ZOR2095, 75BSF1319, 73TL453, 76T683, B-78MI41606, 80IZV1893⟩. The regiospecificity of the reaction has been studied ⟨77AHC(21)207, 75BSF1319, 73TL453, 80IZV1893⟩, as have the rotational barriers to interconversion ⟨77AHC(21)207, 76T683⟩.

A newer method for the preparation of nitronic esters, namely utilizing the O-trimethylsilyl ester, has been reported and these are prepared by the reaction of alkylnitro compounds and N,N-bis(trimethylsilyl)acetamide. These nitronic esters also undergo cycloaddition with alkenes to produce isoxazolidines (equation 54) ⟨74MIP41601, 74DOK109, 78ACS(B)118⟩.

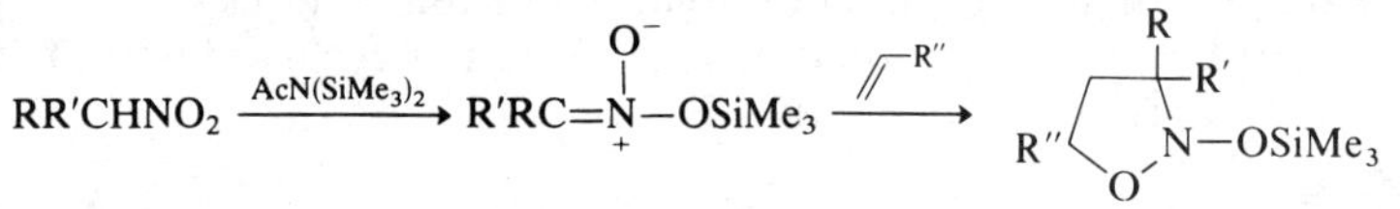

$$\text{RR'CHNO}_2 \xrightarrow{\text{AcN(SiMe}_3)_2} \text{R'RC}=\overset{+}{\text{N}}-\text{OSiMe}_3 \xrightarrow{\;\;\nearrow^{-\text{R''}}\;\;} \underset{\text{R''}}{\overset{R}{\underset{\text{O}}{\bigtriangleup}}}\overset{\text{R'}}{\underset{\text{N}-\text{OSiMe}_3}{}} \quad (54)$$

Steroidal alkene (**531a**) reacted with a nitronic ester at 14 000 atmospheres to produce an isoxazolidine (**532a**) ⟨80IZV1893⟩.

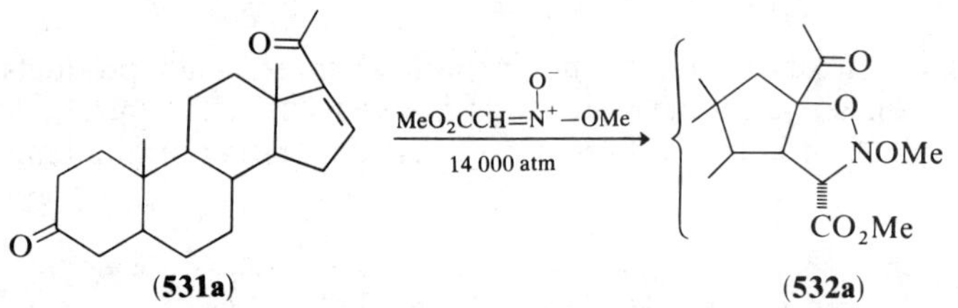

(**531a**)   (**532a**)

A number of studies on the reaction of nitronic esters and alkenes and on the rotational barriers to inversion in the isoxazolidines have been reported. The inversion is a buckling process of the ring ⟨76T683, 75IZV2348, 75BSF1319, 73TL453⟩ and the process was studied by NMR and an inversion barrier of 67 kJ mol^{-1} was recorded ⟨75IZV1451⟩. Resolution of certain N-alkoxyisoxazolidines has been accomplished and the inversion process here also studied ⟨76IZV1903, 77IZV716⟩.

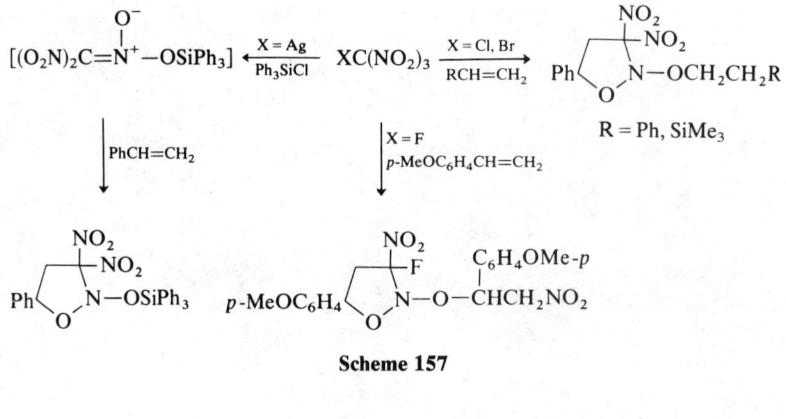

Scheme 157

4.16.4.17.3 By the reaction of tetranitromethane and halotrinitromethanes with alkenes

Halotrinitromethanes decompose to form nitrones which in the presence of alkenes generate isoxazolidines (Scheme 157) ⟨73ZOR269, 72MI41604⟩. The silver salt of trinitromethane reacted with silyl chloride to generate silylnitronic esters which with alkenes likewise generated isoxazolidines (Scheme 157) ⟨73IZV203⟩. When the halogen was fluorine, the halogen was retained in the final product (Scheme 157) ⟨77ZOR2495⟩.

Tetranitromethane was decomposed in the presence of alkenes to generate isoxazolidines (equation 55) ⟨72ZOR1419⟩.

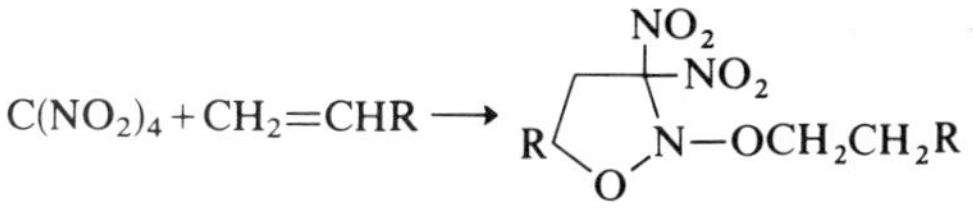

$$C(NO_2)_4 + CH_2=CHR \longrightarrow \underset{\underset{O}{R}}{\overset{NO_2}{\underset{|}{\overset{|}{\underset{N-OCH_2CH_2R}{}}}}}\tag{55}$$

4.16.4.17.4 Other methods

A number of other syntheses were discussed by Takeuchi and Furusaki and the most common involved reaction of hydroxylamine with selected α,β-unsaturated ketones to give isoxazolidine-3- or -5-ols, which exist in equilibrium with an open-chain counterpart ⟨77AHC(21)207⟩. A similar equilibrium was observed in the reaction of α,β-unsaturated ketones with N-hydroxyurea. The geometric orientation of the ring substituents was studied as a dynamic process (Scheme 158) ⟨75TL2337⟩.

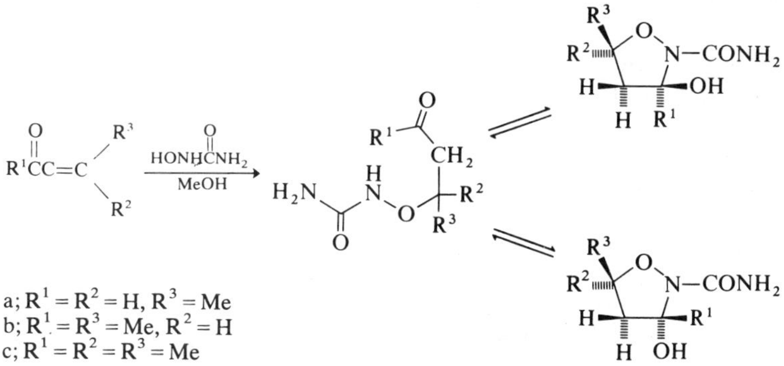

a; $R^1 = R^2 = H$, $R^3 = Me$
b; $R^1 = R^3 = Me$, $R^2 = H$
c; $R^1 = R^2 = R^3 = Me$

Scheme 158

The reaction of hydroxylamine with an α,β-unsaturated pyruvic acid (533) produced (534) ⟨76TL1825⟩. Hydroxylamines reacted with the nonconjugated ketoalkene (535) to produce (536) ⟨73T2683⟩.

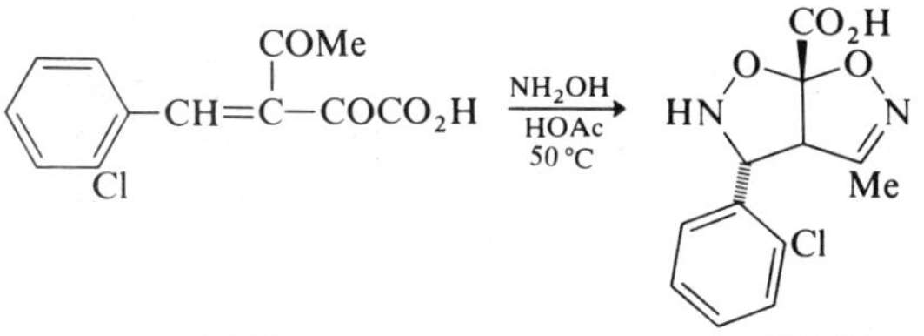

(533)   (534)

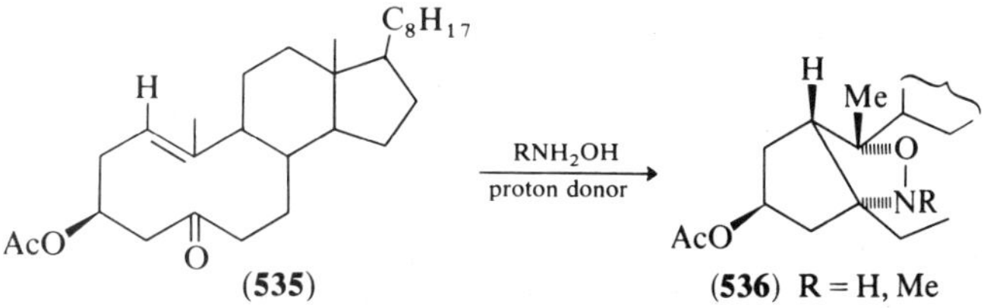

(535)   (536) R = H, Me

Azetidine N-oxides produce isoxazolidines by a thermal ring expansion ⟨77AHC(21)207, 75GEP2365391⟩, and nitrosobenzenes react with alkenes to provide isoxazolidines ⟨77AHC(21)207, 79IZV1059⟩.

The reduction of 3,5-diphenylisoxazoline with sodium cyanoborohydride produced a mixture of isomeric 3,5-diphenylisoxazolidines. The ^{1}H and ^{13}C NMR spectra were utilized to distinguish the isomers ⟨80LA101⟩. Sodium borohydride reductions likewise reduce isoxazolines to isoxazolidines (equation 56) ⟨80JA4265⟩.

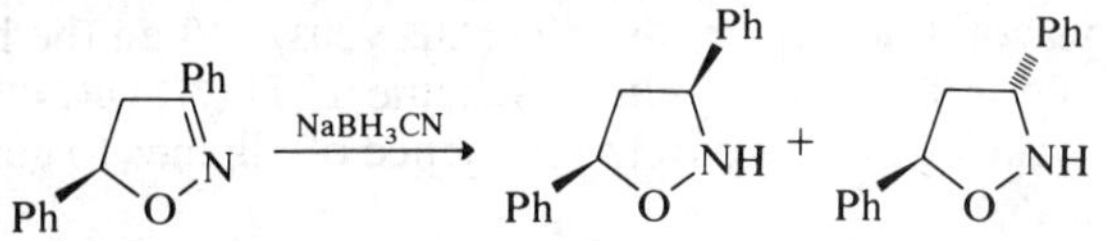

(56)

Treatment of (**537**) with acid chloride (**538**) in the presence of triethylamine produced isoxazolidine (**539**) in 45% yield ⟨80IZV1694⟩.

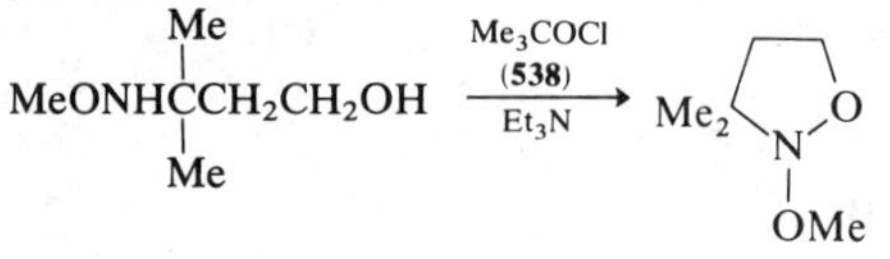

4.16.4.18 Isoxazolidinones

Three isoxazolidinones are possible, the 3-, 4- and 5-isoxazolidinones, with 3-isoxazolidinones being the group containing the most members. Only one synthesis of 4-isoxazolidinones has been reported, although a number of 4-arylmethylene compounds exist. A number of isoxazolidine-3,5-diones have been synthesized. A review by Quilico with nine references appeared in 1962 ⟨62HC(17)1, p. 7⟩ and since then the number of references has slightly more than tripled.

4.16.4.18.1 3-Isoxazolidinones

3-Isoxazolidinones have been prepared by the cyclization of ethyl 3-O-hydroxylaminopropionate and by the cyclization of α-chloropropionhydroxamic acid ⟨62HC(17)1, p. 7⟩. 3-Isoxazolidinones can also be prepared by the hydrolysis of 3-alkoxy- or 3-amino-2-isoxazolines or 2-isoxazolinium salts (Scheme 159) ⟨74BSF1651, 62G501⟩.

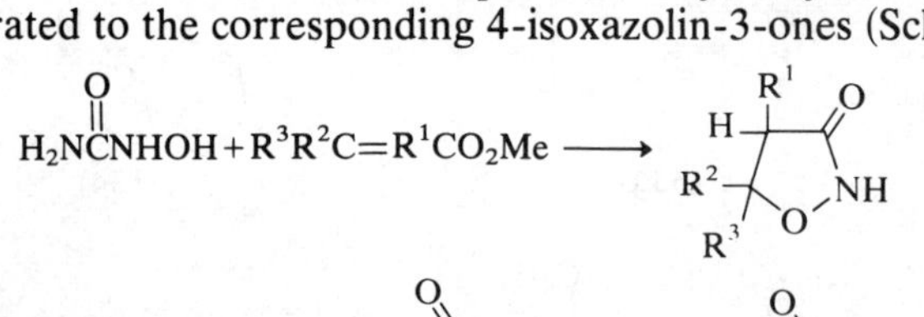

Scheme 159

N-Hydroxyurea and acrylic esters produce 3-isoxazolidinones (equation 57) ⟨76BSF1589⟩. Phenylhydroxylamines react with diketene to provide 5-hydroxyisoxazolidones ⟨80JHC727⟩ which can be dehydrated to the corresponding 4-isoxazolin-3-ones (Scheme 160).

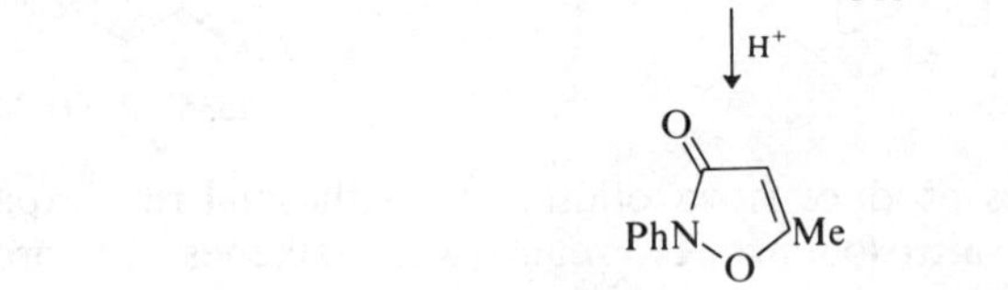

(57)

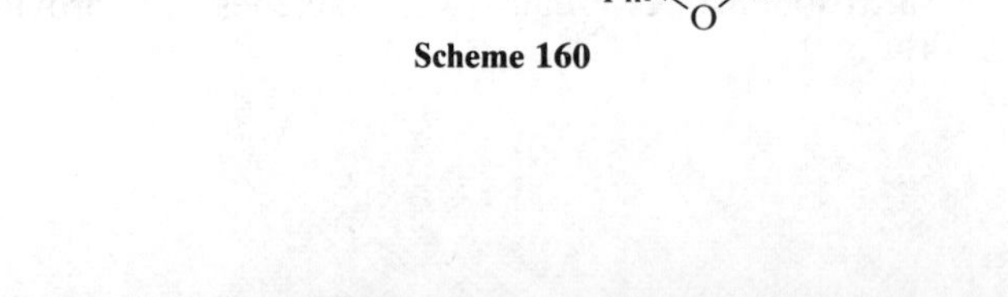

Scheme 160

4.16.4.18.2 4-Isoxazolidinones

The first synthesis of 4-isoxazolidinones (**540**) resulted from the reaction of 2,4-dibromo-2,4-dimethyl-3-pentanone with hydroxylamine ⟨81H(16)1855⟩.

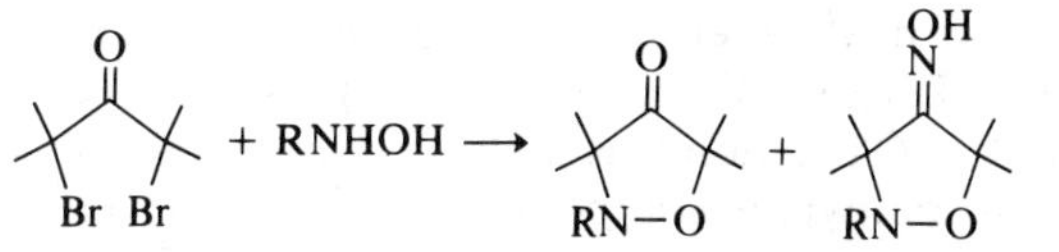

(**540**) R = H, COPh, CO$_2$Et

4.16.4.18.3 5-Isoxazolidinones

The first 5-isoxazolidinone was synthesized in 1912 by the cyclization of α-hydroxylamino-β,β-diphenylpropionic acid ⟨62HC(17)1, p. 7⟩. The ester (**541**) underwent an intramolecular Michael cyclization to produce 2-methyl-3-phenyl-5-isoxazolidinone (**542**) ⟨78TL3985⟩.

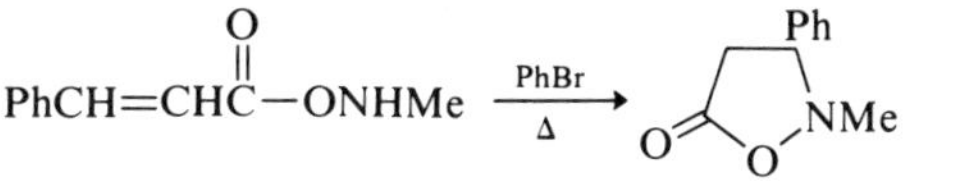

(**541**)   (**542**)

Nitrones react with zinc Reformatskii salts ⟨77AP873, 79T647⟩ or with the anion of dimethyl malonate to generate 5-isoxazolidinones (Scheme 161) ⟨76AP935⟩.

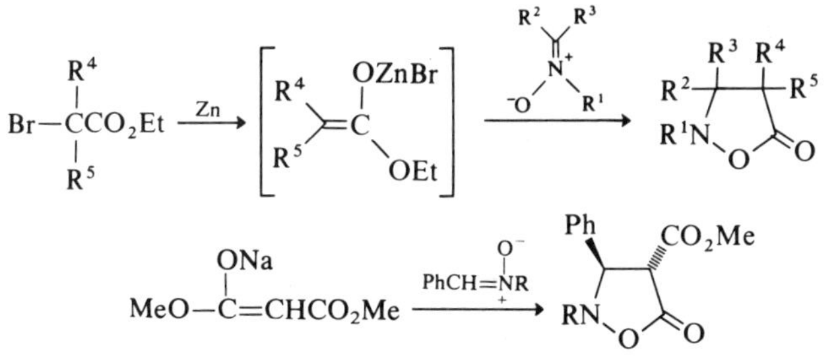

Scheme 161

The cycloaddition of nitrones with ketenes produced 5-isoxazolidinones as well as oxazolones, as shown in Scheme 162 ⟨78H(9)457, 79JOC2961⟩. In a similar fashion, nitrones also react with ketenimines to generate the 5-isoxazolidinone imines ⟨75JHC175, 68JHC881⟩.

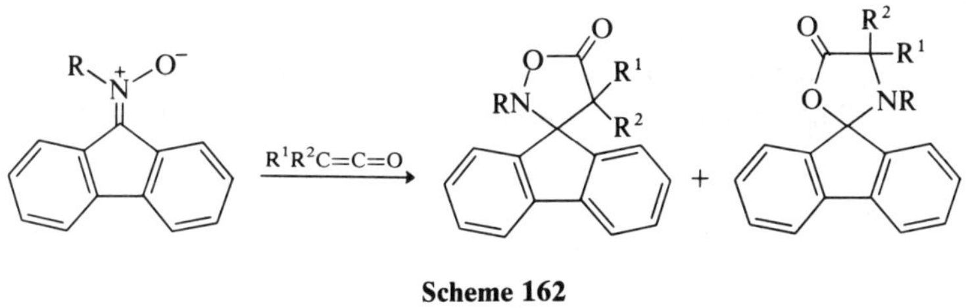

Scheme 162

4.16.4.18.4 Isoxazolidine-3,5-diones

This system was generated by the reaction of hydroxylamine with malonyl dihalides ⟨65HCA1973, 80AP39⟩ or by the saponification of 3-isoxazolin-5-ones as shown in Scheme 163. These compounds are strong acids and can exist in a variety of forms, with the dione form being the most favored (equation 58) ⟨80JHC299⟩.

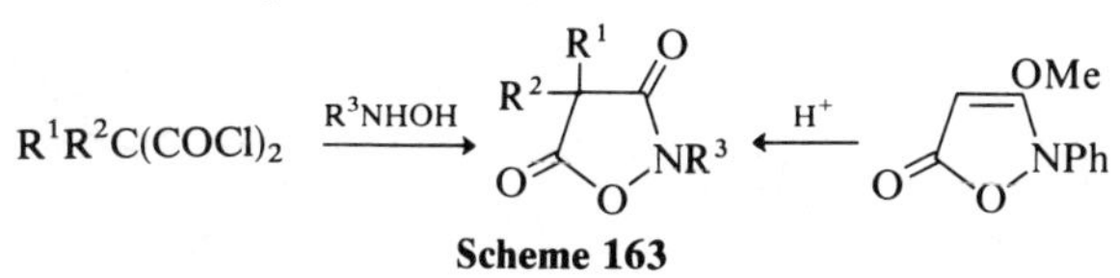

Scheme 163

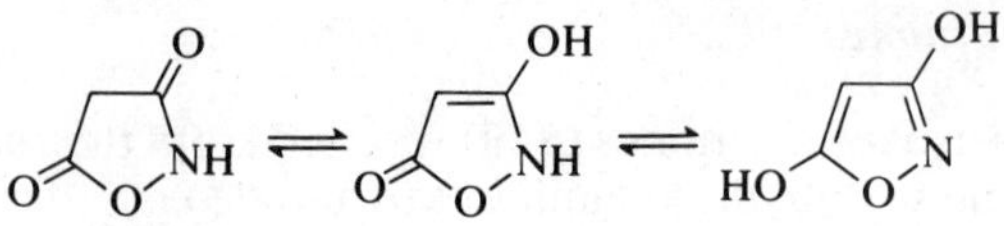

$$\underset{(58)}{\text{[structure: keto-NH tautomer]} \rightleftharpoons \text{[OH-NH tautomer]} \rightleftharpoons \text{[di-OH isoxazole tautomer]}}$$

The reaction of (**543**) with Michael acceptors gave both *C*- and *N*-alkylated products (Scheme 164) ⟨80JHC299⟩. Diazomethane and (**544**) produced *O*-methylation at the C-3 oxygen atom ⟨68M2534⟩. The 4-position was highly active, condensing with aldehydes in the presence of base to give 4-methylene compounds (**545**) which were hydrogenated to (**546**). These β-keto acids (**546**) decarboxylated on heating to give (**547**) ⟨68M2534⟩.

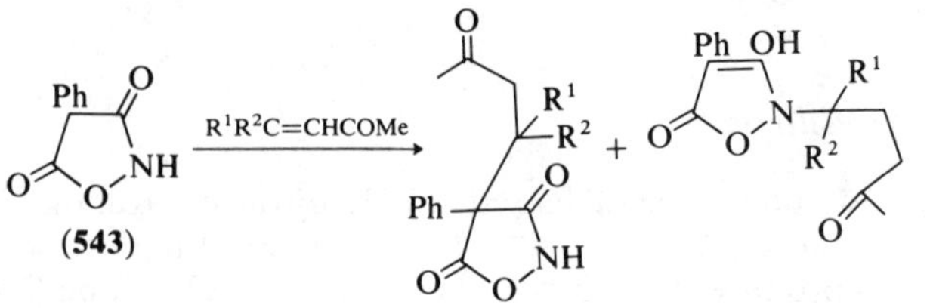

Scheme 164

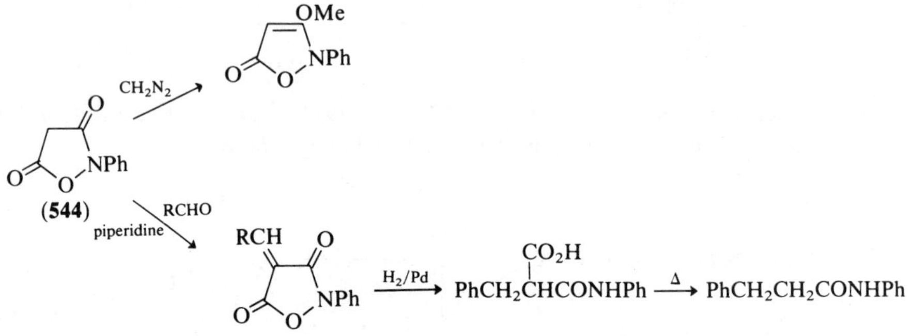

Scheme 165 [illustration of 1,2-benzisoxazole numbering]

4.16.4.19 Benzisoxazoles

4.16.4.19.1 1,2-Benzisoxazoles

(i) By ring closure of oximes

Most synthetic methods involve the formation of bond C(1)—C(7a) [or C(1)—C(7a) and C(2)—C(3)] by the cyclization of a preformed oxime or one generated *in situ*, or by formation of bond C(1)—C(2) [or C(1)—C(2) and C(2)—C(3)] by a variety of transformations (Scheme 165). The remaining methodology involves mainly heterocyclic rearrangements ⟨67AHC(8)277⟩.

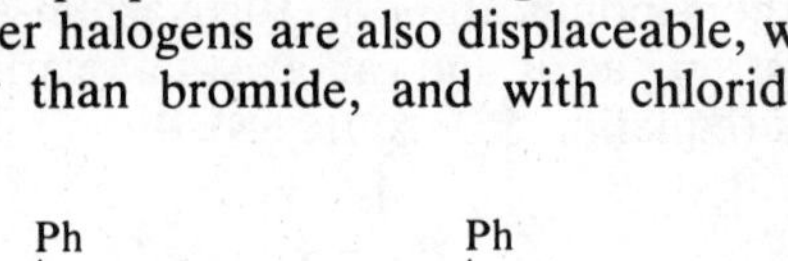

The first reported synthesis of the 1,2-benzisoxazole nucleus was the preparation of 3-phenyl-1,2-benzisoxazole from *o*-bromobenzophenone oxime (equation 59) and this method was extended to the preparation of a large number of derivatives ⟨67AHC(8)277, 1892CB1498, 79EUP2666⟩. Other halogens are also displaceable, with iodide or fluoride being comparable with or better than bromide, and with chloride being the least reactive ⟨39LA(540)83, 41LA(546)273⟩.

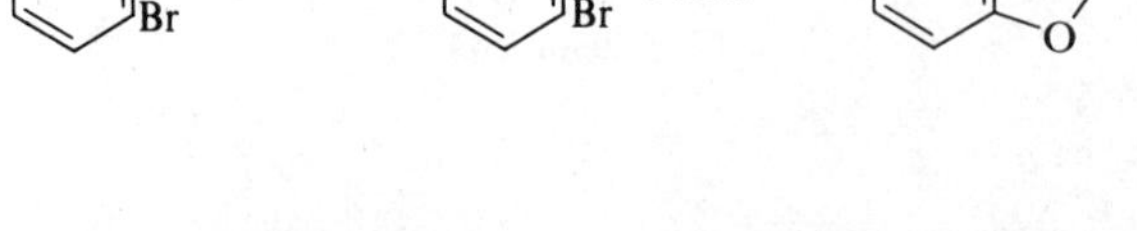

$$\underset{(59)}{\text{o-bromobenzophenone} \xrightarrow{NH_2OH} \text{o-bromobenzophenone oxime} \xrightarrow{\Delta} \text{3-phenyl-1,2-benzisoxazole}}$$

In one example where a 2-Cl and 2'-F were present, only the chloride displacement product (**548**) was reported; also the tetrafluoro-1,2-benzisoxazole (**550**) was prepared by heating the oxime (**549**) (Scheme 166) ⟨79EUP2666, 71MI41600⟩.

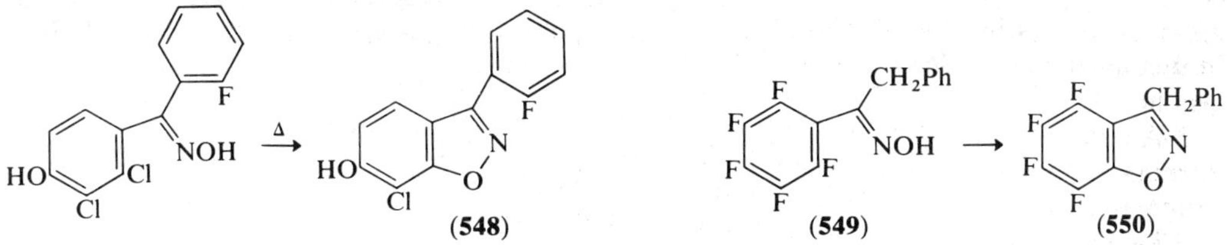

Scheme 166

The course of the reaction is dependent on the configuration of the oxime. The (Z)-oxime gave 1,2-benzisoxazoles as the primary product while the (E)-oxime generally produced a Beckmann rearrangement product with or without subsequent cyclization to a benzisoxazole (Scheme 167) ⟨67AHC(8)277⟩. Bunnett conducted a kinetic study on the reaction shown in Scheme 167 and determined that cyclization to intermediate (**551**) was the rate determining step ⟨61JA3805⟩.

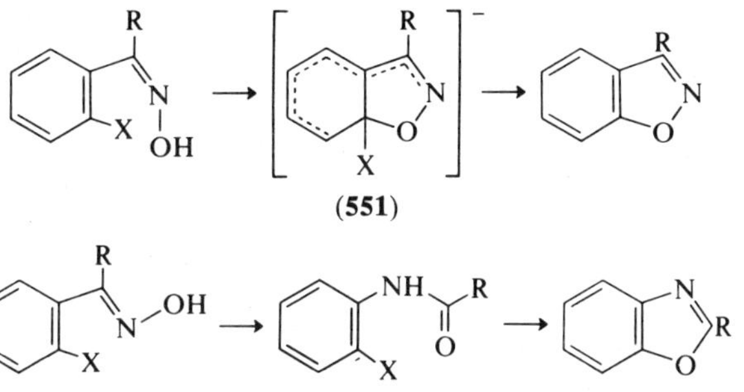

Scheme 167

Other displaceable groups, generally with suitable activation, include MeO, HO, NO_2 and NH_2 ⟨67AHC(8)277⟩. The treatment of o-nitrobenzophenone with hydroxylamine and base generated 1,2-benzisoxazoles. Intermediate oximes generated by nitrosation of o-nitrobenzyl compounds can also be cyclized ⟨67AHC(8)277, 71IJC1311⟩. A novel displacement of MeO occurred in the reaction of the dihydroindazole derivative (**552**) with hydroxylamine to form (**553**) ⟨76MI41608⟩.

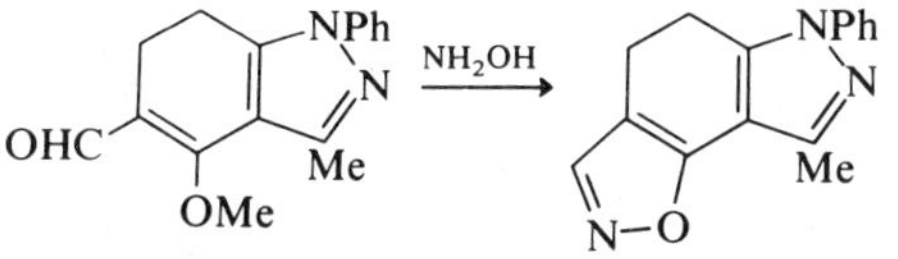

Certain β-hydroxyoximes, unsubstituted in the *ortho* position of an adjacent benzene ring, cyclize when treated with phosphoric acid to give styrylbenzoxazole (equation 60).

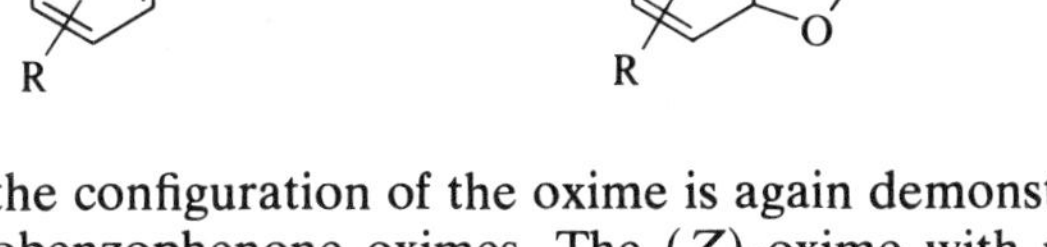

(60)

The importance of the configuration of the oxime is again demonstrated in the oxidative cyclization of o-aminobenzophenone oximes. The (Z)-oxime with nitrous acid produced 3-phenyl-1,2-benzisoxazole, while the (E)-oxime with similar treatment yielded benzotriazine N-oxides ⟨27CB1736⟩.

An unusual replacement of hydroxyl with the concomitant formation of a 1,2-benzisoxazole occurred with the base treatment of 2,6-dihydroxyacetophenone oxime (**554**). The use of 0.5 equivalents of KOH/MeOH gave only a 1,2-benzisoxazole. Increasing the concentration of KOH through 2 equivalents gave the 1,2-benzisoxazole and increasing amounts of a benzoxazole (Scheme 168). The benzoxazole did not arise from the 1,2-benzisoxazole under the reaction conditions ⟨73JCS(P1)2220⟩. This reaction is similar to previously reported base conversions ⟨67AHC(8)277, 70HCA1883⟩. Other replacements of HO or MeO require substantial activation such as the presence of a nitro group. In (555) sufficient activation was present to permit cyclization to form 5,7-dinitro-3-phenyl-1,2-benzisoxazole (556) ⟨67AHC(8)277⟩.

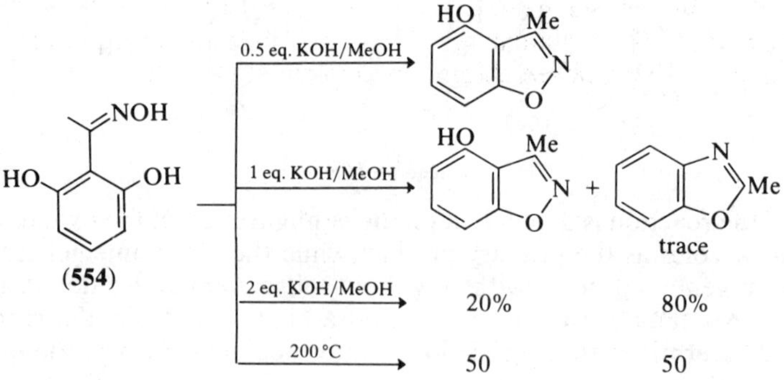

Scheme 168

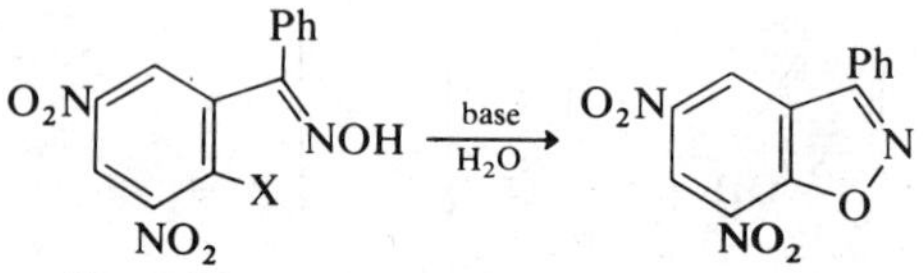

A series of 3-aryl-1,2-benzisoxazoles was prepared by the treatment of 2-hydroxybenzophenone oximes with Na_2CO_3 ⟨79JMC1554⟩. Base induced transformations of ketone *o*-acyloximes lead to a variety of products, the distribution of which depends on the oxime configuration, the base employed and the leaving group ⟨73JCS(P1)2220, 75GEP2450053, 79JMC1554, 72T3295, 77JIC875⟩. Acyl-(*E*)-oximes usually undergo a Beckmann rearrangement with subsequent cyclization to benzoxazoles, and the use of excess aqueous base generated the starting oxime ⟨67AHC(8)277⟩. The use of sodium acetate produced 1,2-benzisoxazoles ⟨71T711, 80IJC(B)571, 77IJC(B)1058⟩.

The sulfonate ester of *o*-hydroxyacetophenone oximes, when treated with pyridine, are similarly converted into a benzoxazole, but cyclize to a 1,2-benzisoxazole when treated with aqueous KOH ⟨73JCS(P1)2220, 71T711⟩.

Heating of the (*E*)-oximes of *o*-hydroxyacetophenones produced 1,2-benzisoxazoles as the predominant product, while thermolysis of the (*Z*)-oximes yielded only minor amounts of 1,2-benzisoxazoles ⟨67AHC(8)277, 73IJC541⟩.

The use of dehydrating agents such as sulfuric or phosphoric acid on (555; X = OH) was also successful, and these closures may proceed *via* mixed anhydrides ⟨67AHC(8)277, 75MIP41600⟩. Carbonyldiimidazole effected the conversion of hydroxamic acid (557) into a 3-hydroxy-1,2-benzisoxazole derivative ⟨79JHC1277⟩. The mixed anhydride (558) where

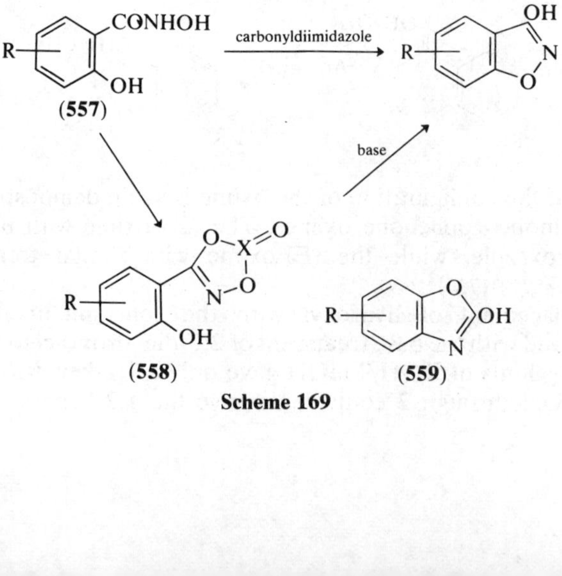

Scheme 169

X = C or S, obtained from (**557**) and thionyl chloride or phosgene, underwent base rearrangement to 3-hydroxy-1,2-benzisoxazoles and some 2-hydroxybenzoxazole (**559**) (Scheme 169) ⟨67AHC(8)277, 69JHC123⟩. Two potential intermediates in the decomposition of (**558**) are an acylnitrene or an aziridinone, with the former being a more likely intermediate ⟨69JHC123, 71DIS(B)4483⟩.

Thionyl chloride/pyridine treatment of oximes (**557**) is believed not to proceed *via* an intermediate like (**558**) in the generation of 1,2-benzisoxazoles, but by a chloramine intermediate (Scheme 170). A similar reaction of an *N*-phenylhydroxamic acid generated a benzisoxazolinone *via* a proposed chloramine intermediate (Scheme 170) ⟨77AJC1847⟩.

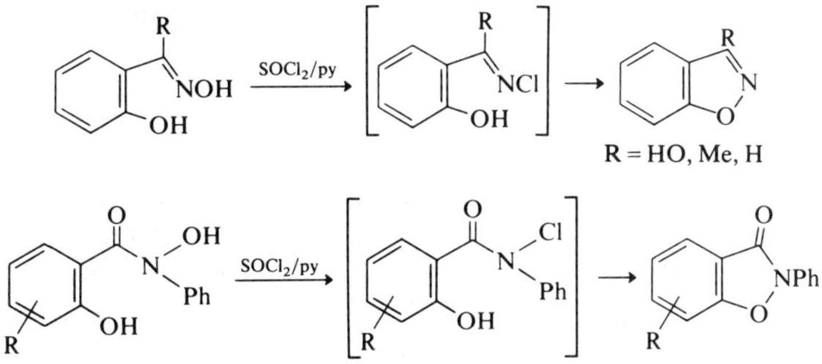

Scheme 170

The treatment of 2-hydroxyacetophenone with hydroxylamine-*O*-sulfonic acid in dilute aqueous base produced 3-methyl-1,2-benzisoxazole. The mechanism was reported to be a C(2)—C(3) ring closure *via* intermediate (**560**) (Scheme 171). Salicylaldehyde failed to cyclize with dilute base, but with 20% KOH and hydroxylamine-*O*-sulfonic acid the transformation to 1,2-benzisoxazole succeeded ⟨76MI41600⟩. Kemp and Woodward isolated an oxime sulfonate (**561**) from salicylaldehyde and hydroxylamine-*O*-sulfonic acid and the subsequent decomposition gave 1,2-benzisoxazole in 95% yield ⟨65T3019⟩.

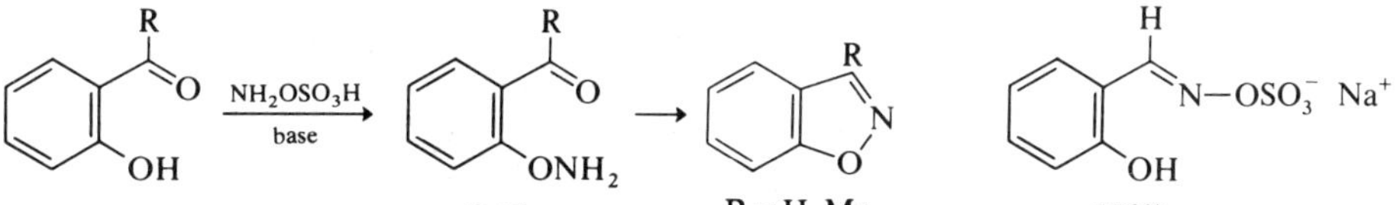

Scheme 171

The synthesis of 3-acyl- and 3-aroyl-1,2-benzisoxazoles was accomplished by the desulfurization of (**562**). The dithioacetal (**562**) was prepared by the addition of the lithium salt of propenedithioacetal to an isocyanate with subsequent base cyclization (equation 61) ⟨B-79MI41609⟩.

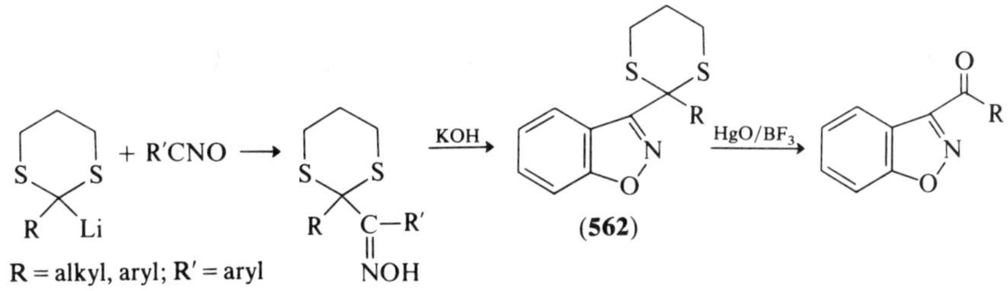

(61)

(ii) Other methods

The utilization of azidyl ion with salicylaldehyde gave 1,2-benzisoxazole accompanied by up to 15% of 2-aminobenzoxazole ⟨67AHC(8)277⟩. Nitrile oxide addition to benzyne gave 3-substituted 1,2-benzisoxazole. Nitrile oxide addition to ketone enamines gave reduced derivatives which were brominated with NBS, and subsequent loss of HBr also generated 3-substituted 1,2-benzisoxazoles ⟨67AHC(8)277, 80IJC(B)571⟩.

1,2-Benzisoxazoles were obtained from quinazolin-3-ols in an ArF displacement activated by MnO_2 (equation 62) ⟨74JHC885⟩. The 1,2-benzisoxazole was produced as a minor side product in the reaction of (**563**) with hydroxylamine ⟨73MI41600⟩.

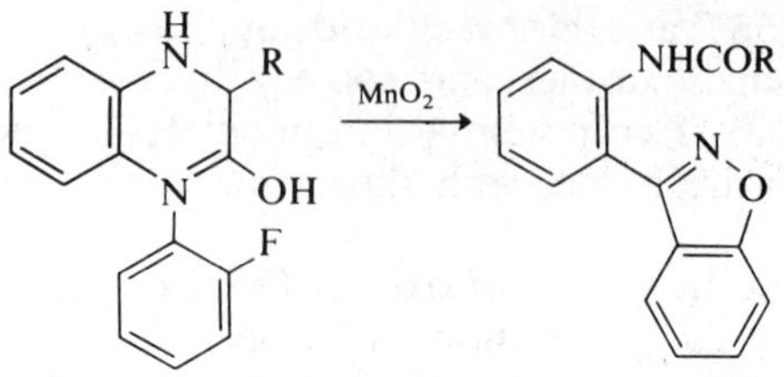

The literature had reported the preparation of a coumarin hydroxylamine by the reaction of 4-hydroxycoumarin with hydroxylamine. A reinvestigation of the reaction showed the product to be 1,2-benzisoxazole-3-acetic acid (Scheme 172) ⟨69JHC279⟩.

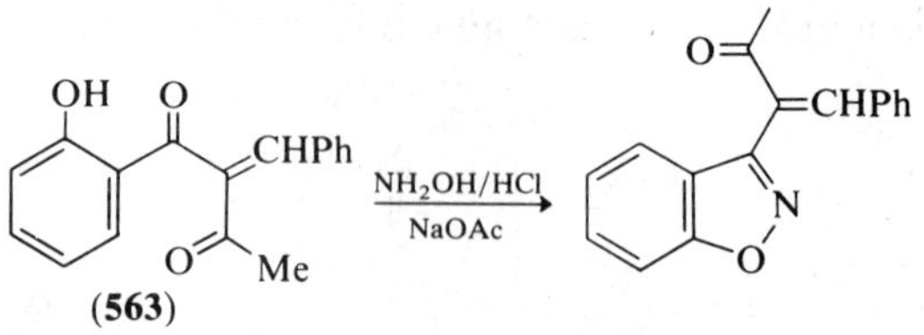

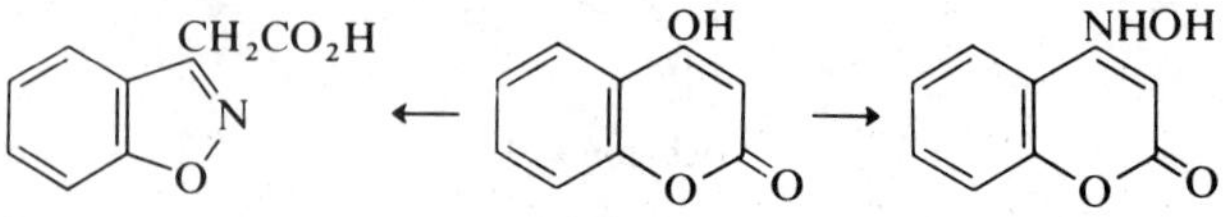

Scheme 172

(iii) 1,2-Benzisoxazole N-oxide

1,2-Benzisoxazole N-oxides have recently been synthesized from 2-hydroxypropiophenone oxime and lead tetraacetate (equation 63) ⟨80CC421⟩.

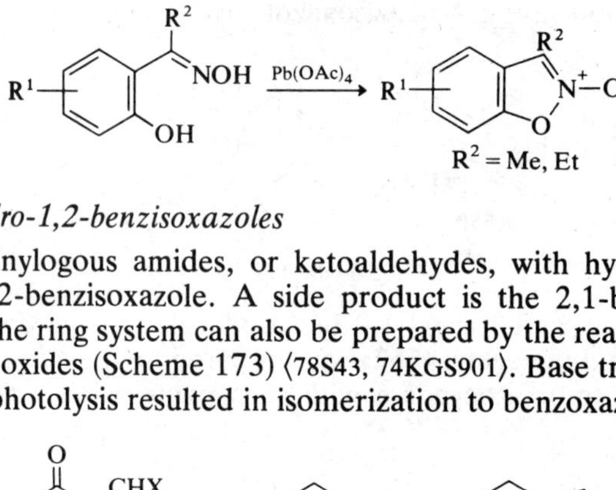

(iv) 4,5,6,7-Tetrahydro-1,2-benzisoxazoles

The reaction of vinylogous amides, or ketoaldehydes, with hydroxylamine produced 4,5,6,7-tetrahydro-1,2-benzisoxazole. A side product is the 2,1-benzisoxazole (Scheme 173) ⟨67AHC(8)277⟩. The ring system can also be prepared by the reaction of cyclohexanone enamines with nitrile oxides (Scheme 173) ⟨78S43, 74KGS901⟩. Base treatment produced ring fission products and photolysis resulted in isomerization to benzoxazoles ⟨76JOC13⟩.

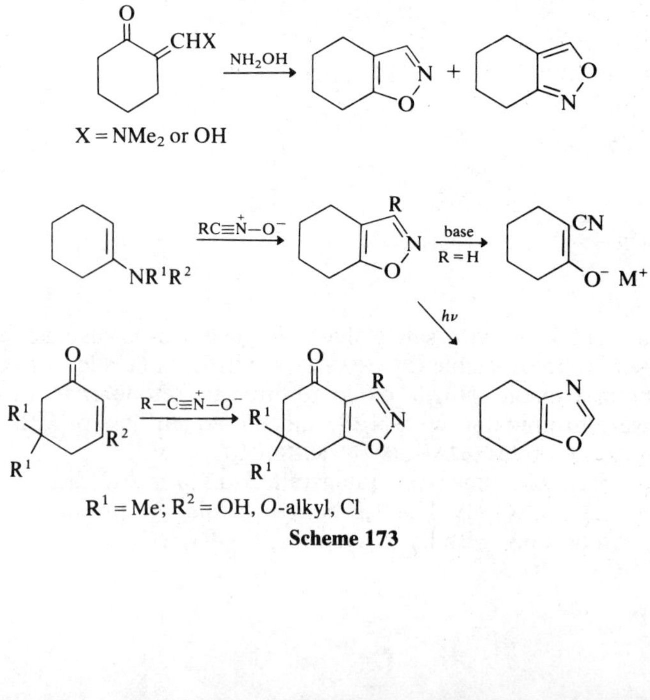

Scheme 173

(v) 4,7-Dihydro-1,2-benzisoxazoles

4,7-Dihydro-1,2-benzisoxazole was prepared by the cycloaddition of benzonitrile N-oxide to 1,4-benzoquinones with subsequent oxidation of the hydroquinone (Scheme 174) ⟨72JAP(K)7242659, 50G140⟩.

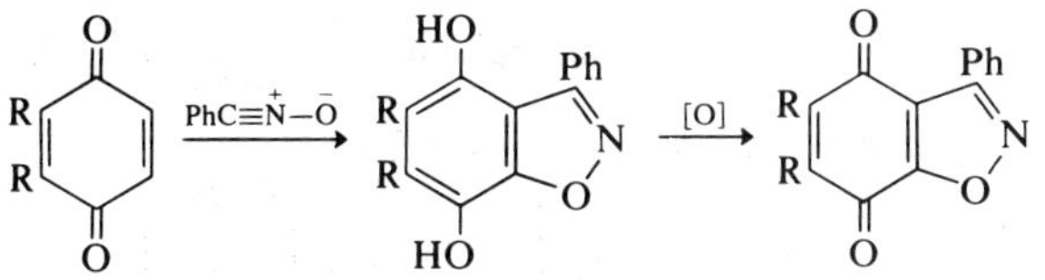

Scheme 174

(vi) 3a,4,5,6,7a-Hexahydro-1,2-benzisoxazoles

This series was prepared by an analogous addition of a species such as benzonitrile N-oxide or 2,2-dimethylpropanenitrile N-oxide to cyclohexanones or cyclohexenes (Scheme 175) ⟨67AHC(8)277⟩.

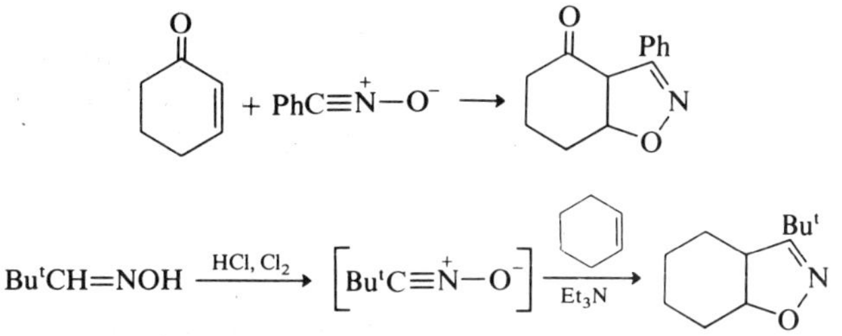

Scheme 175

The addition of the anion of α-bromo-α-nitrotoluene (564) to cyclohexene gave the hexahydro derivative (565) of 3-phenyl-1,2-benzisoxazole ⟨75TL2131⟩. An unusual hexahydro derivative (566) was produced by the bis addition of benzonitrile N-oxide to benzoquinone ⟨67AHC(8)277⟩.

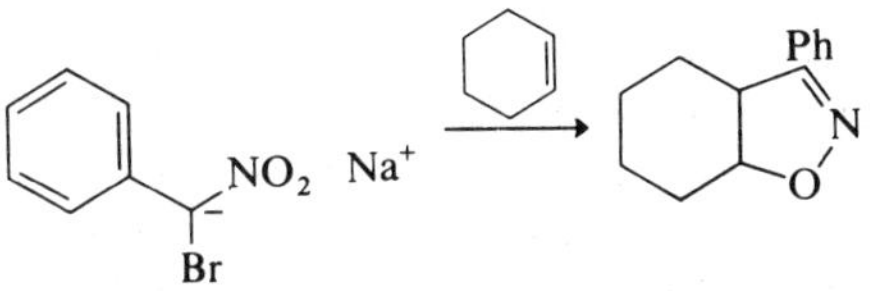

(564)   (565) 30% 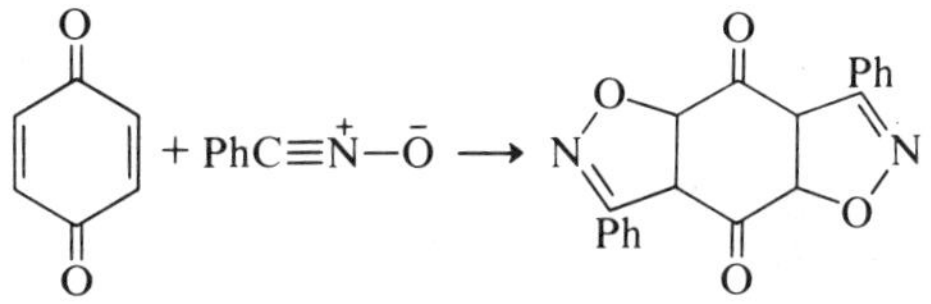   (566)

(vii) 3a,4,7,7a-Tetrahydro-1,2-benzisoxazoles

The only reported synthesis of this system is the synthesis of 3a,4,7,7a-tetrahydro-1,2-benzisoxazol-3-yl dimer by the reaction of 1,4-cyclohexadiene with cyanogen di-N-oxide (equation 64) ⟨65LA(687)191⟩.

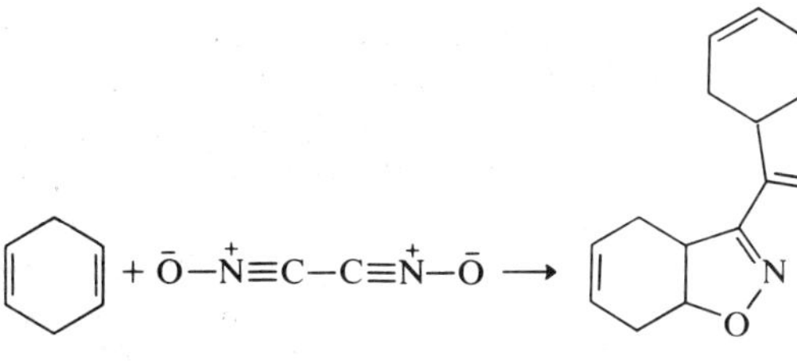

(64)

(viii) 4,5-Dihydro-1,2-benzisoxazoles

The synthesis of this series involved the reaction of disubstituted or benzo fused 6-keto(formyl)-2-cyclohexenones with hydroxylamine (Scheme 176), Base degradation gave α-cyanoketones which can be further degraded to diacids ⟨67AHC(8)277, 80IJC(B)406⟩.

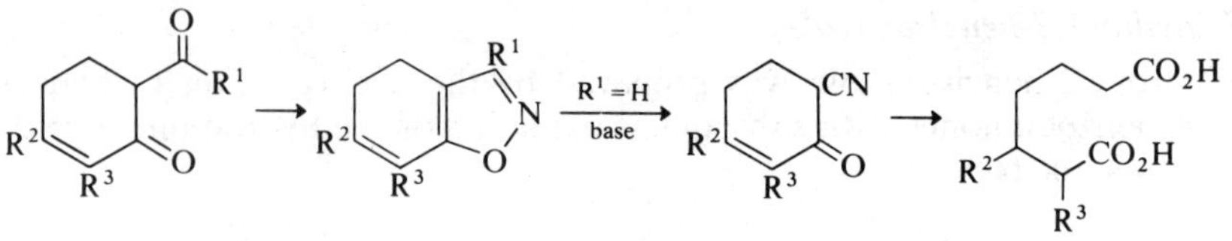

Scheme 176

(ix) 2,3-Dihydro-1,2-benzisoxazoles

This series is punctuated by several unusual synthons, one the reaction of tropone (**567**) with aniline to give (**568**). The cyclization of (**569**) with sodium hydroxide gave (**570**) ⟨67AHC(8)277, 69M602⟩. The reaction of *in situ* generated benzyne with the nitrone (**571**) produced (**572**) ⟨67AHC(8)277⟩.

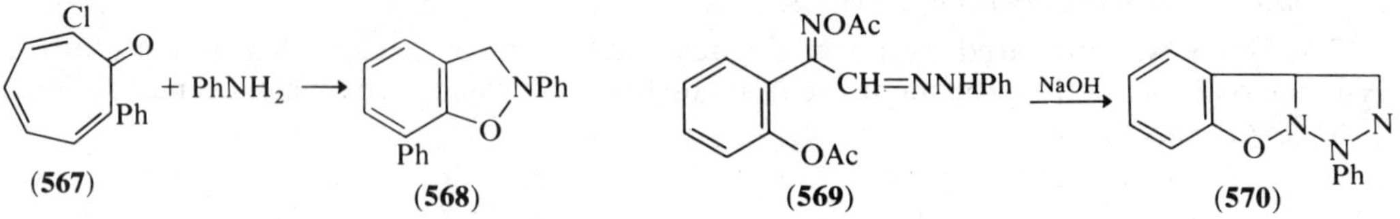

4.16.4.19.2 2,1-Benzisoxazoles

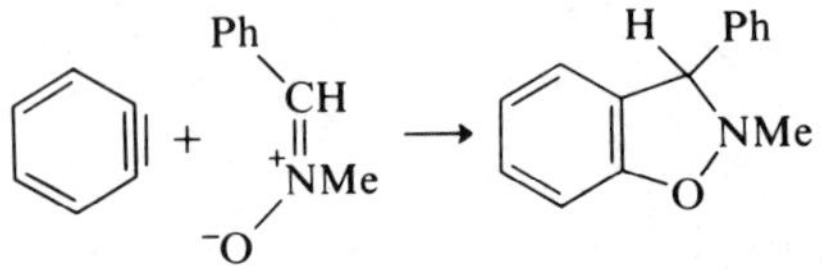

2,1-Benzisoxazoles, also called anthranils as derivatives of anthranilic acid, are most commonly formed by the closure of bonds C(1)—C(2) or C(2)—C(3), or the introduction of atom C(3) resulting in formation of bonds C(2)—C(3) and C(3)—C(3a). As with the 1,2-benzisoxazole series, many early structural ambiguities were present in assignments ⟨67AHC(8)277, 62HC(17)1, 66DIS(B)102⟩. The 3-hydroxy compound is primarily in the keto form and only recently have ethers been reported.

Early structures, proposals and refutations included benzoazetenone, cyclobutadienes ⟨B-65MI41600⟩ or an oxarirane ⟨04CB966, 26LA(437)162, B-35MI41600⟩, and these references should be consulted for this early work.

The structure was established on the basis of substitution at the 3-position ⟨11CB2409⟩, optical investigations ⟨26LA(437)162⟩, dipole moment (3.06 D) ⟨44MI41600⟩ and IR measurements ⟨57MI41600, 66DIS(B)102⟩. The first 2,1-benzisoxazole to be synthesized was 5,6-dimethoxy-2,1-benzisoxazole-4-carboxylic acid in 1881 ⟨1881JPR353⟩.

(i) From nitro compounds

The earliest methods of preparation, and the most utilized, involve the reduction of 1-nitrophenones or *o*-nitrobenzenes that contain a group with an oxygen function on the α-carbon atom. This approach has been discussed in detail previously ⟨67AHC(8)277⟩. Agents used to effect this reductive transformation include Zn ⟨67AHC(8)277, 62MI41600, 65NKK526⟩, Sn ⟨67AHC(8)277, 76GEP2529292, 65RRC1035, 62JOC3683, 76CJC1336, 74RTC139⟩. Na_2S, $Pd/BaSO_4$, *t*-butoxide ⟨67AHC(8)277⟩, $SOCl_2$ ⟨76GEP2529292, 79JHC1249⟩, alkyl phosphites ⟨77JOC1791, 66CC491⟩ and H_2 ⟨67AHC(8)277⟩, and the mechanism of conversion of these 2-nitrobenzyl systems into 2,1-benzisoxazoles has been well studied ⟨66DIS(B)102⟩.

3-Phenyl-1,2-benzisoxazole (**574**) was prepared from (**573**) and sulfuric acid. The inclusion of $NaNO_2$ in the reaction medium yielded the acridone (**575**) in 40% yield ⟨65RRC1035⟩.

3-Aryl-1,2-benzisoxazoles can also be prepared by the reaction of *o*-nitrobenzaldehydes and an aromatic hydrocarbon catalyzed by sulfuric acid ⟨67AHC(8)277⟩. Cyclization of (**576**) by Zn/HOAc gave (**577**); sulfuric acid cyclization gave (**578**) and (**579**) ⟨62JOC3683, 66DIS(B)102⟩.

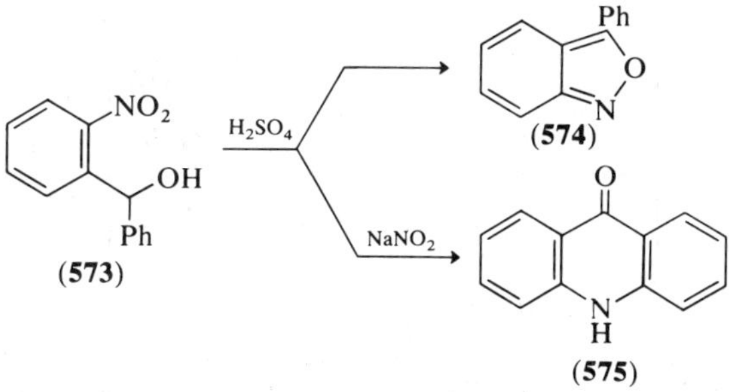

Reductive ring closure with thionyl chloride led to the introduction of a chloro group in the 5-position. When the 5-position was blocked by a substituent, halogen attack occurred in the 7-position. The mechanism is shown in Scheme 177 ⟨67AHC(8)277⟩.

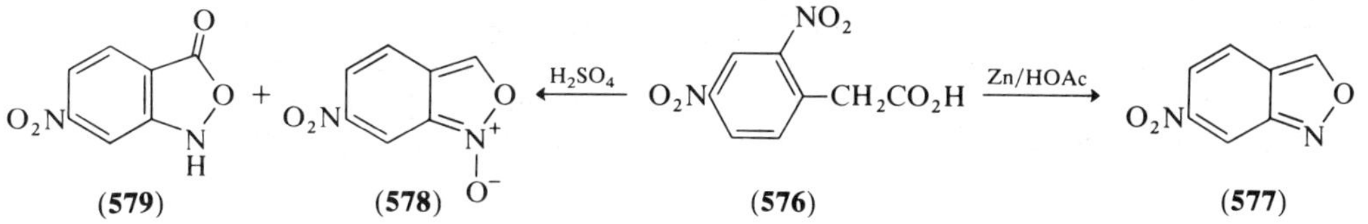

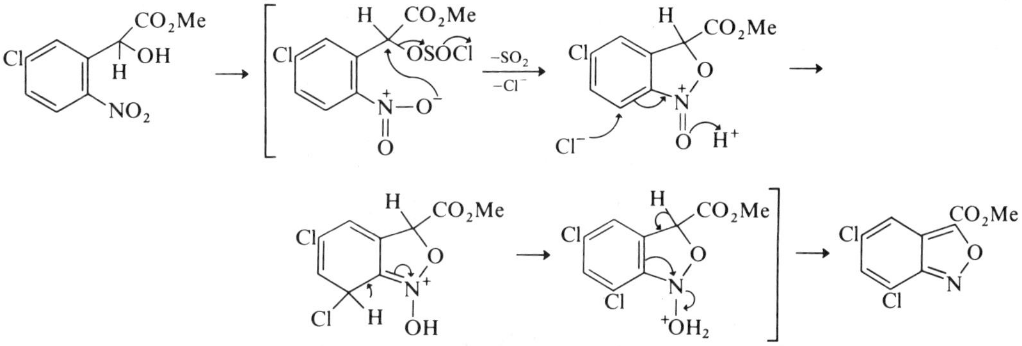

Scheme 177

Certain substituted *o*-nitrotoluenes can be induced to cyclize, forming 2,1-benzisoxazoles. Bis(2-nitrophenyl)methane when irradiated gave 3-(*o*-nitrophenyl)-2,1-benzisoxazole. The possible intermediates including a biradical were discussed ⟨74TL4359⟩. 3-(*o*-Nitrophenyl)-2,1-benzisoxazole was prepared by the acid cyclization of bis(2-nitrophenyl)methanol (Scheme 178) ⟨65RRC1035⟩.

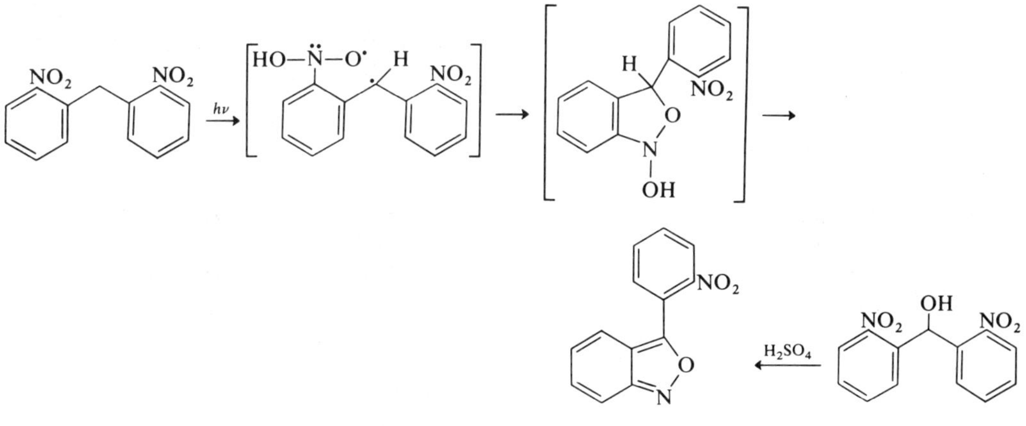

Scheme 178

The intermediacy of an *aci*-nitro compound has been proposed for the sulfuric acid cyclization of *o*-nitrophenylacetic acid to yield a mixture of 2,1-benzisoxazole and 2,1-benzisoxazole-3-carboxylic acid. The acid does not decarboxylate under the reaction conditions. The proposed *aci*-nitro intermediate cyclized to an *N*-hydroxy compound which decomposed to the products (Scheme 179) ⟨70JCS(C)2660⟩.

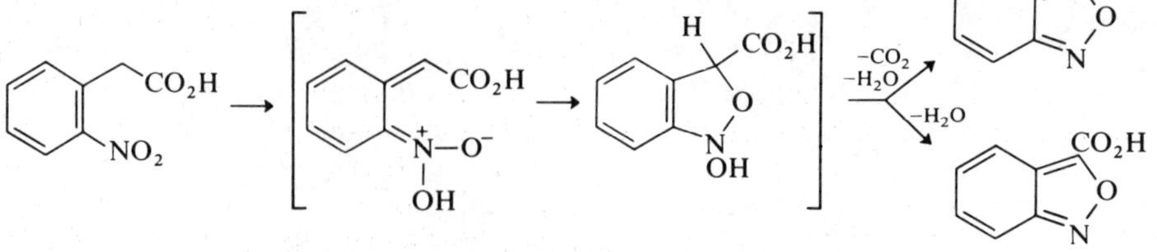

Scheme 179

Bis(2,4,6-trinitrophenyl)methane when treated with NaAc in acetic acid produced (**580**) as a thermostable explosive ⟨80MIP41600⟩. The conversion of o-nitrotoluene into 2,1-benzisoxazole was effected by mercury(II) oxide catalysis. A mercury containing intermediate was isolated and was demonstrated to be converted into 2,1-benzisoxazole ⟨67AHC(8)277⟩. The treatment of o-nitrotoluene derivative (**581**) with sulfuric acid gave (**582**) in 35% yield ⟨72MI41607⟩.

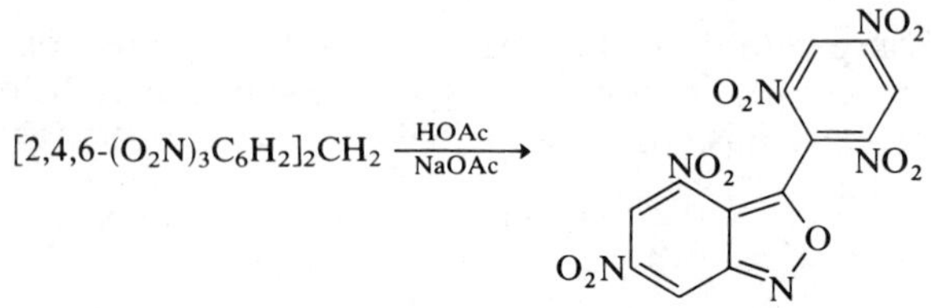

(ii) From nitroso compounds

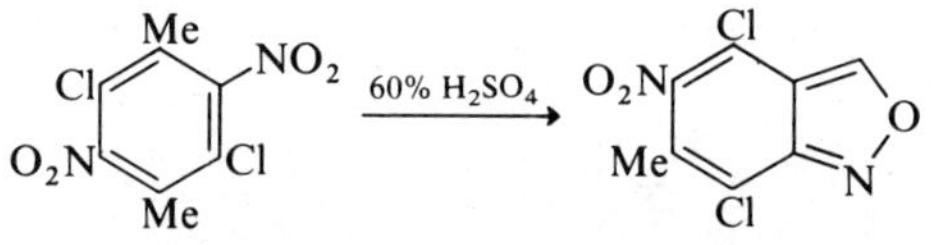

The reaction of o-nitrosoacetophenone with hydrogen gave 3-methyl-1,2-benzisoxazole ⟨67AHC(8)277⟩. Treatment with triphenylphosphine ⟨75KGS1195⟩ or acetic anhydride ⟨73MIP41600⟩ also effected the closure to a 2,1-benzisoxazole (Scheme 180). The ring closure was also achieved by the use of HBr/benzene or HCl/MeOH (Scheme 180) ⟨76KGS886, 73KGS1334⟩.

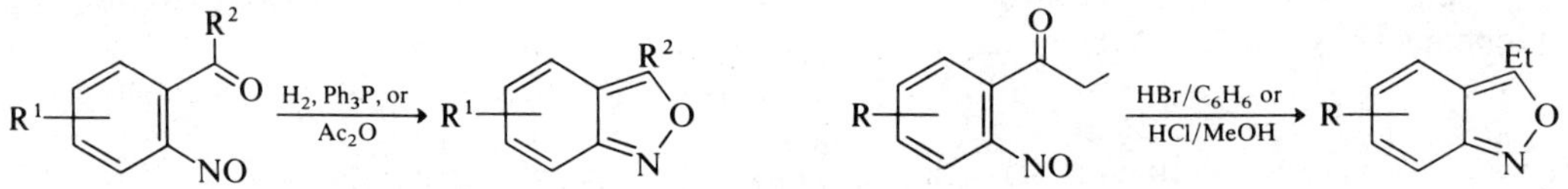

Scheme 180

(iii) By ring closure of anilines

Many methods for the oxidative cyclization of 2-aminophenones to 2,1-benzisoxazoles have been reported and these were extensively discussed by Wünsch and Boulton ⟨67AHC(8)277⟩. An illustrative reaction is the treatment of an aminophenone with hydrogen peroxide or potassium persulfate to produce 2,1-benzisoxazoles (Scheme 181) ⟨64USP3261870⟩.

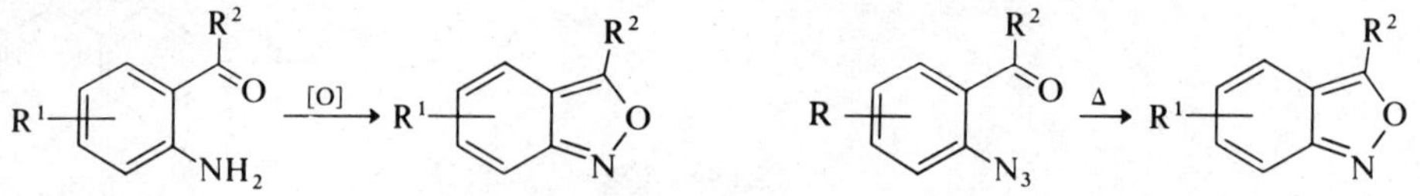

Scheme 181

(iv) By decomposition of azides

Heating of o-azidophenones effected cyclization and production of 2,1-benzisoxazoles (Scheme 181) ⟨67AHC(8)277, 72JA4952, 74JAP7441196, 74JHC125, 78TL2309⟩. The decomposition of azide (**583**) gave the 2,1-benzisoxazole (**584**) as the major product, along with a minor amount of hydroxyquinolone (**585**) ⟨78TL2309⟩.

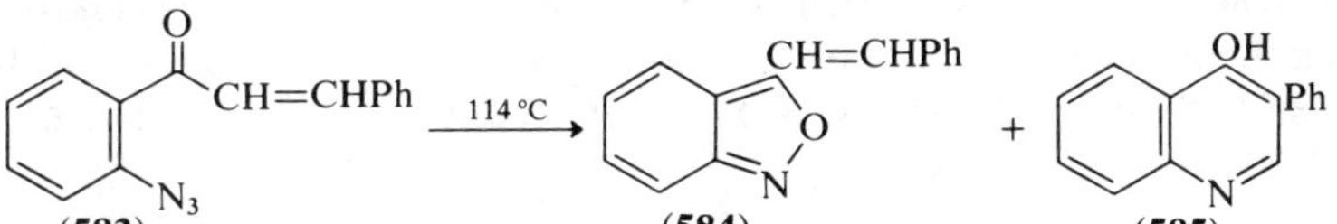

(**583**) (**584**) (**585**)

(v) By the introduction of C-3

This method generally involved the addition to nitrobenzenes of the anions of phenylacetonitrile ⟨67AHC(8)277, 72MI41600, 64NEP6407011, 64USP3156704, 72MI41601, 72MI41603⟩, cyanoacetate ⟨67AHC(8)277⟩ or malonate ⟨67AHC(8)277⟩. An intermediate *o*-quinoid had been isolated and shown to be converted into the 2,1-benzisoxazole (Scheme 182) ⟨67AHC(8)277, 64USP3156704⟩.

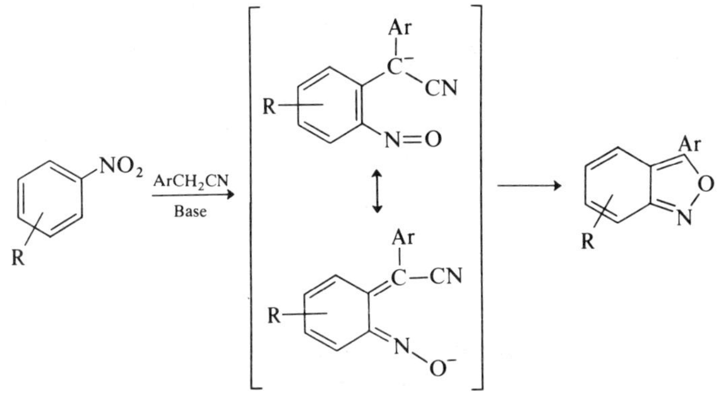

Scheme 182

(vi) Other syntheses

The *N*-oxide of indolone rearranged to produce 3-benzoyl-2,1-benzisoxazole ⟨65JOC1104⟩, and an isoxazolyl-2,1-benzisoxazole (**587**) was prepared from a benzilideneacetylacetone

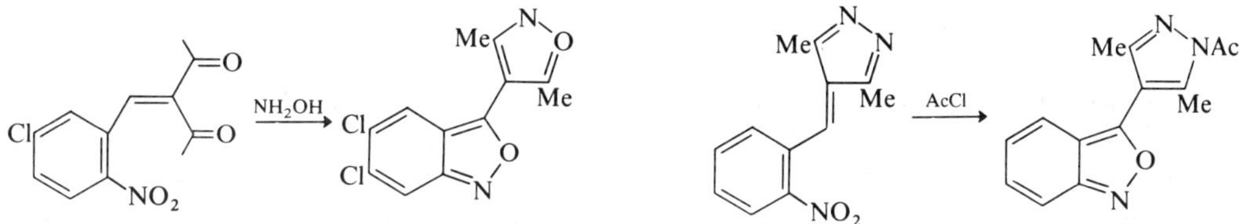

(**586**) (**587**) (**588**) (**589**)

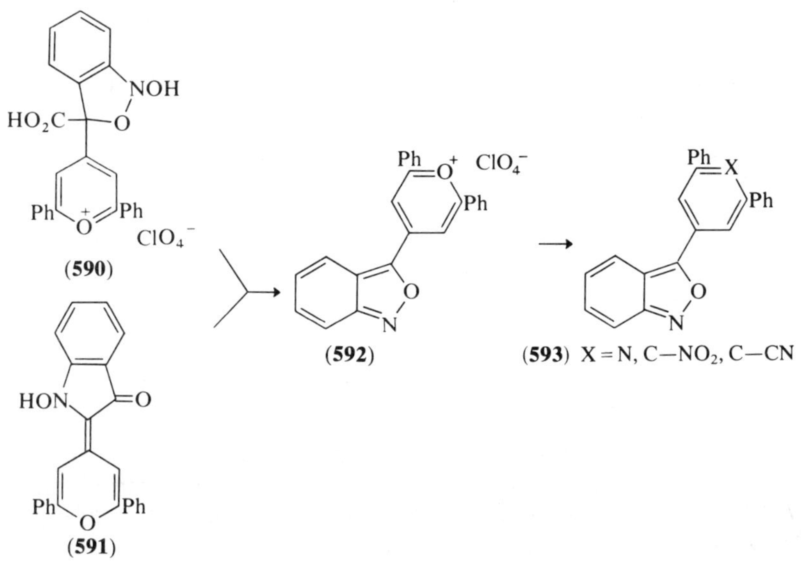

(**590**) (**591**) (**592**) (**593**) X = N, C—NO$_2$, C—CN

(**586**) and hydroxylamine ⟨76JHC661⟩. The reaction of (**588**) with acetyl chloride produced the 2,1-benzisoxazole (**589**) ⟨76CPB1106⟩. Benzisoxazol-3-ylpyrylium salts (**592**) and aryl-2,1-benzisoxazoles of type (**593**) were synthesized from (**590**) or (**591**) ⟨74JHC395⟩.

The photodecomposition of benzotriazine N-oxides produced 3-substituted 2,1-benzisoxazoles (Scheme 183) ⟨73JA2390⟩. The photolysis of cinnoline 1-oxides produced 3-methyl-2,1-benzisoxazoles (Scheme 183) ⟨74TL2643, 74T2645⟩.

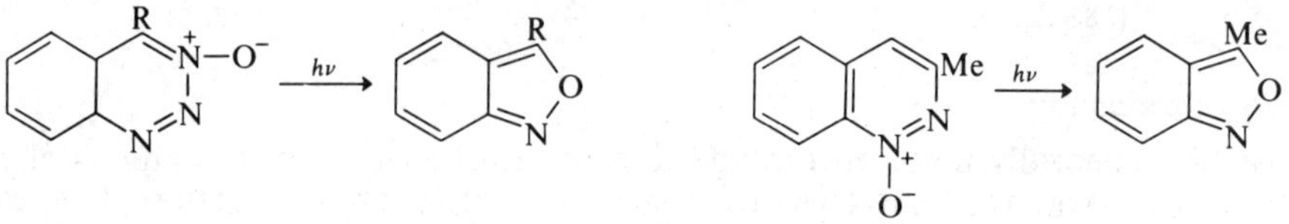

Scheme 183

The hydroxide ion induced rearrangement of N-hydroxyisatin to 2,1-benzisoxazole-3-carboxylic acid has been described. A ^{14}C label at the 3-position produced a 2,1-benzisoxazole with the label in the carboxyl carbon (Scheme 184) ⟨66DIS(B)102⟩.

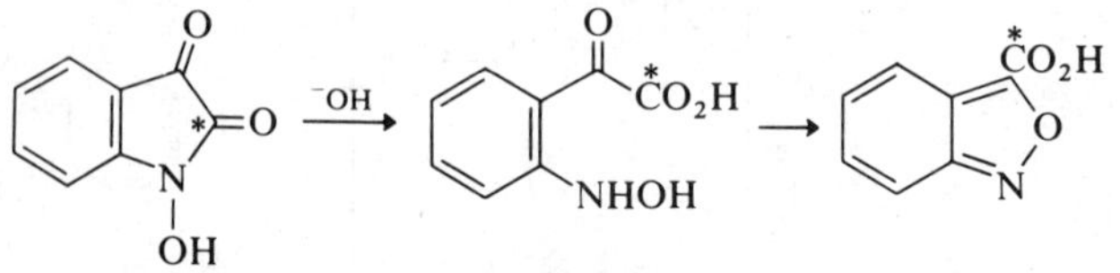

Scheme 184

A novel N-oxide (**595**) was prepared by the solvolysis at room temperature of (**594**) in 1:1 water–dioxane. A kinetic and mechanistic study of this reaction was carried out.

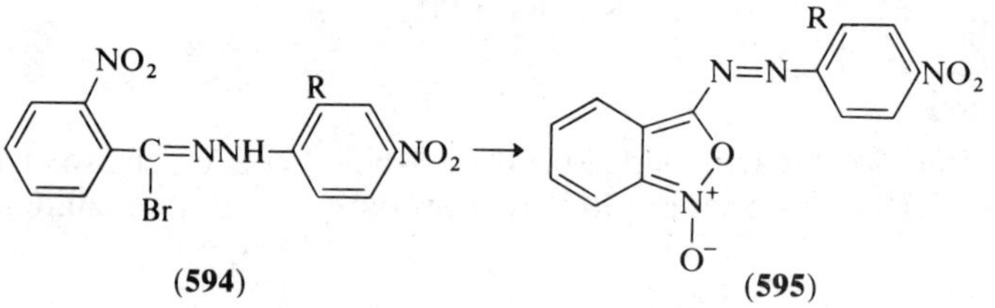

(**594**)   (**595**)

(*vii*) *3-Hydroxy- and 3-amino-2,1-benzisoxazoles*

The synthesis of 3-hydroxy-2,1-benzisoxazoles has been accomplished by the reduction of *o*-nitrobenzoic acids with tin, zinc or tin(II) chloride, by the acid cyclization of *o*-hydroxyaminobenzoic acids ⟨67AHC(8)277⟩ or by the oxidation of anthranilic acids (Scheme 185) ⟨75CJC1336⟩.

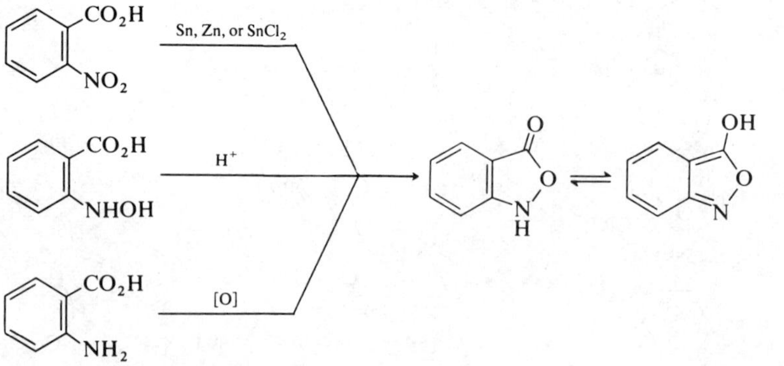

Scheme 185

The electrochemical reduction of 3-nitrophthalic acid at controlled potentials gave 2,1-benzisoxazole-3-carboxylic acid. Cyclization is presumed to proceed *via* an intermediate oxime ⟨67AHC(8)277⟩. Treating 5-iodoanthranilic acid with acetic anhydride gave 3-acetoxy-5-iodo-2,1-benzisoxazole (**596**) ⟨65JMC550⟩.

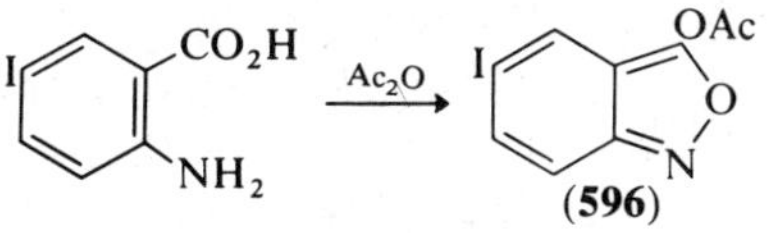

Controlled hydrogenation over Ni or the electrochemical reduction of o-nitrobenzonitriles produced 3-amino-2,1-benzisoxazoles either as the major product or by-product, depending in part on the reaction media and ratio of reactants ⟨72BSF2365, 65CB1562⟩. Reduction of o-nitrobenzonitrile gave either 3-amino-2,1-benzisoxazole or 2-aminobenzonitrile. The benzisoxazole is presumed to arise via an intermediate hydroxylamine. The electrochemical reduction of o-nitrobenzonitrile at acid pH produced the hydroxylamine as the primary product. Reduction at neutral pH gave the amino-2,1-benzisoxazole and the hydroxylamine ⟨72BSF2365⟩.

(viii) 4,5,6,7-Tetrahydro-2,1-benzisoxazoles

The 4,5,6,7-tetrahydro-2,1-benzisoxazole group is the most reported category. The action of hydroxylamine on β-diketones produced a mixture of 2,1- and 1,2-benzisoxazoles (Scheme 186). Ring opening gave β-ketoacids which, under subsequent alkali treatment, degraded to diacids ⟨67AHC(8)277⟩.

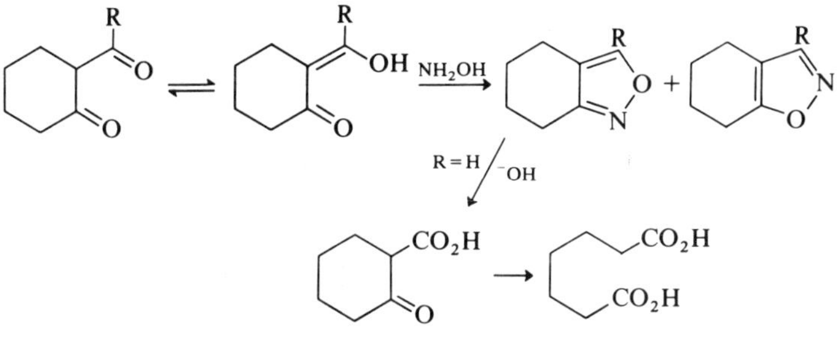

Scheme 186

The reaction of the dilithio salt (**597**) with methyl benzoate and subsequent acid hydrolysis yielded 3-phenyl-4,5,6,7-tetrahydro-2,1-benzisoxazole (**598**) ⟨76JPS1408⟩. The oxime (**599**) was converted into (**600**) by treatment with ethanolic HCl ⟨75JCS(P1)1959⟩.

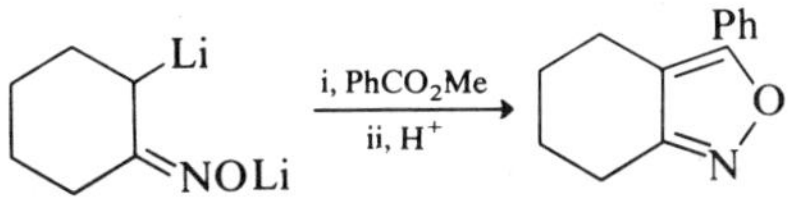

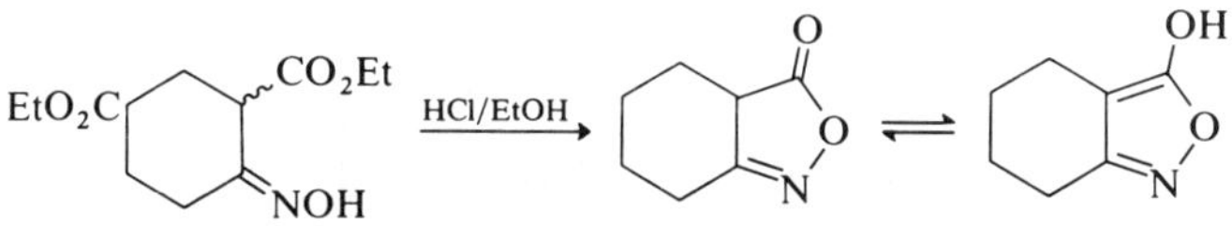

Cyclic α-cyanoketones, when treated with hydroxylamine, yielded 3-amino-4,5,6,7-tetrahydro-2,1-benzisoxazole. This compound could also be obtained by sodium or potassium cyanide interaction with chlorocyclohexanone oxime (Scheme 187) ⟨67AHC(8)277, 66BCJ1125⟩.

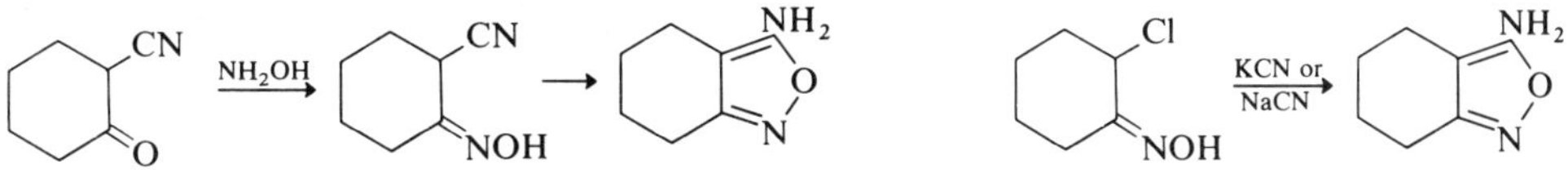

Scheme 187

Conversion of 2-cyano-5,5-dimethyl-3-ethoxy-2-cyclohexene (**601**) into (**602**) by hydroxylamine has been accomplished ⟨73MI41604⟩. Reduction of the homocyclic ring in the 3-substituted 2,1-benzisoxazole occurred with H_2/Pd–C to give the tetrahydro analogs (equation 65) ⟨74JHC395⟩.

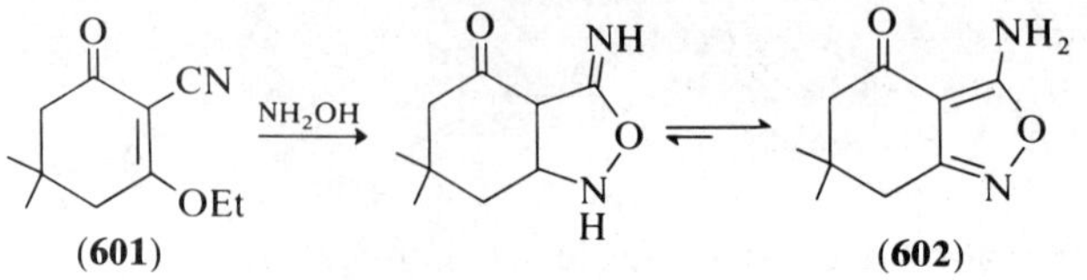

(601) **(602)**

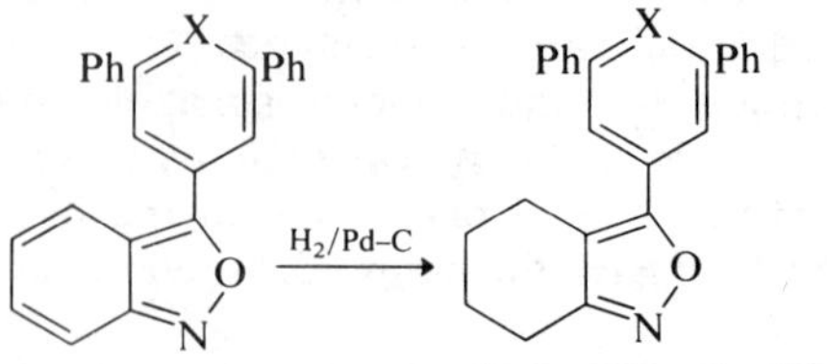

$X = N, C{-}NO_2, C{-}CN$ (65)

(ix) 1,3-Dihydro-2,1-benzisoxazoles

The 1,3-dihydro derivatives were prepared from benzisoxazolium cations by $NaBH_4$ reduction (Scheme 188). Subsequent heating produced an aminobenzaldehyde ⟨75USP3911132, 70JOC2440⟩. Hydrogenation of quinoid, 3,5-dihydro-2,1-benzisoxazole, produced the 1,3-dihydro analog (Scheme 188) ⟨73JCS(P1)4372⟩.

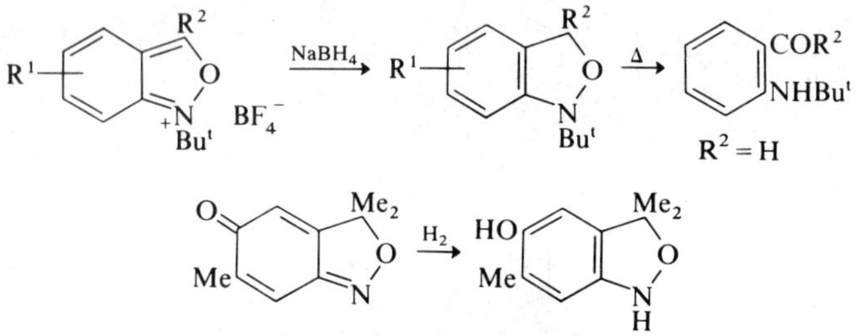

Scheme 188

The 1,3-dihydro-2,1-benzisoxazole-3-carboxylic acid was prepared by either base treatment of *N*-hydroxyisatin or the rearrangement of *N*-hydroxyisatin monooxime with dilute acid ⟨67AHC(8)277⟩. The addition of amines to 3-unsubstituted-2,1-benzisoxazolium salts gave 3-amino-1,3-dihydro-2,1-benzisoxazoles ⟨72CPB2209⟩.

(x) 4,7-Dihydro-2,1-benzisoxazoles

The 4,7-dihydro derivative (**606**) was prepared by the action of azide on (**603**), the action of hydroxylamine on (**604**), or the action of lead tetraacetate on (**605**) ⟨74S30⟩.

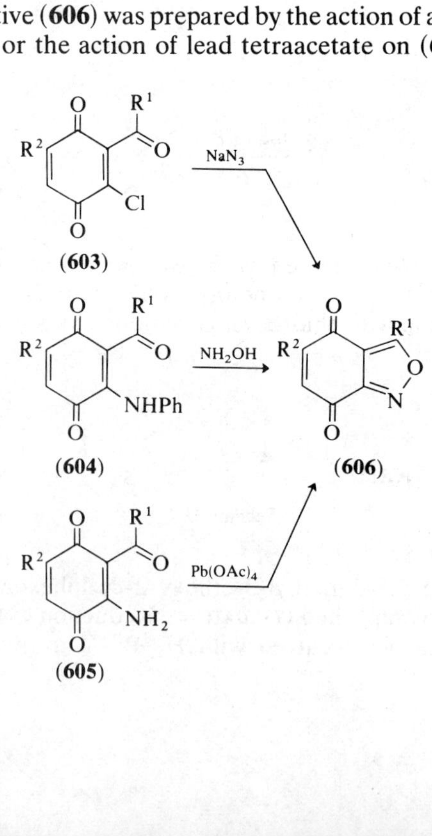

(603) **(604)** **(606)** **(605)**

(xi) 4,5-Dihydro-2,1-benzisoxazoles

A 4,5-dihydro derivative (**608**) was prepared by the intramolecular 1,3-dipolar cyclization of (**607**) ⟨75MI41607⟩.

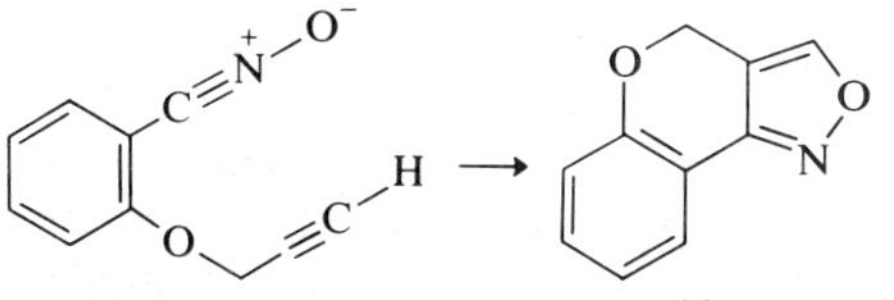

(xii) 1,3,3a,4,5,6,7,7a-Octahydro-2,1-benzisoxazoles

Several 1,3,3a,4,5,6,7,7a-octahydro derivatives have been reported, all by intramolecular nitrone cyclizations. The HgO catalyzed cyclization of (**609**) gave (**610**), which could be subsequently ring opened to (**611**) ⟨67AHC(8)277⟩. Via a similar 1,3-dipolar intramolecular cycloaddition, the nitrone (**612**) yielded (**613**).

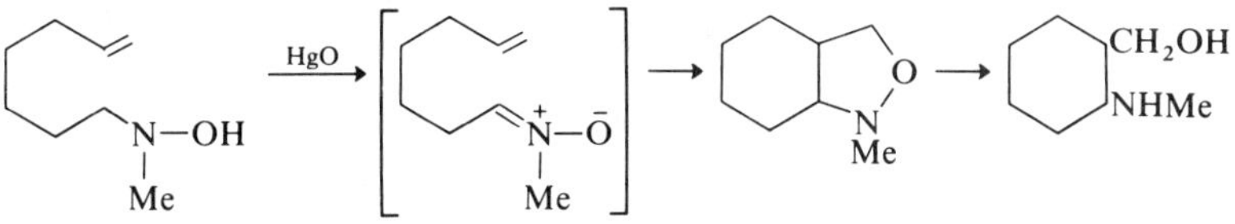

4.16.5 APPLICATIONS

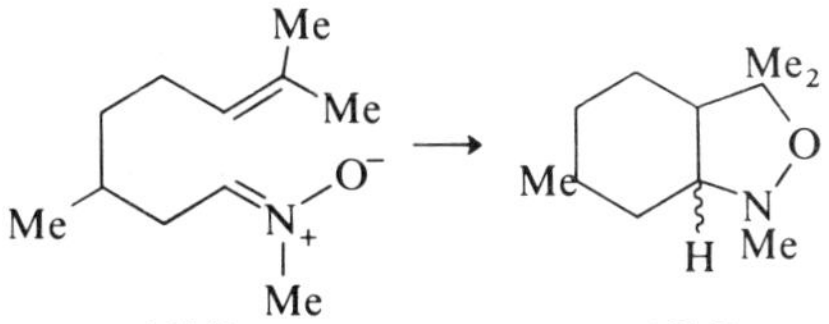

4.16.5.1 Pharmaceutical Applications

Isoxazoles represent a large group of compounds, a number of which display interesting medicinal or agricultural activity or have some other industrial utility.

Pharmacologically useful isoxazoles ⟨B-82MI41600⟩ include antibacterial sulfonamides (**614**), (**615**) and (**616**), semisynthetic penicillins (**617**), (**618**), (**619**) and (**620**), semisynthetic cephalosporin (**621**), anabolic steroid (**622**), the monoamine oxidase inhibitor (**623**) (used in psychotherapy), antiinflammatory agent (**624**) and antitumor agent (**625**).

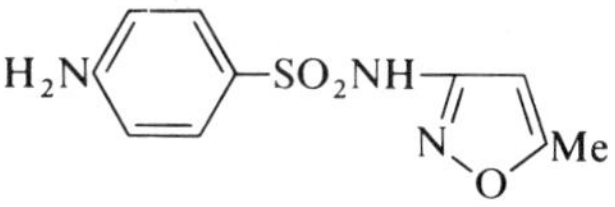

(**614**) Sulfamethoxazole

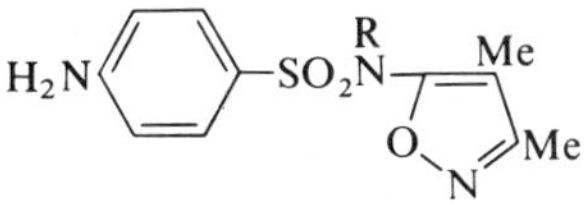

(**615**) R = H; Sulfisoxazole
(**616**) R = Ac; Acetylsulfisoxazole

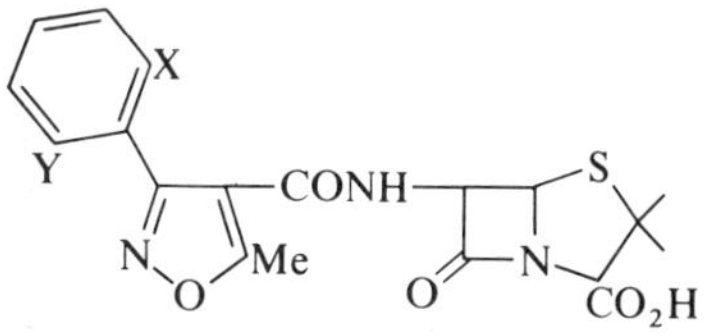

(**617**) X = Y = H; Oxacillin
(**618**) X = H, Y = Cl; Cloxacillin
(**619**) X = Y = Cl; Dicloxacillin
(**620**) X = F, Y = Cl; Floxacillin

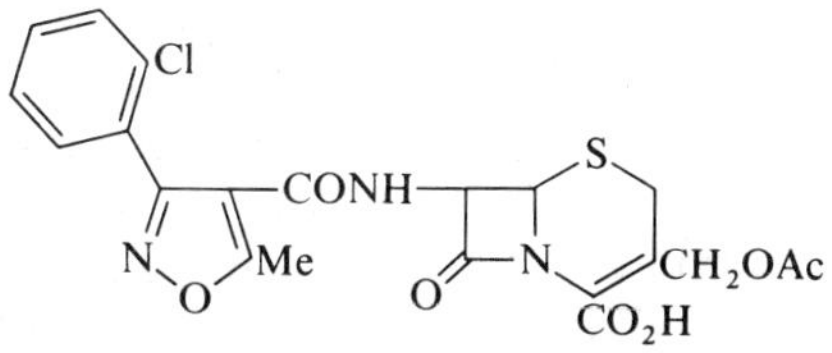

(**621**) Cefoxazole

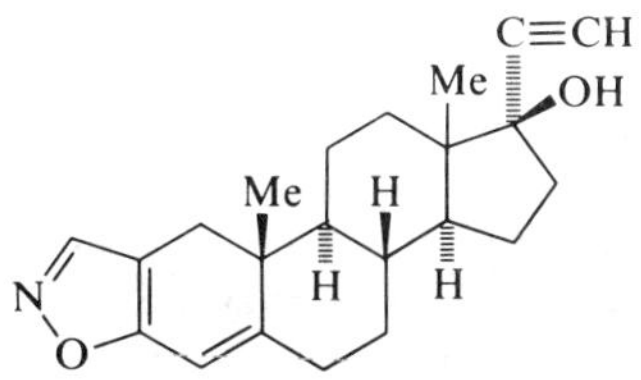

(**622**) Danazol

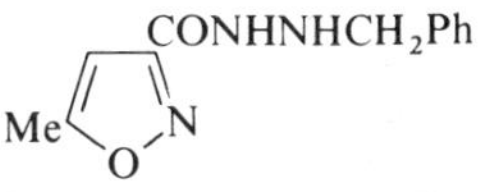

(**623**) Isocarboxazid

(624) Isoxicam

(625) Acivicin

Diverse derivatives of isoxazolecarboxylic acid (**626**) have been reported active as hypoglycemics and hypolipemics ⟨73MI41603, 79JAP(K)7966685, 79JAP(K)7973773⟩, antidiabetics ⟨79JAP(K)7914968, 79JAP(K)7909278⟩, antiinflammatory–analgesic agents ⟨80GEP2854438, 80GEP2854439⟩ and antifungals ⟨76GEP2525023⟩. Other isoxazoles with reported antiinflammatory–analgesic activity include isoxazolylureas (**627**) ⟨75JAP(K)7595272, 79JAP(K)7973774⟩, isoxazolylmethanols (**628**) ⟨76CPB1757, 79JAP(K)7934728, 75MI41606⟩, isoxazolyl ethers (**629**) ⟨75JAP(K)7557975⟩ and other structures containing an isoxazole ⟨79JAP(K)7973774, 79GEP2902025, 79MI41608, 76USP3957805⟩.

(**626**) X = OH, NHR'

(**627**)

(**628**)

(**629**)

Isoxazoles (**630a**) have reported diuretic activity ⟨75JAP(K)7595272⟩, while (**630b**) have reported hypolipemic activity ⟨77USP4032644⟩ and (**630c**) have CNS effects.

Nitroisoxazoles ⟨77MI41609⟩ and isoxazoles linked to a nitroimidazole ⟨77USP4010176⟩ or nitrofuran moiety ⟨72MIP41600⟩ exhibited typanosomal and trichomonacidal activity ⟨75MI41605⟩. Phenylisoxazoles with halogen or multiple halogen substituents are reported as anthelmintics ⟨77JMC934⟩ or are reported effective against endoparasitic nematodes ⟨75USP3879533⟩. Isoxazoles in levamisole molecules also exhibit anthelmintic activity with low toxicity ⟨78GEP2747122⟩.

(**630**) a; X = OH
b; X = CO_2H
c; X = $CONHR^3$

(**631**)

Isoxazole derivatives have reported antiallergic properties by acting as histamine blocking agents ⟨75AP75, 75BRP1395929⟩, by effecting adrenaline release or have a prophylactic antiallergic effect ⟨78USP4066769, 79BRP1548397⟩.

Fused isoxazoles (**631**) were prepared as GABA analogs ⟨75MI41604⟩ and some exhibited CNS depression effects ⟨74JAP(K)7480062⟩ or were effective as minor tranquilizers, muscle relaxants and/or sleep inducers ⟨76USP3966748, 79USP4163057⟩.

Hypolipemic effects were reported for piperidinoisoxazoles ⟨77USP4049813⟩ and hypoglycemic effects were reported for isoxazoles having a guanidino ⟨79JAP(K)7930171⟩, biguanidino ⟨79JAP(K)7912371⟩, s-triazino ⟨79JAP(K)7944686⟩ or acrylic acid ⟨79JAP(K)7944665⟩ residue.

A series of hydroxystyrylisoxazoles have reported antiarrhythmia–antihypertensive properties ⟨79GEP2818998⟩.

Fungicidal activity was reported for isoxazoles containing azo linkages ⟨73AJC2705⟩, sulfonanilamide residues or having a naphthalene ⟨74JAP(K)74134672⟩, iodonium ⟨75USP3896140⟩ or halomethylene ⟨77GEP2723688⟩ residue in the molecular structure.

Antibacterial activity was reported primarily for isoxazole derivatives as part of the acyl side-chain in penams ⟨77MIP41601, 79JAP(K)7984592, 75FES128⟩ and cephams ⟨75GEP2455884, 79FRP2422667, 79BRP2018247⟩ or in association with isoxazolyl sulfa drugs ⟨74GEP2342213, 76JIC181⟩.

Novel nitrothiazolylisoxazole (**632**) ⟨76LA13⟩, diisoxazole (**633**) ⟨77JIC875⟩ and fluorinated isoxazole (**634**) ⟨79MI41620⟩ all have cited antimicrobial activity.

Novel steroids which contain an isoxazole fused ring within the structure have biological activity which is primarily contraceptive ⟨74MI41604, 79MI41607, 79USP4160027⟩ and a variety of other indications ⟨75USP3869467, 75ZOB2090⟩.

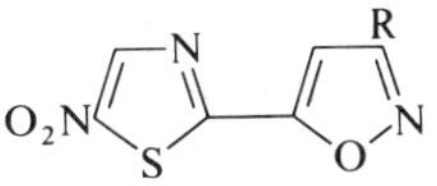

(632) 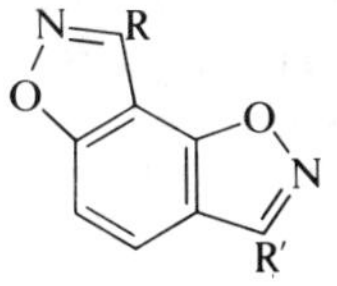   (633) 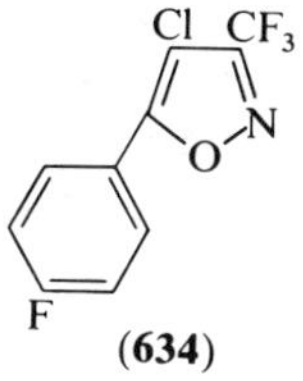   (634)

A number of isoxazoles have antitumor activity where the isoxazole is the base portion of the structure. Acivicin (**625**; AT-125) and (**635**) ⟨74MI41603, 75GEP2514984⟩ are related and (**625**) is currently undergoing clinical evaluation ⟨79MI41606, 79MI41605, 79MI41604, 80JBC(25)6734⟩. Other agents that have reported antitumor activity are glycosylated isoxazoles ⟨76MI41607⟩ and azoisoxazolyl mustards ⟨78GEP2703492, 80JAP(K)8045607⟩.

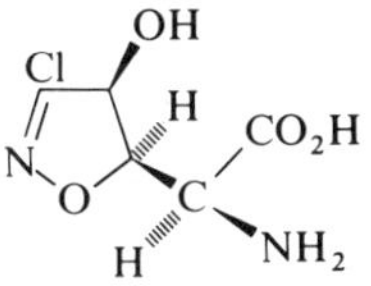

(635)

Weakly absorbed aminoisoxazole (**636**) has reported antidiarrheal effects in pigs without any side effects ⟨77MIP41600⟩. Sulfanilamide isoxazoles have been used to control atrophic rhinitis in newborn swine.

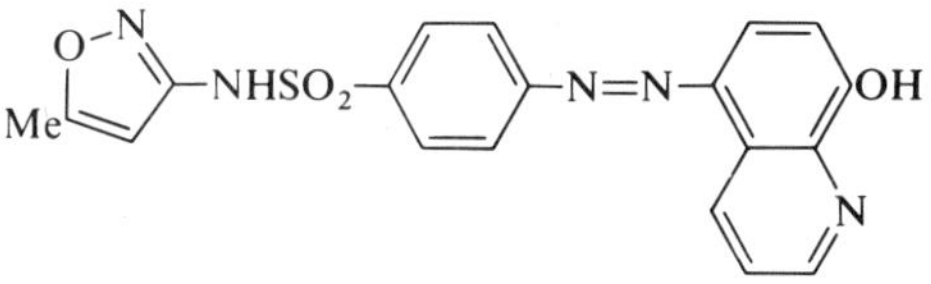

(636)

A significant number of references report therapeutic activity or potential therapeutic activity for a variety of 1,2-benzisoxazole derivatives. These include antibacterial ⟨65FES686, 79JPS1156, 72JAP7242659⟩, monoamine oxidase inhibitors ⟨64MI41600⟩, CNS active agents ⟨69GEP1915644, 74E405, 77GEP2640652, 80AF603, 80AF477, 78MI41611, 76CPB2918, 76CPB2673, 74JAP(K)74116054, 76CPB632⟩, antiparkinsominism ⟨73GEP2313256⟩, β-blockers ⟨77JAP(K)7731070, 78GEP2711382⟩, antiallergics ⟨77IJC(B)1051⟩, antifungal ⟨78JAP(K)7879862, 72JAP7242659⟩, antiinflammatory ⟨75GEP2450053, 79JMC1554⟩, diuretic and antihypertensive ⟨79EUP2666⟩, analogs of biologically active indoles ⟨75MI41603, 76JAP(K)76136666⟩ and antifertility ⟨78MI41605, 78MI41604, 74MI41602, 78MI41603, 77MI41608, 74MI41601⟩.

In comparison to the 1,2-benzisoxazole ring system, only a modest number of derivatives from the 2,1-benzisoxazole series have a patented usage. These include antiinflammatory ⟨72USP3642897⟩, antituberculotic ⟨65JAP6520705⟩, CNS agents ⟨76GEP2529292⟩, lipodemia ⟨75USP3911132⟩ and analogs of psilocene and muscomal ⟨74RTC139⟩.

4.16.5.2 Agricultural Applications

Herbicidal activity has been reported for isoxazolylureas ⟨79JAP(K)7946779, 79JAP(K)7944723⟩, pyridylisoxazolidines ⟨77GEP2639189⟩, thiocarbamates ⟨78FRP2392981⟩, nitrophenylisoxazoles ⟨79JAP(K)7952074⟩ and isoxazolecarboxylic acids ⟨79USP4139366, 80USP4187099⟩.

3-Hydroxy-5-methylisoxazole and its derivatives are active as soil fungicides and have growth promoting and plant regulatory effects predominantly on rice ⟨79GEP2900708, 77MI41601, 76JAP(K)76110032, 75MI41602⟩. 3-Isoxazolinones have stimulatory effects on plant growth ⟨79JAP(K)7952074⟩, as do glycosylisoxazoles ⟨74JAP(K)74107845⟩, quinoid isoxazoles ⟨79JAP(K)7901314⟩ and isoxazolyl diphenyl ethers ⟨79MIP41600⟩. Ureas based on N-isoxazol-3-one have reported plant growth regulation effects ⟨77USP4044018⟩ and selected carboxylic acid derivatives have similar activity ⟨79USP4144047, 77AJC2225⟩.

Haloisoxazolylureas have acardicidal and insecticidal properties ⟨76JAP(K)7659858⟩ and isoxazoles with a phosphorothiolate functionality are broad spectrum insecticides ⟨80USP4212861, 78USP4104375, 78JAP(K)7825566⟩.

1,2-Benzisoxazole derivatives have also found applications as pesticides or insecticides ⟨72GEP2049641⟩.

4.16.5.3 Other Applications

Polyisoxazolines and polyisoxazoles have applications as semiconductors ⟨73MI41602⟩, occur in the backbone of heat resistant polymers ⟨76JCS(P1)570, 74USP3793296⟩ and are present in azo dyes ⟨75BRP1414503⟩. Additional applications have been reported in photographic processes ⟨77GEP2653800, 77GEP2625026, 78USP4092327⟩ and as corrosion inhibitors in fuels and lubricants ⟨79USP4172079⟩. Some 2,1-benzisoxazoles have found applications as thermostable explosives ⟨80MIP41600⟩.

4.17

Isothiazoles and their Benzo Derivatives

D. L. PAIN, B. J. PEART and K. R. H. WOOLDRIDGE
May & Baker Ltd., Dagenham

4.17.1 INTRODUCTION	132
4.17.2 THEORETICAL METHODS	132
4.17.2.1 Charge Densities and Bond Orders	132
4.17.2.2 Energy Levels and Ionization Potentials	133
4.17.3 STRUCTURAL METHODS	134
4.17.3.1 X-Ray Diffraction	134
4.17.3.2 Microwave Spectroscopy	136
4.17.3.3 ^1H NMR Spectroscopy	136
4.17.3.4 ^{13}C NMR Spectroscopy	137
4.17.3.5 Nitrogen NMR Spectroscopy	139
4.17.3.5.1 ^{14}N NMR spectra	139
4.17.3.5.2 ^{15}N NMR spectra	139
4.17.3.6 UV Spectroscopy	140
4.17.3.7 IR Spectroscopy	141
4.17.3.7.1 Isothiazole ring vibrations	141
4.17.3.7.2 Exocyclic and substituent vibrations	142
4.17.3.8 Mass Spectrometry	142
4.17.4 THERMODYNAMIC ASPECTS	143
4.17.4.1 Melting and Boiling Points	143
4.17.4.2 Solubilities and pK_a Values	144
4.17.4.3 Stability and Aromaticity	144
4.17.5 TAUTOMERISM	145
4.17.5.1 Hydroxy–Ketone Tautomerism	145
4.17.5.2 Amine–Imine Tautomerism	146
4.17.5.3 Other Tautomers	146
4.17.6 FULLY CONJUGATED RINGS: REACTIVITY AT RING ATOMS	146
4.17.6.1 General Survey of Reactivity	146
4.17.6.2 Thermal and Photochemical Reactions involving No Other Species	147
4.17.6.3 Reactions with Electrophiles and Oxidants	147
4.17.6.4 Reactions with Nucleophiles and Reducing Agents	149
4.17.6.5 Miscellaneous Reactions at Ring Atoms	152
4.17.7 FULLY CONJUGATED RINGS: REACTIVITY OF SUBSTITUENTS	153
4.17.7.1 General Survey of Reactivity	153
4.17.7.2 Fused Benzene Rings	154
4.17.7.3 C-Linked Substituents	155
4.17.7.4 N-Linked Substituents	157
4.17.7.5 O-Linked Substituents	159
4.17.7.6 S-Linked Substituents	160
4.17.7.7 Halogen Substituents	161
4.17.7.8 Metal Substituents	164
4.17.7.9 Substituents at Ring Heteroatoms	164
4.17.8 SATURATED AND PARTIALLY SATURATED RINGS	164
4.17.8.1 Reactivity at Ring Atoms	165
4.17.8.2 Reactivity of Substituents	166
4.17.9 SYNTHESIS FROM NON-HETEROCYCLIC COMPOUNDS	166
4.17.9.1 Formation of One Bond	166
4.17.9.1.1 Formation of the nitrogen–sulfur bond	166
4.17.9.1.2 Formation of a bond adjacent to a heteroatom	167
4.17.9.1.3 Formation of a carbon–carbon bond	167

4.17.9.2	Formation of Two Bonds	168
4.17.9.2.1	From [4+1] atom fragments	168
4.17.9.2.2	From [3+2] atom fragments	169
4.17.9.3	Formation of Three Bonds	169
4.17.10	SYNTHESIS FROM OTHER HETEROCYCLIC COMPOUNDS	170
4.17.10.1	From Six-membered Heterocycles	170
4.17.10.2	From Other Five-membered Heterocycles	170
4.17.10.3	From Four-membered Heterocycles, Cephalosporins and Penicillins	172
4.17.10.4	From Seven-membered Heterocycles	172
4.17.10.5	From Fused Heterocyclic Systems	172
4.17.11	BEST PRACTICAL METHODS OF SYNTHESIS	173
4.17.12	APPLICATIONS	173
4.17.12.1	Important Compounds	173
4.17.12.2	Chemical Applications	174
4.17.12.3	Biological and Industrial Applications	175

4.17.1 INTRODUCTION

Isothiazole (**1**) was first prepared in 1956 and since then its chemical and physical properties have been extensively studied. The benzisothiazoles, on the other hand, have been known for much longer, the most widely known member of this family, saccharin (**2**), having first been prepared in 1879. There are two isomers of benzisothiazole, 1,2-benzisothiazole (**3**) and 2,1-benzisothiazole (**4**). The numbering system shown in (**3**) and (**4**) will be used throughout this chapter. Several reviews on isothiazoles ⟨65AHC(4)107, 72AHC(14)1, 79RCR289⟩, benzisothiazoles ⟨52HC(4)297, 72AHC(14)43⟩ and saccharin ⟨73AHC(15)233⟩ have been published, and recent advances have been regularly surveyed ⟨75SST(3)541, 77SST(4)339, 79SST(5)345, B-80MI41700⟩.

The possible structures for isothiazoles are discussed in Section 4.01.1, and attention in this chapter will be directed mainly towards the aromatic systems, as defined in Section 4.01.1. The saturated isothiazole 1,1-dioxides (**5**) are known as sultams, and bicyclic compounds of structure (**6**) are called isopenems. Isothiazoles readily coordinate to metals ⟨76MI41703, 78MI41701, 79MI41700, 80MI41701⟩. Coordination usually takes place through the nitrogen atom, but sulfur coordination can occur with 'soft' metals such as cadmium or mercury. Some specific coordination complexes are discussed in later sections.

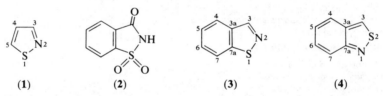

4.17.2 THEORETICAL METHODS

4.17.2.1 Charge Densities and Bond Orders

A number of MO calculations has been carried out, and these have had mixed success in predicting chemical reactivity or spectroscopic parameters such as NMR chemical shifts and coupling constants. Most early calculations did not take into account the contribution of the sulfur 3d-orbitals to the ground state, and this accounts for some of the discrepancies between calculations and experimental observations. Of the MO methods used, CNDO/2 and CNDO/S have been most successful; the INDO approximation cannot be used because of the presence of the sulfur atom.

The theoretical methods agree in placing most of the charge density on the isothiazole nitrogen atom, with sulfur bearing a small net positive charge ⟨65AHC(4)107, 74CJC833, 75CJC596, 79RCR289⟩. The calculated distribution of charge density on the carbon atoms of the heterocyclic ring depends on the approximation used, but in general C(4) is calculated to have the highest electron density and C(3) the least. When attempting to correlate such calculations with chemical reactivities it is important to establish whether ground or excited states are involved, or whether it is the electron distribution of an intermediate, rather than that of the original molecule, which determines the product of a chemical reaction. For example, isothiazoles are nitrated at the 4-position ⟨72BSF162, 75JCS(P2)1620⟩ as expected from MO calculations, indicating that this carbon atom has the highest electron density, but whilst isothiazole and its monomethyl derivatives are nitrated as the free bases even under conditions of high acidity, the 3,5-dimethyl derivative, being more basic, reacts in the form of the conjugate acid. Comparison of partial rate factors with those of benzene and other aromatic compounds ⟨74JCS(P2)399⟩ shows that the 4-position of isothiazole is some 10^4 times more reactive towards electrophiles than would be expected simply on the basis of the π-electron density at the carbon atoms and the electronegativity of the heteroatoms.

Calculated π-bond orders are summarized in Table 1. These calculations are supported by chemical evidence that the S—N bond is the one most easily cleaved. Attempts have been made to relate bond orders and electron densities to NMR coupling constants ⟨74CJC833, 77CJC619⟩ and ^{13}C NMR ⟨75CJC596, 75CJC1677⟩ and ^{15}N NMR ⟨78JOC4693⟩ chemical shifts, with limited success.

Table 1 Calculated π-Bond Orders of Isothiazoles

Isothiazole substituent	Bond					Calculation method	Ref.
	S—N	N—C	C(3)—C(4)	C(4)—C(5)	C—S		
None	0.502	0.705	0.634	0.707	0.594	HMO	65AHC(4)107
None	0.474	0.707	—	—	—	PPP	72T637
None	0.227	0.870	0.410	0.850	0.302	CNDO/2	74CJC833
3-Ph	0.34	0.82	0.43	0.86	0.37	PPP	73BSF1743

Isothiazole acts as a deactivating (electron-withdrawing) substituent on the benzene ring in 3-, 4- and 5-phenylisothiazoles under conditions of electrophilic attack ⟨71BSF4310⟩. The calculated π-bond orders of the isothiazole ring for 3-, 4- and 5-phenylisothiazoles in the ground state are very similar ⟨73BSF1743⟩ and there is little π-electron delocalization between the isothiazole and the benzene rings ⟨81T3627⟩. In the first electronically excited state the π-electrons are more extensively delocalized.

The involvement of the sulfur 3d-orbitals in the ground electronic state is perhaps most obvious in the case of 2,1-benzisothiazole, where X-ray studies, deductions of bond order from NMR coupling constants ⟨72JCS(P2)565⟩, and chemical reactivity suggest that resonance forms such as (7) contribute significantly. This behaviour is not seen with the analogous benzisoxazole, where the oxygen 3d-orbitals are no longer of sufficiently low energy to become involved in bonding.

(7)

4.17.2.2 Energy Levels and Ionization Potentials

The ionization potentials of isothiazole and a number of substituted derivatives have been measured by photoelectron spectroscopy and mass spectrometry. These measurements agree quite well with the results of MO calculations using CNDO/2 ⟨72MI41700⟩ or CNDO/S ⟨B-75MI41703, 78JST205⟩ approximations. The measured and calculated ionization potentials for the five occupied orbitals of isothiazole having highest energy are given in Table 2. For the first ionization the measured vertical and adiabatic potentials in the photoelectron spectrum are the same, implying that this MO is a localized π-orbital. (The vertical potential would be less than the adiabatic if the orbital were strongly involved in bonding as the ionization would result in changes in molecular geometry.) This observation is consistent with MO calculations which show that the HOMO (frontier orbital) is a localized π-orbital concentrated on the sulfur atom. The next highest orbital, also of π-type, is concentrated on nitrogen and C(5), followed by non-bonding σ-orbitals (the lone pairs of electrons) on nitrogen and sulfur, respectively. Electron donor substituents lower the first ionization potential, whilst electron attractors raise it, and the order of ionization potentials, *i.e.* isoxazole > isothiazole > furan > thiophene > cyclopentadiene, corresponds to the electronegativity of the heteroatom (O > N > S > C), a more electronegative heteroatom stabilizing the HOMO and thus increasing the ionization potential. Photoelectron spectra confirm that the HOMO's of 1,2-benzisothiazole ⟨78JST203⟩ and 2,1-benzisothiazole ⟨78JST33⟩ are also π-orbitals. The calculated dipole moments are 2.49 D for 1,2-benzisothiazole and 2.34 D for 2,1-benzisothiazole.

Table 2 Ionization Potentials and Orbital Energies of Isothiazole

Orbital	Ionization potential (eV)			Orbital energy (eV)
	PE^a	MS^b	$Calc.^b$ (CNDO/S)	$(CNDO/S)^a$
π (HOMO)	9.47	9.80	9.62	−10.73
π	10.11	—	10.52	−10.89
$n_N(\sigma)$	10.67	—	11.20	−11.98
$n_S(\sigma)$	12.3	—	—	−13.55
π	13.2	—	—	−15.57

^{a} ⟨78JST205⟩. ^{b} ⟨B-75MI41703⟩.

The polarographic half-wave reduction potential of 4-nitroisothiazole is −0.45 V (pH 7, *vs.* saturated calomel electrode). This potential is related to the electron affinity of the molecule and it provides a measure of the energy of the LUMO. Pulse radiolysis and ESR studies have been carried out on the radical anions arising from one-electron reduction of 4-nitroisothiazole and other nitro-heterocycles ⟨76MI41704⟩.

4.17.3 STRUCTURAL METHODS

4.17.3.1 X-Ray Diffraction

The dimensions of the heterocyclic ring of some typical crystalline derivatives of isothiazole, 1,2- and 2,1-benzisothiazole are listed in Table 3. Bond lengths should be compared with those of single and double bonds given at the foot of the table. All the molecules are more or less planar except (12), where the bulky 3-substituent causes twisting. The S—N bonds are significantly shorter than the single bond values, confirming the MO predictions of π-electron delocalization. In addition, the short N—C bonds in the 1,2-benzisothiazol-3-ones (10) and C—S and S—N bonds in saccharin (7) confirm that π-delocalization occurs in these molecules. In the case of saccharin (which is planar apart from the S=O bonds) this delocalization must involve the empty sulfur 3*d*-orbitals. The benzene ring of saccharin is significantly distorted so that the angles at positions 4 and 7 are reduced to 117.7° and 116.0°, respectively, suggesting an *o*-quinonoid contribution to the structure. The very short S—N bond of (12) can be accounted for if resonance forms such as (13) are taken into account. In the case of 5-chloro-2,1-benzisothiazole the bond lengths and planarity of the molecule show the effect of resonance of the type exemplified in (7) on the ground state of the molecule.

(13)

X-Ray diffraction studies have been carried out on other 1,2-benzisothiazoles ⟨72JCS(P2)2125, 74CSC535, 76G769⟩ and complexes of substituted 1,2-benzisothiazoles with

Structure	Bond lengths (Å)[a]					Heterocyclic ring angles (°)					Ref.
	S–N	N–C	C(3)–C(4)	C(4)–C(5)	C–S	CSN	SNC	NCC	CCC	CCS	
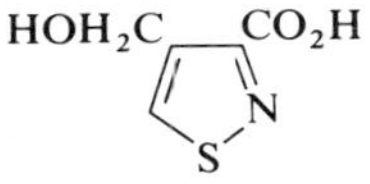 HOH₂C–[CO₂H isothiazole] (8a)	1.665	—	—	—	1.704	—	—	—	—	—	78JOC79
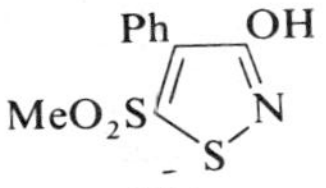 MeO₂S–[Ph, OH isothiazole] (8b)	1.661	1.316	1.397	1.380	1.715	93.3	109.2	118.8	107.7	111.0	77JCS(P2)1332

Structure	S–N	N–C	C(3)–C(3a)	C(3a)–C(7a)	C–S	CSN	SNC	NCC	CCC	CCS	Ref.
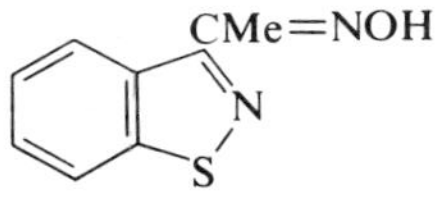 CMe=NOH benzisothiazole (9)	1.664	1.324	1.454	1.402	1.725	94.7	111.9	114.0	110.8	108.6	73AX(B)43
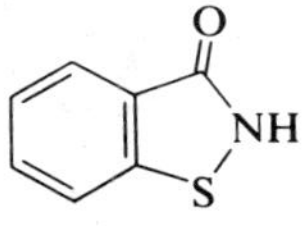 O=, NH benzisothiazolone (10a)	1.706	1.36	1.47	1.38	1.73	90.5	115.5	109.9	111.6	112.4	70G629
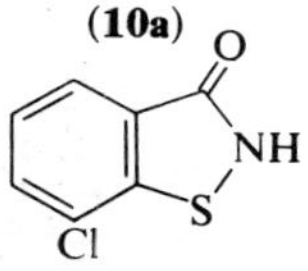 O=, NH, Cl benzisothiazolone (10b)	1.68	1.34	1.48	1.39	1.72	88.9	118.4	108.5	110.2	114.0	69AX(B)2349
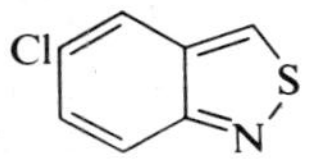 Cl benzisothiazole (11)	1.636	1.359	1.378	1.419	1.664	97.7	107.1	116.3	109.5	109.4	72JCS(P2)565
Saccharin	1.663	1.369	1.474	1.369	1.758	92.2	115.1	109.6	112.9	110.1	69AX(B)2257
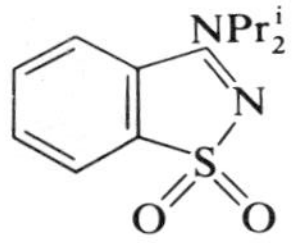 NPr$_2^i$ (12)	1.616	1.320	1.520	1.388	1.753	96.5	111.8	114.7	108.5	108.5	69CB1468

[a] Standard lengths (Å): C–C 1.537, C=C 1.335, S–N 1.735, C–N 1.413, C=N 1.29, C–S 1.812, C=S 1.554.

copper and cobalt ⟨71AX(B)1775, 72AX(B)1207⟩. The copper complex of saccharin ⟨81MI41700⟩ is octahedral, with the two saccharin molecules coordinated *trans* to each other, and water molecules taking up the remaining four coordination sites.

It has been confirmed by X-ray crystallography that dehydromethionine (**14**) exists as a zwitterion, with a pyramidal (sp^3) nitrogen ⟨76JA965⟩. The geometry around the nitrogen atom suggests that p_π–d_π bonding does not occur, though the S—N bond is shorter (1.679 Å) than expected for a single bond. In the solid state, the molecule exists in an envelope conformation with C(3) out of the plane of the other ring atoms. The crystal structure of an isopenem derivative has recently been obtained ⟨80TL619⟩.

(**14**)

4.17.3.2 Microwave Spectroscopy

The microwave spectrum of isothiazole shows that the molecule is planar, and enables rotational constants and NQR hyperfine coupling constants to be determined ⟨67MI41700⟩. The total dipole moment was estimated to be 2.4 ± 0.2 D, which agrees with dielectric measurements. Asymmetry parameters and ^{14}N NQR coupling constants show small differences between the solid and gaseous states ⟨79ZN(A)220⟩, and the principal dipole moment axis approximately bisects the S—N and C(4)—C(5) bonds.

4.17.3.3 ^{1}H NMR Spectroscopy

Representative chemical shifts from the large amount of available data on isothiazoles are included in Table 4. The chemical shifts of the ring hydrogens depend on electron density, ring currents and substituent anisotropies, and substituent effects can usually be predicted, at least qualitatively, by comparison with other aromatic systems. The resonance of H(5) is usually at a lower field than that of H(3) but in some cases this order is reversed. As is discussed later (Section 4.17.3.4) the chemical shift of H(5) is more sensitive to substitution in the 4-position than is that of H(3), and it is also worth noting that the resonance of H(5) is shifted downfield (typically 0.5 p.p.m.) when DMSO is used as solvent, a reflection of the ability of this hydrogen atom to interact with proton acceptors. This matter is discussed again in Section 4.17.3.7.

Table 4 ^{1}H NMR Chemical Shifts of Isothiazoles and Isothiazole Cations

Substituent				Isothiazole chemical shifts δ (p.p.m.)				Isothiazole cation chemical shifts δ (p.p.m.)				
3	4	5	Solvent	3	4	5	Ref.	Solvent	3	4	5	Ref.
H	H	H	CCl$_4$	8.54	7.26	8.72	65CB1111	H$_2$SO$_4$	9.1	7.9	9.6	65CB1111
Me	H	H	CCl$_4$	—	7.00	8.54	65CB1111	D$_2$SO$_4$	—	7.56	9.33	74JCS(P2)399
H	Me	H	CCl$_4$	8.24	—	8.21	65CB1111	H$_2$SO$_4$	8.9	—	9.2	65CB1111
H	H	Me	CCl$_4$	8.24	6.92	—	65CB1111	H$_2$SO$_4$	8.9	7.6	—	65CB1111
H	H	Cl	CCl$_4$	8.37	7.35	—	75CJC1642	—	—	—	—	—
H	H	Br	CCl$_4$	8.32	7.4	—	75CJC1642	—	—	—	—	—
H	Br	H	CCl$_4$	8.33	—	8.57	75CJC596	—	—	—	—	—
H	NH$_2$	H	CDCl$_3$	7.43	—	8.06	76MI41701	TFA	8.92	—	9.22	76MI41701
H	CN	H	CCl$_4$	8.69	—	9.17	75CJC596	—	—	—	—	—
H	NO$_2$	H	CCl$_4$	9.00	—	9.42	75CJC596	H$_2$SO$_4$	9.6	—	10.3	65CB1111
OH	H	H	CDCl$_3$	—	6.48	8.28	71JHC571	—	—	—	—	—

The signal from H(3) is broadened because ^{14}N–H(3) spin coupling is not completely washed out by quadrupolar relaxation of the nitrogen atom ⟨68MI41701⟩. This broadening may be useful in assigning signals in the spectrum and it is usually reduced in solvents of high viscosity or at lower temperatures, or on protonation of the nitrogen. (Protonation of the nitrogen reduces the coupling constant, $J(^{14}N, H\text{-}3)$, from about 10 Hz to 3 Hz.)

Typical coupling constants for isothiazoles are given in Table 5. The electronegative nitrogen atom reduces $^3J_{3,4}$ and $^4J_{3,5}$ from the values of 3.50 Hz and 0.27 Hz, respectively, in thiophene. The ^4J values correlate quite well with π-bond orders calculated by MO methods ⟨74CJC833⟩.

Table 5 Typical ^{1}H Coupling Constants of Isothiazoles ⟨74CJC833⟩

Substituent						Coupling constant (Hz)			
3	4	5	$^3J_{3,4}$	$^3J_{4,5}$	$^4J_{3,5}$	$^4J_{CH,Me}$	$^5J_{CH,Me}$	$^5J_{Me,Me}$	$^6J_{Me,Me}$
H	H	H	1.66	4.66	0.15	—	—	—	—
Me	H	H	—	4.55	—	−0.27	0.51	—	—
H	Me	H	—	—	0.33	−0.44 (3)	—	—	—
						−0.98 (5)			
H	H	Me	1.63	—	—	−1.00	0.55	—	—
Me	H	Me	—	—	—	−0.25 (3)	—	—	−0.23
						−1.00 (5)			
H	Me	Me	—	—	—	−0.45	0.59	0.60	—

Exocyclic conjugation causes a small upfield shift of the ring hydrogen resonances, as can be seen in Table 6. The increase in π-bond fixation also results in an increase in the $^3J_{4,5}$ coupling constant to about 6.0 Hz. The use of coupling constants for the investigation of tautomerism is discussed in Section 4.17.5.

Table 6 ^{1}H NMR Chemical Shifts and Coupling Constants of Isothiazoles (15) with Exocyclic Conjugation

(15)

X	Solvent	δ (p.p.m.) H(4)	H(5)	$J_{4,5}$ (Hz)	Ref.
O	CDCl$_3$	6.05	7.98	6.0	71JHC571
S	CDCl$_3$	6.90	8.25	6.0	80CPB487

The ^{1}H NMR spectrum of 1,2-benzisothiazole has a singlet from H(3) at 8.73 p.p.m. (CCl$_4$ solution), the other hydrogens giving a multiplet at 7.12–8.00 p.p.m. In 2,1-benzisothiazole the signal from H(3) moves downfield to 9.06 p.p.m. (neat liquid), again with a multiplet at 7.09–7.78 p.p.m. ⟨72AHC(14)43⟩. The chemical shift of the signal from H(3) is sometimes very large, e.g. 10.07 p.p.m. (CDCl$_3$ solution) in 4-nitro-2,1-benzisothiazole ⟨69JCS(C)2189⟩. Interproton coupling constants in 2,1-benzisothiazoles correlate with bond orders and are consistent with some degree of o-quinonoid structure. Thus $^3J_{4,5}$ (9 Hz) is substantially larger than $^3J_{5,6}$ (7.5 Hz) ⟨72JCS(P2)565⟩. The existence of a large (0.9 Hz) long range coupling between hydrogens 3 and 7 in 2,1-benzisothiazoles can be useful in structure elucidation.

Under basic conditions the relative rate of deuterium exchange of H(5) and H(3) in isothiazole has been shown by NMR data to be 400:1, with no exchange occurring at H(4). Similarly, for methylisothiazoles, hydrogens of the methyl groups exchange at relative rates of $100:1:10^{-4}$ for the 5-, 4- and 3-methyl derivatives, respectively ⟨66JA4265, 69JHC199⟩. The high reactivity of the 5-position may not be due simply to the electron distribution in the ground state, but to σ–$3d$(S) bonding which could stabilize the anions formed under the conditions used. H(5) also exchanges most rapidly in quaternary isothiazolium cations ⟨66JA4263⟩.

4.17.3.4 ^{13}C NMR Spectroscopy

The ^{13}C NMR chemical shifts of a number of substituted isothiazoles and 2,1-benzisothiazoles are given in Tables 7 and 8. More extensive data may be obtained by

Table 7 ^{13}C NMR Chemical Shifts of Isothiazoles

Substituent			Chemical shifts δ (p.p.m.)			Solvent	Ref.
3	4	5	3	4	5		
H	H	H	157.0	123.4	147.8	CDCl$_3$	75CJC836
Me	H	H	166.7	123.9	148.1	CDCl$_3$	75CJC836
H	Me	H	158.8	134.2	143.6	CDCl$_3$	75CJC836
H	H	Me	157.6	123.3	163.0	CDCl$_3$	75CJC836
H	Ph	H	156.0	139.9	142.6	CDCl$_3$	75CJC836
H	H	Ph	158.1	119.9	167.3	CDCl$_3$	75CJC836
CO$_2$H	H	H	160.4	125.7	151.6	DMSO	75CJC596
H	CO$_2$H	H	158.4	132.7	154.9	DMSO	75CJC596
H	H	CO$_2$H	158.6	127.4	158.3	DMSO	75CJC596
H	H	CHO	159.6	129.4	165.3	Acetone	75CJC1677
H	Br	H	157.9	107.0	147.0	DMSO	75CJC596
H	CN	H	159.2	110.1	158.8	Acetone	75CJC1677
H	NO$_2$	H	152.8	146.8	151.9	DMSO	75CJC596
Ph	H	Ac	168.1	122.5	166.2	CCl$_4$	79JOC510
Ph	Ac	H	168.1	135.5	154.3	CCl$_4$	79JOC510
Ph	OH	Ph	159.2	144.7	144.0	CDCl$_3$	79BSF(2)26
Bu^t	OH	Bu^t	169.6	144.7	155.8	CDCl$_3$	79BSF(2)26

Table 8 ^{13}C NMR Chemical Shifts of 2,1-Benzisothiazoles in CDCl$_3$ ⟨75CJC836⟩

Substituent	Chemical shifts δ (p.p.m.)						
	3	3a	4	5	6	7	7a
None	144.5	134.5	122.1	124.2	128.6	121.6	161.5
4-Me	142.7	135.7	131.3	122.9	128.8	119.0	161.7
7-Me	144.4	134.6	119.7	124.4	127.2	130.9	161.8
7-NO$_2$	148.7	137.6	129.6	127.2	122.8	140.6	151.7

referring to the original papers. As with other aromatic systems, substituents such as OH, NO$_2$ or alkyl cause a downfield shift of the resonance of the ring carbon bearing the substituent, whilst CN or Br cause an upfield shift. The fused benzene substituent in 1,2- and 2,1-benzisothiazoles can be considered to a first approximation as an alkyl substituent on the isothiazole ring, as ring currents have a much smaller effect in ^{13}C than in ^{1}H NMR spectra. Normal substituent increments can be used to calculate the effects of substitution in the benzene ring of 2,1-benzisothiazoles. Of the isothiazole carbons, C(5) is most affected by ring substituents, and the large effect of substituents at the 4-position on the resonance of C(5) has been ascribed to resonance forms of the type (**16b**) or (**17b**) ⟨75CJC596⟩. Within a structurally related group of molecules such as the isothiazoles, it can be assumed that paramagnetic terms alone contribute to ^{13}C NMR shielding differences. Thus, an increase in electron density, as in (**16b**) should result in an increase in the radius of the 2p-electron cloud at C(5) and move the ^{13}C NMR signal to higher field, while the converse applies to (**17b**).

Some results have been published on 2-alkylisothiazole-3- and -5-thiones ⟨75CJC836, 80JCS(P1)2693⟩. As expected, the resonance of the carbon attached to the exocyclic sulfur atom is shifted downfield, to around 185 p.p.m. in the case of the 3-thiones and to over 190 p.p.m. for the 5-thione. It is possible that ^{13}C NMR chemical shifts could be used to investigate tautomerism in related compounds. Saccharin has carbon resonances at 161.0 (3), 127.9 (3a), 125.1 (4), 134.7 and 135.5 (5 and 6), 121.2 (7) and 139.5 (7a) p.p.m. in DMSO solution ⟨82UP41700⟩.

The ^{13}C NMR chemical shifts of non-aromatic isothiazoles can be predicted with reasonable accuracy using standard substituent increments. A particular usefulness of ^{13}C NMR is its ability to distinguish between very similar compounds, and for this reason ^{13}C NMR finds application in pharmaceutical and other analyses. As an example ^{13}C NMR allows ready distinction of the diastereoisomers of dehydromethionine (14) and the possibility of detection of one diastereoisomer in the presence of the other ⟨79JOC2632⟩.

One bond carbon–hydrogen couplings normally fall into the following ranges: $^1J_{C(3),H(3)}$ = 180–196 Hz; $^1J_{C(4),H(4)}$ = 170–176 Hz; $^1J_{C(5),H(5)}$ = 184–197 Hz. Electron-withdrawing groups (e.g. NO_2, CHO) increase the coupling constant of the adjacent C—H bond, whilst electron donors (e.g. alkyl) cause a small reduction. MO calculations based on the assumption that the coupling constant depends on the C(2s)–H(1s) σ-bond order predict that $^1J_{C(3),H(3)}$ > $^1J_{C(5),H(5)}$, whereas in fact the converse is the case ⟨75CJC1677, 77CJC619⟩. However, for coupling along several bonds, $^2J_{C(4),H(3)}$ (ca. 13 Hz) > $^2J_{C(4),H(5)}$ (ca. 5 Hz) and $^3J_{C(3),H(5)}$ (ca. 10 Hz) > $^3J_{C(5),H(3)}$ (ca. 5 Hz), consistent with predictions made on the basis of the electronegativity of the ring heteroatoms ⟨77CJC619⟩.

4.17.3.5 Nitrogen NMR Spectroscopy

4.17.3.5.1 ^{14}N NMR spectra

Despite the broadness of ^{14}N NMR signals because of quadrupolar relaxation, these spectra have found applications in at least two areas of isothiazole chemistry; indeed, measurement of the half-width of the ^{14}N signal can provide structural information in addition to that based on chemical shifts. Analysis of ^{14}N chemical shifts of complexes of isothiazole with metal carbonyls has confirmed that coordination takes place through nitrogen ⟨72JOM(44)325⟩ and ^{14}N NMR offers a means of distinguishing between 1,2- and 2,1-benzisothiazoles ⟨78OMR(11)385⟩ whereas ^{13}C and ^{1}H chemical shifts are sometimes equivocal in this respect. Relevant data are summarized in Table 9. Nitrogen NMR data for a wide range of heterocycles, including isothiazole, are compared in ⟨78MI41702⟩.

Table 9 ^{14}N NMR Data of some Isothiazoles

Structure	Chemical shifts δ (p.p.m.)[a]	Reference signal	Half-width (Hz)	Solvent	Ref.
Isothiazole	−80	$MeNO_2$	108	Neat	72T637
Isothiazole + $Cr(CO)_5$	−145	$NaNO_3$	—	Acetone	72JOM(44)325
Isothiazole + $W(CO)_5$	−167	$NaNO_3$	—	Acetone	72JOM(44)325
1,2-Benzisothiazole	−76	$MeNO_2$	450	Ether	78OMR(11)385
2,1-Benzisothiazole	−121	$MeNO_2$	640	Neat	78OMR(11)385

[a] A minus sign indicates a signal to high field.

4.17.3.5.2 ^{15}N NMR spectra

With improvements in instrument sensitivity and the use of techniques such as enhancement by polarization transfer (INEPT), it can be expected that natural abundance ^{15}N NMR spectra will become increasingly important in heterocyclic chemistry. The chemical shifts given in Table 10 illustrate the large dispersion available in ^{15}N NMR, without the line broadening associated with ^{14}N NMR spectra.

Coupling constants give information about the bonding and hybridization at the nitrogen atom, $^1J_{N,H}$ values being typically about 75, 90 and 135 Hz for pyramidal, trigonal and linear nitrogen atoms, reflecting the increasing s-character of the N—H bond. Longer range ^{15}N–H couplings are much smaller; typical values for isothiazole are: $^2J_{N,H(3)}$ = 14.20 Hz, $^3J_{N,H(4)}$ = 1.86 Hz, $^3J_{N,H(5)}$ = 1.32 Hz. The sign of the coupling constant can be obtained by selective population transfer techniques ⟨81JMR(42)337⟩ and depends on conformation and the nature of any heteroatoms through which the coupling occurs.

Table 10 ^{15}N NMR Chemical Shifts of some Isothiazoles

Structure	Chemical shifts δ (p.p.m.)[a]		Solvent	Ref.
	NH_3	$MeNO_2$		
Isothiazole	298.4	−81.8	Neat	81JMR(42)337
Saccharin	148	−232	DMSO	78JOC4693
(18)	120	−260	DMSO	78JOC4693
(19)	149	−231	CDCl$_3$	78JST227

[a] Downfield shifts positive. Chemical shifts from NH_3 or $MeNO_2$ calculated using approximate values $MeNO_2$ 380 p.p.m., NO_3^- 376 p.p.m., NH_3 0 p.p.m. ⟨B-79MI41701⟩.

4.17.3.6 UV Spectroscopy

Isothiazole has an absorption maximum in ethanol solution at 244 nm, with a molar absorptivity of 5200. This absorption occurs at a longer wavelength than with pyrazole or isoxazole, the displacement being due to the presence of the sulfur atom. A series of approximate additive wavelength shifts has been drawn up in Table 11 and this should enable prediction of UV maxima of isothiazoles with reasonable accuracy, even for multiply substituted compounds. The longest wavelength band results from a $\pi \rightarrow \pi^*$ electronic transition. As with other five-membered heterocyclic compounds no long wavelength $n_N \rightarrow \pi^*$ transition has been detected in the UV spectrum, and this observation is consistent with the MO calculations and photoelectron spectroscopy results discussed in Section 4.17.2.2; the weak $n_N \rightarrow \pi^*$ absorption is likely to be lost under more intense $\pi \rightarrow \pi^*$ bands.

Table 11 Bathochromic Shift caused by Substituents on Isothiazole (Isothiazole λ_{max} = 244 nm, ethanol solvent)

Substituent	Position of substituent and shift (nm)[a]					
	3-	Ref.	4-	Ref.	5-	Ref.
Me	4	63JCS2032	7	64JCS446	−1	64JCS446
CO$_2$H(R)	12	64JCS3114	6	63JCS2032	23	63JCS2032
CONH$_2$	12	64JCS3114	7	64JCS446	—	—
CN	12	64JCS3114	11	63JCS2032	27	63JCS2032
NH$_2$	—	—	39	63JCS2032	14	63JCS2032
NHAc	—	—	26	63JCS2032	11	63JCS2032
NO$_2$	—	—	27	63JCS2032	—	—
Cl	6	75MI41704	10	64JCS446	−1	75CJC1642
Br	0	75MI41704	12	63JCS2032	−1	63JCS2032
I	—	—	18	64JCS446	1	75CJC1642

[a] A minus sign indicates a shift to shorter wavelength.

There is very little published information on the UV spectra of 1,2-benzisothiazoles, though more data are available on the 2,1-isomers. The spectra are complex with as many as six maxima above 200 nm. Representative wavelengths of maxima are collected in Table 12. In all cases the most intense bands ($\varepsilon > 15\,000$) are those at short wavelengths, but all the bands indicated in the table have molar absorptivities greater than 4000, except those of 3-amino-2,1-benzisothiazole. Saccharin absorbs weakly at 350 nm and 277 nm, with intense bands below 230 nm (ethanol solvent) ⟨82UP41700⟩. It exists as the anion except in acid solutions. The UV spectra of cations formed from 3-amino-2,1-benzisothiazole are discussed in ⟨69CB1961⟩. Further applications of UV spectroscopy in studying tautomeric

Table 12 UV Absorption Maxima of Benzisothiazoles (solvent ethanol or methanol)

1,2-Benzisothiazoles			λ_{max} (nm)				Ref.
Unsubstituted	205	222	252	297	302		82UP41700
			261		308		
3-CN, 5-Cl		224	269	276	320		67JCS(C)2364
3-CONH$_2$, 5-Cl		222			320		67JCS(C)2364
2,1-Benzisothiazoles							
Unsubstituted	203	221		288	298	315	69JOC2985
3-CN	206	231		306	318	342	71AJC2405
3-CO$_2$H	207	230		306	315	338	71AJC2405
3-NHAc		226		290	302	351	71AJC2405
3-Br	208	222	281	292	303	331	71AJC2405
3-NH$_2$		230	286	298		373	71AJC2405

equilibria are given in Section 4.17.5, and it is also worth noting that a number of highly conjugated 3-substituted benzisothiazoles are used as dyes (see Section 4.17.12.3).

4.17.3.7 IR Spectroscopy

4.17.3.7.1 Isothiazole ring vibrations

In Table 13 an attempt has been made to correlate data on monosubstituted isothiazoles from the literature and the authors' laboratories. The table is not intended to list every absorption band, and in some cases bands will be weak or absent depending on the substituent, but it should serve as a guide to the regions where absorption might be expected, and the vibrations which do not involve C—H bonds should stay relatively fixed whatever the type of substitution as long as coupling with vibrations of exocyclic bonds does not occur. In many cases the most intense bands in the IR spectra of isothiazoles are to be found near 840 and 750 cm^{-1}. Detailed assignments of the IR and Raman spectra of isothiazoles are given in ⟨70CR(C)(270)1677, 71CJC2254, 73CR(B)(276)31, 75CJC1642, 75MI41704, 81MI41701⟩, and the normal modes of vibration have been calculated using valence force field models ⟨75SA(A)1115, 76CJC3850, 77CJC2302⟩.

Table 13 IR Absorptions of Monosubstituted Isothiazoles[a]

3-Substituted (cm^{-1})	4-Substituted (cm^{-1})	5-Substituted (cm^{-1})	Assignment[b]
3101±9	3109±9	—	ν(C—H(5))
3094±9	—	3094±9	ν(C—H(4))
—	3055±15	3055±15	ν(C—H(3))
1490±20	1500±30	1500±20	ν(C=C)[c]
1375±10	1370±20	1380±20	ν(C=N)[c]
1320±20	1310±25	1280±15	ν(C—C)
—	1210±20	1220±10	δ(C—H(3))
1065±10	—	1045±10	δ(C—H(4))
—	880±15	910±15	γ(C—H(3))
840±10	850±10	820±20	ν(C—S)
810±20	770±20	750±10	ν(S—N)
755±15	(2 bands)	670±20	γ(C—H)

[a] Only the most intense or reproducible bands are listed.
[b] The principal component only of composite bands is shown. ν = stretch; δ = deformation, in plane; γ = deformation, out of plane.
[c] Some substituents (*e.g.* OH, NH$_2$, Me) cause a shift to a higher frequency than indicated.

The relative basicity and acidity of isothiazole and its methyl derivatives have been compared by IR spectroscopy ⟨77MI41702⟩. The isothiazoles, dissolved in inert solvents (*e.g.* CCl$_4$, CS$_2$) containing traces of butanol (a proton donor), interact with the butanol OH group causing small displacements of the O—H stretching frequency. It was concluded that basicity increases in the series isothiazole < 5-methylisothiazole < 4-methylisothiazole < 3-methylisothiazole, in accord with the measured pK_a values. Acidity of the isothiazole ring hydrogen atoms, measured by changes in the isothiazole C—H stretching frequency in proton-accepting solvents (DMSO, HMPT), decreases in the order H(5) > H(4) > H(3). This result is consistent with NMR experiments showing that deuterium exchange occurs most rapidly at H(5) under basic conditions.

4.17.3.7.2 Exocyclic and substituent vibrations

Isothiazole ring substituents usually have IR bands in the regions expected, but some 5-carboxylic acids absorb at frequencies as high as 1750 cm^{-1} ⟨65AHC(4)107⟩. The characteristic out of plane C—H deformations of benzisothiazoles give bands in the region expected for 1,2-disubstituted benzene rings; for example this absorption occurs at 760 cm^{-1} in saccharin. IR studies on OH, NH$_2$ and SH substituents have been used to distinguish between tautomers, as described in Section 4.17.5.

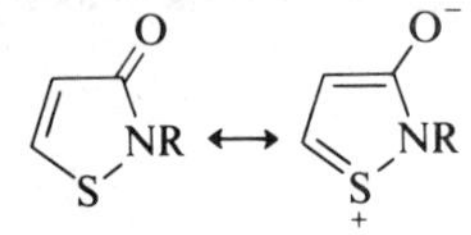

(20a)   (20b)

The exocyclic carbonyl group of isothiazol-3-ones absorbs in the region 1610–1660 cm^{-1} ⟨71JHC591⟩. 2-Methylisothiazol-3-one itself has the C=O and C=C bands at 1660 and 1629 cm^{-1}, respectively, in CCl$_4$ solution ⟨64TL1477⟩. The low carbonyl frequency is due in part to contributions from the resonance form (20b). The carbonyl frequency increases in sulfoxides (1660–1730 cm^{-1}) and 1,1-dioxides (1690–1740 cm^{-1}) where such forms are not favourable. Sulfoxides (1060–1190 cm^{-1}) and sulfones (1330–1360 and 1150–1190 cm^{-1}) absorb in the regions expected (e.g. saccharin, 1353 and 1162 cm^{-1}), but resonance forms related to (13) cause a reduction of the frequency of the asymmetric SO$_2$ vibration to near 1280 cm^{-1} ⟨70CB3166⟩. A similar situation arises in 3-amino-1,2-benzisothiazole 1-oxides.

4.17.3.8 Mass Spectrometry

A characteristic feature of the electron impact mass spectra of isothiazoles is that the molecular ion is intense (often the base peak) and even in the case of molecules with side chains which fragment readily, the most intense fragment ions normally include the isothiazole ring. The fragmentation of many substituted isothiazoles can be related to that of isothiazole itself (Scheme 1) ⟨72AHC(14)1⟩. Metastable and IKE spectra suggest that isothiazole and thiazole isomerize to a common intermediate before fragmentation to the ion at m/e 58 (Scheme 1), though linked scan and appearance potential measurements suggest that this is not necessarily the case for other daughter ions which arise by fragmentation directly from the molecular ions ⟨78OMS119⟩. Intermediates of identical structure are also believed to be involved in the fragmentations of benzisothiazole and benzothiazole ⟨74OMS(9)1161⟩.

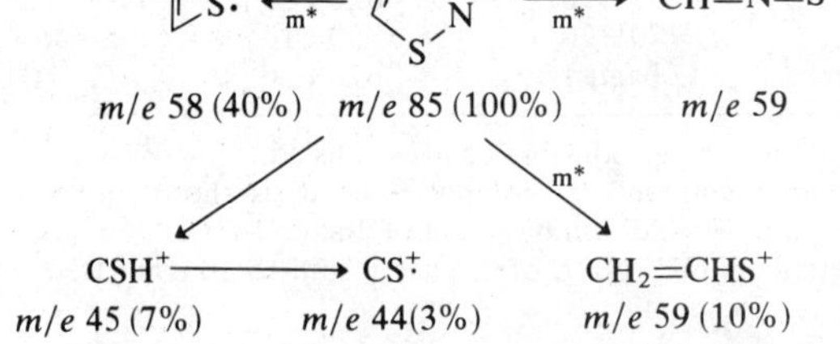

m/e 58 (40%) m/e 85 (100%) m/e 59

CSH$^+$ ⟶ CS$^+$ CH$_2$=CHS$^+$
m/e 45 (7%) m/e 44 (3%) m/e 59 (10%)

m^* indicates a metastable ion

Scheme 1

Ring expansion can occur during the fragmentation of methylisothiazoles (Scheme 2) and alkylthio-substituted isothiazoles ⟨73OMS(7)463, 72CR(C)1871⟩. An interesting feature of the mass spectrum of (**21**) is the simultaneous loss of two atoms of sulfur. This may result from rearrangement of the molecular ion to give the dithiolium radical ion (**22**) which can then lose S_2 ⟨73OMS(7)463⟩.

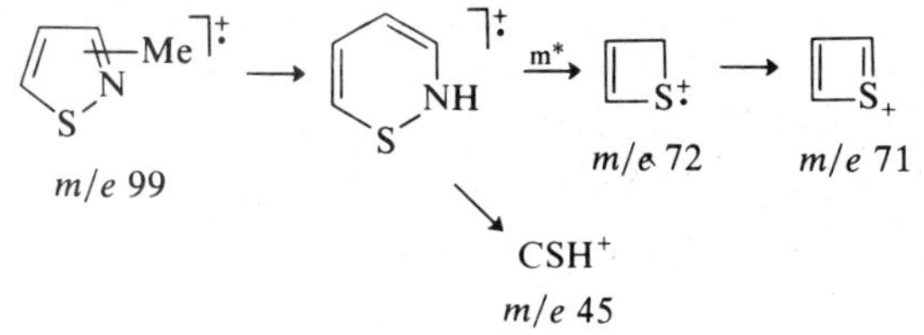

Scheme 2

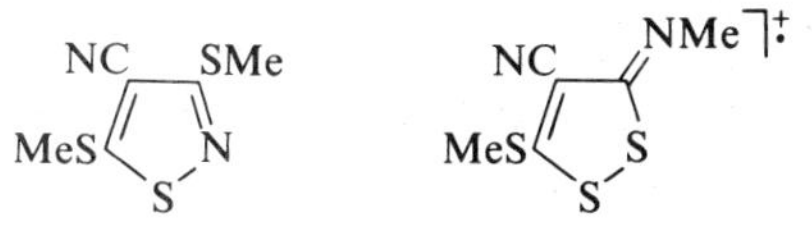

(**21**) (**22**)

The fragmentation of substituted 1,2-benzisothiazoles can usually be related to that of the parent molecule (Scheme 3) ⟨72OMS(6)1321, 73OMS(7)327⟩. Sulfone derivatives readily undergo isomerization with ring expansion resulting in the replacement of C—S bonds by C—O, as frequently occurs with sulfonamides. Saccharin provides an example of this behaviour, the molecular ion rearranging to give (**23**) which then fragments to (**24**) ⟨69OMS(2)1117⟩. The presence of the sulfone group reduces the intensity of the molecular ion, so that the base peak of saccharin, at m/e 76, arises from the $C_6H_4^+$ ion.

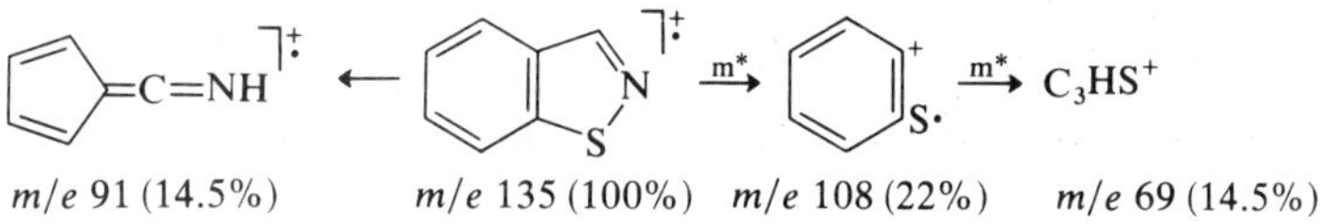

m/e 91 (14.5%) m/e 135 (100%) m/e 108 (22%) m/e 69 (14.5%)

Scheme 3

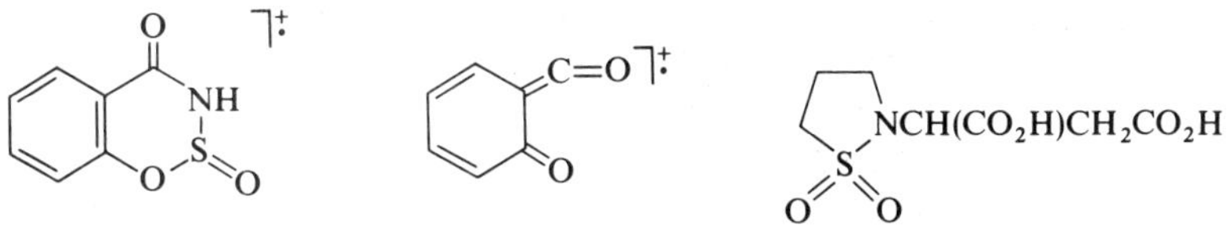

(**23**)   (**24**)   (**25**)

Alternative ('soft') ionization techniques are not usually required for aromatic isothiazoles because of the stability of the molecular ions under electron impact. This is not the case for the fully saturated ring systems, which fragment readily. The sultam (**25**) has no significant molecular ion under electron impact conditions, but using field desorption techniques the $[M+1]^+$ ion is the base peak ⟨73T3861⟩ and enables the molecular weight to be confirmed.

4.17.4 THERMODYNAMIC ASPECTS

4.17.4.1 Melting and Boiling Points

Isothiazoles and benzisothiazoles are usually liquids or low melting solids, polar substituents increasing the melting points because of the possibility of hydrogen bonding and other interactions in the crystalline state. The melting points of 3-, 4- and 5-monosubstituted isothiazoles are usually fairly similar, and data for a range of 4-substituted isothiazoles and 3-substituted benzisothiazoles are given in Table 14. The presence of sulfoxide or sulfone groups increases the melting point of simple isothiazoles; thus the melting point of 3-hydroxyisothiazole, its 1-oxide and 1,1-dioxide are 74, 75 and 118 °C, respectively. Saccharin (228 °C) and 1,2-benzisothiazol-3-one (158 °C) have high melting points because these molecules exist in the crystalline state as dimer units with hydrogen bonding between the N—H and C=O groups ⟨69AX(B)2257, 70G629⟩. The N-methyl analogues, where such intermolecular bonding is not possible, have lower melting points of 131 and 54 °C, respectively ⟨72AHC(14)43⟩.

Table 14 Melting and Boiling Points of some Isothiazoles and Benzisothiazoles[a]

Substituent[a]	Isothiazole m.p. (b.p.) (°C)	Ref.	1,2-Benzisothiazole[c] m.p. (°C)	2,1-Benzisothiazole[c] m.p. (b.p.) (°C)
H	(113)	59JCS3061	37	(242)
Me	(146)	66T2119	—	55
Cl	(143)	70CB112	40	−10
Br	31	64JCS446	59	64
CO_2H	161	59JCS3061	—	212
CN	94	68JMC70	—	104
NH_2	45	59JCS3061	115	178
NO_2	86	64JCS446	—	126
OH[b]	74	65JOC2660	158	—
OMe[b]	(147)	77JHC725	—	—

[a] Data are given for substituents in the 4-position of isothiazole and the 3-position in the benzisothiazoles, except where indicated. Boiling points of liquids (at 760 mmHg) are given in brackets.
[b] 3-Isomer.
[c] All data for benzisothiazoles taken from ⟨72AHC(14)43⟩.

4.17.4.2 Solubilities and pK_a Values

The solubility of isothiazole in water is about 3.5% ⟨65AHC(4)107⟩. Isothiazole is miscible with most organic solvents. 1,2-Benzisothiazole is slightly soluble in water and is volatile in steam. It is soluble in strong acids and most organic solvents. 2,1-Benzisothiazole is soluble in concentrated acids but is reprecipitated on dilution, and is miscible with most organic solvents ⟨72AHC(14)43⟩. The solubilities of substituted isothiazoles and benzisothiazoles follow the general guidelines given in Section 4.01.4.1. The solubility in aqueous acids or bases depends also on the pK_a value, and some relevant pK_a values of substituted isothiazoles are collected in Table 15. 2,1-Benzisothiazole is a very weak base with a pK_a of −0.05 ⟨76AJC1745⟩, whilst 3-amino-2,1-benzisothiazole is, as expected, a stronger base with a pK_a of 3.8 in 50% methanol ⟨65JMC515⟩. Saccharin is a strong acid with a pK_a of 1.30 ⟨73AHC(15)233⟩. 3-Hydroxyisothiazoles are acidic, the OH group giving some phenolic reactions such as the colour reaction with $FeCl_3$.

Table 15 pK_a Values of Isothiazoles

Substituent	pK_a as Base	Ref.	pK_a as Acid	Ref.
H	−0.51	72AHC(14)1	—	—
3-Me	0.48	74JCS(P2)399	—	—
5-Me	0.02	74JCS(P2)399	—	—
3-NH_2	2.49	72AHC(14)1	—	—
4-NH_2	3.58	72AHC(14)1	—	—
5-NH_2	2.70	72AHC(14)1	—	—
3-OH, 5-Me	—	—	8.15	65AHC(4)107
3-OH, 5-Ph	—	—	7.48	65AHC(4)107
3-CO_2H	—	—	3.17	63UP41700
4-CO_2H	—	—	3.45	63UP41700
5-CO_2H	—	—	2.60	63UP41700

Phase equilibria of the isothiazole–water system have been investigated by differential thermal analysis ⟨76BSF1043⟩, and it has been established that a stable crystalline clathrate (isothiazole·34H_2O) forms below 0 °C.

4.17.4.3 Stability and Aromaticity

Isothiazole behaves as a typical stable aromatic molecule. Thermolysis of substituted isothiazoles at 590 °C leads to the formation of thioketenes ⟨80MI41700⟩ and phenylisothiazoles undergo photoisomerism (Section 4.17.6.2) ⟨73BSF1743, 81T3627⟩. 1,2-Benzisothiazole boils at 220 °C without appreciable decomposition, and the 2,1-isomer displays none of the instability associated with *o*-quinonoid systems, for reasons which have already been discussed. Thus, 2,1-benzisothiazole is more stable than 2,1-benzisoxazole ⟨74AJC1221⟩.

The stability of isothiazole derives from the fact that it has an aromatic delocalized π-electron system. The ^{1}H NMR chemical shifts, which depend, *inter alia*, on ring currents, and the high stability of the molecular ions in mass spectrometry, are typical of aromatic compounds, and X-ray measurements confirm the partial double bond character of all the bonds of the ring.

The degree of bond fixation in isothiazole and its 3- and 5-methyl derivatives has been investigated by measurement of ring-hydrogen exchange rates under acidic conditions ⟨74JCS(P2)399⟩. The results suggest that the degree of bond fixation in isothiazole is small, and the aromaticity of the series of heterocycles increases in the order isoxazole < pyrazole < isothiazole. Another measure of aromaticity is the non-local diamagnetic susceptibility perpendicular to the plane of the molecule ⟨80ZN(A)712⟩. This parameter can be measured from the rotational Zeeman effect and offers some advantages over NMR measurements as the non-local contributions are directly related to the ring current, and molecules are examined in the gas phase, so that intermolecular interactions are reduced. Typical values for aromatic systems are -35 to -40×10^{-5} J T^{-2}, and the value for isothiazole (-38.1×10^{-5} J T^{-2}) falls in this range. The non-local susceptibilities for isoxazole and cyclopentadiene (-26.9 and -18.8×10^{-5} J T^{-2}, respectively) are indicative of much smaller ring currents and hence less aromaticity.

Saccharin does not comply with the normal $(4n+2)\pi$-electron rule for aromaticity, but in view of the fact that it has been shown earlier to have a degree of π-electron delocalization through the sulfur atom, and for convenience of classification of its chemical reactions, it will be considered to be aromatic in the subsequent sections dealing with its chemistry.

4.17.5 TAUTOMERISM

Annular tautomerism does not occur in isothiazoles or benzisothiazoles. Substituent tautomers can sometimes be distinguished by chemical methods, but it is important that reaction mechanisms and the relative rates of interconversion of tautomeric starting materials or isomeric reaction products are carefully investigated. Physical methods only will be considered in this section, and references to original publications can be found in a comprehensive review ⟨76AHC(S1)1⟩.

4.17.5.1 Hydroxy–Ketone Tautomerism

3-Hydroxyisothiazole exists in the hydroxy form in non-polar solvents such as cyclohexane or ether. Comparison of UV spectra with those of compounds of fixed structure shows that significant amounts of the keto form are present in aqueous solution ⟨70T2497⟩. The 3-keto and 3-hydroxy forms can also be distinguished by ^{1}H NMR spectroscopy, 3-alkoxyisothiazoles having $^3J_{4,5} = 4.6$ Hz, whilst for 3-keto compounds of fixed structure this coupling constant is about 6 Hz. Coupling constants do not vary significantly from one solvent to another and thus they provide a convenient way of establishing the tautomeric form present. 3-Hydroxyisothiazole has $^3J_{4,5} = 4.6$ Hz in CDCl$_3$ solution, confirming the UV results noted above for non-polar solvents. In polar solvents such as DMSO or methanol $^3J_{4,5}$ increases to about 5.0 Hz, indicating the presence of more keto tautomer in the equilibrium. Chemical shift correlations are generally less clear-cut unless comparisons can be made within a closely related series. 3-Hydroxyisothiazoles substituted in the 4- or 5-positions also exist in the hydroxy form, as does 3-hydroxy-5-phenylisothiazole, even in polar solvents. 5-Hydroxyisothiazoles have been little studied, but UV and IR results suggest that the 4-bromo and 4-nitro compounds are predominantly in the 5-keto form (see Section 4.17.7.5). X-Ray and spectroscopic examination ⟨75AJC129⟩ show that 1,2-benzisothiazol-3-one and its derivatives, 2,1-benzisothiazol-3-one and saccharin all exist in the keto form. Potentiometric methods have recently been used ⟨81G71⟩ to examine the tautomeric and ionization equilibria of 1,2-benzisothiazol-3-one.

4.17.5.2 Amine–Imine Tautomerism

Comparison of UV data for 3-aminoisothiazoles with those of reference compounds confirms that they exist in the 3-amino form. A more recent investigation of 4-aminoisothiazole ⟨76MI41701⟩ using deuterium exchange experiments of the type described in Section 4.01.5.2, and analysis of the symmetric and antisymmetric NH_2 stretching frequencies in its IR spectrum, show that this compound also exists in the 4-amino form.

3-Amino-1,2-benzisothiazole and its derivatives exist in the 3-amino form. In this case comparisons of UV spectra with compounds of known structure, and the observation of coupling between the 3-NH group and alkyl hydrogens in the NMR spectra of 3-ethylamino- and 3-methylamino-1,2-benzisothiazole provide clear evidence of structure ⟨69CB1961⟩. The sulfoxides also exist in the 3-amino tautomeric form ⟨78KGS1632⟩. The NMR coupling patterns and UV spectra of 3-alkylamino-1,2-benzisothiazolium salts in 0.1N HCl indicate that the conjugate acids are protonated on the ring nitrogen. The protonation of 3-amino-1,2-benzisothiazoles has recently been studied by potentiometric measurements ⟨81G71⟩.

4.17.5.3 Other Tautomers

Thiol–thione tautomers have not been extensively studied, but UV and IR evidence show that 5-phenylisothiazole-3-thiol exists in the SH form. Ring–chain tautomerism of 2,3-dihydro derivatives of 1,2-benzisothiazole can occur (**26a** ⇌ **26b**) and the position of equilibrium depends very much on the solvent, physical state and nature of the substituents ⟨69JOC919, 81KGS1209⟩.

(**26a**)   (**26b**) X = O, NH or N-acyl, R = alkyl or aryl

4.17.6 FULLY CONJUGATED RINGS: REACTIVITY AT RING ATOMS

4.17.6.1 General Survey of Reactivity

The isothiazole ring is very stable to moderate heating, but quaternized compounds are dequaternized by this means (see Section 4.02.3.12). Photolysis of isothiazole itself causes loss of hydrogen cyanide, forming thiirene, but substituted compounds are converted into mixtures of thiazoles and isothiazoles isomeric with the starting compound (Section 4.1.1.2). 1,2-Benzisothiazoles suffer straightforward ring cleavage, but their 1,1-dioxides, 2,1-benzisothiazoles and derivatives of saccharin give products containing no sulfur.

Electrophilic substitution occurs at the 4-position, but benzo compounds are substituted only in the benzene ring (Section 4.02.1.4). Isothiazoles need more vigorous nitration conditions than does benzene, but halogenation is easier. All substituted compounds which can potentially exist as tautomers involving the ring nitrogen atom can be alkylated at nitrogen, and benzo fusion has little effect on the rate of reaction (Section 4.02.1.3). The ring sulfur atom can be oxidized to sulfoxides and sulfones, provided an O-linked substituent is present in the molecule.

Isothiazoles and 2,1-benzisothiazoles are stable to most nucleophilic reagents, but quaternized compounds are dequaternized and/or suffer cleavage of the N—S bond. 1,2-Benzisothiazoles also suffer N—S bond cleavage, following attack at sulfur, but 1,2-benzisothiazole 1,1-dioxides are cleaved at the C—N bond. Saccharin derivatives are attacked at the carbonyl function. In cases where N—S bond cleavage occurs, recyclization can sometimes occur, often producing thiophene compounds.

5-Lithioisothiazoles are readily prepared by the action of butyllithium, and the isothiazole ring is desulfurized by Raney nickel (see Section 4.02.1.8). Few cycloaddition reactions are known.

4.17.6.2 Thermal and Photochemical Reactions involving No Other Species

Isothiazole compounds are very stable to moderate heat, but very strong heating can cause breakdown to thioketenes ⟨80MI41700⟩. Flash vacuum pyrolysis of some 3-aryl-1,2-benzisothiazole 1,1-dioxides (27) at 850 °C causes loss of sulfur dioxide ⟨79SST(5)345⟩, the benzoxazole (28) being the main product, but with 3-methyl compounds, 2-hydroxybenzonitrile is the only product. Isothiazolylium and 1,2-benzisothiazolylium salts have been dequaternized by strong heating (see Section 4.02.3.12) ⟨72AHC(14)1, 72AHC(14)43⟩.

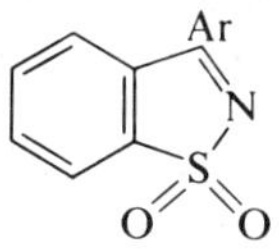

(27) 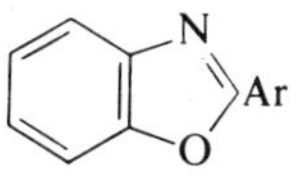   (28)

A considerable amount of work has been reported concerning photochemical reactions. The products include other isothiazoles, thiazoles and cleavage products (see Section 4.02.1.2). Isothiazole itself loses hydrogen cyanide to give thiirene ⟨79SST(5)345⟩. Methylisothiazoles give mixtures of isomeric isothiazoles, together with methylthiazoles ⟨72T3141⟩, and arylisothiazoles behave similarly ⟨72AHC(14)1, 71BSF1103, 73BSF1743, 77SST(4)339⟩. Several possible mechanisms have been proposed ⟨72T3141, 77SST(4)339⟩, all of which are very complex. A simplified version is shown in Scheme 4. 2-Substituted isothiazol-3-ones give high yields of 3-substituted thiazol-2-ones ⟨79CC786⟩. 1,2-Benzisothiazole gives a low yield of 2,2′-dicyanodiphenyl disulfide ⟨77SST(4)339⟩. 2,1-Benzisothiazole, 1-methyl-2,1-benzisothiazole 2,2-dioxide and 2-alkylsaccharins produce decomposition products containing no sulfur ⟨79CL9, 79HCA391⟩.

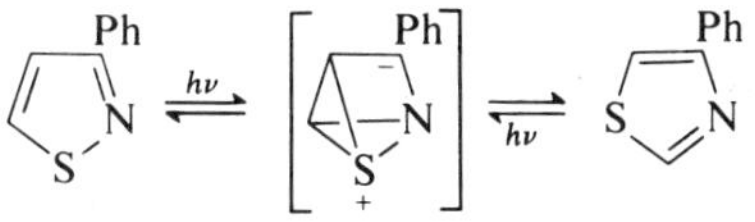

Scheme 4

4.17.6.3 Reactions with Electrophiles and Oxidants

Simple isothiazoles are halogenated at the 4-position ⟨65AHC(4)107, 72AHC(14)1, 69FRP1555414, 79SST(5)345⟩, but yields are often poor, possibly because of the formation of perhalogeno compounds. 2-Substituted isothiazol-3-ones (29) are halogenated much more easily (see Section 4.01.1.4), giving firstly the 4-halogeno compounds (30), then the 4,5-dihalogeno compounds (31), and with excess of chlorine at 60 °C addition of chlorine across the 4,5-double bond occurs to give compounds (32) ⟨77JHC627⟩. When the 4-position is occupied, as in compound (33), bromine can cause bromination at the 5-position in low

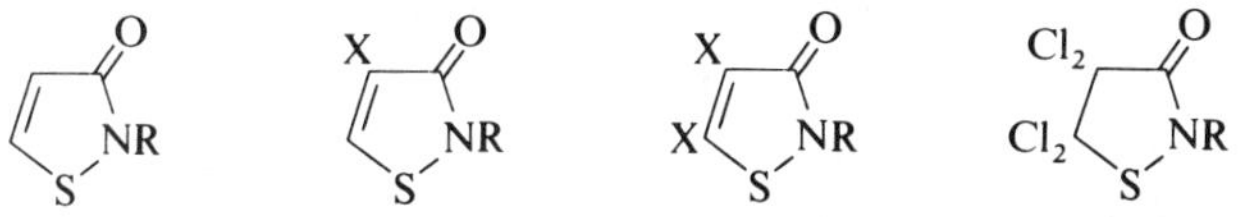

(29)   (30)   (31)   (32)

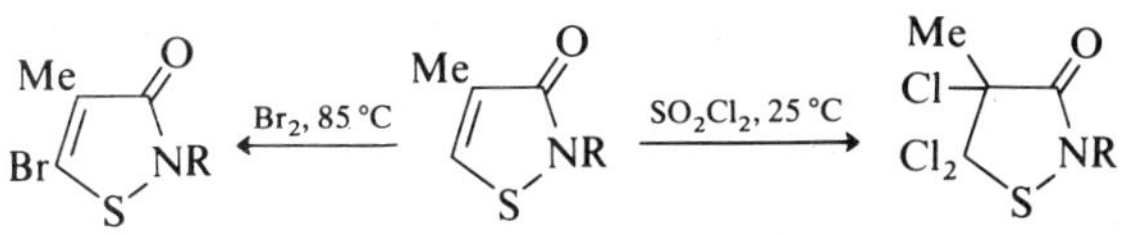

(34)   (33)   (35)

Scheme 5

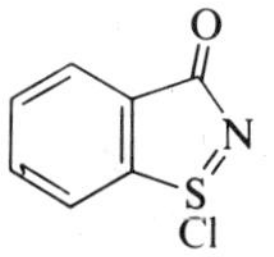

(36)

yield to give compound (**34**), but reaction with sulfuryl chloride at 25 °C gives compound (**35**; Scheme 5) ⟨76JHC1321⟩. Chlorination of 1,2-benzisothiazol-3-one with chlorine gives the S-chloro compound (**36**) ⟨80MI41700⟩. N-Halogenation occurs when the corresponding S-oxide is treated with t-butyl hypochlorite ⟨78ZOR862⟩, and when sodium saccharin reacts with chlorine, bromine or bromine chloride ⟨73AHC(15)233, 76S736⟩.

Nitration occurs exclusively at the 4-position in good yield ⟨65AHC(4)107, 72AHC(14)1⟩, except when easily nitratable substituents are present, when the latter are nitrated instead (see Section 4.02.1.4). 5-Halogeno ⟨80JHC385⟩ and 3-alkoxy ⟨76USP3957808⟩ groups do not interfere. Kinetic studies have shown that isothiazole and the monomethyl compounds are much less reactive than 3,5-dimethylisothiazole ⟨72BSF162⟩. This is because isothiazole and the monomethyl compounds are nitrated solely as free bases, whereas 3,5-dimethylisothiazole is nitrated as the conjugate acid ⟨71JCS(B)2365, 75JCS(P2)1620⟩. 3,5-Dimethylisothiazole is nitrated more easily than the isosteric 1,3,5-trimethylpyrazole ⟨78JCS(P2)613⟩, but less easily than the corresponding thiazole ⟨75SST(3)541⟩.

3-Alkyl-5-acetamidoisothiazoles are nitrosated at the 4-position by nitrosylsulfuric acid ⟨78JOC4154⟩. Isothiazoles are sulfonated at the 4-position using either oleum or sulfur trioxide ⟨72AHC(14)1⟩.

The ring nitrogen atom is readily alkylated in all compounds bearing a 3- or 5-substituent which can tautomerize to a ring NH structure. Alkylating agents include diazomethane ⟨72AHC(14)1, 76ACS(B)781⟩, dimethyl sulfate ⟨75SST(3)541, 79SST(5)345⟩, Meerwein's reagent ⟨79CB1288⟩ and alkyl halides ⟨79JMC237⟩. 1,2-Benzisothiazol-3-one also usually alkylates at nitrogen ⟨72AHC(14)43⟩, but with propargyl bromide in the presence of potassium carbonate both O- and N-propargyl compounds are obtained ⟨79SST(5)345⟩. The corresponding benzisoxazolone behaves similarly, but in that series the N-propargyl compound is unstable and undergoes ring expansion. The saccharin anion is usually alkylated at nitrogen via a S_N2 transition state, but isopropyl and s-butyl halides give O-alkyl products by S_N1 type reactions ⟨73AHC(15)233⟩. N-Glycosides are formed by the action of protected bromo sugars ⟨77SST(4)339⟩. Formaldehyde reacts with isothiazol-3-one to give the 2-hydroxymethyl compound, and Mannich conditions lead to 2-aminomethyl compounds ⟨69FRP1555414⟩. The reaction of sodium saccharin with pentafluoropyridine in the presence of a crown ether gives compound (**37**) ⟨79TL1217⟩. Two rather curious alkylation reactions are the reaction between 3-chloro-1,2-benzisothiazole 1,1-dioxide and pyridine 1-oxide to give compound (**38**), and the reaction between saccharin and vinyl acetate in the presence of a platinum metal complex to give compound (**39**) ⟨74JOC1795, 78GEP2725379⟩.

Early reports suggested that quaternary salts were formed only slowly ⟨72AHC(14)1⟩, and quantitative measurements show that isothiazole reacts less rapidly than 1-methylpyrazole, but more rapidly than isoxazole, in line with the relative basicities of the rings ⟨73AJC1949, 77SST(4)339⟩. As expected, the rate of quaternization with alkyl iodides falls somewhat with increasing size of alkyl group, but steric effects are unimportant ⟨76AJC1745⟩. Benzo fusion has very little effect on the rate (see Section 4.02.1.3), and both 1,2- and 2,1-benzisothiazoles are quaternized at approximately the same rate ⟨79SST(5)345⟩, although very little attention has been given to the 1,2-series. 2,1-Benzisothiazolium iodides are crystalline solids, but other salts tend to be oily ⟨73JCS(P1)1863⟩. The usual range of alkyl halides, sulfates, tosylates and fluorosulfonates has been used for quaternization, as well as triethyloxonium borofluoride. It is interesting to note that 3-substituted 2,1-benzisothiazoles have also been quaternized using alkyl orthoformates in the presence of Lewis acids ⟨75SST(3)541⟩.

C-Acylation is unknown. Attempts to carry out Friedel–Crafts and Vilsmeier–Haack reactions were unsuccessful ⟨65AHC(4)107, 72AHC(14)1, 72AHC(14)43⟩. 1,2-Benzisothiazol-3-one is readily acylated at the nitrogen atom using either carboxylic or sulfonic acid chlorides ⟨72AHC(14)43⟩. When 3-amino-2,1-benzisothiazole is treated with acid chlorides, both nitrogen atoms are acylated to give compound (**40**). The saccharin anion is acylated at the nitrogen atom using either carboxylic or sulfonic acid chlorides, or by reaction with carboxylic acids in the presence of N,N'-dicyclohexylcarbodiimide ⟨73AHC(15)233, 73T3985⟩.

(40) NCOR, COR (benzo-fused with S, N)

(41) Ar, S, NaS, NCSSNa (isothiazole)

(42) N=CHNMe₂, CHO (benzothiophene)

(43) NSAr (saccharin-type, SO₂)

(44) NCH₂SMe (saccharin-type, SO₂)

4-Aryl-3,5-dimercaptoisothiazole reacts with carbon disulfide to give compound (41) ⟨79JPR1021⟩. 3-Methyl-1,2-benzisothiazole, when subjected to the Vilsmeier–Haack reaction, undergoes ring cleavage and recyclization to give the benzothiophene (42) ⟨72JCS(P1)3006⟩. Saccharin reacts at nitrogen with arenesulfenyl chlorides to give compounds (43) ⟨75S165⟩, or with DMSO to give compound (44) ⟨68MI41700⟩.

Compounds having an oxygen-linked substituent can be oxidized to either sulfoxides or sulfones, in both the mononuclear series ⟨73SST(2)556, 75TL4123, 79SST(5)345, 80JCS(P1)2693⟩ and the 1,2-benzisothiazole series ⟨72AHC(14)43, 78MI41700, 81USP4276298⟩. Nitric acid or *m*-chloroperbenzoic acid usually produce sulfoxides, and hydrogen peroxide or chromium trioxide give sulfones. Decarboxylation occurs during the oxidation of 1,2-benzisothiazole-3-acetic acid to give 3-methyl-1,2-benzisothiazole 1,1-dioxide ⟨72JCS(P1)3006⟩.

4.17.6.4 Reactions with Nucleophiles and Reducing Agents

The isothiazole ring is, in general, stable to alkali ⟨65AHC(4)107⟩, although a few examples of ring cleavage at the N—S bond are known ⟨79CB1288, 80JCS(P1)2693⟩ (see Section 4.02.1.6). 2,1-Benzisothiazole is also stable to hot alkali ⟨72AHC(14)43⟩, in marked contrast to 2,1-benzisoxazole, but isothiazolium salts are dequaternized ⟨77IJC(B)886⟩. 1,2-Benzisothiazole 1,1-dioxides suffer cleavage at the C—N bond, and suitable compounds recyclize with ring expansion. For example, the 1,2-benzisothiazole 1,1-dioxide (45) gives the 1,2-benzothiazine 1,1-dioxide (46) ⟨76CC771⟩. Saccharin is hydrolyzed to *o*-carboxybenzenesulfonamide ⟨73AHC(15)233⟩.

(45) BrCMeEt (benzisothiazole SO₂)

(46) Et, Me (benzothiazine SO₂, NH)

(47) ArCO, NR (isothiazol-3-one)

(48) RNHCOCH=⟨S,S⟩=CHCONHR (dithietane)

Sodium alkoxides behave similarly to the hydroxides in that they occasionally cause ring cleavage at the N—S bond ⟨77T1057⟩. In the case of the isothiazol-3-one (47), two molecules of the cleavage product rejoin to form the dithietane (48) ⟨80JHC1645⟩. Sodium hydrogen carbonate also gives dithietanes ⟨80CPB2629⟩. 1,2-Benzisothiazole 1,1-dioxides are cleaved at the C–N bond, ring expansion occurring in suitable compounds ⟨79SST(5)345, 80JHC1281⟩. 2-Methoxycarbonylsaccharin is attacked at the ring carbonyl group by alkoxides, but at the ester carbonyl group by phenoxides ⟨80MI41700⟩.

3-Amino-1,2-benzisothiazole is cleaved by nitrous acid to give di-(*o*-cyanophenyl) disulfide ⟨73SST(2)556⟩.

Amines normally have no effect on the isothiazole ring, but 3-chloro-4-nitroisothiazole (49) reacts with cyclic amines such as morpholine at 0 °C to give the enamine (50) ⟨75JOC955⟩. The mechanism may involve attack either at sulfur or at the 5-position (Scheme 6).

Quaternized compounds are dequaternized in both the mononuclear ⟨74CJC1738⟩ and the 2,1-benzisothiazole series ⟨72CPB2372⟩, but in the latter case only using secondary amines; primary amines or ammonia cause ring cleavage and loss of sulfur ⟨72AHC(14)43, 72CPB2372⟩. 1,2-Benzisothiazole compounds suffer attack at the sulfur atom. Saccharin reacts with amines to give the compounds (**51**), but in the case of arylamines, prolonged heating causes recyclization and formation of compounds (**52**; Scheme 7) ⟨73AHC(15)233⟩.

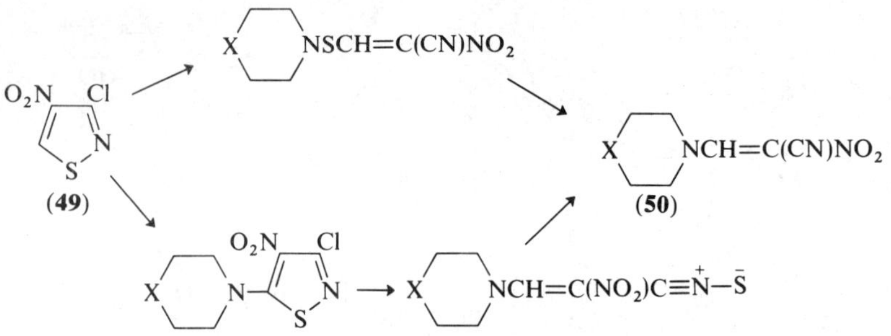

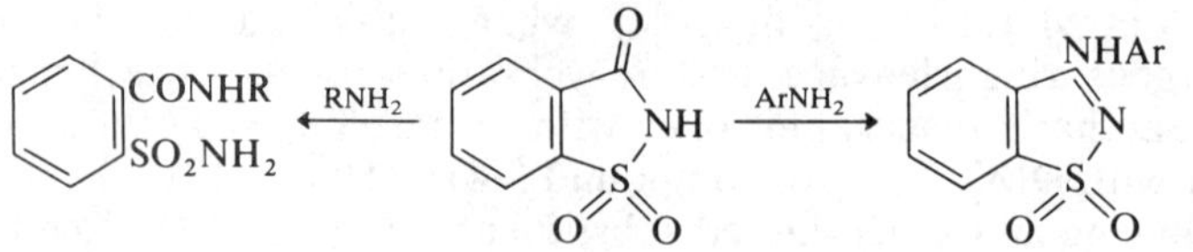

Scheme 7

The 2-phenylsulfonyl compound (**53**) reacts with aniline to give the dibenzothiazepine (**54**), but *N,N*-dialkylanilines lead to *p*-dialkylaminophenylthio compounds (**55**; Scheme 8) ⟨52HC(4)297⟩. 2-Phenacyl-1,2-benzisothiazol-3-one (**56**) gives with triethylamine the benzo-1,3-thiazine (**57**), but the corresponding saccharin derivative gives a benzo-1,2-thiazine 1,1-dioxide (**58**) ⟨76JOC1325⟩. Saccharin itself reacts with amines to give 3-substituted amino-1,2-benzisothiazole 1,1-dioxides ⟨77SST(4)339, 81ZN(B)1640⟩.

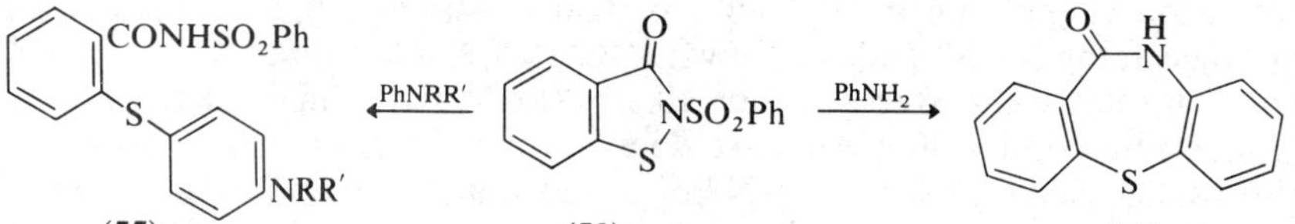

Scheme 8

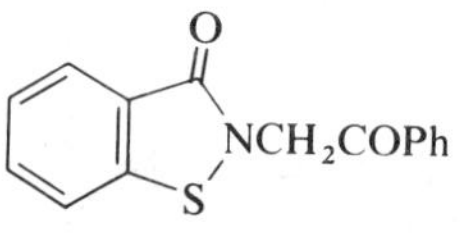

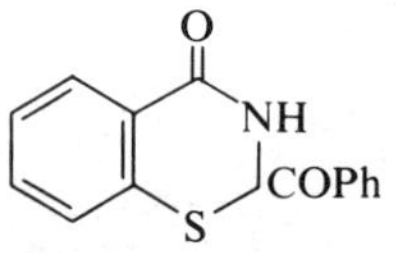

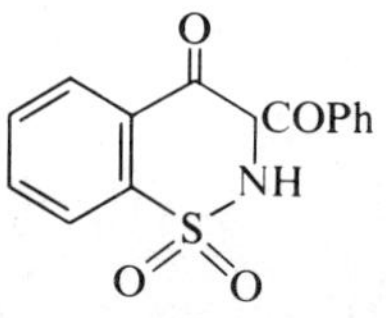

Isothiazolium salts (**59**) react with phenylhydrazine to give pyrazoles (**60**) ⟨72AHC(14)1⟩ (see Section 4.02.1.6). When treated with hydrazine hydrate, 3-chloro-1,2-benzisothiazole gives di-(*o*-cyanophenyl) disulfide ⟨73SST(2)556⟩, but 2,1-benzisothiazole gives *o*-aminobenzaldehyde azine ⟨72AHC(14)43⟩. 2-Substituted saccharins give the expected *o*-sulfamoylbenzohydrazides.

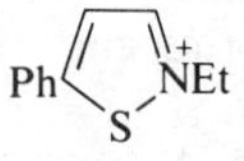

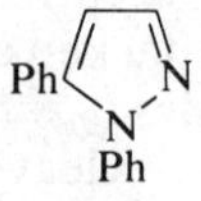

2-Alkylisothiazolium salts (**61**) undergo N—S bond cleavage when treated with hydrogen sulfide or thiophenol to form acyclic products (**62**), but 2-aryl compounds give 1,2-dithioles (**63** or **64**; Scheme 9) ⟨75SST(3)541, 77SST(4)339⟩.

1,2-Benzisothiazol-3-ones form di-(*o*-carbamoylphenyl) disulfides when treated with hydrogen sulfide ⟨52HC(4)297⟩. The corresponding 3-chloro compounds give di-(*o*-cyanophenyl) disulfides or *o*-cyanophenyl aryl disulfides when treated with thiophenols ⟨69BCJ1152,

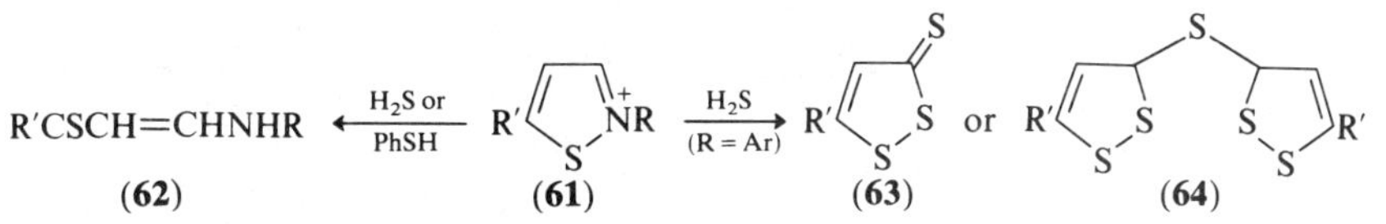

Scheme 9

73SST(2)556⟩. With thioacetic acid, however, benzo-1,2-dithioles are formed ⟨72AHC(14)43⟩. 3-Amino-1,2-benzisothiazoles and their ring quaternized derivatives react with thiophenols to give o-mercaptobenzamidines, which are difficult to obtain by other methods ⟨82LA14⟩. 3-Alkoxy-1,2-benzisothiazole 1,1-dioxides when treated with thiophenoxide ion give saccharin and the phenyl alkyl sulfide ⟨77SST(4)339⟩. 2-Methoxycarbonylsaccharin (65) behaves similarly with thiophenoxide ion, but the compound (66) is obtained using thioalkoxide ion (Scheme 10) ⟨80MI41700⟩.

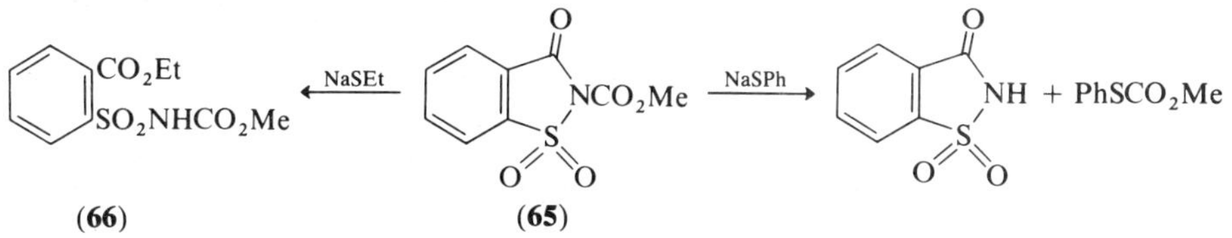

Scheme 10

Most isothiazoles are lithiated at the 5-position by the action of butyllithium or other organolithium compounds, provided that this position is vacant ⟨65AHC(4)107, 72AHC(14)1, 77SST(4)339⟩. 3-Methyl-4-nitroisothiazole, however, is inert ⟨65AHC(4)107⟩. 2,1-Benzisothiazole is lithiated at the 3-position, which corresponds to the 5-position in the mononuclear series ⟨75JHC877⟩. 4-Methylisothiazole forms the 5-lithio derivative, but the presence of by-products produced in subsequent reactions suggests the possibility of lithiation at the 3-position also ⟨72AHC(14)1⟩. 3-Substituted 1,2-benzisothiazoles suffer attack at sulfur and cleavage of the N—S bond ⟨72AHC(14)43, 73SST(2)556⟩.

Saccharin is alkylated or arylated at the 3-position by the action of alkyl- or aryl-lithium at low temperature to give a 3-alkyl- or aryl-1,2-benzisothiazole 1,1-dioxide, which can be further alkylated and acylated to give compound (67) ⟨79USP4178447⟩. At temperatures above 0 °C, ring cleavage and further alkylation (or arylation) occurs giving compound (68) ⟨73AHC(15)233, 74JCS(P1)2589⟩.

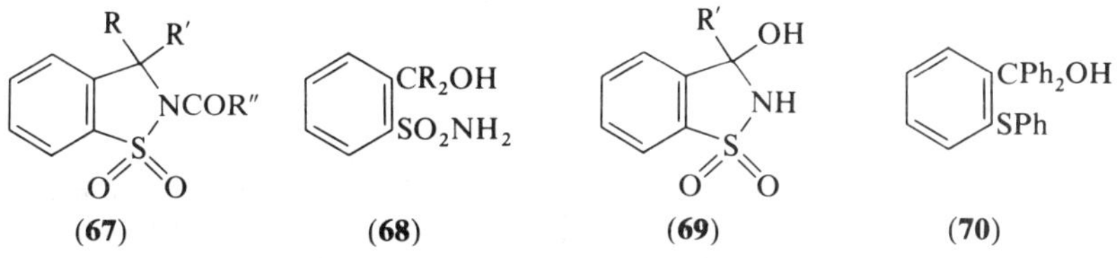

Reaction with Grignard reagents follows the general pattern of the reactions of butyllithium. There is no report of the treatment of mononuclear isothiazoles nor of 2,1-benzisothiazoles, but 3-chloro-1,2-benzisothiazole 1,1-dioxide is dialkylated at the 3-position ⟨79JA6981⟩. Saccharin reacts with alkyl- or aryl-magnesium halides to give 3-alkyl- or aryl-1,2-benzisothiazole 1,1-dioxides, together with the tertiary alcohols (68) ⟨73AHC(15)233, 77SST(4)339⟩. 2-Acylsaccharins are attacked at the ring carbonyl to give compound (69) ⟨58JOC2002⟩, but phenylmagnesium bromide attacks the ring sulfur atom of 2-phenylsulfonyl-1,2-benzisothiazol-3-one to give the ring cleaved compound (70) ⟨72AHC(14)43⟩.

Active methylene compounds react at sulfur, causing cleavage of the N—S bond, and in many cases recyclization produces thiophene compounds. Thus, 2-ethylisothiazol-3-one with diethyl malonate in the presence of sodium ethoxide gives an acyclic compound (71) ⟨72AHC(14)1⟩ (see Section 4.02.1.5), but isothiazolium salts (72) react with sodium benzoylacetate to give 2-benzoylthiophenes (73) ⟨75SST(3)541⟩. Reaction of diethyl malonate on the isothiazolylium salt (74) produces a different type of thiophene compound (75) ⟨77CJC1123⟩. 1,2-Benzisothiazoles also give benzothiophenes ⟨52HC(4)297, 73SST(2)556⟩.

EtNHCOCH=CHSCH(CO₂Et)₂

(71)   (72) 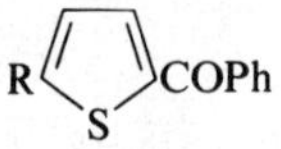   (73)

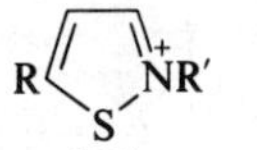

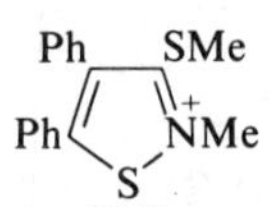

(74) 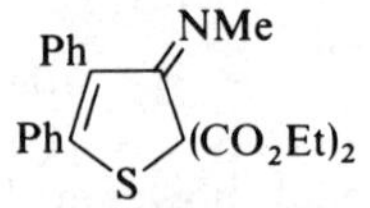   (75)

3-Chloro-1,2-benzisothiazole reacts with sodium cyanide to give a mixture containing mainly *o*-cyanophenyl thiocyanate with some di-(*o*-cyanophenyl) disulfide. An 80% yield of the latter was obtained using cuprous cyanide in DMF ⟨73SST(2)556⟩.

Borohydride reduction of 3-aryl-1,2-benzisothiazole 1,1-dioxides gives the 2,3-dihydro compounds ⟨73JMC1170⟩. Reduction of either 2-methylsaccharin or 2-hydroxymethylsaccharin with lithium aluminum hydride gives the same product, *N*-methyl-*o*-hydroxymethylbenzenesulfonamide ⟨73AHC(15)233⟩.

5-Iminoisothiazoles (76) react with triphenyl- or tributyl-phosphine to give ketenimines ⟨74CB502, 81CB536⟩, and with carbon dioxide and carbon disulfide to give oxathiazoles (77) and dithiazoles (78), respectively ⟨79SST(5)345⟩. A thione group is introduced into 2-methylisothiazolylium salts by the action of sulfur and pyridine ⟨72AHC(14)1, 73CI(L)1162, 75SST(3)541⟩. DMF reacts with 3-hydroxyisothiazoles to form 1,3-thiazines (79) ⟨72JAP(K)7217781⟩. Vilsmeier–Haack conditions applied to 3-methyl-1,2-benzisothiazole causes nucleophilic attack and the formation of compound (80) ⟨72JCS(P1)3006⟩. Isothiazolium salts (81) react with arylidenedimethylsulfuranes to give thiophenes (82) ⟨77CJC1123⟩.

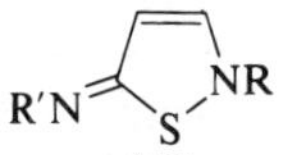

(76) 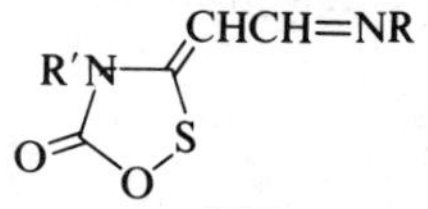   (77) 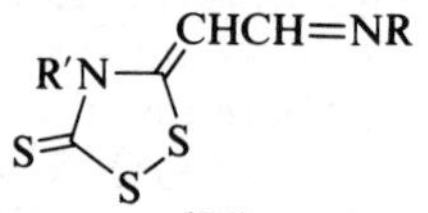   (78)

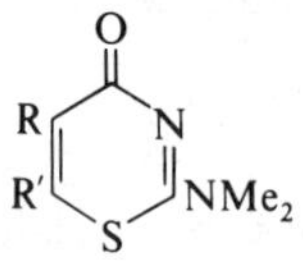

(79) 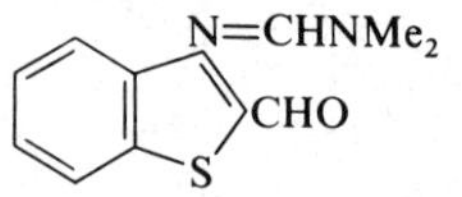   (80) 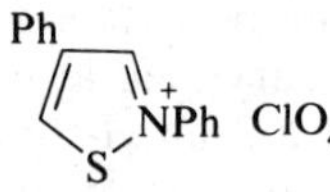   (81) 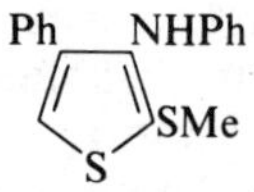   (82)

4.17.6.5 Miscellaneous Reactions at Ring Atoms

Isothiazoles are desulfurized by Raney nickel (see Section 4.01.1.8), forming acyclic products ⟨76TL2163, 80JHC1645⟩. With zinc and hydrochloric acid, 3-chloro-1,2-benzisothiazole gives di-(*o*-cyanophenyl) disulfide ⟨73SST(2)556⟩, but saccharin suffers only reduction of the carbonyl group, giving 2,3-dihydrobenzisothiazole 1,1-dioxide ⟨75MI41701⟩. 2-Benzoyl-1,2-benzisothiazol-3-one 1-oxide follows the former course, but the corresponding 2-acetyl compound gives a recyclized product (83) ⟨52HC(4)297⟩.

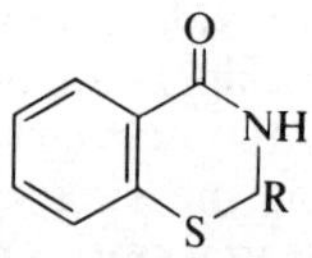

(83) 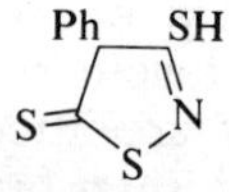   (84) 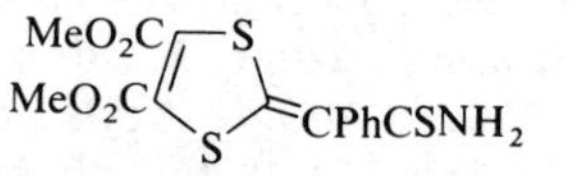   (85)

Few isothiazoles undergo simple cycloaddition reactions. 4-Nitroisothiazoles add to alkynes (see Section 4.17.7.4). With 5-thiones (84) and dimethyl acetylenedicarboxylate, addition to both sulfur atoms leads to 1,3-dithioles (85) ⟨77SST(4)339, 80H(14)785, 81H(16)156, 81H(16)595⟩. Isothiazol-3-one 1-oxide and the corresponding 1,1-dioxide give normal adducts with cyclopentadiene and anthracene ⟨80MI41700⟩, and saccharin forms simple 1:1 or 1:2 adducts with dimethyl acetylenedicarboxylate ⟨72IJC(B)881⟩.

3-Substituted 1,2-benzisothiazole 1,1-dioxides form benzo-1,2-thiazepines (86) with 1-diethylaminoprop-1-yne ⟨76H(5)95⟩. 2,1-Benzisothiazole does not react with maleic anhydride, but the 3-amino and 3-alkylamino derivatives react in different ways giving compounds (**87**) and (**88**) respectively ⟨75SST(3)541⟩. It should be noted that, in contrast, 2,1-benzisoxazole reacts normally.

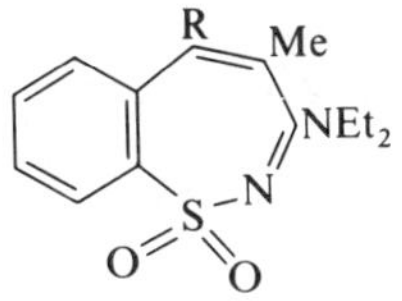

(**86**) 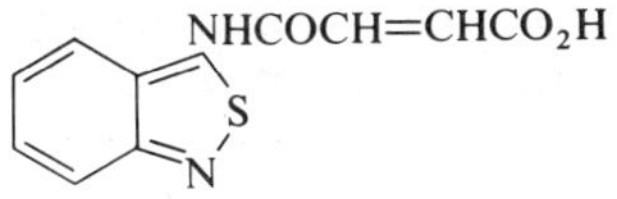 (**87**) 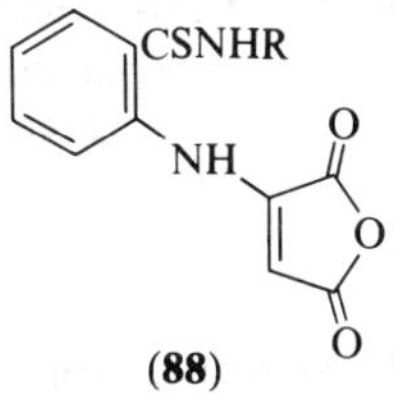 (**88**)

1,2-Benzisothiazoles undergo a variety of photochemical additions (Scheme 11) ⟨81T3377, 81TL525, 81TL529⟩. Isothiazole and its 3- and 4-methyl derivatives undergo radical phenylation on heating with benzoyl peroxide to give mixtures of methylphenylisothiazoles ⟨72BSF1173⟩.

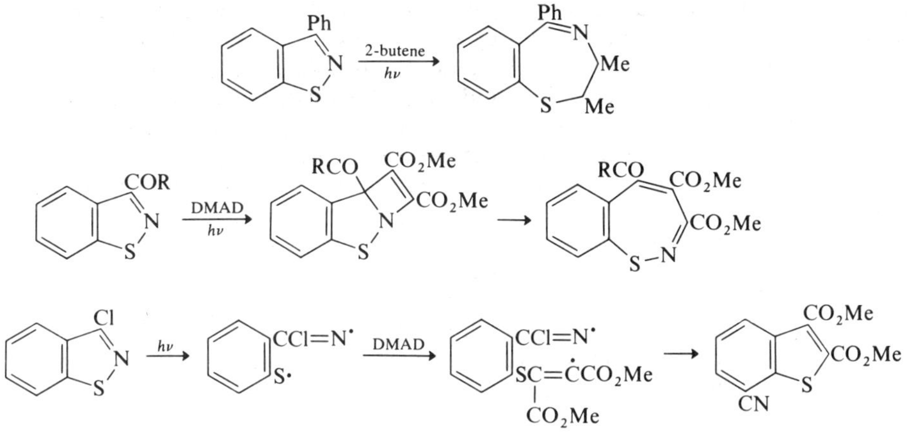

Scheme 11

Isothiazole and its 4- and 5-methyl derivatives react with chromium and tungsten hexacarbonyls under the influence of light to give pentacarbonylmetal compounds, the metal being coordinated *via* the nitrogen atom ⟨72JOM(44)325, 75MI41702⟩.

4.17.7 FULLY CONJUGATED RINGS: REACTIVITY OF SUBSTITUENTS

4.17.7.1 General Survey of Reactivity

Both 1,2- and 2,1-benzisothiazoles react with electrophiles to give 5- and 7-substituted products (see Section 4.02.3.2). The isothiazole ring has little effect on the normal characteristics of the benzene ring. *C*-Linked substituents react almost wholly normally, the isothiazole ring having little effect except that phenyl substituents are deactivated (see Section 4.17.2.1). There are, however, considerable differences in the ease of decarboxylation of the carboxylic acids, the 4-isomer being the most stable (see Section 4.02.3.3).

All amino compounds are normal weak amines, except 3-amino-1,2-benzisothiazole 1,1-dioxide which is acidic. Compounds having 3- or 5-amino substituents can exist in tautomeric forms involving the ring nitrogen atom, but their reactions indicate that they exist largely in the amino form. 3-Amino-2,1-benzisothiazole, however, has a greater tendency to exist in the imino form. All the amines form diazonium salts, which undergo deamination and Sandmeyer reactions (see Section 4.02.3.5). Compounds having an amino group adjacent to a carbamoyl group or other carbonyl function can be cyclized to form condensed ring systems.

Hydroxy and mercapto substituents at the 3- and 5-positions can also exist in tautomeric forms (see Section 4.01.5.2) and can be alkylated at either the substituent or the ring nitrogen atom. 3-Methoxy groups are not replaced by nucleophiles, but both 3- and 5-alkylthio groups react readily, as does 3-methoxy-1,2-benzisothiazole. Alkylthio compounds can be oxidized to sulfoxides and sulfones, and the latter readily undergo nucleophilic replacement. All the hydroxy compounds react with phosphorus pentachloride to give the chloro derivatives.

4-Chloro compounds are aromatic in character, but 5-chloro compounds and 3-chloro-2,1-benzisothiazoles react readily with nucleophiles (see Section 4.02.3.9). 3-Chloroisothiazoles and 3-chloro-1,2-benzisothiazoles are less reactive, and the more vigorous conditions necessary for replacement often cause ring cleavage. Halogens at all positions can be replaced by cyano groups (see Section 4.02.3.9).

5-Lithio compounds react with many different reagents, forming a variety of products (see Section 4.02.3.10).

4.17.7.2 Fused Benzene Rings

Nitration of 1,2-benzisothiazole gives equal quantities of the 5- and 7-nitro compounds ⟨72AHC(14)43⟩ (see Section 4.02.3.2), as also do the 3-methyl ⟨71JCS(C)3994, 80JCR(2)197⟩ and 4-bromo-3-methyl analogues ⟨80JCR(S)197⟩, but 4-chloro-1,2-benzisothiazole surprisingly gives only the 7-nitro derivative ⟨71JCS(C)3994, 80MI41700⟩. 2,1-Benzisothiazole is nitrated predominantly at the 5-position, but lesser amounts of the 7- and 4-nitro compounds are also obtained (see Section 4.02.3.2). In general, the isothiazole ring has little effect on the normal properties of the benzene ring, and the direction of nitration is governed by the substituents in the benzene ring ⟨72AHC(14)43⟩. Thus, 1,3-disubstituted 2,1-benzisothiazoles and their 2,2-dioxides are nitrated at the 5-position ⟨73JHC249, 78JHC529⟩. 6-Methyl- and 5-methyl-2,1-benzisothiazoles, however, are nitrated only at the 7- and 4-positions, respectively ⟨80JHC533⟩.

Bromination follows a similar course, except that 2,1-benzisothiazole produces the 4,5,7-tribromo derivative when excess of reagent is used ⟨72AHC(14)43⟩, and 3,5-disubstituted 1,2-benzisothiazoles are brominated at the 4-position ⟨80JCR(S)197⟩. 7-Amino-4-chloro-1,2-benzisothiazole is brominated at the 6-position ⟨71JCS(C)3994⟩.

Friedel–Crafts acylation usually fails ⟨72AHC(14)43⟩, but 3-substituted 1-methyl-2,1-benzisothiazole 2,2-dioxides can be acetylated at the 5-position ⟨73JHC249⟩. 1-Methyl-2,1-benzisothiazol-3-one can be chlorsulfonated at the 5-position ⟨78JHC529⟩. Vilsmeier–Haack formylation causes cleavage of the isothiazole ring ⟨80JCR(S)197⟩.

Nitro-1,2-benzisothiazoles are readily reduced to the amines ⟨72AHC(14)43, 80JCR(S)197⟩, as is 6-nitrosaccharin ⟨69JHC745⟩. When 5-nitro-2,1-benzisothiazole is treated with hydroxylamine, it is aminated at the 4-position ⟨80JHC537⟩.

5-Amino-1,2-benzisothiazoles undergo normal Sandmeyer conversions, but poor yields are obtained from 7-amino compounds (**89**), the major products being 1,2,3-benzothiadiazoles (**90**; Scheme 12). Abnormal coordination of the metal with the diazonium salt, also involving the sulfur atom, is thought to cause the rearrangement. The proportion of normal Sandmeyer product can be increased by the use of iron(II) chloride in place of copper(I) chloride ⟨71JCS(C)3994⟩. The same rearrangement takes place to a small extent during deamination with hypophosphorous acid ⟨71JCS(C)3994⟩. Deamination of 5-amino-3-methyl-4-nitro-1,2-benzisothiazole (**91**) in the presence of chloride is accompanied by replacement of the nitro group by chlorine to give compound (**92**) ⟨80JCR(S)197⟩.

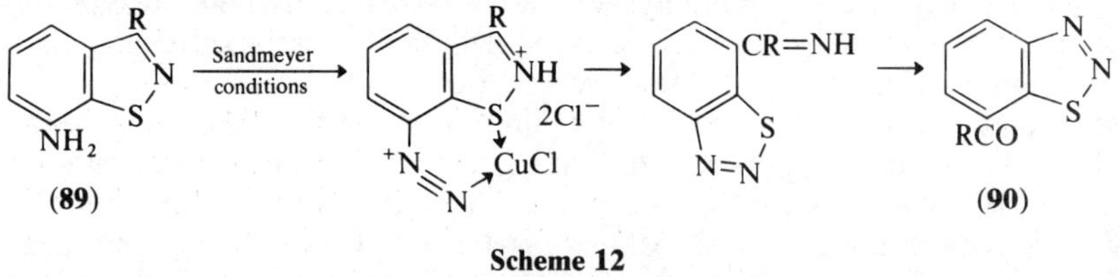

Scheme 12

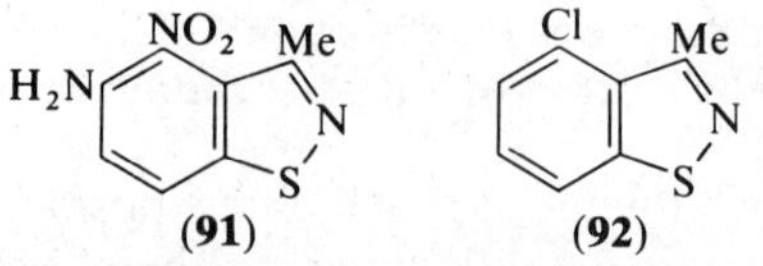

The chlorine atom of 4-chloro-1,2-benzisothiazole is susceptible to nucleophilic attack only when activated, for example by a 7-nitro group ⟨71JCS(C)3994, 75LA1994, 80MI41700⟩. 2-Acetyl-6-chlorosaccharin reacts with anhydrous hydrazine to give 3,6-dihydrazino-1,2-benzisothiazole 1,1-dioxide but the chlorine atom does not react with alcoholic hydrazine ⟨73AHC(15)233⟩. The weakened benzene ring of 5-amino-1,2-benzisothiazole is cleaved by

4.17.7.3 C-Linked Substituents

Methylisothiazoles undergo the expected reactions. Thus, photochemical bromination occurs with either bromine ⟨71MI41700⟩ or *N*-bromosuccinimide ⟨65AHC(4)107, 72AHC(14)1⟩. The latter, in the presence of benzoyl peroxide, reacts with 3-methylisothiazole to give the tribromomethyl compound ⟨75JOC3381⟩, but 3-methyl-1,2-benzisothiazole under the same conditions gives only the monobromomethyl compound ⟨72AHC(14)43, 77JCS(P2)1114, 79JCR(S)395⟩. The corresponding 1,1-dioxide reacts with bromine at room temperature to give the monobromomethyl compound, but on heating in benzene with an excess of bromine it gives the tribromomethyl derivative ⟨76CC771⟩.

Methylisothiazoles do not react with benzaldehyde, but 5-methylisothiazole forms a derivative with *m*-nitrobenzaldehyde, whereas the 4-isomer does not (see Section 4.02.3.3). This can be utilized to separate isomers ⟨65AHC(4)107, 72AHC(14)1⟩. 1,3-Dimethyl-2,1-benzisothiazolylium salts do react with aromatic aldehydes, but more slowly than with the corresponding benzisoxazolium salts ⟨79SST(5)345⟩. 4-Methylisothiazole has been oxidized in the vapour phase to give 4-formylisothiazole, and 3- and 4-methylisothiazoles may be oxidized to the carboxylic acids ⟨72AHC(14)1⟩ (see Section 4.02.3.3). 4-Methylisothiazole reacts with benzophenone under photochemical conditions to give the tertiary alcohol (**93**), whereas other heterocycles give oxetanes ⟨80JHC1777⟩.

(**93**)

Substituted methyl groups attached to the isothiazole ring react as normal aliphatic compounds. Thus, bromomethyl compounds can be converted into iodomethyl ⟨78CPB3888⟩ and cyanomethyl derivatives ⟨72AHC(14)1⟩, and can be oxidized by potassium permanganate to give the carboxylic acids ⟨72AHC(14)1⟩. Isothiazolylalanines are prepared by the acetamidomalonate route ⟨72AHC(14)1⟩. 3-Dibromomethylisothiazole is converted into the aldehyde ⟨65AHC(4)107⟩, and 3-tribromomethylisothiazole reacts with ethanolic silver nitrate to give ethyl isothiazole-3-carboxylate ⟨75JOC3381⟩. The rates of solvolysis in 80% ethanol of the three 2-chloro-2-(isothiazolyl)propanes have been compared with those for cumyl chloride, and the σ^+ values appropriate for the replacement of a benzene ring by an isothiazole ring determined ⟨77SST(4)339⟩. Cyanomethyl compounds form imino ethers, from which amino acids have been prepared ⟨73JMC978, 71MI41700⟩.

Methyl isothiazole-4-acetate is nitrosated by pentyl nitrite, and the oximino ester formed can be hydrolyzed to the α-keto ester ⟨78GEP2745246⟩. Secondary alcohols (**94**) can be oxidized to the ketones ⟨72GEP2223648⟩.

(**94**)

Diethyl 1,2-benzisothiazole-3-malonate does not form a pyrimidine derivative when treated with urea and sodium ethoxide, but gives instead the acetate ester, which can be converted successively into the amide, nitrile and acid. The last is decarboxylated at 165 °C to give 3-methyl-1,2-benzisothiazole ⟨72JCS(P1)3006⟩. 1,2-Benzisothiazole-3-acetic acid can also be reduced to the aldehyde using sodium hydride and ethyl formate, and can be α-brominated. More vigorous bromination conditions cause decarboxylation followed by the formation of the di- and tri-bromomethyl compounds ⟨78CPB3888⟩. 2-Hydroxymethylsaccharin is easily converted into the chloromethyl compound, and this will undergo normal nucleophilic replacement of chlorine with amines and thiols ⟨73AHC(15)233⟩.

The isothiazole ring causes deactivation of phenyl substituents, particularly in 3-phenylisothiazoles. Nitration of arylisothiazoles, however, occurs only in the benzene ring, the pattern of substitution varying greatly with the position of the aryl group on the isothiazole

Table 16 Nitration of Phenylisothiazoles ⟨71BSF4310⟩

Compound	ortho	Isomers formed (%) meta	para	2,4
3-Phenylisothiazole	25	65	10	—
4-Phenylisothiazole	25	3	72	—
5-Phenylisothiazole	20	19	61	5

ring (Table 16). 4-Phenylisothiazole is brominated in both rings to give 5-bromo-4-*p*-bromophenylisothiazole ⟨72AHC(14)1⟩ (see Section 4.02.3.4). Substituents on the benzene ring undergo normal reactions of aromatic compounds ⟨75JAP(K)7512080⟩, including carbonation of 5-*p*-hydroxyphenylisothiazole to give compound (**95**) ⟨78JMC1100⟩.

(**95**)

The reactions of isothiazolyl aldehydes and ketones are almost wholly normal (see Section 4.02.3.4). They form oximes, phenylhydrazones, thiosemicarbazones, acetals and ketals. The aldehydes reduce ammoniacal silver nitrate, form cyanhydrins and undergo the Cannizzaro and Wittig reactions. They can be reduced to the alcohols with borohydride, and oxidized to the acids with silver oxide. They condense with diethyl malonate and with nitromethane ⟨65AHC(4)107, 72AHC(14)1⟩. An anomalous reaction occurs between 4-bromo-5-formyl-3-methylisothiazole and 2-aminopyridine to give compound (**96**) ⟨72AHC(14)1⟩. 3-Formylisothiazole condenses with ammonium carbonate and potassium cyanide to give a poor yield of a hydantoin (**97**), which is hydrolyzed to isothiazol-3-ylglycine, also in poor yield ⟨71USP3579506⟩. The oxime of 4-acetyl-3-methylisothiazole produces 4-amino-3-methylisothiazole by Beckmann rearrangement ⟨72AHC(14)1⟩, and the corresponding phenylhydrazine cyclizes by the action of zinc chloride to give 4-(2-indolyl)-3-methylisothiazole ⟨77IJC(B)473⟩.

(**96**)   (**97**)

There are marked differences in the thermal stability of carboxyl groups at different positions of the ring. Isothiazole-5-carboxylic acids are decarboxylated relatively easily ⟨65AHC(4)107, 72AHC(14)1, 73CC524, 78JOC3736, 79LA1534⟩, the 3-carboxylic acids somewhat less easily ⟨71TL1281, 72AHC(14)1⟩, but the 4-carboxylic acids require high temperatures or heating of the copper salts ⟨65AHC(4)107, 72AHC(14)1, 76IJC(B)394⟩ (see Section 4.02.3.4). All the acids may be converted into acid chlorides and esters by standard methods ⟨65AHC(4)107, 72AHC(14)1, 77JHC725⟩, but 5-acetamido-3-methylisothiazole-4-carboxylic acid cyclizes by the action of thionyl chloride, to give compound (**98**) ⟨73MI41700⟩.

(**98**)   (**99**)   (**100**)

Esters of isothiazolecarboxylic acids undergo normal Claisen condensation to give β-keto esters ⟨65AHC(4)107, 72AHC(14)1⟩, and are reduced to the alcohols by lithium aluminum hydride ⟨72AHC(14)1, 73JMC512⟩. Ethyl isothiazole-5-glyoxylate is reduced by lithium tri-*t*-butoxyaluminum hydride, and the product hydrolyzed to give the corresponding α-hydroxy acid ⟨74JMC34⟩. Methyl 3-methylisothiazole-4-carboxylate does not react at the 5-position with butyllithium as expected, but suffers attack at the ester carbonyl group, giving compound (**99**) ⟨65AHC(4)107⟩. With methyllithium, however, and with Grignard reagents, tertiary alcohols (**100**) are formed ⟨75JOC3381, 79JOC510⟩.

In an attempt to prepare an isothiazolobenzodiazepine, ethyl 5-*o*-aminoanilino-3-methylisothiazole-4-carboxylate was treated with sodium methoxide, but the only reaction was transesterification to the methyl ester ⟨76IJC(B)394⟩. Only the 5-ester group of dimethyl 3-methylisothiazole-4,5-dicarboxylate reacts with *N,N'*-diphenylguanidine, as with the corresponding isoxazole compound, but the product could not be cyclized, even under drastic conditions. This is in marked contrast to the isoxazole compound which cyclized at room temperature ⟨80JCS(P1)1667⟩.

Isothiazole acid chlorides react normally to form amides, hydrazides and azides, are reduced to the aldehydes by lithium trialkoxyaluminum hydrides, and undergo the Arndt–Eistert and Reissert reactions ⟨65AHC(4)107, 72AHC(14)1⟩. Isothiazole-5-carbonyl chloride is converted into the thiol acid by the action of sodium hydrosulfide ⟨74MI41700⟩. 3-Methylisothiazole-4-carbonyl chloride reacts with methylmagnesium iodide in the presence of iron(III) chloride at −20 °C, but aryl Grignard reagents do not react ⟨72AHC(14)1⟩.

Isothiazoleamides undergo normal Hoffman rearrangement ⟨72AHC(14)1, 77JHC725⟩, give carboxylic acids on treatment with nitrous acid, and are converted into nitriles by phosphorus oxychloride ⟨72AHC(14)1⟩. Reduction with diborane gives the aminomethyl compounds ⟨76ACS(B)781⟩. 1,2-Benzisothiazole-3-carboxamide reacts with hydrazine hydrate to give the hydrazide ⟨71FRP2043473⟩. Isothiazole carboxyazides undergo normal Curtius rearrangement, except that in the 5-series, the isocyanate partially trimerizes to give the trioxo-*s*-triazine (**101**) as a by-product ⟨72AHC(14)1, 79JMC237⟩.

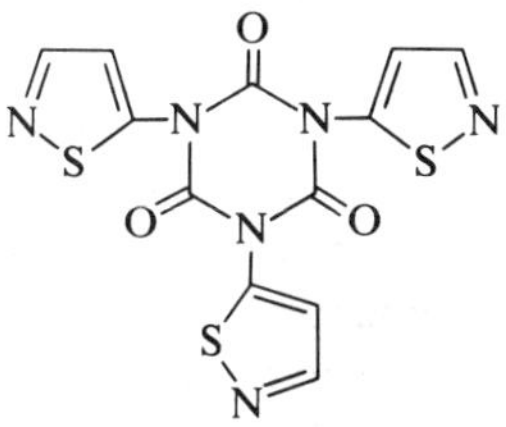

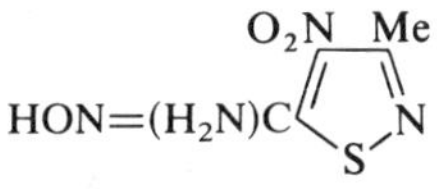

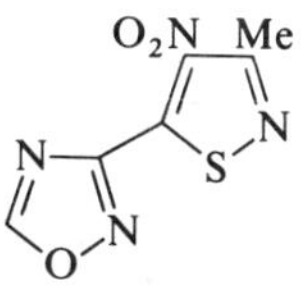

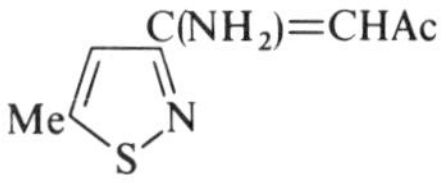

Cyanoisothiazoles can be hydrolyzed to amides and acids ⟨65AHC(4)107, 72AHC(14)1, 77JHC725, 80CPB2629⟩, and 3-cyano-2,1-benzisothiazole has also been hydrolyzed to the acid ⟨72AHC(14)43⟩. Hydrogen sulfide in pyridine gives the 5-thioamide. Grignard reagents react normally, but Stephens reduction of 5-cyano-3-methylisothiazole gives the 5-aminomethyl compound, rather than the aldehyde ⟨65AHC(4)107⟩. 5-Cyano-3-methyl-4-nitroisothiazole reacts with hydroxylamine to give the amidoxime (**102**), which can be cyclized to the 1,2,4-oxadiazole (**103**) by the action of triethyl orthoformate ⟨75MI41700⟩. Acetone adds to the cyano group of 3-cyano-5-methylisothiazole to give compound (**104**) as main product. 5-Cyano-3-methylisothiazole behaves similarly ⟨77H(6)1985⟩.

4.17.7.4 *N*-Linked Substituents

4-Nitroisothiazoles are reduced to the amines by either catalytic hydrogenation or by chemical methods ⟨65AHC(4)107, 72AHC(14)1⟩ (see Section 4.02.3.6). 4-Nitroisothiazole (**105**) reacts with 1-amino-2-phenylacetylenes at room temperature to give both cycloaddition (**106**) and open chain products (**107**), the proportion of the former increasing with increasing polarity of the solvent used. The open chain product gives 4-amino-5-chloroisothiazole (**108**) when treated with hydrochloric acid (Scheme 13) ⟨76RTC67, 77SST(4)339⟩.

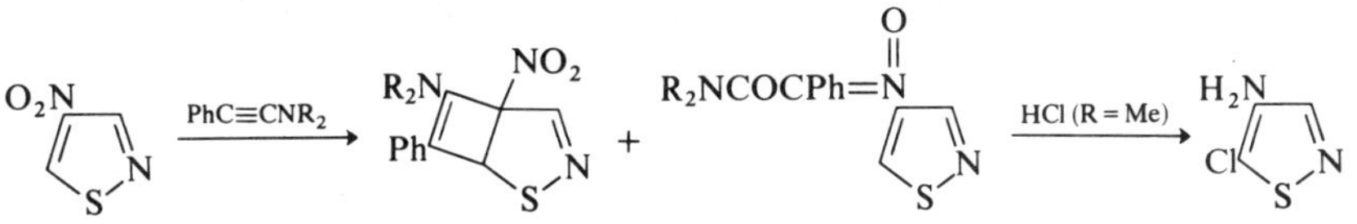

Scheme 13

All the aminoisothiazoles are normal weak amines. The 3- and 5-amino compounds can theoretically exist in tautomeric forms, but their reactions indicate that the amino forms predominate ⟨65AHC(4)107, 72AHC(14)1⟩. 3-Amino-2,1-benzisothiazole has a greater tendency to exist in the imino form. On acylation it gives diacyl derivatives ⟨72AHC(14)43⟩ (see Section 4.02.3.5), whereas acylation of the mononuclear amines is normal. The latter also form sulfonamides ⟨77SST(4)339, 77JHC725⟩, ureas and thioureas ⟨72AHC(14)1⟩.

All three mononuclear amines and 3-amino-2,1-benzisothiazole can be diazotized and coupled to form azo dyes (see Section 4.02.3.5), but 3-amino-1,2-benzisothiazole reacts with nitrous acid to give 3,3′-bis(1,2-benzisothiazoline) ⟨72AHC(14)43⟩, and 5-aminoisothiazoles bearing an electron-withdrawing group at the 4-position form N-nitrosoamines ⟨72AHC(14)1⟩. 4-Aminoisothiazoles having an amide group or a secondary amino group at an adjacent position form cyclic products following diazotization ⟨72AHC(14)1, 75ZC18⟩.

All the diazonium salts undergo the Sandmeyer reaction to give halogen compounds (see Section 4.02.3.5), but 5-cyano compounds cannot be obtained by this means. Isothiazole-3-diazonium borofluoride decomposes to give 3-fluoroisothiazole ⟨72AHC(14)1⟩. 5-Amino-3-methylisothiazole reacts with nitrous acid in the presence of copper to give the 5-nitro compound ⟨75MI41700⟩. 4-Hydroxy compounds cannot be obtained from the diazonium salts, but 3- and 5-hydroxy compounds are available by this means, although the conditions are critical in the 5-series ⟨72AHC(14)1⟩. All three mononuclear amines and 3-amino-2,1-benzisothiazole can be deaminated by treatment of the diazonium salts with hypophosphorous acid (see Section 4.02.3.5). Mononuclear diazonium salts undergo the Meerwein and Gomberg–Hey reactions (see Section 4.02.3.5). With thiourea, 3-methylisothiazole-5-diazonium chloride gives the disulfide, but 3-methylisothiazole-4-diazonium chloride suffers ring cleavage and recyclization to give 4-acetyl-1,2,3-thiadiazole ⟨72AHC(14)1⟩. Isothiazole-4-diazonium salts react with sodium azide to give 4-azidoisothiazole ⟨80TL2995⟩.

5-Amino-3-methylisothiazole undergoes the Skraup reaction to give compound (**109**) in poor yield ⟨72AHC(14)1⟩. 5-Amino-4-carboxy-3-methylisothiazole condenses with aromatic aldehydes to give Schiff bases which can be reduced to the benzylamines ⟨76MI41700⟩, and formaldehyde reacts with 4-aminoisothiazole to give 4-formylaminoisothiazole ⟨76MI41701⟩. 5-Amino-4-cyanoisothiazoles condense with DMF in the presence of thionyl chloride to give the dimethylaminomethylideneamino derivatives ⟨77USP4032322⟩.

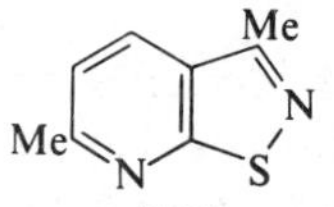

(**109**) 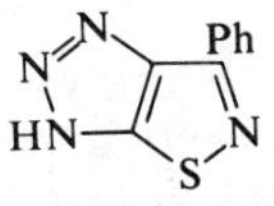   (**110**)

5-Amino-3-phenylisothiazole cyclizes on reaction with 2-azido-3-ethylbenzothiazolylium borofluoride to give compound (**110**) ⟨78HCA108⟩. 4-Aminoisothiazole reacts with 1,4-dibromobutane to give 4-pyrrolidinylisothiazole ⟨80MI41700⟩.

3-(2-Chloroethylamino)isothiazole (**111**) cyclizes to 3-aziridinylisothiazole (**112**) under the influence of potassium t-butoxide, but undergoes intramolecular quaternization if kept at 50 °C for two days, to give compound (**113**; Scheme 14) ⟨79JOC1118⟩.

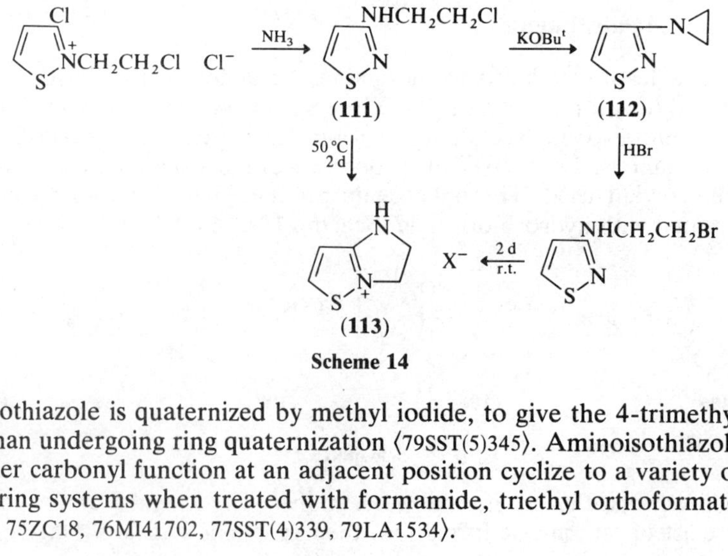

Scheme 14

4-Aminoisothiazole is quaternized by methyl iodide, to give the 4-trimethylammonium salt, rather than undergoing ring quaternization ⟨79SST(5)345⟩. Aminoisothiazoles having an amide or other carbonyl function at an adjacent position cyclize to a variety of condensed heterocyclic ring systems when treated with formamide, triethyl orthoformate or the like ⟨65AHC(4)107, 75ZC18, 76MI41702, 77SST(4)339, 79LA1534⟩.

3-Amino-1,2-benzisothiazole condenses with phosphorus pentachloride, and the chlorine atoms in the product (**114**) can be replaced stepwise by reaction with aniline (Scheme 15) ⟨62ZOB1878⟩. 3-Amino-1,2-benzisothiazole 1,1-dioxide is an acidic compound and shows no reactions of an amine ⟨73AHC(15)233⟩. The dimethylamino analogue, obtained by high temperature reaction of saccharin with hexamethylphosphoramide, reacts with pyrrolidine to give the pyrrolidinyl analogue ⟨77SST(4)339⟩. 3-Hydrazino-1,2-benzisothiazole condenses normally with aromatic aldehydes ⟨76G769⟩. The azide group of 3-azido-1,2-benzisothiazole 1,1-dioxide behaves as a pseudohalogen and is readily replaced by nucleophiles ⟨73AHC(15)233⟩.

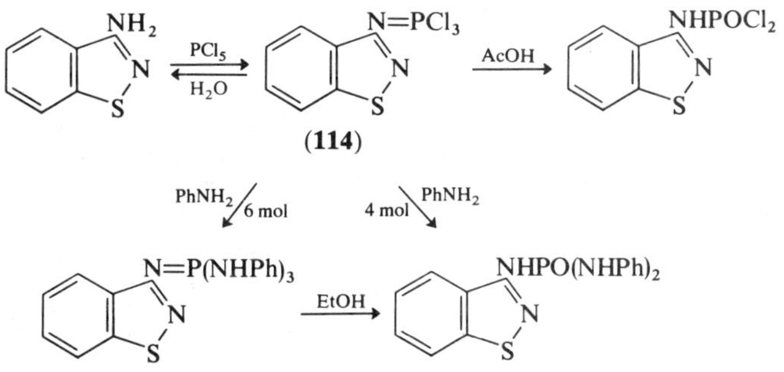

Scheme 15

4.17.7.5 *O*-Linked Substituents

Isothiazole compounds having 3- or 5-hydroxy groups can exist in tautomeric forms involving the ring nitrogen atom (see Section 4.02.3.7). Acylation of 3-hydroxyisothiazole in aprotic solvents gives almost exclusively the 3-acyloxy compound, but migration of the acyl group occurs on heating or storage, the final position of equilibrium depending on the nature of the acyl group ⟨72AHC(14)1⟩. 3-Hydroxy derivatives of benzisothiazoles are indiscriminately described in the literature as '3-hydroxy' and '3-ones', without any evidence of actual structure.

Most 3-hydroxy compounds give mixtures of *O*- and *N*-alkyl derivatives when treated with diazomethane, alkyl halides or dimethyl sulfate ⟨72AHC(14)1, 76ACS(B)781, 77JHC725, 79SST(5)345⟩, but compounds with a 5-phenyl substituent are attacked only at nitrogen ⟨72AHC(14)1, 72JCS(P1)1432⟩ (see Section 4.02.3.7). *N*-Alkylisothiazol-3-ones can be further methylated by methyl fluorosulfonate to give compounds (**115**) ⟨78JHC695, 79JOC1118⟩. 3-Hydroxy-1,2-benzisothiazole and 3-hydroxy-2,1-benzisothiazole also give mixtures of *O*- and *N*-substituted products on alkylation or acylation ⟨72AHC(14)43, 75AJC129⟩. *O*-Substituted organophosphorus compounds have been prepared from both 3-hydroxyisothiazoles ⟨72AHC(14)1, 73GEP2325043⟩ and 3-hydroxy-1,2-benzisothiazoles ⟨72AHC(14)43⟩.

(**115**)

Saccharin reacts with sugars in the presence of diethyl azodicarboxylate and triphenylphosphine to give both *O*- and *N*-substituted products ⟨76JHC1131⟩. 1,2-Benzisothiazol-3-one reacts with acetic anhydride, firstly at nitrogen, and at higher temperatures at oxygen also (Scheme 16) ⟨52HC(4)297⟩. It forms cyanhydrins and bisulfite derivatives, reflecting its ketonic character, and isocyanates react to give *N*-carbamoyl compounds ⟨77SST(4)339⟩.

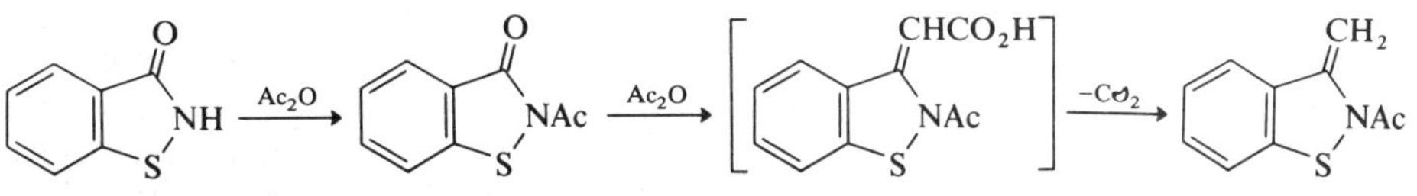

Scheme 16

3-Cyanoacetyl-5-nitro-1,2-benzisothiazole loses carbon dioxide on heating to give the 3-cyanomethyl compound ⟨79JCR(S)395⟩. 3-Hydroxyisothiazole gives 3-chloroisothiazole on treatment with phosphorus pentachloride ⟨79JOC1118⟩, but the 4,5-diphenyl analogue unexpectedly gives a derivative containing phosphorus (**116**) on subsequent reaction with methanol ⟨72JCS(P1)1432⟩. 2-Alkylisothiazol-3-ones ⟨72AHC(14)1, 78JHC695, 79JOC1118⟩, 3-hydroxy-1,2-benzisothiazoles ⟨72AHC(14)43⟩, 3-hydroxy-2,1-benzisothiazoles ⟨73JHC413⟩ and saccharin ⟨77JAP(K)7736663⟩ all give 3-chloro derivatives using a variety of reagents. *N*-Substituted 2,1-benzisothiazol-3-ones, however, give quaternized products (**117**) ⟨79SST(5)345⟩.

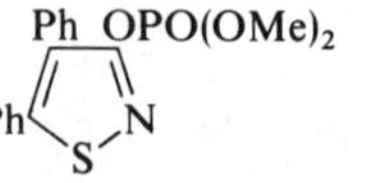

(**116**) 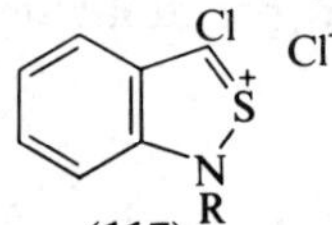 (**117**)

3-Methoxyisothiazoles are demethylated by hydrobromic acid ⟨76ACS(B)781, 69FRP1555414⟩, but nucleophilic displacement of the alkoxy group has not been reported in mononuclear compounds. 3-Alkoxy-1,2-benzisothiazoles, however, react with hydrazine hydrate to give the 3-hydrazino compound ⟨76G769, 79USP4140693⟩. The reaction of 3-alkoxy-1,2-benzisothiazole 1,1-dioxides (**118**) with nucleophiles can take two possible courses, depending upon the nature of the nucleophile (Scheme 17). The action of *t*-amines or pyrolysis at 180 °C causes Chapman–Mumm rearrangement, to give *N*-alkyl-3-hydroxy compounds ⟨73AHC(15)233, 77SST(4)339⟩.

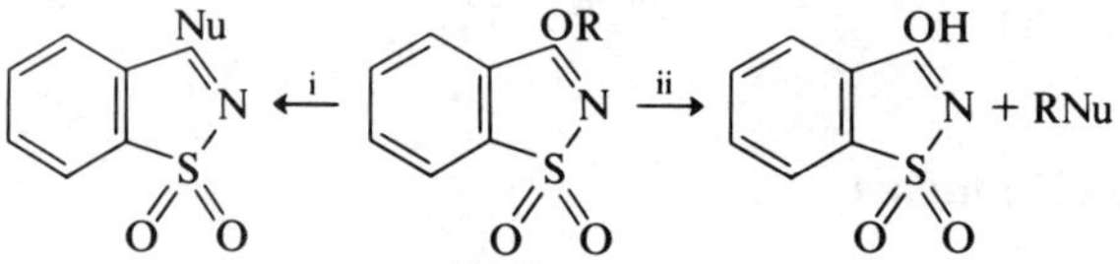

(**118**)

i, Nu⁻ of low leaving reactivity, *e.g.* BuNH$_2$; ii, Nu⁻ of high leaving reactivity, *e.g.* PhO⁻

Scheme 17

When saccharin is treated with diethyl phosphorothiolothionate, the 3-ethylmercapto compound is obtained, rather than the expected organophosphorus compound ⟨77ACS(B)460⟩. Treatment of saccharin with phosphorus pentoxide and amines gives 3-(substituted-amino)-1,2-benzisothiazole 1,1-dioxides, *via* an intermediate phosphate ⟨81ZN(B)1640⟩. Reduction of saccharin with zinc and hydrochloric acid gives 2,3-dihydro-1,2-benzisothiazole 1,1-dioxide, the method being used to estimate saccharin in foodstuffs ⟨75MI41701⟩.

Little study has been made of 4- and 5-hydroxyisothiazoles. 4-Hydroxyisothiazole methylates and acylates normally at oxygen ⟨72AHC(14)1, 79BSF(2)26⟩ (see Section 4.02.3.7). Some 4-substituted 5-hydroxyisothiazoles exist predominantly in the keto form, and it has been suggested that there is some contribution from the zwitterionic species (**119**; Scheme 18). 4-Bromo-3-methyl-5-hydroxyisothiazole (**120**) forms only the methoxy derivative with diazomethane, but the 4-nitro analogue does not react ⟨72AHC(14)1, 76AHC(S1)1⟩.

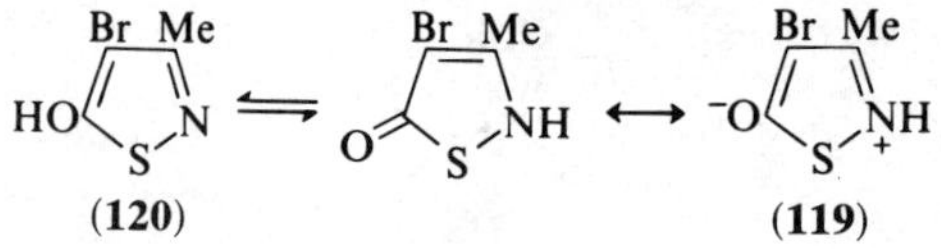

(**120**) (**119**)

Scheme 18

4.17.7.6 *S*-Linked Substituents

As with the hydroxy compounds, 5-mercapto groups exist predominantly in the thione form, but 3-mercapto groups exist in the thiol form. This causes the 3-mercapto compounds to be the more reactive, but both 3- and 5-mercapto compounds are readily alkylated as their sodium salts. 3,5-Disubstituted compounds can thus be made to react stepwise with different alkylating agents ⟨72AHC(14)1, 72BSF2296, 80G233, 80JCS(P1)1029, 80JPR1021⟩. 4-Aryl-3-hydroxy-5-mercaptoisothiazoles are methylated only at sulfur, using dimethyl sulfate ⟨77AJC1815⟩. 2-Alkyl-3- or -5-thiones react with methyl iodide to give methylthioisothiazolium salts ⟨72AHC(14)1, 72AHC(14)43, 79CJC207⟩. Thiosaccharin is methylated at the thione group ⟨73AHC(15)233⟩.

Acylation of 3,5-dimercaptoisothiazoles also occurs preferentially at the 3-substituent. When the product (**121**) is methylated or oxidized, the thiol ester rearranges to give the thione ester (**122**; Scheme 19) ⟨80CL401, 80JCS(P1)2693, 81JCS(P1)1401⟩. 2,4-Diphenylisothiazole-5-thione reacts with ethyl azidoformate to give the 5-ethoxycarbonylimino compound (**123**) ⟨76CJC3879⟩, but triethyl phosphite gives the bis compound (**124**) ⟨81JHC437⟩.

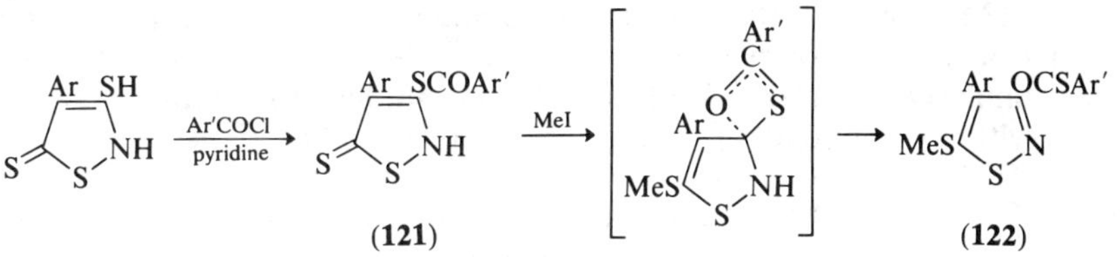

Scheme 19

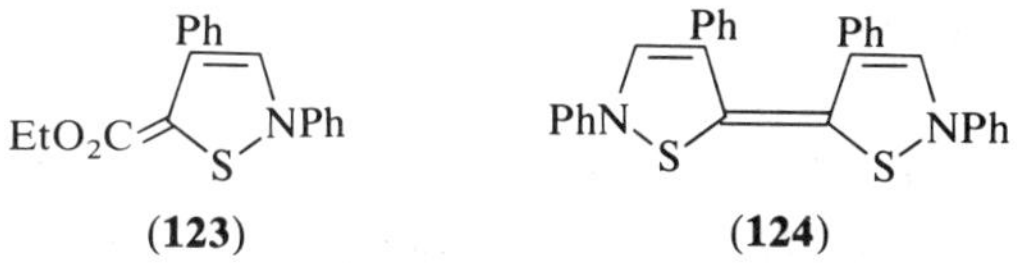

Only one 4-mercapto compound (**125**) is known. This reacts with sulfur monochloride to give the fused heterocyclic ring compound (**126**) ⟨77BEP853648⟩. 5-Alkylthio groups can be displaced by nucleophilic reagents such as hydrazine or alkoxide ⟨72AHC(14)1⟩, or dealkylated using sodium and liquid ammonia ⟨80CPB2629⟩. 3-Methylthio-2,5-diphenylisothiazolium salts are both demethylated and dequaternized by pyridine to give compound (**127**) ⟨74CJC1738⟩.

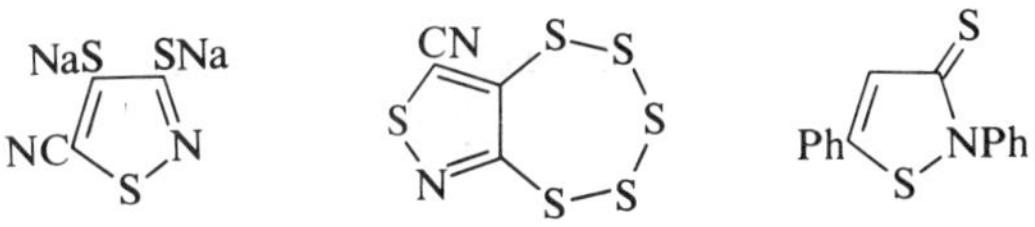

3-Alkylthio-1,2-benzisothiazole 1,1-dioxides do not undergo rearrangement on pyrolysis, as does the corresponding alkoxy compound ⟨73AHC(15)233⟩. 5-Lithiothio compounds react with ethyl bromoacetate to give the expected products (**128**), but with mercury(II) acetate the 3-methyl compound gives compound (**129**), whereas the 4-methyl analogue gives compound (**130**) ⟨77SST(4)339⟩. Thiosaccharin is hydrolyzed to saccharin by water, and other nucleophilic displacements of the 3-thione group occur with amines and with hydrazine ⟨73AHC(15)233⟩.

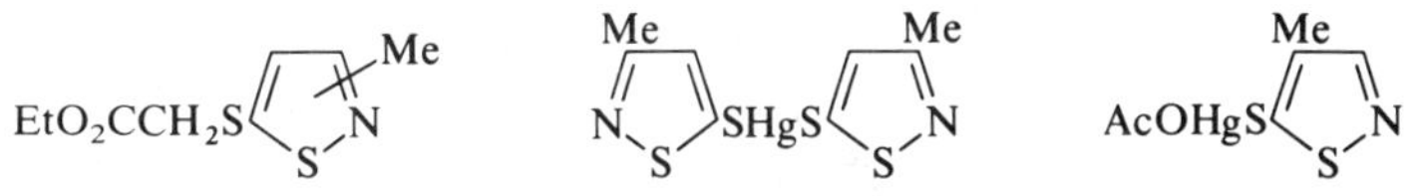

Both 3- and 5-alkylthio compounds can be oxidized to sulfoxides and sulfones ⟨72AHC(14)1, 69FRP1555414, 76JPR127, 77AJC1625⟩. The resulting 5-alkylsulfonyl groups are readily replaced by nucleophiles ⟨79SST(5)345⟩, but there are few reports of replacement of a 3-alkylsulfonyl group ⟨75SST(3)541⟩. 3-Mercapto-5-phenylisothiazole can be oxidized to the sulfonic acid, and 5,5'-sulfides to the sulfoxides and sulfones ⟨72AHC(14)1⟩. 1,2-Benzisothiazole-3-thione and 1-alkyl-2,1-benzisothiazole-3-thiones are oxidized to the corresponding 3-ones ⟨72AHC(14)43⟩.

4.17.7.7 Halogen Substituents

Halogen atoms at the various positions of isothiazole show interesting differences in reactivity (see Section 4.02.3.9). The 4-halogeno compounds are the most benzene-like. Halogens at the 3- and 5-positions both react with nucleophiles, especially when activated by a suitable 4-substituent, but 3-halogeno compounds are less reactive, and their replacement is sometimes accompanied by ring cleavage ⟨72AHC(14)1, 80JCS(P1)1029, 80JHC385⟩. This also applies to 3-halogeno-1,2-benzisothiazoles ⟨72AHC(14)43, 73SST(2)556, 75LA1994,

77MI41700⟩, but 3-halogeno-2,1-benzisothiazoles undergo normal nucleophilic replacement ⟨71MI41701, 77SST(4)339, 80MI41700⟩. 3-Chloro-1,2-benzisothiazole 1,1-dioxide displays the reactivity of a cyclic imidoyl chloride, and the halogen can be replaced by reaction with sodium azide ⟨73AHC(15)233⟩ or with alcohols or thiols ⟨80MI41700⟩.

2-Alkyl-3-chloroisothiazolium chlorides (**131**) react with ammonia to give 3-alkyl-aminoisothiazoles (**132**) ⟨80MI41700⟩, but with amines give mixtures of isomeric isothiazolium salts (**133** and **134**; Scheme 20) ⟨79JOC1118⟩, compound (**134**) presumably being formed by a similar mechanism to compound (**132**). 2-Alkyl-3-chloro-1,2-benzisothiazolium salts (**135**) behave similarly ⟨80CB2490⟩. With phenols and thiophenols, they give the 3-aryloxy (**136**) and 3-arylthio compounds. In some cases, secondary reactions cause the formation of the isomers (**137**) also. The latter can be dequaternized by heating to give the phenols (**138**; Scheme 21) ⟨77SST(4)339⟩. Reaction of compounds (**135**) with N,N-dialkylanilines gives only analogues of compounds (**137**) and (**138**) ⟨79CB3286⟩. With thioacetic acid they give either the thiones (**139**) or the iminobenzodithioles (**140**), depending on the nature of R. Hydroxylamine gives compounds (**141**; Scheme 21) ⟨72AHC(14)43⟩.

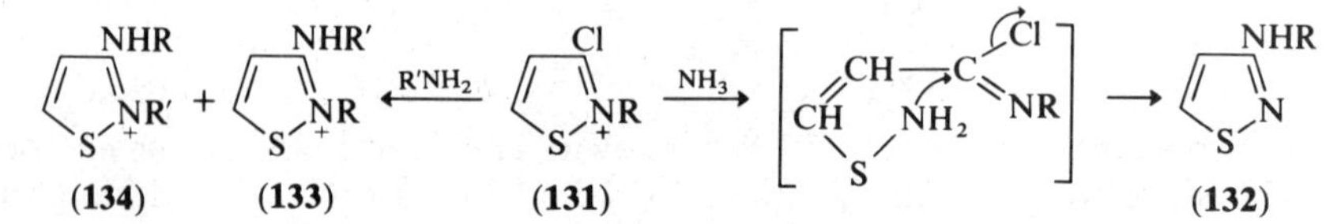

Scheme 20

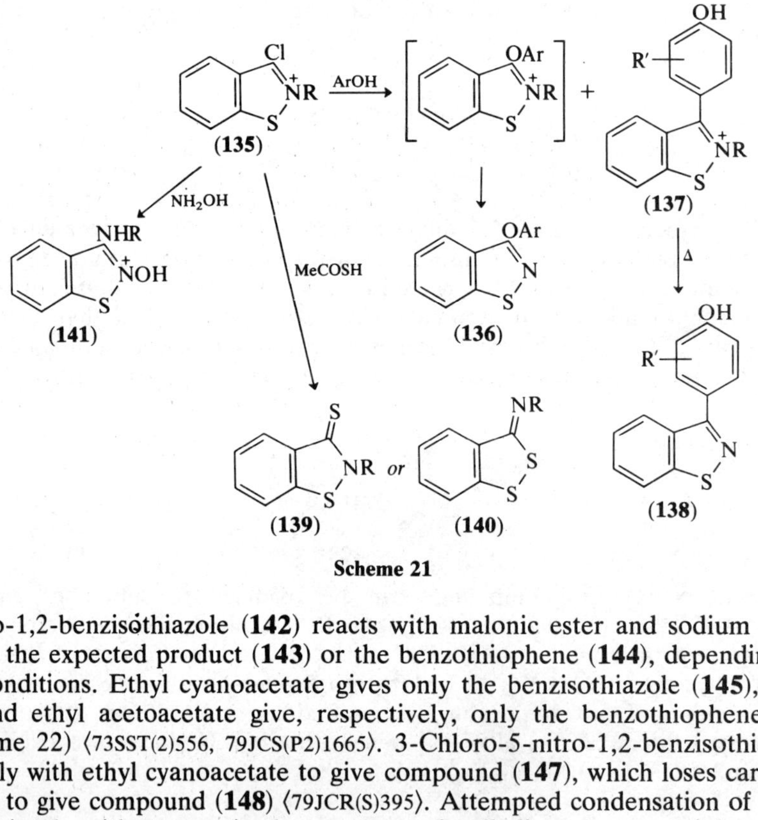

Scheme 21

3-Chloro-1,2-benzisothiazole (**142**) reacts with malonic ester and sodium ethoxide to give either the expected product (**143**) or the benzothiophene (**144**), depending upon the reaction conditions. Ethyl cyanoacetate gives only the benzisothiazole (**145**), but acetylacetone and ethyl acetoacetate give, respectively, only the benzothiophenes (**146** and **144**; Scheme 22) ⟨73SST(2)556, 79JCS(P2)1665⟩. 3-Chloro-5-nitro-1,2-benzisothiazole reacts anomalously with ethyl cyanoacetate to give compound (**147**), which loses carbon dioxide on heating to give compound (**148**) ⟨79JCR(S)395⟩. Attempted condensation of 5-bromo-3-methylisothiazole with styrene in the presence of palladium acetate, triphenylphosphine and base gave only 5,5'-bis(3-methylisothiazolyl) ⟨81CPB3543⟩.

All the halogenoisothiazoles react with copper(I) cyanide to give the corresponding nitriles ⟨65AHC(4)107, 72AHC(14)1, 75JAP(K)7504068, 77JHC725⟩ (see Section 4.02.3.9). 3-Chloro-1,2-benzisothiazole suffers ring cleavage when treated with sodium cyanide, the main product being o-cyanophenyl thiocyanate ⟨71TL1075⟩. 5-Halogenoisothiazoles react readily with butyllithium to give the lithio compound, but halogen atoms at the 3- and 4-positions do not usually react ⟨65AHC(4)107, 72AHC(14)1, 73JMC978⟩. 3-Chloro-2,1-benzisothiazole with a large excess of methylmagnesium iodide gives a 3:1 mixture of 2,1-benzisothiazole and

i, diethyl malonate, NaOEt/EtOH, 30 min, r.t., then 30 min, reflux; ii, diethyl malonate, NaOEt/EtOH, 48 h, r.t., or using phase transfer catalyst; iii, ethyl cyanoacetate, NaOEt; iv, acetylacetone, NaOEt; v, ethyl acetoacetate

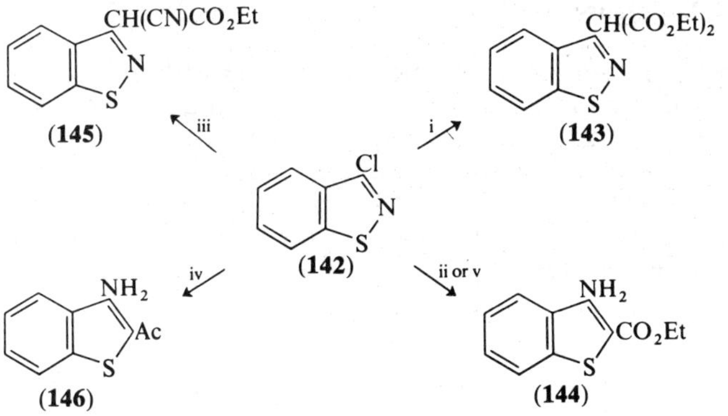

Scheme 22

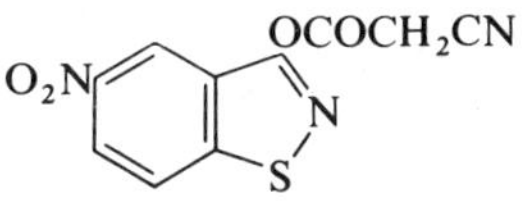

the 3-methyl compound ⟨71MI41701⟩. 3-Chloro-1,2-benzisothiazole 1,1-dioxide gives the compound (149) ⟨79JA6981⟩. Reduction of 3-chloro-1,2-benzisothiazole with zinc dust gives di-o-cyanophenyl disulfide, and not 3,3'-bis(benzisothiazole) as at first thought ⟨72AHC(14)43⟩.

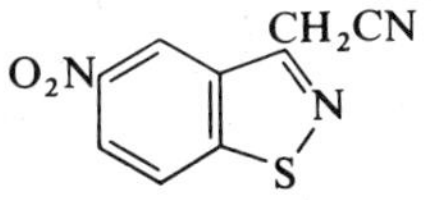

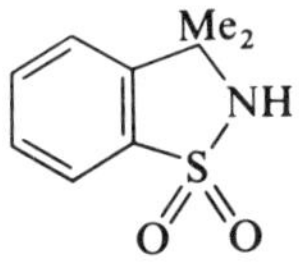

5-Bromo-3-methyl-4-nitroisothiazole can be converted into the 5-iodo analogue by reaction with sodium iodide in acetone ⟨65AHC(4)107⟩. Halogen exchange also takes place when 4-bromo-3-methylisothiazole-5-diazonium chloride is treated with methyl methacrylate and hydrolyzed, giving the chloro compound (150) ⟨72AHC(14)1⟩.

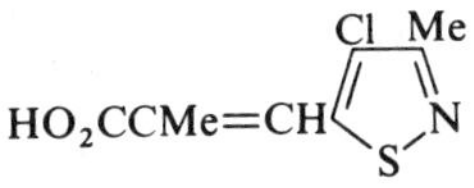

Various iodoisothiazoles on irradiation lose iodine and produce free radicals, which when trapped with benzene compounds give mixtures of o-, m- and p-arylisothiazoles ⟨73BSF1822, 75SST(3)541⟩. 5-Bromo-3-methylisothiazole (151) on treatment with potassium amide gives a high yield of 4-bromo-3-methylisothiazole (152). The mechanism of this reaction is not thought to involve isothiazolyne, but to be as shown in Scheme 23. This mechanism is supported by the fact that an equimolar mixture of compound (153) and 3-methylisothiazole under the same reaction conditions gives a quantitative yield of compound (152) ⟨72AHC(14)1⟩.

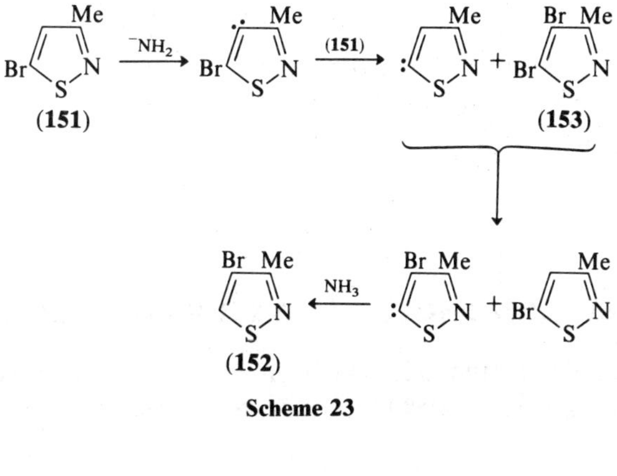

Scheme 23

4.17.7.8 Metal Substituents

The readily available 5-lithioisothiazoles (**154**) react with a wide range of reagents to introduce functional groups at the 5-position (Scheme 24) ⟨65AHC(4)107, 72AHC(14)1, 71OPP167, 73JMC978, 75JHC49, 75JOC3381, 77SST(4)339⟩ (see Section 4.02.3.10). The only known 4-lithio compound, 4-lithio-5-methyl-3-phenylisothiazole, reacts with carbon dioxide to give the acid ⟨72AHC(14)1⟩, and 3-lithio-2,1-benzisothiazole reacts with methyl iodide to give the 3-methyl compound ⟨75JHC877⟩. Isothiazole-4- and -5-mercurioacetate compounds react with bromine in the presence of sodium bromide to give the corresponding bromo compounds ⟨72AHC(14)1⟩.

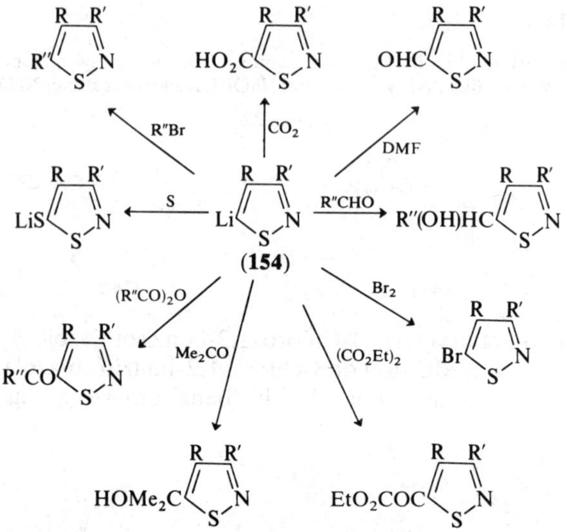

Scheme 24

4.17.7.9 Substituents at Ring Heteroatoms

The hydroxy groups of N-hydroxyalkylisothiazol-3-ones behave as normal alcohols, giving the expected products with isocyanates and thionyl halides, and being replaceable by aromatic amines ⟨76JHC1097⟩. N-Chlorosaccharin reacts with thiophenols by a free radical mechanism to give the N-arylthio compounds ⟨80MI41700⟩, and phosphorus and arsenic derivatives (**155**) and (**156**) have been prepared from N-chloromethylsaccharin ⟨74JPR851⟩. N-Methoxycarbonylsaccharin can be attacked by nucleophiles at either carbonyl group ⟨80MI41700⟩, and loses carbon dioxide on pyrolysis to give N-methylsaccharin. Pyrolysis of higher esters, however, gives saccharin, carbon dioxide and an alkene ⟨74NKK1539⟩. S-Chloro-1,2-benzisothiazol-3-one and related compounds are hydrolyzed to 1,2-benzisothiazol-3-one 1-oxide ⟨75SST(3)541, 78ZOR862⟩.

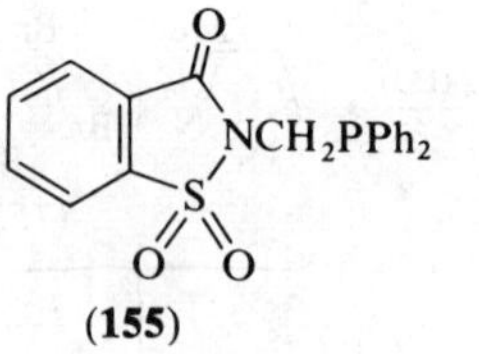

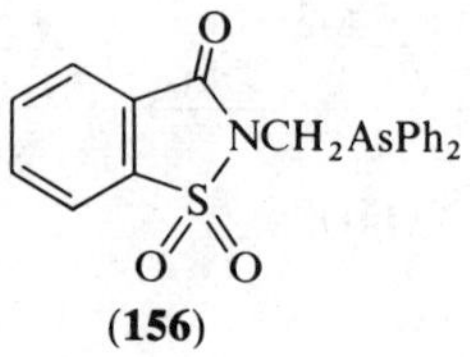

4.17.8 SATURATED AND PARTIALLY SATURATED RINGS

Dihydro and tetrahydro compounds have been little studied. Too few reactions are known to merit full classification. They have therefore been placed in two classes only.

4.17.8.1 Reactivity at Ring Atoms

Esters of 3-aryl-4,5-dihydroisothiazole-4- and -5-carboxylic acids are unstable and aromatize in acetone solution ⟨78JOC3742⟩. The 1-methyl 1-oxide compound (**157**) is converted into 1,2-benzisothiazole on heating at 155 °C ⟨80MI41700⟩. 1-Methyl-3*H*-2,1-benzisothiazole 2,2-dioxide (**158**) loses sulfur dioxide on pyrolysis to give compound (**159**) ⟨80CC471⟩, and condenses with aromatic aldehydes to give compounds (**160**; Scheme 25) ⟨75SST(3)541⟩.

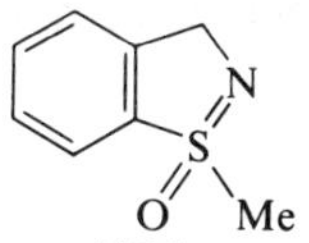

(**157**)

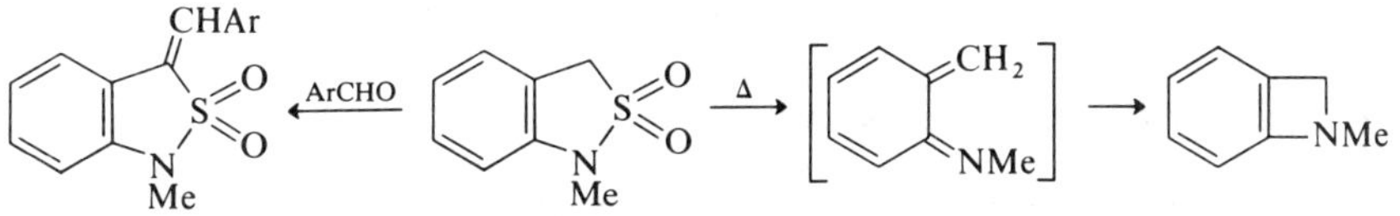

(**160**)   (**158**)   (**159**)

Scheme 25

Chlorination of 2-substituted isothiazolidin-3-one 1-oxides with sulfuryl chloride gives the 4,5-dichloro derivatives ⟨79SST(5)345⟩. 3,3-Dimethyl-1,2-benzisothiazoline 1,1-dioxide can be methylated at the nitrogen atom ⟨79JA6981⟩, and 1-arylisothiazolidine 1-oxide is quaternized using Meerwein's reagent ⟨73JA7692⟩. Certain compounds in the mononuclear and 2,1-benzisothiazole series can be oxidized to give ring sulfoxides and sulfones ⟨73TL5213, 77SST(4)339, 77USP4031227, 81T2181⟩.

Isothiazoline-5-thiones suffer ring cleavage when treated with Grignard reagents ⟨77SST(4)339⟩, but the cycloheptatrieno compound (**161**) gives the 3-lithio derivative on treatment with butyllithium ⟨81JOC3575⟩. The zwitterionic isothiazolidine (**162**), obtained by oxidation of methionine, reverts to a deuterated methionine (**163**) on treatment with deuterated methanol and hydrogen sulfide ⟨79TL2937⟩.

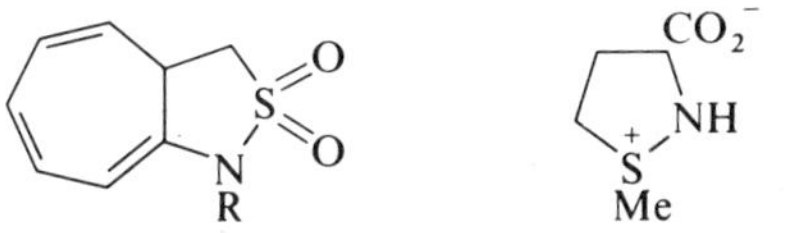

(**161**)   (**162**)   CD$_3$SCH$_2$CH$_2$CH(NH$_2$)CO$_2$H   (**163**)

The quaternized 2,1-benzisothiazoline (**164**) undergoes ring cleavage when treated with water, to give compound (**165**). With sodium hydroxide, however, dequaternization occurs to give compound (**166**), which can be oxidized to the 2-oxide (**167**; Scheme 26) ⟨74TL3319⟩.

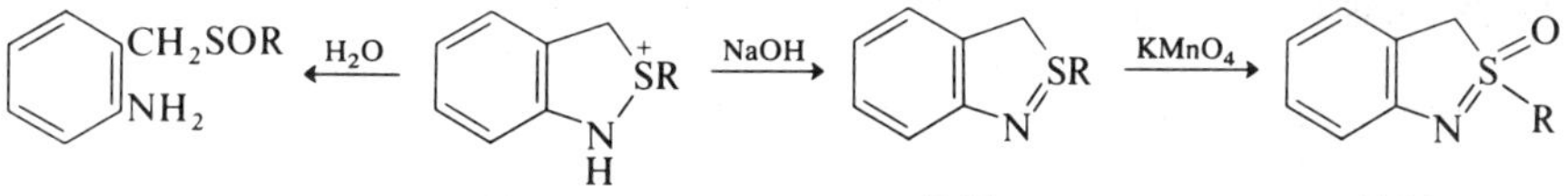

(**165**)   (**164**)   (**166**)   (**167**)

Scheme 26

Isothiazolidinones have been desulfurized by Raney nickel ⟨74JOC1210⟩. The 2,1-benzisothiazoline 2,2-dioxide (**158**) undergoes photocycloaddition with dimethyl acetylenedicarboxylate and loss of sulfur dioxide to give a compound of probable structure (**168**) ⟨80CC471⟩.

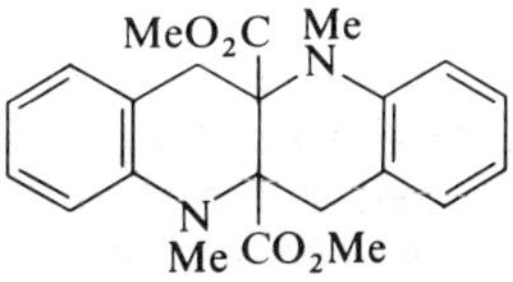

(**168**)

4.17.8.2 Reactivity of Substituents

1-Aryl-4-hydroxy-5H-isothiazole 1-oxide is methylated at the hydroxy group by diazomethane ⟨73JOC20⟩. 2-Alkyl-3-hydroxymethylisothiazolidine 1,1-dioxides are oxidized to the aldehydes, and the latter converted *via* Wittig reactions to analogues of prostaglandins ⟨80MI41700⟩. The acetyl group of 5-acetyl-1-methyl-2,1-benzisothiazoline 2,2-dioxide can be reduced by borohydride to the secondary alcohol, or converted into the carboxylic acid by treatment with sodium hypobromite ⟨73JHC249⟩. The amino group of 3-amino-3-phenyl-1,2-benzisothiazoline 1,1-dioxide condenses with benzaldehyde, but is replaced by hydroxy on treatment with ethanol ⟨69JOC919⟩.

4.17.9 SYNTHESIS FROM NON-HETEROCYCLIC COMPOUNDS

4.17.9.1 Formation of One Bond

4.17.9.1.1 Formation of the nitrogen–sulfur bond

One of the best methods of synthesis of isothiazoles is by direct oxidation of γ-iminothiols (**169**) or their tautomers. The reaction is capable of many ramifications and is represented by the general equation shown in Scheme 27. The substituents represent a wide range of groups. Thus, iminothioamides (**169**; $R^5 = NH_2$) are oxidized to give 3-alkyl-5-aminoisothiazoles (**170**; $R^5 = NH_2$), amidines (**169**; $R^3 = NH_2$) produce 3-amino compounds, iminoethers (**169**; $R^3 = OR$) and their thio analogues give 3-alkoxy and 3-alkylthio compounds (**170**; $R^3 = OR$ or SR), and amides (**169**; $R^3 = OH$) give 3-hydroxy compounds ⟨65AHC(4)107, 72AHC(14)1, 80MI41700⟩. Variations at the other end of the molecule include thioamides (**169**; $R^5 = NH_2$) producing 5-aminoisothiazoles, thioesters (**169**; $R^5 = SMe$) leading to 5-methylthio compounds and thioaldehydes (**169**; $R^5 = H$) and thioketones (**169**; $R^5 = Ar$) giving 5-unsubstituted and 5-aryl compounds ⟨65AHC(4)107, 72AHC(14)1, 72AP902, 74BSF2507, 79SST(5)345, 80MI41700⟩.

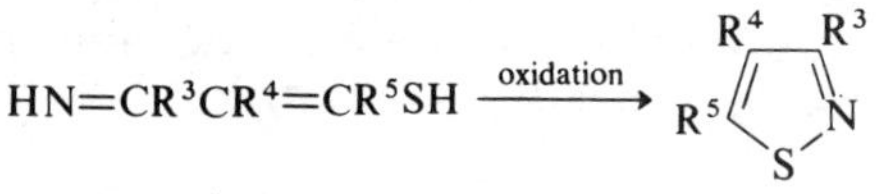

$$HN=CR^3CR^4=CR^5SH \xrightarrow{\text{oxidation}} \underset{(170)}{R^5 \diagdown \underset{S}{\diagup} N}^{R^4 \; R^3}$$

(169) (170)

Scheme 27

The most usual oxidizing agents employed are halogens or hydrogen peroxide. Two special cases are the oxidation of thioamides (**171**), derived from 3-aminobut-2-enoic acid, to give compounds (**172**) ⟨74CB502⟩, and of 2-mercapto-5-methylbenzylamine to give 5-methyl-1,2-benzisothiazole (**173**) ⟨72AHC(14)43⟩. The necessary starting material is sometimes generated *in situ* ⟨72AHC(14)1⟩. A particular case is the conversion of a disulfide into the corresponding sulfenyl halide by the halogen being used for oxidation, prior to ring closure ⟨73SST(2)556, 78JOC1604, 79SST(5)345⟩.

RNHCMe=CR′CSNHR″   (**171**)   (**172**) 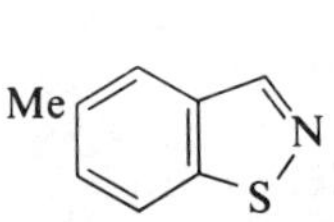   (**173**)

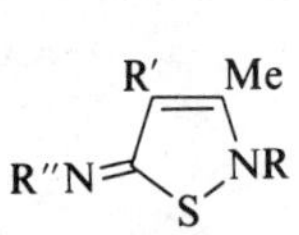

The oxidizing agent may also cause further reaction subsequent to ring closure. Thus, compound (**174**), obtained from diazomethane and carbon disulfide, can be both oxidized and chlorinated by chlorine to give 3,5-dichloro-4-cyanoisothiazole (**175**), but when hydrogen peroxide is employed, compound (**176**) is obtained. Ring closure using sulfur gives the dimercapto compound (**177**; Scheme 28) ⟨65AHC(4)107, 72AHC(14)1⟩. Similarly, treatment of 2-cyano-3-oxothiobutyramide with bromine gives 4-acetyl-5-amino-3-bromoisothiazole ⟨75JPR771⟩. 2-Aryl-2-cyanodithioacetic acid with sulfur gives 4-aryl-3,5-dimercaptoisothiazole ⟨79JPR1021, 80JCS(P1)2693⟩, but with hydrogen peroxide gives 4-aryl-3-hydroxy-5-mercaptoisothiazole ⟨79SST(5)345⟩. The use of chloramine as oxidizing agent can produce 3-amino compounds ⟨75SST(3)541⟩.

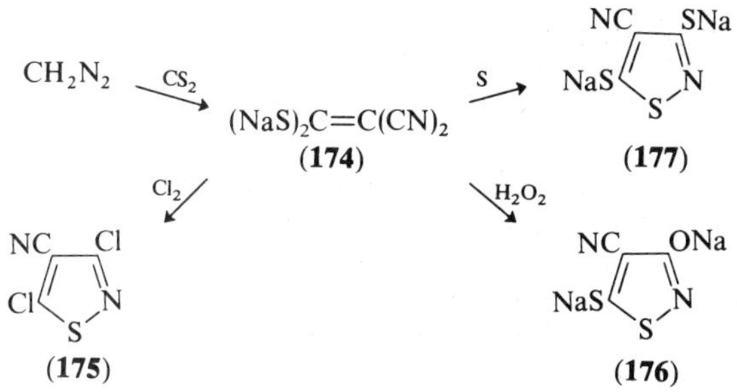

Scheme 28

By suitable choice of starting materials, the above methods can also be used to prepare 2-substituted isothiazol-3-ones ⟨72AHC(14)1, 73TL2159, 73TL5213, 76JHC1321⟩, their dihydro analogues ⟨81T2181⟩ or isothiazolium salts ⟨72AHC(14)1, 75ZC478, 77SST(4)339, 80MI41700⟩.

Although 1,2-benzisothiazoles can be prepared by the oxidation method, they are also available by cyclization of o-mercaptobenzaldoximes and ketoximes with polyphosphoric acid ⟨72AHC(14)43, 73JCS(P1)356, 77JCS(P2)1114⟩. This method has been improved by the use of S-t-butyl analogues, which do not suffer from the instability of the free thiols ⟨79SST(5)345⟩.

Other eliminative cyclizations include thermal loss of hydrogen cyanide from thiocyanatoamides to give isothiazol-3-ones or 1,2-benzisothiazol-3-ones ⟨72AHC(14)1, 77MI41701⟩, thermal loss of ammonia from o-sulfamoylbenzamides to give saccharin derivatives ⟨73AHC(15)233⟩, and the conversion of o-methanesulfinylbenzamides into 1,2-benzisothiazol-3-ones by thionyl chloride or acetyl chloride ⟨81CC510⟩. An androsteno[17,16-d]isothiazole has been prepared by treatment of a 16-mercaptopregnen-20-one with phosphorus oxychloride, followed by dehydration with thionyl chloride ⟨81JHC1485⟩.

2,1-Benzisothiazole compounds are prepared by the oxidation of o-aminophenyl-methanethiol ⟨72AHC(14)43⟩, o-aminothiobenzamides ⟨80JHC65⟩ or o-aminothiobenzoic acid ⟨75SST(3)541, 76JPR161⟩, or by reduction of the corresponding o-nitro compounds with tin(II) chloride ⟨72AHC(14)43, 77SST(4)339⟩. Treatment of o-aminobenzyl methyl sulfide with N-chlorosuccinimide, followed by oxidation gives the 2,1-benzisothiazole 2-oxide (**178**) ⟨79SST(5)345⟩.

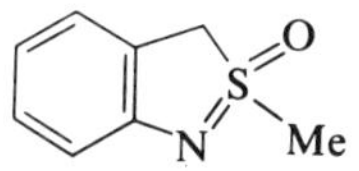

4.17.9.1.2 Formation of a bond adjacent to a heteroatom

This method is not applicable to mononuclear isothiazoles other than tetrahydroisothiazole 1-oxides and 1,1-dioxides. For example, 3-tosyloxypropanesulfonamides cyclize in the presence of base to give sultams ⟨80MI41700⟩. 2,1-Benzisothiazole compounds can be obtained by cyclization of N-sulfinyl-o-toluidines and similar compounds ⟨79SST(5)345⟩. In a related cyclization, treatment of o-methoxycarbonylbenzenesulfenamides with strong base gives 1,2-benzisothiazol-3-ones ⟨77SST(4)339⟩. This method is most useful for the preparation of 1,2-benzisothiazole 1,1-dioxides and saccharin derivatives from o-substituted benzenesulfonamides. Quite a wide range of compounds is available by variation of the group at the o-position and the N-substituent of the sulfonamide. Saccharin is manufactured commercially by the ring closure of o-carboxybenzenesulfonamide ⟨73AHC(15)233, 77SST(4)339⟩. A variant of the method uses the pyrolysis of o-substituted benzenesulfonyl azides, giving poor yields of dihydro compounds ⟨75JOC1525⟩.

4.17.9.1.3 Formation of a carbon–carbon bond

A few isothiazoles, 2,1-benzisothiazoles and monocyclic sultams have been prepared by methods involving formation of the 4,5-bond. Thus, condensation of α-cyanooximes (**179**) with ethyl mercaptoacetate or related compounds gives a compound (**180**) which cyclizes to the isothiazole compound (**181**; Scheme 29). *N*-Methanesulfonyl-*o*-chloroanilines cyclize to dihydro-2,1-benzisothiazole 2,2-dioxides (**182**) by the action of potassium amide in liquid ammonia ⟨77SST(4)339⟩.

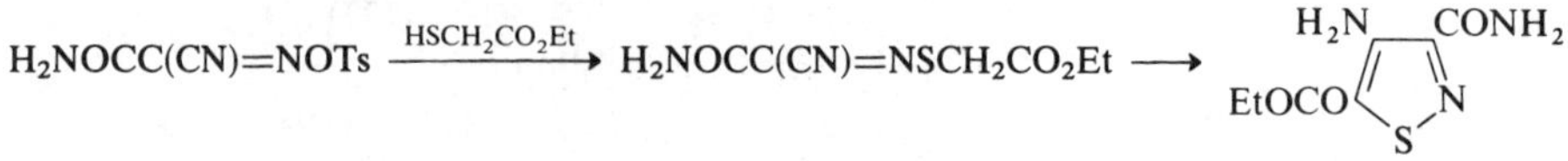

H$_2$NOCC(CN)=NOTs $\xrightarrow{\text{HSCH}_2\text{CO}_2\text{Et}}$ H$_2$NOCC(CN)=NSCH$_2$CO$_2$Et ⟶

(**179**)   (**180**)   (**181**)

Scheme 29

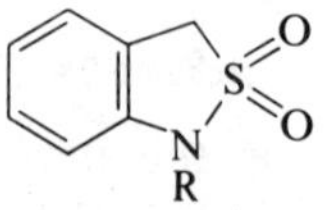

(**182**)

4.17.9.2 Formation of Two Bonds

4.17.9.2.1 From [4+1] atom fragments

The single atom fragment may be either nitrogen or sulfur. α,β-Unsaturated aldehydes having a thiocyanato or thiolsulfonate substituent at the β-position cyclize in liquid ammonia or with amines to give isothiazoles with a free 3-position. The use of analogous ketones gives 3-substituted compounds ⟨65AHC(4)107, 79SST(5)345⟩. Condensation of 3-chlorosulfenylpropionyl chloride with amines gives 2-substituted 4,5-dihydroisothiazol-3-ones ⟨72AHC(14)1⟩.

1,2-Benzisothiazoles are formed by the action of ammonia or amines on benzenesulfenyl halides bearing an *o*-carbonyl function ⟨72AHC(14)43, 80MI41700⟩. Analogous diphenyl disulfides behave similarly ⟨77SST(4)339⟩. Sulfinyl chlorides or phenylsulfoxides produce 1,2-benzisothiazole 1-oxides ⟨70ZOR2273, 77SST(4)339⟩, and sulfonyl chlorides give 1,2-benzisothiazole 1,1-dioxides and saccharin derivatives ⟨73AHC(15)233, 73SST(2)556⟩. *o*-Phenylsulfinylbenzoic esters react similarly with hydrazoic acid to give compound (**183**) ⟨71AG(E)76, 77SST(4)339⟩.

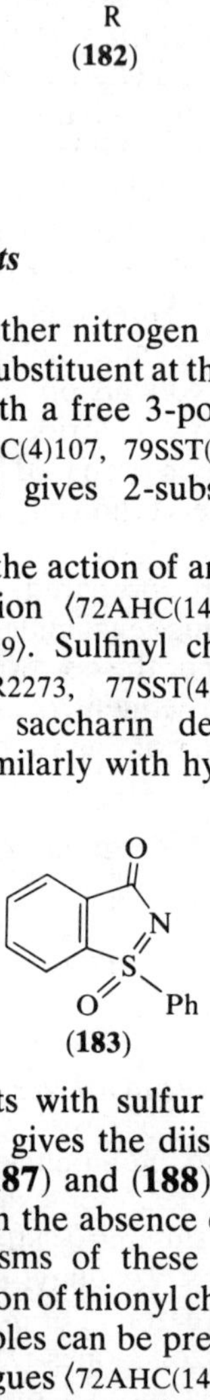

(**183**)

The α-acylbenzylamine (**184**) reacts with sulfur monochloride to give the 5-formylisothiazole (**185**), but thionyl chloride gives the diisothiazolyl disulfide (**186**; Scheme 30) ⟨72AHC(14)1⟩. The polychloronitriles (**187**) and (**188**) both give the 3,4-dichloroisothiazole (**189**) on heating strongly with sulfur in the absence of solvent (Scheme 31). The inventors offer no explanation of the mechanisms of these curious reactions ⟨74GEP2231098⟩. 3-Aroylpropionamides cyclize by the action of thionyl chloride to give 5-aroylisothiazol-3-ones ⟨80JHC1645, 81T3867⟩. 2,1-Benzisothiazoles can be prepared by the reaction between thionyl chloride and *o*-toluidine or its homologues ⟨72AHC(14)43, 73OPP197, 77SST(4)339⟩. Compounds (**190**) and (**191**) react together to give a 2,1-benzisothiazole (**192**; Scheme 32) ⟨80MI41700⟩.

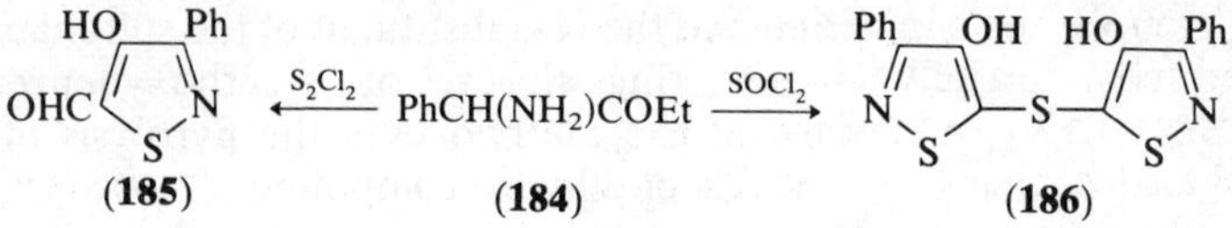

(**185**) $\xleftarrow{\text{S}_2\text{Cl}_2}$ PhCH(NH$_2$)COEt $\xrightarrow{\text{SOCl}_2}$ (**186**)

(**184**)

Scheme 30

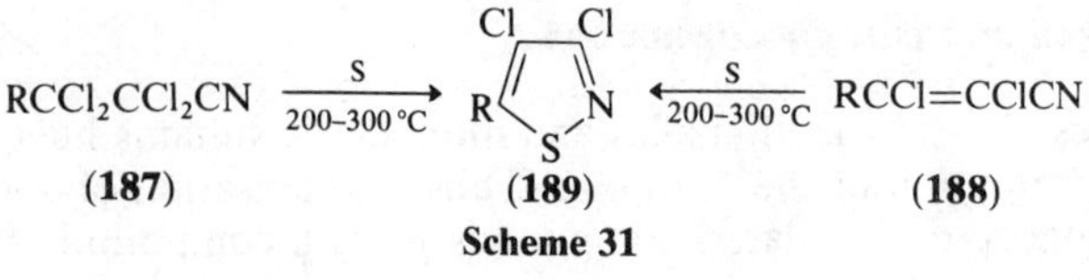

RCCl$_2$CCl$_2$CN $\xrightarrow[200-300\,°C]{\text{S}}$ (**189**) $\xleftarrow[200-300\,°C]{\text{S}}$ RCCl=CClCN

(**187**)   (**188**)

Scheme 31

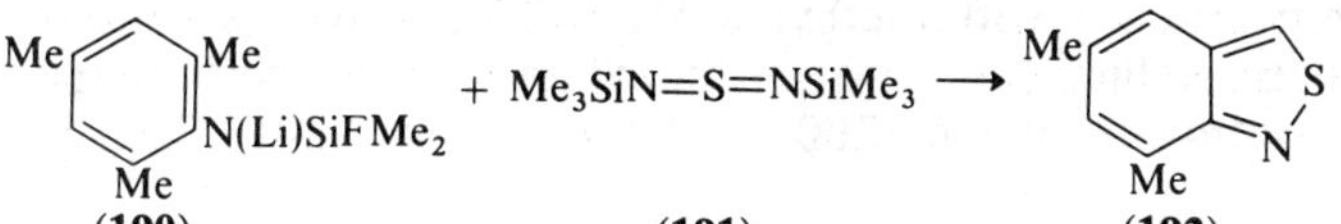

Scheme 32

4.17.9.2.2 From [3 + 2] atom fragments

The enamines (193) react with thiophosgene to give isothiazoles unsubstituted at the 5-position (194) ⟨72AHC(14)1⟩. The use of thiocyanates or carbon disulfide in place of thiophosgene leads respectively to nitrogen and sulfur functions at the 5-position ⟨75SST(3)541, 78JCS(P1)1017⟩. Trichloromethylsulfenyl chloride produces 5-chloro compounds or 5,5-dichloro-4,5-dihydro compounds ⟨75SST(3)541, 80MI41700⟩. Cyclic enamines react analogously. For example, cyclohexanone imines with carbon disulfide and sulfur give 4,5,6,7-tetrahydro-2,1-benzisothiazole-3-thione ⟨72AHC(14)43⟩.

Nitrile sulfides undergo cycloaddition reactions with substituted alkynes to give isothiazoles such as compound (195) ⟨75SST(3)541, 79JOC510, 80JOC3753⟩ (see Section 4.17.10.2). Aryl N-thiocarbonyldiphenylsulfimides react similarly ⟨76BCJ3124⟩. Chlorosulfonyl isocyanate undergoes photocycloaddition with excess of ethylene to give the dihydroisothiazole 1,1-dioxide (196), other conditions giving other products ⟨73SST(2)556⟩.

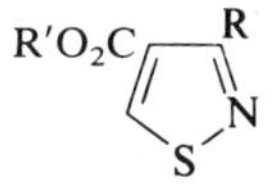

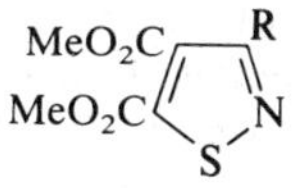

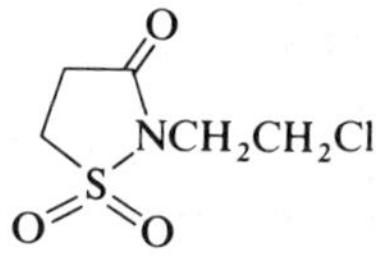

3-Chloro-α,β-unsaturated aldehydes condense with ammonium thiocyanate to give isothiazoles ⟨76EGP122249⟩. 2,3-Diphenylcyclopropenone reacts with N-sulfinylcyclohexylamine in the presence of nickel tetracarbonyl to give the isothiazolin-3-one 1-oxide (197) ⟨79SST(5)345⟩. Cholesteryl acetate reacts with trithiazyl trichloride in pyridine to give the isothiazolo steroid (198) ⟨77JCS(P1)916⟩.

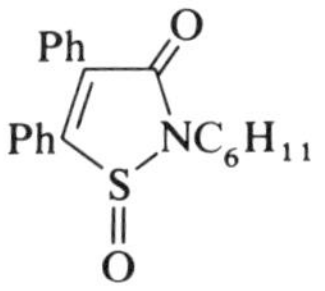

4.17.9.3 Formation of Three Bonds

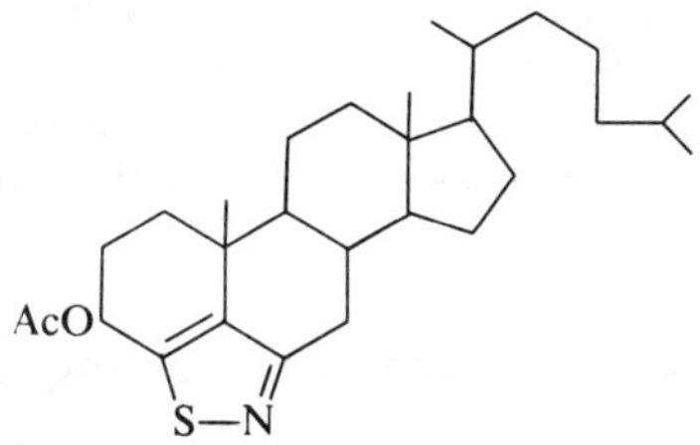

Isothiazole and its phenyl and lower alkyl derivatives can be prepared by passing a mixture of an alkene, sulfur dioxide and ammonia over an oxide catalyst at elevated temperature. Propylene gives isothiazole, isobutene gives 4-methylisothiazole which is not readily available by other methods, but both 1-butene and 2-butene give mixtures of 3- and 5-methylisothiazoles, because of isomerization of the alkene. Higher alkenes give low yields, and increasing amounts of thiophene compounds ⟨65AHC(4)107, 72AHC(14)1, 79SST(5)345⟩. The reaction between propynal, sodium thiosulfate and ammonia is the best laboratory method of preparing isothiazole itself ⟨72AHC(14)1⟩.

1,2-Benzisothiazoles can be prepared by the reaction of aromatic chloro compounds with sulfur and ammonia. Thus, 2,6-dichlorobenzylidene chloride gives 4-chloro-1,2-benzisothiazole ⟨72AHC(14)43⟩, and 2-chlorobenzophenone gives 3-phenyl-1,2-benzisothiazole ⟨79GEP2734866⟩.

Phenylacetonitrile reacts with diethyl sulfite to give 3-hydroxy-4,5-diphenylisothiazole, together with other products ⟨75SST(3)541⟩. Phenylketene reacts with compound (**199**) to give a mixture of the isothiazolidinone (**200**) and the pyrrole (**201**; Scheme 33) ⟨77SST(4)339, 79SST(5)345⟩.

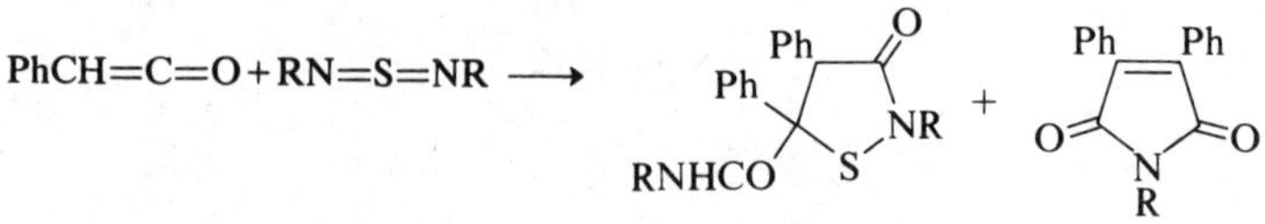

$$PhCH{=}C{=}O + RN{=}S{=}NR \longrightarrow$$

(**199**)   (**200**)   (**201**)

Scheme 33

4.17.10 SYNTHESIS FROM OTHER HETEROCYCLIC COMPOUNDS

4.17.10.1 From Six-membered Heterocycles

1,2-Benzisothiazole-3-acetic acids are prepared in good yield by the action of hydroxylamine on 4-hydroxy-1-thiocoumarins ⟨73SST(2)556⟩. Sulfilimines (**202**), derived from thiochromanones, give the 1,2-benzisothiazoles (**203**) in more modest yield, on treatment with alkali ⟨81JCS(P1)1037⟩. Pyrolysis of 2,4,6-tri-(o-azidophenyl)-2,4,6-trimethyl-1,3,5-trithiane, obtained from o-azidoacetophenone, gives 3-methyl-2,1-benzisothiazole ⟨71TL1315⟩. Sulfur is extruded from 3-methylthio-1,4,2-dithiazines by pyrolysis, to give 3-methylthioisothiazoles ⟨79SST(5)345⟩.

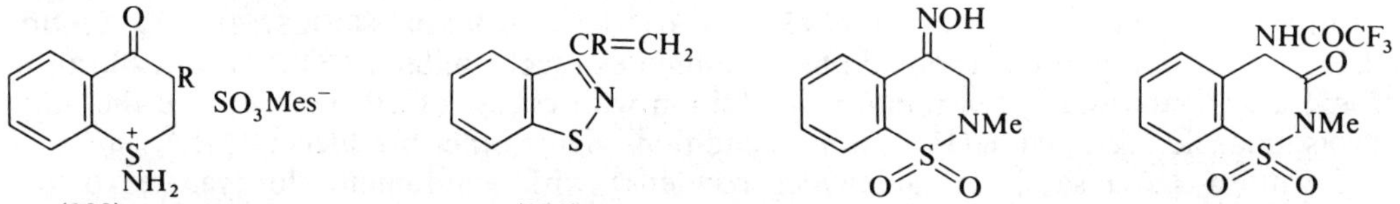

(**202**)   (**203**)   (**204**)   (**205**)

The little known 1-substituted-2,1-benzisothiazol-3-ones can be prepared from N-substituted isatoic anhydrides by successive treatment with potassium hydrosulfide and hydrogen peroxide ⟨78JHC529⟩. Isothiazoles can also be obtained by ring contraction of thiazines. Thus, the 1,2-benzothiazine 1,1-dioxide (**204**) is treated with trifluoroacetic acid to give compound (**205**), which is oxidized by permanganate to give N-methylsaccharin ⟨77JHC1063⟩. 2-Oximino-1,3-thiazines and their benzo analogues produce isothiazoles and 1,2-benzisothiazoles on treatment with polyphosphoric acid ⟨77JCS(P2)1114, 81ZC326⟩.

4.17.10.2 From Other Five-membered Heterocycles

Certain thiophene compounds having nitrogen substituents can be converted into isothiazoles and 1,2-benzisothiazoles. Thus, the azidothiophene (**206**) on refluxing in xylene gives a mixture of the isothiazole (**207**) and the thienopyridazine (**208**; Scheme 34) ⟨81CC550⟩. 3-Acetamido-2-nitrobenzothiophene treated with iron(II) oxalate at 225 °C gives a mixture of 3-cyano-1,2-benzisothiazole and the corresponding amide ⟨72AHC(14)43⟩. The same amide can be obtained by treatment of benzothiophene-2,3-dione with ammonia ⟨71FRP2043473⟩.

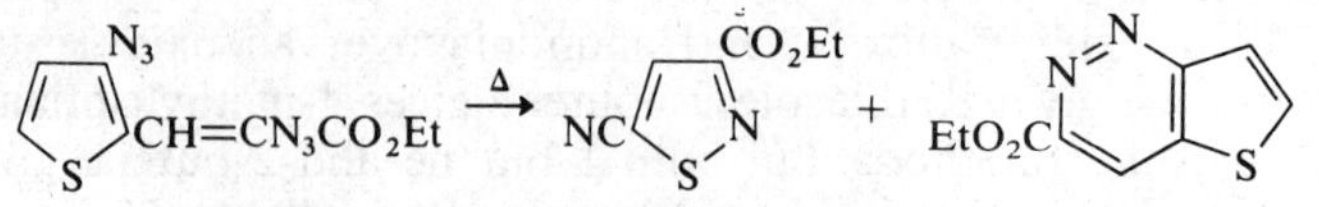

(**206**)   (**207**)   (**208**)

Scheme 34

Isoxazole compounds can be converted into the corresponding isothiazoles by successive catalytic hydrogenation, sulfuration with phosphorus pentasulfide and oxidation with chloranil ⟨72AHC(14)1, 75SST(3)541⟩. 2,1-Benzisoxazoles give the 2,1-benzisothiazoles directly, by the action of phosphorus pentasulfide in either pyridine or molten imidazole ⟨73SST(2)556, 77SST(4)339⟩. (See also Chapter 4.16 for further discussion of these topics.)

Mixtures of isothiazoles having phenyl or alkyl substituents are obtained, together with isomeric thiazoles, by photolysis of thiazole compounds in the presence or absence of iodine. The reactions are to some extent reversible (*cf.* Section 4.17.6.2). The proposed mechanisms are extremely complex, and the original papers need to be consulted ⟨74T879, 75SST(3)541, 80MI41700⟩. 3,4-Dimethyl-1,2,5-thiadiazole reacts with benzyne to give 3-methyl-1,2-benzisothiazole ⟨81AHC(28)183⟩. 2-Phenyl-5-bis(trifluoromethylmethylidene)-1,2,3,4-thiatriazole loses nitrogen on pyrolysis, and the organic fragment cyclizes to give a good yield of 3-hexafluoroisopropyl-2,1-benzisothiazole ⟨78JOC2500⟩.

2,1-Benzoxathioles and their *S*-substituted derivatives react with ammonia or amines to give 1,2-benzisothiazole 1-oxides and saccharin derivatives. Thus, the chloro compound (**209**), obtained by chlorination of *o*-mercaptobenzoic acid or of the corresponding disulfide, reacts with ammonia to give compound (**210**), with urea to give the diimino compound (**211**), and with arylsulfonamides to give the 1-oxides (**212**; Scheme 35) ⟨52HC(4)297, 72AHC(14)43⟩. 2,1-Benzoxathiol-3-one 1-oxide with ammonia gives a mixture of 1,2-benzisothiazol-3-one 1-oxide and the corresponding 3-imino derivative ⟨80MI41700⟩. The corresponding 1,1-dioxide gives *N*-arylsaccharins when treated with aromatic amines ⟨77USP4006007⟩. 3,3-Difluoro-2,1-oxathiole 1,1-dioxide, obtained by pyrolysis of lithium *o*-trifluoromethylbenzenesulfonate, reacts with ammonia to give saccharin ⟨73SST(2)556⟩.

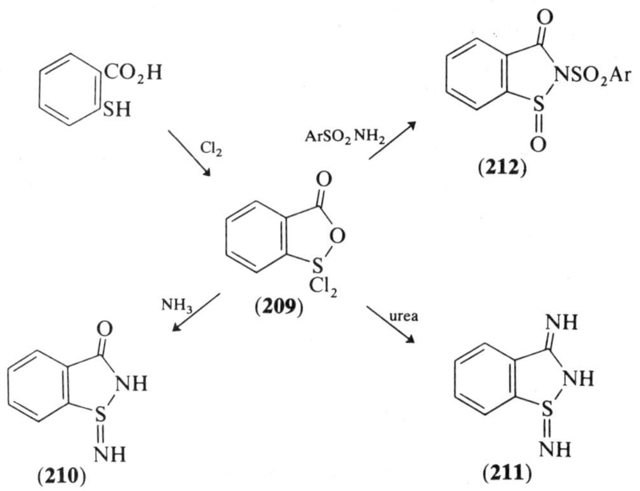

Scheme 35

1,2-Dithiolylium salts, particularly those with aryl substituents, react with ammonia to give isothiazoles (see Section 4.01.1.6). The original substitution pattern remains unchanged, the quaternary sulfur atom being replaced by a nitrogen atom. The same reaction using amines gives either 2-substituted isothiazoles or may also modify the 3-substituent ⟨72AHC(14)1, 75SST(3)541, 77SST(4)339, 80AHC(27)151⟩.

Some unquaternized dithiole compounds also produce isothiazoles. Thus, 3-acetylimino-1,2-dithioles react with hydroxylamine to give 3-acetamidoisothiazoles ⟨79SST(5)345⟩, and the 1,2-dithiolethione (**213**) reacts with amines to give the isothiazolethiones (**214**) ⟨81JHC437⟩. 1,2-Benzodithiol-3-ones and -3-thiones give the corresponding 1,2-benzisothiazol-3-ones and -3-thiones by treatment with amines ⟨52HC(4)297, 72AHC(14)43⟩. (For a further discussion see Chapter 4.30.)

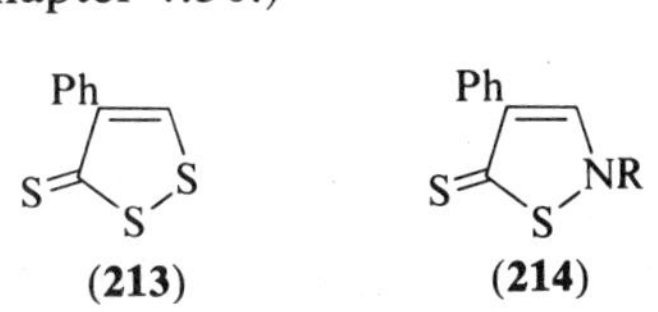

1,3,4-Oxathiazol-2-ones (**215**) on heating with alkenic or alkynic esters give the isothiazoles (**216**). This reaction proceeds by the intermediate formation of the nitrile sulfide (**217**) ⟨72AHC(14)1, 75SST(3)541, 80MI41700⟩. The same intermediate is produced by photolysis of 1,2,3,4-thiatriazoles, leading to the same isothiazole compounds ⟨75JA6197⟩. Photolysis of mixtures of the 1,3,2-oxathiazolylium-5-olate (**218**) and dimethyl acetylenedicarboxylate also gives isothiazoles by the same mechanism. Heating the same reaction mixture in the dark, however, produces the isomeric isothiazole compound (**219**) *via* the cycloadduct (**220**; Scheme 36) ⟨75SST(3)541⟩. (See also Chapter 4.33 for a further discussion of this topic.)

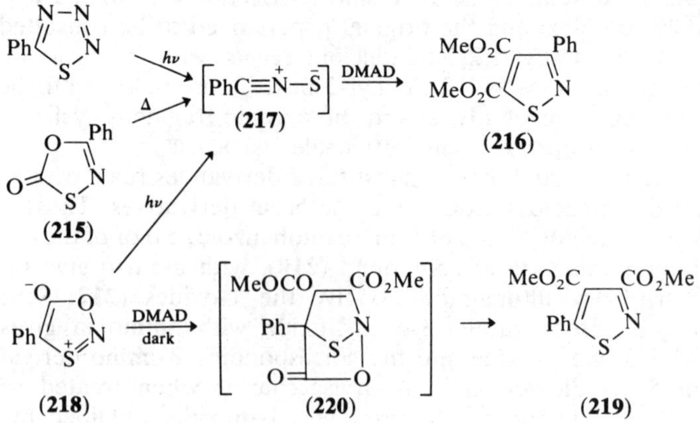

(215) (217) (216) (218) (220) (219)

Scheme 36

4.17.10.3 From Four-membered Heterocycles, Cephalosporins and Penicillins

Cephalosporin *S*-oxides and penicillin *S*-oxides (**221**) can be converted into isothiazol-3-ones (**222**) by the action of bases. These reactions proceed *via* an intermediate azetidinonesulfenic acid (**223**; Scheme 37) ⟨77SST(4)339⟩. Attempts to prepare *β*-lactam compounds from isothiazoles have, as yet, been unsuccessful ⟨81T2181⟩.

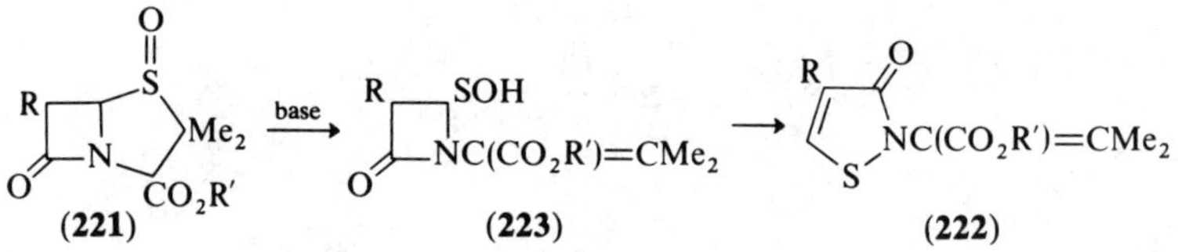

(221) (223) (222)

Scheme 37

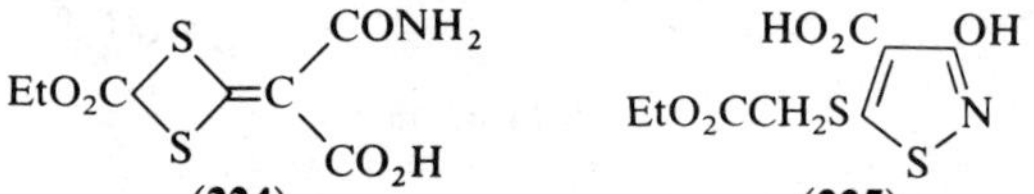

(224) (225)

The dithietane (**224**) reacts reversibly with base to give the isothiazole (**225**) ⟨80CPB2629⟩.

4.17.10.4 From Seven-membered Heterocycles

o-Thiocyanatoacetophenones react with hydroxylamine to give 3,1,4-benzoxathiazepines, which with polyphosphoric acid give 3-methyl-1,2-benzisothiazoles. The latter are also formed by heating *o*-thiocyanatoacetophenones with semicarbazide in ethylene glycol, probably *via* a 1,3,4-benzothiadiazepine intermediate ⟨79JCR(S)395⟩.

4.17.10.5 From Fused Heterocyclic Systems

Chlorination of sodium cyanodithioformate gives 3,4-dichloro-5-cyanoisothiazole (**227**), probably *via* the isothiazolodithiine (**226**) ⟨72AHC(14)1⟩. The thienoisothiazole (**228**) undergoes cycloaddition with alkynic esters to give adducts such as compound (**229**), which

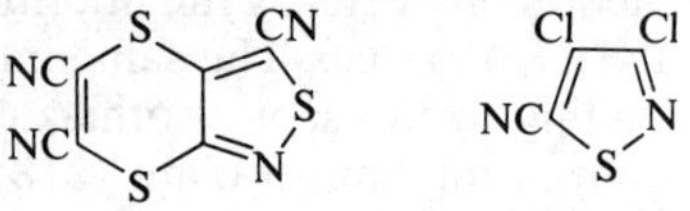

(226) (227)

lose sulfur to give 2,1-benzisothiazoles such as compound (**230**; Scheme 38) ⟨80MI41700⟩.

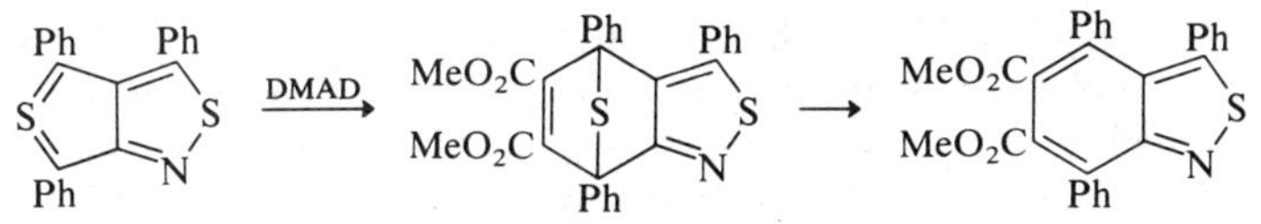

(**228**) (**229**) (**230**)

Scheme 38

4.17.11 BEST PRACTICAL METHODS OF SYNTHESIS

Isothiazole itself is best prepared by the reaction between propynal, ammonia and sodium thiosulfate (see Section 4.17.9.3). A wide range of substituted mononuclear isothiazoles can be obtained by oxidative cyclization of γ-iminothiols and related compounds (see Section 4.17.9.1.1). Substituents at the 3-position need to be in place before cyclization, but 4-substituents are readily introduced by electrophilic reagents (see Section 4.17.6.3), and 5-substituents *via* lithiation (see Section 4.17.6.4).

2,1-Benzisothiazoles are best prepared by oxidative cyclization of *o*-aminothiobenzamides (see Section 4.17.9.1.1), reaction of *o*-toluidines with thionyl chloride (see Section 4.17.9.2.1) or by sulfuration of 2,1-benzisoxazoles (see Section 4.17.10.2). 1,2-Benzisothiazoles can also be prepared from *o*-disubstituted benzene compounds, cyclodehydration of *o*-mercaptobenzaldoximes or oxidative cyclization of *p*-mercaptobenzylamines (see Section 4.17.9.1.1) being the most convenient. Both series of benzo compounds are readily substituted at the 5- and 7-positions by electrophilic reagents.

1,2-Benzisothiazole 1,1-dioxides and saccharin derivatives are best prepared by cyclization of *o*-substituted benzenesulfonamides (see Section 4.17.9.1.2).

4.17.12 APPLICATIONS

4.17.12.1 Important Compounds

The isothiazole ring does not occur in nature. By far the most important synthetic isothiazole derivative is saccharin. This was the first non-carbohydrate sweetening agent to be discovered, as long ago as 1879. It is about 300 times as sweet as sucrose, and is still used in many countries as a non-nutritive sweetener. After it was found that administration of massive doses to rats caused bladder cancer, its use was banned in the New World, but the controversy continues as to whether there is any danger when it is used in small quantity. Saccharin is also used as an additive in electroplating processes ⟨73AHC(15)233⟩.

Various isothiazol-3-ones and their metal salt complexes are good industrial microbicides, particularly the 2-octyl compound (**231**) and mixtures of 2-methylisothiazol-3-one with its 5-chloro analogue ('Kathon-886'). Since they also control the growth of algae, they are used in the paper making industry, especially in Japan, either alone or in combination with other agents. They have been recommended as preservatives to prevent fungal growth in a wide range of manufactured goods, such as emulsion paints, wood varnishes, adhesives, natural and artificial leather, cosmetics, *etc.* They can also be used to protect objects immersed in seawater from attack by algae and marine creatures. A large number of references to these compounds exist, mainly in specialist journals and the patent literature.

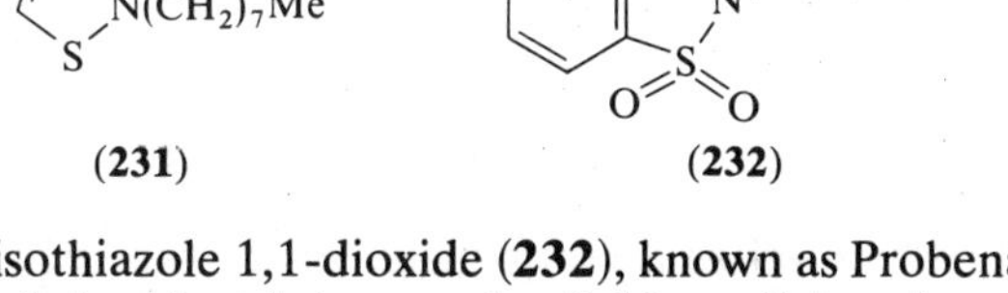

(**231**) (**232**)

2-Allyloxy-1,2-benzisothiazole 1,1-dioxide (**232**), known as Probenazole or Oryzaemate, is useful in the control of rice blast fungus under field conditions ⟨75SST(3)541⟩.

4.17.12.2 Chemical Applications

N-Bromosaccharin ⟨71IJC(B)1355⟩ and *N*-chlorosaccharin ⟨73IJC(B)609⟩ are useful alternatives to *N*-halogenosuccinimides in halogenation and oxidation reactions. *N*-Bromosaccharin gives better yields than *N*-bromosuccinimide for the benzylic bromination of toluene ⟨76S736⟩. *N*-Acylsaccharins are useful acylating agents, particularly for tertiary hydroxy groups. They also selectively acylate hydroxyamines at nitrogen under mild conditions ⟨72IJC(B)1194, 73AHC(15)233⟩. *N*-Arylsulfenylsaccharins have been suggested as sulfenylating agents. Reaction with Grignard reagents or alkyllithiums gives aryl thioethers, and with thiols gives alkyl aryl disulfides ⟨75S165⟩. 3-Bromomethyl-1,2-benzisothiazole affords a means of introducing a phenethyl group, by alkylation and subsequent removal of the heteroatoms ⟨72AHC(14)43⟩.

1-Alkyl-2,1-benzisothiazolium salts are useful intermediates for the preparation of substituted *o*-aminobenzaldehydes, which are often inaccessible by other means. This also affords a good method for the preparation of *o*-aminobenzaldehyde itself (Scheme 39) ⟨73JCS(P1)1863⟩. Reaction of thiophenol with 3-dialkylamino-1,2-benzisothiazoles or their quaternized derivatives provides a method of synthesis of *N*-substituted *o*-mercaptobenzamidines, which are otherwise difficult to obtain ⟨82LA14⟩. Carboxylic acids are reduced to the corresponding aldehydes by reaction with 3-chloro-1,2-benzisothiazole 1,1-dioxide, and reduction of the product (**233**) with sodium di-(2-methoxyethoxy)aluminum hydride (Scheme 40) ⟨77SST(4)339⟩. 3-Chloro-1,2-benzisothiazole 1,1-dioxide can also be used as a safe alternative to the dangerous azidoacetyl chloride in the synthesis of azetidinones from carboxylic acids and ketone imines ⟨77S407⟩.

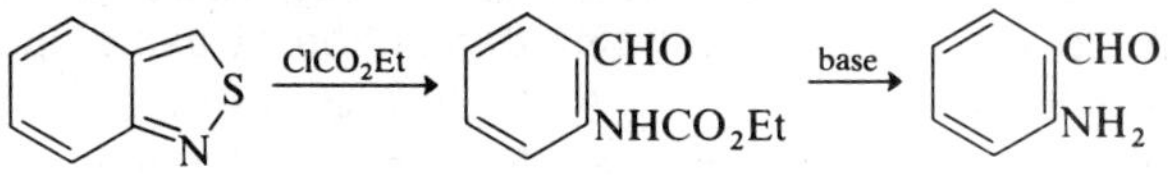

Scheme 39

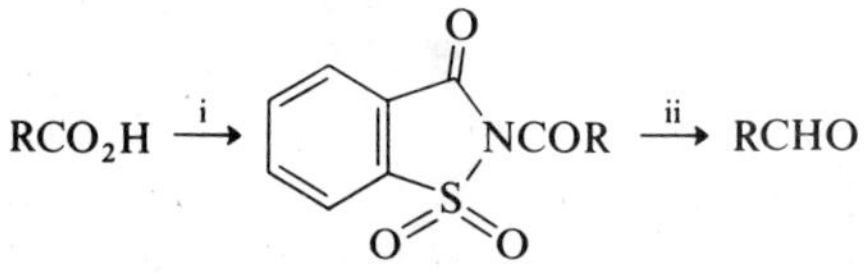

$$RCO_2H \xrightarrow{i} (233) \xrightarrow{ii} RCHO$$

i, 3-chloro-1,2-benzisothiazole 1,1-dioxide; ii, sodium di(2-methoxyethoxy)aluminum hydride

Scheme 40

The succinimide derivative (**234**) can be used in peptide synthesis for conversion of amino acids into their succinimide esters (**235**; Scheme 41) ⟨79CL1265⟩. 3-Substituted mercapto-1,2-benzisothiazole 1,1-dioxides (**236**) have been recommended as an odourless means of storage of thiols. The latter are readily regenerated by the action of piperidine ⟨81CL1457⟩.

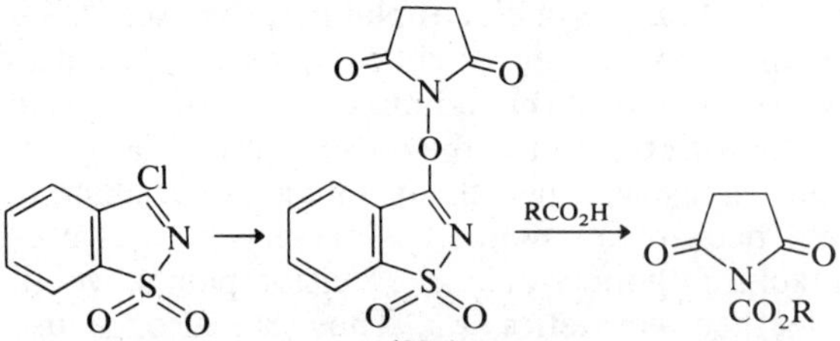

Scheme 41

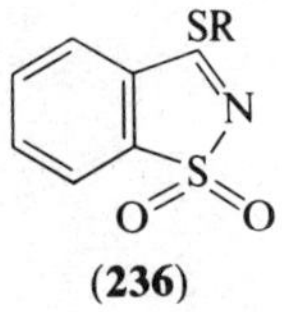

(**236**)

The isothiazole ring was used as a template in Woodward's brilliant total synthesis of colchicine (**237**; Scheme 42) ⟨72AHC(14)1⟩.

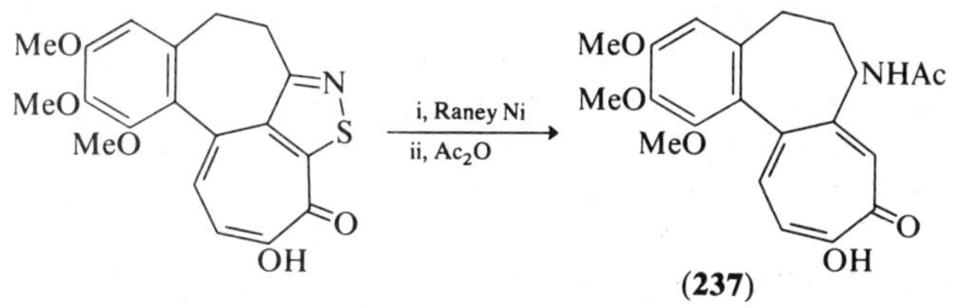

Scheme 42

4.17.12.3 Biological and Industrial Applications

The isothiazole ring has been incorporated into a wide range of known biologically active compounds, either as a substituent group or taking the place of another ring. Although many of these isothiazole compounds exhibit biological activity, only in a few cases has the activity of the model compound been improved. All types of pharmacological activity have been claimed. Among the more interesting are the saccharin derivative (**238**), which has potent sedative, hypnotic and anticonvulsant activity ⟨79SST(5)345⟩, amides and esters of 2,1-benzisothiazole-3-carboxylic acids, some of which are promising local anaesthetics ⟨79SST(5)345⟩, and the adrenergic β-blockers (**239**) ⟨80JMC65⟩. The sultam analogue (**240**) of 11-deoxyprostaglandin-E_2 has been prepared ⟨80MI41700⟩, and the pyridinium salt (**241**) has moderate hypoglycaemic activity ⟨75SST(3)541⟩. The cyclizine analogue (**242**) has appetite suppressant activity, which can be utilized in the management of broiler breeder hens ⟨81UP41700⟩. The amides (**243**) are potent antiinflammatory agents ⟨80MI41702⟩.

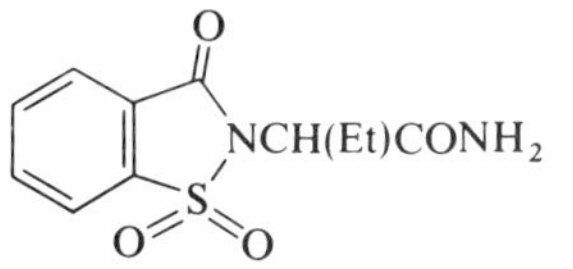

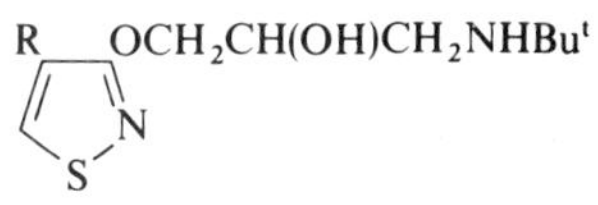

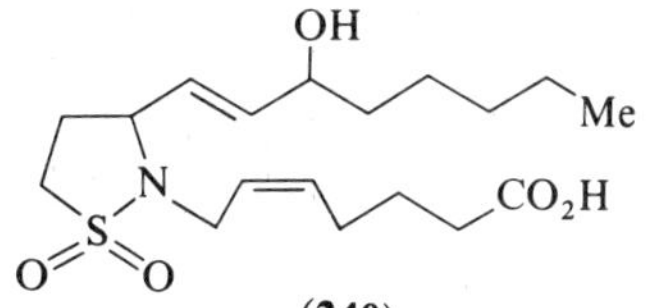

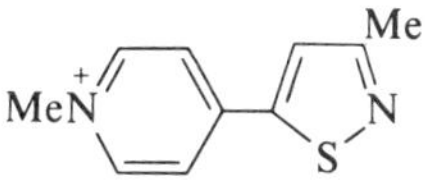

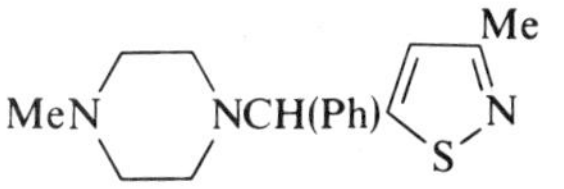

Isothiazole substituents have been attached to β-lactam antibiotics and to macrocyclic antibiotics such as erythromycin. The sulfa drug, Sulfasomizole (**244**) also has good antibacterial activity. Thiosemicarbazones of 5-formyl- and 5-acetylisothiazoles show high activity against the pox group of viruses ⟨65AHC(4)107⟩.

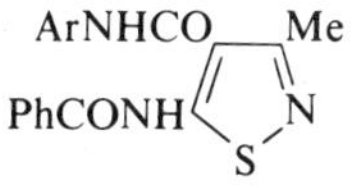

The most interesting compounds in the agrochemical sphere are the acid (**245**) and related compounds, which have high herbicidal activity of the auxin type ⟨77SST(4)339, 79SST(5)345⟩. Organophosphorus insecticides incorporating the isothiazole ring have been prepared ⟨72AHC(14)1, 72AHC(14)43⟩, and the chrysanthemic acid derivatives (**246**) are also claimed as being insecticidal ⟨78JAP(K)7892768⟩.

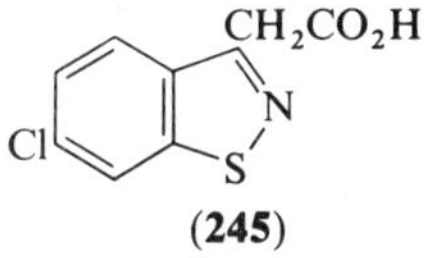

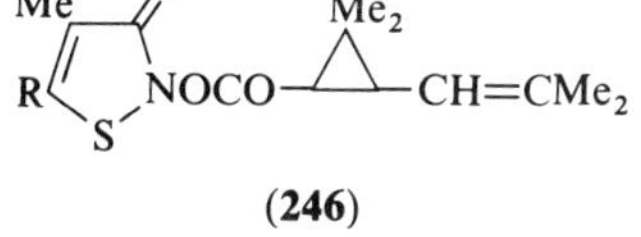

1,2-Benzisothiazole derivatives having biological activity in the medical and agrochemical fields have been reviewed ⟨81MI41702⟩.

There are many references in the patent literature to azo dyes prepared from 4- and 5-aminoisothiazoles, 3-, 5- and 7-amino-1,2-benzisothiazoles, and their quaternized derivatives. These are particularly useful in the dyeing of synthetic fibres. Isothiazole compounds have also been suggested for other industrial purposes, such as corrosion inhibitors, fireproofing agents, additives in rubber vulcanization, photographic chemicals and fluorescent whiteners in detergents.

4.18

Oxazoles and their Benzo Derivatives

G. V. BOYD
Chelsea College, University of London

4.18.1 INTRODUCTION	178
4.18.2 STRUCTURE	179
4.18.2.1 Survey of Structures	179
4.18.2.2 Theoretical Methods	179
4.18.2.3 Structural Methods	180
4.18.2.3.1 X-Ray diffraction	180
4.18.2.3.2 Microwave spectroscopy	181
4.18.2.3.3 NMR spectroscopy	181
4.18.2.3.4 UV spectroscopy	182
4.18.2.3.5 IR spectroscopy	182
4.18.2.3.6 Mass spectrometry	183
4.18.2.3.7 Photoelectron spectroscopy	183
4.18.2.4 Thermodynamic Aspects	184
4.18.2.4.1 Thermochemistry	184
4.18.2.4.2 Aromaticity	184
4.18.2.4.3 Conformation and configuration	185
4.18.2.5 Tautomerism	185
4.18.2.5.1 Substituent tautomerism	185
4.18.2.5.2 Ring–chain tautomerism	186
4.18.3 REACTIVITY	187
4.18.3.1 Reactions at Fully Conjugated Rings	187
4.18.3.1.1 General survey of reactivity	187
4.18.3.1.2 Reactions of oxazoles and oxazolium salts	188
4.18.3.1.3 Reactions of oxazole N-oxides	198
4.18.3.1.4 Reactions of non-aromatic compounds in potential tautomeric equilibrium with oxazoles: oxazolones	199
4.18.3.1.5 Reactions of mesoionic compounds	206
4.18.3.2 Reactions of Partially Saturated and Saturated Rings	211
4.18.3.2.1 Dihydrooxazoles	211
4.18.3.2.2 Oxazolidines and oxo derivatives	213
4.18.3.3 Reactions of Substituents	215
4.18.3.3.1 Fused benzene rings and phenyl substituents	215
4.18.3.3.2 Alkyl groups	215
4.18.3.3.3 Other carbon-linked functional groups	216
4.18.3.3.4 Amino and mercapto groups	216
4.18.4 SYNTHESES	216
4.18.4.1 Synthesis of Oxazoles, Benzoxazoles and Oxazolium Salts	216
4.18.4.1.1 Ring synthesis from non-heterocyclic compounds	216
4.18.4.1.2 Ring synthesis by transformation of heterocyclic compounds	222
4.18.4.2 Synthesis of Oxazole N-Oxides	223
4.18.4.3 Synthesis of Oxazolones and Mesoionic Oxazoles	223
4.18.4.3.1 2(3H)-Oxazolones and benzoxazolones	223
4.18.4.3.2 2(5H)-Oxazolones	224
4.18.4.3.3 4(5H)-Oxazolones and anhydro-4-hydroxyoxazolium hydroxides	225
4.18.4.3.4 5(4H)-Oxazolones and anhydro-5-hydroxyoxazolium hydroxides	225
4.18.4.3.5 5(2H)-Oxazolones	227
4.18.4.4 Synthesis of Dihydrooxazoles	227
4.18.4.4.1 2,3-Dihydrooxazoles and 2,3-dihydrobenzoxazoles	227
4.18.4.4.2 2,5-Dihydrooxazoles	228
4.18.4.4.3 4,5-Dihydrooxazoles	228
4.18.4.5 Synthesis of Oxazolidines and Oxo Derivatives	229
4.18.4.5.1 Oxazolidines	229
4.18.4.5.2 2-Oxazolidinones	229

4.18.4.5.3 4-Oxazolidinones	230
4.18.4.5.4 5-Oxazolidinones	230
4.18.4.5.5 2,4-Oxazolidinediones	231
4.18.4.5.6 2,5-Oxazolidinediones	231
4.18.4.5.7 4,5-Oxazolidinediones	231
4.18.5 OXAZOLES IN NATURE AND APPLICATIONS OF OXAZOLES	232
4.18.5.1 Naturally Occurring Oxazoles	232
4.18.5.2 Uses of Oxazoles	233

4.18.1 INTRODUCTION

The history of oxazole (**1**) begins in 1887 when Hantzsch ⟨1887CB3118⟩ gave that name to a class of compounds then few in number. Eleven years earlier Ladenburg ⟨1876CB1524⟩ had obtained 2-methylbenzoxazole from *o*-aminophenol and acetic anhydride. The parent compound (**2**) was prepared in 1947 by a lengthy route ⟨47JCS96⟩; a more practical synthesis was described in 1962 ⟨62AG(E)662⟩. There was only sporadic interest in oxazoles until the Second World War, when a joint Anglo-American research effort was launched to synthesize penicillin, which was at first thought to contain the oxazole nucleus. This work, which is the foundation of modern oxazole chemistry, is summarized in the monumental treatise 'The Chemistry of Penicillin' ⟨B-49MI41800⟩.

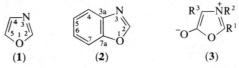

Interest in oxazoles was greatly stimulated by Kondrateva's discovery ⟨57MI41800⟩ that these compounds function as dienes in the Diels–Alder reaction and by Huisgen's work ⟨B-67MI41800⟩ on 1,3-dipolar cycloaddition reactions of mesoionic derivatives (**3**). 5(4*H*)-Oxazolones, because of their close relationship to α-amino acids, have been studied by protein chemists for many years. Naturally occurring oxazoles are rare, but during the past 20 years a few have been found as alkaloids and antibiotics. Oxazoles are applied as scintillators, fluorescent whitening agents and drugs.

There are several reviews on the chemistry of oxazoles. The early history has been summarized by Wiley ⟨45CRV(39)401⟩, and Cornforth has written very readable, comprehensive monographs on oxazoles and benzoxazoles ⟨B-57MI41801, B-57MI41802⟩. Recent reviews are by Lakhan and Ternai ⟨74AHC(17)99⟩, a chapter by Campbell ⟨B-79MI41800⟩ and the very detailed accounts by Turchi and Dewar ⟨75CRV389⟩ and Turchi ⟨81MI41800⟩. Reviews of more specialized topics are mentioned in the appropriate sections. These reviews will be referred to frequently and original papers will, in general, only be cited if they are not quoted in the reviews.

In this chapter, oxazole and its derivatives are named and numbered as in *Chemical Abstracts*. Thus compound (**6**) is called 4,5-dihydrooxazole rather than 2-oxazoline or Δ²-oxazoline, (**7**) is 2,5-dihydrooxazole, the betaines (**3**) are named anhydro-5-hydroxyoxazolium hydroxides and not oxazolium 5-oxides or oxazolium 5-olates, and the oxo derivatives (**4**) and (**5**) are 5(4*H*)-oxazolone and 5(2*H*)-oxazolone, respectively, the position of the 'extra' hydrogen atom being indicated in parentheses. The fully saturated compound (**8**) is oxazolidine; its oxo derivatives are named oxazolidinones and oxazolidinediones, *e.g.* compound (**9**) is 2-oxazolidinone and (**10**) is 4,5-oxazolidinedione. A formula such as (**11**) is not meant to imply that all the substituents are methyl groups; it represents a general oxazolidine derivative and is used in place of the cumbersome expression (**12**; R^1–R^7 = H, alkyl or aryl).

Oxazoles and their Benzo Derivatives

The order in which the various topics are discussed is evident from the table of contents. It should be pointed out that thiones are treated together with oxazolones, while aminooxazoles, which are tautomeric with oxazoleimines, will usually be found in sections dealing with oxazoles.

The chemistry of benzoxazole differs from that of oxazole in many respects. However, the presence of additional benzene rings, as in the naphthooxazoles and phenanthrooxazoles, does not introduce any new features; these compounds are therefore mentioned only incidentally.

4.18.2 STRUCTURE

4.18.2.1 Survey of Structures

Fully conjugated aromatic compounds of the oxazole series are the oxazoles (13), benzoxazoles (14), oxazolium and benzoxazolium salts, e.g. (15), and the oxazole N-oxides (16). There are five systems with exocyclic conjugation: the 2(3H)- and 2(5H)-oxazolones (17) and (18), respectively, the 4(5H)-oxazolones (19), the 5(2H)-oxazolones (20), and, finally, the important 5(4H)-oxazolones (21). Some sulfur analogues of these compounds, such as 2(3H)-oxazolethione (22), are known; in contrast, nitrogen compounds exist predominantly or exclusively as aminooxazole tautomers. Derived from the 2(3H)-oxazolones are the benzoxazolones (23), benzoxazolethiones and benzoxazoleimines (24). There are two betaine structures, the mesoionic anhydro-4-hydroxyoxazolium hydroxides (25) and the 5-hydroxy analogues (26), the 'münchnones'; these are closely related to the 4- and 5-oxazolones, respectively.

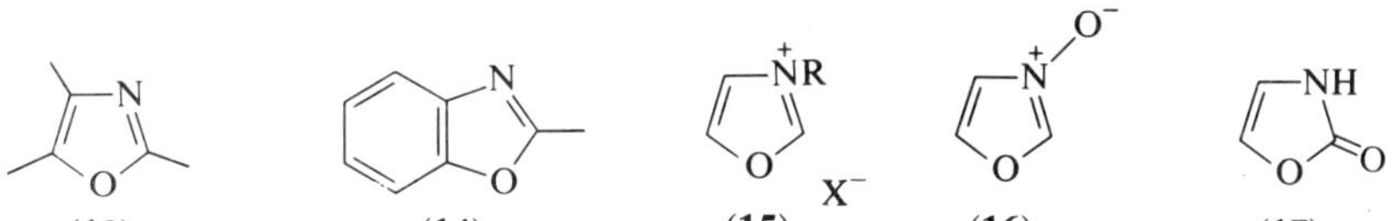

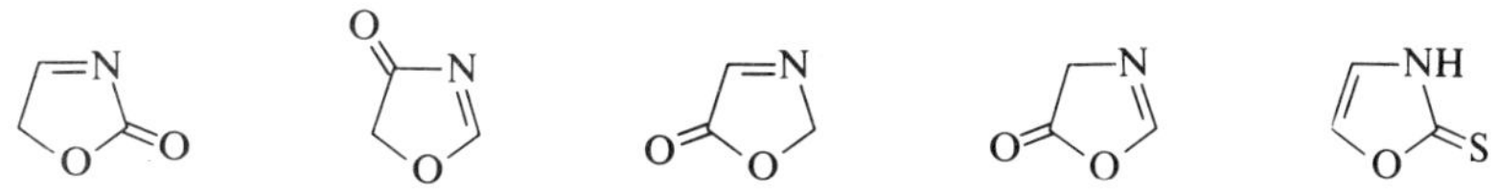

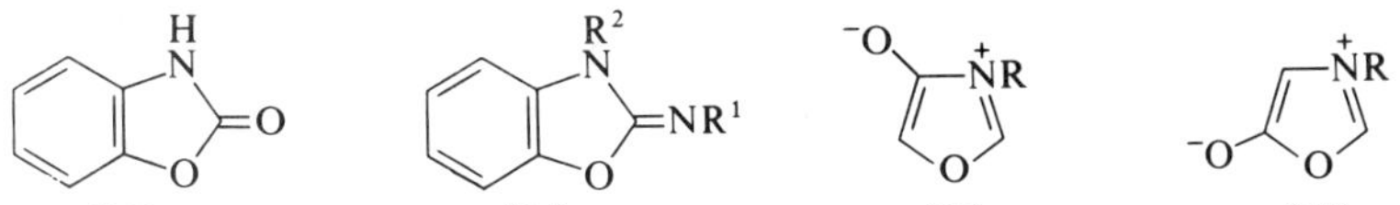

Representatives of all possible dihydrooxazoles, 2,3-dihydrooxazole (27), 2,5-dihydrooxazole (28) and 4,5-dihydrooxazole (29) have been prepared, but surprisingly little is known about 2,3-dihydrobenzoxazoles (30), which are tautomeric with phenolic Schiff's bases (see Section 4.18.2.5.2).

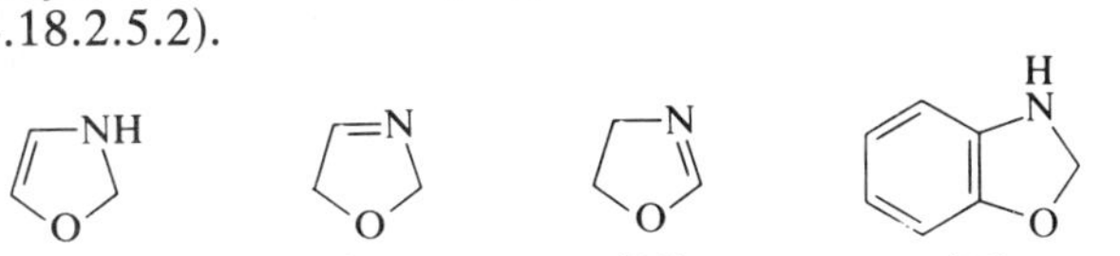

The fully saturated oxazolidine (31), its three monooxo derivatives, 2-oxazolidinone (32), 4-oxazolidinone (33) and 5-oxazolidinone (34), and the three oxazolidinediones (35)–(37) complete the list of oxazole structures. Oxazolidinetriones (38) have not been reported.

4.18.2.2 Theoretical Methods

There have been several molecular orbital treatments of the structure and reactivity of oxazole; these are summarized by Turchi and Dewar ⟨75CRV389⟩. Table 1 lists the σ, π and net charge distributions calculated by the all-valence *ab initio* method, the net charges obtained by the MINDO/3 method and the bond lengths calculated by the latter method.

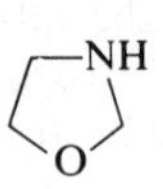

(31) 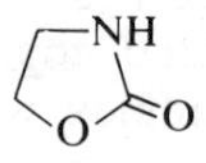 (32) 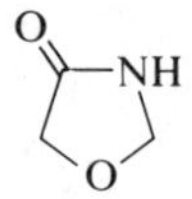 (33) 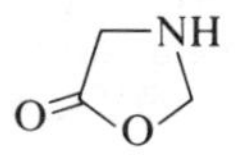 (34)

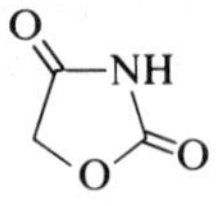

(35) 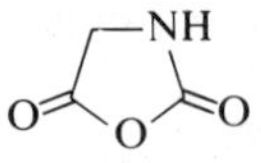 (36) 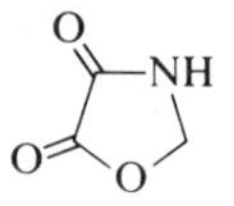 (37) 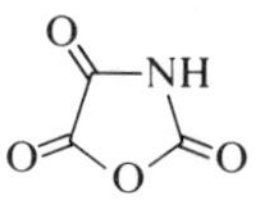 (38)

It is seen that the σ-bond framework of the molecule is polarized to a great extent in accordance with the electronegativities of the ring atoms and the net charge distribution is dominated by this polarization. Of the three carbon atoms, C(2) is predicted to have the lowest net charge and C(4) the highest; these results are in broad agreement with ^{1}H and ^{13}C NMR spectroscopy (see Section 4.18.2.3.3); furthermore, C(2) is the most reactive towards nucleophiles and the least reactive towards electrophiles. Calculated free-valence indices predict that position 5 should be the most susceptible to attack by free radicals.

Table 1 Charge Distribution and Bond Lengths in Oxazole

	Ab initio			MINDO/3		MINDO/3 Bond length (Å)
	q_σ	q_π	q_{net}	q_{net}		
O	−0.63	0.32	−0.31	−0.35	O—C(2)	1.35
C(2)	0.14	−0.06	0.08	0.35	O—C(5)	1.35
N	−0.17	−0.08	−0.25	−0.19	C(2)—N	1.30
C(4)	−0.09	−0.09	−0.18	−0.05	N—C(4)	1.41
C(5)	0.04	−0.09	−0.05	0.19	C(4)—C(5)	1.36

4.18.2.3 Structural Methods

4.18.2.3.1 X-Ray diffraction

The X-ray structures of several aryloxazoles have been determined. The bond lengths (Å) and angles of the oxazole ring in 2,2'-p-phenylenebis(5-phenyloxazole) (39) are given in Figure 1 ⟨65AX942⟩. The oxazole ring is planar; the distances indicate considerable bond fixation in oxazole and, in particular, there is a striking difference in the lengths of the C(2)—N and C(4)—N bonds. The X-ray structure of anhydro-2-m-bromophenyl-5-hydroxy-3-methyl-4-trifluoroacetyloxazolium hydroxide (40) has been determined ⟨75JCS(P2)1280⟩. The significant bond lengths are shown in Figure 2. The exocyclic carbon–oxygen bond C(5)—O(6) is short (1.23 Å), like those in γ-lactones, whereas the C(5)—O(1) bond is much longer than the other bonds in the ring.

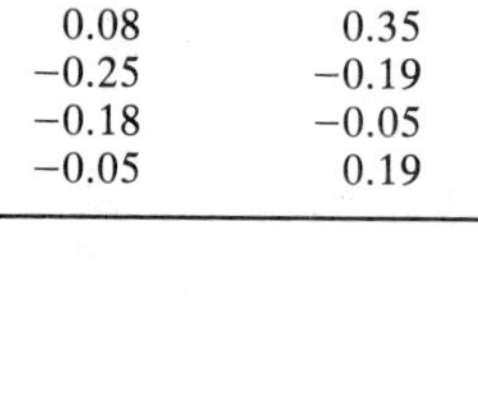

Figure 1 Bond lengths (Å) and bond angles (°) for 2,2'-p-phenylenebis(5-phenyloxazole) (39)

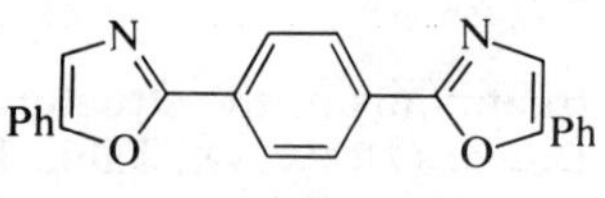

(39) 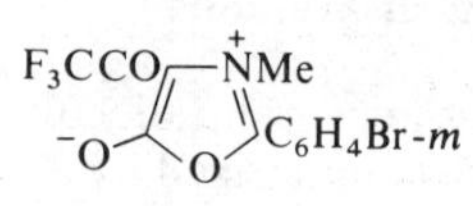 (40)

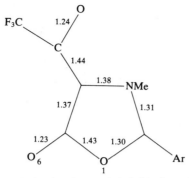

Figure 2 Bond lengths (Å) for anhydro-2-*m*-bromophenyl-5-hydroxy-3-methyl-4-trifluoroacetyloxazolium hydroxide (**40**)

4.18.2.3.2 Microwave spectroscopy

The dipole moment (1.50 D) and quadruple coupling constants of oxazole and its monodeuterio derivatives have been determined ⟨78ZN(A)549⟩. Microwave spectroscopy of oxazole indicates that the molecule has the planar geometry shown in Figure 3 ⟨78ZN(A)145⟩.

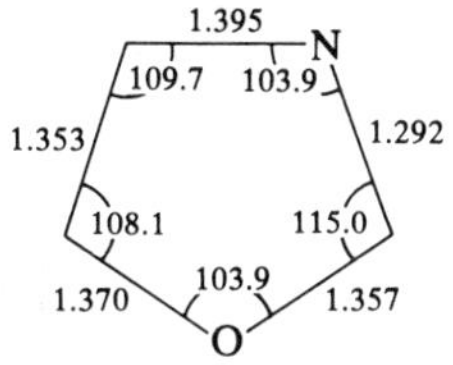

Figure 3 Microwave spectral data for oxazole

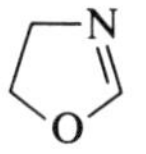

(**41**)

The microwave spectrum of 4,5-dihydrooxazole (**41**) has been reported in detail ⟨74JCP(61)253⟩; the dihydrooxazole ring is planar.

4.18.2.3.3 NMR spectroscopy

The ^{1}H NMR spectra of oxazole, methyloxazoles and benzoxazole are collected in Tables 2 and 3; ^{13}C and ^{14}N NMR spectra of oxazole and benzoxazole are found in Tables 4–6. The ^{13}C NMR spectra of numerous oxazolidines have been reported ⟨79MI41801⟩; for ^{1}H NMR data of 2-oxazolidinones (**42**; R^1 = H or Me, R^2 = Me or Ph), see ⟨72CB2462⟩, and for ^{1}H and ^{13}C NMR spectra of a series of 2,4-oxazolidinediones (**43**; R = H or Me), see ⟨79OMR(12)178⟩.

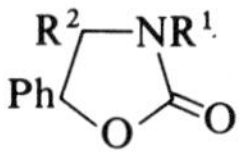

(**42**) 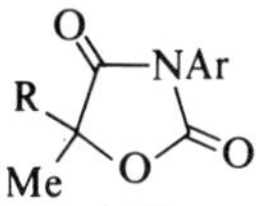   (**43**)

Table 2 ^{1}H NMR Spectra[a] of Ring Hydrogen Atoms in Oxazole and of Methyl Hydrogen Atoms in Methyloxazoles
⟨69JCS(B)270⟩

^{1}H Chemical shifts (p.p.m.)			Coupling constants (Hz)			^{1}H Chemical shifts (p.p.m.)		
2H	4H	5H	2–4	2–5	4–5	2-Me	4-Me	5-Me
7.95	7.09	7.69	—	0.8	0.8	2.2–2.3	1.9–2.2	2.15–2.25

[a] In CCl$_4$.

Table 3 The ^{1}H NMR[a] Spectrum of Benzoxazole ⟨69CPB1815⟩

^{1}H Chemical shifts (p.p.m.)					Coupling constants (Hz)									
2	4	5	6	7	2–4	2–5	2–6	2–7	4–5	4–6	4–7	5–6	5–7	6–7
7.46	7.67	7.80	7.79	7.73	0.3	0	0.3	0	8.0	0.9	0.7	7.4	1.1	8.1

[a] In MeOH.

Table 4 The ^{13}C NMR[a] Spectrum of Oxazole ⟨79CJC3168⟩

^{13}C Chemical shifts (p.p.m.)			C–H coupling constants (Hz)				
2	4	5	2–2	2–4	2–5	4–2	4–4
150.6	125.4	138.1	231.1	10.7	7.9	8.9	195.3

[a] In CDCl$_3$.

Table 5 The ^{13}C NMR Spectrum of Benzoxazole[a,b] ⟨78OMR(11)385⟩

^{13}C Chemical shifts (p.p.m.)						
2	3a	4	5	6	7	7a
152.6	140.1	120.5	125.4	124.4	110.8	150.0

[a] For numbering, see formula (**2**). [b] In CDCl$_3$.

Table 6 The ^{14}N NMR Spectra of Oxazole and Benzoxazole ⟨72T637⟩

Oxazole[a,b]	Benzoxazole[a,c]
+124 ± 1 p.p.m.	+140 ± 3 p.p.m.

[a] Referred to internal MeNO$_2$. [b] In CCl$_4$. [c] Neat.

4.18.2.3.4 UV spectroscopy

The UV absorption maxima of oxazole, benzoxazole and various substituted oxazoles are collected in Table 7.

Table 7 UV Spectral Data for Oxazole Derivatives

Compound	Solvent	λ_{max} (nm)	log ε	Ref.
Oxazole	MeOH	205	3.59	60CB1389
Benzoxazole	Vapour	247	—	78MI41800
4-Methyl-5-phenyloxazole	MeOH	265	4.23	60CB1389
2,4-Dimethyl-5-phenyloxazole	MeOH	264	4.29	60CB1389
4-Phenyloxazole	MeOH	245	4.25	60CB1389
5-Methyl-4-phenyloxazole	MeOH	247	4.18	60CB1389
2,5-Dimethyl-4-phenyloxazole	MeOH	247	4.06	60CB1389
2,5-Diphenyloxazole	C$_6$H$_{12}$	302	2.81	63T169

4.18.2.3.5 IR spectroscopy

The IR stretching modes of oxazoles are at 1650–1610, 1580–1550, 1510–1470, 1485 and 1380–1330 cm^{-1}. Experimental and calculated frequencies of the normal vibrational modes of oxazole have been published ⟨75JOU1737⟩; the Raman and IR spectra of gaseous benzoxazole have been determined ⟨72CR(B)(274)532⟩. Carbonyl absorptions of oxazolones and mesoionic oxazoles are collected in Table 8. The IR spectra of 2,4-diarylmünchnones (**44**) contain intense carbonyl absorptions at *ca.* 1710 cm^{-1} (CH$_2$Cl$_2$) ⟨70CB2581⟩. Münchnones lacking substituents at C(4), which are unstable (see Section 4.18.3.1.5(i)), show carbonyl bands at 1740–1730 cm^{-1}, which gradually decrease in intensity ⟨72JCS(P1)914⟩. A normal coordinate analysis of the IR spectrum of 2-oxazolidinone has been carried out and fundamental frequencies have been assigned ⟨80SA(A)199⟩.

Table 8 $\nu_{C=O}$ Frequencies for Oxazolones and Mesoionic Oxazoles

Class of compound	ν_{max} (cm^{-1})
2(3H)-Oxazolones	1780–1750
2(5H)-Oxazolones	ca. 1780
4(5H)-Oxazolones	1770–1740
5(2H)-Oxazolones	ca. 1820
5(4H)-Oxazolones	ca. 1820
Anhydro-4-hydroxyoxazolium hydroxides	1670–1660
Anhydro-5-hydroxyoxazolium hydroxides	1740–1710

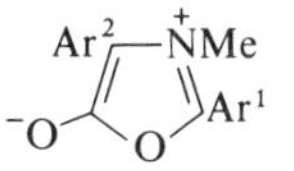

(44)

4.18.2.3.6 Mass spectrometry

The mass spectra of oxazoles show sequential loss of carbon monoxide and a cyanide; the fate of the remaining fragment depends on the nature and position of substituents. In general, the radical cation (45) rearranges to (46) by migration of the 5-substituent and fragmentation ensues (Scheme 1) ⟨76OMS1047⟩. A detailed examination of the mass spectrum of benzoxazole has revealed that the molecular ion first isomerizes to o-hydroxybenzonitrile radical cation (47), which then loses either carbon monoxide or hydrogen cyanide ⟨74OMS(9)149⟩. The mass spectrum of benzoxazolethione shows loss of sulfur ⟨73JOC1356⟩; 2-aminobenzoxazole fragments to carbon monoxide and a radical cation regarded as (48) ⟨70OMS(3)1341⟩.

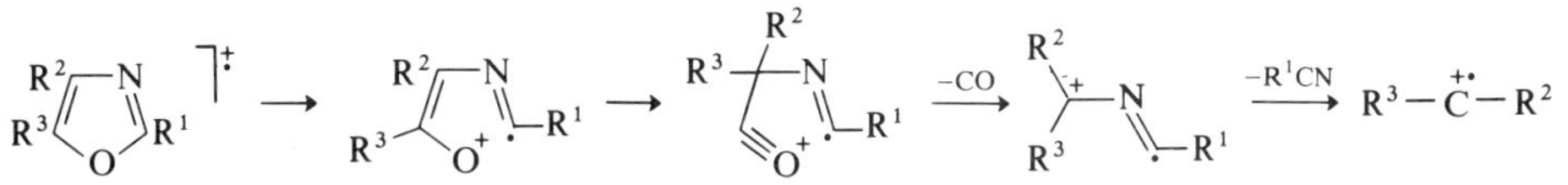

Scheme 1

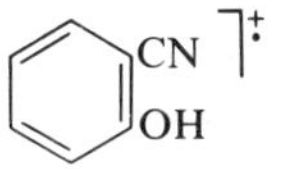

(47) 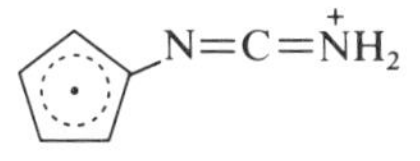   (48)

4.18.2.3.7 Photoelectron spectroscopy

The photoelectron spectra of oxazole, isoxazole, 1,2,5-oxadiazole and 1,3,4-oxadiazole in the He(I) and He(II) regions ⟨77JST(40)191⟩ and those of benzoxazole and other benzazoles in the He(I) region ⟨78JST(43)203⟩ have been measured. The orbital energies were assigned on the basis of *ab initio* calculations. The data for oxazole and benzoxazole are reported in Table 9.

The photoelectron spectra of 2-oxazolidinone and a series of related compounds (49; X = CH$_2$, S or NH, Y = O, S or Se) have been determined and the experimental ionization potentials compared with energy levels calculated by CNDO/2 and *ab initio* methods ⟨80JST(69)151⟩.

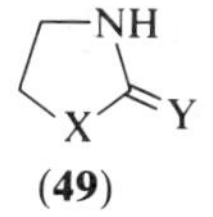

(49)

Table 9 UV Photoelectron Spectra[a] of Oxazole and Benzoxazole ⟨77JST(40)191, 78JST(43)203⟩

Oxazole			Benzoxazole		
Ionization potential	Calculated orbital energy	Assignment	Ionization potential	Calculated orbital energy	Assignment
9.83	10.63	a''	9.13	9.62, 10.11	π,π
11.19	12.59	a'	10.95	12.59	LP_N
11.80	12.90	a''	11.55	13.17	π
14.17	15.76	a'	12.46	14.38, 14.73	π,σ
14.17	16.75	a'	13.23	15.73	σ
14.80	17.73	a'	14.28	16.51	σ
15.63	18.81	a''	14.61	17.26, 17.94	σ,σ
17.80	20.88	a'	16.35	19.10, 19.29	π,σ
19.56	22.50	a'	16.95	20.24, 20.52	σ,σ
20.96	22.92	a'	19.43	22.88, 23.40	σ,σ
24.12	28.81	a'	20.15	24.73	σ
26.5 (?)	33.96	a'	—	—	—

[a] Values in electron volts; 1 eV = 96.485 kJ mol^{-1}.

4.18.2.4 Thermodynamic Aspects

Oxazole is a colourless liquid of b.p. 69–70 °C; benzoxazole has m.p. 31, b.p. 183 °C. Oxazoles are about 10 000 times weaker as bases than the corresponding pyridines; oxazole itself has pK_a 0.8 ± 0.2 ⟨69JCS(B)270⟩.

4.18.2.4.1 Thermochemistry

The heat of formation of oxazole, calculated by the MINDO/3 method, is −1.82 kJ mol^{-1}. The thermodynamic functions of heat capacity, heat content, free energy and entropy have been calculated for gaseous oxazole and compared with those for isoxazole, thiazole, isothiazole and 1,3,4-thiadiazole ⟨68SA(A)361⟩. For calculated and experimental ionization potentials of oxazole and benzoxazole, see Section 4.18.2.3.7

4.18.2.4.2 Aromaticity

Although oxazole possesses a sextet of π-electrons, all its properties indicate that the delocalization is quite incomplete; hence it has but little aromatic character. There is considerable bond fixation (see Section 4.18.2.3.1), hydroxyoxazoles are unstable relative to their oxo tautomers, oxazolediazonium salts are unknown, oxazoles function as dienes in the Diels–Alder reaction (see Section 4.18.3.1.2(vii)) and electrophilic substitution is rare. The chemistry of oxazole is dominated by its tendency to undergo ring-opening rather than preserve its type.

Anhydro-5-hydroxyoxazolium hydroxides (**50**), the münchnones, in common with other mesoionic compounds ⟨76AHC(19)1⟩, can be represented as resonance hybrids of aromatic forms, summarized by the symbol (**51**), and of the non-aromatic 1,3-dipolar forms (**52**) and (**53**). The münchnones function as 1,3-dipoles in cycloaddition reactions (see Section 4.18.3.1.5) and they can react as the open-chain valence tautomers (**54**) although such ketenes have not been detected spectroscopically. While the relative stability of mesoionic compounds is often attributed to their aromaticity, there is no evidence, in the writer's opinion, that the münchnones possess a sextet of delocalized π-electrons. This is most strikingly shown by X-ray crystallography (see Section 4.18.2.3.1); MINDO/3 ⟨76JCS(P2)548⟩ and $\omega\beta$-calculations ⟨75JCS(P2)1280⟩ also indicate that the conjugation does not extend over all the ring atoms. An aromatic formula such as (**51**) or (**40**) does not represent the structure satisfactorily; it is better depicted by the symbol (**55**), which emphasizes that cyclic conjugation is incomplete.

4.18.2.4.3 Conformation and configuration

The ^{13}C NMR ⟨79MI41801⟩ and IR spectra ⟨63CB2029⟩ of numerous oxazolidines have been reported. The conformation of oxazolidines depends on the nature and position of substituents ⟨74MI41800⟩; *ab initio* calculations on the parent molecule indicate that the envelope conformation (**56**) with the nitrogen atom at the tip has the lowest energy ⟨79JST(51)133⟩.

(**56**)   (**57**)   (**58**)

X-Ray crystallography of the 3-oxazolidinone (**57**) shows that the ring is flat. The crystal structure of 2,5-oxazolidinedione (**58**) has been determined; both exocyclic oxygen atoms are slightly displaced to the same side of the central plane ⟨76BCJ954⟩.

The circular dichroism of *cis*- and *trans*-2-oxazolidinones (**59**; R = H or Me) has been measured ⟨79T2009⟩. Geometrical isomerism round the exocyclic double bond in 4-(arylmethylene)-5-(4*H*)-oxazolones (**60**), the unsaturated azlactones, is a well-studied phenomenon ⟨75S749⟩. It has been shown by X-ray crystallography ⟨74T351⟩ that the stable form of 4-benzylidene-2-phenyl-5(4*H*)-oxazolone has the (*Z*) configuration (**61**). It isomerizes to the labile (*E*) form (**62**) in polyphosphoric acid and the change is reversed in pyridine.

(**59**)   (**60**)   (**61**)   (**62**)

4.18.2.5 Tautomerism

4.18.2.5.1 Substituent tautomerism

The oxazolones and their analogues with exocyclic sulfur and nitrogen atoms are in potential tautomeric equilibrium with hydroxyoxazoles, oxazolethiols and aminooxazoles, respectively. It can be stated at the outset that for the oxygen and sulfur compounds the one and thione forms are favoured, whereas the nitrogen compounds exist predominantly as aminooxazoles.

The IR spectra of 2(3*H*)-oxazolones exhibit amide carbonyl bands at 1780–1750 and NH absorptions at 3213 and 3124 cm^{-1}; they therefore exist in the form (**63**) ⟨56CB2825, 71CB2134⟩. The known 2(5*H*)-oxazolones (**64**) contain two substituents at C(5); hence the position of the double bond is fixed. The thione–thiol equilibrium (**65**) ⇌ (**66**) has been investigated for a series of compounds by measuring the acidity constants of the tautomers and their *N*- and *S*-alkyl derivatives ⟨69ACS2888⟩. It was found that the thione forms predominate to the extent of 10^5–10^7. 2-Aminooxazoles ⟨81KGS1011⟩ exist as such both in the solid state and in solution, as shown by pK_a measurements ⟨67BSF2040⟩.

(**63**)   (**64**)   (**65**)   (**66**)   (**67**) X = O or S

Benzoxazolone and benzoxazolethione likewise exist entirely as the one tautomer (**67**), whereas 2-aminobenzoxazole and its derivatives must be formulated as such.

4(5*H*)-Oxazolones (**68**) show carbonyl absorption in the range 1770–1740 cm^{-1}. They are in potential tautomeric equilibrium with 4-hydroxyoxazoles (**69**) and the mesoionic anhydro-4-hydroxyoxazolium hydroxides (**70**). There is no evidence for the former and the latter have likewise not been observed spectroscopically; however, they must be present to a small extent in the equilibrium mixture because they can be trapped as 1,3-cycloadducts with alkynes (see equation 57, Section 4.18.3.1.5) ⟨79JOC626⟩. Significantly, the rates of the cycloaddition reactions and the yields of the products were higher in sulfolane than in the less polar solvent acetic anhydride.

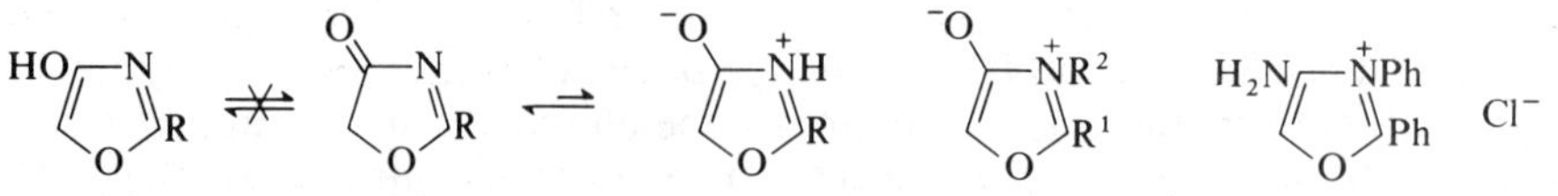

(69)   (68)   (70)   (71)   (72)

Anhydro-4-hydroxyoxazolium hydroxides (71) absorb at 1670–1660 cm^{-1} in the infrared. The conjugate acid (72) of a mesoionic anhydro-4-aminooxazolium hydroxide shows strong IR bands at 3350–3150 (NH) and at 1680–1630 cm^{-1} ⟨69BCJ2310⟩.

5(4H)-Oxazolones (73), the saturated azlactones, have been studied intensively ⟨B-57MI41801, B-57MI41802, 65AHC(4)75, 69MI41800, 77AHC(21)175⟩. They show carbonyl and C=N absorptions in the 1820 and 1660 cm^{-1} regions, respectively. Azlactones derived from chiral α-amino acids, e.g. compound (74), can be obtained in optically active forms which racemize easily. The derived salts (75; R^2 = H, Me or Ph) likewise exhibit optical activity; they show intense carbonyl bands at 1890–1880 and C=N$^+$ absorptions at ca. 1650 cm^{-1}.

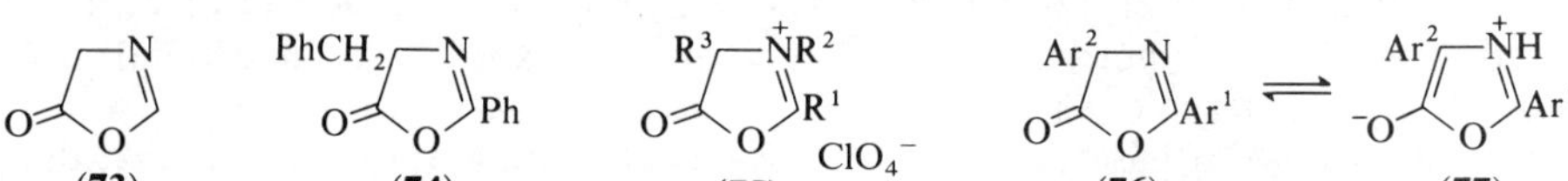

(73)   (74)   (75)   (76)   (77)

There is spectroscopic evidence that saturated azlactones (76) exist in solution in equilibrium with the mesoionic tautomers (77) ⟨68BSF4636, 70JA4340⟩. The bright-yellow diphenyl compound (77; Ar^1 = Ar^2 = Ph) could be isolated by crystallization from hot propan-2-ol; it isomerized to the corresponding colourless azlactone in refluxing toluene ⟨76CB2648⟩. The 2-p-nitrophenyl derivative (77; Ar^1 = p-NO$_2$C$_6$H$_4$, Ar^2 = Ph) exists entirely in the betaine form, as evidenced by its photoelectron spectrum ⟨77TL4587⟩. 5(4H)-Oxazolones react as tautomeric anhydro-5-hydroxyoxazolium hydroxides with dipolarophiles (see Section 4.18.3.1.5(iii)).

5(2H)-Oxazolones (78) are unstable with respect to the 4H-isomers (79); however, the 2-trifluoromethyl derivatives are unique in that the 2H-forms (78; R^1 = CF$_3$) are favoured ⟨62LA(658)128⟩. This was evident from the ^{1}H NMR spectra of numerous derivatives containing phenyl or alkyl groups at C(4) ⟨64CB2023⟩. While 2,2-disubstituted 5(2H)-oxazolones (80) have a fixed structure, the 2-methylene derivatives, such as 2-benzylidene-4-methyl-5(2H)-oxazolone (81), are in potential tautomeric equilibrium with 5(4H)-oxazolones, e.g. (82). However, compound (81) exists entirely in this form, no doubt because it is stabilized by conjugation of the phenyl group with the exocyclic double bond.

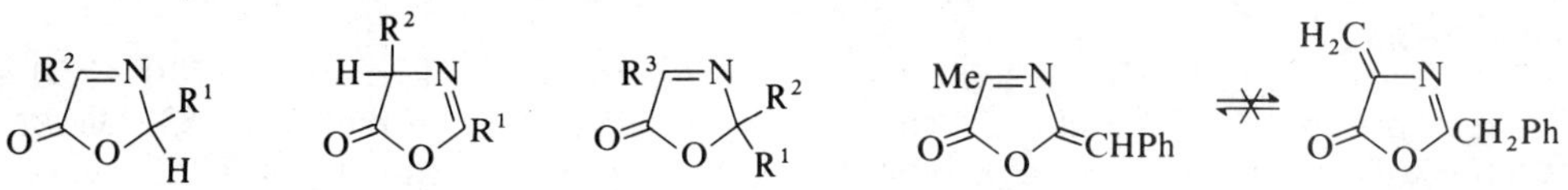

(78)   (79)   (80)   (81)   (82)

4.18.2.5.2 Ring–chain tautomerism

2,3-Dihydrooxazoles (83) are rare and there are few authentic 2,3-dihydrobenzoxazoles (84), most existing as phenolic Schiff's bases (85) ⟨B-57MI41800, B-57MI41801, 61JOC432⟩. The bis(trifluoromethyl) derivative (86) (ν_{NH} 3220 cm^{-1}) is an exception ⟨74BCJ785⟩. It has been reported ⟨80JHC1629⟩ that compounds (87) have the 2,3-dihydrobenzoxazole structure shown and that they exist in equilibrium with the bis(dihydrobenzoxazoles) (88), the proportion of the latter increasing from 0–60% in the series R = 2-thienyl < 2-furyl < 3-thienyl.

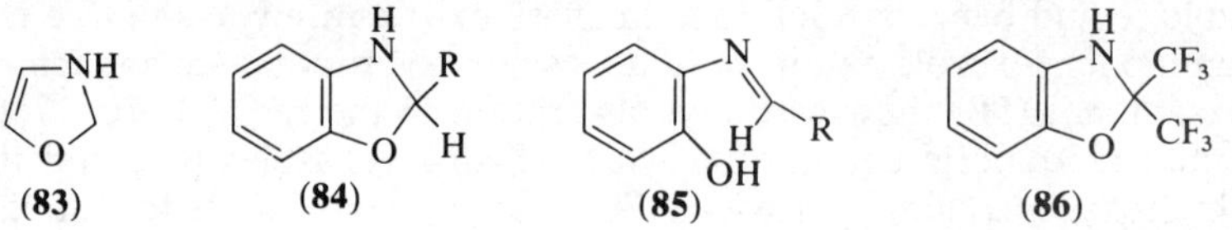

(83)   (84)   (85)   (86)

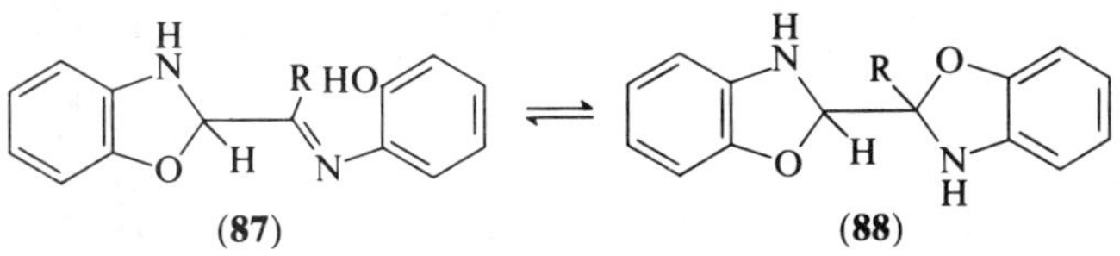

(87) ⇌ (88)

There is evidence for ring–chain tautomerism (89) ⇌ (90) in substituted oxazolidines. The ^{1}H NMR spectra of oxazolidines, determined for the neat compounds, indicate that the proportion of the open form is higher for (90; $R^1 = R^2 = Me$) than for (90; $R^1 = Me$, $R^2 = Bu^t$). In solution the equilibrium shifts in favour of the open isomers in solvents that can form hydrogen bonds ⟨68TL3557⟩. Protonated N-methyloxazolidine (91) exists in trifluoroacetic acid solution as an equilibrium mixture containing 10% of the tautomer (92); the appearance of two singlets in the alkenic region of the ^{1}H NMR spectrum may indicate the presence of a third form (93) ⟨80JA3588⟩. The mass spectra of oxazolidine and 2-aryloxazolidines (94) provide evidence for ring–chain tautomerism in the gas phase: the appearance of $(M-OH)^{+\cdot}$ ions is attributed to the open forms, that of $(M-Ar)^{+\cdot}$ mainly to the ring isomers. The presence of electron-donating substituents in the aryl groups favours the open forms (95) ⟨71T4407⟩.

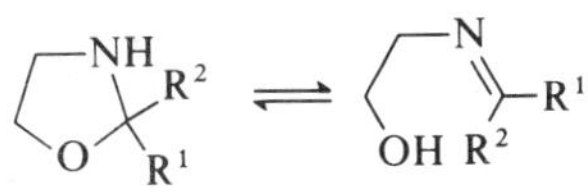

(89) ⇌ (90) 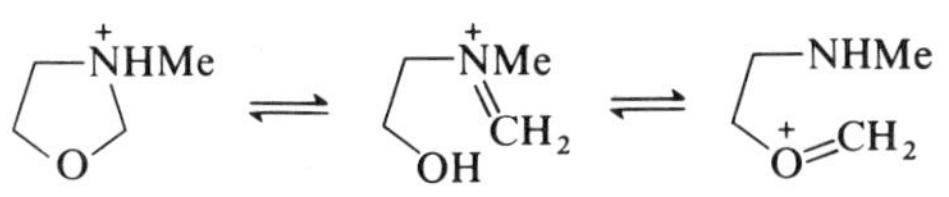   (91) ⇌ (92) ⇌ (93)

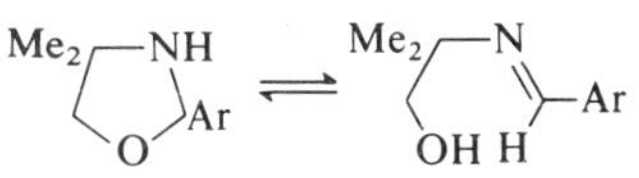

(94) ⇌ (95) 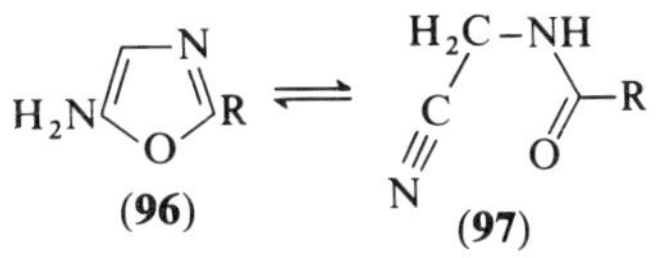   (96) ⇌ (97)

5-Aminooxazoles (96), which are unsubstituted in the 4-position, exist in equilibrium with (acylamino)acetonitriles (97), in which the latter predominate.

4.18.3 REACTIVITY

4.18.3.1 Reactions at Fully Conjugated Rings

4.18.3.1.1 General survey of reactivity

Oxazole is a thermally stable, weakly basic compound. Simple oxazoles are not readily reduced and they resist attack by alkali and other nucleophilic reagents under mild conditions but at elevated temperatures the ring is cleaved. The oxazole nucleus is destabilized by carbonyl substituents at C(4); such compounds rearrange on heating by a process of ring-opening and alternative ring-closure (see Section 4.18.3.1.2). Few electrophilic substitution reactions of neutral oxazoles are known. The chemistry of the oxazole system can be understood in broad terms by regarding it as a hybrid of pyridine and furan. Thus the basic nitrogen atom can be protonated and alkylated and halogen atoms at C(2) are reactive as in 2-halogenopyridines. Considerable bond fixation as in furan is indicated by the ability of oxazoles to undergo Diels–Alder reactions with dienophiles.

Oxazoles are readily attacked by most oxidizing agents, the ring being destroyed. Thus 2,5-diphenyloxazole reacts with peracetic acid to give benzoic acid and benzamide, rather than an N-oxide, and 2-methyl-4-phenyloxazole is oxidized to benzoic acid by potassium permanganate. 4,5-Diaryloxazoles yield the corresponding benzils by oxidation with chlorine or bromine. It appears that oxidizing agents initially attack the 4,5-double bond, since chromic acid converts compound (98) into the keto amide (99).

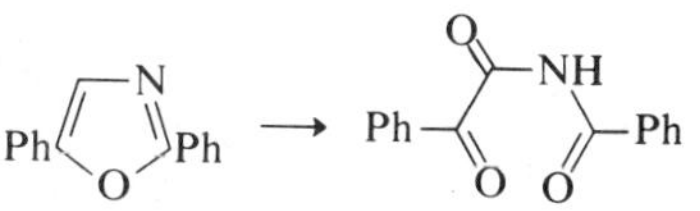

(98) → (99) 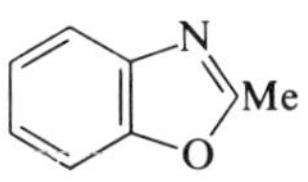   (100) 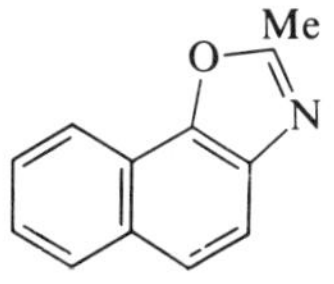   (101)

The oxazole ring of benzoxazoles, in which the 4,5-bond is masked by incorporation in a benzene ring, survives many oxidation reactions. 2-Methylbenzoxazole (**100**), for example, can be oxidized to benzoxazole-2-carboxylic acid, and 2-methylnaphth[2,1-*d*]oxazole (**101**) forms the corresponding aldehyde by the action of selenium dioxide.

The reaction of oxazoles with singlet oxygen is discussed in Section 4.18.3.1.2(vii).

Oxazolium salts are readily attacked by nucleophilic reagents at position 2 (*cf.* **102**); ring fission usually ensues and may be followed by recyclization, as in the transformation into imidazoles (see Section 4.18.3.1.2(iv)). *N*-Alkyl-2-chlorobenzoxazolium salts (**103**), because of the lability of the chlorine atom, are useful for effecting various condensation and dehydration reactions (see Section 4.18.3.1.2(iv).).

(**102**) (**103**)

The chemistry of oxazolones, particularly that of 5(4*H*)-oxazolones (**104**), is full of interest. These compounds are attacked by some nucleophiles at C(2) (*cf.* **105**), but fission of the carbonyl–oxygen bond (*cf.* **106**), leading to α-amino acids or their derivatives, is more usual. 5(4*H*)-Oxazolones react with electrophiles at C(4) or, less commonly, at C(2) by way of the resonance-stabilized anions (**107**), and they can function as tautomeric 1,3-dipoles (**108**) in cycloaddition reactions.

(**104**) (**105**) (**106**)

(**107**) (**108**)

The closely related mesoionic anhydro-5-hydroxyoxazolium hydroxides (**109**) also show a wide range of reactivity. They undergo electrophilic substitution reactions at C(4), ring-cleavage by nucleophilic attack at C(5) (*cf.* **110**), cycloadditions at the 2,4-positions and, moreover, they may behave as valence-tautomeric acylamino ketenes (**111**).

(**109**) (**110**) (**111**)

4.18.3.1.2 Reactions of oxazoles and oxazolium salts

(i) Thermal and photochemical reactions

The mass spectra of oxazoles are discussed in Section 4.18.2.3.6. Oxazole, its simple alkyl and aryl derivatives, and benzoxazoles are stable up to temperatures as high as 400 °C. However, oxazoles carrying a carbonyl substituent at position 4 readily rearrange on heating. This reaction, the Cornforth rearrangement, was discovered when the preparation of the acyl chloride (**112**) was attempted. The product proved to be the rearranged chlorooxazole (**113**). It has since been found that, in general, oxazoles of type (**114**), where R^1 may be alkyl or aryl, R^2 hydroxy, alkoxy, oxide anion or chloro, and COR^3 may be an aldehyde, ester, amide or acyl chloride group, isomerize to compounds (**116**) at 90–120 °C. The change is irreversible except in the case of (**114**; R^1 = Ph, R^2 = R^3 = OEt), where an equilibrium mixture containing 35% of the rearranged compound is produced. The proposal that the stabilized nitrile ylides (**115**) are intermediates is supported by kinetic measurements and the results of MINDO/3 calculations ⟨77JCS(P2)724⟩.

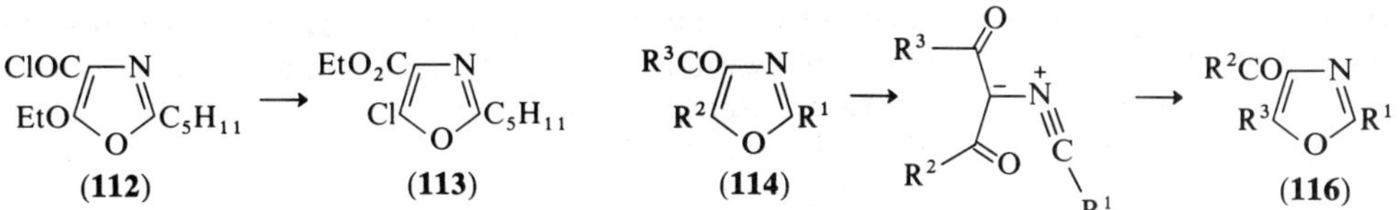

(112) (113) (114) (115) (116)

Oxazoles are generally photostable and are, indeed, produced by light-induced rearrangements of isoxazoles (see Section 4.18.4.1.2) ⟨B-80MI41800⟩. However, irradiation of 2,5-diphenyloxazole (**117**) in ethanol gives a mixture of 3,5-diphenylisoxazole (**118**), 4,5-diphenyloxazole (**119**), the phenanthrooxazole (**120**) and a trace of benzoic acid. These reactions, which proceed by two distinct paths, are rationalized as shown in Scheme 2 ⟨77JCS(P1)239⟩.

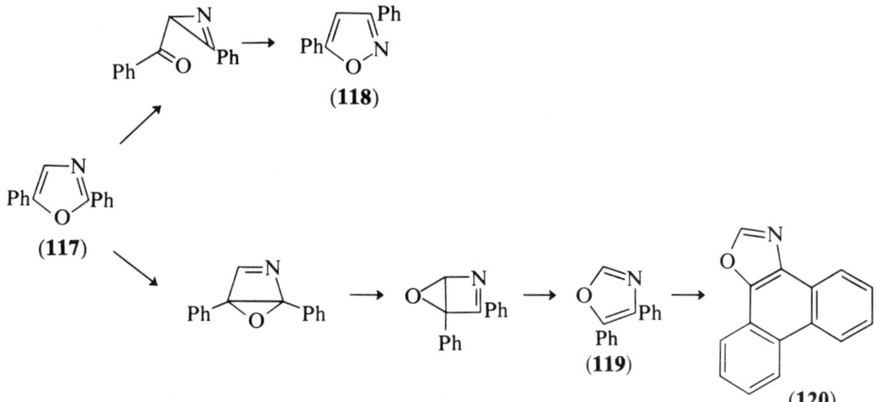

(117) (118) (119) (120)

Scheme 2

2-Phenylbenzoxazole (**121**) forms an amorphous photodimer, either the 1,3-diazetidine (**122**) and/or the isomer (**123**). The dimer reverts to the monomer in solution; the reaction is very fast under acid catalysis and 116 kJ mol^{-1} is released. 2-Phenylbenzoxazole is thus an efficient system for the storage and conversion of light energy ⟨82CC380⟩.

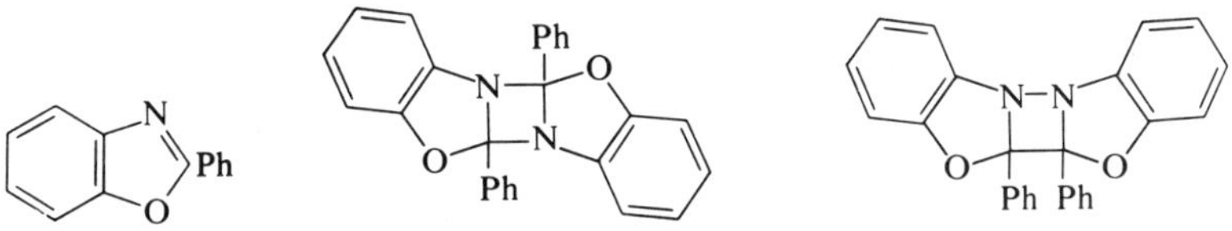

(121) (122) (123)

(ii) Electrophilic attack at nitrogen

Oxazoles and benzoxazoles yield salts with acids, alkyl halides, methyl toluene-*p*-sulfonate, triethyloxonium fluoroborate, *etc.* Quaternization is subject to the well-known steric effect of bulky substituents; 2-phenylbenzoxazole, for instance, does not react with ethyl iodide even at 240 °C. The weakly basic nature of oxazole (see Section 4.18.2.4) is reflected in the instability of its simple salts, such as hydrochlorides and picrates, which are readily hydrolyzed, even in air. 2-Aminooxazoles and 2-aminobenzoxazoles are alkylated at the ring nitrogen atom; the anilinonaphthooxazole (**124**), on the other hand, reacts with methyl iodide to yield a mixture of methyl derivatives (equation 1).

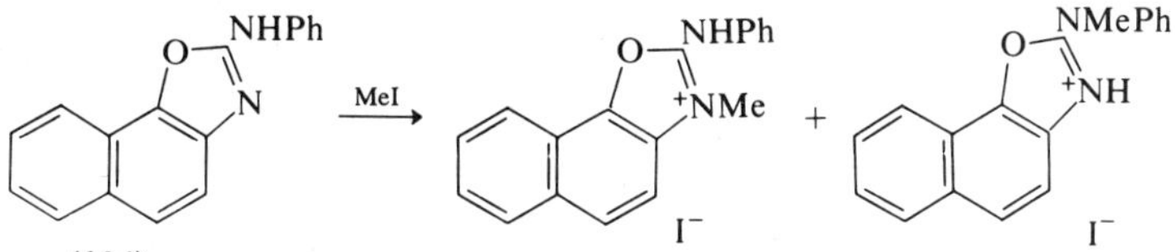

(124) (1)

If a pyridyl substituent is attached to the oxazole ring at C(2), it is alkylated in preference to the oxazole nitrogen atom; 4- and 5-pyridyloxazoles yield bisquaternary salts.

Benzoxazole is attacked by dimethyl acetylenedicarboxylate at the nitrogen atom; the resulting betaine, a 1,4-dipole, adds a second molecule of the ester to form the rearranged pyridobenzoxazole (**125**) ⟨78AHC(23)265⟩. When the reaction is carried out in a solvent, the betaine undergoes ring-opening, recyclization and final cycloaddition with additional ester to yield the pyranobenzoxazine (**126**; Scheme 3) ⟨80H(14)15⟩.

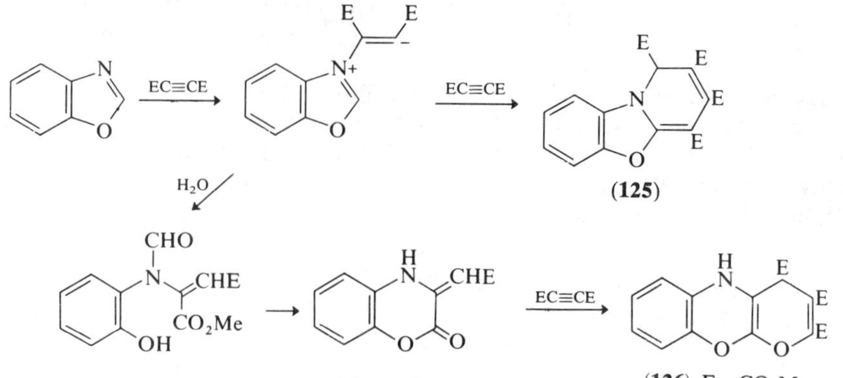

Scheme 3

(iii) Electrophilic attack at carbon

The inductive effect of the nitrogen atom renders the oxazole nucleus electron-deficient. Electrophilic substitution reactions of oxazole and its alkyl and aryl derivatives under essentially neutral conditions would therefore be expected to be difficult and, indeed, few are known. The electron-withdrawing effect of the nitrogen atom should affect position 2 most strongly and position 4 less, while the electron-donating resonance effect of the oxygen atom operates at C(4), which suggests that C(4) should be most reactive towards electrophiles, followed by C(5), with C(2) being least reactive. These simple considerations are borne out by the results of MO calculations of varying degrees of sophistication (see Section 4.18.2.2) which mostly place a higher electron density at position 4 than 5, and the lowest at C(2). These calculations are supported to some extent by experiment.

A systematic study of substitution reactions of oxazole itself has not been reported. Bromination of 2-methyl-4-phenyloxazole or 4-methyl-2-phenyloxazole with either bromine or NBS gave in each case the 5-bromo derivative, while 2-methyl-5-phenyloxazole was brominated at C(4). Mercuration of oxazoles with mercury(II) acetate in acetic acid likewise occurs at C(4) or C(5), depending on which position is unsubstituted; 4,5-diphenyloxazole yields the 2-acetoxymercurio derivative. These mercury compounds react with bromine or iodine to afford the corresponding halogenooxazoles in an electrophilic replacement reaction ⟨81JHC885⟩. Vilsmeier–Haack formylation of 5-methyl-2-phenyloxazole with the DMF–phosphoryl chloride complex yields the 4-aldehyde.

Attempts to carry out substitution reactions on oxazoles in acid media, such as nitration with nitric and sulfuric acids or chlorosulfonation, fail because the highly electron-deficient oxazolium cation is involved. Phenyloxazoles are nitrated at the *para* positions of the phenyl groups; 2,5-diphenyloxazole affords 5-(4-nitrophenyl)-2-phenyloxazole, as expected. Nitrooxazoles are now available by the action of dinitrogen tetroxide on the corresponding iodo compounds ⟨81JHC885⟩. Benzoxazoles are nitrated at the benzene ring; thus 2-methylbenzoxazole gives a mixture of 5- and 6-nitro derivatives.

The presence of amino substituents in the 2- or 5-position activates the oxazole ring (4-aminooxazoles are unknown). Thus 2-amino-3-aryloxazoles yield the derivatives (**127**; X = Cl or Br) on halogenation, and nitration of 2-dimethylamino-4-phenyloxazole afforded 2-dimethylamino-5-nitro-4-(4-nitrophenyl)oxazole, the sole example of direct nitration of the oxazole ring. 5-Dialkylamino-2-phenyloxazoles react with trifluoroacetic anhydride to form the ketones (**128**) and they couple with arene diazonium fluoroborates; the products (**129**) readily rearrange to 1,2,4-triazoles (**130**) in DMSO solution.

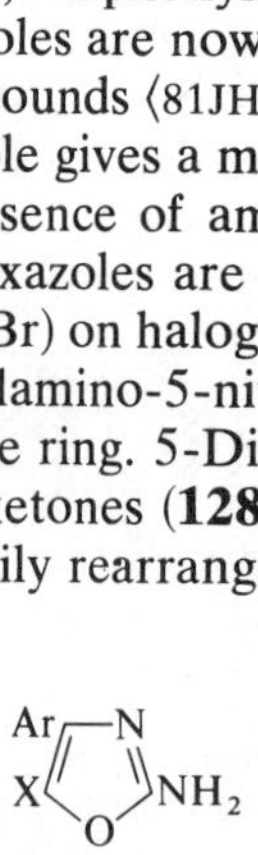

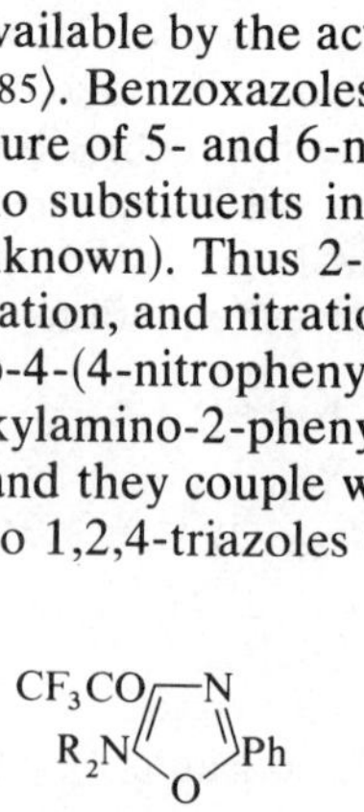

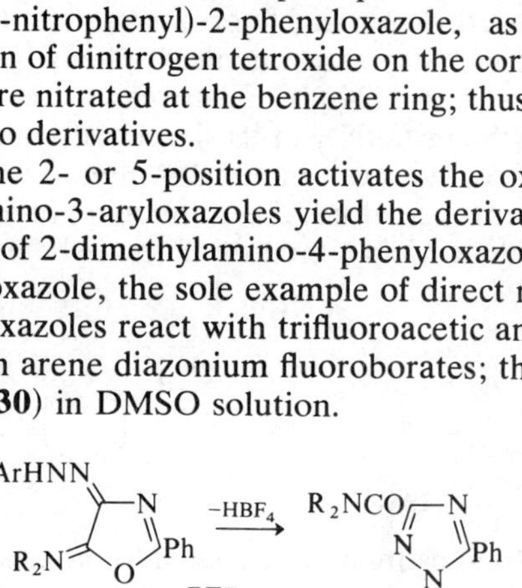

(iv) Nucleophilic attack at carbon, including reactions of halogenooxazoles and oxazolium salts

It is to be expected that the electron-deficient carbon atoms of oxazoles and *a fortiori* of oxazolium ions would be readily attacked by nucleophiles; indeed, many such reactions are known. Nucleophilic substitutions of hydrogen atoms attached to the oxazole ring are, however, rare; an example is the formation of 2-aminobenzoxazole in high yield by the action of hydroxylamine on benzoxazole. More typical is nucleophilic attack at C(2) or C(5), followed by ring-opening. Thus heating alkyl- or aryl-oxazoles with ammonia or amines above 200 °C gives imidazoles (equation 2). Formamide may be used in place of ammonia; it is believed that water or ammonia, which is present in the reaction mixture, cleaves the oxazole ring. Thiazoles are produced in low yields when mixtures of oxazoles and hydrogen sulfide are passed over alumina at 350 °C.

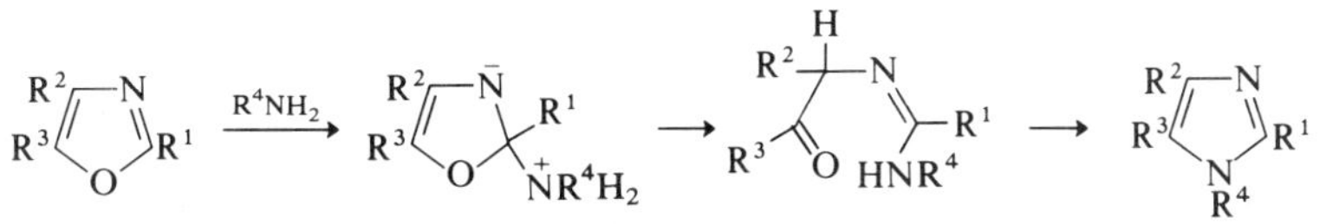

$$\text{(equation 2)}$$

The presence of electron-withdrawing substituents at C(4) facilitates reactions with nucleophiles, which may well be initiated by attack at C(5). Thus oxazole-4-aldehydes undergo ring-fission on treatment with aqueous alkali to form (acylamino)malondialdehydes (equation 3), and 2-pentyloxazole-4-carboxylic acid yields 2-pentylimidazole, with concomitant decarboxylation, when heated with ammonia at 150 °C. An example of a more complex ring transformation is the formation of 3-aminopyridines (**131**) by the action of malononitrile on 4-acetyloxazoles under alkaline conditions (equation 4).

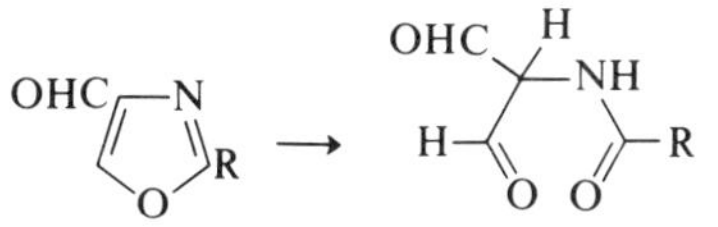

$$\text{(equation 3)}$$

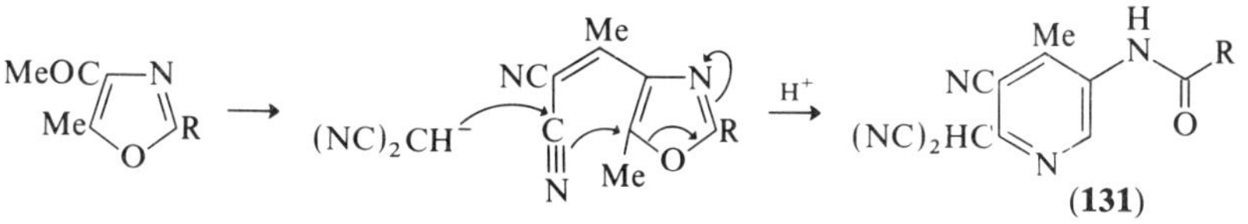

$$\text{(equation 4)}$$

Oxazolium salts are easily attacked by nucleophiles. The hydrolysis of oxazoles and benzoxazoles by hydrochloric acid has long been known. Triphenyloxazole gives benzoic acid, benzaldehyde and ammonium chloride; 2-methylbenzoxazole affords *o*-acetamidophenol, and 5-ethoxy-2-pentyloxazole is rapidly converted into the acylamino ester (**132**) by dilute hydrochloric acid at room temperature (Scheme 4).

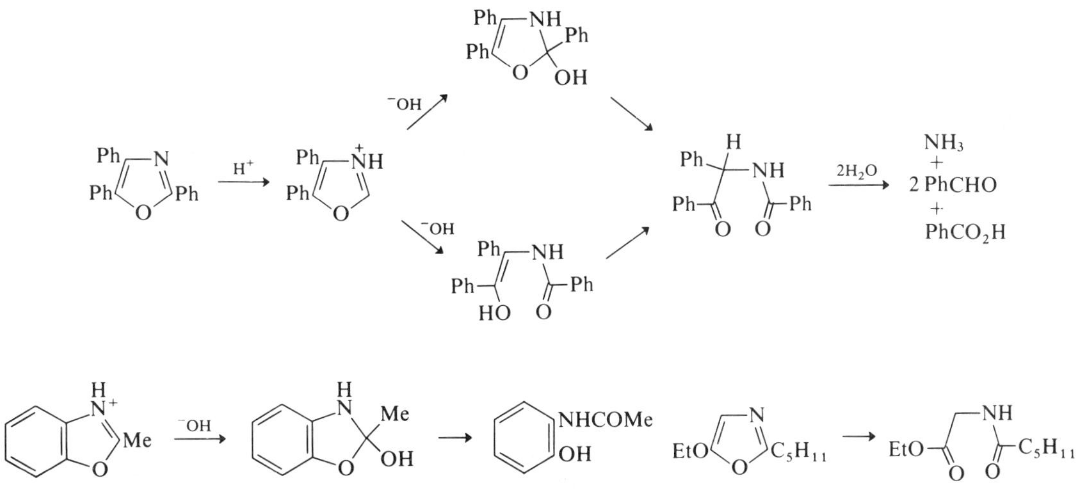

Scheme 4

Esters of oxazole-4-carboxylic acids are easily hydrolyzed by hydrochloric acid to give salts of α-amino ketones (equation 5); the action of 2,4-dinitrophenylhydrazine on the ester (**133**) in the presence of hydrochloric acid leads to the 2,4-dinitrophenylhydrazone (**134**).

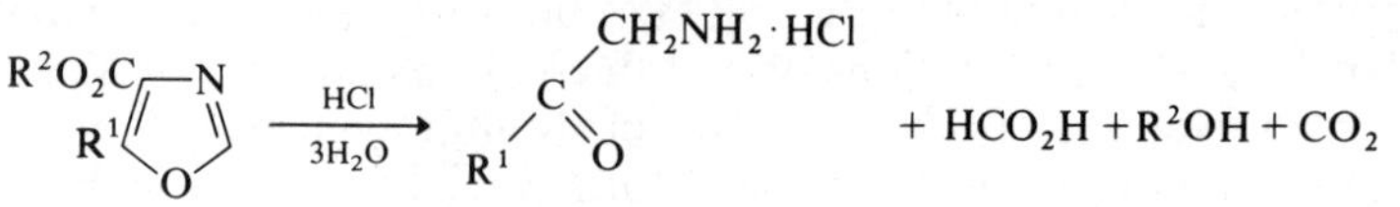

$$\text{R}^2\text{O}_2\text{C-oxazole-R}^1 \xrightarrow[3\text{H}_2\text{O}]{\text{HCl}} \text{R}^1\text{COCH}_2\text{NH}_2\cdot\text{HCl} + \text{HCO}_2\text{H} + \text{R}^2\text{OH} + \text{CO}_2 \quad (5)$$

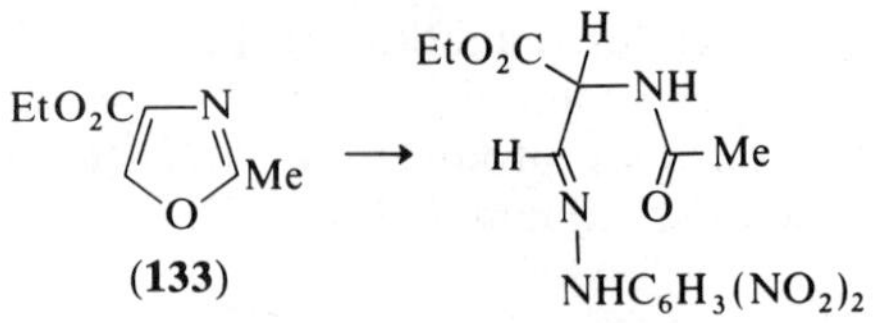

(**133**) → (**134**)

Quaternary oxazolium salts are easily cleaved by alkali; N-methylbenzoxazolium iodide affords N-methyl-2-formamidophenol (**135**). N-Methyloxazolium fluorosulfonates react with ethanolic ammonia to give imidazoles; with sodium hydrogen sulfide, thiazolium salts are formed. Hydrolysis or aminolysis of 2-amino-N-phenacyloxazolium salts (**136**) is followed by cyclization to yield imidazolones (**137**) or imidazoimidazoles (**138**), respectively (Scheme 5).

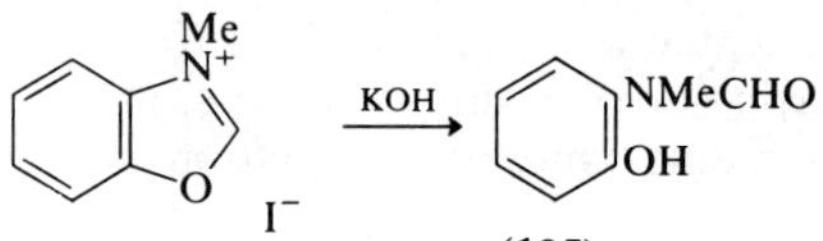

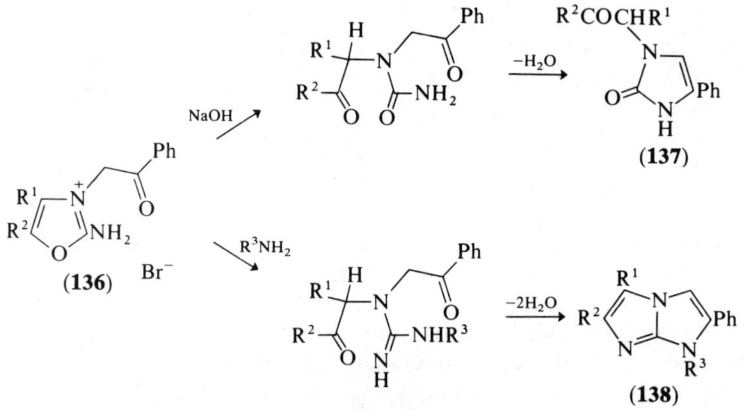

Scheme 5

The ease of replacement of halogen atoms attached to the oxazole ring by nucleophiles follows the order Hal-2 ≫ Hal-4 > Hal-5. Halogen–metal exchange of 4- or 5-bromooxazoles with butyllithium, followed by quenching with D_2O, affords the corresponding deuterated oxazoles, but other replacement reactions are difficult. The 4-bromooxazole (**139**) gives a low yield of mandelic acid on alkaline hydrolysis (equation 6). The substitution of all three chlorine atoms in (**140**) by piperidine has been observed.

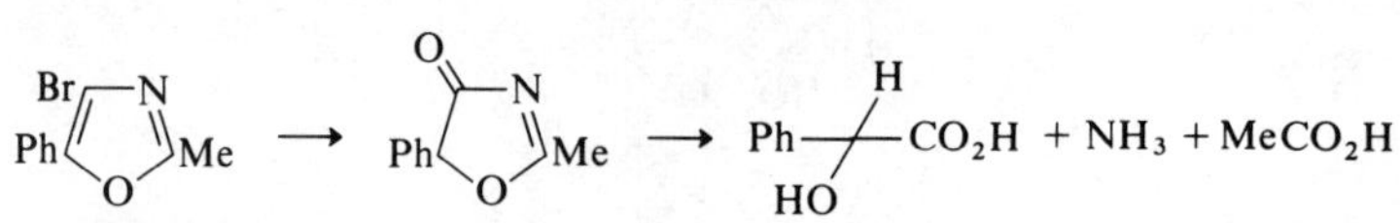

$$\text{(139)} \rightarrow \rightarrow \text{Ph-CH(OH)-CO}_2\text{H} + \text{NH}_3 + \text{MeCO}_2\text{H} \quad (6)$$

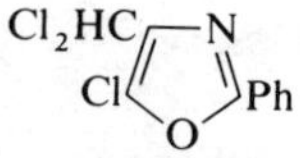

(**140**)

Halogen atoms in position 2 of oxazoles and benzoxazoles are easily replaced by hydroxide, alkoxide, sulfide and carbon anions and by amines; the products, if suitably constituted, may undergo secondary reactions (equations 7 and 8).

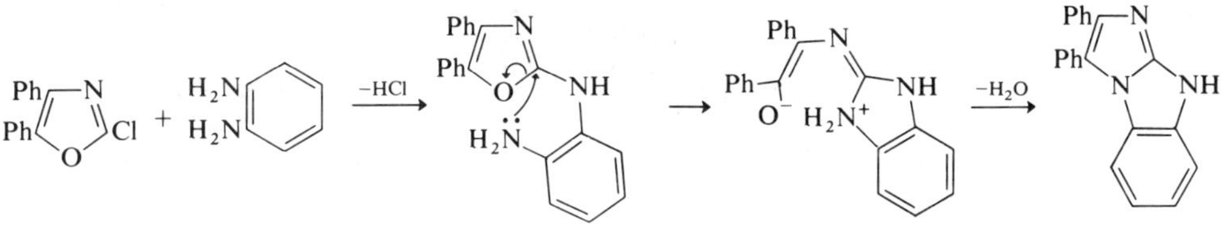

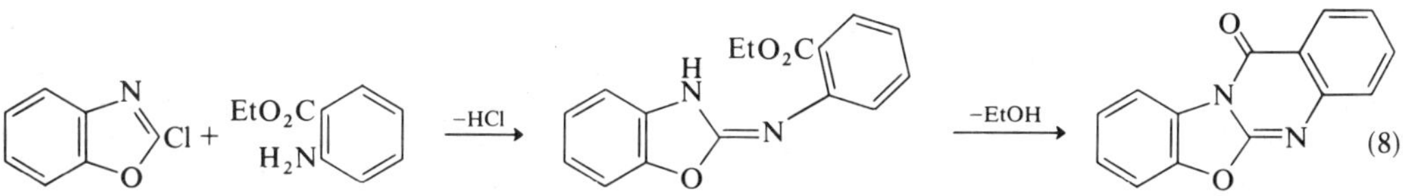

The reactivity of halogen atoms in position 2 of oxazoles is considerably enhanced by quaternization. This is best illustrated by the synthetic utility of the salt (141) which, like other N-alkyl-2-halogeno-azolium and -azinium salts, effects a number of condensation reactions; but it far surpasses these other salts in activity ⟨79AG(E)707⟩. Thus it activates carboxylic acids and it dehydrates formamides to isocyanides (Scheme 6). Amides are similarly converted into cyanides, ketones (RCH_2COAr) into alkynes ($RC\equiv CAr$), and cyanohydrins ($RCH(OH)CN$) are transformed into the corresponding chloro compounds ⟨79CL1117⟩.

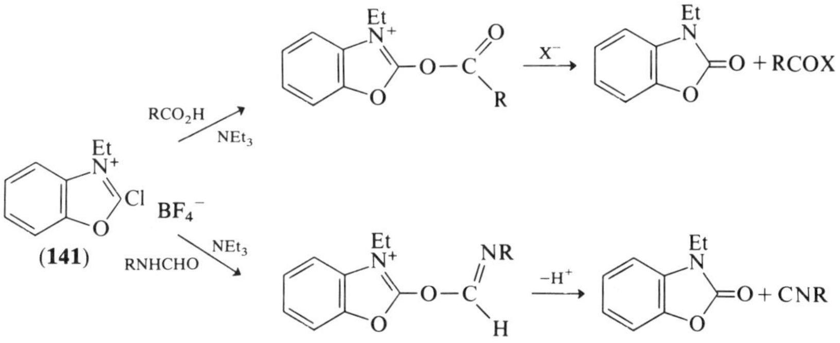

Scheme 6

(v) Nucleophilic attack at hydrogen

Oxazoles with a free 2-position are metallated thereat by n-butyllithium, while 2,4,5-trimethyloxazole forms the anion (142) ⟨81JOC1410⟩. In the presence of a carboxyl or methoxycarbonyl group at C(4) the metallation is directed to the 5-position, even in the absence of a 2-substituent, to give, for example, the lithio compounds (143) and (144) ⟨81TL3163⟩. 2-Lithiooxazoles can equilibrate with open-chain isocyanides and the diphenyl compound has been trapped as a mixture of cis- and trans-trimethylsilyl ethers (equation 9).

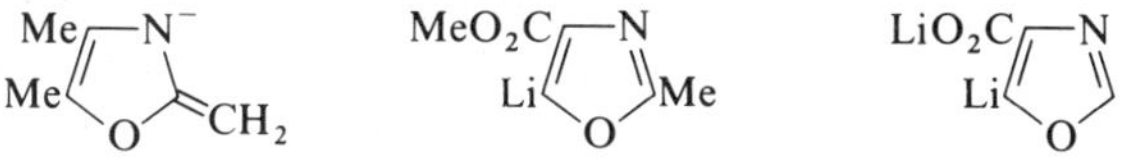

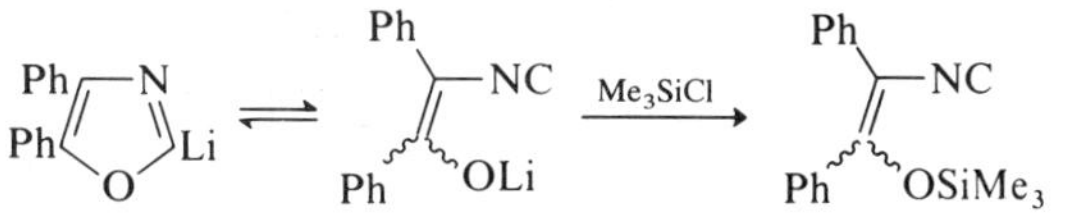

The proton at C(2) in oxazole is acidic, that at C(5) less so. Instantaneous deuteration at C(2) was observed when oxazole was dissolved in DMSO-d_6 containing sodium methoxide; a slower exchange occurred at C(5). The effect is enhanced by the presence of electron-attracting substituents at C(4) and even more in oxazolium ions, which owe their reactivity to the stability of ylides such as (145). Unlike thiamine and other thiazolium salts, oxazolium ions do not catalyze biochemical reactions or the benzoin condensation (see Chapter 4.19). This is due to a rapid ring-opening of the ylides, *e.g.* (**145**) → (**146**), at pH 7.4 or above, which is several thousand times faster than that of the corresponding thiazolium ylides.

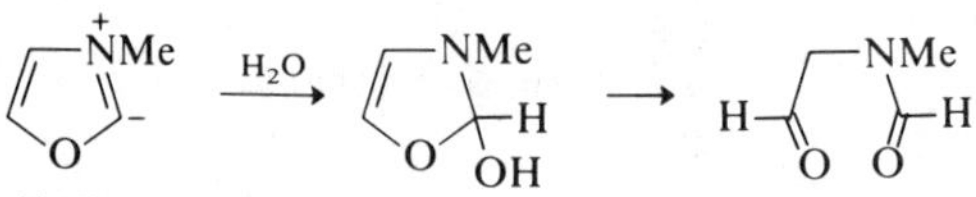

The oxazoliumcarboxylic acid (**147**) is easily decarboxylated *via* the ylide (**148**); the neutral compound (**149**) is much more stable due to the low equilibrium concentration of the zwitterionic tautomer (**150**; Scheme 7). Oxazolium salts lacking substituents at the 2-position react with dialkyl acylphosphonates in the presence of triethylamine to give mixtures of 1,4-oxazin-3-ones and 2-azetidinones; the reaction (see Scheme 8) proceeds by electrophilic attack of the phosphonate on an oxazolium ylide, *e.g.* (**151**), followed by insertion of oxygen into the carbon–phosphorus bond, ring-opening, and formation of the enolate anion (**152**) which can cyclize in two alternative ways with expulsion of the phosphonate group.

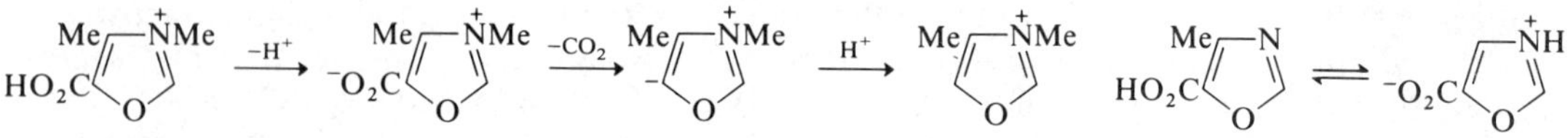

Scheme 7

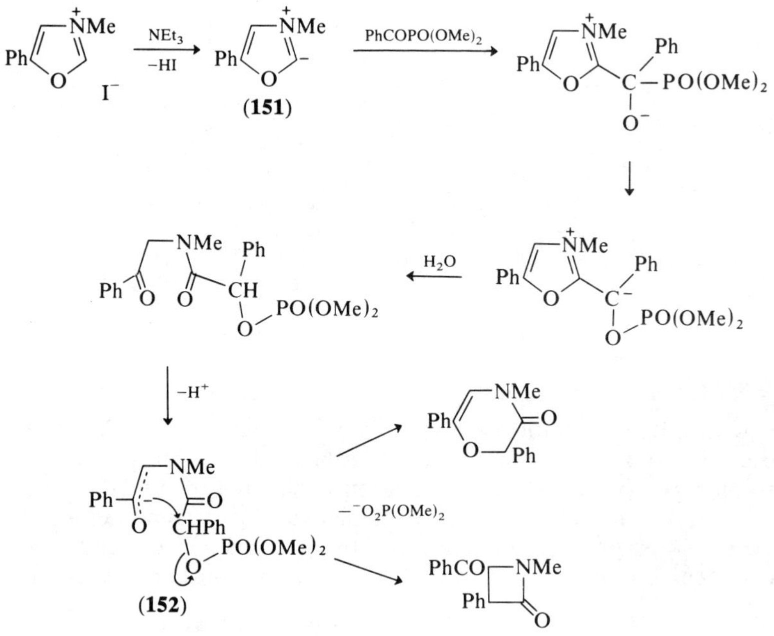

Scheme 8

(vi) Electrochemical reactions and reduction

The polarographic reduction of 2,5-diaryloxazoles occurs in two steps: the first produces a stable radical anion, the second gives a dianion which is rapidly protonated to a 2,3-dihydrooxazole. Further reduction and hydrogenolysis eventually yields an unstable amino alcohol, $Ar^1CH(OH)NHCH_2CH_2Ar^2$. Chemical or catalytic reduction of oxazoles invariably results in ring cleavage. The radical anion produced from oxazole has been assigned structure (**153**) on the basis of its ESR spectrum. 2-Substituted 4,5-diphenyloxazoles are reduced by sodium and ethanol to amides (**154**) if the 2-substituent is alkyl, whereas amino alcohols (**155**) are produced if it is aryl. Catalytic hydrogenation of aryloxazoles over platinum results in hydrogenation of the aryl groups prior to ring fission to yield amides. 2,5-Diphenyloxazole is cleaved by sodium and ethanol or lithium aluminum hydride to the

(153) (154) (155)

alcohol PhCHOHCH$_2$NHCH$_2$Ph. Reduction of benzoxazole with sodium in liquid ammonia leads to o-(methylamino)phenol.

(vii) Reactions with cyclic transition states

(a) Diels–Alder reactions. Of the oxygen-containing azoles, only oxazoles function as dienes in the Diels–Alder reaction. Since the observation ⟨57MI41800⟩ that alkyloxazoles combine with maleic anhydride there have been more than 100 papers and patents on the scope, mechanism and synthetic utility of this reaction ⟨69RCR540⟩.

Few reactions of the parent oxazole with the usual alkenic and alkynic dienophiles have been reported. Most oxazoles which yield Diels–Alder adducts contain electron-releasing substituents, the order of reactivity being alkoxy > alkyl ~ 4-phenyl > acetyl > ethoxycarbonyl. This sequence suggests that the oxazole functions as the electron-rich component and that the reaction is governed by interaction of the highest occupied molecular orbital of the oxazole and the lowest unoccupied orbital of the dienophile. Cycloadditions with 'inverse electron demand' of electron-deficient oxazoles with electron-rich dienophiles can be envisaged.

The primary adducts (**156**) and (**157**) of oxazoles with alkenes and alkynes, respectively, are usually too unstable to be isolated. An exception is compound (**158**), obtained from 5-ethoxy-4-methyloxazole and 4,7-dihydro-1,3-dioxepin, which has been separated into its *endo* and *exo* components. If the dienophile is unsymmetrical the cycloaddition can take place in two senses. This is usually the case in the reactions of oxazoles with monosubstituted alkynes; with alkenes on the other hand, regioselectivity is observed. Attempts to rationalize the orientation of the major adducts by the use of various MO indices, such as π-electron densities or localization energies and by Frontier MO theory ⟨80KGS1255⟩ have not been uniformly successful. A general rule for the reactions of alkyl- and alkoxy-substituted oxazoles is that in the adducts the more electronegative substituent R^4 of the dienophile occupies the position shown in formula (**156**). The primary adducts undergo a spontaneous decomposition, whose outcome depends on the nature of the groups R and on whether alkenes or alkynes have been employed.

(156) (157)

(158)

The reaction of oxazoles with alkenic dienophiles leads to pyridines. The intermediates (**156**) cleave to unstable dihydropyridinols (**159**), which aromatize in four ways (Scheme 9): pyridines are formed by dehydration (path A), while 3-hydroxypyridines results from elimination of the molecule R^3H (path B), elimination of R^4H if R^3 is hydrogen (path C), or dehydrogenation if R^3 is hydrogen (path D). Generally, more than one path is followed and a mixture of products results. If the group R^3 has no tendency to leave as an anion, *e.g.* if it is alkyl, pyridines are produced by path A, as in the reaction of 2,4,5-trimethyloxazole with fumaronitrile to yield the dicyanopyridine (**160**; equation 10). Path B becomes dominant if R^3 is a good leaving group, such as alkoxy or cyano. This constitutes an important method for the synthesis of vitamin B$_6$ (pyridoxin) (**162**) and its analogues. Thus treatment of 5-ethoxy-4-methyloxazole with diethyl fumarate gives the diester (**161**), which is converted into pyridoxin by reduction (equation 11). An even shorter route (equation

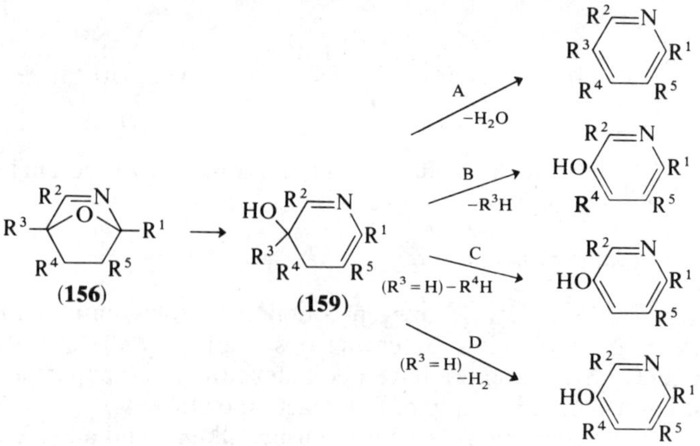

Scheme 9

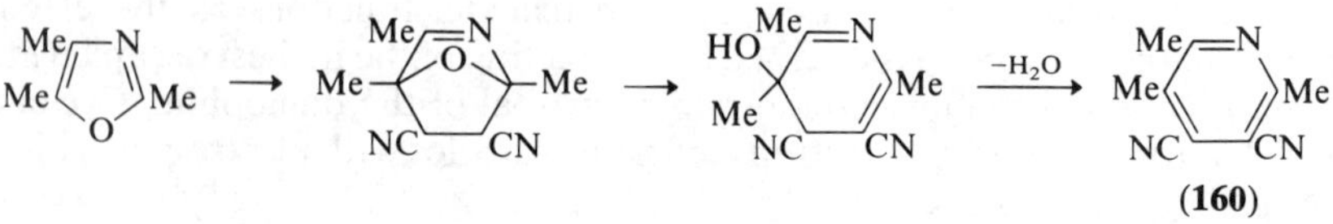

12) employs 5-cyano-4-methyloxazole and the acetal (**163**) in a cycloaddition reaction with inverse electron-demand.

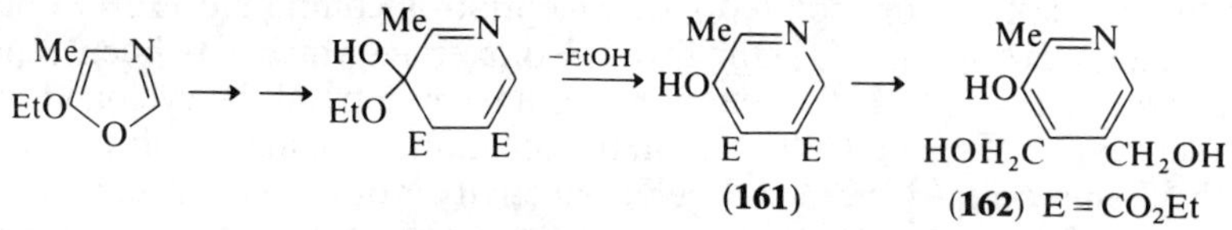

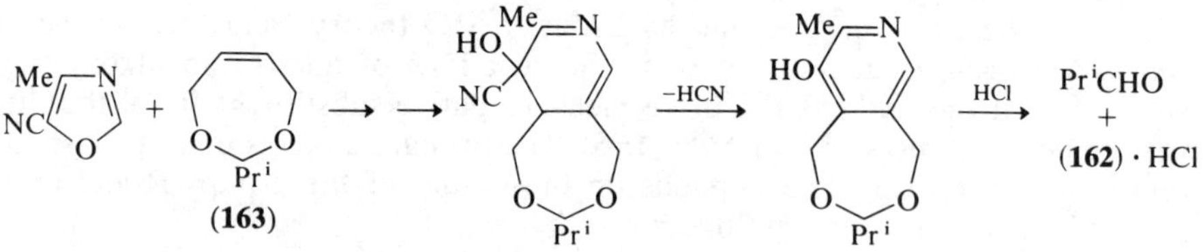

Pathways C and D are less well documented, mainly because oxazoles lacking a 5-alkoxy substituent show reduced reactivity towards the usual dienophiles. An example of reaction C is the formation of 3-hydroxy-2-methylpyridine from 4-methyloxazole and acrylonitrile (equation 13). Dehydrogenation (path D) is rare but proceeds in the presence of a hydride acceptor. Thus 4-methyloxazole reacts with ethyl acrylate in the presence of hydrogen peroxide to give ethyl 3-hydroxy-2-methylpyridine-5-carboxylate in 27% yield (equation 14).

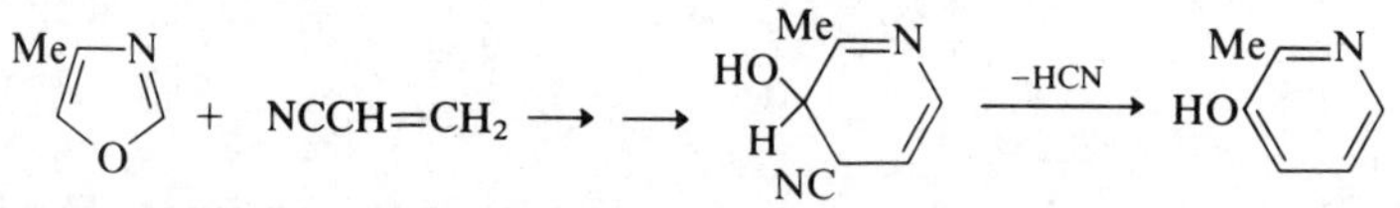

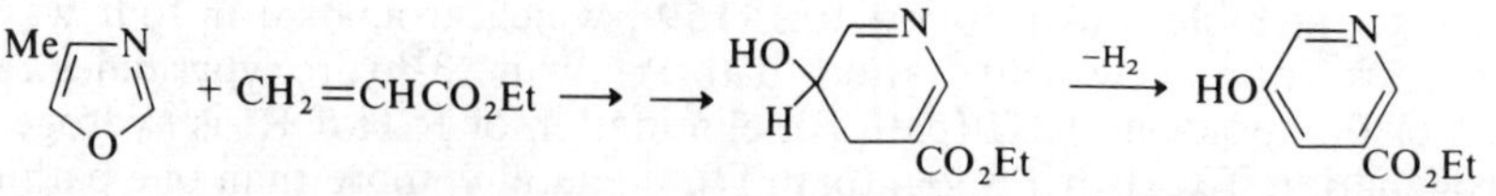

The reaction of oxazoles with alkynes is entirely different, leading to furans. The adducts (**157**) eliminate a cyanide in a retro-Diels–Alder process (equation 15). A typical example is the formation of the ester (**164**) from 5-ethoxy-4-methyloxazole and dimethyl acetylenedicarboxylate (equation 16); equation (17) illustrates the production of two regioisomers in this reaction ⟨79MI41802⟩; a more elaborate case is the twofold addition of benzyne to 4-methyl-2,5-diphenyloxazole to give the bridged dihydroanthracene shown in equation (18) ⟨80TL3627⟩.

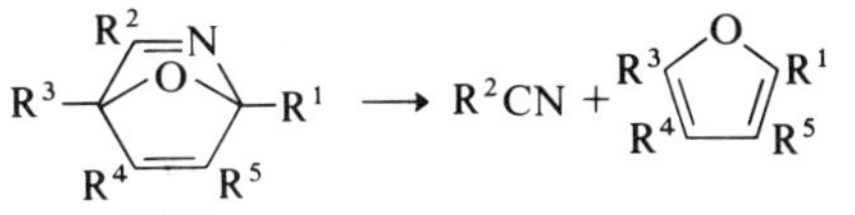

$$\text{(157)} \quad \longrightarrow \quad R^2CN + \text{product} \tag{15}$$

$$\text{oxazole} + EC\equiv CE \xrightarrow{-MeCN} \text{furan derivative} \tag{16}$$

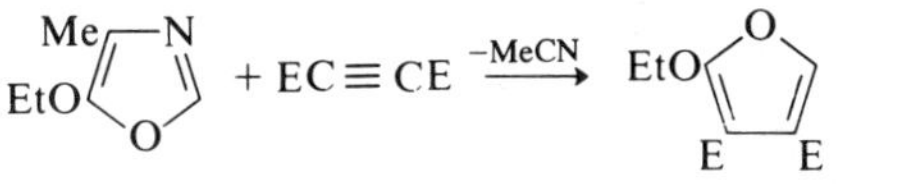

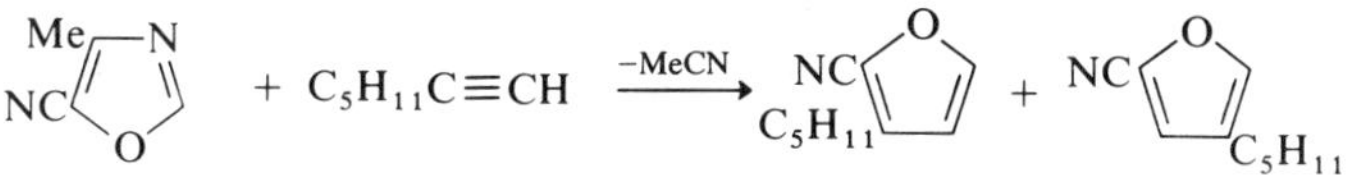

(164) E = CO$_2$Me

$$\text{oxazole} + C_5H_{11}C\equiv CH \xrightarrow{-MeCN} \text{products} \tag{17}$$

$$\text{oxazole} + \text{benzene} \xrightarrow{-MeCN} \text{intermediate} \xrightarrow{benzyne} \text{product} \tag{18}$$

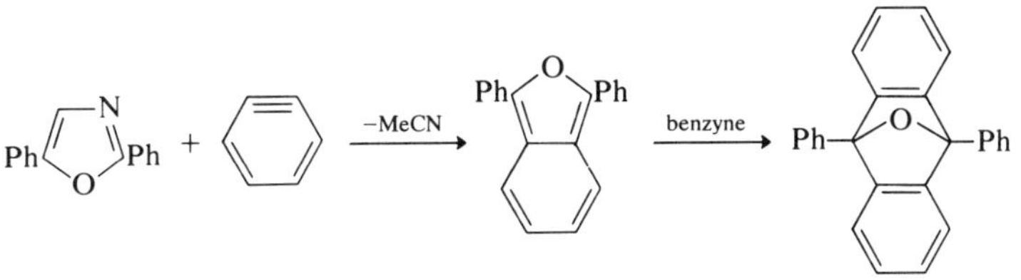

A very different type of Diels–Alder reaction is that of the Schiff's base (165) with pyrrolidinocyclohexene (equation 19) ⟨76CPB2889⟩; it is a rare example of a cycloaddition to the 1,3-diaza-1,4-diene system.

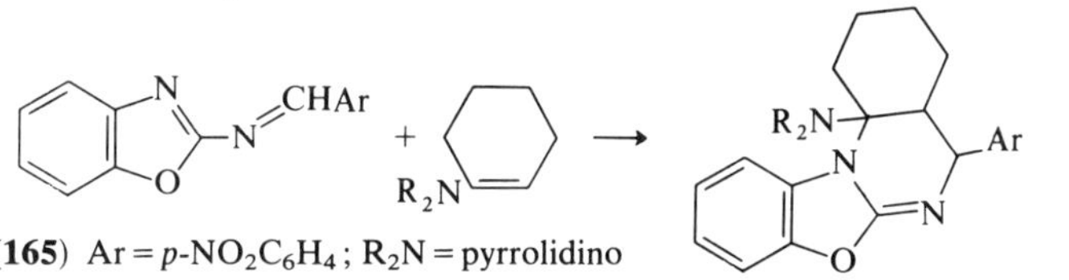

$$\text{(165)} + \text{cyclohexene} \longrightarrow \text{product} \tag{19}$$

(165) Ar = p-NO$_2$C$_6$H$_4$; R$_2$N = pyrrolidino

(b) *Photooxygenations.* Oxazoles are susceptible to attack by singlet oxygen to yield a variety of products arising from ring fission of the initial adduct. Oxazoles lacking alkoxy substituents yield triacylamines, acid anhydrides and/or cyanides. When 2,4,5-triphenyl oxazole is irradiated in ether or benzene in the presence of oxygen, a mixture of tribenzoylamine, benzoic anhydride and benzonitrile is produced; when the photooxygenation is conducted in methanol, only the amide is obtained. These products arise from a common intermediate which decomposes in two ways (Scheme 10). Singlet oxygen combines with the oxazole in a Diels–Alder reaction to give the adduct (166), which either suffers fission to benzonitrile and benzoic anhydride or it rearranges to the isoimide (167), which in turn isomerizes to tribenzoylamine. Exclusive formation of triacylamines occurs in alcoholic solvents, *e.g.* both 2-methyl-4,5-diphenyloxazole and 4-methyl-2,5-diphenyloxazole afford acetyldibenzoylamine when exposed to air and visible light in methanol containing Methylene Blue as a sensitizer. Cycloalkenooxazoles can be made to yield either cyanides or isoimides, depending on the choice of solvent. Thus photooxygenation of compound (168) in dichloromethane gives the mixed anhydride (169), whereas irradiation in methanol results in a mixture of the isoimide (170) and its rearrangement product (171).

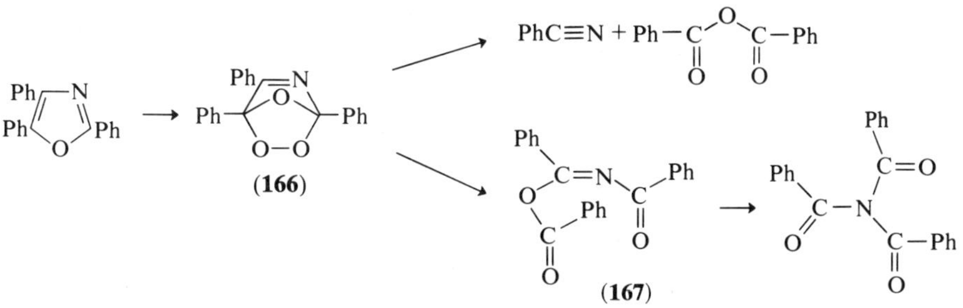

Scheme 10

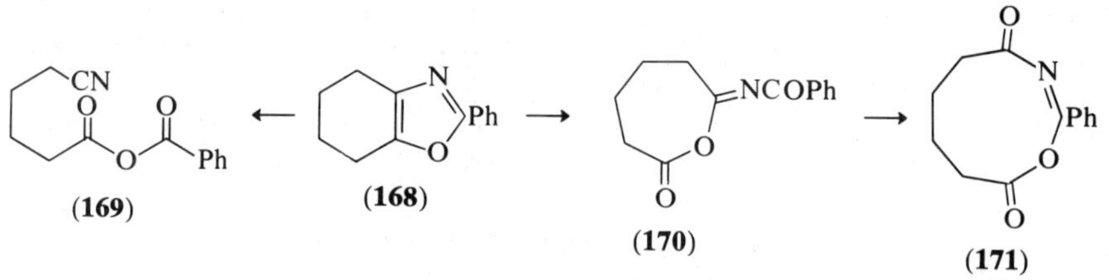

The participation of singlet oxygen in these reactions was confirmed by the use of 9,10-diphenylanthracene peroxide (**172**), a thermal source of $^1O_2$, which, when heated with oxazoles, gave the same products as were formed in the photooxidations. The reaction of the peroxide (**172**), doubly labelled with ^{18}O, with unsymmetrically substituted oxazoles gave triacylamines labelled in the acyl groups R^1CO and R^2CO, which proves that the initial peroxide was the Diels–Alder adduct (**173**); addition of oxygen at the 4,5-double bond (*cf.* **174**) would have produced a different distribution of the labels (Scheme 11) ⟨72JA7180⟩.

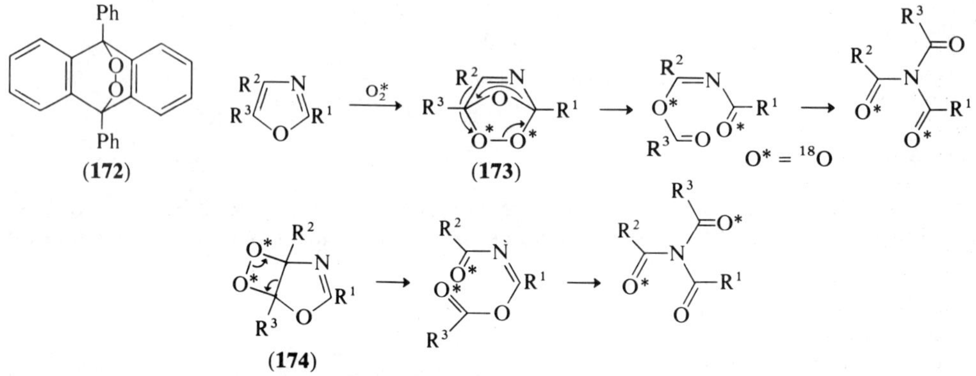

Scheme 11

The dye-sensitized photooxygenation of oxazoles with alkoxy substituents in positions 2, 4 or 5 gives 1,2,4-dioxazoles, *e.g.* (**176**). These novel heterocycles are believed to arise by addition of oxygen to the 4,5-double bond leading to peroxiranes, such as (**175**), as the primary intermediates (equation 20).

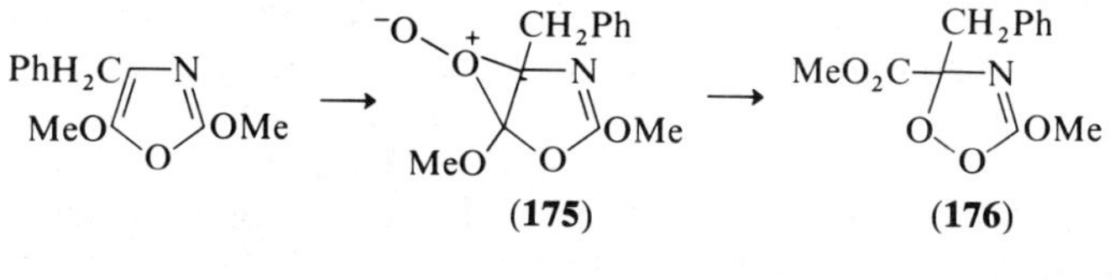

4.18.3.1.3 Reactions of oxazole N-oxides

The chemistry of the *N*-oxides of oxazole has not been explored as thoroughly as that of the pyridine analogues. Both types of compounds are reversibly protonated to give *N*-hydroxy cations, form complexes with iodine and with tetracyanoethylene, and both are deoxygenated by triphenylphosphine and other tervalent phosphorus compounds.

Oxazole *N*-oxides having a 4-methyl substituent are attacked by acetic anhydride to yield 4-acetoxyoxazoles (equation 21). The combined action of benzoyl chloride and potassium cyanide leads to compounds of the Reissert type, *e.g.* (**177**). The reaction of 4-methyloxazole *N*-oxides with phenyl isocyanate gives 5-hydroxy-4-methylene-1-phenyl-4,5-dihydroimidazoles by cycloaddition, extrusion of carbon dioxide and recyclization (Scheme 12); with 4-phenyloxazole *N*-oxides the reaction takes a different course, yielding imidazooxazolidinones (Scheme 13).

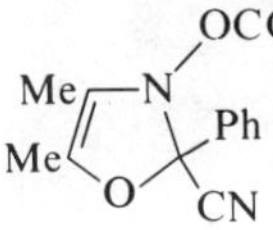

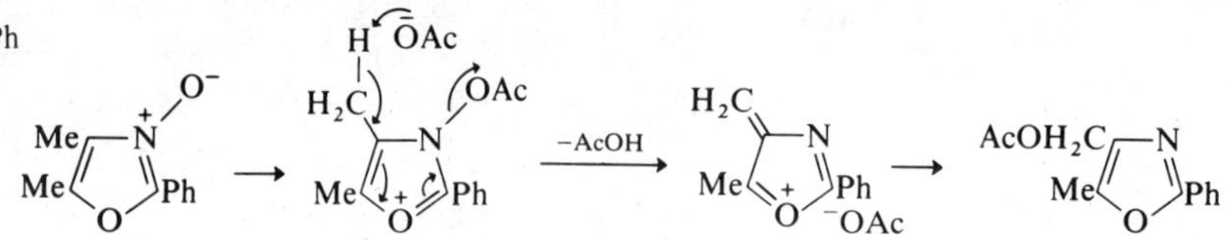

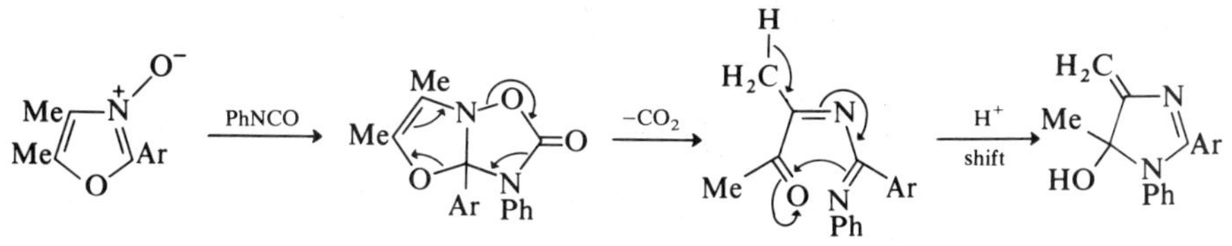

Scheme 12

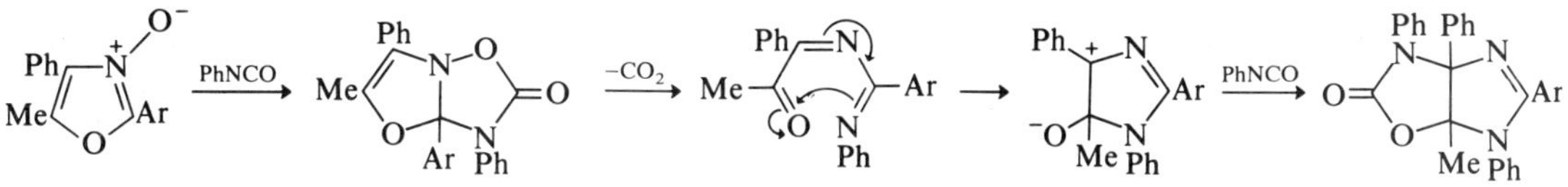

Scheme 13

4.18.3.1.4 Reactions of non-aromatic compounds in potential tautomeric equilibrium with oxazoles: oxazolones

The rich and varied chemistry of oxazolones is summarized in this section ⟨46OR(3)198, 55QR150, B-57MI41801, B-57MI41802, 65AHC(4)75, 69MI41800, 77AHC(21)175⟩. Little is known about oxazolethiones; oxazoleimines exist in the amino form. Of the five possible oxazolones (see Section 4.18.2.1), all of which are now known, the 5(4H)-oxazolones (**178**), the azlactones, are the oldest and most familiar. 5(2H)-Oxazolones (**179**), the pseudooxazolones, follow and representatives of the other three systems have become available in recent years. 4-Methylene-5(4H)oxazolones (**180**), the unsaturated azlactones, and 2-methylene-5(2H)oxazolones (**181**) are not tautomeric with hydroxyoxazoles; they are nevertheless included here because of their close relationship to the azlactones and pseudooxazolones. Cycloaddition reactions of 5(4H)- and 4(5H)-oxazolones proceed by way of mesoionic tautomers and these mesoionic compounds are described in the following section.

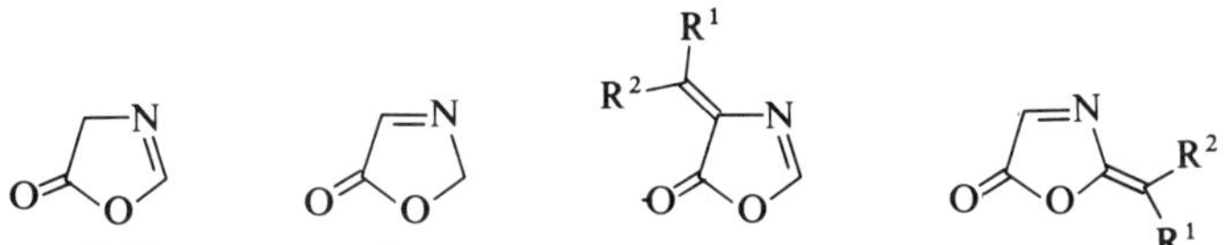

(**178**) (**179**) (**180**) (**181**)

(i) Thermal and photochemical reactions

When the ethoxy-2(5H)-oxazolone (**182**) is heated it isomerizes to an oxazolidinedione by migration of the ethyl group (equation 22).

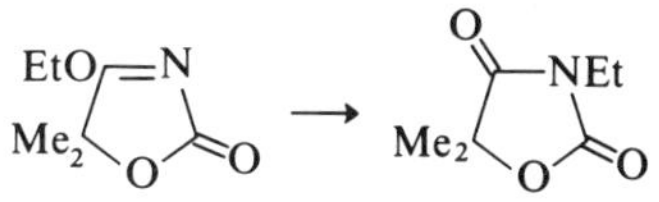

(**182**)   (22)

Simple 5(4H)-oxazolones are stable at moderate temperatures. The diphenyl derivative (**183**) yields the acylimine (**184**) on flash vacuum pyrolysis (equation 23) ⟨80AG(E)564⟩. The presence of acyl substituents at C(4) destabilizes the ring; such compounds rearrange at 180 °C to oxazoles by a process akin to that described in Section 4.18.3.1.2(i) (equation 24). 4-Allyl-5(4H)-oxazolones undergo a thermal 3,3 sigmatropic rearrangement (equation 25) ⟨81TL2435⟩.

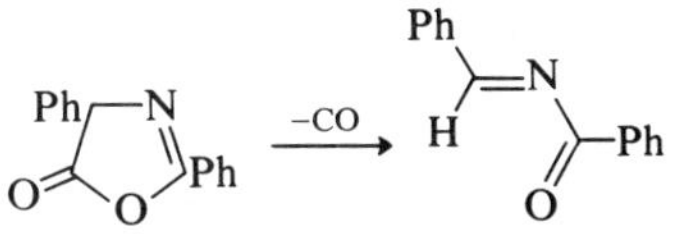

(**183**)   (**184**)   (23)

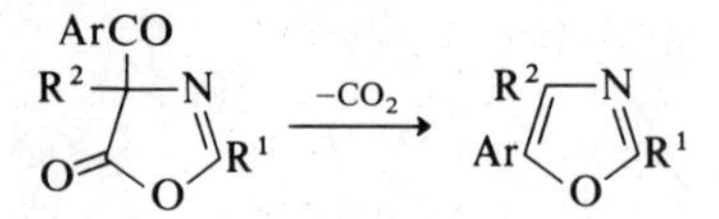

R^1, R^2 = alkyl

(24)

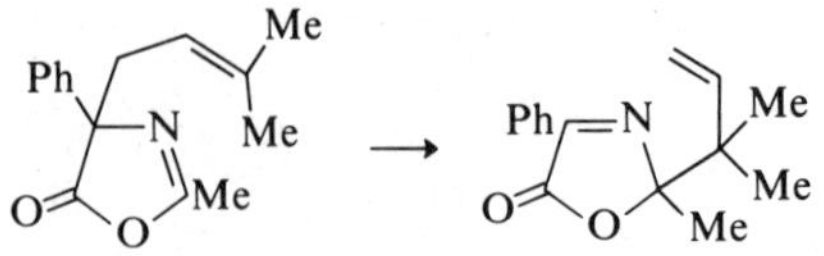

(25)

Flash vacuum pyrolysis of the 5(2*H*)-oxazolone (**185**) gives the imine (**186**) *via* a transient nitrile ylide (equation 26) ⟨80CC503⟩. The nitrile ylide (**188**), produced by thermolysis of the 5(2*H*)-oxazolone (**187**; Ar = *p*-ClC_6H_4), undergoes an intramolecular 1,2-cycloaddition with retention of configuration (Scheme 14) ⟨79AG(E)167⟩. 2,4-Disubstituted 2-acyl-5(2*H*)-oxazolones lose carbon dioxide on heating to form trisubstituted oxazoles, in which the groups at C(2) and C(4) are interchanged (equation 27).

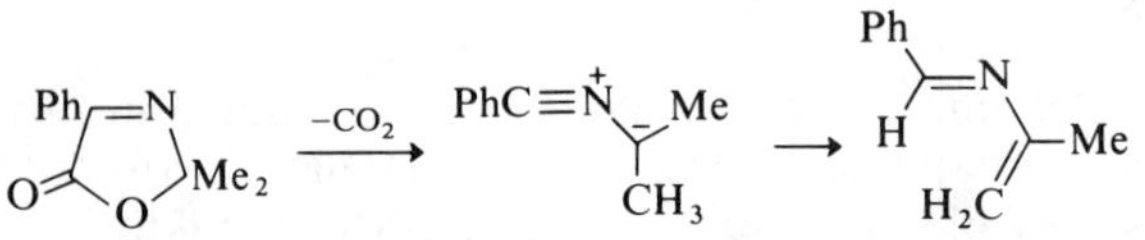

(**185**) (**186**)

(26)

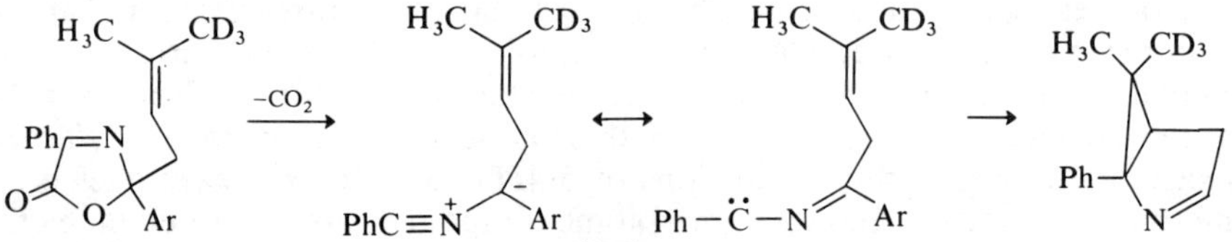

(**187**) (**188**)

Scheme 14

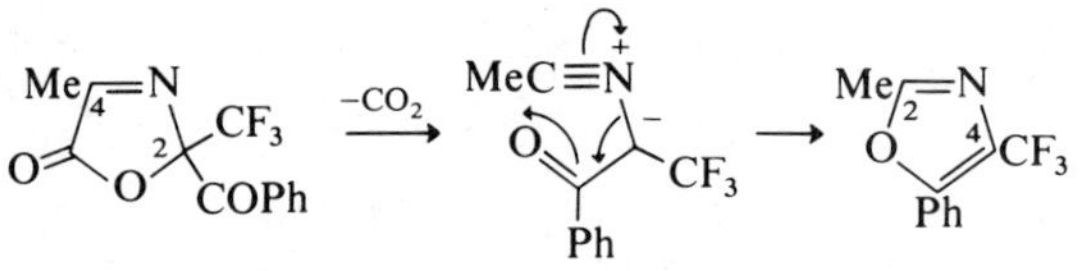

(27)

The photorearrangement of 2-phenyl-4(5*H*)-oxazolone to a 2(3*H*)-oxazolone proceeds by way of an isocyanatooxirane (equation 28). Prolonged exposure of 5(4*H*)-oxazolones to UV light causes decarbonylation and formation of acylimines (*cf.* **184**) ⟨81TL2435⟩. The photochemical 1,3 rearrangement of the allyl-5(2*H*)-oxazolone shown in equation (29) proceeds in accordance with the rules of orbital symmetry ⟨81TL2435⟩. Photolysis of 5(2*H*)-oxazolones leads to nitrile ylides, which can be isolated in special cases (equation 30).

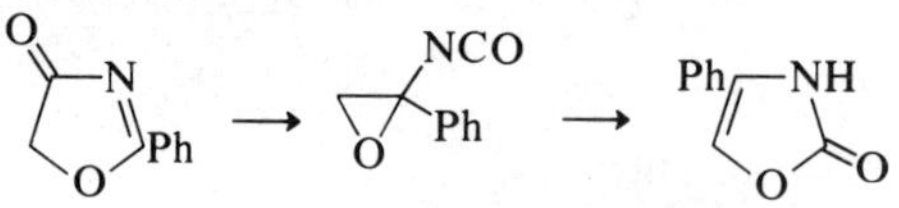

(28)

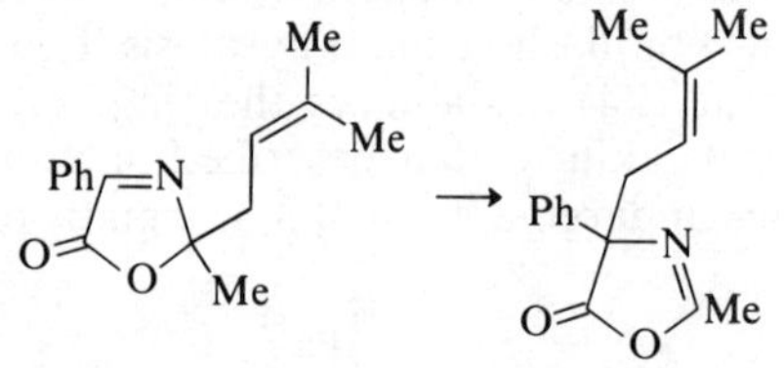

(29)

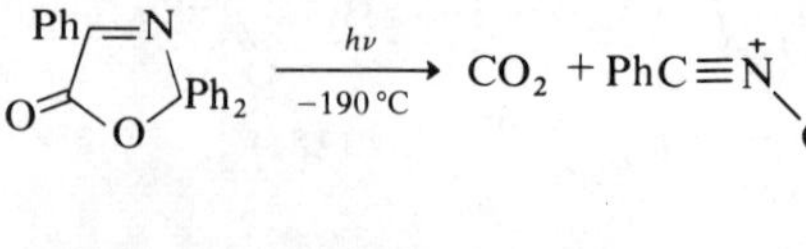

(30)

The (Z) isomer (**189**) of 4-benzylidene-2-phenyl-5(4H)-oxazolone rearranges to the (E) form on irradiation. The photodimerization of compound (**190**) in solution results in the centrosymmetric cyclobutane derivative (**191**). A different kind of photodimerization occurs when the oxazolone (**192**) is irradiated in the solid state, compound (**193**) being formed ⟨79TL2461, 79TL3139⟩.

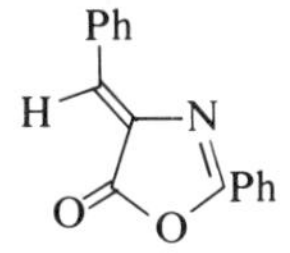

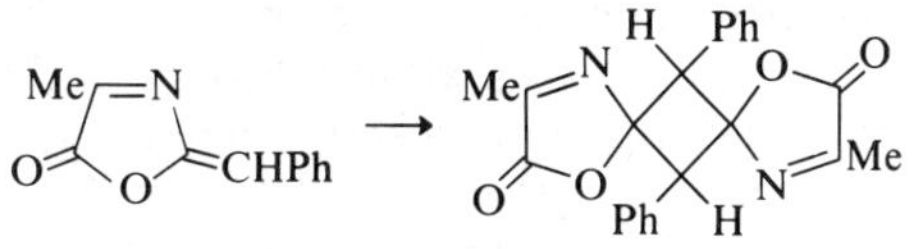

(ii) Electrophilic attack

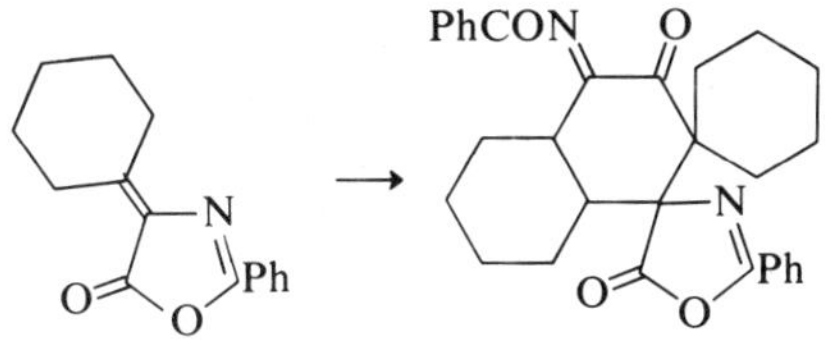

Neutral oxazolones and oxazolethiones are inert towards most electrophiles. 5(4H)-Oxazolones are reversibly protonated at the nitrogen atom. When the 2-oxazolone (**194**) is treated with sulfuric acid, 2-methyl-4,5-diphenyloxazole is produced (equation 31). 2(3H)-Oxazolones and benzoxazolones react with phosphoryl chloride to yield the corresponding 2-chloro-oxazoles and -benzoxazoles (equation 32).

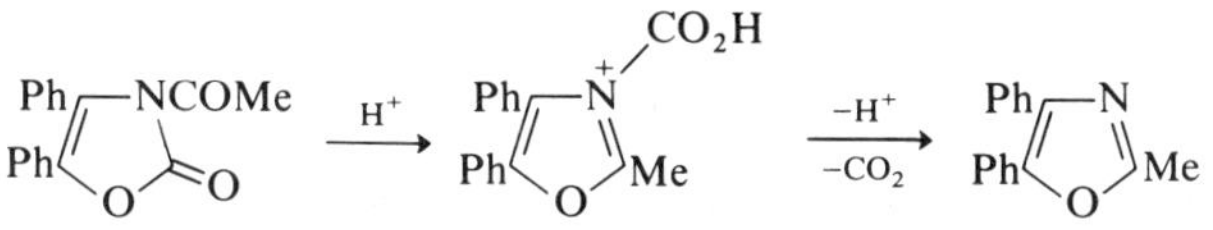

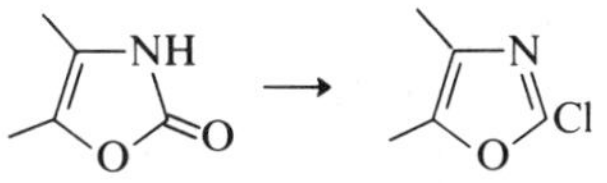

Alkylation of benzoxazolone and benzoxazolethione in alkaline media occurs by way of the delocalized anions (**195**) and results in N-alkylbenzoxazolones and 2-alkylthiobenzoxazoles, respectively. Treatment of benzoxazolethione with benzoyl chloride in the presence of a base yields N-benzoylbenzoxazolethione; however, under carefully controlled conditions (triethylamine/THF at −30 °C) the acylation is directed to the sulfur atom. Both the N- and the S-benzoyl derivatives are benzoylating agents for alcohols and amines ⟨81S991⟩. Benzoxazolone is N-aminated by hydroxylamine-O-sulfonic acid; the product (**196**) reacts with lead tetraacetate to yield a nitrene, which is stabilized as a singlet. It undergoes stereospecific *cis* addition to alkenes to give aziridines (equation 33) ⟨69JCS(C)772⟩. Acylation of 2-trifluoromethyl-5(2H)-oxazolones yields mainly the O-acyl derivatives, which rearrange to C-acyl compounds in the presence of pyridine (equation 34).

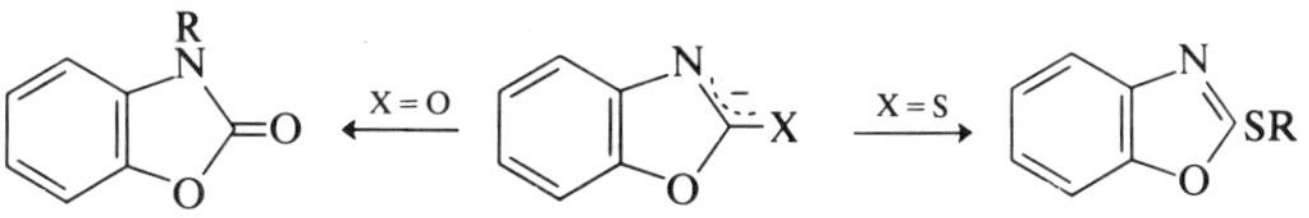

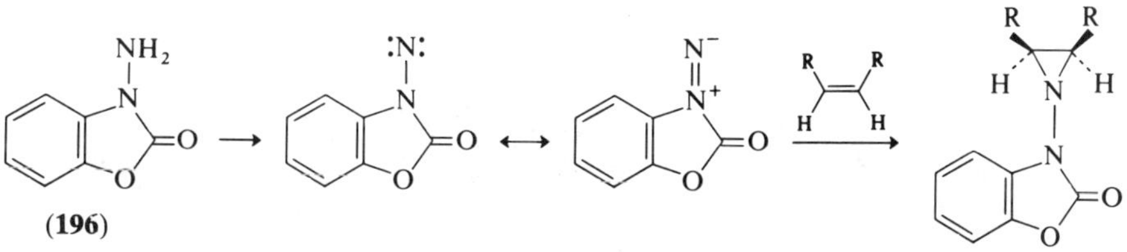

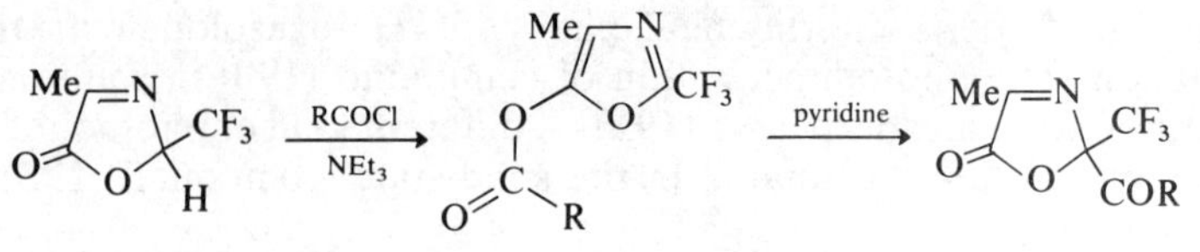

$$\text{(34)}$$

5(4H)-Oxazolones are attacked by a variety of electrophiles at C(4); these reactions, which require the presence of bases, proceed through the enolate anions (**197**). This type of anion adds to carbonyl compounds, a key step in the Erlenmeyer synthesis of unsaturated azlactones (equation 35) (see Section 4.18.4.3.4). The anions are intermediates in the formation of the amides (**198**) when oxazolones are treated with enamines (Scheme 15) ⟨71JCS(C)598⟩.

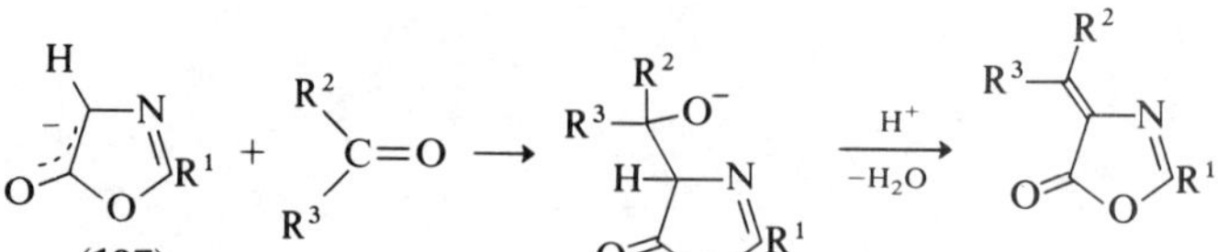

$$\text{(35)}$$

Scheme 15

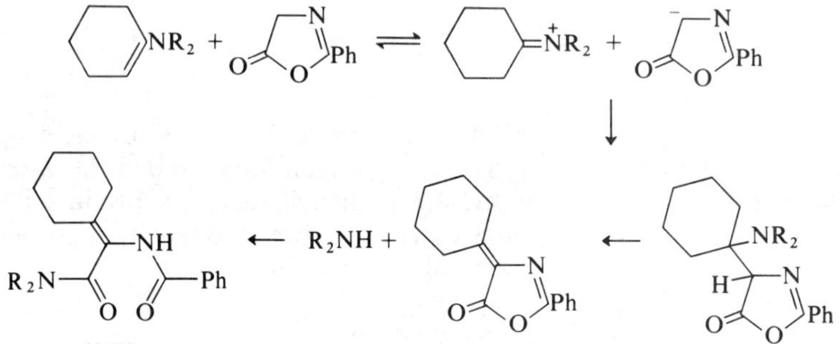

5(4H)-Oxazolones are alkylated at position 4 by alkyl halides, allyl halides and electrophilic alkynes, such as methyl propiolate (equation 36). In contrast, 2-phenyloxazolones react with methyl vinyl ketone at both C(4) and C(2) to yield a mixture of Michael adducts (equation 37). If the phenyl substituent is replaced by the bulky 2,4,6-trimethylphenyl group the addition is directed exclusively to C(4) ⟨81CB2580⟩. Alkylation of 5(4H)-oxazolones is a key step in the synthesis of ketones from α-amino acids (Scheme 16). The outcome of this sequence is the union of the electrophilic fragment R^3 with the group R^2CO; the amino acid thus functions as the equivalent of an acyl anion ⟨78AG(E)450⟩.

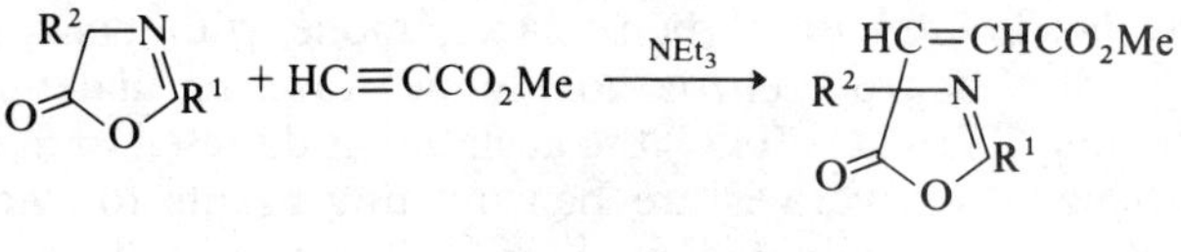

$$\text{(36)}$$

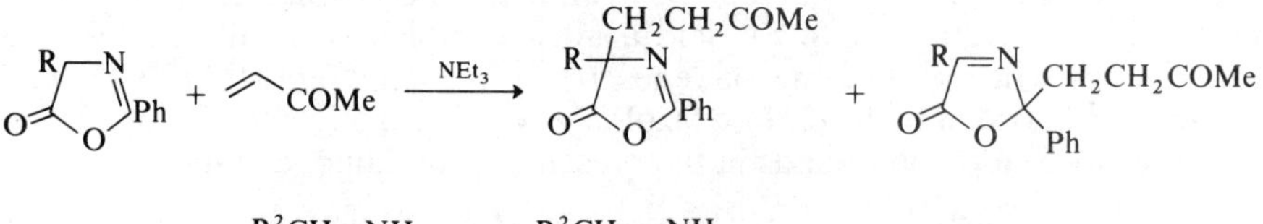

$$\text{(37)}$$

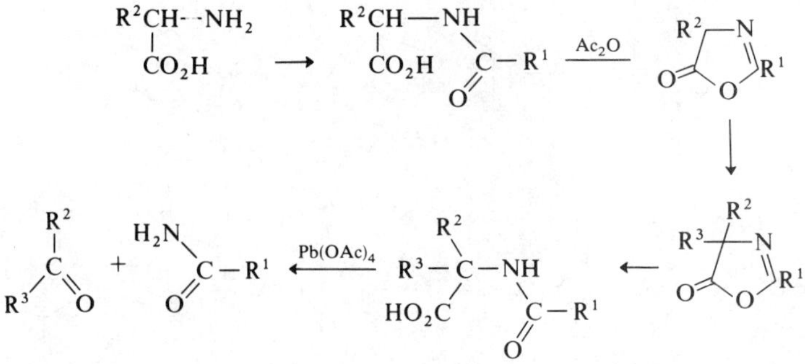

Scheme 16

(iii) Nucleophilic attack

2(3H)-Oxazolones are fairly inert towards hydrolysis; benzoxazolone yields o-aminophenol and carbon dioxide on prolonged heating with acid or alkali. 3,4-Diphenyl-2(3H)-oxazolone reacts with two molecules of a secondary amine, by initial attack at the carbonyl group, to afford an enamine (equation 38) ⟨66TL6009⟩. Organometallic reagents convert 2(3H)-oxazolones into α-acylamino ketones (equation 39); similarly, benzoxazolone yields o-benzamidophenol by the action of phenylmagnesium bromide.

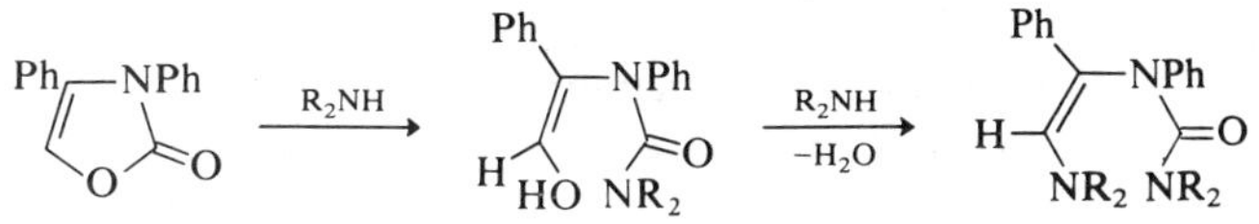

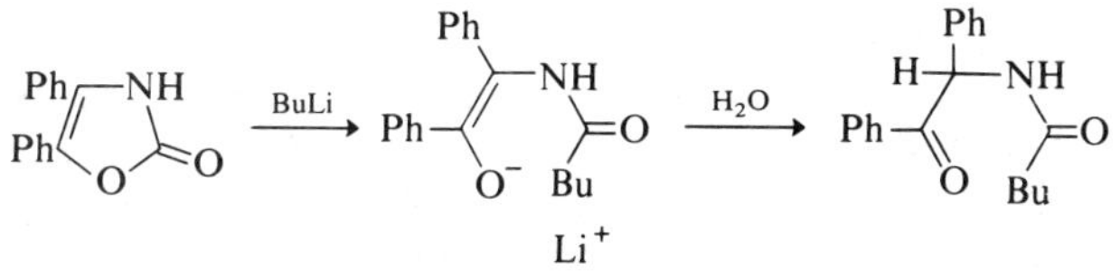

2(5H)-Oxazolones give hydroxy ketones on alkaline hydrolysis; lithium aluminum hydride reacts at the C=N bond to form the corresponding 2-oxazolidinones (Scheme 17).

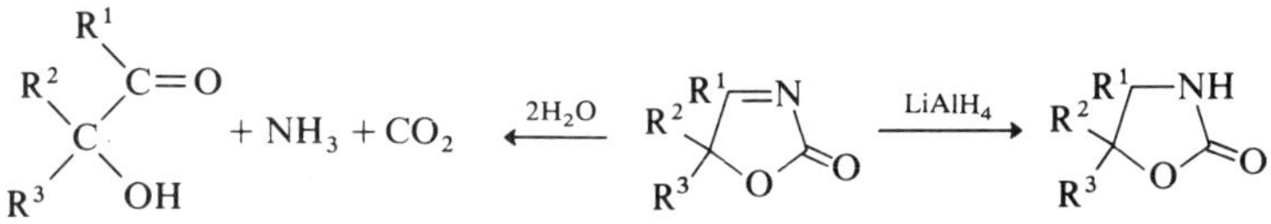

Scheme 17

The ring of 4(5H)-oxazolones is readily opened under mild hydrolytic conditions, amides being produced (equation 40). Compounds possessing a 2-alkoxy substituent undergo hydration rather than hydrolysis to yield eventually 2,4-oxazolidinediones (equation 41).

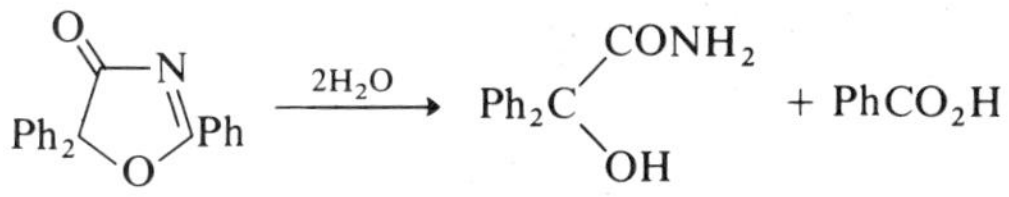

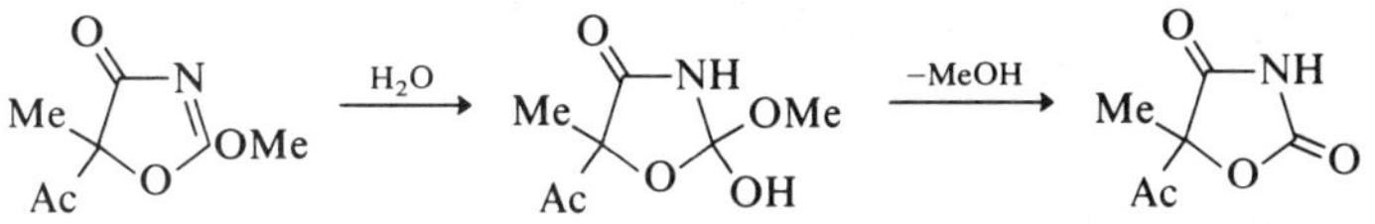

5(4H)-Oxazolones react readily with nucleophiles, C(5) and C(2) being possible sites for attack, and, in the case of unsaturated azlactones, C(α) as well (see **199–201**). It has been proved by using water labelled with ^{18}O that the acid-catalyzed hydrolysis of unsaturated azlactones proceeds by alkyl–oxygen fission (equation 42). The formation, hydrolysis and reduction of 4-methylene-5(4H)-oxazolones is a well-established method for the synthesis of α-amino acids, e.g. phenylalanine (equation 43). The addition of hydrazoic acid to 5(4H)-oxazolones without methylene groups at C(4) likewise occurs exclusively at C(2) to yield tetrazoles by ring-opening and recyclization (equation 44).

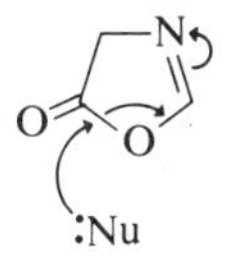

(199) 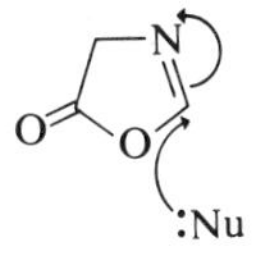   (200) 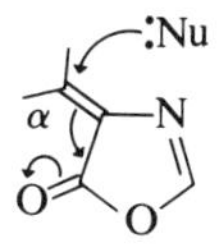   (201)

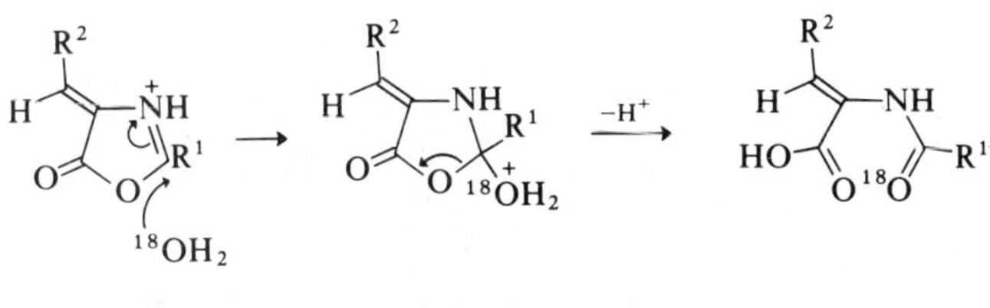

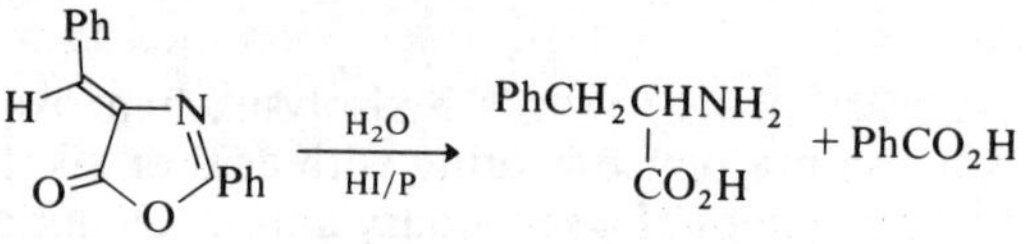

However, most nucleophiles attack 5-oxazolones at the carbonyl group and the products are derivatives of α-amino acids formed by acyl–oxygen fission. Thus the action of alcohols, thiols, ammonia and amines leads, respectively, to esters, thioesters and amides; orthophosphate anion gives acyl phosphates (Scheme 18). The use of α-amino acids in this reaction results in the establishment of a peptide link. Cysteine is acylated at the nitrogen atom in preference to the sulfur atom. Enzymes, *e.g.* α-chymotrypsin and papain, also readily combine with both saturated and unsaturated azlactones. A useful reagent for the introduction of an α-methylalanine residue is compound (**202**). Both the trifluoroacetamido and ester groups in the product are hydrolyzed by alkali to give a dipeptide. The alkaline hydrolyzate may be converted into the benzyloxycarbonyl derivative, which forms a new oxazolone on dehydration. Reaction with an ester of an amino acid then yields a protected tripeptide (equation 45).

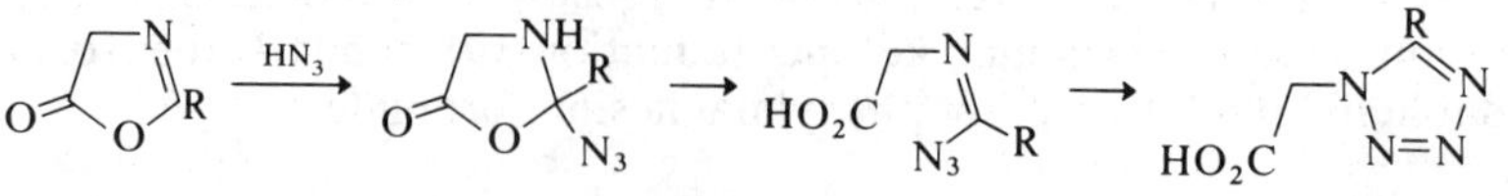

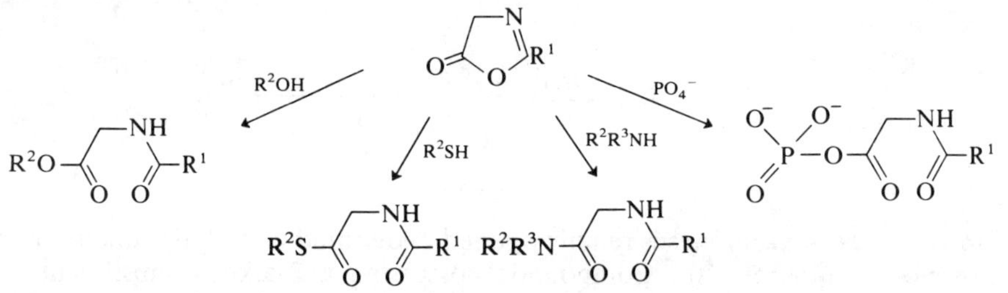

The benzylideneoxazolone (**189**) reacts with amines to give α-benzamidocinnamides, which can be induced to cyclize in different ways by a choice of conditions (Scheme 19). The *N*-benzyl compound (**203**), for instance, yields an imidazolinone under the influence of bases, whereas the action of perchloric acid produces the oxazole derivative (**204**). The latter gives the stable oxazoleimine (**205**) on deprotonation. Treatment of the ternary imonium perchlorate (**206**) with ammonia, followed by sodium hydroxide, affords the orange diazafulvene shown ⟨72JCS(P1)1140⟩.

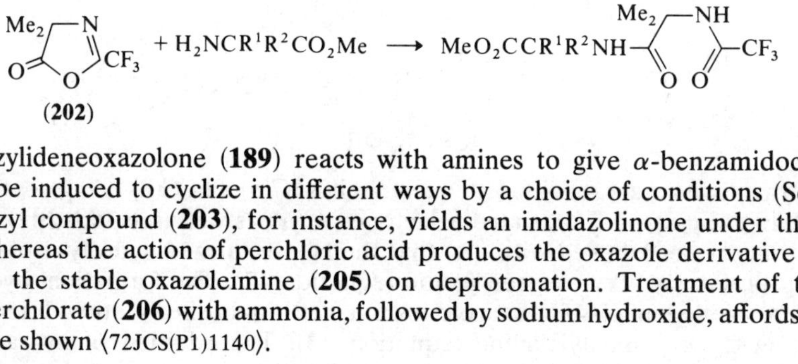

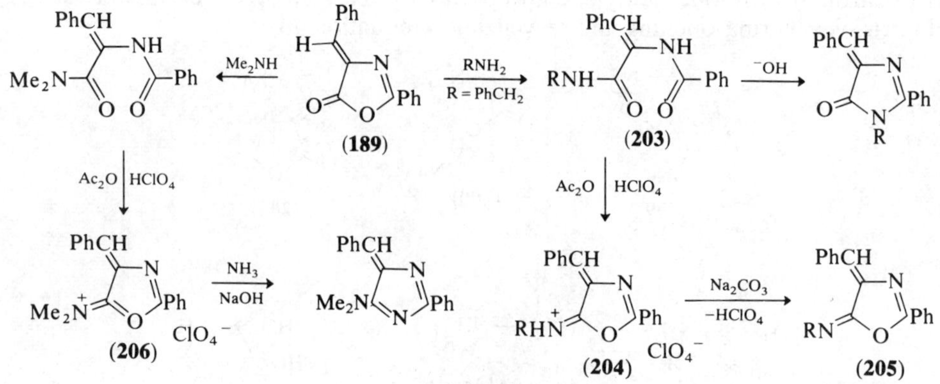

Treatment of saturated azlactones with aromatic compounds under Friedel–Crafts conditions gives acylamino ketones in high yield (equation 46). 4-Benzyl-2-methyl-5(4H)-oxazolone undergoes an intramolecular reaction to yield an acetamidoindanone (equation 47). Friedel–Crafts reactions of 4-(arylmethylene)-5(4H)-oxazolones are complicated by the presence of an additional electrophilic centre (*cf.* **201**) and may follow three courses. The unsaturated azlactone (**189**) adds benzene under the influence of aluminum chloride to form the saturated azlactone (**207**); in inert solvents (**189**) undergoes an intramolecular acylation to yield a mixture of the indenone (**208**) and the isoquinoline (**209**; Scheme 20).

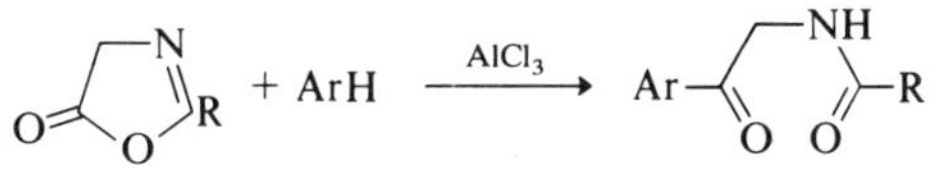

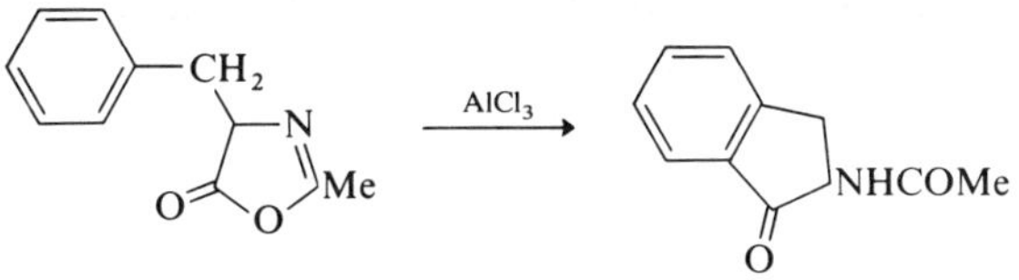

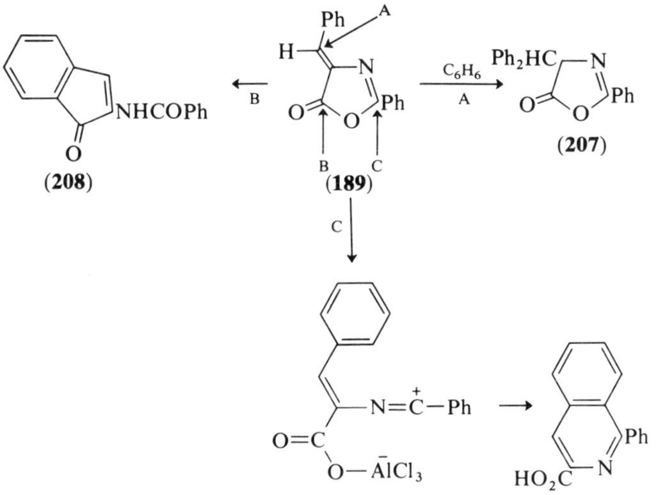

Scheme 20

The outcome of Grignard reactions on 5(4H)-oxazolones likewise depends on the nature of the substituent at C(4). The saturated azlactone (**210**) is converted by the action of phenylmagnesium bromide into the tertiary acylamino alcohol shown in equation (48); on the other hand, saturated azlactones are obtained by the action of alkyl Grignard reagents on (arylmethylene)oxazolones (equation 49). This type of oxazolone reacts with arylmagnesium halides at the carbonyl group; in the presence of copper(I) chloride, however, the addition is directed to the methylene carbon atom.

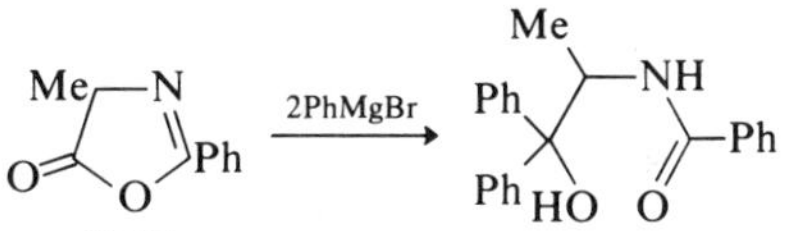

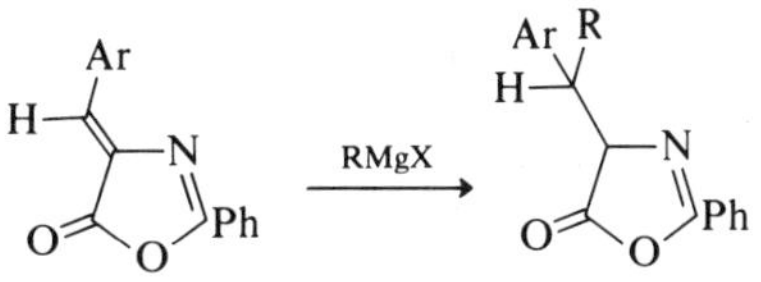

Hydride donors, such as calcium borohydride, reduce compound (**189**) to 2-benzamidocinnamyl alcohol, $PhCH=CH(NHCOPh)CH_2OH$.

Nucleophilic attack at the exocyclic double bond occurs in the reactions of the benzylidene derivative (**189**) with diazomethane and the sulfurane PhCOCH=SMe$_2$, which yield the spirocyclopropanes (**211**) and (**212**), respectively ⟨81H(16)209⟩.

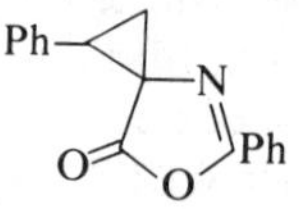

(**211**) 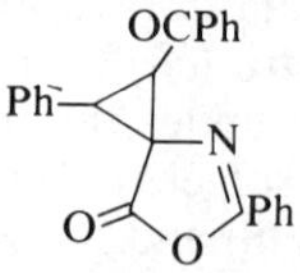 (**212**) 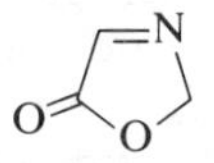 (**213**) 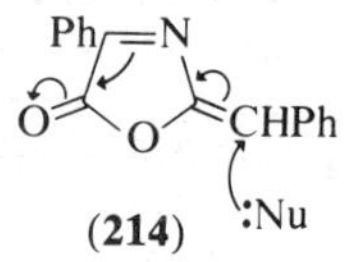 (**214**)

The 5(2*H*)-oxazolones (**213**) present two sites, C(4) and C(5), to nucleophilic attack; they usually react at the latter. The benzylidene derivative (**214**), the most thoroughly studied member of this class, possesses an additional electrophilic centre at the exocyclic carbon atom. However, alkaline hydrolysis of this compound affords phenylacetamide and benzoylformic acid by acyl–oxygen fission (equation 50). α-Keto acids are also obtained when 2-trifluoromethyl-5(4*H*)-oxazolones are hydrolyzed, the reaction involving preliminary isomerization to a 5(2*H*)-oxazolone. The example shown in equation (51) represents the first non-enzymatic synthesis of an optically active α-keto acid. An instance of nucleophilic attack at C(4) of a 5(2*H*)-oxazolone is the formation of the oxazolidinone (**215**) in a Grignard reaction (equation 52). However, the typical behaviour of unsaturated pseudooxazolones like (**214**) is conjugate addition of a nucleophile, followed by further transformations of the resulting 5(4*H*)-oxazolones. This is illustrated by the reaction of compound (**214**) with benzene in the presence of aluminum chloride to yield, after aqueous work-up, the acylamino acid (**216**; equation 53).

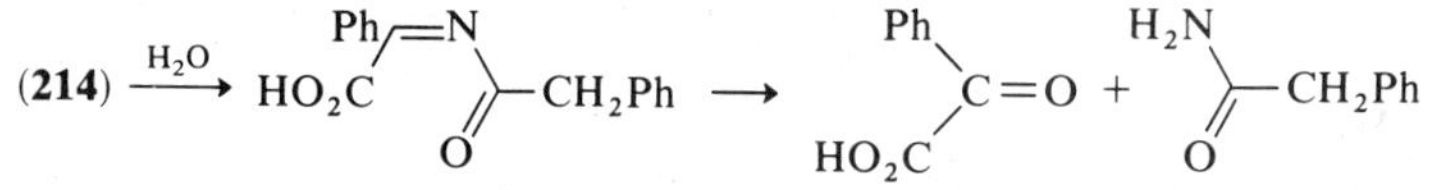

$$(214) \xrightarrow{H_2O} \quad \quad \rightarrow \quad \quad \quad \quad \quad \quad \quad \quad \quad \quad (50)$$

$$\text{(+)-form} \rightleftharpoons \quad \xrightarrow{H_2O} \quad \quad C=O + NH_3 + CF_3CHO \quad (51)$$
$$\text{(+)-form}$$

$$(214) \xrightarrow{PhMgBr} (215) \quad \quad \quad \quad \quad \quad \quad \quad \quad (52)$$

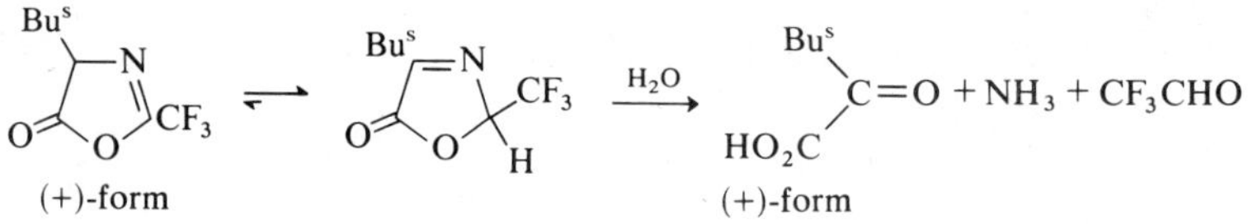

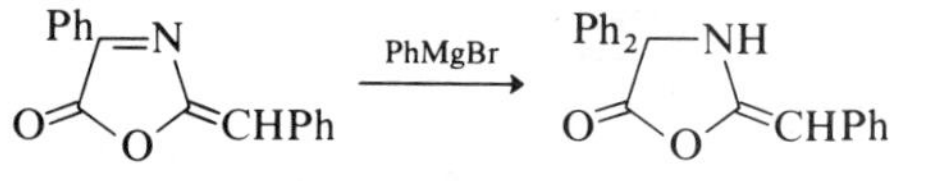

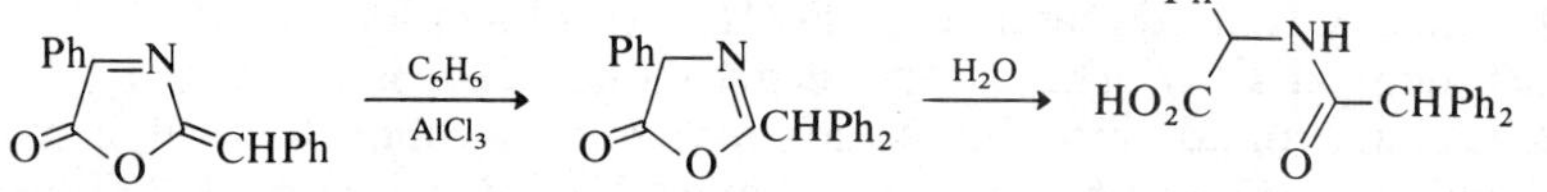

$$(214) \xrightarrow[AlCl_3]{C_6H_6} \quad \xrightarrow{H_2O} \quad (216) \quad \quad \quad \quad \quad (53)$$

4.18.3.1.5 Reactions of mesoionic compounds

This section deals with the comparatively new anhydro-4-hydroxyoxazolium hydroxides (**217**), the isomeric 5-hydroxy compounds (**218**) and the amino analogues (**219**). The chemistry of the münchnones (**218**) has been intensively studied as a result of the pioneering research of Huisgen's school in Munich on 1,3-dipolar cycloaddition reactions of these compounds ⟨B-67MI41800⟩. There have been so many publications in this area that only the bare outlines can be presented here; the reader is referred to recent reviews ⟨76AHC(19)1, 81MI41800⟩ for fuller accounts.

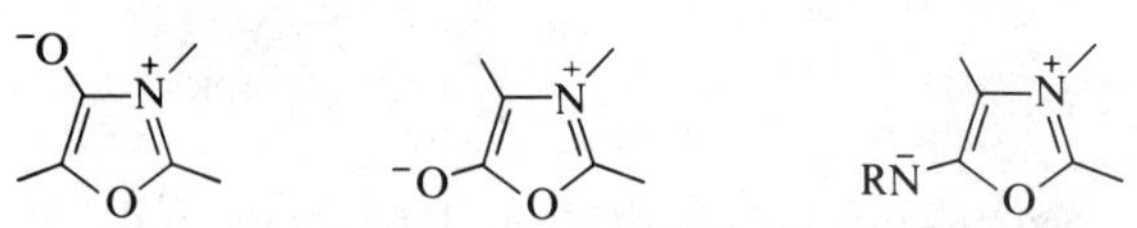

(**217**) (**218**) (**219**)

(i) Thermal and photochemical reactions

Anhydro-5-hydroxyoxazolium hydroxides lacking substituents at C(4) dimerize spontaneously by a process in which one molecule acts as an electrophile and the other as a nucleophile (Scheme 21). This accounts for the fact that dimeric products of this type are obtained by the action of dicyclohexylcarbodiimide on acylamino acids of the general formula $R^1CONR^2CH_2CO_2H$. Substituents at position 4 stabilize the mesoionic system: the first compounds to be prepared were the acetyl derivatives (**220**) ⟨B-49MI41800⟩ and (**221**) ⟨58CI(L)461⟩ and much of the more recent work has been carried out with the relatively stable methyldiphenyl compound (**222**). This münchnone decomposes above 115 °C to yield the allene (**225**) with loss of carbon dioxide. The mechanism proposed for this remarkable reaction (Scheme 22) involves valence isomerization to the ketene (**223**), which undergoes a 1,3-dipolar cycloaddition with the münchnone. The product loses carbon dioxide to form a new betaine (**224**), which collapses to the allene as shown.

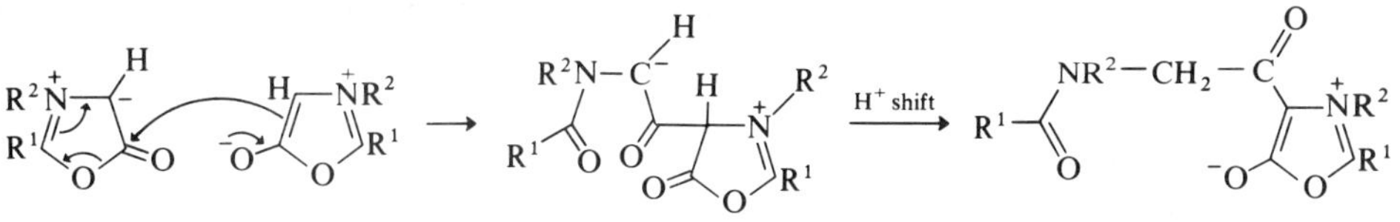

Scheme 21

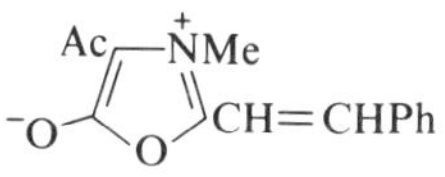

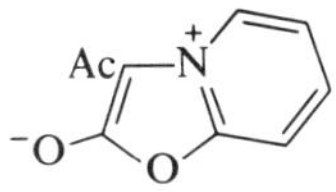

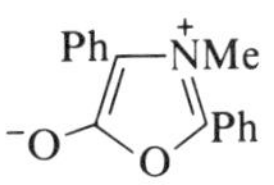

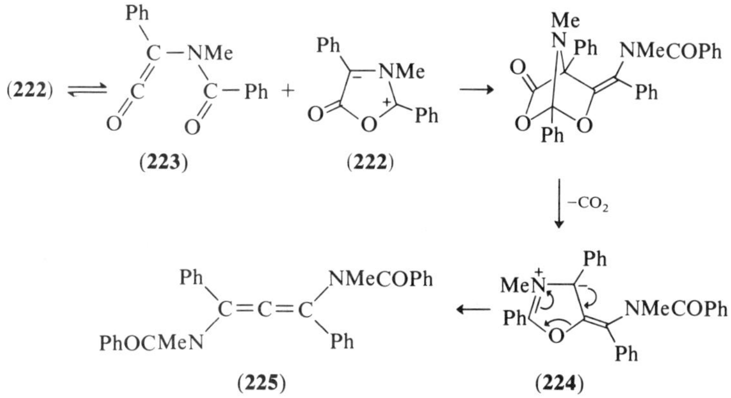

Scheme 22

Compound (**226**) is stable at −50 °C; when its solution in ethanol–chloroform is irradiated at this temperature, the ketene (**227**) is formed and is trapped as the ester (**228**) by reaction with the ethanol (equation 54). No reaction occurs in the dark.

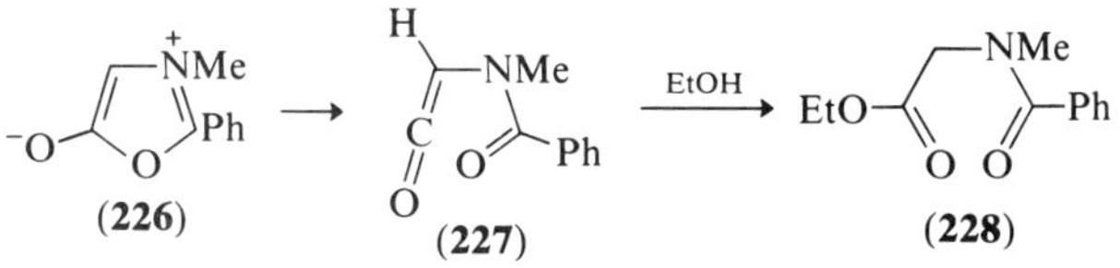

(54)

(ii) Electrophilic attack

Anhydro-5-hydroxyoxazolium hydroxides are attacked by electrophiles at position 4 (cf. **229**). If this position is free, substitution ensues, as in the dimerization reaction of Scheme 21. Other examples are uncatalyzed acylations (equation 55) and diazo coupling (equation 56). The key step in the Dakin–West reaction, i.e. the formation of α-acetamido ketones from α-amino acids and acetic anhydride in the presence of pyridine, is acetylation of an intermediate betaine (**230**; Scheme 23).

(55)

(56)

Scheme 23

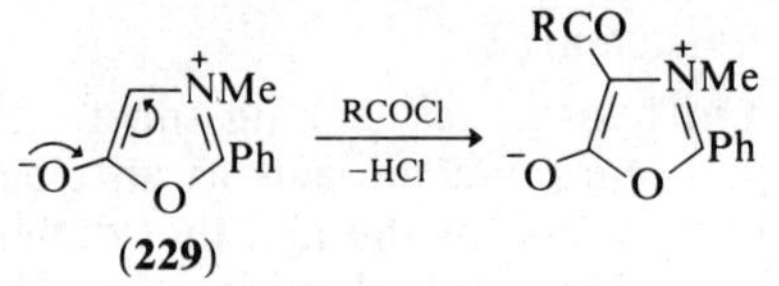

(*iii*) *Cycloaddition reactions*

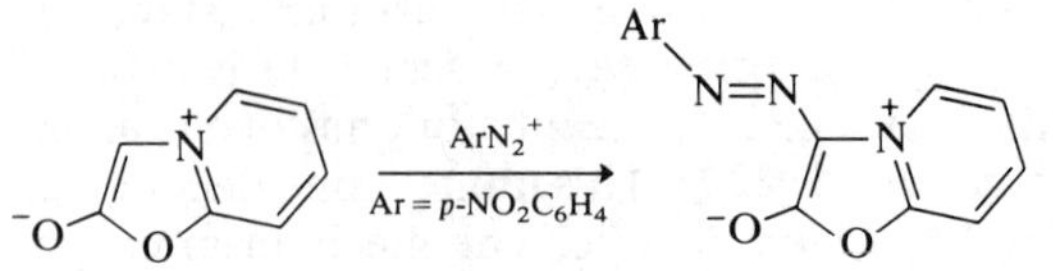

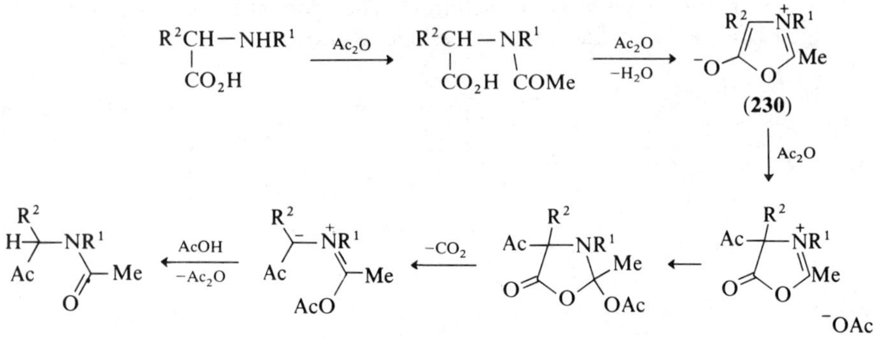

Anhydro-4-hydroxyoxazolium hydroxides, such as compound (**231**), behave as carbonyl ylides (**232**) in cycloaddition reactions, yielding bicyclic adducts with alkenes and carbonyl compounds (Scheme 24). The adducts produced by combination with alkynes fragment spontaneously in a retro-Diels–Alder reaction, giving furans (equation 57). The formation of a furan by the action of DMAD on the 4(5*H*)-oxazolone (**233**) shows that the latter exists in equilibrium with the mesoionic tautomer (**234**; equation 58) ⟨79JOC626⟩.

Scheme 24

(57)

(58)

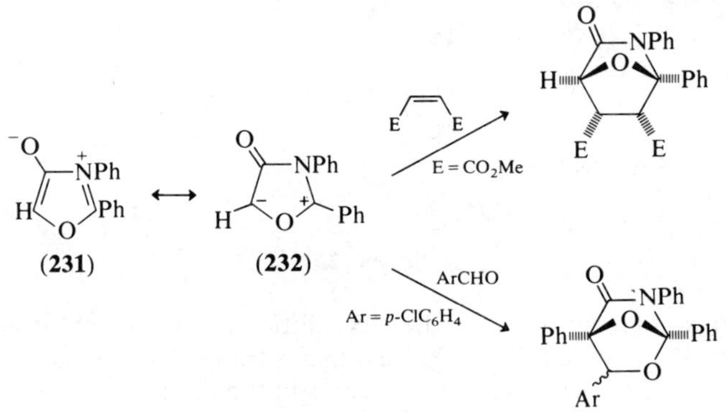

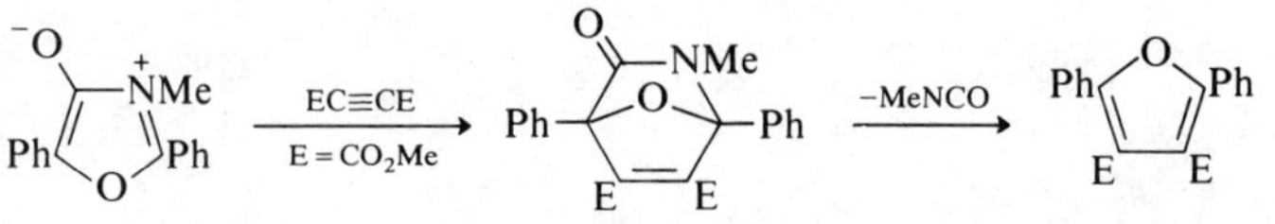

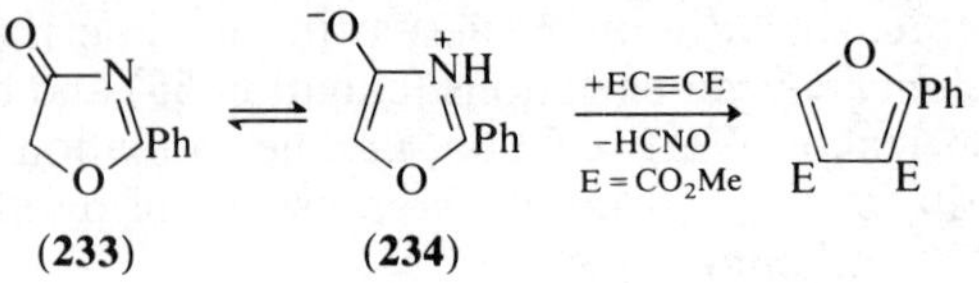

Cycloaddition reactions of anhydro-5-hydroxyoxazolium hydroxides, *e.g.* (235), have been investigated thoroughly but no satisfactory comprehensive theoretical treatment to account for the orientation of the products has yet been presented. While the münchnones usually function as 1,3-dipoles (236), several reactions involving the intermediacy of the valence-isomeric ketenes (237) have come to light.

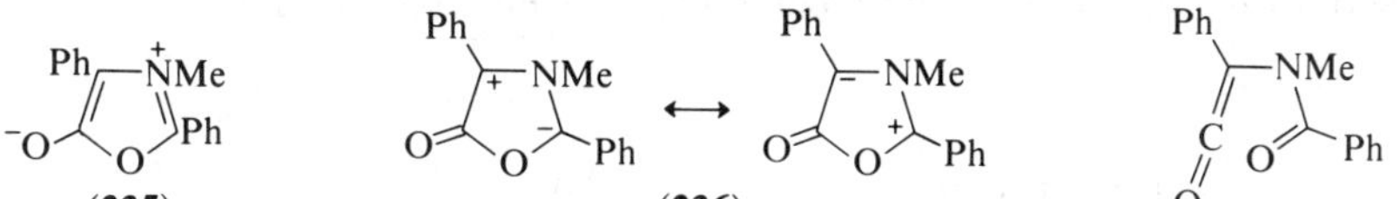

The 1,3-dipolar cycloaddition of alkenes and other double-bond compounds to the münchnones results in the adducts (238) which usually cannot be isolated. Carbon dioxide is extruded in a cycloreversion to give new 1,3-dipoles (239) whose subsequent fate is determined by the nature of the dipolarophile X=Y (equation 59). Alkenes react with compound (235) to give intermediate azomethine ylides which yield stable products by tautomerization, as in the formation of the 2,3-dihydropyrrole (240) in the reaction with styrene (equation 60). If the alkene possesses a suitable leaving group pyrroles are produced (equation 61). 5(4H)-Oxazolones are in tautomeric equilibrium with münchnones (see Section 4.18.2.5.1); this accounts for the formation of 1-pyrrolines, *e.g.* (241), when azlactones are treated with alkenes (equation 62). Cyclopropenes and cyclobutenes add to anhydro-5-hydroxyoxazolium hydroxides to yield ring-expanded products, as illustrated in equation (63). The intramolecular cycloaddition of compound (242; R = Ph) results in a mixture of the regioisomers (243; R = Ph) and (244); in contrast, the methyl analogue (242; R = Me) yields solely compound (243; R = Me). These reactions are also notable for the thermal stability of the primary adducts (equation 64) ⟨80TL3419⟩.

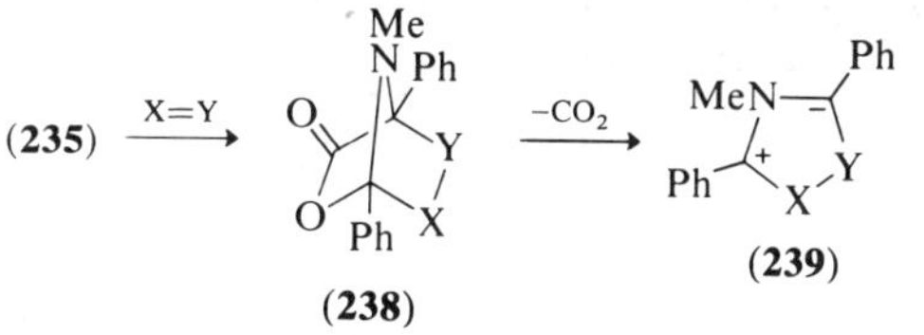

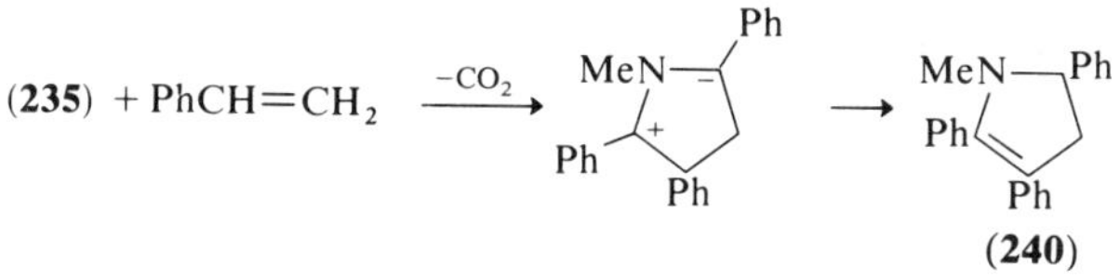

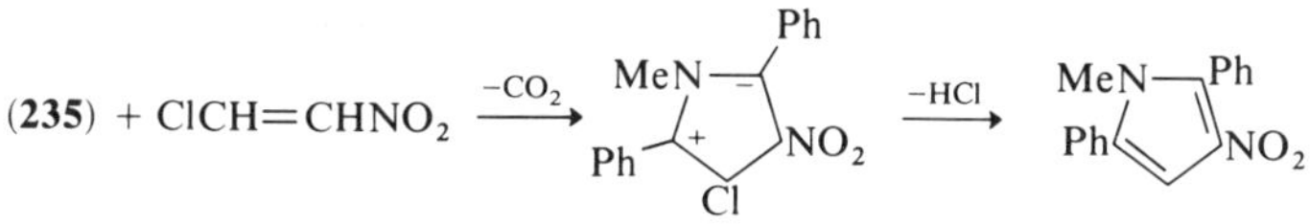

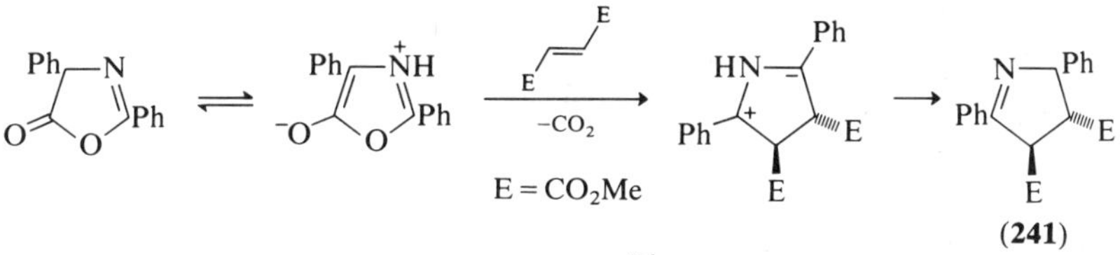

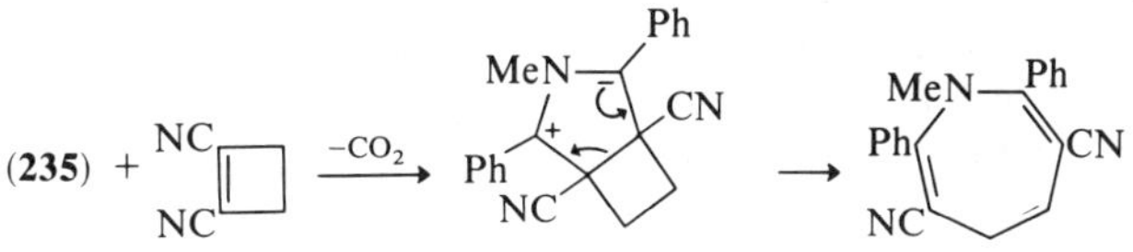

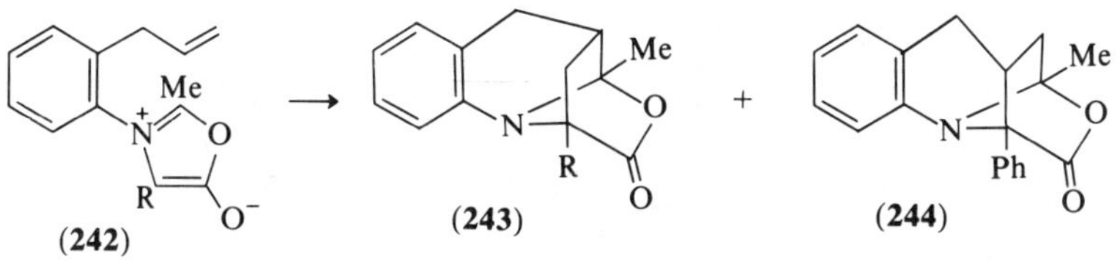

Carbonyl compounds and other dipolarophiles containing heteroatoms react with münchnones. Thus benzaldehyde forms the betaine (**245**) which suffers ring-cleavage to yield the enamine (**246**; equation 65). Thiocarbonyl compounds and nitrosobenzene behave in an analogous manner. The action of dipolarophiles containing cumulative double bonds results in the formation of new mesoionic systems. Thus carbon disulfide and phenyl isothiocyanate afford a zwitterionic thiazole and imidazole, respectively (Scheme 25).

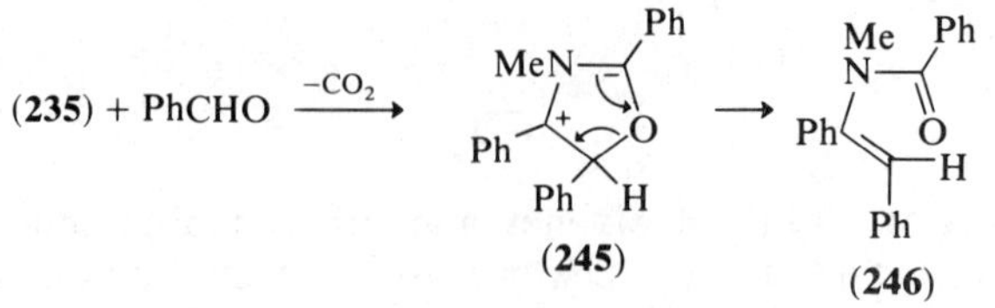

(65)

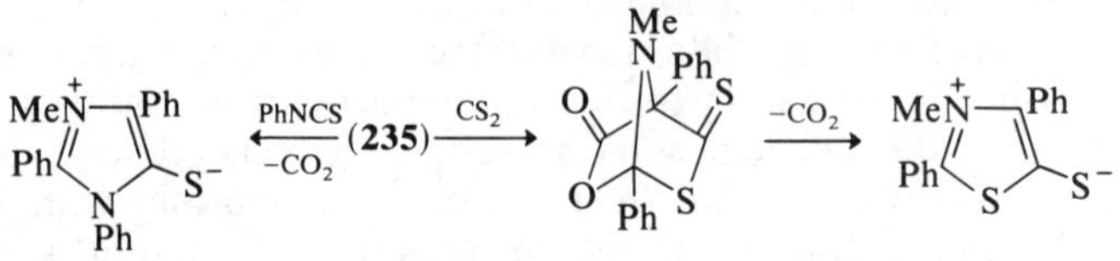

Scheme 25

The 1,3-dipolar cycloaddition of alkynes to anhydro-5-hydroxyoxazolium hydroxides leads to pyrroles (equation 66). Electron-withdrawing substituents in the alkyne are activating and DMAD reacts 48 times faster with compound (**235**) than does methyl propiolate, which in turn is 7000 times more reactive than oct-1-yne. The reaction has many ramifications and it is frequently used to trap unisolable münchnones with or without substituents on the nitrogen atom by an *in situ* reaction in which the betaine is generated in the presence of the dipolarophile (equation 67). It has become an important method for constructing the pyrrole ring of fused heterocyclic systems, as in the example of equation (68).

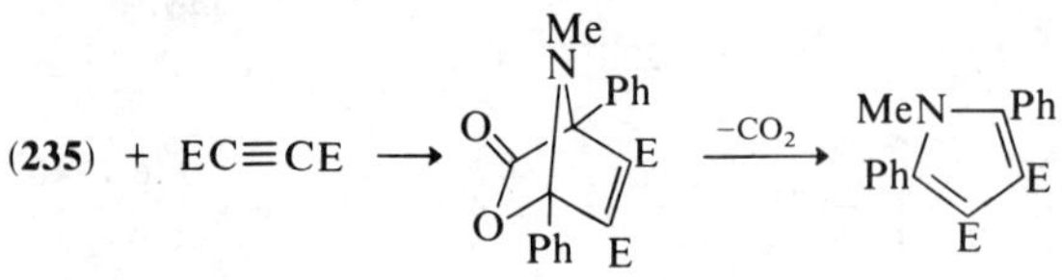

(66)

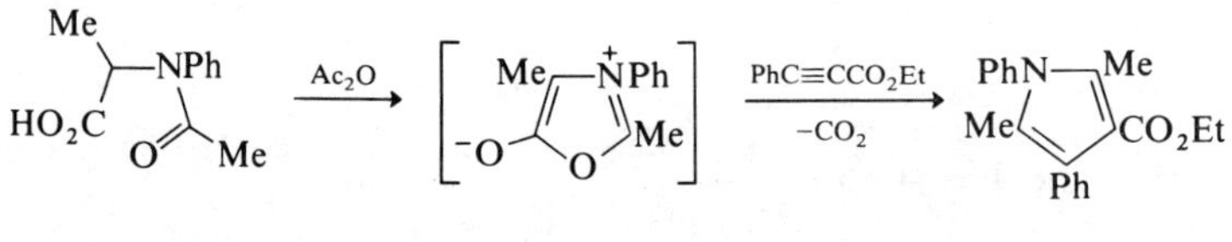

(67)

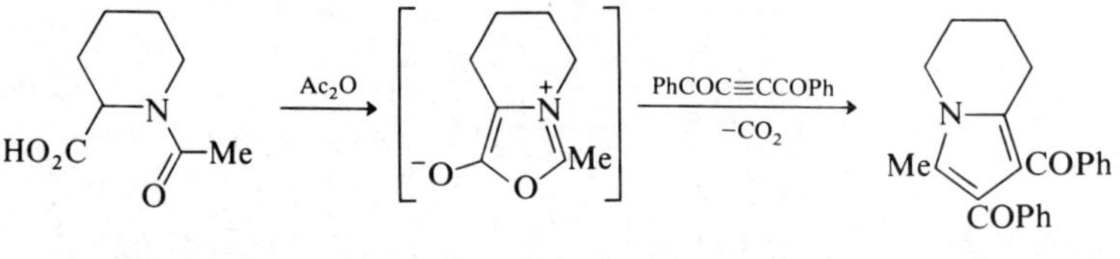

(68)

Some reactions of münchnones occur *via* acylamino ketenes, the covalent valence tautomers of the betaines. The ketenes are intermediates in the thermolysis (see Scheme 22) and in the formation of azetidinones from imines (equation 69); they are thought to be involved in the aminolysis of the mesoionic compounds, which results in amides of α-acylamino acids, and in the formation of the benzodioxin (**247**) by the combined action of acetic anhydride and tetrachloro-*o*-benzoquinone on *N*-benzoylalanine (equation 70).

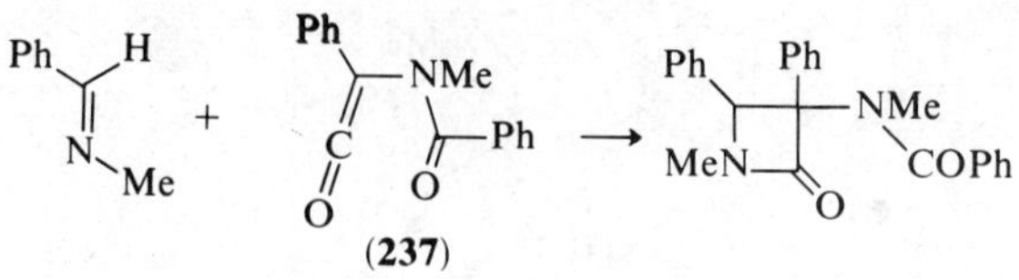

(69)

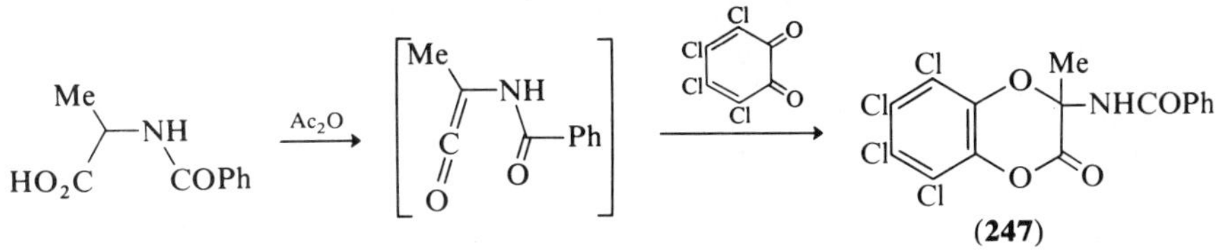

(247) (70)

Anhydro-5-aminooxazolium hydroxides (**249**), generated from Reissert compounds (**248**), combine with alkynes to afford pyrroloisoquinolines; the reaction (Scheme 26) is analogous to that of münchnones with alkynes.

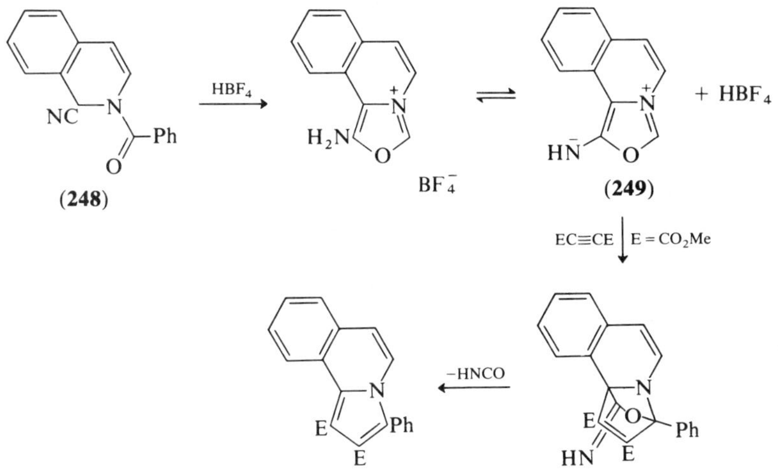

(248) (249)

Scheme 26

4.18.3.2 Reactions of Partially Saturated and Saturated Rings

4.18.3.2.1 Dihydrooxazoles

While representatives of all three dihydrooxazoles (**250**)–(**252**) are known, only the chemistry of the last has been thoroughly explored and the present discussion is confined to these compounds ⟨71CRV483⟩.

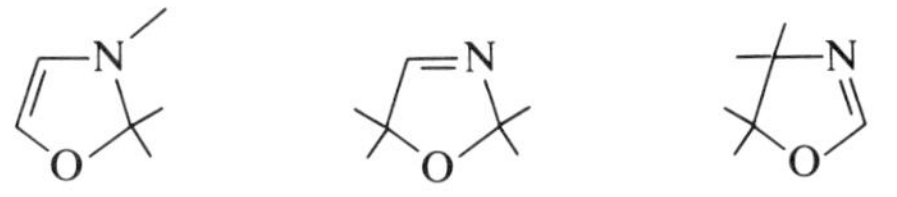

(**250**) (**251**) (**252**)

The 4,5-dihydrooxazoles (**252**) are weakly basic, their salts being readily hydrolyzed to yield esters of β-amino alcohols (equation 71). A different mode of ring-fission is observed when dihydrooxazoles are treated with concentrated hydrochloric acid; *N*-(β-chloroethyl)amides are formed (equation 72). The same type of reaction takes place when acid chlorides are used (equation 73). However, if position 2 is free, *N*-acyl-2-chlorooxazolidines are produced (equation 74) ⟨81JOC3742⟩. 4,5-Dihydrooxazoles undergo cationic polymerization induced by dimethyl sulfate or ethyl trifluoromethanesulfonate (equation 75) ⟨78MI41801⟩; alternating copolymers, *e.g.* (**253**), are obtained with acrylic acid ⟨77MI41800⟩.

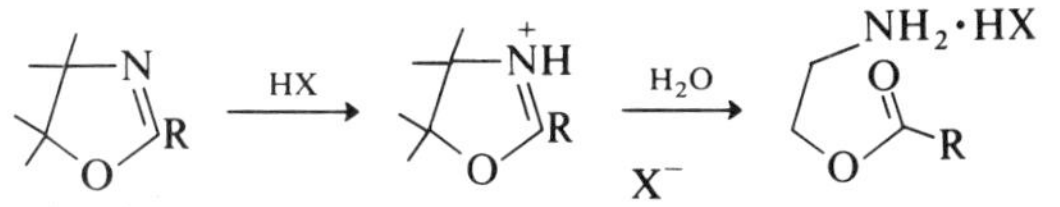

(**252**) (71)

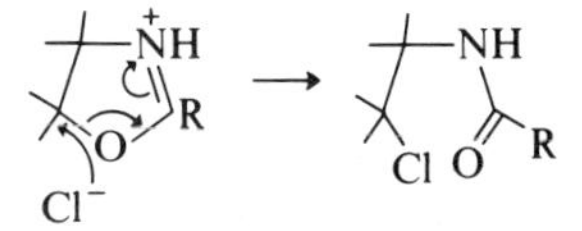

(72)

 (73)

 (74)

 (75)

R = H, Me or Ph

(253)

(254) (255)

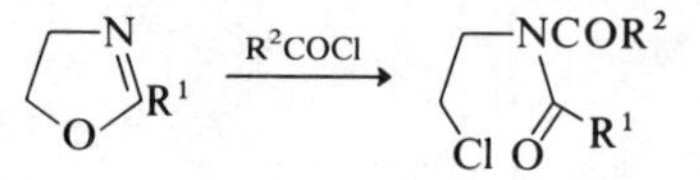

2-Alkyl-4,5-dihydrooxazoles (**254**) are metallated by butyllithium *etc.* at the alkyl group and the resulting lithium derivatives (**255**) undergo a variety of useful transformations. Meyer's work on asymmetric synthesis ⟨78ACR375⟩ is based on the reactions of the chiral dihydrooxazoles (**257**), which are prepared from the readily available optically active aminoglycol (**256**; equation 76). Lithiation of (**257**), followed by treatment with an alkyl halide R^2X, yields mainly one of the two diastereoisomeric dihydrooxazoles (**258**) which on hydrolysis affords an optically active dialkylacetic acid (equation 77). The (4*S*,5*S*)-dihydrooxazoles (**257**) have also been used for the stereoselective synthesis of α- and β-hydroxy acids. Lithiation of the benzyl derivative (**259**) and subsequent oxidation with oxygen gives the benzoyldihydrooxazole (**260**). The latter reacts with organometallic compounds RM to yield the alcohols (**261**) as mixtures of diastereoisomers which can be separated by medium-pressure liquid chromatography (MPLC). Hydrolysis then affords optically pure (*S*)-α-hydroxy acids (equation 78) ⟨80JOC2912, 80JOC2785⟩. The chiral ethyl derivative (**262**) reacts with 9-borabicyclononane trifluoromethanesulfonate in the presence of ethyldiisopropylamine to form the ethylideneoxazolidine (**263**). Addition of an aliphatic aldehyde RCHO, followed by treatment with methanolic hydrogen peroxide, gives the alkylated dihydrooxazoles (**264**) which are hydrolyzed to β-hydroxy acids in which the *erythro* isomers predominate to the extent of 98% (equation 79) ⟨81JA4278⟩.

 (76)

(256) (257)

 (77)

(258)

 (78)

(259) (260) (261)

R = Me, Et, Pr, Pr^i, *etc.*

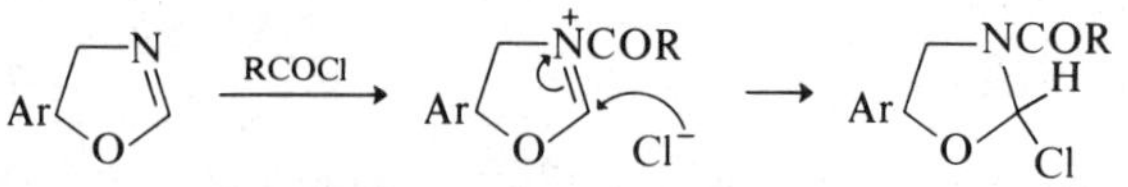

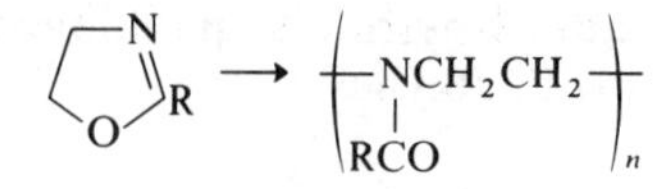

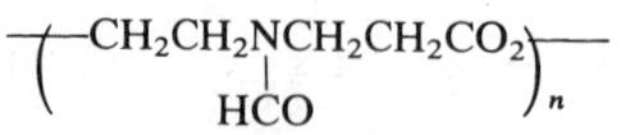

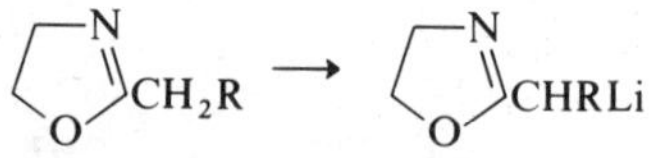

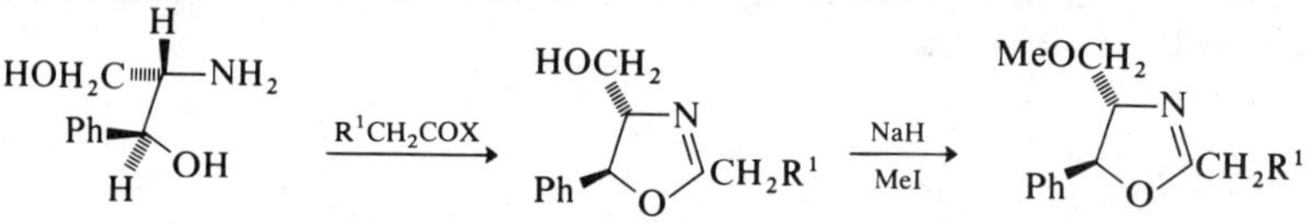

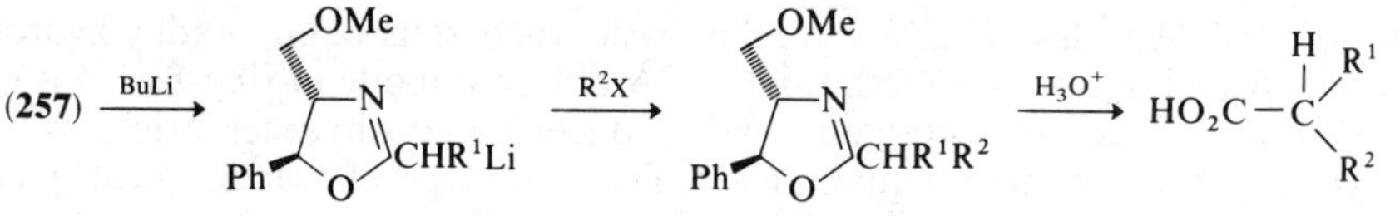

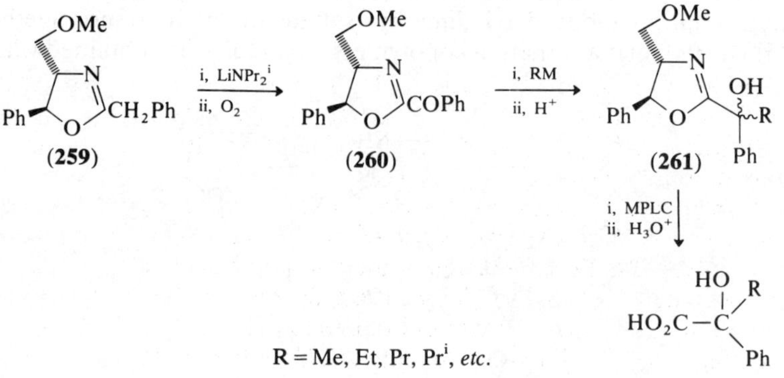

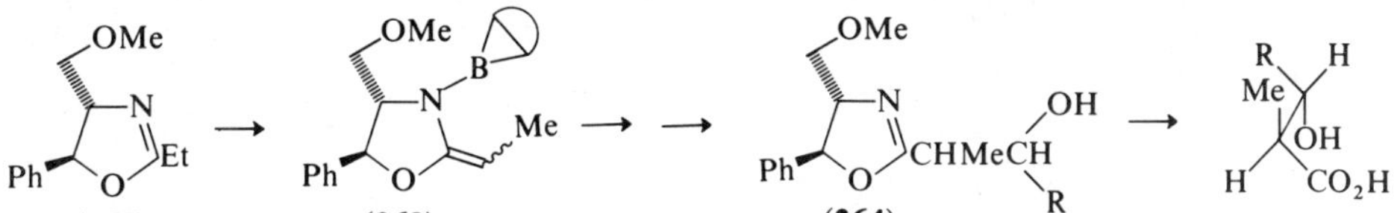

(262) → (263) → (264) → (79)

4.18.3.2.2 Oxazolidines and oxo derivatives ⟨B-57MI41801⟩

(i) Oxazolidines ⟨53CRV(53)309⟩

The ring–chain tautomerism of oxazolidines is discussed in Section 4.18.2.5.2. Oxazolidines form unstable salts which are easily hydrolyzed (equation 80) ⟨77RRC1413⟩. N-Substituted oxazolidines react with enamines in the presence of trifluoroacetic acid to yield 1,4-oxazepines in a ring-opening 1,5-cycloaddition reaction (equation 81) ⟨80LA1573⟩.

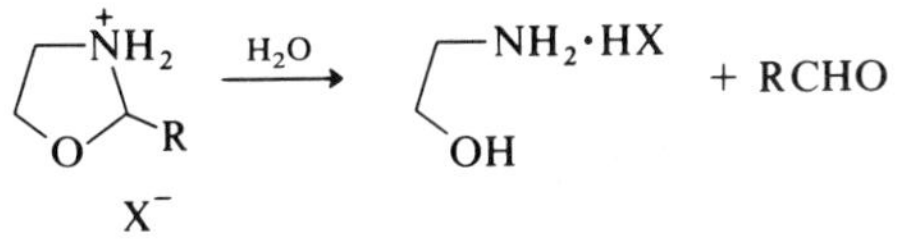

(80)

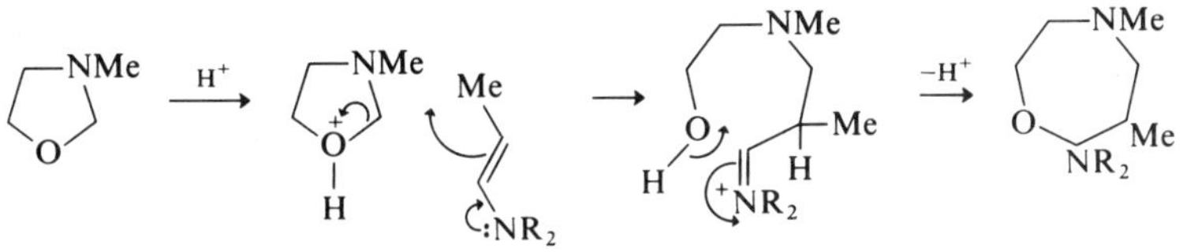

(81)

Oxidation of 2,2,4,4-tetraalkyloxazolidines gives stable nitroxide radicals (**265**) which are used as spin labels for probing biomolecular structures ⟨69ACR17, B-76MI41800⟩. The stearic acid derivative 12-doxylstearic acid (**266**) is commercially available.

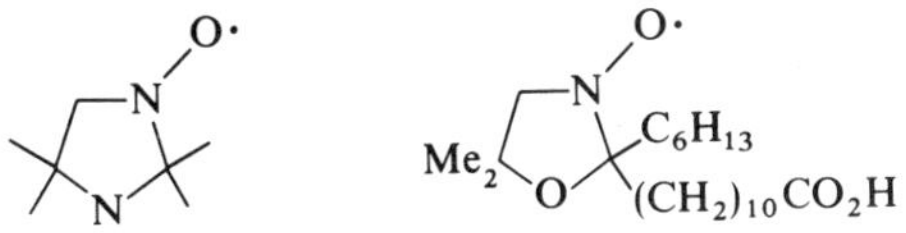

(265)   (266)

(ii) Oxazolidinones ⟨67CRV197⟩

2-Oxazolidinones are virtually neutral; they do not give stable salts with acids but they can be alkylated at the nitrogen atom under basic conditions. They are rather resistant to hydrolysis; vigorous treatment with alkali yields β-amino alcohols whereas hydrochloric acid gives β-chloroalkylamines (equation 82). The combined action of aromatic hydrocarbons and aluminum chloride on 2-oxazolidinones leads to β-arylethylamines which are formed by alkyl–oxygen fission (equation 83) ⟨80TL1719⟩.

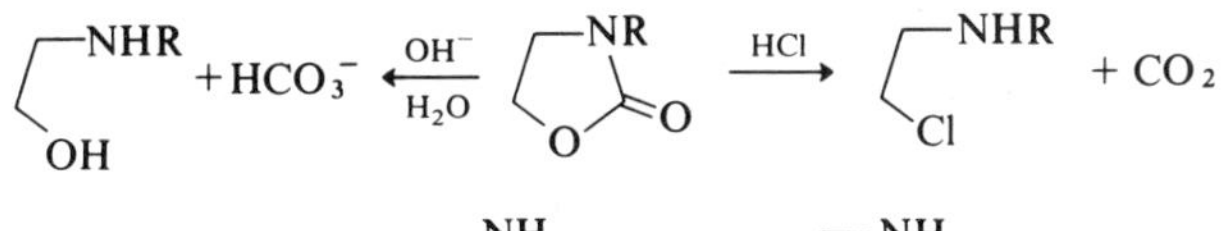

(82)

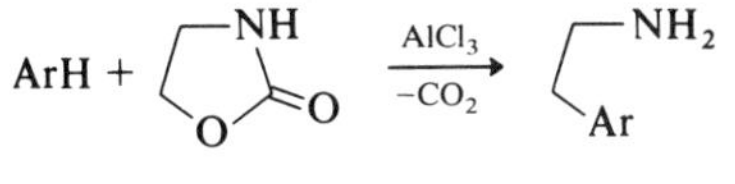

(83)

2-Oxazolidinones react with nitrous acid, the resulting N-nitroso compounds being decomposed by alkali to yield a variety of products: alkynes, aldehydes and vinyl ethers. These transformations are rationalized by postulating the intermediacy of transient alkenediazonium ions (Scheme 27) ⟨79CB2120⟩.

Little is known about the chemistry of 4- and 5-oxazolidinones. The former are hydrolyzed by dilute acids to give α-hydroxyamides and carbonyl compounds (equation 84); they form O-methyl derivatives (**267**) by the action of methyl iodide and silver oxide.

The bis(trifluoromethyl)-5-oxazolidinone (**268**) decomposes on heating to carbon dioxide and the azomethine ylide (**269**). The latter cyclizes to the aziridine (**270**) but in the presence of DMAD it forms the 1,3-dipolar cycloadduct (**271**; Scheme 28) ⟨77MI41801⟩.

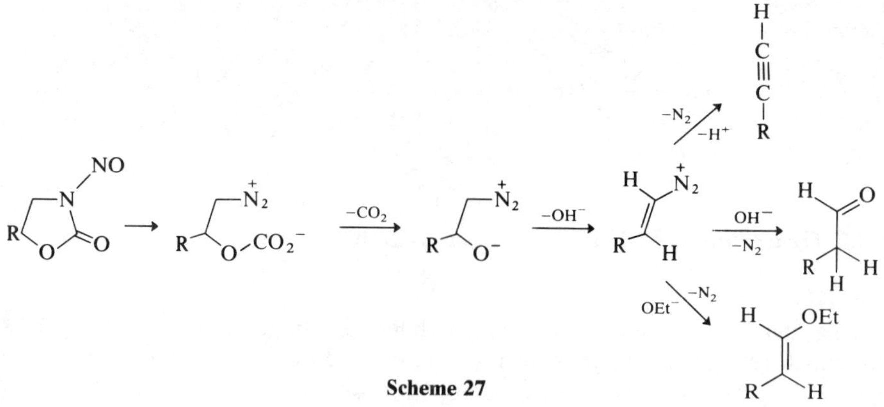

Scheme 27

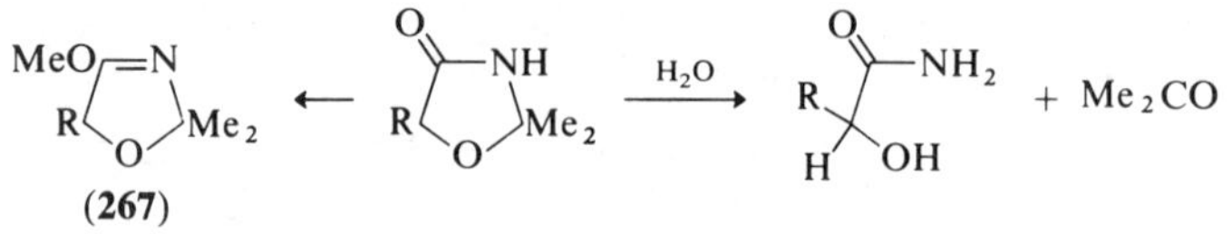

$$(84)$$

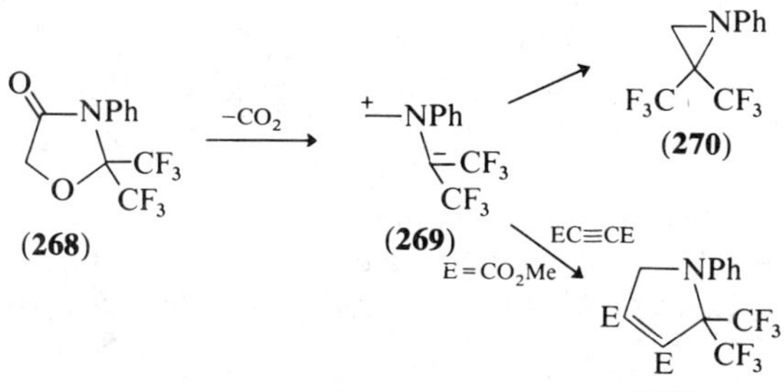

Scheme 28

(iii) Oxazolidinediones

Oxazolidine-2,4-diones are stable compounds. Those without a substituent at the nitrogen atom are acidic, dissolving in sodium hydroxide solution and being regenerated on acidification. The salts react with alkylating agents to form *N*-alkyloxazolidine-2,4-diones, *e.g.* (**272**), which are more easily hydrolyzed than the parent compounds. Sodium hydroxide yields a mixture of the hydroxyamide (**273**) and a carboxylic acid, probably (**274**; equation 85). The silver salt of 5,5-dimethyloxazolidine-2,4-dione reacts with ethyl iodide to give the *O*-ethyl derivative (**275**) (see Section 4.18.4.3.2) which rearranges to the *N*-ethyl isomer on heating (equation 86).

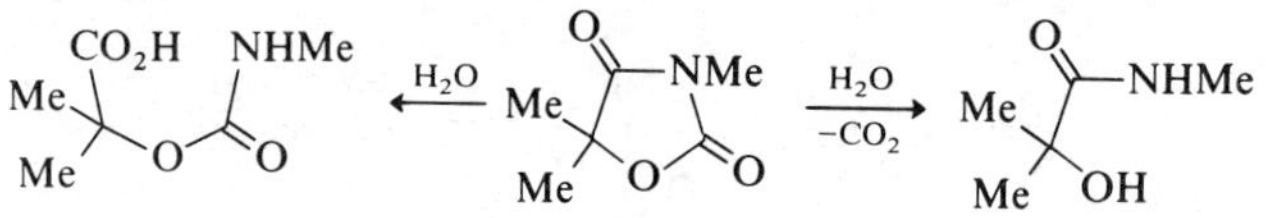

$$(85)$$

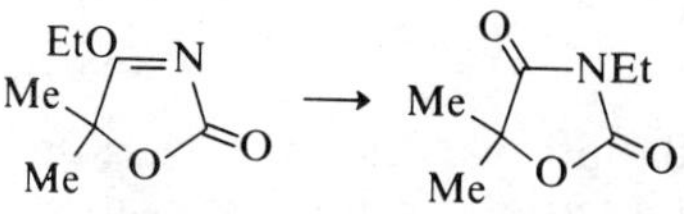

$$(86)$$

Oxazolidine-2,5-diones (**276**) are the anhydrides of *N*-carboxy-α-amino acids. They react with nucleophiles, such as water, alcohols or amines, to give unstable *N*-carboxy acids, esters or amides which lose carbon dioxide (equation 87). The products (**277**) are themselves nucleophilic reagents and may react with another molecule of the oxazolidinedione to yield a dipeptide (equation 88). The process may continue, leading to a polypeptide. Chiral oxazolidine-2,5-diones derived from optically active α-amino acids undergo stereospecific 'polymerization' ⟨72C501⟩.

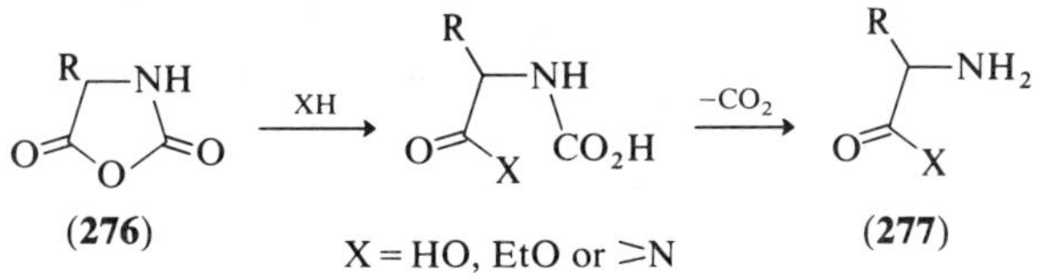

$$\text{(276)} \xrightarrow{XH} \underset{X = HO, EtO \text{ or } >N}{\text{intermediate}} \xrightarrow{-CO_2} \text{(277)} \quad (87)$$

$$\text{(276)} + \text{(277)} \xrightarrow{-CO_2} \text{product} \quad (88)$$

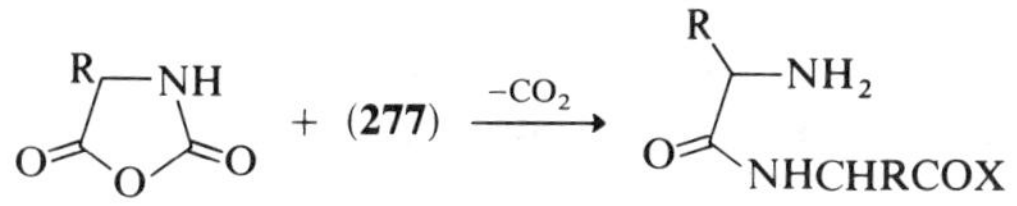

4.18.3.3 Reactions of Substituents

The chemical transformations of substituents attached to the oxazole nucleus present few surprises. Most undergo standard reactions, the sole exception being the activity of 2-alkyl groups in oxazoles and benzoxazoles ⟨B-57MI41801, B-57MI41802, 74AHC(17)99, 75CRV389⟩.

4.18.3.3.1 Fused benzene rings and phenyl substituents

Nitration of 2-methylbenzoxazole yields a 4:1 mixture of 6- and 5-nitro derivatives (cf. 2). Phenyl groups attached to oxazole at C(2), C(4), or C(5) are nitrated in the *para*-positions and the resulting (nitrophenyl)oxazoles can be reduced chemically to the corresponding amino compounds.

4.18.3.3.2 Alkyl groups

Both 4- and 5-methyloxazoles are normal compounds rather like toluene, but if the alkyl group is at position 2 it is capable of undergoing a variety of addition and condensation reactions.

The reactivity of 2-alkyloxazoles in base-catalyzed additions and condensations is due to the intermediacy of delocalized anions analogous to (**142**). 2-Methyl-4,5-diphenyloxazole forms the adduct (**278**) by the action of benzophenone in liquid ammonia containing lithium amide (equation 89), and 2-methylbenzoxazole condenses with diethyl oxalate in the presence of potassium ethoxide to yield the keto ester (**279**). Anions with more extended conjugation are involved in the condensation of 2-(propen-1-yl)benzoxazole with diethyl oxalate (equation 90) and in the formation of the stilbene derivative (**280**) from 5-phenyl-2-*p*-tolyloxazole and benzylideneaniline ⟨78AHC(23)171⟩.

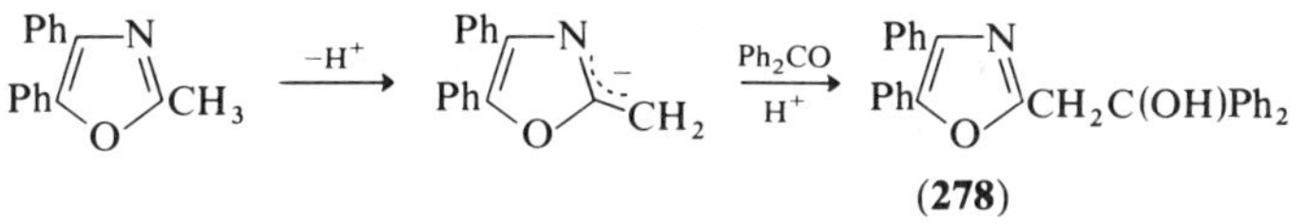

$$\text{Ph-oxazole-CH}_3 \xrightarrow{-H^+} \text{Ph-oxazole-CH}_2^- \xrightarrow[H^+]{Ph_2CO} \text{Ph-oxazole-CH}_2C(OH)Ph_2 \quad (89)$$
(**278**)

(**279**) benzoxazole-CH$_2$COCO$_2$Et

(**280**) Ph-oxazole-C$_6$H$_4$-CH=CHPh

$$\text{benzoxazole-CH=CHMe} \rightarrow \text{benzoxazole-CH=CHCH}_2\text{COCO}_2\text{Et} \quad (90)$$

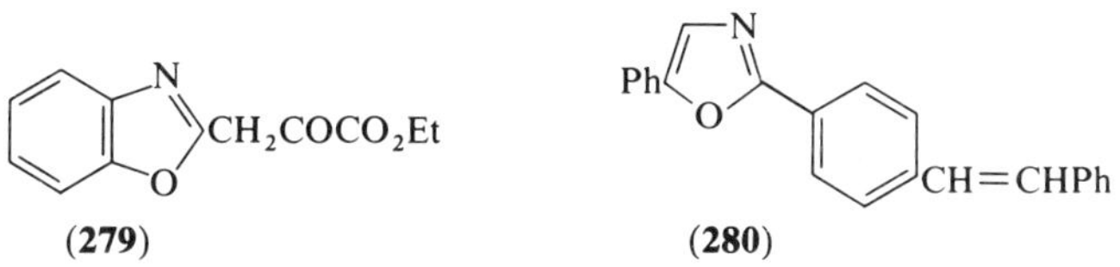

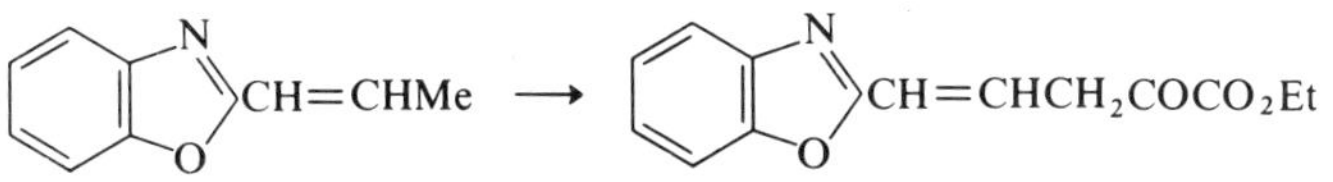

Deprotonation of 2-alkyloxazolium salts generates reactive 2-methylenedihydrooxazoles which behave as enamines towards a variety of electrophiles (equation 91). Thus the cyanine dyes (**281**) and (**282**) are obtained, respectively, by condensation of 2,3-dimethylnaphth[2,1-*d*]oxazolium iodide with 1-ethylquinolinium iodide and of 2,3-dimethylbenzoxazolium iodide with triethyl orthoformate. 2-Methylbenzoxazole reacts with benzaldehyde in the presence of zinc chloride to give 2-styrylbenzoxazole *via* the methylene tautomer shown in equation (92).

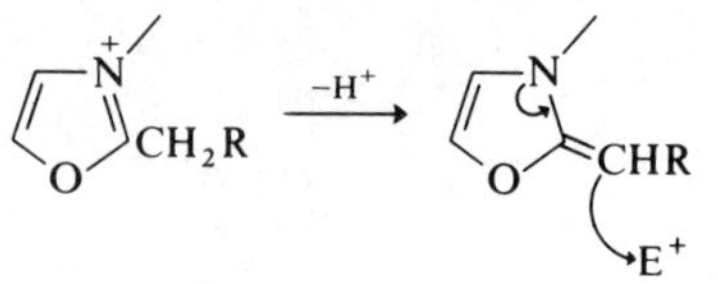

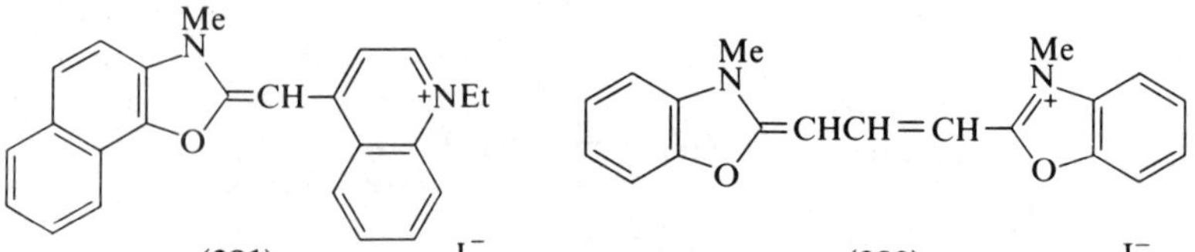

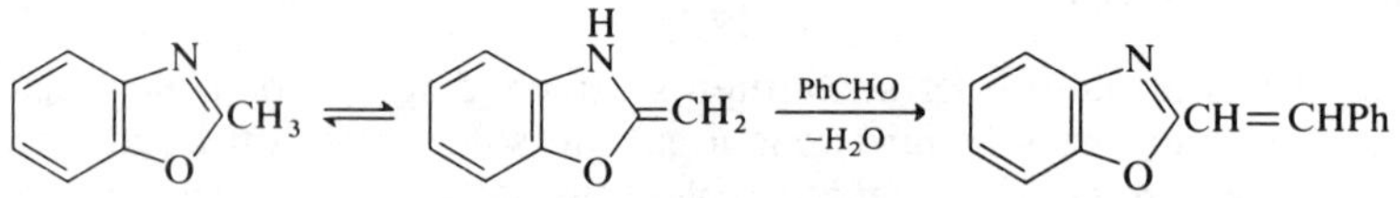

For the oxidation of 2-methylbenzoxazole and 2-methylnaphth[2,1-*d*]oxazole, see Section 4.18.3.1.1.

4.18.3.3.3 Other carbon-linked functional groups

Hydroxyalkyl groups attached to the oxazole nucleus can be converted into haloalkyl groups by means of thionyl chloride or phosphorus pentabromide and the resulting haloalkyloxazoles undergo the usual replacement reactions with hydroxides, alkoxides, cyanides, amines and thiolates. Oxazole-4-aldehydes behave normally towards sodium hydrogen sulfite, hydroxylamine and hydrazines; they condense with nitroethane to give 1-(oxazol-4-yl)-2-nitropropenes. Oxazolecarboxylic acids form acid chlorides, amides and esters and they can be reduced to the corresponding primary alcohols. Ethyl 4-methyloxazole-5-carboxylate has been transformed into the hydrazide and thence into the acyl azide, which underwent Curtius rearrangement to 5-isocyanato-4-methyloxazole; azides derived from oxazole-2-carboxylic acids behave similarly. Cyanooxazoles, obtained by the dehydration of amides of oxazolecarboxylic acids, can be reduced to the corresponding aldehydes by the Stephen method.

4.18.3.3.4 Amino and mercapto groups

The 2- and 5-aminooxazoles afford normal acyl and arenesulfonyl derivatives but attempts to diazotize the amino group have been unsuccessful. Oxazoles with a free amino group in position 4 are unknown, as are hydroxyoxazoles. 2-Methylthiooxazoles are produced when oxazole-2-thiones are methylated in alkaline solution. For the alkylation and benzoylation of benzoxazole-2-thione, see Section 4.18.3.1.4(ii).

4.18.4 SYNTHESES

4.18.4.1 Synthesis of Oxazoles, Benzoxazoles and Oxazolium Salts

4.18.4.1.1 Ring synthesis from non-heterocyclic compounds

(i) *Formation of one bond*

One of the most reliable methods for constructing the oxazole ring is the cyclodehydration of α-acylamino ketones, the Robinson–Gabriel synthesis (1909/1910) (equation 93). The reaction is usually conducted in the presence of sulfuric acid or phosphorus pentachloride and more recently polyphosphoric acid, phosgene or anhydrous hydrogen fluoride have been used. The required starting compounds are available from the Dakin–West reaction (see Section 4.18.3.1.5(ii)) or from the Friedel–Crafts reaction of saturated azlactones (see Section 4.18.3.1.4(iii)). Two separate experiments, in which compound (**283**) was labelled with ^{18}O, first at the ketone group and then at the amide carbonyl group, have demonstrated that the oxygen atom of the oxazole ring is derived from that of the acyl group (equation 94) ⟨73JOC2407⟩. The synthesis has been extended in various ways. Esters of α-amino acids are converted into 5-alkoxyoxazoles (equation 95); *N*-formamido esters yield 5-alkoxyoxazoles unsubstituted in position 2 which are valuable for the synthesis of analogues of pyridoxin. The method has also been used for preparing 5-dialkylaminooxazoles, *e.g.* $R^3 = NMe_2$ in equation (93).

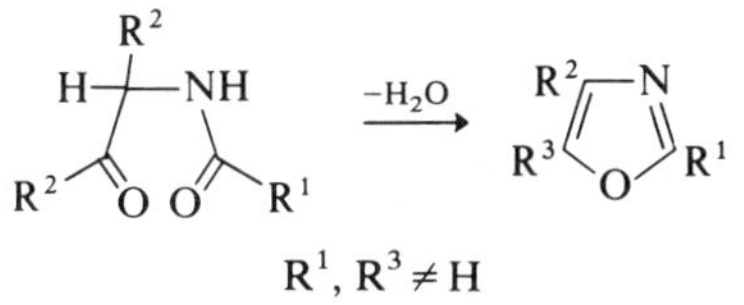

$$(93)$$

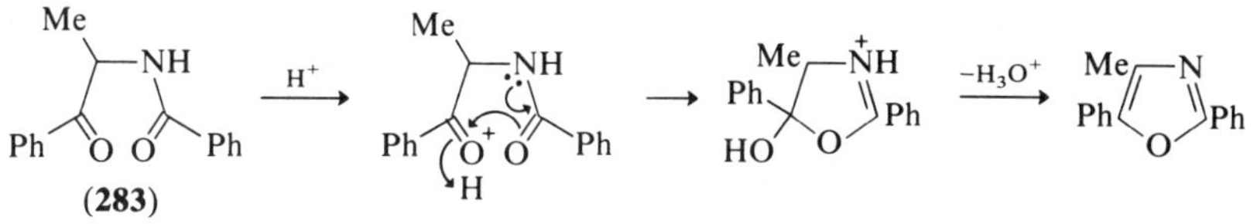

$$R^1, R^3 \neq H$$

$$(94)$$

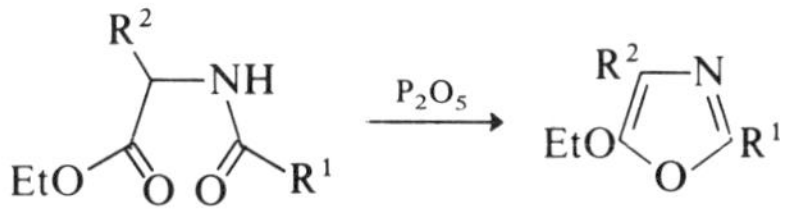

$$(95)$$

Thermal dehydration of *o*-(acylamino)phenols is the method of choice for the preparation of benzoxazoles (equation 96) and other annulated oxazoles. *O,N*-Diacyl derivatives of *o*-aminophenols cyclize at lower temperatures than do the monoacyl compounds. The synthesis is often carried out by heating the aminophenol with the carboxylic acid or a derivative, such as the acid chloride, anhydride, an ester, amide or nitrile. The Beckmann rearrangement of oximes of *o*-hydroxybenzophenones leads directly to benzoxazoles (equation 97).

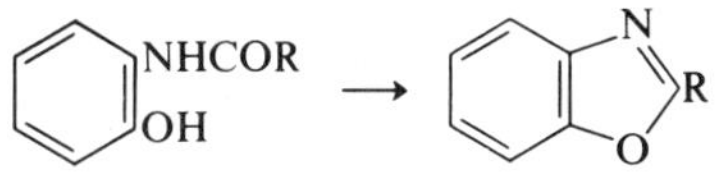

$$(96)$$

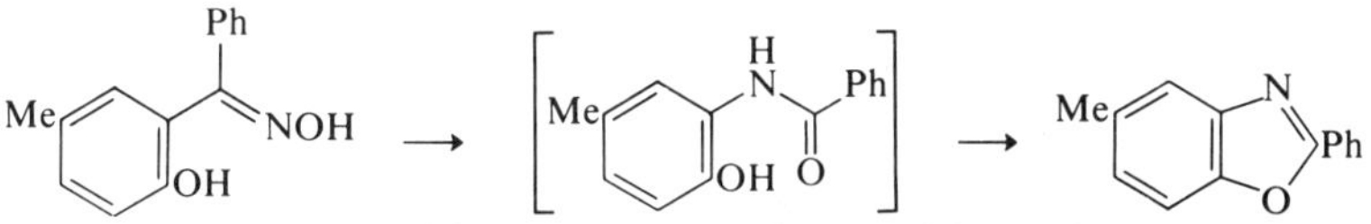

$$(97)$$

Benzoxazoles are formed by the action of potassium amide in liquid ammonia on *N*-aroyl derivatives of both *o*- and *m*-chloroaniline. The reaction does not proceed directly from an intermediate aryne (**284**), as was once thought, but *via* isolable *o*-hydroxyphenylamidines (**285**; Scheme 29) ⟨81JOC3256⟩.

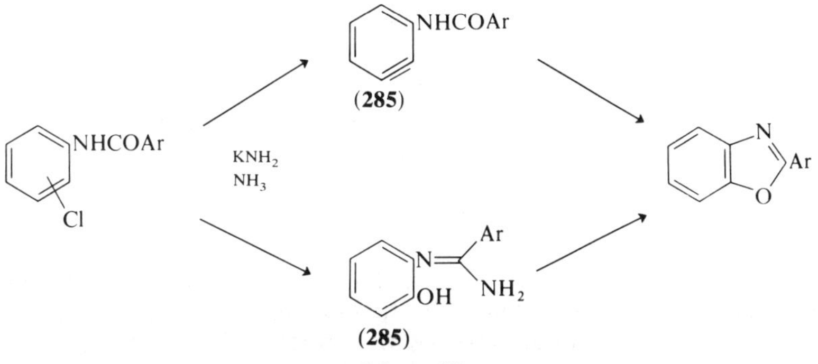

Scheme 29

Quaternary oxazolium and benzoxazolium salts are prepared by alkylation (see Section 4.18.3.1.2.2); they are also produced by the action of acetic anhydride and an inorganic acid (perchloric or fluoroboric) on the appropriate acylamino ketone (equation 98). *N*-Aryl derivatives are obtained in this way.

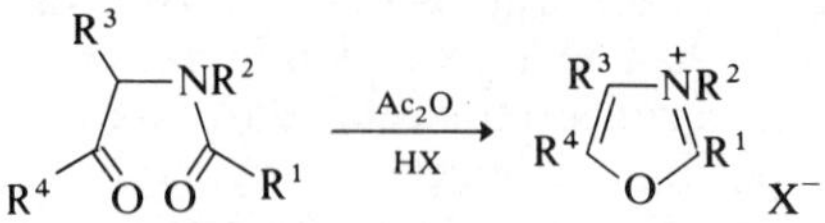

$$\text{(98)}$$

There are several ring-closure reactions in which a carbon–oxygen link of the oxazole ring is formed by a process other than dehydration. Carbonylnitrile ylides undergo 1,5-dipolar cyclization to yield oxazoles (equation 99) ⟨79CRV181⟩; the Cornforth rearrangement (see Section 4.18.3.1.2(i)) is of this type. The ylides can also be generated by thermolysis of 2-acyl-5(2*H*)-oxazolones (see Section 4.18.3.1.4(i)). *N*-Aroyl derivatives of propargylamines cyclize to 2,5-disubstituted oxazoles in the presence of mercury(II) acetate, sulfuric acid or sodium hydride (equation 100). The first fluorooxazoles were obtained by heating *N*-acylimines of hexafluoroacetone with tin(II) chloride (equation 101).

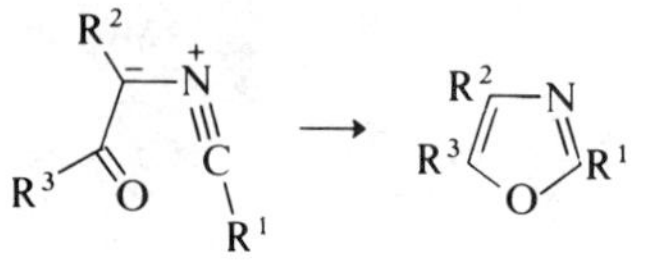

$$\text{(99)}$$

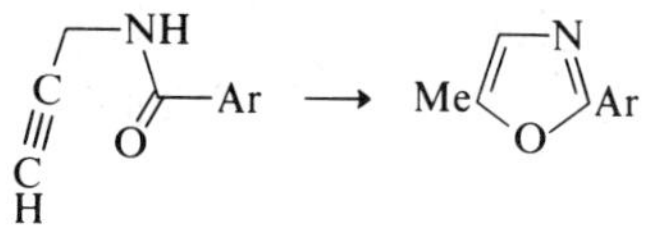

$$\text{(100)}$$

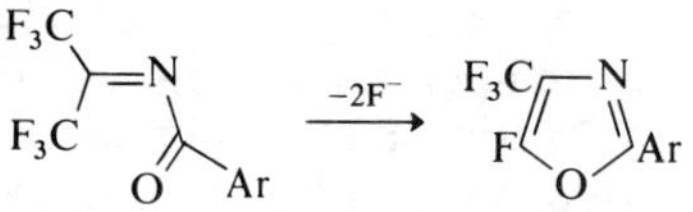

$$\text{(101)}$$

The Pomeranz–Fritsch synthesis of isoquinolines from the acetals (**286**) often gives oxazoles as by-products. Isoquinoline formation is favoured by the presence of electron-releasing groups in the benzene ring and suppressed when the ring is deactivated. Nitroarylmethylene compounds (**286**; X = NO$_2$), for example, give only oxazoles (equation 102).

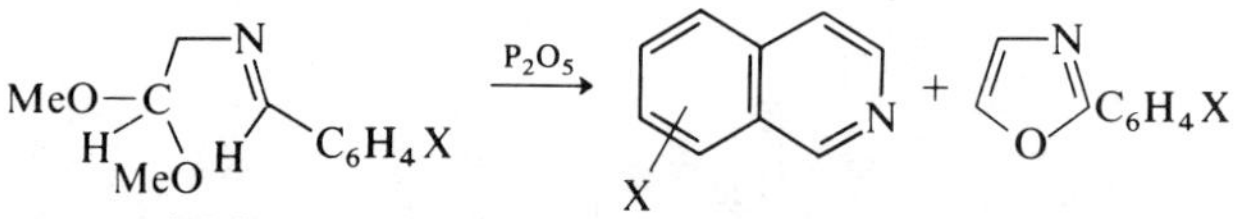

$$\text{(102)}$$

(**286**)

Benzoxazoles and other condensed oxazoles are obtained by the oxidative ring-closure of phenolic Schiff's bases (equation 103) ⟨78H(10)57⟩. The formation of phenanthrooxazoles by the action of benzylamine or other amines on phenanthraquinone (equation 104) is a reaction of this type.

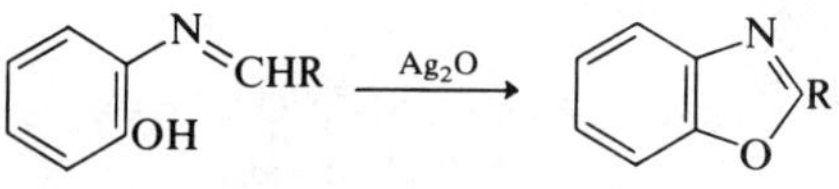

$$\text{(103)}$$

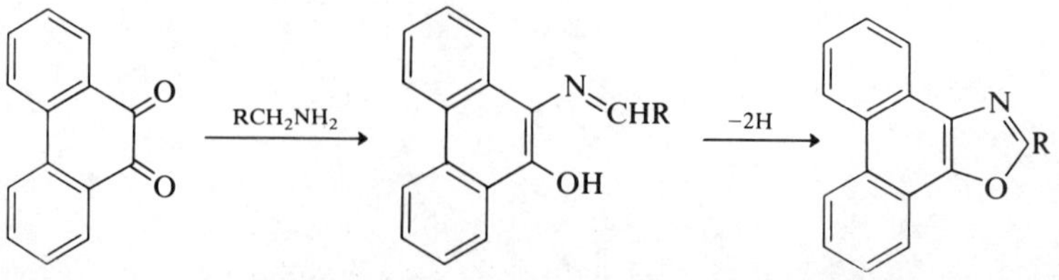

$$\text{(104)}$$

Aminooxazoles are produced by the cyclization of suitable precursors. *N,N*-Dialkyl-*N'*-propargylureas form 2-dialkylaminooxazoles (*cf.* equation 100). The action of inorganic acids on α-acylaminoacetonitriles leads to salts of 5-aminooxazoles (equation 105; R^2 = H, alkyl or aryl). Treatment of α-acylamino-β,β-dichloroacrylonitriles with secondary amines gives α-chloroenamines (**287**) which undergo spontaneous ring-closure, presumably by way of transient keteneimmonium salts, to afford 5-dialkylaminooxazoles in good yields (equation 106).

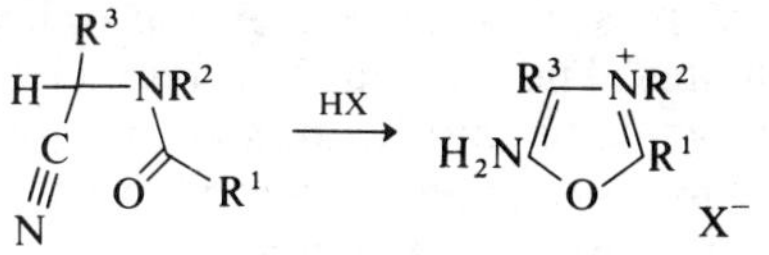

(105)

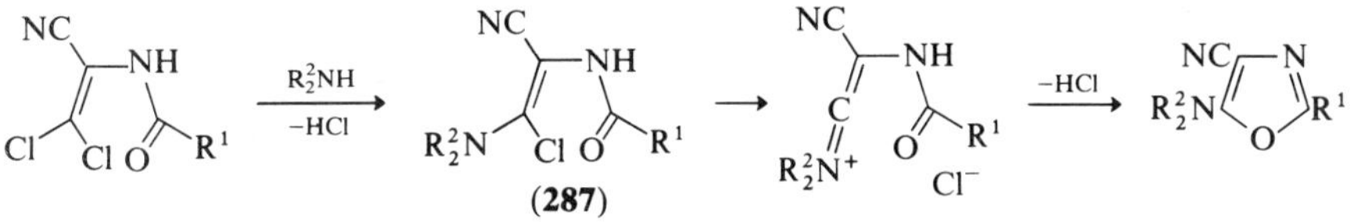

(106)

(ii) Formation of two bonds

(a) From [4 + 1] atom fragments: $O-C-C-N+C$. The formation of 2,4-disubstituted oxazoles from aldehyde cyanhydrins and aldehydes in the presence of hydrogen chloride (equation 107) is known as the Fischer synthesis (1896). The method is only suitable for preparing diaryloxazoles and it suffers from the disadvantage that if the aryl residues are different a mixture of four isomers may be produced because of an exchange of hydrogen cyanide from one aldehyde to the other. The mechanism of this classical reaction has not been elucidated. The formation of 2-aryl-4-chloro-5-phenyloxazoles from aromatic aldehydes and benzoyl cyanide (equation 108) ⟨77JHC317⟩ is a related process.

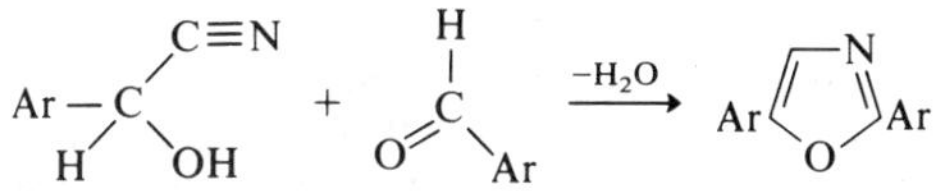

(107)

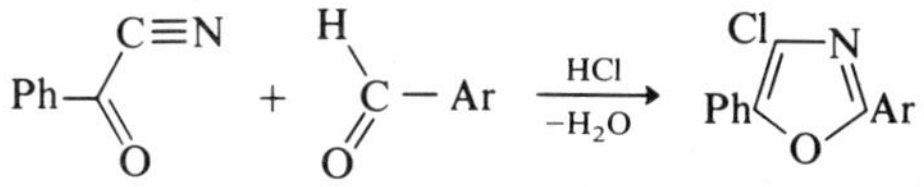

(108)

Carbon atom 2 of the oxazole ring is also supplied by aldehydes in their reaction with α-(hydroxylamino) ketones, which proceeds in the presence of sulfuric acid and acetic anhydride (equation 109). Three further oxazole syntheses involving incorporation of a C(2) fragment are the condensation of triethyl orthoformate with the hydrochlorides of α-aminoacetophenones (equation 110), the reaction of acyl chlorides with α-azido ketones or α-azido esters in the presence of triphenylphosphine (equation 111), and the preparation of 2-aminobenzoxazole and benzoxazoleimines from *o*-aminophenols and cyanogen bromide (equation 112).

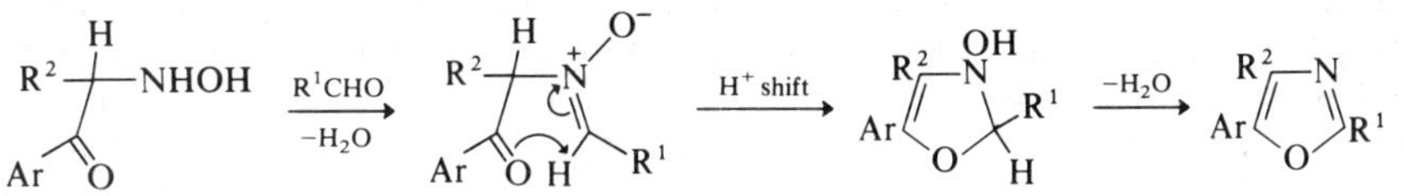

(109)

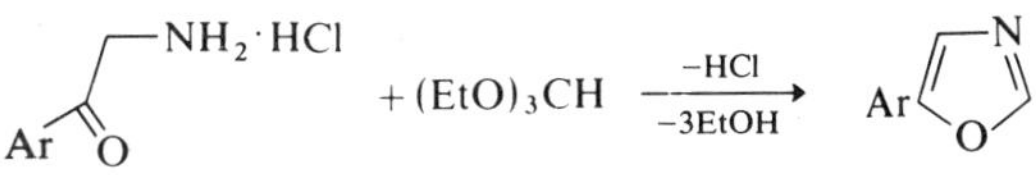

(110)

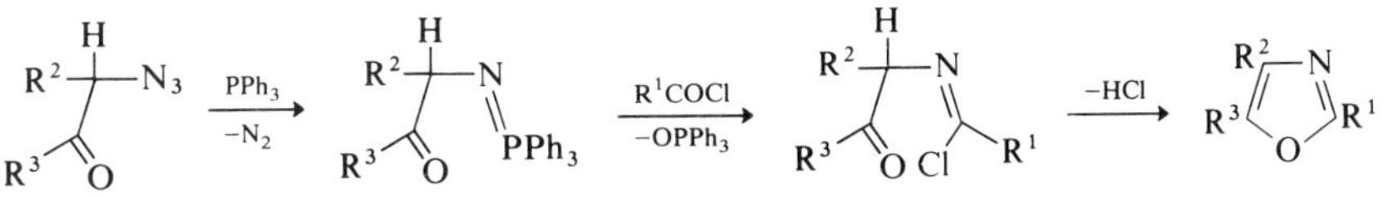

(111)

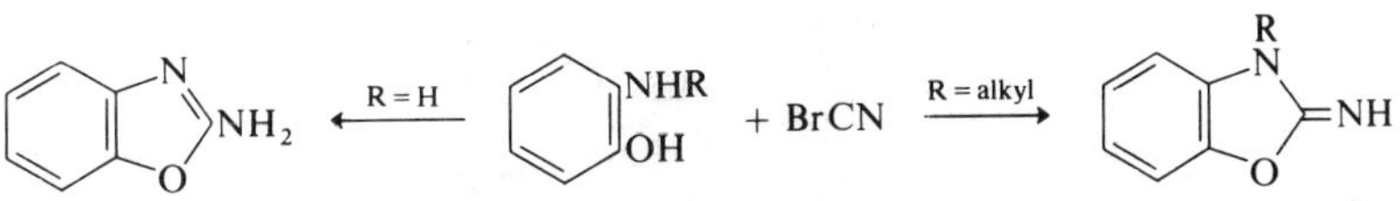

(112)

$C-C-O-C+N$. The formation of oxazoles from α-acyloxy ketones and ammonium salts was discovered in 1937 when it was found that treatment of benzoin benzoate with ammonium acetate in hot acetic acid gave triphenyloxazole in excellent yield. It has been shown that the reaction proceeds by way of intermediate enamines (equation 113). The synthesis is quite general and it is only limited by the difficulty of obtaining the starting keto esters, particularly formates. The latter are probably intermediates in the preparation of cycloalkenooxazoles from acyloins and formamide in hot sulfuric acid (equation 114). Another variation is to heat a mixture of an α-bromo ketone, the sodium salt of a carboxylic acid and ammonium acetate in acetic acid (equation 115).

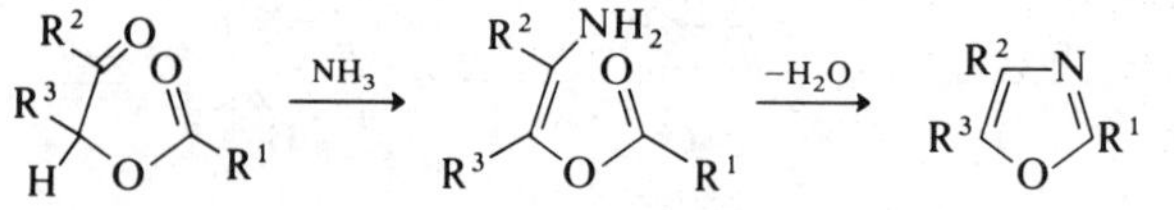

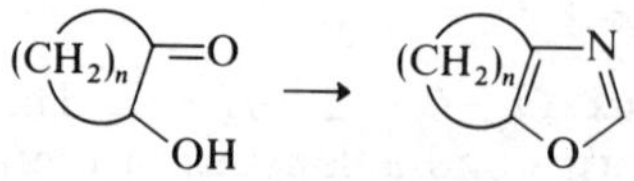

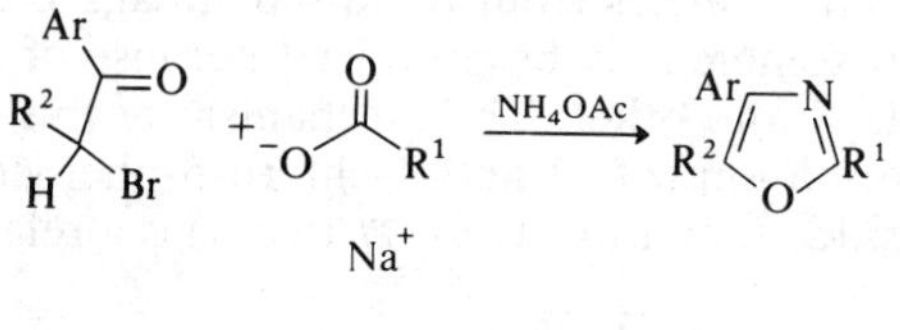

$C-N-C-O+C$. An oxazole synthesis, in which C(5) is introduced in the form of an anion, is the reaction of aroyl isocyanates with stabilized sulfonium ylides. The initial adducts eliminate dimethyl sulfide on heating to give hydroxyoxazole ketones or esters (equation 116).

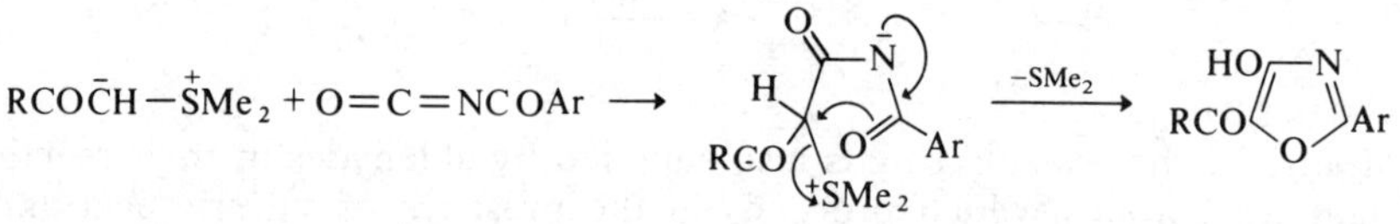

(b) *From [3+2] atom fragments*: $C-N-C+C-O$. An important development in the synthesis of 2-unsubstituted oxazoles has been the use of tosylmethyl isocyanides by van Leusen ⟨80MI41801⟩ and of α-metallated isocyanides by Schöllkopf ⟨79PAC1347⟩ and their respective coworkers.

The tosyl compound reacts with aldehydes in the presence of potassium carbonate to yield 5-alkyl- or 5-aryl-oxazoles, the intermediate dihydrooxazoles (which can be isolated) eliminating toluene-*p*-sulfinic acid (Scheme 30). Use of acyl chlorides in place of aldehydes leads to 4-tosyloxazoles (**288**). Furthermore, alkylation of tosylmethyl isocyanide with an alkyl halide R^1X, followed by treatment with an aldehyde R^2CHO, yields a 4,5-disubstituted oxazole (**289**). A related reaction is that of *N*-tosylmethyl-*N'*-tritylcarbodiimide with aromatic aldehydes under phase-transfer catalysis to yield 2-tritylaminooxazoles which are readily converted into 2-amino-5-aryloxazoles (equation 117) ⟨81JOC2069⟩.

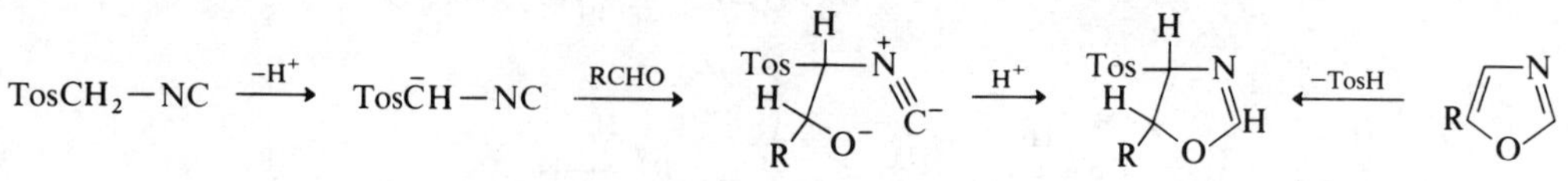

Scheme 30

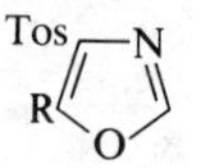

(**288**) 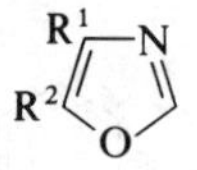   (**289**)

$$ArCHO + TosCN=C=NTr \longrightarrow Ar\underset{O}{\overset{N}{\diagup}}NHTr \xrightarrow{HCl} Ar\underset{O}{\overset{N}{\diagup}}NH_2 \quad (117)$$

Tr = Ph₃C

α-Metallated isocyanides, produced by the action of butyllithium or potassium *t*-butoxide on isocyanides, react with acyl chlorides, carboxylic esters or amides to form oxazoles in yields of up to 80% (equation 118). The substituents R^1, R^2 may be varied widely; 4-alkoxycarbonyloxazoles are obtained from alkyl isocyanoacetates. The reaction of α-metallated isocyanides with derivatives of dibasic acids, for example dimethyl oxalate, affords bioxazolyls (equation 119), and polyoxazolyls have been prepared from isonitriles and suitably substituted oxazoles, *e.g.* equation (120).

$$R^1H\bar{C}-NC \xrightarrow{R^2COX} \underset{R^2}{\overset{R^1}{\diagdown}}\underset{O}{\overset{H}{\underset{|}{N^+}}}_{C^-} \longrightarrow \underset{R^2}{\overset{R^1}{\diagup}}\underset{O}{\overset{N}{\diagdown}} \quad (118)$$

$$2\,MeNC + MeO_2C-CO_2Me \xrightarrow{2BuLi} \text{bioxazolyl} \quad (119)$$

$$2\,PhCH_2NC + \underset{MeO_2C}{\overset{MeO_2C}{\diagdown}}\underset{O}{\overset{N}{\diagup}} \xrightarrow{2BuLi} \text{tris-oxazole with Ph substituents} \quad (120)$$

Another synthesis based on the combination of these fragments is the 1,3-dipolar cycloaddition of carbonyl compounds to nitrile ylides. The latter are generated by photolysis of 1-azirines; when the irradiation is carried out in the presence of an acyl chloride and triethylamine, oxazoles are formed in moderate yields (equation 121).

$$\underset{H\,\,R^1}{\overset{Ph\diagdown N}{\diagup}} \xrightarrow{h\nu} PhC\equiv\overset{+}{N}\diagdown\underset{H}{\overset{}{C}}-R^1 \xrightarrow{R^2COCl} \underset{R^2\,\,O\,\,H}{\overset{Ph\diagdown N}{\diagup}}R^1 \xrightarrow{-HCl} \underset{R^2}{\overset{Ph\diagdown N}{\diagup}}\underset{O}{\overset{}{\diagdown}}R^1 \quad (121)$$

$R^1, R^2 = Ph\ or\ Bu^t$

N—C—O + C—C. The construction of the oxazole ring by the condensation of α-halogeno ketones with primary amides (equation 122) is the Blümlein–Lewy synthesis (1884/1888). The method succeeds best when the resulting oxazole contains one or more aryl substituents. The use of formamide leads to oxazoles with a free 2-position and in this case it is possible that the reaction proceeds as in equation (113). 2-Aminooxazoles are produced by the action of α-halogeno ketones on urea and its derivatives (equation 123) or on cyanamide ⟨80ZOR2185⟩. The mercury(II) sulfate-catalyzed condensation of alkynic alcohols or their esters with primary amides leads to trisubstituted oxazoles (equation 124).

$$\underset{H\,\,X}{\overset{Ar}{\underset{R^2}{\diagup}}}=O + \underset{O}{\overset{H_2N}{\diagdown}}R^1 \longrightarrow \underset{H\,\,O\,\,R^1}{\overset{Ar}{\underset{R^2}{\diagup}}}\overset{O^+NH_2}{\underset{X^-}{\diagdown}} \xrightarrow[-HX]{-H_2O} \underset{R^2}{\overset{Ar\diagdown N}{\diagup}}\underset{O}{\overset{}{\diagdown}}R^1 \quad (122)$$

$$\underset{R^4\,H\,X}{\overset{R^3\,\,O}{\diagup}} + \underset{O}{\overset{H_2N}{\diagdown}}-NR^1R^2 \longrightarrow \underset{R^4}{\overset{R^3\diagdown N}{\diagup}}\underset{O}{\overset{}{\diagdown}}NR^1R^2 \quad (123)$$

$$\underset{R^2\,\,OH\,H}{\overset{HC\equiv C}{\diagup}} + \underset{O}{\overset{H_2N}{\diagdown}}R^1 \longrightarrow \underset{R^2}{\overset{Me\diagdown N}{\diagup}}\underset{O}{\overset{}{\diagdown}}R^1 \quad (124)$$

$C-C-O+N-C$. α-Hydroxy ketones react with monosubstituted cyanamides under the influence of sodium hydroxide to yield derivatives of 2-aminooxazole (equation 125) ⟨76S591⟩. Oxazoles are obtained by the action of nitriles on α-diazo carbonyl compounds in the presence of Lewis acids, such as aluminum chloride or boron trifluoride, and the reaction is thought to involve the intermediacy of nitrilium salts (equation 126). Nitrilium salts are also the effective agents in the formation of oxazoles from α-chloro ketones and nitriles in the presence of tin(IV) chloride (equation 127).

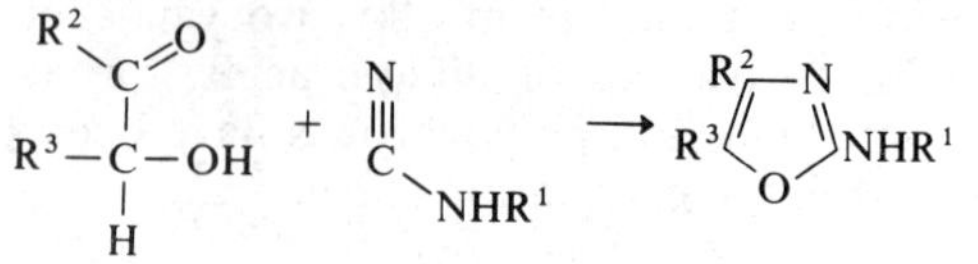

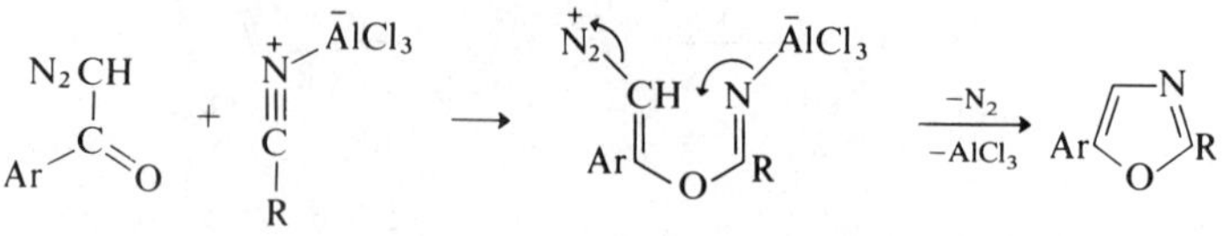

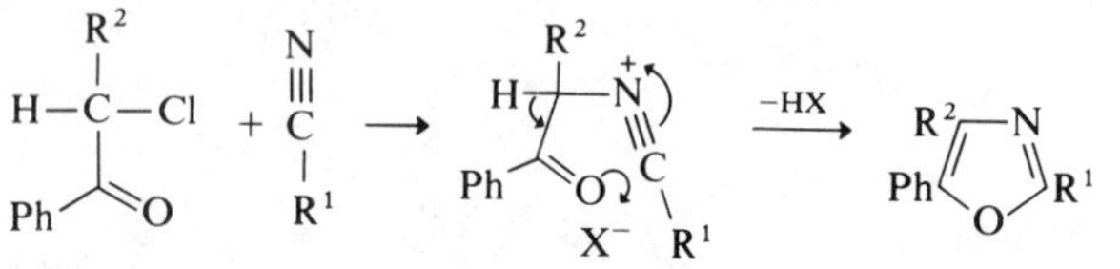

$C-C-N+O-C$. A general synthesis of oxazoles is the reaction of oximes of α-methylene ketones with aliphatic or aromatic acyl chlorides (equation 128) ⟨81IJC(B)322⟩.

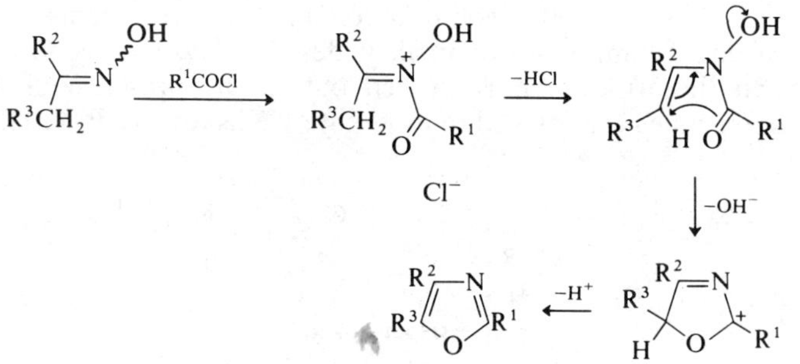

4.18.4.1.2 Ring synthesis by transformation of heterocyclic compounds

The formation of oxazoles by the Cornforth rearrangement, which is a useful synthetic procedure, is described in Section 4.18.3.1.2(i). Oxazoles are also satisfactorily prepared by reduction of their N-oxides (see next section) with zinc and acetic acid. Treatment of oxazoles and benzoxazoles with alkylating agents gives quaternary salts.

C-Acylaziridines rearrange thermally to oxazoles by way of azomethine ylides which undergo 1,5-dipolar cyclization (equation 129). Both cis- and trans- 2-benzoyl-N-t-butyl-3-phenylaziridine afford 2,5-diphenyloxazole. When the N-substituent was phthalimido the intermediate betaines could be trapped as 1,3-dipolar cycloadducts with DMAD. N-Acylaziridines undergo scission of the C—N bond, followed by recyclization; thus the action of aroyl chlorides on aziridines leads to oxazoles (equation 130).

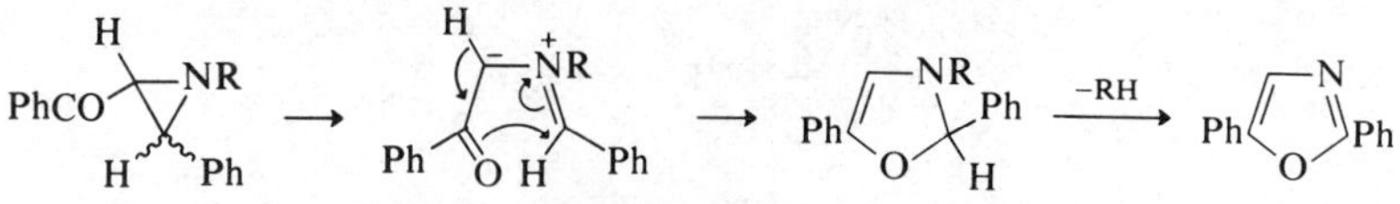

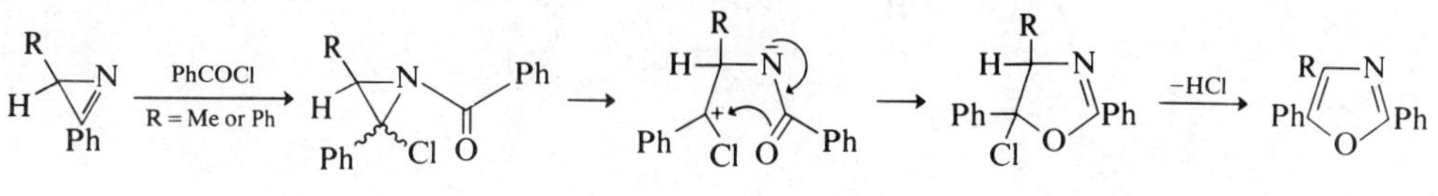

Oxazoles are the products of the rearrangement of isoxazoles and of C-acylazirines, these processes being intimately related ⟨81MI41800, 76ACR371, 77H(6)143⟩. Photolysis of 3-acylazirines gives nitrile ylides which cyclize thermally to oxazoles (equation 131). Flash vacuum pyrolysis of 3,5-diphenylisoxazole yields as major products 2,5-diphenyloxazole, 1,2-diphenylazirine, 2-phenylindole and benzamide. ^{13}C labelling experiments indicate that the first two compounds are formed as shown in Scheme 31.

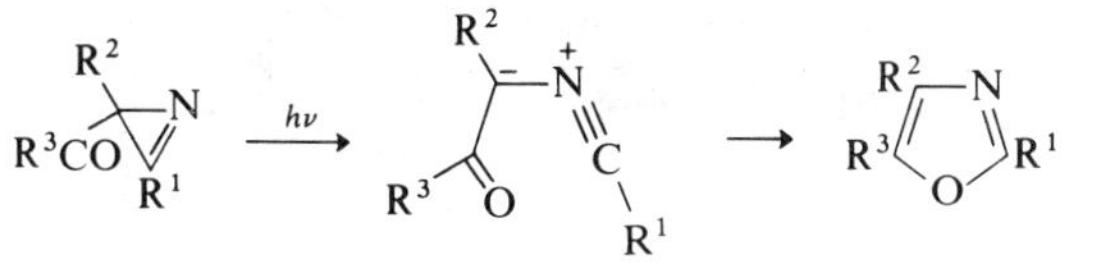

(131)

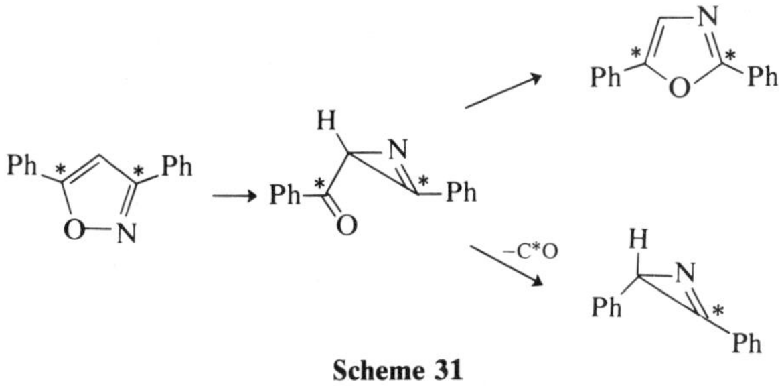

Scheme 31

4.18.4.2 Synthesis of Oxazole N-Oxides

Oxazole N-oxides cannot be made by oxygenation of oxazoles. The only method of synthesis remains the condensation of monooximes of 1,2-dicarbonyl compounds with aldehydes in the presence of hydrogen chloride (equation 132) ⟨15CB897⟩. The aldehyde may be aromatic or aliphatic (including formaldehyde) and the oxime may be derived from an aromatic diketone or it may be an α-keto aldoxime, leading to a 2,5-disubstituted oxazole N-oxide. It may also contain an additional carbonyl group as in equation (133).

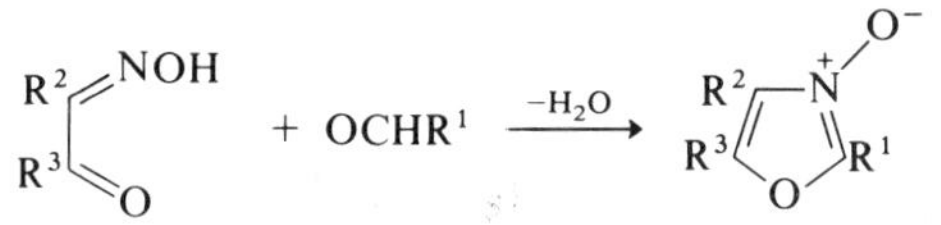

(132)

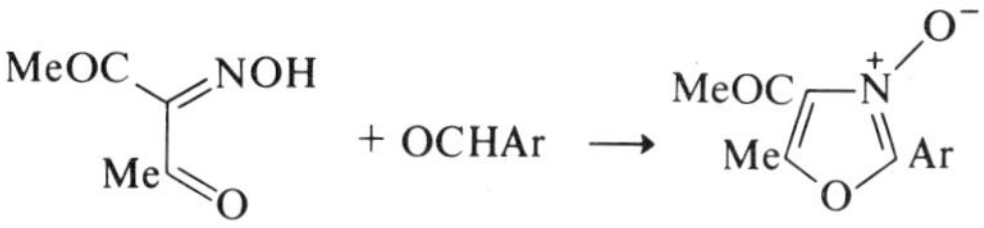

(133)

4.18.4.3 Synthesis of Oxazolones and Mesoionic Oxazoles

Synthetic methods for these classes of compounds are conveniently discussed together.

4.18.4.3.1 2(3H)-Oxazolones and benzoxazolones

2(3H)-Oxazolones are formed by the spontaneous cyclization of β-oxo isocyanates (equation 134). Similarly, o-hydroxyphenyl isocyanate, produced by the Curtius rearrangement of the azide of salicylic acid or by the action of sodium hypochlorite on salicylamide, forms benzoxazolone (equation 135). An analogous reaction is the formation of N-phenylbenzoxazolone by the action of thionyl chloride on the hydroxamic acid shown in equation (136) ⟨78TL2325⟩. Pyrolysis of aryl azidoformates affords benzoxazolones by nitrene insertion (equation 137) ⟨81CC241⟩.

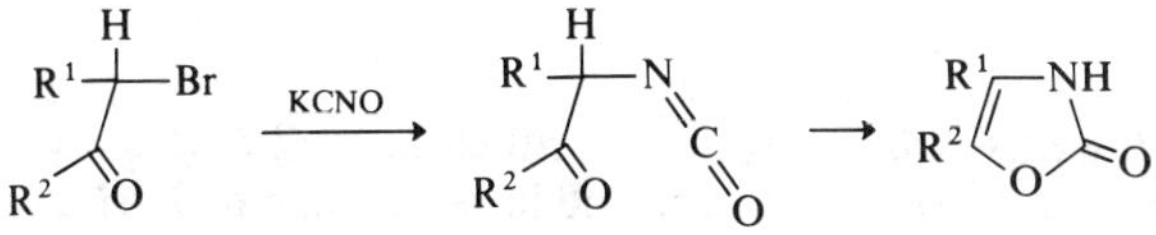

(134)

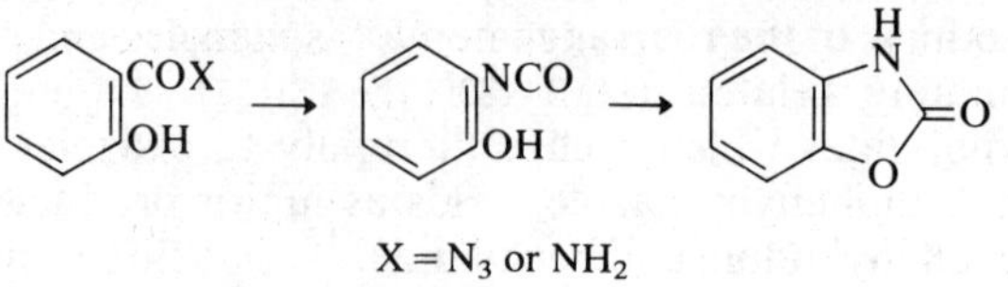

(135), 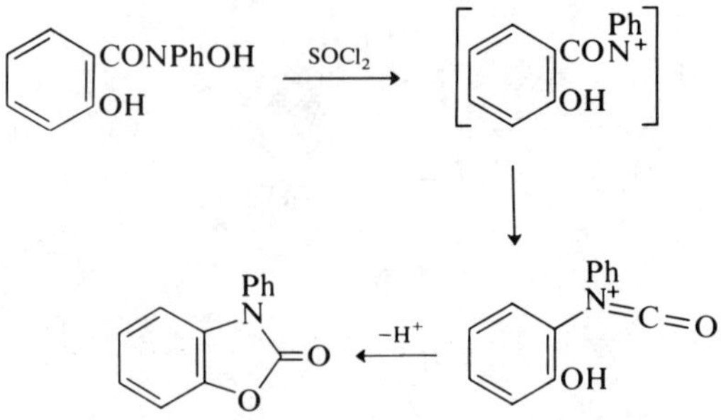 (136), 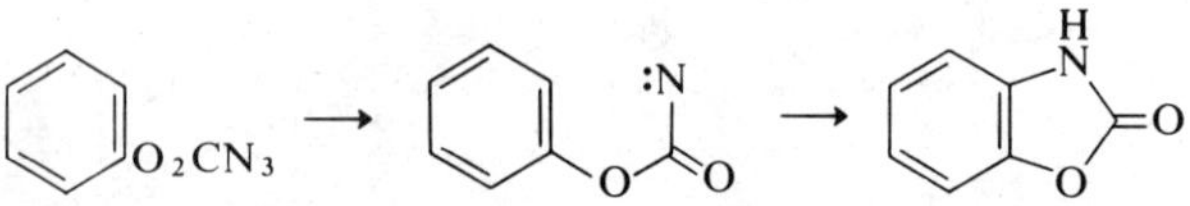 (137)

The synthesis of 2(3H)-oxazolones by incorporation of carbon atom 2 into a four-atom chain is exemplified by the condensation of phenacylaniline with ethyl chloroformate (equation 138). Benzoxazolones are similarly prepared from o-aminophenols and derivatives of carbonic acid, such as phosgene, diethyl carbonate, urethane (EtO_2CNH_2) or urea (equation 139). The analogous condensation of carbon disulfide or cyanogen bromide with o-aminophenols leads to benzoxazolethiones or 2-aminobenzoxazoles, respectively (equation 140).

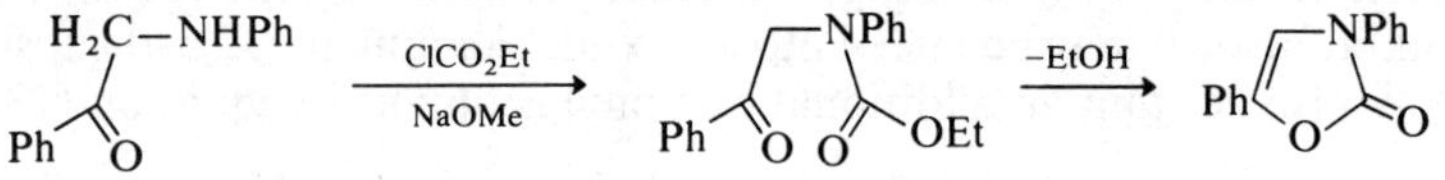

(138), 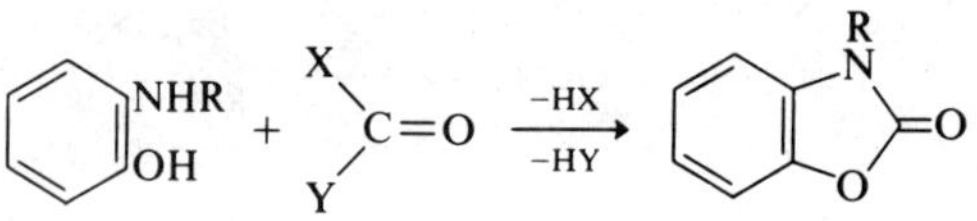 (139), 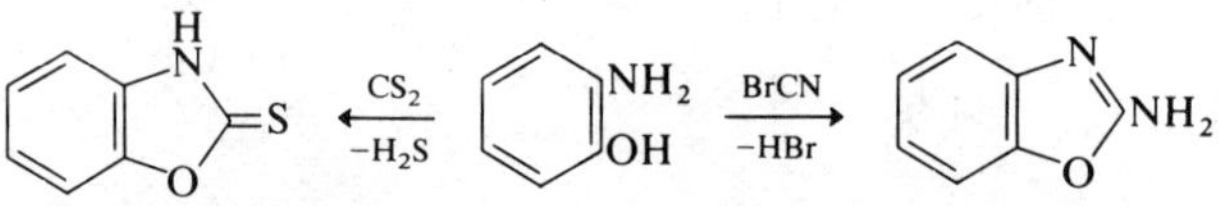 (140)

Cyclic carbonates react with primary amines to give β-oxo carbamates which are dehydrated to 2(3H)-oxazolones (equation 141). Thermal rearrangement of the isoxazolone (**290**) results in the oxazolone (**291**; equation 142).

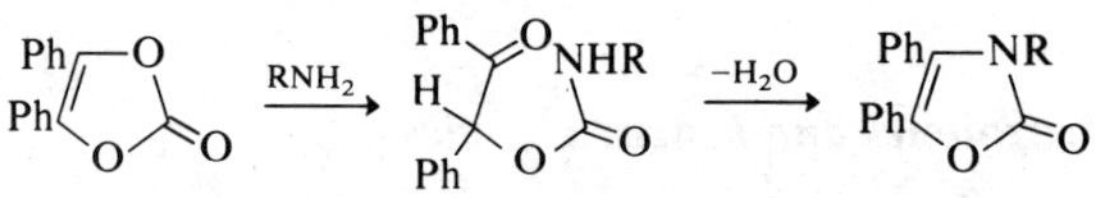

(141), 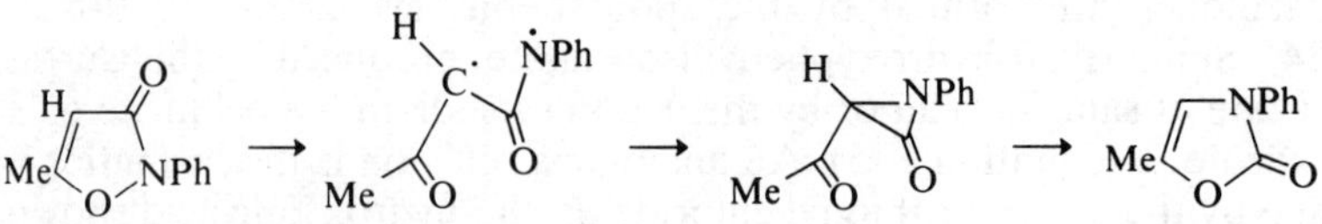 (142)

(**290**)   (**291**)

4.18.4.3.2 2(5H)-Oxazolones

Treatment of tertiary α-hydroxy ketones with chlorosulfonyl isocyanate yields intermediates which are hydrolyzed to carbamates. The latter cyclize to 2(5H)-oxazolones on heating (equation 143). Acetone cyanohydrin is converted into the chloro derivative (**292**) by the action of phosgene (equation 144). The first 2(5H)-oxazolone was obtained by alkylation of an oxazolidinedione (see Section 4.18.3.2.2 (iii)).

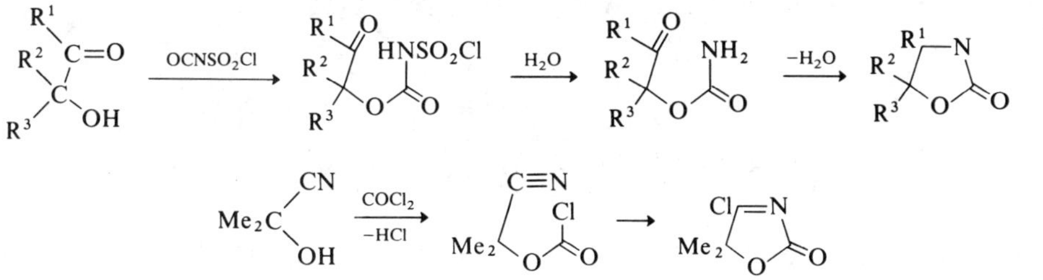

(144)

4.18.4.3.3 4(5H)-Oxazolones and anhydro-4-hydroxyoxazolium hydroxides

4(5H)-Oxazolones were unknown until 1949 when the 2-phenyl derivative was prepared by the action of diazomethane on benzoyl isocyanate (equation 145) ⟨49JA4059⟩. Treatment of the phospholene (**293**) with acyl isocyanates (**294**; R = Ph, MeO or Ph$_2$N) affords 5-acetyl-5-methyl-4(5H)-oxazolones (equation 146).

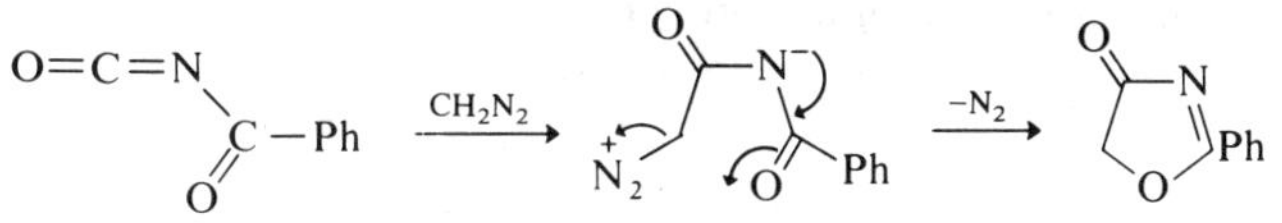

(145)

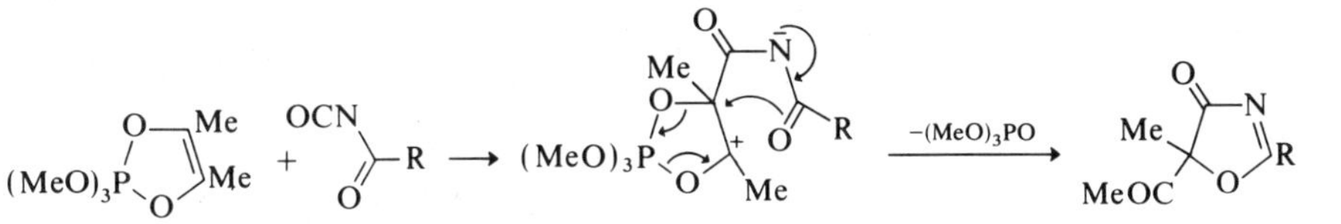

(146)

Anhydro-4-hydroxyoxazolium hydroxides were first obtained in 1974 by the decomposition of the α-diazoimides (**295**) induced by copper(II) acetylacetonate. The reaction proceeds by way of a carbene or a carbenoid species (equation 147). The triphenyl derivative is formed when the imide shown in equation (148) is treated with triethyl phosphite ⟨82JOC723⟩.

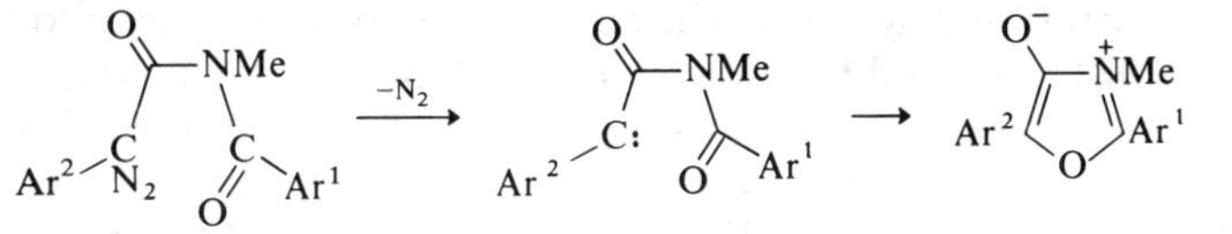

(147)

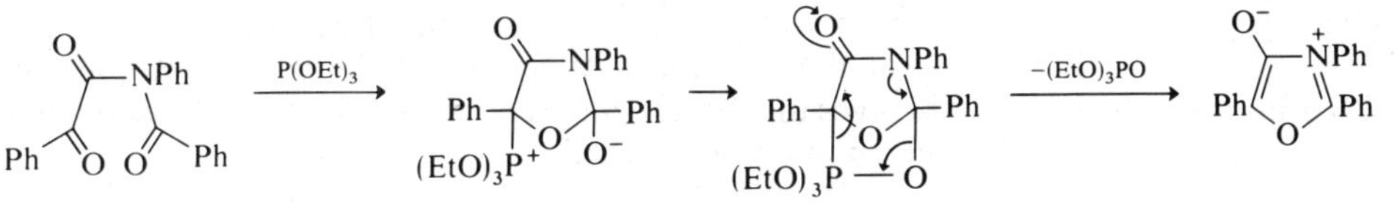

(148)

A 4-amino analogue is produced by the action of N-phenylbenzimidoyl chloride on hydroxyacetonitrile in the presence of sodium hydride (equation 149).

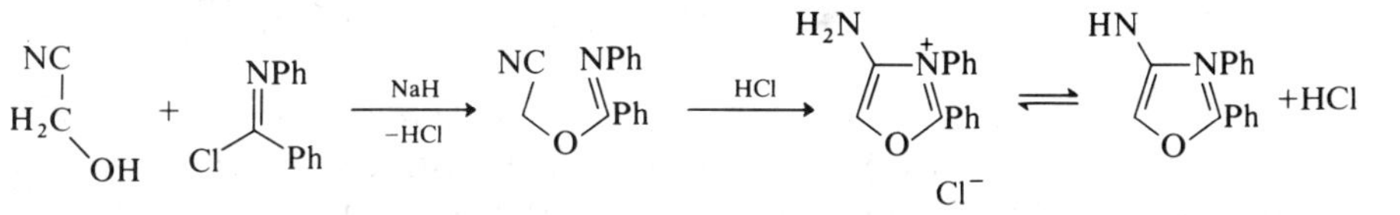

(149)

4.18.4.3.4 5(4H)-Oxazolones and anhydro-5-hydroxyoxazolium hydroxides

The standard procedure for the preparation of 5(4H)-oxazolones is dehydration of α-acylamino acids with acetic anhydride (equation 150). Saturated azlactones lacking a substituent at C(4) are rather sensitive and difficult to isolate and in such cases it is advantageous to carry out the dehydration in the presence of perchloric acid. The resulting stable salts can be deprotonated under mild conditions (equation 151).

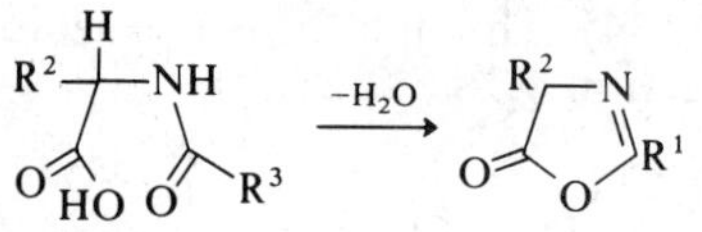

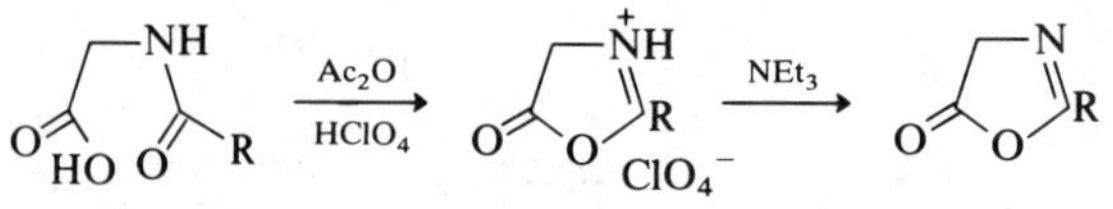

The condensation of acylaminoacetic acids with aldehydes under the influence of acetic anhydride and sodium acetate, the classical Erlenmeyer synthesis (1893) ⟨46OR(3)198, 55QR150, B-57MI41801⟩ leads to 4-methyleneoxazolones, the unsaturated azlactones (equation 152) (see Section 4.18.3.1.4(ii)). There are some limitations to the method: it is not generally applicable to ketones, and hydroxy groups in the carbonyl component, e.g. in salicylaldehyde, are acetylated in the process. These disadvantages have been overcome by the use of polyphosphoric acid as the condensing agent which is effective even with alkyl aryl ketones. Moreover, aromatic aldehydes yield the labile (E) isomers (**296**), whereas the (Z) isomers are obtained by the older procedure.

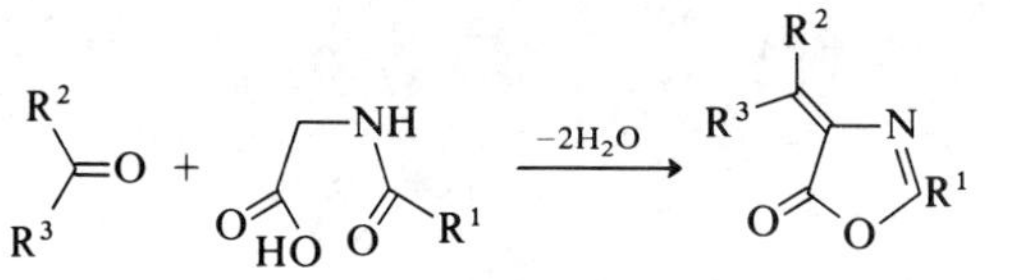

An extension of the Erlenmeyer synthesis is the condensation of acylamino acids with triethyl orthoformate which leads to the ethoxymethylene derivatives (**297**). These can be hydrolyzed to the corresponding enols, which in turn can be converted into chloromethylene compounds, e.g. (**298**). The lability of the chlorine atom in this compound (see arrows) can be put to good account; reaction with organometallic compounds or with benzene and its derivatives under Friedel–Crafts conditions yields unsaturated lactones (**298**; aryl or heteroaryl in place of Cl). The method is especially valuable in cases where the aldehyde is not readily available.

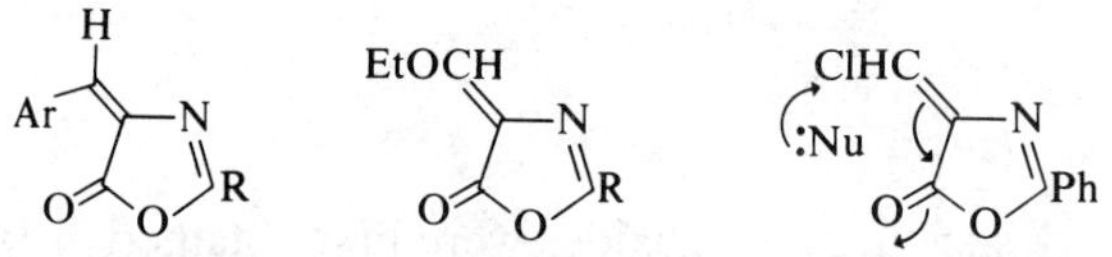

Bergmann's synthesis (1926) is still used to prepare unsaturated azlactones containing only alkyl substituents. It consists of the treatment of α-alkyl α-(α-halogenoacyl)amino acids with acetic anhydride and it involves the isomerization of an intermediate 2-methylene-5(2H)-oxazolone. An example is given in equation (153).

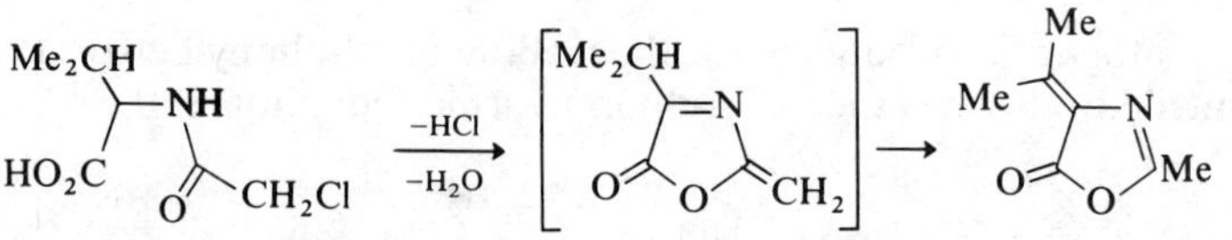

The standard method of preparing anhydro-5-hydroxyoxazolium hydroxides (**300**) is by dehydration of N-substituted α-acylamino acids, usually with acetic anhydride (equation 154); dicyclohexylcarbodiimide and trifluoroacetic anhydride have also been used. N-Acylglycines (**299**; R^3=H) give rise to unstable mesoionic oxazoles which tend to dimerize (see Section 4.18.3.1.5(i)); if such acids are treated with trifluoroacetic anhydride, stable trifluoroacetyl derivatives (**300**; R^3 = CF$_3$CO) are isolated. Numerous unstable betaines (**300**; R^1, R^2 = alkyl or aryl, R^3 = H) have been generated by the action of triethylamine on the corresponding hydroperchlorates, prepared by the method shown in equation (151).

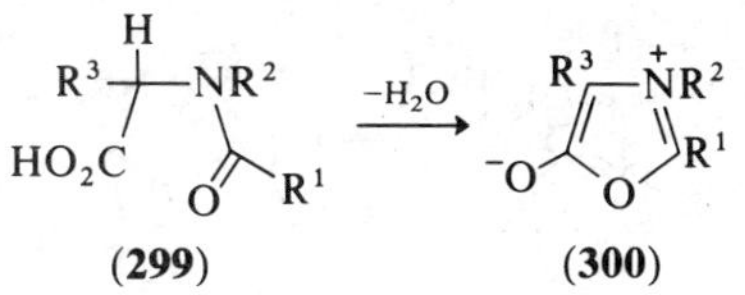

$$\text{(299)} \xrightarrow{-H_2O} \text{(300)} \quad (154)$$

The generation of the condensed anhydro-5-aminooxazolium hydroxide (**301**) is described in Section 4.18.3.1.5(iii). Monocyclic compounds of this type are produced in a similar manner from *N*-substituted acylaminoacetonitriles (equation 155).

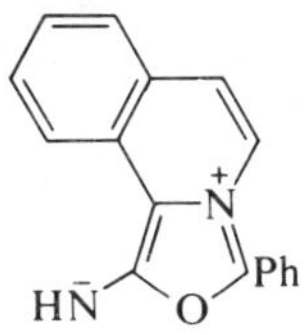

(**301**)

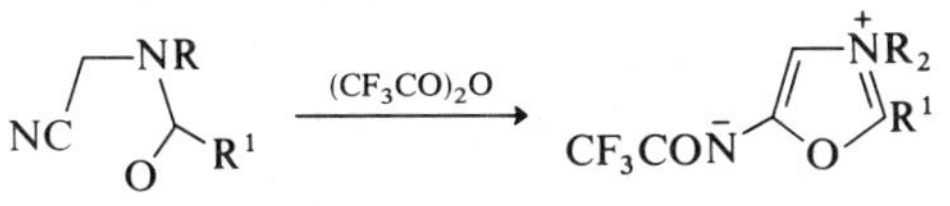

$$\xrightarrow{(CF_3CO)_2O} \quad (155)$$

4.18.4.3.5 5(2H)-Oxazolones

The first simple 5(2*H*)-oxazolones were the trifluoromethyl derivatives obtained by treating α-amino acids with trifluoroacetic anhydride (equation 156). A general synthesis of 2,2-disubstituted 5(2*H*)-oxazolones is the condensation of α-(hydroxylamino) acids with ketones (equation 157) ⟨80S55⟩. The dehydrogenation of 2,2-bis(trifluoromethyl)-5-oxazolidinones affords bis(trifluoromethyl) derivatives (equation 158) ⟨79LA1547⟩.

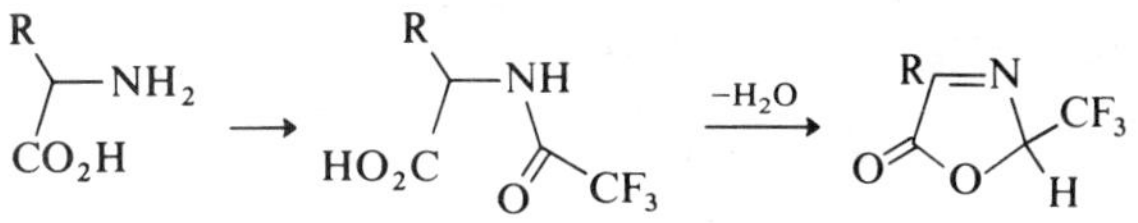

$$\xrightarrow{-H_2O} \quad (156)$$

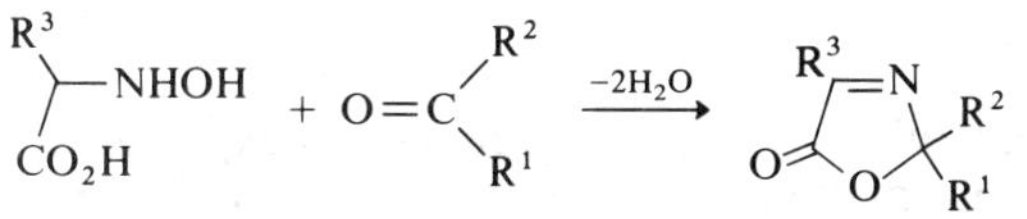

$$+ O=C \xrightarrow{-2H_2O} \quad (157)$$

$$\xrightarrow[h\nu]{Br_2} \quad (158)$$

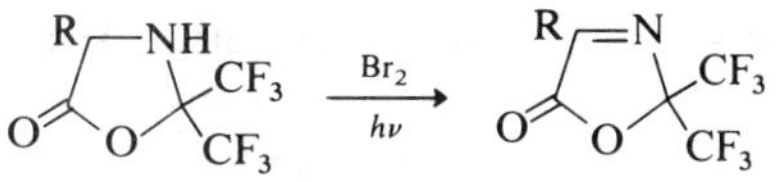

2-Arylmethylene-5(2*H*)-oxazolones, 'unsaturated pseudooxazolones', are prepared by the combined action of acetic anhydride and pyridine on α-(β-chloroarylacetyl)amino acids (equation 159) or on α-acryloylamino acids (equation 160).

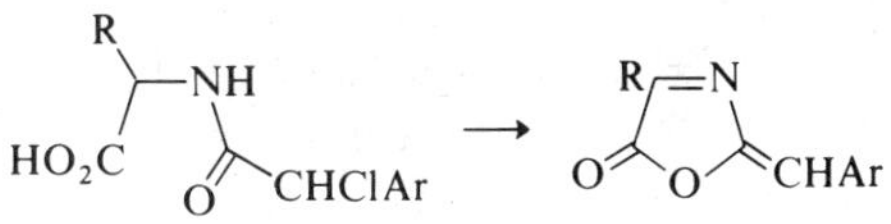

$$\rightarrow \quad (159)$$

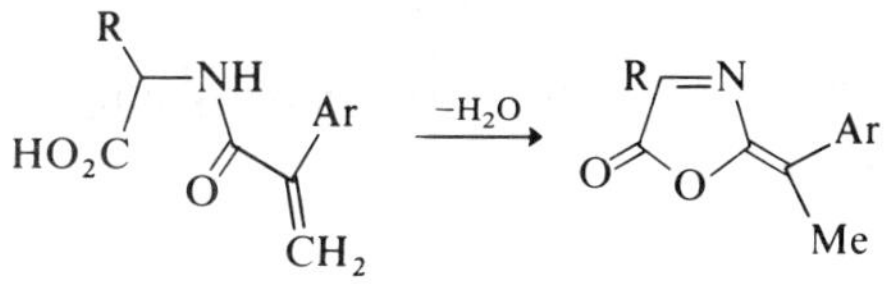

$$\xrightarrow{-H_2O} \quad (160)$$

4.18.4.4 Synthesis of Dihydrooxazoles ⟨71CRV483⟩

4.18.4.4.1 2,3-Dihydrooxazoles and 2,3-dihydrobenzoxazoles

2,3-Dihydrooxazoles are quite rare; they are produced by the thermal rearrangement of *C*-acylaziridines (equation 161). Both *cis*- and *trans*-2-benzoyl-1-cyclohexyl-3-phenylaziridine react with diphenylcyclopropenone to yield compound (**302**; equation 162). The

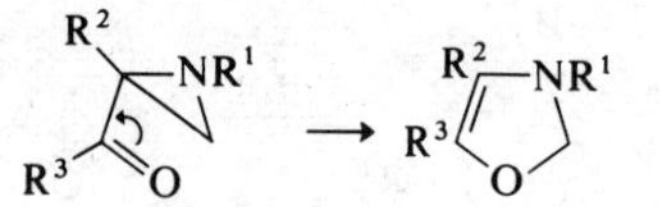

$$\underset{R^3}{\overset{R^2}{\bigvee}}\underset{O}{\overset{NR^1}{\bigvee}} \longrightarrow R^3\underset{O}{\overset{R^2}{\bigvee}}NR^1 \qquad (161)$$

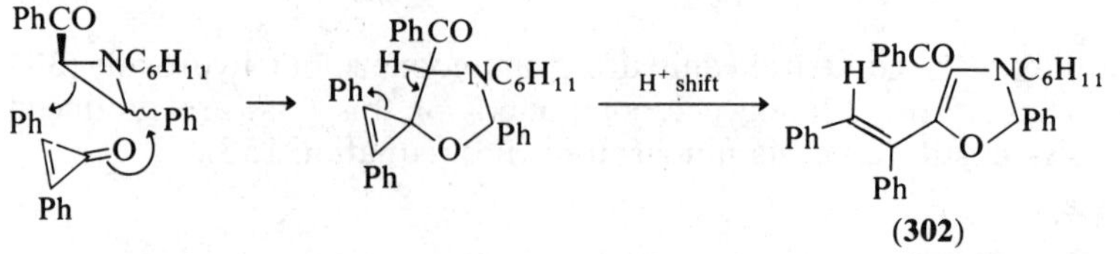

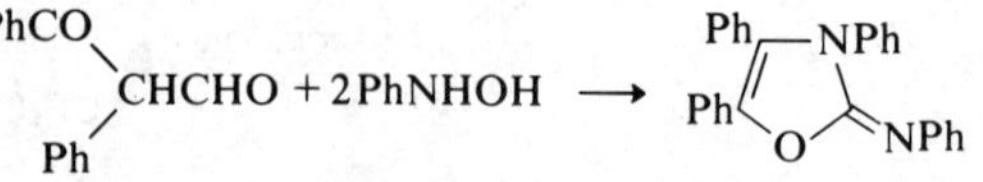

(162) — formation of (302)

(163) — (303) + 2PhNHOH → (304)

phenylimino derivative (**304**) is formed by the action of *N*-phenylhydroxylamine on the keto aldehyde (**303**; equation 163) ⟨80JCS(P1)385⟩.

The tautomeric equilibrium between phenolic Schiff's bases and 2,3-dihydrobenzoxazoles is discussed in Section 4.18.2.5.2. An authentic 2,3-dihydrobenzoxazole, the 2,2-bis(trifluoromethyl) derivative, is obtained by the action of the dithietane (**305**) on *o*-aminophenol (equation 164) ⟨74BCJ785⟩.

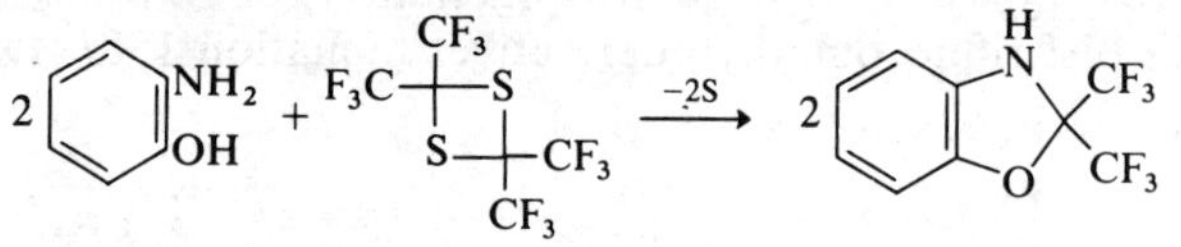

(164) — with (305)

4.18.4.4.2 2,5-Dihydrooxazoles

The preparation of 2,5-dihydrooxazoles from α-hydroxy ketones, aldehydes and gaseous ammonia is illustrated in equation (165). A derivative of 2,5-dihydrooxazole is obtained by the 1,3-dipolar cycloaddition of benzaldehyde to the nitrilium ylide (**306**), generated from α-chlorobenzylidene-*p*-nitrobenzylamine (equation 166).

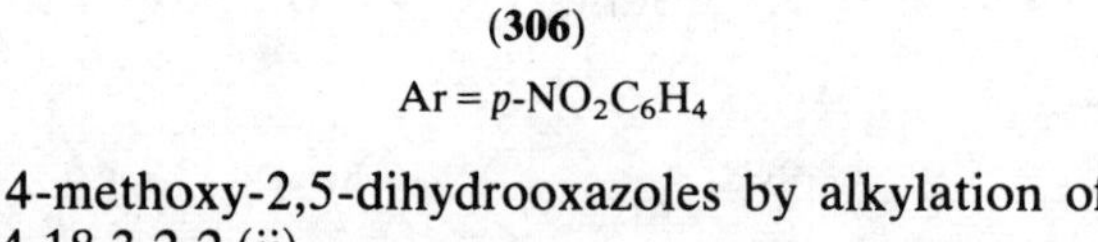

(165)

(166) — via (306), Ar = *p*-NO$_2$C$_6$H$_4$

The formation of 4-methoxy-2,5-dihydrooxazoles by alkylation of 4-oxazolidinones is described in Section 4.18.3.2.2 (ii).

4.18.4.4.3 4,5-Dihydrooxazoles

4,5-Dihydrooxazoles are readily prepared from *N*-acyl derivatives of β-hydroxylamines by heating or by the action of thionyl chloride, sulfuric acid or phosphorus pentoxide (equation 167) ⟨79JCS(P1)539⟩. The direct condensation of carboxylic acids with β-hydroxylamines succeeds best with substituted compounds, such as norephedrine (equation 168). There are many variations of this general method, such as the use of imino ether hydrochlorides (equation 169) ⟨78ACR375⟩ and of cyanides (equation 170) ⟨74LA996⟩.

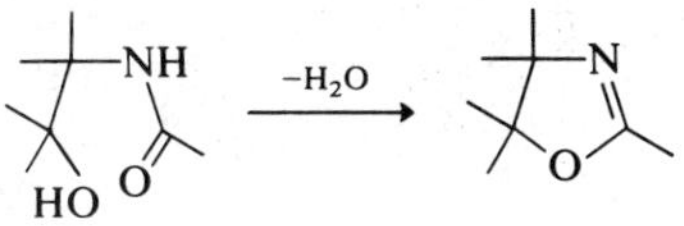

Oxazoles and their Benzo Derivatives

(167)

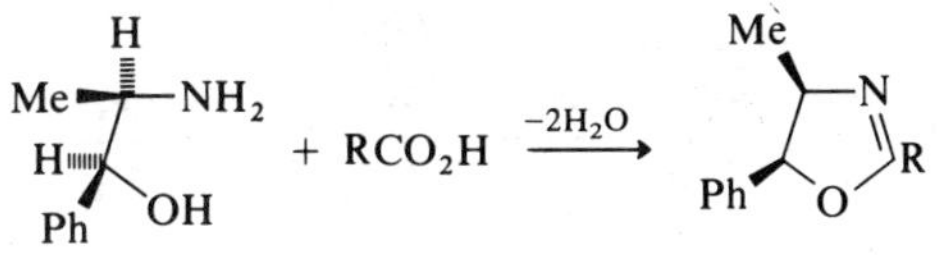

(168)

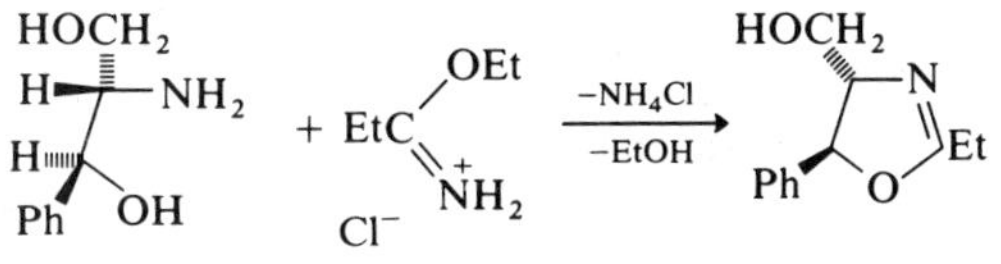

(169)

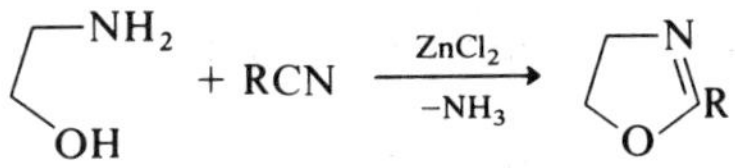

(170)

Oxiranes are converted into 4,5-dihydrooxazoles by treatment with cyanides in the presence of sulfuric acid (equation 171). α-Lithiated isocyanides react with aromatic aldehydes to furnish 2-lithiodihydrooxazoles (equation 172) ⟨79JCS(P1)652⟩.

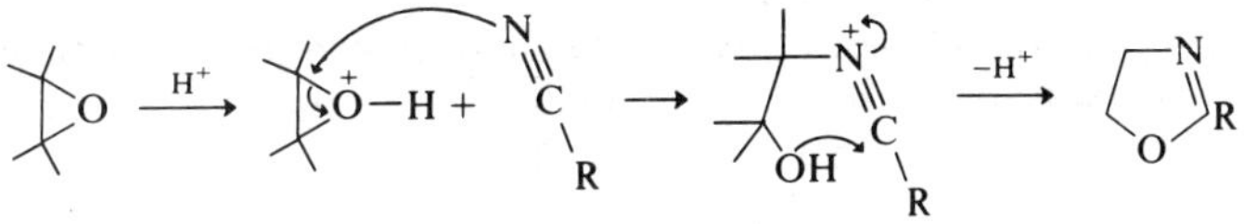

(171)

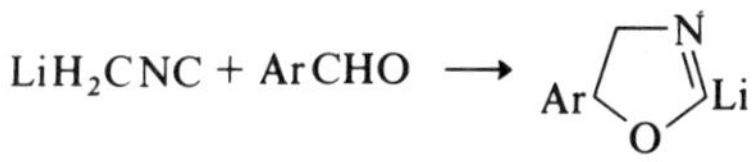

$$LiH_2CNC + ArCHO \longrightarrow$$ (172)

4.18.4.5 Synthesis of Oxazolidines and Oxo Derivatives ⟨53CRV(53)309, 67CRV197⟩

4.18.4.5.1 Oxazolidines

Oxazolidines are commonly prepared from β-amino alcohols and carbonyl compounds (equation 173). The condensation is conveniently conducted in boiling benzene with continuous removal of water. Treatment of 2-aminoethanol with formaldehyde yields the parent compound ⟨67BSF571⟩. An interesting reaction is the formation of the enantiomerically pure oxazolidine (**307**) from (−)-ephedrine and the alkynic sulfone $PhSO_2C{\equiv}CMe$ ⟨79JCS(P1)1430⟩.

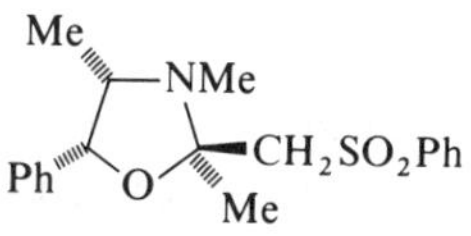

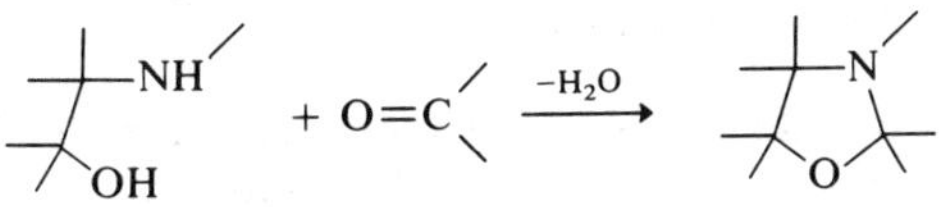

(173)

(**307**)

4.18.4.5.2 2-Oxazolidinones

The standard preparation of 2-oxazolidinones is by treatment of β-amino alcohols with phosgene or its synthetic equivalents ethyl chloroformate, dialkyl carbonates or even urea (equation 174). Isocyanates react with oxiranes in the presence of amines to yield 2-oxazolidinones (equation 175) ⟨79LA200⟩. The action of isocyanates on cyclic carbonates results in 2-oxazolidinones (equation 176) and photolysis of alkyl azidoformates affords oxazolidinones *via* intermediate nitrenes (equation 177).

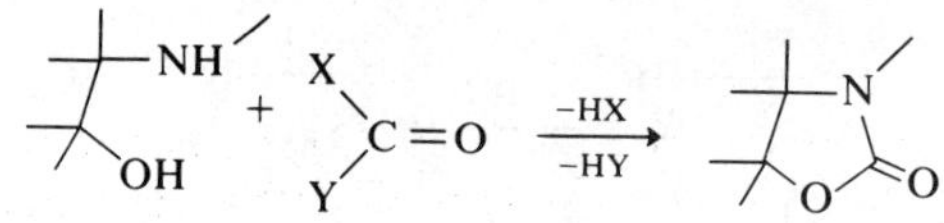

$$\text{(174)}$$

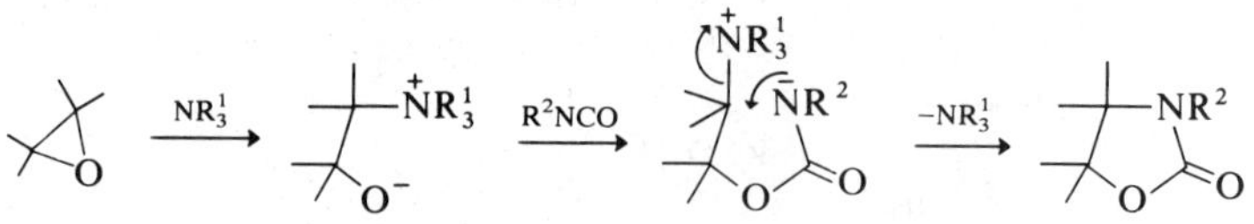

$$\text{(175)}$$

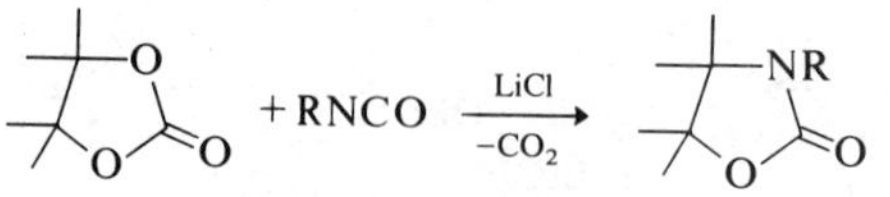

$$\text{(176)}$$

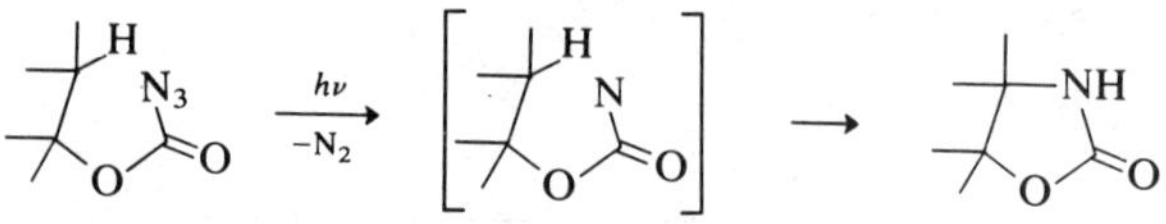

$$\text{(177)}$$

4.18.4.5.3 4-Oxazolidinones

4-Oxazolidinones are the products of the acid-catalyzed condensation of α-hydroxyamides with aldehydes and ketones (equation 178). Tertiary amides derived from pyruvic acid undergo intramolecular cyclization when irradiated (equation 179) ⟨78JOC419⟩. Treatment of the α-bromo amide (**308**) with sodium hydride yields *inter alia* the 'dimeric' oxazolidinone (**309**), presumably by way of an α-lactam, which adds to the carbonyl group of a second molecule of the amide (equation 180) ⟨80JCS(P1)2249⟩.

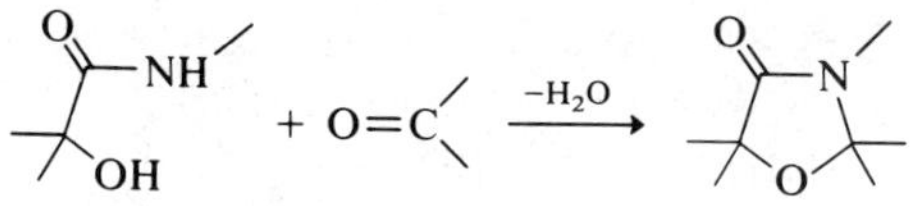

$$\text{(178)}$$

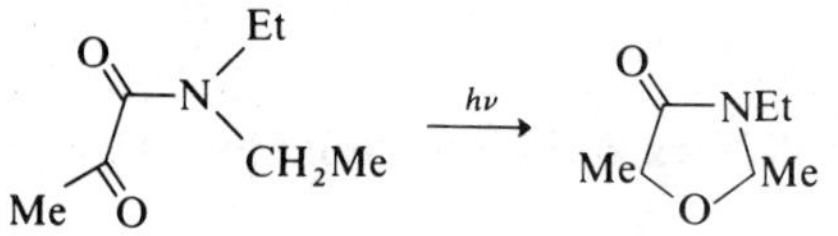

$$\text{(179)}$$

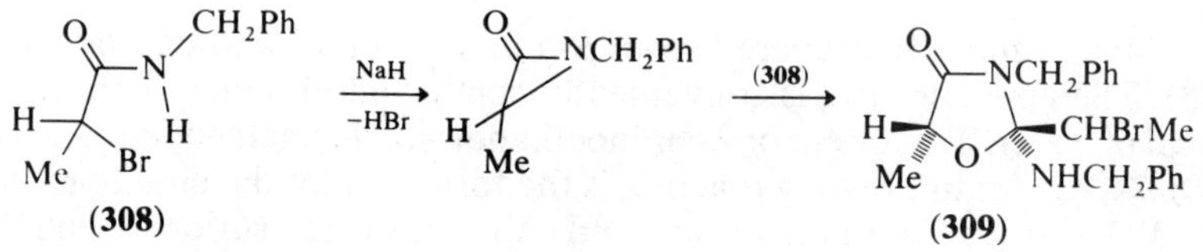

$$\text{(180)}$$

4.18.4.5.4 5-Oxazolidinones

The sodium salt of *N*-benzylideneglycine reacts with acyl chlorides to yield *N*-acyl-5-oxazolidinones (equation 181). Fluorinated 5-oxazolidinones are obtained by condensing α-amino acids with hexafluoroacetone (equation 182) ⟨77MI41801⟩.

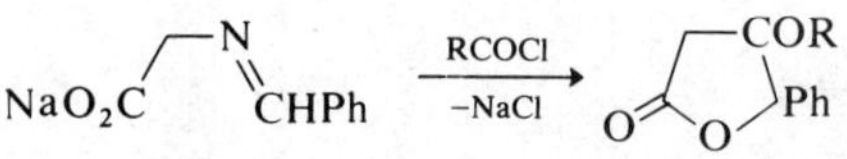

$$\text{(181)}$$

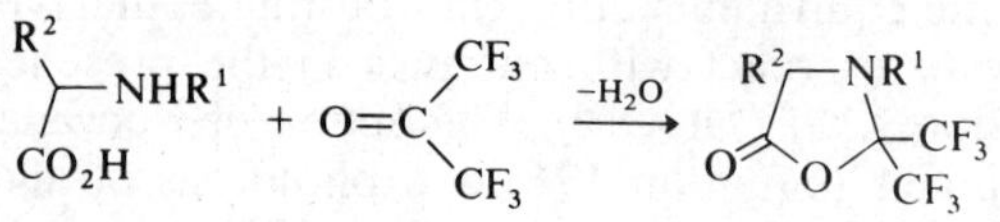

$$\text{(182)}$$

4.18.4.5.5 2,4-Oxazolidinediones

A logical construction of the 2,4-oxazolidinedione system is by condensation of an α-hydroxyacylamide with ethyl chloroformate or dimethyl carbonate under basic conditions (equation 183). Two other general syntheses are by the action of isocyanates on α-hydroxy acids in the presence of sodium (equation 184) and by treatment of esters of α-hydroxy acids with urea and sodium ethoxide (equation 185).

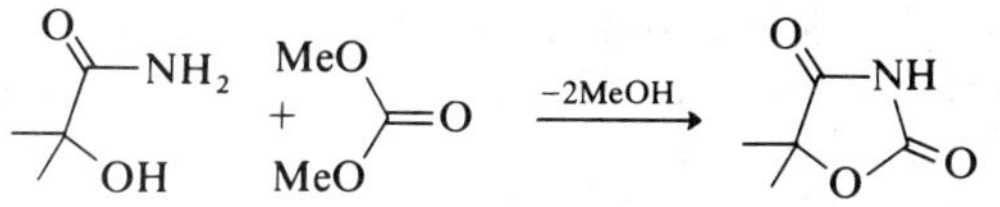

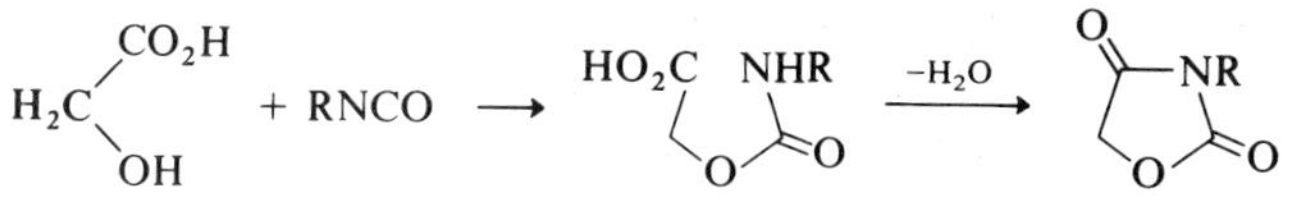

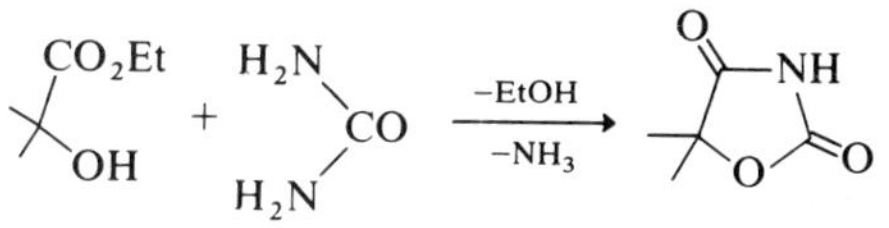

4.18.4.5.6 2,5-Oxazolidinediones

2,5-Oxazolidinediones are important in peptide synthesis (see Section 4.18.3.2.2(iii)); several methods for preparing these compounds have been devised.

The most general synthesis is the reaction of phosgene with α-amino acids (equation 186). Another method, which is mainly of historical interest, is the action of heat on α-(methoxycarbonylamino)acyl chlorides (equation 187).

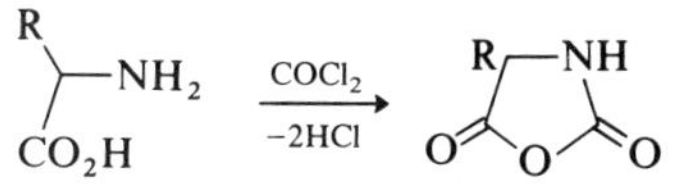

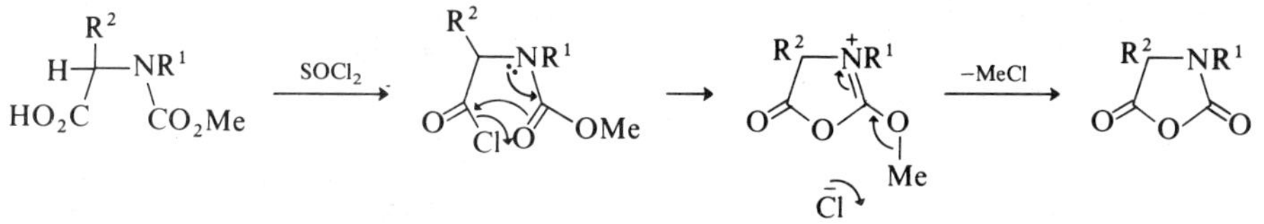

4.18.4.5.7 4,5-Oxazolidinediones

Treatment of *N*-(hydroxymethyl)benzamide with oxalyl chloride in the presence of potassium carbonate gives *N*-benzoyloxazolidine-4,5-dione in poor yield (equation 188). An efficient general synthesis of *N*-acyl-2-trichloromethyloxazolidine-3,4-diones from amides and chloral is outlined in equation (189) ⟨77AP242⟩.

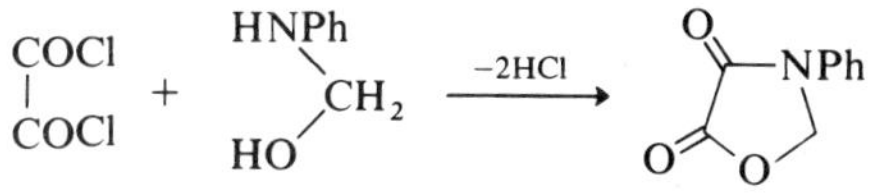

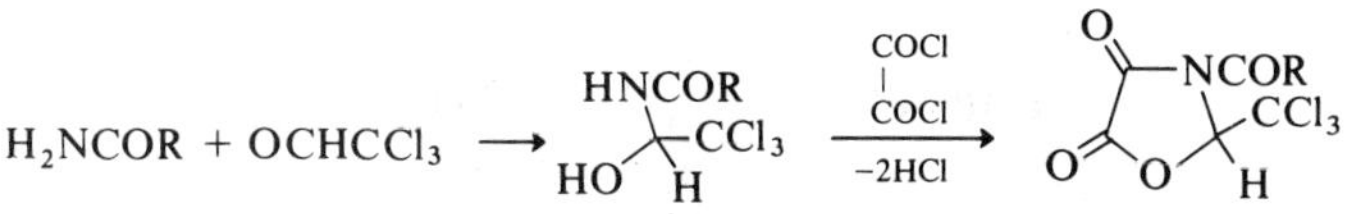

4.18.5 OXAZOLES IN NATURE AND APPLICATIONS OF OXAZOLES
⟨75CRV389, 81MI41800⟩

4.18.5.1 Naturally Occurring Oxazoles

The oxazole nucleus is rarely found in nature. During the past 20 years some alkaloids have proved to be relatively simple 2,5-disubstituted oxazoles. The first was annuloline (310) from *Lolium multiflorum*; another is pimprinine (311) from *Streptomyces pimprina*. Several oxazoles have been isolated from *Halfordia scleroxyla*: halfordine (312), halfordinone (313), halfordinol (314; R = H) and N-methylhalfordinium chloride (315) ⟨64AJC119⟩. The halfordinol isopentenyl ether (314; R = CH$_2$CH=CMe$_2$) and balsoxin [2-phenyl-5(3,4-dimethoxyphenyl)oxazole] occur in the Jamaican amyris plants *Amyris plumieri* and *Amyris balsamifera*, respectively. The isopentenyl ether has also been isolated from the seeds of the hard-shell citrus fruit *Aeglopsis chevalieri*. Annuloline, pimprine and halfordinol have all been synthesized by standard methods.

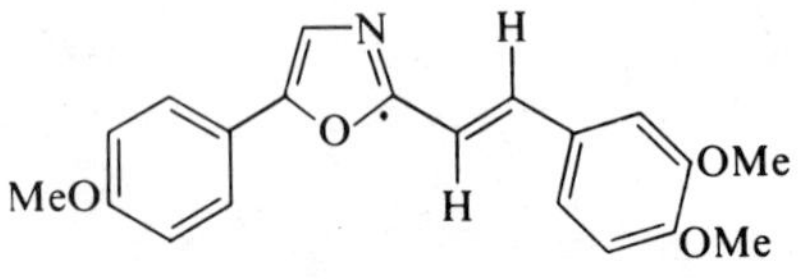

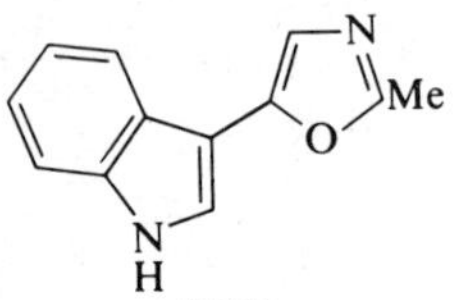

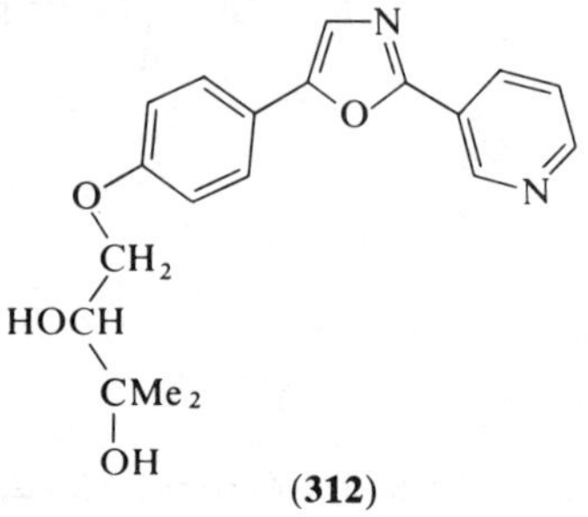

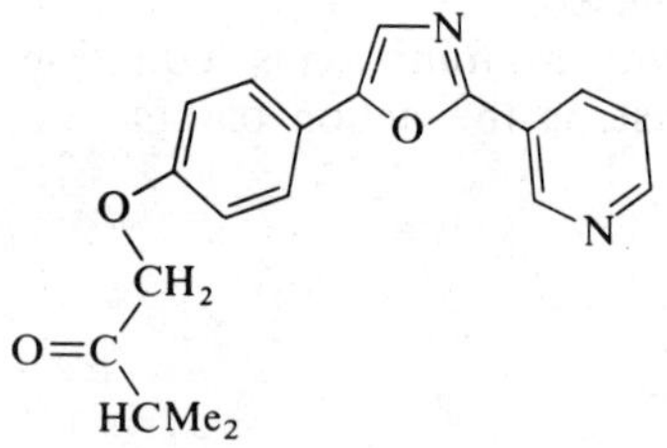

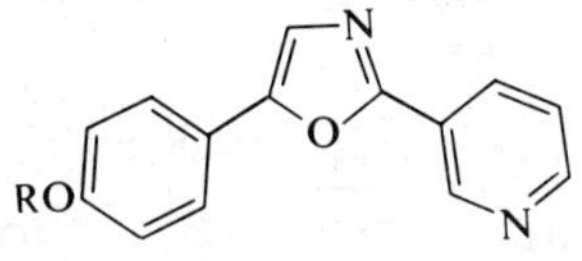

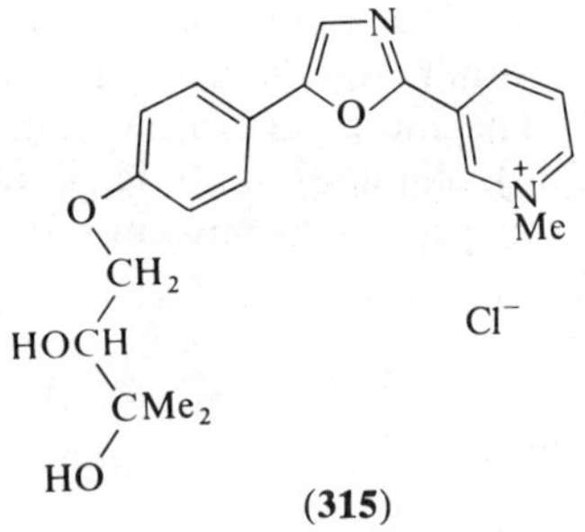

Some macrocyclic antibiotics contain oxazole rings. Ostreogrycin A, isolated from the soil organism *Streptomyces ostreogriseus*, is identical with virginiamycin A and its structure (316) has been confirmed by X-ray analysis. Griseoviridin (317) is a broad-spectrum antibiotic; the dilactone conglobatin (318) was isolated from fermented cultures of *Streptomyces conglobatus*. The X-ray structures of both antibiotics have been reported. The most complex antibiotic discovered so far is bernimamycin A (319), an inhibitor of bacterial protein synthesis.

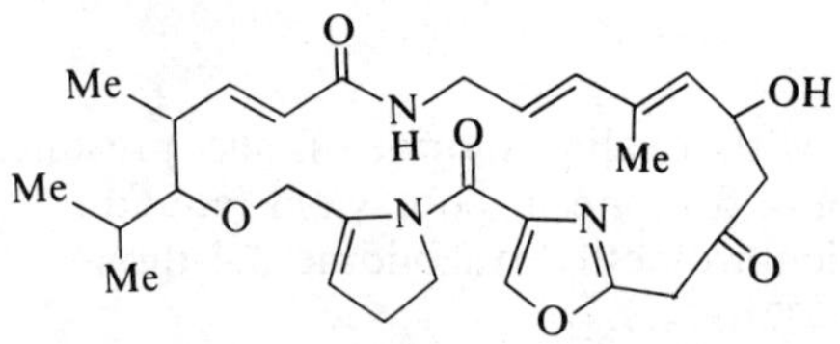

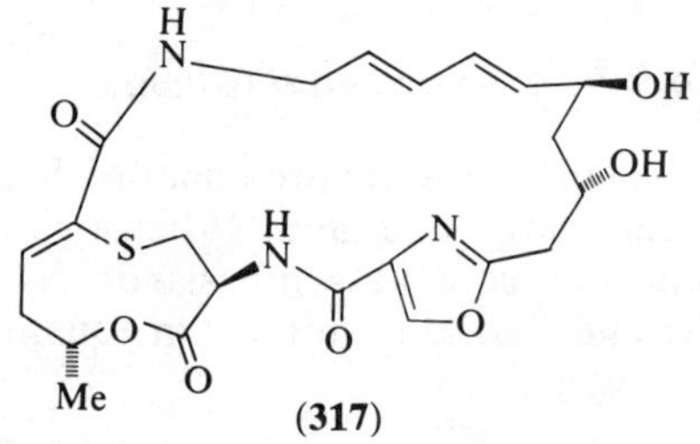

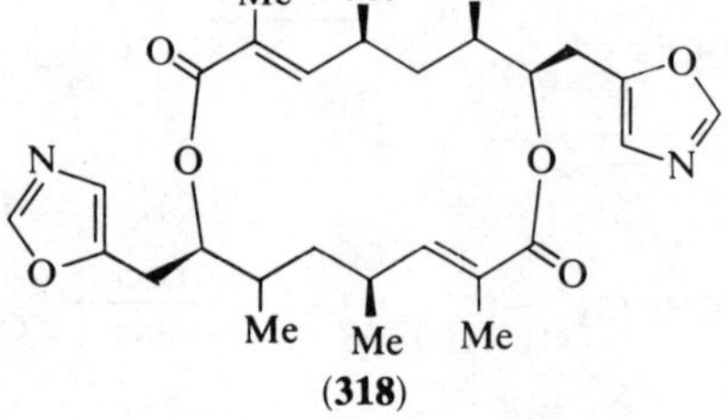

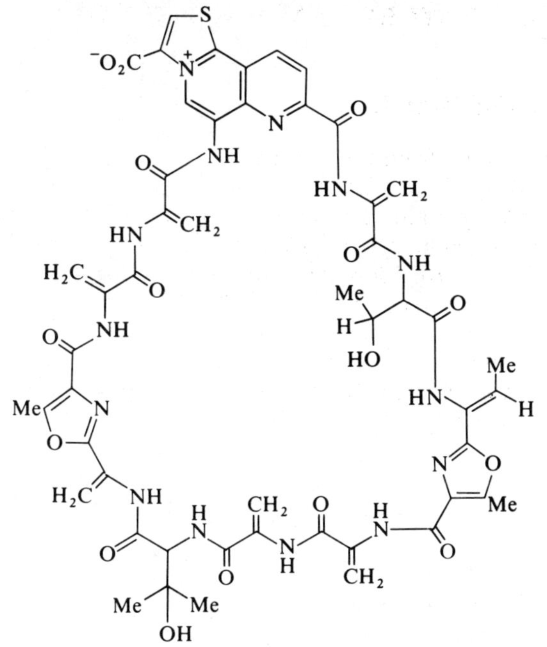

(319)

4.18.5.2 Uses of Oxazoles

By far the most important application of oxazoles is based on the scintillator properties of 2,5-diaryloxazoles ⟨B-70MI41800⟩. Diaryl- and arylstryryl-oxazoles are fluorescent whitening agents, and 2-arylbenzoxazoles are used as dye lasers ⟨B-75MI41800⟩.

Medicinals derived from oxazole include the antiepileptic drugs trimethadione (**320**; R = Me) and paramethadione (**320**; R = Et) ⟨80MI41802⟩, the sedative and muscle-relaxant zoxazolamine (**321**), and furazolidone (**322**), which is effective against a wide range of enteric infections. Derivatives of oxazolidine show promise as appetite depressants ⟨65MI41800⟩.

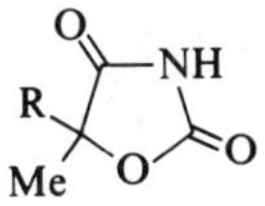

(320) 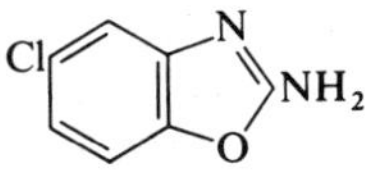   (321) 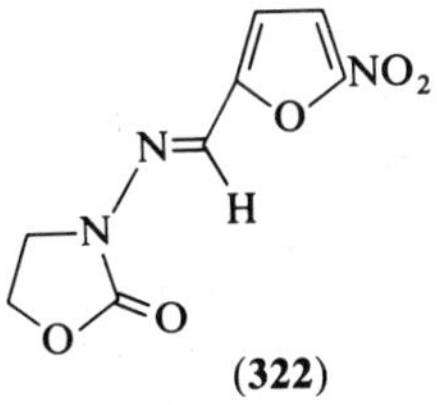   (322)

4.19

Thiazoles and their Benzo Derivatives

J. V. METZGER
Université d'Aix-Marseille

4.19.1	STRUCTURE OF THIAZOLE	236
4.19.1.1	Theoretical Methods	236
4.19.1.2	Molecular Dimensions	238
4.19.1.2.1	X-Ray diffraction	238
4.19.1.2.2	Microwave spectroscopy	239
4.19.1.3	Molecular Spectra	239
4.19.1.3.1	UV spectroscopy	239
4.19.1.3.2	Photoelectron spectroscopy	240
4.19.1.3.3	IR and Raman spectroscopy	241
4.19.1.3.4	NMR spectroscopy	242
4.19.1.3.5	Mass spectrometry	244
4.19.1.3.6	EPR spectroscopy	244
4.19.1.4	Thermodynamic Aspects	244
4.19.1.4.1	Intermolecular forces	244
4.19.1.4.2	Stability and stabilization	246
4.19.1.4.3	Conformation	247
4.19.1.5	Tautomerism	247
4.19.1.5.1	2-Substituted thiazoles	247
4.19.1.5.2	4- and 5-substituted thiazoles	248
4.19.1.5.3	Ring–chain tautomerism	249
4.19.2	REACTIVITY OF THIAZOLE	249
4.19.2.1	Reactions at the Aromatic Ring	249
4.19.2.1.1	General survey of reactivity	249
4.19.2.1.2	Thermal and photochemical reactions formally involving no other species	250
4.19.2.1.3	Electrophilic attack at nitrogen	251
4.19.2.1.4	Electrophilic attack at carbon	254
4.19.2.1.5	Attack at sulfur	257
4.19.2.1.6	Nucleophilic attack at carbon	257
4.19.2.1.7	Nucleophilic attack at hydrogen	261
4.19.2.1.8	Reaction with radicals and electron-deficient species: reactions at surfaces	263
4.19.2.1.9	Reactions with cyclic transition states	266
4.19.2.2	Reactions of Non-aromatic Compounds	269
4.19.2.2.1	Isomers of aromatic derivatives	269
4.19.2.2.2	Dihydro compounds	270
4.19.2.2.3	Tetrahydro compounds	273
4.19.2.3	Reactions of Substituents	273
4.19.2.3.1	General survey of substituents on carbon	273
4.19.2.3.2	Benzenoid rings	274
4.19.2.3.3	Other C-linked substituents	274
4.19.2.3.4	N-linked substituents	281
4.19.2.3.5	O-linked substituents	285
4.19.2.3.6	S-linked substituents	289
4.19.2.3.7	Halogen atoms	292
4.19.2.3.8	Metal- and metalloid-linked substituents	292
4.19.2.3.9	Substituents attached to ring nitrogen atoms	293
4.19.3	SYNTHESES	293
4.19.3.1	Thiazoles	294
4.19.3.1.1	Synthesis from α-halocarbonyl compounds (Type A): Hantzsch's synthesis	294
4.19.3.1.2	Synthesis by rearrangement of α-thiocyano ketones (Type A): Tcherniac's synthesis	299
4.19.3.1.3	Synthesis from α-aminonitriles (Type B): Cook–Heilbron's synthesis	300
4.19.3.1.4	Synthesis from acylaminocarbonyl compounds and phosphorus pentasulfide and related condensations (Type C): Gabriel's synthesis	302

4.19.3.1.5	Synthesis from nitriles and α-mercapto ketones or acids: 2,4-disubstituted and 4-hydroxythiazole derivatives (Type D)	302
4.19.3.1.6	Synthesis from thioacids, thiocarbamates and thioureas (Type F)	303
4.19.3.1.7	Synthesis from O,S-heterocycles (Type G)	304
4.19.3.1.8	Type H synthesis of thiazoles	305
4.19.3.1.9	Type J synthesis of thiazoles	305
4.19.3.1.10	Type K synthesis of thiazoles	306
4.19.3.2	Δ^2-Thiazolines	306
4.19.3.2.1	Type A synthesis of Δ^2-thiazolines	306
4.19.3.2.1	Type B synthesis of Δ^2-thiazolines	307
4.19.3.2.3	Type C synthesis of Δ^2-thiazolines	307
4.19.3.2.4	Type D synthesis of Δ^2-thiazolines	308
4.19.3.2.5	Type E synthesis of Δ^2-thiazolines	308
4.19.3.2.6	Type F synthesis of Δ^2-thiazolines	309
4.19.3.2.7	Type G synthesis of Δ^2-thiazolines	309
4.19.3.2.8	Type H synthesis of Δ^2-thiazolines	309
4.19.3.2.9	Type K synthesis of Δ^2-thiazolines	310
4.19.3.2.10	Type N synthesis of Δ^2-thiazolines	312
4.19.3.3	Δ^3-Thiazolines	312
4.19.3.4	Δ^4-Thiazolines	314
4.19.3.4.1	Type A synthesis of Δ^4-thiazolines	314
4.19.3.4.2	Type B synthesis of Δ^4-thiazolines	314
4.19.3.4.3	Type D synthesis of Δ^4-thiazolines	315
4.19.3.4.4	Type G synthesis of Δ^4-thiazolines	315
4.19.3.4.5	Type H synthesis of Δ^4-thiazolines	316
4.19.3.4.6	Type N synthesis of Δ^4-thiazolines	316
4.19.3.4.7	Type O synthesis of Δ^4-thiazolines	316
4.19.3.5	Thiazolidines	316
4.19.3.5.1	Type A synthesis of thiazolidines	316
4.19.3.5.2	Type B synthesis of thiazolidines	317
4.19.3.5.3	Type D synthesis of thiazolidines	319
4.19.3.5.4	Type E synthesis of thiazolidines	320
4.19.3.5.5	Type G synthesis of thiazolidines	320
4.19.3.5.6	Type K synthesis of thiazolidines	320
4.19.3.5.7	Type O synthesis of thiazolidines	320
4.19.3.6	Benzothiazoles	321
4.19.3.6.1	Type A synthesis of benzothiazoles	321
4.19.3.6.2	Type B synthesis of benzothiazoles	322
4.19.3.6.3	Type C synthesis of benzothiazoles	324
4.19.3.6.4	Type D synthesis of benzothiazoles	324
4.19.3.6.5	Type F synthesis of benzothiazoles	326
4.19.3.6.6	Type K synthesis of benzothiazoles	326
4.19.4	APPLICATIONS	326
4.19.4.1	Natural Occurrence	326
4.19.4.2	Pharmaceutical and Phytosanitary Uses	328
4.19.4.3	Analytical and Synthetic Uses	328
4.19.4.4	Industrial Uses	330

4.19.1 STRUCTURE OF THIAZOLE

4.19.1.1 Theoretical Methods

The electronic structure of thiazole has been approached using various theoretical methods ⟨79HC(34-1)1⟩. The best set of numerical values introduced into simple Hückel-type MO calculations is, for Coulomb integrals, $\alpha_N = \alpha + 0.5\beta$, $\alpha_S = \alpha_C$, and for resonance integrals, $\beta_{CN} = \beta_{CS} = \beta_{CC} = 0.5$ ⟨61CCC156⟩. The resulting net π-charges, π-bond orders and free valence numbers are given in Figure 1 and the calculated localization energies are given in Table 1. More elaborate treatments have also been applied to thiazole including *ab initio* methods and all-valence electron methods (PPP and CNDO). Figure 2 reproduces the total net charges ($\sigma + \pi$) given by an *ab initio* calculation ⟨79HC(34-1)1, p. 29⟩. Whatever the method employed, common trends can be noted in the electronic properties of the thiazole molecule. In all cases the net π-charge on sulfur is positive, whereas its net σ-charge is sometimes positive, sometimes negative. The all-electrons *ab initio* method gives a positive total net charge. In all cases the total net charge on nitrogen is negative. Of the three carbon atoms of the ring the situation is clear for C-2 only; its total net charge is either positive or very close to zero, while its net π-charge is positive. The charges at C-4 and

C-5 are not uniform; the net π-charge at C-4 is very low, whereas that at C-5 is slightly negative. From their net σ-charges the three hydrogen atoms are predicted to have decreasing acidity in the order H-2 ≥ H-5 > H-4, which is consistent with experimental results. The distribution of the π-bond orders gives a picture of the distribution of π-electrons along the σ-frame of the ring and, therefore, of its aromaticity, which can be confirmed by NMR and UV spectroscopic methods. Thiazole is slightly aromatic with a certain dienic character. The two bonds to nitrogen are different; N—C(2), having more π-character than N—C(4), should be shorter, which is confirmed by microwave spectroscopy. The free valencies, calculated by simple HMO methods, in accordance with the radical localization energies, predict a decreasing order of free radical reactivity of the three positions of the ring as 2 > 5 > 4 and the ionic localization energies, in accordance with the net π-charges diagram, predict decreasing orders of electrophilic and nucleophilic reactivities, respectively, as 5 > 2 > 4 and 2 > 5 > 4. All of these predictions have been confirmed by experiment. The calculated electronic distribution can be used to evaluate the dipole moment of thiazole: all-valence electron methods predict a moment of 1.7–1.8 D with an orientation of $-63°$ relative to the C(2)—C(5) axis, whereas the experimental values are 1.61 D and an angle of $-53°$.

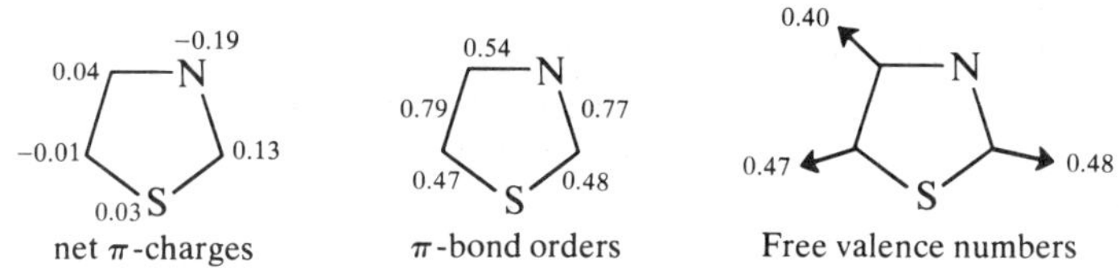

Figure 1 Electronic diagrams for thiazole by HMO method

Table 1 Localization Energies (β units) for Thiazole by HMO Method

	Position 2	4	5	Reactivity
Radical	1.99	2.51	2.01	2 > 5 > 4
Electrophilic	2.19	2.55	2.01	5 > 2 > 4
Nucleophilic	1.79	2.47	2.01	2 > 5 > 4

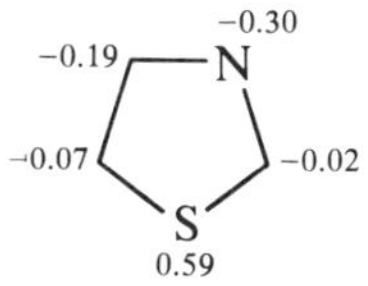

Figure 2 Total electronic net charges of thiazole by *ab initio* method

Thiazole and benzothiazole derivatives have also been studied by various theoretical methods, the predictions of which are usually in accordance with the experimental data. The relatively high thermal stability of thiazole and benzothiazole correlates with calculated bond-breaking energies for hydrogen atoms bound to these rings ⟨67MI41900⟩; theoretical calculations predict that the most probable conformation of 2-, 4- and 5-phenylthiazole is a non-planar one and that the angle of twist about the inter-ring linkage increases in the order 4 < 2 < 5 ⟨72T2799⟩; various NMR data can be successfully correlated with various theoretical calculations, for example ^{13}C chemical shifts and J_{CH} of alkyl-substituted thiazoles ⟨72BSF2679⟩, ^{1}H chemical shifts of the three phenylthiazoles ⟨72BSF1040⟩, ^{13}C chemical shifts of 2-methylbenzothiazole ⟨73JPR765⟩. An extensive theoretical study of the thiazolium, benzothiazolium and benzoselenazolium ions allowed an excellent correlation between their absorption spectra, their polarographic reduction potentials, their reactivity and the calculated values ⟨72BSF3862⟩. The stabilization energies of the donor–acceptor molecular complexes of thiazoles and phenylthiazoles with TCNE have been theoretically estimated and these values correlate well with the experimental values ⟨73MI41900⟩. MO calculations have been used for the interpretation of the quenching or sensitizing of the photo-Fries rearrangement by several heterocycles including thiazole ⟨78T481⟩. The HMO

4.19.1.2 Molecular Dimensions

Several methods have been applied to the determination of the size, shape, configuration and conformation of thiazole and its derivatives.

4.19.1.2.1 X-Ray diffraction

The X-ray structures of 18 thiazole derivatives are mentioned in ⟨72PMH(5)1⟩. Since the publication of this review, further interesting structures have been determined by this technique: ring geometry and deformations related to steric overcrowding have been determined by low-temperature X-ray measurements on a set of polyalkyl-Δ^4-thiazoline-2-thiones; the geometry of the ring is shown in Figure 3 ⟨76AX(B)1317, 76AX(B)1321⟩. Contrary to previous spectroscopic evidence, X-ray analysis has shown that 2-aryliminothiazolidines exist predominantly in this form in the solid state with a Z configuration (Scheme 1) ⟨77AX(B)99⟩. A crystallographic study of the anhydro base of 2,3-dimethylbenzothiazolium has confirmed the structure (**1**), deduced previously from NMR and chemical reactivity data (see Section 4.19.2.3.3(i)(d)). The configuration is that of an asymmetrical dimer of 2-methylene N-methylbenzothiazoline with the sulfur atoms in the *trans* positions. The planes of the benzothiazole and benzothiazoline rings subtend an angle of 91° ⟨69BSF3970⟩. X-ray analysis of the structure of 2-(o-hydroxyphenyl)benzothiazole, a compound which forms insoluble chelates with bivalent metal ions, reveals a strong intramolecular chelating hydrogen bond O—H···N with an O—N distance of 2.605 Å (**2**) ⟨70ACS3729⟩. Among other natural products the structure of the antibiotic althiomycin has been established by chemical and spectroscopic methods and by the unequivocal formulation of a derivative, as bisanhydroalthiomycin, by X-ray crystallographic analysis (**3**) ⟨74MI41900⟩.

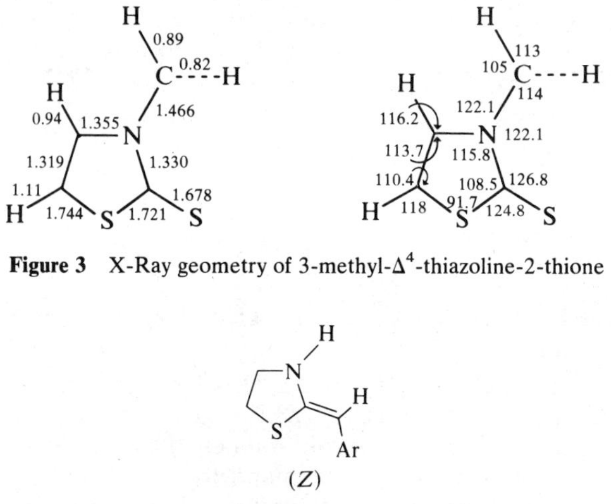

Figure 3 X-Ray geometry of 3-methyl-Δ^4-thiazoline-2-thione

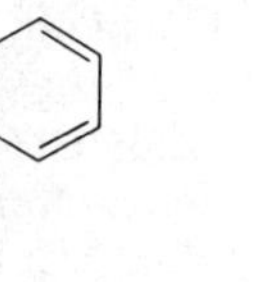

Scheme 1

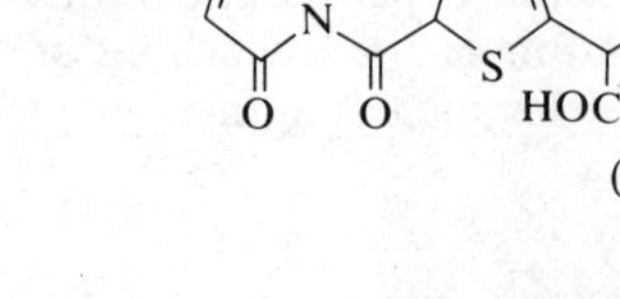

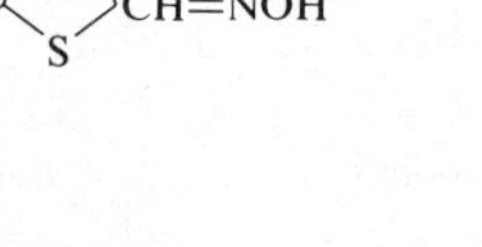

4.19.1.2.2 Microwave spectroscopy

The complete determination of the geometrical parameters of the thiazole molecule has been made by a microwave spectroscopic study of thiazole itself and of eight isotopically labelled isomers (Figure 4) ⟨71JST(9)222⟩. The structure obtained for thiazole is surprisingly close to an average of the structures of thiophene and 1,3,4-thiadiazole. From the direction of the quadrupole axis of nitrogen it is concluded that its lone pair is symmetrically placed outside the ring, along the bisector of angle C(2)—N—C(4).

Figure 4 Molecular structure of thiazole; bond lengths (Å) and bond angles (°)

4.19.1.3 Molecular Spectra

4.19.1.3.1 UV spectroscopy

The vapor phase UV absorption spectrum of thiazole exhibits two bands at (A) 228 and (B) 206.5 nm. Both bands undergo a red shift of about 3 nm when thiazole is dissolved in hydrocarbon solvents, while the red shift of band (A) increases when the solvent is hydroxylic. This bathochromic shift is typical of $\pi \rightarrow \pi^*$ transitions and is confirmed by PPP-CI calculations ⟨67BSF3283⟩. In acidic solution the red shift of band (A) is 9 nm. Methyl substitution does not alter fundamentally the absorption spectrum of thiazole, inducing a slight bathochromic shift both in the vapor phase and in dichloromethane solution, the effect decreasing in the order 4- > 5- > 2-, with a significant increase in intensity (Table 2) ⟨75CR(C)(281)1⟩. Benzothiazole and 2-methylbenzothiazole exhibit similar behavior (Table 3). Introduction of a functional group is characterized by a more pronounced bathochromic effect (Table 4). Quantum chemical calculations predict UV transitions which are in good agreement with the experimental values in the case of thiazole and its three methyl derivatives. A very weak absorption has been detected at 269.5 nm that could correspond to an $n \rightarrow \pi^*$ transition, predicted by calculation at 280 nm. UV absorption spectroscopy has been investigated in connection with steric interactions in the Δ^4-thiazoline-2-thione series (4) ⟨74JST(22)389⟩. It was earlier demonstrated, by NMR techniques, that 4-alkyl-3-isopropyl-Δ^4-thiazoline-2-thiones exist in solution as equilibrium mixtures of two conformers A (5) and B (6), the relative populations of which vary with the size of R^4: for $R^4 = Bu^t$ the population of rotamer A is 100%, whereas for $R^4 = Me$ it is only 28%. Starting from the observed absorption wavelength for extreme cases and making allowance for electronic effects, it was possible to estimate a standard wavelength of absorption for (5; $R^4 = H$) at 323.8 nm and for (6; $R^4 = H$) at 320.0 nm. The steric compression of the thiocarbonyl chromophore is stronger for (5) than for (6) and results in a red shift of its absorption. This is in agreement with PPP π and CNDO/2 calculations ⟨72BSF1055⟩ that associate the 320 nm absorption of these compounds with an electronic $\pi \rightarrow \pi^*$ transition strongly localized in the C=S π-bond. All the results concerning the compounds (4) could be rationalized by the assumption that the C=S bond is bulkier in the ground state than in the excited state.

Table 2 UV Absorption Spectra of Thiazole and Monomethyl Derivatives

	Vapour phase[a] λ_{max} (nm)	Solution[b] (CH_2Cl_2) λ_{max} (nm)	$\log \varepsilon$
Thiazole	228	234	3.05
2-Methyl	231.6	236	3.60
4-Methyl	237.4	242.5	3.53
5-Methyl	234.7	241.8	3.48

[a] ⟨67BSF3283⟩. [b] ⟨75CRC(281)1⟩.

Table 3 UV Absorption Spectra (alcohol) of Benzothiazole and 2-Methylbenzothiazole

	λ (nm)	log ε	λ (nm)	log ε	λ (nm)	log ε
Benzothiazole	294	3.13	285	3.22	252	3.76
2-Methyl	293	3.14	283	3.20	252	3.87

Table 4 UV Absorption Spectra (ethanol) of Some Typical Derivatives of Thiazoles ⟨70CR(C)(271)17⟩

	λ_{max} (nm)	log ε		λ_{max} (nm)	log ε
Thiazole	233	3.59	2-Bromo	247.6	3.69
2-Methoxy	235	3.67	5-Bromo	247.4	3.61
2-Ethoxy	237.5	3.68	2-Iodo	253	—
5-Ethoxy	255.2	3.62	2-Nitro	308	3.68
2-Chloro	244.5	3.72	5-Nitro	298.5	3.68
4-Chloro	245.6	3.46	2-Dimethylamino	265.3	3.96

(4)   (5)   (6)

As in the case of pyridine, the quaternization of thiazole induces a bathochromic shift of the UV absorption spectrum. In ethanol the long wavelength maximum at 233 nm (log ε = 3.59) for thiazole moves to 240 nm (log ε = 3.62) for 3-methylthiazolium tosylate. As in the case of the free bases, the substitution of a nuclear hydrogen atom by a methyl group induces a bathochromic shift that decreases in the order of the position substituted, 4- > 5- > 2-, and this perturbation could be accounted for by a theoretical model based on the PPP π method, using the fractional core charge approximation ⟨72BSF3862⟩.

Thiazole exhibits phosphorescence in the 380 to 500 nm region, lifetimes range from 70 ms in a xenon matrix to 2 s in a neon matrix. This region coincides with the region of the $T_1 \rightarrow S_0$ transition predicted from calculations ⟨67BSF3283⟩.

Charge transfer complexes have been observed in CCl_4 solutions of benzothiazole and iodine. They have an intense absorption band at 282 nm, corresponding to an ionization potential of 8.65 eV ⟨61BBA(46)576⟩.

4.19.1.3.2 Photoelectron spectroscopy

Ionization potentials have been determined for thiazole and some derivatives by photoelectron spectroscopy, the assignments of the bands being made on the basis of CNDO/2 or PPP calculations (Table 5a) ⟨79HC(34.1)1⟩. Δ^4-Thiazoline-2-thione has a photoelectronic

Table 5a Ionization Potentials of Various Thiazoles (eV) ⟨79HC(34-1)1, p. 52⟩

	π_3	π_2	σ_N	σ_S	π_1
Thiazole	9.50	10.25	10.40	12.70	13.4
2-Methyl	9.20	10	10.3	12.7	13.15
2-Ethyl	9	9.95	10.1	—	—
2-i-Propyl	8.95	9.85	10.1	—	—
2-t-Butyl	8.95	9.80	10.10	—	—
2,4-Dimethyl	8.71	9.9	10.0	—	—
2-Chloro	9.37	10	67	13.1	14.5
2-Bromo	9.30	10.52	10.62	13.06	14.02
4-Bromo	9.23	10	64·	12.9	14.1
5-Bromo	9.26	10.35	10.45	13.3	14.0
2-Amino	8.45	10.1		12.5	13.6
2-Dimethylamino	7.8	9.7		12.1	13.1
2-Methoxy	8.8	10.15	10.3	12.8	13.8
2-Nitro	10.1	11.1		13.5	14
Benzothiazole	8.67	9.00	10.5	10.59	11.34

Table 5b Ionization Potentials of Δ^4-Thiazoline-2-thione ⟨74JA6217⟩

	π	n	π	π	σ
Δ^4-Thiazoline-2-thione	7.74	8.12	10.32	10.82	12.48

spectrum which shows a clean separation between the two highest molecular orbitals and the others (Table 5b) ⟨74JA6217⟩. The highest MO of π-symmetry (7.74 eV) is essentially localized on the dithiocarbamic part of the structure whereas the second one (8.12 eV) is highly localized on the exocyclic sulfur atom. This peculiarity allowed the first correlation between n-orbital energy and sulfur nucleophilicity to be made.

4.19.1.3.3 IR and Raman spectroscopy

The study of the IR spectrum of thiazole in various physical states (solid, liquid, vapor, in solution) and of isotopically labeled isomers, established the symmetry properties of the main vibrations of the molecule, and the calculation of its normal modes of vibration defined a force field for it and confirmed quantitatively the assignments given in Table 6 ⟨79HC(34-1)1⟩. The Raman spectrum of thiazole has been interpreted in the same way. The IR and Raman studies of a large number of thiazole derivatives allowed the designing of a table giving the mean vibration frequencies characteristic of C—H bonds [ν(CH), δ(CH), γ(CH)] as a function of the substitution pattern (Table 7) ⟨79HC(34-1)1⟩. The IR spectra of benzothiazole and of numerous mono-, di- and tri-substituted derivatives have also been studied and the main fundamental vibrations assigned ⟨71CJC956⟩. Many problems of tautomerism relative to amino-, hydroxy- or mercapto-thiazoles or benzothiazoles have been studied using IR spectroscopy (see Section 4.19.1.5). Similarly, IR studies have shown that in the series of thiazoline-2-one, -thione and -selenone, the zwitterionic form increases on passing from oxygen to selenium (**7**).

Table 6 Thiazole Vibration Frequencies and Assignments ⟨79HC(34-1)1, p. 59⟩

Symmetry mode	Experimental frequencies ν (cm^{-1})	Theoretical frequencies	Theoretical assignment[a]	Experimental assignment
A'	3126	3125	97ν(C^{5}H)	ν(C^{5}H)
		3091	97ν(C^{4}H)	ν(C^{4}H)
	3092	3087	99ν(C^{2}H)	ν(C^{2}H)
	1478	1477	58ν(C=C)+28ν(C—N)	ω_1
	1378	1393	60ν(C=N)+25ν(C=C)	ω_2
	1319	1329	34ν(C—N)+23ν(C=C) + 22ν(C=N)	ω_3
	1326	1248	27δ(C^{2}H)+27δ(C^{4}H)+17δ(C^{5}H)	δ(CH)
	1122	1105	41δ(C^{4}H)+7δ(C^{5}H)+7δ(C^{2}H)	δ(CH)
	1041	1047	72δ(C^{5}H)+24δ(C^{4}H)	δ(CH)
	862	875	60δ(N)+38ν(C—S)	ω_4
	811	817	93ν(C—S)	ω_5
	753	749	61ν(C—S)+32δ(N)	ω_6
	610	607	65δ(N)+33ν(C—S)	ω_7
A"	877	877	86γ(C^{4}H)+11γ(C^{2}H)	γ(C^{4}H)
	795	804	68γ(C^{2}H)+28γ(C^{4}H)	γ(C^{2}H)
	718	704	98γ(C^{5}H)	γ(C^{5}H)
	602	603	22τ(C=C)+44τ(C—N)+22τ(C=N)	Γ_1
	467	474	46τ(C—S)+12τ(C=C)	Γ_2

[a] δ(N) = angular bending of the ring; τ(C—X) = torsion of the C—X bond.

Table 7 Mean Positions of the C—H Vibrations of Thiazole Derivatives ⟨79HC(34-1)1, p. 65⟩

Position of substituent	ν(C^{5}H)	ν(C^{4}H)	ν(C^{2}H)	δ(C^{2}H)	δ(C^{4}H)	δ(C^{5}H)	γ(C^{4}H)	γ(C^{2}H)	γ(C^{5}H)
2	3118±8	3087±7	—	—	1164±37	1057±11	874±8	—	702±21
4	3121±11	—	3085±9	1218±20	—	1091±46	—	808±6	722±4
5	—	3089±9	—	1226±8	1109±7	—	851±4	785±2	—
2,4	3120±12	—	—	—	—	1083±55	—	—	707±32
2,5	—	3088±13	—	—	1149±12	—	839±2	—	—
4,5	—	—	3085±7	1215±29	—	—	—	785±3	—

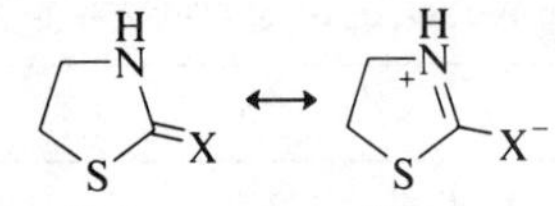

(7a) ⇌ (7b)

Raman and IR spectroscopic studies of Δ^2-thiazoline and its 2-alkyl derivatives indicate that the ring is almost planar, a result in agreement with theoretical calculations by the EHT method of the energy of the molecule as a function of its conformation ⟨78JST(50)247⟩. A complete assignment of the vibrations of benzothiazoline has been proposed ⟨77HCA215⟩ and has been used to characterize the structural modifications resulting from the photochemical opening of the pyran ring of photochromic benzothiazolinic spiropyrans (Scheme 2) ⟨78HCA1072⟩. The IR spectra of Δ^2-thiazolin-5-ones and thiazolidine-2,5-diones feature carbonyl absorption bands that are displaced by ca. 100 cm^{-1} towards lower frequencies in comparison with the corresponding oxazole derivatives. Raman and IR spectroscopic studies of thiazolidine and its N-deuterio derivative are compatible with the existence of only one half-chair conformation whilst similar studies of the three monomethylthiazolidines suggest the existence, at ambient temperature, of two conformations corresponding to pseudo axial and pseudo equatorial methyl groups (Figure 5) ⟨75JST(25)329, 75JST(25)343⟩. A valence force field has been determined for thiazolidine and five deuterated species. The experimental (IR and Raman) and theoretical results (normal mode study) permit, for all the compounds, a complete assignment of the various vibrations ⟨78MI41900⟩. The study of the normal modes of vibration of the three monomethylthiazolidines permitted a detailed assignment of their vibration spectra. This study confirms a conformational equilibrium, at ambient temperature, between the two half-chair conformers of the monomethylthiazolidines ⟨78JST(50)233⟩, whereas photoelectron spectroscopy, combined with MO calculations, suggests the predominance of certain conformers (Figure 6) ⟨77MI41900⟩.

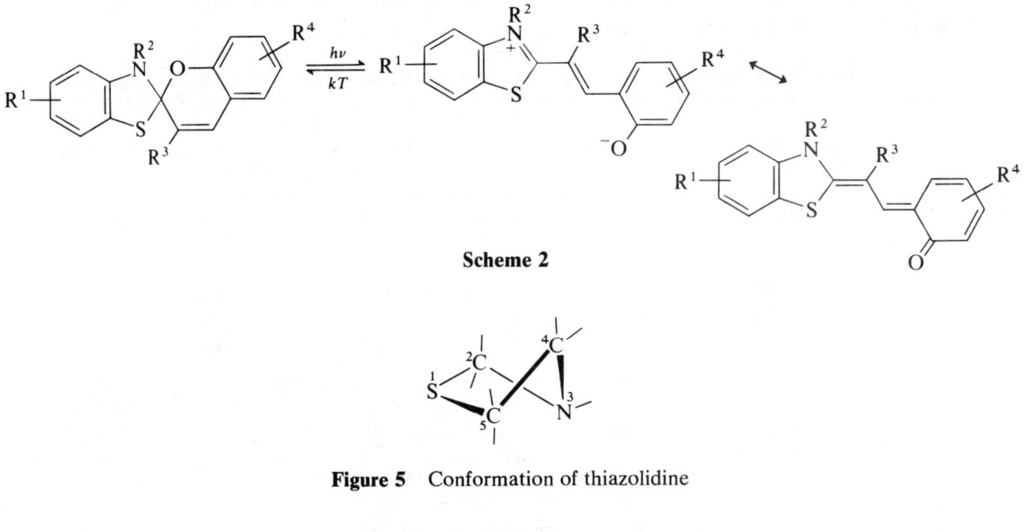

Scheme 2

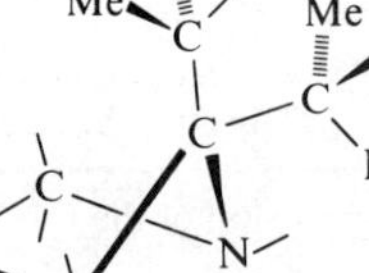

Figure 5 Conformation of thiazolidine

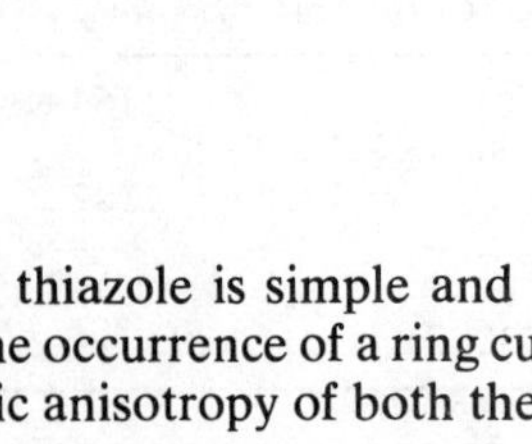

Figure 6 Conformation of 4,4-diisopropylthiazolidine

4.19.1.3.4 NMR spectroscopy

(i) ^1H NMR spectroscopy

The proton NMR spectrum of thiazole is simple and first order. The large deshielding observed can be correlated with the occurrence of a ring current connected with the aromatic sextet of thiazole and the magnetic anisotropy of both the sulfur and nitrogen heteroatoms.

Spin-coupling results in first-order coupling patterns indicating normal H(2)–H(5) and H(4)–H(5) coupling with J_{HH} values of 1.95 and 3.15, respectively, but negligible H(2)–H(4) coupling. Quaternization enhances the deshielding of the three protons and affords a small H(2)–H(4) coupling with J_{HH} of 0.70. 2-Alkyl-substituted 3-methylthiazolium salts have their N-methyl group chemical shift linearly correlated with Taft's E_s values for the o-alkyl substituent. Chemical shifts of 2-methyl protons of a large number of 5- and 6-substituted 2-methylbenzothiazoles are linearly correlated with the Hammett substituent constants, using σ_m for the 5- and σ_p for the 6-substituents. This result indicates that electronic effects are transmitted to the 2-position predominantly through the nitrogen atom. The methylene proton spectrum of 2-methyl-Δ^2-thiazoline analyzes as an A_2B_2 system with equal J_{AB} values (8.22 Hz), the low-field half being further split by coupling with the methyl protons ($J_{CH_3-(CH_2)_4}$ = 1.61 Hz) indicating that the molecule is probably planar. The thiocarbonyl group of Δ^4-thiazoline-2-thiones exhibits a large magnetic anisotropy which has led to the experimental discovery of a new kind of conformational transmission, the gear effect ⟨76JA2847⟩. In 4-alkyl-3-isopropyl-Δ^4-thiazoline-2-thiones, the conformational state of the 3-substituent (which is clearly visible due to the thiocarbonyl anisotropy) is strongly dependent on the alkyl group at C-4 and the 5-substituent (Figure 7). When the 4-substituent is an isopropyl group, the conformational equilibria cannot be explained by the classical steric effect. In this geared system substitution at one side of the structural block of two interacting groups affects the reactivity at the other end of the block. The barriers to the rotation about the N—C bond have been determined by dynamic ^{1}H NMR for various N-alkyl substituents and for the first time evidence was given for restricted rotation of the ethyl group ⟨75TL1985⟩.

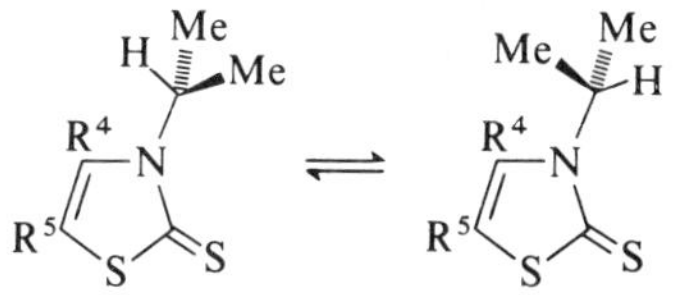

Figure 7 Conformational equilibrium of 4,5-disubstituted 3-isopropyl-Δ^4-thiazoline-2-thiones

(ii) ^{13}C NMR spectroscopy

The ^{13}C–H spin coupling constants of thiazole have been determined on samples with ^{13}C natural abundance or on samples specifically enriched in the 2-, 4- or 5-positions (Table 8). The ^{13}C chemical shifts of benzothiazole are reported in Table 9.

Table 8 J_{CH} (Hz) Coupling Constants in Thiazole ⟨71ACS2739⟩

$^1J_{CH}$		$^2J_{CH}$		$^3J_{CH}$	
C^2H^2	210.0	C^4H^5	14.9	C^2H^4	15.4
C^4H^4	184.2	C^5H^4	16.4	C^2H^5	6.2
C^5H^5	189.0	—	—	C^4H^2	7.2
—	—	—	—	C^5H^2	3.6

Table 9 ^{13}C Chemical Shifts of Benzothiazole[a] ⟨77BSB95⟩

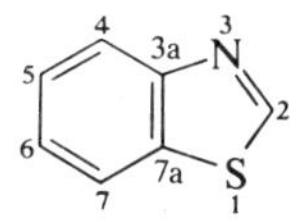

Position	$\delta(^{13}C)$ (p.p.m.)	Position	$\delta(^{13}C)$ (p.p.m.)
C^2	155.2	C^6	125.2
C^{3a}	153.2	C^7	122.1
C^4	123.1	C^{7a}	133.7
C^5	125.9		

[a] Relative to TMS.

(iii) ^{15}N NMR spectroscopy

^{15}N–H coupling constants have been determined on a sample of thiazole enriched in this isotope (Table 10).

Table 10 ^{15}N–H (Hz) Coupling Constants in Thiazole
⟨71ACS2739⟩

	$^2J_{NH}$		$^3J_{NH}$
NH^2	10.56	NH^5	1.97
NH^4	10.6		

4.19.1.3.5 Mass spectrometry

The first characteristic of the mass spectra of thiazoles is the high intensity of the molecular ion, resulting from the aromatic character of the ring. The main fragmentations of the molecular ions are described in Figure 8 ⟨66JCS(B)339⟩. The most important is the scission of the S—C(2) and N—C(4) bonds (process *a*) with elimination of HCN and retention of the positive charge by the sulfur-containing fragment which appears as a thiirenium radical cation. The mass spectra of alkyl, aryl and functional derivatives of thiazole follow the same trends for molecular ions and fragmentation patterns, making this technique particularly useful for the detection and identification of thiazole compounds.

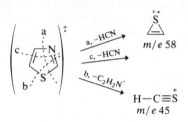

Figure 8 Fragmentation processes of thiazole

4.19.1.3.6 EPR spectroscopy

The 2-thiazolyl radical could be indirectly observed by EPR using the method of spin trapping in the photolysis of 2-iodothiazole (Scheme 3) ⟨70T2759⟩. Coupling constants for this 2-thiazolyl-*t*-butyl nitroxide radical (in gauss) are as follows: $a(NO) = 9.90$, $a(N-3) = 2.35$, $a(H-4) = 0.48$, $a(H-5) = 3.15$.

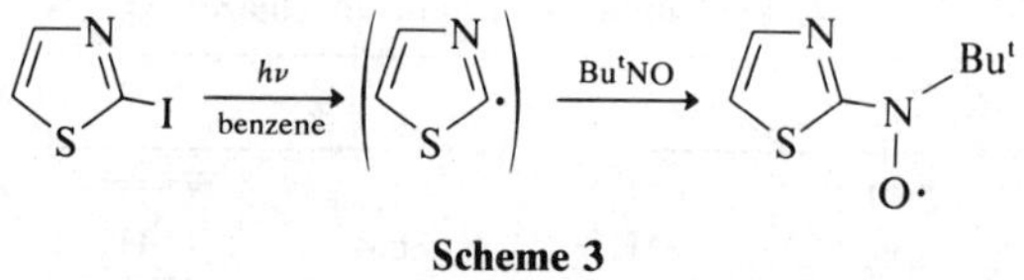

Scheme 3

4.19.1.4 Thermodynamic Aspects

4.19.1.4.1 Intermolecular forces

(i) *Melting and boiling points*

The thermodynamic study of thiazole has led not only to the determination of important practical data, but also to the discovery of thiazole self-association in the liquid state.

The vapor–liquid equilibrium of highly purified thiazole has been determined from 100 to 760 mmHg and the experimental data correlated by the Antoine equation ⟨69MI41900⟩, with P in mmHg and t in °C (equation 1).

$$\log_{10} P = 7.14112 - 1424.800/(t + 216.194) \qquad (1)$$

2-Methylthiazole behaves similarly fitting the Antoine equation (2).

$$\log_{10} P = 7.04109 - 1406.419/(t + 209.257) \qquad (2)$$

The normal boiling points of pure thiazole and 2-methylthiazole are respectively $t_{760} = 118.24$ and 128.49 °C. From the diagram of crystallization of thiazole and its mono-, di-

Table 11 Cryoscopic Data for Various Thiazoles ⟨66CR(C)(263)1333⟩

Thiazole	T_{fo} (°C)[a]	ΔH (J mol^{-1})[b]	Thiazole	T_{fo} (°C)[a]	ΔH (J mol^{-1})[b]
Thiazole	−33.6	9614	2,4-Dimethylthiazole	−50.3	—
2-Methylthiazole	−34.7	12 122	4,5-Dimethylthiazole	−17.5	—
4-Methylthiazole	−44.1	8778	2,4,5-Trimethylthiazole	−32.5	8778
5-Methylthiazole	−40.4	7524			

[a] True temperature of crystallization of ideally pure sample.
[b] Molar enthalpy of fusion of ideally pure sample.

and tri-methyl derivatives, it was possible to obtain the true temperatures of crystallization and molar enthalpies of fusion of ideally pure thiazoles (Table 11) ⟨66CR(C)(263)1333⟩.

The heat capacity and thermodynamic functions of thiazole, 2-methylthiazole and benzothiazole have been determined by adiabatic calorimetry from 5 to 340 K ⟨68MI41900, 68MI41902, 69MI41901⟩. Table 12 reproduces the heat capacity (C_p), entropy ($S°$) and Gibbs function ($G° - H°/T$) of the liquid phase at 298.15 K. The variation of C_p for crystalline thiazole between 145 and 175 K reveals a marked inflection that has been attributed to a gain in molecular freedom within the crystal lattice. The heat capacity of the liquid phase varies nearly linearly with temperature to 310 K, at which temperature it rises more rapidly. This thermal behavior, which is not uncommon for nitrogen compounds, has been attributed to weak intermolecular association. The remarkable agreement of the third-law ideal-gas entropy at 298.15 K (277.8 J mol^{-1} K^{-1}) with the spectroscopic value calculated from experimental data (277.6 J mol^{-1} K^{-1}) ⟨68MI41901, 68MI41902, 69MI41901⟩ indicates that the crystal is in an ordered form at 0 K.

Table 12 Thermodynamic Properties of Thiazole Derivatives ⟨68MI41900, 69MI41901⟩

Thiazole	C_p[a]	$S°$[a]	$(G° - H°)/T$[a]
Thiazole	120.89	169.79	−75.28
2-Methylthiazole	150.52	211.68	−94.05
Benzothiazole	189.35	209.63	−86.02

[a] All in J mol^{-1} K^{-1} at 298.15 K.

(ii) *Solubility*

Thiazole is entirely soluble in water at room temperature but when distilled exhibits an azeotropic mixture having, under a pressure of 750 mmHg, a molar fraction of water of 0.72 and an equilibrium temperature of 65 °C. Thus, dehydration of thiazole cannot usually be performed by simple distillation; one must use dehydrating agents. Thiazole is entirely miscible with most organic solvents. At lower temperatures the binary mixtures exhibit varied behavior; with cyclohexane, carbon tetrachloride and benzene for example, one observes eutectic mixtures having, respectively, the following characteristics (m.p./°C, molar fraction of thiazole): −38.4, 0.82; −60.8, 0.46; −48.5, 0.70 ⟨68CR(C)(267)114⟩. The first system is characterized by a partial miscibility of the liquid phases, the second one is unstable with incongruent melting points at −54 and −52.8 °C and the third is a simple eutectic mixture. Observed deviation from ideality for the solutions are attributable to thiazole self-association. The constants of self-association have been estimated for cyclohexane solutions of thiazole ($K_{assoc} = 5$ at 6.5 °C) and for benzene solutions of thiazole ($K_{assoc} = 3$ at 5.5 °C). Similarly, molar excess functions have been determined for various thiazole–solvent binary mixtures (Table 13). For cyclohexane the excess enthalpy (H^E) is positive and large whereas for solvents with aromatic character it is low, and even negative in the case of pyridine.

Table 13 Partial Molal Excess Enthalpy at Infinite Dilution (H^E) of Thiazole in Various Solvents at 318.15 K ⟨79HC(34-1)1, p. 88⟩

Solvent	H^E (J mol^{-1})	Solvent	H^E (J mol^{-1})
C_6H_{12}	10 199	C_6F_6	2801
C_6H_6	1200	Thiophene	401
CCl_4	3386	Pyridine	−301

A conclusion of all these thermodynamic studies is the existence of thiazole–solvent and thiazole–thiazole associations. These associations are confirmed by the results of viscosimetric and diffusiometric studies on thiazole and binary mixtures of thiazole and cyclohexane or CCl_4. In the case of cyclohexane, the solvent seems to destroy some thiazole self-associations (aggregates) existing in the pure liquid, whereas in the case of carbon tetrachloride there is association of two thiazole molecules with one solvent molecule ⟨67BSF4465⟩. The most probable mode of self-association of thiazole is of the $n–\pi$ type from the lone electron pair of the nitrogen of one molecule to the LUMO of the other ⟨73MI41901⟩. The same methods applied to Δ^2-thiazolines reveal self-association constants of 3 to 5 ⟨67BSF4583⟩.

(iii) *Chromatography*

Thin layer chromatography and gas-liquid chromatography have been widely applied for the separation and for the identification of thiazoles in reaction mixtures. From systematic studies it appears that thiazole, alkyl- and aryl-thiazoles and benzothiazoles are best separated on stationary phase of low polarity in GLC and with eluents of low polarity in TLC. It has been possible to correlate, for these series of compounds, the R_F of TLC or the specific volume of retention in GLC with the number of carbon atoms in the aliphatic side chain, and also with the rate constants of quaternization of the cyclic nitrogen atom. This last observation indicates a significant participation of the nitrogen atom in the chromatographic processes ⟨67BSF846⟩.

4.19.1.4.2 *Stability and stabilization*

(i) *Thermochemistry*

The thermal stabilities of thiazole and of some aryl and benzo derivatives have been determined ⟨66BSF2857⟩ and the pyrolysis temperatures are reported in Table 14. They have been correlated with the energy of the LUMO whereas there is no correlation with calculated resonance energies.

Table 14 Pyrolysis Temperature of Some Thiazoles and Benzothiazoles ⟨66BSF2857⟩

Thiazole	T (°C)	Thiazole	T (°C)
Thiazole	530	2,4-Bis(p-biphenylthiazole)	510
2,4-Diphenylthiazole	431	Benzothiazole	556
2-Phenyl-4-biphenylthiazole	442	2-Methylbenzothiazole	446
4-Phenyl-2-biphenylthiazole	452	2-Phenylbenzothiazole	504

(ii) *Aromaticity*

Many physicochemical properties of thiazole and benzothiazole are consistent with the aromatic character of the heterocyclic ring. The geometrical structure of thiazole, discussed in Section 4.19.1.2.2, is very close to the average of that of thiophene and 1,3,4-thiadiazole suggesting an aromaticity of the same order for the three molecules. The π-bond orders calculated either by HMO or by CNDO/2 methods are indicative of a thiazole molecule situated between an aromatic system and a dienic system (see Section 4.19.1.1). The microwave spectrum of thiazole exhibits practically no inertial defect, which implies entire coplanarity of the eight atoms of the ring, in accordance with an important cyclic conjugation of their p_z orbitals (see Section 4.19.1.2.2). The important ^{1}H NMR deshielding of ring protons for thiazole and benzothiazole has been correlated with the occurrence of a ring current connected with the aromatic character of the thiazole ring (and also with the magnetic anisotropy of both sulfur and nitrogen heteroatoms). Ring currents for thiazole and thiophene are estimated to be very similar, as a result of both empirical calculations ⟨66BSF3537⟩ and experimental data ⟨66JCS(B)127⟩. The important magnetic susceptibility anisotropy of thiazole is also compatible with its aromatic character ⟨74TL253⟩. Finally the large intensity of the molecular ion in the mass spectrum of thiazoles and benzothiazoles is associated with the aromatic character of these rings (see Section 4.19.1.3.5).

4.19.1.4.3 Conformation

As already mentioned, many conformational studies have been performed in the thiazole series. IR and Raman studies of Δ²-thiazoline and its 2-alkyl derivatives have shown that the ring is almost planar, a result in agreement with theoretical calculations (see Section 4.19.1.3.3). The same types of studies on thiazolidine are compatible with the existence of only one half-chair conformation for the ring, whilst in the case of the three monomethyl-thiazolidines they suggest the existence, at ambient temperature, of two conformations corresponding to pseudoaxial and pseudoequatorial methyl groups (Figure 5); in the same series PE spectroscopy combined with MO calculations suggest the predominance of certain conformers (see Section 4.19.1.3.2). The population analysis of the conformers of geared molecules in the series 3,4- and 3,4,5-alkyl-substituted Δ⁴-thiazoline-2-thiones was possible by ¹H NMR techniques, owing to the large magnetic anisotropy of the 2-thiocarbonyl group (see Section 4.19.1.3.4).

4.19.1.5 Tautomerism

Thiazole substituted in the 2-, 4- or 5-position by XH groups (XH = NHR, OH, SH) is susceptible to protomerism (Scheme 4).

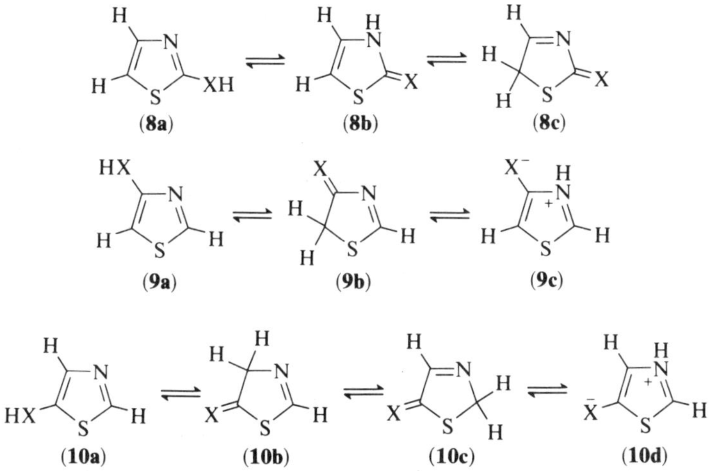

[Scheme 4 showing tautomeric structures (8a)–(8c), (9a)–(9c), (10a)–(10d)]

Scheme 4

4.19.1.5.1 2-Substituted thiazoles

The position of the protomeric equilibrium may be determined by spectroscopic methods.

A simple HMO treatment of 2-substituted thiazoles, limited to protomeric forms (**8a**) and (**8b**), gives their relative energies of formation as a function of the heteroatom X (Table 15). In agreement with the experimental evidence, the equilibrium is displaced towards form (**8a**) only when X = NR. This simple treatment also allows an estimation of the general effects to be expected when substituents are introduced in the 4- or 5-positions or on the exocyclic nitrogen atom: this is done by varying the coulomb integral α of the introduced substituent. The calculations show interesting trends: when X = O or S electronic substituent effects are not expected to invert the protomeric equilibrium position. In the case of X = NR, the equilibrium would be shifted towards (**8a**) when an electron-donor substituent is present in any position, and toward (**8b**) when electron-acceptor substituents are present. One must take into account that this simple treatment does not include the medium effects, which are known to be important, nor the existence of associated species ⟨79HC(34-2)1⟩.

Both 2-hydroxythiazole and 2-mercaptothiazole have been studied to determine the position of the protomeric equilibrium (**8a** ⇌ **8b** ⇌ **8c**). Most studies indicate that (**8b**) is largely predominant in neutral solution for X = O or S. The equilibrium constant K_T of the protomeric system (**12a**) ⇌ (**12b**) for 4,5-dimethyl-Δ⁴-thiazoline-2-thione in water (Scheme 5) has been evaluated as $K_T = 10^6$ ⟨69ACS2879⟩. These compounds are strongly

Table 15 HMO Calculated Formation Energies for 2-XH-Substituted Thiazoles (**8a**) and 2-X-Substituted Δ^4-Thiazolines (**8b**) ⟨79HC(34-2)1, p. 2⟩

X	$FE(\mathbf{8a}) - FE(\mathbf{8b})^a$
O	−0.616
S	−0.140
N	+0.075

[a] In β units; the more negative values are those for which structure (**8b**) is the more stable.

associated in dilute hexane and in carbon tetrachloride solutions and even stronger association occurs in the solid state. The protomeric equilibrium, therefore, is part of a more complex system which involves species such as (**13** and **14**; Scheme 6). The hydrogen-bonding abilities of thiazole-2-thiols and related azoles toward DMSO and p-chlorophenol has led to a better understanding of the factors which affect the self-association of mercapto-azoles ⟨77JCS(P2)1015⟩. An IR study of benzothiazoline-2(3H)-thiones has shown that hetero associations between different heterocyclic thioamides result in greater stabilities than the homo associated species of the parent compounds ⟨77TL251⟩.

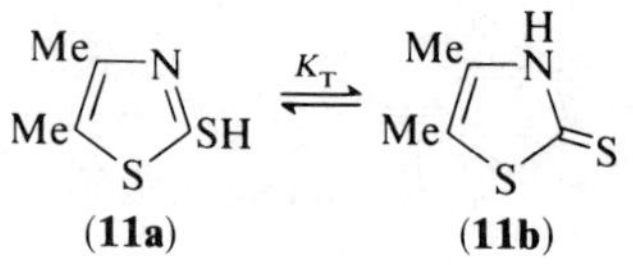

Scheme 5

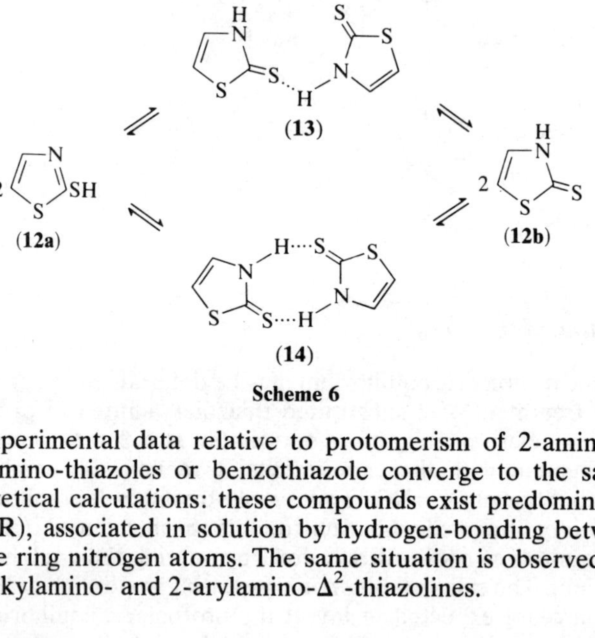

Scheme 6

All available experimental data relative to protomerism of 2-amino-, 2-alkylamino-, 2-aryl or heterylamino-thiazoles or benzothiazole converge to the same result as that predicted by theoretical calculations: these compounds exist predominantly in the amino form (**8a**; X = NHR), associated in solution by hydrogen-bonding between the exocyclic NH groups and the ring nitrogen atoms. The same situation is observed in the case of the corresponding 2-alkylamino- and 2-arylamino-Δ^2-thiazolines.

4.19.1.5.2 4- and 5-substituted thiazoles

In 4- and 5-substituted thiazoles the character of the protomerism is not so sharply defined. IR and ^{1}H NMR data indicate that in non-polar solvents, as in the solid state, Δ^2-thiazolin-4-ones exist predominantly in the keto form (**15b**) but that polar solvents such as DMSO shift the keto–enol equilibrium towards the enol form (**15a**): in acetone equal amounts of (**15**; R^2 = Ph, R^5 = H) enol and keto form were found, whereas in DMSO more than 90% of the enol form exists at equilibrium ⟨65ACS1215⟩. The third form (**15c**) corresponds to the mesoionic 4-hydroxythiazole and its existence has been suggested from reactivity experiments but it could not be established spectroscopically.

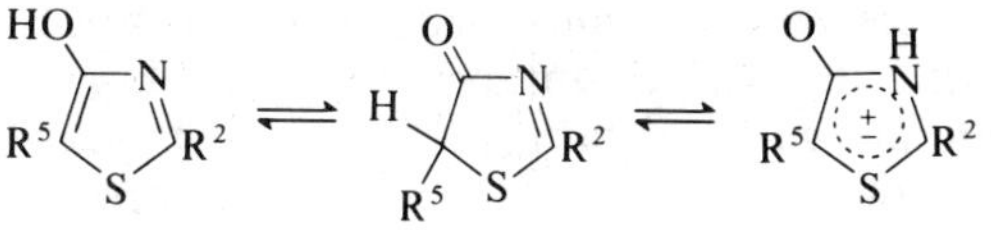

(15a) (15b) (15c)

Scheme 7

Δ^2-Thiazolin-5-one may exist in three tautomeric forms (**16a, b, c**): the protomeric equilibrium has been studied by ^{1}H NMR, IR and UV spectroscopies. Polar solvents favor the enolic (**16a**) and the mesoionic (**16c**) forms (Table 16), this latter form being detected by the yellow color (428 nm) which appears in polar solvents on high dilution ⟨70TL169⟩.

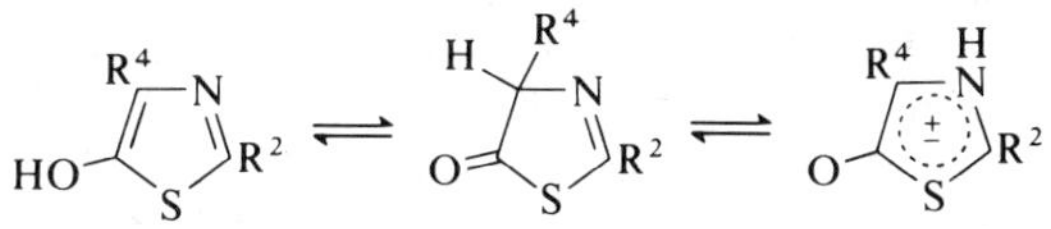

(16a) (16b) (16c)

Scheme 8

Table 16 Protomeric Equilibrium for 4-Methyl-2-phenyl-Δ^2-thiazolin-5-one ⟨70TL169⟩

Solvent	(**16b**)	(**16a**)	*Solvent*	(**16b**)	(**16a**)
$CDCl_3$	100	0	Cyclohexanone	50	50
CD_2Cl_2	100	0	CF_3CH_2OH	45	55
CD_3NO_2	100	0	PhOH	30	70
$PhNO_2$	95	5	MeOH	20	80
PhCN	80	10	DMF	10	90
MeCN	85	15	DMSO/HMPA	0	100
Acetone	65	35			

Tautomerism of Δ^2-thiazoline-5-thiones has not been investigated intensively. 2-Phenylthiazole-5-thione exists in the thiol form in both polar and non-polar solvents ⟨76T579⟩. This behavior is in contrast with that of the corresponding thiazolones.

4.19.1.5.3 Ring–chain tautomerism

2-Azido-thiazole and -benzothiazole exist in equilibrium with a corresponding thiazolo[2,3-*e*]tetrazole form (Scheme 9). In the case of 2-azidobenzothiazole, at 20 °C in dioxane about the same proportion of both tautomers exist in equilibrium ⟨77CJC1728⟩.

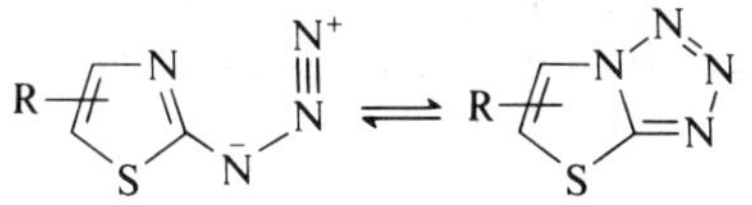

(17) (18)

Scheme 9

4.19.2 REACTIVITY OF THIAZOLE

4.19.2.1 Reactions at the Aromatic Ring

4.19.2.1.1 General survey of reactivity

(i) *Reactivity of neutral thiazole*

As discussed in the general chapter (Section 4.01.1.1) thiazole may be considered to be derived from benzene by replacing a CH group with a nitrogen atom and a CH=CH group at position 3,4 of the corresponding pyridine with a sulfur atom. The chemistry of thiazole therefore shows similarities to those of both pyridine and thiophene. Thus electrophilic reagents attack the lone electron pair on nitrogen but do not attack that on sulfur. The three carbon atoms of the ring behave quite differently, taking into account the complete dissymmetry of the structure which is mirrored in its electronic distribution. As discussed in Section 4.19.1.1 the whole reactivity of the ring in nucleophilic, electrophilic and radical reactions is well explained by data from MO calculations, and the aromatic character of the molecule, which appears in its chemical behavior, is also demonstrated by many physicochemical properties (see Section 4.19.1.4.2(ii)). The fused ring of benzothiazole exhibits a similar chemical reactivity in good agreement with theoretical calculations (see Section 4.19.2.3.2(i)).

(ii) *Thiazolylium ions*

Thiazolylium quaternary salts show great reactivity toward nucleophiles. The unsubstituted 2-position of thiazolylium and benzothiazolylium cations is exceptionally reactive and its hydrogen atom is removed by bases affording an ylide which is particularly stabilized by the adjacent sulfur atom. This stabilization is for a large part responsible for the natural selection of thiazolylium, among other azolyliums, as the active part of vitamin B_1. 2-Alkylthiazolylium and benzothiazolylium quaternary salts are transformed by bases into exceptionally reactive anhydro bases which have wide application in synthesis.

(iii) *Thiazolones, thiazolethiones and thiazolimines*

These compounds are characterized by their tautomerism, and they exhibit ambident electrophilic reactivity.

4.19.2.1.2 Thermal and photochemical reactions formally involving no other species

(i) *Fragmentations*

It has been shown (Section 4.19.1.3.5) that the large intensity of the molecular ion in the mass spectrum of thiazoles results from the aromatic character of the ring. The stability, as measured by the relative intensity of the molecular ion, is enhanced by the presence on ring carbon atoms of aryl substituents or of an electron-releasing group (NMe_2); it is lowered by the presence of an electron-attracting group (NO_2). More generally, the thiazole ring is reputed stable enough to be incorporated into thermally stable polymers, some of which contain benzothiazole or benzobisthiazole units (see Section 4.19.4.4).

(ii) *Rearrangements*

The UV photochemical rearrangement of alkyl- and aryl-thiazoles into isomeric thiazoles and isothiazoles has been studied with the aid of isotopic labelling ⟨74T879⟩. Among the possible mechanisms two are compatible with the observed products: the first implies 180° rotations around the bonds adjacent to the sulfur atom ⟨72TL1429⟩, the second corresponds to a valence-bond isomerization (Scheme 10) ⟨70CC386⟩. The latter is in agreement with the variation of the π-bond orders (calculated by the PPP method) between the ground state and the first excited state in the case of 2-phenylthiazole and with the free-valence numbers of the intermediate conjugated diradical (Scheme 11) ⟨73BSF1743⟩. In some cases, the yield is large enough to allow the method to be preparative (*e.g.* 32% of 3,4-diphenylisothiazole from 2,5-diphenylthiazole). The reaction is reversible and the thermodynamically more stable isomer forms predominantly. Another photochemically-induced important rearrangement concerns certain thiazolinic spiropyrans (**19**). These molecules exhibit photochromism by their reversible conversion into 'open form' structures of the merocyanine type (**20**). During this rearrangement the C-2 carbon atom of the thiazoline ring rehybridizes from sp^3 to sp^2 and the π-system extends over the whole framework of the molecule. The nature and position of the substituents in both heterocyclic moieties influence the UV–visible absorption and the first order thermal decolorization rate constants of the individual merocyanines. A Hammett type correlation could be established between this rate constant and the σ_p^+ parameters of the 6-substituent (**19**, **20**; R^3); on the other hand the steric effect of the 3'-substituent (**19**, **20**; R^1) has a very large influence on the half-life of the merocyanine ⟨68MI41900⟩.

Certain mesoionic derivatives of thiazole (**21**) rearrange into thiazolone (**22**) under irradiation. When the photolysis is conducted in methanol, the corresponding β-lactam (**23**) is obtained.

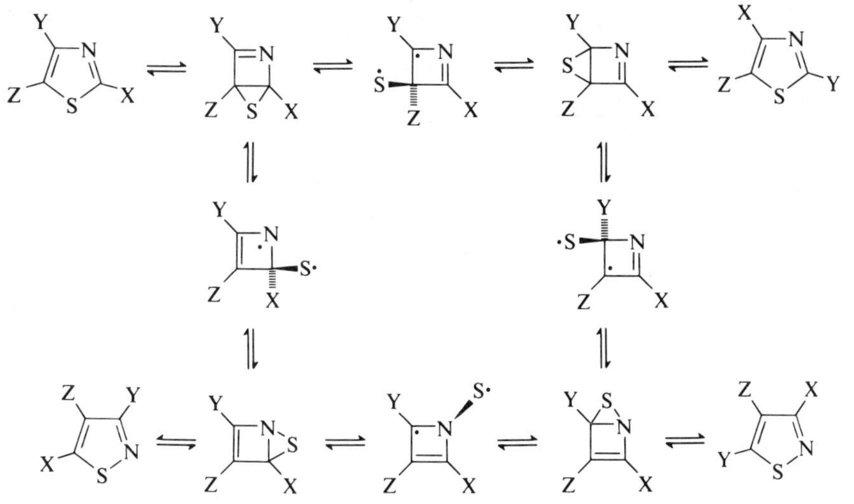

Scheme 10

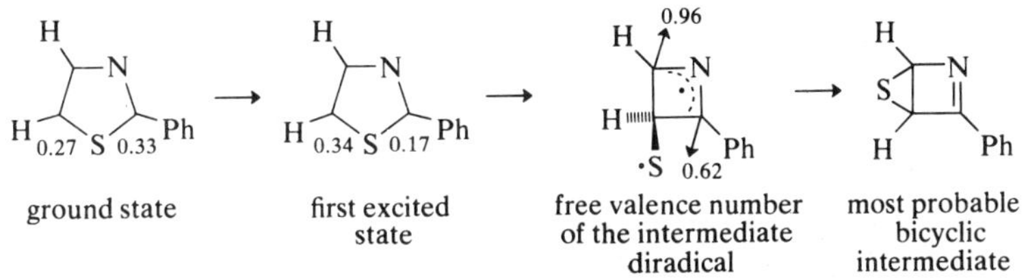

Scheme 11 π-Bond order

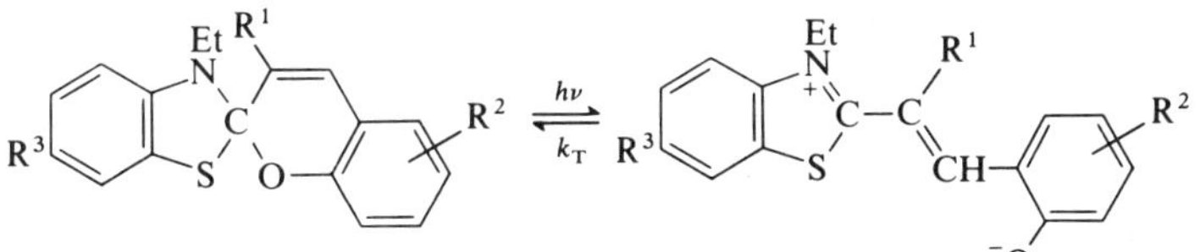

(19) uncolored (20) colored

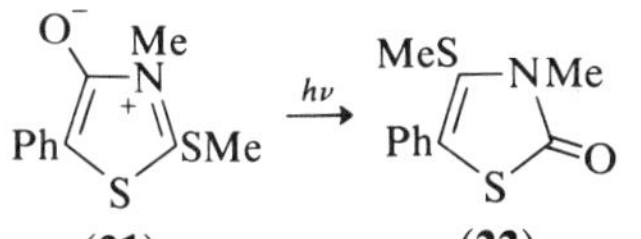

(21) (22) 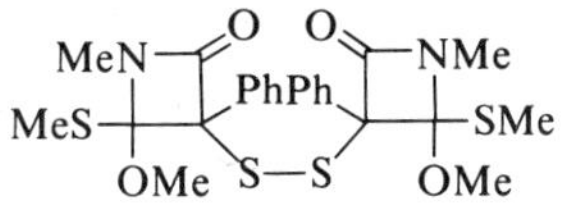 (23)

4.19.2.1.3 Electrophilic attack at nitrogen

(i) *Introduction*

The most striking analogy between thiazole and pyridine results from the presence, in both molecules, of a ring N atom, contributing two σ-bonds to the σ-framework of the molecule and one $2p_z$ electron to the π-system, while a lone pair remains localized on the nitrogen atom, described by an sp^2-hybrid orbital, the axis of which lies in the σ-plane and is directed along the bisector of the C—N—C angle. The lone pair of thiazole is, however, less reactive than that of pyridine, due to the enhanced aromatic character of the latter ring which is responsible for a more effective stabilization of the (incipient to integral) positive charge developing on the nitrogen atom along the reaction path (Scheme 12). Table 17 shows three sets of physicochemical data that illustrate this difference. These are: (i) the thermodynamic basicity, which is three orders of magnitude lower for thiazole than for pyridine; (ii) the enthalpy of reaction with BF_3 in nitrobenzene solution, which is 10% lower for thiazole; (iii) the specific rate of reaction with methyl iodide in acetone at 40 °C, which is about 50% lower for thiazole and which in DMSO at 25 °C is of the order of $\frac{1}{21}$ that of pyridine.

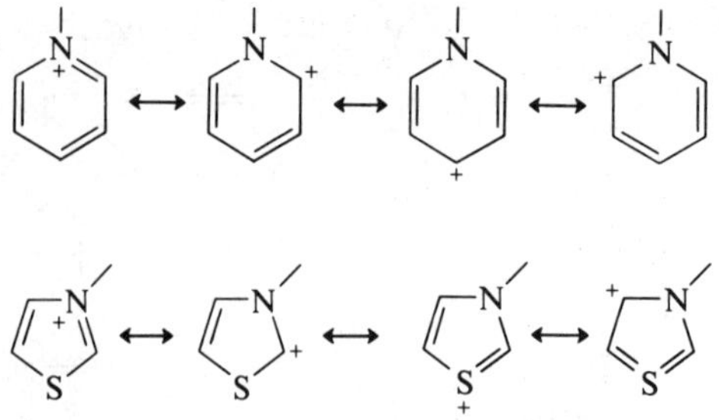

Scheme 12

Table 17 Physicochemical Comparative Data for the Electrophilic Reactivity of Nitrogen of Pyridine and Thiazole ⟨79HC(34-1)1, p. 125⟩

	Thiazole	Pyridine
pK_a	2.52	5.27
$\Delta H(BF_3)^a$	123.93	139.32
$k_2(MeI)^b$	15.8×10^{-6}	35.0×10^{-6}

^{a} Enthalpy of reaction with BF_3 in $PhNO_2$ at 25 °C, in J mol^{-1}.
^{b} Specific rate of methylation by MeI in acetone at 40 °C, in l mol^{-1} s^{-1}.

(ii) *Basicity of thiazoles*

Table 18 shows the pK_a values of some representative thiazoles and the free enthalpy of dissociation of their conjugate acids. Alkyl groups are weakly base-strengthening due to their +I effect, which decreases in the order 2-, 4-, 5-alkyl, in agreement with the net π-charge on nitrogen calculated by various theoretical methods ⟨79HC(34-1)1, p. 40⟩. A conjugated +M 2-amino group markedly enhances the basicity while the introduction into the 5-position of a strongly −M nitro group results in a large decrease of the pK_a. A fused benzene ring has little effect on the basicity of the nitrogen atom. The variation of pK_a in both 2- and 4-alkyl-substituted thiazoles after the α-series of Ingold (Me, Et, Pr^i, Bu^t) shows that both a +I activating effect and a steric deactivating *ortho* effect have to be taken into account.

Table 18 pK_a and Free Enthalpy of Dissociation of Some Representative Thiazoles ⟨79HC(34-1)1, p. 92, 79HC(34-2)1, p. 18⟩

Thiazole	pK_a	$\Delta G°$ (kJ mol^{-1})	Thiazole	pK_a	$\Delta G°$ (kJ mol^{-1})
Thiazole	2.52	14.55	4-Ethylthiazole	3.20	—
2-Methylthiazole	3.43	19.40	4-t-Butylthiazole	3.06	—
4-Methylthiazole	3.15	18.02	Benzothiazole	1.2	—
5-Methylthiazole	3.12	17.26	2-Aminothiazole	5.39	—
2-Ethylthiazole	3.37	—	2-Acetylamino-4-methylthiazole	3.75	—
2-i-Propylthiazole	3.28	—	2-Amino-5-nitrothiazole	0.93	—
2-t-Butylthiazole	3.15	—	4-Methyl-2-carboxythiazole	1.20	—

Water-insoluble, crystallizable thiazole salts have long been used for characterization purposes. For example, the picrate of thiazole has a melting point of 159 °C, that of 2-methylthiazole melts at 153 °C and that of 2-benzylbenzothiazole at 139 °C.

(iii) *Metal ions*

Like pyridines, thiazoles form complex salts by addition to aqueous solutions of metal salts and these complexes have long been used for the identification of the heterocycle. The most common complexes utilized are platinum double salts such as, for example, (4-methylthiazole·HCl)$_2$ PtCl$_4$ (m.p. dec. 204 °C).

Thiazole, benzothiazole and their alkyl homologues (L) form well-characterized tetrahedral complexes ML_2X_2 with Zn(II) and Co(II) salts and tetrahedral or octahedral complexes with Ni(II) and Cu(II) salts; Pt(III) complexes are square-planar.

Chelate complexes of the same metal salts are formed with 2-(2-pyridyl)thiazole, 4-(2-pyridyl)thiazole and 2,4-bis(2-pyridyl)thiazole. All these complexes are invariably metal–nitrogen coordinated.

(iv) *Alkyl halides and related compounds*

Thiazoles readily undergo *N*-alkylation by alkyl halides or tosylates (Menshutkin reaction). The sensitivity of this S_N2 quaternization reaction to the molecular environment of the nitrogen atom has been used to evaluate, in a quantitative way, steric and electronic effects of ring substituents. The electronic effect of alkyl substituents (unperturbed by any steric effect) may be evaluated from the rate constants for the reaction of 5-alkylthiazoles with methyl iodide (in nitrobenzene at 30 °C): Table 19 shows that introducing a methyl group at the 5-position corresponds to an acceleration of the rate by a factor of 2 but that each addition of a methyl ramification to the 5-alkyl group enhances the rate only by a factor of 1.1. The data in Table 19 fit well with a Hammett–Taft equation (3):

$$\log \frac{k}{k_{Me}} = -0.5\sigma^* \tag{3}$$

which confirms the inductive nature of the electronic influence of 5-alkyl groups on the rate of reaction.

Table 19 Rate Constants for the Reaction of 2- and 4-Alkylthiazoles with MeI in PhNO$_2$ at 25 °C (l mol^{-1} s^{-1})

R	2-Alkylthiazoles			4-Alkylthiazoles		
	$k \times 10^6$	$\log k/k_{Me}$	k/k_{Me}	$k \times 10^6$	$\log k/k_{Me}$	k/k_{Me}
Me	11.83	0	1	9.78	0	1
Et	9.25	−0.108	0.78	7.85	−0.096	0.80
Pr^i	4.32	−0.436	0.37	2.58	−0.582	0.26
Bu^t	0.29	−1.611	0.02$_4$	0.12	−1.910	0.01$_2$

The same alkyl substituents introduced at the 2- and 4-positions, *i.e. ortho* to the nitrogen atom, induce both inductive acceleration and steric retardation of the reaction.

The observed deceleration (Table 20) is then essentially attributed to steric interaction in the transition state between the 2- or 4-alkyl group and the entering methylating agent. A good correlation between $\log k/k_{Me}$ and the Taft parameter E_s has been found for both series (equation 4).

$$\log \frac{k}{k_{Me}} = \delta E_s \tag{4}$$

Table 20 Rate Constants for the Reaction of 5-Alkylthiazoles with MeI in PhNO$_2$ at 30 °C (l mol^{-1} s^{-1})

R	$k \times 10^6$	$\log k/k_{Me}$	k/k_{Me}
H	21.95	−0.286	0.52
Me	42.48	0	1
Et	46.41	0.039	1.09
Pr^i	52.23	0.095	1.24
Bu^t	55.17	0.117	1.31

The value of δ is 0.96 for 2-alkylthiazoles and 1.14 for 4-alkylthiazoles, whereas for 2-alkylpyridines it is 2.03. These interesting results show that although 2- and 4-alkyl substituents contribute similarly to the basicity of the thiazole ring, the dissymmetric structure of the ring makes the 4-position more sensitive than the 2-position to steric effects, the bond angle R(2)—C(2)—N (123.6°) being greater than that of R(4)—C(4)—N (119.4°). The 2-alkylpyridines, with R(2)—C(2)—N still smaller (115°), are twice as sensitive to steric effects as their thiazole analogues.

The results of kinetic studies of the quaternization of 4-alkyl-2,5-dimethylthiazoles do not give the same type of linear correlations with E_s as 2- and 4-monoalkyl- or 2,4-dialkylthiazoles: 4-isopropyl-2,5-dimethylthiazole, for example, reacts 6.5 times slower than predicted by a linear correlation. These results can be explained by the existence, for 4-isopropyl, of preferred conformations, induced by the presence of the 5-methyl group which apparently makes it as bulky as a *t*-butyl group and slows down the rate of reaction. This result leads to a limitation in the use of Taft's E_s parameter to cases where the environment of a substituent does not induce preferential conformation for the latter ⟨73JA3807⟩.

Because of its dissymetrical geometry around the nucleophilic nitrogen centre, thiazole affords a unique model for the study of the steric implications of S_N2 reactions. As we have seen, in the case of the Menshutkin quaternization reaction, both the 2- and 4-positions allow the possibility of introducing, near the electrophilic carbon centre, various substituents whose electronic and steric effects could be finely adjusted. Thus it was possible to obtain experimental evidence for the geometrical variations in the transition state of S_N2 reactions, induced by changing the leaving group. From a comparative study of quaternization rates of *ortho*-substituted thiazoles and pyridines with various methylating agents (MeI, MeSO$_3$F, MeOTs) it was established experimentally that an increase in the basicity of the leaving group moves the transition state of this S_N2 reaction closer to the products ⟨76JA1260⟩. Annulation of thiazole to a benzene ring leads to retardation of the rate by a factor of 5 to 10.

(v) *Peroxy acids*

Thiazole *N*-oxides may result from the direct oxidation of thiazoles with hydrogen peroxide or with tungstic acid or peracetic acid. These products are unstable and break down by autoxidation: 2,4-dimethylthiazole 3-oxide rearranges in acetic anhydride into 2-acetoxy-4-methylthiazole and 4-acetoxymethyl-2-methylthiazole. The same products result in low yields from the rearrangement, under the same conditions, of 4-methylthiazole 3-oxide ⟨64CJC2375⟩. 2-Methylbenzothiazole affords the *N*-oxide, in low yield, by treatment with monoperoxymaleic or trifluoroperacetic acids. *N*-Oxide formation from 6-substituted 2-methylbenzothiazoles (Scheme 13) is favored by electron-releasing substituents (Me, OMe) and inhibited by electron-withdrawing substituents (Cl, NO$_2$). The assigned *N*-, rather than *S*-, oxide structure of the products agrees with their UV and IR spectra, and is confirmed by a comparison of their calculated and observed dipole moments ⟨69CPB1598⟩.

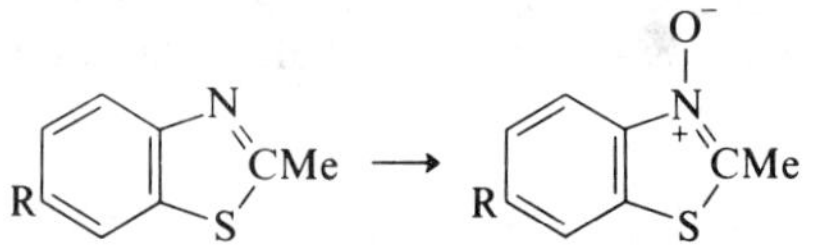

Scheme 13

(vi) *Other Lewis acids*

Benzothiazole and 2-methylbenzothiazole form with diborane a 1:1 adduct which rearranges into *N*-substituted benzothiazaborole (Scheme 14) ⟨78CC971⟩.

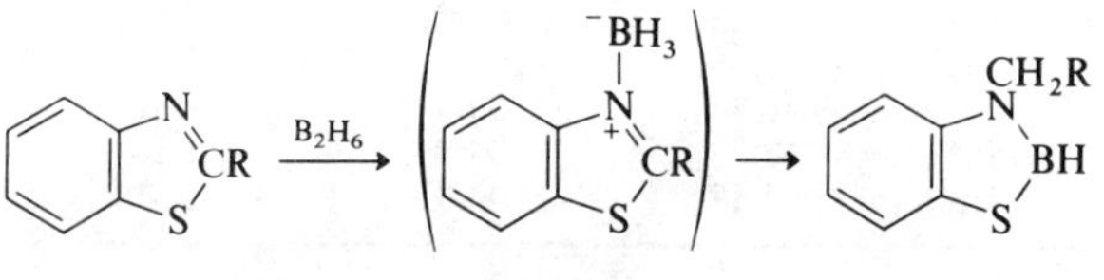

Scheme 14

4.19.2.1.4 Electrophilic attack at carbon

(i) *Reactivity and orientation*

Despite its 'π-excessive character', thiazole, just as with pyridine, is resistant to electrophilic substitution, the ring nitrogen atom deactivating the heterocyclic nucleus towards electrophilic attack. Moreover, most electrophilic substitutions which proceed in strongly acid media (nitration, sulfonation, Friedel–Crafts reactions) involve the protonated form of thiazole or some quaternary thiazole derivative whose reactivity toward electrophiles is still lower than that of the free base. Thiazole derivatives undergo electrophilic substitution only when the ring is activated by electron-releasing groups (essentially hydroxy and amino groups). Electrophilic substitution takes place at the 5-position if it is unsubstituted, or at the 4-position if the latter is already occupied. This orientation in electrophilic substitution is consistent with the calculated net π-charges of the three carbon atoms of the ring ⟨48BSF1021⟩ and the electrophilic localization energies of the three positions ⟨61CCC156⟩. In the case of benzothiazole no electrophilic substitution takes place at the 2-position, although the benzene ring may react as will be discussed in Section 4.19.2.3.2.

(ii) *Nitration*

No nitration of thiazole occurs with the classical nitration reagents, even under forcing conditions. A correlation between the ability of thiazole derivatives to be nitrated and the ^{1}H NMR chemical shifts of their hydrogen atoms shows that only those thiazoles that present ^{1}H chemical shifts lower than 476 Hz can be nitrated. From the lowest field signal of thiazole (497 Hz) one can infer that its nitration is quite unlikely ⟨66BSF2395⟩. Competition experiments have allowed the determination of orientation and reactivity relative to benzene in the nitration of alkylthiazoles (70 °C, sulfonitric acid). As can be seen from Tables 21 and 22, the 5-position of thiazole is two to three times more reactive than the 4-position and alkylation of the 2-position enhances the rate of nitration by a factor of 15 in the 5-position and of 8 to 14 in the 4-position. 2-Aminothiazoles and their *N*- or 4-substituted derivatives undergo nitration predominantly in the 5-position. A detailed kinetic study of nitration of thiazoles shows that 2,4- and 2,5-dialkylthiazoles are nitrated *via* their conjugate acids at all acidities and that the 5-position is more reactive by a factor of about two (Table 23). Despite the deactivating influence of the 2-thiazolyl substituent, the nitration of 2- and 3-(2-thiazolyl)thiophene occurs exclusively in the vacant α-position of the thiophene ring which is more reactive than the 5-position of thiazole.

Table 21 Orientation and Reactivity Relative to Benzene in the Nitration of Alkylthiazoles at 70 °C by Sulfonitric Acid ⟨68CR(C)(266)714⟩

Thiazole	Position	Reactivity ($\times 10^3$)	Thiazole	Position	Reactivity ($\times 10^3$)
4-Me	5	9	2-Me-5-Et	4	59
5-Me	4	6	2-Et-5-Me	4	63
2,4-DiMe	5	150	2-Bu^t-5-Me	4	83
2,5-DiMe	4	46			

Table 22 Orientation in the Nitration of 2-Substituted Thiazoles

Thiazole	4-NO_2 (%)	5-NO_2 (%)	Ref.
2-Me	31	69	68JOC2544
2-*n*-Pr	29	71	69JHC575
2-MeO	25	75	69CR(C)(269)1560

Table 23 Standard Rate Constants for the Nitration of Thiazoles ⟨75JCS(P2)1614⟩

Thiazole	Position	log $k_2^{\circ\,a}$	Thiazole	Position	log $k_2^{\circ\,a}$
2,4-Dimethyl	5	−6.90	5-Ethyl-2-*t*-butyl	4	−6.96
2,5-Dimethyl	4	−7.60	5-*i*-Propyl-2-*t*-butyl	4	−7.43
2,3,4-Trimethyl	5	−7.52	2-Methoxy-4-methyl	5	−4.62
2,3,5-Trimethyl	4	−7.72			

^{a} Calculated at $H_0 = 6.60$ and 25 °C.

(iii) *Sulfonation*

Thiazole sulfonation occurs only under forcing conditions: the action of oleum at 250 °C for 3 hours in the presence of mercury(II) sulfate leads to 65% formation of 5-thiazole sulfonic acid. 4-Methyl- and 2,4-dimethyl-thiazoles undergo sulfonation at the 5-position under equally forcing conditions (oleum, 200 °C). 2-Aminothiazole is sulfonated at low temperature (0 °C) affording at first the 2-sulfamic acid which on heating rearranges to 2-amino-5-sulfonic acid (Scheme 15).

(iv) Acid-catalyzed hydrogen exchange

Thiazole heated at 118 °C for 8 days with SO_4D_2 exchanges at all three positions but the total deuterium content is only 25%.

(v) Halogenation

Unsubstituted thiazole does not react with chlorine or bromine in an inert solvent. Of the monomethylthiazoles only the 2-isomer undergoes bromination giving the 5-bromo derivative. When the 5-position is not free, as in 2,5-dimethylthiazole, no reaction occurs. 2-Hydroxy and 2-amino groups strongly activate the 5-position. However, bromination of 2-amino-4-(2-furyl)thiazole under mild conditions occurs successively on the furan and thiazole rings (Scheme 16).

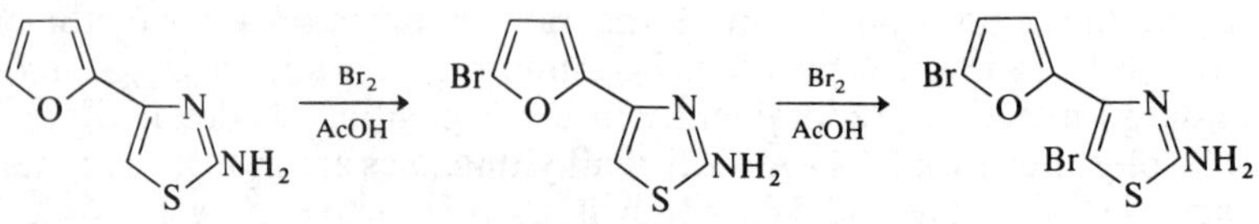

Scheme 16

(vi) Acylation and alkylation

Under appropriate conditions activated thiazoles are alkylated at the 5-position: 2-amino-4-methylthiazole is alkylated in the 5-position by heating with t-butyl alcohol in sulfuric acid (**24**). Under similar conditions 4-methyl-2-phenylthiazole is alkylated by cyclohexanol. 2-Acetylamino-4-methylthiazole reacts with dimethylamine and formaldehyde to afford the corresponding Mannich base (**25**). 2-Hydroxy-4-methylthiazole fails to react when submitted to Friedel–Crafts benzoylation conditions whereas it reacts normally in Gattermann and in Reimer–Tiemann formylation reactions yielding the 5-formyl derivative (**26**). 2,4-Dimethylthiazole undergoes perfluoroalkylation when heated at 200 °C for 8 h in a sealed tube with perfluoropropyl iodide and sodium acetate (**27**).

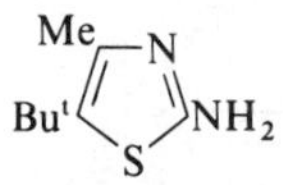

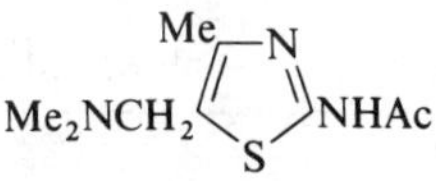

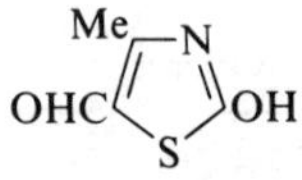

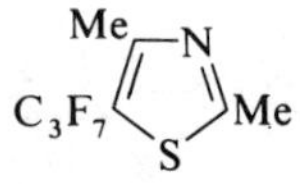

(vii) Mercuration

Thiazole reacts on heating with mercury(II) acetate in aqueous acetic acid yielding 2,4,5-tris(acetoxymercurio)thiazole (**28**), but the rate of the reaction decreases in the order C-5 > C-4 > C-2. 2-Methyl-, 4-methyl- and 2,4-dimethyl- and 4-methyl-2-phenyl-thiazole afford, under the same conditions, the 5-acetoxymercurio derivatives. In the case of 2,4-diphenylthiazole, mercuration at room temperature occurs in the 5-position giving (**29**) whereas at 60–70 °C both the 5-position and the para-position of the 2-phenyl ring are mercurated (**30**). The mercurio derivatives are reconverted into the starting material by hydrochloric acid and in the corresponding 5-bromo analogue by bromine in carbon tetrachloride. Mercuration may thus be used as an extractive technique to separate thiazole from complex mixtures such as crude petroleum.

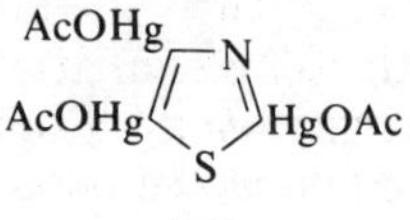

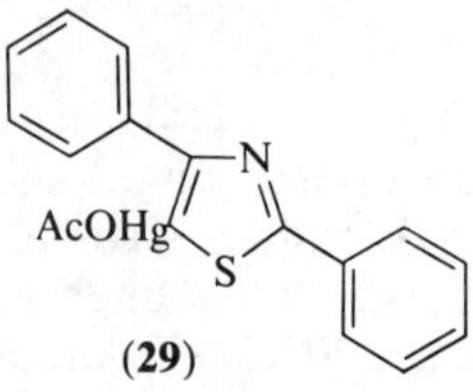

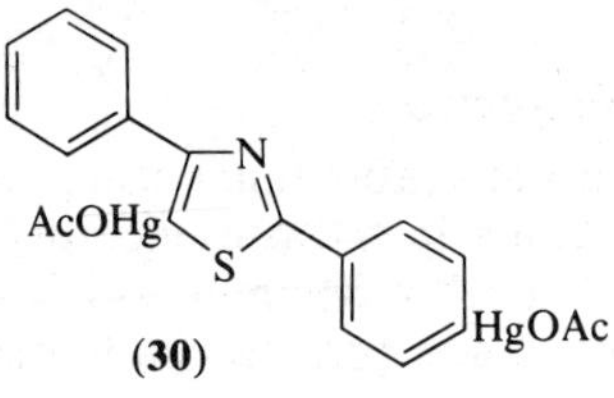

(viii) *Diazo coupling*

Only 2-aminothiazole derivatives are reactive enough toward diazonium salts to undergo the diazo coupling reaction. The azo group substitutes exclusively at the 5-position when the latter is unsubstituted (**31**).

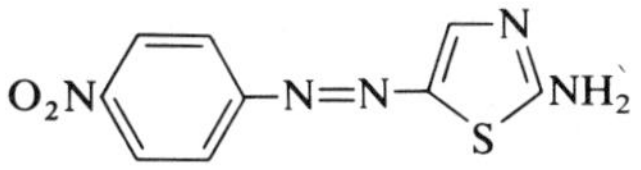

$$O_2N-C_6H_4-N=N-\text{[thiazole-5-yl]}-NH_2$$

(**31**)

(ix) *Oxidation*

Thiazole is relatively resistant to oxidation. Under photosensitized oxygenation, triphenylthiazole affords various products of ring cleavage, depending on the sensitizer and the solvent employed (Scheme 17) ⟨69BCJ2973⟩.

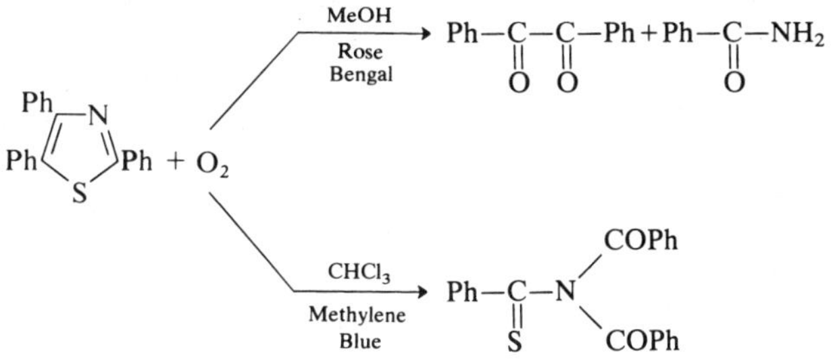

Scheme 17

4.19.2.1.5 Attack at sulfur

Active Raney nickel induces desulfurization of thiazoles. The reaction apparently occurs by two competing mechanisms ⟨57JCS1652⟩. In the first, under alkaline conditions, ring cleavage occurs before desulfurization, whereas in the second, using a neutral catalyst, the initial desulfurization is followed by fission of the C—N bond and formation of carbonyl derivatives (Scheme 18). Because of the easy and versatile syntheses of thiazoles, this reaction could have interesting synthetic applications.

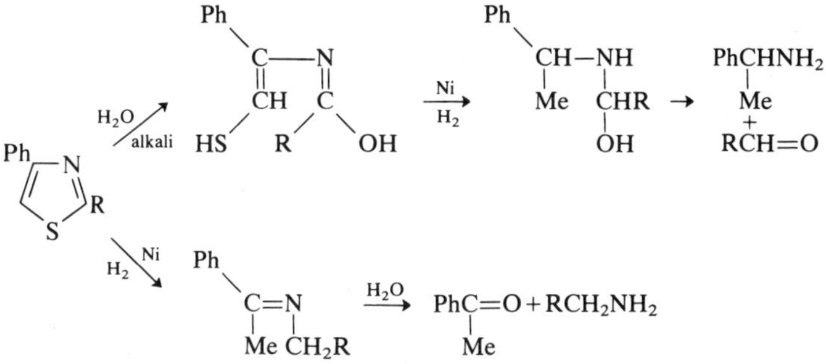

Scheme 18

4.19.2.1.6 Nucleophilic attack at carbon

According to electronic theory and to MO calculations (see Section 4.19.1.1), the carbon atom of the neutral thiazole ring that is the most sensitive to nucleophilic attack must be C-2 for which the π-electron density is lowest and the nucleophilic polarization energy a minimum. The same prediction is true for benzothiazole. However the reaction of nucleophiles is not observed unless strong electron-withdrawing substituents activate the C-2 position or very strong nucleophiles are employed. Quaternization of the ring nitrogen considerably enhances this reactivity and allows the attack at C-2 by mild nucleophiles. Simultaneously the acidity of H-2 is greatly increased making the thiazolylium salts very reactive species in a variety of condensation reactions.

(i) Hydroxide ion and other O-nucleophiles

(a) *Neutral thiazoles*. Thiazole and benzothiazole are resistant to attack by hydroxide and alkoxide ions. However 5- or 6-nitrobenzothiazole adds methoxide ion at C-2 yielding an unstable anionic adduct which undergoes ring opening (Scheme 19) to an *o*-aminothiophenol derivative.

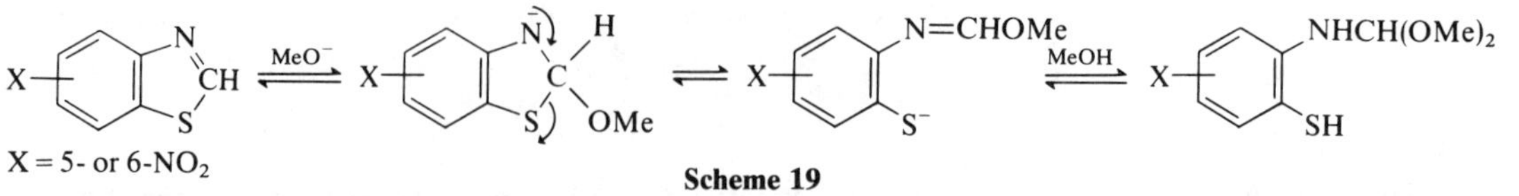

X = 5- or 6-NO$_2$

Scheme 19

(b) *The thiazolylium ring*. 2-Unsubstituted or 2-aryl-*N*-methyl-thiazolylium and benzothiazolylium cations react reversibly with water yielding pseudobases or carbinolamines (**32**; Scheme 20). The solution shows an acidic pH and this hydrolysis develops by progressive addition of sodium hydroxide, and leads to the consumption of two equivalents of the base for complete neutralization. This one step titration of thiazolylium salts corresponds to a midpoint in the vicinity of pH 9–11 at room temperature which is usually designated as pK_{av}. The second equivalent of base is consumed for the neutralization of the thiol function which results from the ring opening of the pseudobase (**32**). The accepted mechanism (Scheme 21) involves a triple equilibrium: formation of the pseudobase (K_{R^+}), ring–chain tautomerism of the pseudobase (K_T), and transformation of the open-chain pseudobase into its conjugate base (K_R). At the midpoint of the titration (pH = pK_{av}), equal concentrations of the thiazolylium cation (**33**) and the ring-opened thiolate anion (**36**) are present in equilibrium with only minor amounts of the pseudobase (**34**) and neutral thiol (**35**). Values of pK_{R^+} are in the range 10–12 for thiazolylium cations and 8–9 for benzothiazolylium cations ⟨79AHC(25)46⟩. Kinetic studies of this reversible ring-opening of thiazolylium salts have shown that in dilute solutions the rate-determining step for ring opening is the formation of the pseudobase (**34**). The presence at C-2 of a substituent slows the rate of ring opening by steric effect: a factor of 100 is observed for the introduction of a 2-methyl group.

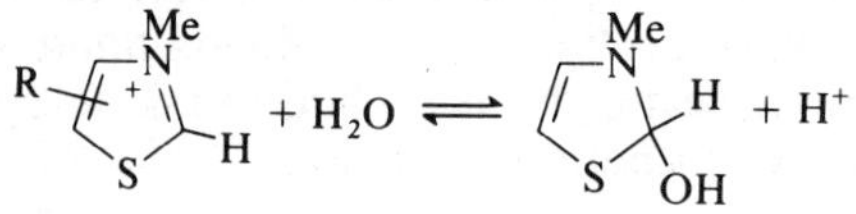

(**32**)

Scheme 20

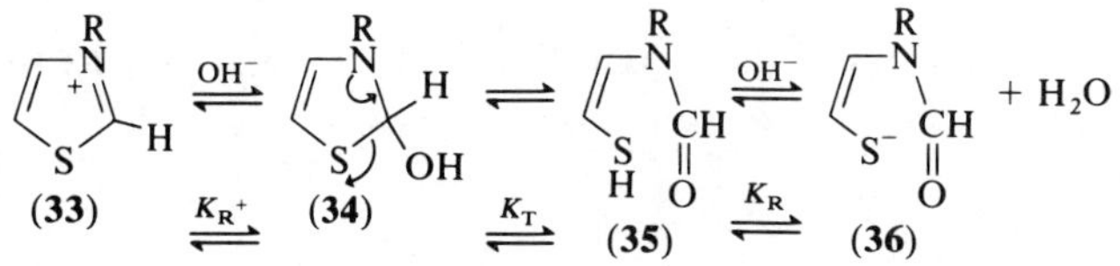

(**33**) (**34**) (**35**) (**36**)

Scheme 21

When the *O*-nucleophile is an alcohol, attack on thiazolylium and benzothiazolylium cations yields an adduct very similar to a pseudobase but which resists further attack by alcoholate base (Scheme 22): a 10^{-4} molar solution of 3-methylbenzothiazolylium salt (**37**; R = Me, R' = H, R" = benzo) in methanol or ethanol contains, in equilibrium with the quaternary salt, 8% and 54% respectively of the benzothiazolinic ether (**38**; R = Me, R' = H, R" = benzo, R''' = Me or Et). When the 2-substituent R' is a primary or a secondary alkyl group, some other very important reactions occur which are discussed in Section 4.19.2.3.3(i)(d)

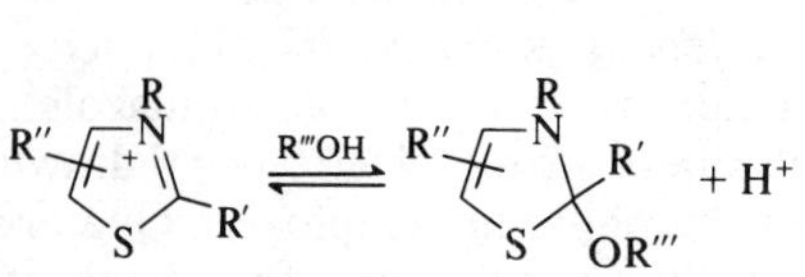

(**37**) (**38**)

Scheme 22

(ii) Amines and amide ions

(a) *Neutral thiazoles.* Sodium amide, a very strong nucleophile, reacts at 150 °C with 4-methylthiazole or benzothiazole, yielding the corresponding 2-amino derivative. In a similar way benzothiazole and its 6-nitro derivative react with hydroxylamine in aqueous sodium hydroxide, affording the corresponding 2-amino derivative (Scheme 23).

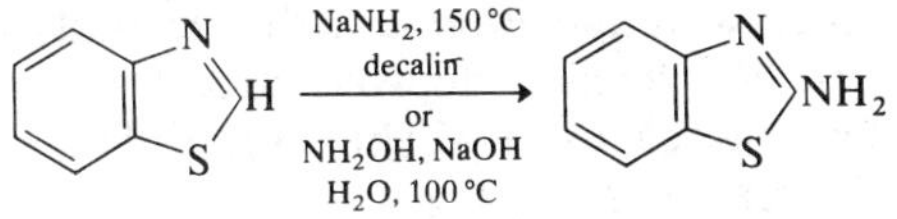

Scheme 23

(b) *The thiazolylium ring.* Thiazolylium salts are sufficiently activated to react smoothly with amines affording the corresponding adducts at C-2 (Scheme 24).

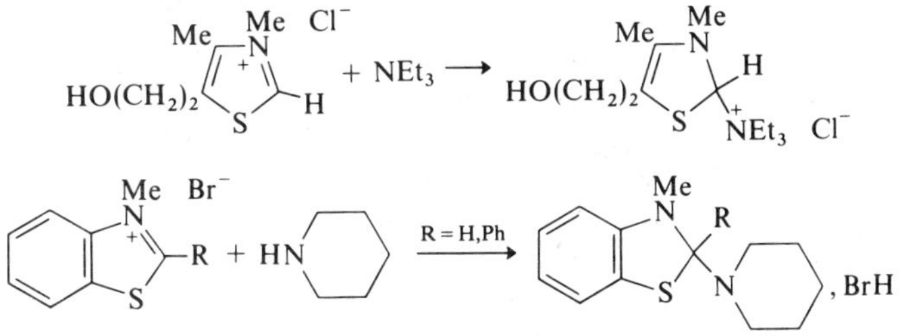

Scheme 24

(iii) Carbanions

Organolithium and organomagnesium reagents attack thiazoles and benzothiazoles as strong bases, abstracting ring hydrogen atoms (see Section 4.19.2.1.7(i)). Phenylmagnesium bromide reacts with 3-methylbenzothiazolylium iodide, under nitrogen, to give 3-methyl-2-phenylbenzothiazoline (**39**) which, in the presence of air, is oxidized to the disulfide (**40**; Scheme 25). In contrast, 3-methyl-2-phenylbenzothiazolylium iodide and phenyl magnesium bromide give only 2,2-diphenylbenzothiazoline (**41**) in nitrogen or in air (Scheme 26). 3-Methylbenzothiazolylium halides react with phenacyl chloride in the presence of a base affording the expected C-2 adduct (**42**) which rearranges into 2-benzoyl-4-methyl-4-*H*-1,4-benzothiazine (**43**; Scheme 27) ⟨75JHC1005⟩. Chloracetone reacts similarly.

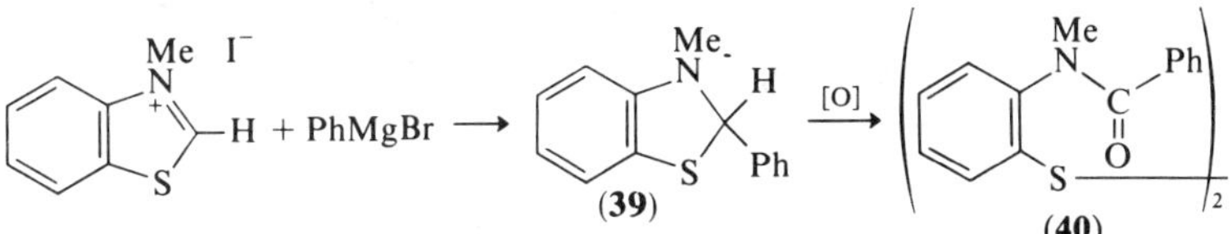

Scheme 25

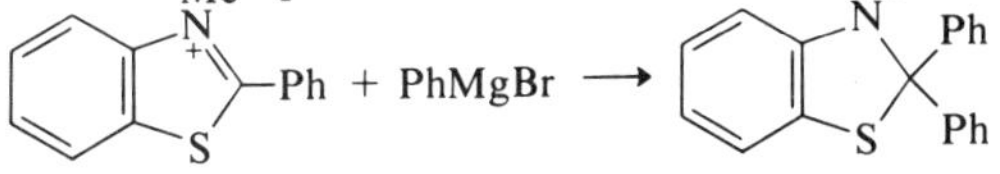

Scheme 26

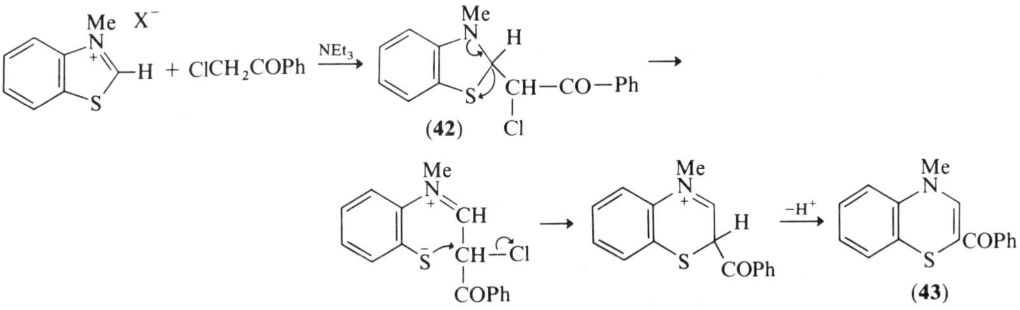

Scheme 27

Anhydro bases resulting from the proton abstraction by a base at an activated α-methyl group of a quaternary salt (see Section 4.19.2.3.3(iv)(a)) are active *C*-nucleophiles. These attack the C-2 position of a thiazolium salt affording adducts whose further reaction may lead to thiacyanines. Scheme 28 summarizes the successive steps in the reaction resulting from the addition of sodium ethoxide to a fairly concentrated ethanolic solution of 2,3-dimethylbenzothiazolylium salt (**45**) ($c = 0.1$ mol l^{-1}). The initially formed anhydro base (**46**) cannot be isolated, it reacts as a nucleophile with a second molecule of benzothiazolylium salt yielding an adduct (**47**) which is deprotonated by ethoxide anion affording the 'dimeric anhydro base' (**48**) whose reactivity will be discussed later (see Section 4.19.2.3.3.i). Monocyclic thiazolylium salts react similarly.

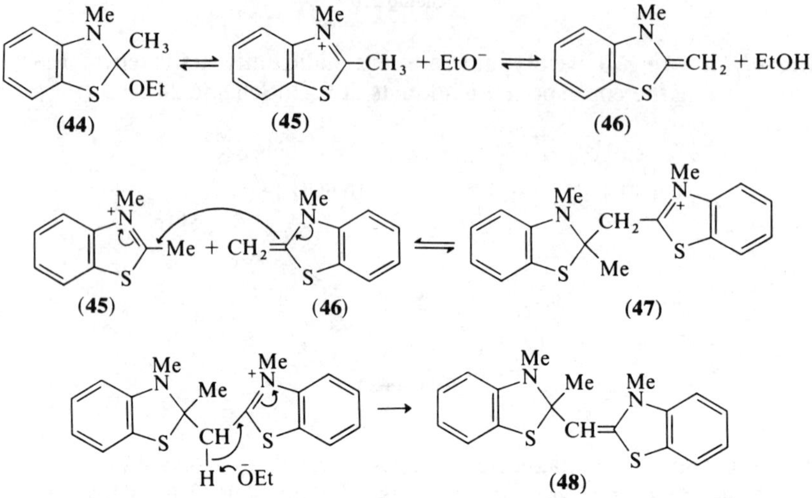

Scheme 28

(iv) Reduction with complex hydrides

Reduction of benzothiazole with lithium aluminum hydride produces *o*-(methylamino)thiophenol. Applied to 2,2'-bibenzothiazole the method affords the stable diamine (**49**; Scheme 29).

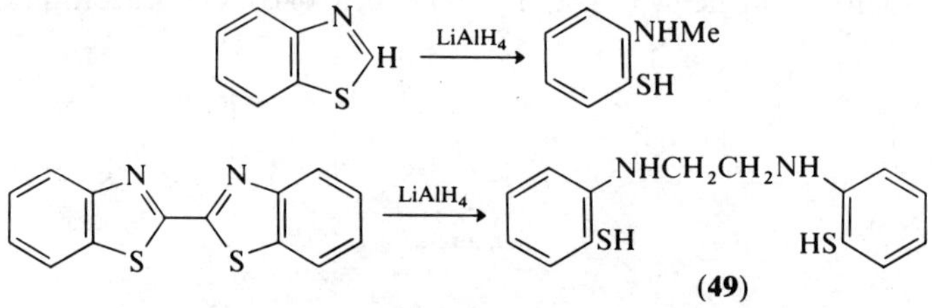

Scheme 29

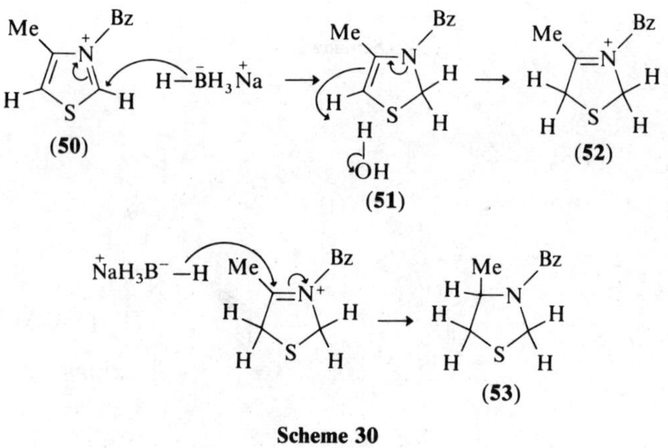

Scheme 30

Thiazolylium and benzothiazolylium salts react normally with aqueous sodium borohydride yielding the corresponding thiazolidine or benzothiazoline. The mechanism and the stereochemistry of the reaction have been studied with thiazolylium salts chosen as models for thiamine, using borodeuteride/hydride and deuterium/protium oxide. The pathway described in Scheme 30 was suggested; it involves the addition of a nucleophilic hydride at C-2 (**50** → **51**), the addition of an electrophilic proton at C-5 (**51** → **52**) and the addition of a second nucleophilic hydride at C-4 (**52** → **53**).

4.19.2.1.7 Nucleophilic attack at hydrogen

(i) *Metallation at a ring carbon atom*

Thiazole and benzothiazole exchange the C-2 hydrogen for lithium or magnesium when treated with an ethereal solution of phenyl- or butyl-lithium at −60 °C or ethylmagnesium bromide at 0 °C then at 25 °C (Scheme 31). When the 2-position of thiazole is occupied by a methyl group, the reaction of butyl-lithium at low temperature (−100 °C) affords three independent lithio salts (**54**, **55** and **56**; Scheme 32) in the approximate ratio 52:3:45. As the temperature is increased, the 2-lithiomethyl derivative (**54**), which is less stable than the 5-lithio isomer (**56**), decomposes up to +5 °C at which point it has almost entirely disappeared.

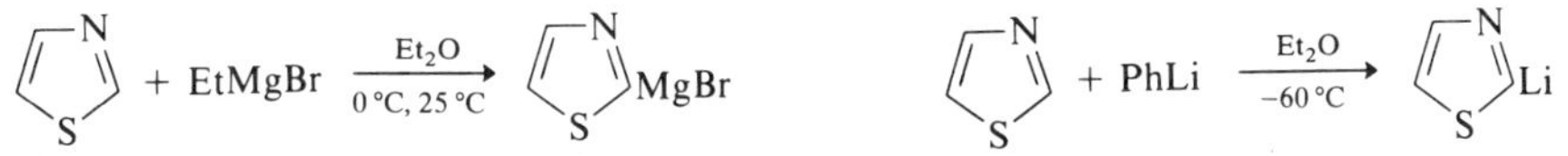

Scheme 31

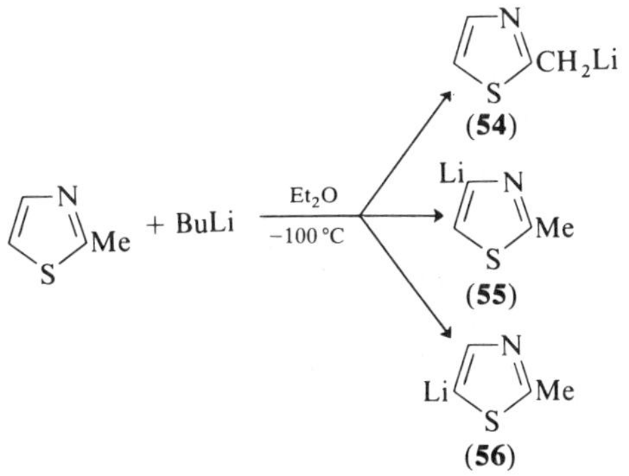

Scheme 32

(ii) *Hydrogen exchange at ring carbon in neutral thiazoles*

Over a wide range of pH (or pD) thiazole undergoes deuterodeprotonation by attack of DONa in D_2O or MeONa in MeOD ⟨74AHC(16)10⟩. At low values of pD (0 to 11) only the C-2 hydrogen exchanges and at a rate close to that observed for 3-ethylthiazolylium iodide. For higher pD (12 to 14) both C-2 and C-5 hydrogen atoms exchange and at a similar rate (in the pD interval, the ratio k_2/k_5 decreases from 12 to 1.1). The pD rate profile of exchange of both C-2 and C-5 protons is consistent with two competitive mechanisms. The first one (Scheme 33) operates essentially between pD 0 and pD 11. Its rate determining step corresponds to the ylide (**58**) formation by base abstraction of H-2 from the ND^+ conjugate acid of thiazole (**57**) which is formed in an acid–base preequilibrium step. This is followed by the *trans* deuteration of the ylide (**58**) (elimination–addition mechanism). In such acidic conditions, the conjugate acid of thiazole (**57**) exchanges only its H-2 proton to give exclusively 2-deuterothiazole (**59**), a situation quite similar to that observed for the *N*-ethylthiazolylium cation. The second mechanism, operating at higher pD, corresponds to a rate determining attack on H-2 and H-5 by the base (DO^- or RO^-), followed by a fast deuteration of the carbanion formed (**60a, b**) yielding D-2 or D-5 thiazole (**61a, b**; Scheme 34). Under the same conditions, benzothiazole exchanges its C-2 hydrogen atom at a rate that is three times slower than for thiazole. Phase transfer catalysis accelerates H/D exchange by NaOD in benzene for both thiazole and benzothiazole.

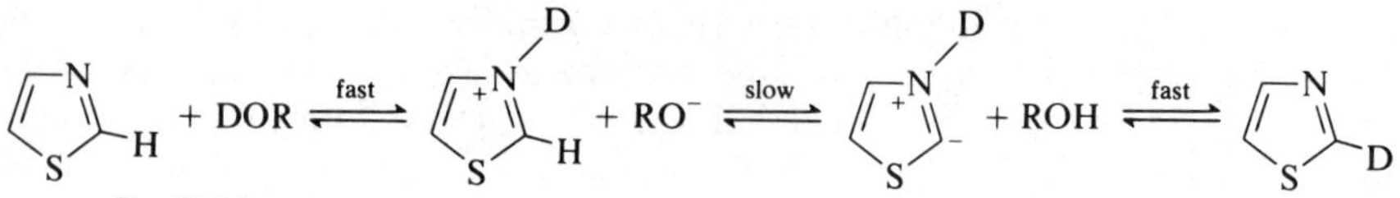

Scheme 33

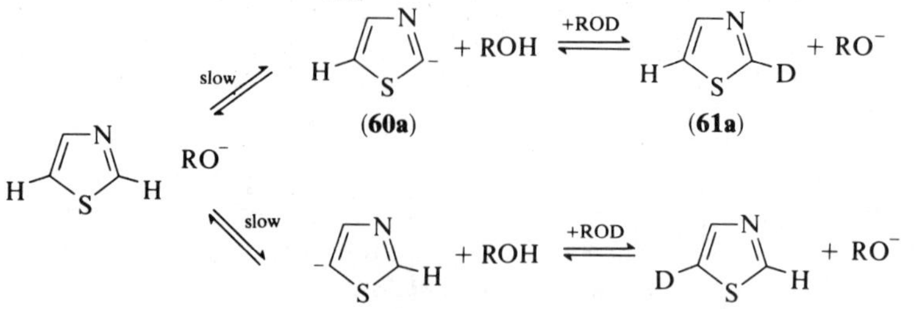

Scheme 34

(iii) Hydrogen exchange at ring carbon in thiazolylium cations

Hydrogen kinetic acidity of the thiazolylium ion has received much attention since the discovery by Breslow ⟨56CI(L)R28, 57CI(L)893⟩ and Ingraham and Westheimer ⟨56CI(L)846⟩ that the catalytic action of thiamine (**62**) results from the base-induced abstraction of the 2-hydrogen atom leading to the formation of the very reactive thiazolylium ylide (**63**; Scheme 35). Deuterium exchange takes place very readily at the C-2 position of thiazolylium quaternary salts with D_2O, even in the absence of any basic catalyst, whereas neither H-5 nor H-4 exchange. The postulated mechanism (Scheme 36) involves the abstraction of H-2 by the base followed by the deuteration of the ylide (**65**). The ease of formation of the 2-ylide results from the combined effect of the following factors: the high s-character of the C(2)—H σ-bond (as shown by its unusually high ^{13}C–H coupling constant); the withdrawing inductive effect of the heterocyclic cationic ring (analogous to HCN acidity); stabilization of the ylide (**65**) by a resonance contribution from the carbene-like structure (**65b**); d–σ overlap of the electron pair of the anion with an empty d-orbital of sulfur ⟨79HC(34-1)1, p. 113⟩. The ratio of the rates of deuterodeprotonation for the thiazolylium cation and neutral thiazole is of the order 10^7, accounting for the major influence of the positive charge on nitrogen on the proton's lability.

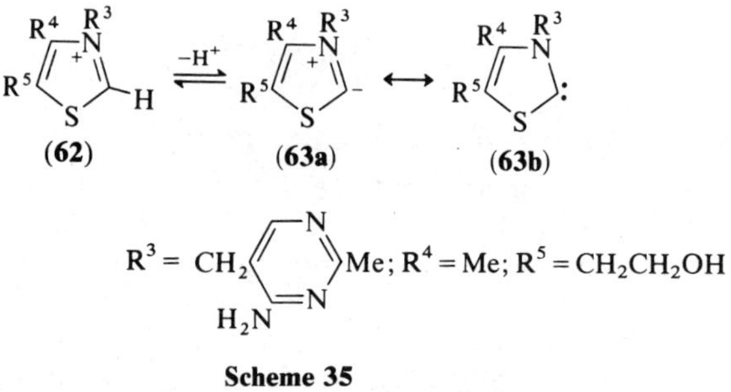

$R^3 = CH_2$![pyrimidine]Me; $R^4 = Me$; $R^5 = CH_2CH_2OH$

Scheme 35

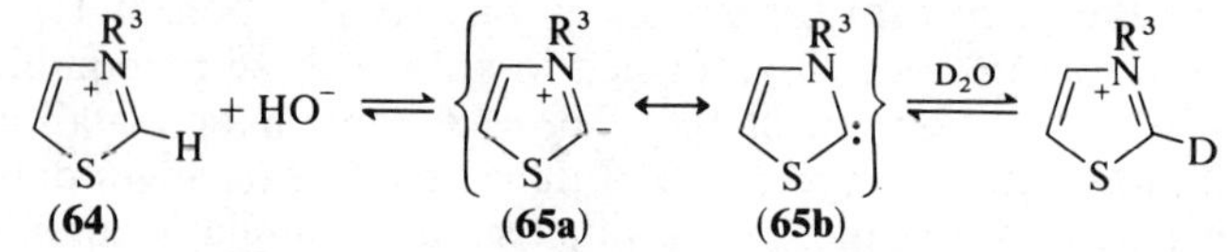

Scheme 36

In aprotic medium (anhydrous acetone at reflux) triethylamine deprotonates 3-alkylbenzothiazolylium salts yielding an intermediate carbene (**66**) which can form a dimer (**67**; Scheme 37), or be trapped giving adducts with tetracyanoethylene (**68**), azides (**69**), or giving insertion derivatives with benzaldehyde (**70**) or activated methylene compounds (**71**).

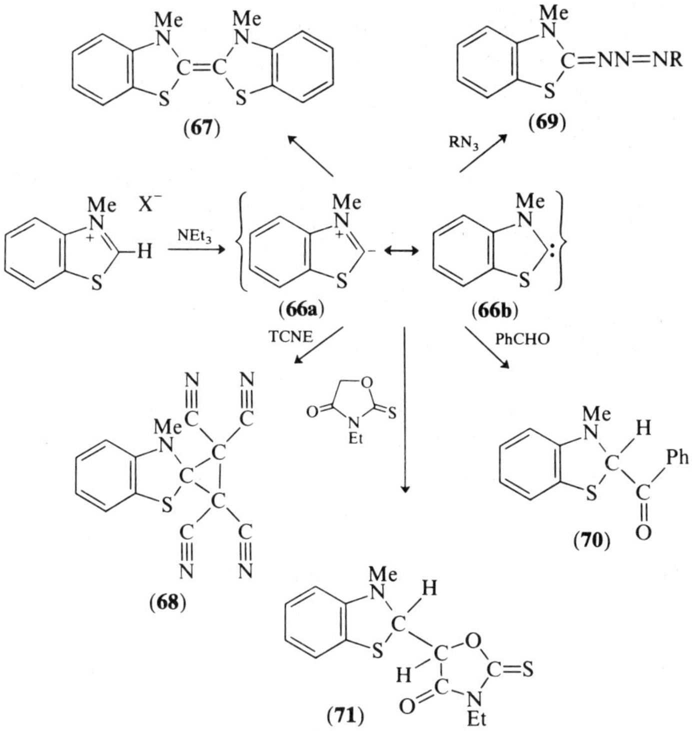

Scheme 37

(iv) *C-Acylation* via *deprotonation*

Attempts to acylate 2-unsubstituted thiazolylium salts with dialkylacylphosphonates under basic conditions yield an intermediate (**72**) which rearranges with ring expansion affording a 1,4-thiazine (**73**; Scheme 38). Under the same reaction conditions, 2-unsubstituted benzothiazolylium salts give 1,4-benzothiazines of type (**74**; Scheme 39).

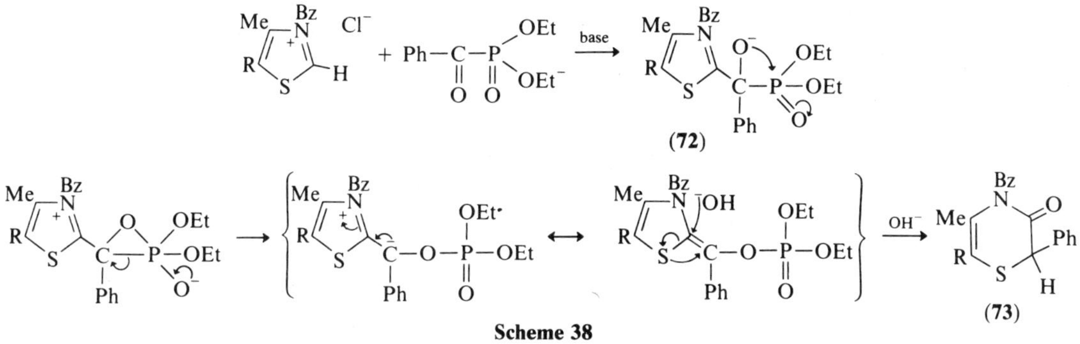

Scheme 38

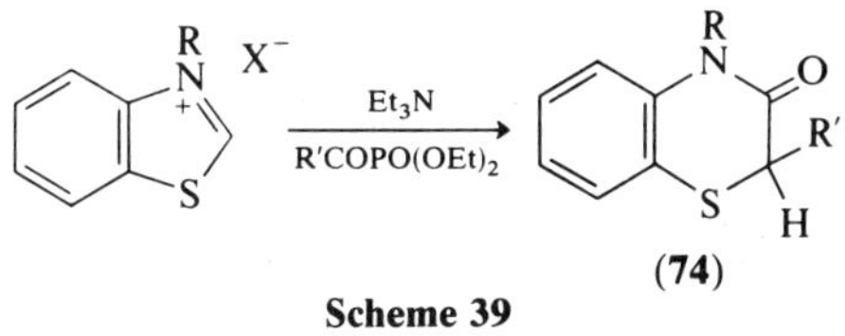

Scheme 39

4.19.2.1.8 Reactions with radicals and electron-deficient species: reactions at surfaces

(i) *Free radical attack at the ring carbon atoms*

(a) *Halogen atoms.* Vapor-phase bromination of thiazole at 250–400 °C on pumice gives 2-bromothiazole when equimolecular quantities of reactants are mixed and a low yield of 2,5-dibromothiazole when an excess of bromine is used. This preferential orientation to the 2-position is interpreted as an indication of the free-radical nature of the reaction, a conclusion that is in agreement with the free-valence indices of the three positions of thiazole which decrease in the order $2>5>4$. Bromine in chloroform solution produces no nuclear bromination but only perbromides are formed (Scheme 40).

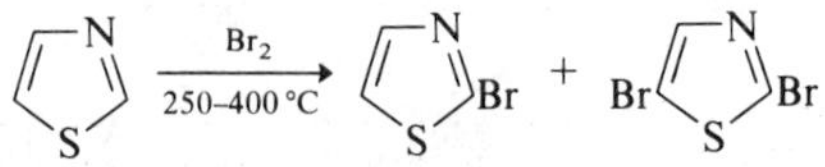

Scheme 40

Vapor-phase bromination of benzothiazole affords the 2-bromo derivative whereas attempts to perform the bromination by bromine in acetic acid solution yield only perbromides which on heating in alcoholic solution give substitution products containing bromine in the benzo ring. Chlorination in the 2-position results when benzothiazole is heated at 160 °C with phosphorus pentachloride in a sealed tube.

(b) *Aryl radicals.* Free radical arylation of thiazole has been performed either by the Gomberg–Bachmann decomposition of aryldiazonium chlorides, by the thermal decomposition of benzoyl peroxide or *N*-nitrosoacetanilide, or by the photolysis of benzyl peroxide or iodobenzene.

The three monophenylthiazoles are obtained in the practically constant ratio $2:5:4 \equiv 60:30:10\%$. Competing reactions with nitrobenzene give an estimation of the overall reactivity of thiazole relative to benzene of 0.75 with the partial rate factors $f_2 = 2.2 : f_5 = 1.9 : f_4 = 0.5$. When phenylation is performed in acidic medium (thermal decomposition of benzoyl peroxide in acetic acid solution), the substrate entering the reaction is the conjugate acid of thiazole. The overall reactivity is enhanced to 1.25 and the partial rate factors become $f_2 = 6 : f_5 = 1 : f_4 = 0.3$. These experimental observations are in agreement with the free valencies and radical localization energies calculated for thiazole and its conjugate acid using the HMO method (Scheme 41) ⟨61CCC156⟩. Alkyl substituents slightly activate the substrate toward free radical phenylation, but they also induce some steric hindrance to the approach of the aryl radical from the *ortho* positions (Scheme 42).

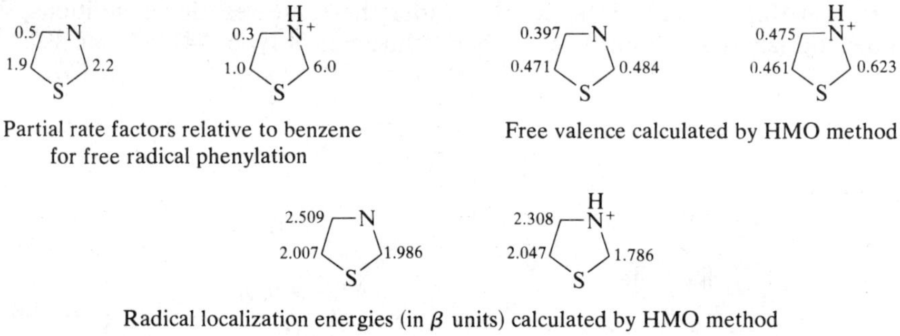

Partial rate factors relative to benzene for free radical phenylation

Free valence calculated by HMO method

Radical localization energies (in β units) calculated by HMO method

Scheme 41

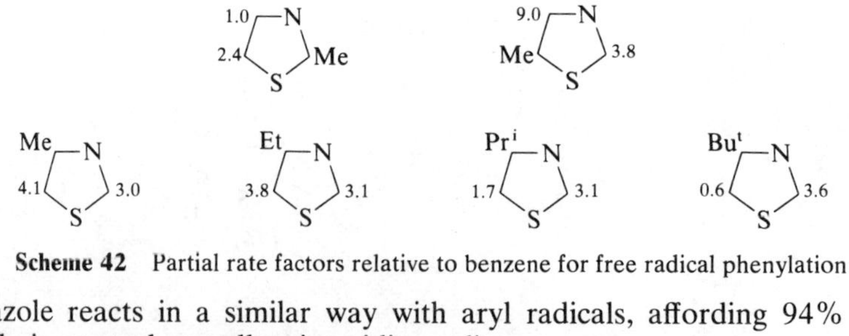

Scheme 42 Partial rate factors relative to benzene for free radical phenylation

Benzothiazole reacts in a similar way with aryl radicals, affording 94% of 2-phenylbenzothiazole in neutral as well as in acidic medium.

(c) *Alkyl and acyl radicals.* Methyl free-radicals, generated either by thermolysis of lead tetraacetate in acetic acid solution or by radical cleavage of DMSO by H_2O_2 and iron(II) salts, afford 2- and 5-methylthiazoles in the ratio 86:14 respectively, in agreement with the nucleophilic character of alkyl free radicals. Cyclohexyl radicals, generated by photolysis of *t*-butyl peroxide in excess cyclohexane, attack thiazole in positions 2,4,5 yielding the three monocyclohexylthiazoles in the ratio 73:14:13, respectively, showing an increase of the 2-isomer and a decrease of the 5-isomer, relative to phenylation reactions. This is in agreement with the electronic charges at both positions of the nucleus.

Similar reactions are observed in the thermolysis at 200 °C, in the presence of powdered copper, of N-benzylthiazolylium chloride. The main products formed are 2- and 5-benzylthiazole.

Heating thiazole with phenylazotriphenylmethane at 75 °C for 24 hours affords 2-phenyl-5-triphenylmethylthiazole (**75**; Scheme 43). The 1-adamantyl radical and other nucleophilic alkyl and acyl radicals react with 2-substituted benzothiazoles in a homolytic *ipso* substitution yielding the corresponding 2-alkyl or 2-acylbenzothiazole (Scheme 44) ⟨77CC316⟩.

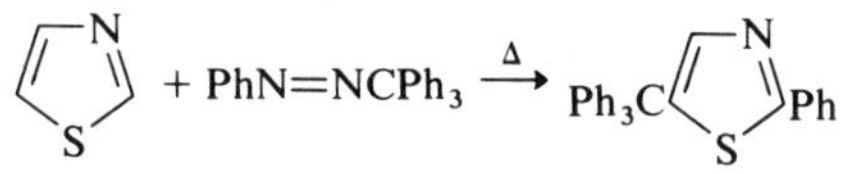

Scheme 43

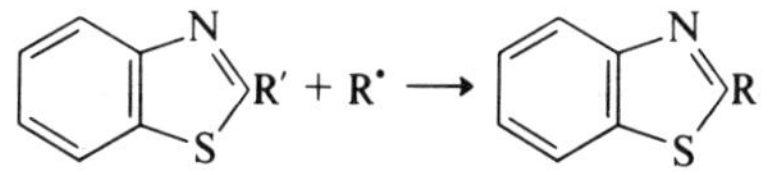

R′ = COMe, CO$_2$Et, Cl, F, SPh, SOPh, SO$_2$Ph; R˙ = alkyl, aryl

Scheme 44

Benzothiazole is acylated selectively in the 2-position by acyl radicals generated from a variety of aldehydes under the influence of the redox system *t*-butylhydroperoxide–ferrous sulfate. The nucleophilicity of acyl radicals is confirmed by the higher reactivity of 6-nitrobenzothiazole. The reaction, which could detect acyl radicals in the oxidation of aldehydes by various oxidizing agents, could serve as a diagnostic test for the presence of such radicals ⟨71JCS(C)1747⟩.

(ii) *Electrochemical reactions*

Neutral thiazole and its simple alkyl or aryl derivatives are not reduced polarographically but introduction in the ring of electron-attracting groups may cause the nucleus to undergo reduction ⟨70AHC(12)213⟩.

Many thiazolylium salts are reduced polarographically: 3-methylbenzothiazolylium iodide (**76**), for example, is reduced in a two-electron wave affording 3-methylbenzothiazoline (**77**) but, on a preparative scale, part of the intermediate radical dimerizes giving (**78**). Aryl groups in the 2-position hinder dimerization and the polarographic reduction of 2-aryl-3-methylbenzothiazolylium perchlorate (**79**) yields the corresponding benzothiazoline (**80**; Scheme 46).

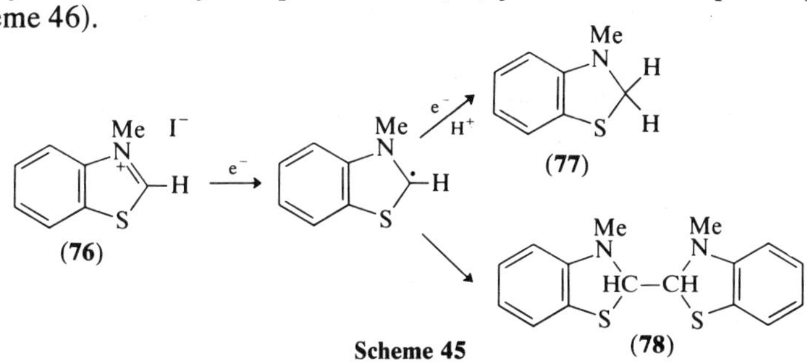

Scheme 45

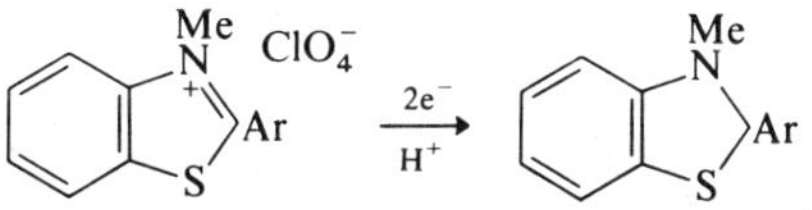

Scheme 46

The effect of substituents in the 2-aryl group on the reduction potential may be correlated satisfactorily with their Hammett σ-values ⟨69ZOB54⟩.

(iii) Thiazolyl radicals

Thiazol-2-yl radicals have been generated by silver oxide oxidation of thiazol-2-ylhydrazine in various aromatic solvents. The thermal decomposition of thiazol-2-ylcarbonyl peroxide in aromatic solvents affords thiazole with good yields of 2-arylthiazoles. Thiazol-4-ylcarbonyl peroxide gives fair yields of 4-arylthiazoles but thiazol-5-ylcarbonyl peroxide gives no product indicative of thiazol-5-yl radical formation ⟨70CJC3554⟩. The most general way of producing thiazol-2-yl radicals is the thermal decomposition, in a variety of aromatic solvents, at 0 °C and in the presence of alkali, of 2-thiazole diazonium salts resulting from the aprotic diazotization of the corresponding 2-aminothiazole by an alkyl nitrite, the yield of 2-arylthiazole obtained being around 40%. The strong electrophilic character of thiazol-2-yl radicals is shown by their ability to substitute at positions of high π-electron density of the aromatic substrate in which they are generated (Scheme 47) ⟨73JCS(P2)1093⟩.

Scheme 47 Partial rate factors for the free radical substitution of aromatic substrates: (phenylation)/thiazol-2-ylation

In agreement with the theory of polarized radicals, the presence of substituents on heteroatomic free radicals can slightly affect their polarity. Introduction in the 5-position of electron-withdrawing substituents slightly enhances the electrophilic character of thiazol-2-yl radicals (Table 24) ⟨74T4171⟩.

Table 24 Reactivity of 5-Substituted Thiazol-2-yl Radicals toward Toluene and Nitrobenzene Relative to Benzene ⟨74T4171⟩

Substituent	$\dfrac{PhMe}{PhH}K$	$\dfrac{PhNO_2}{PhH}K$	Substituent	$\dfrac{PhMe}{PhH}K$	$\dfrac{PhNO_2}{PhH}K$
H	2.2	1.5	SCN	3.25	—
Br	3.2	0.52	CN	3.0	0.35

4.19.2.1.9 Reactions with cyclic transition states

(i) Diels–Alder reactions and 1,3-dipolar cycloadditions

In contrast to oxazole, thiazole does not undergo the Diels–Alder cycloaddition reaction. This behavior can be correlated with the more dienic character of oxazole relative to thiazole.

Unexpectedly, the 1,3-dipolar cycloaddition of dimethyl acetylenedicarboxylate (DMAD) with 3-dicyanomethyl ylide of thiazole (**81**) yields a fused six-membered ring (**84**) rather than the five-membered ring (**82**). The initially formed thiazolopyrrole derivative (**82**) is strongly polarized by the *gem*-dicyano group and its pyrrole ring cleaves spontaneously with proton elimination. The ring closure of the intermediate (**83**) leads to the final stable derivative of 5*H*-thiazolo[3,2-*a*]pyridine (**84**; Scheme 48) ⟨68JA3830⟩.

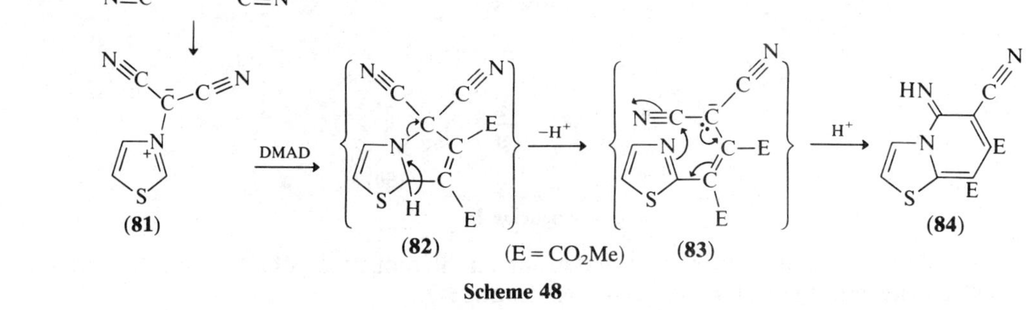

Scheme 48

Dipolarophiles (DMAD, dibenzoylacetylene, ethyl propiolate) condense with ylides (**85**) resulting from quaternization of 4-methylthiazole with an α-bromo ketone or ester and subsequent deprotonation. The 1:1 molar adduct obtained (**86**) rearranges to a pyrrolothiazine (**87**; Scheme 49) ⟨76JOC187⟩.

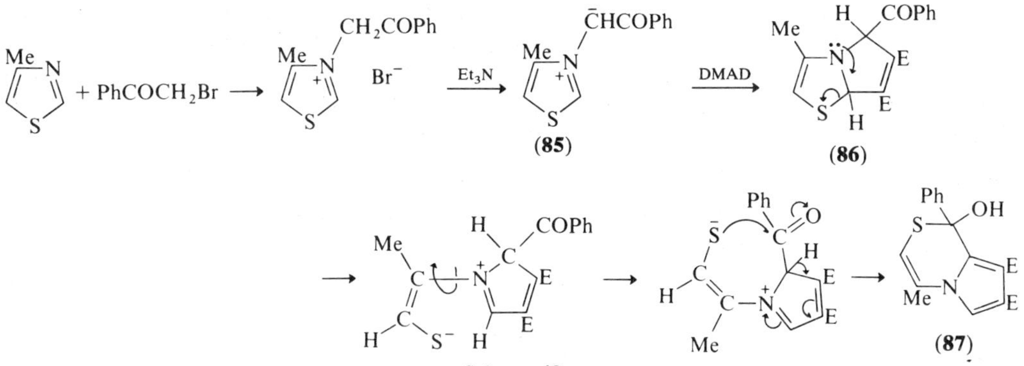

Scheme 49

Thiazole itself reacts with DMAD at room temperature in DMF, the initially formed adduct (**88**) rearranging either *via* a concerted suprafacial 1,5-sigmatropic shift or by a non-concerted pathway proceeding *via* zwitterion (**89**) to (**90**; R = R' = H; Scheme 50) ⟨79HC(34-1)1, p. 97⟩. The three monomethylthiazoles and 2,5-dimethylthiazole undergo the same type of cycloaddition with rearrangement when condensed with DMAD in DMF (Scheme 50; R,R' = Me, H; H, Me-4; H, Me-5; Me, Me-5).

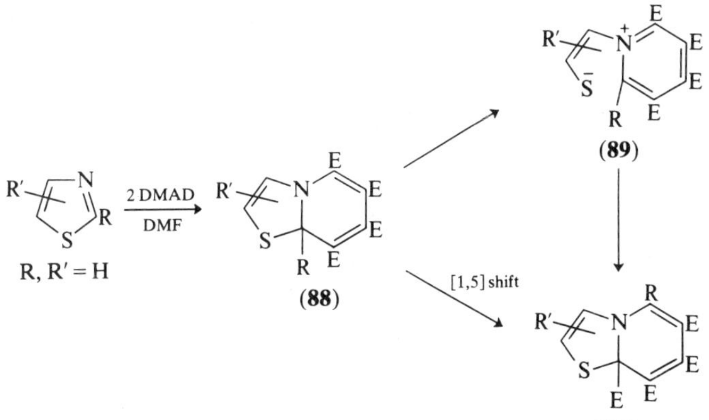

Scheme 50

A quite different course is observed when the reaction of 2-alkyl substituted thiazoles is performed in methanol or acetonitrile. In this case 2:1 adducts with seven-membered azepine rings (**91**) are formed in which two of the original activated hydrogen atoms have essentially altered positions (Scheme 51).

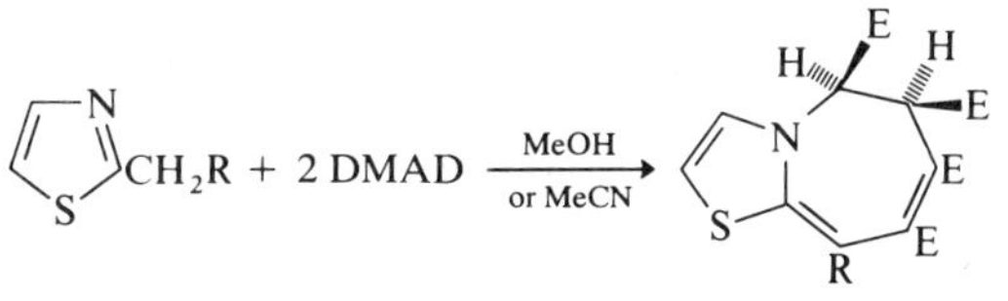

Scheme 51

Benzothiazole undergoes the same type of cycloaddition with DMAD. In DMF the final product is (**92**), whereas in aqueous methanol (**93**) and (**94**) are formed (Scheme 52) ⟨79SST(5)358, p. 401⟩.

Benzothiazole 3-oxide reacts with some dipolarophiles yielding 2-substituted benzothiazoles (Scheme 53) ⟨70CPB1176⟩. 1,4-Cycloadditions of dipolarophiles are observed with mesoionic thiazoles. The 4- and 5-hydroxythiazolylium hydroxide inner salts (**95**) and (**98**) react with dipolarophiles, such as electron-deficient alkynes, to give cycloadducts which react further by extrusion either of sulfur (**96**) or of a simple molecule (**97**, **99**; Scheme 54). An example is given for the reaction of DMAD with 4-hydroxy-2,3-diphenylthiazolylium

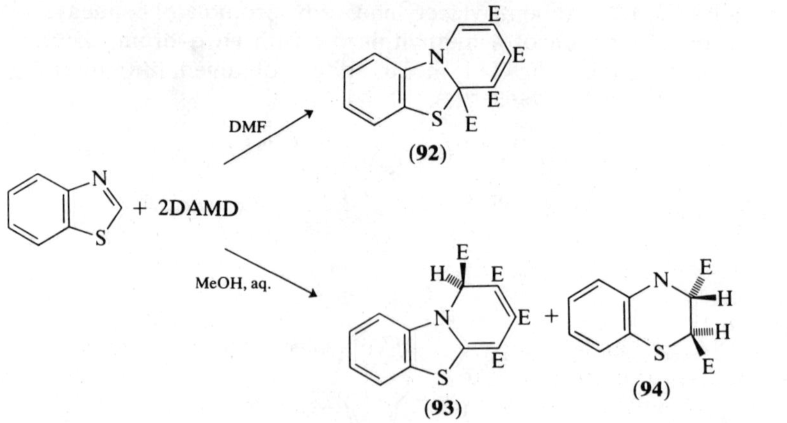

Scheme 52

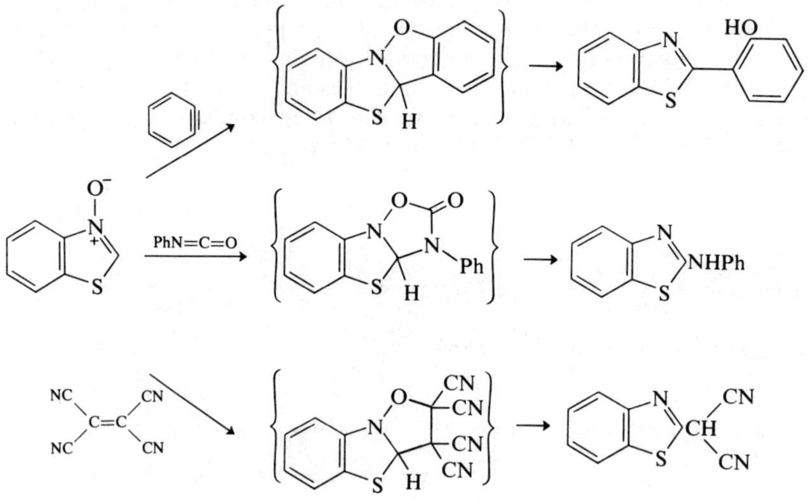

Scheme 53

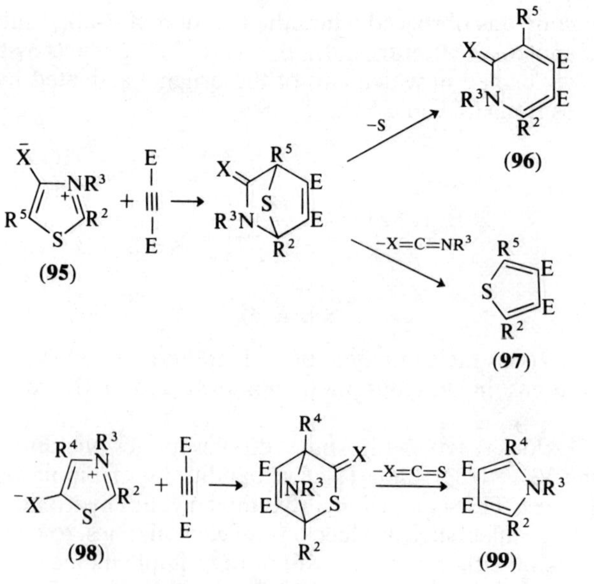

Scheme 54

Scheme 55

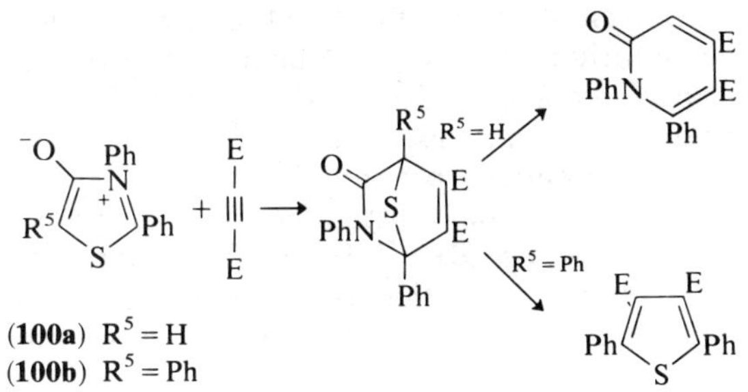

inner salt (**95**; $R^5 = H$) and 4-hydroxy-2,3,5-triphenylthiazolylium inner salt (**95**; $R^5 = Ph$). In the first case sulfur extrusion leads to a substituted pyrid-2-one whereas in the second case, phenyl isocyanate extrusion affords a tetrasubstituted thiophene (Scheme 55) ⟨69CC1129⟩. 2,4,5-Triphenylthiazole (**101**) consumes 0.45 mol of (singlet) oxygen in methanol, in the presence of a sensitizer, to form benzil (**104**) and benzamide (**105**) (11 and 18%, respectively) presumably *via* the cyclic peroxide (**102**) and the Schiff's base (**103**; Scheme 56) ⟨69BCJ2973⟩.

Scheme 56

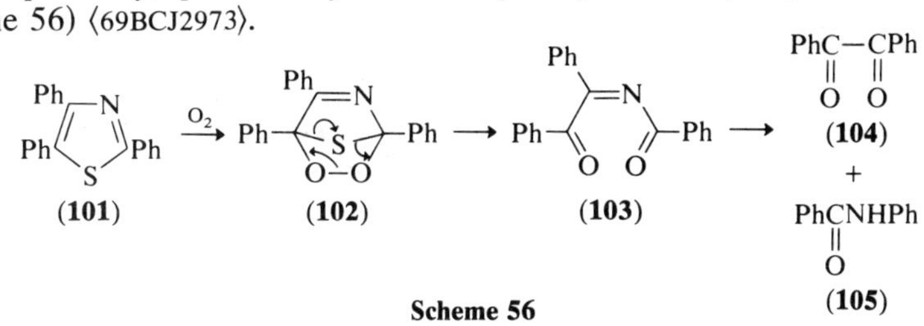

4.19.2.2 Reactions of Non-aromatic Compounds

4.19.2.2.1 Isomers of aromatic derivatives

Thiazoles substituted in the 2-, 4- or 5-position with an amino, hydroxy or mercapto group are in tautomeric equilibrium with the corresponding imino-, oxo- or thioxo-thiazolines (Scheme 57). A similar protomerism has been established for (4-phenylthiazol-2-yl)acetone (**109**; Scheme 58).

Scheme 57

Scheme 58

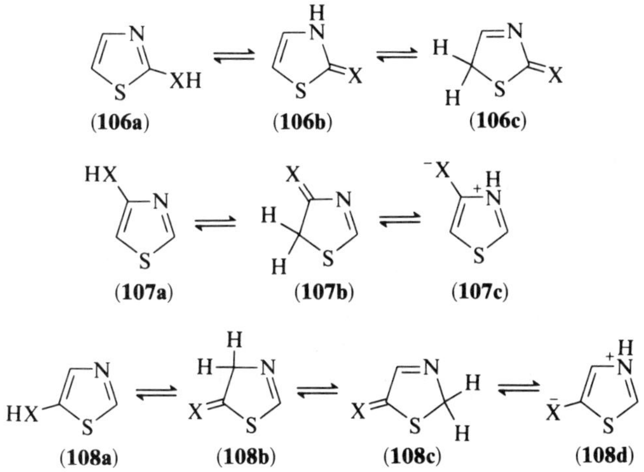

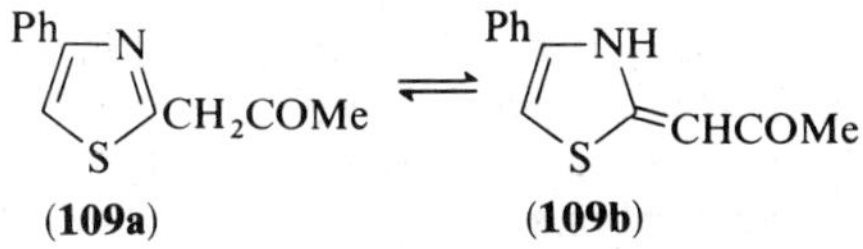

For this class of thiazole, most of the chemical and physicochemical studies are centered around the protomeric equilibrium and its consequences. The position of this equilibrium has been determined by spectroscopic methods and is discussed in the next sections dealing with the aromatic nature of these compounds.

Thiazoles disubstituted by potentially protomeric groups (NHR, OH, SH) also exhibit tautomerism (Scheme 59). Among the most important compounds of this series are thiazolidine-2,4-dione (**110**, XH = OH, YH = OH), 2-thioxothiazolidine-4-one or rhodanine (**110**; XH = SH, YH = OH), 4-thioxothiazolidine-2-one or isorhodanine (**110**; XH = OH, YH = SH) and thiazolidine-2,4-dithione or thiorhodanine (**110**; XH = SH, YH = SH).

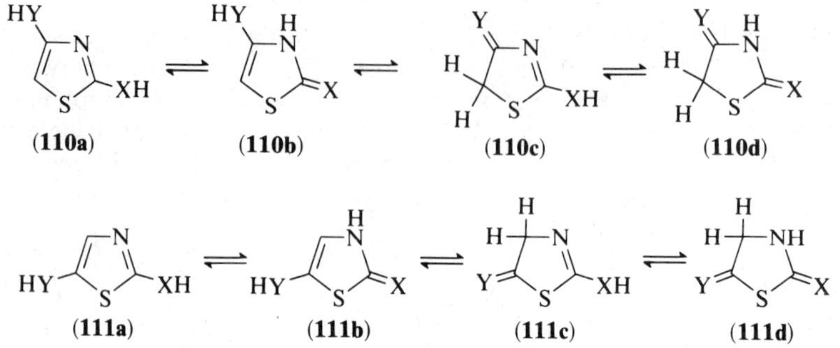

Scheme 59

4.19.2.2.2 Dihydro compounds

(i) *Tautomerism*

Dihydrothiazoles bearing protomeric amino, hydroxy or mercapto groups in a position corresponding to enamine, enol or enethiol functions are in tautomeric equilibrium (Scheme 60) with these functional groups.

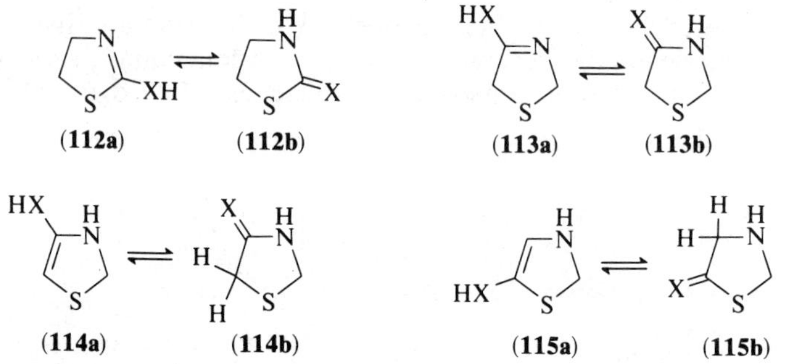

Scheme 60

(ii) *Aromatization*

Δ^2-Thiazolines are converted into the corresponding thiazoles on treatment with nickel peroxide at room temperature. Benzothiazolines are also very easily oxidized to the corresponding benzothiazoles, the aromatization of the heteroring being the driving force of the reaction which corresponds to hydride transfer.

(iii) *Other reactions*

Δ^2-Thiazolines are readily quaternized by alkyl halides yielding 3-alkyl-Δ^2-thiazolinium salts. *p*-Nitrobenzyl chloride in benzene acylates 2-methyl-Δ^2-thiazoline (**116**) affording, initially, 2-chloro-*N*-acylthiazolidine (**117**) which is dehydrohalogenated by triethylamine to 2-methylene-*N*-acylthiazolidine (**118**). 2-Methyl-Δ^2-thiazoline (**116**) is acetylated by acetyl chloride in anhydrous acetonitrile to give *N*,*S*-diacetylcysteamine (**120**) accompanied by 2-[β-2'-acetamidoethylthio-β-methyl]vinyl-Δ^2-thiazoline (**121**) and the cyclic trimer (**123**) of the 2-methylene-*N*-acetylthiazolidine (**122**; Scheme 61) ⟨71T19⟩. 2-Phenylthiazoline (**124**) condenses with substituted acetyl chlorides in the presence of triethylamine to yield bicyclic lactams (**125**). Similarly diphenylketene adds to 2-phenyl-Δ^2-thiazoline affording a β-lactam (**126**; Scheme 62). Nitration of 2-phenyl-Δ^2-thiazoline results in *meta* substitution of the nitro group into the phenyl nucleus, which is in contrast to the behavior of 2-, 4- or 5-phenylthiazole where nitration occurs only in the *para* position. The Δ^2-thiazoline ring is not reduced by lithium aluminum hydride at room temperature whereas aluminum amalgam in moist ether affords thiazolidine. This reduction is a step in an interesting sequence for the preparation of aldehydes (**129**; Scheme 63) ⟨75JOC2025⟩. A major interest of this process is that the hydrolysis of the thiazolidine (**128**) occurs under neutral conditions, thus enabling acid-sensitive aldehydes to be obtained. Hydrolysis of Δ^2-thiazoline occurs readily in acidic medium and each derivative exhibits a characteristic bell-shaped dependence of observed first-order rate constant for initial rate of hydrolysis upon pH. The reaction proceeds *via* attack of water on the thiazolinium ion (**131**) to yield an uncharged 2-hydroxythiazolidine intermediate (**132**) which is converted into a mixture of *N*- (**133**) and *S*-acetylcysteamine (**134**; Scheme 64) ⟨65JA2743⟩. Dimethyl acetylenedicarboxylate condenses with Δ^2-thiazolines yielding adducts similar to those obtained with thiazole (Scheme 65) ⟨69CR(C)(268)870⟩. The photoisomerization of Δ^2-thiazolines (**137**) leads to *N*-alkenylthioamides (**139**; Scheme 66) ⟨72CC896⟩.

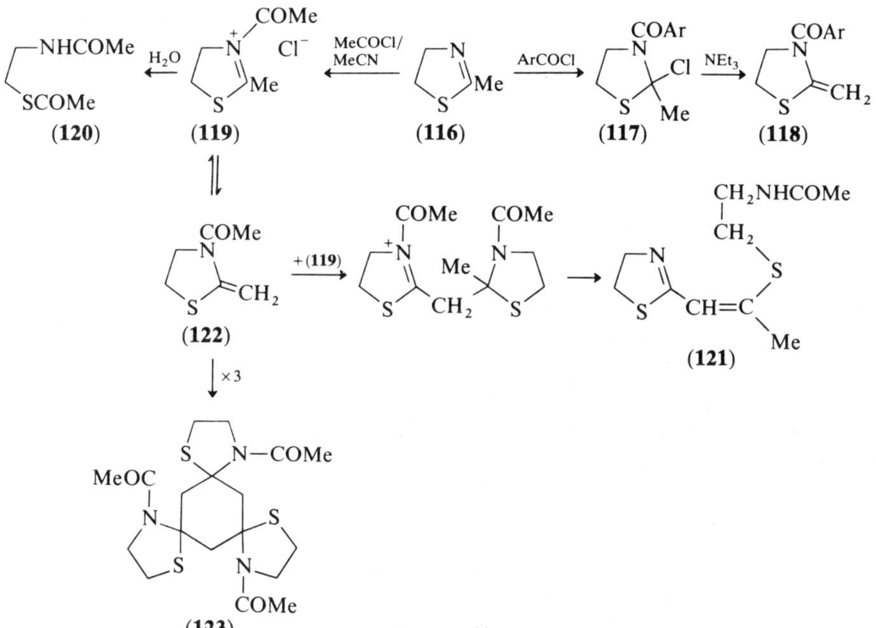

Scheme 61

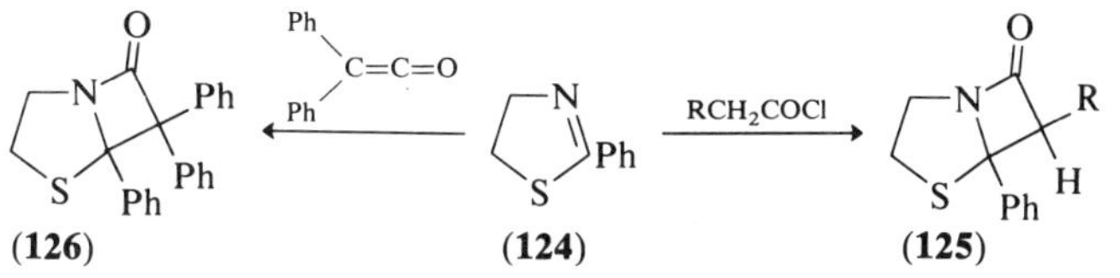

Scheme 62

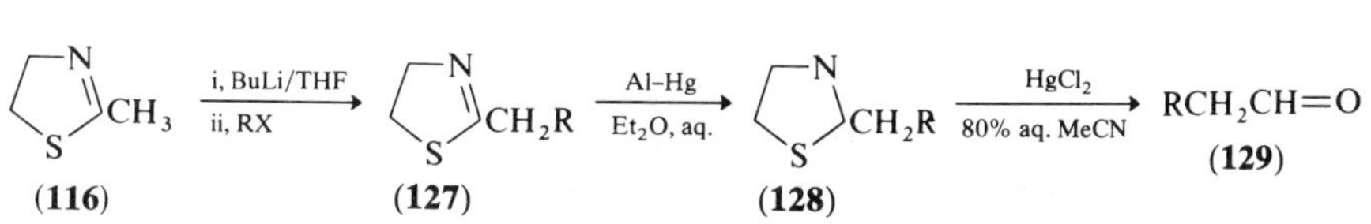

Scheme 63

The simple 2-amino-Δ^2-thiazolines are strong bases, readily soluble in water. Hot concentrated hydrochloric acid opens the ring affording mercapto-alkylamines. 2-Amino-Δ^2-thiazolines are tautomeric; alkylation of 2-amino-Δ^2-thiazoline occurs at the ring nitrogen atom (**140**), whereas alkylation of 2-monoalkylamino-Δ^2-thiazolines occurs at the exocyclic nitrogen atom (**141**; Scheme 67). Reaction of 2-amino-Δ^2-thiazoline with malonic esters

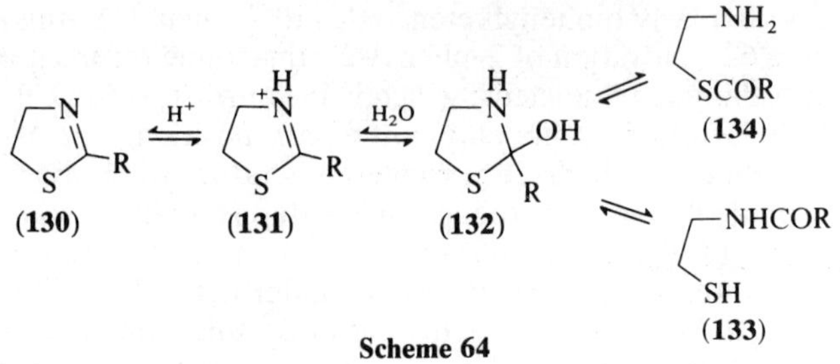

Scheme 64

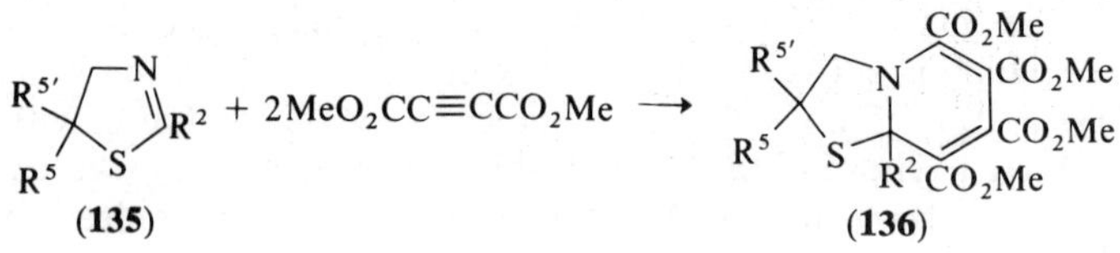

(130) ⇌ (131) ⇌ (132) ⇌ (134) ⇌ (133)

Scheme 65

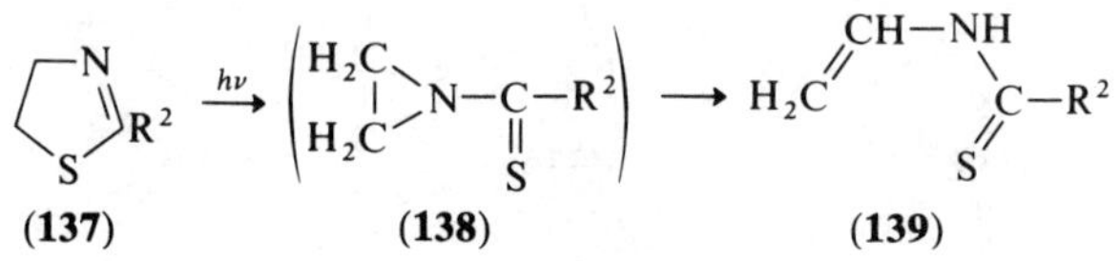

(135) + 2MeO$_2$CC≡CCO$_2$Me → (136)

Scheme 66

(137) →^{hν} (138) → (139)

affords a bicyclic product (**142**) whereas 2-aminothiazole, under the same reaction conditions, gives the 2-acylamino derivative (Scheme 68). Like 2-amino-Δ^2-thiazoline, 2-mercaptothiazolines are readily cleaved on acid hydrolysis to give β-mercaptoalkylamines. The sodium or potassium salt of 2-mercaptothiazoline may be obtained by the action of dilute alkaline solutions and is alkylated, yielding the corresponding 2-alkylmercapto-Δ^2-thiazoline. Oxidation of 2-mercapto-Δ^2-thiazoline with hydrogen peroxide in neutral medium affords the disulfide, whereas in alkaline media desulfurization takes place and 2-hydroxy-Δ^2-thiazolines are formed. Δ^3-Thiazolines are readily aromatized to thiazoles with oxidants (Scheme 69).

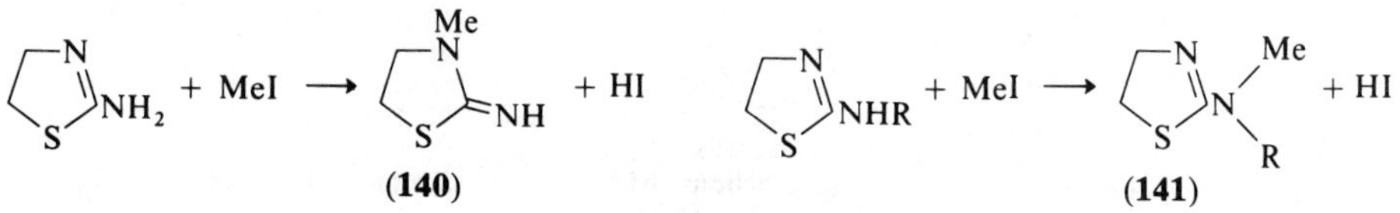

Scheme 67

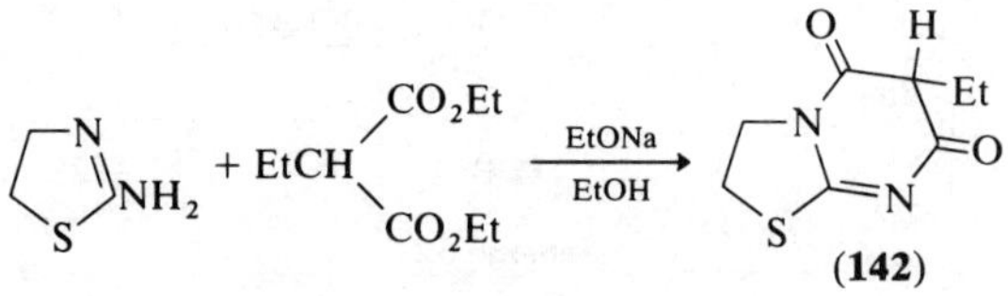

(140)   (141)

Scheme 68

→ (142)

Most Δ^4-thiazolines bear a protomeric group (=NR, =O, =S) and their reactivity is discussed in Sections 4.19.2.3.4, 5 and 6. 2,3-Dihydrobenzothiazoles or benzothiazolines are readily oxidized by iodine or ferric chloride to the corresponding benzothiazoles (Scheme 70).

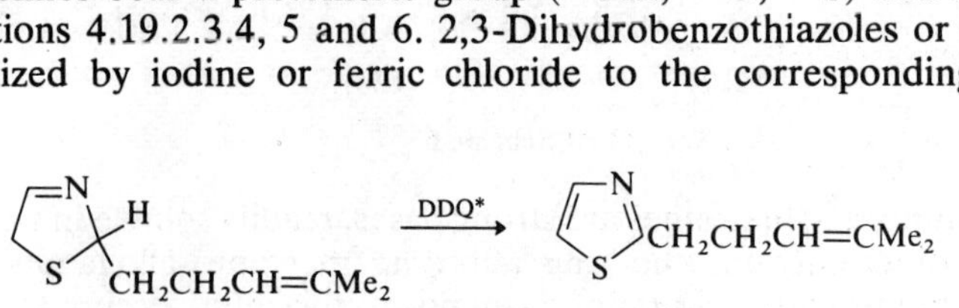

$\xrightarrow{\text{DDQ*}}$

* 2,3-dichloro-5,6-dicyano-1,4-dibenzoquinone

Scheme 69

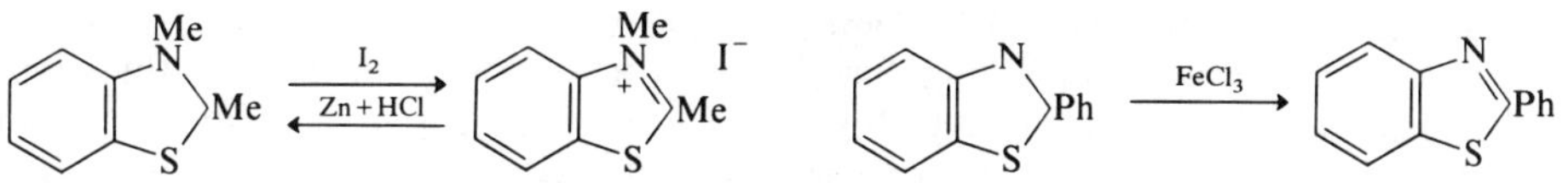

Scheme 70

4.19.2.2.3 Tetrahydro compounds

(i) Tautomerism

As discussed in Section 4.19.2.2.2(i), imino, one or thione derivatives of thiazolidine are in tautomeric equilibrium (Scheme 60). Furthermore, many thiazolidines are in equilibrium with an acyclic thiol form or, in aqueous media, with the β-mercaptoalkylamine and the carbonyl compound from which they were formed (Scheme 71). The position of the equilibrium depends on the nature of the ring substituents.

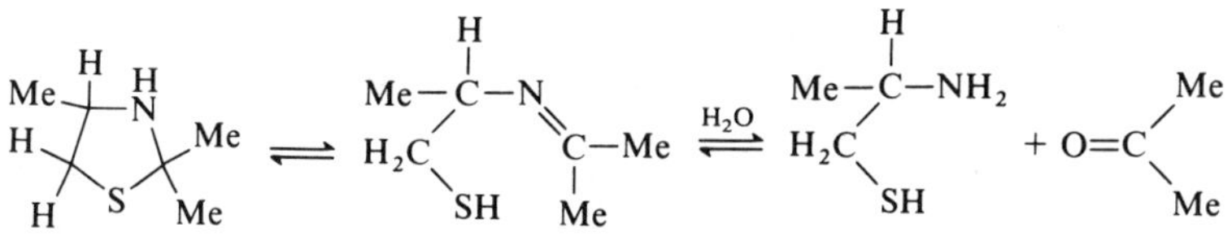

Scheme 71

(ii) Other reactions

Thiazolidines unsubstituted on the nitrogen atom are readily hydrolyzed by boiling aqueous solutions of acids or bases and the dissociation is complete in the presence of a compound reacting with one of the cleavage products. Thus phenylhydrazine gives the hydrazone of the carbonyl compound corresponding to carbon-2; mercury(II) and silver salts cleave the ring leading to metal sulfides of the corresponding mercaptoamine and the carbonyl component may be isolated (see Scheme 63, Section 4.19.2.2.2(iii)). The presence of substituents on the nitrogen atom stabilizes the thiazolidine ring. In its application to the problem of the structure of penicillin, the cleavage of thiazolidines or their N-acyl derivatives with Raney nickel has been of great importance: desulfurization occurs and the product is a substituted α-amino acid (Scheme 72). Similarly, sodium in liquid ammonia cleaves the thiazolidine ring affording an N-alkylaminothiol (Scheme 73).

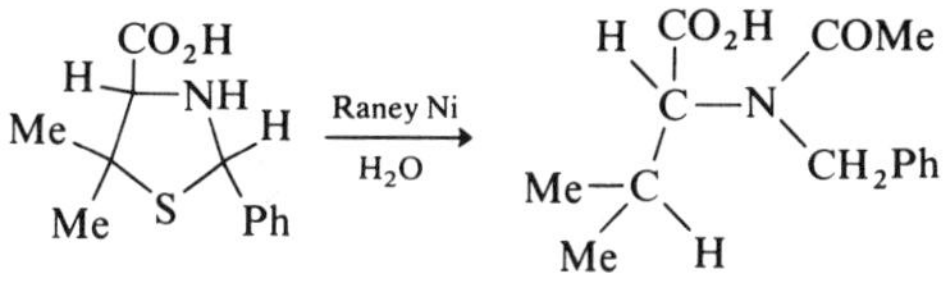

Scheme 72

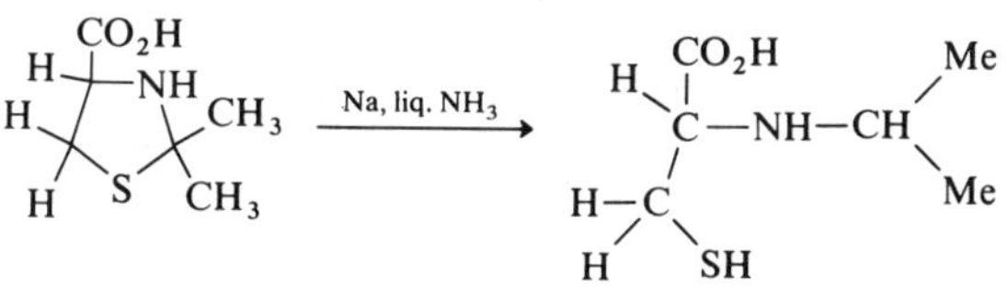

Scheme 73

4.19.2.3 Reactions of Substituents

4.19.2.3.1 General survey of substituents on carbon

When compared to benzenoid substituents, those attached to carbon atoms of the thiazole ring behave very differently according to their position. As discussed in Section 4.19.1.1, the electron distribution in the thiazole molecule corresponds to position 2 being electron-deficient, position 4 almost neutral and position 5 slightly electron-rich. Thus 2-methylthiazole is appreciably more reactive than toluene in reactions involving a trace of its reactive anion, whereas the 4- and 5-isomers do not exhibit such an enhanced reactivity. Similarly the three isomeric thiazolecarboxylic acids are stronger acids than their analogs in the benzene and pyridine series, the strongest acid being the 2-isomer. Isomeric halogenothiazoles exhibit a more complex pattern of nucleophilic substitution; nevertheless 2-halogenothiazole is more reactive than 2-halogenopyridine. Protic substituents such as NH_2, OH and SH are characterized by the tautomerism of the molecule.

4.19.2.3.2 Benzenoid rings

(i) *Fused benzene rings: unsubstituted*

Electrophilic substitution of benzothiazole occurs on the fused benzene ring, essentially at the 4 and 6 positions. Mononitration with HNO_3–H_2SO_4 at room temperature yields a mixture of 6-, 7-, 4- and 5-nitrobenzothiazoles in the ratio 50%, 23%, 20% and 7%, respectively ⟨61JCS2825⟩. Sulfonation with oleum at 100 °C affords 4-, 6- and 7-benzothiazole sulfonic acids in the ratio 70% : 25% : 5%, respectively ⟨16JPR(93)183⟩. Bromination in acetic acid at 100 °C gives 4,6-dibromobenzothiazole. The orientation in these electrophilic substitutions agrees with the net π-charges estimated by the HMO method. As already discussed, nucleophilic substitution of benzothiazole occurs at the C-2 position. Radical phenylation realized in competition with nitrobenzene gives partial rate factors, relative to benzene, of 7 : 4.4 : 3 : 1 for the 2-, 4-, 7- and 5,6-positions. Acidification of the reaction medium enhances the reactivity of the C-2 position by a factor of 20 ⟨70BSF2705⟩. Radical localization energies calculated by the ω method are in agreement with these results only in the hypothesis of an early transition state.

(ii) *Fused benzene rings: substituted*

Bromination of 4-, 5-, 6- and 7-aminobenzothiazole in chloroform yields mixtures of mono- and di-bromo derivatives according to Scheme 74 ⟨65JCS2248⟩. Claisen rearrangement of 6-allyloxy-2-methylbenzothiazole affords 7- and 5-allyl-6-hydroxy-2-methylbenzothiazole in a ratio of about 20 : 1. This ratio is changed to 2 : 1 by replacing the 2-methyl group by a substituent with greater $+M$ effect, such as diethylamino ⟨41CB(B)1407⟩.

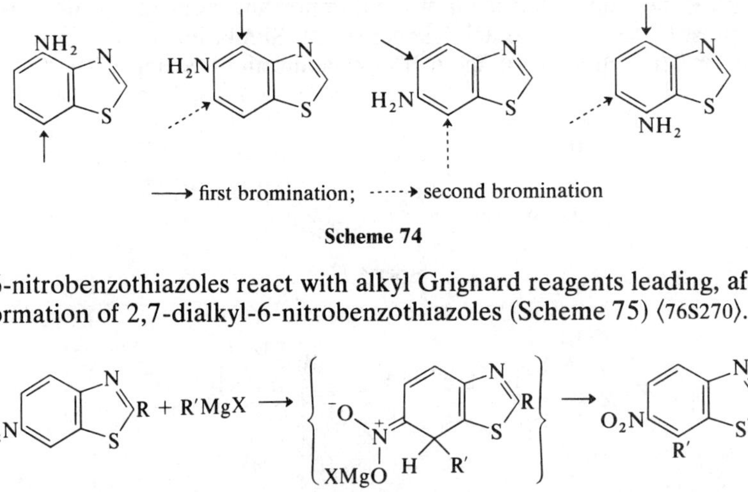

⟶ first bromination; ┄┄▸ second bromination

Scheme 74

2-Alkyl-6-nitrobenzothiazoles react with alkyl Grignard reagents leading, after hydrolysis, to the formation of 2,7-dialkyl-6-nitrobenzothiazoles (Scheme 75) ⟨76S270⟩.

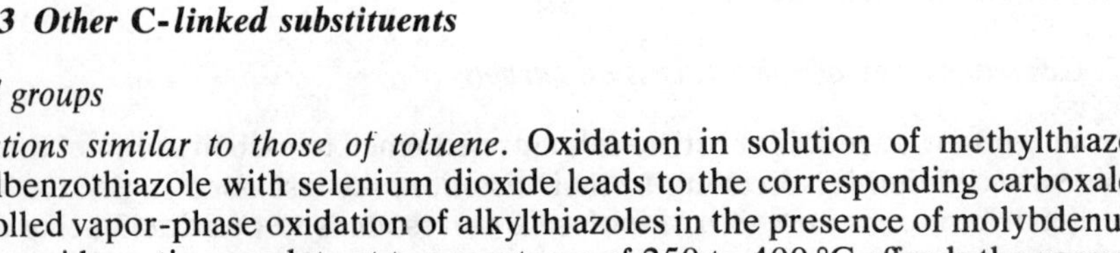

Scheme 75

4.19.2.3.3 Other C-linked substituents

(i) *Alkyl groups*

(a) *Reactions similar to those of toluene.* Oxidation in solution of methylthiazoles and 2-methylbenzothiazole with selenium dioxide leads to the corresponding carboxaldehydes.

Controlled vapor-phase oxidation of alkylthiazoles in the presence of molybdenum oxide, vanadium oxide or tin vanadate at temperatures of 250 to 400 °C affords the corresponding formyl derivative in moderate yields.

Chlorination of 2-methylbenzothiazole occurs when this compound reacts with chlorine gas at 80 °C in acetic acid–sodium acetate solution. The hydrolysis of the 2-trichloromethyl derivative thus formed gives 2-benzothiazolecarboxylic acid.

(b) *Alkylthiazoles: reactions involving essentially complete anion formation.* Under the action of strong organometallic bases, 2-methylthiazole and benzothiazole undergo hydrogen–metal exchange converting them into the corresponding pseudobenzylic metal derivatives (Scheme 76). Butyllithium reacts at −78 °C with 2-methylthiazole giving a mixture of 2-lithiomethyl and 5-lithio derivatives (with a minute quantity of the 4-lithio salt) in a ratio of 1:4, observable by deuterolysis and isotopic analysis of the 2-methylthiazole obtained. Starting from the mixture formed at −78 °C and allowing the temperature to increase results in progressive decomposition of the 2-lithiomethyl derivative (benzylic type) whereas the 5-lithio compound (aromatic type) is stable under these conditions (Scheme 77).

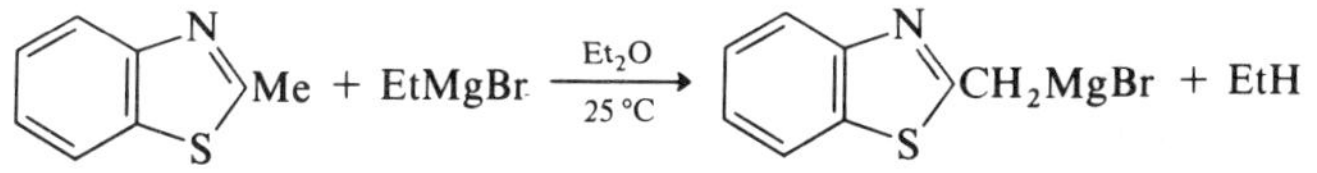

Scheme 76

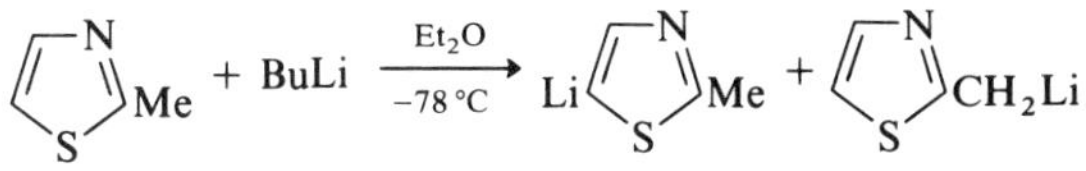

Scheme 77

A possible mechanism for the formation of both lithio derivatives is described in Scheme 78.

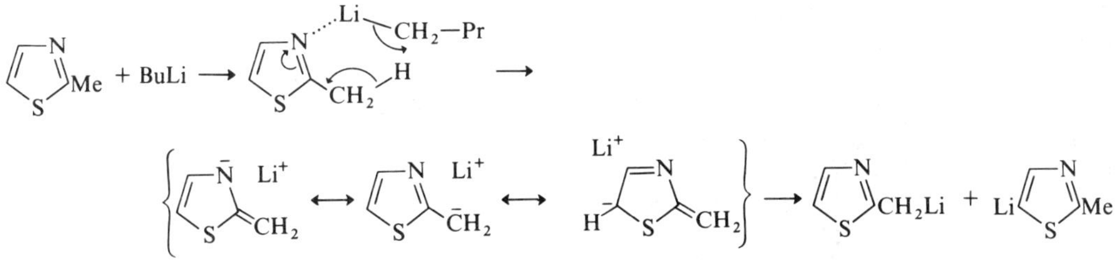

Scheme 78

Further reaction with D_2SO_4 or alkyl halides is normal ⟨67BSF4134⟩.

Meyers ⟨73JA3408⟩ has identified the product resulting from a temperature increase from −78 °C to room temperature of the lithio salt of 2,4-dimethylthiazole (Scheme 79). 2-Lithiomethylthiazole (144) adds to unmetalated 2,4-dimethylthiazole (143) which needs to be present in only trace amounts, generating adduct (145). Rearrangement to an open-chain imine (146) provides an intermediate whose acidity toward lithiomethylthiazole (144a) is rather large. Proton abstraction by (144a) gives the dilithio intermediate (147) and regenerates 2-methylthiazole (143) for further reaction. During the final hydrolysis (147) gives the dimer (148) which could be isolated. Crossover experiments at −78 °C and 25 °C

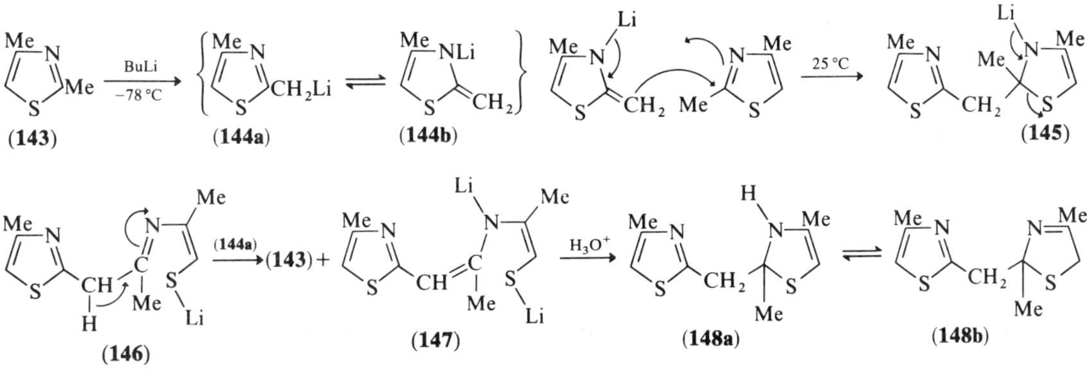

Scheme 79

using thiazoles bearing different substituents (**149**; R = Me, Ph) proved that, at low temperature, the lithio derivatives (**150**) and (**151**) do not exchange H/Li and that the product ratios (**153/154**) observed are the result of independent metalation of the 2-methyl and 5-positions in a kinetically controlled process (Scheme 80) ⟨74JOC1192⟩.

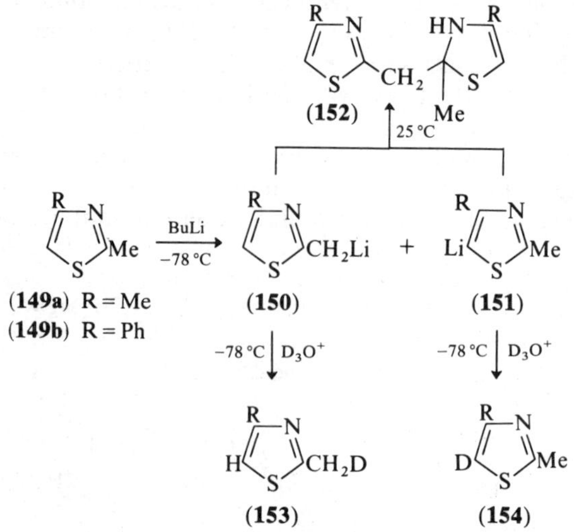

Scheme 80

(c) *Alkylthiazoles: reactions involving traces of reactive anions.* The proton attached to the carbon atom of a 2-alkyl group of thiazole, benzothiazole or Δ^2-thiazoline is easily abstracted by bases. Such a pseudo-acidity is connected with the resonance-stabilization of the resulting carbanion (Scheme 81). Isotopic H/D exchange is observed for 2-methylbenzothiazole when dissolved in D_2O/ethanol in the presence of triethylamine. A rate constant of $2.1 \times 10^{-5} s^{-1}$ is obtained at 180 °C ⟨61ZOB1962⟩. 2-Methyl- and 2-ethyl-benzothiazoles condense with aromatic aldehydes at room temperature in 50% aqueous sodium hydroxide and in the presence of triethylbenzylammonium chloride leading to secondary carbinols (Scheme 82) ⟨75TL3519⟩. The same type of condensation results when a 2-methylthiazole or benzothiazole is heated at 150 °C with benzaldehyde in the presence of zinc chloride. On heating the formed carbinol dehydrates to the styryl derivative (Scheme 83). Neither 4-methyl- nor 5-methyl-thiazoles undergo such a condensation, which is in agreement with the particular stabilization of the carbanion attached to the C-2 atom of the thiazole ring. Numerous similar condensations have been described, for example the bisformylation of 2-methylbenzothiazole in the Vilsmeyer–Haack reaction (Scheme 84) ⟨72CCC2273⟩. Hydroxyalkylation of 4-substituted 2-methylthiazole occurs when it is treated with lithium amide in liquid ammonia followed by a carbonyl compound; the tertiary carbinol readily dehydrates to the styryl derivative (Scheme 85).

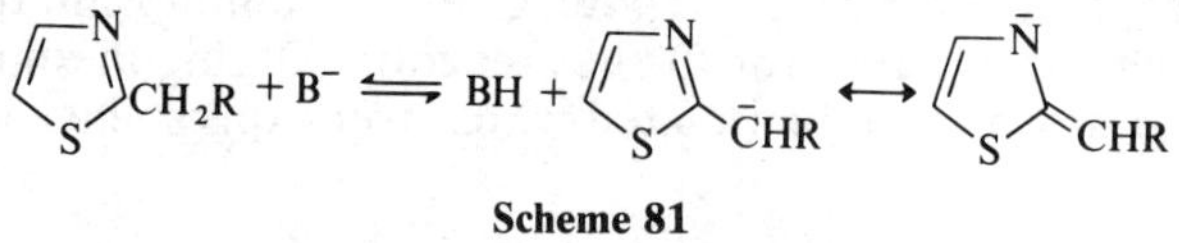

Scheme 81

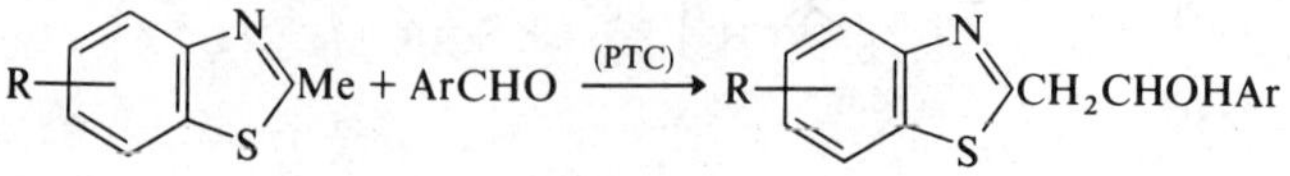

Scheme 82

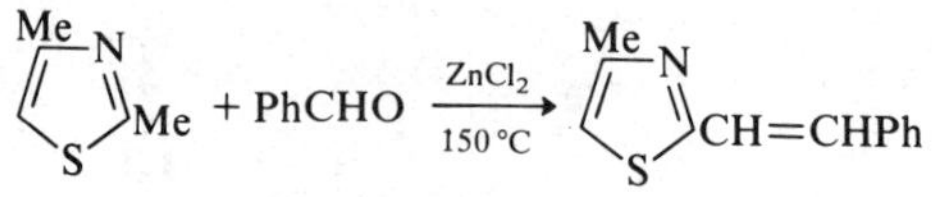

Scheme 83

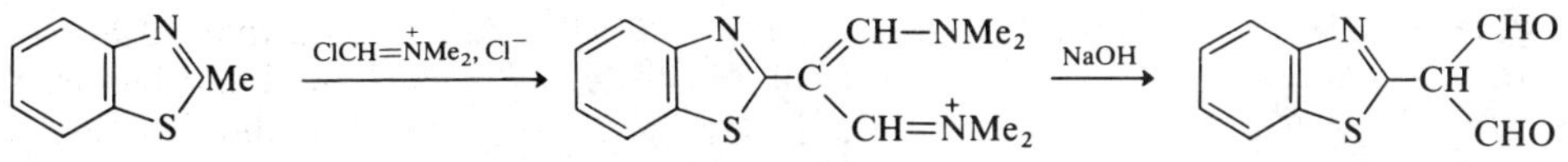

Scheme 84

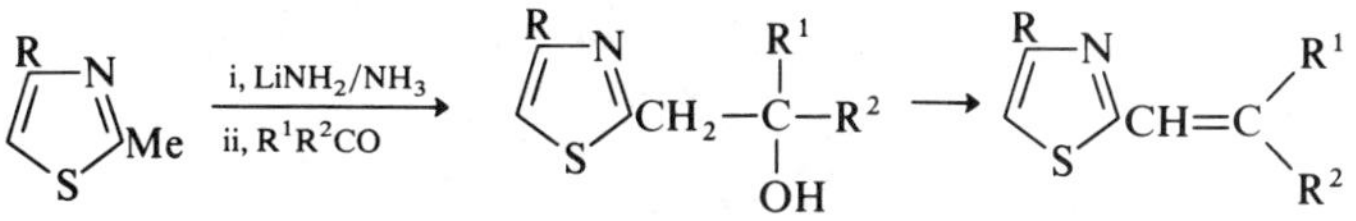

Scheme 85

(d) *C-Alkylthiazolylium cations.* Under basic conditions 2-methylthiazolylium salts (**155**) form a pseudobase (**156**) which reversibly loses a proton leading to the neutral anhydro base (**157**) if the base is sufficiently concentrated. If dilute base is used, ring opening occurs affording (**158**). The anhydro base (**157**) cannot be isolated; it reacts with a second molecule of the thiazolylium salt (**155**), giving an adduct (**159**) which by proton loss affords the dimeric anhydro base (**160**) in equilibrium with its zwitterionic open-chain form (**161**). Both monomeric and dimeric anhydro bases possess enaminic structures which makes them highly reactive. They are the key intermediates in the synthesis of thiazolocyanines. 2-Methylbenzothiazolylium salts undergo the same type of deprotonation in basic medium. The monomeric anhydro base (**163**) could be obtained by treating the 2-methylbenzothiazolylium salt (**162**) with a large excess of a strong base (tetramethylguanidine) in a solvent in which the quaternary salt is only sparingly soluble, thus avoiding further reaction with the latter ⟨69TL2709⟩. The isolated dimeric anhydro base of 2-methylbenzothiazolylium (**164**) reacts with other quaternary salts (**165**) affording 'mixed' anhydro bases (**166**; Scheme 88).

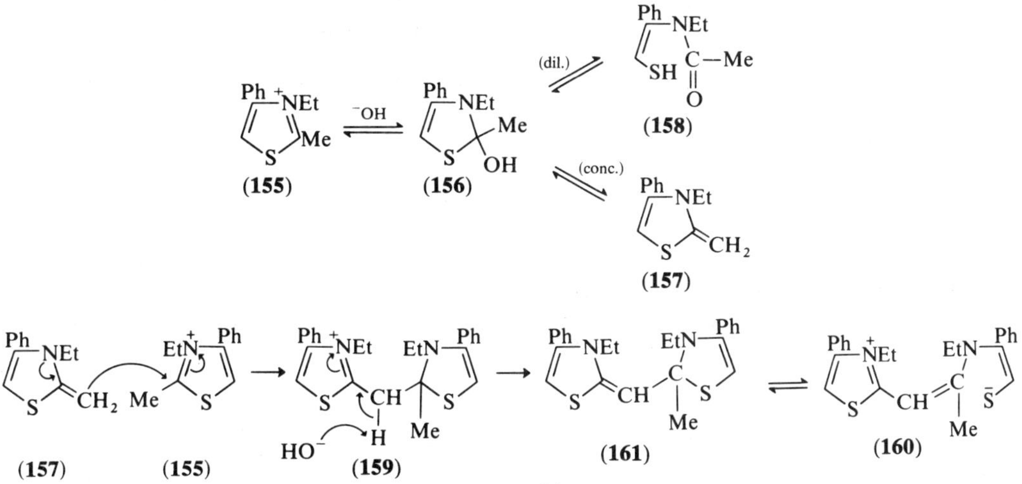

Scheme 86

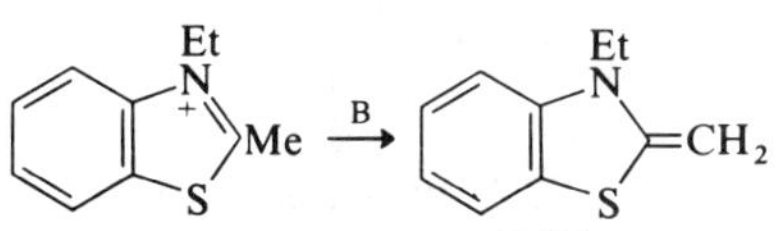

Scheme 87

Heating a pyridine solution of a quaternary salt of 2-methylbenzothiazolylium (**167**) with triethylamine affords a *meso* substituted thiacarbocyanine (**168**). The mechanism of this reaction implies an initial formation of the dimeric anhydro base (**169**) which then reacts as a base with a third molecule of (**167**) transforming it into a monomeric anhydro base (**170**) and an anilinovinyl derivative of benzothiazolylium (**171**). These then add together to form the precursor (**172**) of the *meso* substituted thiacarbocyanine (**168**; Scheme 89) ⟨64BSF2888⟩. A large number of cyanine dyes derived from thiazole, benzothiazole and thiazoline rings have been synthesized and many of them are used as chromatic sensitizing dyes in photographic emulsions ⟨79HC(34-3)1, p. 23⟩.

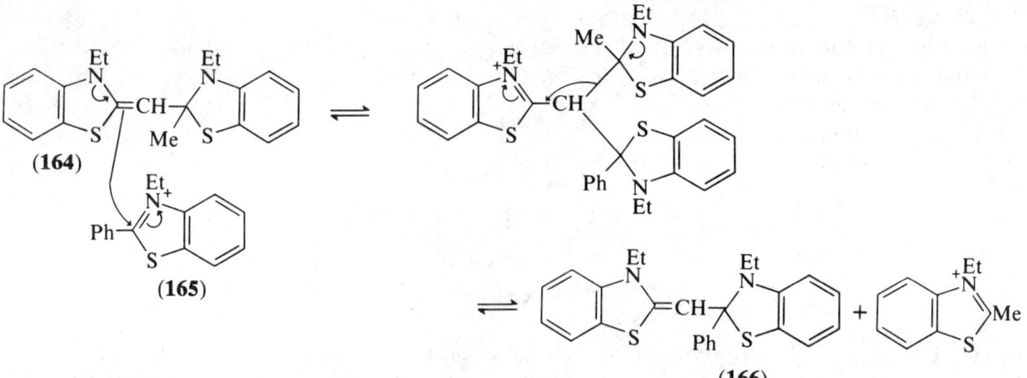

Scheme 88

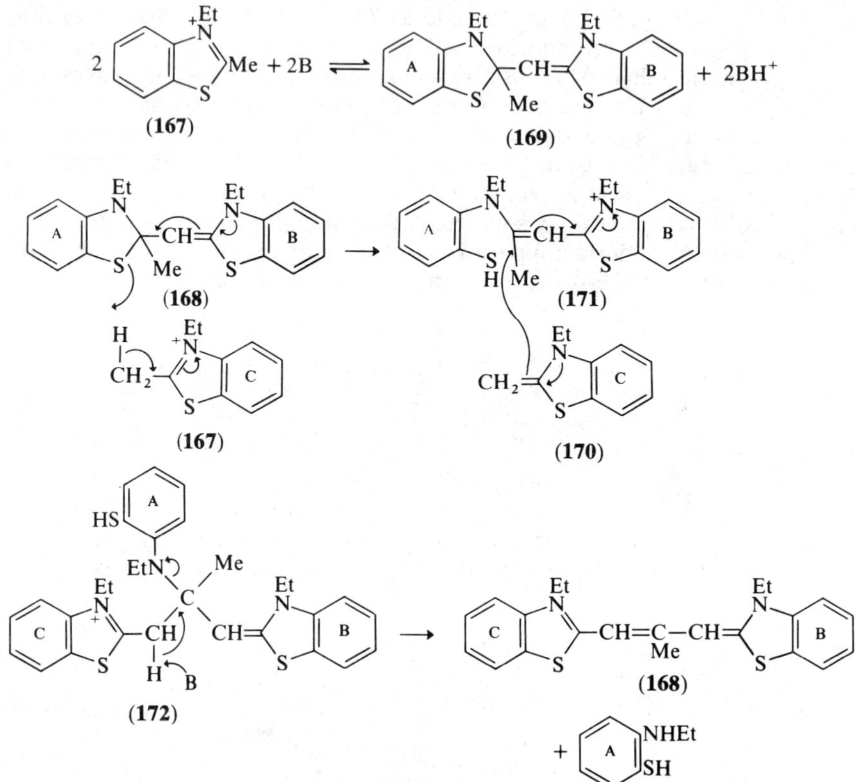

Scheme 89

(e) *Free radical reactions of alkyl substituents.* In addition to their substitution into the heterocyclic ring, cyclohexyl radicals may induce hydrogen abstraction from the side-chain of alkyl thiazoles, producing benzylic type radicals which react further by the sequence shown in Scheme 90 ⟨71CR(C)(272)854⟩. Vinyl- and isopropenyl-thiazoles undergo radical polymerization and copolymerization with styrene, 2-vinylthiazole being more reactive than its 4-isomer.

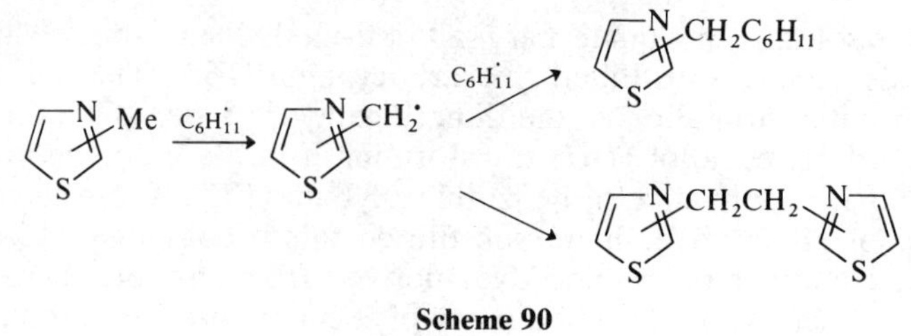

Scheme 90

(ii) *Aryl groups*

Nitration of the three phenylthiazoles gives the corresponding *p*-nitrophenylthiazoles. This behavior is to be contrasted with that of 2-phenyl-Δ^2-thiazoline which gives only the *m*-nitrophenyl derivative (see Section 4.19.2.2.2(i)). In the case of 2-phenylbenzothiazoles, electrophilic substitution reactions such as chlorination, bromination and nitration occur exclusively in the fused benzene ring, especially at the 6-position.

(iii) *Other C-linked groups*

(a) *Carboxylic acids.* Thiazolecarboxylic acids are stronger acids than their analogs in the benzene and pyridine series. Their pK_a values, in increasing order, are 2-thiazole < 2.5, 5-thiazole 2.9, 4-thiazole 3.5, benzoic 4.2, 3-pyridine 4.8. The three thiazoleacetic acids are slightly stronger acids than phenylacetic acid, their pK_a values in increasing order being 2- and 5-thiazole 3.9, 4-thiazole 4.1, phenyl 4.3. Because of their amphoteric character all these thiazolecarboxylic acids are difficult to study in aqueous solution. 2-Thiazolecarboxylic acids decarboxylate very easily, as do 2-benzothiazole and 2-naphthothiazole analogs. The 5-carboxylic acids also decarboxylate readily but 4-carboxylic acids are relatively stable. These differences are illustrated in Scheme 91. The decarboxylation very probably involves the zwitterionic or betaine form of the acid (Scheme 92), the positive charge of which enhances the rate of decarboxylation of the carboxylate anion. The kinetics of decarboxylation of 5-carboxylic acids derived from oxazole, thiazole and imidazole have been measured (Scheme 93) and suggest that the 5-ylides (**164a**) are generated at approximately relative rates of $10^{5.4}:10^3:1$ for oxazole, thiazole and imidazole respectively, exemplifying the special bonding properties of sulfur in the azolylium ring ⟨71JA7045⟩. *N*-Oxidation of benzothiazole enhances the ease of decarboxylation of the 2-carboxylic acid.

$$\text{2,5-diacid} \xrightarrow{85\,°C} \text{5-acid} \xrightarrow{140\,°C} \text{thiazole}$$

$$\text{2,4-diacid} \xrightarrow{140\,°C} \text{4-acid}$$

Scheme 91

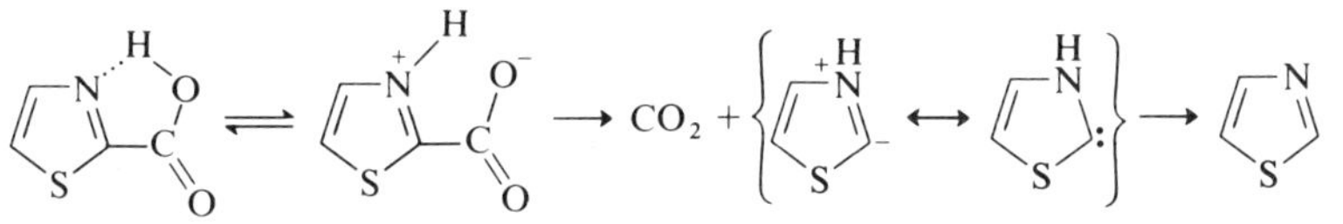

Scheme 92

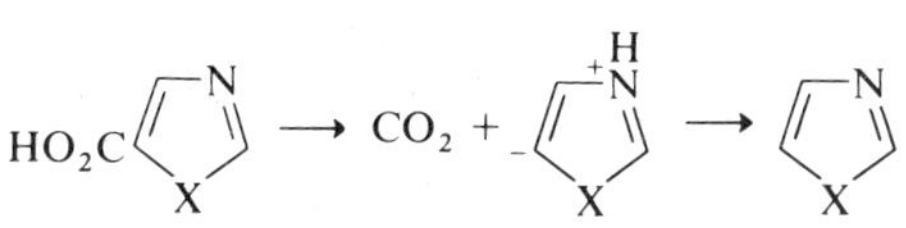

Scheme 93

The three thiazoleacetic acids also decarboxylate easily, the 2-isomer still being the least stable. The higher homologs of the thiazoleacetic acids are stable. The methyl esters of thiazolecarboxylic acids are readily obtained by the action of diazomethane and, for the more stable, by the common acid-catalyzed procedure. Under mild conditions the saponification of diethyl 4,5-thiazoledicarboxylate leads to the 4-monoester (Scheme 94). In the presence of sodium ethoxide, ethyl 4-methylthiazole-5-carboxylate undergoes the Claisen ester condensation with ethyl acetate and the hydrolysis of the resultant β-keto ester affords 5-acetyl-4-methylthiazole (Scheme 95). Reduction of diethyl 2-amino-4,5-thiazoledicarboxylate with lithium aluminum hydride leads, with a small excess of reducing agent, to ethyl 2-amino-4-hydroxymethyl-5-thiazolecarboxylate whereas a large excess of the reducing agent affords 2-amino-4-hydroxymethyl-5-methylthiazole (Scheme 96). Thiazole-4- and -5-carboxylic acids react with thionyl chloride to give the corresponding acid chlorides. Under the same conditions the potassium salt of 2-ethylthiazole-4,5-dicarboxylic acid affords the anhydride (Scheme 97). Thiazole acid chlorides react with diazomethane to give the corresponding diazoketones (**165a**). These react with alcoholic hydrogen chloride to give chloroacetylthiazole (Scheme 98) and heated with alcohol in the

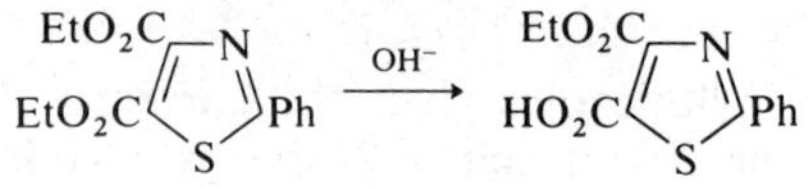

Scheme 94

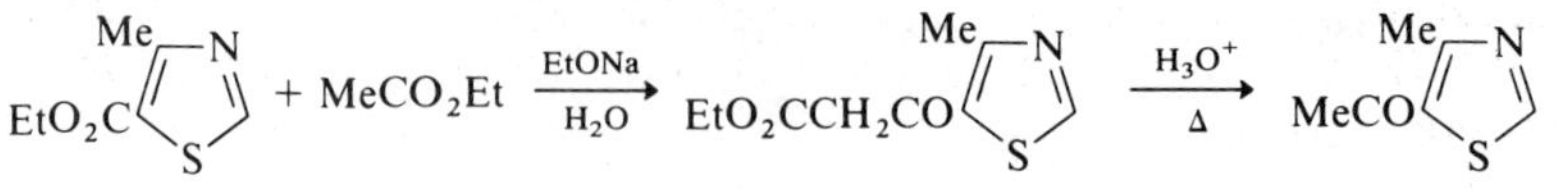

Scheme 95

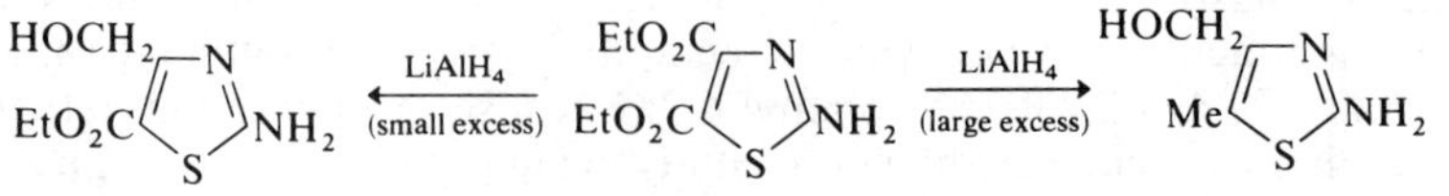

Scheme 96

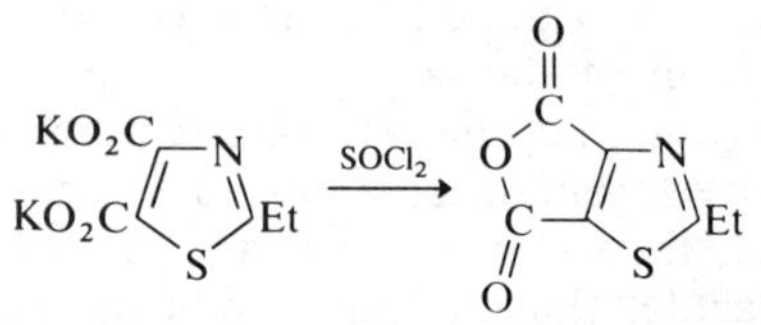

Scheme 97

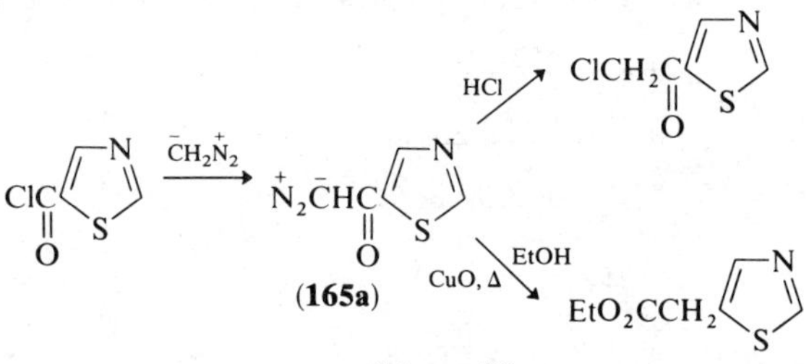

Scheme 98

presence of copper oxide, the 5-diazomethylketone (**165a**) gives ethyl 5-thiazoleacetate (the Wolff rearrangement).

Thiazolecarboxylic acids, esters or acid chlorides react readily with ammonia or various amines, affording the corresponding carboxamides. Dehydration of the amides with phosphorus pentoxide or phosphoryl chloride occurs readily and gives the corresponding nitriles (Scheme 99). Thiazolecarboxylic acid hydrazides are obtained in a similar way, using hydrazine or substituted hydrazines instead of ammonia or amines. The Raney nickel reduction of cyanothiazoles leads to the corresponding amino compounds, the 4-cyano derivative being the isomer most readily reduced.

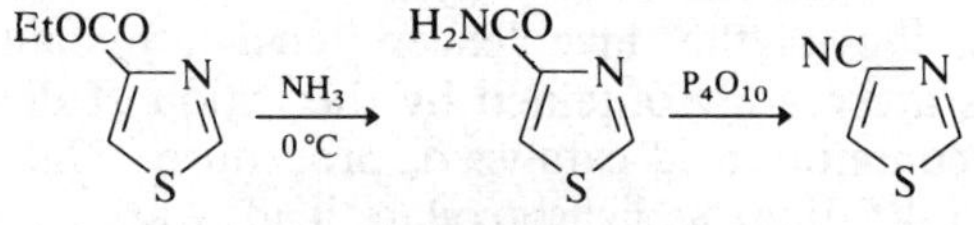

Scheme 99

(b) *Aldehydes and ketones.* Thiazole aldehydes behave in a standard way, disproportionating in the presence of a strong base, giving equal amounts of the corresponding alcohol and carboxylic acid (Cannizzaro reaction); reduction with isopropanol and aluminum isopropoxide gives the corresponding primary alcohols (Meerwein, Pondorf, Verley reaction), oxidation with most oxidizing agents forms the corresponding acids, and they undergo the benzoin condensation in the presence of potassium cyanide.

Thiazolyl ketones also exhibit the standard reactivity of arylketones.

(c) *Chloroalkyl derivatives.* A quantitative study of the solvolysis of α-thiazolylethyl chlorides has been carried out which provides a useful set of σ^+ constants for the thiazolyl group: -0.18, -0.01 and $+0.26$ for 5-, 4- and 2-thiazolyl, respectively ⟨73JOC3316⟩. This result is in agreement with the qualitative order of decreasing electrophilic reactivity observed for the ring (Section 4.19.2.1.4) (Scheme 100). The introduction on the same substrates of various substituents (Scheme 101) provides information on the transmission of substituent effects in the thiazole system. In the series 1-(2-X-5-thiazolyl)ethyl (**168a**) and 1-(5-X-2-thiazolyl)ethyl (**169a**) chlorides, reaction rates are well correlated by Brown's ⟨58JA4079⟩ substituent constants σ_p^+ for the substituent X. In contrast, no correlation exists in the case of 1-(2-X-4-thiazolyl)ethyl and 1-(4-X-2-thiazolyl)ethyl chlorides.

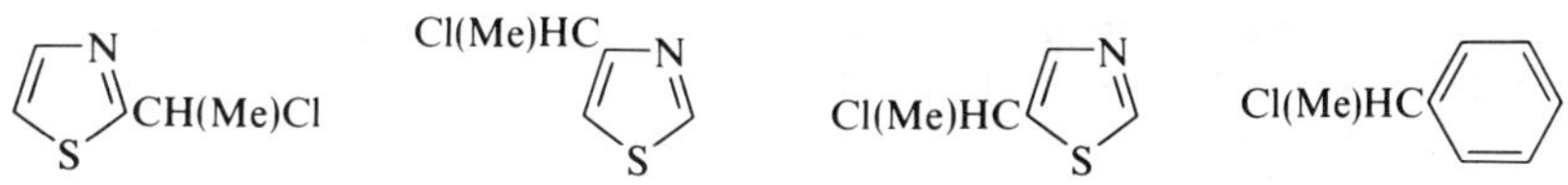

1 25 174 21

Scheme 100 Relative rates of solvolysis in 80% ethanol at 45 °C

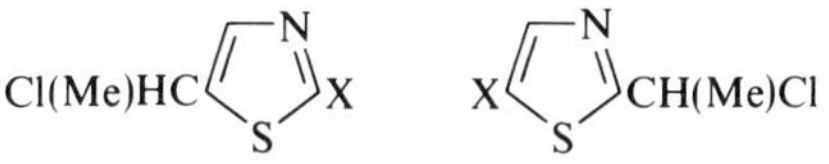

(**168a**) (**169a**)

X = OMe, SMe, Me, H, Cl

Scheme 101

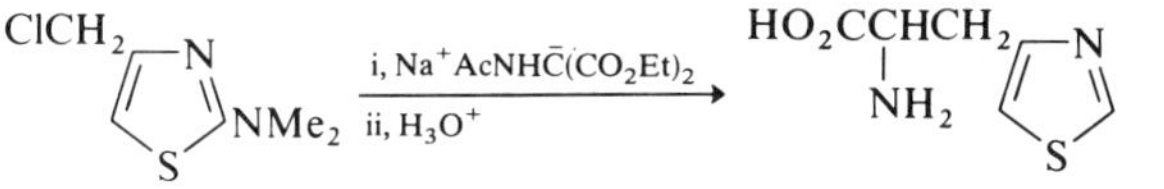

Scheme 102

Chloromethylthiazoles are fairly reactive and are useful intermediates in the synthesis of functional derivatives (Scheme 102).

(d) *Vinyl groups.* Alkynic cyanine dye analogues (**172a**) have been prepared by reacting 3-alkyl-2-chloropropenylbenzothiazolylium salts (**170a**) with the betaine (**171a**; Scheme 103) ⟨77JOC1035⟩.

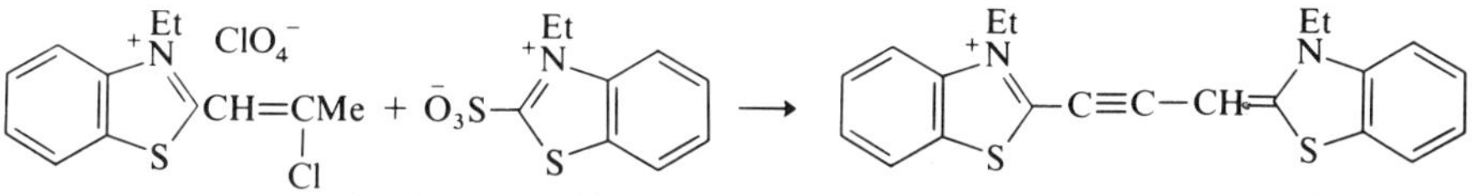

(**170a**) (**171a**) (**172a**)

Scheme 103

4.19.2.3.4 ***N-Linked substituents***

(i) *Aminothiazoles*

As expected, 2-aminothiazole is more basic (pK_a = 5.28) than thiazole (pK_a = 2.52), protonation occurring on the ring nitrogen atom. Electron-withdrawing substituents lower the basicity whereas electron donating substituents increase it. 2-Amino-4-*p*-R-phenylthiazoles (**173**) and 2-amino-5-*p*-R-phenylthiazoles (**174**) have had their pK_a values correlated with the σ_p values of the substituents in the phenyl ring. The tautomeric imino derivatives are more basic than their amino analogs: 2-aminotetrahydrobenzothiazole (**175**; pK_a = 6.11) is less basic than 2-imino-3-methyltetrahydrobenzothiazole (**176**; pK_a = 10), indicating that in a protomeric equilibrium the amino form is predominant (Scheme 104).

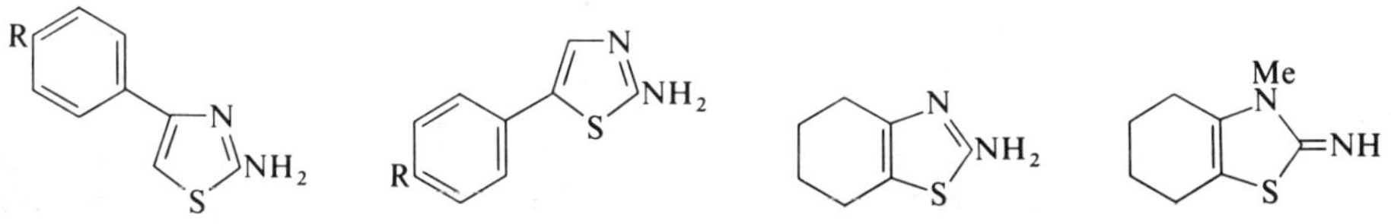

(**173**) (**174**) (**176**)

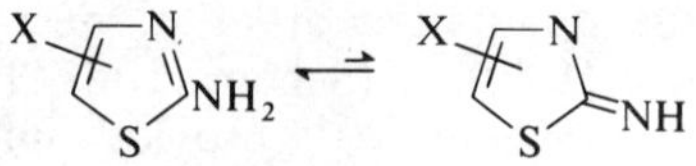

Scheme 104

(a) *Reaction with saturated electrophilic reagents*. The reactivity of 2-aminothiazole towards reactants bearing an sp^3 C electrophilic center follows the general pattern (Scheme 105). If the thiazole reacts in its neutral form, the ring nitrogen atom is the more reactive center, except when bulky substituents are present at the C-4 position. If the thiazole reacts in the form of its conjugate base, the ambident anion leads to mixture of products resulting from *N*-ring and *N*-exocyclic reactivity. The 5-aminothiazoles react at the ring nitrogen atom with alkylating agents, affording stable thiazolylium salts.

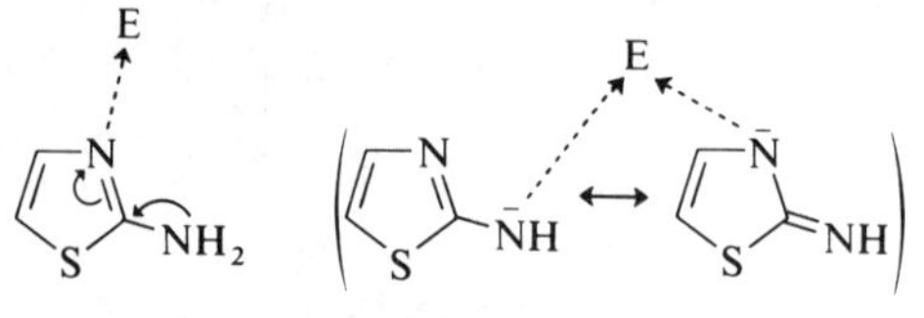

Scheme 105

2-Aminobenzothiazoles are alkylated on the ring nitrogen atom with a certain amount of *N*-exocyclic alkylation whereas both 2-aminonaphtho-[1,2]- and -[2,1]-thiazoles give exclusively ring nitrogen methylation.

(b) *Reaction with unsaturated electrophilic reagents*. Under mild conditions, 2-aminothiazoles react at their exocyclic nitrogen atom with aromatic aldehydes, yielding Schiff bases (Scheme 106). These Schiff bases, under stronger conditions, may afford benzylidenebisaminothiazoles (Scheme 107). 2-Amino-4-phenylthiazole reacts with salicylaldehyde at both the exocyclic nitrogen atom and at the C-5 position (Scheme 108).

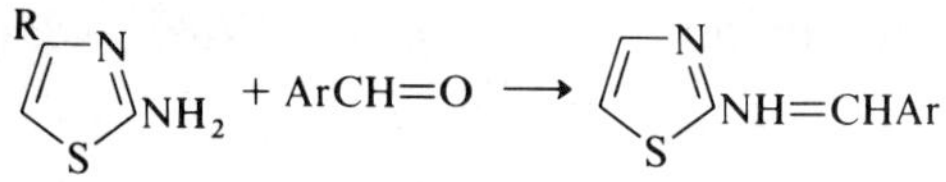

Scheme 106

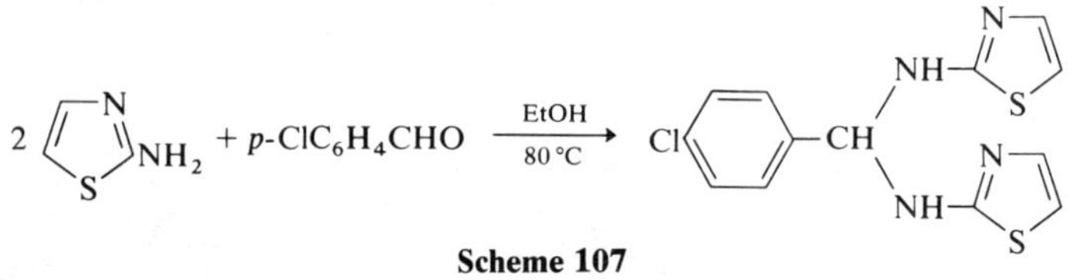

Scheme 107

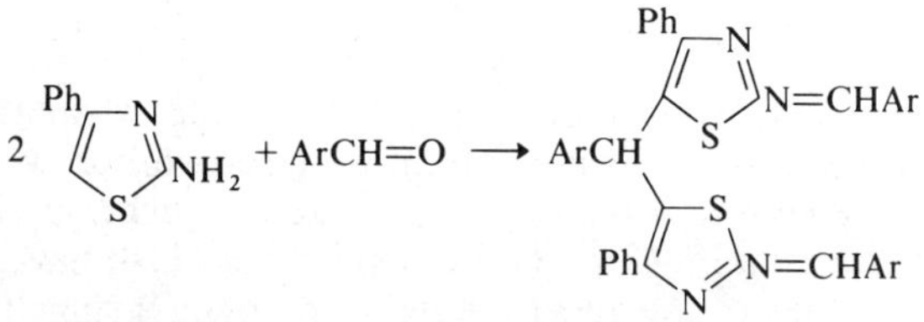

Scheme 108

This nucleophilic reactivity of 2-aminothiazoles has been used to prepare biheterocyclic compounds. Thus 2-aminothiazole reacts with chlorovinyl methyl ketone yielding 5-methylthiazolo[3,2-*a*]pyrimidinium chloride (Scheme 109). In the presence of formaldehyde, aminothiazoles react with enols affording condensation products, through the intermediacy of Mannich bases (Scheme 110). In acidic medium and at higher temperatures, condensation of 4-aryl-2-aminothiazole with benzaldehyde takes place at the C-5 position (Scheme 111). The same orientation is observed when 2-amino-4-methylthiazole is alkylated by secondary or tertiary alcohols in 85% sulfuric acid (Scheme 112).

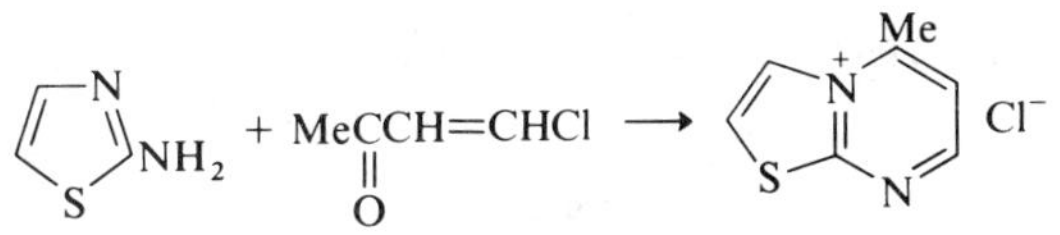

Scheme 109

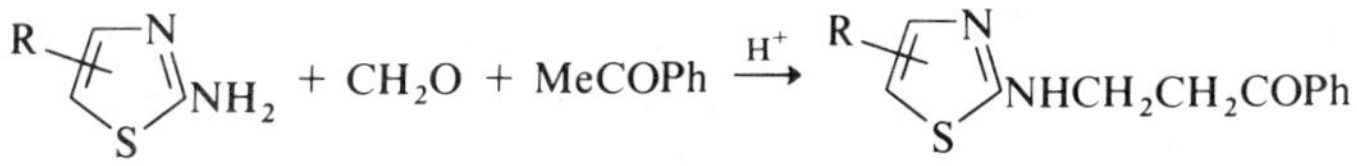

Scheme 110

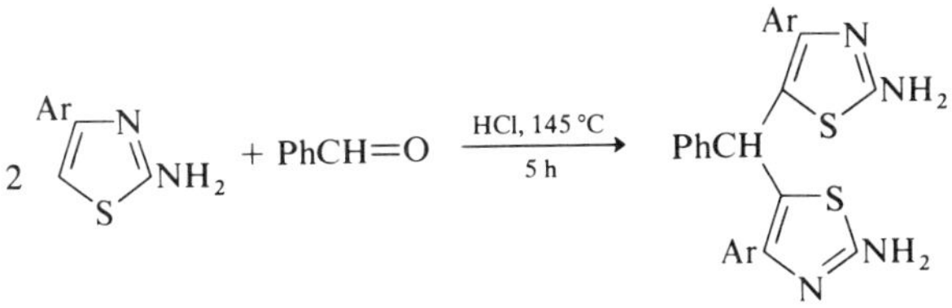

Scheme 111

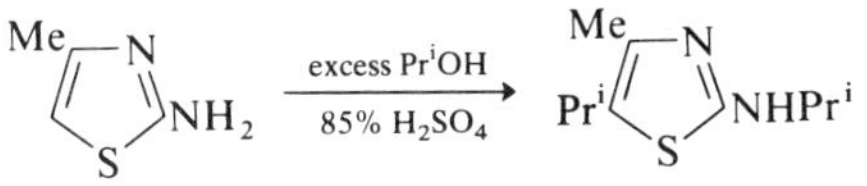

Scheme 112

Acylation of 2-, 4- and 5-aminothiazoles takes place on the exocyclic nitrogen atom. Acyl halides are the most commonly used acylating agents and α- or β-halogenated acyl chlorides lead to biheterocyclic derivatives (Scheme 113). Acetic anhydride acetylates aminothiazoles and benzothiazoles on the exocyclic nitrogen atom. Esters react also with 2-aminothiazoles in the presence of acidic catalysts and α,β-unsaturated esters afford thiazolopyrimidine derivatives (Scheme 114). The reaction of 2-aminothiazoles with alkyl or aryl isocyanates or isothiocyanates gives the corresponding thiazolylureas or thioureas (Scheme 115). Similarly the sodium salt of N-2-thiazolyldithiocarbamic acid is obtained by reaction in NaOH between 2-aminothiazole and carbon disulfide (Scheme 116). Sulfonyl halides react with 2-aminothiazoles, affording sulfonamido derivatives, some of which have been used as antibiotics (Scheme 117).

Scheme 113

Scheme 114

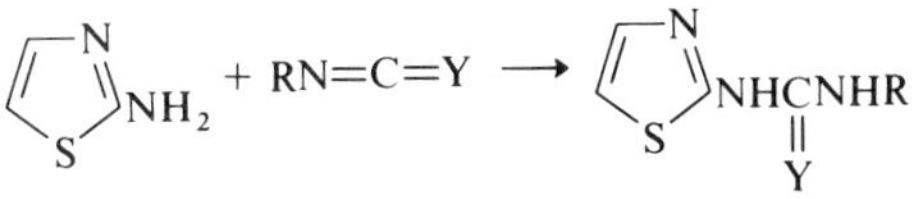

Scheme 115

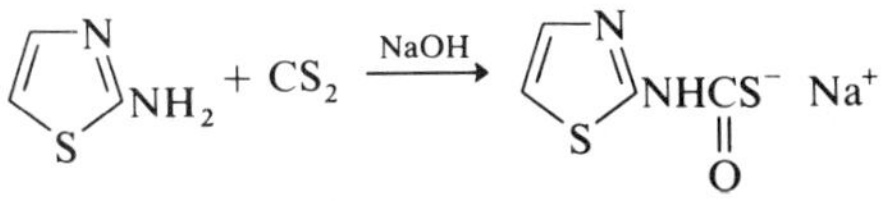

Scheme 116

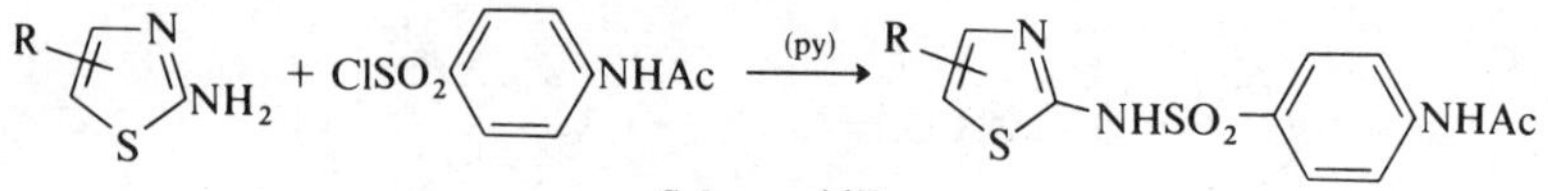

Scheme 117

(c) *Diazotization.* 2-Amino- and 5-amino-thiazoles are most readily diazotized in concentrated solutions of oxygen-containing acids such as 50% sulfuric acid, fluoroboric–phosphoric acid, or mixtures of nitric and phosphoric acids. The diazotization of 2-aminobenzothiazole is more difficult. The diazonium salts undergo normal Sandmeyer reactions to give halogeno- or cyano-thiazoles. Reduction of the 2-thiazole diazonium salt by hypophosphorus acid affords the 2-unsubstituted thiazole, which is a standard method of preparation of thiazole starting from commercial 2-aminothiazole (Scheme 118).

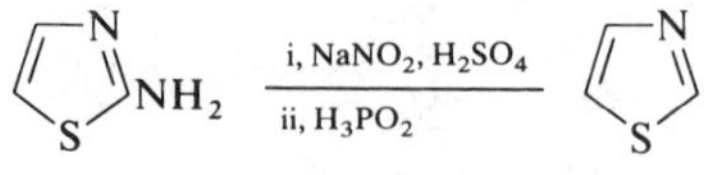

Scheme 118

(d) *Reaction with other elements.* 2-Aminothiazole reacts with trimethylsilyl chloride and triethylamine affording the *N*-bis(trimethylsilyl) derivative (Scheme 119). Some phosphonamidothiazolyl derivatives exhibiting pesticidal properties are obtained by reacting 2-aminothiazoles with reactive phosphorus derivatives (Scheme 120).

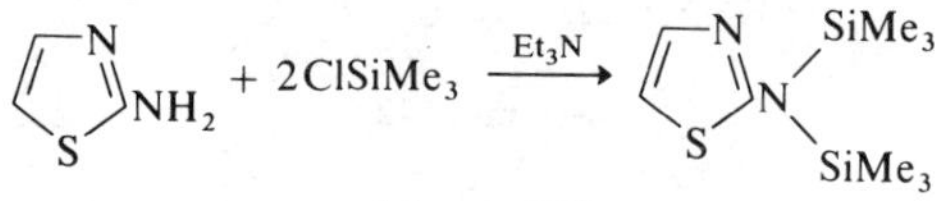

Scheme 119

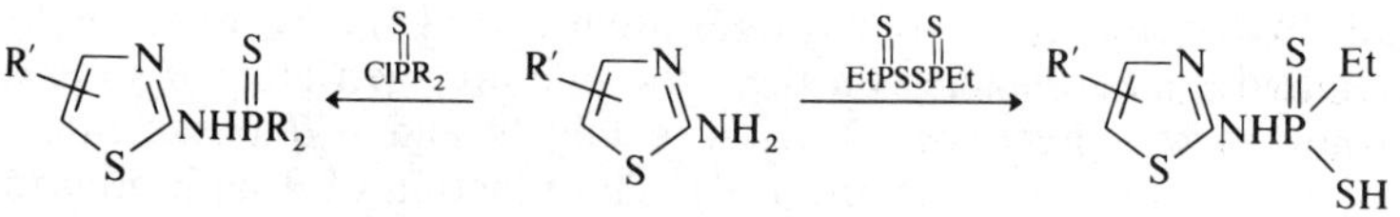

Scheme 120

(ii) *Imines*

The alkylation of 2-acetamido-5-nitrothiazole with alkyl halides or with bromoacetic acid in DMF, occurs predominantly on the ring nitrogen atom, affording iminothiazolines (Scheme 121). 2-Imino-3-alkylbenzothiazoline (**177**) reacts with acrylic acid yielding the *N*-substituted derivative (**178**; Scheme 122). 3-Substituted 2-nitrosoiminothiazolines (**179**) or benzothiazolines (**180**) are relatively stable by virtue of their high resonance stabilization. Their photolysis occurs with the loss of nitric oxide and leads to open-chain disulfides (Scheme 123). The reversible oxidation of the yellow thiazol-2-one azine (**181**) reveals two reversible one-electron exchanges, through the very stable radical cation (**182**) to the red azo dication (**183**; Scheme 124).

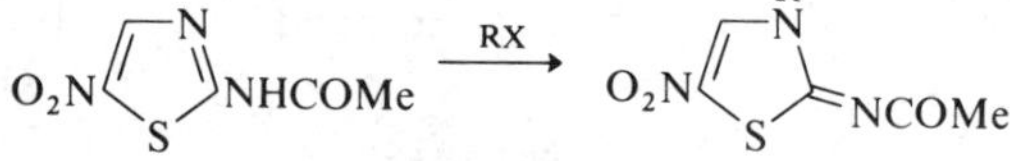

Scheme 121

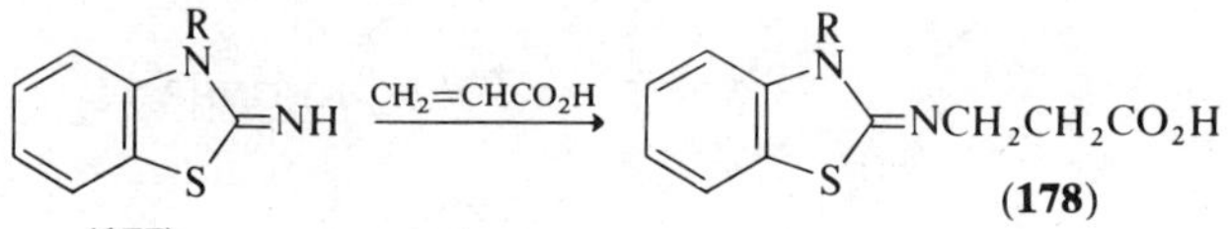

Scheme 122

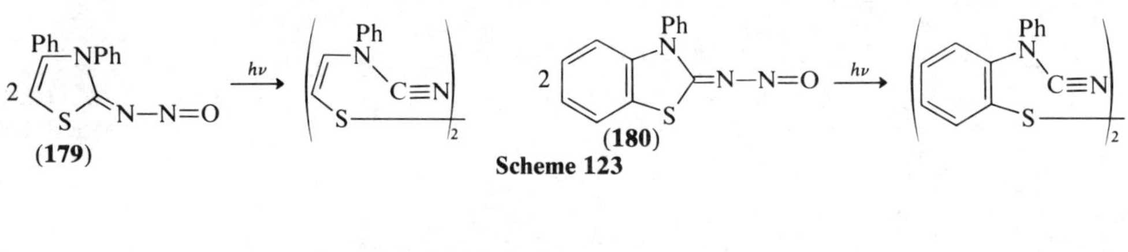

Scheme 123

(181) (182) (183)

Scheme 124

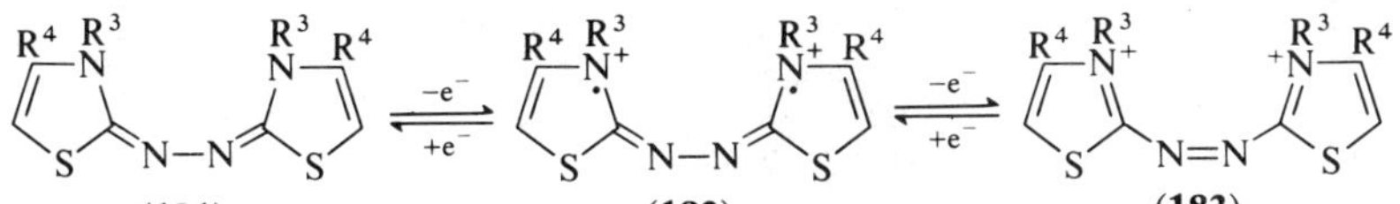

(iii) *Formazans*

2-Benzothiazolylformazans form complexes of composition L_2M^{2+} with nickel, copper and cobalt cations (**184**).

(**184**)

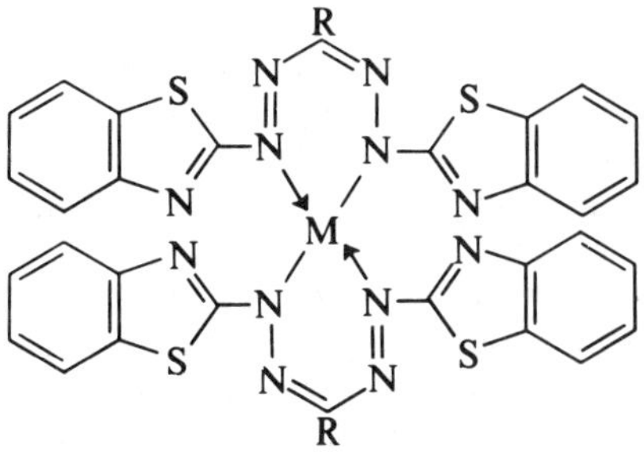

(iv) *Nitro groups*

2-Nitrothiazole reacts with nucleophiles, such as MeO^- or F^-, yielding the normal *ipso* substitution product (Scheme 125). The 5-nitrothiazole, under similar conditions, decomposes.

2-Nitrothiazoles are reduced to 2-aminothiazoles, either catalytically or chemically, whereas the reduction of 5-nitrothiazole gives poor yields of impure 5-aminothiazole. 2-Nitrobenzothiazole is reduced to the azo derivative by tin(II) chloride in basic medium (Scheme 126).

Scheme 125

Scheme 126

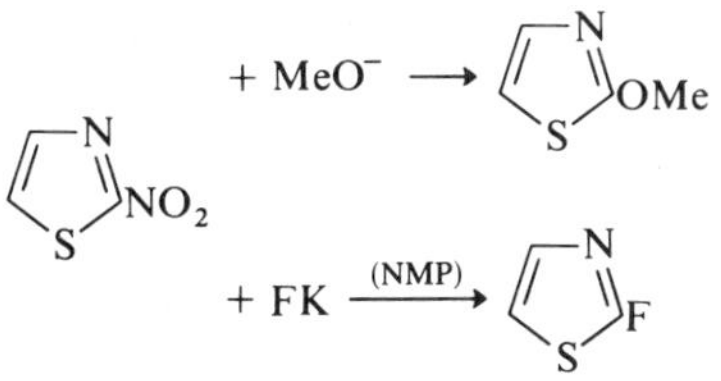

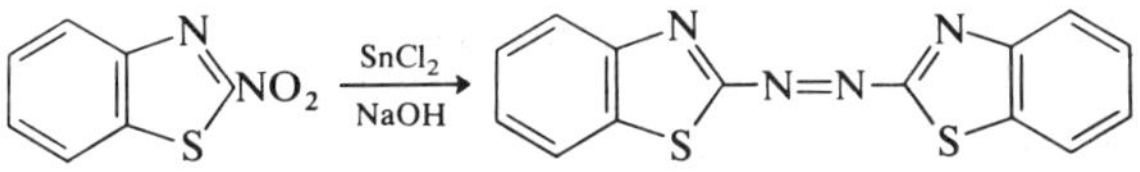

4.19.2.3.5 O-Linked substituents

As already discussed in Section 4.19.2.2.1 hydroxyl derivatives of thiazole are tautomeric, according to Scheme 127, the oxo forms usually being the most stable. The interconversion of the hydroxyl and carbonyl forms proceeds under acid or base catalysis. This protomeric behavior of the thiazolinones corresponds to an ambident nucleophilic reactivity which is most marked for the conjugate anion arising in basic medium.

(i) *2-Hydroxythiazoles*

2-Hydroxythiazole and benzothiazole exist mainly in the Δ^4-thiazolin-2-one form (**185b**). They are quite stable and react mostly in addition–elimination reactions rather than undergoing ring opening reactions.

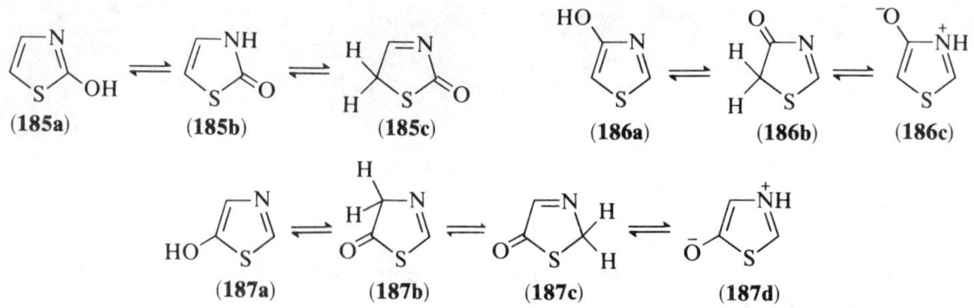

(185a) (185b) (185c) (186a) (186b) (186c)

(187a) (187b) (187c) (187d)

Scheme 127

(a) *Electrophilic attack at oxygen.* 3,4-Dimethyl-Δ^4-thiazolin-2-one (**188**) is a weak base which does not undergo protonation in media of acidity up to $H_0 = -3.99$. The protonation occurs at the carbonyl oxygen atom (Scheme 128), the conjugate acid (**189**) being mesomeric. Alkylation of Δ^4-thiazolin-2-ones may yield O—R or N—R derivatives according to experimental conditions. With diazomethane in ethanol, *O*-methylation takes place whereas *N*-methylation occurs when a basic solution of the thiazolinone reacts with methyl iodide (Scheme 129). Similarly benzothiazolin-2-one is *N*-alkylated by alkyl halides in the presence of sodium ethoxide. More probably the site of reaction, in basic medium, depends on the size of the R^4 substituent: the sodium salt of 4-methyl-Δ^4-thiazolin-2-one reacts with $ClPO(OEt)_2$ or $ClPS(OEt)_2$ in dry acetone providing the corresponding diethyl 4-methylthiazolyl 2-phosphate or thiophosphate (Scheme 130). When phosphoryl chloride is used, the reaction goes further and 2-chlorothiazole is obtained (Scheme 131).

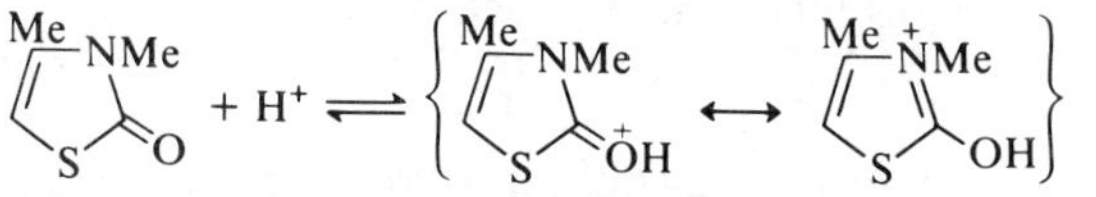

(188) (189a) (189b)

Scheme 128

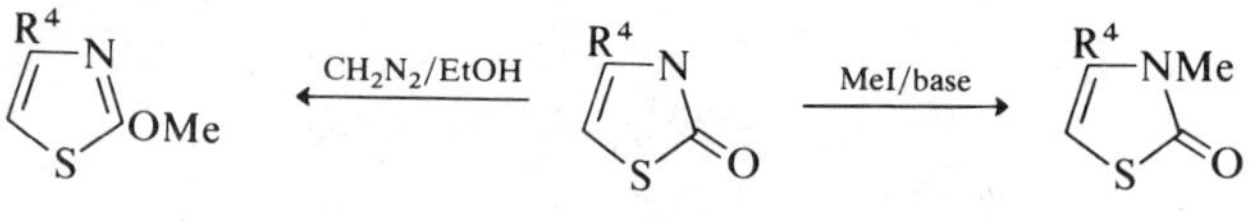

Scheme 129

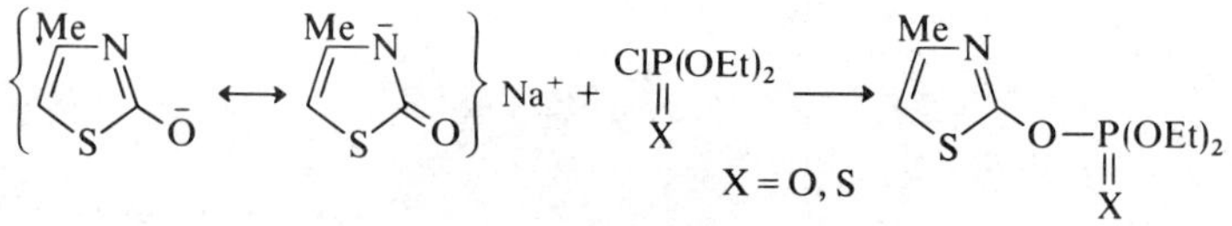

Scheme 130

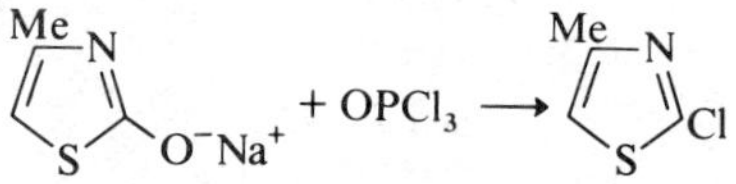

Scheme 131

(b) *Nucleophilic displacements.* The last reaction is typical of the attack on the 2-carbonyl function by a strong nucleophile. In a similar way dry HCl in ether converts 4-aryl-Δ^4-thiazolin-2-ones to the corresponding 2-chlorothiazoles (Scheme 132). Phosphorus pentasulfide attacks the carbonyl group affording the corresponding thiocarbonyl derivative (Scheme 133). The formation of thiochrome (**191**) results from an intramolecular nucleophilic attack on the carbonyl group of the 2-oxo derivative of thiamine (**190**; Scheme 134).

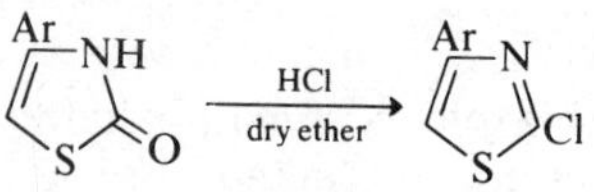

Scheme 132

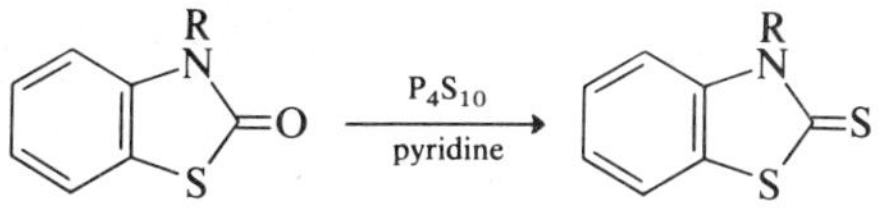

Thiazoles and their Benzo Derivatives

Scheme 133

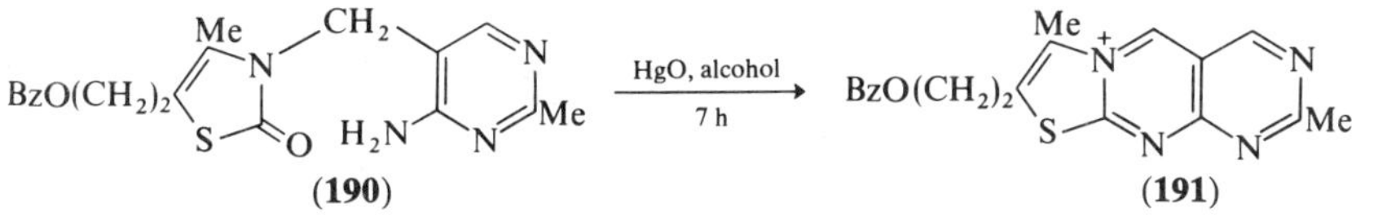

Scheme 134

(ii) 4- and 5-Hydroxythiazoles

Although 4- and 5-hydroxythiazoles may also exist in a tautomeric equilibrium (Scheme 127), their oxo forms (**186b**, **187b**, **187c**) are not cyclic conjugated structures as are those of 2-hydroxythiazole (**185b**). Complementary zwitterionic structures (**186c**, **187d**) have to be considered which correspond to the mesoionic 4- and 5-hydroxythiazole derivatives (**192**, **193**). These mesoionic thiazol-4- and -5-ones are hybrids of the dipolar structures shown in Scheme 135. The 4- and 5-hydroxythiazoles exist mainly under the non-aromatic Δ^2-thiazolin-4- and -5-one structures (**186b**, **187b**). They are less stable than the Δ^4-thiazolin-2-ones and some of their reactions proceed with ring opening. The pK_a values of 2-substituted 4-hydroxythiazoles range between 6.6 and 6.9 pK_a units.

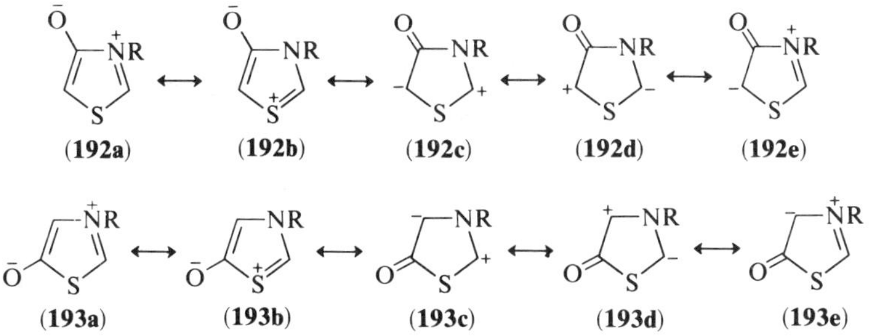

Scheme 135

Δ^2-Thiazolin-4-one presents three nucleophilic centers (C-5, oxygen and nitrogen) and two electrophilic centers (C-4 and C-2). Catalyzed by acids, dimerization yields compound (**194**) ⟨65ACS1215⟩ whereas dilute sodium hydroxide induces the formation of compound (**195**) ⟨66CR(C)(262)1017⟩. The nucleophilic reactivity of the oxygen atom has been observed in the acetylation by acetic anhydride of 2-aryl- and 2-heteryl-Δ^2-thiazolin-4-ones (Scheme 136). 2-Alkoxy and 2-methyl derivatives of Δ^2-thiazolin-4-one (**196**) react with OPCl$_3$ to yield thiazolylphosphoric esters (**197**) which have insecticidal uses (Scheme 137). An example of the electrophilic reactivity of the C-4 atom is the easy formation of oxime and phenylhydrazone derivatives. 5-Aryl-Δ^2-thiazolin-4-one (**198**) gives the 1,3-dipolar cycloaddition product (**199**) with methyl fumarate and methyl maleate (Scheme 138). Under similar conditions, treatment of (**198**) with dimethyl acetylenedicarboxylate (DMAD) yields a thiophene derivative (**202**) when R = Ph and a pyridone derivative (**203**) when R = H (Scheme 139). The proposed mechanism involves the formation of a mesoionic intermediate (**200**) which reacts in a cycloaddition with a second molecule of DMAD, yielding (**201**), the decomposition of which depends on the R substituent.

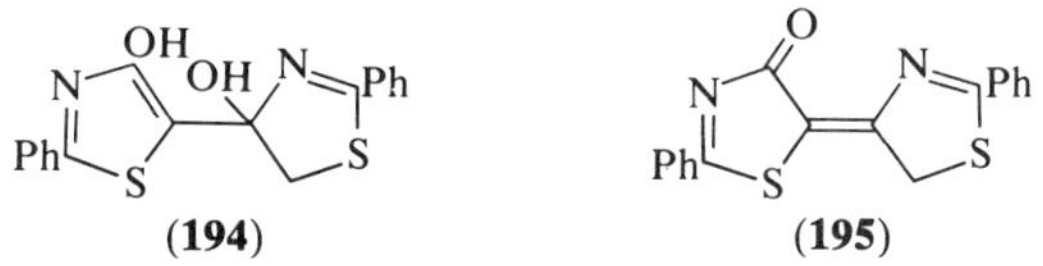

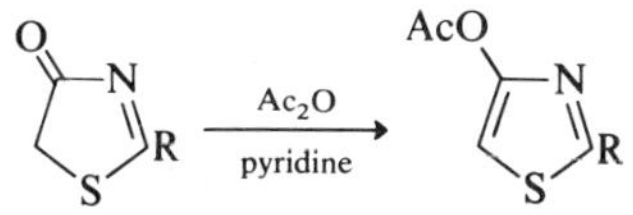

Scheme 136

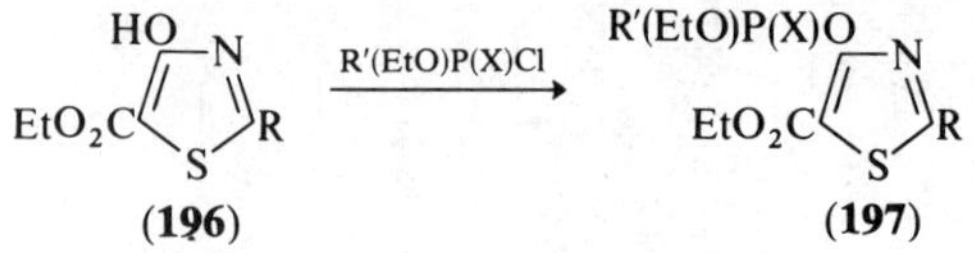

288   *Thiazoles and their Benzo Derivatives*

Scheme 137

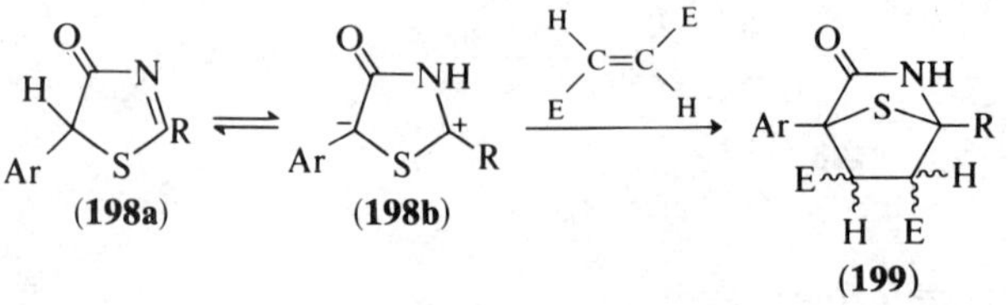

Scheme 138

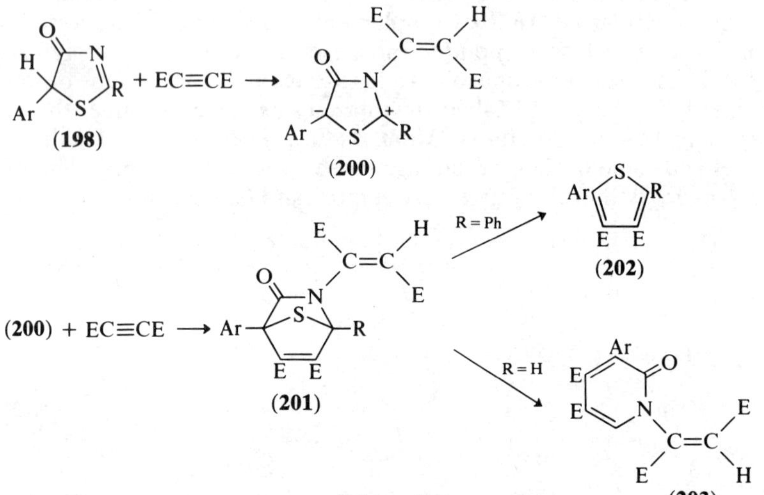

Scheme 139

The reactivity of Δ²-thiazolin-5-ones is closely related to the protomeric equilibrium shown in Scheme 127. The nucleophilic activity of the C-5 oxygen atom is clearly demonstrated by its reaction with acetyl chloride, acetic anhydride, benzoyl chloride, methyl or phenyl isocyanate, carbamoyl chloride or phosphorus derivatives in the presence of bases, to give (**205–208**; Scheme 140).

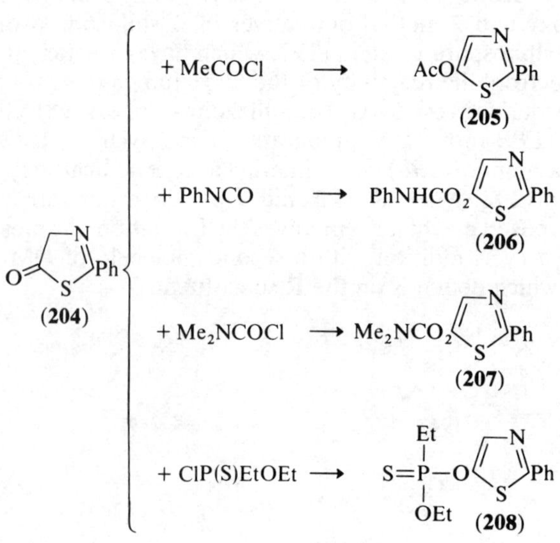

Scheme 140

(iii) Alkoxyl groups

2-Methoxythiazoles (**209**) are converted into the corresponding N-methyl-Δ^4-thiazolin-2-ones (**210**) by heating with excess methyl iodide. The reaction mechanism involves the formation of a quaternary salt followed by nucleophilic attack on the methyl carbon atom of the 2-methoxy group. The reaction is reversible (Scheme 141). On heating, 2-allyloxythiazole (**211**) undergoes a thermal rearrangement to N-allyl-Δ^4-thiazolin-2-one (**212**) in excellent yield (Scheme 142). A first order rate law is found and activation parameters show a negative entropy in accord with that measured in most Claisen rearrangements. 2-Alkoxy-3-methylbenzothiazolylium salts can be used to alkylate various nucleophiles including thiols, phenols, halide ions and cyanide ions (Scheme 143). This reactivity shows that 3-methylbenzothiazol-2-one is a good leaving group. 2-Aryloxybenzothiazoles undergo the Fries rearrangement under the influence of Lewis acids, or on photolysis (Scheme 144).

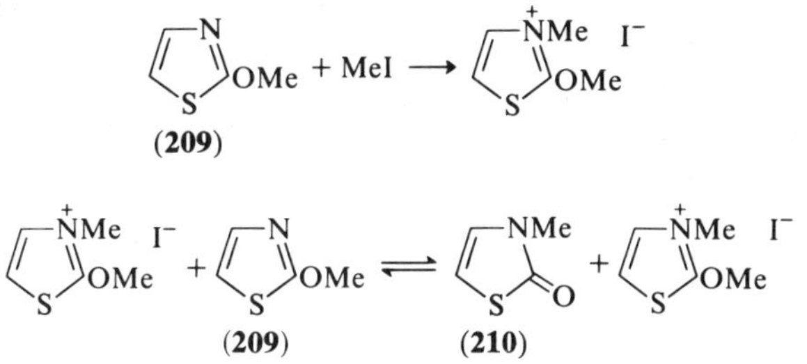

Scheme 141

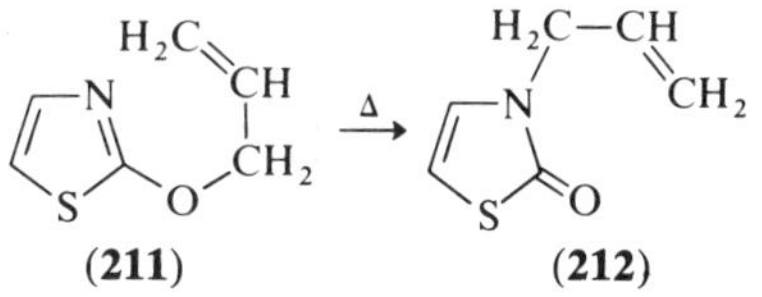

Scheme 142

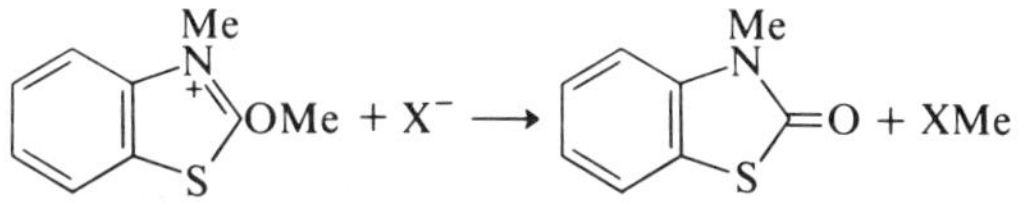

Scheme 143

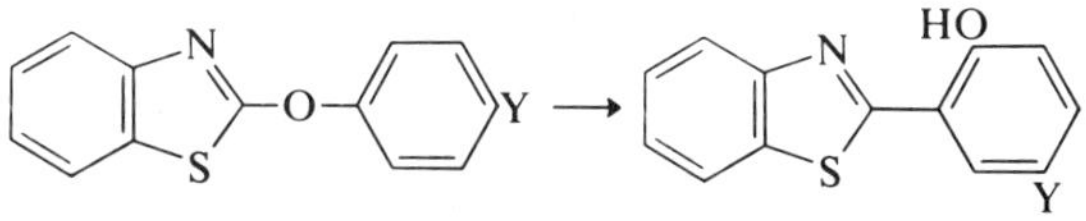

Scheme 144

5-Alkoxythiazoles are easily cleaved by acids to give the 5-hydroxy derivatives.

4.19.2.3.6 S-Linked substituents

(i) Mercapto compounds: tautomerism

The same tautomerism as that described for hydroxythiazoles is observed for mercaptothiazoles (see Sections 4.19.2.3.5(i) and (ii)). 2-Mercaptothiazole exists predominantly in the Δ^4-thiazoline-2-thione form (**213**) whereas 2-phenylthiazole-5-thiol exists in the thiol form (**214**).

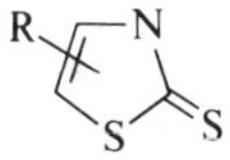

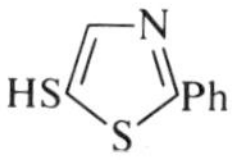

(ii) *Thiones*

Reactivity of Δ^4-thiazoline-2-thiones and derivatives involves four main possibilities: nucleophilic reactivity of the exocyclic sulfur atom or of the ring nitrogen atom, electrophilic reactivity of carbon 2, and electrophilic substitution on carbon 5.

Nucleophilic reactivity of the sulfur atom is observed in neutral or acidic medium. Rate constants for methylation by methyl iodide are 200 times greater for 3-methyl-Δ^4-thiazoline-2-thiones (**215**) than for the isomeric 2-(methylthio)thiazoles (**216**). Bulky substituents on the nitrogen atom increase the nucleophilic reactivity of the exocyclic sulfur atom, which is related to a steric decompression of the thiocarbonyl group in the transition state. Oxidation of Δ^4-thiazoline-2-thione by N_2O_4 in chloroform yields 76% of the 2-sulfonic acid whereas reaction with H_2O_2 in acid medium yields the unstable sulfinic acid, which then loses SO_2 to give good yields of 2-unsubstituted thiazoles (Scheme 146). Under basic conditions the reactive species is the ambident anion (**217**). Reaction takes place on the nitrogen atom when the electrophilic center is an sp^2 carbon atom. Acrylonitrile reacts with the sodium salt of 4,5-dimethyl-Δ^4-thiazoline-2-thione (**218**) to yield 3-(2-cyanoethyl)-4,5-dimethyl-Δ^4-thiazoline-2-thione (**219**) but the size of the R^4 substituent is a determining factor in the S *versus* N ratio.

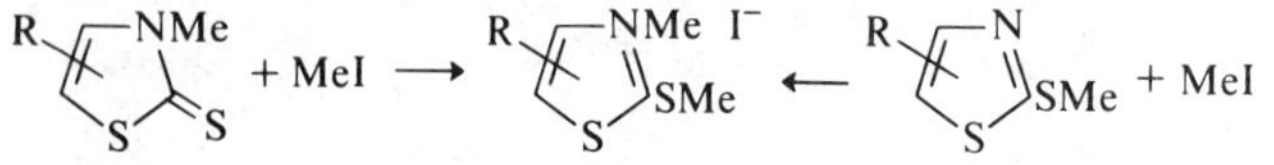

Scheme 145

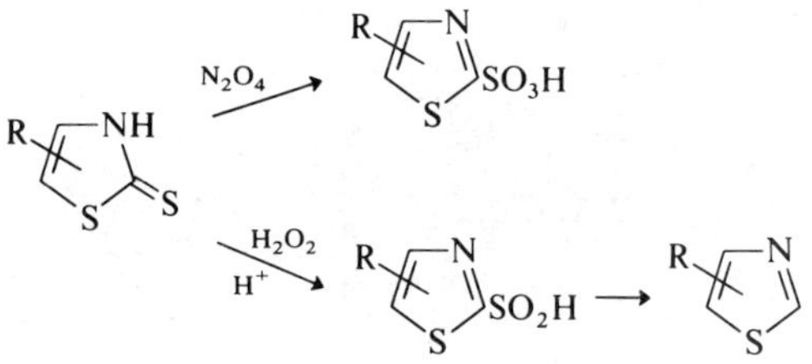

Scheme 146

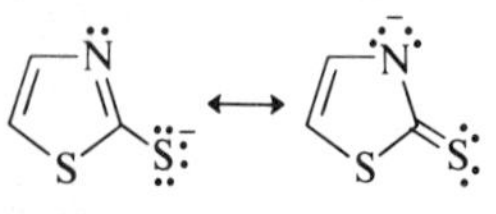

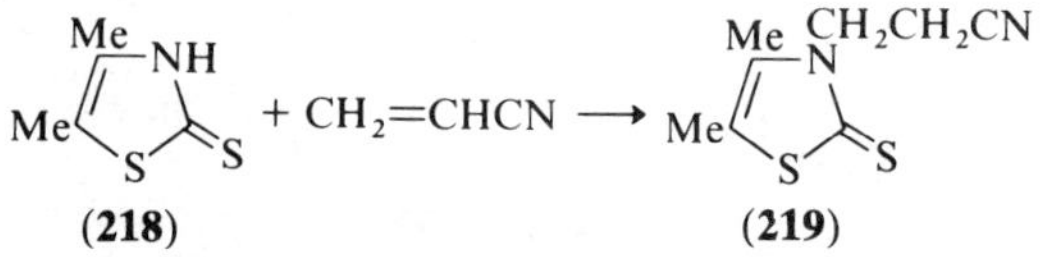

The electrophilic character of the C(2) atom in Δ^4-thiazoline-2-thiones is demonstrated in the Cook rearrangement of the 5-amino derivative ⟨47JCS1594⟩ (Scheme 147).

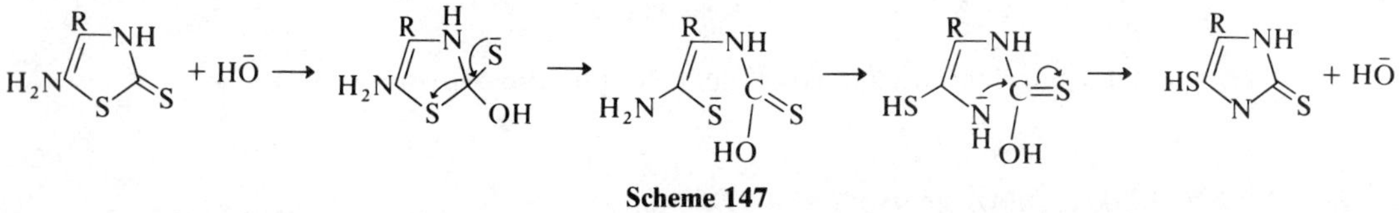

Scheme 147

Electrophilic substitution at C-5 occurs when Δ^4-thiazoline-2-thiones are treated with KSCN and bromine or monochlorourea. The same reactivity of C-5 is observed when 3-(morpholinomethyl)-Δ^4-thiazoline-2-thione is heated in ethanol giving after 7 hours heating the 5-(morpholinomethyl)-Δ^4-thiazoline-2-thione (Scheme 148) ⟨63ZOB3667⟩.

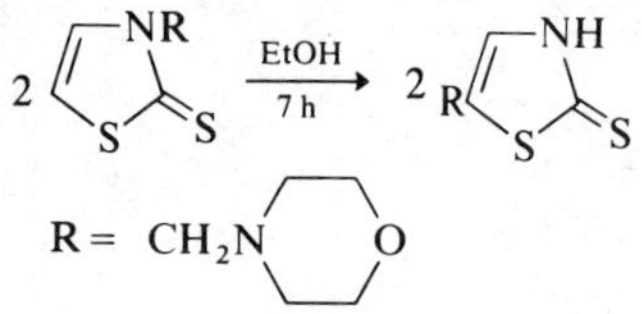

Scheme 148

2-Phenylthiazole-5-thiol exists in the thiol form in both polar and non-polar solvents, in contrast to the behavior of the corresponding thiazolinone. Its alkylation in alkaline medium takes place on the exocyclic sulfur atom. 5-Thiazolyl sulfides are reactive toward oxidizing agents, yielding the corresponding 5-sulfones (Scheme 149).

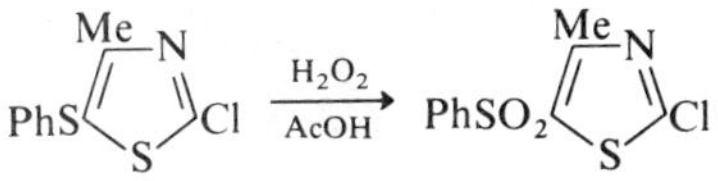

Scheme 149

(iii) Alkylthio groups

Like their oxygen analogues, 2-alkylthiothiazoles (**220**) rearrange to 3-alkyl-Δ^4-thiazoline-2-thiones (**222**) when heated with iodine, or preferably, with catalytic amounts of the corresponding quaternary salt (**221**; Scheme 150). The values of ΔH, ΔS and ΔG associated with this equilibrium measure the importance of steric interactions between R^3 and R^4 substituents in the Δ^4-thiazoline-2-thione ring (Table 25). If 3-methyl-4-t-butyl-Δ^4-thiazoline-2-thione (**223**) is heated with the catalyst, the isomeric thioether (**224**) is formed in quantitative yield.

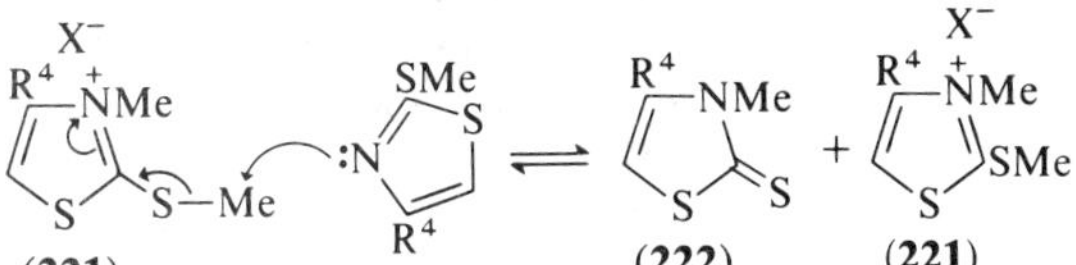

Scheme 150

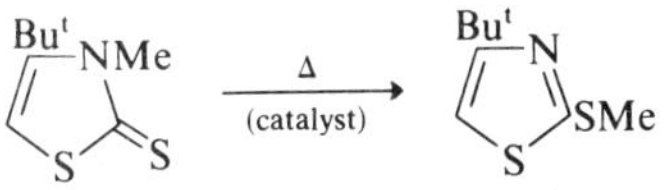

Table 25 Thermodynamic Values Associated with the Equilibrium (**220**) ⇌ (**222**)
⟨79HC(34-2)1, p. 407⟩

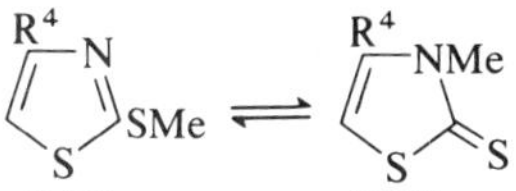

R^4	ΔH (J mol^{-1})	ΔS (J mol^{-1} K^{-1})	ΔG (J mol^{-1})
H	21 318	117	8360
Me	28 424	192	6688
Pr^i	29 260	280	−1672

2-Methylthio groups in thiazolylium cations are readily substituted by nucleophilic carbanions generated in the medium, this reaction being used in a classical synthesis of merocyanines dyes (Scheme 151).

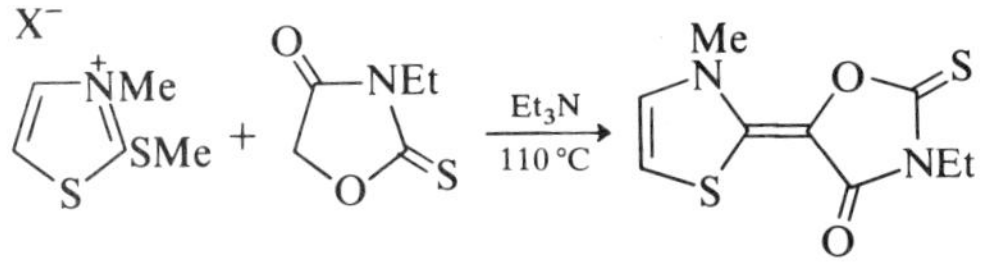

Scheme 151

(iv) Sulfonic acid groups

2-Thiazolesulfonic acid reacts with nucleophiles leading to the corresponding 2-substituted compounds (Scheme 152).

Thiazole-4- or -5-sulfonic acids are relatively stable substances.

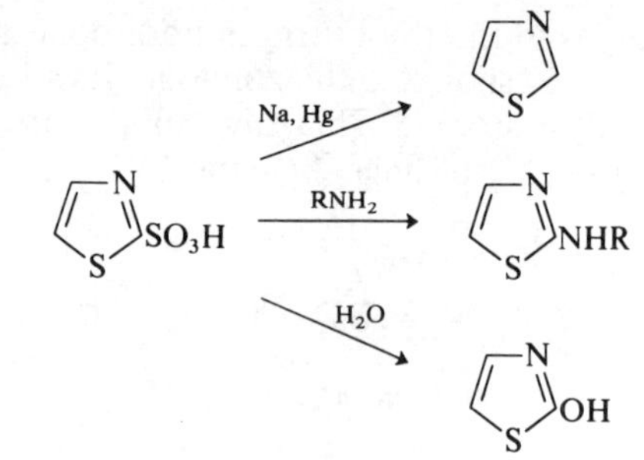

Scheme 152

4.19.2.3.7 Halogen atoms

(i) *Nucleophilic substitution reactions: neutral thiazoles*

Halogens bonded to the thiazole ring exhibit different degrees of reactivity depending on the position of bonding and on the reagents employed. 2-Halogenothiazoles easily react with nucleophiles leading to the expected products of nucleophilic substitution. They are more reactive than the corresponding 2-halogenopyridines (Table 26). Halogen atoms in the 4- and 5-positions of thiazoles exhibit different reactivities depending on the nucleophile: with methoxide ion in methanol the reactivity sequence is 5-chlorothiazole > 2-chlorothiazole > 4-chlorothiazole, while with thiophenoxide ion or piperidine in methanol, the sequence 2 > 4 > 5 is observed. With the series of alkoxide ions methoxide, ethoxide, isopropoxide and *t*-butoxide in the corresponding alcohol, the reactivity ratios of 4-chlorothiazole *versus* 2-chlorothiazole are: 0.07, 0.6, 4, 42. The nucleophilic reactivity of 2-halogenothiazoles is strongly affected by substituents: 2-chloro-5-nitrothiazole is more reactive than 2-chlorothiazole by a factor of 10^8.

Table 26 Relative Reactivities of Some Halogenothiazoles and Pyridines ⟨79HC(34-1)1, p. 568⟩

	MeO^- in MeOH	PhS^- in MeOH		MeO^- in MeOH	PhS^- in MeOH
2-Fluorothiazole	2220	27	2-Iodothiazole	0.5	2.5
2-Chlorothiazole	1	1	2-Chlorobenzothiazole	416	422
2-Bromothiazole	1.3	3.9	2-Chloropyridine	4×10^{-3}	—

(ii) *Other reactions*

Halogen atoms in position 2 can be easily replaced by hydrogen using zinc and acetic acid or tin and hydrochloric acid. Catalytic reduction can also be realized with Raney nickel in the presence of potassium hydroxide or triethylamine. 2-Bromothiazole can be reduced electrochemically.

Halogenothiazoles react with *n*-butyllithium leading to the corresponding thiazolyllithium. Similarly the Grignard reagents can be prepared in all thiazole positions using magnesium in the presence of an excess of magnesium bromide while the normal procedure fails.

Treating 2-bromothiazole with copper at 170 °C in cumene as solvent affords the 2,2'-bithiazole.

4.19.2.3.8 Metals and metalloid-linked substituents

Metallated thiazoles show expected properties. Thus 2-lithiothiazole reacts normally, as shown in Scheme 153. Acetoxymercuriothiazoles are hydrodemetalled by hydrochloric acid and converted into the corresponding bromo derivative by bromine in carbon tetrachloride (see Section 4.19.2.1.4(vii); Scheme 154).

The silicon–heterocyclic bond of 2-trimethylsilylbenzothiazole (**225**) is unusually labile. It is cleaved smoothly by benzaldehyde, with formation of the silyl ether (**226**) which is hydrolyzed to the alcohol (**227**); the cleavage by benzoyl chloride affords the corresponding ketone (**228**; Scheme 155).

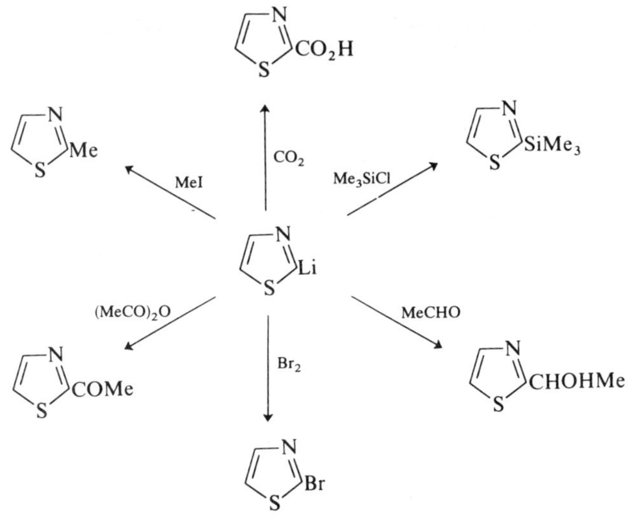

Scheme 153

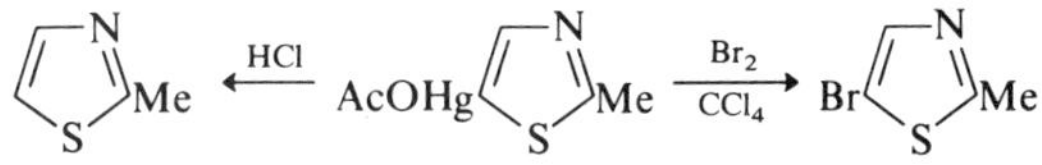

Scheme 154

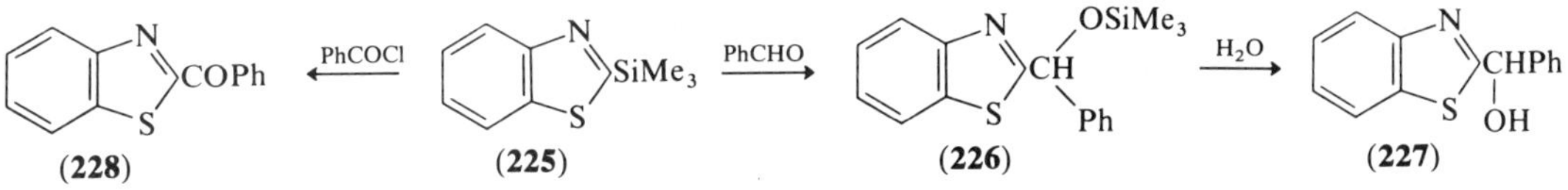

Scheme 155

4.19.2.3.9 Substituents attached to ring nitrogen atoms

(i) *Alkyl groups*

N-Alkyl groups in thiazole quaternary salts can be removed by nucleophilic S_N2 reactions. This reaction which is the reverse of quaternization (see Section 4.19.2.1.3(iv)), is best achieved with soft nucleophiles like triphenylphosphine in an aprotic solvent (Scheme 156).

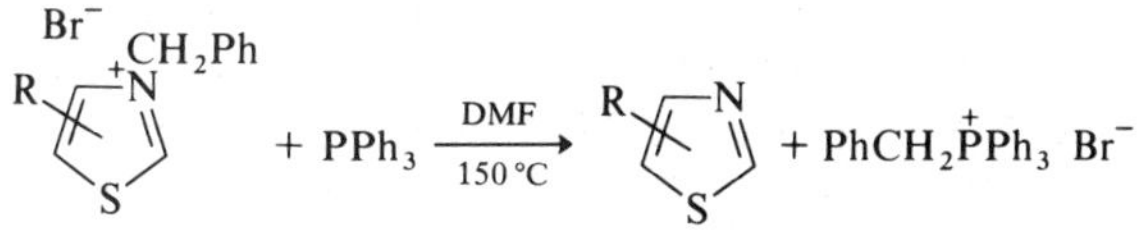

Scheme 156

N-Phenacyl-2-methylthiazolylium salts (**229**) react under aprotic conditions with sodium acetate in acetic anhydride to form a pyrrolo[2,1-*b*]thiazole (**230**). When the reaction is carried out in an aqueous solution of sodium hydrogen carbonate, extension of the ring occurs with formation of the dihydro-1,4-thiazine derivative (**231**).

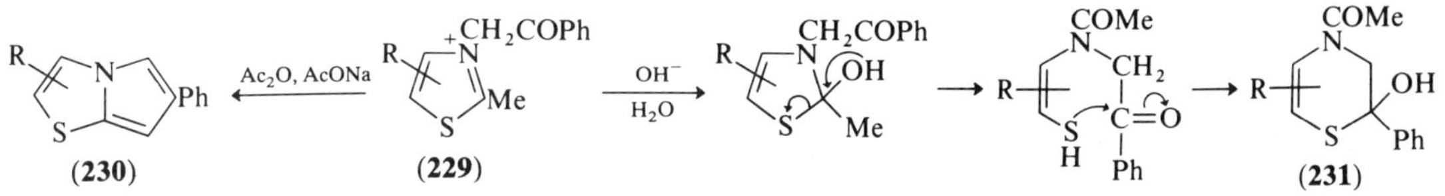

4.19.3 SYNTHESES

The numerous syntheses of thiazoles and benzothiazoles are classified according to the nature of the components which join to form the ring system, following the scheme proposed by Sprague and Land ⟨B-57MI41900⟩.

In the case of monocyclic compounds (thiazoles, thiazolines, thiazolidines) the syntheses are classified as shown in Scheme 157. For the benzothiazoles they are classified according to Scheme 158.

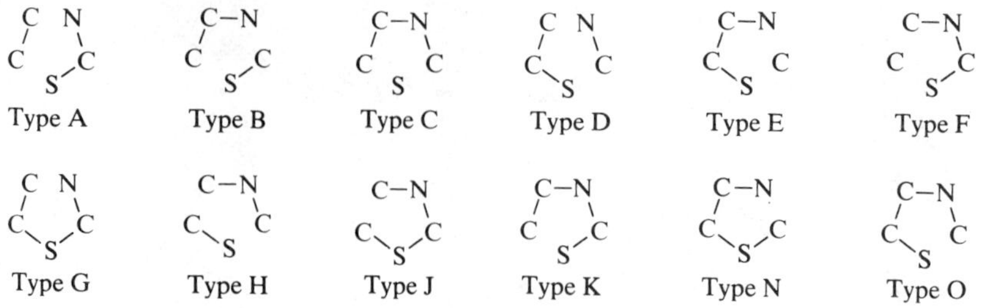

Type A   Type B   Type C   Type D   Type E   Type F

Type G   Type H   Type J   Type K   Type N   Type O

Scheme 157

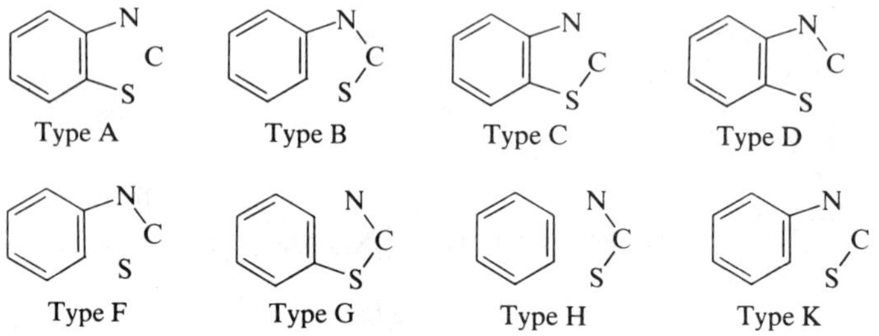

Type A   Type B   Type C   Type D

Type F   Type G   Type H   Type K

Scheme 158

4.19.3.1 Thiazoles

4.19.3.1.1 *Synthesis from α-halocarbonyl compounds (Type A): Hantzsch's synthesis*

First described in 1887 ⟨1887CB3118⟩ by Hantzsch, the cyclization of α-halocarbonyl compounds by a great variety of reactants bearing the N—C—S fragment of the ring is still the most widely used method of synthesis of thiazoles.

(i) *Reaction with thioamides*

Thiazole itself can be obtained by condensing chloroacetaldehyde and thioformamide (Scheme 159). The reaction is explosive and proceeds in low yield because of the instability of the thioformamide under acid conditions. Higher thioamides are more stable and react under milder conditions with chloroacetaldehyde, affording 2-substituted thiazoles in moderate yields. It is possible, and often preferable, to prepare the thioamide *in situ* in dioxane solution by the action of phosphorus pentasulfide on the corresponding amide and in the presence of solid $MgCO_3$ (Scheme 160). With arylthioamides, except for some nitrothiobenzamides, yields are usually higher and the cyclization is carried out over several hours in boiling absolute alcohol. Chloroacetaldehyde can be replaced in these reaction by derivatives such as 1,2-dichloro- or dibromo-ethyl methyl or ethyl ether, 1,2-dichloro- or dibromo-ethyl acetate, 2-chloro- or dibromo-ethyl acetate, and 2-chloro or bromo-diethyl-acetal.

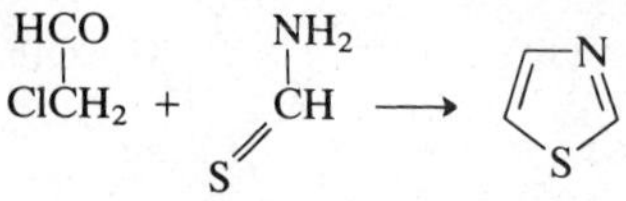

Scheme 159

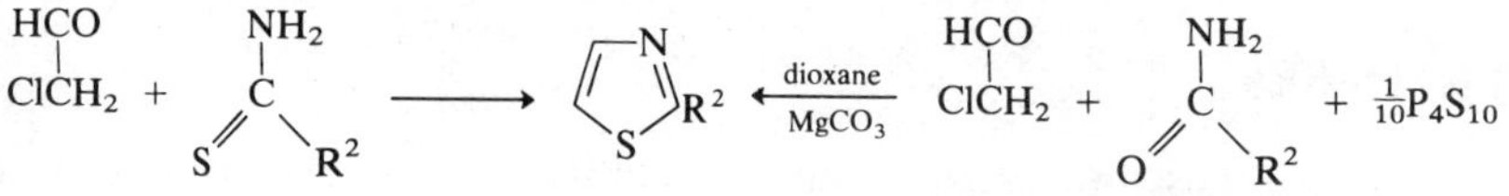

Scheme 160

5-Mono- and 2,5-di-substituted thiazoles are obtained in a similar way by the cyclization of higher α-halo aldehydes with thioformamide or higher thioamides (Scheme 161). The yields usually decrease with the size of the R^2 chain of the thioamide.

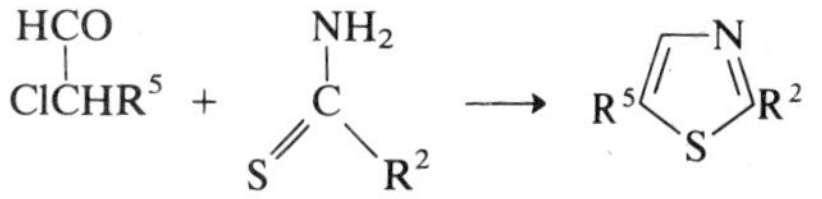

Scheme 161

One of the causes of the moderate yields obtained in this group of reactions is the instability of the α-halo aldehydes used.

α-Halo ketones condense with thioformamide and higher thioamides giving 4-, 2,4-, 4,5- and 2,4,5-substituted thiazoles (Scheme 162). Bromo ketones lead to better yields than chloro ketones and, similarly, generation of the thioformamide *in situ* leads to better results. 1-Chloro-1,2-epoxides can replace the halo ketones in this reaction (Scheme 163).

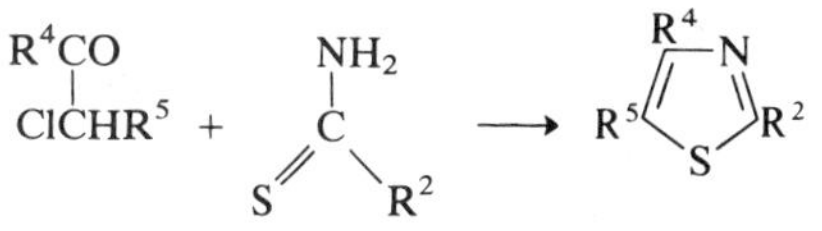

Scheme 162

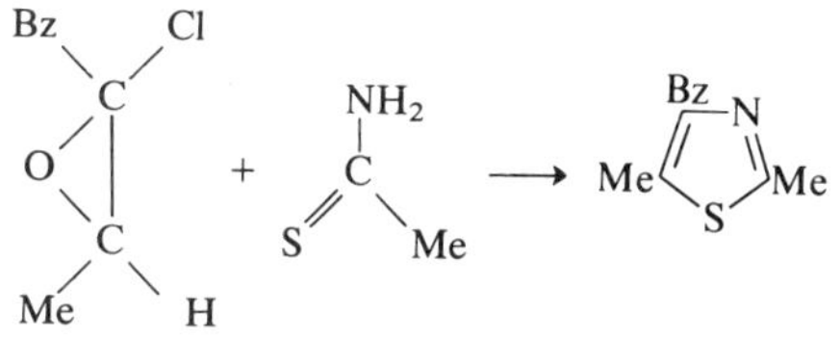

Scheme 163

Many functional derivatives incorporating a thiazole ring in their structure have been produced by this method. An illustrative example is the condensation of pentaacetyl-D-galactonic acid thioamide with chloroacetone to give 4-methyl-2-(D-galactopentaacetoxypentyl)thiazole which is deacetylated in the usual way (Scheme 164) ⟨69ACH(62)179⟩.

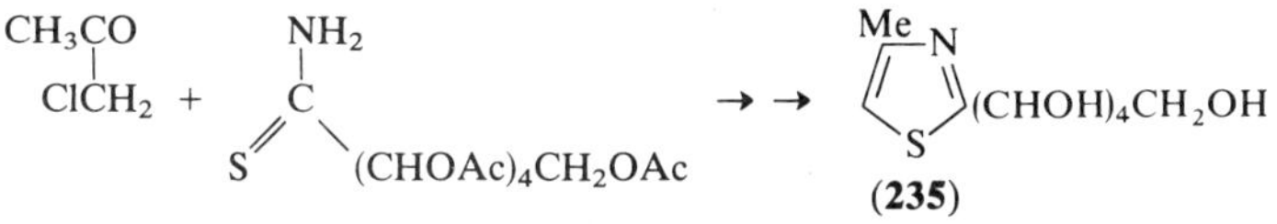

Scheme 164

The mechanism of the Hantzsch synthesis has been established and is shown in Scheme 165. Substitution of the halogen atom of the α-halo ketone by the sulfur atom of the thioamide occurs first to give an open-chain α-thioketone (232), which under transprotonation proceeds to give a 4-hydroxy-Δ²-thiazoline (233) in aprotic solvents, or a thiazole (234) by acid-catalyzed dehydration of the intermediate thiazoline in protic solvents.

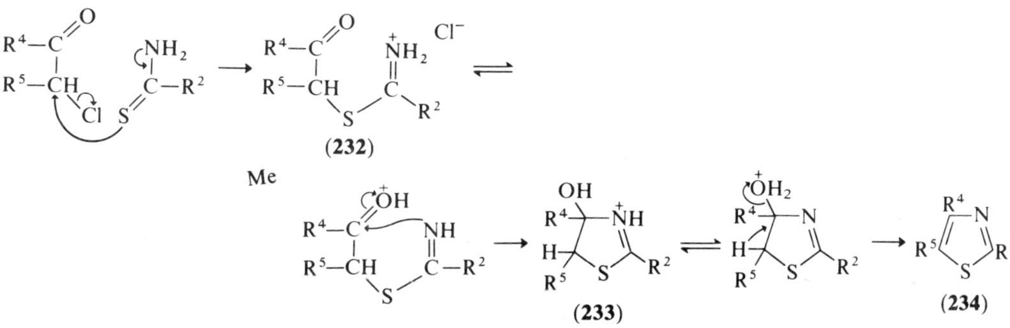

Scheme 165

(ii) Reaction with N-substituted thioamides (thiazolylium salts)

The application of the Hantzsch synthesis to monosubstituted thioamides affords the corresponding N-substituted thiazolylium salts. This method is particularly valuable for the preparation of those thiazolylium compounds in which the substituent on the ring nitrogen atom cannot be introduced by direct quaternization, for example the synthesis of aryl or heteryl thiazolylium salts. In that case the intermediate α-thioketone can usually be isolated, its conversion into the thiazolylium salt occurring by simple heating (Scheme 166).

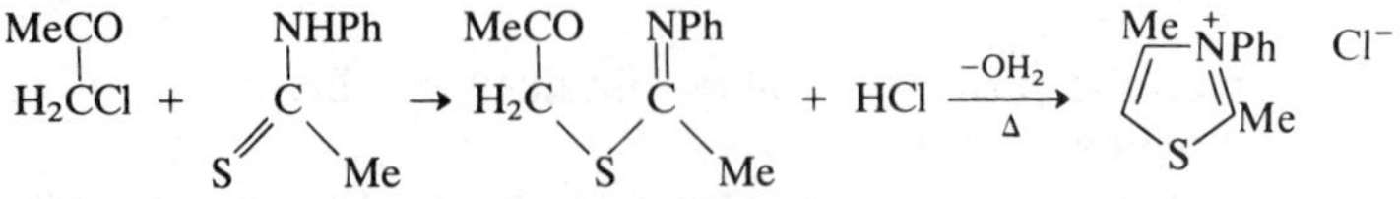

Scheme 166

(iii) Reaction with thiourea (2-aminothiazoles)

Of all the methods described for the synthesis of thiazole compounds, the most efficient involves the condensation of equimolar quantities of thiourea and α-halo ketones or aldehydes to yield the corresponding 2-aminothiazoles (Scheme 167) ⟨1888LA(249)31⟩. The reaction occurs more readily than that of thioamides and can be carried out in aqueous or alcoholic solution, even in a distinctly acid medium, an advantage not shared by thioamides which are often unstable in acids. The yields are usually excellent. A derived method condenses the thiourea (2 mol) with the non-halogenated methylene ketone (1 mol) in the presence of iodine (1 mol) or another oxidizing agent (chlorine, bromine, sulfuryl chloride, chlorosulfonic acid or sulfur monochloride) (Scheme 168) ⟨45JA2242⟩.

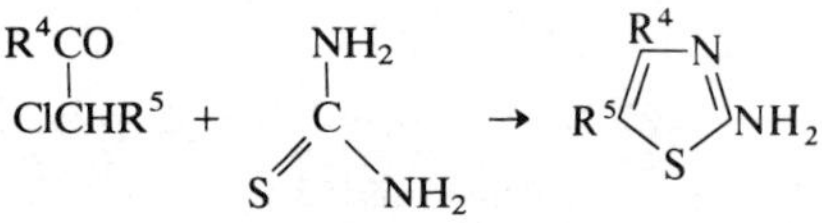

Scheme 167

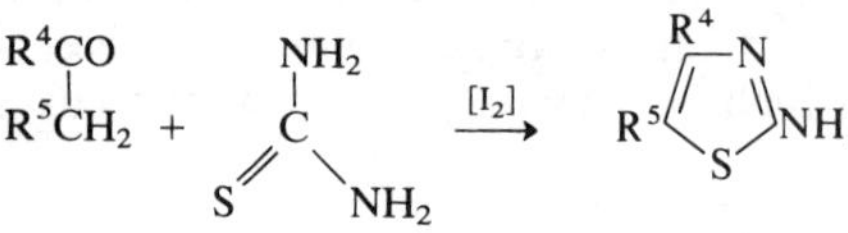

Scheme 168

The α-halo acids and their esters also react with thiourea to give 2-amino-4-hydroxythiazoles. Pseudothiohydantoin (**237**) is thus obtained by condensation of ethyl α-chloroacetate with thiourea and in the course of the reaction it is possible to isolate the acyclic intermediate (**236**; Scheme 169). Similarly 2,4-diaminothiazole hydrochloride is readily prepared by the reaction of thiourea with chloroacetonitrile in warm alcoholic solution (Scheme 170).

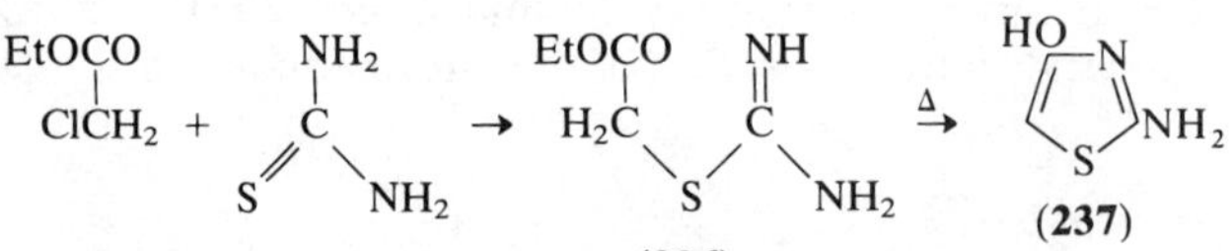

Scheme 169

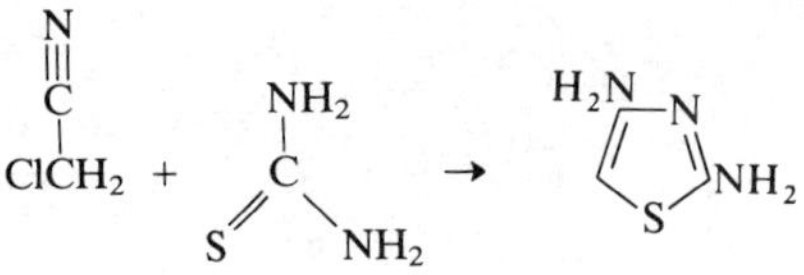

Scheme 170

A variation of Hantzsch's synthesis, using thiourea in conjunction with α-diazo ketones in place of α-halo ketones, has proved to be generally applicable and is shown in Scheme 171.

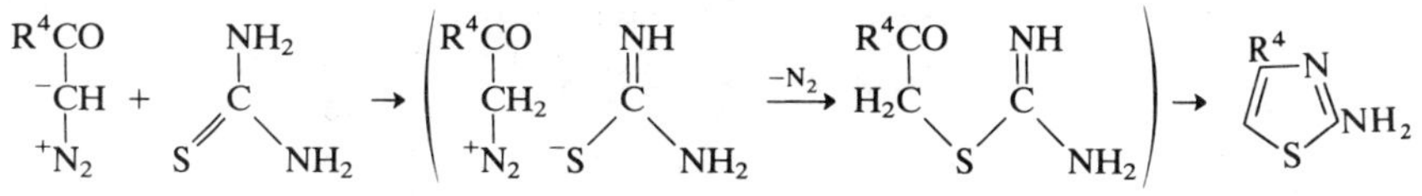

Scheme 171

2-Aminothiazoles are invaluable intermediates for the preparation of 2-unsubstituted thiazoles by diazotization followed by reduction with hypophosphorous acid. Application of the Sandmeyer reaction leads to 2-halogeno- and 2-cyano-thiazoles.

(iv) Reaction with N-substituted thioureas

The cyclization of *N*-substituted thioureas with halocarbonyl compounds gives 2-monosubstituted aminothiazoles (Scheme 172). When the α-halocarbonyl compound is derived from an acid or an ester, the *N*-aryl or *N*-acylthioureas afford the 2-monoaryl- or acylamino-4-hydroxythiazole (**238**) whereas the *N*-allylthiourea gives the isomeric 3-allyl-4-hydroxy-2-iminothiazole (**239**; Scheme 173). *N,N*-Disubstituted thioureas condense with α-halocarbonyl compounds to give 2-disubstituted aminothiazoles (Scheme 174) whereas *N,N'*-dialkylthioureas lead to 3-alkyl-2-alkylimino-Δ^4-thiazoline (Scheme 175). Thiosemicarbazide reacts with α-halocarbonyl compounds yielding the corresponding 2-hydrazinothiazoles (**241**). In neutral medium (alcohol) the acyclic intermediate (**240**), analogous to those obtained with thiourea and thioamides can be isolated, whereas under acid conditions a thiosemicarbazone intermediate (**242**) is obtained. This can be cyclized in alcohol either into the corresponding 2-hydrazinothiazole (**241**) in the presence of benzaldehyde or into the 1,3,4-thiadiazine (**243**) in the absence of benzaldehyde (Scheme 176) ⟨52CB1122⟩.

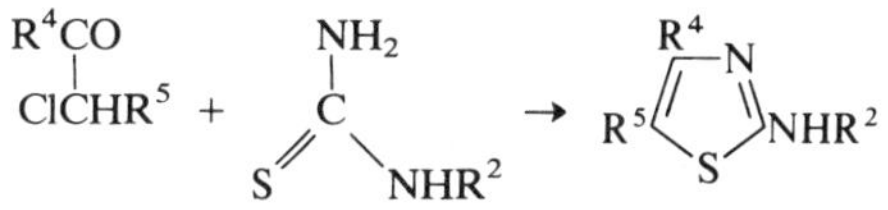

Scheme 172

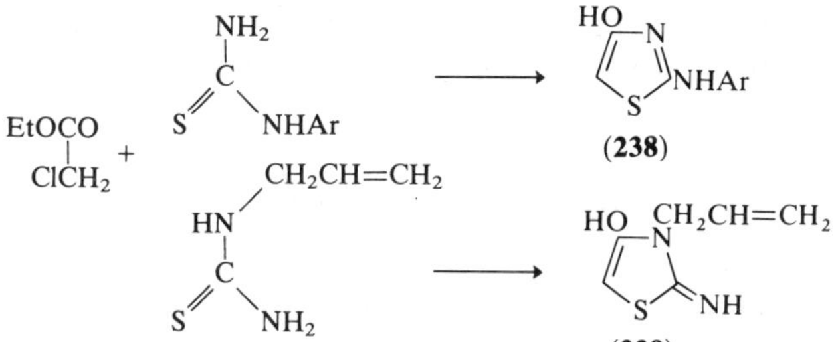

Scheme 173

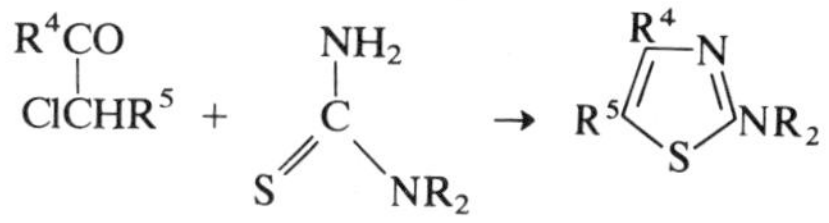

Scheme 174

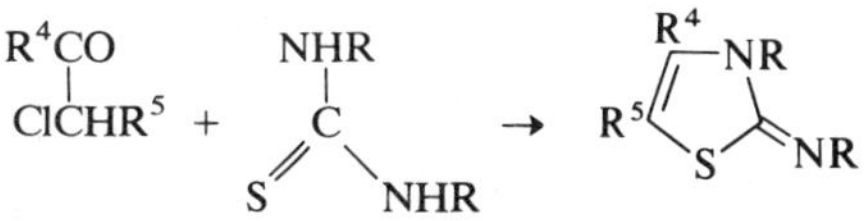

Scheme 175

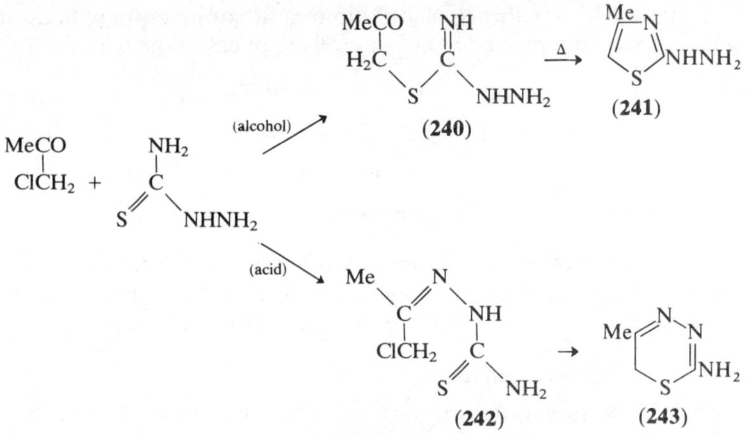

Scheme 176

(v) *Reaction with salts and esters of thiocarbamic acid (2-hydroxythiazoles)*

This method, initiated by Marchesini in 1893 ⟨1893G(24)65⟩ (Scheme 177), consists of the condensation of an α-halocarbonyl compound with ammonium thiocarbamate (**244**; $R^2 =$ NH$_4$) or its esters (**244**; $R^2 =$ alkyl). The reaction is carried out at low temperature in aqueous medium and then allowed to stand overnight. 2-Hydroxythiazoles (**244**; $R^2 =$ H) give 2-chlorothiazole derivatives almost quantitatively upon treatment with phosphorus oxychloride. This constitutes a convenient synthetic method for these compounds when the conversion of 2-amino- into 2-chloro-thiazole fails.

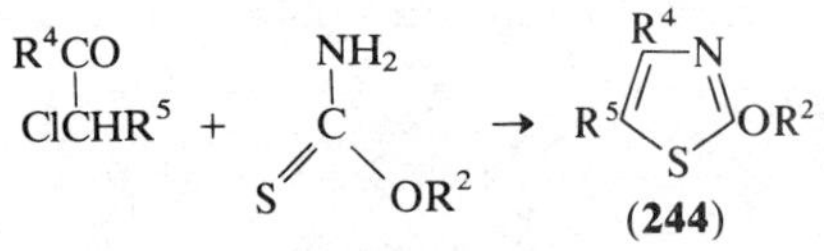

Scheme 177

(vi) *Reaction with salts or esters of dithiocarbamic acids (2-mercaptothiazoles)*

First discovered by Miolati in 1893 ⟨1893G(23)437⟩, the reaction of ammonium dithiocarbamate with α-halocarbonyl compounds gives 2-mercaptothiazole derivatives, usually in good yields. These compounds have found wide use as accelerators in the vulcanization of rubber (Scheme 178).

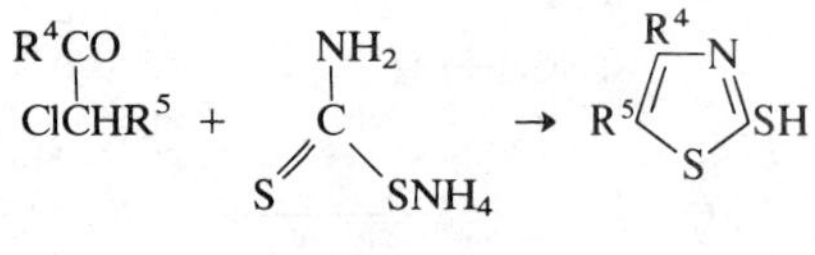

Scheme 178

α-Bromoalkynes also give 2-mercaptothiazoles by condensation with ammonium dithiocarbamate in alcohol solution. They are not isolated in their free form (**245**) but treated immediately in the reaction mixture with methyl iodide to obtain their methyl derivatives which are easier to separate (Scheme 179). 2-Mercaptothiazoles with heterocyclic substituents in the 4-position have also been prepared, for example, N-alkyl-2-benzimidazolyl chloromethyl ketones (**246**) give the corresponding 2-mercaptothiazole derivatives (**247**; Scheme 180).

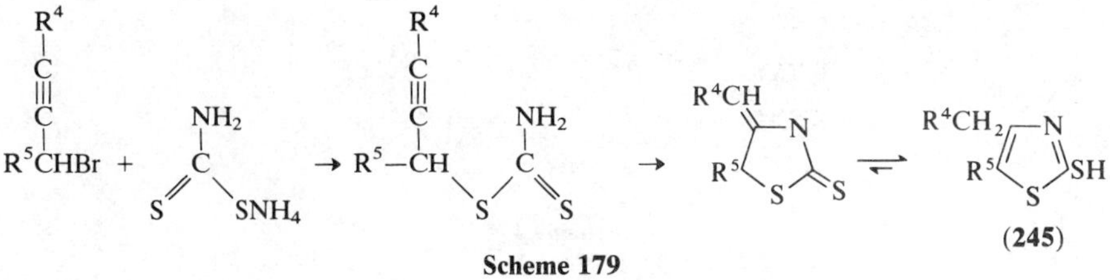

Scheme 179

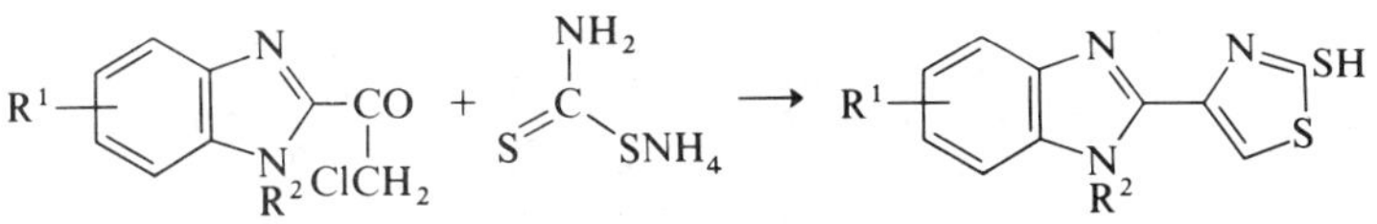

[Scheme 180 structures: (246) → (247)]

Substitution of an ester of dithiocarbamic acid or its salts leads directly to 2-substituted mercaptothiazoles (Scheme 181).

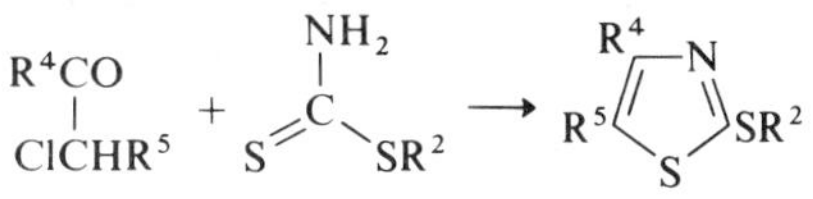

[Scheme 181]

The mechanism of this cyclization is similar to that proposed for the Hantzsch synthesis. Hydroxythiazoline intermediates have been isolated which, on heating with dilute HCl, lead to Δ^4-thiazoline-2-thione. The precise control of the pH during the reaction permits the isolation of an acyclic intermediate (248) which is in equilibrium with its cyclic isomer (249; Scheme 182). The large influence of steric effects on the 3-position has been demonstrated: with R^4 = Me and R^3 = Bu^t dealkylation at nitrogen is observed upon dehydration of the intermediate thiazoline (Scheme 183) ⟨67BSF1948⟩.

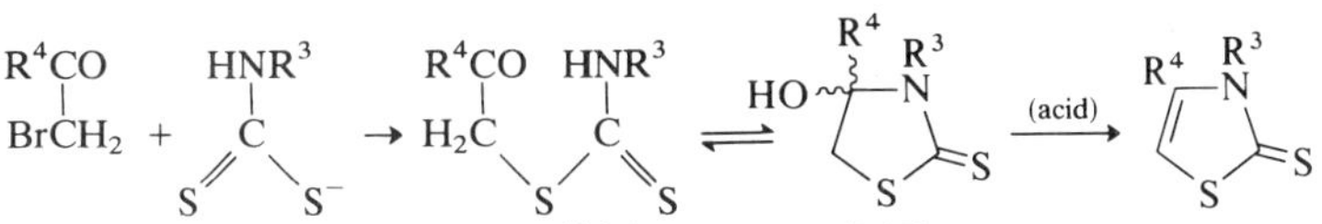

[Scheme 182]

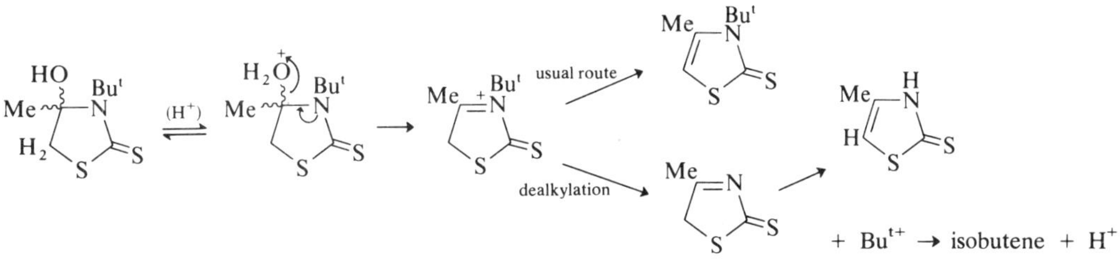

[Scheme 183]

4.19.3.1.2 Synthesis by rearrangement of α-thiocyano ketones (Type A): Tcherniac's synthesis

By variation of the experimental conditions, a fairly large variety of thiazoles, variously substituted at the 2-position, can be obtained from α-thiocyano ketones. This method was first developed by Tcherniac ⟨1892CB2607⟩. The α-thiocyano ketones are easily obtained from α-halocarbonyl compounds and a metal thiocyanate in alcohol solution. They are very sensitive compounds and isomerize when treated with acids, bases or labile hydrogen and sulfur compounds.

(i) *Acid or alkaline hydrolysis*

The cyclization of α-thiocyano ketones (250) in aqueous acid or alkaline solution leads to 2-hydroxythiazoles (251) after dilution with water (Scheme 184). The yields are moderate. With α-thiocyanoacetophenones (252), 4-aryl-2-hydroxythiazoles (254) can be obtained in high yields (Scheme 185). Very early, Arapides ⟨1888LA(249)27⟩ was able to demonstrate the formation of the acyclic compound (253) in the course of this reaction.

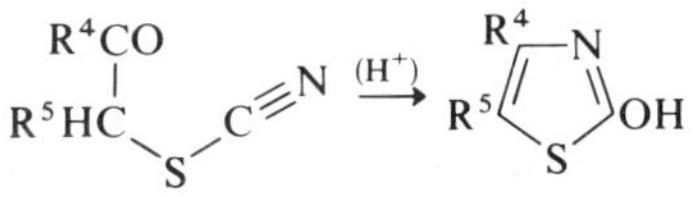

[Scheme 184: (250) → (251)]

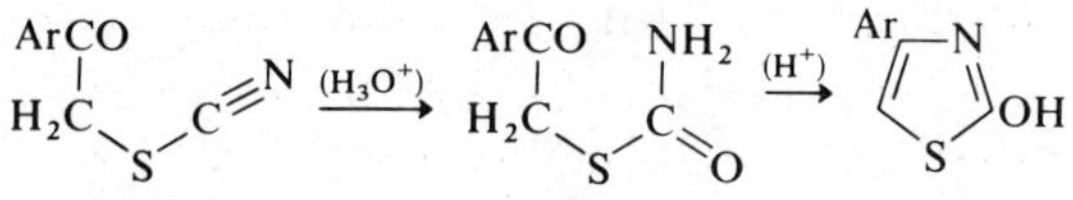

Scheme 185

(ii) *Action of dry hydrogen halides*

Treatment of α-thiocyano ketones at low temperature with dry HCl in ether solution gives satisfactory yields of 2-chlorothiazole derivatives (Scheme 186). Under the same conditions hydrogen bromide leads to the corresponding 2-bromothiazoles.

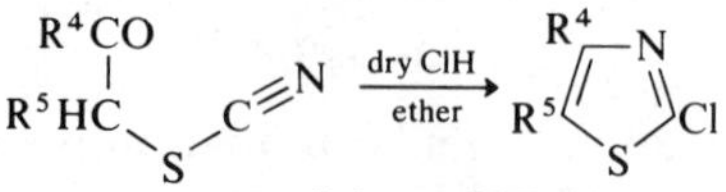

Scheme 186

(iii) *Action of labile sulfur*

Thioacids react with α-thiocyanoacetophenone to produce 2-mercapto-4-phenylthiazole (Scheme 187). Other sulfur compounds such as thiourea, ammonium dithiocarbamate or H_2S also lead to 2-mercaptothiazoles (Scheme 188) with two by-products (**257**) and (**258**).

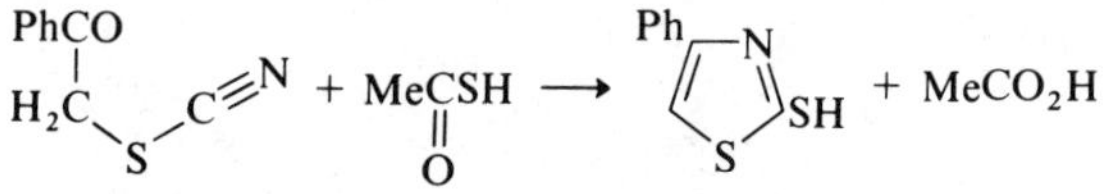

Scheme 187

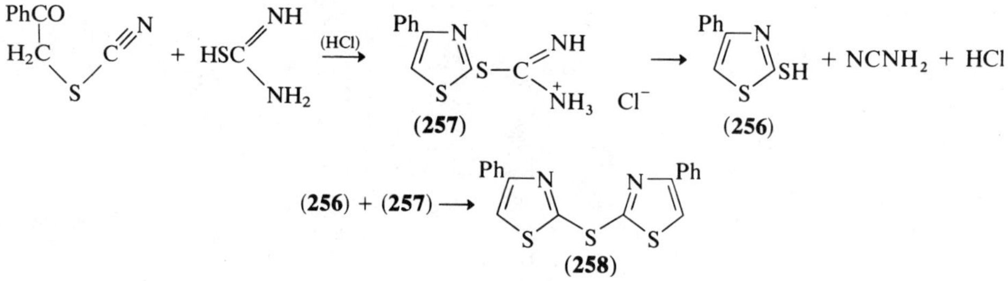

Scheme 188

(iv) *Action of labile nitrogen*

α-Thiocyano ketones react with ammonium chloride or amine hydrochlorides to give 2-aminothiazoles or their *N*-substituted derivatives. For example, the action on α-thiocyanoacetone of ammonia in ether solution gives 2-amino-4-methylthiazole, but in poor yield. However, methylamine in ether at 0 °C gives the initial adduct *S*-acetonyl-*N*-methylisothiourea (**259**) in 80% yield (Scheme 189). The cyclization of this intermediate occurs after prolonged standing at room temperature, by fusion or by heating with dilute HCl, to afford the 4-methyl-2-methylaminothiazole (**260**).

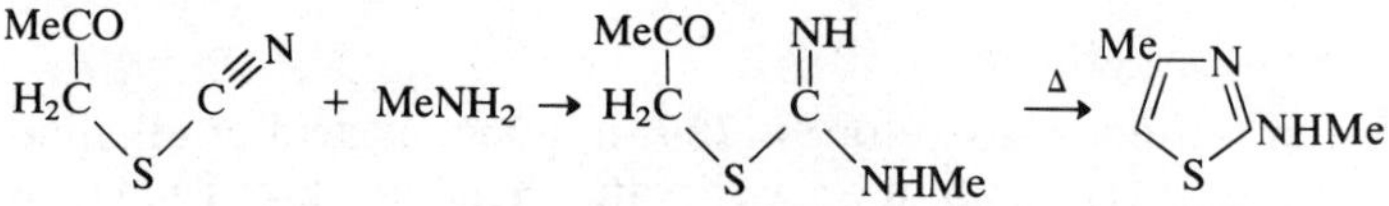

Scheme 189

4.19.3.1.3 *Synthesis from α-aminonitriles (Type B): Cook–Heilbron's synthesis*

This type of synthesis, which was investigated initially by Cook and Heilbron in 1947, gives 5-aminothiazoles with various substituents in the 2-position by reacting an α-aminonitrile with salts and esters of dithioacids, carbon disulfide, carbon oxysulfide, and isothiocyanates under exceptionally mild conditions ⟨47JCS1594⟩.

(i) *Salts and esters of dithioacids: 5-aminothiazole derivatives and related condensations*

By condensing the salts or the esters of either dithioformic (**261**) or dithiophenacetic acids with α-aminonitriles (**262**), 5-aminothiazoles (**264**) in which $R^2 = H$, benzyl and $R^4 =$ phenyl, ethoxycarbonyl or phenoxycarbonyl, are obtained in fairly good yields (Scheme 190). Acyclic thioamides (**263**) have been isolated in certain cases as intermediates in this reaction. Another method of synthesis of 2-unsubstituted 5-aminothiazoles (**267**) from ethyl formate is shown in Scheme 191. The α-aminonitrile reacts with ethyl formate to give the intermediate α-amidonitrile (**265**) which when treated with H_2S is converted into the corresponding thioamide (**266**). Reflux with acetic anhydride affords the 5-aminothiazole (**267**).

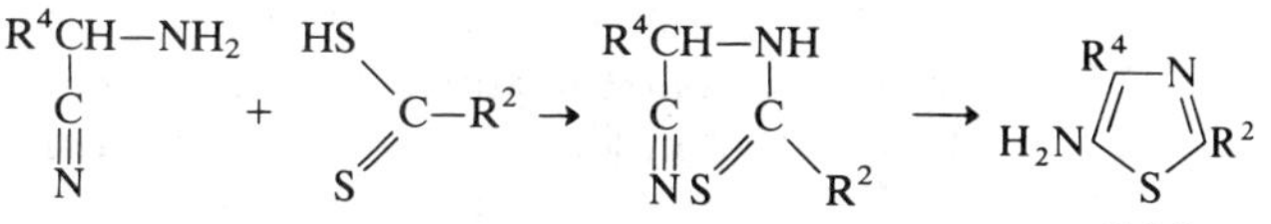

Scheme 190

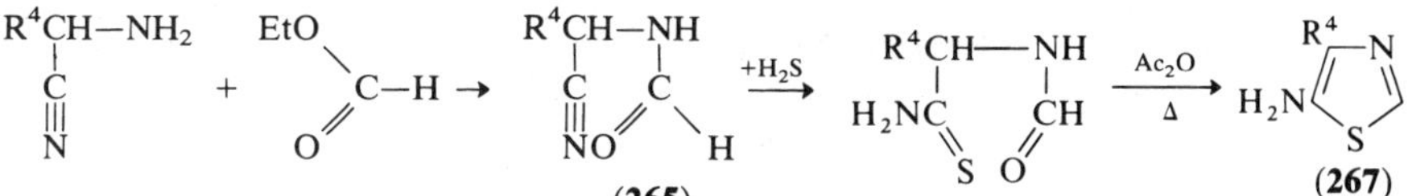

Scheme 191

(ii) *Carbon disulfide: 2-mercapto-5-aminothiazole derivatives*

Carbon disulfide readily reacts with α-aminonitriles giving 5-amino-2-mercaptothiazoles (**268**) which can be converted into 5-aminothiazoles unsubstituted in the 2-position (Scheme 192). Under the same conditions methylaminoacetonitrile (**269a**) reacts with carbon disulfide in the presence of acetic anhydride with ethyl acetate as solvent to give 3-methyl-Δ^4-thiazoline-2-thione (**270**; Scheme 193).

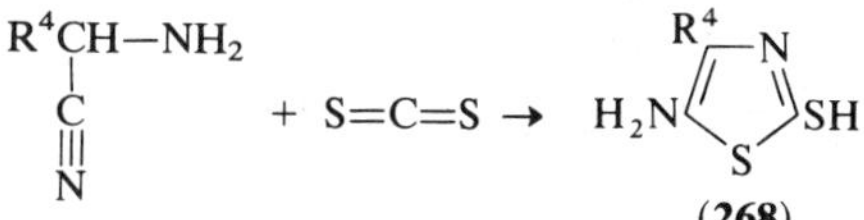

Scheme 192

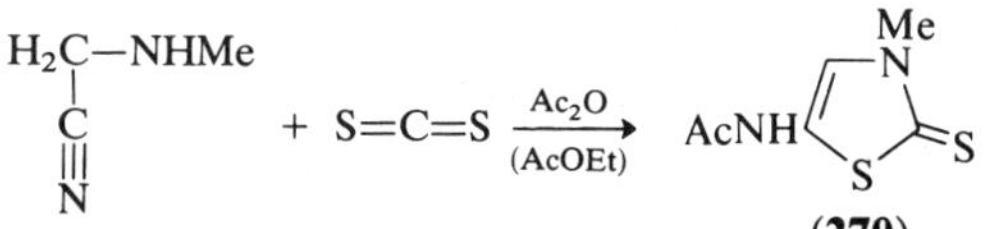

Scheme 193

(iii) *Carbon oxysulfide: 2-hydroxy-5-aminothiazole derivatives*

By condensing carbon oxysulfide with α-aminonitriles, the corresponding 5-amino-2-hydroxythiazoles can be obtained (Scheme 194). In the presence of benzaldehyde the reaction leads to 5-benzylideneaminothiazole derivatives in good yields.

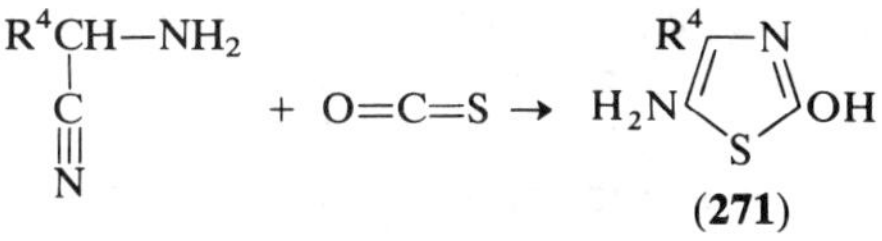

Scheme 194

(iv) *Isothiocyanates: 2,5-diaminothiazole derivatives*

Isothiocyanates (**272**) condense with α-aminonitriles affording 2-substituted 5-aminothiazoles (**274**) through an acyclic intermediate (**273**). In some cases the 2,5-disubstituted aminothiazole (**275**) was isolated as a by-product.

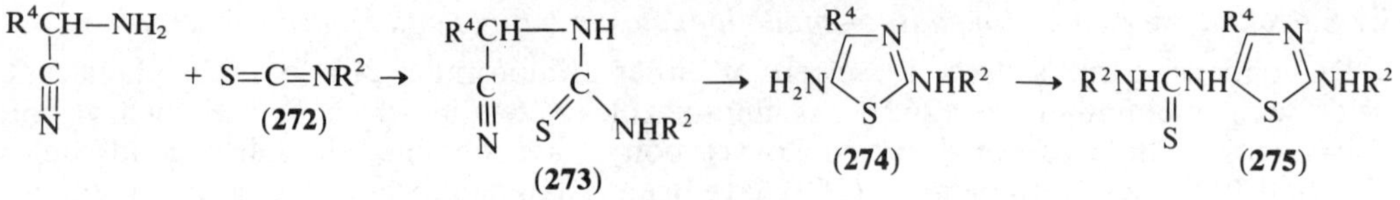

302   *Thiazoles and their Benzo Derivatives*

$$R^4CH-NH_2 \quad + \quad S=C=NR^2 \quad \rightarrow \quad \underset{(273)}{\overset{R^4CH-NH}{\underset{N}{\overset{|}{C}}\underset{S}{\overset{|}{C}}\diagdown NHR^2}} \quad \rightarrow \quad \underset{(274)}{H_2N\underset{S}{\overset{R^4}{\diagup}N}NHR^2} \quad \rightarrow \quad \underset{(275)}{R^2NHCNH\underset{S}{\overset{R^4}{\diagup}N}NHR^2}$$
                         (272)

Scheme 195

4.19.3.1.4 Synthesis from acylaminocarbonyl compounds and phosphorus pentasulfide and related condensations (Type C): Gabriel's synthesis

The reaction was first described by Gabriel in 1910 ⟨10CB134, 10CB1283⟩ when he warmed an acylamino ketone (**276**) with a stoichiometric amount of phosphorus pentasulfide. The reaction (Scheme 196) is similar to that of cyclization of other five-membered oxygen- and sulfur-containing rings from 1,4-dicarbonyl compounds. The method has not been studied extensively and is restricted to the preparation of alkyl-, aryl-, or alkoxy-thiazoles substituted mostly in the 2-, 5-, or 2,5-positions. Yields range from 45 to 80%. Thiazole itself was obtained in a 62% yield from formylaminoacetal (Scheme 197) ⟨14CB3163⟩.

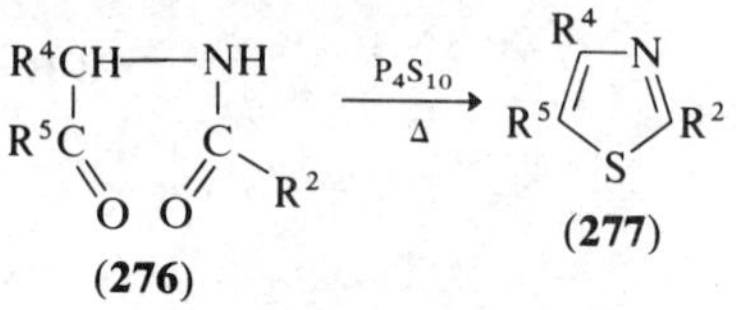

Scheme 196

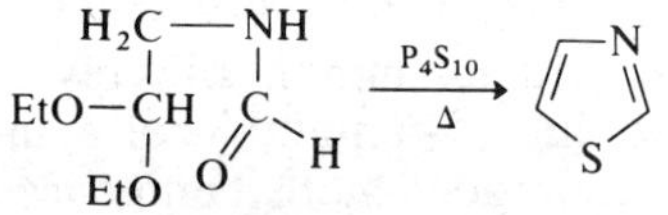

Scheme 197

The Gabriel synthesis has been used for the preparation of 5-alkoxythiazoles. The formyl derivative of an α-amino ester is heated with P_4S_{10} in an inert solvent on a steam bath for 8 hours (Scheme 198), the alkoxythiazole being isolated by ether extraction of the aqueous alkali-treated reaction mixture.

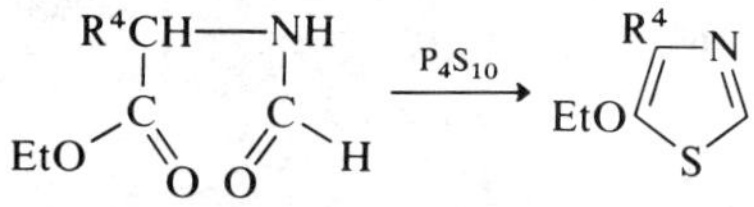

Scheme 198

4.19.3.1.5 Synthesis from nitriles and α-mercapto ketones or acids: 2,4-disubstituted and 4-hydroxythiazole derivatives (Type D)

Besides α-halocarbonyl compounds, α-mercapto ketones and acids are also used for the preparation of thiazoles from nitriles and aldoximes.

(i) *α-Mercapto ketones*

2,4-Disubstituted thiazoles are readily obtained by the action of α-mercapto ketones (**278**) on nitriles (**279**). The reaction is carried out in benzene solution at 0 °C by passing a current of dry hydrogen chloride through the mixture (Scheme 199). In a variation of this reaction, by refluxing α-mercapto ketones (**278**) with aldoximes (**280**) for 2 hours at 100 °C, 2,4-di- or 2,4,5-tri-substituted thiazoles are obtained in good yields (Scheme 200).

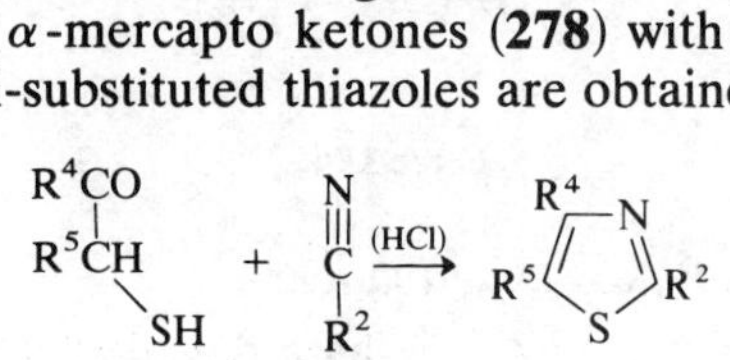

Scheme 199

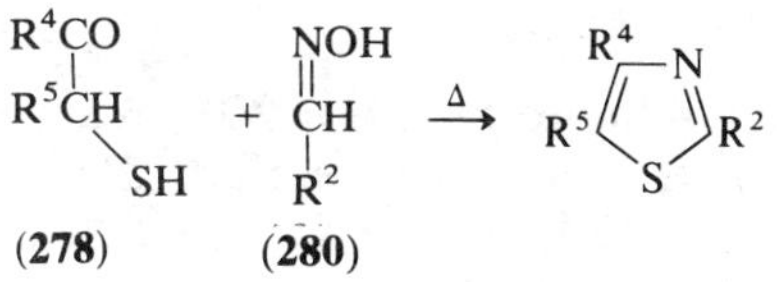

Scheme 200

Cyanamide (**279**; $R^2 = NH_2$) condenses with α-mercapto ketones to give the corresponding 2-aminothiazoles.

(ii) α-Mercapto acids: 4-hydroxythiazole derivatives

α-Mercapto acids (**281**) condense with nitriles to afford 4-hydroxythiazoles (**283**), the reaction being carried out in ethereal or alcoholic solution saturated with dry HCl, for several hours at 0 °C (Scheme 201). In some cases, the acyclic intermediate (**282**) has been isolated, and it undergoes cyclization to thiazole upon heating in toluene.

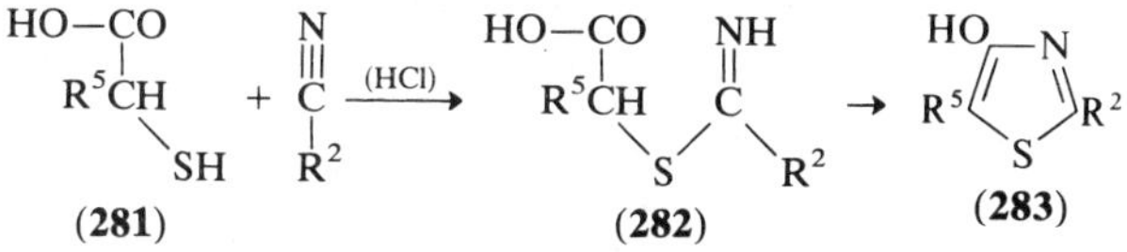

Scheme 201

(iii) α-Aminothioketones

Treatment of enamines of type (**284**) with sulfur and cyanamide at room temperature in ethanol produces a range of 2-aminothiazoles (**287**) in 30–70% yield. Initial formation of an α-aminoenethiol intermediate (**285a**), which is a tautomer of an α-aminothioketone (**285b**), is probably followed by addition of cyanamide, yielding (**286**). Elimination of amine finally produces the observed thiazole (**287**; Scheme 202) ⟨70JPR776⟩.

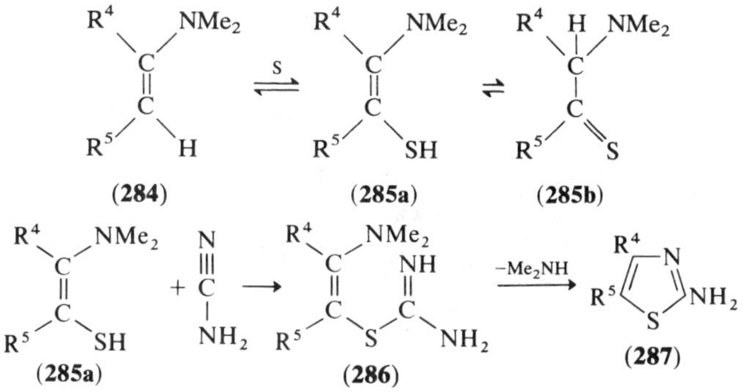

Scheme 202

4.19.3.1.6 Synthesis from thioacids, thiocarbamates and thioureas (Type F)

In this type of synthesis, one of the reactants supplies only the carbon at position 5 of the resulting thiazole.

(i) Thiocarboxylates

2-Substituted 4-amino-5-cyanothiazole (**290**) can be prepared by the following sequence of reactions (Scheme 203). A dithioester reacts with cyanamide to give the cyanoimidothiocarboxylate (**288**) which condenses with chloroacetonitrile (**289**) affording the thiazole (**290**) ⟨70AP625⟩.

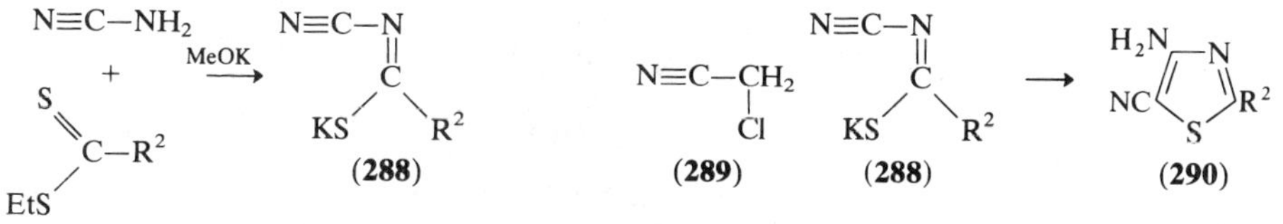

Scheme 203

(ii) *Thiocarbamates*

Potassium salts of cyanoimidodithiocarbonates (**291**) react successively with an alkyl halide and α-halo esters, nitriles, amides or ketones (**293**) in basic medium to yield 2,5-disubstituted 4-aminothiazoles (**294**; Scheme 204) ⟨67JPR(309)97⟩.

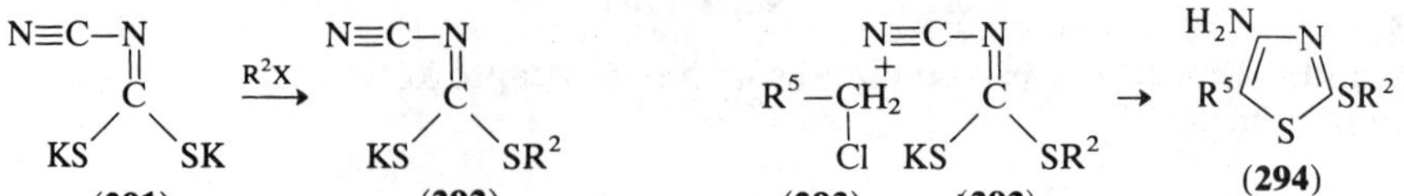

(**291**) (**292**) (**293**) (**292**) (**294**)

$R^5 = CO_2R$, C≡N, $CONH_2$

Scheme 204

In a similar way, N^2-ethoxythiocarbonyl-N^1-phenylbenzamidine (**295**) reacts with R^5CH_2Br (R^5 = COMe, COAr, CO_2Et, $COCH_2Br$) to give the corresponding 2,5-disubstituted 4-phenylthiazole (**296**; Scheme 205) ⟨76S403⟩.

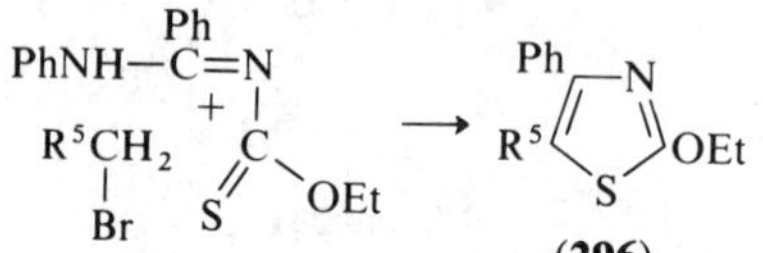

(**295**) (**296**)

R^5 = COMe, COAr, CO_2Et, $COCH_2Br$

Scheme 205

(iii) N*-Acylthiourea*

A synthesis of 2-aminothiazoles based on a corresponding thiophene synthesis involves the reaction of an alkyl halide R^5CH_2X (**297**) with an *N*-acylthiourea (**298**) in basic medium (Scheme 206) ⟨74ZC470⟩.

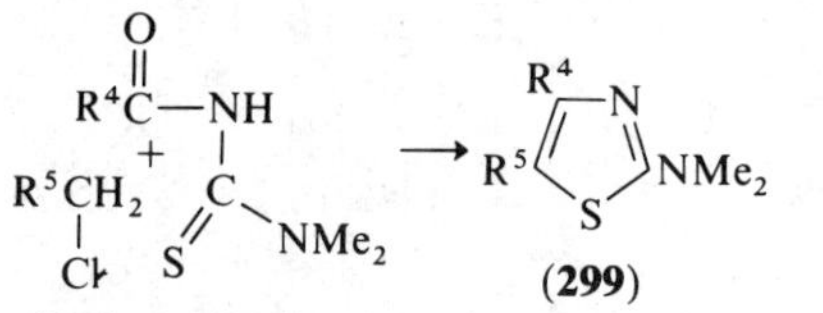

(**297**) (**298**) (**299**)

Scheme 206

4.19.3.1.7 Synthesis from O,S-heterocycles (Type G)

N-(5-Aryl-1,3-oxathiol-2-ylidene) tertiary iminium salts (**300**) react with ammonia affording an open-chain intermediate (**301**) in equilibrium with a 4-hydroxy-Δ^2-thiazoline (**302**) which dehydrates to 2-amino-4-arylthiazole (**303**; Scheme 207) ⟨71CC1318⟩. The reaction of hydroxylamine with (**300**) followed by that of 10% methanolic HCl gives the corresponding thiazole *N*-oxide chloride (**304**) ⟨78H(9)1223⟩.

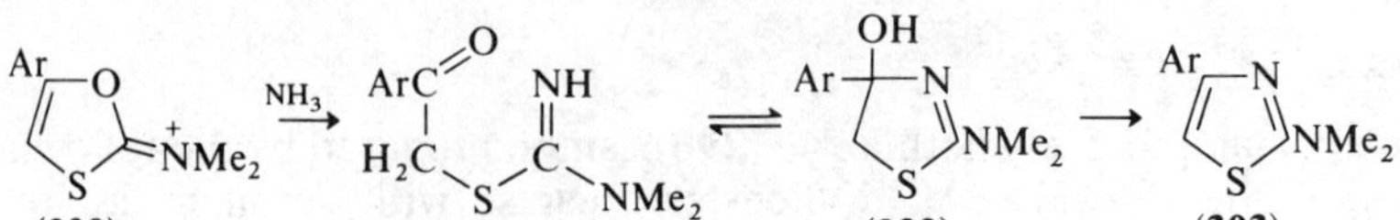

(**300**) (**301**) (**302**) (**303**)

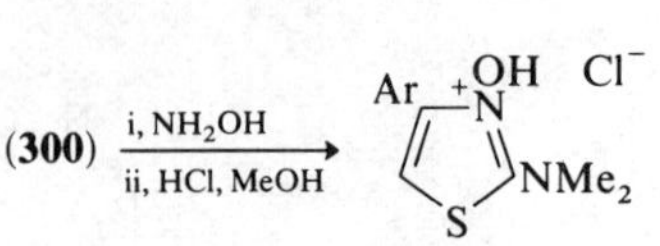

(**304**)

Scheme 207

4.19.3.1.8 Type H synthesis of thiazoles

(i) *Cycloaddition of nitrile ylides*

Benzonitrilo-4-nitrophenylmethide (305; Ar = p-$C_6H_4NO_2$) interacts with an excess of methyl mono- or di-thiobenzoate (306a, b) in triethylamine to give high yields of (307). The same ylide (305) reacts with dimethyl trithiocarbonate (308) in the cold in a direction opposite to the preceding, yielding (309; Scheme 208) ⟨72CB1307⟩.

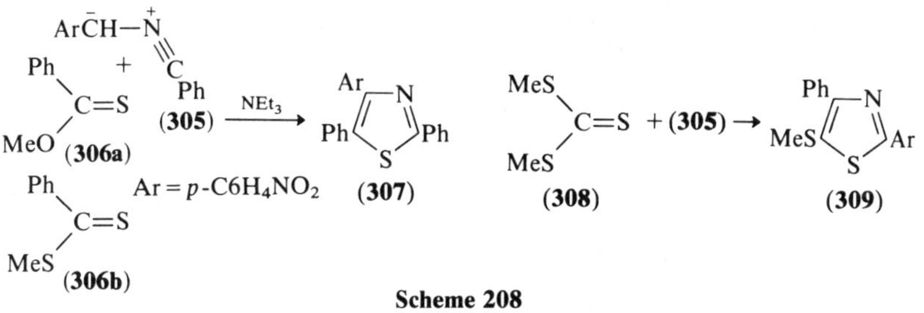

Scheme 208

(ii) *Cycloaddition of metal isocyanides*

α-Metallated isocyanides (310) react with carbon disulfide yielding the metal salts of the corresponding 4-substituted thiazole-5-thiols (311; Scheme 209) ⟨76LA2122⟩. Thionoesters (312) react similarly with (310) giving 4-mono- or 4,5-di-substituted thiazoles (313; Scheme 209) ⟨76S681⟩.

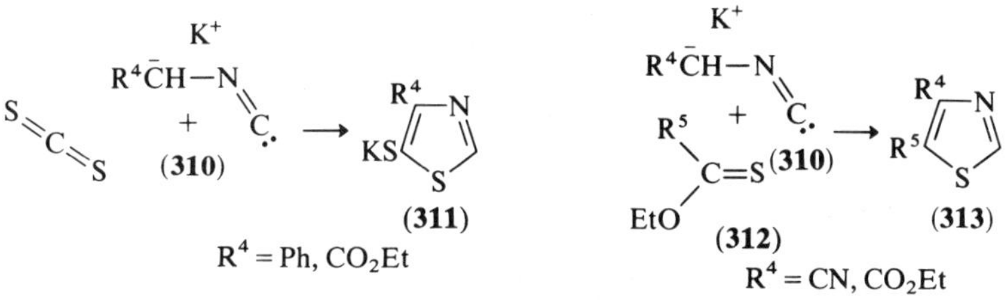

Scheme 209

(iii) *Mesoionic thiazoles from mesoionic oxazoles*

5-Mercaptothiazolylium hydroxide inner salts (316) can be prepared in a 'one pot' reaction from 5-hydroxyoxazolylium hydroxide inner salts (314) and CS_2 (Scheme 210) ⟨76T583⟩.

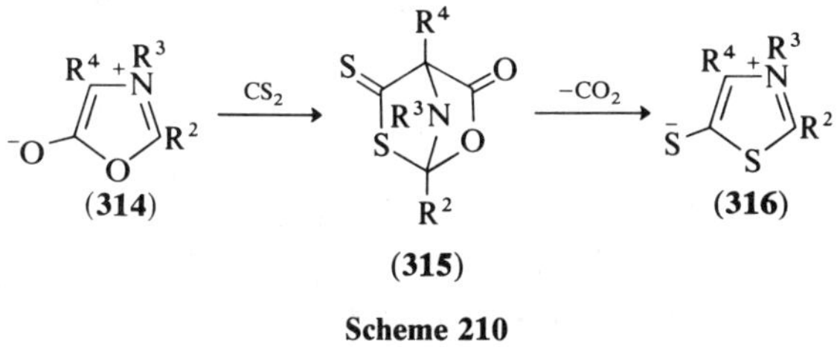

Scheme 210

4.19.3.1.9 Type J synthesis of thiazoles

1,3-Oxathiolylium salts (317) condense with cyanamide in the presence of sodium ethoxide to give an open-chain intermediate (318) which cyclizes to substituted 4-aminothiazoles (319; Scheme 211) ⟨73JPR497⟩.

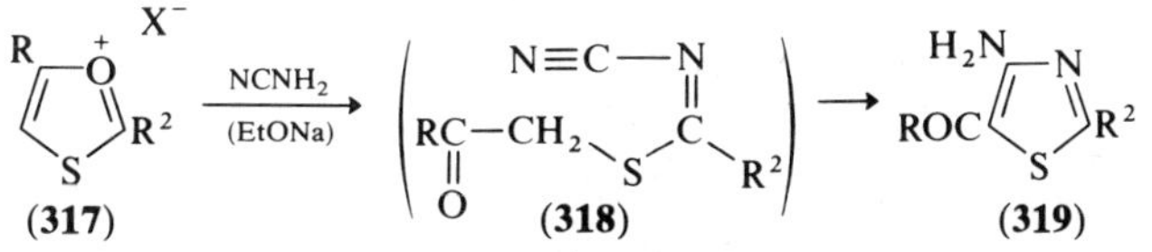

Scheme 211

4.19.3.1.10 Type K synthesis of thiazoles

This type of cyclization corresponds to the formation of the C(5)—S bond.

(i) *Thioacylamino compounds*

2-Phenylthiazole (**321**) can be prepared by cyclization, under mild conditions, of thiobenzamidoacetal (**321**; Scheme 212) ⟨57JCS1556⟩. Similarly *N*-thiobenzoyl-α-amino acids or their amides are cyclized by dissolution in trifluoroacetic anhydride to thiazoles.

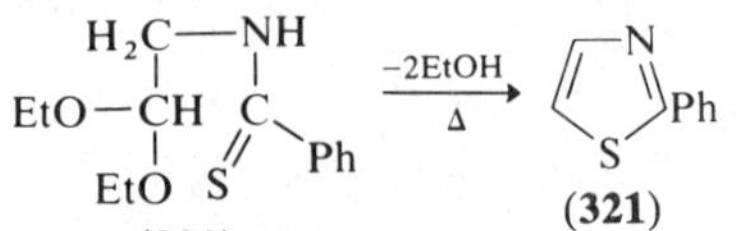

(**320**) (**321**)

Scheme 212

(ii) *Acylaminothioacetamides*

Heated with $POCl_3$, 2-formylaminothioacetamide (**322**; R^2, $R^4 = H$) affords 5-aminothiazole (**323**; R^2, $R^4 = H$). In a similar manner, ethyl 2-acetamido-2-thiocarbamoylacetate (**322**; $R^2 = Me$, $R^4 = CO_2Et$) gives 5-amino-4-ethoxycarbonyl-2-methylthiazole (**323**; $R^2 = Me$, $R^4 = CO_2Et$; Scheme 213).

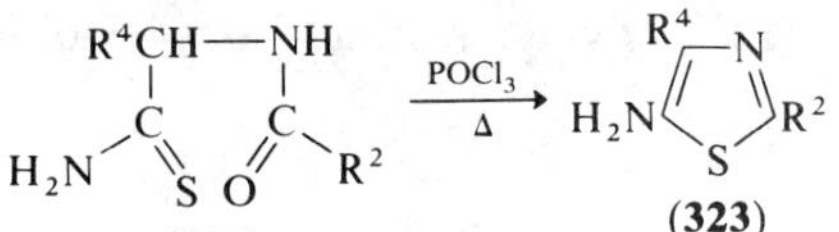

(**322**) (**323**)

Scheme 213

4.19.3.2 Δ^2-Thiazolines

4.19.3.2.1 Type A synthesis of Δ^2-thiazolines

The Hantzsch synthesis is applicable to α-halo acids and their derivatives, (**324**) leading to 2-substituted Δ^2-thiazolin-4-ones (**325**; Scheme 214).

Non-functionalized Δ^2-thiazolines (**328**) are obtained by the classical Gabriel's synthesis which condenses a thioamide (**326**) with 1,2-dibromoalkanes (**327**; Scheme 215) ⟨1890CB157⟩. When an *N*-arylthioamide is employed, the product (**329**) is a 3-arylthiazolinium salt. This constitutes the only route to these salts with an aryl group substituted on the nitrogen atom.

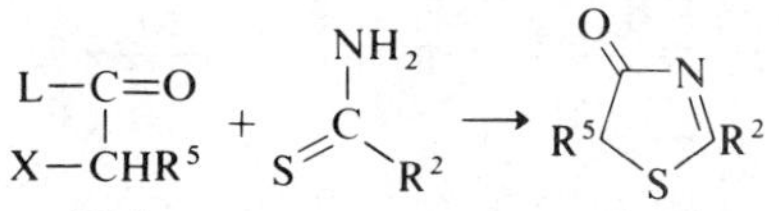

(**324**) (**325**)

L = OH, OEt; R^2 = alkyl, aryl, NR_2, OR, SR

Scheme 214

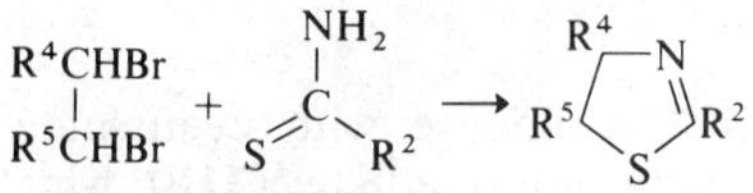

(**327**) (**326**) (**328**)

R^2 = alkyl, aryl

Scheme 215

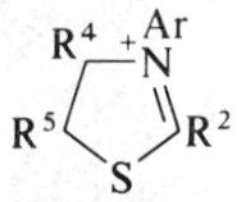

(**329**)

Thioamides and thioureas (**326**; R^2 = alkyl, NR_2) condense with dimethyl acetylenedicarboxylate (**330**), yielding a 2-substituted 5-methoxycarbonylmethylene-Δ^2-thiazolin-4-one (**331**; Scheme 216).

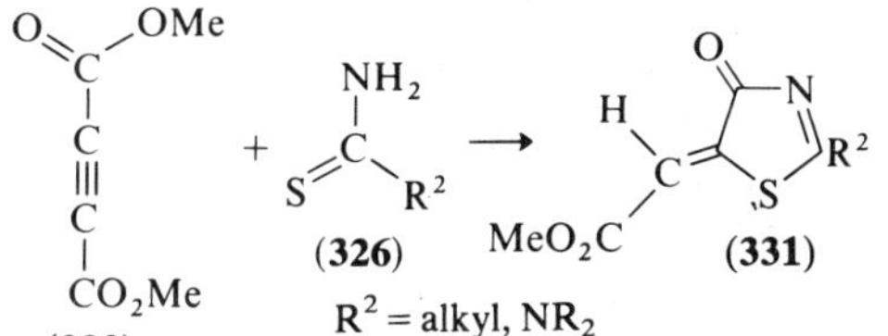

(**330**) (**326**) R^2 = alkyl, NR_2 (**331**)

Scheme 216

Thiourea and phenyldithiocarbamate (**326**; $R^2 = NR_2$, SPh) react with glycidic esters (**332**) to give Δ^2-thiazolin-4-ones (**333**; Scheme 217) ⟨71BSF4021⟩.

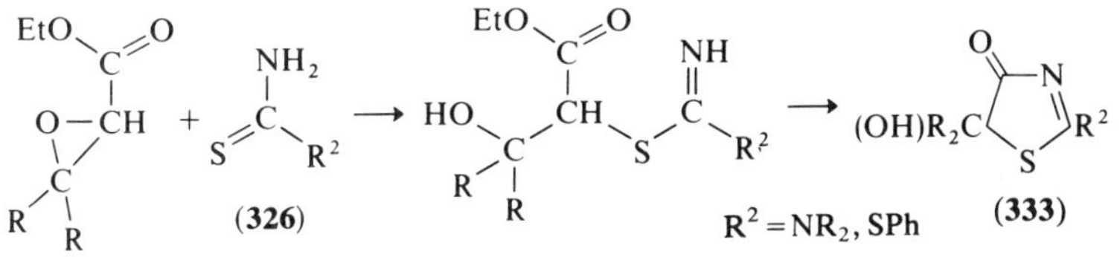

(**332**) (**326**) $R^2 = NR_2$, SPh (**333**)

Scheme 217

4.19.3.2.2 *Type B synthesis of Δ^2-thiazolines*

One of the most valuable methods for the preparation of Δ^2-thiazolines is of this class. The reaction of 2-haloalkylamines (**334**) with thioamides, metal thiocyanates or carbon disulfide give 2-alkyl- or -aryl- (**335**), 2-amino- (**336**), and 2-mercapto- (**337**) Δ^2-thiazolines, respectively (Scheme 218) ⟨17CB804⟩. A method derived from the last procedure leads to a convenient synthesis of 2-phenyl-Δ^2-thiazolines (**340**) under very mild conditions and consists of the condensation between α-aminothiols (**338**) and thiobenzoylmercaptoacetic acid (**339**; Scheme 219) ⟨74TL1863⟩ (this method could be better classified under Type E, Section 4.19.3.2.5).

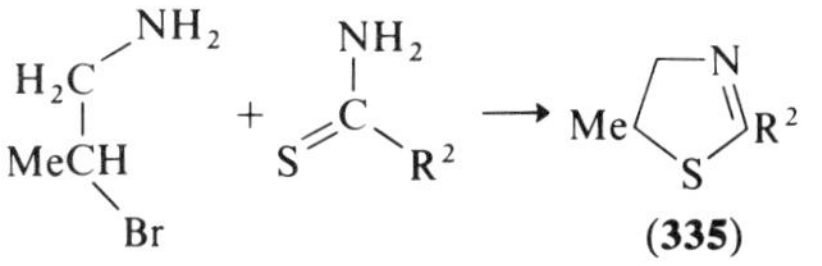

(**334**) (**335**)

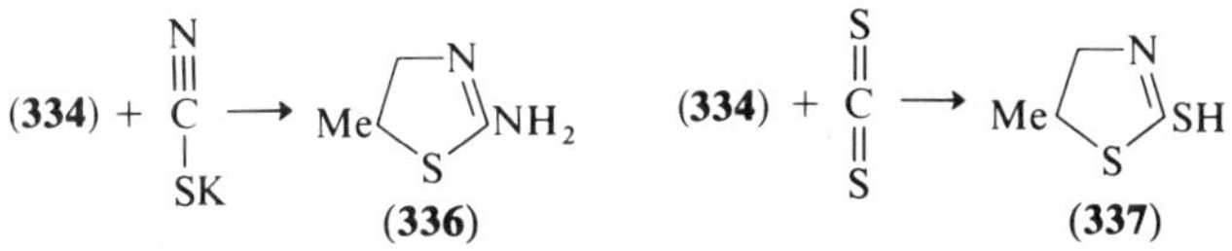

(**336**) (**337**)

Scheme 218

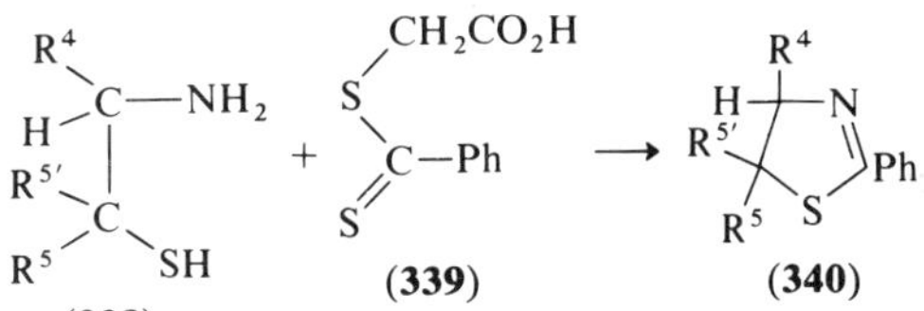

(**338**) (**339**) (**340**)

Scheme 219

4.19.3.2.3 *Type C synthesis of Δ^2-thiazolines*

This type of ring closure has been widely applied in the synthesis of Δ^2-thiazolines. *N*-(β-Substituted alkyl)amides (**341**; X = halogen, OH, SH, OAc) yield thiazolines (**342**) on treatment with phosphorus pentasulfide (Scheme 220) ⟨35JA1079⟩. Δ^2-Thiazoline itself

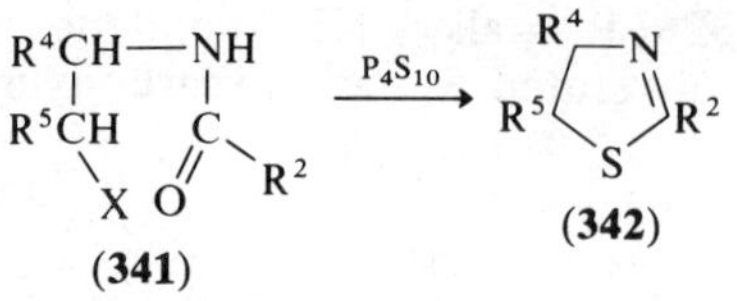

Scheme 220

is best prepared by this method from formylamino-2-ethanol, and the 2-alkyl- and 2-aryl-Δ^2-thiazolines from the corresponding acylamino-2-ethanols.

Another interesting and general method starts from an alkene which provides the C(4)—C(5) fragment of the ring, by reaction with a nitrile R^2CN in the presence of a halogen X_2 giving an *N*(-*β*-haloalkyl)iminohalide (**344**). This intermediate reacts with alkaline sulfide solutions yielding stereospecifically in one step the thiazoline (**346**; Scheme 221) ⟨69JPR408⟩. Replacing the nitrile by thiocyanate esters or dialkylcyanamides affords the corresponding 2-alkylthio- or 2-dialkylamino-thiazolines.

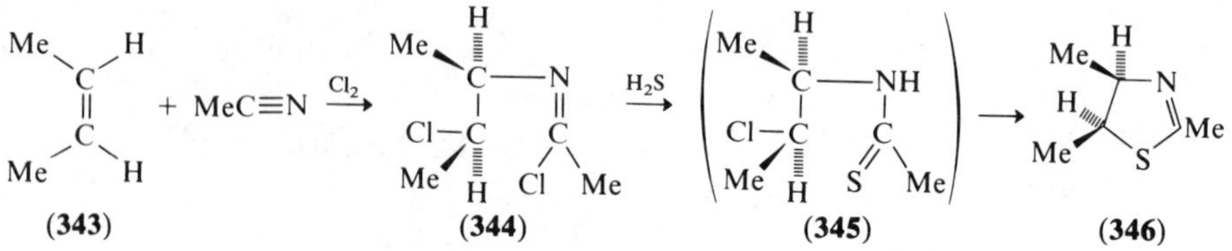

Scheme 221

4.19.3.2.4 Type D synthesis of Δ^2-thiazolines

The reaction of *α*-mercaptocarboxylic acids (**347**) with dicyanogen affords good yields of 4,4′-dioxo-2-bithiazolinyl (**349**; Scheme 222) ⟨69JOC2053⟩.

Cysteamine (**350**) treated with nitriles (**351**; $R = CO_2Me$, CN, Br) in ethanol yields the corresponding Δ^2-thiazoline (Scheme 223) ⟨77CR(C)(285)257⟩.

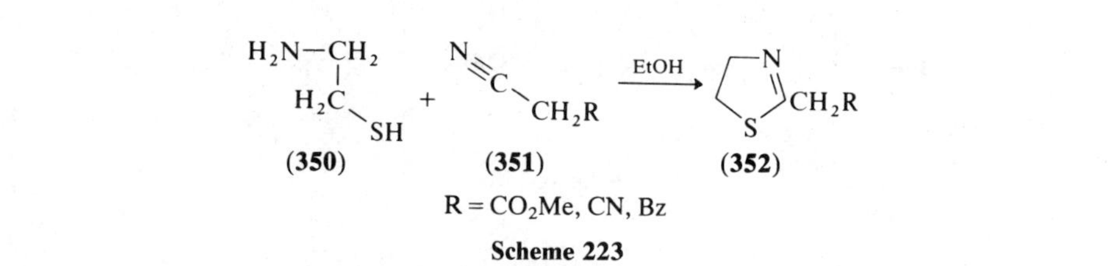

Scheme 222

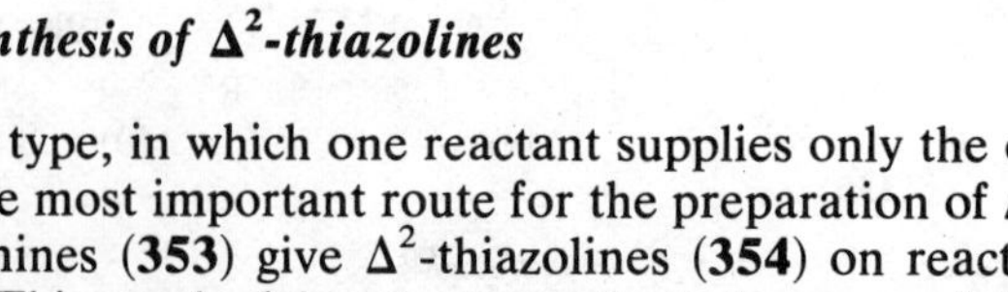

$R = CO_2Me$, CN, Bz

Scheme 223

4.19.3.2.5 Type E synthesis of Δ^2-thiazolines

Cyclizations of this type, in which one reactant supplies only the carbon at position 2 of the ring, constitute the most important route for the preparation of Δ^2-thiazolines.

β-Mercaptoalkylamines (**353**) give Δ^2-thiazolines (**354**) on reaction with esters, imino esters or thioesters. This method has been particularly successful with cysteine and its derivatives (Scheme 224) ⟨16B1110⟩.

2-Acyl-Δ^2-thiazolines (**357**) are similarly prepared by the condensation of cysteamine (**350**) and the 2,2-dialkoxyalkanenitrile (**355**), followed by acid hydrolysis of the intermediate (**356**; Scheme 225) ⟨72RTC711⟩.

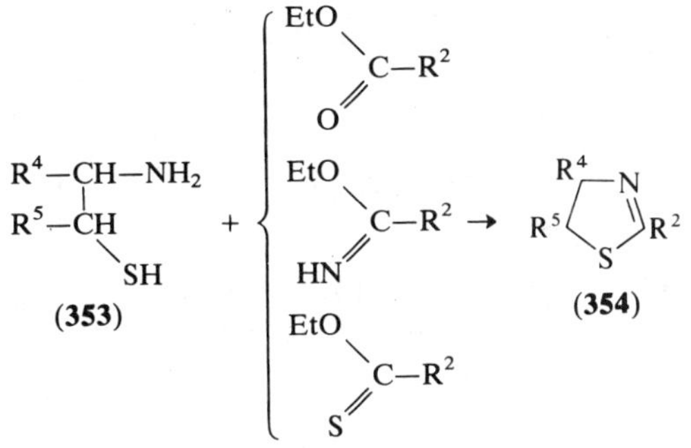

Scheme 224

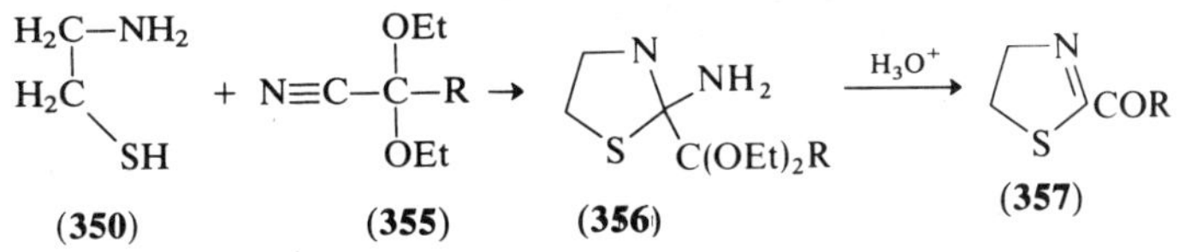

Scheme 225

4.19.3.2.6 Type F synthesis of Δ²-thiazolines

Cyclizations of this class, in which one reactant supplies only the carbon at position 5 of the ring, are rare. Diazoalkanes (**358**) condense with thiobenzoylisocyanate (**359**) affording Δ²-thiazolin-4-ones (**360**; Scheme 226) ⟨73CB1496⟩.

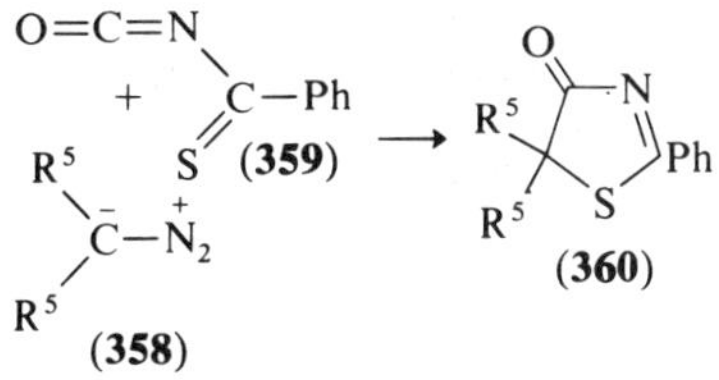

Scheme 226

4.19.3.2.7 Type G synthesis of Δ²-thiazolines

Ring closure between nitrogen and a suitably placed electron-deficient carbon atom occurs when the intermediate (**363**) resulting from the Michael addition of the N-alkylthioamide (**362**) to the α-bromo-α,β-unsaturated acid (**361**) cyclizes with displacement of bromide ion. The resulting 2-substituted 4-carboxy-Δ²-thiazolinium bromide (**364**) is highly reactive, owing to the important steric interaction between the substituents R^2 and R^3, and its facile hydrolysis affords a convenient preparation of cysteine derivatives (**365**; Scheme 227) ⟨71ACS1⟩.

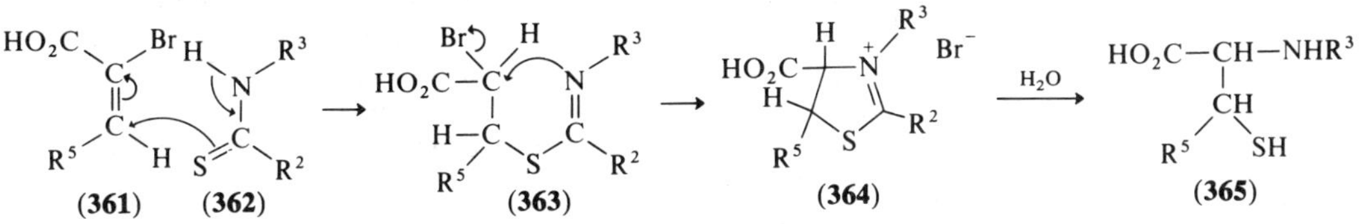

Scheme 227

4.19.3.2.8 Type H synthesis of Δ²-thiazolines

The ring expansion of aziridines and azirines corresponds to this type of ring closure leading to Δ²-thiazolines. N-Thioacylaziridines (**366**) rearrange under acid conditions to 2-aryl-Δ²-thiazolines (**368**) either by an S_N1 or S_N2 mechanism, depending on the nature of the acid catalyst and the solvent used ⟨69JA5835, 69JA5841⟩.

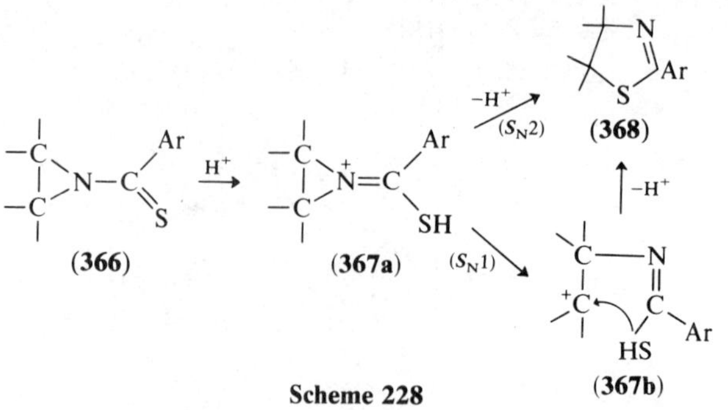

Scheme 228

Similarly, treatment of *N*-aroylaziridines with phosphorus pentasulfide in boiling toluene affords good yields of 2-aryl-Δ^2-thiazolines ⟨71CC329⟩.

N-Unsubstituted aziridines react with thiocyanic acid with ring enlargement, producing 2-amino-Δ^2-thiazolines. The reaction proceeds stereospecifically with 100% Walden inversion as illustrated in Scheme 229. *cis*-2,3-Dimethylaziridine (**369**) gives exclusively *trans*-2-amino-4,5-dimethyl-Δ^2-thiazoline (**371**) ⟨72JOC4401⟩.

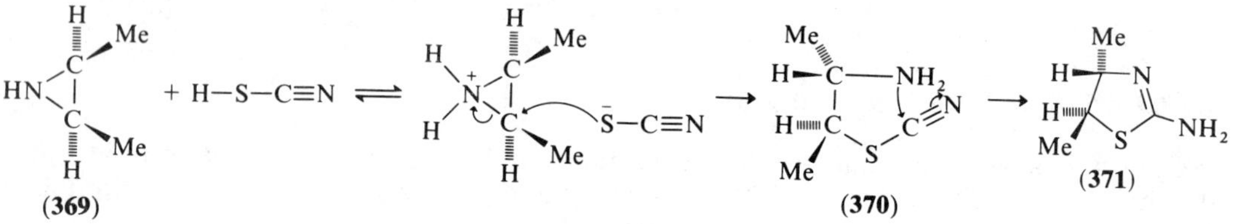

Scheme 229

2,3-Diphenylazirine (**372**) undergoes a photochemically induced addition of methyldithiobenzoate in benzene, *via* the nitrile ylide (**373**), affording a mixture of the two stereoisomeric Δ^2-thiazolines (**375**; Scheme 230) ⟨72TL4087⟩.

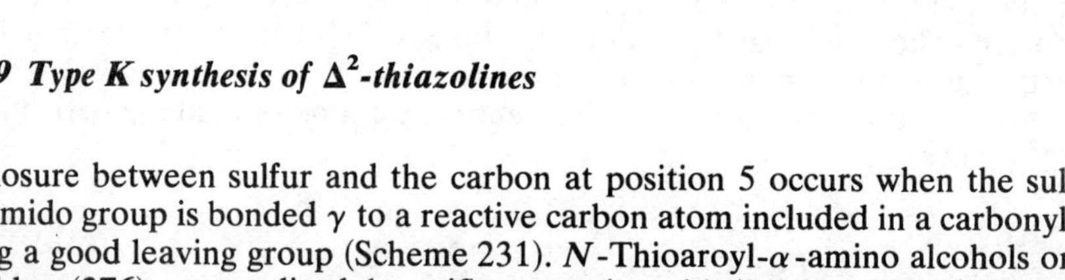

Scheme 230

4.19.3.2.9 Type K synthesis of Δ^2-thiazolines

Ring closure between sulfur and the carbon at position 5 occurs when the sulfur atom of a thioamido group is bonded γ to a reactive carbon atom included in a carbonyl function or bearing a good leaving group (Scheme 231). *N*-Thioaroyl-α-amino alcohols or acids or their amides (**376**) are cyclized by trifluoroacetic acid (Scheme 232) ⟨69JCS(C)1117⟩ to Δ^2-thiazolin-5-ones (**377**). β-Chloroethyl or acetyl isothiocyanate (**378**) react with arylamines to give β-chloroethyl or acylthioamides which undergo cyclization to 2-arylamino-Δ^2-thiazolin-4-ones (**379**; Scheme 233) ⟨68MI41903⟩.

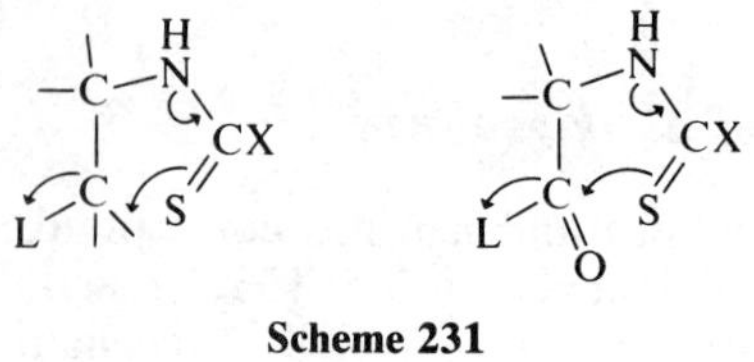

Scheme 231

Scheme 232 (376) → (377)

Scheme 233 (378) + ArNH₂ → (379)

A similar type of reaction is observed with the intermediate (**381**) resulting from the reaction of a secondary amine, a thiol or an alcohol, with β-chloroethyl isothiocyanate (**380**; Scheme 234) ⟨75SC143⟩.

Scheme 234 (380) + XH → (381) → (382)

A biogenetic-type synthesis of the bicyclic penicillin–cephalosporin antibiotics from an acyclic tripeptide equivalent involves a double cyclization of compound (**383**) to form both the thiazoline and the β-lactam rings in one step (**384**; Scheme 235) ⟨75JA5008⟩.

Scheme 235 (383) → (384)

N-Allylthioureas and thioamides (**385**) undergo an acid-catalyzed cyclization to Δ²-thiazolines (**387**). The reaction proceeds through the carbenium intermediate (**386**; Scheme 236) ⟨70JOC3768⟩. In a similar way, α-isothiocyanatoacrylates (**388**) add alcohols or thiols affording intermediate α-thioacylaminoacrylates (**389**) which cyclize under acid or basic conditions to 2-substituted Δ²-thiazoline-4-carboxylates (**390**; Scheme 237) ⟨76LA1997⟩.

Scheme 236 (385) → (386) → (387)

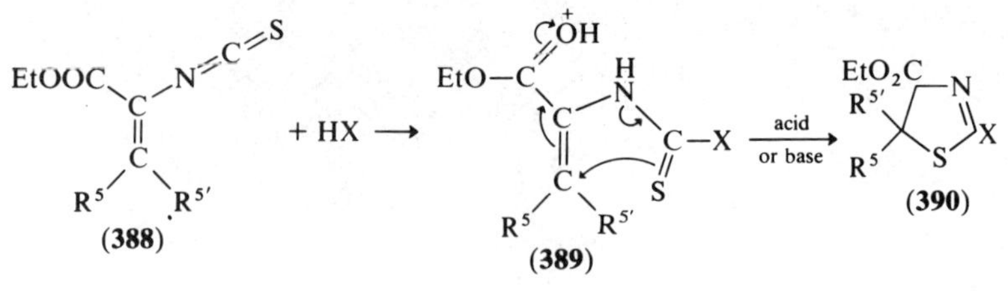

4.19.3.2.10 Type N synthesis of Δ² thiazolines

The well-known cyclization of the radioprotective S-aminoethylisothiouronium bromide to 2-amino-Δ²-thiazoline is dependent on this type of ring closure (Scheme 238) ⟨57JA5667⟩.

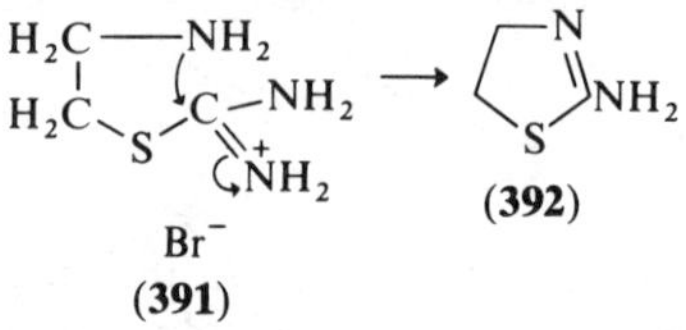

Scheme 238

4.19.3.3 Δ³-Thiazolines

Δ³-Thiazolines were only recently discovered by Asinger ⟨58AG667⟩ as products of a general reaction of ketones or aldehydes with a mixture of sulfur and ammonia. The reaction gives a complex mixture of products, depending on the temperature of the reaction and the ketone to sulfur ratio. In the case of diethyl ketone, the main product is 2,2,4-triethyl-5-methyl-Δ³-thiazoline (394) which results from the condensation of the α-mercapto ketone with a second molecule of ketone before or after its reaction with ammonia (Scheme 239). This reaction, which corresponds to the type D synthesis of the thiazoline ring (mercapto ketone + ketoimine) has been extended to the condensation of an α-mercapto ketone with an aldehyde or a ketone, affording a large variety of mono- to tetra-alkyl-substituted Δ³-thiazolines (397; Scheme 240). Thus 4-methyl-Δ³-thiazoline (397; $R^2 = R^{2'} = H$, $R^4 =$ Me, $R^5 = H$) could be obtained in a yield of 62% from mercaptoacetone (395; $R^4 = $ Me, $R^5 = H$) and formaldehyde (396; R^2, $R^{2'} = H$) ⟨58AG667⟩. Similarly 2-alkyl-substituted Δ³-thiazolines (399) may be prepared by the reaction of an aldehyde with ammonia and 2-mercaptoacetaldehyde or its disulfide or dimer (398; Scheme 241) ⟨74S294⟩. According to an H type synthesis, Δ³-thiazolines (403) result from the thermal cyclization and ring expansion of 2,2-dimethyl-3-dimethylaminoazirine (400) with isothiocyanates (401). The

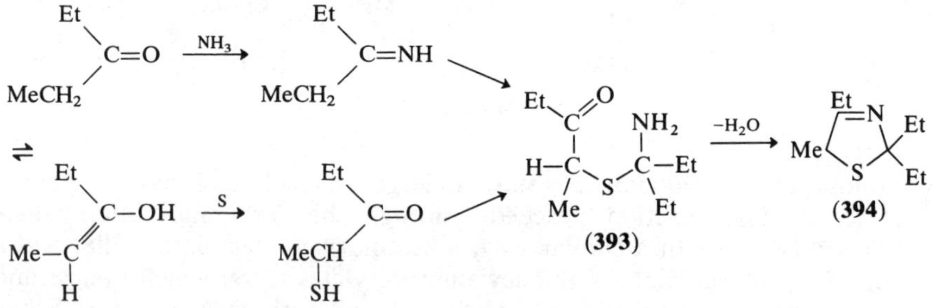

Scheme 239

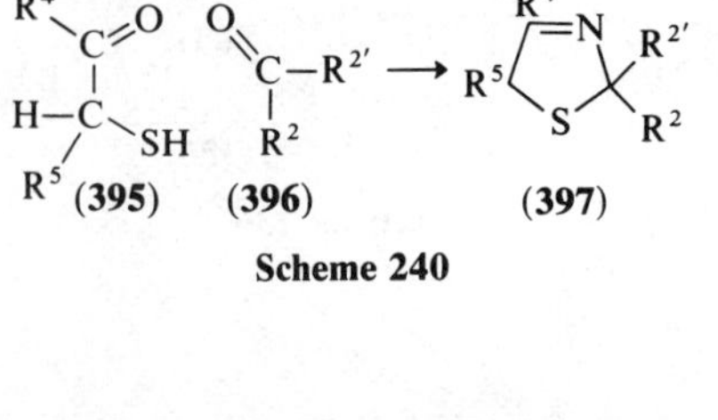

Scheme 240

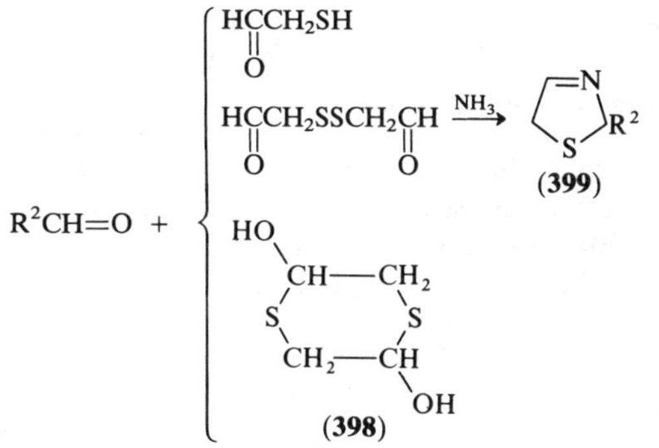

Scheme 241

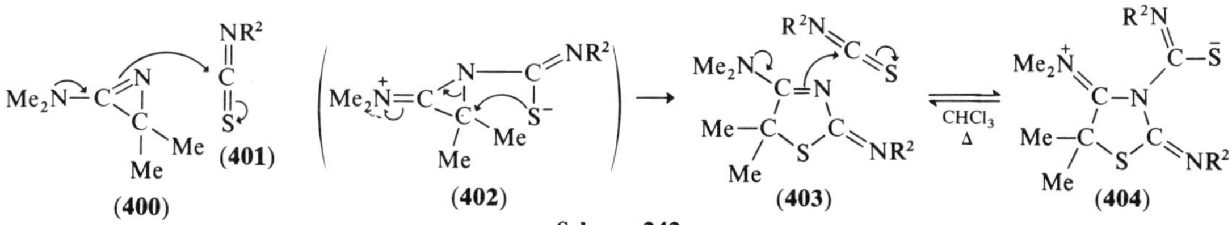

Scheme 242

4-dimethylamino-Δ^3-thiazoline (**403**) may react further with isothiocyanate to give dipolar thiazolidines (**404**). On being heated in chloroform, the thiazolidine (**404**; $R^2 = Ph$) loses phenyl isothiocyanate to revert to the Δ^3-thiazoline (**403**; $R^2 = Ph$; Scheme 242) ⟨74CB3574⟩.

In a similar way 3-dimethylamino-2-ethyl-2-phenylazirine (**405**) reacts with carbon disulfide affording a dipolar adduct (**406**) which may undergo two different types of ring-opening. A 1,2-cleavage of the azirine ring, which is favored by delocalization of the positive charge over the phenyl group, leads to intermediate (**407**) which may cyclize either to 4-dimethylamino-5-ethyl-5-phenyl-Δ^3-thiazoline-2-thione (**409**) or to the isomeric six-membered dihydro-Δ^3-1,3-thiazine-2-thione (**410**). A 1,3-cleavage of dipolar adduct (**406**) or its bicyclic isomer (**411**) gives the 4,4-disubstituted 2-thioxo-thiazolidin-5-one (**413**; Scheme 243) ⟨77TL1351⟩.

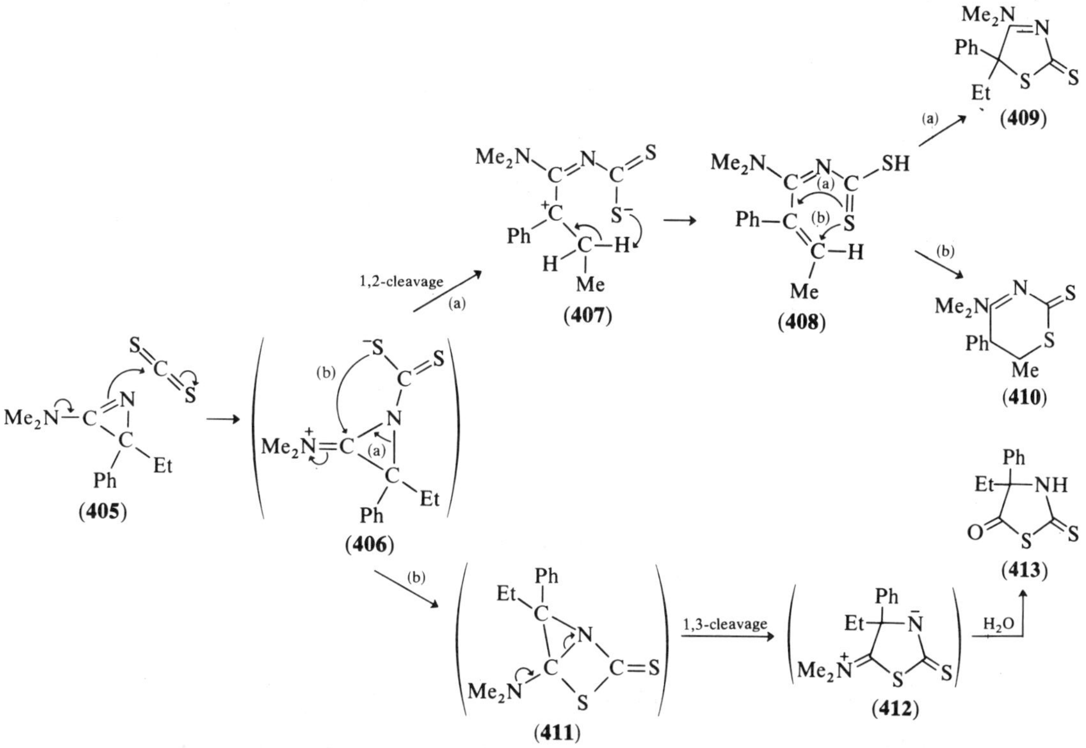

Scheme 243

4.19.3.4 Δ⁴-Thiazolines

4.19.3.4.1 Type A synthesis of Δ⁴-thiazolines

A great variety of Δ⁴-thiazolines are produced by Hantzsch's synthesis in which α-halogeno aldehydes or ketones condense with monothio- or dithio-carbamates or thioureas or thiosemicarbazides affording derivatives of Δ⁴-thiazolin-2-one or -thione or -imine, respectively (Scheme 244). The mechanism of this reaction has been studied in the cyclization leading to Δ⁴-thiazoline-2-thione ⟨71BSF1902⟩ and involves an isolable intermediate 4-hydroxythiazolidine-2-thione (**420**). The dehydration of this compound under acidic conditions is assisted by the adjacent nitrogen atom and the intermediate Δ³-thiazolinylium cation (**421**) may lose a trimethylcarbenium ion when R^3 is *t*-butyl and affords a Δ⁴-thiazoline-2-thione unsubstituted at the ring nitrogen atom (**423**) ⟨70ZC432⟩. On treatment with *N*-bromosuccinimide, *N,N'*-diphenylthiourea condenses with acetone or acetophenone to give 4-methyl- or 4-phenyl-2-phenylimino-3-phenyl-Δ²-thiazoline (**424**; Scheme 246).

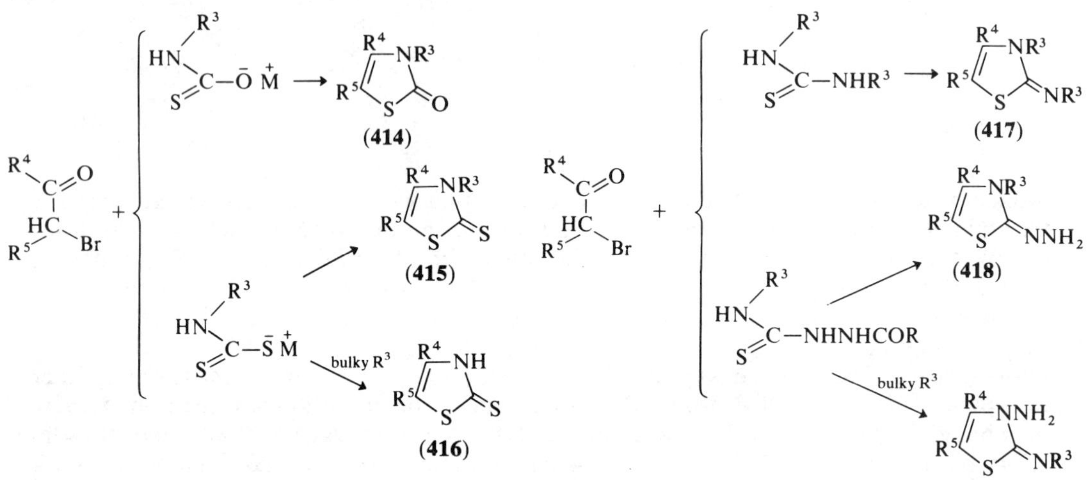

Scheme 244

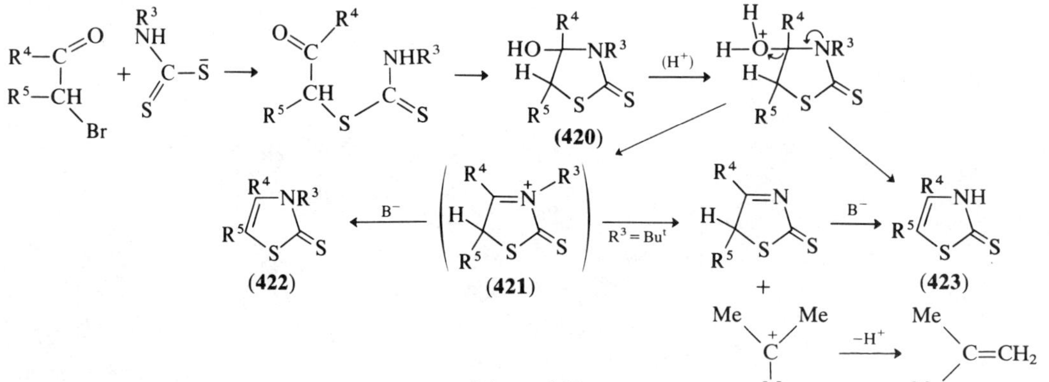

Scheme 245

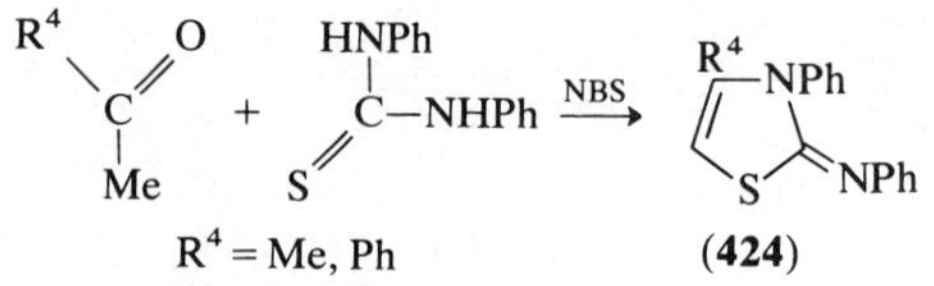

Scheme 246

4.19.3.4.2 Type B synthesis of Δ⁴-thiazolines

Aryl isothiocyanates (**425**) condense with alkylaminoalkynes (**426**) to give 3-alkyl-2-arylimino-5-methyl-Δ⁴-thiazolines (**427**; Scheme 247) ⟨77IJC(B)133⟩.

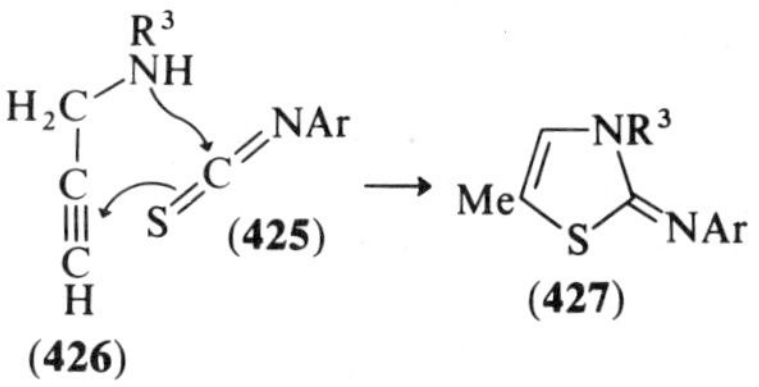

Scheme 247

4.19.3.4.3 *Type D synthesis of Δ^4-thiazolines*

2,3-Dihydrothiamine (**430**) has been prepared by condensation of 5-hydroxy-3-mercaptopentan-2-one (**428**) with formaldehyde and 4-amino-5-aminomethyl-2-methylpyrimidine (**429**; Scheme 248) ⟨55YZ677⟩. This reaction has been developed as a general method of synthesis of Δ^4-thiazolines of the D type. α-Mercapto ketones (**431**) react with primary aliphatic amines (**432**) and an aldehyde (**433**), or the corresponding aldimine (**434**), to give *N*-substituted Δ^4-thiazolines (**435**; Scheme 249) ⟨58AG667⟩. This last method has been extended to include the use of amino acids (**432**; $R^3 = CRCO_2Et$), affording Δ^4-thiazolin-3-ylacetic ester derivatives (**435**; $R^3 = CRCO_2Et$).

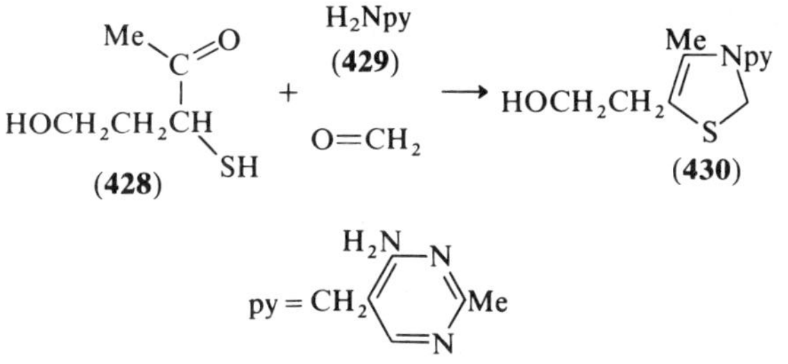

Scheme 248

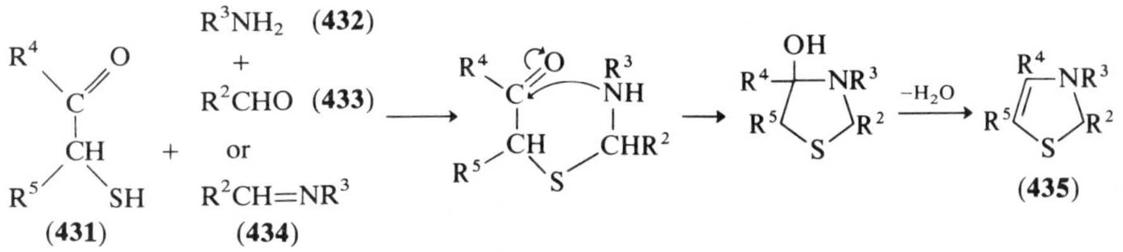

Scheme 249

4.19.3.4.4 *Type G synthesis of Δ^4-thiazolines*

The treatment of an α-thiocyano ketone (**436**) with sulfuric acid affords a 4,5-disubstituted Δ^4-thiazolin-2-one (**437**; Scheme 250) ⟨75IJC532⟩. In a similar way, dithiocarbazic esters (**438**) cyclize giving *N*-substituted Δ^4-thiazoline-2-thiones (**439**; Scheme 251).

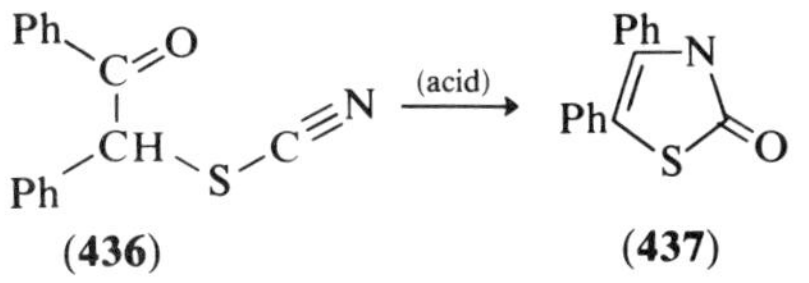

Scheme 250

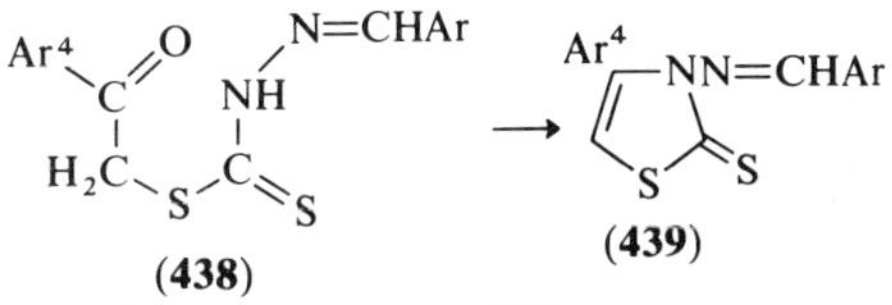

Scheme 251

4.19.3.4.5 Type H synthesis of Δ^4-thiazolines

The condensation of 3-aroylaziridines (**440**) and aryl isothiocyanates (**441**) in boiling benzene is a general one-step synthesis for 4-aroyl-Δ^4-thiazolines (**442a, b**). Two isomers are formed: 4-aroyl-5-aroylamino-Δ^4-thiazoline (**442a**) results from a concerted [2+3] cycloaddition whilst the formation of 2-arylamino-4-aroyl-Δ^4-thiazoline (**442b**) depends on nucleophilic attack of the sulfur of the isothiocyanate group at the 2-position of the aziridine ring (Scheme 252) ⟨69CJC3557⟩.

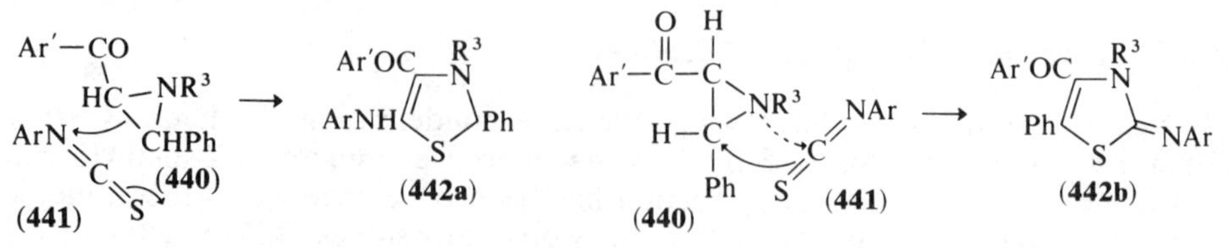

Scheme 252

4.19.3.4.6 Type N synthesis of Δ^4-thiazolines

The condensation of arylamines with phenacyl thiocyanate (**443**) affords 3-aryl-2-imino-4-phenyl-Δ^4-thiazolines (**444**; Scheme 253) ⟨72IJC318⟩.

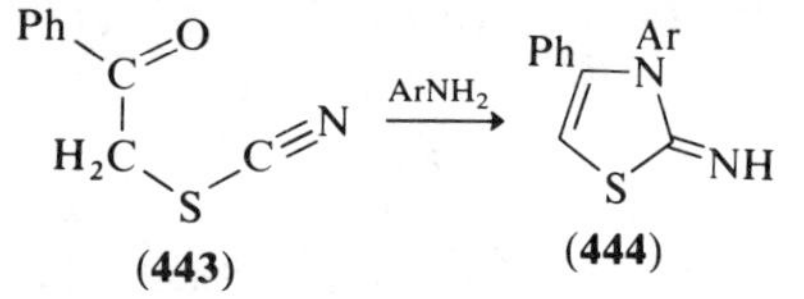

Scheme 253

4.19.3.4.7 Type O synthesis of Δ^4-thiazolines

Heating thiamine disulfide (**445**) in triethylamine yields the Δ^4-thiazoline-2-thione (**446**) as the main product. The reaction is probably initiated by the nucleophilic scission of the disulfide bond followed by liberation from one of the fragments of the sulfur which is responsible for the introduction of the thione group at the 2-position (Scheme 254) ⟨68JCS(C)2871⟩.

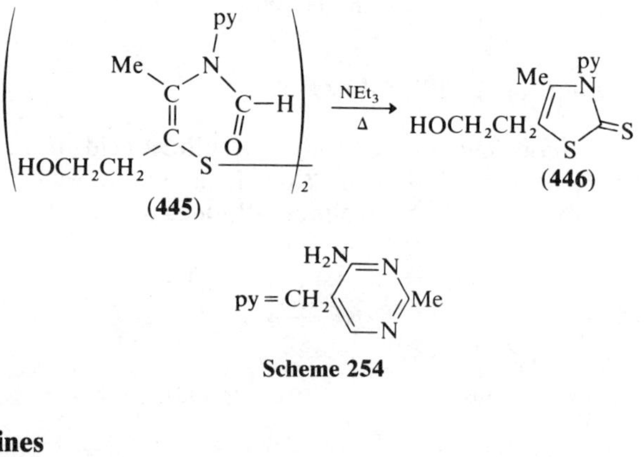

Scheme 254

4.19.3.5 Thiazolidines

4.19.3.5.1 Type A synthesis of thiazolidines

2-Substituted thiazolidin-4-ones (**449**) are easily prepared by the reaction between chloracetic acid, or its derivatives (**447**), and a thiourea, a thiosemicarbazide, or a mono- or di-thiocarbamate (**448**; Scheme 255). The same α-chloro acid derivatives react with metallic thiocyanates yielding an intermediate (**450**), the cyclization of which gives a 2-iminothiazolidin-4-one (**451**). In these reactions α-chloracetic acid may be replaced by other two-carbon components such as oxalyl chloride (**452**) to give substituted thiazolidine-4,5-diones (**453**), with maleic anhydride (**454**) to give substituted 5-carboxymethyl-thiazolidin-4-ones (**455**), with acetylenedicarboxylic esters (**456**) to give substituted 5-carbalkoxymethylenethiazolidin-4-ones (**457**), or with 1,2-dibromoalkanes (**458**) to give 2-substituted thiazolidines (**459**).

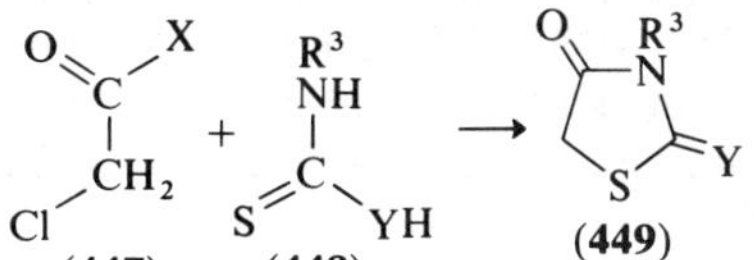

(**447**) (**448**) (**449**)

X = OH, OR, NHAr, $NHCONH_2$;
YH = NHR, $NHNH_2$, OH, SH

Scheme 255

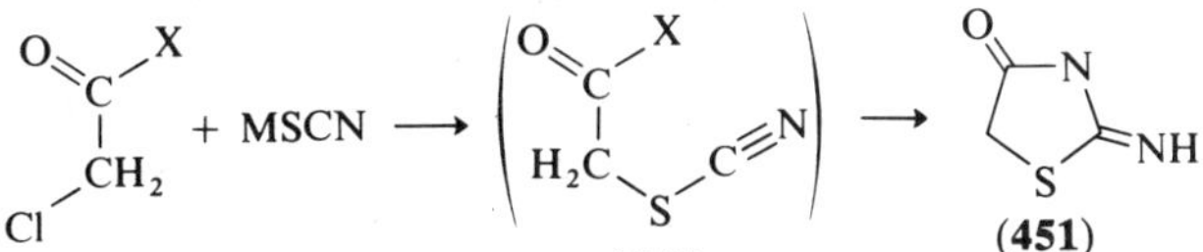

(**450**) (**451**)

Scheme 256

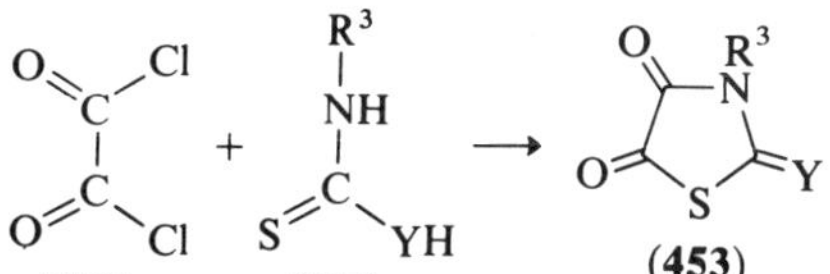

(**452**) (**448**) (**453**) YH = SH

Scheme 257

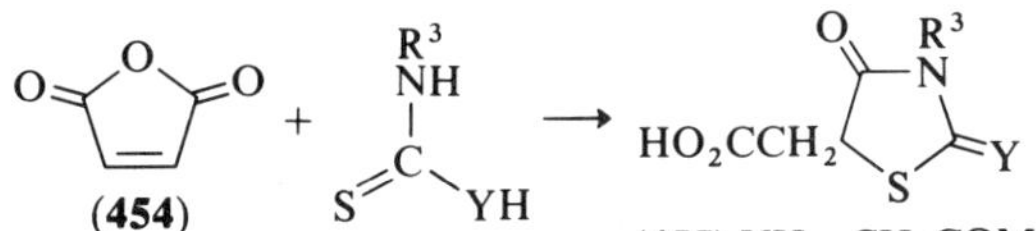

(**454**) (**455**) YH = CH_2COMe

Scheme 258

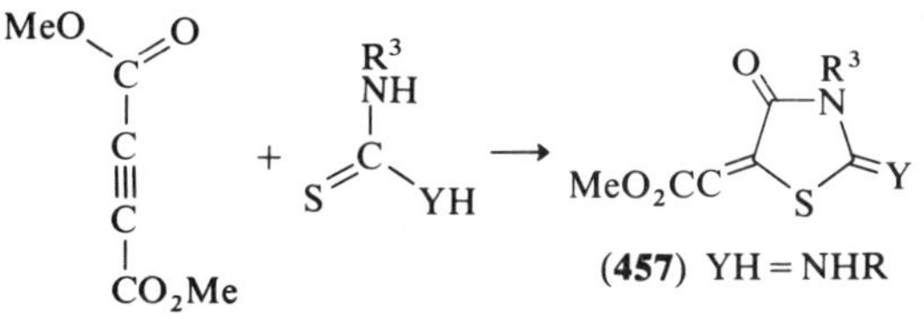

(**456**) (**457**) YH = NHR

Scheme 259

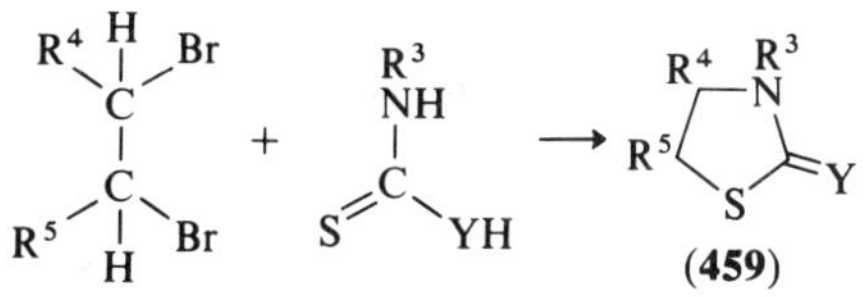

(**458**) (**459**) YH = NHR

Scheme 260

4.19.3.5.2 Type B synthesis of thiazolidines

2-Bromoethylamine (**460**) reacts with methyl isothiocyanate (**461**) giving an intermediate thiazolidine (**462**) which adds further to methyl isothiocyanate yielding the 3-*N*-methyl-thiocarbamoyl-2-methyliminothiazolidine (**463**; Scheme 261). The interaction of 2-

substituted 2-amino alcohols (**464**) with carbon disulfide, in sodium hydroxide solution, affords mixtures of 4-substituted oxazolidine-2-thiones (**466**) and thiazolidine-2-thiones (**465**; Scheme 262). Like amino alcohols, aminoalkyl hydrogen sulfates (**467**) react with carbon disulfide, in alkaline media, providing an interesting route to thiazolidine-2-thiones (**468**; Scheme 263) ⟨65JOC491, 65JOC495⟩. α-Amino ketones (**469**) react similarly with carbon disulfide to give thiazolidine-2-thiones (**471**) which equilibrate in solution with the acyclic tautomer (**470**; Scheme 264) ⟨76CB139⟩. Ethyleneimine reacts with aldehydes and ketones yielding intermediates which undergo thiolation and cyclization under the action of sulfur or hydrogen sulfide, the complex mixture of products containing thiazolidine (Scheme 265) ⟨71M321⟩. The use of *cis* and *trans*-2,3-dialkylaziridines in conjunction with CS_2 affords fairly good yields of 4,5-disubstituted thiazolidine-2-thiones (**477**) of opposite geometrical configuration. The reaction is stereospecific for *cis*-aziridines but only stereoselective for *trans* isomers. A proposed mechanism which accounts for this observed selectivity of inversion in the aziridine ring is shown in Scheme 266 ⟨71JOC1068⟩. Applied to 2-alkyl-aziridines this reaction leads to 4-alkylthiazolidine-2-thiones indicating that the attack of the sulfur nucleophile occurred on the least substituted carbon atom of the aziridine. The steric course and kinetics of the ring-cleavage of 1,2-disubstituted aziridines (**478**) during their cycloaddition with methyl dithiobenzoate have been studied and indicated that all four theoretically possible racemates of thiazolidines (**479**) bearing three asymmetric centers are obtained (Scheme 267) ⟨71TL477⟩.

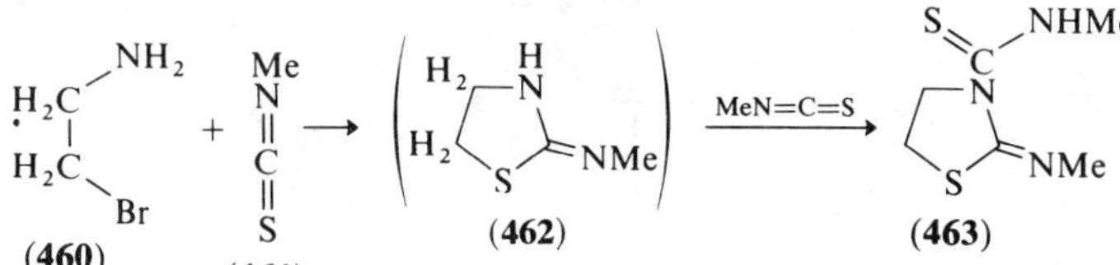

Scheme 261

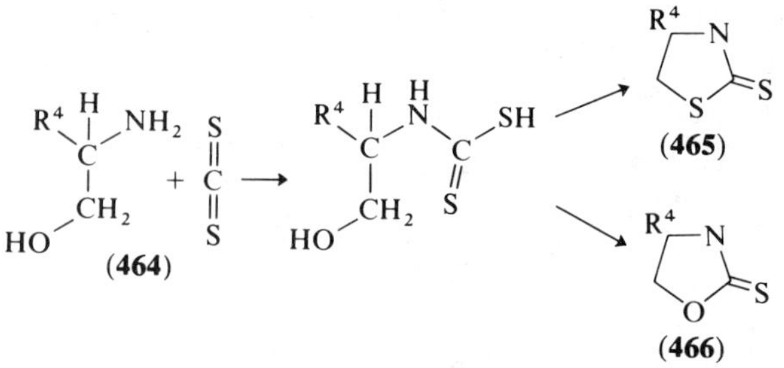

Scheme 262

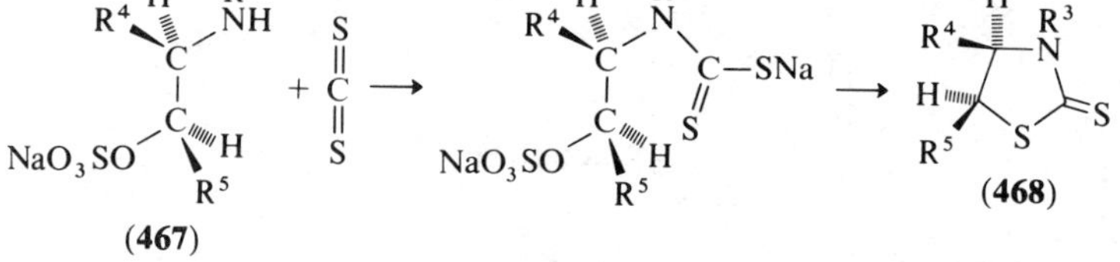

Scheme 263

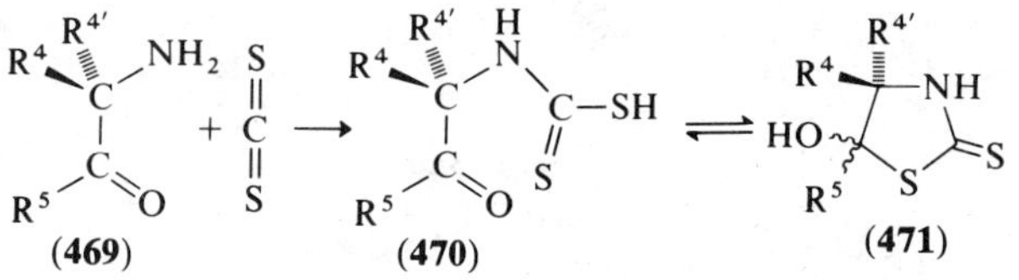

Scheme 264

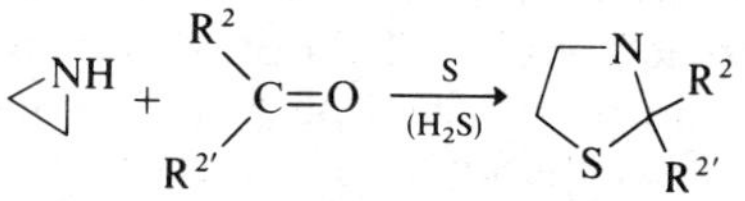

Scheme 265

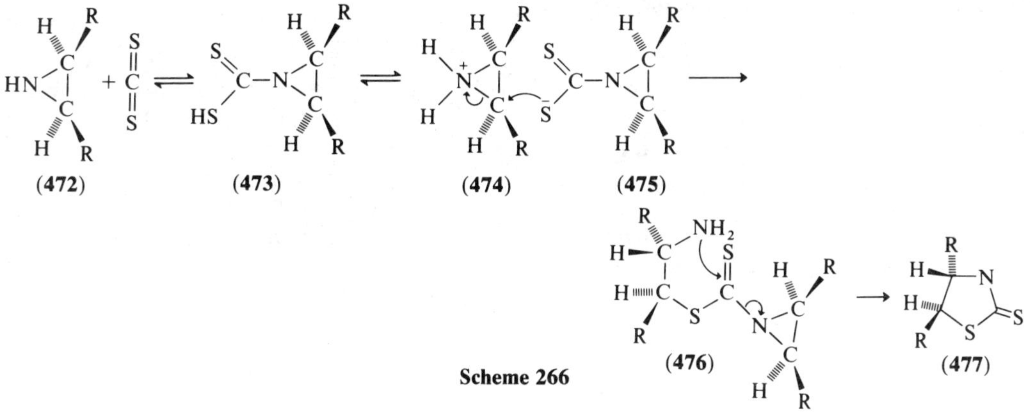

(472) (473) (474) (475) (476) (477)

Scheme 266

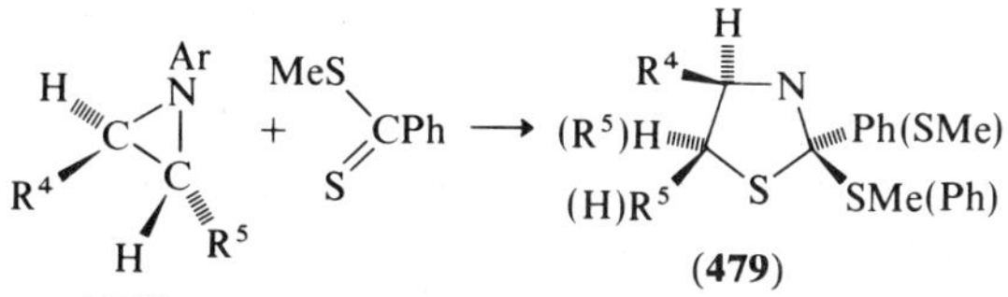

(478) (479)

Scheme 267

4.19.3.5.3 Type D synthesis of thiazolidines

α-Mercaptocarboxylic acids (**480**) react with enamines derived from ethyl acetoacetate (**481**) to give thiazolidin-4-ones (**482**; Scheme 268) ⟨70T4641⟩. Similarly, β-aryl-α-mercaptoacrylic acids (**483**) add to the C=N bond of Schiff's bases (**484**) affording thiazolidin-4-ones (**485**; Scheme 269) ⟨73IJC128⟩. The addition of the same acid (**483**) to phenyl isothiocyanate yields 5-arylidene-3-phenylrhodanines (**486**). An interesting synthesis of substituted thiazolidin-4-one 1,1-dioxides (**489**) results from 1,3-dipolar addition of the adduct of ketene and sulfur dioxide (**487**) with benzylidene aniline (**488**). The ketene–sulfur dioxide adduct may be represented by a mesomeric structure which corresponds to a linear 1,3-dipolar species (**487b**; Scheme 270) ⟨69JHC729⟩.

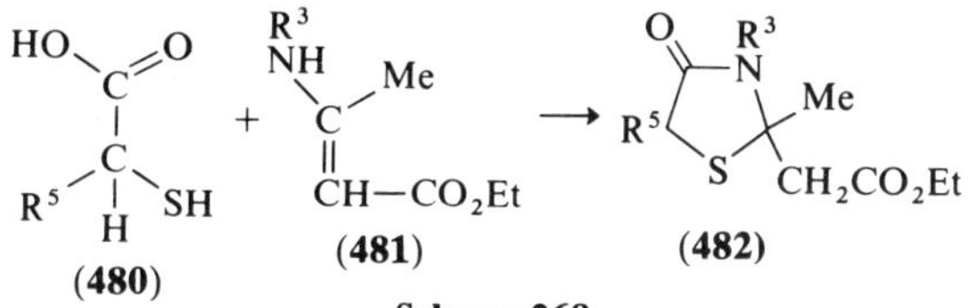

(480) (481) (482)

Scheme 268

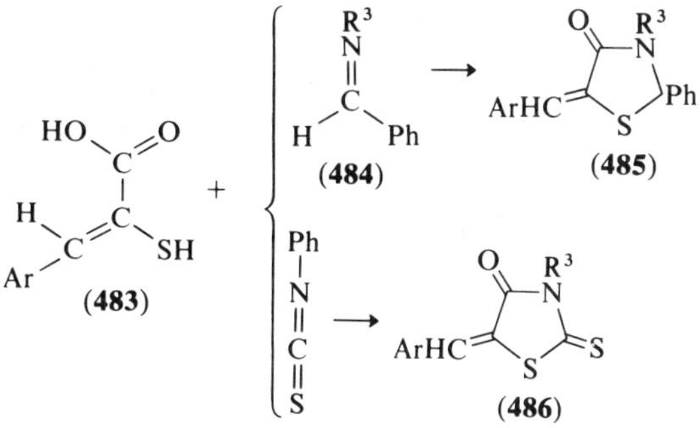

(483) (484) (485) (486)

Scheme 269

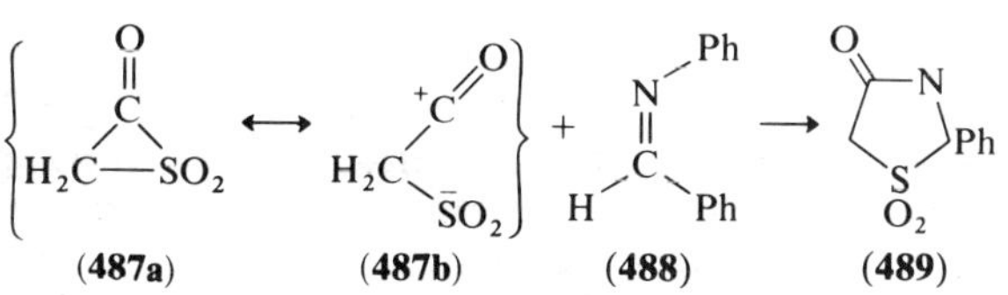

(487a) (487b) (488) (489)

Scheme 270

4.19.3.5.4 Type E synthesis of thiazolidines

This is the most general method for the preparation of thiazolidine and its homologs and involves the reaction of carbonyl compounds with α-aminothiols ⟨B-49MI41900⟩. Aldehydes, ketones or their derivatives such as acetals, thioacetals or Schiff's bases can be used, and among α-aminothiols cysteine and its decarboxylation product cysteamine, play an important role. The parent compound, thiazolidine (**491**) is obtained in very high yield as the hydrochloride, by the reaction of cysteamine hydrochloride (**490**) with aqueous formaldehyde at ambient temperature (Scheme 271). In this reaction ethyleneimine and hydrogen sulfide can be substituted for the aminothiol. Similarly, 4-substituted thiazolidin-2-ones are formed in high yields when α-aminothiols are treated successively with carbon monoxide and oxygen in the presence of triethylamine and a catalyst such as selenium ⟨74TL2899⟩.

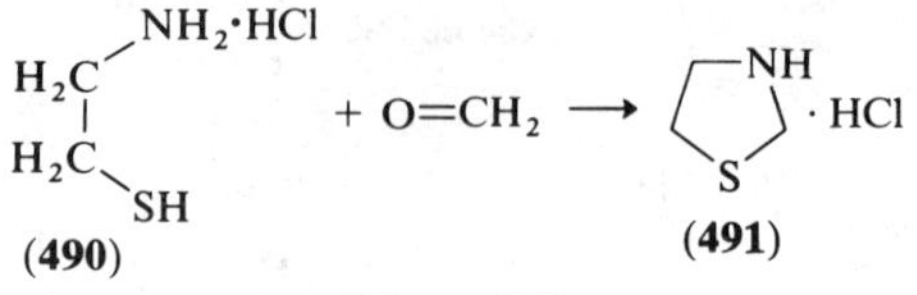

Scheme 271

4.19.3.5.5 Type G synthesis of thiazolidines

The reaction of ethyl cyanothioacetate (**492**) with an amine and aromatic aldehydes affords diversely substituted 5-arylidene-2-iminothiazolidin-4-one derivatives (**493**; Scheme 272) ⟨71NKZ867⟩. Thiazolidinium salts (**496**) are formed by reaction of thiiranes (**494**) with dimethylchloromethylamine (**495**) in acetonitrile (Scheme 273) ⟨70CB3058⟩.

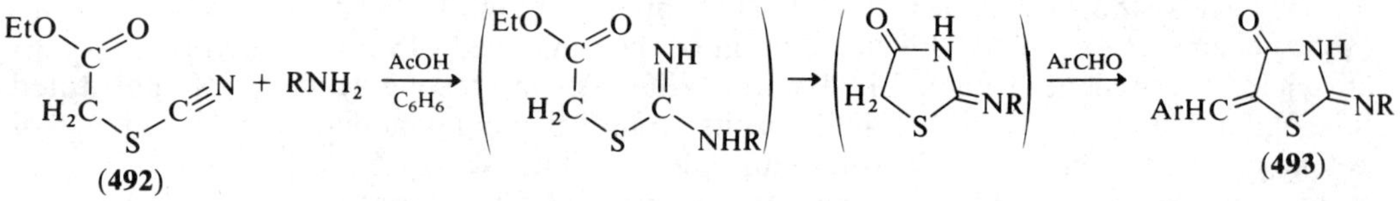

Scheme 272

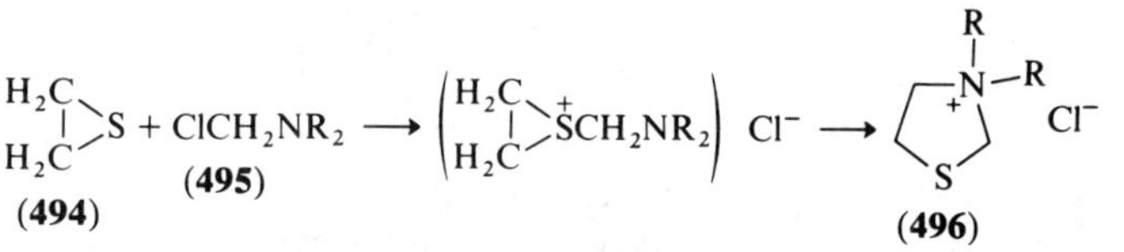

Scheme 273

4.19.3.5.6 Type K synthesis of thiazolidines

The adduct of allyl isothiocyanate and dimethyl malonate (**497**) cyclizes in the presence of halogen to give the corresponding thiazolidine (**498**; Scheme 274) ⟨72JOC318⟩.

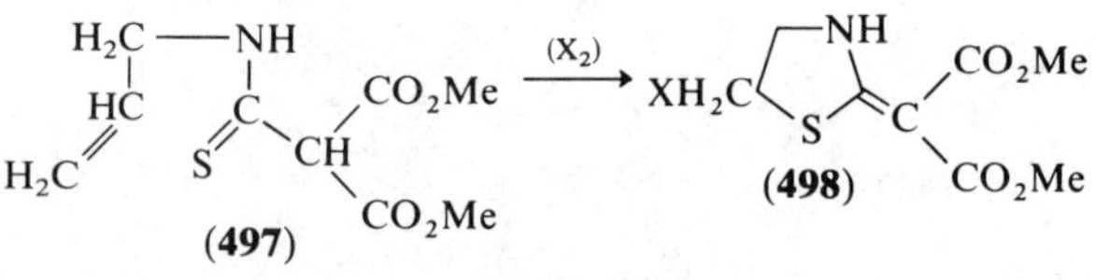

Scheme 274

4.19.3.5.7 Type O synthesis of thiazolidines

Free radical initiation or photolysis of N-allyl-2-aminothiols (**499**) leads essentially to a thiazolidine especially when the reaction is performed at higher temperature with a high aminothiol concentration in a good hydrogen-donating solvent and with starting compounds containing a secondary amino group. Thiazolidines (**501**) result from intramolecular allylic hydrogen abstraction by the thiyl radical, followed by double bond migration and final cyclization of the intermediate enamine (**500**; Scheme 275) ⟨81BSF(2)449⟩.

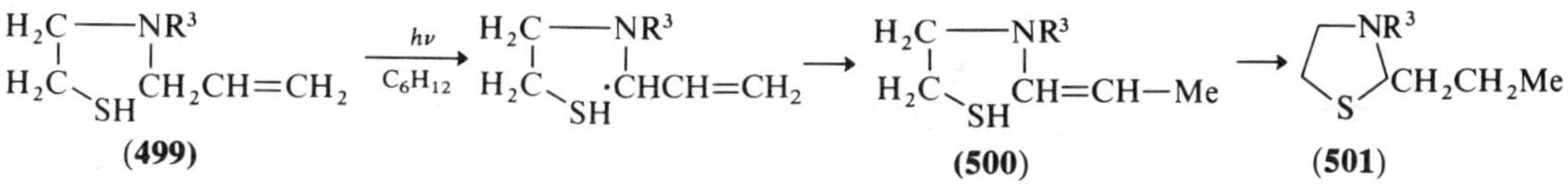

4.19.3.6 Benzothiazoles

4.19.3.6.1 Type A synthesis of benzothiazoles

This mode of preparation of benzothiazoles is one of the earliest and most valuable methods for the cyclization of the heterocyclic ring on an o-difunctionalized benzene (or naphthalene). The starting material is an o-aminothiophenol (or naphthol) or an o-aminoarylthiosulfuric acid.

(i) Synthesis from o-aminothiophenols

o-Aminothiophenols (**502**) are very oxygen sensitive and are employed in the form of derivatives such as acid salts, alkaline salts, zinc salts or disulfides. The last precursors generate the free thiophenol under reducing conditions. The appreciable reactivity of o-aminothiophenol allows this compound to react with a wide variety of reagents.

(ii) Synthesis from aldehydes

Aldehydes (**503**) condense in acid or basic medium giving an intermediate Schiff's base (**504**) which could be isolated in certain cases, and whose cyclization affords the benzothiazoline (**505**) which dehydrogenates giving the 2-substituted benzothiazole (**506**; Scheme 276). In certain cases this dehydrogenation must be assisted by an oxidant (iron(III) chloride). Ketones react similarly but more slowly and give lower yields of benzothiazolines ⟨56MI41900, 56MI41901⟩. o-Aminothiophenols (**507**) react with aromatic 1,2-diketones (**508**) in boiling

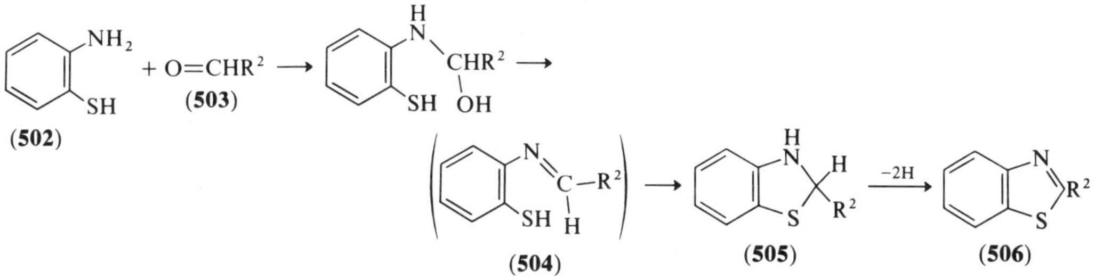

methanol yielding 2-aroyl-2-arylbenzothiazolines (**509**) which are converted in boiling acetic acid into the corresponding 2-arylbenzothiazoles (**510**). Benzothiazolin-2-one is obtained in good yield by treating o-aminothiophenol successively with carbon monoxide and oxygen in the presence of triethylamine and a catalyst such as selenium in DMF ⟨75TL1969⟩.

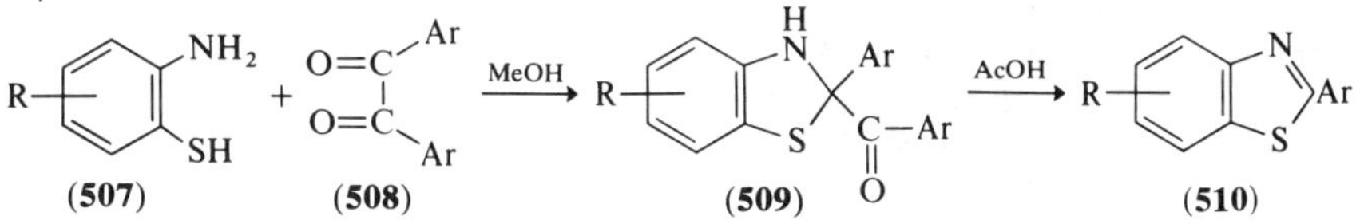

(iii) Synthesis from carboxylic acids

Carboxylic acids (**511**) and their chlorides, anhydrides, amides, esters and nitriles, condense with o-aminothiophenols (**507**) or their zinc salts, affording 2-substituted benzothiazoles (**513**), the only exception being acetic acid. The intermediate o-acylaminothiophenols (**512**) could be isolated and readily cyclized. The activation of carboxylic acids in

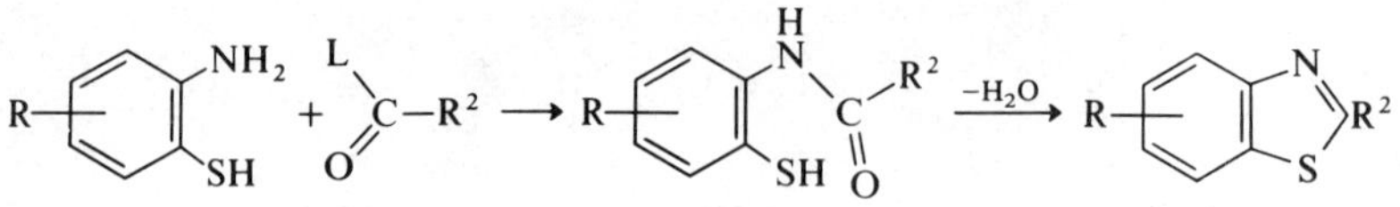

Scheme 278

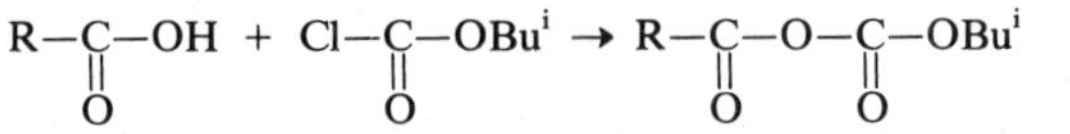

Scheme 279

their condensation with o-aminothiophenols can be realized by an initial treatment with isobutyl chloroformate which transforms the acid into a more reactive mixed anhydride (Scheme 279) ⟨69CB568⟩.

Bisbenzothiazolylalkanes could be obtained by cyclization of o-aminothiophenol with aromatic and certain aliphatic diacids. Acetic anhydride is used to obtain the most important 2-methylbenzothiazole. The reaction of an o-aminothiophenol with phosgene, and with thiophosgene or carbon disulfide, gives a 2-hydroxy- and a 2-mercapto-benzothiazole, respectively. o-Aminothiophenol adds to alkynecarbonitriles (514) forming vinylamines (515) which cyclize in the presence of sodium ethoxide to an intermediate benzothiazoline (516). This loses acetonitrile on distillation at atmospheric pressure to give 2-substituted benzothiazole (517). The practical preparation of benzothiazoles by this general type of ring closure requires satisfactory methods for the preparation of o-aminothiophenols. These intermediates are accessible by reduction of di(o-nitrophenyl) disulfides by zinc and acetic acid, affording the zinc salt which can be used directly for cyclization. They may be prepared also by the Herz process which treats the Herz bases, obtained by the action of S_2Cl_2 on aromatic amines having a free *ortho* position, with basic reducing agents, or by reduction of o-aminoarylthiosulfuric acids.

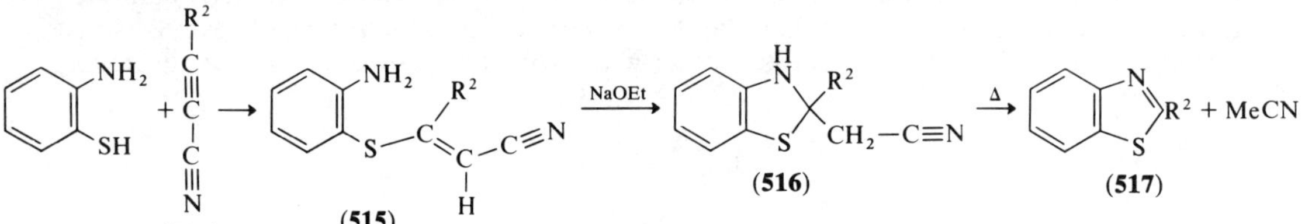

Scheme 280

4.19.3.6.2 Type B synthesis of benzothiazoles

This second mode of preparation of benzothiazoles is also well established and represents a general method for the cyclization of derivatives of arylamines. The main derivatives adapted to the formation of the thiazole ring include N-arylthioamides, arylthioureas, aryl isothiocyanates, esters of arylthiocarbamic acids and esters of aryldithiocarbamic acids. Their cyclization results from the action of an appropriate oxidizing agent.

(i) *Synthesis from* N-*arylthioamides or thioanilides*

N-Arylthioamides or thioanilides (518) which are usually prepared from the corresponding amides by sulfurization, are oxidized in basic medium by potassium ferricyanide to 2-substituted benzothiazoles (519; Scheme 281) (Jacobson's method). The best yields are obtained for arylamines bearing electron-donating groups which enhance the electron density *ortho* to the nitrogen atom.

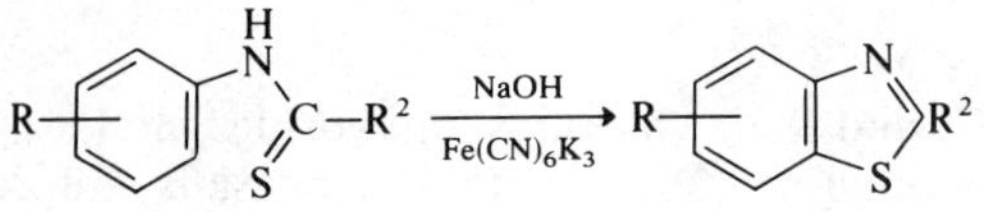

Scheme 281

The thioanilides of α-cyanomalonic ester (**520**), which exist preferentially in their enethiol form, are cyclized by bromine to benzothiazolines (**521**) which can be decarboxylated to the corresponding 2-methylbenzothiazoles (**522**; Scheme 282) ⟨69CB351⟩. 2-Bromo- or 2-chloro-thioacetanilides (**523**) undergo an intramolecular displacement of halogen on treatment with sodium hydride in *N*-methylpyrrolidone or sodium methoxide in DMF, to give a correspondingly substituted 2-methylbenzothiazole (**522**; Scheme 283) ⟨76S730⟩.

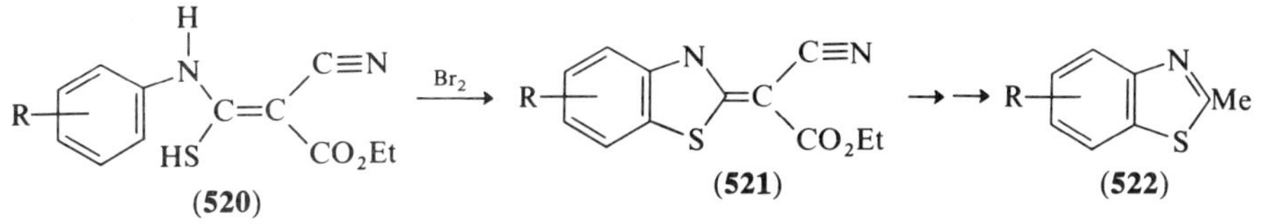

Scheme 282

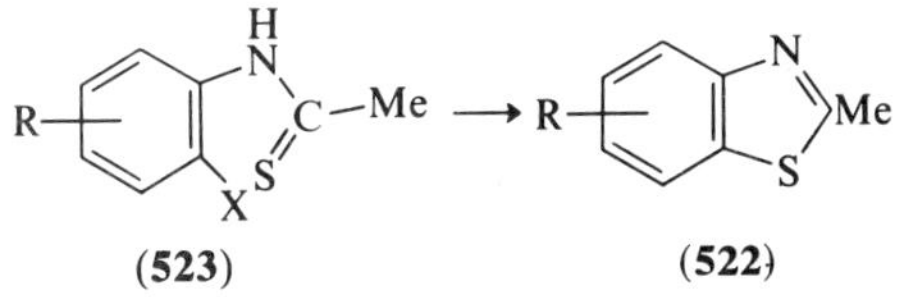

Scheme 283

(ii) Synthesis from arylthioureas

Mono-, di- and tri-substituted arylthioureas (**524**) are very easily cyclized to 2-aminobenzothiazoles (**525**) by the action of bromine in a solvent such as $CHCl_3$, CCl_4, CS_2, S_2Cl_2 (Hugershoff's method) followed by a treatment with SO_2 and with a base (Scheme 284). In the case of unsymmetrical *N*,*N*′-diarylthioureas both possible 2-arylaminobenzothiazoles are obtained, and that which bears on the exocyclic nitrogen atom the least *ortho* activated aryl group is obtained in greater yield ⟨56MI41900, 56MI41901⟩.

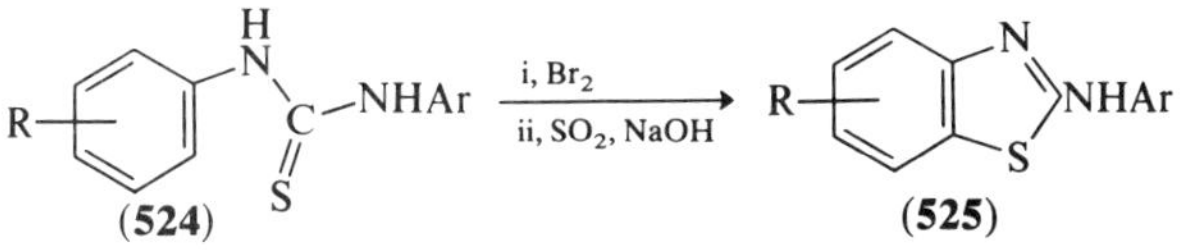

Scheme 284

(iii) Synthesis from aryl isothiocyanates

Aryl isothiocyanates (**526**) can be cyclized by heating with PCl_5 to 2-chlorobenzothiazoles (**527**) (Hunter's method) or with sulfur to 2-mercaptobenzothiazoles (**528**; Scheme 285).

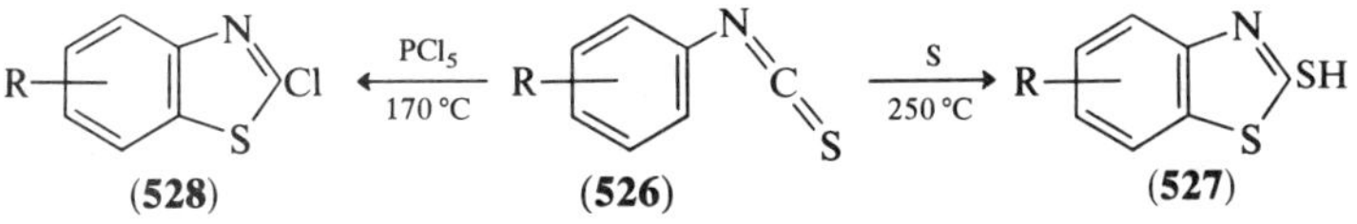

Scheme 285

(iv) Synthesis from arylmonothiocarbamates or arylthiourethanes

Arylmonothiocarbamates or arylthiourethanes (**529**), easily prepared from aryl isothiocyanates and alcohols, are oxidized by potassium ferricyanide, in basic medium, to 2-alkoxybenzothiazoles (**530**; Scheme 286) (Jacobson–Hunter method).

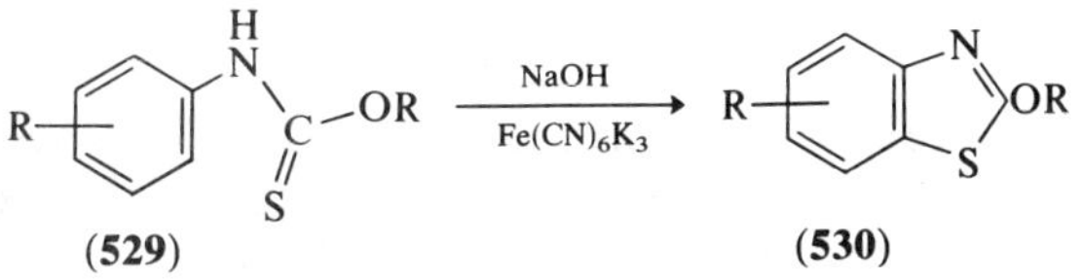

Scheme 286

(v) Synthesis from aryldithiocarbamates

Aryldithiocarbamates (**531**), easily obtained by the reaction of ammonia and carbon disulfide on arylamines, are cyclized, with dehydrogenation, by heating with sulfur at 200 °C to 2-thiobenzothiazoles (**532**; Scheme 287).

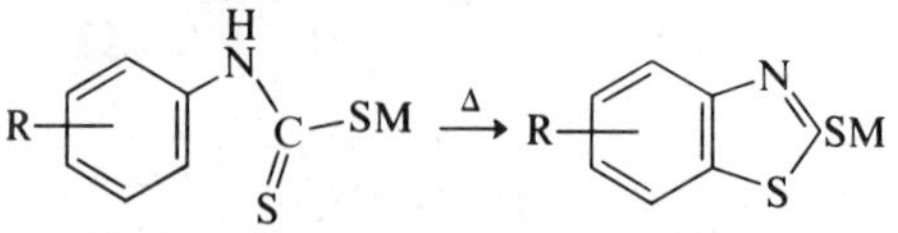

Scheme 287

4.19.3.6.3 Type C synthesis of benzothiazoles

Among the compounds bearing an S—C group are the thiocyanates, the S-acylthiophenols and the arylthioesters.

(i) Synthesis from o-nitroaryl thiocyanates

o-Nitroaryl thiocyanates (**533**), readily obtained by the action of potassium thiocyanate on o-nitrochloroarenes, are reduced to the amino derivatives (**534**) which undergo cyclization to 2-aminobenzothiazoles (**535**; Scheme 288). The o-aminoaryl thiocyanates may be prepared directly by the action of thiocyanogen (N≡C—S)$_2$ on arylamines.

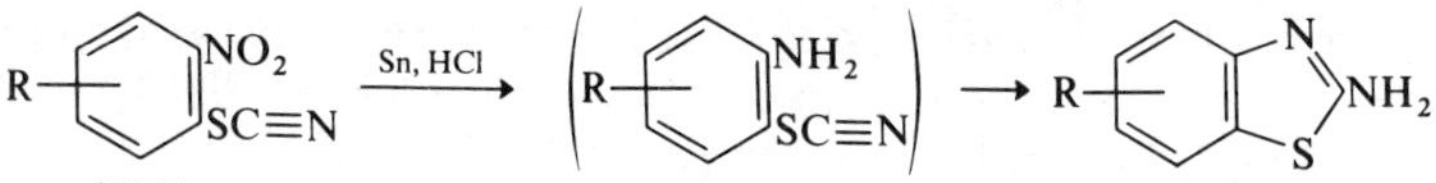

Scheme 288

(ii) Synthesis from S-acyl-o-aminothiophenols

S-Acyl-o-aminothiophenols (**537**), generally formed by the reduction of S-acyl-o-nitrothiophenols (**536**) readily undergo cyclization to 2-hydroxybenzothiazoles (**538**; Scheme 289).

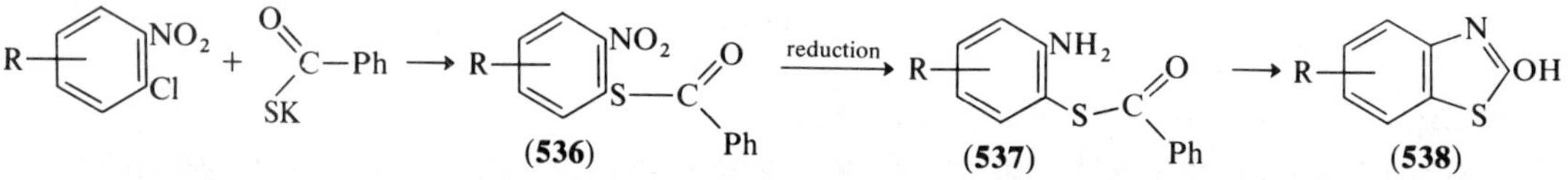

Scheme 289

(iii) Synthesis from o-nitroaryl thioethers

o-Nitrochlorobenzene (**539**) reacts with the sodium salt of α-mercaptoacetic acid affording o-nitrophenylthioacetic acid (**540**) which, in the presence of acetic anhydride and further treatment with base or acid, gives the 2-hydroxybenzothiazole (**541**; Scheme 290).

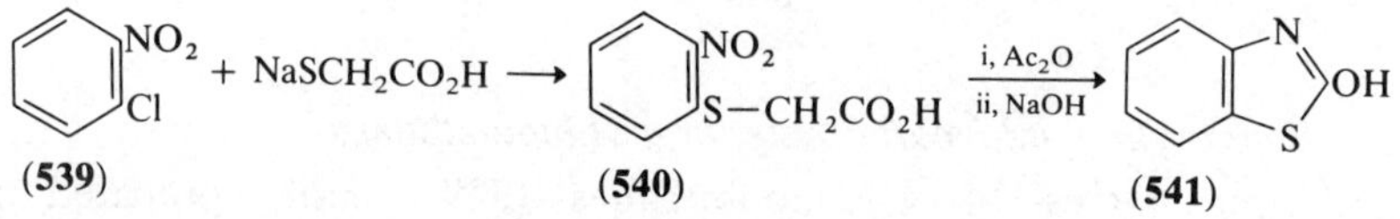

Scheme 290

4.19.3.6.4 Type D synthesis of benzothiazoles

Benzothiazines (**542**) may rearrange into benzothiazoles (**543**) by oxidation in the presence of a base (Scheme 291). This type of synthesis was prompted by the suggestion that the biosynthesis of firefly luciferin (**544**) could result from a similar rearrangement of the benzothiazine formed by the condensation of p-benzoquinone and cysteine ⟨75CC42⟩. The reaction of *p*-benzoquinone (**545**) with ethyl cysteinate hydrochloride (**546**) affords an intermediate (**547**) which on acetylation rearranges to 7-acetoxy-1,4-benzothiazine ester (**548**). Oxidation of the corresponding amide (**549**) with potassium ferricyanide in the presence of NaOEt gives 2-ethoxycarbonyl-6-hydroxybenzothiazole (**550**; Scheme 292).

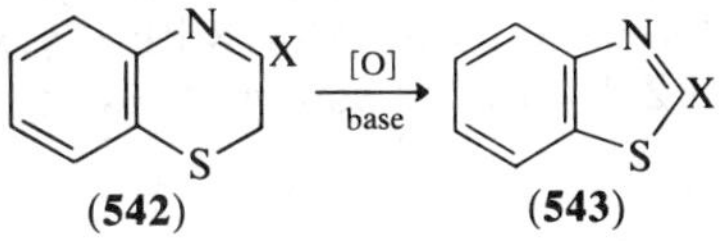

Scheme 291

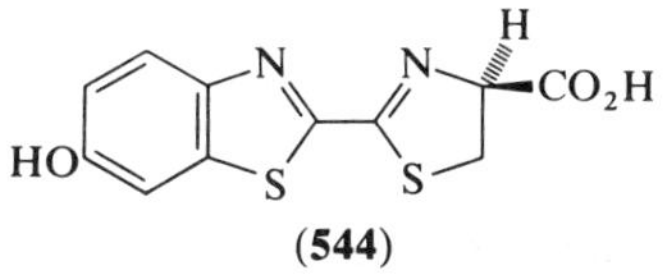

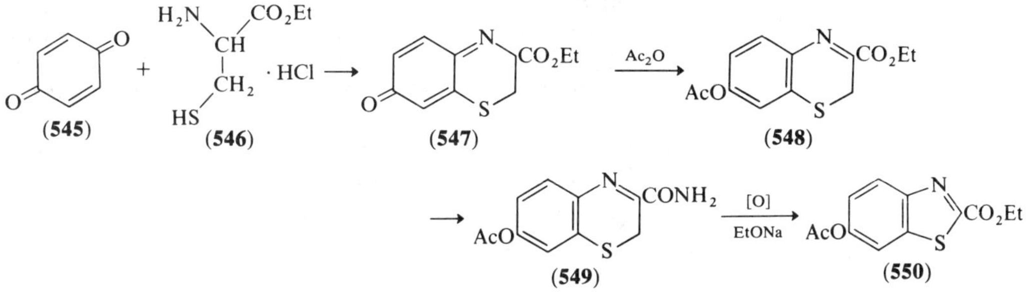

Scheme 292

Catalytic hydrogenation of 3-phenyl-2*H*-1,4-benzothiazine (**551**) in the presence of palladium–charcoal or Raney nickel affords 2-methyl-2-phenyl-2,3-dihydrobenzothiazole (**553**). The reaction is probably initiated by the hydrogenolysis of the C—S bond of (**551**) resulting in the intermediate thiophenol (**552**) which cyclizes to the final benzothiazoline (**553**) ⟨69JHC635⟩.

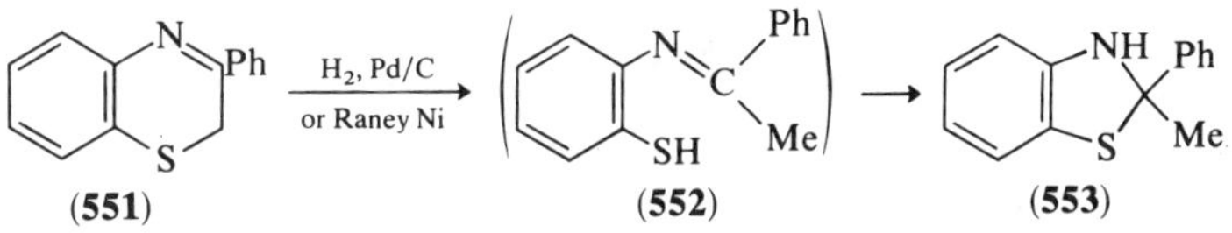

Scheme 293

2-Acylaminothiophenols (**554**) dehydrate on heating in an acidic medium affording 2-substituted benzothiazoles (**555**; Scheme 294). 2-Acylaminothiophenols (**558**) are prepared by acylation of 2-aminothiophenols (**556**) or by reduction of di(*o*-acylaminobenzene) disulfides (**557**) by various reducing agents (Scheme 295).

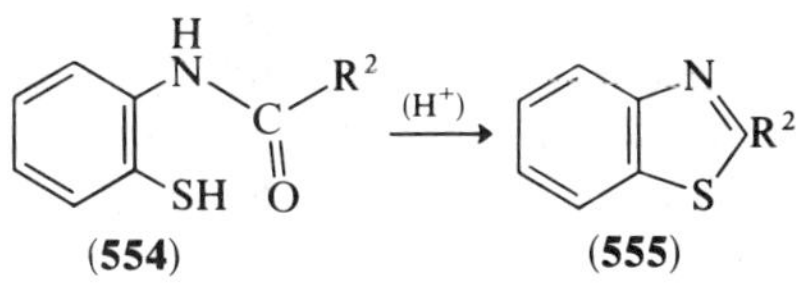

Scheme 294

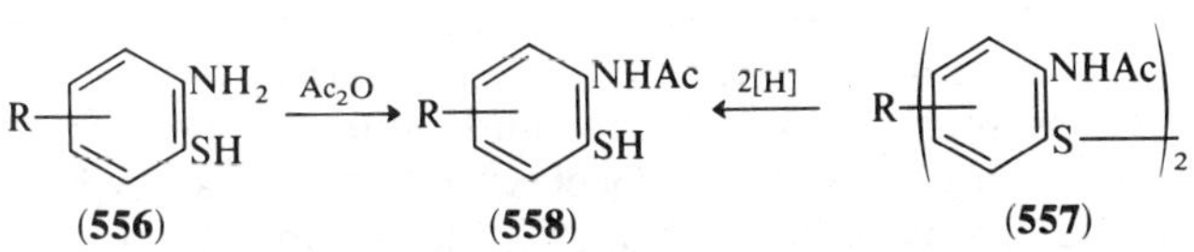

Scheme 295

4.19.3.6.5 Type F synthesis of benzothiazoles

Benzothiazoles are formed by the action of sulfur on alkyl- and acyl-anilines. The reaction is realized at elevated temperatures and is accompanied by secondary reactions. In certain cases (aryl and aroyl derivatives) the yields are satisfactory. Thus benzanilide (**559**) reacts with sulfur affording 2-phenylbenzothiazole (**560**) with a yield of 70%. Benzothiazole is obtained in a similar way (yield 21%) by the action of sulfur on N,N-dimethylaniline.

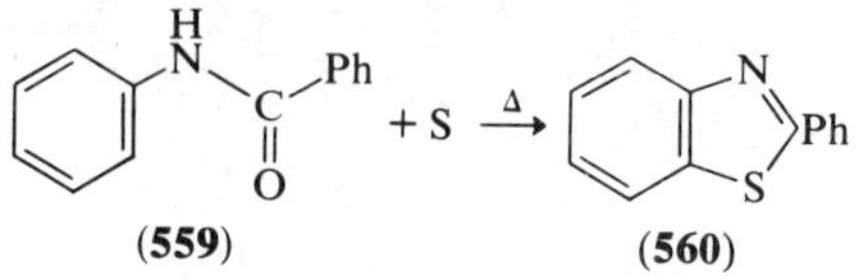

Scheme 296

4.19.3.6.6 Type K synthesis of benzothiazoles

An industrial process for the preparation of 2-p-aminophenyl-6-methylbenzothiazole (**562**) (dehydrothiotoluidine) is based on the high temperature reaction of sulfur with p-toluidine (**561**; Scheme 297). A similar process involves the simultaneous action of sulfur and carbon disulfide on a primary arylamine. Benzothiazole-2-thione (**564**) (2-mercaptobenzothiazole) is synthesized industrially by reaction of aniline (**563**) with sulfur and carbon disulfide (Scheme 298). In this synthesis, sulfur plays the role of a dehydrogenating reagent and may be replaced by nitrobenzene.

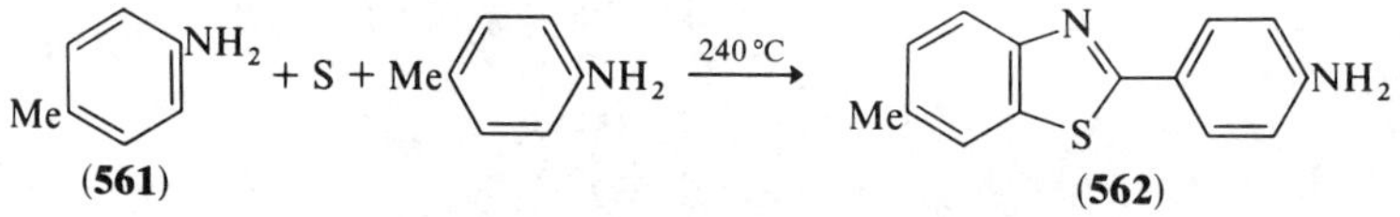

Scheme 297

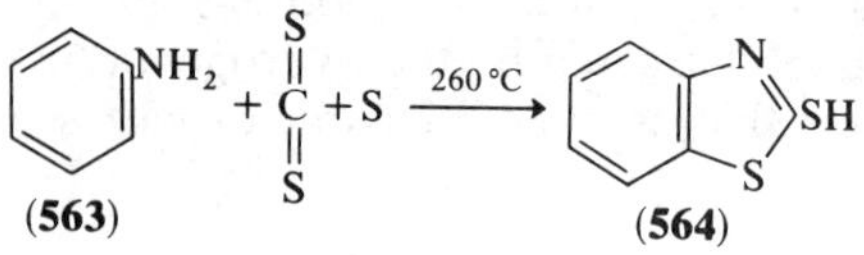

Scheme 298

4.19.4 APPLICATIONS

4.19.4.1 Natural Occurrence

The most important naturally occurring thiazole derivative is thiamine (vitamin B1). Thiamine pyrophosphate (**565**) is the essential coenzyme in the enzymatic decarboxylation of pyruvate to acetaldehyde. As mentioned in Section 4.19.2.1.7(iii) Breslow ⟨58JA3719⟩ proposed the thiazolium ylide (**566**) as the actual catalyst for this reaction. The zwitterionic adduct (**568**) which it forms with pyruvate (**567**) decarboxylates readily, in a way similar to that of β-keto acids and the resulting enamine (**569**) is rapidly protonated to (**570**) which loses a molecule of acetaldehyde with regeneration of the ylide (**566**). Model experiments involving (**568**; R^3 = Me, R^4 = Me, R^5 = H) have shown that the first order rate constant for the decarboxylation reaction, which is pH dependent, is greatly increased in solvents less polar than water. The observed rate of the naturally occurring reaction seems to indicate that the enzymatic catalysis is effected through the binding of the thiazolylium moiety of (**565**) in a region of the enzyme less polar than water ⟨70JA5707⟩. A reason for the natural selection of the thiazolylium heterocycle as the active part of thiamine, as compared with other azolylium systems such as oxazolylium and imidazolylium, may be found in the observation that, of the three azolylium species, only the thiazolylium ion is suited for function in cocarboxylase. It is thermodynamically stable at physiological pH and the rate of formation of the active thiazolylium ylide enables it to be an effective catalyst. The oxazolylium and the imidazolylium heterosystems fail in one of the two criteria: the oxazolylium ion is not stable to ring opening at physiological pH whereas the imidazolylium ion, although stable to ring opening, exhibits too slow a rate of ylide formation, which renders it a poor thiazolylium substitute ⟨74B5358⟩.

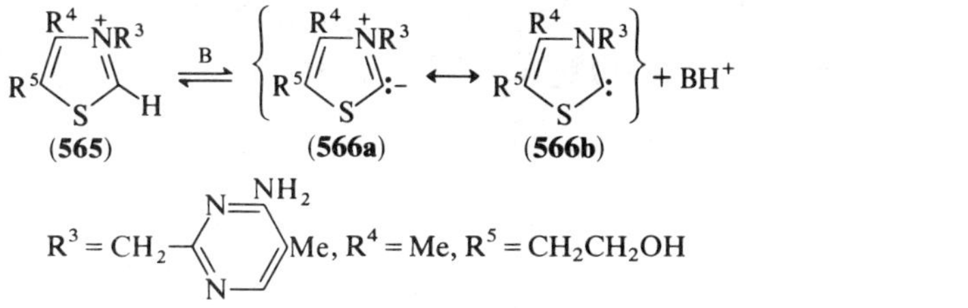

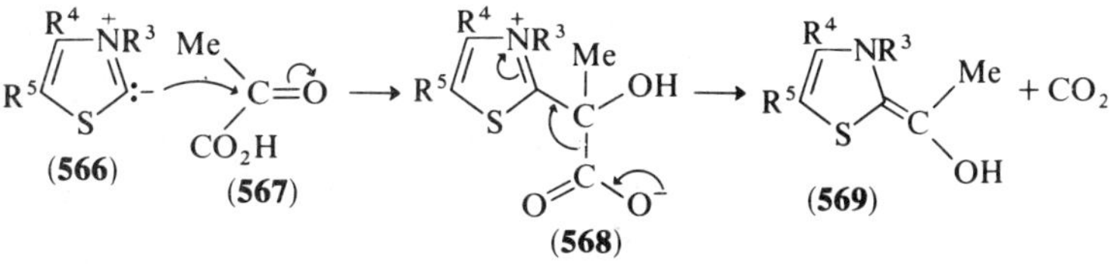

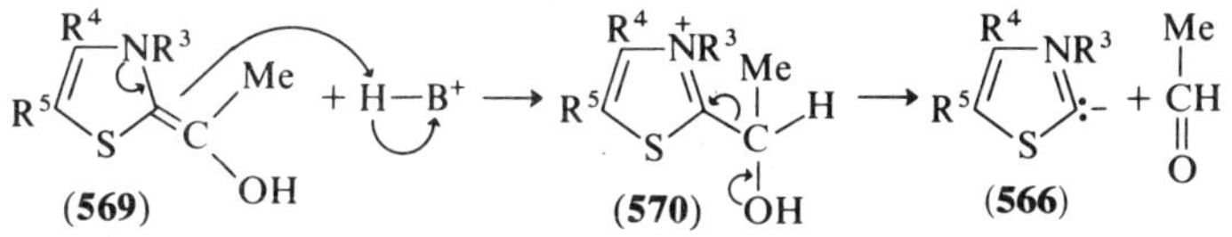

$R^3 = CH_2-\text{(pyrimidinyl)}Me, R^4 = Me, R^5 = CH_2CH_2OH$

Penicillins are also very important naturally occurring thiazolidine derivatives. They are produced by isolation from cultures of mutant strains of the mold *Penicillium chrysogenum*. The most important is penicillin G (**571**; $R = CH_2C_6H_5$) but other natural, biosynthetic or semisynthetic penicillins are produced and used as antibiotics. Other natural antibiotics such as althiomycin (**3**) ⟨75MI41900⟩ or micrococcin ⟨66JCS(C)1361⟩ contain thiazole rings as do many metabolic products of living organisms such as 2-amino-4-(4-carboxythiazol-2-yl)butyric acid (**572**) which has been isolated from the fungus *Xerocomus subtomentosus* ⟨69BSB299⟩, or aeruginoic acid which has been isolated from the culture medium of *Pseudomonas aeruginosa* and has the structure 2-o-hydroxyphenylthiazole-4-carboxylic acid (**573**) ⟨70ABC780⟩. Luciferin, a natural product responsible for the bioluminescence and chemiluminescence of fireflies, has a structure involving both a benzothiazole and a Δ^2-thiazoline ring (**544**) ⟨71MI41900⟩. The reddish-brown macromolecular pigments of bird feathers contain structural units of the type (**574**) ⟨70G461⟩. A large number of natural flavors and aromas of foods contain the thiazole nucleus ⟨81MI41900⟩. 4-Methyl-5-vinylthiazole (**575**) is present in cocoa and passion fruit aromas, benzothiazole (**576**) is a constituent of roast walnuts, coconut, cocoa, beer, and roast pork liver aromas while 2-acetylthiazole (**577**) and 2-acetyl-5-methyl-Δ^2-thiazoline (**578**) are present in beef aroma. 2-Isobutylthiazole (**579**) is the most important flavoring constituent of tomatoes. 2,4,5-Trimethylthiazole and certain Δ^3-thiazolines (**580**), di- or tri-substituted by alkyl, acyl or alkoxy groups exhibit walnut and roast hazel nut flavors. Associated with pyrazines and cyclohexanones, 2-alkylthiazolidines (**581**) enhance cocoa and chocolate aromas. Thiazoles have been detected in certain petroleums in small quantities.

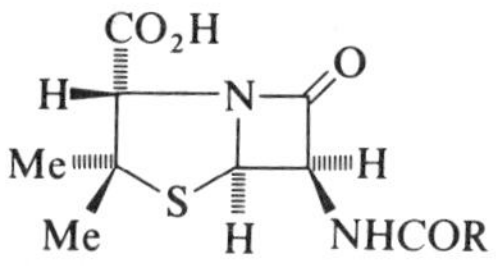

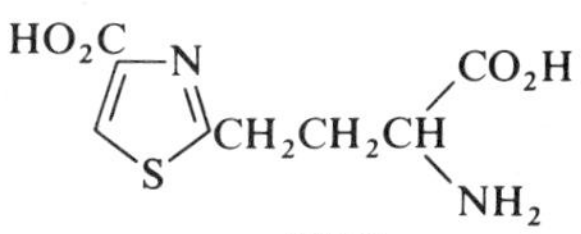

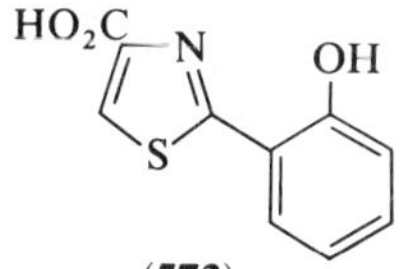

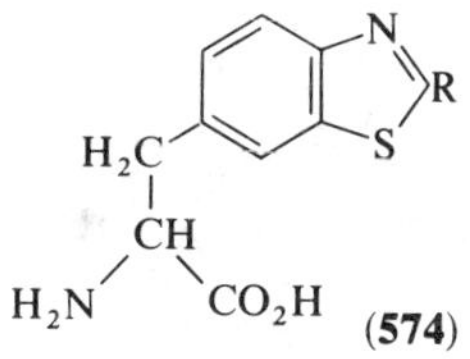

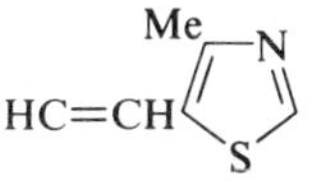

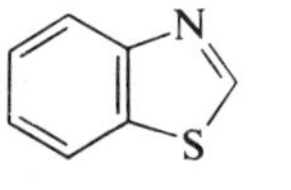

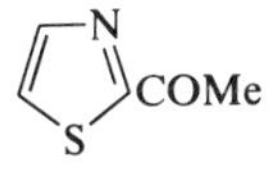

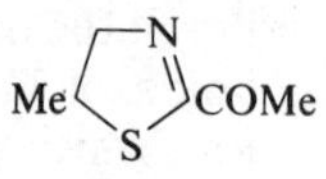

4.19.4.2 Pharmaceutical and Phytosanitary Uses

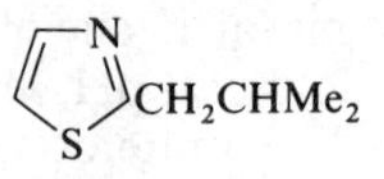

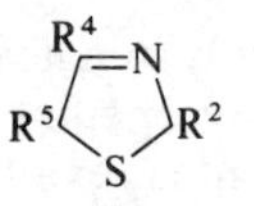

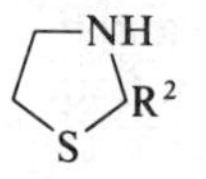

One of the first commercial synthetic drugs containing thiazole was 'Sulphathiazole', a simple sulfamide antibiotic derived from 2-aminothiazole (582). More recently a large number of thiazole derivatives have been found to exhibit pharmacological activity ⟨70SST(1)378, 70SST(1)410, 73SST(2)587, 73SST(2)653, 75SST(3)566, 75SST(3)617, 77SST(4)354, 77SST(4)386, 79SST(5)358, 79SST(5)393, 79HC(34-2)1, B-80MI41900⟩. More specifically, 2-(p-chlorophenyl)thiazol-4-ylacetic acid (583) possesses anti-inflammatory properties ⟨69MI41903⟩, 'Thiabendazol' or 2-(4-thiazolyl)benzimidazole (584) is widely used as an anthelmintic and fungicide. Other derivatives such as 3-substituted 4-aminothiazoline-2-thiones (585) possess antifungal activity, inhibiting *in vivo* the growth of *Xanthomonas oryzae* ⟨69YZ699⟩, or 4-chloro-2-oxobenzothiazolin-3-ylacetic acid (586) ('Benazolin') which is used as an ingredient of herbicides ⟨69MI41902⟩. One of most active schistosomicidal drugs to date is 1-(5-nitro-2-thiazolyl)-2-imidazolidinone (587) ('Niridazole') ⟨71JMC10⟩. Finally a very large number of functional derivatives of thiazole exhibit interesting biological properties, they are summarized in ⟨79HC(34-2)1, pp. 132, 436⟩.

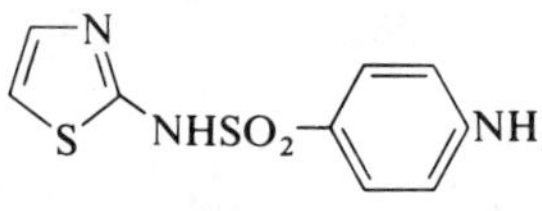

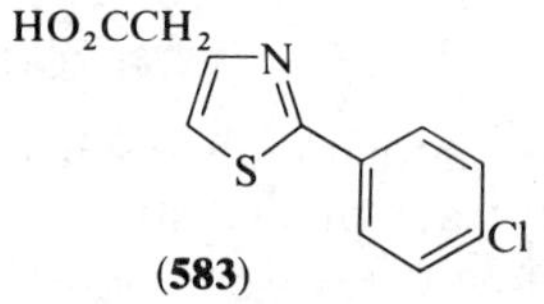

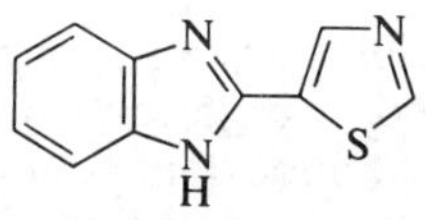

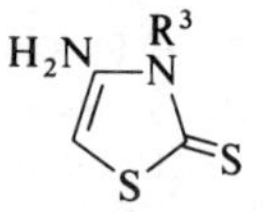

4.19.4.3 Analytical and Synthetic Uses

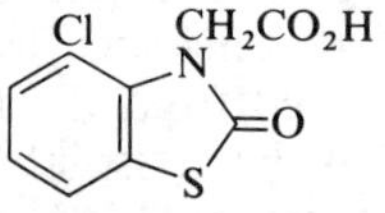

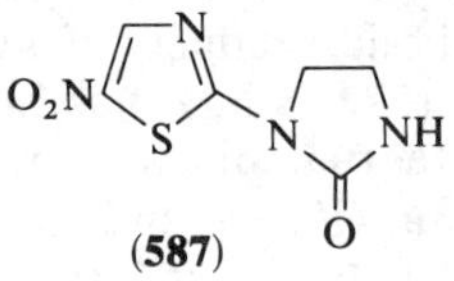

3-Methylbenzothiazolin-2-one hydrazone (588) is known to yield blue solutions with phenols and carbonyl compounds in oxidizing media and is used in the colorimetric estimation of phenols, probably by the formation of colored compounds such as (589) ⟨57LA(609)143⟩. 1-(2-Thiazolylazo)-2-hydroxy-3-naphthoic acid ('TAHN') (590) is used as a complexing indicator in alkaline solution. It forms pink-violet chelates with certain metals and gives a sharp end-point in the titration of Cu or Ni with H_4-EDTA. 1-(2-Thiazolylazo)-2-naphthol ('TAN') (591) is used similarly in the spectrophotometric determination of Ni ⟨68NKK951⟩. 2-(2-Thiazolylazo)-1,8-dihydroxynaphthalene-3,6-disulfonic acid (592a) derivatives and their benzothiazole analogs (592b) also give complexes of contrasting color with Al, Zn, Zr, Th and Ga and are therefore useful reagents in the photometric analysis of these elements ⟨68IZV2666⟩. A large number of analytical applications of thiazole derivatives are described in ⟨79HC(34-2)1, p. 155⟩.

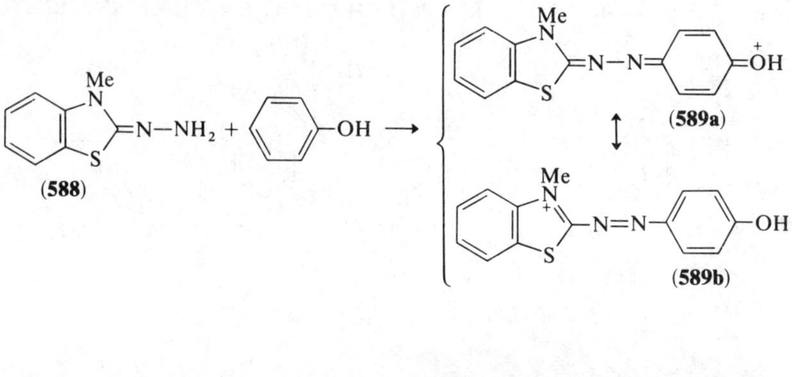

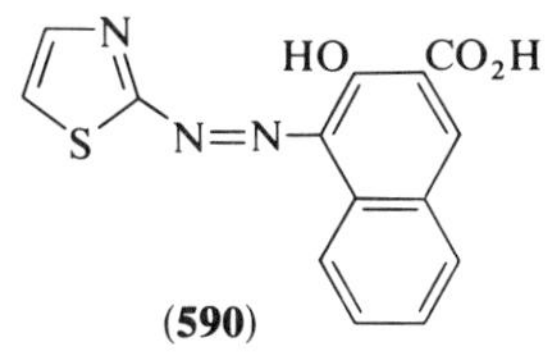

(590), 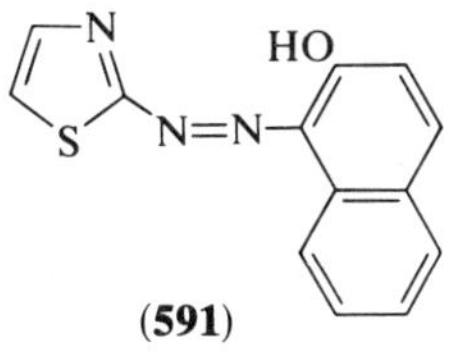 (591), 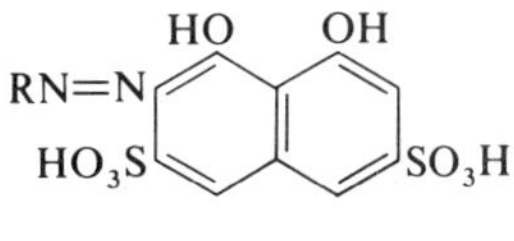 (592a), (592b)

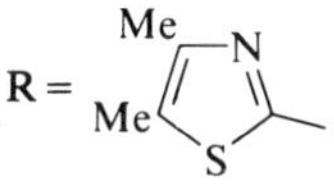

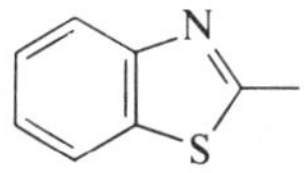

Synthetic uses of thiazole derivatives are not uncommon. As already discussed in Section 4.19.2.2.2(iii), Δ²-thiazolines may serve as synthons for aldehydes, the 2-carbon atom being introduced as the formyl center (Scheme 63) ⟨75JOC2021⟩. Similarly the benzothiazole nucleus may serve as a useful carbonyl equivalent for the synthesis, in high yields, of vinyl aldehydes or ketones. Benzothiazol-2-yllithium (593) reacts with a ketone to yield the corresponding carbinol (594) which on dehydration gives the corresponding 2-vinylbenzothiazole (595). Compound (595) is readily quaternized and the quaternary salt (596) is either alkylated in the 2-position with alkyllithium (597) or reduced with NaBH₄ (599). The resulting benzothiazolines are hydrolyzed to yield a vinyl ketone (598) or a vinyl aldehyde (600), respectively (Scheme 299) ⟨78TL13⟩.

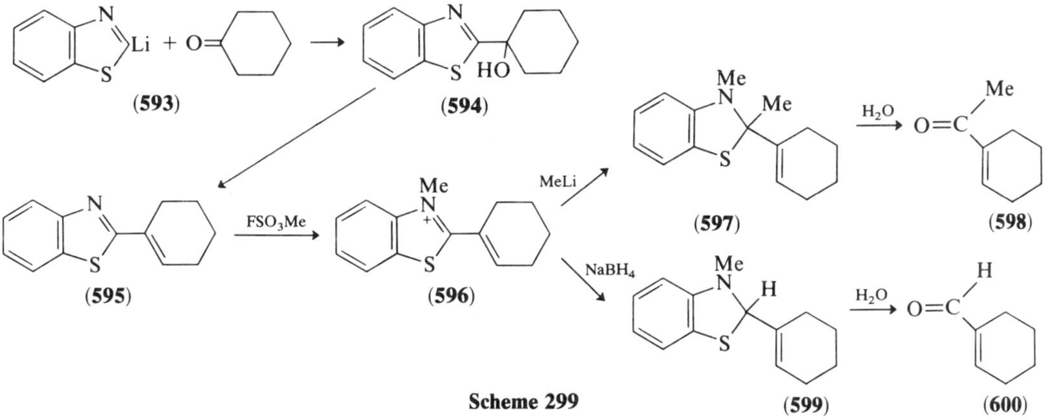

Scheme 299

3-Methylbenzothiazoline-2-thione (601) converts oxiranes (602) into thiiranes (604) in trifluoroacetic acid, stereospecifically and in high yield, by a mechanism involving nucleophilic ring opening by the thione sulfur atom and transfer of the benzothiazolylium moiety to the oxirane oxygen atom (Scheme 300) ⟨75CC621⟩. This reaction is based on the good leaving ability of benzothiazolinone (603).

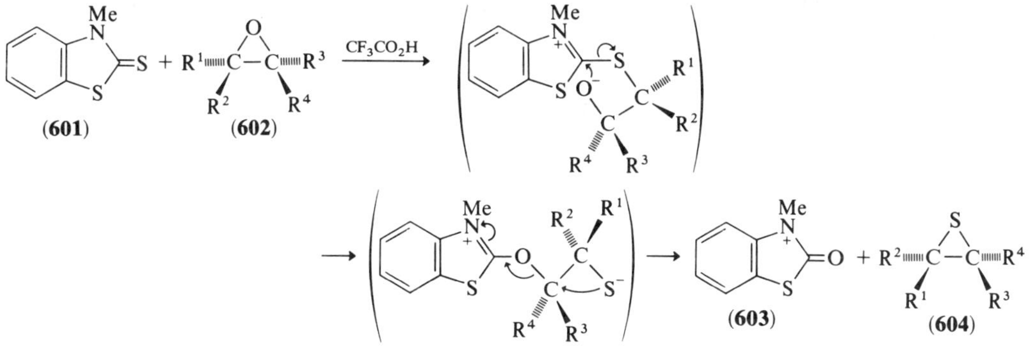

Scheme 300

The N-alkylation of L-cysteine results from a series of reactions involving the condensation of L-cysteine methyl ester (605) with ethyl iminoacetate (606), followed by N-alkylation of the intermediate 4-methoxycarbonyl-2-methylthiazoline (607). Acid hydrolysis of the thiazolinium salt (608; Scheme 301) then gives (609) ⟨70ACS(B)3129⟩.

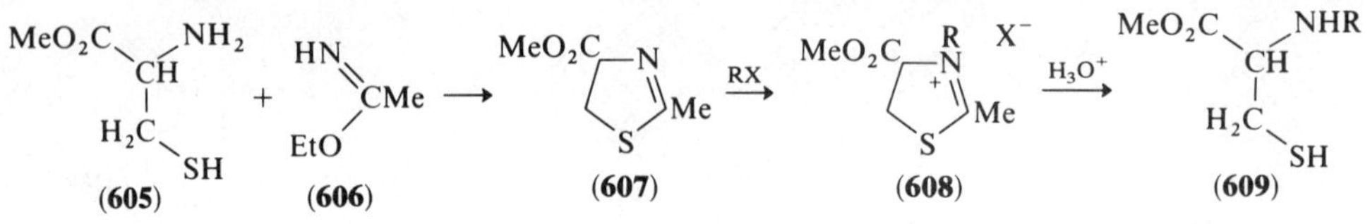

4.19.4.4 Industrial Uses

2-Mercaptothiazoles and benzothiazoles have found wide use as accelators in rubber vulcanization and as antioxidants ⟨73SST(2)587, 73SST(2)653⟩. A large number of cyanine dyes, derived from thiazolylium and benzothiazolylium salts (**610**) has been synthesized and many of them have been used in silver photography, essentially as sensitizing dyes ⟨79HC(34-3)1, p. 23⟩. Other derivatives resulting from the condensation of 2-ethylbenzothiazolylium salts with substituted salicylaldehydes exhibit the phenomenon of photochromism. These benzothiazolinospiropyrans are colorless in their cyclized form (**611**) but become colored under UV irradiation. The colored merocyanine (**612**) thus obtained bleaches on heating, giving back the colorless spiropyran. Such compounds have found uses in photographic and reprographic systems (Scheme 302) ⟨70SST(1)410, p. 416⟩.

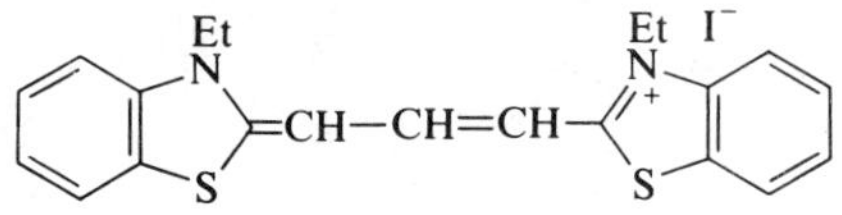

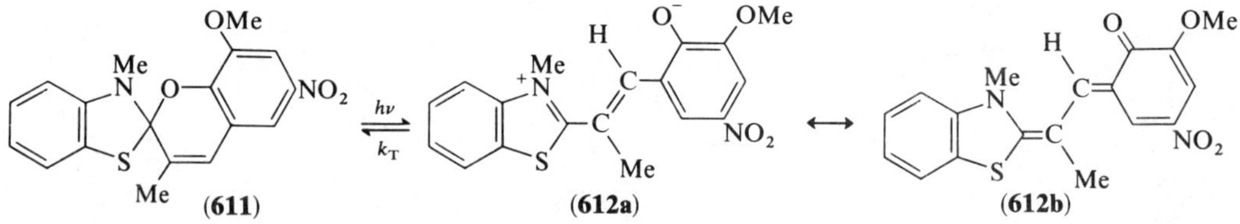

The high thermal stability of the thiazole nucleus has prompted the synthesis of polymers incorporating this structure and which have good thermal stability. Thermoplastic polymers have been prepared by the interaction of benzothiazolediamines (**613**) and aromatic dianhydrides (**614**). These ordered heterocyclic copolymers (**615**) show interesting thermal stability, resistance to oxidation and may, in suitable cases, be spun into excellent fibers (Scheme 303) ⟨70SST(1)410, p. 417⟩. Other thermoresistant polymers incorporating benzothiazole residues (**618**) are obtained by the condensation of 2,6-dichloro- or 2,6-diphenoxy-benzobisthiazole (**616**) with aromatic diamines (**617**). They are sparingly soluble brown powders which show only slight softening below their decomposition temperature (Scheme 304) ⟨70MI41900⟩.

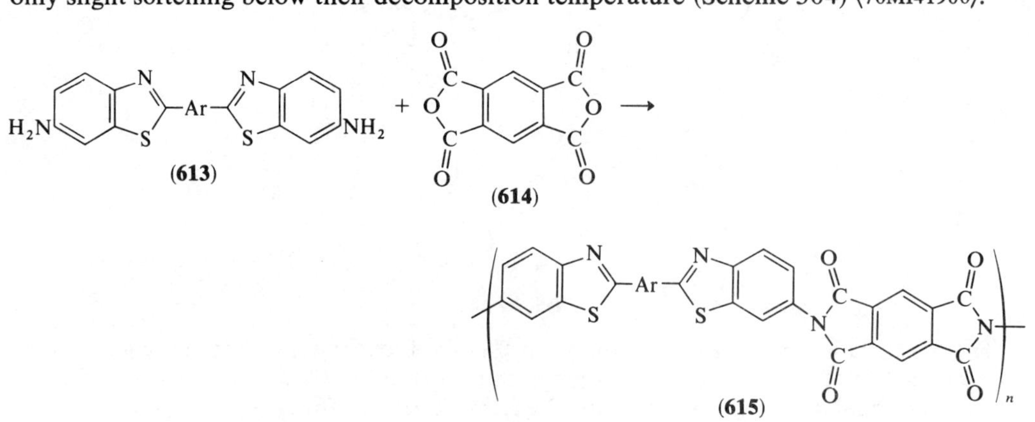

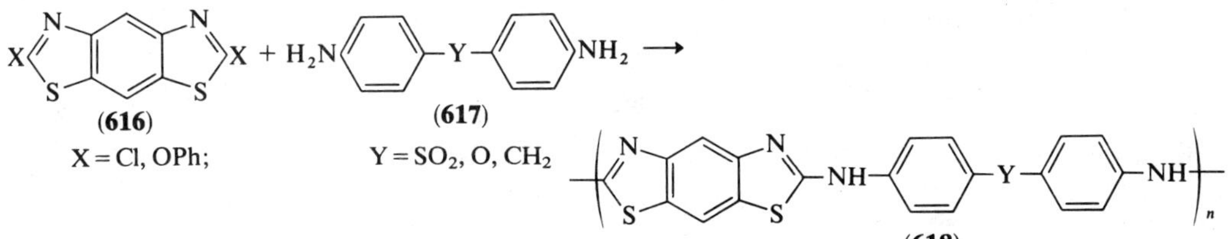

Scheme 304

(616) X = Cl, OPh;
(617) Y = SO$_2$, O, CH$_2$
(618)

4.20

Five-membered Selenium–Nitrogen Heterocycles

I. LALEZARI

Montefiore Medical Center and Albert Einstein College of Medicine, New York

4.20.1	INTRODUCTION	334
4.20.2	ISOSELENAZOLES (1,2-SELENAZOLES)	334
4.20.2.1	*Structure*	334
	4.20.2.1.1 Theoretical methods and molecular dimensions	334
	4.20.2.1.2 Molecular spectra	334
4.20.2.2	*Reactivity*	336
	4.20.2.2.1 Ozonolysis	336
	4.20.2.2.2 Amination	336
	4.20.2.2.3 Ring opening	336
	4.20.2.2.4 N-Alkylation	337
	4.20.2.2.5 Reactions involving the benzene ring	337
	4.20.2.2.6 Ring transformation reactions	337
4.20.2.3	*Synthesis*	337
4.20.3	SELENAZOLES, SELENAZOLINES AND SELENAZOLIDINES	339
4.20.3.1	*Structure*	339
	4.20.3.1.1 Theoretical methods	339
	4.20.3.1.2 Molecular dimensions	339
	4.20.3.1.3 Molecular spectra	340
4.20.3.2	*Reactivity*	340
	4.20.3.2.1 Electrophilic substitution	340
	4.20.3.2.2 Reactivity of the 2-methyl group of selenazoles	341
	4.20.3.2.3 Ring expansion	341
4.20.3.3	*Synthesis*	342
	4.20.3.3.1 Selenazoles	342
	4.20.3.3.2 Mesoionic selenazoles	343
	4.20.3.3.3 Benzoselenazoles and hetero-fused selenazoles	344
	4.20.3.3.4 Selenazolines	345
	4.20.3.3.5 Selenazolidines	346
4.20.4	1,2,3-SELENADIAZOLES	347
4.20.4.1	*Structure*	347
	4.20.4.1.1 General	347
	4.20.4.1.2 Molecular spectra	347
	4.20.4.1.3 Thermodynamic aspects	349
4.20.4.2	*Reactivity*	349
	4.20.4.2.1 Reactivity at the ring atoms	349
	4.20.4.2.2 Reactivity of substituents	351
4.20.4.3	*Synthesis*	352
4.20.5	1,2,4-SELENADIAZOLES	354
4.20.5.1	*Structure*	354
4.20.5.2	*UV Spectroscopy*	355
4.20.5.3	*Synthesis*	355
4.20.6	1,3,4-SELENADIAZOLES	356
4.20.6.1	*Structure*	356
	4.20.6.1.1 Molecular dimensions	356
	4.20.6.1.2 Molecular spectra	356
4.20.6.2	*Reactivity*	357
4.20.6.3	*Synthesis*	357
	4.20.6.3.1 Fused 1,3,4-Selenadiazoles	359

4.20.7 HYPERVALENT FIVE-MEMBERED SELENIUM- AND TELLURIUM-
NITROGEN HETEROCYCLES 359
 4.20.7.1 Benzo-2,1,3-thiaselenazolylium Chlorides, Benzo-1,2,3-thiaselenazolylium Chlorides and
    Benzo-1,2,3-diselenazolylium Chlorides 359
  4.20.7.1.1 Reactivity 359
  4.20.7.1.2 Synthesis 360
 4.20.7.2 $7\lambda^4$-[1,2,5]Oxaselenazolo[2,3-b][1,2,5]oxaselenazoles and $7\lambda^4$-[1,2,5]Oxatellurazolo[2,3-
b][1,2,5]oxatellurazoles 360
  4.20.7.2.1 Structural methods 360
  4.20.7.2.2 Reactivity 361
  4.20.7.2.3 Synthesis 361
 4.20.7.3 $6a\lambda^4$-Selena- and $6a\lambda^4$-Tellura^{IV}-1,2,5,6-tetraazapentalenes 361
 4.20.7.4 1-Oxa-6,6$a\lambda^4$-diselena-2-azapentalenes 361
  4.20.7.4.1 Structure 361
  4.20.7.4.2 Synthesis 362
 4.20.7.5 6,6$a\lambda^4$-Diselena-1,2-diazapentalenes 362
  4.20.7.5.1 Structure 362
  4.20.7.5.2 Synthesis 363

4.20.1 INTRODUCTION

Although selenazoles have been known for many years, it is only in the last decade that they have been the subject of relatively intense investigation with numerous systems being synthesized and characterized. In contrast, five-membered tellurium-nitrogen heterocycles have been represented by only a few examples in the literature. However, both selenium– and tellurium–nitrogen heterocycles have received little attention with regard to theoretical studies and there is a similar lack of information concerning their thermodynamic properties, aromatic character, solubilities, *etc.* These areas show considerable promise for future development, especially as both selenium and tellurium intermediates are becoming more readily available. In this chapter, different classes of five-membered selenium- and tellurium–nitrogen heterocycles are discussed. 1,2,5-Selenadiazoles, described in Chapter 4.26, are not included in this discussion.

4.20.2 ISOSELENAZOLES (1,2-SELENAZOLES)

4.20.2.1 Structure

4.20.2.1.1 Theoretical methods and molecular dimensions

X-Ray diffraction of the 5-(1,2-benzoselenazol-3-yl)pentadienonitrile derivative (**1**) indicates a relatively localized double bond as in formaldoxime. The heterocyclic C—N bond distance (1.29 Å) is significantly shorter than that in 1,2-benzoselenazole (1.41 Å) ⟨81JCS(P1)607⟩. Although the C—Se—N bond angle (91°) found for (**1**) is larger than that of selenium in some other five-membered rings (in 1,3,4-selenadiazole the C—Se—C angle was found to be 81.8°, see Section 4.20.6), it is smaller than the values (95°) for the angle of sulfur in the 1,2-benzothiazoles studied. This is consistent with the observation that the bond angle at the heteroatom decreases with increase in atomic number.

(**1**)

4.20.2.1.2 Molecular spectra

(i) *^{1}H NMR spectroscopy*

The ^{1}H NMR spectral parameters for the parent compound, isoselenazole (**4**) are as follows: δ 9.26 (H-5), 7.35 (H-4; $J_{5,4}$ = 5.3 Hz), 8.98 (H-3; $J_{4,3}$ = 1.8 Hz). Comparison of these chemical shifts and coupling constants with those reported for isothiazole indicates a deshielding of the order of 0.3 p.p.m. for H-3 and 0.7 p.p.m. for H-5 in the selenium system. 1,2-Benzoselenazole (2) and 1,2-benzotellurazole (3) also exhibit similarities. The H-3 proton in compound (2) appears at δ 9.15 and in compound (3) at 10.15 (in the sulfur analog it is at δ 8.73). Coupling constants of $J(^{77}Se-H-3) = 24$ Hz in (2) and $J(^{125}Te-H-3) = 35$ Hz in (3) support their structures ⟨73JHC267⟩.

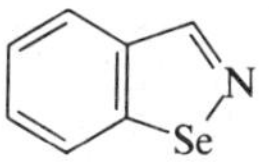

(2) 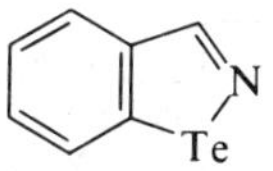   (3) 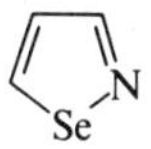   (4)

The spectra of N-methyl-1,2-benzoselenazolium iodide (5) and its tellurium analog (6) are very similar. In compound (5) the 3-proton appears at δ 9.5 and in compound (6) it is at δ 9.65. The methyl groups resonate at δ 4.10 [J(H-3–NMe) = 1.4 Hz] and δ 3.67 [J(H-3–NMe) = 1.5 Hz], respectively ⟨78JHC865⟩.

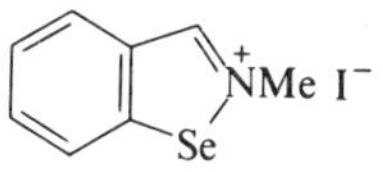

(5) 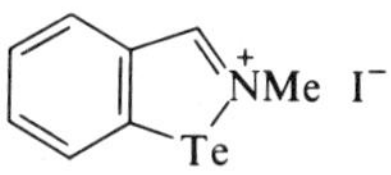   (6)

(ii) ^{13}C NMR spectroscopy

In their ^{13}C NMR spectra, the C-3 signals for 1,2-benzoselenazole (2) and its tellurium analog (3) appear at δ 160.7 and 168.3, respectively. This shift may be related to the decrease in electron–withdrawing activity in going from selenium to tellurium ⟨78JHC865⟩.

(iii) ^{77}Se NMR spectroscopy

The ^{77}Se NMR spectral data for five isoselenazoles *ortho* fused with benzene, thiophene and selenophene (2, 7–10) are reported in Table 1 ⟨80BSB773⟩.

Table 1 ^{77}Se NMR Data

Compound	Chemical shift (p.p.m.)[a]	Compound	Chemical shift (p.p.m.)[a]
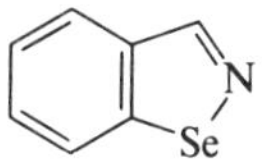 (2)	1013	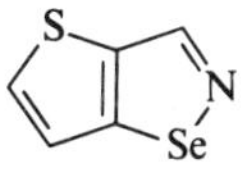 (9)	979
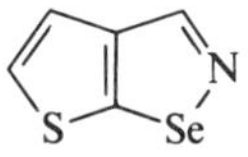 (7)	996.2	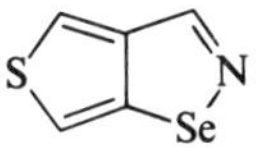 (10)	926.9
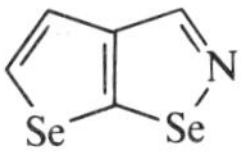 (8)	1027, 589.2[b]		

[a] Reference, Me$_2$Se; solvent, CDCl$_3$.   [b] Selenophene 526 p.p.m.

(iv) Mass spectrometry

The mass spectral data for (2) and (3) are consistent with the structures. In both compounds the most intense ions are the molecular ions. Among the major fragments are those corresponding to loss of HCN ⟨78JHC865⟩.

(v) UV spectroscopy

The UV spectral data for isoselenazoles and 1,2-benzoselenazoles are very similar to those of their sulfur analogs. As expected, moderate bathochromic shifts are observed in the selenium compounds (Table 2) ⟨73JHC267⟩. For isoselenazole (4), λ_{max} (log ε) in ethanol was found to be 212 (3.29), 266 nm (3.69).

Table 2 UV Spectral Data (EtOH) for 1,2-Benzoselenazole and Analogs ⟨78JHC865⟩

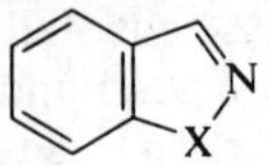

X	λ_{max} (nm) ($\log \varepsilon$)
S	204 (4.19), 222 (4.36), 303 (3.57)
Se	203 (4.15), 228 (4.30), 318 (3.67)
Te	207 (4.27), 242 (4.25), 341 (3.65)

4.20.2.2 Reactivity

4.20.2.2.1 Ozonolysis

As a result of the ring stability of 1,2-benzoselenazoles, ozonolysis of 3-styryl-1,2-benzoselenazole (**11**) at −30 °C afforded 57% of 1,2-benzoselenazole-3-carbaldehyde (**12**). Silver oxide oxidation of (**12**) gave a high yield of 1,2-benzoselenazole-3-carboxylic acid (**13**) which was converted into the amide (**14**). The amide (**14**) can be dehydrated to give the nitrile (**15**). Formation of 3-amino-1,2-benzoselenazole (**16**) through a Curtius reaction has been reported. The 1,2-benzoselenazole derivative (**1**) on ozonolysis affords a low yield of the aldehyde (**12**).

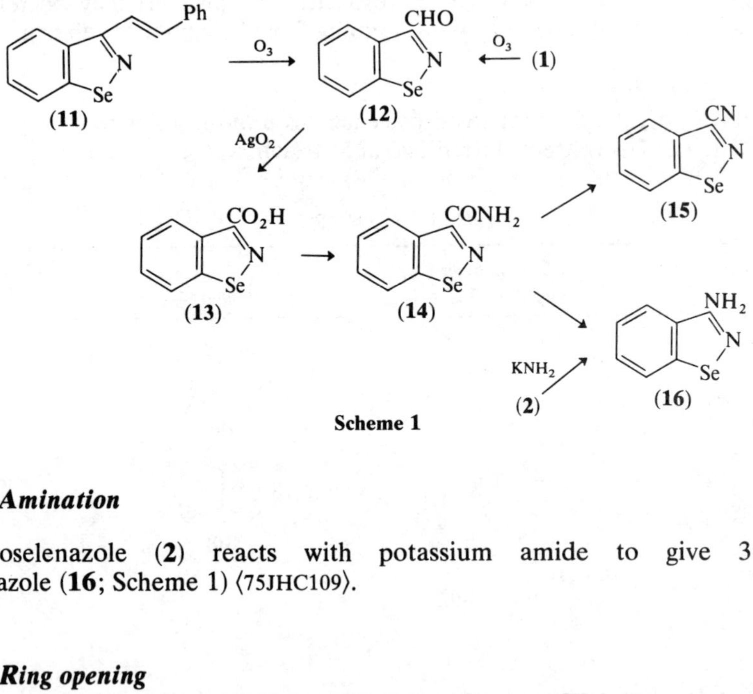

Scheme 1

4.20.2.2.2 Amination

1,2-Benzoselenazole (**2**) reacts with potassium amide to give 3-amino-1,2-benzoselenazole (**16**; Scheme 1) ⟨75JHC109⟩.

4.20.2.2.3 Ring opening

The reaction of 1,2-benzoselenazole (**2**) and *n*-butyllithium results in the formation of the ring-opened selenium compounds shown in Scheme 2 ⟨75JHC1091⟩.

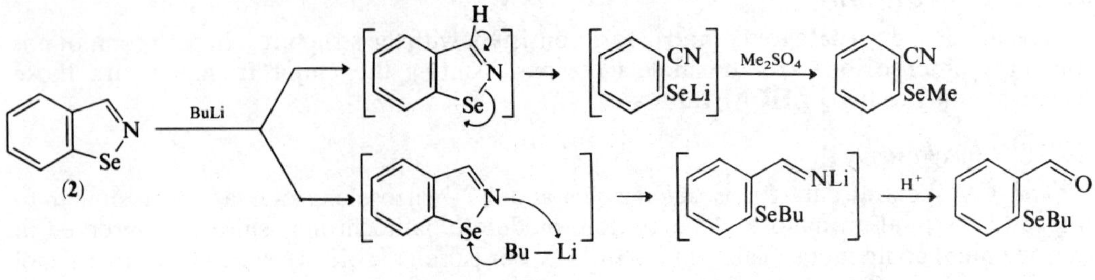

Scheme 2

4.20.2.2.4 N-Alkylation

1,2-Benzoselenazole (**2**) and 1,2-benzotellurazole (**3**) are readily quaternized by methyl iodide (Scheme 3).

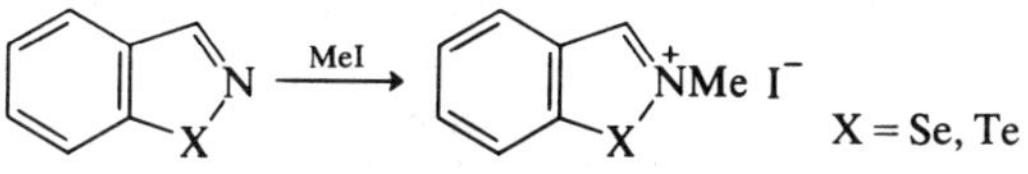

Scheme 3

4.20.2.2.5 Reactions involving the benzene ring

Nitration of 1,2-benzoselenazole affords a 55/45 mixture of the 5-nitro- (**17**) and 7-nitro-1,2-benzoselenazole (**18**) in an overall quantitative yield. The structures of the two isomers were confirmed by independent synthesis of the nitro compounds (see Section 4.20.2.3). Bromination of (**2**) results in the formation of a mixture of five mono-, di- and tri-bromo compounds and bromination does not occur at position 3 of the molecule ⟨78JHC865⟩.

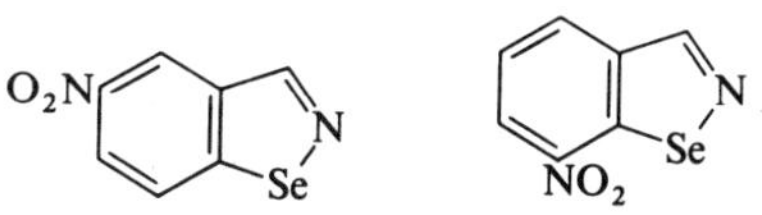

4.20.2.2.6 Ring transformation reactions

1,2-Benzoselenazolium salts (**19**) react with ammonia to give 3-alkylamino-1,2-benzoselenazole (**20**). The reaction product of (**19**) with thiolacetic acid is shown in Scheme 4. In contrast to 1,2-benzoselenazolethione (**23**) which in polar solvents is in equilibrium with the imine (**22**), 1,2-benzoselenazolone (**25**) is quite stable ⟨76BSF1124⟩.

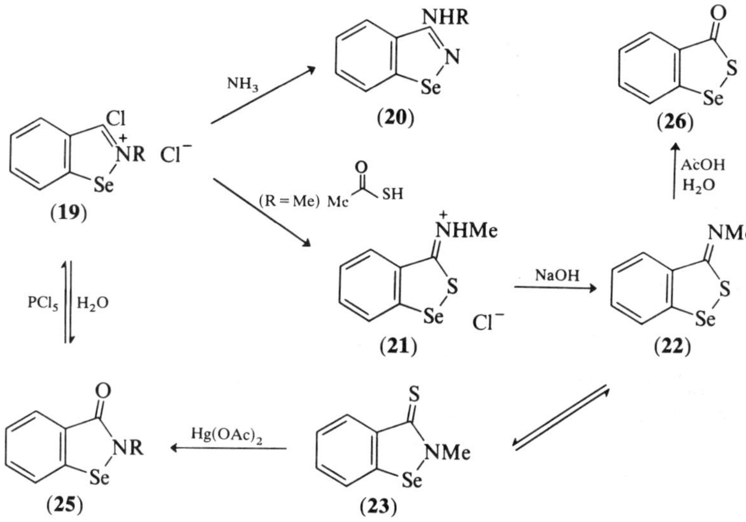

Scheme 4

4.20.2.3 Synthesis

Unsubstituted isoselenazole (**4**) is prepared by ring closure of 3-cyanoselenoacrolein with liquid ammonia. Using the appropriate butenone, 3-methylisoselenazole (**27**) is obtained (Scheme 5) ⟨62AG753⟩. Weber and Renson ⟨73JHC267⟩ have prepared a series of isoselenazoles by the addition of bromine followed by ammonia to *cis*-β-methylseleno-2-propanals (Scheme 6). 1,2-Benzoselenazoles were prepared by a similar reaction (Scheme 7). Using the basic features of the above reaction, isoselenazoles fused with thiophene or selenophene rings, represented by compounds (**7–10**), were prepared (Scheme 8) ⟨80BSB773⟩.

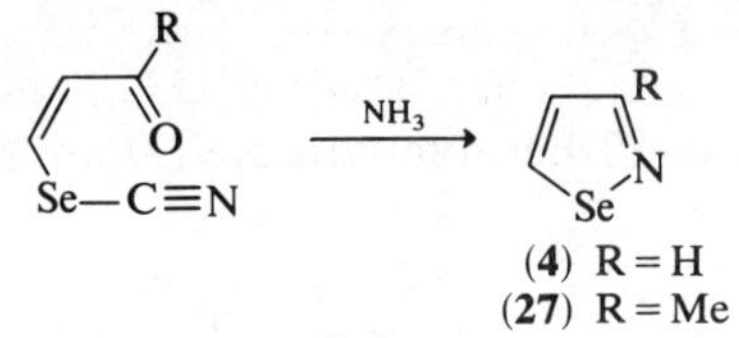

Scheme 5

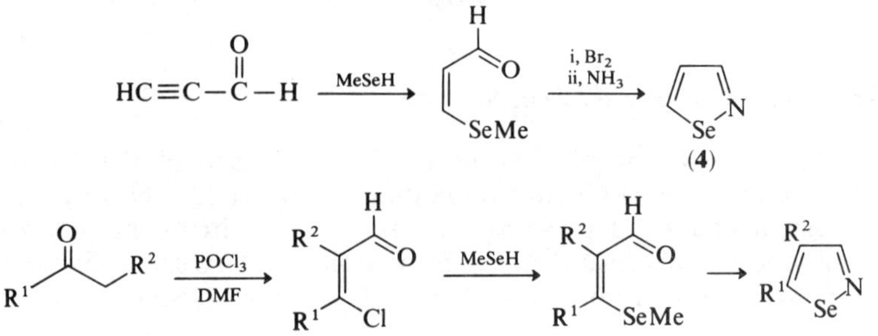

Scheme 6

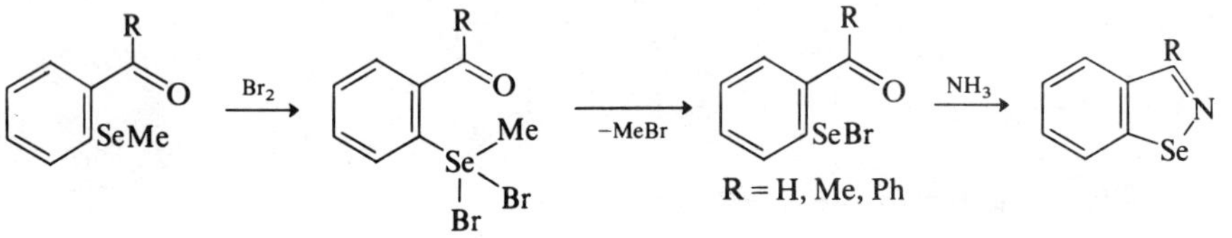

Scheme 7

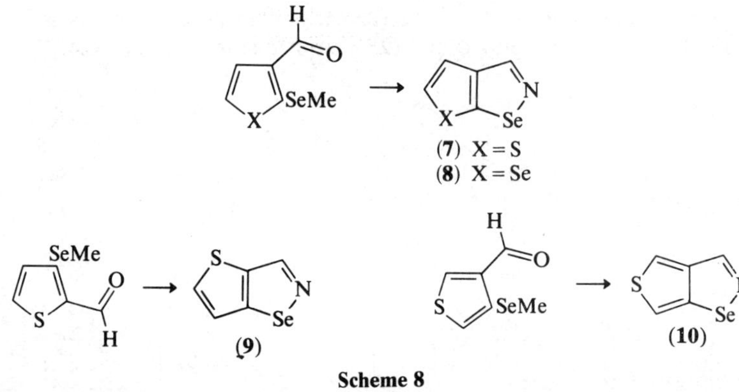

Scheme 8

Cyclic dehydration of *o*-methylselenobenzaldehyde oxime with PPA affords 1,2-benzoselenazole (2). Interestingly, the corresponding acetophenone derivative gives 2-methyl-1,3-benzoselenazole (28a). 1,2-Benzotellurazole (3), the only known compound in the isotellurazole series, is synthesized by cyclic dehydration of *o*-butyltellurobenzaldehyde oxime (29) as well as by the reaction of ammonia with *o*-bromotellurobenzaldehyde (30; Scheme 9) ⟨78JHC865⟩.

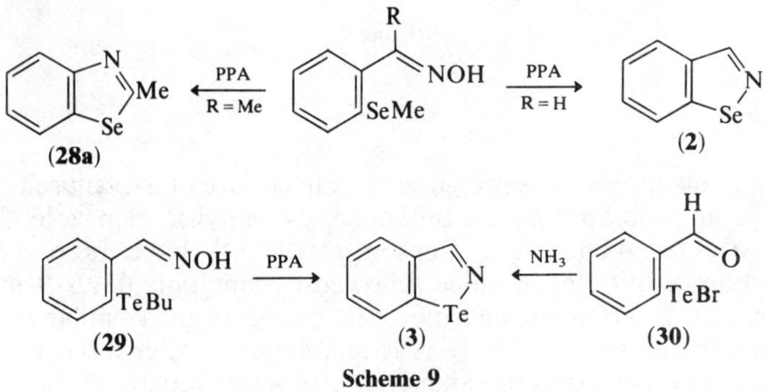

Scheme 9

3-Substituted 1,2-benzoselenazoles are synthesized by the reaction of 6H, 7H-benzo[b]selenophene-6,7-dione (31) with bromine followed by ammonia. The amide (14; Scheme 10) may be used for the preparation of the nitrile (15) or the amino compound (16) ⟨75JHC1091⟩.

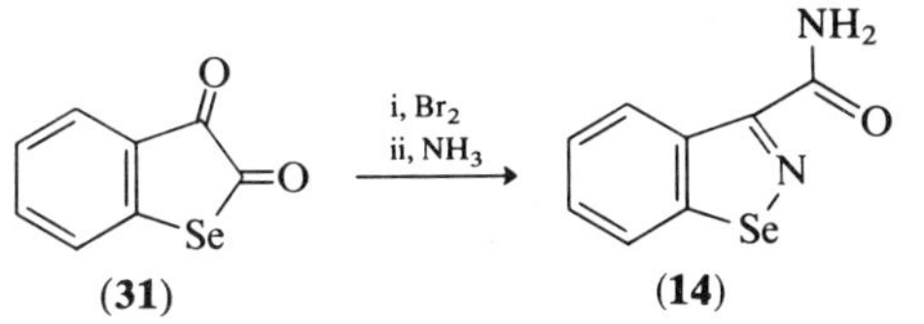

Scheme 10

Cyclocondensation of the amide (32) with phosphorus pentachloride affords the 1,2-benzoselenazolium salt (19; Scheme 11). Isoselenazolium perchlorates (33) are synthesized as shown in Scheme 12 ⟨76S273⟩.

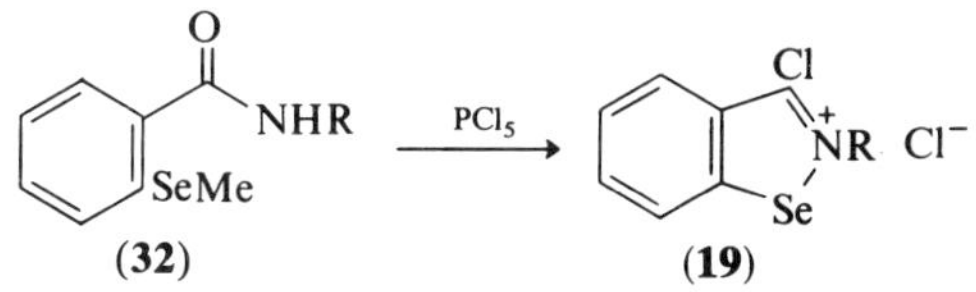

Scheme 11

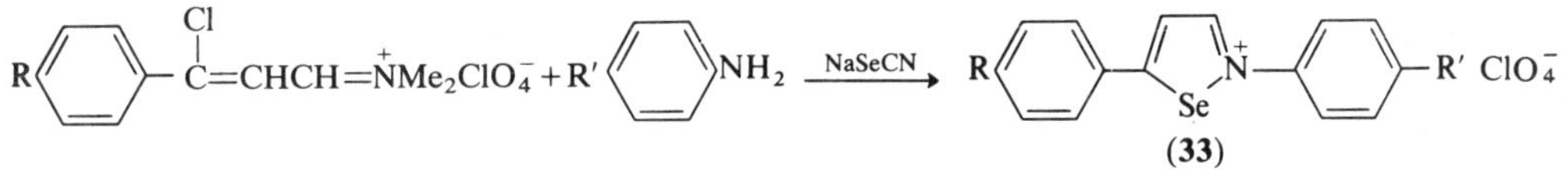

Scheme 12

4.20.3 SELENAZOLES, SELENAZOLINES AND SELENAZOLIDINES

4.20.3.1 Structure

4.20.3.1.1 Theoretical methods

The NMR and UV spectra of some benzoselenazolium salts were examined and interpreted with respect to π-electron density and $(N - V_1)$ transition energies obtained by Hückel molecular orbital (HMO) calculations ⟨73JPR587⟩.

N-Methyl chemical shifts in quaternized azoles and their benzo analogs appear to be determined by resonance interaction from the heteroatom, with the order of donating ability of the heteroatom being $NMe > O > S \approx Se$ ⟨74JHC1011⟩.

4.20.3.1.2 Molecular dimensions

(i) *X-Ray diffraction*

An X-ray diffraction study of 2-amino-1,3-selenazol-2-inium bromide (34) reveals that the molecule is planar except for C-5 which is displaced 0.43 Å from the plane. Both C⋯N distances are short [C(2)⋯N(6), 1.28; C(2)⋯N(3), 1.35 Å]. It is concluded that the positive charge concentration is more pronounced on the amine nitrogen atom C(2)—Se and C(5)—Se bond distances are 1.99 and 1.88 Å, respectively with a C—Se—C bond angle of 86° ⟨68AG(E)811⟩.

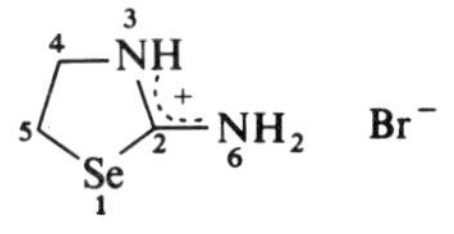

4.20.3.1.3 Molecular spectra

(i) ^1H NMR spectroscopy

The chemical shift of H-2 of benzoselenazole has been compared with those of its thia and oxa analogs. The observed deshielding effect of the heteroatom is opposite to the order of electronegativity. $^{13}C-^1H$ coupling constants have also been measured (Table 3) ⟨67JHC139⟩.

Table 3 ^1H NMR Data for Benzoazoles

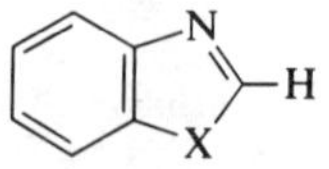

X	Chemical shift (δ, p.p.m.)		Coupling constant (H-2) (Hz)	
	CCl_4	CF_3CO_2H	Neat	CF_3CO_2H
Se	9.85	11.10	3.56	3.60
S	8.93	10.19	3.51	3.50
O	8.05	9.83	3.85	4.05

(ii) ^{13}C NMR spectroscopy

The ^{13}C NMR spectral parameters of benzoselenazole (28) and its thione derivative (35) compared with their benzothiazole analogs show that substitution of ring sulfur by selenium results in a marked deshielding effect (3–5 p.p.m.) at C-2. This deshielding effect has been attributed to a decrease in total charge density at the α-carbon atom resulting from a greater tendency for the d-orbitals of the heteroatom to accept electrons from the ring for selenium than for sulfur. This effect apparently outweighs the opposite trend anticipated from the differences in electronegativity of sulfur and selenium ⟨77JOC3725⟩. The ^{13}C NMR chemical shifts of compounds (28) and (35) are given in Table 4.

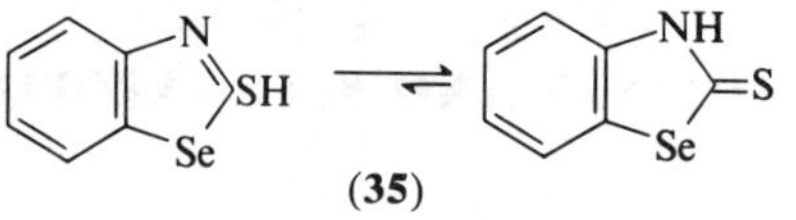

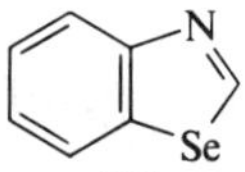

Table 4 ^{13}C NMR Chemical Shifts[a] of Compounds (28) and (35)

Compound	Chemical Shifts (δ, p.p.m.)		
	C-2	C-3	C-7a
(28)	160.6[b]	154.8	137.3
(35)	193.5	142.7	129.9

[a] DMSO-d_6 solvent. [b] $^{13}C-^{77}Se$ coupling constant for C-2 = −136.7 Hz ⟨77JOC3725⟩.

(iii) Mass spectrometry

The electron impact mass spectra of a series of selenazoles have been compared with those of their thiazole analogs. The results indicate that, in general, the mass spectrometric fragmentation patterns of the selenium heterocycles can be predicted by an examination of the corresponding sulfur analog. One of the characteristics of the mass spectra of selenium compounds is that they are rich in ions due to the presence of the six relatively abundant stable selenium isotopes ⟨81JHC1335⟩.

4.20.3.2 Reactivity

4.20.3.2.1 Electrophilic substitution

The 5-position of selenazoles is very reactive towards electrophilic reagents. Nitration of 2-alkylselenazoles affords 5-nitro derivatives under relatively mild conditions. 2-

Acetamidoselenazole (**36**) under the same reaction conditions gives the nitro derivative (**37**). 2-Diethylaminoselenazoles (**38**) and nitrous acid afford 5-nitroso derivatives (**39**; Scheme 13) ⟨75CS(8A)39⟩. The chemical behavior of (**39**) is comparable with that of aromatic nitroso compounds.

Scheme 13

Sulfonation of 2-alkylselenazoles yields the corresponding sulfonic acids.

Direct bromination of 2-aminoselenazoles affords the corresponding 5-bromo derivatives, stable only as hydrobromide salts. Diazonium salts also react with selenazoles to give the 5-azo derivatives.

Compounds of type (**40**) react with *p*-nitrosodialkylanilines to give deeply-colored azomethines (**41**; Scheme 14) ⟨63ZC388⟩.

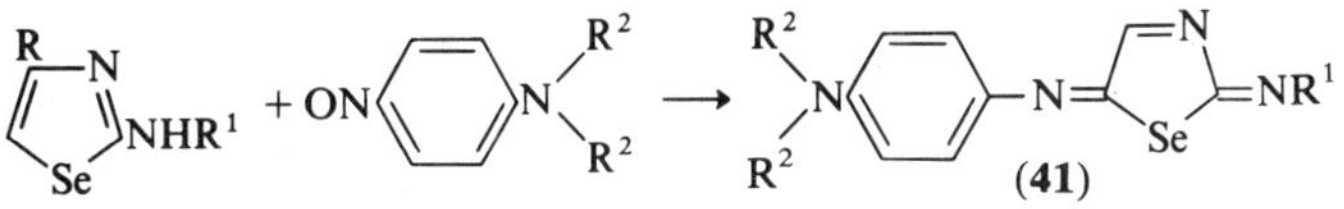

(**40**) R^1 = aryl, N=CHR^3

Scheme 14

4.20.3.2.2 Reactivity of the 2-methyl group of selenazoles

Selenium dioxide oxidation of 2,4-dimethylselenazole (**42**) affords 4-methylselenazole (**43**; Scheme 15) *via* the intermediate 2-carboxylic acid ⟨49YZ566⟩.

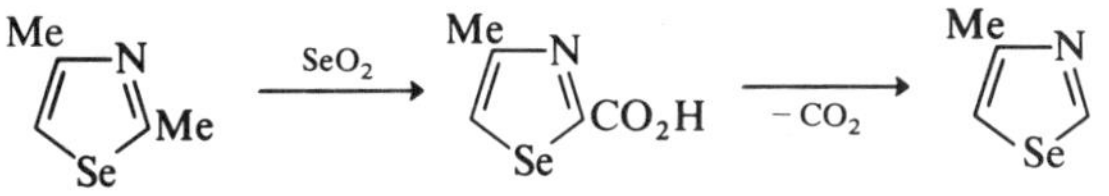

Scheme 15

2-Methylbenzoselenazole (**28a**) and chloral affords the alcohol (**44**). Hydrolysis of the latter yields β-(2-benzoselenazolyl)acrylic acid (**45**) which can be oxidized to benzoselenazole-2-carbaldehyde (**46**) along with benzoselenazole-2-carboxylic acid (**47**; Scheme 16) ⟨52JCS3197⟩. Aldehyde (**46**) was obtained by selenium dioxide oxidation of (**28a**) ⟨53CB888⟩.

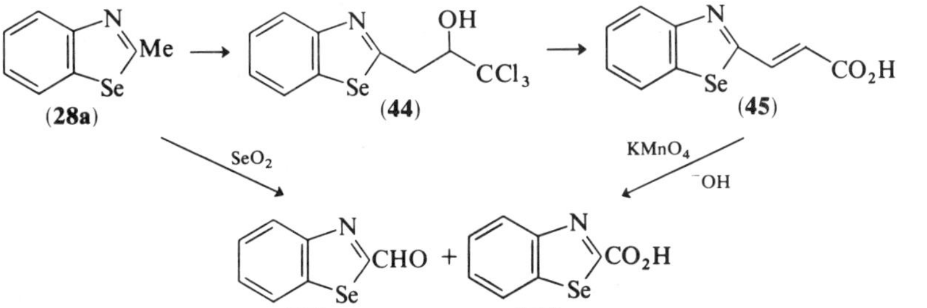

Scheme 16

2-Methylbenzoselenazole (**28a**) reacts with benzoyl chloride in the presence of triethylamine to give unsaturated derivatives shown in Scheme 17 ⟨79JPR320⟩.

4.20.3.2.3 Ring expansion

Quaternization of benzoselenazole (**38**) with α,α'-dibromo-*o*-xylene affords bromide (**48**) which by treatment with sodium hydroxide solution rearranges to *N*-formyl-6,11-dihydrodibenzo[*b,f*][1,4]selenazocine (**49**; Scheme 18) ⟨80TL2429⟩.

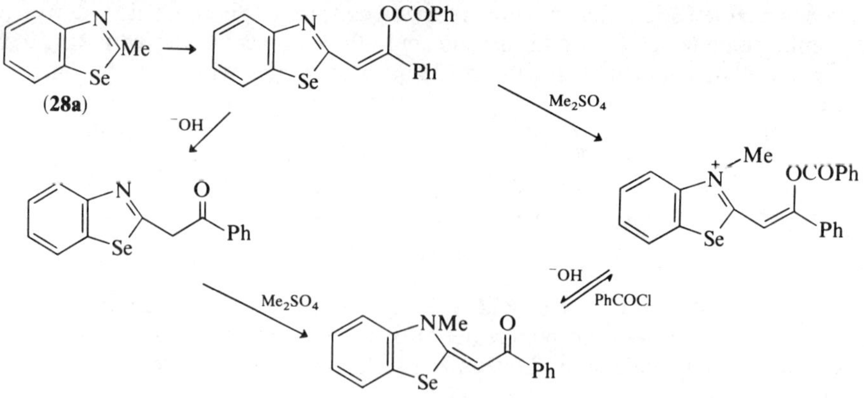

Scheme 17

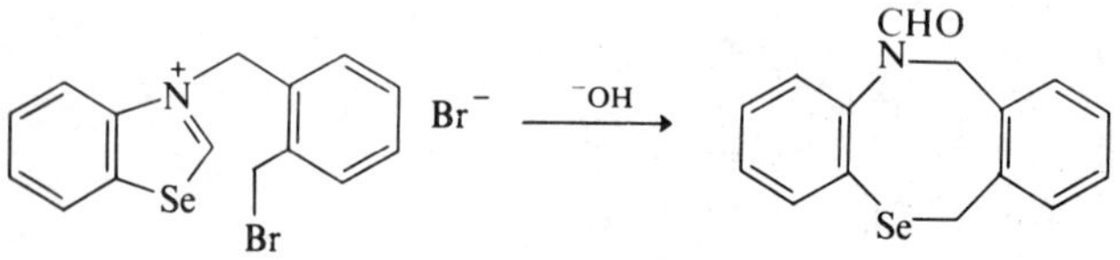

Scheme 18

4.20.3.3 Synthesis

4.20.3.3.1 Selenazoles

Hofmann ⟨1889LA(250)294⟩ was the first to report the synthesis of selenazoles in a modified Hantzsch reaction in which derivatives of selenocarboxamides are allowed to react with α-chlorocarboxyl compounds. Other substituted selenazoles are prepared similarly (Scheme 19).

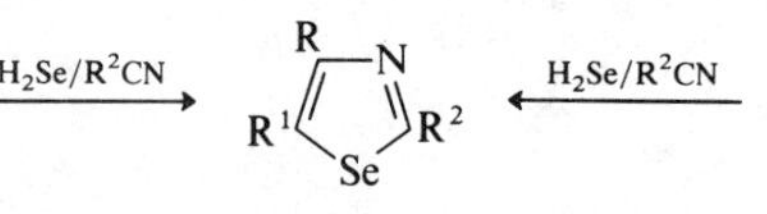

Scheme 19

Hydrogen selenide reacts with a mixture of a nitrile and an α-halo ketone or 2-halo aldehyde acetal in the presence of a condensation catalyst to give selenazoles (Scheme 20) ⟨48YZ191, 79S66⟩.

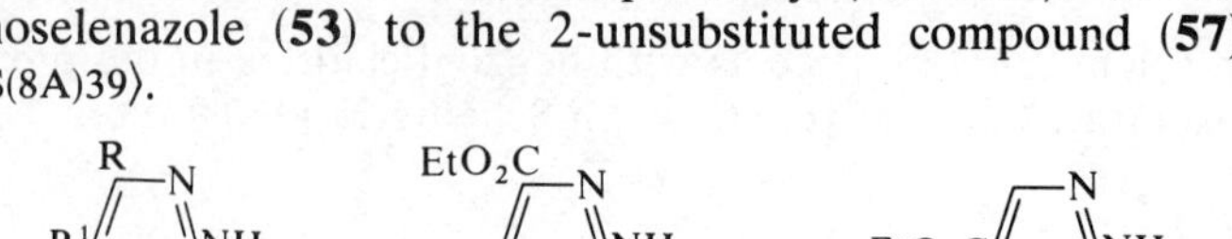

Scheme 20

Selenourea and α-halocarboxyl compounds react similarly to give 2-aminoselenazoles (**50**) ⟨53CR(237)906⟩. Selenourea and ethyl bromopyruvate or methyl formylchloroacetate afford aminoselenazoles (**51**) and (**52**), respectively ⟨75JHC675⟩. Deamination of 4,5-diphenyl-2-aminoselenazole (**53**) to the 2-unsubstituted compound (**57**) is outlined in Scheme 21 ⟨75CS(8A)39⟩.

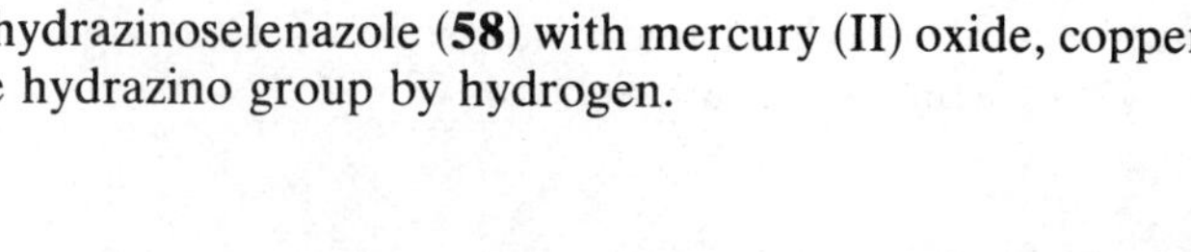

Oxidation of 2-hydrazinoselenazole (**58**) with mercury (II) oxide, copper(II) salts or silver oxide replaces the hydrazino group by hydrogen.

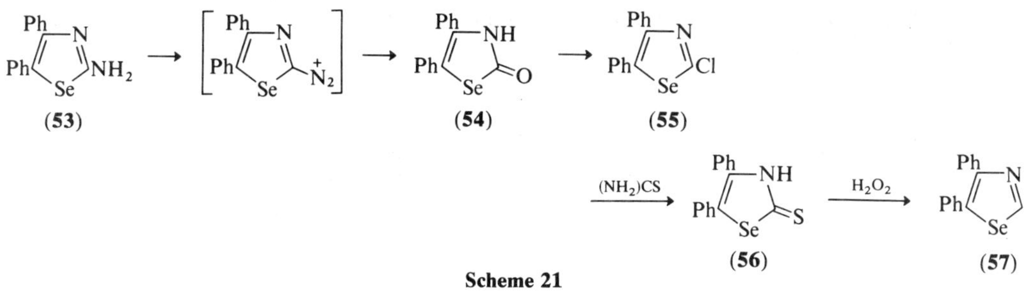

Scheme 21

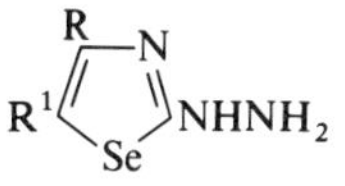

(53) → (54) → (55)

(55) —(NH₂)CS→ (56) —H₂O₂→ (57)

(58)

Selenourea and 1,3-dichloropropan-2-one yield 2-amino-4-chloromethylselenazole hydrochloride (59) which was subsequently acetylated to give (60). Sodium diethyl formamidomalonate reacts with the acetamido compound (60) to give the protected amino acid derivative (61). Hydrolysis of (61) affords (±)-β-(2-amino-1,3-selenazol-4-yl)alanine (62; Scheme 22) ⟨81JHC205⟩.

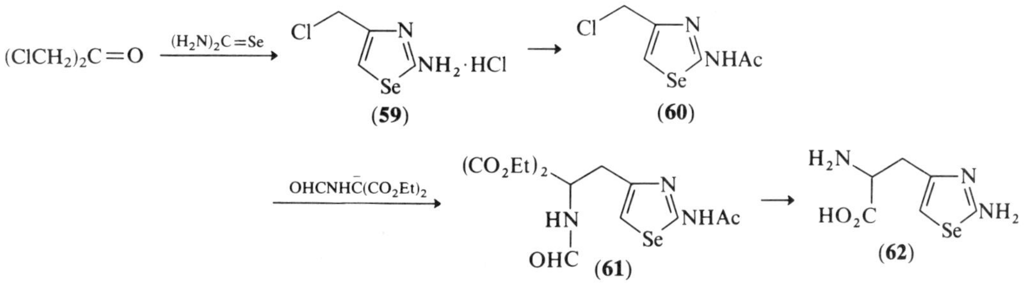

$(ClCH_2)_2C=O$ —$(H_2N)_2C=Se$→ (59) → (60) —$OHCNHC(CO_2Et)_2$→ (61) → (62)

Scheme 22

4.20.3.3.2 Mesoionic selenazoles

Anhydro-4-hydroxy-2,3,5-triphenyl-1,3-selenazolium hydroxide (65) is prepared by a multistep reaction: selenobenzanilide (63) reacts with α-bromophenylacetic acid to give the α-seleno acid (64); this acid readily cyclizes to the mesoionic selenazole (65), which upon reaction with DMAD affords an unstable adduct which by extrusion of selenium gives the pyridone diester (66; Scheme 23) ⟨75CC617⟩.

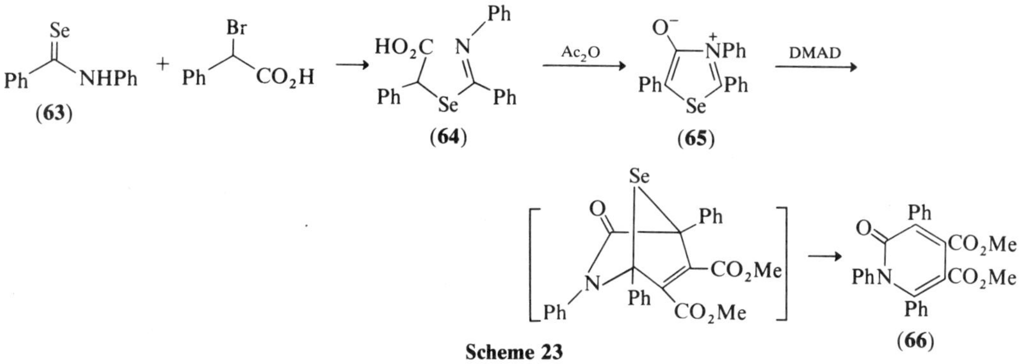

(63) + (Br compound) → (64) —Ac₂O→ (65) —DMAD→ [adduct] → (66)

Scheme 23

In independent work, a variety of selenoamides and 1,2-bielectrophiles such as α-bromophenylacetyl chloride and 2-bromo-2-ethoxycarbonylacetyl chloride gave mesoionic selenazoles (65). The adduct formed from the mesoionic selenazole (67) and phenyl isocyanate leads to anhydro-4-mercapto-2-p-methoxyphenyl-6-oxo-1,3,5-triphenylpyrimidinium hydroxide (68; R = p-OMeC₆H₄). This constitutes the first example of conversion of a five-membered mesoionic system into a six-membered heteroaromatic betaine of this type (Scheme 24) ⟨77JOC1644⟩.

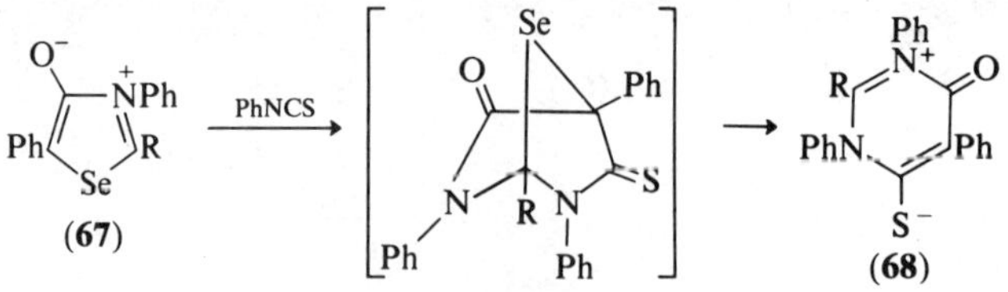

(67) (68)

Scheme 24

4.20.3.3.3 Benzoselenazoles and hetero-fused selenazoles

2-Phenylbenzoselenazole, the first compound in this series, was reported by Bauer in 1913 ⟨13CB92⟩. Zinc bis(*o*-aminophenylselenolate) and benzoyl chloride affords (**35a**), whereas with phosgene or thiophosgene it gives 2-hydroxybenzoselenazole (**35b**) and 2-mercaptobenzoselenazole (**35**), respectively (Scheme 25) ⟨35JCS1762⟩.

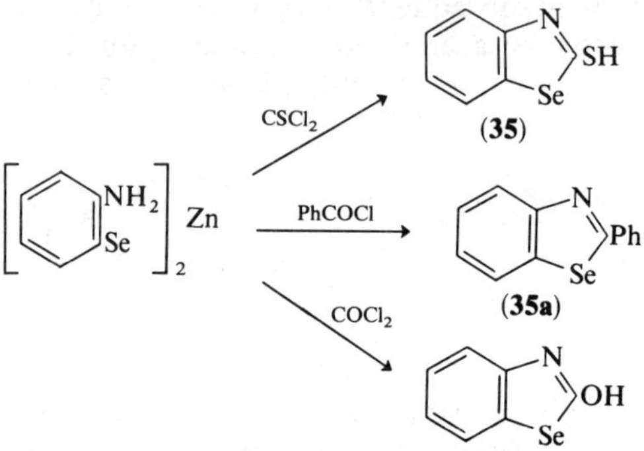

(35) (35a) (35b)

Scheme 25

Oxidation of 1-phenyl-2-selenourea with bromine affords 2-aminobenzoselenazole (**69**).

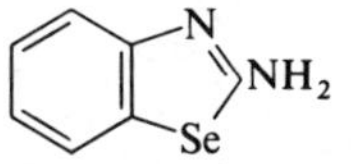

(69)

Cyclic condensation of *o*-methylselenoacetophenone oxime with PPA affords 2-methylbenzoselenazole (**28a**) ⟨78JHC865⟩.

2-Methylthieno[3,2-*d*]selenazole (**71**) is prepared by thermal cyclization of the thiophene derivative (**70**; Scheme 26) ⟨80BSF(2)151⟩.

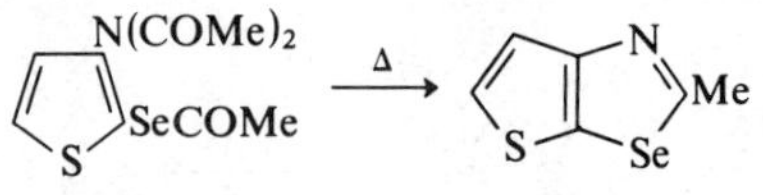

(70) (71)

Scheme 26

2-Aryl-4-chloromethylselenazoles (**72**) are hydrolyzed to alcohols (**73**). Manganese dioxide oxidation of (**73**) affords 4-formyl derivatives (**74**) which by condensation with ethyl azidoacetate give azidocrylates (**75**). Thermal cyclization of compounds (**75**) give pyrrolo[3,2-*d*]selenazoles (**76**; Scheme 27) ⟨79JHC1563⟩.

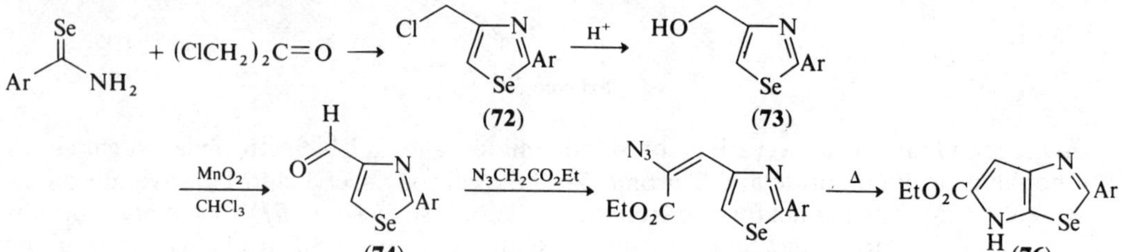

(72) (73) (74) (75) (76)

Scheme 27

Selenazoles (**72**) and methyl salicylate or methyl-*o*-mercaptobenzoate lead to ethers (**77**) or thioethers (**78**). The acid chlorides of (**77**) and (**78**) through a Friedel-Craft cyclization afford 1-benzoxepino[3,4-*d*]selenazoles (**79**) and 1–benzothiepino[3,4-*d*]selenazoles (**80**), respectively (Scheme 28) ⟨81JHC789⟩.

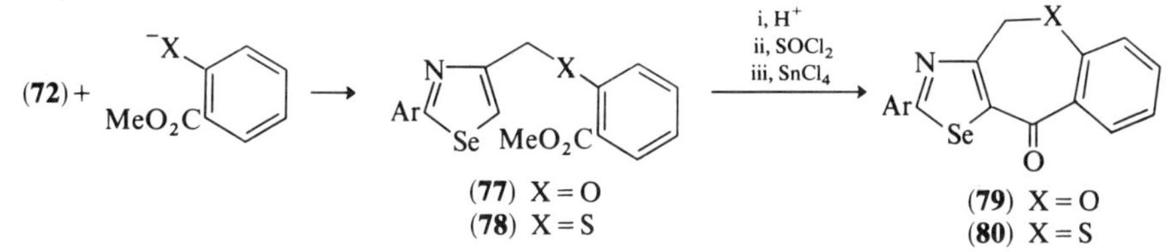

Scheme 28

4.20.3.3.4 Selenazolines

In 1892 Michels ⟨1892CB3048⟩ reported the synthesis of 2-methyl-2-selenazoline (**82**) by ring closure of the acetyl derivative of bis(2-aminoethyl)diselenide (**81**; Scheme 29).

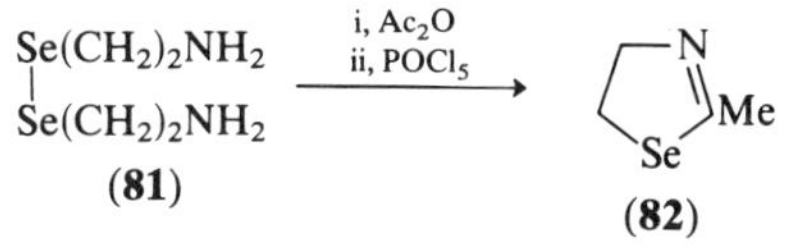

Scheme 29

2-Amino-2-selenazoline (**34**) is prepared by two methods outlined in Scheme 30 ⟨62JOC2899, 65JOC2454⟩. Cyclocondensation of substituted propargylamines and aryl selenocyanates affords high yields of 2-arylaminoselenazoline (**83**; Scheme 31) ⟨78KGS917⟩. Substituted selenosemicarbazides (**84**) react with chloroacetone under neutral conditions to give selenazolines (**85**) and (**86**). The reaction carried out under basic conditions affords substituted 1,3,4-oxadiazoles (Scheme 32) ⟨80MI42000⟩.

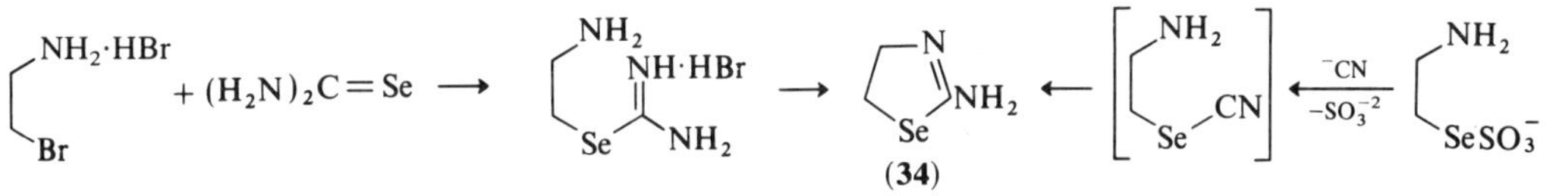

Scheme 30

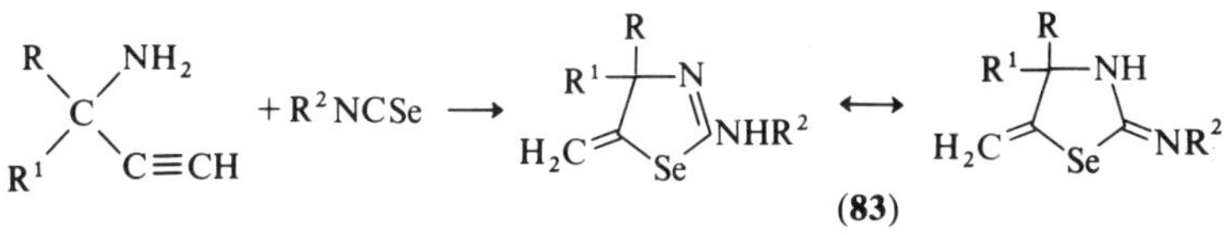

Scheme 31

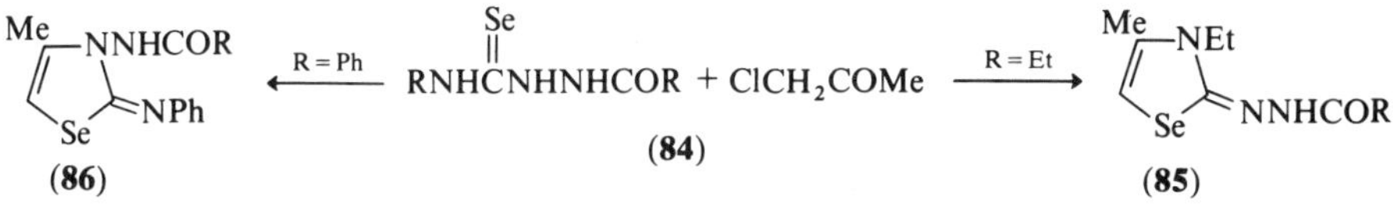

Scheme 32

α-Halocarboxylic esters and selenoamides give 4-oxoselenazoline (**87**; Scheme 33) ⟨75CS(8A)39⟩. Cyclic condensation of 2-aminoselenazoles (**88**) with ethyl propiolate affords, in addition to 7*H*-selenazolo[3,2-*a*]pyrimidin-7-ones (**89**), the adducts (**90** and **91**; Scheme 34) ⟨75JHC675⟩.

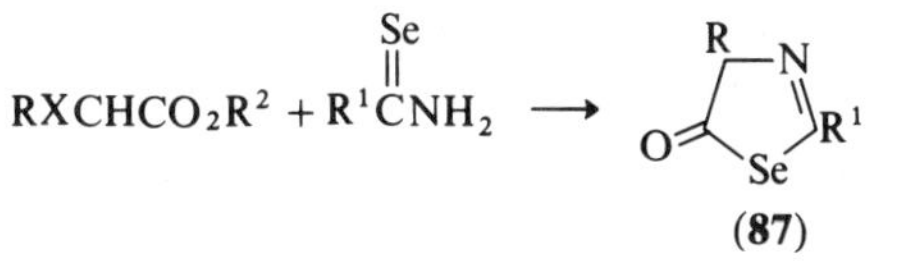

Scheme 33

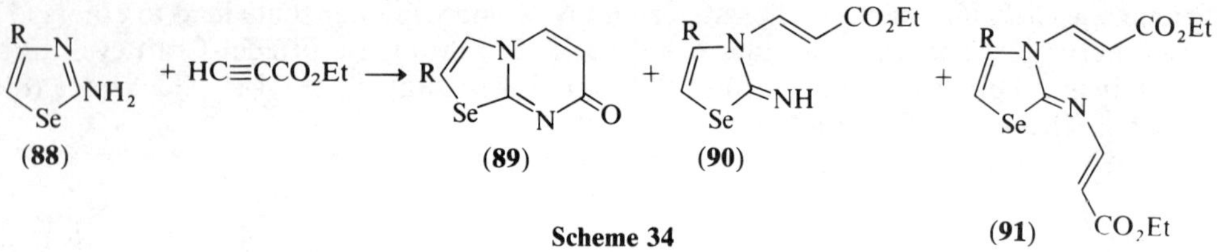

Scheme 34

4.20.3.3.5 Selenazolidines

Aziridines and aldehydes or ketones react with hydrogen selenide to give selenazolidines (**92**; Scheme 35) ⟨66BSB243, 72BSB279, 72BSB295⟩. Using selenopencillamines instead of the aziridines in the above reaction, selenazolidines (**93**; Scheme 36) are obtained ⟨72BSB303⟩. Selenazolidine-2-thiones (**94**) are obtained from sodium hydrogen selenide and 2-haloethyl isothiocyanates (Scheme 37) ⟨78CCC2298⟩.

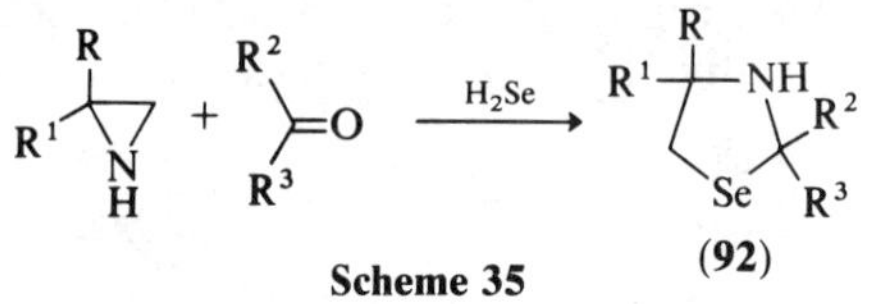

Scheme 35

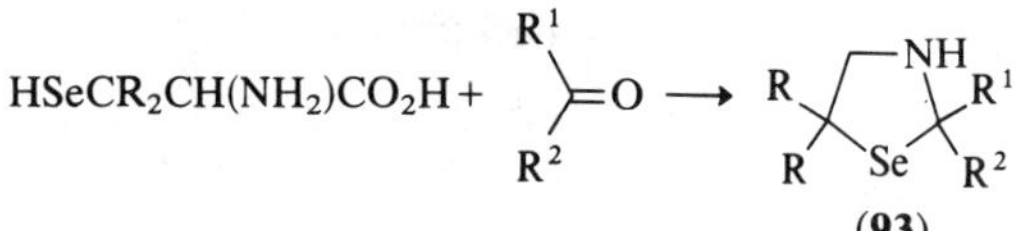

Scheme 36

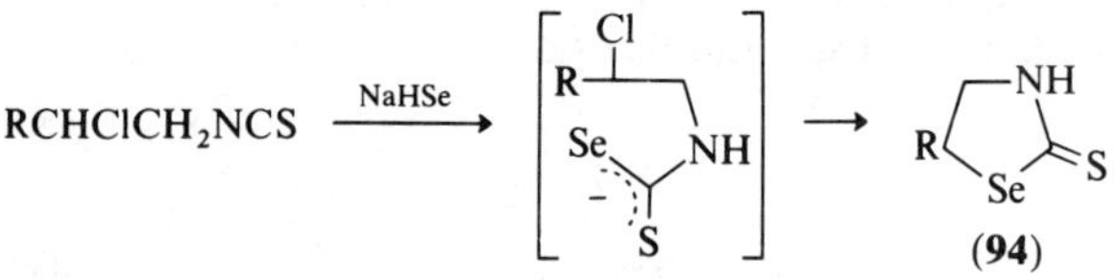

Scheme 37

(±)-Selenazolidine-2-carboxylic acid (**95**) is prepared by reduction of selenocysteamine and subsequent treatment with sodium glyoxylate (Scheme 38) ⟨79MI42000⟩. (±)-Selenazolidine-4-carboxylic acid (**96**) is synthesized in good yield by cyclic condensation of DL-selenocysteine with formaldehyde (Scheme 39) ⟨77MI42000⟩.

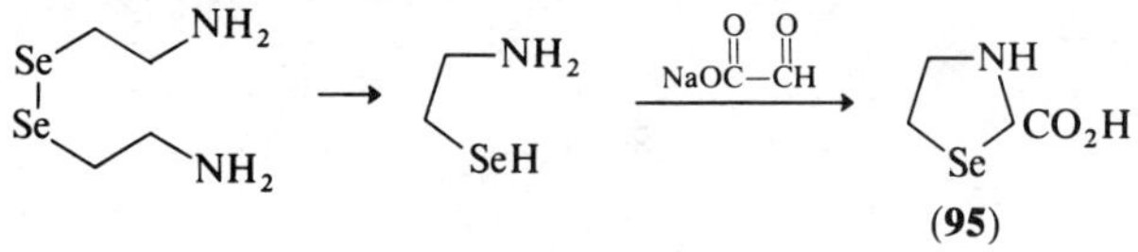

Scheme 38

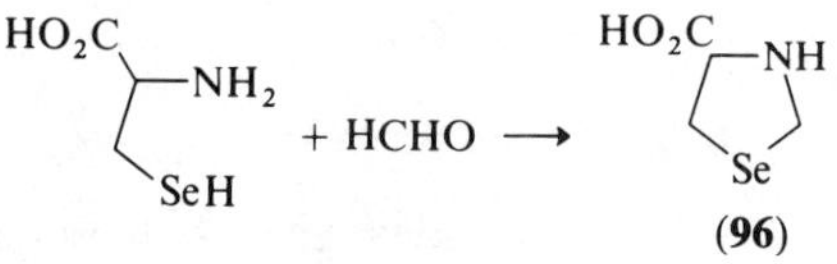

Scheme 39

The selenazolidines are unstable compounds. Their reactions with aldehydes, resulting in a ring-opening–ring-closure sequence, have been studied (Scheme 40) ⟨72BSB289⟩. Ring opening has also been observed with 4-thioxoselenazolid-2-one upon treatment with hydrazine (Scheme 41) ⟨73KGS930⟩.

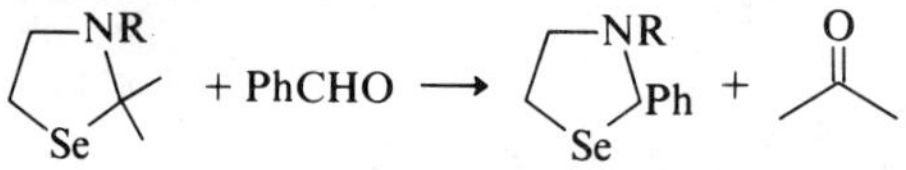

Scheme 40

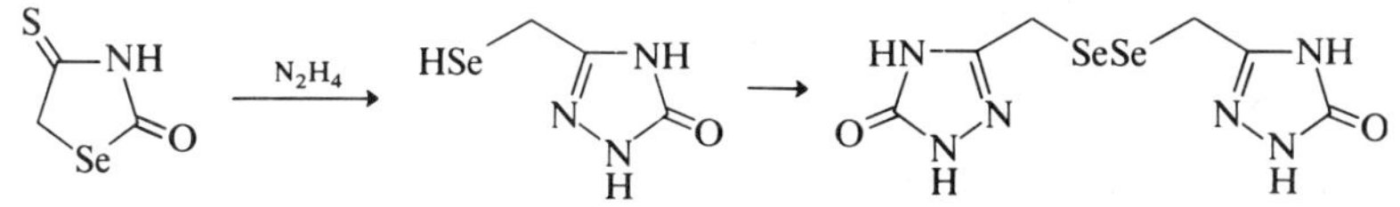

Scheme 41

(±)-6-Phenyl-2,3,5,6-tetrahydroimidazo[2,1-*b*]selenazole (selenotetramizole; **97**) is prepared by a three-step reaction sequence shown in Scheme 42. The biological activity of the (−)-isomer is comparable with that of (−)-tetramizole ⟨78JMC496⟩. The selenazolo[3,2-*b*][1,2,4]triazine system (**98**) is accessible from 6-substituted 3-selenoxo-2,3,4,5-tetrahydro-1,2,4-triazin-5-one by reaction with bromoacetic acid followed by treatment with acetic anhydride (Scheme 43) ⟨71JHC1011⟩.

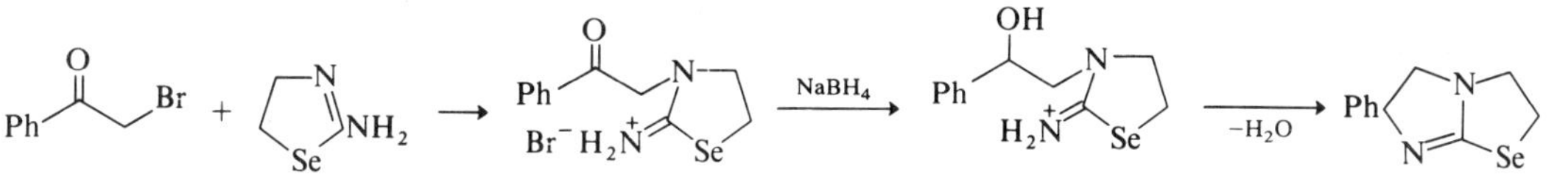

Scheme 42

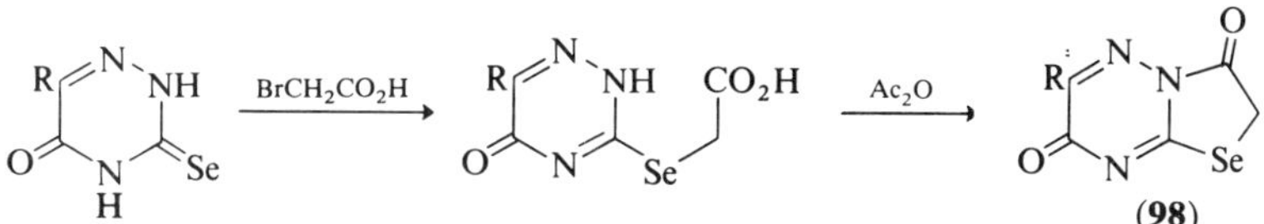

Scheme 43

The reaction of 2-amino-2-selenazoline (**34**) and ethoxycarbonyl isothiocyanate affords 4-thioxo-2,3,6,7-tetrahydro-4*H*-selenazolo[3,2-*a*][1,3,5]triazin-2-one (**99**). Phenyl isothiocyanate and (**34**) give the 1,3,5-triazine derivative (**100**; Scheme 44) ⟨74JOC1819, 69T191⟩.

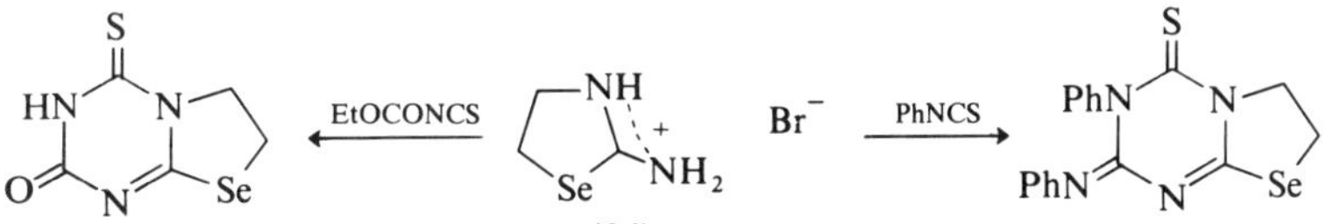

Scheme 44

4.20.4 1,2,3-SELENADIAZOLES

4.20.4.1 Structure

4.20.4.1.1 General

The heat and light sensitivity and high chemical reactivity of 1,2,3-selenadiazoles towards strong base have made these easily available heterocyclic compounds the subject of numerous investigations during the last decade. 1,2,3-Selenadiazoles are readily decomposed by extrusion of nitrogen, selenium or both of these elements. On the basis of these properties, 1,2,3-selenadiazoles are remarkable chemical reagents for the synthesis of different types of compounds such as alkynes, 1,3-diselenafulvenes, thiaselenols, selenophenes, 1,4-diselenines, selenoesters, selenoamides and alkynylselenocarboxylic acids.

4.20.4.1.2 Molecular spectra

(i) *¹H NMR spectroscopy*

The ¹H NMR spectrum of 1,2,3-selenadiazole (**101**) exhibits two doublets at δ 8.81 and 9.47 for C-4 and C-5 ($J = 3.8$ Hz). These chemical shifts are consistent with the aromaticity of the molecule. In compound (**101**) and its 4-substituted derivatives, H-5 appears as a singlet and a weak doublet around the singlet. This doublet is assigned to the ^{1}H–^{77}Se

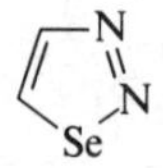

(101) 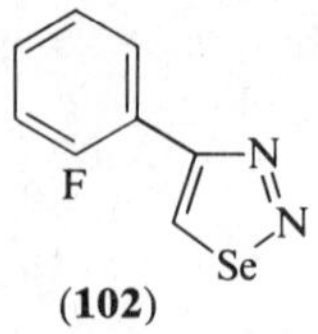 **(102)**

coupling, $J = 40 \pm 2$ Hz. The relative intensities of the doublet to the singlet is $7.2 \pm 0.3\%$, consistent with 7.5% of natural abundance of ^{77}Se isotope ⟨71JOC2836, 80JOC2632⟩.

Interestingly, the H-5 chemical shift of 4-(2-fluorophenyl)-1,2,3-selenadiazole (**102**) appears as a doublet ($J = 0.9$ Hz), arising from coupling with the *ortho* fluorine atom. No coupling is observed for the *para*-fluoro derivative ⟨74JCS(P1)30⟩.

The H-4 of 5-(*p*-toluensulfonyl)-1,2,3-selenadiazole (**103**) appears as a sharp singlet at δ 8.9 ⟨78JPS1336⟩. In the cycloalkeno-1,2,3-selenadiazole series, the chemical shifts are as expected. In 10,11-dihydrophenanthro[1,2-*d*][1,2,3]selenadiazole (**104**) the H-4 appears as a doublet at δ 9.50 ($J = 8$ Hz) ⟨78JHC501⟩.

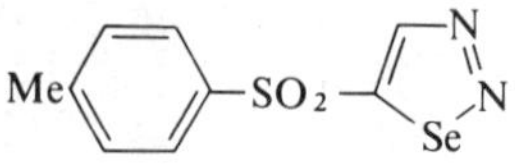

(103) 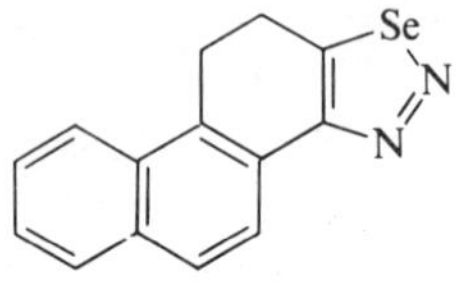 **(104)**

(ii) ^{13}C NMR spectroscopy

^{13}C NMR spectra of a number of 1,2,3-selenadiazoles including cycloalkeno-1,2,3-selenadiazoles have been measured and interpreted. Chemical shifts and coupling constants of ^{1}H–^{13}C, ^{1}H–^{77}Se and ^{13}C–^{77}Se are given and discussed. The coupling constants for unsubstituted 1,2,3-selenadiazole are as follows: (values in Hz), ^{1}H(4)–^{1}H(5) 3.8; ^{1}H(4)–^{13}C(4), 191.1; ^{1}H(4)–^{77}Se ≤ 6.0; ^{1}H(5)–^{13}C(4), 9.2; ^{1}H(5)–^{13}C(5), 191.7; ^{1}H(5)–^{77}Se, 39.3; ^{13}C(4)–^{77}Se, 27.9; ^{13}C(5)–^{77}Se, 136.8 ⟨81ZN(B)1017⟩.

(iii) UV spectroscopy

UV spectra of the parent ring (**101**) and some substituted 1,2,3-selenadiazoles are reported. For compound (**101**), λ_{max} (ethanol) occurs at 282 nm (log $\varepsilon = 2.95$). All other 1,2,3-selenadiazoles studied exhibit UV absorptions at 282–328 nm (log $\varepsilon = 3.01$–3.47).

(iv) Mass spectrometry

Mass spectral fragmentation patterns in the spectra of these compounds are in accord with the formation of alkynes. The first step in the fragmentation of 1,2,3-selenadiazoles is the loss of N$_2$ followed by extrusion of selenium and formation of the corresponding alkyne. The abundance of the alkynic ion in the mass spectrum appears to be dependent on the nature of the substituent group present in the selenadiazole. When the alkynic ion cannot be stabilized by the formation of a cation on the adjacent carbon atoms, the abundance of the alkynic ion decreases (10% in the parent compound and zero for 4-*t*-butyl-1,2,3-selenadiazole). On the basis of the mass spectral pattern it is possible to predict the yield of the alkynic compound formed through pyrolysis or photolysis of a given 1,2,3-selenadiazole ⟨71JOC2836, 78JPS1336⟩.

(v) PE spectroscopy

In the variable temperature photoelectron spectroscopy of 1,2,3-selenadiazole (**101**) at 5×10^{-2} mbar between 450 and 750 °C, acetylene is always formed along with selenoketene (**105**). At 800 °C, acetylene is the sole product. 1,2,3-Benzoselenadiazole (**106**) does not produce benzyne even at 800 °C. The reaction product was found to be the highly reactive 6-fulveneselone (**107**). The PE spectra of (**101**), (**106**) and their decomposition products are very similar to those of their corresponding sulfur analogs (Scheme 45) ⟨80CB3187⟩

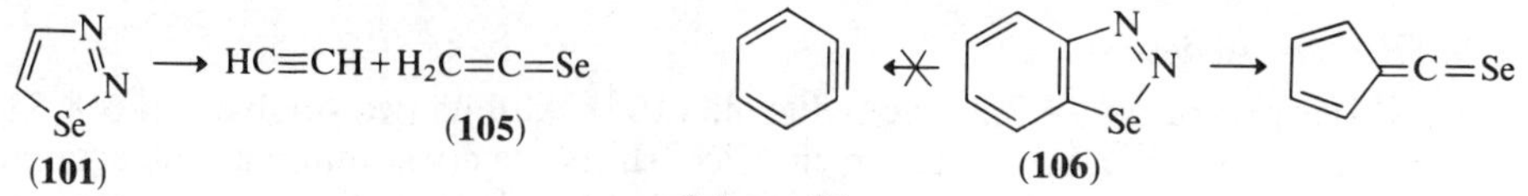

(**101**) → HC≡CH + H$_2$C=C=Se (**106**)
            (**105**)

Scheme 45

4.20.4.1.3 Thermodynamic aspects

The heats of combustion and sublimation of crystalline 4-phenyl-1,2,3-selenadiazole (**108**) is found to be 4586 and 94 kJ mol^{-1} respectively. Heats of formation and a C—Se bond energy of 238 kJ mol^{-1} are also calculated ⟨73JCS(P2)1732⟩.

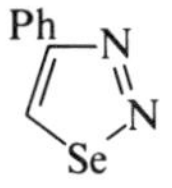

(**108**)

4.20.4.2 Reactivity

4.20.4.2.1 Reactivity at the ring atoms

(i) *Thermal and photochemical reactions*

1,2,3-Selenadiazoles are thermally or photochemically decomposed to give alkynes, selenium and nitrogen (Scheme 46) ⟨70AG(E)464, 71JOC2836⟩. The utility of this reaction is shown by the various types of alkynic compounds prepared by controlled thermolysis of 1,2,3-selenadiazoles (Scheme 47) ⟨75JHC801, 77JHC567, 77JHC745, 78JPS1336⟩. Similarly, diacetylenes are prepared from bis(selenadiazoles) (**109**; Scheme 48) ⟨73JHC655⟩.

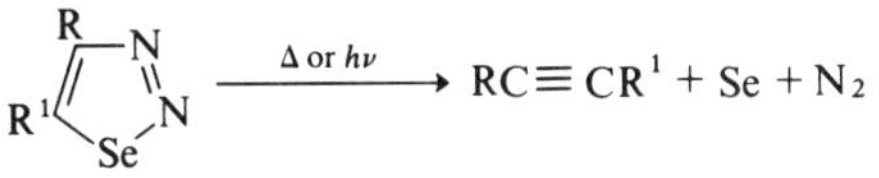

$$RC\equiv CR^1 + Se + N_2$$

Scheme 46

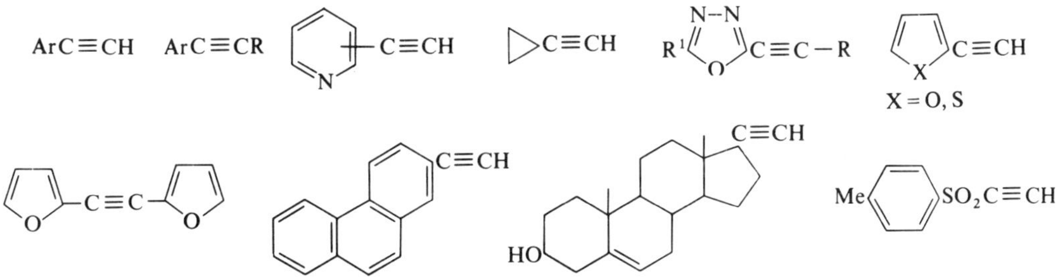

ArC≡CH   ArC≡CR   Scheme 47

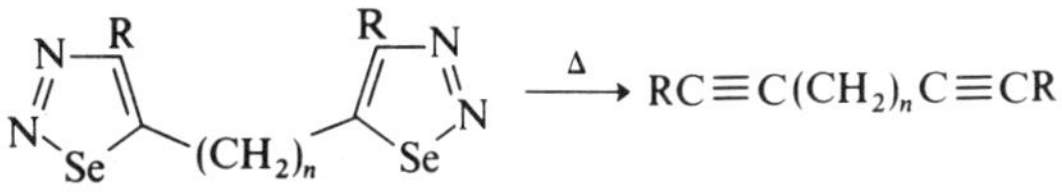

Scheme 48: $RC\equiv C(CH_2)_n C\equiv CR$

4-Aryl-1,2,3-selenadiazoles upon prolonged heating are converted into selenophenes. Since the latter compounds have also been obtained from the reaction of arylalkynes with selenium, 1,2,3-selenadiazoles seem to be indirect precursors of selenophenes (Scheme 49) ⟨73JHC953, 79JHC1405⟩.

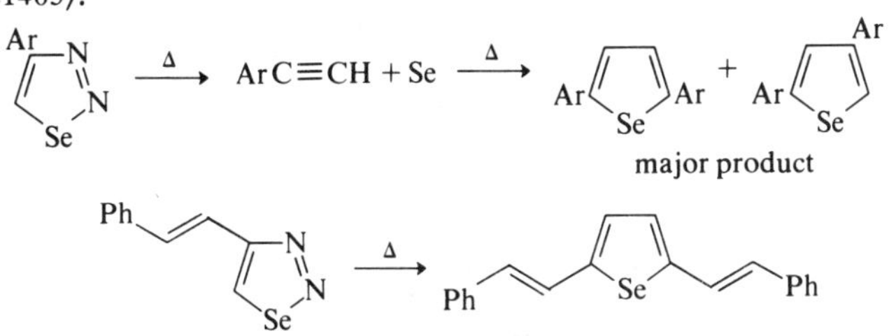

Scheme 49

To stabilize the antiaromatic 4π-electron selenirenes (**115**), considered as possible intermediates in the thermolysis and photolysis of 1,2,3-selenadizoles, Rees and coworkers ⟨77CC418, 72JCS(P1)2165⟩ studied the iron carbonyl complex formation of 1,2,3-selenadiazoles.

Nitrogen extrusion by diiron nonacarbonyl affords a mixture of two isomeric carbene iron complexes (**111**) rather than (**110**). The fact that a crossover product is formed is taken as evidence for the intermediacy of selenirene and carbenes (**112a,b**).

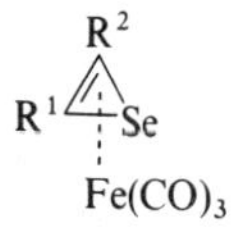

Gas-phase irradiation (λ = 2350–2800 Å) of 1,2,3-selenadiazole (**101**) produces selenoketene (**113**), acetylene, ethynyl selenol (**114**) and selenirene (**115**). Irradiation (λ = 2750–3250 Å) of the reaction mixture leads to the destruction of selenirene. Further irradiation using light of λ = 2350–2800 Å resulted in the selenoketene and acetylene bands being reduced in intensity as monitored by IR spectroscopy. The bands assigned to selenirene reappear and those belonging to ethynyl selenol are enhanced at this stage (Scheme 50) ⟨77JA4842⟩.

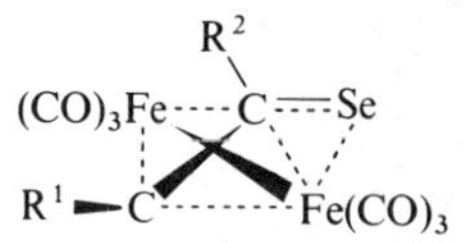

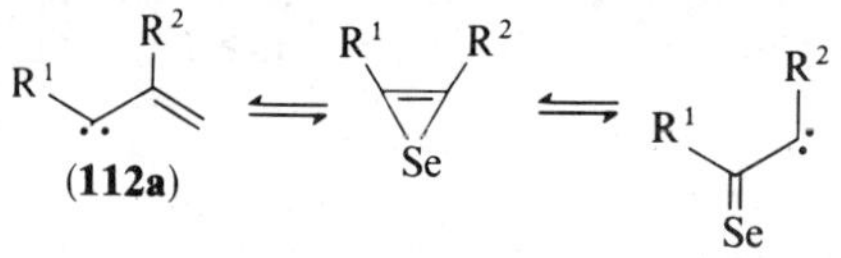

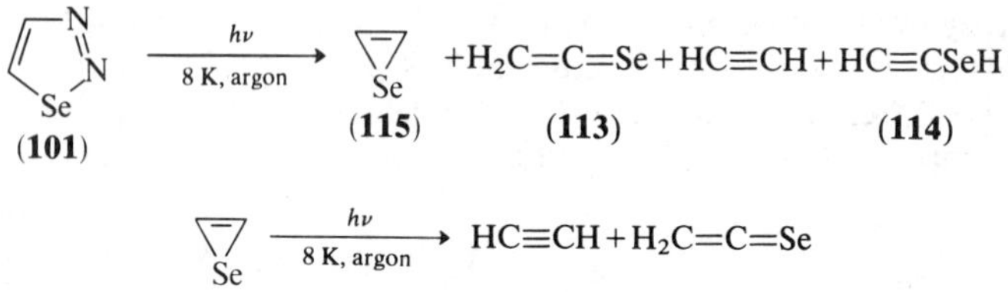

Scheme 50

Selenoketene produced by pyrolysis of 1,2,3-selenadiazole was identified by observation of the microwave absorption of the isotopic species including monodeuterated selenoketene (**116**). An r_s structure has been derived with distances of C=Se 1.706 Å, C=C 1.303 Å and an HCH bond angle of 119.7° ⟨78MI42000⟩.

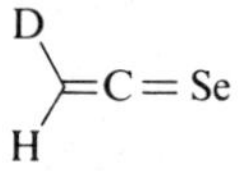

The structural assignment of selenoketene was also based on an *ab initio* SCF calculation of its PE spectrum as well as on a radical cation state comparison with ketene and thioketene by mass spectrometry ⟨80CB3187⟩.

Selenoketenes formed by pyrolysis of 4-alkyl-1,2,3-selenadiazoles in the vapour phase are isolated at −196 °C as highly colored materials which dimerize rapidly.

When selenoketene vapour is mixed with gaseous dimethylamine, the expected formation of *N,N*-dimethylselenoamide takes place (Scheme 51) ⟨79CC99⟩.

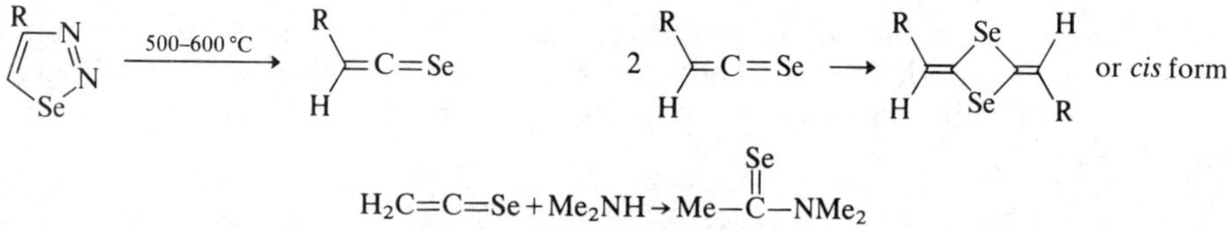

Scheme 51

(ii) Base-catalyzed decomposition of 1,2,3-selenadiazoles

4-Substituted 1,2,3-selenadiazoles react with strong base to form $\beta.\omega$-disubstituted-1,4-diselenafulvenes (**117**). The *cis* isomer (**117**) is formed initially and is then slowly isomerized to the *trans* form. The rate of isomerization is enhanced by the addition of a trace of an acid, *via* the diselenolylium ion (Scheme 52) ⟨73JOC338⟩. The kinetics and mechanism of the base-catalyzed decomposition of 4-aryl-1,2,3-selenadiazoles have been studied. It has been shown that arylethynylselenolates (**118**) and arylselenoketenes (**119**) are intermediates in the diselenafulvene formation ⟨74JOC3906⟩.

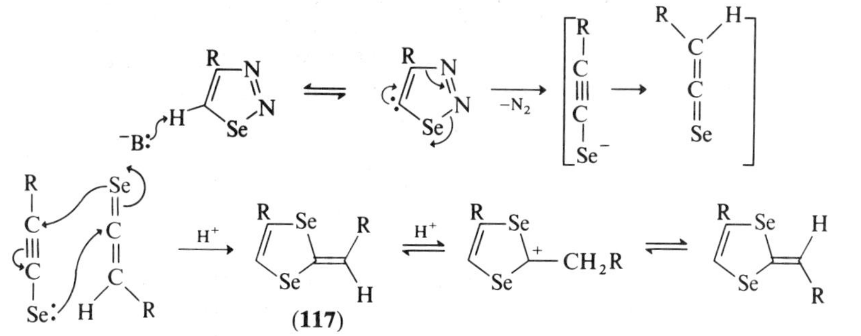

(117)

Scheme 52

$ArC{\equiv}CSe^-$
(118)

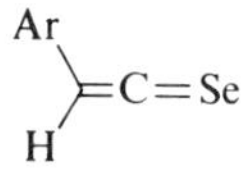

(119)

Photolysis of 4-ethoxycarbonyl-1,2,3-selenadiazoles (**120**) affords, in addition to the corresponding alkynes as major products, small amounts of mixtures of *cis*- and *trans*-1,3-diselenafulvenes (**121**; Scheme 53) ⟨72TL445⟩.

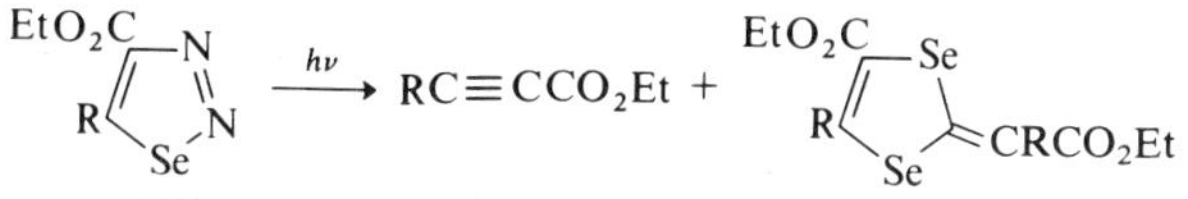

(120) **(121)**

Scheme 53

Cycloalkeno-1,2,3-selenadiazoles (**122**) are converted into thiaselenoles (**123**). Subsequent treatment of the latter compounds with triethylphosphine or triethyl phosphate affords *sym*-diselenadithiafulvalenes of type (**124**). 4,6-Dihydrothieno[3,4-*d*]-[1,2,3]selenadiazole (**125**) undergoes a similar ring transformation reaction to give compounds (**126**) and (**127**; Scheme 54) ⟨76JA3916⟩.

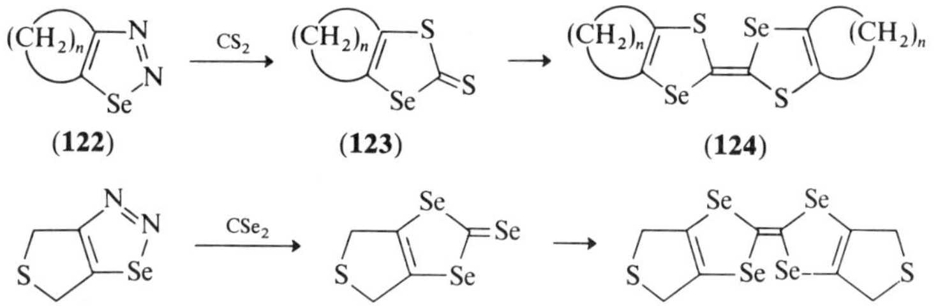

(122) **(123)** **(124)**

(125) **(126)** **(127)**

Scheme 54

2-Thioxo-1,3-thiaselenoles (**129**) are prepared from 4-substituted 1,2,3-selenadiazoles *via* reaction of selenolates (**128**) and carbon disulfide at room temperature ⟨77S764, 80JHC545, 80JOC2632⟩. Selenoamides (**130**) are prepared from the reaction of 4-substituted 1,2,3-selenadiazoles with amines containing potassium hydroxide ⟨77S328⟩. The seleno esters (**131**) are obtained similarly using alcohols instead of amines ⟨76JOC729⟩. The reaction of selenolates (**128**) and DMAD affords selenophene derivatives (**132**) as well as alkynic compounds (**133**) and (**134**) ⟨77S765⟩. Similar results were obtained with methyl phenylpropiolate ⟨81ZOR667⟩. Alkylation of selenolates (**128**) gives selenoethers (**137**). Phenyl isothiocyanate and selenolate (**128**) afford 2-phenylimino-1,3-thiaselenoles (**138**) ⟨81MIP859361⟩. Selenolates (**128**) and halo esters afford alkynylseleno esters (**139**) which are hydrolyzed to give the corresponding acids (**140**; Scheme 55) ⟨82UP42000⟩.

4.20.4.2.2 Reactivity of substituents

In contrast to the ring-fission reaction of 1,2,3-selenadiazoles in the presence of strong base, ethyl 4-methyl-1,2,3-selenadiazole-5-carboxylate (**141**) is hydrolyzed smoothly to the carboxylic acid (**142**) which in turn is converted into the ester *via* its acid chloride (**143**). The corresponding carbamates (**146**) are prepared by hydrazinolysis of esters (**141**) to give the hydrazides (**144**), conversion to azides (**145**), and finally the Curtius reaction. 2-(4-Substituted 1,2,3-selenadiazol-5-yl)-1,3,4-oxadiazole derivatives (**147–149**), as well as alkynes (**150**), were also synthesized as shown in Scheme 56 ⟨71JOC2836, 76JPS304, 77JHC567⟩.

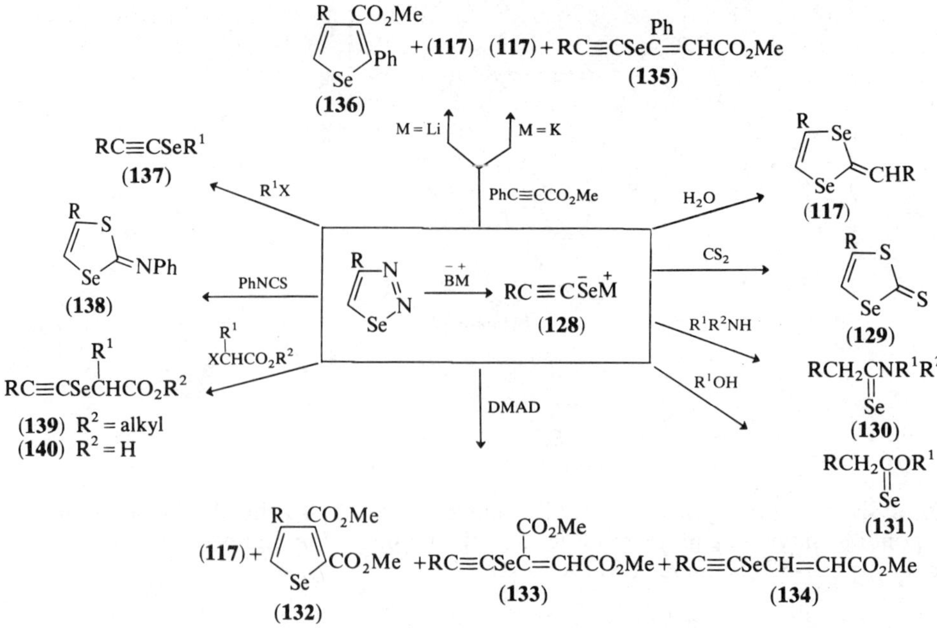

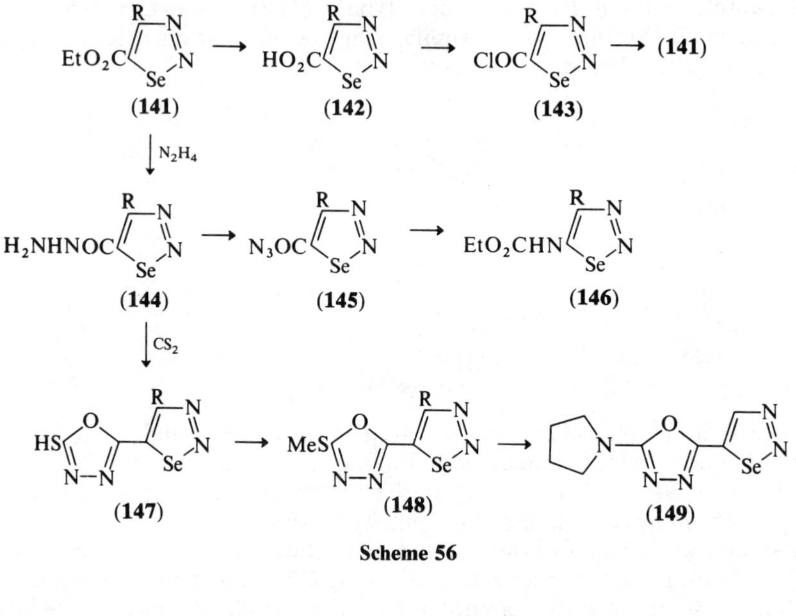

Scheme 56

4.20.4.3 Synthesis

1,2,3-Selenadiazoles, including the parent ring, are prepared by selenium dioxide oxidative ring closure of semicarbazones as described in 1969 by Lalezari *et al* (Scheme 57) ⟨69TL5105⟩.

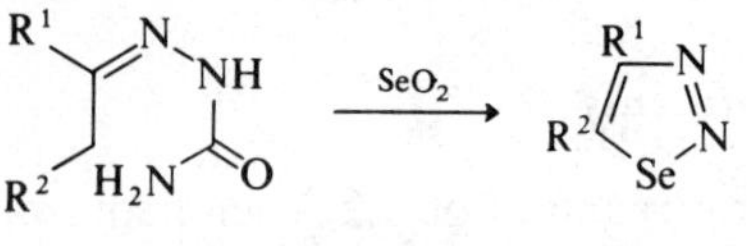

Scheme 57

R^1 and R^2 are H or different types of substituents including cyclic or heterocyclic systems. Thus, semicarbazones of aldehydes or ketones including cyclic ketones with two geminal α-hydrogen atoms are reacted in acetic acid or dioxane with selenium dioxide 〈70AG(E)464, 71JOC2836, 72JHC1411, 74JCS(P1)30, 78JHC501, 79JHC1405, 80JOC2632〉. The proposed mechanism for this reaction (Scheme 58) is supported by the formation of ammonium bicarbonate when dioxane is used as solvent. Using 4-phenylsemicarbazones, sym-diphenylurea is formed as a result of phenyl isocyanate hydrolysis.

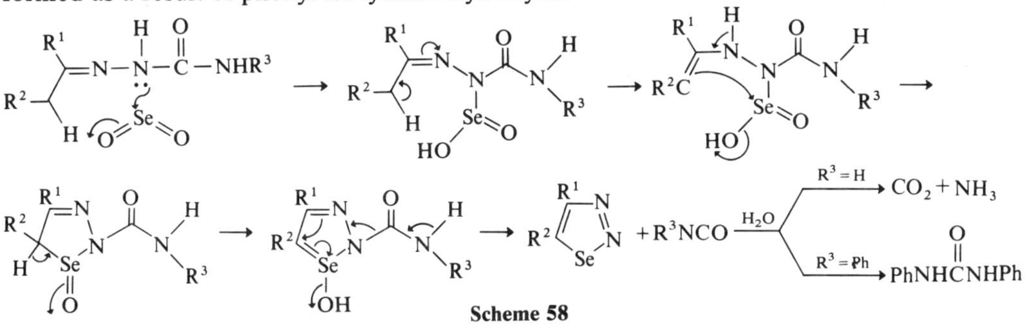

Scheme 58

Ketones possessing two different α-methylene groups accessible to the reaction afford a mixture of two isomeric 1,2,3-selenadiazoles. Comparison of the reaction products derived from ethyl methyl ketone and phenylacetone semicarbazones shows that the direction of ring closure depends on the acidity of the α-hydrogens as determined by the substituent effect (Scheme 59) 〈71JOC2836〉.

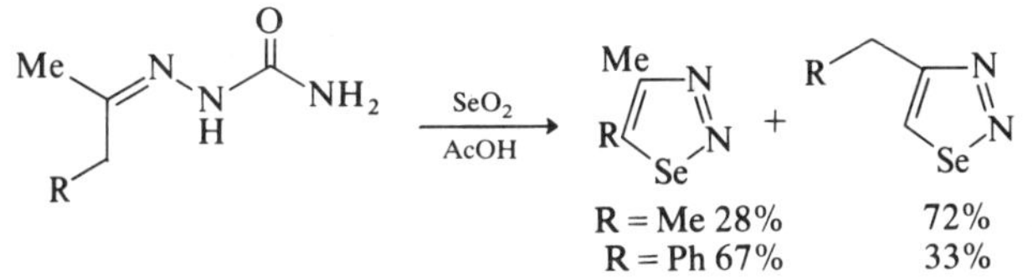

R = Me 28%   72%
R = Ph 67%   33%

Scheme 59

The Z/E equilibrium in asymmetrical semicarbazones and related compounds also affects the regioselectivity of the ring-closure reaction (Scheme 60) 〈81CB2938〉. 1,2,3-Selenadiazoles (**151**) and (**152**) as well as ^{75}Se radiolabelled (**151**) were prepared for biological evaluation 〈79MI42001, 80JHC1245〉.

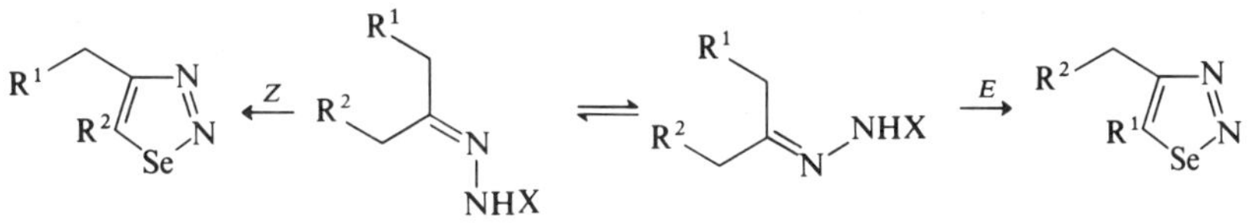

X = $CONH_2$, CO_2Et, $SO_2C_6H_4Me$-p

Scheme 60

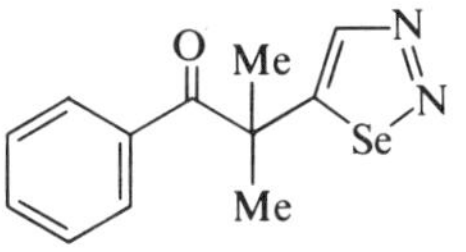

(**151**) 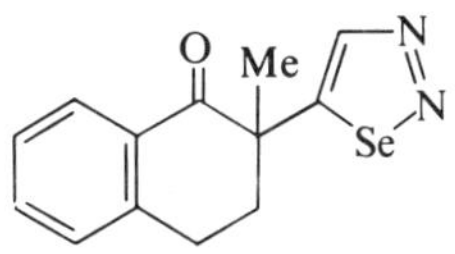   (**152**)

Starting from cycloalkanones, cycloalkeno-1,2,3-selenadiazoles (**122**) are obtained. When the cycloalkene moiety consist of eight or more carbon atoms, pyrolysis or photolysis will convert the selenadiazoles to the corresponding cycloalkynes (**153**). Smaller ring selenadiazoles on pyrolysis afford dicycloalkeno-1,4-diselenines (**154**). In the case of cyclohexeno-1,2,3-selenadiazole (**155**) the octahydroselenanthrene (**156**) is obtained. When pyrolysis is conducted at 250 °C for a prolonged period, some selenanthrene (**157**) and dibenzoselenophene (**158**) result (Scheme 61)〈72JHC1411〉. Some of the alkynes prepared by pryolysis of cycloalkeno-1,2,3-selenadiazoles have been trapped by tetraphenylcyclopentadienone 〈71T187, 81CB2382〉. 10,11-Dihydrophenanthro[1,2-d][1,2,3]selenadiazoles (**159**) are prepared as shown in Scheme 62 〈78JHC501〉.

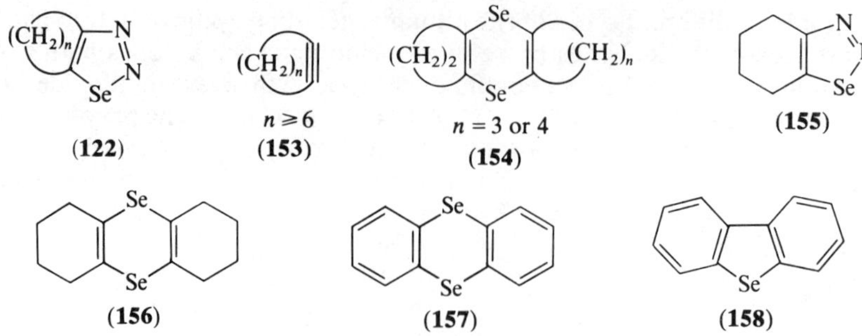

(122) (153) (154) (155)
n ≥ 6   n = 3 or 4

(156) (157) (158)

Scheme 61

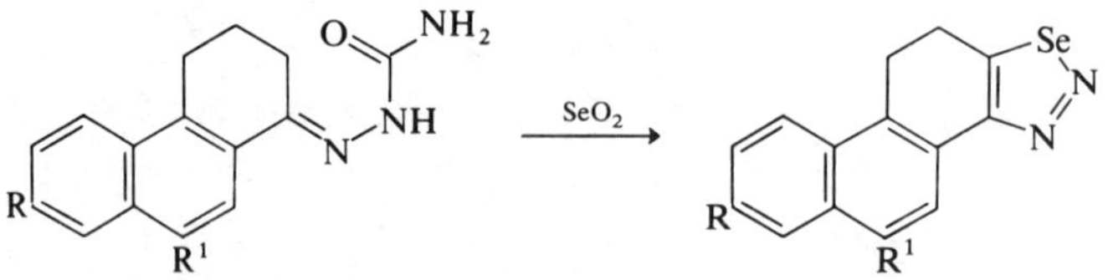

Scheme 62 (159)

4,6-Dihydrothieno[3,4-d][1,2,3]selenadiazoles (**125**) ⟨76JA3916⟩ and [1]-benzoselenopheno[3,4-d][1,2,3]selenadiazole (**160**) ⟨79MI42001⟩ are prepared from their corresponding ketone semicarbazones (Scheme 63).

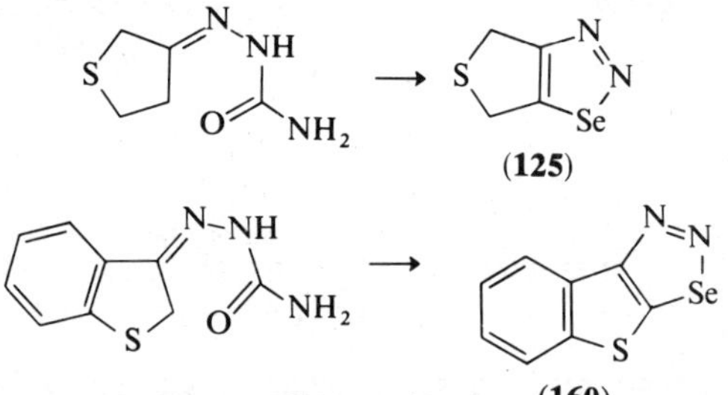

(125)

(160)

Scheme 63

Base-catalyzed reaction of selenadiazole (**161**) with acid chlorides affords tricyclic selenadiazoles (**162**; Scheme 64) ⟨78CB3423⟩. 1,2,3-Benzoselenadiazole (**106**) and its 4- and 5-methyl derivatives reported by Keimatsu and Satoda in 1935 ⟨35YZ233⟩ are prepared by the reaction of nitrous acid and 2-aminobenzeneselenols (Scheme 65).

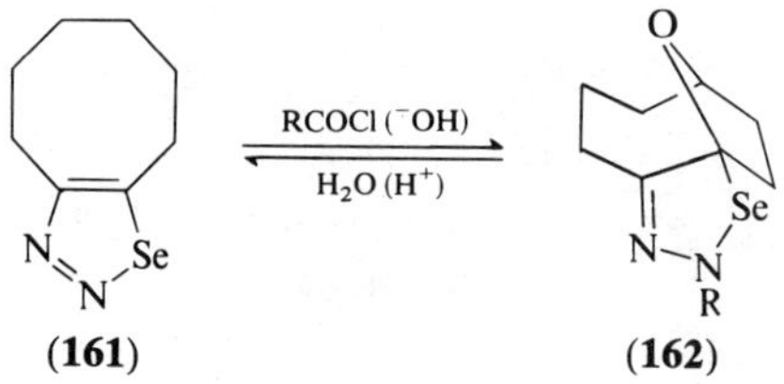

(161) (162)

Scheme 64

(106)

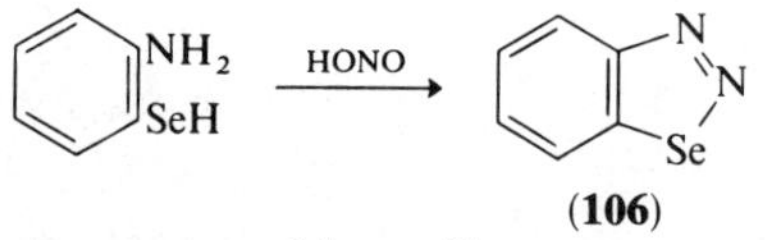

Scheme 65

4.20.5 1,2,4-SELENADIAZOLES

4.20.5.1 Structure

Very little is known about this class of compounds. Photolysis of 3,5-diphenyl-1,2,4-selenadiazole (**163**) at room temperature affords benzonitrile and selenium on the basis of spectroscopic evidence, in particular the IR band at 2200 cm^{-1} (solid nitrogen). The transient photolysis product at 20 and 85 K has been identified as benzonitrile selenide. Prolonged photolysis or warming of the samples from cryogenic temperatures produces exclusively benzonitrile. Photolysis at room temperature is *via* the singlet state, with a quantum yield of 0.085 ± 0.05 (Scheme 66) ⟨77TL3981; 77ACS(B)848⟩.

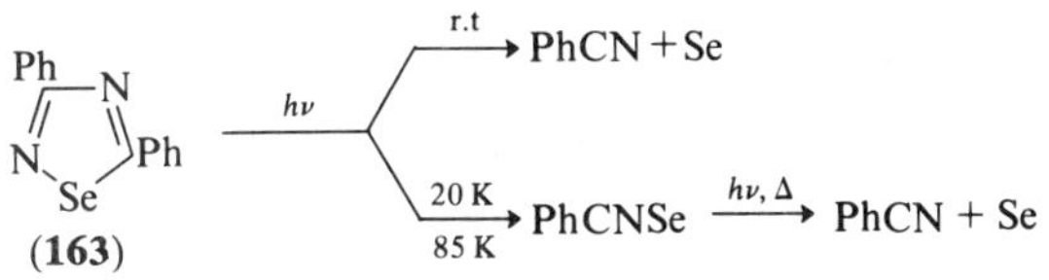

Scheme 66

4.20.5.2 UV Spectroscopy

The UV spectrum of 5-amino-3-phenyl-1,2,4-selenadiazole (**165b**) is very similar to that of its thia analog. In methanol, (**165b**) exhibits two maxima at 280 (log ε 3.78) and 232 nm (log ε 4.36).

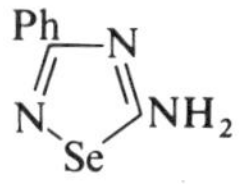

(**165**)

4.20.5.3 Synthesis

Diphenyl- and di-*p*-tolyl-1,2,4-selenadiazoles (**163**) are prepared by iodine cyclooxidation of the corresponding arylselenoamides (Scheme 67) ⟨04CB2550⟩. Some diaryl-1,2,4-selenadiazoles including 3,5-di(2-thienyl)-1,2,4-selenadiazole (**164**) are prepared by the same method ⟨78S768⟩. Diaryl-1,2,4-selenadiazoles are crystalline compounds. Compound (**163b**) is reduced by sodium in ethanol to give *p*-tolylmethylamine.

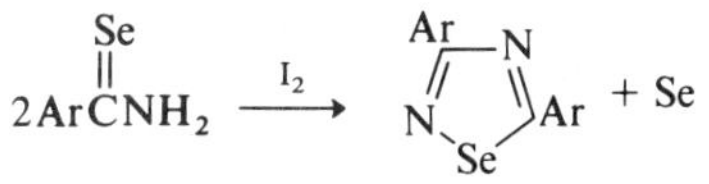

(**163a**) Ar = Ph
(**163b**) Ar = *p*-tolyl

Scheme 67

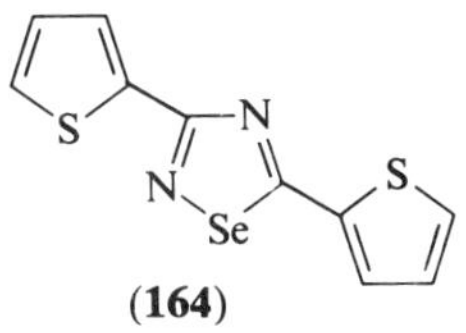

(**164**)

5-Amino-3-methyl- or 5-amino-3-phenyl-1,2,4-selenadiazoles (**165**) are prepared by reaction of potassium selenocyanate with *N*-haloamidines (Scheme 68) ⟨63CB1289⟩.

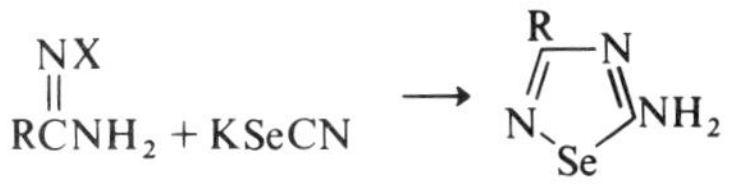

(**165a**) R = Me
(**165b**) R = Ph

Scheme 68

Compounds (**165**) are unstable, the 3-methyl derivative being completely decomposed after one day standing at room temperature. They are also very unstable towards strong bases. The acetyl derivatives, however, are more stable. Nitrosamine derivatives of (**165a**) have also been prepared.

4.20.6 1,3,4-SELENADIAZOLES

4.20.6.1 Structure

4.20.6.1.1 Molecular dimensions

(i) *Microwave spectroscopy*

Analysis of the microwave spectrum of 1,3,4-selenadiazole (**166**) shows the molecule to be planar with a C—Se bond length of 1.868 Å, C=N 1.302 Å and N—N 1.37 Å. The C—Se—C angle is found to be 81.8° in this molecule and is the smallest known angle in a planar five-membered ring reported to date.

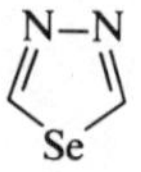

(**166**)

Rotational constants of $C_2H_2N_2{}^{80}Se$ in MHz are: $A = 8479.55 \pm 2.06$, $B = 3372.69 \pm 0.03$ and $C = 2411.80 \pm 0.03$ 〈69JSP459〉.

4.20.6.1.2 Molecular spectra

(i) *UV spectroscopy*

The UV absorption spectrum of (**166**) and of 2-amino-1,3,4-selenodiazole (**167**) in alcohol show λ_{max} 232 (log ε 3.37) and 261 nm (2.59), respectively 〈70JOC806, 71JHC835〉.

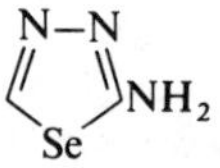

(**167**)

(ii) *^1H NMR spectroscopy*

The chemical shifts of the protons of 1,3,4-selenadiazole (**166**) in CCl_4 appear at δ 9.94 with $J(^{77}Se-{}^1H) = 55.3$ Hz. In compound (**167**) the ring proton appears at δ 9.23 [$J(^{77}Se-{}^1H) = 55$ Hz] and the protons of the amino group resonate at δ 2.68 (DMSO-d_6).

(iii) *^{13}C NMR spectroscopy*

^{13}C NMR spectra of 5-substituted 3H-1,3,4-selenadiazoline-2-thiones and -2-selones (**168–170**) have been reported. Data suggest that substitution of selenium for exocyclic sulfur at C-2 leads to a shielding effect at the carbon of attachment and for selones a deshielding effect at C-5. Substitution of selenium for the ring sulfur of 1,3,4-thiadiazoles results in deshielding at both C-2 and C-5. This effect may be accounted for by a decrease in the total charge density on the α-carbon resulting from a reduced tendency for the d-orbitals of the heteroatom of the ring to overlap with the π-orbitals of the ring system. This effect would be expected to be greater for selenium than for sulfur (Table 5) 〈77JOC3725〉.

Table 5 ^{13}C NMR Data[a] for Selenadiazole Derivatives (**168**)–(**170**) 〈77JOC3725〉

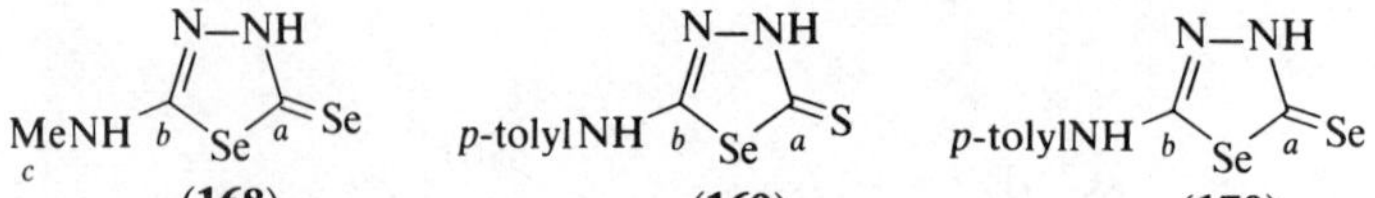

Compound	Chemical Shift (p.p.m.)		
	a	b	c
(**168**)	171.7	166.1	308
(**169**)	186.4	157.2	—
(**170**)	175.2	161.1	—

[a] Reference TMS, solvent DMSO-d_6.

(iv) Mass spectrometry

The molecular ions for (**166**) and its dideuterio derivatives (**174**), as well as ions of selenium and its satellites, comprise the major ions in the mass spectra of these compounds. Evidence for the following major ions has been obtained: $SeCHN^+$, $CHSe^+$, Se^+ and HCN^+.

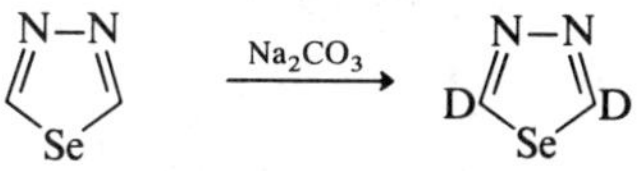

(**166**) +D_2O (**174**)

Scheme 69

Among the major fragment ions of 2,5-diaryl-1,3,4-selenadiazoles (**171**) are ArC≡N^+ and the ion (**171a**) ⟨70JOC806, 79JHC365⟩.

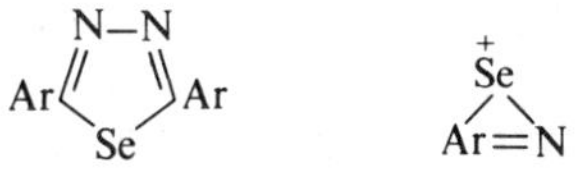

(**171**) (**171a**)

4.20.6.2 Reactivity

Thermolabile 2,2,5,5-tetrasubstituted Δ^3-1,3,4-selenadiazolines are useful in the preparation of sterically very crowded alkenes. Thermal decomposition of (**172**) generates episelenide (**173**) and provides access to the alkene by ready loss of selenium (Scheme 70) ⟨76JCS(P1)2079⟩.

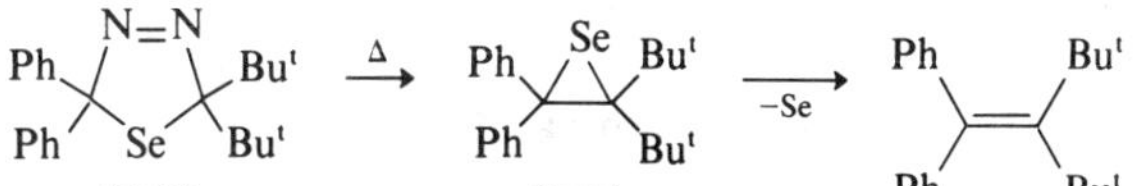

(**172**) (**173**)

Scheme 70

4.20.6.3 Synthesis

2,5-Dialkyl- or 2,5-diaryl-1,3,4-selenadiazoles (**176**) are prepared from N^1,N^2-diacylhydrazines (**175**; Scheme 71) ⟨04JPR(69)509⟩. Compounds (**171**) are prepared in low yields by hydrazinolysis of arylselenoamides (Scheme 72) ⟨79JHC365⟩. The parent compound (**166**) is prepared by the reaction of hydrogen selenide and *N,N*-dimethylformamide azine (Scheme 73) ⟨70JOC806⟩.

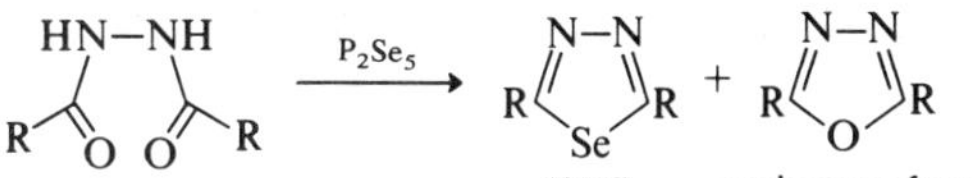

(**175**) (**176**) major product

R = Me, Ph

Scheme 71

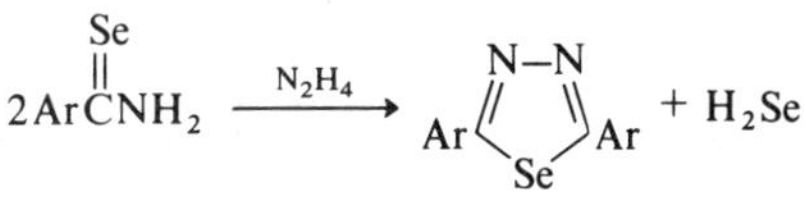

(**171**)

Scheme 72

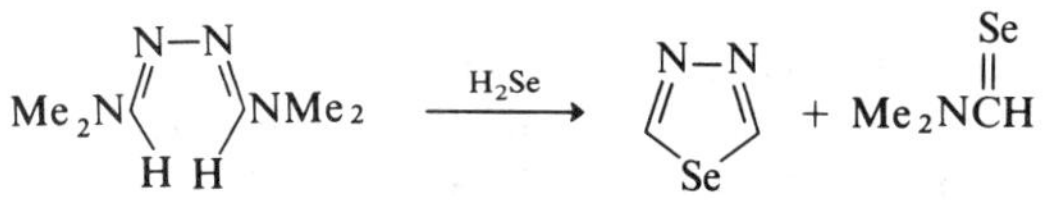

(**166**)

Scheme 73

α-Halohydrazones (**177**) react with potassium selenocyanate to give 3,5-disubstituted 2,3-dihydro-2-imino-1,3,4-selenadiazoles (**178**). Compound (**178**) reacts with nitrous acid and affords the 2-oxo compound (**179**) through heating (Scheme 74) ⟨38G665, 80JHC1185⟩.

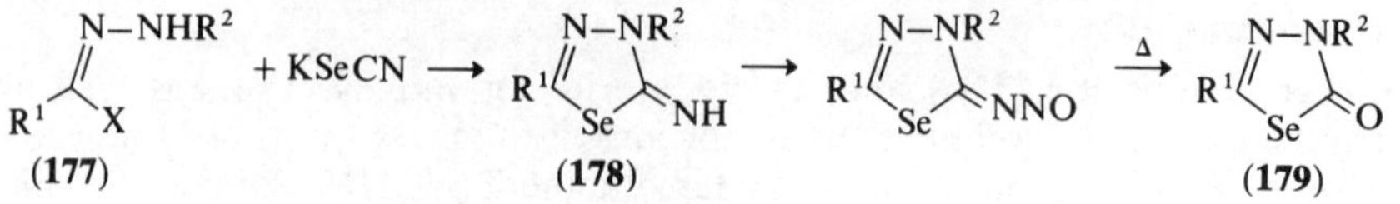

(177) (178) (179)

Scheme 74

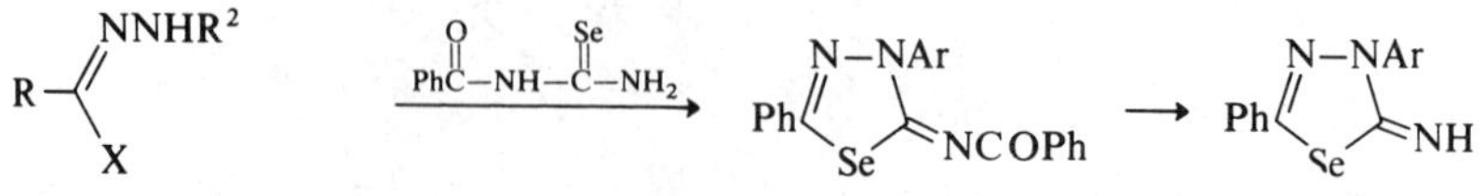

(177) $R^1 = Ph, R^2 = Ar$ (180) (181)

Scheme 75

α-Halohydrazones (**177**) and benzoylselenourea afford the cyclic benzoylimines (**180**) which in turn give the 1,34-selenadiazolineimines (**181**; Scheme 75) ⟨73JPR510⟩.

Phenacyl selenocyanate and arenediazonium chlorides afford 5-benzoyl-2-imino-2,3-dihydro-1,3,4-selenadiazoles (**182**; Scheme 76) ⟨80BCJ1185⟩. 4-Substituted 3-selenosemicarbazides (**183**), by a cyclocondensation reaction with carbon diselenide, afford 5-substituted 3*H*-1,3,4-selenadiazoline-2-selones (**168**) and (**170**). Compound (**183**; R = *p*-tolyl) and thiophosgene give (**169**; Scheme 77) ⟨77JOC3725⟩.

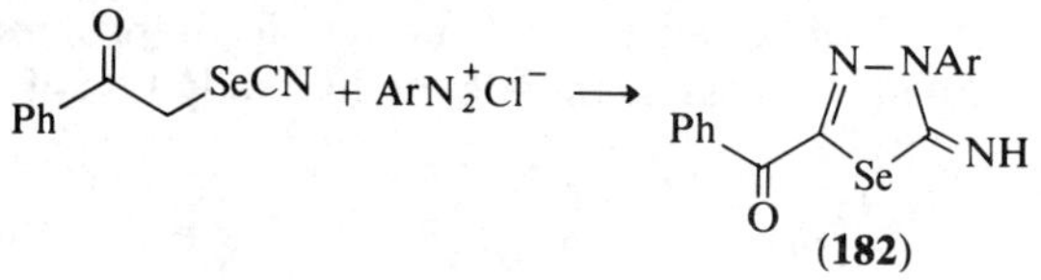

(182)

Scheme 76

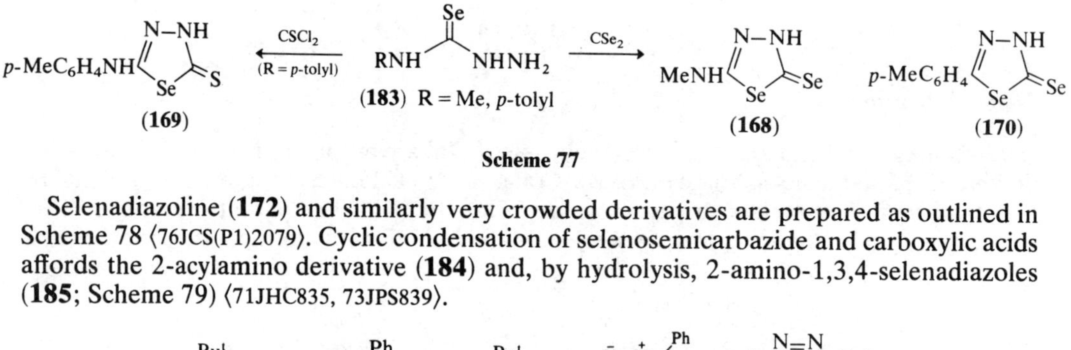

(169) (183) R = Me, *p*-tolyl (168) (170)

Scheme 77

Selenadiazoline (**172**) and similarly very crowded derivatives are prepared as outlined in Scheme 78 ⟨76JCS(P1)2079⟩. Cyclic condensation of selenosemicarbazide and carboxylic acids affords the 2-acylamino derivative (**184**) and, by hydrolysis, 2-amino-1,3,4-selenadiazoles (**185**; Scheme 79) ⟨71JHC835, 73JPS839⟩.

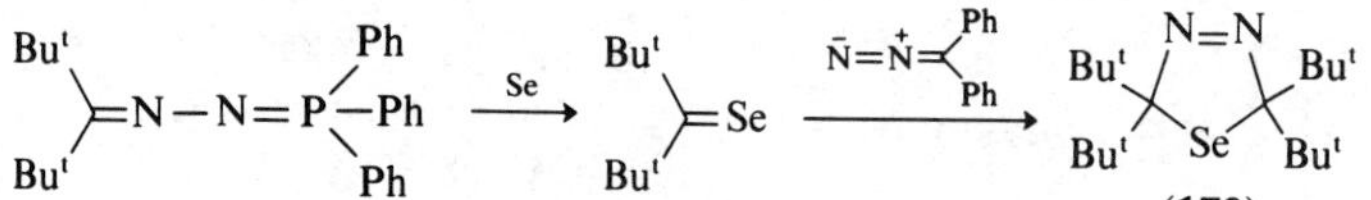

(172)

Scheme 78

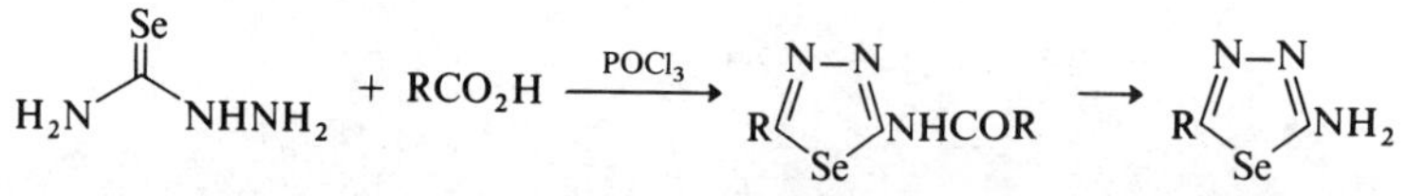

(184) (185)

Scheme 79

Another variation of this synthesis involves cyclization of 4-substituted 1-acylselenosemicarbazides (Scheme 80) ⟨73JPR164⟩. 2,5-Diamino-1,3,4-selenadiazoles of type (**187**) are prepared from bis(selenourea) (**186**; Scheme 81) ⟨73JPR155⟩.

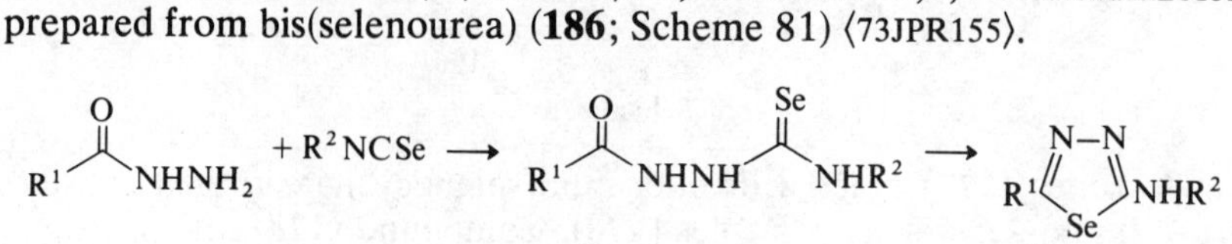

Scheme 80

(186) (187)

Scheme 81

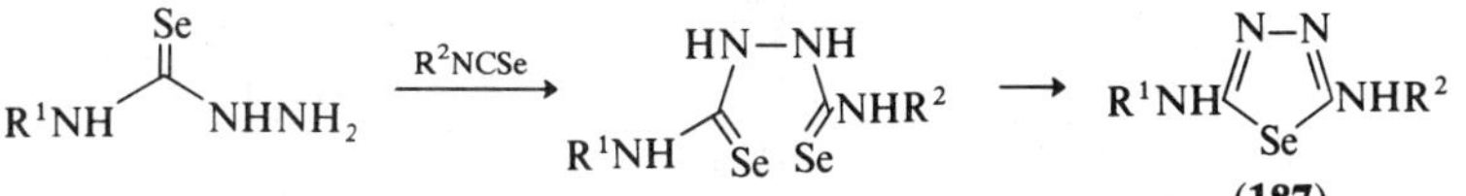

4.20.6.3.1 Fused 1,3,4-selenadiazoles

6-Phenylimidazo[2,1-*b*][1,3,4]selenadiazoles (**188**) are prepared by the reaction of (**185**) and 2-chloroacetophenone followed by cyclic dehydration (Scheme 82) ⟨71JHC835⟩.

(185) (188)

Scheme 82

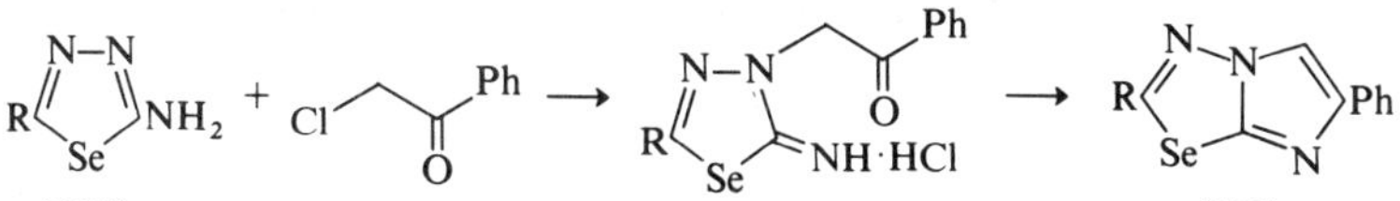

5-Ethyl derivatives of (**185**) and ethyl propiolate afford 2-ethyl-7*H*-1,3,4-selenadiazolo[3,2-*a*]pyrimidin-7-one (**189**) and 2- or 3-ethoxycarbonyl-5*H*-selenanzolo[3,2-*a*]pyrimidin-5-one (**190**). The reaction of (**185**) and DAMD afford the bicyclic compounds (**191**; Scheme 83) ⟨75JHC675⟩.

(191) (185) (189) (190) Major product

Scheme 83

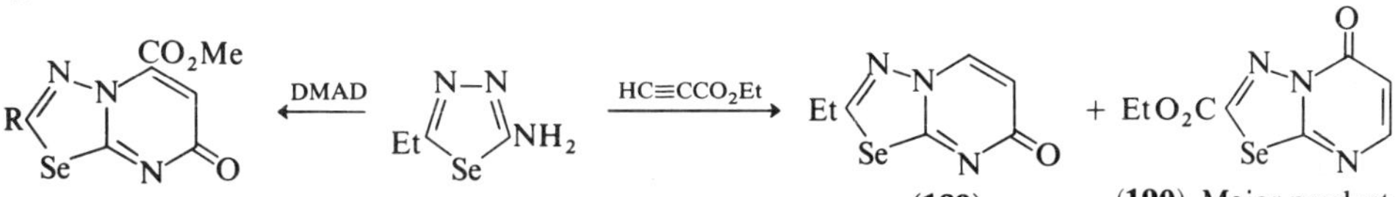

1,4-Cycloaddition of diphenylketene to some Schiff bases (**192**) derived from 2-amino-5-phenyl-1,3,4-selenadiazole (**185**; R = Ph) affords 7-(4-substituted phenyl)-2,6,6-triphenyl-6,7-dihydro-5*H*-1,3,4-selenadiazolo[3,2-*a*]pyrimidin-5-one (**193**; Scheme 84) ⟨76CPB2532⟩. The selenadiazolopyrimidinium perchlorates of type (**194**) are prepared by the reaction of (**185**; R = Ph) and dicarbonyl compounds ⟨74KGS1435⟩.

(192) (193)

Scheme 84

(194)

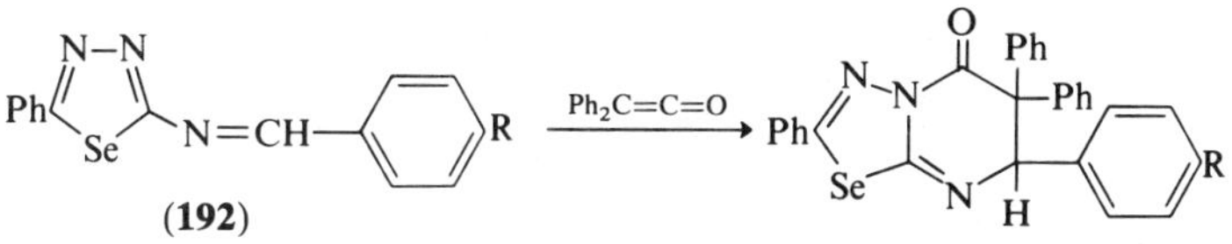

4.20.7 HYPERVALENT FIVE-MEMBERED SELENIUM- AND TELLURIUM-NITROGEN HETEROCYCLES

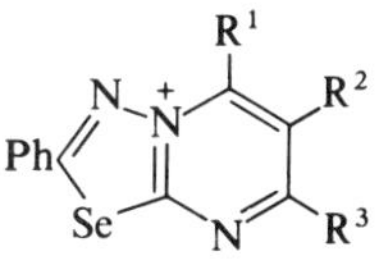

4.20.7.1 Benzo-2,1,3-thiaselenazoylium Chlorides (195) Benzo-1,2,3-thiaselenazolylium Chlorides (196) and Benzo-1,2,3-diselenazolylium Chlorides (197)

4.20.7.1.1 Reactivity

6-Chloro or 6-methoxy derivatives of the compounds (**195–197**) afford the corresponding 6-amino derivatives by a nucleophilic displacement reaction using anilines, dimethylamine and piperidine (Scheme 85) ⟨77KGS1499⟩.

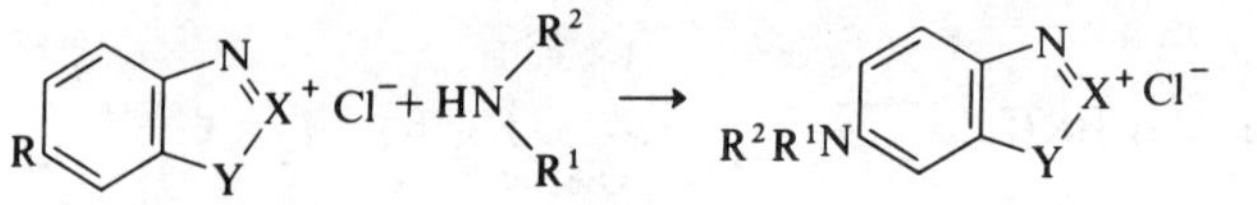

Scheme 85

R = Cl, OMe   X, Y = S, Se

4.20.7.1.2 Synthesis

Compounds (**195–197**) are prepared starting with o-aminothiophenols or o-aminoselenophenols and thionyl chloride. Mono- and di-selenium compounds (**196–197**) are synthesized by reaction of selenous acid and the corresponding sulfur derivatives (Scheme 86) ⟨75KGS275, 76KGS1361⟩.

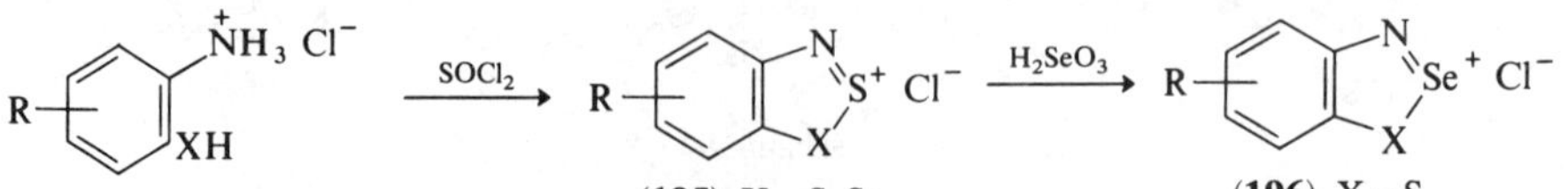

(**195**) X = S, Se

(**196**) X = S
(**197**) X = Se

Scheme 86

4.20.7.2 $7\lambda^4$-[1,2,5]Oxaseleno[2,3-b][1,2,5]oxaselenazoles and $7\lambda^4$-[1,2,5]Oxatellurazolo[2,3-b][1,2,5]oxatellurazoles

4.20.7.2.1 Structural methods

(i) Molecular dimensions

King and Felton ⟨49JCS274⟩ proposed a bicyclic structure (**200**) for the compounds prepared from tris(oxime)derivatives. Although the structures (**202–205**) probably contribute to the overall resonance hybrid, the X-ray diffraction results, however, support the structure (**201**). The Se—O distances were 2.017 and 2.030 Å, thus implying a very strong Se—O interaction. The single bond covalent radii for Se—O is reported to be 1.83 Å and that of the corresponding van der Waals' radii is 3.40 Å. The O—Se—O bond angle was found to be 199.2°. The molecule is almost planar (maximum deviation 0.036 Å) ⟨71CC594, 72JCS(P2)2001, 73TL1565⟩.

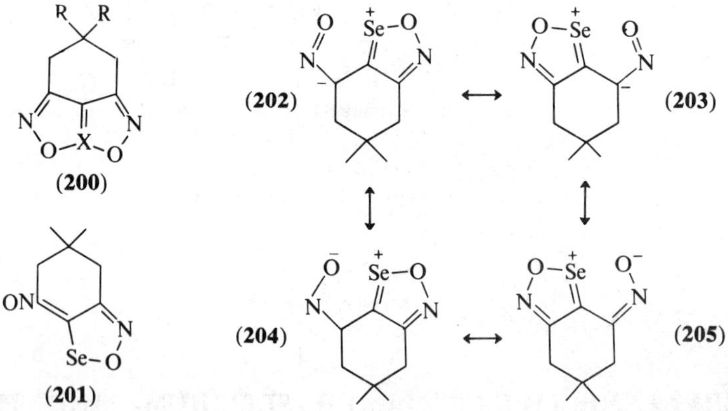

(**200**)

(**201**)

(**202**) ↔ (**203**)

↕   ↕

(**204**) ↔ (**205**)

(ii) Molecular spectra

(a) *NMR spectroscopy.* The NMR spectral data for a series of compounds (**198**) and (**199**) (see Scheme 88) consistent with a symmetrical structure, have been reported ⟨70BSF4517, 71BSF4591⟩.

(b) *Mass spectrometry.* The mass spectral data for compounds (**198**) and (**199**) are compared with their sulfur analogs and the spectral data show the high degree of stability of the three analogs studied. The fragmentation pattern is independent of the nature of the heteroatoms sulfur, selenium or tellurium. The molecular ion is the most intense ion and its fragmentation starts with loss of one NO group followed by a second one. The structures assigned to the remaining fragments are cyclopropene-thione, -selone and -tellurone (**206**) ⟨75JHC639⟩.

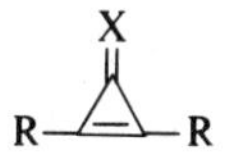

(**206**) X = S, Se, Te

4.20.7.2.2 Reactivity

Compound (**200a**) is thermally stable and is unreactive towards bromine and concentrated sulfuric acid at room temperature. The lack of reactivity towards aniline is consistent with the absence of a normal nitroso group. The compound is sensitive, however, to alkaline reagents and to reducing agents (Scheme 87) ⟨71CC594, 79BSF(2)199⟩.

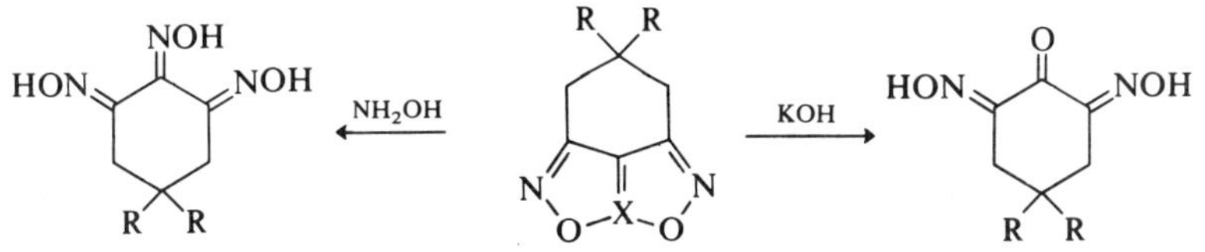

(**200a**) R = H, X = Se
(**200b**) R = H, X = Te

Scheme 87

4.20.7.2.3 Synthesis

Selenium dioxide and tellurium dioxide react with 1,3-diketone dioximes unsubstituted at position 2, to give compounds (**198**) and (**199**), respectively. Selenium monochloride and tellurium tetrachloride give identical results (Scheme 88) ⟨49JCS274, 79BSF(2)199⟩.

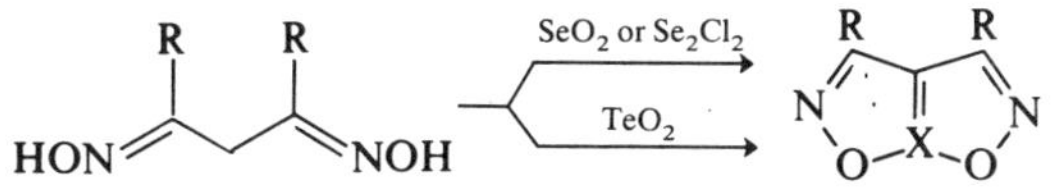

(**198**) X = Se
(**199**) X = Te

Scheme 88

4.20.7.3 6aλ^4-Selena- and 6aλ^4-Tellura-1,2,5,6-tetraazapentalenes

Very little is known about these hypervalent selenium and tellurium compounds.

The NMR spectrum of compound (**207**) (stable green crystals) in $CDBr_3$ at 110 °C consist of bands at δ 3.06 (s, 6H, CH_3) and 7.64–8.60 (m, 6H, aromatics) ⟨71BSF4591⟩. In the tellurium series the bis(2,4-dinitrophenylhydrazone) of 1,3-cyclopentadione and tellurium dioxide give only 2% of (**208**).

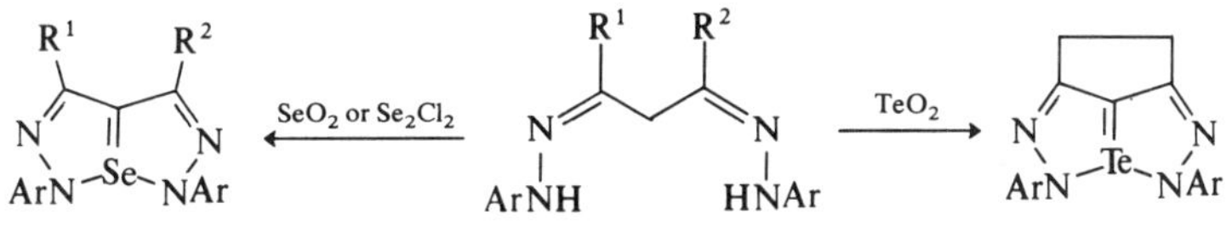

(**207**) Ar = 2,4$(O_2N)_2C_6H_3$, $R^1 = R^2$ = Me
(**208**) Ar = 2,4$(O_2N)_2C_6H_3$

Scheme 89

4.20.7.4 1-Oxa-6,6aλ^4-diselena-2-azapentalenes

4.20.7.4.1 Structure

Three relatively stable compounds (**210a–210c**) of this class of hypervalent selenium heterocycles are known. The assignment of a bicyclic structure is based on physical and spectroscopic similarities with their sulfur analogs. The UV, NMR and mass spectral data are reported in Table 6.

Table 6 Spectral Data for some Heteropentalene Derivatives (210) ⟨72JCS(P1)1360⟩

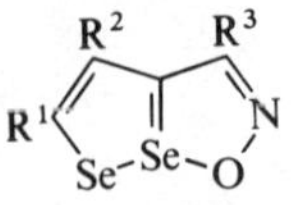

(210)

Compound	UV λ_{max}(nm)(log ε)	$^1H\ NMR$ δ (p.p.m.), J (Hz) (CDCl$_3$)	Mass Spectra principal ions
(210a)	446 (3.78), 315sh (3.50) 253 (4.22), 218 (4.07)	2.78 (d, $J_{4-Me(5)}$ = 1.0, Me) 8.40 (q, $J_{4-Me(5)}$ =1.0, H-4), 9.20 (H-3)	251, 253, 255
(210b)	458 (3.91), 354 (4.01) 242 (4.41).	7.42–7.92 (m, aromatics), 8.82 (H-4), 9.32 (H-5).	313, 315, 317
(210c)	459 (3.82), 305br (3.40) 252 (4.33), 217 (4.32)	10.56 (q, $J_{5-Me(4)}$ = 0.7), 3.06 (d, $J_{5-Me(4)}$ = 0.7), 3.01 (Me-3)	265, 267, 269

4.20.7.4.2 Synthesis

Brief treatment of diselenolylium salts (209) with sodium nitrite in acetic acid–acetonitrile affords 1-oxa-6,6aλ^4-diselena-2-azapentalenes (Scheme 90) ⟨72JCS(P1)1360⟩.

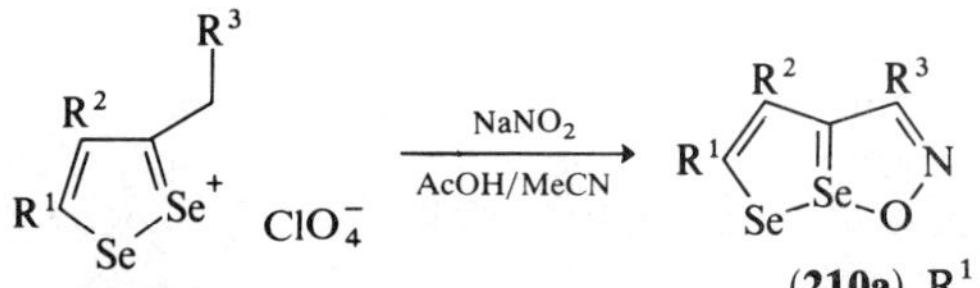

(209) $\xrightarrow{\text{NaNO}_2}{\text{AcOH/MeCN}}$ (210)

(210a) R^1 = Me, R^2 = R^3 = H
(210b) R^1 = Ph, R^2 = R^3 = H
(210c) R^1 = H, R^2 = R^3 = Me

Scheme 90

4.20.7.5 6,6aλ-Diselena-1,2-diazapentalenes

4.20.7.5.1 Structure

Structure elucidation of this class of hypervalent selenium compounds (211) is based on spectroscopic methods and on a comparison of the results with those obtained for their established sulfur analogs. The compounds exist as stable, purple or violet crystals. UV, NMR and mass spectral data are reported in Table 7 ⟨76JCS(P1)228,76ACS(B)600⟩.

Table 7 Spectral Data for some Diselenadiazapentalene Derivatives (211) ⟨76JCS(P1)228⟩

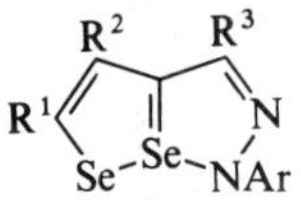

(211)

Compound	UV max (nm) (log ε)	$^1H\ NMR$ δ(p.p.m.), J (Hz) (CDCl$_3$)	Mass spectra principal ions
(211a)	5.29 (4.17), 340vbr (3.54), 2.80 (4.19), 250 (4.47), 212 (4.41)	2.72(D, $J_{Me(5)-H(4)}$ = 0.9) 7.75 (q, H-4), 8.48 (H-3) 7.30–7.84 (m, aromatics)	328, 330
(211b)	542 (4.14), 345 (3.39), 286 (4.22), 251 (4.46), 202 (4.48)	9.72 (br, H-5), 2.94 (d, $J_{Me(4)-H(5)}$ = 0.8), 2.96 (Me-3), 7.23–7.90 (m, aromatics)	342, 344
(211c)	543 (4.34), 346br (4.20), 277 (4.00), 244 (4.35), 220 (4.36), 203 (4.50)	2.78(Me-5), 8.13 (H-4), 8.51 (H-3), 7.85–8.29 (m, aromatics)	373, 375

4.20.7.5.2 Synthesis

Arenediazonium fluoroborates react with 3-methyl- or 3-methylene-1,2-diselenolylium salts (**209**) in water or alcohol to afford 6,6aλ^4-diselena-1,2-diazapentalenes (**211**). The polar solvents used appear to assist the reaction as shown in Scheme 91 ⟨76JCS(P1)228⟩.

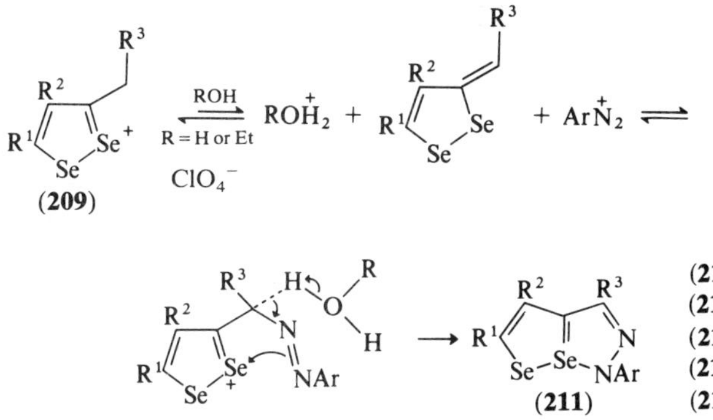

(**211a**) R^1 = Me, $R^2 = R^3$ = H, Ar = Ph
(**211b**) R^1 = H, $R^2 = R^3$ = Me, Ar = Ph
(**211c**) R^1 = Me, $R^2 = R^3$ = H, Ar = p-$O_2NC_6H_4$
(**211d**) R^1 = H, $R^2 = R^3$ = Me, Ar = p-$O_2NC_6H_4$
(**211e**) R^1 = Me, $R^2 = R^3$ = H, Ar = p-$MeOC_6H_4$

Scheme 91

4.21
1,2,3- and 1,2,4-Oxadiazoles

L. B. CLAPP
Brown University, Providence

4.21.1	STRUCTURE OF 1,2,3-OXADIAZOLES	366
	4.21.1.1 1,2,3-Oxadiazoles as Intermediates	366
	4.21.1.2 1,2,3-Oxadiazolines as Intermediates	367
	4.21.1.3 Sydnones—Theoretical Methods	367
	4.21.1.3.1 Dipole moments and dielectric constants	368
	4.21.1.3.2 Electron densities	368
	4.21.1.4 Molecular Dimensions	369
	4.21.1.4.1 X-Ray diffraction	369
	4.21.1.5 Molecular Spectra	369
	4.21.1.5.1 UV spectra	369
	4.21.1.5.2 IR spectra	370
	4.21.1.5.3 NMR spectra (^1H, ^{13}C, ^{15}N)	370
	4.21.1.5.4 Mass spectra	370
	4.21.1.6 Thermodynamic Aspects	371
	4.21.1.6.1 Intermolecular forces	371
	4.21.1.6.2 Stability of the ring	371
4.21.2	REACTIVITY OF 1,2,3-OXADIAZOLES	371
	4.21.2.1 General Survey of Reactivity	371
	4.21.2.2 Thermal and Photochemical Decomposition	372
	4.21.2.3 Electrophilic Substitution at C-4	372
	4.21.2.4 Nucleophilic Substitution at C-4	373
	4.21.2.5 Reactions with Cyclic Intermediates	373
	4.21.2.5.1 Thermally and photochemically induced additions	373
	4.21.2.5.2 Singlet oxygen	375
	4.21.2.5.3 Other photochemical reactions	375
	4.21.2.5.4 Pyrazoles and pyrazolines from sydnones	376
	4.21.2.6 Reactions of Substituents	376
4.21.3	SYNTHESIS OF 1,2,3-OXADIAZOLES	376
	4.21.3.1 Sydnones	376
	4.21.3.2 Sydnonimines	377
	4.21.3.3 1,2,3-Oxadiazolines	377
4.21.4	APPLICATIONS	378
4.21.5	STRUCTURE OF 1,2,4-OXADIAZOLES	378
	4.21.5.1 Theoretical Methods: Dipole Moments and Electron Densities	378
	4.21.5.2 Molecular Dimensions: X-Ray Diffraction	378
	4.21.5.3 Molecular Spectra	379
	4.21.5.3.1 UV spectra	379
	4.21.5.3.2 IR spectra	379
	4.21.5.3.3 NMR spectra (^1H)	379
	4.21.5.3.4 Mass spectra	379
	4.21.5.3.5 Prototropy	380
	4.21.5.4 Thermodynamic Aspects	381
	4.21.5.4.1 Intermolecular forces	381
	4.21.5.4.2 Stability of the ring	382
4.21.6	REACTIVITY OF 1,2,4-OXADIAZOLES	382
	4.21.6.1 General Survey of Reactivity	382
	4.21.6.2 Thermal and Photochemical Decomposition	382
	4.21.6.3 Electrophilic Substitution	382
	4.21.6.4 Nucleophilic Substitution	383
	4.21.6.5 Reactions of Substituents	383
	4.21.6.6 Reduction	384
	4.21.6.7 Rearrangements	385

4.21.7 SYNTHESIS OF 1,2,4-OXADIAZOLES 386
   4.21.7.1 Ring Closure on Carbon 386
      4.21.7.1.1 1,2,4-Oxadiazoles (from amidoximes) 386
      4.21.7.1.2 Δ^2-Oxadiazolines 388
      4.21.7.1.3 Δ^3-Oxadiazolines 388
      4.21.7.1.4 From N-acylimino ethers 388
   4.21.7.2 1,3-Dipolar Cycloadditions 389
      4.21.7.2.1 1,2,4-Oxadiazoles 389
      4.21.7.2.2 Δ^2-Oxadiazolines 389
      4.21.7.2.3 Δ^4-Oxadiazolines 389
      4.21.7.2.4 Completely reduced 1,2,4-oxadiaza rings 390
      4.21.7.2.5 Oxadiazolidinediones 390
   4.21.7.3 From Oxidations 391
      4.21.7.3.1 Of non-ring compounds 391
      4.21.7.3.2 Of oxadiazolines 391
4.21.8 APPLICATIONS 391

Oxadiazoles are numbered by designating the heteroatoms as shown in Scheme 1. The position of the double bond in partially reduced rings is designated as Δ^2- or Δ^4- with the terminal ending -oline. The fully saturated ring is described by the terminal ending -olidine. Substituents may be referred to as occupying positions C-3 (R), N-4 (R') or C-5 (R'').

Scheme 1

4.21.1 STRUCTURE OF 1,2,3-OXADIAZOLES

1,2,3-Oxadiazoles are unknown. The importance of the 1,2,3-oxadiaza ring system lies in the stabilization offered in the derivatives of higher oxidation state known as sydnones (**1**) and sydnonimines (**2**). Fusion with an aromatic ring does not stabilize the five-membered 1,2,3-oxadiazole ring so that what would be called 1,2,3-benzoxadiazole (**3**) actually is more stable as an o-quinone diazide (**4**) ↔ (**5**) ⟨66RCR388⟩.

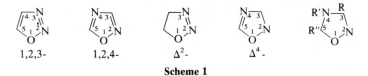

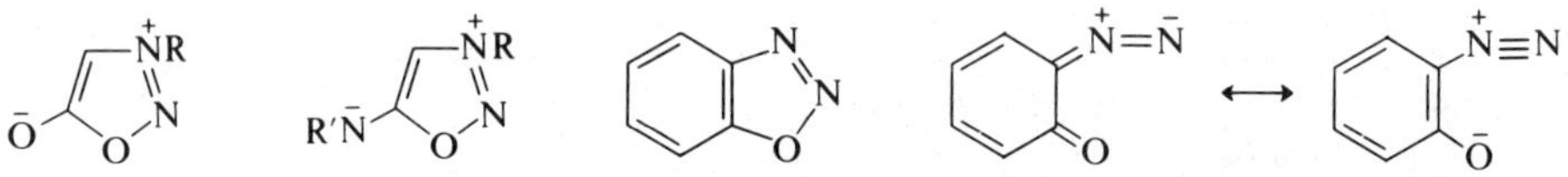

The ionization potentials of (**4**) and (**5**) and the known analogs (**6**) and (**7**) have been measured by mass spectrometry ⟨74CS(6)222⟩ and that of the unknown (**3**) has been estimated; they are included with the formulae. The stabilizing influence of the zwitterionic structures (**4**) and (**5**) is apparently more important than what might be gained by aromatization as in (**3**). IR spectral evidence for the quinone diazide structure, absorption in the 2015–2157 cm^{-1} region, is also convincing ⟨66RCR388⟩.

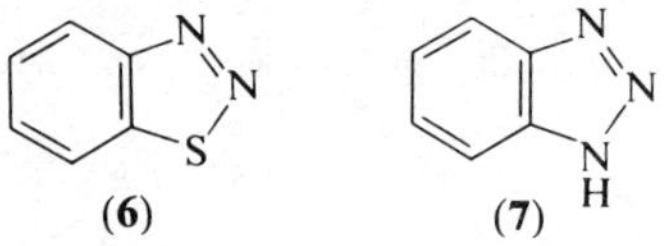

4.21.1.1 1,2,3-Oxadiazoles as Intermediates

The 1,2,3-oxadiazole structure surfaced early ⟨1899JPR(39)107⟩ as the formulation of diazoesters (**8**) until it was laid to rest in 1921 as noncyclic (**9**) ⟨21HCA239⟩. The ring structure was revived again ⟨51JCS3016⟩ as an intermediate (**10**) in the high pressure addition of nitrous oxide to alkynes where the identifiable product was diphenylketene (11). The ring has also been proposed as a transient intermediate (13) in the catalytic reduction of 3-benzyl-sydnonimine (12) ⟨63HCA805⟩. An aromatic 1,2,3-oxadiazole 3-oxide has been suggested once in similar vein ⟨59JCS257⟩.

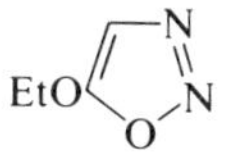

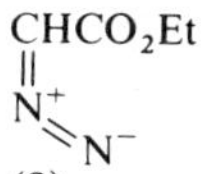

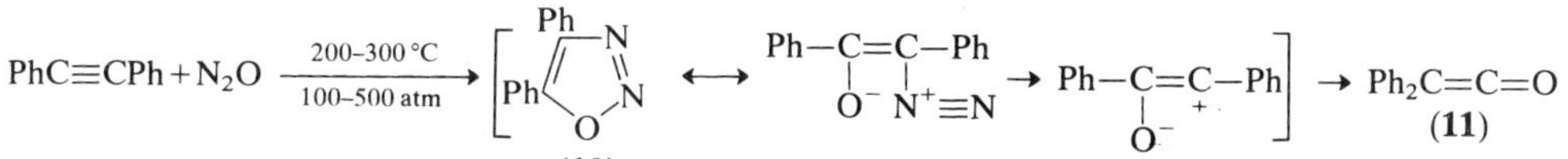

4.21.1.2 1,2,3-Oxadiazolines as Intermediates

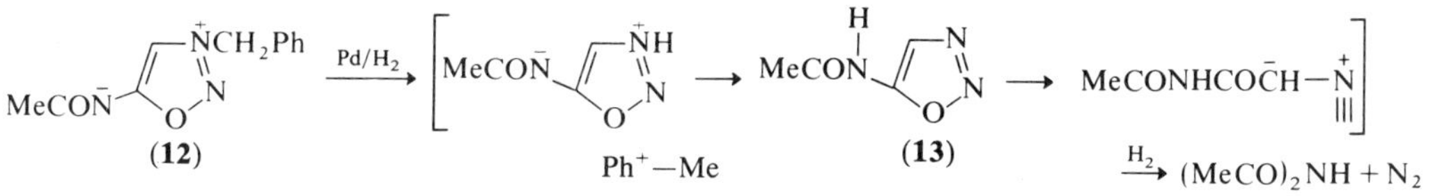

The partially reduced (oxadiazoline) ring was suggested as an intermediate in the addition of nitrous oxide to 2-methyl-2-butene under high pressure ⟨51JCS2999⟩ by analogy with the alkyne addition intermediate (10). However, the first mention of this reduced ring system came in 1907 ⟨07CB479⟩ in a study of the reaction products of aldehydes with diazomethane ⟨28CB1118⟩. The latest report of the 1,2,3-oxadiazoline (the partially reduced ring; 14) is in the decomposition of 1,3-bis(2-chloroethyl)-1-nitrosourea, an active antitumor agent ⟨80JMC1245⟩.

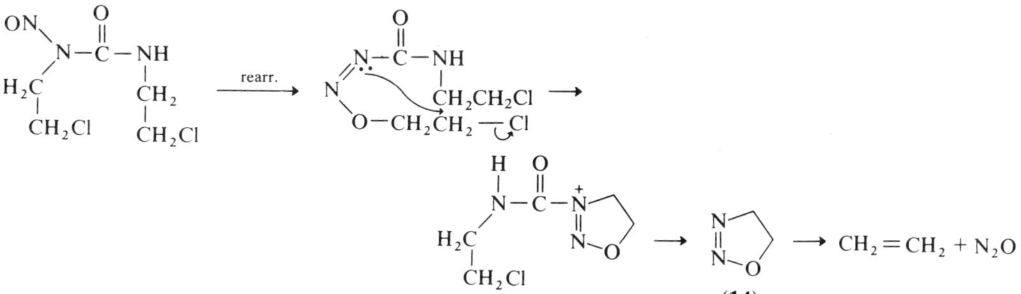

4.21.1.3 Sydnones—Theoretical Methods

The structure of sydnones was not immediately agreed upon by chemists after the original discovery by Earl and Mackney at the University of Sydney ⟨35JCS899⟩. The difficulty in assigning a structure to sydnones arose because no satisfactory covalent structure could be written. The original structure (15) was too strained to be plausible and many of the properties could be better accounted for by an open chain nitrosoamino acid (16) or a mixed anhydride ring (17). The bicyclic structure (15) was not at once discarded but the Chemical Society ⟨59JCS55⟩ earlier recommended the mesoionic form (18) rather than the mesomeric betaine structure (19). Structure (18) is intended as a summary of the mesomers (19)–(21) ⟨64CRV129⟩ and in this chapter the sydnones are represented by structure (19). X-Ray evidence has shown that the valence tautomer (22) should also be considered in any discussion of the structure of the sydnones ⟨67JA5977⟩ (Section 4.21.1.4.1).

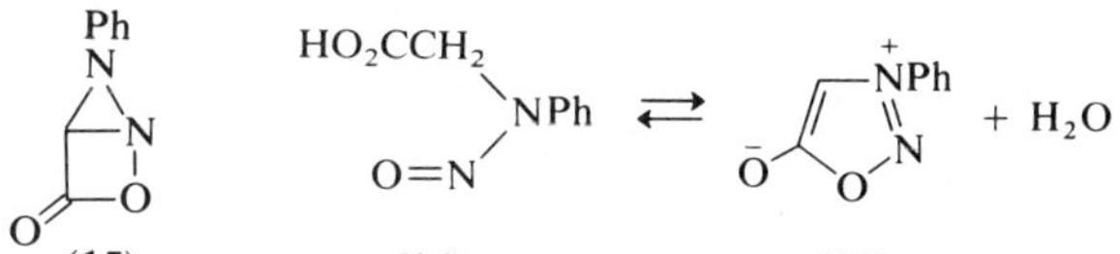

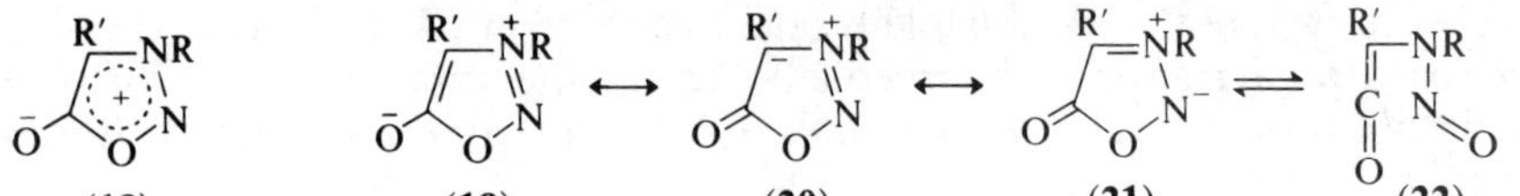

(18)   (19)   (20)   (21)   (22)

4.21.1.3.1 Dipole moments and dielectric constants

The strongly polar character of the sydnones needs the charge separations shown in (**18**)–(**21**) to account for the large dipole moments, generally greater than 6 D. This is comparable to the dipole moment associated with p-nitrodimethylaniline (6.87 D), portrayed as due largely to the dipolar form (**23**) since addition of the moments of dimethylaniline (1.58 D) and nitrobenzene (3.95 D) still falls short of 6.87 D. Some measurements are given in Table 1 ⟨see also 71PMH(4)239⟩.

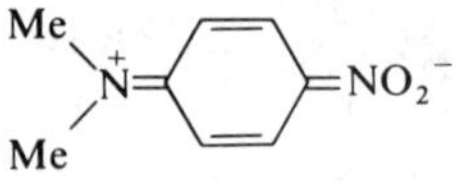

(23)

Table 1 Dipole Moments of Sydnones in Benzene

Substituent	Dipole moment (D)	Ref.	Substituent	Dipole moment (D)	Ref.
3-Ph	6.47	48JCS2269	3-PhCH$_2$	6.28	48JCS2269
	6.48	49JCS746		6.27	49JCS746
3-p-C$_6$H$_4$Cl	5.00	48JCS2269	3-Cy	6.70	48JCS2269
	5.01	49JCS746	3-Me	7.31	70JST(5)236
3-p-Tolyl	6.86	48JCS2269		6.33	71JST(9)321
	6.89	49JCS746			

The dielectric constant is very high, e.g. $\varepsilon = 144.0$ at 40 °C for 3-methylsydnone ⟨77MI42100⟩, a value comparable to that of N-methylacetamide ($\varepsilon = 165.0$ at 40 °C). Water ($\varepsilon = 80$) and acetone ($\varepsilon = 20$), though polar compounds, have much lower dielectric constants.

4.21.1.3.2 Electron densities

Electron densities calculated by quantum mechanical methods (Hückel assumptions and later ones) are shown in Table 2. Early calculations suggested marked charge separations within the ring, O-1 and N-3 being positive while N-2 and O-6 are negative. Later calculations with more refined methods have tended to reduce these differences and even to reverse the signs of the excess charges. Houk has commented ⟨75ACR361⟩, 'Although in principle quantum mechanics can lay bare all the naked truths of chemistry, many of us are still revelling in fondly opening [pushing] one button at a time and savoring each new insight.' Perhaps the next calculation will further modify the results shown in Table 2 but the dipole moments must still be accounted for ⟨71PMH(4)239⟩.

Table 2 π-Electron Densities in Sydnones

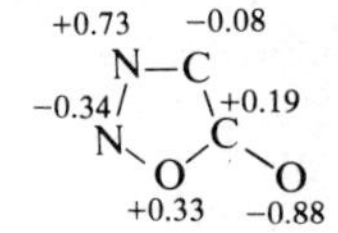

Hückel ⟨49JCS746⟩	Hückel ⟨51MI42100⟩	ω-empirical HMO technique ⟨66JPS807⟩

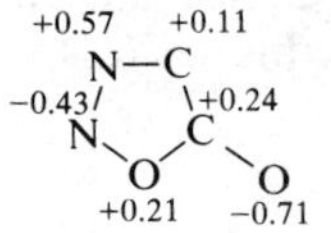

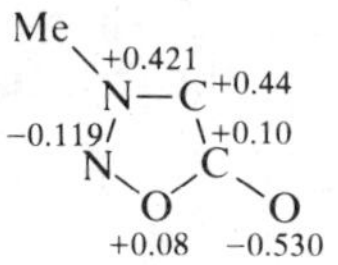

Table 2 (*cont.*)

CNDO/2 ⟨70JST(5)236⟩

iterative extended Hückel ⟨71JST(9)321⟩

ab initio ⟨72JST(12)197⟩

4.21.1.4 Molecular Dimensions

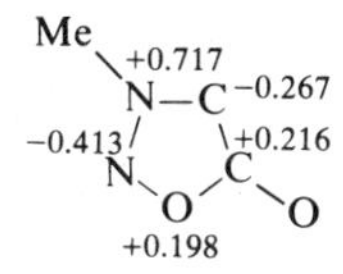

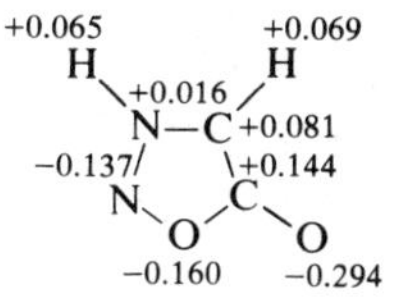

4.21.1.4.1 X-Ray diffraction

The observed bond distances in sydnones can scarcely be called 'normal' since no two bonds are alike. The aromatic C—C distance in benzene is 1.397 Å, somewhat shorter than the only C—C bond length in the sydnones (**24**) ⟨79AX(B)2449⟩, (**25**) ⟨79JCS(P2)1751⟩ and (**26**) ⟨67JA5977⟩. It has been observed that the bond lengths in the sydnone ring are not strongly influenced by fusion with another ring system.

(**24**)

(**25**)

(**26**)

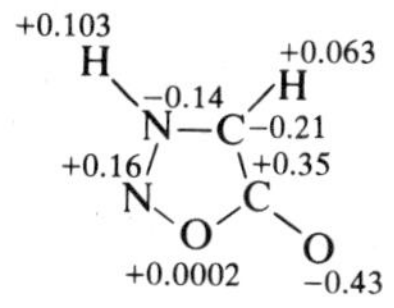

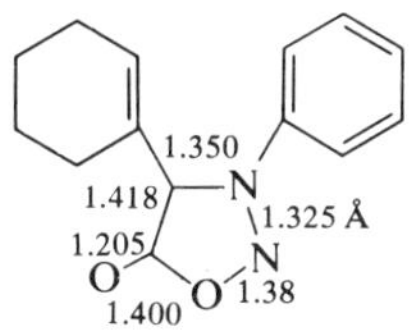

The exocyclic carbonyl bond length is 1.20 Å, apparently not shortened by any ionic character and not appreciably varying in any sydnone since the first X-ray measurements were made ⟨63AX471, 72PMH(5)18, 19⟩. The deformation of the bond angles about the carbonyl group in (**26**) has been attributed to a contribution from structure (**22**) ⟨67JA5977⟩.

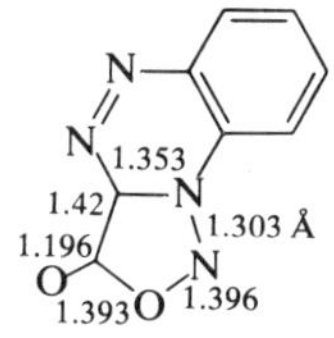

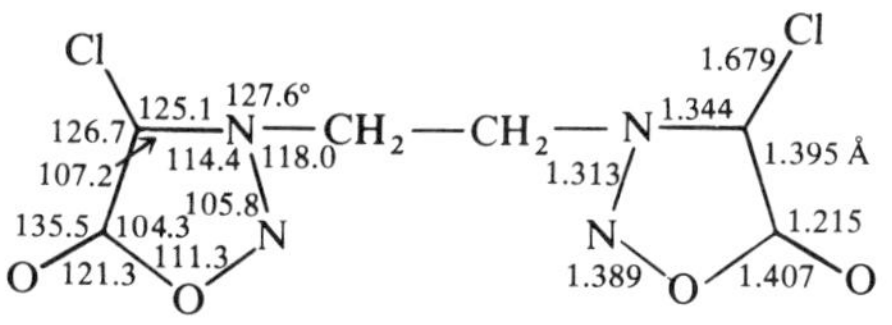

4.21.1.5 Molecular Spectra

4.21.1.5.1 UV spectra

In general alkylsydnones show a single absorption maximum at 290 nm in the UV region with an extinction coefficient of *ca.* 8000. Aryl substituents at N-3 result in a bathochromic shift to about 310 nm ($\varepsilon = 5600$) but aryl groups at C-4 add a further shift, perhaps due to more effective conjugation, with a resultant λ_{max} of 317 nm ($\varepsilon = 7700$). Some recent measurements are given in Table 3 (see also Chapter 4.01, Table 23).

Table 3 IR and UV Spectra of Some Sydnones ⟨79LA63⟩

Substituent 3-R	4-R′	*IR (KBr)* ν(C=O) (cm^{-1})	λ_{max} (nm) in CH_2Cl_2 (log ε)
$4\text{-}NO_2C_6H_4$	H	1773, 1730	335 (3.48)
$4\text{-}NO_2C_6H_4$	Ph	1740	309 (3.94)
$3\text{-}MeOC_6H_4$	MeCO	1790, 1690	324 (3.97)
Me_2N	Br	1757	312 (3.93)
Morpholino	PhS	1750	315 (3.86)
Me_2N	CN	1790, 1770	307 (3.94)

4.21.1.5.2 IR spectra

The IR absorption frequency for the exocyclic carbonyl group in the sydnones in the range 1750–1770 cm^{-1} has suggested a similarity to ν-lactones (1770 cm^{-1}) rather than to tropones (1638 cm^{-1}). This is somewhat disturbing if the sydnone is considered to have the ionic character of structure (1). However, some Δ^3-oxazolones absorb in the 1760 cm^{-1} region. IR absorption data for some recently prepared sydnones are given in Table 3 ⟨see also 63PMH(2)229⟩.

4.21.1.5.3 NMR spectra (^{1}H, ^{13}C, ^{15}N)

As shown in Table 4 the 3-alkyl protons in the NMR spectra of sydnones are considerably further downfield than those of the 4-alkyl protons. The strong deshielding effect of the positively charged N-3 atom accounts for this shift. Protons in other five-membered rings, e.g. thiophene or 1,2,4-oxadiazole, have chemical shifts near those of six-membered aromatic rings (δ 7–8 p.p.m.) but the sydnone protons are upfield (δ 6.2–6.8 p.p.m.) ⟨63CI(L)1926⟩ (see also Chapter 4.01, Tables 9 and 17).

Table 4 NMR Spectra of Some Sydnones ⟨64CRV129⟩

Substituent	Chemical shifts (δ, p.p.m.)			
	Sydnone proton	3-Alkyl protons	4-Alkyl protons	Aryl protons
3-Ph	6.78	—	—	7.70
3-PhCH$_2$	6.19	5.36	—	7.44
3-n-C$_{16}$H$_{33}$	6.33	4.27	—	—
3-Me-4-Ph	—	4.13	—	7.50
4-Me-3-Ph	—	—	2.17	7.68
3-PhCH$_2$-4-CO$_2$Me	—	5.50	3.54	7.36
3-Bu^n	6.34	—	—	—

The ^{13}C spectra of several sydnones and sydnonimines have been determined ⟨74JCS(P2)875⟩. In 3-methylsydnone the N-methyl carbon resonance is at 39.8 p.p.m. and the ring carbons 4 and 5 resonate at 96.8 and 169.2 p.p.m., respectively. These data and those for other sydnones have been correlated with ring structure.

The nitrogen atoms in sydnones and sydnonimines have been independently labelled with ^{15}N ⟨77T2571⟩ and the NMR spectra of the labelled products have been determined ⟨79JMR(36)227⟩ (see also Chapter 4.01, Table 22).

4.21.1.5.4 Mass spectra

In the mass spectrometer the loss of mass in a sydnone ring due to NO and CO ($M-30$ and $M-58$, respectively) may occur consecutively or simultaneously. Mass $[M-58]$ is commonly prominent as the ion of 100% relative intensity, although $M^{\ddot{+}}$ may often be distinguished. The fragmentation pattern of 3-p-nitrophenyl-4-phenylsydnone (27) is shown in Scheme 2 with fragments of high relative intensity ⟨79LA63⟩. In the fused ring sydnone

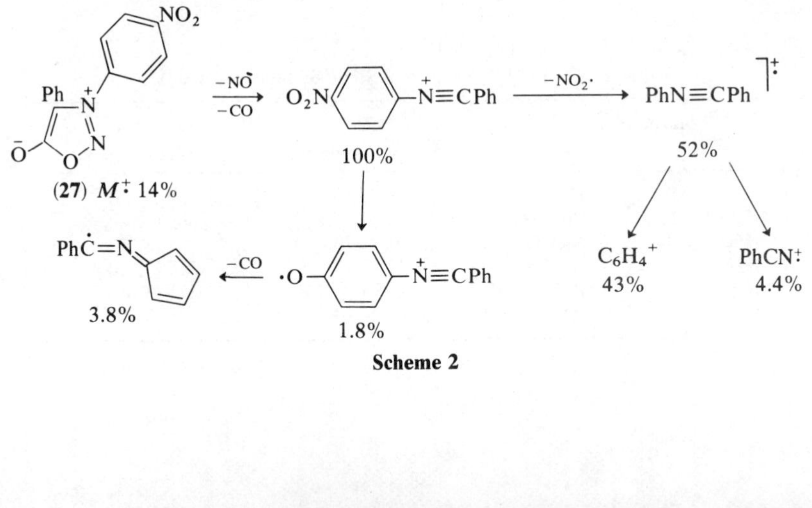

Scheme 2

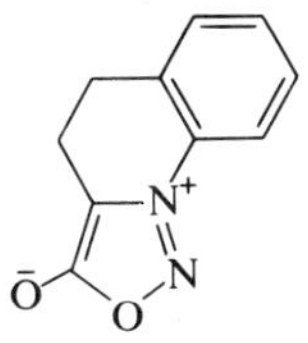

(28) initial loss of NO was followed consecutively by CO, HCN, HC≡CH and finally Ph^+ (m/e 77) as the principal fragment ions ⟨68AJC1665; see also 80MI42110⟩.

4.21.1.6 Thermodynamic Aspects

4.21.1.6.1 Intermolecular forces

Alkylsydnones are mainly liquids (or low-melting solids) that can be distilled at reduced pressure without decomposition. The polar character of the molecules suggests some solubility in water but they are also soluble in most organic solvents except ligroin. Arylsydnones are crystalline solids with melting points of normal range for organic substances. Stewart has compiled an extensive list of sydnones and sydnonimines prepared before 1964 ⟨64CRV129⟩.

Recently viscosity, density and index of refraction measurements have been recorded for some liquid alkylsydnones ⟨77MI42100⟩. For example, for 3-methylsydnone at 40 °C the viscosity is 5.501 cP, the density is 1.3085 g cm^{-3} and the index of refraction n_D is 1.515. Measurements on the conductance of salt solutions in 3-methylsydnone are also available ⟨80MI42111⟩.

4.21.1.6.2 Stability of the ring

The stability of the sydnone ring is shown by its resistance to catalytic hydrogenation. For example, 3-phenylsydnone is only reduced to N-phenylglycine with the very reactive Adams catalyst ⟨57QR15⟩.

For thermal and photochemical stabilities of sydnones, see Sections 4.21.2.1 and 4.21.2.5.

4.21.2 REACTIVITY OF 1,2,3-OXADIAZOLES

4.21.2.1 General Survey of Reactivity

Sydnones and sydnonimines are stable in acid solution at room temperature but are subject to rapid hydrolysis in basic solution. The kinetics of alkaline hydrolysis of 3-alkylsydnonimines was found to be third order, first order in sydnonimine and second order in hydroxide ion at pH ~8 ⟨63ZOB3699⟩. The mechanism shown in Scheme 3 rationalizes the kinetics of the hydrolysis. The nitrosonitrile (29) is hydrolyzed to the nitrosoamide at higher pH ⟨65KGS328⟩. The rate law for 3-arylsydnonimines is k[syd][OH$^-$].

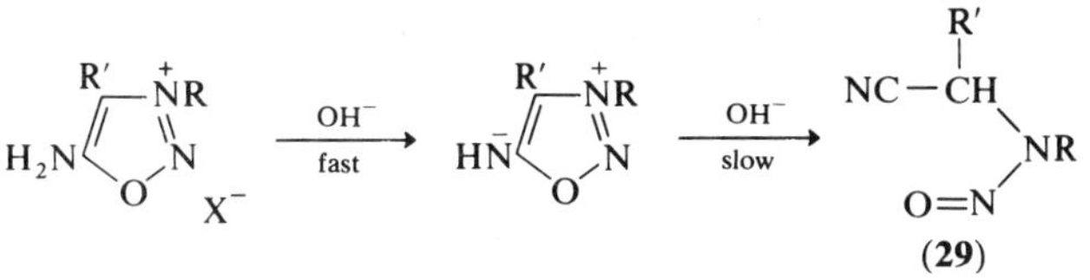

Scheme 3

In acid solution sydnones and sydnonimines are hydrolyzed at elevated temperatures. In view of the demonstration ⟨70JA3133⟩ that the proton is located on the exocyclic oxygen atom (or nitrogen ⟨80OMR(13)274⟩), the mechanism shown in Scheme 4 appears plausible, although other mechanisms were suggested earlier ⟨57QR15, 67ZOR942, 67JPS149⟩. Acid hydrolysis of sydnones was once advocated as a synthetic path to alkylhydrazines (30) ⟨55JA1843⟩.

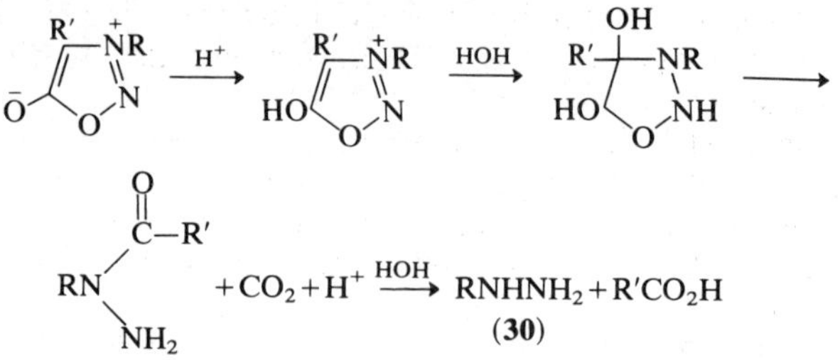

4.21.2.2 Thermal and Photochemical Decomposition

Thermal and photochemical decompositions of sydnones differ depending on the substrate. The reactions are addition–elimination reactions involving various cyclic intermediates (Section 4.21.2.5).

4.21.2.3 Electrophilic Substitution at C-4

The sydnone ring opens readily in base but much less readily in acid so it is possible to effect electrophilic substitutions under carefully controlled acidic conditions. Only the C-4 carbon atom has excess π-electron density (Table 2) so that electrophilic substitutions, summarized in Table 5, occur there ⟨64CRV129⟩.

Bromination can be carried out with bromine in acetic anhydride at 0 °C, with NBS and with other reagents but the most efficient reagent is bromine in ether in the presence of sodium hydrogen carbonate.

A nitro group may be substituted at C-4 in 3-phenylsydnone but when phenyl groups occupy both the N-3 and C-4 positions there is a duality of electronic effects. The sydnone ring is electron-attracting at N-3 but electron-releasing at C-4. Thus nitration gave the 4-*p*-nitrophenyl derivative with 3,4-diphenylsydnone. With a benzyl group at N-3 in (31) nitration gave a mixture of products, compound (32) predominating when 1–2 moles of nitric acid were used and compound (33) being in excess when 6 moles of nitric acid were used. In 3-methyl-4-phenylsydnone, nitration in acetic acid at 25 °C gives 4-nitrophenyl substitution, and a 2,4-dinitrophenyl derivative results under more vigorous reaction conditions.

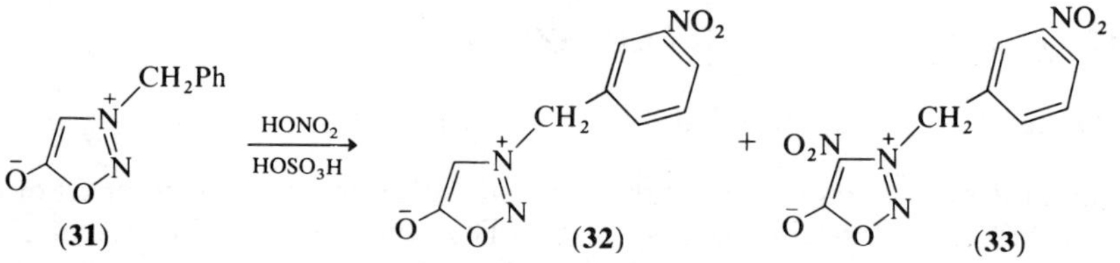

Table 5 Electrophilic Substitutions in Sydnones

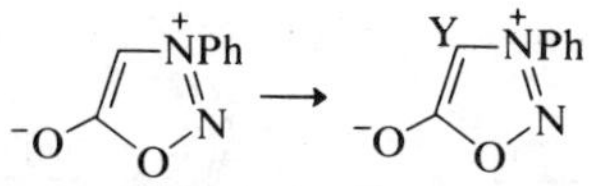

Reagents	Y
(a) Br$_2$, ether, NaHCO$_3$	Br
(b) HONO$_2$ + HOSO$_3$H, 0 °C	NO$_2$
(c) SO$_3$ (dioxane)	SO$_3$H
(d) ClSO$_3$H + H$_3$PO$_4$	SO$_2$Cl
(e) Ac$_2$O + BF$_3$ (ether)	COMe
(f) HCONMe$_2$ + POCl$_3$	CHO
(g) Hg(OAc)$_2$	HgOAc
(h) DMSO + AcCl	SMe

Sulfonation at C-4 can be accomplished with sulfur trioxide complexed with dioxane. Chlorosulfonic acid introduces the sulfonyl chloride group into the ring from which esters and sulfonamides can be made ⟨79AP977, 81AP470, 503⟩.

The Friedel–Crafts acetylation of 3-phenylsydnone was accomplished with boron trifluoride etherate as catalyst. Formylation at C-4 by the Vilsmeier procedure occurred with 3-phenylsydnone. Mercuration is easily afforded with mercury(II) acetate or mercury(II) chloride and thioethers can be made directly with DMSO in acetyl chloride ⟨74T409⟩. At least one fused ring as in compound (35) has been made by a coupling reaction on the sydnone (34) at C-4 ⟨79JCS(P2)1751⟩.

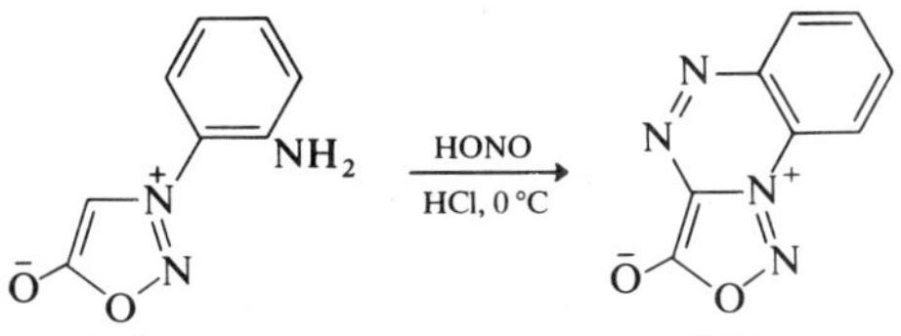

4.21.2.4 Nucleophilic Substitution at C-4

Several replacement reactions at C-4 in sydnones may be carried out but aqueous bases must be avoided. Butyllithium can be used to displace bromine from a 3-phenylsydnone; the resulting organolithium salt can be carbonylated, will add to ketones, and forms a silyl derivative ⟨80CB1830⟩. A sydnone Grignard derivative can also be made and will add ketones in the normal way ⟨80JCS(P1)20⟩. Sodium borohydride will reduce a sydnone sulfone, formed by oxidation of a thioether (Table 5) with hydrogen peroxide, back to the unsubstituted sydnone ⟨74T409⟩.

Table 6 Nucleophilic Substitutions in Sydnones

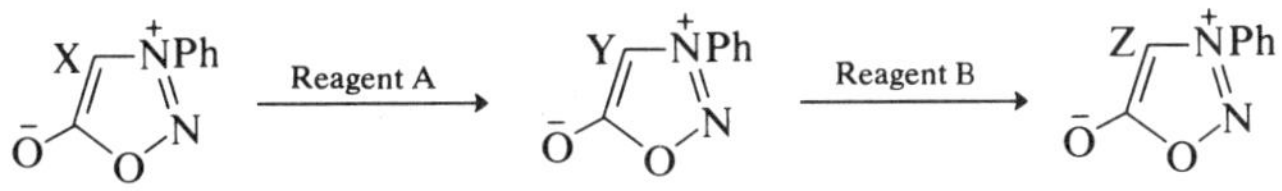

X	Reagent A	Y	Reagent B	Z	Ref.
Br, H	Bu^{n}Li	Li	CO_2 $COCl_2$ MeCOCHMe$_2$	CO_2H $(CO)_{1/2}$ MeC(OH)CHMe$_2$	B-69MI42100
Br	Mg, ether, MeI	MgBr	I_2 Ac$_2$O RCHO	I COMe CH(OH)R	B-69MI42100
SMe (Table 5)	H_2O_2	SO$_2$Me	NaBH$_4$	H	74T409

4.21.2.5 Reactions with Cyclic Intermediates

4.21.2.5.1 Thermally and photochemically induced additions

Thermochemical and photoinduced decomposition of sydnones gave different products when an alkyne was available to trap the transient intermediate ⟨71JOC1589⟩. For example, when 3,4-diphenylsydnone (36; Scheme 5) was heated in the presence of DMAD, addition occurred to give (37) followed by elimination of carbon dioxide resulting in dimethyl 1,5-diphenylpyrazole-3,4-dicarboxylate (38). In the photochemically induced reaction loss of carbon dioxide occurred first, followed by addition to the dipolar intermediate (40) to give dimethyl 1,3-diphenylpyrazole-4,5-dicarboxylate (41) as the final product.

The phenylnitrilimine (40) or (45) was first suggested as an intermediate in the photochemical decomposition of the sydnone system when (46) was isolated from (42; Scheme 6) ⟨66TL4043⟩. Four laboratories agreed almost simultaneously in 1971 that (45) was the key intermediate by various trapping and labelling experiments ⟨71JOC1589, 71HCA1275,

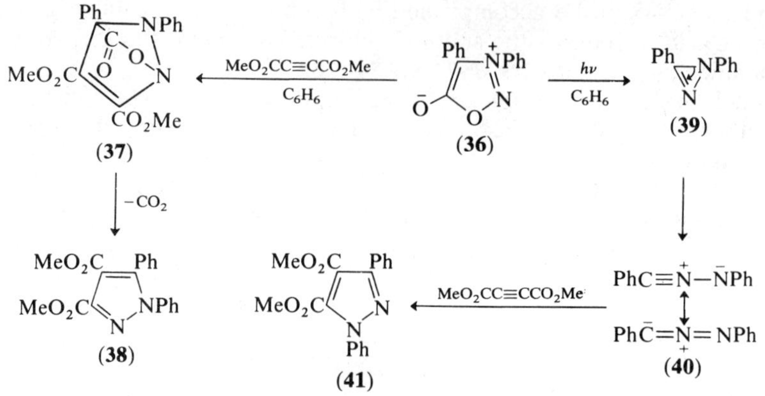

Scheme 5

71BCJ1667, 71TL2749⟩. Other fragments such as (**43**) and (**44**) have been suggested as precursors to (**45**), and (**47**) and (**48**) have also been considered as intermediates ⟨72CC498⟩. Besides DMAD, alkenes ⟨78HCA1477⟩ and other trapping agents ⟨79CB1635, 80HCA653⟩ have been used to catch the dipolar moiety ⟨see also 81CB2450, 1737⟩. When 1,3-butadiene was used as a trap with (**44**) and (**45**; Ar = Ph), the product was 1,3-diphenyl-5-vinylpyrazole, suggesting (**45**) as the trapped fragment. A [2+4] addition would be expected for (**44**) to give a six-membered ring ⟨80MI42100⟩.

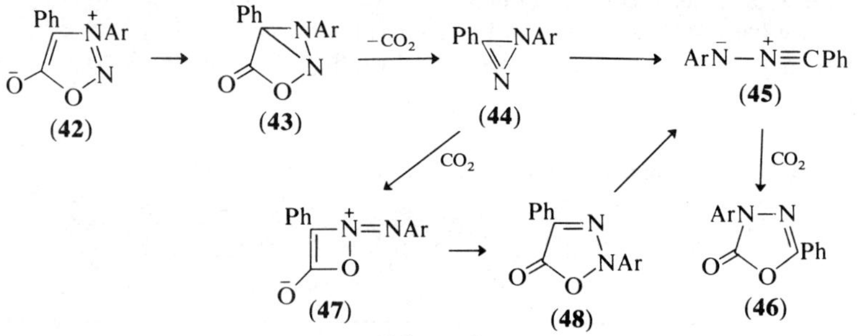

Scheme 6

Another product formed in the photochemical decomposition of a sydnone in dioxane was a triazole. By labelling the nitrogen in (**49**) as shown, the triazole (**53**) was found to contain two ^{15}N atoms (Scheme 7). The argument for (**50**) as the key in the mechanism was verified with a crossover experiment in which an equimolar mixture of 3-*p*-tolyl-4-phenylsydnone (**54**) and 3,4-di-*p*-tolylsydnone (**55**) was irradiated to give the three triazoles (**56**), (**57**) and (**58**) in the ratios given in Scheme 8.

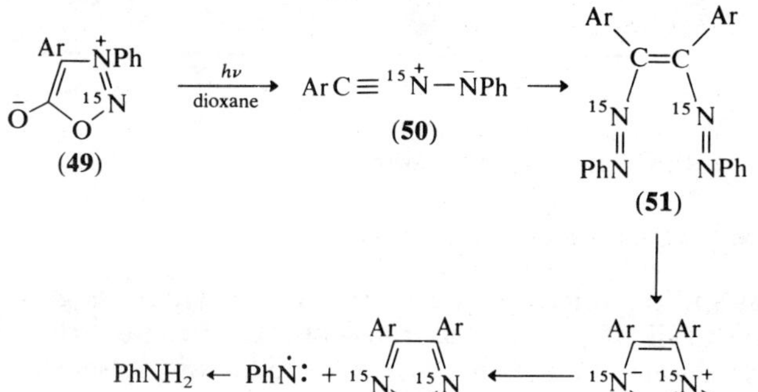

Scheme 7

The same mechanism was used to explain the isolation of approximately the same percentages of the three products (**60**), (**61**) and (**62**) from the photolysis of 3-cyclohexyl-4-phenylsydnone (**59**; Scheme 9) ⟨72BCJ3202⟩.

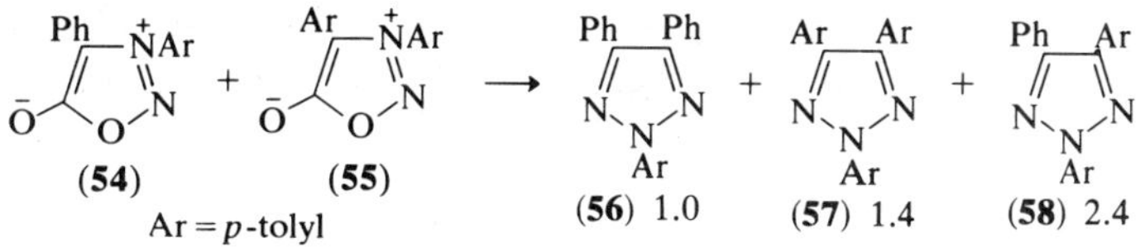

1,2,3- and 1,2,4-Oxadiazoles

Scheme 8

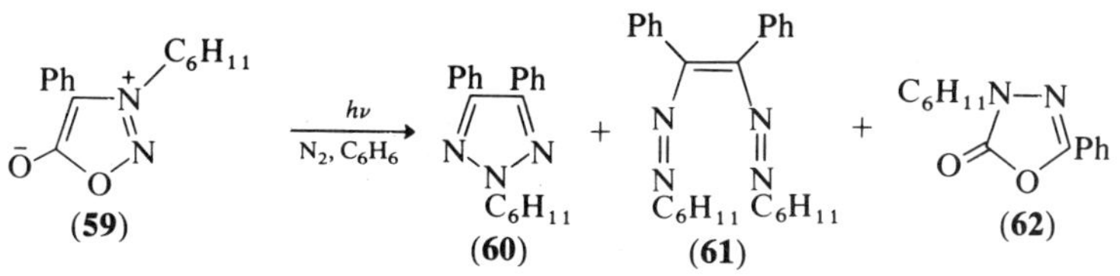

Scheme 9

4.21.2.5.2 Singlet oxygen

Singlet oxygen will also add to a sydnone (**63**) as a dipolarophile in the presence of Rose Bengal as a sensitizer. The products suggest that two pathways are followed simultaneously since benzoic acid (**64**) and dibenzoylphenylhydrazine (**65**) were identified among the reaction products (Scheme 10) ⟨79JOC2957⟩.

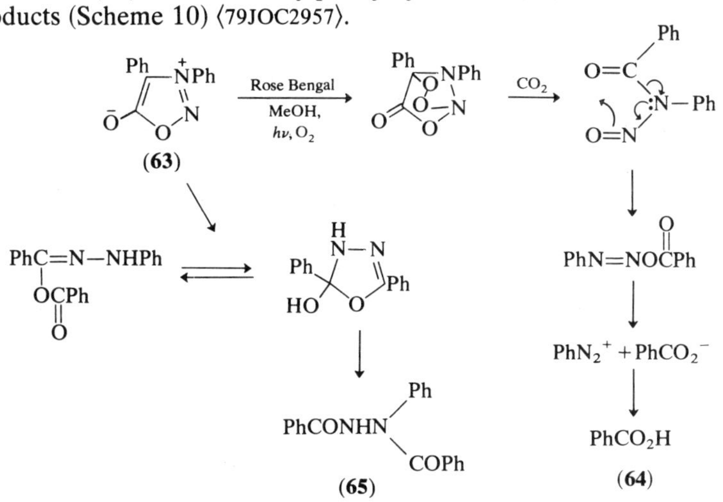

Scheme 10

4.21.2.5.3 Other photochemical reactions

Several sydnones have been discovered that develop color when irradiated ⟨77CCC811, 77-BCJ3268, 79JHC1059⟩ while some develop color only in solid glasses ⟨79AX(B)2256, 79AX(B)437⟩. The original observation ⟨55CI(L)119, 55JA6604⟩ concerned the development of a blue color in irradiated 3-(3-pyridyl)sydnone. The phenomenon has been interpreted by analogy to photochromism in aziridines but in sydnones the photochemical reaction is not reversible. Trozzolo's two-step process to explain the phenomenon (Scheme 11) involves an orbital symmetry-allowed conrotatory ring closure in (**66**) to give the bicyclic (**67**) which is colorless ⟨79PAC261⟩. As (**67**) warms, relief of strain by a reverse conrotatory thermal process is not possible because of steric constraint. Instead, relief occurs through ring opening with low

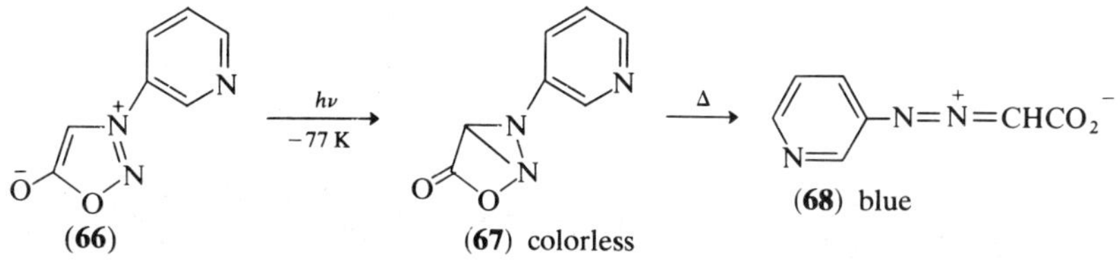

Scheme 11

activation energy to give the colored ylide (**68**). The argument is consistent with the photochemical reactions just described (Section 4.21.2.5.1) although the blue ylide has not been trapped.

4.21.2.5.4 Pyrazoles and pyrazolines from sydnones

The thermal addition–elimination of unsaturated compounds to sydnones was fully exploited after the original discovery as a synthetic route to pyrazoles and pyrazolines ⟨60ZOB698, 81H(16)35⟩. The rate of the thermal reaction was found to be greatest for alkynes in the order $MeO_2CC{\equiv}CCO_2Me > HC{\equiv}CCO_2Me > PhC{\equiv}CCOR > HC{\equiv}CCH_2OH > RC{\equiv}CH$ ⟨62AG(E)48, 68CB536, 77HCA1087, 79CB1193⟩. Alkenes added under more severe conditions but the intermediate pyrazolines often aromatized readily to pyrazoles ⟨62AG(E)49, 68CB552⟩. The kinetics of the reaction have been studied for numerous examples ⟨68CB1059⟩ and a wide range of dipolarophiles has been utilized: acrylonitrile ⟨68CB829⟩, alkenes ⟨68CB839⟩, TCNE ⟨79JOC2395⟩, bis-alkynes ⟨66JHC155⟩, bicyclic alkenes ⟨80JOC479, B-69MI42100⟩, benzyne ⟨68CB1056⟩ and trialkylborane ⟨80ZN(B)568⟩. Substituted sydnones also take part in thermal addition–eliminations; *e.g.* 3-aminosydnones ⟨76CPB3001⟩ and halo-substituted sydnones ⟨75CJC913, 74CB3036⟩. This cycloaddition route is often the preferred synthetic method for pyrazoles since the yields are high and the reaction is often regiospecific.

4.21.2.6 Reactions of Substituents

Various functional groups on the sydnone ring undergo reactions commonly associated with this functionality. For example, amides, esters and hydrazides can be made from 3-phenylsydonone-4-carboxylic acid. The tertiary alcohol (Table 6) can be dehydrated to alkenes with phosphoric anhydride and the ketone (Table 6) will condense with benzaldehyde ⟨73BCJ3304⟩. The aldehyde (**69**) can be converted with Wittig reagents into a mixture of *E* and *Z* alkenes (**70**) ⟨76JPR823, 78ZC262⟩. Arylimino derivatives of sydnone aldehydes are also attainable by azeotropic loss of water from the aldehyde and aniline ⟨78JPR206⟩.

A side chain at C-4 cannot be oxidized to the carboxylic acid ⟨B-69MI42100⟩.

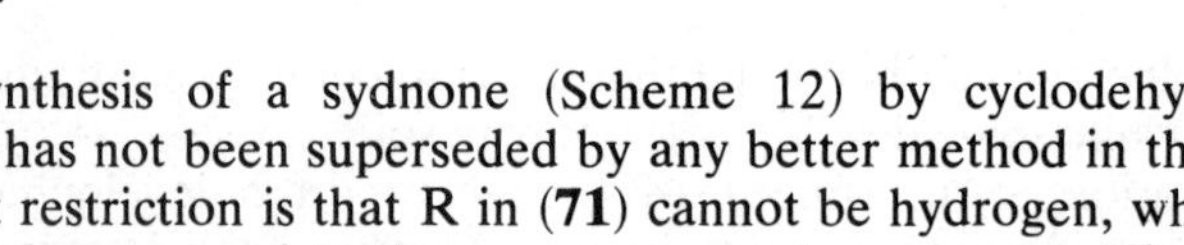

4.21.3 SYNTHESIS OF 1,2,3-OXADIAZOLES

4.21.3.1 Sydnones

The original synthesis of a sydnone (Scheme 12) by cyclodehydration of an *N*-nitrosoamino acid has not been superseded by any better method in the intervening time. The only apparent restriction is that R in (**71**) cannot be hydrogen, which means that the parent compound does not exist. However, no one has succeeded in placing a strongly negative group, *e.g.* picryl, at N-3.

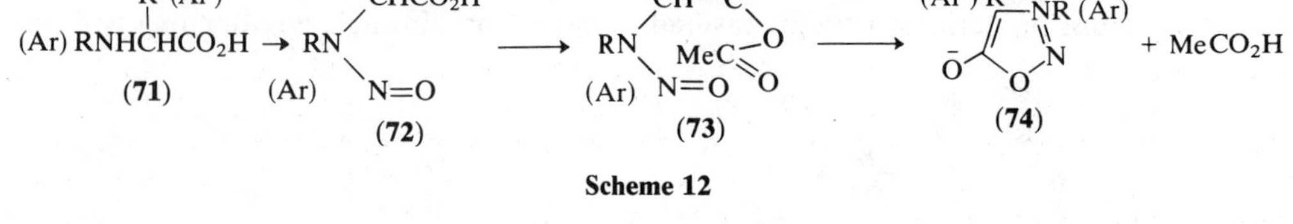

Scheme 12

In general there are three steps in the synthesis of a sydnone: (a) nitrosation of the amino acid (**71**), (b) mixed anhydride formation (**73**), and (c) cyclization to (**74**). The mixed anhydride is not generally isolated although it was shown early that it could be. In a few cases the method fails when the nitroso derivative cannot be made or when a different reaction occurs during the dehydration step. Among dehydrating agents trifluoroacetic anhydride gives the most rapid results, product formation often occurring in seconds ⟨73OSC(5)962⟩. Thionyl chloride, phosphorus oxychloride, phosphoric anhydride and carbodiimides have all been used successfully as dehydrating agents.

A summary of sydnones prepared before 1964 and their physical properties is given by Stewart ⟨64CRV129⟩.

4.21.3.2 Sydnonimines

Sydnonimines can be made by a similar, general method (Section 4.21.3.1) involving nitrosating the corresponding substituted aminonitrile (75) and cyclizing the intermediate (76). Substituents may be placed on the exocyclic nitrogen atom by ordinary methods in acidic or buffered solution ⟨62HCA2441⟩, for example, carbamates and urea derivatives. As might be expected, sydnonimines are more stable in acid and less stable in base than sydnones.

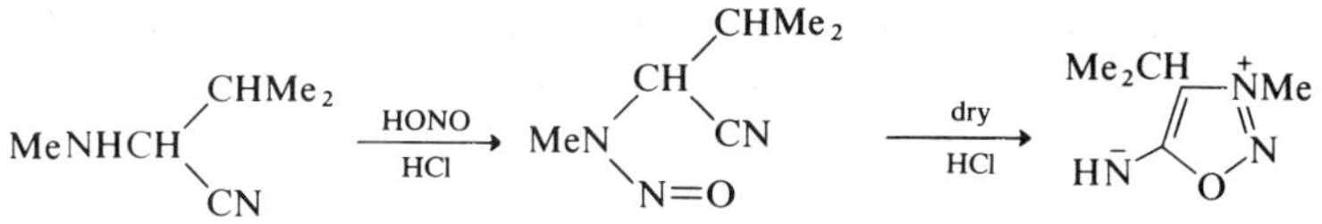

Sydnonimines and sydnones can be readily synthesized in a three-component reaction of the Mannich type (Scheme 13) ⟨71CCC2640⟩, for example to prepare 3-N-hydroxy- (78) or 3-N-amino- (79) substituted sydnonimines ⟨70CPB128, 70JHC123, 71T4449⟩.

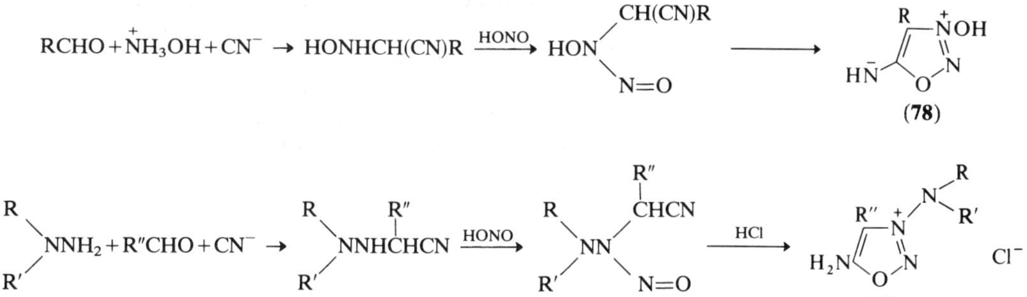

Scheme 13

Since the nitrogen atom in each of the three components in the Mannich sydnonimine synthesis can be labelled independently with ^{15}N, e.g. with $^{15}NH_2OH$, $Na^{15}NO_2$ and $KC^{15}N$, the NMR spectrum due to each ^{15}N can be determined unambiguously ⟨77T2571, 79JMR(36)227⟩.

4.21.3.3 1,2,3-Oxadiazolines

Besides the stabilization attained by the oxidation state represented by the sydnones, it was recently discovered ⟨77DIS(B)(38)1721⟩ that the oxidation state represented by an N-oxide (81) also lends stability. Thus far only one example is known, although other dienes related to (80) were evaluated as starting compounds (Scheme 14).

The diene (80) when reacted with nitrosyl chloride gave a compound (81), 4,4-dimethyl-5-(2-methylpropyl)-Δ^2-1,2,3-oxadiazoline 3-oxide. The structure of (81) was established by its chemical properties, IR and NMR spectra. The alkene linkage remaining in the molecule underwent additions of bromine, hydrogen chloride and hydrogen bromide without disrupting the ring and it was ozonized and oxidized to the carboxylic acid (82). However, deoxygenation with triphenylphosphine gave a mixture of (83; 72%) and a rearranged aldehyde (84; 28%). Reduction with LAH gave a rearranged primary alcohol (85; 86%) and hydrazine (65%). It is not clear why the double bond remaining in (81) does not react further with nitrosyl chloride.

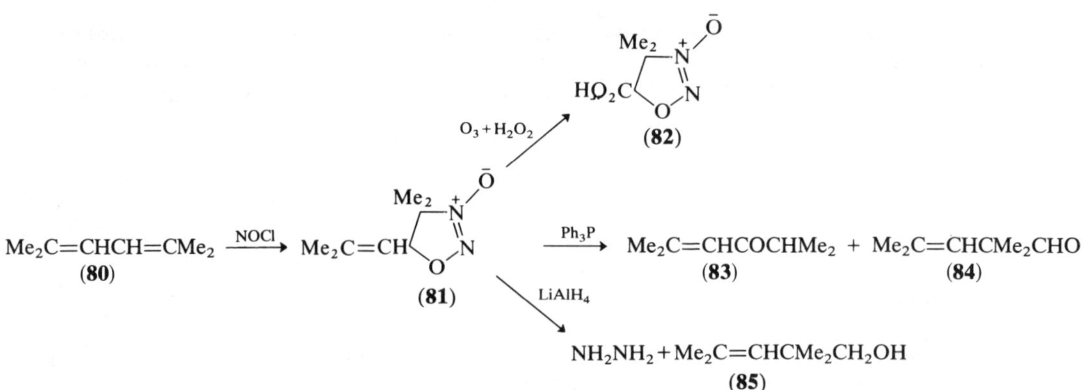

378   1,2,3- and 1,2,4-Oxadiazoles

Me₂C=CHCH=CMe₂ →NOCl→ Me₂C=CH–[ring (81)] →O₃+H₂O₂→ HO₂C–[ring (82)]

(81) →Ph₃P→ Me₂C=CHCOCHMe₂ + Me₂C=CHCMe₂CHO
                    (83)   (84)

(81) →LiAlH₄→ NH₂NH₂ + Me₂C=CHCMe₂CH₂OH
                              (85)

Scheme 14

4.21.4 APPLICATIONS

Among 1,2,3-oxadiazole derivatives two compounds stand out as having received the most attention in tests for biological activity. Both are sydnonimines: Molsidomine (**86**) and Sydnocarb (**87**).

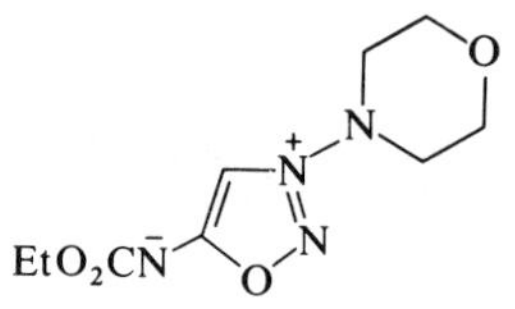

(**86**) Molsidomine (SLN-10, Morial, Corvaton Morsydomine)

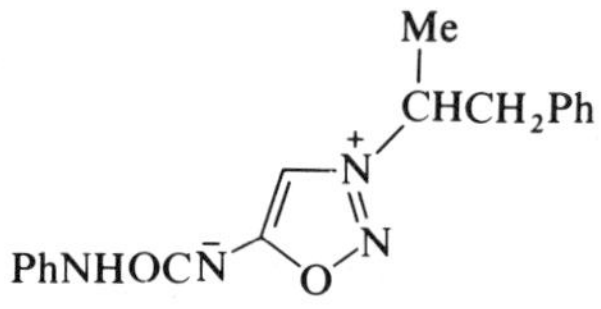

(**87**) Sydnocarb (Mesocarb)

Molsidomine has a long-term effect in vasodilation and so diminishes the work of the heart in cases of ischemic heart disease. The toxicity is very low. So much attention has been given to it since it was developed in 1967 by Takeda Chemical Industries in Japan that a symposium was held in Munich in 1978 devoted entirely to its properties and chemotherapy ⟨B-79MI42101⟩. Molsidomine acts like nitroglycerine in treating angina pectoris; onset is slower but it retains its activity for a much longer period ⟨80MI42108, 80MI42101⟩. It is currently being tested for use in the U.S.

Sydnocarb acts on the central nervous system and has been patented in many countries as a psychostimulant, an antidepressant and as a stimulant to motor activities ⟨80RCR28⟩. Sydnocarb has been tested extensively in the U.S.S.R. ⟨80MI42102-7⟩.

Many other sydnones and sydnonimines have been tested for antiinflammatory, antitumor, antibacterial, analgesic and antipyretic activity ⟨67JPS149, 79MI42100, 80MI42102⟩.

4.21.5 STRUCTURE OF 1,2,4-OXADIAZOLES

4.21.5.1 Theoretical Methods: Dipole Moments and Electron Densities

The structure of 1,2,4-oxadiazoles has not been in dispute since covalent structures can be drawn from the methods of synthesis. The geometry of the ring was estimated for theoretical calculations years before X-ray measurements were available. Microwave spectra and recently fluorescence spectra have been used to determine the dipole moments of oxadiazoles and oxadiazolines, the latter being high by comparison (Table 7). Electron densities were calculated for 5-methyl-3-phenyl-1,2,4-oxadiazole (**88**) and its 2,3-dihydro derivative (**89**) ⟨77JOC1555⟩.

4.21.5.2 Molecular Dimensions: X-Ray Diffraction

The oxadiazole ring has little aromatic character. Both C—N bond distances suggest conjugated double bond character in the only oxadiazole (**90**) that appears to have been subjected to X-ray analysis ⟨79AX(B)2256⟩. The C—N bond lengths in the aminopyridine

Table 7 Dipole Moments of Oxadiazoles and Oxadiazolines

1,2,4-Oxadiazole	Dipole moment (D)	Method	Ref.
Unsubstituted	1.2 ± 0.3	Microwave spectra, Stark effect	76ACH(20)65
	1.34	CNDO/2	76AHC(20)65
3,5-Diphenyl	1.56	Measured in dioxane	76AHC(20)65
3-Methyl-5-phenyl	1.63	Measured in dioxane	76AHC(20)65
5-Ethyl-3-phenyl	1.55	Fluorescence spectra	77JOC1555
4,5-Dihydro-5-methyl-3-phenyl	4.90	Fluorescence spectra	77JOC1555

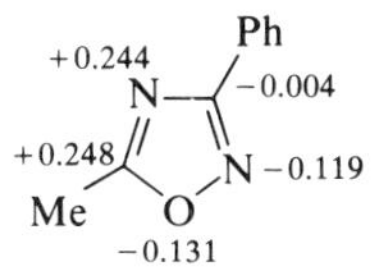

moiety (**90**), in contrast, were normal aromatic pyridine bonds (1.333 and 1.348 Å). The bond angles in this same molecule (**91**) are not those of a regular pentagon.

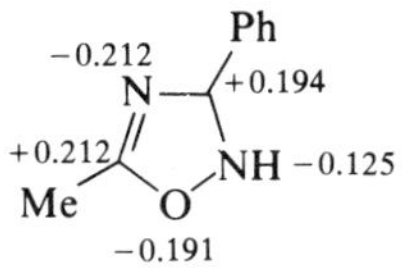

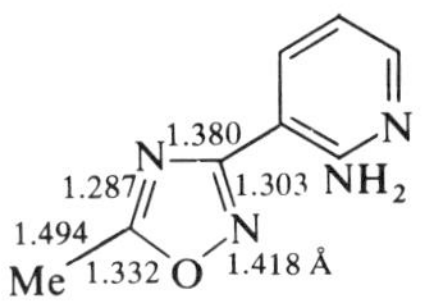

4.21.5.3 Molecular Spectra

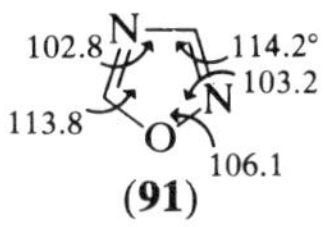

4.21.5.3.1 UV spectra

UV spectra also suggest that the oxadiazole ring is better described as a conjugated system rather than as an aromatic system. In the 3-phenyl derivative λ_{max} was 233 nm and in the 5-phenyl analog it was 250 nm, but the 3,5-diphenyl derivative absorbed at a maximum between the two (245 nm) ⟨76AHC(20)65⟩.

4.21.5.3.2 IR spectra

Enough 1,2,4-oxadiazoles and their reduced derivatives have been made that there is general agreement on where to expect the C=N absorption in the IR spectrum, *i.e.* 1560–1590 cm^{-1}. In the partially reduced (oxadiazoline) ring the Δ^2 double bond appears at 1550–1565 cm^{-1} and the NH bands at 3320 and 3220–3250 cm^{-1}. In an amino derivative, 5-amino-3-phenyl-1,2,4-oxadiazole, the C=N bond absorbs at higher frequency (1660 cm^{-1}) ⟨63PMH(2)229⟩.

4.21.5.3.3 NMR spectra (^{1}H)

Protons on the oxadiazole ring are shifted downfield with respect to protons in benzene. At C-3 the chemical shift is δ 8.0–8.2 and at C-5 it is δ 8.3–8.7 p.p.m.

Protons in the aromatic ring at C-3 or C-5 are not coupled with the remaining hydrogen at the conjugate position in the oxadiazole ring. NMR spectra were used to distinguish between Δ^2- and Δ^3-oxadiazolines ⟨76AHC(20)65⟩.

4.21.5.3.4 Mass spectra

In oxadiazoles, 1,3-retroaddition (Section 4.21.7.2.1) of the nitrile oxide and nitrile (split *a* + *b* in **92**; Scheme 15) does not compete well with the break at *c*, the weakest bond, followed by the break at *d*, except when X is NH$_2$ ⟨76AC(R)57, 77AC(R)371, 77AC(R)621, 80OMS573⟩.

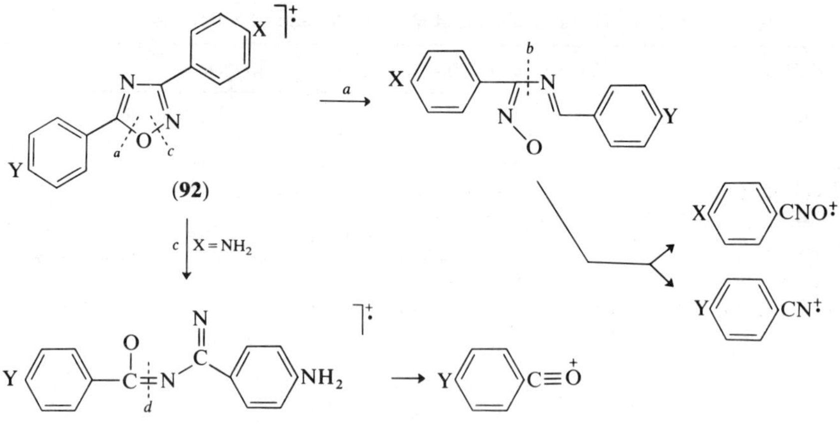

Scheme 15

Oxadiazolines (Scheme 16) show a more complex fragmentation pattern, although as expected the cleavage involves the weak N—O bond (patterns *a* and *c* in **93**) more frequently than the C—O bond (pattern *b*). However, the 4,5-dihydro derivative (**94**) loses hydrogen concurrently with an ethyl group in accord with the sequences (**94**) → (**95**) and (**94**) → (**96**) ⟨78OMS14⟩.

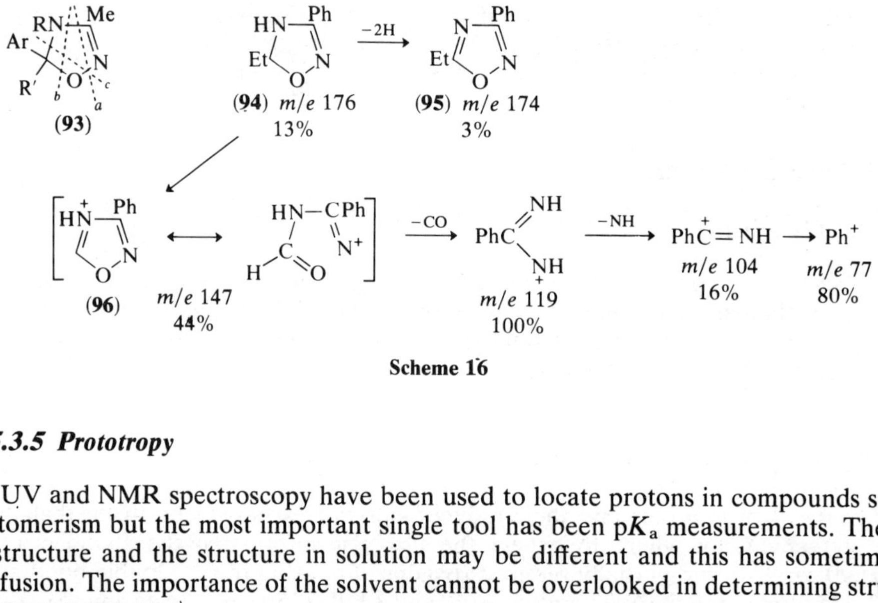

Scheme 16

4.21.5.3.5 Prototropy

IR, UV and NMR spectroscopy have been used to locate protons in compounds subject to tautomerism but the most important single tool has been pK_a measurements. The solid state structure and the structure in solution may be different and this has sometimes led to confusion. The importance of the solvent cannot be overlooked in determining structure ⟨76AHC(20)65, 80NJC527⟩.

In 5-hydroxy-3-phenyl-1,2,4-oxadiazole (**97a**) the keto forms (**97b, c**) predominate due to the proximity of the electronegative oxygens at the sp^2 carbon. Of these three tautomers (**97c**) is favored by IR, UV and NMR spectral data ⟨76BSB35⟩. The tautomer (**98b**) is more important in the 5-phenyl isomer in *solution*. Nevertheless the position of tautomeric equilibrium in the two cases is dependent on solvent. For example, in the solid (**98b, c**) predominates but in acetone and other oxygen solvents (**98a**) allows for an effective hydrogen bonded dimer (**99**).

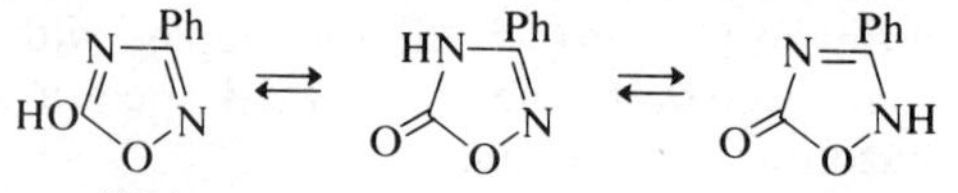

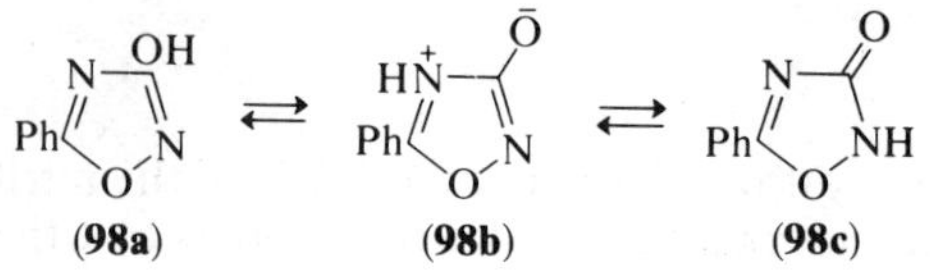

(**99a**) (**99b**)

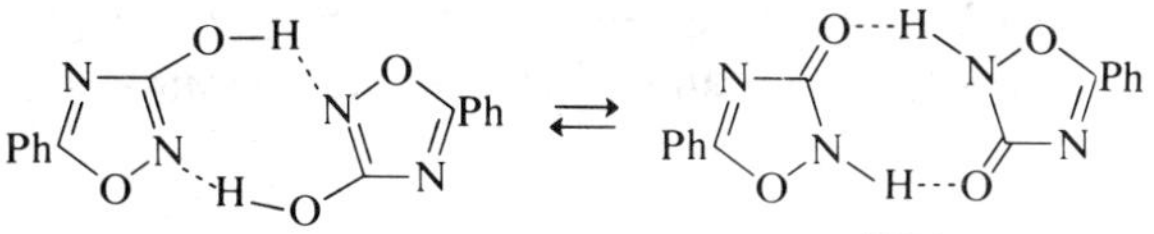

The silver salt of (**100**) gives a 1:1 mixture of *N*- and *O*-ethyl derivatives (**101**) and (**102**) on reaction with ethyl iodide. The more stable *N*-ethyl derivative is obtained in excess when the reaction mixture is allowed to stand. Diazomethane also gives a mixture of *O*- and *N*-methyl derivatives. Both N-4 (**104**) and N-2 (**105**) derivatives have been positively identified only in a synthesis from 5-methyl-3-phenyl-Δ^2-1,2,4-oxadiazoline (**103**).

(**100**) (**101**) (**102**)

(**103**) (**104**) (**105**)

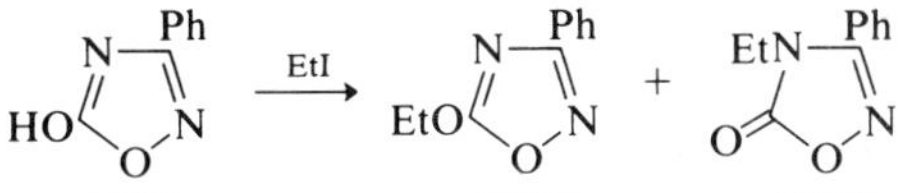

An oxadiazolium inner salt was found to be an intermediate in the rearrangement of a nitrone (**106**) to an amide (**108**). The red salt (**107**) was isolated. In accord with the mechanism as shown, only when the nitrone function was substituted at C-3 did the rearrangement occur ⟨79JHC1477⟩.

(**106**) (**107**) (**108**)

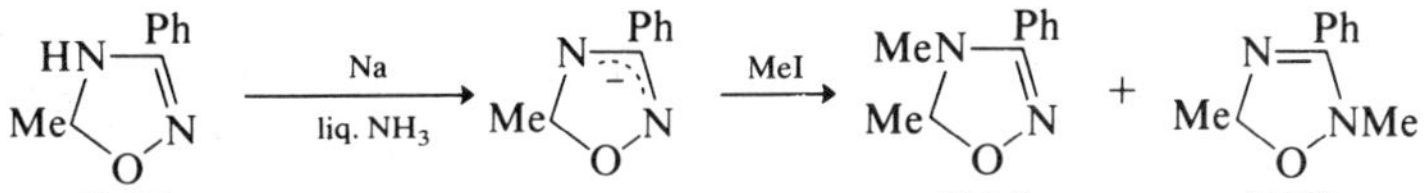

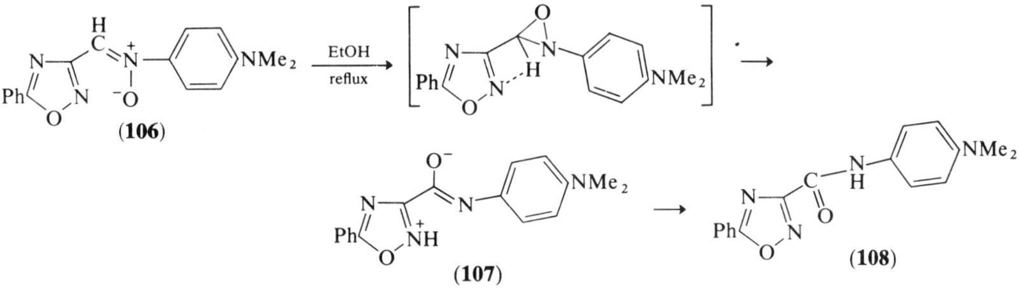

In aminooxadiazole derivatives the tautomeric imine form is less significant since (**109**) is more basic. In the corresponding sulfur analog to (**110**) there is only evidence for the thione form (**111**) with the hydrogen at N-2. Some recently synthesized oxadiazoles with aldehyde functions at C-3 or C-5 were unexpectedly found to be stabilized in the hydrated form (**112**) ⟨79JHC1469⟩. This suggests that the oxadiazole ring is a strong electron-withdrawing group (see Section 4.21.6.3).

(**109**) (**110**) (**111**) (**112**)

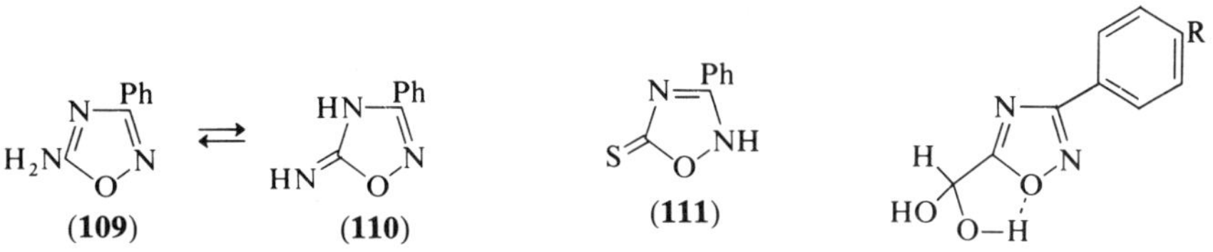

4.21.5.4 Thermodynamic Aspects

4.21.5.4.1 Intermolecular forces

The physical properties of 1,2,4-oxadiazoles are unexceptional. The boiling point of the parent compound is 87 °C, within the the normal range of other two-carbon compounds. The heats of vaporization of the first three homologs (39.5–42 kJ mol^{-1}) are comparable to that of ethanol. Thermodynamic functions have been calculated over the range

4.21.5.4.2 Stability of the ring

3,5-Diphenyl-1,2,4-oxadiazole can be heated in concentrated sulfuric acid or recrystallized from fuming nitric acid and appears to decompose only slowly at 250 °C. This inertness disappears when there is a hydrogen atom at C-3 or C-5 (see also Sections 4.21.6.1–4.21.6.6 and 4.21.5.3.5) ⟨1884CB1685⟩.

4.21.6 REACTIVITY OF 1,2,4-OXADIAZOLES

4.21.6.1 General Survey of Reactivity

The stability of 3,5-diphenyl-1,2,4-oxadiazole (Section 4.21.5.4.2) contrasts with that of the 3-methyl derivative (**113**) which hydrolyzes in acid to acetamidoxime. In basic solution, acetonitrile and ammonia are formed. The ring is stable to mild reducing agents (Section 4.21.6.6).

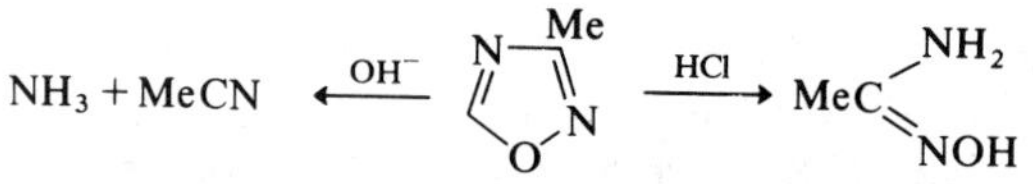

$$NH_3 + MeCN \xleftarrow{OH^-} \text{(113)} \xrightarrow{HCl} MeC(NH_2)=NOH$$

(**113**)

4.21.6.2 Thermal and Photochemical Decomposition

Although many oxadiazoles can be distilled (<300 °C) without noticeable decomposition, an open flame will cause slow decomposition. For example, compound (**114**) gave *p*-chlorophenyl isocyanate and *p*-methoxybenzonitrile as products.

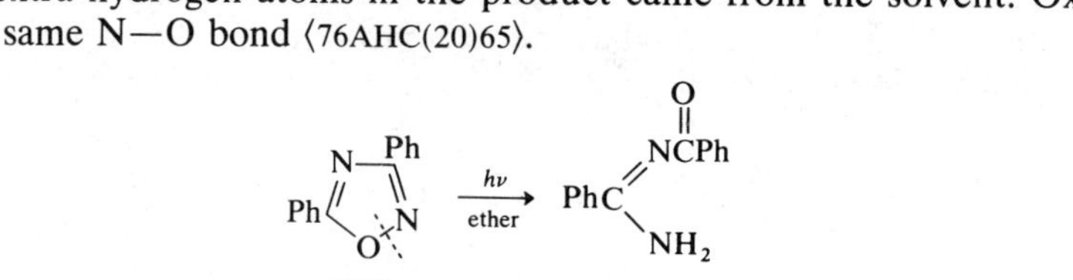

(**114**) $\xrightarrow{250\,°C}$ MeO–C$_6$H$_4$–CN + Cl–C$_6$H$_4$–NCO

In contrast, photochemical decomposition of (**115**) broke the ring at the weakest bond, N—O. The extra hydrogen atoms in the product came from the solvent. Oxadiazolidines break at the same N—O bond ⟨76AHC(20)65⟩.

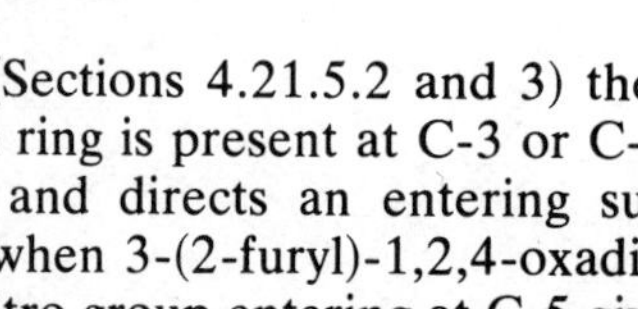

(**115**) $\xrightarrow[\text{ether}]{h\nu}$ PhC(NH$_2$)=NC(=O)Ph

4.21.6.3 Electrophilic Substitution

As mentioned previously (Sections 4.21.5.2 and 3) the oxadiazole ring lacks aromatic character. When an aromatic ring is present at C-3 or C-5, the oxadiazole ring acts as an electron-withdrawing group and directs an entering substituent to the *meta* position ⟨76AHC(20)65⟩. For example, when 3-(2-furyl)-1,2,4-oxadiazole (**116**) is nitrated the furan ring is substituted, with the nitro group entering at C-5 giving (**117**) ⟨75MI42100⟩.

Mercury(II) chloride results in introduction of the —HgCl group at C-5 in 3-phenyl-1,2,4-oxadiazole and the group can be displaced by Cl with chlorine gas. Other electrophilic substitutions are not successful.

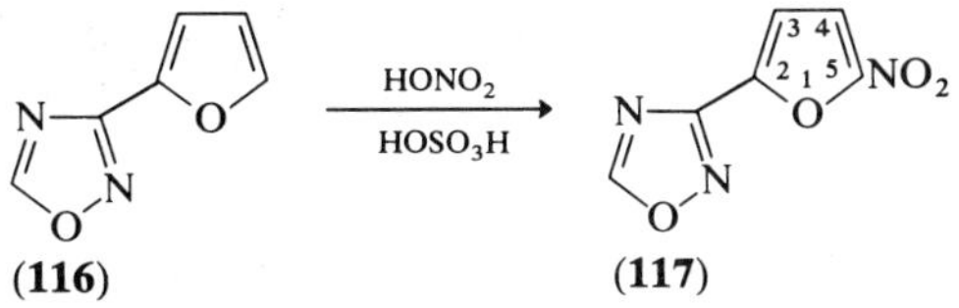

4.21.6.4 Nucleophilic Substitution

Nucleophilic displacements that have been carried out on oxadiazoles are summarized in Table 8. Halogen at C-5 may be displaced by alkoxy, hydroxy or amino groups. The ethoxy group at C-5 may be replaced by hydroxy or amino groups. These reactions are characteristic of those occurring at an aliphatic sp^2 carbon atom and not aromatic nucleophilic displacements. A difference in lability of groups at C-3 and C-5 is shown by the CCl_3 group at these two positions (Table 8).

By deuterium exchange in basic solution the relative ease of displacement of hydrogen at C-5 and of hydrogen in the methyl group at C-5 has been determined (Table 8).

Table 8 Nucleophilic Displacements in 1,2,4-Oxadiazoles

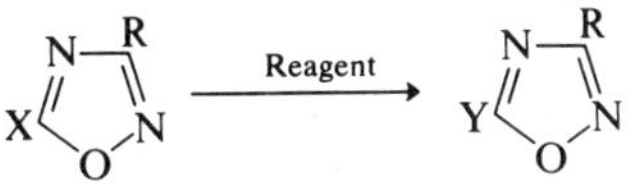

Substrate			Product
R	X	Reagent	Y
Me	Cl	OH⁻	OH
Me	Cl (OEt)	RNH₂	NHR
Me	OH	Ag⁺	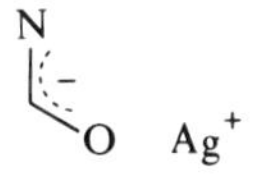 N⁻...O Ag⁺
Me	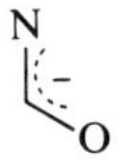 N⁻...O	EtI	OEt + NEt (1:1)
Ph	CCl₃	(H₂N)₂C=NH	NHC(=NH)NH₂
Ph	CCl₃	OH⁻ (alc.)	OH
CCl₃	Me	OH⁻ (alc.)	Me (CO₂⁻ at C-3)
Ph	H	OD⁻	D (rel. rate, 80)
Ph	Me	OD⁻	CH₂D (rel. rate, 1)
Me	Me	OD⁻	CH₂D (rel. rate, 20)

4.21.6.5 Reactions of Substituents

The greater reactivity of the methyl group at C-5 compared to that at C-3 (Table 8) is due to the stabilization by mesomerism in the anion (**118**) compared to (**119**).

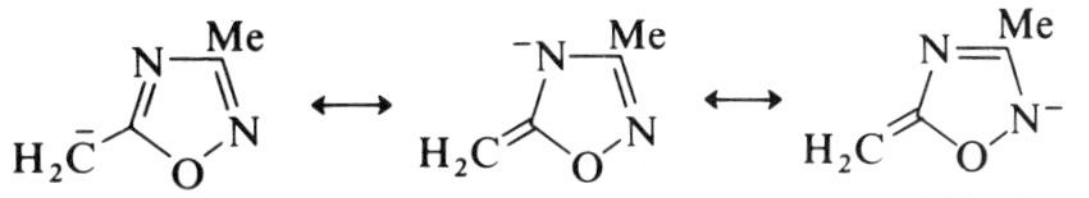

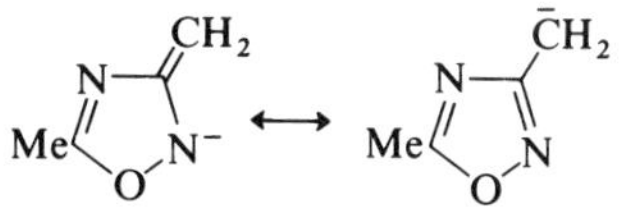

Reactivity of the C-5 methyl group in (**120**) is exhibited in aldol-type condensations, e.g. benzaldehyde with an acid catalyst to give (**121**) or ethyl oxalate with a strong base to give (**122**). The greater reactivity of a methyl group at C-5 is also shown in the results of lithiation of (**123**) and (**125**), followed by treatment with carbon dioxide and water. However, compound (**124**) loses carbon dioxide upon heating to revert to (**123**), analogous to the behavior of malonic acids.

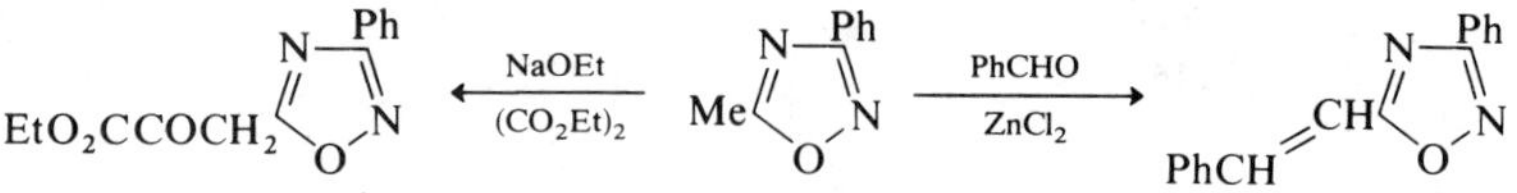

An ester group at C-3 or C-5 is not different in function from any ester in aliphatic or aromatic compounds. The ester can be converted into an amide or a hydrazide, or it may be hydrolyzed. The hydrazide can be converted into an acid azide and rearranged to an isocyanate which in turn will form a carbamate or can be hydrolyzed to an amine. However, the amine group can be replaced with chlorine by diazotization in hydrochloric acid, not an ordinary pattern of behavior ⟨76AHC(20)65⟩ although similar reactions do occur in the 1,2,4-triazole ring system.

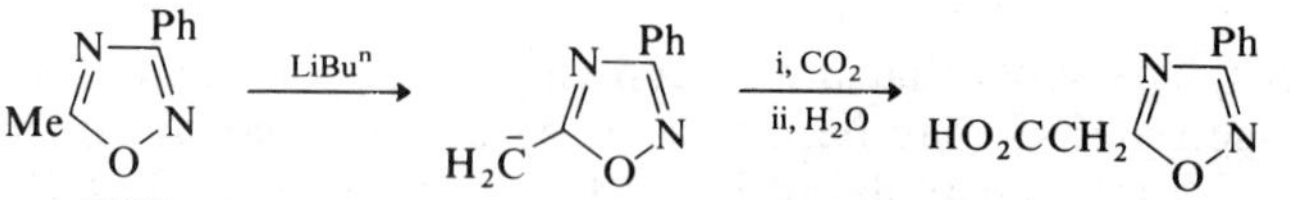

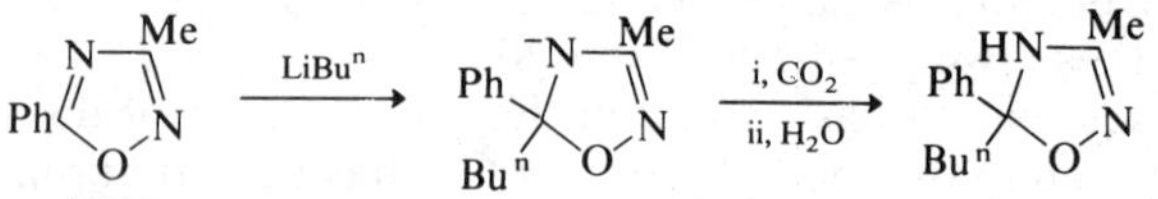

4.21.6.6 Reduction

Reagents vigorous enough to alter double bonds in oxadiazoles or oxadiazolines always break the ring with the notable exceptions of diborane and sodium borohydride. Diborane in THF reduces the ring (**126**) at Δ^4 giving the Δ^2-oxadiazoline ⟨78JHC1373⟩. Sodium borohydride likewise does not break the ring and has been used to reduce a carbonyl group adjacent to the ring to give alcohols used as CNS depressants ⟨76FES393⟩. Neutral catalytic reduction, e.g. with PtO_2, Raney Ni or Pd/C, always appears to break the N—O bond in either oxadiazoles (**127**) or Δ^4-oxadiazolines. LAH, a strong base in contrast, breaks the C—O bond in (**128**) to give substituted amidoximes ⟨76AHC(20)65⟩.

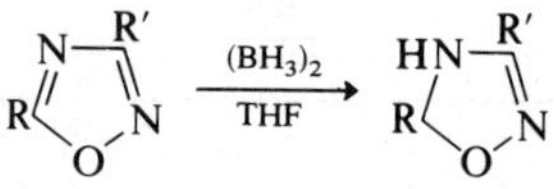

In the catalytic hydrogenolysis of an oxadiazole (**129**) bearing an o-nitrophenyl (or amino) group at C-5 the ring was opened and then closed again to give a quinazolone (**130**) ⟨76CPB1197, 77H(6)107⟩. However, the appropriately substituted quinazolone (**131**) in acid solution hydrolyzed back to the oxadiazole (**132**) ⟨79H(12)239⟩.

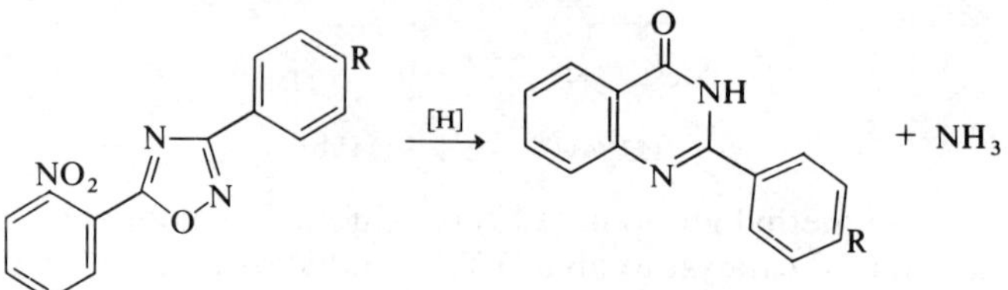

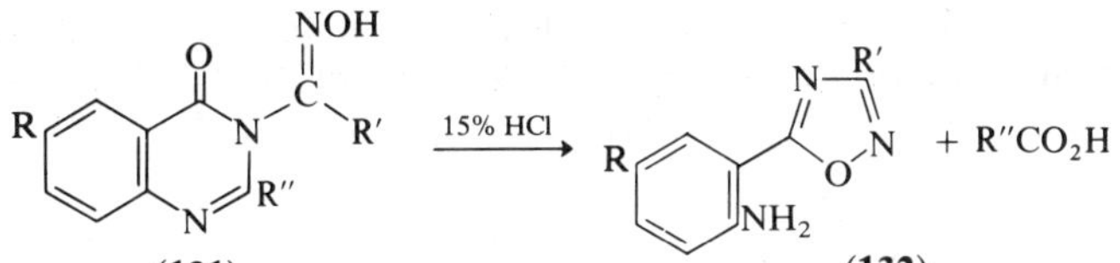

4.21.6.7 Rearrangements

The generalized formulation of rearrangements ⟨81AHC(29)141⟩ in heterocyclic compounds has been found to have some restrictions, mainly from the extensive work of Italian chemists. In (**133**) D must be more electronegative than N, oxygen in presently known examples. It follows that D—N is a single bond. The second restriction is that Z must represent a good nucleophile but polar, aprotic solvents help. Z can be O, S, Se, C or N (Table 9) as the bonding A=B—D in the five-membered ring of (**133**) gives way to the new bonding X=Y—Z in (**134**) where C=N is common to both rings. The rearrangements are either base-catalyzed or occur at the melting point of the solid (**133**).

Table 9 Rearranged Products of 1,2,4-Oxadiazole Rings
⟨76AHC(20)65, 75JHC985⟩

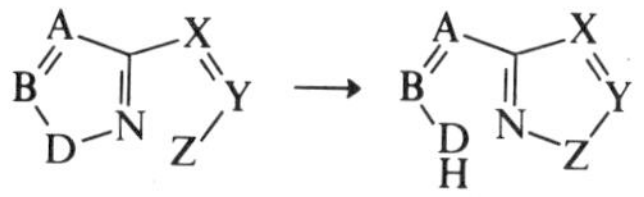

XYZ	XYZ
NCC imidazole	CNN 1,2,3-triazole
CNO 1,2,5-oxadiazole	NCN 1,2,4-triazole
NCS 1,2,4-thiadiazole	CCO benzisoxazole
NCSe 1,2,4-selenadiazole	

Recently it was shown that the *E* isomer (**135**) must first rearrange to the *Z* isomer (**136**) before the second rearrangement of the 1,2,4-oxadiazole to the 1,2,3-triazole (**137**) could occur. Where an internal nucleophilic displacement of N on N occurred, both *p*-methoxy and *m*-nitro groups on Ar in (**138**) were found to speed up the rearrangement to (**140**) in comparison with phenyl itself in buffered solvents. This means a change of mechanism. In the first case, the methoxy group on Ar helps the S_N1-type transition state (N—N bond forming in **139**) more than H (in phenyl) assists at low pS^+ values (~3.8–6). In the second case the *m*-NO_2 group helps the ionization (N—O bond breaking in **139**) more than H in phenyl assists this part of the mechanism at higher pS^+ values (6.5–8.0). At constant pS^+ of 3.8, the Hammett ρ-value was −1.33 for 11 compounds in this rearrangement where Ar was phenyl ⟨81AHC(29)141, 81JHC723⟩.

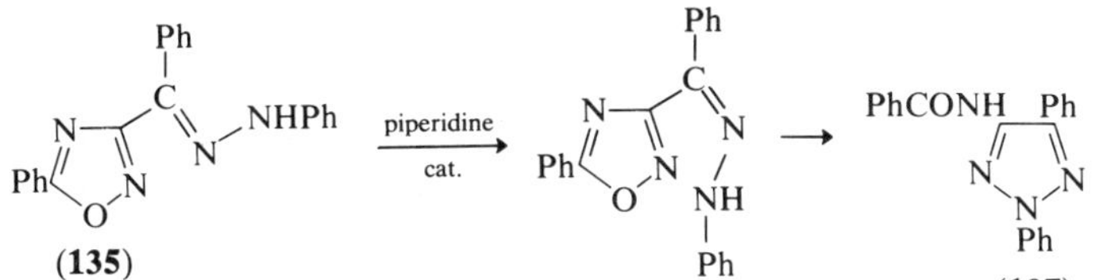

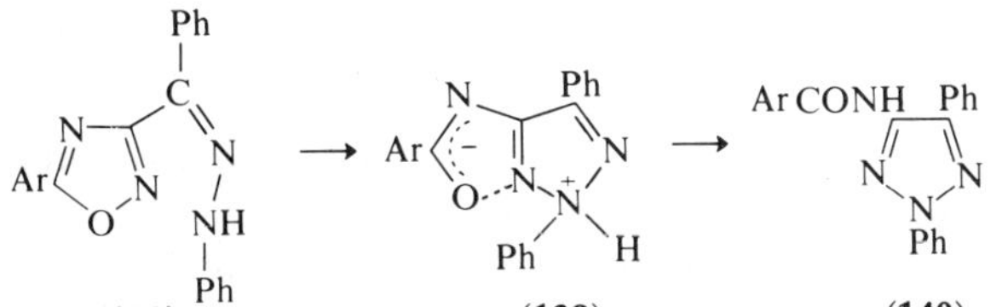

An equilibrium between two oxadiazoles may exist in certain cases if the substituents are of the right character. For example, the equilibrium between (**141**) and (**142**) was established very rapidly in DMSO but slowly in less polar solvents.

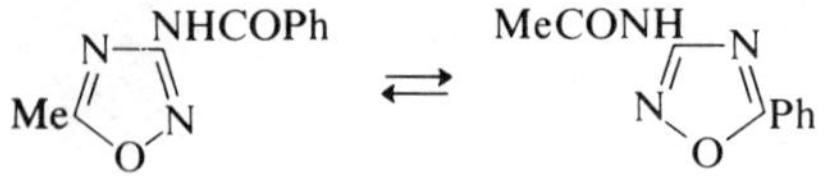

(**141**) ⇌ (**142**)

The enamine (**143**) was rearranged in DMF with sodium ethoxide to an aminoimidazole (**145**) through the intermediate (**144**). Caution must be observed in using the generalized rearrangement as a guide since other rearrangements may interpose. For example, the closely related enamine (**146**) rearranged to a mixture of the expected imidazole (**147**) and the pyrazole (**148**), but the latter predominated in the ratio 3:1 ⟨78JA4208⟩. The explanation involved a postulated diazirine as the source of the pyrazole (**148**). With a phenylhydrazino group at C-5 a rearrangement occurs in which a 1,2,4-oxadiazol-5-one is formed ⟨81JCS(P1)1703⟩. A rearrangement not fitting the generalized pattern is that of an oxime of a 3-phenacyl-1,2,4-oxadiazole to an isoxazole ⟨81T1415⟩.

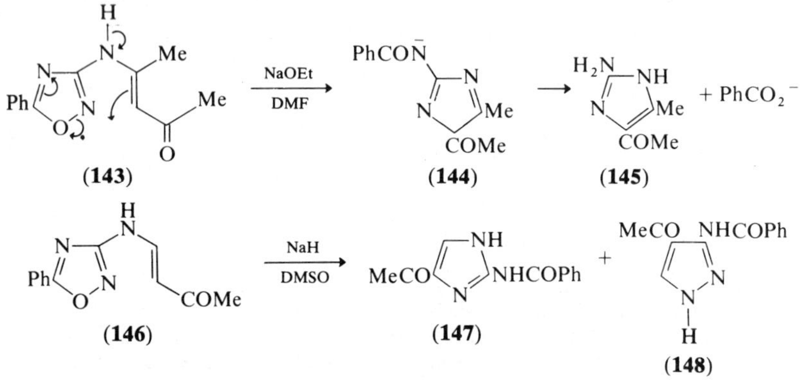

(**143**) →[NaOEt, DMF] (**144**) → (**145**) + $PhCO_2^-$

(**146**) →[NaH, DMSO] (**147**) + (**148**)

4.21.7 SYNTHESIS OF 1,2,4-OXADIAZOLES

Two general methods of synthesizing 1,2,4-oxadiazoles can be pictured as joining of the skeletons shown in Scheme 17 ⟨76AHC(20)65⟩. The fragments making up the moieties (**149a**), (**149b**), (**150c**) and (**150d**) can be disguised in a variety of ways. The carbon C-5 in (**149b**) can be in oxidation state +4, +3 or +2 to give oxadiazoles or the reduced ring oxadiazolines. The sp^2 carbon at C-3 is commonly in an amidoxime or an iminoether. The fragments (**150c**) and (**150d**) may be triple bonds as shown, but double bonds in either moiety give Δ^2- or Δ^4-oxadiazolines, or oxadiazolidine derivatives. These two methods of ring formation comprise at least 95% of the successful syntheses of 1,2,4-oxadiaza compounds.

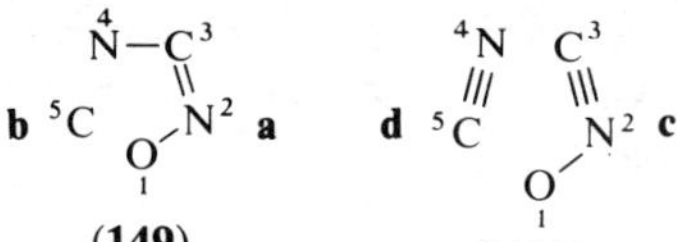

(**149**)   (**150**)

Scheme 17

4.21.7.1 Ring Closure on Carbon

4.21.7.1.1 1,2,4-Oxadiazoles (from amidoximes)

An amidoxime (**151**) is a good starting point for a suitable reagent containing an sp^2 carbon (C-3) with two nitrogen atoms attached. The carbon to end up as C-5 in the +3 oxidation state is furnished by an anhydride (**152**). The O-acyl derivative (**153**) first formed is ordinarily not isolated. A kinetic study suggests that the rate-determining step is the proton transfer (**153a**) → (**153b**) following cyclization ⟨80JCS(P2)1792⟩. Occasionally mistakes in structure have been made by assuming that the anhydride carbon atom always takes the C-5 position as opposed to the C-3 position ⟨81JHC37⟩.

(152) (151) (153) (153a) (153b)

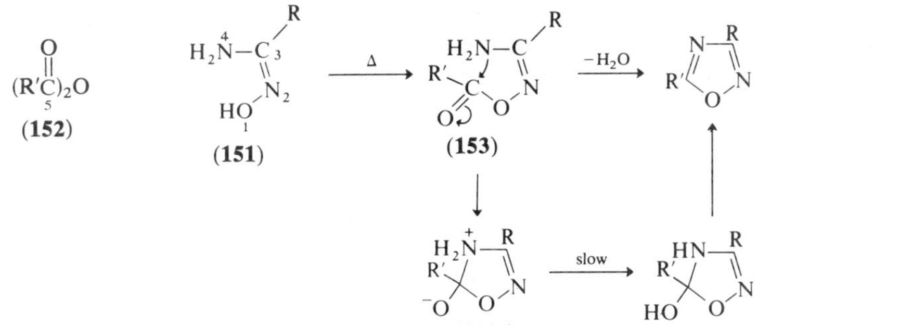

Amidoximes (**155**) may be made by heating a nitrile (**154**) with hydroxylamine in acid solution. The sulfone function does not interfere with subsequent ring formation ⟨79JHC1197⟩. Amidoximes (**158**) may also be of an aldoxime (**156**) followed by ammonolysis of the oximinochloride (**157**).

(154) (155) (156) (157) (158)

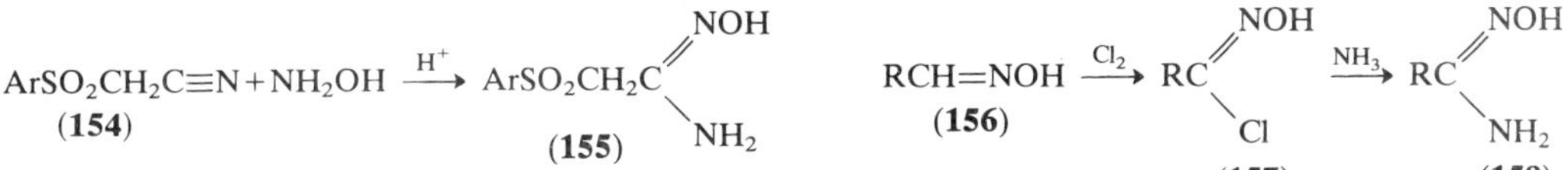

The source of the C-5 carbon in (**149b**) can also be acid chlorides, ketenes, esters, ortho esters and amides. Trifluoroacetic anhydride is particularly efficient in the dehydration step, often acting in minutes.

Synthesis of oxadiazole (**159**) by heating an amide with an amidoxime salt is practical since separation and recovery are simple. A successful innovation of this method, with yields of oxadiazoles (**161**) of 81–95%, is the disguise of the (**149a**) fragment in an *N,N*-dimethylalkanamide dimethyl acetal. The available R groups in (**160**) at present are methyl and hydrogen ⟨79JOC4160⟩.

(159)

(160) (161)

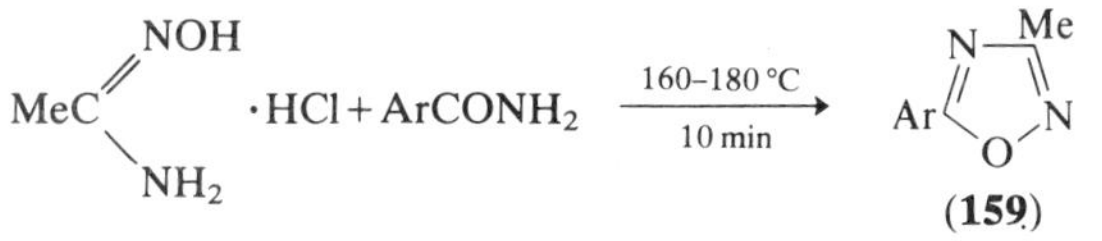

Still another variation of this method is the reaction of a mixed imide with hydroxylamine in which the R group takes the C-3 position as in (**162**) ⟨81JHC1197⟩.

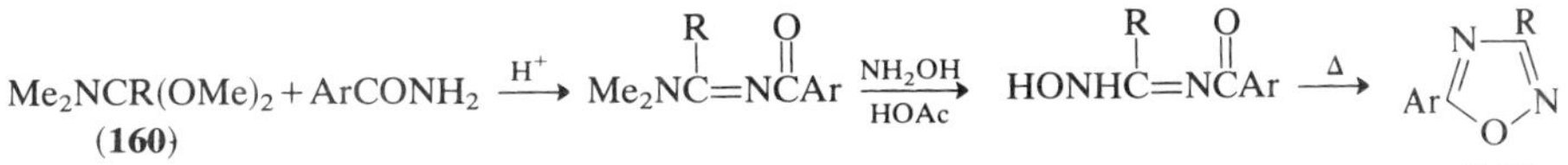

The oximino group in (**149**) can be introduced through nitrosation on carbon if there is an active hydrogen at that carbon atom. For example, nitrosation of the acylamino half ester of malonic acid gives an intermediate (**163**) that can isomerize to an oxime (**164**) and then cyclize to a 3-ethoxycarbonyloxadiazole (**165**).

(162)

(163) (164) (165)

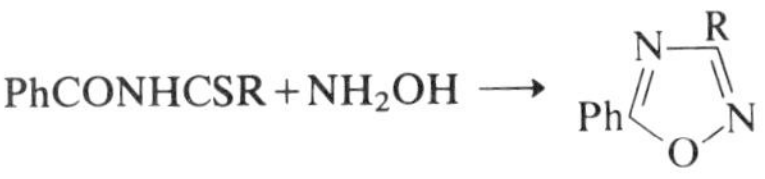

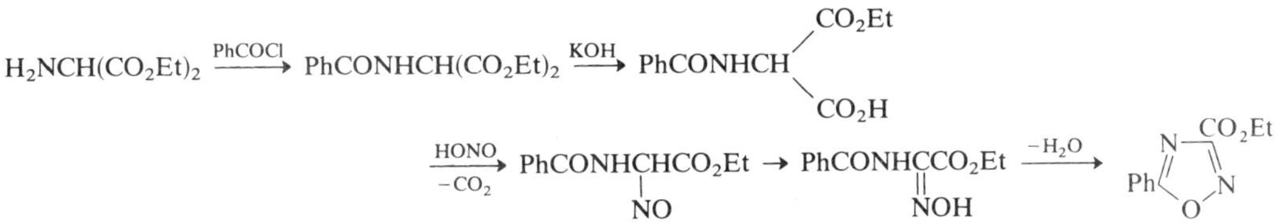

When the (**149b**) fragment is a carbonic acid derivative (**166**), *i.e.* of oxidation state +4, a hydroxy or an amino group may be introduced at C-5. If chloroethyl formate is used with an amidoxime, a 5-hydroxyoxadiazole (**167**) is obtained.

A 3-hydroxyoxadiazole (**168**) was obtained by hydrolysis of the product obtained by adding trimethylsilylazide to an acyl isocyanate.

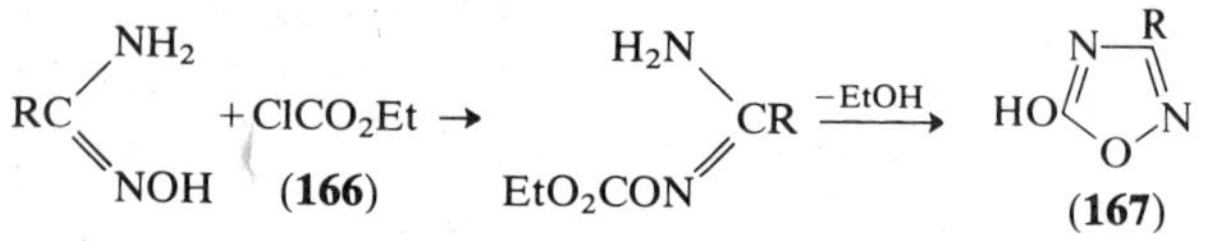

$$RC(NH_2)=NOH + ClCO_2Et \rightarrow EtO_2CON=CR(NH_2) \xrightarrow{-EtOH} \text{HO-oxadiazole-R (167)}$$

$$ArCNCO + Me_3SiN_3 \longrightarrow \text{Ar-oxadiazole-OSiMe}_3 \xrightarrow{H_2O} \text{Ar-oxadiazole-OH (168)}$$

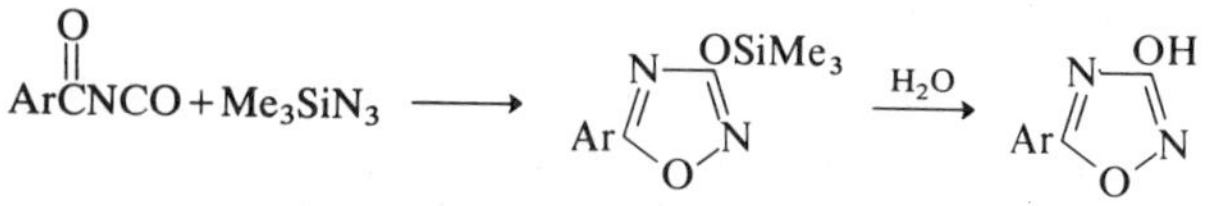

Cyanogen bromide on the amidoxime (**169**) gives the 5-amino derivative (**170**) ⟨75ZC57⟩. 5-Amino-3-methylthiooxadiazole (**172**) ⟨75JHC37⟩ and 3,5-diamino derivatives (**173**) and (**174**) ⟨80HCA832⟩ have been synthesized from the cyanoimino compound (**171**). When the amine was *p*-nitroaniline, (**173**) was obtained in 68% yield and with *t*-butylamine (**174**) was formed in 52% yield.

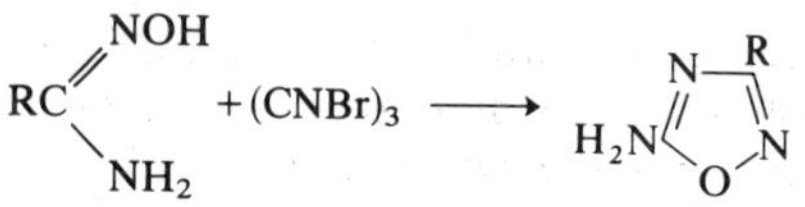

$$RC(NOH)(NH_2) + (CNBr)_3 \longrightarrow H_2N\text{-oxadiazole-}R \text{ (170)}$$

$$\text{(172) } H_2N\text{-oxadiazole-SMe} \xleftarrow{NH_2OH} NCN=C(SMe)_2 \text{ (171)} \xrightarrow{RNH_2} NCN=C(SMe)(NHR) \xrightarrow{NH_2OH} H_2N\text{-oxadiazole-NHR (173)} \quad RNH\text{-oxadiazole-NH}_2 \text{ (174)}$$

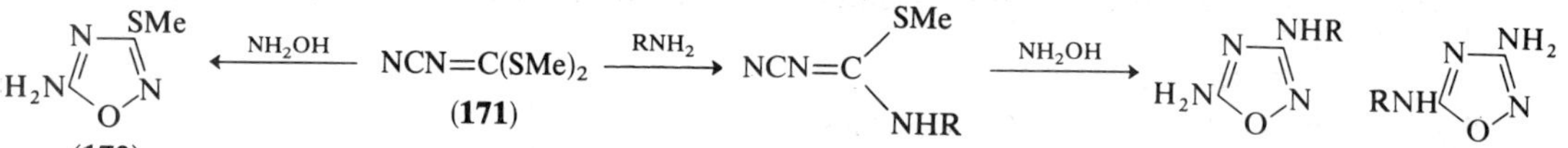

A new variation is the activation of the oximino group in the amidoxime by replacement with =N—Cl. This was accomplished by oxidation of an imino group in a guanidine derivative (**175**) with sodium hypochlorite. A nitrene (**176**) was suggested as an intermediate ⟨80HCA841⟩.

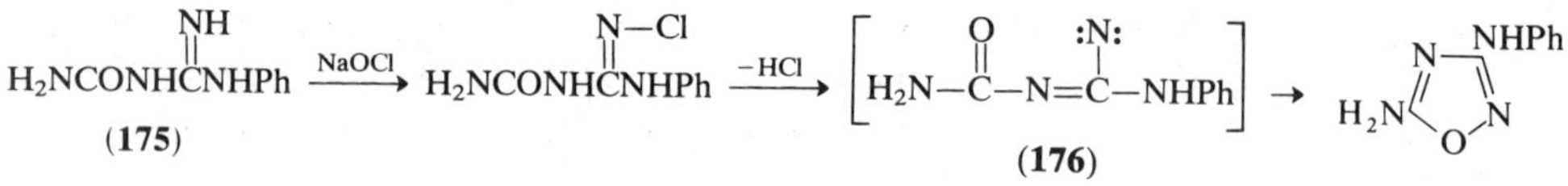

$$H_2NCONHCNHPh(=NH) \xrightarrow{NaOCl} H_2NCONHCNHPh(=NCl) \xrightarrow{-HCl} [H_2N-C(O)-N=C-NHPh, :N:] \rightarrow H_2N\text{-oxadiazole-NHPh}$$
(175)   (176)

4.21.7.1.2 Δ²-Oxadiazolines

When the sp^2 center in (**149b**) is an aldehyde, its addition to an amidoxime gives a Δ²-oxadiazoline (**177**).

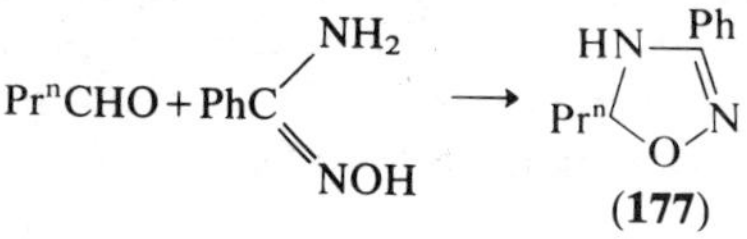

$$Pr^nCHO + PhC(NH_2)(NOH) \rightarrow Pr^n\text{-oxadiazoline-Ph (177)}$$

4.21.7.1.3 Δ³-Oxadiazolines

One synthesis of a Δ³-oxadiazoline (**179**) was reported from a substituted guanidine (**178**, not isolated) by condensation with acetone (see also Section 4.21.6.6).

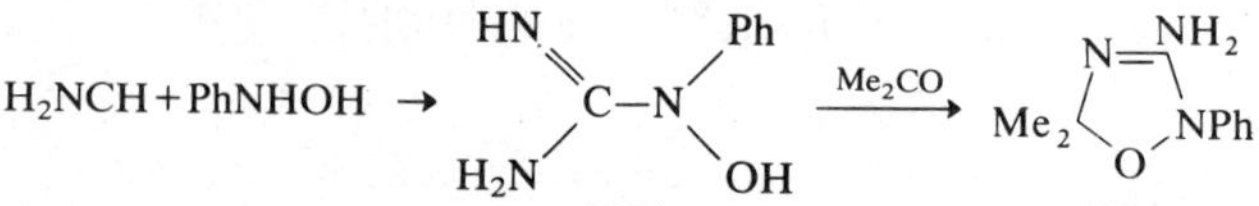

$$H_2NCH + PhNHOH \rightarrow (HN=)C(NH_2)N(Ph)(OH) \text{ (178)} \xrightarrow{Me_2CO} Me_2\text{-oxadiazoline-NPh, NH}_2 \text{ (179)}$$

4.21.7.1.4 From N-acylimino ethers

Another source of the sp^2 carbon at C-3 is an *N*-acylimino ether (**180**). The second nitrogen at C-3 is obtained from hydroxylamine and dehydration gives the oxadiazole (**181**).

An imino ether (**182**) can also be used as the source of the C-5 carbon if an amidoxime furnishes the one at C-3.

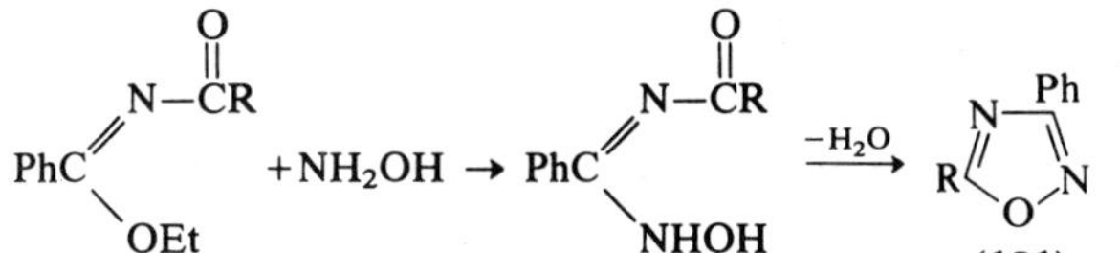

[Reaction scheme showing PhC(OEt)=N-CR(=O) (**180**) + NH$_2$OH → PhC(NHOH)=N-CR(=O) → (−H$_2$O) oxadiazole (**181**)]

[Reaction scheme showing PhC(NH$_2$)=NOH + MeC(OEt)=NH·HCl → (Δ) 3-methyl-5-phenyl-1,2,4-oxadiazole (**182**)]

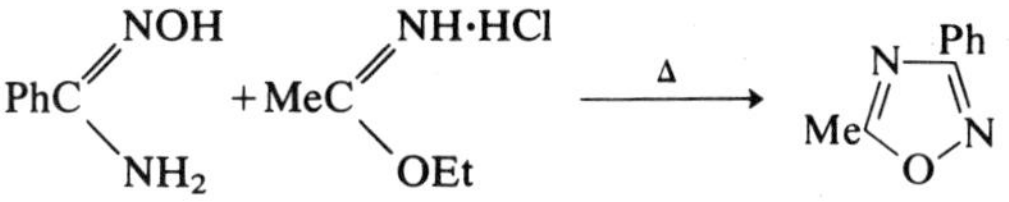

4.21.7.2 1,3-Dipolar Cycloadditions

4.21.7.2.1 1,2,4-Oxadiazoles

The second general method of generating the 1,2,4-oxadiazole ring (**183**) is to add the fragments (**150c**) and (**150d**) in a 1,3-dipolar cycloaddition. Aliphatic nitrile oxides are generated *in situ* from a nitroalkane with phenyl isocyanate as dehydrating agent ⟨B-71MI42100, 80JCS(P1)1635⟩ or in other ways ⟨80CPB3296⟩. Aromatic nitrile oxides are much less reactive and can be isolated but again are more often generated in the presence of the nitrile.

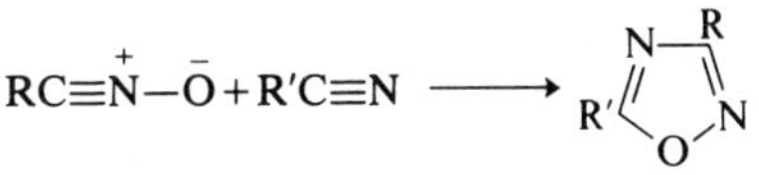

$$RC\equiv\overset{+}{N}-\overset{-}{O} + R'C\equiv N \longrightarrow \text{(183)}$$

$$ArC(NOH)Cl \xrightarrow{Et_3N} ArC\equiv\overset{+}{N}-\overset{-}{O} + Et_3\overset{+}{N}H\,Cl^-$$

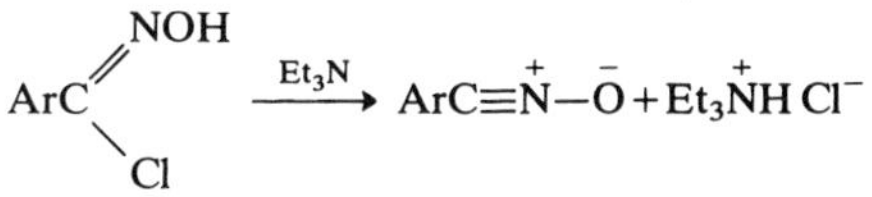

A novel use of this general method is the addition of an aromatic nitrile oxide to the nitrile equivalent tied up in the triazine (**184**). The reaction occurs to give a good yield only in the presence of boron trifluoride etherate whose role is illustrated in Scheme 18 ⟨78BCJ1484⟩.

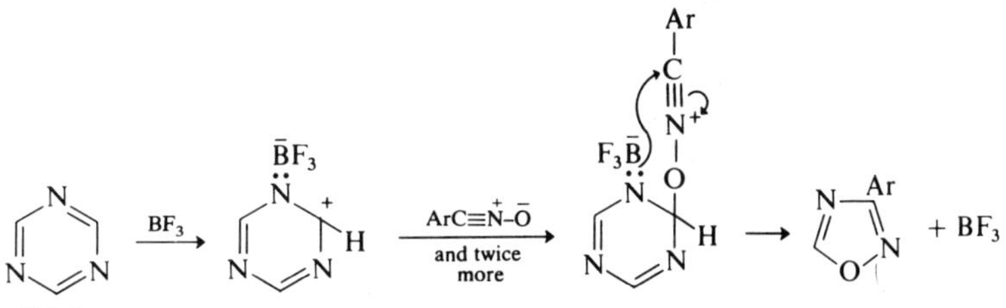

[Scheme 18: triazine (**184**) + BF$_3$ → BF$_3$-adduct → ArC≡N-O (and twice more) → intermediate → 1,2,4-oxadiazole + BF$_3$]

Scheme 18

4.21.7.2.2 Δ^2-Oxadiazolines

A large number of Δ^2-oxadiazolines (**187**) have been synthesized by the direct addition of a nitrile oxide (**186**) to a Schiff base (**185**), analogous to the corresponding synthesis of oxadiazoles (above) ⟨77IJC(B)848, 77JOC1555⟩.

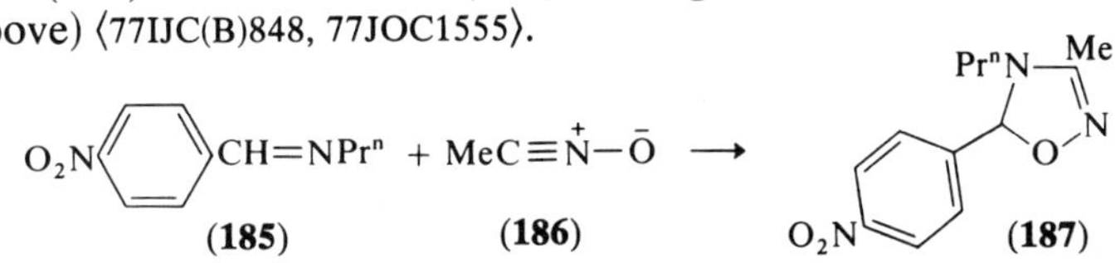

$$O_2N\text{-C}_6H_4\text{-CH=NPr}^n\;(\mathbf{185}) + MeC\equiv\overset{+}{N}-\overset{-}{O}\;(\mathbf{186}) \longrightarrow (\mathbf{187})$$

4.21.7.2.3 Δ^4-Oxadiazolines

The 1,3-dipolar cycloaddition resulting in a Δ^2-oxadiazoline may be inverted to give a Δ^4-oxadiazoline (**190**). The starting compounds are nitrones (**188**) and nitriles (**189**) for example. Even when a C=C bond might be expected to compete with C=N in dipolar addition to nitrones, the C=N wins as is shown in the reaction of a nitrone (191) with TCNE to give (192) ⟨80JGU117⟩.

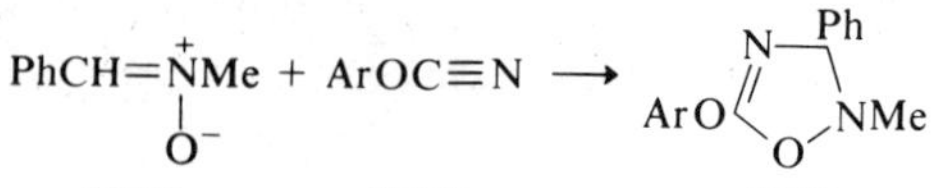

$$PhCH=\overset{+}{N}Me + ArOC\equiv N \longrightarrow $$
(188) (189) (190)

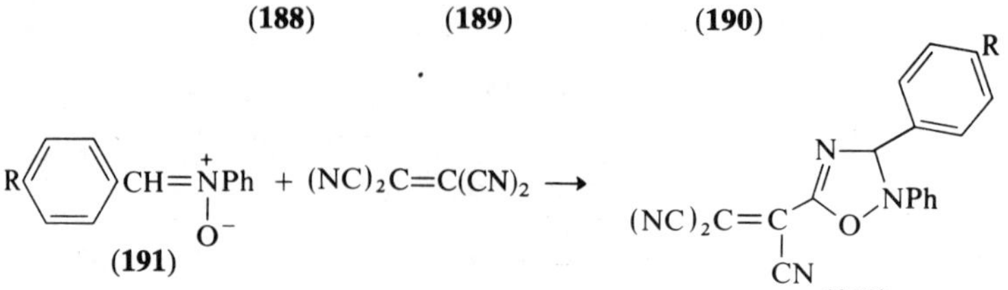

$$R\text{-}C_6H_4\text{-}CH=\overset{+}{N}Ph + (NC)_2C=C(CN)_2 \longrightarrow$$
(191) (192)

4.21.7.2.4 Completely reduced 1,2,4-oxadiaza rings

The completely reduced oxadiaza ring may be obtained by suitable dipolar additions involving potential double bonds rather than triple bonds as in (150c) and (150d). For example, a nitrone (193) added to a substituted ketenimine (194) gives an oxadiazolidine (195). An oxaziridine (196), an isomer of the corresponding nitrone (and sometimes interchangeable with it), gives a reduced ring, a substituted oxadiazolidinone (198) with phenyl isocyanate (197). The cumulative double bond system of (194) may be replaced by diphenylcarbodiimide, PhN=C=NPh, to give the same type of 1,3-dipolar addition product.

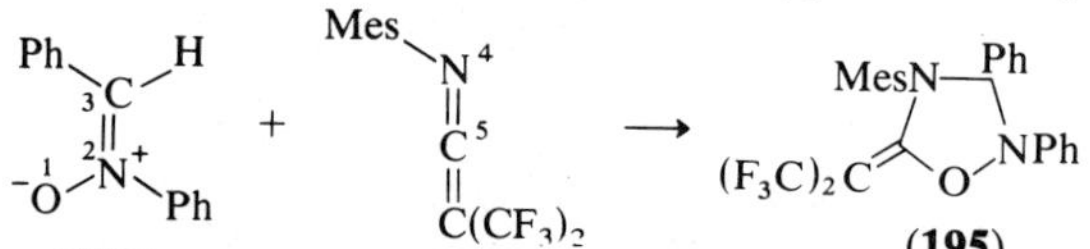

(193) (194) (195)

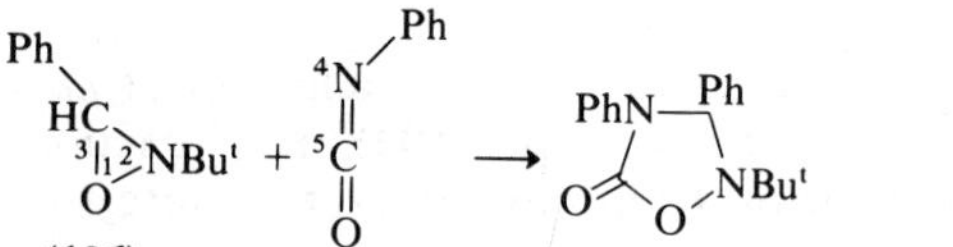

(196) (197) (198)

In another example an amine oxide (199) adds to phenyl isocyanate (200), resulting in a fused ring system (201) along with a smaller amount of the isomer (202) ⟨74T3723, 77CZ154; see also 81AP10, 81LA191, 80RTC278⟩.

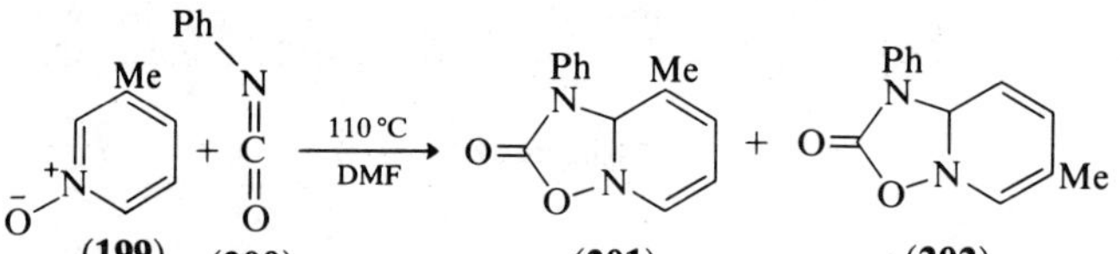

(199) (200) (201) (202)

4.21.7.2.5 Oxadiazolidinediones

A few examples of oxadiazolidinediones have been synthesized where the carbons at both C-3 and C-5 are in the +4 oxidation state ⟨81AP294⟩. Perhaps the most interesting ones are two natural products (203) and (204) isolated from *Quisqualic fructus*. Both are amino acid derivatives and have been synthesized ⟨75YZ236, 76AHC(20)65⟩. A comparison is being made of the biological function of the quisqualates with the related glutamic acid analog ⟨81MI42100-3⟩.

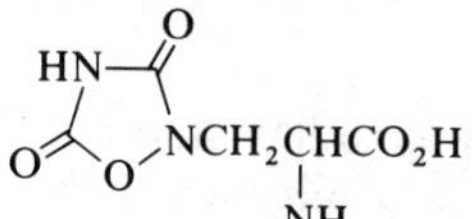

(203)

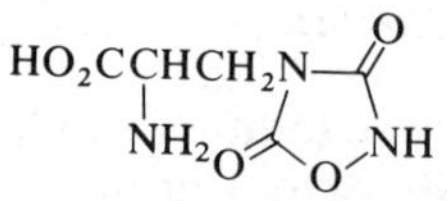

4.21.7.3 From Oxidations

4.21.7.3.1 Of non-ring compounds

Less common, and in general less satisfactory, are oxidation reactions to synthesize oxadiazoles. Mild oxidation of benzamidoxime (**205**) leads to 3,5-diphenyl-1,2,4-oxadiazole (**206**).

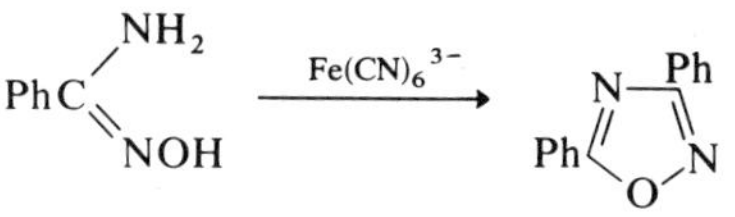

Benzaldoxime can be oxidized with nitrogen dioxide in ether to a dimer that is in turn oxidized to an oxadiazole ⟨73S156⟩. Some imidazoles have also been oxidized to oxadiazoles ⟨69RTC204⟩.

4.21.7.3.2 Of oxadiazolines

The partially reduced oxadiazoline ring (**207**; Section 4.21.7.2.2) can be oxidized to an oxadiazole (**208**) by various oxidizing agents ⟨76AHC(20)65, 77JOC1555⟩.

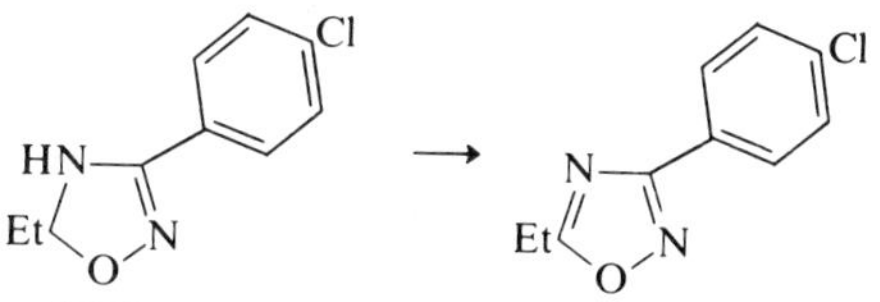

4.21.8 APPLICATIONS

A number of 1,2,4-oxadiazoles have been found useful as chemotherapeutic agents. The most effective ones are collected in Table 10. Others have been tested for various properties ⟨80JMC690⟩. Methazole (**209**), an oxazolidinedione, has been used to control various weeds ⟨73MI42100, 80MI42109, 81MI42104⟩.

Table 10 Chemotherapeutic Agents Containing a 1,2,4-Oxadiazole Nucleus ⟨76AHC(20)65⟩

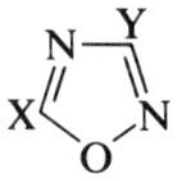

X	Y	Trade name	Activity[a]
$Et_2N(CH_2)_2$	Ph	Oxalmine, Bredon, Perebron	d, e
$Et_2N(CH_2)_2NH$	Ph	Irrigor	a, f
$C_5H_{10}N$	CH_2CHPh_2	Libexin	d
Et_2NCH_2	Ph		a, b
$C_5H_{10}N(CH_2)_3$	Ph		a, e
$Et_2N(CH_2)_2$	o-C_6H_4OH		b, c, e
$Bu^n_2N(CH_2)_2$	Ph		a, c, e

[a] a, anaesthetic; b, analgesic; c, antispasmodic; d, antitussive; e, antiinflammatory; f, vasodilator; g, anthelmintic.

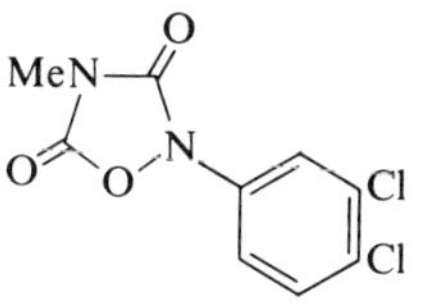

4.22

1,2,5-Oxadiazoles and their Benzo Derivatives

R. M. PATON
University of Edinburgh

4.22.1 INTRODUCTION	394
4.22.2 STRUCTURE	394
4.22.2.1 Theoretical Methods	395
4.22.2.2 Molecular Dimensions	395
4.22.2.2.1 Furazans	395
4.22.2.2.2 Furoxans	396
4.22.2.3 Molecular Spectra	397
4.22.2.3.1 NMR spectra	397
4.22.2.3.2 IR and UV spectra	398
4.22.2.3.3 Mass spectra	399
4.22.2.3.4 Miscellaneous spectroscopic and physical properties	399
4.22.3 REACTIVITY	400
4.22.3.1 Reactions of the Heterocyclic Ring of Furazans and Benzofurazans	400
4.22.3.1.1 Thermal and photochemical ring cleavage	400
4.22.3.1.2 Reaction with electrophiles and oxidizing agents	401
4.22.3.1.3 Reaction with nucleophiles and reducing agents	402
4.22.3.1.4 Heterocyclic ring transformations and miscellaneous reactions	403
4.22.3.2 Reactions of the Heterocyclic Ring of Furoxans and Benzofuroxans	403
4.22.3.2.1 Thermal and photochemical isomerization	403
4.22.3.2.2 Thermal and photochemical ring cleavage	404
4.22.3.2.3 Reaction with electrophiles and oxidizing agents	405
4.22.3.2.4 Reaction with nucleophiles and reducing agents	405
4.22.3.2.5 Heterocyclic ring transformations	407
4.22.3.3 Benzofurazans and Benzofuroxans: Reactions of the Homocyclic Ring	409
4.22.3.3.1 Electrophilic attack	409
4.22.3.2.2 Nucleophilic attack	410
4.22.3.3.3 Miscellaneous reactions and transformations involving substituents	411
4.22.3.4 Monocyclic Furazans and Furoxans: Reactions of Substituents	412
4.22.3.4.1 Alkyl- and aryl-furazans and -furoxans	412
4.22.3.4.2 Acyl-substituted furazans and furoxans	413
4.22.3.4.3 Furazan- and furoxan-carboxylic acids and their derivatives	413
4.22.3.4.4 Amino- and nitro-furazans and -furoxans	413
4.22.3.4.5 Furazan and furoxan alcohols and thiols, and their derivatives	414
4.22.4 SYNTHESES	415
4.22.4.1 Furazans	415
4.22.4.1.1 Dehydration of α-dioximes	415
4.22.4.1.2 Deoxygenation of furoxans	416
4.22.4.1.3 Rearrangements of other heterocyclic systems	417
4.22.4.1.4 Miscellaneous furazan syntheses	417
4.22.4.2 Benzofurazans	418
4.22.4.2.1 Dehydration of o-quinone dioximes	418
4.22.4.2.2 Synthesis from o-disubstituted arenes	418
4.22.4.2.3 Deoxygenation of benzofuroxans	419
4.22.4.2.4 Miscellaneous benzofurazan syntheses	419
4.22.4.3 Furoxans	420
4.22.4.3.1 Oxidation of α-dioximes	420
4.22.4.3.2 Dehydration of α-nitroketone oximes	420
4.22.4.3.3 Dimerization of nitrile oxides	421
4.22.4.3.4 Miscellaneous furoxan syntheses	423

4.22.4.4 Benzofuroxans	424
4.22.4.4.1 Oxidation of o-quinone dioximes	424
4.22.4.4.2 Decomposition of o-nitroaryl azides	424
4.22.4.4.3 Oxidation of o-nitroanilines	425
4.22.4.4.4 Miscellaneous benzofuroxan syntheses	425
4.22.5 APPLICATIONS	425

4.22.1 INTRODUCTION

From the time they were first reported during the second half of the 19th century the 1,2,5-oxadiazoles and their derivatives, notably the N-oxides, have been the subject of continuous and often intensive investigation. Much of the early work was concerned with structure determination, particularly of the N-oxides for which over the years a variety of formulations were proposed (see Section 4.22.2), while more recently it has been synthetic applications of their chemistry that have attracted most attention. Natural products incorporating such oxadiazoles are unknown.

The term furazan originally proposed by Wolff ⟨1890LA(260)79⟩ for the 1,2,5-oxadiazole skeleton (1) finds widespread use and is favoured by *Chemical Abstracts*; other names found in the early literature include: glyoxime anhydrides, furo[aa_1]diazoles and azoxazoles. For largely historical reasons associated with their uncertain structure, the class of heterocycles now known to be the N-oxides of the furazans posses their own nomenclature based on the name furoxan (2). Although they are currently designated furazan oxides in *Chemical Abstracts*, the term furoxan is still in common use. In view of their distinctive chemistry the name seems worthy of retention and will be employed throughout this review. In the years before their structure was fully established they were also called glyoxime peroxides and dioxadiazetidines.

Furazans and furoxans with a benzene ring fused at the 3- and 4-positions are generally referred to as benzofurazans (3) and benzofuroxans (4); other names include 3,4-benzo-1,2,5-oxadiazole or 2,1,3-benzoxadiazole for (3) and 3,4-benzo-1,2,5-oxadiazole 2-oxide or 2,1,3-benzoxadiazole 1-oxide or (in some pre-1960 reports) *o*-dinitrosobenzene for the corresponding furoxan (4). Three acronyms are to be found, particularly in the biochemical and pharmacological literature: BFZ and BFX (prefixed by the sites and names of substituents) for benzofurazan and benzofuroxan, respectively, and Nbf-Cl for the fluorogenic label, 7-chloro-4-nitrobenzofurazan.

The 1,2,5-oxadiazole ring system has been the subject of several reviews. Both furazans and furoxans were covered by Boyer ⟨B-61MI42200⟩, Behr ⟨62HC(17)283⟩ and more recently by Stuart ⟨75H(3)651⟩, while Kaufman and Picard ⟨59CRV429⟩ and Boulton and Ghosh ⟨69AHC(10)1⟩ dealt solely with furoxans and benzofuroxans; of particular value is a recent comprehensive account by Gasco and Boulton ⟨81AHC(29)251⟩ covering the literature up to the middle of 1980. There are also two reviews restricted to one distinctive and intensively studied reaction of benzofuroxans (see Section 4.22.3.2.5.i) ⟨75S415, 76H(4)767⟩, and on covering the biochemical and pharmacological aspects of benzofurazans and benzofuroxans ⟨81MI42200⟩.

4.22.2 STRUCTURE

The 1,2,5-oxadiazole framework for monocyclic furazans (1) was correctly deduced at an early stage from analytical data and their chemical reactions, and has since been confirmed by X-ray crystallography and microwave spectroscopy (see Section 4.22.2.2). Likewise for benzofurazans, benzenoid structures such as (5) and (6), which were originally put forward to explain their stability and resistance to oxidation, have been rejected in favor of the *o*-quinonoid formulation (3).

(5) **(6)** **(7)** **(8)** **(9)**

In marked contrast the structure of furoxans was for many years a matter of some controversy. Among the formulations proposed, and thus frequently to be found in the early literature, were the dioxadiazine (or glyoxime peroxide) (7) and the bicyclic arrangements (8) and (9). It was not until the isomerism characteristic of asymmetrically substituted furoxans was fully appreciated (see Section 4.22.3.2.1) that the N-oxide structure (2), originally proposed by Wieland ⟨03LA(329)225⟩ and by Werner ⟨B-04MI42200⟩ some 60 years previously, finally became universally accepted.

Similar uncertainty surrounded the benzofuroxan series. Indeed as late as 1961 the mesomeric system (10) ↔ (12) was being considered and at that time frequent reference was made to 'o-dinitrosobenzene'. Thereafter incontrovertible evidence for a rapid interconversion of isomeric N-oxides was provided by NMR spectroscopy (see Section 4.22.2.3) and the benzofuroxan formula (4) was finally confirmed by X-ray crystallography. The history of the problem is contained in previous reviews and a summary of the development of the major theories through to final proof of the Wieland–Werner structure is presented in the recent account by Gasco and Boulton ⟨81AHC(29)251⟩.

(10) **(11)** **(12)**

4.22.2.1 Theoretical Methods

Molecular orbital (MO) calculations at various levels of approximation (see Section 4.01.2) have been applied to furazans and to a lesser extent furoxans. The Hückel method has been used for a wide range of N—O-containing compounds including the parent furazan (**1**; $R^1 = R^2 = H$) and furoxans for which the molecular geometries were known ⟨70CB3370⟩; fair correlations are obtained between calculated bond orders and experimental bond lengths. MO calculations have also been performed on dimethylfurazan and the results compared with observed IR and NMR spectra ⟨77KGS1110⟩. The equilibria between the benzofurazan 1- and 3-oxides have been the subject of several theoretical investigations ⟨81AHC(29)251⟩. Reasonable activation energies (ca. 80 kJ mol^{-1}) are obtained by assuming the reaction proceeds via a ψ-o-dinitrosobenzene transition state (**10–12**). The Pariser–Pople–Parr procedure which gives adequate correlation with UV spectra is less satisfactory for predicting the equilibrium constants. CNDO/2 calculations have also been utilized. The photoelectron spectrum for benzofurazan has been interpreted with the aid of semiempirical ⟨73T3085⟩ and more recently ab initio ⟨78JST(43)33⟩ MO calculations. The latter method has also been used for unsubstituted furazan. A high level of agreement was obtained between observed nuclear quadrupole coupling constants and calculated values, both in magnitude and direction ⟨79ZN(A)220⟩; other calculated molecular properties include its dipole moment, diamagnetic susceptibilities and photoelectron spectrum ⟨79JST(51)87⟩.

4.22.2.2 Molecular Dimensions

4.22.2.2.1 Furazans

X-Ray crystal structures have been reported for a variety of furazans, mostly in the monocyclic series. Typical bond lengths and angles are presented in Table 1. The heterocyclic ring is planar and usually symmetrical. π-Bond orders of 70–80% for N(2)—C(3) and C(4)—N(5), and of 40–50% for C(3)—C(4) are commonplace, suggesting significant π-electron delocalization; in contrast O(1)—N(2) and N(5)—O(1) are essentially single bonds. For diphenylfurazan (**1**; $R^1 = R^2 = Ph$) the two phenyl rings have different dispositions relative to the oxadiazole with dihedral angles of 19° and 62°. This results in distortion of the heterocycle as evidenced by dissimilar values for the O—N (1.386 and 1.412 Å) and C—N (1.330 and 1.364 Å) bonds; likewise the bond angles ONC (103° and 106°) and NCC (108° and 111°) also differ ⟨76AX(B)1079⟩. The molecular geometry of the parent furazan (**1**; $R^1 = R^2 = H$) has been determined by microwave spectroscopy ⟨65JCP(43)166⟩; the reported bond lengths and angles are broadly similar to those reported for substituted analogues.

Table 1 Bond Lengths and Angles in Furazans and Benzofurazans[a]

Bond	Length (Å)	Bonds	Angle (°)
a, e	1.36–1.41	ab, de	103–108
b, d	1.29–1.36	bc, cd	107–110
c	1.41–1.44	ea	110–113

[a] Data taken from ⟨65JCP(43)166, 71AX(B)1388, 75CSC561, 76CSC113, 76CSC329, 78AX(B)2953⟩.

4.22.2.2.2 Furoxans

A greater variety of both monocyclic and benzofuroxans has been examined ⟨81AHC(29)251⟩. The bond lengths and angles listed in Table 2 show that the effect of the fused benzo moiety on the geometry of the oxadiazole ring is minimal. As for furazans the heterocyclic ring is nearly planar but the exocyclic oxygen atom at N(2) causes substantial distorton. A common feature of the cases studied is π-electron delocalization with C(3)—C(4) showing significant (*ca.* 30%) double bond character. Other noteworthy features are the long O(1)—N(2) and short N(2)—O(*exo*) bonds. The exocyclic oxygen invariably lies within 0.05 Å of the plane of the heterocycle. Intramolecular hydrogen bonding to the *N*-oxide group is evident in furoxan-3-carboxamides and -3-carbohydrazides.

Table 2 Bond Lengths and Angles in Furoxans and Benzofuroxans[a]

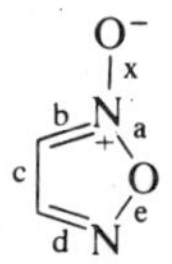

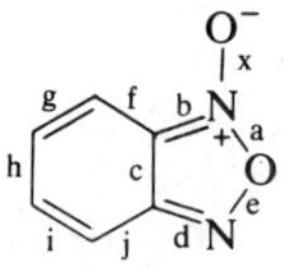

	Furoxans				Benzofuroxans		
Bond	Length (Å)	Bonds	Angle (°)	Bond	Length (Å)	Bonds	Angle (°)
a	1.42–1.46	ab	105–109	a	1.42–1.46	ab	106–107
b	1.28–1.33	bc	105–110	b	1.31–1.33	bc	107–108
c	1.40–1.43	cd	109–115	c	1.40–1.42	cd	111–113
d	1.29–1.32	de	106–108	d	1.32–1.33	de	104–106
e	1.35–1.39	ea	106–109	e	1.38–1.39	ea	108–109
x	1.18–1.25	ax	115–118	x	1.22–1.24	ax	116–118
				f	1.41–1.43	cf	122–124
				g	1.34–1.36	fg	115–117
				h	1.44–1.45	gh	121–123
				i	1.34–1.36	hi	121–123
				j	1.42–1.43	ij	117–118
						jc	119–120

[a] Data taken from ⟨81AHC(29)251, 79AX(B)3076, 80LA1557, 81JCS(P2)1240⟩.

The same overall pattern of bond lengths and angles is found for benzofuroxans, with short N(1)—O(*exo*) and O(2)—N(3) and longer N(1)—O(2) distances. It has been suggested ⟨72AX(B)1116⟩ that these trends indicate appreciable contributions from forms (**13**) and (**14**) to the resonance hybrid; however, the short C(4)—C(5) and C(6)—C(7) bond lengths, implying substantial bond localization in the fused benzo ring, argue in favour of the conventional N-oxide formulation (4) being the major component.

Benzotrifuroxan (15; $n = 1$) and the thiadiazolo compound (16) have also been studied; they both have longer than average N(1)—O(2) bonds, 1.48 and 1.47 Å, respectively. In contrast the two endocyclic N—O distances in 4,6-dinitrobenzofuroxan are almost the same. Benzofuroxans usually crystallize in the tautomeric form which predominantes in solution.

Trimethylenefuroxan (17; $n = 3$) and the acenaphtho compound (18) have exceptionally long O(1)—N(2) bonds (1.49 and 1.50 Å, respectively), a fact that may be associated with ring strain and their tendency to undergo ring-cleavage reactions (see Section 4.22.3.2.1). One monosubstituted furoxan (2; $R^1 = p$-HOC$_6$H$_4$, $R^2 =$ H) has been examined; its structural parameters are similar to those of disubstituted analogues ⟨80LA1557⟩. The parent unsubstituted furoxan is unknown.

4.22.2.3 Molecular Spectra

4.22.2.3.1 NMR spectra

The NMR spectra of ^{1}H, ^{13}C, ^{14}N, ^{15}N and ^{17}O nuclei have all been used for the characterization of furazans ⟨81AHC(29)251⟩ and to a lesser extent furazans.

Examples with protons directly attached to the oxadiazole ring are few and no systematic study has been made of their ^{1}H NMR spectra. The parent unsubstituted furazan shows the expected single peak with δ_H 8.66 (neat) or 8.19 p.p.m. (CDCl$_3$) ⟨65JOC1854⟩. Similar sensitivity to solvent is exhibited by phenylfurazan (δ_H 8.60 in CDCl$_3$ and 8.42 in CCl$_4$) ⟨71JOC5⟩ and by p-hydroxyphenylfurazan (δ_H 9.06 in acetone-d_6 and 8.47 in CDCl$_3$) ⟨80LA1557⟩. In the corresponding monosubstituted furoxan series the 3-H and 4-H of 4-phenyl- and 3-phenyl-furoxans absorb at δ_H 7.26 and 8.55 p.p.m. (CDCl$_3$), respectively; similarly the 4-H of 3-(p-hydroxyphenyl)furoxan absorbs at δ_H 8.55 p.p.m.

For monocyclic furoxans and furoxans fused to non-aromatic rings the signals due to the substituent groups can be used for structure identification. For dimethylfuroxan the upfield peak (at 2.16 p.p.m.) is assigned to the 3-methyl (MeC=NO$_2$) and that at lower field (2.38 p.p.m.) to the 4-methyl group (MeC=NO) by comparison with dimethylfurazan (2.31 p.p.m.). Structural assignments for numerous isomeric pairs of furoxans have been made on this basis; in several cases confirmation has been provided by X-ray crystallography (see Section 4.22.2.2).

Similar chemical shift differences (ca. 0.25 p.p.m.) are observed for the protons at the 4- and 7-positions of benzofuroxans; 5-H and 6-H show a smaller separation (ca. 0.15 p.p.m.). In each case the proton conjugated to the N-oxide is at higher field. For benzofurazan the vicinal proton–proton coupling constants ($J_{4,5} = J_{6,7} = 9.3$ and $J_{5,6} = 6.4$ Hz) indicate substantial bond localization, in accord with the results of X-ray crystal structure analysis. A similar pattern emerges for benzofuroxan ⟨63JCS197⟩.

Proton resonance techniques provided the first clear-cut evidence for the unsymmetrical structure of benzofuroxans and for their tautomerism. The unsymmetrical (ABCD) spectrum observed at low temperature (−40 °C) becomes well-resolved and symmetrical (A$_2$B$_2$) above

100 °C. Equilibrium constants and activation energies have been estimated for a wide range of benzofuroxans, both by detailed line-shape analysis and more qualitatively from coalescence temperatures. That the interconversion of the 1- and 3-oxide isomers is more facile for benzofuroxans than for monocyclic compounds is readily demonstrable. Both the early work and more recent results have been reviewed ⟨69AHC(10)1, 81AHC(29)251⟩.

Greater attention has recently been focused on the ^{13}C NMR spectra. Not only can the non-equivalence of the two ring carbons and the tautomerism characteristic of furoxans be demonstrated, but conformational features such as ring strain can be examined. For a series of 3-substituted 4-phenylfurazans (1; R^1 = Ph, R^2 = H, Me, Ph, SPh, SO$_2$Ph, NO$_2$, NH$_2$, NMe$_2$) $\delta_{C(3)}$ is found to vary from 139 to 160 p.p.m. ⟨82JHC427⟩. The most significant feature for the corresponding furoxans is the large upfield displacement of the C(3) resonance. While the C(4) peak remains in the range 143-160 p.p.m. depending on the nature of the substituent, that for C(3) is shifted to 102-126 p.p.m. For monosubstituted furazans and furoxans ^{13}C-^{1}H couplings of *ca.* 200 Hz are found for the ring carbon and attached proton. At low temperatures benzofuroxan and its derivatives show two peaks, at *ca.* 152 and 114 p.p.m.; on heating these coalesce. For ring-strained furazans such as trimethylenefurazan (19; n = 3) and -furoxan (17; n = 3) and the norbornane and camphor compounds (20) and (21), the C(4) peak is displaced downfield to 165-168 p.p.m.; in the last case the presence of two isomers (21a) and (21b) is easily discernible ⟨82UP42200⟩.

(19) (20) (21a) (21b)

The two dissimilar nitrogens in furoxans can be detected by ^{15}N NMR spectroscopy. Solvent dependent δ_N values in the ranges +18 to +26 p.p.m. from MeNO$_2$ are found for N(2) and -3 to +14 for N(5); for benzofuroxans the two peaks coalesce on heating. Dimethylfurazan and benzofurazan each show the expected single absorptions, at -31 and -42 p.p.m., respectively ⟨78JOC2542⟩. ^{14}N Spectra have also been reported for the two furazans above and several monocyclic and benzofuroxans; in all cases only a single broad signal was observed. Benzotrifuroxan has received particular attention; two signals in each of the ^{13}C and ^{15}N spectra confirm the structure as (15; n = 1) rather than hexanitrosobenzene ⟨80OMR(14)356⟩.

The ^{17}O NMR spectrum of dimethylfuroxan consists of two distinct peaks at δ_O -460 and -350 p.p.m. from H$_2$O; the former can be assigned to O(1) by comparison with dimethylfurazan (δ_O -475 p.p.m.) ⟨61HCA865⟩. Dynamic resonance effects are observed for benzofuroxan ⟨62HCA504⟩.

4.22.2.3.2 IR and UV spectra

The IR spectra of several furazans and furoxans have been investigated systematically and there are also numerous other less detailed reports ⟨71PMH(4)265⟩. Among the useful absorption bands for furazans are: 1625-1560 (C=N—O), 1430-1385 (N—O), and 1040-1030 and 890-880 cm^{-1} (heterocyclic ring). Examination of the IR and Raman spectra of the parent unsubstituted furazan and the mono- and di-deuterated compounds has allowed a complete assignment of the fundamental frequencies to be made ⟨73SA(A)1393⟩. Furoxans have characteristic peaks at 1625-1600 (C=N—O), 1475-1410 (=N$^+$(O$^-$)—O), 1360-1300 cm^{-1} (N—O), and 1190-1150, 1030-1000 and 890-825 cm^{-1} (heterocyclic ring); for strained furoxans, *e.g.* the norbornane and camphor compounds (20) and (21), the C=N—O peak occurs at higher frequency (1675-1655 cm^{-1}). Benzofuroxans have four strong diagnostic bands at or near 1630, 1600, 1545 and 1500 cm^{-1} ⟨69AHC(10)1⟩.

Although there have been few systematic investigations, UV absorption data are available for a wide range of furazans and furoxans. The spectrum of furazan in the region 230-250 nm has been studied in detail and rationalized in terms of a π–π* transition from a planar ground state to a non-planar excited state ⟨72CJC2088⟩. Dimethylfurazan shows a band at 228-241 nm and its *N*-oxide one at *ca.* 260 nm; a similar pattern emerges for other monocyclic compounds. The more extended conjugation in the chromophores of benzofurazan and benzofuroxan results in the expected shift of λ_{max} to longer wavelength (usually 350–410 nm). There are two, often overlapping, band systems with some fine structure for substituted analogues. When electron-donating groups are present the lowest energy band extends well into the visible region; e.g. 7-amino- and 7-alkylthio-4-nitrobenzofurazans absorb at ca., 465 and 425 nm, respectively. The λ_{max} values are often strongly pH and solvent dependent. The formation of Meisenheimer complexes of nitrobenzofurazans and benzofuroxans is accompanied by substantial changes in their UV spectra, thereby allowing the kinetics and equilibria of these processes to be examined.

The 4-nitro compounds also exhibit visible fluorescence dependent on the electron-donating ability of substituents conjugated with the nitro group and on steric interference with the conjugation between the two groups ⟨81MI42200⟩. The fluorescence is particularly intense for 7-alkylamino-4-nitrobenzofurazans in the region 525–545 nm; as these compounds are readily formed by nucleophilic attack of amines on 7-chloro-4-nitrobenzofurazan (Nbf-Cl), the latter has found widespread use as a fluorogenic label (see Section 4.22.5).

4.22.2.3.3 Mass spectra

There are two or three general modes of fragmentation for furazans under ionizing radiation (Scheme 1). Initial cleavage of the labile O(1)—N(2) bond can be followed by expulsion of either nitrile (path a) or NO (path b); ions attributable to RC$^+$ (path c) are also common ⟨B-71MS526⟩. Both positive and negative ion spectra have been recorded for benzofurazan; pathways involving loss of NO are observed in each case ⟨78OMS379⟩.

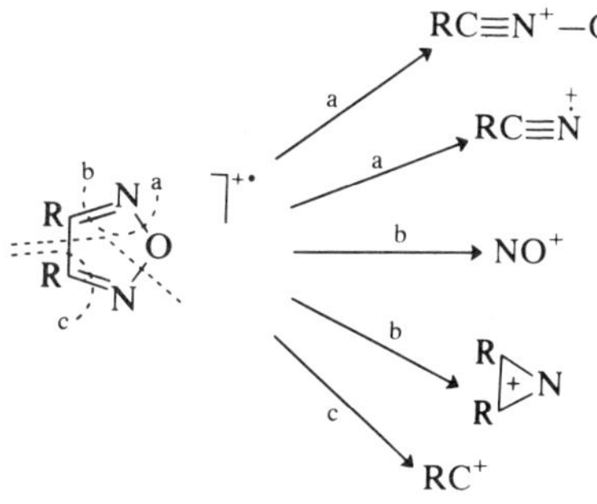

Scheme 1

The mass spectra of furoxans differ markedly from those of most other aromatic N-oxides. The $(M-16)^+$ ion is often weak, oxygen loss rarely being a significant fragmentation process. Prominent ions usually occur at $(M-NO)^+$, $(M-N_2O_2)^+$ and 30 (NO$^+$) a.m.u. These can be rationalized by assuming ring opening at the weak O(1)—N(2) bond to give the 1,2-dinitroso tautomer, followed by breaking of one or both of the C—N bonds (Scheme 2). As for the furazans, cleavage can also occur at C(3)—C(4) yielding nitrile oxides; this behaviour closely parallels the thermal 1,3-dipolar cycloreversion process (see Section 4.22.3.2.2).

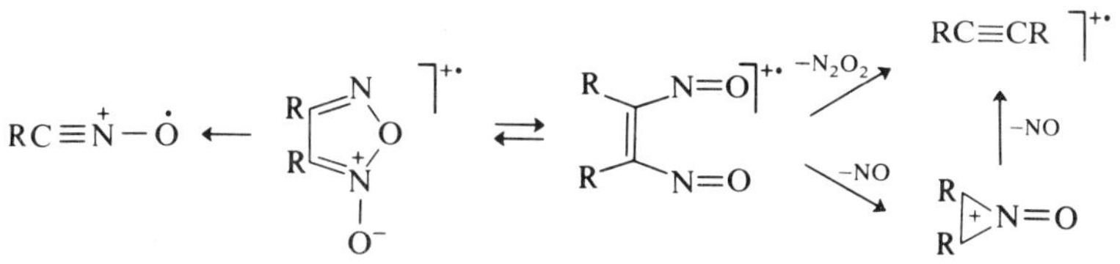

Scheme 2

4.22.2.3.4 Miscellaneous spectroscopic and physical properties

The UV PE spectra of furazan (He(I) and He(II) regions) and of benzofurazan (He(I)) have been recorded ⟨77JST(40)191, 78JST(43)33⟩ and assigned on the basis of *ab initio* MO calculations (see Section 4.22.2.1). For benzotrifuroxan (**15**; n = 1) the X-ray PE spectrum shows that there are two different types of carbon, nitrogen and oxygen atoms, confirming that it is a trifuroxan rather than a hexanitroso compound ⟨72RTC552⟩.

The radical anions produced by alkali metal or electrochemical reduction of benzofurazans can be studied by ESR spectroscopy ⟨68KGS360⟩ and the results interpreted with the aid of MO calculations of the spin density distribution ⟨75KGS1055⟩. A spectrum consisting of thirteen equally spaced lines is found for the anion of (15; $n = 0$), consistent with coupling to the six equivalent nitrogens ⟨69JCS(B)681⟩. Similar reduction of benzofuroxan affords the radical anion of benzofurazan. The ^{14}N NQR spectrum of furazan has been recorded at 77 K ⟨79ZN(A)220⟩ and the observed coupling constants compared with the results from microwave spectroscopy and *ab initio* MO calculations.

Furazan is a stable liquid with b.p. 98 °C; the melting points and/or boiling points of numerous furazans and furoxans are listed in earlier reviews. Heats of combustion are the most widely investigated of other thermochemical properties; comparison of several furazans with their N-oxides gives a value of 247 ± 4 kJ mol^{-1} for the energy of the $N^+—O^-$ bond. The heats of combustion, sublimation and formation of benzofurazan have been compared with those of phenazine and an estimate made of the strengths of the N—C (439 kJ mol^{-1}) and O—N (294 kJ mol^{-1}) bonds ⟨80MI42200⟩. The gas-phase thermal decomposition of furazans has been studied and kinetic parameters related to the thermal stability indices, calculated by the extended Hückel method ⟨78IZV313⟩. The detonation properties of (15; $n = 1$) and some nitrofuroxans have also been measured.

The dipole moment of furazan (3.38 D) is markedly greater than that for the corresponding 1,2,5-thiadiazole (1.57 D); typical values for other furazans fall in the range 4.0–4.8 D. Benzofurazan (3) (4.04 D) has been compared with its sulfur (1.79 D) and selenium (1.19 D) analogues ⟨73JHC773⟩; it was concluded that the mesomeric charge transfer increases through the series and that (3) is *o*-quinone-like with resonance occurring mainly in the heterocyclic ring. For benzofuroxan the dipole moment is 5.29 D; values between 4.6 and 5.5 D are found for most alkyl- and aryl-substituted furoxans.

4.22.3 REACTIVITY

The reactions of furoxans and benzofuroxans have recently been comprehensively surveyed ⟨81AHC(29)251⟩; earlier reviews are available for furazans and benzofurazans ⟨B-61MI42200, 62HC(17)283, 75H(3)651⟩. Unless otherwise stated the material contained in this section is cited therein. Seperate sections are devoted to the reactions at the heterocyclic rings of furazans (including benzofurazans) and of their furoxan counterparts in view of their distinctive chemistries. Transformations involving the homocyclic rings of benzofurazans and benzofuroxans, and of the substituents at the 3- and 4-positions of monocyclic furazans and furoxans are treated in a unified manner.

4.22.3.1 Reactions of the Heterocyclic Ring of Furazans and Benzofurazans

4.22.3.1.1 Thermal and photochemical ring cleavage

The heterocyclic ring of 1,2,5-oxadiazoles undergoes thermal and photochemical cleavage at the 1–2 and 3–4 bonds yielding nitrile- and nitrile oxide-derived products in a process which appears to closely parallel their electron impact-induced fragmentation ⟨75ACS(B)483⟩. In most cases temperatures in excess of 200 °C are required, but for ring-strained analogues such as acenaphthofurazan (22) and the thianorbornane-based compound (23) much less forcing conditions are needed. At 130 °C reaction of (23) with DMAD gives (24), presumably by way of the nitrile–nitrile oxide (25) ⟨80H(14)423⟩; 8-cyanonaphtho-1-nitrile oxide is generated from (22) under similar conditions ⟨73JOC1054⟩. In contrast, only at *ca.* 250 °C does diphenylfurazan decompose giving benzonitrile, phenyl isocyanate and 3,5-diphenyl-1,2,4-oxadiazole, the last two products being attributed to rearrangement and 1,3-dipolar cycloaddition with benzonitrile of the initially-formed benzonitrile oxide. It has been proposed that the O(1)—N(2) bond is the first to break generating an iminyl–iminoxyl diradical intermediate ⟨80DOK(255)917⟩. The kinetics of the gas-phase thermolysis of dimethylfurazan to methyl isocyanate and 3,5-dimethyl-1,2,4-oxadiazole have also been examined ⟨81MI42201⟩. When phosphites are present, deoxygenation leads to two nitrile fragments; *e.g.* diphenylfurazan with triphenyl phosphite at 270 °C, and (**22**) with trimethyl phosphite at 80 °C, give benzonitrile and 1,8-dicyanonaphthalene, respectively ⟨72JOC1842⟩.

The photochemical breakdown proceeds in like manner. On irradiation in benzene diphenylfurazan affords benzonitrile, diphenylfuroxan and 3,5-diphenyl-1,2,4-oxadiazole. In alcohols solvent-incorporated products are also formed.

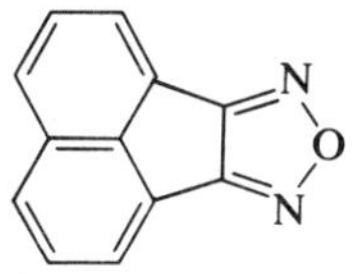

(**22**)

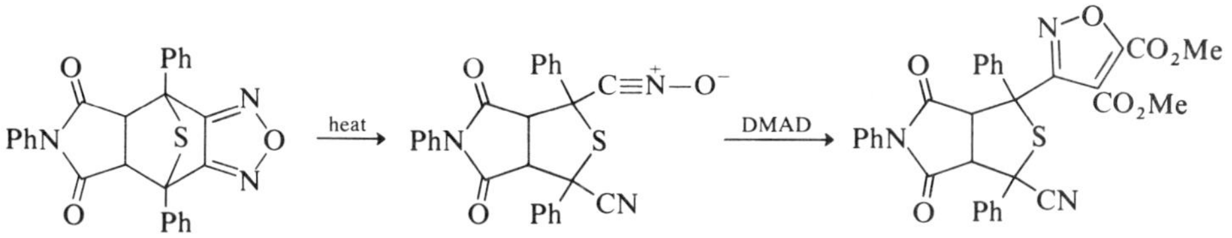

(**23**) → heat → (**25**) → DMAD → (**24**)

Benzofurazans show greater thermal stability but may be cleaved photochemically. Irradiation of benzofurazan in benzene and in methanol gives the azepine (**26**) and the urethane (**27**), respectively; in the presence of triethyl phosphite (Z,Z)-1,4-dicyanobuta-1,3-diene is formed. The proposed mechanism (Scheme 3) involves nitrile oxide, oxazirene, acyl nitrene and isocyanate intermediates, and is supported by spectrophotometric studies ⟨76HCA2727⟩ and by trapping of the nitrile oxide as its isoxazole cycloadduct with DMAD ⟨75JOC2880⟩.

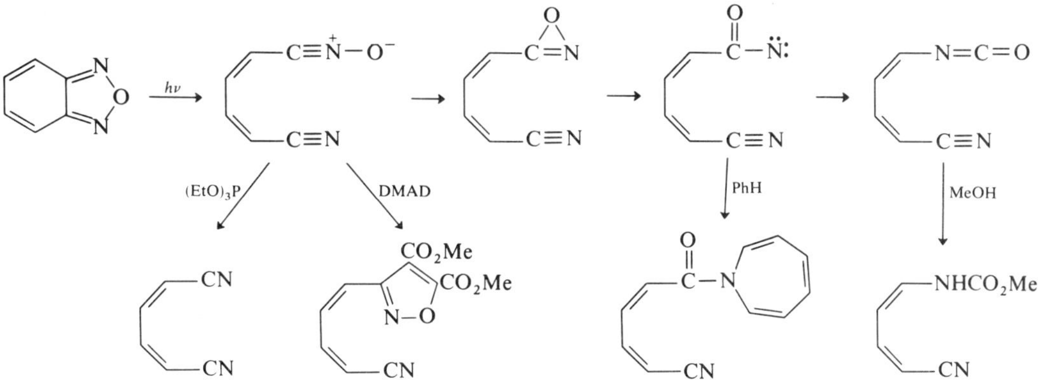

Scheme 3

4.22.3.1.2 Reaction with electrophiles and oxidizing agents

The heterocyclic ring is notably resistant to attack by electrophilic reagents. Nitration and halogenation of arylfurazans occurs only at the aryl groups; even 3-phenylfurazan fails to react at the unsubstituted 4-position ⟨78MI42200⟩ (see Section 4.22.3.4.1). Benzofurazan is nitrated in the homocyclic ring (see Section 4.22.3.3.1). Reaction with acid is not favoured; benzofurazan dissolves unchanged in concentrated sulfuric acid, and diphenylfurazan is not attacked by hydrochloric acid at 200 °C. ^{1}H NMR spectroscopy has been used to examine the acid–base properties of some disubstituted furazans ⟨81KGS35⟩; methylphenylfurazan has a pK_a value of -4.9 ± 0.2, compared with -8.4 ± 0.1 for benzofurazan. Quaternization of the ring nitrogens is difficult but has been accomplished using dimethyl sulfate in sulfolane ⟨74AJC1917⟩. This reaction is approximately 8 and 75 times slower, respectively, than the corresponding process for 1,2,5-thia- and 1,2,5-selena-diazoles. Some aminofurazans react with 2-chloro-1-acylalkenes in the presence of perchloric acid to give oxadiazolopyrimidinium salts, *e.g.* (**28**) ⟨80UKZ637⟩. The cyclization is reversed by treatment with sodium hydroxide, whereas ammonia generates the enaminofurazan (**29**).

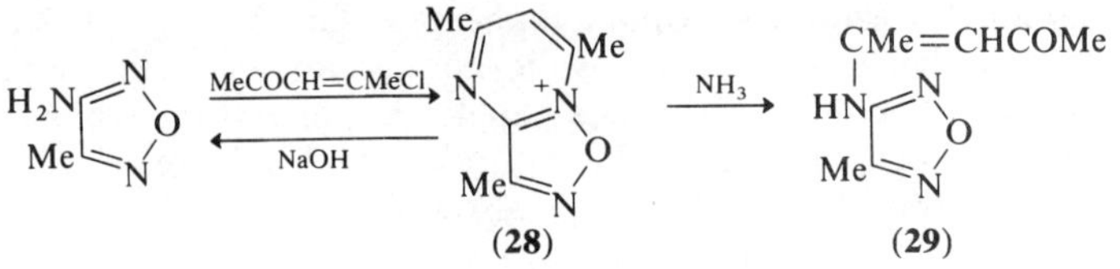

Direct oxidation of furazans to furoxans has not been achieved. The resistance of the oxadiazole ring to oxidizing agents is illustrated by the preferential attack of potassium permanganate at alkyl substituents; dimethylfurazan is converted into furazandicarboxylic acid *via* the methylfurazanmonocarboxylic acid (see Section 4.22.3.4.1). Furazans appear to be less susceptible to ozonolysis than furoxans.

4.22.3.1.3 Reaction with nucleophiles and reducing agents

The furazan ring is generally resistant to attack by nucleophiles. Benzofurazans tend to react at the homocyclic ring, halides being displaced by, for example, amines (see Section 4.22.3.3.2). Monocyclic analogues can also undergo reaction with nucleophiles at the substituents without disturbing the heterocyclic ring (see Section 4.22.3.4). Furazans are much more susceptible to reduction than oxidation. Benzofurazans on treatment with tin and hydrochloric acid are converted into *o*-diaminoarenes, but catalytic hydrogenation usually takes place on the homocyclic ring; the major product, tetramethylenefurazan (**19**; $n = 4$) is accompanied by only traces of *o*-phenylenediamine ⟨72T3271⟩. It has been reported that at temperatures above 200 °C phosphorus with hydrogen iodide reduces diphenylfurazan to 1,2-diphenylethane, whereas lithium aluminum hydride causes ring cleavage to two primary amine fragments.

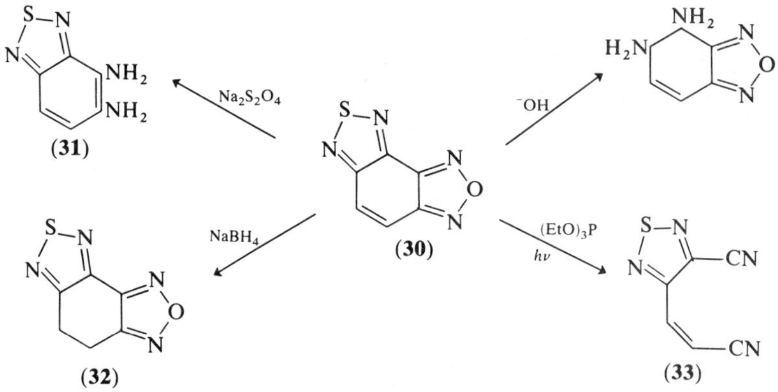

Scheme 4

The oxadiazole can be selectively reduced in the presence of other heterocycles. Furazano[3,4-*d*]pyrimidines and furazano[3,4-*e*]pyrazines have been converted into the corresponding *o*-diamino compounds using zinc and acetic acid. The thiadiazolobenzofurazan (**30**) with sodium hydrosulfite gives the diaminobenzothiadiazole (**31**); in contrast sodium hydroxide at 135–140 °C cleaves the sulfur-containing ring, while sodium borohydride and triethyl phosphite (with photolysis) yield the dihydro and dicyano compounds (**32**) and (**33**), respectively (Scheme 4) ⟨78CPB3896⟩. Polarographic reduction of disubstituted furazans shows 1–3 waves, the first of which corresponds to transfer of two electrons and N—O bond cleavage ⟨79KGS38⟩. The redox behavior of benzofurazans has also been examined and compared with their thio and seleno analogues ⟨72ZOB2049; 74JHC763⟩.

The parent furazan is a stable liquid which in the presence of base undergoes ring scission to α-oximinoacetonitrile ⟨65JOC1854⟩. Monosubstituted furazans, irrespective of the nature of the substituent, are likewise isomerized by alkali to the oximes of α-ketonitriles. Aminoacyl- and diacyl-furazans are susceptible to attack, but dialkyl, diaryl and alkylaryl analogues are comparatively inert. Treatment of benzofurazan with phenyllithium results in cleavage of the heterocyclic ring and formation of 2-anilinotriphenylamines. On the other hand the oxadiazole remains intact when butyllithium reacts with dimethylfurazan, lithiation taking place at one of the methyl groups (see Section 4.22.3.4.1). Furazans appear to show low reactivity towards Grignard reagents.

4.22.3.1.4 Heterocyclic ring transformations and miscellaneous reactions

A variety of rearrangements has been reported ⟨81AHC(29)141⟩ in which the oxadiazole is transformed into a new heterocyclic system bearing a hydroximino substituent (Scheme 5). The role of X=Y—Z can be filled by N=C—N, N=C—S, C=N—N, and C=N—O. 3-Arylformamidinofurazans on heating with potassium *t*-butoxide rearrange to the oximes of 3-acyl-1,2,4-triazoles. Phenyl isothiocyanate converts aminofurazans to their thiourea derivatives, which cyclize to 5-anilino-3-hydroximino-1,2,4-thiadiazoles ⟨77JCS(P1)1616⟩. On heating or treatment with alkali dibenzoylfurazan forms phenyl (4-phenylfurazan-3-yl)glyoxime; in the benzo fused series a similar rearrangement (**34**) → (**35**) has been reported. The sodium salt of 3-acetyl-4-methylfurazan is converted on heating into the oxime of 4-acetyl-5-methyl-1,2,3-triazole; similarly with alkali the phenylhydrazone (**36**) affords (**37**) ⟨76T1277⟩. Treatment of 3-amino-4-benzoylfurazan with base gives 5-phenyl-1,2,4-oxadiazole-3-carboxamide ⟨80JCS(P2)1096⟩. Similar rearrangements are reported for related heterocyclic systems (see Section 4.02.3.2).

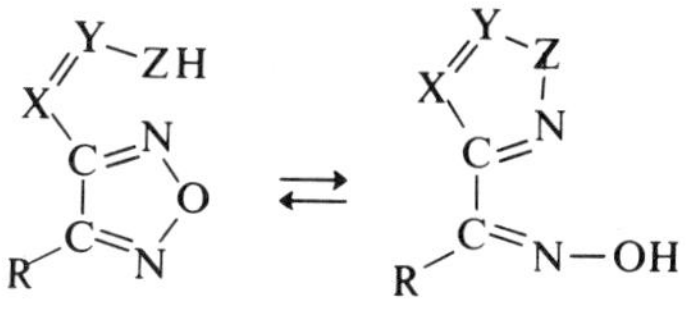

Scheme 5

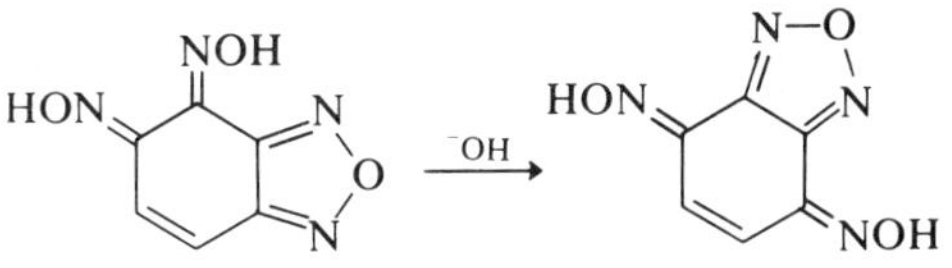

(34)   (35) 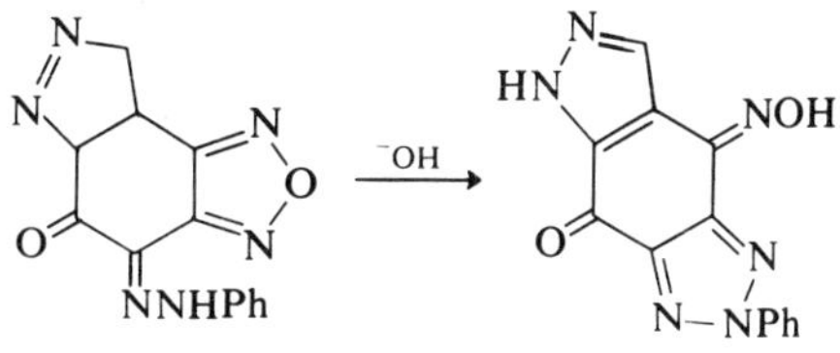   (36)   (37)

Coordination compounds have been prepared which incorporate 1,2,5-oxadiazoles as ligands, *e.g.* $[M(C_2N_2O)_3](SbCl_6)_2$ where M = Mg(II), Mn(II), Fe(II), Co(II), Ni(II) and Zn(II). The metal ions are in a regular octahedral environment of six nitrogen atoms with the ligands forming bidentate bridges ⟨78ZN(B)1120⟩. Similar complexes are also formed from benzofurazans and potassium tetrachloropalladate ⟨81ZOB1192⟩.

4.22.3.2 Reactions of the Heterocyclic Ring of Furoxans and Benzofuroxans

4.22.3.2.1 Thermal and photochemical isomerization

Tautomerism between the 2-oxides and 5-oxides is a consistent feature of furoxan chemistry. *cis*-Dinitrosoalkenes and *o*-dinitrosoarenes are likely intermediates, although definitive evidence has not yet been presented. A number of reactions have, however, been rationalized in terms of such dinitroso species; *e.g.* the formation of the bisazoxy compound (**39**) from benzofuroxan and *p*-anisyl azide ⟨75TL3577⟩.

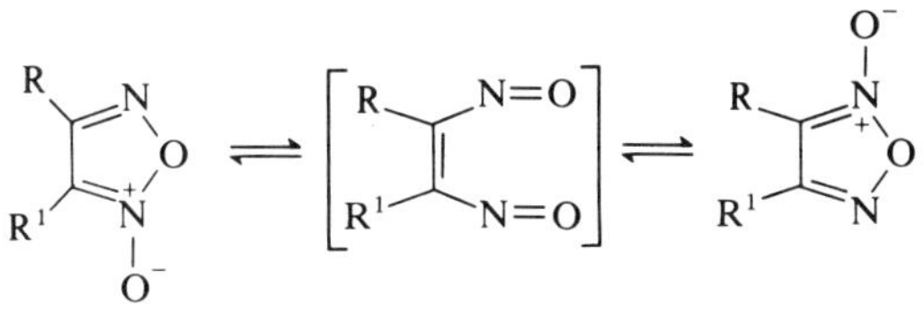

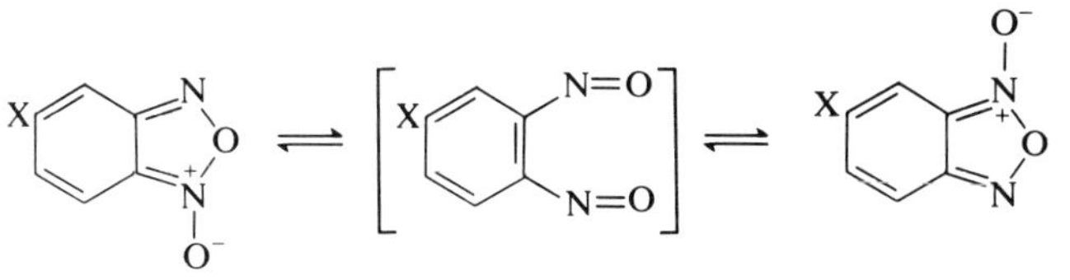

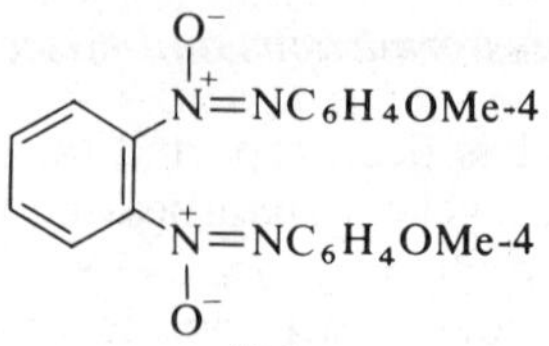

(39)

The rate of the process depends critically on the nature of the substituents. It is slow for unfused examples such as 3-methyl-4-phenylfuroxan, temperatures of ca. 100 °C being required ($\Delta G^{\ddagger}$ 133 kJ mol^{-1}). The energy barrier is somewhat reduced when an electron-donor group is present. Fusion to an aliphatic 6-membered ring also produces a small acceleration, attributable to strain; a larger effect might be expected for more strained analogues, e.g. trimethylenefuroxan (17; $n = 3$), but in these cases decomposition to nitrile oxides invariably intervenes (see Section 4.22.3.2.2). There is a further dramatic decrease in the energy barrier when there is a fused aromatic ring. For benzofuroxan the isomerization occurs at ambient temperature ($\Delta G^{\ddagger}$ 58 kJ mol^{-1}). Within the benzofuroxan series the electronic influence of substituents is negligible, but some steric effects have been observed.

The position of the equilibrium is influenced by a number of factors. In the monocyclic series electron-donor substituents favour the 4-position; 3-amino-4-arylfuroxans are not detected in the presence of their 4-amino-3-aryl isomers. On the other hand a corresponding preference for electron-attracting groups in the 3-position is not usually observed. Steric and hydrogen-bonding effects can also be significant. For benzofuroxans electronic effects operate for 5-/6-substituents. Carboxy, carbalkoxy, cyano and nitro substituents favour the 6-substituted tautomer, but less uniform behaviour is shown by donor groups. Additional steric factors are evident for 4-/7-substituted derivatives, and more heavily biased equilibria are generally observed for furoxans fused to heterocyclic rings.

The equilibrium constants for these furoxan interconversions can usually be measured from the ^{1}H NMR spectra (see Section 4.22.2.3.1); for rapidly equilibrating systems, such as benzofuroxans, low temperatures are necessary. The factors influencing both the position of the equilibrium and the rate of the process are thoroughly discussed in a recent review ⟨81AHC(29)251⟩.

Photochemical isomerization of furoxans has also been reported. For example, irradiation of 7-chloro-4-nitrobenzofuroxan generates the thermally less favoured 4-chloro-7-nitro isomer.

4.22.3.2.2 Thermal and photochemical ring cleavage

Under forcing conditions disubstituted furoxans cleave at the O(1)—N(2) and C(3)—C(4) bonds to give two nitrile oxide fragments. This process, which is the reverse of one of the principal preparative routes, can be regarded as a retro-1,3-dipolar cycloaddition reaction. Temperatures in excess of 200 °C are usually necessary, but when the substituents are bulky or there is ring strain the process is more facile. For example, diphenylfuroxan at 245 °C affords benzonitrile oxide (Scheme 6; R = Ph), which either rearranges to phenyl isocyanate or in the presence of a dipolarophile (X=Y) is trapped as its 1,3-dipolar cycloadduct. Using the FVP technique the nitrile oxides can be isolated ⟨79TL2443⟩. The kinetics of the gas-phase pyrolysis (220–300 °C) have also been examined ⟨81MI42201⟩. In contrast, more sterically hindered diadamant-1-ylfuroxan fragments in refluxing carbon tetrachloride.

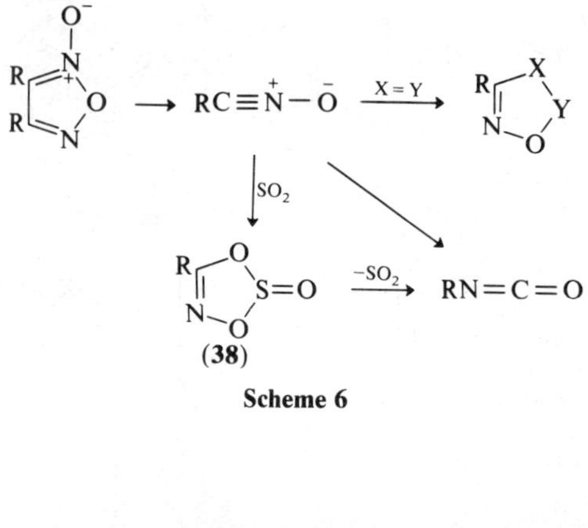

(38)

Scheme 6

Whereas disubstituted furoxans produce monofunctional nitrile oxides and isocyanates, bicyclic analogues such as decamethylenefuroxan (**17**; $n = 10$) give rise to bisnitrile oxides and diisocyanates. This reaction is particularly sensitive to ring strain; compounds (**17**; $n = 10$) and (**17**; $n = 3$) require temperatures of *ca.* 220 °C and 130 °C, respectively, while compounds (**20**), (**21**) and the dicyclopentadiene-derived furoxan (**40**) react at 80–100 °C. Products from the nitrile oxides are readily obtained at these lower temperatures, but the rearrangement to isocyanate is slow, polyfuroxans being formed instead. The process can, however, be adapted for the synthesis of diisocyanates by the addition of catalysts such as sulfur dioxide ⟨78CC113⟩; dioxathiazoles (**38**) are intermediates. The ready availability of (**40**) from dicyclopentadiene (see Section 4.22.4.3.2) therefore provides access to the diisocyanate (**41**), which is suitable for polyurethane formation. The use of sulfur dioxide likewise allows monoisocyanates to be prepared from monocyclic furoxans under FVP conditions ⟨80CI(L)665⟩. Diacyl- and disulfonyl-furoxans also give nitrile oxide-derived products ⟨81JOC316, 81TL3371⟩, but benzofuroxans are much less susceptible to decomposition. Bisiminoxyl biradicals, resulting from initial cleavage of the O(1)—N(2) bond, have been postulated as intermediates in furoxan fragmentations ⟨80DOK(255)917⟩.

(**40**) (**41**)

4.22.3.2.3 Reaction with electrophiles and oxidizing agents

Furoxans, like furazans, show low reactivity towards electrophiles; aryl substituents can be nitrated and halogenated, and benzofuroxans can undergo substitution in the homocyclic ring without disturbing the heterocyclic portion. Reaction with acid is also slow. Even the monosubstituted compound 3-phenylfuroxan can be steam distilled in the presence of acid without decomposition; with concentrated hydrochloric acid at 160 °C cleavage to benzoic acid occurs, whereas nitric acid leads to *p*-nitrobenzoic acid. Benzofuroxan has a pK_a value of -8.3 ± 0.1, similar to that for benzofurazan. Quaternization is difficult for all furoxans; benzofuroxan does not react with triethyloxonium tetrafluoroborate, but an *N*-quaternized intermediate may be involved in reactions such as the formation of 1-hydroxybenzimidazole 3-oxide from benzofuroxan and methyl trifluoromethanesulfonate. Benzofuroxan also reacts with 2,4-dianisyl-1,3,2,4-dithiaphosphetane 2,4-disulfide, 2,1,3-benzothiadiazole being formed, presumably *via* rearrangement of benzofurazan *N*-sulfide and deoxygenation ⟨80BSB247⟩.

The furoxan ring is generally resistant to attack by oxidizing agents, reaction often taking place preferentially at the substituent groups (see Section 4.22.3.4.1). Benzofuroxans may however be converted into the corresponding *o*-dinitrosoarenes using trifluoroperacetic or persulfuric acid ⟨81CC365⟩; milder reagents, *e.g.* performic and peracetic acids, appear to give no reaction. Some diarylfuroxans have been subjected to ozonolysis; carboxylic acid salts are formed on alkaline hydrolysis of the first-formed adducts.

4.22.3.2.4 Reaction with nucleophiles and reducing agents

Grignard reagents react with monocyclic furoxans to give mainly nitriles and ketones. The proposed mechanism, a modified version of which is also applicable to their reduction with lithium aluminum hydride, involves attack by the Grignard carbanion at C(3), followed by collapse of the resulting adduct to nitrile and nitronate fragments (Scheme 7); further reaction with excess Grignard reagent and hydrolysis yields the ketones. In contrast monosubstituted furoxans afford glyoximes.

In the benzofuroxan series nucleophilic attack occurs preferentially at the homocyclic ring if activating nitro groups are present, halides being readily displaced (see Section 4.22.3.3.2). Otherwise reaction generally takes place at N(5) of the oxadiazole; thus

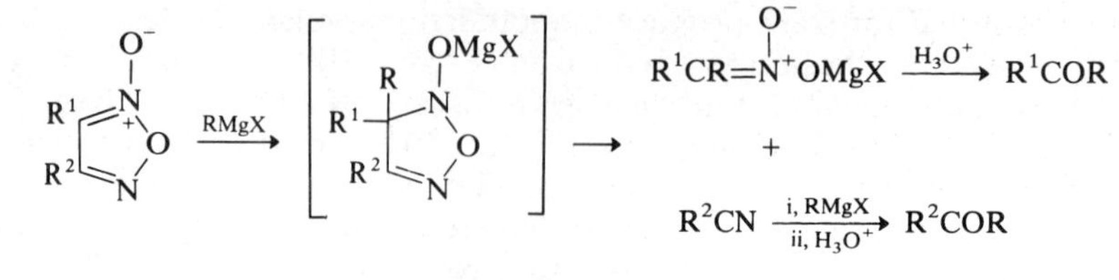

Scheme 7

secondary amines give mainly o-nitroarylhydrazines, together with other reduction products. Similarly pyrido[2,3-c]furoxan (**42**) with DMSO gives 2-dimethylsulfoximino-3-nitropyridine. 5-Dimethylaminopyrimidinol[4,5-c]furoxan reacts with nucleophiles (water, alcohols, amines, hydride ion) to give covalent adducts (**43**; Nu = OH, OR, NR$_2$, H) in which the furoxan ring remains intact ⟨82CC60⟩. In marked contrast the 7-dimethylamino-5-methyl analogue (**44**) undergoes aminative opening of the oxadiazole ring affording 4-amino-2-methyl-5-nitrosopyrimidine; under similar conditions it is the pyrimidine ring of (**45**) that is cleaved yielding the furoxancarboxamide (**46**) ⟨82CC267⟩. Monosubstituted derivatives, such as 3- and 4-phenylfuroxans, are particularly susceptible to ring opening by base, giving α-nitrophenylacetonitrile and α-oximinophenylacetonitrile oxide, respectively. Various substituents directly attached to the heterocyclic ring are displaced by nucleophiles. 3-Nitro-4-phenylfuroxan, prepared by peroxide oxidation of the amino derivative, reacts with sodium methoxide to give 3-ethoxy-4-phenylfuroxan ⟨81JCS(P2)1240⟩. Similar substitutions have been reported for 3-methyl-4-nitrofuroxan with phenol, ethoxide, pyrrolidine, thiophenol and ethyl sulfide ⟨73JHC587⟩.

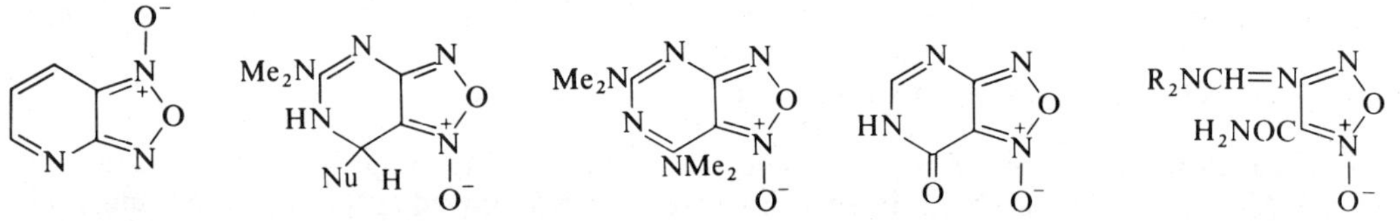

Displacement of halides by secondary amines and of sulfonyl groups by alkoxides can also take place. Furoxancarboxylic acids are attacked by base to give acyclic products, but their derivatives can undergo nucleophilic acyl substitutions. Likewise nucleophilic addition reactions can be accomplished for ketofuroxans, although ring cleavage is also commonplace. The generation of new heterocyclic systems by reaction with nucleophiles is dealt with in Section 4.22.3.2.5.

The reduction of furoxans has been the subject of extensive investigation and proves to be more facile than their oxidation. Furazans, α-dioximes, nitriles and 1,2-diamines are formed depending on the conditions.

Sodium borohydride reduces dialkyl-, diaryl-, and benzo-furoxans to dioximes, whereas the use of lithium aluminum hydride results in cleavage at both O(1)—N(2) and C(3)—C(4) bonds yielding primary amines. Rupture of the carbon–carbon bond may also occur during catalytic hydrogenation, but dioximes are the more usual products; compound (**17**; n = 4) gives cyclohexane-1,2-dione dioxime and hexamethylenediamine with, respectively, palladium/charcoal at 20 °C and Raney nickel at 100 °C. Reduction with dissolving metals generally affords dioximes but furazans may also be formed, presumably as secondary dehydration products. Commonly used reagents include zinc and acetic acid, tin and hydrochloric acid, and tin(II) chloride in acid. Similar behavior is shown by monosubstituted furoxans. Reduction of benzofuroxans to o-quinone dioximes has been reported using methanol and potassium hydroxide, hydroxylamine and alkali, and by thiols; this last reaction has been proposed as the basis for assaying thiol groups in enzymes (see Section 4.22.5). Copper and hydrochloric acid reduce benzofuroxan to o-nitroaniline.

Tervalent phosphorus compounds, such as trialkyl and triaryl phosphites and phosphines, deoxygenate furoxans to furazans. For simple dialkyl-, diaryl- and alkylaryl-furoxans forcing conditions (e.g. in refluxing triethyl phosphite) are required. At still higher temperatures (refluxing triphenyl phosphite) fragmentation and further reduction to nitriles takes place. In contrast more strained analogues and benzofuroxans react much more readily; acenaphthofuroxan (**18**) gives acenaphthofurazan (**22**) at room temperature and 1,8-dicyanonaphthalene on warming. Benzofuroxan can be deoxygenated by heating with sodium azide in ethylene glycol or carboxylic acid solvents. Thermal deoxygenation, *e.g.* of imidazo[4,5-*f*]benzofuroxans, has also been reported. Polarographic studies reveal that the products of electrochemical reduction of furoxans are pH dependent, with dioxime anions being formed initially. Other reducing agents include hydrazine hydrate, sulfur dioxide, ammonium and alkali metal sulfides, and iodide ion.

4.22.3.2.5 *Heterocyclic ring transformations*

The generation from furoxans of other heterocyclic systems, some of which show useful biological activity (see Section 4.22.5), has been the subject of intensive investigation. The following subsections summarize the conversion of benzofuroxans into quinoxaline and benzimidazole oxides, the rearrangement of 4-substituted benzofuroxans, and the transformation of monocyclic furoxans into isoxazoles and isoxazolines, furazans, and pyrazolines. More detailed discussion is to be found in recent comprehensive reviews ⟨75S415, 76H(4)767, 81AHC(29)251⟩.

(i) *Transformation of benzofuroxans into quinoxaline oxides and benzimidazole oxides*

Enamines and enolate anions react readily with benzofuroxan to form quinoxaline dioxides (**47**). This process, which is frequently referred to as the Beirut reaction, can be adapted to accommodate a variety of substituents: for example the use of ynamines and β-diketones leads to 2-amino- and 2-acyl-substituted products, respectively. Polycyclic compounds can also be synthesized: benzo[*a*]phenazine and indolo[2,3-*b*]quinoxaline dioxides are formed from β-naphthol and indole. In the absence of an α-hydrogen atom for elimination (*e.g.* with Me$_2$NCH=CMe$_2$) the corresponding dihydroquinoxaline dioxide (**48**) is obtained. 1-Hydroxyquinoxalin-2-one 4-oxides are the major products from β-ketonitriles, β-keto esters or α-(methylthio) ketones.

(**47**)   (**48**)   (**49**)   (**50**)

Accompanying elimination reactions can lead to quinoxaline mono-*N*-oxides: benzofuroxan and morpholine with cinnamaldehyde (or benzylideneacetone) yield 3-phenylquinoxaline 1-oxide, the formyl (or acetyl) group being removed by base. The reaction takes a different course when leaving groups such as CN$^-$, RSO$_2^-$ and NO$_2^-$ are present. Cyanoacetamides, primary nitroalkanes and β-ketosulfones yield 1-hydroxybenzimidazole 3-oxides (**49**), whereas 2*H*-benzimidazole 1,3-dioxides (**50**) are formed from secondary nitro compounds. In the absence of such leaving groups the oxidation level of the products may be lower: benzofuroxan and ethyl acetoacetate give ethyl 1-hydroxybenzimidazole-3-carboxylate, while 1,3-dihydroxybenzimidazol-2-one results from treatment with formaldehyde and alkali.

These reactions apply to a wide range of 5-substituted benzofuroxans and heterocyclic ring-fused analogues; 4-substituted benzofuroxans have received much less attention. The mechanism of the reaction has been the subject of some debate ⟨76H(4)767, 81AHC(29)251⟩. While not all aspects have yet been fully explained, it is generally accepted that it involves nucleophilic attack at N(3) of the ground state of benzofuroxan, or less likely at one of the nitrogens of *o*-dinitrosobenzene (Scheme 8). Subsequent cyclization to 5- or 6-membered rings is dependent on the nature of the substituents.

(ii) *Rearrangement of 4-substituted benzofuroxans*

Benzofuroxans (**51**) bearing a variety of unsaturated substituents (X=Y) in the 4-position rearrange to nitroheterocycles (**52**; Scheme 9); the role of X=Y can be filled by C=N, C=O, N=N, and N=O. Nitroindazoles (**52**); X=C, Y=NR) are formed from imines, oximes and hydrazones of 4-acylbenzofuroxans; equilibrium between the parent 4-formylbenzofuroxan and 7-nitroanthranil (**52**; X=CH, Y=O) has also been demonstrated ⟨77JOC897⟩, the proportions of the isomers being dependent on the substituent in the

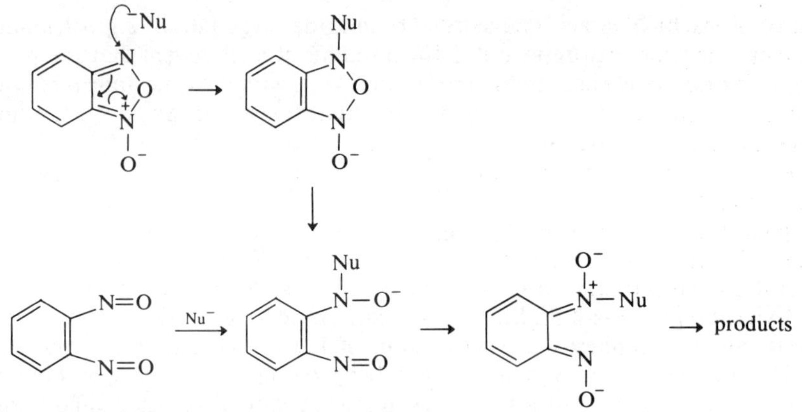

7-position. Likewise the 4-arylazo- and 4-nitroso-5-dimethylamino analogues, generated *in situ* from 5-dimethylaminobenzofuroxan by reaction with arenediazonium chlorides and nitrous acid, respectively, isomerize to nitrobenzotriazoles (**52**; X = N, Y = NAr) and 4-nitrobenzofurazan (**52**; X = N, Y = O). An unusual intramolecular redox process accounts for the formation of 7-benzenesulfonyl-4-nitrobenzofurazan on heating the corresponding benzofuroxan sulfoxide (see Scheme 14, Section 4.22.3.3.3).

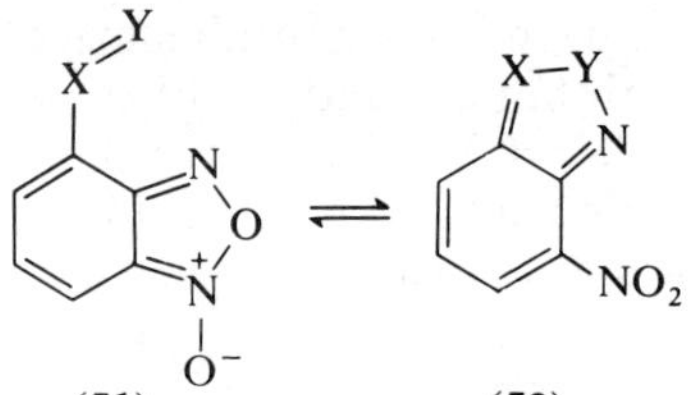

Nitro groups in the 4-position may also participate in the rearrangement. For example, 5-chloro- and 5-methyl-4-nitrobenzofuroxans rearrange to their 7-chloro- and 7-methyl-4-nitro isomers, the ease of the process being attributed to steric inhibition of resonance for the nitro group in the starting materials. Similarly, 4-ethoxycarbonylamino-5,7-dinitrobenzofuroxan equilibrates with its 4,6-dinitro-5-urethane isomer.

(iii) *Rearrangements of monocyclic furoxans*

Monocyclic furoxans can be transformed into a variety of other heterocyclic systems, including isoxazolines, isoxazoles, pyrazolines and furazans. Much of the early literature dealing with these reactions has required substantial revision; an up-to-date assessment is provided in a recent review ⟨81AHC(29)251⟩, and there follows hereafter only a summary of the better-substantiated transformations.

4-Aryl-3-methylfuroxans rearrange on heating in alcoholic alkali hydroxides or alkoxides in a process which has become known as the isoxazoline transposition or Angeli's rearrangement (Scheme 10). The products have recently been firmly established ⟨81G167⟩ to be oximes of 3-arylisoxazolin-4-ones (**53**), as proposed by Angeli, rather than as 3-aryl-5-hydroxisoxazolin-4-imines which have sometimes been favoured. A mechanism involving cleavage of O(1)—N(2) and nitrosoalkene intermediates accounts for the observed transformations.

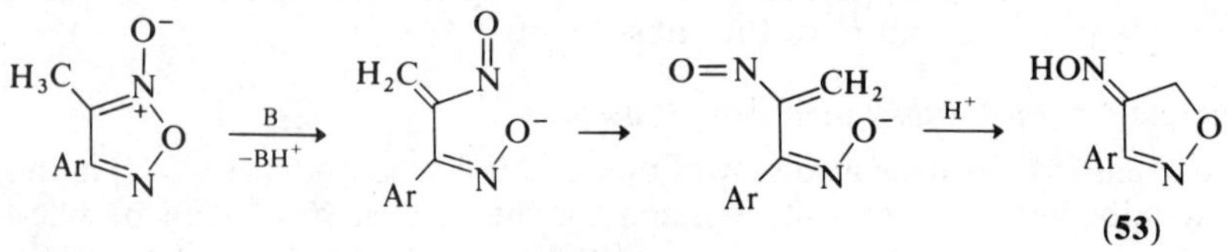

Heating furoxans with alkenes may also yield 2-isoxazolines as a result of initial fragmentation to nitrile oxides and subsequent 1,3-dipolar cycloaddition (see Section 4.22.3.2.2);

similarly, thermolysis in alkynes and nitriles leads to isoxazoles and 1,2,4-oxadiazoles. Nitrile oxide cleavage products also account for the formation of isoxazolines and isoxazoles from monosubstituted furoxans with alkenes and alkynes; for example decarboxylation of 4-methylfuroxan-3-carboxylic acid in the presence of alkynes yields 3-acetohydroximinoisoxazoles *via* 2-hydroximinopropionitrile oxide, and furoxan-3-carbaldoxime apparently rearranges to 5-amino-4-nitroisoxazole which ring opens to fulminuric acid (H$_2$NCOCH(CN)NO$_2$).

Ammonia and primary amines react with dibenzoylfuroxan to give 3-amino-4-nitrosophenylisoxazole (**54**), which may rearrange thermally to 3-amino-4-benzoylfurazans. Earlier suggestions that the products are 3-acyl-5-amino-1,2,4-oxadiazoles have been discounted ⟨69JHC317⟩. A mechanism involving nitrile oxide and glyoxime intermediates has been proposed (Scheme 11) and it is apparent that other diacylfuroxans behave similarly.

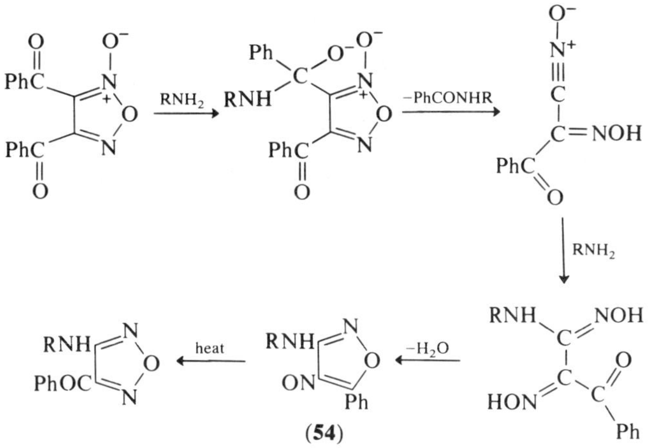

Scheme 11

Furoxans bearing oxime groups in the 4-position rearrange to α-nitroalkylfurazans in a process analogous to that observed for benzofuroxans (Scheme 9); *e.g.* furoxan-3,4-dicarbaldoxime (α-isocyanilic acid) gives 2-(furazan-3-yl)-2-nitroacetaldoxime (β-isocyanilic acid), while the oxime of 4-benzoyl-3-methylfuroxan is converted into a furazan, which was originally believed to be methyl-α-nitrobenzylfurazan ⟨36G819⟩ but has subsequently been assigned as its phenyl-α-nitroethyl isomer ⟨81AHC(29)251⟩.

4.22.3.3 Benzofurazans and Benzofuroxans: Reactions of the Homocyclic Ring

The reactions of the homocyclic ring are considered in three parts: attack by electrophiles; attack by nucleophiles; and transformations involving substituent groups, together with miscellaneous reactions. Unless otherwise stated common behaviour is assumed for benzofurazans and benzofuroxans.

4.22.3.3.1 Electrophilic attack

The most facile electrophilic substitution is nitration. Both benzofurazan and benzofuroxan react preferentially at the 4-position, and a second nitro group can sometimes be inserted at C(6); 4- and 5-substituted analogues are nitrated at C(7) and C(4), respectively. 5-Substituted 4-nitrobenzofuroxans frequently rearrange to their 7-substituted 4-nitro isomers (see Section 4.22.3.2.5.ii). Other electrophiles generally react less readily, nitrosation and diazo coupling occurring only in the presence of activating groups. Bromine adds to benzofurazan to give a 4,5,6,7-tetrabromo adduct (**55**; X = Br) whereas 5-bromo- and

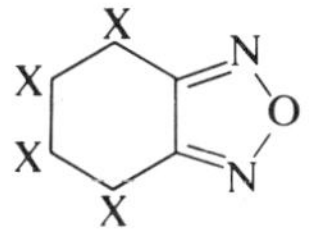

(**55**)

5-methyl-benzofurazan undergo substitution (Br$_2$/HBr) in the 4-position; treatment of the resulting 4-bromo compounds with cuprous cyanide in DMF affords the corresponding cyano derivatives ⟨74JHC813⟩. Naphtho[1,2-c]-furazans and -furoxans also undergo electrophilic substitutions, the main site of attack being the 4-position.

4.22.3.3.2 Nucleophilic attack

The susceptibility of the homocyclic ring to attack by nucleophiles has been the subject of intensive investigation, particularly in the last decade. Halides can be displaced by, for example, amines, alkoxides, phenoxides and thiolates. 4-Halogenobenzofurazans yield products with the incoming nucleophile located at either the 4- or 5-positions, resulting from 'normal' and 'cine' substitution, respectively; similar behavior is shown by 5-halogeno derivatives. The relative amounts depend on the polarity of the solvent and on steric factors; bulkier nucleophiles favor normal products, whereas the proportion of cine substitution increases as the halogen is varied from fluorine to bromine. The mechanisms involved have been studied in detail and the results recently reviewed ⟨81MI42200⟩. Addition–elimination (AE_a) is favored ⟨74T863⟩ for cine products from thiomethoxy dehalogenations, while both AE_a and S_NAr pathways operate for normal substitution (Scheme 12). Elimination–addition has also been considered, e.g. for methoxy dehalogenations ⟨74JCS(P2)1171⟩.

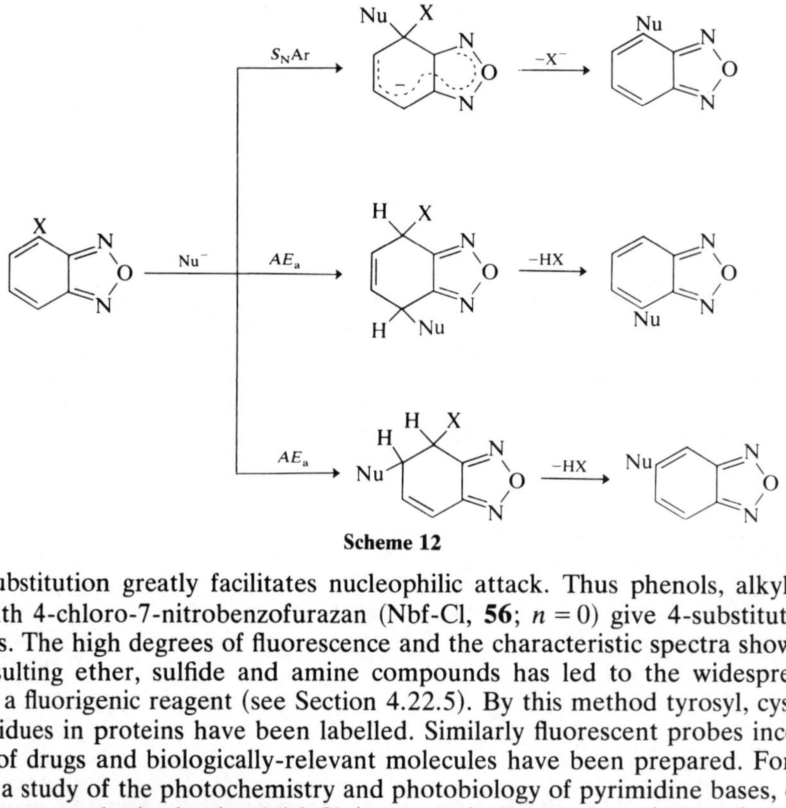

Scheme 12

Nitro substitution greatly facilitates nucleophilic attack. Thus phenols, alkylthiols and amines with 4-chloro-7-nitrobenzofurazan (Nbf-Cl, **56**; $n = 0$) give 4-substituted 7-nitro derivatives. The high degrees of fluorescence and the characteristic spectra shown by each of the resulting ether, sulfide and amine compounds has led to the widespread use of Nbf-Cl as a fluorigenic reagent (see Section 4.22.5). By this method tyrosyl, cysteinyl and amino residues in proteins have been labelled. Similarly fluorescent probes incorporating a variety of drugs and biologically-relevant molecules have been prepared. For example, as part of a study of the photochemistry and photobiology of pyrimidine bases, compound (**57**) has been synthesized using Nbf-Cl ⟨75JOC1559⟩; light energy (270 nm) absorbed by the thymine is efficiently transferred to the fluorescent nitrobenzofurazan fragment ⟨75JA245⟩. For benzofuroxans additional reactivity results from the presence of the N-oxide moiety. 7-Chloro-4-nitrobenzofuroxan (**56**; $n = 1$) reacts readily with nucleophiles, the displacement of halide sometimes being accompanied by isomerization (see Section 4.22.3.2.5.ii).

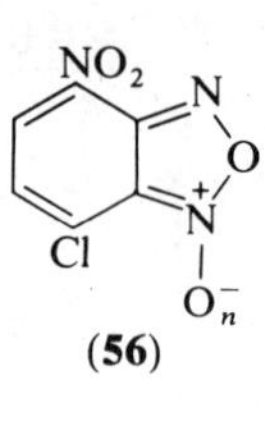

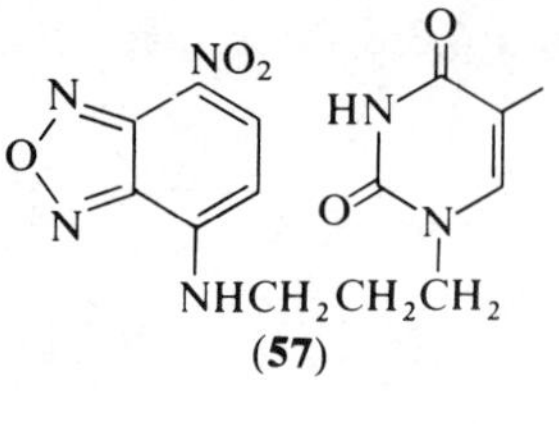

The formation of Meisenheimer complexes, all of which appear to be highly colored ⟨81MI42200⟩, has also attracted close attention, partly due to the possibility they may be associated with high levels of biological activity. ^{13}C NMR spectroscopy ⟨76OMR(8)56⟩ and stopped-flow techniques ⟨75JCS(P2)1469⟩ have proved of particular value for determining their structure and examining the kinetics of the system.

4,6-Dinitrobenzofuroxan is slightly acidic, forming salts (**58**) by addition of hydroxide ion; at higher pH the dianion (**59**) is produced. Hydroxide ion can also displace methoxide from 7-methoxy-4-nitrobenzofuroxan to yield the 7-hydroxy derivative, which may subsequently rearrange to the 5-hydroxy isomer. Substitution occurs when 5,6-dinitrobenzofuroxan is treated with primary arylamines, but heterocyclic ring opening is observed when benzofuroxan is heated with aniline, 2,5-dianilino-1,4-benzoquinone dianil being the major product. Azide ion replaces chloride from chloronitro derivatives; for the 5-chloro-4-nitro compound spontaneous loss of nitrogen from the product affords furoxanobenzofuroxan (**60**). Both reduction and substitution take place when iodobenzofuroxans are treated with lithium dimethylcuprate ⟨79JOM(166)265⟩; methylation predominates when iodomethane is added as quenching agent.

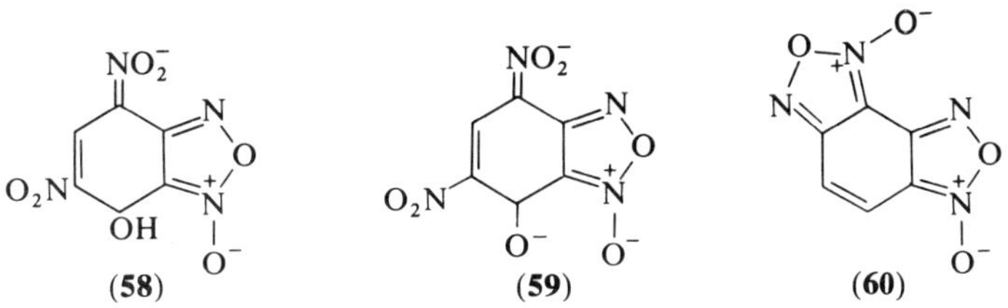

Displacement of a nitro group has also been observed. 4,7-Bis(arylsulfonyl)benzofuroxans are formed from (**56**; $n = 1$) and arylsulfinates; furthermore, the arylsulfonyl groups in the product may themselves be displaced by other nucleophiles.

Substitution in the homocyclic ring may be accompanied by deoxygenation of the furoxan. 4-Nitrobenzofuroxan with dialkylamines gives 4-dialkylamino-7-nitrobenzofurazan; 5-trifluoromethyl and 5-ethoxycarbonyl compounds are reported to react similarly at C(6).

4.22.3.3.3 Miscellaneous reactions and transformations involving substituents

A variety of transformations involving groups attached to the homocyclic ring has been reported. 5-Chlorocarbonylbenzofuroxans react readily with amino acids to give mixtures of the 5- and 6-amides; the chiroptical properties of these have been examined in detail ⟨80JHC213⟩. Decarboxylation of carboxylic acid derivatives has also been reported ⟨1893CB2897⟩, *e.g.* 1,2-naphthofurazan-4-carboxylic acid is converted into the parent naphthofurazan on boiling with alkali or sulfuric acid. Early attempts to prepare 5-aminobenzofuroxans directly from the corresponding anilines failed; an alternative approach was therefore developed which involves Curtius rearrangement of the acyl azide (**61**) formed from benzofuroxan-5-carboxylic acid, and hydrolysis of the urethane derivative (**62**) of the resulting isocyanate (Scheme 13).

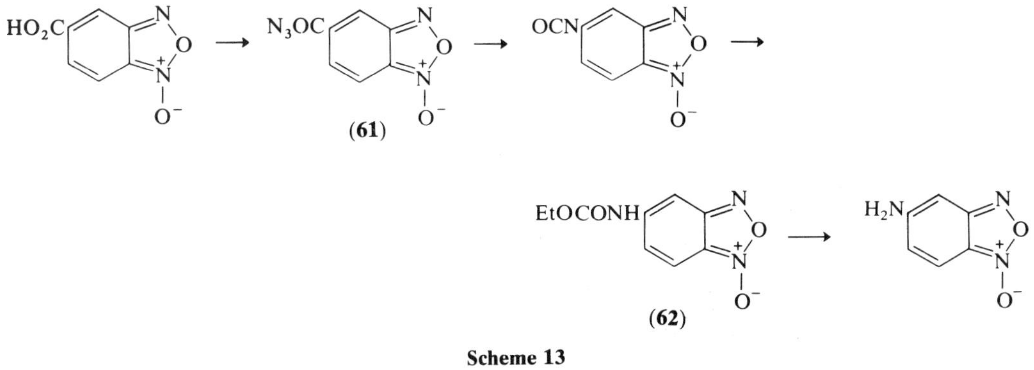

Scheme 13

The ability of aminobenzofurazans to form imine derivatives has been exploited for the construction of fused heterocyclic systems; 4,5-diaminobenzofurazan on treatment with benzil yields 7,8-diphenylfurazano[3,4-*f*]quinoxaline ⟨79T241⟩. The conversion, using *m*-chloroperbenzoic acid, of 4-nitro-7-phenylthiobenzofurazan (**63**) to the sulfone (**64**) apparently involves oxidation of the sulfide to sulfoxide (**65**), followed by intramolecular oxygen transfer (Scheme 14).

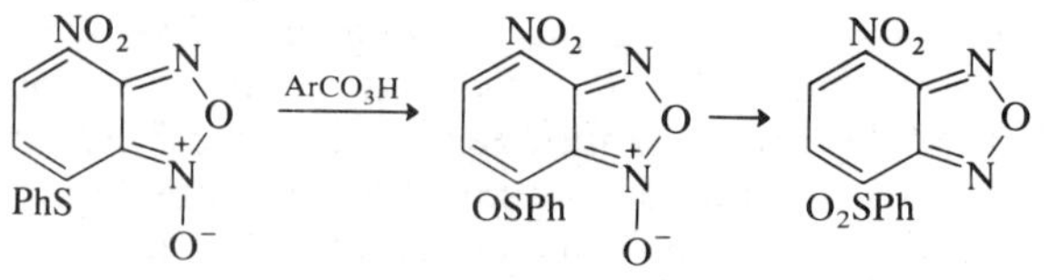

Scheme 14

The double bonds formulated at the 4,5- and 6,7-positions are sufficiently localized and activated to undergo cycloaddition reactions. 1,3-Dienes such as isoprene and butadiene give 1:1 and 2:1 Diels–Alder adducts with 4-nitrobenzofuroxan; similarly, diazomethane reacts at C(5)—C(6) in benzofurazans and furoxans forming pyrazolino 1,3-dipolar cycloadducts, which may rearrange or fragment with loss of nitrogen yielding 5,6-cyclopropa-fused derivatives ⟨76T1277⟩.

A rare example of conversion of benzofurazans to their monocyclic analogues is provided by Beckmann fragmentation reactions; both (**66**) and (**67**) undergo homocyclic ring cleavage to the nitrilecarboxylic acids (**68**) and (**69**), respectively ⟨75G723⟩. Fused heterocyclic rings can also be cleaved to monocyclic derivatives; 4-aminopyrimidino[4,5-*c*]furazan reacts with methylamine to give 3-aminofurazan-4-*N*-methylcarboxamide ⟨71JOC3211⟩. Transformations involving substituents attached to such fused heterocycles can also be accomplished ⟨80JOC3827⟩.

4.22.3.4 Monocyclic Furazans and Furoxans: Reactions of Substituents

4.22.3.4.1 Alkyl- and aryl-furazans and -furoxans

Although numerous dialkyl, diaryl and alkylaryl furazans are known, the reactions of the substituents have not been thoroughly investigated. Even less attention has been paid to the corresponding furoxans.

Alkyl groups, whether attached directly to the heterocycle or to aryl substituents, are oxidized by potassium permanganate to carboxylic acids. Both methylfurazancarboxylic acid and furazandicarboxylic acid can be formed from dimethylfurazan; similarly di(*p*-tolyl)-furazan yields the di(*p*-carboxyphenyl) derivative ⟨73MI42200⟩.

Aryl groups can be nitrated, brominated and chlorinated, particularly when electron-withdrawing substituents are present; no reaction occurs at the heterocyclic ring. Even monophenylfurazan reacts exclusively at the phenyl group; the furazan is *ortho/para*-directing and proves to be somewhat less activating than the corresponding 1,2,5-thiadiazole which exhibits greater π-delocalization ⟨78MI42200⟩. Dimethylfurazan can be brominated using NBS with AIBN ⟨80GEP2919293⟩; five chlorinated products have been detected for the corresponding free radical chlorination ⟨77KGS30⟩ and the kinetics of this process have been examined ⟨78ZOR1255⟩.

Dimethylfurazan also undergoes lateral lithiation with butyllithium and affords (methylfurazanyl)acetic acid after carboxylation. The scope this process presents for achieving transformations involving pendant groups without disruption of the heterocycle is illustrated by the synthesis (Scheme 15) of methylvinylfurazan ⟨81JHC1247⟩.

4-Substituted 3-methylfuroxans undergo the isoxazolinic transposition in which the methyl becomes incorporated into a new heterocycle (see Section 4.22.3.2.5.iii).

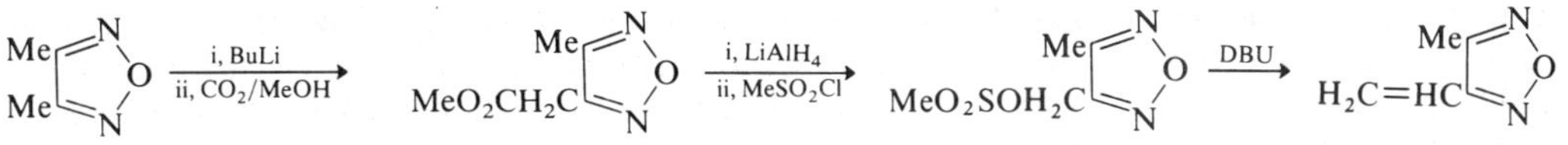

4.22.3.4.2 Acyl-substituted furazans and furoxans

Even though acyl furazans and furoxans are more susceptible to heterocyclic ring cleavage than their alkyl and aryl analogues, many of the usual reactions of the carbonyl group are possible. Borohydride reduction yields the expected alcohols, and oxidation of acetyl-furazans to the carboxylic acids can be accomplished using acidic permanganate. With diazomethane epoxides are produced. Oxime, hydrazone and dichloro derivatives are formed by treatment with hydroxylamine, arylhydrazines and phosphorus pentachloride, respectively. The oximes and hydrazones may undergo rearrangements involving the oxadiazole ring (see Sections 4.22.3.1.4 and 4.22.3.2.5.iii). Condensation of diaroyl-furazans and -furoxans with hydrazine provides a route to fused pyridazines (**70**; $n = 0, 1$) ⟨77MI42200⟩; similarly furazanopyridines (**71**) may be prepared using aminoalkanes and DBU ⟨79S687⟩.

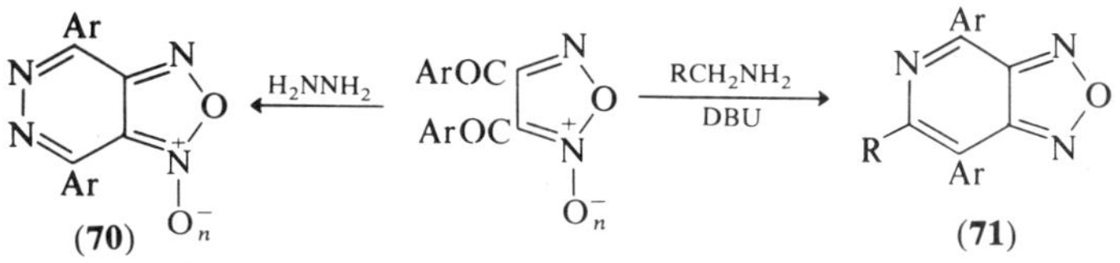

4.22.3.4.3 Furazan- and furoxan-carboxylic acids and their derivatives

The dicarboxylic acid compounds are apt to undergo ring cleavage reactions, particularly in the presence of alkali; boiling water converts furazandicarboxylic acid to cyano-oximinoacetic acid, presumably *via* initial decarboxylation to the monoacid.

Acid derivatives including esters, amides, halides and nitriles are readily accessible. Dicyanofuroxan (**72**) shows in its reactions some similarities to phthalonitrile ⟨75LA1029⟩; it also provides a source of fused pyridazino- and oxazino-furoxans (**73**) and (**74**) *via* addition with hydrazine and hydroxylamine, respectively ⟨82H(19)1063⟩. The tetronic acid compound (**75**) yields a hydroxyamide (**76**) on aminolysis ⟨79S977⟩.

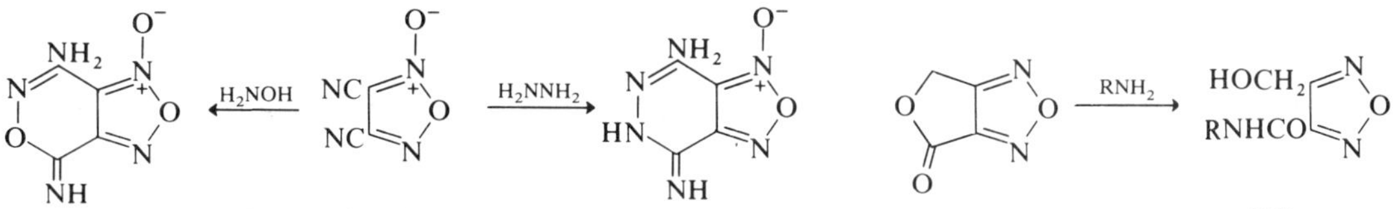

Arylfurazancarboxamides on treatment with alkaline hypochlorite undergo Hofmann degradation to the amines; likewise carbamates result from Curtius rearrangement of furazanylacyl azides in the presence of alcohols.

4.22.3.4.4 Amino- and nitro-furazans and -furoxans

Aminofurazans have received much greater attention than the corresponding furoxans. Both methyl- and aryl-aminofurazans have been studied in detail ⟨73JPR791⟩ and it has been shown that a variety of reactions are possible without affecting the oxadiazole ring. Urea, acetyl and enamino derivatives can be prepared using aryl isocyanates, acetic anhydride and vinyl ethers. Schiff's bases are formed with aldehydes and acetals; benzaldehyde can also react with two molecules of amine to give methylenediamino compounds, *e.g.* (**77**). With phenyl isothiocyanate the initial thiourea adduct undergoes rearrangement to a 1,2,4-thiadiazole ⟨77JCS(P1)1616⟩ (see Section 4.22.3.1.4). The triazene (**78**) is formed on diazotization, while permanganate oxidation affords the azo(methylfurazan) (**79**), which in turn can be reduced to its hydrazo analogue with phenylhydrazine. Peroxide oxidation yields nitro derivatives ⟨81JCS(P2)1240⟩. Treatment of 3-amino-4-phenylazofurazan with lead tetraacetate ⟨74BCJ1493⟩ yields heteropentalene (**80**), which can also be prepared by reaction of nitrosobenzene with aminoazidofurazan or by thermolysis of the azidophenylazo compound ⟨81ZOR1123⟩. Diaminofurazan reacts with α-dicarbonyl compounds to form furazano[3,4-b]pyrazines ⟨78JOC341⟩ and with β-keto esters to give fused diazepinones, e.g. (**81**), and pyrazinones ⟨78KGS1196⟩; oxidation with Caro's acid, ammonium persulfate or hydrogen peroxide yields aminonitro, diaminoazo and azoxy compounds ⟨81ZOR861⟩.

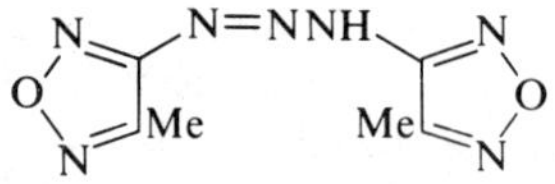

Some aryl- and aroyl-aminofuroxans and bis(dialkylamino)furoxans have been reported. 4-Amino-3-methylfuroxan has been prepared by hydrolysis of its *O*-benzylurethane. The 4-amino-3-aryl compounds can be acylated but the 3-amino-4-aryl isomers appear to be less reactive. On oxidation with hydrogen peroxide 3-aryl-4-nitrofuroxans are formed. The nitro group can be displaced by nucleophiles (see Section 4.22.3.2.4) or reduced back to the amino group using tin(II) chloride.

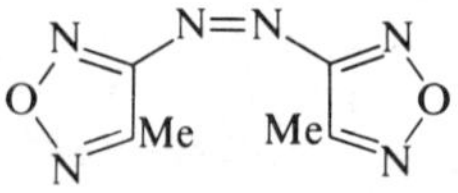

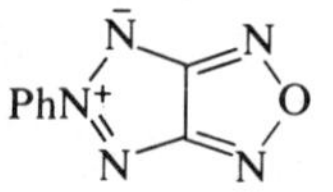

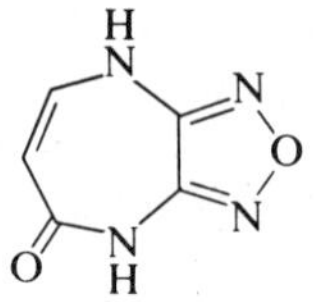

4.22.3.4.5 *Furazan and furoxan alcohols and thiols, and their derivatives*

While several furazans and furoxans bearing oxygen- and sulfur-containing substituents are known, few reactions involving these groups have been described. A rare example of direct attachment of hydroxyl to the oxadiazole is provided by the formation of quinol-2-ylhydroxyfurazan during the thermolysis of α-oximino-α-quinolylacetyl azide; acetoxy and methoxy derivatives were formed from acetic anhydride and diazomethane, respectively ⟨79JHC689⟩. Furazans bearing more remote alcohol groups are more commonplace and undergo the expected reactions; with ketones dihydroxymethylfurazan forms cyclic ketals, which may themselves be converted into ester, chloro and vinyl ether derivatives ⟨81ZOR1047⟩; its dinitrate ester has also been prepared ⟨75IZV1870⟩.

Furazan and furoxan sulfides and sulfones have been reported. The sulfides, formed by displacement of nitro from nitrofuroxans by thiophenols, can be oxidized to the corresponding sulfones using hydrogen peroxide.

The diversity of reactions that can be performed on substituents attached to the furazan ring is well illustrated by the five-stage synthesis of the thieno[3,4-c][1,2,5]oxadiazole system (**82**) from the dibenzoyl compound (**83**; Scheme 16): borohydride reduction affords the

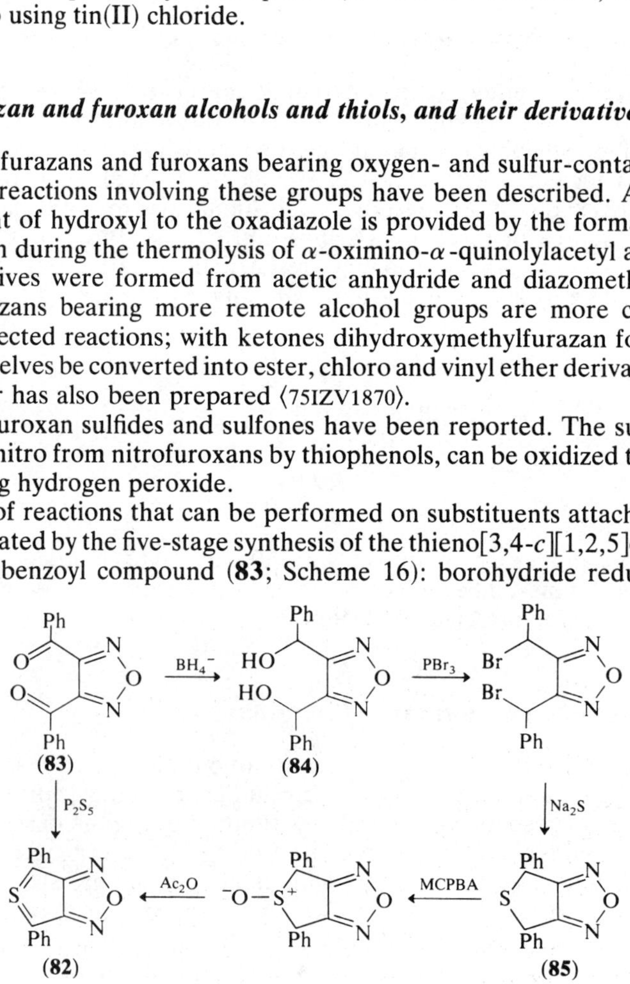

Scheme 16

diol (**84**) which is converted into its dibromo derivative using phosphorus tribromide; treatment with sodium sulfide, peracid oxidation of the resulting cyclic sulfide (**85**) followed by dehydration yields (**82**). Each step occurs without disruption of the furazan nucleus ⟨77H(6)1173⟩. Direct conversion of (**83**) into (**82**) using phosphorus pentasulfide has also been reported, as have cycloadditions of the thiocarbonyl ylide dipole in this non-classical 10π-electron system with activated alkenes such as N-phenylmaleimide ⟨80H(14)423⟩. The resulting cycloadduct (**86**) can undergo thermal fragmentation of the furazan (see Section 4.22.3.1.1) or desulfurization with methoxide ion to the benzofurazan derivative (**87**).

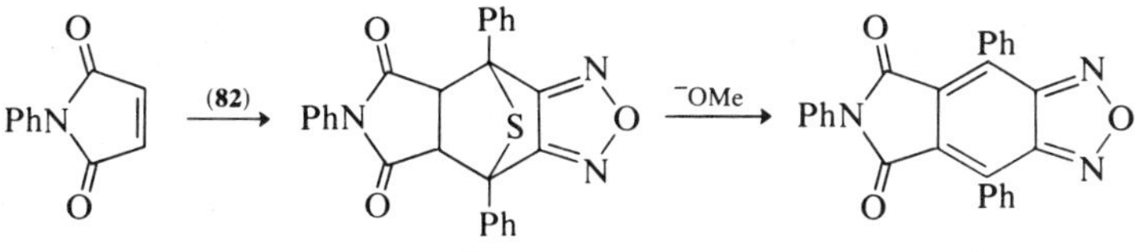

4.22.4 SYNTHESES

In view of the diversity of methods utilized, separate subsections are devoted to the synthesis of furazans, benzofurazans, furoxans and benzofuroxans.

4.22.4.1 Furazans

Earlier reviews ⟨B-61MI42200, 62HC(17)283, 75H(3)651⟩ examined the synthesis of furazans up to 1975, and unless stated, the material in the following sections is cited therein. Few new methods of preparative significance have emerged since then. The two main routes comprise: the dehydrative cyclization of α-dioximes and the deoxygenation of furoxans.

4.22.4.1.1 Dehydration of α-dioximes

The most frequently used method for the synthesis of monocyclic furazans involves the dehydration of the appropriately substituted α-dioxime. Its utility stems from the ready availability of the precursor. Reduction of furoxans, which may themselves be formed from a variety of sources (Scheme 17) including glyoximes (see Section 4.22.4.2.1), oximation of 1,2-diones, and α-nitrosation–oximation of ketones are among the most favored approaches.

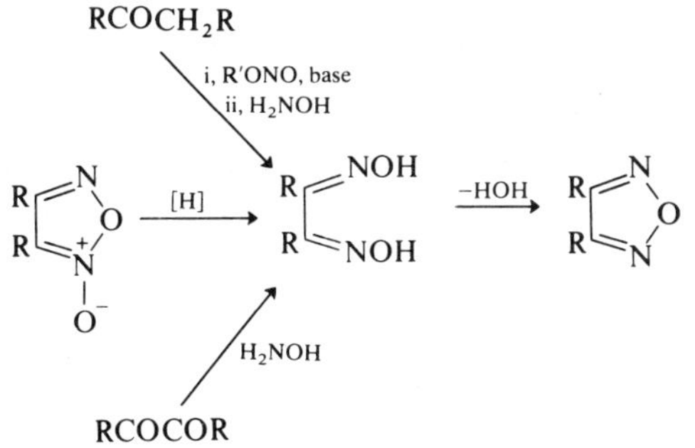

Scheme 17

The conditions used to achieve this transformation vary widely, depending on the nature of the substituent groups. The parent furazan, which had proved elusive for so many years, was first prepared in 1965 by treatment of glyoxime with succinic anhydride at 150–170 °C, followed by distillation ⟨65JOC1854⟩. Monosubstituted furazans are likewise best made by dehydration of the corresponding α-dioximes with acetic anhydride or sulfuric acid. For unsubstituted and all monosubstituted furazans the presence of alkali must be avoided due to their susceptibility to isomerization under these conditions.

In marked contrast disubstituted furazans are generally stable to both heat and chemical attack and a wide range of dehydrating agents may be utilized. Thus 3,4-diphenylfurazan has been reported to be formed from α- or β-benzil dioxime on heating at 200–210 °C, and on warming with aqueous ammonia, sodium hydroxide and copper sulfate. Other reagents which have been successfully used include urea, acetic anhydride, phenyl isocyanate and phosphorus oxychloride, although in some cases the last reagent also gives rise to 1,2,4-oxadiazole by-products. Similarly, thionyl chloride with benzil dioxime yields 3,5-diphenyl-1,2,4-oxadiazole. On the other hand, for α-dioximes for which such Beckmann rearrangements are inhibited the use of thionyl chloride provides a mild and effective route. It has proved of particular value for strained furazans which fragment on prolonged heating (Section 4.22.3.1.1). The technique is illustrated by the preparation of furofurazan derivatives (**88**), suitable for subsequent conversion to 3-acylamido-4-hydroxyalkylfurazans (**89**), from 3-unsubstituted tetronic acids (Scheme 18) ⟨79S977⟩.

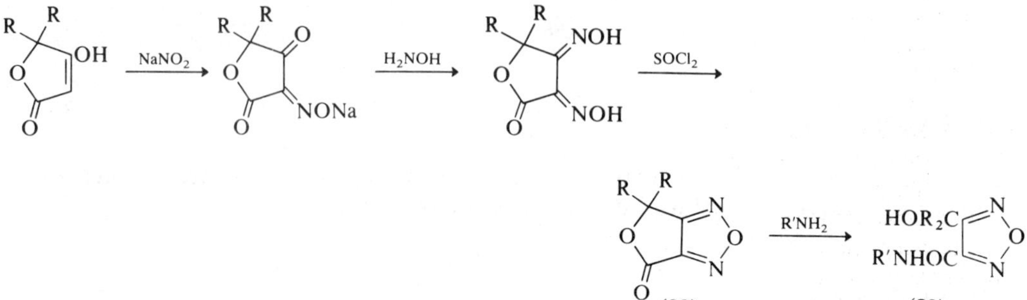

Scheme 18

The α-dioxime dehydration route is compatible with a wide range of substituents; not only is it suitable for alkyl-, aryl- (including heteroaryl) and acyl-furazans, but cyclization in the presence of carboxyl, amino and arylamino groups can also be achieved. It is not limited to monofurazans; formation of bisfurazanocyclohexane derivatives, such as (**90**) and (**91**), via dehydration of the corresponding tetraoximes has also been reported.

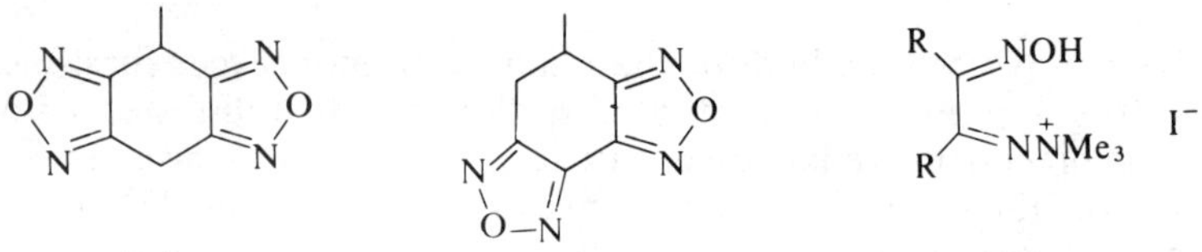

A modification involving prior conversion to α-dioxime diesters, followed by cyclization via steam distillation or by reaction with alkali is claimed to result in high yields. Similarly furazans result from base treatment of (**92**), generated from methyl iodide and *N,N*-dimethylhydrazones of α-isonitroso ketones ⟨79MI42200⟩. Amination as well as dehydration is apparently involved in the transformation, using alkaline hydroxylamine, of pyruvaldoxime into 3-amino-4-methylfurazan, and of isonitrosoacetophenone into the 4-phenyl analogue.

4.22.4.1.2 Deoxygenation of furoxans

The reduction of furoxans, which is described in detail in Section 4.22.3.2.4, offers a valuable alternative route, particularly when the initial furoxan is available from sources other than α-dioximes. While care must be taken to avoid further reduction to glyoximes, amino compounds, nitriles and hydrocarbons, a variety of reagents may be used. Trialkyl phosphites have proved most generally applicable. Simple dialkyl-, alkylaryl- and diarylfurazans usually require more vigorous conditions, *e.g.* refluxing triethyl phosphite, than do strained derivatives, prolonged heating in these cases leading to ring cleavage and nitrile formation. Other reagents used include triarylphosphines, phosphorus pentachloride, and red phosphorus with hydriodic acid. Reductions with dissolving metals and metal ions, such as zinc/acetic acid and tin(II) chloride, have also found widespread use, the furazans probably arising as secondary products *via* dehydration of the initially formed α-dioximes.

The method is suitable for furazans bearing diverse substituents including alkyl, aryl, acyl, cyano and amino groups.

4.22.4.1.3 Rearrangements of other heterocyclic systems

Furazans are available *via* the mononuclear rearrangement of isoxazoles and 1,2,4-oxadiazoles bearing oxime groups in the 3-position ⟨81AHC(29)141⟩. The approach is illustrated in Scheme 5, with X=Y—Z being N=C—O or C=C—O.

3-Acylisoxazoles yield furazans on heating with alkaline hydroxylamine, presumably by oximation and subsequent rearrangement. Likewise 3-benzohydroximino-4-benzoyl-5-phenylisoxazole when treated with 20% potassium hydroxide is reported to give 3-dibenzoylmethyl-4-phenylfurazan, which was subsequently converted to the 3-phenacyl-4-phenyl derivative on further reaction with alkali or acid.

Similar behavior is shown by 3-acyl-1,2,4-oxadiazoles and their oximes. Thus 3-benzoyl-5-phenyl-1,2,4-oxadiazole is transformed by hydroxylamine into 3-benzamido-4-phenylfurazan, while the oximes of 3-aroyl- and 3-acetyl-5-methyl-1,2,4-oxadiazoles afford 3-aryl- and 3-methyl-4-aminofurazans in the presence of 6M hydrochloric acid.

Hydroxylamine has also been used to generate furazans from 3-isonitrosopyrroles, and it has been claimed that 5-aryl-4-nitrosoimidazoles may be converted into 3-amino-4-arylfurazans using the same reagent.

Thermolysis of 3-anilino-4-nitroso-5-phenylisoxazole, obtained from aniline and dibenzoylfuroxan, produces 3-anilino-4-benzoylfurazan in high yield ⟨69JHC317⟩; other acylanilino derivatives have been prepared similarly. Further examples of furazan formation resulting from the rearrangement of furoxans are described in Section 4.22.3.2.5(ii).

4.22.4.1.4 Miscellaneous furazan syntheses

Reaction of vinyl azides with nitrosyl tetrafluoroborate gives furazans in good yield, particularly when the alkene is 1,2-dialkyl-substituted. For example, 1-azidocyclooctene and 2-azidobut-2-enes yield (**19**; $n=6$) and (**1**, $R^1=R^2=Me$), respectively; 1,2,4-oxadiazoles may also be formed as by-products.

The general resistance of the 1,2,5-oxadiazole ring to chemical attack and the consequent ability of its substituent groups to undergo chemical transformations, which were described in Section 4.22.3.4, provides wide scope for the synthesis of furazans. For example, tetramethylenefurazan (**19**; $n=4$) and its tetrachloro derivatives (**55**; X = Cl) are formed in good yield by hydrogenation and chlorination of the corresponding benzofurazans. Furazancarboxylic acids may be prepared by oxidation with potassium permanganate of alkyl- and acyl-substituted furazans, and by hydrolysis of its amide, ester, and halide derivatives. Conversely, these carboxamides and esters are obtainable from the acids *via* its acyl halides.

Halogenation and metallation of alkyl substituents may also be exploited. Chloroalkyl groups are formed by direct chlorination, thus providing a route to hydroxyalkylfurazans and their derivatives, while lithiation is the first step in the synthesis (Scheme 15) of alkylvinylfurazans suitable for polymer applications.

The availability of amino-substituted furazans, both by direct synthesis from aminoglyoximes and *via* Hofmann degradation of the corresponding carboxamides, allows urea, thiourea, imino and acylamido derivatives, for example, to be prepared by reaction with isocyanates, isothiocyanates, aldehydes and acyl halides, respectively ⟨73JPR791, 77JCS(P1)1616⟩. Alkoxy and acyloxy derivatives are likewise formed from hydroxyfurazans ⟨79JHC689⟩.

By employing furazans bearing reactive groups attached at both 3- and 4-positions bicyclic derivatives may be constructed. For instance the furazanothiophene (**85**) and diazepine (**93**) are obtained by treatment of di(α-chlorobenzyl)- and diamino-furazans with sodium sulfide and ethyl acetoacetate, respectively ⟨77H(6)1173, 78KGS1196⟩. Further examples are described in Section 4.22.3.4.

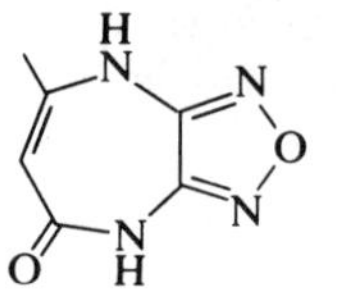

4.22.4.2 Benzofurazans

As for the monocyclic furazans, most of the synthetic approaches have been covered in previous reviews ⟨B-61MI42200, 62HC(17)283, 75H(3)651⟩. The principal routes comprise: the formation of the oxadiazole ring by cyclization of o-quinone dioximes or o-disubstituted arenes; and the deoxygenation of the appropriate benzofuroxan.

4.22.4.2.1 Dehydration of o-quinone dioximes

A variety of conditions have been employed to convert o-quinone dioximes into benzofurazans, including dehydration with acetic anhydride, thionyl chloride, sulfuric acid, and phenyl isocyanate, and steam distillation in the presence of alkali. Alternatively, cyclization may be achieved by boiling with aqueous alkali or thermolysis of the corresponding dioxime diacetates or dibenzoates.

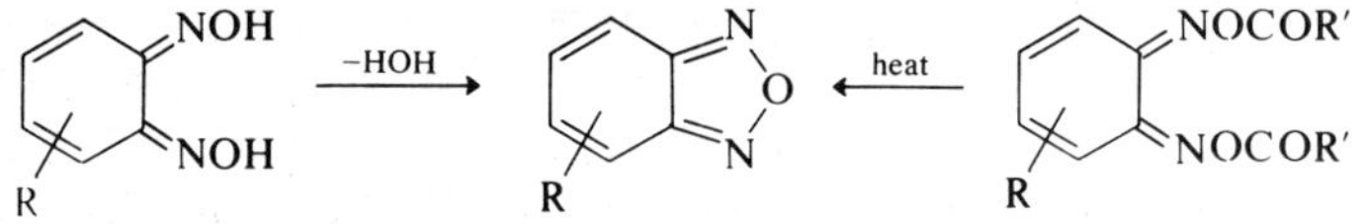

The utility of the method depends on the availability of the dioxime precursors. When the parent quinone is readily accessible, direct oximation provides a straightforward approach which has been used for the acenaphtho- and phenanthro-furazans, (**22**) and (**94**; $n=0$), although in the former case mild conditions ($SOCl_2$, 20 °C) are required for the subsequent cyclodehydration in order to avoid fragmentation of the product ⟨73JOC1054⟩ (see Section 4.22.3.1.1). o-Quinone dioximes can be generated by reduction of benzofuroxans, but in many such cases direct deoxygenation to the benzofurazan is also possible.

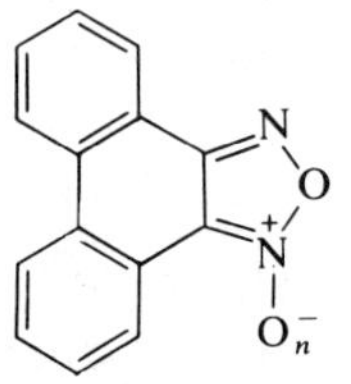

This route can accommodate a wide range of substituents in the homocyclic ring, and it is applicable to both mono- and bis-furazans. It may also be used for furazans with fused heterocyclic rings.

4.22.4.2.2 Synthesis from o-disubstituted arenes

Heating o-nitrosophenols with hydroxylamine is reported to give furazans, naphtho[1,2-c]furazan (**95**) being formed from both 1-nitroso-2-naphthol and 2-nitroso-1-naphthol, presumably by oximation of the tautomeric o-naphthoquinone monooximes and subsequent dehydration. Compound (**95**) has also been prepared by oxidation, using alkaline ferricyanide or hypochlorite, of 1-amino-2-nitroso- and 2-amino-1-nitroso-naphthalene. This latter approach is suitable for heterocyclic fused furazans; thus 4,6-diamino-5-nitrosopyrimidine is converted into the furazanopyrimidine (**96**) by oxidation with lead tetraacetate ⟨71JOC3211⟩. In a similar reaction alkaline hypochlorite oxidizes o-nitrosoacetanilide to benzofurazan in quantitative yield.

(95) (96)

o-Halonitrosoarenes provide an alternative source; 1,3-dichloro-2-nitrosobenzene on heating in DMSO at 100 °C generates 1-azido-3-chloro-2-nitrosobenzene, which spontaneously loses nitrogen with ring closure to give 4-chlorobenzofurazan ⟨66TL2887⟩. *o*-Dinitroarenes behave similarly and offer the advantage of being more readily accessible ⟨73JCS(P1)1954⟩. The reduction of *o*-dinitroarenes with sodium borohydride ⟨73AJC1683⟩ and thermolysis of *o*-nitroanilines, *o*-nitroarylcarbamates and 1-(*o*-nitroaryl)-1,2,5-triphenylphospholes are also reported to afford benzofurazans in low to moderate yields ⟨75MI42200, 70CJC3059, 74JCS(P1)1694⟩.

4.22.4.2.3 Deoxygenation of benzofuroxans

A valuable route to benzofurazans is provided by deoxygenation of the corresponding benzofuroxan. This may be accomplished either directly using trialkyl phosphites, tributyl- or triphenyl-phosphine, or indirectly *via* the quinone dioxime using, for example, methanol and potassium hydroxide, hydroxylamine and alkali, sodium azide in DMSO or ethylene glycol, sodium borohydride, and occasionally thermolysis alone. More detailed discussion of these reactions is included in Section 4.22.3.2.4.

4.22.4.2.4 Miscellaneous benzofurazan syntheses

Photolysis of 2,1,3-benzoselenadiazole (97) affords benzofurazan (96%) in a process which is assumed to involve cleavage of the heterocyclic ring to give the nitrososelenonitroso isomer (98), followed by extrusion of selenium and ring closure ⟨79TL745⟩. A much lower yield (37%) was obtained on thermal treatment, 2,1,3-benzoselenadiazole being the main coproduct. The isomerization of certain 7-substituted benzofuroxans, which was described in Section 4.22.3.2.5(ii), results in deoxygenation and furazan formation; although satisfactory yields have been achieved in individual cases, this general approach does not appear to offer widespread synthetic scope.

(97) (98) Nbf-Cl $\xrightarrow{+Nu^-}_{-Cl^-}$ (99)

As for the monocyclic furazans the stability of the heterocyclic ring permits a wide range of derivatives to be prepared by chemical modification of the substituents. For benzofurazans both electrophilic substitutions (*e.g.*, nitration, sulfonation) and nucleophilic displacement of halides are of synthetic value, with the latter process being especially favored when activating nitro groups are present. Thus (99) is readily obtained from Nbf-Cl (56; $n = 0$). Section 4.22.3.3.2 gives a fuller account of the scope and mechanism of this process. In basic media some benzofuroxans bearing electron-withdrawing substituents undergo nucleophilic substitution in the homocyclic ring with concomitant deoxygenation of the furoxan portion. Still further synthetic opportunities result from the ability of the substituents to undergo modification without disruption of the benzofurazan framework (see Section 4.22.3.3).

The condensation reactions of 3,4-disubstituted 1,2,5-oxadiazoles can lead to heterosubstituted analogues of benzofurazan (see Section 4.22.3.4). Recent attention has focussed on the preparation of furazanopyrazines by reaction of diaminofurazan with α-diones and of furazano-pyridazines and -pyridines from diacylfurazans with hydrazine and primary amines, respectively.

4.22.4.3 Furoxans

The synthesis of furoxans has been discussed in a recent review ⟨81AHC(29)251⟩, and the material summarized in the following sections is cited therein unless otherwise stated. The principal routes comprise: oxidative ring closure of α-dioximes; dehydration of α-nitro ketone oximes; and dimerization of nitrile oxides. In choosing the method the possibility of equilibration between the 2- and 5-oxides of asymmetrically substituted furoxans, which was described in Section 4.22.3.2.1, must be taken into account. There has been no case of direct oxidation of a furazan to a furoxan.

4.22.4.3.1 Oxidation of α-dioximes

The availability of α-dioximes, which makes them the most common source of furazans also accounts for their widespread use in furoxan preparations. Divers oxidizing agents have been employed, including alkaline hypohalites, potassium ferricyanide, lead tetraacetate, oxides of nitrogen, cerium(IV) ion, N-iodosuccinimide ⟨80H(14)1279⟩, and phenyliodine(III) bistrifluoroacetate. Electrochemical oxidation has also been reported.

The ring closure can be achieved stereospecifically, thus allowing the individual isomers of asymmetrically substituted furoxans to be prepared. For example, the two *amphi* forms (**100**) and (**101**) of *p*-methoxybenzil dioxime are specifically oxidized by ferricyanide to (**102**) and (**103**), whereas the *syn* and *anti* isomers (**104**) and (**105**) give mixtures of the two furoxans. With some oxidants the process is non-stereospecific, either due to a change of mechanism, or as a result of isomerization of the dioxime prior to cyclization.

A variety of substituents can be accommodated, including alkyl, aryl, acyl, amino and halogens. It is the method of choice for monosubstituted furoxans, which are susceptible to thermal and hydrolytic ring cleavage. It is also suitable for polycyclic analogues such as hexamethylene- and camphor-furoxans, (**17**; *n* = 6) and (**21**), some of which are thermally unstable and must therefore be generated and stored at low temperatures. Earlier reports that the trimethylene analogue (**17**; *n* = 3) cannot be made this way are incorrect; aqueous alkaline hypochlorite oxidizes 1,2-dioximinocyclopentane to the furoxan at 0 °C ⟨79JCR(S)314⟩. It has also been reported that α-hydroximino oximes are converted to furoxans in high yield on treatment with sodium hypobromite.

4.22.4.3.2 Dehydration of α-nitro ketone oximes

Among the earliest reported furoxans were the arylmethyl derivatives resulting from the reaction of nitrous acid with naturally occurring arylpropenes such as isosafrole and anethole.

Pseudonitrosites and α-nitro ketone oximes were identified as intermediates in this process. This route has since been developed to provide access to mono- and poly-cyclic furoxans from readily-available alkenes and cycloalkenes. Treatment of the alkene with dinitrogen trioxide affords the nitro-nitroso adduct, which is readily isolated as its nitroso dimer (pseudonitrosite); thermal isomerization to the α-nitro ketone oxime, followed by dehydration with cyclization leads to the furoxan (Scheme 19).

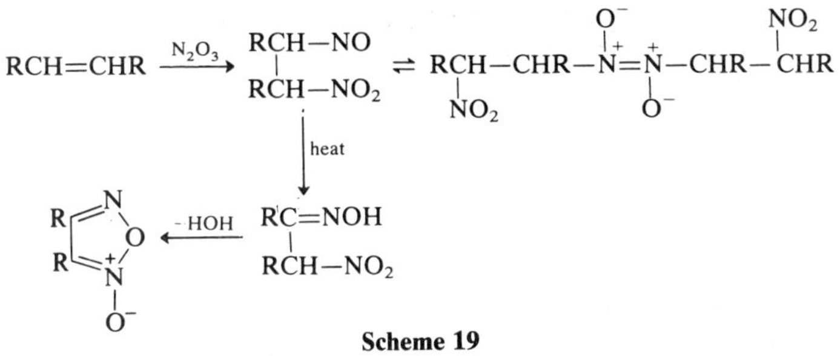

$$RCH=CHR \xrightarrow{N_2O_3} \begin{array}{c} RCH-NO \\ | \\ RCH-NO_2 \end{array} \rightleftharpoons RCH-CHR-\overset{\overset{O^-}{|}}{\underset{\underset{NO_2}{|}}{N}}=\overset{\overset{NO_2}{|}}{\underset{\underset{O^-}{|}}{N}}-CHR-CHR$$

$$\downarrow \text{heat}$$

$$\underset{\underset{O^-}{|}}{\overset{R}{\underset{R}{\diagup}}}\overset{N}{\underset{\overset{+}{N}}{\diagdown}}O \xleftarrow{-HOH} \begin{array}{c} RC=NOH \\ | \\ RCH-NO_2 \end{array}$$

Scheme 19

Sulfuric and phosphoric acids at 110–120 °C proved to be effective dehydrating agents, but limited the method to thermally stable furoxans. Compounds (**2**; $R^1 = R^2 = Me$) and (**17**; $n = 4, 6, 10$) are among the examples which may be obtained under these reaction conditions ⟨65CB1831, 71GEP2062928⟩. More recently it has been established that sulfur trioxide or chlorosulfonic acid in DMF is equally efficient, enabling strained furoxans which are prone to thermal ring cleavage to be prepared. Dicyclopentadiene and norbornane furoxans (**40**) and (**20**) have been synthesized in this way ⟨78CC113⟩. With terminal alkenes nitrofuroxans are sometimes produced, *e.g.* propene with dinitrogen trioxide or tetroxide and methacrylic acid with nitrous acid give 3-methyl-4-nitrofuroxan.

An alternative and complementary approach, also proceeding *via* α-nitrooximes, which has been used to generate furoxans of the sugar series but which should be more generally applicable, involves reaction of α-tosyloxy ketones with hydroxylamine hydrochloride in aqueous pyridine followed by treatment of the resulting oximinopyridine salt with sodium nitrite. The latter step is understood to proceed *via* nucleophilic displacement of the pyridine by nitrite ion, followed by cyclization of the nitrooxime ⟨72CC1117⟩.

Treatment of some α-ethynyl acetates with a mixture of nitrosyl fluoride and nitrosonium tetrafluoroborate affords bis[(acetoxyalkyl)carbonyl]furoxans; the presence of the α-acetoxy group appears to be essential ⟨81JOC312⟩. Monosubstituted furoxans have been obtained by nitrosation of unsaturated hydrocarbons; thus the reaction of nitrous acid with *p*-hydroxycinnamic acid affords 3-(*p*-hydroxyphenyl)furoxan, its structure being confirmed by X-ray crystallography ⟨80LA1557⟩.

4.22.4.3.3 Dimerization of nitrile oxides

Nitrile oxides, whether isolated or generated *in situ* by one of the methods outlined below (Scheme 20), dimerize spontaneously to furoxans. The yields are high and it has proved a very effective and general route to symmetrically disubstituted derivatives; asymmetric substitution can be achieved using a mixture of nitrile oxides, but it is preparatively inconvenient to separate the products and alternative methods are usually employed.

While the precise mechanism has been a matter of some debate, recent frontier orbital treatment, taken with earlier kinetic studies determining the entropy (negative) and the effect of solvent (low) and substituents (Hammett $\rho = +0.87$), favor a one-step concerted pathway, one nitrile oxide moiety fulfilling the role of 1,3-dipole and the other that of the dipolarophile ⟨B-77MI42201⟩. The rate is also critically dependent on the size of the substituent, bulky groups generally slowing the process. Thus at room temperature acetonitrile oxide (**106**; R = Me) has a lifetime of less than one minute, whereas the sterically hindered mesitonitrile oxide (**106**; R = 2,4,6-$Me_3C_6H_2$) can be isolated and stored, only forming dimesitylfuroxan on warming (60–65 °C). The biomolecular dimerization competes with unimolecular rearrangement to the isomeric isocyanate, the latter becoming dominant at higher temperatures; on heating rapidly to >100 °C mesitonitrile oxide forms mesityl isocyanate in near quantitative yields ⟨65JOC2809⟩. The preparation of furoxans from nitrile oxides is therefore carried out in concentrated solution at ambient temperature. Under carefully controlled conditions other modes of dimerization have been observed, leading to 1,4,2,5-dioxadiazines (**107**), 1,2,4-oxadiazole 4-oxides (**108**) and nitrile oxide-derived polymers. In the same circumstances fulminic acid (**106**; R = H) polymerizes and the parent furoxan is not formed.

The principal sources (Scheme 20) of nitrile oxides comprise: hydroximoyl halides and oximes; nitrolic acids and their precursors; and primary nitroalkanes. In addition there is a range of miscellaneous methods of generation which can be applied in particular instances.

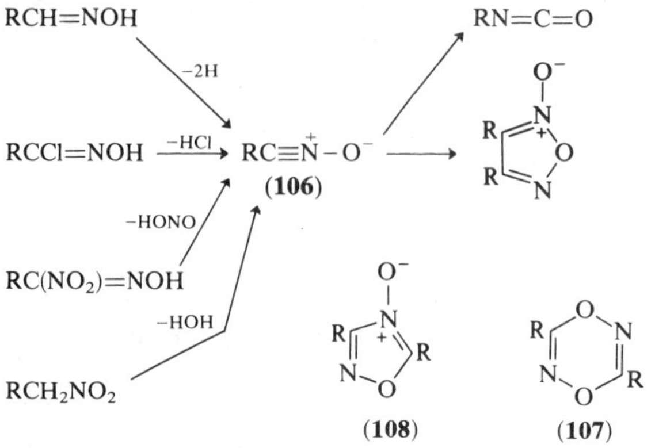

Scheme 20

(i) *From hydroximoyl halides and oximes*

Hydroximoyl halides, which are readily available ⟨B-71MI42200⟩ *via* the halogenation of oximes (*e.g.* with chlorine, NBS, nitrosyl chloride) or from hydrogen halides with nitrolic acids, are the most widely used source of nitrile oxides. They are easily dehydrohalogenated by treatment with bases, particularly trialkylamines, sodium carbonate or acetate, and alkali hydroxides. Concentration of the resulting solution facilitates furoxan formation. The dehydrochlorination can also be achieved by heating in an inert solvent (*e.g.* refluxing xylene). While this last technique is appropriate for the formation of many of the 1,3-dipolar cycloadducts, it is not suitable for the synthesis of furoxans. The yields are low and by-products, including isocyanates, are prevalent ⟨71JOC2155⟩.

Direct dehydrogenation of oximes is also possible, particularly for sterically hindered nitrile oxides. The most successful oxidizing agents are alkaline hypobromite and NBS in the presence of base, the latter being valuable when basic groups are present. Lead tetraacetate and alkaline hypochlorite have also been employed, but the yields are lower. The formation of furoxans from the reaction of alkoximes with nitrogen oxides is believed to proceed *via* the corresponding nitrolic acids.

(ii) *From nitrolic acids and their precursors*

Nitrile oxides are generated from nitrolic acids [$RC(NO_2)=NOH$] by elimination of the elements of nitrous acid. This can occur spontaneously, on heating, or by the action of sodium bicarbonate, and it thereby provides access to symmetrically substituted dialkyl-, diaryl-, and diacyl-furoxans.

Nitrolic acids, which can be isolated from sources such as nitroalkanes and aldoximes, are also probable intermediates in a wide variety of furoxan preparations involving nitric and nitrous acid, and nitrogen oxides. Diacylfuroxans have been synthesized *via* nitrosation of methyl ketones; for example acetone reacts with anhydrous dinitrogen tetroxide or nitric acid/sodium nitrite to give diacetylfuroxan.

(iii) *From nitroalkanes*

Dehydration of primary nitroalkanes with phenyl isocyanate or acetic anhydride in the presence of catalytic triethylamine affords nitrile oxides, which may be trapped as their 1,3-dipolar cycloadducts or allowed to dimerize to the corresponding furoxans. Other dehydrating agents that have been used include diketene, sulfuric acid and, when the α-methylene group is activated by electron-withdrawing groups, boron trifluoride in acetic anhydride, trifluoroacetic anhydride with triethylamine, and nitric acid in acetic acid.

The capacity of the method to generate nitrile oxides bearing diverse substituents, which has led to its recent use in natural product synthesis, *e.g.* of ergot alkaloids such as chanoclavine ⟨80JA4265⟩, is also valuable for the preparation of furoxans. Dialkoxycarbonyl- and bisarylsulfonyl-furoxans have been obtained in this way, while the formation of the bicyclic furoxan (**109**) from bis(β-nitroethyl) sulfide represents a rare case of intramolecular dimerization of a bis(nitrile oxide) ⟨77JA6754⟩. It is the method of choice for the lower alkanonitrile oxides, and hence their furoxan dimers, for which the hydroximoyl chloride-based route is preparatively inconvenient. For example, diethylfuroxan is readily obtained (93%) from nitroethane using phenyl isocyanate and a trace of triethylamine ⟨60JA5339⟩.

(**109**)

(iv) From miscellaneous sources of nitrile oxides

Acyl and perfluoroalkyl diazomethanes react with nitrogen oxides, and α-diazosulfones with nitrosyl chloride or dinitrogen trioxide, to give furoxans. The process is believed to involve nitrosation, followed by loss of nitrogen to form the nitrile oxide which subsequently dimerizes. Nitrosation of dimethylphenacylsulfonium bromide with nitric acid/sodium nitrite gives dibenzoylfuroxan.

Dihalofuroxans are formed on dehydrohalogenation of dihaloximes and when halogens react with mercury or silver fulminates. Iodine with the silver salt of phenylnitrosolic acid gives diphenylfuroxan. Boron trifluoride in acetic anhydride reacts with 2-(ethoxycarbonyl-nitromethylene)thiazolidine to give diethoxycarbonyl furoxan. In each case nitrile oxides are likely or proven intermediates.

4.22.4.3.4 Miscellaneous furoxan syntheses

Nitration of 1,2-dialkylvinyl azides with nitryl fluoroborate yields 1-azido-2-nitroalkene (**110**), which spontaneously decomposes with loss of nitrogen to the dialkylfuroxan. Compound (**110**) is also formed on treatment of 1,2-dinitroalkenes with azide ion and from 2-nitroacrylic ester with bromine azide.

(**110**)

Nitrosyl chloride with metal acetylides produces nitrosoalkynes which may dimerize and rearrange to furoxans; 1-nitrosohex-1-yne (**111**) gives butylhexynylfuroxan (**112**), while arylaminobutylfuroxans (**113**) are formed from the same nitrosoalkyne with anilines and nitrosyl chloride.

(**112**)   (**111**)   (**113**)

The resistance of the furoxan ring to chemical attack allows derivatives to be prepared via the reactions of the substituents (Section 4.22.3.4). Carboxylic acids are available by permanganate oxidation of methyl derivatives or by hydrolysis of the corresponding esters; reaction with ammonia affords carboxamides. Acylfuroxans provide a source of hydroxyalkyl compounds by reduction, and oximes, for example, via nucleophilic addition. Acylation and oxidation of aminofuroxans allows the amide and nitro derivatives to be prepared. Nucleophilic displacements of nitro substituents can take place, but can be somewhat hazardous on account of the explosive nature of these compounds. Alkoxy derivatives are formed with sodium alkoxide, while reaction with thiolate anions yields sulfides, from which sulfones can be synthesized by peracid oxidation. Nitrofuroxans have also been reduced to the corresponding amines. Other transformations of preparative significance include: nitration and halogenation of aryl substituents; nucleophilic substitution of sulfonyl or halogen by alkoxide ion; and reduction of nitro to amino using tin(II) chloride.

4.22.4.4 Benzofuroxans

Most of the main synthetic routes (Scheme 21) have been summarized in previous reviews ⟨69AHC(10)1, 81AHC(29)251⟩. They involve: oxidation of *o*-quinone dioximes; decomposition of *o*-nitroaryl azides; and oxidation of *o*-nitroanilines. The interconversion of the 2- and 5-oxides for asymmetrically substituted derivatives is more facile compared with the monocyclic analogues, and mixtures of isomers are commonly formed. In no case has a benzofuroxan been prepared by direct oxidation of the corresponding benzofurazan.

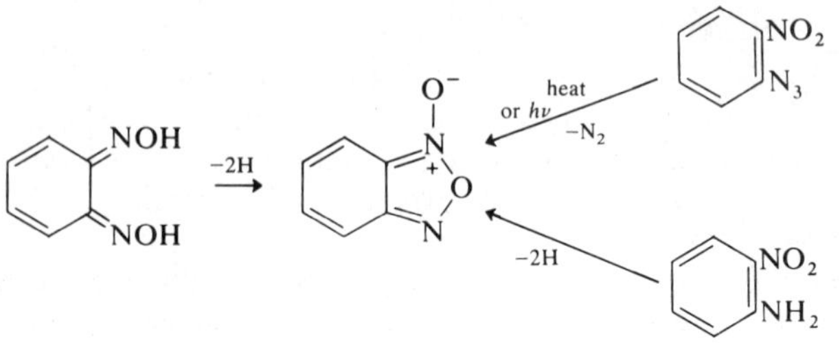

Scheme 21

4.22.4.4.1 Oxidation of o-quinone dioximes

Benzofuroxans are formed from *o*-quinone dioximes by oxidation with, for example, alkaline ferricyanide, nitric acid, bromine water and chlorine, While the reaction is usually straightforward and high yielding the method is not generally applicable since the dioximes themselves are not readily obtainable and are often best prepared *via* reduction of the furoxan (see Section 4.22.3.1.3). However it can be used when the parent quinones or their monooximes (*o*-nitrosophenols) are available from other sources. Thus it is the method of choice for the acenaphtho- and phenanthro-furoxans, (**18**) and (**94**; $n=1$), respectively. In other cases alternative routes, such as the oxidation of *o*-nitroanilines or the thermolysis of *o*-nitroaryl azides, are more commonly utilized.

4.22.4.4.2 Decomposition of o-nitroaryl azides

The most reliable method for the preparation of benzofuroxans involves the thermal decomposition of *o*-nitroaryl azides ⟨51OS(31)14⟩. On heating at 100–150 °C in, for example, glacial acetic acid ring closure is accomplished with loss of nitrogen. More forcing conditions are often required when the participating groups are severely crowded. The reaction appears to involve a simple concerted cyclization and elimination mechanism. Photolysis provides a viable, although rarely used, alternative to the thermolytic procedure. Numerous substituted benzofuroxans have been made by this route, and it has also been utilized for a variety of hetero-substituted analogues including thieno, imidazo, oxadiazolo, thiadiazolo, pyrido, pyrimidino, pyridazino and quinolino derivatives. Di- and tri-furoxans such as (**15**; $n=1$) have been prepared by this method.

Nitrotetrazolo compounds are an additional source; pyrido[2,3-*c*]furoxan (**42**) is formed from the nitropyridotetrazole (**114**), presumably *via* the isomeric azidonitropyridine (Scheme 22).

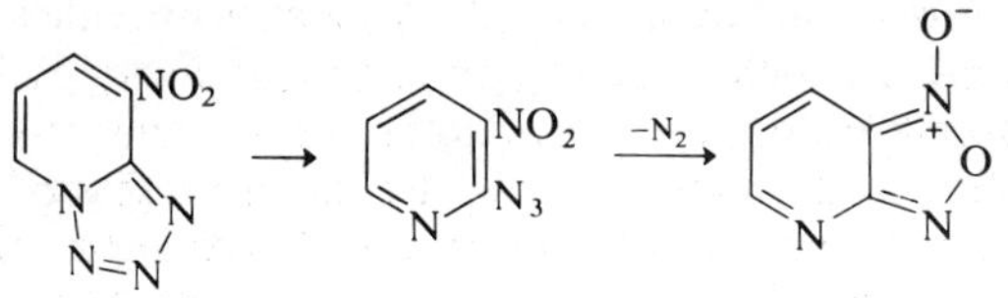

Scheme 22

4.22.4.4.3 Oxidation of o-nitroanilines

Extensive use has been made of the ease with which o-nitroanilines undergo oxidative ring closure to benzofuroxans ⟨57OS(37)1⟩. Alkaline hypochlorite is the most common reagent, and the reaction probably involves N-chlorination, deprotonation and ring closure of the N-chloroacetanilide anion (Scheme 23). Phenyliodosodiacetate is an alternative oxidizing agent, which proves of particular value in cases where alkaline hypochlorite reacts with the product. Hofmann degradation of o-nitroarylcarboxamides also affords benzofuroxans, presumably *via* an o-nitroarylamine intermediate.

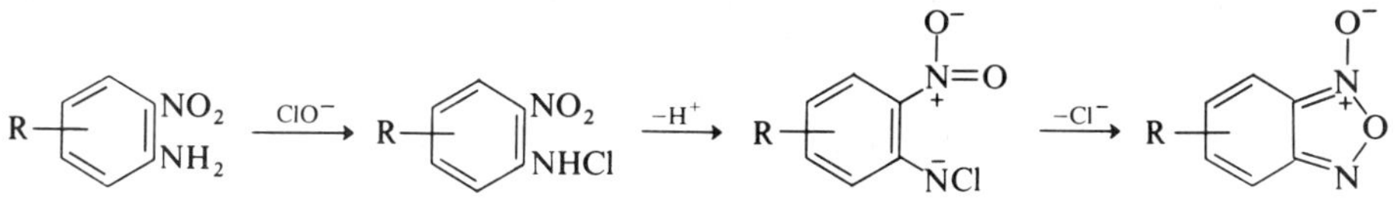

Scheme 23

4.22.4.4.4 Miscellaneous benzofuroxan syntheses

The reactions of the homocyclic ring of benzofuroxans, which are described in detail in Section 4.22.3.3, provide access to numerous derivatives. Nucleophilic displacement of halides is facile when activating nitro groups are present, allowing alkoxy, aryloxy, thio and amino groups to be introduced. Electrophilic substitutions, *e.g.* nitration, are also valuable. Further transformations may also be performed on benzo-ring substituents. Such modifications include: acetoxy to hydroxy; acetamido to amino; and acyl halides to esters and amides. Some reactions of the substituents of monocyclic furoxans allow heterosubstituted analogues of benzofuroxans to be prepared. For example, pyridazinofuroxans are formed by condensation of diacylfuroxans with hydrazine.

4.22.5 APPLICATIONS

Various reactions of furazans and furoxans find use in organic synthesis. Reduction of benzofuroxans is often the most convenient route to o-quinone dioximes. Under forcing conditions monocyclic furoxans yield nitriles and amines as the products of ring cleavage. For non-aromatic fused furoxans, some of which can be readily prepared from cycloalkenes (see Section 4.22.4.3.2), this reaction provides a route to dinitriles and diamines; by this means hexamethylenediamine is accessible from cyclohexene *via* (**17**; $n = 4$) ⟨67MI42200⟩. Such furoxans are also a source of diisocyanates, *e.g.* (**41**), suitable for polyurethane applications ⟨74BRP1521690⟩. The reactions of benzofuroxans have attracted particular attention in the search for compounds of medicinal value. Extensive use has been made of the Beirut reaction (see Section 4.22.3.2.5.i) by which quinoxaline and benzimidazole oxides are formed; high levels of antibacterial activity have been detected in some of the products. Benzofuroxan is readily reduced by thiols and this reaction has been proposed as a biochemical assay for enzyme-SH groups.

A wide spectrum of biological activity is claimed for compounds incorporating the 1,2,5-oxadiazole framework. Penicillin and cephalasporin derivatives have been prepared and tested as antibacterials. Fungistatic, antimicrobial and anthelmintic activities have been reported, and various examples have been recommended as plant-growth regulators, pesticides, algicides, herbicides, rodenticides, radioprotectants, vasodilators, anticancer agents, anticonvulsants and muscle relaxants. The hepatic and intestinal toxicity of some furoxan and furoxan derivatives have also been investigated ⟨80MI42201⟩. Benzofurazans and benzofuroxans have been studied in great detail ⟨81MI42200⟩. Their antiproliferative and immunosuppressant properties have been examined. Nucleic acid and protein biosynthesis in various mammalian cells is inhibited and structure–activity relationships have been obtained. A large number of derivatives have also been screened by the Ames method, and it was concluded that the presence of nitro and N-oxide functions confers potent *in vivo* antitumor activity, but also favors mutagenicity. The effects of diverse oxadiazoles on the peripheral and central nervous systems have been investigated. Tests for inhibition of monoamineoxidase indicate that polycyclic compounds such as (**60**) are highly active, whereas monocyclic and benzofuroxans are less potent. The patent literature contains many references to benzofurazans and furoxans as herbicides, fungicides and antimicrobials, but in general they appear to be less effective than the corresponding thiadiazoles.

The ease with which chloride is displaced by amines, phenols and thiols from Nbf-Cl (**56**; $n = 0$) (see Section 4.22.3.3.2), and the distinctive fluorescence shown by the resulting —NR—, —O—, and —S— derivatives has led to their use as fluorigenic agents. Numerous applications have been described ⟨81MI42200⟩ including: exploring the structure and conformation of the active sites of enzymes; and monitoring active transport across, and lateral diffusion of phospholipids in, red cell and liposome membranes ⟨80MI42202, 81MI42202⟩. Nbf-labelled disulfides have also been employed to selectively label protein thiol groups; in this case the problems associated with colabelling tyrosyl and amino groups are avoided ⟨75BJ(151)417⟩. Nbf-Cl is often regarded as the reagent of choice for detecting and quantifying amine drugs and pollutants in foodstuffs, soils, water supplies, *etc.* The technique is highly sensitive; for example, nitrosamines in food ⟨80MI42203⟩ and amphetamines in urine ⟨76MI42200⟩ being detectable in 25 ng ml^{-1} and 0.1 μg ml^{-1} quantities, respectively. A further advance is provided ⟨81MI42200⟩ by the recent development of ammonium 4-chloro-7-sulfobenzofurazan (Sbf-Cl), which is reported to be a stable non-mutagenic, water-soluble reagent which reacts specifically with protein thiols. The products display fluorescent intensities far greater than the corresponding Nbf compounds, and the Sbf group can be removed from the protein by thiolysis.

The facile ring-opening of strained bicyclic furoxans to dinitrile oxides has been exploited for the modification of polymers; Compounds (**20**) and (**40**) react with polybutadienes and polyisoprenes forming bis(isoxazoline) crosslinks ⟨74BRP1474691⟩. Furoxan-containing polymers ('polyfuroxans'), *e.g.* (**115**), can be prepared by intermolecular dimerization of bisnitrile oxides) and are generally stable up to 200 °C ⟨79MI42201⟩. Various polymers incorporating the furazan moiety prove to be heat and hydrolysis resistant. Polyamides such as (**116**), prepared from bis(chlorocarbonyl)furoxan and *trans*-2,5-dimethylpiperazine ⟨75MI42201⟩, have been used for water desalination by reverse osmosis; high water permeability and good sodium chloride rejection are claimed during extended use ⟨73GEP2263774⟩. Resins prepared from diaminofurazan and formaldehyde have been used as a film forming material ⟨80MI42204⟩. The various applications of furazans and furoxans have been summarized ⟨75H(3)651, 81AHC(29)251⟩.

(**115**)   (**116**)

Nitrofuroxans have been examined as explosives and propellants. Miscellaneous uses include the inhibition of polymerization and corrosion. Metallurgical, photographic, and electrochemical applications have also been considered.

4.23
1,3,4-Oxadiazoles

J. HILL
University of Salford

4.23.1 INTRODUCTION	427
4.23.2 STRUCTURE	428
4.23.2.1 Theoretical Methods	428
4.23.2.2 Structural Methods	428
4.23.2.2.1 ^1H NMR spectroscopy	428
4.23.2.2.2 UV spectroscopy	429
4.23.2.2.3 IR and Raman spectroscopy	429
4.23.2.2.4 Mass spectrometry	429
4.23.2.2.5 Other spectroscopic methods and physical properties	429
4.23.2.3 Thermodynamic Aspects (Melting and Boiling Points, Solubility, Aromaticity and Conformation)	430
4.23.2.4 Tautomerism	430
4.23.3 REACTIVITY	430
4.23.3.1 Introduction	430
4.23.3.2 Fully Conjugated Rings; Reactivity at Ring Atoms	431
4.23.3.2.1 Thermal and photochemical reactions, formally involving no other species	431
4.23.3.2.2 Electrophilic attack at nitrogen	431
4.23.3.2.3 Electrophilic substitution at carbon	432
4.23.3.2.4 Reaction of nucleophiles at carbon	432
4.23.3.2.5 Nucleophilic attack at hydrogen; acidity	435
4.23.3.2.6 Photoreactions and cycloadditions	435
4.23.3.2.7 Reaction with metals	436
4.23.3.3 Non-aromatic Compounds	436
4.23.3.3.1 2,3-Dihydro-1,3,4-oxadiazoles (Δ^2-1,3,4-oxadiazolines)	436
4.23.3.3.2 2,5-Dihydro-1,3,4-oxadiazoles (Δ^3-1,3,4-oxadiazolines)	437
4.23.3.3.3 2,3,4,5-Tetrahydro-1,3,4-oxadiazoles (1,3,4-oxadiazolidines)	438
4.23.3.4 Reactivity of Substituents	438
4.23.3.4.1 C-Linked substituents	438
4.23.3.4.2 N-Linked substituents	439
4.23.3.4.3 O- and S-Linked substituents	439
4.23.4 SYNTHESIS	440
4.23.4.1 Ring Synthesis	440
4.23.4.1.1 Cyclization with the formation of one bond	440
4.23.4.1.2 Cyclization with the formation of two bonds	441
4.23.4.1.3 Cyclization with the formation of three bonds	444
4.23.4.2 Ring Transformations	444
4.23.5 APPLICATIONS	445
4.23.5.1 Biologically Active 1,3,4-Oxadiazoles	445
4.23.5.2 Other Applications	445

4.23.1 INTRODUCTION

1,3,4-Oxadiazole (**1**) is a thermally stable neutral aromatic molecule ⟨65JA5800⟩. Other aromatic systems are 1,3,4-oxadiazolium cations (**2**) and the exocyclic-conjugated mesoionic 1,3,4-oxadiazoles (**3**) and 1,3,4-oxadiazolines (**4**). Also known are derivatives of the non-aromatic reduced systems, 2,3-dihydro-1,3,4-oxadiazole (Δ^2-1,3,4-oxadiazoline; **5**),

2,5-dihydro-1,3,4-oxadiazole (Δ^3-1,3,4-oxadiazoline; **6**) and 2,3,4,5-tetrahydro-1,3,4-oxadiazole (1,3,4-oxadiazolidine; **7**). The literature prior to 1965 is surveyed in a comprehensive review ⟨66AHC(7)183⟩ and other useful reviews are available ⟨B-61MI42300, 62HC(17)263, 64RCR508⟩.

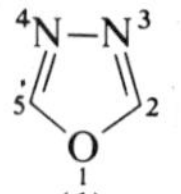

(1) 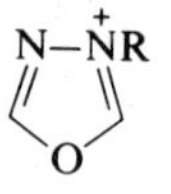 (2) 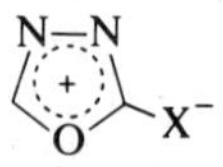 (3) X = O, S, NR 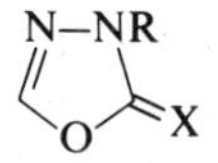 (4) X = O, S, NR

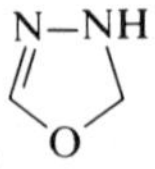

(5) 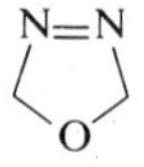 (6) 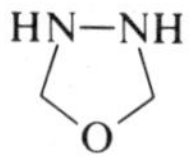 (7)

4.23.2 STRUCTURE

4.23.2.1 Theoretical Methods

Various SCF-MO methods have been used to calculate the electron distribution in 1,3,4-oxadiazole. Structural parameters, dipole moment and data relating to its UV (λ_{max} calculated to be in the region 193–203 nm), NMR, NQR and microwave spectra have been derived ⟨77MI42300, 77CHE954, 79JST(51)87, 73IJC1017⟩. Studies on 1,3,4-oxadiazole and its cation indicate a maximum positive charge in the 2-position. Molecular diagrams for 1,3,4-oxadiazole, 2-phenyl- and 2,5-diphenyl-1,3,4-oxadiazole, and oligomeric oxadiazoles have been derived and conjugation between the rings is found to be similar to that in polyphenyls ⟨72CHE943⟩. Calculated ionization potentials for 1,3,4-oxadiazole ⟨79JST(51)87⟩ and Δ^3-1,3,4-oxadiazolin-5-ones ⟨78CJC1319⟩ have been compared with values obtained from PE spectra.

4.23.2.2 Structural Methods

4.23.2.2.1 ^1H NMR spectroscopy

Chemical shifts for hydrogen attached to the ring in simple 1,3,4-oxadiazoles are given in Table 1.

Table 1 Chemical Shifts of Ring Protons in 1,3,4-Oxadiazoles

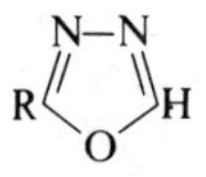

R	Solvent	δ (p.p.m.)	Ref.
H	CDCl$_3$	8.73	66JOC3442
Me	CDCl$_3$	8.53	66JOC3442
Et	CDCl$_3$	8.48	66JOC3442
PhCH$_2$	CDCl$_3$	8.26	66JOC3442
MeS	d_6-DMSO	9.42	72CJC3079
HS[a]	d_6-DMSO	8.88	72CJC3079

[a] Δ^2-1,3,4-Oxadiazoline-5-thione.

The chemical shifts of the ring protons are lowered by quaternization as shown by the values of δ (TFA) 10.7 and 9.3 p.p.m. for the 2- and 5-protons in 3-phenyl-1,3,4-oxadiazolium perchlorate. A δ (CDCl$_3$) value of 2.58 p.p.m. for the methyl protons in 2-methyl-1,3,4-oxadiazole reflects the low electron density at the ring carbon atom. The value of the chemical shift of the proton or protons in the O-alkyl or N-alkyl group assists in differentiating between isomeric 2-alkoxy-1,3,4-oxadiazoles and 4-N-alkyl-Δ^2-1,3,4-oxadiazolin-5-ones.

4.23.2.2.2 UV spectroscopy

UV spectra of substituted 1,3,4-oxadiazoles are similar to those of similarly substituted benzenes, particularly in the case of 2-phenyl- and 2,5-diphenyl-1,3,4-oxadiazole (λ_{max} (EtOH): 247.5 nm, log ε 4.26, and 280 nm, log ε 4.44 respectively). However, no absorption above 200 nm is shown by 1,3,4-oxadiazole itself and calculated values (Section 4.23.2.1) for its long wavelength absorption are in the region of 200 nm compared with λ_{max} ca. 260 nm for benzene. 2-Methyl- and 2-ethoxycarbonyl-1,3,4-oxadiazole, and Δ^2-1,3,4-oxadiazoline-5-thione have the following λ_{max} (log ε) values respectively: 206 nm (2.62) (methanol), 243 nm (3.2) and 260 nm (4.12) (ethanol).

UV spectra of 1,3,4-oxadiazoles have been described ⟨66AHC(7)183⟩ and the spectra of isosydnones (1,3,4-oxadiazolium-2-olates) have been recorded ⟨69JCS(B)1185⟩.

4.23.2.2.3 IR and Raman spectroscopy

Aspects of the IR spectra of 1,3,4-oxadiazoles have been reviewed ⟨66AHC(7)183⟩ and the spectrum of 1,3,4-oxadiazole itself has been recorded ⟨65JA5800⟩. The spectra are generally characterized by bands at 1640–1560 ($\nu_{C=N}$), 1030–1020 (ν_{C-O}) and 970 cm^{-1}. Typical for 2-substituted oxadiazoles are bands at 3140 (ν_{CH}), 1640–1560 ($\nu_{C=N}$), 1120–1100 (probably ring deformation) and 645–635 cm^{-1} ⟨69JPR646⟩. The band for C=N stretching is useful in distinguishing between 2-amino-1,3,4-oxadiazoles (1640–1610 cm^{-1}) and Δ^2-1,3,4-oxadiazolin-5-imines (1710–1680 cm^{-1}). 2-Methyl-Δ^2-1,3,4-oxadiazolin-5-one appears to exist largely as tautomer (**9a**; R = Me) in chloroform as indicated by bands at 3472 and 3268 (NH), 1786 (CO), 1650 (C=N), 1321 (C—O—C) and 935 cm^{-1}. 5,5-Disubstituted Δ^3-1,3,4-oxadiazolin-2-ones exhibit significant bands (CCl$_4$) at 1840–1820 (CO) and ca. 1540 cm^{-1}(C=N) and isosydnones (1,3,4-oxadiazolium-2-olates) show a prominent band at 1770–1758 cm^{-1}. Ring stretching modes for 1,3,4-oxadiazoles are given in Table 26 (Section 4.01.3.7).

A complete assignment of fundamental vibrations of 1,3,4-oxadiazole has been made and thermodynamic functions have been calculated ⟨76SA(A)971⟩. Raman spectra of several 1,3,4-oxadiazoles have been determined ⟨62HC(17)263⟩.

4.23.2.2.4 Mass spectrometry

The mass spectra of 1,3,4-oxadiazole, 2,5-diphenyl- and 2-perfluoroalkyl-1,3,4-oxadiazoles, including suggested fragmentation pathways, have been discussed ⟨B-71MS528⟩. The molecular ion is the base peak in the spectra of 1,3,4-oxadiazole and 2-amino-5-phenyl-1,3,4-oxadiazole, and it is a major peak in the spectra of 2-methyl-1,3,4-oxadiazole ⟨80HCA588⟩ and 2,5-diphenyl-1,3,4-oxadiazole. A characteristic fragment is RCO$^+$ and loss of CO or HCO may be typical for 2-substituted 1,3,4-oxadiazoles. Loss of HNCO is significant in the spectrum of 2-amino-5-phenyl-1,3,4-oxadiazole ⟨76OMS304⟩. The mode of fragmentation of 1,3,4-oxadiazolium-2-aminides may be used to distinguish them from the isomeric 1,3,4-triazolium-2-olates ⟨74JCS(P1)645⟩.

4.23.2.2.5 Other spectroscopic methods and physical properties

Bond lengths and angles, dipole moments (Table 2, Section 4.01.3.2) and ^{14}N nuclear coupling have been determined and the ^{14}N hyperfine structure has been analyzed for several 1,3,4-oxadiazoles ⟨74PMH(6)53⟩. Observed ^{14}N quadrupole coupling constants have been compared with values obtained from microwave spectra ⟨74JCP(61)1494⟩.

Fluorescence, which is strong in diaryl-1,3,4-oxadiazoles, has been discussed ⟨66AHC(7)183⟩ and has been used in studies of aminooxadiazole tautomeric systems (Section 4.23.3.4).

The ESR spectrum of the anion radical derived from 2,5-diphenyl-1,3,4-oxadiazole has been determined and the hyperfine splittings are based on MO calculations ⟨80AHC(27)31⟩. In the same review, hyperfine splittings are given for the radicals produced on oxidation by lead oxide of 3-substituted oxadiazolidine-2,5-diones. ESR spectra of radical cations derived from 3,4-disubstituted 1,3,4-oxadiazolidines have been described ⟨74JA2916⟩.

PE spectra of 1,3,4-oxadiazolidines (Section 4.23.2.3), 1,3,4-oxadiazole and Δ^3-1,3,4-oxadiazolin-2-ones (Section 4.23.2.1) have been recorded. X-Ray diffraction data are available for several substituted 1,3,4-oxadiazoles and dipole moments have been measured ⟨62HC(17)263⟩.

4.23.2.3 Thermodynamic Aspects (Melting and Boiling Points, Solubility, Aromaticity and Conformation)

1,3,4-Oxadiazole (b.p. 150 °C) and its lower alkyl derivatives are liquids at room temperature whereas aryl derivatives are solids (2-phenyl- and 2,5-diphenyl-1,3,4-oxadiazole melting at 34–35 and 138 °C respectively). In general, melting points increase in the series: Δ^2-1,3,4-oxadiazoline-5-thiones < Δ^2-1,3,4-oxadiazolin-5-ones < 2-amino-1,3,4-oxadiazoles (Δ^2-1,3,4-oxadiazolin-5-one and 2-amino-1,3,4-oxadiazole melting at 120 and 156 °C respectively).

Lower-alkyl-substituted 1,3,4-oxadiazoles are soluble in water, solubility decreasing with increasing MW. Solubility in water decreases in the 1,3,4-oxadiazole series: 2,5-dialkyl > 2-aryl > 2-alkyl-5-aryl > 2,5-diaryl. A similar trend is shown by aminooxadiazoles.

1,3,4-Oxadiazole is aromatic, having an estimated resonance energy of 167.4 kJ mol^{-1}. The ring is stable to heat, a property which has been exploited in the production of heat-resistant poly-1,3,4-oxadiazoles (Section 4.23.5.2).

PE spectroscopy has been used to identify gas phase conformations of 1,3,4-oxadiazolidines. The ring adopts a half-chair conformation ⟨80CB221⟩. Inversion at nitrogen in 3,4-disubstituted 1,3,4-oxadiazolidines has been studied using variable temperature NMR spectroscopy ⟨75JCS(P2)1191⟩.

4.23.2.4 Tautomerism

2-Hydroxy- (**8a**), 2-mercapto- (**8b**) and 2-amino-1,3,4-oxadiazoles (**8c**) are in equilibrium with the tautomeric oxadiazolines (**9a**), (**9b**) and (**9c**) respectively. Evidence from UV ⟨72CJC3079⟩ and IR (Section 4.23.2.2.3) spectra supports structure (**9a**) for Δ^2-1,3,4-oxadiazolin-5-ones and structure (**9b**) for Δ^2-1,3,4-oxadiazoline-5-thiones. The UV and IR spectra, fluorescence and pK values of 2-amino-1,3,4-oxadiazoles indicate that the amine tautomer (**8c** or **d**) rather than the imine tautomer (**9c** or **d**) predominates ⟨69BSF870, 874, 64CR(258)4579⟩.

a; X = O
b; X = S
c; X = NH
d; X = NR

4.23.3 REACTIVITY

4.23.3.1 Introduction

As 1,3,4-oxadiazoles have a relatively low electron density at carbon (positions 2 and 5) and a relatively high electron density at nitrogen (positions 3 and 4), the major reactions are nucleophilic attack at carbon, generally followed by ring cleavage, and electrophilic attack at nitrogen. This reactivity towards nucleophiles, also catalyzed by acid, causes difficulties when carrying out reactions which involve basic or acidic conditions. The ring is more stable when substituted by one or more aryl groups. Tautomeric oxadiazoles (**8**) or (**9**) react with electrophiles at ring nitrogen, at the exocyclic heteroatom (Section 4.23.3.4) or at both centers. Reactions in the substituent group of alkyl- or aryl-1,3,4-oxadiazoles are possible but they are limited by the sensitivity of the ring to the reagents used.

4.23.3.2 Fully Conjugated Rings; Reactivity at Ring Atoms

4.23.3.2.1 Thermal and photochemical reactions, formally involving no other species

1,3,4-Oxadiazole is thermally stable and this stability is increased on substitution, particularly by aryl and perfluoroalkyl groups. Oxadiazolin-5-ones and -5-thiones are somewhat less stable and undergo selective pyrolysis at high temperatures.

Oxadiazolinones (**10**) lose carbon dioxide at high temperatures to give nitrilimines (**11**) which react further (Scheme 1). Recyclization in the nitrilimine, formed at 210–230 °C from oxadiazolinone (**10a**), yields a 2-alkoxy-1,3,4-oxadiazole which, at the reaction temperature, undergoes a further rearrangement to oxadiazolinone (**10c**). A similar rearrangement occurs on thermolysis of oxadiazolinone (**10b**) ⟨74TL3875⟩. 2,5-Diphenyl-1,3,4-oxadiazole is produced on heating oxadiazolinone (**10d**) at 250 °C ⟨78IJC(B)146⟩. Oxadiazolinones (**10e**) and (**10f**) are more stable but undergo flash vacuum pyrolysis at 500 °C to give indazoles. At higher temperatures, loss of nitrogen also occurs, and styrene or fluorene respectively is produced in high yield ⟨78JOC2037⟩. A variety of hydrocarbons are formed on flash vacuum pyrolysis of oxadiazolinones (**10**; R^2 = alkenyl) ⟨80JOC4065⟩.

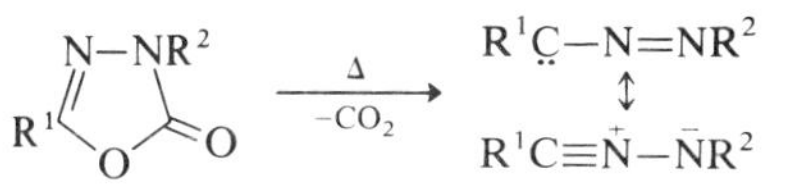

	R^1	R^2
a;	Ph	CO_2R
b;	Ph	COSEt
c;	Ph	R
d;	Ph	COPh
e;	Me	Ph
f;	Ph	Ph

Scheme 1

Carbon dioxide is lost on thermolysis of 4-ethoxycarbonyl-2-phenyl-Δ^2-1,3,4-oxadiazoline-5-thione. Migration of the ethyl group from oxygen to sulfur leads to the product, 5-ethylthio-2-phenyl-1,3,4-oxadiazole ⟨78IJC(B)146⟩.

Photolysis of oxadiazolinone (**10f**) yields a nitrilimine (**11f**) which, in the presence of alkenes, undergoes cycloaddition to give pyrazoles ⟨68TL325⟩. Azobenzene is produced on irradiation of 3,4-diphenyl-1,3,4-oxadiazolidine-2,5-dione.

4.23.3.2.2 Electrophilic attack at nitrogen

Alkyl- and aryl-1,3,4-oxadiazoles are neutral compounds and 2-amino derivatives are weak bases. Protonation is believed to occur at ring nitrogen in position 3 which facilitates ring cleavage in aqueous acid. The pK_a (water) value of 2-amino-5-methyl-1,3,4-oxadiazole is 2.37 and values in the range 2.3–2.7 have been recorded for 2-amino-5-phenyl- and 2-N-methylamino-5-phenyl-1,3,4-oxadiazole in 50% aqueous ethanol. In the same solvent pK_a values of the more basic imines, 4-methyl-2-phenyl- and 4,N-dimethyl-2-phenyl-Δ^2-1,3,4-oxadiazolin-5-imine, are 6.31 and 6.38 respectively ⟨64CR(258)4579⟩.

Few examples of the alkylation of 1,3,4-oxadiazoles have been reported. The chemical shifts of the 2-methyl protons in the salts formed by protonation or alkylation of 2-methyl-5-phenyl-1,3,4-oxadiazole indicate that reaction takes place at ring nitrogen in position 3 ⟨70JCS(C)1397⟩. Alkylation of oxadiazolines (**12**) and aminooxadiazole (**13**) generally results in substitution at ring nitrogen in position 4 or 3 respectively, particularly under neutral conditions. In alkaline media, alkylation at the exocyclic heteroatom may occur. 2-Aryl-Δ^2-1,3,4-oxadiazolin-5-ones react with alkyl iodides in the presence of silver oxide to give O- and/or N-alkyl derivatives, depending on the alkyl iodide used ⟨73BSF254⟩. The Mannich reaction converts oxadiazolinones (**12a**) and oxadiazolinethiones (**12b**) into 4-aminoalkyloxadiazolines (**14a**) and (**14b**) respectively. 2-Amino-5-aryl-1,3,4-oxadiazoles react with methyl iodide or benzyl chloride to give 3-alkyl derivatives ⟨70CHE141⟩. N-Substituted 2-amino-5-phenyl-1,3,4-oxadiazoles are methylated by dimethyl sulfate at the amino group

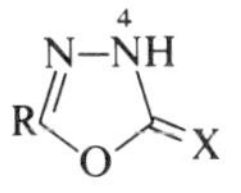

(**12a**) X = O
(**12b**) X = S

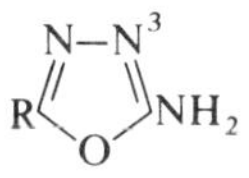

(**13**)

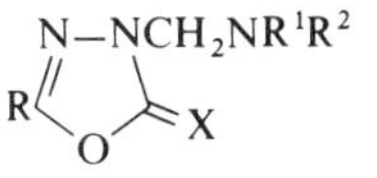

(**14a**) X = O
(**14b**) X = S

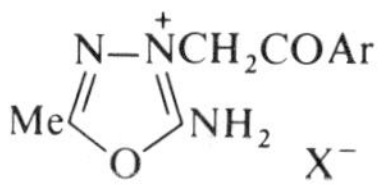

(**15**)

or at ring nitrogen, depending on the nature of the substituent ⟨72JHC107⟩. Aminooxadiazolium salts (**15**) are formed when amine (**13**; R=Me) is treated with phenacyl halides.

Heating 2-alkoxyoxadiazoles (**16**) with alkyl halides produces labile salts which react further to yield mixtures of two 4-alkyloxadiazolin-5-ones (Scheme 2) ⟨74TL3871⟩.

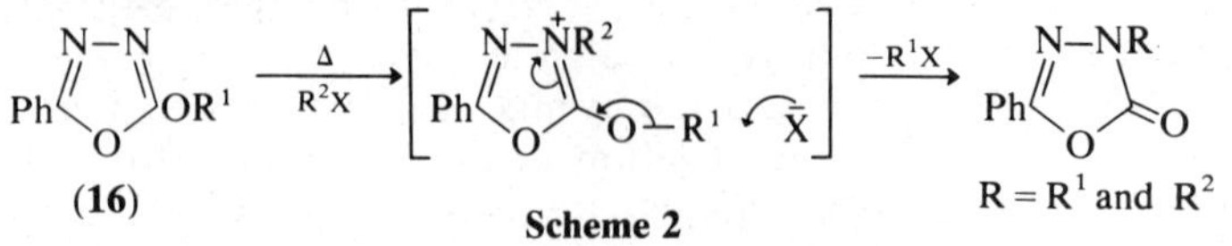

Scheme 2

Oxadiazolinones (**12a**) and thiones (**12b**) acylate, for example with acetic anhydride, benzoyl chloride or alkyl chloroformates ⟨74TL3875⟩, at ring nitrogen in position 4. 4-Acetyl derivatives have been described as 'new acylating reagents' ⟨76NKK315⟩. In contrast, acetylation of 2-methylamino-5-phenyl-1,3,4-oxadiazole takes place at the methylamino group ⟨72JHC107⟩.

2,5-Disubstituted 1,3,4-oxadiazoles form double salts with mercury(II) chloride or with silver nitrate, which, in the latter case, are used to purify or characterize the oxadiazole.

4.23.3.2.3 Electrophilic substitution at carbon

The relatively low electron density at carbon, coupled with the possibility of protonation at nitrogen, makes electrophilic substitution at carbon difficult. A further problem is acid-catalyzed ring cleavage, particularly with alkyloxadiazoles. No examples of nitration or sulfonation of the oxadiazole ring have been reported and attempted brominations were unsuccessful. A low yield of 2-(2-furoyl)-5-phenyl-1,3,4-oxadiazole is obtained when 2-phenyl-1,3,4-oxadiazole is treated with 2-furoyl chloride in the presence of triethylamine ⟨77LA159⟩.

4.23.3.2.4 Reaction of nucleophiles at carbon

The attack of a nucleophile at carbon (Scheme 3) leads either to nucleophilic displacement (path a) or ring cleavage (path b), the latter being the most common result. Attempts to reduce 1,3,4-oxadiazoles to dihydro- and tetrahydro-1,3,4-oxadiazoles have failed.

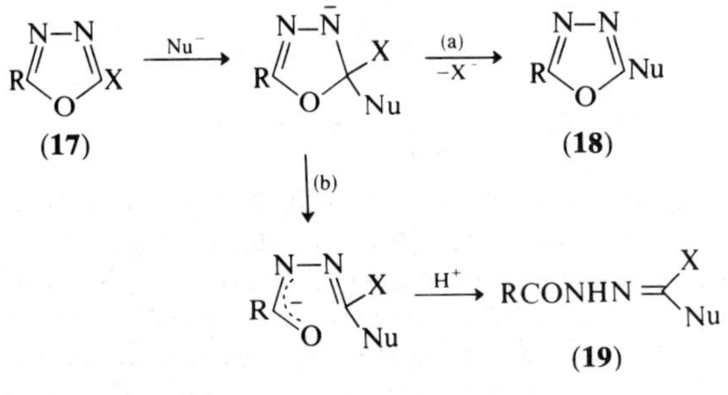

Scheme 3

(i) Nucleophilic displacement

Treatment of 2-chloro- (**17**; X=Cl) or 2-methylsulfonyl-oxadiazoles (**17**; X=SO$_2$Me) with amines, thiourea or azide ion yields the corresponding 2-substituted oxadiazoles (**18**; Nu=NHR or NR^{1}R^2, SH or N$_3$ respectively). Conversion into the hydroxyoxadiazole (**18**; Nu=OH) (an oxadiazolin-5-one) is effected using aqueous acid or alkali. A low yield of the 2-chloro compound (**17**; X=Cl, R=Ph) is obtained by heating the corresponding 2-hydroxyoxadiazole in phosphorus oxychloride with phosphorus pentachloride. Oxadiazoles (**17**; X=Cl or SO$_2$Me) react with hydrazine to give 1,2-bis(oxadiazol-2-yl)-hydrazines ⟨69IJC760⟩.

(ii) Nucleophilic attack with ring cleavage

The most frequently encountered result of the reaction of a 1,3,4-oxadiazole with a nucleophile is ring opening (Scheme 3) to a hydrazine derivative (**19**). This may undergo further reaction such as hydrolysis, or recyclization to a 1,2,4-triazole (**23**) where X or Nu is an amino group ⟨66AHC(7)183⟩.

Alkyl- and aryl-1,3,4-oxadiazoles (**20**) undergo acid- or base-catalyzed ring opening in water. Susceptibility to hydrolysis increases with solubility (Section 4.23.2.3). Hence alkyloxadiazoles ring-open more readily than aryloxadiazoles and 2,5-diaryl-1,3,4-oxadiazoles are fairly stable in dilute acid or alkali at 100 °C. The initial product of hydrolysis is a diacylhydrazine (**21**) which suffers further hydrolysis under more vigorous conditions.

(**20**)   (**21**) R^{1}CONHNHCOR^2 (**22**) R^3 = H, R, Ar, NHR   (**23a**) X = H
                                                                                    (**23b**) X = Ph
                                                                                    (**23c**) X = NHR^3

The reaction of 1,3,4-oxadiazoles with ammonia, primary amines or hydrazines provides a useful synthesis of 1,2,4-triazoles (Chapter 4.12). In some cases, the initial ring cleavage product (**22**) may be isolated. Good yields of triazoles (**23a**), (**23b**) and (**23c**) are obtained on heating oxadiazoles (**20**) with formamide in ethylene glycol ⟨75JAP(K)7535165⟩, with aniline ⟨75FRP2244462⟩ or with hydrazines ⟨76CHE711⟩. In contrast, 2,5-bis(trifluoromethyl)-1,3,4-oxadiazole reacted with hydrazine to form an s-tetrazine derivative ⟨78JCS(P1)378⟩. 2-Aryl-1,3,4-thiadiazoles (Chapter 4.27) are also produced by a ring opening, cyclization sequence when 2-aryl-1,3,4-oxadiazoles are treated with phosphorus pentasulfide.

2-Phenyl- (**24a**) and 2,5-diphenyl-1,3,4-oxadiazole (**24b**), in their electronic excited states, undergo nucleophilic attack by lower MW alcohols to give adducts which ring-open (path a) or undergo cycloelimination (path b) with subsequent formation of a triazole (Scheme 4) ⟨77CL1207⟩.

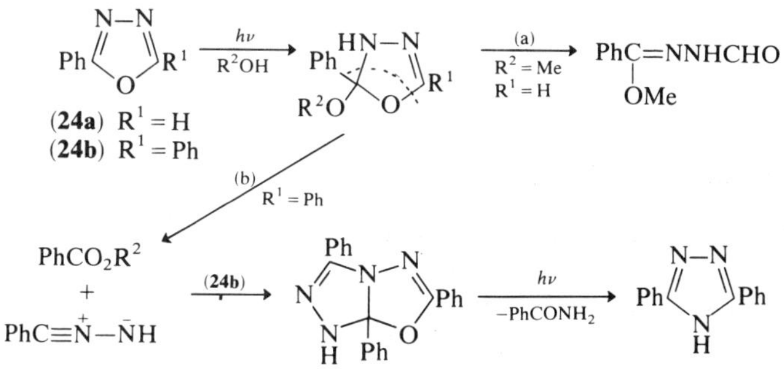

(**24a**) R^1 = H
(**24b**) R^1 = Ph

Scheme 4

Oxadiazolinones (**25a**) undergo ring opening in hot water to form hydrazinecarboxylic acids (**21**; R^2 = OH) which decarboxylate to acylhydrazines. These acylhydrazines may subsequently attack the ring of the starting oxadiazole causing cleavage to 1,5-diacylcarbonohydrazides. Oxadiazolinethiones (**25b**) are more resistant to nucleophilic attack and thione (**25b**; R = 5-nitro-2-furyl) is stable in hot water. 4-Substituted oxadiazolines (**26a**) and (**26b**) undergo similar ring cleavage reactions ⟨66AHC(7)183⟩. Oxadiazolinone (**26a**; R^1 = R^2 = Me) is converted into the corresponding thione by the action of phosphorus pentasulfide ⟨66ACS57⟩. Ring opening of 2-methyl-Δ^2-1,3,4-oxadiazolin-5-one on treatment with a tertiary amine or a Lewis acid catalyst is followed by polymerization.

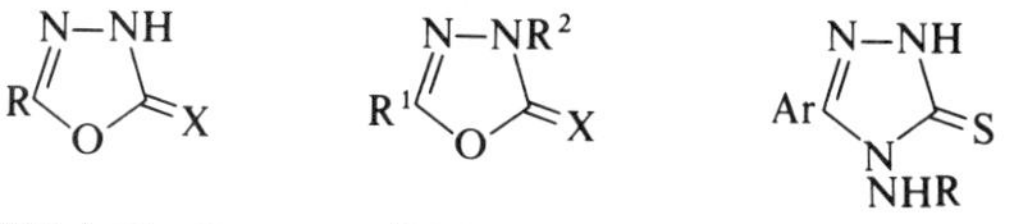

(**25a**) X = O   (**26a**) X = O   (**27**)
(**25b**) X = S   (**26b**) X = S

Ring cleavage of oxadiazolines (**25**) with ammonia, amines or hydrazines yields acylsemicarbazides. Thione (**25b**; R = 5-nitro-2-furyl) forms stable salts with amines which, in some cases, suffer ring opening on heating. 2-Aryloxadiazolinethiones (**25b**; R = aryl) react with hydrazines RNHNH$_2$ to give triazolinethiones (**27**) ⟨71KGS905⟩.

Mannich bases (**26**; $R^2 = CH_2NR^3R^4$) are dealkylated in acid or in alkali to the corresponding oxadiazolines (**25**).

Acid-catalyzed hydrolytic cleavage of 2-aminooxadiazoles (**28**) generally leads to extensive decomposition. Hydrazine derivatives (**30**) may be isolated after heating diamines (**29**) in hydrochloric acid, whereas 2,5-diamino-1,3,4-oxadiazole is stable in hot 6M hydrochloric acid.

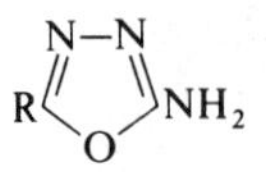

(**28**) 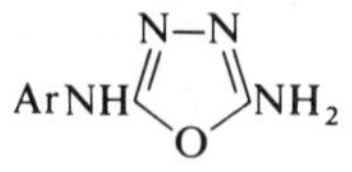   (**29**)   ArNHCONHNHCONH$_2$ (**30**)

Heating aminooxadiazoles (**28**) with aqueous sodium hydroxide usually results in ring cleavage followed by cyclization to a triazolinone (**31a**) (Chapter 4.12). In a similar manner, 2,5-dianilino-1,3,4-oxadiazole is converted into the aminotriazolinone (**31b**) ⟨65LA(685)176⟩. 1,2,4-Triazole derivatives are also formed when aminooxadiazoles (**28**) are treated with primary amines or hydrazine. In some cases the initial cleavage products are isolated and converted into triazoles by heating in alkali ⟨66AHC(7)183⟩. The dione monoimine (**32**) rearranges to the triazolinedione (**33**) in ethanolic potassium hydroxide. Prolonged heating results in ring cleavage with loss of carbon dioxide. Loss of the *t*-butyl group in imine (**32**) occurs in concentrated hydrochloric acid ⟨75JOC3112⟩.

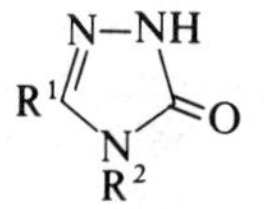

(**31a**) $R^2 = H$
(**31b**) $R^1 = PhNH, R^2 = Ph$
(**32**) $R = Pr^i$ or Bu^t 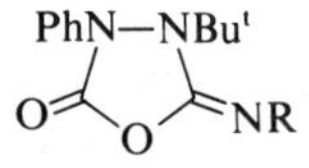
(**33**) $R = Pr^i$ or Bu^t 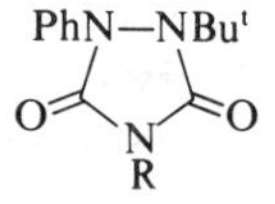

Nucleophilic attack on 1,3,4-oxadiazolium salts occurs under mild conditions and is usually followed by ring cleavage, often with subsequent recyclization to another heterocycle. Typical reactions are shown by the triphenyl oxadiazolium salt (**34**; Scheme 5) ⟨71JCS(C)409⟩. With hydrogen sulfide, salt (**34**) cleaves to a thioacylhydrazine and with cyclopentadienyl anion, ring opening to the hdyrazine derivative (**35**) takes place. A similar reaction occurs with ethyl cyanoacetate in the presence of triethylamine, but the ring cleavage product is unstable and reacts further to give pyrazole (**36**) ⟨71JCS(C)225⟩. 2-Amino-3-phenacyl-1,3,4-oxadiazolium salts rearrange to imidazolinones in alkali and yield imidazoles (Chapter 4.08) on treatment with amines in liquid ammonia ⟨66AHC(7)183⟩.

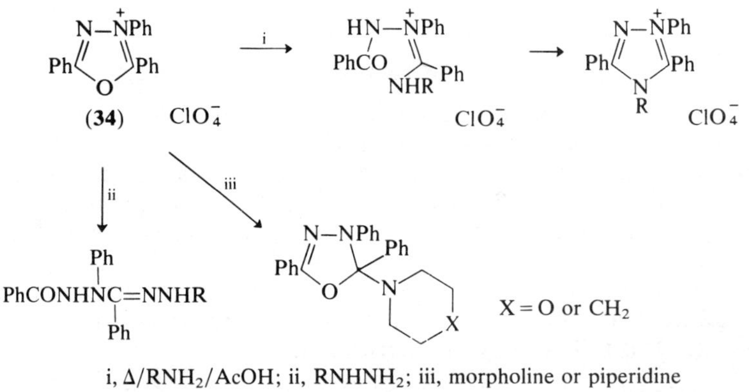

X = O or CH$_2$

i, Δ/RNH$_2$/AcOH; ii, RNHNH$_2$; iii, morpholine or piperidine

Scheme 5

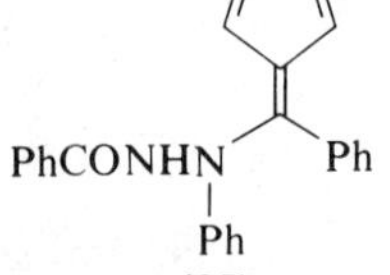

(**35**) 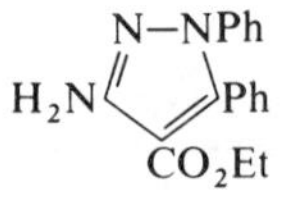   (**36**)

On heating with benzoic anhydride, 2-alkoxy-5-phenyl-1,3,4-oxadiazoles yield 2,5-diphenyl-1,3,4-oxadiazole. The mechanism for the reaction probably involves nucleophilic attack by benzoate anion on the initially formed 3-benzoyl-1,3,4-oxadiazolium benzoate ⟨78IJC(B)146⟩.

4,5-Diphenylisosydnone (4,5-diphenyl-1,3,4-oxadiazolium-2-olate; **37**) undergoes nucleophilic attack at both carbon atoms 2 and 5 to give ring cleavage products ⟨66AHC(7)183⟩. Oxadiazolium-2-thiolates (**38**) rearrange to thiadiazolium-2-olates (**39**) on heating in ethanol, by a mechanism which probably involves nucleophilic attack by ethanol. Treatment with aniline yields the triazolium-2-thiolates (**40**) and with diethylamine, ring cleavage to an acylthiosemicarbazide takes place ⟨74JCS(P1)627⟩. The oxadiazolium-2-aminide (**41**) cleaves to a semicarbazide derivative in acid and is converted into 2,3-diphenyl-Δ^3-1,2,4-triazolin-5-one in alkali ⟨73CHE1216⟩.

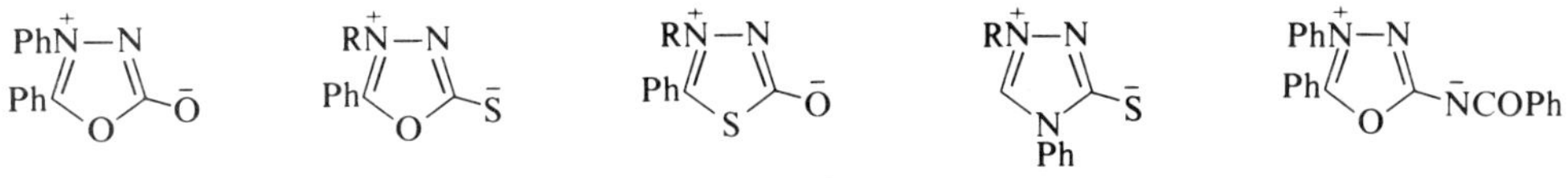

4.23.3.2.5 Nucleophilic attack at hydrogen; acidity

Deprotonation at C-2 or C-5 has not been reported. Oxadiazolin-5-ones (**25a**) and -5-thiones (**25b**) are weak acids and formally undergo deprotonation at the NH group. pK_a values (aqueous DMF) in the range 6.6–7.8 have been quoted for oxadiazolinones (**25a**; R = aryl or hetaryl) and a slightly higher pK_a (water) of 7.93 has been quoted for 2-methyl-Δ^2-1,3,4-oxadiazolin-5-one. Oxadiazolinethiones (**25b**; R = H, Me or Ph) are stronger acids, having pK_a values (water) of 3.85, 4.45 and 4.27 respectively.

4.23.3.2.6 Photoreactions and cycloadditions

UV irradiation of 2,5-diphenyloxadiazole (**43**) together with benzo[*b*]thiophene (**42**) yields an oxadiazepine (**44**) as the initial and major product, along with the 3-substituted benzo[*b*]thiophenes (**45a**) and (**45b**). These latter products are also obtained on photolysis of the oxadiazepine (Scheme 6). Prolonged irradiation gives products (**45**) and what is formally a *trans* [2+2] cycloadduct (**46**). This cycloadduct is also formed from oxadiazole (**43**) and benzo[*b*]thiophene on irradiation with benzophenone as a sensitizer and the reaction presumably involves the triplet state of the oxadiazole. Somewhat similar photoreactions have been observed between oxadiazole (**43**) and methylbenzo[*b*]thiophenes, indazoles, furan and indene ⟨77BCJ3281⟩.

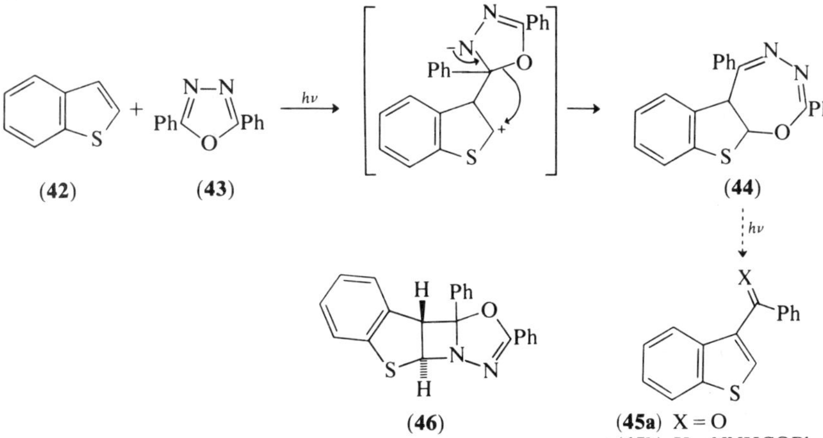

Scheme 6

Oxadiazolium salts (**47**; X = ClO$_4$, BF$_4$ or SbCl$_6$) undergo [4$^+$+2] cycloaddition to 1-diethylaminoprop-1-yne (path a) to yield oxadiazoloquinolines (**48**; Scheme 7). In some cases, nucleophilic attack (path b) occurs instead of cycloaddition and the initial ring cleavage product cyclizes to a pyrazole (**49**) with loss of RCOX. Alternatively, a [3$^+$+2] cycloaddition (path c) followed by cycloreversion produces the pyrazoles (**50**). A reaction similar to path (c) also takes place with 1,1-diethoxyethylene ⟨79CB3623⟩.

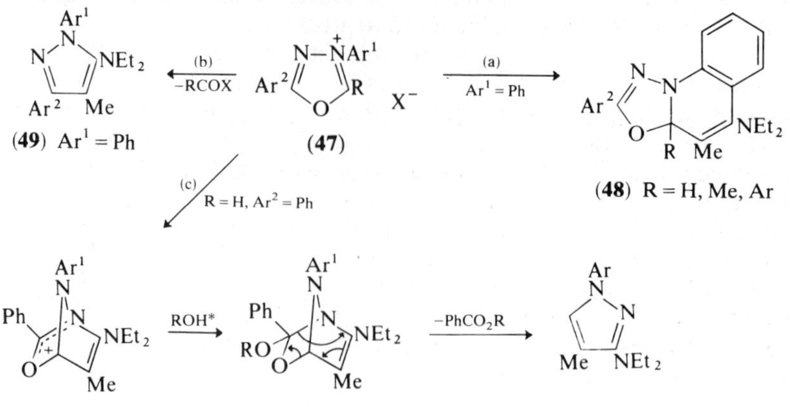

(**49**) $Ar^1 = Ph$ (**47**) (**48**) R = H, Me, Ar

* During chromatographic work-up (**50**) Ar = 2,4-dinitrophenyl

Scheme 7

4.23.3.2.7 Reaction with metals

2,5-Diphenyl-1,3,4-oxadiazole reacts with sodium to yield a radical anion which dimerizes ⟨70KGS(S2)303⟩. 2-Substituted Δ^2-1,3,4-oxadiazoline-5-thiones yield the corresponding 2-substituted 1,3,4-oxadiazoles on heating with Raney nickel. With an excess of nickel, ring cleavage to an amide results ⟨B-61MI42300⟩.

4.23.3.3 Non-aromatic Compounds

4.23.3.3.1 2,3-Dihydro-1,3,4-oxadiazoles (Δ^2-1,3,4-oxadiazolines)

Ring opening is the most common reaction of Δ^2-oxadiazolines, either thermally or on attack by nucleophiles. A rare example of oxidation to a 1,3,4-oxadiazole is the conversion of 4-acetyl-2,5-diphenyl-Δ^2-1,3,4-oxadiazoline into 2,5-diphenyl-1,3,4-oxadiazole on treatment with LTA ⟨67JOC3318⟩.

4-Acyl-2-ethoxyoxadiazolines (**51a**) exist in equilibrium with the corresponding ring-opened azomethinimine (**52**) at room temperature, whereas the 2-methyloxadiazolines (**51b**) only ring-open on heating. Depending on the nature of the substituents, the ring-opened form (**52**) rearranges reversibly or irreversibly to a diacylhydrazone ⟨66AHC(7)183⟩. An unusual fragmentation of the N—N bond occurs in the thermolysis of the 2-ethoxy-oxadiazolines (**53**) to an aryl cyanide Ar^1CN and a urethane Ar^2NHCO_2Et ⟨72TL3169⟩.

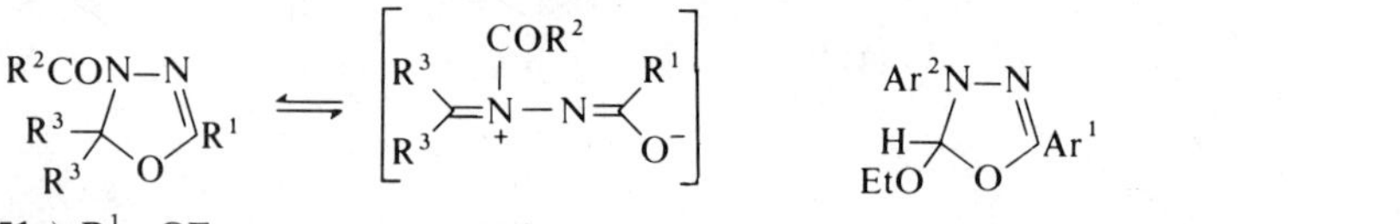

(**51a**) $R^1 = OEt$
(**51b**) $R^1 = Me$
(**52**) (**53**)

Δ^2-1,3,4-Oxadiazolines readily undergo ring cleavage in acid to carbonyl compounds and hydrazines ⟨B-61MI42300⟩. The acyloxadiazolines (**54**) form hydrazine derivatives in acid, and on treatment with alkali yield acylhydrazones (**55**) which are useful intermediates in the preparation of α-hydroxyketones (Scheme 8) ⟨77CL245⟩.

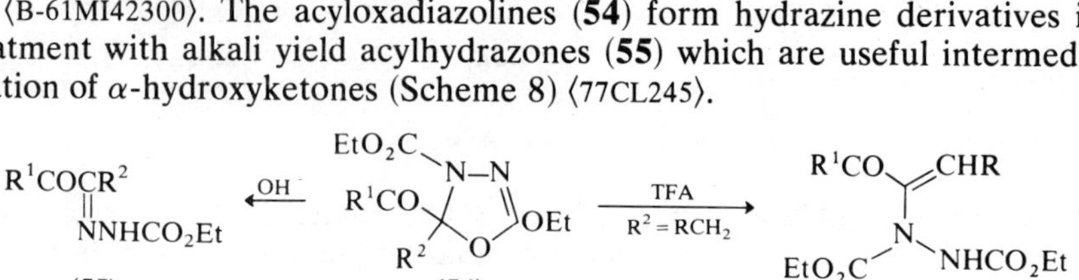

(**55**) (**54**)

Scheme 8

5,5-Dichloro-2-methyl-4-phenyl-Δ^2-1,3,4-oxadiazoline reacts with ammonia, water or hydrogen sulfide to give the corresponding oxadiazolin-5-imine, oxadiazolin-5-one or oxadiazoline-5-thione respectively. Reduction with tin or zinc and hydrochloric acid yields 2-methyl-4-phenyl-Δ^2-1,3,4-oxadiazoline ⟨B-61MI42300⟩.

4.23.3.3.2 2,5-Dihydro-1,3,4-oxadiazoles (Δ^3-1,3,4-oxadiazolines)

The most studied reactions of Δ^3-1,3,4-oxadiazolines are thermal and photochemical ring cleavage, generally with loss of a molecule of nitrogen.

Fluoromethyl-Δ^3-oxadiazolines (**56**) lose nitrogen on heating in benzene to give a carbonyl ylide (**57**) which reacts further to form an epoxide and a ketone. In some cases rearrangement to an enol ether (**58**) occurs (Scheme 9) ⟨78JA4260⟩. 2-Acetoxy-Δ^3-oxadiazolines are, in general, thermally unstable and decompose, often at room temperature, in a similar manner. Oxadiazoline (**59**) decomposes at 150 °C, with loss of nitrogen and diphenylketene, to give diphenylmethylene. Trapping with an alkene produces a cyclopropane derivative ⟨68TL2281⟩.

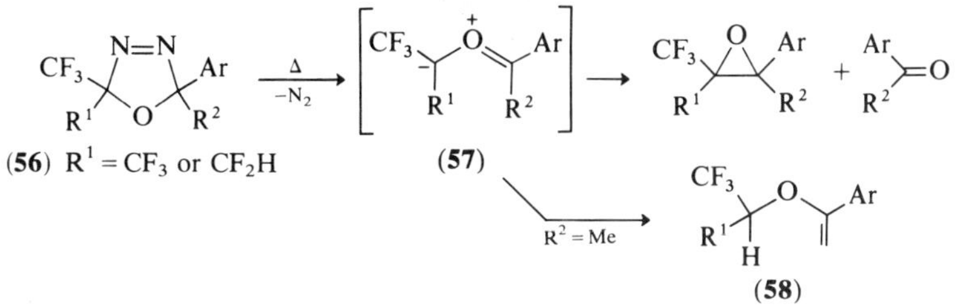

Scheme 9

Oxadiazolinones (**60a**) ⟨72CJC2326⟩ and the corresponding N-phenylimines (**60b**) ⟨69JOC3233⟩ decompose on heating by two distinct pathways (Scheme 10). An unusual 'three-piece' fragmentation to stable molecules (path a) is the preferred route in non-polar solvents, whereas in polar solvents the dominant reaction is formation of a diazoalkane (path b). Application of this type of reactivity to synthesis is illustrated by the thermolysis of the epoxy compound (**61**) in acetonitrile to give the epoxyketone (**62a**) and 4,4-dimethylhept-6-yn-2-one. The latter product is formed by further reaction of the diazoalkane (**62b**) produced via path (b) ⟨78CJC308⟩.

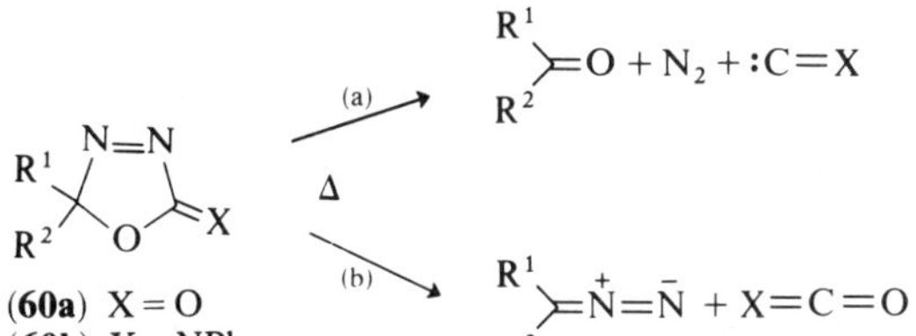

(**60a**) X = O
(**60b**) X = NPh
(**60c**) X = NCH$_2$Ph, $R^1 = R^2$ = Me

Scheme 10

Oxadiazoline (**59**), on irradiation in benzene, forms an equilibrium mixture with its synthetic precursors diphenylketene and diphenyldiazomethane. Ring opening occurs on irradiation in the presence of alcohols ⟨66AHC(7)183⟩. Photolysis of oxadiazolinones (**60a**) proceeds similarly to thermolysis and a ketone R^1COR^2 is produced concurrently with loss of nitrogen and carbon dioxide. In the presence of cyclohexene the major product is the hydrazone (**63**) which is believed to arise from a bimolecular photoreaction of the alkene with the oxadiazolinone. Similar photoreactions occur with 2-phenylpropene and diethyl ether ⟨78S535⟩.

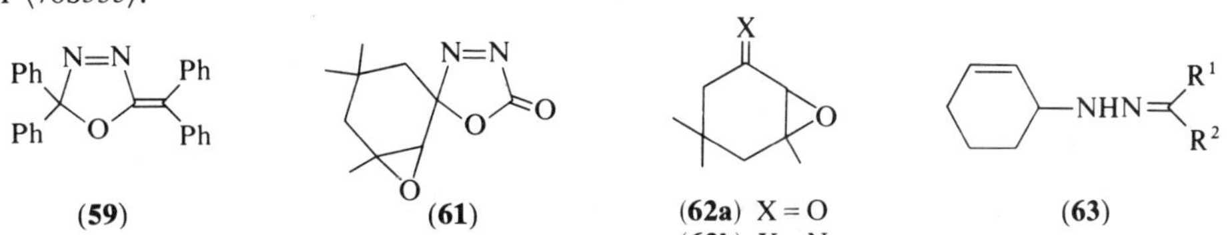

(**59**)   (**61**)   (**62a**) X = O
                        (**62b**) X = N$_2$
            (**63**)

Hydrogenation of oxadiazoline (**59**) over Raney nickel and of oxadiazolinimine (**60c**) over palladium charcoal yields hydrazones, probably by reduction to oxadiazolidines followed by ring opening. Δ^3-1,3,4-Oxadiazolin-2-imines may be hydrolyzed to the corresponding oxadiazolinones in cold acid.

4.23.3.3.3 2,3,4,5-Tetrahydro-1,3,4-oxadiazoles (1,3,4-oxadiazolidines)

Few reactions have been reported for 1,3,4-oxadiazolidines (**64**). The ring is cleaved by amino compounds to give a tetrahydro-s-tetrazine and an imine (Scheme 11) ⟨73AP134⟩.

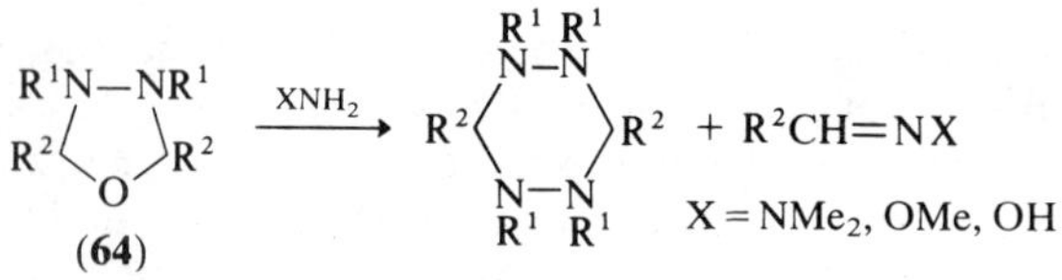

Scheme 11

4.23.3.4 Reactivity of Substituents

2-Alkyl- and 2-aryl-1,3,4-oxadiazoles undergo reactions similar to those of benzene derivatives, such as halogenation in an alkyl side-chain and electrophilic substitution in an aryl group. The main limitation is that ring cleavage may occur, particularly under acid or alkaline conditions (Section 4.23.3.1). Tautomeric oxadiazoles (**65**) are alkylated or acylated at a ring nitrogen or at the exocyclic heteroatom as indicated by the arrows.

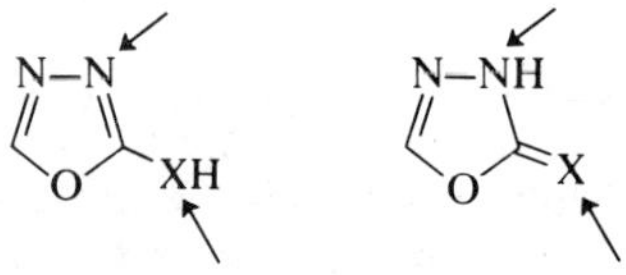

(**65**) X = O, S, NR

4.23.3.4.1 C-Linked substituents

Electrophilic substitution into the aryl nucleus of a 2,5-diaryl-1,3,4-oxadiazole can be carried out but such reactions are difficult with the more acid sensitive monoaryl-1,3,4-oxadiazoles. A variety of transformations of the functional group on an aryl ring in a mono- or di-aryl-1,3,4-oxadiazole have been performed ⟨66AHC(7)183⟩.

1,3,4-Oxadiazol-2-ylpropenes yield polymers on treatment with azoisobutyrylnitrile ⟨72MI42300⟩.

2-Diphenylmethyl-1,3,4-oxadiazoles undergo halogenation and oxidation at the methyl group. Typical reactions of an active methylene group, such as methylation, halogenation and condensation with benzaldehyde, take place with the ester (**66a**) ⟨67MI42300⟩. Treatment of the ester with ammonia or hydrazine yields the corresponding amide or hydrazide respectively. The latter, in nitrous acid, forms an azide which is converted into the amine (**66b**) on heating. 2-Chloromethyl- and 2-iodomethyl-1,3,4-oxadiazoles undergo nucleophilic displacement reactions with toluene-4-thiol and with amines ⟨73IJC732⟩. With triethyl phosphite, a phosphonate is produced which condenses with benzaldehyde to give a 2-styryl-1,3,4-oxadiazole ⟨79BRP1550440⟩. A new route to heterocyclic nitriles involves a three-step conversion of oxadiazole (**66c**) into the nitrile (**66d**) ⟨79S102⟩.

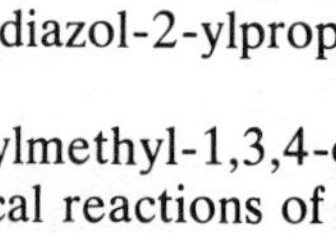

(**66a**) X = CO₂Et
(**66b**) X = NH₂
(**66c**) X = Cl
(**66d**) X = CN

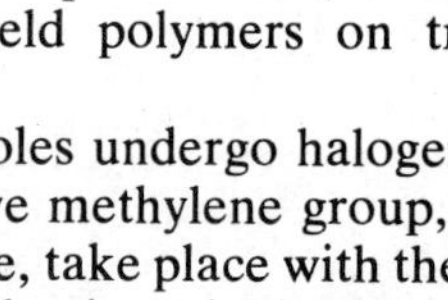

(**67a**) R = aryl
(**67b**) R = H or alkyl

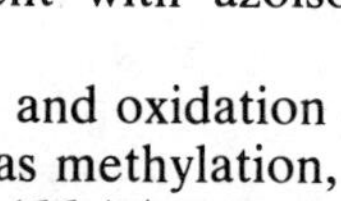

(**68**)

Metallation at the methyl group of 2-methyl-5-phenyl-1,3,4-oxadiazole takes place on treatment with butyllithium at −78 °C. On warming to room temperature a dimeric product is formed ⟨74JOC1189⟩.

2-Benzyloxadiazolium salts (**67a**) deprotonate in alkali to form coloured anhydro bases (**68**). Treatment of salts (**67a**) with acetyl chloride, or salts (**67b**) with triethylamine, results in a bimolecular reaction, with rearrangement, to give 3-(2,2-diacylhydrazino)pyrazoles ⟨71JCS(C)2314⟩. 2-Methyl-3-aryl-1,3,4-oxadiazolium salts condense with aldehydes to yield 2-styryl-1,3,4-oxadiazolium salts but no reaction occurs between aldehydes and the isomeric 5-methyl-3-aryl-1,3,4-oxadiazolium salts.

If care is taken to avoid ring cleavage, 5-aryl-1,3,4-oxadiazole-2-carboxylic acids will undergo typical reactions such as the formation of acid chlorides, amides and esters. Decarboxylation may occur on heating, for example with 5-amino-1,3,4-oxadiazole-2-carboxylic acids ⟨77JHC1385⟩, and an amide has been dehydrated to a nitrile ⟨78GEP2808842⟩.

4.23.3.4.2 N-Linked substituents

2-Amino-1,3,4-oxadiazoles alkylate either at ring nitrogen in position 3 or at the amino group, the latter predominating under alkaline conditions (Section 4.23.3.2.2). The anion derived from acetylaminooxadiazole (**69a**) reacts with ethyl bromoacetate and phenacyl bromide to give the N-alkyl derivatives (**69b**) and (**69c**) respectively ⟨77H(7)73⟩. 2-Amino-5-aryl-1,3,4-oxadiazoles add to alkynic esters and subsequent cyclization yields pyrimidooxadiazoles (**70**) (Chapter 4.29) ⟨73T2937⟩.

(**69a**) $R^1 = H, R^2 = Ac$
(**69b**) $R^1 = H, R^2 = CH_2CO_2Et$
(**69c**) $R^1 = Ac, R^2 = CH_2COPh$

(**70**)

2-Amino-1,3,4-oxadiazoles form amides with acetic anhydride, benzoyl chloride and arenesulfonyl chlorides. With phenyl isocyanate, 5-substituted 2-amino-1,3,4-oxadiazoles yield urea derivatives (**71a**). Using an excess of phenyl isocyanate an isocyanate (**71b**) is produced, and with three equivalents of phenyl isocyanate 1,3,4-oxadiazolo[3,2-a]-1,3,5-triazine-5,7-diones are obtained (Chapter 4.29) ⟨74T221⟩.

2-Amino-5-aryl-1,3,4-oxadiazoles form stable N-nitrosamines (**71c**) which are converted into hydrazines (**71d**) with zinc in acetic acid. Treatment of the hydrazine (**71d**) with nitrous acid yields the corresponding azide and with benzaldehyde a hydrazone is formed ⟨B-61MI42300⟩. On heating in benzene or in nitrobenzene, the nitrosamine (**71c**) undergoes the Gomberg–Bachmann reaction to yield 2,5-diaryl-1,3,4-oxadiazoles ⟨73JCS(P1)1357⟩.

Imidophosphoranes (**71e**) ⟨80JIC1243⟩ or phosphorimidic trichlorides (**71f**) ⟨79JHC1097⟩ are formed when 2-aminooxadiazoles react with dibromotriphenylphosphorane or phosphorus pentachloride respectively. 2-Amino-1,3,4-oxadiazoles form silver and copper salts and yield complexes with heavy metal salts ⟨62ZC69⟩.

(**71a**) X = NHCONHPh
(**71b**) X = NCO
(**71c**) X = NHNO, R = aryl
(**71d**) X = NHNH$_2$, R = aryl
(**71e**) X = N=PPh$_3$
(**71f**) X = N=PCl$_3$

(**72**)

Treatment of the imine (**72**; R = H) with phenyl isocyanate, acetic anhydride or an arenesulfonyl chloride ArSO$_2$Cl gives the corresponding N-substituted imines (**72**; R = CONHPh, Ac or SO$_2$Ar respectively) ⟨70CHE141⟩.

4.23.3.4.3 O- and S-Linked substituents

Alkylation and acylation of Δ^2-1,3,4-oxadiazolin-5-ones (**73a**) generally occurs at ring nitrogen in position 3 (Section 4.23.3.2.2).

Δ^2-1,3,4-Oxadiazoline-5-thiones (**73b**) react with a variety of alkyl halides, in the presence of base, to give *S*-alkyl ethers (**74a**). 2-Phenyl-Δ^2-1,3,4-oxadiazolin-5-one and the corresponding 5-thione yield a 4-alkylated product on reaction with diphenyldiazomethane, whereas other diazoalkanes form *S*-alkyl derivatives with the thione ⟨77MI42301⟩. Thiones (**73b**; R = aryl) are converted into the corresponding oxadiazolin-5-ones on treatment with hydrogen peroxide, and on oxidation with chlorine, a sulfonyl chloride (**74b**) is formed. Disulfides are produced on oxidation of thiones with bromine ⟨79MI42300⟩ and thioethers (**74a**; R^1 = Me) are oxidized to sulfones (**74c**) with permanganate ⟨73JPR185⟩.

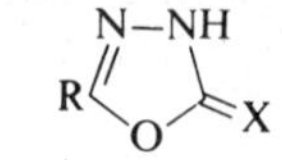

(**73a**) X = O
(**73b**) X = S

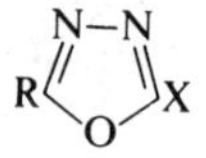

(**74a**) X = SR^1
(**74b**) X = SO_2Cl, R = aryl
(**74c**) X = SO_2Me

4.23.4 SYNTHESIS

Most 1,3,4-oxadiazoles are best obtained by synthesis from acylic precursors. Such reactions are mainly 'one-bond' or 'two-bond' cyclizations. For convenience, cyclizations of intermediates formed from two reactants are classed as 'one-bond' cyclizations if the intermediate can be isolated. A few minor routes to 1,3,4-oxadiazoles which are described in the reviews listed earlier (Section 4.23.1) are omitted from this section and no references are given for reactions covered by these reviews.

4.23.4.1 Ring Synthesis

4.23.4.1.1 Cyclization with the formation of one bond

The only common mode of cyclization is formation of the O—C(2) bond, usually by nucleophilic attack of the carbonyl oxygen of an amide group at the carbon atom which becomes C-2 in the 1,3,4-oxadiazole ring (Scheme 12).

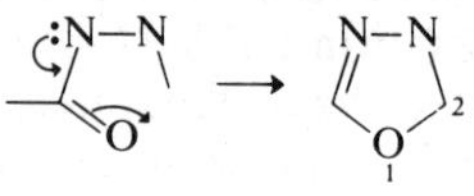

Scheme 12

(i) From 1,2-diacylhydrazines and related compounds

The most widely applicable route to 2,5-dialkyl-, 2-alkyl-5-aryl and 2,5-diaryl-1,3,4-oxadiazoles (**76**) is the thermal or acid catalyzed cyclization of 1,2-diacylhydrazines (**75**; Scheme 13). The method may also be used for monosubstituted oxadiazoles (**76**; R^1 = H) and esters (**76**; $R^2 = CO_2R$) ⟨78GEP2808842⟩.

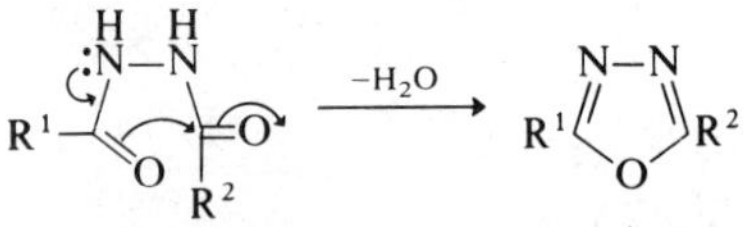

(**75**)

(**76**) R^1 = H, alkyl, aryl, hetaryl
R^2 = alkyl, aryl, hetaryl, CO_2R

Scheme 13

1-Phenyl-1,2-diacylhydrazines (**77**) cyclize in acetic acid in the presence of acid HX to form oxadiazolium salts (**78**) ⟨70JCS(C)1397⟩.

$R^1CONHNCOR^2$
|
Ph

(**77**)

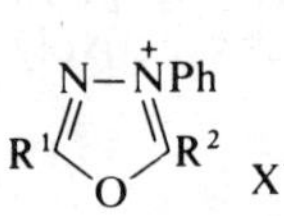

(**78**) X = ClO_4 or BF_4
R^1 = H, alkyl, aryl, OEt
R^2 = H, alkyl, Ph

The major route to Δ^2-1,3,4-oxadiazolin-5-ones (**80**) is the thermal cyclization of acylhydrazinecarboxylic acid derivatives (**79**; Scheme 14). Acid chlorides (**79**; X = Cl), conveniently prepared from acylhydrazines $R^1CONHNHR^2$ and phosgene, are usually cyclized *in situ*. Δ^2-1,3,4-Oxadiazolin-5-one (**80**; $R^1 = R^2 = H$) has been prepared in this manner from formylhydrazine and phosgene.

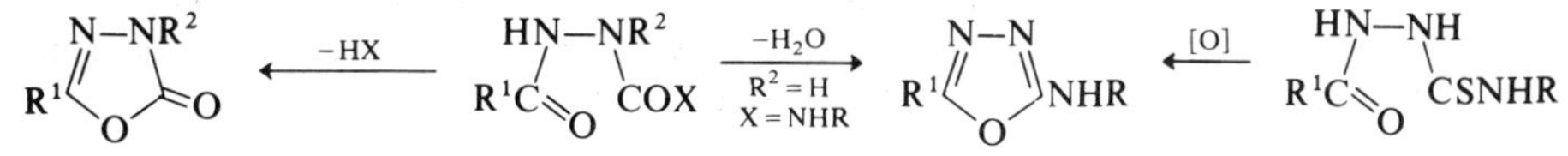

(**80**) (**79**) X = Cl, OR, NHR (**81**) (**82**)

R^2 = H, alkyl, aryl

Scheme 14

Thiophosgene reacts with acylhydrazines in a manner similar to that of phosgene (above) to produce oxadiazoline-5-thiones (**83**) and the reaction represents an important route to these compounds.

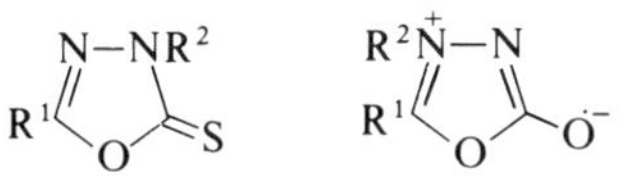

(**83**) (**84**)

Cyclization in phosphorus oxychloride of semicarbazides (**79**; X = NHR) yields aminooxadiazoles (**81**) whereas thermolysis leads to loss of ammonia (when X = NH_2) and formation of an oxadiazolinone (**80**). Cyclization to aminooxadiazoles (**81**) occurs when thiosemicarbazides (**82**) are heated with an oxidizing agent such as lead oxide. This reaction has been widely applied to the synthesis of aminooxadiazoles, sometimes in low yields, and has been used to prepare 2-amino-1,3,4-oxadiazole (**81**; $R^1 = R^2 = H$). *S*-Methyl ethers of thiosemicarbazides (**82**) cyclize, with loss of methanethiol, to aminooxadiazoles (**81**) on heating, but in PPA cyclization to 2-methylthio-1,3,4-oxadiazoles occurs.

1-Substituted 1-acylhydrazines react with phosgene to form isosydnones (**84**) ⟨69JCS(B)1185⟩.

Isocyanide dichlorides, formally imines derived from phosgene, react in a manner similar to that of phosgene (above) except that 2-amino-1,3,4-oxadiazole derivatives, instead of oxadiazolin-5-ones, are produced (Scheme 15) ⟨71CC1223, 81AP193, 73CHE1216⟩.

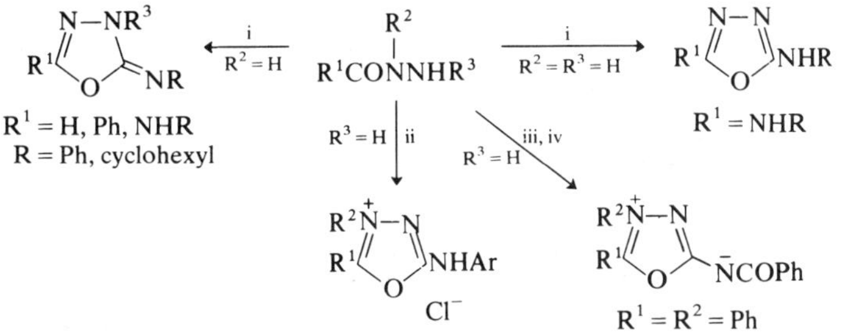

i, $RN{=}CCl_2$; ii, $ArN{=}CCl_2$; iii, $PhCON{=}CCl_2$; iv, NEt_3/Δ

Scheme 15

Hydrazine-1,2-dicarboxylic acid derivatives (**85**) cyclize thermally, or on treatment with base, to alkoxyoxadiazolinones (**86**) or oxadiazolidinediones (**87**) ⟨76JOC3763⟩.

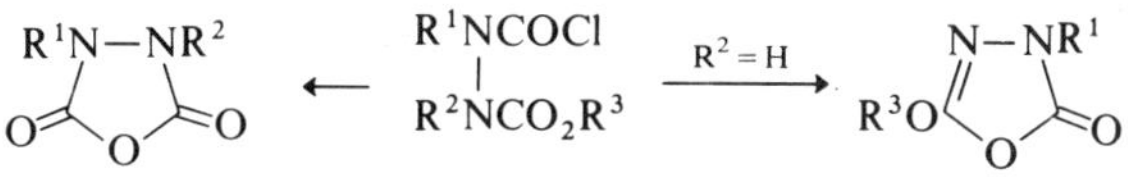

(**87**) (**85**) (**86**)

The main route to Δ^2-oxadiazoline-5-thiones (**83**; R^1 = alkyl, aryl, CO_2Et, $R^2 = H$) is the cyclization of dithiocarbazates (**88**) prepared from acylhydrazines (**89**; $R^2 = H$) and carbon disulfide. The ester (**83**; $R^1 = CO_2Et$, $R^2 = H$), obtained by this method, has been hydrolyzed

$R^1CONHNHCS_2^-M^+$ $R^1CON(R^2)NH_2$ 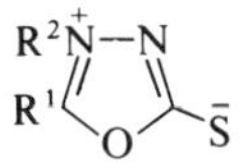

(**88**) M = K or NH_4 (**89**) (**90**) R^2 = Ph, Me

and decarboxylated to Δ^2-1,3,4-oxadiazoline-5-thione (**83**; $R^1 = R^2 = H$) ⟨72CJC3079⟩. Acylhydrazines (**89**; R^2 = Me or Ph) react with carbon disulfide to give oxadiazolium-2-thiolates (**90**) ⟨74JCS(P1)627⟩.

A versatile route to 2-aminooxadiazoles (**92**), which generally proceeds in higher yield than oxidation of thiosemicarbazides (**82**), is the cyclization in aqueous or alcoholic solution of 1-acyl-2-cyanohydrazines (**91**) prepared from acylhydrazines and cyanogen bromide (Scheme 16). Alternatively, hydrazines (**91**) are obtained by acylation of cyanohydrazines ⟨66EGP52668⟩.

$$\text{RCONHNH}_2 \xrightarrow{\text{CNBr}} \text{RCONHNHCN} \longrightarrow \underset{(\textbf{92})}{\underset{O}{\overset{N-N}{R\diagup\!\!\!\diagdown}}\text{NH}_2}$$
$$(\textbf{91})$$

Scheme 16

Acylureas react with hypohalite in alkali to yield N-haloureas which are converted into oxadiazolinones (**94**). Rearrangement of the N-halourea to an isocyanate presumably occurs during the reaction (Scheme 17).

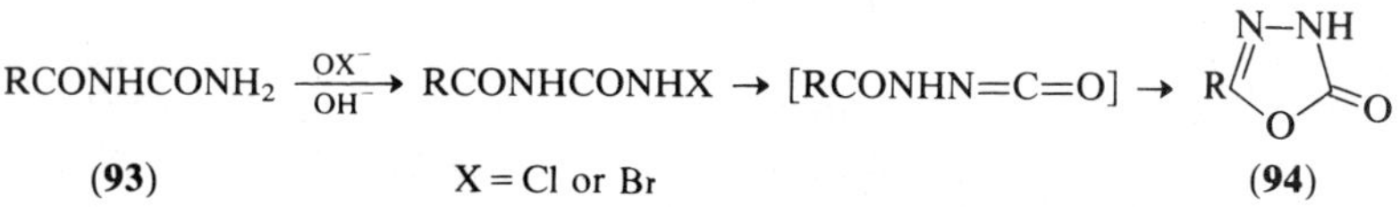

$$\underset{(\textbf{93})}{\text{RCONHCONH}_2} \xrightarrow[\text{OH}^-]{\text{OX}^-} \underset{X=\text{Cl or Br}}{\text{RCONHCONHX}} \rightarrow [\text{RCONHN}=\text{C}=\text{O}] \rightarrow \underset{(\textbf{94})}{\underset{O}{\overset{N-\text{NH}}{R\diagup\!\!\!\diagdown}}\!\!\!=\!\!O}$$

Scheme 17

(ii) From acylhydrazones

Substituted acylhydrazones (**95**), formally derivatives of 1,2-diacylhydrazines (**75**), also cyclize to oxadiazoles (**76**), with loss of HX. Imidol ethers (**95a**; $R^2 = H$), prepared from acylhydrazines $R^1\text{CONHNH}_2$ and ortho esters HC(OR)_3, cyclize on heating to monosubstituted oxadiazoles (**76**; $R^1 = H$). This reaction is an important synthesis of monosubstituted 1,3,4-oxadiazoles, usually with ethyl orthoformate as the ortho ester, and it has been used to prepare the parent 1,3,4-oxadiazole (**76**; $R^1 = R^2 = H$) ⟨65JA5800⟩. 2,5-Disubstituted 1,3,4-oxadiazoles are obtained using ortho esters $R^1\text{C(OR}^2)_3$ or imido ester salts $R^1\text{C(OR}^2)=\overset{+}{\text{N}}\text{H}_2\text{Cl}^-$. A similar reaction between an acylhydrazine $R^1\text{CONHNH}_2$ and dimethyl monothionemalonate yields oxadiazoles (**76**; $R^2 = \text{CH}_2\text{CO}_2\text{Me}$) *via* imidol ethers (**95a**; R = Me, $R^2 = \text{CH}_2\text{CO}_2\text{Me}$) ⟨68CJC2255⟩.

Halohydrazones (**95b**; R^1 = alkyl or aryl) or halosemicarbazones (**95b**; $R^1 = \text{NH}_2$, NHR or NR_2), prepared from hydrazones or semicarbazones respectively by the action of halogen in alkali or, preferably, by treatment with bromine in acetic acid in the presence of sodium acetate ⟨71AC(R)587⟩, cyclize to 1,3,4-oxadiazoles. A dipolar intermediate (**96**) has been proposed. This method is suitable for the preparation of 2-aminooxadiazoles (**81**) or aminoesters (**81**; $R^1 = \text{CO}_2\text{-alkyl}$) ⟨77JHC1385⟩ from halosemicarbazones, or for the preparation of oxadiazole-2-carboxylic acid esters (**76**; $R^2 = \text{CO}_2R$) from halohydrazones (**95b**; $R^2 = \text{CO}_2R$). Thermolysis of the dihalohydrazones (**97**) yields 2-halooxadiazolinones (**98**) ⟨74JOC2336⟩.

$$\underset{\begin{array}{c}(\textbf{95a}) \text{ X}=\text{OR}\\(\textbf{95b}) \text{ X}=\text{halogen}\\(\textbf{95c}) \text{ X}=\text{NHR}\end{array}}{R^1\!\!\diagup\!\!\!\overset{\text{HN-N}}{\underset{O}{\diagdown}}\!\!\!\overset{\displaystyle X}{\underset{R^2}{\diagdown}}} \qquad \underset{(\textbf{96})}{R^1\overset{O}{\overset{\|}{C}}\!-\!\bar{\text{N}}\!-\!\text{N}\!=\!\overset{+}{\text{C}}R^2} \qquad \underset{\begin{array}{c}(\textbf{97}) \text{ X}=\text{Cl, Br}\\R=\text{Me, Et}\end{array}}{X_2\text{C}=\text{NN(CO}_2R)_2} \qquad \underset{\begin{array}{c}(\textbf{98}) \text{ X}=\text{Cl, Br}\\R=\text{Me, Et}\end{array}}{X\!\!\diagup\!\!\overset{N-\text{NCO}_2R}{\underset{O}{\diagdown}}\!\!=\!\!O}$$

Aldehyde or ketone acylhydrazones (**99**) undergo oxidative cyclization to oxadiazoles (**100**) ⟨72CPB1663⟩ or to the thermally unstable Δ^3-oxadiazolines (**101**; Scheme 18) ⟨67TL3501⟩. Lead tetraacetate is generally the preferred oxidizing agent. Ketone semicarbazones (**99**; $R^3 = \text{NHR}$) react with LTA to yield oxadiazolinimines (**102**) which, in acid, are converted into oxadiazolinones (**103**) ⟨72CJC3248⟩. Cyclization of ketone hydrazones (**99**) in acetic anhydride gives 4-acetyl-Δ^2-oxadiazolines (**104**) ⟨76JOU1102⟩.

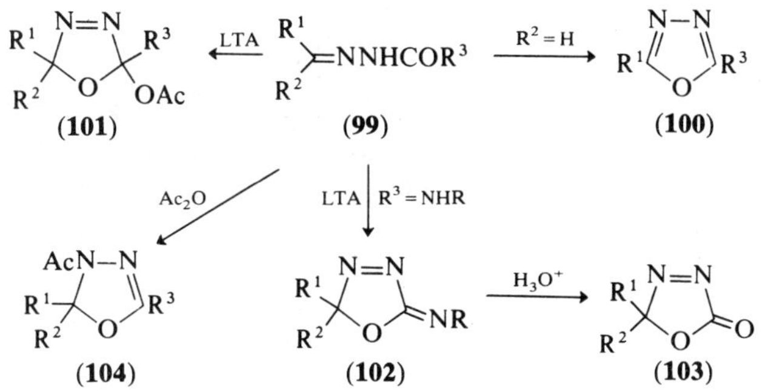

1,3,4-Oxadiazoles

Scheme 18

4.23.4.1.2 Cyclization with the formation of two bonds

A useful preparation of 5-aryloxadiazolines (**106a**) is the reaction of acylhydrazines (**105a**) with aromatic aldehydes ⟨81ZC182⟩. Treatment of acylhydrazines (**105a**) with ethyl orthoformate yields 5-ethoxyoxadiazolines (**106b**) ⟨72TL3169⟩.

$Ar^1CONHNHR$

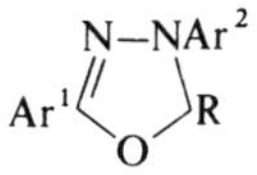

(**105a**) R = Ar^2
(**105b**) R = H

(**106a**) R = aryl
(**106b**) R = OEt

Azo compound (**107a**) reacts with carbene :CXY (X = H, Y = H or CO_2Et) derived from diazoalkane N_2CXY to yield an oxadiazoline (**108a**; Scheme 19). The carbene :CXY (X = Y = Cl or Br), formed by loss of PhHgBr from the mercury compound $PhHgCX_2Br$ (X = Cl or Br) ⟨74JOC2336⟩, reacts similarly to produce an unstable oxadiazoline (**108b**) which loses benzoyl halide to form a 2-halooxadiazole (**109**). A related reaction of azo compound (**107b**) with ylides (**110**) provides a route to oxadiazolines (**108c**) ⟨77CL245⟩.

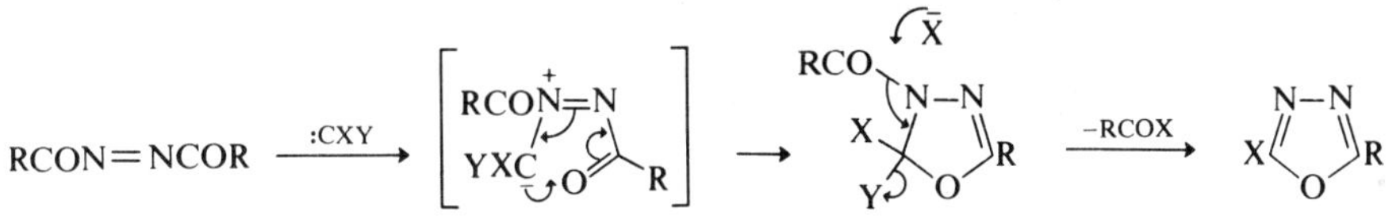

(**107a**) R = Ph
(**107b**) R = OEt

(**108a**) R = Ph, X = H
Y = H or CO_2Et
(**108b**) R = Ph
X = Y = Cl or Br
(**108c**) R = OEt
X = R^1, Y = COR^2

(**109**) R = Ph
X = Cl or Br

Scheme 19

1,3-Dipoles such as diazoalkanes $PhC(R^1)N_2$ undergo cycloaddition to fluoroalkyl ketones CF_3COR^2 to give Δ^2-oxadiazolines (**111**) ⟨78JA4260⟩. Diphenylketene reacts in a similar manner with diphenyldiazomethane.

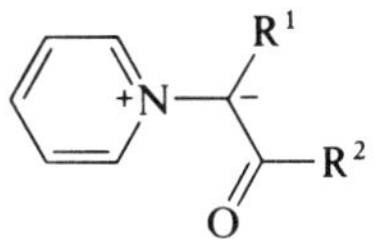

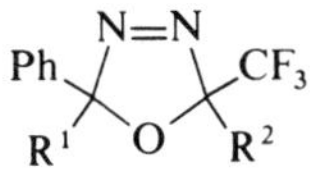

(**110**) R^1 = Me or OEt
R^2 = H or alkyl

(**111**) R^1 = Me or Ar
R^2 = CF_3 or CF_2H

Silver salts of acylhydrazones (**99**) yield 4-acetyloxadiazolines (**104**) on treatment with acetyl chloride. With benzoyl chloride the corresponding 4-benzoyl-Δ^2-oxadiazolines are produced ⟨67T1379⟩. Aldazines $(RCH=N)_2$ react with LTA to form Δ^2-oxadiazolines (**104**; $R^1 = R^3 = R$, $R^2 = H$), which, on prolonged reaction time, lose acetic acid to give 2,5-disubstituted 1,3,4-oxadiazoles ⟨67JOC3318⟩.

4.23.4.1.3 Cyclization with the formation of three bonds

Of the few examples of 1,3,4-oxadiazole synthesis from three components, the most important is the preparation of oxadiazolidines (112) from hydrazines R^1NHNHR^1 and aldehydes R^2CHO. Tetrahydro-s-tetrazines are usually formed as by-products ⟨73AP134⟩. The best results are achieved using a mild dehydrating agent such as magnesium sulfate ⟨64RTC877⟩ or a molecular sieve ⟨77CZ302⟩.

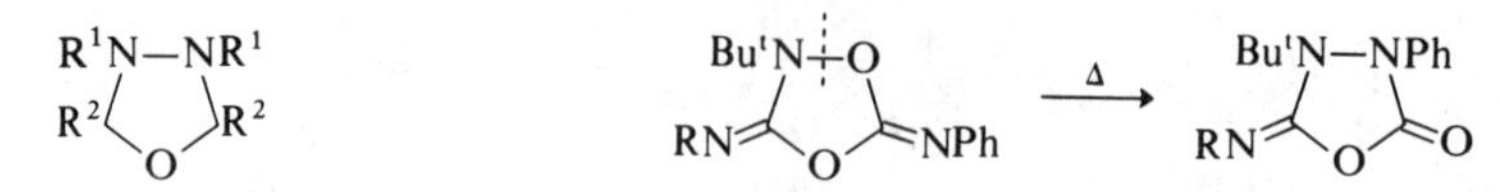

(112)   (113)   (114) R = Bu^t or Pr^i

2-Methyl-2-nitrosopropane, alkyl isocyanide RNC and phenyl isocyanate form a 1:1:1 adduct (113) which on heating at 80 °C rearranges to the oxadiazolidine (114) ⟨75JOC3112⟩.

4.23.4.2 Ring Transformations

Tetrazoles (115a) lose nitrogen on heating to give nitrilimines (116) which undergo cycloaddition to isocyanates or ketones with the subsequent formation of 1,3,4-oxadiazole derivatives (Scheme 20). The nitrilimines from 3-acyltetrazoles (115b) cyclize directly to 2,5-disubstituted 1,3,4-oxadiazoles. This reaction provides a useful synthetic route to 1,3,4-oxadiazoles and it has been applied to the synthesis of poly-1,3,4-oxadiazoles and macrocyclic systems containing four oxadiazole rings ⟨64RCR508⟩. 3-Acyltetrazoles (115b; $R^1 = H$) yield 2-alkyloxadiazoles (117; $R^1 = H$) on irradiation or on heating ⟨80HCA588⟩. 5-Aminotetrazoles are converted into 2-amino-1,3,4-oxadiazoles on heating in acetic anhydride by a reaction which involves migration of an acetylamino group.

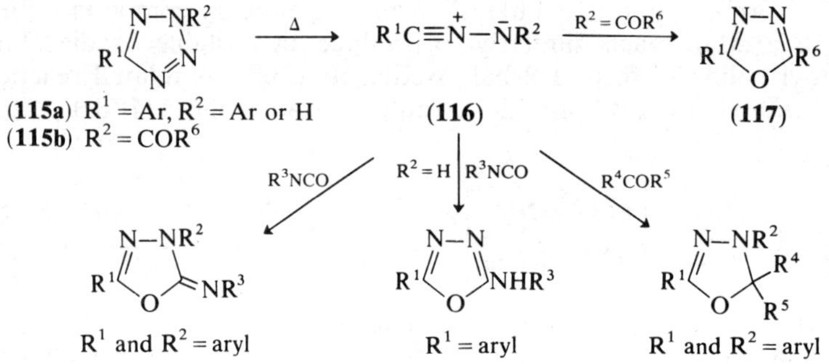

(115a) $R^1 = Ar$, $R^2 = Ar$ or H   (116)   (117)
(115b) $R^2 = COR^6$

R^1 and R^2 = aryl R^1 = aryl R^1 and R^2 = aryl

Scheme 20

Monosubstituted oxadiazoles (119; R = alkyl, aryl, hetaryl, styryl or CO_2Et) are formed in a 'relay' synthesis from triazolylbenzamidines (118) on treatment with acid chlorides RCOCl or anhydrides $(RCO)_2O$ ⟨69JPR646⟩.

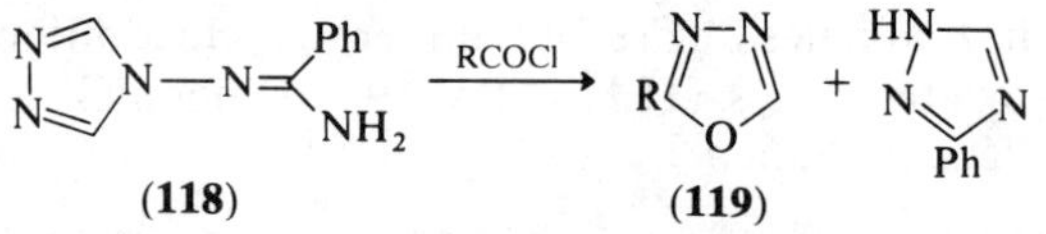

(118)   (119)

Other routes to 1,3,4-oxadiazoles from five-membered heterocycles include ozonolysis of pyrazoles and rearrangement of hydantoins in the presence of hypohalite.

The conversion of diaryl-s-tetrazines into oxadiazoles is shown in Scheme 21.

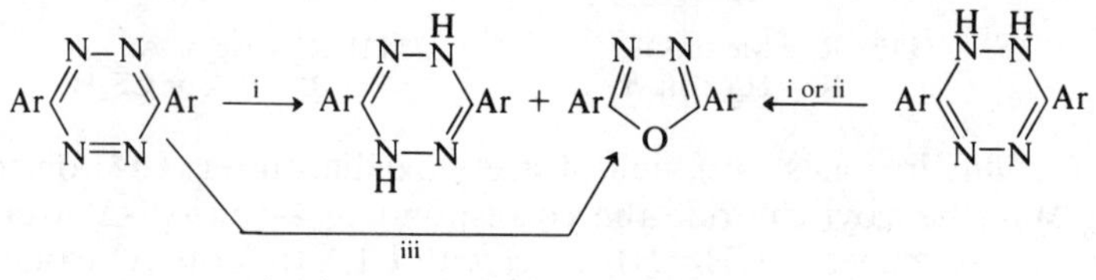

i, Δ/HCl; ii, KOH/EtOH; iii, Δ/AcO$_2$H

Scheme 21

The action of bromine in acetic acid, in the presence of sodium acetate, on triazinones (**120**) provides access to rare oxadiazolyl ketones (**121**) ⟨79JHC145⟩.

Examples of syntheses from three-membered heterocycles are the synthesis of the oxadiazolylpropylamine (**122**) from acylhydrazine ArCONHNH$_2$ and 3,3-dimethyl-2-dimethylaminoazirine ⟨78HCA2419⟩, and the conversion of 3,3-dialkyldiaziridines into oxadiazolines (**123**) on heating above 150 °C with an anhydride (R^{3}CO)$_2$O ⟨76MIP42300⟩.

4.23.5 APPLICATIONS

1,3,4-Oxadiazoles have a wide variety of uses, in particular as biologically active compounds in medicine and in agriculture, as dyestuffs, UV absorbing and fluorescent materials, heat-resistant polymers and scintillators. Reviews of the relevant literature prior to 1965 are available ⟨66AHC(7)183, 62HC(17)263⟩.

4.23.5.1 Biologically Active 1,3,4-Oxadiazoles

4-Alkyl-2-(4-pyridyl)-Δ^2-1,3,4-oxadiazolin-5-ones and the corresponding 5-thiones show leprostatic and tuberculostatic properties ⟨66AF1034⟩. The antitubercular activity of 2-(4-pyridyl)-Δ^2-1,3,4-oxadiazolin-5-ones may be partly related to their metabolism to isonicotinoylhydrazide. 4-Arylaminomethyl-Δ^2-1,3,4-oxadiazoline-5-thiones (Mannich bases) exhibit antibacterial ⟨78JIC108⟩, antiproteolytic and anticonvulsant activity ⟨78JPS1507, 78MI42300⟩. Analgesic, antipyretic and antiphlogistic properties have also been described for Δ^2-oxadiazolin-5-ones and Δ^2-oxadiazoline-5-thiones. 2-Amino-1,3,4-oxadiazoles act as muscle relaxants ⟨66JMC478⟩ and show antimitotic activity ⟨74MI42300⟩. Analgesic, antiinflammatory, anticonvulsive, diuretic and antiemetic properties are exhibited by 5-aryl-2-hydroxymethyl-1,3,4-oxadiazole derivatives ⟨74GEP2403357⟩, 2-hydroxyphenyl-1,3,4-oxadiazole acts as a hypnotic and as a sedative, and 2-(1,1-diphenylalkyl)-1,3,4-oxadiazoles are antidiarrheal agents ⟨76JMC1221⟩.

A wide variety of Mannich bases derived from Δ^2-1,3,4-oxadiazoline-5-thiones show fungicidal activity ⟨73MI42300, 77JIC1143⟩ and in some cases also act as bactericides and insecticides ⟨79JIC1230⟩. 2-(1,2,4-Triazol-4-yl)- and 2-amino-5-aryl-1,3,4-oxadiazoles ⟨76MI42300⟩, disulfides and thioethers derived from 1,3,4-oxadiazole-2-thiols ⟨79MI42300⟩ and 1,3,4-oxadiazolyl-2-sulfones ⟨76MI42301⟩ also show fungicidal activity.

A variety of 2,5-diaryl-, 2,5-dialkyl- and 2-alkyl-5-aryl-1,3,4-oxadiazoles show herbicidal activity, particularly against broad-leafed weeds and grasses in crops such as rice and corn ⟨80JAP8027042, 76MIP42301⟩. Several reports on a wide selection of herbicidal uses for oxadiazolinone (**124**) appear yearly in the recent literature.

4-Aryl-Δ^2-1,3,4-oxadiazolin-5-ones stimulate root formation in beans ⟨77MI42302⟩ and together with 2-(2-carboxyphenyl)-5-aryl-1,3,4-oxadiazoles ⟨77AJC2225⟩ act as plant growth regulators.

4.23.5.2 Other Applications

2,5-Diaryl-1,3,4-oxadiazoles, particularly those containing 4-biphenylyl or 1-naphthyl substituents, have been described for use as liquid and plastic scintillators for the measurement of weak β-rays, γ-rays, fast neutrons and ionized particles in cosmic radiation. 2,5-Di-*m*-tolyl-1,3,4-oxadiazole is reported to be a cheap scintillator with properties comparable to those of commercial products ⟨77USP4017738⟩.

Conjugated systems containing a 1,3,4-oxadiazole ring find use as fluorescent whiteners in detergents or in cotton or polyamide fibres. Typical are the 2,5-diaryloxadiazoles (**125**) ⟨79MIP42300⟩. Bis-oxadiazoles (**126**) are useful as fluorescent whiteners in polymer fibres and films rather than for textiles ⟨76MIP42302⟩. 4-[2-(1,3,4-Oxadiazol-2-yl)vinyl]stilbenes are optical brighteners for cotton, polyester, polyamide and PVC fabrics ⟨79BRP1550440⟩ and 2,5-diaryl-1,3,4-oxadiazoles act as UV absorbers and light-screening agents.

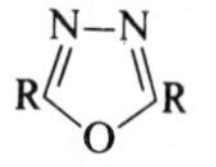

(**125**)

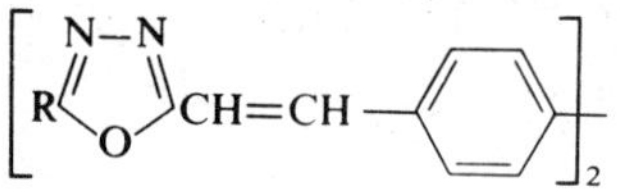

(**126**) R = Ph or *m*-Tol

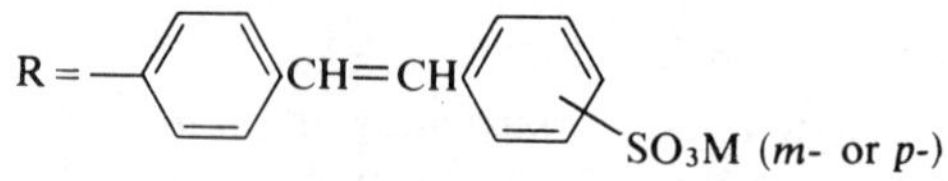

R = —⟨phenyl⟩CH=CH⟨phenyl⟩—SO₃M (*m*- or *p*-)

M = H or Na

1,3,4-Oxadiazoles are used in coatings in electrographic reproduction processes and as tone improvers and development accelerators in photography. Aminooxadiazoles are described as photographic sensitizers and 2-methyl-5-phenyl-1,3,4-oxadiazole acts as a photographic supersensitizer when used in conjunction with a cyanine dye ⟨71USP3615633⟩. Halogenated 2-alkyl-5-styryl-1,3,4-oxadiazoles are used in photosensitive compositions suitable for lithographic printing plates ⟨79GEP2851471⟩.

Poly-1,3,4-oxadiazoles have a high thermal stability and have been used in film materials. Heat-resistant polymers have been prepared by incorporating 1,3,4-oxadiazole units into polymers ⟨70MI42300⟩ and heat-resistant polyamides have been synthesized from 3,5-bis(4-aminophenyl)-1,3,4-oxadiazole and isophthaloyl chloride ⟨73MI42301⟩.

Several examples of anthraquinone and azo dyes containing a 1,3,4-oxadiazole ring have been reported although less frequently in recent years. The use of 2-alkyl-5-aryl-1,3,4-oxadiazoles as photographic colour formers has been described.

2,5-Bis(perfluoroalkyl)-1,3,4-oxadiazoles have a high thermal stability and are suitable for use as heating-bath liquids and solvents for fluorinated polymers.

4.24

1,2,3-Thiadiazoles and their Benzo Derivatives

E. W. THOMAS

The Upjohn Company, Kalamazoo

4.24.1	INTRODUCTION	448
4.24.2	STRUCTURE OF 1,2,3-THIADIAZOLES	448
4.24.2.1	*Survey of Possible Structures*	448
4.24.2.1.1	Neutral aromatic systems	448
4.24.2.1.2	Mesoionic compounds	448
4.24.2.1.3	Nonaromatic compounds	448
4.24.2.2	*Theoretical Methods*	449
4.24.2.2.1	π-Electron density	449
4.24.2.2.2	Frontier electron densities	449
4.24.2.3	*Methods for Structure Determination*	449
4.24.2.3.1	X-ray diffraction	449
4.24.2.3.2	Microwave spectroscopy	450
4.24.2.3.3	^{1}H NMR spectroscopy	451
4.24.2.3.4	^{13}C NMR spectroscopy	451
4.24.2.3.5	Mass spectral fragmentation	451
4.24.2.3.6	IR spectroscopy	452
4.24.2.3.7	UV spectroscopy	452
4.24.2.4	*Thermodynamic Aspects*	452
4.24.2.4.1	Melting and boiling points	452
4.24.2.4.2	Solubility	453
4.24.2.4.3	Chromatography	453
4.24.2.4.4	Thermochemistry	453
4.24.2.4.5	Aromaticity	453
4.24.3	REACTIVITY	453
4.24.3.1	*Thermal and Photochemical Reactions*	453
4.24.3.1.1	Fragmentations	453
4.24.3.1.2	Rearrangements	455
4.24.3.2	*Electrophilic Attack at Nitrogen*	456
4.24.3.2.1	Introduction	456
4.24.3.2.2	Proton acids: basicity of 1,2,3-thiadiazoles	456
4.24.3.2.3	Alkyl halides and related compounds (mesoionic compounds)	456
4.24.3.2.4	Halogens	456
4.24.3.2.5	Peracids	456
4.24.3.3	*Attack at Sulfur*	457
4.24.3.3.1	Electrophilic	457
4.24.3.3.2	Nucleophilic	457
4.24.3.4	*Nucleophilic Attack at Carbon*	457
4.24.3.5	*Reactions at Hydrogen*	457
4.24.3.6	*Reaction with Radicals*	458
4.24.3.7	*Reaction of Substrates*	458
4.24.3.7.1	Benzenoid rings	458
4.24.3.7.2	S-Linked substituents	458
4.24.3.7.3	C-Linked substituents	459
4.24.4	SYNTHESIS	459
4.24.4.1	*Introduction*	459
4.24.4.2	*Dipolar Cycloadditions*	459
4.24.4.3	*From Hydrazones and Thionyl Chloride*	460
4.24.4.4	*1,2,3-Benzothiadiazoles*	460
4.24.4.5	*From Other Heterocycles*	461

4.24.5	IMPORTANT COMPOUNDS — APPLICATIONS	461
4.24.5.1	Insecticide Synergist	461
4.24.5.2	Sedative	462
4.24.5.3	Herbicide	462
4.24.5.4	Antibacterial	462
4.24.5.5	Polymers	462

4.24.1 INTRODUCTION

For nearly a century chemists have studied 1,2,3-thiadiazoles. 1,2,3-Benzothiadiazoles were the first in this series to be synthesized and through 1950 most of the citations referred to them. Hodgson and Dodgson's review ⟨48MI42400⟩ and Bambas' review ⟨52HC(4)3⟩ cover this early period.

Hurd and Mori's synthesis of 1,2,3-thiadiazoles in 1955 ushered in the modern era of research for this heterocycle. The review by Sherman ⟨B-61MI42400⟩ discusses various syntheses and covers the literature through 1960.

During the past decade an excellent series, *Organic Compounds of Sulphur, Selenium, and Tellurium*, has reviewed advances in this area. Most recently Barton's compendium ⟨B-79MI42400⟩ includes a brief review of the thiadiazole literature through 1977. This review covers the literature through 1981.

Although thiadiazoles have been known for a long time, of the four possible isomers, 1,2,3-thiadiazoles account for the fewest literature citations. A substantial amount of this literature has dealt with thermal and photochemical reactions of 1,2,3-thiadiazoles, and most recently attention has been focused on mesoionic derivatives. Still, the many gaps that exist in our knowledge leave this field a fertile area for further research.

4.24.2 STRUCTURE OF 1,2,3-THIADIAZOLES

4.24.2.1 Survey of Possible Structures

4.24.2.1.1 Neutral aromatic systems

The 1,2,3-thiadiazole (**1**) possesses three contiguous heteroatoms in a five-membered ring and exists as a remarkably stable neutral aromatic compound.

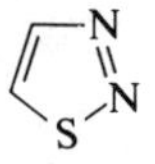

(**1**)

4.24.2.1.2 Mesoionic compounds

More recently fully aromatic mesoionic compounds such as (**2**) have been generated. Since this is a new area, information in previous reviews is limited ⟨76AHC(19)1, 77SST(4)417⟩.

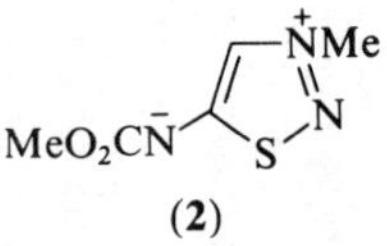

(**2**)

4.24.2.1.3 Nonaromatic compounds

Examples of nonaromatic 1,2,3-thiadiazole derivatives such as (**3**) and (**4**) are rare. A previous review cites few references to these compounds due to the paucity of work in this area ⟨70CRV593⟩.

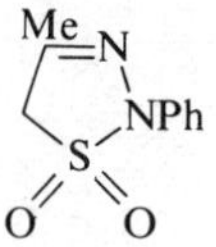

(**3**) 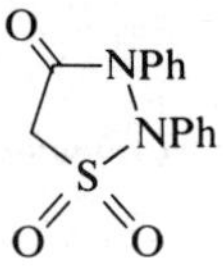   (**4**)

4.24.2.2 Theoretical Methods

4.24.2.2.1 π-Electron density

Two articles concerning MO calculations for 1,2,3-thiadiazoles have been reported. The earlier paper employs a linear combination of atomic orbitals (LCAO) with the Hückel approximation to calculate the π-electron densities illustrated in Figure 1a ⟨61CCC156⟩. These calculations do not accurately predict the site of reaction with electrophiles. The more recent paper employed a linear combination of gaussian orbitals (LCGO) to the Hartree–Fock method to calculate the π-electron densities illustrated in Figure 1b ⟨77JST(39)189⟩. These calculations include the contribution of sulfur d-orbitals and demonstrate a higher electron density at N-3 which is consistent with most electrophilic reactions. As predicted by electron density, nucleophiles attack at sulfur.

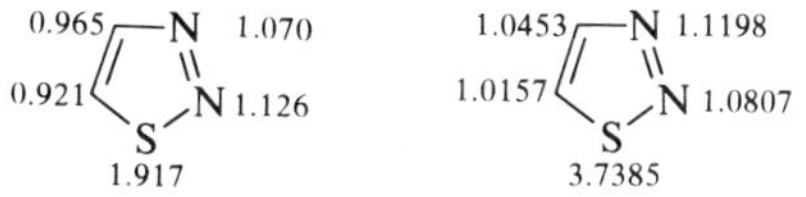

Figure 1 Molecular diagrams for the π-electron densities in 1,2,3-thiadiazole

4.24.2.2.2 Frontier electron densities

Frontier molecular orbital calculations have been employed successfully to predict reactions by looking at the LUMO in nucleophilic substitution reactions and the HOMO in electrophilic reactions. Although these calculations have become quite routine, they have unfortunately not been applied to 1,2,3-thiadiazoles.

4.24.2.3 Methods for Structure Determination

4.24.2.3.1 X-Ray diffraction

An X-ray diffraction study, when possible, is the ultimate structure proof. Bond lengths are known for certainty to 0.001 Å and bond angles to 0.1°. Although the literature abounds with X-ray data for heterocycles there are no X-ray spectra for a 4,5-dialkyl substituted, nonionic 1,2,3-thiadiazole. However, an X-ray structure has been reported for a 1,2,3-benzothiadiazole (5), and the bond lengths and bond angles are listed in Table 1 ⟨79AX(B)3114⟩.

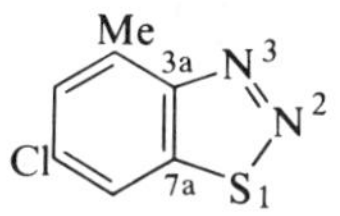

(5)

Table 1 Molecular Dimensions for 1,2,3-Benzothiadiazole (5)

Bond lengths (Å)		Bond angles (°)	
S—C(7a)	1.715	C(3a)—S—N(2)	92.1
S—N(2)	1.708	S—N(2)—N(3)	112.1
N(2)—N(3)	1.284	N(2)—N(3)—C(3a)	114.3
N(3)—C(3a)	1.371	N(3)—C(3a)—C(7a)	113.7
C(3a)—C(7a)	1.401	C(3a)—C(7a)—S	107.7

Analysis of the data indicates the bond lengths are in the normal range for this essentially planar molecule and the benzene portion exhibits considerable π-conjugation. The best resonance structure for the thiadiazole is a partial double bond for S—C(7a) and S—N(2), and a nearly complete double bond for N(2)—N(3) and C(7a)—C(3a). One might expect the bond angles and lengths for thiadiazole to vary if it were not fused to the benzene ring.

X-Ray data have also been reported for a mesoionic compound, 5-acylamino-3-methyl-1,2,3-thiadiazole (**2**) ⟨80TL2101⟩, and for 5-phenyl-1,2,3-thiadiazole 3-*N*-oxide (**6**) ⟨78ZN(B)316⟩. Molecular dimensions for (**2**) and (**6**) are in Tables 2 and 3 respectively.

Table 2 Molecular Dimensions for 1,2,3-Thiadiazole (**2**)

Bond lengths (Å)		Bond angles (°)	
S—C(5)	1.736	C(5)—S—N(2)	94
S—N(2)	1.657	S—N(2)—N(3)	109
N(2)—N(3)	1.309	N(2)—N(3)—C(4)	119
N(3)—C(4)	1.331	N(3)—C(4)—C(5)	112
C(4)—C(5)	1.390	C(4)—C(5)—S	106
N(3)—Me	1.474		

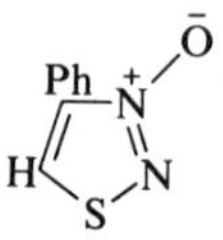

(**6**)

Table 3 Molecular Dimensions for 1,2,3-Thiadiazole (**6**)

Bond lengths (Å)		Bond angles (°)	
S—C(5)	1.685	C(5)—S—N(2)	94.0
S—N(2)	1.645	S—N(2)—N(3)	109.5
N(2)—N(3)	1.313	N(2)—N(3)—C(4)	116.4
N(3)—C(4)	1.401	N(3)—C(4)—C(5)	109.4
C(4)—C(5)	1.351	C(4)—C(5)—S	110.6
C(5)—H	1.02		
N(3)—O	1.270		

These few examples represent a major share of literature citations for 1,2,3-thiadiazole X-ray determinations. One clearly sees that an X-ray of a simple 1,2,3-thiadiazole would be worthwhile as a reference.

4.24.2.3.2 Microwave spectroscopy

Microwave spectroscopy is a powerful tool for the determination of molecular structure. Thiazoles and thiadiazoles have been studied by this technique, but it was not until 1976 that a paper on the microwave spectrum of 1,2,3-thiadiazole appeared. Bond distances and angles for 1,2,3-thiadiazole (**7**) are listed in Table 4 ⟨76MI42400⟩. The success of this project is owed in part to the development of double resonance modulated (DRM) microwave spectroscopy which allows for quick analysis of an individual spectrum.

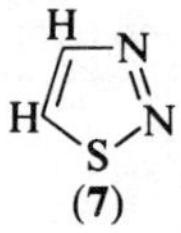

(**7**)

Table 4 Molecular Dimensions for 1,2,3-Thiadiazole (**7**)

Bond lengths (Å)		Bond angles (°)	
S—N(2)	1.6917	C(5)—S—N(2)	92.91
N(2)—N(3)	1.2897	S—N(2)—N(3)	111.21
N(3)—C(4)	1.3662	N(2)—N(3)—C(4)	113.95
C(4)—C(5)	1.3686	N(3)—C(4)—C(5)	114.15
C(5)—S	1.6888	C(4)—C(5)—S	107.79

A structural feature of note for this planar molecule is the near equal bond length of S—N(2) (1.692 Å) and S—C(5) (1.689 Å). Another interesting feature is the N(2)—N(3) bond length (1.290 Å), which is the shortest bond observed so far in an aromatic heterocyclic ring.

4.24.2.3.3 ^{1}H NMR spectroscopy

^{1}H NMR chemical shifts of 1,2,3-thiadiazoles give another indication as to the aromatic character of these compounds. Compiled in Table 5 are a number of examples of proton chemical shifts for ring substituted 1,2,3-thiadiazoles.

Table 5 ^{1}H NMR Spectral Data for Ring Hydrogens of 1,2,3-Thiadiazoles

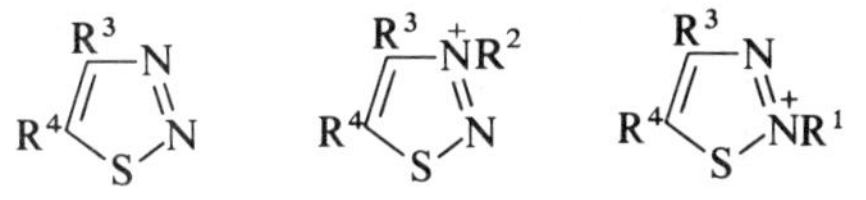

R^1	R^2	R^3	R^4	$\delta(^1H)$ for ring hydrogens (p.p.m.) 4	5	Solvent	Ref.
—	—	H	H	Ab centered at 8.80		CCl_4	78JOC2487
—	—	Me	H	—	8.20	CCl_4	78JOC2487
—	—	H	Me	8.35	—	CCl_4	78JOC2487
—	—	H	Ph	8.70	—	$CDCl_3$	75LA1257
—	—	Ph	H	—	8.60	$CDCl_3$	75LA1257
—	Me	Ph	H	—	9.93	DMSO	80JHC1217
Me	—	Ph	H	—	10.17	DMSO	80JHC1217

4.24.2.3.4 ^{13}C NMR spectroscopy

^{13}C NMR data can be quite diagnostic for heterocyclic structures. With reasonable certainty one can usually expect the carbon atom adjacent to the nitrogen atom in a 1,2,3-thiadiazole to have a lower field chemical shift than the carbon atom adjacent to the sulfur. Several examples are illustrated in Table 6.

Table 6 ^{13}C NMR Spectral Data for Ring Carbons of 1,2,3-Thiadiazoles

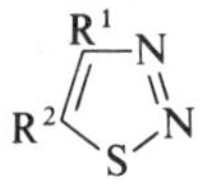

R^1	R^2	$\delta(^{13}C)$ for ring carbons (p.p.m.) 4	5	Solvent	Ref.
H	H	147.3	135.8	$CDCl_3$	78JHC1383
Ph	H	163.9	130.9	$CDCl_3$	78JHC1383
Ph	Ph	157.5	150.8	$CDCl_3$	75LA1257
—(CH$_2$)$_4$—		157.8	147.4	$CDCl_3$	75LA1257

4.24.2.3.5 Mass spectral fragmentation

In their mass spectra 1,2,3-thiadiazoles commonly fragment initially by expulsion of nitrogen, and the fragmentation then proceeds to form thiirene ion radical (**8**; Scheme 1). These studies have been reviewed ⟨73SST(2)717⟩ and additional studies have shown, depending on substituents, that thioketene radical (**9**) can be formed in the mass spectrum ⟨72T1353⟩.

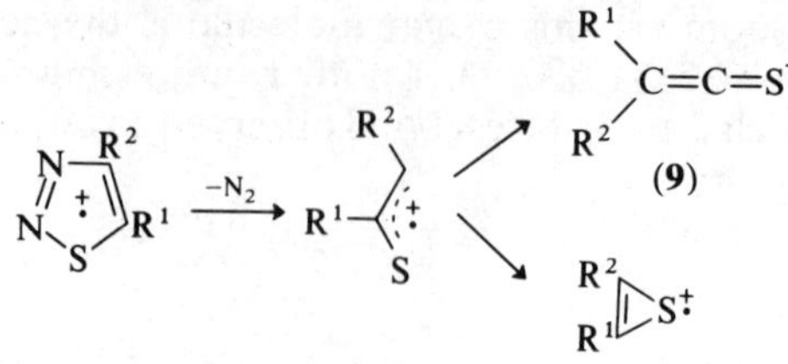

Scheme 1

4.24.2.3.6 IR spectroscopy

The IR spectra of a number of five-membered N and N,S-heterocyclic compounds have been compared ⟨64CJC43⟩, and the regions of absorption are very similar for the thiadiazole isomers. Characteristic vibrations for 1,2,3-thiadiazoles are: 1560–1475, 1350–1280 cm^{-1} (ring skeletal); 1265–1200, 1190–1175, 1150–950 cm^{-1} (ring breathing and CH in-plane deformations); and 910–890, 705–670 cm^{-1} (CH out-of-plane deformations).

4.24.2.3.7 UV spectroscopy

The UV spectra of a number of 1,2,3-thiadiazoles have been compiled along with those of other heterocycles ⟨71PMH(3)180⟩. Several examples are listed in Table 7.

Table 7 UV Spectral Data for 1,2,3-Thiadiazoles

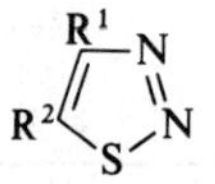

R^1	R^2	λ_{max}^{a} (ε_{max})	Ref.
H	H	211 (4380), 249 (1460), 294 (195)	78JOC2487
Me	H	213 (4820), 258 (1850), 290 (220)	78JOC2487
H	Me	217 (5300), 253 (2100), 293 (245)	78JOC2487

^{a} nm, in cyclohexane.

4.24.2.4 Thermodynamic Aspects

4.24.2.4.1 Melting and boiling points

There has been no attempt to correlate melting and boiling points of 1,2,3-thiadiazoles with structural parameters. Table 8 lists several examples taken from the literature for

Table 8 Melting and Boiling Points for 1,2,3-Thiadiazoles

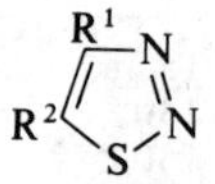

R^1	R^2	M.p./b.p. (°C)	R^1	R^2	M.p. (°C)
H	H	157	Ph	Bu^t	120–122
Me	H	87–89	H	CO$_2$H	104–106
Bu^n	H	210^a	CO$_2$H	H	200–205 (dec)
H	Ph	52–56	CO$_2$H	Me	113–114
Ph	H	77–79	CO$_2$H	SEt	179–181
Ph	Ph	93–94	NH$_2$	H	44–46
Me	Ph	300^a	H	NH$_2$	145–147 (dec)
Bu^t	Ph	106–108	Br	Br	79–80

^{a} Corrected to atmospheric pressure.

derivatives containing a variety of substituents. The parent compound, 1,2,3-thiadiazole, is a yellow liquid and its boiling point is 157 °C at atmospheric pressure. When 1,2,3-thiadiazoles are heated above 200 °C they usually decompose (Section 4.24.3.1).

4.24.2.4.2 Solubility

The parent compound, 1,2,3-thiadiazole, is soluble in alcohol, ether and water. Most 1,2,3-thiadiazoles are freely soluble in methylene chloride and chloroform. Solvents such as ether and alcohol have been used to recrystallize 1,2,3-thiadiazoles. No systematic study of the solubility characteristics of 1,2,3-thiadiazoles has been undertaken.

4.24.2.4.3 Chromatography

No systematic study of 1,2,3-thiadiazoles by TLC or GC has appeared. As long as the temperatures for GC analysis are below 200 °C (the approximate decomposition temperature for 1,2,3-thiadiazole) this method should be satisfactory. Most simple thiadiazoles are isolated by distillation but more complex structures are quite amenable to column chromatography or TLC ⟨79JMC1214⟩.

4.24.2.4.4 Thermochemistry

There have been no papers with experimental thermochemical data for 1,2,3-thiadiazoles. Thermochemical data derived from quantum mechanical calculations for 1,2,3-thiadiazoles are also not available. These areas clearly are worthy of greater research attention.

4.24.2.4.5 Aromaticity

There has been no work published assessing the aromatic character of 1,2,3-thiadiazole. From the Hückel definition of aromaticity, $(4n+2)$ π-electrons in a ring constitute an aromatic compound and by this criterion 1,2,3-thiadiazoles should be considered aromatic compounds. Chemical shifts for 1,2,3-thiadiazoles in both ^{13}C and ^{1}H NMR, which are related to the electronic environment, are in accord with an aromatic heterocyclic ring.

4.24.3 REACTIVITY

4.24.3.1 Thermal and Photochemical Reactions

4.24.3.1.1 Fragmentations

The majority of 1,2,3-thiadiazole chemistry has been directed toward the study of thermal and photochemical reactions. Upon thermolysis 1,2,3-thiadiazoles afford a number of varied products ⟨77T449, 77JOC575⟩ as illustrated in equation (1). The mechanism of formation of the myriad of products has received considerable attention.

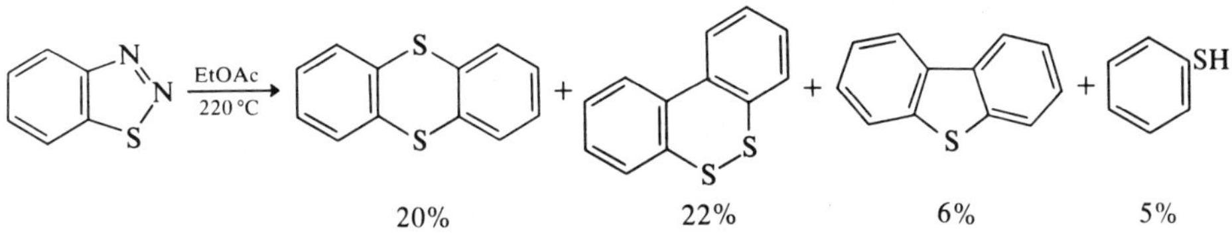

The course of the thermal as well as photochemical reactions of 1,2,3-thiadiazole is believed to proceed, as in the case of the mass spectrum, with the expulsion of nitrogen forming a diradical intermediate which can react to form products or generate additional intermediates ⟨B-79MI42400⟩. The latter have been pursued with great interest.

As early as 1916, Staudinger and Siegwart predicted that nitrogen extrusion from thermally excited 4,5-diphenyl-1,2,3-thiadiazole would be facile leading to thioketenes ⟨B-61MI42400⟩. Their experiments only led to the isolation of tetraphenylthiophene.

More recently researchers have been successful in generating thioketenes from thiadiazoles ⟨77T449, 75JHC605, 79LA1734⟩. Thioketenes are very reactive compounds forming deeply colored solutions which can be handled in solution at low temperature ⟨77CB1225⟩. They react with alcohols and amines quantitatively (Scheme 2) ⟨77SST(4)417, 75AG(E)248⟩.

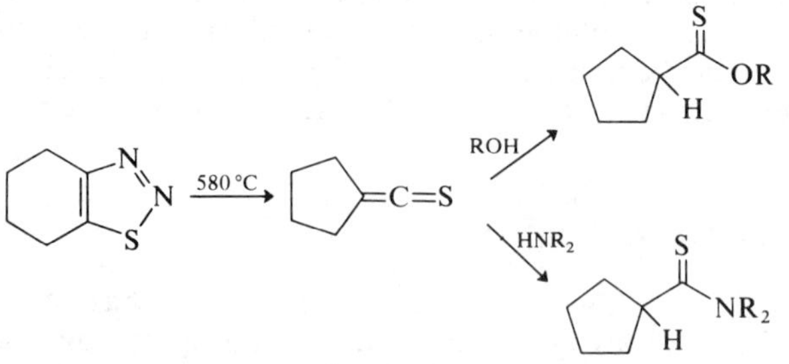

Scheme 2

Support for another intermediate, the thiirene (**10**), has been generated by experiments in which one thiadiazole (**11**) forms two isomers as illustrated in Scheme 3 ⟨77TL2643⟩. Others have also proposed the existence of a thiirene intermediate ⟨79CB1769⟩. Attempts to trap the thiirene as an iron complex (**12**) failed; instead, reaction of nonacarbonyldiiron with thiadiazoles affords red crystalline carbene complexes (**13**) ⟨75SST(3)670⟩.

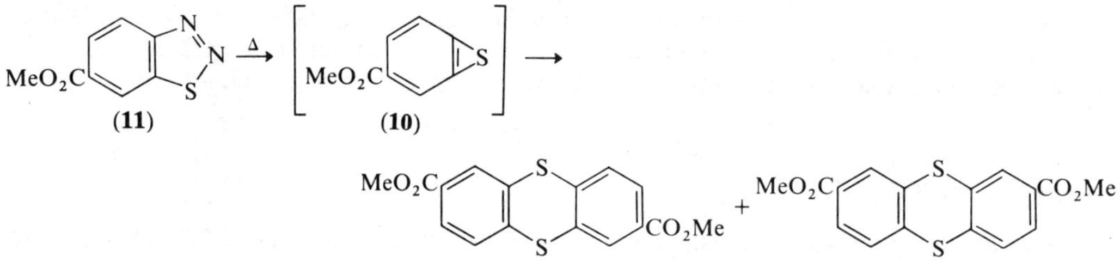

Scheme 3

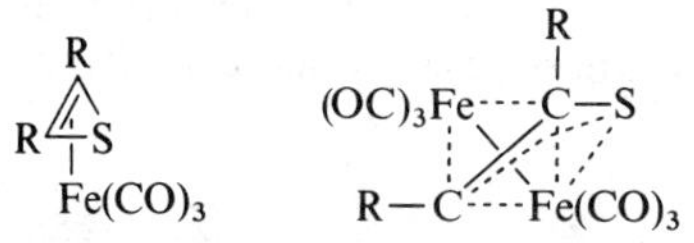

For similar systems, results have been reported ⟨77JOC575⟩ which do not require a thiirene intermediate (equation 2). Further, researchers monitoring the decomposition of 1,2,3-thiadiazoles by PE spectroscopy saw no evidence of thiirenes ⟨77JA1663⟩, only thioketenes (equation 3).

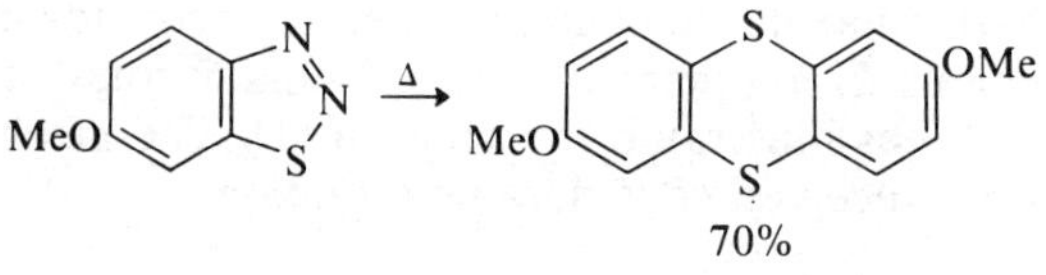

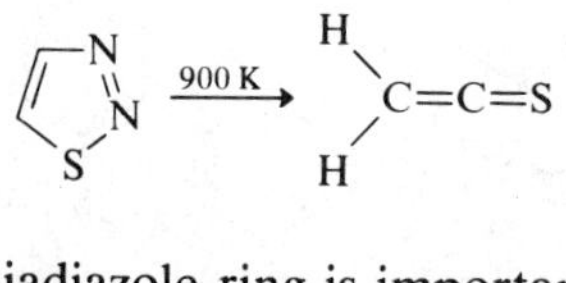

Clearly, substitution on the thiadiazole ring is important in this study ⟨79TL2785⟩. Additional evidence for the existence of thiirenes is presented in the following section.

Like thermal reactions, photochemical reactions of 1,2,3-thiadiazoles afford many products ⟨B-79MI42400, 73SST(2)717⟩. Subtle changes, such as ring size, can have a marked effect on product distribution for these reactions (equations 4 and 5) ⟨79JHC1295, 79CB3728⟩. Intermediates proposed for the thermal reaction of 1,2,3-thiadiazoles have also been proposed as photochemical intermediates. The ease of low temperature work with photochemical reactions has led to spectroscopic characterization of these molecules.

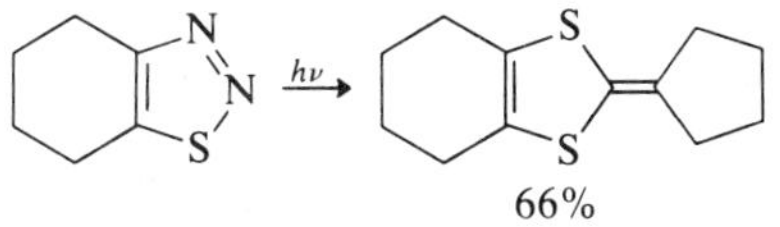

$$\text{(4)}$$

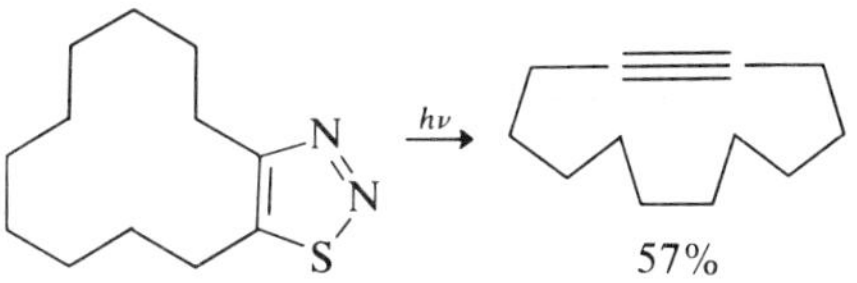

$$\text{(5)}$$

Evidence for the formation of diradical (**14**; equation 6) from the photolysis of 1,2,3-thiadiazoles has been gained by ESR work ⟨79JA3976⟩. Photolysis of thiadiazoles at low temperature led to the characterization of thioketenes by IR and a thiirene was postulated as a reaction intermediate ⟨74JA6768⟩.

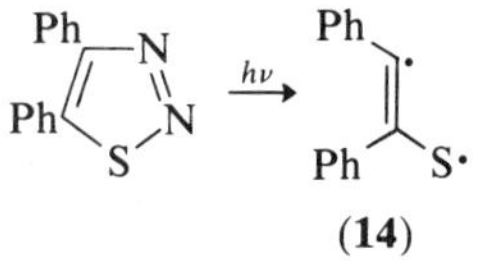

$$\text{(6)}$$

(**14**)

Strong evidence for a thiirene intermediate in a photochemical reaction was gained by examining reaction products. Photolysis of either (**15**) or (**16**) in the presence of hexafluoro-2-butyne leads only to the thiophene (**17**; equation 7) ⟨78JOC2487⟩.

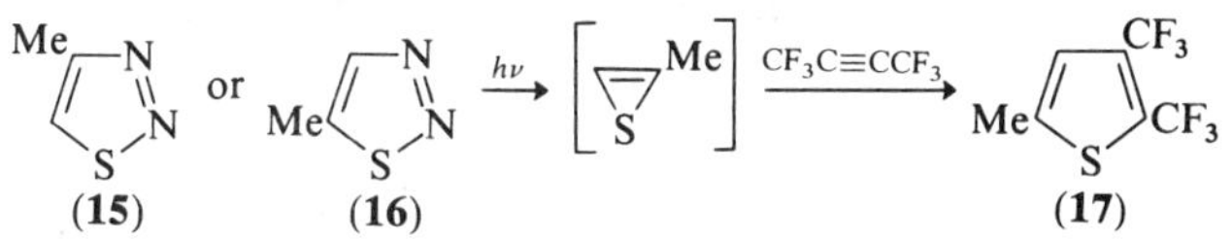

$$\text{(7)}$$

(**15**) (**16**) (**17**)

Thiirene (**19**) was first characterized by irradiating (**18**) in an argon matrix at 8 K and then observing the appearance of its IR spectrum ⟨77JA4842, 81AHC(28)231⟩. Upon further irradiation (**19**) affords thioketene (**20**; equation 8). A carbonyl group substituted on the thiirene adds stability, making detection easier ⟨78JOC2490⟩.

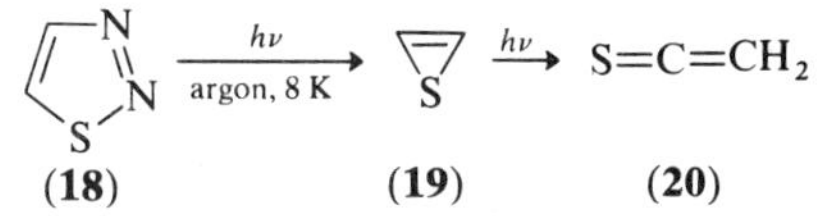

$$\text{(8)}$$

(**18**) (**19**) (**20**)

1,2,3-Thiadiazole trioxides fragment when irradiated ⟨77LA1347⟩, yet the products are quite different from photolysis products of the parent 1,2,3-thiadiazole (equation 9).

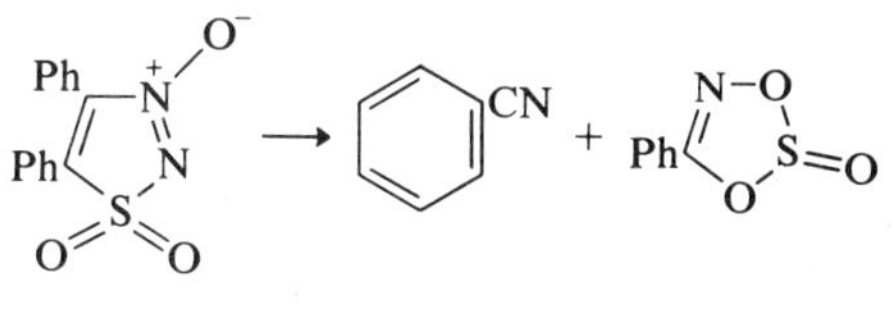

$$\text{(9)}$$

4.24.3.1.2 Rearrangements

1,2,3-Thiadiazole N-oxides isomerize, when irradiated, from one N-oxide isomer to the other. An X-ray crystal structure of (**21**) established the N-oxide is at the 3 position. The authors ⟨78ZN(B)316⟩ then concluded that the product from photolysis of (**21**) must be the N-oxide (**22**; equation 10).

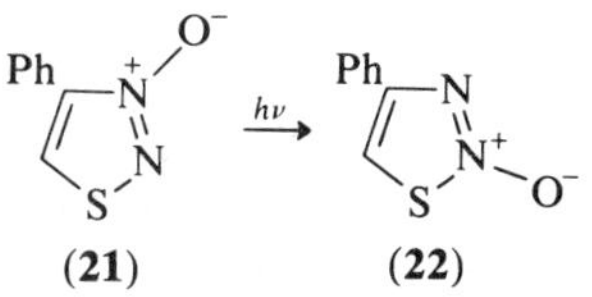

$$\text{(10)}$$

(**21**) (**22**)

Amino-1,2,3-thiadiazoles, when heated ⟨79S470⟩, undergo rearrangement to 1,2,3-triazoles (equation 11).

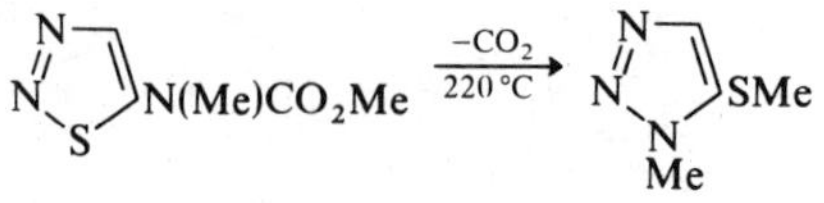

$$\underset{S}{\overset{N}{\underset{\|}{N}}}\!\!\diagdown\!\!N(Me)CO_2Me \xrightarrow[220\,°C]{-CO_2} \underset{\underset{Me}{N}}{\overset{N}{\underset{\|}{N}}}\!\!\diagdown\!\!SMe \qquad (11)$$

4.24.3.2 Electrophilic Attack at Nitrogen

4.24.3.2.1 Introduction

Reaction of 1,2,3-thiadiazoles with electrophiles has not been extensively studied, so effects of ring substituents have not been tabulated. Reaction of 1,2,3-thiadiazoles with acyl halides or metal ions has not even been reported.

4.24.3.2.2 Proton acids: basicity of 1,2,3-thiadiazoles

1,2,3-Thiadiazoles are weak bases and form a deliquescent hydrochloride salt which is decomposed in water ⟨B-61MI42400⟩.

4.24.3.2.3 Alkyl halides and related compounds (mesoionic compounds)

For a number of years it has been known that 1,2,3-thiadiazole reacts with methyl iodide ⟨B-61MI42400⟩, yet the site of alkylation was unknown. Owing to the increased interest in mesoionic compounds, the alkylation of 1,2,3-thiadiazoles has received renewed attention.

Reaction of (**23**) with dimethyl sulfate affords two *N*-alkyl isomers (equation 12) ⟨80JHC1217⟩. Others have shown that for substrates like (**24**) alkylation occurs exclusively at the 3-position (equation 13) ⟨80TL2101, 81JCS(P1)1591, 81CPB1743⟩.

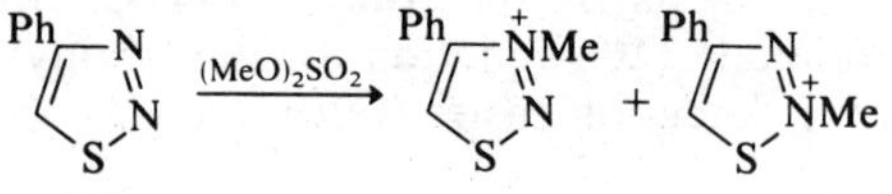

(12)

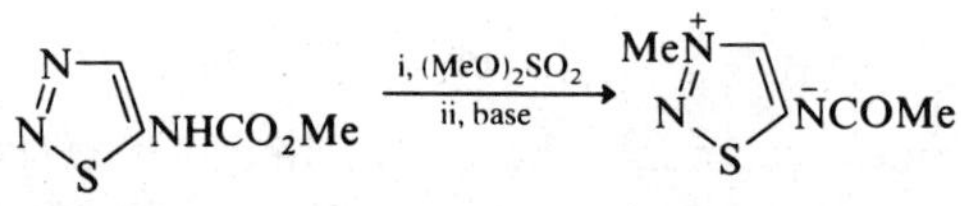

(**23**)

(13)

(**24**)

4.24.3.2.4 Halogens

Attempts to brominate thiadiazoles (**25**) and (**26**) were unsuccessful ⟨65JCS5166⟩. No reports of successful halogenation of thiadiazoles have appeared.

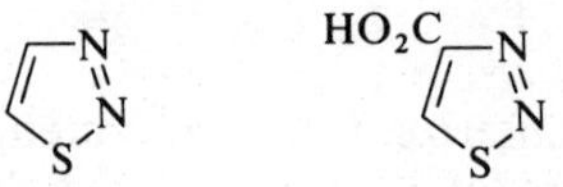

(**25**)   (**26**)

4.24.3.2.5 Peracids

Reaction of 1,2,3-thiadiazole with peracids has been reported to give an *N*-oxide in the 2-position ⟨75SST(3)670⟩. An X-ray analysis of *N*-oxide (**21**) reveals, however, that peracids react with 1,2,3-thiadiazoles at the 3-position ⟨78ZN(B)316⟩.

4.24.3.3 Attack at Sulfur

4.24.3.3.1 Electrophilic

Addition of more than one equivalent of peracid to a 1,2,3-thiadiazole results first in oxidation of the 3-position, followed by oxidation of the sulfur atom (equation 14). ⟨77LA1347⟩.

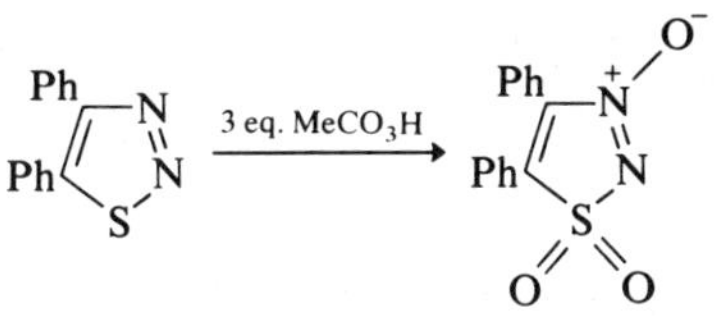

$$\text{Ph-thiadiazole} \xrightarrow{\text{3 eq. MeCO}_3\text{H}} \text{Ph-thiadiazole N-oxide SO}_2 \quad (14)$$

4.24.3.3.2 Nucleophilic

n-Butyllithium reacts with 1,2,3-thiadiazoles and nitrogen is evolved. By quenching the reaction with methyl iodide and analyzing the products (equation 15), the authors concluded fragmentation is initiated by attack at sulfur ⟨71OPP163⟩.

$$\text{Ph,Ph-thiadiazole} \xrightarrow[\text{ii, MeI}]{\text{i, Bu}^n\text{Li}} \text{PhC≡CPh} + \text{N}_2 + \text{Bu}^n\text{SMe} + \text{LiI} \quad (15)$$

4.24.3.4 Nucleophilic Attack at Carbon

Kinetic studies have shown sodium methoxide will displace the chlorine substituent from 5-chloro-1,2,3-thiadiazole (equation 16) ⟨74JHC343⟩. The products were not characterized in the paper.

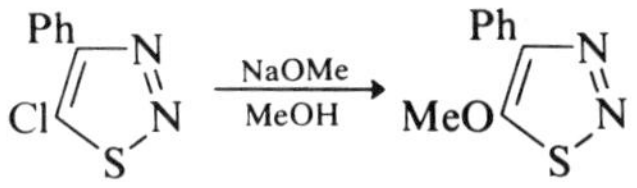

$$\text{Ph,Cl-thiadiazole} \xrightarrow[\text{MeOH}]{\text{NaOMe}} \text{Ph,MeO-thiadiazole} \quad (16)$$

4.24.3.5 Reactions at Hydrogen

Ring protons of 1,2,3-thiadiazoles are known to undergo rapid deuterium exchange under basic conditions ⟨74AHC(16)1⟩. Practically, this led to the conversion of 4-substituted thiadiazoles into thioethers and thiafulvenes (Scheme 4) ⟨68CJC2251, 68CJC1057⟩. Thiadiazoles substituted in the 5-position can be deprotonated and then alkylated (equation 17) ⟨68CJC1057⟩.

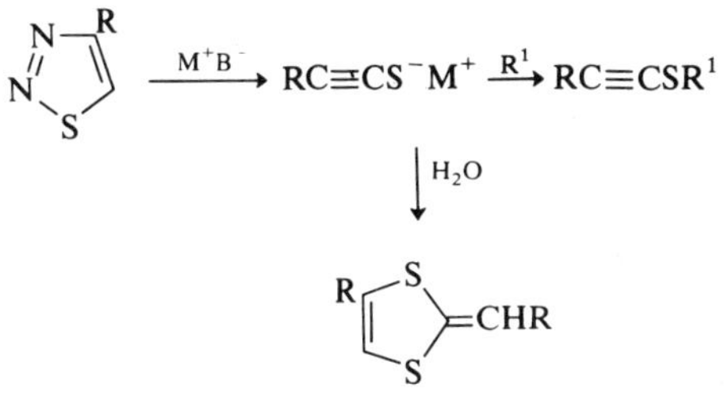

$$\text{R-thiadiazole} \xrightarrow{\text{M}^+\text{B}^-} \text{RC≡CS}^-\text{M}^+ \xrightarrow{\text{R}^1} \text{RC≡CSR}^1$$
$$\downarrow \text{H}_2\text{O}$$
$$\text{R-dithiole=CHR}$$

Scheme 4

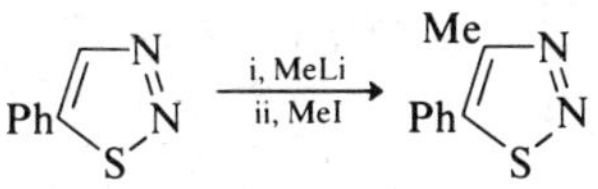

$$\text{Ph-thiadiazole} \xrightarrow[\text{ii, MeI}]{\text{i, MeLi}} \text{Ph,Me-thiadiazole} \quad (17)$$

4.24.3.6 Reaction with Radicals

Montevecchi and Tundo have actively explored the chemistry of radicals reacting with 1,2,3-thiadiazoles. Recently they reported the synthesis of a number of thianthrenes in high yield from 1,2,3-thiadiazoles ⟨81JOC4998⟩, where *t*-butoxy radicals initiated the reaction (equation 18).

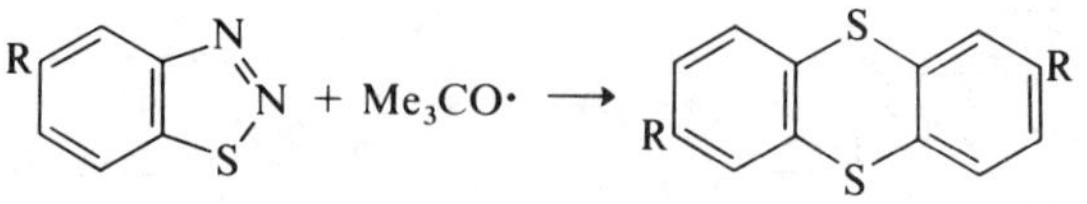

$$\text{benzothiadiazole-R} + \text{Me}_3\text{CO} \cdot \longrightarrow \text{thianthrene-R,R} \tag{18}$$

4.24.3.7 Reaction of Substrates

4.24.3.7.1 Benzenoid rings

1,2,3-Benzothiadiazole (**27**) has been known since 1888. The rich chemistry of this class of compounds has been the subject of many reviews (see Section 4.24.1), the most comprehensive of early work being those by Hodgson and Dodgson ⟨48MI42400⟩ and Bambas ⟨52HC(4)3⟩.

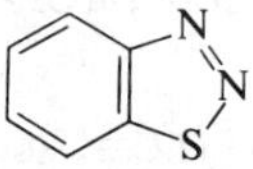

(**27**)

Owing to its aromatic structure which has been compared to naphthalene, 1,2,3-benzothiadiazole survives exposure to 20% potassium hydroxide at 150 °C or 27% sulfuric acid at 200 °C. Potassium permanganate, potassium ferricyanide, chromic acid or dilute nitric acid failed to oxidize the heterocycle ⟨B-61MI42400⟩.

Like naphthalene, 1,2,3-benzothiadiazoles react with electrophiles. Nitration of 1,2,3-benzothiadiazole affords predominately the 4-isomer (equation 19). Bromination of compound (**28**) yields the expected *ortho-para* product, directed by the hydroxy group (equation 20) ⟨B-61MI42400⟩.

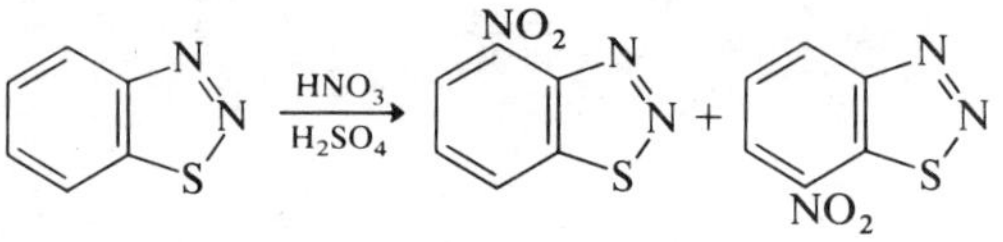

$$\xrightarrow{\text{HNO}_3 / \text{H}_2\text{SO}_4} \tag{19}$$

$$(\mathbf{28}) \xrightarrow{\text{Br}_2 / \text{HOAc}} \tag{20}$$

Substituted benzothiadiazoles may undergo nucleophilic attack. For example, in equation (21) the chlorine atom is displaced by a variety of nucleophiles ⟨75SST(3)670⟩.

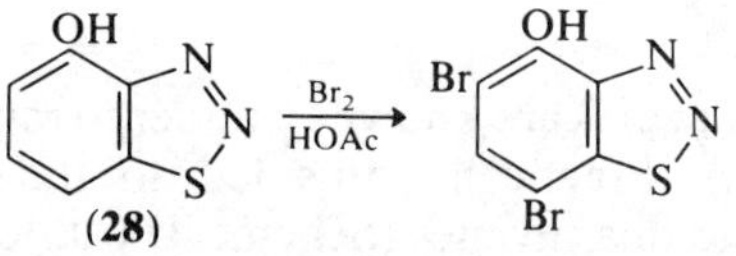

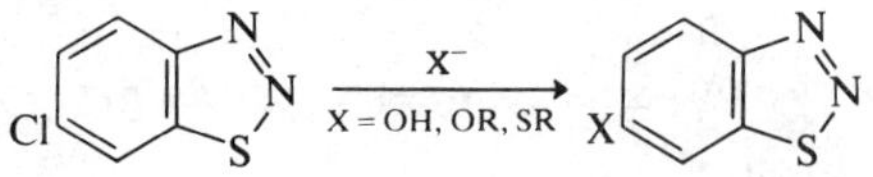

$$\xrightarrow[X = \text{OH, OR, SR}]{X^-} \tag{21}$$

4.24.3.7.2 S-Linked substituents

Recently potassium 1,2,3-thiadiazole-5-thiolate (**29**) has been generated and employed to displace an acetate group ⟨79JMC1214⟩ as in equation (22).

$$\text{(29)} + \text{cephem-OCOMe} \xrightarrow{\text{NaHCO}_3, \text{H}_2\text{O, acetone}} \tag{22}$$

4.24.3.7.3 C-linked substituents

The aromatic nature of 1,2,3-thiadiazoles enhances their stability and affects the chemistry of substituents. The thiadiazole nucleus survives reducing conditions (equation 23) ⟨75JHC1191⟩. Owing to the aromatic ring, α-carbons are easily brominated (equation 24) ⟨80JHC1639⟩.

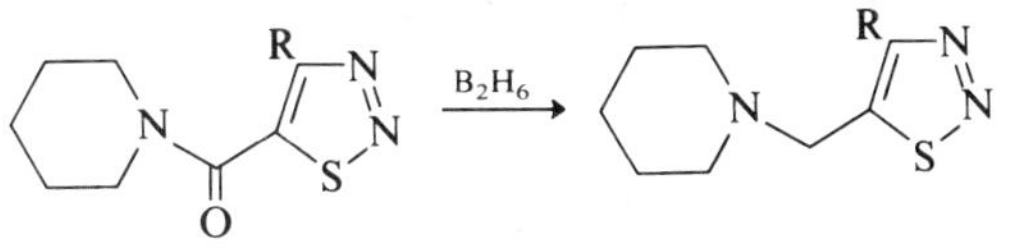

$$\text{piperidine-CO-thiadiazole} \xrightarrow{B_2H_6} \text{piperidine-CH}_2\text{-thiadiazole} \tag{23}$$

$$\text{Pr-thiadiazole} \xrightarrow{NBS} \text{CHBr(Et)-thiadiazole} \tag{24}$$

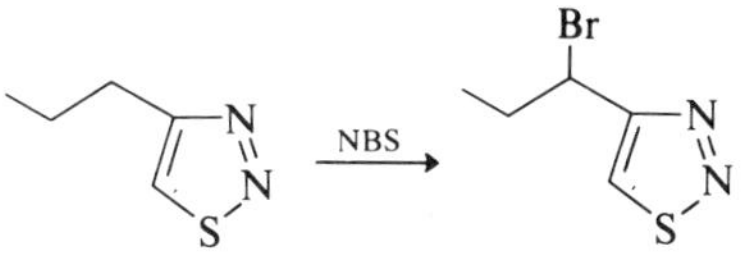

4.24.4 SYNTHESIS

4.24.4.1 Introduction

No new method for the synthesis of 1,2,3-thiadiazoles has appeared since Hurd and Mori's work in 1955 (see below). This and prior syntheses have been reviewed ⟨B-61MI42400⟩, so the purpose of this section is to present an overview with emphasis on recent modifications of existing methods.

4.24.4.2 Dipolar Cycloadditions

One of the earliest methods for synthesizing 1,2,3-thiadiazoles is that of Pechmann and Nold ⟨B-61MI42400⟩, who reacted diazomethane with phenyl isothiocyanate (equation 25). Of the four possible isomers only 1,2,3-thiadiazole (**30**) was formed. This method is somewhat restricted as methyl isothiocyanate does not react with diazomethane at room temperature.

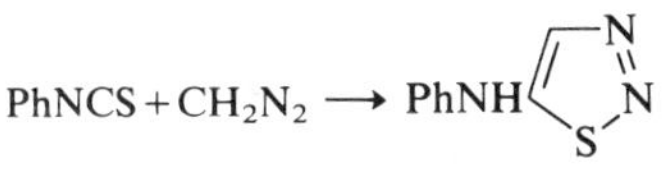

$$\text{PhNCS} + \text{CH}_2\text{N}_2 \longrightarrow \text{PhNH-thiadiazole} \tag{25}$$
(**30**)

Reaction of diazoalkanes with thiocarbonyl compounds, which may be considered an extension of Pechmann's synthesis, forms 1,2,3-thiadiazole derivatives ⟨75SST(3)670⟩. For example, methyl dithioacetate reacts with diazomethane to form a mixture of thiadiazole isomers and thiirane; the latter often is the main product of this type of reaction (equation 26). O-Ethyl thionoacetate reacts with ethyl diazoacetate to form a substituted thiadiazole (equation 27). However, an aryl thionoester reacts with diazomethane to yield a 1,2,3-thiadiazole derivative which is only formed as an intermediate (**31**) and then rapidly decomposes to alkene (Scheme 5).

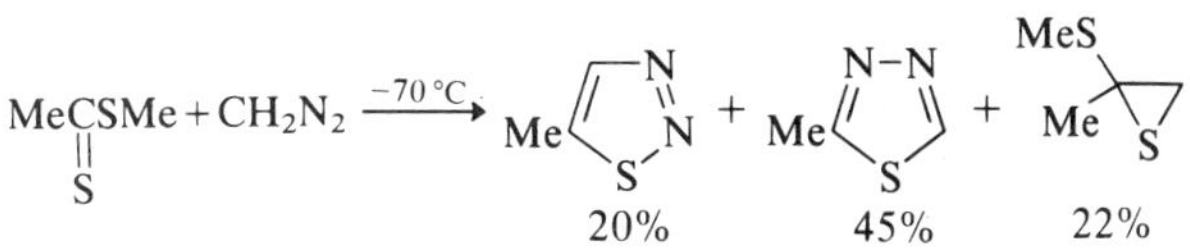

$$\text{MeCSMe} + \text{CH}_2\text{N}_2 \xrightarrow{-70\,°C} \text{Me-thiadiazole (1,2,3)} + \text{Me-thiadiazole (1,2,4)} + \text{Me-thiirane(SMe)} \tag{26}$$
20%   45%   22%

$$\text{MeCOEt} + \text{N}_2\text{CHCO}_2\text{Et} \longrightarrow \text{thiadiazole(CO}_2\text{Et)(Me)} \tag{27}$$

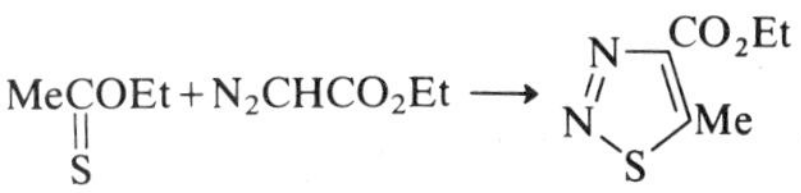

Another variation of the dipolar cycloaddition approach is reaction of carbamoyl isothiocyanates with diazomethane (equation 28) ⟨73SST(2)717⟩. Yields in these reactions are in the range 40–60%, making this route most worthy of synthetic note.

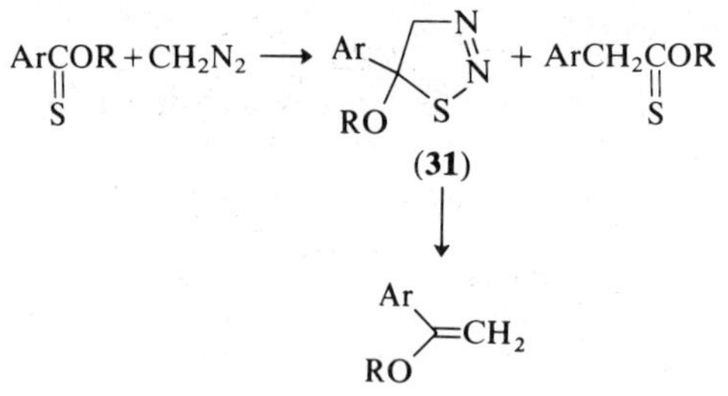

$$ArCOR + CH_2N_2 \longrightarrow \underset{\underset{RO}{|}}{Ar}\diagdown\!\!\!\overset{\overset{N}{\|}}{\underset{S}{\diagup}}\!\!\!N + ArCH_2COR$$

(31)

$$\downarrow$$

$$\underset{RO}{\overset{Ar}{\diagdown}}\!\!=\!CH_2$$

Scheme 5

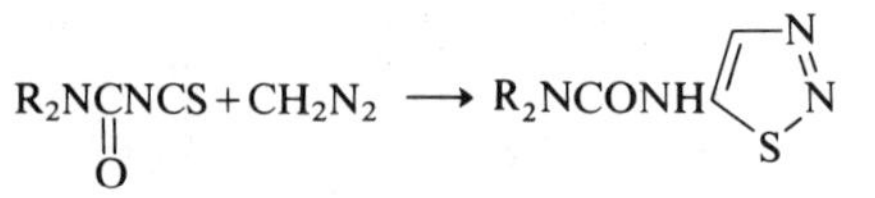

$$R_2N\overset{\|}{\underset{O}{C}}NCS + CH_2N_2 \longrightarrow R_2NCONH\diagdown\!\!\!\overset{N}{\underset{S}{\diagup}}\!\!\!N \tag{28}$$

4.24.4.3 From Hydrazones and Thionyl Chloride

The most general and widely employed synthesis of 1,2,3-thiadiazoles is Hurd and Mori's ⟨55JA5359⟩ now classic synthesis from α-methylene ketones. A variety of ketones (aromatic, cyclic and acyclic) have been converted into their corresponding hydrazones (tosyl and acyl), and reaction of the hydrazones with thionyl chloride produces 1,2,3-thiadiazoles in high yield (equation 29). Along with Hurd and Mori's original paper, several articles have appeared which furnish complete experimental details for the synthesis of a number of 1,2,3-thiadiazoles ⟨68CJC1057, 75JHC1191⟩.

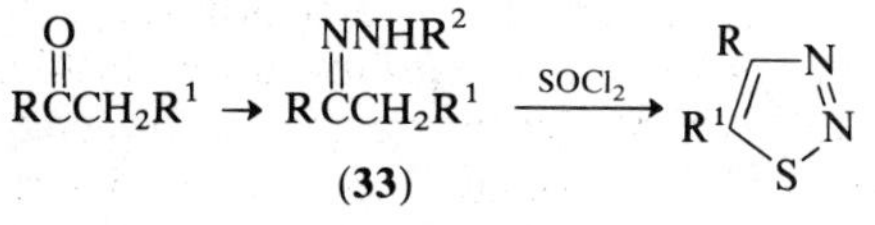

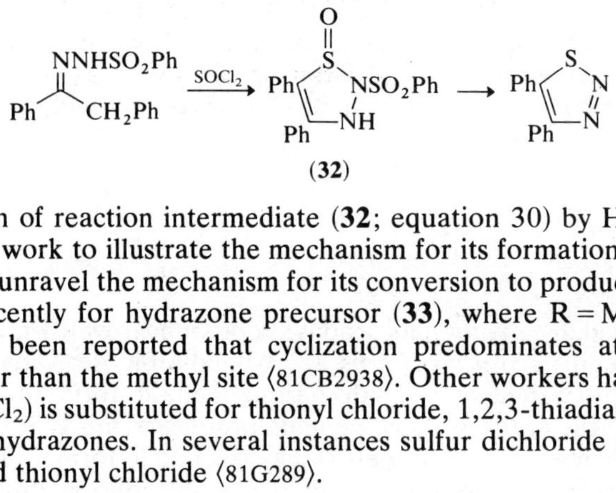

$$R\overset{O}{\overset{\|}{C}}CH_2R^1 \longrightarrow R\overset{NNHR^2}{\overset{\|}{C}}CH_2R^1 \xrightarrow{SOCl_2} \underset{R^1}{\overset{R}{\diagdown}}\!\!\!\overset{N}{\underset{S}{\diagup}}\!\!\!N \tag{29}$$

(33)

$$\underset{Ph}{\overset{NNHSO_2Ph}{\overset{\|}{C}}}\!\!\!CH_2Ph \xrightarrow{SOCl_2} Ph\diagdown\!\!\!\overset{\overset{O}{\overset{\|}{S}}}{\underset{Ph}{\diagup}}\!\!\!\overset{NSO_2Ph}{\underset{NH}{|}} \longrightarrow \underset{Ph}{\overset{Ph}{\diagdown}}\!\!\!\overset{S}{\underset{N}{\diagup}}\!\!\!\overset{}{\underset{}{}}N \tag{30}$$

(32)

Since the isolation of reaction intermediate (**32**; equation 30) by Hurd and Mori there has been no further work to illustrate the mechanism for its formation, nor has there been any further work to unravel the mechanism for its conversion to product under the reaction conditions. Only recently for hydrazone precursor (**33**), where R = Me and R^1 = alkyl or aryl groups, it has been reported that cyclization predominates at the more reactive methylene site rather than the methyl site ⟨81CB2938⟩. Other workers have shown that when sulfur dichloride (SCl_2) is substituted for thionyl chloride, 1,2,3-thiadiazoles are still formed from α-methylene hydrazones. In several instances sulfur dichloride afforded products in higher yield than did thionyl chloride ⟨81G289⟩.

4.24.4.4 1,2,3-Benzothiadiazoles

The first 1,2,3-thiadiazole synthesized, 1,2,3-benzothiadiazole, was prepared by diazotization of o-aminothiophenol with nitrous acid (equation 31) ⟨B-61MI42400⟩, and recently sodium nitrite–acetic acid has been substituted for nitrous acid ⟨B-79MI42400⟩. Another modification, thermal decomposition of diazonium acetate (**34**), affords benzothiadiazole in good yield in contrast to the variable yields usually experienced in the diazotization of o-aminothiophenols (equation 32) ⟨78SST(5)431⟩. Benzothiadiazoles are also available directly from aromatic amines (equation 33) ⟨70JCS(C)2250⟩. Sulfur monochloride reacts with the amine to form a benzothiazothiolium salt which reacts with nitrous acid to yield a chlorinated 1,2,3-benzothiadiazole (**35**). This process, depending on the aromatic ring substitution, may afford a number of products, and yields are variable.

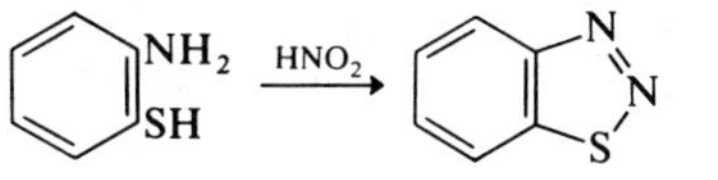

1,2,3-Thiadiazoles and their Benzo Derivatives

(31)

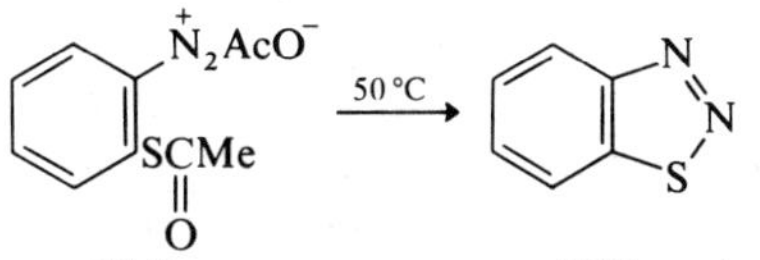

(32)

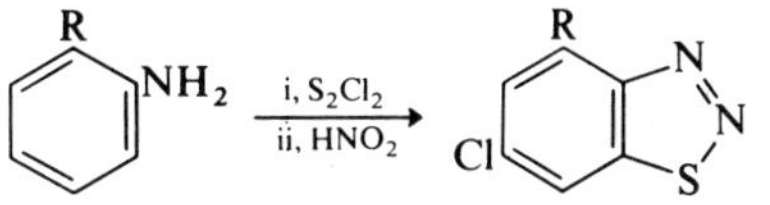

(33)

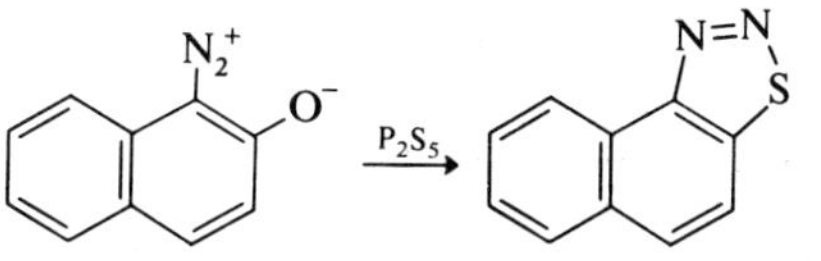

(34)

Diazo oxides when treated with phosphorus pentasulfide also form benzothiadiazoles (equation 34) ⟨B-61MI42400⟩.

4.24.4.5 From Other Heterocycles

Heterocycles also serve as precursors to 1,2,3-thiadiazoles ⟨B-79MI42400⟩. For example, mercapto-substituted triazole (36) exists in equilibrium with thiadiazole (37; equation 35), as does (38) with (39; equation 36). Isothiazoles may react to form benzothiadiazoles (equation 37). Treatment of 1,2,3-oxadiazoles with ammonium hydrogen sulfide ⟨B-61MI42400⟩ provides yet another route to thiadiazoles (equation 38).

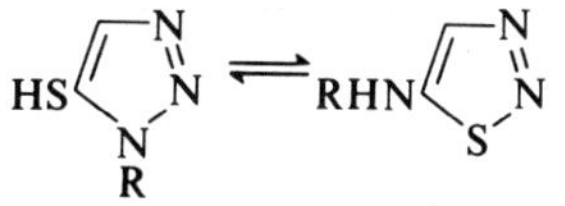

(35)

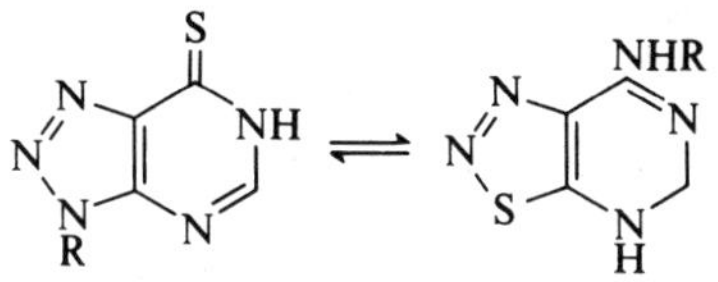

(36)

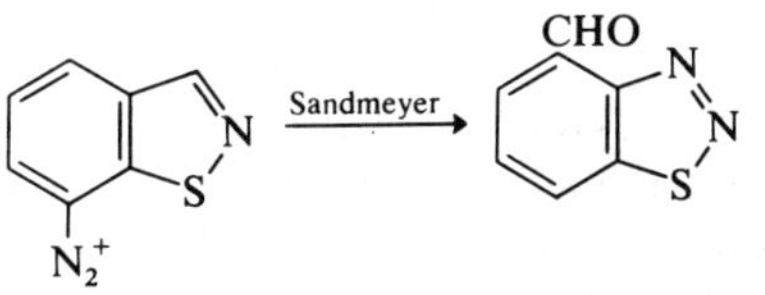

(37)

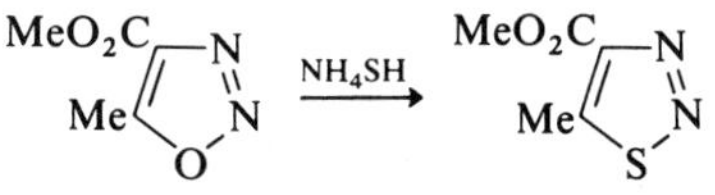

(38)

4.24.5 IMPORTANT COMPOUNDS — APPLICATIONS

4.24.5.1 Insecticide Synergist

Although 1,2,3-benzothiadiazoles are not themselves insecticides, together with carbaryl, several are excellent insecticide synergists against the house fly ⟨76MI42401⟩. Many 1,2,3-benzothiadiazole derivatives were found to be inhibitors of mammalian and insect microsomal oxidation ⟨77MI42400⟩, and recently 6-chloro-1,2,3-benzothiadiazole has been studied for its activity on the coupling site I in rat liver mitochondria ⟨80MI42400⟩.

4.24.5.2 Sedative

Analogs of 1,2,3-thiadiazole-4-carboxylic acid (**40**) were prepared and a number of these compounds have sedative and hypnotic actions comparable to the benzodiazepines ⟨73MI42400⟩. Since this report no further work has appeared in this area.

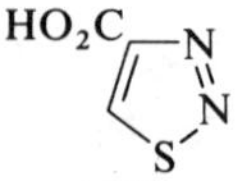

(**40**)

4.24.5.3 Herbicide

Generic structures (**41**) and (**42**) are examples of thiadiazoles with herbicidal activity ⟨75USP3874873, 79GEP2728523⟩.

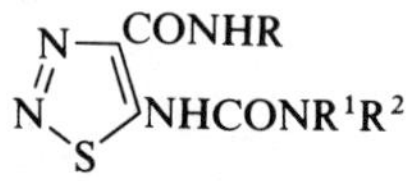

(**41**)

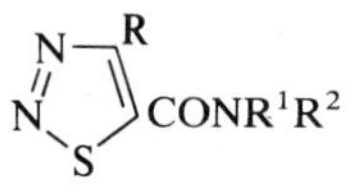

(**42**)

4.24.5.4 Antibacterial

Several examples of 1,2,3-thiadiazoles possessing antibacterial activity have been reported. β-Lactam (**43**) ⟨79JMC1214⟩ and sulfonamide (**44**) ⟨65JCS5166⟩ exhibit antibacterial activity, which one might expect, yet pyridylthiadiazole (**45**) also exhibits antibacterial activity ⟨74JPS628⟩.

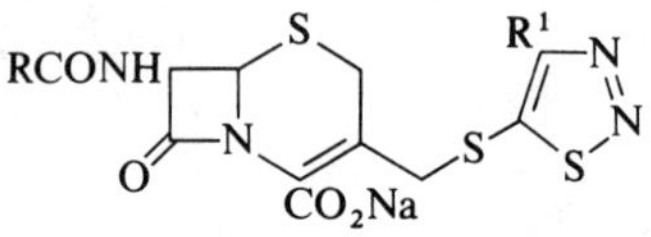

(**43**)

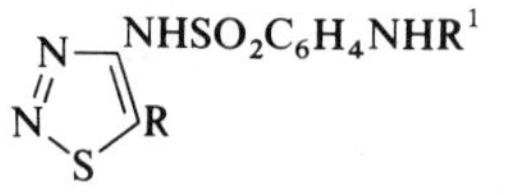

(**44**)

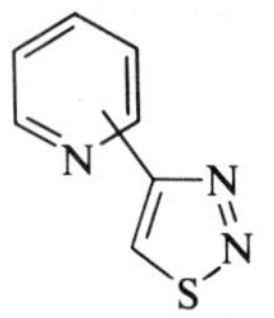

(**45**)

4.24.5.5 Polymers

1,2,3-Thiadiazoles readily undergo photochemical reactions (see Section 4.24.3.1) which have been utilized in polymer chemistry. For instance, polymers of type (**46**), formed by condensation of bisphenol epichlorohydrin copolymers and a 1,2,3-thiadiazole-4-carboxylic acid, undergo photochemically induced crosslinking ⟨70SST(1)444⟩. This type of process is useful in the printing and electronic industries.

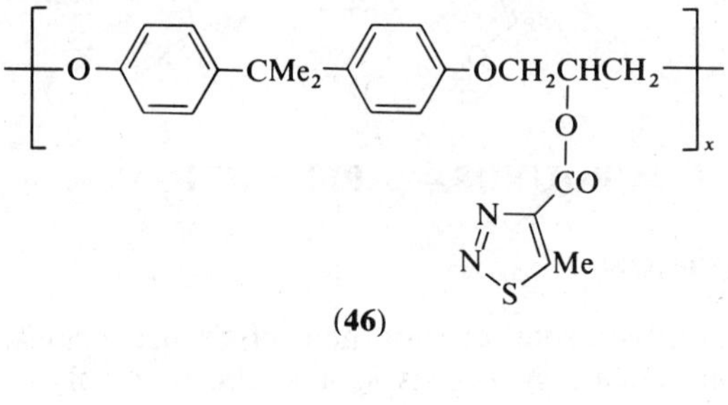

(**46**)

4.25
1,2,4-Thiadiazoles

J. E. FRANZ and O. P. DHINGRA
Monsanto Agricultural Products Co., St. Louis

4.25.1 INTRODUCTION	464
4.25.2 STRUCTURE	464
4.25.2.1 Introduction	464
4.25.2.2 Theoretical Methods	464
4.25.2.3 Molecular Dimensions	465
4.25.2.4 Molecular Spectra	465
4.25.2.4.1 UV spectra	465
4.25.2.4.2 IR spectra	465
4.25.2.4.3 NMR spectra	465
4.25.2.4.4 Mass spectra	466
4.25.2.5 Thermodynamic Aspects	467
4.25.2.6 Tautomerism	467
4.25.3 REACTIVITY	468
4.25.3.1 Reactions at Aromatic Rings	468
4.25.3.1.1 General survey	468
4.25.3.1.2 Thermal and photochemical reactions	469
4.25.3.1.3 Electrophilic attack at nitrogen	469
4.25.3.1.4 Electrophilic attack at carbon	470
4.25.3.1.5 Electrophilic and nucleophilic attack at sulfur	470
4.25.3.1.6 Nucleophilic attack at carbon	470
4.25.3.1.7 Nucleophilic attack at hydrogen	472
4.25.3.1.8 Reactions with radicals and electron deficient species: reactions at surfaces	473
4.25.3.1.9 Reactions with cyclic transition states	473
4.25.3.2 Reactions of Non-aromatic Compounds	473
4.25.3.2.1 Isomers of aromatic compounds	473
4.25.3.2.2 Dihydro compounds	477
4.25.3.2.3 Tetrahydro compounds	478
4.25.3.3 Reactions of Substituents	478
4.25.3.3.1 General survey of substituents on carbon	478
4.25.3.3.2 Benzenoid rings	478
4.25.3.3.3 Other C-linked substituents	478
4.25.3.3.4 N-Linked substituents	480
4.25.3.3.5 O-Linked substituents	485
4.25.3.3.6 S-Linked substituents	486
4.25.3.3.7 Halogen atoms	487
4.25.3.3.8 Metals and metalloid-linked substituents	489
4.25.3.3.9 Fused heterocyclic rings	489
4.25.4 SYNTHESIS	492
4.25.4.1 Ring Synthesis from Non-heterocyclic Compounds	492
4.25.4.1.1 Introduction	492
4.25.4.1.2 Type A syntheses	492
4.25.4.1.3 Type B syntheses	494
4.25.4.1.4 Type C syntheses	496
4.25.4.1.5 Type D syntheses	498
4.25.4.1.6 Type E syntheses	498
4.25.4.1.7 Type F syntheses	499
4.25.4.1.8 Type G syntheses	500
4.25.4.1.9 Miscellaneous synthetic methods	500
4.25.4.2 Ring Syntheses by Transformation of Other Heterocycles	501
4.25.4.2.1 Introduction	501
4.25.4.2.2 Dithiazolidine rearrangements	501
4.25.4.2.3 Oxazole and oxadiazole rearrangements	504
4.25.4.2.4 Thiatriazoline rearrangements	505

4.25.4.3 Practical Methods for Synthesis of Derivatives	505
4.25.4.3.1 General survey	505
4.25.4.3.2 Parent compound	506
4.25.4.3.3 C-Linked derivatives	506
4.25.4.3.4 N-Linked derivatives	507
4.25.4.3.5 O-Linked derivatives	507
4.25.4.3.6 S-Linked derivatives	507
4.25.4.3.7 Halogen-linked derivatives	508
4.25.5 APPLICATIONS AND IMPORTANT COMPOUNDS	509
4.25.5.1 Introduction	509
4.25.5.2 Fungicides	509
4.25.5.3 Herbicides	509
4.25.5.4 Insecticides	510
4.25.5.5 Bactericides	510
4.25.5.6 Dyes	510
4.25.5.7 Lubricant Additives	511
4.25.5.8 Vulcanization Agents	511

4.25.1 INTRODUCTION

1,2,4-Thiadiazole (**1**) was first prepared and characterized in 1955 but products containing this ring system were described as early as 1821. A natural product containing the 1,2,4-thiadiazole nucleus was not reported until 1980. Dendrodoine (**2**) is a cytotoxic material isolated from the marine Tunicate *Dendrodoa grasssularia* ⟨80TL1457⟩.

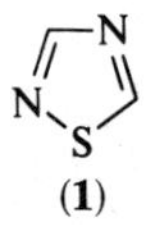

In this chapter the extensive review on 1,2,4-thiadiazoles by Kurzer ⟨65AHC(5)119⟩ was utilized for information published prior to 1965. For the period 1965–1981, many literature references were obtained by computer methods and from short heterocyclic reviews ⟨B-79MI42501, B-79MI42502, 77SST(4)422⟩.

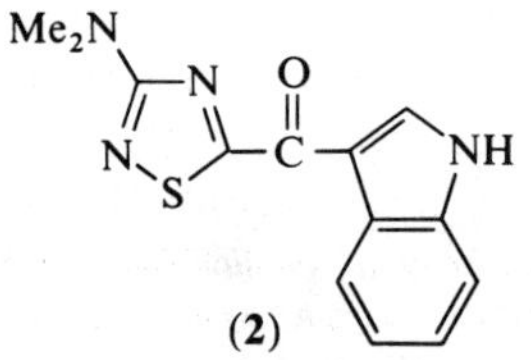

4.25.2 STRUCTURE

4.25.2.1 Introduction

1,2,4-Thiadiazole is classified as a π-excessive sulfur-containing heteroaromatic compound ⟨B-59MI42500⟩. The 1,2,4-thiadiazole nucleus is numbered as in (**1**). The double bonds in the partially reduced rings are designated Δ^2, Δ^3, Δ^4, respectively (Scheme 1) and these compounds are called thiadiazolines. The fully reduced system is termed a thiadiazolidine.

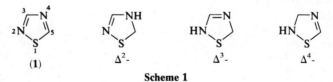

Scheme 1

4.25.2.2 Theoretical Methods

Theoretical calculations using the simple MO method in the LCAO approximation have provided molecular diagrams of the free base and the conjugate acid ⟨61CCC156⟩. According to these calculations, the 5-position of the non-protonated 1,2,4-thiadiazole system should be the most reactive in nucleophilic substitution reactions.

4.25.2.3 Molecular Dimensions

X-Ray diffraction measurements have not been reported on 1,2,4-thiadiazole. Double resonance modulation microwave spectroscopy has recently been used to study the structure of four isomeric thiadiazoles ⟨76ZN(A)1681⟩. The bond lengths and bond angles reported for 1,2,4-thiadiazole are shown in Scheme 2.

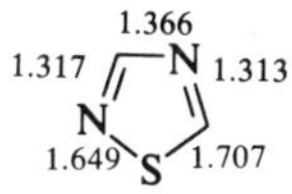

Scheme 2 Bond lengths (Å) and angles (°)

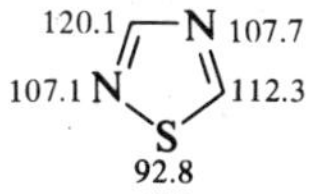

The crystal structure of Hector's base (3) ⟨78CC652⟩ and a thiadiazolinedithione (4) ⟨74JCS(P2)1096⟩ were recently determined.

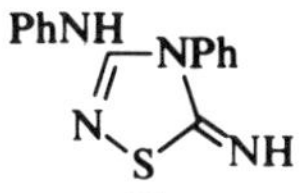

4.25.2.4 Molecular Spectra

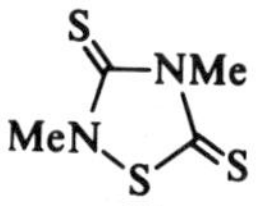

4.25.2.4.1 UV spectra

1,2,4-Thiadiazole has an absorption maximum at 229 nm (log ε 3.7). The introduction of amino groups into the heteroaromatic nucleus results in a bathochromic shift. Thus, the maximum due to 1,2,4-thiadiazole is moved to 247 nm in 5-amino and to 256 nm in 3,5-diamino-1,2,4-thiadiazole. The UV spectra of a variety of substituted 1,2,4-thiadiazoles have been determined, but the information so far available does not provide any broad correlation between structure and absorption characteristics ⟨65AHC(5)119⟩.

4.25.2.4.2 IR spectra

Vibrational assignments for IR spectral absorptions of thiadiazoles were made by Rao and Venkataraghavan ⟨64CJC43⟩. Spectra published until 1965 are tabulated in a review by Kurzer ⟨65AHC(5)119⟩ and data since 1965 are listed in Table 1.

Table 1 IR Absorption Spectral Data for 1,2,4-Thiadiazoles

Compound	Ref.
5-Alkoxy-3-hydroxy-1,2,4-thiadiazole	79JHC961
5-Benzyloxy-3-hydroxy-1,2,4-thiadiazole	79JHC961
5-Dimethylamino-3-hydroxy-1,2,4-thiadiazole	79JHC961
3,5-Diaryl-1,2,4-thiadiazole	74JOC962
3-Aryl-1,2,4-thiadiazole	74JOC962
3-Methyl-5-phenyl-1,2,4-thiadiazole	74JOC962
3,5-Bis(alkoxycarbonylamino)-1,2,4-thiadiazole	73CPB2396
3-Halo-1,2,4-thiadiazole	73JOC465
3-Methyl-5-phenylamino-1,2,4-thiadiazole	80JOC3750
5-Amino-3-arylamino-1,2,4-thiadiazoles	77IJC(B)490
1,2,4-Thiadiazole-5-thione	73CJC2353

4.25.2.4.3 NMR spectra

The chemical shift of the protons in 1,2,4-thiadiazoles are downfield from benzene, the C-3 proton being farther downfield than the C-5 proton (Table 2).

Table 2 ^{1}H NMR Chemical Shifts of 1,2,4-Thiadiazoles

Compound	δ(H-5)	δ(H-3)	Ref.
3-Phenyl-1,2,4-thiadiazole	9.90	—	74JOC962
5-Phenyl-1,2,4-thiadiazole	—	8.66	80JOC3750
3-Methylthio-1,2,4-thiadiazole	9.88	—	73CJC2353
5-Methylthio-1,2,4-thiadiazole	—	8.58	73CJC2353

A good correlation between solvent shifts ($\nu_{CCl_4} - \nu_{C_6D_6}$) for methyl groups attached to sp^2-hybridized nitrogen atoms and the corresponding Hammett σ constants was observed for compounds (**5 → 8**). The solvent shifts for the methyl groups attached to sp^3-hybridized nitrogen atoms have been shown to correlate with the corresponding Taft $\sigma°$ constants ⟨78NKK1256⟩.

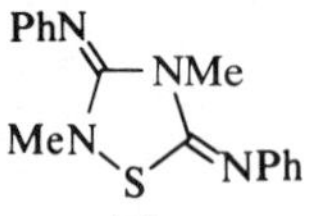

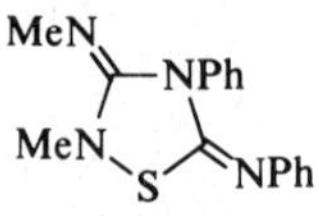

Press *et al.* recently studied the course of acylation and carbamidization reactions of 3-amino-5-methylthio-1,2,4-thiadiazole (**9**) using ^{1}H, ^{13}C and ^{15}N NMR spectroscopy. It had been known for a long time that 3- and 5-amino-1,2,4-thiadiazoles react with 'hard' electrophiles such as acid chlorides at the amino substituent to give the corresponding amides. Using ^{15}N labelled compounds, Preiss *et al.* have demonstrated that the reaction (**9 → 10**; Scheme 3) is not as straightforward as it looks. Instead, the 'hard' electrophile first acylates the ring nitrogen atom at the 2-position and the dihydro heteroaromatic compound (**11**) first formed then rearranges into the exocyclic acylation product (**10**) *via* a Dimroth rearrangement (Scheme 3) ⟨81JPR279⟩.

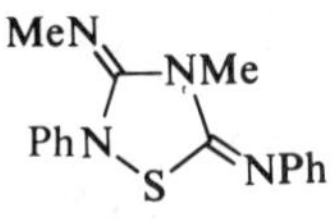

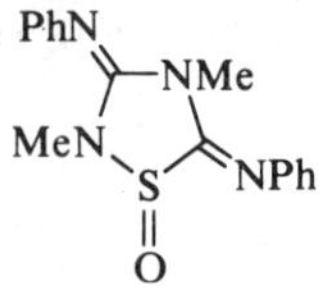

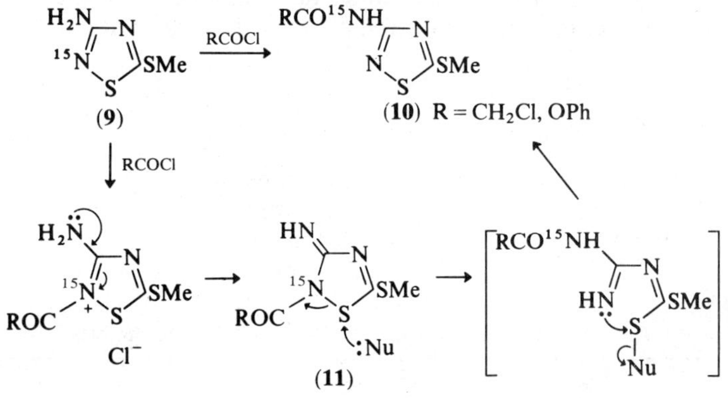

Scheme 3

In the case of softer electrophiles, such as methyl isocyanate, reaction is on the exocyclic nitrogen atom and the resultant urea (**12**) is stable and does not undergo Dimroth rearrangement (Scheme 4) ⟨81JPR279⟩.

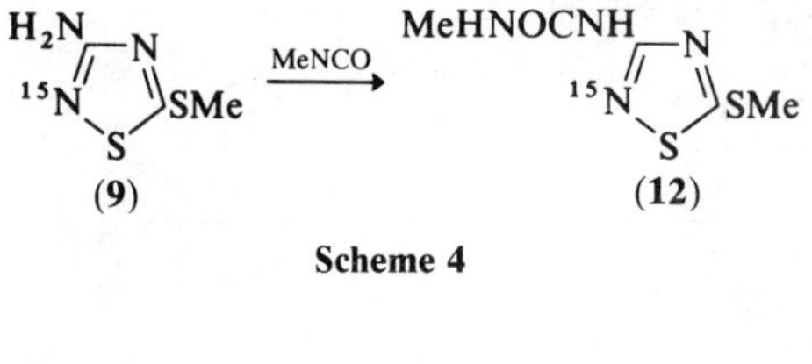

Scheme 4

4.25.2.4.4 Mass spectra

Mass spectra of 1,2,4-thiadiazoles have been determined in increasing number in recent years. Two general fragmentation pathways (a and b) are illustrated in Scheme 5 ⟨72JHC651⟩.

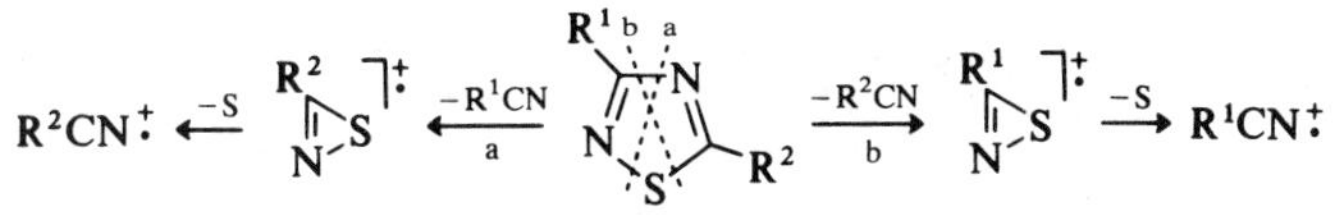

Scheme 5

In the case of compounds containing a hydrogen-labile functional group at the 3-position a significant new fragmentation (Schemes 6 and 7), in addition to the two general pathways (Scheme 5) is observed ⟨73OMS(7)555⟩.

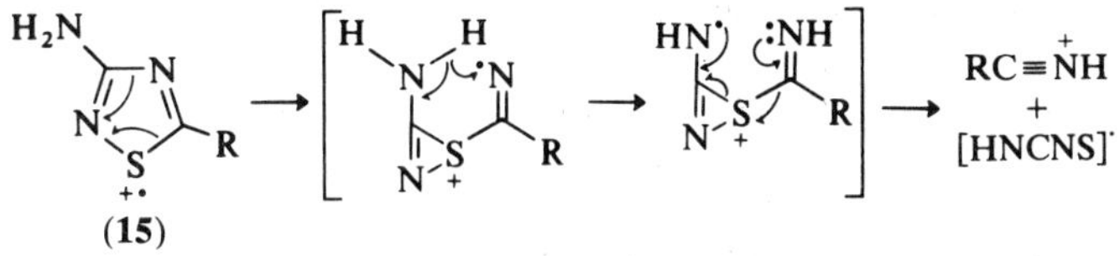

Scheme 6

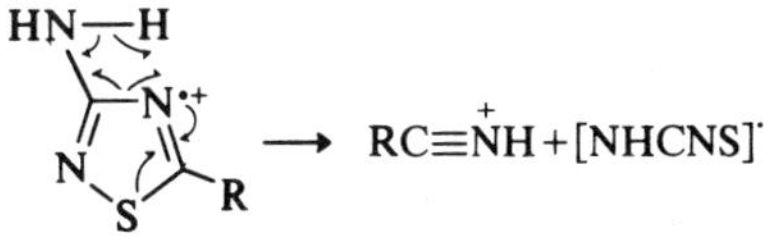

Scheme 7

Mass spectra of 5-amino-3-arylamino-1,2,4-thiadiazoles exhibit a new fragmentation process involving two consecutive hydrogen shifts and ring cleavage leading to the elimination of the NHSH radical (Scheme 8) ⟨77AJC563⟩.

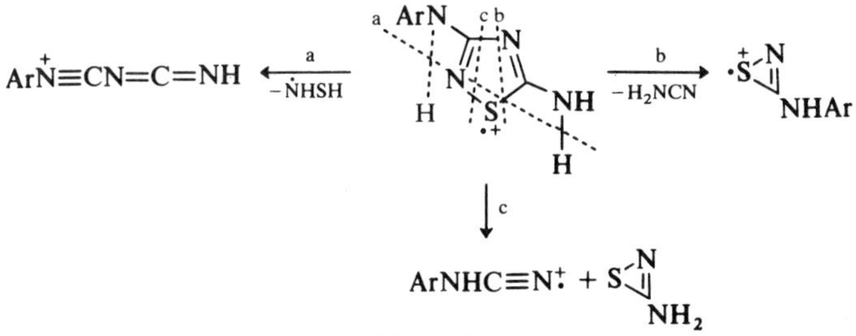

Scheme 8

4.25.2.5 Thermodynamic Aspects

Thermodynamic functions (entropy, heat capacity, enthalpy and free energy functions) have not been reported for 1,2,4-thiadiazoles. The ionization constants of a number of 1,2,4-thiadiazoles have been determined potentiometrically or by Hammett's method ⟨65AHC(5)119⟩. Polarographic measurements of a series of methylated 5-amino-1,2,4-thiadiazoles show that thiadiazoles are not reducible in methanolic lithium chloride solution, while thiadiazolines are uniformly reduced at $E_{0.5} = -1.6 \pm 0.02$ V. This technique has been used to assign structures to compounds which may exist theoretically as either thiadiazoles or thiadiazolines ⟨65AHC(5)119⟩.

The photoelectron spectra for 1,2,4-thiadiazole and other isomeric thiadiazoles have recently been assigned ⟨77JST(39)189⟩.

4.25.2.6 Tautomerism

Tautomerism in heterocyclic five-membered rings was recently reviewed by Elguero *et al.* ⟨76AHC(S1)266⟩.

The position of equilibrium in the tautomers of 3-hydroxy-1,2,4-thiadiazoles (Scheme 9) is not conclusively known. The existing chemical evidence suggests that the OH form (13) predominates. However, a UV spectral study was interpreted to suggest that the lactam NH form (14a) contributed substantially to the tautomeric equilibrium in ethanol ⟨76AHC(S1)266, p. 377⟩.

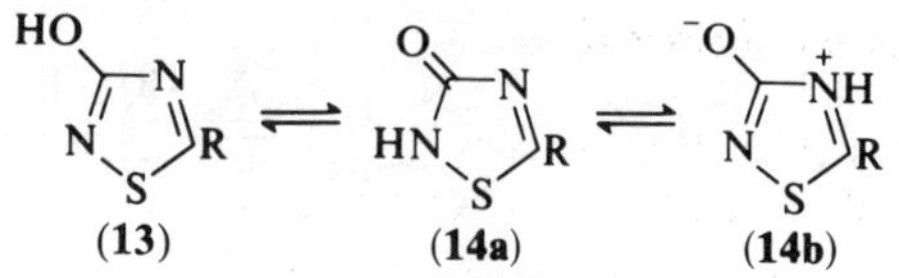

Scheme 9

The tautomeric situation is simpler in 3- and 5-amino derivatives. Here the equilibrium lies predominantly on the side of the amino forms (**15**) and (**16**), respectively ⟨76AHC(S1)266, p. 437⟩.

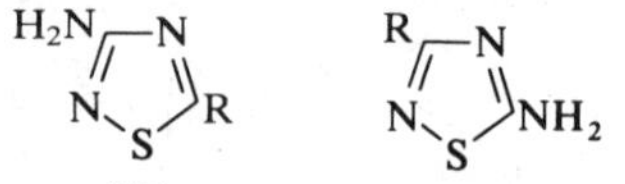

In the case of 5-thio derivatives, the IR spectra show no clear SH absorption as would be anticipated for (**17a**), and the thione structure (**17b**) or (**17c**) was tentatively assigned (Scheme 10) ⟨73CJC2353⟩. The IR spectrum in Nujol of perthiocyanic acid is considered to indicate that the dithione structure (**18a**) is favored over (**18b**) ⟨63JCS3168⟩.

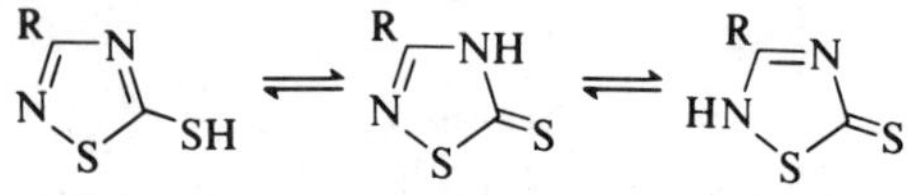

Scheme 10

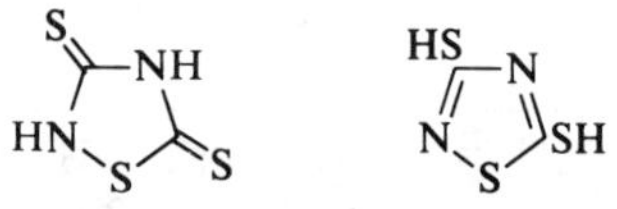

4.25.3 REACTIVITY

4.25.3.1 Reactions at Aromatic Rings

4.25.3.1.1 General survey

The parent compound is quite sensitive to acid and alkali, as well as oxidizing and reducing agents. Aqueous alkali decomposes it to ammonia, hydrogen sulfide and sulfur; hydrogen peroxide oxidizes it to sulfate ion, and Zn–HCl leads to reductive ring cleavage. Substituents at the 3- and 5-positions stabilize the nucleus toward acid, alkali, oxidizing agents and reducing agents ⟨65AHC(5)119⟩.

The 5-position in 1,2,4-thiadiazoles is most reactive in nucleophilic substitution reactions. Chlorine, for example, may be displaced by nucleophiles (Nu) such as fluoride, hydroxide, thiol, amino, hydrazino, sulfite and azido groups (Scheme 11). Active methylene compounds such as malonic, acetoacetic and cyanoactic esters as their sodio derivatives also displace the 5-halo substituent ⟨65AHC(5)119⟩. The reaction follows second-order kinetics, the rate determining step being addition of the nucleophile at C-5 followed by rapid elimination of X.

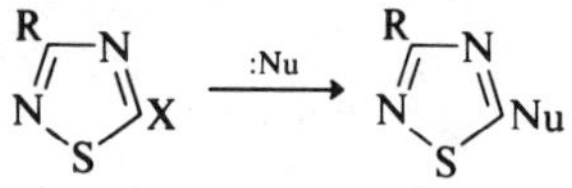

Scheme 11

In contrast, halogen in the 3-position of 1,2,4-thiadiazoles is inert toward most nucleophilic reagents.

Electrophilic reactions of 1,2,4-thiadiazoles are very limited; the parent compound forms salts with mineral acids, forms a methiodide, and also gives addition compounds with heavy-metal salts.

4.25.3.1.2 Thermal and photochemical reactions

Due to the aromatic nature of the nucleus, 1,2,4-thiadiazoles are generally quite stable to heat. Their thermal stability, solubility and resistance to acids are influenced by the nature of the 3- and 5-substituents ⟨65AHC(5)119⟩. Mass spectral decomposition patterns of various substituted 1,2,4-thiadiazoles are discussed in Section 4.25.2.4.4.

Surprisingly, photochemical behavior of 1,2,4-thiadiazoles has not been studied to date.

4.25.3.1.3 Electrophilic attack at nitrogen

1,2,4-Thiadiazoles are weak bases. They form readily hydrolyzable salts with mineral acids and addition compounds with heavy-metal salts. Amino-1,2,4-thiadiazoles also form salts with picric acid and picrolonic acid ⟨65AHC(5)119⟩.

Methylation of 5-amino-1,2,4-thiadiazoles (**16**; R = H, Ph), leads to N-4 derivatives (**19**) which on warming in ethanol undergo a Dimroth rearrangement to 5-methylamino-1,2,4-thiadiazoles (**20**) when R = H. Available evidence suggests that the isomerization proceeds by the sequence shown in Scheme 12. Ethylation of (**16**) proceeds in low yield and no reaction is observed when R = Ph. Introduction of higher alkyl, allyl and benzyl groups is also unsuccessful.

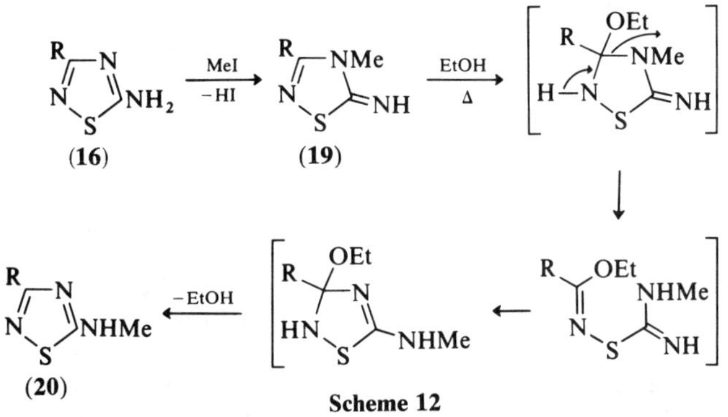

Scheme 12

The initial N-4 derivative (**22**) obtained by reacting phenacyl bromide with 5-amino-1,2,4-thiadiazole (**21**) undergoes cyclization to the fused imidazolothiadiazole heterocycle (**23**; Scheme 13) ⟨65AHC(5)119⟩. The diquaternary salt (**24**) of 1,2,4-thiadiazole (**1**) is obtained when trimethyloxonium tetrafluoroborate is used as an alkylating agent (Scheme 14) ⟨72JOC2259⟩. A bicyclic compound (**27**) is formed when 5-amino-3-methyl-1,2,4-thiadiazole (**25**) and diethylethoxymethylene malonate (**26**) are heated in trichlorobenzene (Scheme 15) ⟨65AHC(5)119⟩.

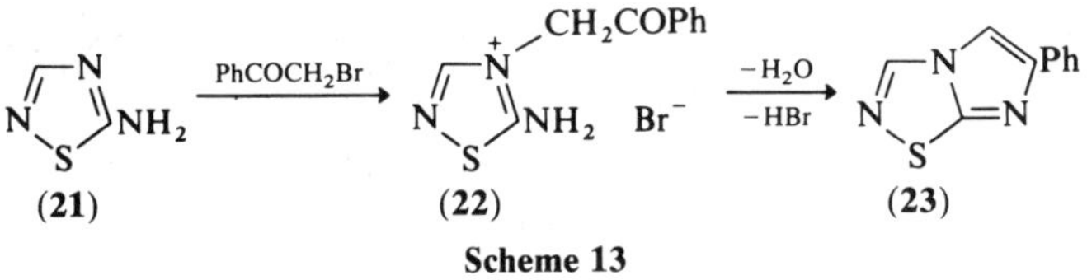

Scheme 13

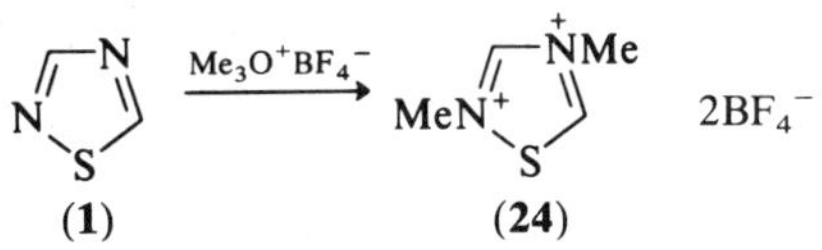

Scheme 14

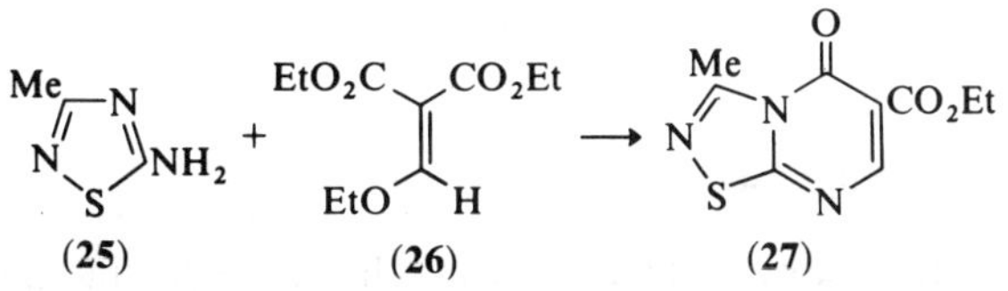

Scheme 15

Acylation of 3-amino-5-methylamino-1,2,4-thiadiazole (**28**) with benzoyl chloride (or arenesulfonyl chlorides) introduces one acyl group when one equivalent is employed, and three when a large excess is used to produce (**29**) and (**30**), respectively (Scheme 16). With acetic anhydride, however, acylation terminates with the formation of the 3-monoacyl derivative ⟨65AHC(5)119⟩. For the mechanism of acylation, see Scheme 3. In the case of 3,5-diarylamino-1,2,4-thiadiazoles, no triacyl derivative is obtained even when excess of acylating agent is employed. Acetyl and benzoyl chlorides give monoacyl derivatives and p-toluenesulfonyl chloride forms 3,5-di(p-toluenesulfonyl) derivatives ⟨65AHC(5)119⟩.

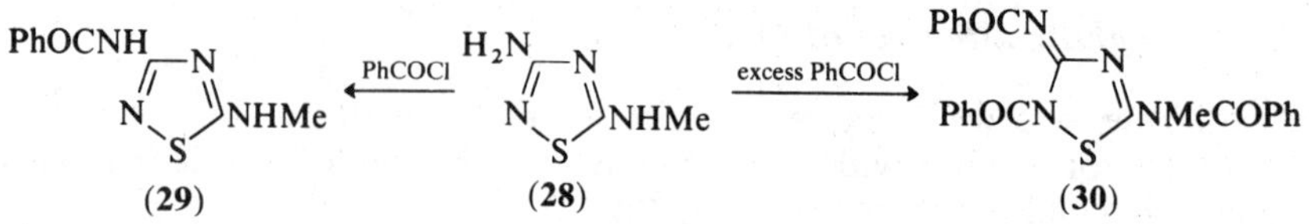

Scheme 16

4.25.3.1.4 Electrophilic attack at carbon

Electrophilic reactions at the ring carbons of 1,2,4-thiadiazole are virtually unknown.

4.25.3.1.5 Electrophilic and nucleophilic attack at sulfur

The S,S-dioxides of 1,2,4-thiadiazoles are only accessible by cyclization of precursors already incorporating the oxidized sulfur functions. Attempts to prepare them by oxidation of the preformed heteroaromatic ring have led to ring cleavage giving sulfate ion ⟨65AHC(5)119⟩.

Nucleophilic attack at the sulfur of 1,2,4-thiadiazoles has been implicated in many of the ring transformations of 1,2,4-thiadiazoles described throughout this chapter (*e.g.* see Schemes 3, 17, 18, 37, 39, 43, 44, 54, 127, 130). In general, 'soft' nucleophiles attack at sulfur and 'hard' nucleophiles at the C-5 carbon atom of 1,2,4-thiadiazoles. For example, the salt (**31**) reacts with sodium sulfide, sodium thiosulfate and sodium benzenethiolate giving, in each case, the thiobenzoylamidine (**32**; Scheme 17) ⟨77JCS(P1)1791⟩. Borohydride also attacks, at the sulfur of (**31**) to give (**32**) as the only detectable product. The initial product obtained from the reaction of cyanide ion with (**31**) is the thiocyanate (**33**) but on attempted recrystallization this material undergoes isomerization to the isothiocyanate (**34**), followed by cyclization to the triazinethione (**35**; Scheme 18) ⟨77JCS(P1)1791⟩.

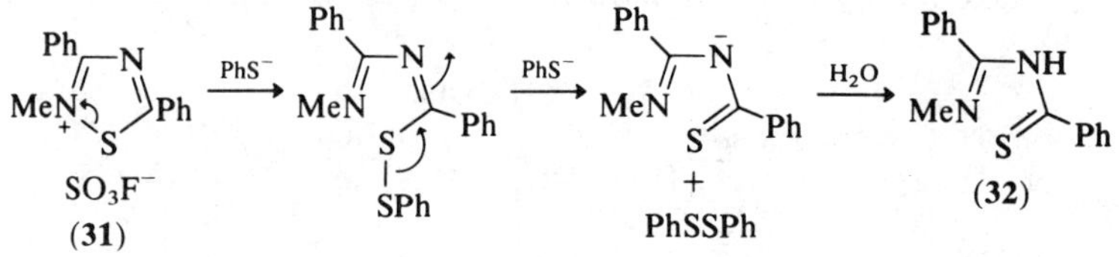

Scheme 17

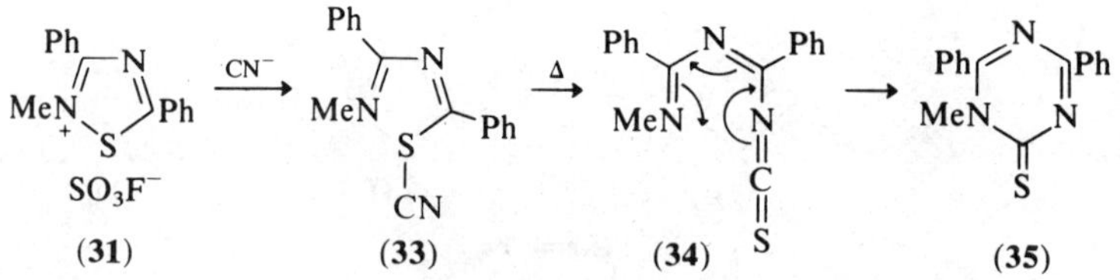

Scheme 18

4.25.3.1.6 Nucleophilic attack at carbon

The parent compound is quite sensitive to alkali and reducing agents. Aqueous alkali decomposes it to ammonia, hydrogen sulfide and sulfur. Substituents at the 3- and 5-positions exert a marked stabilizing influence on the heterocyclic nucleus. For example, 3,5-diphenyl-1,2,4-thiadiazole resists the action of hot mineral acids and prolonged boiling is required for attack by alkalis ⟨65AHC(5)119⟩.

The resistance of 3,5-diamino-1,2,4-thiadiazoles to alkaline reagents depends upon the degree of substitution. Thus, the compound (36) is stable to cold alkali, but is rapidly degraded by hot alkali to amidinoureas (37; Scheme 19). The stability of 5-alkylamino and 5-arylamino homologs is considerably higher; thus, aryl homologs are unaffected after several hours refluxing in 3N sodium hydroxide solution. Nucleophilic attack at the C-5 carbon atom of 1,2,4-thiadiazoles has been proposed as a reaction mechanism for many of the ring transformations described throughout this chapter. Thus, salt (31) reacts with alkoxide and hydroxide ions at room temperature to give ring opened products (38) and (39), respectively (Scheme 20). In the case of hydroxide ion, in addition to forming (39) as the major product, small amounts of N-methylbenzamide and benzonitrile are also formed. These products are not obtained by the action of the aqueous base on the initially formed benzoylamidine (39), thus suggesting some nucleophilic attack at the C-3 position (Scheme 21) in addition to that occurring at C-5 ⟨77JCS(P1)1791⟩.

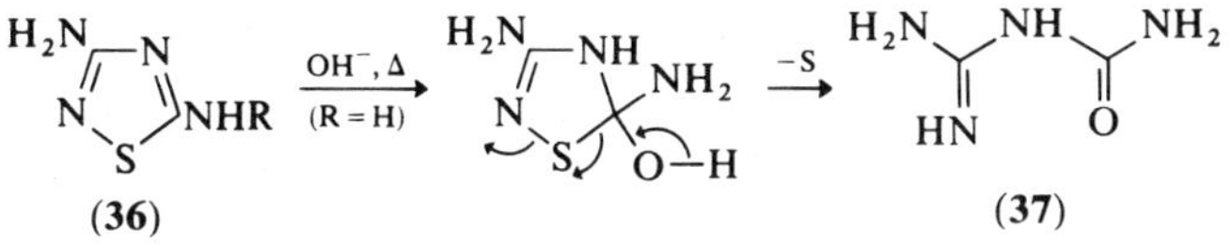

Scheme 19

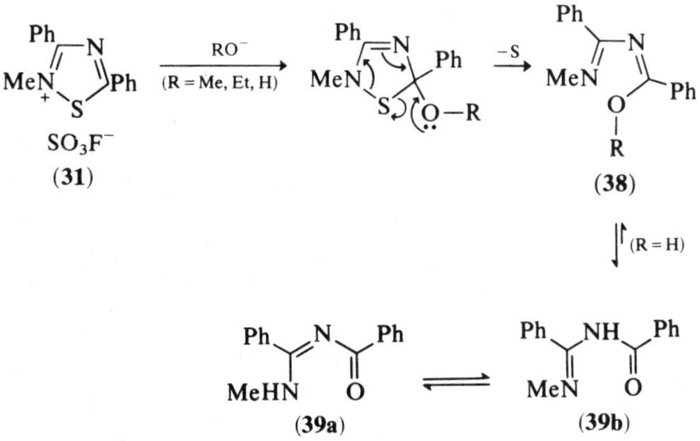

Scheme 20

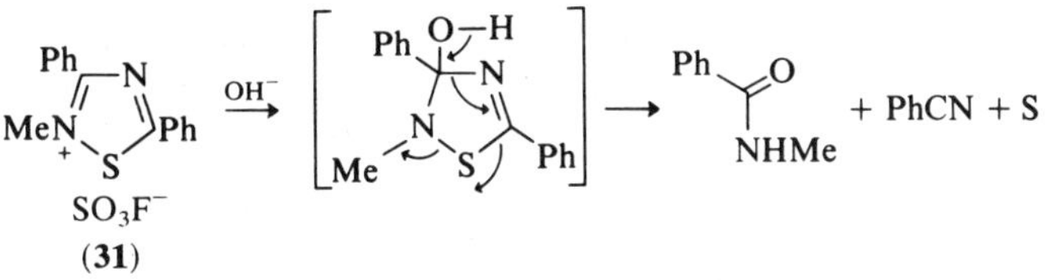

Scheme 21

Primary and secondary amines also function as hard nucleophiles and attack at the C-5 position of (40) to yield salts (41; Scheme 22). When the amines are hydrazines and hydroxylamine, the initially formed salts (42) cyclize to give a 1,2,4-triazole or oxadiazole (43; Scheme 23) ⟨77JCS(P1)1791⟩. A similar mechanism probably operates in the transformation (44 → 46; Scheme 24). However, the original authors proposed a mechanism involving nucleophilic displacement of the methylthio group to produce (45) followed by attack of the hydrazine group at C-5 with the extrusion of sulfur ⟨61CB1682⟩. This mechanism appears to be improbable for both electronic and steric reasons. A carbon nucleophile, the dicyanomethanide ion, also attacks at the C-5 position of (31) and the initial product (47) cyclizes to yield the dihydropyrimidine (48). On treatment with further dicyanomethanide ion or aqueous base, compound (48) is converted via a Dimroth rearrangement into the pyrimidine (49; Scheme 25) ⟨77JCS(P1)1791⟩.

R^1, R^2=H, Ph; H, Bz; $-(CH_2)_5-$

Scheme 22

Scheme 23

Scheme 24

Scheme 25

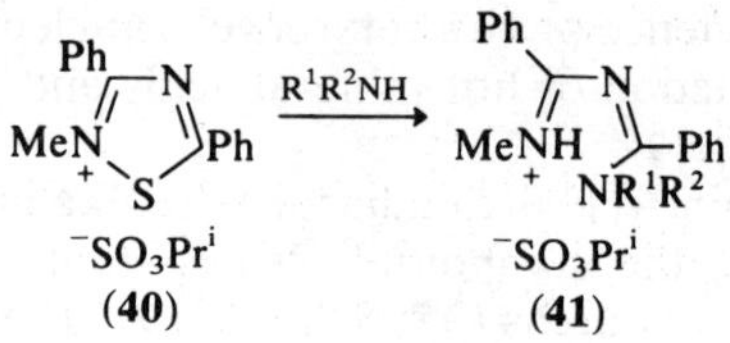

4.25.3.1.7 Nucleophilic attack at hydrogen

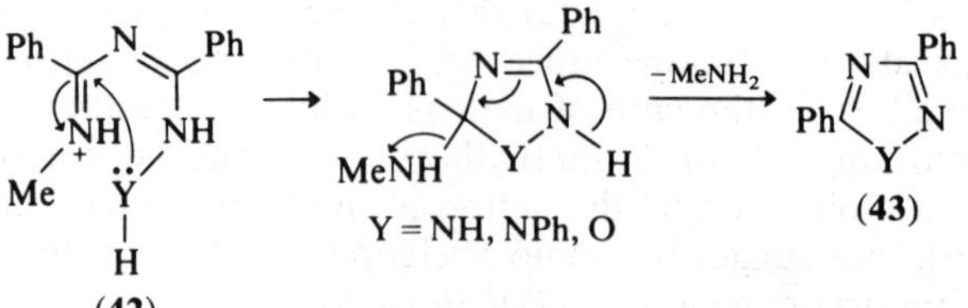

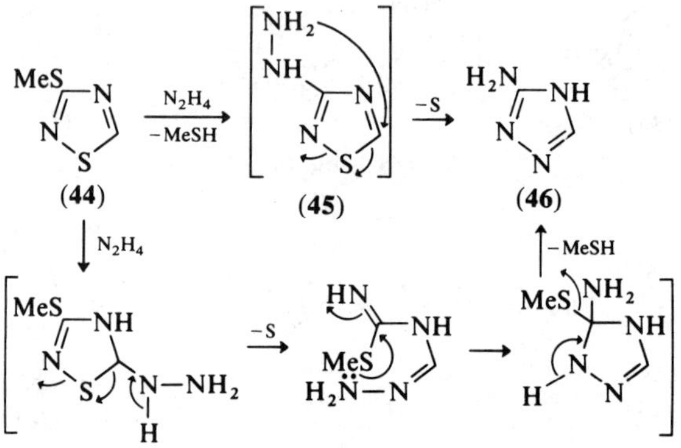

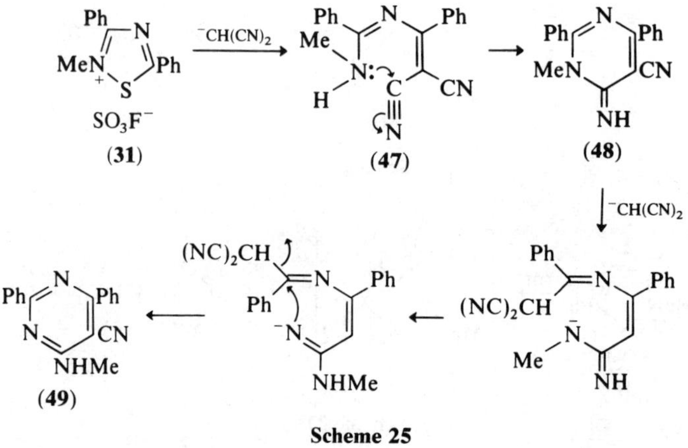

1,2,4-Thiadiazole (**1**) on treatment with a weak base (K_2CO_3) in D_2O is deuterated in the 5-position to yield the monodeutero derivative (**50**; Scheme 26). Under strongly alkaline conditions, the molecule is completely decomposed ⟨76ZN(A)1681⟩.

Scheme 26

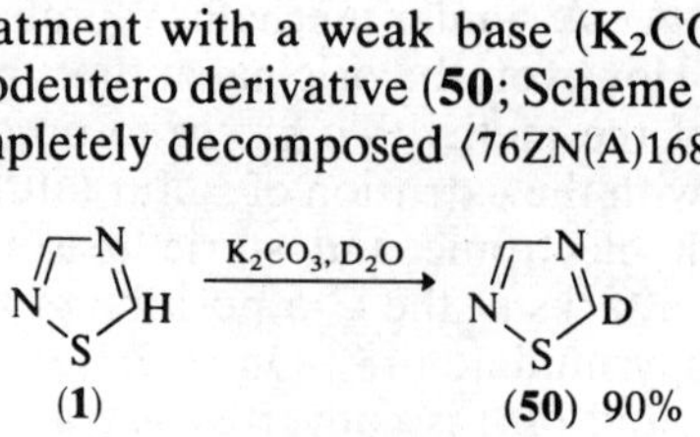

4.25.3.1.8 Reactions with radicals and electron deficient species: reactions at surfaces

Reactions of 1,2,4-thiadiazoles with radicals and electron deficient species are virtually unknown. Catalytic and dissolving metal reductions usually result in S—N bond cleavage. For example, the 5-anilino-3-hydroxy derivative (**51**) gives a good yield of 1-phenyl-2-thiobiuret (**52**) on Zn–HCl reduction (Scheme 27). Reduction of the diamino derivative (**53**) gives amidinothiourea (**54**) from which it may be prepared by oxidation (Scheme 28). Under similar conditions, cleavage of the 3,5-diphenyl derivative (**55**) results in loss of sulfur and formation of benzylbenzamidine (**56**; Scheme 29). Reduction of 5-alkylamino- or 5-arylamino-3-alkylthio derivatives (**57**) with H_2S in pyridine–triethylamine or sodium in liquid ammonia yields 1-substituted dithiobiurets (**58**; Scheme 30).

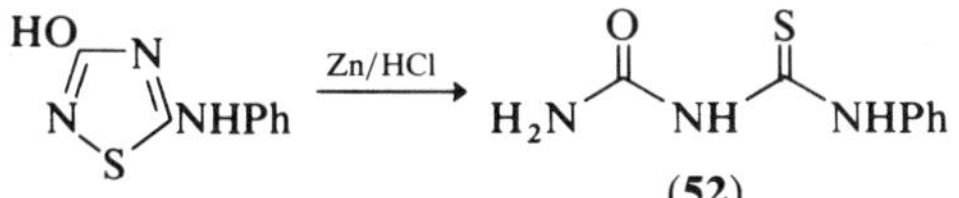

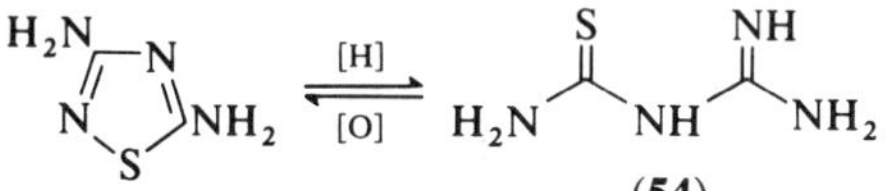

4.25.3.1.9 Reactions with cyclic transition states

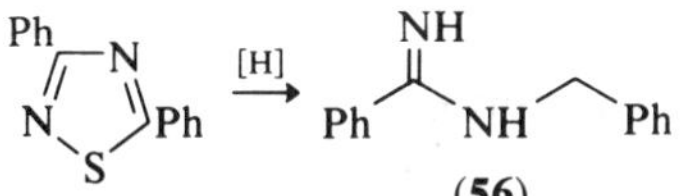

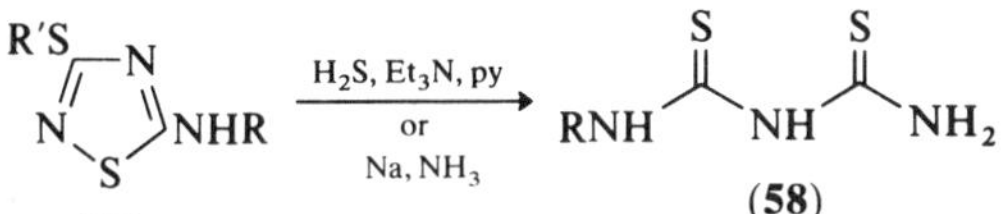

No authentic 1,3-dipolar cycloaddition reactions of 1,2,4-thiadiazoles are known. For proposals related to these transition states see Section 4.25.3.2.1 and Schemes 35, 36 and 38.

4.25.3.2 Reactions of Non-aromatic Compounds

4.25.3.2.1 Isomers of aromatic compounds

The Δ^4-1,2,4-thiadiazoline derivative (**59**) undergoes reductive cleavage at the N—S bond by H_2S in ethanol, forming 1,3,5-triazines (**60**) and (**61**; Scheme 31). Treatment of (**59**) with aqueous ethanolic sodium hydroxide also gives a triazine (**62**) ⟨73CL917⟩. Apparently the last reaction proceeds by nucleophilic attack of hydroxide ion at the 5-position followed by ring opening, loss of sulfur and recyclization. The Δ^4-1,2,4-thiadiazoline derivative (**63**) on treatment with dry HCl in methanol is converted into the 2-carbamoyl derivative (**64**). When (**63**) is treated with a mixture of free amines and amine hydrochlorides in the ratio 2 : 1 in refluxing ethanol, 2-amidino-3-imino-5-methylthio derivatives (**65**) are formed in 80–97% yields (Scheme 32).

In the case of aniline, a weak base, no substitution occurs and (**64**) is obtained ⟨75BCJ310⟩. Oxidation of 1-phenylthiourea (**66**) with H_2O_2 gives Hector's base (**3a**; Scheme 33). This

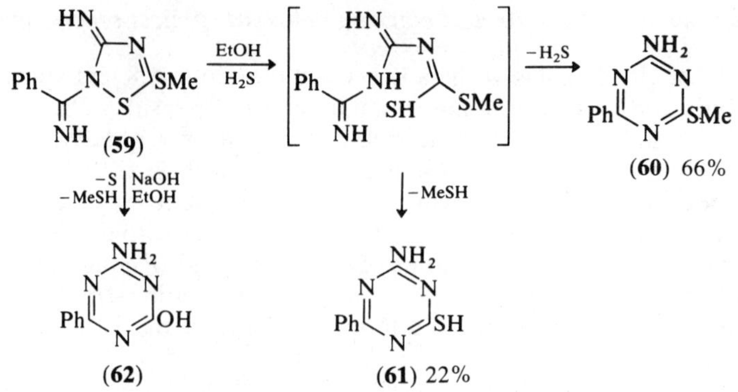

Scheme 31

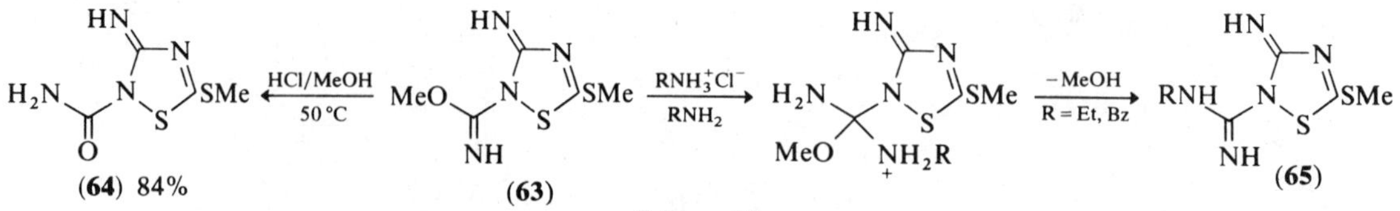

Scheme 32

base was first prepared in 1889 and until very recently its structure was considered to be (**3b**), when an X-ray analysis showed it to be (**3a**) ⟨78CC652⟩. For a detailed analysis of arguments in favor of structure (**3b**), see the review by Kurzer ⟨65AHC(5)119⟩. Hector's base (**3a**) undergoes a base-catalyzed Dimroth rearrangement to give Dost's base (**67**). This isomerization proceeds by the mechanism outlined in Scheme 12. Reduction of (**3a**) under mild conditions (H$_2$S, 25 °C) gives amidinothiourea (**68**) which on mild hydrolysis forms diphenylguanidine (**69**). The guanidine (**69**) can also be obtained directly from (**3a**) by reduction under more drastic conditions ⟨65AHC(5)119⟩.

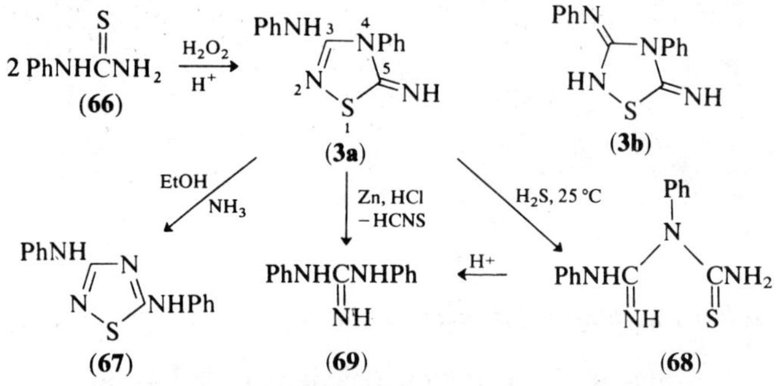

Scheme 33

Hector's base (**3a**) forms a 1:1 adduct with carbon disulfide for which the alternative structures (**70**) and (**71**) were suggested by Butler (Scheme 34) ⟨78JCR(M)0855⟩. A recent X-ray analysis study has shown the correct structure of the adduct to be (**72**), readily explained in terms of an attack by the dithiocarbamate sulfur of (**71**) on the ring-sulfur atom.

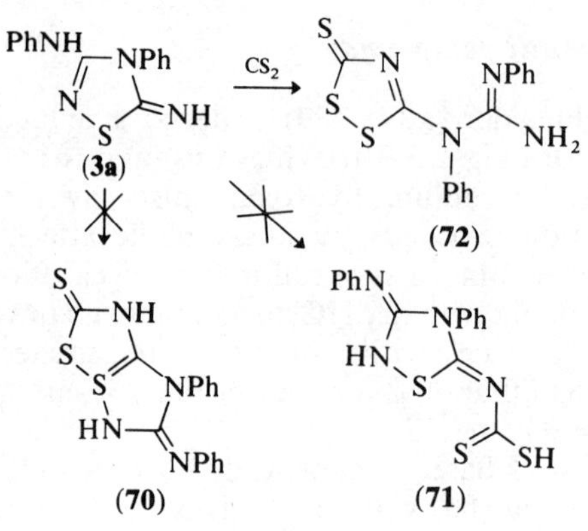

Scheme 34

Hector's base (**3a**) and (**73**; Ar = C₆H₄Me-4) react with electron deficient alkynes (**74**) to give 2-arylaminothiazoles (**76**). The intermediacy of a 1,3-dipolar cycloadduct (**75**) has been suggested ⟨75TL459⟩ to explain this result (Scheme 35). In order to provide support for the intermediate formation of (**75**), the reactions of 2-imino-3,4-diphenylthiazoline (**77**) with alkynes were studied. No 1,3-dipolar addition product (**78**) could be isolated; instead, simple addition products (**79**) were obtained in high yields (Scheme 36).

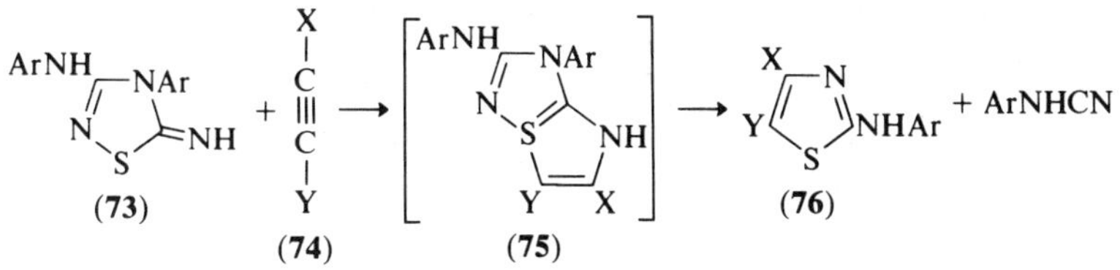

Scheme 35

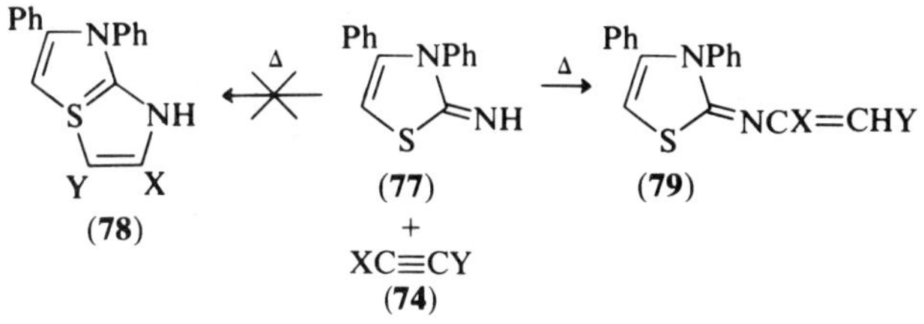

Scheme 36

The authors of this review article suggest that the results indicated in Schemes 34, 35 and 36 are consistent with a stepwise nucleophile substitution mechanism ⟨64JA107⟩ as outlined in Scheme 37. When Z is nitrogen the driving force for the reaction is the ease of cleavage of the N—S bond by a soft nucleophile with a concerted elimination of phenylcyanamide. In the thiazole (Z = CH), however, the stable C—S bond precludes ring fission by a nucleophilic attack at sulfur and the initial adduct (**79**) (presumably *trans*) is obtained as a stable product.

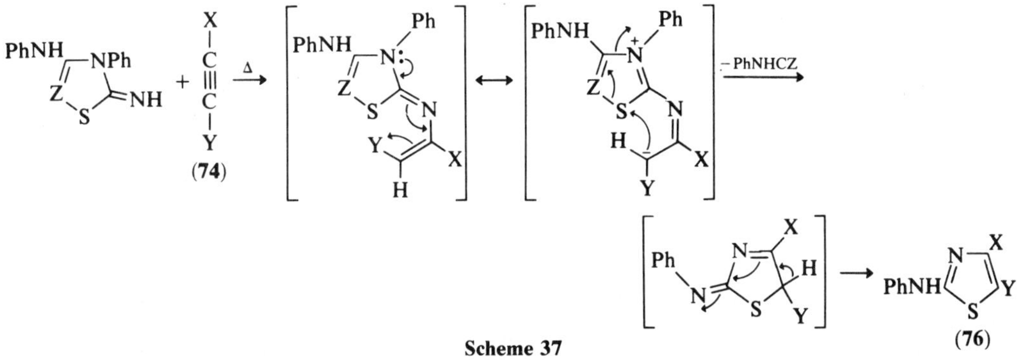

Scheme 37

Reactions of the Hector's bases with arylcyanamides (**80**) have also been studied and 1:1 adducts were obtained. Structures (**81**), based on spectral data, were initially suggested for these products which were formed in fair to good yields ⟨75TL455⟩. A recent X-ray analysis of the adduct (Ar¹ = Ar² = 4-BrC₆H₄) has shown the correct structure to be (**82**; Scheme 38). The molecule is planar with an S—N(1) distance of 2.538 Å compared with the sum of van der Waals' radii of 3.35 Å, suggesting a strong interaction between sulfur and the N(1) atoms ⟨76CL723⟩. Here again, as in the case of the reactions between alkynes and Hector's bases (Scheme 35), a 1,3-dipolar addition mechanism has been postulated without supporting evidence. The authors of this review article suggest that a simple nucleophilic substitution mechanism as shown in Scheme 39 cannot be eliminated. When (**82**) is treated with sodium ethoxide in dry ethanol under reflux, followed by acid work-up, Dost's bases (**83**) are formed in fair yield (Scheme 40). Thermal decomposition of Hector's bases, *e.g.* (**73**), gives a mixture of (**82**) and (**84**) as shown in Scheme 41. Compound (**82**) could arise as indicated in Scheme 39 and Compound (**84**) may be formed by isomerization of (**73**) or by loss of aryl cyanamide from (**82**) (*e.g.* see Scheme 40).

In contrast to the formation of Hector's base (**3a**) by the oxidation of 1-phenylthiourea, oxidation of 1-phenyl-3-methylthiourea and 1,3-diphenylthiourea give thiadiazole (**86**;

Scheme 38

Scheme 39

Scheme 40

Scheme 41

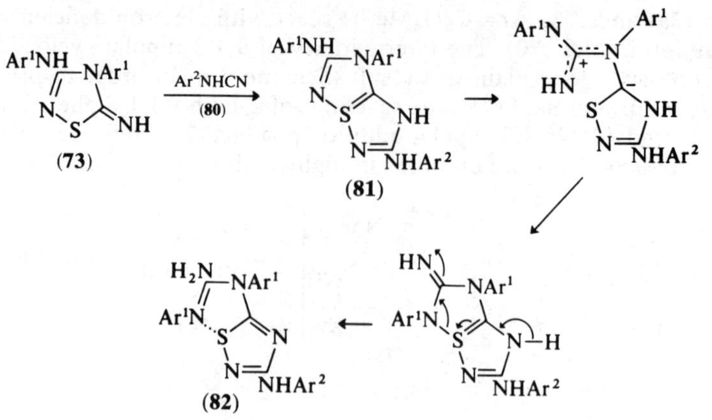

R = Me) and Hugerschoff's base (**87**; R = Ph), respectively (Scheme 42). The products obtained are independent of the oxidizing agent used but (**86**; R = Me) can be converted into (**87**; R = Me) by prolonged acid treatment. The preferential formation of (**86**) and also of (**87**) can be explained by the mechanism illustrated in Scheme 43.

Scheme 42

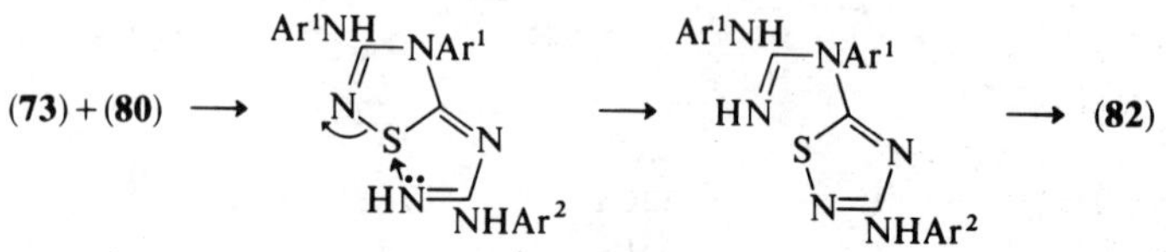

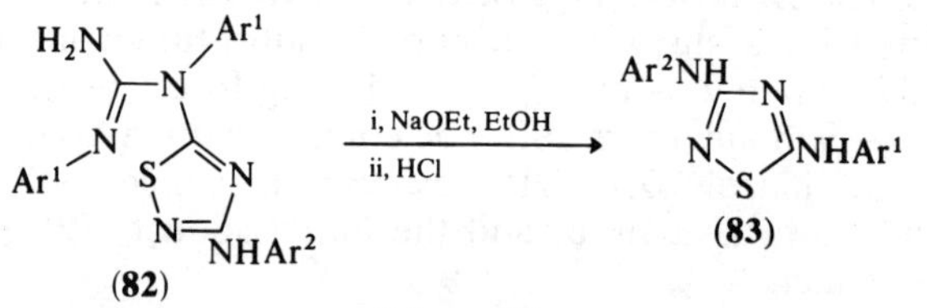

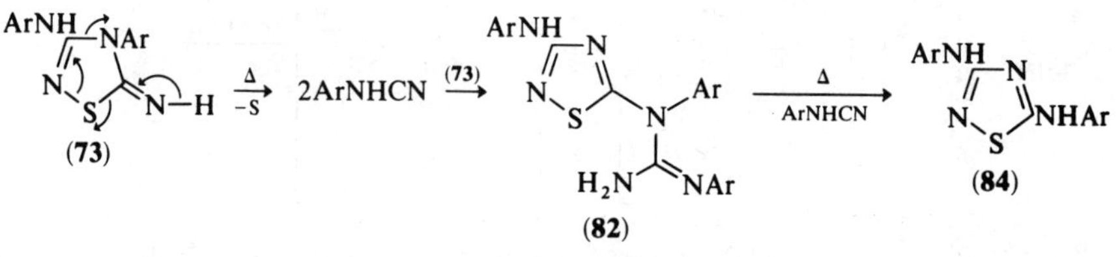

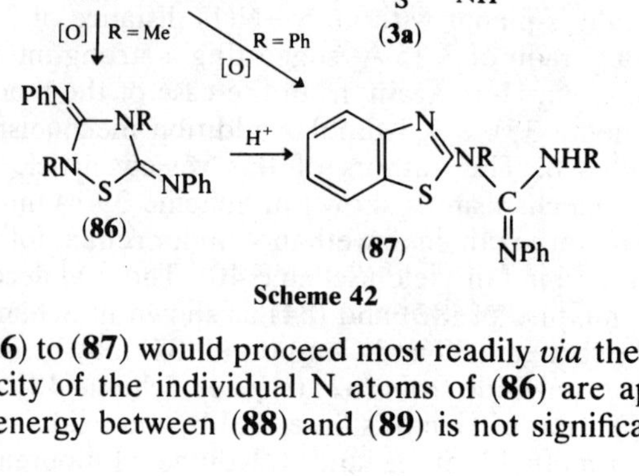

Rearrangement of (**86**) to (**87**) would proceed most readily *via* the protonated form (**89**). When R = Ph the basicity of the individual N atoms of (**86**) are approximately equal so that the difference in energy between (**88**) and (**89**) is not significantly different and the concentration of (**89**) is large enough to result in rapid rearrangement. When R is methyl, however, the basicity of the ring nitrogen atoms would be considerably greater than the exocyclic nitrogen atoms and protonated form (**88**) would predominate over (**89**). As a result, the concentration of (**89**) would be low and rearrangement to (**87**; R = Me) would require prolonged heating in acid media. 3,5-Diimino-1,2,4-thiadiazolidines (**86**) and 5-amino-3-imino-Δ^4-1,2,4-thiadiazoline (**90**) lack the resonance stabilization of the isomeric thiadiazoles and are readily cleaved at the N—S bond by hydrogen sulfide ⟨65AHC(5)119, p. 175⟩. The free base (**90**) is relatively stable in the solid state but is readily isomerized to a Hungerschoff's base (**91**) when refluxed in ethanol for one hour, as indicated in Scheme 44.

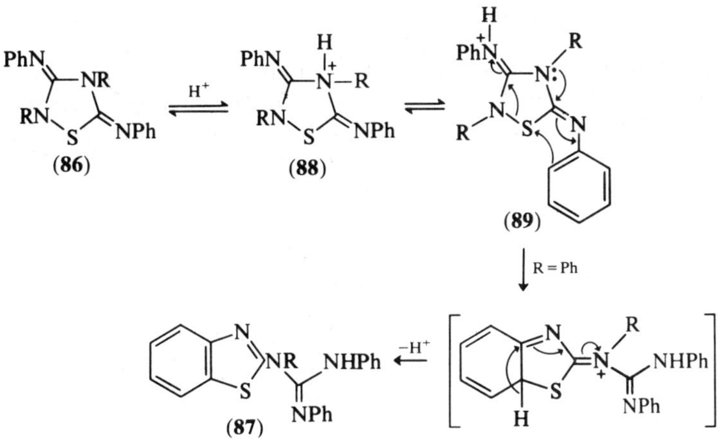

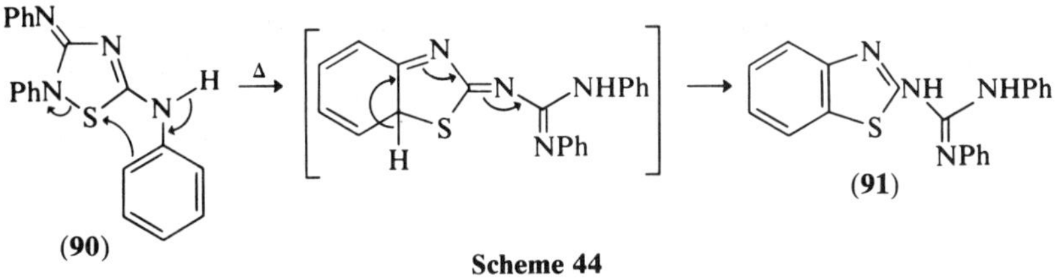

4.25.3.2.2 Dihydro compounds

The 3-alkylthio group in 3,4-disubstituted Δ^2-1,2,4-thiadiazoline 1,1-dioxide (**92**; X = S) is displaced by ammonia under mild conditions to give the corresponding amines (**95**; R = H). However, ring opening to aminomethylsulfonylurea (**94**) occurs under more vigorous conditions. The alkoxy or alkylthio group in (**92**) can also be displaced by primary and secondary amines to yield products of type (**95**). When R' is phenyl, the amine also displaces OR' giving (**96**). Chlorination of (**92**; X = O) gives the 3-chloro derivative (**93**) which provides access to a variety of functional derivatives of this ring system (Scheme 45) ⟨74BSF1580⟩.

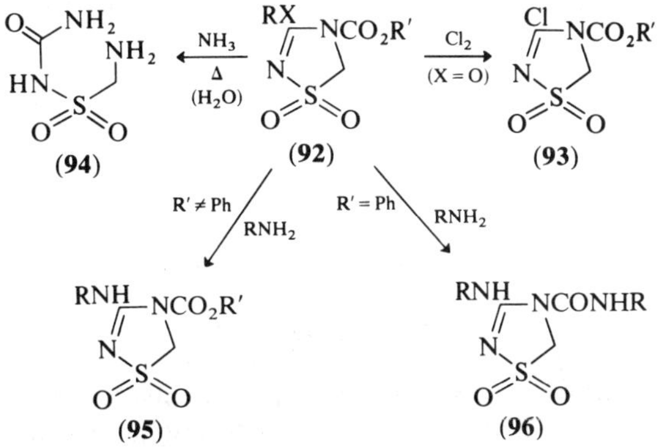

4.25.3.2.3 Tetrahydro compounds

1,2,4-Thiadiazolidines cannot be prepared by catalytic or chemical reductions. Reagents vigorous enough to reduce the double bonds in a 1,2,4-thiadiazole ring also cleave the ring (see Section 4.25.3.1.8). No other synthesis of 1,2,4-thiadiazolidines has been reported.

4.25.3.3 Reactions of Substituents

4.25.3.3.1 General survey of substituents on carbon

Substituents at the 3- and 5-positions of the 1,2,4-thiadiazole ring exert a marked stabilizing influence on the reactions of the substituted compounds with acid, alkali, oxidizing agents and reducing agents. By contrast, 1,2,4-thiadiazole is sensitive to dilute acid and cold dilute alkali, is easily cleaved by reduction, and is oxidized to sulfate by 30% hydrogen peroxide ⟨65AHC(5)119⟩. Many of the properties of the substituted derivatives appear to be related to the lower electron density at the 5-position of the heterocyclic ring relative to the 3-position. Thus, 5-methyl-1,2,4-thiadiazoles are sufficiently acidic to be metalated by butyllithium ⟨70CJC2006⟩ and 5-carboxy-1,2,4-thiadiazoles are readily decarboxylated at 100 °C ⟨74JOC962⟩. Likewise, 3-halo-1,2,4-thiadiazoles are relatively inert compared to the 5-substituted isomers which approach 2-halothiazoles and 4-halopyrimidines in susceptibility to nucleophilic displacement of halogen ⟨65AHC(5)119⟩. Both 3- and 5-hydroxy-1,2,4-thiadiazoles are generally acidic compounds, comparable in acidity to phenol and nitrophenol, respectively. The mercapto-1,2,4-thiadiazoles are even more acidic. The pK_a of 5-mercapto-3-methyl-1,2,4-thiadiazole at 25 °C is 5.18 ⟨65AHC(5)119⟩. Aminothiadiazoles are weak bases but form salts with strong acids. The greater basicity of the 5-amino compound relative to the 3-isomer appears to be anomalous. The diazonium salts of 5-amino-1,2,4-thiadiazoles are considerably more stable and reactive than those of the 3-isomers. In fact, the former diazonium salts incorporating the 1,2,4-thiadiazole ring are among the most reactive of their kind among heterocycles of all types ⟨65AHC(5)119⟩.

4.25.3.3.2 Benzenoid rings

See fused heterocyclic rings in Section 4.25.3.3.9.

4.25.3.3.3 Other C-linked substituents

3,5-Dimethyl-1,2,4-thiadiazole (**97**) is selectively converted into the 5-methyllithio derivative (**98**) on treatment with butyllithium. Subsequent carbonation of (**98**) yields the corresponding carboxylate salt (**99**) in fair yield (Scheme 46) ⟨70CJC2006⟩. The conversion of 5-chloro-3-trichloromethyl-1,2,4-thiadiazole (**100**) to fluorinated analogs (**101**), (**102**) and (**103**) is illustrated in Scheme 47 ⟨65AHC(5)119⟩. In these reactions, silver fluoride is most efficient in the replacement of nuclear chlorine whereas Swarts mixture (SbF_3, $SbCl_3$, Cl_2) preferentially exchanges aliphatic halogen. One method used for the synthesis of (**100**) in 90% yield involves the chlorination of (**104**) in the presence of UV light.

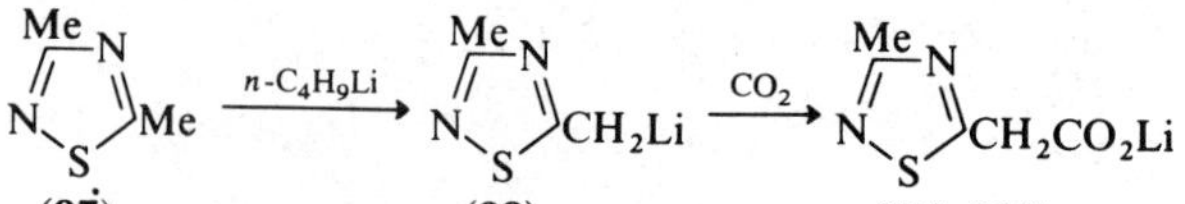

Scheme 46

5-Substituted derivatives of type (**105**) have been reported to undergo cleavage in the presence of acid to yield the acetic acid derivative (**106**; Scheme 48) ⟨65AHC(5)119⟩. The nitrile functions of the isomeric compounds (**107**) and (**108**) have been converted into the corresponding amides, carboxylic acids and methyl esters in the usual fashion ⟨81JOC771⟩.

Tricyanoethylene derivatives such as (**109**) undergo solvolysis in 95% ethanol to yield predominantly (85%) the carboxylate esters (**110**) and smaller amounts of the nitriles (**111**; Scheme 49) ⟨76JOC620⟩.

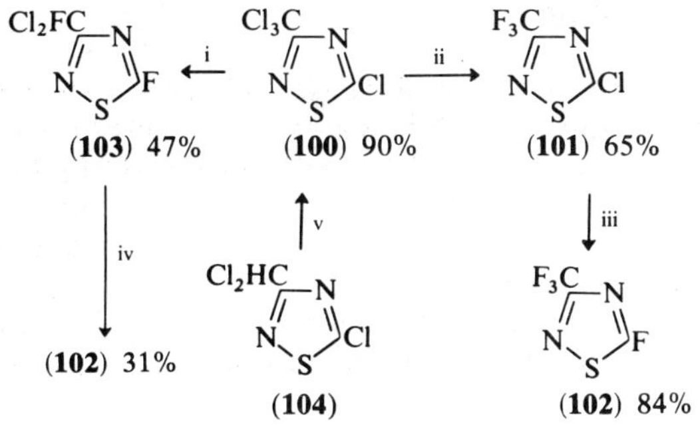

1,2,4-Thiadiazoles

Scheme 47

i, AgF, 140 °C, 10 min; ii, SbF$_3$, SbCl$_3$, Cl$_2$, 150 °C, 4 h; iii, AgF, reflux, 4 h; iv, AgF, reflux, 24 h; v, Cl$_2$, CCl$_4$, *hν*

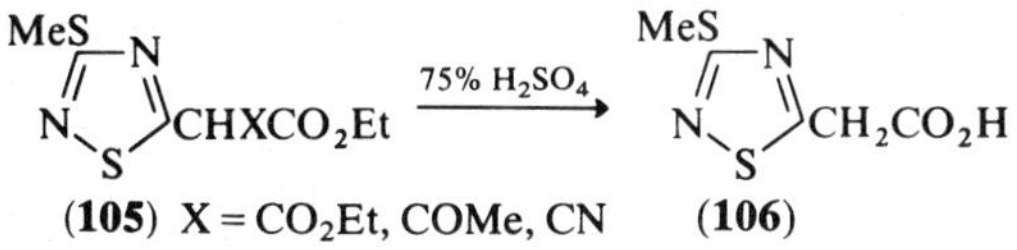

Scheme 48

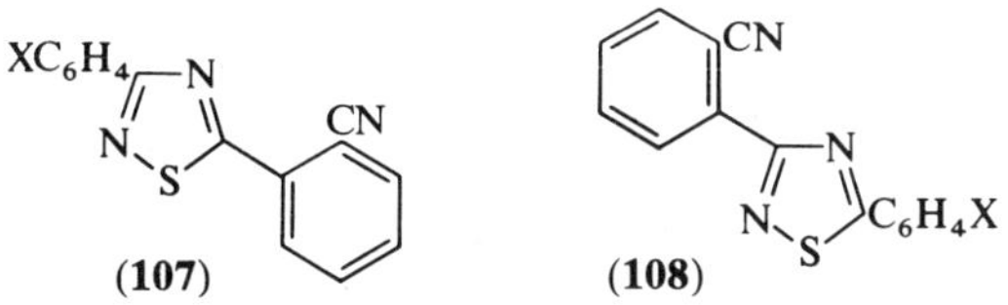

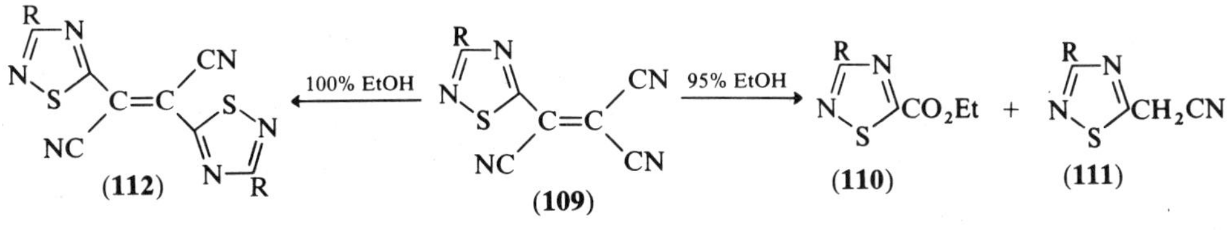

Scheme 49

By contrast, solvolysis of (**109**) in absolute ethanol produces dicyanoethylenes (**112**) as the primary products. The phenyl derivative (**112**; R = Ph) is too insoluble to react with hot 95% ethanol but the methyl derivative (**113**) yields a 1:1 mixture of nitrile (**114**) and ester (**116**), presumably *via* acylnitrile (**115**), in about 90% yield as indicated in Scheme 50. The carboxylate esters (**110**) are readily saponified to the acids (**117**) which readily undergo decarboxylation at 100 °C to produce 3-substituted 1,2,4-thiadiazoles (**118**) in good yields (Scheme 51) ⟨74JOC962⟩.

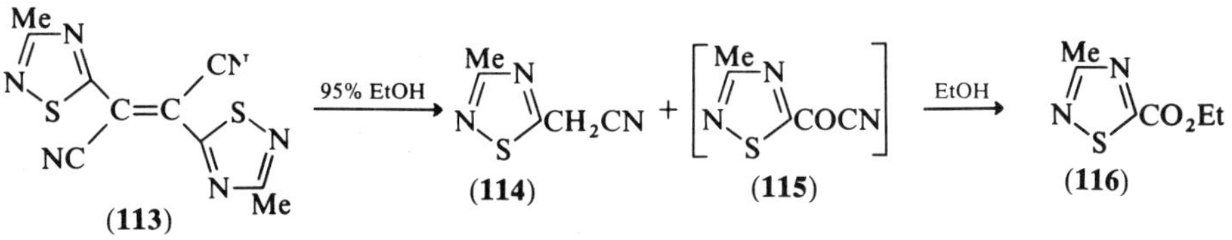

Scheme 50

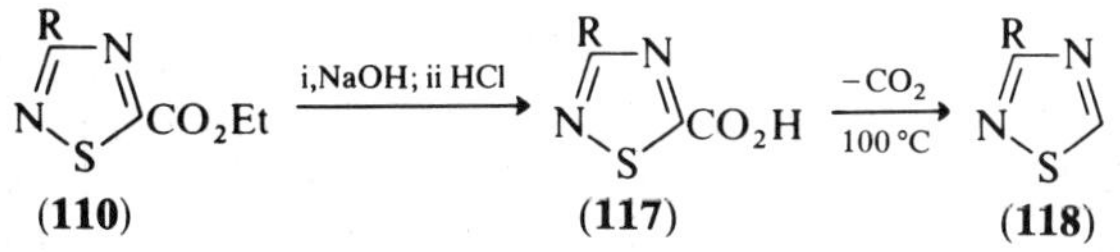

Scheme 51

4.25.3.3.4 N-Linked substituents

5-Amino-1,2,4-thiadiazoles are weak bases. The basicity is further decreased by 3-alkylsulfonyl and 3-alkylthio substitution but increased to a slight extent in compounds containing the 3-dialkylamino group. 3-Amino-1,2,4-thiadiazoles are generally much weaker bases than the 5-amino isomers ⟨65AHC(5)119⟩. Crystalline picrate and *p*-toluenesulfonate salts of amino-1,2,4-thiadiazoles are readily prepared and the formation of insoluble 1:1 complexes with silver nitrate has been reported.

Under mild conditions, amino-1,2,4-thiadiazoles are resistant to the action of acids but prolonged treatment with hot mineral acid can lead to decomposition with liberation of sulfur ⟨65AHC(5)119⟩. 5-Amino-1,2,4-thiadiazoles generally decompose in hot alkali but the rate of degradation is dependent on the nature of the 3-substituent. Under similar conditions, 3-amino-1,2,4-thiadiazoles are much more stable. Thus, 3-amino-5-arylamino-1,2,4-thiadiazoles are unaffected after being refluxed in 3N sodium hydroxide ⟨65AHC(5)119⟩.

As a general rule, reducing agents promote ring cleavage of amino-1,2,4-thiadiazoles by initial scission of the N—S bond. The rate of reduction and the occurrence of subsequent reactions are dependent on the specific conditions used and the nature of the substituents elsewhere in the molecule ⟨65AHC(5)119⟩. 5-Amino-1,2,4-thiadiazoles are readily cleaved by acidic iodide or hydrogen sulfide whereas the 3-amino isomers are much more resistant to these reagents. Zinc in hydrochloric acid, however, will effectively reduce essentially all types of amino-1,2,4-thiadiazoles. 5-Anilino-3-phenyl-1,2,4-thiadiazole (**119**) is slowly converted by the latter reagent into *N*-benzyl-*N'*-phenylthiourea (**120**) as indicated in Scheme 52.

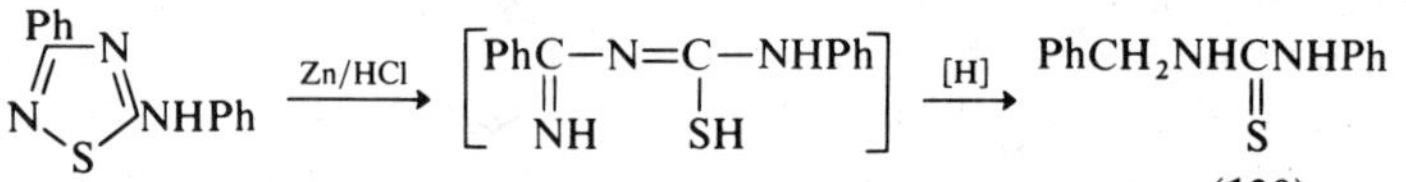

Scheme 52

5-Amino-1,2,4-thiadiazoles (**16**) are alkylated at the 4-position of the heterocyclic ring to produce salts of type (**121**) when heated with methyl iodide or reactive halides such as phenacyl bromide ⟨77G1⟩. The reaction is hindered when 3-substituents are present on the ring and when higher molecular weight alkyl, allyl or benzyl halides are used. By contrast, benzhydryl and trityl chlorides (which are 'harder' electrophiles) alkylate (**16**) at the 5-amino function (presumably by an S_N1 reaction) to produce products of type (**122**) as indicated in Scheme 53 ⟨65AHC(5)119⟩.

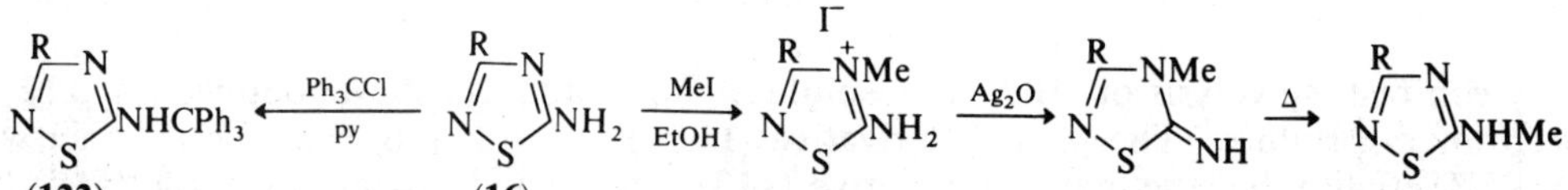

Scheme 53

The free bases (**19**), obtained by neutralization of (**121**), undergo a Dimroth rearrangement to (**20**) when heated or on prolonged storage (see Section 4.25.3.1.3). Formation of (**20**) is rapid and occurs in good yield when R is hydrogen but the reaction is severely hindered when the 3-substituent is methyl, benzyl or phenyl ⟨63CB534⟩. Alkylation of 3-amino-5-phenyl-1,2,4-thiadiazole (**15**; R = Ph) with methyl iodide was reported to require several months at room temperature. The product of this reaction was claimed to be (**123**) which on treatment with silver oxide was believed to form the zwitterionic compound (**124**) ⟨63CB534⟩ as indicated in Scheme 54. Attempts to reproduce these results by the method described, however, have failed ⟨70CB1805⟩. Treatment of 3-amino-5-phenyl-1,2,4-thiadiazole (**15**; R = Ph) with trimethyloxonium tetrafluoroborate readily produces the 2-alkylated salt (**125**) in good yield. Since the picrate salts prepared from (**125**) and (**123**) are identical ⟨70CB1805⟩, the structure of (**124**) is probably incorrect. Neutralization of (**125**) with dilute alkali results in a Dimroth type rearrangement to (**127**), possibly *via* (**126**) as indicated in Scheme 54 (where X is a soft nucleophile). By analogy with 5-amino-1,2,4-thiadiazoles (**16**), the 3-amino isomer (**15**) is alkylated by benzhydryl chloride (and trityl chloride) at the primary amino function to produce (**128**).

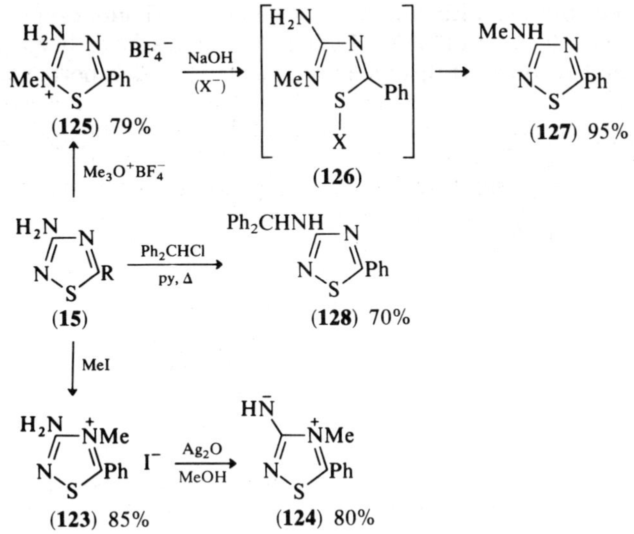

1,2,4-Thiadiazoles   481

Scheme 54

5-Amino-1,2,4-thiadiazoles (**16**) readily yield monoacylated derivatives of type (**129**) under the usual conditions whereas both monoacyl and diacyl derivatives (**130** and **131**, respectively) are easily obtained from the 3-amino isomers ⟨65AHC(5)119, 70CB1805⟩. In a similar manner the 3-amino-1,2,4-thiadiazoles produce mono- and di-sulfenamide derivatives when treated with sulfenyl halides. In general, the reactions of 3-amino- and 5-amino-1,2,4-thiadiazoles with sulfonyl halides under basic conditions produce only low yields of the desired derivatives. Sulfonamides of type (**132**) are best prepared by ring synthesis methods as illustrated in Scheme 55 ⟨75LA1961⟩.

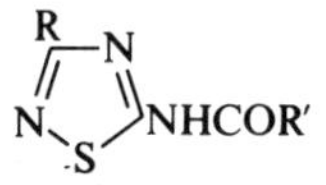

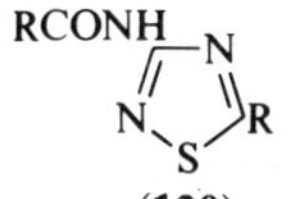

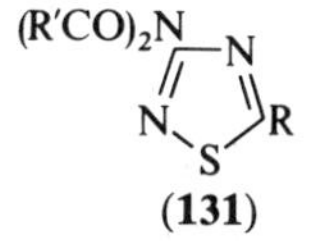

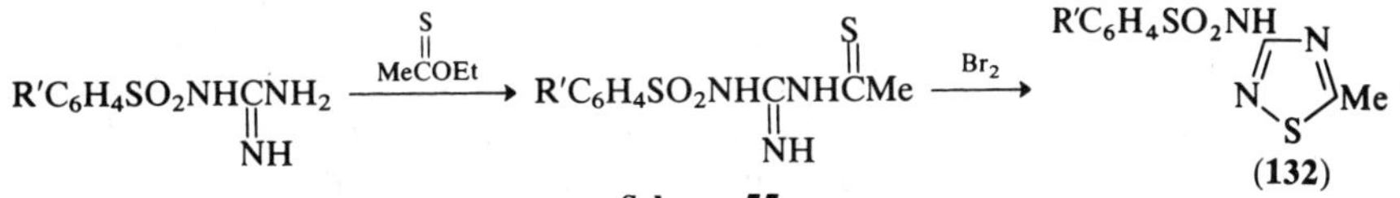

Scheme 55

By contrast, however, 3,5-diamino-1,2,4-thiadiazoles are selectively converted into amides and sulfonamides in good yields. Thus, acetylation of 3,5-diamino-1,2,4-thiadiazole (**53**) with excess acetic anhydride yields the symmetrical derivative (**133**) whereas under similar conditions 3-amino-5-methylamino-1,2,4-thiadiazole and 3,5-diarylamino-1,2,4-thiadiazoles ⟨65AHC(5)119⟩ produce the monoacylated derivatives (**134**) and (**135**), respectively. Sulfonylation also results in the preferential formation of the 3-sulfonamides but reaction with excess reagent eventually produces the 3,5-disubstituted derivatives (**136**) and (**137**) in good yields ⟨65AHC(5)119⟩.

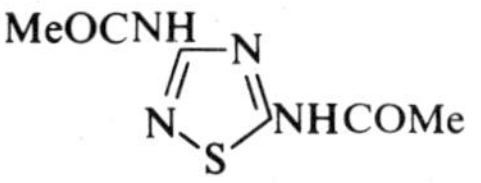

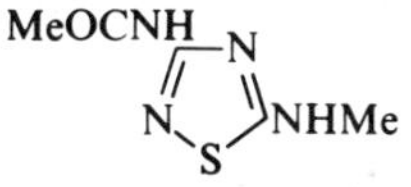

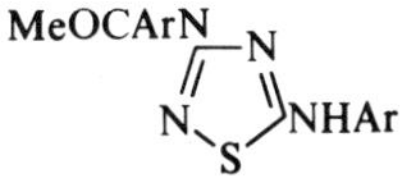

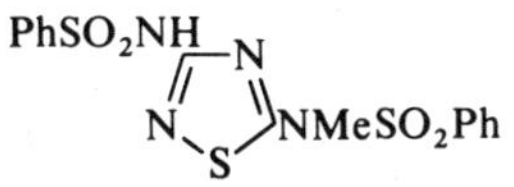

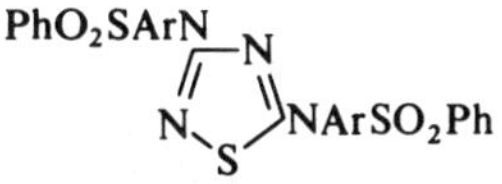

Acetylation of 5-amino-3-dimethylamino-1,2,4-thiadiazole yields the expected 5-acetylamino derivative but no product was obtained when sulfonylation was attempted. 3-Amino-5-phenyl-1,2,4-thiadiazole (**15**; R = Ph) forms the expected derivatives (**138**), (**139**) and

(140) when allowed to react with suitable isocyanates, carbamates and chloroformates as indicated in Scheme 56 ⟨65AHC(5)119⟩. Similar derivatives have been prepared from the 5-amino isomers and products of this type have been claimed as herbicides and bactericides (see Section 4.25.5).

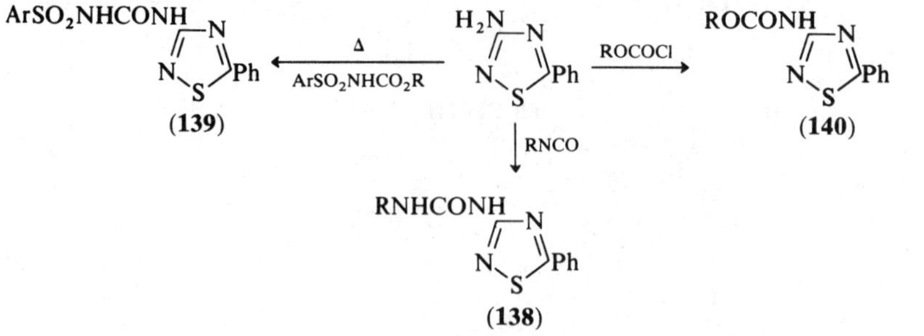

Scheme 56

The 5-imino isomers (19) also react in the normal manner with anhydrides, sulfonyl halides, isocyanates and chloroformates as indicated in Scheme 57 to afford (141)–(144) ⟨54CB68⟩.

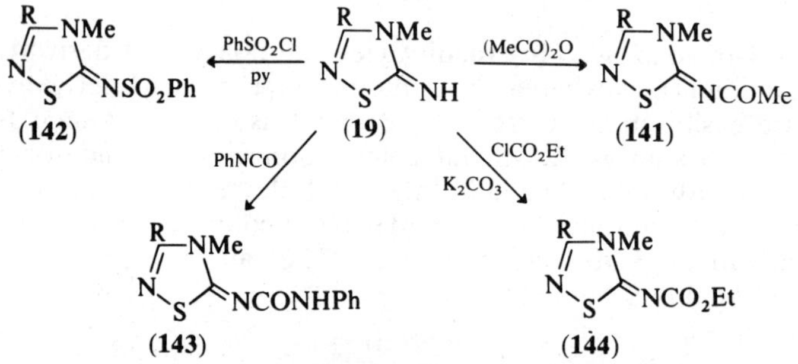

Scheme 57

The reaction of 5-amino-1,2,4-thiadiazoles (16) with aldehydes produces α-amino alcohols (145), azomethines (146) or bis-amines (147) depending on the reactants and reaction conditions, as illustrated in Scheme 58. The nitration of 5-amino-3-methyl-1,2,4-thiadiazole (25) in a mixture of 98% sulfuric acid and 95% nitric acid at 0 °C is reported to produce (148) in good yield ⟨65AHC(5)119⟩, but a similar reaction with 3-amino-1,2,4-thiadiazoles has not been described. Products of type (149) can be obtained, however, by the ring closure of N-nitro-N'-thiocarbamylguanidines with alkaline hydrogen peroxide.

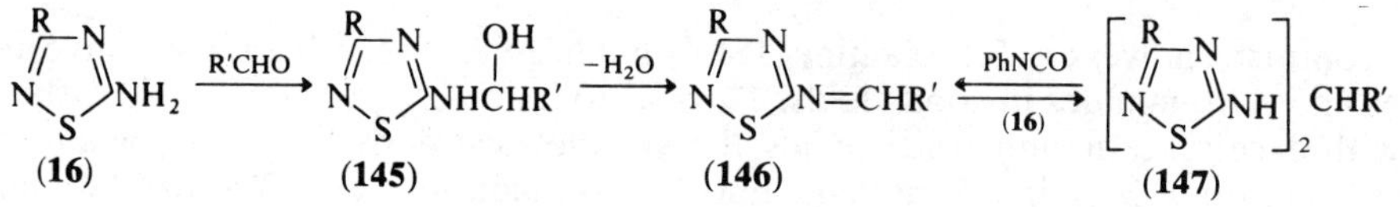

Scheme 58

3-Amino-1,2,4-thiadiazoles fail to yield nitrosamines under the usual conditions but relatively stable 5-nitrosamines (150) are easily prepared as indicated in Scheme 59 ⟨65AHC(5)119⟩.

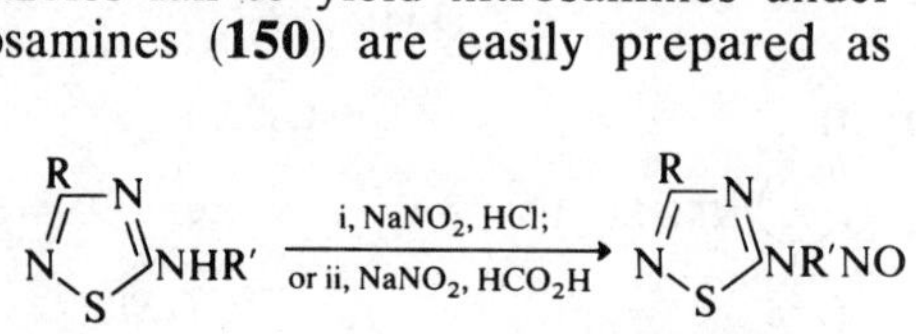

R = OR″, SR″, NR″$_2$, SO$_2$R″, Ar; R′ = H, Ar, R″

Scheme 59

Molecular weight determinations indicate that the 5-nitrosamines associate as dimers in non-polar media. The site of alkylation of (**150**; R'=H) by diazomethane depends on the nature of the 3-substituent. Thus, the 3-phenyl derivative is alkylated on the exocyclic NH group to produce (**151**) (also obtained by nitrosation of the *N*-methyl derivative (**20**)) whereas the 3-benzyl derivative undergoes alkylation at the 4 position of the ring. Reduction of nitrosamines to the corresponding hydrazines (**152**) can be accomplished in good yields using lithium aluminum hydride as indicated in Scheme 60. Owing to their amphoteric character, 5-nitrosamines dissolve in both alkalis and acids. In acidic media the corresponding diazonium salt (**153**), which is in equilibrium with (**150**), may be precipitated under certain conditions whereas decomposition to hydroxyl- or halogen-substituted derivatives (**154**) and (**155**), respectively, may occur under other conditions ⟨65AHC(5)119⟩. Addition of suitable phenols to acidic solutions of (**150**) produces azo dyes directly so that isolation of (**153**) is not necessary. In aqueous or alcoholic solutions the dimerization of (**150**) to triazines (**156**) (60–90% yields) is catalyzed by acids. The triazines (**156**) are very stable to acids and alkali under the usual conditions and are not converted into (**153**) on treatment with nitrous acid. Alkylation of (**156**) occurs on the exocyclic nitrogen to yield (**157**). The influence of reaction conditions and the mechanism of triazine formation have been discussed ⟨60CB963⟩. In benzene the dimerization of (**150**; R = Ph) takes a different course and the azoxythiadiazole (**158**) is obtained in 40% yield ⟨65AHC(5)119⟩. The azoxy derivative is also obtained in excellent yield when the hydroxylamine (**159**) is oxidized with chromic–acetic acid or when the nitrosamine (**150**) is coupled with (**159**). The hydroxylamine (**159**) dissolves in base to produce the azo derivative (**160**) which is precipitated by addition of acid. Reduction of (**160**) with hydrogen sulfide produces the hydrazo compound and in alkaline media this product is reconverted to (**160**) by air oxidation.

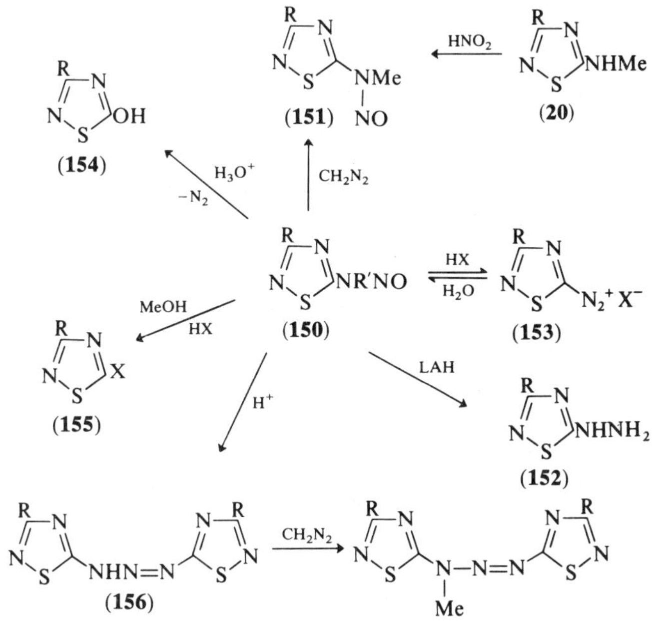

Scheme 60

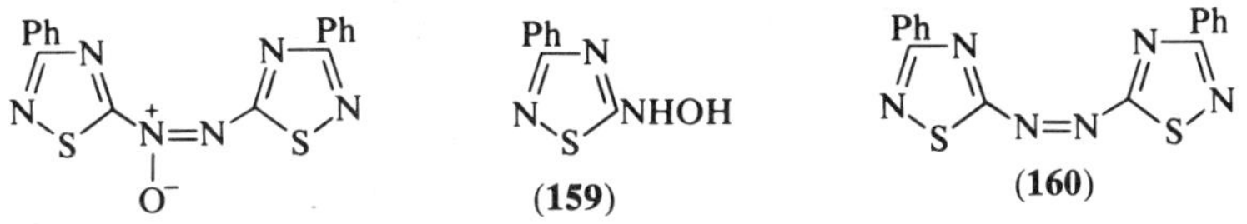

The 5-hydrazino-1,2,4-thiadiazoles (**152**) are stable in acid and base and readily form hydrazone derivatives on reaction with suitable carbonyl compounds. Thus, (**152**; R = Me, Ph, *m*-O$_2$NC$_6$H$_4$) reacts with β-keto esters to form 3-alkyl-1-(1,2,4-thiadiazol-5-yl)pyrazolin-5-ones (**161**) which undergo coupling with *p*-nitrosodialkylanilines to form azomethine dyes (**162**) ⟨65AHC(5)119⟩.

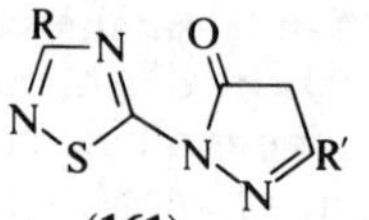

(161) 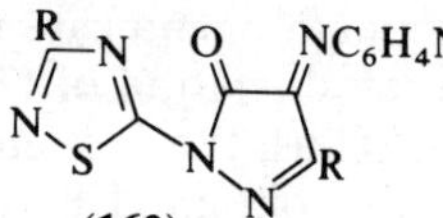   (162)

By contrast, 3-hydrazino-1,2,4-thiadiazoles (163), which are prepared by ring closure methods, are very sensitive to acids and undergo elimination of sulfur with formation of 5-arylamino-1,2,4-triazoles (164) as indicated in Scheme 61 ⟨63JCS4566⟩. The mechanism of this conversion is not known but presumably involves the attack of a hard nucleophile at the 5-position followed by ring opening, loss of sulfur and recyclization as has been observed with the quaternary salts ⟨77JCS(P1)1791⟩ (see Schemes 22, 23, and 24). Hydrazine (163) reacts with aldehydes to produce hydrazones (165) and with dimethyl malonate to yield pyrazoles (166). Diazonium salts of type (153) have been prepared by the methods

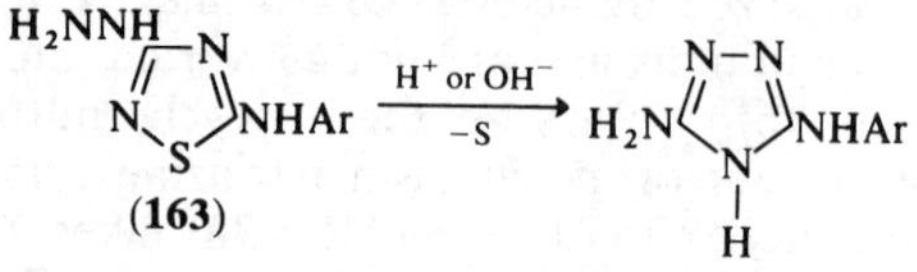

(163)   (164)

Scheme 61

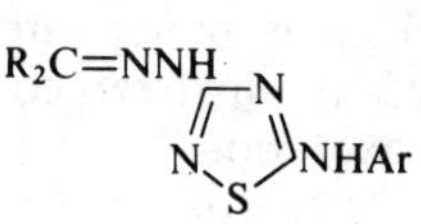

(165) 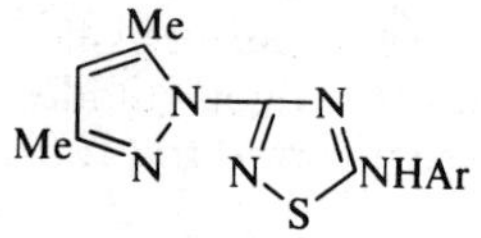   (166)

indicated in Scheme 62 and often can be isolated as crystalline tetrafluoroborate or perchlorate salts. Certain electron-donating groups in the 3-position, such as dialkylamino, can make diazotization difficult ⟨65AHC(5)119⟩. The unstable diazonium salts of 3-amino-1,2,4-thiadiazoles are not often isolated but have been obtained in a few cases using the nitrosyl tetrafluoroborate procedure. Some of the more common reactions of 1,2,4-thiadiazole-5-diazonium salts are illustrated in Scheme 63.

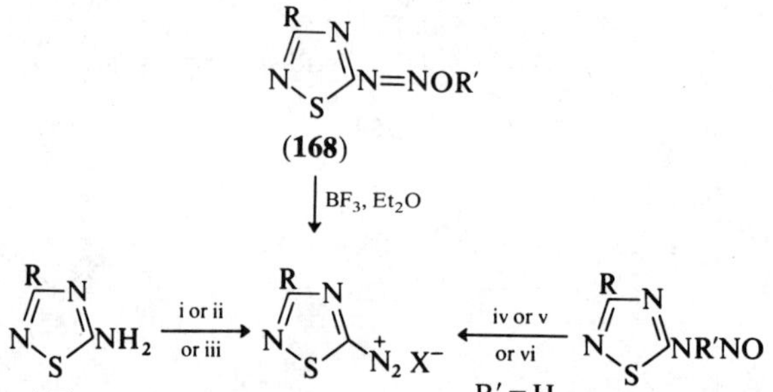

(168)

$\downarrow$ BF$_3$, Et$_2$O

(16)   (153)   (150)
         R' = H

i, conc. H$_2$SO$_4$ or H$_3$PO$_4$, NaNO$_2$; ii, NO$^+$BF$_4^-$, HOAc, 0 °C; iii, HOAc, NaNO$_2$; iv, BF$_3$, Et$_2$O; v, H$_2$SO$_4$ or HClO$_4$; vi, HBF$_4$, HOAc

Scheme 62

Solutions of thiadiazole diazonium salts (153) readily couple with a variety of aromatic compounds to form monoazo dyes of type (167). Diazonium salts incorporating the 1,2,4-thiadiazole ring are especially reactive coupling reagents due to the electronegative nature of the ring and the high positive charge localized on the terminal nitrogen of the diazo group. Thus, the diazonium salts of 3-alkylsulfonyl-5-amino-1,2,4-thiadiazoles even couple with *m*-xylene to form azo dyes (167; R=3-alkylsulfonyl; Ar = *m*-xylyl) in ~75% yields ⟨65AHC(5)119⟩. Owing to their stability, high reactivity and ready accessibility, thiadiazole diazonium salts have been extensively used to prepare monoazo dyes for use in polymers.

Treatment of (153) with a large excess of alcohol yields the diazo ethers (168). Similar products are not obtained in the benzene series. In the presence of low concentrations of acid, (168) condenses with nitrosamines (150) to yield triazines of type (169). The triazines (169) also are readily obtained from the reaction of the diazonium salts (153) with primary

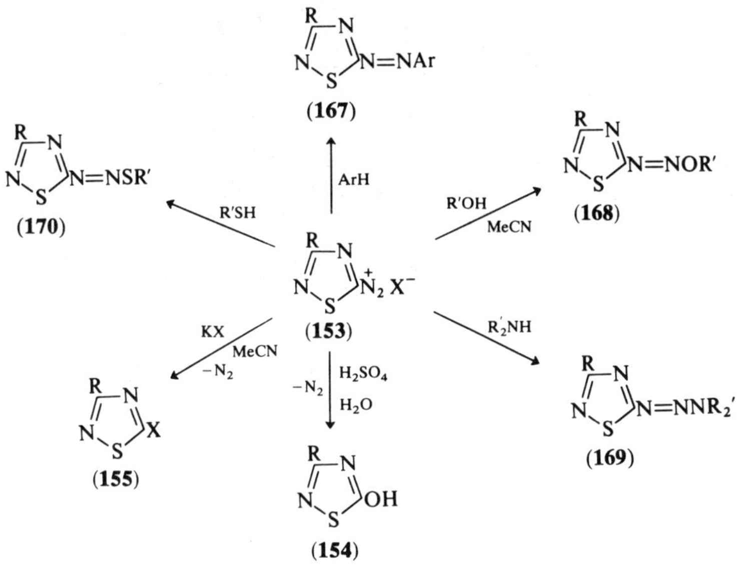

Scheme 63

or secondary amines in acetonitrile. Products of type (**170**) are obtained when certain divalent sulfur compounds are allowed to react with the 5-diazonium salts (**153**). Thus, 3-phenyl-1,2,4-thiadiazole-5-thiol (**17**; R = Ph) and thiourea react with 3-phenyl-1,2,4-thiadiazol-5-yl diazonium tetrafluoroborate (**153**; R = Ph, X = BF_4^-) to produce (**171**) and (**172**), respectively. Treatment of (**172**) with alkali results in degradation with formation of the mercapto derivative (**17**). The reaction of (**153**; R = Ph) with diethyl sulfide proceeds with nitrogen evolution to produce a sulfonium salt (**173**). Hydrolytic cleavage of (**173**) in water at neutral pH results in the formation of (**174**) in 66% yield ⟨65AHC(5)119⟩. 3-Substituted 1,2,4-thiadiazole-5-diazonium tetrafluoroborates (**153**) react with potassium halides in acetonitrile in the absence of catalysts to produce the corresponding 5-halo derivatives (**155**). With the exception of iodide salts, the latter reaction does not proceed with the isomeric 3-diazonium compounds. In general, however, the 3-halides may be obtained by treatment of diazotized 5-substituted 3-amino-1,2,4-thiadiazoles with halogen acids and copper powder or with cuprous halide salts in acetonitrile. Hydrolysis of (**153**) in warm aqueous sulfuric acid produces the 5-hydroxy derivatives (**154**) in moderate to good yields ⟨65AHC(5)119⟩.

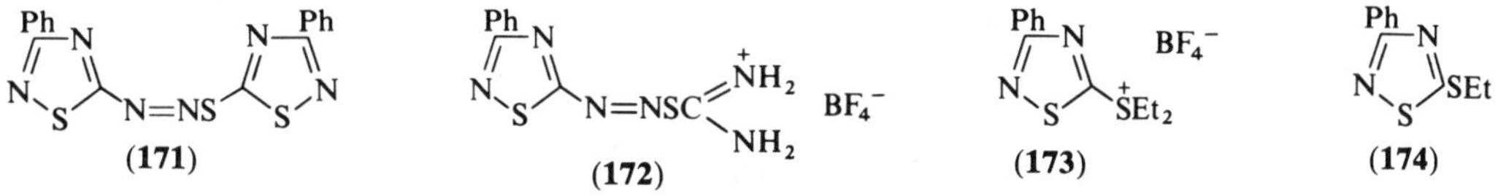

4.25.3.3.5 O-Linked substituents

Hydroxy-1,2,4-thiadiazoles are generally quite acidic. 3-Ethyl-5-hydroxy-1,2,4-thiadiazole, for example, is more acidic than 4-nitrophenol and 4-hydroxypyrimidine, but less so than 2,4-dinitrophenol. 3-Hydroxy-1,2,4-thiadiazoles give purple colors with iron(III) chloride, indicating their phenolic nature. Their IR spectra in a chloroform solution or in a suspension in Nujol suggest that the hydroxy form (**13**) is mostly present (absorptions *ca.* 1530–1550 and 1670–75 cm^{-1}) (Scheme 64) ⟨79JHC961⟩. Electrophilic reagents react with 3-hydroxy-1,2,4-thiadiazoles either at OH or with the N-2 atom depending upon the nature of the electrophile. Reagents which can be considered as 'hard' acids (acyl ions and sulfonyl ions) react at oxygen. Trimethylsilyl chloride, the dialkoxyphosphoryl and thionophosphoryl chlorides also react at oxygen and thus may be considered as generating 'hard' acids. On the other hand, isocyanates and anhydrides react exclusively at nitrogen and thus behave as 'soft' acids (Scheme 65) ⟨79JHC(16)961⟩. Carbocations are also soft electrophiles and alkylate on nitrogen. Thus, the bis(trimethylsilyl) derivative (**175**) reacts with 1,2,3,5-tetra-*O*-acetyl-β-D-ribofuranose (**176**) in the presence of tin(IV) chloride to produce the 2-substituted derivative (**177**) in 60% yield (Scheme 66) ⟨76JHC169⟩.

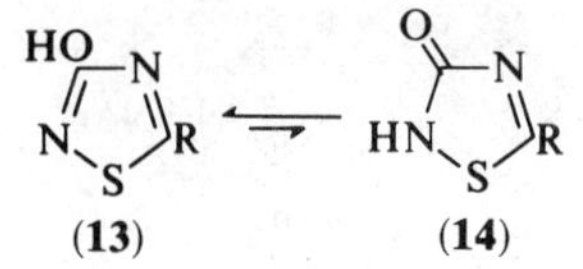

Scheme 64

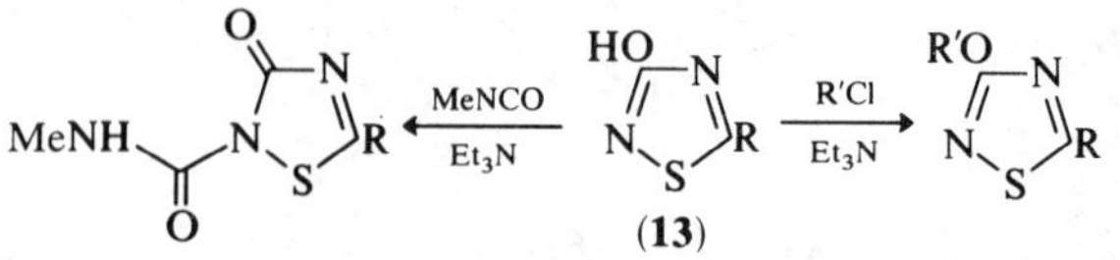

Scheme 65

R = OR″, SR″, NMe₂; R′ = CONMe₂, SO₂Me, PO(OEt)₂, PS(OEt)₂, SiMe₃

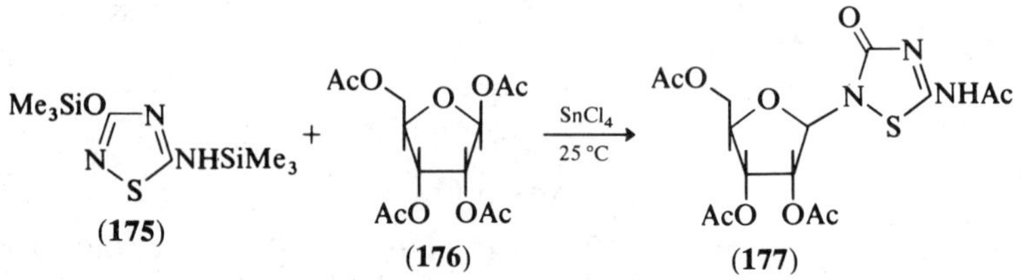

Scheme 66

4.25.3.3.6 S-Linked substituents

The position of equilibrium in the tautomers of 1,2,4-thiadiazole-5-thiol (**178**) appears to be on the thione side (**178b**) or (**178c**). The solid state IR spectrum of (**178**) shows no signs of SH absorption. On treatment with potassium hydroxide, (**178**) is converted into a stable potassium salt (**179**) which reacts with methyl iodide yielding the 5-methyl ether (**180**; Scheme 67) ⟨73CJC2353⟩.

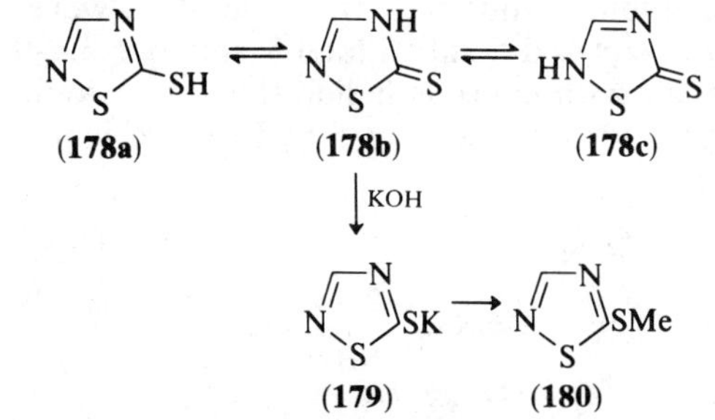

Scheme 67

The disulfide (**181**) on treatment with mercury forms a stable mercury salt (**182**) which reacts with sodium sulfide to give the corresponding sodium salt (**183**). The latter salt cannot be isolated, but its presence was confirmed by transformation to the S-methyl ether (**184**; Scheme 68) ⟨73CJC2353⟩. 3-Halo-1,2,4-thiadiazole sulfides (**185**) can be readily oxidized to the corresponding sulfoxides (**186**) and sulfones (**187**). m-Chloroperbenzoic acid is the reagent of choice for the preparation of sulfoxides and hydrogen peroxide in acetic anhydride–acetic acid gives a good yield of sulfones. Attempts to convert (**185**; R=Me) to the corresponding 5-chloromethyl sulfide (**188**) using sulfuryl chloride were unsuccessful (Scheme 69) ⟨73JOC465⟩.

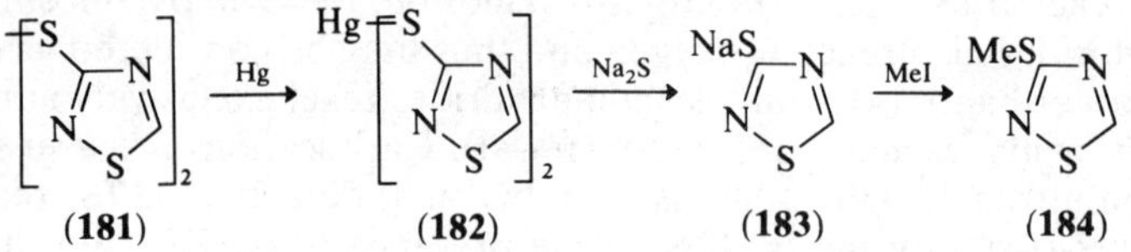

Scheme 68

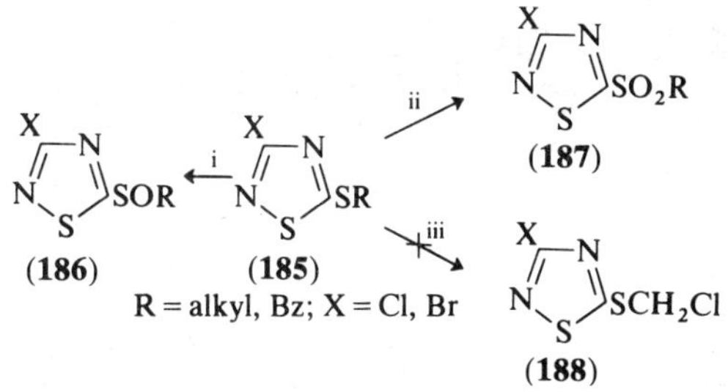

(187)

(186) ← i — (185) — iii → (188)

R = alkyl, Bz; X = Cl, Br

i, 3-ClC$_6$H$_4$CO$_3$H, CHCl$_3$; ii, H$_2$O$_2$, MeCO$_2$H, (MeCO)$_2$O; iii, SO$_2$Cl$_2$, Δ (R = Me)

Scheme 69

Chlorine in aqueous media oxidizes 3-alkylthio-1,2,4-thiadiazoles (**189**) successively to the sulfoxide (**190**) and sulfone (**191**). However, in glacial acetic acid, chlorine displaces the 3-benzylthio group in (**192**) giving 5-amino-3-chloro-1,2,4-thiadiazole (**193**; Scheme 70) ⟨65AHC(5)119⟩. The 3-alkylthio groups in 1,2,4-thiadiazoles are difficult to replace. Thus 3-alkylthio-1,2,4-thiadiazoles resist the action of aniline, ammonia, molten urea and ammonium acetate. However, hydrazine attacks 3-methylthio-1,2,4-thiadiazole (**44**) forming 3-amino-1,2,4-triazole (**46**; see Scheme 24). Hydrazine also reacts with the quaternary salt (**195**), prepared by methylation of (**194**), forming the hydrazone (**196**; Scheme 71) ⟨65AHC(5)119⟩.

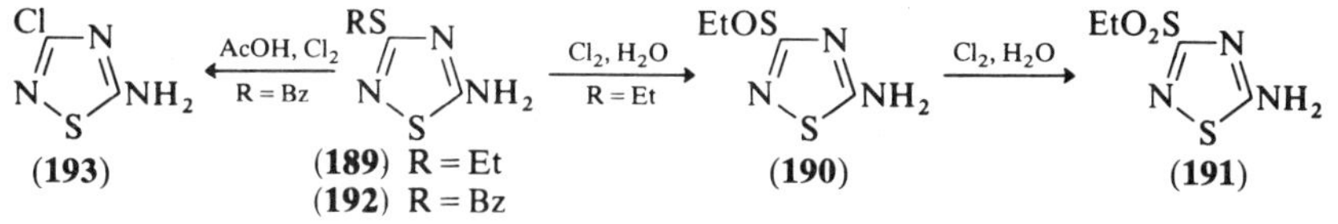

(193) ← AcOH, Cl$_2$, R = Bz — (189) R = Et / (192) R = Bz — Cl$_2$, H$_2$O, R = Et → (190) — Cl$_2$, H$_2$O → (191)

Scheme 70

(194) — (MeO)$_2$SO$_2$ → (195) — N$_2$H$_4$ → (196)

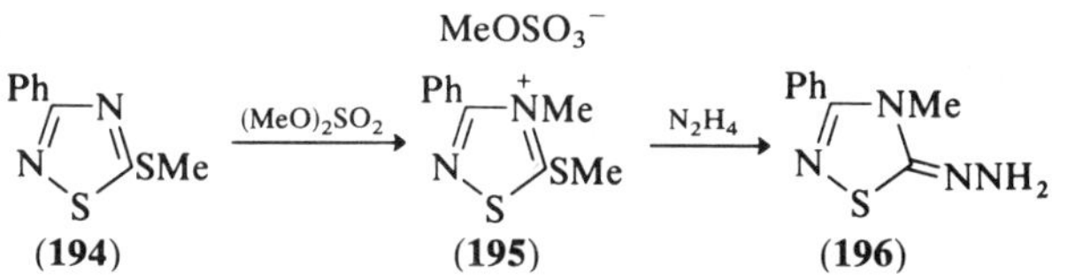

Scheme 71

Mercapto-1,2,4-thiadiazole (**197**) is readily oxidized to the corresponding disulfide (**198**) which can be reduced back to the starting thiol by sodium amalgam in alcohol. The disulfide link can also be cleaved by metallic mercury to give the corresponding mercaptide (**199**). The latter product on reaction with sodium sulfide forms an unstable sodium salt (**200**) whose presence was confirmed by various transformations shown in Scheme 72 ⟨65AHC(5)119⟩. Thus, treatment of (**200**) with hydrobromic acid presumably results in the formation of an intermediate (**201**) which undergoes attack on the ring sulfur by the soft nucleophile bromide ion to eventually produce the dithiazole (**202**). The 5-chlorine of (**200**) is readily replaced by sulfide to yield the dithiol (**203**) but treatment with alkali produces a mixture of (**204**) and (**205**). Cyanogen bromide converts (**204**) and (**200**) into the corresponding thiocyanates (**206**) and (**207**), respectively.

The IR spectrum in Nujol of 1,2,4-thiadiazolidine-3,5-dithione indicates that the dithione structure (**18a**) predominates. However, the dithione can be readily dialkylated on sulfur. Thus, treatment with 2,3-dichloropropene in aqueous sodium hydroxide forms (**208**; Scheme 73) ⟨65AHC(5)119⟩. The disulfenamide (**209**) condenses with aldehydes and ketones to give the corresponding imines (**210**) and (**211**; Scheme 74) ⟨74ZOB2553⟩. 5-Mercapto-3-methylthio-1,2,4-thiadiazole (**212**) reacts with di-n-butyltin chloride in THF to give di-n-butyltin derivative (**213**; Scheme 75) ⟨72USP3634442⟩.

4.25.3.3.7 Halogen atoms

(i) Nucleophilic substitution reactions

Nucleophilic displacements on the sp^2 carbon atom at C-5 of 1,2,4-thiadiazole (**155**; X = Cl) are plentiful. Chlorine, for example, may be displaced by fluoride, hydroxy, amino,

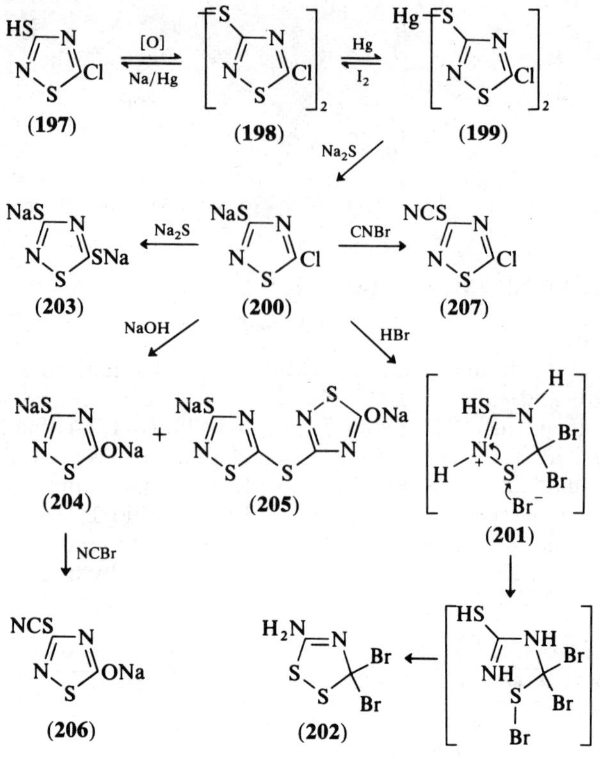

Scheme 72

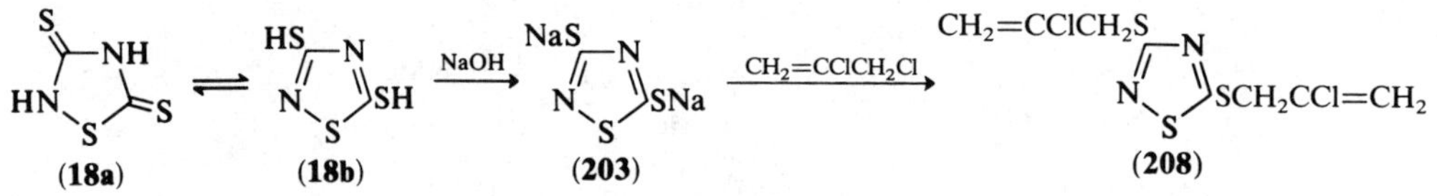

Scheme 73

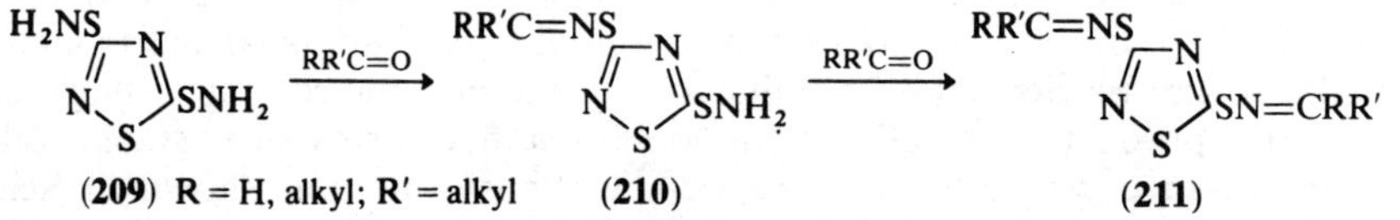

Scheme 74

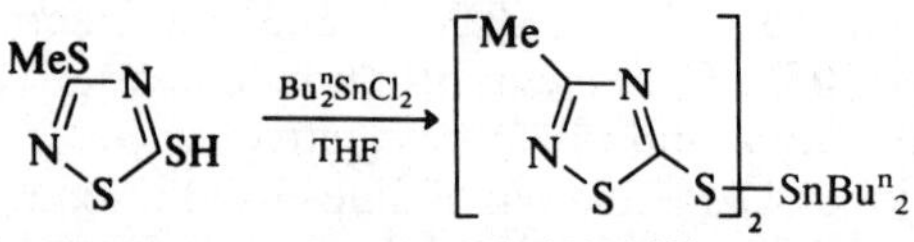

Scheme 75

mercapto, hydrazino, sulfonamido, sulfinato and azido groups (Scheme 76) ⟨65AHC(5)119⟩ to produce (**214**), (**154**), (**215**), (**17**), (**216**), (**217**), (**218**) and (**219**), respectively. Active methylene anions also displace the 5-halogen substituent; for example, the 5-chloro derivative (**220**) reacts with the sodium salt of malonic ester to give (**221**) which is decarboxylated to (**222**) with 75% H_2SO_4 (Scheme 77). The nucleophilic displacement reactions follow second order kinetics, the rate-determining step being the addition at C-5 of the nucleophile, followed by rapid elimination of halogen ⟨65AHC(5)119⟩. In contrast, 3-halo-1,2,4-thiadiazoles (**223**) are notoriously inert to nucleophilic displacement. Displacement with the alkoxide group has been reported (Scheme 78) to yield (**224**) ⟨65AHC(5)119⟩.

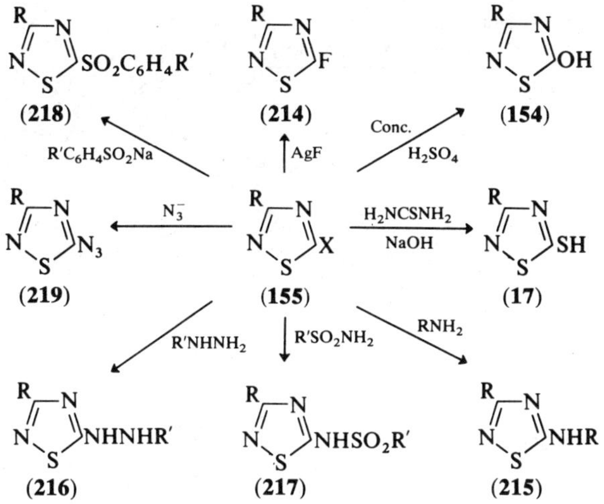

Scheme 76

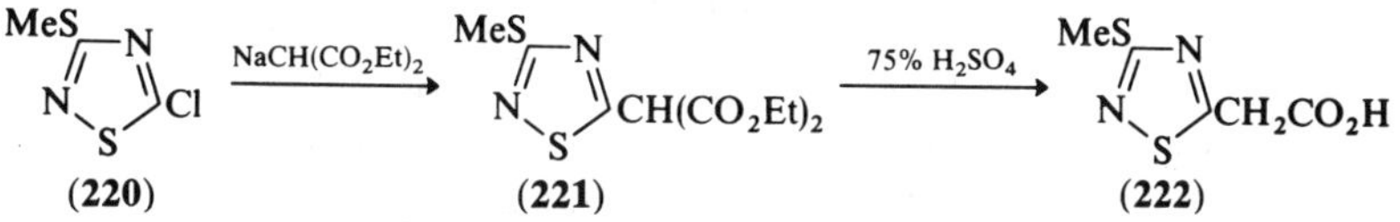

Scheme 77

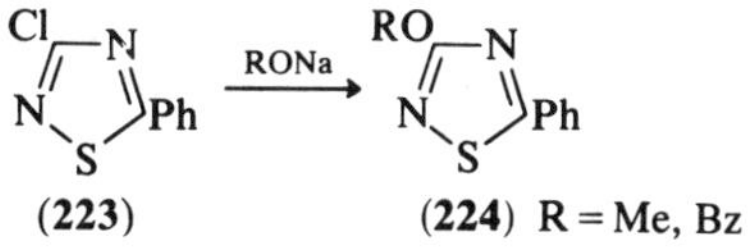

Scheme 78

(ii) Other reactions

5-Bromo-1,2,4-thiadiazole (**225**) is reduced to the parent heterocycle by H_2 and Raney nickel (Scheme 79) ⟨56CB1534⟩.

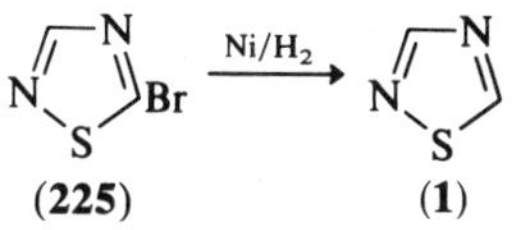

Scheme 79

4.25.3.3.8 Metals and metalloid-linked substituents

Metalation of the 1,2,4-thiadiazole nucleus is not known to date, presumably because of its instability to base.

4.25.3.3.9 Fused heterocyclic rings

Dilute hydrochloric acid in ethanol cleaves the thiadiazoline ring of (**226**) yielding the carbamate (**227**) as indicated in Scheme 80. The 5,5-ring fused systems (**228**) and (**229**) are completely stable under the above reaction conditions ⟨75JCS(P1)375⟩. The 5,6-ring fused system (**230**) is quite sensitive to acid or alkali like the parent 1,2,4-thiadiazole. Thus, compound (**230**) on treatment with aqueous sodium hydroxide solution is transformed into 2-cyanoaminopyridine (**231**). Catalytic hydrogenation of (**230**) results in cleavage of the S—N bond leading to the thiourea derivative (**232**) from which (**230**) is formed by oxidation (Scheme 81) ⟨75OPP55⟩.

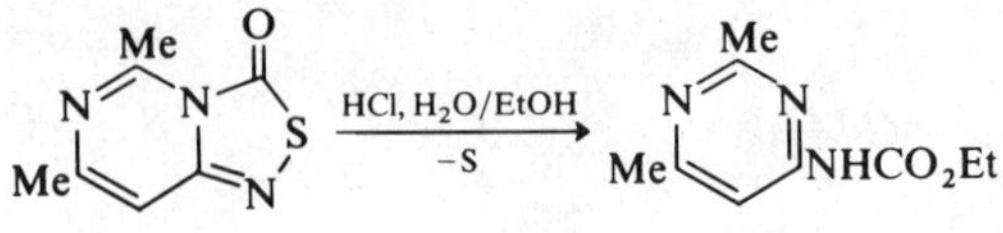

(226) →[HCl, H₂O/EtOH, −S] (227)

Scheme 80

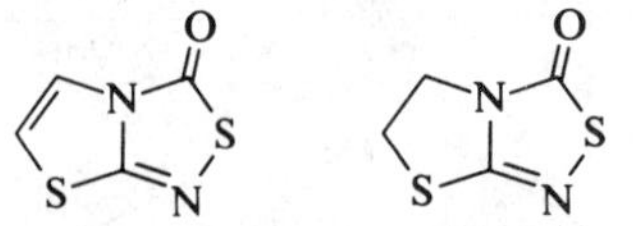

(228) (229)

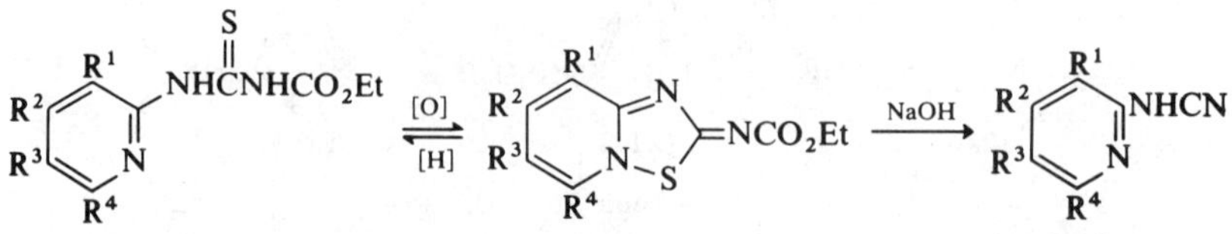

(232) R = H, Me, CO₂Et ⇌[[O]/[H]] (230) →[NaOH] (231)

Scheme 81

3*H*-1,2,4-Thiadiazolopyrimidine (**233**) undergoes a Dimroth-type rearrangement in 10% ethanolic HCl or sodium hydroxide solutions forming 2*H*-1,2,4-thiadiazolopyrimidine (**234**; Scheme 82). In contrast, substituted 3*H*-1,2,4-thiadiazolopyrimidine (**235**) does not give the rearranged product (**236**), but undergoes cleavage to the 3-acetyl derivative (**237**) on treatment with 10% ethanolic HCl, and to (**238**) on treatment with alkali. Reaction of (**238**) with POCl₃ under reflux for 1 h gives the cyclized product (**236**). Interestingly, the NMR spectrum of (**236**) shows only two signals for the four methyl groups and a single signal for the two aromatic protons, suggesting that (**236**) may be best represented as a resonance hybrid of the hypervalent sulfur compound (**236b**; Scheme 83) ⟨74JOC3783⟩. The substituted 3*H*-1,2,4-thiadiazolopyridine system (**239**) is stable to hot dilute acid or alkali, 30% H₂O₂ and sodium metaperiodate. Dienophiles such as maleic anhydride and dimethyl acetylenedicarboxylate do not add to it and quaternization with methyl iodide is slow, forming stable colored salts (**240**) in 85–95% yield (Scheme 84). Bromination with bromine in acetic acid results in halogenation of the 3-pyridyl substituent to produce (**241**) ⟨70JOC1965⟩.

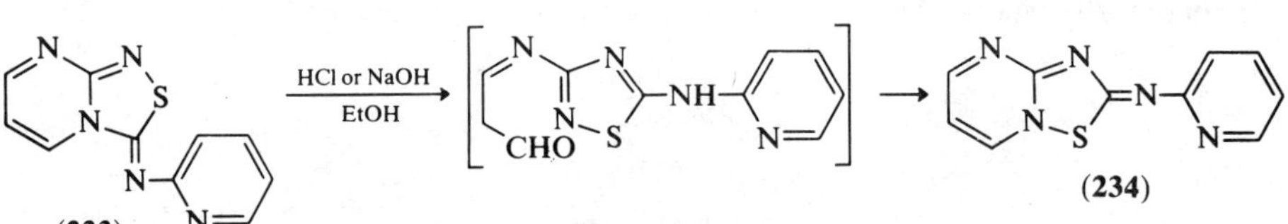

(233) →[HCl or NaOH, EtOH] [...] → (234)

Scheme 82

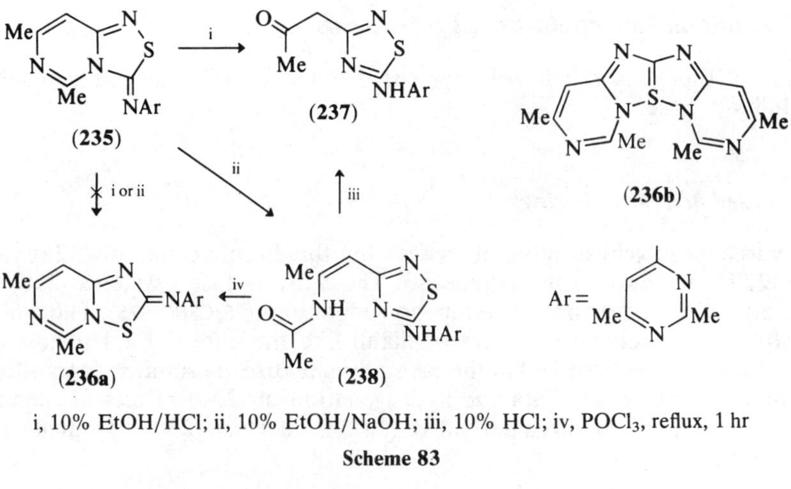

(235) →[i] (237)

(236b)

(235) ↛[i or ii], →[ii] (238) →[iii] (237); (238) →[iv] (236a)

Ar = 4,6-dimethylpyrimidin-2-yl (Me–pyrimidine–Me)

i, 10% EtOH/HCl; ii, 10% EtOH/NaOH; iii, 10% HCl; iv, POCl₃, reflux, 1 hr

Scheme 83

(241) (239) R = Me, H, Cl, Br (240)

Scheme 84

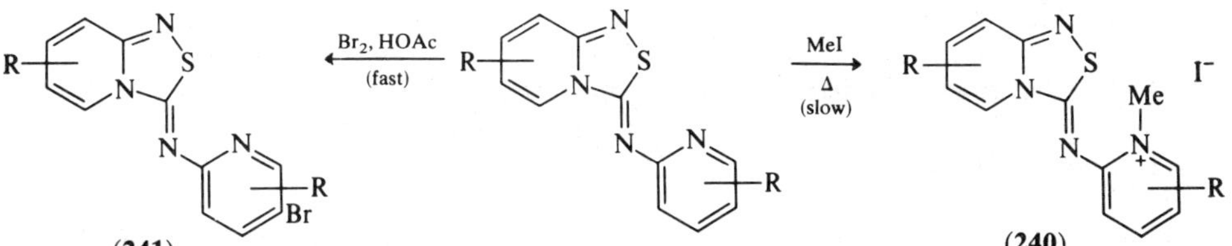

3*H*-1,2,4-Thiadiazolopyridines (**242**) and (**243**) are stable to acid hydrolysis and form unstable salts when reacted with methyl iodide ⟨71JOC1846⟩. The 3*H*-isoxazolothiadiazole system (**244**) is unstable to light, 10% HCl and P_4S_{10} treatment (Scheme 85). Similarly, acid hydrolysis of the 3*H*-[1,3,4]thiadiazolo[1,2,4]thiadiazole system (**245**) gave 3-nitroaniline, 2-amino-5-methyl-1,3,4-thiadiazole (**246**) and sulfur (Scheme 86) ⟨75JOC2600⟩. Oxidation of the 6*H*-[1,3,4]thiadiazolo[3,2-*b*][1,2,4]thiadiazol-6-one system (**247**) with *m*-chloroperbenzoic acid gives the sulfone (**248**) in 22% yield (Scheme 87). Chlorine in (**249**) can be displaced by a methylthio group to give (**250**) which on oxidation with peracetic acid gives sulfone (**251**). Alkaline hydrolysis of (**250**) gives 3-amino-6-methylthiopyridazine (**252**; Scheme 88) ⟨73JOC1575⟩.

(242) (243)

(244)

Scheme 85

(245) (246)

Scheme 86

(247) (248)

Scheme 87

(249) (250) (251) (252)

Scheme 88

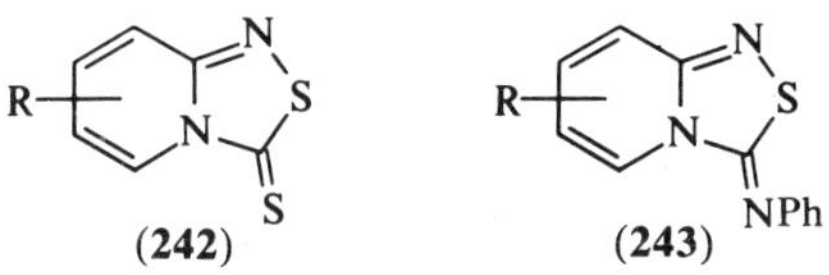

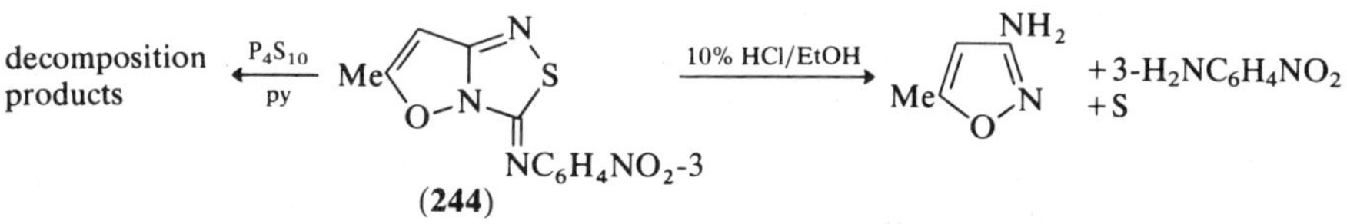

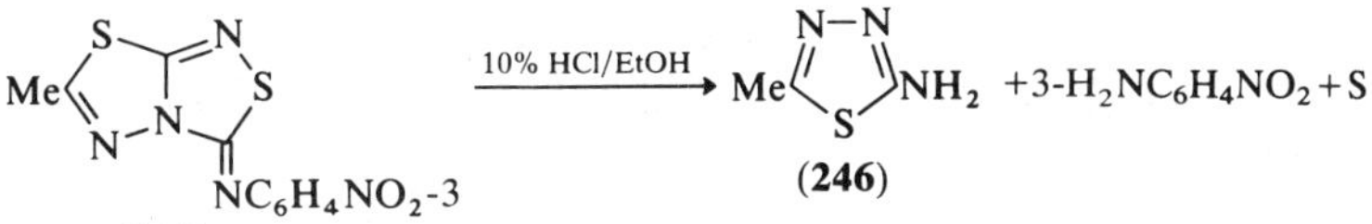

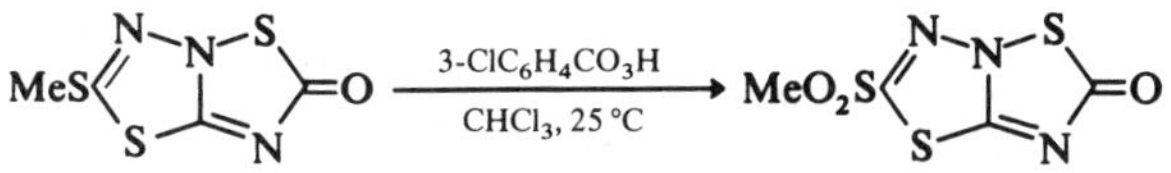

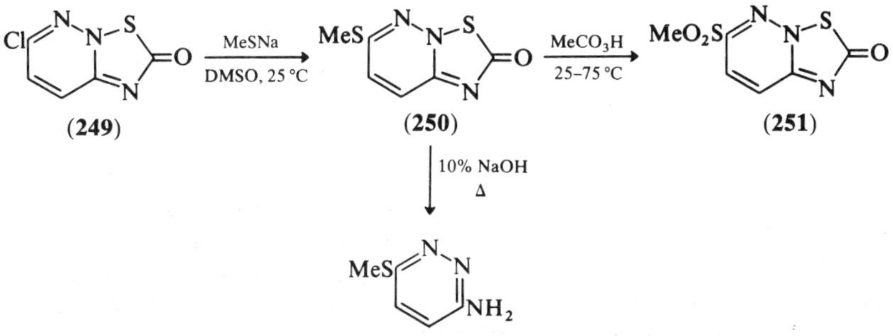

4.25.4 SYNTHESIS

4.25.4.1 Ring Syntheses from Non-heterocyclic Compounds

4.25.4.1.1 Introduction

Although the first 1,2,4-thiadiazole derivative was prepared by Wohler in 1821, the parent compound was not reported until 1955 ⟨56CB1534⟩. In general, 1,2,4-thiadiazole derivatives are prepared by appropriate intra- or inter-molecular ring closure reactions and the substituents are then modified as required. Seven potential ring closure routes (denoted by dotted lines), which are designated by the types A to G, are discussed in this chapter. Presumed intermediates in these reactions are not considered sufficient criteria for assignment of reaction type unless they have been isolated and their conversion to 1,2,4-thiadiazoles demonstrated. For this reason a few synthetic routes assigned to specific classes in earlier reviews have been reassigned.

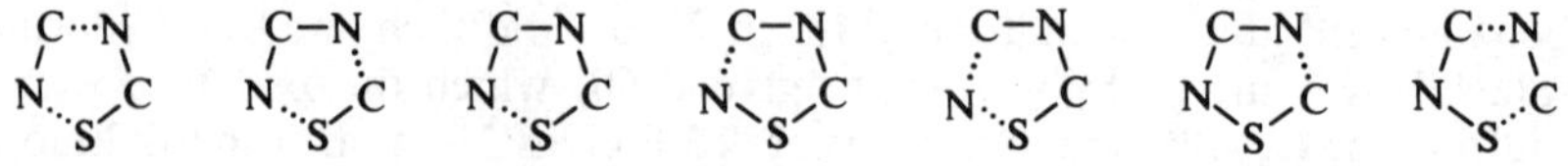

4.25.4.1.2 Type A syntheses

Many thioamides (**253**) are converted into 1,2,4-thiadiazoles (**255**) on treatment with a variety of oxidizing agents including halogens, hydrogen peroxide, sulfur monochloride, thionyl chloride, sulfuryl chloride, ammonium persulfate, nitrous acid, ozone, N-chloromethylcarbamates, phosphorus pentachloride, 2-bromoindane-1,3-dione and N-sulfinylsulfonamides ⟨65AHC(5)119⟩. The yields of the 1,2,4-thiadiazoles obtained are widely variable, depending on the thioamide, the oxidizing agent and the conditions used (Scheme 89). The mechanisms of these reactions are not known in most cases although initial oxidation at the sulfur atom to an intermediate (**254**) is usually implied. Thioamide S-oxides (**254**; OX = O) have been isolated in good yields from reactions utilizing hydrogen peroxide as oxidant but their role in the formation of 1,2,4-thiadiazoles is open to question. As indicated in Scheme 90, a mechanism in which the S-oxide (**257**) functions as an imidoacylating agent of thiobenzamide (**256**) to form (**258**) has been proposed ⟨60LA(633)49⟩ but other pathways have not been excluded.

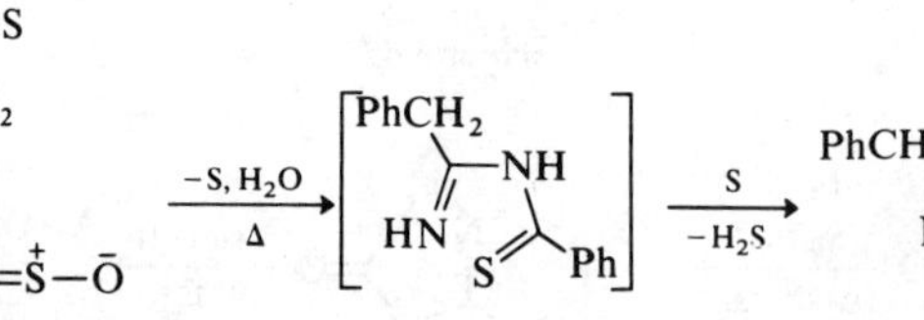

R = alkyl, aryl, heterocyclic, NHCOMe, OR'

Scheme 89

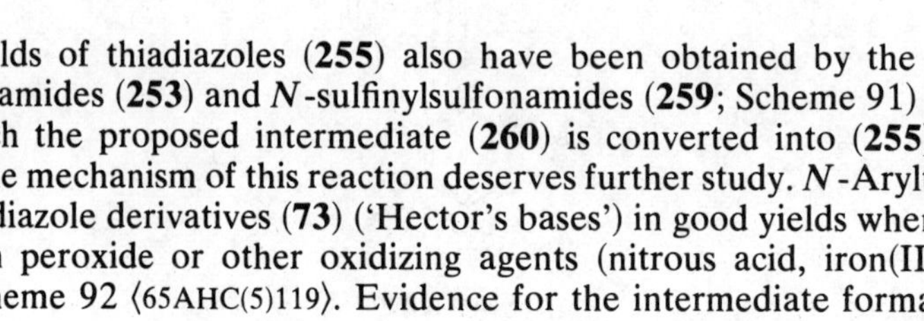

Scheme 90

Moderate yields of thiadiazoles (**255**) also have been obtained by the thermolysis of mixtures of thioamides (**253**) and N-sulfinylsulfonamides (**259**; Scheme 91) ⟨62AG135⟩. The manner in which the proposed intermediate (**260**) is converted into (**255**) has not been discussed and the mechanism of this reaction deserves further study. N-Arylthioureas (**261**) form 1,2,4-thiadiazole derivatives (**73**) ('Hector's bases') in good yields when oxidized with acidic hydrogen peroxide or other oxidizing agents (nitrous acid, iron(III) chloride) as indicated in Scheme 92 ⟨65AHC(5)119⟩. Evidence for the intermediate formation of dithio-

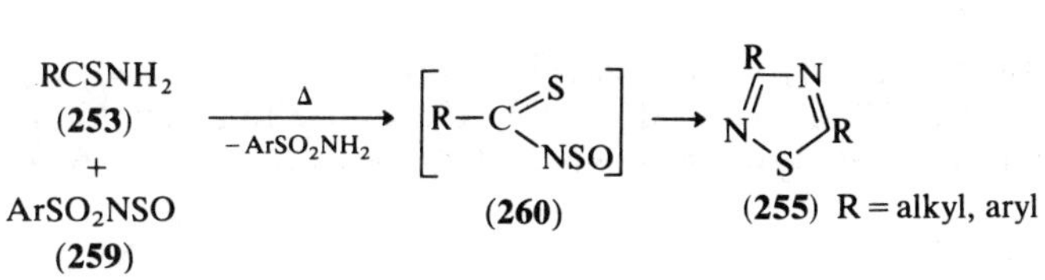

1,2,4-Thiadiazoles

Scheme 91

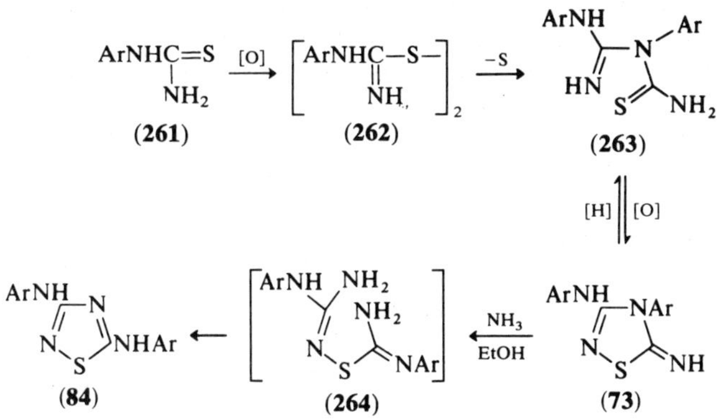

Scheme 92

formamidines (**262**) and N^1-aryl-N^3-arylamidinothioureas (**263**) in this conversion has been reported.

The 'Hector's bases' (**73**) undergo a Dimroth rearrangement, presumably via (**264**), to the symmetrical 'Dost's bases' (**84**) when heated under alkaline conditions. Oxidation of N-alkyl-N-arylthioureas affords the symmetrical thiadiazoles (**265**) directly whereas the products obtained from N,N'-disubstituted thioureas (**266**) depend on the nature of the substituents ⟨80JCR(S)407⟩.

Oxidation of (**266**; Ar = Ph) with nitrous acid produces the thiadiazole (**86**) which can be isolated when R is alkyl. Treatment of (**86**) with acid produces 'Hugerschoff bases' (**87**; Scheme 93). When R of (**266**) is aryl, isomerization is rapid and only (**87**) is obtained when acidic peroxide is used as the oxidant ⟨65AHC(5)119⟩. Recently the reaction has been studied using mixtures of thiourea and N,N'-disubstituted thioureas. Products of type (**267**) and (**268**) were isolated from N-aryl-N'-alkylthioureas but only (**268**) was obtained from the N-aryl-N'-aryl and N-alkyl-N'-alkyl derivatives (Scheme 94) ⟨80IJC(B)667, 75IJC(B)241⟩.

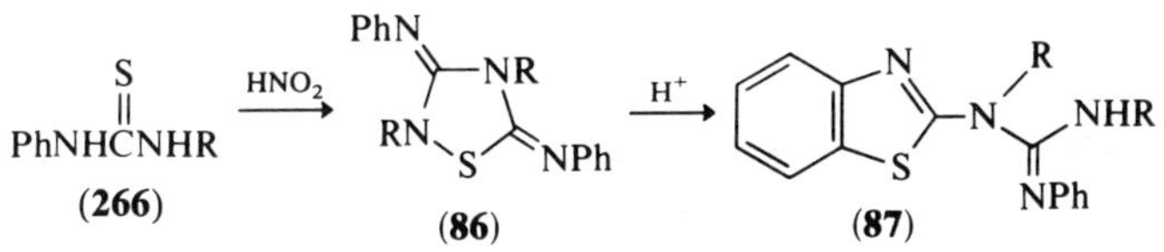

Scheme 93

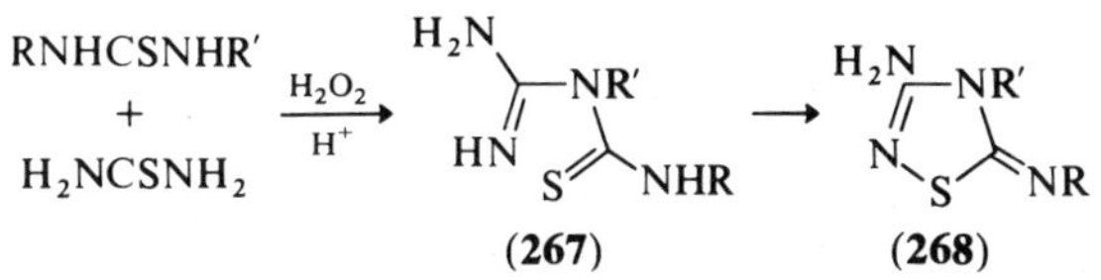

Scheme 94

The chlorination of mixtures of isothiocyanates and isocyanates produces sparingly soluble 3-oxathiadiazolium salts (**270**) via the intermediate iminochloromethylsulfenyl chloride derivatives (**269**; Scheme 95) ⟨70AG(E)62⟩. The salts (**270**) can easily be converted into derivatives (**271**), (**272**) and (**273**) as indicated in Scheme 95. When R is methoxymethyl, compound (**270**) spontaneously forms (**274**) in good yields ⟨76JHC169⟩. Cyanates also react with iminochloromethylsulfenyl chlorides (**269**) to produce thiadiazolium salts (**275**) which undergo hydrolysis and aminolysis to produce (**276**) and (**277**), respectively ⟨70AG(E)62⟩.

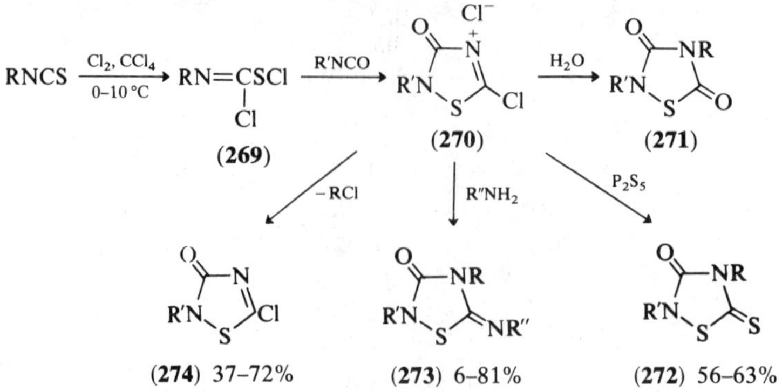

Scheme 95

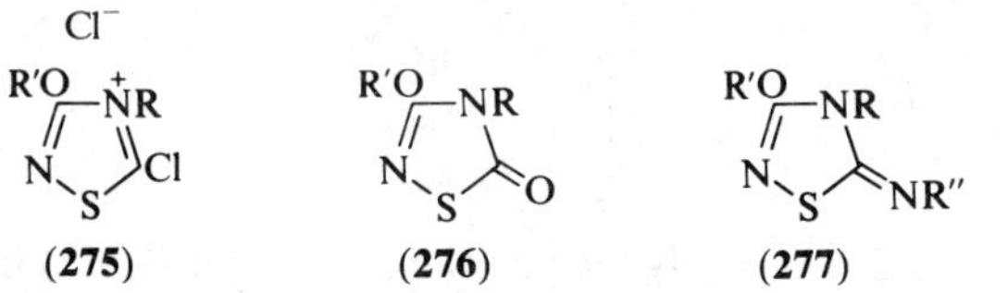

4.25.4.1.3 Type B syntheses

Amidoximes (**278**) yield 5-mercapto-1,2,4-thiadiazoles (**279**) when treated with carbon disulfide whereas the 5-amino derivatives (**280**) are obtained from isothiocyanates, as indicated in Scheme 96 ⟨65AHC(5)119⟩. Products of type (**280**) are also produced in the reaction between N-sulfenylamidines (**281**) and isothiocyanates (Scheme 97).

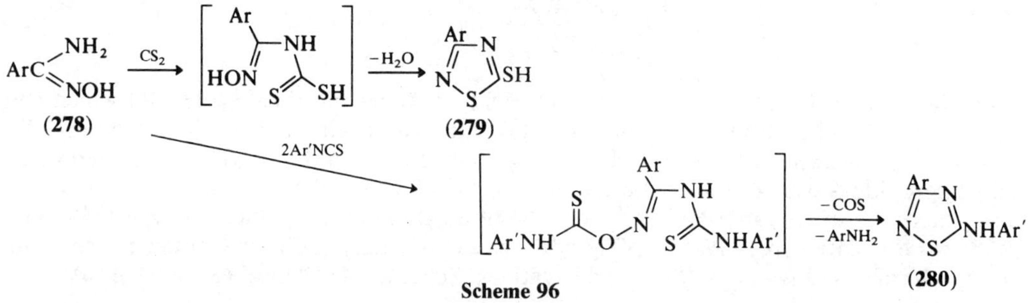

Scheme 96

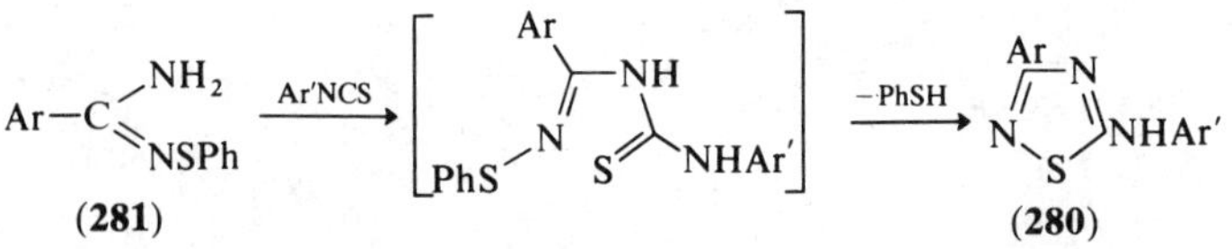

Scheme 97

A wide variety of reagents have been used to convert amidines (**282**) into 1,2,4-thiadiazoles. Treatment of (**282**) with sodium hypochlorite followed by potassium thiocyanate at pH 3 produces 5-amino derivatives (**284**) in moderate to good yields (Scheme 98). In one example an unstable intermediate (**283**; R = OMe, R' = H) was isolated but not fully characterized ⟨65AHC(5)119⟩. Products of type (**285**) are obtained when amidines are treated with iminochloromethylsulfenyl chlorides (**269**; Scheme 99) ⟨71T4117⟩. When R' is aryl (**285**) is the exclusive product whereas a mixture of isomers is obtained when R' is aroyl. 5-Amino-1,2,4-thiadiazoles related to (**284**) and (**285**) also are formed in good yields when N-haloamidines are treated with isothiocyanates. Thus, 5-methylamino-3-trichloromethyl-1,2,4-thiadiazole (**287**) is obtained in 73% yield from the reaction of N-bromotrichloroacetamidine (**286**) with methyl isothiocyanate (Scheme 100) ⟨78USP4107377⟩.

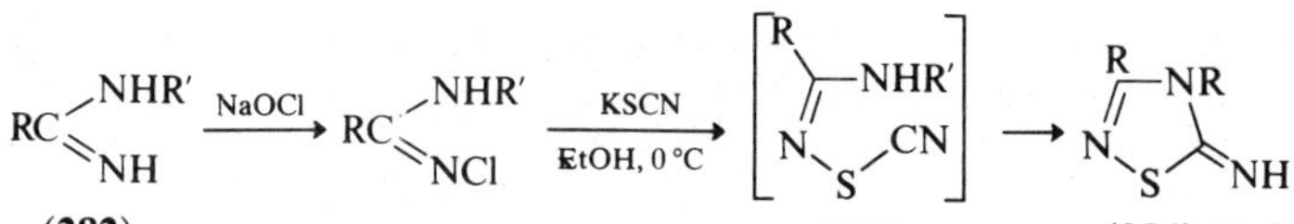

(282) (283) (284) 40–90%

R = alkyl, aryl, OR″, SR″, NR''_2, H; R′ = alkyl, aryl, H

Scheme 98

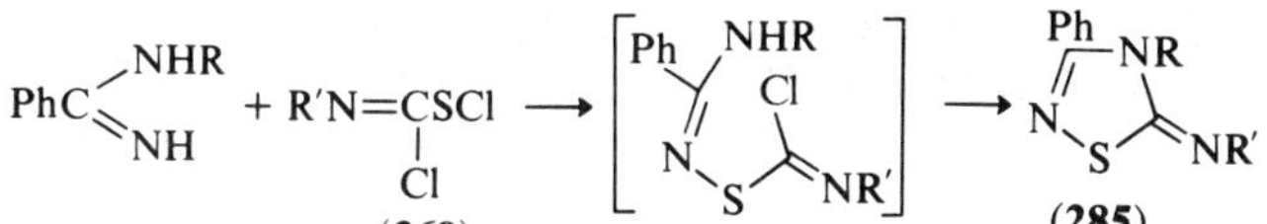

(269) (285)

R = alkyl, aryl; R′ = aryl, ArCO

Scheme 99

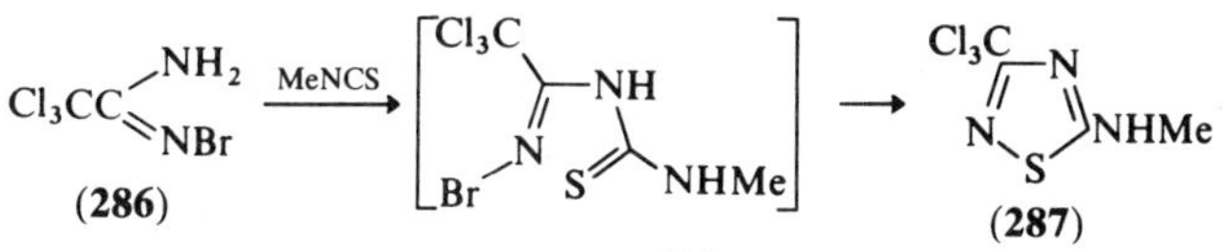

(286) (287)

Scheme 100

Amidines (**282**; R′ = H) also condense with carbon disulfide in the presence of sulfur to produce 5-mercapto-1,2,4-thiadiazoles (**17**) in reasonable yields, as indicated in Scheme 101 ⟨68BRP1116198⟩. A general method for the synthesis of 5-chloro-1,2,4-thiadiazoles (**289**; X = Cl) involves the reaction of amidines with trichloromethylsulfenyl chloride (**288**; X = Cl) in the presence of a base under mild conditions (Scheme 102) ⟨65AHC(5)119⟩. A variety of 3-substituted thiadiazoles (**118**) are obtained when dichloromethylsulfenyl chloride (**288**; X = H) is used in this reaction.

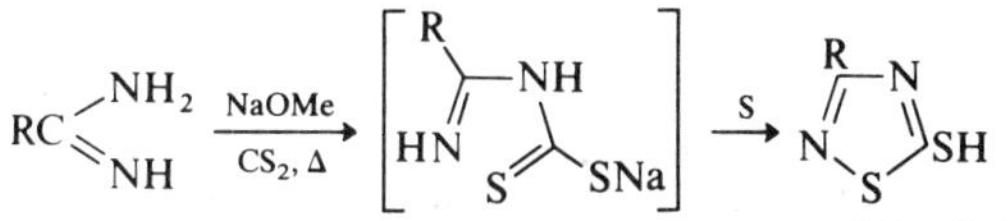

(17) 40–75%

R = MeS, Ph, 4-$O_2NC_6H_4$, 3-MeC_6H_4, 4-MeC_6H_4

Scheme 101

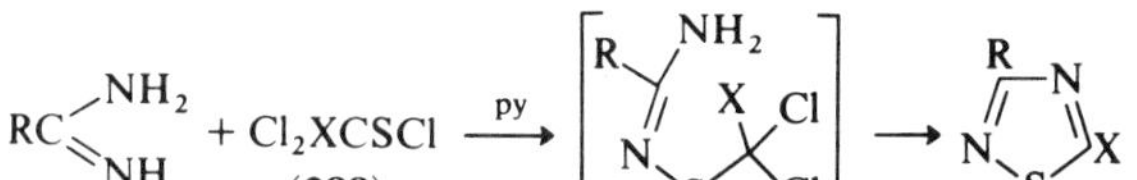

(288)

R = alkyl, haloalkyl, aryl, SR′; X = H, Cl (289) 35–70%

Scheme 102

Bicyclic 1,2,4-thiadiazolones (**293**) are formed in moderate yields when cyclic amidines (**290**) (α-amino heterocycles) are allowed to react with chlorocarbonylsulfenyl chloride (**291**) in the presence of an acid acceptor (Scheme 103) ⟨73JOC1575⟩.

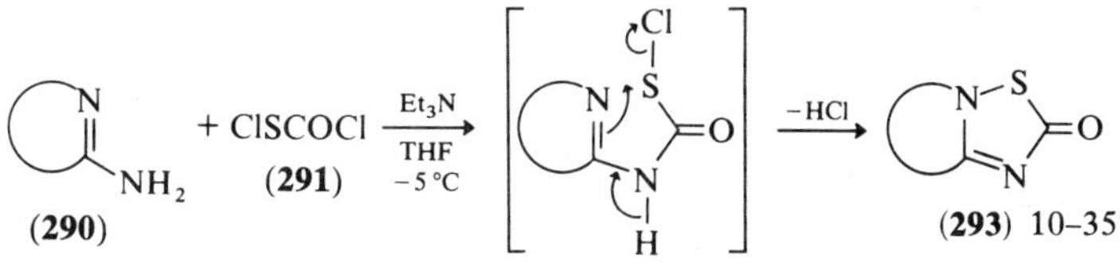

(290) (291) (292) (293) 10–35%

Scheme 103

Heterocycles (**290**) which were found to undergo this reaction include 2-aminopyridines, 3-aminopyridazines and 2-amino-5,6-dihydro-4*H*-1,3-thiazine. Sulfenyl chlorides of type (**292**) were postulated as intermediates but were not isolated or detected. By contrast, it also has been reported ⟨75JCS375⟩ that some closely related heterocycles (**290**) react with the same reagent to yield thiadiazolones of type (**295**; Scheme **104**). Heterocycles (**290**) which undergo this reaction include 2-aminothiazole, 2-aminopyrimidine, 4-amino-2,6-dimethylpyrimidine and 2-amino-4,5-dihydrothiazole. It was postulated that the formation of intermediate (**294**) would be favored relative to (**292**) and that the degree of hydrogen bonding by the solvents (chloroform *vs.* THF) determines whether (**293**) or (**295**) is formed. It is interesting, however, that reverse modes of addition were used in the formation of (**293**) and (**295**). The slow addition of (**290**) to (**291**) would be expected to produce (**293**) *via* (**292**) but the reverse order of addition, used to prepare (**295**), may initially result in the formation of a bis derivative (**296**) which could undergo base-catalyzed cyclization to (**295**; Scheme 105). In this case the initial intermediate would again be (**292**) and not (**294**). The formation of imine derivatives of (**295**) is discussed under type F syntheses.

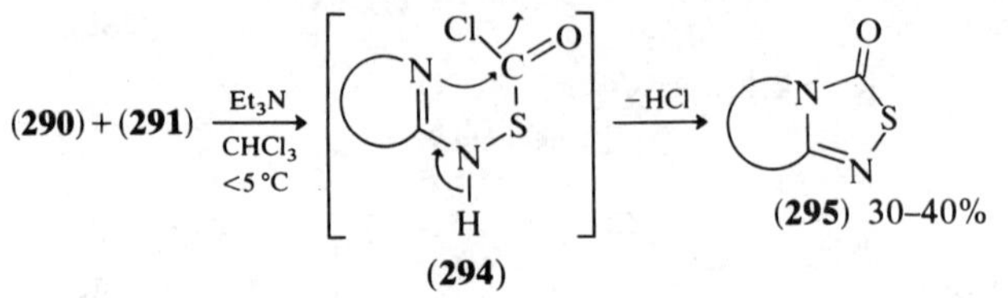

Scheme 104

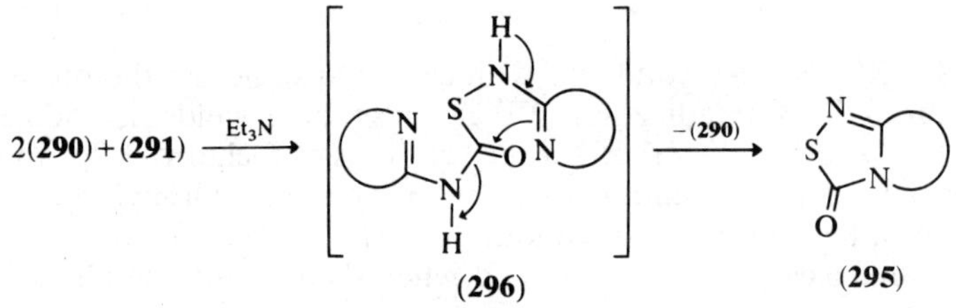

Scheme 105

Chlorocarbonylsulfenyl chloride (**291**) also condenses with a variety of ureas, thioureas, guanidines, carbodiimides and cyanamides to produce 1,2,4-thiadiazol-2-one derivatives of types (**298**), (**297**) and (**299**) as indicated in Scheme 106 ⟨70AG(E)54, 73CB3391⟩.

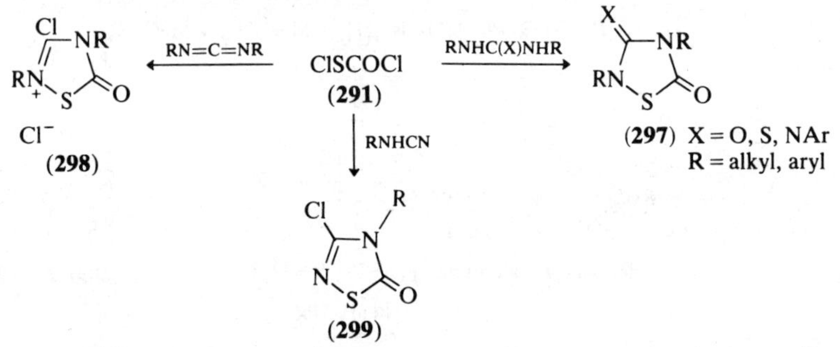

(**297**) X = O, S, NAr
R = alkyl, aryl

Scheme 106

4.25.4.1.4 Type C syntheses

A very general method for the synthesis of 1,2,4-thiadiazoles (**301**) in reasonable yields is based on the oxidation of thioacylamidines (**300**; Scheme 107) ⟨65AHC(5)119⟩. Typical oxidizing agents include bromine, iodine, nitric acid, acidic hydrogen peroxide and arylsulfonyl halides in the presence of pyridine. Products of type (**303**) are made in a similar manner from thiocarbamylguanidines (**302**; Scheme 108). The thioacylguanidines and thioacylamidines have been prepared by a variety of procedures. Thus, acylguanidines (**304**) have been sulfurized with phosphorus pentasulfide to the thioacyl derivatives (**305**) which on oxidation afford 3-amino-1,2,4-thiadiazoles (**306**) in variable yields (Scheme 109) ⟨70CB1805⟩. Acyl- and sulfonylthioacyl-guanidines (**307**) are prepared by acylation of the appropriate guanidine with a thioester (Scheme 110). Oxidation of (**307**) with hydrogen peroxide in pyridine or with bromine produces the products of type (**308**) ⟨75LA1961⟩. The

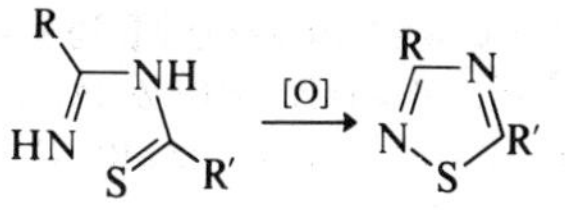

Scheme 107

R = alkyl, aryl, NHR″, NHAr, NHCOAr, NHSO$_2$R″, NHSO$_2$Ar, NR″$_2$, OR″, SR″, NHN=CR″$_2$, NHNO$_2$, NH$_2$
R′ = alkyl, aryl, NHR″, NHAr, NR″$_2$, OR″, NH$_2$

Scheme 108

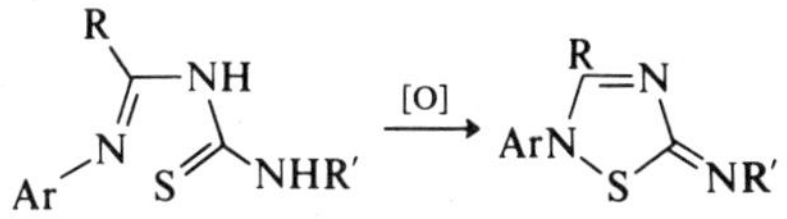

(302) R = NHAr, NR″$_2$; (303)
R′ = H, alkyl, aryl

Scheme 109

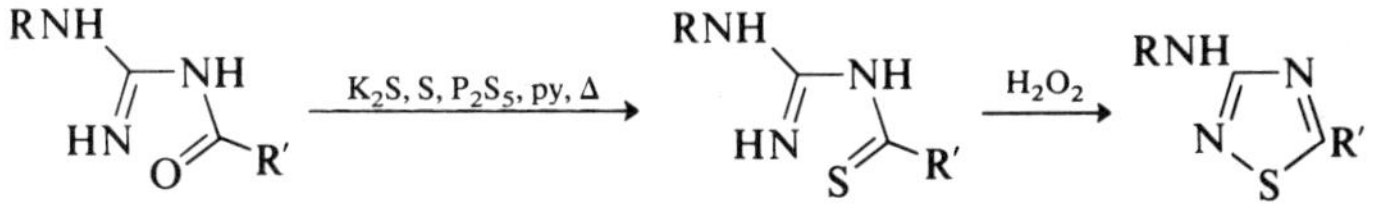

(304) R = H, Me; R′ = alkyl, aryl (305)   (306) 1–76%

Scheme 110

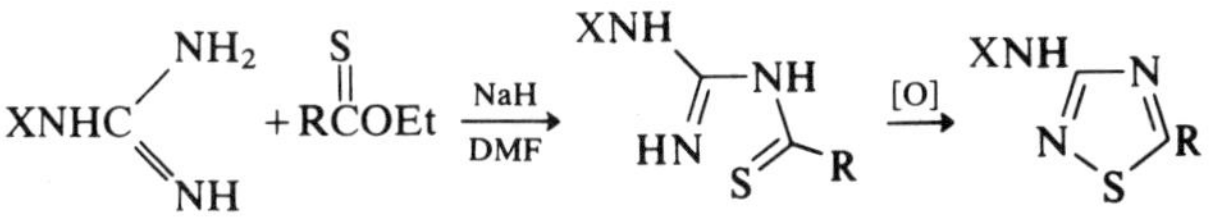

R = alkyl, aryl, ArCH$_2$; X = R′CO, R′SO$_2$ (307)   (308)

unsubstituted 3-amino derivatives (**308**; X = H) can be readily obtained in yields of 25–65% by basic hydrolysis of the corresponding *N*-acetyl-1,2,4-thiadiazoles.

Similar reactions have been carried out with amidines ⟨67NEP6610627⟩. Thus trichloroacetamidine (**309**) reacts with chlorothioformate esters to produce thioacyl products (**310**) which are readily oxidized to 5-alkoxy-3-trichloromethyl-1,2,4-thiadiazoles (**311**). The yields in both steps are about 80% (Scheme 111). The bromine oxidation of *N*-thiocarbamyl derivatives (**312**) of cyclic amidines (2-aminopyridine, 2-aminothiazole, 2-aminopyrimidine) yields thiadiazolium salts (**313**) ⟨71JPR1148⟩. In the 2-aminopyridine series, products of type (**230**; Scheme 81) are obtained in 20–73% yield when R is an ethoxycarbonyl group ⟨75OPP55⟩.

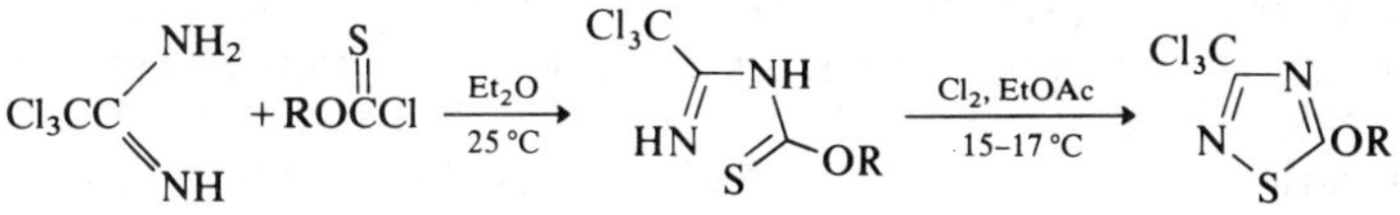

Scheme 111

(309)   (310)   (311)

(312)   (313)

Cyanodithioimidocarbonate salts have become very useful reagents for the preparation of many 3,5-disubstituted 1,2,4-thiadiazoles. Dipotassium cyanodithioimidocarbonate (**314**) is readily prepared from cyanamide and carbon disulfide under alkaline conditions and is readily converted into the reactive thiadiazoles (**315**), (**316**), (**317**), (**318**) and (**319**), as shown in Scheme 112 ⟨71JOC14⟩. Alkylation of (**314**) produces derivatives of type (**320**) which react with chlorine or sulfuryl chloride to yield 5-alkylthio-3-chloro-1,2,4-thiadiazoles

(**185**; X = Cl) ⟨67JOC1566⟩. The alkyl cyanodithioimidocarbonates (**320**) and the oxygen and nitrogen analogs (**322**; X = O, NR′) have also been prepared from cyanamide and esters of type (**321**; Scheme 113) ⟨79JHC961⟩.

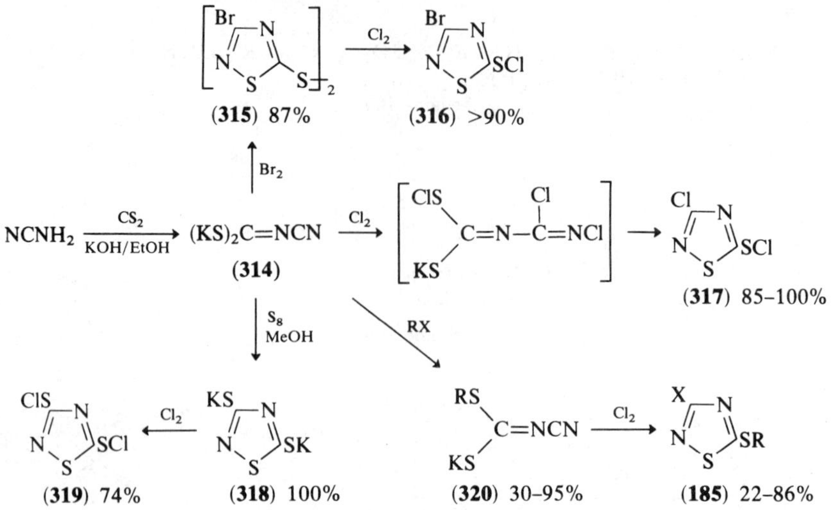

Scheme 112

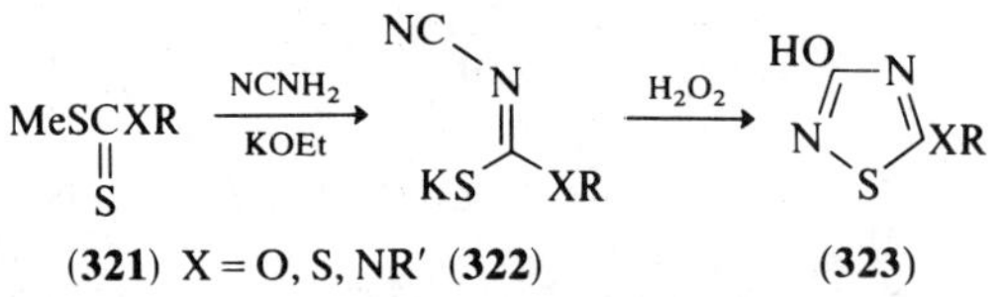

Scheme 113

4.25.4.1.5 Type D syntheses

A few examples previously assigned to this class on the basis of postulated intermediates have been reassigned to type E syntheses ⟨75BCJ310, 73CL917⟩. At the present time there are no authenticated examples of this type of ring closure.

4.25.4.1.6 Type E syntheses

N-Chloro derivatives of amidines, isoureas and guanidines react with potassium *S*-methyl cyanodithioiminocarbonate (**320**; R = Me) to yield thiadiazole systems of type (**324**; Scheme 114) ⟨75BCJ310; 73CL917⟩. A related transformation occurs when the cyanothioiminocarbonates (**322**) and chloramine are allowed to condense at low temperature ⟨76EGP119791⟩. In this manner a variety of 5-substituted 3-amino-1,2,4-thiadiazoles (**325**) are obtained in good yields (Scheme 115). The use of ethoxycarbonylthioiminocarbonates (**326**) in this

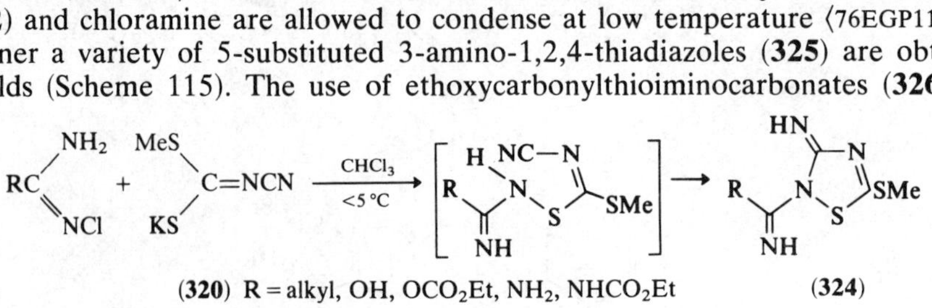

Scheme 114

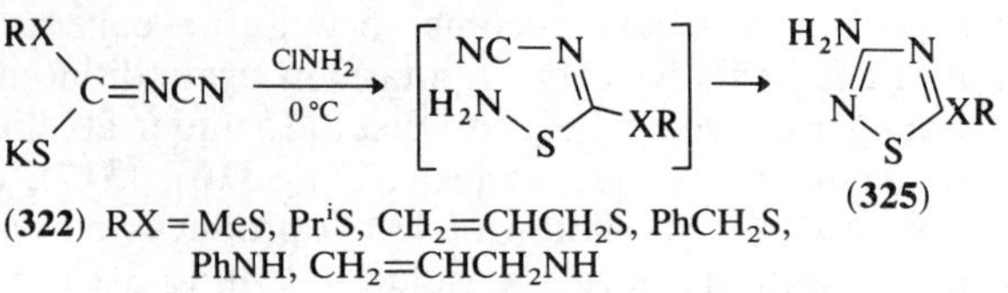

Scheme 115

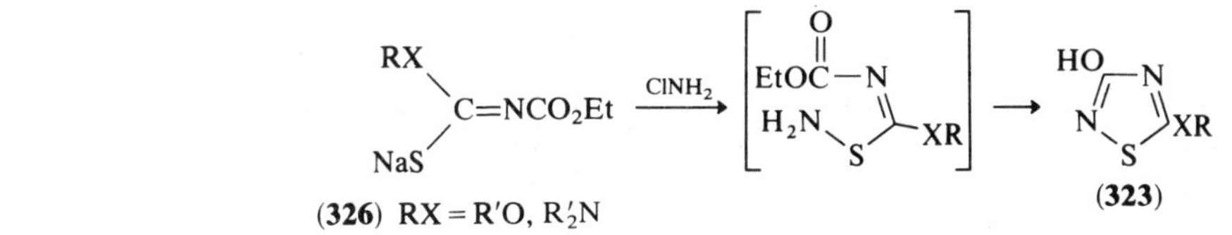

reaction results in the formation of the 3-hydroxy analogs (**323**; Scheme 116) ⟨79JHC961⟩. A third high yield synthesis of 1,2,4-thiadiazoles (**301**; Scheme 117), has been reported recently ⟨80JOC3750⟩.

The reaction of a thioamide or thiourea and an *N,N*-dimethylacylamide dimethyl acetal at room temperatures produces imino derivatives of type (**327**). The latter compounds readily condense with amino transfer reagents such as hydroxylamine-*O*-sulfonic acid and mesitylsulfonyloxyamine (MSH) to afford (**301**; R = H, Me; R′ = aryl, ArNH, 4-pyridyl) in excellent yields. An unusual reaction of the E category is the condensation of nitrosobenzene with thiobenzoyl isocyanate, which produces the thiadiazolone (**328**) in 58% yield (Scheme 118) ⟨73CB1496⟩. Presumably, nitrobenzene is another reaction product, although this has not been characterized.

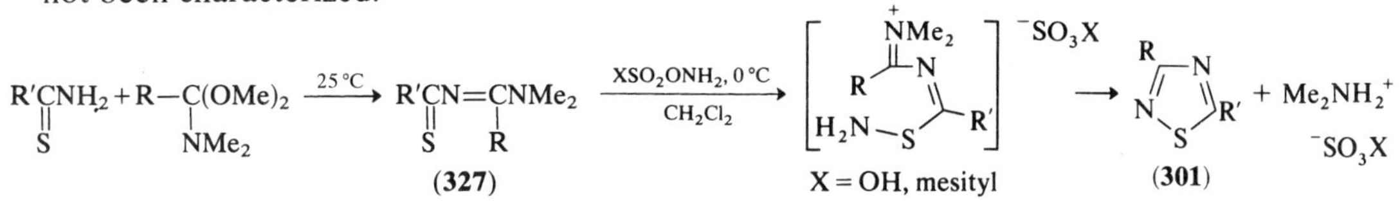

4.25.4.1.7 *Type F syntheses*

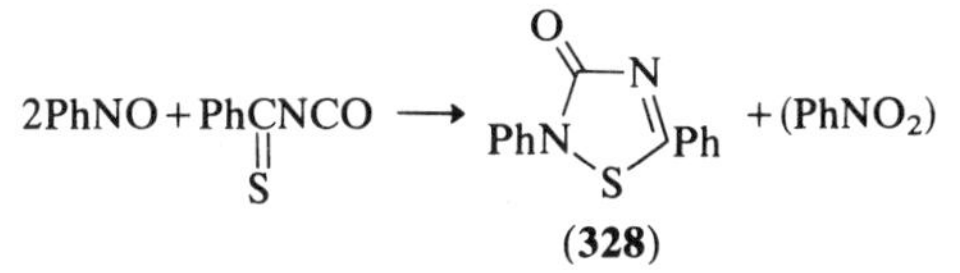

In a series of publications ⟨75JOC2600, 70JOC1965, 73JOC3087⟩, Potts and coworkers have reported that cyclic amidines (**290**) readily condense with trichloromethylsulfenyl chloride (**329**) to yield the sulfenamides (**330**; Scheme 119). Treatment of the latter compounds with aromatic amines in the presence of triethylamine results in cyclization, possibly via an intermediate such as (**331**), to produce bicyclic products of type (**332**). Heterocycles (**290**) which have been used successfully in this reaction include 2-amino-1,3,4-thiadiazoles, 3-aminopyridazines, 2-aminopyrimidines, 2-aminopyrazines, 2-aminopyridines, 3-aminoisoxazoles and 5-amino-1,2,4-thiadiazoles. The sulfenamide derivative (**330**) of 2-aminopyridine also was found to react with sodium sulfide and with diethyl malonate to produce (**333**) and (**334**) respectively. Attempts to hydrolyze (**332**) to (**295**) under acidic conditions failed.

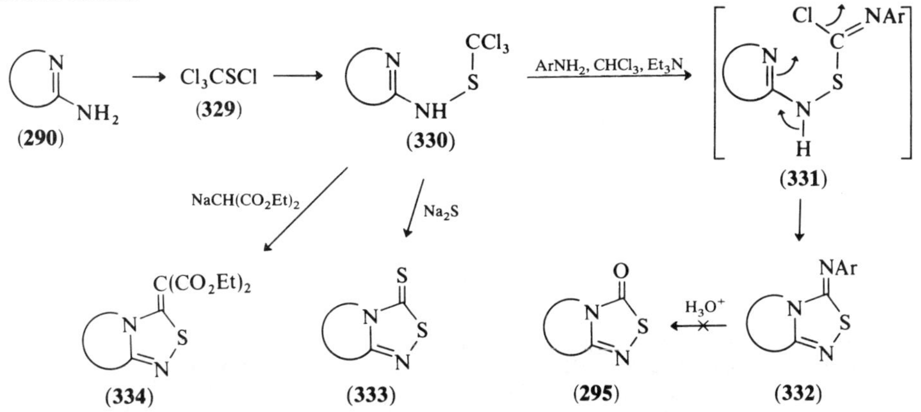

1,2,4-Thiadiazole *S,S*-dioxides (**336**) have been prepared by the cyclization of *N*-halomethylsulfonyl-amidines and -guanidines (**335**; Scheme 120) ⟨70JCS(C)1429⟩. *N*-chloromethylsulfonyl-isoureas and -isothioureas (**337**) cyclize in the presence of base and chloroformate esters to yield the related *S,S*-dioxides (**92**) which easily undergo chlorinolysis to produce (**93**; Scheme 121) (see also Scheme 45) ⟨74BSF1580⟩.

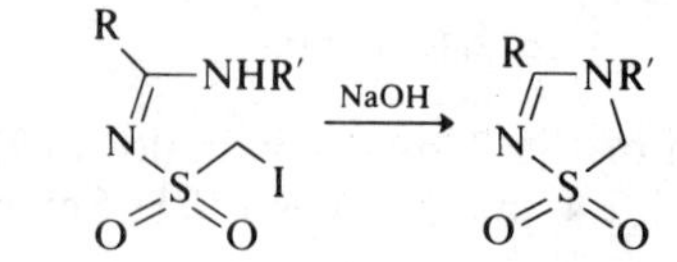

(**335**) R = Ar, R″NH; R′ = alkyl (**336**)

Scheme 120

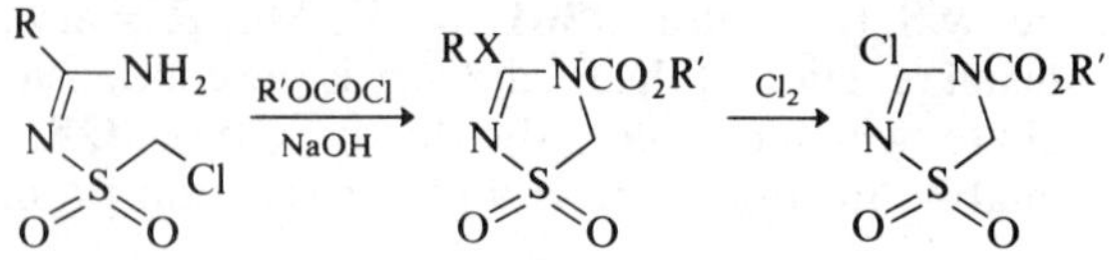

(**337**) X = O, S (**92**) (**93**)

Scheme 121

4.25.4.1.8 Type G syntheses

Since the discovery of nitrile sulfides (**339**) as reactive 1,3-dipoles ⟨70TL1381⟩, Franz and coworkers have trapped intermediates of this type with a variety of dipolarophiles ⟨78JOC3742, 78JOC3736⟩. The use of nitriles as acceptors in this reaction has resulted in a general method for the synthesis of 3,5-disubstituted 1,2,4-thiadiazoles (**301**) of unambiguous structure ⟨81JOC771, 77JOC1813, 74JOC962⟩. A side reaction which competes with 1,3-dipolar cycloaddition is the decomposition of (**339**) to the corresponding nitrile and sulfur (Scheme 122). Best yields of (**301**) are obtained when an electrophilic nitrile is used and/or the nitrile is employed in considerable excess. Thus, decomposition of the oxathiazolones (**338**; R = aryl) in a 35 molar excess of aryl nitriles at 190 °C produces a variety of 3,5-diaryl-1,2,4-thiadiazoles (**301**) in 16–74% yields. When ethyl cyanoacetate is used as the acceptor, high yields (76–94%) of (**301**; R′ = CO_2Et) are obtained at lower mole ratios of reactants (1 : 4) and at lower temperatures (130 °C). The reactions of (**339**) with tetracyanoethylene are especially facile and either mono adducts (**109**) (50–70% yields) or bis adducts (**112**) (47–80% yields) may be produced as major products (Scheme 49) depending on the mole ratios of reactants used ⟨76JOC620⟩. The bis adduct (**112**) is also formed in good yields when (**109**) is refluxed in anhydrous ethanol. A mechanism for this transformation was proposed. Nitrile oxides were found to react with tetracyanoethylene in a similar manner.

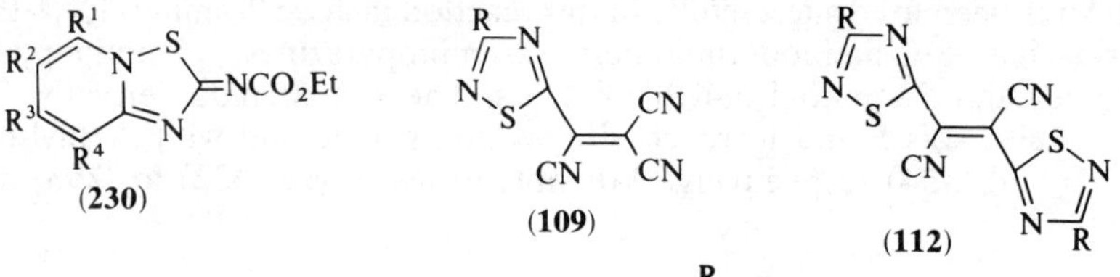

(**230**) (**109**) (**112**)

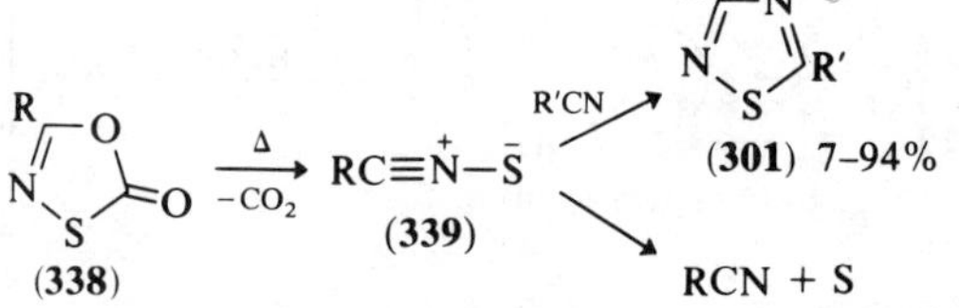

(**338**) (**339**) (**301**) 7–94%

R = aryl, alkyl, CO_2Et; R′ = aryl, CO_2Et, CH_2CO_2Et, C(CN)=C(CN)$_2$

Scheme 122

4.25.4.1.9 Miscellaneous synthetic methods

3,5-Diphenyl-1,2,4-thiadiazole (**55**) is formed in 25% yield when a mixture of benzonitrile, sulfur and hydrogen sulfide is heated under pressure at 190–215 °C. Since no reaction occurred in the absence of hydrogen sulfide, a type A synthesis involving the intermediate formation of thiobenzamide was suggested, as indicated in Scheme 123 ⟨62JOC869⟩. It was reported later, however, that this thiadiazole is formed in 75% yield when the reaction is carried out at 250 °C in the absence of hydrogen sulfide but in the presence of a catalytic amount of trioctylamine ⟨67AG(E)1084⟩. A mechanism consistent with this observation is indicated in Scheme 124. It is also possible that a type G synthesis occurs through the intermediate formation of benzonitrile sulfide. Apparently, little effort has been made to develop this simple procedure as a general method for the synthesis of 1,2,4-thiadiazoles.

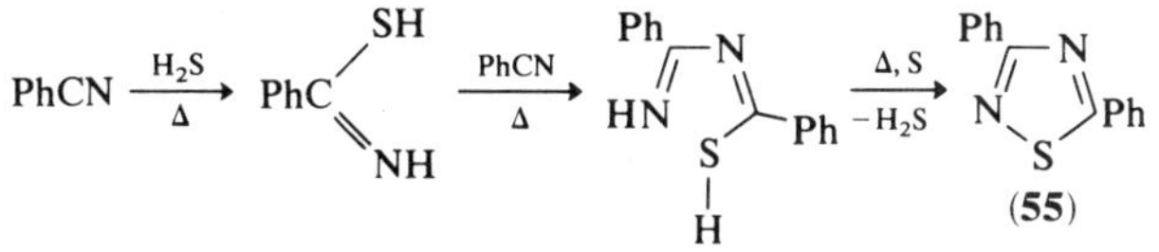

Scheme 123

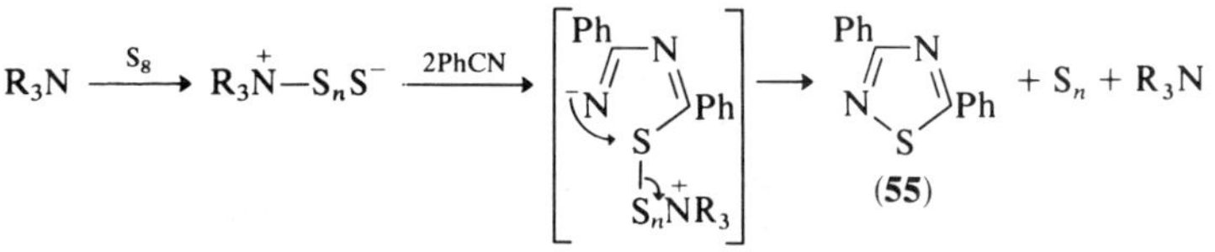

Scheme 124

A one-step procedure for the synthesis of 5-arylimino-1,2,4-thiadiazolin-3-ones (**273**) from a mixture of alkyl azide, substituted phenyl isothiocyanate and aryl or alkyl isocyanate has been reported ⟨76AG(E)489⟩. Evidence was presented to support the proposal that a 1,2,3,4-thiatriazoline (**340**) is an intermediate in this reaction (Scheme 125).

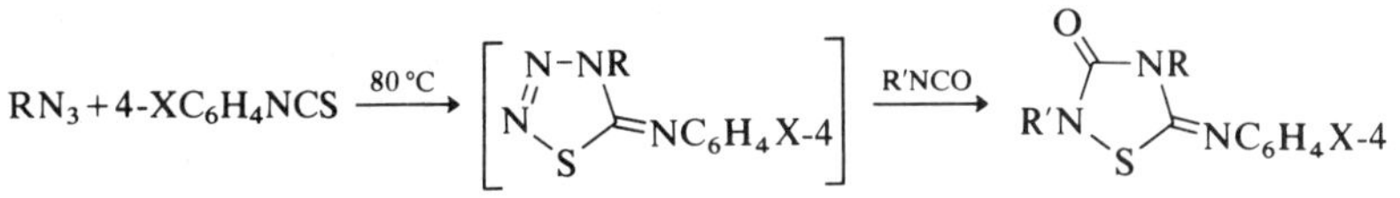

R = alkyl; R' = alkyl, aryl; X = H, Cl, Me (**340**)   (**273**) 60–95%

Scheme 125

4.25.4.2 Ring Synthesis by Transformation of Other Heterocycles

4.25.4.2.1 Introduction

Ring systems which have been converted into 1,2,4-thiadiazole derivatives include 3,5-diimino-1,2,4-dithiazolidines (**341**), 5-imino-1,2,4-dithiazolidinones (**342**), isoxazoles (**343**), oxadiazoles (**344**) and (**345**), and 5-imino-1,2,3,4-thiatriazolines (**346**). In general a ring opening reaction followed by rotation and ring closure occurs in these transformations.

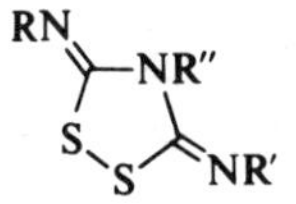

(**341**) 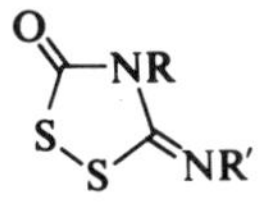   (**342**) 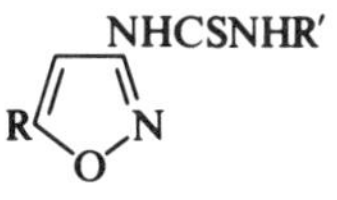   (**343**)

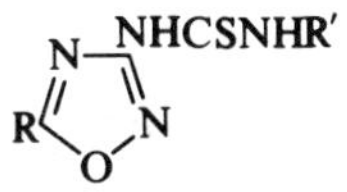

(**344**) 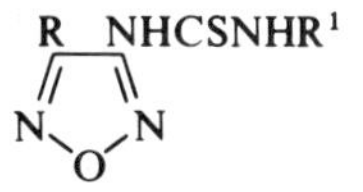   (**345**) 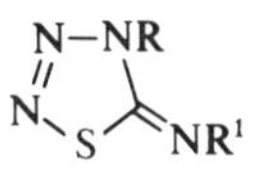   (**346**)

4.25.4.2.2 Dithiazolidine rearrangements

The oxidation of dithiobiurets (**58**) in acid media produces diimino-1,2,4-dithiazolidines (**347**), commonly known as thiurets. Treatment of thiurets with aromatic amines results in ring opening to form thiocarbamylguanidines (**348**) and sulfur. The latter reaction mixtures sometimes spontaneously produce 3,5-diamino-1,2,4-thiadiazole derivatives (**349**; Scheme 126) ⟨65AHC(5)119⟩. The formation of products of type (**349**) does not occur when aliphatic amines are used in place of the aromatic amines.

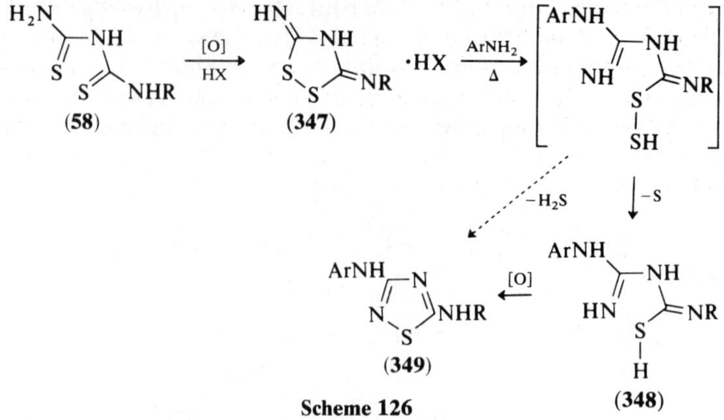

Scheme 126

The alkylation of 'thiurets' (**347**) in basic media produces the imino derivatives (**352**) possibly *via* intermediates such as (**350**) and (**351**) (Nu = soft nucleophile), as indicated in Scheme 127 ⟨74IJC134⟩. The acid-catalyzed dealkylation of (**352**; R' = *t*-butyl) results in a reverse isomerization to the thiuret (**347**). The reaction takes a different course when *N*-alkyl thiurets of type (**353**) are alkylated in the presence of an amine. In this case intermediate (**354**) is postulated to exchange with the amine to form a guanidine which undergoes ring closure to yield (**355**; Scheme 128) ⟨74JOC2235⟩. Thermolysis of (**356**) also produces (**355**) presumably *via* a spiran intermediate.

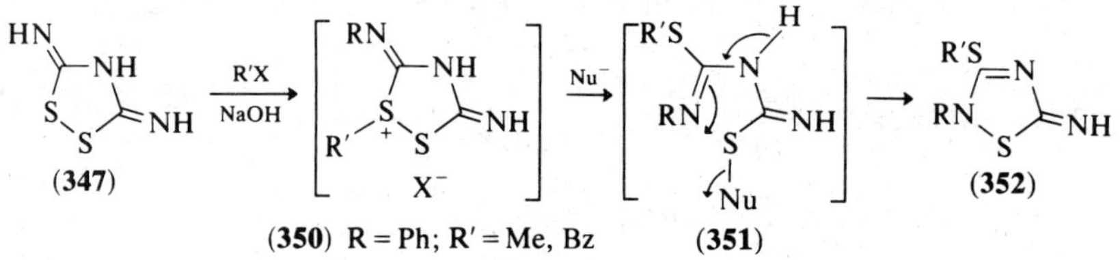

Scheme 127

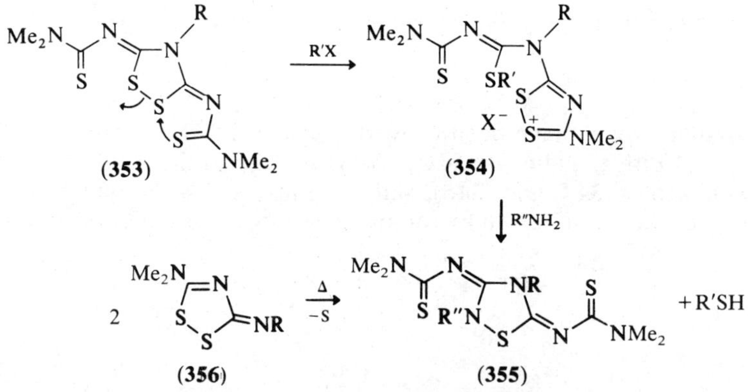

Scheme 128

The alkylated derivative (**357**) or (**356**; R = Me) reacts with sodium azide with elimination of sulfur and formation of 3,5-bis(dimethylamino)-1,2,4-thiadiazole (**358**) in 75% yield (Scheme 129) ⟨71JOC3465⟩.

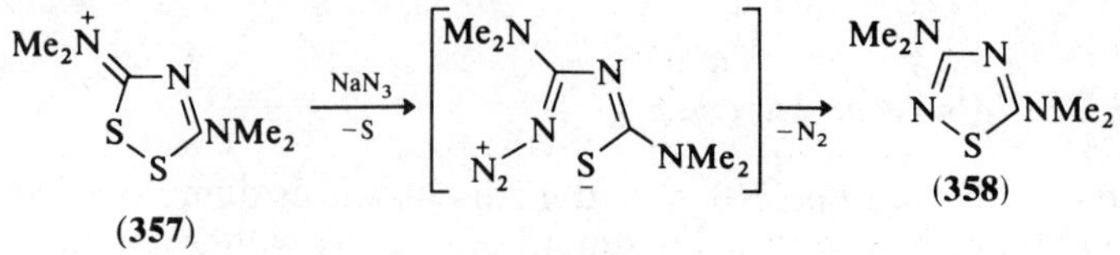

Scheme 129

In 1821 Wöhler discovered that a solid deposited from concentrated aqueous solutions of thiocyanic acid. The solid, which was called 'isoperthiocyanic acid' (3-imino-5-mercapto-1,2,4-dithiazole) (361), formed a new product 'perthiocyanic acid' (3,5-dimercapto-1,2,4-thiadiazole) (18) when treated with alkali and then acid. On storage 'perthiocyanic acid' (18) readily reverted to 'isoperthiocyanic acid' (361) ⟨65AHC(5)119⟩. The mechanisms of these interconversions are still not known with certainty but the transformations outlined in Scheme 130 are suggested. Wöhler proposed the initial formation of a dimer of thiocyanic acid for which structure (359) appears resonable. Addition of the imine function of (359) to the nitrile function of HSCN would produce the trimer (360) which could readily eliminate hydrogen cyanide to produce 'isoperthiocyanic acid' (361).

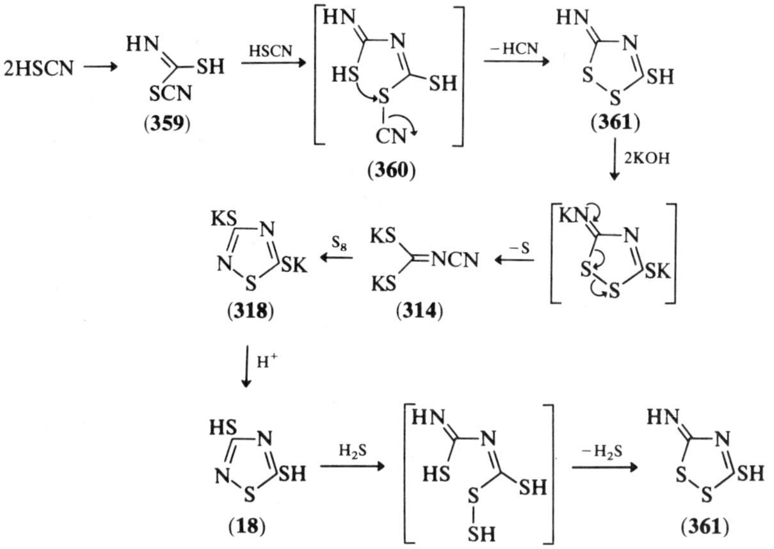

Scheme 130

Ring cleavage of dianion (361) in the manner indicated affords sulfur and the dipotassium salt of cyanodithioimidocarbonate (314). The latter product has been isolated from reaction mixtures of this type in the form of its barium salt. More recently it has been demonstrated that (314) and sulfur form (318) in essentially quantitative yield when refluxed in methanol for 15 minutes ⟨71JOC14⟩. Careful acidification of (318) produces the free acid (18) which has been postulated to revert to (361) by the pathway shown in Scheme 130 ⟨65AHC(5)119⟩. The latter isomerization is catalyzed by soft nucleophiles such as H_2S, sulfurous acid, thiols and probably sodium cyanide, e.g. via (360). 3,5-Dimercapto-1,2,4-thiadiazole (18) is soluble in ether and water whereas the isomeric compound (361) is insoluble. In a similar manner the S-methyl ether (362) was found to undergo disproportionation on treatment with alkali to produce (363) in good yield (Scheme 131) ⟨69JOC2562⟩. A mechanism to account for the conversion of sulfur and (314) to (318) has not been proposed. One possibility is illustrated in Scheme 132. Dithiirane imines (364) are unknown but dithiiranes ⟨79AG(E)941⟩ and dithiiranethiones ⟨80JCP(72)4242⟩ have been proposed as reaction intermediates [also see Section 4.25.4.2.4 for thiaziridinimines (379)].

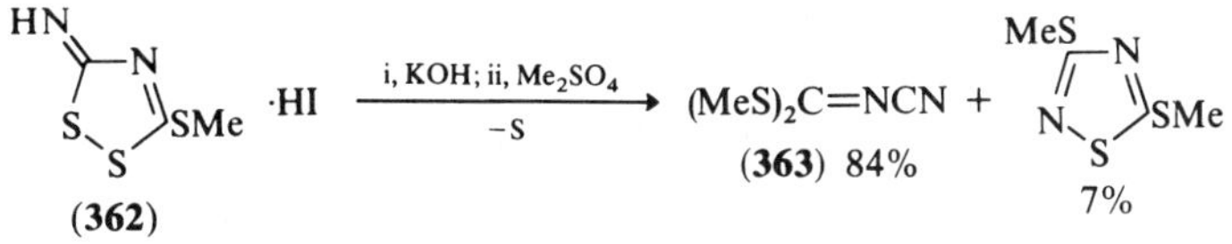

Scheme 131

Chlorination of isothiocyanates produces iminohalomethylsulfenyl halides (269) which react further with isothiocyanates to yield dithiazolium salts of type (365). Treatment of (365) with 95% ethanol leads to the formation of salts of 'isothiocyanate oxides' (366) whereas treatment with hydrogen sulfide produces salts of 'isothiocyanate sulfides' (367), as indicated in Scheme 133 ⟨65AHC(5)119⟩. Neutralization of salts (367) affords the free base 'isothiocyanate sulfides' (368) which are stable with one exception. Melting or recrystallization of N,N'-dimethyliminothiono-1,2,4-dithiazolidine (368; R = Me) results in a Dimroth

Scheme 132

rearrangement to *N,N'*-dimethyldithiono-1,2,4-thiadiazole (**369**) which can be reconverted to (**368**) by treatment with acid followed by neutralization. Neutralization of the 'isothiocyanate oxide' salts (**366**) yields the corresponding dithiazole derivatives (**370**) when the substituents are aryl or arylalkyl, but thiadiazolidines (**371**) are obtained by a Dimroth type rearrangement when R is lower alkyl. Treatment of (**370**) and (**371**) with acid results in the regeneration of (**366**). The mechanism of these interconversions are presumably related to those illustrated in Scheme 130.

Scheme 133

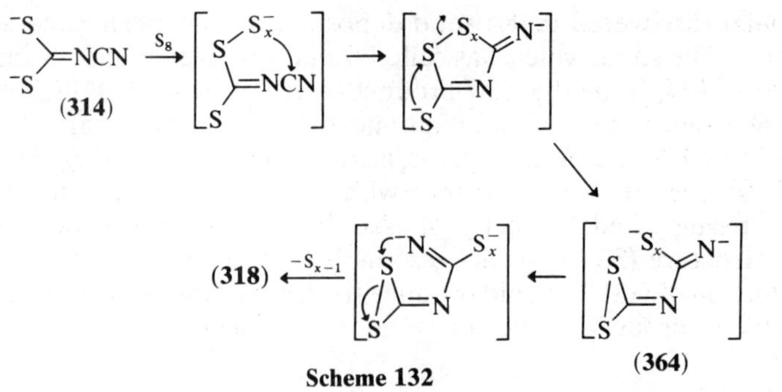

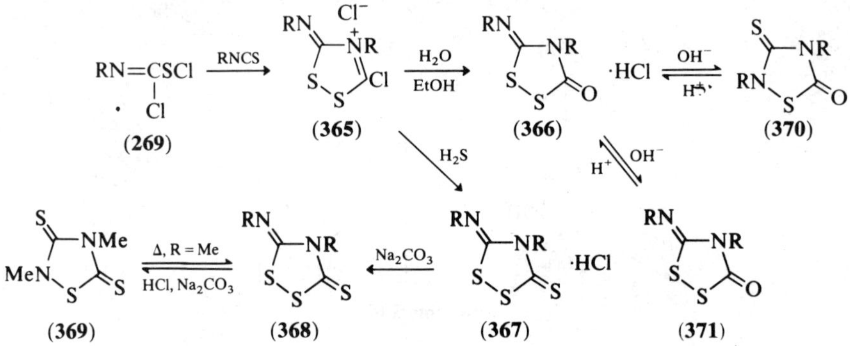

4.25.4.2.3 Oxazole and oxadiazole rearrangements

Heating 5-alkyl-3-amino-1,2,4-oxadiazoles, 5-alkyl-3-aminoisoxazoles and 4-alkyl-3-amino-1,2,5-oxadiazoles with phenyl isothiocyanate produces the corresponding thioureas (**344**), (**343**) and (**345**) which rearrange by a common mechanism to the 1,2,4-thiadiazoles (**372**), (**373**) and (**374**), respectively (Schemes **134, 135, 136**) ⟨74CC358, 77JCS(P1)1616⟩.

Scheme 134

Scheme 135

Scheme 136

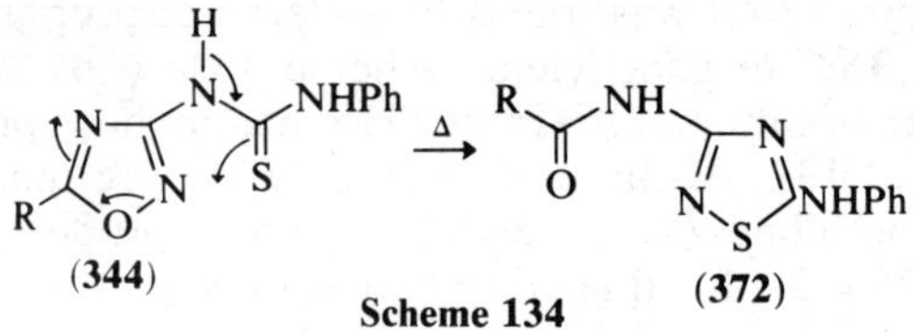

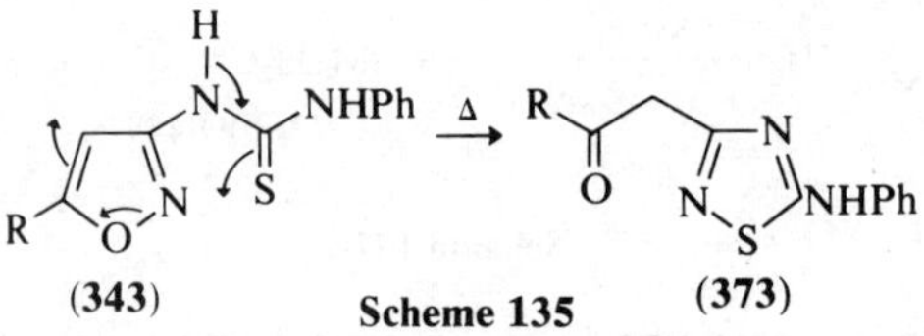

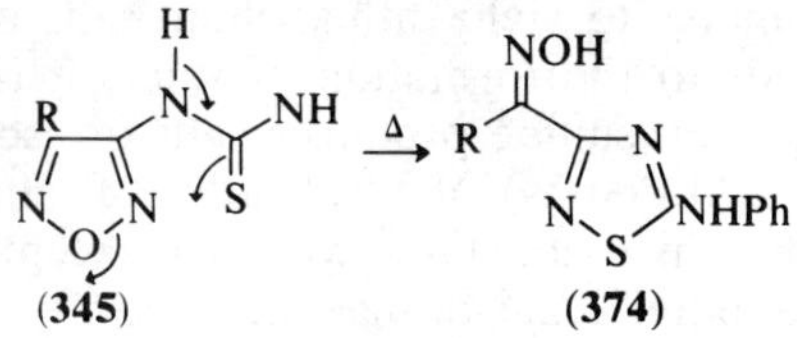

4.25.4.2.4 Thiatriazoline rearrangements

Alkyl azides readily undergo 1,3-dipolar cycloaddition to arylsulfonyl isothiocyanates (375) to yield thiatriazolines (376). Thermolysis of (376) in the presence of isocyanates or carbodiimides produces 1,2,4-thiadiazole derivatives (378) and (379), respectively. The intermediate formation of a thiaziridinimine (377) has been postulated as indicated in Scheme 137 ⟨75JOC1728, 75S52⟩. The use of isothiocyanates as dipolarophiles produces dithiazolidines (380) instead of the thiadiazole derivatives. In these reactions the intermediate thiazirine (377) functions as a 1,3-dipole with the positive charge primarily localized on sulfur. It was recently proposed that the reaction of oxaziridines (381) with isothiocyanates produces a similar thiazirine intermediate (382) which reacts in a different regiospecific manner with isothiocyanates to produce 1,2,4-thiadiazole derivatives (383) and (384; Scheme 138) ⟨74JOC957⟩.

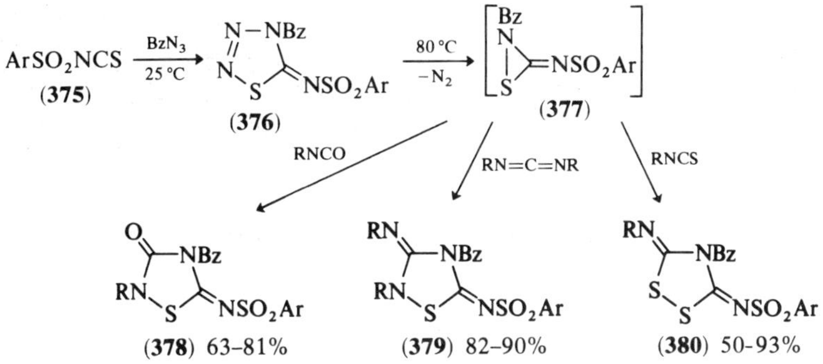

Scheme 137

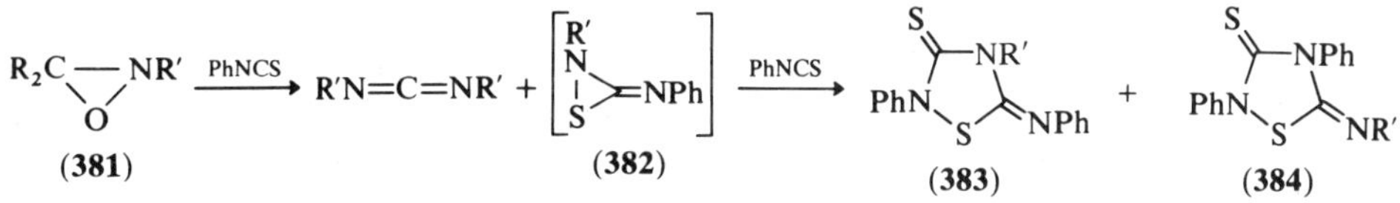

Scheme 138

4.25.4.3 Practical Methods for Synthesis of Derivatives

4.25.4.3.1 General survey

Practical methods of synthesis include both ring forming reactions and the transformation of reactive 1,2,4-thiadiazole derivatives. The most useful 1,2,4-thiadiazole intermediates are those substituted with an amino or halogen atom in the 5-position, i.e. (16) and (155), respectively. The latter materials can be readily interconverted and utilized to produce a wide variety of substituted compounds (385) as indicated in Scheme 139 ⟨65AHC(5)119⟩. The most versatile ring forming method for the synthesis of 1,2,4-thiadiazoles (301) is the oxidation of thioacyl amidines and related compounds of general structure (300), as indicated in Scheme 140 ⟨65AHC(5)119⟩ (also see Scheme 107). The oxidizing agents most commonly used are bromine, acidic hydrogen peroxide and arylsulfonyl halides in pyridine. Other

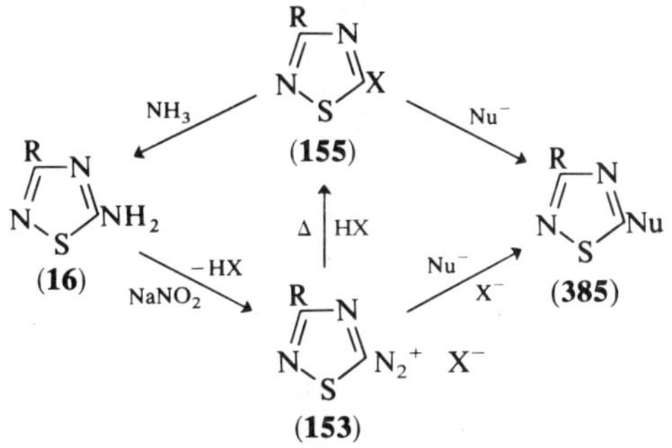

Nu = F, Br, OH, OR, OAr, SH, SR, SAr, NH_2, NHR, NR_2, NHAr, $NHSO_2R$, $NHNR_2$, NHOR, N_3, SO_2R, SO_2Ar, R, H, Ar

Scheme 139

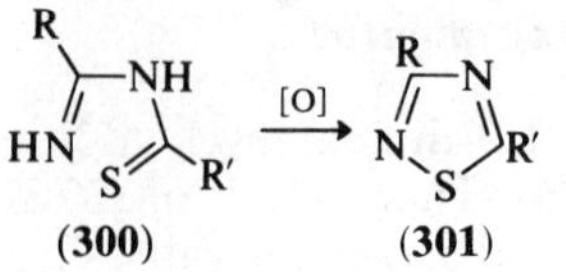

R = alkyl, aryl, NHR, NHAr, NHCOAr, NHSO$_2$R, NR$_2$, NH$_2$, NHNO$_2$, NHN=CR$_2$, OR, SR; R' = alkyl, aryl, NH$_2$, NHR, NR$_2$, OR, SR

Scheme 140

versatile ring forming reactions used in the synthesis of derivatives of this heterocyclic system are discussed in the following section.

4.25.4.3.2 Parent compound

The only synthesis of 1,2,4-thiadiazole (**1**) is the one reported by Goerdler *et al.* in 1955 and is illustrated in Scheme 141 ⟨56CB1534⟩. Owing to the sensitivity of (**1**) to ring opening reactions, it is not a suitable starting material for the preparation of other derivatives.

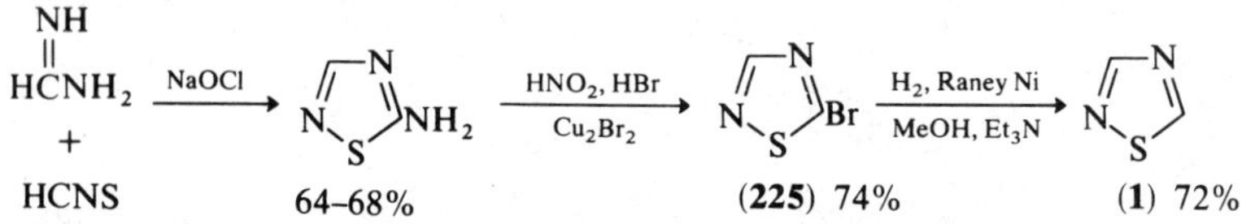

Scheme 141

4.25.4.3.3 C-Linked derivatives

The coupling of thioamides (**253**) by use of *N*-sulfinylsulfonamides (Scheme 142) appears to be a practical method for the synthesis of a variety of 3,5-diaryl- and 3,5-dialkyl-1,2,4-thiadiazoles (**301**) in moderate yields. However, the procedure has not been widely utilized (see Scheme 91) ⟨62AG135⟩. Thioamides also have been coupled by the use of numerous oxidizing agents ⟨65AHC(5)119⟩ but the reported yields of 1,2,4-thiadiazoles have been quite variable (see Scheme 89).

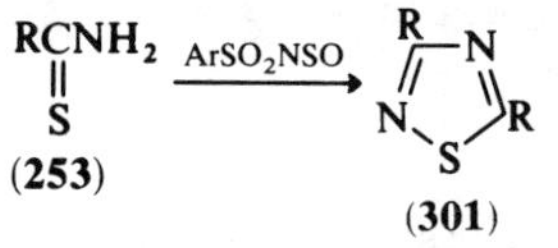

Scheme 142

A more recent method which provides 3,5-disubstituted products (**301**) of unequivocal structure involves the thermolysis of a 5-substituted 1,3,4-oxathiazol-2-one (**338**) in the presence of a nitrile (usually in excess) (Scheme 143) ⟨77JOC1813⟩. Best yields are obtained when the intermediate nitrile sulfides ⟨70TL1381⟩ are generated in the presence of nitriles which are electron deficient (see Scheme 122). Another excellent method for the synthesis of *C*-linked 1,2,4-thiadiazoles involves the amidation of thioacylamidines (**353**) as indicated in Scheme 144 ⟨80JOC3750⟩ (also see Scheme 117).

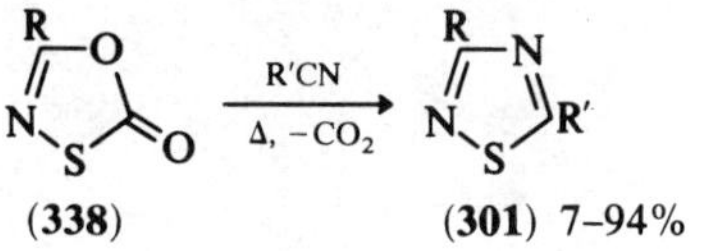

Scheme 143

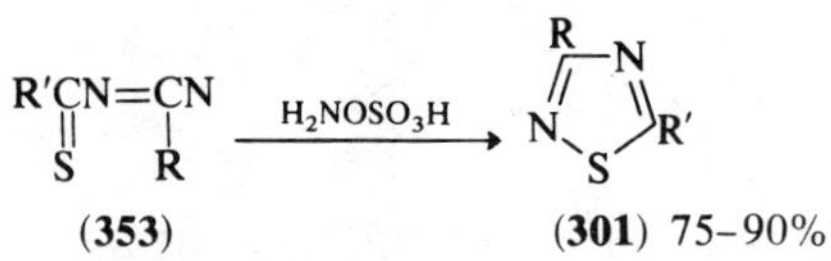

Scheme 144

4.25.4.3.4 N-Linked derivatives

The reaction of amidines with sodium hypochlorite and potassium thiocyanate (Scheme 145) is a mild method for producing a variety of 5-amino- and 3,5-diamino-1,2,4-thiadiazoles (**16**) in good yields (also see Scheme 98) ⟨65AHC(5)119⟩. N-substituted derivatives are obtained when isothiocyanates (RCNS) are used in place of potassium thiocyanate ⟨78USP4107377⟩.

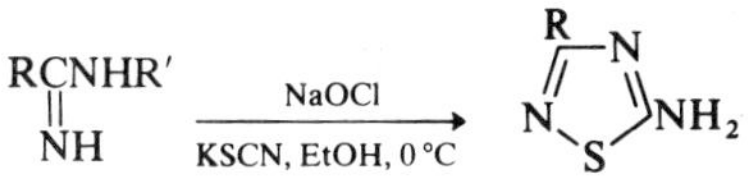

$$\underset{\underset{\text{NH}}{\overset{\text{RCNHR'}}{\|}}}{(282)\ R'=H} \xrightarrow[\text{KSCN, EtOH, 0 °C}]{\text{NaOCl}} \underset{(16)\ 40\text{–}90\%}{\text{R-N=N-NH}_2\ (\text{thiadiazole ring})}$$

Scheme 145

Oxidation of N-arylthioureas (**261**) with acidic hydrogen peroxide produces 'Hector's bases' (**73**) which readily isomerize under basic conditions to afford 3,5-diamino-1,2,4-thiadiazoles (**84**; Scheme 146) ⟨65AHC(5)119⟩ (also see Scheme 92). A variety of 3-amino- and 3,5-diamino-1,2,4-thiadiazoles (**325**) also have been obtained by the treatment of iminocarbonates (**322**) with chloramine at low temperature, as illustrated in Scheme 147 (also see Scheme 115) ⟨76EGP119791⟩.

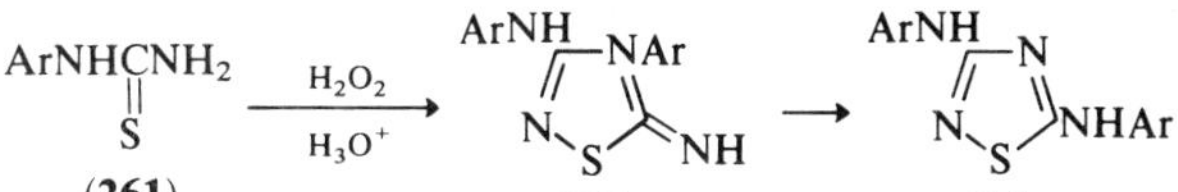

$$\underset{(261)}{\text{ArNHCNH}_2\ (\text{C=S})} \xrightarrow[\text{H}_3\text{O}^+]{\text{H}_2\text{O}_2} \underset{(73)}{\text{ArNH-NAr thiadiazole =NH}} \longrightarrow \underset{(84)}{\text{ArNH thiadiazole NHAr}}$$

Scheme 146

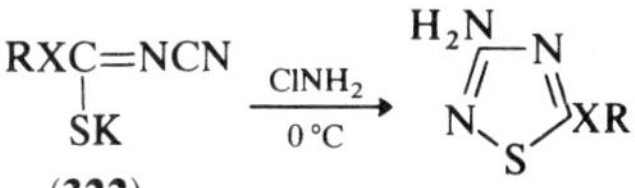

$$\underset{\underset{\text{SK}}{\overset{\text{RXC=NCN}}{|}}}{(322)} \xrightarrow[0\ °\text{C}]{\text{ClNH}_2} \underset{(325)}{\text{H}_2\text{N thiadiazole XR}}$$

Scheme 147

4.25.4.3.5 O-Linked derivatives

5-Alkoxy-1,2,4-thiadiazoles (**311**; R = alkyl) have been prepared in good yields by the oxidation of N-alkoxythiocarbonylamidines (**310**), as indicated in Scheme 148 (also see Scheme 111) ⟨67NEP6610627⟩.

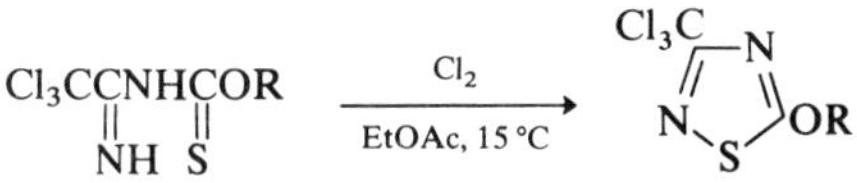

$$\underset{\underset{\text{NH S}}{\overset{\text{Cl}_3\text{CCNHCOR}}{\|\quad\|}}}{(310)} \xrightarrow[\text{EtOAc, 15 °C}]{\text{Cl}_2} \underset{(311)\ 80\%}{\text{Cl}_3\text{C thiadiazole OR}}$$

Scheme 148

3-Hydroxy and 3-hydroxy-5-alkoxy derivatives (**323**) are obtained on oxidation of iminodicarbonates of type (**322**; X = O, S, N) with hydrogen peroxide (Scheme 149) (see Scheme 113) ⟨79JHC961⟩. Products of type (**323**) also are obtained from the reaction of iminodicarbonates of type (**326**) with chloramine (see Scheme 116).

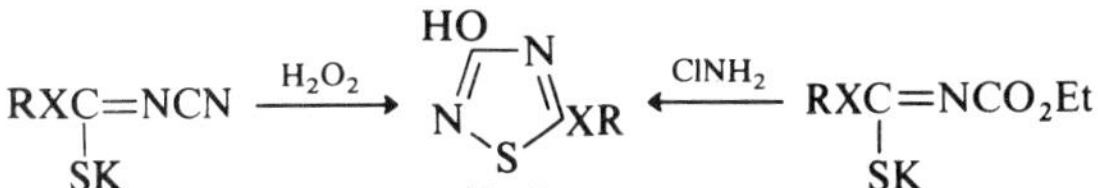

$$\underset{\underset{\text{SK}}{\overset{\text{RXC=NCN}}{|}}}{(322)} \xrightarrow{\text{H}_2\text{O}_2} \underset{(323)}{\text{HO thiadiazole XR}} \xleftarrow{\text{ClNH}_2} \underset{\underset{\text{SK}}{\overset{\text{RXC=NCO}_2\text{Et}}{|}}}{(326)}$$

Scheme 149

4.25.4.3.6 S-Linked derivatives

The most convenient method of preparing thio derivatives of 1,2,4-thiadiazole is *via* dipotassium cyanodithioiminocarbonate (**314**), as outlined in Scheme 150 ⟨71JOC14⟩ (also

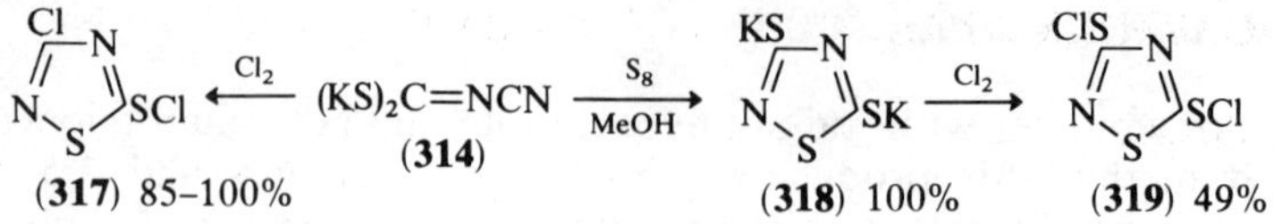

(317) 85–100% (314) (318) 100% (319) 49%

Scheme 150

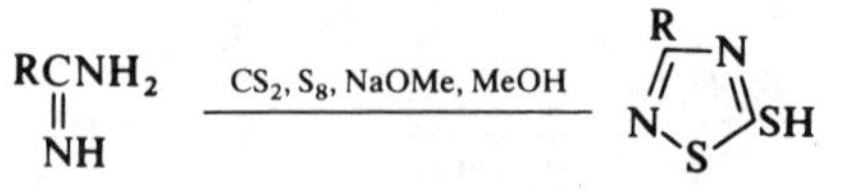

(17) 40–75%

Scheme 151

see Scheme 112). Another convenient procedure is the reaction of amidines with carbon disulfide and sulfur (Scheme 151) ⟨68BRP1116198⟩ (also see Scheme 101).

4.25.4.3.7 Halogen-linked derivatives

A general procedure for the synthesis of 5-chloro-1,2,4-thiadiazoles (**289**) is the reaction of amidines with trichloromethylsulfenyl chloride (**329**) in the presence of an acid acceptor (Scheme 152) (also see Scheme 102). 3-Halo derivatives (**387**; X = Cl, Br, I; Scheme 153) have been obtained in moderate yields from the corresponding amines (**386**) using the Sandmeyer–Gatterman reaction (Scheme 153) ⟨60JCS3234⟩ (see Section 4.25.3.3.4). 3-Chloro-1,2,4-thiadiazolin-2-ones of types (**298**) and (**299**) can be prepared by treatment of chlorosulfenylcarbonyl chloride (**291**) with carbodiimides and cyanamides, respectively (Scheme 154) ⟨70AG(E)54⟩. The isomeric products (**270**) and (**274**) are produced by the chlorination of mixtures of isocyanates and isothiocyanates (Scheme 155) (see also Scheme 95).

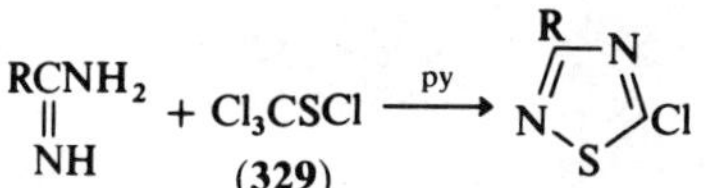

(329) (289)

Scheme 152

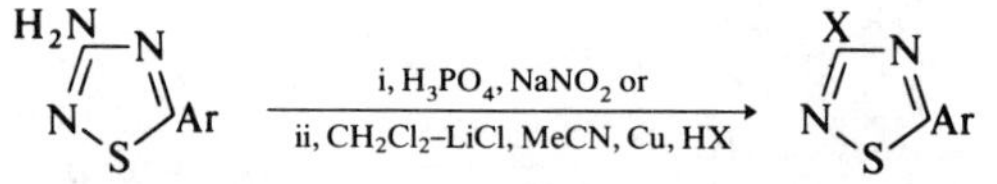

(386) (387)

Scheme 153

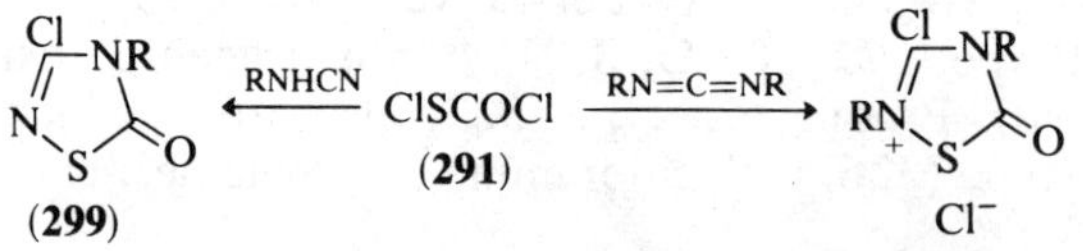

(299) (291) (298)

Scheme 154

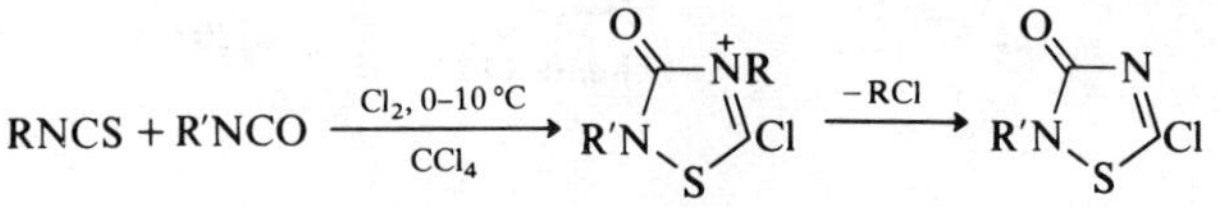

(270) (274)

Scheme 155

4.25.5 APPLICATIONS AND IMPORTANT COMPOUNDS

4.25.5.1 Introduction

During the past 15 years numerous patents have issued on the synthesis and use of 1,2,4-thiadiazoles as fungicides, herbicides, insecticides, bactericides, dyes, lubricant additives and vulcanization accelerators. The largest number of patents appear to be concerned with fungicides.

4.25.5.2 Fungicides

Etridiazole (5-ethoxy-3-trichloromethyl-1,2,4-thiadiazole) (**388**), is a commercial product used as a soil fungicide and as a component of seed dressings ⟨B-79MI42500⟩. Several formulations of this compound are sold under a variety of names such as Terrazole®, Pansoil®, Truban®, Aaterra®, etc. The product is useful for the control of pythium and phytophthora spp and other fungal diseases of turf, vegetables, fruit, cotton, groundnuts, ornamentals and tobacco ⟨76MI42500⟩. Terrazole® was found to be taken up by the roots of tomato seedlings and was translocated unaltered into stems and leaves but was not detected in plant tissue after one day. By contrast, the fungicide was persistent in soil for about five weeks ⟨72MI42500⟩. Several synergistic combinations of (**388**) with other fungicides have been claimed ⟨71MI42500⟩. One metabolite of Etridiazole ⟨74JAP74117630⟩ is a 3-carboxy-5-ethoxy-1,2,4-thiadiazole (**389**). A process patent describes the synthesis of (**388**) from N-chlorotrichloroacetamidine and potassium ethyl xanthate ⟨79USP4143044⟩. Another method of synthesis involves the treatment of 5-chloro-3-trichloromethyl-1,2,4-thiadiazole with ethanol under basic conditions.

(**388**)   (**389**)

Several 5-amino-3-trichloromethyl-derivatives (**390**) of 1,2,4-thiadiazole also have been claimed as fungicides ⟨81USP4254265, 67USP3324141⟩. 3-Halo-5-thio derivatives such as (**391**) and (**392**) are patented as fungicides for use on rice ⟨75JAP7589535, 72USP3691183⟩. Many additional 5-thio-substituted derivatives represented by (**393**) and (**394**) have also been claimed for this utility ⟨72USP3692794, 72USP3692794⟩.

(**390**)   (**391**)   (**392**)   (**393**)   (**394**)

4.25.5.3 Herbicides

A wide variety of 1,2,4-thiadiazole derivatives have been patented as herbicides and/or plant growth regulators. The majority of the products appear to be 5-amino substituted. Thus, products such as (**395**), (**396**) and (**397**) are claimed to exhibit selective preemergent and/or postemergent activity ⟨81FRP2457289, 80USP4207089, 80USP4207090⟩. Other 5-substituted derivatives patented as herbicides include (**398**), (**399**) and (**400**) ⟨73USP3736328⟩. Amides of type (**401**) are claimed to be pre- and post-emergent herbicides at 2–25 lb acre^{-1} with selective action on morning glory and barnyard grass ⟨75USP3859296⟩. The 5-thio derivative (**402**) is a plant growth regulator ⟨80GEP1574430⟩ and is claimed to increase branching rate, fruit set and the yields of soybean and corn. Products of type (**403**) have been patented as safeners (antidotes) of other herbicides for use in crops ⟨78USP4115095⟩.

(**395**)   (**396**)   (**397**)   (**398**)

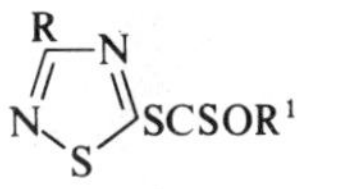

(399) R-N=...-SCSOR¹
(400) Cl₃C-...-OAr
(401) R¹-...-CONR²CHMe₂
(402) Ph-...-SCH₂CO₂R
(403) RO₂C-...-C₆H₄R'

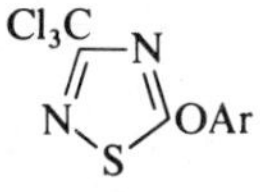

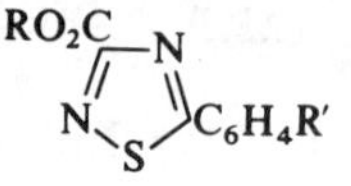

4.25.5.4 Insecticides

The fungicide Etridiazole (**388**) also has been patented as an insecticide and has been claimed to inhibit locomotion of Southern armyworm and potato beetle larva to a greater extent than methyl parathion ⟨77USP4057639⟩. Thiophosphate derivatives (**404**) and (**405**) have been patented as insecticides, acaricides and nematocides ⟨75GEP2500485, 73USP3755571⟩. The 5-chloro derivatives (**406**) and (**407**) are claimed to be effective against 2-spotted spider mites and nematodes, respectively ⟨73USP3720684, 72JAP7230833⟩.

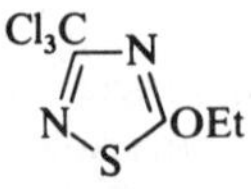

(388) Cl₃C-...-OEt
(404) Et-O-P(R)(=O)-SCH₂-...-XMe
(405) MeS-...-NH-CO-CH₂-S-P(R)(OR')(=O)
(406) RR'N-...-Cl
(407) EtS-...-Cl

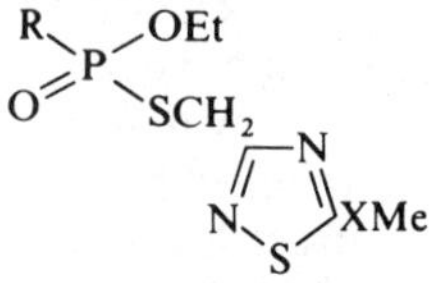

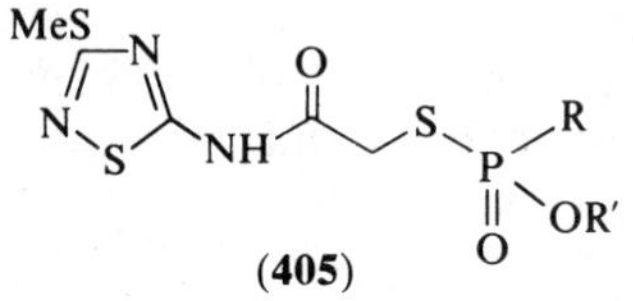

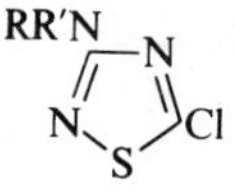

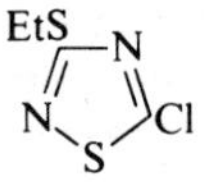

4.25.5.5 Bactericides

Several 3,5-diamino derivatives (**408**) are claimed to be agricultural bactericides whereas (**409**) is said to be effective in the control of intestinal parasites ⟨74JAP7418899, 67BRP1083607⟩. The 1,2,4-thiadiazoles (**410**) and (**411**) have exhibited activity as antimalarials especially against resistant strains of *Plasmodium berghei* ⟨73JHC611⟩. Derivatives (**412**) inhibit the growth of *E. coli*, *Shingella dysenteraie*, *Salmonella typhosa* and *Staphococcus aureas*. General antimicrobial activity is claimed for derivatives of the general class (**413**) ⟨67JAP6708028, 69YZ1437⟩.

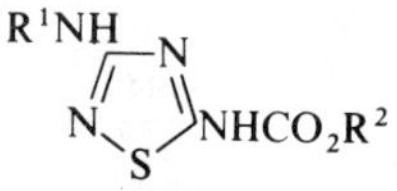

(408) R¹NH-...-NHCO₂R²
(409) Et-...-NHSO₂-C₆H₄-NHCO-C₆H₄-CO₂H
(410) Cl₃C-...-N(CCl₃...)-(CH₂)₃NMe₂
(411) Cl₃C-...-NH-C₆H₃(CH₂NEt₂)(OMe)
(412) O₂N-furan-CH=CH-...-NR₂
(413) MeS(O)ₙ-...-SO₂Ar

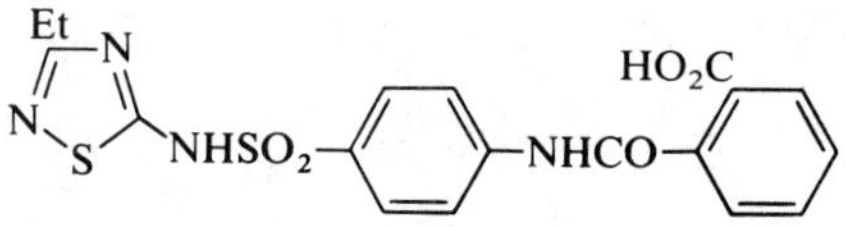

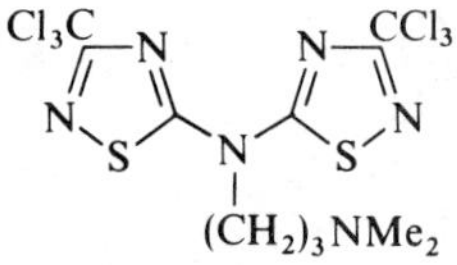

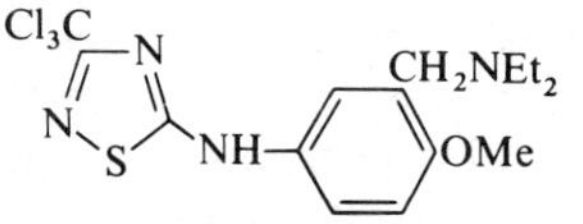

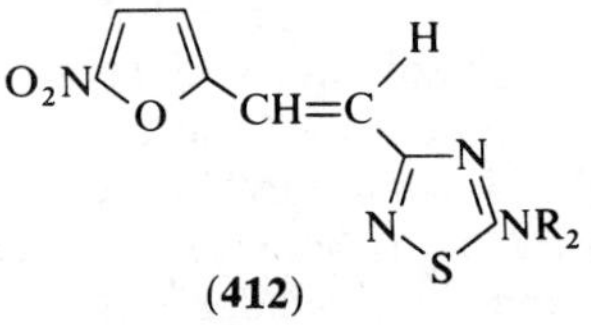

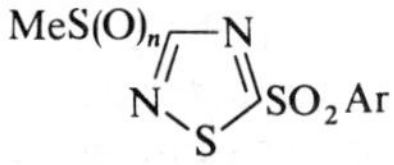

4.25.5.6 Dyes

Many dyes based on 3-substituted 5-diazonium derivatives of 1,2,4-thiadiazoles have been patented. For example (**414**) dyes polyesters red with good light fastness, wash fastness and thermal stability whereas (**415**) is claimed to be more suitable for dyeing polyacrylonitrile fibers red ⟨69FRP1581417, 69BRP1147546⟩. The 5-chloro derivatives (**416**) have been patented as red to violet dyes of good light fastness for polyesters, polyamides and Ni-containing polypropylene ⟨70GEP2006131⟩. The anthraquinone derivative (**417**) is claimed to be suitable as a vat dye for cotton ⟨70GEP1907407⟩.

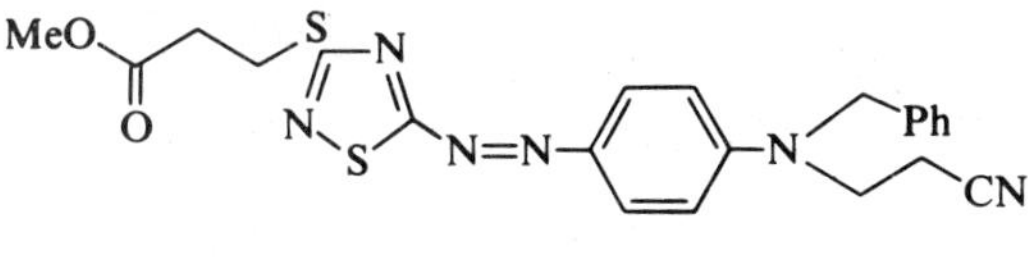

(414) 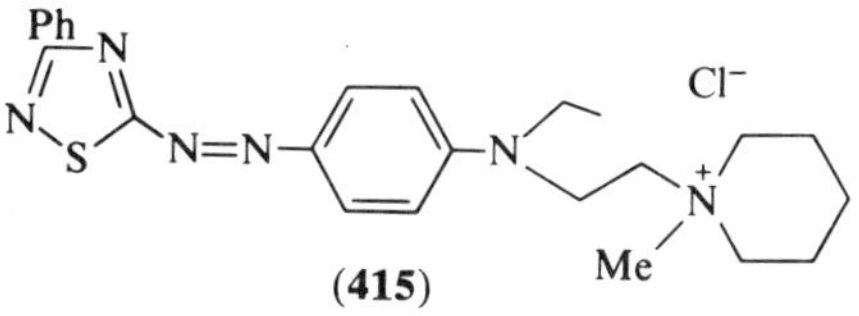 (415)

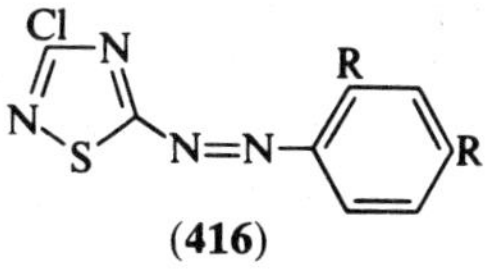

(416) 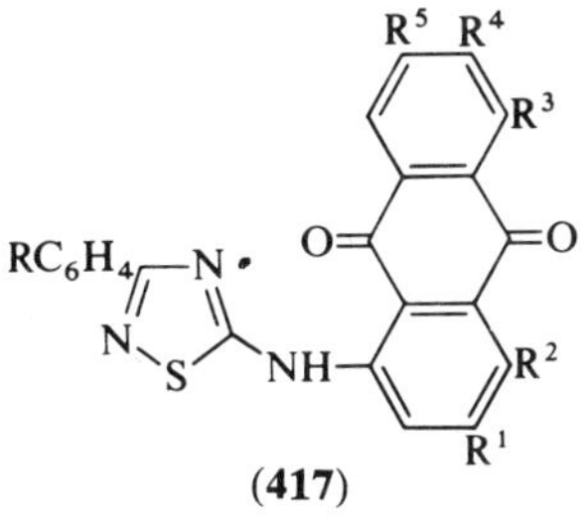 (417)

4.25.5.7 Lubricant Additives

A number of thio-substituted derivatives such as (**418**), (**419**), (**420**) and (**421**) have been claimed to be useful as corrosion inhibitors and extreme pressure additives for greases ⟨75USP3904537, 78USP4107059, 66MI42500⟩.

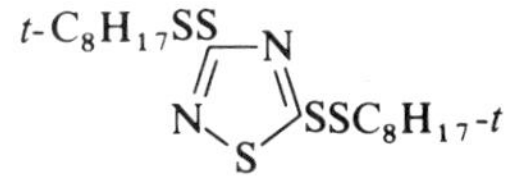

(418) 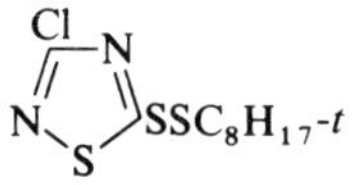 (419) 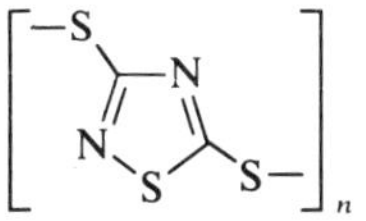 (420) 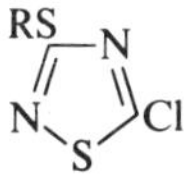 (421)

4.25.5.8 Vulcanization Agents

Certain thio-substituted 1,2,4-thiadiazoles (**422**), (**423**), (**424**) have been patented as efficient vulcanization accelerators ⟨76MI42501, 75USP3899502, 75USP3904619⟩.

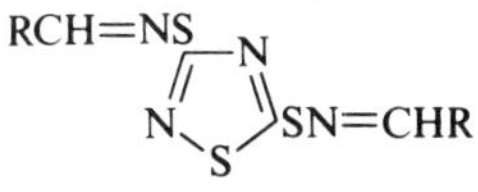

(422) 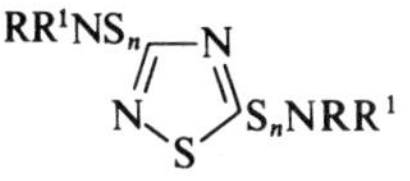 (423) 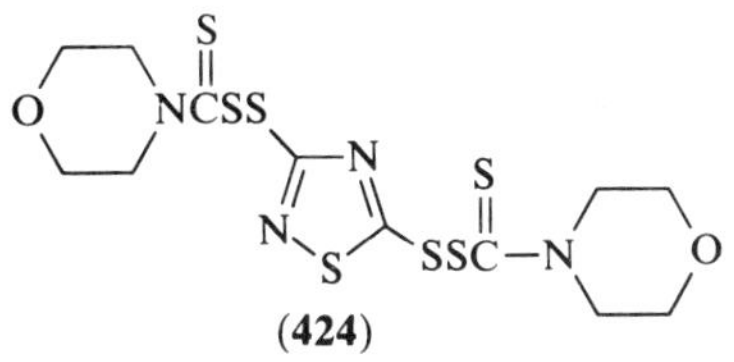 (424)

4.26

1,2,5-Thiadiazoles and their Benzo Derivatives

L. M. WEINSTOCK and I. SHINKAI

Merck Sharp and Dohme Research Laboratories, Rahway

4.26.1 INTRODUCTION	513
4.26.2 STRUCTURE AND PHYSICAL PROPERTIES	514
4.26.2.1 Molecular Geometry	514
4.26.2.2 MO Calculations	516
4.26.2.3 NMR Spectroscopy	517
4.26.2.3.1 ^1H NMR spectroscopy	517
4.26.2.3.2 ^{13}C NMR spectroscopy	518
4.26.2.3.3 ^{14}N and ^{15}N NMR spectroscopy	519
4.26.2.4 UV Spectroscopy	519
4.26.2.5 IR and Raman Spectroscopy	521
4.26.2.6 Mass Spectrometry	522
4.26.2.7 ESR Spectroscopy	522
4.26.2.8 Ionization Constants	523
4.26.2.9 Electronic Structure of Benzo-2,1,3-thia- and -selena-diazoles	523
4.26.3 REACTIVITY	524
4.26.3.1 Reactions at Heterocyclic Rings	524
4.26.3.1.1 General survey of reactivity	524
4.26.3.1.2 Thermal and photochemical reactions	525
4.26.3.1.3 Electrophilic attack at carbon, nitrogen or sulfur	525
4.26.3.1.4 Nucleophilic attack at carbon, hydrogen or sulfur	526
4.26.3.1.5 Reactions with radicals (reduction)	528
4.26.3.1.6 Cyclic transition state reactions	529
4.26.3.2 Reactions of Non-aromatic Compounds	529
4.26.3.2.1 S-Monoxides and dioxides	529
4.26.3.2.2 Reduced compounds	531
4.26.3.3 Reactions of Substituents	531
4.26.3.3.1 Benzenoid rings	531
4.26.3.3.2 Heterocyclic rings	533
4.26.3.3.3 Other C-linked substituents	535
4.26.3.3.4 N-Linked substituents	536
4.26.3.3.5 O-Linked substituents	536
4.26.3.3.6 Halogens	537
4.26.4 SYNTHESES	537
4.26.4.1 Synthesis of 1,2,5-Thiadiazole and 1,2,5-Selenadiazole Rings	537
4.26.4.1.1 Synthesis from [4+1] atom fragments	537
4.26.4.1.2 Synthesis from [3+2] atom fragments	539
4.26.4.1.3 Miscellaneous syntheses	541
4.26.4.2 Synthesis of Substituted 1,2,5-Thiadiazoles	541
4.26.5 APPLICATIONS OF 1,2,5-THIADIAZOLES	542

4.26.1 INTRODUCTION

For almost a century, since 2,1,3-benzothiadiazole (**1**) was discovered by Hinsburg in 1889, the chemistry of 1,2,5-thia- and -selena-diazoles has held considerable interest for organic and physical chemists. Three periods of progress can be distinguished; during the first, 1890–1958, interest focused on the synthesis and substitution of benzo derivatives (**1**) and (2) and other polycyclic analogs. During the second period, 1958–1972, the parent systems (3) and (4) were reported and synthetic thiadiazole chemistry was greatly advanced. A wide range of physical and theoretical methods was applied in studying the aromaticity, chemical bonding, substituent effects and isosterism of (1)–(4). Since 1972, medicinal research on 1,2,5-thiadiazoles has been stimulated by the discovery of the potent β-adrenergic blocking agent timolol. Other thiadiazoles with exceptional pharmacological properties and with potential industrial applications, such as polymers, plant growth regulators and fungicides, have also received increasing attention in recent years.

(1) X = S
(2) X = Se
(3) X = S
(4) X = Se

Aspects of the earlier thiadiazole and selenadiazole chemistry have been previously reviewed ⟨68AHC(9)107, 70RCR923, B-61MI42600⟩. In this chapter the literature through 1981 is surveyed with emphasis on developments of the past 20 years.

4.26.2 STRUCTURE AND PHYSICAL PROPERTIES

4.26.2.1 Molecular Geometry

In its ground state, 1,2,5-thiadiazole (3) has been well characterized by a number of spectroscopic studies. 1,2,5-Thiadiazole is a planar, cyclic molecule of C_{2v} symmetry to within 0.1 Å by microwave spectroscopy ⟨63JA3553⟩ (see Section 4.01.3.2) and gas electron diffraction ⟨64JA162, 61JA4475⟩. The structural parameters for (3) and related compounds were previously discussed in detail ⟨68AHC(9)107⟩.

Recently, double resonance modulation microwave spectroscopy has been used to redetermine the complete substitution structures of (3) and [3,4-^{2}H$_2$]-1,2,5-thiadiazole ⟨78ZN(A)1511⟩. The results are summarized in Table 1. The C—C distance (1.417 Å) is very nearly the same as that found in 1,2,5-oxadiazole (1.421 Å) ⟨65JCP(43)166⟩ and the C(3)—C(4) bond in thiophene (1.423 Å) ⟨61JSP(7)58⟩. It is longer than in benzene (1.397 Å) but shorter than the C(2)—C(3) single bond distance in cyclopentadiene (1.46 Å). These data indicate that the C—C bond in (3) possesses more single than double bond character. The C—N bond distance is intermediate between that of pyridine and oxadiazole and displays ca. 70% double bond chracter. The S—N bonds of (3) possess about 30% double bond character, indicating that substitution of two nitrogen atoms at the 2- and 5-positions of thiophene has little effect on the overall aromaticity.

Table 1 Structural Parameters for 1,2,5-Thiadiazole

Coordinates (Å)		Bond lengths (Å)		Bond angles (°)	
S	+1.04966	S—N	1.62956 ± 0.003	NSN	99.44 ± 0.2
N	−0.00387	C=N	1.32702	SNC	106.51
C	−1.21834	C—C	1.41669	NCC	113.77
H	−2.08948	C—H/D	1.08051	NCH/D	119.96
				CCH/D	126.27

The nitrogen quadrupole coupling (NQR) constants of (3) were determined and the dipole moment confirmed as $\mu = 1.58 \pm 0.02$ D ⟨78ZN(A)1518⟩. The NQR spectra for three isomeric thiadiazoles have been obtained at 77 K. The asymmetry parameters and coupling constants from NQR and microwave spectra are in good agreement in both magnitude and direction with those obtained by *ab initio* calculations ⟨79ZN(A)220⟩. The difference in energy with the two basis sets (*sp/spd*) is very small. The $3d$-orbitals are of no importance in the bonding of these molecules but merely add variational flexibility to the calculations. The NQR spectra of (**1**) ⟨70JCP(52)2787, 78OMR(11)385⟩ and (**2**) ⟨73MI42601, 78JPC463⟩ have also been reported.

X-ray diffraction data (see Section 4.01.3.1) are not available for (**3**). The first X-ray work on a 1,2,5-thiadiazole was performed on the 3,4-dicarboxamide by McDonald ⟨68AHC(9)107⟩. Recently, the crystal structures of 3,4-diphenyl-1,2,5-thiadiazole and 3,4-diphenyl-1,2,5-selenadiazole were determined ⟨76AX(B)1074⟩. Selected bond lengths and bond angles (Table 2) show that the heterocyclic rings have C_{2v} symmetry. The geometrical features of 3,4-diphenyl-1,2,5-thiadiazole (Table 3) show good agreement with those of the parent ring (**3**) derived from microwave spectroscopy or electron diffraction. The C—N bond distance of 3,4-diphenyl-1,2,5-thiadiazole (1.335 Å) is lengthened relative to 1.300 Å found in the corresponding selenadiazole and 1,2,5-oxadiazole ⟨65JCP(43)166⟩, indicating a greater electron delocalization in the sulfur heterocycle. This is probably a consequence of d,π-bonding. In both the sulfur and selenium derivatives, little conjugation exists between the heteroaromatic rings and the phenyl groups.

Table 2 Selected Bond Lengths (Å) and Bond Angles (°) for 3,4-Diphenyl-1,2,5-thiadiazole and 3,4-Diphenyl-1,2,5-selenadiazole

Bond lengths	Thiadiazole ($X = S$)	Selenadiazole ($X = Se$)
X—N(1)	1.635	1.807
X—N(2)	1.630	1.814
N(1)—C(1)	1.327	1.298
N(2)—C(2)	1.342	1.303
C(1)—C(2)	1.435	1.442
Bond angles		
N(1)—X—N(2)	99.2	92.1
X—N(1)—C(1)	107.2	107.9
X—N(2)—C(2)	107.2	106.8
N(1)—C(1)—C(2)	113.7	116.0
N(2)—C(2)—C(1)	112.8	117.1

Table 3 Geometrical Features of 1,2,5-Thiadiazoles

	Microwave spectroscopy[a]	Electron diffraction[b]	X-Ray diffraction[c]
S—N	1.631Å	1.632Å	1.635Å
N—C	1.328	1.329	1.333
C—C	1.420	1.413	1.435
C—H	1.079	1.080	—
N—S—N	99.33°	99.24°	99.12°
S—N—C	—	106.30	107.60

[a] ⟨63JA3553⟩. [b] ⟨64JA162⟩. [c] ⟨76AX(B)1074⟩

Recently, the X-ray analysis of 3,4-bis(methylthio)-1,2,5-thiadiazole 1-oxide demonstrated that the oxidized form of the ring is essentially non-aromatic and shows a pyramidal sulfoxide structure. Interaction between the sulfur lone pair of electrons and the diene is small, the C(3)—C(4) bond length lying closer to that of cyclopentadiene than of thiophene or (**3**). Theoretical calculations indicate that aromaticity effects lower the inversion barrier nearly equally in the thiophene and thiadiazole 1-oxides by stabilizing the planar transition state and destabilizing the pyramidal structure ⟨82JA1375⟩.

The X-ray crystal structures for [1,2,5]thiadiazolo[3,4-*g*]benzofurazan 1-oxide ⟨77AX(B)3685⟩, 2,5-diphenyl-(1-phenylimino)-1,2,5-thiadiazolidine-3,4-dione ⟨77CB3149⟩, 7-amino-[1,2,5]thiadiazolo[3,4-*d*]pyrimidine ⟨71JA7281⟩, 2,3,3,5-tetraphenyl-1,2,5-thiadiazolidin-4-one ⟨70BCJ1905⟩, acenaphtho[1,2-*c*][1,2,5]thiadiazole ⟨71CC1623⟩, timolol maleate ⟨76CR(C)(283)401⟩, [1,2,5]thiadiazolo[3,4-c][1,2,5]thiadiazole ⟨75JOC2749⟩ and anthra[1,2-c][1,2,5]selenadiazole-6,11-dione ⟨73MI42602⟩ have also been reported. The crystal structure of (**1**), (**2**) and benzo-2,1,3-oxadiazole (**5**) will be discussed in Section 4.26.2.9 in connection with the mesomeric properties of these derivatives.

4.26.2.2 MO Calculations

The electronic structure of the monocyclic thiadiazoles is of considerable interest ⟨61CCC156⟩ because they are isoelectronic with thiophenes, thiazoles, pyrazines and related five- and six-membered heterocycles. The comparative study of the π-electron structures of the ground states of the symmetrical isomers, 1,3,4-thiadiazole (**6**) and 1,2,5-thiadiazole (**3**), using the Pariser–Parr–Pople approximation ⟨53JCP(21)767, 53MI42600, 67MI42600⟩ of the SCF–MO theory (see Section 4.01.2.2) has been reported ⟨68ACS(B)2051⟩. The calculated bond orders and π-electron charges, summarized in Table 4, indicate a high degree of polarity in both isomers. The value obtained for the π-electron charge distributions permits the conclusion that (**3**) should be more sensitive to both nucleophilic and electrophilic attack than (**6**). The derived bond distances are presented in Table 5, together with experimental values obtained by microwave spectroscopy ⟨63JA3553⟩. Another SCF–CI calculation has been made on (**3**), assuming the sulfur atom to be in the $s^2 p_x p_y p_z^2$ electronic configuration and donating two p_z electrons to the π-electron system ⟨69JIC779⟩. Microwave spectroscopic data of (**3**) show that the bond angle at the sulfur atom is close to 90°, indicating that the π-bonds are formed from p_x and p_y orbitals of sulfur. Moreover, even under drastic conditions, (**3**) cannot be quaternized on sulfur, suggesting that the two p_z electrons enter into conjugation with the π-electron system. Calculated and experimental spectral data were in excellent agreement.

Table 4 Bond Orders and π-Electron Charges for 1,2,5-Thiadiazole (**3**) and 1,3,4-Thiadiazole (**6**)

Bond/Atom	(**3**)		(**6**)	
	SCF	HMO[a]	SCF	HMO[b]
1–2	0.3109	0.468	0.3392	0.464
2–3	0.7452	0.780	0.8902	0.787
3–4	0.3830	0.541	0.3099	0.526
1	+0.2652	—	+0.1647	—
2	−0.5227	—	+0.1702	—
3	+0.3899	—	−0.2522	—

[a] ⟨61CCC156⟩. [b] ⟨66MI42600⟩.

Table 5 Calculated and Observed Bond Distances (Å) for 1,2,5-Thiadiazole (**3**) and 1,3,4-Thiadiazole (**6**)

Bond	(**3**)		(**6**)	
	R_{calcd}	R_{obs}	R_{calcd}	R_{obs}
1–2	—	1.631	—	1.7205
2–3	1.324	1.328	1.298	1.3024
3–4	1.448	1.420	1.368	1.3710

The semiempirical CNDO/2 method indicates that (**3**) loses its aromaticity on excitation, becoming non-planar ⟨77CCC2060⟩. Perturbation theory and the graph–theoretical definition of resonance energy have been combined and applied to heterocyclic π-systems ⟨78T3419⟩. Thus, the prediction that 1,2,4-thiadiazole should possess less resonance energy than (**3**) or the 1,3,4-derivative is in agreement with an analysis of bond lengths derived from microwave spectra of thiadiazoles and oxadiazoles ⟨66JSP(19)283⟩. Complete sets of harmonic symmetry force constants have been developed for (**3**) and (**4**) ⟨69ACS3139⟩. Based on harmonic force fields, spectroscopic calculations, such as mean amplitudes of vibration and atomic vibration mean-square amplitudes, were performed for (**3**) and (**4**) ⟨69ACS3407⟩. Some mean amplitudes of vibration for (**3**) have been observed by electron diffraction ⟨64JA162, 61JA4475⟩, which are in satisfactory agreement with the calculated values (Table 6).

Table 6 Mean Amplitudes of Vibration (Å) for Thia- and Selena-diazoles (3), (4) and (6)

Bond	(3)	(4)	(6)
C—H	0.0768 (0.078)[a]	0.0773	0.0772
S—C	—	—	0.0435
Se—N	—	0.0459	—
S—N	0.0438 (0.048 ± 0.008)	—	—
N—C	0.0470 (0.040 ± 0.0012)	0.0472	0.0462
N—N	—	—	0.0462
C—C	0.0518 (0.045 ± 0.026)	0.0516	—

[a] Obtained by electron diffraction ⟨61JA4475⟩.

The results of CNDO/2–SCF–MO calculations have been used to evaluate the effect of d-orbitals on the absorption spectra of thiolester and sulfur–nitrogen compounds including (3) ⟨73JCS(P2)1542⟩. The principal effects of the inclusion of d-orbitals in the basis set is to lower the energy of the HOMO. In general, the predicted absorption maxima for $\pi-\pi^*$ transitions are too large. The major source of error is the large energy for the occupied molecular orbitals. In view of this error in the orbital energies, it is impossible to make a meaningful judgement as to whether or not the inclusion of d-orbitals in the calculation is necessary to explain $\pi-\pi^*$ electronic transition energies.

The LCGO calculations of dipole and second moments for a number of five-membered heterocycles agreed with the experimental absolute dipole moment ⟨74JCS(P2)420⟩. The dipole moments of 4-, 5- and 6-substituted 2,1,3-benzazoles (1), (2) and (5), determined in benzene at 25 °C ⟨73JHC773⟩, show that the 2,1,3-benzoxadiazole (5) exists primarily in the *ortho*-quinonoid form, while (1) and (2) are more in the benzenoid form. The greatest amount of mesomeric contribution occurs in the selenium compound.

The assignment of the photoelectron spectra of complex molecules by comparison with MO calculations of ordering is now a well-established technique in heterocyclic chemistry (see Section 4.01.3.9). Thiophene has been the subject of a number of recent non-empirical studies, all of which lead to substantial agreement about the valency shell orbital ordering ⟨75JCS(P2)974, 75JCS(P2)1223, 76MI42600⟩. The photoelectron spectra and *ab initio* MO calculations for the isomeric thiadiazoles and related heterocycles suggest that 1,3,4-thiadiazole is less aromatic than the other isomers ⟨77JST(39)189, 77JST(40)191⟩. The inclusion of $3d$-orbitals leads to much better agreement with the observed dipole moments, both in magnitude and direction. Calculated and measured ionization potentials suggest that $3d$-orbital participation is very small ⟨73T3085⟩. Similarly, *ab initio* calculations and photoelectron spectra for (1) and related compounds were also reported ⟨78JST(43)33, 78MI42600, 77MI42600, 73JCS(F2)1155⟩.

4.26.2.3 NMR Spectroscopy

4.26.2.3.1 *¹H NMR spectroscopy*

The chemical shifts of the protons in (3) and related heterocycles are shown in Table 7 ⟨68AHC(9)107⟩. The electron deficiency of the 1,2,5-thiadiazole nucleus is apparent. Substituents on the ring affect the remaining proton in the expected manner, *i.e.* electron withdrawing substituents show downfield shifts ($\Delta\delta$ −0.4 to −0.6), while substituents with small inductive effects exhibit very slight shifts ($\Delta\delta$ 0 ± 0.1). Electron-donor substituents show a significant shielding effect ($\Delta\delta$ 0.7–0.9) (see also Section 4.01.3.3).

NMR spectroscopy of molecules partially oriented in nematic solvents has been used to study internal potential and preferred conformations of medium-sized molecules. The NMR spectra of 3-phenyl-1,2,5-oxadiazole, -thiadiazole and -selenadiazole in the nematic phase have been analyzed ⟨77JCS(P2)561⟩. A twisted conformation of the bond between the two rings was found. The angle of twist for the selenium, sulfur and oxygen derivatives varies over a range of only a few degrees. The sulfur derivative has the lowest twist angle, suggesting that the conjugation between the phenyl and heterocyclic rings is the strongest of this series. NMR studies of (1) in nematic phases show that the preferred molecular orientation is with the C_2-axis along the direction of the magnetic field. This resembles the way most aromatic molecules are oriented under such conditions ⟨72MI42600⟩.

Table 7 ^{1}H NMR Spectra of 1,2,5-Thiadiazole and Related Heterocycles

Compound	Solvent	Chemical shift^a δ(p.p.m.)
(3)	CCl$_4$	8.70
(4)	CCl$_4$	9.28^b
(6)	CCl$_4$	9.12^c
1,2,5-Oxadiazole	CCl$_4$	8.66^d
Pyrazine	CCl$_4$	8.60
Pyridine	CDCl$_3$	8.60 (H-2)
Benzene	CDCl$_3$	7.37
Thiophene	CDCl$_3$	7.3 (H-2)
Furan	CDCl$_3$	7.1 (H-2)
		7.42 (H-2)
		6.37 (H-3)

^{a} TMS as internal standard. ^{b} ⟨67AG(E)364⟩. ^{c} ⟨67DIS(B)(27)4330⟩. ^{d} ⟨65JOC1854⟩.

The ^{1}H NMR spectra of the benzenoid ring system in (1), (2) and (5) have been fully analyzed as AA'BB' spin systems and their chemical shifts and coupling constants compared ⟨69CPB1821, 69SA(A)1027, 67KGS297⟩. The chemical shifts and coupling constants exhibit interesting features related to the nature of the Group VIA heteroatoms. The chemical shift $\nu_{\delta AB}$ decreases with more delocalized π-electron distribution, suggesting a d-orbital model for sulfur. Substituent effects on coupling constants in (2) have been documented ⟨72JCS(P2)1682⟩. Calculated and measured spectra of (1) and (2) are in good agreement ⟨69KGS62⟩. The *ortho* coupling constant $J_{A'B'}$ in 5-substituted benzo-2,1,3-selenadiazole is larger than the corresponding coupling ($J_{AB} = J_{A'B'}$) in (4) and this is particularly marked in the 5-methoxy derivative. Likewise, the *meta* coupling constant $J_{AB'}$ is increased and the *para* coupling $J_{AA'}$ is reduced by substitution. This increase in the *meta* coupling constant in the order H < Me < Cl < MeO < Me$_2$N has also been found in a number of substituted quinolines ⟨65JCS6004⟩.

The N-methyl chemical shifts of quaternized (1), (2) and (5) in sulfolane ⟨74JHC1011⟩ and ^{1}H NMR spectra of selenolo[3,2-*e*]benzo-2,1,3-selenadiazole ⟨73BSF677⟩, 5H-5-phenyl-[1,2,3]triazolo[4,5-*c*][1,2,5]-oxadiazole, -thiadiazole and -selenadiazole ⟨74BCJ1493, 81JST(74)343⟩, 5-oximinobenzo-2,1,3-thiadiazole derivatives ⟨72T303, 74T3839, 76T1277⟩, [1,2,5]selenadiazolo[3,4-*b*]pyridine and [1,2,5]selenadiazolo[3,4-*c*]pyridine ⟨68T6577⟩ have been reported.

4.26.2.3.2 ^{13}C NMR spectroscopy

The ^{13}C NMR chemical shifts and one-bond coupling constants $J(^{13}$C–H) have been obtained for (1), (2) and (5) in carbon disulfide (Table 8) ⟨74OMR(6)430⟩ (see also Section 4.01.3.4). The chemical shifts for the 3a-position in the title compounds are the reverse of what would be expected on the basis of a purely inductive effect of the Group VIA heteroatoms. If the chemical shifts are controlled by through-bond effects, then mesomeric interaction of (2) is of considerable importance in these systems. Thus, the ^{13}C NMR

Table 8 ^{13}C NMR Spectra of

Chemical shifts (p.p.m.) and coupling constants (Hz)	(1) $X = S$	(2) $X = Se$	(5) $X = O$
$\delta(3a) = \delta(7a)$	155.2 ± 0.2	161.0	144.4
$\delta(4) = \delta(7)$	121.6	124.8	111.2
$\delta(5) = \delta(6)$	129.0	129.6	127.2
$^1J_{C(4)-H} = {}^1J_{C(7)-H}$	166 ± 3 Hz	173	176
$^1J_{C(5)-H} = {}^1J_{C(6)-H}$	159	166	163

measurements show that selenium is the most electron withdrawing towards the five-membered ring, with sulfur showing a slightly smaller effect. Both are more deshielding than oxygen. The ^{13}C NMR spectrum of the interesting ring system [1,2,5]thiadiazolo[3,4-c][1,2,5]thiadiazole has also been reported ⟨75JOC2749⟩.

4.26.2.3.3 ^{14}N and ^{15}N NMR spectroscopy

The ^{15}N NMR chemical shifts for (1), (2) and (5) (see Section 4.01.3.5) show that replacement of oxygen in (5) with sulfur leads to a large upfield shift of ca. 85 p.p.m. ⟨78JOC2542⟩. Interestingly, the ^{15}N chemical shift of (2) appears only 43 p.p.m. upfield from that of (5). The unexpectedly high-field position of the ^{15}N resonance of (5) indicates that the type of bonding between sulfur and nitrogen may differ from that between oxygen and nitrogen. The participation of 3d-orbitals of sulfur to enter sulfur diimide-type resonance was used to explain these effects. Similar resonance is highly unlikely for (5) and the 4d-orbitals of selenium may not be easily involved in this type of π-bonding.

The ^{14}N chemical shifts obtained in a continuous-wave system at 4.3346 MHz for (1), (3) and their N-oxide derivatives are summarized in Table 9 ⟨78SA(A)877⟩. The 1,2,5-thiadiazole system shows a considerable increase in shielding of the nitrogen nuclei compared with those of 1,2,5-oxadiazole. 2,1,3-Benzothiadiazole 1-oxide gives two nitrogen resonance signals, one close to that of (3) and the other at about 100 p.p.m. to lower frequency, also characteristic of N-oxides. ^{14}N NMR is a strong tool for determination of benzenoid or quinonoid structures of isomeric benzazoles ⟨78OMR(11)385, 78JPC463, 78BAP291⟩, which are discussed in Section 4.26.2.9.

Table 9 ^{14}N NMR Spectra of 1,2,5-Thiadiazole and Related Heterocycles

Compound	Chemical shift δ (p.p.m.)^c	Half-height width H(Hz)
(1)	+50±2	190±10
(2)	342^a	495^b
(1)	308^a	297^b
(3)	+34±1	78±4
1,2,5-Oxadiazole	−32±1	150±5
Isothiazole	275^a	—
(5)	−36±2	490±30
2,1,3-Benzothiadiazole 1-oxide	+26±1 (N)	47±2
	+122±1 (NO)	210±20
2,1,3-Benzoxadiazole 2-oxide	+27±3	370±20

^{a} Relative to NH$_4^+$ (aqueous), positive downfield.
^{b} Peak-to-peak width of the derivative line.
^{c} External reference MeNO$_2$.

4.26.2.4 UV Spectroscopy

The similarity of the UV absorption spectra (see Section 4.01.3.6) of 1,2,5-thiadiazoles and the corresponding pyrazine analogs (Table 10) reveals their isoelectronic relationship ⟨68AHC(9)107⟩. The pyrazine molecules have much better developed $n-\pi^*$ bands at longer wavelength. When one considers that the π-electron density in the five-membered ring is formally 1.2 electrons per atom, while it is 1.0 for a six-membered ring, the reduced tendency of the former to promote a non-bonding electron into the π-system is understandable ⟨63PMH(2)61⟩. The substitution of a —CH=CH— group by a sulfur atom has little effect on the energy of the electronic transitions. The band assignments are readily inferred from shifts to shorter wavelengths in solvents of increasing polarity and the disappearance of the band in strong acid solution. The shorter wavelength band, due to the $\pi-\pi^*$ excitation, is relatively insensitive to changes in solvent polarity. No evidence of an $n-\pi^*$ transition could be found in the spectrum of 1,2,5-thiadiazole. From these observations it can be inferred that the nitrogen lone-pairs are involved in the molecular orbital of the ring and that significant changes in symmetry of the molecular orbital take place on excitation. While (3) shows a marked displacement in UV absorption maxima compared to the aliphatic

	(3)	(6)	(7)	(8)
λ_{max}(log ε)	253 (3.93)	210 (3.57)	206 (4.23)	205 (4.11)
	—	—	209 (4.27)	208 (4.06)

1,4-diazene (**7**), indicating that sulfur is able to conjugate with the chromophore, there is virtually no difference in the absorption maxima of (**6**) compared to the aliphatic 1,2-diazene (**8**) ⟨67DIS(B)(27)4330⟩. The UV absorption maxima of various 1,2,5-thiadiazoles are given in Table 11 ⟨68AHC(9)107⟩.

Table 10 Comparison of UV Absorption Spectra of 1,2,5-Thiadiazoles and Analogous Pyrazines ⟨59DIS(19)3136⟩

Compound	Solvent	λ_{max} (nm) (log ε)	
		$n-\pi^*$	$\pi-\pi^*$
(**3**)	H$_2$O	—	253 (3.89)
	conc. HCl	—	255 (3.89)
	96% H$_2$SO$_4$	—	261 (3.99)
	Isooctane	—	260 (3.68)
			257 (2.83)
			253 (3.87)
			250 (3.86)
Pyrazine	H$_2$O	301 (2.98)	261 (3.82)
			267 (3.72)
	conc. HCl	—	272 (3.78)
	96% H$_2$SO$_4$	—	284 (3.88)
	Isooctane	309 (2.81)	253 (3.83)
		315 (2.91)	260 (3.82)
			267 (3.56)
1,2,5-Thiadiazole-3-carboxylic acid	H$_2$O	—	263 (4.01)
Pyrazine-2-carboxylic acid	H$_2$O	310 (2.87)	268 (3.95)
1,2,5-Thiadiazole-3,4-dicarbonitrile	MeOH	—	272 (4.01)
		—	225 (3.85)
2,3-Dicyanopyrazine	MeOH	—	275 (3.82)
			230 (4.02)

Table 11 UV Absorption Maxima of 3-Substituted 1,2,5-Thiadiazoles ⟨68AHC(9)107⟩

3-Substituent	Solvent	λ_{max} (nm)	log ε
H	Methanol	253	3.89
Me	Methanol	257	3.93
Cl	Methanol	263	3.94
CO$_2$H	Isopropanol	263	4.02
CN	Methanol	263	4.00
Ph	Isooctane	280	4.13
OMe	Methanol	272	3.93
OH	Methanol	273	3.87
NH$_2$	Methanol	298	4.08
NHCO$_2$R	Isopropanol	279	4.07
COR	Methanol	265	4.12

The vapour-phase UV absorption spectrum of (**3**) near 2600 Å has been shown to be $\pi-\pi^*$ in nature and to involve a planar ground state and a non-planar excited state having a barrier to inversion of ca. 460 cm^{-1}. This appears to be the first reported observation of a cyclic molecule having a planar ground state and a non-planar excited state. In its ground electronic state (**3**) has been shown to have considerable aromatic character, the result of sulfur p- and d-orbital participation as pd-hybrids in the π-bonding of the ring. Vapor-phase UV studies show clearly that excitation of an electron from a π- to a π^*-orbital destroys the aromatic π-bonding in (**3**) and may be useful in theoretical studies of π-bonding in sulfur-containing molecules ⟨71CC1368⟩. Investigations of the vapor-phase UV absorption spectra of (**3**) in the 3050–2600 Å region reveal that the excited state is non-planar ⟨74CJC100⟩. The barrier to planarity has been calculated to be in the order of 115 ± 30 cm^{-1}.

4.26.2.5 IR and Raman Spectroscopy

The IR and Raman spectra (see Section 4.01.3.7) of (3) were discussed extensively earlier ⟨68AHC(9)107, 67G379⟩. Recently, the assignments of the fundamental vibrations of 3-chloro- and 3,4-dichloro-1,2,5-thiadiazole derivatives have been obtained and are, as expected, consistent with a planar molecule ⟨66SA1417, 72CJC627, 73CJC680⟩. These assignments for the ring modes are compared with those of the parent molecule (3) in Table 12. Good correlations were obtained for three ring stretching vibrations, the ring in-plane bending and the ring out-of-plane bending modes. The polarized and depolarized Raman spectra of (3) have been measured as a function of temperature ⟨76JCP(64)4475⟩. Due to the small depolarization ratios of totally symmetric vibrations, the contribution of the polarizability anisotropy to the polarized spectra is small and the widths of polarized spectra are determined mainly by non-reorientational isotopic relaxation processes.

Table 12 Fundamental Vibrations of 1,2,5-Thiadiazoles

Symmetry species no.	Unsubstituted	1,2,5-Thiadiazole 3-Chloro	3,4-Dichloro
1	A_1 3106[a]	A_1 3111[a]	A_1 1334[b]
2	1350[b]	1325[b]	1278[b]
3	1251[b]	1287[b]	814[b]
4	1041[c]	1150[c]	738[g]
5	806[d]	825[d]	471[d]
6	688[d]	430[d]	190[h,k]
7	A_2 (908)[j]	A_2 868[e]	A_2 451[f,k]
8	(500)[j]	649[f]	178[i,k]
9	B_1 3108[a]	B_1 518[f]	B_1 525[f]
10	1461[b]	233[i]	271[i]
11	1227[c]	B_2 1460[b]	B_2 1451[b]
12	895[d]	949[d]	1031[d]
13	780[d]	786[d]	874[d]
14	B_2 838[e]	701[g]	833[g]
15	520[f]	297[h]	358[h]

[a] C—H stretching. [b] Ring stretching. [c] C—H in-plane bending. [d] Ring in-plane bending. [e] C—H out-of-plane bending. [f] Ring out-of-plane bending. [g] C—Cl stretching. [h] C—Cl in-plane bending. [i] C—Cl out-of-plane bending. [j] Non-observed frequencies. [k] Measured by Raman in neat liquid.

The IR spectra of (1) and its derivatives show a well-defined aromatic system, evidenced by the correlation of the C—C and C—H stretching frequencies of (1) and its 4- and 5-substituted derivatives (Me, Cl and Br) with the corresponding frequencies for benzene derivatives (Table 13) ⟨69CHE180, 69KGS235⟩.

Table 13 Comparison of IR Frequencies of Substituted 2,1,3-Benzothiadiazoles and Benzene Derivatives ⟨69CHE180⟩

	C—C stretching	C—H Out-of-plane bending			C—H In-plane bending		
Position of substituent Substituted benzene	—	1,2	1,2,3	1,2,4	1,2	1,2,3	1,2,4
	1620–1560	770–735	810–750	860–800	1225–1175	1175–1125	1225–1175
	1600–1560		725–680	750–700	1125–1090	1110–1070	1175–1125
	1525–1475			900–860	1070–1000	1070–1000	1125–1090
	1470–1440				1000– 960	1000– 960	1070–1000
							1000– 960
Position of substituent Substituted 2,1,3-benzothiadiazole	—	—	4	5	—	4	5
	1610–1583	744	800	816–802	1230	1150–1140	1250–1220
	1570–1535	755	750	750	1130	1085–1075	1130
	1546–1506			860	1043		1045–1014
	1486–1455				982		930

4.26.2.6 Mass Spectrometry

The major fragmentation pathways in the mass spectra (see Section 4.01.3.8) of (**3**), (**4**) and 1,2,5-oxadiazole are shown in Figure 1 ⟨75ACS(B)483⟩. Direct loss of the heteroatom X from the molecular ions was not observed in any case. Application of the metastable defocusing technique revealed that elimination of X (O, S and Se) takes place from $RCNX^{\ddagger}$ or the rearranged ion $RNCX^{\ddagger}$, with formation of a fragment with m/e 103 ($PhCN^{\ddagger}$ or $PhNC^{\ddagger}$). Since this process appears to be the predominant pathway for the formation of these ions, the abundance of the m/e 103 fragment in the three spectra (R = Ph, X = O, S and Se) indicates that the order of elimination of X is Se > S > O. The mass spectra of the parent compounds indicate the loss of HCN from the parent ions. Further fragmentation of $RCNX^{\ddagger}$, however, is clearly affected by the nature of the substituent. Elimination of selenium takes place from $PhCNSe^{\ddagger}$, while $HCNSe^{\ddagger}$ leads to abundant $Se^{\ddagger}$ with elimination of HCN. Similar effects are found in the spectra of the thiadiazoles, but the sulfur ion is less abundant. This mode of fragmentation is negligible in the case of 1,2,5-oxadiazole. The mass spectral behavior of the three series of compounds show a close analogy to their thermal and photochemistry (see Section 4.26.3.1.2).

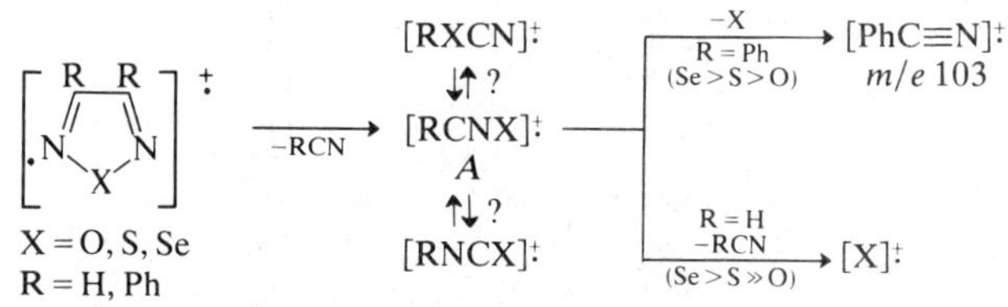

Figure 1 Fragmentation scheme of 1,2,5-thiadiazoles and related heterocycles

The mass spectra of the benzo derivatives (**1**), (**2**) and (**5**) indicate that stability of NO *versus* NS and NSe strongly influences fragmentation ⟨78OMS379⟩. Thus, while $M^{\ddagger}$ – NO is the major fragmentation pathway in (**5**), the corresponding loss of NX (X = S or Se) is much less important in (**1**) and not observed in the selenium analog (**2**). The m/e 76 ion in the spectrum of (**2**) has previously been suggested to be the benzyne ion ⟨65CC624⟩. Metastable defocusing studies and exact mass measurements, however, have established its elemental composition as C_5H_2N. Similarly, the composition of the m/e 77 and 78 fragments was found to be C_5H_3N and C_5H_4N, respectively. Intense parent ions are dominating features of the mass spectra of a number of substituted benzo-2,1,3-thiadiazoles. The fragmentation of the heterocyclic ring (loss of H, HCN and NS) is far less important than fragmentation of the substituents ⟨75ACS(B)622⟩. Mass spectral studies played an important role in the structural assignments of 3-methyl-4-phenyl-1,2,5-selenadiazole 2-oxide ⟨76ACS(B)675⟩ and 2,5-diaryl-1-(arylimino)-1λ^4,2,5-thiadiazolidine-3,4-diones ⟨77OMS628⟩.

4.26.2.7 ESR Spectroscopy

The ESR spectra of the radical anions of (**1**), (**2**) and (**5**), generated by potassium metal reduction ⟨67JCS(A)771⟩ or by electrochemical reduction ⟨67KGS811⟩, have been interpreted using MO theory. A p-orbital model gives a satisfactory description of the bonding of sulfur. Solvent effects on the nitrogen hyperfine splitting constants of the anion radicals of (**1**) and (**2**) are discussed in terms of a mesoionic structure. The ESR spectrum of benzoselenadiazole anion radical has a pattern similar to that of 2-methyl-1,2,3-benzotriazole, indicating that replacement of selenium with an N-methyl group has little effect on the spin distribution ⟨70BCJ268⟩. The unrestricted Hartree–Fock method has been employed in the study of spin density distributions in the radical anion of (**1**). The results are in agreement with experimental ESR data after annihilation of the quartet spin resonance ⟨69IJC97⟩.

MO calculations for electronic configuration, spin density and bond length have been reported for anion radicals of (**1**)–(**5**) ⟨75KGS1055⟩.

ESR parameters of the anion radical of [1,2,5]thiadiazolo[3,4-c][1,2,5]thiadiazole show good agreement with those calculated by McLachlan's perturbation method using two models for the sulfur atom (sp/spd). The HMO method does not differentiate between the p-orbital and d-orbital models for the thiadiazole anion radicals. Due to the presence of a plane of symmetry through the sulfur atom, the contribution of the $3d$-orbital is absorbed into that of the $3p$-orbital for the totally symmetric odd-electron orbital. The same MO parameters can be used to predict hyperfine splitting in a series of aromatic anion radicals of related thiadiazole analogs including (**1**), thiadiazolopyridine and thiadiazoloquinoxaline ⟨76JPC1786⟩. The ESR spectral data for CuL_2X_2 (X = F, Cl and Br, L = **1** or **2**) indicate that L is coordinated to Cu through the sulfur or selenium and the halo groups are in the inner coordination sphere ⟨76MI42601⟩.

4.26.2.8 Ionization Constants

The ionization constants of isomeric thiadiazoles and related heterocycles, obtained by spectrophotometric methods, are given in Table 14. Diprotonation of these very weakly basic ring systems seems unlikely. In ring systems where a sulfur atom replaces an isoelectronic vinyl group α to a nitrogen atom, the pK_a is decreased by *ca.* 6 log units [pK_a of (**3**) $(-4.9) - pK_a$ of pyrazine $(0.6) = -5.5$]. When the same replacement is made β to an azine nitrogen atom, the observed decrease is *ca.* 2.6 log units [pK_a of thiazole $(2.53) - pK_a$ of pyridine $(5.23) = -2.70$]. The difference in magnitudes of pK_a values can be explained by the inductive effect of the ring sulfur atom which more effectively withdraws electrons from the nitrogen lone pair when located in the α-position. The extremely low basicity of the 1,2,5-isomer indicates that the π-orbital in the ring is of high electron density and that the lone pair electron density on the nitrogen atoms is correspondingly low ⟨67DIS(B)(27)4330⟩.

Table 14 Ionization Constants of Isomeric Thiadiazoles and Related Heterocycles

Compound	pK_{a1}	pK_{a2}	Compound	pK_{a1}	pK_{a2}
1,2,5-Thiadiazole	−4.90	—	Pyrazine	0.70	−5.8
1,2,3-Thiadiazole	−2.00	—	Thiazole	2.33	—
1,2,4-Thiadiazole	−1.29	—	Isothiazole	−0.51	—
1,3,4-Thiadiazole	−0.90	—			

The electronegative influences of the thiadiazole system greatly increases the acidity of substituents (Table 15). The effect is further enhanced in the 1-monoxide and 1,1-dioxide analogs. The ionization constants of 4- and 5-hydroxy-2,1,3-benzothiadiazoles were determined in water ⟨70AC(R)801⟩.

Table 15 Ionization Constants of 1,2,5-Thiadiazoles

R^1	R^2	n	pK_{a1}	pK_{a2}	Ref.
CO_2H	H	0	2.47	—	68AHC(9)107
CO_2H	CO_2H	0	1.59	4.14	—
OH	H	0	5.10	—	68AHC(9)107
OH	Me	0	5.10	—	68AHC(9)107
OH	OEt	0	4.40	—	68AHC(9)107
OH	Cl	0	3.65	—	68AHC(9)107
OH	CN	0	2.97	—	68AHC(9)107
OH	OH	0	4.68	7.50	69OPP255
OH	OH	1	2.62	6.58	69OPP255
OH	OH	2	2.20	5.50	75JOC2743
NH_3^+	H	0	2.90	—	68AHC(9)107

4.26.2.9 Electronic structure of Benzo-2,1,3-thia- and -selena-diazoles

Although a large body of physical and chemical data is available related to the electronic structure and double bond fixation of (**1**), (**2**) and (**5**), no single structure, quinonoid (a) or benzenoid (c) (Figure 2), satisfies all the results (see Section 4.01.4.2.2). The *ortho* coupling constants in the ^{1}H NMR spectra of (**1**) (8.95 Hz), (**2**) (9.08 Hz) and (**5**) (9.20 Hz) vary significantly with the degree of fixation of the intervening bonds ⟨68SA(A)1869⟩ and parallel the double bond fixation determined by X-ray diffraction ⟨51AX193⟩. There is a considerable shortening of the C(4)—C(5) bond compared to the C(5)—C(6) bond. The non-bonding electrons associated with the tertiary nitrogen atoms are least firmly held in (**2**) and are increasingly bound in (**1**) and (**5**), in order of electronegativity effects. Similarly, the application of ^{1}H NMR spectroscopy to structural problems such as partial bond fixation in bicyclic heterocycles (**1**), (**5**) and 2-methyl-1,2,3-benzotriazole has been reported ⟨69OMR(1)311⟩. The differences between $J_{4,5}$ and $J_{5,6}$ are 2.68, 2.23 and 1.9 for (**5**), (**1**) and N-methyl derivatives, respectively. This indicates that bond fixation of the sequence shown in (c) decreases in the order of O > S ≫ NMe. Charge separated canonical forms of type (b) will be the most important for the N-Me derivative. Canonical form (c) is evidently not of great importance. Charge-separation ion forms of type (b) probably increase in importance for oxygen, sulfur and N-methyl, respectively.

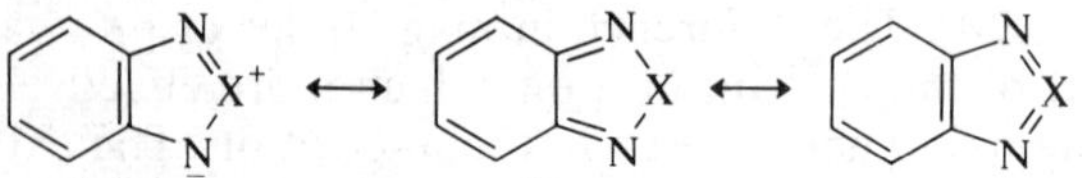

(**1**) X = S
(**2**) X = Se
(**5**) X = O

Figure 2 Canonical forms

^{13}C NMR studies of (**1**), (**2**) and (**5**) indicate that selenium is the most electron withdrawing towards the five-membered ring, with sulfur showing a slightly smaller effect. The mesomeric interaction of type (b) is of considerable importance in these systems ⟨74OMR(6)430⟩. The large chemical shift difference (*ca.* 85 p.p.m.) in the ^{14}N NMR spectra of (**1**) and (**7**) clearly suggest that (**5**) has a more quinonoid-like structure than (**1**). On the other hand, the data obtained by microwave spectra indicate that there is double bond fixation in the six-membered ring in both (**1**) and (**5**).

The chemical reactivity of (**1**) and (**2**) is a dichotomy. Benzenoid character is indicated in normal electrophilic substitution (nitration, sulfonation and catalyzed halogenation, Section 4.26.4.1) and in the lack of dienophile reactivity in the Diels–Alder reaction. On the other hand, typical dienoid character is exhibited in the facile ozonolysis of the benzene ring of (**1**) and in the easy, non-catalyzed tetra-addition of halogen (see Section 4.26.4.1).

Kinetic studies of reaction rates of the 4- and 5-chloro derivatives of all three benzo-2,1,3-diazoles with methoxide anion show that the 5-chloro compounds are more reactive than the corresponding 4-chloro derivatives (see Section 4.26.3.3). The rates of reaction for the former were in the order O ≫ Se > S and for the latter the order was O > Se ≫ S. This again underlines the high bond order between C(4) and C(5) and the higher electron density at C(4) compared to that at C(5) (see Section 4.26.4.1) ⟨68SA(A)1869⟩.

4.26.3 REACTIVITY

4.26.3.1 Reactions at Heterocyclic Rings

4.26.3.1.1 General survey of reactivity

1,2,5-Thiadiazole is a planar, thermally stable and weakly basic aromatic ring system. Aromatic forms of the 1,2,5-thiadiazole nucleus are generally stable to concentrated mineral acids and are only slightly sensitive to base. Base-catalyzed deuterium exchange of the ring protons can be effected ⟨67G1870⟩. Reduced and oxidized forms of the ring, on the other hand, readily undergo hydrolytic fission. The 1,2,5-thiadiazoles are reactive toward nucleophilic attack at carbon, sulfur or a ring proton, the site depending on the structure of the thiadiazole substrate and the nature of the nucleophile. Reductive cleavage and desulfurization also takes place with vigorous reducing agents. The system is very electron poor and is relatively inert in electrophilic substitution reactions, quaternizations, complexation with metal salts and oxidation. The pronounced acidity of 3-hydroxy-1,2,5-thiadiazole (pK_a 5.1) and the 3-carboxylic acid (pK_a 2.47) reflect the magnitude of the electron-withdrawing effect on substituents.

4.26.3.1.2 Thermal and photochemical reactions

Simply-substituted 1,2,5-thiadiazoles, e.g. the parent ring and alkyl derivatives, withstand heating to at least 250–300 °C. The mass spectrum of 1,2,5-thiadiazole (see Section 4.26.2.6), which displays a major molecular ion, is further indication of the inherent stability of this ring system. Under UV photolytic conditions the aromatic thiadiazoles slowly fragment to nitriles and sulfur. Attempts to trap the postulated nitrile sulfide intermediate were not successful ⟨68CC977⟩.

Photolysis of derivatives of (4) leads to quantitative formation of nitriles and selenium via a nitrile selenide intermediate, detected and characterized by low temperature spectrophotometric methods. Benzo derivative (2) was slowly converted to selenium and a mixture of cis,cis, cis,trans and trans,trans isomers of 2,4-hexadienedinitriles. The cis,cis isomer is the probable kinetic product ⟨77ACS(B)848⟩.

Photolysis of 2,1,3-benzothiadiazole 1-oxide produces 1,3-dihydro-2,1,3-benzothiadiazole 2,2-dioxide, shown by flash photolysis to be formed via hydration of the 2-oxide intermediate ⟨78ACS(B)625⟩. Independent of this process 2-thionitrosobenzene is generated reversibly as a short-lived intermediate, analogous to the thermal and photochemical formation of a 1,2-dinitroso intermediate from benzofuroxans. Preliminary flash photolysis and spectrometric results point to a nitroselenanitroso pathway in the photolysis of 2,1,3-benzoselenadiazole 2-oxide to benzofurazan ⟨76ACS(B)675⟩.

4.26.3.1.3 Electrophilic attack at carbon, nitrogen or sulfur

The 1,2,5-thiadiazoles were relatively inert in electrophilic substitution reactions (see also Section 4.02.1.4). The parent ring does not react with bromine, either under irradiation or in the presence of iron(III) salts, nor does it enter the Friedel–Crafts or nitration reactions. Electrophilic deuteration has been effected in low yield under drastic conditions (250 °C in trideuterophosphoric acid). If the thiadiazole ring bears an activating group, e.g. amino or methyl, electrophilic substitution can be achieved ⟨68AHC(9)107, 70RCR923⟩.

In light of the failure of other common electrophilic substitution methods, the facile bischloromethylation of the parent 1,2,5-thiadiazole is noteworthy ⟨74CC585⟩. This reaction takes place in refluxing acetic acid, formaldehyde and hydrogen chloride, even in the absence of a Lewis acid catalyst. These results suggest that a mechanism differing from that of 'normal' aromatic chloromethylation is operating in the thiadiazole reaction, possibly the ylide mechanism ⟨70T685⟩ observed in the deuterium exchange of thiazole, pyridine, imidazole and other nitrogen heterocycles.

The nitrogen and sulfur heteroatoms in 1,2,5-thiadiazoles have low electron densities, corresponding low basicity and nucleophilicity, and relatively low reactivity with acids, alkylating agents, metal salts and oxidizing agents. 1,2,5-Thiadiazole, for example, is an extremely weak base, exhibiting a pK_a value of -4.90 ⟨67DIS(B)(27)4330⟩ compared to pK_a 0.6 for the isoelectronic pyrazine system (see Section 4.26.2.8).

The 1,2,5-thiadiazole system is relatively resistant to oxidation. Depending on the reagents and structure of the thiadiazole it is possible to oxidize substituents even under fairly vigorous conditions without affecting the thiadiazole ring. When oxidation of the ring does occur, the site of attack is usually the sulfur atom leading to thiadiazole 1-oxides, 1,1-dioxides ⟨81H(16)1561⟩ or ring destruction. Thus 3-[(2-methoxy-4-nitrophenyl)]-1,2,5-thiadiazole is oxidized to 2-methoxy-4-nitrobenzoic acid by permanganate ⟨72IJS(A)25⟩, and the peracetic acid oxidation of (1) and several substituted analogs gives ammonium sulfate and glycolaldehyde ⟨70RCR923⟩. Similar results are obtained with unsubstituted thiadiazole ⟨68AHC(9)107⟩. Diethoxy-1,2,5-thiadiazole, on the other hand, can be smoothly converted to the 1-oxide and 1,1-dioxide with m-chloroperbenzoic acid ⟨81H(16)1561⟩. Dihydroxy-1,2,5-thiadiazole was oxidized to 1,2,5-thiadiazole-3,4-dione 1-oxide with 2,3-dichloro-5,6-dicyano-1,4-benzoquinone via the possible intermediacy of a tetravalent sulfur species ⟨69OPP255⟩, and N,N-di-t-butyl-1,2,5-thiadiazole-3,4-dione produces the S-imino compounds on reaction with sulfonyl azides ⟨80CZ111⟩.

Quaternary salts of 2,1,3-benzothiadiazole (**1**) can be prepared only under forcing conditions and with reactive alkylating agents. Alkyl iodides are not suitable for quaternization of (**1**) but methyl sulfate at 100 °C slowly forms the monomethyl salt ⟨65JCS6769⟩. Benzoselenadiazoles, however, react rapidly with dimethyl sulfate, producing quaternary salts in high yield. Ethylation of (**2**) with ethyl *p*-toluenesulfonate failed but succeeded with the more reactive ethyl 2,4-dinitrobenzenesulfonate ⟨70RCR923⟩. Quaternization of the 5(6)-substituted derivatives of (**1**) and (**2**) with dimethyl sulfate leads to a nearly equimolar mixture of isomeric salts ⟨76CHE289⟩. Unambiguous confirmation of the structures of these salts was obtained by their independent syntheses from appropriately substituted *N*-methyl-*o*-phenylenediamines.

A kinetic study of the quaternization of (**1**), (**2**) and (**3**) and related heterocycles with dimethyl sulfate ⟨74AJC1917, 65JCS6769⟩ showed the order of activation by the heteroatom in benzazoles to be Se > NMe > S > O ⟨74AJC1917⟩. Benzo fusion has very little effect on the rate of quaternization of thiadiazole, implying that development of benzenoid character in the product (**9**) is relatively unimportant as a driving force for the reaction (see Section 4.26.2.9).

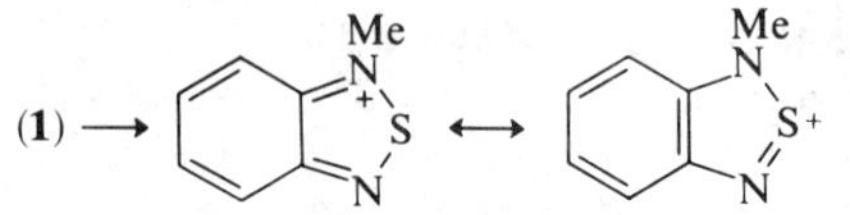

Compounds (**1**) and (**2**) form very insoluble complexes with a variety of metals: Cd^{2+}, Zn^{2+}, Co^{2+}, Ni^{2+}, Fe^{2+}, Hg^{2+}, Cu^{2+}, Mn^{2+}, Ag^{+}, Ce^{4+}, Ti^{4+} ⟨70RCR923⟩. Differential thermal analysis and thermogravimetric analysis ⟨71MI42600, 71MI42601⟩ show that the complexes of (**2**) are substantially more stable than those of (**1**), consistent with other data concerning the relative basicity of the donor nitrogen atoms. In some cases, *e.g.* the Cd^{+} complex, a substituent in the 5(6)-position has a predictable effect on the thermal stability of the benzoselenadiazole complexes, but the $CuCl_2$ complexes did not follow the expected trend. The IR spectra of these complexes have been studied in detail ⟨77ZOB1888, 73ZOB1179⟩ but other structural data, particularly X-ray diffraction, are lacking.

The naphtho[2,3-*d*][2,1,3]selenadiazole forms two stable complexes with $PdCl_2$ which serve as the basis of five analytical procedures (gravimetric, spectrophotometric, radiometric, fluorometric and atomic absorption) for the specific determination of macro to submicro quantities of palladium ⟨70MI42600⟩.

4.26.3.1.4 *Nucleophilic attack at carbon, hydrogen or sulfur*

As expected from the electronic structure, the 1,2,5-thiadiazoles readily undergo nucleophilic attack (see also Section 4.02.1.6). The site of attack, however, is variable and can take place either at carbon, sulfur or a ring proton (Scheme 1; also see Section 4.26.3.3.6).

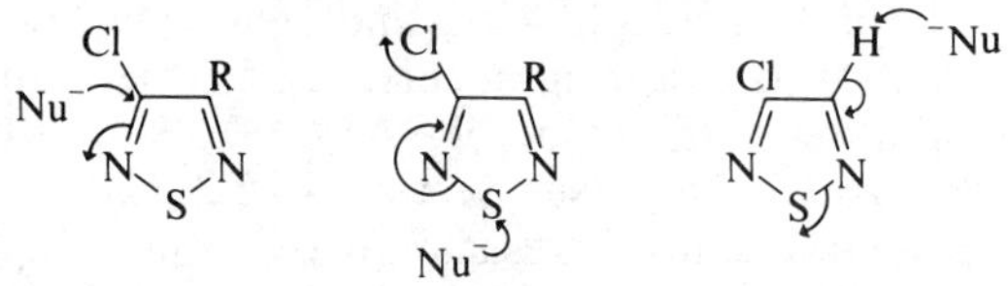

Scheme 1

Attack at carbon leads to production of the expected displacement product while attack at sulfur can result in either rearrangement products or total ring destruction. Abstraction of the ring proton proceeds to either ring destruction or simple proton exchange ⟨68AHC(9)107⟩. This is in contrast to the chemistry of the furazans, where proton abstraction and ring cleavage to α-cyanoximes are synchronous ⟨65JOC1854⟩.

The primary site of attack in the reactions of the 1,2,5-thiadiazoles and selenadiazoles with lithium alkyls or Grignard reagents is usually at the ring sulfur or selenium (Scheme 2, path A) but attack on nitrogen (path B) or the ring carbon (path C) also occurs. *N*-Phenyl-*o*-phenylenediamine is produced in the reaction of (**1**) and (**2**) with phenyllithium, probably *via* path B ⟨55JCS1468⟩. The driving force for this mode of addition could depend on the stable benzenoid character of the addition intermediate.

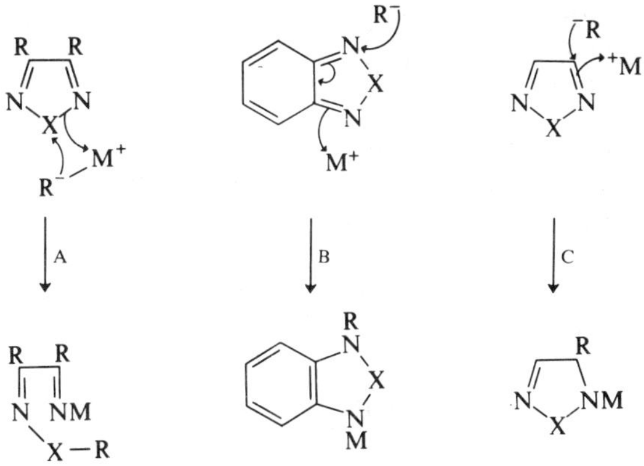

Scheme 2

Monocyclic thiadiazole and selenadiazoles bearing alkyl or aryl substituents at the 3,4-positions usually react at sulfur with the organometallic reagents *via* path A to yield, after hydrolysis, a thio or seleno ether, ammonia and a 1,2-dicarbonyl compound ⟨74JOC2294, 70CJC2006⟩. Unsubstituted thiadiazole, on the other hand, reacts with Grignard reagents, at least in part, *via* path C. An intermediate forms which is converted to a *C*-alkylated 1,2,5-thiadiazole ⟨80T1245⟩ or -selenadiazole ⟨79S979⟩ on treatment with sulfur dichloride or selenium tetrachloride. The intermediate was originally formulated as arising from Grignard attack at the sulfur atom of (**3**) ⟨80T1245⟩.

Reaction of the quaternary compound (**10**) with ammonia in acetonitrile gives the rearrangement product (**11**), indicating attack at sulfur at some stage of the process ⟨79JOC1118⟩. Two mechanistic pathways could account for these results (path A and path B, Scheme 3). The reaction of (**10**) with CD_3NH_2 produces an approximately 1:1 mixture of isomers (**14**) and (**15**). This has been interpreted as a process proceeding by path B and the product-forming intermediate (**13**) ⟨82PC42600⟩. The intermediate (**12**), stemming from reaction of CD_3NH_2 and (**10**) *via* path A, would produce (**14**) exclusively. Path A predominates in the reactions of the corresponding isothiazole cations.

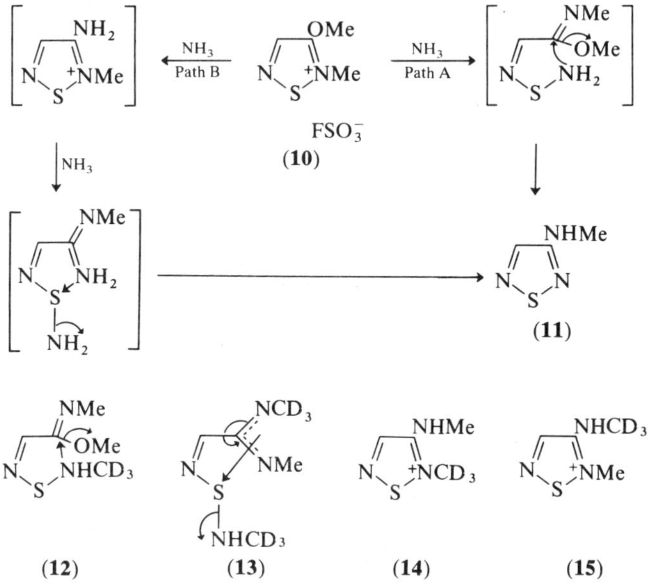

Scheme 3

The reaction of the quaternary compound (**16**) with carbon nucleophiles results in attack at sulfur leading to an open-chain intermediate that recyclizes to ring-expanded compounds ⟨79TL1281⟩. Reaction of (**16**) with one equivalent of cyanide ion proceeds according to Scheme 4 to produce (**17**) in high yield. When an excess of cyanide ion is employed, (**17**) reacts further to form 80% of a compound whose structure is represented by either (**18**) or (**19**). The reaction of (**16**) with methyl propiolate anion produces (**20**) *via* a similar process.

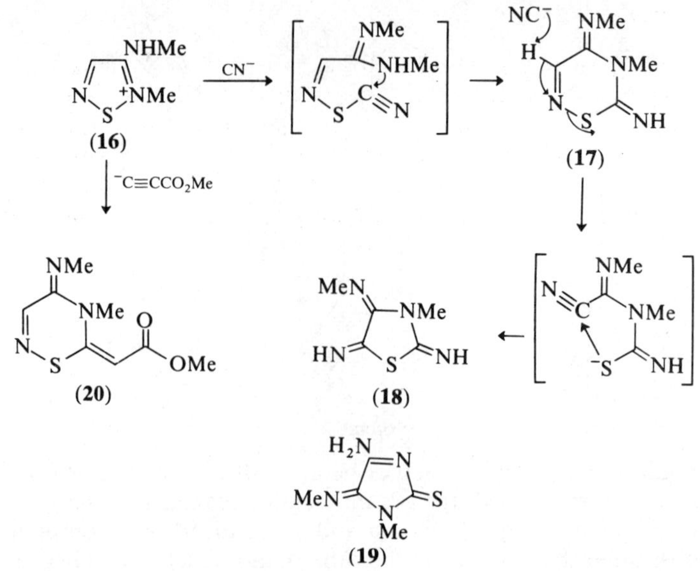

Scheme 4

In certain bicyclic and polycyclic thiadiazoles the ring can be hydrolytically cleaved to an o-diamine and sulfur dioxide. This process is a reversal of the reaction pathway involved in the synthesis of thiadiazoles from o-diamines and thionyl chloride or N-sulfinylaniline under anhydrous conditions (see Section 4.26.5.1.1), and is analogous to, but much slower than, the hydrolysis of sulfurdiimides. In contrast to (1), which steam distilled without decomposition, steam distillation of [1,2,5]thiadiazolo[3,4-c][1,2,5]thiadiazole (21) produced only oxamide. Under milder conditions (75 °C, 12 h) a mixture of the diamine (22), oxamide and elemental sulfur was obtained ⟨75JOC2749⟩. Similar hydrolytic instability was observed in thiadiazole rings angularly joined to an anthraquinone ⟨70RCR923⟩ and benzofuroxan ⟨78CPB3896⟩.

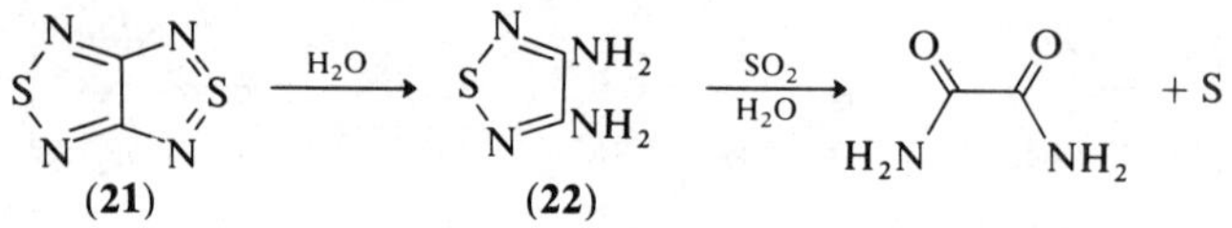

4.26.3.1.5 Reactions with radicals (reduction)

The 1,2,5-thiadiazole ring is susceptible to reduction with the more vigorous reducing agents including zinc and tin in acids, sodium and alcohol, lithium aluminium hydride, and Raney nickel ⟨68AHC(9)107⟩ (see also Section 4.02.1.8). Reduction occurs at the N—S bond with regeneration of the NCCN substructure and formation of hydrogen sulfide, a property which has been employed as a method of structural proof for 1,2,5-thiadiazoles. Reduction of bicyclic thiadiazoles also provides a convenient method for preparing heterocyclic and aromatic o-diamines which are difficult to obtain by other methods ⟨70RCR923⟩. For example, nitration of (1) or (2) followed by reduction produces 1,2,3-triaminobenzene.

The thiadiazole ring was employed as a diamine protecting group in a synthesis which led to the unambiguous formation of the 9-substituted adenine arprinocid (24) ⟨78JOC960⟩. The thiadiazole (23) readily underwent amine replacement, formylation, and reduction with Raney nickel to produce (24; Scheme 5).

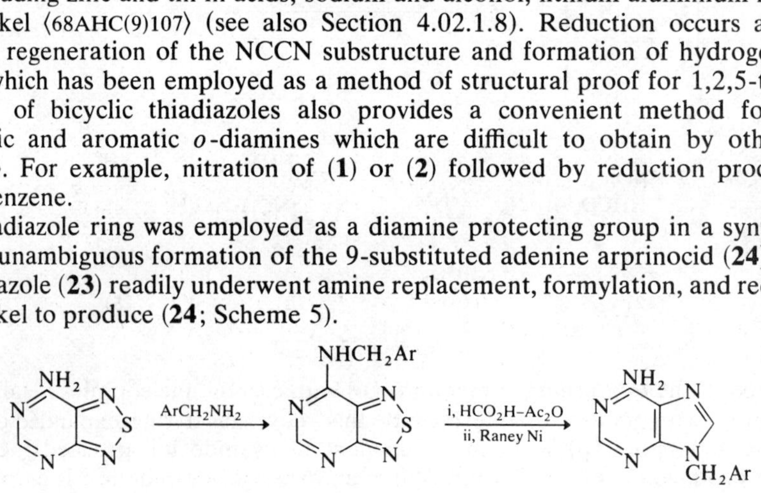

Scheme 5

4.26.3.1.6 Cyclic transition state reactions

Attack at the heteroatoms was observed in the reactions of (**1**) and (**2**) with dimethyl acetylenedicarboxylate and with benzyne ⟨78JCS(P1)1006⟩. Compound (**2**) behaves as a heterodiene toward the acetylene ester, producing the quinoxaline (**25**) and selenium. The diester (**25**) was obtained only in trace amounts from (**1**). These reactions, and the failure of (**1**) and (**2**) to react with maleic anhydride, contrast with the behavior of naphtho[2,3-c][1,2,5]thiadiazole which reacts with maleic anhydride across the center ring ⟨64TL3815⟩.

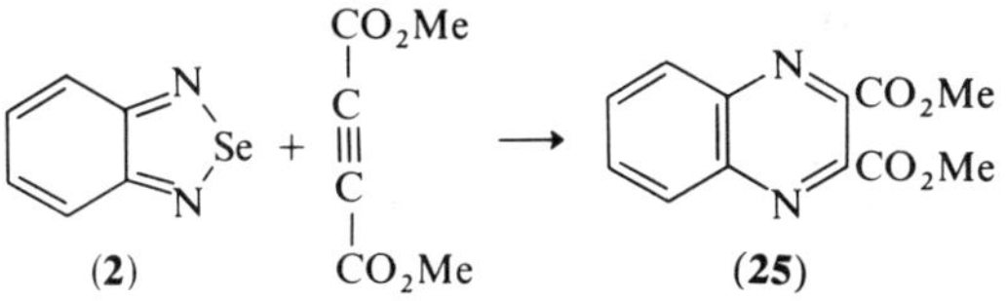

Benzyne, generated either by oxidation of 1-aminobenzotriazole with lead tetraacetate or by decomposition of benzenediazonium-2-carboxylate, was efficiently trapped by (**2**) to give 88% of the 1,2-benzisoselenazole (**26**); however, it was trapped in only 5–10% yield by (**1**). The series of adducts analogous to (**26**), prepared from benzyne and substituted benzoselenadiazoles ⟨81JCS(P1)607⟩, occurred *via* attack of the benzyne at the selenium atom followed by reorganization of the intermediate according to Scheme 6. Benzyne addition to dimethylthiadiazole produced methyl derivatives of quinoline and 1,2-benzisothiazole ⟨82CC299⟩.

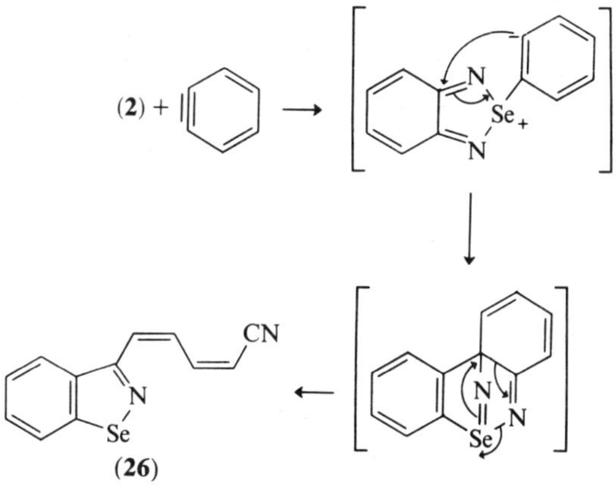

Scheme 6

4.26.3.2 Reactions of Non-aromatic Compounds

4.26.3.2.1 S-Monoxides and dioxides

1,2,5-Thiadiazole 1-oxides and 1,1-dioxides are non-aromatic (see Section 4.26.2.2), less thermally stable than the aromatic form ⟨68AHC(9)107, 74S22⟩ and exhibit greatly enhanced reactivity toward nucleophiles. The observation has been made that these systems behave like strong acyl functions and that the chloro, alkoxy and amino derivatives have chemical behavior resembling an acyl chloride, ester and amide of a strong acid ⟨75JOC2743, 70CRV593⟩. The pK_a values of the hydroxy derivatives are in accord with these properties (see Section 4.26.2.8). Dichlorothiadiazole 1,1-dioxide (**27**) is extremely reactive with nucleophiles and can be hydrolyzed and undergoes reaction with methanol without base catalysis (Scheme 7). The dimethoxy derivative (**28**) retains the reactivity of an active ester and is readily converted into the diamines (**29**) and (**30**). The diamine (**29**) forms a disodium salt which condenses with (**27**), producing the resonance-stabilized dianion (**31**). The dichloride (**27**) also condenses with anthranilic acid followed by acetic anhydride to form the pentacyclic system (**32**).

Thiadiazole 1-monoxides react analogously ⟨81H(16)1561, 81H(16)1565⟩. When alkylthio is the leaving group, substitution with carbon nucleophiles can be accomplished. Thus, the methylthio analog (**33**) reacts smoothly with malonate anion to form (**34**). Analogous

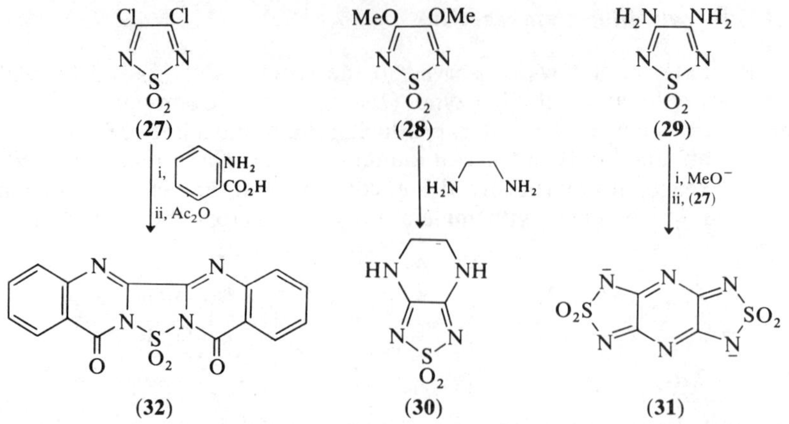

Scheme 7

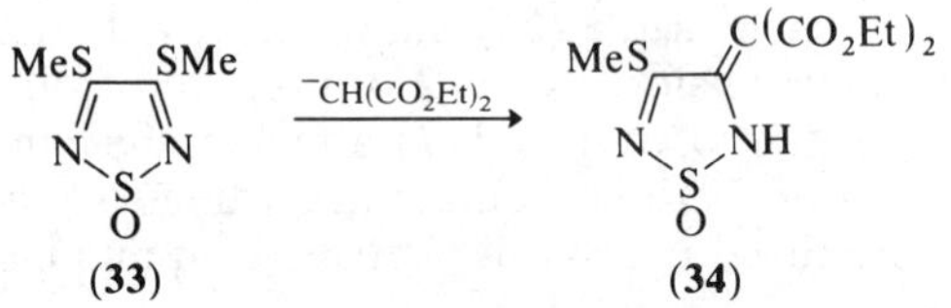

products were obtained with acetophenone and benzyl anions ⟨81H(16)1561⟩. There are two interesting features in the chemistry of 1,2,5-thiadiazole 1-monoxides. First, unsymmetrically substituted thiadiazole monoxides are chiral and, by virtue of the high sulfoxide inversion barrier, can be separated into two stable isomers ⟨82JA1375⟩. Second, suitably substituted thiadiazole monoxides produce reactive N-sulfinylamine intermediates which recyclize by nucleophilic addition and cycloaddition to affect a variety of facile ring interconversions (Scheme 8). The thiadiazole 1-oxide (**35**) reacts with phenylhydrazine to form (**36**) which is spontaneously converted into the thiatriazole (**38**) *via* (**37**). By a similar pathway, (**35**) forms (**39**) on reaction with amidines. The N-sulfinylamine intermediates are also trapped intramolecularly by cycloaddition, as in the conversion of (**40**) into (**41**), and by cyclocondensation, *e.g.* (**42**) → (**43**) ⟨82UP42600⟩.

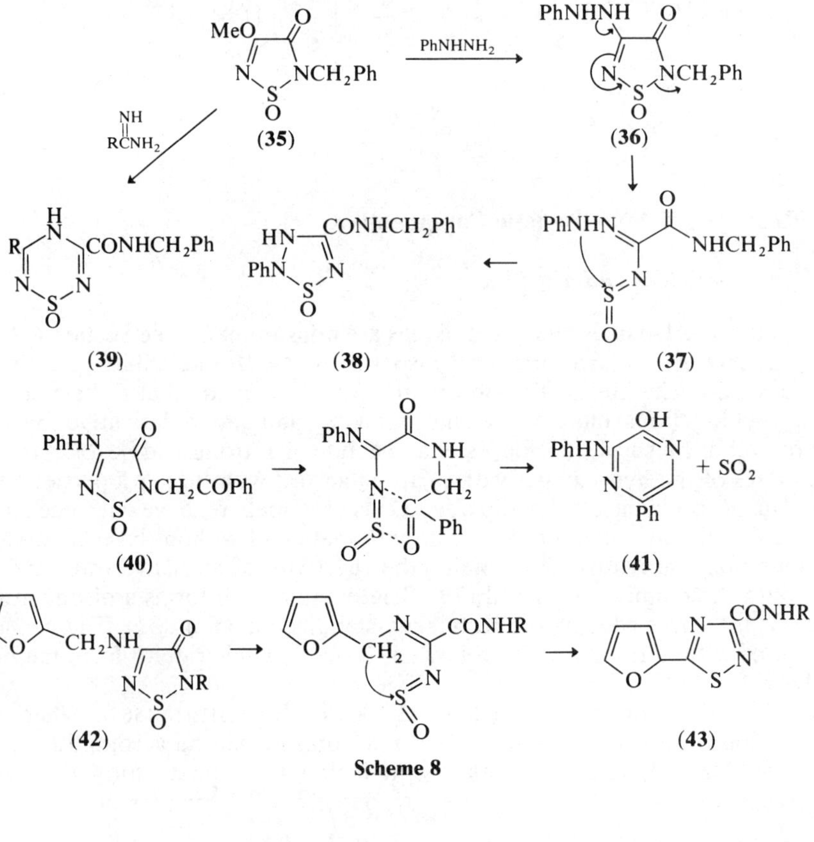

Scheme 8

Thiadiazole 1-monoxide and 1,1-dioxides react with electrophiles at the ring nitrogen atom (Scheme 9). Alkylation of the disilver salt of 3,4-dihydroxy-1,2,5-thiadiazole 1,1-dioxide with methyl iodide gives (44) and (45) ⟨75JOC2743⟩, while the potassium salt of the thiadiazole monoxide produces the *N*-alkylated and *N*-acetylated products ⟨81H(16)1561⟩. Thermal treatment of alkoxy derivatives of thiadiazole monoxides and dioxides results in O→N alkyl migration to produce 2-alkyl-3-ones, *e.g.* (47)→(46). The amino derivatives of thiadiazole monoxides also react with a wide variety of electrophilic reagents on the ring nitrogen atom ⟨81H(16)1565⟩.

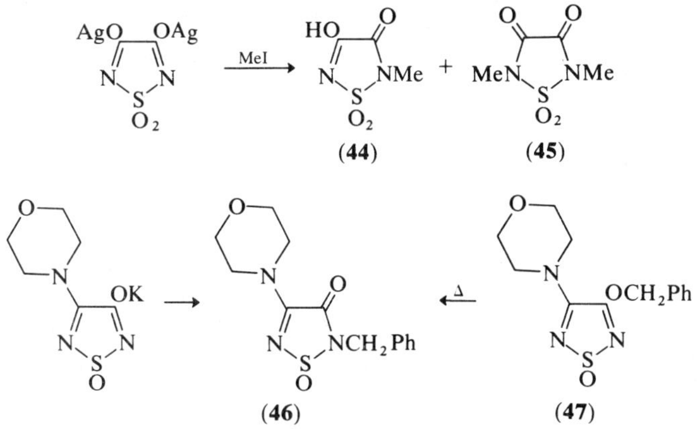

Scheme 9

4.26.3.2.2 Reduced compounds

The nitrogen–sulfur bond in reduced forms of 1,2,5-thiadiazole is much less stable than in the aromatic form. The reduced systems are readily hydrolytically desulfurized to the open-chain NCCN portion of the ring ⟨80JPR273, 69OPP255, 69T4277⟩. Mesoionic thiadiazoles are also sensitive to basic hydrolysis ⟨81JCS(P1)1033⟩.

4.26.3.3 Reactions of Substituents

4.26.3.3.1 Benzenoid rings

The 2,1,3-benzothiadiazoles and benzoselenadiazoles readily enter into most classical electrophilic aromatic substitution reactions, *e.g.* nitration, sulfonation, halogenation and chloromethylation, but are relatively inert under Friedel–Crafts condition ⟨70RCR923⟩ (see also Section 4.02.3.2). Orientation in these reactions is similar to the naphthalenes, but the reactivity of the heterocycle is lower. Electrophiles attack (1) and (2) at the C-4 (C-7) position. Substituents in the benzene ring exert a greater orientation effect than the fused thiadiazole ring. The deactivation of position 5(6) to electrophilic attack is overcome by the presence of a donor substituent at position 4. For example, mononitration of 4-methyl- and 4-ethoxy-2,1,3-benzothiadiazole yields a mixture of 5- and 7-nitro derivatives in the proportions 35:65 and 50:50, respectively, while the 4-hydroxy derivative is nitrated exclusively in position 5. Most of the other electrophilic reactions (sulfonation, halogenation and chloromethylation) of benzothiadiazoles bearing an electron donor at position 4 take place at position 7 and also readily yield 5,7-disubstituted products. Donor substituents in position 5 of (1) and (2) always direct electrophiles to position 4. The presence of electron-withdrawing substituents exerts the predicted deactivating and *meta*-directing effect, *e.g.* nitration of either 4- or 5-nitro-2,1,3-benzothiadiazole produces the same 4,6-dinitro derivative ⟨63JGU1714⟩.

When the benzene ring of (1) is strongly activated, the milder electrophilic substitution reactions (nitrosation, azo coupling and thiocyanation) are readily effected in a manner analogous to the naphthalenes ⟨70RCR923⟩. Nitrosation of the 4- and 5-hydroxy derivatives of (1) occurs at positions 5 and 4, leading to nitroso derivatives that exist predominantly in the quinone monoxime form ⟨74T3839⟩.

Chlorination of 2,1,3-benzothiadiazoles can take place either by addition or substitution, depending on the reaction conditions, the nature of the substituents in the benzene ring and the catalyst employed. The uncatalyzed reaction of (1) with chlorine is exothermic and produces an isomeric mixture of tetrachloro addition products, which form 4,7-dichloro-2,1,3-benzothiadiazole on treatment with base ⟨70RCR923⟩. In the presence of an iron catalyst, chlorine substitution in the 4,7-positions predominates.

Bromination of (1) leads to the tetrabromide which, when treated with base, forms 4,7-dibromo-2,1,3-benzothiadiazole ⟨70RCR923⟩. It was subsequently demonstrated that the 4,6-dibromo (20%) and a trace of the 4,5-dibromo derivatives are also produced in this reaction ⟨70JHC629⟩. In the presence of iron, bromination of a melt of (1) yields almost exclusively the 4,7-dibromo derivative even when equimolar quantities of bromine are used. As in the chlorination of (1), the process under these conditions cannot be limited to monosubstitution ⟨70RCR923⟩. Monobromination and subsequent dibromination can be achieved, however, when the reaction is carried out by dropwise addition of bromine to (1) in 47% hydrobromic acid at 130 °C ⟨70JHC629⟩.

Ipso-substitution can result if an electron withdrawing substituent is present at position 4 of benzothiadiazole. Several examples of *ipso*-substitution are known including chlorodenitration, chloro and bromodesulfonation and nitrodebromination ⟨70RCR923⟩. Transbromination products ⟨71JOC207⟩ have been detected in some nitrodebrominations (Scheme 10). Treatment of 4,7-dibromo-2,1,3-benzothiadiazole (48) with 70% nitric acid under reflux produces 4-bromo-7-nitro-2,1,3-benzothiadiazole (50) together with the 4,5,7-tribromo (51) and tetrabromo (52) analogs. The polybromination products probably result from attack of (48) by a cationic bromine species derived from the *ipso*-substitution intermediate (49). Since the nitronium ion concentration in 70% nitric acid is very low, the attacking species in this process is probably the nitrosonium ion (NO$^+$) and the reaction best characterized as a nitrosodebromination, followed by rapid oxidation to the nitro compound. Normal nitronium ion attack on (48) with sodium nitrate in concentrated sulfuric acid ⟨67CHE662⟩ gives exclusively the 6-nitro compound (53). The reactions summarized in Scheme 10 are analogous to nitrosodebrominations that have been observed in the benzene series ⟨76ACR287⟩.

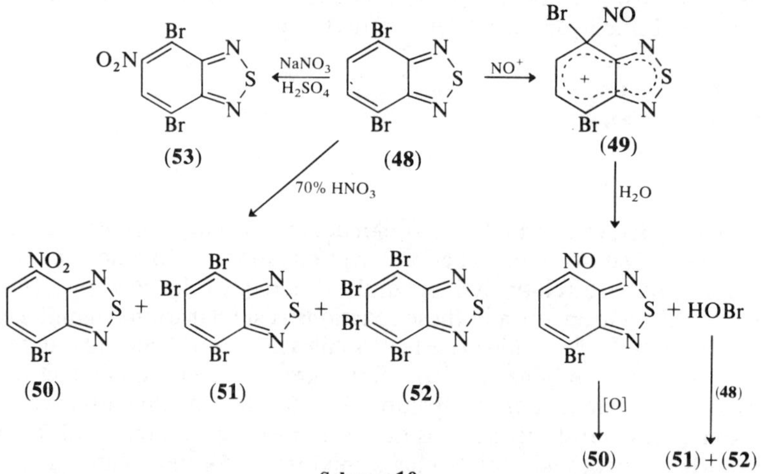

Scheme 10

The direct amination of (1) and (2) by hydroxylamine in concentrated sulfuric acid in the presence of vanadium pentoxide ⟨71CHE561, 72CHE297⟩ gives equal amounts of the 4- and 5-amino derivatives, indicating a probable radical mechanism. Similar results were obtained on amination of the 4- and 5-methyl derivatives of (1) and (2).

The electron-acceptor character of the annulated thiadiazole and selenadiazole rings favor nucleophilic substitution in the benzene rings of (1) and (2). In the absence of other activating groups the preferred site of nucleophilic attack is the 5-position, although differences in reactivity with the 4-position are not large. The ratio $k(5\text{-Cl})/k(4\text{-Cl})$ in the reaction with sodium methoxide in methanol is 5 in the thiadiazole and 13 in the selenadiazole series ⟨63AC(R)1697⟩. The major differences in the stability of transition states (54) and (55) may reflect the fact that attack by a negatively charged nucleophile at the 4-position is repelled by the lone pair of electrons of the *peri* aza group ⟨71JCS(B)2209⟩. The thiadiazole and selenadiazole analogs are $>10^3$ times faster than chlorobenzene but are ca. 10^2 times slower than o- and p-nitrochlorobenzene in the reaction with methoxide ion. The 4-bromo-5-methyl-2,1,3-benzothiadiazole reacts with morpholine, ethanolic potassium hydroxide and cuprous cyanide to produce fair yields of the corresponding displacement products ⟨76IJC(B)1001, 74JHC777⟩.

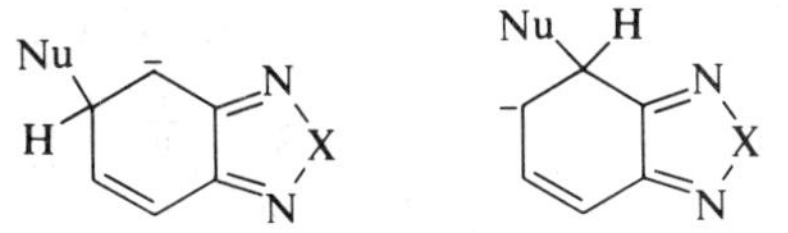

(54)   (55)

The introduction of an activating nitro group into the halogenobenzothiadiazole and selenadiazole causes dramatic increases in reactivity by a factor of 10^5–10^7, which exceeds the reactivity of mononitro-activated chlorobenzenes by a factor of 10^3 ⟨69MI42600, 71JCS(B)2209⟩. Detailed kinetic studies of the methoxydechlorination of 5-chloro-4-nitro-2,1,3-benzothiadiazole ⟨71JCS(B)2209⟩ and 7-chloro-4-nitro-2,1,3-benzothiadiazole ⟨77T855⟩ indicate that these reactions proceed by a 'normal' overall second order, 2-step, S_NAr2-type mechanism. The chloro group in 5-chloro-4-nitro-2,1,3-benzothiadiazole is readily replaced by ammonia in boiling ethylene glycol, conditions which do not effect replacement in o- and p-chloronitrobenzene ⟨68CHE186⟩. Substitution reactions employing other nucleophiles, e.g. aniline, diethylamine, hydrazine, phenylhydrazine, sodium ethoxide, sodium hydroxide, potassium thiocyanate, sodium diethylthiophosphate and sodium azide are all readily carried out in the 2,1,3-benzothiadiazole series under fairly mild conditions ⟨70RCR923, 80CPB1909⟩.

The oxidation of 2,1,3-benzothiadiazole (**1**) with aqueous potassium permanganate was very thoroughly investigated and developed into a procedure which produces good yields of 1,2,5-thiadiazole-3,4-dicarboxylic acid (**58**) ⟨68AHC(9)107⟩. The detection of polyols (**56**) and (**57**), oxalate and carbonate among the reaction products led to the interpretation of the oxidation as proceeding *via* hydroxylation of the benzene ring as shown in Scheme 11 ⟨71MI42603⟩. The dioxide (**59**) is the major product of forced oxidation of (**1**) or (**58**) at 80–100 °C ⟨75JOC2743⟩.

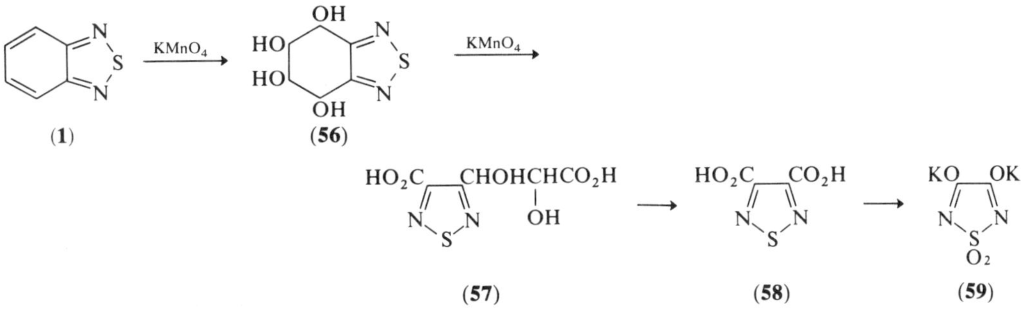

Scheme 11

4.26.3.3.2 Heterocyclic rings

The fusion of the 1,2,5-thiadiazole or selenadiazole system to a heterocyclic nucleus greatly enhances the electrophilicity of the substituted ring. These properties are manifested in increased rates of hydrolysis and more facile displacement, Diels–Alder and cycloaddition reactions. The synthetically useful nucleophilic cleavage of the pyrimidine ring of [1,2,5]thiadiazolo[3,4-*d*]pyrimidines was studied in detail by Shealy and coworkers and previously discussed ⟨68AHC(9)107⟩.

Leaving groups present in heterocyclic rings fused to the 1,2,5-thiadiazole and selenadiazole systems are highly mobile and are displaced readily under substitution conditions ⟨70RCR923, 70RTC5⟩. The chloro group in (**60**) is extremely reactive and hydrolyzed by aqueous acid much more readily than the corresponding imidazopyridine. Similarly, the substituents Cl, SR and SH in the 7-position of the pyrimidothiadiazole (**61**) are easily replaced. The amino group in (**23**) can be substituted by heating with secondary amines ⟨68AHC(9)107⟩ and substituted benzylamines ⟨78JOC960⟩. Nucleophilic substitution of (**62**)

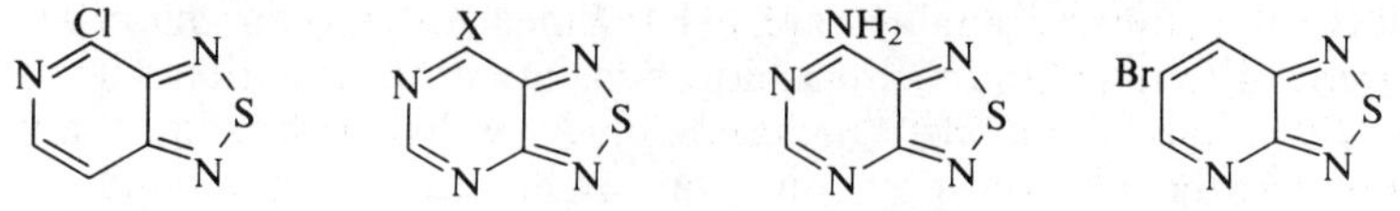

(60) (61) X = Cl, SH, SR (23) (62)

with morpholine resulted in rearrangement to the 7-substitution product *via* a 6,7-dehydro intermediate ⟨79IJC(B)13⟩.

The nonclassical 10π-electron system, thieno[3,4-c][1,2,5]thiadiazole (63), containing tetravalent sulfur, behaves like a thiocarbonyl ylide in cycloadditions (Scheme 12). No reaction is observed at the N—S=N dipole. The unsubstituted analog is not stable enough to be isolated but is gradually converted to the dimer (64) ⟨79TL4493⟩. Generation of (63) in the presence of dimethyl acetylenedicarboxylate or N-phenylmaleimide gives rise to dimethyl 2,1,3-benzothiadiazole-5,6-dicarboxylate (66) and the adduct (65), respectively. The 2,5-diphenyl analog (67; Scheme 13), on the other hand, is stable and readily isolated as brilliant purple needles ⟨69JA6891⟩. Compound (67) undergoes a Diels–Alder reaction with N-phenylmaleimide at 140 °C, producing a mixture of the two thermally interconvertible *endo* and *exo* adducts (68) and (69) ⟨79CL1029⟩. The cycloaddition reaction of (67) with 6,6-diphenylfulvene formed two stereoisomeric [3+2] adducts (70) and (71), also subject to retrocycloaddition and recombination. This is indicated by the time dependence of the product ratio and the facile thermal conversion of (71) into the benzothiadiazole (72) *via* (67) by refluxing with dimethyl acetylenedicarboxylate, a consequence of the rapid desulfurization of the strained intermediate cycloadduct ⟨80CL1031, 80JOC2956⟩.

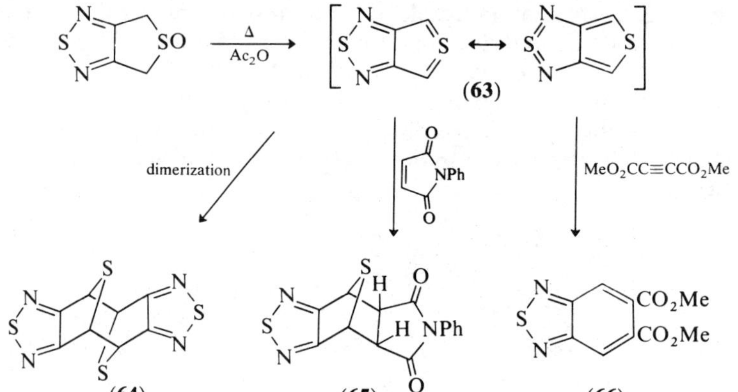

Scheme 12

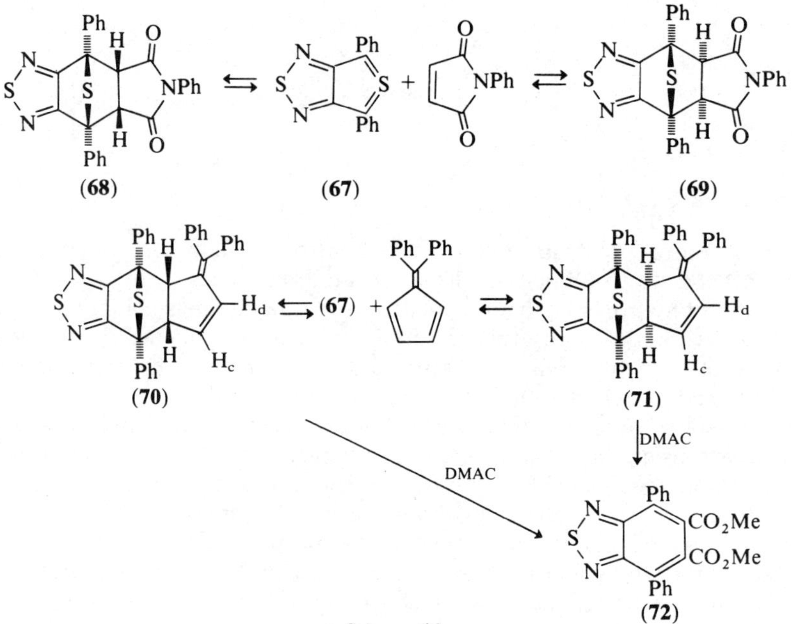

Scheme 13

All attempts to isolate the nonclassical 14π-electron tricyclic system (**73**) resulted only in the formation of dimer (**73a**), condensed at the benzene ring (Scheme 14). Reaction of (**73**) with N-phenylmaleimide, on the other hand, takes place at both the benzene and thiophene rings ⟨69JA6891⟩.

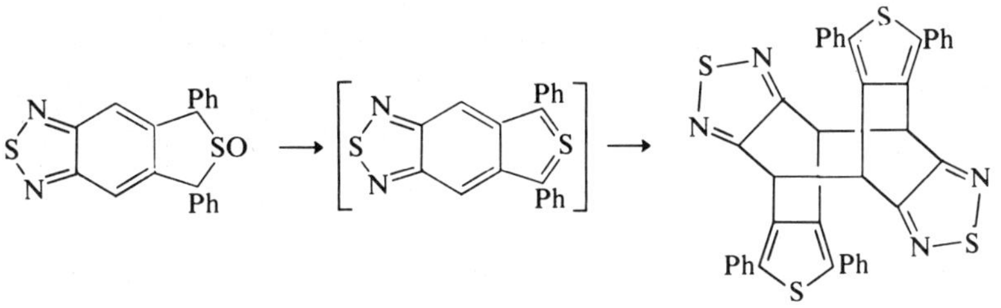

Scheme 14

The [1,2,5]thiadiazolo[3,4-c][1,2,5]thiadiazole system (**21**), being heteroaromatic according to bond lengths, does not undergo any of the cycloaddition reactions described above ⟨75JOC2749⟩. It is interesting to note that 3,7-diphenyl-1,5-dithia-2,4,6,8-tetrazocine (**74**) is planar and also has molecular parameters and spectral properties suggestive of a delocalized, 10π aromatic system ⟨81JA1540⟩. The bis(dimethylamino) analog (**75**), however, is folded along an axis drawn through the sulfur atoms.

(**74**) R = Ph
(**75**) R = NMe$_2$

4.26.3.3.3 Other C-linked substituents

The effect of the 1,2,5-thiadiazole system on carbon-bound substituents (see Section 4.02.3.3.1) can be summarized as follows: (1) stabilization of carbanions; (2) destabilization of carbenium ions; (3) enhanced S_N2 reactivity and repressed S_N1 reactivity ⟨68AHC(9)107⟩. Aryl substituents are rendered more reactive to nucleophiles ⟨72IJS(A)25⟩ and deactivated in reactions with electrophiles, which are directed to the *ortho/para* positions by the thiadiazole ring ⟨78MI42601, 72IJS(A)25⟩.

Metallation of 3,4-dimethyl-1,2,5-thiadiazole (**76**) to the anion (**77**) was accomplished with the use of a non-nucleophilic base, lithium diisopropylamide ⟨81JHC1247⟩. Nucleophilic attack at sulfur (see Section 4.26.3.2) resulted in an alkyllithium reagent ⟨70CJC2006⟩. The lithiomethyl derivative (**77**) was carboxylated to (**78**) with carbon dioxide and converted to the vinyl derivative (**79**) *via* an esterification, reduction, mesylation and base elimination sequence.

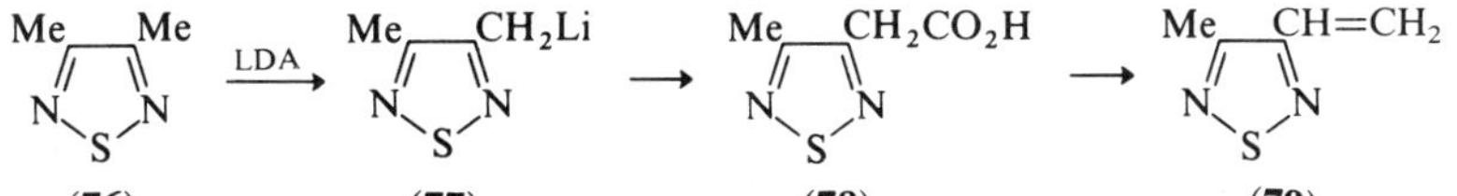

(**76**)   (**77**)   (**78**)   (**79**)

The pronounced electron-withdrawing nature of the 1,2,5-thiadiazole system is also evidenced by strong carbonyl electrophilic activation and by enhancement of carboxy acidity. The acid dissociation constants of thiadiazole acids, discussed in Section 4.26.2.8, fall in the range of 1.5–2.5. The 1,2,5-thiadiazole carboxylic acids are easily decarboxylated at 160–200 °C. This reaction has been used for the synthesis of monosubstituted derivatives as well as the parent ring and deuterated derivatives ⟨68AHC(9)107⟩. An efficient bromodecarboxylation of 3-amino-1,2,5-thiadiazolecarboxylic acid has also been reported ⟨70BRP1190359⟩.

The enhanced acidity of the thiadiazole acids is reflected in the reactivity of the corresponding esters which undergo hydrolysis, aminolysis and hydrazinolysis with extreme ease. Thiadiazole-amides and -hydrazides behave normally in the Hofmann and Curtius reactions.

The acid chlorides have served as useful synthetic intermediates leading to ketones via the malonic acid synthesis and Friedel–Crafts reaction, thiadiazole acetic acid derivatives and halo ketones via the Arndt–Eistert synthesis and carbinols by hydride reduction ⟨68AHC(9)107⟩. The dialkylcadmium conversion of acid chlorides into ketones fails in the 1,2,5-thiadiazole series. The major product is either a tertiary carbinol or the corresponding dehydration product, by virtue of the high reactivity of the intermediate ketone.

Acetyl thiadiazole also displays enhanced reactivity in both the carbonyl and the methyl groups and enters the aldol condensation, the Mannich reaction, cyanoethylation reaction, the Willgerodt reaction, the Leukart reaction and borohydride reduction, all in high yield under very mild conditions ⟨68AHC(9)107⟩.

Addition of nucleophiles to cyanothiadiazole under basic conditions takes place with unusual ease. Hydrolysis to the amide, for example, can be effected at 0 °C in the presence of a catalytic amount of sodium hydroxide or basic ion-exchange resin. At reflux temperature, hydrazine and monosubstituted hydrazines convert 3,4-dicyano-1,2,5-thiadiazole into the [1,2,5]thiadiazolo[3,4-d]pyridazines ⟨71JHC441⟩. The base-catalyzed addition of acetone to cyanothiadiazole forms an enamino ketone, used as a key intermediate for the synthesis of a number of heterocyclic ring systems, e.g. isothiazole, isoxazole, pyrazole, pyrimidine and thiazole ⟨77H(6)1985⟩.

4.26.3.3.4 N-Linked substituents

3-Amino-1,2,5-thiadiazole is a weak base (see Section 4.26.2.8) but is sufficiently nucleophilic to readily form acyl, aroyl, carbamyl, sulfanilyl and sulfinyl derivatives on the exocyclic nitrogen atom. Similarly, all the observed reactions of diaminothiadiazole ⟨76JHC13⟩ involve the exocyclic nitrogen atoms. IR, UV and NMR spectroscopic evidence clearly indicate, however, that aminothiadiazole protonates on the ring nitrogen atom ⟨68AHC(9)107⟩.

Diazotization of the monoamine in acidic aqueous media formed only the diazoamino compound in high yield. In an attempt to generate and trap 1,2,5-thiadiazyne via diazotization of the 3-amino-4-carboxylic acid in the presence of anthracene, a mixture comprising 9-nitroanthracene and a small amount of 9-thiocyanoanthracene was obtained ⟨71TL2143⟩.

Despite the tendency of the thiadiazole ring to undergo reductive cleavage, milder reducing agents selectively reduce nitro, nitroso, azo, acyl chlorides and even ester functions without rupture of the ring system. Selective reducing agents include sodium borohydride ⟨67JHC445⟩, diisobutylaluminum hydride ⟨81JHC1247⟩, iron in acetic acid, sodium dithionite, sodium sulfide, zinc in calcium chloride, alkaline iron(II) sulfate ⟨70RCR923, 69CHE53⟩, sodium hydrosulfite ⟨75JHC829⟩, palladium and sodium phosphinite ⟨78T213⟩, and palladium on carbon ⟨74JHC835⟩.

4.26.3.3.5 O-Linked substituents

Hydroxythiadiazoles exhibit marked acid properties (see Section 4.26.2.8) and exist essentially in the hydroxy form ⟨68AHC(9)107⟩. Alkylation with alkyl halides or sulfates and base in dipolar aprotic solvents usually forms ethers. In media where the oxygen atom is screened by solvation or by formation of ion pairs, mixtures of N- and O-alkylation products are formed. The reaction of 3-chloro-4-hydroxy-1,2,5-thiadiazole with epichlorohydrin catalyzed by piperidine in the absence of a solvent gave a mixture of N- and O-alkylated products in about equal amounts ⟨72JMC651⟩. Treatment of 3-hydroxy-4-(N-morpholino)-1,2,5-thiadiazole with chloroacetic acid in aqueous sodium hydroxide produced very low yields of O-alkylated product ⟨78CJC722⟩. An attempt to prepare a mesoionic compound by the alkylation of 3-hydroxy-4-phenyl-1,2,5-thiadiazole with dimethyl sulfate and methyl fluorosulfonate in the absence of base produced 2-methyl-4-phenyl-1,2,5-thiadiazol-3-one ⟨81JCS(P1)1033⟩. Diazomethane gave a mixture of N- and O-methylation products.

Hydroxythiadiazoles can be converted into chloro compounds by the usual methods, e.g. with $POCl_3$. The O-acylation of the sodium salt of hydroxythiadiazoles with dimethylthiocarbamoyl chloride followed by thermal rearrangement of the thiolcarbamate and hydrolysis is a convenient route to 1,2,5-thiadiazole-3-thiones ⟨73CJC2349⟩.

4.26.3.3.6 Halogens

The halogens in 1,2,5-thiadiazoles are reactive but 3-chloro-1,2,5-thiadiazole usually does not produce high yields in displacement reactions, probably because of ring decomposition *via* proton abstractions (see Section 4.26.3.1.4) ⟨68AHC(9)107⟩.

A detailed study of the conversion of 3,4-dichloro-1,2,5-thiadiazole into 3,4-diamino-1,2,5-thiadiazole has been carried out ⟨76JHC13⟩. Reaction with lithium or sodium amide produces only 4% of the diamine together with cyano-containing by-products, a consequence of direct attack on sulfur. Use of a less powerful nucleophile, ammonia or potassium phthalimide, resulted in an increased attack on carbon and produced the diamine in 24% and 66% yields, respectively. Secondary amines, *e.g.* morpholine ⟨76JOC3121⟩ and dimethylamine ⟨72JMC315⟩, produce the normal displacement products. The reaction of dichlorothiadiazole with potassium fluoride in sulfolane gives a mixture of 3-chloro-4-fluoro- and 3,4-difluoro-1,2,5-thiadiazole ⟨82CB2135⟩.

The presence of an electron-donating group on the thiadiazole ring promotes attack on carbon and good yields of the displacement products. In contrast to 3,4-dichloro-1,2,5-thiadiazole, which is destroyed by caustic hydrolysis, the 3-chloro-4-*N*-morpholino analog can be hydrolyzed to 3-hydroxy-4-*N*-morpholino-1,2,5-thiadiazole in nearly quantitative yield ⟨76JOC3121⟩. Alkoxy group behaves similarly; 4-chloro-3-ethoxy-1,2,5-thiadiazole produces 62% of the corresponding sulfanilamido derivative on reaction with sulfanilamide and base ⟨67JOC2823⟩.

4.26.4 SYNTHESES

4.26.4.1 Synthesis of 1,2,5-Thiadiazole and 1,2,5-Selenadiazole Rings

There are three principal methods for the preparation of the 1,2,5-thiadiazole and 1,2,5-selenadiazole ring systems. These are the [4+1], [3+2] (type A), and [3+2] (type B) methods plus a few miscellaneous syntheses.

4.26.4.1.1 Synthesis from [4+1] atom fragments

The syntheses from [4+1] atom fragments, in which the Group VIA heteroatom is introduced between two nitrogen atoms, are the most widely applicable and versatile methods available for construction of the 1,2,5-thiadiazole and 1,2,5-selenadiazole ring systems. These methods have been applied to the synthesis of monocyclic and polycyclic aromatic forms of these ring systems in addition to the direct formation of 1-oxides and 1,1-dioxides, 2-oxides, quaternary salts and reduced forms. The earliest use of the [4+1] synthesis dates back to 1889 when Hinsburg prepared 2,1,3-benzothiadiazole (**1**) and 2,1,3-benzoselenadiazole (**2**) from *o*-phenylenediamine and sodium bisulfite, or selenium dioxide, respectively.

These reactions have undergone extensive development and have proven to be general for the synthesis of these ring systems from any substituted *o*-diamine ⟨70RCR923, B-61MI42600⟩. The choice of the sulfur reagent, thionyl chloride, sulfur dioxide, sulfinylaniline or sulfur monochloride, depends upon the nature of substituents present but selenium dioxide is usually the reagent of choice for fused selenadiazoles ⟨70RCR923⟩. The reaction of *o*-phenylenediamine with selenium(IV) is so rapid and complete that it has been suggested as the basis for spectrophotometric ⟨79MI42600⟩ and gas chromatographic ⟨81MI42600⟩ determination of selenium.

The mechanism for the formation of 2,1,3-benzothiadiazole from an *o*-diamine and thionyl chloride involves an *o*-sulfinylaniline ⟨59DIS(19)3136, 67CB2170, 68CB371⟩ and not an *o*-disulfinylamine intermediate as proposed by early workers.

1,2,5-Thiadiazoles and 1,2,5-selenadiazoles fused to other heterocyclic systems have attracted considerable attention. These polycyclic systems, obtained either via [4+1] syntheses from the corresponding o-diamines or via annulation of a heterocycle to a compound already containing a thiadiazole or selenadiazole ring, involve fusion to pyridine ⟨70RCR923, 78JOC960, 79JMC944, 70RTC5, 80S842, 79S687⟩, pyridazine ⟨71JHC441, 71MI42602⟩, pyrazine ⟨76JHC13, 75JHC451, 74JOC1235⟩, pyrazole ⟨81JOC4065⟩, 1,2,5-thiadiazole ⟨75JOC2749, 76JHC13⟩, triazole ⟨74BCJ1493⟩, thiadiazine ⟨79JMC944⟩, thiophene ⟨69JA6891⟩, quinoline ⟨81S316, 76CHE61⟩, quinoxaline ⟨76JHC13, 79JHC1617⟩, benzimidazole ⟨81IJC(B)111⟩, benzoxadiazole ⟨80CPB1909, 78CPB3896⟩, benzisothiazole ⟨80JHC537⟩, benzothiadiazole ⟨75JHC829⟩, acridine ⟨72JHC1109⟩ and coumarin ⟨80KGS853⟩.

Due to the lack of suitable aliphatic starting materials, the synthesis of aromatic forms of monocyclic 1,2,5-thiadiazoles was not accomplished for more than 70 years after the first report of the benzo analogs. The reaction of cis-diaminomaleonitrile (HCN tetramer) and thionyl chloride to produce 3,4-dicyanothiadiazole and the oxidative degradation of 2,1,3-benzothiadiazole (**1**) to thiadiazole dicarboxylic acid (**49**) (see Section 4.26.3.3) were the earliest methods available for monocyclic derivatives. Later, general synthetic methodology for monocyclic 1,2,5-thiadiazoles, not requiring unsaturated amines, was devised which readily leads to a variety of substituted derivatives, alkyl, aryl, halo, hydroxy, alkoxy, cyano and amino as well as the parent system. A general model for substrates applicable in these syntheses can be described in terms of an acyclic NCCN system in which the N—C groups exist at any hybridization, sp, sp^2 and sp^3. Compounds containing these C—N functionalities in any combination react with sulfur monochloride or sulfur dichloride to form an appropriately substituted thiadiazole ⟨68AHC(9)107, 67JOC2823⟩. These syntheses are summarized in Table 16. In a few limited cases, thionyl chloride ⟨70RCR923⟩, tetrasulfur tetranitride (see below) or sulfur tetrafluoride ⟨76CB2442⟩ served as the thio cyclizing agent but the sulfur chlorides are the most versatile and generally applicable reagents for this purpose. For example, the sulfur chlorides are strongly electrophilic and capable of forming N—S bonds with a wide range of nucleophiles, e.g. amines, imines, amides, nitriles and oximes, and can also serve as the oxidant, as required in the syntheses of thiadiazoles from

Table 16 Synthesis of 1,2,5-Thiadiazoles from Aliphatic Compounds and Sulfur Chlorides

Starting Material	R^1	R^2	Yield (%)	Ref.
$R^1CH(NH_2)CH(NH_2)R$	H	H, alkyl, aryl	50–70	68AHC(9)107, 67JOC2823
	H	CO_2H, vinyl		80T1245
$(NH_2)(CN)C=C(CN)(NH_2)$ or $NH_2(CN)C=C(CN)NH_2$	CN	CN	60	68AHC(9)107, 59DIS(20)1593
$R^1CN(NH_2)CONH_2$	H, alkyl, aryl	OH	40–60	68AHC(9)107, 67JOC2823, 68CPB544
$NH_2CH_2CONHR^1$	a	—	40–80	79JOC1118
$R^2NHCH(R^1)CONH_2$	b	—	50–65	81JCS(P1)1033
$NH_2CH_2C(=NH)NH_2$	H	NH_2	66	67JOC2823
$NH_2CH(R^1)CN$	H, alkyl, aryl	Cl	50–70	67JOC2823
$NCC(=NH)C(=NH)CN$	CN	CN	93	72JOC4136
$R^1C(=NSiMe_3)-C(=NSiMe_3)R^2$	Aryl	Aryl	80–100	68LA(711)174, 79JOM(166)25
$R^1C(=NH)C(=NH)R^2$	Alkoxy	H, alkoxy	50–90	67JOC2823
NH_2COCN	OH	Cl	88	67JOC2823
$R^1C(=NH)CN$	Alkoxy	Cl	89–90	67JOC2823
NCCN	Cl	Cl	90	68AHC(9)107, 67JOC2823
$NCC(=NOH)CONH_2$	CN	OH	68	64JA2861
$R^1C(=NOH)CN$	Aryl	Cl	15	67JOC2823
$R^1C(=NOH)C(=NOH)R^2$	Alkyl, aryl[c]	Alkyl, aryl	40–45	67JOC2823, 70JOC1165

[a] 2-Substituted 2H-1,2,5-thiadiazol-3-ones are formed. [b] The products are mesoionic 5-alkyl-4-aryl-1,2,5-thiadiazolium-3-olates. [c] N-Oxides have also been detected.

reduced substrates (*e.g.* α-diamines). The sulfur chlorides also readily add to nitrile groups, providing chloro-substituted thiadiazoles. 1,2,5-Selenadiazoles in the aromatic form have been similarly prepared, usually utilizing selenium monochloride as the cyclizing agent, but the literature is less extensive in this area. α-Diamines have been employed in the synthesis of the parent selenadiazole system ⟨67AG(E)364, 67G1870⟩ as well as alkyl and aryl derivatives ⟨79S979⟩. α-Dioximes produce a mixture of the selenadiazole and the corresponding 2-oxide ⟨76ACS(B)675⟩.

Thiadiazole oxides and reduced forms are readily prepared *via* [4 + 1] cyclizations (Table 17). The reagents employed for this purpose are thionyl chloride, sulfuryl chloride, sulfur tetrafluoride, *N*,*N*′-bis(*p*-toluenesulfonyl)sulfur diimide, cyclic diimides, pentafluoroethyliminosulfur difluoride and sulfamide.

Table 17 Synthesis of 1-Oxides and Reduced Forms of 1,2,5-Thiadiazole from [4+1] Atom Fragments

Starting material	Cyclizing reagent	Product	Ref.
EtOC(=NH)C(=NH)OEt	SOCl$_2$	3,4-Diethoxy-1,2,5-thiadiazole 1-oxide	81H(16)1561, 68LA(711)174
H$_2$NCMe$_2$CN	SOCl$_2$ or SO$_2$Cl$_2$	3-Chloro-4,4-dimethyl-Δ^2-1,2,5-thiadiazoline 1-oxide or 1,1-dioxide	75MI42600
Bu^{t}NHCOCH$_2$NHPh	SOCl$_2$	2-*t*-Butyl-5-phenyl-1,2,5-thiadiazolidin-3-one 1-oxide	77JOC1015
Me$_3$SiNHCH$_2$CH$_2$NHSiMe$_3$	SF$_4$	3,4-Dihydro-1λ^4,2,5-thiadiazole	77CB3205
H$_2$NCHMeCH$_2$NH$_2$	TosN=S=NTos	3,4-Dihydro-3-methyl-1λ^4,2,5-thiadiazole	69TL4117
MeNHCH$_2$CH$_2$NHMe	CF$_3$CF$_2$N=SF$_2$	1-Pentafluoroethylimino-2,5-dimethyl-1,2,5-thiadiazolidine	79IC213
MeNHCH$_2$CH$_2$NHMe	SO$_2$Cl$_2$	2,5-Dimethyl-1,2,5-thiadiazolidine 1,1-dioxide	78CB1915
o-C$_6$H$_4$(NHTos)$_2$	SOCl$_2$	1,3-Bis(*p*-toluenesulfonyl)-1,3-dihydro-2,1,3-benzothiadiazole 1-oxide	59DIS(19)3136, 79CB1012
o-C$_6$H$_4$(NH$_2$)$_2$	NH$_2$SO$_2$NH$_2$	1,3-Dihydro-2,1,3-benzothiadiazole 1,1-dioxide	71JCS(C)993
PhNHC(=NPh)C(=NPh)Ph	SOCl$_2$	2,5-Diphenyl-3,4-diphenylimino-1,2,5-thiadiazolidine 1-oxide	80JPR273

4.26.4.1.2 Synthesis from [3+2] atom fragments

The N—S—N grouping in the [3+2] (type A) thiadiazole syntheses can be derived from either an *N*,*N*′- or *S*,*S*-disubstituted sulfur diimide, diaminosulfane, sulfamide or tetrasulfur tetranitride. The cycloaddition reaction of *N*,*N*′-disubstituted sulfur diimides with diphenylketene has received considerable attention. The product of the reaction of diphenylsulfur diimide (**80**) and diphenylketene in refluxing benzene was originally assumed to be the 1,2-thiazetidinone (**81**; Scheme 15) on the basis of spectral and chemical evidence ⟨69TL447⟩, but later shown by an X-ray study ⟨70BCJ1905⟩ to bear the thiadiazole structure (**82**). A product analogous to (**82**) was also obtained from the reaction of diphenylketene with di-*p*-ethoxycarbonylphenylsulfur diimide ⟨69T4277⟩. It was further demonstrated that the products of these reactions are temperature dependent ⟨72JOC3810⟩ and that at 8 °C the 2:1 cycloadduct (**84**) is the principal product. Refluxing (**84**) in benzene in the presence of (**80**) forms (**82**). Of several possible pathways for these reactions, one in which (**82**) and (**84**) are competitively formed *via* the same intermediate, (**83**), was considered the most reasonable.

In contrast to (**80**), the low temperature reaction of diphenylketene with di-*t*-butylsulfur diimide (**85a**) and with bis(*p*-toluenesulfonyl)sulfur diimide (**85b**) ⟨70TL1427⟩ at low temperatures affords the isolable [2+2] cycloadducts (**86a**) and (**86b**) which are converted into the [3+2] adducts (**87a**) and (**87b**) at 70 °C.

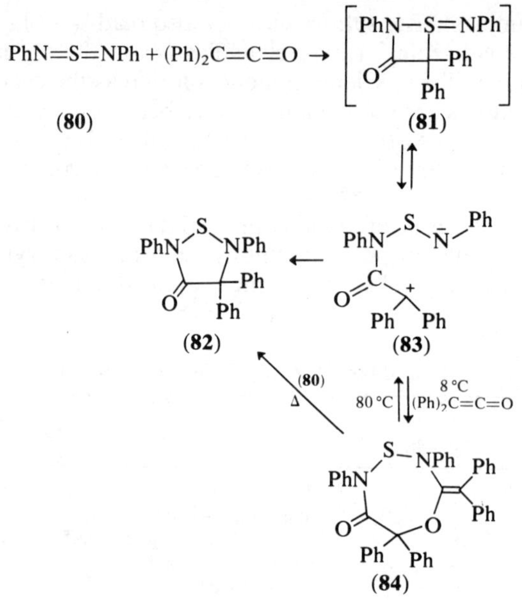

Scheme 15

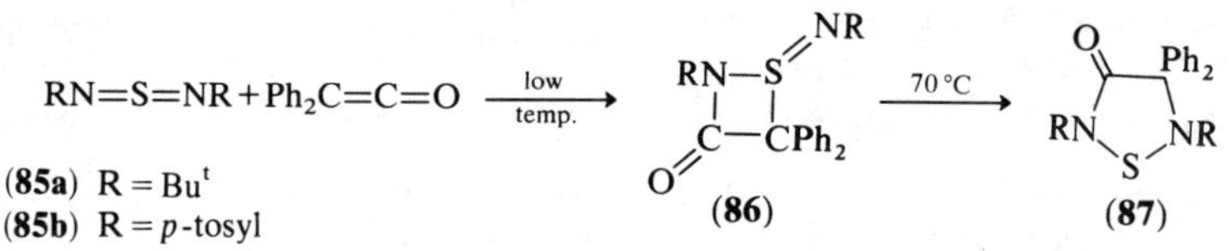

Scheme 16

Cyclocondensation of *N,N'*-diarylsulfur diimides with oxalyl chloride produces the thiadiazolediones (**89**) ⟨75AG(E)762, 77CZ35, 77CB3149⟩ but the 1,1-dichloro analogs (**90**) are the products derived from the *N,N*-bis(*t*-butyl) analog (Scheme 16) ⟨77CZ35, 77CB3149⟩. Reduction of (**90**) produces the thiadiazole dione (**91**) ⟨77S63⟩, which can also be prepared *via* cyclization of the diaminosulfane (**92**) ⟨81CB80⟩. The 1,2,5-thiadiazole (**88**) can also be derived from sulfur diimides by reaction with 2-phenylmalonyl chloride ⟨69M959⟩.

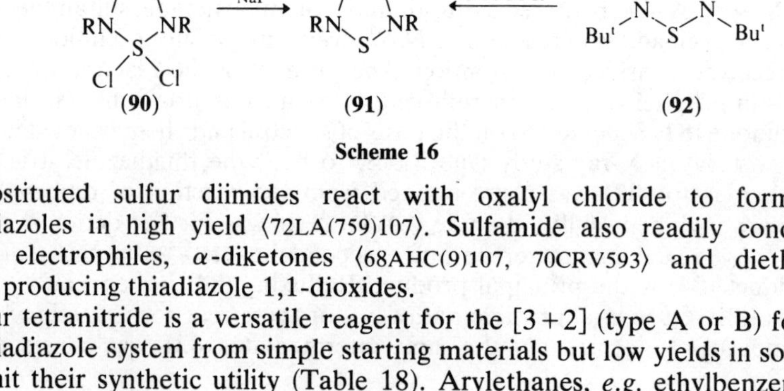

S,S-Disubstituted sulfur diimides react with oxalyl chloride to form the 1,1-dialkylthiadiazoles in high yield ⟨72LA(759)107⟩. Sulfamide also readily condenses with bifunctional electrophiles, α-diketones ⟨68AHC(9)107, 70CRV593⟩ and diethyl oxalate ⟨75JOC2743⟩, producing thiadiazole 1,1-dioxides.

Tetrasulfur tetranitride is a versatile reagent for the [3+2] (type A or B) formation of the 1,2,5-thiadiazole system from simple starting materials but low yields in some of these methods limit their synthetic utility (Table 18). Arylethanes, *e.g.* ethylbenzene, 2-ethylnapthalene and 1,2-diphenylethane, react with S_4N_4 in refluxing xylene to give low yields of aryl-substituted thiadiazoles together with elemental sulfur and ammonia ⟨67G1614, 65AG(E)239⟩. Cyclic analogs, *e.g.* 9,10-dihydrophenanthrene and tetrahydronaphthalene, react with S_4N_4 to provide low yields of some polycyclic thiadiazoles which are difficult to

Table 18 Synthesis of 1,2,5-Thiadiazole from [3+2] Atom Fragments Using S_2N_4

Starting material	R^1	R^2	Yield (%)	Ref.
$R^1CH_2CH_2R^2$	Phenyl	H, phenyl	10–20	67G1614, 65AG(E)239
$R^1CH_2CH(NH_2)R^2$	Phenyl	H, phenyl	60	71G259
$R^1CH_2COR^2$	Aryl	Aryl, alkyl	10–60	79S524, 80JHC1681
$R^1C(=NOH)COR^2$	Phenyl	Phenyl	4	79JCS(P1)2905
$R^1C\equiv CR^2$	Aryl, CO_2Me, CN, H	Aryl, CO_2Me, CN, phenyl	16 7	79JHC1009, B-68MI42600, 77H(6)933
α- or β-naphthyl-amines	a	a	20–60	59DIS(19)3136

^a $R^1-R^2=$1,2-Naphtho.

obtain ⟨72JOC2587⟩. When the ethylene units in the starting materials are partially or fully oxidized, yields are substantially higher. Thus, arylethylamines form the corresponding thiadiazole in 60% yields ⟨71G259⟩.

The general reaction of S_4N_4 with readily available benzyl ketones provides a convenient method for the preparation of a wide variety of disubstituted 1,2,5-thiadiazoles in 20–60% yields ⟨79S524, 80JHC1681⟩. Benzil monoximes, however, result in very poor yields ⟨79JCS(P1)2905⟩. The reaction of negatively substituted alkynes with S_4N_4 is the basis of another general preparative method for 1,2,5-thiadiazole from [3+2] atom fragments ⟨B-68MI42600, 79JHC1009⟩. The reaction is fairly efficient (50–80%) when disubstituted alkynes are employed but yields drop to less than 20% with monosubstituted alkynes ⟨77H(6)933⟩. In the latter case 3-aminothiadiazoles are also formed as by-products. A novel reaction of S_4N_4 with alkoxybenzenes directly produces benzothiadiazole analogs, but yields are less than 5% ⟨77JHC963⟩. In the first reported synthesis of a thiadiazole from [3+2] fragments, napthothiadiazoles were prepared in 40% yield by the action of S_4N_4 on napthylamines ⟨59DIS(19)3136, 70RCR923⟩. Attempts to extend the method to anilines were unsuccessful.

Trithiazyl trichloride ($S_3N_3Cl_3$) was shown to be much more electrophilic than S_4N_4. The trichloride reacts with acenaphthylene and *trans*-stilbene to produce moderate yields of acenaphtho[1,2-c][1,2,5]thiadiazole and 3,4-diphenylthiadiazole, respectively ⟨77JCS(P1)916⟩.

4.26.4.1.3 Miscellaneous syntheses

The unique synthesis of the potassium salt of 4-cyano-3-hydroxy-1,2,5-thiadiazole from sulfur dioxide and potassium cyanide is of potential industrial importance ⟨64JA2861⟩. The principal products (*ca.* 50%) of the reaction of *N*-sulfinylanilines and aryl isocyanides are 2,5-diaryl-3,4-diarylimino-1,2,5-thiadiazolidine 1-oxides ⟨80JPR273⟩. In an attempted Strecker degradation of ethyl glycinate the unexpected product was 30% of diethyl 1,2,5-thiadiazole-3,4-dicarboxylate ⟨75CPB2654⟩. 1-Cyanocyclopentyliminosulfur dichloride isomerizes thermally or under the action of hydrogen chloride into 3-chloro-4-(4-chlorobutyl)-1,2,5-thiadiazole ⟨73ZOR2502⟩.

Thiadiazolines have also been formed *via* 1,5-cyclization of vinyl-substituted sulfur diimide intermediates which form when 2,2-diphenylvinylamine and sulfur di-*p*-toluenesulfonyldiimide are kept in benzene ⟨66AG(E)1042⟩, and by the base-catalyzed 1,5-cyclization of 2-propynylsulfonamides ⟨80JOC482⟩.

4.26.4.2 Synthesis of Substituted 1,2,5-Thiadiazoles

The best practical methods for the preparation of various types of derivatives of 1,2,5-thiadiazoles are summarized in Table 19 with cross-referencing to earlier sections in this review.

Table 19 Preparation of Derivatives of 1,2,5-Thiadiazole

Derivative	Method	Section or Ref.
Unsubstituted ring	Ring formation	4.26.4.1.1
	Decarboxylation of dicarboxylic acid	4.26.3.1.1
Deutero	Decarboxylation of deuterated acids	4.26.3.3.3
	Base-catalyzed exchange	4.26.3.1.1
Fused carboxylic and heterocyclic rings	Cyclization of o-diamines	4.26.4.1.1
C-Linked substituents		
Alkyl	Ring formation	4.26.4.1.1
Aryl	Ring formation	4.26.4.1.1
Alkenyl	Wittig synthesis	68AHC(9)107
	Elimination reactions	4.26.3.3.3
Chloromethyl	Chloromethylation	4.26.3.1.3
	Chlorination of methyl derivatives	68AHC(9)107
	Replacement of hydroxy group	4.26.3.3.5
Hydroxymethyl	Reduction of esters, acyl chlorides and ketones	4.26.3.3.3
Carboxylic acids	Oxidation of benzothiadiazole and methylthiadiazoles	4.26.3.3.1
	Hydrolysis of pyrimidothiadiazoles	4.26.3.3.2
	Hydrolysis of cyanothiadiazoles	4.26.3.3.3
Carboxymethyl	Arndt–Eistert synthesis	68AHC(9)107
	Carboxylation of α-lithiomethyl derivatives	4.26.3.3.3
Cyano	Ring synthesis	4.26.4.1.1
	Dehydration of amides	4.26.3.3.3
Keto	Friedel–Crafts reaction of thiadiazole	4.26.3.3.3
	Acid chlorides	4.26.3.3.3
	Malonic ester synthesis	
O-Linked substituents		
Hydroxy	Ring synthesis	4.26.4.1.1
Alkoxy	Ring synthesis	4.26.4.1.1
	Alkylation of hydroxythiadiazoles	4.26.3.3.5
	Displacement reactions	4.26.3.3.6
S-Monoxides	Ring synthesis	4.26.4.1.1
	Oxidation of thiadiazoles	4.26.4.1.1
S-Dioxides	Ring synthesis	4.26.3.1.3
	Oxidation of thiadiazoles	4.26.4.1.1
N-Oxides	Ring synthesis	4.26.4.1.1
N-Linked substituents		
Amino	Ring synthesis	4.26.4.1.1
	Displacement of chlorothiadiazoles	4.26.3.3.6
	Hydrolysis of pyrimidothiadiazoles	4.26.3.3.2
	Hofmann or Curtius reactions	68AHC(9)107
Chloro	Ring synthesis	4.26.4.1.1
	Replacement of hydroxy	4.26.3.3.5
	Chlorination of activated thiadiazoles	4.26.3.1.3
Fluoro	Replacement of chloro	4.26.3.3.6
Mercapto	Replacement of hydroxy	4.26.3.3.5

4.26.5 APPLICATIONS OF 1,2,5-THIADIAZOLES

The 1,2,5-thiadiazoles have been broadly applied in the areas of pharmaceutical, agricultural, industrial and polymer chemistry. The potential pharmaceutical applications include antibiotics (**93**) and (**94**) ⟨67USP3322749⟩, histamine H_2-receptor antagonists (**95**) and (**96**) ⟨81GEP3033169, 82JMC207, 82JMC210⟩ and β-adrenergic blocking agents, e.g. timolol (**97**) ⟨72JMC651, 76JOC3121⟩. The hemimaleate salt of timolol (**97**), the most important 1,2,5-thiadiazole thus far synthesized, is the active chemotherapeutic agent in Blocadren® and Timoptic®. Blocadren® is in world-wide use for the treatment of high blood pressure and in reducing the incidence of subsequent myocardial infarction (heart attack). Timoptic® is the preferred drug in the treatment of the eye disease glaucoma. Two major urinary metabolites of (**97**) in man involve oxidation and hydrolytic cleavage of the morpholine ring ⟨80MI42600⟩.

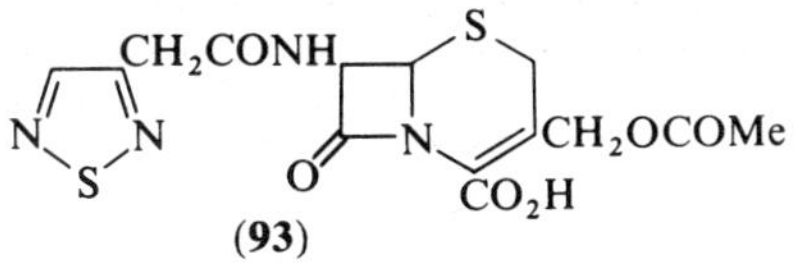

1,2,5-Thiadiazoles and their Benzo Derivatives

(93) 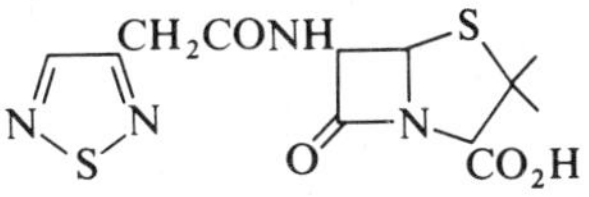 **(94)**

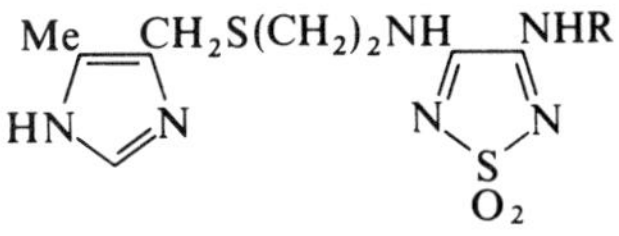

(95) 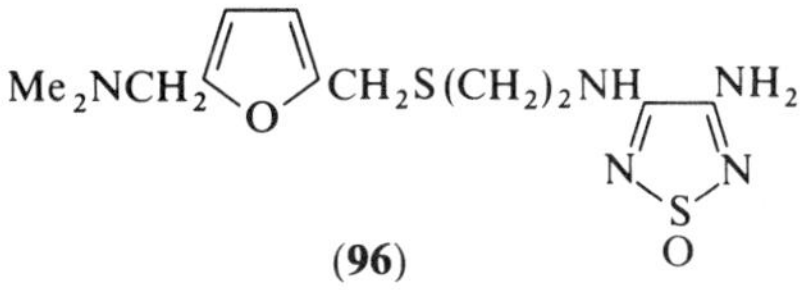 **(96)**

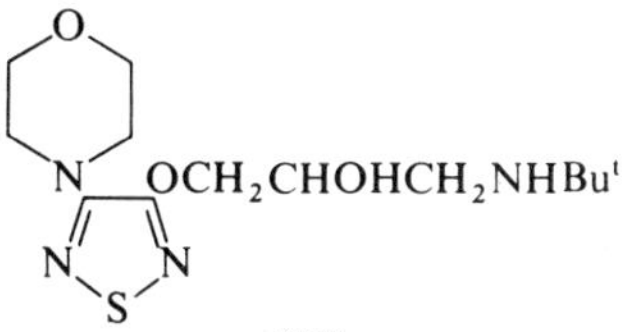

(97)

A number of other thiadiazoles have been found that are active as fungicides (**98**) ⟨80GEP2852869⟩ and (**99**) ⟨78USP4075205⟩, bactericides (**100**) and (**101**) ⟨78USP4094986, 74USP3854000, 78USP4127584⟩, herbicides (**102**) ⟨75MI42601⟩, plant-growth regulators (**103**) ⟨70N395⟩, insecticides (**104**) ⟨74USP3819354⟩, coccidiostats (**105**) ⟨62USP3066147⟩ and antihelmintics (**106**) ⟨62USP3055907⟩.

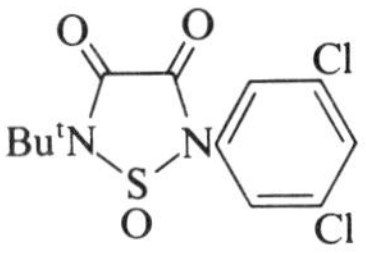

(98) 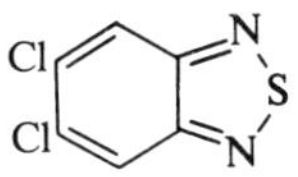 **(99)** 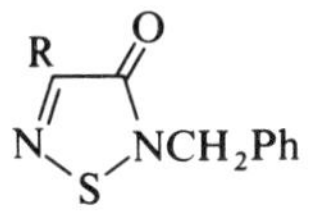 **(100)** 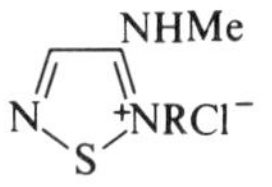 **(101)** 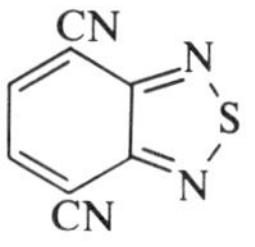 **(102)**

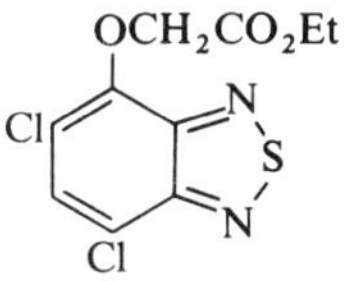

(103) 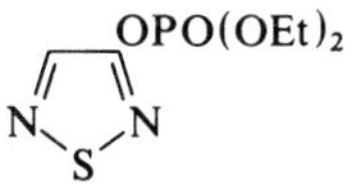 **(104)** 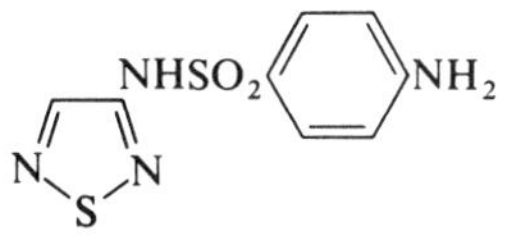 **(105)** 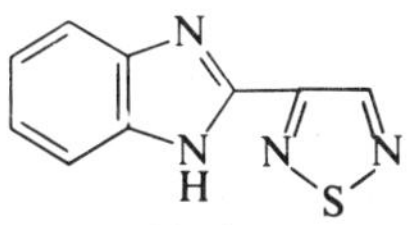 **(106)**

The 1,2,5-thiadiazole ring has been incorporated into the backbone of a number of polymers with desirable chemical and thermal stability ⟨79MI42601⟩. Thiadiazole polyamides, polyesters, polyethers, polyurethanes and polyhydrazides are useful as packaging materials, fibers, electrical insulators and membranes for reverse osmosis. The polyamide (**107**), obtained from the reaction of *trans*-dimethylpiperazine with 1,2,5-thiadiazole-3,4-dicarbonylchloride, produced fibers of high tensile strength, moduli and work recovery values ⟨74MI42600, 73MI42600⟩ and membranes with high water permeability and high salt rejection ⟨73GEP2263774⟩. The thiadiazole polyesters (**108**) form strong films which require a 50% oxygen atmosphere to sustain combustion ⟨73MI42600, 74USP3786028⟩.

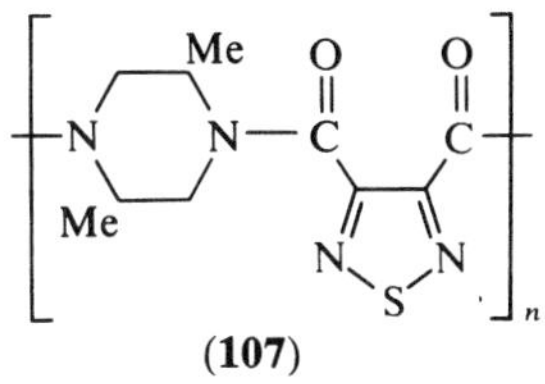

(107) 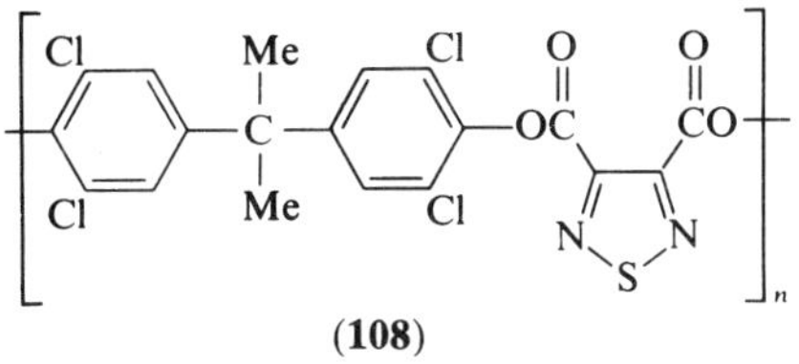 **(108)**

4.27
1,3,4-Thiadiazoles

G. KORNIS
The Upjohn Company, Kalamazoo

4.27.1	INTRODUCTION	546
4.27.2	STRUCTURE	546
4.27.2.1	Survey of Possible Structures	546
4.27.2.2	Theoretical Methods	547
4.27.2.2.1	MO approximation	547
4.27.2.2.2	Self consistent field, configuration interaction (SCF, CI)	547
4.27.2.3	Structural Methods	547
4.27.2.3.1	X-Ray diffraction	547
4.27.2.3.2	Microwave spectroscopy	548
4.27.2.3.3	Dipole moments	549
4.27.2.3.4	^1H NMR spectroscopy	549
4.27.2.3.5	^{13}C, ^{14}N and ^{15}N NMR spectroscopy	549
4.27.2.3.6	Combined spectroscopic methods	551
4.27.2.3.7	UV spectroscopy	551
4.27.2.3.8	IR spectroscopy	552
4.27.2.3.9	Mass spectrometry	552
4.27.2.3.10	Photoelectron spectroscopy	553
4.27.2.3.11	Spectral data for non-aromatic systems	553
4.27.2.4	Thermodynamic Aspects	555
4.27.2.4.1	Melting and boiling points	555
4.27.2.4.2	Solubility	555
4.27.2.4.3	Chromatography	555
4.27.2.4.4	Thermochemistry	556
4.27.2.4.5	Aromaticity	556
4.27.2.4.6	Conformation and Configuration	556
4.27.2.5	Tautomerism	557
4.27.3	REACTIVITY	558
4.27.3.1	Survey of Reactivity	558
4.27.3.2	Thermal and Photochemical Reactions Formally Involving No Other Species	558
4.27.3.3	Electrophilic Attack at Nitrogen	560
4.27.3.4	Electrophilic Attack at Carbon	560
4.27.3.5	Electrophilic Attack at Sulfur	561
4.27.3.6	Nucleophilic Attack	561
4.27.3.6.1	Ring opening	561
4.27.3.6.2	Amination	562
4.27.3.6.3	Reduction	562
4.27.3.6.4	Hydrogen exchange at ring carbon	562
4.27.3.7	Reactions with Radicals and Electron-Deficient Species	562
4.27.3.8	Cycloaddition Reactions	563
4.27.3.9	Reactions of Substituents on Carbon	563
4.27.3.9.1	Carbon substituents	563
4.27.3.9.2	Halogen substituents	564
4.27.3.9.3	Sulfur substituents	565
4.27.3.9.4	Amino substituents	565
4.27.4	SYNTHESES	568
4.27.4.1	Survey of Synthetic Procedures	568
4.27.4.2	Formation of One Bond	568
4.27.4.3	Formation of Two Bonds	569
4.27.4.3.1	Cyclizations	569
4.27.4.3.2	Dipolar cycloadditions	571
4.27.4.3.3	Thiadiazolidines	573
4.27.4.3.4	Mesoionic compounds	573
4.27.4.4	Formation of Three Bonds	574
4.27.4.5	Formation by Ring Transformations	575

4.27.5 APPLICATIONS	575
4.27.5.1 Survey of Applications	575
4.27.5.2 Selected Commercial Compounds	575
4.27.5.3 Biological Effects	576
4.27.5.4 Patents	577

4.27.1 INTRODUCTION

The thiadiazole system contains the following members: the 1,2,3-thiadiazoles (**1**) and their benzo derivatives (**2**), the 1,2,4-thiadiazoles (**3**), the 1,3,4-thiadiazoles (**4**), and the 1,2,5-thiadiazoles (**5**) and their benzo derivatives (**6**) ⟨B-79MI42700⟩. Most of the published work, by far, is on 1,3,4-thiadiazoles. Between 1967 and March 1, 1982 *Chemical Abstracts* lists 724 references for this ring system. This includes the 1,3,4-thiadiazolines (**7**) and (**8**) and the 1,3,4-thiadiazolidines (**9**).

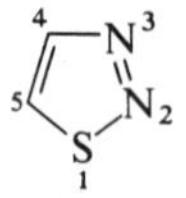

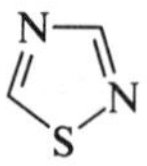

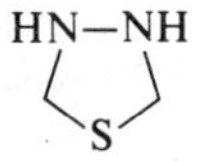

1,3,4-Thiadiazoles were first described in 1882 by Fischer and further developed by Busch and his coworkers. The advent of sulfur drugs and the later discovery of mesoionic compounds greatly accelerated the rate of progress in the field of thiadiazoles. Thiadiazoles carrying mercapto, hydroxy and amino substituents can exist in many tautomeric forms and this property is being intensively studied using modern instrumental methods.

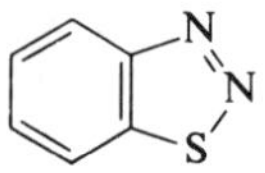

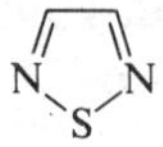

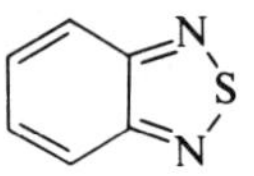

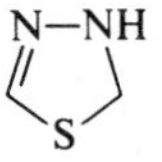

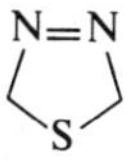

The literature of 1,3,4-thiadiazoles has been extensively reviewed. The 120 page survey by Bambas ⟨52HC(4)81⟩ is the definitive work up to 1952, and is followed by Sherman's review in 1961 ⟨B-61MI42700⟩ and then by Sandstrom's more modern treatment up to 1967 ⟨68AHC(9)165⟩. The Specialist Periodical Reports of the Chemical Society on Organic Compounds of Sulfur partially cover the literature between 1970 and 1981 ⟨70SST(1)449, 73SST(2)725, 75SST(3)687, 77SST(4)431, 79SST(5)440⟩.

A brief review appeared recently ⟨B-79MI42700⟩ and much useful information can also be obtained from the reference series 'Advances in Heterocyclic Chemistry' and 'Physical Methods in Heterocyclic Chemistry'. This chapter will discuss some of the earlier work but will emphasize developments that took place after 1967.

4.27.2 STRUCTURE

4.27.2.1 Survey of Possible Structures

The 1,3,4-thiadiazoles are conveniently divided into three subclasses:

(a) Aromatic systems which include the neutral thiadiazoles (**4**) and constitute a major part of this review.

(b) Mesoionic systems (**10**) which are defined as five-membered heterocycles which are not covalent or polar and possess a sextet of electrons in association with the five atoms comprising the ring ⟨B-79MI42701⟩.

(c) Non-aromatic systems such as the 1,3,4-thiadiazolines (**7**) and (**8**) and the tetrahydro-1,3,4-thiadiazolidines (**9**). In the partially reduced systems the position of the double bond is denoted by the prefix Δ, with (**7**) being a Δ^2-1,3,4-thiadiazoline.

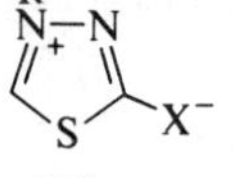

4.27.2.2 Theoretical Methods

Literature references to the application of theoretical methods to 1,3,4-thiadiazoles are limited.

4.27.2.2.1 MO approximation

The reaction of the acid chloride phenylhydrazone (**11**) with base gives the nitrile-imine 1,3-dipolar compound (**12**) which reacts with potassium thiocyanate to give the Δ^2-thiadiazoline (**13**; Scheme 1). Thus the cycloaddition occurs at the C=S and not the C=N bond. This regioselectivity can be explained in terms of the frontier orbital treatment. Due to the electron rich nature of the thiocyanate anion, its reaction with (**12**) is expected to be controlled by the LUMO and HOMO of (**12**) and the thiocyanate respectively. As the HOMO of the thiocyanate anion has the larger orbital coefficient on the sulfur atom, it can be concluded that the larger orbital coefficient in the LUMO of (**12**) is on the carbon atom. This is also in agreement with other dipolar cycloadditions ⟨82H(19)57⟩.

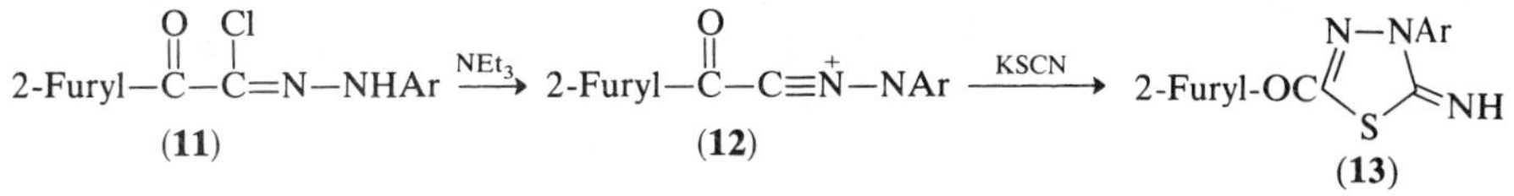

Scheme 1

2-(*N*-Benzylhydrazino)-1,3,4-thiadiazole (**14**) rearranges quantitatively in the presence of HCl to 4-amino-2-benzyl-1,2,4-triazoline-3-thione (**15**). In HCl–HOAc, (**14**) gives a mixture of (**15**) and (**16**). MO calculations indicate that the π-electron stabilization is -0.61β units greater for the 4-amino-1,2,4-triazoline-3-thione system than for the 2-hydrazino-1,3,4-thiadiazole system ⟨70ZC406⟩.

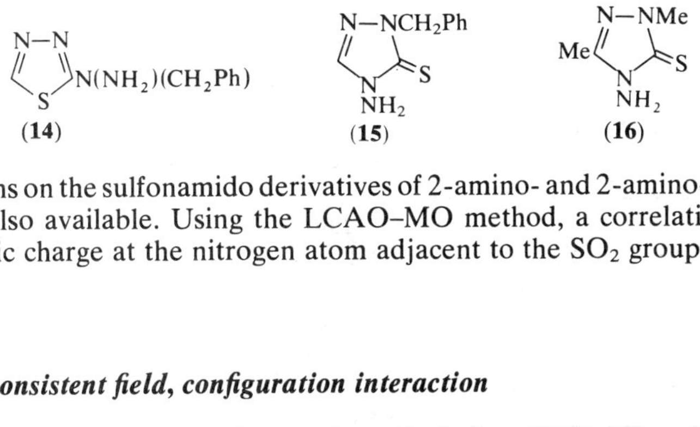

MO calculations on the sulfonamido derivatives of 2-amino- and 2-amino-5-methyl-1,3,4-thiadiazole are also available. Using the LCAO–MO method, a correlation between pK_a and the electronic charge at the nitrogen atom adjacent to the SO$_2$ group can be obtained ⟨67JPS608⟩.

4.27.2.2.2 Self consistent field, configuration interaction

Self consistent field, configuration interaction calculations (SCF, CI) on 1,3,4-thiadiazoles indicate that the electronic state of the sulfur atom is described by $s^2 p_x p_y p_z^2$ in which the two p_z electrons are conjugated with the π-electron system of the rest of the molecule. The SCF eigenvalues and the molecular geometries are available ⟨69JIC779⟩.

4.27.2.3 Structural Methods

4.27.2.3.1 X-Ray diffraction

The structural features of many 1,3,4-thiadiazoles have been determined by X-ray diffraction techniques and several examples are presented below. 1,3,4-Thiadiazole in the vapor phase is a planar molecule having C_{2v} symmetry. Bond distances and angles are in agreement with those determined by microwave spectroscopy. The bond lengths and bond angles are as follows with values in parentheses being estimated error limits: (C—H) = 1.081 (±0.028), (N—C) = 1.304 (±0.010), (N—N) = 1.381 (±0.016), (S—C) = 1.722 (±0.006) Å; CSC = 86.4 (±0.4), SCN = 114.8 (±0.5), CNN = 112.0 (±0.4), SCH = 124.1 (±3.0), HCN = 121.1 (±3.0)°. Bond distances and angles are those consistent with the average structure (r_α) ⟨70ACS2525, 74AX(B)1642⟩.

5-Amino-1,3,4-thiadiazole-2-thiol is also planar. Bond lengths suggest that the compound exists in the solid state in the thione rather than the thiol form. The ring is aromatic with a localized C(1)=N(1) double bond of length 1.305 Å. Other C—N bond lengths are 1.334 and 1.339 Å and the N—N bond length is 1.382 Å. The cyclic C—S distances are 1.746 and 1.748 Å and the exocyclic C—S bond length is 1.678 Å. Hydrogen bonding of the type N—H···S is also present, joining pairs of molecules across centers of symmetry ⟨72AX(B)1584⟩.

The crystal structure of acetazolamide (5-acetamido-1,3,4-thiadiazole-2-sulfonamide), a potent inhibitor of carbonic anhydrase, is known. Bond distances are shown in (17) ⟨74JCS(P2)532⟩.

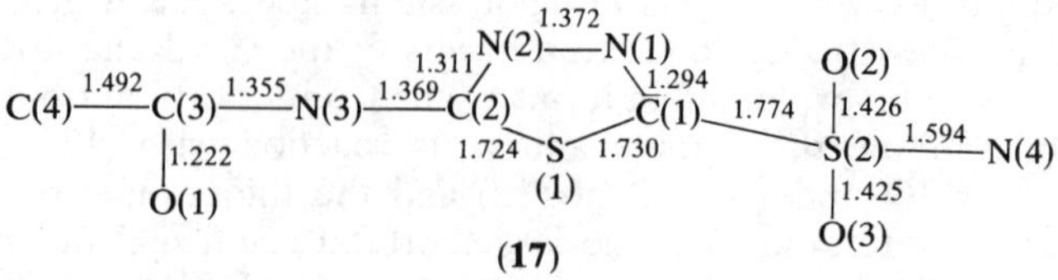

(17)

The crystal structure of a highly insoluble thiadiazole derivative is shown in (18). There is an intramolecular hydrogen bond between the hydroxyl and the carbonyl groups, and there is one intermolecular hydrogen bond between N(6) in the open chain and N(3) in the thiadiazole ring ⟨80JOC4860⟩ (see also Section 4.27.3.9.4).

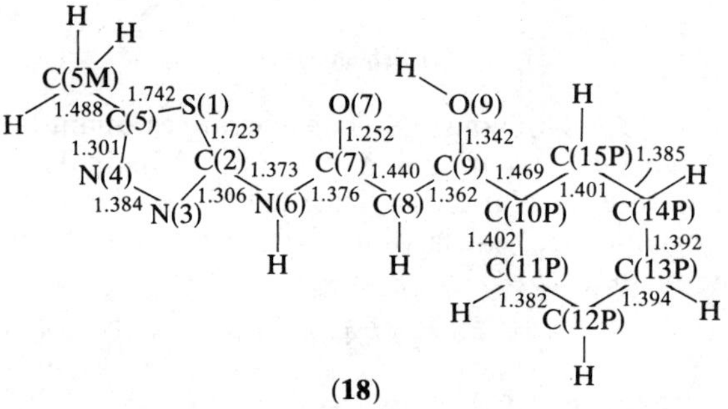

(18)

2-Methyl-3-phenyl-5-phenylazo-1,3,4-thiadiazoline is a condensation product obtained from dithiazone, which is an important analytical reagent in trace analysis, and acetaldehyde, present in minute quantities, even in highly purified solvents. Its structural proof rests on X-ray crystallography; it is planar due to electron delocalization within the Ph—N=N—C=N—N—Ph chain. The refinement is to a final R value of 0.049 ⟨80AX(B)1626⟩.

The structure of 5,5'-dithiobis(3-methyl-1,3,4-thiadiazoline-2-thione) (19) is known from X-ray crystallographic studies; the thiadiazoline ring is planar and the substituents are only slightly displaced from the plane. Bond lengths and bond angles are available and demonstrate extensive conjugation in the ring ⟨78AX(B)3803⟩.

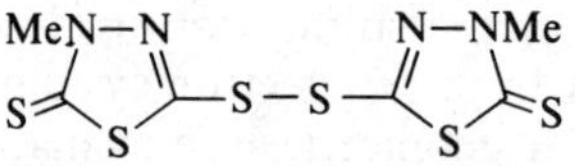

(19)

4.27.2.3.2 Microwave spectroscopy

The technique is useful for determining the structure of 1,3,4-thiadiazole with an uncertainty of 0.03 Å for hydrogen atoms and less than 0.003 Å for other atoms ⟨68AHC(9)165, p. 199⟩.

In a second investigation of the microwave spectrum of 1,3,4-thiadiazole, ^{14}N quadrupole coupling and centrifugal distortion for lines with $J \leq 5$ and $J \leq 50$, respectively, are determined. These are compared with results obtained from 1,3,4-oxadiazoles and pyridazine. Comprehensive tables showing microwave transitions, the microwave spectrum and the σ and π population from ^{14}N quadrupole coupling data for 1,3,4-thiadiazole are also given ⟨71JST(9)163⟩ (see also Sections 4.01.3.2, 4.01.4.2.2.(i) and (iii), and 4.01 Table 2).

4.27.2.3.3 Dipole moments

1,3,4-Thiadiazole has a dipole moment of 3.0 D directed from the S atom toward the center of the N—N bond, as calculated by a combination of techniques ⟨68AHC(9)165, p. 200⟩. Dipole measurements have been utilized to substantiate the mesoionic structure of certain 1,3,4-thiadiazoles by comparing the results with those obtained for similar isosydnones. They also indicate that an exocyclic sulfur atom carries a larger negative charge then does an exocyclic oxygen atom ⟨69JCS(B)1194⟩.

4.27.2.3.4 ¹H NMR spectroscopy

A comprehensive study on the proton magnetic resonance spectrum of 2-amino-1,3,4-thiadiazoles with alkyl and aryl substituents in the 5-position has been published. The chemical shifts are compared with those of other five-membered heterocycles such as selenadiazoles and triazoles and the signals of the ring protons for the selenadiazoles are at a lower field than for the thiadiazoles, a difference opposite to that expected from the electronegativity of the Group VI elements. This may be explained by mesomeric changes in charge distribution and increases from S to Se in the deshielding effect of the diamagnetic anisotropy of the ring and in the paramagnetic contribution from the heteroatom. The ring protons of thiadiazoles are less shielded than those of the triazoles. When the substituent in the 5-position is a phenyl group, the *ortho* protons can be distinguished from the *meta* and *para* protons. The NMR spectra of thiadiazoles are solvent dependent. When the substituent in the 5-position is H or Me the signals are shifted towards higher fields on changing from CCl_4 to benzene, suggesting the formation of a π-type complex between the thiadiazole and the benzene molecule with the 5-substituent placed in the shielding area of the diamagnetic anisotropy of the benzene molecule. The signals from the NH group exhibit a shift towards higher fields on dilution with $CDCl_3$, suggesting intermolecular hydrogen bonding (solute–solute). In DMSO, however, the shifts of the NH protons are concentration independent, but are shifted towards higher fields with increasing temperature ⟨72ACS459⟩.

The chemical shifts for a large number of 2,5-dialkyl-1,3,4-thiadiazoles are known ⟨73ACS391⟩ and selected compounds are shown in Table 1 (for the proton NMR spectra of thiadiazolidines see ⟨80JCS(P1)574⟩).

Table 1 2,5-Dialkyl-1,3,4-thiadiazoles: Proton NMR Chemical Shifts ⟨73ACS391⟩

R^2	R^5	C—C—	—C—	(ring)	—C—	—C—	—C	Solvent
CH_3	CH_3	—	2.72		2.72	—	—	$CDCl_3$
CH_3	C_2H_5	—	2.72		3.08	1.39	—	None
CH_3	$CH(CH_3)_2$	—	2.70		3.37	1.34	—	None
CH_3	$CH_2CH_2CH_3$	—	2.69		3.02	1.78	0.96	None
CH_3	$CH_2C(CH_3)_3$	—	2.75		2.97	—	1.02	None
CH_3	$CH_2C_6H_5$	—	2.62		4.32	7.25	—	$CDCl_3$
C_2H_5	$CH_2CH_2CH_3$	1.40	3.10		3.02	1.81	1.02	$CDCl_3$
C_2H_5	$CH_2C_6H_5$	1.31	3.00		4.32	7.27	—	$CDCl_3$
$CH(CH_3)_2$	$CH_2CH_2CH_3$	1.37	3.38		3.02	1.79	0.98	None

4.27.2.3.5 ¹³C, ¹⁴N and ¹⁵N NMR spectroscopy

¹³C spectroscopy is becoming more common in the field of thiadiazoles. For example, it can be utilized to distinguish between thione and thiol tautomers. The compounds shown in Table 2 exist as the thione tautomer except for the disulfide (**20**). Substitution of selenium for ring sulfur leads to deshielding at both C-2 and C-5; this is essentially independent of the number of ring nitrogen atoms ⟨77JOC3725⟩ (see also Sections 4.27.2.3.6 and 4.27.4.3.1).

Table 2 ^{13}C NMR Chemical Shifts^a of 1,3,4-Thiadiazoles (**20**) and (**20a**) ⟨77JOC3725⟩

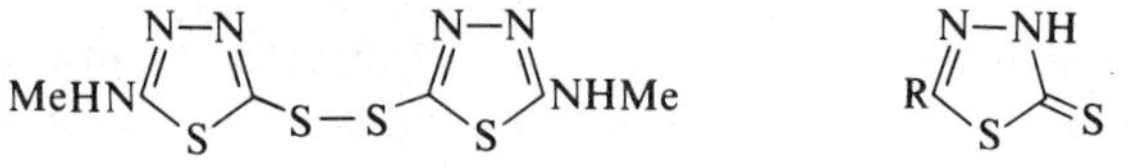

Compound	C-2	C-5	Other	$\Delta\delta^b$ C-2	C-5
(**20a**) R = MeNH	180.6	161.6	29.9 (Me)	—	—
(**20**)	148.6	173.0	31.2 (Me)	−32	11.4
(**20a**) R = Me-C$_6$H$_4$-NH— 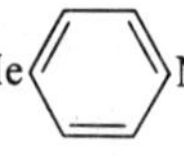	181.1	156.6	137.4 (C-1'); 117.6 (C-2'); 129.4 (C-3'); 131.2 (C-4'); 20.3 (Me)	−28.5	7.9
(**20a**) R 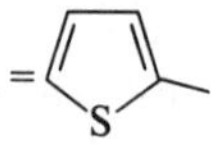 = 2-thienyl	187.2	154.2	—	—	—

^{a} Solvent is DMSO-d_6. Chemical shift values are in p.p.m. downfield from internal TMS.
^{b} $\Delta\delta$ value for thione to disulfide change.

^{13}C NMR is also helpful in determining whether substitution of (**21**) takes place on the *endo* nitrogen atom as in (**22**) or on the *exo* nitrogen atom as in (**23**; Scheme 2). This is a very common problem due to the ambident nucleophilic nature of 2-amino-1,3,4-thiadiazoles. Thus, tosylation of (**21a, b**) in the presence of triethylamine yields (**22a, b**) with the tosyl group on the endocyclic nitrogen atom. On standing in chloroform, however, isomerization takes place to give (**23a, b**), with the tosyl group now on the exocyclic nitrogen atom. Treatment with acetyl chloride and triethylamine gives (**24a**) directly. Table 3 shows the effect an imino structure (**22**) has on the chemical shifts of C-2 and C-5 ⟨77JHC515⟩.

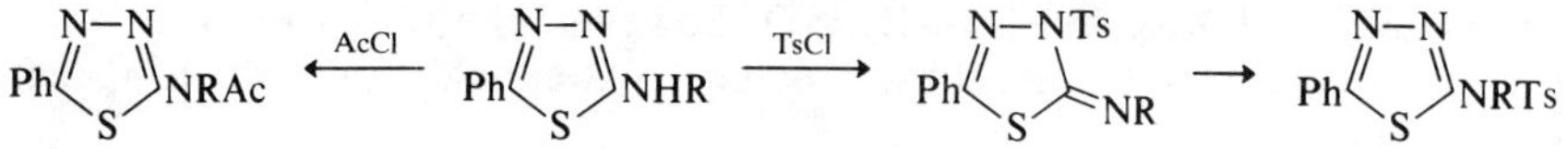

Scheme 2

a, R = PhCH$_2$; b, R = Ph

Table 3 Carbon Chemical Shifts^a in ^{13}C NMR Spectra of Aminothiadiazoles ⟨77JHC515⟩

Compound	C-2	C-5	Others
(**21a**)	168.8	156.6	48 (CH$_2$, ^1J(C–H) = 139 Hz)
(**21b**)	164.5^b	157.9^b	140.9 (PhN, C-1); 117.8 (PhN, C$_o$); 122.3 (PhN, C$_p$)
(**22a**)	151.5^b	149.5^b	21.6 (Me); 62.3 (CH$_2$, ^1J(C–H) = 135 Hz)
(**22b**)	152.1^c	149.7^c	21.8 (Me); 151.2 (PhN, C-1); 120.6 (PhN, C$_o$); 125.3 (PhN, C$_p$)
(**23a**)	161.7	166.5	21.6 (Me); 53.3 (CH$_2$, ^1J(C–H) = 142 Hz)
(**23b**)	163.6^d	166.7^d	138.5 (PhN, C-1); 127–130 (PhN, C$_{o,m,p}$); 21.7 (Me)
(**24a**)	160.1^d	165.4^d	22.4 (Me); 51.6 (CH$_2$, ^1J(C–H) = 141.5 Hz); 170.5 (CO)
(**24b**)	160.8^d	164.6^d	23.2 (Me); 169.9 (CO); 139.5 (PhN, C-1); 127–130 (PhN, C$_{o,m,p}$)

^{a} Solvent for compounds (**21a, b**) is DMSO-d_6. The rest are in CDCl$_3$. Reference standard is TMS in each case.
^{b} Established by selective decoupling.
^{c} Values are tentative.
^{d} Obtained from the undecoupled spectrum.

^{15}N NMR spectroscopy has not been applied widely to 1,3,4-thiadiazoles due to the low natural abundance and low sensitivity of detection of ^{15}N. Recent technological advances, however, have facilitated the recording of such spectra.

Organic nitrogen compounds have a chemical shift range of over 900 p.p.m., which is four times the range for ^{13}C and 90 times the range for ^{1}H. This wide range aids the detection of small changes in the chemical structure of molecules, such as in distinguishing the prevalent tautomeric form of 2-amino- and 2-thio-1,3,4-thiadiazole. In Table 4 the most likely structure of each tautomer is shown, and the appropriate ^{15}N chemical shifts are given ⟨78H(11)121⟩.

Table 4 ^{15}N and ^{14}N NMR Chemical Shifts for 1,3,4-Thiadiazoles

Compound	^{15}N Chemical shifts (p.p.m.)[a,c,e]	Solvent	^{14}N Chemical shifts (p.p.m.)[b,d,e]	Solvent
H₂N-[N-N / S]-NH₂	81.7 (N-3, N-4) 317.8 (N-2', N-5')	—	87.9 (N-3, N-4) 324.0 (N-2', N-5')	1 M DMSO
Me-[N-N / S]-NH₂	72.8 (N-3, N-4) 312.6 (N-2')	1 M DMSO	79.0 (N-3, N-4) 318.8 (N-2')	1 M DMSO
HS-[N-N / S]-SH	108.0 (N-3, N-4)	1.2 M EtOH	114.2 (N-3, N-4)	1 M EtOH
H₂N-[N-NH / S]=S	110.6 (N-3) 160.5 (N-4) 308.7 (N-2')	2 M DMSO	116.8 (N-3) 166.7 (N-4) 314.9 (N-2')	2 M DMSO

[a] ⟨78H(11)121⟩. [b] ⟨81MI42701⟩. [c] Upfield from external 1 M H^{15}NO₃ in D₂O. [d] Nitromethane reference standard. [e] Primed numbers refer to the exocyclic nitrogen atoms.

^{14}N NMR is useful in distinguishing between isomers of heterocyclic entities, such as oxadiazoles and thiadiazoles. The ^{13}C chemical shifts for 1,2,5- and 1,3,4-thiadiazoles are almost the same, and the ^{1}H chemical shifts for 1,2,3- and 1,2,5-thiadiazoles are identical. In contrast, the nitrogen chemical shifts are very different. Thus, in 1,2,3-thiadiazoles the N-3 resonates at −59 p.p.m. and the N-2 at −33 p.p.m., while in 1,2,5-thiadiazole both nitrogen atoms resonate at +35 p.p.m. Finally, in 1,3,4-thiadiazole both nitrogen atoms resonate at +10 p.p.m. (a plus value means upfield from nitromethane) ⟨78MI42701⟩. ^{14}N values for other thiadiazole derivatives are presented in Table 4 ⟨81MI42701⟩.

4.27.2.3.6 Combined spectroscopic methods

The combination of several spectroscopic techniques can be exceedingly useful in the structural elucidation of thiadiazoles. Thus, the heteroaromatic thiadiazole (**25**) and the nonaromatic thiadiazoline (**26**) can be easily distinguished. The IR absorption of the carbonyl stretch is 1660–1675 cm^{-1} for (**25**) and 1610–1620 cm^{-1} for (**26**), while in the UV spectra the λ_{max} value is 260 ± 3 for (**25**) and 301 ± 10 for (**26**). ^{13}C NMR with broad-band and off-resonance proton decoupling techniques has also been applied to this problem. The carbonyl carbon (difference of 3.8–10.8 p.p.m.), C-5 (5.3–5.9 p.p.m.), C-2 (3.2–3.9 p.p.m.), the benzylic carbon (5.7 p.p.m.) and the 5'-methyl carbon (1.2 p.p.m.) are the characteristic signals for differentiating between the two isomeric pairs ⟨80OMR(14)515⟩. A further example is provided by the diazomethane treatment of acylamido-1,3,4-thiadiazoles (**27**), which gives an isomeric mixture of the *exo* (**28**) and *endo* (**29**) methylated products. These can be identified by the characteristic carbonyl absorption in the IR spectrum [(**28**), 1667 cm^{-1}; (**29**), 1612 cm^{-1}] and by the differences in the ^{13}C NMR chemical shift at C-2 and C-5 ⟨80JHC1469⟩.

(**25**) R¹ = COPh, COCH₂Ph; R² = CH₂CH=CH₂, CH₂Ph   (**26**)

(**27**) →[CH₂N₂] (**28**) + (**29**)

4.27.2.3.7 UV spectroscopy

Unconjugated 1,3,4-thiadiazoles have no selective absorption above 220 nm. Substituents with lone pairs cause bathochromic shifts; the effect of an alkylthio group is greater than that of an amino group. *p*-Nitrophenyl groups cause large bathochromic shifts, while *m*-nitrophenyl groups cause hypsochromic shifts.

The UV spectra of 2-aminoarylidine-1,3,4-thiadiazole derivatives (**30**) show four bands. Two of these are in the 204-265 nm region and are due to the $\pi-\pi^*$ transition state of the benzene ring. The third band is in the 265–305 nm region and is attributed to the excitation of the π-electrons within the central C=N and that within the C=N bond of the heterering. The fourth band is due to a transition within the whole molecule (intramolecular charge transfer) and its position is sensitive to the nature of X. The acidity constants (pK_1) for these compounds can be determined from spectral shifts in buffer solutions and it is concluded that the pK_1 value increases as the electron-donating character of the substituent on the phenyl group increases. This shows the electron-withdrawing character of the 1,3,4-thiadiazole ring and the coplanarity between the ring and the rest of the molecule ⟨81MI42700⟩. In some earlier work, UV spectroscopy has also been utilized to investigate the tautomerism and acidity of thiadiazolyl hydrazones and thiadiazoline-5-thiones ⟨64ACS871, 66ACS57⟩. Further information on UV work ⟨71PMH(3)180⟩ and pK_a values ⟨68AHC(9)165, p. 203⟩ is available.

4.27.2.3.8 IR spectroscopy

Characteristic vibration bands are presented in Chapter 4.01.3.7, Tables 26 and 27. Further information is also available in Section 4.27.2.3.6 ⟨68AHC(9)165, p. 202, 70SA(A)2057⟩. The IR and electronic spectra of the metal complexes of bismuthiol I and II (thiadiazole derivatives) have been published ⟨80BSF(1)451⟩.

4.27.2.3.9 Mass spectrometry

The electron impact mass spectral fragmentation of 1,3,4-thiadiazoles was studied on a series of 2- and 2,5-substituted derivatives. The substituents (R^1) at the 5-position of (**31**) were H, alkyl and aryl groups and those at position 2 were amino ($R^2 = R^3 = H$), ureido ($R^2 = H$, $R^3 = CONHR$), carbamoyl ($R^2 = H$, $R^3 = CO_2R$) and alkylamido ($R^2 = H$, $R^3 = COR$) groups. The fragmentation of these compounds usually starts with the elimination of the ureido or carbamoyl functions with hydrogen rearrangement to give the 5-substituted 2-amino-1,3,4-thiadiazoles. The fragmentation of the 1,3,4-thiadiazole ring system is depicted in (**31**), (**32**) and (**33**). When R^1 is H the ions resulting from cleavages *a* and *b* predominate. However, in those cases where R^1 = aryl, fragmentations along *c*, *d*, *e*, *f* and *g* compete favorably with those along *a* and *b*. This is explained by the stabilizing effect of the aryl group on the resulting ions ⟨77JHC401⟩.

The mass spectrometry of 1,3,4-thiadiazolines substituted with aryl groups in the 2- and 4-positions and with H, alkyl, aryl, alkoxy and dialkylamino groups in the 5-position has also been investigated. The major fragmentation reactions occur by cleavages of the ring bonds by pathways *a*, *b*, and *c*, and by loss of substituents from the thiadiazoline ring (especially from the 5-position). Thus, if R' and R" in (**34**) are alkyl or dialkylamino, the resulting thiadiazolium ion (**35**) is the most intense ion. For the other cases the intensity of this fragment ion is reduced. An interesting rearrangement is observed when the S—C(5) bond is cleaved and a new bond between the aromatic system at N(4) and the sulfur radical is formed to give (**36**). This structure fragments to yield ions as indicated by *d* as well as an ion Ar_2S^+ ⟨74OMS(9)181⟩.

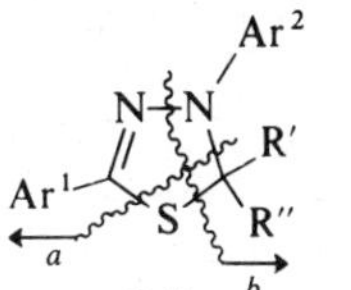

The fragmentation of 2-phenyl-1,3,4-thiadiazolin-5-one and -5-thione begins with the breaking of the C—N bond followed by the ejection of the S=C=Y molecule (Y = O or S). The major part of ΔH_R° is associated with the stabilization of the neutral fragment ⟨81OMS29⟩.

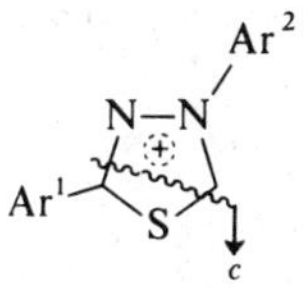

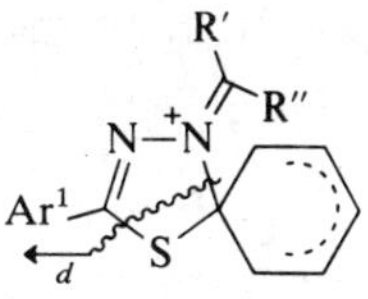

4.27.2.3.10 Photoelectron spectroscopy

Although this technique is not often mentioned in the field of 1,3,4-thiadiazoles, it can be useful for structural elucidation. For example, the sulfur $2p$ bond energy in (37) is considerably greater than in (38) (see Table 5). Also the nitrogen $1s$ electrons of the exocyclic nitrogen atom have a bond energy of 396.8 versus 398.2 for (38) ⟨72LA(764)94⟩.

Table 5 X-Ray Photoelectron Data for Nitrogen and Sulfur in Compounds (37) and (38) ⟨72LA(764)94⟩

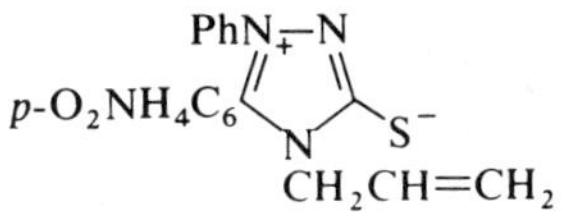

Compound	N, 1s (eV)	S, 2p (eV)	Compound	N, 1s (eV)	S, 2p (eV)
(37)	396.8	163.7	(38)	398.2	160.6
	398.3			399.9	
	400.2			400.1	
	404.6 (NO$_2$)			405.0 (NO$_2$)	

4.27.2.3.11 Spectral data for non-aromatic systems

Thiadiazolines and thiadiazolidines have been investigated by many spectroscopic techniques. In Sections 4.27.2.3.5 and 4.27.2.3.6 the use of UV, IR, ^{1}H NMR and ^{13}C NMR for distinguishing isomeric structures is discussed ⟨80OMR(14)515, 80JHC1469, 77JHC515⟩. Further examples of the application of NMR methods are given in Section 4.27.4.3.1 ⟨81JCS(P1)360⟩, while mass spectral information can be found in Section 4.27.2.3.9 ⟨74OMS(9)181⟩. In Tables 6–10 spectroscopic values for selected thiadiazolines and thiadiazolidines are summarized.

Table 6 Spectral Data for 2,4-Disubstituted-5-phenylimino Derivatives (203) and (204) ⟨80JCS(P1)574⟩

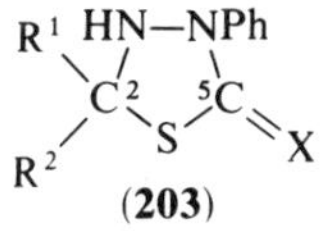

	^{13}C Chemical shifts (δ, p.p.m.)					^{1}H Chemical shifts (δ, p.p.m.)				ν (NH)
Compound	R^1	R^2	X	C-2	C-5	Solvent	R^1	R^2	NH	$M^{\ddot{+}}$	(cm^{-1})
(203)	Me	Me	NPh	79.2	176.3	DMSO-d_6	1.4 (s)	1.4 (s)	5.1 (s)	283	3160
	Me	Et	NPh	82.2	177.4	CDCl$_3$	1.4 (s)	1.0 (t)	5.0 (s)	297	3120
	—(CH$_2$)$_5$—		NPh	81.8	177.6	CDCl$_3$	1.2–2.1 (m)		4.8 (s)	323	3160
	Me	Ph	NPh	82.4	177.0	CDCl$_3$	1.7 (s)	6.8–7.3 (m)	5.4 (s)	345	3100
	Ph	Ph	NPh	88.5	177.1	CDCl$_3$	7.2–7.6 (m)		5.5 (s)	407	3190
	Pr	H	NPh	76.6	175.8	DMSO-d_6	—	—	—	—	—
	Me	Me	S	73.1	194.2	DMSO-d_6	—	—	—	—	—
(204)	—	—	—	149.9	168.6	CDCl$_3$	—	—	—	—	—

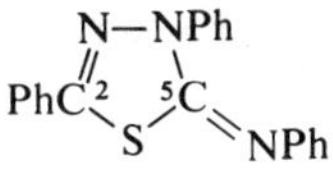

Table 7 ^{13}C and ^{1}H NMR Chemical Shifts for 2,5-Disubstituted-Δ^2-1,3,4-thiadiazoline Derivatives ⟨82CC188⟩

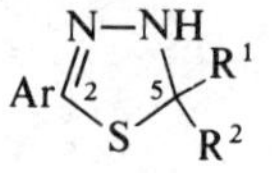

Ar	Substituents R^1	R^2	^{13}C Chemical shifts (C-5) (p.p.m.)^a	^{1}H Chemical shifts (C-5—H) (p.p.m.)^b
Ph	H	Ph	74.65	6.3
Ph	H	4-MeOC$_6$H$_4$	74.35	6.3
Ph	Me	Me	79.94	—
4-MeOC$_6$H$_4$	H	4-MeOC$_6$H$_4$	74.59	6.3

^{a} 20 MHz, CDCl$_3$. ^{b} 90 MHz, CDCl$_3$.

Table 8 ^{1}H NMR Chemical Shifts for 5,5-Dialkyl-1,3,4-thiadiazolidine-5-thiones and Their Precursor ⟨70ACS179⟩

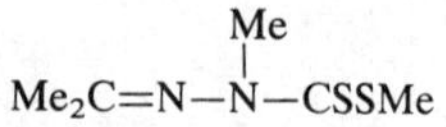

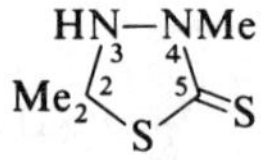

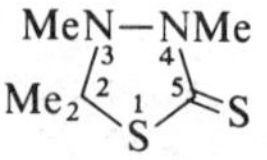

Me$_2$C=N—N(Me)—CSSMe		HN—NMe ring		MeN—NMe ring		
Chemical shifts (p.p.m.)		Chemical shifts (p.p.m.)		Chemical shifts (p.p.m.)		
Me$_2$C	N—Me	Me$_2$C	N—Me	Me$_2$C	N(3)—Me	N(4)—Me
1.94	3.67	1.69	3.59	1.64	2.72	3.51
2.18	—	—	—	—	—	—

Table 9 ^{1}H NMR Chemical Shifts of Non-aromatic Protons for 3-Aryl-5-phenyl-2-thioacyl-methylene Derivatives ⟨81JCS(P1)2952⟩^{a}

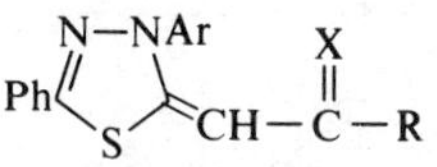

X	Compound Ar	R	Chemical shifts Me (p.p.m.)	Chemical shifts H (p.p.m.)
S	Ph	Ph	—	7.13–8.03 (under aromatic region)
O	Ph	Ph	—	6.70
S	2,4-Br$_2$C$_6$H$_3$	Me	2.70	6.60
O	2,4-Br$_2$C$_6$H$_3$	Me	2.12	5.45
S	2,4-Br$_2$C$_6$H$_3$	Ph	—	7.03
O	2,4-Br$_2$C$_6$H$_3$	Ph	—	6.16
S	Ph	Me	2.67	7.01
O	Ph	Me	2.17	6.10

^{a} Mass spectral data are also presented in this ref.

Table 10 ^{13}C NMR Chemical Shifts for 2-Phenyl-1,3,4-thiadiazolin-5-ones and -thiones ⟨82OMR(18)159⟩

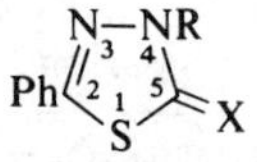

Substituents R	X	C-2	C-5	Chemical shifts (p.p.m.) Ph	Others
H	O	153.4	173.4	126.3; 129.1; 131.0	—
H	S	161.5	189.1	126.8; 129.3; 129.5; 131.6	—
Et	O	149.8	169.1	125.8; 129.0; 130.5	13.9 (Me); 42.5 (CH$_2$)

4.27.2.4 Thermodynamic aspects

4.27.2.4.1 Melting and boiling points

The melting and boiling points of a very large number of thiadiazoles have been summarized ⟨52HC(4)81⟩. 1,3,4-Thiadiazole melts at 43 °C while the 2-methyl homologue melts with decomposition at 201 °C. 2-Amino-1,3,4-thiadiazole has a melting point of 193 °C and the 2-aminomethyl homologue melts at 165 °C; 2-amino-5-methyl-1,3,4-thiadiazole melts at 224 °C with decomposition. The unusually high melting point of 255 °C displayed by (**18**) is explained by inter- and intra-molecular hydrogen bonding. Table 11 shows the melting and boiling points of a selected number of thiadiazole derivatives.

Table 11 Melting and Boiling Points of Thiadiazoles (**A**), (**B**) and (**C**) ⟨52HC(4)81⟩

Compound	Substituents R^1	R^2	m.p. (°C)	Compound	Substituents R^1	R^2	m.p. (°C) (b.p. °C/mmHg)
(A)	H	H	43	(A)	Me	Me	64
(A)	H	Me	201	(A)	Pr	Pr	(127/13)
(A)	H	NH_2	193	(A)	Ph	Ph	141
(A)	Me	NH_2	224	(A)	NH_2	NH_2	221
(A)	Et	NH_2	201	(A)	NH_2	$NHNH_2$	207
(A)	CMe_3	NH_2	184	(A)	Br	Br	111
(A)	CF_3	NH_2	225	(A)	NHMe	NHMe	175
(A)	Ph	NH_2	229	(A)	NH_2	SCH_2Ph	157
				(A)	MeCONH	SH	205

Compound	Substituents R^1	R^2	R^3	m.p. (°C)	Compound	Substituents R^1	R^2	R^3	m.p. (°C)
(B)	Me	Ph	Me	80	(C)	Me	H	Ph	240
(B)	$PhCH_2$	Ph	H	68	(C)	Ph	H	Ph	173
(B)	$PhCH_2$	Ph	Ph	92	(C)	Ph	Ph	Me	92
(B)	Ph	Ph	Ph	97	(C)	Ph	Me	Ph	102

The melting and clearing points of many 2,5-diphenyl-1,3,4-thiadiazole derivatives are available. They form crystalline liquid phases to various degrees ⟨80JPR933⟩.

4.27.2.4.2 Solubility

1,3,4-Thiadiazole as well as the following derivatives are soluble in water: 2-amino, 2,5-dimethyl, 2,5-diethyl, 2-methylamino and 2-aminomethyl-5-methyl. Generally, as the substituents in the 2- and 5-position increase in size, water solubility decreases, while solubility in organic solvents increases. Solubilities for many 1,3,4-thiadiazoles are tabulated in ⟨52HC(4)81⟩.

4.27.2.4.3 Chromatography

1,3,4-Thiadiazoles, like most other organic molecules, have been extensively chromatographed on silica gel using a variety of solvents ⟨74JOC2951, 80JOC4860⟩.

Acetazolamide (**225**) can be separated from other drugs by TLC on silica gel or cellulose using 40:15:15:15 BuOH:MeOH:$CHCl_3$:25% NH_3 as the solvent system. Detection is by fluorescence after spraying with Rhodamine B ⟨73MI42700⟩.

Good separation of isomeric 5-methyl-2-nitroanilino-1,3,4-thiadiazoles can be achieved on $Al_2O_3/CaSO_4$ thin layer plates ⟨68MI42700⟩.

In paper chromatography, 5-anilino-2-mercapto-1,3,4-thiadiazole is used as a reagent in the ascending chromatography of various cations ⟨68RRC909⟩, while 2-amino-1,3,4-

thiadiazole-5-sulfonamide is used for the affinity chromatography of plant carbonic anhydrase ⟨76MI42701⟩.

2-Methylamino-5-(1-methylallylamino)-1,3,4-thiadiazole is an oxidation product of methallibure, a veterinary oestrus regulator. Separation can be achieved by chromatography on alkylated polydextran gel and elution with mixed organic solvents. Metal salts can be removed by complexation with diphenylthiocarbazone ⟨70JPS835⟩.

4.27.2.4.4 Thermochemistry

The combined translational, rotational and vibrational contributions to the molar heat capacity, heat content, free energy and entropy for 1,3,4-thiadiazoles are available between 50 and 2000 K. They are derived from the principal moments of inertia and the vapor-phase fundamental vibration frequencies ⟨68SA(A)361⟩.

Thermal studies of bis(2-amino-5-phenyl-1,3,4-thiadiazole)copper(II) sulfate indicate that decomposition is a nucleation-controlled process and starts at the site of defects ⟨78MI42702⟩.

Based on thermogravimetric studies, the metal chelates of 5-amino-1,3,4-thiadiazole-2-thiol are shown to be stable up to 200 °C and undergo one-step decomposition to the free metal ⟨82MI42700⟩.

4.27.2.4.5 Aromaticity

The aromatic character of 1,3,4-thiadiazole can be demonstrated with the aid of microwave spectroscopy. Using the differences between the measured bond lengths and covalent radii, aromaticity, as shown by π-electron delocalization, diminishes in the order: 1,2,5-thiadiazoles > thiophene > 1,3,4-thiadiazole > 1,2,5-oxadiazole ⟨66JSP(19)283⟩. The microwave spectrum was further refined by later workers ⟨71JST(9)163⟩.

4.27.2.4.6 Conformation and configuration

The energy barrier to internal rotation of the dimethylamino group in a number of heterocyclic molecules is known. For 1,3,4-thiadiazoles the rotational barriers are greater by 1.7–2.9 kJ mol^{-1} than for the corresponding 1,3,4-oxadiazoles. This is attributed to a difference in bonding interaction between the dimethylamino group and the heterocyclic ring system ⟨77SST(4)431⟩.

The difference in reactivity between *cis*- and *trans*-dihydro-1,3,4-thiadiazole derivatives is presented in Section 4.27.3.2 ⟨81CB802⟩.

Treatment of geometrical isomers of sulfines with 2-diazopropane gives in high yield the desired 1:1 adducts, Δ^3-1,3,4-thiadiazoline oxides. Each of the geometrical isomers yields a single product, distinctly different from the one obtained from the other isomer. The spatial arrangement of the S=O group and that of the substituents is retained in the product, indicating that the cycloaddition is stereospecific and that the products are formed in a concerted manner (see also Section 4.27.4.3.2.ii) ⟨73TL3589⟩.

The stereoisomeric formation of bis(Δ^3-1,3,4-thiadiazolines) from dithiones and diazomethane is also discussed in Section 4.27.4.3.2(ii) ⟨71JOC3885⟩.

The reaction of 1-(4-methoxyphenyl)-2,2-dimethyldiazopropane with sulfur dioxide results in the formation of 2,5-di-*t*-butyl-2,5-bis(4-methoxyphenyl)-Δ^3-1,3,4-thiadiazoline 1,1-dioxide (**191**). In the 100 MHz NMR spectrum the aromatic protons appear as an unexpectedly complex pattern. A computer analysis of the spectrum yields the parameters shown in Table 12, indicating that the complexity is due to slow rotation about the sp^2–sp^3 carbon–carbon single bond as shown by the arrows, with the pairs H-2'/H-6' and H-3'/H-5' exchanging environments. This was confirmed by the high-temperature NMR spectrum in various solvents, when the complex multiplet collapsed to the expected symmetrical multiplet. Utilizing the nuclear Overhauser effect, the preferred conformation of (**191**) can be deduced and is shown in Table 12. The highly populated conformation suggests a value for ϕ other than 30°, with the *t*-butyl group equidistant from H-2' and H-6', H-6' being

Table 12 NMR Chemical Shifts for Aromatic Protons of (**191**) ⟨75AJC151⟩

(**191**)

Conformation of (**191**)

H-Position	Chemical shifts (p.p.m.)[a]	Coupling constants (Hz)
2'	763.22	J (2'-3') 8.82, J (3'-5') 2.93
3'	689.13	J (2'-5') 0.23, J (3'-6') 0.30
5'	698.04	J (2'-6') 2.49, J (5'-6') 8.88
6'	748.25	— —

[a] At 100 MHz.

closer to the *t*-butyl group. The relative assignments of H-3' and H-5' are obtained from the magnitudes of the coupling constants ⟨75AJC151⟩.

An extensive discussion and leading references on the stereochemistry of thiadiazolines and their pyrolysis to alkenes by a two-fold extrusion process can be found in ⟨78JCS(P1)45⟩ (see also Section 4.27.4, Scheme 33).

4.27.2.5 Tautomerism

The 1,3,4-thiadiazole ring system, with three heteroatoms, does not exhibit tautomerism in its fully conjugated form. However, when certain substituents are present, tautomerism is possible. 1,3,4-Thiadiazolin-2-ones (**39**; X = O) and -2-thiones (**39**; X = S) exist in the oxo and thione forms, respectively, as shown by spectroscopic and LCAO–MO calculations. 2-Amino-1,3,4-thiadiazoles exist in the amino form in solution and in the solid state; the K_T value is 10^5 as shown by basicity measurements. UV spectroscopy and LCAO–MO calculations show that the amino tautomer is also the main species when there is an alkoxy group in the 5-position (**40**; R^1 = alkoxy, R^2 = NH$_2$), or if the exocyclic nitrogen atom is part of a hydrazone group (**40**; R^2 = NHN=CR^2).

(**39**)   (**40**)   (**41**)

While the nitroso group has no influence on the tautomeric equilibrium the sulfonamido group shifts the equilibrium towards the imido tautomer (**39**; X = NSO$_2$Ph). Aminoalkoxythiadiazoles exist in the amino form (**40**; R^1 = NHR; R^2 = OR) as shown by IR studies ⟨76AHC(S1)380, 76AHC(S1)380, p. 407, 76AHC(S1)380, p. 482⟩. The structure of 2,5-dimercapto-1,3,4-thiadiazole is controversial; IR, UV and pK_a measurements suggest (**41**) as the most likely tautomeric form. ^{15}N NMR evidence, however, indicates tautomerism between (**41**) and (**40**; $R^1 = R^2$ = SH) with the equilibrium strongly favoring the latter ⟨78H(11)121⟩. 2-Amino-5-mercapto-1,3,4-thiadiazole can exist in four forms, but it is likely to exist as shown in (**39**; R = NH$_2$; X = S) ⟨68AHC(9)165, p. 205⟩.

1,3,4-Thiadiazoline-2-thiones (**42**) are in equilibrium with the open-chain hydrazone tautomer (**43**); under basic conditions the hydrazone salt (**44**) is the predominant form (Scheme 3). The reverse also occurs. Compounds of type (**45**) which were thought to be linear are in equilibrium with the cyclic Δ^2-1,3,4-thiadiazolines (**46**). These results are based on NMR evidence but are also confirmed chemically by reactions such as the alkylation of (**47**) to give high yields of (**48**). The mechanism for the cyclization–ring cleavage involves the electron displacement (**49** ⇌ **50**) ⟨70LA(731)142⟩. Later workers were unable to reproduce the spectroscopic data for (**46**; R = H; R' = 4-MeOC$_6$H$_4$; R'' = Ph) which were claimed to prove the presence of the acyclic isomer ⟨82CC188⟩.

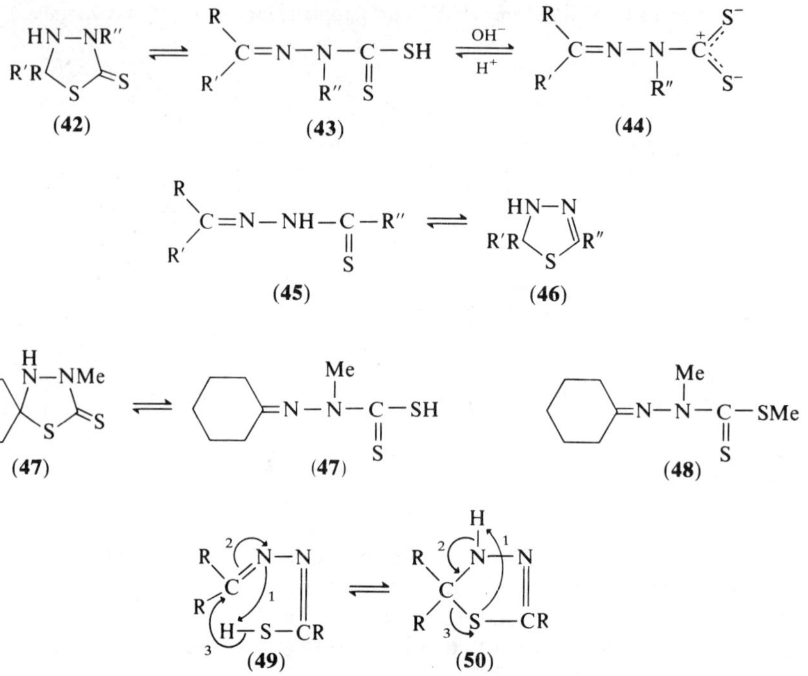

Scheme 3

4.27.3 REACTIVITY

4.27.3.1 Survey of Reactivity

Some of the characteristic reactions of the 1,3,4-thiadiazole nucleus are ring opening by strong base, ease of nucleophilic attack and the formation of mesoionic compounds by quaternization. The substituents in the 2- and 5-positions have a large effect in determining the reactivity of the molecule as a whole. Thus the ambident nucleophilicity of 2-aminothiadiazoles gives rise to electrophilic attack on both the amino group and the nuclear nitrogen atom. Ring formation between these two nitrogen atoms is also a common reaction. 2-Mercaptothiadiazoles react similarly to arenethiols while a methyl group on the thiadiazole ring has a reactivity similar to that in picoline. Nucleophiles easily displace halogen atoms from the thiadiazole nucleus. This is due to the electronegativity of the two nuclear nitrogen atoms which impart a low electron density to the carbon atom of the nucleus. The ease of decarboxylation of carboxythiadiazoles as well as the lack of electrophilic attack on unsubstituted thiadiazoles can also be explained by the low electron density on the nuclear carbon atoms. Selected examples in this section illustrate the general reactivity of 1,3,4-thiadiazoles.

4.27.3.2 Thermal and Photochemical Reactions Formally Involving No Other Species

The thermal and photochemical fragmentation of 1,3,4-thiadiazoles often follows the fragmentation pattern observed in the mass spectrometer. The thermolysis of 2,5-dihydro-1,3,4-thiadiazoles (**51**) gives (**52**), while photolysis yields the ketazine (**53**; Scheme 4). The *cis*-2,5-di-*t*-butyl-2,5-dihydro-1,3,4-thiadiazole 1,1-dioxide (*cis*-**54**) undergoes thermolysis at 50 °C to give the azine (**55**). The *trans* isomer (*trans*-**54**), however, undergoes thermolysis only above 145 °C to give the alkylidenehydrazide (**56**) plus sulfur monoxide which disproportionates into sulfur and sulfur dioxide. Compound (**56**) was independently prepared from dimethylpropanal (**57**) and the hydrazide (**58**). Thermolysis of the isomeric 2,3-dihydro-1,3,4-thiadiazole 1,1-dioxide (**59**) also gives rise to (**56**). The concerted [4+1] cycloelimination of sulfur dioxide from the *trans* isomer (**54**) cannot take place due to steric hindrance ⟨81CB802⟩.

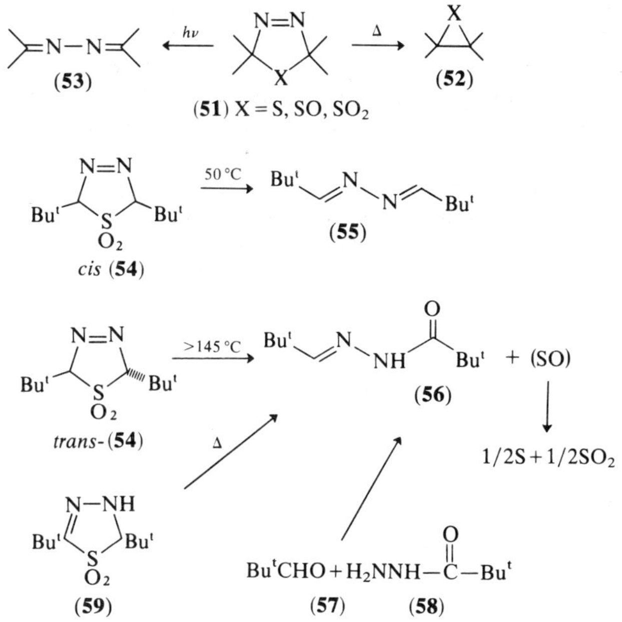

1,3,4-Thiadiazoles

Scheme 4

Photolysis of mesoionic anhydro-2,3-diaryl-5-mercapto-1,3,4-thiadiazolium hydroxide (**60**) in acetonitrile at >3000 Å yields, through oxidative cyclization, the tetracyclic (**61**; Scheme 5). In contrast, irradiation at 2537 Å gives *N*-phenylthiobenzamide (**62**) in 27% yield. Similarly, irradiation of (**63a**) in 0.01 M MeCN at 2537 Å gives (**64a**; 21%) and elemental sulfur (40%). Photolysis of (**63a**) in MeCN:MeOH (1:4) yields (**64a**; 33%) and sulfur (26%). Further, photolysis of the isomeric (**63b**) leads to (**64b**; 20%) and sulfur (35%). This photofragmentation proceeds through a valence tautomerization to yield the *N*-isothiocyanatothioamide (**65**) in which the N—N bond is cleaved to yield the radical precursor of the thioamide and the isothiocyanate radical. The thioamide is then formed by hydrogen abstraction from the solvent. The photochemical decomposition of the isothiocyanate group is known to proceed with loss of elemental sulfur ⟨69TL4627⟩.

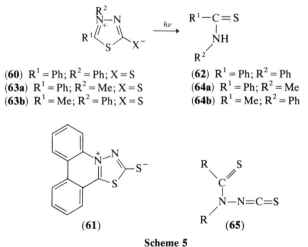

(**60**) $R^1 = Ph$; $R^2 = Ph$; $X = S$
(**63a**) $R^1 = Ph$; $R^2 = Me$; $X = S$
(**63b**) $R^1 = Me$; $R^2 = Ph$; $X = S$

(**62**) $R^1 = Ph$; $R^2 = Ph$
(**64a**) $R^1 = Ph$; $R^2 = Me$
(**64b**) $R^1 = Me$; $R^2 = Ph$

Scheme 5

5-Amino-2-imino-3-phenacyl-1,3,4-thiadiazoline (**66**) undergoes the Dimroth rearrangement to yield the 1,2,4-triazole derivative (**67**; Scheme 6) ⟨75M1291⟩.

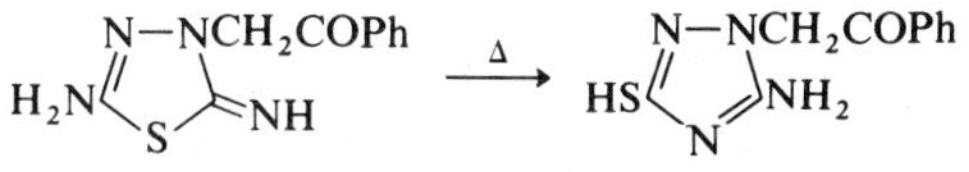

Scheme 6

4.27.3.3 Electrophilic Attack at Nitrogen

1,3,4-Thiadiazoles undergo a facile reaction with electrophiles on the annular nitrogen atoms. In an extensive study, the quaternization of 2,5-disubstituted 1,3,4-thiadiazoles with methyl iodide was compared with quaternization in the pyridazine series. The influence of the alkyl substituents on product distribution was less in the thiadiazole series than in the pyridazine series, probably due to the difference in geometry of the two compounds. The conditions chosen for the quaternization were such that the reaction was irreversible and the product distribution therefore kinetically controlled. Some of the results obtained are summarized in Table 13 ⟨73ACS391⟩. NMR data on these compounds are presented in Section 4.27.2.3.4.

Table 13 The Quaternization of Selected 1,3,4-Thiadiazoles ⟨73ACS391⟩[a]

2-Substituent	5-Substituent	Quaternization at $N(3)$ (%)	Quaternization at $N(4)$ (%)	$\log k_{N(3)}/k_{N(4)}$
H	Me	55	45	+0.09
Et	Me	43.5	56.5	−0.11
Bu^t	Me	2.5	97.5	−1.60
Pr^i	Et	38	62	−0.21
Bu^t	Pr^i	8.5	91.5	−1.03
Pr^n	Me	42	58	−0.14
Bu^s	Me	39	61	−0.19
Bu^s	Et	45.5	54.5	−0.08
Bz	Et	48	52	−0.04

[a] Product distribution obtained by reaction with methyl iodide at 50 °C in acetonitrile.

The reaction of 2-amino-5-benzoyl-1,3,4-thiadiazole with various methylating agents occurs at both the annular nitrogen in position 3 and at the exocyclic nitrogen atom. Methyl iodide in methanol gives methylation on the annular nitrogen, as does dimethyl sulfate/potassium carbonate in acetone; the latter, however, leads to some methylation on the exocyclic nitrogen as well. With diazomethane, methylation occurs on both the annular nitrogen and on the acylamino group ⟨75JHC841⟩.

α-Halo ketones react at N-3 of 2,5-diamino-1,3,4-thiadiazoles to give the thiadiazoline (**68**), which undergoes facile cyclization to the imidazothiadiazole (**69**) ⟨77M665⟩.

Bromo monosaccharides react with 1,3,4-thiadiazoline-2,5-dithiones on both the exocyclic sulfur and the annular nitrogen atoms. Treatment with 2 moles of 2,3,4,6-tetra-O-acetyl-α-D-glucopyranosyl bromide (α-acetobromoglucose, ABG) yields a mixture of (**70**) and (**71**). Using an excess of the aglycone, the S-monoglucose (**72**) is obtained. Compound (**71**) can also be prepared by reacting ABG with the monomercury(II) salt of the 2,5-dithione. Treatment with mercury(II) bromide of (**70**) and (**72**) causes transglycosidation to (**71**) in high yield ⟨73JPR915, 77SST(4)431, p. 434⟩.

(**70**) GL = β-D-glucopyranosyl

4.27.3.4 Electrophilic Attack at Carbon

Due to the low electron density at the carbon atoms in 1,3,4-thiadiazole, such reactions as nitration, sulfonation, acetylation, halogenation, *etc.* normally do not take place. However, 2-amino-substituted 1,3,4-thiadiazoles (**73a–i**) react with bromine in acetic acid to give the 5-bromo derivatives (**74a–i**). Similarly, the thiadiazolines (**75b**) and (**75d–f**) yield the corresponding 5-bromo derivatives (**76b**) and (**76d–f**). The thiadiazoline (**75a**), however, when brominated in acetic acid in the presence of sodium acetate gives a mixture of (**76f**), (**75f**) and (**76a**). Nitrosation, acetylation and benzoylation of (**74a**) yields the expected (**74c**), (**74d**) and (**74e**), respectively ⟨77JHC823⟩. The 5-chlorothiadiazole (**77**; R = Cl) is obtained in high yield on heating (**77**; R = SO$_2$Cl), probably through a concerted mechanism. See also Section 4.02.1.4.

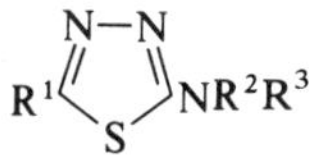

a; R^2 = R^3 = H
b; R^2 = H; R^3 = Me
c; R^2 = H; R^3 = NO
d; R^2 = H; R^3 = COMe
e; R^2 = H; R^3 = COPh
f; R^2 = R^3 = Me
g; R^2 = Me; R^3 = NO
h; R^2 = Me; R^3 = COMe
i; R^2 = Me; R^3 = COPh

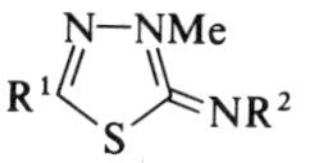

(**73**) R^1 = H
(**74**) R^1 = Br

a; R^2 = H
b; R^2 = Me
c; R^2 = NO
d; R^2 = COMe
e; R^2 = COPh
f; R^2 = Br

(**75**) R^1 = H
(**76**) R^1 = Br

Scheme 7

(**77**)

4.27.3.5 Electrophilic Attack at Sulfur

Although direct oxidation of the annular sulfur atom in 1,3,4-thiadiazoles has not been reported, Δ^3-1,3,4-thiadiazoline 1-oxides and 1,1-dioxides can be obtained by indirect means (see Section 4.27.4.3, Schemes 26 and 27).

4.27.3.6 Nucleophilic Attack

4.27.3.6.1 Ring opening

The 1,3,4-thiadiazole ring reacts with strong nucleophiles. Treatment of the parent compound with base leads to ring fission. 2-Amino- and 2-methylamino-1,3,4-thiadiazole (**78**; R^1 = H, R^2 = H, Me) rearrange in the presence of methylamine to the triazolinethione (**79**; R^1 = H, R^2 = H, Me; Scheme 8).

2-Amino-1,3,4-thiadiazole (**78**, R^1 = H, R^2 = H) on heating with benzylamine gives a mixture of (**78**; R^1 = H, R^2 = CH$_2$Ph) and (**79**; R^1 = H, R^2 = CH$_2$Ph). The rearrangement to (**79**) probably proceeds through the amidrazone (**80**). On treatment with hydrazine hydrate, (**78**; R^1 = Cl, R^2 = H) rearranges to a mixture of (**81**) and (**82**), probably via the hydrazide (**83**). 2,5-Dichloro- and 2,5-dimercapto-1,3,4-thiadiazole, under the same conditions, give only (**82**). These rearrangements can also take place under acidic conditions. Treatment of (**84**) with hydrochloric acid gives the triazolinethione (**85**; R = H), while with hydrochloric acid containing acetic acid a mixture of (**85**; R = H) and (**85**; R = Me) is obtained, presumably through intermediate (**86**) ⟨68AHC(9)165, p. 194⟩.

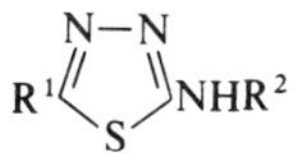

(**78**) 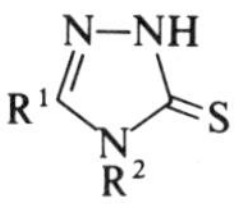 (**79**) 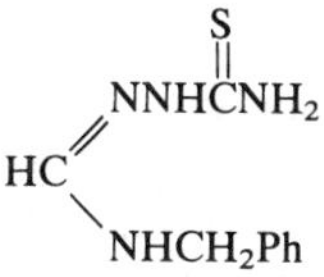 (**80**) 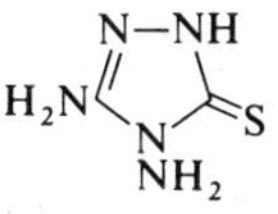 (**81**) 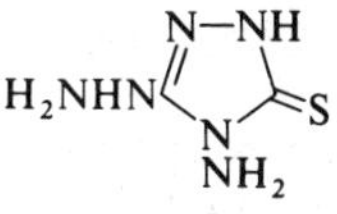 (**82**)

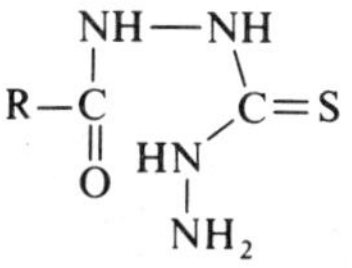

(**83**) 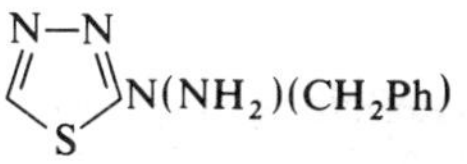 (**84**) 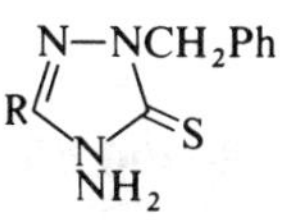 (**85**) 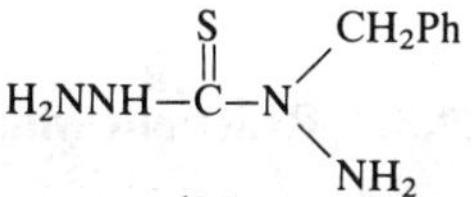 (**86**)

Scheme 8

The mesoionic compound (**87**) undergoes nucleophilic attack by aniline at C-2 to yield (**88**), which further reacts with excess aniline to give (**89**) and (**90**). Lithium aluminum hydride reduction of (**87**) removes the sulfur atom to give (**91**; Scheme 9) ⟨70SST(1)449, p. 452⟩.

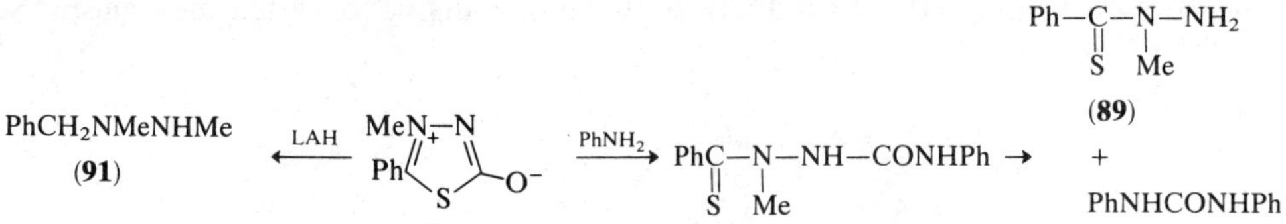

Scheme 9

The thiadiazolium iodide (**92**) hydrolyzes to *N*-benzoyl-*N*-phenyldithiocarbamate (**93**). Above pH 4, the reaction is of first order and almost independent of pH, in accord with the rate determining formation of the neutral intermediate (**94**), *via* a cationic transition state ⟨77SST(4)431, p. 435⟩.

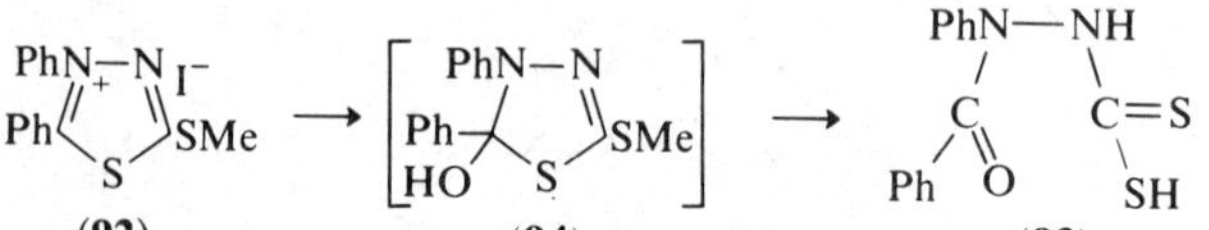

4.27.3.6.2 Amination

Direct nuclear amination of certain thiadiazoles is possible, illustrating the ease of nucleophilic attack. Thus, (**95**; R = H) reacts with hydroxylamine in the presence of base to give (**95**; R=NH$_2$) presumably *via* the imine (**96**) ⟨73SST(2)725, p. 741⟩.

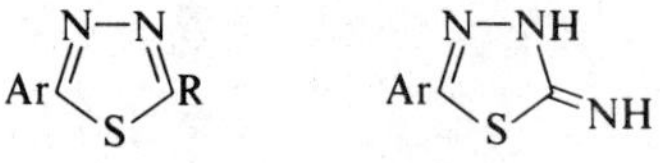

2-Amino-1,3,4-thiadiazolium salts (**97**) on treatment with alkali rearrange to mesoionic 1,2,4-triazole-3-thiones (**98**) probably by deprotonation of (**97**), nucleophilic attack at C-5, followed by ring cleavage and recyclization to (**98**) ⟨75SST(3)687, p. 690⟩.

4.27.3.6.3 Reduction

2-Amino-5-phenyl-1,3,4-thiadiazole is reduced by sodium amalgam to benzaldehyde thiosemicarbazone, but the 5-H and 5-methyl analogs do not react ⟨68AHC(9)165, p. 197⟩ (see also Scheme 9 for LAH reduction).

4.27.3.6.4 Hydrogen exchange at ring carbon

The deprotonation rate of 1,3,4-thiadiazolium salts is about 10^4 times faster than for thiazolium salts. The rate equation is first order in both substrate and OD$^-$ and the mechanism involves rate determining proton abstraction by OD$^-$ followed by fast reaction with D$_2$O to give the exchanged product ⟨74AHC(16)14⟩.

4.27.3.7 Reactions with Radicals and Electron-deficient Species

α-Alkoxythiadiazolines react with diethyl acetylenedicarboxylate *via* a carbene ylide and a 1 → 4 migration of the *p*-nitrophenyl group to give the thiadiazole derivative shown in Scheme 10 ⟨77TL2095⟩.

Scheme 10

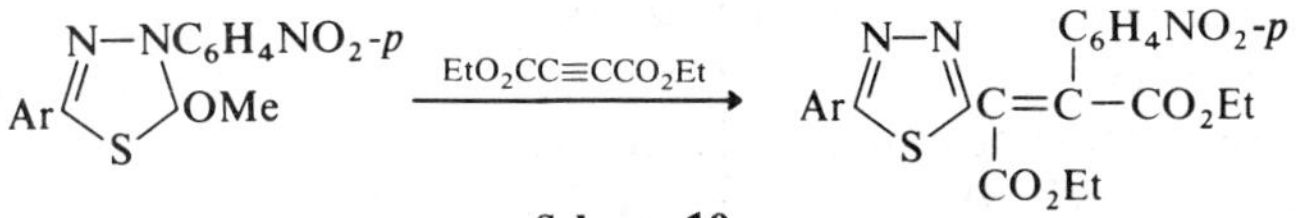

4.27.3.8 Cycloaddition Reactions

Iminobenzodithioles (**100**) are produced when 1,3,4-thiadiazolinethiones (**99**; $R^1 = H$, Me, Ac; $R^2 = Me$, Ph) are reacted with benzyne, most likely by a 1,3-dipolar addition mechanism followed by elimination of RSCN ⟨79SST(5)440, p. 444⟩.

(**99**) (**100**)

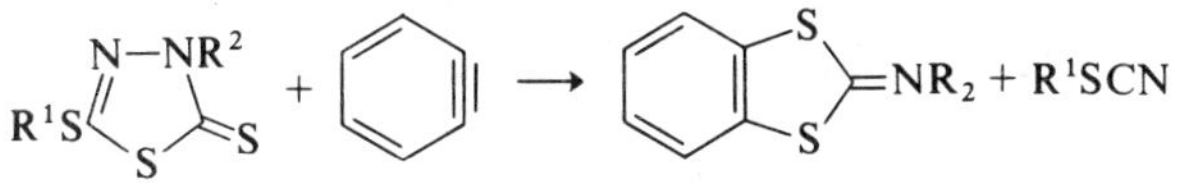

Oxidation of 1,3,4-thiadiazolidine-2,5-dione (**101**) or its monopotassium salt (**103**) with various oxidizing agents leads to 1,3,4-thiadiazoline-2,5-dione (**102**) which, though unstable, can be trapped with reactive dienes to give Diels–Alder adducts. Reaction of (**101**) with dienes and copper(II) chloride in DMF gives the Diels–Alder adduct (**104a**) from cyclopentadiene (65%), (**104b**) from 1,3-cyclohexadiene (43%), (**104c**) from 2,3-dimethylbutadiene (23%) and (**104d**) from isoprene (4%). In the absence of a diene, (**102**) decomposes to nitrogen, carbon monoxide and carbon oxysulfide, a fragmentation analogous to that of 1,2,4-triazoline-3,5-dione. Polar solvents increase the rate of oxidation of (**101**) and (**103**) to give (**102**); methylene chloride, however, slows it down and enhances decomposition. The Diels–Alder reactivity of (**102**) and 4-phenyl-1,2,4-triazoline-3,5-dione, a very potent dienophile, is about the same as measured by their competition for cyclopentadiene (Scheme 11).

(**104a**) $R^1, R^2 = CH_2$; $R^3 = R^4 = H$
(**104b**) $R^1, R^2 = CH_2CH_2$; $R^3 = R^4 = H$
(**104c**) $R^1 = R^2 = H$; $R^3 = R^4 = Me$
(**104d**) $R^1 = R^2 = R^3 = H$; $R^4 = Me$
[O] = $CuCl_2$, $Pb(OAc)_4$, Bu^tOCl

(**106a**) $R = PhCH_2$
(**106b**) $R = Me$
(**106c**) $R = CH_2{=}CHCH_2$
(**106d**) $R = Bu^t$

Scheme 11

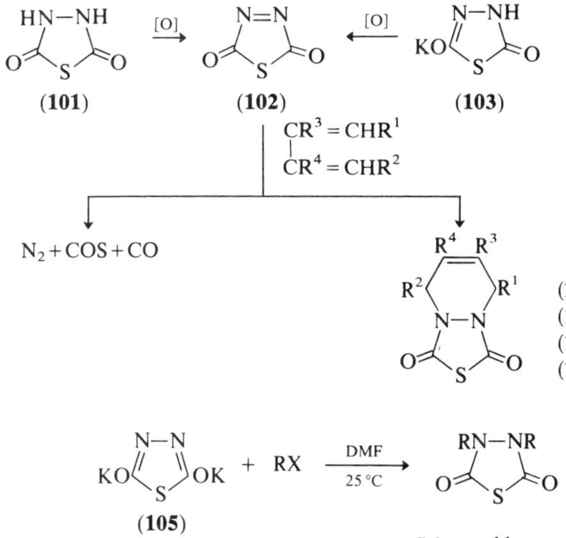

The monopotassium salt (**103**) and the dipotassium salt (**105**) can be prepared in quantitative yield from (**101**) and potassium hydroxide. Reaction of (**105**) in DMF with alkyl halides gives dialkylation products: (**106a**) from benzyl bromide, (**106b**) from methyl iodide, (**106c**) from allyl bromide, and (**106d**) from *t*-butyl bromide in low yield.

A previously claimed synthesis of (**101**) has been shown to be erroneous ⟨74JOC2951⟩.

4.27.3.9 Reactions of Substituents on Carbon

4.27.3.9.1 Carbon substituents

Treatment of 2-methyl-5-phenyl-1,3,4-thiadiazole (**107**; R = H) with *n*-butyllithium at −78 °C gives the lithio derivative (**107**; R = Li) which, on treatment with methyl iodide, gives the expected 2-ethyl homologue (**107**; R = Me; Scheme 12). However, if (**107**; R = Li) is allowed to warm from −78 °C to 25 °C the dimer (**108**) is formed, which on heating above 150 °C reverts to starting material. The dimerization proceeds *via* intermediate (**109**) and not the ketene-imine (**110**) as previously suggested ⟨74JOC1189⟩.

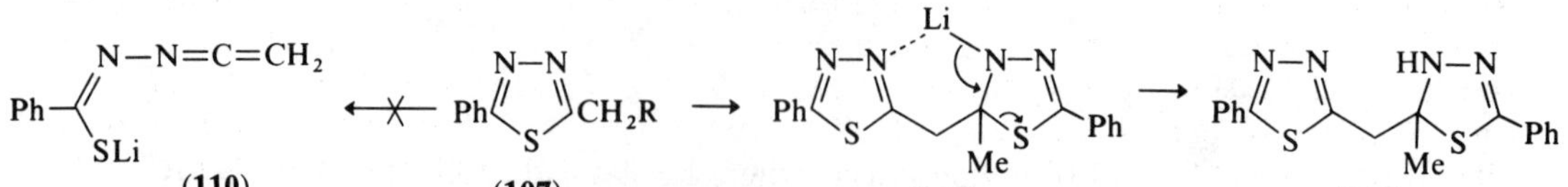

Scheme 12

4,4′-(1,3,4-Thiadiazole-2,5-diyl)bis(1-methylpyridinium) diiodide is reduced by a one-electron transfer to a radical cation (**111**). The reduction is done in the pH range 5.4–8.0 and the half-wave potential is −0.39 independent of pH and concentration. The stability of the diiodide was determined by UV data and from proton NMR data it was evident that the one-electron transfer was not totally reversed by air ⟨81JHC409⟩.

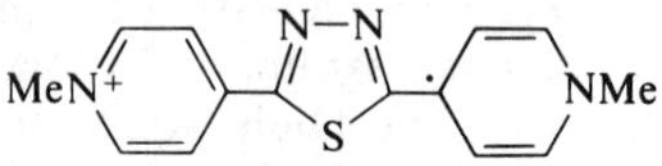

(**111**)

5-Amino-1,3,4-thiadiazole-2-carboxylic acid decarboxylates in water over a pH range 0.25–3.91. Rate constant measurements suggest a decarboxyprotonation mechanism of the zwitterion ⟨78JOC4042⟩.

4.27.3.9.2 Halogen substituents

Yield data obtained from the reaction of *p*-substituted 2-aryl-5-chloro-1,3,4-thiadiazoles with nucleophiles such as ethoxide, cyanide, phenoxide, hydroxide and thiophenoxide ions, and aromatic amines and piperidine, show that piperidine is the best substrate for investigating the nucleophilic reactivity of these thiadiazoles. Kinetic studies in ethanol and in benzene in the temperature range 30–60 °C are first order with respect to both reagents in ethanol; however, a third-order term appears in benzene. *p*-Substituents on the phenyl ring affect the rate of nucleophilic exchange by a Hammett ρ value of 1.47 at 50 °C in ethanol. The reactivity order in ethanol is OMe < H < Cl ≪ NO_2. A comparison of the reactivity towards piperidine of three isomeric chlorophenyl thiadiazoles is shown in Table 14. Nucleophilic displacement of chlorine is favored when a sulfur atom is in the α-position to the reactive center. As expected, solvent effects are important. When the *para* substituent is H or NO_2, nucleophilic substitution of chlorine by piperidine is faster in ethanol than in benzene by factors of 17.6 and 22.9, respectively. This suggests a process between neutral molecules giving rise to a polar transition state ⟨68T3209⟩.

Table 14 Reactivity of Three Isomeric Thiadiazoles with Piperidine at 30 °C in Ethanol ⟨68T3209⟩

Structure	Rate constant ($mol^{-1} s^{-1}$)	Structure	Rate constant ($mol^{-1} s^{-1}$)	Structure	Rate constant ($mol^{-1} s^{-1}$)
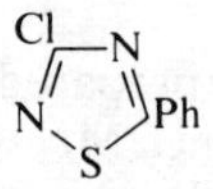	$10^4 K = 0.028$	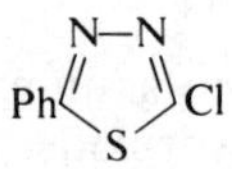	$10^4 K = 1.79$	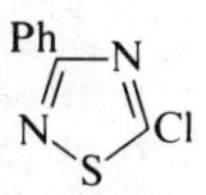	$10^4 K = 198$

2-Amino-5-phenyl-1,3,4-thiadiazole (**112**; R = NH_2) on diazotization gives the 2-chloro derivative (**112**; R = Cl), which on treatment with hydrazine followed by benzoylation gives (**112**; R = NHNHCOPh). Cyclization with phosphorus oxychloride yields the triazolo derivative (**113**) in moderate yield ⟨66JOC3528⟩.

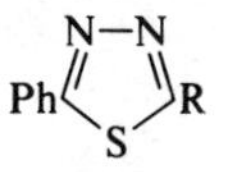

(**112**) 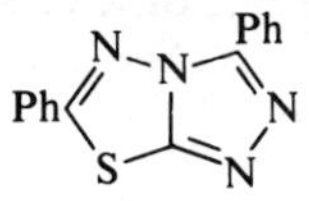   (**113**)

4.27.3.9.3 Sulfur substituents

2,5-Dimercapto-1,3,4-thiadiazole (Bismuthiol I) and 5-mercapto-3-phenyl-1,3,4-thiadiazoline-2-thione (Bismuthiol II) form complexes with many metals and this has led to their use as antioxidants in lubricating oils and in paints. Structural elucidation of these complexes involves spectroscopic and magnetic moment measurements. The complexes consist of bidentate bridging ligands through the thiocarbonyl and deprotonated thiol group. The complexes formed with cobalt, nickel and copper are tetrahedral, octahedral and square planar, respectively; zinc and cadmium complexes have tetrahedral structures ⟨80BSF(1)451⟩. Some workers claim that the cobalt and nickel complexes are square planar ⟨77MI42700⟩. 2-Amino-5-phenyl-1,3,4-thiadiazole also forms complexes with Cu(II). The ligand is bonded through nitrogen and sulfur and has a distorted octahedral geometry ⟨79IJC(A)191⟩.

The mesoionic 4,5-diphenyl-1,3,4-thiadiazolium-2-thiolate (**114**) reacts with methyl azodicarboxylate to yield the azothiadiazole (**115**) and not the tetrazine betaine (**116**) as previously claimed. Compound (**115**) can also be prepared from 2-amino-5-phenyl-1,3,4-thiadiazole and nitrosobenzene ⟨71CC837⟩.

Another mesoionic thiadiazole (**117**), with a thio substituent, reacts with dimethyl acetylenedicarboxylate to give the thiophene (**119**) and the S-cyanothioimidate (**120**), presumably through the intermediate (**118**) ⟨72CC1300⟩.

4.27.3.9.4 Amino substituents

Aminothiadiazoles react similarly to other aromatic amines; however, the presence of a nitrogen atom in the 3-position facilitates the formation of bicyclic compounds. Examples of the reactivity of the amino group are given in Scheme 12a. Treatment of 2-aminothiadiazole (**121**) with nitrous acid yields the N-nitroso compound (**122**), which acetylates on the *exo* nitrogen atom with concomitant loss of the nitroso group to give (**123**). Acetylation or benzoylation of (**121**) gives (**124**; R = Me or Ph); acetylation of (**124**; R = Ph) leads to loss of the benzoyl group to give (**123**). Methylation and base treatment of (**121**) gives the Δ^2-1,3,4-thiadiazoline (**125**) due to the greater nucleophilicity of the N-3 atom when compared with the exocyclic nitrogen atom. Nitrosation, acetylation, benzoylation or methylation of (**125**) leads, as expected, to the corresponding nitroso (**126**), acetyl or benzoyl (**127**; R = Me, Ph) and methyl (**128**) derivatives. Aminothiadiazoles with a nitroso (**122**) or acetyl (**124**; R = Me) substituent on the exocyclic nitrogen atom methylate on both nitrogen atoms, to give mixtures of (**126**) and (**129**), and (**127**; R = Me) and (**130**; R = Me), respectively ⟨77JHC1263⟩.

The reaction of unsubstituted 2-amino-1,3,4-thiadiazole (**131**) with 1,3-dicarbonyl compounds is dependent on the nature of the dicarbonyl compound (Scheme 13). Thus, reaction of (**131**) with pentane-2,4-dione gives the 4,6-dimethyl-2-thiocyanatopyrimidine (**132**). The formation of (**132**) may proceed *via* the cation (**132a**). With ethyl acetoacetate, however, a mixture of (**133**) and (**134**) is formed, (**133**) also being converted into (**134**) on heating.

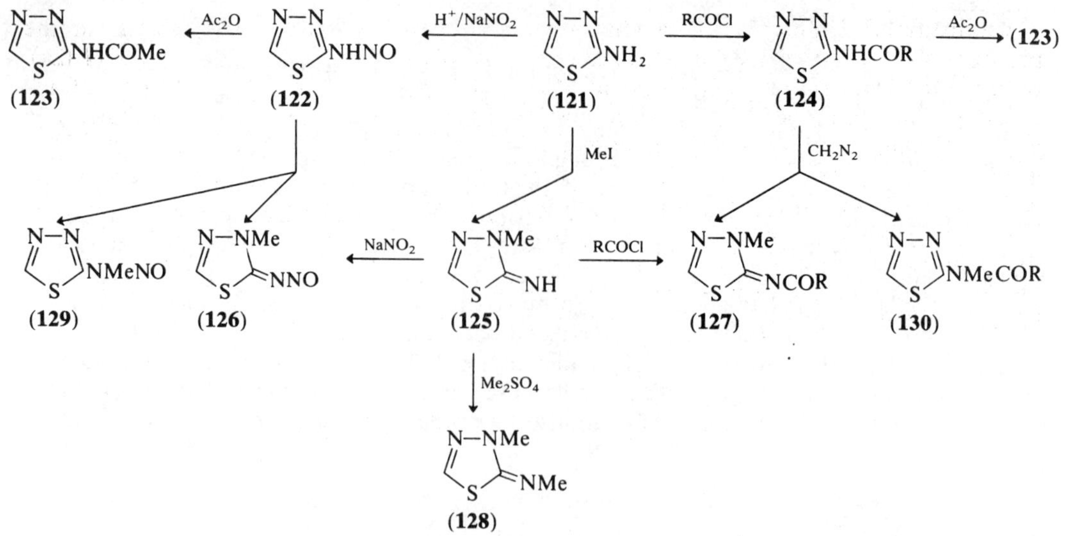

When diethyl malonate is the diketone component, the ester (**135**) is first formed, which then cyclizes to the expected (**136**). Reaction of (**131**) with diketene gives a mixture of (**137**) and (**138**), the latter on acid cyclization also yields (**134**). This rearrangement is by a 1,3-shift of the acetoacetyl group to give the transient imine followed by cyclization. Similar rearrangements are known. Compound (**139**), the 7-one isomer of (**134**), is obtained from the reaction of (**131**) with α-bromocrotonic acid ⟨76JHC291⟩.

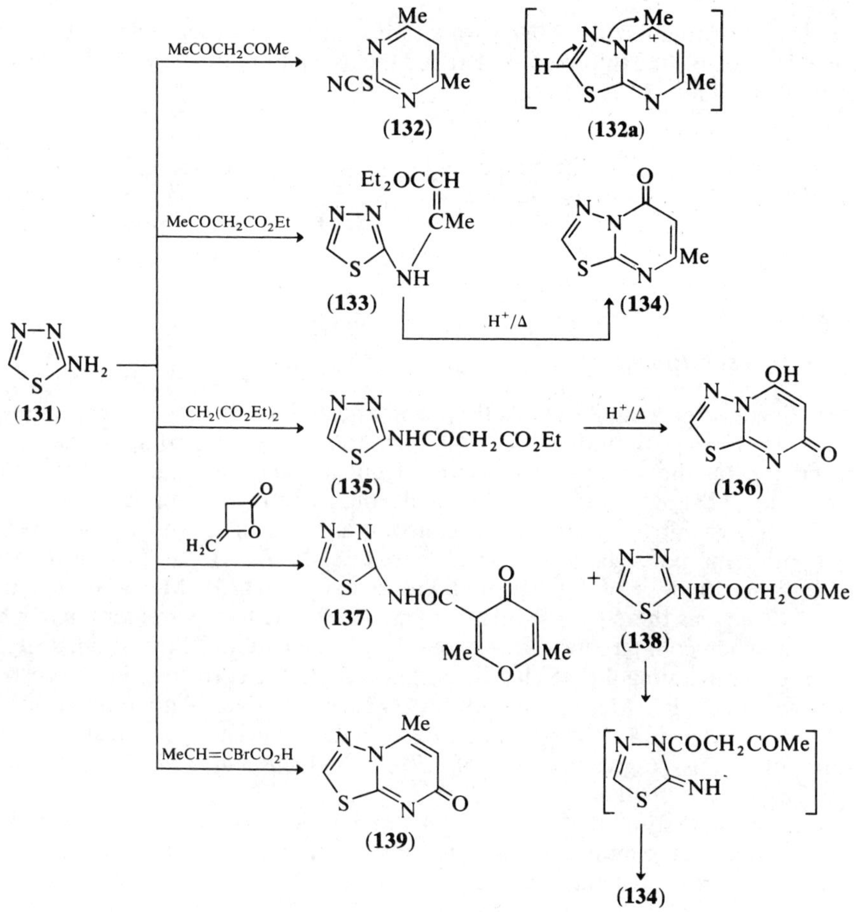

Scheme 13

5-Substituted 2-amino-1,3,4-thiadiazoles (140) react with ethyl acetoacetate, in the presence of TsOH, on the ketonic function to give (141) which cyclizes exclusively to (142a), on treatment with sulfuric acid (Scheme 14). In the presence of sodium methoxide, attack occurs on the ethoxycarbonyl group to give the butanamide (143a), which in sulfuric acid cyclizes through partial rearrangement to a mixture of mainly (144a) and some (142a). This is in contrast to the reaction of (140) with ethyl benzoylacetate, where attack occurs exclusively on the ethoxycarbonyl group, with TsOH or sodium methoxide, to give the highly insoluble inter- and intra-molecularly hydrogen bonded intermediate (143b; enol form) whose structure was ascertained by X-ray crystallography (see Section 4.27.2.3.1). On acid cyclization, (143b) also partially rearranges to a mixture of (144b) and (142b); the major constituent, however, is the rearranged (142b) ⟨80JOC4860⟩.

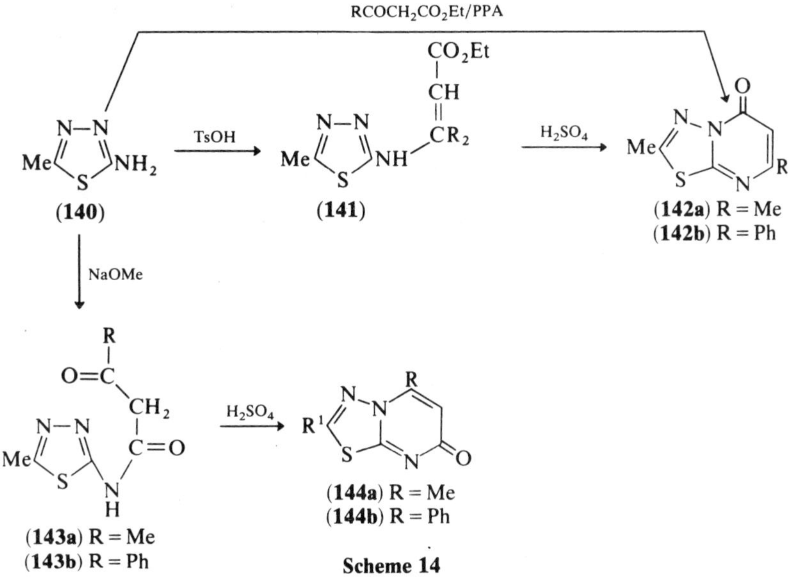

Scheme 14

Aminothiadiazoles also react with halo aldehydes and halo ketones in a bidentate fashion to give imidazo[2,1-b][1,3,4]thiadiazoles (145). The NMR properties, aromatic character, basicity and crystal structure data are available ⟨80JCS(P2)421⟩. Aminothiadiazoles also react with trichloromethanesulfenyl chloride to give the sulfenamide (146) which in the presence of an aromatic amine cyclizes to 3H-[1,3,4]thiadiazolo[2,3-c][1,2,4]thiadiazole (147) ⟨75JOC2600⟩.

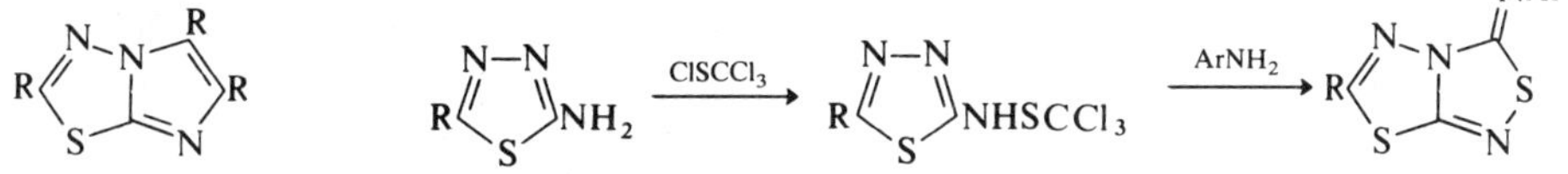

2,5-Diamino-1,3,4-thiadiazole also reacts with bifunctional groups such as ethyl acetoacetate to give the thiadiazolopyrimidone (148); however, with esters of malonic acid only (149) is obtained which does not cyclize under the conditions which yield (148) ⟨71M550⟩.

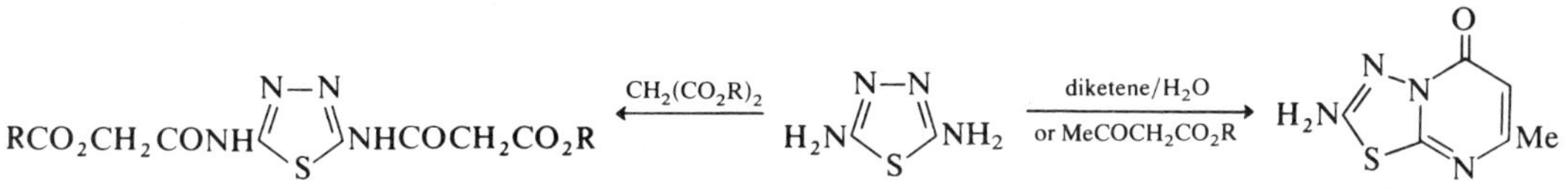

With diethyl oxalate the diamide (150) is produced which easily cyclizes to the imidazothiadiazole (151) ⟨81ZC185⟩.

568   1,3,4-Thiadiazoles

5-Substituted 1,3,4-thiadiazol-2-ylhydrazines (**152**) react with α-keto acids to give the hydrazones (**153**) which cyclize in acid to 1,3,4-thiadiazolotriazines (**154**; Scheme 15). Treatment of (**152**) with α-cyanoacetophenone gives the pyrazolyl[1,3,4]thiadiazole (**155**) ⟨76JHC117⟩.

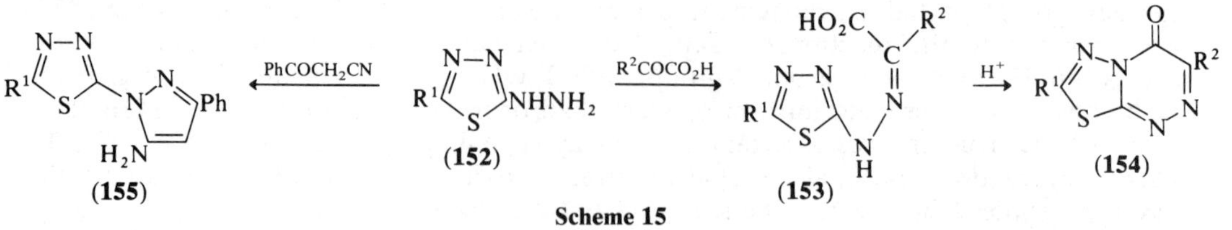

Scheme 15

4.27.4 SYNTHESES

4.27.4.1 Survey of Synthetic Procedures

The synthesis of 1,3,4-thiadiazoles is discussed in terms of the number of bonds being formed and by ring transformation. Thiadiazole synthesis by one-bond formation is exemplified by the cyclization of an acylated thiosemicarbazide as shown in Scheme 16. The most common two bond formation takes place via 1,3-dipolar cycloadditions presented in Scheme 23.

Thiadiazolidines can be obtained from aliphatic aldehydes or ketones and disubstituted hydrazine derivatives (Scheme 29). A typical preparation of a mesoionic compound consists in the reaction of 1-methylthioacylhydrazine and phosgene (Scheme 31a). Syntheses by three-bond formation are rare; for example, a one-pot reaction of an aldehyde with hydrazine and sulfur. A typical ring transformation reaction is the irradiation of 1,3,4-oxadiazoles to yield 1,3,4-thiadiazoles.

Detailed examples of these procedures are discussed below. For excellent reviews of the older literature see ⟨52HC(4)81, B-61MI42700, 68AHC(9)165⟩.

4.27.4.2 Formation of One Bond

The most common procedure for the synthesis of 5-substituted 2-amino-1,3,4-thiadiazoles is the acylation of a thiosemicarbazide followed by dehydration. Sulfuric acid, polyphosphoric acid and phosphorus halides are some of the reagents used. The most recent procedure utilizes 1.5 moles of methanesulfonic acid as the dehydrating agent, and the thiadiazoles are obtained in high yield and good purity (Scheme 16a) ⟨80JHC607⟩. 5-Alkyl-2-methylamino-1,3,4-thiadiazoles are prepared from a suitable carboxylic acid and 4-methylthiosemicarbazide in the presence of three parts of polyphosphoric acid and one part of concentrated sulfuric acid. Again the yields and purity are quite good ⟨75OPP179⟩. 2-Alkylamino-1,3,4-thiadiazoles unsubstituted in the 5-position can be prepared in high yields by the reaction of 4-alkylthiosemicarbazides with orthoformate esters in the presence of small amounts of concentrated hydrochloric acid. None of the isomeric triazolethione is formed under these conditions (Scheme 16b) ⟨73JOC3947⟩. The nature of the substituent in the 4-position of the thiosemicarbazide has a profound effect on its reactivity. Thus 4-methylthiosemicarbazide (**156a**) and phenyl isocyanide give only the triazole-3-thione (**158a**), while thiosemicarbazide (**156b**) yields 2-amino-1,3,4-thiadiazole (**157b**). 4-Phenylthiosemicarbazide (**156c**) and phenyl isocyanide give a mixture of 1:1.3 of (**157c**) and (**158c**). The reaction is catalyzed by copper(II) chloride and aniline is the side product

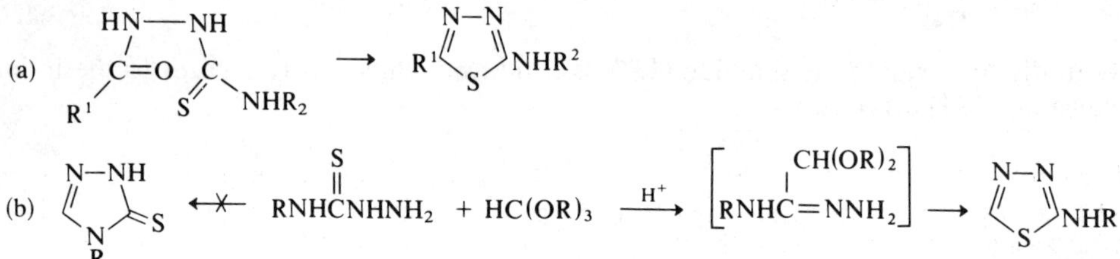

Scheme 16

⟨77ACS(B)264⟩. Thiosemicarbazones can also furnish thiadiazoles. Iron(III) chloride oxidation of (159) gives the acylated (160) which on treatment with base yields (161; Scheme 17). Thiazolines can be made in the same way ⟨77JHC853⟩.

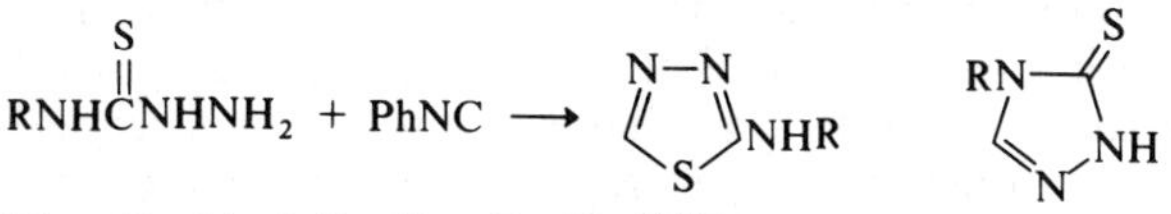

(156) a, R = Me; b, R = H; c, R = Ph (157)   (158)

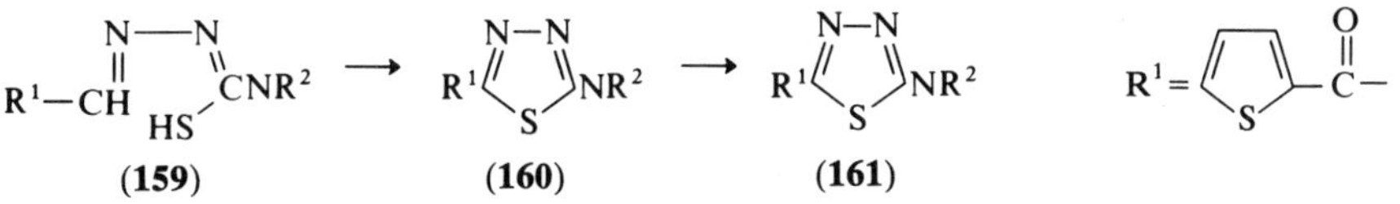

(159) → (160) → (161) $R^1 =$ furyl-C(O)-

Scheme 17

The reaction of oxalyl hydrazide with aryl isothiocyanate gives the bis(thiosemicarbazide) (162) which under acidic conditions cyclizes to the bis(1,3,4-thiadiazole) (163). Under alkaline conditions the 5-mercapto-1,2,4-triazole (164) is obtained ⟨77BSB399⟩.

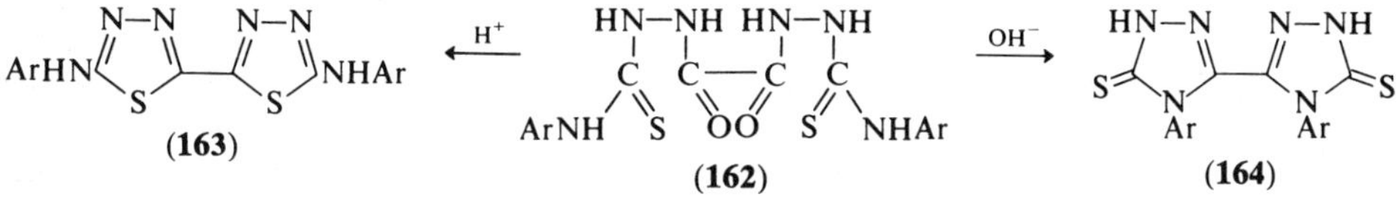

(163) ← H^+ — (162) — OH^- → (164)

Acetylation of benzaldehyde thiosemicarbazone (165) yields the thiadiazoline (166) and not the diacyl derivative (167) as previously suggested (Scheme 18). Structural proof rests on spectroscopic data and on oxidation to the thiadiazole (168) followed by deacetylation with hydrazine to (169). Mild acid or base hydrolysis of (166) furnishes the starting (165) while methylation gives (170), a product identical to that obtained from the acetylation of (171) ⟨80JOC1473⟩.

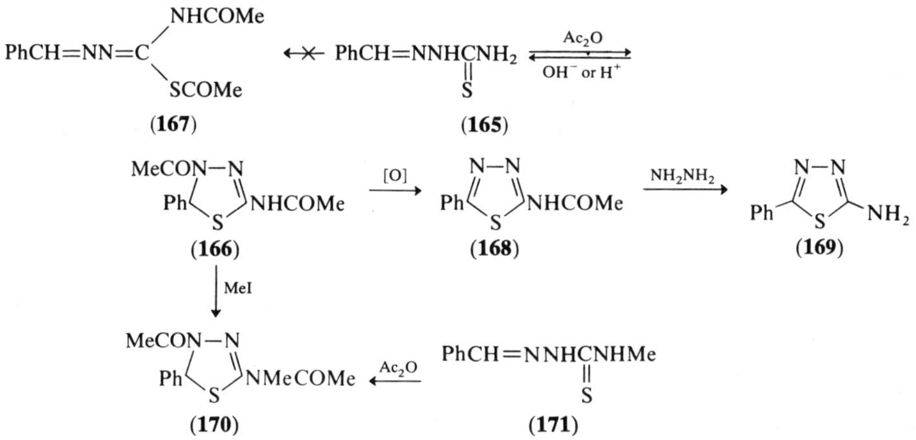

(167) ⇽╳⇾ PhCH=NNHC(S)NH$_2$ (165) ⇌ Ac_2O / OH^- or H^+

(166) — [O] → (168) — NH_2NH_2 → (169)

(166) ↓ MeI

(170) ← Ac_2O — PhCH=NNHC(S)NHMe (171)

Scheme 18

4.27.4.3 Formation of Two Bonds

This is the most widely used procedure for the synthesis of thiadiazoles, thiadiazolines, thiadiazolidines and mesoionic thiadiazoles. It is classified into the following subsections.

4.27.4.3.1 Cyclizations

The parent molecule, 1,3,4-thiadiazole (174), was first synthesized in 1956 by a four-step reaction sequence starting from thiosemicarbazide. A second procedure utilizes hydrazine and potassium dithioformate. A later procedure is shown in Scheme 19. Dehydration of DMF with thionyl chloride or phosgene gives the formamidoyl chloride (172) which on treatment with *N,N'*-diformylhydrazine gives the dihydrochloride of (**173**). The free base (**173**) is liberated with sodium ethoxide, which then cyclizes to thiadiazole in the presence of hydrogen sulfide in an overall yield of 65% ⟨67AG(E)361⟩. 2-Amino-1,3,4-thiadiazole is also prepared from thiosemicarbazide and a mixture of formic and hydrochloric acids in a tedious procedure with an overall yield of 65% (Scheme 20a) ⟨76JHC291⟩. The synthesis of other simple thiadiazoles is outlined in Schemes 20b–d. A general method for the synthesis of 1,3,4-thiadiazolines consists in the treatment of aldehydes or ketones with N(2)-substituted thiohydrazides (Scheme 20e).

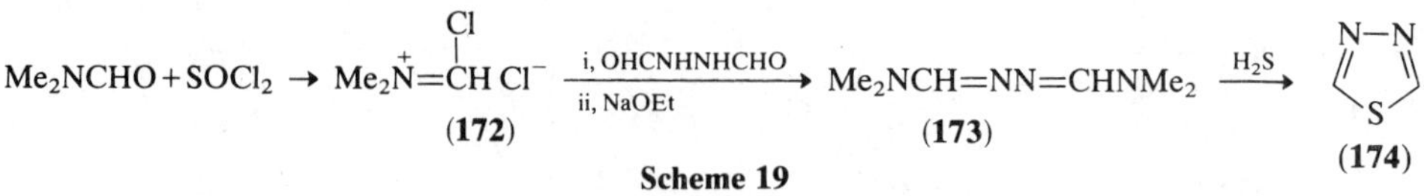

Scheme 19

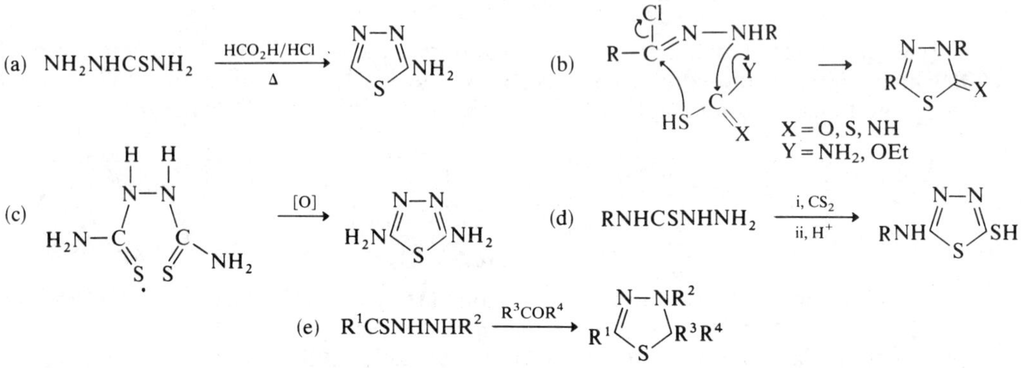

Scheme 20

Various 2-substituted 1,3,4-thiadiazoline-5-thiones can be prepared in good yield by the reaction of amidrazones with CS_2 in ethanol at room temperature (Scheme 21a) ⟨70CPB1696⟩. 4-Aryl-2-benzoyl-5-imino-Δ^2-1,3,4-thiadiazolines are prepared from activated thiocyanates and benzenediazonium chloride, presumably *via* a hydrazone (Scheme 21b) ⟨75TL163⟩. 2-Thio- and 2-amino-1,3,4-thiadiazoles carrying a carbamate in the 5-position are obtained in one step by the reaction of a dithioimidate with a hydrazide. The yields are quite high (Scheme 22a) ⟨80ZC413⟩.

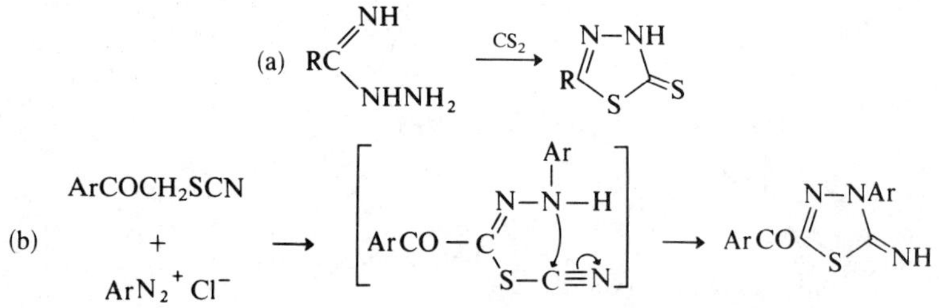

Scheme 21

A general synthesis of 4-amidino-Δ^2-thiadiazolines consists in the reaction of 1-chloro-2,3-diazabutadiene with thiourea at room temperature (Scheme 22b). The characteristic spectroscopic features of Δ^2-1,3,4-thiadiazolines are associated with the sp^3-hybridized C-5 atom and its methine proton. The C-5 atom resonates in the ^{13}C NMR spectrum at about $\delta 70.5$, the signal appearing as a doublet in the off-resonance spectrum, while the methine proton resonates in the ^1H spectrum at about $\delta 7$, displaying a strongly deshielded environment ⟨81JCS(P1)360⟩.

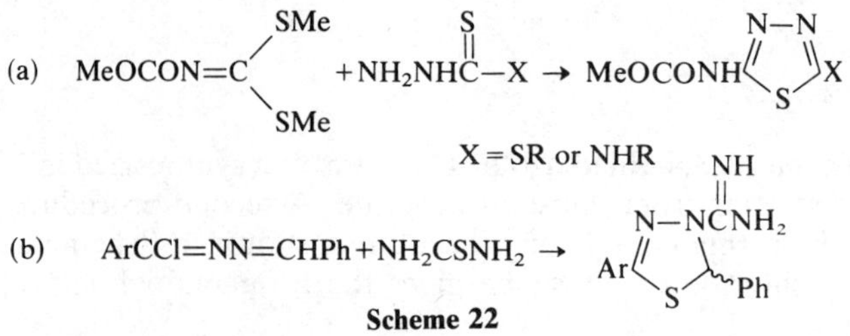

Scheme 22

4.27.4.3.2 Dipolar cycloadditions

This procedure has been widely used during the last decade for both synthetic and mechanistic reasons. Selected examples are given here, and many more may be found in the literature.

(i) *From acid chloride phenylhydrazones*

N-Aryl- and *N,N*-dialkyl-thiobenzamides react with acid chloride phenylhydrazones to give 5-anilino- and 5-dialkylamino-2,4,5-triaryl-1,3,4-thiadiazolines, respectively. Treatment with alcohol gives the corresponding 5-alkoxy-1,3,4-thiadiazoline (Scheme 23a) ⟨76ACS(B)837⟩. A second example of the use of acid chloride phenylhydrazone is its reaction with a thioketene to provide thiadiazolines with a methylene group in the 5-position (Scheme 23b) ⟨66AG(E)970⟩.

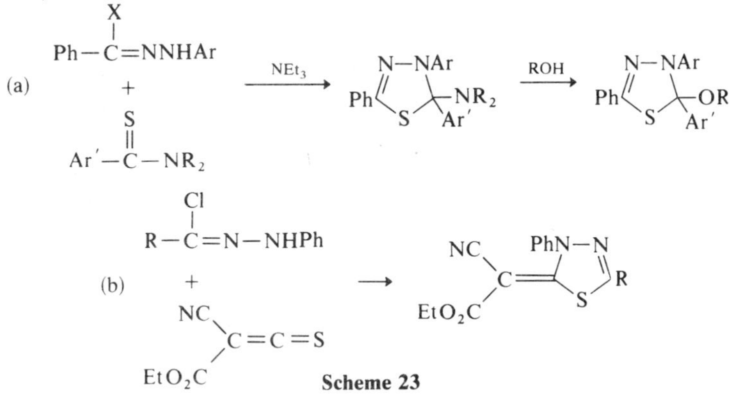

Scheme 23

(ii) *From diazo compounds*

The reaction of arylsulfonyl-substituted sulfines (**175**) with diazomethane gives Δ^3-1,3,4-thiadiazoline 1-oxides (**176**) which, however, are unstable and rearrange *via* an isomerization of the Δ^3- to the Δ^2-thiadiazoline oxide. This is followed by an elimination and readdition of sulfinic acid and loss of water in a Pummerer-type aromatization to give the rearranged thiadiazole (**177**; Scheme 24) ⟨73TL5009⟩.

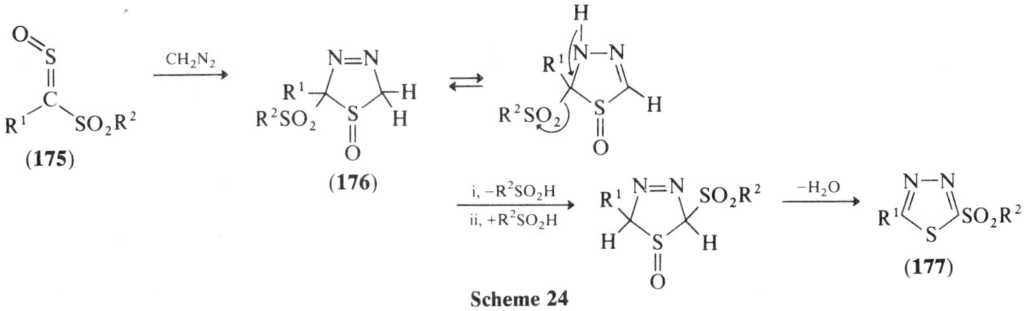

Scheme 24

The reaction of diazomethane with thiobenzophenone at −78 °C yields 2,2-diphenyl-1,3,4-thiadiazoline (**178**) which, however, decomposes at −30 °C to give 2,2,3,3-tetraphenyl-1,4-dithiane (**180**) in 95% yield. At higher dilution, (**178**) yields a mixture consisting of (**180**; 41%) and (**181**+**182**; 38%; Scheme 25). The thiocarbonyl ylide (**179**)

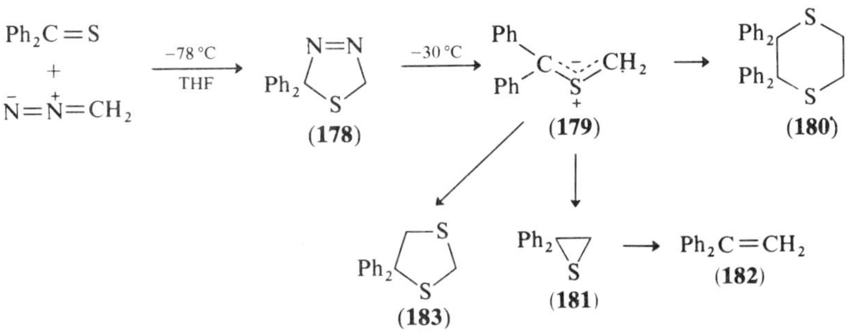

Scheme 25

can be trapped by dipolarophiles to give compounds of type (183). The isolation of (183) from diazomethane and thiobenzophenone can be explained by two 1,3-dipolar cycloadditions separated by a 1,3-dipolar cycloreversion ⟨81JA7032⟩.

Dithiones and diazomethane undergo dipolar additions at 0 °C to give stereoisomeric bis(Δ^3-1,3,4-thiadiazolines). Thus, (184), (185; $n = 5$) and (185; $n = 6$) give (186), (187; $n = 5$) and (187; $n = 6$). On thermolysis the bis adducts lose nitrogen to give stereoisomeric mixtures of diepisulfides (188), (189; $n = 5$) and (189; $n = 6$) ⟨71JOC3885⟩.

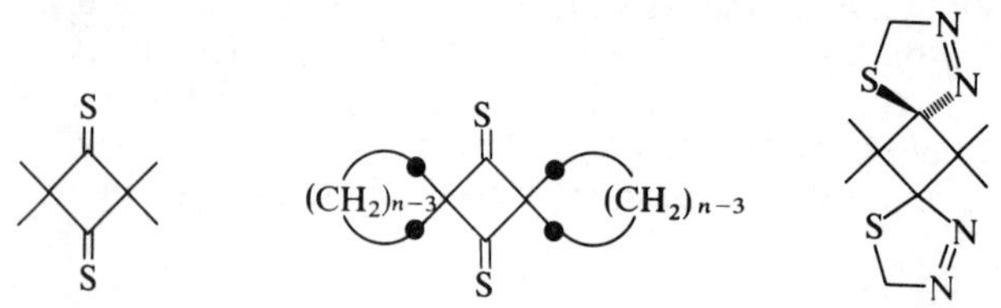

(184)   (185)   (186)

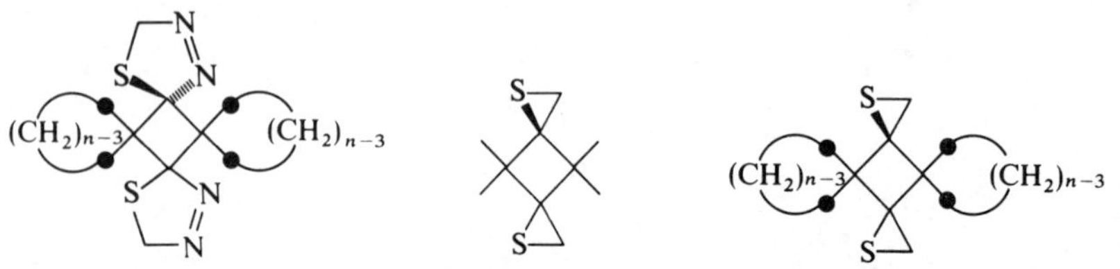

(187)   (188)   (189)

The reaction of 2-diazopropane with geometrical isomers of different types of sulfines yields Δ^3-1,3,4-thiadiazoline 1-oxides as the single product by a concerted regiospecific cycloaddition process. Introduction of bulky substituents in either of the reactants will sterically hinder the cyclization to a five-membered ring, and give rise to non-stereospecific formation of episulfoxides (Scheme 26) ⟨73TL3589⟩.

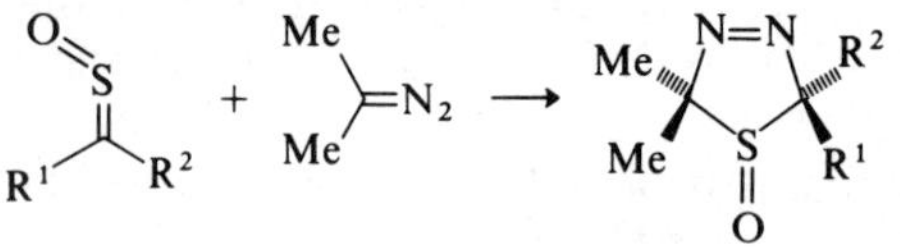

Scheme 26

An extension of this work is treatment of diazopropane derivatives with SO_2 in benzene to give a 50% yield of the Δ^3-1,3,4-thiadiazoline 1,1-dioxide (191; Scheme 27). The reaction proceeds probably via the sulfene (190) and the main side product is the ketone (192). None of the episulfone (193) or stilbene (194) could be found. The NMR spectra of compounds of type (191) show temperature dependence due to the restricted rotation of the aryl group ⟨75AJC151⟩. These findings differ from results obtained on very similar compounds by later workers who obtained a mixture of the thiadiazoline and the episulfone ⟨81CB787⟩. Δ^3-1,3,4-Thiadiazolines (195; Scheme 28) are prepared by dehydrogenation of 1,3,4-thiadiazolidines (196), or by the action of H_2S on the addition product (197) of chlorine and azine (198) to give (199). On pyrolysis, (195) gives the episulfide (200) by conrotatory ring closure in agreement with orbital symmetry considerations. Compound (195) undergoes a stereospecific cycloaddition with diethyl azodicarboxylate to give (201) with complete retention of configuration. Cycloaddition of (199) with dimethyl acetylenedicarboxylate gives the dihydrothiophene (202) ⟨72JOC4045⟩.

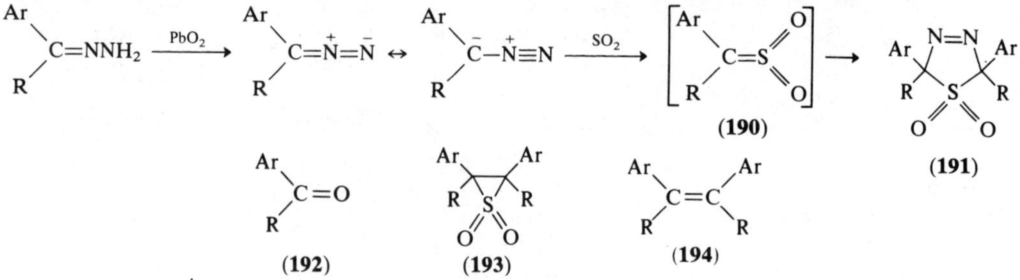

Scheme 27

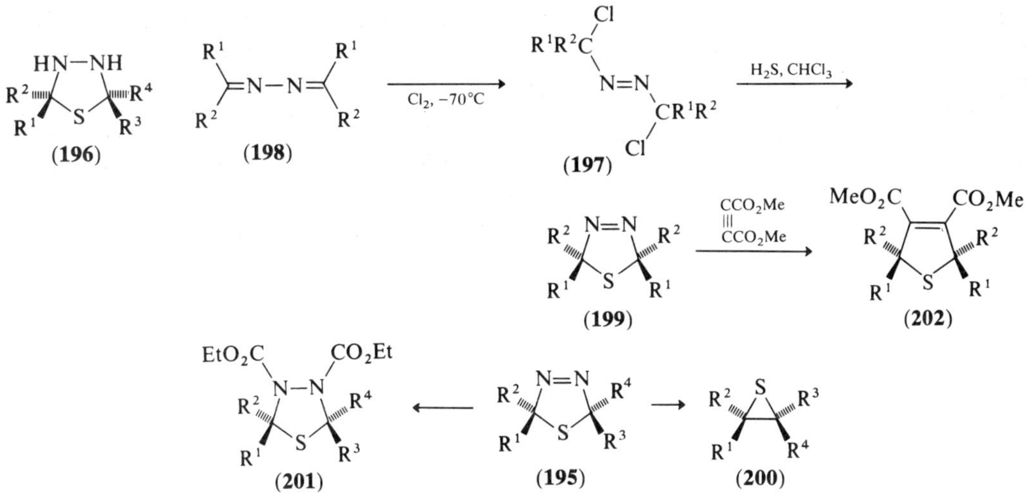

Scheme 28

4.27.4.3.3 Thiadiazolidines

The general procedure for the preparation of 1,3,4-thiadiazolidine-2-thiones with aliphatic substituents in the 3-, 4- and 5-positions is the reaction between an aliphatic aldehyde or ketone and a hydrazinium dithiocarbazate (Scheme 29). Ring opening to the salts or methyl esters of alkylidenedithiocarbazic acid occurs in the presence of hydroxide ion or methyl iodide, respectively ⟨70ACS179⟩.

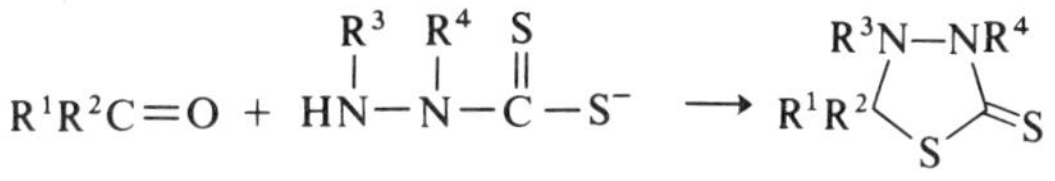

Scheme 29

2,2-Disubstituted 4-phenyl-5-phenylimino-1,3,4-thiadiazolines (**203**; X = NPh) are prepared by the reaction of phenyl isothiocyanate with ketonic phenylhydrazones in DMF in the presence of sodium hydride. Replacing the isothiocyanate with carbon disulfide leads to (**203**; X = S). Aliphatic aldehydes react similarly; however, benzaldehyde phenylhydrazone furnishes the thiosemicarbazone which on thermolysis yields the thiadiazoline (**204**). The structures of these compounds are supported by detailed spectroscopic measurements ⟨80JCS(P1)574⟩.

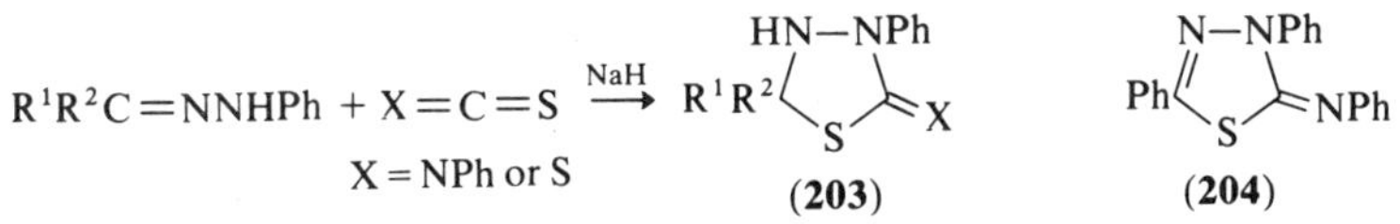

Scheme 30

4.27.4.3.4 Mesoionic compounds

The synthesis and some properties of mesoionic 1,3,4-thiadiazoles are nicely summarized in a recent review ⟨B-79MI42701⟩. The most common starting materials are thioacylated hydrazine derivatives. Some of the methods utilized are shown in Scheme 31. 1-Methyl-1-thioacylhydrazine and thiophosgene or phosgene react in the presence of potassium carbonate to give the mesoionic thiadiazole shown in Scheme 31a. These compounds do not react with alkenic and alkynic dipolarophiles ⟨68CC672⟩. When X = NR the thiadiazole rearranges to the mesoionic triazole ⟨71CC1222⟩. Mesoionic 2-methylene-1,3,4-thiadiazoles are prepared from thiocarboxylic acid hydrazides and 3,3-dichloroacrylonitrile (Scheme 31b) ⟨70TL5083⟩. CS_2 can also be used in the synthesis of mesoionic compounds (Schemes 31b, c) ⟨68TL5881⟩. Very similar procedures are discussed in ⟨68JHC277, 71CC1436, 77IJC(B)499⟩. A recent publication describes the synthesis of mesoionic [1,3,4]thiadiazolo[3,2-*a*]pyridine derivatives ⟨82JCS(P1)351⟩.

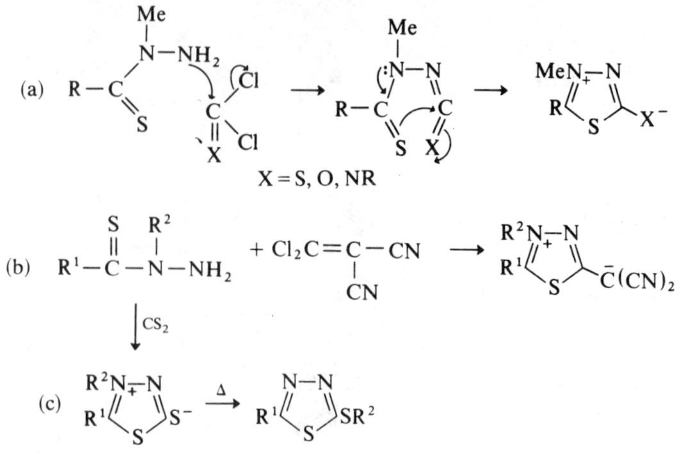

4.27.4.4 Formation of Three Bonds

2,5-Dialkyl-1,3,4-thiadiazoles are prepared in high yield by the one-pot reaction between an aldehyde, hydrazine hydrate and elemental sulfur (Scheme 32). Chlorination of the 2,5-dimethylthiadiazole yields the bis(trichloromethyl) derivative which undergoes a wealth of reactions with various nucleophiles ⟨80LA1216⟩.

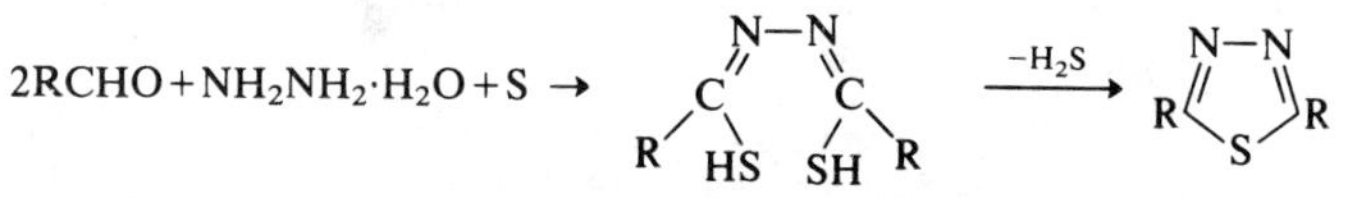

Scheme 32

A similar reaction provides thiadiazolines. Treatment of cyclohexanones with hydrazine and H_2S in ethanol yields (**205**), also obtainable from the azine (**206**; Scheme 33). Oxidation of (**205**) with manganese dioxide gives the thiadiazoline (**207**), a compound which undergoes the 'two-fold extrusion process', *i.e.* the removal of nitrogen and sulfur to form an alkene. Thus, pyrolysis of (**207**) gives the thiirane (**208**) which on treatment with zinc and acetic acid gives good yields of the alkene (**209**). These reactions have extensive stereochemical implications ⟨78JCS(P1)45⟩.

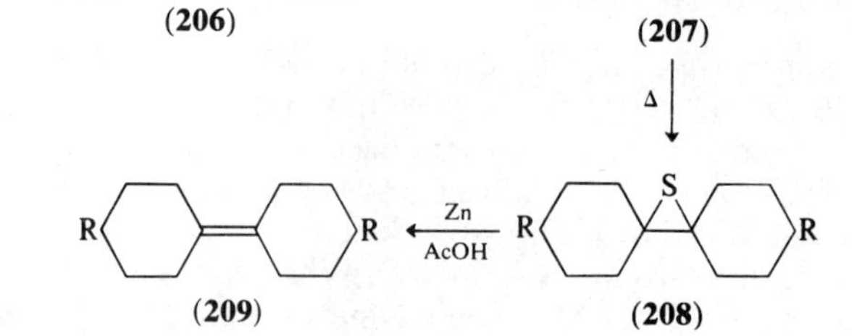

Scheme 33

2,5-Dihydro-1,3,4-thiadiazole 1,1-dioxides (*cis* and *trans*) are obtained by the Staudinger–Pfenninger reaction of *t*-butyldiazomethane and sulfur dioxide. Some thiirane is also formed (Scheme 34) ⟨81CB802⟩.

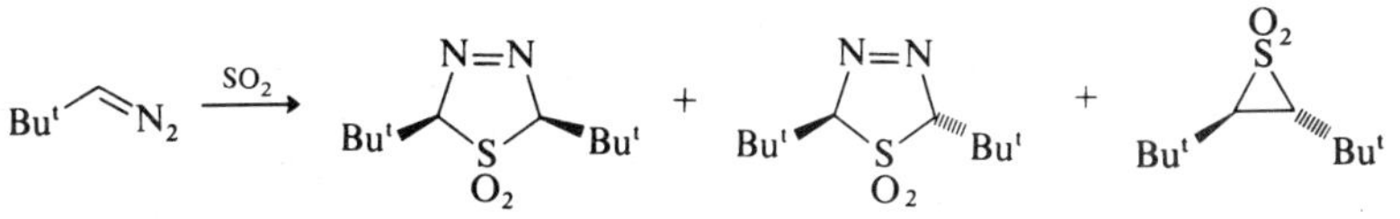

4.27.4.5 Formation by Ring Transformations

1,3,4-Thiadiazoles can easily be obtained from 1,3,4-oxadiazoles. Thus (210) in refluxing ethanolic HCl rearranges to (211) ⟨67MI42700⟩. Similarly, the mesoionic oxadiazole (212) gives (213) when heated in ethanol or ethanethiol ⟨68CC499⟩. Thiadiazolium salts (214) on treatment with aldehydes give thiadiazolines (215) ⟨78JHC1515⟩. Finally, the reaction of benzothiazinedithione (216) with benzoyl hydrazides yields 2,5-disubstituted thiadiazoles (217), presumably via the intermediate (218) ⟨74ZC305⟩.

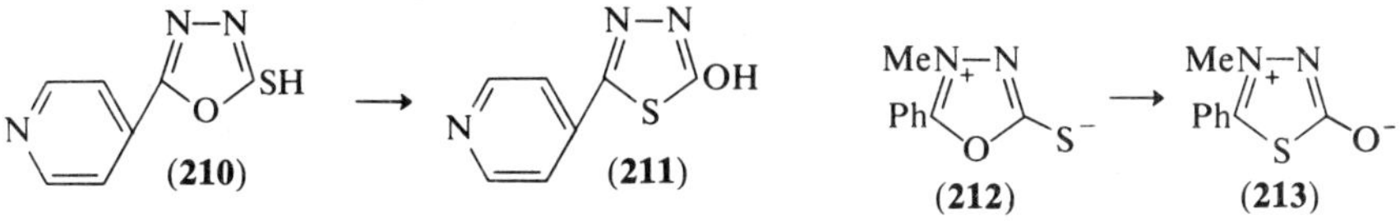

4.27.5 APPLICATION

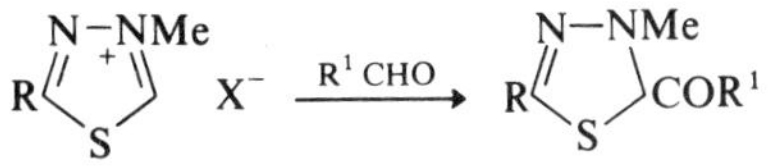

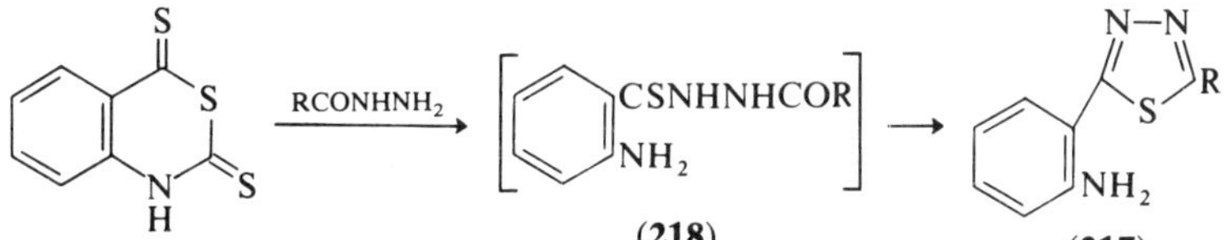

4.27.5.1 Survey of Applications

1,3,4-Thiadiazoles have applications in many fields. The earliest uses were in the pharmaceutical area as antibacterials with properties similar to those of the well-known sulfonamide drugs. Some of the later uses are as antitumor and anti-inflammatory agents, pesticides, dyes, lubricants and analytical reagents.

Between 1967 and March 1982, *Chemical Abstracts* lists 215 patents for 1,3,4-thiadiazole derivatives, 21 patents for thiadiazolines and 6 patents for thiadiazolidines.

4.27.5.2 Selected Commercial Compounds

The early antimicrobial compounds such as lucosit (219; R = Me) and globucid (219; R = Et) can be prepared by treatment of the 2-amino-1,3,4-thiadiazole with N-acetylsulfonyl chloride in pyridine followed by deacetylation. A more recent antimicrobial compound (224) can be efficiently synthesized by the route shown in Scheme 35. The thiadiazole (220) is obtained by the usual cyclization procedure and then oximated to (221). Dehydration and the Pinner amidine synthesis followed by treatment with aminoacetaldehyde dimethyl acetal yields (222). Sulfuric acid cyclization furnishes the thiadiazoloimidazole derivative (223) which is further methylated, acetylated, nitrated and hydrolyzed to the desired (224) ⟨69JHC(6)835⟩.

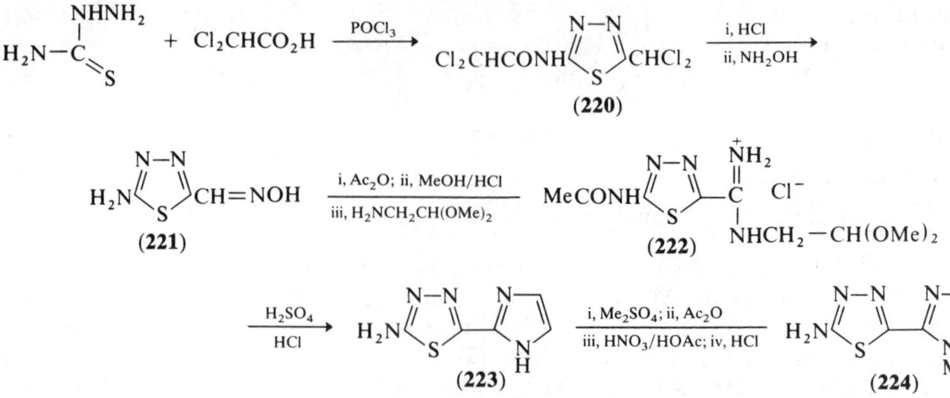

Scheme 35

Acetazolamide (**225**) and methazolamide (**226**) are two carbonic anhydrase inhibitors which can be derivatized by Mannich-type derivatives to increase their solubility. Under physiological conditions the parent compound is regenerated ⟨71JMC458⟩. Cefazolin (**227**) is a thiadiazole analog of cephalosporanic acid useful as an antibacterial ⟨70MI42701⟩.

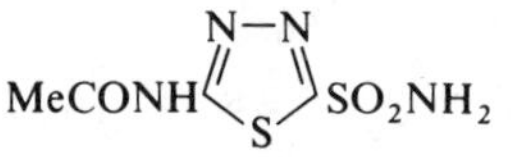

The well-known herbicide spike is prepared in high yield by heating in toluene a mixture of pivalic acid, N-methylthiosemicarbazide, sulfuric acid and polyphosphoric acid to give (**228**; R = H), which on further treatment with methyl isocyanate yields (**228**; R = CONHMe) ⟨81USP4283543⟩. Other patented processes are known ⟨76GEP2541115⟩. Details on the structure–activity relationship of compounds in this class are available ⟨70MI42700⟩. A thiadiazole-containing insecticide is methidathion (**229**).

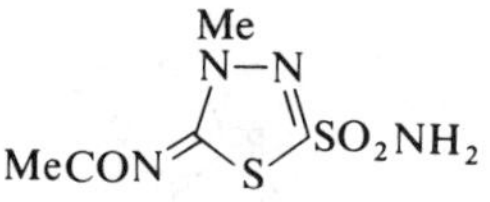

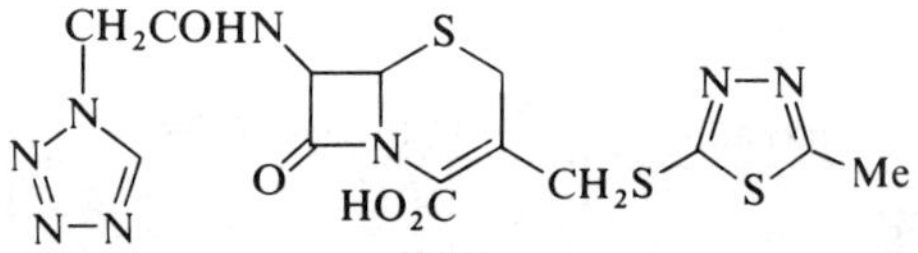

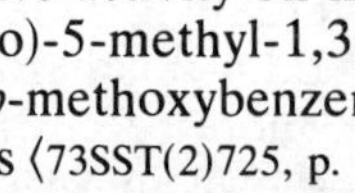

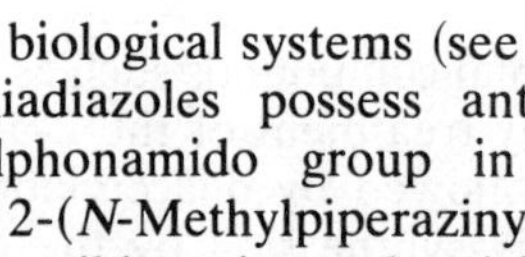

4.27.5.3 Biological Effects

1,3,4-Thiadiazoles have activity on many biological systems (see also Section 4.27.5.4). 2-(Halogenoacetylamino)-5-methyl-1,3,4-thiadiazoles possess antitumor activity while thiadiazoles with a p-methoxybenzenesulphonamido group in the 2-position have hypoglycemic properties ⟨73SST(2)725, p. 744⟩. 2-(N-Methylpiperazinyl)- and 2-(N,N-dialkylaminoethyl)amino-1,3,4-thiadiazoles display antihistamine and anticholinergic activity while 2,2'-methylenediiminobis(1,3,4-thiadiazole) is effective against tumors in mice ⟨77SST(4)431, p. 437⟩. 2-Amino-1,3,4-thiadiazole-5-thiol is an effective radioprotective agent ⟨79SST(5)440, p. 445⟩.

The pharmacological fate of the antitumor agent 2-amino-1,3,4-thiadiazole in dogs is known ⟨80MI42700⟩. Its distribution and metabolism in mice and monkeys has also been investigated ⟨78MI42700⟩. The most pronounced effect of aminothiadiazole on ribonucleotide pools of leukemia L 1210 cells is the lowering of guanine ribonucleotide pools and the elevation of IMP. A marked inhibition of the incorporation of 8-^{14}C labeled inosine into guanine nucleotides also occurs ⟨76MI42700⟩.

5-Substituted 2-anilino-1,3,4-thiadiazoles uncouple oxidative phosphorylation in rat liver mitochondria and photosynthetic phosphorylation in spinach chloroplasts ⟨75ZN(C)183⟩.

4.27.5.4 Patents

The purpose of this subsection is to further outline the many applications of 1,3,4-thiadiazoles. Although the vast majority of patented compounds never become commercial products, patents are a measure of the industrial importance of a particular class of compounds. Table 15 presents selected examples of the many areas in which 1,3,4-thiadiazoles have been patented.

Table 15 Selected 1,3,4-Thiadiazole Patents

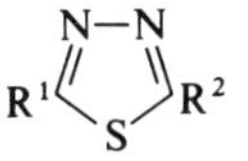

R^1	R^2	Utility	Ref.
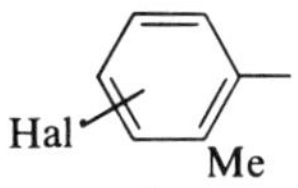 Hal-C₆H₃(Me)-	—NHNHR	Muscle relaxant	78GEP2727146
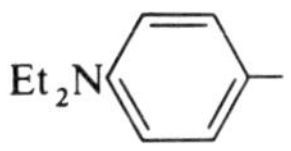 Et₂N-C₆H₄-	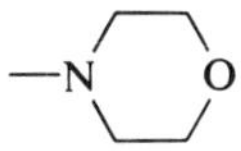 —N(morpholino)O	Antiviral	79EGP136963
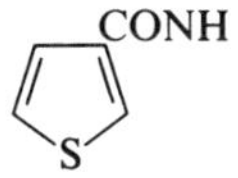 2-(CONH)-thienyl	H	Anticarcinogenic	80JAP(K)8028946
CF₃	—SO-heterocycle or phenyl	Antiparasitic	77GEP2533605
 Cl-C₆H₄-	—NHCHO	Bactericidal	72JAP7207549
Cl	 -C₆H₃(Hal, Me)-	Algicidal	76USP3959301
EtSO—	—CONHEt	Nematicidal	80GEP2853196
H, alkyl	—SOCH₂CO₂H	Fungicidal	77JAP(K)7725028
Alkyl	—NHC(O)NHMe	Herbicidal	72BRP1266542
MeS	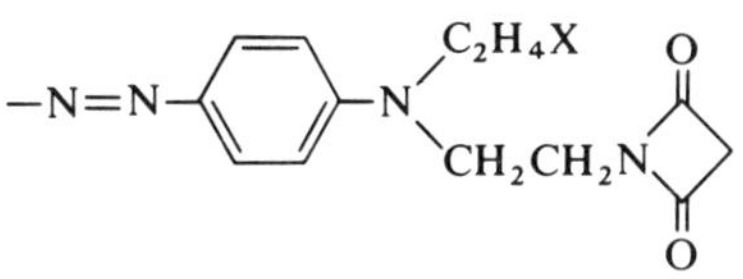 —N=N—C₆H₄—N(C₂H₄X)(CH₂CH₂N(succinimido))	Dyes	70USP3493556
Oxidized—S—	—S—Oxidized	Lubricant	81USP4246126
Polymer—S—	—S—Polymer	Cross linkers for polymers	81USP4288576

Structure: 1,3,4-thiadiazole with R^1 at C-5 and R^2 at C-2 (N—N / S ring).

4.28
Oxatriazoles and Thiatriazoles

A. HOLM
University of Copenhagen

4.28.1 INTRODUCTION	580
4.28.2 1,2,3,4-OXATRIAZOLES AND THIATRIAZOLES	580
4.28.2.1 Introduction	580
4.28.2.2 Structure	581
4.28.2.2.1 General survey	581
4.28.2.2.2 Theoretical methods	581
4.28.2.2.3 Structural methods	581
4.28.2.2.4 Tautomerism: prototropic	583
4.28.2.3 Reactivity of the Ring System	584
4.28.2.3.1 Reactivity at the ring atoms	584
4.28.2.4 Reactivity of Substituents	590
4.28.2.4.1 1,2,3,4-Thiatriazoles	590
4.28.2.4.2 3,5-Disubstituted 1,2,3,4-oxa- and -thia-triazolium salts	592
4.28.2.5 Syntheses	593
4.28.2.5.1 1,2,3,4-Thiatriazoles	593
4.28.2.5.2 Δ^2-1,2,3,4-Thiatriazolines	596
4.28.2.5.3 Δ^3-1,2,3,4-Thiatriazolines	597
4.28.2.5.4 3,5-Disubstituted 1,2,3,4-oxa- and -thia-triazolium salts	597
4.28.2.5.5 5-Amino-1,2,3,4-selenatriazoles	597
4.28.2.6 Applications	598
4.28.2.6.1 Analytical uses	598
4.28.2.6.2 Miscellaneous	598
4.28.3 MESOIONIC 1,2,3,4-OXATRIAZOLES AND -THIATRIAZOLES	598
4.28.3.1 Introduction	598
4.28.3.2 Structure	599
4.28.3.2.1 Theoretical methods	599
4.28.3.2.2 Structural methods	599
4.28.3.2.3 Molecular spectra	600
4.28.3.3 Thermodynamic Aspects	601
4.28.3.3.1 Interconversion of mesoionic isomers	601
4.28.3.4 Reactivity of the Ring Systems	601
4.28.3.4.1 Reactivity at the ring atoms	601
4.28.3.5 Reactivity of Substituents	603
4.28.3.5.1 Salt formation	603
4.28.3.5.2 Other side-chain reactions	603
4.28.3.6 Syntheses	603
4.28.3.6.1 Mesoionic 1,2,3,4-oxatriazol-5-ones	603
4.28.3.6.2 Mesoionic 1,2,3,4-oxatriazole-5-thiones	604
4.28.3.6.3 Mesoionic 1,2,3,4-oxatriazol-5-imines	604
4.28.3.6.4 Mesoionic 1,2,3,4-thiatriazol-5-ones, -5-thiones, -5-imines and -5-alkenes	605
4.28.3.7 Applications	605
4.28.4 1,2,3,5-OXATRIAZOLES AND -THIATRIAZOLES	605
4.28.4.1 Introduction	605
4.28.4.2 Structure	606
4.28.4.2.1 Δ^2- and Δ^4-1,2,3,5-oxatriazolines	606
4.38.4.2.2 Δ^3- and Δ^4-1,2,3,5-thiatriazolines and 1,2,3,5-thiatriazolidines	606
4.28.4.2.3 1,2,3,5-Thiatriazole and -selenatriazole heteropentalenes	607
4.28.4.3 Reactivity of the Ring Systems	607
4.28.4.3.1 Δ^2- and Δ^4-1,2,3,5-oxatriazolines	607
4.28.4.3.2 Δ^3- and Δ^4-1,2,3,5-thiatriazolines	607
4.28.4.3.3 1,2,3,5-Thiatriazolidines	608
4.28.4.3.4 1,2,3,5-Thiatriazole and -selenatriazole heteropentalenes	609

4.28.4.4 Syntheses	609
4.28.4.4.1 Δ^2- and Δ^4-1,2,3,5-oxatriazolines	609
4.28.4.4.2 Δ^3- and Δ^4-1,2,3,5-thiatriazolines	609
4.28.4.4.3 1,2,3,5-Thiatriazolidines	611
4.28.4.4.4 1,2,3,5-Thiatriazole and -selenatriazole heteropentalenes	612
4.28.4.5 Applications	612

4.28.1 INTRODUCTION

This chapter discusses the chemistry of the known representatives of five-membered ring systems containing three nitrogen atoms and an oxygen or sulfur atom. Several examples of such ring systems containing a selenium atom are now known, and these are also included because of their close chemical resemblance to their sulfur analogs. Although the peripheral heteroatom composition precludes benzo derivatives of these ring systems, several ring-fused systems have been obtained and these are also described.

The diverse nature of the known oxatriazoles and thiatriazoles and their extremely varied chemistry make it more convenient to arrange a discussion of their chemistry in the three groups below.

4.28.2 1,2,3,4-OXATRIAZOLES AND -THIATRIAZOLES

4.28.2.1 Introduction

This section deals with the chemistry of 5-substituted 1,2,3,4-thiatriazoles (**1**), Δ^2- (**2**) and Δ^3-1,2,3,4-thiatriazolines (**3**) (4,5- and 2,5-dihydro-1,2,3,4-thiatriazoles, respectively), 3,5-disubstituted 1,2,3,4-oxa- (**4**) and -thia-triazolium salts (**5**) and the novel 1,2,3,4-selenatriazoles (**6**). This last group has a close chemical resemblance to 1,2,3,4-thiatriazoles. 5-Substituted 1,2,3,4-oxatriazoles have not been isolated. In contrast to the 5-substituted 1,2,3,4-thiatriazoles, which are known only in the thermodynamically stable, heteroaromatic ring form (except thiobenzoyl azide *S*-oxide (**8**) ⟨76AHC(20)145⟩), 5-substituted 1,2,3,4-oxatriazoles are observed only as the open-chain acyl azides (**7**; X = O), the treatment of which falls outside the scope of this review (see also Section 4.01.1.2). However, 3,5-disubstituted 1,2,3,4-oxatriazolium salts (**4**) are known. Like their thia analogues (**5**), they may be regarded as salts of the corresponding 1,2,3,4-oxa- and -thia-triazoles or as mesoionic 1,2,3,4-oxa- and -thia-triazole derivatives. The last two types of compounds are discussed below in Section 4.28.3.

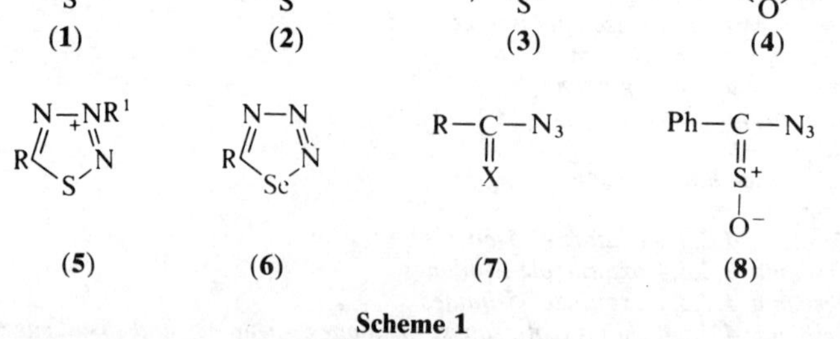

Scheme 1

The chemistry of 1,2,3,4-thiatriazoles was reviewed by Jensen and Pedersen in 1964 ⟨64AHC(3)263⟩ and by Holm in 1976 ⟨76AHC(20)145⟩. Certain aspects of the chemistry of thiatriazoles have been reviewed by Jensen ⟨69ZC121⟩. Recent progress lies mainly within the chemistry of Δ^2-1,2,3,4-thiatriazolines (**2**).

$1\lambda^4$-1,5-Dihydro-1,2,3,4-thiatriazoles (**9**) were suggested as products in the reaction between *S*-alkylated *N,N*-disubstituted thioamides and azide ion. The correctness of the structural assignment has been questioned, however ⟨76AHC(20)145⟩, and the alternative triazene structure (**10**) was proposed. An X-ray crystal structure determination has now settled the problem and has definitely shown that the product possesses the triazene structure ⟨79BSB107⟩.

$$\underset{(9)}{\underset{R^2}{R^3R^4N}}\overset{N=N}{\underset{S}{\bigvee}}\overset{N}{\underset{}{\rightleftarrows}} \quad \xcancel{\quad} \quad \underset{R^3R^4N}{\overset{R^1}{\bigvee}}\overset{}{\underset{+}{C-S-R^2}} + N_3^- \longrightarrow \underset{R^3R^4N}{\overset{R^1}{\bigvee}}C=N-N=N-S-R^2$$
$$(10)$$

4.28.2.2 Structure

4.28.2.2.1 General survey

5-Amino- and 5-alkylamino-1,2,3,4-thiatriazoles were prepared in 1896 by Freund and coworkers from thiosemicarbazides and nitrous acid as the first members of this class of compounds ⟨1896CB2500, 1896CB2491⟩. For a long time it was a matter of controversy whether these compounds should be regarded as thiatriazoles (**1**) or thioacyl azides (**7**; X = S). It was not until 1957 that the question was settled by the IR investigations of Lieber *et al.*, who concluded that the original formulation as thiatriazoles was correct ⟨64AHC(3)263⟩. All later investigations on thiatriazoles have confirmed this result and in fact the open-chain thioacyl azides have never been observed except for thiobenzoyl azide *S*-oxide (**8**) ⟨76AHC(20)145⟩. Differential thermal analysis shows that on heating to room temperature, (**8**) decomposes to give benzonitrile, SO and N_2 without formation of thiatriazole *S*-oxide ⟨77T2231⟩. The structures of several 1,2,3,4-thiatriazoles have been investigated by X-ray diffraction (see Section 4.28.2.2.3).

The heteroaromatic selenatriazole ring system (**6**) represents a new class of compounds isolated so far only as 5-amino-1,2,3,4-selenatriazoles. IR and mass spectra are in agreement with the selenatriazole structure and an alternative open-chain selenoacyl azide structure (**7**; X = Se) can be rejected because of the lack of an azide absorption in the IR heterocumulene region (*ca.* 2000–2300 cm^{-1}) ⟨82UP42800⟩.

The structure of Δ^2-1,2,3,4-thiatriazolines (**2**) has been discussed previously ⟨76AHC(20)145⟩ and the structural assignment of Δ^3-1,2,3,4-thiatriazolines (**3**) is discussed in Section 4.28.2.3.1(i).

4.28.2.2.2 Theoretical methods

A series of MO calculations has been performed on the structure, formation and nucleophilic reactivity of 1,2,3,4-thiatriazoles (**1**) ⟨76AHC(20)145⟩. However, these calculations have been criticized (see Section 4.28.2.2.3 for a detailed discussion). MO calculations have been carried out for the hypothetical ring closure of thioformyl azide *S*-oxide (**8**; H substituted for Ph) to thiatriazole *S*-oxide. It is suggested that the variable chemical behavior of thioacyl azides and their *S*-oxides is due to disruption of the aromatic character of the thiatriazole *S*-oxide ⟨77T2231⟩.

4.28.2.2.3 Structural methods

(i) X-Ray diffraction

An X-ray structural analysis of 5-phenyl-1,2,3,4-thiatriazole, 5-phenyl-1,2,3,4-thiatriazole 3-oxide and 3-ethyl-5-phenyl-1,2,3,4-thiatriazolium tetrafluoroborate has been performed (Figure 1) ⟨76ACS(A)351⟩. This work confirmed the sites of oxidation and alkylation (see Section 4.28.2.3.1(ii)) of the thiatriazole ring which previously were based on ^{1}H, ^{15}N and ^{13}C NMR spectroscopy ⟨75JOC431, 76AHC(20)145⟩. It further showed the five-membered rings to be planar and the bond lengths indicate a considerable resonance stabilization. However, compared to the corresponding resonance stabilized 1,2,5- ⟨63JA3553, 61JA4475⟩ and 1,3,4-thiadiazoles ⟨66JSP(19)283, 70ACS2525⟩, a noteworthy difference is indicated. The S(1)—C(5) bond length is shortened by 0.02 Å compared to that of 1,3,4-thiadiazole and the S(1)—N(2) bond length is increased by 0.05 Å compared to that of 1,2,5-thiadiazole. The planes of the benzene rings are tilted to a small degree (3.6–18.0°) and the C(5)—Ph bond lengths (1.46–1.47 Å) indicate conjugation between the rings. The introduction of oxygen at N(3) results in significant lengthening of the N(2)—N(3) and N(3)—N(4) bonds.

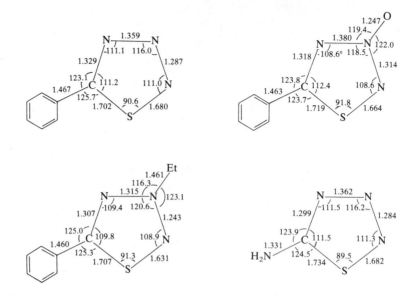

Figure 1 Bond lengths (Å) and bond angles (°) of 1,2,3,4-thiatriazoles

As pointed out ⟨76AHC(20)145⟩, the single bond character of the S(1)—N(2) bond should be reflected in the chemistry of thiatriazoles. In fact the thermal decomposition of thiatriazoles leading to formation of nitrogen, sulfur and an organic fragment is assumed to proceed *via* an initial ring-opening at the S(1)—N(2) bond (see Section 4.28.2.3.1(i)).

The crystal and molecular structure of 5-amino-1,2,3,4-thiatriazole has been determined (Figure 1) ⟨79MI42801⟩. Some important features of this study concerning the discussion of the positional reactivity of the thiatriazole ring are pointed out by the authors. Susceptibility for alkylation has been predicted to be the highest at the N-4 position (N-4 ⩾ N-2 > N-3) based on CNDO and SCF MO–CNDO calculations ⟨76AHC(20)145⟩, in disagreement with the experimentally observed alkylation, oxidation and complex formation of phenylthiatriazole at the 3-position (see Section 4.28.2.3.1(ii)). Zaworotko *et al.* suggest that the reason for the failure of the theoretical model may result from differences between the observed and calculated structures. The ring substituent participates in resonance interaction with the heteroaromatic ring as borne out by the following: sp^2-hybridization of the nitrogen of the amino substituent, the closeness to planarity of the entire molecule, and the short C(5)—NH$_2$ bond which indicates strong π-bonding between C(5) and NH$_2$. It seems likely on this evidence that the observed reactivity of position 3 is caused by electronic changes imposed when the substituent at the 5-position participates in the π-system.

(ii) Molecular spectra

IR and mass spectra of 5-substituted 1,2,3,4-thiatriazoles have been reviewed ⟨76AHC(20)145⟩. Recently, in-plane normal vibration modes of 5-amino-1,2,3,4-thiatriazole and 5-mercapto-1,2,3,4-thiatriazole have been tentatively assigned in the solid state on the basis of a normal coordinate analysis, transferring force constants from related molecules ⟨77PIA(A)(86)265⟩.

The ^{1}H, ^{15}N and ^{13}C NMR spectra of phenyl- and alkylthio-thiatriazoles and of the corresponding 3-ethylthiatriazolium tetrafluoroborates have been recorded ⟨76AHC(20)145⟩. Spectroscopic data (IR, NMR, MS and UV) have been reported for thiatriazolium (**79**; X = S, Y = O, R = Ph, 4-MeC$_6$H$_4$, 4-ClC$_6$H$_4$) and oxatriazolium tetrafluoroborates (**79**; X = O, Y = S, R = Ph) (see Section 4.28.2.5.4) ⟨79JCS(P1)732⟩. ^{13}C(5) shifts in 5-substituted thiatriazoles have been compiled in Table 1 (see also Section 4.28.2.2.4).

UV spectra of 5-phenyl-1,2,3,4-thiatriazole (λ_{max} = 280 nm, ε = 10 800) ⟨60CB2353⟩ and of some 5-amino-1,2,3,4-thiatriazoles have been recorded ⟨77JOC3725, 80CCC2329⟩.

ESCA and IR spectra of 3-phenyl-1,2,3,4-thiatriazole 3-oxide have been published ⟨75T1783⟩. The UV spectrum exhibits maxima at 254 (ε = 19 300) and 320 nm (ε = 3000) ⟨75T1783, 76AHC(20)145⟩.

Table 1 ^{13}C NMR Chemical Shifts^a of C(5)-substituted Thiatriazoles

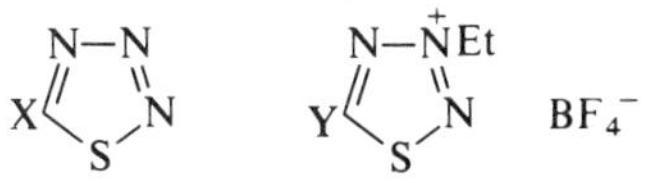

Substituent		Chemical shift	
X	Y	(δ, p.p.m.)	Ref.
Ph	—	178.46	76AHC(20)145, p. 154
PhNH	—	173.8	77JOC 3725
BzS	—	179.8	74JOC 3770
PhCOCH$_2$S	—	179	74JOC 3770
PhCOS	—	171.5	74JOC 3770
—	Ph	186.42	76AHC(20)145, p. 154

^{a} In CDCl$_3$.

The ^{1}H, ^{13}C NMR, IR and MS data of 4-alkyl-5-sulfonylimino-Δ^2-1,2,3,4-thiatriazolines (**17**) have been published ⟨73JOC2916, 74JA3973, 77JHC1417⟩ (see also Section 4.28.2.2.4).

4.28.2.2.4 Tautomerism: prototropic

For 5-(monosubstituted)amino-1,2,3,4-thiatriazoles previous evidence, based on dipole moments, IR and other properties ⟨64AHC(3)263⟩, favored the amino form (**11**) over the imino form (**12**). L'abbé and coworkers have compared ^{13}C chemical shifts in model compounds (**13**) and (**14**) (Table 2; see also Table 1) with seven monosubstituted 5-aminothiatriazoles. The latter are all found to resonate from 173.1–179.4 p.p.m. (DMSO-d_6 or CDCl$_3$, δ values) and it is therefore concluded that they all, including 5-aminothiatriazole, exist exclusively in the amino form (at least under the given conditions). ^{1}H NMR measurements confirm this conclusion ⟨77JHC1417⟩.

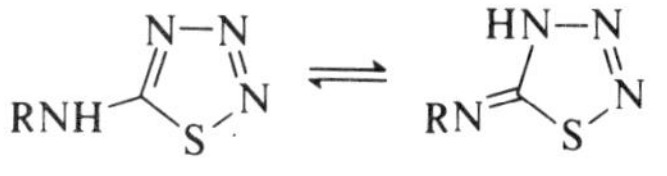

Table 2 ^{13}C NMR Chemical Shifts^a of Thiatriazoles (**13**) and (**14**) ⟨77JHC1417⟩

Compound	R	Chemical shift (C-5) (δ, p.p.m.)	Compound	R	Chemical shift (C-5) (δ, p.p.m.)
(**13**)	Ph	180.4	(**14**)	Ph	156
(**13**)	Tos	173.1	(**14**)	Tos	166.3

^{a} In CDCl$_3$.

This satisfactory method was also used to settle the similar thione–thiol (**15–16**) problem of 5-mercaptothiatriazole which was claimed by Jensen and Pedersen to exist in the thiol form ⟨64AHC(3)263⟩. However, in the ^{13}C NMR spectrum of 5-mercaptothiatriazole (0 °C, acetone-d_6) the C(5) resonates at 193.1 p.p.m., corresponding to the thione structure, whereas the carbon in the thiol form would be expected to resonate at 175–180 p.p.m. ⟨77JHC1417⟩.

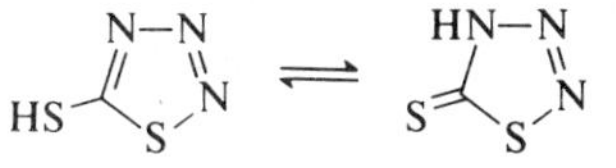

4.28.2.3 Reactivity of the Ring System

4.28.2.3.1 Reactivity at the ring atoms

(i) *Thermal and photochemical reactions involving no other species*

(a) *1,2,3,4-Thiatriazoles.* It is characteristic of all known 1,2,3,4-thiatriazoles that they undergo thermal decomposition forming nitrogen, sulfur and an organic fragment (equation 1). The relative stability of the thiatriazole is highly dependent on the substituent. Very few alkylthiatriazoles have been isolated as they decompose around 0 °C except for the remarkably stable fluorinated thiatriazole (66) (m.p. 60.5–61 °C ⟨70JOC3470⟩; see Section 4.28.2.5). Arylthiatriazoles are much more stable and may be stored at room temperature without noticeable decomposition over a period of years. Thiatriazoles attached at the 5-position to N-, S- or O-linkages are more or less stable. From qualitative observations on many of these types of thiatriazoles, Solanki and Trevedi suggest the following order of stability: NH > O > S ⟨71JIC843⟩.

$$X \underset{S}{\overset{N-N}{\diagdown}} \!\!\!\! N \xrightarrow{\Delta} XCN + N_2 + S \qquad (1)$$

The decomposition of thiatriazoles has been used for the preparation of certain types of compounds. 5-Alkoxy- and 5-aryloxy-1,2,3,4-thiatriazoles decompose smoothly in ethereal solution at 20 °C with formation of alkyl and aryl cyanates (equation 1, X = RO). A thermally unstable N-cyanatodiphenylmethanimine (X = Ph$_2$C=NO—) has been generated from the corresponding thiatriazole ⟨77ACS(B)687⟩. Although other methods are available for the syntheses of cyanates, the thiatriazole method is the most generally applicable for the thermally and chemically sensitive alkyl cyanates ⟨76AHC(20)145, p. 166, B77MI42803⟩.

The similarly smooth decomposition of S-substituted 5-mercaptothiatriazoles has been used for the preparation of acylated thiocyanates which are difficult to obtain otherwise (equation 1; X = RC(=O)S—) ⟨76AHC(20)145, p. 164, 76T745⟩. 5-Mercaptothiatriazole, the free acid (X = SH), is reported to be relatively stable below 10 °C but decomposes thermally with the expected formation of (polymeric) thiocyanic acid. Its oxidation product the disulfide is sensitive to impact and may explode even under water. Further details are found in the 1964 review ⟨64AHC(3)263⟩.

5-Aminothiatriazoles decompose more or less violently on heating to their melting point ⟨64AHC(3)263⟩. When heated in aqueous solution the expected cyanamide (R^{1}R^{2}NCN) is formed (equation 1; X = NR^{1}R^2) and is isolated as the trimeric melamine. In the case of the unsubstituted 5-aminothiatriazole, dicyandiamide is the final product ⟨64AHC(3)263⟩. The thermal properties of 5-aminothiatriazoles have been studied using TG (thermal gravimetry) and DTA (differential thermal analysis). Substitution of the protons in 5-aminothiatriazole with phenyl or benzyl groups increases the stability, and a further increase in stability is observed by introducing ethoxycarbonyl substituents ⟨77MI42802⟩.

Early results concerning the mechanism of the thermal decomposition of thiatriazoles were covered in the 1976 review ⟨76AHC(20)145⟩. In the meantime new evidence has accumulated and a decomposition mechanism has been proposed, based on kinetic measurements ⟨78JOC4816⟩. The first-order decomposition of 5-phenylthiatriazole in bromobenzene was followed by UV spectroscopy ($E_a = 118.0$ kJ mol^{-1}, $\Delta S^{\ddagger} = 11.7$ J K^{-1} mol^{-1}) and a linear Hammett correlation was obtained with substituted phenylthiatriazoles ($\rho \approx 0$). Isotopic labeling in the 2-position with ^{15}N had no measurable effect on the rate, while labeling in the 4-position led to a kinetic isotope effect around 1.04 which is about the effect calculated for breaking of the N(3)—N(4) bond in the rate determining step. This result is *a priori* in agreement with retro-cycloaddition of N$_2$S but a careful search for it in the mass spectrum of decomposing 5-phenylthiatriazole was unsuccessful. Attempts to trap thiobenzoyl nitrene with various dipolarophiles were similarly unsuccessful and a high yield of benzonitrile was obtained in all cases (92–100%). An ion in the mass spectrum at m/e 135, of the same stoichiometric composition as thiobenzoylnitrene, was shown to be produced primarily by electron impact rather than being a thermally formed species. However, by varying the ion source temperature it was observed that increasing temperature resulted in increasing, although small, amounts of phenyl isothiocyanate (up to 6%), together with benzonitrile. This result was confirmed by pyrolysis in a quartz tube in combination with GLC–MS. In short, a mechanism is suggested involving formation of hitherto unknown thiobenzoyl azide (in equilibrium with the thiatriazole) which undergoes synchronous fragmentation without an intermediate thioacylnitrene (equation 2). Only indirect evidence for thiobenzoyl azide was obtained. Benzonitrile may be formed by a Grob-type fragmentation ⟨69AG543⟩ from the *E*-conformer of thiobenzoyl azide (equation 3), and phenyl isothiocyanate may be formed from the *Z*-conformer in a Curtius rearrangement (equation 4). A DTA investigation of the decomposition of phenylthiatriazole in bromobenzene supports the hypothesis of a complex process ⟨79MI42802⟩.

$$Ph\underset{S}{\overset{N-N}{\underset{\diagdown}{\diagup}}}N \rightleftarrows Ph-\underset{S}{\overset{\|}{C}}-N=\overset{+}{N}=\bar{N} \rightleftarrows PhCN + S + N_2 \quad (2)$$

$$\underset{Ph}{\overset{^-S}{\diagdown}}C\overset{\cdot}{=}N\underset{\overset{+}{N_2}}{\diagdown} \rightarrow PhCN + S + N_2 \quad (3)$$

$$\underset{Ph}{\overset{^-S}{\diagdown}}C=N\overset{\overset{+}{N_2}}{} \rightarrow PhN=C=S + N_2 \quad (4)$$

The results of kinetic measurements on the thermal decomposition of 5-alkoxythiatriazoles, studied by Jensen *et al.*, are consistent with this scheme ⟨76AHC(20)145, p. 149⟩. The formation of pentyl cyanide and pentyl isothiocyanate in a 4:1 molar ratio from thermally decomposing 5-pentylthiatriazole ⟨76AHC(20)145, p. 163⟩ can apparently be rationalized along the same lines.

The thiatriazole photochemistry published prior to 1976 has been reviewed ⟨76AHC(20)145⟩. Briefly summarized, the photoreactions of arylthiatriazoles leading to nitriles, sulfur, nitrogen and small amounts of isothiocyanates (equation 5) take place from an excited singlet state. The nitrogen formed originates exclusively from the N-2 and N-3 positions. Benzonitrile sulfide was trapped as a cycloaddition product with dimethyl acetylenedicarboxylate (DMAD) in 9% yield (equation 6) and its photogeneration from phenylthiatriazole and decay were examined by UV spectroscopy in an EPA glass at cryogenic temperatures. No indication of the intermediacy of thiobenzoylnitrene or thiobenzoyl azide in the chemical or spectroscopic investigations was found. Isothiocyanate is suggested to be formed directly from an excited singlet state although precursors could not be rigorously excluded ⟨76AHC(20)145⟩.

$$Ar\underset{S}{\overset{N-N}{\underset{\diagdown}{\diagup}}}N \xrightarrow{h\nu} ArCN + ArN=C=S + S + N_2 \quad (5)$$

$$Ph\underset{S}{\overset{N\dot{-}N}{\underset{\diagdown}{\diagup}}}N \xrightarrow{h\nu} PhC\overset{=N}{\underset{S}{\diagdown}} \xrightarrow[>20\,K]{\Delta} Ph\overset{+}{C}N - \bar{S} \xrightarrow[>220\,K]{\Delta} PhCN + S$$

$$\Big\downarrow MeO_2CC\equiv CCO_2Me$$

$$Ph\underset{CO_2Me}{\overset{N-S}{\underset{\diagdown}{\diagup}}}CO_2Me \quad (6)$$

Irradiation of phenylthiatriazole and other phenyl-substituted C-, N-, and S-containing heterocycles (see Section 4.01.1.2, Scheme 1) in PVC films at 10 K has been monitored using UV spectroscopy. An intermediate intervening between the heterocycle and benzonitrile sulfide was detected, which is considered to be phenylthiazirine. On heating to 20 K and higher temperatures, phenylthiazirine undergoes electrocyclic ring-opening with formation of benzonitrile sulfide, which decomposes rapidly to benzonitrile and sulfur above 220 K (equation 6) ⟨78JCS(P1)746⟩.

Scrutiny of a number of other five-membered C-, N- and S-containing heterocycles ⟨78JCS(P1)1445⟩ with the framework depicted in Scheme 2 has revealed that facile photofragmentation takes place when extrusion of a small inorganic fragment such as N_2 (thiatriazole), N_2O (thiatriazole 3-oxide), CO, CO_2, *etc.* is possible. In these cases formation of nitrile sulfides appears to be general, as demonstrated by isolation of the cycloaddition product between the nitrile sulfide and DMAD (equation 6) in 5–21% yield depending upon the starting material (see also Chapters 4.17 and 4.34).

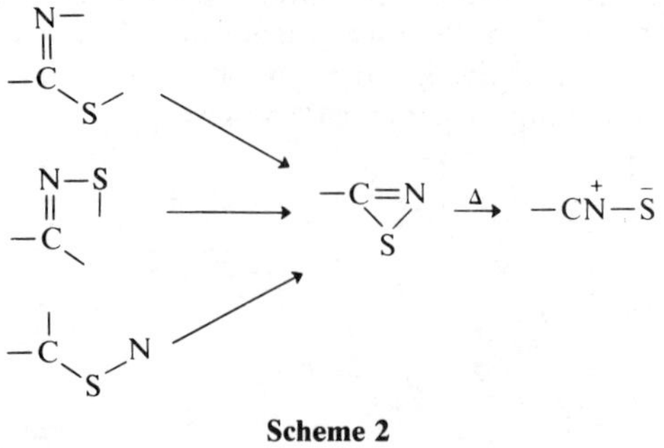

Scheme 2

Rate constants for the first-order decay of nitrile sulfides have been estimated by means of flash photolysis of thiatriazoles and mesoionic oxathiazoles and the results used to calculate activation parameters ⟨79JCS(P1)960⟩.

(b) Δ^2-1,2,3,4-Thiatriazolines. Δ^2-Thiatriazolines decompose at slightly elevated temperatures. 4-Alkyl-5-sulfonylimino-1,2,3,4-thiatriazolines (**17**; $R^1 = 4\text{-}XC_6H_4SO_2$, $R^2 =$ alkyl or aralkyl) give carbodiimides at 45–80 °C, trapped with either water or $(COCl)_2$ (50% yield) ⟨73JOC2916⟩. Likewise 4-alkyl-5-alkylimino-1,2,3,4-thiatriazolines (**17**; $R^1 =$ alkyl, $R^2 =$ alkyl) give carbodiimides (7–52% yield), decomposing slowly around 40–60 °C but rapidly at 125 °C (equation 7) ⟨78JCS(P1)1440⟩. Both types of thiatriazolines decompose in the presence of dipolarophiles with formation of characteristic cycloaddition products (e.g. **20**). 4-Alkyl-5-alkyliminothiatriazolines react with the dipolarophile in a bimolecular reaction, as discussed further in Section 4.28.2.3.1(iii). Kinetic experiments show that the rate of thermal decomposition of 4-benzyl-5-tosyliminothiatriazoline is not influenced by the presence of enamines, excluding direct attack of the enamine on the thiatriazoline. Instead, this result indicates formation of a discrete intermediate which apparently is either a thiaziridine (**18**) and/or the isomeric open-chain 1,3-dipolar species (**19**; equation 8) ⟨75BSF1127, 76AHC(20)145, 80AG277⟩.

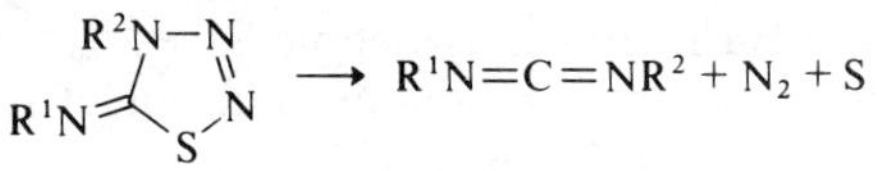

$$R^1N=C=NR^2 + N_2 + S \quad (7)$$

(**17**) $R^1 = 4\text{-}XC_6H_4SO_2$ or alkyl; $R^2 =$ alkyl or aralkyl

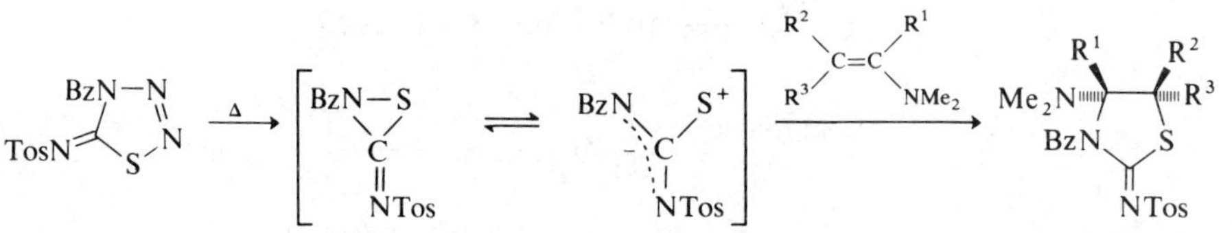

(8)

(**18**) (**19**) (**20**)

In the absence of dipolarophiles the intermediate loses sulfur to give the carbodiimide; however, the intermediate may be trapped with a number of alkenes, heterocumulenes and other reagents such as ynamines, ketenes, isocyanates, isothiocyanates, carbodiimides, ketenimines, sulfonylimines, imines, nitriles, thiocarbonyl compounds, Wittig reagents and the already mentioned enamines ⟨80AG277⟩. Contemporary knowledge of the chemistry of these sulfonyliminothiatriazolines is mainly due to the meticulous work of L'abbé and his coworkers ⟨80AG277⟩.

4-Alkyl-5-aryliminothiatriazolines (**21**) do not produce carbodiimides but give rise to benzothiazoles (**22**) when heated in toluene at 120–130 °C for several hours (equation 9). With the 3-chloro compound, two isomeric benzothiazoles are formed resulting from attack

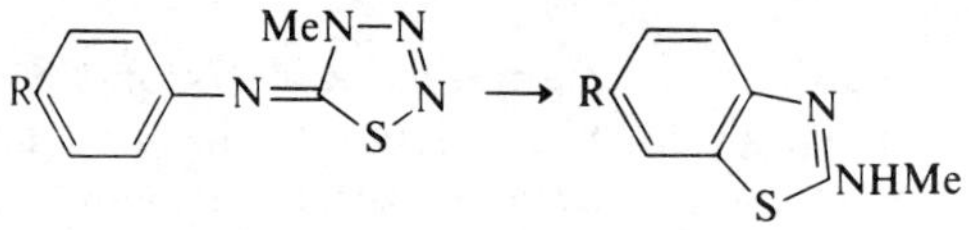

(9)

(**21**) R = H, Me, Cl   (**22**)

either *ortho* or *para* to the chlorine substituent ⟨71AP687⟩. The mechanism of this process was not discussed but it may be rationalized by assuming an electrophilic attack on the benzene ring of the 1,3-dipole formed by loss of nitrogen.

(c) *Δ³-1,2,3,4-Thiatriazolines.* The 1:1 adducts (**23**) formed from bis-(trifluoromethyl)thioketene and aryl azides (see Section 4.28.2.5.3) are believed to be the new Δ³-1,2,3,4-thiatriazolines ⟨78JOC2500⟩. Structural assignment is based on ¹⁹F NMR and IR spectra, and also on thermolysis leading to a product believed to be 2,1-benzisothiazole (**24**; equation 10). The latter gives 2-EtNHC₆H₄CH₂CH(CF₃)₂ on reductive desulfurization with Raney nickel in ethanol (*N*-ethylation of amines with ethanol in the presence of Raney nickel is a known reaction). The alternative benzothiazole structure (**25**) was excluded by independent synthesis of this compound from 2-aminobenzenethiol and bis(trifluoromethyl)thioketene.

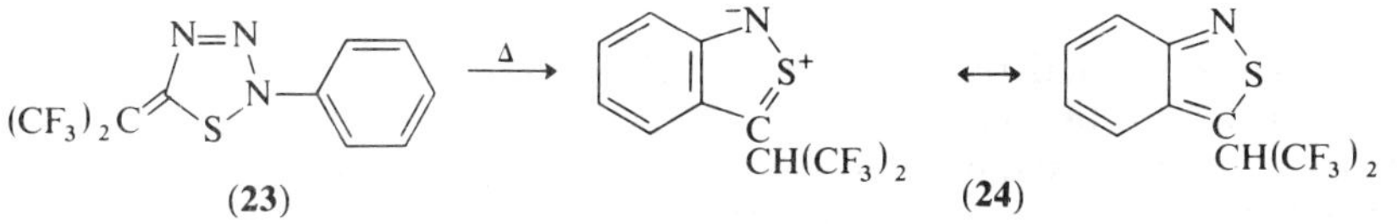

$$(10)$$

A surprising reaction has been observed on thermolysis of an arylthiatriazoline (**26**) blocked in the *ortho* positions with methyl groups. The 1,2-benzisothiazole (**27**), in which a methyl group is lost according to NMR analysis, is formed in 24% yield along with 'tar' (equation 11).

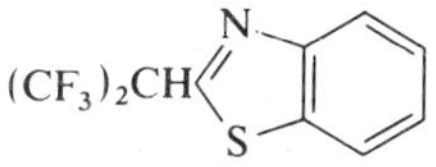

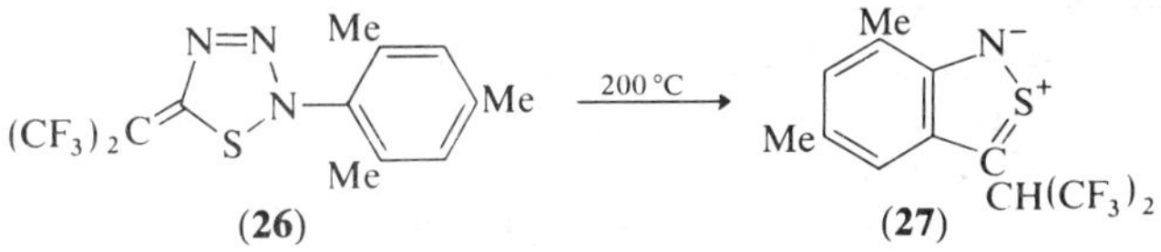

$$(11)$$

(*d*) *3,5-Disubstituted 1,2,3,4-thiatriazol.um salts.* 3-Ethyl-5-phenylthiatriazolium tetrafluoroborate (Figure 1) is found to be thermally stable up to 180–200 °C in the solid state. The alkyl group in the 3-position blocks the ability of the system to eject nitrogen under mild conditions ⟨75JOC431⟩.

5-Phenyl-1,2,3,4-thiatriazole 3-oxide (Figure 1) is likewise thermally quite stable, decomposing at around 185 °C ⟨75T1783⟩.

(*e*) *1,2,3,4-Selenatriazoles.* 5-(*N,N*-Disubstituted amino)-1,2,3,4-selenatriazoles are thermally unstable, decomposing slowly at room temperature with quantitative formation of the corresponding cyanamides, nitrogen and selenium. Thus they resemble the corresponding sulfur compounds which are, however, significantly more stable. A half-life of 8 days at room temperature in CHCl₃ is estimated for 5-(diethylamino)selenatriazole (equation 12) ⟨82UP42800⟩.

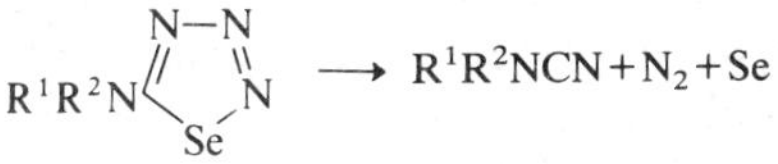

$$R^1R^2NCN + N_2 + Se \quad (12)$$

5-Cyclohexylamino-1,2,3,4-selenatriazole (**29**) dissociated into hydrazoic acid and isoselenocyanate on attempted isolation (equation 13) ⟨82UP42800⟩. Monosubstituted 5-amino-1,2,3,4-thiatriazoles also undergo this type of ring cleavage, although only in the presence of base (see Section 4.28.2.4.1).

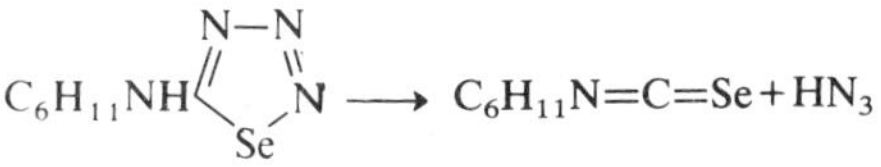

$$C_6H_{11}N=C=Se + HN_3 \quad (13)$$

(*ii*) *Electrophilic and nucleophilic attack*

(*a*) *1,2,3,4-Thiatriazoles.* 5-Phenyl-1,2,3,4-thiatriazole is attacked in the 3-position by potent oxidizing (3-ClC₆H₄CO₃H) and alkylating (Et₃O⁺BF₄⁻) agents to give the 3-oxide and the 3,5-disubstituted thiatriazolium salt, respectively (Figure 1). These reactions, and the physical and chemical properties of the products, have been discussed in detail elsewhere ⟨76AHC(20)145⟩. The reader's attention may be directed to the fact that 3,5-disubstituted thiatriazolium as well as oxatriazolium salts are also obtained by alkylation of mesoionic 1,2,3,4-thia- and oxa-triazoles (see Section 4.28.2.5.4).

The reaction of alkoxythiatriazoles with trivalent phosphorus compounds to give 2,2-dihydro-1,3,4,5,2-thiatriazaphosphorines (**30**; equation 14) has been reviewed ⟨76AHC(20)145⟩.

$$RO\underset{S}{\overset{N-N}{\diagdown N}} \xrightarrow{R_3P} RO\underset{N\diagdown N}{\overset{S}{\diagdown}PR_3} \xrightarrow{\Delta} ROCN + N_2 + S{=}PR_3 \qquad (14)$$

(**30**)

5-Phenylthiatriazole and titanium tetrachloride form an orange, thermally stable 1:1 complex, the structure of which has not been investigated ⟨76AHC(20)145⟩. $W(CO)_6$ reacts similarly, and X-ray crystallography on the red-orange complex has shown W to be coordinated to the 3-position of the thiatriazole ring (thiatriazole: $W(CO)_5$, 1:1) ⟨82UP42801⟩. However, stronger Lewis acids like $AlCl_3$ and BBr_3 cause immediate thermal decomposition of thiatriazoles ⟨76AHC(20)145, p. 151⟩.

In general, the thiatriazole ring is rather inert, as pointed out by Jensen and Pedersen. Thus it is stable towards chlorine and nitric acid. 5-Phenylthiatriazole is a very weak base, dissolving in concentrated hydrochloric or nitric acid. It is reported to withstand warm, alcoholic alkali and Grignard reagents. However, phenylmethanethiol is formed on reduction with LAH ⟨64AHC(3)263, p. 269⟩.

(iii) Reactions with cyclic transition states

(*a*) Δ^2-*1,2,3,4-Thiatriazolines.* 4-Alkyl-5-arylimino-1,2,3,4-thiatriazoline (**31**) is thermally stable to about 110 °C, at which temperature it decomposes with formation of benzothiazole (**22**; R = H; equation 9). However, when mixed with phenyl isothiocyanate at room temperature, nitrogen is evolved and compounds (**32**), (**33**) and (**22**) are formed (equation 15). This clearly demonstrates that a bimolecular mechanism is operating. Compound (**32**) is assumed to be formed by addition of the isothiocyanate C=N double bond across the ring, while (**33**) may arise from (**32**) by a Dimroth rearrangement (see below) ⟨77JOC1159⟩.

$$\underset{\textbf{(31)}\ R=Ph,\ Bz}{\underset{PhN}{\overset{MeN-N}{\diagdown}}\underset{S}{\diagup}N} \xrightarrow{RNCS} \underset{\textbf{(32)}}{\underset{PhN}{\overset{MeN-\diagup S}{\diagdown}}\underset{S-NR}{\diagup}} + \underset{\textbf{(33)}}{\underset{PhN}{\overset{MeN-\diagup NR}{\diagdown}}\underset{S}{\diagup}} + (\textbf{22}) \qquad (15)$$

4-Alkyl-5-alkylimino-1,2,3,4-thiatriazolines react along the same lines ⟨78JCS(P1)1440⟩, since addition of electron-rich alkenes or heterocumulenes such as enamines, carbodiimides, isocyanates or styrene causes immediate evolution of nitrogen at room temperature. The reaction with β-*N,N*-dimethylaminostyrene with formation of (**34**) and (**35**) in almost equal amounts is shown as an example in equation (16). Compound (**34**) is assumed to be formed by cycloaddition–elimination across the ring while compound (**35**) may be formed by 1,3-dipolar cycloaddition across the imino group and S-1. Diphenylketene adds in highly diluted solutions across the ring of (**36**) to form (**37**; equation 17). In more concentrated

$$\underset{R,\ R^1=Et,\ Me;\ Me,\ Et}{\underset{RN}{\overset{R^1N-N}{\diagdown}}\underset{S}{\diagup}N} + \underset{Ph}{\overset{H}{\diagdown}}C{=}C\underset{H}{\overset{NMe_2}{\diagup}} \xrightarrow{-N_2} \underset{\textbf{(34)}}{\underset{RN}{\overset{R^1N{-}NMe_2}{\diagdown}}\underset{S}{\diagup}\underset{Ph}{\overset{H}{\diagdown}}} + \underset{\textbf{(35)}}{\underset{R^1N}{\overset{RN{-}NMe_2}{\diagdown}}\underset{S}{\diagup}\underset{Ph}{\overset{H}{\diagdown}}} \qquad (16)$$

$$\underset{\textbf{(36)}}{\underset{MeN}{\overset{EtN-N}{\diagdown}}\underset{S}{\diagup}N} \xrightarrow{Ph_2C{=}C{=}O} \underset{\textbf{(37)}}{\underset{MeN}{\overset{EtN{-}\overset{O}{\diagdown}}{\diagdown}}\underset{S}{\diagup}\underset{Ph}{\overset{Ph}{\diagdown}}} + \underset{\textbf{(38)}}{\underset{EtN}{\overset{MeN{-}\overset{O}{\diagdown}}{\diagdown}}\underset{S}{\diagup}\underset{Ph}{\overset{Ph}{\diagdown}}} \qquad (17)$$

solution (10^{-2} M), an approximately 1:1 mixture of (**37**) and (**38**) is formed, the latter apparently arising by addition across the imino group and S-1 but the dilution effect may indicate a more complex reaction.

Only in the case of sulfonyliminothiatriazolines has a unimolecular mechanism involving primary formation of a reactive 1,3-dipole or thiaziridine been demonstrated. This reactive intermediate may be trapped with dipolarophiles (see Section 4.28.2.3.1(i)).

In two independent communications, Revitt ⟨75CC24⟩ and L'abbé *et al.* ⟨75JHC607⟩ describe formation of heterocycles from alkyl azides and aryl isothiocyanates. The reaction was shortly afterwards investigated in detail, as illustrated by the reactions of benzyl azide with two equivalents of aryl isothiocyanate ⟨77JOC1159⟩. At 60 °C five products are obtained, a tetrazolinethione (**40**), two thiadiazolidines (**41**) and (**42**), and two dithiazolidines (**43**) and (**44**). Compound (**41**) is formed as the major product at the beginning of the reaction, whereas (**43**) predominates at the end. To explain this complicated reaction mixture, it is assumed that (**40**) and (**39**) are formed by addition to the C=N and to the C=S bond, respectively (equation 18), with the latter as the predominant reaction. Thiatriazoline (**39**), which cannot be isolated, is assumed to undergo further reaction with the isothiocyanate (equation 19). Independent experiments proved (**41**) to undergo the Dimroth ⟨69ZC241⟩ rearrangement in a quantitative manner to yield (**43**) under mild conditions, in agreement with its build up during the reaction. The by-products (**42**) and (**44**) were only isolated in one case (Ar = 4-$NO_2C_6H_4$) and in small amounts (see iminothiatriazolines as masked dipoles below).

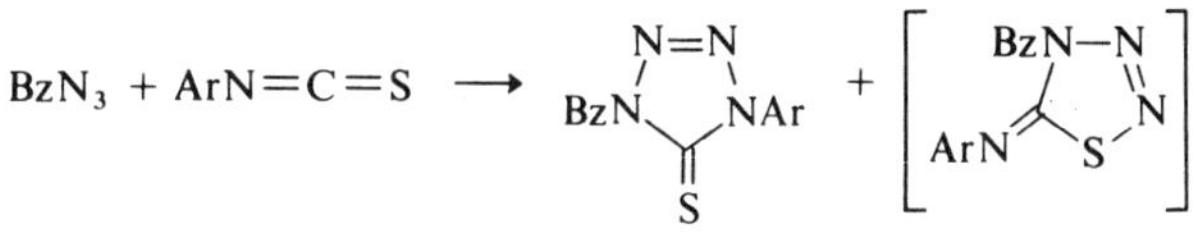

$$BzN_3 + ArN=C=S \longrightarrow \underset{(\mathbf{40})\ Ar = Ph, 4-NO_2C_6H_4,}{\underset{4-ClC_6H_4, 4-MeC_6H_4}{\text{[structure]}}} + [\text{structure (39)}] \quad (18)$$

$$(\mathbf{39}) \xrightarrow{\text{major path}} (\mathbf{41}) \longrightarrow (\mathbf{43}) \quad (19)$$

Acyl isothiocyanates react with alkyl azides, presumably to give the primary products (**45**) which have not been isolated so far (equation 20). Reaction of (**45**) with a second mole of acyl isothiocyanate affords 1,2,4-dithiazolidine (**46**) and 1,2,4-thiadiazole (**47**). The formation of these products may be rationalized along the same principles as discussed above ⟨79BSB245⟩.

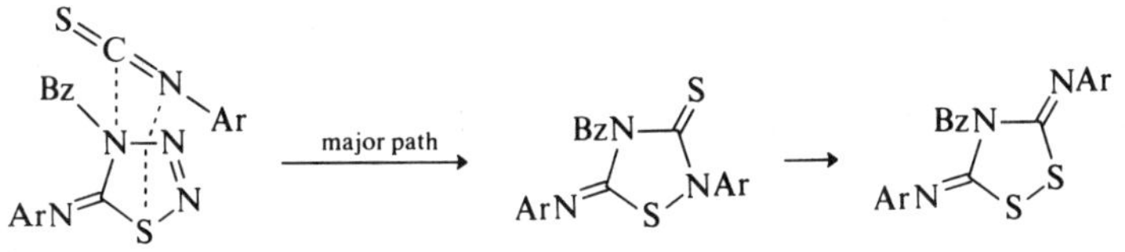

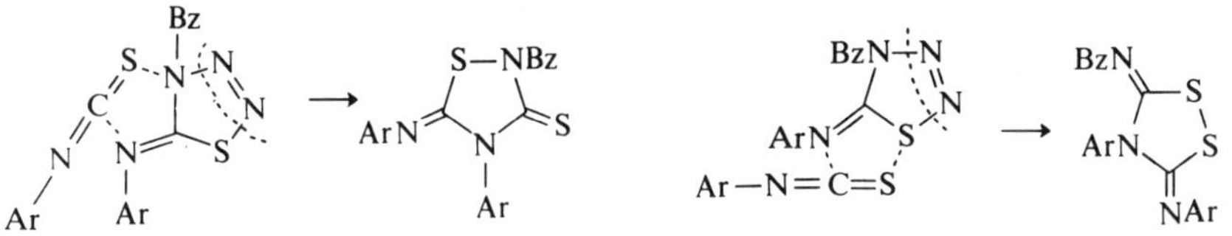

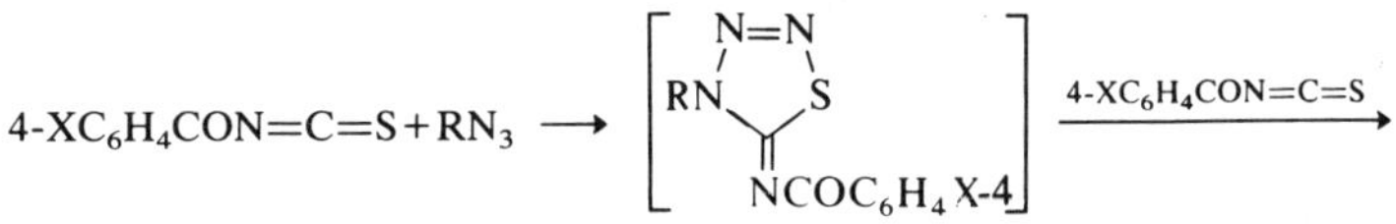

$$4\text{-}XC_6H_4CON=C=S + RN_3 \longrightarrow [\text{structure (45)}] \xrightarrow{4\text{-}XC_6H_4CON=C=S}$$

$$[\text{structure (46)}] + [\text{structure (47)}] \quad (20)$$

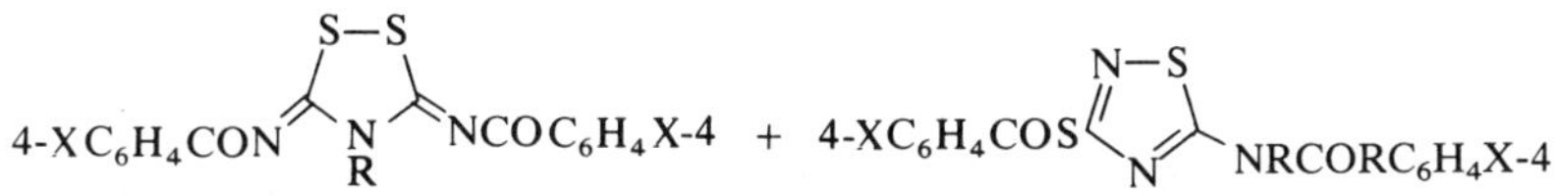

The reaction of isocyanates with iminothiatriazolines generated *in situ* has been employed in the preparation of 5-arylimino-1,2,4-thiadiazolidin-3-ones (**49**; equation 21). The utility of this principle is based on the observation that most isocyanates do not react with azides. A direct reaction between 4-methyl-5-phenyliminothiatriazoline and butyl isocyanate gives 95% yield of the thiazolidin-3-one ⟨76AG510⟩.

$$RN_3 + R^1NCO + 4\text{-}XC_6H_4NCS \longrightarrow \left[\begin{array}{c} \text{(intermediate)} \end{array} \right] \longrightarrow \text{(49)} \quad (21)$$

R, R¹, X = cyclo-C₆H₁₁, Bu, Cl (84%); Bu, Ph, Me (82%); Bz, Bu, H (60%); Bu, Bu, H (73%) (yields are based on isothiocyanate)

(b) *Δ²-1,2,3,4-Thiatriazolines.* Iminothiatriazolines undergo in certain cases cycloaddition across S(1)—C=N(5). In this respect they may be considered as masked 1,3-dipoles (*e.g.* **50**), as pointed out by L'abbé *et al.* ⟨78JOC4951⟩. This is a characteristic reaction mode of the electrophilic acyl isothiocyanates and of sulfenes with aryliminothiatriazolines. Reaction of iminothiatriazoline (**50**) with one equivalent of aroyl isothiocyanate or ethoxycarbonyl isothiocyanate thus gives a single product (**51**) (equation 22). A kinetic study of the reaction with benzoyl isothiocyanate in benzene and in acetonitrile was undertaken by these authors. Second-order kinetics, moderately negative entropies of activation and small solvent effects indicate a concerted cycloaddition–elimination mechanism through a dithiapentalene-like transition state, although the data do not rigorously exclude the alternative stepwise mechanism. In the absence of solvent and with a threefold excess of benzoyl isothiocyanate, compound (**52**) was isolated; this was formed from (**51**) by a Dimroth rearrangement (equation 23).

$$(50) \xrightarrow{\text{RCONCS}} \left[\text{intermediate} \right] \xrightarrow{-N_2} (51) \quad (22)$$

$$(51) \xrightarrow{\text{R}^1\text{NCS}} \left[\text{intermediate} \right] \xrightarrow{-\text{R}^1\text{NCS}} (52) \quad (23)$$

Sulfenes, generated *in situ* from alkylsulfonyl chlorides and triethylamine, react with thiatriazoline (**50**) to give sultams (**53**). The sultams isomerize on heating at 60 °C in the presence of benzoyl chloride, or AlCl₃ and methanesulfonyl chloride (equation 24).

$$(50) \xrightarrow[-N_2]{\text{RCH}=SO_2} (53) \xrightarrow{\text{catalyst}} \text{product} \quad (24)$$

4.28.2.4 Reactivity of Substituents

4.28.2.4.1 1,2,3,4-Thiatriazoles

(i) *Alkylation and acylation of 5-aminothiatriazoles without ring cleavage*

Alkylation or acylation of 5-aminothiatriazoles may in principle lead to several isomeric products by attack at the different sites on the ring, or side chain. From ¹H NMR data, Lippmann *et al.* concluded that only 5-(*N*-acylphenylamino)thiatriazoles are formed upon acylation of 5-(phenylamino)thiatriazole ⟨73ZC134⟩. Methylation of 5-(arylamino)thiatriazoles with dimethyl sulfate gives only 5-(*N,N*-arylmethylamino)thiatriazoles but treatment with diazomethane gives 4-methylated 5-iminothiatriazolines or a mixture of 4- and 5-methylated products (see Section 4.28.2.5.2). Alkylation with trialkyloxonium tetrafluoroborates affords exclusively 4-alkylated 5-iminothiatriazolines (see Section 4.28.2.5.2). In contrast, alkylation takes place at the 3-position when the 5-amino group is absent (see Section 4.28.2.3.1(ii)).

(ii) Acid–base induced cleavage and Dimroth rearrangement of 5-aminothiatriazoles

The 5-alkylaminothiatriazoles are cleaved by alkali to give azide ion and an isothiocyanate (equation 25). Similarly, unsubstituted 5-aminothiatriazole gives isothiocyanate ion ⟨64AHC(3)263, p. 280⟩. The same reaction takes place to a minor extent with arylaminothiatriazoles ⟨64AHC(3)263⟩ but they generally undergo isomerization to 5-mercaptotetrazoles (Dimroth rearrangement, see Section 4.02.3.5.1). This reaction is reversible ⟨76AHC(20)145, p. 167⟩.

$$\text{RNH-}\underset{S}{\underset{|}{\overset{N-N}{\diagdown}}}\text{N} \xrightarrow{OH^-} RNCS + N_3^- \tag{25}$$

On treatment of 5-amino- and 5-alkylamino-thiatriazoles with concentrated hydrochloric acid, formamidine disulfide dihydrochlorides (**54**) are formed, along with 50% hydrolysis leading to hydrazoic acid, carbon dioxide and hydrogen sulfide. It is postulated that hydrazoic acid reduces the positive ion presumably formed by ring cleavage of the thiatriazole in the presence of the strong acid (equation 26) ⟨64AHC(3)263⟩.

$$\text{RNH-}\underset{S}{\underset{|}{\overset{N-N}{\diagdown}}}\text{N} \xrightarrow[-N_2]{H^+} \underset{NH}{\overset{RNHC-\ddot{S}:}{\underset{+}{\|}}} \xrightarrow{HN_3} \underset{+NH_2}{\overset{(RNHC-S)_2 + N_2}{\|}} \tag{26}$$

(**54**)

The reaction takes a different course with 5-phenylamino-1,2,3,4-thiatriazole, affording 2-aminobenzothiazole by electrophilic attack of the sulfenium ion on the aryl group (equation 27) ⟨64AHC(3)263⟩.

$$\text{PhNH-}\underset{S}{\underset{|}{\overset{N-N}{\diagdown}}}\text{N} \xrightarrow{H^+} \underset{PhN}{\overset{+S}{\diagdown}}C-NH_2 \rightarrow \text{[benzothiazole]}-NH_2 + H^+ \tag{27}$$

(iii) Acylation and reaction with isocyanates of 5-aminothiatriazoles with ring cleavage: heteropentalenes

Beer and Hart ⟨77CC143⟩ and the group of L'abbé ⟨77AG(89)420⟩ observed independently the interesting fact that 1,6-oxa-6aλ^4-thia-3,4-diazapentalenes (**57**) are formed by aroylation or acylation, in the presence of base, of the easily accessible 5-aminothiatriazole. Further studies ⟨81BSB89⟩ led to a mechanism involving formation of a reactive inner salt (**55**) as the key step (equation 28).

$$H_2N-\underset{S}{\underset{|}{\overset{N-N}{\diagdown}}}\text{N} \xrightarrow[-N_2, -HCl]{RCOCl, \text{ base}} \left[\underset{O}{\overset{R}{\underset{\|}{C}}}\underset{S^+}{\overset{N\cdots NH}{\underset{|}{C}}}\right] \rightarrow R\underset{O-S}{\overset{N\diagup NH}{\diagdown}} \xrightarrow[-N_2, -HCl]{RCOCl, \text{ base}} R\underset{O-S-O}{\overset{N-N}{\diagdown}}R \tag{28}$$

(**55**)　　　　　(**56**)　　　　　(**57**)

R = alkyl, aryl

When R = Me, a mixture of the pentalene and of 1,2,4-thiadiazole (**58**) was obtained (equation 29). In the absence of base, HCl was liberated which caused decomposition of 5-aminothiatriazole to formamidine disulfide (equation 26).

$$H_2N-\underset{S}{\underset{|}{\overset{N-N}{\diagdown}}}\text{N} \xrightarrow[-N_2, -HCl]{2\,RCOCl, \text{ base}} \left[\underset{O}{\overset{R}{\underset{\|}{C}}}\underset{S^+}{\overset{N\cdots NH}{\underset{|}{C}}}\right] \xrightarrow{-S} \left[\underset{O}{\overset{RC-NHCN}{\|}}\right] \xrightarrow{(55)} RCONH\underset{N-S}{\overset{N}{\diagdown}}NHCOR \tag{29}$$

(**55**)　　　　　　　　　　　(**58**)

Likewise, thiobenzoyl chloride gives 3,4-diaza-1,6,6aλ^4-trithiapentalene (**59**) in low yield (11%), which may also be prepared from the above diazapentalene and P_4S_{10} ⟨81BSB89⟩.

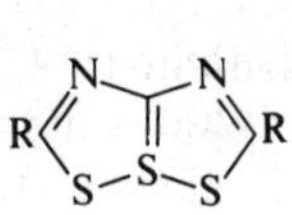

(59)

5-Imino-Δ^3-1,2,4-thiazolidines (61) are obtained (77–87% yield) when equimolar amounts of imidoyl chlorides (60) react with 5-aminothiatriazole in the absence of base (equation 30). The free bases may be liberated by treatement with $NaHCO_3$ and these react with a second mole of imidoyl chloride [$R^2C(=NR^3)Cl$] to give a 1,3,4,6-tetraaza-6aλ^4-thiapentalene (62), or with dibenzylcarbodiimide to give the corresponding amino-substituted pentalene (R^2 = NHBz, R^3 = Bz). It is not practical to treat the aminothiatriazole directly with two moles of imidoyl chloride as a mixture of (61) and (62) is formed ⟨81BSB89⟩. Yamamoto and Akiba have extended this method with other examples ⟨79H(13)297⟩.

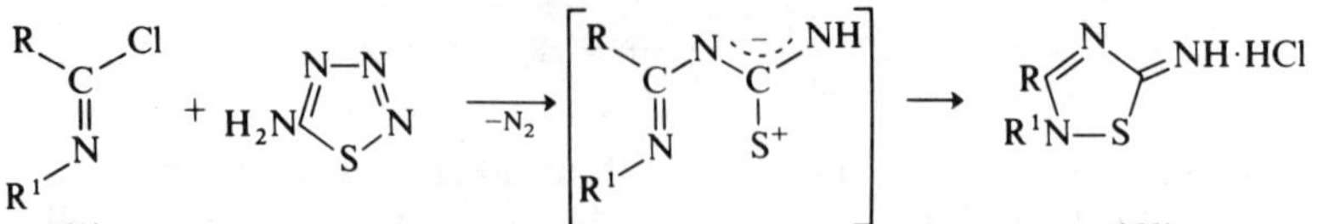

(60) (61) (30)

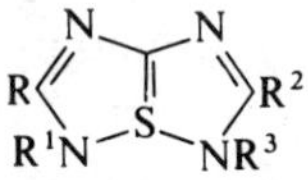

(62)

Kaugars and Rizzo have found that 5-alkylamino- or 5-arylamino-1,2,3,4-thiatriazoles react with isocyanates to give 3-oxo-Δ^4-1,2,4-thiadiazolin-5-ylureas (65) ⟨79JOC3840⟩. The structures were verified by independent synthesis and by ^1H and ^{13}C NMR spectroscopy. In the presence of triethylamine the exothermic reaction went to completion in a few hours. The reaction may either take place by attack at the amino group to give (63), followed by N_2 extrusion to form the reactive dipole (64; equation 31), or the reaction may be initiated by attack at the ring in the 4-position to give an intermediate thiatriazoline, which, as discussed above, reacts with heterocumulenes (see Section 4.28.2.3.1(iii)). As expected, 5-(dialkylamino)thiatriazoles were not found to react with isocyanates.

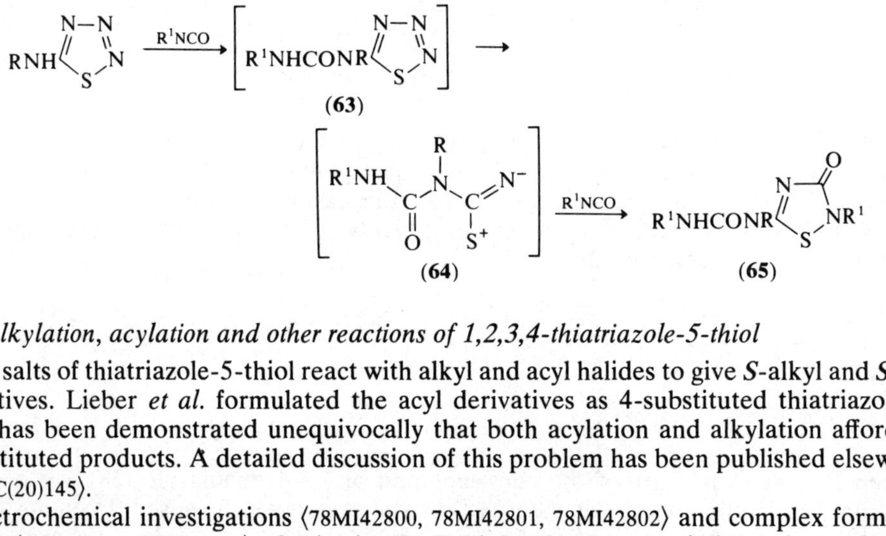

(63)

(64) (65) (31)

(iv) Alkylation, acylation and other reactions of 1,2,3,4-thiatriazole-5-thiol

The salts of thiatriazole-5-thiol react with alkyl and acyl halides to give S-alkyl and S-acyl derivatives. Lieber et al. formulated the acyl derivatives as 4-substituted thiatriazolines, but it has been demonstrated unequivocally that both acylation and alkylation afford the 5-substituted products. A detailed discussion of this problem has been published elsewhere ⟨76AHC(20)145⟩.

Electrochemical investigations ⟨78MI42800, 78MI42801, 78MI42802⟩ and complex formation studies ⟨80MI42800, 75MI42800⟩ of thiatriazole-5-thiol have been carried out, but a detailed discussion of the results is beyond the scope of this chapter. For oxidation and further reactions of thiatriazole-5-thiol see Sections 4.28.2.5.1 and 4.28.2.6.

4.28.2.4.2 *3,5-Disubstituted 1,2,3,4-oxa- and thia-triazolium salts*

The alkoxy group in 5-ethoxy-3-phenylthiatriazolium tetrafluoroborate undergoes facile substitution reactions with certain nucleophiles to form mesoionic compounds (see Section

4.28.3.6.4). In contrast, 3-ethyl-5-phenylthiatriazolium tetrafluoroborate undergoes dealkylation on treatment with nucleophiles (equation 32) ⟨75JOC431⟩.

$$\text{Ph}\underset{S}{\overset{N-NEt}{\diagdown N}}^+ BF_4^- \xrightarrow{OH^-} \text{Ph}\underset{S}{\overset{N-N}{\diagdown N}} + EtOH + BF_4^- \qquad (32)$$

3-Ethoxy-5-phenylthiatriazolium tetrafluoroborate similarly undergoes reaction at the 3-substituent. In aqueous solution it is thus rapidly transferred into a mixture of thiatriazole and the 3-oxide (equation 33) ⟨75JOC431⟩.

$$\text{Ph}\underset{S}{\overset{N-N}{\diagdown N}} + MeCHO + H^+ \xleftarrow{H_2O} \underset{(82)}{\text{Ph}\underset{S}{\overset{N-NOEt}{\diagdown N}}^+ BF_4^-} \xrightarrow{H_2O} \text{Ph}\underset{S}{\overset{N-N}{\diagdown N}}^+\!\!\!{}^{O^-} + EtOH + H^+ \qquad (33)$$

4.28.2.5 Syntheses

4.28.2.5.1 1,2,3,4-Thiatriazoles

(i) 1,2,3,4-Thiatriazole substituted with alkyl and aryl groups and with chlorine

Thiatriazoles with an aromatic or heteroaromatic substituent in the 5-position are quite stable and may be prepared from thiohydrazides and nitrous acid or by the reaction of aromatic dithioesters with sodium azide (equation 34) ⟨64AHC(3)263⟩. The highly unstable 5-*t*-butyl- ⟨61ACS1104⟩ and cyclohexyl-thiatriazoles ⟨61JOC5221⟩ are prepared from the corresponding thiohydrazides and nitrous acid but the method is not generally applicable because of difficulties in obtaining the required aliphatic thioacyl hydrazides. The alkylthiatriazoles cannot be isolated in a pure state as they decompose rapidly at 0 °C. Thiatriazoles cannot be obtained from aliphatic dithioacids and sodium azide. A compound isolated from the reaction between 1-acetylthio-1-hexyne and ammonium azide (equation 35) is believed to be 5-pentylthiatriazole. It is an oil that solidifies at about −16 °C and could not be analyzed because of its explosive character and poor stability at room temperature. The structure is partially deduced from its spectra and thermal properties (see Section 4.28.2.3.1(i)) ⟨67RTC670⟩. Raasch prepared 5-[2,2,2-trifluoro-1-(trifluoromethyl)ethyl]thiatriazole (**66**) from bis(trifluoromethyl)thioketene and hydrazoic acid (equation 36). It is remarkable by being a 5-alkylthiatriazole stable at room temperature ⟨70JOC3470⟩.

$$RCSNHNH_2 + HNO_2 \longrightarrow R\underset{S}{\overset{N-N}{\diagdown N}} \longleftarrow RCSSR^1 + NaN_3 \qquad (34)$$

$$BuC\equiv CSCOMe \xrightarrow{NH_3} BuC\equiv CSH + MeCONH_2$$

$$BuC\equiv CSH \rightleftarrows BuCH=C=S \xrightarrow{HN_3} C_5H_{11}CSN_3 \longrightarrow H_{11}C_5\underset{S}{\overset{N-N}{\diagdown N}} \qquad (35)$$

$$(CF_3)_2C=C=S \xrightarrow{HN_3} (CF_3)_2CH\underset{S}{\overset{N-N}{\diagdown N}} \qquad (36)$$
$$(\mathbf{66})$$

Aralkylthiatriazoles are slightly more stable than the alkyl derivatives and 5-benzylthiatriazole was obtained in a crystalline state from phenylthioacethydrazide and nitrous acid ⟨60CB2353⟩. A compilation of references to 5-alkyl-, aryl- and heteroaryl-1,2,3,4-thiatriazoles has been published ⟨64AHC(3)263⟩.

The highly explosive 5-chloro-1,2,3,4-thiatriazole is prepared from thiophosgene and sodium azide (94% yield) (equation 37). It has been used for the preparation of 5-substituted

$$CSCl_2 + N_3^- \longrightarrow Cl\underset{S}{\overset{N-N}{\diagdown N}} \qquad (37)$$
$$(\mathbf{67})$$

1,2,3,4-aminothiatriazoles. The authors stress that *great care* should be exercised in handling *any reaction product* of thiophosgene and azide ion ⟨61JOC1644⟩.

(ii) 5-Amino-1,2,3,4-thiatriazoles

5-Amino- and 5-substituted amino-1,2,3,4-thiatriazoles are obtained from the reaction of thiosemicarbazides with nitrous acid (equation 38). In some cases this method fails for the disubstituted aminothiatriazoles but they can be prepared from the corresponding thiocarbamoyl chloride and sodium azide. A few mono- and di-substituted aminothiatriazoles have been synthesized by reacting 5-chloro-1,2,3,4-thiatriazole (67) with an amine ⟨61JOC1644⟩. Some selected aminothiatriazoles, prepared according to these methods, have been listed ⟨64AHC(3)263⟩. A number of benzylaminothiatriazoles have been obtained by the nitrosation method ⟨71JIC843⟩. Isothiocyanates and hydrazoic acid react to give 5-alkyl- or 5-aryl-aminothiatriazoles (equation 39) ⟨64AHC(3)263⟩ but this reaction is less convenient and less efficient than the reaction of thiosemicarbazides and nitrous acid. The method has, however, been used recently to prepare some 5-(arylsulfonylamino)thiatriazoles from sulfonylisothiocyanates ⟨77CCC1557⟩, 5-(1-imidazolyl)thiatriazoles from thiocarbonyldiimidazoles ⟨78T453⟩ and 2-, 3- and 4-pyridylaminothiatriazoles ⟨80CCC2329⟩. The isomeric mercaptotetrazoles are formed when sodium azide is used instead of the free acid ⟨64AHC(3)263, p. 280⟩.

$$RR^1NCSNHNH_2 + HNO_2 \longrightarrow RR^1N\underset{S}{\overset{N-N}{\underset{\diagdown}{\diagup}}}N + 2H_2O \qquad (38)$$

$$RN=C=S + HN_3 \longrightarrow RNH\underset{S}{\overset{N-N}{\underset{\diagdown}{\diagup}}}N \qquad (39)$$

Trimethylsilyl azide (TMSA) also reacts with isothiocyanates to give thiatriazoles ⟨77CCC2945⟩. The yields with this reagent are approximately the same as with hydrazoic acid but the reaction is much faster (~2 days) and working with the more unpleasant hydrazoic acid is avoided (equation 40). The first step in this reaction is claimed to be the cycloaddition of trimethylsilyl azide to isothiocyanate leading to a 1:1 cycloadduct which in aqueous methanolic medium decomposes into the thiatriazole and hexamethyldisiloxane. A much lower reactivity between isothiocyanate and TMSA in dry benzene is reported ⟨80JOC5130⟩, stressing the need for a hydroxylic solvent in the thiatriazole forming process. Trimethylsilyl azide can be prepared *in situ*. Thus 5-phenylaminothiatriazole was obtained in a single step from chlorotrimethylsilane, sodium azide and phenyl isothiocyanate in DMF (83%) ⟨79S35⟩.

$$RN=C=S + Me_3SiN_3 \xrightarrow{EtOH, H_2O} RNH\underset{S}{\overset{N-N}{\underset{\diagdown}{\diagup}}}N \qquad (40)$$

Alkylidenaminothiatriazoles have been prepared from NaN_3 and α-bromoisothiocyanates ⟨79CB1102, 79CB1956⟩.

It may be mentioned that bis(imino)thiazetidines (68), which are prepared by the reaction of carbodiimides with isothiocyanates, react with hydrazoic acid to give an aminothiatriazole. Although this reaction may be of interest from other points of view, it has no general synthetic application (equation 41) ⟨81BSB63⟩.

$$\underset{(68)}{\overset{NBu^t}{\underset{NTos}{Me N\diagup S}}} \xrightarrow{HN_3} MeNH\underset{N}{\overset{N-N}{\underset{\diagdown}{\diagup}}}N\text{-}Bu^t + TosNH\underset{S}{\overset{N-N}{\underset{\diagdown}{\diagup}}}N \qquad (41)$$

An alternative method for the preparation of 5-aminothiatriazoles by way of an aza transfer reaction has been reported ⟨78OPP59⟩. Benzenediazonium tetrafluoroborate or diazotized sulfanilic acid react rapidly at room temperature with thiosemicarbazides to give substituted aminothiatriazoles (equation 42). The procedure may be of importance with sensitive substrates since nitrous acid is avoided and the reaction proceeds under almost neutral conditions.

$$\text{RNHCSNHNH}_2 + \text{ArN}_2^+ \xrightarrow[-\text{ArNH}_3^+]{} \left[\text{RNHC}(=\text{S})-\text{N}_3 \right] \longrightarrow \text{RNH-[thiatriazole]} \quad (42)$$

(iii) 1,2,3,4-Thiatriazole-5-thiol and its derivatives

Thiatriazole-5-thiol is a fairly strong acid and its salts are readily obtained from the water-soluble alkali and alkaline earth azides with CS_2 at 40 °C (equation 43) ⟨64AHC(3)263⟩. These should be handled with care. The potassium salt may explode violently when spread on a porous plate and the slightly soluble heavy-metal salts are very sensitive to shock even under water. An improved method for the preparation and storage of sodium thiatriazole-5-thiolate has been reported ⟨74MI42800⟩. 1,2,3,4-Thiatriazole-5-thiol is prepared by addition of concentrated hydrochloric acid to a chilled solution of the sodium salt, obtained as a white or slightly yellow crystalline compound. The free acid can also be prepared from hydrazoic acid and carbon disulfide (equation 43) ⟨64AHC(3)263⟩.

$$N_3^- + CS_2 \longrightarrow {}^-S\text{-[thiatriazole]} \xrightarrow{H^+} HS\text{-[thiatriazole]} \longleftarrow HN_3 + CS_2 \quad (43)$$

The salts of thiatriazole-5-thiol react with alkyl and acyl halides to give *S*-alkyl and *S*-acyl derivatives. A table with references to 1,2,3,4-thiatriazolethiols was published in the 1964 review ⟨64AHC(3)263⟩. A number of 5-(acylthio)thiatriazoles has been prepared by Christophersen ⟨71ACS1160, 71ACS1162, 76T745⟩. Nitrosation of a thiohydrazide can also be used for the preparation of mercaptothiatriazoles (*e.g.* equation 44) ⟨71ACS2015, 71JIC843⟩.

$$\text{Ph}_2\text{CHS} - \underset{\underset{S}{\|}}{C} - \text{NHNH}_2 \xrightarrow{\text{HNO}_2} \text{Ph}_2\text{CHS-[thiatriazole]} \quad (44)$$

Formation of *S*-substituted 5-mercaptothiatriazoles from trithiocarbonate *S,S*-dioxides ⟨71CC314⟩ or alkylthio- or arylthio-thiocarbonyl *p*-toluenesulfonyl disulfides reacting with azide ion has been reported ⟨79CC1135⟩ but these methods have not found any synthetic application.

The free acid and its salts are transformed into bis(5-thiatriazolyl) disulfide (**69**) by various oxidizing agents. Chemical and physical properties of this highly explosive solid have been reviewed elsewhere ⟨64AHC(3)263⟩.

[thiatriazole]–S–S–[thiatriazole]   NCS–[thiatriazole]
(**69**)   (**70**)

5-Thiocyanato-1,2,3,4-thiatriazole (**70**) is formed from the potassium salt of the thiol and cyanogen bromide. The spectral properties are in agreement with the structural assignment and, on heating, $S(CN)_2$ is formed as expected ⟨64AHC(3)263⟩.

Extremely shock-sensitive alkylene bis(5-thiatriazolyl) sulfides were prepared by Pilgram *et al.* by reacting sodium thiatriazole-5-thiolate with dihaloalkanes ⟨65AG348, 71JHC899⟩. Dichloromaleimide reacted in an analogous manner ⟨71JHC899⟩.

The reaction of carbon disulfide with tributyltin azide and with $[\pi\text{-}C_5H_5\text{Ni}(PBu_3)_2]^+N_3^-$ to give thiatriazoles or thiatriazolinethiones has been discussed ⟨76AHC(20)145, p. 165⟩. 5-Mercaptothiatriazole complexes are believed to be formed by reaction of CS_2 with azides of Rh, Pd, Pt, Co, Cu, Ag and Au. They may be thermally or photochemically converted into thiocyanato complexes ⟨79JCS(D)371⟩.

(iv) 5-Alkoxy- and 5-aryloxy-1,2,3,4-thiatriazoles

A number of 5-alkoxy, 5-aryloxy and 5-aralkoxythiatriazoles have been prepared ⟨76AHC(20)145, p. 166⟩ from alkoxythiocarbonylhydrazines and nitrous acid ⟨64ACS825, 65ACS438, 66ACS2107, 69ACS1567, 70ACS1512⟩, or from *O*-alkyl or *O*-aryl chlorothioformates and sodium azide (equation 45) ⟨64AG303, 64TL2829, 65CB2059, 65CB2689, 65CB2063, 65CR(260)2839, 67BSF422, 71JIC843⟩.

$$\text{ROCSCH}_2\text{CO}_2\text{H} \xrightarrow{\text{NH}_2\text{NH}_2} \text{ROCNHNH}_2 \xrightarrow{\text{HNO}_2} \text{RO-[thiatriazole]} \xleftarrow{N_3^-} \text{ROCCl} \quad (45)$$
(with C=S groups)

4.28.2.5.2 Δ^2-1,2,3,4-Thiatriazolines

Neidlein and Tauber found that 5-(arylamino)thiatriazoles can be alkylated to give either 5-(alkylarylamino)thiatriazoles and/or 4-alkyl-5-(arylimino)thiatriazolines depending on the alkylating agent and the aryl group. With dimethyl sulfate in alkaline solution, all compounds investigated gave 5-substituted thiatriazoles, while alkylation with diazomethane gave either the iminothiatriazoline or a mixture of the thiatriazoline and the thiatriazole (equation 46). 5-Phenylamino- and 5-(4-methylphenylamino)-thiatriazoles gave exclusively compound (71) with diazomethane, whereas 4-chloro- and 3-chloro-phenylthiatriazoles gave a mixture of compounds (71) and (72) ⟨71AP687, 77JHC1417⟩.

$$XC_6H_4N\underset{(71)}{\overset{MeN-N}{\underset{S}{\bigvee}}}N \xleftarrow{CH_2N_2} XC_6H_4NH\underset{S}{\overset{N-N}{\bigvee}}N \xrightarrow[\text{or } CH_2N_2]{(MeO)_2SO_2} XC_6H_4N\underset{Me\ S}{\overset{N-N}{\bigvee}}N \quad (46)$$
(72)

4,5-Disubstituted iminothiatriazolines may be formed by reaction of alkyl azides and aryl isothiocyanates but the reaction is relatively slow and the thiatriazoline formed undergoes further reaction with the isothiocyanate present (Section 4.28.2.3.1(iii)). 4-Alkyl-5-sulfonyliminothiatriazolines (73) are readily obtained when the more reactive sulfonyl isothiocyanates are treated with alkyl azides at room temperature (equation 47) ⟨73JOC2916, 74JA3973⟩.

$$RN_3 + 4\text{-}XC_6H_4SO_2NCS \rightarrow 4\text{-}XC_6H_4SO_2N\underset{S}{\overset{RN-N}{\bigvee}}N \quad (47)$$
(73)

R, X (yield %, m.p. °C): Bu, H (49, 60–61); Bu, Me (59, 74–75 dec.); Bu, Cl (76, 102–4 dec.); Bz, H (—, oil); Bz, Me (70, 101–103 dec.); Bz, Cl (62, 116–118)

Both types of thiatriazoline mentioned above and 4,5-dialkyliminothiatriazolines may be prepared in moderate to high yield by alkylation of 5-alkylamino-, 5-arylamino- and 5-sulfonylamino-thiatriazoles with trialkyloxonium tetrafluoroborates (equation 48). Thiatriazolines are formed exclusively with this reagent ⟨78JCS(P1)1440⟩. The thiatriazoline tetrafluoroborates are crystalline solids which may be stored at room temperature and serve as convenient sources for the thermally labile free bases (75). These may be liberated with aqueous $NaHCO_3$ ⟨78JCS(P1)1440⟩.

$$RNH\underset{S}{\overset{N-N}{\bigvee}}N \xrightarrow{R^1_3O\ BF_4^-} RN\underset{S}{\overset{R^1N-N}{\bigvee}}N\cdot HBF_4 \xrightarrow{NaHCO_3} RN\underset{S}{\overset{R^1N-N}{\bigvee}}N \quad (48)$$
(74) (75)

R, R¹ (yield %, m.p. °C) (74): Me, Me (56, 115); Et, Me (73, 133); Me, Et (29, 87–88); Et, Et (38, 142); Ph, Me (100, 162); Ph, Et (91, 164); (75): 4-$MeC_6H_4SO_2$, Me (57, 117–118)

Thiatriazolines are reported as possible transient intermediates in reactions of azide ion with thiobenzophenone (equation 49a) and thiobenzophenone S-oxides (equation 49b) ⟨76ACS(B)997⟩. 1,3-Dithietane-2,4-diylidene bis(cyanoacetic ester) (76) reacts with sodium azide to give a compound assumed to be the sodium salt of alkylidene-1,2,3,4-thiatriazolidine (77). Nitrogen is evolved on acidification of the sodium salt and a hitherto unknown perhydro-1,4,2,5-dithiadiazine (78) is formed (equation 49c) ⟨81ZC102⟩.

$$Ph_2C=S + N_3^- \rightarrow \left[\underset{Ph}{\overset{Ph}{\bigvee}}\underset{S}{\overset{N=N}{\bigvee}}N\right] \xrightarrow[-N_2,-S]{H^+} Ph_2C=NH \quad (49a)$$

$$Ar_2C=\overset{O^-}{\underset{\parallel}{S}} + N_3^- \rightarrow \left[\underset{Ar}{\overset{Ar}{\bigvee}}\underset{\underset{\parallel}{S}}{\overset{N=N}{\bigvee}}N\right] \rightarrow Ar_2C=N_2 + NSO^- \quad (49b)$$

$$\underset{NC}{\overset{RO_2C}{\bigvee}}C=C\underset{S}{\overset{S}{\bigvee}}C=C\underset{CO_2R}{\overset{CN}{\bigvee}} + N_3^- \rightarrow \left[\underset{NC}{\overset{RO_2C}{\bigvee}}C=C\underset{S}{\overset{N=N}{\bigvee}}N\right] \xrightarrow{H^+} \underset{NC}{\overset{RO_2C}{\bigvee}}C=C\underset{S-N}{\overset{H}{\underset{H}{\bigvee}}}C=C\underset{CO_2R}{\overset{CN}{\bigvee}} \quad (49c)$$
(76) (77) (78)

4.28.2.5.3 Δ^3-1,2,3,4-Thiatriazolines

1:1 Adducts are formed from bis(trifluoromethyl)thioketene and aryl azides ⟨78JOC2500⟩. The yellow, crystalline products are believed to be the so far unknown Δ^3-1,2,3,4-thiatriazolines (**23**; equation 50), as indicated by ^{19}F and IR spectra and the formation of 2,1-benzisothiazole on thermolysis (Section 4.28.2.3.1(i)).

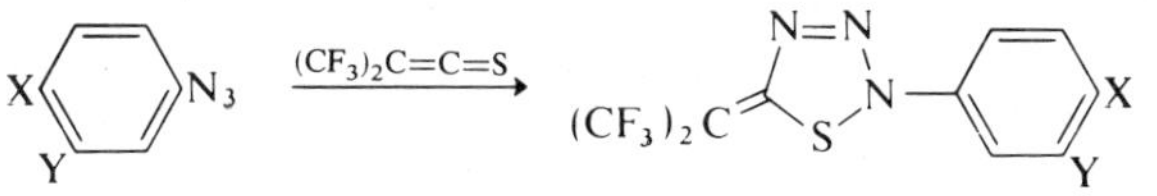

(50)

(**23**)

X	H	Me	OMe	Cl	Cl	2,4,6 Me$_3$C$_6$H$_2$
Y	H	H	H	H	Cl	
Yield (%)	27	35	57	14	20	41
m.p. (°C)	125–125.5	124.5–125	113–113.7	119–119.5	96.5–97	126.7–127.2

Bis(trifluoromethyl)thioketene and aryl azides form orange 2:1 adducts, which can also be obtained from the thiatriazolines and one equivalent of bis(trifluoromethyl)thioketene. The structure of these adducts is unknown.

4.28.2.5.4 3,5-Disubstituted 1,2,3,4-oxa- and -thia-triazolium salts

Alkylation of mesoionic 3-aryl-1,2,3,4-thiatriazol-5-ones (**79**; X = S, Y = O, R = Ph, 4-MeC$_6$H$_4$, 4-ClC$_6$H$_4$) and of mesoionic 3-phenyl-1,2,3,4-oxatriazole-5-thione (**79**; X = O, Y = S, R = Ph) with triethyloxonium tetrafluoroborate yields thiatriazolium salts (**80**; equation 51) ⟨79JCS(P1)732⟩.

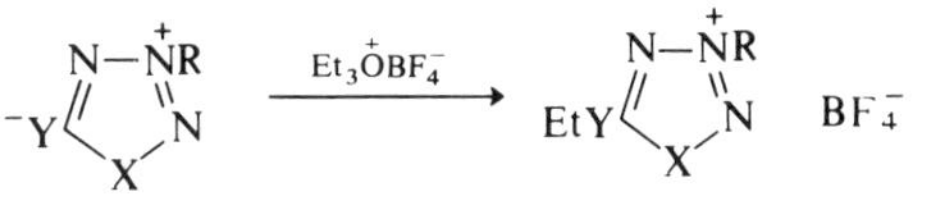

(51)

(**79**)   (**80**)

Alkylation with triethyloxonium tetrafluoroborate of a 1,2,3,4-thiatriazole substituted in the 5-position with an aryl or an alkylthio group gives a thiatriazolium salt (**81**; equation 52) ⟨76AHC(20)145⟩. With an alkyl- or aryl-amino group in the 5-position, however, a 4,5-disubstituted thiatriazoline is formed (see Section 4.28.2.5.2). Under similar alkylation conditions 5-ethoxythiatriazoles decompose completely to nitrogen, sulfur and ethyl cyanate. This is the normal thermal decomposition reaction of 5-alkoxythiatriazoles (see Section 4.28.2.3.1(i)) ⟨76AHC(20)145⟩.

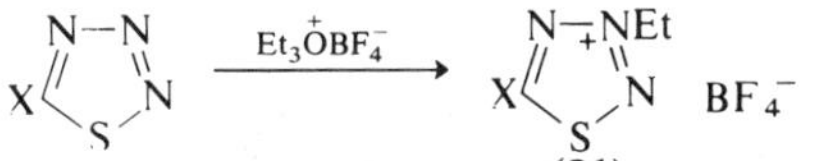

(52)

X = SEt, SMe, Ph   (**81**)

A thiatriazolium salt (**82**) is formed on alkylation of 5-phenyl-1,2,3,4-thiatriazole 3-oxide (equation 53) ⟨75T1783⟩.

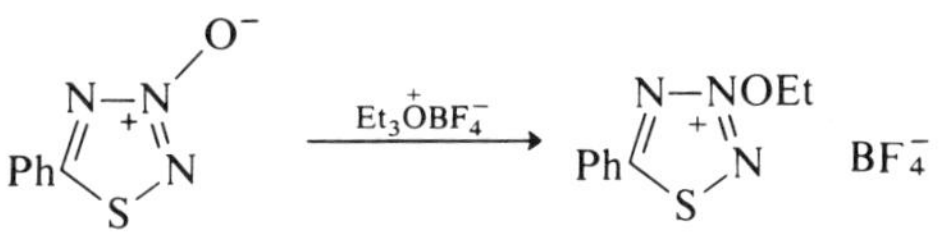

(53)

(**82**)

4.28.2.5.5 5-Amino-1,2,3,4-selenatriazoles

Treatment of thiohydrazides with nitrous acid leads to formation of thiatriazoles. 5-Amino-1,2,3,4-selenatriazoles may be prepared in the same manner from the corresponding selenosemicarbazides (**83**; R^1 = R^2 = Et, ca. 60%) but the reaction is accompanied by the formation of elemental selenium. A comparable yield is achieved by use of an aza transfer reagent (55–70%) (equation 54). The best yield and purest material are obtained from the reaction between the azide ion and bis(*N,N*-dialkylselenocarbamoyl) selenide (**84**; 80% with $R^1 = R^2 = Et$; equation 55) ⟨82UP42800⟩.

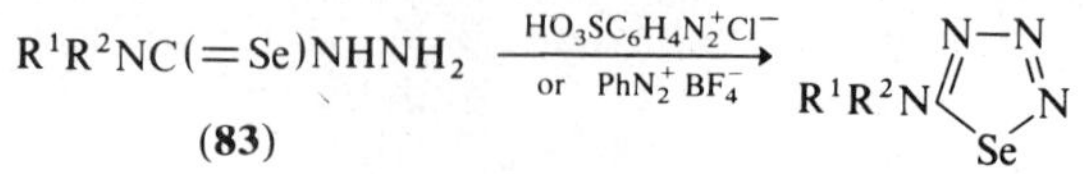

$$R^1R^2NC(=Se)NHNH_2 \xrightarrow[\text{or } PhN_2^+ BF_4^-]{HO_3SC_6H_4N_2^+Cl^-} R^1R^2N\underset{Se}{\overset{N-N}{\diagdown\diagup}}N \qquad (54)$$
$$(83)$$

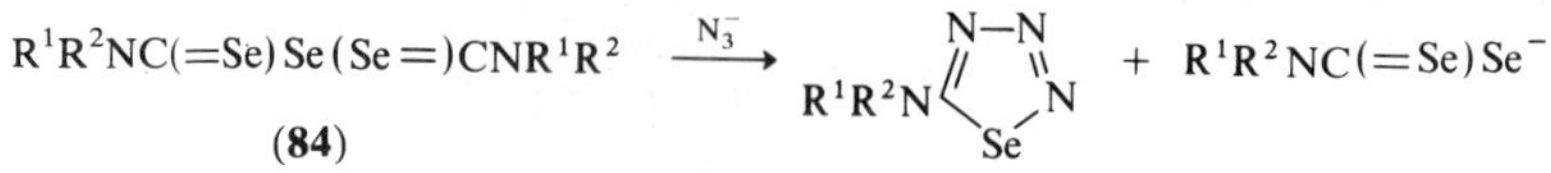

$$R^1R^2NC(=Se)Se(Se=)CNR^1R^2 \xrightarrow{N_3^-} R^1R^2N\underset{Se}{\overset{N-N}{\diagdown\diagup}}N + R^1R^2NC(=Se)Se^- \qquad (55)$$
$$(84)$$

4.28.2.6 Applications

4.28.2.6.1 Analytical uses

The reaction between CS_2 and the azide ion to give 1,2,3,4-thiatriazole-5-thiolate ($CS_2N_3^-$) has been used to develop a spectrophotometric method for the quantitative determination of CS_2 ⟨77MI42800, 79MI42803⟩ as well as azide ion ⟨77MI42801⟩. The thiatriazole-5-thiolate concentration may also be estimated quantitatively by permanganate titration ⟨79MI42800⟩. The oxidation of $CS_2N_3^-$ to $(CS_2N_3)_2$ with iodine (CS_2-catalyzed iodine–azide reaction ⟨64AHC(3)263⟩) has been elaborated into a method for the quantitative determination of CS_2 ⟨67NI42800⟩.

Sodium thiatriazole-5-thiolate gives colored precipitates with copper(II) or bismuth salts which may be used as a test for soluble inorganic azides by way of their reaction with CS_2 ⟨72MI42800⟩.

4.28.2.6.2 Miscellaneous

A series of 4-pyridinylacetamidocephalosporins attached to heterocyclic systems including the thiatriazole ring has been prepared and investigated for antibacterial activity ⟨79AF362⟩. Eight new and seven known 5-arylaminothiatriazoles have been evaluated for their anticancer activity ⟨79AF728⟩. 5-Aminothiatriazoles have been investigated as chemosterilants ⟨76MI42800⟩ and fungicides ⟨76ABC759⟩ and patented for use as lubricants, additives to paints, *etc.*

4.28.3 MESOIONIC 1,2,3,4-OXATRIAZOLES AND -THIATRIAZOLES

4.28.3.1 Introduction

The known systems comprise mesoionic 1,2,3,4-oxatriazoles and -thiatriazoles (see Chapter 4.01, Scheme 5) of the structures (**85**)–(**91**). Preparative aspects of the chemistry of five-membered mesoionic heterocycles have been extensively reviewed by Ollis and Ramsden covering the literature until 1974 ⟨76AHC(19)1⟩. The review contains a detailed discussion of the mesoionic concept and the representation of mesoionic heterocycles and other systems. Since then a number of investigations of the chemistry of mesoionic 1,2,3,4-oxatriazoles and -thiatriazoles have been carried out, in particular by the authors mentioned and their coworkers. Mesoionic thiatriazoles were first described in 1976.

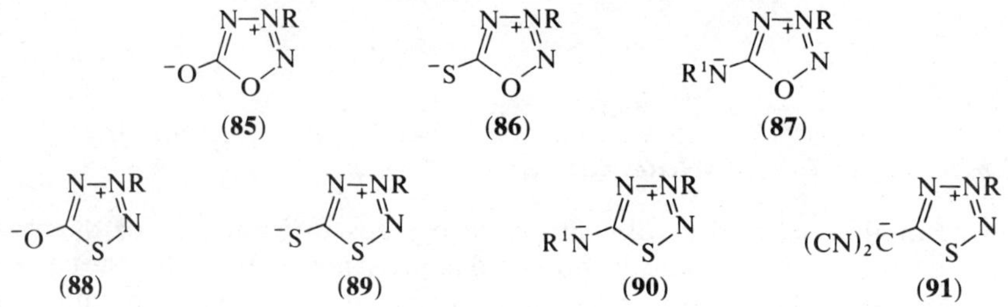

Scheme 3

Heterocycle (**85**) may be systematically described as 3-substituted anhydro-5-hydroxy-1,2,3,4-oxatriazolium hydroxide. However, this nomenclature is somewhat cumbersome and the terminology mesoionic has been used throughout this review. Thus compounds (**85**)–(**87**) and (**91**) are named mesoionic 1,2,3,4-oxatriazol-5-ones (**85**), -5-thiones (**86**), -5-imines (**87**) and mesoionic 1,2,3,4-thiatriazole-5-alkenes (**91**), respectively. Using IUPAC Rule C-87 (**85**) is named 3-substituted 1,2,3,4-oxatriazol-5-ylio oxide or 3-substituted 1,2,3,4-oxathiazolylium-5-olate and (**87**; R = R^1 = Ph) is named N-[3-phenyl-5-(1,2,3,4-oxatriazolio)]anilide.

4.28.3.2 Structure

The formulation of a mesoionic structure for these types of compounds is strongly supported by spectral data, dipole moments and X-ray crystallography.

4.28.3.2.1 Theoretical methods

MO calculations by Sundaram and Purcell ⟨68MI42800⟩ give a value of 5.3 D for the mesoionic 1,2,3,4-oxatriazol-5-one ring moment. This may be compared with the dipole moment of the N-phenyl derivative measured to be 6.1 D (**85**; R = Ph) in benzene solution (see below). Further discussions of these MO calculations do not seem justified considering the development in this area since the work was published.

4.28.3.2.2 Structural methods

The structures of mesoionic 3-phenyl-5-phenylimino-1,2,3,4-oxatriazole imine (**92**) ⟨75ACS(A)45⟩ and mesoionic 3-phenyl-1,2,3,4-oxatriazol-5-one (**93**) ⟨75ACS(A)799⟩ have been investigated by X-ray diffraction methods. The results obtained show the same structural features as those found for the isoelectronic sydnones ⟨76AHC(19)1, p. 90⟩. Thus the central five-membered rings are planar, the ring CO bond length is relatively long (1.42–1.43 Å), the exocyclic CN bond length (1.271 Å) in (**92**) is a 'pure' double bond, and the exocyclic CO bond length in (**93**) is a 'pure' double bond. The two N-phenyl bond lengths in (**92**) (1.437 Å and 1.402 Å, respectively) indicate that there is only small, if any, coupling between the conjugation in the two phenyl rings and in the mesoionic ring.

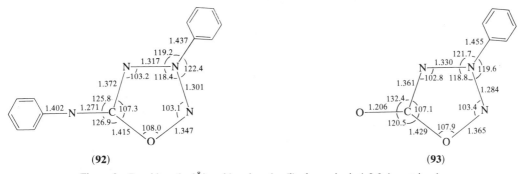

Figure 2 Bond lengths (Å) and bond angles (°) of mesoionic 1,2,3,4-oxatriazoles

The electronic dipole moments of mesoionic compounds of structure (**94**) depend on the heteroatoms or groups X and Y. This relationship has been discussed by Ollis *et al.* ⟨78JCS(P1)600, 79JCS(P1)736⟩ on the basis of a number of measured values for mesoionic oxa- and thia-triazoles as well as oxa- and thia-diazoles. The largest moments are associated with an exocyclic dicyanomethylene group (**91**; R = Ph; 8.8 D) and Y = S (**86**; R = aryl, 5.1–7.9 D). Replacement of the exocyclic sulfur atom by oxygen (**85**; R = Ph, 6.1 D) and (**88**; R = aryl, 4.4–6.1 D) results in a reduction of the magnitude of the dipole moment; replacement by a nitrogen atom (**87**; R and R^1 = aryl, 4.2–6.3 D) and (**90**; R^1 = R = Ph, 3.7 D) gives a further lowering of the dipole moment. The relatively small dipole moment in the last case is still consistent with the mesoionic formulation.

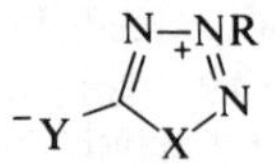

(94)

4.28.3.2.3 Molecular spectra

(i) NMR spectra

^{1}H NMR chemical shifts for substituents in mesoionic oxatriazole-5-thiones (**86**) ⟨79JCS(P1)732⟩, oxatriazol-5-imines (**87**) ⟨79JCS(P1)736⟩, mesoionic thiatriazol-5-ones (**88**) ⟨79JCS(P1)732⟩, thiatriazole-5-thiones (**89**) ⟨79JCS(P1)732⟩, thiatriazol-5-imines (**90**) ⟨79JCS(P1)741⟩, and 5-alkylidenethiatriazoles (**91**) ⟨79JCS(P1)744⟩ have been reported.

(ii) IR spectra

Mesoionic oxatriazol-5-ones exhibit a strong absorption in the C=O stretching region around 1780–1800 cm^{-1} ⟨79JCS(P2)533, 80JCS(P2)1437, 64JCS906⟩. In the case of mesoionic thiatriazol-5-ones, ν(CO) is shifted to 1675–1680 cm^{-1} ⟨79JCS(P1)732⟩. Mesoionic oxatriazole-5-thiones show a strong C=S absorption at 1365–1375 cm^{-1} ⟨74JCS(P1)645⟩. This is also the case with mesoionic thiatriazole-5-thiones, although it appears at lower frequency (1265–1270 cm^{-1}) ⟨79JCS(P1)732⟩. Mesoionic oxatriazol-5-imines show a strong C=N absorption at 1675–1680 cm^{-1} ⟨79JCS(P1)741⟩. In the corresponding thiatriazoles this appears at 1580–1590 cm^{-1} ⟨79JCS(P1)741⟩.

(iii) UV spectra

UV and visible spectral data for the various mesoionic systems have been reported. Typical values are selected ($\lambda_{max}, \varepsilon$): colorless oxatriazol-5-ones (**85**; R = 4-ClC$_6$H$_4$) (ethanol) 218 (9900), 275 (13 500) ⟨64JCS906⟩; yellow oxatriazole-5-thiones (**86**; R = Ph) 259 (21 650), 394 (2615) ⟨79JCS(P1)732⟩; deep red oxatriazol-5-imines (**87**; R^1 = R = Ph) 265 (31 120), 410 (2539) ⟨79JCS(P1)732⟩; colorless or pale yellow thiatriazol-5-ones (**88**; R = Ph) 215sh (9850), 280 (11 000), 330 (4250) ⟨79JCS(P1)732⟩; orange thiatriazole-5-thiones (**89**; R = Ph) 233 (3490), 292 (20 200), 450 (1715) ⟨79JCS(P1)732⟩; red thiatriazol-5-imines (**90**; R^1 = R = Ph) 279 (29 300), 441 (1830) ⟨79JCS(P1)732⟩; orange 5-alkylidenethiatriazoles (**91**; R = Ph) 238 (5880), 298 (31 800), 471 (3620) ⟨79JCS(P1)744⟩.

(iv) Mass spectra

Several accounts of the mass spectral fragmentation pattern of mesoionic compounds have been published ⟨76AHC(19)1, p. 88⟩. Recently the usefulness of mass spectrometry to distinguish between pairs of isomers (see next section) has been extended to mesoionic 1,2,3,4-oxatriazoles and -thiatriazoles. A detailed account of the individual types of compounds will not be given. The general fragmentation pattern according to Ollis and Ramsden is depicted in Scheme 4 ⟨79JCS(P1)747⟩.

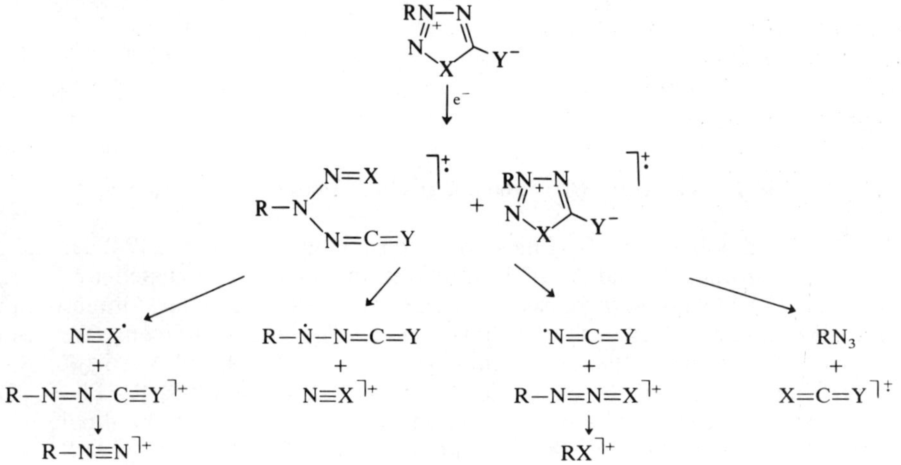

X = O or S; Y = O, S, NR or C(CN)$_2$

Scheme 4

4.28.3.3 Thermodynamic Aspects

4.28.3.3.1 Interconversion of mesoionic isomers

Oxatriazole (**87**) has been shown to undergo rearrangement in ethanolic alkali solution to give tetrazole (**95**) ⟨1896CB1686, 29CB1449, 71ACS625⟩. Oxatriazole (**86**) similarly gives the thermodynamically more stable thiatriazole (**88**) when heated in aqueous ethanolic ammonia ⟨76CC306, 79JCS(P1)732⟩, but all attempts to effect the transformation (**90**) to (**96**) or (**96**) to (**90**) have failed (Scheme 5) ⟨76CC307, 79JCS(P1)741⟩.

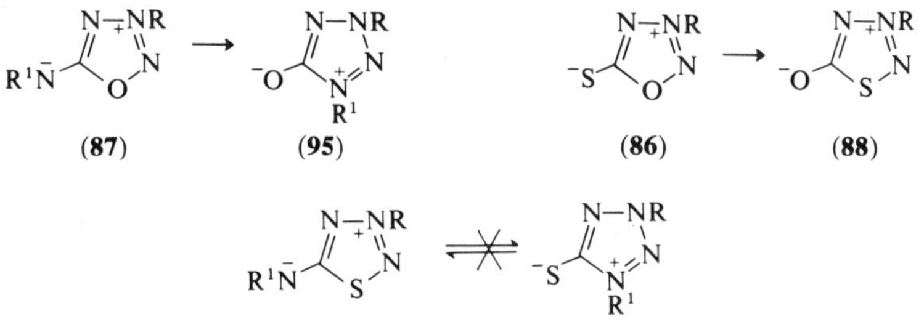

(**87**) (**95**) (**86**) (**88**)

(**90**) (**96**)

Scheme 5

The mechanisms of the rearrangements have not been established but they may well proceed *via* a betaine intermediate similar to those considered for similar transformations of mesoionic isomers (Section 4.01.1.2(ii)) ⟨79JCS(P1)732⟩. Ollis and Ramsden have treated this type of rearrangement in more detail in their review ⟨76AHC(19)1⟩.

4.28.3.4 Reactivity of the Ring Systems

4.28.3.4.1 Reactivity at the ring atoms

(i) Thermal reactions and 1,3-dipolar cycloaddition reactions

Thermally mesoionic oxatriazoles are relatively stable. Heating of (**85**) in tolane (**97**) at 200 °C for 20 d in the presence of LiCl induces CO_2 fragmentation and formation of cycloadduct (**99**) in 37% yield. There is no bimolecular reaction as found in the case of the sydnones and isosydnones (equation 56) ⟨68CB536⟩.

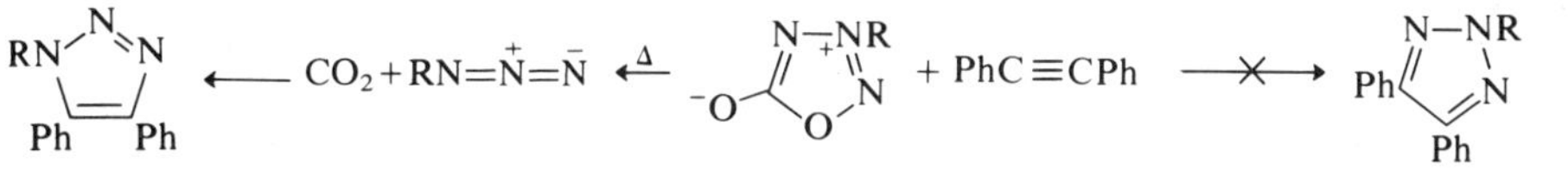

(56)

(**99**) R = cyclohexyl (**85**) (**97**) (**98**)

Oxatriazole (**100**), on the other hand, undergoes an interesting exchange reaction with *p*-chlorophenyl isocyanate to form a mixture of oxatriazoles (equation 57). It is suggested that 1,3–dipolar cycloaddition leads to a bicyclic adduct (**101**) which may either lose phenyl isocyanate or *p*-chlorophenyl isocyanate to produce an equilibrium mixture. This is the only known example of a mesoionic compound of the general type (**94**; X or Y = NR, O or S) participating in a 1,3-dipolar cycloaddition ⟨79JCS(P1)736⟩. Thus oxatriazole-5-thiones as well as thiatriazol-5-ones are found to be unreactive towards 1,3-dipolarophiles such as alkenes, alkynes and heterocumulenes ⟨79JCS(P1)732⟩.

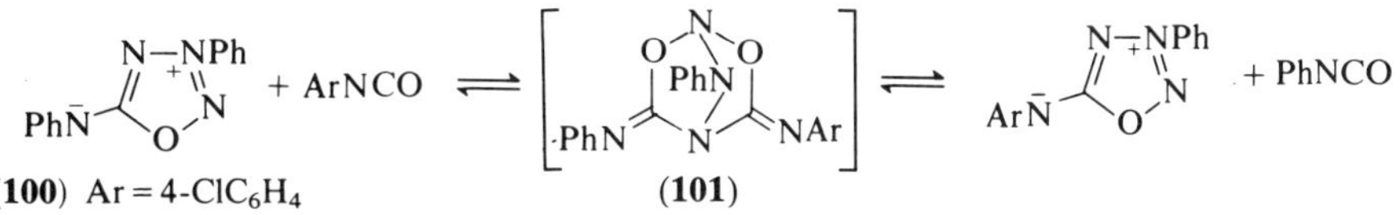

(57)

(**100**) Ar = 4-ClC_6H_4 (**101**)

(ii) Photochemical reactions

Photolysis of (**102**) in the presence of water yielded cyclohexanone (27%) which was suggested to be formed *via* cyclohexyl azide and the antiaromatic, unknown triazirine (equation 58) ⟨70CC1591⟩. This interesting possibility spurred another group to investigate the photochemistry of mesoionic 3-phenyl-1,2,3,4-oxatriazol-5-one and the corresponding 2-^{15}N-labelled compound. Small amounts of phenyl azide are formed (phenyl azide itself is further photolyzed under the given conditions) along with phenyl isocyanate (equation 59). IR analysis proved that the phenyl azide formed is ^{15}N-labelled exclusively in the 3-position. Therefore, any symmetric intermediate like the triazirene can be ruled out. The phenyl group must migrate during or before ring fragmentation ⟨79T409⟩.

$$\underset{(102)}{H_{11}C_6N\underset{N}{\overset{N}{\diagdown}}\overset{O^-}{\underset{O}{\diagup}}} \xrightarrow{h\nu} [C_6H_{11}N\underset{N}{\overset{N}{\diagdown}}] \rightarrow [C_6H_{11}N=\overset{+}{N}=\bar{N}] \xrightarrow[H_2O]{h\nu} \underset{}{\bigcirc}=O \quad (58)$$

$$PhN=C=O + {}^{14}N^{15}NO \xleftarrow{h\nu} \underset{Ph}{\overset{N\diagdown}{\underset{N^{15}}{\diagup}}\overset{O^-}{\underset{O}{\diagup}}} \xrightarrow{h\nu} PhN=\overset{+}{N}={}^{15}\bar{N}+CO_2 \quad (59)$$
$$(66\%)$$

(iii) Hydrolysis, reduction and other reactions

Mesoionic 3-cyclohexyl-1,2,3,4-oxatriazol-5-one decomposes in strong acid to give cyclohexanol, carbon dioxide and hydrazoic acid ⟨56JA5124⟩ and 3-phenyl-1,2,3,4-oxatriazol-5-one gives phenyl azide and carbon dioxide ⟨32G912⟩. Treatment of mesoionic 3-phenyl-1,2,3,4-oxatriazole-5-thione with hot aqueous alkali gives phenyl azide; with hot dilute sulfuric acid, phenol is formed in low yield ⟨79JCS(P1)732⟩. Mesoionic 3-phenyl-1,2,3,4-thiatriazol-5-one is hydrolyzed to phenyl azide by aqueous sodium hydroxide, but was recovered unchanged from hot 80% sulfuric acid ⟨79JCS(P1)732⟩. Busch and Schmidt observed the formation of small amounts (1–2%) of a colorless compound on treatment of oxatriazole (**87**; R = R^1 = Ph) with boiling mineral acid ⟨29CB1449⟩. It was later identified as oxatriazole (**85**, R = Ph; equation 60) ⟨71ACS625⟩. Among the other hydrolysis products, azoformamide (**103**) was identified ⟨29CB1449⟩.

$$PhN=NCONHPh \xleftarrow{H^+, H_2O} \underset{(87)\ R=R^1=Ph}{\overset{N-NPh}{\underset{PhN\diagdown O\diagup N}{}}} \xrightarrow{H^+, H_2O} \underset{(85)\ R=Ph}{\overset{N-NPh}{\underset{-O\diagdown O\diagup N}{}}} \quad (60)$$
(**103**)

More detailed studies of the acid-catalyzed hydrolysis of mesoionic oxatriazoles have been carried out by Tillett and coworkers. The acid-catalyzed ring opening of mesoionic 3-isopropyl- and 3-methyl-1,2,3,4-oxatriazol-5-one was studied in aqueous solutions of mineral acids. Analysis of the different data is consistent with an *A*-1 mechanism for the hydrolysis of isopropyloxatriazole (equation 61) ⟨76JOC3040, 79JCS(P2)533⟩, while methyloxatriazole hydrolyzes by an *A*-2 mechanism in which proton transfer seems to be partially rate determining (equation 62) ⟨79JCS(P2)533⟩. Analyses of the data of the acid-catalyzed hydrolysis of mesoionic 3-ethyl-1,2,3,4-oxatriazol-5-one are consistent with a gradual change-over from an *A*-2 mechanism at low acid concentration to a predominantly *A*-1 mechanism at high acidities ⟨80JCS(P2)1437⟩.

$$\overset{N-NPr^i}{\underset{-O\diagdown O\diagup N}{}} + H_3O^+ \rightleftharpoons \overset{N-NPr^i}{\underset{O\diagdown O\diagup NH}{}} + H_2O \xrightarrow{slow}$$
$$Pr^{i+} + HN_3 + CO_2 \quad (Pr^{i+} + H_2O \rightarrow Pr^iOH + H^+) \quad (61)$$

$$\overset{N-NMe}{\underset{-O\diagdown O\diagup N}{}} + H_3O^+ \rightleftharpoons \overset{N-NMe}{\underset{O\diagdown O\diagup NH}{}} + H_2O$$

$$\underset{O\diagdown O\diagup NH}{\overset{N-N\overset{Me}{\diagdown}\overset{H}{\underset{\diagup}{O}}\ \cdot OH_2}{}} \rightarrow MeOH + HN_3 + CO_2 + H_3O^+ \quad (62)$$

The mesoionic oxatriazole ring may also undergo opening on reduction or treatment with hydrogen sulfide (equation 63) ⟨29CB1449, 71ACS625⟩.

$$\text{PhNHNHCONHPh} + \text{PhN}=\text{NCONHPh} \xleftarrow[\text{HOAc}]{\text{Zn}} \underset{(103)}{\overset{\text{N}\!\!-\!\!\text{NPh}}{\underset{\text{PhN}^-\diagdown_{\text{O}}\diagup^{\text{N}}}{\bigtriangleup}}} \xrightarrow{\text{H}_2\text{S}} \text{PhNHNHCSNHPh} + \text{S} + \text{NH}_3 \quad (63)$$

4.28.3.5 Reactivity of Substituents

4.28.3.5.1 Salt formation

Mesoionic 1,2,3,4-oxatriazol- (**87**) and 1,2,3,4-thiatriazol-5-imines (**90**) form salts with HCl or HBF$_4$ from which the bases may be regenerated by neutralization with NH$_3$ or NaHCO$_3$ giving (**87**) ⟨72ACS(26)858, 79JCS(P1)736⟩, or with sodium hydroxide, giving (**90**) ⟨79JCS(P1)741⟩.

Oxatriazol-5-imines unsubstituted at the exocyclic nitrogen are amphoteric. Thus treatment of (**87**; R^1 = H, R = Ph) with potassium hydroxide in ethanol gives a colorless crystalline salt (equation 64) ⟨29CB1449⟩. Christophersen and Treppendahl later found this to have the remarkable open-chain structure (**104**) and showed that the reaction is reversible. A pK_a value of 6 has been estimated for compound (**87**; R^1 = H, R = Ph) ⟨72ACS858⟩.

$$\underset{\text{HN}}{\overset{\text{N}\!-\!\text{NPh}}{\diagdown_{\text{O}}\diagup^{\text{N}}}} \xrightleftharpoons[\text{H}^+, -\text{K}^+]{\text{KOH}, -\text{H}_2\text{O}} \underset{(104)}{\text{Ph}-\overset{\text{N}=\text{O}}{\underset{|}{\text{N}}}-\text{N}=\text{C}=\bar{\text{N}}\ \text{K}^+} \quad (64)$$

Mesoionic 1,2,3,4-oxatriazole-5-thiones (**86**) and thiatriazol-5-ones (**88**) are alkylated with triethyloxonium tetrafluoroborate to give 3,5-disubstituted oxa- and thia-triazolium tetrafluoroborates (**80**; equation 51), while oxatriazolethiones (**86**) were found to be inert towards methyl iodide ⟨76CC306, 79JCS(P1)732⟩.

4.28.3.5.2 Other side-chain reactions

A benzoyl derivative of oxatriazolimine (**87**; R^1 = H, R = Ph) has been prepared by direct benzoylation ⟨29CB1449, 71CPB559⟩ and by nitrosation of 4-benzoyl-1-phenylthiosemicarbazide (equation 65) ⟨72ACS858⟩. Compound (**105**) absorbs at 1647 cm^{-1} in the IR, indicating a polarized carbonyl function, and it forms a salt with HCl in MeOH ⟨72ACS858⟩.

$$\underset{\text{HN}}{\overset{\text{N}\!-\!\text{NPh}}{\diagdown_{\text{O}}\diagup^{\text{N}}}} \xrightarrow{\text{PhCOCl}} \underset{(105)}{\text{Ph}-\overset{-\text{O}}{\underset{|}{\text{C}}}=\text{N}\diagdown_{\text{O}}\diagup^{\text{N}\!-\!\text{NPh}}} \xleftarrow{\text{HNO}_2} \text{PhNHNHCSNHCOPh} \quad (65)$$

An *N*-nitroso derivative of (**87**; R^1 = H, R = Ph) has been prepared ⟨71CPB559⟩. It is thermally unstable, decomposing to give a high yield of mesoionic 3-phenyl-1,2,3,4-oxatriazol-5-one (equation 66).

$$\text{O}=\text{N}-\bar{\text{N}}\diagdown \overset{\text{N}\!-\!\text{NPh}}{\underset{\text{O}}{\diagdown_{\text{O}}\diagup^{\text{N}}}} \rightarrow {}^-\text{O}\diagdown \overset{\text{N}\!-\!\text{NPh}}{\underset{\text{O}}{\diagdown_{\text{O}}\diagup^{\text{N}}}} \quad (66)$$

4.28.3.6 Syntheses

4.28.3.6.1 Mesoionic 1,2,3,4-oxatriazol-5-ones

3-Alkyloxatriazolones are obtained by the action of nitrous acid upon the corresponding semicarbazides ⟨55JA1280, 65JOC567, 70JMC196, 71JHC89⟩. At low temperatures the intermediate *N*-nitrososemicarbazides may be isolated. They cyclize on heating in acid solutions (equation 67) ⟨56JA5124⟩. 3-Methyloxatriazole has been prepared by another method from

$$\text{RNHNHCONH}_2 \xrightarrow{\text{HNO}_2} \underset{\underset{\text{NO}}{|}}{\text{RNNHCONH}_2} \xrightarrow{\Delta, \text{H}^+} {}^-\text{O}\diagdown \overset{\text{N}\!-\!\text{NR}}{\underset{\text{O}}{\diagdown_{\text{O}}\diagup^{\text{N}}}} \leftarrow \left[\underset{\underset{\text{NO}}{|}}{\text{RN}-\text{N}=\text{C}=\text{O}}\right] \xleftarrow{\text{COCl}_2} \underset{\underset{\text{NO}}{|}}{\text{RNNH}_2}$$

(67)

N-nitroso-N-methylhydrazine and phosgene. The reaction is suggested to proceed through an intermediate isocyanate (equation 67) ⟨62BCJ766⟩.

3-Aryloxatriazolones cannot be prepared by this method since the N-nitrosoarylsemicarbazides readily undergo loss of nitroxyl with the resulting formation of an arylazomethanamide ⟨1895CB1925⟩. They can be formed by nitrosation of potassium arylhydrazonomethane disulfonates, a reaction first described by von Pechmann in 1896 ⟨1896CB2161⟩. Recognition of the mesoionic structure is due to Farrar ⟨64JCS906⟩. The reaction, which involves potassium diazomethane disulfonate (**106**), is rather complicated but has been used for specific introduction of ^{15}N into the ring (equation 68) ⟨79T409⟩. 3-Aryloxatriazolones are prepared more conveniently by coupling diazonium salts with the ammonium salt of trinitromethane to give (**107**) ⟨56JA5124, 32G503, 32G912, 33G269, 33G862, 35G201, 46G259, 15G12, 16G56, 33G471⟩ or the potassium salt of dinitromethane to give (**108**) ⟨53LA(579)28⟩ and decomposing the coupling products (equation 69). The addition of nitrous acid is not required because intermediate N-nitroso derivatives are presumably formed by hydrolysis of the nitro compounds.

$$(SO_3K)_2C=\overset{+}{N}=\overset{-}{N} \xrightarrow[MeCO_2H]{K_2SO_3} (SO_3K)_2CHN_2^+ KSO_3^- \xrightarrow[-N_2]{2PhN_2^+OH^-}$$
(**106**)

$$(SO_3K)_2C=N-N-N=NPh \xrightarrow[HCl]{Na^{15}NO_2} \underset{(68)}{\text{mesoionic ring}} + PhN_2^+Cl^-$$
with Ph substituent

$$ArN=NC(NO_2)_3 \xrightarrow[\Delta]{H_2O} \underset{(107)}{\text{ring}} \xleftarrow[\Delta]{H_2O} ArNHN=C(NO_2)_2 \quad (69)$$
(**108**)

4.28.3.6.2 Mesoionic 1,2,3,4-oxatriazole-5-thiones

Arylhydrazines and carbon disulfide in ethanol give arylhydrazinium dithiocarbamate salts. Oxatriazolethiones are formed on subsequent nitrosation at 0 °C. Attempts to isolate nitrosoaryldithiocarbazates (**109**) have not been successful (equation 70) ⟨76CC306, 79JCS(P1)732⟩.

$$ArNHNH_3^+ \; ArNHNHCSS^- \xrightarrow{HNO_2} \left[\begin{array}{c} NO \quad S \\ | \quad \quad \| \\ Ar-N \quad C \\ \quad NH \quad SH \end{array} \right] \rightarrow \underset{(109)}{\text{mesoionic ring}} \quad (70)$$

4.28.3.6.3 Mesoionic 1,2,3,4-oxatriazol-5-imines

Busch and Becker isolated a mesoionic oxatriazolimine as early as in 1896 from the reaction of 1,4-diphenylthiosemicarbazide with nitrous acid. It was formulated as a bicyclic compound named 'isotetrazolone' ⟨1896CB1686⟩. Further studies were published in 1929 ⟨29CB1449⟩ but the mesoionic structure was first recognized by Christophersen and Treppendahl ⟨71ACS625, 72ACS858⟩ and confirmed by Hanley, Ollis and Ramsden ⟨79JCS(P1)736⟩. Oxatriazolimines (**87**) with R = aryl or alkyl and R^1 = aryl, alkyl or H have been prepared by this method (equation 71). This is in noticeable contrast to the corresponding reaction with semicarbazides where alkyl-substituted but not aryl-substituted semicarbazides give mesoionic 1,2,3,4-oxatriazolones (see Section 4.28.3.6.1).

$$RNHNHCSNHR^1 \xrightarrow{HNO_2} RN-NHCSNHR^1 \xrightarrow{-H_2S} \underset{(87)}{\text{mesoionic ring}} \quad (71)$$
with NO on N

Other routes leading to oxatriazolimines have been described. Treatment of compound (**110**; R = cyclohexyl) ⟨65JOC567⟩ and of compound (**111**; R = Ph) ⟨71CPB559⟩ with hydrogen chloride gives the corresponding oxatriazolimines isolated as the hydrochlorides. Sodium bicarbonate or ammonia liberates the free base.

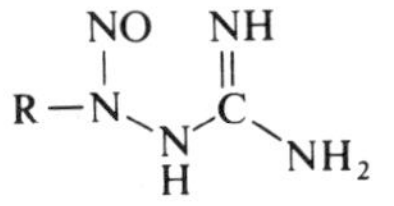

(110) 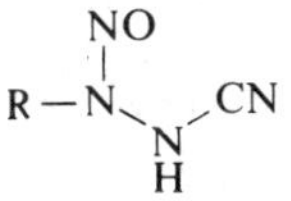   (111)

4.28.3.6.4 Mesoionic 1,2,3,4-thiatriazol-5-ones, -5-thiones, -5-imines and -5-alkenes

Mesoionic thiatriazolones (88) are prepared by isomerization of the corresponding oxatriazolethiones (86; Scheme 5) ⟨79JCS(P1)732⟩.

Mesoionic thiatriazolethiones (89) ⟨79JCS(P1)732⟩, thiatriazolimines (90) ⟨79JCS(P1)741⟩ and thiatriazolebiscyanomethylides (91) ⟨79JCS(P1)744⟩ are formed from 3,5-disubstituted 1,2,3,4-thiatriazolium tetrafluoroborates as shown in Scheme 6.

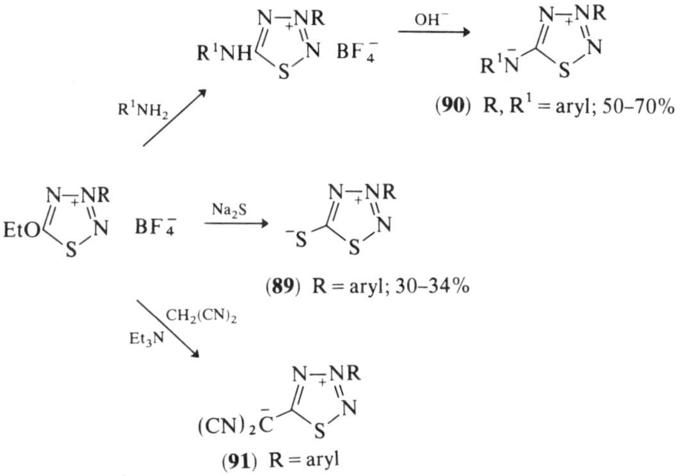

(90) R, R^1 = aryl; 50-70%

(89) R = aryl; 30-34%

(91) R = aryl

Scheme 6

4.28.3.7 Applications

A number of mesoionic 1,2,3,4-oxa- and -thia-triazoles, prepared by standard methods, were assessed for pesticidal and herbicidal activity ⟨79BRP2015878⟩. Pharmacological activity of mesoionic compounds has been reviewed ⟨76AHC(19)1⟩.

4.28.4 1,2,3,5-OXATRIAZOLES AND -THIATRIAZOLES

4.28.4.1 Introduction

The 1,2,3,5-oxatriazoles are known as oxatriazolines (112, 113; X = O) and the thiatriazoles as thiatriazolines (114, 113; X = S) and thiatriazolidines (115) often as S-oxides. Recently the so-called non-classical heteropentalenes containing the 1,2,3,5-thia- and -selena-triazole ring system (116) have been prepared.

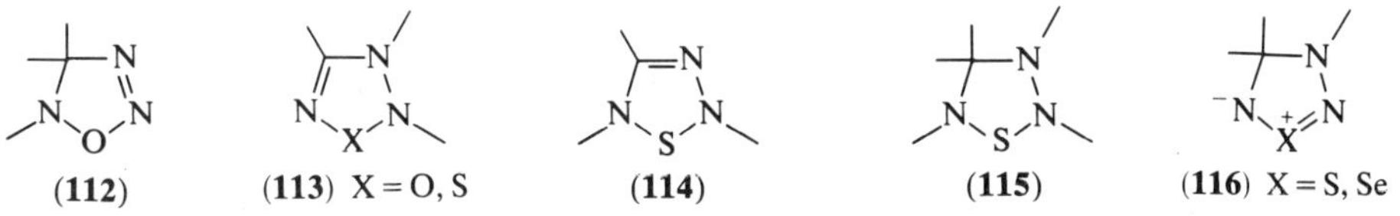

(112)   (113) X = O, S   (114)   (115)   (116) X = S, Se

Scheme 7

4.28.4.2 Structure

4.28.4.2.1 Δ^2- and Δ^4-1,2,3,5-oxatriazolines

The structures of (112) and (113; X = O) are inferrred from the mode of synthesis. The spectroscopic properties of (113; X = O; IR, ^{1}H NMR) are in agreement with the structure assignment ⟨72CB2841⟩.

4.28.4.2.2 Δ^3- and Δ^4-1,2,3,5-thiatriazolines and 1,2,3,5-thiatriazolidines

Generally, the structures of the 1,2,3,5-thiatriazoles (114) and (113; X = S) are inferred from the mode of synthesis. The structure of 1,2,3,5-thiatriazolo[5,4-a]pyridine 3-oxide (117) has been discussed in more detail. 2-Hydrazinopyridine is formed upon hydrolysis and the IR spectrum exhibits an NH stretching vibration at 3280 cm^{-1}. This suggests the interesting possibility of tautomeric and zwitterionic structures (Scheme 8). Methylation with diazomethane gives the 1-methyl derivative which apparently is best formulated as the zwitterionic Δ^4-thiatriazoline (118). UV spectroscopic properties of (117) (λ_{max} 235, 290 and 342) and (118) (λ_{max} 235, 290 and 345 nm) reveal a close structural similarity ⟨63CB2519⟩. A comparative NMR spectroscopic study of (117) and the Δ^3-1,2,3,5-thiatriazoline (119; R^1 = H, R^2 = Me) supports the description of (117) as a pyridinium salt (Scheme 9) ⟨70CB1918⟩.

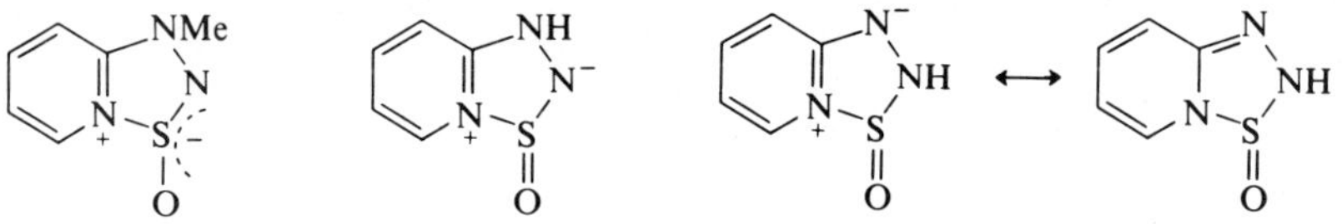

Scheme 8

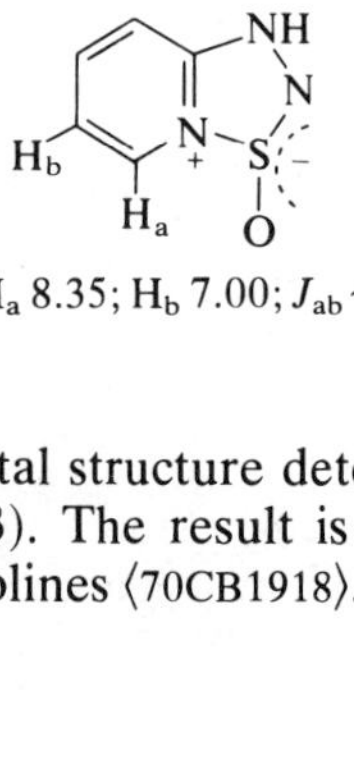

(117) H$_a$ 8.35; H$_b$ 7.00; J_{ab} ~ 4.5 Hz

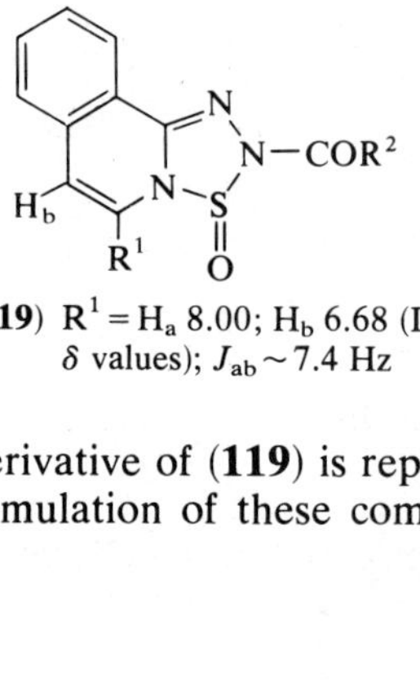

(119) R^1 = H$_a$ 8.00; H$_b$ 6.68 (DMSO, δ values); J_{ab} ~ 7.4 Hz

Scheme 9

A crystal structure determination of the chloro derivative of (119) is reproduced below (Figure 3). The result is in agreement with the formulation of these compounds as Δ^3-thiatriazolines ⟨70CB1918⟩.

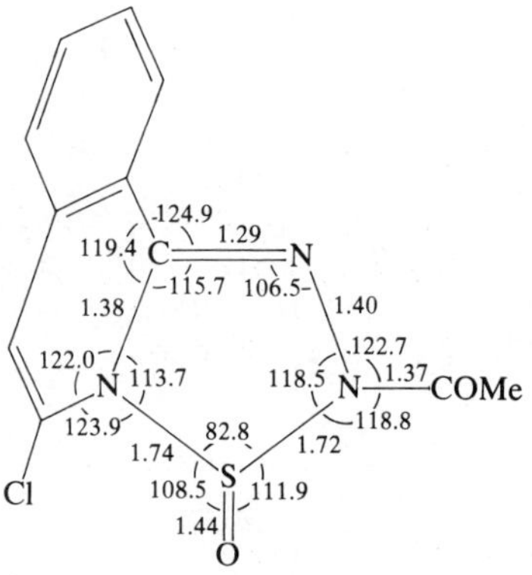

Figure 3 Bond lengths (Å) and bond angles (°) of 2-acetyl-5-chloro-2H-1,2,3,5-thiatriazolo[4,5-a]isoquinoline S-oxide

4.28.4.2.3 1,2,3,5-Thiatriazole and -selenatriazole heteropentalenes

The structure is inferred from the mode of synthesis and from spectral data ⟨81JOC4065⟩.

4.28.4.3 Reactivity of the Ring Systems

4.28.4.3.1 Δ^2- and Δ^4-1,2,3,5-oxatriazolines

Δ^2-Oxatriazoline (**121**) is reported to undergo a facile nitrogen extrusion reaction with formation of a nitrone (equation 72) ⟨76BCJ3173⟩.

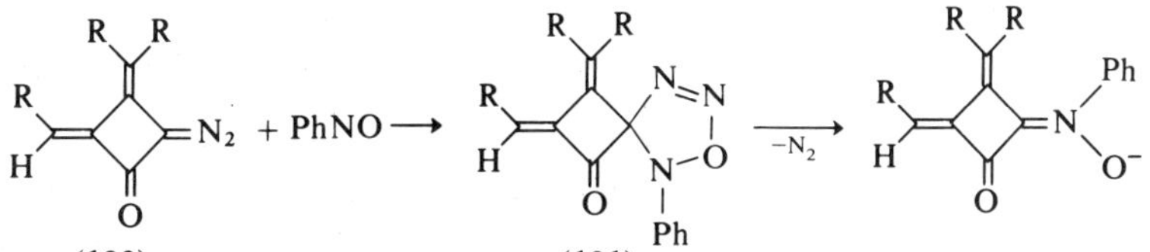

Δ^4-Oxatriazolines (**122**) are thermally unstable, undergoing rapid isomerization in solution with formation of (**123**) ⟨72CB2841, 66TL405, 64TL887⟩. Betaines are considered likely intermediates (equation 73).

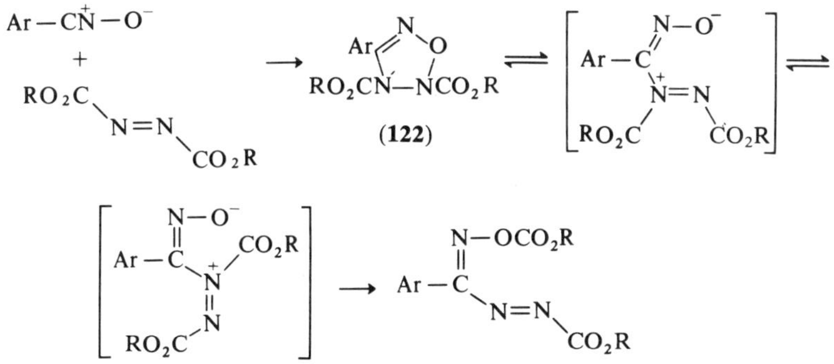

(**123**) R = Me, Et; Ar = Ph, 4-ClC$_6$H$_4$, 4-NO$_2$C$_6$H$_4$, 3-NO$_2$C$_6$H$_4$

The orange azo compound (**123**) may undergo a further reaction with nitrile oxides (used for the oxatriazoline formation), most likely to form a new oxatriazoline which isomerizes to (**124**; equation 74) ⟨72CB2841⟩.

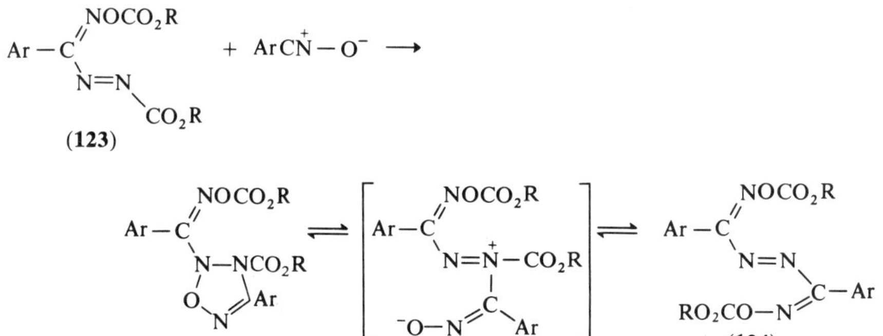

4.28.4.3.2 Δ^3- and Δ^4-1,2,3,5-thiatriazolines

As discussed in Section 4.28.4.2.2, the spectroscopic properties of 1,2,3,5-thiatriazolopyridine 3-oxide (**117**) suggest a zwitterionic character. This is reflected in its facile reactions with nucleophiles shown in Scheme 10. The products formed by addition of methanol or HN$_3$ to decomposing (**117**) may be interpreted as trapping of a reactive intermediate hetarenium ion (**125**) ⟨63CB2519⟩. Thiatriazolines of structure (**119**) can be recrystallized unchanged from ethanol and are thermally stable to about 250 °C ⟨70CB1918⟩.

Thiatriazoline (**133**) was unchanged after 16 h in boiling xylene ⟨71CB639⟩. Thiatriazoline (**135**) is likewise reported to be thermostable ⟨70CB1934⟩ and the same appears to be the case with thiatriazolines (**134**) ⟨80LA1376⟩. Note added in proof: see also ⟨81JOC4567⟩.

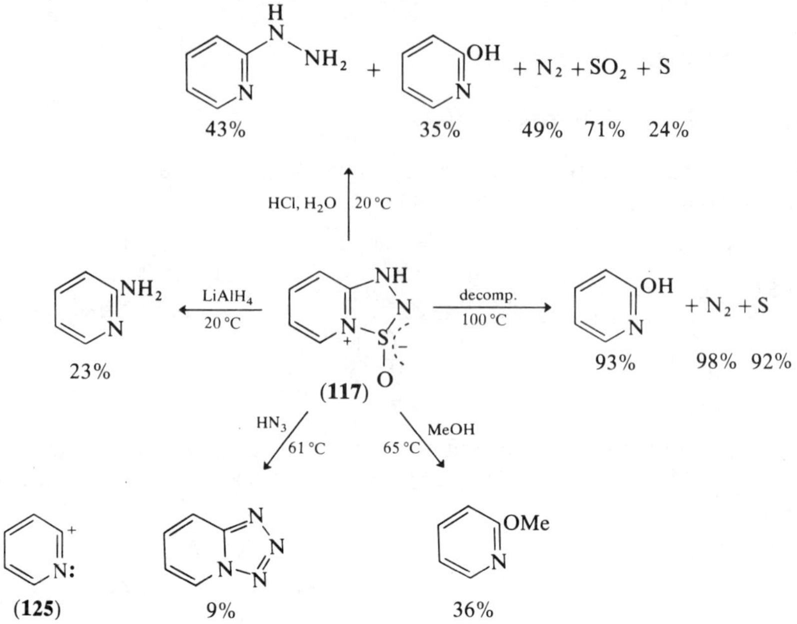

Scheme 10

4.28.4.3.3 1,2,3,5-Thiatriazolidines

Fully substituted thiatriazolidine *S*-oxides are generally reported to be thermally stable. The thermal stability of the 2-unsubstituted thiatriazolidine *S*-oxides (**126**) depends on the substituents R^1 and R^2. The transformation of (**126**) into (**127**) either on heating or on treatment with thionyl chloride is shown in equation (75) ⟨78JOC1677⟩.

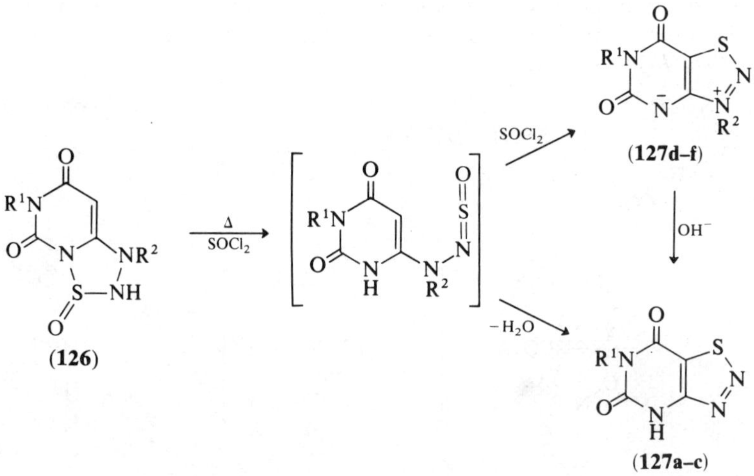

(75)

R^1, R^2: **a**, Me, H; **b**, Pr, H; **c**, Bz, H; **d**, Me, Me; **e**, Pr, Me; **f**, Bz, Me

2-Unsubstituted 1,2,3,5-thiatriazolidinone 1,1-dioxides (**128**) are thermally stable and moderately acidic compounds. Their anions readily undergo methylation and benzoylation to give 2-substituted products. On cyanomethylation a ring expansion takes place to give a 1,2,4,6-thiatriazinone (equation 76). Phenacyl chloride (2 equiv.) and phenacylidine dichloride also give ring expanded products (equation 77). Introduction of the second phenacyl group agrees with an intermediate anion (equation 76) ⟨77JCR(S)239, 77JCR(M)2826⟩.

[Equation (76) scheme: compound (128) with ClCH(R³)CN/base gives intermediate, then loses CN⁻ to give product]

[Equation (77) scheme: compound (128) with PhCOCH₂Cl/base gives products]

3-Unsubstituted 1,2,3,5-thiatriazolidinone 1,1-dioxides are thermally unstable, decomposing with SO₂ extrusion ⟨77JCR(S)238, 77JCR(M)2813⟩.

4.28.4.3.4 1,2,3,5-Thiatriazole and -selenatriazole heteropentalenes

None of the ring-fused thiatriazole and selenatriazole derivatives undergo cycloaddition with electron deficient dipolarophiles (see Section 4.28.4.4.4) ⟨81JOC4065⟩.

4.28.4.4 Syntheses

4.28.4.4.1 Δ^2- and Δ^4-1,2,3,5-oxatriazolines

The Δ^2-oxatriazoline (121) is obtained from the reaction of nitrosobenzene with diazoketone (120). It is thermally labile (equation 72) ⟨76BCJ3173⟩. Δ^4-Oxatriazolines have been prepared by reaction of nitrile oxides with azodicarboxylates. Thus benzonitrile oxides (Ar = Ph, 4-ClC₆H₄, 4-NO₂C₆H₄) and dimethyl azodicarboxylate react in ether at −15 °C to give a precipitate of the colorless crystalline dimethyl 4-phenyl-Δ^4-1,2,3,5-oxatriazoline-2,3-dicarboxylates (122, equation 73; see Section 4.28.4.3.1) ⟨72CB2841⟩. The oxatriazoline is unstable in solution and a successful preparation depends on its rapid crystallization from the solution. The diethyl carboxylates are soluble under the experimental conditions and rearrange before isolation. The reaction between 3-nitrobenzonitrile oxide and diethyl azodicarboxylate gives (123) ⟨72CB2841⟩ and not 1,2,3,5-oxatriazoline (121), as originally proposed ⟨64TL887⟩.

Reaction between nitrile oxides and other types of diazo compounds has not led to the expected oxatriazolines. Phenyl benzoyl diimide and ethyl benzoylazocarbonate gave rise to some substituted hydrazines by an unknown mechanism. Azobenzene did not react with nitrile oxides and 4-chlorobenzenediazocyanide reacted at the nitrile group instead of at the N=N bond ⟨72CB2841⟩.

4.28.4.4.2 Δ^3- and Δ^4-1,2,3,5-thiatriazolines

This class of compounds has been prepared only as the S-oxides. In most cases thionyl chloride serves as the cyclization agent. Thus the first Δ^3-thiatriazoline (129) was prepared by Huisgen et al. from diphenylbenzamidrazone and SOCl₂, and from 1,3-dipolar cycloaddition between diphenylnitrileimine and thionylaniline (equation 78) ⟨62LA(658)169⟩.

$$PhC\equiv\overset{+}{N}-\overset{-}{N}-Ph + PhN=S=O \longrightarrow \underset{(129)}{\begin{array}{c}PhC=N\\ \mid \quad \quad \backslash\\ PhN \quad NPh\\ \backslash \; / \\ S \\ \parallel \\ O\end{array}} \xleftarrow[\text{pyridine}]{SOCl_2} PhNHC=NNHPh \quad | \quad Ph \quad (78)$$

Other types of amidrazones behave similarly. Thus compounds (130), (131) and (132) give rise to the respective thiatriazolines (133) ⟨71CB639⟩, (134) ⟨80LA1376⟩ and (135) ⟨70CB1934⟩ on treatment with SOCl₂.

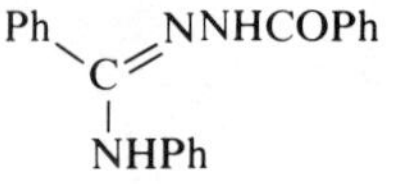

(130), 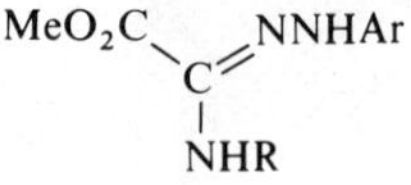 (131), 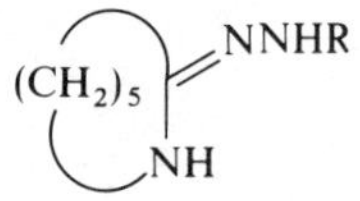 (132)

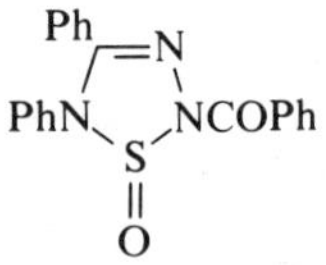

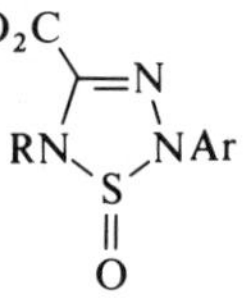

(134) R = alkyl or aryl 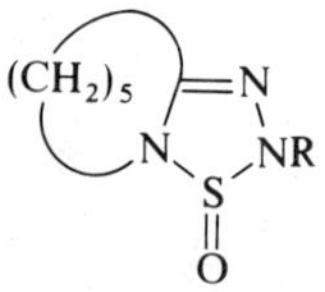   (135) R = PhCO, 4-MeOC$_6$H$_4$CO, 4-ClC$_6$H$_4$CO, pyridyl-CO

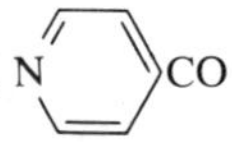

Certain heterocycles may be considered 'masked' amidrazones. Thus 1-(2-acetylhydrazino)isoquinolines react with SOCl$_2$ to give 1,2,3,5-thiatriazoline S-oxides (119; equation 79). Formation of 1,2,3,4-oxathiadiazoles (136) from reaction at the acetylhydrazino group is also possible and was observed in a single case (equation 79) ⟨70CB1918⟩. Diisoquinolylhydrazine (137) reacts in the same manner to give thiatriazole (138) ⟨70CB1918⟩. Compound (139) gives 1,2,3,5-thiatriazolobenzodiazepine derivatives (140) having properties as tranquilizers and sedatives ⟨73USP3737434⟩.

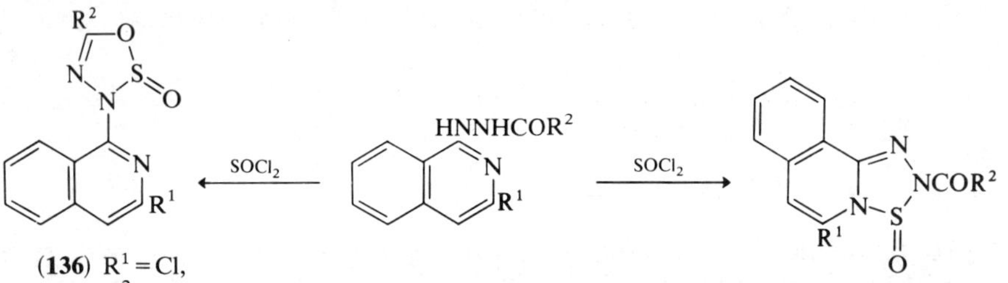

(136) R^1 = Cl, R^2 = 2-NO$_2$C$_6$H$_4$

(119) R^1, R^2 = Cl, Me; H, Ph; H, C≡CPh; H, Me; H, 2-NO$_2$C$_6$H$_4$

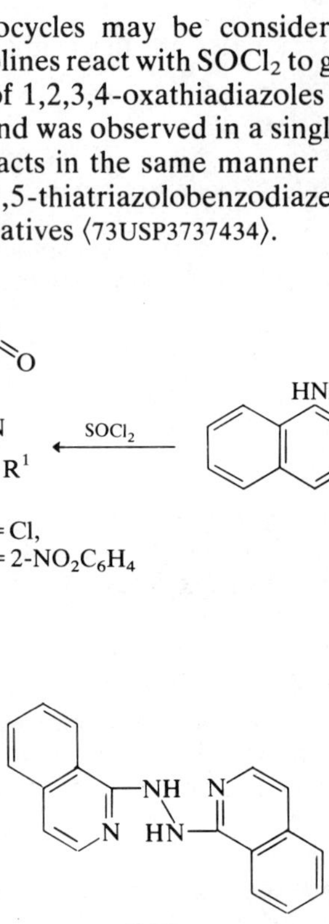

(137), 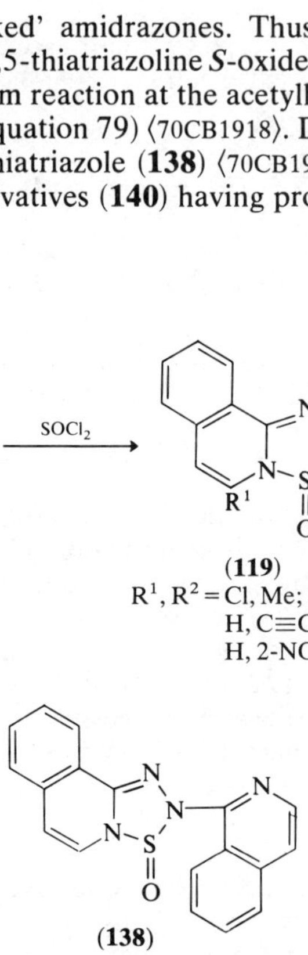 (138), 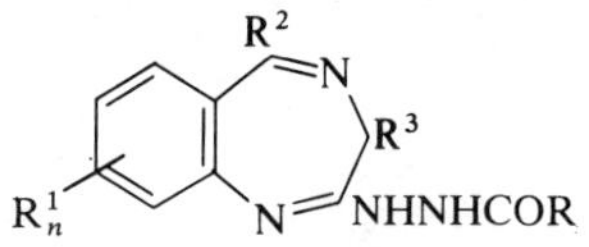 (139), 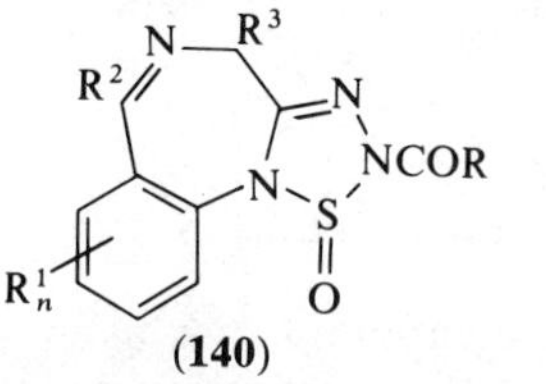 (140)

2-Hydrazinopyridine reacts with SOCl$_2$ to give the pale yellow hydrochloride of 1,2,3,5-thiatriazolo[5,4-a]pyridine 3-oxide (117), which on treatment with NaHCO$_3$ gives the free base (equation 80). This, however, is sufficiently acidic to react with diazomethane and gives a methyl derivative (118; equation 81). The position of the methyl group is inferred from the results of hydrolysis (equation (81)) ⟨63CB2519⟩.

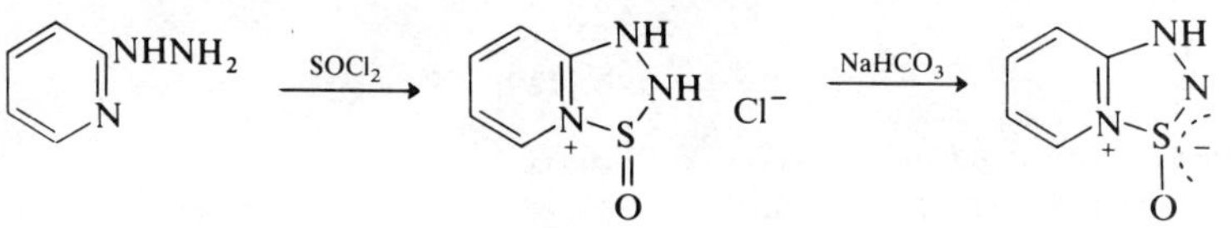

(117)

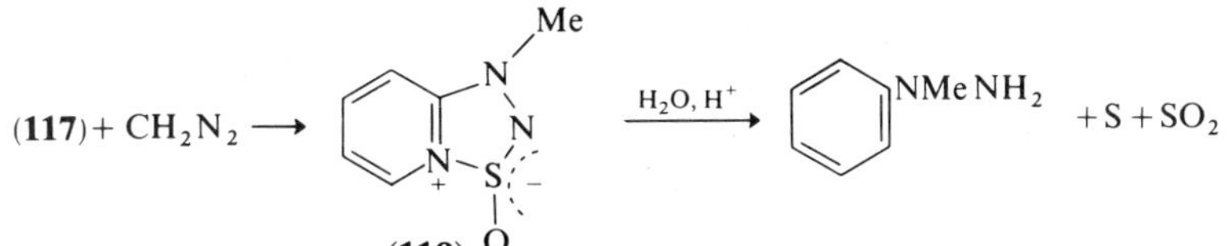

4.28.4.4.3 1,2,3,5-Thiatriazolidines

1,2,4,3-Triazasilole-5-thione reacts with SCl_2, $COCl_2$, etc. to give heterocyclic compounds. In the case of SCl_2, 1,2,3,5-thiatriazolidine-4-thiones (**141**) are formed (equation 82) ⟨78ZC336⟩.

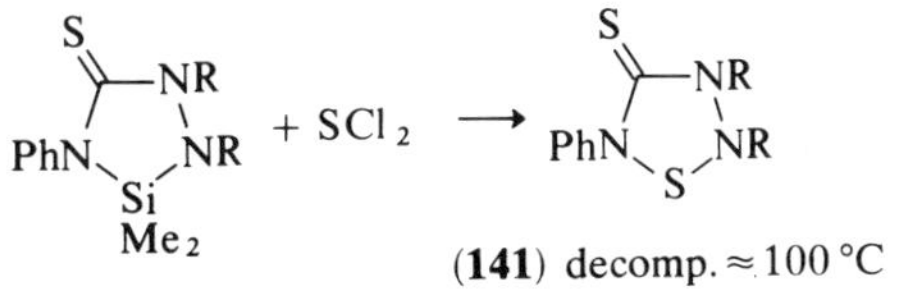

$$\text{(141) decomp.} \approx 100\,°C$$
$$R = Me\,(94\%),\,Ph\,(67\%)$$

The same ring system can be prepared from 1,2,4-trisubstituted semicarbazides or thiosemicarbazides and $SOCl_2$ to give 1,2,3,5-thiatriazolidine S-oxides (**142**; equation 83). Formation of isomeric 5-iminooxathiadiazolidines is ruled out, since the IR spectrum shows a strong carbonyl absorption at 1710–1720 cm^{-1} but no imino band ⟨79JHC895⟩. 1,2,4-Trisubstitution is a necessity for cyclization. 1,4-Disubstituted semicarbazides form acyclic azo compounds and 2,4-disubstituted semicarbazides form acyclic sulfinyl amines. It is of interest to point out that attempts have failed to produce five-membered rings by cyclization of semicarbazides with either sulfur dichloride or sulfuryl chloride ⟨79JHC895⟩. Thiatriazolidines (**144**) and (**126**) are formed from $SOCl_2$ and the corresponding indazolones (**143**) ⟨81LA1361⟩ and uracils (**145**) ⟨78JOC1677⟩, respectively.

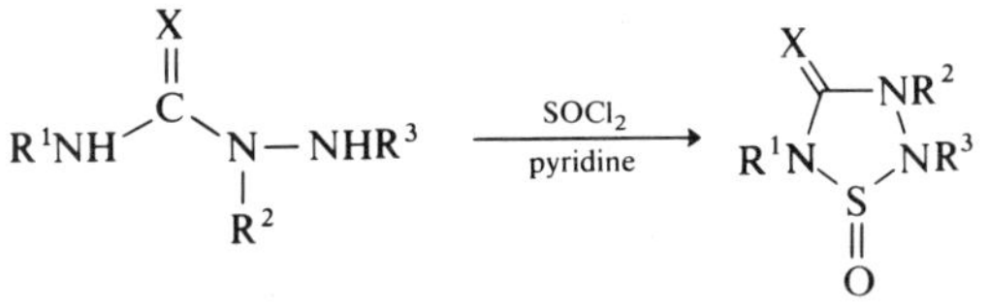

(83)

(**142**) 25–85%

$R^1 = Bu^t$ or Ar; $R^2 = Me$; $R^3 = $ alkyl; $X = O$ or S

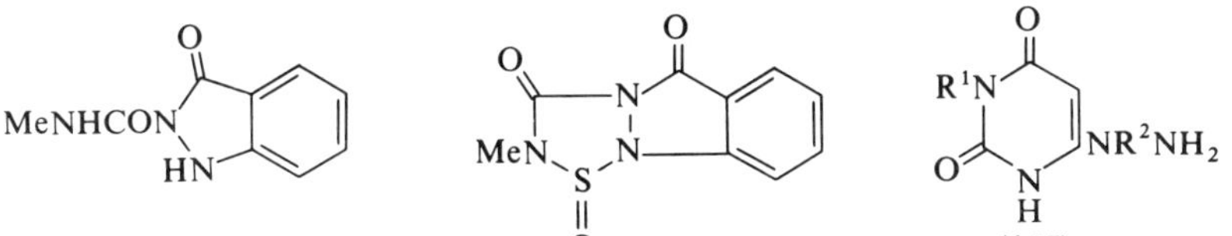

(**143**)   (**144**)   (**145**)

Heterocyclic synthesis using 2-alkyl-2-chloro- and -fluoro-sulfonylcarbamoyl chlorides has been exploited. 1,2,3,5-Thiatriazolidine 1,1-dioxides are formed with the reagents shown in Scheme 11. The reaction is initiated by attack of the carbamoyl chloride at the most basic nitrogen followed by base-induced cyclization. With 4-methyl-1-phenylsemicarbazide the protecting methylcarbamoyl group is eliminated under the alkaline conditions. In a similar reaction with phenylhydrazine, the unsubstituted nitrogen being the most basic, phenylazocarboxamide (**147**) was isolated. This indicated rapid decomposition of the presumed intermediate 3-unsubstituted 1,2,3,5-thiatriazolidinone 1,1-dioxide (**146**; equation

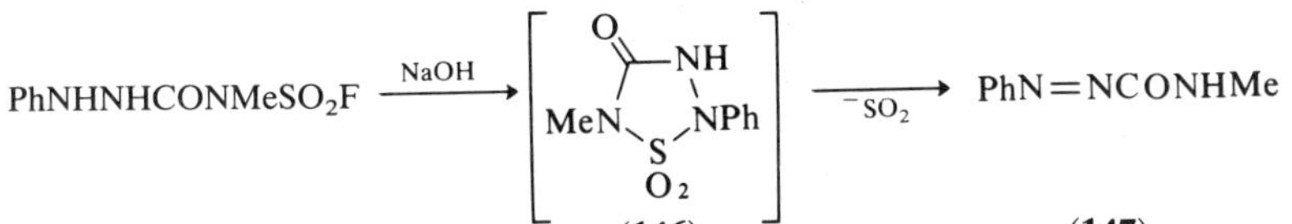

$$PhNHNHCONMeSO_2F \xrightarrow{NaOH} \left[\text{(146)} \right] \xrightarrow{-SO_2} PhN=NCONHMe \quad (84)$$

(**146**)   (**147**)

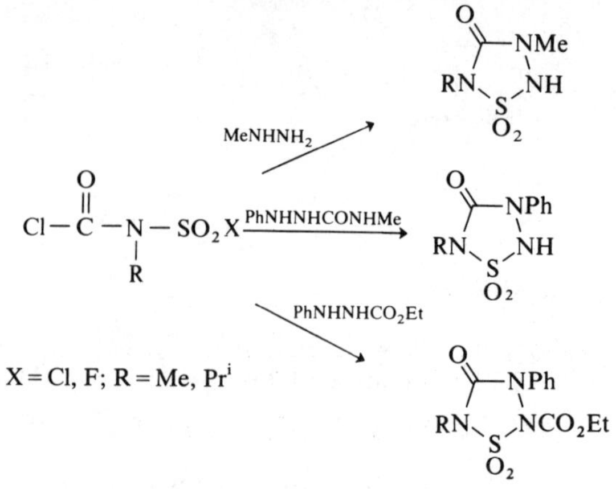

Scheme 11

84). Small amounts of thiatriazolidine were also formed from initial reaction at the phenyl-substituted nitrogen ⟨77JCR(S)238, 77JCR(M)2813⟩.

4.28.4.4.4 1,2,3,5-Thiatriazole and -selenatriazole heteropentalenes

The imidazolo[1,2-c]-thia- and -selena-triazole (**116**) ring systems are prepared from the corresponding 1,2-diaminoimidazoles as shown in equation (85). Benzimidazolo[1,2-c]-thia- and -selena-triazoles are prepared in an analogous fashion in 47 and 5% yield, respectively ⟨81JOC4065⟩.

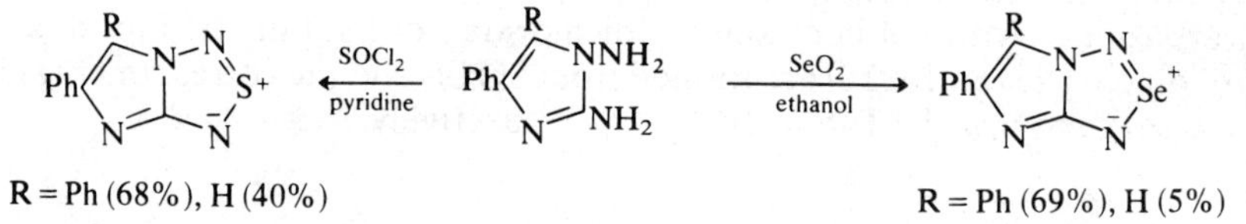

4.28.4.5 Applications

1,2,3,5-Thiatriazole derivatives have been patented for various uses, *e.g.* as sedatives and tranquilizers (**139**), as herbicides (**142**), and as chemosterilants and lubricant additives.

4.29

Five-membered Rings (One Oxygen or Sulfur and At Least One Nitrogen Atom) Fused with Six-membered Rings (At Least One Nitrogen Atom)

K. UNDHEIM
University of Oslo

4.29.1	INTRODUCTION	616
4.29.2	ISOXAZOLE AND ISOTHIAZOLE FUSED WITH AZINES: INTRODUCTION	616
4.29.2.1	*Survey of Possible Ring Systems*	*616*
4.29.2.2	*General Survey of Reactivity*	*617*
4.29.3	ISOXAZOLE FUSED WITH AZINES	617
4.29.3.1	*Introduction*	*617*
4.29.3.2	*Structure*	*618*
4.29.3.2.1	*X-Ray diffraction*	*618*
4.29.3.2.2	*Molecular spectra*	*618*
4.29.3.3	*Reactivity*	*621*
4.29.3.3.1	*Survey of reactivity*	*621*
4.29.3.3.2	*Thermal and photochemical reactions involving no other species*	*621*
4.29.3.3.3	*Electrophilic attack at nitrogen*	*622*
4.29.3.3.4	*Electrophilic attack at carbon*	*622*
4.29.3.3.5	*Nucleophilic attack at carbon*	*623*
4.29.3.3.6	*Reductive ring opening*	*624*
4.29.3.3.7	*Nucleophilic attack at hydrogen*	*624*
4.29.3.3.8	*Other C-linked substituents*	*624*
4.29.3.4	*Synthesis: Fully Conjugated Rings*	*625*
4.29.3.4.1	*2H-Isoxazolo[2,3-a]pyridine*	*625*
4.29.3.4.2	*Isoxazolo[4,5-b]pyridine*	*626*
4.29.3.4.3	*Isoxazolo[4,5-c]pyridine*	*626*
4.29.3.4.4	*Isoxazolo[5,4-c]pyridine*	*627*
4.29.3.4.5	*Isoxazolo[5,4-b]pyridine*	*627*
4.29.3.4.6	*Isoxazolo[4,3-b]pyridine*	*628*
4.29.3.4.7	*Isoxazolo[4,3-c]pyridine*	*628*
4.29.3.4.8	*Isoxazolo[3,4-c]pyridine*	*629*
4.29.3.4.9	*Isoxazolo[3,4-b]pyridine*	*630*
4.29.3.5	*Saturated and Partially Saturated Rings: Reactivity*	*630*
4.29.3.6	*Saturated and Partially Saturated Rings: Synthesis*	*631*
4.29.3.6.1	*2H-Isoxazolo[2,3-a]pyridine*	*631*
4.29.3.6.2	*Isoxazolo[4,5-c]pyridine*	*633*
4.29.3.6.3	*Isoxazolo[5,4-c]pyridine*	*633*
4.29.3.6.4	*Isoxazolo[5,4-b]pyridine*	*633*
4.29.3.6.5	*Isoxazolo[4,3-c]pyridine*	*633*
4.29.3.6.6	*Isoxazolo[3,4-c]pyridine*	*634*
4.29.3.7	*Applications*	*634*
4.29.4	ISOTHIAZOLE FUSED WITH AZINES	635
4.29.4.1	*Introduction*	*635*
4.29.4.2	*Structure*	*635*
4.29.4.2.1	*Molecular spectra*	*635*
4.29.4.3	*Reactivity*	*636*
4.29.4.3.1	*Survey of reactivity*	*636*
4.29.4.3.2	*Electrophilic attack at nitrogen*	*636*
4.29.4.3.3	*Electrophilic attack at carbon*	*637*
4.29.4.3.4	*Nucleophilic attack at carbon*	*637*

4.29.4.3.5 Nucleophilic attack at sulfur	638
4.29.4.3.6 Desulfurization	638
4.29.4.3.7 Nucleophilic attack at hydrogen	638
4.29.4.3.8 Other C-linked substituents	639
4.29.4.3.9 Rearrangement	639
4.29.4.4 Synthesis	639
4.29.4.4.1 2H-Isothiazolo[2,3-a]pyridine	639
4.29.4.4.2 Isothiazolo[4,5-b]pyridine	640
4.29.4.4.3 Isothiazolo[4,5-c]pyridine	641
4.29.4.4.4 Isothiazolo[5,4-c]pyridine	641
4.29.4.4.5 Isothiazolo[5,4-b]pyridine	642
4.29.4.4.6 Isothiazolo[4,3-b]pyridine	642
4.29.4.4.7 Isothiazolo[4,3-c]pyridine	643
4.29.4.4.8 Isothiazolo[3,4-c]pyridine	643
4.29.4.4.9 Isothiazolo[3,4-b]pyridine	644
4.29.4.5 Applications	644
4.29.5 OXAZOLE AND THIAZOLE FUSED WITH AZINES: INTRODUCTION	**644**
4.29.5.1 Survey of Possible Ring Systems	644
4.29.5.2 General Survey of Reactivity	645
4.29.6 OXAZOLE FUSED WITH AZINES	**645**
4.29.6.1 Introduction	645
4.29.6.2 Structure	645
4.29.6.2.1 X-Ray diffraction	645
4.29.6.2.2 Molecular spectra	648
4.29.6.3 Reactivity	648
4.29.6.3.1 Survey of reactivity	648
4.29.6.3.2 Electrophilic attack at nitrogen	653
4.29.6.3.3 Electrophilic attack at carbon	654
4.29.6.3.4 Nucleophilic attack at carbon	654
4.29.6.3.5 Nucleophilic attack with ring opening	655
4.29.6.3.6 Other C-linked substituents	657
4.29.6.4 Synthesis: Fully Conjugated Systems	657
4.29.6.4.1 5H-Oxazolo[3,2-a]pyridine	657
4.29.6.4.2 3H-Oxazolo[3,4-a]pyridine	659
4.29.6.4.3 Oxazolo[4,5-b]pyridine	659
4.29.6.4.4 Oxazolo[4,5-c]pyridine	660
4.29.6.4.5 Oxazolo[5,4-c]pyridine	661
4.29.6.4.6 Oxazolo[5,4-b]pyridine	661
4.29.6.5 Saturated and Partially Saturated Rings: Introduction	662
4.29.6.6 Saturated and Partially Saturated Rings: Structure	662
4.29.6.6.1 Thermodynamic aspects	662
4.29.6.7 Saturated and Partially Saturated Rings: Synthesis	663
4.29.6.7.1 5H-Oxazolo[3,2-a]pyridine	663
4.29.6.7.2 3H-Oxazolo[3,4-a]pyridine	666
4.29.6.7.3 Oxazolo[4,5-b]pyridine	667
4.29.6.8 Applications	667
4.29.7 THIAZOLE FUSED WITH AZINES	**668**
4.29.7.1 Introduction	668
4.29.7.2 Structure	672
4.29.7.2.1 X-Ray diffraction	672
4.29.7.2.2 Molecular spectra	672
4.29.7.2.3 Thermodynamic aspects	682
4.29.7.3 Reactivity	682
4.29.7.3.1 Survey of reactivity	682
4.29.7.3.2 Photochemical reactions involving no other species	682
4.29.7.3.3 Electrophilic attack at nitrogen	683
4.29.7.3.4 Electrophilic attack at carbon	684
4.29.7.3.5 Nucleophilic attack at carbon	686
4.29.7.3.6 Nucleophilic attack with ring opening	688
4.29.7.3.7 Nucleophilic attack at hydrogen	689
4.29.7.3.8 Other C-linked substituents	689
4.29.7.3.9 N-Linked substituents	690
4.29.7.3.10 O-Linked substituents	691
4.29.7.3.11 S-Linked substituents	691
4.29.7.3.12 Reactions with cyclic transition states	691
4.29.7.4 Synthesis: Fully Conjugated Rings	692
4.29.7.4.1 5H-Thiazolo[3,2-a]pyridine	692
4.29.7.4.2 3H-Thiazolo[3,4-a]pyridine	696
4.29.7.4.3 Thiazolo[4,5-b]pyridine	696
4.29.7.4.4 Thiazolo[4,5-c]pyridine	698
4.29.7.4.5 Thiazolo[5,4-c]pyridine	699
4.29.7.4.6 Thiazolo[5,4-b]pyridine	699

4.29.7.5 Saturated and Partially Saturated Rings: Introduction	700
4.29.7.6 Saturated and Partially Saturated Rings: Structure	700
4.29.7.6.1 Thermodynamic aspects	700
4.29.7.7 Saturated and Partially Saturated Rings: Reactivity	700
4.29.7.7.1 Electrophilic attack at sulfur	700
4.29.7.7.2 Nucleophilic attack with ring opening	701
4.29.7.7.3 Nucleophilic attack at hydrogen	702
4.29.7.8 Saturated and Partially Saturated Rings: Synthesis	703
4.29.7.8.1 5H-Thiazolo[3,2-a]pyridine	703
4.29.7.8.2 3H-Thiazolo[3,4-a]pyridine	708
4.29.7.9 Applications	709
4.29.8 OXADIAZOLES AND THIADIAZOLES FUSED WITH AZINES: INTRODUCTION	**711**
4.29.8.1 Survey of Possible Ring Systems	711
4.29.8.2 General Survey of Reactivity	712
4.29.9 OXADIAZOLES FUSED WITH AZINES	**712**
4.29.9.1 Introduction	712
4.29.9.2 Structure	712
4.29.9.2.1 X-Ray diffraction	712
4.29.9.2.2 Molecular spectra	713
4.29.9.3 Reactivity	713
4.29.9.3.1 Survey of reactivity	713
4.29.9.3.2 Thermal rearrangements	716
4.29.9.3.3 Electrophilic attack at nitrogen	718
4.29.9.3.4 Electrophilic attack at carbon	718
4.29.9.3.5 Nucleophilic attack at carbon	719
4.29.9.3.6 Nucleophilic attack with ring opening	720
4.29.9.3.7 Reductive ring opening	722
4.29.9.3.8 Nucleophilic attack at hydrogen	722
4.29.9.3.9 Other C-linked substituents	722
4.29.9.4 Synthesis	722
4.29.9.4.1 1H-[1,2,3]Oxadiazolo[3,4-a]pyridine	722
4.29.9.4.2 2H-[1,2,4]Oxadiazolo[2,3-a]pyridine	723
4.29.9.4.3 3H-[1,2,4]Oxadiazolo[4,3-a]pyridine	725
4.29.9.4.4 5H-[1,2,4]Oxadiazolo[4,5-a]pyridine	725
4.29.9.4.5 5H-[1,3,4]Oxadiazolo[3,2-a]pyridine	726
4.29.9.4.6 1H,3H-[1,3,4]Oxadiazolo[3,4-a]pyridazine	727
4.29.9.4.7 2H-[1,2,5]Oxadiazolo[2,3-a]pyridine	727
4.29.9.4.8 [1,2,3]Oxadiazolo[4,5-b]pyridine	728
4.29.9.4.9 [1,2,3]Oxadiazolo[4,5-c]pyridine	728
4.29.9.4.10 [1,2,3]Oxadiazolo[5,4-b]pyridine	728
4.29.9.4.11 [1,2,5]Oxadiazolo[3,4-b]pyridine	728
4.29.9.4.12 [1,2,5]Oxadiazolo[3,4-c]pyridine	730
4.29.9.5 Applications	730
4.29.10 THIADIAZOLES FUSED WITH AZINES	**731**
4.29.10.1 Introduction	731
4.29.10.2 Structure	731
4.29.10.2.1 X-Ray diffraction	731
4.29.10.2.2 Molecular spectra	731
4.29.10.3 Reactivity	732
4.29.10.3.1 Survey of reactivity	732
4.29.10.3.2 Thermal rearrangements	732
4.29.10.3.3 Electrophilic attack at nitrogen	735
4.29.10.3.4 Electrophilic attack at carbon	736
4.29.10.3.5 Nucleophilic attack at carbon	736
4.29.10.3.6 Nucleophilic attack with ring opening	737
4.29.10.3.7 Nucleophilic attack at hydrogen	740
4.29.10.3.8 Other C-linked substituents	740
4.29.10.3.9 Desulfurization	740
4.29.10.4 Synthesis	741
4.29.10.4.1 1H-[1,2,3]Thiadiazolo[3,4-a]pyridine	741
4.29.10.4.2 2H-[1,2,4]Thiadiazolo[2,3-a]pyridine	741
4.29.10.4.3 3H-[1,2,4]Thiadiazolo[4,3-a]pyridine	742
4.29.10.4.4 5H-[1,2,4]Thiadiazolo[4,5-a]pyridine	742
4.29.10.4.5 5H-[1,3,4]Thiadiazolo[3,2-a]pyridine	743
4.29.10.4.6 1H,3H-[1,3,4]Thiadiazolo[3,4-a]pyridazine	744
4.29.10.4.7 2H-[1,2,5]Thiadiazolo[3,4-a]pyridine	745
4.29.10.4.8 [1,2,3]Thiadiazolo[4,5-b]pyridine	745
4.29.10.4.9 [1,2,3]Thiadiazolo[4,5-c]pyridine	745
4.29.10.4.10 [1,2,3]Thiadiazolo[5,4-b]pyridine	746
4.29.10.4.11 [1,2,5]Thiadiazolo[3,4-b]pyridine	747
4.29.10.4.12 [1,2,5]Thiadiazolo[3,4-c]pyridine	748
4.29.10.5 Applications	748

4.29.1 INTRODUCTION

Heterocyclic systems defined by the title of this chapter represent an important, new and expansive area of heterocyclic chemistry. The importance is underlined by a prolific number of papers in the literature in the past few years.

An attempt has been made in this chapter to bring some order into the organization and systematic treatment of the great number of different *ortho*-fused ring systems which can be constructed with the heteroelements as defined by the title. Our main systematic scheme is based on the number, the nature and the relative locations of the azole heteroatoms when the azine ring contains the minimum number of nitrogens. The azine ring is pyridine except for two *ortho*-fused oxadiazole and thiadiazole systems where the two nitrogens are at bridgehead positions and hence the azine ring is pyridazine.

With one nitrogen and one oxygen or sulfur as azole 1,2-locants, *ortho*-fused derivatives are classified as isoxazolo- and isothiazolo-azines; with the same heteroelements as azole 1,3-locants the fused derivatives are classified as oxazolo- and thiazolo-azines. With two nitrogens the compounds are oxadiazolo- or thiadiazolo-azines which are subdivided further according to the relative locations of the azole heteroatoms. This approach can also be extended to fused oxatriazole and thiatriazole. These systems, however, are omitted from this discussion since little work has hitherto been reported.

Each azole ring can be *ortho*-fused to the azine ring in a characteristic and fixed number of ways; the fused azole systems are subgrouped accordingly. Further organization is based on the relative position of the azine nitrogen and on the relative positions of additional heteroatoms in the azine ring.

The numbering of the skeleton in a ring system is generally a consequence of the way in which the skeleton is named. In order to avoid confusion as to the peripheral numbering sequence and locations of substituents in the fused systems, structural formulae are drawn as in *Chemical Abstracts* in accordance with the rules of heterocyclic nomenclature. Thereby the reader is not required to possess detailed knowledge of the elaborate series of priorities used in the structural presentation and numbering of ring systems.

Besides the challenge of new chemistry represented by these fused systems and the desire for mapping of their physicochemical properties, much of the interest in these systems is caused by their close structural analogies to purines and the essential biological functions of the latter; hence a great number of patent applications have been filed for potential biological applications. A further stimulus has been the discovery of some of these ring systems in natural products.

The most intensively investigated fused ring systems have been thiazoloazines and to a lesser degree oxazoloazines. In general most work has been done on fully conjugated ring systems although many investigations have been reported on fully saturated and, in particular, on partly saturated isoxazolo-, oxazolo- and thiazolo-azines.

The properties of the various fused systems can be readily understood and rationalized from the fundamental physicochemical properties of the parent heterocyclic rings as discussed in the ensuing sections.

4.29.2 ISOXAZOLE AND ISOTHIAZOLE FUSED WITH AZINES: INTRODUCTION

4.29.2.1 Survey of Possible Ring Systems

There are nine different ways of constructing fused isoxazolo- or isothiazolo-pyridines. These are the main groups into which such compounds have been classified (Chart 1).

The nitrogen is common to both rings in group (*a*) since it is in a bridgehead position. The fully conjugated (*a*)-system is cationic (group *A*). In the series (*b*)–(*e*) the oxygen (or sulfur) is attached to the azine ring, and in the series (*f*)–(*i*) it is the azole nitrogen which is attached to the azine ring.

The relative position of the azine nitrogen can be changed. The system with bridgehead nitrogen (group *a*) contains only one nitrogen whereas the other systems contain two nitrogens.

The pyridine ring may contain additional heteroatoms, *e.g.* as diazines and triazines. These systems are subgrouped under the pyridine group according to the position of the

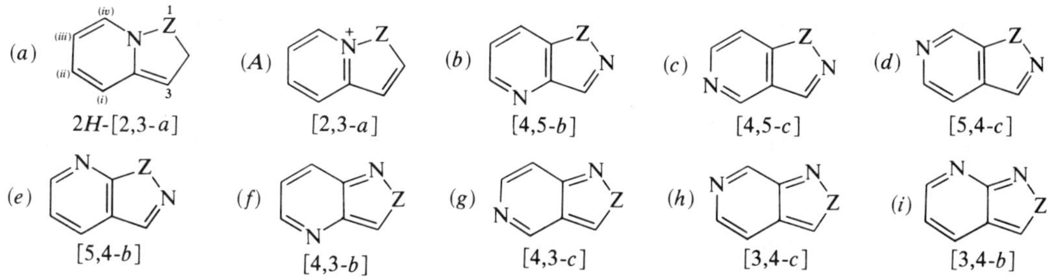

Chart 1 Ring systems formed by fusion of isoxazole or isothiazole with pyridine (Z = O, S)

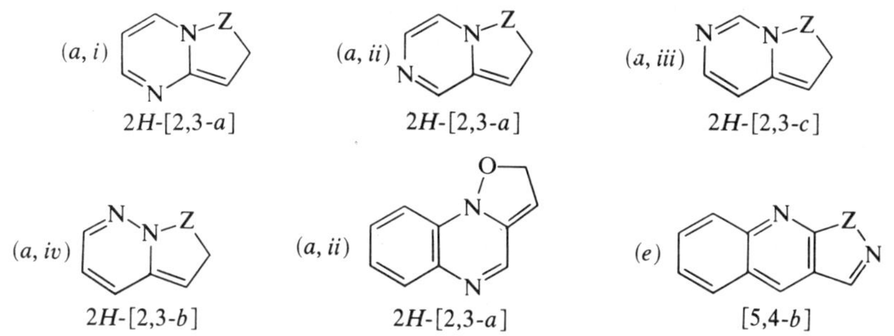

Chart 2 Ring systems formed by fusion of isoxazole or isothiazole with pyridine containing additional heteroatoms and/or rings (Z = O, S)

second heteroatom. They are treated in the order of increasing number for the position of the second heteroatom (Chart 2).

Polycyclic azines, *e.g.* quinoline and quinoxaline, are grouped under the parent azine ring to which the azole is fused.

Fused ring systems can be prepared either from an azine precursor or from an isoxazole or isothiazole precursor. In the sections on synthesis of these compounds the former method is grouped under 'Azine approach', the latter under 'Azole approach'.

4.29.2.2 General Survey of Reactivity

Electrophilic attack at nitrogen occurs preferentially in the azine ring. Electrophilic attack at carbon occurs in the azine ring when activated by electron-releasing substituents.

The ease of nucleophilic substitution in the azine ring is increased after fusion with the azole. Substitution is also possible in the azole ring, especially in the fused isothiazole systems.

The N—O bond in isoxazoloazines is cleaved by hydrogenolysis. Besides reductive desulfurization, the N—S bond in isothiazoloazines is cleaved by nucleophilic attack at sulfur. Both the N—O and N—S bonds are cleaved on proton abstraction in the azole ring.

Fused azolium derivatives readily suffer ring opening reactions by nucleophilic attack. The tendency for ring opening of the azole ring is generally increased in fused dihydroazoles.

4.29.3 ISOXAZOLE FUSED WITH AZINES

4.29.3.1 Introduction

Isoxazole can be fused with pyridine in nine different ways; the section on synthesis has been organized accordingly. Further fused systems arise from the diazines, the triazines and their areno and heteroareno homologues.

Most of the work described so far deals with the synthesis of members of the many possible ring systems. Both saturated and partially saturated ring systems have been prepared, but most information comes from the fully conjugated ring systems.

Chemical properties are discussed in the section on reactivity. There are several claims in the patent literature for potential use in medicine of isoxazoloazines; a review is given in the section on applications.

4.29.3.2 Structure

4.29.3.2.1 X-Ray diffraction

Bond lengths and bond angles are listed in Table 1. The compounds are fully conjugated and as such are planar.

Table 1 X-Ray Diffraction Data for Fused Isoxazoloazines

Compound	Bond lengths (Å) and angles (°)			
	Ring A		Ring B	
7-Bromo-5-chloro-3-phenyl-isoxazolo[4,5-d]pyrimidine ⟨72MI42901⟩	a^1 1.42 a^1a^3 103	a^3 1.40 a^1b^1 112	a^2 1.37 a^2a^3 120	a^3 1.40 a^2b^5 120
	b^1 1.27 a^3c^1 110	c^1 1.35 c^1h^1 106	b^2 1.32 a^3b^2 121	b^3 1.30 b^2b^3 114
	h^1 1.44 b^1h^1 108		b^4 1.38 b^3b^4 130	b^5 1.32 b^4b^5 115
7-Bromo-3-phenylisoxazolo[4,5-d]pyridazin-4(5H)-one ⟨72MI42900⟩	a^1 1.47 a^1a^4 105	a^4 1.36 a^1b^1 109	a^2 1.50 a^2a^4 118	a^3 1.40 a^2b^2 111
	b^1 1.32 a^4c^1 111	c^1 1.34 b^1h^1 108	a^4 1.36 a^3a^4 125	b^2 1.39 a^3b^3 119
	h^1 1.44 c^1h^1 108		b^3 1.32 b^2g^1 129	g^1 1.37 b^3g^1 118
1-Acetyl-4,6-dimethylisoxazolo[3,4-b]pyridin-3(1H)-one ⟨79AX(B)470⟩	a^1 1.49 a^1a^5 108	a^5 1.36 a^1c^1 104	a^2 1.41 a^2a^3 113	a^3 1.39 a^2a^5 120
	b^1 1.44 a^5b^1 109	c^1 1.40 b^1h^1 106	a^4 1.36 a^3a^4 123	a^5 1.36 a^4b^2 125
	h^1 1.40 c^1h^1 111		b^2 1.37 a^5b^3 129	b^3 1.35 b^2b^3 111
5-Bromo-4,6-dimethyl[3,4-b]pyridin-3(1H)-one ⟨79AX(B)468⟩	a^1 1.46 a^1a^5 105	a^5 1.37 a^1c^1 105	a^2 1.39 a^2a^3 114	a^3 1.43 a^2a^5 123
	b^1 1.31 a^5b^1 117	c^1 1.36 b^1h^1 103	a^4 1.35 a^3a^4 123	a^5 1.37 a^4b^2 122
	h^1 1.46 c^1h^1 111		b^2 1.36 a^5b^3 121	b^3 1.39 b^2b^3 118

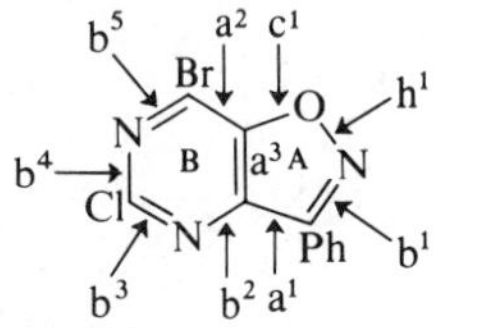

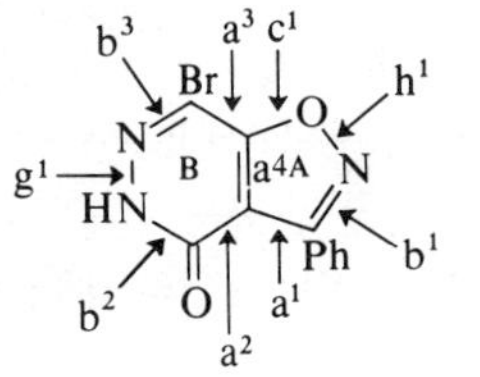

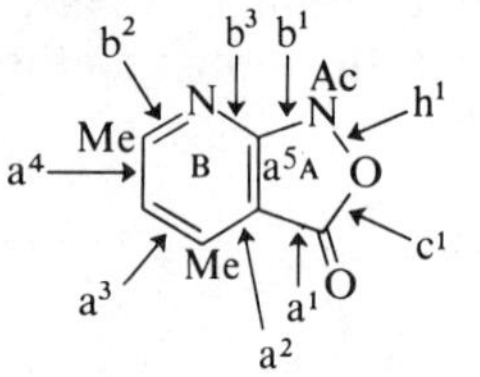

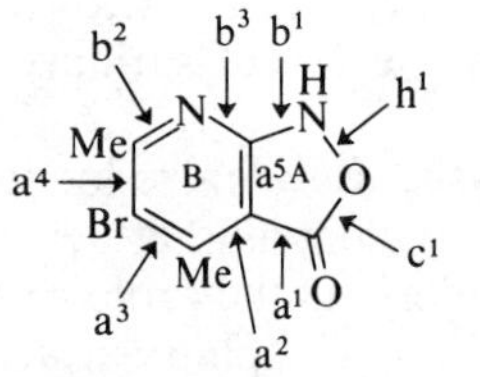

4.29.3.2.2 Molecular spectra

(a) *¹H NMR spectroscopy*. Proton chemical shifts and coupling constants for ring CH of fully conjugated and partially saturated systems are listed in Table 2.

(b) *¹³C NMR spectroscopy*. Few ¹³C NMR data have been reported and the available data are listed in Table 3. The chemical shifts and the $^1J_{CH}$ coupling in the six-membered ring are as in azines.

(c) *UV spectroscopy*. UV data for fully conjugated systems are listed in Table 4.

(d) *IR spectroscopy*. Assignments of characteristic frequencies are not possible due to the diversity of the systems.

CO bands are expected to have absorption frequencies as in the parent monocycle or in their benzo homologues, *e.g.* (**1**). The tautomerism of hydroxy and amino compounds largely

Table 2 ^{1}H NMR Data for Fused Isoxazoloazines

Compound	Solvent	2	3	4	5	6	7	Coupling constants (Hz)	Ref.
R^1 Ph, R^2 H, R^3 Br	DMSO-d_6	8.85	—	—	—	—	10.07		77JOC1364
R^1 Ph, R^2 OH, R^3 H	TFA	—	7.74	—	—	8.14	8.95	$J_{2,3}$ 2.0, $J_{6,7}$ 8.0	71JCS(C)1196
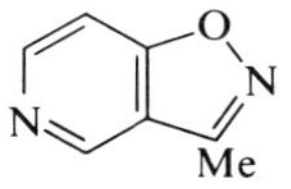 (isoxazolo, Me)	CDCl$_3$	—	2.62	8.95	—	8.60	7.42	$J_{4,7}$ 1.0, $J_{6,7}$ 6.0	75JCS(P1)2190
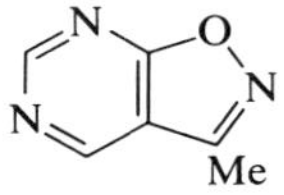 (isoxazolo, Me)	CDCl$_3$	—	2.58	8.08	7.33	8.60	—	$J_{4,5}$ 8.0, $J_{4,6}$ 1.7, $J_{5,6}$ 4.8	77JHC435
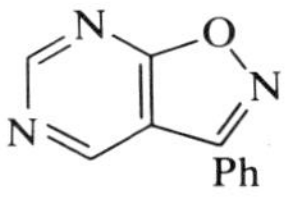 (isoxazolo, Ph)	CDCl$_3$	—	—	9.40	—	9.16	—		77H(7)51
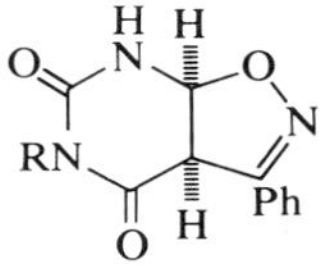 Reduced, R = H	DMSO-d_6	5.15 (H-3a), 5.64 (H-7a)						$J_{3a,7a}$ 10.0	72JOC2983
R = Me	DMSO-d_6	5.27 (H-3a), 5.65 (H-7a)						$J_{3a,7a}$ 10.0	72JOC2983

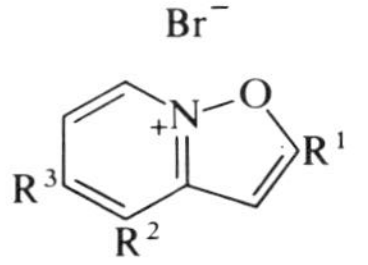

Table 3 ^{13}C NMR Data for Fused Isoxazoloazines

Compound	Solvent	^{13}C chemical shifts (p.p.m.)						$^1J_{CH}$ coupling constants (Hz)	Ref.
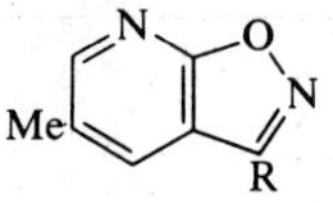		3	3a	4	5	6	7a		
R^1 = Me	CDCl$_3$	155.4	113.4	131.0	129.2	151.4	168.6	18.1 (5-Me), 10.7 (3-Me) J_4 163.8, J_6 179.4	79S449
R^1 = Ph	CDCl$_3$	157.1	111.7	131.8	129.8	151.5	169.2	18.2 (5-Me)	79S449

Table 4 UV Absorption Data for Fused Isoxazoloazines

Compound			Solvent	λ_{max}^a ($\log \varepsilon$)	Ref.
R^1	R^2	R^3			
Ph	H	H	EtOH	322 (4.30), 272 (4.22)	77JOC1364
H	OH	Br	EtOH	335 (3.6), 285 (4.02), 223^b	71JCS(C)1196
H	OH	Br	NaOH/EtOH	384^b (3.48), 312 (4.13), 245 (4.17)	71JCS(C)1196
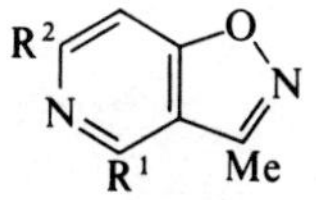 H	H	—	MeOH	265^b (2.93), 232 (3.82)	75JCS(P1)2190
OH	OH	—	MeOH	285 (4.11), 265^b (3.93)	75JCS(P1)2190
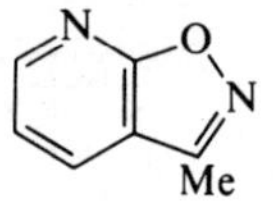			MeOH	291 (3.74), 284 (3.86), 280 (3.84), 232 (3.71)	77JHC435
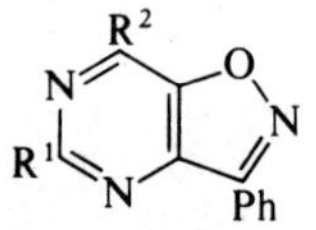 Cl	Cl	—	EtOH	313 (3.94), 228 (4.21)	67T675
OMe	OMe	—	EtOH	307 (4.00), 218 (4.27)	67T675
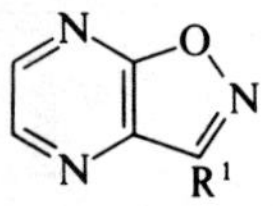 Me	—	—	Cyclohexane	293 (4.00), 287 (3.98), 241 (3.33)	73JHC181
Ph	—	—	EtOH	301 (4.07), 230 (4.04)	68T4907
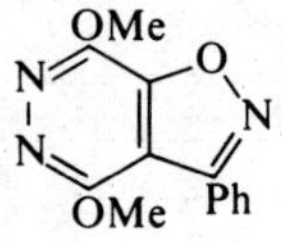			EtOH	247 (4.07), 223 (4.25)	67T681
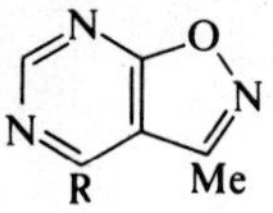 OH	—	—	EtOH	267 (3.73), 242^b (3.78), 235 (3.81)	64JOC2116
NHMe	—	—	EtOH	283 (3.96), 252 (3.99)	64JOC2116

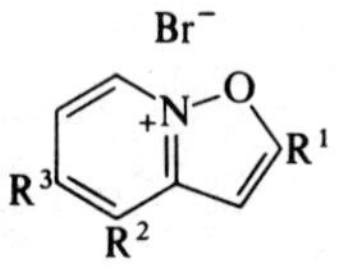

Five-membered Rings Fused with Six-membered Rings 621

Table 4 (cont.)

R^1	Compound R^2	R^3	Solvent	$\lambda_{max}{}^{a}$ $(\log \varepsilon)$	Ref.
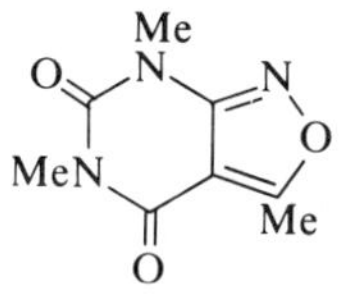	(structure: Me-O-N fused isoxazole with MeN, Me, O substituents)		EtOH	236 (4.04)	77CPB2974

^a nm. ^b Shoulder.

follows the same pattern. In the examples (**2**) and (**3**) the isoxazole ring takes the place of the phenyl ring in isoquinoline; 3-hydroxyisoquinoline exists predominantly in the hydroxy form and so does the isoxazole analogue (**2**), whereas 1-hydroxyisoquinoline exists in the oxo form which is also the case for the isoxazole analogue (**3**; Table 5).

Table 5 IR Absorption Characteristics for Fused Isoxazoloazines

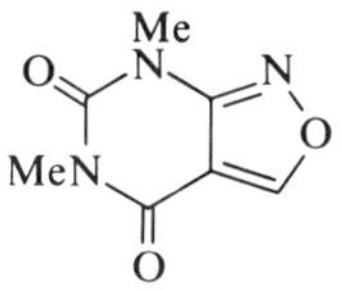 (**1**)	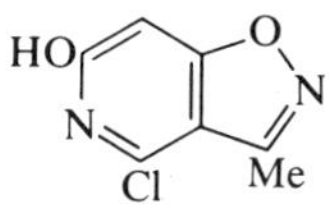 (**2**)	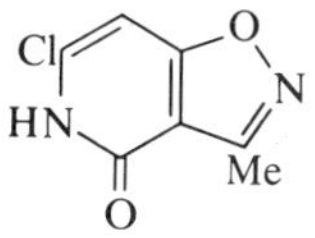 (**3**)
1720, 1670 cm^{-1} (CO)	3200–3300 cm^{-1} (OH)	1670 cm^{-1} (CO)
Nujol	KBr	KBr
78CPB2497	75JCS(P1)2190	75JCS(P1)2190

(*e*) *Mass spectrometry.* Owing to the variety of structures and hence the many modes of fragmentation, the reader should consult the original literature for information on the type of compound of interest.

4.29.3.3 Reactivity

4.29.3.3.1 Survey of reactivity

N-Alkylation and *N*-oxidation occur preferably in the azine ring. Electrophilic substitution at carbon also takes place in the azine ring when activated by strongly electron-releasing groups. The ease of nucleophilic substitution in the azine ring is increased after fusion with the azole ring. In dihydro systems where the azine ring is aromatic the reactivity in the azine ring towards electrophiles and nucleophiles is similar to that of the fully conjugated system.

The N—O bond in isoxazoles is easily cleaved by hydrogenolysis and also as a result of proton abstraction from an unsubstituted C-3 or C-5 position in the isoxazole ring. Reduced isoxazoles having the oxygen attached to an activated azine position may readily undergo nucleophilic opening of the ring at this point.

4.29.3.3.2 Thermal and photochemical reactions involving no other species

Isoxazoles are photolabile. The first step in the reaction is believed to be fission of the N—O bond with formation of a diradical which may recyclize. In the case of the isoxazolopyridine (**4**) the reaction is rationalized as taking place *via* a 2*H*-azirine (**5**) and a carbenoid structure (**6**) before formation of an oxazole ring (**7**). A high pressure mercury lamp is used, and the reaction can be run on a preparative scale ⟨79CB3282⟩. Photolysis of the isoxazole (**8**), which is fused to an azine ring with a different order of its azole heteroatoms, also leads to oxazole formation (**10**) ⟨78CPB2497⟩, but in this instance another type of intermediate is postulated (**9**).

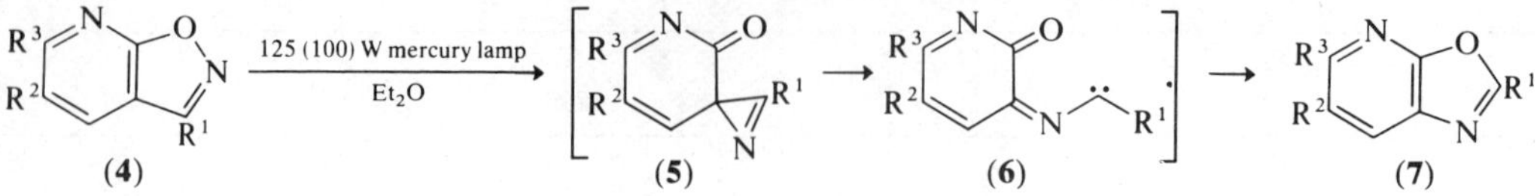

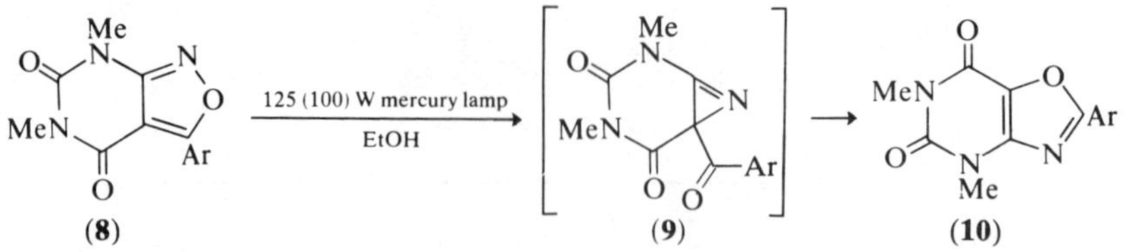

4.29.3.3.3 Electrophilic attack at nitrogen

The low pK_a value of isoxazole (-2.03) indicates that the nitrogen is not a good target for electrophiles. When fused to azine systems selective electrophilic attack on the azine nitrogen(s) is therefore to be expected.

Alkylation of hydroxyazines follows the usual azine behavior with preference for *N*-alkylation as in the fused hydroxypyridine (**11**) ⟨72AP833⟩. In the fused uracil (**12**), using diazomethane or dimethyl sulfate, a mixture of *N,N*-dimethylated (**13**) and *N,O*-dimethylated (**14**) products is obtained. On heating the mixture in acetylacetone in the presence of sodium iodide, the *O*-methyl isomer is converted into its *N*-isomer, a type of rearrangement well established in azines ⟨67T675⟩.

N-Oxidation is an electrophilic reaction which therefore occurs preferentially in the azine ring. In the fused pyrazine (**15**) it is N-7 which is oxidized, access to N-4 being sterically retarded by the 3-methyl group ⟨73JHC181⟩.

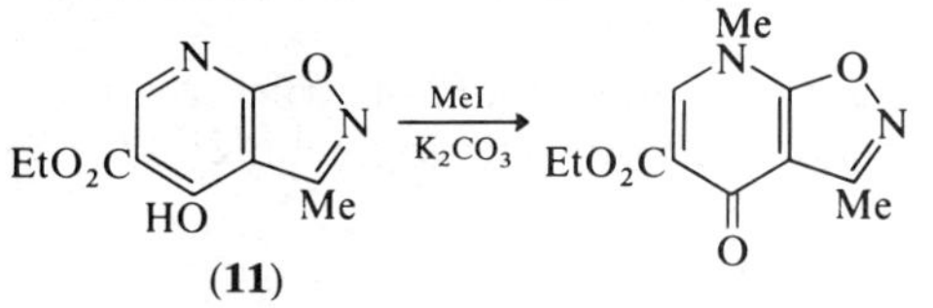

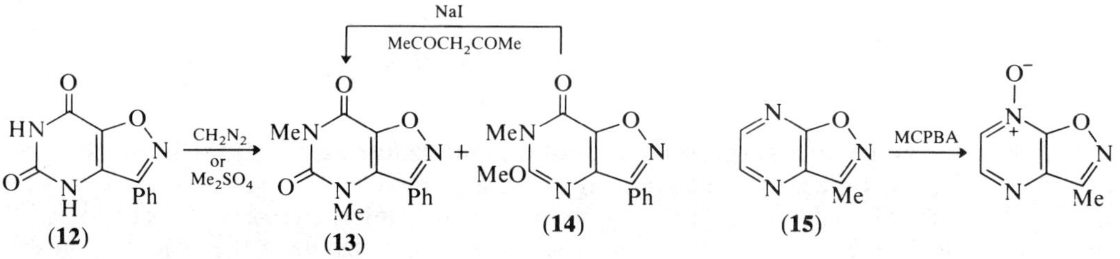

4.29.3.3.4 Electrophilic attack at carbon

Electrophilic substitution in isoxazoles occurs most readily in the 4-position. The latter is part of the ring junction in the fused systems and is only free in the N-bridgehead series.

In general, electrophilic substitution is to be expected in the azine ring when this is substituted by one or more strongly electron-donating groups in the usual manner of azine chemistry. Thus *N*-oxidation promotes electrophilic substitution. Nitration occurs preferentially in the *para* position to the *N*-oxide function, and when this is occupied the nitro group goes into an *ortho* position. Phosphorus oxychloride causes *ortho* chlorination and deoxygenation. Both reactions are observed for the pyrazine *N*-oxide (**16**) ⟨73JHC181⟩.

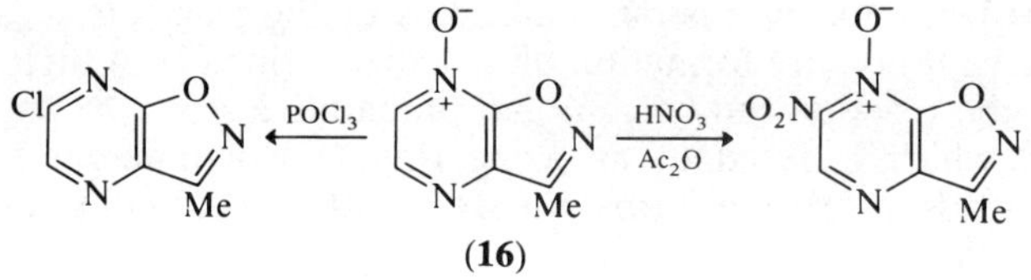

4.29.3.3.5 Nucleophilic attack at carbon

The ease of nucleophilic substitution in an azine ring is increased after fusion with isoxazole, since the charge of the transition state is distributed over a larger electron-accepting aromatic system. The stabilization is most efficient for substitution on the carbon next to the isoxazolo ring. Kinetic studies with the chloro substituted isoxazolopyridines (17) and (18) using methoxide ion substitution show that the ring fusion increases the rate constant 10^2–10^5-fold in the 4- and 6-positions as compared with the corresponding chloropyridines. The 6-chlorine in (18) is more readily displaced than in (17), the ratio between the rate constants being 229. In (18) the chlorine at C-4 reacts eight times faster than at C-6 ⟨79JHC49⟩. These properties compare well with the relative reactivities of haloquinolines and isoquinolines; in isoquinoline a chlorine at C-1 is very reactive, whereas at C-3 fairly vigorous conditions are required for displacement. In (17; R = Cl) methoxide substitution is possible at C-4 or at both C-4 and C-6.

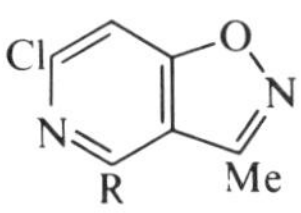

(17) R = H, Cl

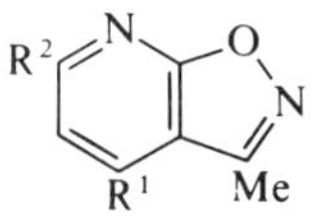

(18) $R^1 = R^2 = Cl$; $R' \neq R^2 = H$, Cl

The reaction of the fused 5,7-dihalopyrimidine (19) to form the 7-methoxy derivative (20) also shows that the more readily displaced chlorine is next to the azole ring ⟨67T675⟩. Hydrazinolysis of the isoquinoline analogue (21) is stepwise with initial 4-substitution (22). The difference in reactivity at C-4 and C-6 in (22) is sufficient for selective acid hydrolysis of the hydrazino group in (22). Reductive cleavage of the hydrazino C—N bond is achieved in the usual manner by the reaction of its tosyl derivative with alkali ⟨75JCS(P1)2190⟩.

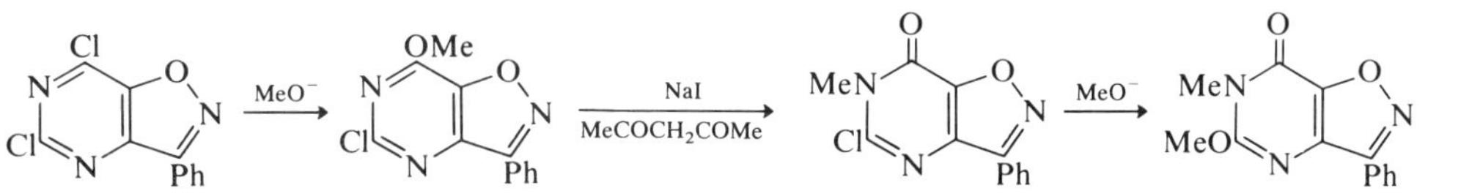

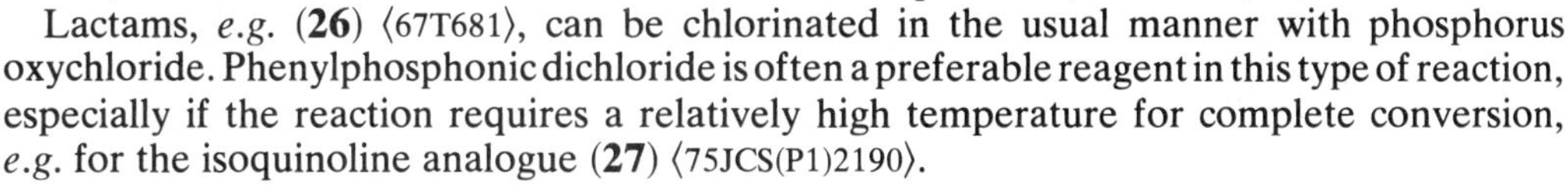

Halogens are displaced by sulfur nucleophiles as in the phthalazine analogue (24) ⟨67T681⟩. In the thiation of the fused uracil (25) the higher reactivity at C-4, for the reasons discussed above, can be used to effect selective thiation in this position ⟨78CPB2497⟩.

Lactams, e.g. (26) ⟨67T681⟩, can be chlorinated in the usual manner with phosphorus oxychloride. Phenylphosphonic dichloride is often a preferable reagent in this type of reaction, especially if the reaction requires a relatively high temperature for complete conversion, e.g. for the isoquinoline analogue (27) ⟨75JCS(P1)2190⟩.

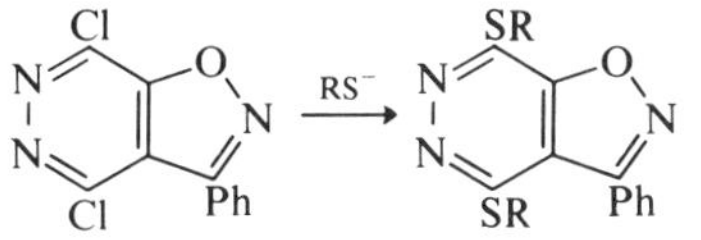

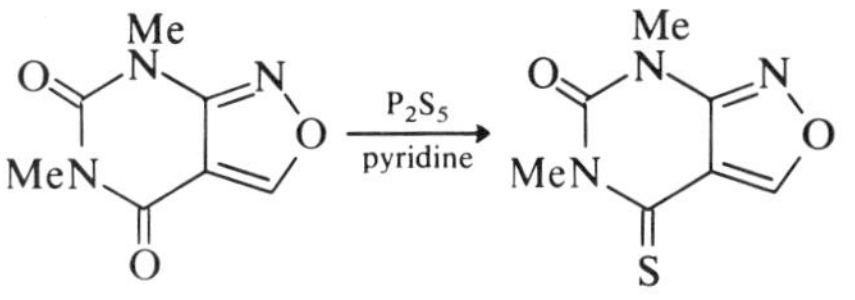

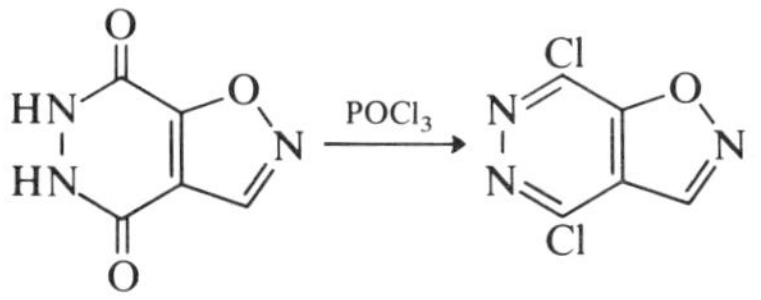

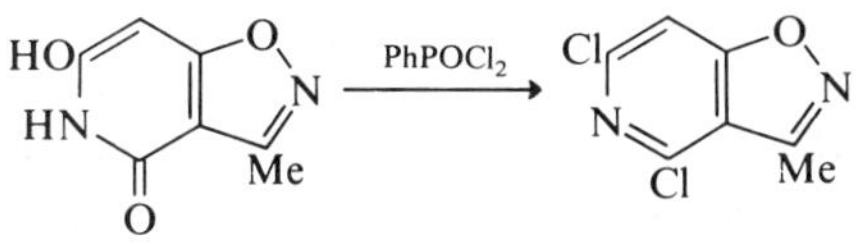

4.29.3.3.6 Reductive ring opening

The N—O bond in isoxazoles is very readily cleaved by hydrogenolysis. Reductive reactions in the substituents may therefore be accompanied by a ring opening reaction through the cleavage of this bond. A number of reductive methods can be used. The isoxazole ring in the fused pyridazine (**28**) is cleaved by hydrogenolysis over Raney nickel ⟨68AC(R)121⟩, whereas the most common catalyst palladium-on-charcoal is used for the fused pyridine (**29**) ⟨73AP746⟩. The 3-oxo derivative (**30**) gives the corresponding amide with zinc metal in alkali ⟨75JCS(P1)693⟩.

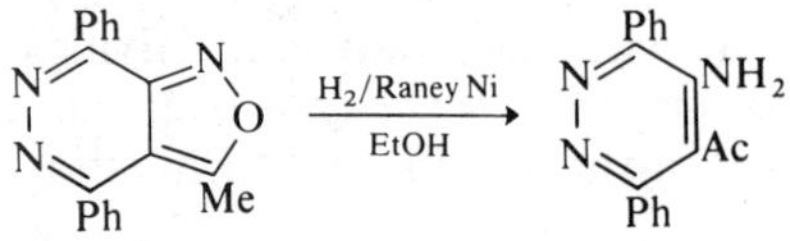

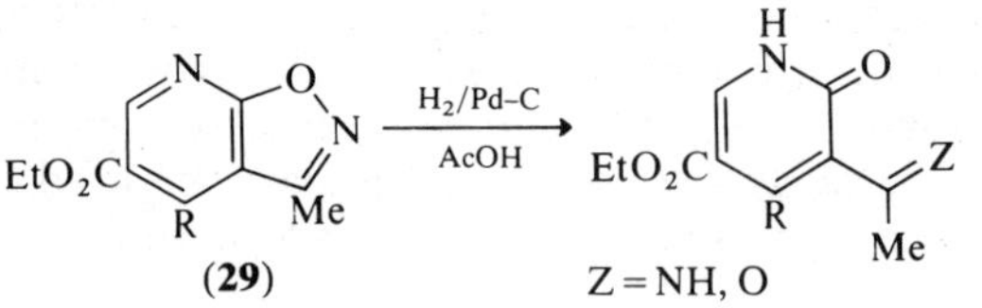

4.29.3.3.7 Nucleophilic attack at hydrogen

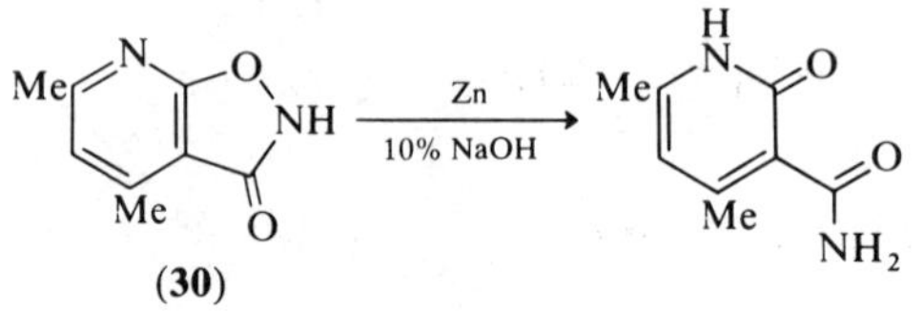

Isoxazoles with no substituent in the 3-position undergo proton abstraction by base which leads to cleavage of the N—O bond and the generation of a cyano group. If it is the 5-position which is unsubstituted in the isoxazole, proton elimination again leads to ring opening, this time with the formation of an intermediate ketene. The latter type of reaction is observed for the fused uracil (**31**); the basicity of piperidine is sufficient for removal of H-3 ⟨78CPB2497⟩. The former type of reaction is seen for the fused 10,11-dihydrophenanthridine (**32**) using sodium methoxide ⟨74JHC455⟩. Carbanion formation at C-1 can equally well be effected by a decarboxylation reaction under alkaline conditions as in the case of the fused quinoxaline (**33**) ⟨72JOC2498⟩. A more detailed discussion of the ring opening of isoxoles may be found in Chapter 4.16.

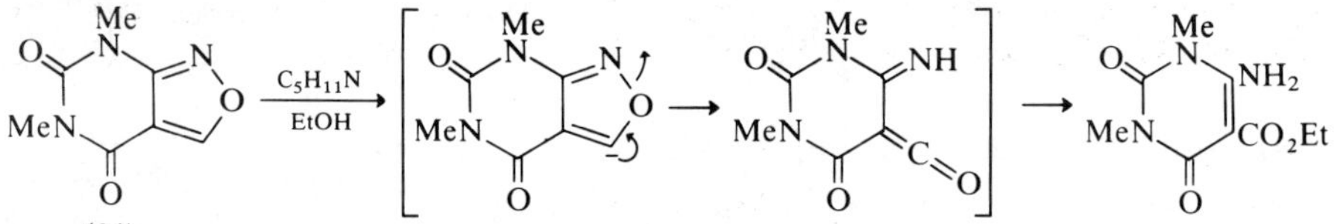

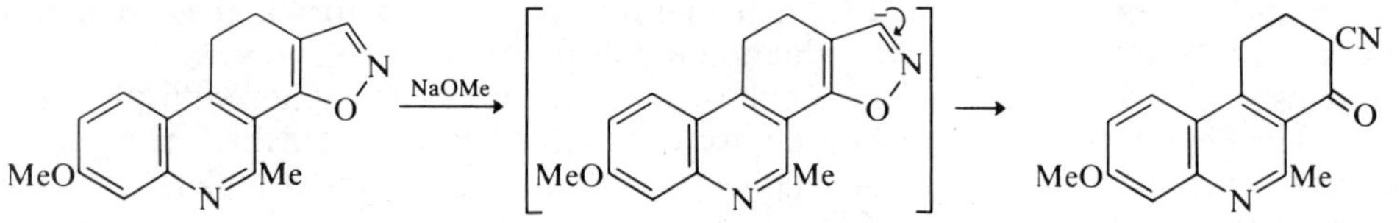

4.29.3.3.8 Other C-linked substituents

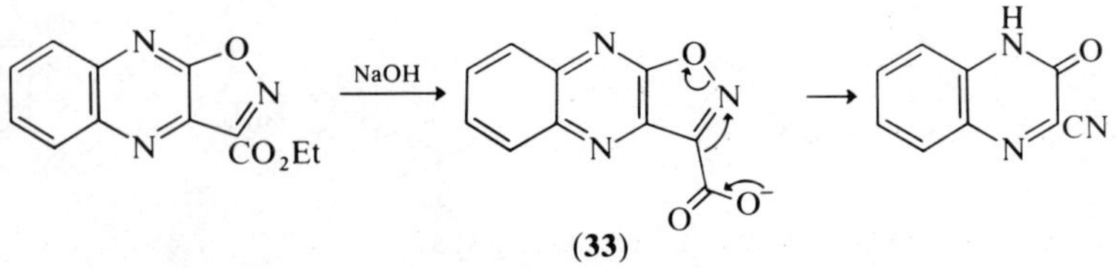

A methyl group attached to the carbon of the isoxazolo ring is activated for electrophilic attack. Bromination of the fused pyridine 3-methyl derivative (**34**) occurs in the methyl substituent and the bromine can be displaced by nucleophiles ⟨75FES992⟩. With oxo reactants or their equivalents 3-vinyl derivatives result, *e.g.* from (35) ⟨77CPB2974⟩. In the 3,4-dimethyl derivative (36) the methyl group in the isoxazolo ring is selectively attacked with sodium methoxide as base ⟨76JHC409⟩.

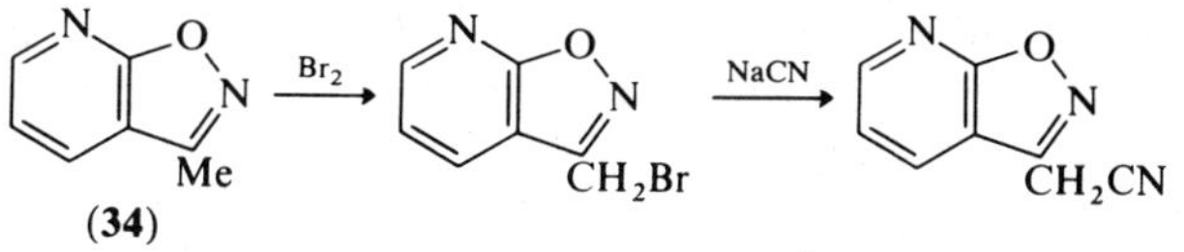

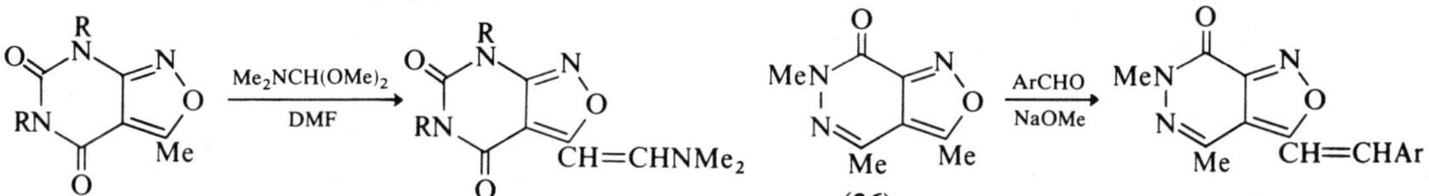

4.29.3.4 Synthesis: Fully Conjugated Rings

4.29.3.4.1 2H-Isoxazolo[2,3-a]pyridine

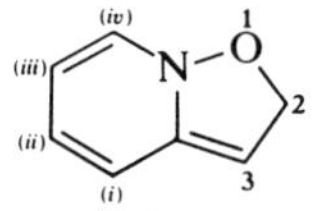

Azine approach. Acid-catalyzed cyclization of 2-β-oxoalkylpyridine 1-oxides (37) constitutes a general method for the preparation of the isoxazolo[2,3-*a*]pyridinium ring system ⟨77JOC1364⟩. By analogy to the synthesis of isoxazoles from nitrile oxides, the oxo chain should be replaceable by an alkynyl or ethenyl chain, the latter containing a substituent which can readily be eliminated from the cyclic product.

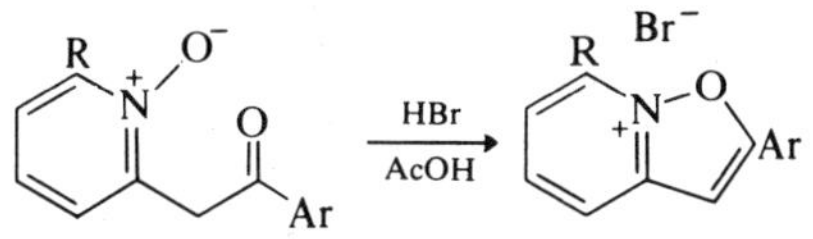

Azole approach. 3-(4-Oxo-1-alkenyl)isoxazoles on acid catalysis will undergo cyclization at the nitrogen atom; if the oxo group is part of an ester, a lactam is formed, *e.g.* (38) ⟨80LA542⟩. The hydrobromide of the betaine (40) has been prepared by ionic dibromination of the ketone (39), followed by elimination of HBr ⟨71JCS(C)1196⟩.

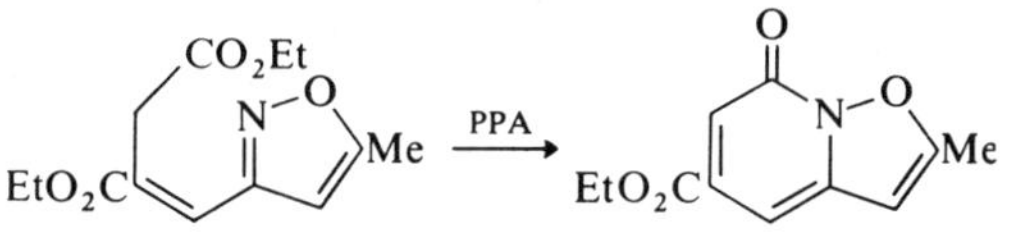

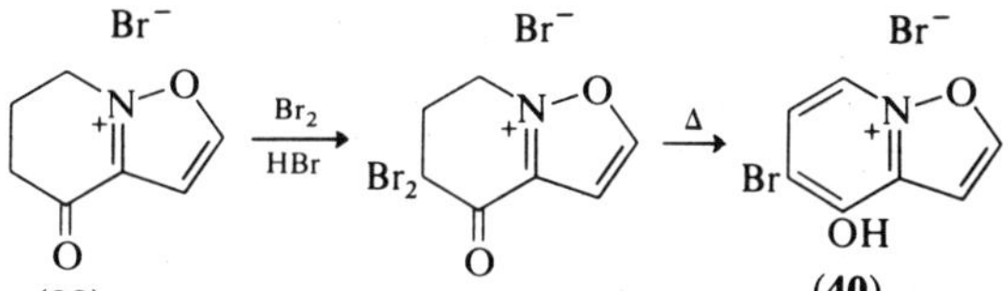

(i) *2H-Isoxazolo[2,3-a]pyrimidine*

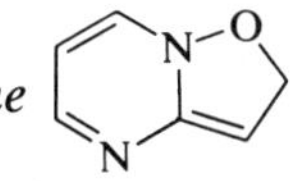

Azole approach. 3-Aminoisoxazoles, which are readily available, form fused pyrimidines (41) when treated with 1,3-dicarbonyl compounds ⟨76UKZ261⟩.

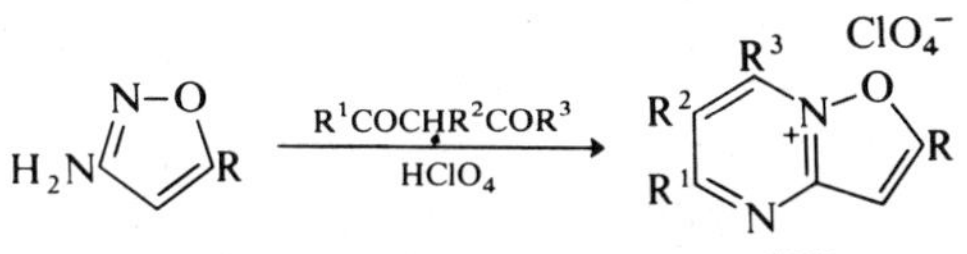

4.29.3.4.2 Isoxazolo[4,5-b]pyridine

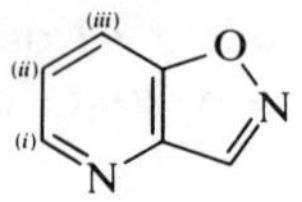

(ii) Isoxazolo[4,5-d]pyrimidine

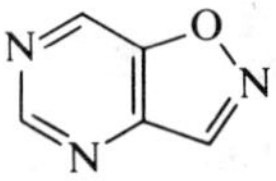

Azole approach. Hofmann degradation of isoxazole-4,5-dicarboxamide goes faster for the 4-substituent, which is converted into an isocyanate group. The latter reacts with the remaining carbamoyl substituent to form the fused uracil (**42**) ⟨67T675⟩.

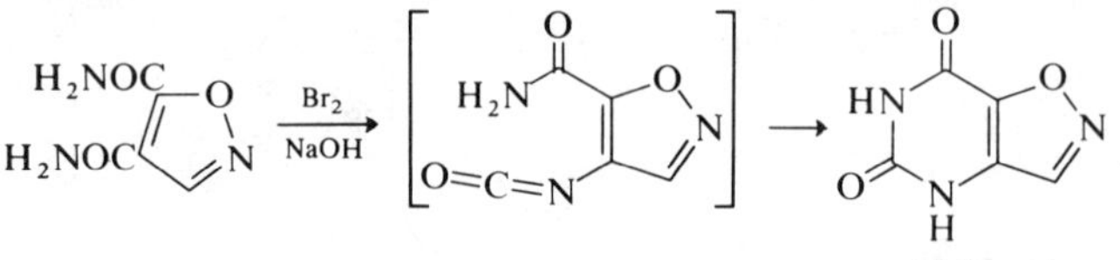

(iii) Isoxazolo[4,5-b]pyrazine

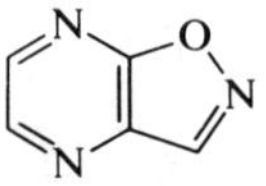

Azine approach. Cyclization of β-oxo oximes yields isoxazoles. In an adaptation of this method to the preparation of fused isoxazoles, 3-ethoxycarbonylmethylene-2(1*H*)-quinoxalinone is nitrosated and the resulting oxime (**43**) cyclized to the fused isoxazole (**44**) ⟨72JOC2498⟩.

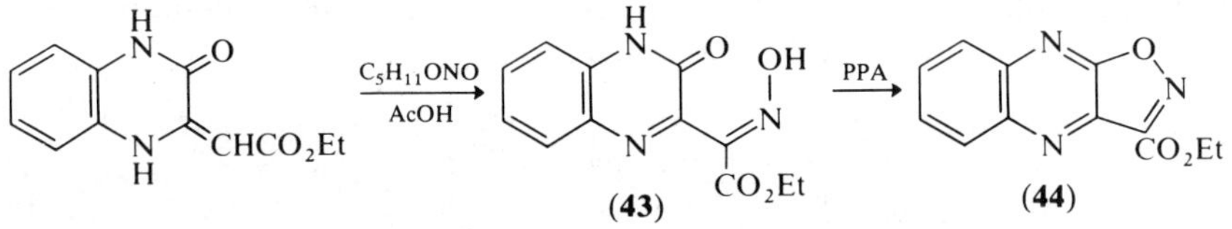

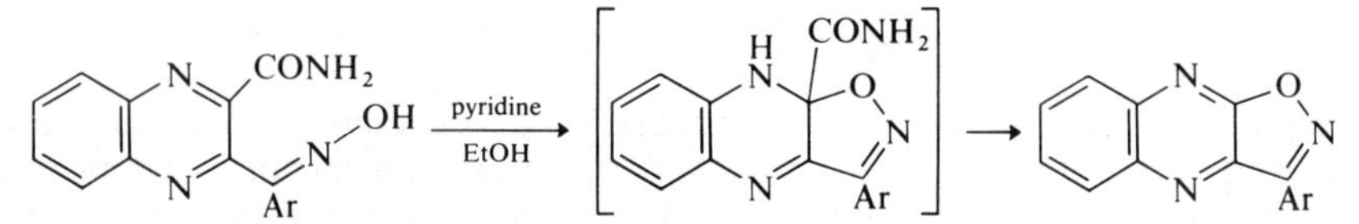

Quinoxaline-3-ketoximes in the presence of pyridine can be cyclized; the reaction proceeds by addition of the oxime hydroxy group to the electron-deficient diazine to form a cyclic dihydro derivative which expels the ring function substituent to regain aromaticity, *e.g.* (**45**) ⟨66HCA2426⟩.

Azole approach. The reaction of 4,5-diaminoisoxazoles (**46**) with 1,2-dicarbonyl compounds represents a convenient route to pyrazines ⟨73JHC181, 68T4907⟩; the reaction is analogous to the formation of quinoxalines from *o*-phenylenediamine.

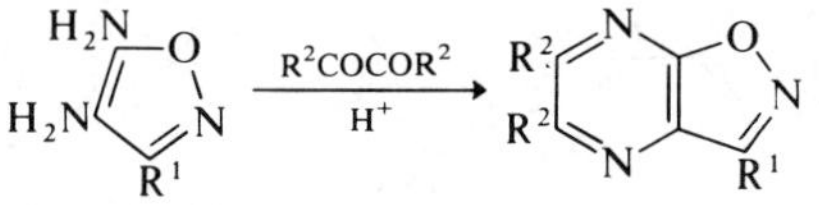

4.29.3.4.3 Isoxazolo[4,5-c]pyridine

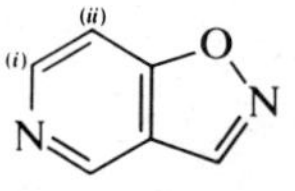

Azole approach. The cyclization of the 4-ethyloxycarbonyl-5-bis(ethyloxycarbonyl)methylisoxazole (**47**) using ammonia has its analogy in the synthesis of 1,3-dihydroxyisoquinolines ⟨75JCS(P1)2190⟩.

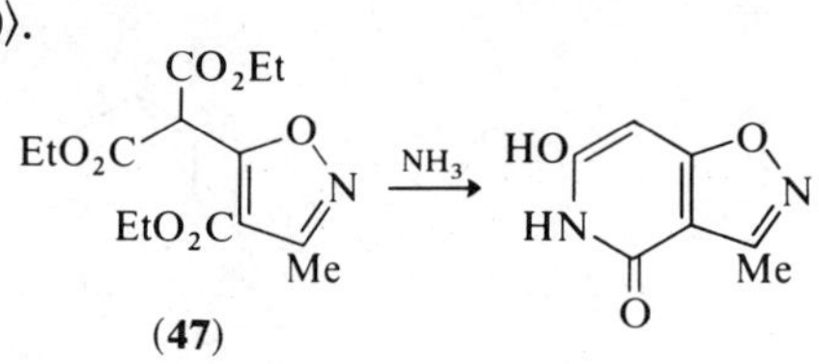

(ii) Isoxazolo[5,4-d]pyrimidine

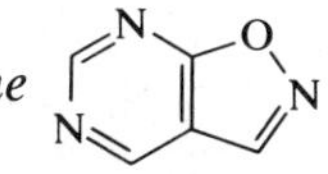

Azine approach. Cyclization of the oxime (**48**) ⟨73G219⟩ corresponds to a common method for the preparation of isoxazoles, *viz.* the reaction between 1,3-dicarbonyl compounds and hydroxylamine. Similarly a cyano group reacts with hydroxylamine to form an *N*-hydroxyamidine which can be cyclized to an isoxazole if the vicinal carbon is activated and carries a leaving group. The 5-cyanopyrimidine (**49**) perhaps behaves unexpectedly in that it is a trichloromethyl substituent which is the leaving group in the cyclization ⟨79JHC1109⟩. This behavior may be attributed to its location in the activated pyrimidine 4-position.

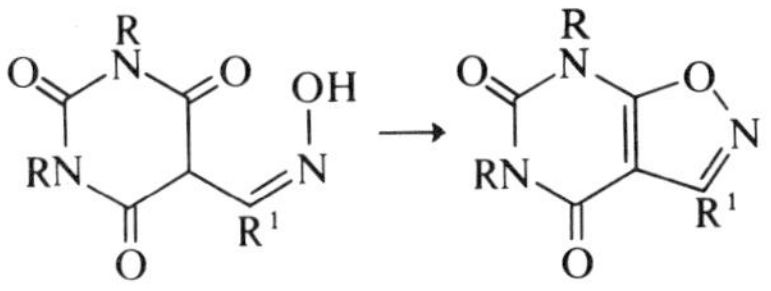

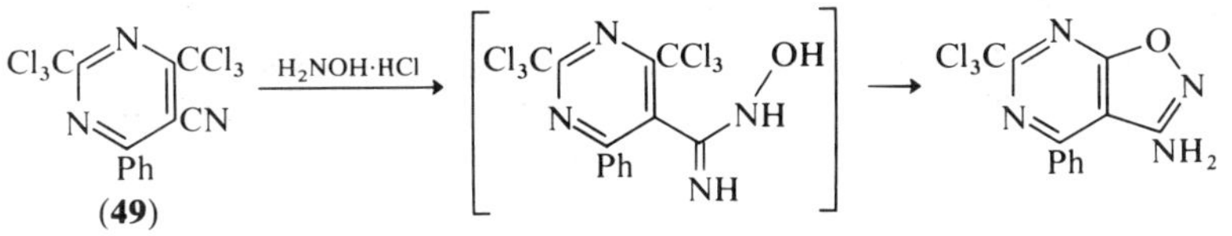

Azole approach. The isoxazole ring can be regarded as analogous to the phenyl ring in the synthesis of quinazolines. One method is to react a vicinal aminocarboxamide with an activated carboxylic acid derivative to form the fused pyrimidine; the isoxazole (**50**) is reacted in this manner ⟨67T687⟩. Sometimes it is advantageous to carry out this type of reaction in two steps, such as in the reaction *via* the 5-ethoxymethyleneamino intermediate (**51**) ⟨64JOC2116⟩.

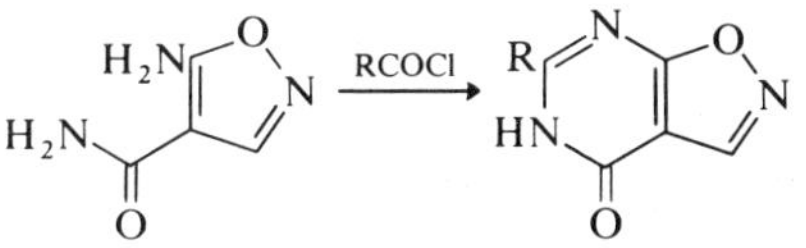

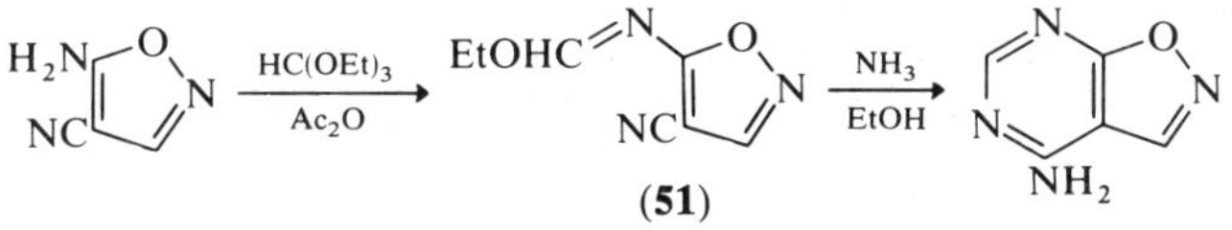

4.29.3.4.4 Isoxazolo[5,4-c]pyridine

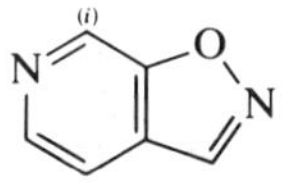

Members of this ring system are described in the section dealing with saturated or partially saturated compounds.

4.29.3.4.5 Isoxazolo[5,4-b]pyridine

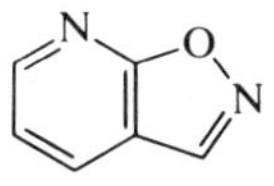

Azole approach. Two main routes lead to fused isoxazoles belonging to this system. Both are adaptations of well-established quinoline syntheses, where the phenyl ring has been replaced by the isoxazole ring. In the first method a carbonyl derivative with at least one hydrogen at an α-carbon atom is condensed with a 4-acyl-5-aminoisoxazole as for the formyl derivative (**52**) ⟨77H(7)51⟩ and the ester (**53**) ⟨77JHC435⟩. In the second method cyclization is effected by means of a carbonyl group which is part of the 5-amino side-chain.

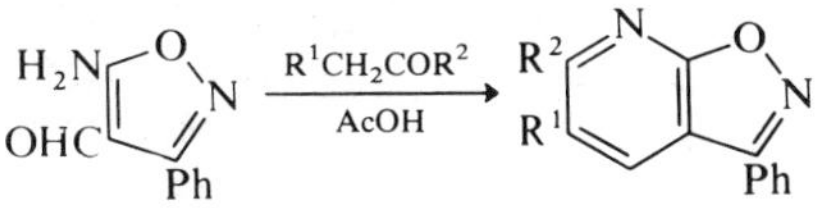

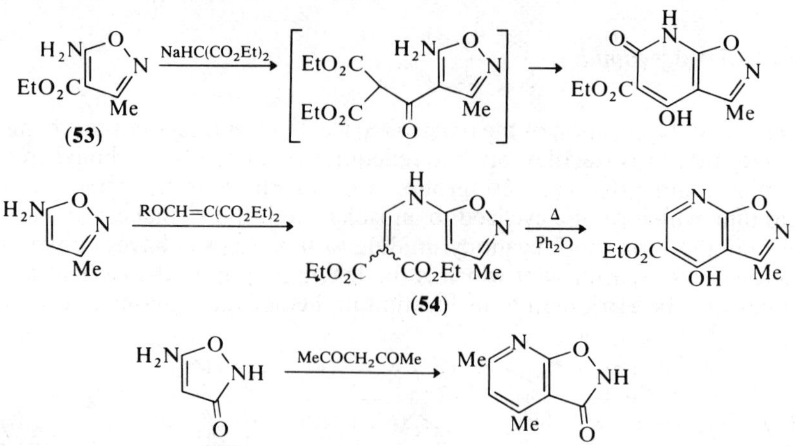

This is illustrated by the cyclization of the ester (**54**) ⟨72AP833⟩ and by the reaction of the 5-aminoisoxazole (**55**) with acetylacetone ⟨75JCS(P1)693⟩.

Cyclization involving the 5-amino group is also possible. An intermediate (**56**) for this reaction has been prepared from dimeric malononitrile and benzohydroximoyl chloride; the intermediate (**56**) is cyclized under alkaline conditions ⟨80CB1195⟩.

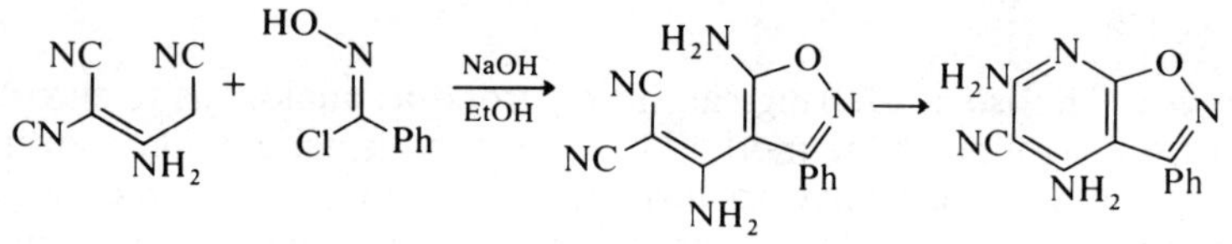

4.29.3.4.6 Isoxazolo[4,3-b]pyridine

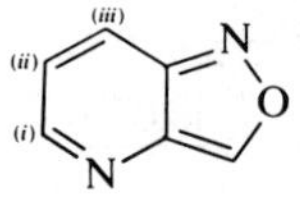

(ii) *Isoxazolo[4,3-d]pyrimidine*

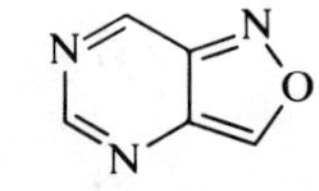

Azine approach. The fused uracil (**58**) has been prepared from 5-amino-1,3,6-trimethyluracil; diazotization of the latter yields a fused triazine *N*-oxide (**57**), which has its triazine ring transformed into an isoxazole ring by tin(II) chloride treatment ⟨62USP3056781⟩.

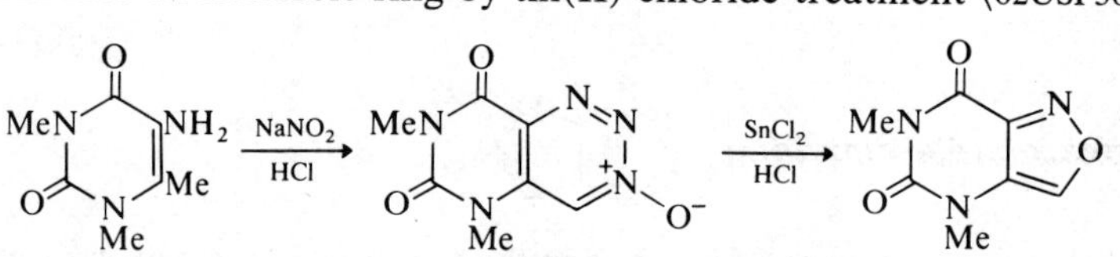

4.29.3.4.7 Isoxazolo[4,3-c]pyridine

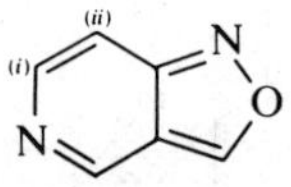

(i) *Isoxazolo[3,4-d]pyridazine*

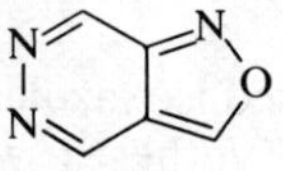

Azole approach. As usual, treatment of a vicinal diacyl derivative with hydrazine gives the corresponding pyridazine ring, product as in the case of (**59**) ⟨65G1478⟩.

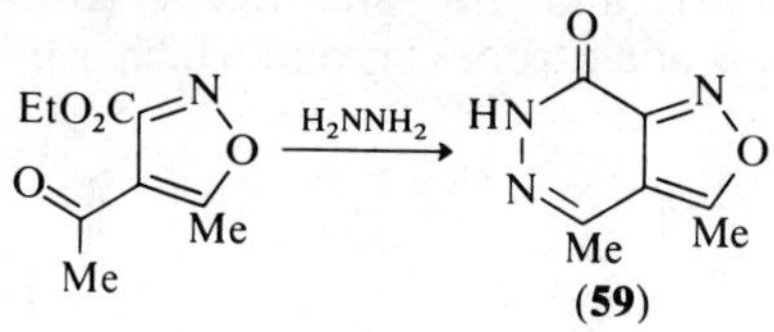

(ii) Isoxazolo[3,4-d]pyrimidine

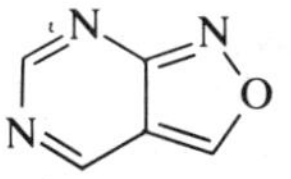

Azine approach. 6-Aminouracils react readily with electrophiles. The 6-hydroxylamino analogue (**60**) on heating with arylaldehydes in DMF undergoes cyclocondensation with formation of the fused isoxazole (**62**). The reaction is thought to involve an arylidene intermediate which is cyclized by a Michael-type addition of the hydroxyl group to the arylidene function ⟨78CPB2497⟩. The same type of product (**62**) is obtained when the hydroxylamine (**60**) is reacted with an acid anhydride. In this case cyclization of the intermediate (**63**) corresponds to nucleophilic substitution on the oxime nitrogen. By analogy, cycloinsertion of the azole carbon into (**60**) can be achieved by using ortho esters; with triethyl orthoformate the compound (**64**) without a 3-substituent is formed ⟨77CPB2974⟩. The latter is also available from (**60**) by the Vilsmeier reaction which proceeds via an intermediate corresponding to the arylidene (**61**).

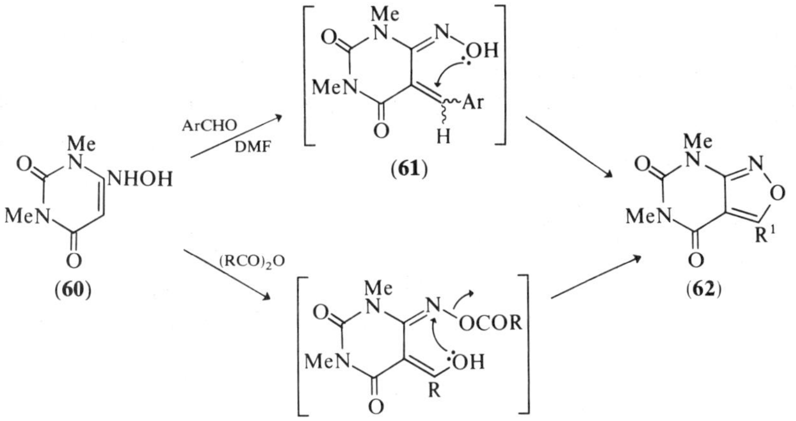

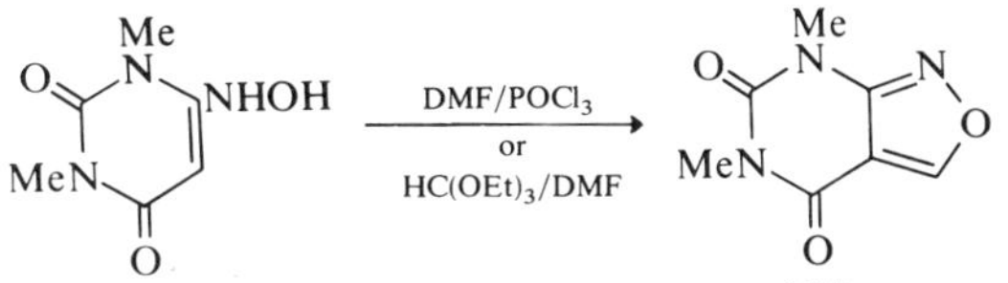

Azole approach. Vicinal amino–cyano groups react with acid chlorides or anhydrides, the product being a pyrimidinone, as in the reaction of the substituted isoxazole (**65**) ⟨73GEP2249163⟩.

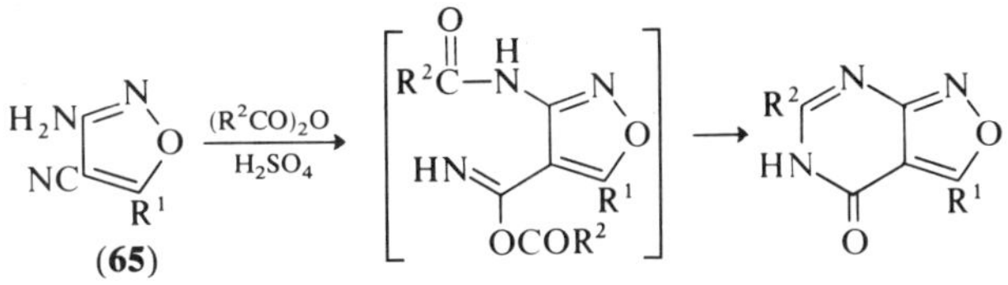

4.29.3.4.8 Isoxazolo[3,4-c]pyridine

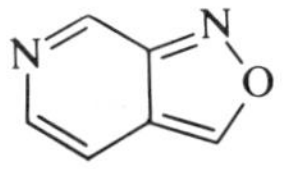

Azine approach. The reduction of a nitro group may be carried out to yield a hydroxylamino group. With a vicinal cyano group under appropriate reaction conditions, cyclization to an isoxazole may result as shown for the formation of (**66**) ⟨68CPB1700⟩.

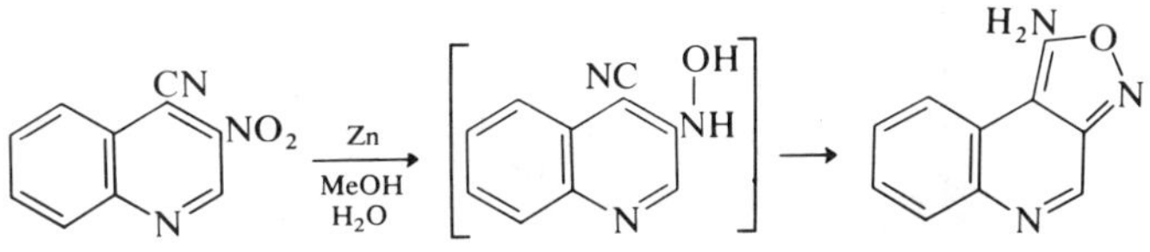

4.29.3.4.9 Isoxazolo[3,4-b]pyridine

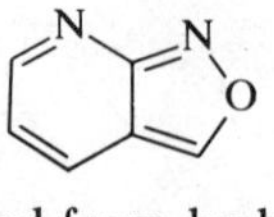

Azine approach. The systems derived from hydroxylaminoazines can be cyclized on to a vicinal carboxy group, in a reaction corresponding to α-lactone formation as shown for (**67**) ⟨75JCS(P1)693⟩.

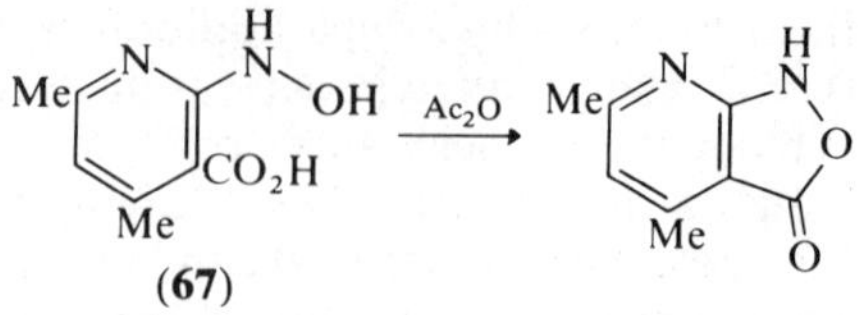

4.29.3.5 Saturated and Partially Saturated Rings: Reactivity

Fused isoxazoline systems where the azine ring remains aromatic show similar behavior towards electrophilic and nucleophilic substitution in the azine ring as the corresponding fully conjugated system. The reduced isoxazole ring, however, may be less resistant towards nucleophilic ring opening reactions than the aromatic isoxazole ring.

Ring opening reactions. 2,3-Dihydroisoxazolo[2,3-a]pyridinium salts undergo ring cleavage by nucleophile attack at C-2 whereby the N-oxide function becomes the leaving group. Alternatively the nucleophile may add to an activated azine position to form a pseudobase. Ring cleavage is shown for (**68**) in its reaction with secondary amines; the pseudobase formed by addition at C-8 is sometimes a minor product ⟨61JOC3802⟩.

In the discussion of the fully conjugated systems, it was pointed out that the N—O bond is very easily cleaved by hydrogenolysis. Therefore fused isoxazolidine systems find use in chiral syntheses. Examples of cleavage reactions are demonstrated with LAH for the N-methylpiperidine derivative (**69**) ⟨78TL4647⟩, hydrogenolysis over Raney nickel for the isoquinoline (**70**) ⟨69CB736⟩ and metal in acid solution for the piperidine (**71**) ⟨78TL2753⟩.

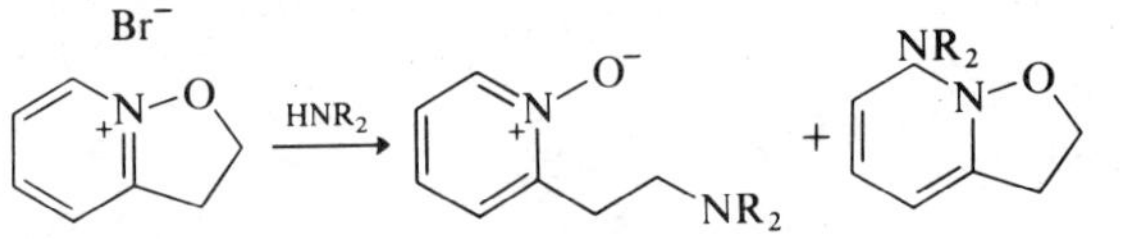

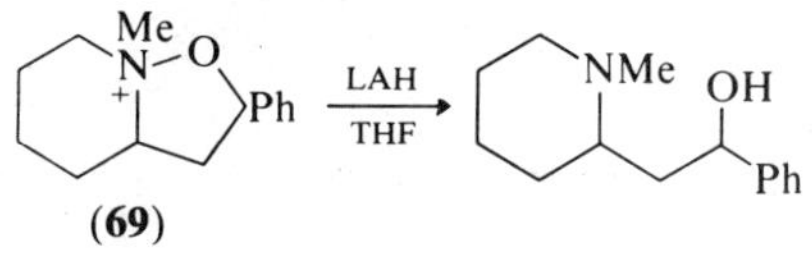

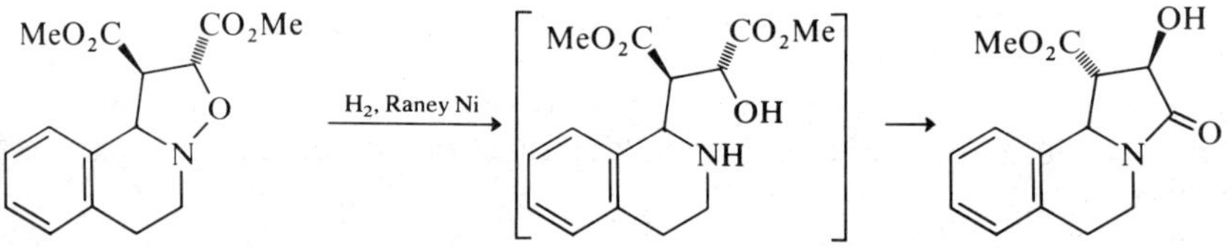

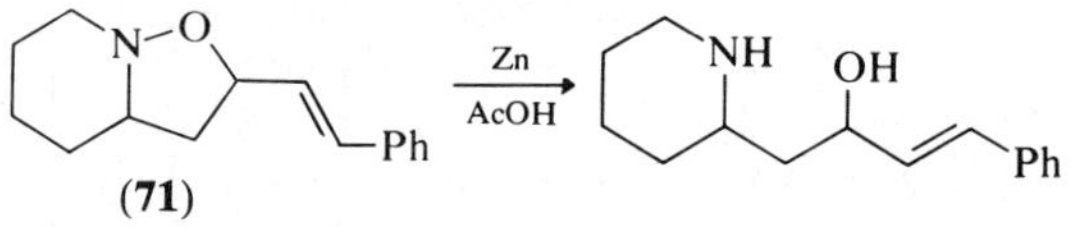

In partially reduced systems where the isoxazole ring has remained aromatic, proton abstraction by base from a vacant position next to one of the azole heteroatoms will occur in the same way as in the fully conjugated system. This is shown for the isoxazolylium salt (**72**) ⟨72JCS(P1)2441⟩, where a consecutive ring closure involving the ketene function and the vicinal oxo group in its hydroxy form (**73**) leads to lactone formation. When the oxime of (**72**) is used in this reaction, the corresponding 4H-pyrrolo[3,2-b]pyridin-2(1H)-one is formed ⟨74JCS(P1)158⟩.

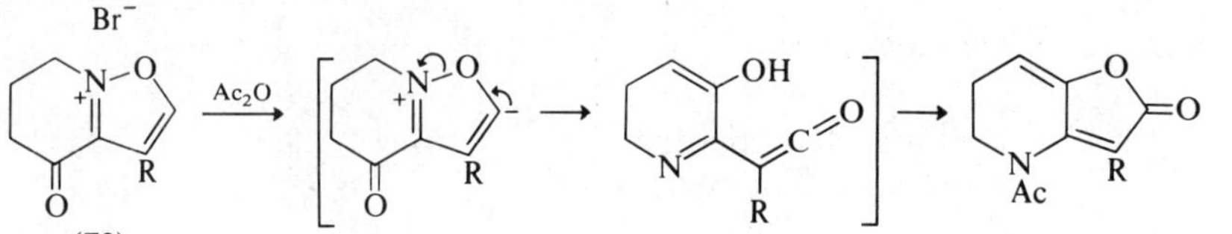

4.29.3.6 Saturated and Partially Saturated Rings: Synthesis

4.29.3.6.1 2H-Isoxazolo[2,3-a]pyridine

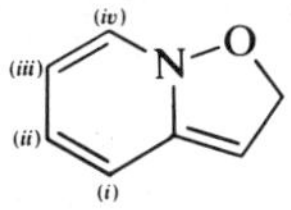

Azine approach. Saturated ring systems with a bridgehead nitrogen atom are best prepared from 3,4,5,6-tetrahydroazine 1-oxides. The latter can be classified as nitrones and as such will undergo 1,3-dipolar cycloaddition reactions.

Studies of the reaction between 3,4,5,6-tetrahydropyridine 1-oxide (74) and *trans*-1-phenyl-1,3-butadiene show addition to the less substituted double bond and regiochemical integrity of the product (75), which is a diastereoisomeric mixture in accordance with frontier orbital calculations for 1-substituted 1,3-butadienes with a conjugated substituent ⟨78TL2753⟩. The frontier orbital treatment predicts that the regioselectivity in the reaction with dipolarophiles decreases as the IP of the dipolarophile is increased. Experimentally it is found that methyl vinyl ketone gives high regioselectivity in its reaction with (74), forming (76), whereas methyl acrylate gives a mixture of the regioisomers (76) and (77) ⟨79TL4167⟩.

Monosubstituted alkenes such as styrene and methyl acrylate, which are capable of secondary orbital interactions with the nitrone (74), give a stereochemical mixture (77a) from *endo* and *exo* addition, but the latter predominates. The unactivated, monosubstituted alkenes give *exo* addition overwhelmingly with this nitrone. The *exo* preference can largely be attributed to steric interactions between ring hydrogens and the substituent R in the disfavored *endo* mode of addition ⟨78TL4647⟩.

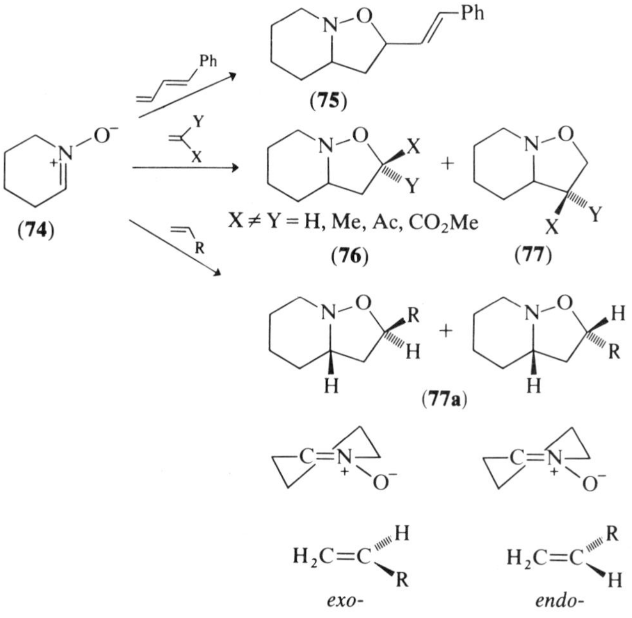

The stereochemistry of maleic and fumaric acid dimethyl esters is retained in their cycloaddition reactions with 3,4-dihydroisoquinoline 2-oxide (78) ⟨69CB736⟩.

Cycloaddition reactions with alkynic compounds are illustrated by the reaction of methyl propiolate with (78) to form (79), and by the reaction of benzyne with (78) to form (80) ⟨69CB904⟩.

In another type of reaction, CH-acidic compounds in the presence of a strong base react with aldonitrones by C—C bond formation to form a hydroxylamine. Formed from the reaction of (78) and dimethyl malonate, the adduct (81) undergoes cyclization to an azlactone ⟨71LA(748)143⟩.

(i) 2H-Isoxazolo[3,2-a]pyrimidine

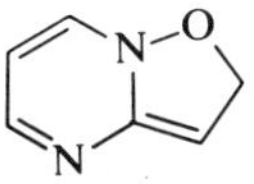

Azine approach. Partially saturated isoxazolo[3,2-b][1,3]benzoxazines (83) are of interest as antiinflammatory agents. They can be synthesized from 2-hydroxybenzohydroxamic acids and α,β-unsaturated carbonyl compounds; the last step involves simple cycloalkylation of the intermediate *N*-oxide (82) ⟨77AF760⟩.

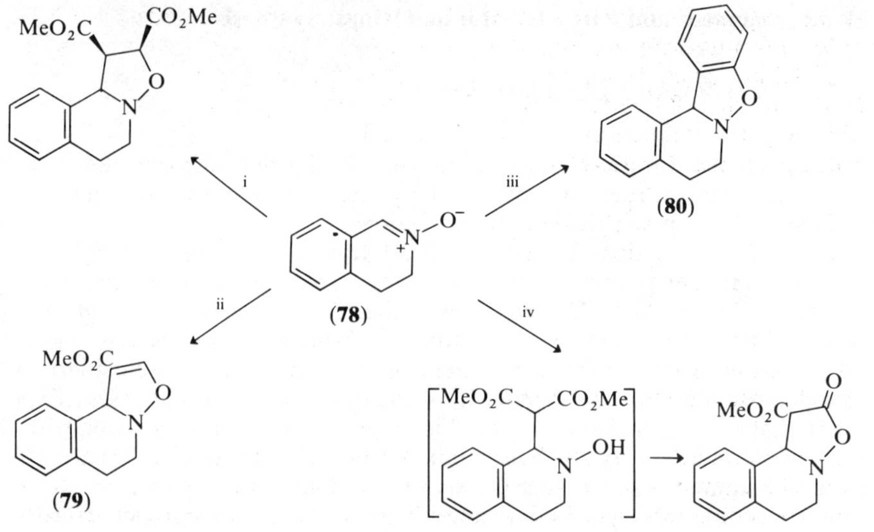

i, MeO₂CCH=CHCO₂Me; ii, HC≡CCO₂Me; iii, o-NH₂C₆H₄CO₂H, NO⁺; iv, NaCH(CO₂Me)₂, MeOH

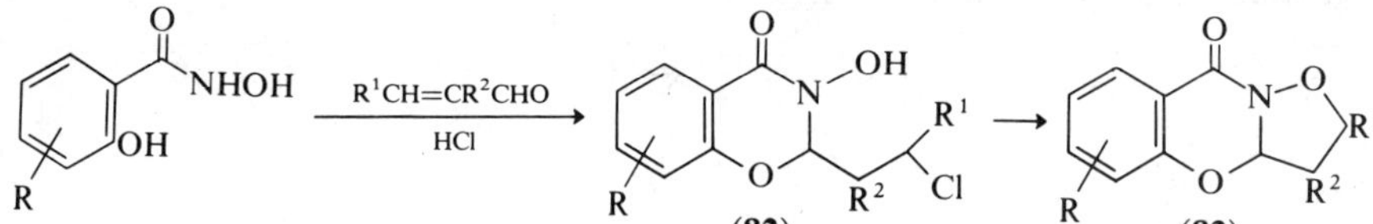

Azole approach. Quinazolines can be prepared from isatoic anhydride and an amide. When the amide function is part of a cyclic structure such as in a 3-isoxazolidinone, the 2,3-dihydro-9*H*-isoxazolo[3,2-*b*]quinazolin-9-one (**84**) is formed ⟨77AF766⟩.

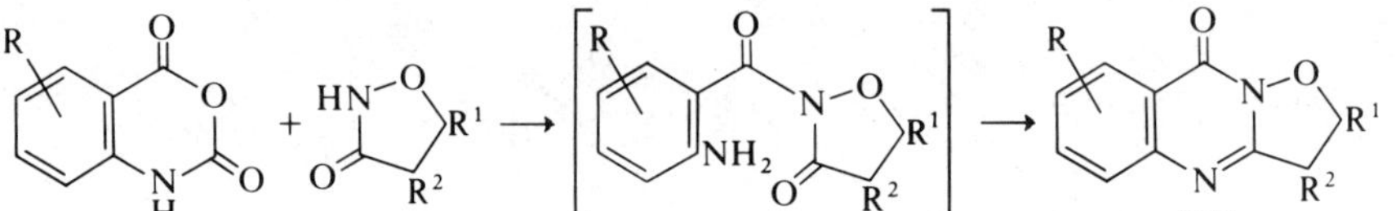

(ii) *2H-Isoxazolo[2,3-a]pyrazine*

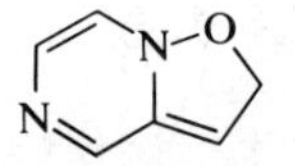

Azine approach. Heteroaromatic *N*-oxides will undergo 1,3-dipolar cycloaddition as a nitrone. Thus quinoxaline 1-oxide reacts with a variety of dipolarophiles, *e.g.* in the formation of the adduct (**85**). The 1,4-dioxide can form cycloadducts at both *N*-oxide functions. The adduct from 2*H*-1,4-benzoxazine 4-oxide and acrylonitrile is the isoxazolo[3,2-*c*][1,4]benzoxazine (**86**) ⟨79T1771⟩.

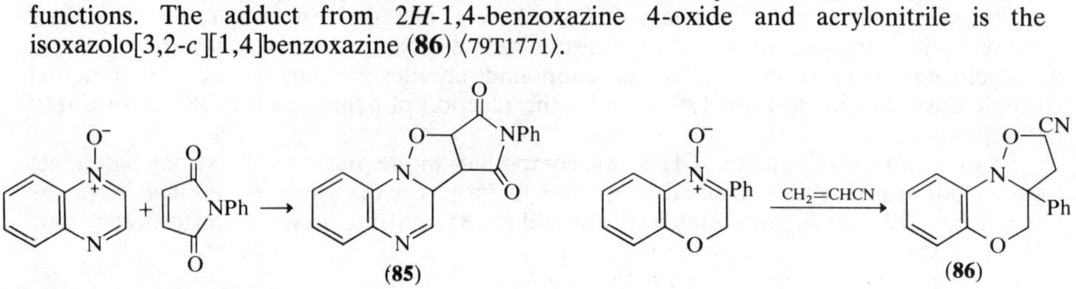

(iv) *2H-Isoxazolo[2,3-b]pyridazine*

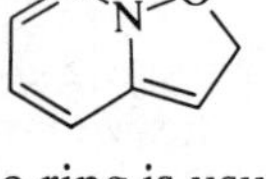

The second heteroatom in the azine ring is usually oxygen, the compounds being derivatives of 1,2-oxazines.

Azine approach. 4,5-Dihydro-6H-1,2-oxazine 2-oxides undergo 1,3-dipolar cycloaddition reacting with appropriately substituted alkenes and alkynes to form isoxazolo[2,3-b][1,2]oxazines. With styrene as the dipolarophile in the reaction with the oxazine (**87**), the product (**88**) with *cis* methyl and phenyl groups is formed. With acrylonitrile and methyl acrylate, some *trans* isomer is formed, but the *cis* isomer is predominant. The rings are always *cis*-fused ⟨77IZV211⟩.

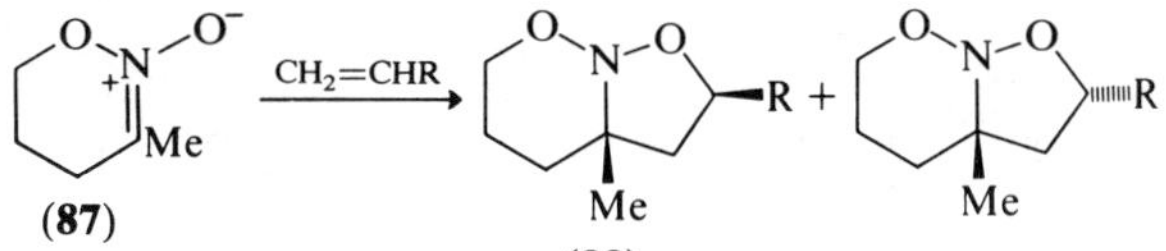

4.29.3.6.2 Isoxazolo[4,5-c]pyridine

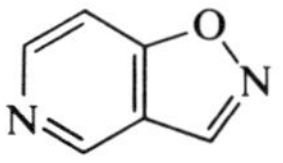

Azine approach. Enamines when used as dipolarophiles with nitrile oxides form isoxazoles. With cyclic enamines such as (**89**) fused isoxazoles are formed; the nitrile oxide in this reaction is generated *in situ* from benzhydroxamoyl chloride ⟨64JOC1582⟩.

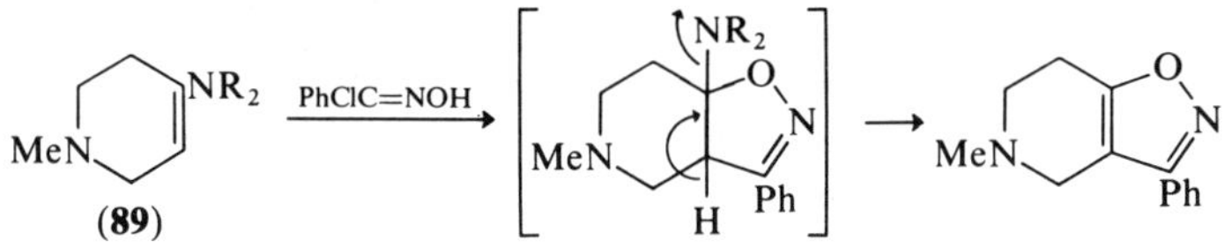

4.29.3.6.3 Isoxazolo[5,4-c]pyridine

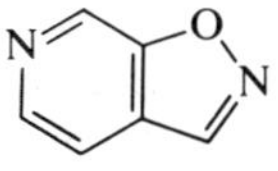

Azine approach. THIP (4,5,6,7-tetrahydroisoxazolo[5,4-c]pyridin-3-ol; **91**) is an active GABA antagonist. It is formed by cyclization of 3-oxopiperidine-4-carbohydroxamic acid in the protected form (**90**). Analogues are similarly prepared. THIP exists as a zwitterion, the pK_a values being 4.4 and 8.5 ⟨79MI42900⟩.

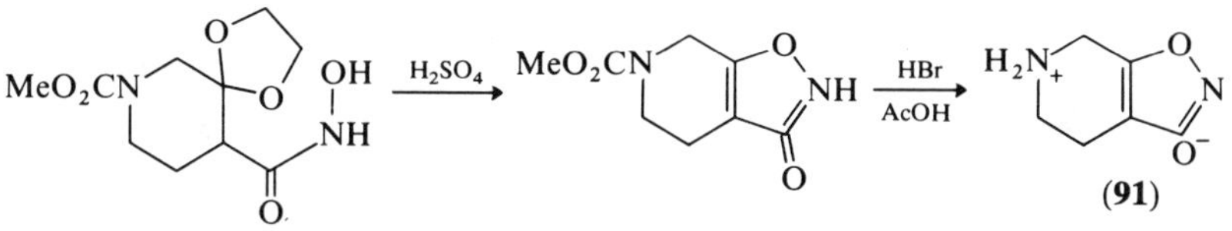

4.29.3.6.4 Isoxazolo[5,4-b]pyridine

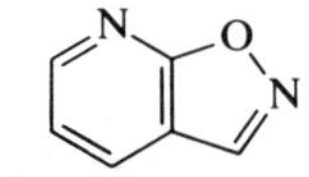

Azine approach. Bicyclic isoxazolines can be prepared by 1,3-dipolar cycloaddition of nitrile oxides to heterocyclic alkenes such as the *N*-acetyl tetrahydropyridine (**92**) ⟨68CPB117⟩.

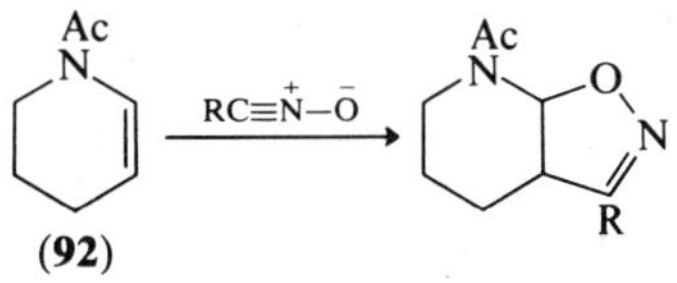

4.29.3.6.5 Isoxazolo[4,3-c]pyridine

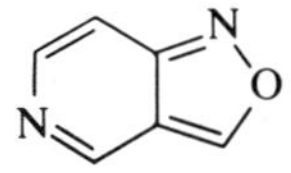

Azine approach. Bicyclic ring formation results if the 1,3-dipolar nitrone–alkene cycloaddition is intramolecular; such a reaction is shown for (**93**) ⟨70TL1117⟩.

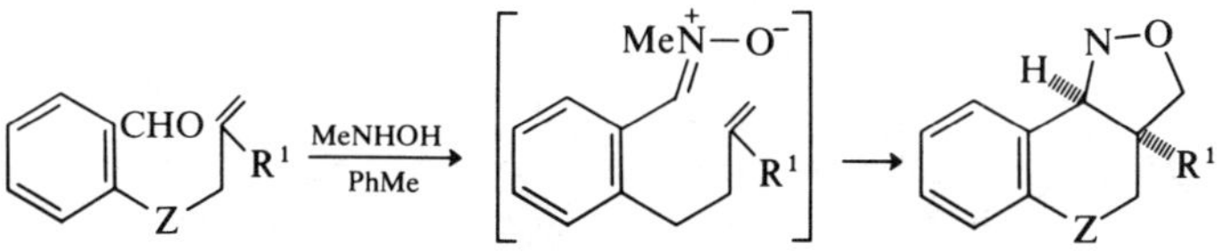

(**93**) Z = O, NCHO

3-Amino-4,5,6,7-tetrahydro derivatives (**95**) are available by the reaction between cyclic α,β-unsaturated β-aminonitriles (**94**) and hydroxylamine. The first step is an exchange of the β-substituent with hydroxylamine and this is followed by cyclization ⟨76JAP(K)7663192⟩. When the nitrile group is replaced by a thioamide group as in (**96**), cyclization under alkaline conditions yields the 3-amino product (**97**; R = NHPh), whereas in acid solution the latter is formed together with the 3-mercapto analogue ⟨70LA(738)60⟩.

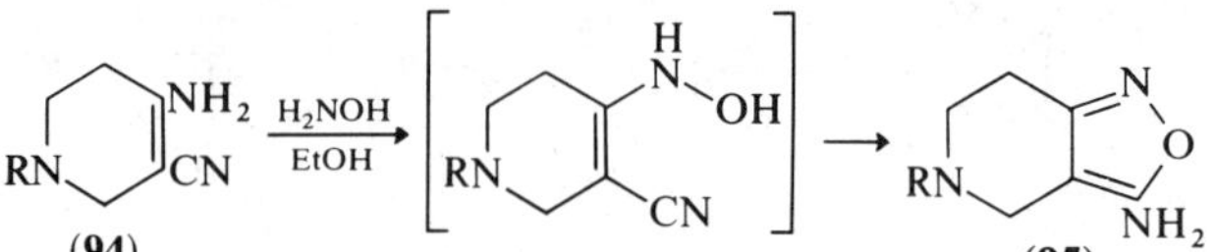

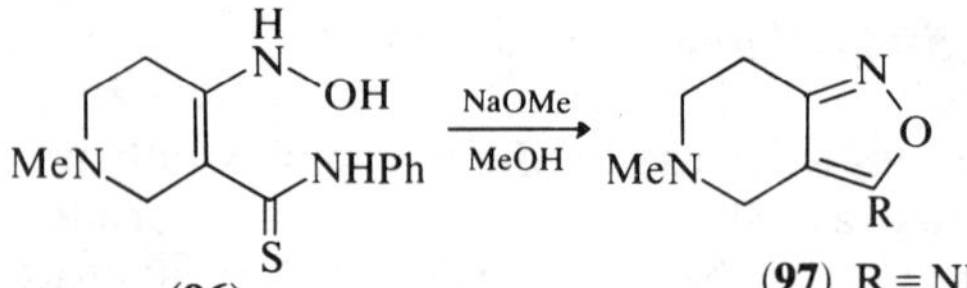

4.29.3.6.6 Isoxazolo[3,4-c]pyridine

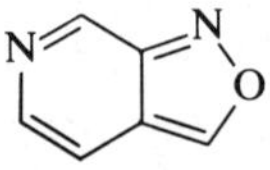

Azine approach. The zwitterionic compound (**99**) with pK_a values of 3.0 and 9.1 is called isoTHIP, being an analogue of the GABA antagonist THIP (**91**) where the order of the azole heteroatoms has been reversed. It can be prepared by the usual isoxazole synthesis from a 1,3-dicarbonyl derivative and hydroxylamine. In the reaction of the oxopiperidine (**98**) with hydroxylamine the regioselectivity is secured by the keto function being more reactive than the ester group ⟨79MI42900⟩.

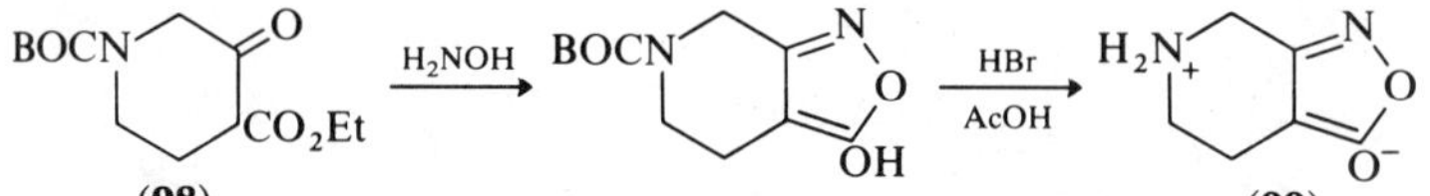

4.29.3.7 Applications

Several patents exist for the potential application of compounds with fused isoxazoloazine structures in human medicine or as agrochemicals. 2-Aryl derivatives of the cation (**100**) are claimed to be useful as inflammation inhibitors and in treatment of gastric hyperacidity ⟨77USP4018781⟩. Substituted isoxazolo[5,4-*b*]pyridines such as (**101**) are claimed to be effective as antiasthmatics and anxiolytics, and for increasing AMP concentrations in the blood ⟨76USP3933823⟩. Isoxazolo[5,4-*d*]pyrimidines with substitution as in (**102**) are claimed to have analgesic properties ⟨79JAP(K)7944696⟩ whereas the substitution pattern in (**103**) results in pesticidal properties ⟨78GEP2812367⟩. Compounds having the general formula (**104**) are protected for their herbicidal properties ⟨73GEP2249163⟩. Meseclazone (**105**) is useful for its antiinflammatory properties ⟨77AF760⟩. The pyrimidine analogues (**106**) have similar properties ⟨77AF766⟩. THIP (4,5,6,7-tetrahydroisoxazolo[5,4-*c*]pyridin-3-ol; **107**) is a very potent GABA (γ-aminobutyric acid) receptor agonist ⟨79MI42901⟩. Compounds belonging to the [4,5-*c*] system (**108**) are claimed to be useful as hypolipidemics. 3-Amino-4,5,6,7-tetrahydroisoxazolo[4,3-*c*]pyridines (**109**) are claimed to possess analgesic, antiinflammatory and psychotropic activities ⟨76JAP(K)7663192⟩.

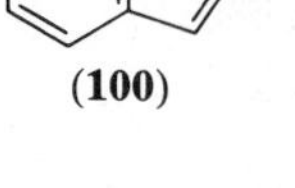

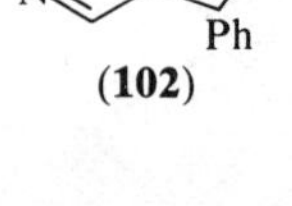

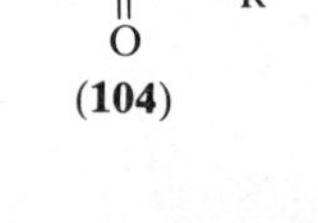

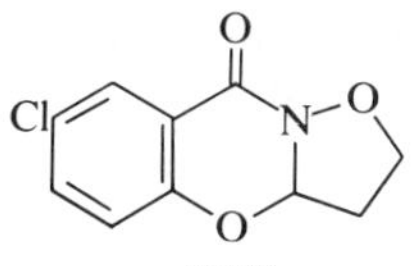

(105) 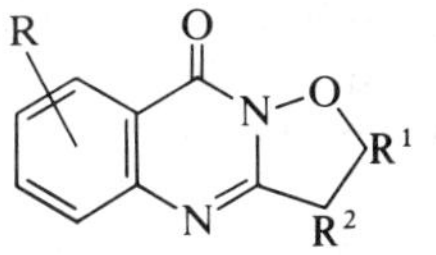 (106) 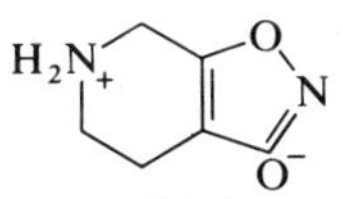 (107) 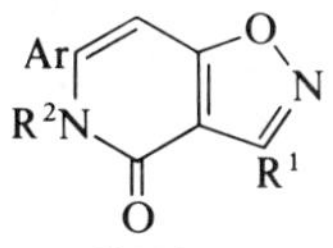 (108) 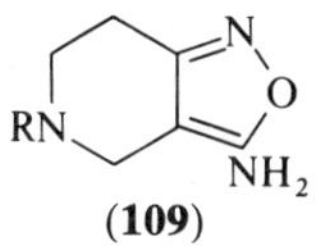 (109)

4.29.4 ISOTHIAZOLE FUSED WITH AZINES

4.29.4.1 Introduction

Isothiazole can be fused with pyridine in nine different ways; the section on synthesis has been organized accordingly. Further fused systems arise from the diazines, the triazines and their areno and heteroareno homologues.

Isothiazoloazine systems have been relatively little explored. Most work deals with the preparation of members of the many ring systems and especially of fully conjugated ring systems.

Chemical properties are discussed in the section on reactivity. Isothiazoloazines are claimed as potential dyes or may have useful biological properties; a review is given in the section on applications.

4.29.4.2 Structure

4.29.4.2.1 Molecular spectra

(a) ^{1}H NMR spectroscopy. ^{1}H NMR data available refer mainly to fused pyridines (Table 6). It should be noted that the chemical shifts for H-2 and H-7 in the parent isothiazolo[2,3-a]pyridinium cation are at very low fields.

Table 6 ^{1}H NMR Data for Fused Isothiazoloazines

Compound	Solvent	^{1}H chemical shifts (p.p.m.)						Coupling constants (Hz)	Ref.
		2	3	4	5	6	7		
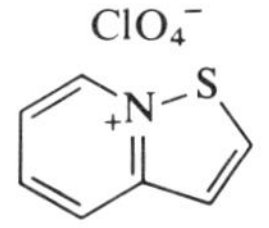 ClO$_4^-$ isothiazolo[2,3-a]pyridinium	DMSO-d_6	9.17	7.78	8.43	8.27	7.79	9.69	$J_{2,3}$ 6.4	73CC150
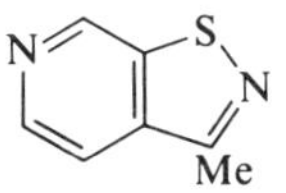 N-Me pyridoisothiazole	CDCl$_3$	—	2.75	7.76	8.60	—	9.31	$J_{4,5}$ 5.5, $J_{5,7}$ 1.0	73CJC1741
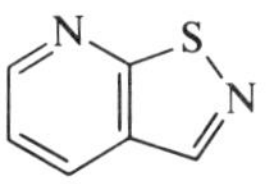 pyridoisothiazole	CDCl$_3$	—	8.93	8.32	7.37	8.75	—	$J_{4,5}$ 8.4, $J_{4,6}$ 1.6, $J_{5,6}$ 4.4	73CJC1741, 78T989
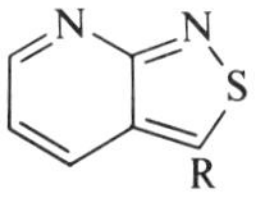 R = H	CDCl$_3$	—	9.44	8.17	7.19	8.90	—	$J_{4,5}$ 8.8, $J_{4,6}$ 2.0, $J_{5,6}$ 4.0	73CJC1741
R = Cl	CDCl$_3$	—	—	8.04	7.23	8.91	—	$J_{4,5}$ 8.5, $J_{4,6}$ 1.8, $J_{5,6}$ 4.0	73CJC1741
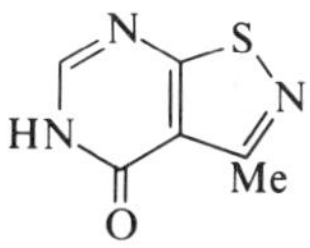 Me-pyrimido isothiazolone	DMSO-d_6	—	2.62	—	—	8.20	—		75JHC883

(b) *UV spectroscopy*. UV data for simple ring systems are largely confined to the fused pyridines in Table 7. The UV absorption of the parent isothiazolo[2,3-*a*]pyridinium cation is very similar to that of the iso-π-electronic quinolizinium ion ⟨73CC150⟩.

Table 7 UV Absorption Data for Fused Isothiazoloazines

Compound	Solvent	λ_{max}[a] ($\log \varepsilon$)	Ref.
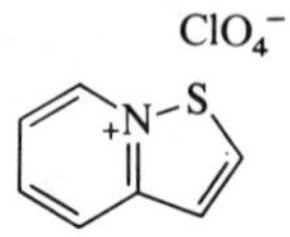 [ClO$_4^-$ salt, isothiazolo-pyridinium]	EtOH	331 (4.11), 318 (4.07), 230 (4.25), 212 (3.85)	73CC150
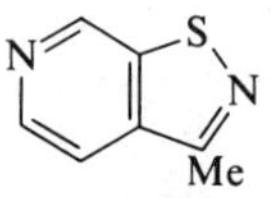 [N-Me isothiazolopyridine]	EtOH	326[b] (3.69), 313 (3.74), 224[b] (4.01), 209 (4.15)	73CJC1741
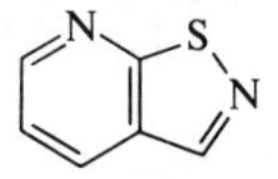 [isothiazolopyridine]	EtOH	306 (3.46), 295 (3.53), 226 (4.24)	73CJC1741, 78T989
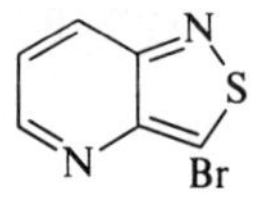 [Br-substituted isothiazolopyridine]	EtOH	324 (4.00), 306 (4.14), 213 (4.29)	73CJC1741
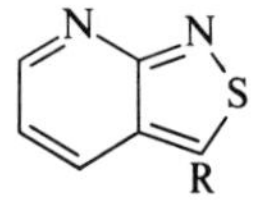 R = H	EtOH	296 (3.86), 213 (4.03)	73CJC1741
R = Cl	EtOH	328[b] (3.67), 303 (3.83), 216 (4.06)	73CJC1741
R = OMe	EtOH	352 (3.58), 297[b] (3.62), 289 (3.66), 219 (4.18)	73CJC1741
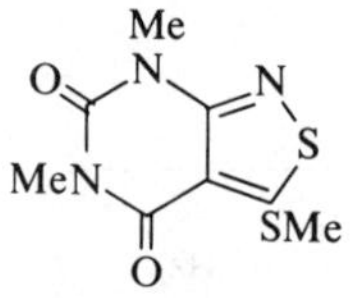 [uracil-fused isothiazole, SMe]	EtOH	294 (4.26)	79H(12)485

[a] nm. [b] Shoulder.

4.29.4.3 Reactivity

4.29.4.3.1 Survey of reactivity

So far little information is available on electrophilic substitution reactions; these are mainly expected to occur in the azine ring when activated by electron-releasing substituents. Nucleophilic substitution reactions, however, occur readily in either ring. The N—S bond may be cleaved by nucleophilic attack at sulfur and this may be the preferential reaction path in some cases. The N—S bond may also be cleaved as a result of proton abstraction from the azole ring.

4.29.4.3.2 Electrophilic attack at nitrogen

In principle *N*-alkylation could occur on the nitrogen in either the azole or the azine ring in the isothiazolo[5,4-*b*]pyridine (**110**). Since *N*-alkylation in 1,2-benzisothiazoles is difficult, preference for azine alkylation is to be expected as observed in the methylation of (**110**) ⟨76JPR779⟩. 2,1-Benzisothiazoles, however, which have a quinonoid electronic arrangement, are readily *N*-alkylated. Similar reactions might be expected in fused heterocyclic analogues. In the methylation of the uracil (**111**) under alkaline conditions it is the usual preference for alkylation of the lactam nitrogen which gives the first product (**112**; $R^3 = H$); the latter reacts further under alkaline conditions with excess methylating agent on the 3-amino group (**112**; $R^3 = Me$), presumably as the anion ⟨76CPB979⟩.

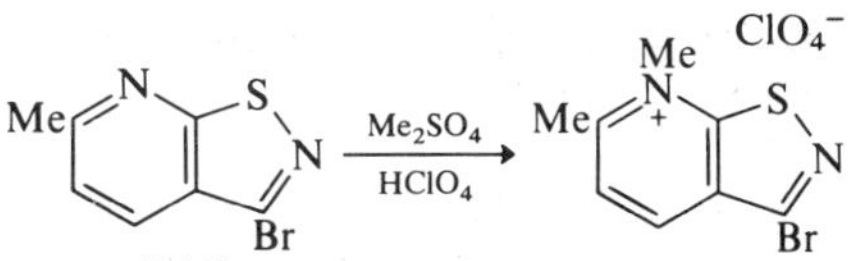

4.29.4.3.3 Electrophilic attack at carbon

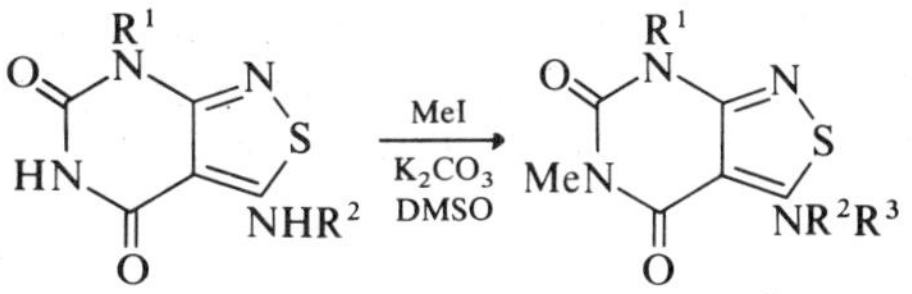

Both the 1,2- and 2,1-benzisothiazoles are attacked by electrophiles in the carbocyclic ring. Electrophilic substitution in analogous azine fused systems is therefore dependent on the presence of strongly electron donating substituents.

4.29.4.3.4 Nucleophilic attack at carbon

Nucleophilic substitution occurs readily in either ring. The bromine at C-3 in the thiazole (113) is substituted by a methoxy group ⟨73CJC1741⟩, and in the uracil derivative (114) it is a 3-methylthio substituent which is replaced by an amino nucleophile ⟨79H(12)485⟩. Aminolysis in an activated azine position proceeds in the usual manner as shown for the 6-chloride (115) ⟨70GEP1950990⟩.

In the thiation of the 3-bromide (113) thiourea is applicable ⟨73CJC1741⟩. Thiation of an oxo group in the azine ring using phosphorus pentasulfide in the manner of azines is illustrated for the fused uracil (116) ⟨65JCS7277⟩; it is the 7-oxo group which is the more reactive as is usually the case in quinazoline like systems.

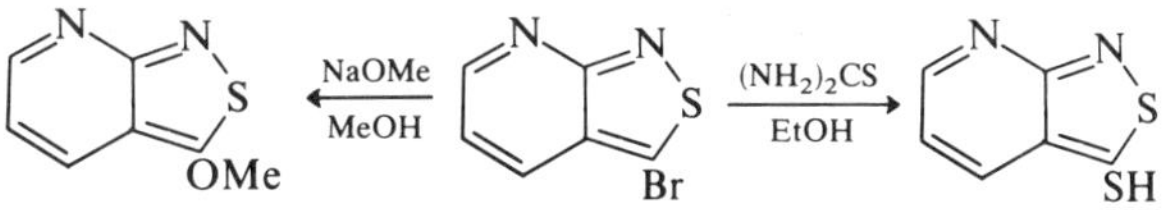

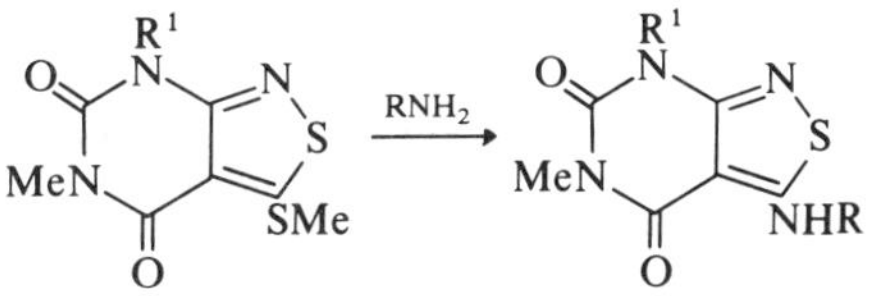

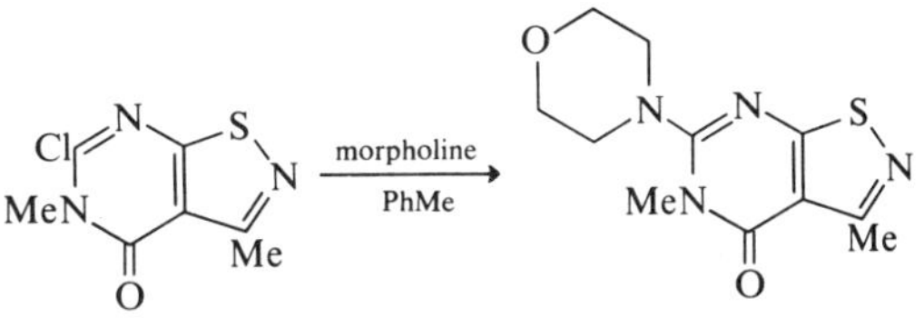

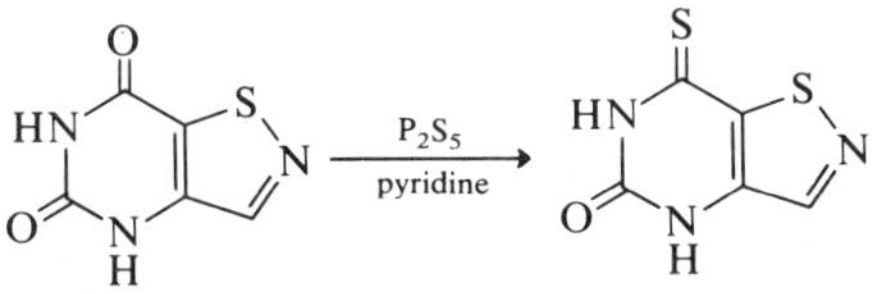

Lactams react with phosphorus chlorides to yield the corresponding chloro compounds as in the uracil (117) ⟨70GEP1950990⟩.

As pointed out, the 3-position in fused isothiazoles is activated for nucleophilic substitution. Amino groups in activated azine positions after diazotization are frequently displaced by the anion of the acid used in the diazotization reaction. Similarly, this type of reaction can be used to substitute the 3-amino group in (118) with a bromine substituent ⟨73CJC1741⟩.

A carbanion can be used in the substitution of a 3-substituent, *e.g.* as in the 3-methylthio derivative (119) ⟨79H(12)485⟩.

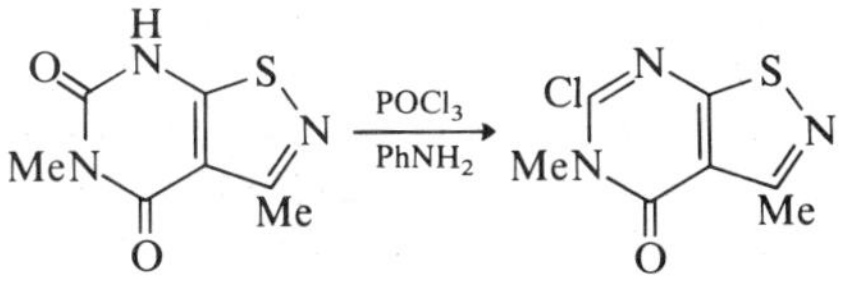

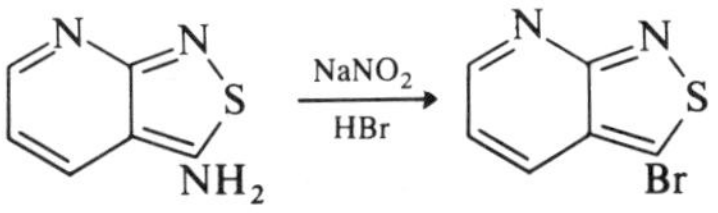

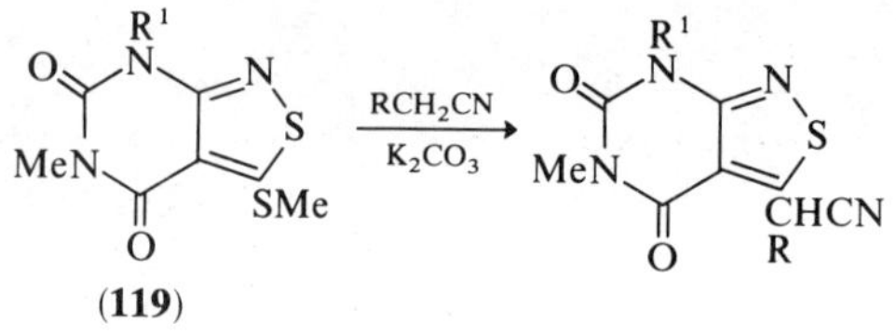

4.29.4.3.5 Nucleophilic attack at sulfur

The N—S bond can be cleaved by nucleophilic attack at the sulfur as a competitive or preferable reaction path to nucleophilic substitution at carbon.

The parent cation (**120**) is cleaved by sodium cyanide ⟨73CC150⟩. In the 3-chloride (**121**) preferential attack is on the ring sulfur. The 3-bromide (**122**) shows the same preferences in its reactions with both nitrogen and carbon nucleophiles ⟨76JPR779⟩.

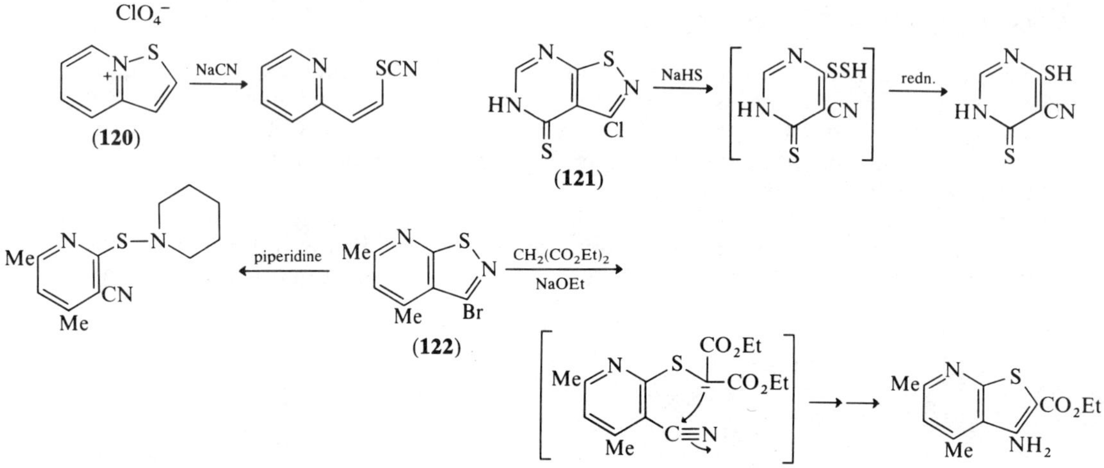

4.29.4.3.6 Desulfurization

A number of methods are available for desulfurization of sulfur-containing heterocycles. The use of Raney nickel is a common method for reductive desulfurization which has been employed in the cleavage reaction of the fused uracil (**123**) ⟨76CPB979⟩.

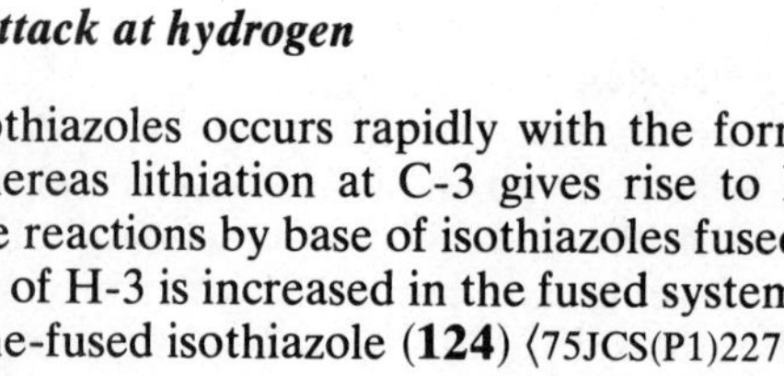

4.29.4.3.7 Nucleophilic attack at hydrogen

Lithiation at C-5 in isothiazoles occurs rapidly with the formation of relatively stable metallated derivatives whereas lithiation at C-3 gives rise to N—S cleavage and nitrile formation. Hence cleavage reactions by base of isothiazoles fused to azines are predictable, especially since the acidity of H-3 is increased in the fused systems. Such cleavage reactions are shown for the quinoline-fused isothiazole (**124**) ⟨75JCS(P1)2271, 78JHC1527⟩.

4.29.4.3.8 Other C-linked substituents

α-Hydrogen-containing carbon substituents in activated ring positions are attacked by electrophiles. In the fused pyridinium derivative (125) it is the 6-methyl group which preferentially reacts with *p*-dimethylaminobenzaldehyde ⟨76JPR779⟩.

By choice of reaction conditions so as to avoid nucleophilic substitution, the diazonium salt from the 3-amine (126) will undergo diazo coupling to form azo dyes ⟨79BRP1550828⟩.

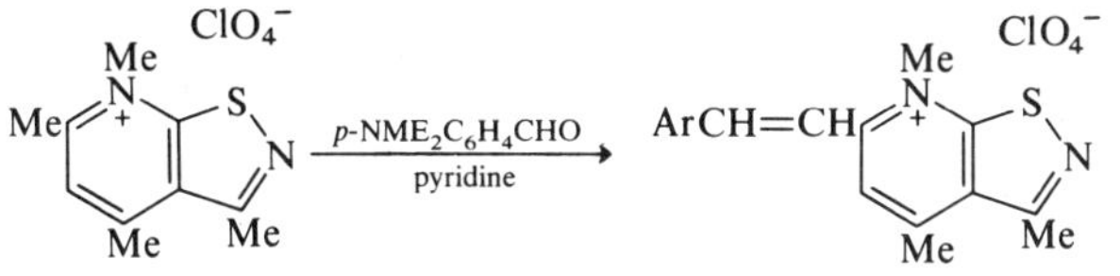

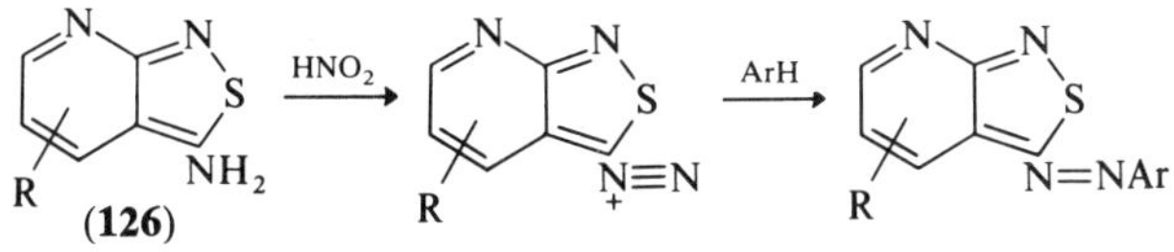

4.29.4.3.9 Rearrangement

Rearrangements involving the isothiazole ring as found in benzisothiazoles are to be expected. The Dimroth-type rearrangement of fused pyrimidine systems has been reported for the imine (127), which on heating with water alone, or in the presence of a base, is rearranged to the fully conjugated structure (128) ⟨68AP611⟩.

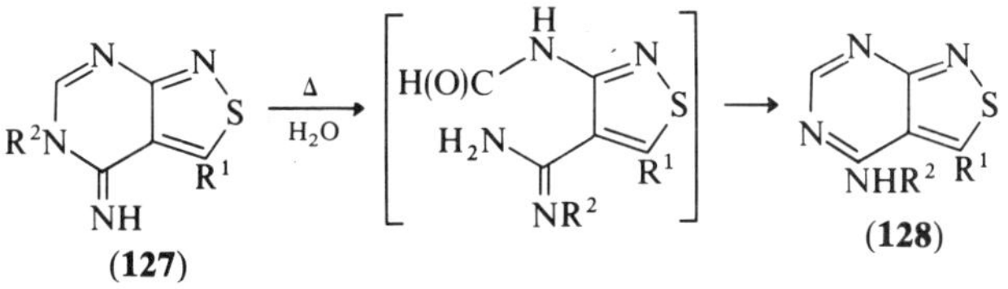

4.29.4.4 Synthesis

4.29.4.4.1 2H-Isothiazolo[2,3-a]pyridine

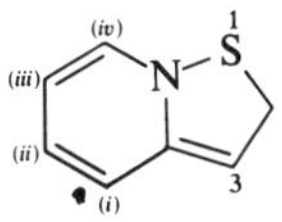

Azine approach. Cyclization by oxidative formation of the S—N bond is frequently the final step in isothiazole syntheses. This approach can be used in the synthesis of the parent cation from the thioformyl precursor (129). 2-Substituted derivatives would require thioketone precursors and such compounds are more readily available as vinyl thioethers, *e.g.* (130). Bromine is used for the oxidative cyclization in this instance giving (131) ⟨73CC150⟩.

Partially reduced derivatives can be prepared in the same way. Substitution in the nitroalkene (132) by an isothiocyanate, followed by oxidation, leads to cyclization ⟨77IJC(B)886⟩.

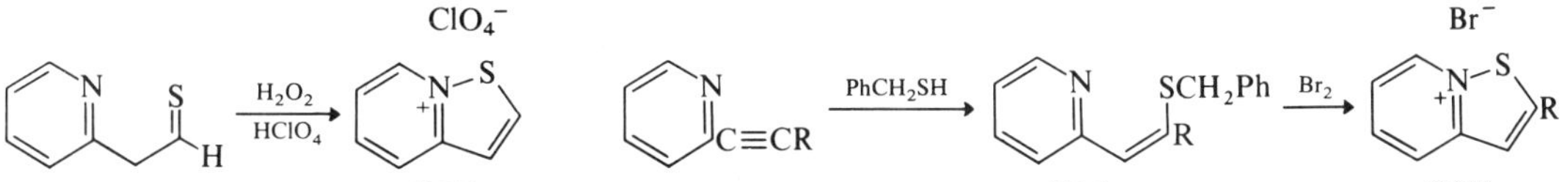

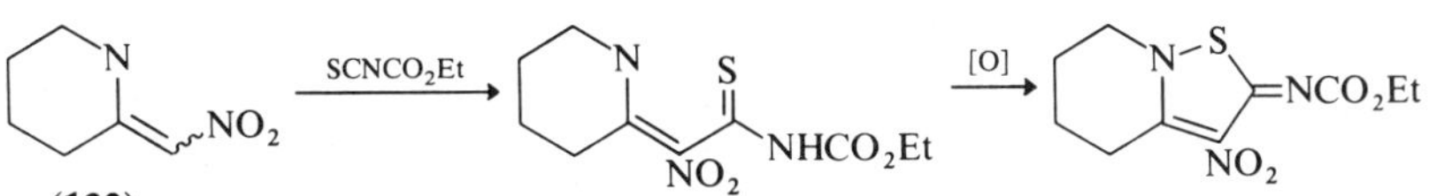

Azole approach. Compounds saturated in the azine ring are available from 3-alkylisothiazoles functionalized in the alkyl chain by an intramolecular alkylation as for (**133**) ⟨69JCS(C)707⟩.

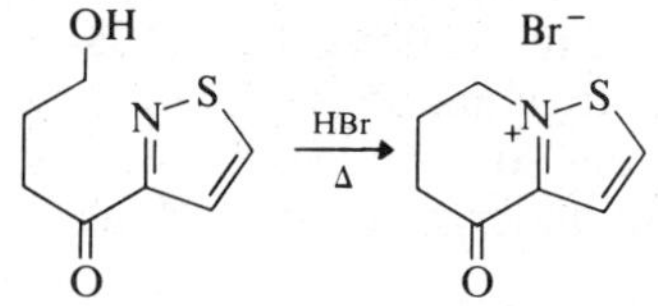

(**133**)

(i) *2H-Isothiazolo[2,3-a]pyrimidine*

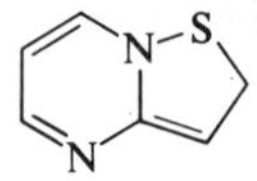

Azole approach. Aminolysis of isatoic anhydride using 3-hydroxyisothiazoles as the amino nucleophile provides a convenient method for preparing 9H-isothiazolo[3,2-b]quinazolin-9-ones (**134**) ⟨69AJC2497⟩.

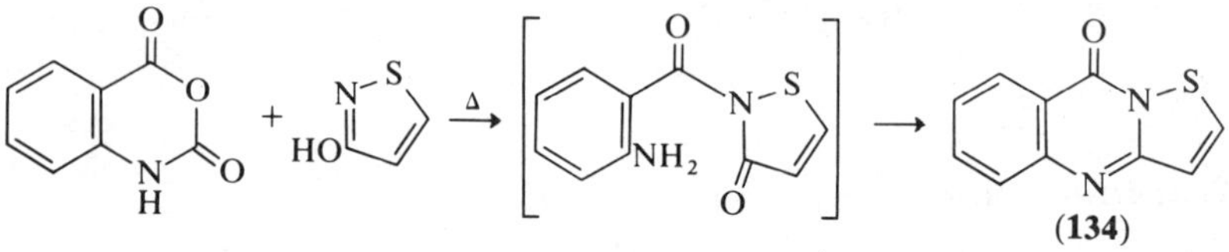

(**134**)

4.29.4.4.2 Isothiazolo[4,5-b]pyridine

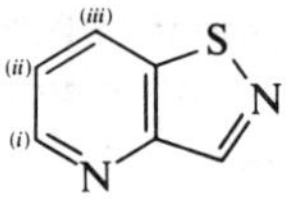

(ii) *Isothiazolo[4,5-d]pyrimidine*

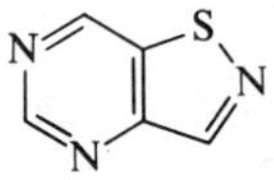

Azole approach. Condensation reactions of the 4-amino-5-cyanoisothiazole (**135**) or its 5-carbamoyl analogue (**136**) are classical methods for the fusion of the pyrimidine ring on to other ring systems ⟨65JCS7277⟩.

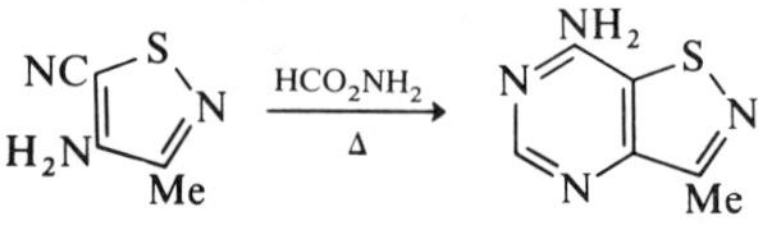

(**135**) 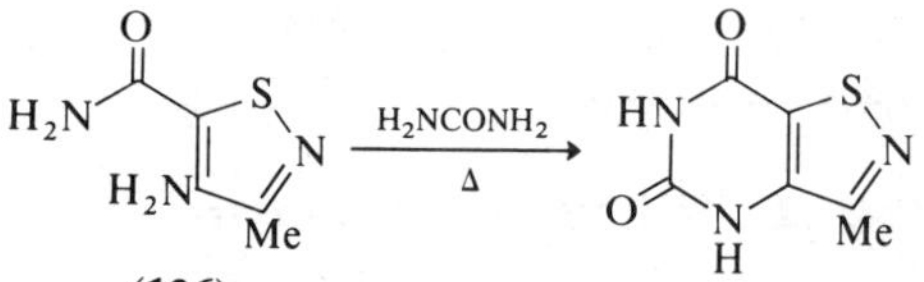 (**136**)

(iii) *Isothiazolo[4,5-b]pyrazine*

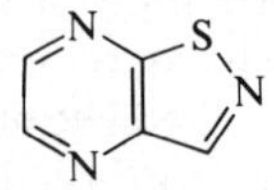

Azole approach. The 4,5-diaminoisothiazole (**137**) reacts with a 1,2-dioxo derivative in a quinoxaline-type synthesis; the difference in the nucleophilicity of the amino groups leads to regioselectivity in product formation ⟨65JCS7277⟩. In the vicinal nitrosoamide (**138**) the reaction with malononitrile provides a functionalized pyrazine for further pteridine (**139**) synthesis ⟨78JOC4154⟩.

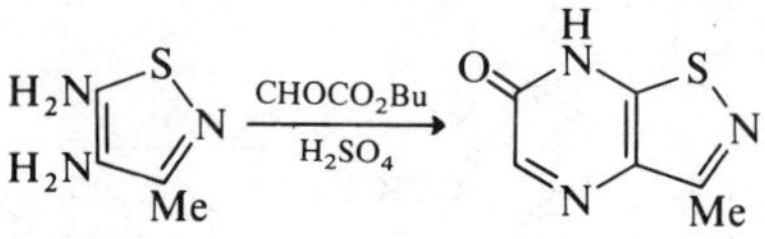

(**137**)

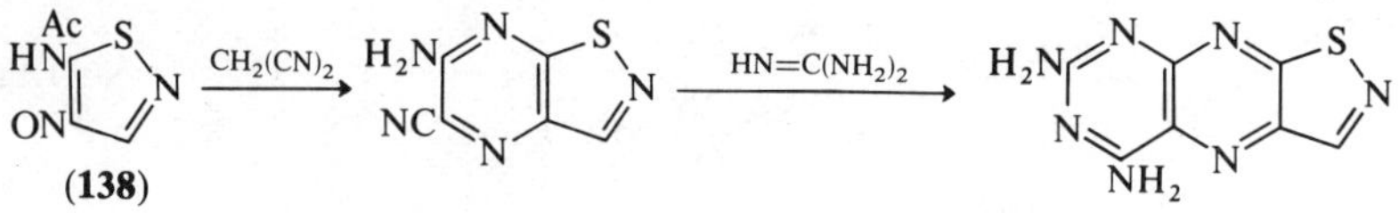

(**138**) (**139**)

4.29.4.4.3 Isothiazolo[4,5-c]pyridine

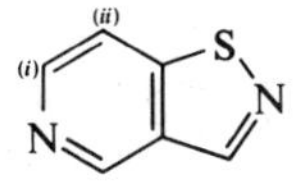

Azine approach. Cyclization of sulfenamides by addition of the amino nitrogen to an electrophilic carbon is an important route to isothiazoles. The sulfenamide substituent may be generated *in situ* from a thiol and chloramine. When the electrophilic carbon is a cyano group, an aminoisothiazole is formed, *e.g.* as in the synthesis from (**140**) ⟨73LA1644⟩.

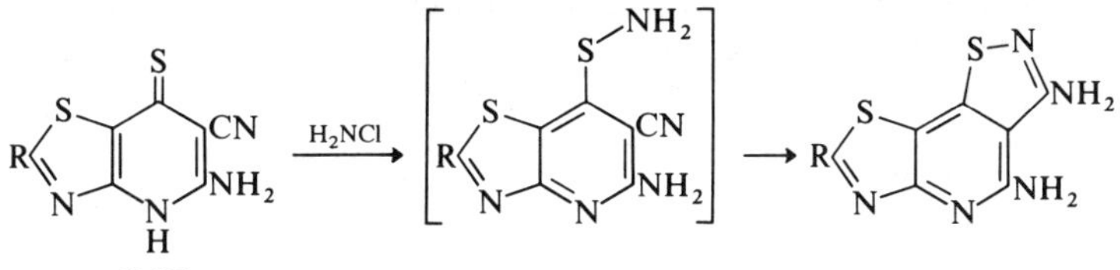

(i) *Isothiazolo[4,5-d]pyridazine*

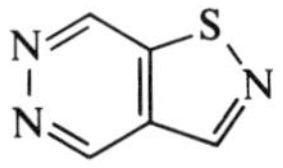

Azole approach. In a phthalazine type synthesis, a 4,5-diacylisothiazole (**141**) is reacted with hydrazine ⟨59JCS3061⟩.

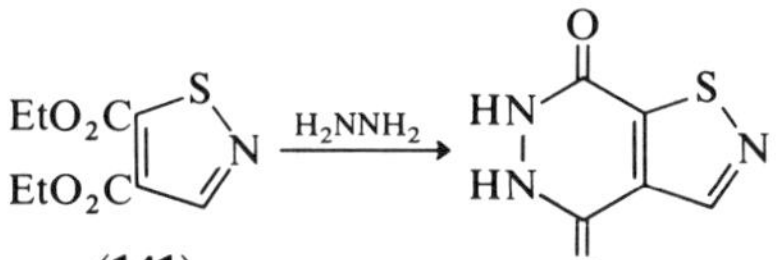

(ii) *Isothiazolo[5,4-d]pyrimidine*

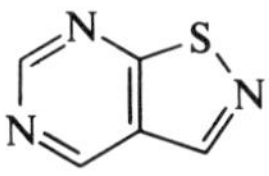

Azole approach. The substituted isothiazole (**142**) can be reacted with an ortho ester to form a methyleneamine which reacts with hydroxylamine to yield the 5-oxide (**143**). With amidines, (**142**) yields 4-amino derivatives ⟨75JHC883⟩. The reaction of 5-amino-4-ethoxycarbonylisothiazole with iminoethers results in pyrimidine annulation, as in the formation of (**144**) ⟨76IJC(B)391⟩.

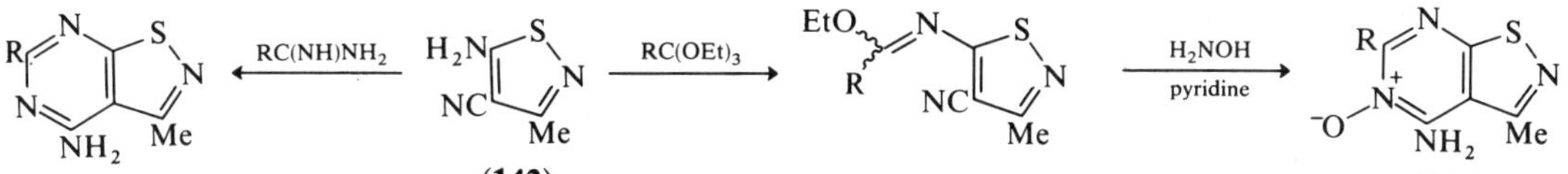

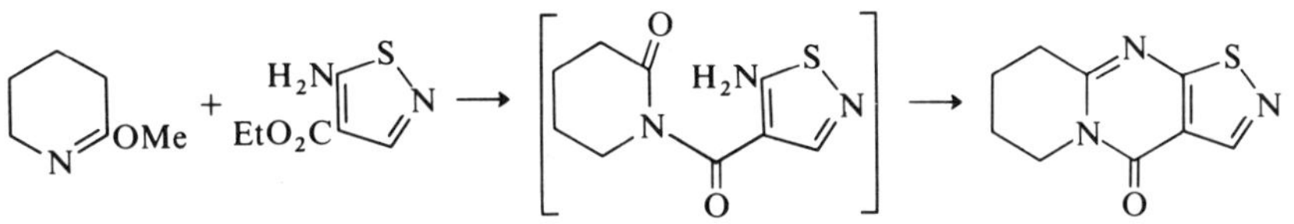

4.29.4.4.4 Isothiazolo[5,4-c]pyridine

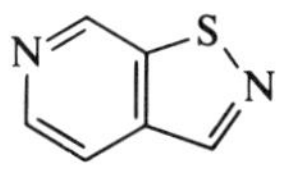

Azine approach. Addition of ammonia to compounds with vicinal acyl–cyanothio groups as in (**145**) results in an intermediate adduct where the amino group is in a position to attack the cyanothio group on the sulfur atom, resulting in isothiazole formation ⟨73CJC1741⟩.

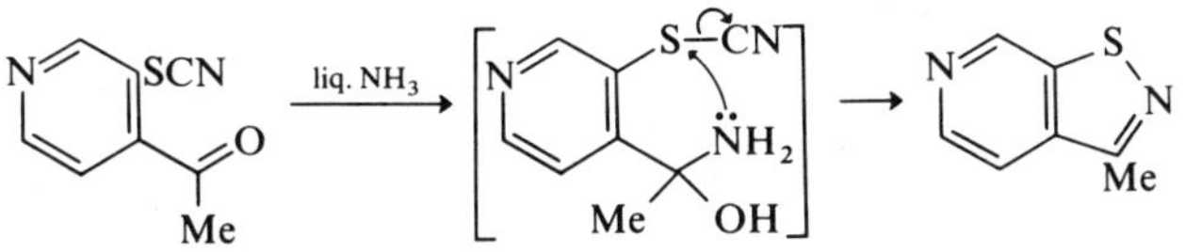

4.29.4.4.5 Isothiazolo[5,4-b]pyridine

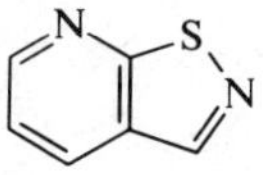

Azine approach. The acyl–thiocyanate cyclization method can also be used in the synthesis of the [5,4-b]-system represented by (**146**) ⟨73CJC1741⟩. Bromination of 3-cyano-2-mercaptopyridines leads to oxidative cyclization *via* the intermediate (**147**). A bromine is introduced at C-3 in the isothiazole moiety by this procedure ⟨76JPR779⟩. If the cyano group is replaced by a formyl group and the reagent is hydroxylamine, the cyclization of the intermediate oxime (**148**) corresponds to nucleophilic attack on nitrogen with cleavage of the N—O bond to furnish the unsubstituted ring system (**146**) ⟨78T989⟩. The same ring system can be prepared from a 3-formyl-2-mercaptopyridine which is converted *in situ* into its sulfenamide (**149**) for cyclization to the quinoline analogue (**150**) ⟨78JHC1527⟩. S-Amidation of the 3-cyano-2-mercaptopyridine (**151**) gives a fused 3-aminoisothiazole ⟨75JPR959⟩.

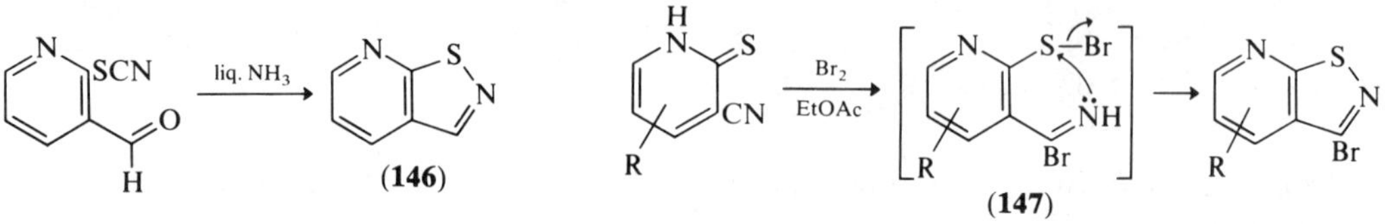

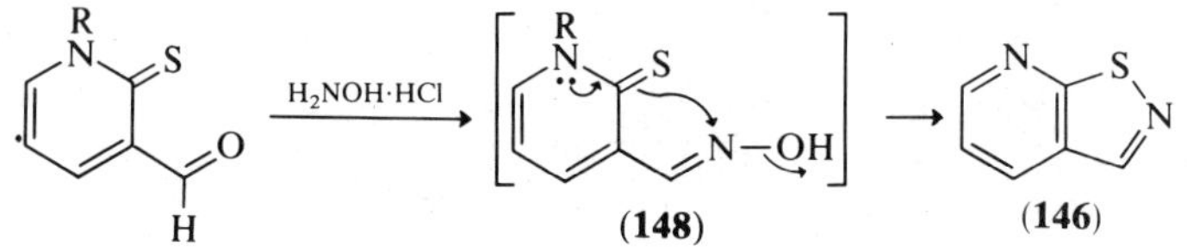

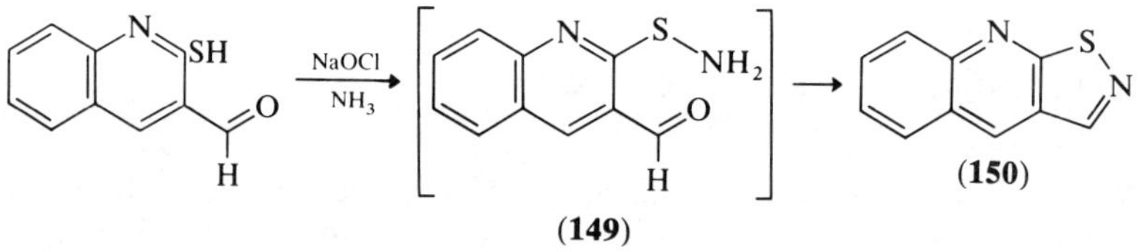

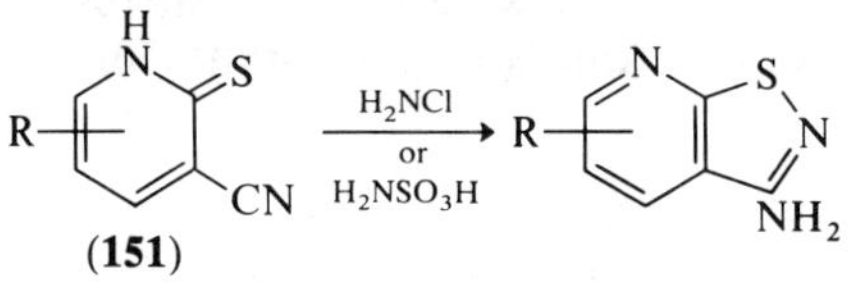

Azole approach. Cyclocondensation of 5-aminoisothiazoles with β-ethoxyacroleins is a simple method for preparing the fused ring system (**153**); in some cases the intermediate (**152**) can be isolated ⟨79S370⟩.

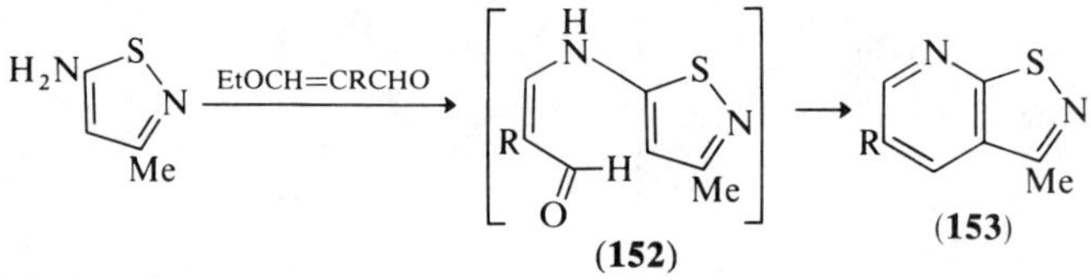

4.29.4.4.6 Isothiazolo[4,3-b]pyridine

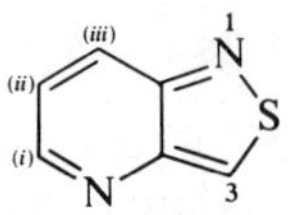

Azine approach. The 3-amino derivative (**154**) has been prepared by the oxidative procedure using the appropriately substituted pyridine ⟨73CJC1741⟩.

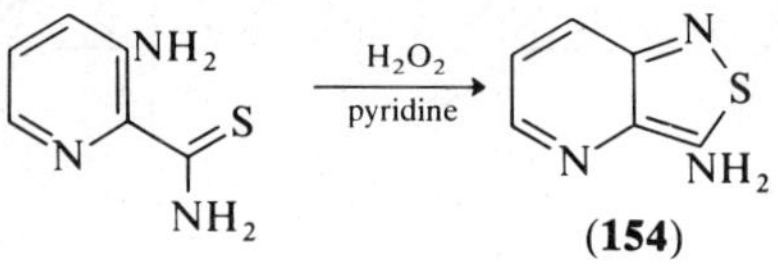

(ii) *Isothiazolo[4,3-d]pyrimidine*

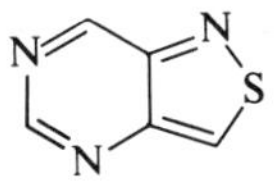

Azole approach. Classical methods for pyrimidine fusions have been employed in the synthesis of (**155**) ⟨65JCS7277⟩.

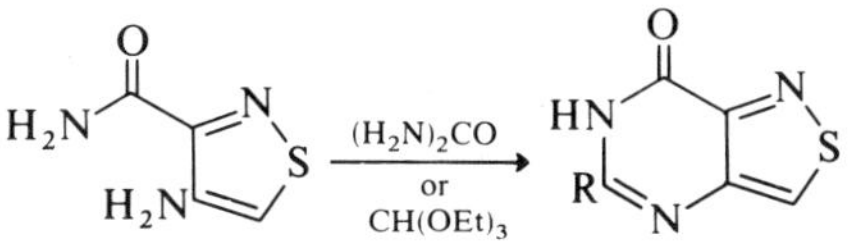

(**155**) R = OH, H

4.29.4.4.7 Isothiazolo[4,3-c]pyridine

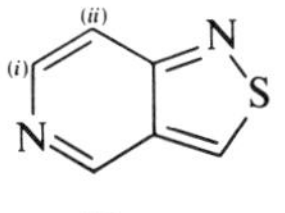

(ii) *Isothiazolo[4,3-d]pyrimidine*

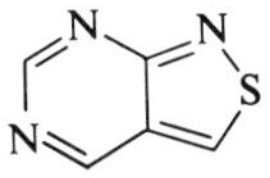

Azine approach. 6-Aminouracils are attacked by electrophiles at C-5. With an isothiocyanate as electrophile a thiocarbamoyl derivative (**156**) is formed which undergoes oxidative cyclization. Sometimes a minor product, the thiazole (**158**), may be formed *via* the thiourea intermediate (**157**). The course of the reaction may perhaps be rationalized on the HSAB principle; the 'hard' isothiocyanate electrophile has to choose between the 'hard' nucleophilic centre at C-5 or the 'soft' nucleophilic centre at 6-NH_2. Isothiocyanates of aminosugars have also been used in these reactions ⟨79CPB1147⟩.

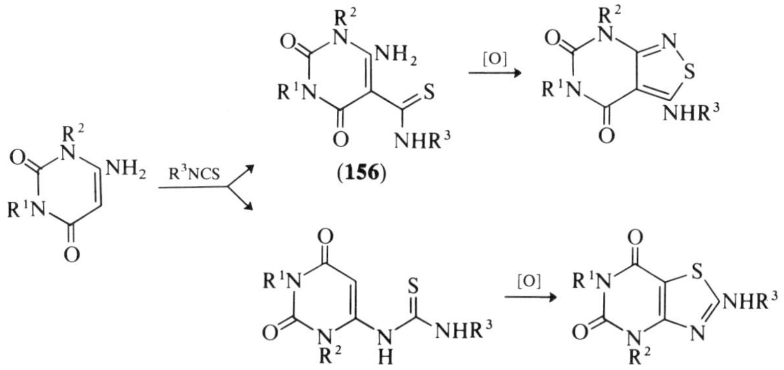

6-Aminouracils as nucleophiles will attack carbon disulfide; cyclization of the dithioester (**159**) by halogen oxidation introduces a thio substituent at C-3 ⟨79H(12)485⟩.

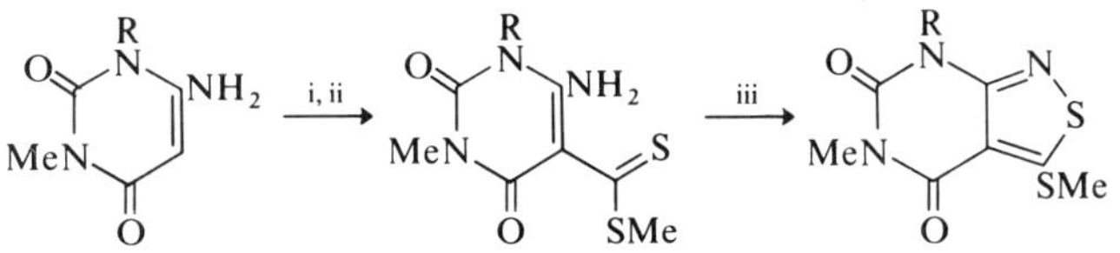

i, CS_2/NaOH; ii, Me_2SO_4/DMSO; iii, I_2

4.29.4.4.8 Isothiazolo[3,4-c]pyridine

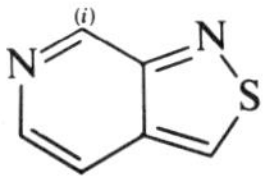

(i) *Isothiazolo[3,4-c]pyridazine*

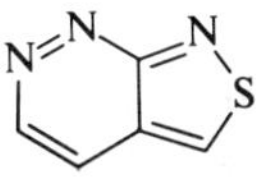

3-Amino-4-cyanopyridazines with H_2S yield the corresponding thioamides which undergo oxidative cyclization to derivatives of this system such as (**160**) ⟨79JPR71⟩.

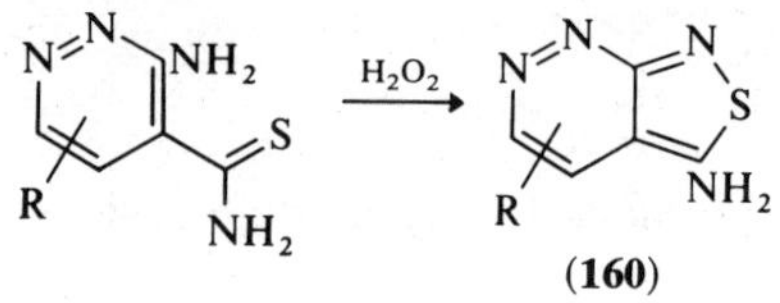

4.29.4.4.9 Isothiazolo[3,4-b]pyridine

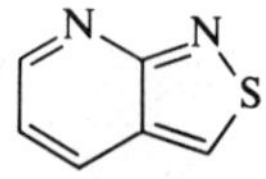

Azine approach. 3-Amino derivatives (**161**) are prepared by oxidative cyclization of 2-amino-3-thiocarbamoylpyridines ⟨73CJC1741⟩.

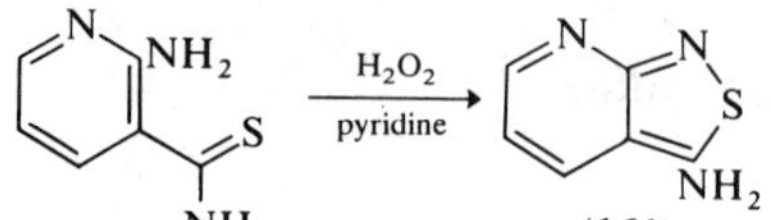

4.29.4.5 Applications

Coupling diazotized 3-aminoisothiazolo[2,3-b]pyridines with arenes, especially anilines, gives dyes (**162**) for polyester fibers ⟨79BRP1550828⟩. Pyrimidine analogues and pyrimidinedione analogues (**163**) are similarly claimed as dyes ⟨75GEP2336978⟩.

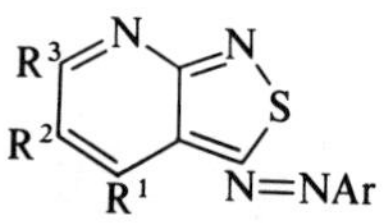

A group of 3-oxoisothiazolopyrimidines (**164**) have been patented for their ability to inhibit human blood platelet aggregation ⟨77GEP2718707⟩. 3,6-Dialkylisothiazolo[3,4-d]pyrimidin-4(5H)-ones (**165**) show useful herbicidal activity ⟨73GEP2249099⟩.

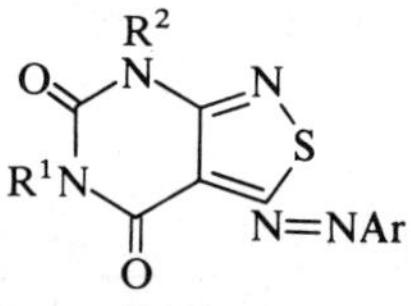

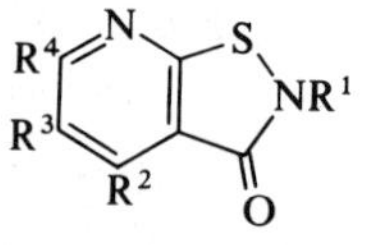

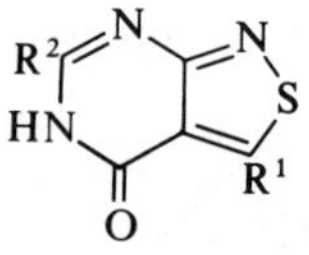

4.29.5 OXAZOLE AND THIAZOLE FUSED WITH AZINES: INTRODUCTION

4.29.5.1 Survey of Possible Structures

There are six different ways of constructing fused oxazolo- or thiazolo-pyridines. These are the main groups into which such compounds have been classified (Chart 3).

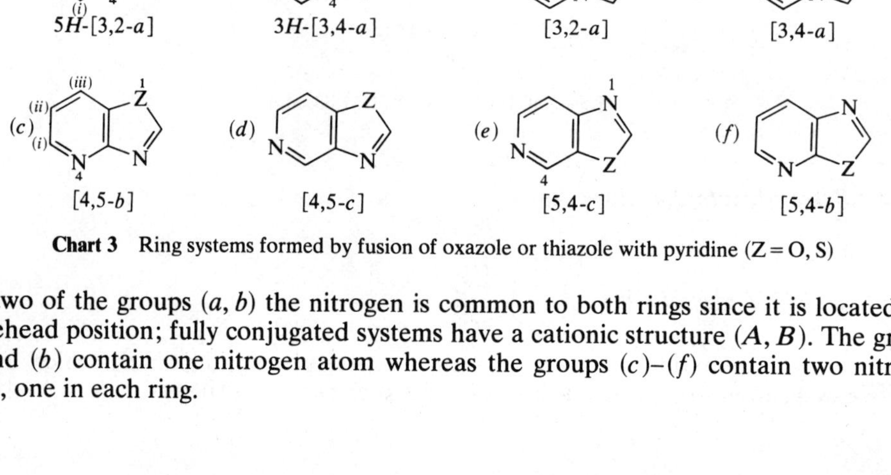

Chart 3 Ring systems formed by fusion of oxazole or thiazole with pyridine (Z = O, S)

In two of the groups (*a*, *b*) the nitrogen is common to both rings since it is located in a bridgehead position; fully conjugated systems have a cationic structure (*A*, *B*). The groups (*a*) and (*b*) contain one nitrogen atom whereas the groups (*c*)–(*f*) contain two nitrogen atoms, one in each ring.

The pyridine ring may contain additional heteroatoms. Most systems hitherto described are derived from a diazine or triazine. These systems are subdivided under the pyridine group according to the position of the second heteroatom. They are treated in the order of increasing number for the position of the second heteroatom (Chart 4).

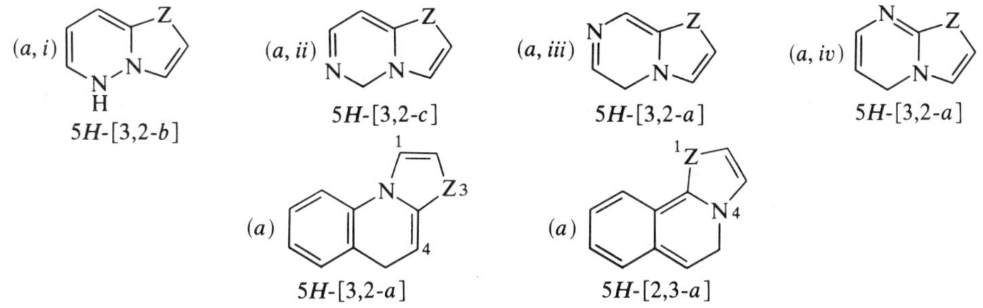

Chart 4 Ring systems formed by fusion of oxazole or thiazole with pyridine containing additional heteroatoms and rings (Z = O, S)

Polycyclic azines, *e.g.* quinoline, pteridine, are grouped under the parent azine ring to which the azole is fused.

Fused ring systems can be prepared either from an azine precursor or from an oxazole or thiazole precursor. In the sections on the synthesis of these compounds the former method is grouped under 'Azine approach', the latter under 'Azole approach'.

4.29.5.2 Survey of Reactivity

The fully conjugated oxazole and thiazole ring systems possess aromatic stability, which is higher in the latter due to better electron-releasing properties of sulfur. In fused dihydroazole systems the resistance towards nucleophilic opening of the azole ring is less.

Electrophilic substitution at nitrogen can occur in either ring, and the preference may depend on the nature of the substituents. Electrophilic substitution at carbon proceeds readily in the presence of electron-releasing substituents.

Nucleophilic substitution reactions are general. Reactions in side-chains are comparable with the same type of reaction in azines. Dimroth-type rearrangements in appropriately substituted azine rings may occur.

The sulfur in saturated or partly saturated thiazolo derivatives may be oxidized by peracids.

4.29.6 OXAZOLE FUSED WITH AZINES

4.29.6.1 Introduction

Oxazole can be fused with pyridine in six different ways and the section on synthesis has been organized accordingly. Further fused systems arise from the diazines and the triazines and their areno and heteroareno homologues.

Considerable knowledge on oxazoloazine chemistry has been built up over the last 15 or so years. A major emphasis has been on the preparation of compounds for biological purposes; oxazolopyrimidines may be regarded as purine analogues. Oxazoloazine ring systems are also found in antibiotics and alkaloids. Fluorescent whitening properties have also stimulated interest in azine-fused oxazoles.

Chemical and physicochemical properties for saturated, partially saturated and fully conjugated ring systems are discussed under the appropriate section on structure, reactivity and synthesis.

4.29.6.2 Structure

4.29.6.2.1 X-Ray diffraction

Bond lengths and bond angles are listed in Table 8.

Table 8 X-Ray Diffraction Data for Fused Oxazoloazines

Compound	Ring A		Bond lengths (Å) and angles (°) Ring B		Ring C	
2-Phenyl-6,7,8,8a-tetrahydro-5H-oxazolo[3,2-a]pyridin-3(2H)-one ⟨77CSC553⟩	a^1 1.506 b^1 1.332 b^2 1.440 c^1 1.420 c^2 1.422	a^1b^1 106.6 a^1c^2 105.1 b^1b^2 113.1 b^2c^1 104.4 c^1c^2 110.5	a^2 1.504 a^3 1.510 a^4 1.516 a^5 1.496 b^2 1.440 b^3 1.452	a^2a^3 110.6 a^2b^3 110.1 a^3a^4 110.3 a^4a^5 110.4 a^5b^2 110.4 b^2b^3 118.9		
1-Substituted 1H,3H,5H-oxazolo-[3,4-a]quinolin-3-one[a] ⟨79JCS(P1)1013⟩	a^1 1.45 (1.48) b^1 1.44 (1.46) b^2 1.37 (1.35) c^1 1.44 (1.46) c^2 1.36 (1.35)	a^1b^2 106.6 (106.3) a^1c^2 107.8 (107.9) b^1b^2 111.0 (111.4) b^1c^1 103.5 (102.2) c^1c^2 110.8 (110.9)	a^2 1.33 (1.34) a^3 1.53 (1.50) a^4 1.50 (1.53) a^5 1.40 (1.40) b^2 1.37 (1.35) b^3 1.40 (1.43)	a^2a^3 119.3 (117.7) a^2b^2 125.3 (128.5) a^3a^4 112.8 (112.4) a^4a^5 122.3 (122.1) a^5b^3 119.0 (118.5) b^2b^3 120.2 (118.0)	a^5 1.40 (1.40) a^6 1.40 (1.38) a^7 1.39 (1.37) a^8 1.38 (1.36) a^9 1.41 (1.40) a^{10} 1.38 (1.38)	a^5a^6 117.7 (118.2) a^6a^7 122.0 (122.2) a^7a^8 120.0 (119.3) a^8a^9 118.3 (120.3) a^9a^{10} 121.3 (119.9) a^5a^{10} 120.5 (119.8)

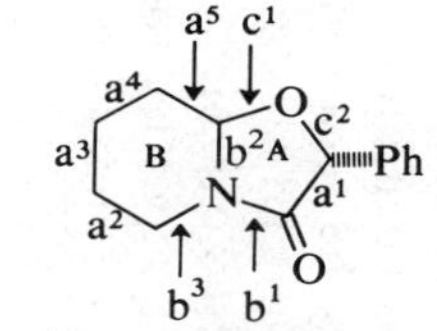

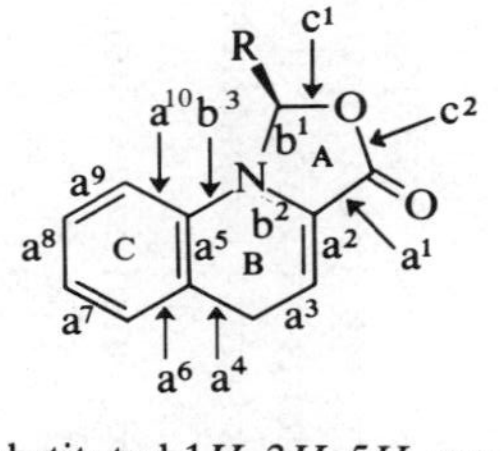

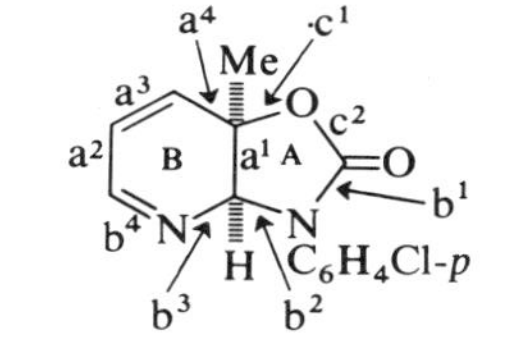

a^1	1.544	a^1b^2	100.84	a^1	1.544	a^1a^4	113.26
b^1	1.363	a^1c^1	104.00	a^2	1.459	a^1b^3	119.74
b^2	1.465	b^1b^2	112.01	a^3	1.311	a^2a^3	121.04
c^1	1.466	b^1c^2	109.66	a^4	1.497	a^2b^4	125.41
c^2	1.358	c^1c^2	109.56	b^3	1.452	a^3a^4	120.99
				b^4	1.282	b^3b^4	117.44

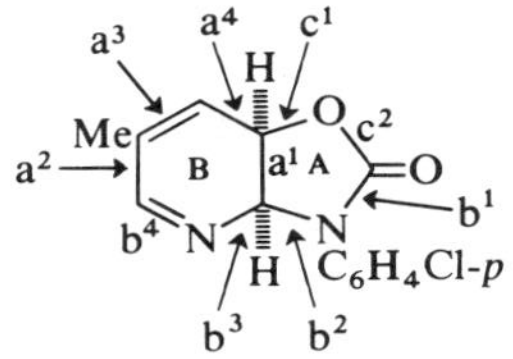

3-(4-Chlorophenyl)-3a,7a-dihydro-7a-methyloxazolo[4,5-b]pyridin-2(3H)-one ⟨79CPB2261⟩

a^1	1.523	a^1b^2	100.69	a^1	1.523	a^1a^4	115.07
b^1	1.364	a^1c^1	103.88	a^2	1.475	a^1b^3	116.93
b^2	1.462	b^1b^2	111.21	a^3	1.311	a^2a^3	117.80
c^1	1.463	b^1c^2	109.35	a^4	1.482	a^2b^4	127.57
c^2	1.361	c^1c^2	108.88	b^3	1.458	a^3a^4	121.08
				b^4	1.269	b^3b^4	116.77

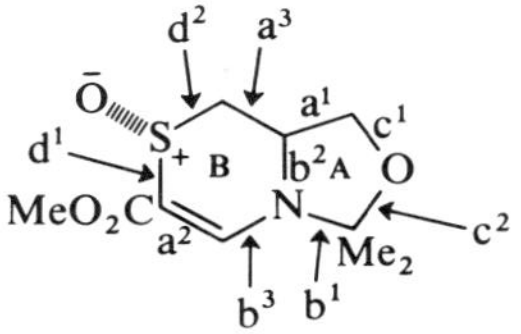

3-(4-Chlorophenyl)-3a,7a-dihydro-6-methyloxazolo[4,5-b]pyridin-2(3H)-one ⟨79CPB2261⟩

a^1	1.538	a^1b^2	99.9	a^2	1.390	a^2b^3	124.6
b^1	1.486	a^1c^1	103.6	a^3	1.506	a^2d^1	123.0
b^2	1.463	b^1b^2	111.8	b^2	1.463	a^3b^2	109.9
c^1	1.433	b^1c^2	101.9	b^3	1.316	a^3d^2	108.4
c^2	1.417	c^1c^2	108.0	d^1	1.770	b^2b^3	122.2
				d^2	1.830	d^1d^2	96.5

Methyl 8,8a-dihydro-3,3-dimethyl-1H,3H-oxazolo[4,3-c]-[1,4]thiazine-6-carboxylate 7-oxide ⟨74JCS(P2)1132⟩

[a] Two unique molecules in the crystal lattice.

4.29.6.2.2 Molecular spectra

(a) *¹H NMR spectroscopy*. Proton chemical shifts and coupling constants for ring CH of some fully conjugated and partially saturated systems are listed in Table 9.

(b) *¹³C NMR spectroscopy*. ¹³C NMR chemical shift data are listed in Table 10. Very few data have been published.

(c) *UV spectroscopy*. UV data for some systems are collected in Table 11. Both fully conjugated and partially saturated systems are included.

(d) *IR spectroscopy*. Some CO stretching frequencies are listed in Table 12. Derivatives of the tautomeric 2-hydroxypyridine system (**166**) all absorb in the region 1795–1820 cm^{-1} ⟨76HCA1593⟩, which is comparable to the absorption at 1790 cm^{-1} for the dihydroquinoline lactone (**167**) ⟨79JHC1589⟩. Compounds (**166**), therefore, exist in the oxo form. The 2-oxopyridinium salt (**168**) like monocyclic oxazolium salts exhibits characteristic high frequency absorption at 1874 cm^{-1} ⟨70JCS(C)1485⟩. The mesoion generated from (**168**) is highly unstable whereas 3-acyl derivatives of the mesoion can be studied. These compounds all exhibit high frequency absorption at 1750–1800 cm^{-1} due to the oxazolone ring. The exocyclic polarized carbonyl group absorbs in the region 1600 cm^{-1} ⟨70JCS(C)1485⟩. In the tautomeric 7-hydroxypyrimidine system (**170**; R = H) there is a strong absorption band at *ca.* 1700 cm^{-1} ⟨70CPB1233⟩ and this is where the *N*-methyl derivative (**170**; R = Me) absorbs ⟨70CPB2242⟩; the hydroxy derivative therefore exists in the usual oxo form of the parent azine.

(e) *Mass spectrometry*. The mass spectral fragmentation patterns for the many different systems can hardly be generalized. The behaviour of salts during mass spectrometry, however, needs some clarification. On electron impact or chemical ionization analysis the salts must initially be thermally converted to volatile species which enter into the gas phase before being broken down by ionization reactions.

The mass spectrum from the pyridinium salt (**171**) has a molecular ion corresponding to loss of perchloric acid from the salt ⟨79JA3607⟩. The volatile species may be an ylide or more likely a vinylic elimination product since thermolysis of (**171**) is known to lead to a rapid generation of allylic pyridones (**172**).

The mass spectrum from the pyrimidinium salt (**173**) shows a molecular ion which corresponds to an adduct between the cation and the anion ⟨75JOC1713⟩. Since the oxazoline ring is very readily opened by attack from the mesylate counterion at C-2, this transformation accounts for the volatile species.

The mesoionic 3-benzoyloxazolopyridine (**174**) and analogues show a molecular ion corresponding to the mass of (**174**) ⟨70JCS(C)1485⟩. In this case the molecule may enter the gas phase in the mass spectrometer as a mesoion or it may isomerize to a ketene (**175**).

Transformations as above are discussed in more detail in the section on thiazoloazines.

4.29.6.3 Reactivity

4.29.6.3.1 Survey of reactivity

Electrophilic attack at nitrogen can occur in either ring, the preference being influenced by the nature of the substituents present. For electrophilic substitution at carbon the presence of electron-donating substituents is normally required.

Table 9 ^{1}H NMR Data for Fused Oxazoloazines

Compound	Solvent	1	2	3	4	5	6	7	8	Coupling constants (Hz)	Ref.
R = M	TFA	—	2.71	8.12	—	8.86	7.84	8.43	8.04	$J_{5,6}$ 7.0, $J_{5,7}$ 1.0, $J_{6,7}$ 7.0, $J_{6,8}$ 1.0, $J_{7,8}$ 8.0	70JCS(C)1938
R = Ph	D$_2$O	—	—	8.52	—	8.72	7.1–7.8	8.20	8.0	$J_{5,6}$ 7.0, $J_{5,7}$ 1.0, $J_{6,7}$ 7.0, $J_{6,8}$ 1.0, $J_{7,8}$ 8.0	70JCS(C)1938
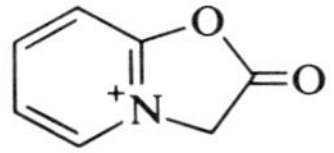	TFA	—	—	5.65	—	—	—	—	—		70JCS(C)1485
R = H	D$_2$O	—	2.1, 6.0	5.50	—	8.80	7.93	8.80	7.85		79JA3607
R = Me	D$_2$O	—	2.2	5.25	—	8.80	7.93	8.80	7.85		79JA3607
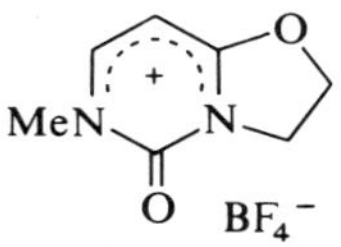	D$_2$O	—	4.57	5.32	—	—	3.69	8.60	6.65	$J_{7,8}$ 8.0	75JOC1713
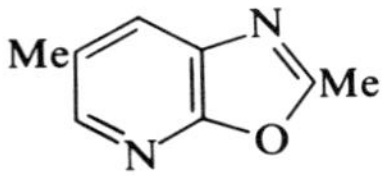	CDCl$_3$	—	2.65	—	—	8.17	2.4	7.8	—	$J_{5,7}$ 3.0	79CB3282
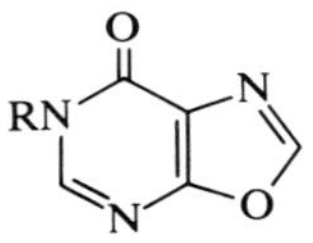 R = H	TFA	—	8.63	—	—	8.77	—	—	—		70CPB1233
R = Me	TFA	—	8.68	—	—	8.83	3.97	—	—		70CPB2242

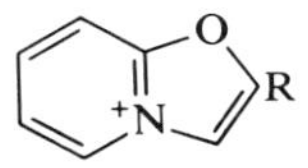

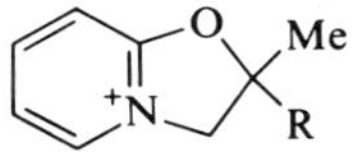

Table 9 ^{1}H NMR Data for Fused Oxazoloazines

Compound	Solvent	\[^{1}H chemical shifts (p.p.m.)\]							Coupling constants (Hz)	Ref.
		1	2	3	4	5	6	7	8		
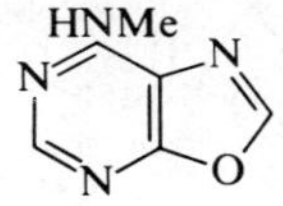 HNMe-structure	DMSO-d_6	—	8.33	—	—	8.61	—	3.03	—		70BCJ3909
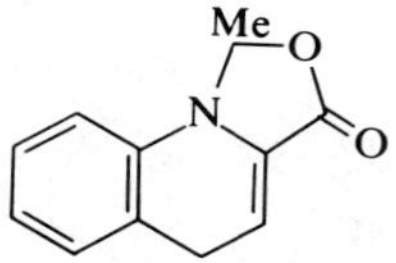 Me-oxazoloquinolinone	CDCl$_3$	1.73, 5.71	—	—	5.6	3.68	—	6.9–7.3 6.5(C-9)	—		79JHC1589
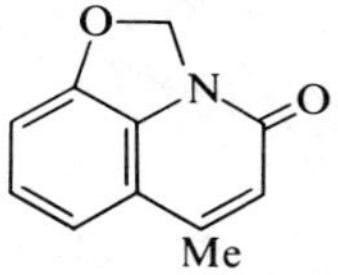 Me-oxazolonaphthyridinone	TFA	—	6.66	—	—	—	2.90	—	—		73JA5003

Table 10 ^{13}C NMR Data for Fused Oxazoloazines

Compound	Solvent	^{13}C chemical shifts (p.p.m.)							J_{CH} coupling constants (Hz)	Ref.
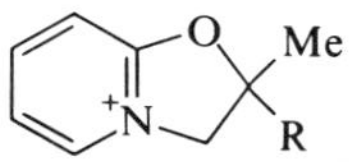		2	3	5	6	7	8	9		
R = H	D$_2$O	106.0	56.8	147.9	128.5	138.3	140.0	—	160.2 (8a), 19.6 (2-Me)	79JA3607
R = Me	CDCl$_3$	92.0	61.2	147.9	139.5	140.2	140.8	—	159.3 (8a), 26.6 (2-Me)	79JA3607
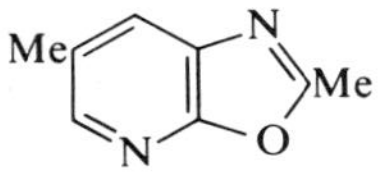	CDCl$_3$	164.5	—	144.3	130.3	127.9	—	—	158.7 (3a), 133.1 (7a), 14.9 (2-Me), 18.3 (6-Me) J_5 179.4, J_7 163.8	79CB3282
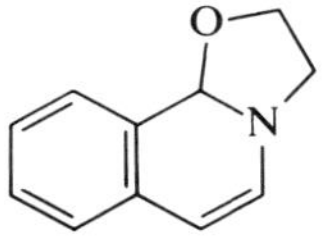	CDCl$_3$	62.3	59.3	130.5	102.5	123.7	128.9	124.8	133.9 (10), 134.4 (6a), 121.3 (10a), 76.1 (10b)	78JOC672
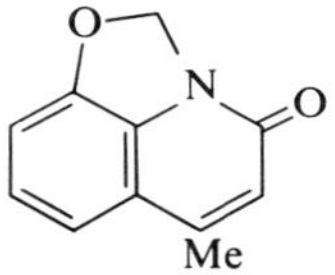		85.7	—	122.6	—	—	—	—		77JA4647

Table 11 UV Absorption Data for Fused Oxazolozines

Compound	Solvent	λ_{max}[a] (log ε)	Ref.
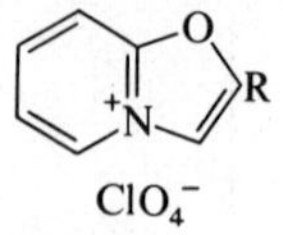 R = Me	EtOH	277[b] (3.91), 270[b] (4.04), 262 (4.08)	67JHC66
R = Ph	EtOH	297 (4.21), 285[b] (4.18), 238 (3.97)	67JHC66
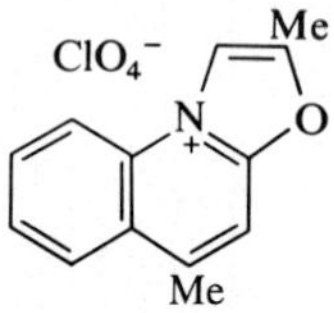	EtOH	308 (3.93), 282 (3.78), 270[b] (3.70), 247[b] (3.82), 230 (4.52)	67JHC66
	CH$_2$Cl$_2$	361, 320[b]	70JCS(C)1485
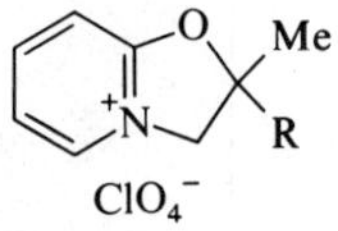 R = H	H$_2$O	285 (3.68)	79JA3607
R = Me	H$_2$O	285 (3.71)	79JA3607
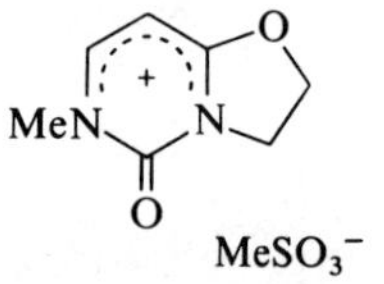	MeCN	290 (4.00), 202 (4.06)	75JOC1713
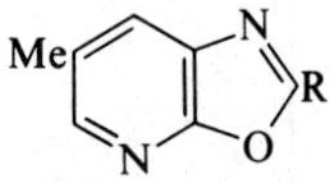 R = Me	Et$_2$O	292 (3.82), 286 (3.93), 283 (3.93), 244 (3.35)	79CB3282
R = Ph	Et$_2$O	324 (4.32), 310 (4.51), 306 (4.43), 267 (4.18), 260 (4.16)	79CB3282
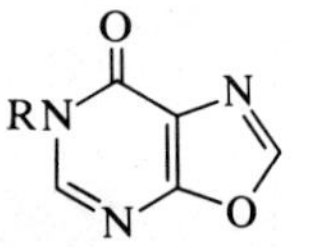 R = H	EtOH	269 (3.69), 244[b] (3.93), 238 (4.02), 234[b] (3.98)	70CPB1233
R = Me	EtOH	272 (3.73), 245 (3.89), 239 (3.97), 233[b] (3.91)	70CPB2242
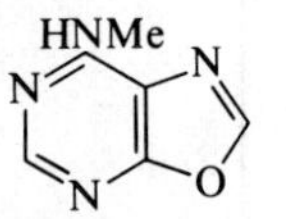	MeOH	283[b] (3.71), 258 (4.20), 207 (4.20)	70BCJ3909
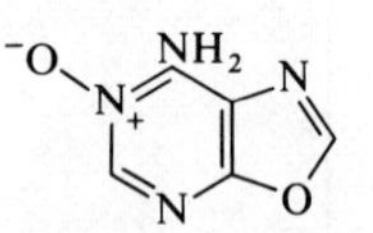	pH 8.5	301 (3.40), 259 (3.96), 230 (4.52)	70BCJ3305
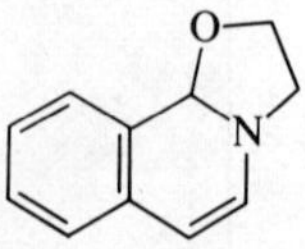	EtOH	246 (3.45), 231 (3.67), 208 (3.97)	78JOC672
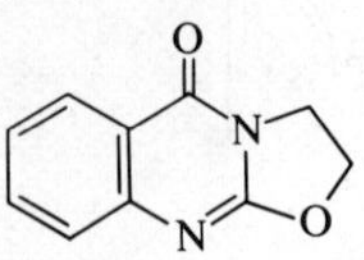	MeOH	318[b], 308 (3.56), 262 (3.79), 254 (3.83), 245 (3.83), 226[b] (4.55), 221 (4.61)	76S469

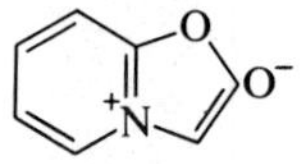

[a] nm. [b] Shoulder.

Five-membered Rings Fused with Six-membered Rings 653

Table 12 IR Absorption Data for CO Groups in Fused Oxazoloazines

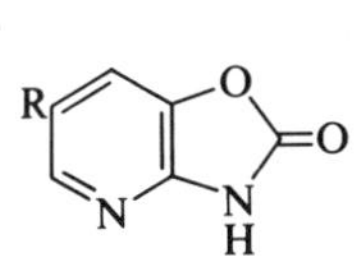

(166)
R = H, Cl, Br, NO$_2$
1795–1820 cm^{-1}
76HCA1593

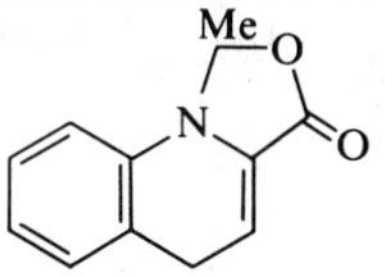

(167)
1790, 1690 cm^{-1}
KBr
79JHC1589

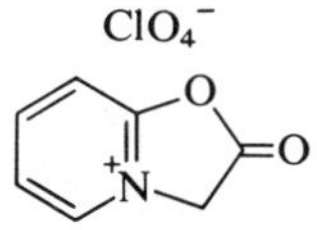

(168)
1874 cm^{-1}
Nujol
70JCS(C)1485

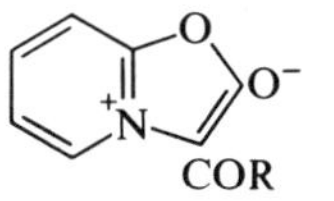

(169)
R = CF$_3$; 1802, 1601 cm^{-1}
R = Ph; 1787, 1582, 1563 cm^{-1}
Nujol
70JCS(C)1485

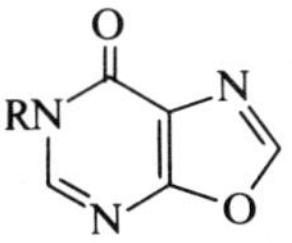

(170)
R = H; 1699 cm^{-1}
R = Me; 1700 cm^{-1}
KBr
70CPB1233, 70CPB2242

Nucleophilic substitution in the azine ring is a normal reaction. The fused oxazole ring, however, is susceptible to nucleophilic ring opening reactions, especially in oxazolium salts and dihydro derivatives. When fused to pyrimidine, substituents in the pyrimidine ring may promote a Dimroth rearrangement of the latter.

4.29.6.3.2 Electrophilic attack at nitrogen

As mentioned above, the relative susceptibility of the two rings to electrophilic attack at nitrogen is largely influenced by the nature of the substituents present. Compounds with a hydroxyl group in positions α or γ to a ring nitrogen are as a rule *N*-alkylated preferentially under alkaline conditions or under the influence of metal catalysts. 3-Hydroxymethylation results when the 2-oxazolinone (**176**) is treated with aqueous formaldehyde. Under Mannich conditions (**176**) reacts to form the 3-*N*,*N*-dimethylaminomethyl derivative. With vinyl acetate and Hg(II) catalysis (**176**) reacts to form the 3-vinyl compound (**177**) ⟨76HCA1593⟩. Metal catalysis is also used in the condensation of the pyrimidine analogue (**178**) with sugars; in the presence of mercury(II) cyanide (**178**) reacts with 1-bromo-2,3,5-tri-*O*-acetyl-D-ribofuranose to form the correspondingly acetylated nucleoside. The latter can be deprotected in the sugar moiety by ethanolic ammonia to form the nucleoside (**179**). Benzoylated sugars react equally well and nucleosides from glucopyranose, arabopyranose and xylopyranose have been prepared in this way and have been found to be the β-anomers ⟨75CPB2104⟩.

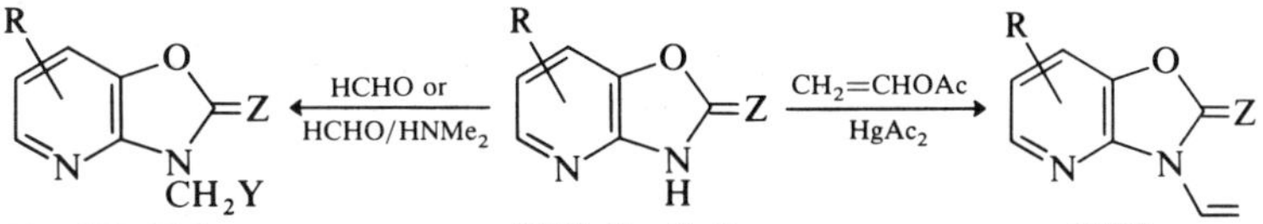

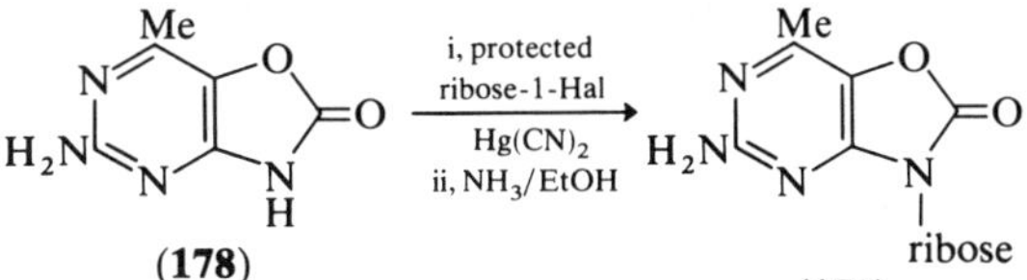

N-Alkylation in the azine ring is demonstrated by the preparation of 2-substituted 6-(β-D-ribofuranosyl)oxazolo[5,4-*d*]pyrimidin-7(6*H*)-ones (**181**). In this case the silyl ether variant of the Hilbert–Johnson method for the synthesis of pyrimidine nucleosides has been used; the pyrimidinone (**180**) is converted into its silyl ether using HMDS and the ether reacted with 2,3,5-tri-*O*-acetyl-D-ribofuranosyl bromide under Hg(II) catalysis. Deacetylation using methanolic ammonia gives the nucleoside with a β-anomeric configuration ⟨74JMC1282⟩.

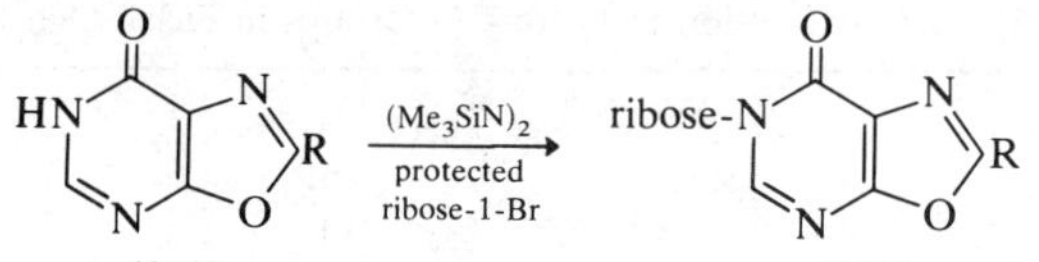

Nitrogen heterocycles are attacked by cyanogen halides at the nitrogen. In the reaction of cyanogen bromide with the saturated, fused oxazole (**182**), a fused medium-ring heterocycle (**183**) is formed when the reaction is carried out under solvolytic conditions ⟨79CI(L)319⟩.

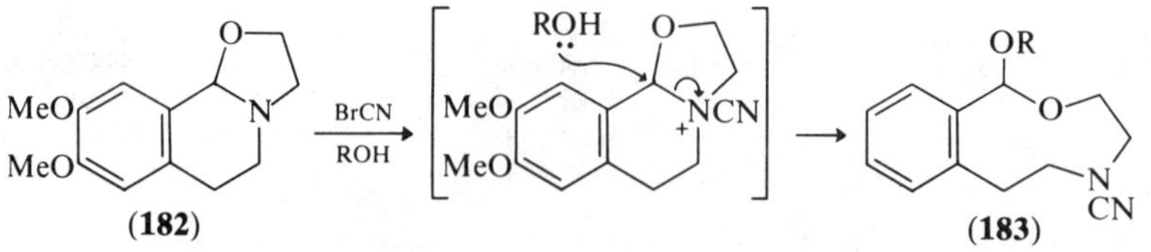

4.29.6.3.3 Electrophilic attack at carbon

Activation by substituents is normally required for electrophilic substitution. The mesoionic oxazole (**184**) is highly reactive. It is stored as its hydroperchlorate and can be generated in solution by addition of triethylamine. It is rapidly substituted by electrophiles at C-3. The 3-acyl derivatives (**185**) are stable compounds because of the stabilization of the negative charge by the exocyclic carbonyl group. Coupling with diazonium salts to form diazo compounds (**186**) is another substitution reaction which leads to stabilized derivatives. The stability is increased by electron-withdrawing substituents in the aryl ring. 3-Benzylidene derivatives are formed with arenealdehydes. Salicylaldehyde reacts rapidly to form a coumarin (**187**); the first formed 3-benzylideneoxazolone is attacked intramolecularly by the phenolic oxygen in a translactonization reaction ⟨70JCS(C)1485⟩. The quinoline homologue (**188**) has similar properties and is stabilized by substitution at C-1 ⟨78TL1887⟩.

In the 2-oxo derivative (**189**), which has no vacant C-position in the azole ring, both halogenation and nitration occur at C-6 in the six-membered ring; C-6 is a pyridine *meta*-position and is *para* to the azole nitrogen ⟨76HCA1593⟩.

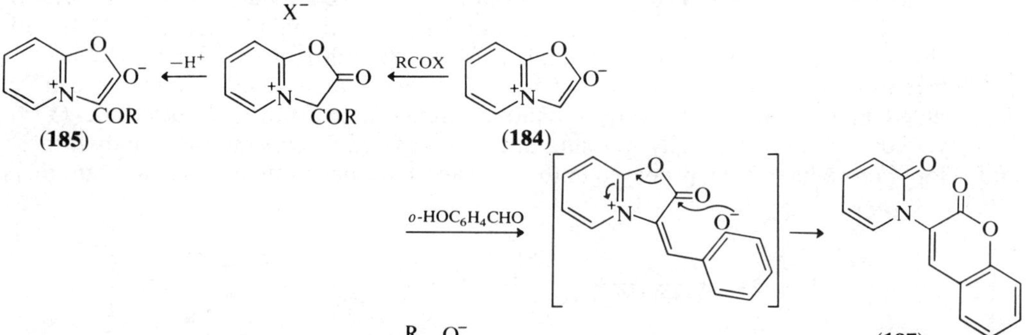

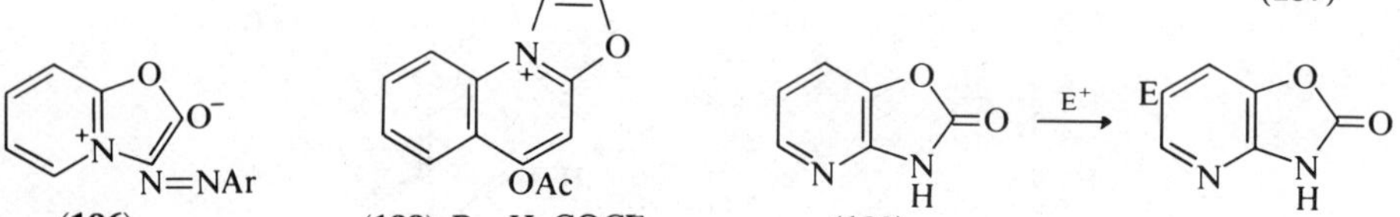

4.29.6.3.4 Nucleophilic attack at carbon

Fused oxazoles are susceptible to ring opening reactions. In appropriately substituted systems, however, nucleophilic substitutions in the azine ring are possible. Thus, in the 7-pyrimidinone (**190**) chlorination is effected in the usual manner by means of phosphorus oxychloride. The chlorine can be substituted by amino or sulfur nucleophiles ⟨71JHC503⟩.

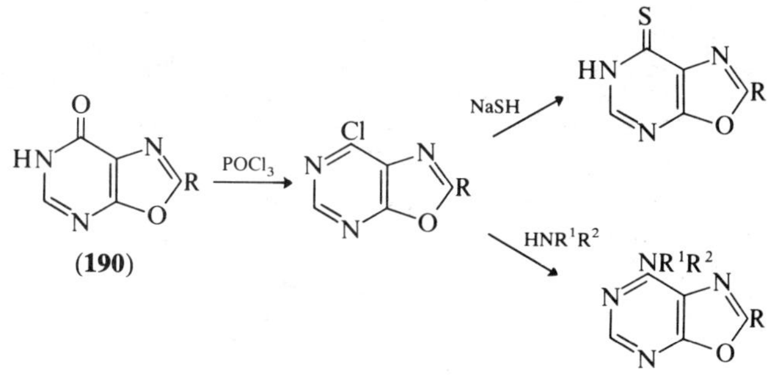

4.29.6.3.5 Nucleophilic attack with ring opening

The oxazole ring possesses less aromatic stabilization than a thiazole ring and is readily opened in many of its fused derivatives, especially in oxazolium salts. In the pyridazinium derivative (**191**) the oxazole ring is opened by oxygen, sulfur or carbon nucleophilic attack at the C-8a ring junction ⟨77YZ422⟩. In the mesoionic pyridine (**192**) an amine attacks at C-2, which is a pseudocarbonyl carbon atom ⟨70JCS(C)1485⟩.

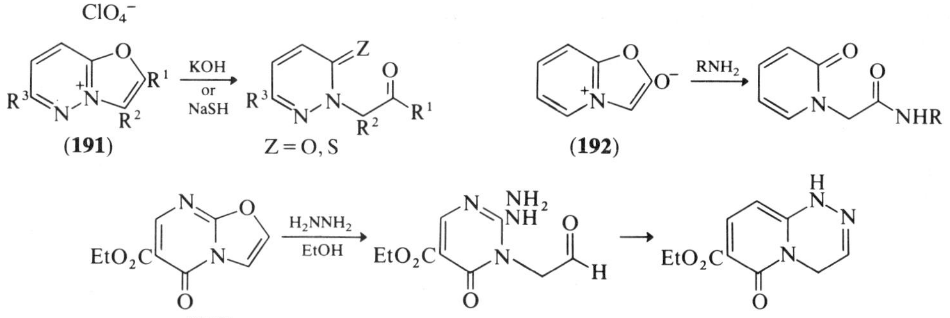

The 5-oxo-5H-oxazolo[3,2-a]pyrimidine-6-carboxylate (**193**) reacts faster than the corresponding thiazolo analogue with hydrazine, which is explicable on the basis of the greater electronegativity of oxygen. The initial attack is at C-2 or more likely at C-8a; the ring-opened product is subsequently cyclized to a triazine ⟨71JCS(C)1615⟩.

The 7-iminopyrimidines (**194**) are chemically unstable because of their great tendency to undergo the Dimroth rearrangement to the more stable 7-aminopyrimidines (**195**) ⟨70BCJ3909⟩. The latter in aqueous alkali or on heating in formamide are converted into hypoxanthines (**196**). This reaction involves opening of the oxazolo ring at C-2 and recyclization with the vicinal amino group ⟨73BCJ506⟩.

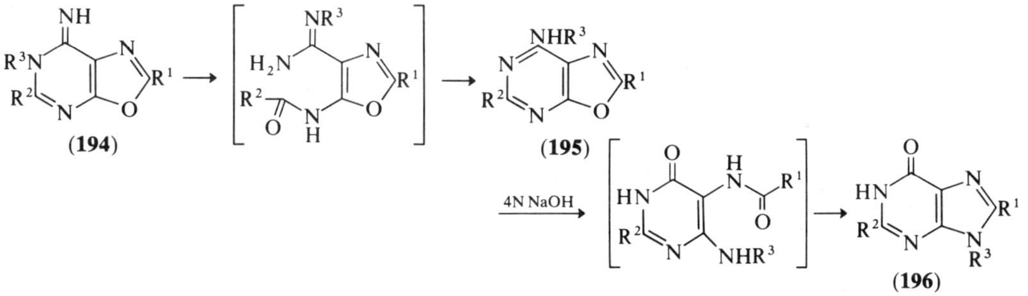

Thiation of the uracil (**197**) by means of phosphorus pentasulfide in pyridine leads to the formation of a fused thiazole (**198**). Presumably the rearrangement, by opening of the oxazole ring and recyclization, occurs after dithiation of the original oxo groups ⟨78CPB765⟩. The conversion of the pyrimidine 6-oxide (**199**) to the hypoxanthine 3-oxide is another example of the same rearrangement involving opening of the oxazolo ring ⟨70BCJ3305⟩. The rearrangement in the oxidation of the 7-aminopyrimidine (**200**) may be similar; (**200**) is transformed into the purine (**202**) with hydrogen peroxide in acetic acid; the first formed product may be the 1-oxide (**201**) ⟨70BCJ2281⟩.

(197) **(198)**

(199)

(200) **(201)** **(202)**

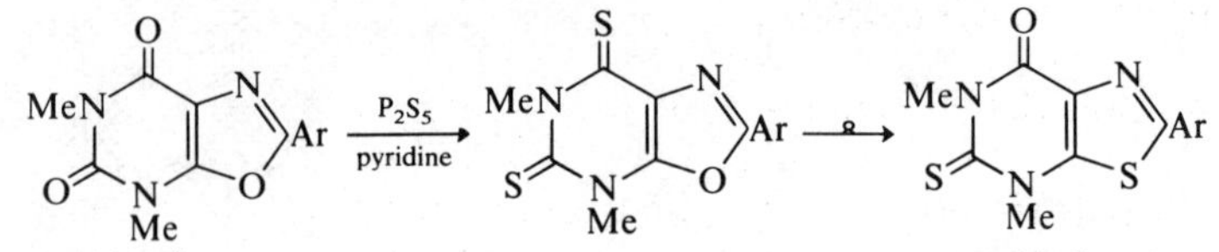

The methylene bridge in the fused oxazoline (**203**), 6-methyl-2*H*,6*H*-oxazolo[5,4,3-*ij*]quinolin-4-one, shows great stability towards acid cleavage, such as heating in 47% hydriodic acid. It is cleaved by oxidation, however, when treated with activated manganese dioxide in acetic acid ⟨73JA5003⟩.

(203)

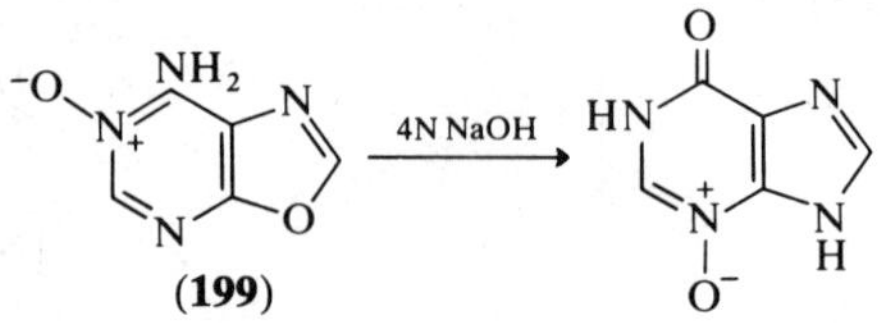

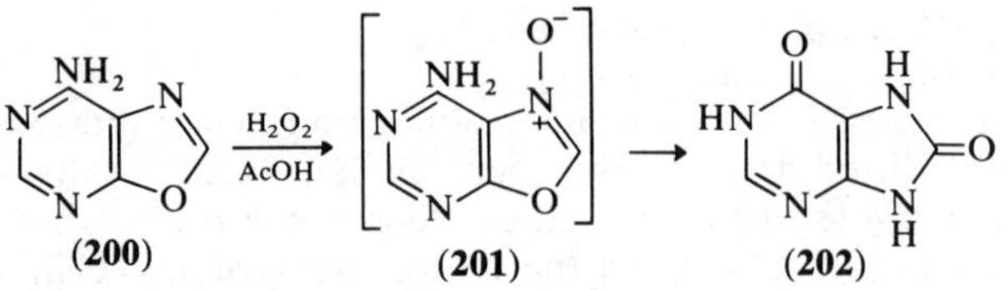

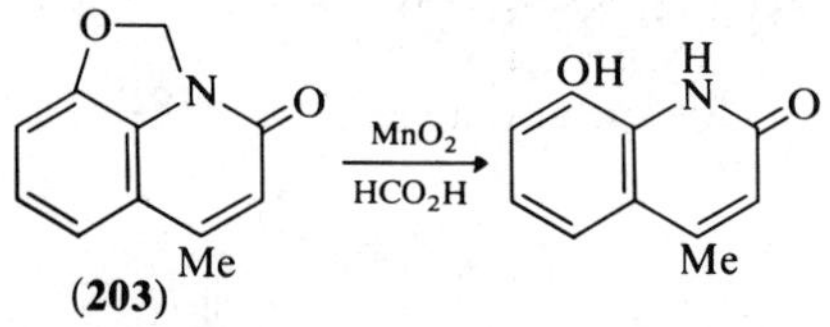

Fused dihydrooxazolium compounds react readily by nucleophilic displacement of the oxygen from the azine ring. Thus, 2-carboxy-2,3-dihydrooxazolo[2,3-*a*]pyridinium bromide reacts with oxygen or nitrogen nucleophiles to form 2-oxo- or 2-imino-1(2*H*)-pyridinelactic acids (**204**) ⟨52JA4906⟩. Similarly, the pyridinium salt (**205**) reacts by substitution in potassium hydroxide solution, whereas potassium *t*-butoxide causes proton abstraction and formation of the *N*-vinyl derivative (**206**) ⟨79JA3607⟩.

(204) Z = O, NH

(205) **(206)**

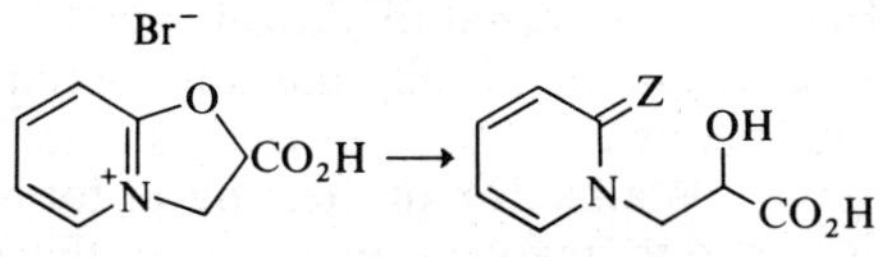

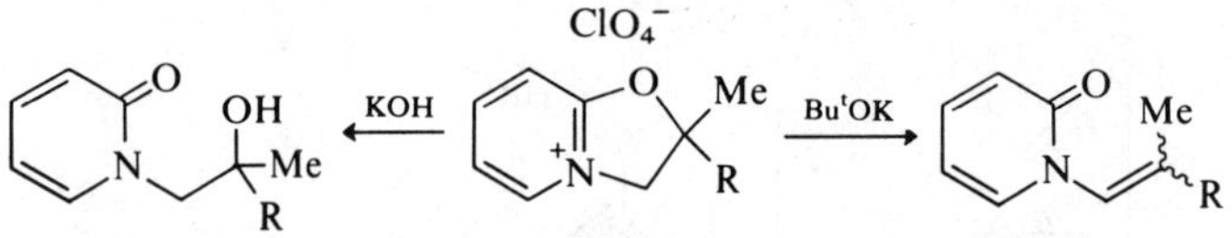

In the 2,3-dihydro-5-oxo-5*H*-oxazolo[3,2-*c*]pyrimidinium salt (**207**) there are three sites for reactions with nucleophilic reagents, *viz*. C-2, C-8a and C-7. Products resulting from attack at C-2 are observed with DMSO, water, alcohols, benzoate, chloride, diethylamine and pyridine. Products resulting from attack at C-8a are observed with water, hydroxide, alcohols, alkoxide and isopropylamine. Diethylamine also causes attack at C-7 of the cation, which results in cleavage of the pyrimidine ring ⟨75JOC1713⟩.

X = OR, Cl, $PhCO_2$, NEt_2, NC_5H_{11} **(207)**

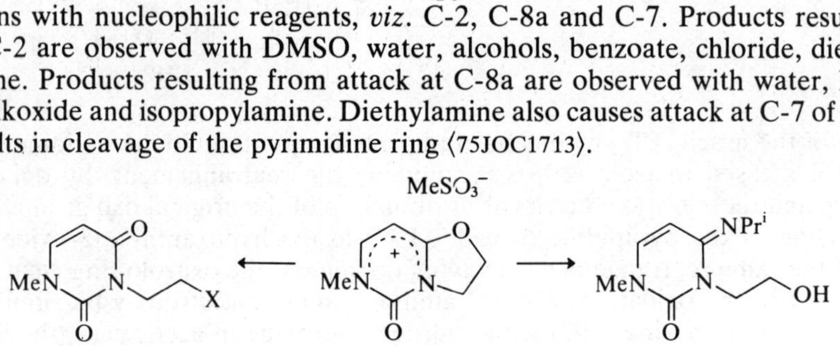

4.29.6.3.6 Other C-linked substituents

Most work on side-chain reactions deals with carbohydrate chemistry in connection with the preparation of nucleoside analogues. In the more simple case of the hydroxymethyl derivative (**208**), thionyl chloride treatment gives the corresponding chloromethyl derivative and the chlorine is replaced by sulfur nucleophiles in the preparation of dithiophosphoric acid derivatives (**209**) ⟨76HCA1593⟩.

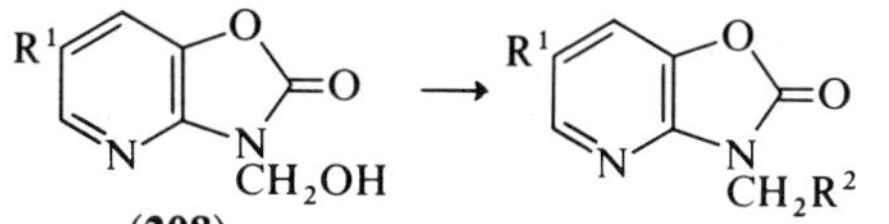

(**208**) → (**209**) $R^2 = Cl, SP(S)(OR)_2$

4.29.6.4 Synthesis: Fully Conjugated Systems

4.29.6.4.1 5H-Oxazolo[3,2-a]pyridine

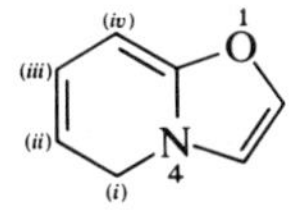

Azine approach. Oxazolo[3,2-a]pyridinium salts (**210**) were first obtained from the cyclodehydration reaction of 1-phenacyl-2(1H)-pyridinone in sulfuric acid ⟨67JHC66⟩. These salts can also be prepared from 2-halo-1-phenacylpyridinium derivatives (**211**) by treatment with a base which causes ylide formation and hence cyclization by intramolecular substitution ⟨69JOC2129, 76CB3646⟩. It is recommended that a bulky tertiary amine is used as base in order to avoid opening of the ring or substitution of the 2-halo substituent in the starting material (**211**). Isoquinoline and quinoline analogues have also been prepared by these methods.

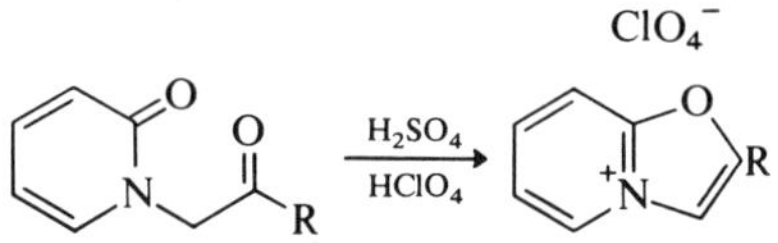

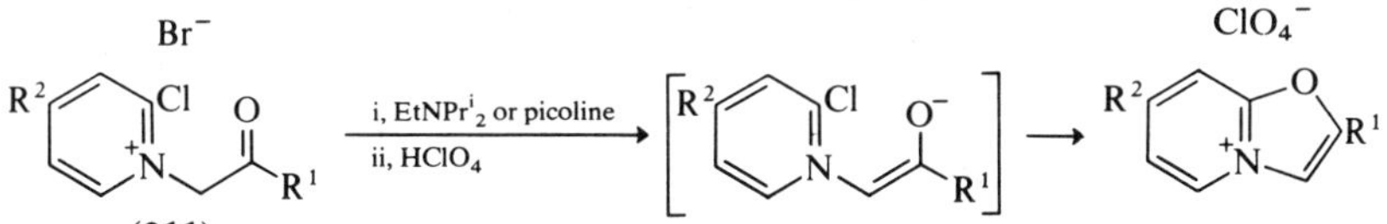

The pyridinium perchlorate (**212**), which is the precursor of the mesoionic oxazolone (**213**), is formed by the action of acetic anhydride and perchloric acid on 1,2-dihydro-2-oxopyridin-1-acetic acid. The mesoion (**213**) is chemically unstable and is generated in solution for its reactions by treatment with a tertiary amine. Compound (**213**) is easily substituted at C-3, and with trifluoroacetic anhydride forms the stable 3-trifluoroacetyl mesoion ⟨70JCS(C)1485⟩.

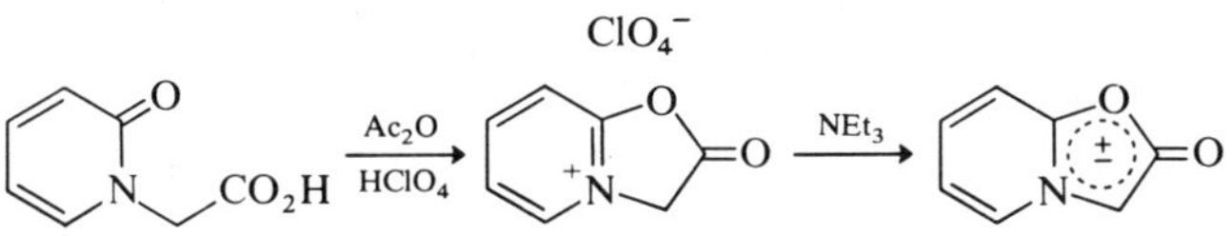

N-Phenacylanthranilic acid reacts with acetic anhydride to form an oxazolo[3,2-a]quinolin-5-one (**214**) ⟨71JOC222⟩. N-Acylation probably accounts for the product first formed and this is followed by cyclocondensations leading to (**214**).

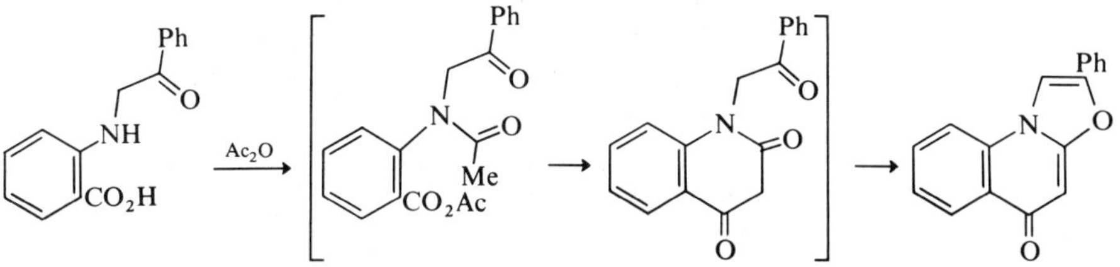

Azole approach. Acid-catalyzed cyclization of 2-(4-ethoxybutyroyl)oxazoles gives 8-oxo-5,6,7,8-tetrahydropyridine (**215**) which is aromatized in boiling acetic anhydride. This reaction has its analogy in the preparation of quinolizinium salts ⟨70JCS(C)1938⟩.

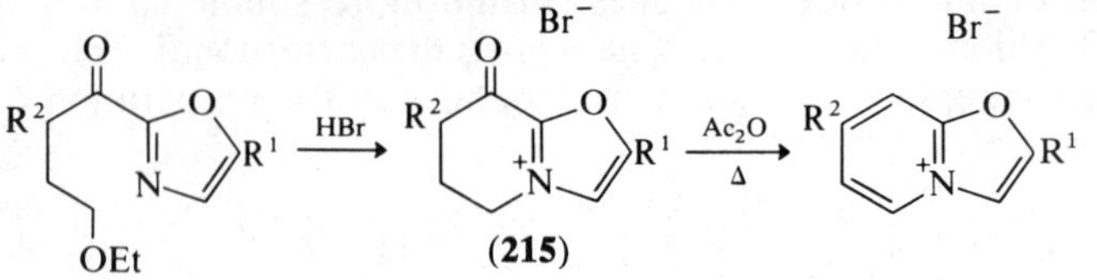

(i) *5H-Oxazolo[3,2-b]pyridazine*

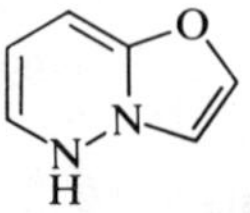

Azine approach. Pyridazinium salts (**216**) can be prepared by acid-catalyzed cyclodehydration of 2-β-oxoalkyl-3(2H)-pyridazinones ⟨77YZ422⟩. The same cyclization of 3,4-dihydro-1,2,3-benzotriazines (**217**) gives the corresponding benzotriazinium salts (**218**). The latter exist in equilibrium with their diazo isomers (**219**); such equilibria are not uncommon for 1,2,3-triazines ⟨67CC1272⟩.

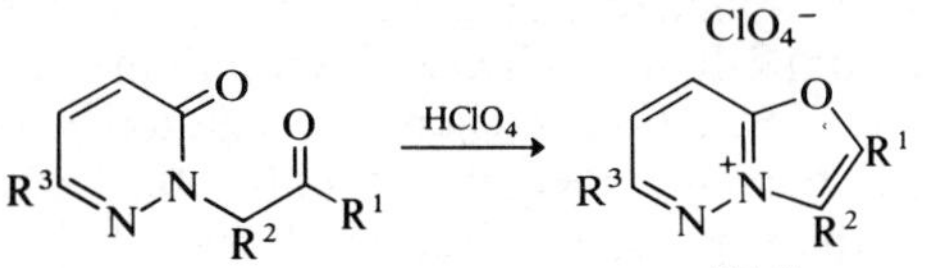

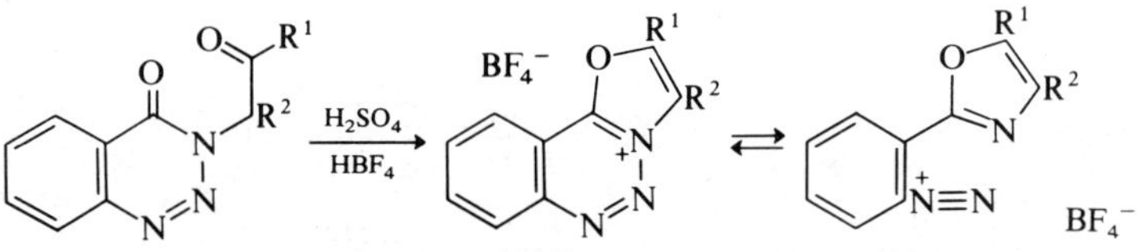

(ii) *5H-Oxazolo[3,2-c]pyrimidine*

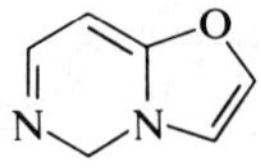

Azole approach. 2-Aminooxazoles react with arylisocyanates on heating to form 2H-oxazolo[3,2-a][1,3,5]triazine-2,4(3H)-diones (**220**). The bicyclic system can also be formed by treatment of the oxazole acetamide or the oxazole urea with two moles of an isocyanate under the same reaction conditions ⟨71JMC1075⟩.

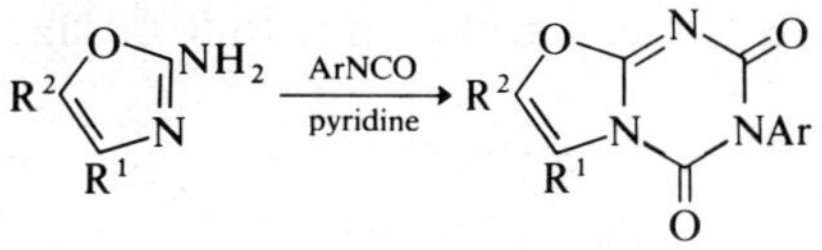

(iv) *5H-Oxazolo[3,2-a]pyrimidine*

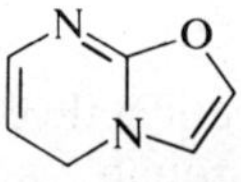

Azole approach. Condensation of 2-aminooxazoles with 1,3-dicarbonyl compounds or β-halo α,β-unsaturated carbonyl compounds yields oxazolo[3,2-a]pyrimidinium salts (**221**). 2-Aminobenzoxazoles in the same way yield benzo homologues ⟨74UKZ633⟩.

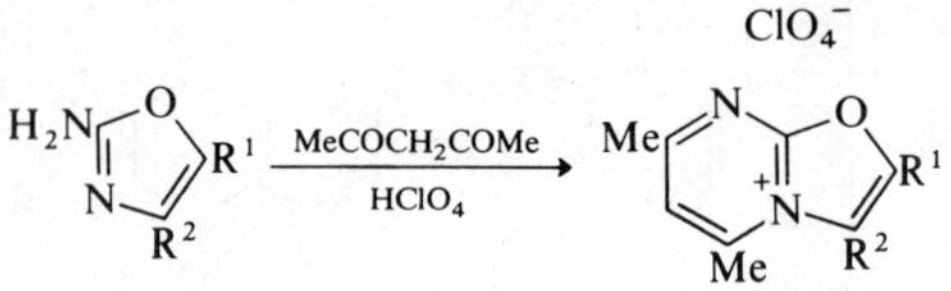

4.29.6.4.2 3H-Oxazolo[3,4-a]pyridine

Azine approach. Reissert compounds such as (**222**) on treatment with acid form transient cyclic intermediates containing the 3,4-fused oxazole ring. When the acid is tetrafluoroboric acid, the Reissert salt (**223**) can be isolated ⟨73JA2392⟩.

If the Reissert compound from the reaction with ethyl chloroformate (**224**) is lithiated at C-2 and subsequently treated with benzaldehyde, the final product is the 1*H*-oxazolo[3,4-*a*]quinolin-1-one (**225**) ⟨73JHC99⟩.

Derivatives with an oxo group between the heteroatoms are formed with phosgene from precursors with structural features as in the isoquinoline (**226**) ⟨70TL5261⟩.

4.29.6.4.3 Oxazolo[4,5-b]pyridine

Azine approach. 2-Substituted oxazolo[4,5-*b*]pyridines are obtained from the reaction between 2-amino-3-hydroxypyridine and a carboxylic acid in the presence of an agent to absorb the water eliminated. Acid anhydrides can be used ⟨57JCS4625⟩. In the presence of PPA, 2-aryl derivatives (**227**) can be made ⟨78JMC1158⟩. Phosgene treatment gives 2-oxo derivatives (**228**; Z=O), cyanogen chloride gives 2-imines (**228**; Z=NH), and with carbon disulfide 2-thiones (**228**; Z=S) are formed. The 2-oxo derivative is also formed by Hofmann degradation of the 2-carbamoylpyridine (**229**) ⟨76HCA1593, 75JHC775⟩.

(ii) Oxazolo[4,5-d]pyrimidine

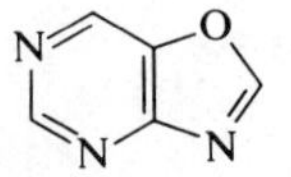

Azine approach. 4-Amino-5-hydroxypyrimidine condenses with acid anhydrides or esters to form the corresponding 2-substituted oxazolopyrimidine system ⟨77CPB491⟩. Under relatively mild reaction conditions the nucleoside 6-amino-5-hydroxyuridine will undergo cyclocondensation with formamide under the influence of polyphosphate ester to form the oxazolo-fused nucleoside (**230**); both protected and non-protected pyrimidine nucleosides have been used in this reaction ⟨73CPB1327⟩.

Photochemically induced rearrangement of the azole ring in the fused isoxazole (**231**) leads to the corresponding oxazole (**232**) in low to moderate yields ⟨78CPB2497⟩.

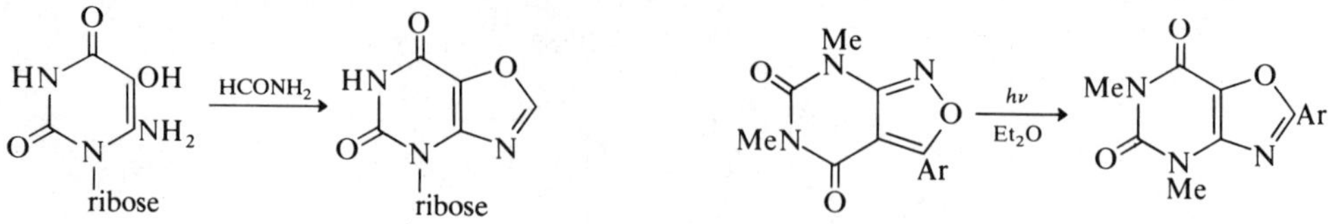

4.29.6.4.4 Oxazolo[4,5-c]pyridine

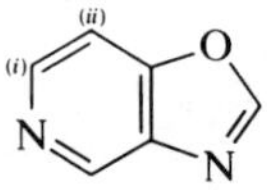

Azine approach. Cyclocondensation between 3-amino-4-hydroxypyridines and carboxylic acids leads to oxazole fusion as in (**233**) ⟨77CR(C)(284)73⟩. The 2-phenyl derivative (**235**) has been obtained *via* a pyridyne-type reaction; 3-benzamido-5-bromopyridine reacts with lithium piperidide *via* an intermediate which can be visualized as the pyridyne enolate (**234**) ⟨73CB220⟩.

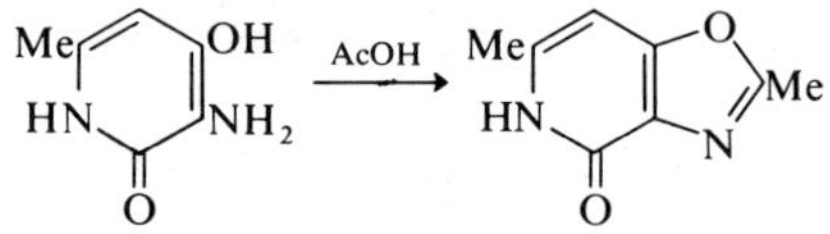

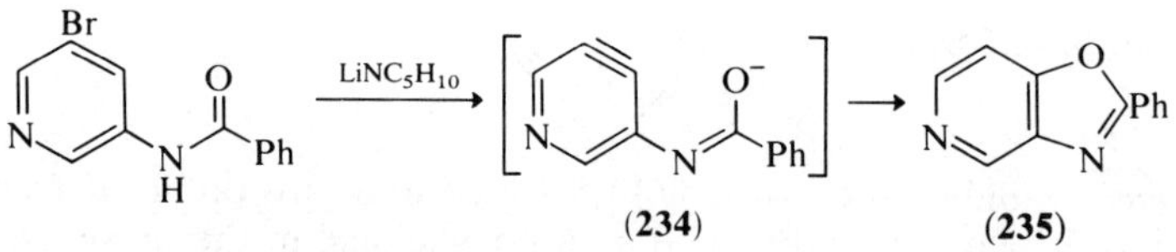

(ii) Oxazolo[5,4-d]pyrimidine

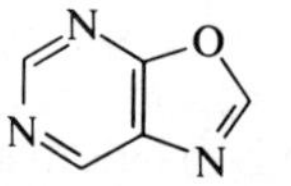

Azine approach. 2-Substituted oxazolo[5,4-*d*]pyrimidin-7-ones (**236**) are formed when 5-amino-4,6-dihydroxypyrimidines are heated with acid anhydrides ⟨71JHC503, 75JOC3141⟩.

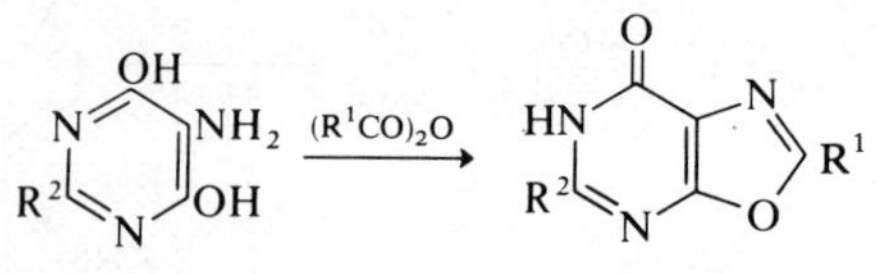

5-Benzylideneamino-6-hydroxyuracils (**237**) can be cyclized under mild reaction conditions; thionyl chloride or NBS induces cyclization ⟨77H(6)1919⟩. The same products can be prepared by heating 6-hydrazone derivatives of uracil (**238**) with sodium nitrite in acetic acid ⟨77H(6)1925⟩.

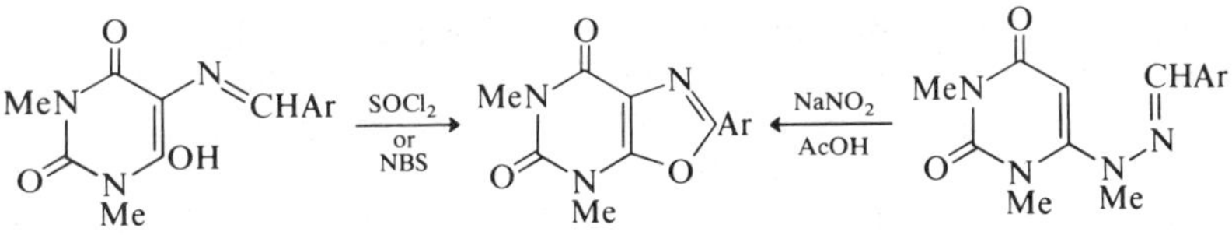

(237) → (238)

Azole approach. 5-Ethoxymethyleneaminooxazole-4-carboxamides (**239**), formed from ethyl orthoformate and the amine, are cyclized to oxazolo[5,4-*d*]pyrimidines when heated in acetic anhydride. This method is superior to the cyclization of the corresponding 5-acylamino derivatives ⟨70CPB2242⟩. Another modification is the use of 4-cyano derivatives (**240**); fused pyrimidines are conveniently formed from heteroarenes which have a cyano and an ethoxymethyleneamino group at adjacent positions. With ammonia or alkylamines, (**240**) reacts in the cold to form the pyrimidine (**241**); with aromatic amines heating is required. The 7-imino derivatives formed from primary amines may be isolated, but they very readily undergo a Dimroth rearrangement under the basic conditions of the reaction to form the more stable 7-aminopyrimidines (**241a**) ⟨70BCJ3909, 73BCJ506⟩. When hydroxylamine is the amine used in the reaction, the 6-oxide (**242**) is formed ⟨70BCJ3305⟩.

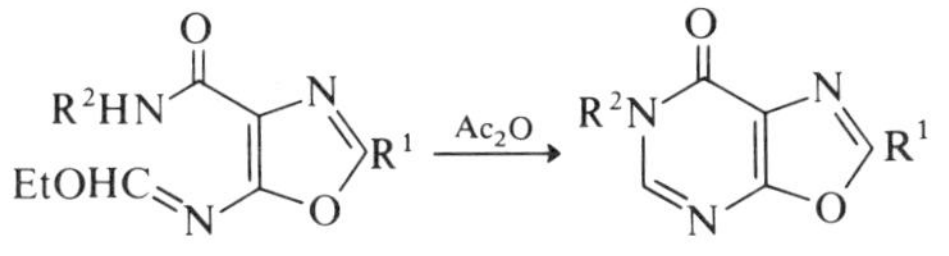

(239) 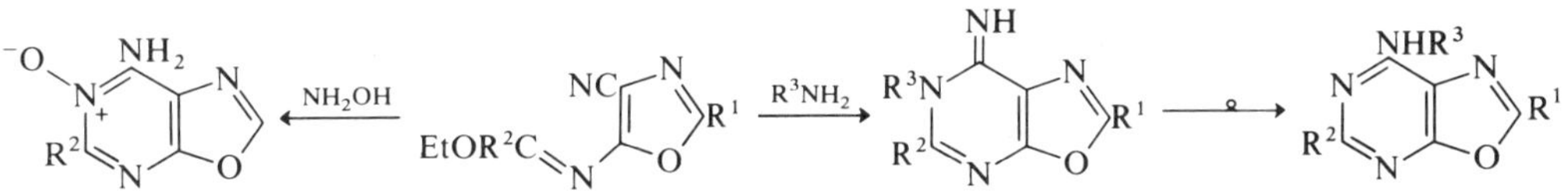 (242) (240) (241) (241a)

Benzoyl isocyanate trimerizes to form 7-benzoylimino-2,5-diphenyloxazolo[5,4-*d*]-[1,3]oxazine (**243**). The latter in the presence of sodium methoxide undergoes a Dimroth-type rearrangement to yield the fused pyrimidine (**244**) after debenzoylation ⟨72JOC2583⟩.

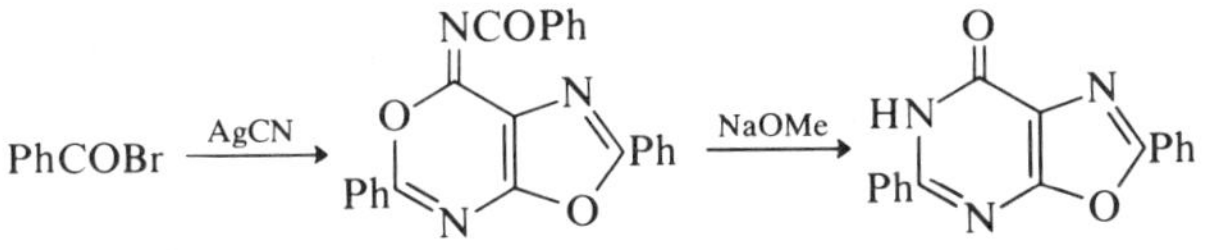

(243) (244)

4.29.6.4.5 Oxazolo[5,4-c]pyridine

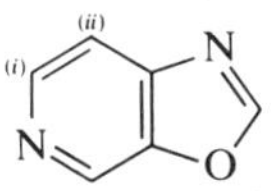

Azine approach. Condensation of 4-amino-3-hydroxypyridines with carboxylic acids yields 2-substituted oxazolo[5,4-*c*]pyridines (**245**) ⟨74GEP2330109⟩.

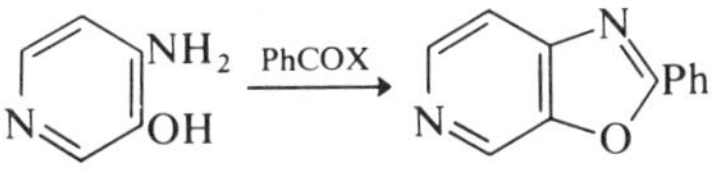

(245)

4.29.6.4.6 Oxazolo[5,4-b]pyridine

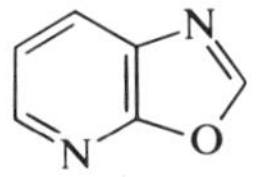

Azine approach. It is claimed that a stepwise procedure is required for the preparation of oxazolo[5,4-*b*]pyridines (**246**) from 3-amino-2(1*H*)-pyridinones; firstly the amino group is acylated and secondly the product is cyclized using a dehydrating agent. Phosphorus pentoxide has been used for the dehydration ⟨59CPB725⟩, but phosphorus oxychloride is said to be a more convenient reagent ⟨78JMC1158⟩. In the cyclization of the naphthyridine (**247**) acetic anhydride has been used ⟨80CPB761⟩.

Oxazolo[5,4-*b*]pyridines can also be prepared in *ca.* 50% yields by photochemical rearrangement of the azole ring in isoxazolo[5,4-*b*]pyridines (**248**); the latter are readily available from 5-aminoisoxazoles and 3-ethoxyacroleins ⟨79CB3282⟩.

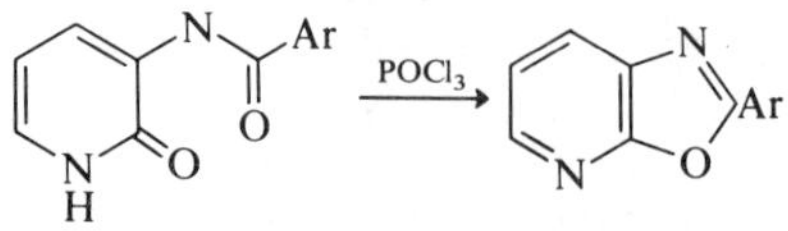

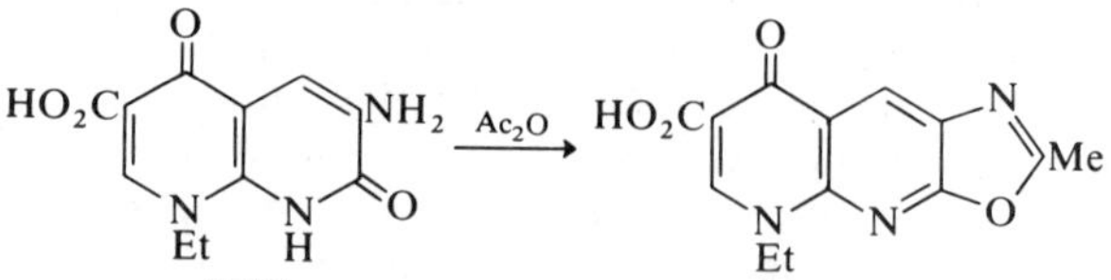

Azole approach. Application of the commonly used method for reductive cyclization of aromatic nitro compounds by means of triethyl phosphite to the 4-*o*-nitrobenzylidene oxazoles (**249**) leads to the fused quinolines (**250**) ⟨69JCS(C)385⟩.

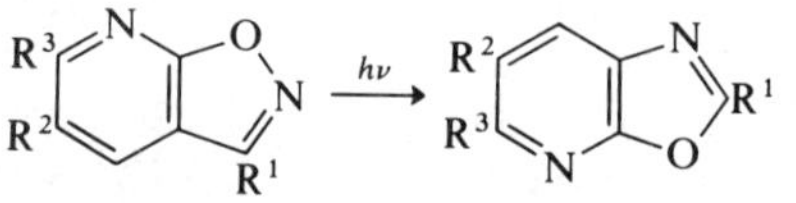

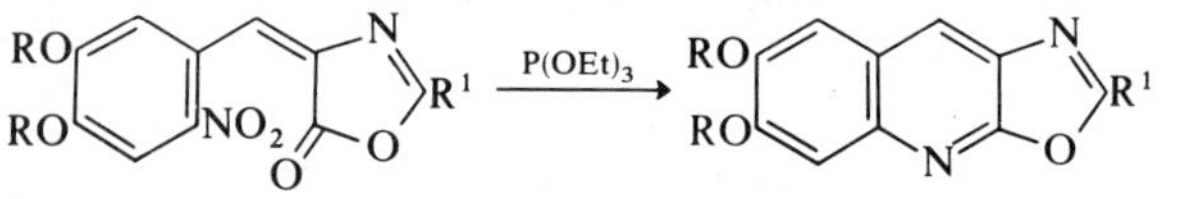

4.29.6.5 Saturated and Partially Saturated Rings: Introduction

The main interest in saturated or partially saturated derivatives has been on compounds having a bridgehead nitrogen atom, *viz.* on members of the 5*H*-oxazolo[3,2-*a*]pyridine and 3*H*-oxazolo[3,4-*a*]pyridine series.

Substitution reactions leading to ring opening for partially saturated systems are discussed in the appropriate sections under fully conjugated rings.

Structural features (X-ray and molecular spectra) are also dealt with under fully conjugated rings.

4.29.6.6 Saturated and Partially Saturated Rings: Structure

4.29.6.6.1 Thermodynamic aspects

Perhydrooxazolo[3,4-*a*]pyridines may exist in solution as an equilibrium mixture of the *trans*-fused conformer (**251a**) and the two *cis*-fused conformers (**251b**) and (**251c**). The conformers are interconvertible by nitrogen inversion and ring inversion. Dominant among the interactions influencing the position of the conformational equilibrium is that arising from the 1,3-arrangement of the heteroatoms in the oxazolidine ring. This interaction is unfavourable in conformers (**a**) and (**c**), which have nearly parallel arrangements of lone pairs, but is minimized in conformer (**b**). The *cis* conformer (**c**) is destabilized by two *gauche*-butane interactions and one *gauche*-propanol-type interaction relative to the *trans* conformer and therefore does not make any significant contribution to the equilibrium mixture. Thus, the equilibrium mixture of (**251**) in solution at room temperature contains from *ca.* 68% (CCl$_4$) to *ca.* 76% (CDCl$_3$) of the *trans* isomer, the remaining being the *cis* isomer (**b**). Perhydro[4,3-*c*][1,4]oxazine (**252**) exists under the same conditions as *ca.* 86% *cis*-fused conformer in equilibrium with *ca.* 14% of the *trans*-fused conformer (**252a**). The difference in the conformational preference in these systems may be ascribed to an increase in ring fusion strain because of the smaller morpholine ring and the replacement of a *gauche*-

butane interaction in the *cis*-fused conformer (**251b**) by an energetically more favoured *gauche*-propanol-type interaction in (**252b**) ⟨74JCS(P2)1419⟩. In perhydrooxazolo[4,3-*c*]-[1,4]thiazine (**253**), however, *ca.* 84% of the equilibrium mixture consists of the *trans* conformer (**253a**). The larger thiomorpholine ring (C—S = 1.82, C—O = 1.43 and C—C = 1.54 Å) is more flexible resulting in decreased strain associated with the ring fusion in the *trans* conformer ⟨76JCS(P2)203⟩.

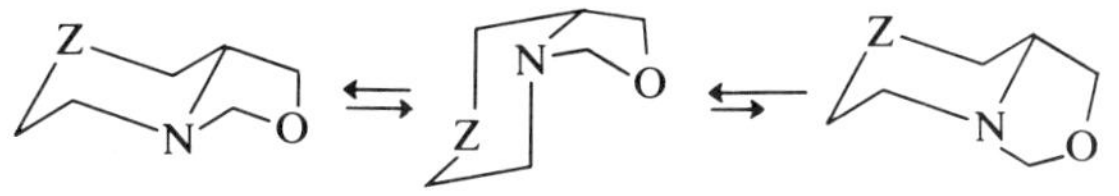

(a) (b) (c)

(**251**) Z = CH
(**252**) Z = O
(**253**) Z = S

4.29.6.7 Saturated and Partially Saturated Rings: Synthesis

4.29.6.7.1 5H-Oxazolo[3,2-a]pyridine

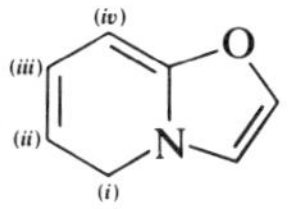

Azine approach. When 2(1*H*)-pyridinone is reacted with 2-bromopropenoic acid, 2-carboxy- -dihydrooxazolo[3,2-*a*]pyridinium bromide is formed by cyclization of the intermediate Michael adduct (**254**) ⟨52JA4906⟩.

Irradiation of aqueous solutions of 1-isobutenyl- and 1-propenyl-2-pyridinones containing perchloric acid leads to formation of 2,3-dihydrooxazolo[3,2-*a*]pyridinium salts. It is suggested that the mechanism involves an initial excited state electrocyclization to generate a pyridinium ylide (**255**). The latter is rapidly trapped by proton transfer from water. In the absence of perchloric acid the hydroxide counterion acts as a nucleophile and opens the ring to the monocyclic alcohol (**256**) ⟨79JA3607⟩.

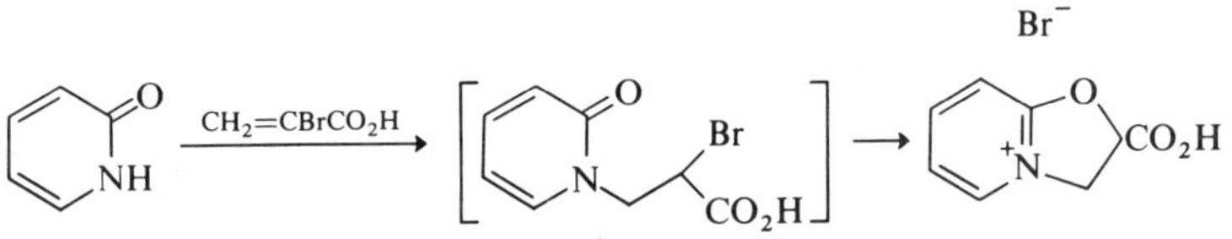

(**254**)

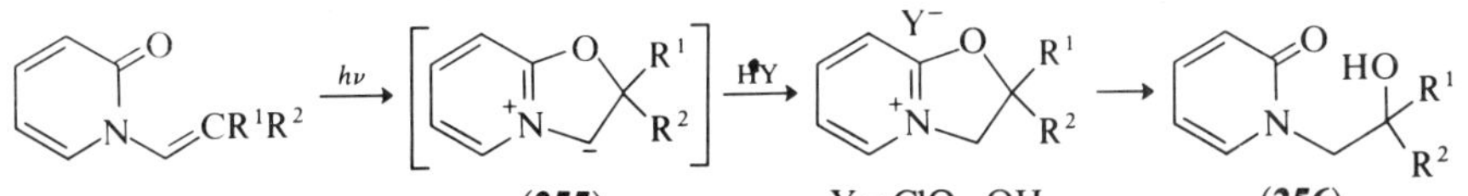

(**255**)   Y = ClO$_4$, OH   (**256**)

Isoquinolines react with ethylene oxide in acetic acid at room temperature to afford 2,3-dihydro-10b*H*-oxazolo[2,3-*a*]isoquinolines (**257**). Quinoline reacts in the same way and both adducts are unstable. Studies of isoquinolines with a good leaving group (Br, CN) at C-1 show spontaneous elimination in the adduct (**257**) with formation of the corresponding isoquinolinium salt. The latter may suffer solvolysis by attack at C-2 and ring opening to the corresponding 1(2*H*)-isoquinolinone ⟨79JOC285⟩. A similar reaction results when an isoquinoline α-ylide is treated with an aryl aldehyde; the first formed adduct (**258**) undergoes cyclization by addition of the oxygen to C-1 in the isoquinoline ⟨67TL3653⟩. Hexahydro derivatives (**259**) are available from 3,4-dihydroisoquinoline by the same hydroxyethylation reaction and cyclization of the adduct under alkaline conditions ⟨66AP817⟩.

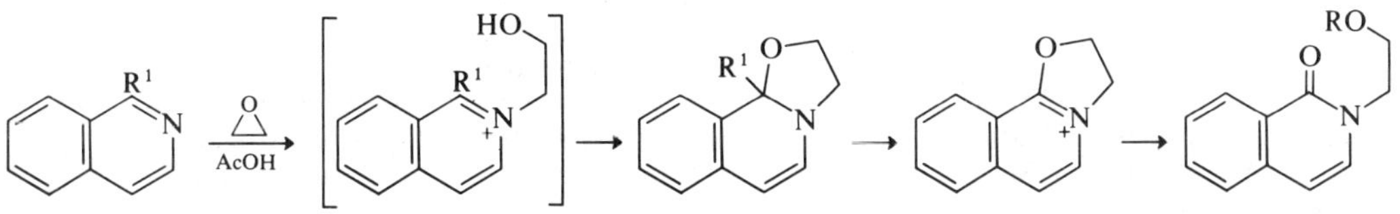

(**257**)   R = OAc, H

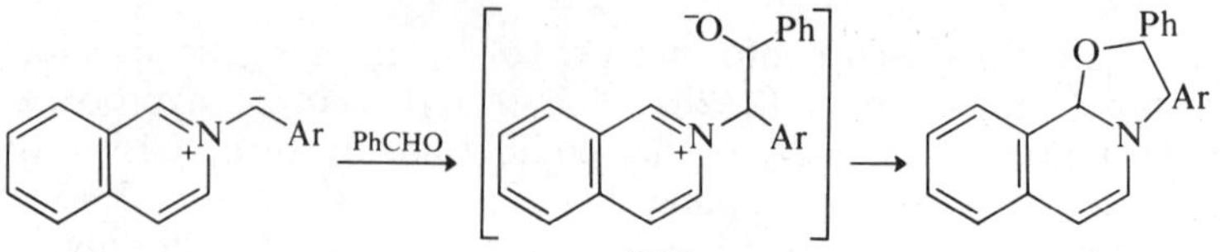

Compounds with two oxo groups in the oxazolidine ring have been prepared. Thus, 3,4-dihydro-2(1H)-quinolinones react with oxalyl chloride to form 5H-oxazolo[3,2-a]quinoline-1,2-diones (**260**) ⟨70M383⟩.

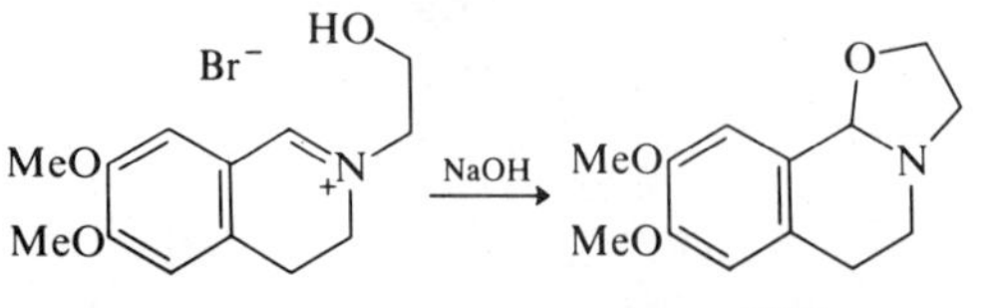

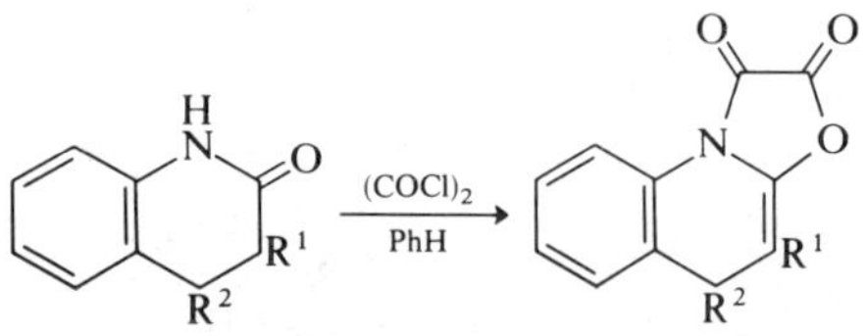

(ii) *5H-Oxazolo[3,2-c]pyrimidine*

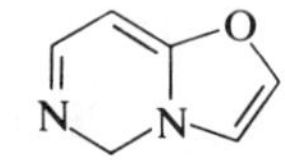

Azine approach. Mesylation of 4-(β-hydroxyethyl)-1-methyluracil gives the corresponding ester (**261**) which is unstable at room temperature. Either neat or in solution in polar solvents such as water or alcohol, the ester is transformed into the bicyclic pyrimidinium salt (**262**) ⟨75JOC1713⟩.

Cyclization of 3-allylquinazoline under the influence of PPA leads to the 2,3-dihydrooxazolo[3,2-c]quinazolinone (**263**) ⟨74JIC453⟩.

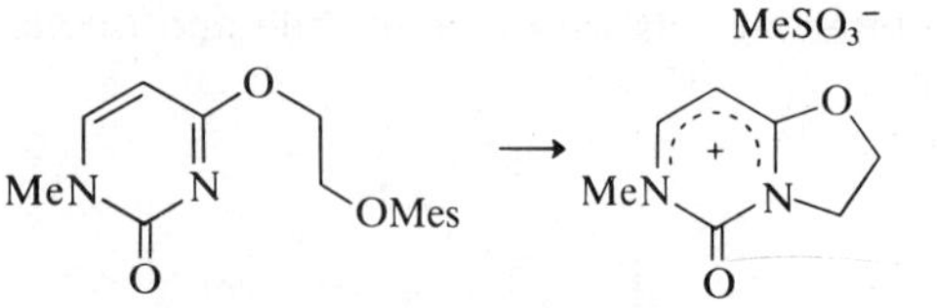

Azole approach. 2-Aminooxazoline forms a cycloadduct with ethoxycarbonyl isothiocyanate. The orientation of the substituents in the adduct shows that the reaction is initiated by isothiocyanate addition to the exocyclic nitrogen (**264**) ⟨71CB3039⟩.

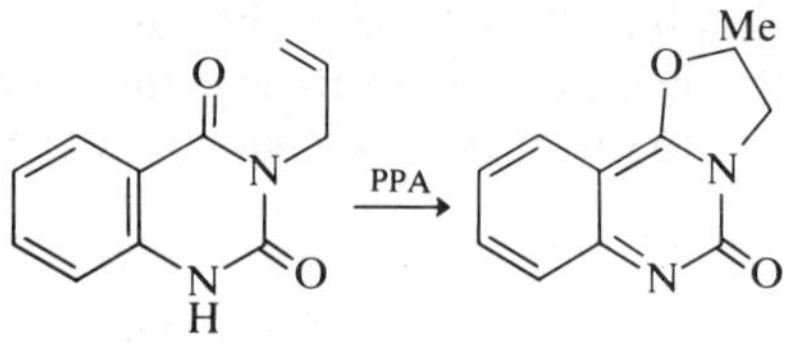

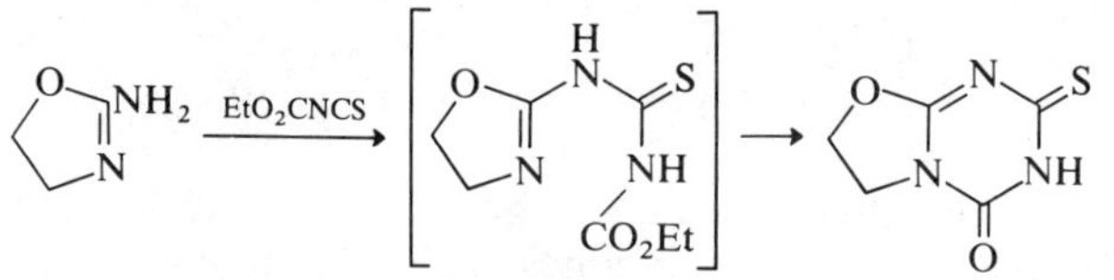

(iii) *5H-Oxazolo[3,2-a]pyrazine*

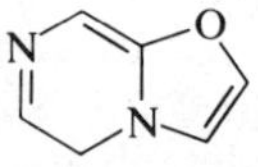

This ring system is embedded in relatively complex ergot alkaloids; total syntheses have been carried out ⟨B-75MI42900, 76M763⟩.

(iv) *5H-Oxazolo[2,3-a]pyrimidine*

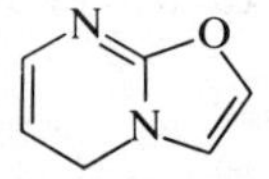

Azine approach. 1-(2-Chloroethyl)uracil is cyclized to the corresponding oxazoline (**266**) in the presence of DBU ⟨77MI42901⟩. By analogy, treatment of 2,4-(1H,3H)quinazolinedione with 1,2-dibromoethane or reacting the 3-(2-chloroethyl) homologue (**267**) with sodium ethoxide leads to oxazoline fusion. The same product is formed from 2-chloro-3-(2-

hydroxyethyl)-4(3*H*)-quinazolone under alkaline conditions. The 5*H*-2,3-dihydroquinazoline (**268**) is also formed by cycloalkylation ⟨79CPB880⟩. Bromination of the 3-allylquinazoline (**269**) leads to the 2-bromomethyloxazoline (**269a**) ⟨74JIC453⟩.

Perhydro derivatives (**270**) are prepared from 1-substituted 1,4,5,6-tetrahydropyrimidines which form cycloadducts with epoxides ⟨70LA(742)128⟩. 4*H*-5,6-Dihydro-1,3-oxazines and -thiazines react in the same way to form products of type (**270**) ⟨66AG(E)875⟩.

(**266**)

(**267**)

(**268**)

(**269**) (**269a**)

(**270**)

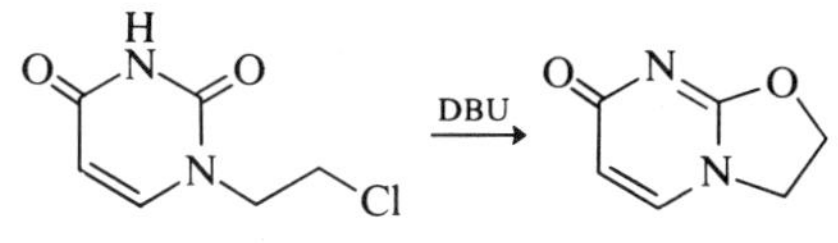

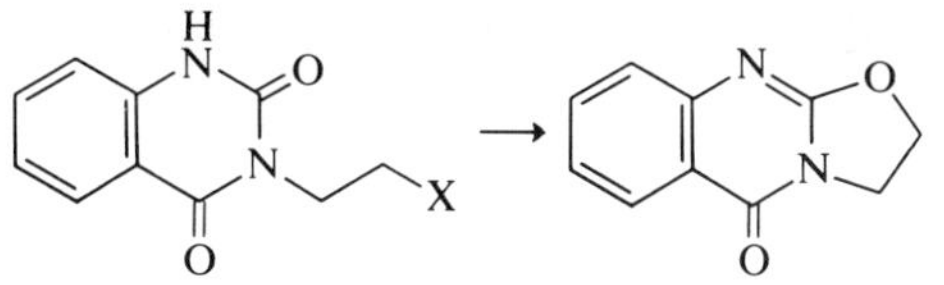

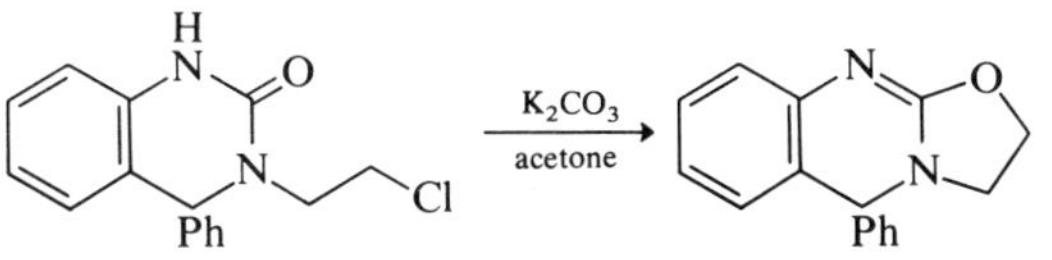

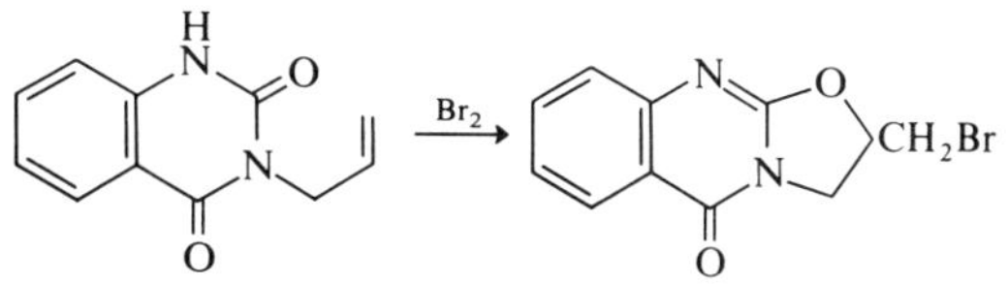

Azole approach. A 'one-pot' synthesis of substituted 2,3-dihydro-5-oxo-5*H*-oxazolo[2,3-*b*]quinazolines (**273**) consists of reacting esters of anthranilic acid with a 2-chloro- or 2-bromoalkyl-isocyanate; the initially formed urea (**271**) is cyclized to the acid salt of the oxazoline (**272**), which on addition of piperidine as base forms the pyrimidine ring ⟨76S469⟩. In a similar reaction the aziridinyl urea (**274**), which is available from aziridinyl isocyanate

(**271**) (**272**) (**273**) (**274**)

and the ester of anthranilic acid, in the presence of sodium iodide in acetone is quantitatively converted into the quinazolone (**273**) ⟨79JHC877⟩.

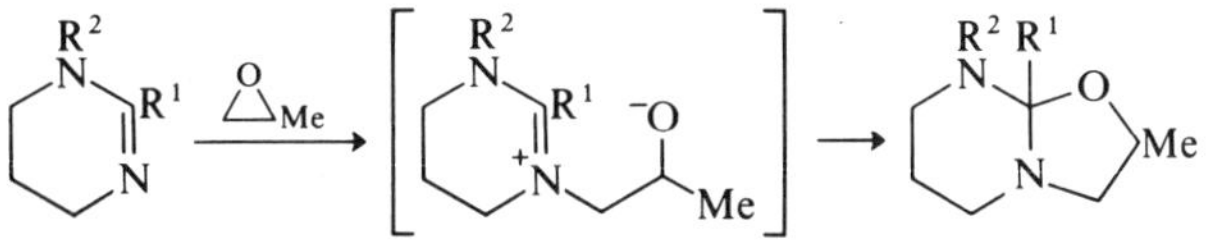

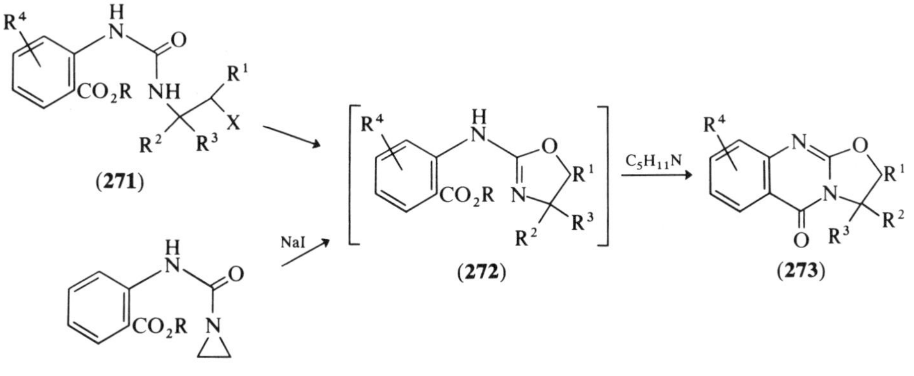

2,3,5,6-Tetrahydro-7*H*-oxazolo[3,2-*a*]pyrimidin-7-ones (**275**) can be prepared from 2-amino-2-oxazolines and *α,β*-unsaturated carboxylic esters. The orientation of the substituents in the reaction product corresponds to Michael addition of the oxazoline nitrogen as the first step ⟨74LA593⟩.

(**275**)

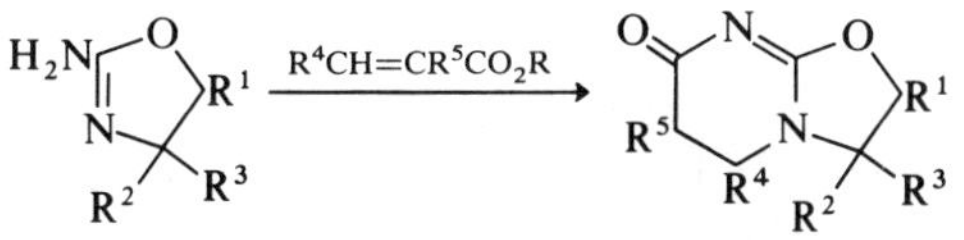

4.29.6.7.2 3H-Oxazolo[3,4-a]pyridine

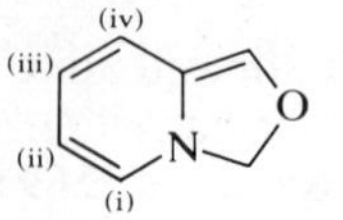

Azine approach. Perhydrooxazolo[3,4-*a*]pyridines (**276**) are available by the action of formaldehyde on the corresponding piperidyl carbinols ⟨66JHC418⟩. Benzo homologues are similarly prepared ⟨63AP389⟩.

In the 8-hydroxy-2(1*H*)quinolinone (**277**) the bridge between the oxygen and the nitrogen is formed from methylene bromide giving the angular fused oxazoline ring ⟨73JA5003⟩.

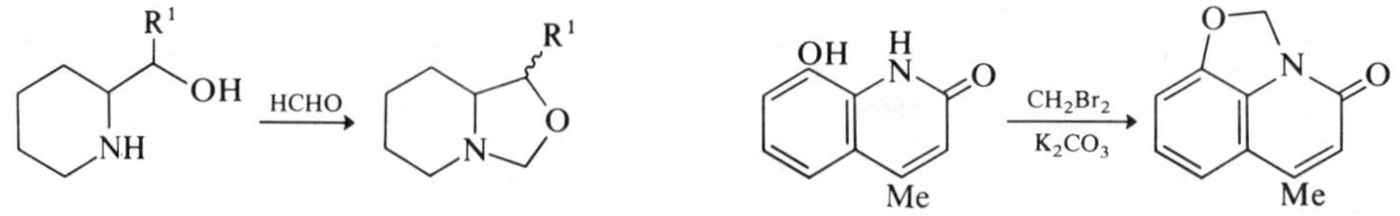

Cyanogen bromide reacts with *threo*-2-piperidyl carbinol (**278**) to form the *trans* imine (**279**). The latter is also formed from the *erythro* 1-carbamate (**280**) using thionyl chloride for the cyclodehydration reaction. The latter reaction therefore proceeds by inversion of the configuration at the carbinol carbon atom ⟨73AP284⟩.

A 3-oxo group is introduced by the action of phosgene, carbonates or as in the reaction of ethyl chloroformate with piperidyl-2-carbinol to furnish (**281**) ⟨73OMR(5)295⟩.

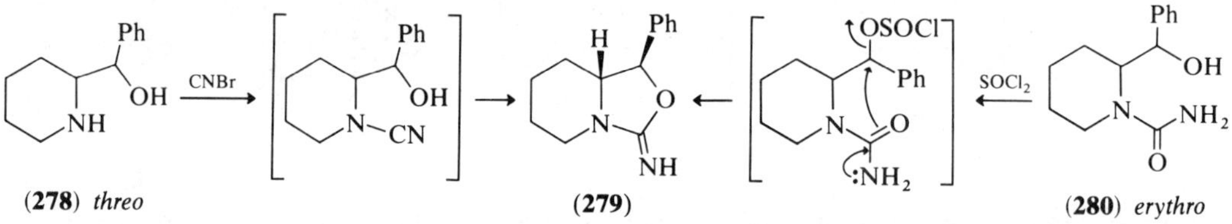

1*H*,3*H*,5*H*-Oxazolo[3,4-*a*]quinolin-3-ones (**283**) can be prepared from 1-acyl-1,2,3,4-tetrahydroquinoline-2-carboxylic acids by treatment with acetic anhydride. The chiral centre at C-2 in the quinoline (**282**) is lost in the reaction and a new chiral centre is created at C-1 in the product. The Dakin–West reaction may be a competitive reaction depending on the nature of the 1-acyl substituent in (**282**). When the acyl group is a peptide chain only the fused oxazole (**283**) is formed ⟨79JHC1589⟩.

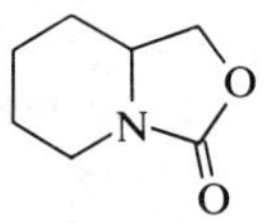

Intramolecular imino Diels–Alder cyclization is a useful method in alkaloid synthesis. The reaction is highly stereoselective but gives results opposite to those usually found in closely analogous 'all-carbon' systems. Thus, the diene (**284**) on pyrolysis gives the single *trans* cycloadduct (**286**). The reaction is rationalized by assuming initial formation of the

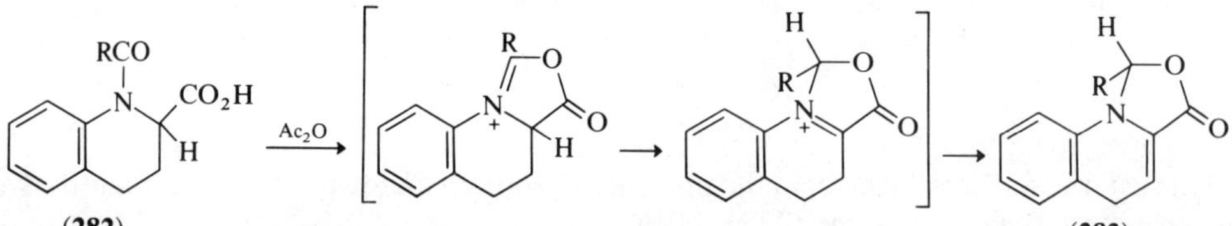

(*E*)-acylamine (**285**), which in the Diels–Alder reaction must have a transition state where the *N*-acyl group is *endo* and the ester group is *exo* in order to account for the formation of the *trans* product (**286**) ⟨80JA1153⟩.

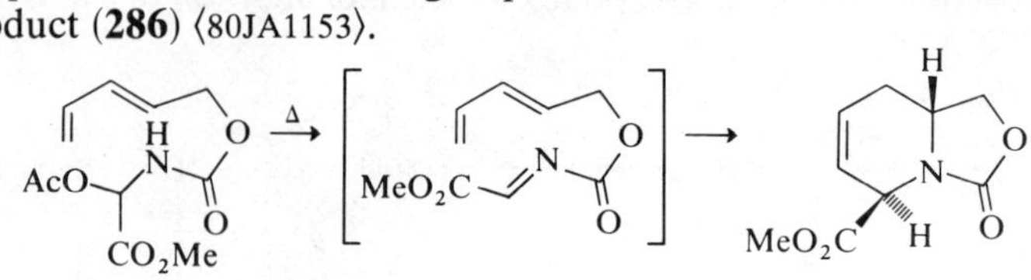

(iii) *3H-Oxazolo[3,4-a]pyrazine*

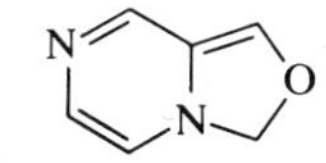

1,4-Thiazine analogues have received a great deal of attention partly because they may arise by chemical transformations from penicillins ⟨73CC226, 74JCS(P1)1572⟩. A simple synthesis of this ring system is shown below.

Azine approach. 8,8a-Dihydro-1-oxo-1*H*,3*H*-oxazolo[4,3-*c*][1,4]thiazines (**265**) are available from the corresponding 1,4-thiazine-3-carboxylic acids and aldehydes or ketones; 2,2-dimethoxypropane with acid catalysis yields the 3,3-dimethyl derivative (**265**; R=Me), and acetaldehyde with a water absorbent yields the monomethyl derivative as a diastereoisomeric mixture ⟨76JCS(P1)584⟩.

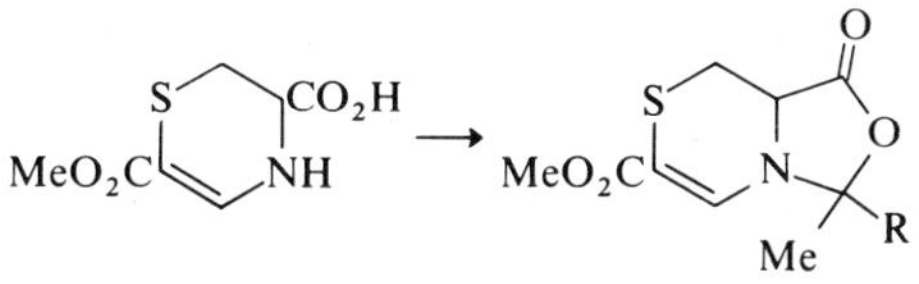

(**265**)

4.29.6.7.3 *Oxazolo[4,5-b]pyridine*

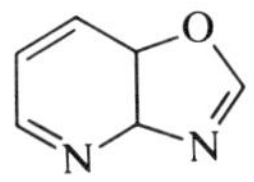

Azine approach. Pyridine *N*-oxides undergo 1,3-dipolar cycloaddition reactions with aryl isocyanates. The first formed 1,2-dihydropyridines are unstable and rearrange at once to the 2,3-dihydropyridines (**287**). From 3-picoline 1-oxide both the 6-methyl and 7a-methyl isomers are formed. The adduct (**288**) from 3,5-dibromopyridine 1-oxide reacts further to eliminate HBr, thereby giving the fully aromatic system ⟨79CPB2261⟩.

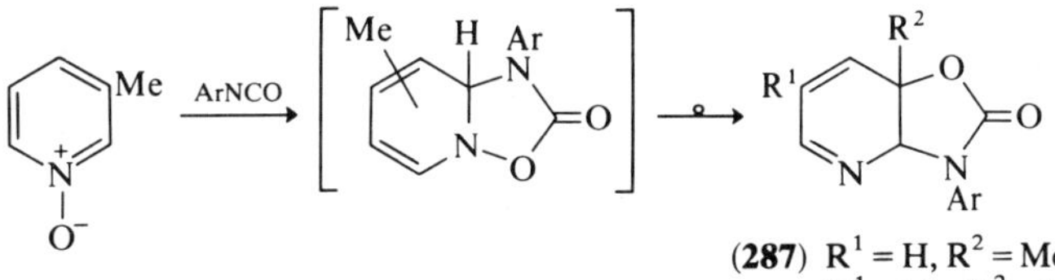

(**287**) R^1 = H, R^2 = Me
R^1 = Me, R^2 = H

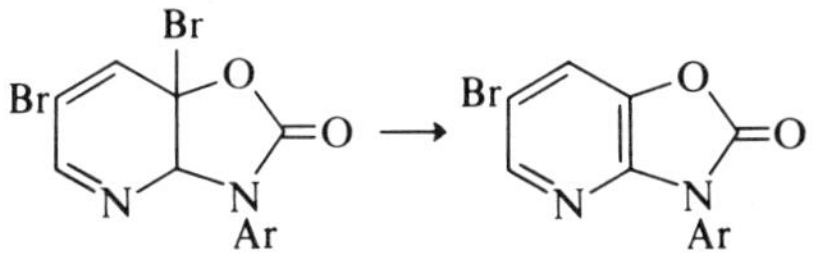

(**288**)

4.29.6.8 Applications

A medically very important group of alkaloids, *viz.* the peptide ergot alkaloids (**289**), contain the fused oxazoloazine ring system, the peptide derived skeleton being 8*H*-oxazolo[3,2-*a*]pyrrolo[2,1-*c*]pyrazine. They are attached to lysergic acid by an amide linkage. The ergot alkaloids originate principally from parasitic fungi, the main alkaloid being ergotamine (**289**; R^1=Me, R^2 = CH$_2$Ph). They show a variety of physiological effects. Both naturally occurring and chemically modified compounds have important medical applications as drugs ⟨B-75MI42900⟩.

The antibiotic nybomycin (**290**; R = OH), 8-hydroxymethyl-6-methyl-2*H*,4*H*-oxazolo[5,4,3-*ij*]pyrido[3,2-*g*]quinolin-4-one, is active against Gram-positive and some Gram-negative bacteria. It was isolated from streptomycete cultures and together with deoxynybomycin (**290**; R=H) it is available by synthesis ⟨73JA5003⟩.

Several of the nucleoside analogues of purine have been tested for anticancer activity. One compound, *viz.* (**291**), was active *in vivo* against leukemia L 1200 in mice ⟨74JMC1282⟩.

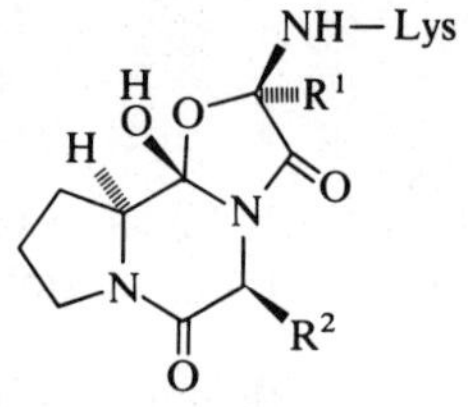

(289) 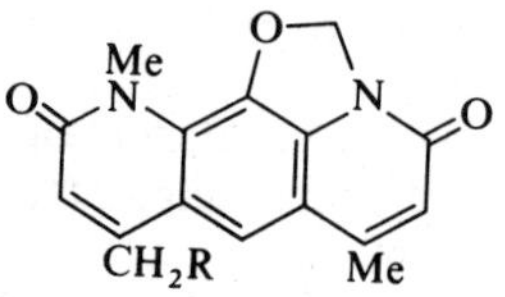 **(290)** 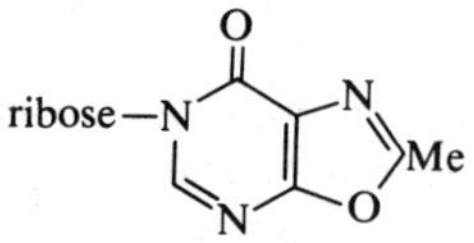 **(291)**

Hexahydro-1-phenyl-3H-oxazolo[3,4-a]pyridin-3-imines (**292**) have antihypotensive and central nervous system stimulant activity which is stereochemically dependent. Most active is the (−)-*trans* form of the parent compound (**292**; R = H, Ar = Ph) ⟨80MI42902⟩. Useful stimulant activity for the central nervous system has also been claimed for 2,3,5,6-tetrahydro-7H-oxazolo[3,2-a]pyrimidin-7-ones ⟨74GEP2253555⟩ and quinazoline homologues (**293**) ⟨74GEP2252122⟩.

2-Phenyloxazolo[4,5-b]pyridine (**294**) and 2-phenyloxazolo[5,4-b]pyridine substituted in the phenyl ring have good antiinflammatory and analgesic activity. Some compounds possess activity comparable to phenylbutazone or indomethacin without producing the irritation in the gastrointestinal tract that acidic antiinflammatory compounds cause ⟨78JMC1158⟩.

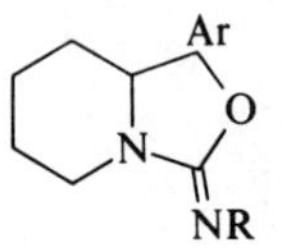

(292) 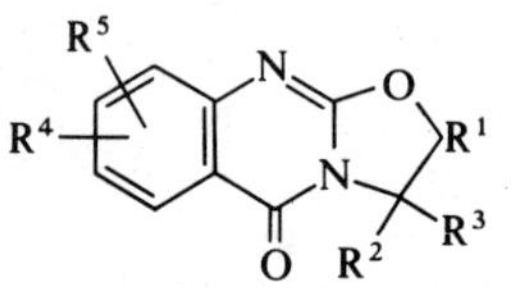 **(293)** 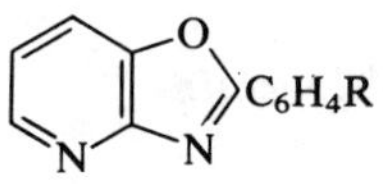 **(294)** 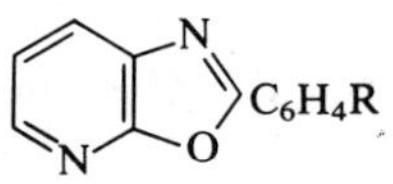 **(295)**

Thiophosphoric and thiophosphonic acid ester aza analogues (**297**) of the benzoxazolone insecticide Phosalone (**296**) have high insecticidal, acaricidal and anthelmintic activities with very low mammalian toxicities ⟨76HCA1593⟩.

A number of patents exist for the application of fused oxazoloazines in fluorescent whiteners, e.g. the use of 4,4'-bis(oxazolopyridin-2-yl)stilbenes (**298**) as fluorescent whiteners for polyesters ⟨70GEP2003575⟩. The highly unsymmetrical compound (**299**) is a fluorescent whitener for polyamide and polyester fibers ⟨75USP3873531⟩.

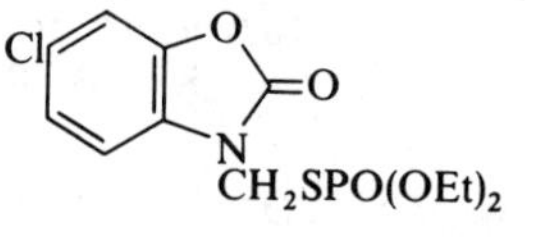

(296) 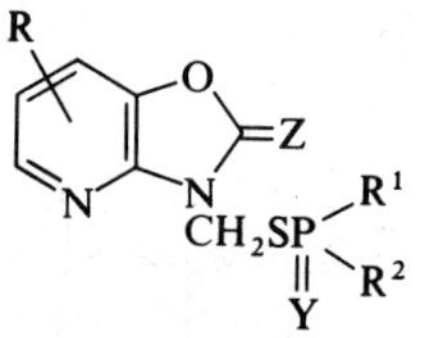 **(297)** Y ≠ Z = O, S

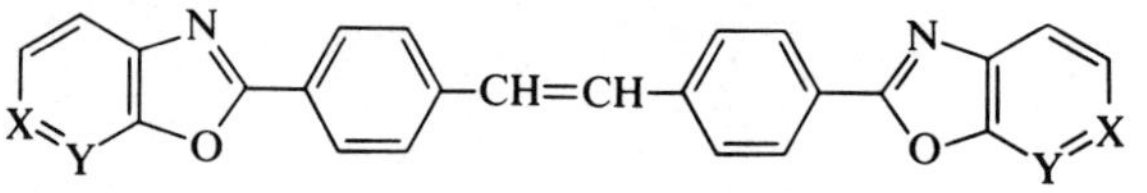

(298) X ≠ Y = N, CH

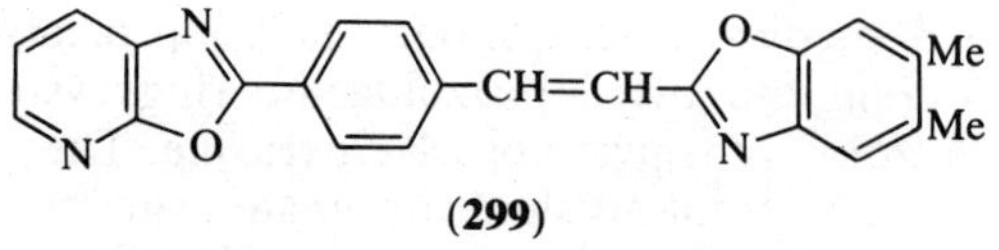

(299)

4.29.7 THIAZOLE FUSED WITH AZINES

4.29.7.1 Introduction

Thiazole can be fused with pyridine in six different ways; the sections on synthesis have been organized accordingly. Further fused systems arise from the diazines and the triazines and their areno and heteroareno homologues.

Table 13 X-Ray Diffraction Data for Fused Thiazoloazines

Compound	Ring A		Bond lengths (Å) and angles (°) Ring B		Ring C	
3-(4-Bromophenyl)thiazolo[3,2-a]pyridinium tetrafluoroborate ⟨80AX(B)1229⟩	a^1 1.358 b^1 1.423 b^3 1.366 d^1 1.723 d^2 1.726	a^1b^1 110.5 a^1d^1 113.6 b^1b^3 113.9 b^3d^2 111.2 d^1d^2 90.7	a^2 1.368 a^3 1.406 a^4 1.372 a^5 1.392 b^2 1.381 b^3 1.366	a^2a^3 119.9 a^2b^2 119.5 a^3a^4 120.8 a^4a^5 117.8 a^5b^3 121.7 b^2b^3 120.2		
Anhydro-3-(4-bromophenyl)-2-mercaptothiazolo[2,3-a]isoquinolinium hydroxide ⟨67JCS(B)1117⟩	a^1 1.37 b^2 1.41 b^5 1.32 d^1 1.67 d^2 1.78	a^1b^1 113 a^1d^1 108 b^1b^3 114 b^3d^2 113 d^1d^2 92	a^2 1.39 a^3 1.43 a^9 1.45 a^{10} 1.39 b^2 1.39 b^3 1.32	a^2a^3 122 a^2b^2 117 a^3a^{10} 118 a^9a^{10} 119 a^9b^3 119 b^2b^3 124	a^4 1.41 a^5 1.41 a^6 1.38 a^7 1.40 a^8 1.43 a^{10} 1.39	a^4a^5 116 a^4a^{10} 121 a^5a^6 124 a^6a^7 119 a^7a^8 119 a^8a^{10} 121
Berninamycinic acid (anhydro-3,8-dicarboxy-6-hydroxythiazolo[2,3-f]-[1,6]naphthyridin-4-ium hydroxide) ⟨76JA299⟩	a^1 1.341 b^1 1.421 b^5 1.347 d^1 1.711 d^2 1.694	a^1b^1 110.5 a^1d^1 113.7 b^1b^5 112.7 b^5d^2 112.8 d^1d^2 90.2	a^2 1.369 a^3 1.437 a^7 1.432 a^8 1.398 b^2 1.393 b^5 1.347	a^2a^3 119.6 a^2b^2 119.8 a^3a^8 119.7 a^7a^8 118.7 a^7b^5 119.9 b^2b^5 122.3	a^4 1.390 a^5 1.370 a^6 1.397 a^8 1.398 b^3 1.360 b^4 1.328	a^4a^5 119.3 a^4a^4 124.0 a^5a^6 118.1 a^6a^8 119.2 a^8b^3 122.3 b^3b^4 117.0

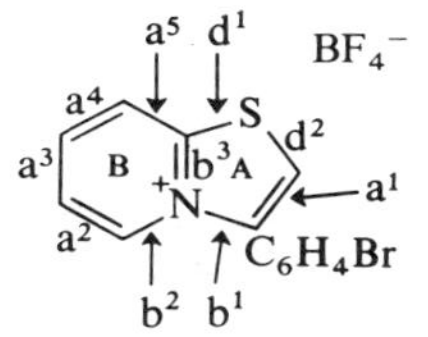

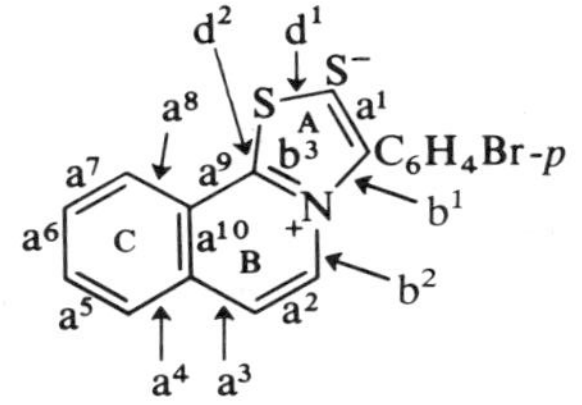

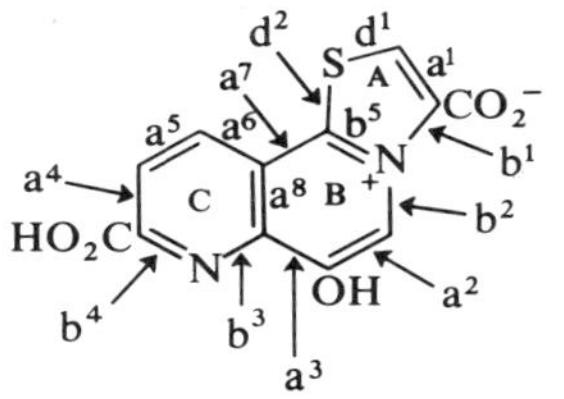

Table 14 X-Ray Data For Partially Saturated Fused Thiazoloazines

Compound	Ring A		Bond lengths (Å) and angles (°) Ring B		Ring C	
Anhydro-2α,3β-dicarboxy-2,3-dihydro-5-methylthiazolo[3,2-a]pyridinium hydroxide ⟨71ACS118⟩	a^1 1.519 b^1 1.481 b^3 1.356 d^1 1.837 d^2 1.734	a^1b^1 105.4 a^1d^1 104.2 b^1b^3 113.5 b^3d^2 113.3 d^1d^2 90.6	a^2 1.379 a^3 1.376 a^4 1.371 a^5 1.385 b^2 1.357 b^3 1.357	a^2a^3 120.9 a^2b^2 117.7 a^3a^4 120.1 a^4a^5 119.0 a^5b^3 119.5 b^2b^3 122.7		
Anhydro-7-bromo-3α-carboxy-2,3-dihydro-8-hydroxy-2β,5-dimethyl-1α-oxothiazolo[3,2-a]pyridinium hydroxide ⟨71ACS1353⟩	a^1 1.515 b^1 1.496 b^3 1.361 d^1 1.830 d^2 1.803	a^1b^1 105.1 a^1d^1 105.3 b^1b^3 113.3 b^3d^2 112.6 d^1d^2 86.9	a^2 1.391 a^3 1.379 a^4 1.431 a^5 1.399 b^2 1.351 b^3 1.361	a^2a^3 120.8 a^2b^2 117.6 a^3a^4 122.7 a^4a^5 112.5 a^5b^3 124.4 b^2b^3 121.9		
4-Ethoxy-5a,6,7,8,9,9a-hexahydro-1-methyl-9-oxopyrido[2,1-b]benzothiazolium bromide ⟨72ACS3131⟩	a^1 1.362 a^2 1.397 a^3 1.360 a^4 1.397 b^2 1.375 b^3 1.343	a^1a^2 122.3 a^1b^2 116.2 a^2a^3 119.5 a^3a^4 118.7 a^4b^3 119.8 b^2b^3 123.5	a^{10} 1.511 b^1 1.494 b^3 1.343 d^1 1.722 d^2 1.844	$a^{10}b^1$ 105.2 $a^{10}d^2$ 103.5 b^1b^3 112.0 b^3d^1 114.7 d^1d^2 90.0	a^5 1.536 a^6 1.525 a^7 1.592 a^8 1.470 a^9 1.523 a^{10} 1.511	a^5a^6 113.7 a^5a^{10} 113.7 a^6a^7 108.9 a^7a^8 108.1 a^8a^9 114.9 a^9a^{10} 112.9

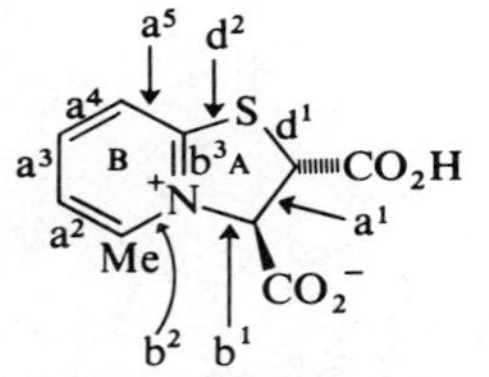

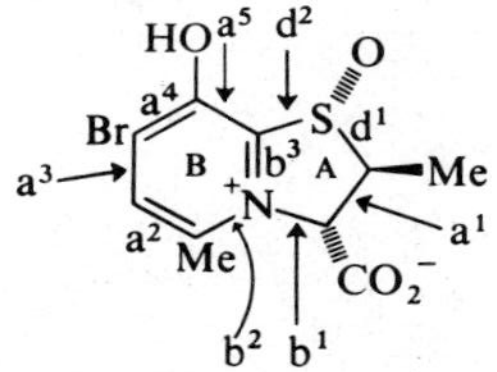

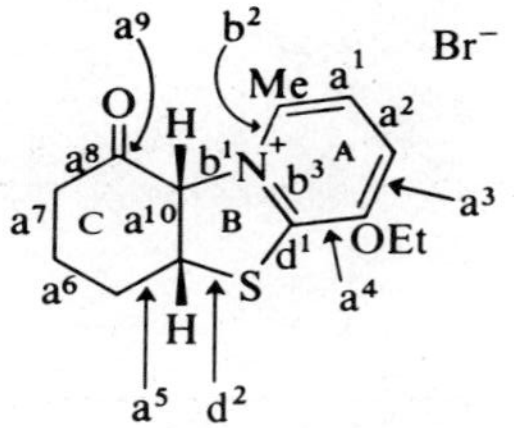

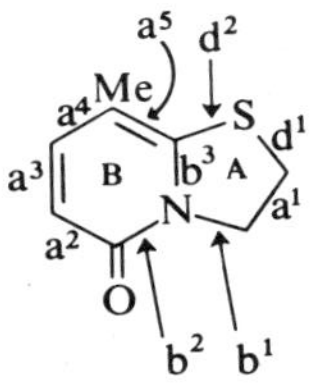

a^1	1.518	a^1b^1	106.4	a^2	1.413	a^2a^3 121.3
b^1	1.482	a^1d^1	106.6	a^3	1.371	a^2b^2 115.2
b^3	1.372	b^1b^3	115.3	a^4	1.408	a^3a^4 122.1
d^1	1.820	b^3d^2	112.2	a^5	1.366	a^4a^5 116.2
d^2	1.749	d^1d^2	91.4	b^2	1.383	a^5b^3 122.0
				b^3	1.372	b^2b^3 —

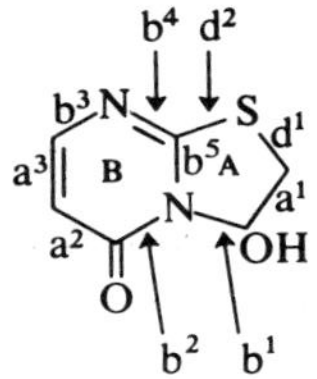

2,3-Dihydro-8-methyl-5H-thiazolo[3,2-a]pyridin-5-one ⟨77ACS(A)340⟩

a^1	1.517	a^1b^1	104.6	a^2	1.431	a^2a^3 121.1
b^1	1.482	a^1d^1	107.3	a^3	1.345	a^2b^2 112.6
b^5	1.361	b^1b^5	115.6	b^2	1.407	a^3b^3 124.0
d^1	1.809	b^5d^2	112.5	b^3	1.371	b^3b^4 115.4
d^2	1.736	d^1d^2	91.0	b^4	1.302	b^2b^5 122.0
				b^5	1.361	b^4b^5 124.8

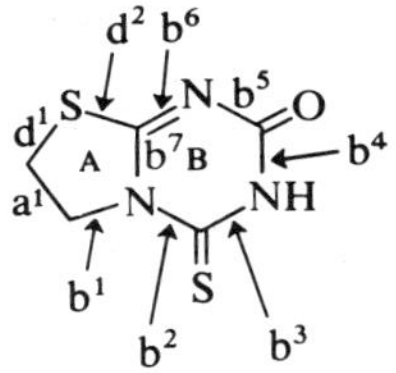

2,3-Dihydro-3-hydroxy-5H-thiazolo[3,2-a]pyrimidin-5-one ⟨80AX(B)1971⟩

a^1	1.500	a^1b^1	108.6	b^2	1.364	b^2b^3 113.4
b^1	1.458	a^1d^1	109.1	b^3	1.357	b^2b^7 120.7
b^7	1.372	b^1b^7	116.6	b^4	1.382	b^3b^4 125.4
d^1	1.799	b^7d^2	112.4	b^5	1.361	b^4b^5 118.3
d^2	1.716	d^1d^2	93.3	b^6	1.292	b^5b^6 117.0
				b^7	1.372	b^6b^7 125.2

3,4,6,7-Tetrahydro-4-thioxo-2H-thiazolo[3,2-a][1,3,5]triazin-2-one ⟨74AX(B)1123⟩

Thiazoloazine chemistry is rapidly becoming an important area of heterocyclic chemistry. The thiazoloazine systems are in general readily accessible and most compounds are chemically stable. The fully conjugated systems, especially when pyrimidine is the azine ring, may be regarded as purine analogues. Hence the emphasis on the synthesis of such compounds for biological purposes, *e.g.* in the preparation of nucleoside and nucleotide analogues. Thiazoloazine ring systems are also found in antibiotics.

Chemical stability and light absorption properties have stimulated the search for new dyes containing thiazoloazine ring systems.

Saturated or some partially saturated systems, *e.g.* for stereochemical or biological work, are accessible from chiral precursors such as the amino acid cysteine.

4.29.7.2 Structure

4.29.7.2.1 X-Ray diffraction

Details of bond lengths and bond angles are listed in Table 13. The fused, fully conjugated ring systems are planar.

In the dihydrothiazolo-azinium or -azinone systems the six-membered ring is planar, whereas the five-membered ring has the envelope conformation. Thus in anhydro-$2\alpha,3\beta$-dicarboxy-2,3-dihydro-5-methylthiazolo[3,2-*a*]pyridinium hydroxide, S-1 and C-3 are coplanar (within 0.05 Å) with the pyridine ring, whereas C-2 is 0.60 Å out of the plane (Table 14) ⟨71ACS118⟩.

4.29.7.2.2 Molecular spectra

(a) *^{1}H NMR spectroscopy*. Proton chemical shifts and coupling constants for ring CH of some fully aromatic systems are listed in Table 15. Partially saturated systems are listed in Table 16. In the azinium derivatives the most deshielded protons are in positions α to the charged nitrogen. The coupling constants do not deviate significantly from the accepted values in azines and thiazoles.

Most of the data for the partially reduced systems refer to 2,3-dihydrothiazolo derivatives.

(b) *^{13}C NMR spectroscopy*. The data available refer to the thiazolo[3,2-*a*]pyridine system. The chemical shifts for the fully aromatic system are recorded in Table 17, the chemical shifts for the 2,3-dihydrothiazolo derivatives in Table 18.

The carbon deshielding is most pronounced in positions α to the nitrogen and this is enhanced in the pyridinium salts. The one-bond carbon–hydrogen coupling $^1J_{CH}$ at sp^2 carbon is also highest in those positions.

(c) *UV spectroscopy*. UV data for fully conjugated and partially saturated systems are listed in Tables 19 and 20, respectively.

(d) *IR spectroscopy*. IR data can be useful in differentiating between isomeric structures by comparison of CO stretching frequencies (Table 21). Thus, pyridinones which can be said to have an *ortho*-quinonoid structure (**300, 302**) display higher carbonyl stretching frequencies than their isomers with a *para*-quinonoid structure (**301, 303**).

The pseudocarbonyl groups in mesoionic compounds absorb in the region 1600–1700 cm^{-1}. The absorption intensities of the pseudocarbonyl groups can be used as measures of the polarity of the carbonyl group. Two bands are generally observed in the stretching region at 1680–1690 and 1630–1655 cm^{-1} in mesoionic thiazolo[3,2-*a*]pyrimidine-5,7-diones (**305, 306**). The average integrated intensities for the carbonyl bands are in the same range as those determined for sydnones and isosydnones and much higher than for covalent carbonyl groups, which indicates a highly polar nature for these pseudocarbonyl groups. Their carbonyl stretching frequencies, however, suggest bond orders similar to covalent models ⟨73JHC487⟩.

(e) *Mass spectrometry*. The nature of the fused systems varies greatly; the fragmentation patterns vary accordingly, especially after electron bombardment. Almost all the data in the literature have been obtained by electron bombardment. Heterocycles in general are sufficiently volatile in the mass spectrometer for analysis but with their salts, however, some thermally induced transformation may have to take place before a volatile species is formed. Favourable for cationic species is deprotonation to an anhydro base (**310 → 311**). Other

Table 15 ^{1}H NMR Data for Fused Thiazoloazines

Compound					Solvent	^{1}H chemical shifts (p.p.m.)							Coupling constants (Hz)	Ref.
						1	2	3	4	5	6	7	8		

Structure 1: R⁴-substituted thiazolo-pyridinium with Y⁻ counterion (R¹, R², R³, R⁴ positions)

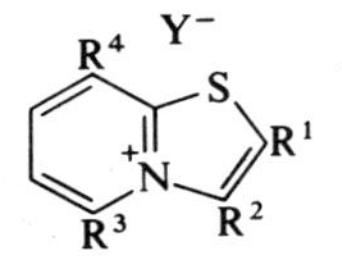

R^1	R^2	R^3	R^4	Y											
H	H	H	H	Br	TFA	—	8.8	8.37	—	9.4	8.02	8.4	8.75	$J_{5,6}$ 7.0, $J_{6,7}$ 7.0, $J_{7,8}$ 9.0	67JCS(C)515
Me	H	H	H	ClO₄	DMSO-d_6	—	2.78	8.36	—	9.10	7.89	8.23	8.49		76CB3653
H	OH	H	H	Br	TFA	—	8.48	—	—	9.17	—	—	—		71ACS5
Me	OH	H	H	Br	TFA	—	2.65	—	—	3.14	7.74	8.07	—		71ACS5
H	H	Me	OH	—	D₂O	—	7.92	8.30	—	8.24	7.46	6.94	—		78ACS(B)651
H	Me	H	O⁻	—	TFA	—	7.85	2.90	—	8.59	7.86	7.80	—		70ACS2956
H	H	Me	O⁻	—	TFA	—	2.81	8.11	—	2.90	7.59	7.59	—		70ACS2956

Structure 2: thiazolo-pyridinone with Me

| 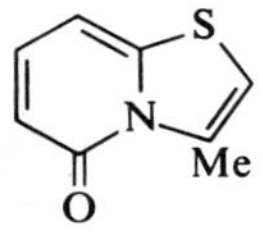 | CDCl₃ | — | 6.32 | 2.88 | — | — | 6.21 | 7.32 | 6.51 | $J_{2,\text{Me}}$ 1.0, $J_{6,7}$ 7.5, $J_{7,8}$ 8.5, $J_{6,8}$ 1.0 | 79JCS(P1)1150 |

Structure 3: thiazolo-pyridinium ClO₄, R = H or Me

| 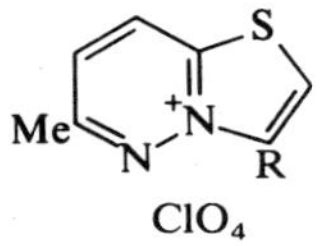 R = H | DMSO-d_6 | — | 8.74 | 9.40 | — | — | 2.74 | 8.09 | 9.15 | $J_{2,3}$ 4.5, $J_{7,8}$ 9.8 | 77CPB299 |
| R = Me | DMSO-d_6 | — | 8.51 | 2.75 | — | — | 2.84 | 8.08 | 9.15 | $J_{7,8}$ 9.8 | 77CPB299 |

Structure 4: thiazolo-pyridinone

| 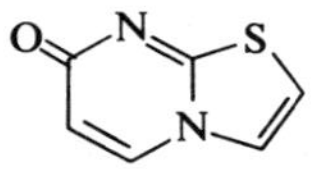 | DMSO-d_6 | — | 7.28 | 7.72 | — | 8.35 | 6.27 | — | — | | 71JCS(C)2094 |

Structure 5: Br-OH thiazolo-pyridinium Me, Br⁻

| 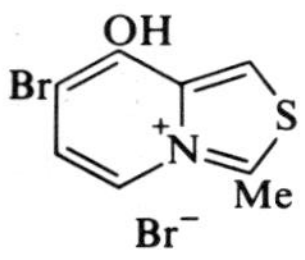 | TFA | 8.12 | — | 2.3 | — | 8.15 | 7.7 | — | — | $J_{5,7}$ 6.6 | 69JCS(C)707 |

Table 15 (cont.)

Compound	Solvent	^1H chemical shifts (p.p.m.)								Coupling constants (Hz)	Ref.
		1	2	3	4	5	6	7	8		
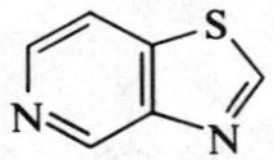 (thiazolo-pyridine)	DMSO-d_6	—	9.57	—	9.42	—	8.60	8.27	—	$J_{6,7}$ 5.3	76T399
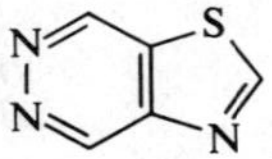 (thiazolo-pyridine)	DMSO-d_6	—	9.89	—	9.98	—	—	10.16	—	$J_{4,7}$ 1.5	71BSF1491
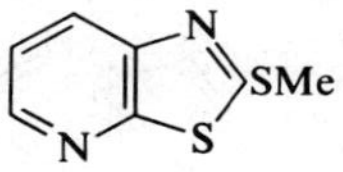 (SMe-thiazolo-pyridine)	CDCl$_3$	—	2.70	—	—	8.28	7.17	7.90	—	$J_{5,6}$ 5.0, $J_{5,7}$ 1.5, $J_{6,7}$ 7.9	76JHC491
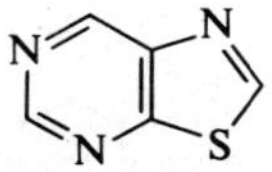 (thiazolo-pyridine)	CDCl$_3$	—	9.19	—	—	9.21	—	9.47	—		69TL1011

Table 16 ^{1}H NMR Data for Partially Saturated Fused Thiazoloazines

Compound	Solvent	1	2	3	^{1}H chemical shifts (p.p.m.) 5	6	7	8	Coupling constants (Hz)	Ref.
![structure] R^1 Me, R^2 H, R^3 Me, R^4 H, Y Br	TFA	—	1.70, 4.5	5.0	2.78	—	—	—		73ACS1749
R^1 H, R^2 H, R^3 H, R^4 O$^-$	TFA	—	3.92	5.27	8.22	7.53	7.75	—	$J_{2,3}$ 8.0, $J_{5,6}$ 6.0, $J_{6,7}$ 8.5, $J_{5,7}$ 1.0	69ACS1704
R^1 H, R^2 H, R^3 Me, R^4 O$^-$	TFA	—	3.88	5.10	2.75	7.30	7.70	—	$J_{2,3}$ 8.0, $J_{6,7}$ 8.5	69ACS1704
R^1 H, R^2 H, R^3 Cl, R^4 O$^-$	TFA	—	3.9	5.3	—	7.4	7.7	—	$J_{6,7}$ 8.5	77ACS(B)919
R^1 H, R^2 CO$_2^-$, R^3 Me, R^4 OH	TFA	—	4.3	6.33	2.78	7.42	7.85	—	$J_{6,7}$ 8.5	69ACS1966
![structure] R^1 H, R^2 CO$_2^-$, R^3 Me, R^4 OH	TFA	—	4.52 (cis), 4.12 (trans)	6.68	2.98	8.10	8.40	—	$J_{2,2}$ 14.5, $J_{2,3}$(cis) 7.0, $J_{2,3}$(trans) 1.0, $J_{6,7}$ 9.0	69ACS1966
R^1 H, R^2 H, R^3 Me, R^4 O$^-$	TFA	—	4.0	5.62	2.93	8.00	8.32	—	$J_{6,7}$ 9.0	69ACS1966
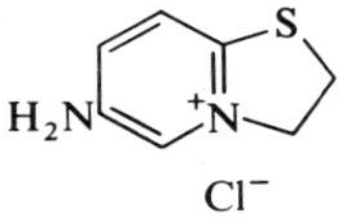 ![structure] H$_2$N— (Cl$^-$)	TFA	—	3.96	5.30	9.05	—	8.43	7.90	$J_{7,8}$ 9.0	78ACS(B)70
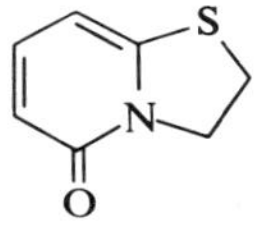 ![structure] lactam	CDCl$_3$	—	3.35	4.46	—	6.20	7.22	6.07	$J_{6,7}$ 8.5, $J_{7,8}$ 7.5	78ACS(B)70

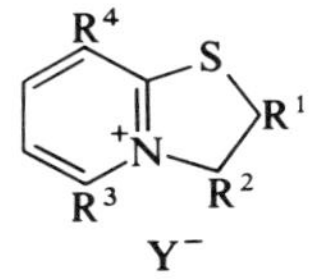

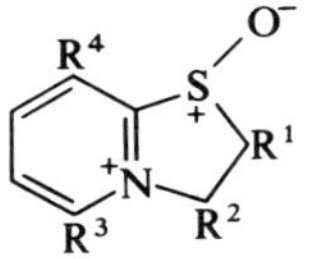

Table 16 (cont.)

Compound	Solvent	^{1}H chemical shifts (p.p.m.)							Coupling constants (Hz)	Ref.
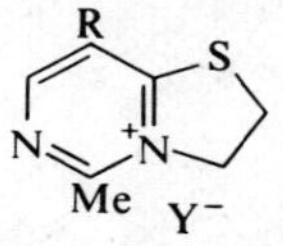		1	2	3	5	6	7	8		
R = O$^-$ (Y absent)	TFA	—	4.07	5.48	3.05	—	8.68	—		69ACS2437
R = OEt, Y = Br	TFA	—	4.16	5.44	3.10	—	8.49	4.53, 1.62		69ACS2437
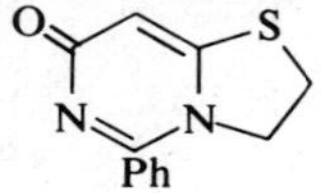 (thiazolopyrimidinone, Ph)	TFA	—	3.73	4.83	7.82	—	—	6.87		72T4737
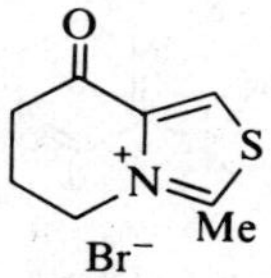 (thiazolo Me Br$^-$)	D$_2$O	9.05	—	3.15	—	—	—	—		69JCS(C)707

Table 17 ^{13}C NMR Data for Fused Thiazoloazines

Compound	Solvent	^{13}C chemical shifts (p.p.m.)							$^1J_{CH}$ coupling constants (Hz)	Ref.
		2	3	5	6	7	8	9 (8a)		
[structure: thiazolo-pyridinium with O⁻, S, N⁺, R]										
R = H	D$_2$O	124.1	133.0	124.6	126.4	120.9	161.9	149.6	J_2 199, J_3 199, J_5 200, J_6 170, J_7 167	78ACS(B)651
R = Br	D$_2$O	114.9	133.7	122.7	127.0	122.6	163.2	149.9		78ACS(B)651
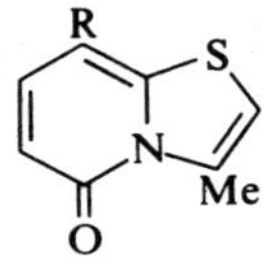 [structure: thiazolo-pyridinone with S, N-Me, =O, R]										
R = H	CDCl$_3$	106.0	139.0	163.2	110.9	137.4	99.7	150.1	18.7 (3-Me)	79JCS(P1)1150
R = Me	CDCl$_3$	106.2	140.0	162.7	111.2	139.1	107.8	147.5	18.8 (3-Me), 17.5 (8-Me)	79JCS(P1)1150

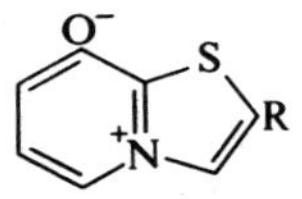

Table 18 ^{13}C NMR Data for Partially Saturated Fused Thiazoloazines

Compound	Solvent	2	3	5	6	7	8	9 (8a)	$^1J_{CH}$ coupling constants (Hz)	Ref.
		^{13}C chemical shift (p.p.m.)								
Y^-										
R^1 R^2 R^3 Y										
H H H Br	D_2O	32.3	63.0	144.2	125.1	147.0	125.8	162.9	J_5 192, J_6 176, J_7 172, J_8 178	82UP42900
H H O^- —	D_2O	31.9	63.9	135.5	125.9	129.1	154.3	152.6	J_5 194, J_6 176, J_7 169	82UP42900
CO_2H Me OH Br	TFA	34.0	72.3	147.9	126.4	131.2	152.6	152.1	171.7 (3-CO_2H), 20.3 (5-Me), J_6 174, J_7 167	82UP42900
R = H	$CDCl_3$	28.2	50.8	162.4	114.8	139.9	100.3	147.9		79JCS(P1)1150
R = Me	$CDCl_3$	28.2	51.3	161.8	114.9	142.5	109.0	145.0	17.7 (8-Me)	79JCS(P1)1150

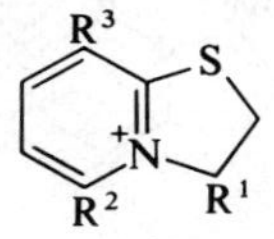

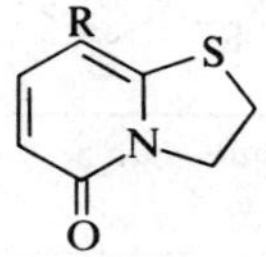

Table 19 UV Absorption Data for Fused Thiazoloazines

Compound					Solvent	λ_{max}[a] (log ε)	Ref.
R^1	R^2	R^3	R^4	Y			
H	H	H	H	Br	H$_2$O	306 (4.24), 295 (4.11), 228[b] (4.02), 224 (4.08), 208 (4.00)	66JHC27
H	O$^-$	H	H	—	MeOH	422 (4.10), 255 (3.80)	71ACS5
Ph	O$^-$	H	H	—	MeOH	418 (4.08), 260 (3.74)	71ACS5
H	Me	H	O$^-$	—	1N HCl	327 (4.18), 317 (4.08), 274 (3.26), 265 (3.27), 239 (3.95)	70ACS2956
H	Me	H	O$^-$	—	1N NaOH	344 (4.00), 328 (3.51), 274 (3.58), 256 (3.76)	70ACS2956
H	H	Me	O$^-$	—	1N HCl	333 (4.11), 324[b] (4.05), 271 (3.91), 263 (4.06), 243 (4.33)	70ACS2956
H	H	Me	O$^-$	—	1N NaOH	354 (3.99), 282 (3.40), 245 (3.84)	70ACS2956

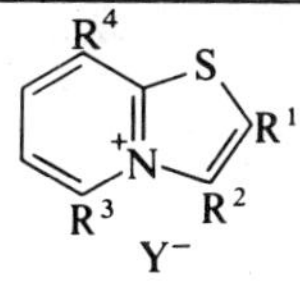

Compound	Solvent	λ_{max}[a] (log ε)	Ref.
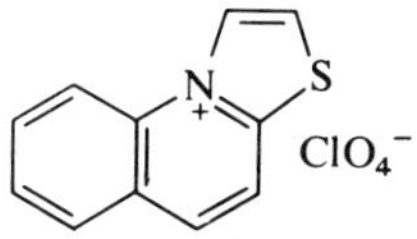 (benzo-fused thiazoloquinolinium ClO$_4^-$)	EtOH	342 (4.38), 327 (4.23), 313[b] (3.98), 260[b] (3.89), 249 (4.12), 237[b] (4.33), 227 (4.34)	67JHC71
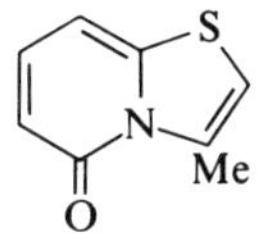 (thiazolopyridinone, Me substituent)	EtOH	382 (4.08), 364 (4.09), 277[b] (3.60), 268 (3.72), 235 (3.89)	79JCS(P1)1150

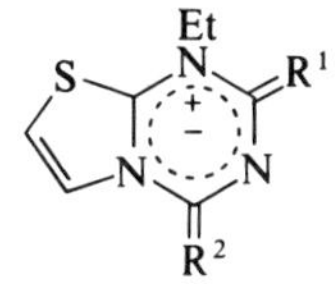			
$R^1 = S, R^2 = O$	H$_2$O	298 (4.02), 276 (4.01), 239[b] (4.05), 216 (4.23)	73JOC3868
$R^1 = O, R^2 = S$	EtOH	320 (4.11), 274 (3.96), 225 (3.92)	73JOC3868

Compound	Solvent	λ_{max}[a] (log ε)	Ref.
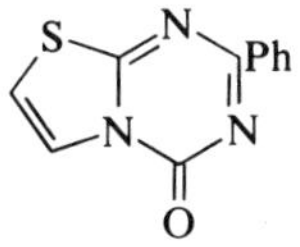 (thiazolotriazinone, Ph)	EtOH	330 (4.18), 273 (4.21)	74S346
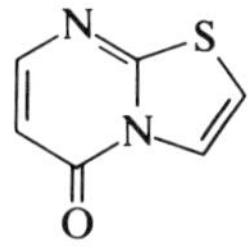 (thiazolopyrimidinone)	MeOH	322 (4.04), 319 (4.11), 259 (3.71), 252 (3.73), 222 (3.96)	71JCS(C)2094
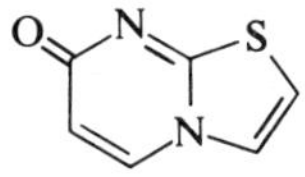 (thiazolopyrimidinone isomer)	MeOH	271 (4.04), 215 (4.14)	71JCS(C)2094
(Br-oxo-thiazolopyridinium)	NaOH	405, 318, 308	69JCS(C)707
	HCl	354, 304, 292, 282[b]	
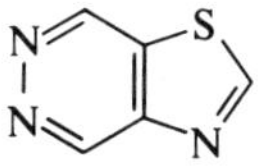 (thiazolopyridazine)	H$_2$O	270 (3.48), 237[b] 213 (3.39)	71BSF1491

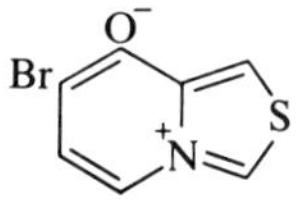

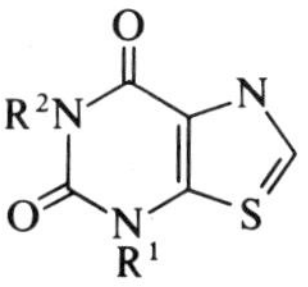

R^1	R^2			
Me	H	pH 1	261 (3.99)	
Me	H	pH 11	266 (4.02)	
H	Me	pH 1	257 (3.96)	75JOC2476
H	Me	pH 11	282 (4.04)	

[a] nm. [b] Shoulder.

Table 20 UV Absorption Data for Partially Saturated Fused Thiazolozines

Compound					Solvent	λ_{max}[a] ($\log \varepsilon$)	Ref.
R^1	R^2	R^3	R^4	Y			
Me	H	Me	H	Br	0.1N HCl	327 (3.99), 252 (3.88), 210 (4.13)	73ACS1749
Me	CO_2^-	Me	H	—	0.1N HCl	329 (4.08), 242 (4.04), 210 (4.10)	73ACS1749
H	H	H	O^-	—	0.1N HCl	330 (3.91), 230 (3.68)	69ACS371
H	H	H	O^-	—	0.1N NaOH	350 (3.98), 245 (3.84)	69ACS371
H	CO_2^-	Me	OH	—	0.1N HCl	340 (4.01), 240 (3.85)	69ACS371

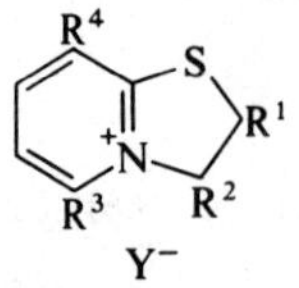

	Solvent	λ_{max} ($\log \varepsilon$)	Ref.
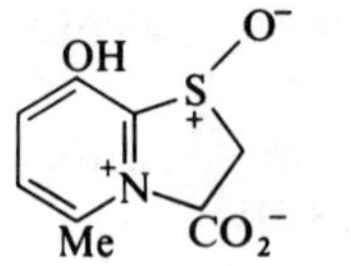 (OH, S, O^-, Me, CO_2^- structure)	0.1N HCl	312 (4.00), 230[b] (3.85), 213 (4.25)	72T1223
	0.1N NaOH	352 (3.85), 255[b] (3.91), 232 (4.12)	72T1223
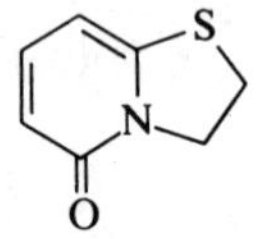 (bicyclic thiazolo-pyridinone)	EtOH	328 (4.11), 243 (3.92)	79JCS(P1)1150
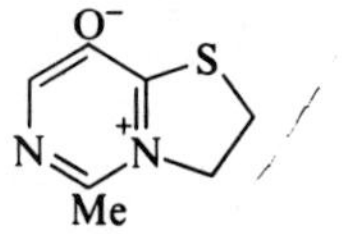 (O^-, N, Me structure)	0.1N HCl	335 (3.79), 240 (3.66)	69ACS2427
	0.1N NaOH	360 (3.84), 265 (3.78)	69ACS2427
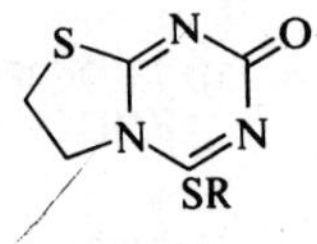 (triazine-thione, R=H)	EtOH	282[b] (4.02), 262 (4.17), 245[b] (4.08), 206 (4.06)	74JOC1819
(triazine-thione, R=BR)	EtOH	236[b] (4.32), 231 (4.32)	74JOC1819
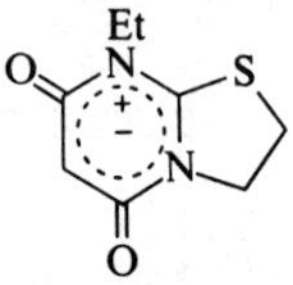 (Et-N mesoionic)	H_2O	248[b] (3.46), 265 (3.88), 222 (4.28)	73JHC487
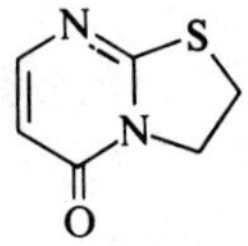 (pyrimidone-thiazoline)	H_2O	290.5 (3.84), 207.5 (4.29)	71JCS(C)1527
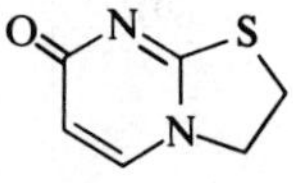 (pyrimidone-thiazoline isomer)	H_2O	260[b] (3.88), 232 (4.30)	71JCS(C)1527

[a] nm. [b] Shoulder.

likely processes are intramolecular Hofmann elimination to form a non-charged isomeric structure (**312 → 313**) or elimination of a substituent (**314 → 316**). Adduct formation between the cation and anion to give a pseudobase, or a nucleophilic attack by the anion leading to ring opening reaction (**317 → 318**) should also be considered. In a fully aromatic system without the possibility for an elimination or substitution reaction, redox processes are likely to prevail; the cation is oxidized with concurrent reduction of the anion to radicals which are the volatile species. The carbon radical after ionization in the gas phase gives back the parent cation.

Mesoionic structures and other internal salts may display a molecular ion of an ambiguous structure which is isobaric with the original molecule. Since the fragmentation patterns of

Table 21 IR Absorption Data for CO Groups in Fused Thiazoloazines

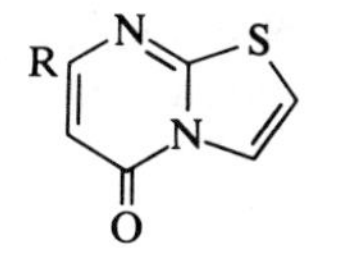

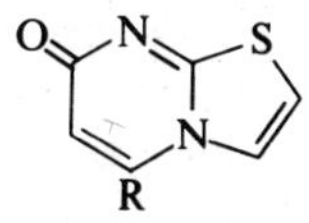

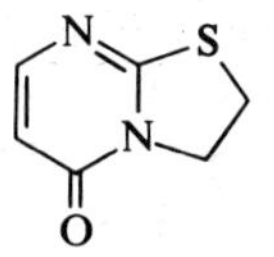

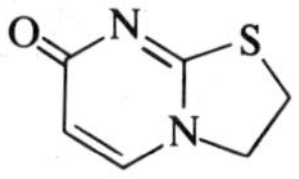

(300)
R = H; 1660 cm^{-1}
R = Me; 1670 cm^{-1}
KBr
71JCS(C)2094

(301)
R = H; 1630 cm^{-1}
R = Me; 1641 cm^{-1}
KBr
71JCS(C)2094

(302)
1670 cm^{-1}
Nujol
71JCS(C)1527

(303)
1640 cm^{-1}
Nujol
71JCS(C)1527

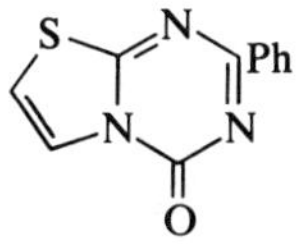

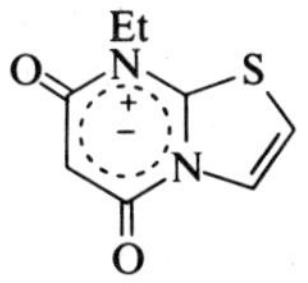

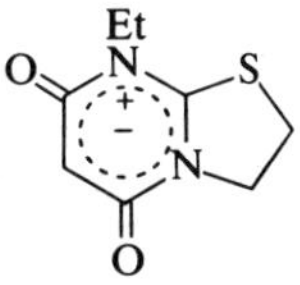

(304)
1695 cm^{-1}
Nujol
74S346

(305)
1690 cm^{-1}
CHCl$_3$
73JHC487

(306)
1670 cm^{-1}
CHCl$_3$
73JHC487

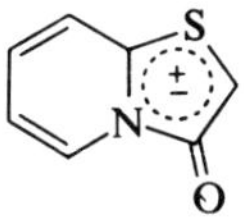

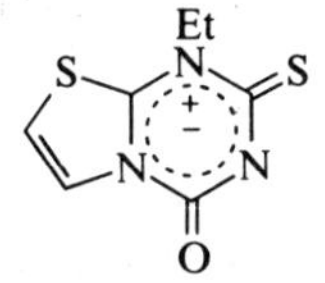

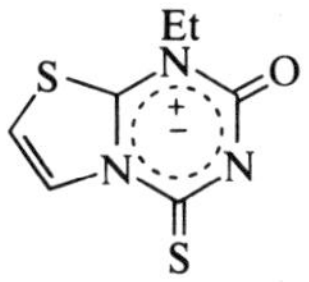

(307)
1630 cm^{-1}
KBr
71ACS5

(308)
1675 cm^{-1}
KBr
73JOC3868

(309)
1690 cm^{-1}
KBr
73JOC3868

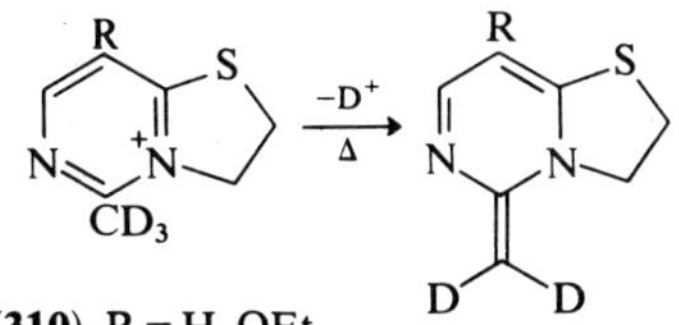

(310) R = H, OEt
(311)
(312) 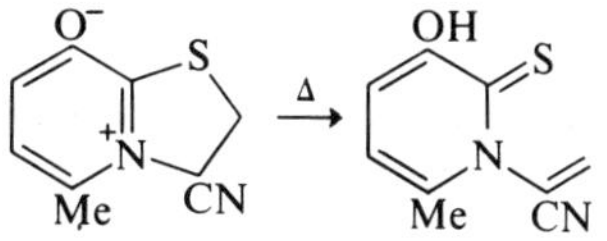
(313)
(314) 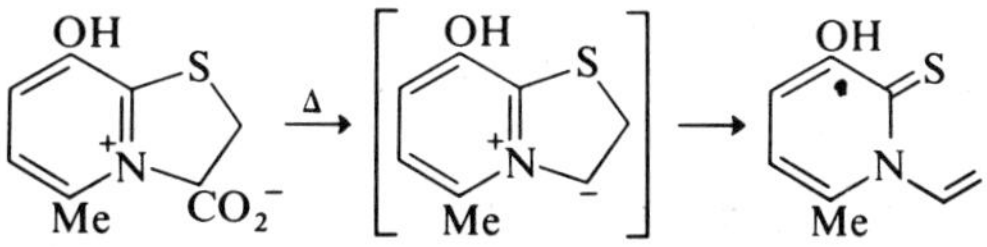
(315)
(316)
(317) 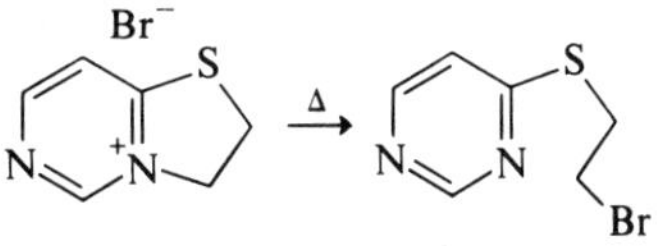
(318)
(319) $IP = 6.93$ eV 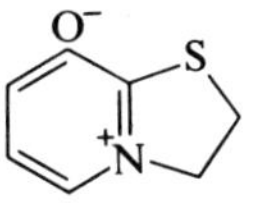
(320) $IP = 6.80$ eV 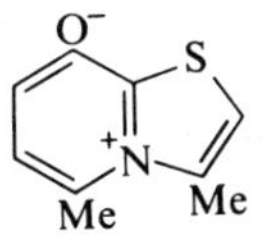
(321) $[M] = 20\%$ (r.i.) 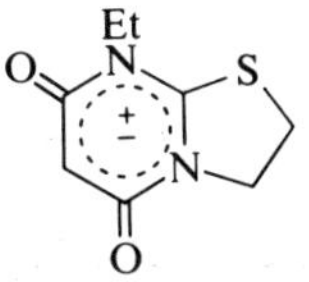
(322) X = O, Y = S; $[M] = 23\%$ (r.i.) 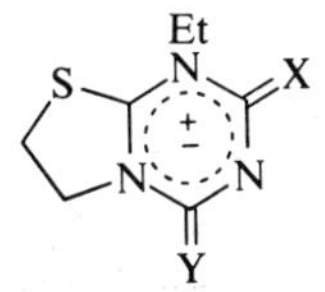
(323) X = S, Y = O; $[M] = 20\%$ (r.i.)
(324) R^1 = Me, R^2 = H; $IP = 6.80$ eV
(324)
(326)
(325) $IP = 8.60$ eV

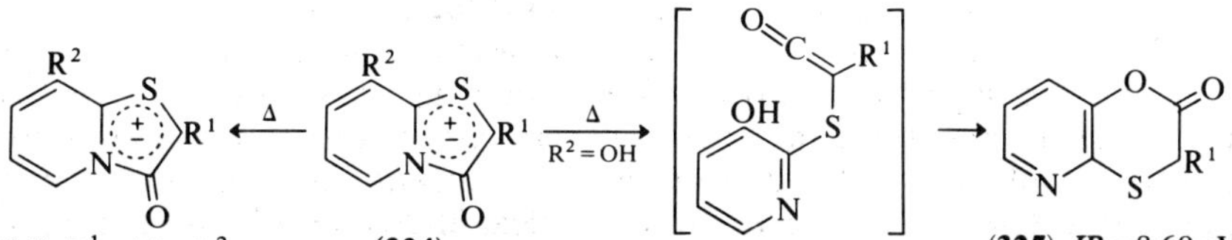

Pyrolytic reactions on mass spectrometry (IP values given as ±0.05 eV)

isomeric molecules may be very similar, other methods must be used to analyze the nature of the gaseous species. It has been found that radicals, or species with internal charge separation, are characterized by exceptionally low appearance (ionization) potentials. Using this technique it has been shown that the betaines (**319**) are volatilized as such without any structure change, whereas mesoions of type (**324**) in the gas phase may exist as mesoions except for derivatives with a vicinal hydroxy group, which are isomerized to the lactones (**325**); ketenes (**326**) seem likely intermediates. In other cases ketenes rather than mesoions are the species in the gas phase ⟨80MI42900⟩.

4.29.7.2.3 Thermodynamic aspects

(*a*) *Acid dissociation constants.* The acid dissociation constants for some anhydro-8-hydroxy-2,3-dihydrothiazolo[3,2-*a*]pyridinium hydroxides are given in Table 22. The pK_a values for the hydroxy group in the bicyclic compounds (4.5–5.0) are comparable with that of 4.96 found for anhydro-3-hydroxy-1-methylpyridinium hydroxide. The high acidity of the carboxy group is caused by the pyridinium ring; its electron deficiency is transmitted more efficiently to C-3 than to C-2, and hence the 3-carboxy group is the more acidic. This correlates with the higher ^{1}H and ^{13}C NMR chemical shift values in the 3-position ⟨81H(15)1349⟩.

Table 22 Acid Dissociation Constants in Water at 20 °C ⟨72ACS1847⟩

Structure	R^1	R^2	R^3	pK_a 1 (*CO$_2$H*)	pK_a 2 (*OH*)
	H	H	Me	—	4.70
	H	CO$_2$H	H	1.2	4.47
	H	CO$_2$H	Me	1.5	4.97
	CO$_2$H	H	Me	2.0	4.96

4.29.7.3 Reactivity

4.29.7.3.1 Survey of reactivity

Electrophilic substitution at nitrogen can occur in either ring depending on the nature of the substituents present. Electrophilic substitution at carbon proceeds readily in the presence of electron-releasing substituents. A wide variety of electrophiles have been studied. Mesoionic systems are readily substituted and cationic structures may become very reactive when conditions are chosen so as to promote intermediate formation of a pseudo- or anhydro-base. Sulfoxides are formed in the peracid oxidation of fused dihydrothiazoles.

Nucleophilic substitution is a general reaction. Fully conjugated ring systems are relatively resistant to ring opening reactions because of aromatic stabilization. Dihydrothiazolo systems are more reactive. Proton abstraction in either ring can be used in regioselective deuterium labelling. Proton abstraction in fused dihydrothiazoles may be used in *N*- or *S*-vinylation reactions of the azines by opening of the thiazole ring.

Mesoionic compounds of this series may participate in 1,3-dipolar cycloaddition reactions; they may also undergo photochemical rearrangements.

4.29.7.3.2 Photochemical reactions involving no other species

Anhydro-8-hydroxythiazolo[3,2-*a*]pyridinium hydroxides (**321**) on photolysis at 350 nm undergo valence isomerism. The product formed (**327**) corresponds to an interchange of the 5- and 8-substituents. It is the pyridine ring which undergoes the transformation and this type of rearrangement is also observed for 1-substituted anhydro-3-hydroxypyridinium hydroxides. In the 2,3-dihydrothiazole analogues (**320**) the unstable intermediate, the 5-aza-2-thiatricyclo[4.3.0.0$^{1.5}$]non-7-en-9-one (**328**), can be isolated; it is further rearranged to the fused pyridinone (**329**) on illumination with a medium pressure mercury lamp. The photochemically labile valence isomer (**328**) is formed by a photochemically allowed ring closure. A 1,2-photochemical shift as indicated in (**328**) probably accounts for the second transformation ⟨81H(15)1349⟩.

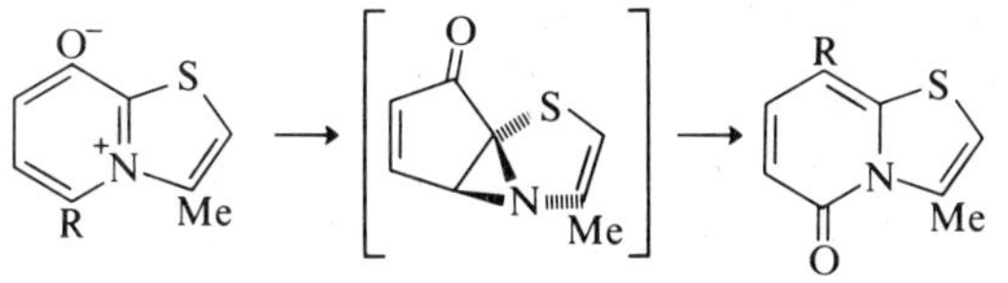

(**321**) R = H, Me   (**327**)

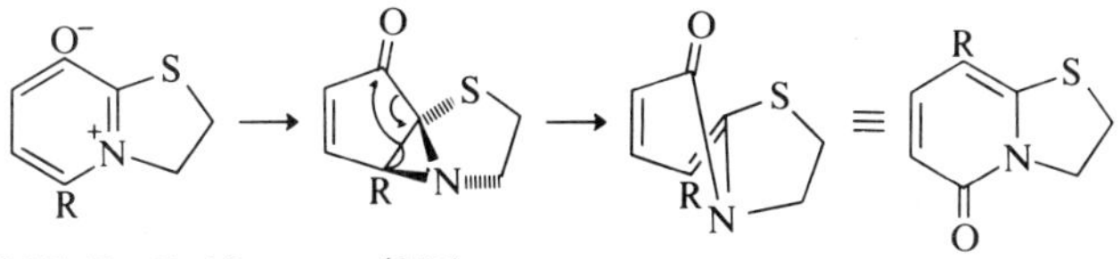

(**320**) R = D, Me   (**328**)   (**329**)

4.29.7.3.3 Electrophilic attack at nitrogen

Fused 1,3-azoles without a common nitrogen at a bridgehead position may be attacked by an electrophile at a nitrogen in either ring. The nature of the substituents present usually controls the selectivity in the attack. Thus, in the naphthyridine (**330**) it is the thiazole nitrogen, which is activated by the 2-methylthio group, which is preferentially methylated ⟨79CPB410⟩. On the other hand, the 2,3-dihydrothiozolo derivative (**331**) is readily methylated on the pyrimidine nitrogen ⟨71JCS(C)1527⟩. A further selective alkylation of activated pyrimidine nitrogen is illustrated by the ribosylation of the bis(trimethylsilyl) ether of the thiazolo[5,4-*d*]pyrimidine (**332**); the protected nucleoside (**333**; R = PhCO) is obtained as an anomeric mixture (ratio $\beta : \alpha = 2 : 1$). In this reaction the same regioselectivity is observed as in the coupling of purines. The protecting groups can be removed in the usual way with methanolic ammonia or sodium methoxide to furnish the nucleoside (**333**; R = H) ⟨75JOC2476⟩.

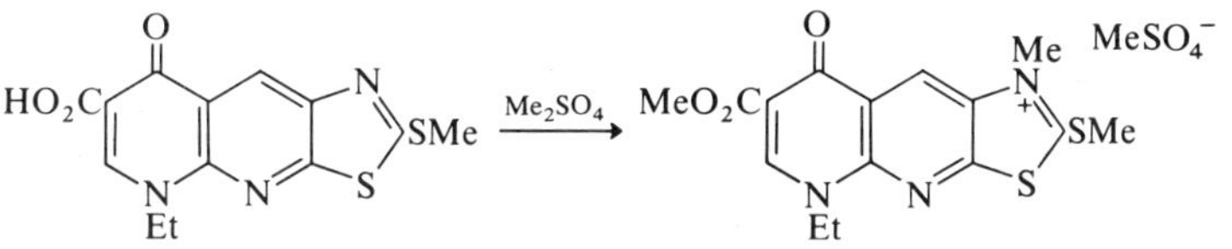

(**330**)

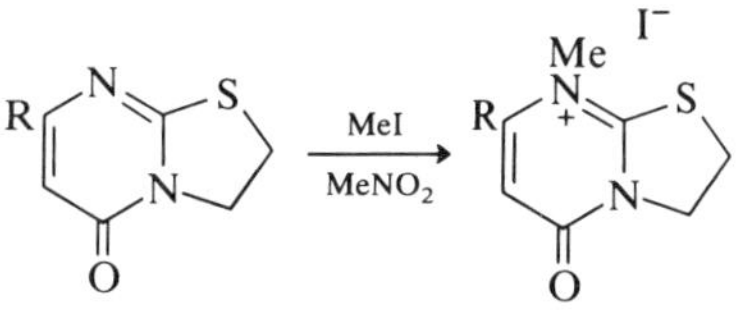

(**331**)

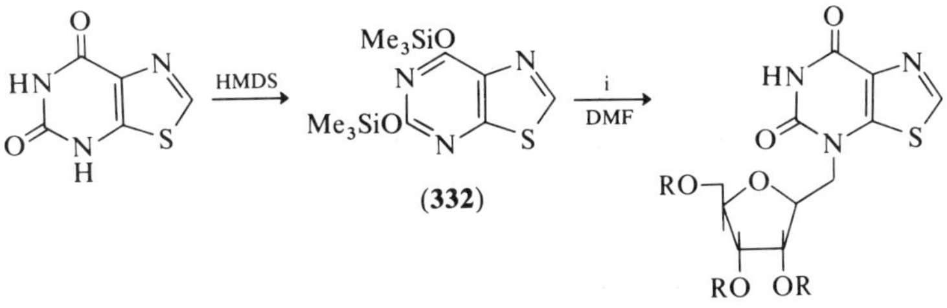

(**332**)

(**333**) R = PhCO, H

i, 2,3,5-tri-*O*-benzoyl-D-ribofuranosyl bromide

In the alkylation of lactams the choice of conditions and reagents is important for controlling the reaction towards *N*- or *O*-alkylation. Methylation of the isoquinoline derivative (**334**) under alkaline conditions gives a mixture from alkylation of both lactam and isoquinoline nitrogen atoms. Under the same conditions the [5,4-c] isomer (**335**) yields the lactam (**335**; R = Me) ⟨68CJC691⟩.

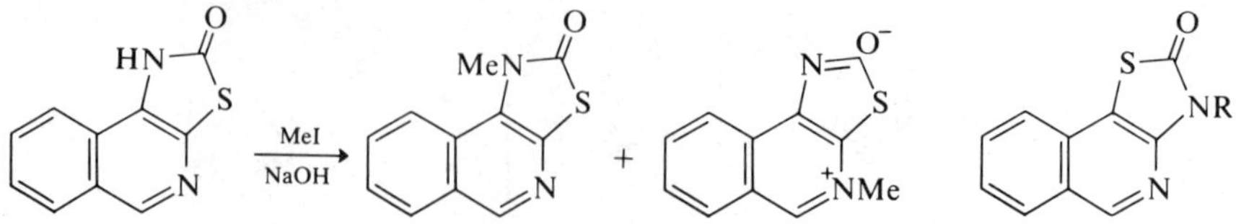

4.29.7.3.4 Electrophilic attack at carbon

Heterocyclic cations will not easily react with electrophiles unless substituted by strongly electron-releasing groups. Alternatively, the reaction conditions are chosen in such a way that a reactive anhydro base or pseudobase intermediate is formed. Thus the thiazolo cation (**336**) can be nitrated at C-3 ⟨78ZOR216⟩ possibly via the ylide (**337**). The 6(8)-nitro-2,3-dihydrothiazolo[3,2-a]pyridinium salts (**338**) and (**340**) are readily brominated in hydroxylic solvents; the regioselectivity and the ease of reaction are consistent with pseudobase intermediates (**339**) and (**341**) ⟨81H(15)1349⟩.

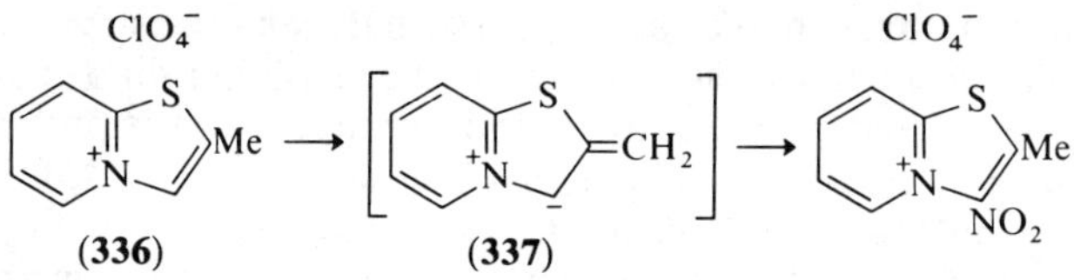

Mesoionic systems may be readily substituted by electrophiles. Thus the thiazolo mesoion (**342**) will couple with diazonium salts despite their relatively weak electrophilicity ⟨80KGS621⟩. Substitution in a fused heteroaromatic betaine azine ring, e.g. (**343**), also takes place with ease. The resonance form (**344**) of the mesoion (**343**) shows that the electrophile will attack at C-6. The substitution in this position is also predicted by MO calculations ⟨73JHC487⟩. Similarly the pyridine ring in pyridinium olates is active towards electrophiles and is substituted in the positions ortho and para to the olate function. Bromination of the 5-methyl derivative (**321**; R = Me) occurs exclusively in the 7-position which is rationalized via the intermediate (**345**). In the absence of a 5-substituent, attack in either the 5- or 7-position occurs; the dibromide is readily formed. No bromination in the thiazole ring is observed. The 2-bromo derivative (**346**) has been made, however, by condensation between the appropriate mercaptopyridine and 1,1,2,2-tetrabromoethane.

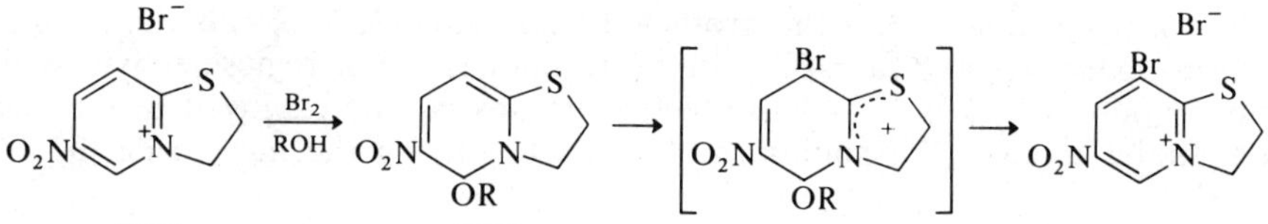

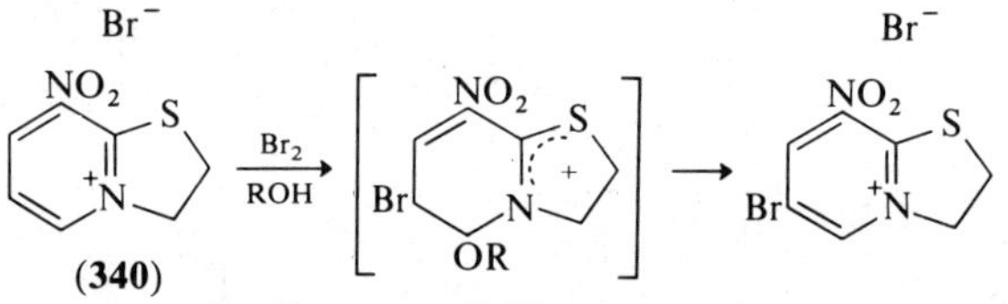

Further studies of electrophilic substitution using the 2,3-dihydrothiazolo analogue (**320**) show that regioselective bromination can be effected in the 7-position (**347**) at low temperature. Chlorination with sulfuryl chloride, however, occurs exclusively in the 5-position. The 5-chloro derivative (**348**) can be further nitrated in the 7-position. By analogy, bromination of quinoline analogues (**349**) occurs in the azine ring, in the 5-position. The [3,2-a]pyrimidine analogue (**350**; R = H) similarly yields the 7-bromide. The activation by

(**342**)

(**343**) (**344**)

(**321**) R = H, Me (**345**) (**346**)

an amino group may also be sufficient for electrophilic substitution to occur in an azinium ring; the 6-amine (**351**; R = H) is selectively brominated at C-5.

(**347**) R = H, Br (**320**) (**348**)

(**349**) (**350**) R = H, Br (**351**) R = H, Br

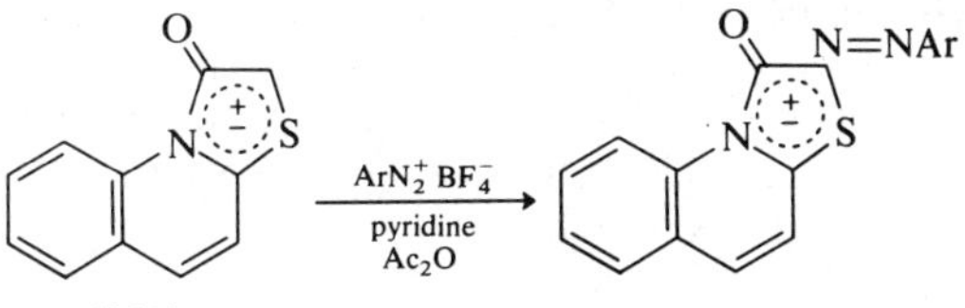

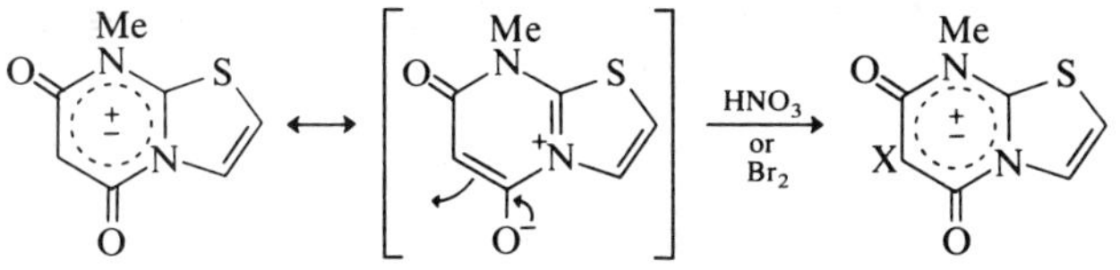

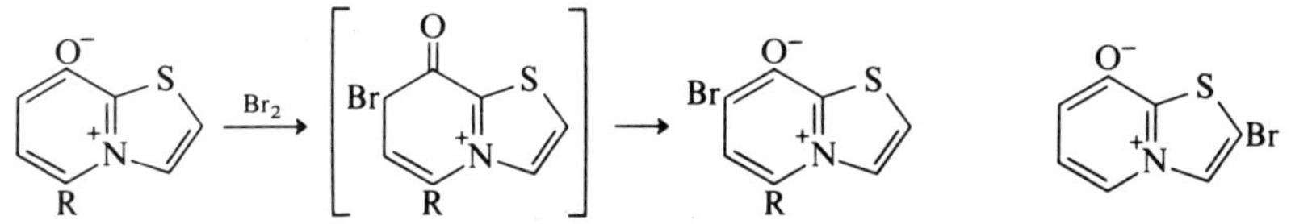

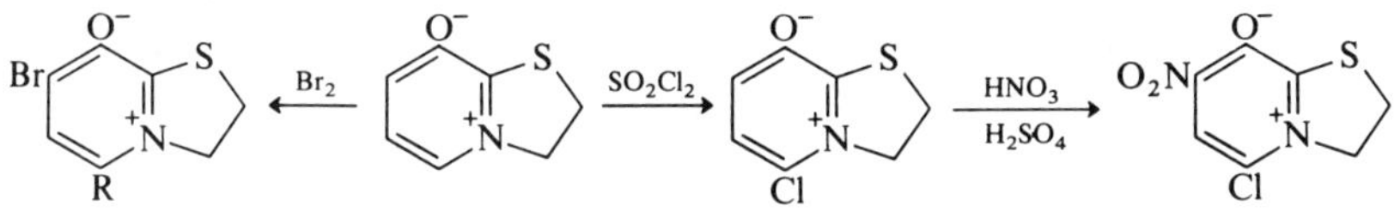

The coupling of the betaines (**352**) with diazonium salts, however, may take an unusual course in that arylation results. The reaction may be rationalized by the formation of a covalent intermediate (**353**) which subsequently suffers a pericyclic rearrangement involving expulsion of nitrogen ⟨81H(15)1349⟩.

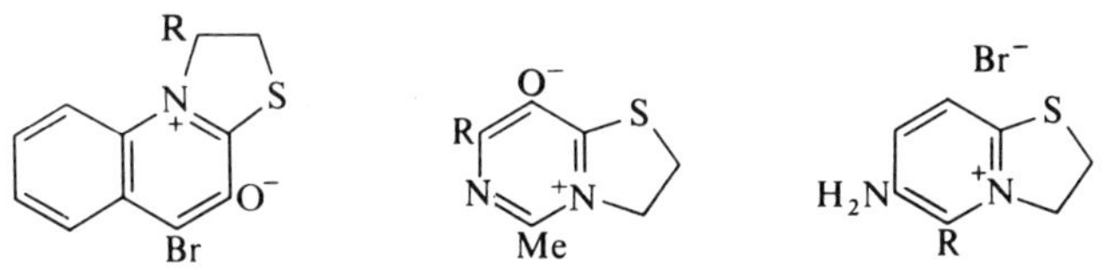

In fused thiazoloazines which carry no positive charge the reactivity towards electrophiles will be as normally experienced for fused azines. In such systems a strongly electron-donating substituent usually more than counteracts the deactivating effect from a nuclear nitrogen. Hence the ready nitrosation of the fused pyrimidinone (**354**) in the 6-position ⟨77ACS(B)167⟩ is readily understood. Even without the 7-amino group, as in (**302**), the electrophile enters the 6-position ⟨71JCS(C)1527⟩.

(**352**) (**353**) R = H, CO_2H

(**354**)

(**302**) X = Br, Cl

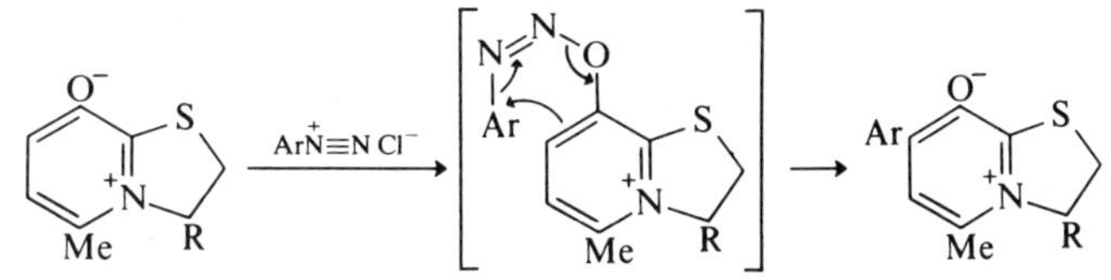

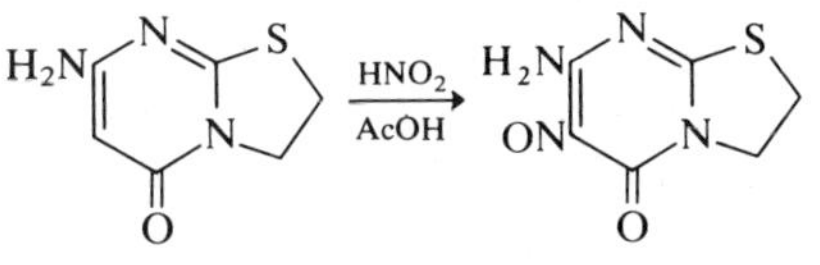

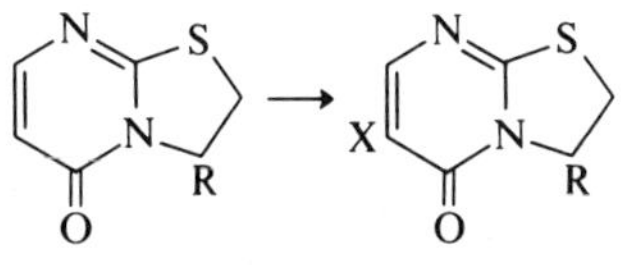

4.29.7.3.5 Nucleophilic attack at carbon

Nucleophilic substitution in fused thiazoloazines proceeds with ease in the activated positions in either ring. N-Quaternization further enhances the ease of substitution, especially in the ring which carries the charge.

(a) *Hydroxide ion and other oxygen nucleophiles.* The 2-methylthio substituent in the thiazolium derivative (355) is hydrolyzed in water ⟨66JPR(304)26⟩. Sulfides are more readily substituted after oxidation. The methoxide ion will displace the 2-methylsulfonyl substituent in (356) ⟨78MI42900⟩. Selective displacement of one of two or more of the same substituents in an azine can sometimes be achieved. Such a case is the selective displacement of the 5-substituent in the dichloropyridazine (357) ⟨71BSF3537⟩. In the dichloropyrimidines (358) and (359) it is the chlorine in the activated azine positions which is hydrolyzed in water ⟨72ACS947⟩. Oxidation of an azine thiolactam is frequently accompanied by hydrolysis whereby the lactam is formed. The peracid oxidation of (360) falls into this pattern ⟨78CPB765⟩.

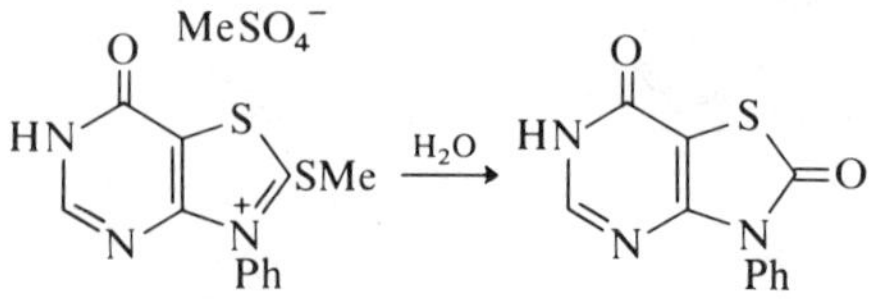

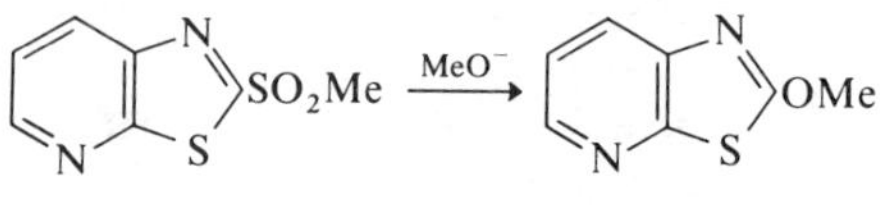

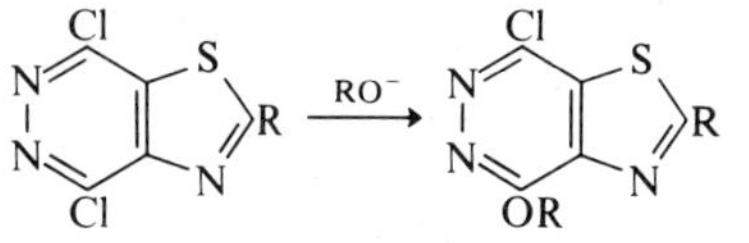

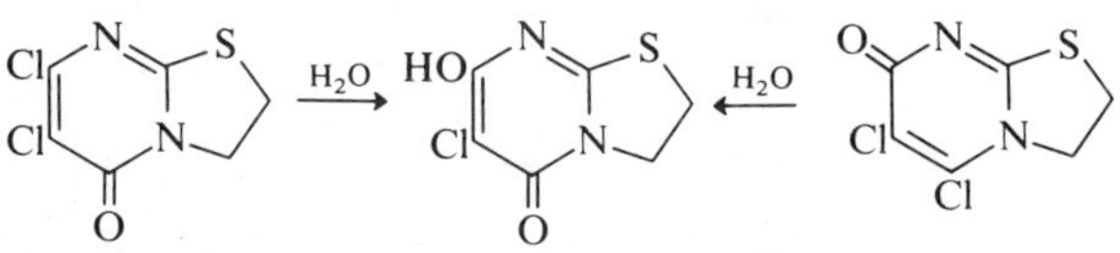

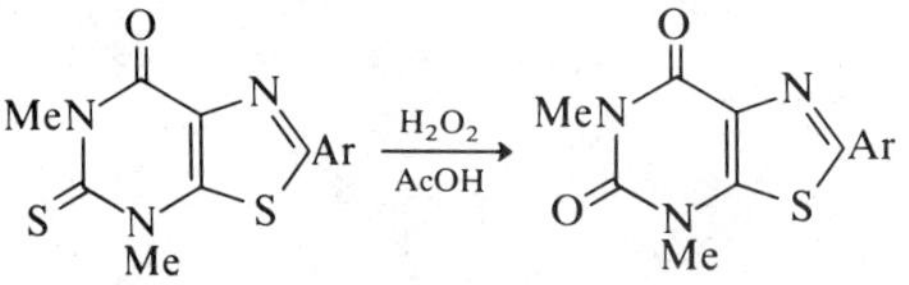

(b) *Amines and amide ions.* In agreement with the ready displacement of a 2-substituent as discussed above, the 2-methylthio derivative (361) reacts with hydrazine by substitution ⟨77JHC1045⟩. In 2,5,7-trichlorothiazolo[5,4-d]pyrimidine it is the 7-chloro substituent which is first displaced. In fused pyrimidines this corresponds to the most active position. In the subsequent nucleophilic substitutions the 2-chloro and finally the 5-chloro substituents are replaced ⟨68CPB750⟩; in trichloropurines the reactivity order is 7-, 5-, 2-. Reduction of the thiazole ring has little effect on substitution in the azine ring. Remarkable, however, is the behaviour of the 6-bromo analogue (363), which reacts with secondary amines to form the 7-amino derivative (364), possibly via a (Michael) addition–elimination process ⟨71JCS(C)1527⟩.

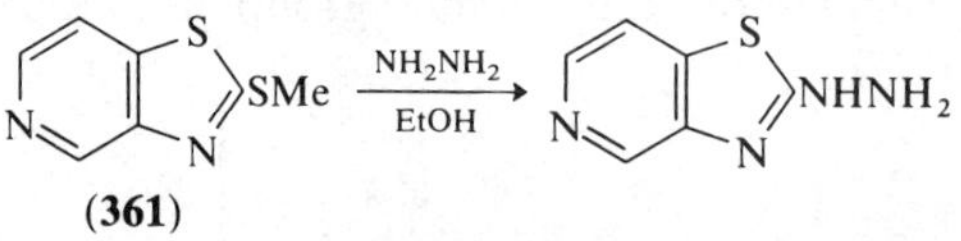

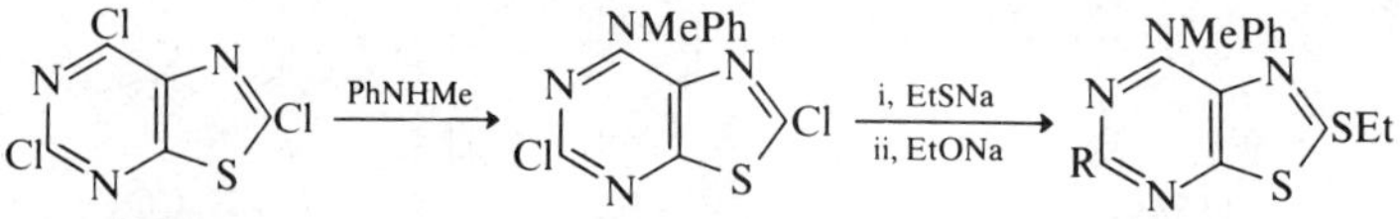

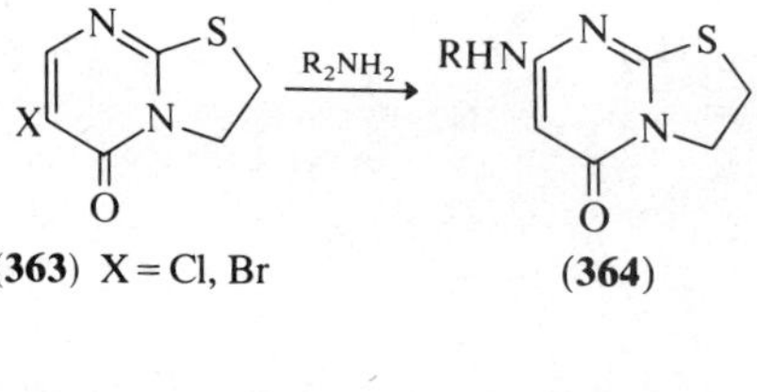

(*c*) *Sulfur nucleophiles.* Azine lactams are usually converted into thiolactams by reaction with phosphorus pentasulfide in an inert solvent or in the presence of a tertiary amine. This also holds for the fused systems (**301**) ⟨78JHC849⟩ and (**302**) ⟨71JCS(C)1527⟩. Selective thiation at C-7 in the 5,7-dione (**365**) ⟨75JOC2476⟩ is in accordance with the higher reactivity of this position in fused pyrimidines. Normal displacement reactions of halogens in the reaction of (**366**) with thiourea ⟨71BSF3537⟩ and of (**367**) with hydrosulfides ⟨81H(15)1349⟩ take place.

(**301**) Z = O, S

(**302**) Z = O, S

(**365**) R = H, β-D-(PhCO)$_3$-ribose

(**366**)

(**367**)

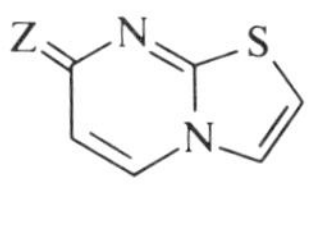

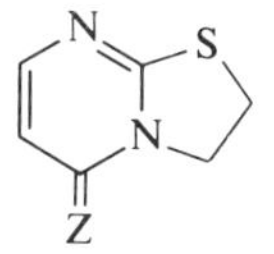

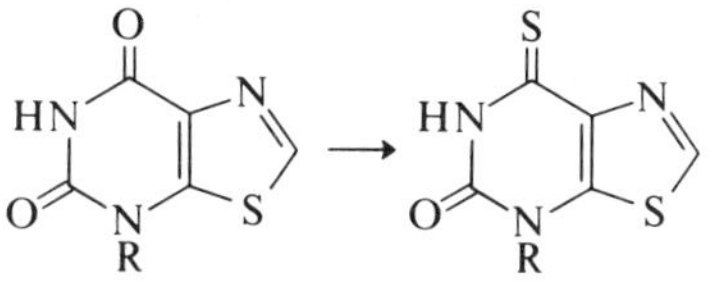

(*d*) *Halide ions.* Oxo groups in both the five- and six-membered rings are converted into chloro substituents in the usual way, using phosphorus halides, *e.g.* (**368**) ⟨68CPB750⟩. A 2-methylthio substituent as in (**369**) is not affected by the phosphorus halide ⟨79JPR260⟩. Diazotization in strong HCl of the 2-amino group in (**370**) gives the 2-chloro compound as is normally the case for diazotization of activated amino groups under these conditions ⟨70JCS2478⟩. A chloro substituent can also be introduced by the usual sulfuryl chloride reaction with thiols in which case an oxo function is not affected, as shown for the 2-mercaptothiazolonaphthyridine (**371**) ⟨79CPB410⟩. A halogen in an activated azine position may be substituted by another halogen (**372**), and a nitro group in an activated position is also readily displaced by a halide ion ⟨81H(15)1349⟩.

(**368**)

(**369**)

(**370**)

(**371**)

(**372**) R^1 = H, R^2 = Br R^1 = H, Cl
R^1 = Cl, R^2 = NO$_2$

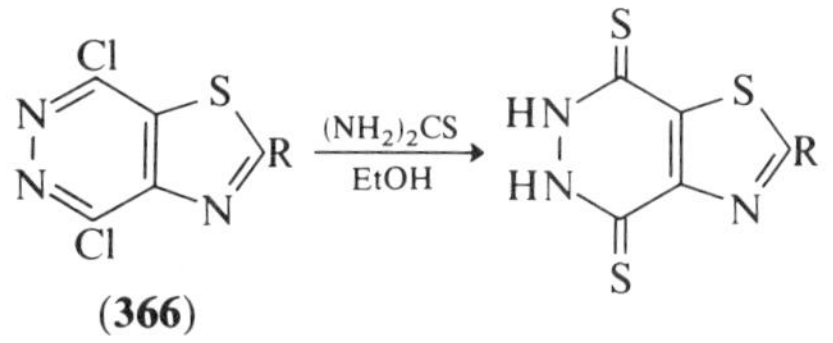

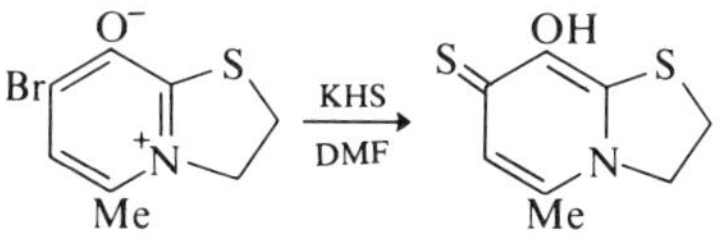

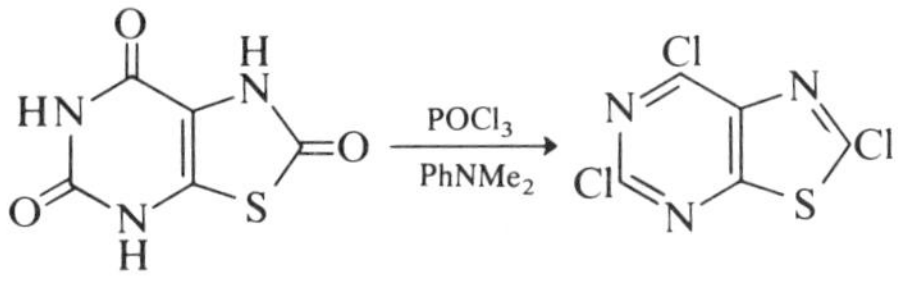

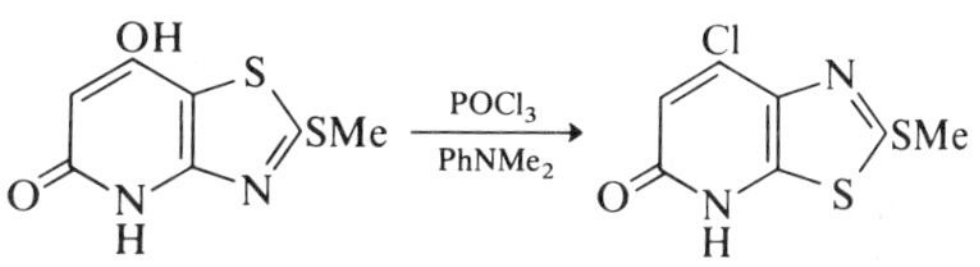

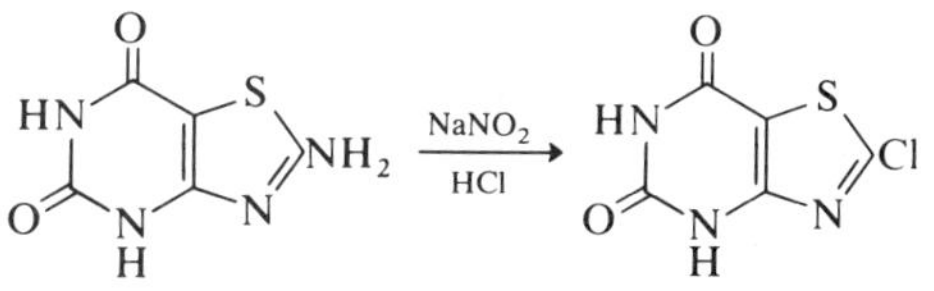

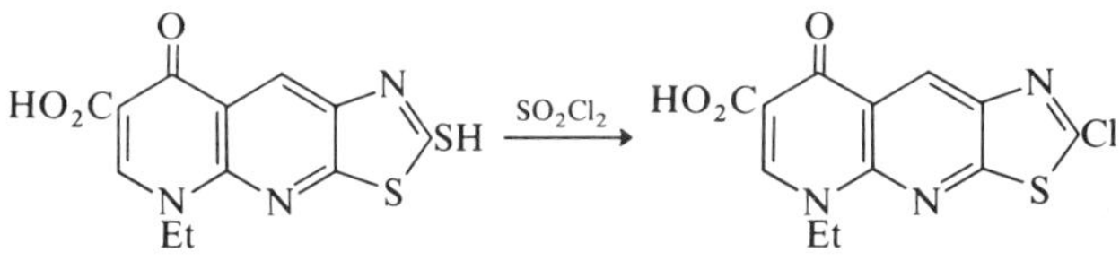

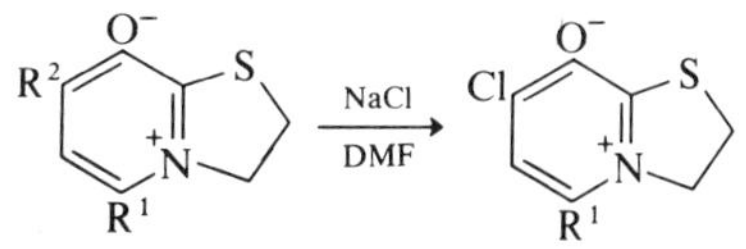

(*e*) *Chemical reduction.* In general, azinium compounds are reduced to di- or tetrahydroazines by metal hydrides. This is also true for fused systems as shown for the isoquinolinium derivative (**373**) ⟨79IJC(B)4⟩. A hydroxy group in the azinium ring, as part of an internal salt, helps to protect this ring towards reduction. Thus Raney nickel desulfurization in the thiazolonaphthyridium derivative (**374**) leads to reduction of the pyridine ring, but the pyridinium ring as an internal salt is not reduced ⟨77TL735⟩. Desulfurization of optically active 2,3-dihydro-8-hydroxy-5-methylthiazolo[3,2-*a*]pyridinium-3-carboxylate gives the corresponding propionic acid (**375**) with retention of optical activity ⟨81H(15)1349⟩. Raney nickel desulfurization of the riboside (**376**) removes the thiazole ring completely ⟨74CPB342⟩. Presumably the first product after the desulfurization is the corresponding *N*-vinyl derivative which is hydrolyzed under the conditions of the reaction.

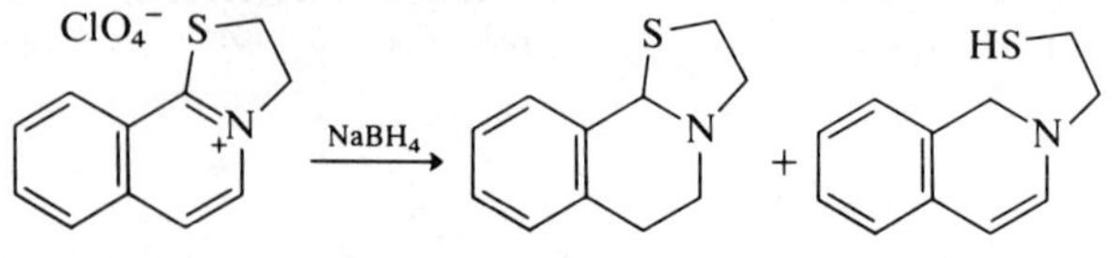

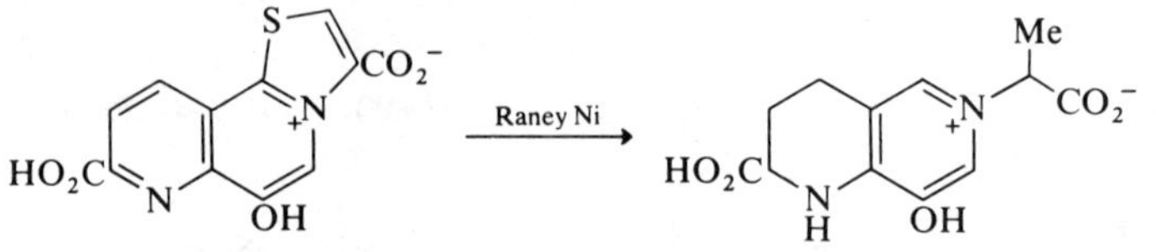

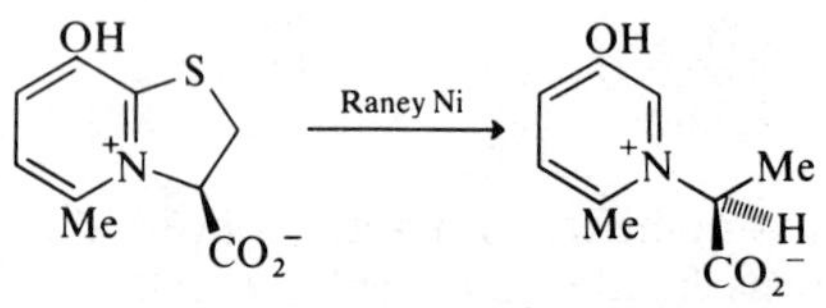

4.29.7.3.6 Nucleophilic attack with ring opening

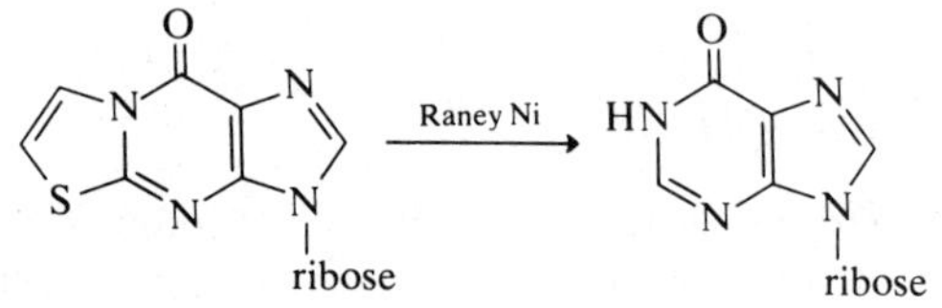

In nucleophilic substitution reactions competition may exist between the normal substitution and attack at a carbon, which eventually leads to opening of a ring. In the absence of suitable leaving groups, ring opening reactions will eventually result. Cations are especially susceptible. Thus the isoquinolinium derivative (**377**) is attacked by active methyl or methylene compounds under carbanionic conditions at the activated isoquinoline 1-position ⟨76CPB1299⟩. The same type of reaction with a hydroxide nucleophile is shown for the pyridazinium derivative (**378**). In the pyrimidine betaine (**379**) ⟨73JHC487⟩ and the triazine betaine (**380**) ⟨73JOC3868⟩ it is the pseudocarbonyl carbon attached to the bridgehead nitrogen which is attacked.

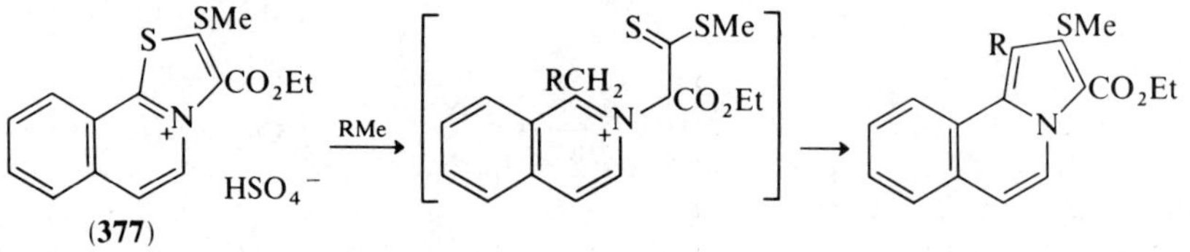

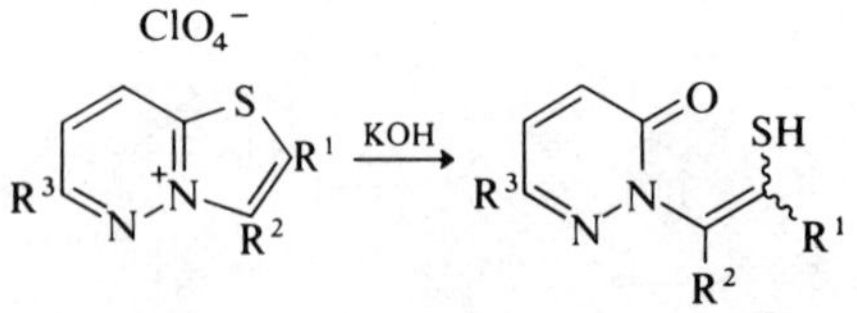

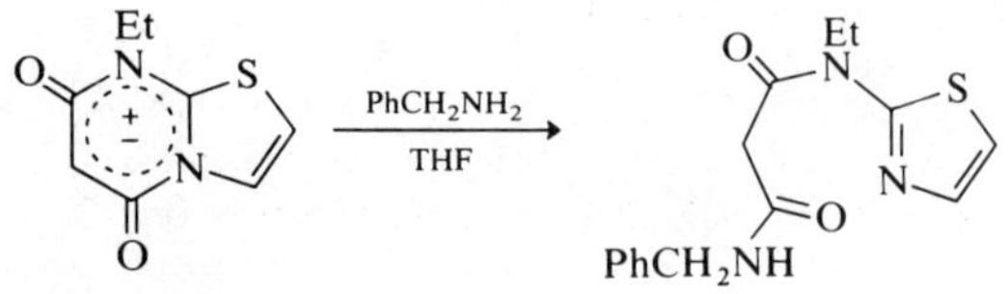

The rather unusual ring opening of the pyrimidinone (**381**) can be rationalized by an intermediate adduct between the amine and the electron-deficient pyrimidinone ring

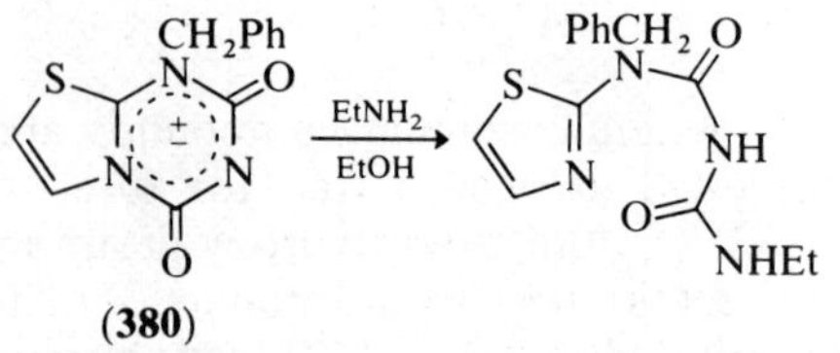

⟨71JCS(C)2094⟩. In the thiazolopyrimidine (**382**) the hydroxy group protects the pyrimidine ring, and it is the thiazole ring which is cleaved with loss of C-2 ⟨70JCS(C)2478⟩.

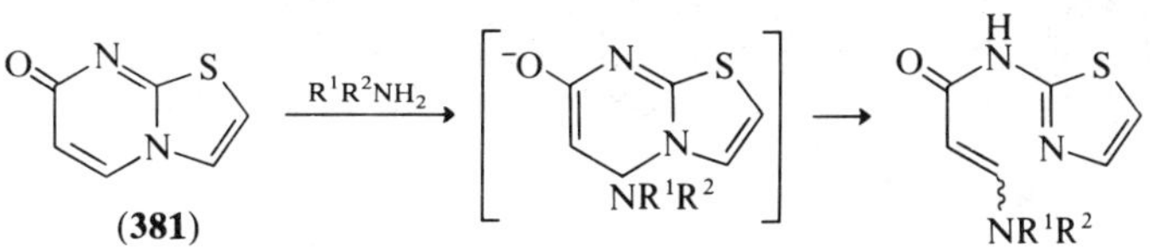

4.29.7.3.7 Nucleophilic attack at hydrogen

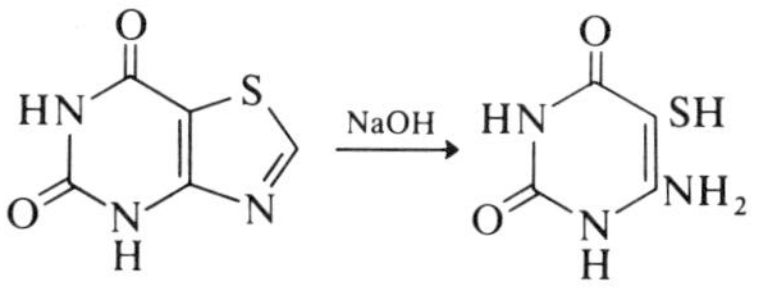

Protons are abstracted with ease in both azinium and azolium rings. In the fused betaine (**321**) it is the protons in the thiazole ring which are the more easily abstracted; deuteration under weakly alkaline conditions proceeds rapidly at both C-2 and C-3. Monodeuteration at C-3 is achieved by deuteration under mildly alkaline conditions of the 2-bromide (**383**) and subsequent hydrogenolysis using zinc in 10% acetic acid. Hydrogenolysis of (**383**) in 10% acetic acid-d_1 gives the 2-deuterated derivative. Dilute and weak d_1-acid is used in order to avoid electrophilic deuteration in the pyridine ring. In 2-methyl or 3-methyl homologues of (**321**) the remaining thiazole proton is selectively exchanged under the above conditions ⟨81H(15)1349⟩.

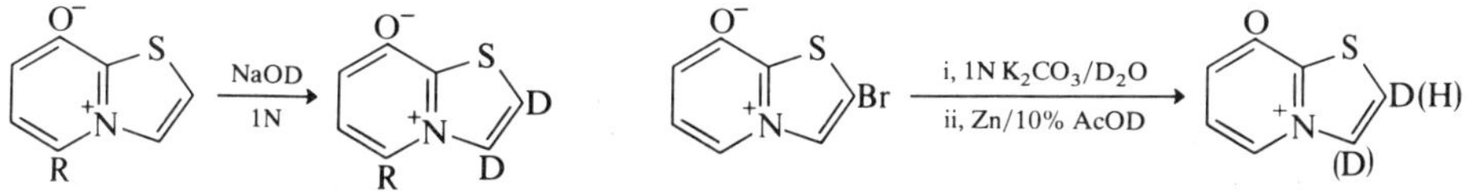

Rate studies of hydrogen–deuterium exchange in thiazolo[4,5-*c*]pyridine (**384**) in MeOH-d_1 show that the proton at C-2, which is in the most active position, is exchanged at almost the same rate as in 5-nitrobenzothiazole and about 10 times as fast as in 5-methylbenzothiazole. The polar effects are said to be transmitted primarily through N-3 ⟨76T399⟩. This suggests that azines with a nitrogen atom in a position *ortho* or *para* to N-3 should be even more activated at C-2.

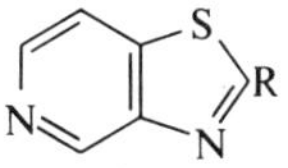

4.29.7.3.8 Other C-linked substituents

The exocyclic reactivity of substituents in fully conjugated rings will be comparable to the behaviour in other azine systems. Thus, an alkyl group in an activated azine position is open for electrophilic attack after deprotonation, especially in a cationic ring. Thus deprotonation of the 5- but not of the 1-methyl group in the isoquinolinium derivative (**385**) leads to an anhydro base as the most likely intermediate for electrophilic attack by a carbonyl electrophile ⟨67JHC71⟩. Nucleophilic substitution in the side-chain proceeds in the normal manner as in (**386**) ⟨79CPB410⟩. Oxidation of the nucleoside (**387**) by metaperiodate in the usual way cleaves the carbohydrate between vicinal hydroxy groups; the resultant dicarbonyl compound can be reduced to the corresponding hydroxy compound (**388**) ⟨75JOC2476⟩.

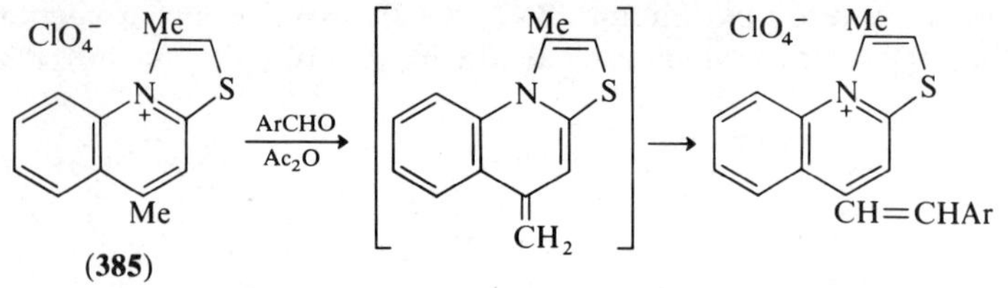

4.29.7.3.9 N-Linked substituents

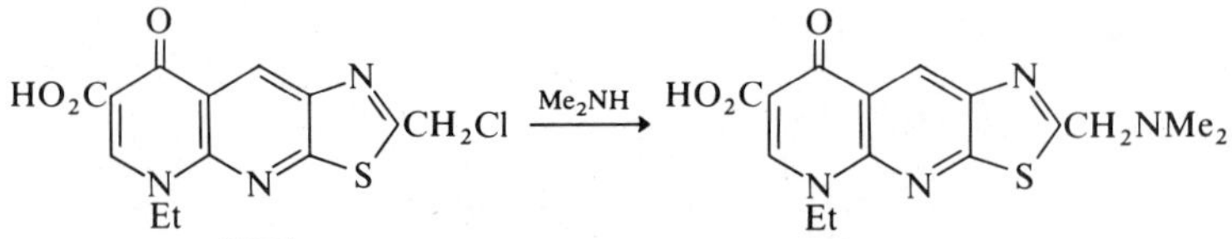

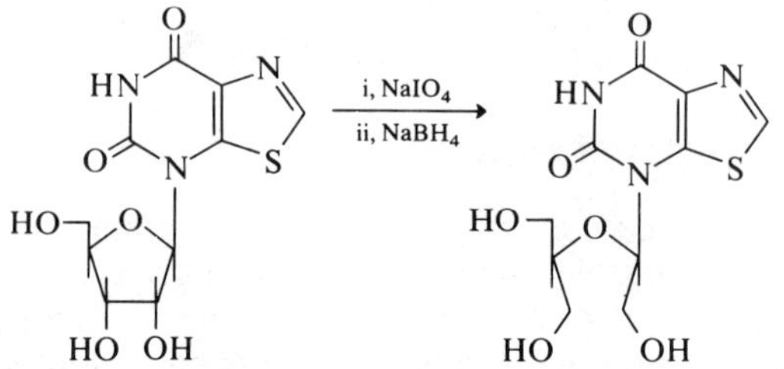

The amino form is the dominating tautomer in the amines in the same way as in the parent monocyclic rings.

The N—O bond in N-oxides is normally readily cleaved by hydrogenolysis. In the case of the thiazolopyrimidine N-oxide (**389**) zinc in acetic acid has been employed ⟨68JHC331⟩.

In azidoazoles and azidobenzazoles equilibria exist between the azido and the tetrazolo forms; the position will depend on the properties of the azole as well as on temperature and solvent. Such equilibria with preference for the tetrazole form also exist in 2-azidothiazolo[5,4-b]pyridine (**390**) and its [4,5-c] isomer ⟨77JHC1045⟩.

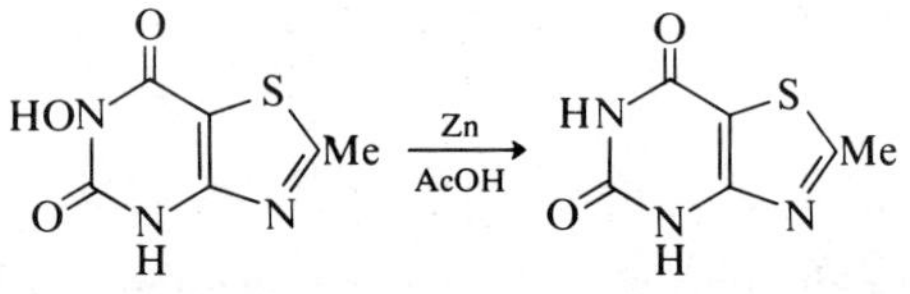

Diazotization of the 8-amino group in the 2,3-dihydrothiazolo derivative (**391**) leads to rearrangement and formation of a fused triazole. The reaction is rationalized in terms of an intermediate pseudobase to overcome the charge in the pyridine ring in the diazonium salt. Ring opening with subsequent cyclization by the diazo group accounts for formation of the fused triazole (**392**). With the amino group in the 6-position as in (**393**), a 1,2-hydride shift in the intermediate pseudobase leads to expulsion of nitrogen and formation of the lactam (**394**) ⟨81H(15)1349⟩.

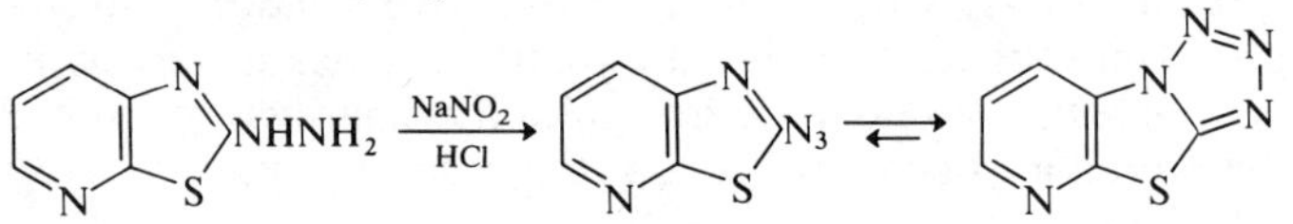

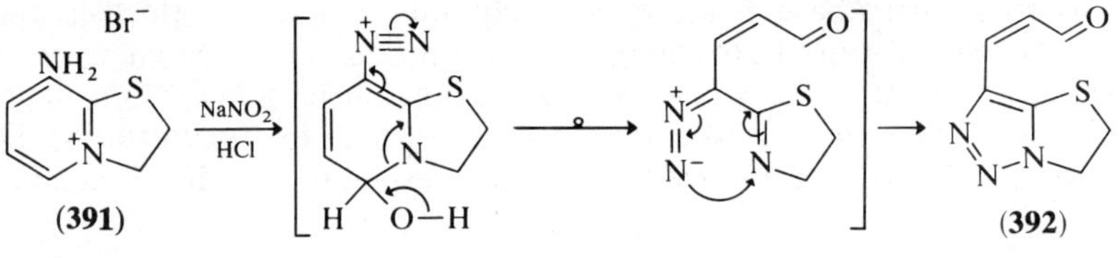

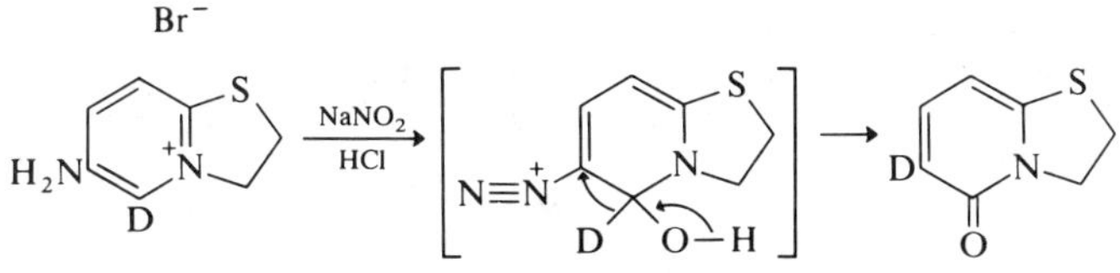

(393) → [intermediate] → (394)

4.29.7.3.10 O-Linked substituents

Hydroxy derivatives are expected to show the same tautomeric preferences for the oxo form in *ortho* and *para* positions to the nitrogen as in the parent monocycle. Thus, for the 2-hydroxythiazole (395) comparison of the spectroscopic data with the data for its *N*-methyl and *O*-methyl derivatives is consistent with the oxo form; the same preference has been shown for its [5,4-*c*] isomer ⟨68CJC691⟩.

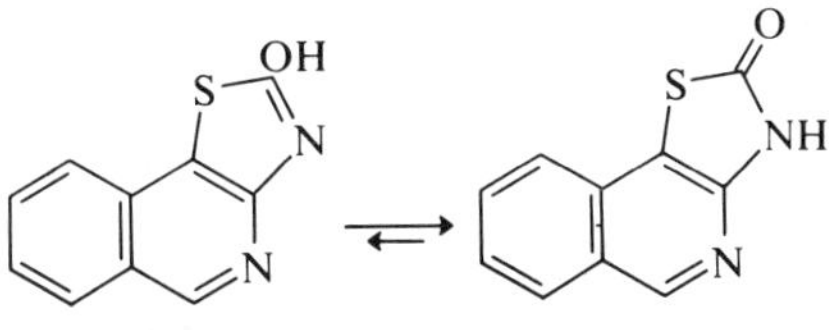

(395)

4.29.7.3.11 S-Linked substituents

The tautomeric preferences of mercapto compounds are to be compared with the behaviour in the respective parent heteromonocycle.

4.29.7.3.12 Reactions with cyclic transition states

Mesoionic compounds may undergo 1,3-dipolar cycloaddition reactions. Thus anhydro-1-hydroxythiazolo[3,2-*a*]quinolinium hydroxide (396) is a substrate for the reaction with DMAD. The formation of the pyrrolo[1,2-*a*]quinoline (397) from this reaction involves COS elimination from the initial adduct. Ethyl propiolate also reacts in the same fashion. The orientation in the cycloadduct can be arrived at from the ylide form (396a). With fumaronitrile, however, the fused pyridinone (398) is formed by loss of sulfur from the primary cycloadduct ⟨78JOC2700⟩.

Fusion of a thiazole to pyrimidine betaines does not change the tendency of the latter for cycloaddition reactions, *e.g.* (306) forms adducts with alkynes ⟨73JHC487⟩. Similarly 1,3-thiazine betaines (399) react as 1,4-dipoles with aryl isocyanate with elimination of COS to produce pyrimidine betaines (400) ⟨76CB3668⟩.

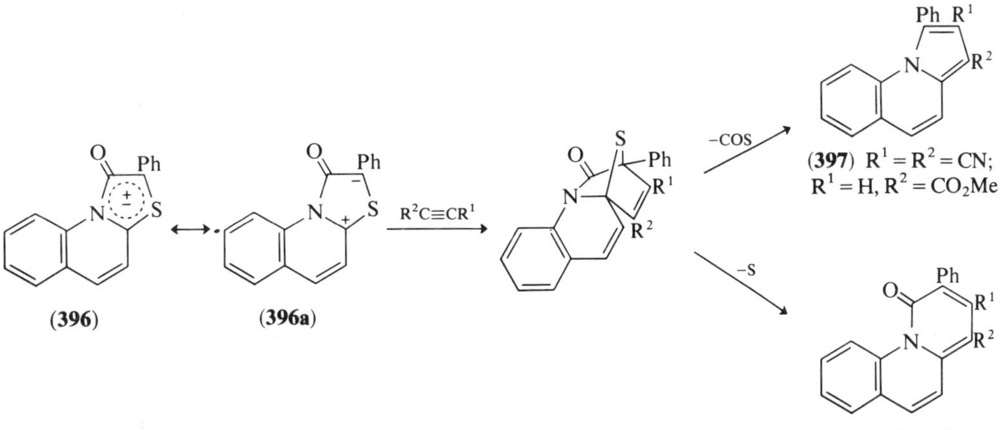

(396) ↔ (396a) → [cycloadduct] → (397) $R^1 = R^2 = CN$; $R^1 = H, R^2 = CO_2Me$

(398) $R^1 = R^2 = CN$

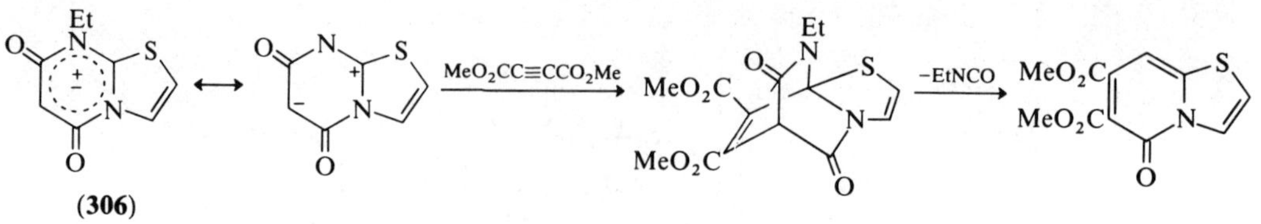

4.29.7.4 Synthesis: Fully Conjugated Rings

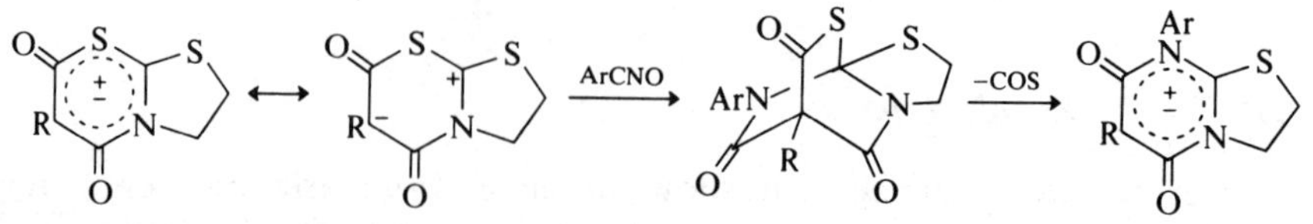

4.29.7.4.1 5H-Thiazolo[3,2-a]pyridine

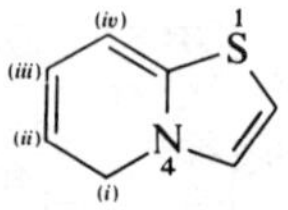

Azine approach. The parent cation and substituted derivatives are available by acid-catalyzed cyclization of 2-β-oxoalkylthiopyridines (**401**) using an acid such as sulfuric, phosphoric or PPA. Chloro or nitro substituents in the pyridine ring do not seriously interfere ⟨66JHC27⟩. The cyclization of 3-hydroxypyridine analogues (**402**) is also at the nitrogen to yield the thiazole derivatives. The cyclization, however, is sensitive to the *peri* interaction between 3- and 5-substituents. In 3,5-dimethyl derivatives (**403**; $R^2 = R^3 = Me$) the steric repulsion is apparent by the unusually low field signals for the methyl protons ⟨81H(15)1349⟩.

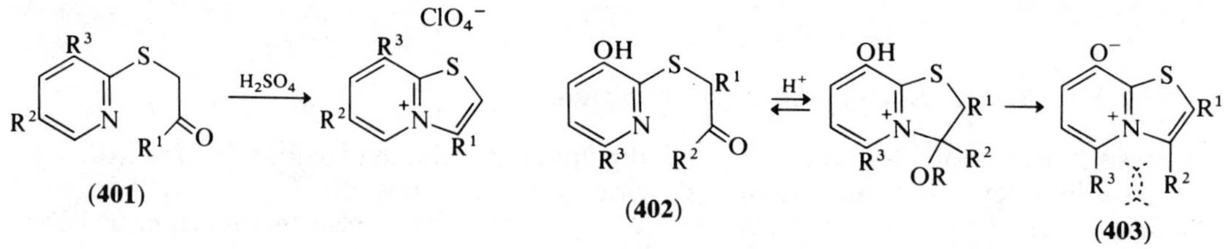

The cyclization can also be effected in the opposite direction, *viz.* at the sulfur atom as in (**404**). For this reaction the sulfur is introduced after the β-oxoalkyl substituent has been attached to the nuclear nitrogen ⟨78JMC489⟩. The intermediate in this cyclization corresponds to the product from the acid-catalyzed Pummerer reaction of the 2,3-dihydrothiazolo sulfoxide (**405**) ⟨81H(15)1349⟩. When the β-oxo group is part of a carboxy group or its derivatives, 3-hydroxythiazolylium derivatives are formed (**408**) ⟨81H(15)1349⟩.

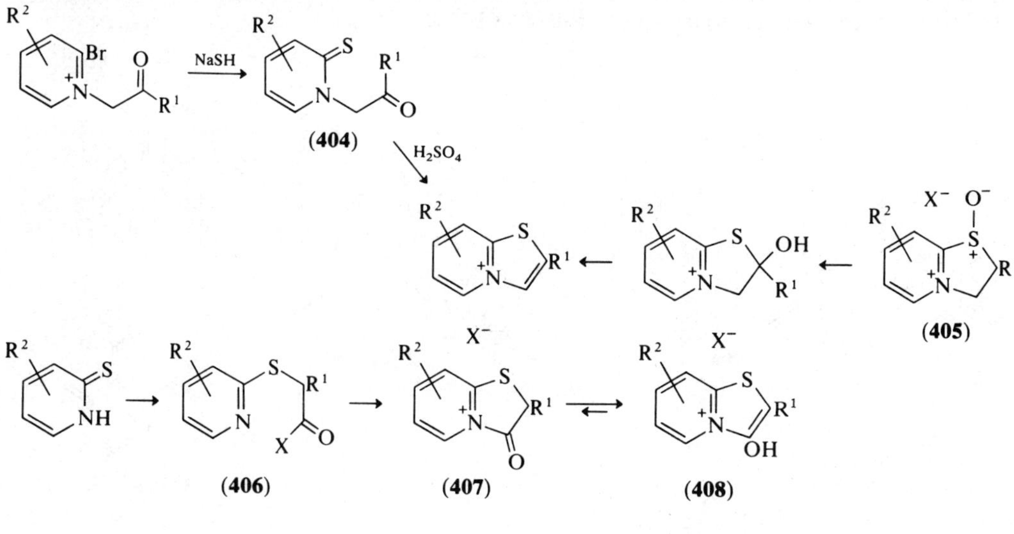

Cyclization on to a nitrile group gives the same type of product (**409**) ⟨78JCR(S)407⟩. The parent mesoion from (**406**; $R^1 = R^2 = H$) was originally synthesized from pyridine-2(1H)-thione and α-bromacetyl bromide, and by cyclization of 2-(2-pyridinethio)acetic acid in acetic anhydride ⟨66BCJ1248⟩.

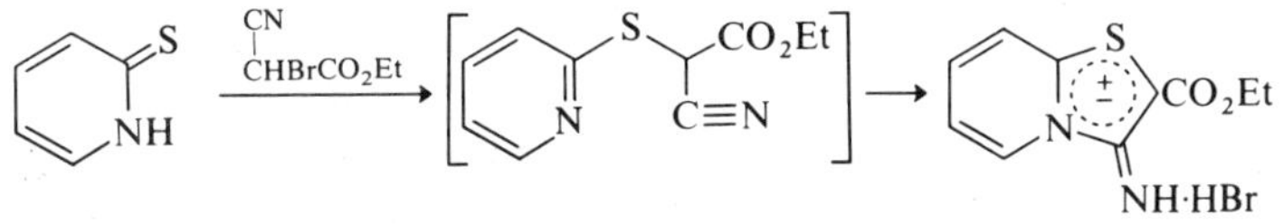

Cyclization occurs directly through catalysis by the acid liberated when a pyridine-2(1H)-thione is heated with an α-halo acid ester. The most convenient method for preparing the thiazole, however, seems to be the cyclization of (2-pyridinethio)acetic acids in acetic anhydride in the presence of pyridine. Without base catalysis the reaction is slow, which suggests a mixed anhydride intermediate. Mixed anhydride formation with ethyl chloroformate in pyridine, or carboxyl activation by DCC in pyridine, gives the mesoionic product. The cyclization reaction and the chemical stability of the thiazole are adversely affected by a pyridine 6-substituent. The initially formed acylpyridinium salt (**407**) undergoes rapid tautomerization to the aromatic thiazole form; equilibrium between the forms (**407**) and (**408**) is verified by rapid deuteration at C-2 ($R^1 = H$) in AcOH-d ⟨81H(15)1349⟩.

Mesoionic sulfur analogues, with the sulfur on C-2, can be prepared by heating a 2-oxo-1(2H)-azine-1-acetic acid, e.g. (**410**), with phosphorus pentasulfide ⟨77JOC2525⟩.

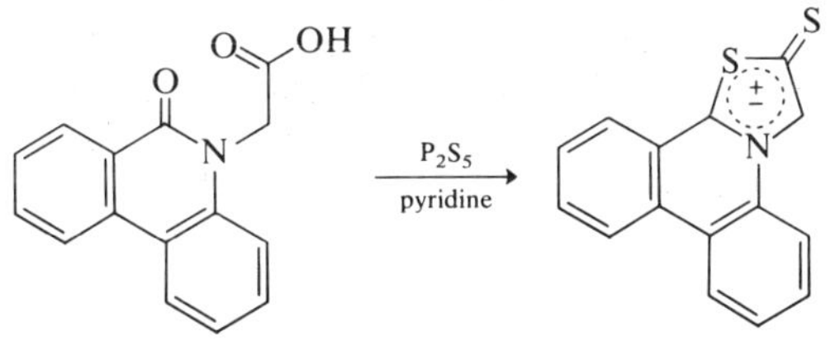

Azole approach. 2,3-Dialkylthiazolium salts (**411**) where the α-carbon atoms carry at least two hydrogens can be made to condense with 1,2-dioxo compounds to form thiazolo[3,2-a]pyridines. A diketone or an oxoaldehyde will yield the cation (**412**) ⟨69AG(E)74⟩. Related is the condensation between a 2-oxo ester and a 2-alkylthiazole as in the synthesis of the 5-lactam (**413**) ⟨74JPR684⟩.

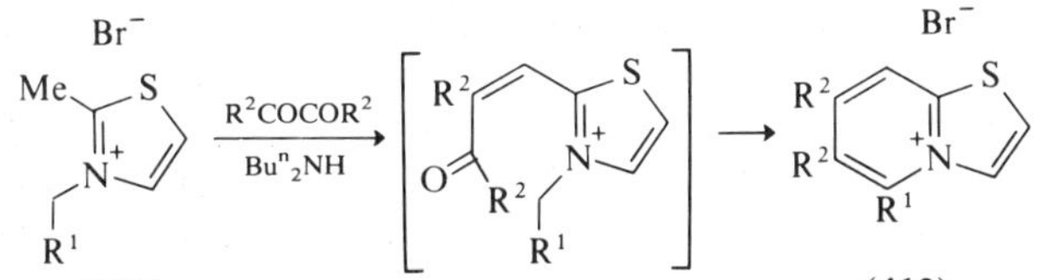

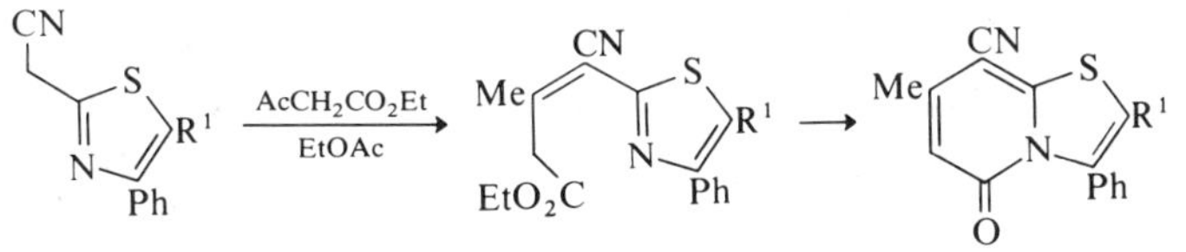

Cycloaddition reactions between thiazoles and DMAD give 1,2-adducts ⟨76TL3463⟩. Three isomers (**414**)–(**416**) are formed in the reaction with 5-phenylthiazole, which has been rationalized as shown.

(i) *5H-Thiazolo[3,2-b]pyridazine*

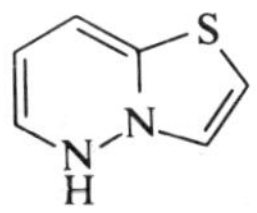

Azine approach. The pyridazines and their cinnoline and phthalazine benzologues can be fused with thiazoles in the same way as pyridine derivatives. The cyclization of the acetonitrile (**417**) is catalyzed by sulfuric acid ⟨77CPB299⟩. The 1,2,4-triazine-3(2*H*)-thione (**418**) reacts with bromoacetone to form the fused thiazole (**419**); cyclization at N-4 apparently is of no importance ⟨79MI42901⟩.

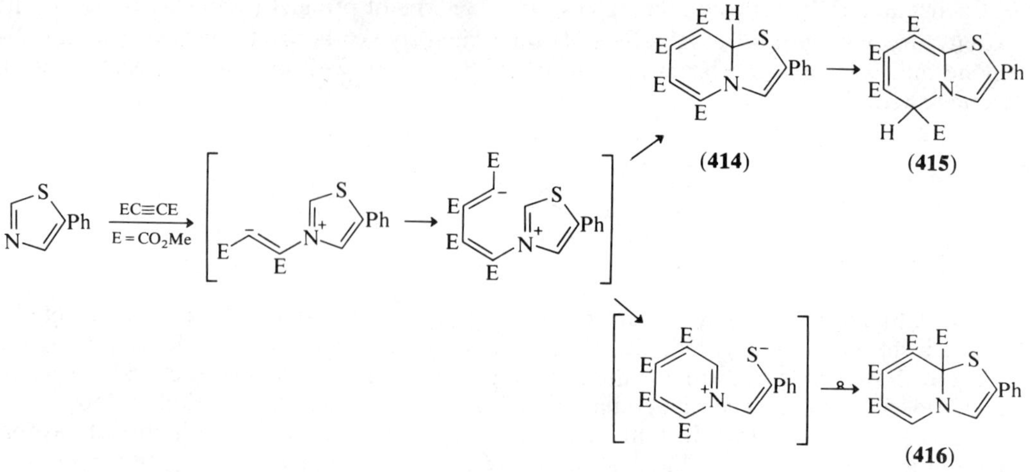

Fused thiazolium betaines can be prepared from dimercaptoazines, *e.g.* as in the reaction of the phthalazine (**420**) ⟨73JAP(K)7391092⟩.

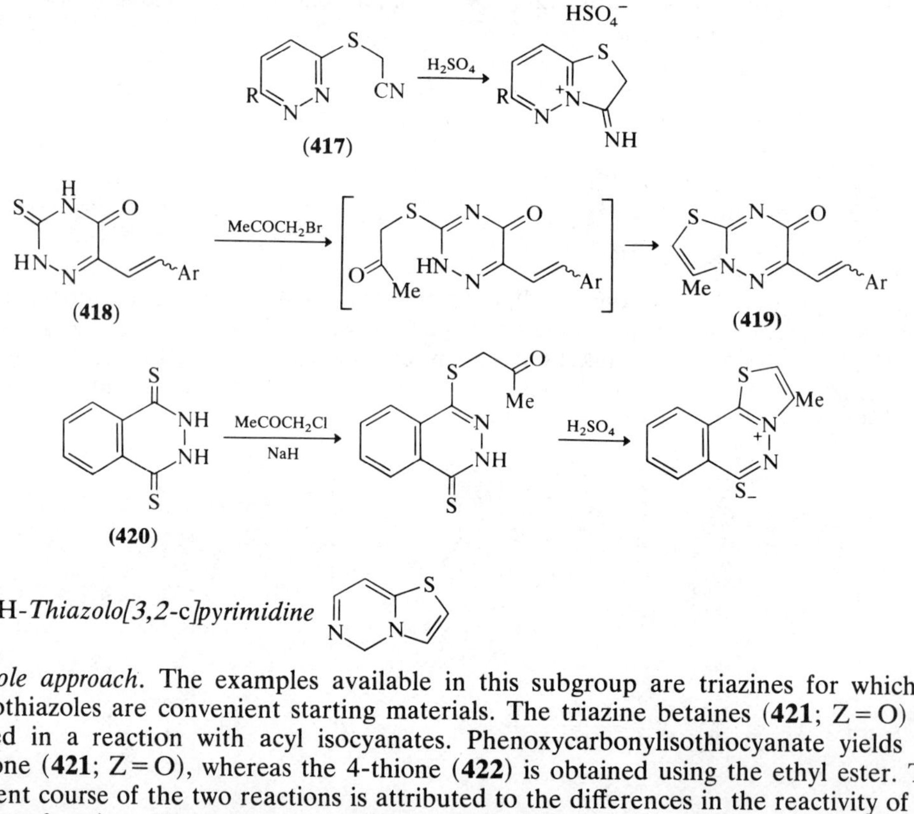

(ii) *5H-Thiazolo[3,2-c]pyrimidine*

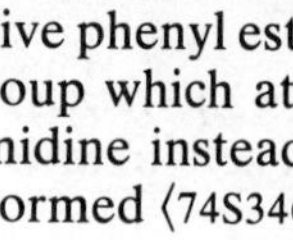

Azole approach. The examples available in this subgroup are triazines for which 2-aminothiazoles are convenient starting materials. The triazine betaines (**421**; Z = O) are formed in a reaction with acyl isocyanates. Phenoxycarbonylisothiocyanate yields the 2-thione (**421**; Z = O), whereas the 4-thione (**422**) is obtained using the ethyl ester. The different course of the two reactions is attributed to the differences in the reactivity of the two ester functions. The more reactive phenyl ester initially reacts with the thiazole nitrogen, whereas it is the isothiocyanate group which attacks the thiazole nitrogen using the ethyl ester reagent ⟨73JOC3868⟩. If an amidine instead is cyclized using a carbonyl reagent such as phosgene, a triazinone (**423**) is formed ⟨74S346⟩.

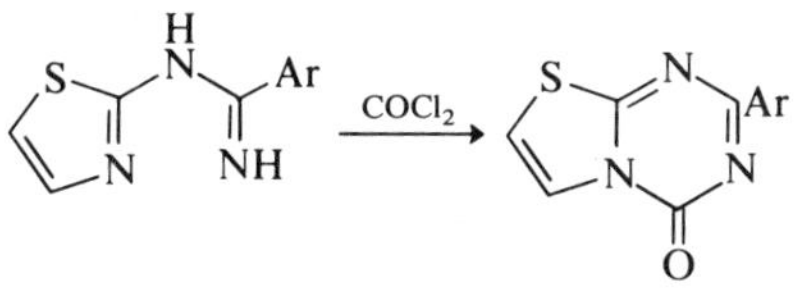

Five-membered Rings Fused with Six-membered Rings

(423)

(iii) *5H-Thiazolo[3,2-a]pyrazine*

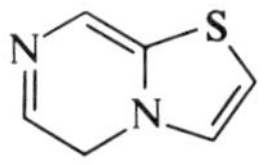

Azole approach. Cyclization of the thiazolo oxime (424) with acid catalysis gives the fused pyridazine 7-oxide (425) in which the oxide bond is cleaved in the usual way by the reaction with phosphorus bromide ⟨69JCS(C)2270⟩. Fused 1,2,4-triazines can be constructed from appropriately substituted thiazole-2-hydrazones as shown for (426) ⟨79MI42901⟩.

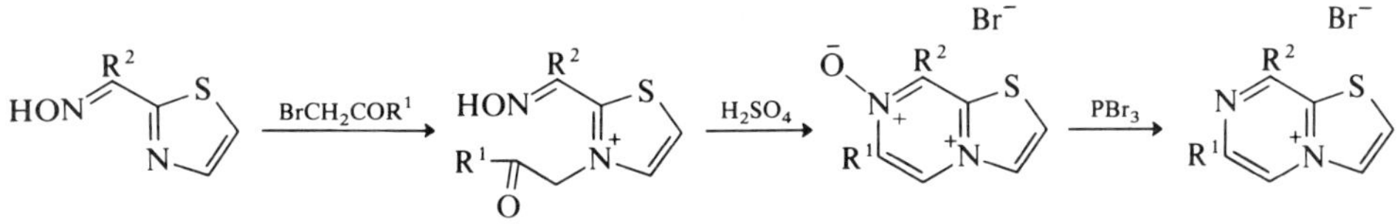

(424)   (425)

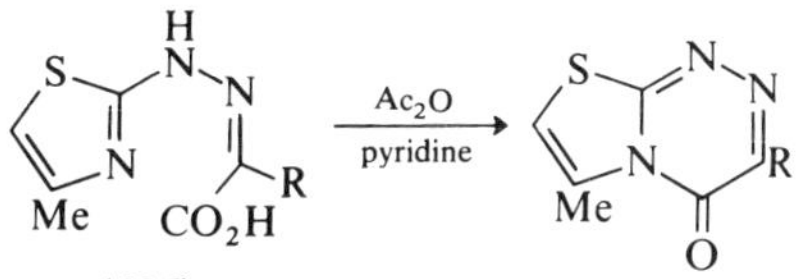

(426)

(iv) *5H-Thiazolo[3,2-a]pyrimidine*

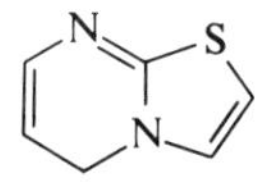

Azine approach. This is the most widely studied pyrimidine system containing a bridgehead nitrogen.

In the cyclization of unsymmetrically substituted 2-(β-oxoalkylthio)pyrimidines a mixture of two isomers may be obtained. Sometimes the choice of reaction conditions coupled with substituent effects may lead to exclusive or almost exclusive formation of the one isomer. Thus in the simple case of (427) preferential cyclization at N-3 results in the absence of a 4-substituent as in (428), whereas almost exclusive formation of the other isomer (429) results in the methyl derivative (R^1 = Me) ⟨67JHC577⟩. In the dichloro analogue (430) both the steric and electronic effects from the vicinal chlorine account for the formation of the 5-oxo product ⟨75ACS(B)1092⟩.

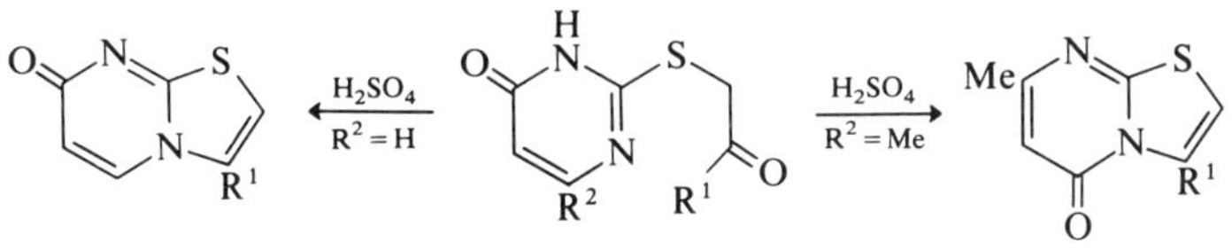

(428)   (427)   (429)

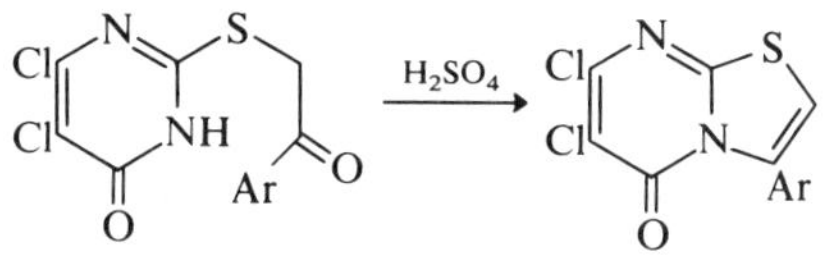

(430)

Benzo homologues react in the same manner. Thus the perimidinethioacetic acid (432) is cyclized to the fused 10-oxo derivative (433) using acetic anhydride in the usual manner ⟨69IJC767⟩. Michael addition of the thiolactam (431) to the activated triple bond in DMAD and subsequent cyclization of the ester group again give a thiazolo[3,2-*a*]perimidin-10(9*H*)-one ⟨76AP928⟩.

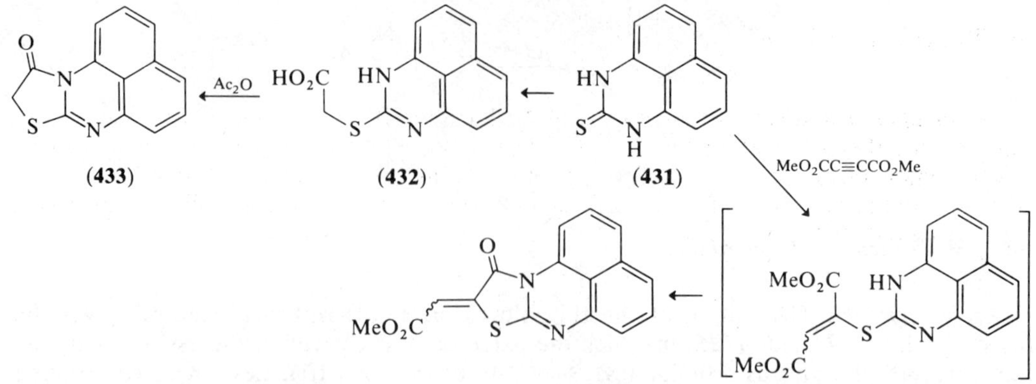

Azole approach. 2-Aminothiazoles are convenient starting materials for the condensation with the missing three-carbon unit in pyrimidine. Examples are found in the literature for β-diketones, malonodialdehyde diethyl acetal and β-halovinylketones or aldehydes ⟨70UKZ483⟩. Equivalent reagents suggest themselves, *e.g.* (ethoxymethylene)malonic esters, which undergo the Michael addition–elimination reaction with 2-aminothiazoles. The resultant products (**434**) are cyclized simply by heating ⟨72JMC1203⟩.

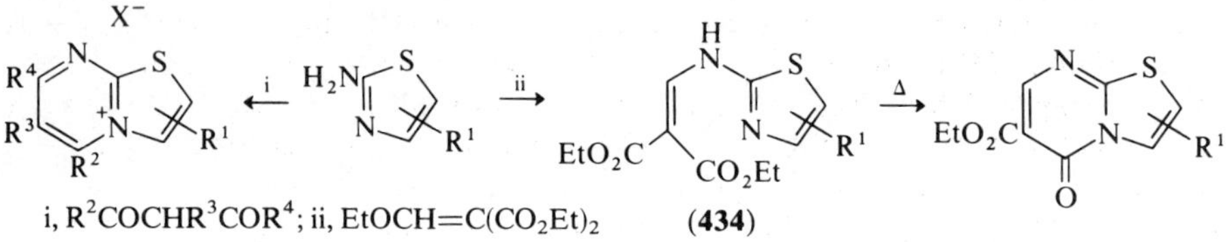

i, $R^2COCHR^3COR^4$; ii, $EtOCH=C(CO_2Et)_2$

4.29.7.4.2 3H-Thiazolo[3,4-a]pyridine

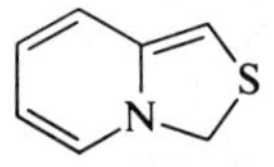

Most compounds within this group are partially saturated and are dealt with in that section.

Azine approach. In the reaction between ethyl bromo(pyridin-2-yl)acetate (**435**) and tetramethylthiourea the first formed tetramethylthiouronium salt is cyclized to the thiazole (**436**). The 3-dimethylamino group is readily hydrolyzed yielding the corresponding lactam ⟨79IJC(B)486⟩.

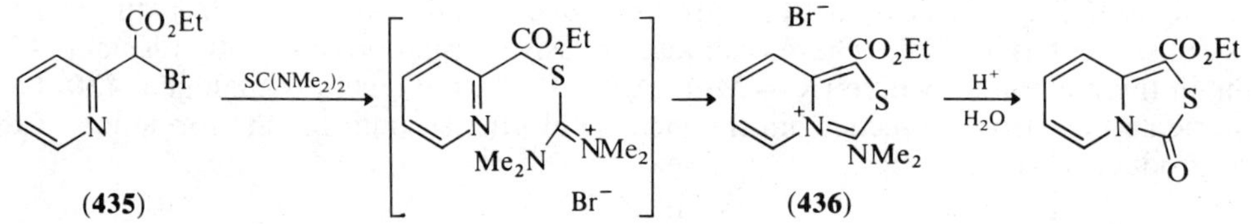

4.29.7.4.3 Thiazolo[4,5-b]pyridine

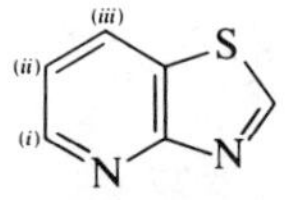

Azine approach. With the heteroatoms in the appropriate vicinal positions in the pyridine nucleus, cyclization with the missing one-carbon unit is by analogy to the preparation of benzothiazoles effected by using a carboxylic acid or its equivalents as for example in the preparation of (**437**) ⟨77USP4038496⟩. The use of a carbonate ester or its equivalents will yield 2-oxy, 2-amino or 2-thio substituted derivatives. 2-Amino derivatives are also available by cyclization of vicinal aminothiocyanates such as (**438**) ⟨66CJC2465⟩.

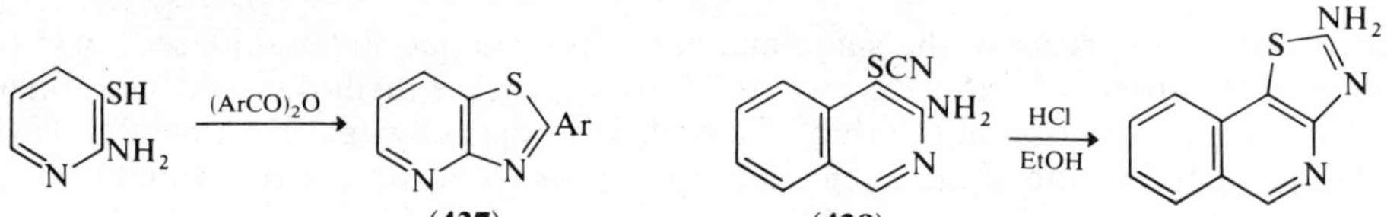

(i) *Thiazolo[4,5-c]pyridazine*

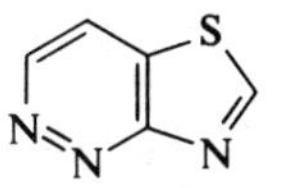

Azine approach. Oxidative cyclization of *N*-thioacetyl derivatives of anilines is a common method for the preparation of benzothiazoles. The reaction may also be applied to the azine analogues, but may proceed less readily because of the decreased nucleophilicity of the ring. In the cyclization of the pyridazine (**439**) the oxidizing agent is alkaline potassium ferricyanide ⟨75JHC337⟩.

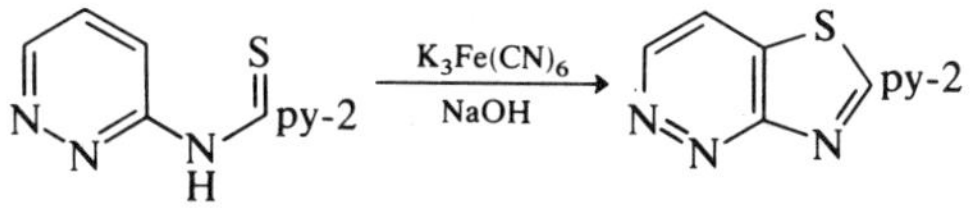

(**439**)

(ii) *Thiazolo[4,5-d]pyrimidine*

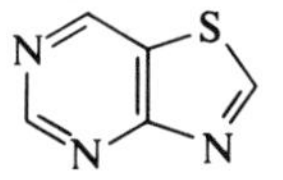

Azine approach. The fused pyrimidines can be synthesized in the same way as the pyridines, *e.g.* by the cyclization of vicinal aminothiocyanates ⟨70JCS(C)2478⟩. Another useful method for aminoazines is the reaction with chlorocarbonylsulfenyl chloride, *e.g.* with the aminopyrimidine (**440**) ⟨73LA1018⟩. The reaction can be rationalized by initial acylation of the amino group which is then cyclized with formation of the 2(3*H*)-one (**441**). Another case is the reaction of the 6-aminouracil (**442**) with thionyl chloride ⟨69JOC3285⟩. The reaction is rationalized as an initial electrophilic substitution at the 5-position of the activated pyrimidine. Subsequently the chlorosulfinyl derivative (**443**) is cyclized to a thiazoline *S*-oxide which loses water to yield the thiazole.

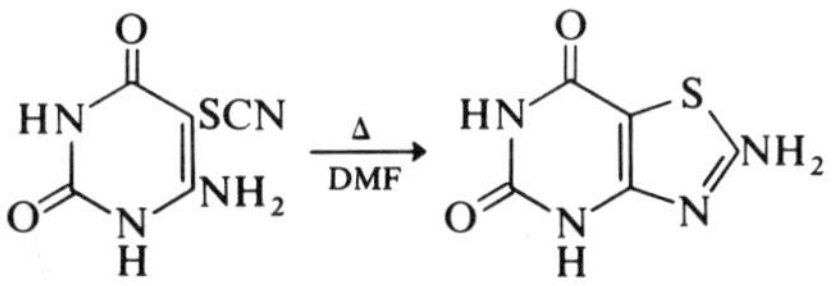

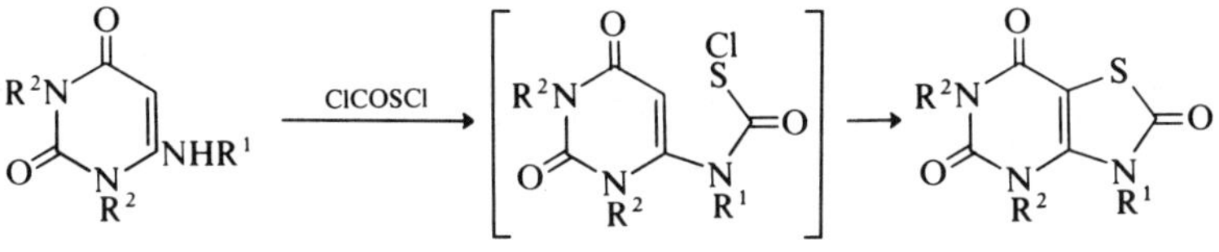

(**440**)   (**441**)

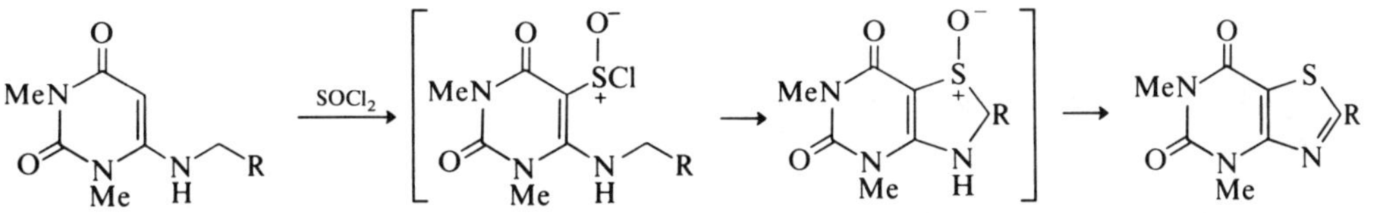

(**442**) R = H, CO$_2$R, CF$_3$ (**443**)

Azole approach. Fused pyrimidines are generally prepared from vicinal aminocarboxamides of carbocyclic or heterocyclic rings by insertion of the missing carbon at the oxidation level of a carboxylic acid. In the synthesis of (**444**) an ortho ester is used ⟨66JPR(304)26⟩. In general, Hofmann degradation of vicinal dicarboxamides leads to fused pyrimidines. In the thiazole (**445**) the 4-amido group undergoes selective degradation and hence the formation of the isomer (**446**).

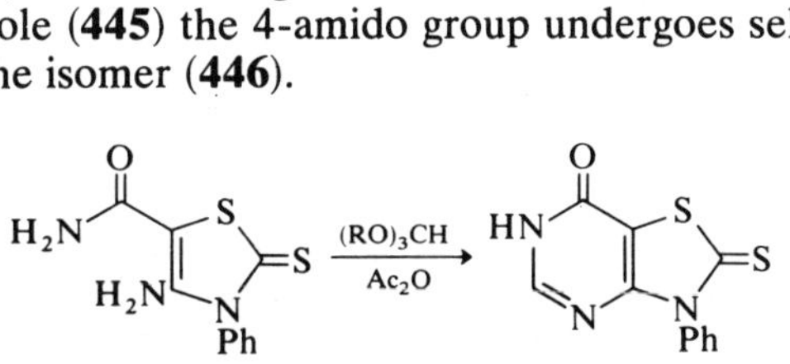

(**444**)

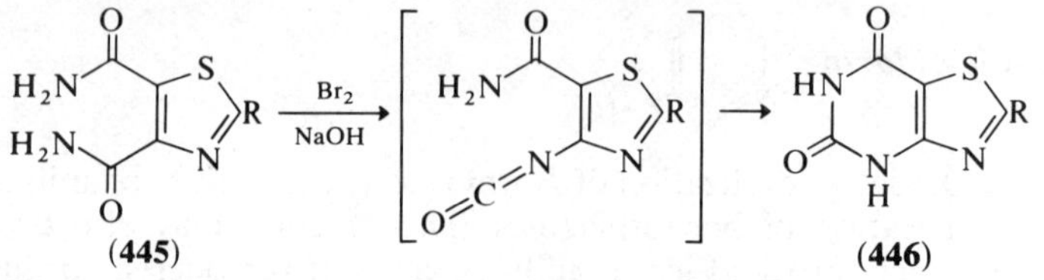

1-Substituted imidazole-4,5-dicarbohydroxamic acids on Lossen rearrangement form 1-hydroxyxanthines. Thiazole-4,5-dicarbohydroxamic acids in their partial Lossen degradation, however, show little differentiation between the 4- and 5-position and hence mixtures of the [4,5-*d*] and the [5,4-*d*] structures (**447**) and (**448**) are formed. The isomer distribution appears to be affected by the solvent as well as by the sulfonyl chloride utilized in the reaction ⟨68JHC331⟩.

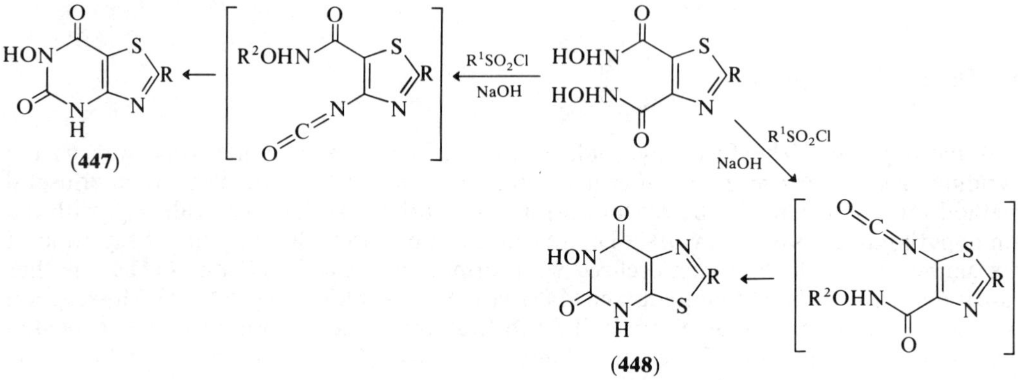

(iii) *Thiazolo[4,5-b]pyrazine*

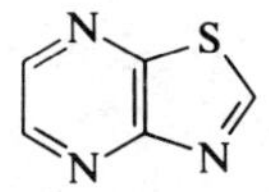

Azine approach. Vicinal dihalides with the halogens in activated azine positions will react with thiourea or thiosemicarbazone to form thiazoles as in the reactions of 2,3-dichloroquinoxaline ⟨77ZC15⟩.

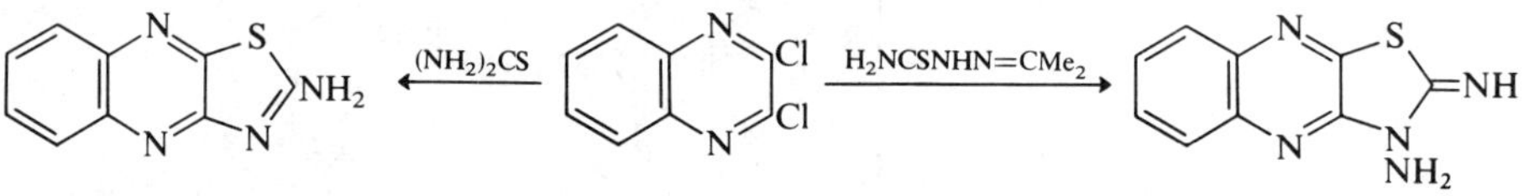

4.29.7.4.4 Thiazolo[4,5-c]pyridine

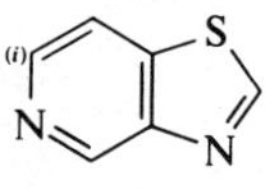

Azine approach. Vicinal amino halides in which the halogen is in an activated azine position, *e.g.* (**449**), can form a thiocarbamic acid ⟨77JHC1045⟩, or an equivalent derivative, *e.g.* (**450**) ⟨77JHC129⟩. This intermediate subsequently undergoes cyclization by nucleophilic displacement of the halogen substituent.

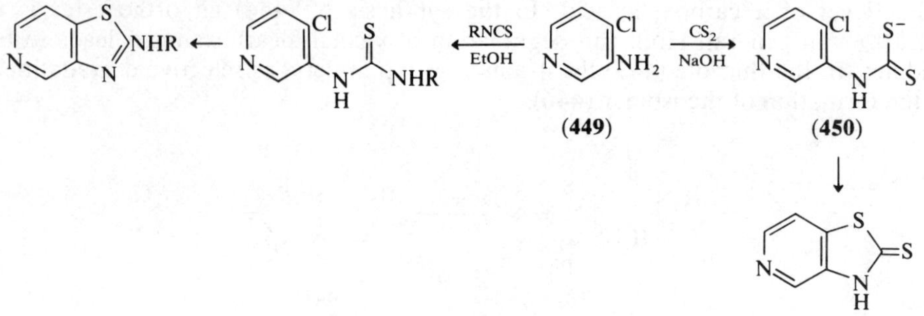

(i) *Thiazolo[4,5-d]pyridazine*

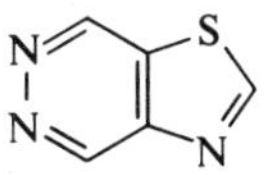

Azole approach. Like phthalazinones, which can be formed from the reaction between phthalic acids and hydrazines, the thiazole analogues can be formed from thiazole-4,5-dicarboxylic acid derivatives. Other 4,5-substituted oxo groups react similarly to give a fused thiazole (**451**) ⟨71BSF1491⟩.

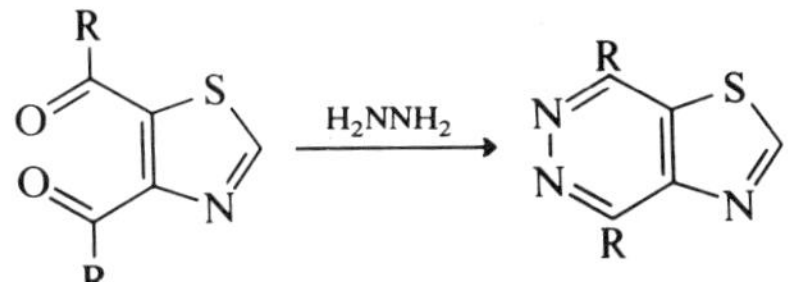

(**451**)

4.29.7.4.5 Thiazolo[5,4-c]pyridine

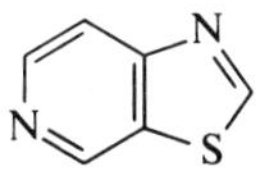

Azine approach. In principle several of the methods used in the previous series can be applied to the preparation of representatives of this system. In the example given, *in situ* generation of a 4-acylamino-3-mercaptopyridine from the protected sulfide (**452**) is followed by cyclization ⟨77JMC1572⟩.

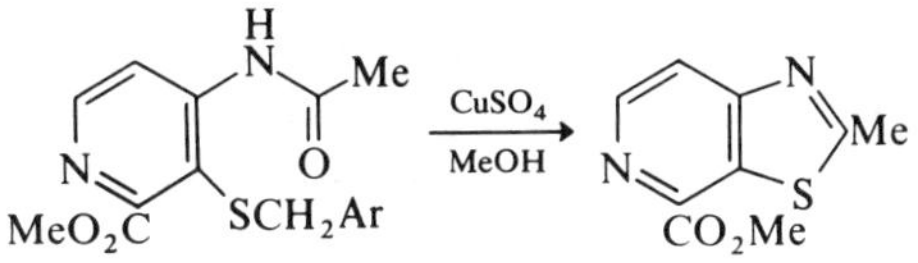

(**452**)

4.29.7.4.6 Thiazolo[5,4-b]pyridine

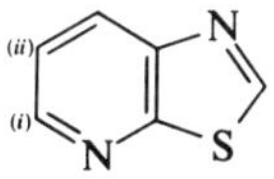

Azine approach. As in the other series, insertion of a one-carbon unit between vicinal amino–thiol groups leads to fused thiazoles. An acid derivative or its equivalent such as an ortho ester may be used; for derivatives without a 2-substituent N,N-dimethylformamide dimethyl acetal is successful ⟨74S120⟩. If a xanthate is the reagent employed, a 2-mercapto derivative, *e.g.* (**453**), results ⟨79CPB410⟩. A 2-mercapto derivative, *e.g.* (**454**), is also available by the reaction of a 3-amino-2-chloroazine with carbon disulfide *via* nucleophilic cyclization of the intermediate thiocarbamic acid ⟨75JPS1371⟩. 2-Amino derivatives can be prepared by reacting an aminoazine, *e.g.* (**455**), with thiocyanogen; the latter attacks as an electrophile at the position next to the amino group, and the cyanothio derivative subsequently undergoes cyclization ⟨73JOC4383⟩.

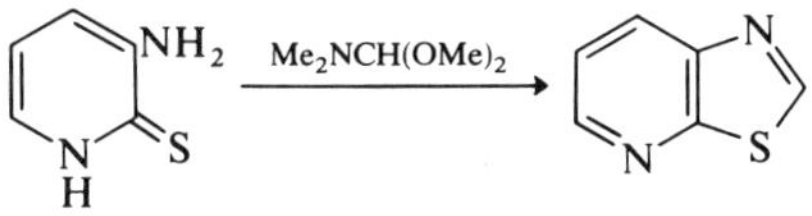

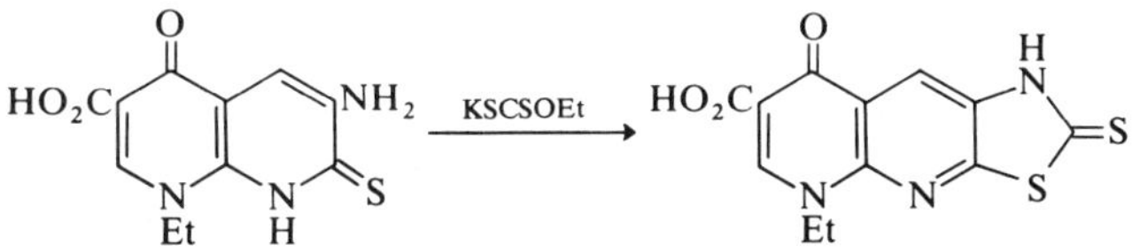

(**453**) 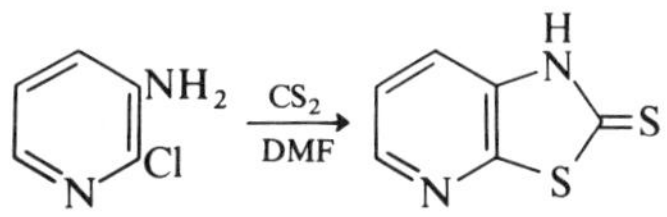   (**454**)

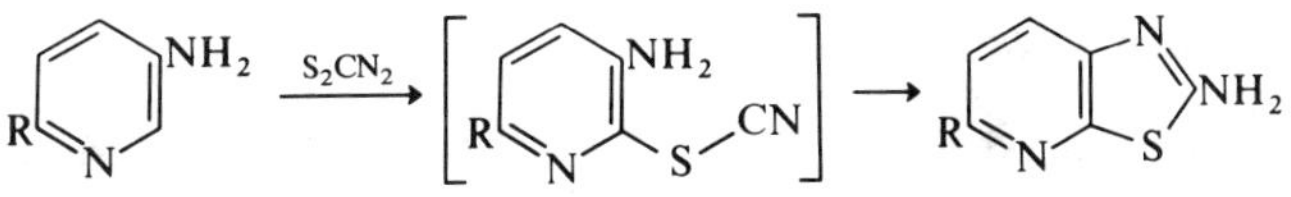

(**455**)

(ii) *Thiazolo[5,4-d]pyrimidine*

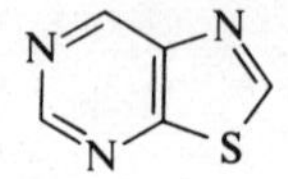

Azine approach. Cyclization of 5-amino-4(6)-pyrimidinethiones (**456**) proceeds in the usual way for this type of substituent arrangement. With phosgene, 2-oxo derivatives result and with xanthates, 2-thio derivates are formed ⟨68CPB741⟩. In (**457**), which has an amino and a mercapto group in the vicinal position to the 5-benzamido group, exclusive thiazole formation results with PPA ⟨65JOC1916⟩.

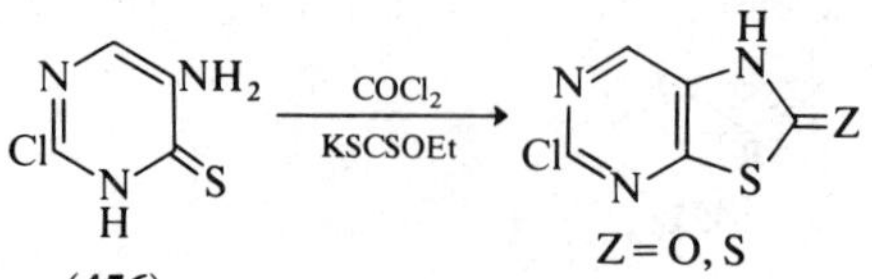

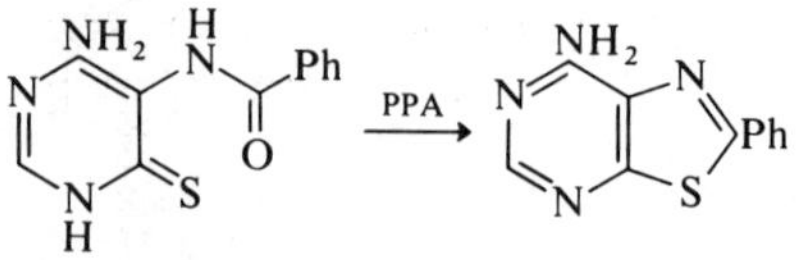

Azole approach. Compounds of this type may be formed in the Lossen rearrangement of thiazole-4,5-dicarbohydroxamic acids to give compounds (**447**) and (**448**).

4.29.7.5 Saturated and Partially Saturated Rings: Introduction

The main interest in saturated or partially saturated compounds has been in systems having a bridgehead nitrogen atom, in particular, reduced forms of 5*H*-thiazolo[3,2-*a*]pyridines.

'Aromatic' reactions in partially reduced systems where the azine ring is fully conjugated are discussed in the appropriate subsections of 'Reactivity' (Section 4.29.7.3). Only the topics characteristic of the reduced systems are surveyed below.

Structural features (X-ray and molecular spectra) are also described in Section 4.29.7.2 for fully conjugated rings.

4.29.7.6 Saturated and Partially Saturated Rings: Structure

4.29.7.6.1 Thermodynamic aspects

Saturated bicyclic hetero systems possessing a nitrogen atom at the bridgehead position exist in solution as an equilibrium mixture of one *trans*-fused and two *cis*-fused conformers interconvertible by nitrogen inversion and ring inversion. This is illustrated for perhydrothiazolo[3,4-*a*]pyridine, which is capable of existing as an equilibrium mixture between the three forms (**458**). The *cis* conformer (**c**) is destabilized relative to the *trans* conformer by two *gauche*-butane interactions, one *gauche*-propanol-type interaction and by the unfavourable generalized anomeric effect. The latter also destabilizes (**a**) relative to the *cis* conformer (**b**). In ^{1}H NMR studies conformer (**c**) is not seen, the equilibrium being between conformer (**a**; 64%) and conformer (**b**; 36%). The NMR studies also show that the conformational equilibria are little affected by solvent changes ⟨80OMR(13)159⟩.

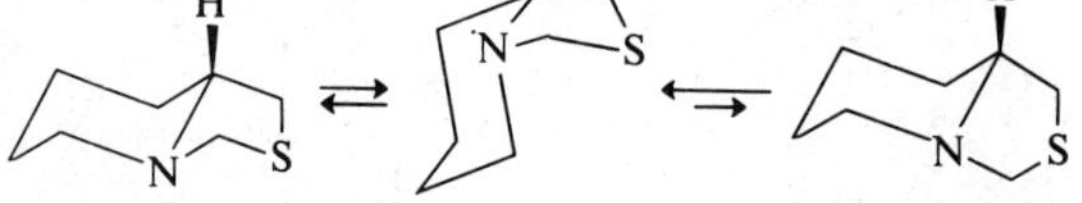

4.29.7.7 Saturated and Partially Saturated Rings: Reactivity

4.29.7.7.1 Electrophilic attack at sulfur

Peracid oxidation of fused dihydrothiazoles yields the corresponding sulfoxides or sulfones. In the oxidation of anhydro-8-hydroxy-2,3-dihydrothiazolo[3,2-*a*]pyridinium hydroxides, sulfoxides are formed. The 3-carboxy derivative (**459**) gives the sulfoxide diastereoisomers in the ratio 9:1. The stereoselectivity in the oxidation is further enhanced in the 2-methyl-3-carboxy analogue. The dominating stereoisomer has the sulfoxide group *cis* with respect to the carboxy group. Therefore the 3-carboxy group and the peracid interact in such a way that the oxidation preferentially takes place from the carboxy side of the ring. The stereoselectivity in the oxidation of 3-carboxy derivatives can be utilized in the preparation of enantiomeric sulfoxides. Thus the performic acid oxidation of (*R*)-(**459**) gives mainly the sulfoxide (1*S*,3*R*)-(**460**), which is readily decarboxylated in acetic acid, due to the activating effect of the pyridinium nitrogen, to the (*S*)-sulfoxide (**461**) ⟨81H(15)1349⟩. In the oxidation of the penicillamine-derived analogue (**462**), in which there is no formal charge separation, stereoselectivity in the oxidation was not reported ⟨71T3447⟩.

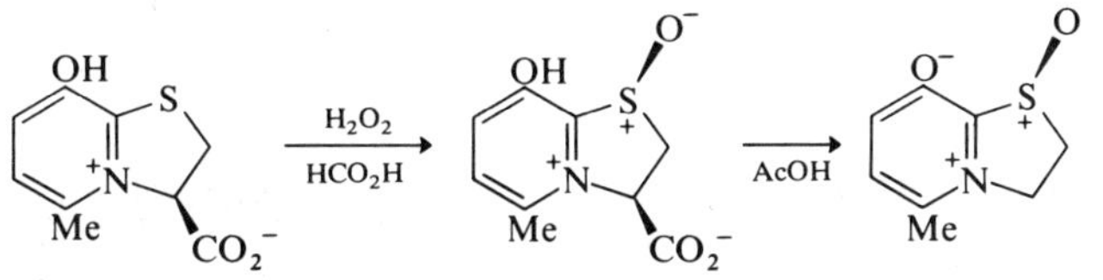

The ease of oxidation depends on the electron availability on the sulfur. In quinoline and pyrimidine analogues of (**459**) the rate of the reaction is decreased, and in these betaine systems sulfone formation is not observed using peracids in the oxidations. A 2-carboxy group as in (**463**; R = H) promotes the Pummerer-type rearrangement. The initially formed hemimercaptal mainly eliminates water to give the thiazole (**464**); a minor product (**465**) may be formed by ring opening ⟨81H(15)1349⟩.

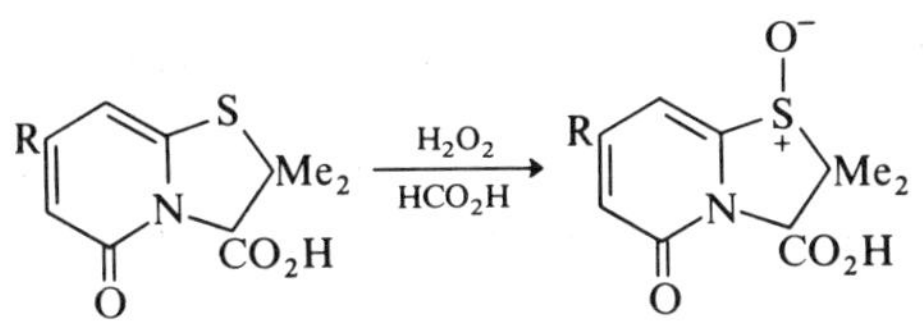

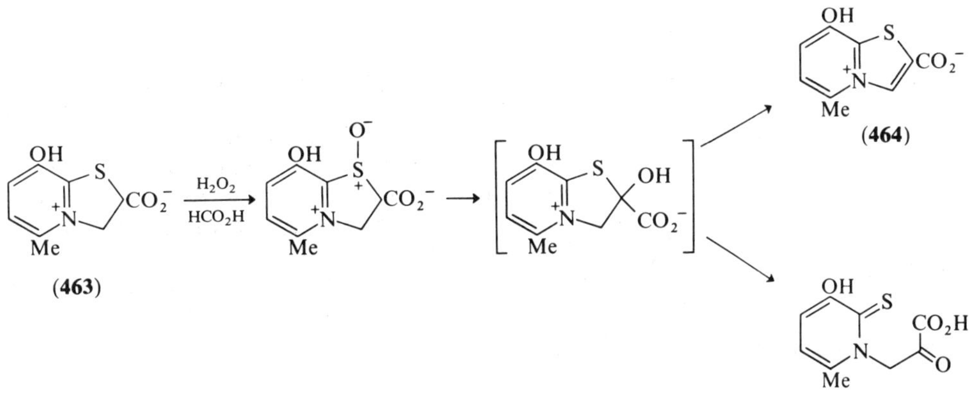

Thioethers of azines are generally oxidizable to sulfones. By analogy (**466**) can be oxidized to its sulfone and the intermediate sulfoxide may be isolated. The latter, when subjected to the Pummerer rearrangement, yields the 2-acetoxy derivative (**467**), which on acid catalysis eliminates acetic acid to form the thiazole ⟨79CPB1207⟩.

4.29.7.7.2 Nucleophilic attack with ring opening

Thiazolo[3,2-*a*]pyridines and their higher azine analogues are normally resistant to opening of the thiazole ring because of aromatic stabilization. 2,3-Dihydrothiazolo derivatives, however, have properties reminiscent of alkyl thioethers in an activated azine position; therefore they may be vulnerable to nucleophilic ring openings as illustrated with the fused

(466)

(467)

pyrimidines (**468**) and (**469**) ⟨70ACS1423⟩ and the fused pteridine (**470**) ⟨77ACS(B)167⟩. With a thiolate as nucleophile the preferential attack may be at the sp^3-hybridized carbon rather than at the azine position, an example being the reaction of the 2,3-dihydrothiazolo cation (**471**) with thiophenolates ⟨65JHC97⟩.

(468)

(469)

(470)

(471)

(472)

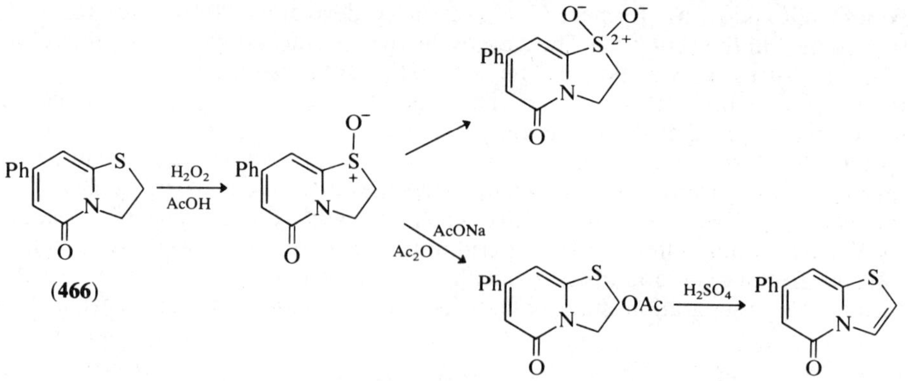

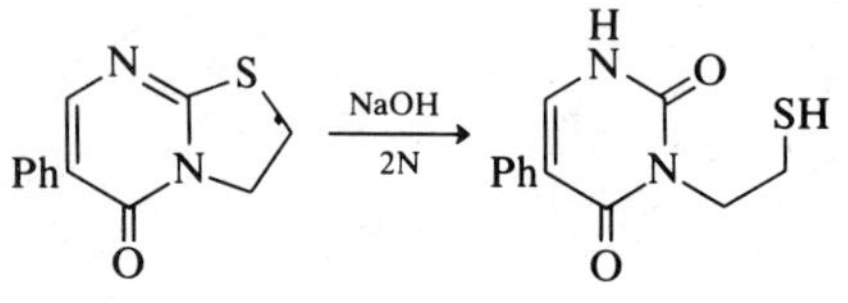

When aromatic stabilization in the azine is blocked, as in the 6,6-disubstituted pyrimidine (**472**), preferential azine cleavage may result ⟨79JHC903⟩.

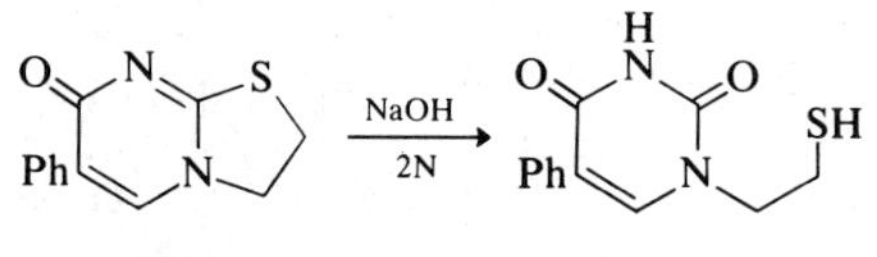

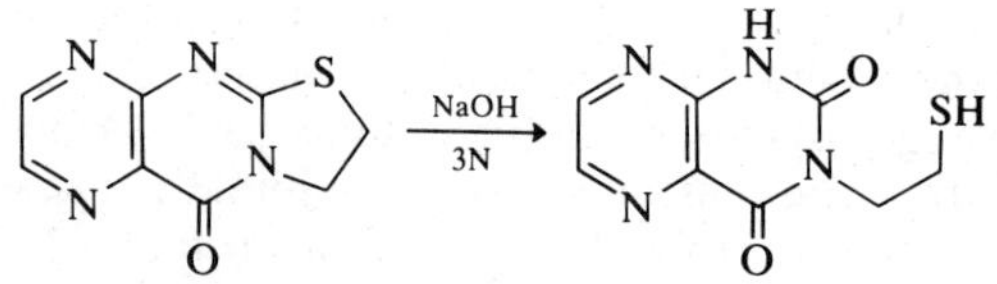

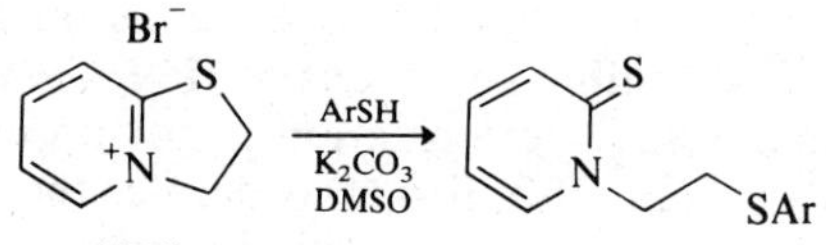

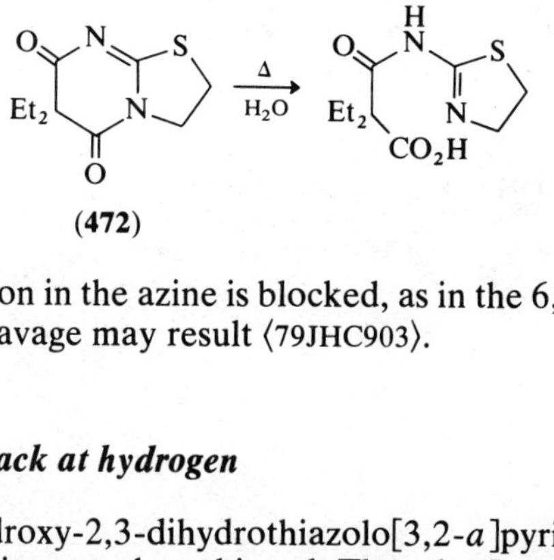

4.29.7.7.3 Nucleophilic attack at hydrogen

Studies of anhydro-8-hydroxy-2,3-dihydrothiazolo[3,2-*a*]pyridinium hydroxide systems show that selective deuterations can be achieved. Thus the 5-methyl group in (**473**) can be selectively deuterated *via* its anhydro base, the difficulty being to avoid deuteration at C-3. A proton at C-5 (**474**) is exchanged under mild alkaline conditions. Deuteration at C-3 is best effected *via* the 3-carboxy derivative (**475**) which readily exchanges the remaining H-3 and is subsequently decarboxylated in the presence of deuterium oxide. Deuteration at C-2, which is the least reactive position, is best carried out before cyclization ⟨81H(15)1349⟩.

(473)

(**474**) R = O⁻, NH⁻

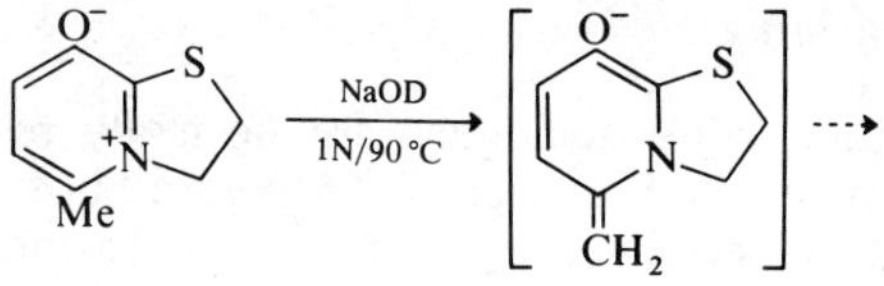

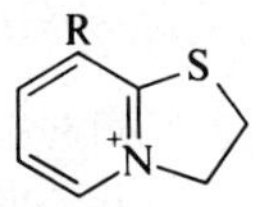

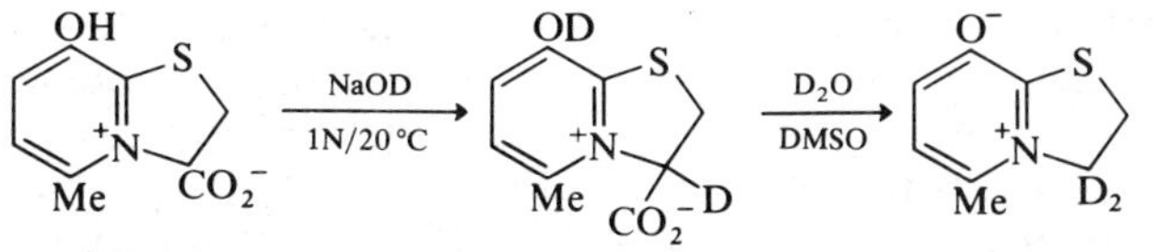

An anion at C-3 (**480**) which is the intermediate in the deuterium labelling at this carbon, in the absence of a proton source, will give rise to a ring opening reaction with the formation of an *N*-vinyl-2-pyridinethione (**481**). Also the protons at C-2 are activated by the sulfur which partly carries the positive charge of the ring system. Proton elimination from C-2 as in (**477**) and ring opening lead to a 2-vinylthiopyridine (**478**). The preferential route of elimination is sensitive to the steric and electronic effects of the substituent. Thus, from the parent betaine (**476**; $R^1 = R^2 = R^3 = H$) 98% relative yield of the *N*-vinyl isomer is formed. The 2-methyl derivative gives only the *N*-vinyl isomer, whereas the 3-methyl derivative gives 70% relative yield of the *S*-vinyl isomer (**478**). A carbanion at C-3 can be specifically generated from a 3-carboxy derivative (**479**), in which case exclusive formation of the *N*-vinyl isomer (**481**) can be achieved ⟨81H(15)1349⟩.

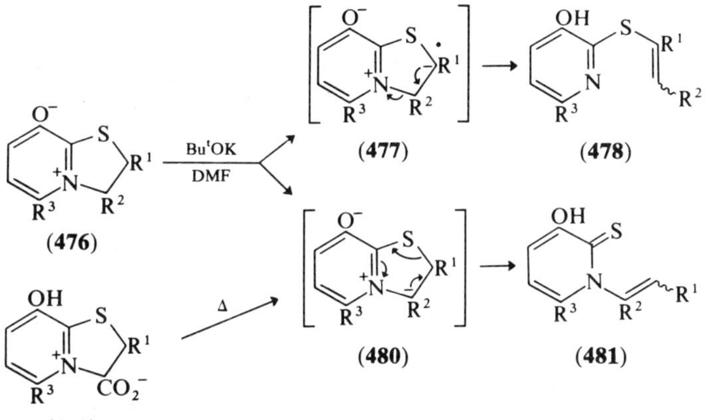

The 3-bromomethyl derivative (**482**), after anion formation at C-3, has the choice of the ring opening reaction as above or elimination of the bromide. The latter pathway is preferred, giving the 3-methylene derivative (**483**) which very readily isomerizes to its aromatic thiazole isomer (**484**) ⟨81H(15)1349⟩. Similarly, the 2-bromomethyl derivative (**485**) on treatment with morpholine undergoes elimination to give the 2-methylene derivative, which with acid catalysis undergoes a prototropic shift to the 2-methyl thiazole (**486**) ⟨74M882⟩.

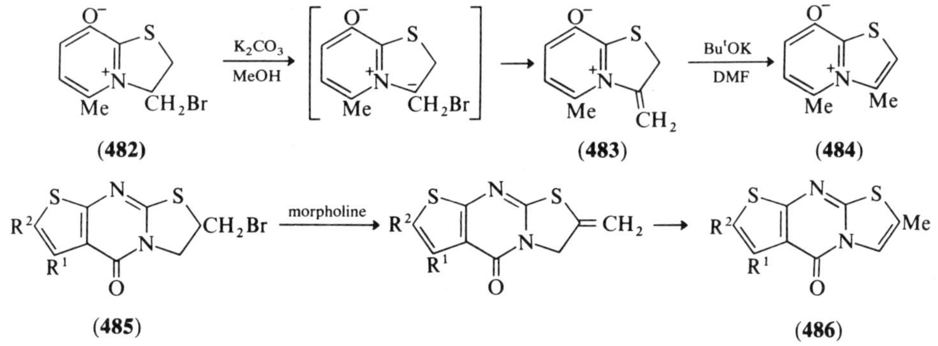

4.29.7.8 Saturated and Partially Saturated Rings: Synthesis

4.29.7.8.1 5H-Thiazolo[3,2-a]pyridine

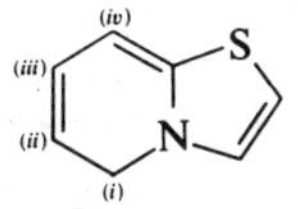

Azine approach. Fused 2,3-dihydrothiazoles are available by cyclization of 2-alkylthiopyridines carrying a leaving group such as a halogen, a hydroxy group or an ester thereof, or some amino function on the β-alkyl carbon atom ⟨81H(15)1349⟩. The parent cation (**487**) was first obtained by the reaction between 2-mercaptopyridine and 1-bromo-2-chloroethane ⟨65JHC97⟩. The reaction of a 1,2-difunctional ethane with a 2-mercaptopyridine initially takes place at the more nucleophilic sulfur. The sulfur is now in a position to exert anchimeric assistance in the cyclization reaction (**489**). Since the pyridine nitrogen atom can attack at either carbon in an episulfonium intermediate, rearranged products can be obtained. The course of the reaction is largely controlled by the choice of reaction conditions. Usually the cyclization occurs directly at the nitrogen atom and therefore does not involve an episulfonium intermediate. An aryl group, however, which stabilizes a carbenium ion, has a tendency to end up at C-3 in the cyclized product, irrespective of its position before the cyclization reaction.

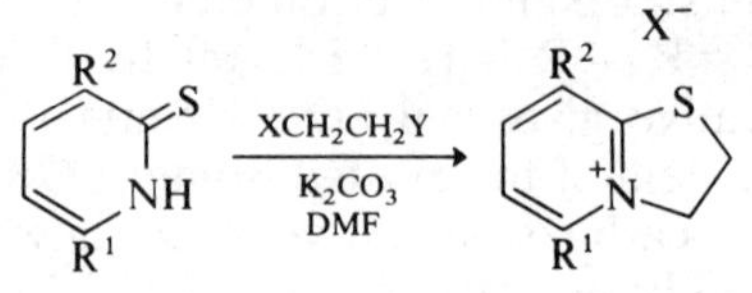

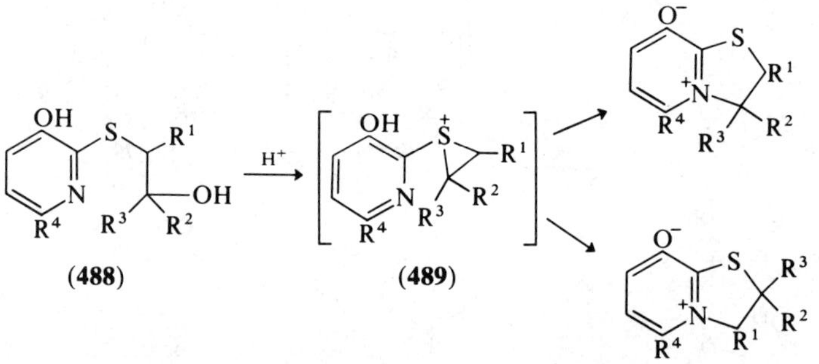

Inversion of the configuration in the cyclization reaction has been demonstrated for the (*R*)-bromo acid (**490**) which can be made from (*R*)-cysteine. The product (**491**) from the cyclization has the *S* configuration. By choice of reaction conditions to promote rearrangement, some of the enantiomeric 2-carboxy isomer (**492**) can be obtained ⟨81H(15)1349⟩.

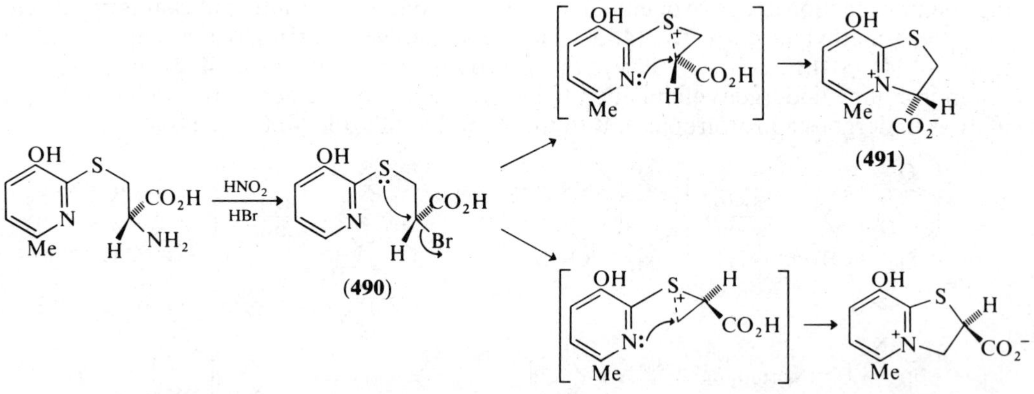

Michael addition of pyridine-2-thiones, *e.g.* (**493**), to α-haloalkenes which are α-substituted by strongly electron-attracting substituents (COR, CO_2R, $CONR^1R^2$, NO_2, CN) introduces the electron-attracting substituent into the 3-position. In the initial adduct (**494**) the halogen-bearing carbon atom has become rehybridized from sp^2 to sp^3, and hence the halogen becomes displaceable for cyclization. Adduct formation is the rate determining step. The electron density or availability on the sulfur is therefore the most important factor for the overall rate of the reaction. Thus a 3-hydroxy-2-mercaptopyridine, which is intramolecularly hydrogen bonded by pseudoring formation between the hydroxy group and the sulfur atom, is significantly less reactive than its 3-ethoxy analogue. Similarly a 3-acetamido analogue is much less reactive than its 5-isomer where this type of hydrogen bonding is excluded.

In the reaction of (Z)-2-bromo-2-butenoic acid the cyclic product (**495**) has the *trans* configuration. Since the cyclization reaction proceeds with inversion, the adduct formation is suprafacial. This is further confirmed by the reaction between 2-bromocyclohex-2-enone and mercaptopyridines where it has been shown that the dihydrothiazole and the cyclohexane rings are *cis*-fused.

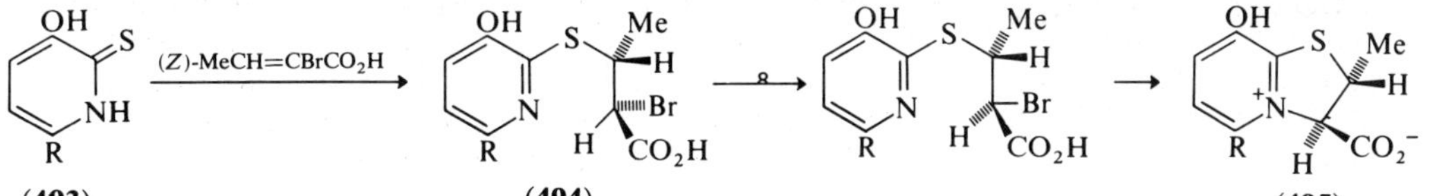

2-Vinylthiopyridines, *e.g.* (**496**) and (**497**), available by various methods, on acid catalysis can be cyclized to dihydrothiazoles. For allyl derivatives (**498**) cyclization by acid or electrophilic catalysis yields 3-methyl derivatives ⟨81H(15)1349⟩.

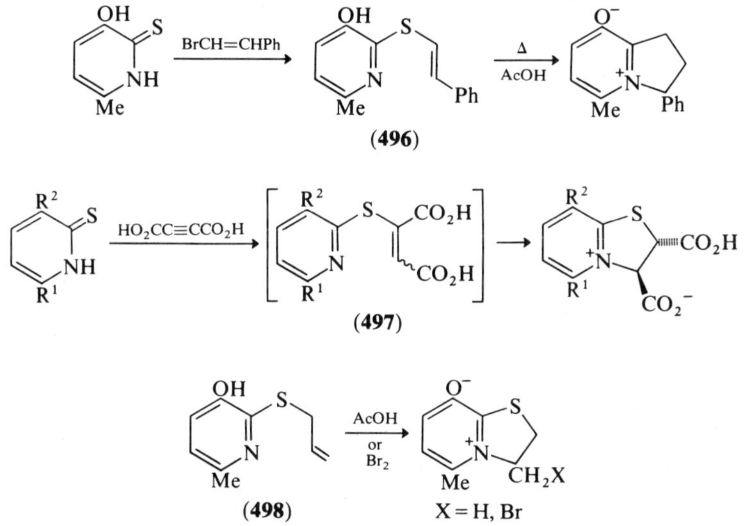

Azole approach. An important route for the synthesis of 3-hydroxypyridines is the reaction between a substituted furan and ammonia, or a primary amine. If the amino acid cysteine is used, the corresponding fused dihydrothiazole (**479**) can be obtained. Starting from optically active cysteine the corresponding cyclic enantiomer is obtained provided the fused product carries a 5-substituent, which for steric reasons prevents the otherwise ready racemization. A number of steps are involved in the overall reaction which may perhaps be rationalized by thiazolidine intermediates. The desired oxidation level in the substituted furan can either be reached by α-oxidation in an alkyl side-chain (**499**) or more generally by raising the furan ring itself to a higher oxidation level as in (**500**) ⟨81H(15)1349⟩.

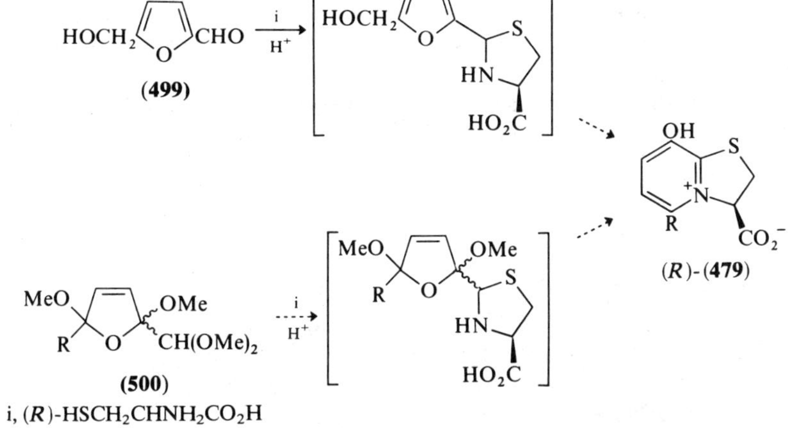

i, (R)-HSCH$_2$CHNH$_2$CO$_2$H

Similarly, 2,3-dihydro-5-oxo-thiazolo[3,2-a]pyridine-3,7-dicarboxylic acids are formed in the reaction between α-amino-β-mercaptocarboxylic acids with citric acid. In the reaction of (S)-penicillamine the cyclic product (**501**) was reported to be racemized ⟨71T3447⟩.

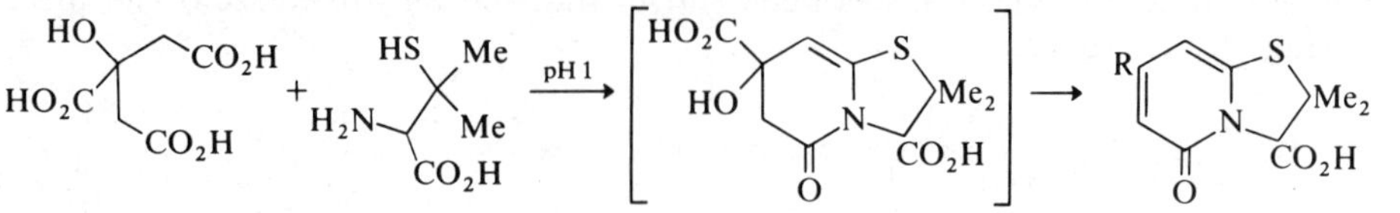

(**501**) R = CO_2H, H

The Hantzsch pyridine synthesis gives initially a dihydropyridine from the cyclization reaction. Adaptation of this reaction to the use of a 2-methylenethiazolidine yields the fused tetrahydro derivative (**502**) ⟨77LA1888⟩. Perhydro derivatives are simply prepared from 2-substituted thiazolidines by cycloalkylation as for (**503**) ⟨80S387⟩. The thiazolidine may also be generated *in situ* as in the reaction between γ-benzoylbutyric acid and 2-mercaptoethylamines under azeotropic conditions to yield (**504**) ⟨65JOC1506⟩.

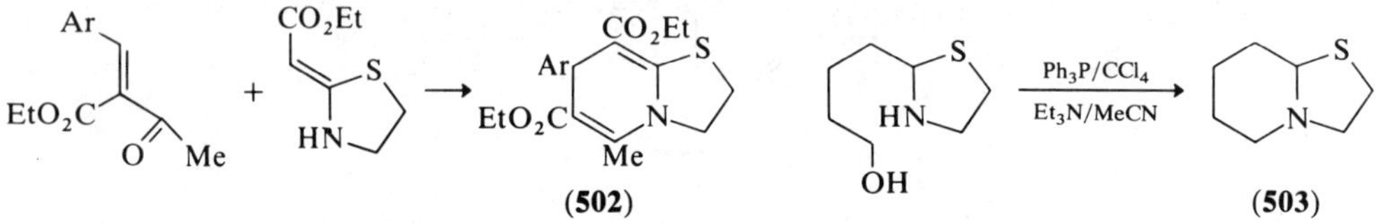

(ii) 5H-Thiazolo[3,2-c]pyrimidine

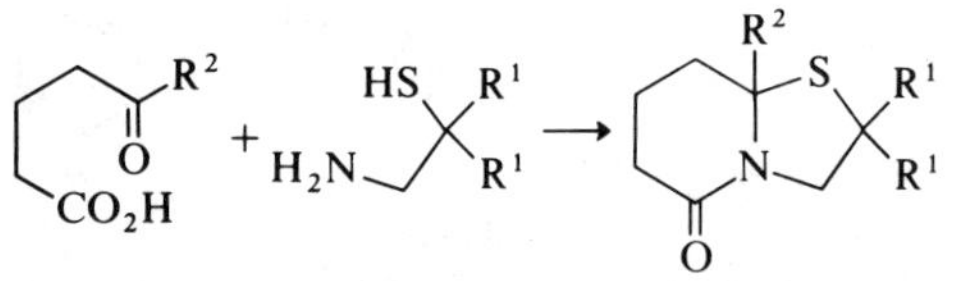

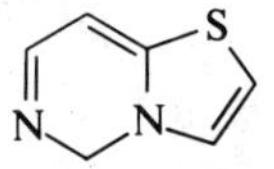

Azine approach. In the same way as in the pyridine series, a pyrimidine-4-thione (**505**) will react with 1,2-difunctional ethanes to form fused dihydrothiazoles. Michael addition to α-bromopropenoic acid and cyclization occur less readily than with comparable pyridines because of the lower nucleophilicity of the system ⟨81H(15)1349⟩.

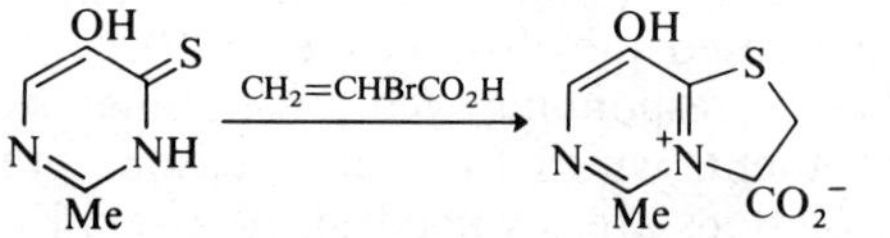

(**505**)

Azole approach. The β-lactam ring in penicillanic acid derivatives can be expanded to a pyrimidine ring. Thus methyl 6β-phthalimidopenicillanate (**506**) is transformed by CSI into the corresponding fused pyrimidine (**507**) with retention of the stereochemistry ⟨78JCS(P1)817⟩. Treatment of penicillanate S-oxides (**508**) with acyl isocyanates in a similar manner yields fused pyrimidines which subsequently eliminate water from the rearranged sulfoxide ⟨74USP3850933⟩.

Ethoxycarbonyl isothiocyanate, as well as benzoyl isothiocyanate, interacts with 2-amino-2-thiazoline at the ring nitrogen atom (**509**). The initial adducts react further to form fused 1,3,5-triazines ⟨75JOC2000⟩.

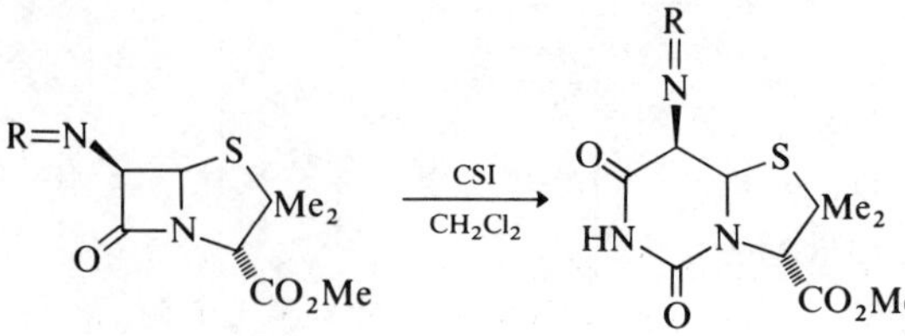

(**506**) R = phthaloyl   (**507**)

(508)

(509) R = OH, Ph

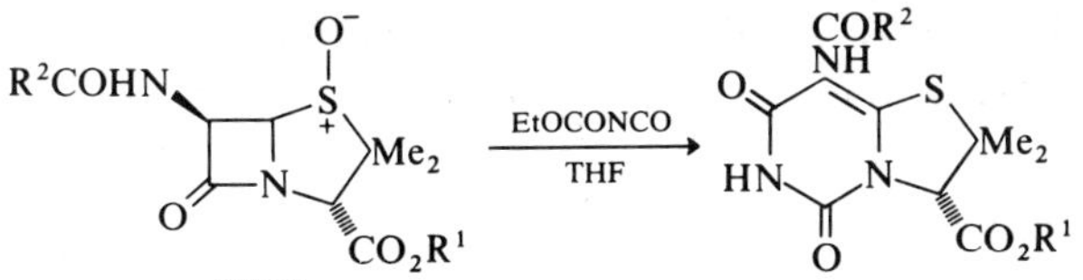

(iii) 5H-*Thiazolo[3,2-*a*]pyrazine*

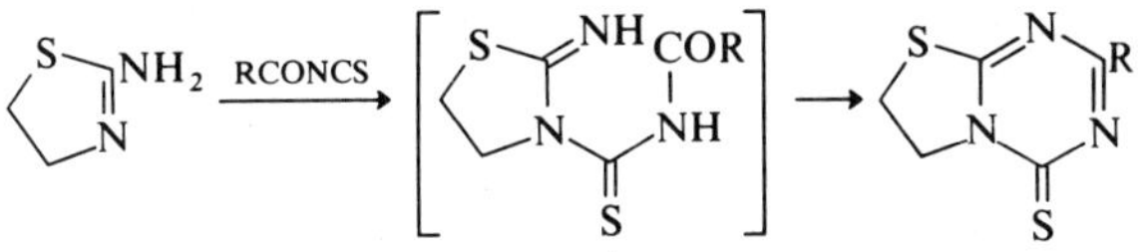

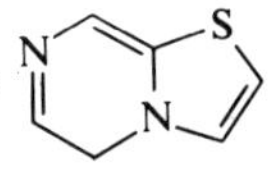

Azole approach. In the synthesis of the fused dioxopiperazine (**510**) a 2-butyloxycarbonylthiazolidine is *N*-acylated by glycine and then cyclized under weakly alkaline conditions ⟨80JHC39⟩.

(510)

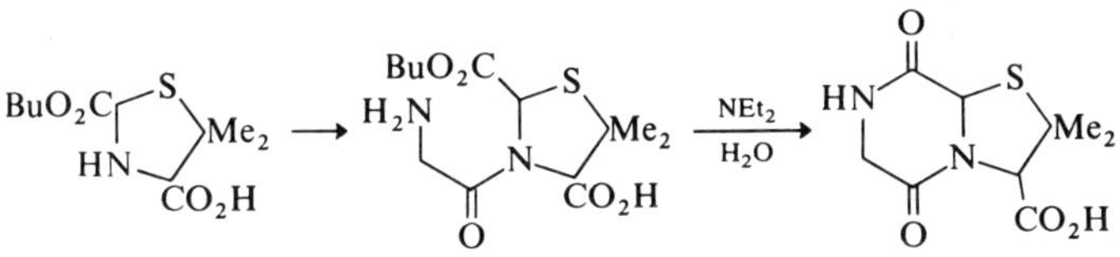

(iv) 5H-*Thiazolo[3,2-*a*]pyrimidine*

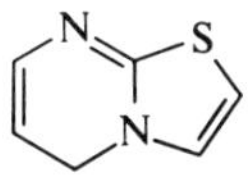

Azine approach. Reactions of unsymmetrically substituted pyrimidine-2-thiones with 1,2-difunctional ethanes may yield two isomeric products on ring fusion, *e.g.* from 2-thiouracil (**511**) ⟨71JCS(C)1527⟩. For the 6-amino homologue, however, only the 5-oxo isomer (**512**) was obtained ⟨77ACS(B)167⟩. In the absence of a base, cyclization of the dichloro-2-thiouracil (**513**) is solvent dependent. Heating (**513**) in DMF, when the HBr from the reaction is boiled off, gives the 5-oxo product (**514**), whereas in acetone the HBr salt of the 7-oxo isomer (**515**) is precipitated ⟨72ACS947⟩.

(511)

(512)

(514) **(513)** **(515)**

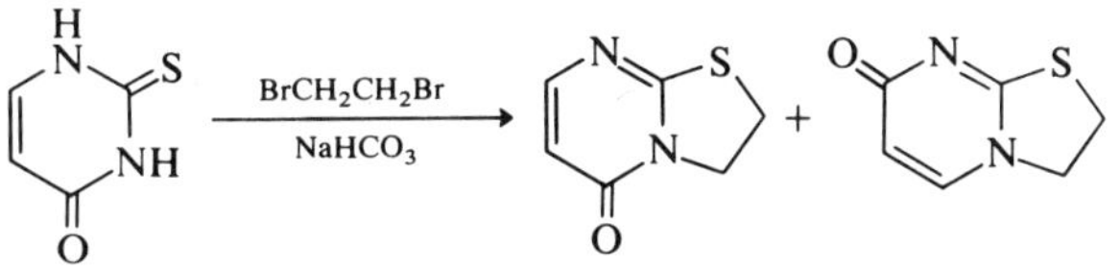

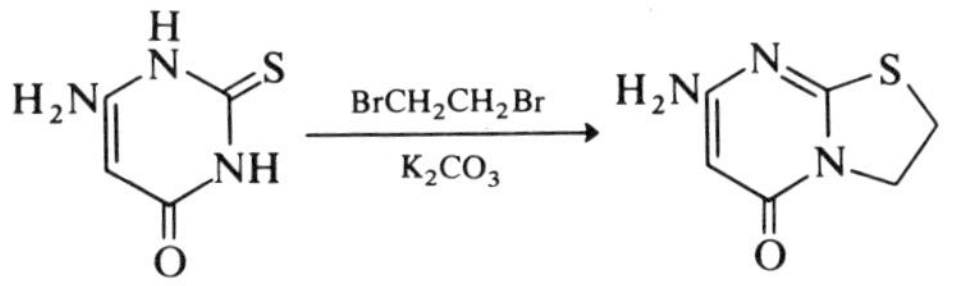

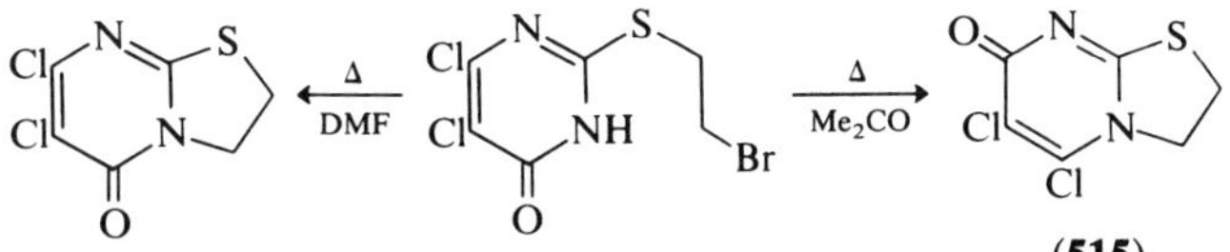

In the reaction of the isothiocyanate (**516**) the initially formed thiourea is cyclized to a pyrimidine-2-thione ring, the quinazoline (**517**), which is unambiguously N-3 substituted and therefore on acid catalysis yields the thiazolidine (**518**). The true configuration of

(S)-serine is retained in the reaction, the product therefore having the R configuration ⟨68HCA241⟩.

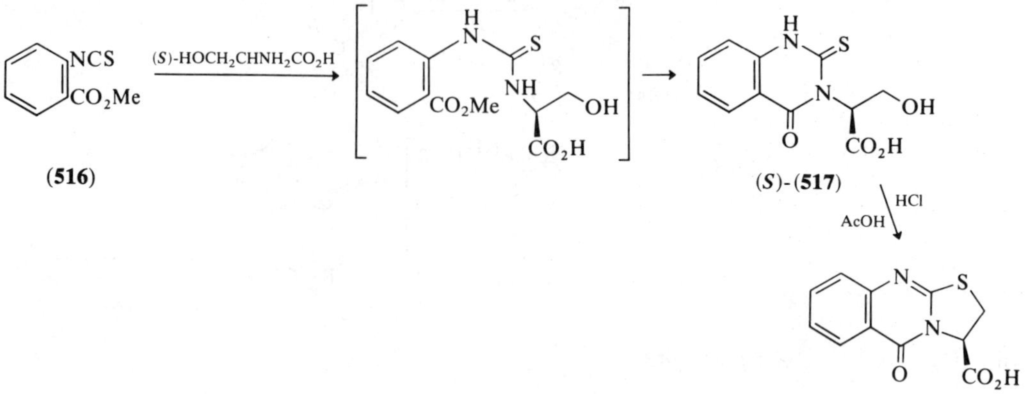

Azole approach. 2-Aminothiazolines as well as 2-aminothiazoles will react with ethoxymethylenemalonic ester to form an enamine which on heating is cyclized to give (**519**).

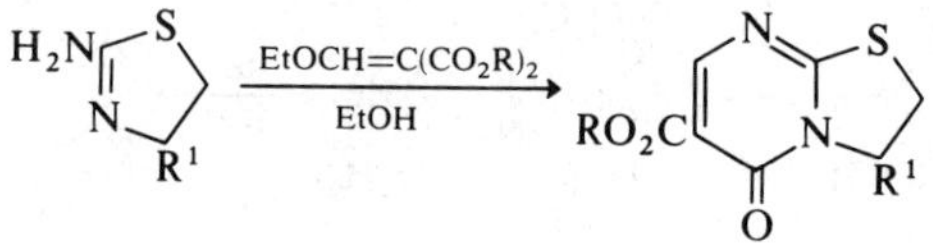

4.29.7.8.2 3H-Thiazolo[3,4-a]pyridine

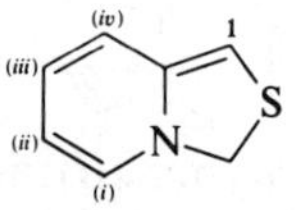

Azine approach. 3-Amino derivatives are readily prepared by the reaction of a 2-hydroxymethylpiperidine with an isothiocyanate with subsequent acid-catalyzed cyclization of the intermediate thiourea. With an optically active alcohol (**520**) the cyclization proceeds with inversion of the configuration ⟨78MI42901⟩.

Carbon–sulfur bond formation by oxidative cyclization of a thiourea derivative using bromine will in the case of (**521**) furnish a thiazolo[*i,j*]quinoline (**522**; Z = NH). The same imine is obtained from 8-mercapto-1,2,3,4-tetrahydroquinoline on treatment with cyanogen bromide; with phosgene the 2-oxo derivative (**522**; Z = O) is formed ⟨63JOC2581⟩.

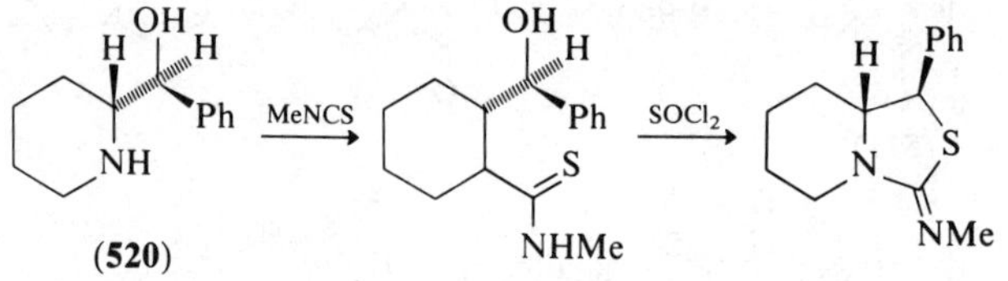

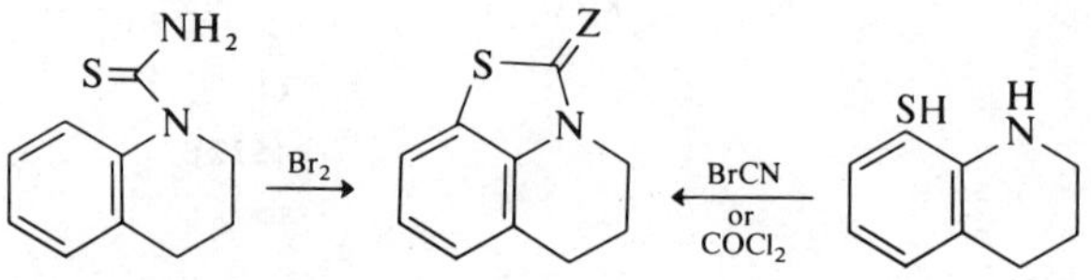

(iii) *Thiazolo[3,4-a]pyrazine*

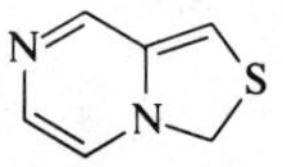

Azole approach. Sulfomycimine (**523**; R = H), from acid hydrolysis of the antibiotic sulfomycine, can be synthesized by the condensation of thiazole-4-carboxylic acid with DL-threonine ethyl ester, followed by cyclization and ester hydrolysis ⟨77H(8)461⟩.

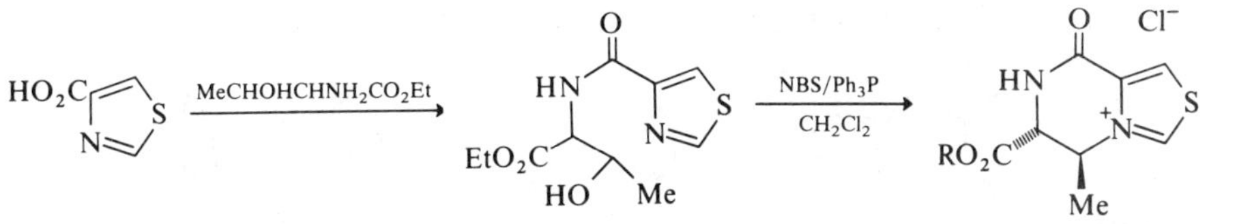

(523) R = Et, H

4.29.7.9 Applications

Berninamycinic acid is one of the products from acid hydrolysis of the cyclic peptide antibiotic berninamycin A, which is a potent inhibitor of bacterial protein synthesis. Berninamycinic acid has been assigned the structure (524), anhydro-3,8-dicarboxy-6-hydroxythiazolo[2,3-f][1,6]naphthyridin-4-ium hydroxide. The 6-hydroxy group arises during hydrolysis from a peptide-bonded amino group ⟨77JA1645⟩.

Berninamycinic acid is also a hydrolysis product from the antibiotic sulfomycin. Another hydrolysis product from this antibiotic is racemic sulfomycimine (523; R absent), anhydro-6-carboxy-5,6,7,8-tetrahydro-5-methyl-8-oxothiazolo[3,4-a]pyrazin-4-ium hydroxide ⟨77TL735⟩.

Acid hydrolysis of bovine liver yields (R)-anhydro-3-carboxy-2,3-dihydro-8-hydroxythiazolo[3,2-a]pyridinium hydroxide (479; R^1 = H, R^3 = Me). This last is formed from the amino acid L-cysteine and carbohydrates present ⟨81H(15)1349⟩.

Compounds with the general pyridazinium formula (525) and S-alkylated derivatives are claimed to have useful bactericidal and fungicidal properties ⟨73JAP(K)7391092⟩, whereas the thiazolo[3,2-a]pyridinium salt (526) is reported to have hypoglycemic activity ⟨78JMC489⟩.

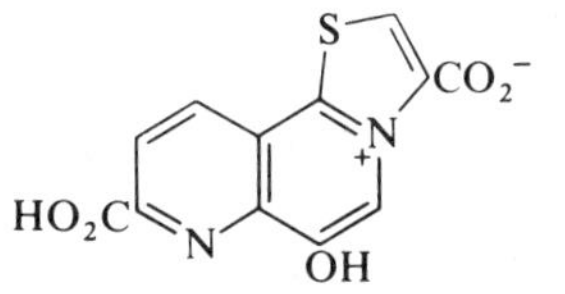

(524) 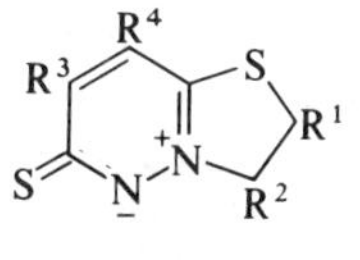   (525) 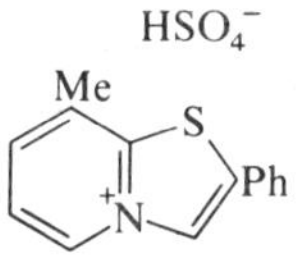   (526)

The betaine derivatives (527), 6,7-dihydro-5,7-dioxo-5H-thiazolo[3,2-a]pyrimidinium hydroxides, have been tested as inhibitors of cyclic AMP phosphodiesterase and found to have theophylline-like activity ⟨78JPS1762⟩.

Pyrimidine thiophosphates (528) are useful insecticides and acaricides ⟨74FRP2197513⟩, whereas the pyrimidine (529) was active against virulent Lewis lung tumour in mice ⟨76JMC524⟩. 2,3,6,7-Tetrahydropyrimidine derivatives (530) can be used in antimicrobial compositions for plants ⟨74FRP2223371⟩; close analogues are claimed to have useful anti-inflammatory properties ⟨77GEP2701853⟩.

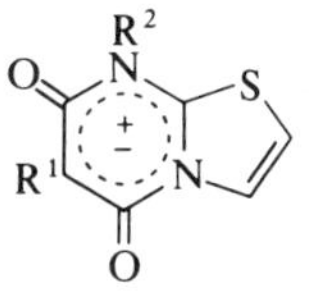

(527) 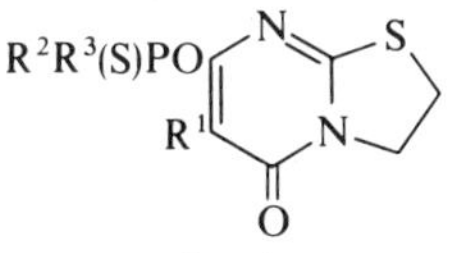   (528) 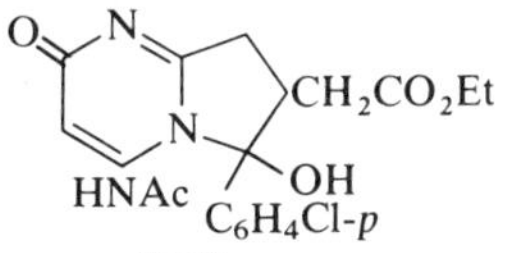   (529) 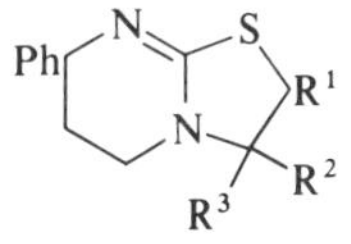   (530)

Thiopegans, fused thiazoloquinazolines, e.g. the 10,11-thiopegan (531), have antibacterial and antimalarial activities ⟨66JIC585⟩. Several patents cover the use of substituted 2H,4H-thiazolo[5,4,3-ij]quinolines as fungicides, e.g. the 5,6-dihydro-2-oxo derivative (532) is highly effective in protecting rice seedlings ⟨78JAP(K)7891134⟩. Triazines (533) are claimed to be bactericides and fungicides for phytopathogenic fungi besides having insecticidal properties; S-alkyl homologues have similar properties ⟨72JAP7212350, 78JAP(K)7884992⟩.

1,4-Dihydropyridine derivatives (534) are claimed to be useful as coronary dilators and antihypertensives ⟨73GEP2210633⟩. Compounds with formula (535) are patent protected as analgesics ⟨76GEP2264979⟩. Several patents cover the use of 1,5,10,10a-tetrahydro-3H-thiazolo[3,4-b]isoquinolines, e.g. (536), as ulcer inhibitors, acting by inhibition of gastric juice secretion ⟨78GEP2706398⟩. The 5-phenyl quinazoline derivative (537) and analogues

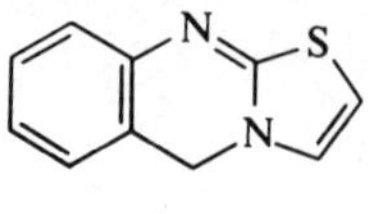

(531) 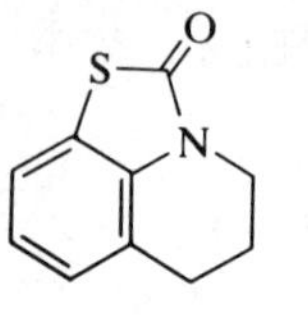 (532) 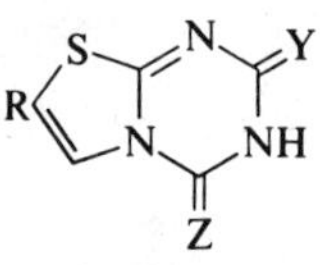 (533) Z ≠ Y = O, S

have hypotensive and blood platelet aggregation inhibiting activities ⟨78JAP(K)7844592⟩, whereas the quinolines (538) show antiinflammatory activities ⟨78BEP866987⟩.

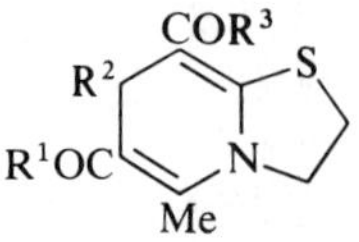

(534) 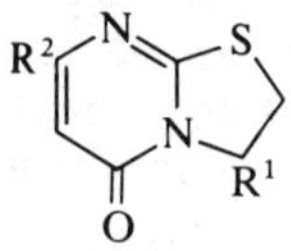 (535) 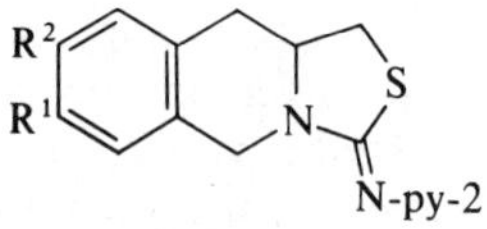 (536)

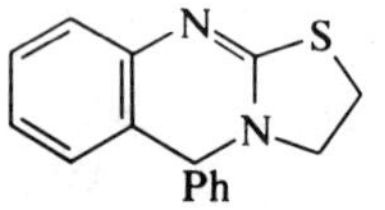

(537) 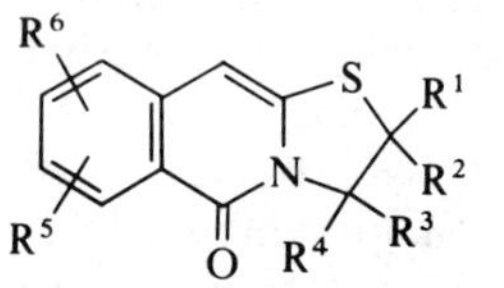 (538)

Thiazolo[4,5-b]pyrazines (539) show bactericidal, fungicidal, herbicidal, insecticidal and nematocidal activities ⟨78USP4075207⟩. Thiazolo[5,4-b][1,8]naphthyridines (540) are active against both Gram-negative and Gram-positive bacterio *in vitro* ⟨79CPB410⟩. The 2,5,7(3H,4H,6H)-triones (541) are claimed to have useful fungicidal activity in rice cultures ⟨73GEP2223421⟩ whereas the 2-amino analogues (542) have andidepressant activities ⟨73USP3772290⟩. Pyridazine derivatives (543) have analgesic properties ⟨79JAP(K)79119490⟩, whereas the 7-amines (544) are patent protected for their ability to inhibit thrombocyte aggregation and for antihypertensive activity ⟨73GEP2155963⟩. The tetrahydropyridines (545) are claimed as antitussives and antihypertensives ⟨73GEP2205065⟩.

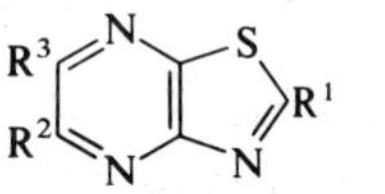

(539) 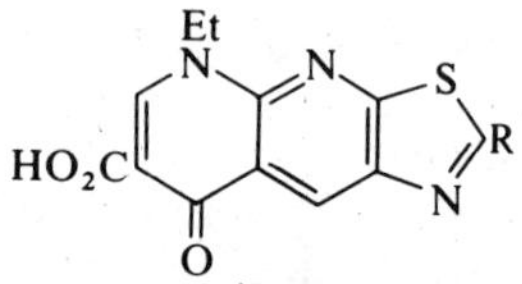 (540) 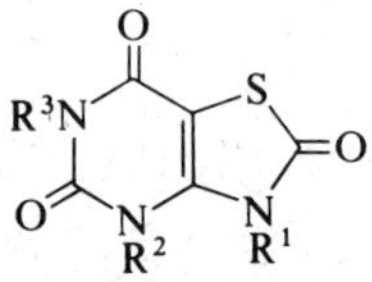 (541) 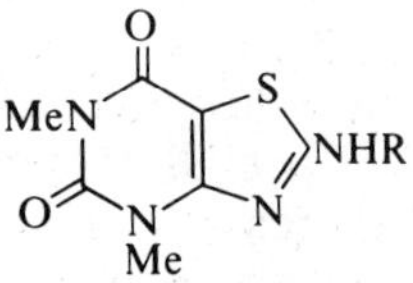 (542)

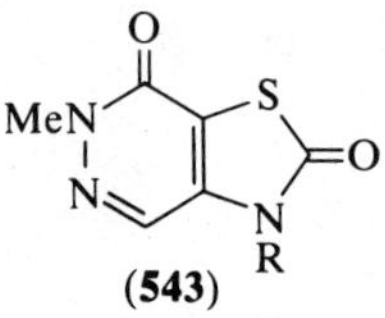

(543) 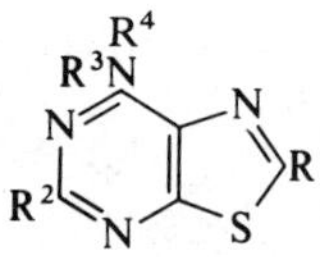 (544) 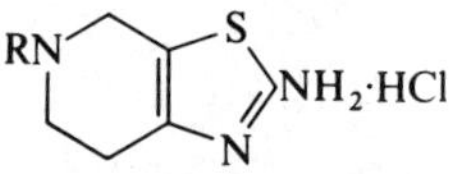 (545)

Thiazolo[5,4-c]isoquinolines (546) and their [4,5-c] isomers can be used as antifoggants for photographic layers ⟨71USP3630745⟩.

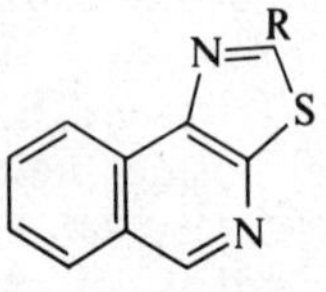

(546) 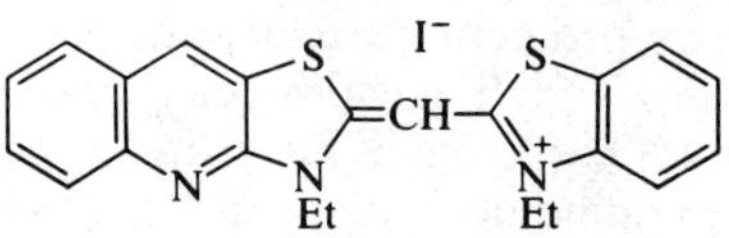 (547)

Benzothiazoles are involved in many cyanine dyes. Aza analogues seem a logical extension *e.g.* with quinoline nuclei as in (547) ⟨65BEP659236⟩.

4.29.8 OXADIAZOLES AND THIADIAZOLES FUSED WITH AZINES: INTRODUCTION

4.29.8.1 Survey of Possible Ring Systems

The heteroatoms in oxadiazoles or thiadiazoles can be arranged in a [1,2,3], a [1,2,4], a [1,3,4] or a [1,2,5] manner. The azoles can be fused to a six-membered ring containing the minimum number of nitrogens in 14 different ways. The possible systems have been classified accordingly as shown in Chart 5.

(a) 1H-[1,2,3]-[3,4-a]

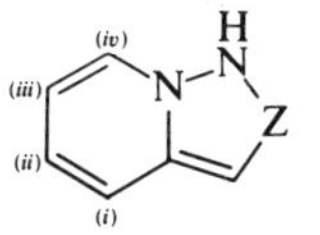

(b) [1,2,3]-[2,3-a]

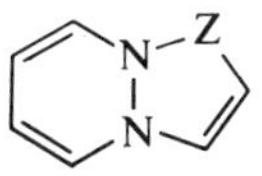

(c) 2H-[1,2,4]-[2,3-a]

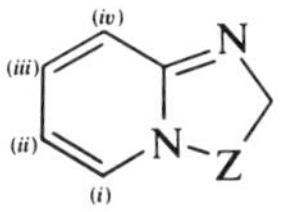

(d) 3H-[1,2,4]-[4,3-a]

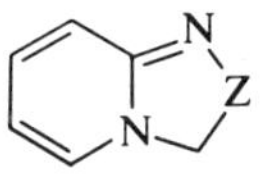

(e) 5H-[1,2,4]-[4,5-a]

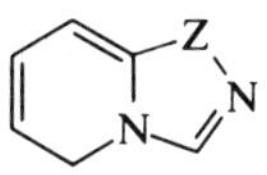

(f) 5H-[1,3,4]-[3,2-a]

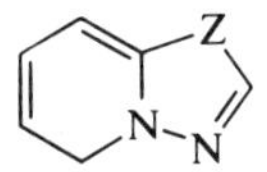

(g) 1H,3H-[1,3,4]-[3,4-a]

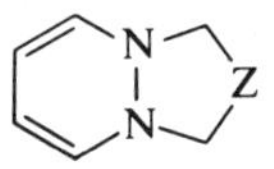

(h) 2H-[1,2,5]-[2,3-a]

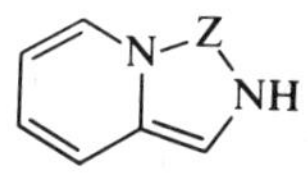

(i) [1,2,3]-[4,5-b]

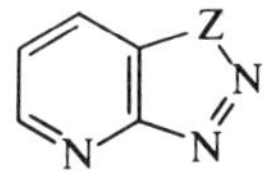

(j) [1,2,3]-[4,5-c]

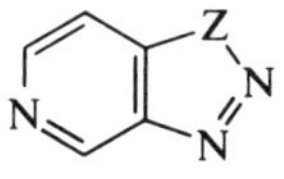

(k) [1,2,3]-[5,4-c]

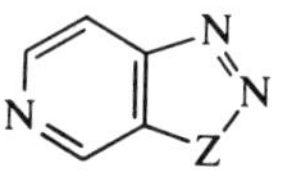

(l) [1,2,3]-[5,4-b]

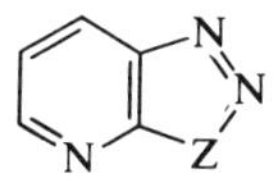

(m) [1,2,5]-[3,4-b]

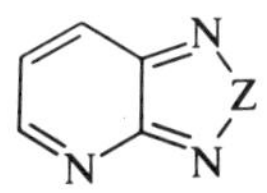

(n) [1,2,5]-[3,4-c]

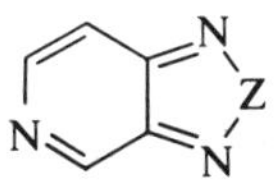

Chart 5 Ring systems formed by fusion of oxadiazole or thiadiazole with pyridine or pyridazine containing the minimum number of nitrogen atoms (Z = O, S)

Two of the systems, *viz.* groups (*b*) and (*g*), are formally diaza-bridged pyridazines, whereas the other systems are pyridine derivatives.

The eight systems (*a*)–(*h*) contain at least one nitrogen atom at a bridgehead position. In the remaining six systems (*i*)–(*n*), no azole nitrogen is part of the six-membered ring. These systems contain an additional nitrogen located in the azine ring.

The pyridine ring or the pyridazine ring for groups (*b*) and (*g*) may also contain additional heteroatoms. Such systems are subgrouped under the pyridine or the pyridazine group according to the position of the second heteroatom. They are treated in the order of increasing number for the position of the second heteroatom in the fused system as shown in Chart 6.

Polycyclic azines are grouped under the parent azine ring to which the azole ring is fused.

Fused ring systems can be prepared either from an azine precursor or from an oxadiazole or thiadiazole precursor. In the sections on the synthesis of these compounds, the former method is grouped under 'Azine approach', the latter under 'Azole approach'.

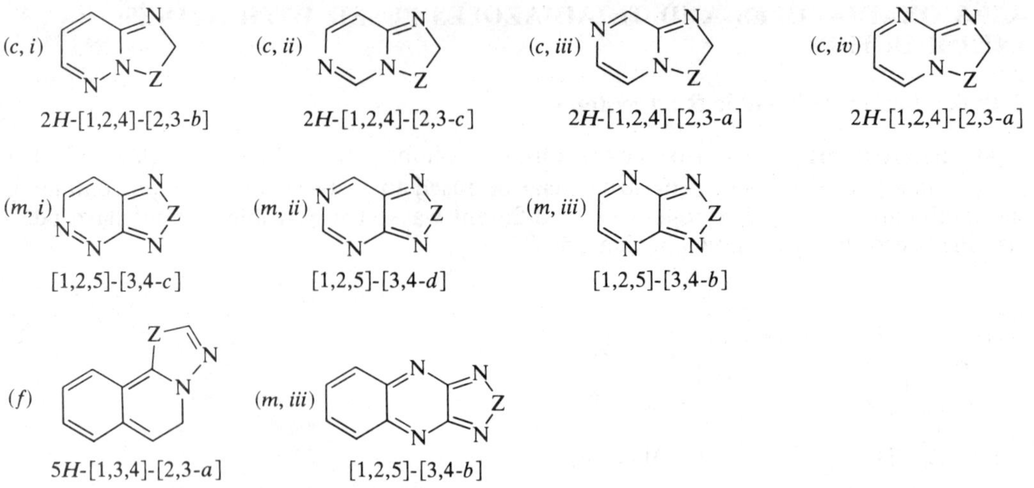

(c, i) 2H-[1,2,4]-[2,3-b] (c, ii) 2H-[1,2,4]-[2,3-c] (c, iii) 2H-[1,2,4]-[2,3-a] (c, iv) 2H-[1,2,4]-[2,3-a]

(m, i) [1,2,5]-[3,4-c] (m, ii) [1,2,5]-[3,4-d] (m, iii) [1,2,5]-[3,4-b]

(f) 5H-[1,3,4]-[2,3-a] (m, iii) [1,2,5]-[3,4-b]

Chart 6 Ring systems formed by fusion of oxadiazole or thiadiazole with pyridine or pyridazine containing additional heteroatoms and rings (Z = O, S)

4.29.8.2 General Survey of Reactivity

The oxadiazole and thiadiazole rings are π-electron deficient and hence do not readily react with electrophiles at nitrogen or at carbon. Electrophilic attack in the azine ring proceeds in the presence of electron-releasing substituents.

Substitution chemistry in these systems is mainly nucleophilic. Substitution in the azine ring is facilitated by the π-electron-deficient azole ring; the effect from the oxadiazole ring is generally stronger than from the thiadiazole ring, which possesses a higher degree of aromatic stabilization.

In some systems, especially in nitrogen bridgehead systems and azolium salts, nucleophiles open the azole ring; oxadiazole systems are the more reactive. Ring opening and Dimroth rearrangements in the azine ring can be observed.

4.29.9 OXADIAZOLES FUSED WITH AZINES

4.29.9.1 Introduction

1,2,3-, 1,2,4-, 1,3,4- and 1,2,5-oxadiazoles can be fused to a six-membered ring containing the minimum number of nitrogens in 14 different ways; the section on synthesis has been organized accordingly. Further fused systems arise from the diazines and the triazines and their areno and heteroareno homologues.

The known chemistry is almost entirely limited to the fully conjugated systems which have received a great deal of attention in recent years. Most systems are readily available and a number of these have been prepared for biological purposes. Others like the furazano[3,4-d]pyrimidines can be used as intermediates in the preparation of purine or pteridine ring systems.

4.29.9.2 Structure

4.29.9.2.1 X-Ray diffraction

The data for anhydro-3-hydroxy[1,2,3]oxadiazolo[4,3-c][1,2,4]benzotriazin-10-ium hydroxide show the molecule to be essentially planar. The O(2)—C(3) bond is relatively long, the exocyclic C(3)—O bond is almost of double bond character, and there is a marked deformation of bond angles around C(3). From the bond angles and bond lengths it appears that a ketene type of structure contributes substantially to the overall bonding situation ⟨79JCS(P2)1751⟩.

The data for 7-amino[1,2,5]oxadiazolo[3,4-d]pyrimidine show that the five- and six-membered rings are slightly tilted from coplanarity about their common bond which has been attributed to distortion from the packing forces in the crystal. The N—O bond lengths are quite similar to those in the parent azole. From comparison of bond lengths it has been estimated that the C—N bonds have 75% double bond character. The fused furazan ring may perhaps be thought of as constituting a pair of strongly electronegative substituents (oximes) on a monocyclic pyrimidine. The extreme ease of nucleophilic substitution at C-7 in this ring system is attributable to the high electron deficiency introduced by the furazan ring ⟨71JA7281⟩.

The data for 7-amino-4H-[1,2,5]oxadiazolo[3,4-d][1,2,6]thiadiazine 5,5-dioxide show planarity of the furazan ring which is quasi-coplanar with the thiadiazine ring. The latter has an envelope conformation ⟨75AX(B)2310⟩.

4.29.9.2.2 Molecular spectra

(a) *^{1}H NMR spectroscopy.* Proton NMR data for some relatively simple systems are listed in Table 24. Worth noting is the deshielding of H-5 in the 2H-[1,2,4]oxadiazolo[2,3-a]pyridine-2-thiones, the shift for this proton being almost the same as for H-5 in the [1,3,4]oxadiazolo[3,2-a]pyridin-4-ium ion. A large deshielding is also seen for the corresponding α-pyridine proton (H-4) in the 7-nitrofurazano[3,4-c]pyridine 3-oxide.

In the two sets of isomeric furoxanopyridines it is seen that the *N*-oxide oxygen causes shielding by ca. 0.2 p.p.m. of H-7 in the 1-oxide with respect to H-7 in the 3-oxide, and this effect is almost the same as in benzofuroxanes.

(b) *UV spectroscopy.* UV data for some simple systems are listed in Table 25. In the furazanopyrimidines it is seen that oxidation to the 1-oxide gives a bathochromic shift for the longwave UV absorption band.

The UV spectra for the 7-aminofurazano[3,4-d]pyrimidines are very similar to those of the sulfur analogues, 7-amino[1,2,5]thiadiazolo[3,4-d]pyrimidines and also to the correspondingly substituted pteridines. This similarity suggests an isoelectronic relationship between these systems ⟨71JOC3211⟩.

(c) *Infrared spectroscopy.* The carbonyl absorption band at 1773 cm^{-1} for 2H-[1,2,4]oxadiazolo[2,3-a]pyridin-2-one (**548**) is in the region for γ-lactone absorption whereas the carbonyl stretching of its isomeric mesoion (**550**) appears as two bands at 1745 and 1785 cm^{-1} ⟨77CJC3736⟩. In the diphenyl substituted homologue (**551**; Z = O) the pseudocarbonyl band is at 1770 cm^{-1} which is in the frequency region for isosydnones; in the sulfur analogue (**551**; Z = S) the absorption band at 1430 cm^{-1} is attributed to pseudo C=S stretching; in the 2-imine (**551**; Z = NPh) the absorption at 1630 cm^{-1} is attributed to exocyclic C=N stretching ⟨80TL4025⟩.

The pseudocarbonyl stretching in the sydno[3,4-a]quinoxalines (**552**) appears at ca. 1790 cm^{-1} whereas the absorption in simple sydnones lies at ca. 1740 cm^{-1}. The band at 1675 cm^{-1} (**552**; R = Me) is due to the 4-oxo group. In the parent compound (**552**; R = H) there is a strong band at 1675 cm^{-1} consistent with the normal preference for the lactam form in azines ⟨72JOC1707⟩.

4.29.9.3 Reactivity

4.29.9.3.1 Survey of reactivity

The oxadiazole rings are π-electron deficient and hence do not readily react with electrophiles at nitrogen or at carbon. The azine ring may be attacked by electrophiles at a nitrogen, especially if the ring contains a strongly electron-releasing substituent. Electrophilic attack at carbon requires the presence of electron-releasing substituents.

Since both the oxadiazole and the azine rings are π-electron-deficient systems, nucleophilic substitution reactions are important. The oxadiazole ring may be regarded as a strongly electronegative substituent which promotes substitution in the azine ring. Most widely studied is the furazano[3,4-d]pyrimidine system, which at C-7 is so strongly activated that amines will readily interchange in this position.

Table 23 X-Ray Diffraction Data for Fused Oxadiazoloazines

Compound	Ring A		Bond lengths (Å) and angles (°) Ring B		Ring C	
5-Methyl-2H-[1,2,4]oxadiazolo[2,3-c]quinazolin-2-one ⟨76TL3615⟩	b^1 1.318 b^2 1.370 b^3 1.354 c^1 1.446 h^1 1.388 c^2 1.198	b^1b^2 105.8 b^1b^3 112.9 b^2c^1 110.0 b^3h^1 108.4 c^1h^1 103.0	a^1 1.406 a^2 1.438 b^3 1.354 b^4 1.371 b^5 1.296 b^6 1.397	a^1a^2 115.8 a^1b^6 123.5 a^2b^3 115.6 b^3b^4 127.4 b^4b^5 118.2 b^5b^6 119.5	a^1 1.406 a^3 1.409 a^4 1.373 a^5 1.396 a^6 1.374 a^7 1.406	a^1a^3 118.3 a^1a^7 121.1 a^3a^4 120.1 a^4a^5 121.4 a^5a^6 120.0 a^6a^7 119.4
Anhydro-3-hydroxy[1,2,3]oxadiazolo[4,3-c][1,2,4]benzotriazin-10-ium hydroxide ⟨79JCS(P2)1751⟩	a^1 1.422 b^1 1.353 c^1 1.393 g^1 1.303 h^1 1.396 c^2 1.196	a^1b^1 105.8 a^1c^1 103.4 b^1g^1 115.7 c^1h^1 112.0 g^1h^1 103.1	a^2 1.388 b^1 1.353 b^2 1.342 b^3 1.387 b^4 1.397 g^2 1.297	a^2b^3 124.3 a^2b^4 113.9 b^1b^2 124.9 b^1b^4 118.8 b^2g^2 118.9 b^3g^2 119.0	a^2 1.388 a^3 1.404 a^4 1.358 a^5 1.391 a^6 1.365 a^7 1.385	a^2a^3 117.4 a^2a^7 123.1 a^3a^4 120.0 a^4a^5 121.0 a^5a^6 121.0 a^6a^7 117.6

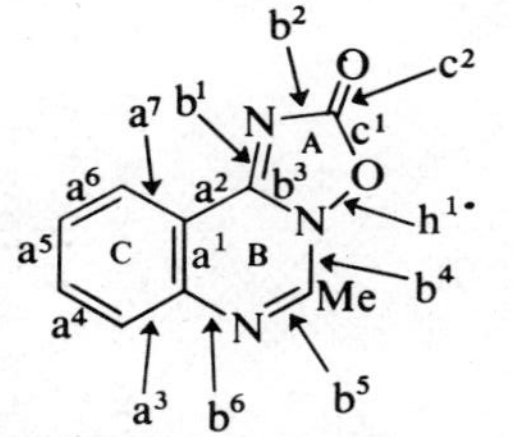

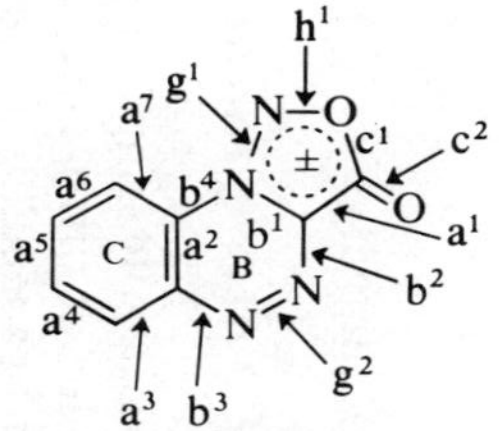

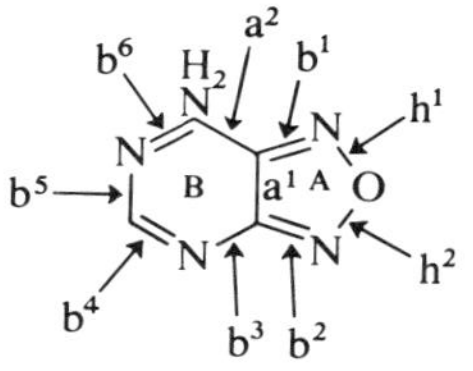

a^1	1.423	a^1b^1	110.7	a^1	1.423	a^1a^2 117.1
b^1	1.301	a^1b^2	108.4	a^2	1.465	a^1b^3 124.7
b^2	1.320	b^1h^1	104.3	b^3	1.366	a^2b^6 116.7
h^1	1.376	b^2h^2	104.4	b^4	1.320	b^3b^4 112.1
h^2	1.393	h^1h^2	112.3	b^5	1.368	b^4b^5 129.5
				b^6	1.330	b^5b^6 119.9

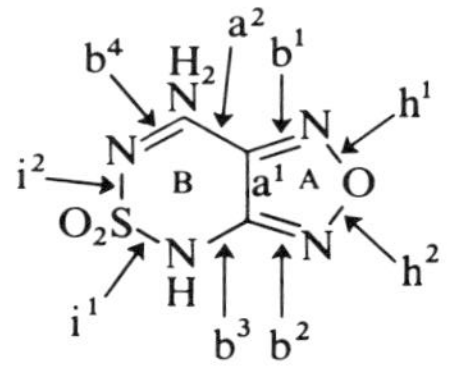

7-Amino[1,2,5]oxadiazolo[3,4-d]pyrimidine ⟨71JA7281⟩

a^1	1.418	a^1b^1	109.8	a^1	1.418	a^1a^2 122.3
b^1	1.299	a^1b^2	109.5	a^2	1.456	a^1b^3 122.4
b^2	1.307	b^1h^1	105.4	b^3	1.362	a^2b^4 119.3
h^1	1.366	b^2h^2	104.0	b^4	1.316	b^3i^1 117.3
h^2	1.408	h^1h^2	111.5	i^1	1.675	b^4i^2 124.6
				i^2	1.592	i^1i^2 106.4

7-Amino-4H-[1,2,5]oxadiazolo[3,4-d]-[1,2,6]thiadiazine 5,5-dioxide ⟨75AX(B)2310⟩

Table 24 ^{1}H NMR Data for Fused Oxadiazoloazines

Compound	Solvent	^{1}H chemical shifts (p.p.m.)					Coupling constants (Hz)	Ref.
		4	5	6	7	8		
[structure: pyridinium-fused oxadiazole with C$_6$H$_4$R-p, ClO$_4^-$]								
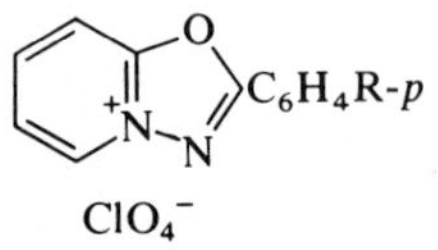 R = H	DMSO-d_6	—	9.7	8.3	8.8	8.8	$J_{5,6}$ 6.0	80JOC5095
R = Cl	DMSO-d_6	—	9.8	8.4	8.8	8.8	$J_{5,6}$ 7.0	80JOC5095
[structure: R^1, R^2-substituted oxadiazole-thione fused to pyridine]								
R^1 = Me, R^2 = H	DMSO-d_6	—	9.50	2.77	8.33	8.00		77CJC3736
R^1 = H, R^2 = Me	DMSO-d_6	—	9.47	7.60	2.88	7.91		77CJC3736
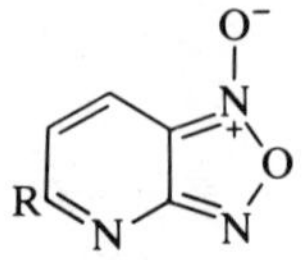 [structure: furoxano-pyridine, N-oxide on top]								
R = H	CDCl$_3$	—	8.94	7.40	7.96	—		70JCS(B)636
R = Me	CDCl$_3$	—	2.77	7.21	7.78	—	$J_{6,7}$ 9.2	70JCS(B)636
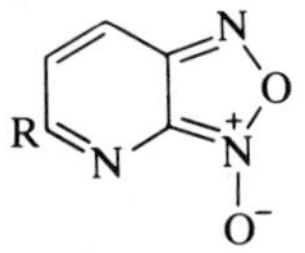 [structure: furoxano-pyridine, N-oxide on bottom]								
R = H	CDCl$_3$	—	8.75	7.56	8.20	—	$J_{5,6}$ 3.7, $J_{5,7}$ 1.4, $J_{6,7}$ 9.2	70JCS(B)636
R = Me	CDCl$_3$	—	2.76	7.37	8.03	—	$J_{6,7}$ 9.3	70JCS(B)636
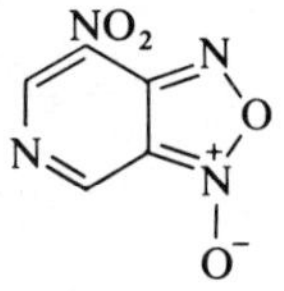 [structure: NO$_2$-substituted furoxano-pyridine]	CDCl$_3$	9.31	—	9.46	—	—		71JCS(C)1211

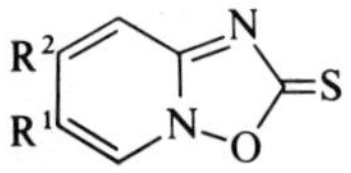

In some cases, depending on the system and the nature of the substituents, opening of a ring rather than substitution may result, *e.g.* opening of the azole ring in nitrogen bridgehead systems.

Opening of the azole ring with recyclization over a vicinal substituent gives access to other ring systems, *e.g.* purines and pteridines. The easy reductive cleavage of the N—O bond is used in the preparation of the latter systems from 7-aminofurazano[3,4-*d*]pyrimidines.

Furoxanoazines exhibit tautomeric rearrangements in the furoxane ring, and Dimroth rearrangements in the azine ring are to be expected.

4.29.9.3.2 *Thermal rearrangements*

Benzofuroxanes exhibit rapid tautomeric rearrangements in the furoxane ring. Fused furoxanoazines behave in the same manner. In [1,2,5]oxadiazolo[3,4-*b*]pyridine 1-oxide (1-furoxano[3,4-*b*]pyridine) an equilibrium exists between the 1-oxide (**553**) and the 3-oxide (**555**), the intermediate being the 2,3-dinitrosopyridine (**554**). ^{1}H NMR data show the coalescence point to be below room temperature; at −45 °C the spectrum is resolved into two well-defined patterns. The oxide group exhibits a considerable preference (3.97 kJ mol^{-1}) for the 1-position which may partly be caused by electronic repulsion between lone pairs on the oxygen and the aza groups, but also by charge delocalization which gives negative charge concentration on the aza group.

Table 25 UV Absorption Data for Fused Oxadiazoloazines

Compound	Solvent	λ_{max}[a] ($\log \varepsilon$)	Ref.

Structure 1:

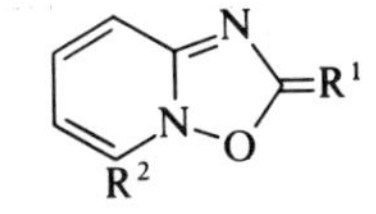 R^1	R^2			
O	H	MeOH	310 (3.60), 245 (4.23)	56JCS2063
O	Me	MeOH	312.5 (3.75), 244 (4.28)	56JCS2063
S	H	EtOH	331.5 (3.72), 289 (4.27)	77CJC3736
S	Me	EtOH	333.5 (3.61), 287 (4.20)	77CJC3736

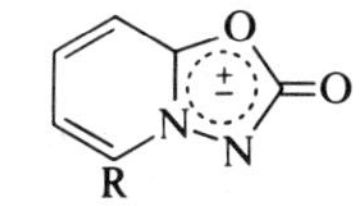

Structure 2:

R			
R = H	EtOH	297 (3.86), 262.5 (3.87), 220 (4.11)	58HCA548
R = H	EtOH	290 (3.80), 268 (3.95), 222 (4.21)	58HCA548

Structure 3:

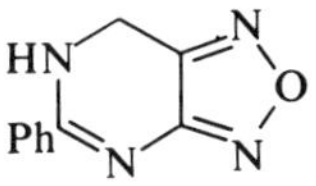	EtOH	302 (4.05), 240 (4.50)	76CPB235

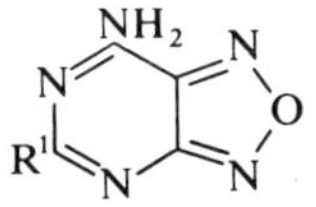

Structure 4:

R			
R = H	EtOH	340 (3.83), 253[b] (3.08), 210 (4.18)	71JOC3211
R = H	EtOH	337 (3.80), 210 (4.24)	71JOC3211

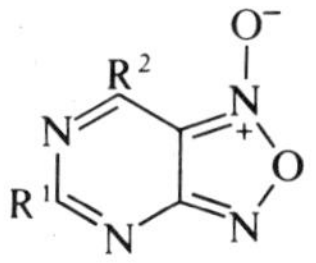

Structure 5:

R^1	R^2			
H	NH_2	pH 7	365 (3.74), 297[b] (3.57), 290 (3.58), 270 (3.45)	68JOC2086
H	OH	pH 7	359 (3.62), 282 (3.64), 255[b] (3.50)	68JOC2086
Me	OMe	MeOH	350 (3.45), 283 (3.54), 245 (3.75), 210 (4.03)	76JCS(P1)1327

[a] nm. [b] Shoulder.

Table 26 IR Absorption Data for CO Groups in Fused Oxadiazoloazines

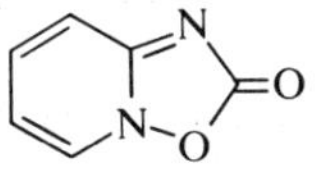

(548)
1773 cm^{-1}
KBr
77CJC3736

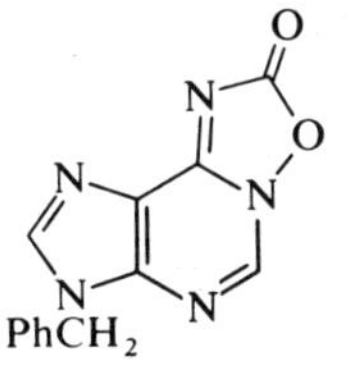

(549)
1815 cm^{-1}
KBr
76CJC2804

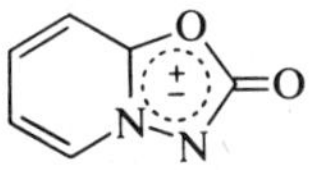

(550)
1785, 1745 cm^{-1}
KBr
77CJC3736

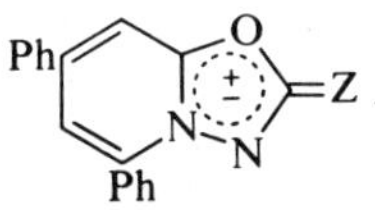

(551)
Z = O; 1770 cm^{-1}
Z = S; 1430 cm^{-1}
Z = NPh; 1630 cm^{-1}
80TL4025

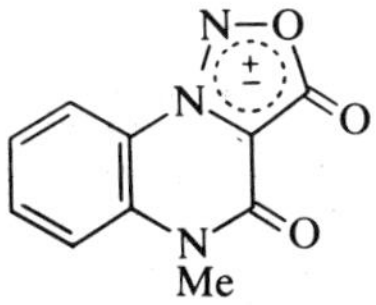

(552)
1790, 1670 cm^{-1}
KBr
72JOC1707

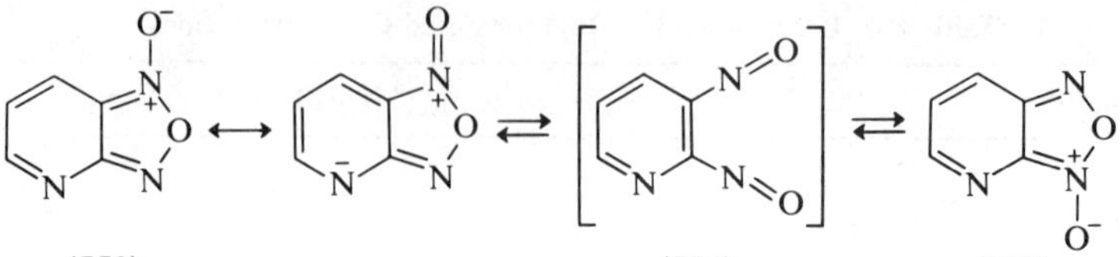

(553) (554) (555)

In benzofuroxanes the nature and position of the substituents have a marked effect on the equilibrium position. In the pyridine analogue the influence of the aza group overrides the effects of the methyl substituents; at −45 °C 80–90% of the equilibrium pair from monomethyl and dimethyl derivatives consists of the 1-oxide isomer (553) ⟨70JCS(B)636⟩.

In the [c]-fused quinoline 60% of the equilibrium mixture at 35 °C consists of the 3-oxide (556). This reversal of preference may be explained by invoking resonance between the N-oxide group and the quinoline nitrogen which can only occur in the 3-oxide ⟨68JCS(B)1516⟩.

In the [d]-fused pyrimidines the tautomerism of the furoxane ring favours the 1-oxide structures (557) to the virtual exclusion of the 3-oxides ⟨76JCS(P1)1327⟩.

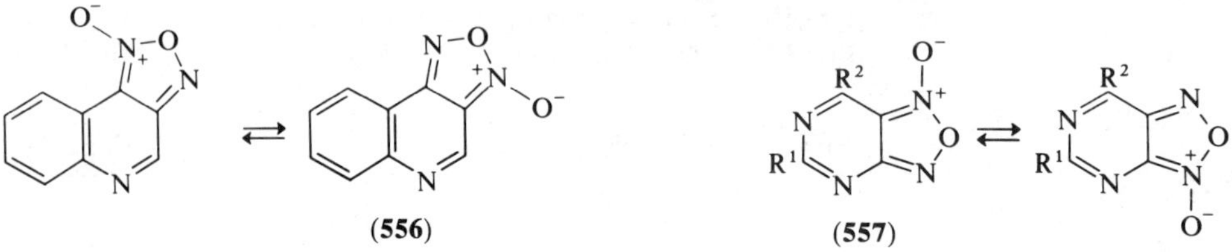

(556) (557)

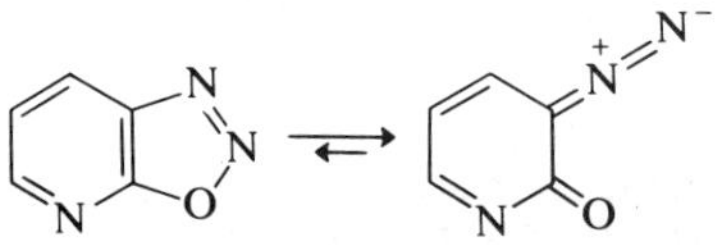

(558)

Another type of problem concerns 1,2,3-oxadiazoles. All spectroscopic evidence for benz[1,2,3]oxadiazoles is consistent with an *o*-quinonoid benzodiazooxide structure ⟨66RCR388⟩. Pyridine analogues are chemically highly unstable and show spectroscopic properties consistent with the pyridodiazooxide structure (558) ⟨74CS(6)222⟩. However, their fused sulfur and nitrogen analogues, the 1,2,3-thiadiazolo and 1,2,3-triazolo derivatives, are perfectly stable as ring compounds.

4.29.9.3.3 Electrophilic attack at nitrogen

The oxadiazole rings are π-electron deficient. Electrophilic attack on an azole nitrogen is therefore difficult to effect. Reactions, however, may take place at an azine nitrogen. In the uracil (559) N-4 alkylation is carried out on the lithium salt ⟨76JCS(P1)1327⟩. The sodium salt of the mesoionic compound (560), sydno[3,4-*a*]quinoxaline, can be exclusively N-5 alkylated using dimethyl sulfate; with diazomethane selective *O*-methylation of the lactam oxygen occurs ⟨72JOC1707⟩. Acylation follows the general rule in azine chemistry which states that it is an amino group and not a nuclear nitrogen atom which is acylated in the final product ⟨71JOC3211⟩.

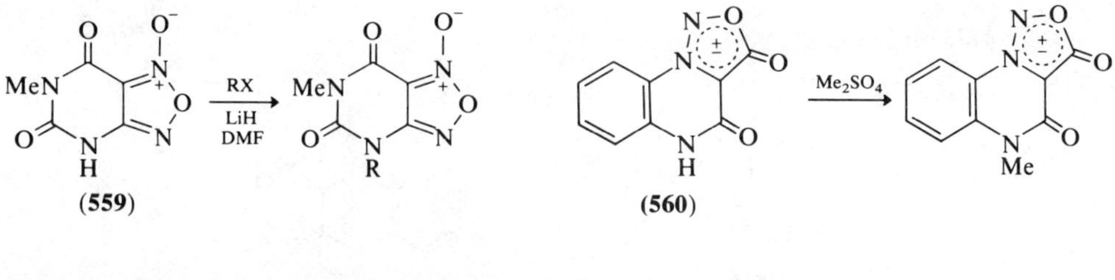

(559) (560)

4.29.9.3.4 Electrophilic attack at carbon

Electrophilic substitution at carbon is largely limited to the azine when this ring is activated by strongly electron-donating substituents.

4.29.9.3.5 Nucleophilic attack at carbon

Since both the oxadiazole and the azine rings are π-electron-deficient systems, nucleophilic substitution reactions are an important part of the chemistry of the oxadiazoloazine ring system. Difficulties, however, may be encountered in some cases when attempting substitution reactions because the nucleophile may cause preferential opening of one of the rings. This is especially true for nitrogen bridgehead systems. Nucleophilic substitutions, therefore, have mainly been studied in [3,4]-fused oxadiazoles, the 1,2,3-oxadiazole system being excluded because of instability as previously pointed out.

The most extensive investigations have been carried out on furazano[3,4-d]pyrimidines. The latter have become important synthons in the preparation of purines; after introduction of the desired substituents in the pyrimidine ring the furazano ring is opened by reductive cleavage whereby a 5,6-diaminopyrimidine is formed.

A 7-substituent in furazano[3,4-d]pyrimidine is replaced very easily by nucleophiles; even an amino group readily exchanges with another amino group, e.g. as in (**561**). The extraordinary ease with which this substitution takes place contrasts sharply with the relatively vigorous conditions required for the analogous displacement of an amino group from the 4-position of the isoelectronic pteridines. Perhaps even more remarkable is the difference in reaction conditions required for the substitution of the 7-amino group in the sulfur analogue, the [1,2,5]thiadiazolo[3,4-d]pyrimidine. X-Ray analysis shows that the C—N bonds in the furazan ring have ca. 75% double bond character whereas these bonds in the sulfur analogue have ca. 50% double bond character. The furazan ring may thus be regarded as a pair of strongly electronegative substituents which will activate the fused system for substitution at C-7 ⟨71JA7281⟩. MO calculations indicate considerable charge separation in the ground state and in particular an unusually low charge density (0.5711) at C-7 in accordance with the high nucleophilic reactivity at this position ⟨76CPB235⟩.

The reaction starts with an equilibrium between the amine and the amine adduct (**562**). The amino group exchange can proceed directly over the sp^3-hybridized intermediate (**562**), or may involve a ring opening to (**563**) which is in equilibrium with (**562**). The pathway depends on the system under investigation ⟨71JOC3211⟩.

The ease of amine substitution is enhanced in 7-acylamino derivatives and also depends on the leaving properties of the amide. Benzamides are more reactive than acetamides and are commonly used in the substitution reactions.

The difference in the reactivity between a 5- and 7-substituent is shown for the dibenzamide (**564**); with ethyl glycinate substitution takes place at C-7, whereas it is the benzoyl group which is split off from the amino group at C-5 ⟨79JOC302⟩.

The ease of the reaction makes it suitable for introduction of aminosugars at C-7; subsequent reductive cleavage of the furazan ring yields important intermediates for purine nucleoside syntheses ⟨80MI42901⟩. Carbon nucleophiles will also displace an acylamino substituent at C-7, e.g. as in (**565**) ⟨80JOC3827⟩.

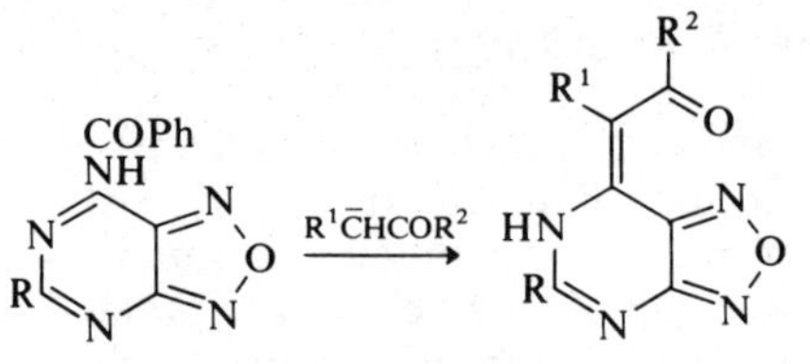

The high π-deficiency at C-7 is also apparent from the nucleophilic addition reaction with sodium borohydride; selective formation of the 6,7-dihydro derivative (566) results with this reagent. The reaction may also proceed further by reductive loss of the 7-substituent ⟨76CPB235⟩. In the pyrazine analogue (567) both LAH and sodium borohydride treatment lead to saturation of the pyrazine ring ⟨78JOC341⟩.

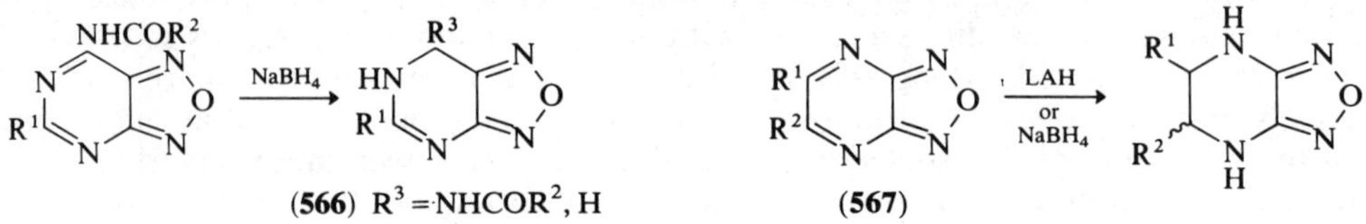

4.29.9.3.6 Nucleophilic attack with ring opening

The anion in the salts (568) and (569) is of low nucleophilic strength in order to avoid adduct formation with the cation and thereby decompositions. Solvolysis is to be expected as in the reaction of (568) with water. With sodium methoxide the 2-methoxypyridinium ylide (570) and its α-pyridone isomer (571) are formed. The former product arises from substitution of the oxygen, which may be regarded as an ether function attached to an activated pyridinium position. The other product (571) arises from competitive attack at C-2. The preferred reaction path can be influenced by the nature of the 2-substituent in (568). When the substituent is a p-methoxy or p-dimethylaminophenyl group, the preferential attack is at C-8a; a p-chloro or p-cyanophenyl substituent favours attack at C-2 ⟨80JOC5095⟩. Further activation of the pyridinium ring as in the 8-cyano derivative (569) gives exclusive attack in the pyridinium ring; the product from the reaction with cyanamide is recycled to a fused triazole (572) ⟨71JCS(C)3873⟩.

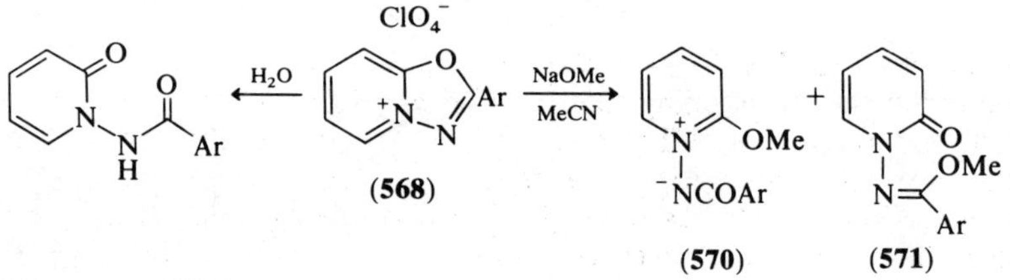

In non-charged systems where the azole oxygen is attached to the azine ring, nucleophilic attack may lead to opening of the ring at the point of oxygen attachment as shown for the reaction between the pyrimidine (573) and benzylamine ⟨78CI(L)92⟩. The final product is a fused triazole (574) due to recyclization of the first formed product. In the mesoion (575), however, the attack is at the oxo carbon at C-2 giving 1-amino-2-pyridinones. In the isomeric lactone (576) the attack is also at the oxo carbon at C-2, the product being a 2-aminopyridine 1-oxide ⟨58HCA548⟩.

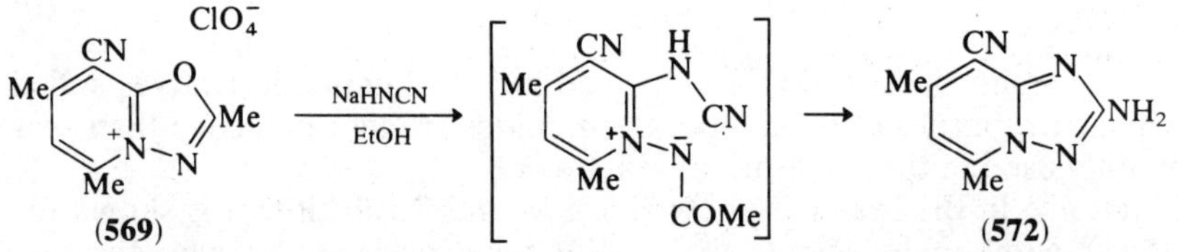

When a nitrogen of the azole is attached to the azine, reaction occurs at C-2 with opening of the azole ring, or in the azine ring. The pyrimidine (577) undergoes the former reaction with potassium hydroxide ⟨77H(6)107⟩, whereas the quinazoline (578) is opened by attack

(573) (574)

(575) X = OEt, NH$_2$

(576) X = OEt, NH$_2$

at C-5; recyclization gives the isomeric structure (**579**). The purine (**580**) is also opened between the azine nitrogens when treated with dilute acid ⟨76CJC2804⟩.

(577) R = COPh, H

(578) (579)

(580)

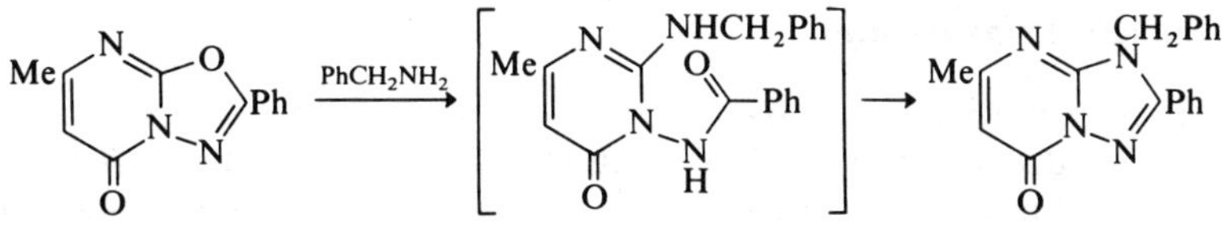

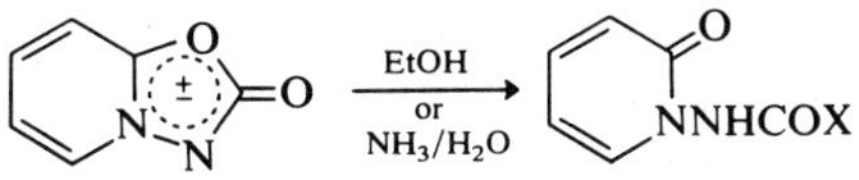

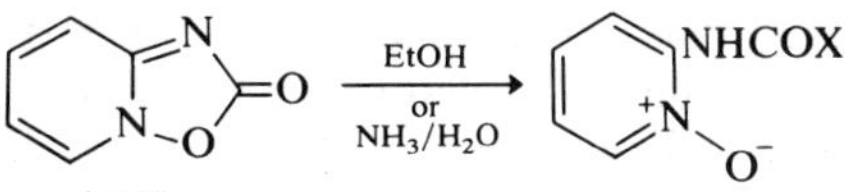

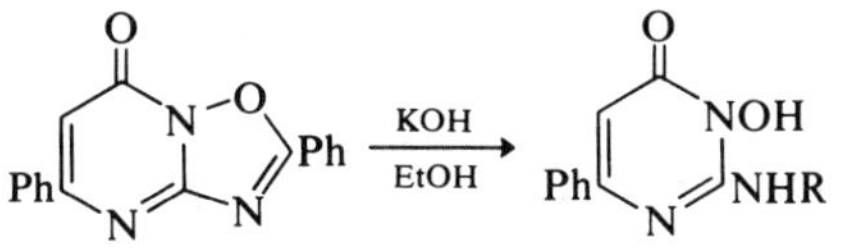

The furoxanouracil (**581**) reacts with benzylamine under vigorous reaction conditions with formation of the corresponding fused imidazole (**583**). It has been suggested that the reaction is initiated by a nucleophilic attack at C-3a leading to the 6-amino-5-nitrosouracil (**582**) which subsequently undergoes cyclization ⟨81H(15)341⟩.

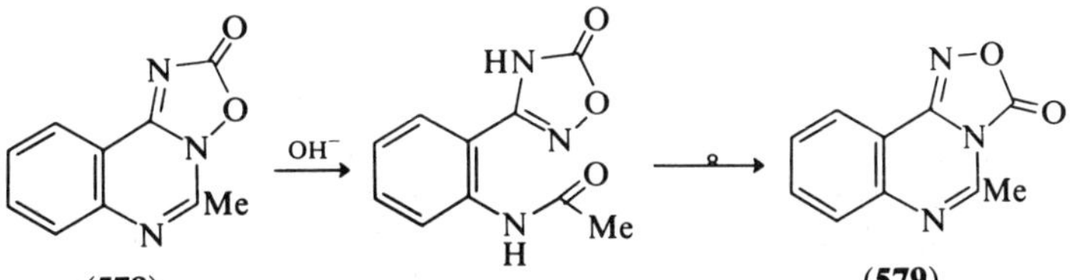

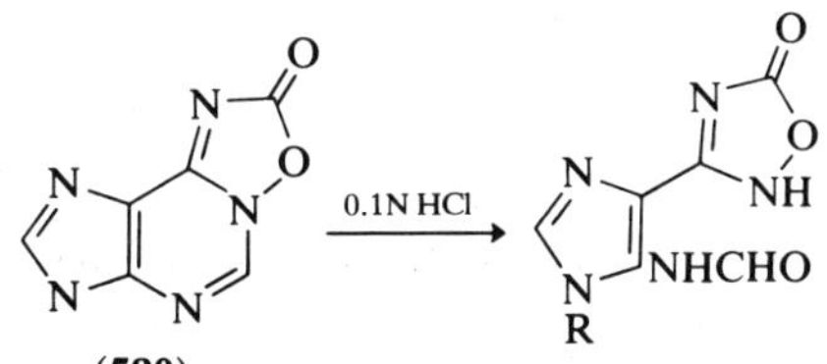

The furoxanopyridine (**584**) has its azole ring expanded into a pyrazine when reacted with enamines. The reactive species is assumed to be the 2,3-dinitrosopyridine (**585**) ⟨80M407⟩. Recent review literature on benzofuroxanes suggests extension of the enamine reaction to enolate anions and related species ⟨75S415, 76H(4)767⟩.

(581) (582) (583)

(584) (585)

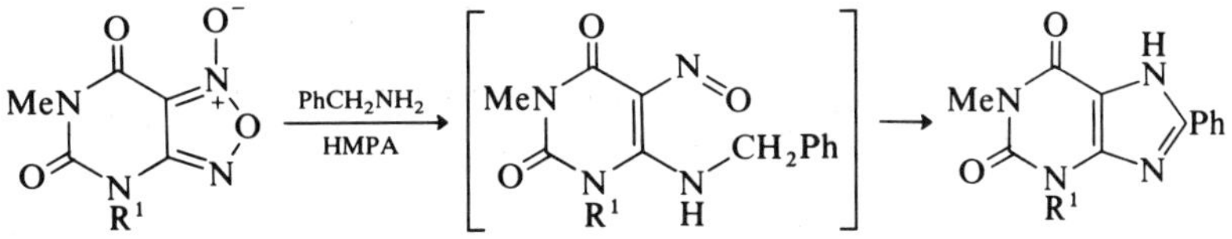

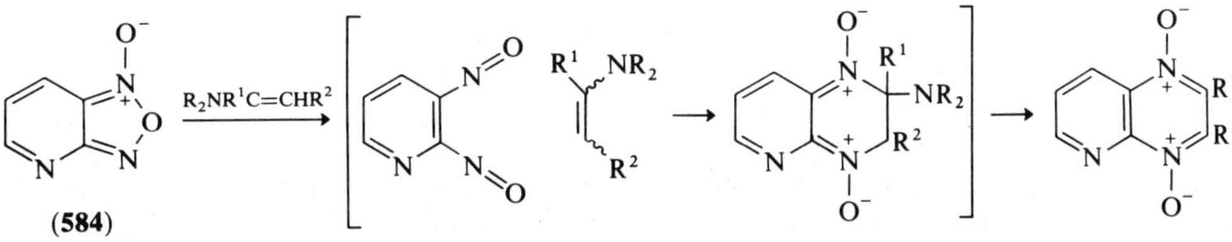

4.29.9.3.7 Reductive ring opening

The N—O bond can easily be cleaved by hydrogenolysis. In heterocyclic rings containing the N—O bond, hydrogenolysis leads to ring cleavage between these heteroatoms. The reaction is of importance in the chemistry of fused furazans since the reductive cleavage gives vicinal diaminoazines. Catalytic hydrogenolysis or dissolving metal reduction in acid solutions is used.

Hydrogenolysis of 7-acylamino pyrimidines (**586**) gives adenine derivatives (**587**); the intermediate 4-acylamino-5,6-diaminopyrimidine is rapidly recyclized because of the nucleophilic 5-amino group. Appropriately 7-substituted furazano[3,4-*d*]pyrimidines, which are readily accessible by substitution reactions at C-7, are thus important intermediates for the preparation of other fused ring systems, *e.g.* pyrrolo[3,2-*d*]pyrimidines (**589**) from (**588**) ⟨80JOC3827⟩, pteridines (**591**) from (**590**) and similarly 7-azapteridines ⟨73JOC2238⟩. Catalytic hydrogenolysis of the furazan ring in the pyrazine (**592**) proceeds similarly ⟨78JOC341⟩.

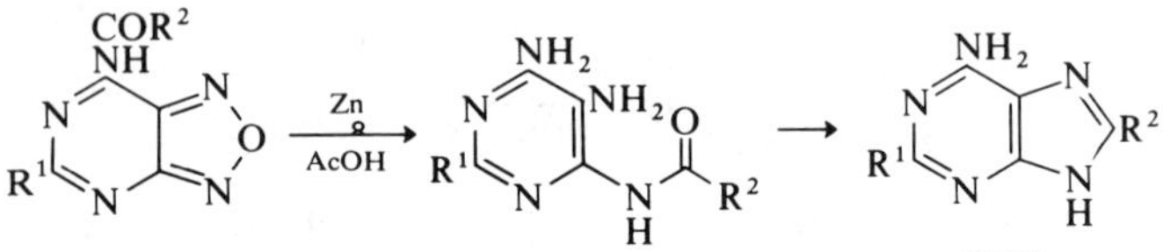

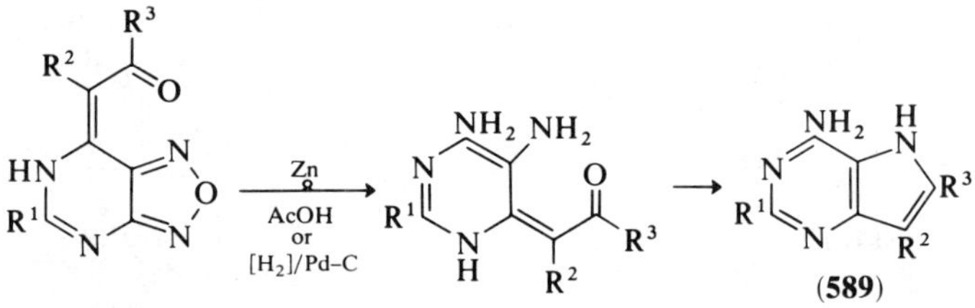

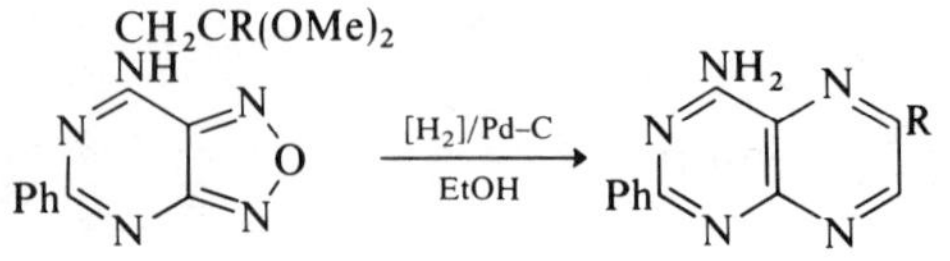

4.29.9.3.8 Nucleophilic attack at hydrogen

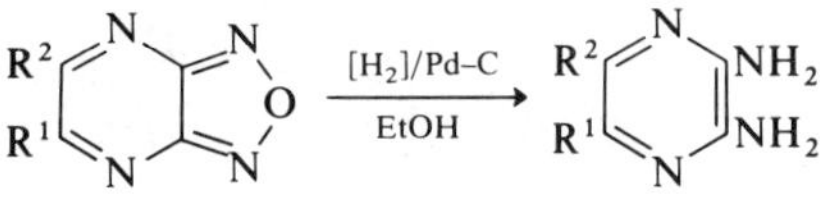

In systems with a methine carbon in the azole ring, proton abstraction by base may lead to ring cleavage as observed in related thiadiazoles. This is a common reaction in isoxazole analogues.

4.29.9.3.9 Other C-linked substituents

Alkyl groups attached to either π-electron-deficient ring will be activated for electrophilic attack after deprotonation as in aldol-type condensations.

4.29.9.4 Synthesis

4.29.9.4.1 1H-[1,2,3]Oxadiazolo[3,4-a]pyridine

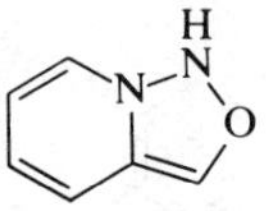

Azine approach. Sydnones are prepared by cyclodehydration of *N*-nitroso-α-aminocarboxylic acids. Fused sydnones can be prepared in essentially the same manner as shown for the pyridoindole (**593**) ⟨70USP3524859⟩ and the thienopyridine (**594**) ⟨81H(16)35⟩.

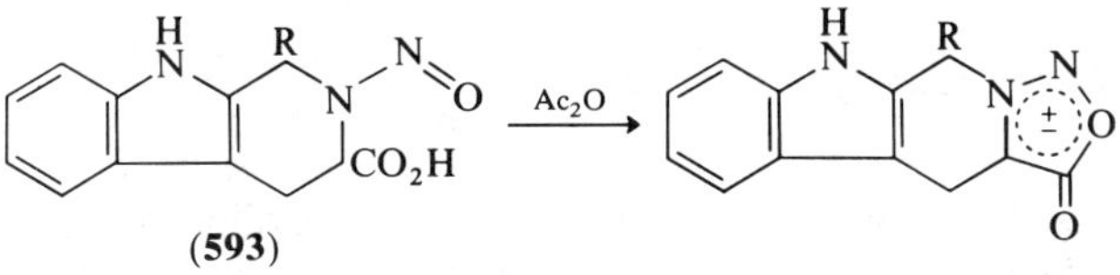

Five-membered Rings Fused with Six-membered Rings

(593)

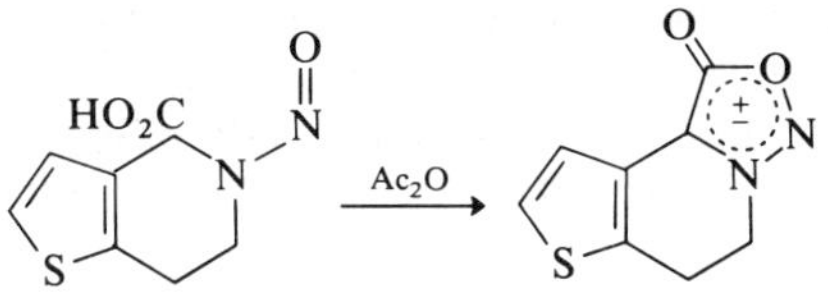

(594)

Azole approach. A sydnone may be the starting material for the fusion of another ring. In (595) the strong base methyllithium abstracts a proton at the amino group and a proton in the sydnone ring; treatment of metallated (595) with CO_2 results in cyclization ⟨72JOC1707⟩. The sydno[3,4-a]pyrazine (596) is formed by a typical isoquinoline-type synthesis ⟨73JAP7329328⟩ in which the sydnone ring takes the place of the phenyl ring.

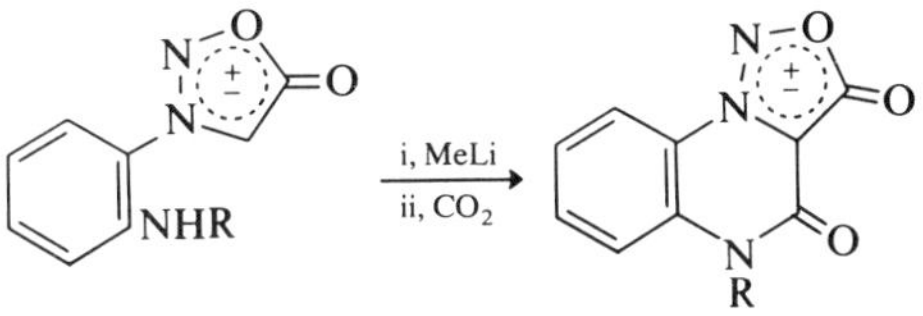

(595)

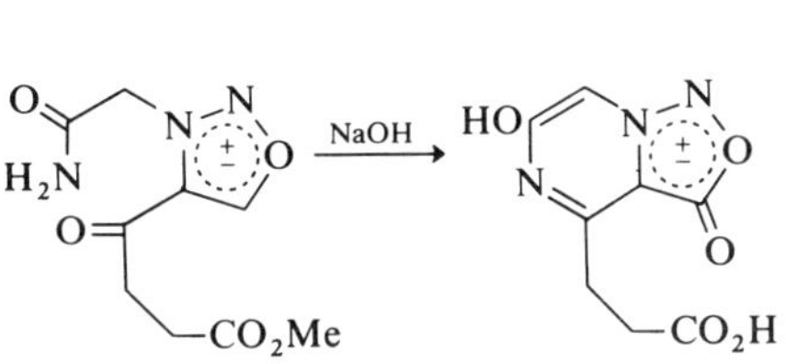

(596)

4.29.9.4.2 2H-[1,2,4]Oxadiazolo[2,3-a]pyridine

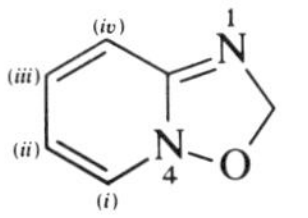

Azine approach. 2-Substituted derivatives are readily accessible from the *N*-oxide of an α-aminoazine. Thus heating 2-ethoxycarbonylamino- or ureido-pyridine 1-oxide (597) leads to cyclocondensation ⟨56JCS2063⟩. The same 2-oxo derivative is formed from 2-aminopyridine 1-oxide and phosgene. The analogous reaction with thiophosgene yields 2-thiones (598) ⟨77CJC3736⟩. Cyanogen bromide similarly yields 2-imines (599); benzo homologues react in the same way ⟨75YZ87⟩.

Aromatic *N*-oxides undergo 1,3-dipolar cycloaddition reactions. Phenyl isocyanate with pyridine 1-oxides involves a reaction sequence which in the end gives a 2-anilinopyridine. The initial product is the ring-fused 1,2-dihydropyridine 1-oxide (600), which rapidly undergoes a 1,5-sigmatropic shift to its more stable 2,3-dihydro counterpart (601) which is then aromatized by CO_2 expulsion ⟨76JHC171, 80H(14)19⟩.

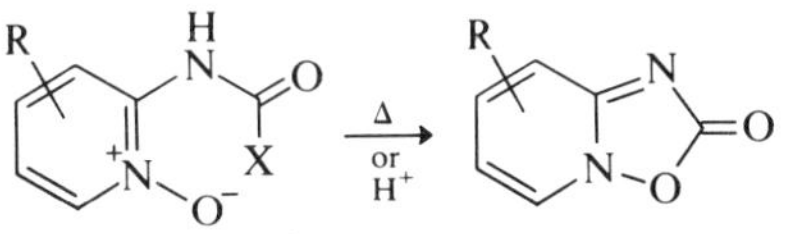

(597) $X = OR^1$, NR^1R^2, Cl

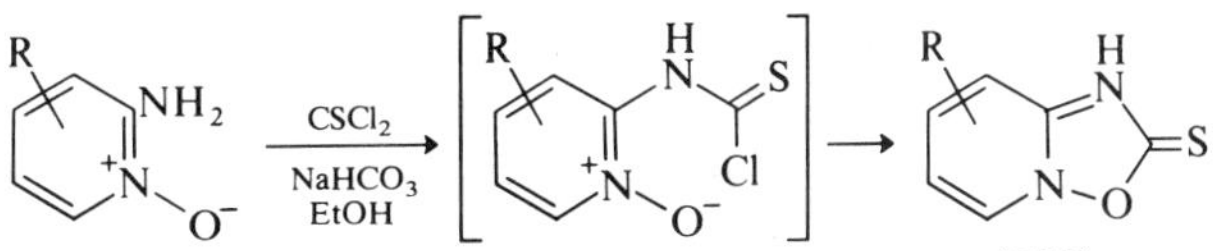

(598)

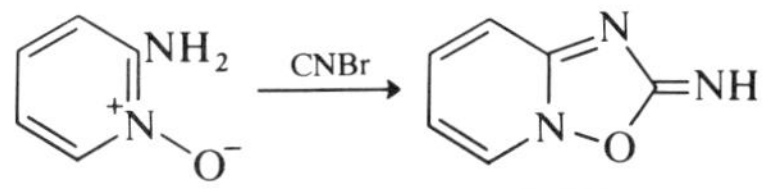

(599)

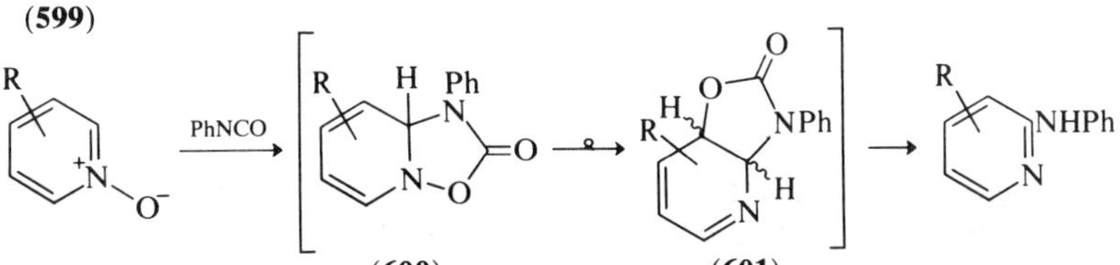

(600) (601)

(ii) *2H-[1,2,4]Oxadiazolo[2,3-c]pyrimidine*

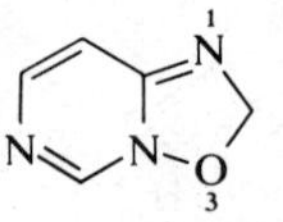

Azine approach. DCC dehydration of the 3-oxoquinazoline-4-hydroxamic acid (**602**) gives an isocyanate (**603**) *via* a Lossen rearrangement; addition of the *N*-oxide oxygen to the isocyanate group effects the cyclization. The same product is formed by the phosgene reaction with 4-amino-2-methylquinazoline 3-oxide ⟨76TL3615⟩.

The cyanogen bromide reaction with the nucleoside adenosine 1-oxide (**604**) gives the 2-imino-7-β-D-ribofuranosyl[1,2,4]oxadiazolo[2,3-*f*]purine (**605**) as the HBr salt. The latter on neutralization tautomerizes to N^6-cyanoadenosine 1-oxide ⟨78CPB2122⟩.

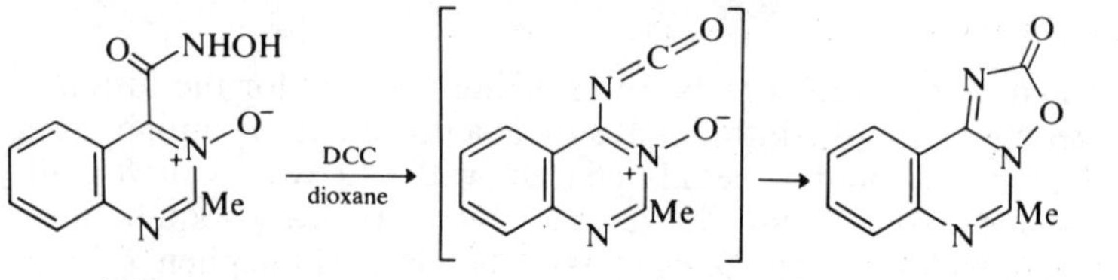

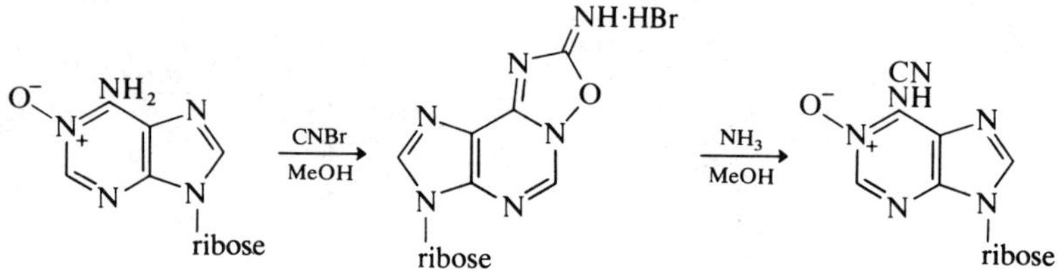

(iii) *2H-[1,2,4]Oxadiazolo[2,3-a]pyrazine*

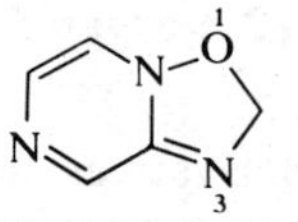

Azine approach. The reaction of phosgene with 2-aminoquinazoline 1-oxides leads to the tricyclic oxadiazole (**606**) ⟨74GEP2232468⟩.

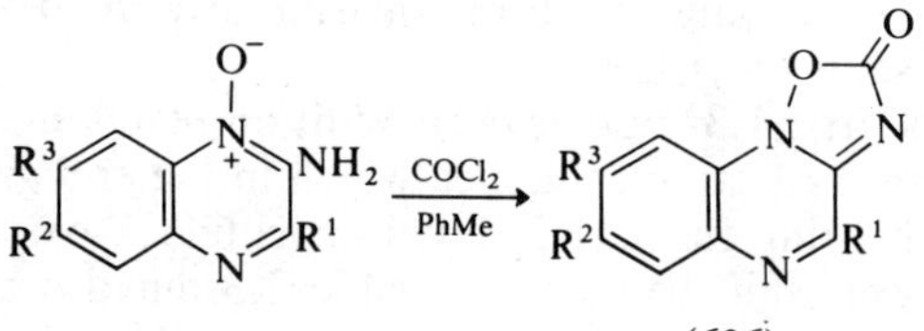

(iv) *2H-[1,2,4]Oxadiazole[2,3-a]pyrimidine*

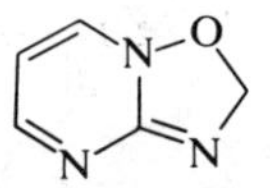

Azine approach. Phosgene ⟨80JAP(K)8017386⟩ or ethyl chloroformate reacts with 2-aminopyrimidine 1-oxide to form (**607**) ⟨78GEP2804519⟩; the triazine (**608**) is similarly formed ⟨80EUP7643⟩.

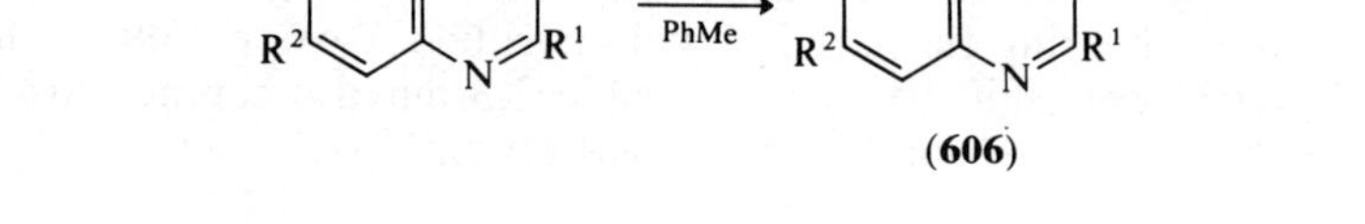

Azole approach. In the cyclocondensation of the 1,2,4-oxadiazole-3-enamine (**609**) with PTSA catalysis the N-2 nitrogen of the azole moiety becomes the bridgehead nitrogen atom ⟨77H(6)107⟩.

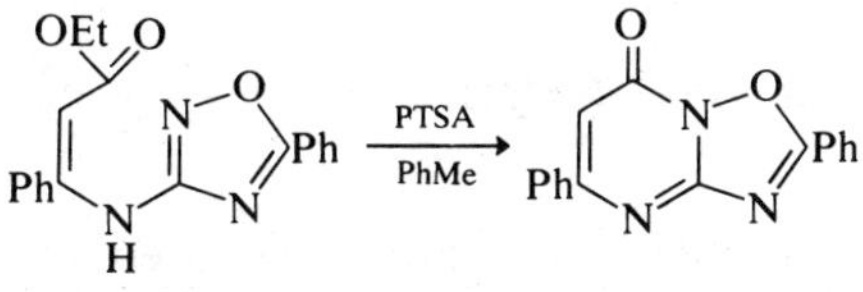

(609)

4.29.9.4.3 3H-[1,2,4]Oxadiazolo[4,3-a]pyridine

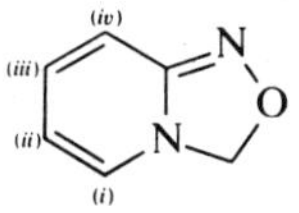

Most syntheses employ hydroxylamino derivatives.

(i) *3H-[1,2,4]Oxadiazolo[4,3-b]pyridazine*

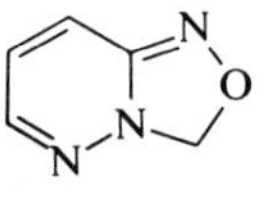

Azine approach. Cyclocondensation of the hydroxylamine ester (610) yields the corresponding fused phthalazine ⟨72BRP1285333⟩.

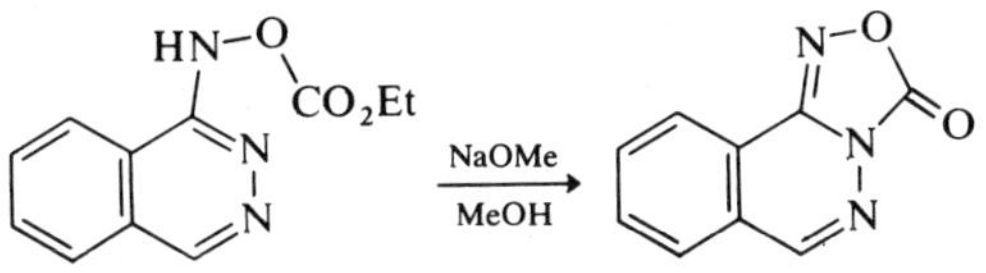

(610)

(iii) *3H-[1,2,4]Oxadiazolo[4,3-a]pyrazine*

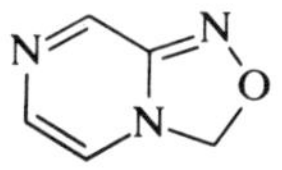

The oxime (611) when reacted with ethyl chloroformate undergoes cyclocondensation to form the dihydroquinoxaline (612); the latter is aromatized on treatment with diethyl azodicarboxylate ⟨73LA190⟩.

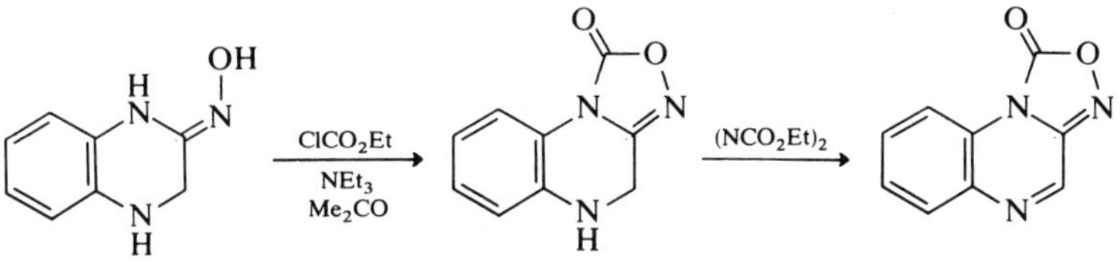

(611)　(612)

4.29.9.4.4 5H-[1,2,4]Oxadiazolo[4,5-a]pyridine

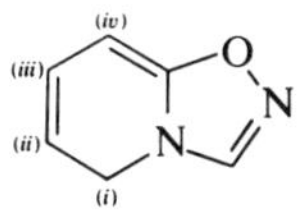

(iv) *5H-[1,2,4]Oxadiazolo[4,5-a]pyrimidine*

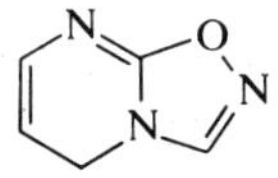

Azine approach. Substituted 6,7,8,8a-tetrahydro-5H-[1,2,4]oxadiazolo[4,5-a]-pyrimidines (613) have been prepared by cycloaddition of nitrile oxides or their chlorooxime equivalents to a tetrahydro pyrimidine ⟨73GEP2155753⟩.

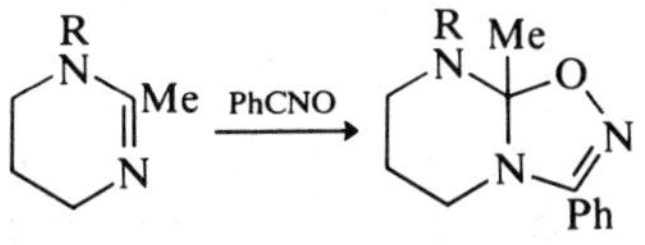

(613)

4.29.9.4.5 5H-[1,3,4]Oxadiazolo[3,2-a]pyridine

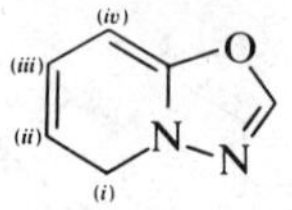

Azine approach. *N*-Acylaminopyridines are often used as starting materials for the preparation of members of this ring system. In the pyridinium derivative (**614**) it is the carbamoyl oxygen which is the nucleophile for the chlorine substitution whereby the azole ring of the oxadiazolopyridinium salt (**615**) is formed ⟨80JOC5095⟩. When the starting material is a 1-acylamino-2(1*H*)-pyridinone, the cyclodehydration is effected by an acid of low nucleophilicity to prevent an anion-induced opening of the azolylium ring (**616**) ⟨70JCS(C)1397⟩. Mesoionic compounds are similarly prepared. Thus the methoxycarbonyl or arylaminocarbonyl derivatives (**617**) are cyclized by heating to the 2-oxo mesoions (**618**) ⟨80TL4025, 74CC621⟩. Thiophosgene on 1-amino-2(2*H*)-pyridinones gives the 2-thioxo mesoions (**619**). The latter are also available from the iminophosphoranes (**620**) and carbon disulfide. Iminophosphoranes are versatile intermediates for many purposes. Thus (**620**) reacts with isocyanate to form the imino mesoion (**621**) ⟨80TL4025⟩.

(**614**) (**615**) (**616**)

(**617**) X = OMe, NHAr (**618**) (**619**)

(**620**) (**621**)

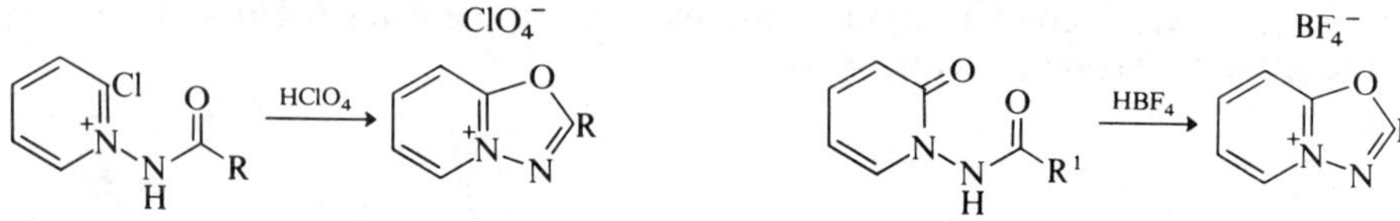

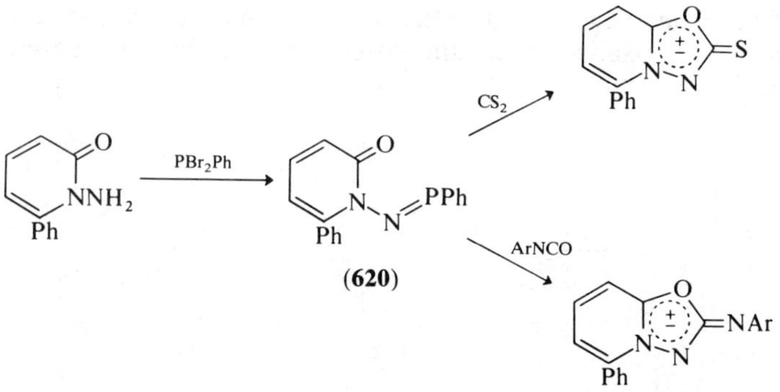

(iv) *5*H-*[1,3,4]Oxadiazolo[2,3-*a*]pyrimidine*

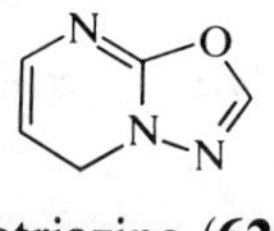

Azine approach. Acylation of the 3-aminotriazine (**622**) results in cyclocondensation to the corresponding oxadiazole ⟨67CB2585⟩. The facile cyclization is caused by the high activation of the 3-position in a 1,2,4-triazine for nucleophilic substitution.

(**622**) R = Ph, OH

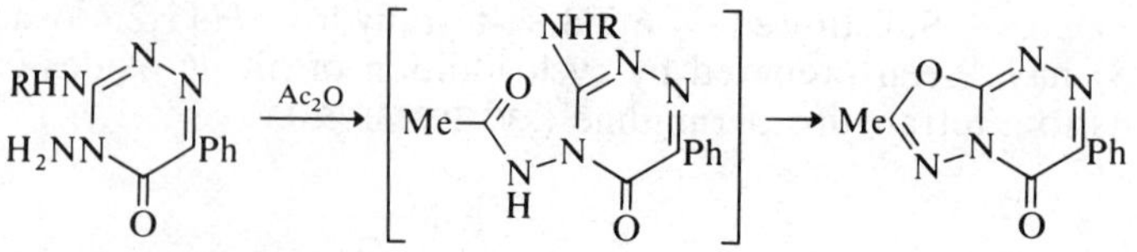

Azole approach. Because 2-amino-1,3,4-oxadiazoles are readily available, they are commonly used as starting materials in the preparation of fused pyrimidines. The 2-amines react with 1,3-dicarbonyl compounds to form the fused pyrimidinium salt (**623**) ⟨75KGS1493⟩. With carbon suboxide the dihydroxypyrimidine (**624**) is formed ⟨62M1441⟩. The enamine (**625**) from the reaction with ethoxymethylenemalonate is cyclized by heating. Enamines from the reactions with β-ketoesters or acetylenecarboxylic acid esters (**626**) are also cyclized by heating ⟨70AP501, 73T2937⟩. With isocyanate a triazine (**627**) can be formed ⟨74T221⟩.

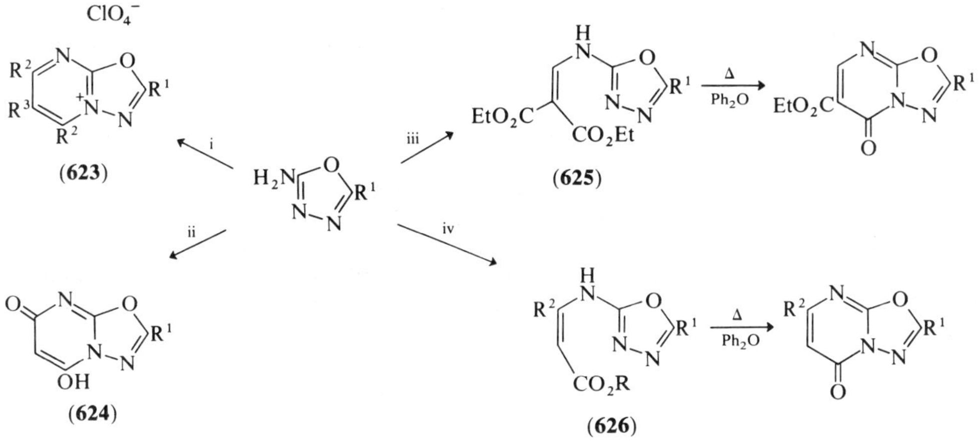

i, $(R^2CO)_2CHR^3$; ii, C_3O_2; iii, $EtOCH=C(CO_2Et)_2$; iv, $AcCH_2CO_2Et$ or $R^2C\equiv CCO_2R$

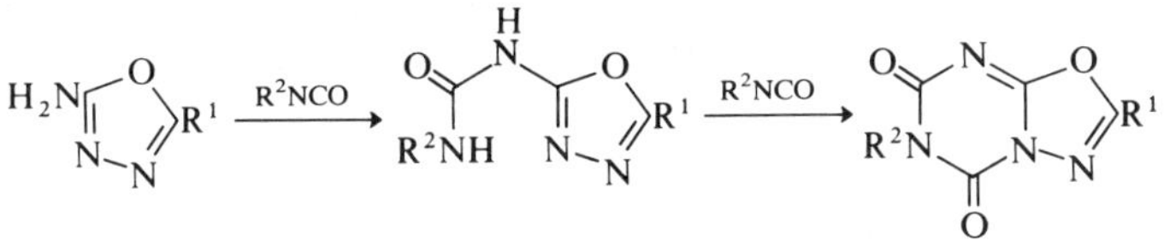

4.29.9.4.6 1H,3H-[1,3,4]Oxadiazolo[3,4-a]pyridazine

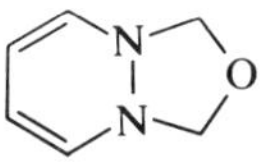

Azine approach. C-1 and C-3 in the ring system can be regarded as aminal carbons; the ring system is formed accordingly. The perhydro derivative (**628**) can be formed from acetaldehyde and perhydropyridazine ⟨72JA7108⟩.

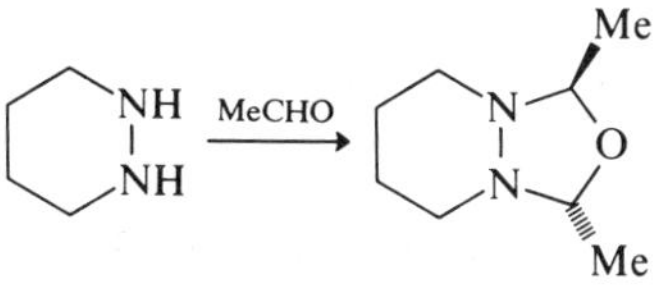

4.29.9.4.7 2H-[1,2,5]Oxadiazolo[2,3-a]pyridine

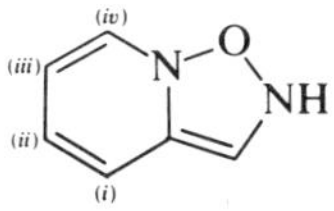

(i) 2H-[1,2,5]Oxadiazolo[2,3-a]pyrimidine

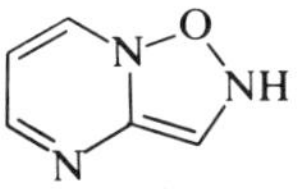

Azole approach. Salts are prepared by treatment of 4-amino-1,2,5-oxadiazoles with α,β-unsaturated carbonyl compounds and perchloric acid. The orientation of the substituents in the product suggests that it is the nuclear nitrogen (N-2) which is involved in the Michael addition giving (**629**) before cyclization ⟨80UKZ637⟩.

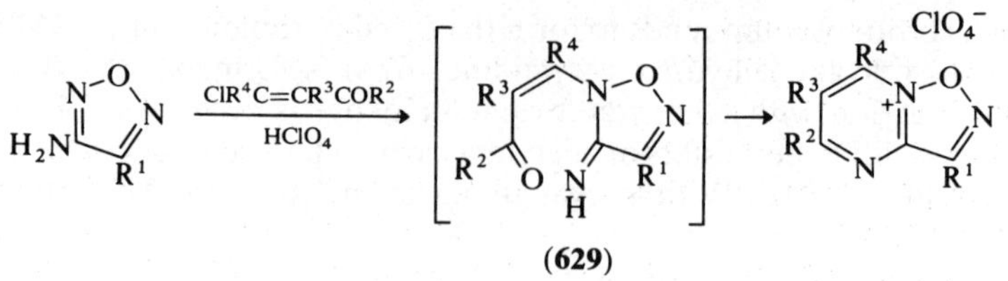

(629)

4.29.9.4.8 [1,2,3]Oxadiazolo[4,5-b]pyridine

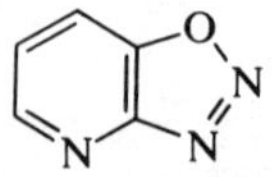

Azine approach. Diazotization of 2-amino-3-hydroxypyridines might be expected to yield the bicyclic ring system (**631**) by bond formation between the oxygen and the diazonium group in the first formed diazonium salt; fused 1,2,3-thiadiazoles and -triazoles are formed by this type of reaction. Compounds prepared by this method, however, are highly unstable and have spectroscopic properties consistent with the pyridodiazooxide structure (**630**) ⟨74CS(6)222⟩.

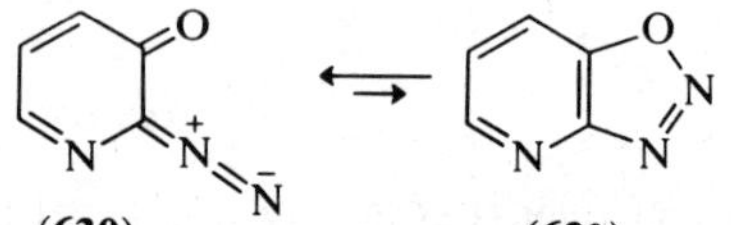

(630)   (631)

All spectroscopic evidence for benzo analogues shows that these compounds exist predominantly in the quinonoid form as benzodiazooxides ⟨66RCR388⟩.

4.29.9.4.9 [1,2,3]Oxadiazolo[4,5-c]pyridine

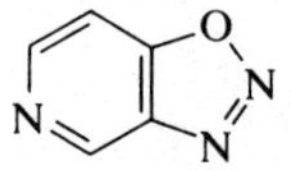

Azine approach. The diazotization reaction of 3-amino-4-hydroxypyridine gives an unstable compound with spectroscopic properties consistent with the pyridodiazooxide form (**632**) ⟨74CS(6)222⟩.

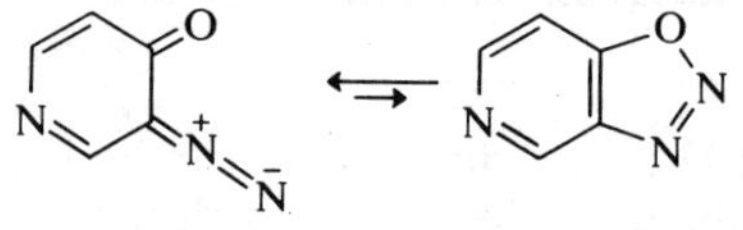

(632)

4.29.9.4.10 [1,2,3]Oxadiazolo[5,4-b]pyridine

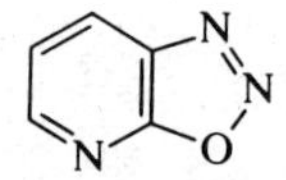

Azine approach. The diazotization reaction of 3-amino-2-hydroxypridine gives an unstable product with spectroscopic properties consistent with a pyridodiazooxide structure (**633**) ⟨74CS(6)222⟩; the 7-nitro analogue behaves in the same way ⟨62MI42900⟩.

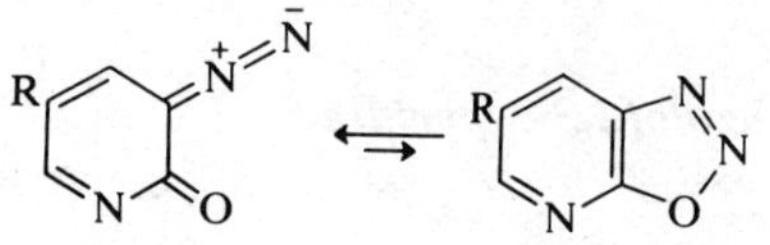

(633)

4.29.9.4.11 [1,2,5]Oxadiazolo[3,4-b]pyridine

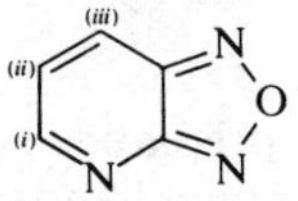

Azine approach. Benzofurazans are conveniently prepared by oxidative cyclization of o-nitrosoanilines. Benzofuroxanes, the N-oxide derivatives of benzofurazans, are similarly prepared from *o*-nitroanilines. The same principle is used in the preparation of azine analogues. Thus, the furoxanopyridines (**634**) are available by oxidative cyclization of 2-amino-3-nitropyridines using iodobenzene diacetate ⟨70JCS(B)636⟩; tautomerism to the 3-oxide isomer should be kept in mind.

2-Nitro-1-azidobenzenes are cyclized to benzofuroxanes on heating. Similarly, 2-azido-3-nitropyridine is smoothly cyclized to furoxanopyridines (**635**) on heating in a dioxane solution. The azidopyridine is in a reversible equilibrium with the dominating tetrazole form (**636**) from which it is liberated for the cyclization reaction ⟨80M407⟩.

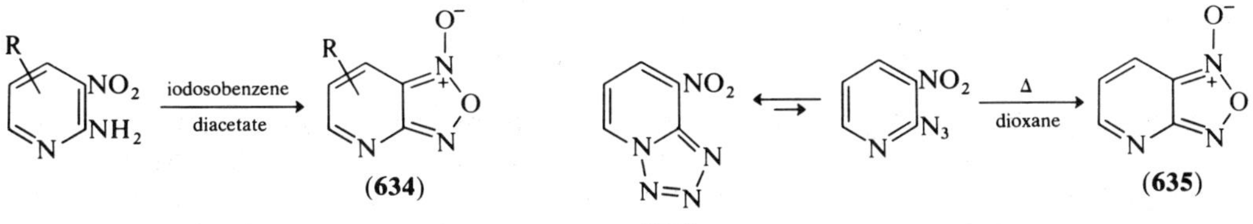

(ii) *[1,2,5]Oxadiazolo[3,4-d]pyrimidine*

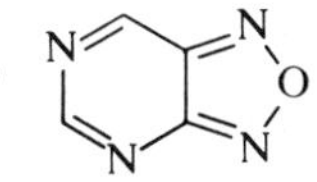

Azine approach. A number of 2-substituted 4,6-diamino-5-nitropyrimidines have been converted into 7-aminofurazano[3,4-*d*]pyrimidines (**637**) upon treatment with LTA in acetic acid. 6-Amino-5-nitrosouracil reacts similarly ⟨71JOC3211⟩. In principle the same reaction can be applied to the preparation of 7-amino-4*H*-[1,2,5]oxadiazolo[3,4-*c*][1,2,6]-thiadiazine 5,5-dioxide (**638**) ⟨76MI42900⟩.

Furoxano derivatives can be prepared by the thermal azido–nitro reaction as shown for (**639**) ⟨68JOC2086⟩.

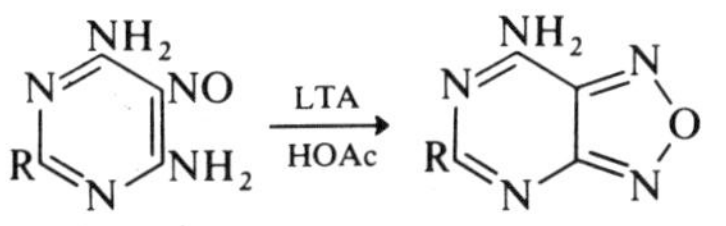

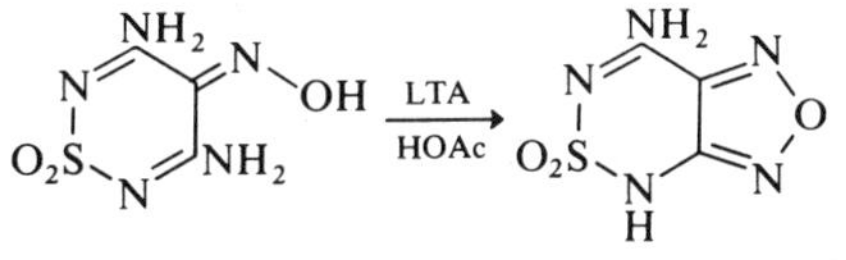

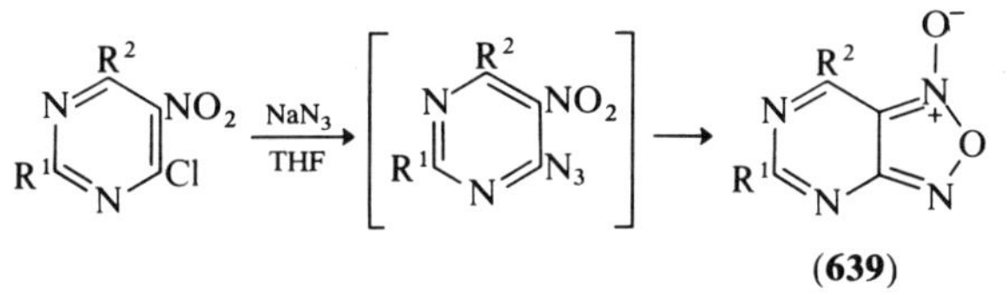

Azole approach. From the amidoxime (**640**) a typical pyrimidine fusion reaction yields the bicyclic product (**641**). The 7-hydroxyimine as first formed is transformed by the formic acid of the reaction to the 7-oxo derivative (**641**) ⟨65JHC253⟩.

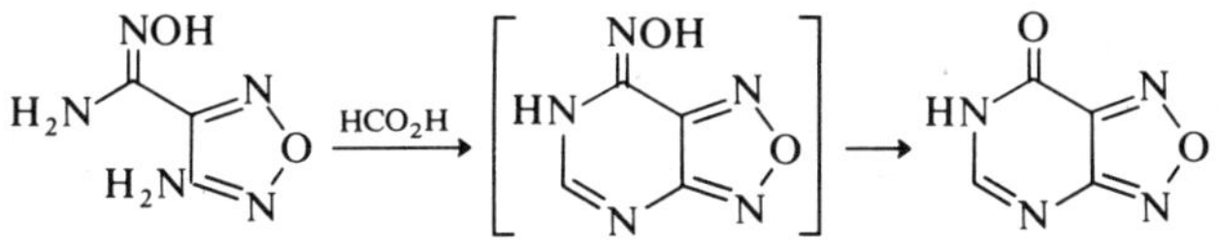

(iii) *[1,2,5]Oxadiazolo[3,4-b]pyrazine*

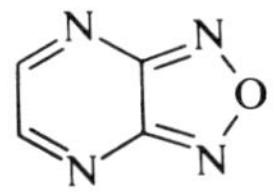

Azole approach. Condensation reactions between 3,4-diamino-1,2,5-oxadiazole and 1,2-dicarbonyl derivatives give pyrazines (**642**) ⟨78JOC341⟩.

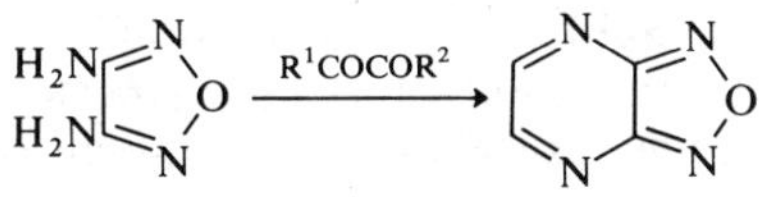

4.29.9.4.12 [1,2,5]Oxadiazolo[3,4-c]pyridine

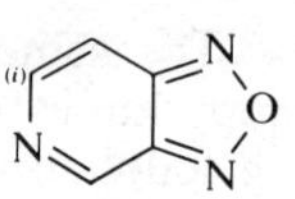

Azine approach. Thermal reactions between the vicinal nitro–azido groups in the pyridine (**643**) ⟨71JCS(C)1211⟩ and in the quinoline (**644**) ⟨68JCS(B)1516⟩ yield the fused furoxanes.

Nitrosation of 1-aminoquinolizinium salts gives furazano[3,4-a]quinolizines (**645**). The first formed product is probably a *C*-nitroso derivative which is subsequently oxidatively cyclized ⟨68JCS(C)1088⟩.

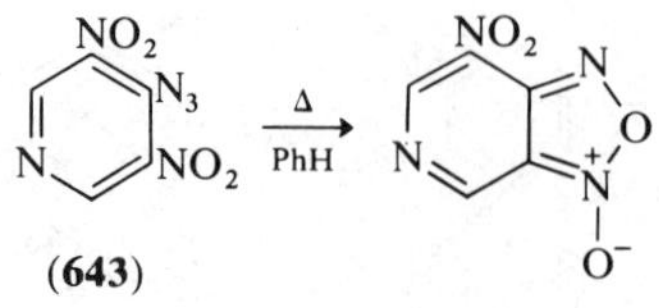

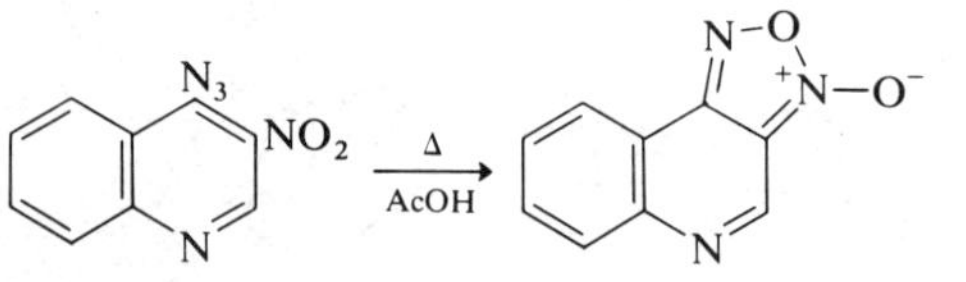

Azole approach. The reaction of 3,4-diacyl-1,2,5-oxadiazole with methylamine derivatives under acid or base catalysis provides a convenient method for the preparation of furazano[3,4-c]pyridines (**646**). In the presence of acid- or base-labile groups, DBU can be used to effect the reaction ⟨80S842, 79S687⟩.

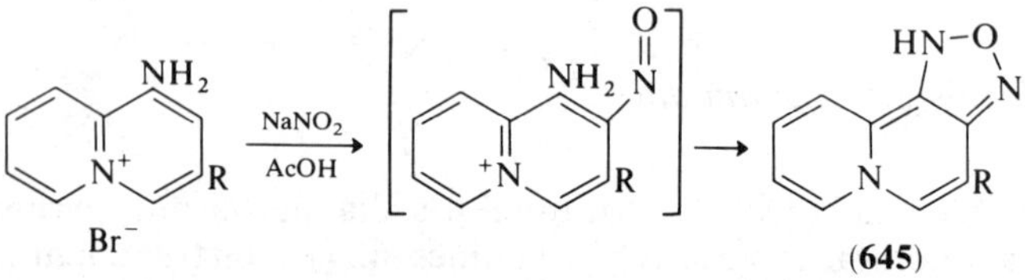

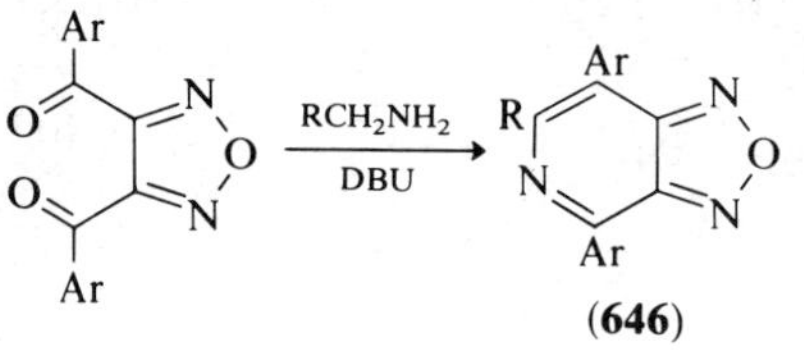

(i) *[1,2,5]Oxadiazolo[3,4-d]pyridazine*

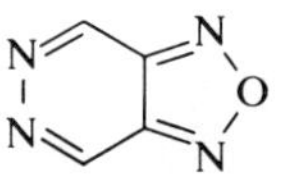

Azine approach. Tributylphosphine can be used for partial deoxygenation of *o*-dinitrosoarenes; with the 4,5-dinitrosopyridazine (**647**) this reagent leads to the formation of a fused furazan ⟨61JOC4684⟩.

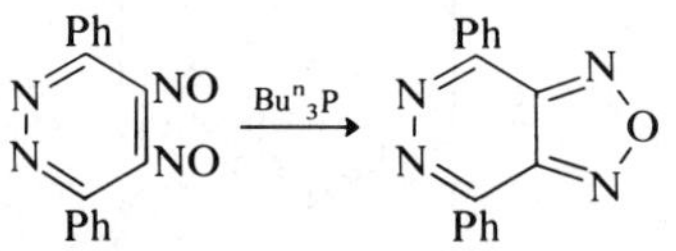

4.29.9.5 Applications

[1,2,3]Oxadiazolo[3,4-a]pyrazine (**648**) and analogues are claimed to have useful antiinflammatory and analgesic activities ⟨73JAP7329329⟩. The same claim has been made for compounds under formula (**649**) ⟨80JAP(K)8085590⟩. Compounds described by formula (**650**) are of interest for use in vasodilatory antihypertensive therapy ⟨78GEP2804518, 81MI42900⟩. The pyridazine derivatives (**651**) are said to be useful as virucides and bronchodilators ⟨72BRP1285333⟩.

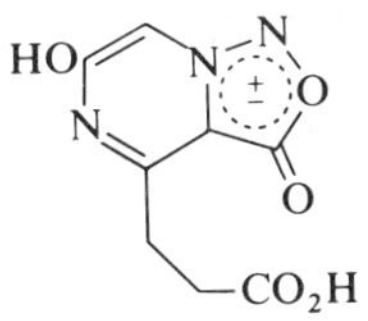

(648) 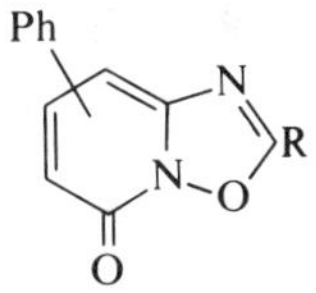 (649) 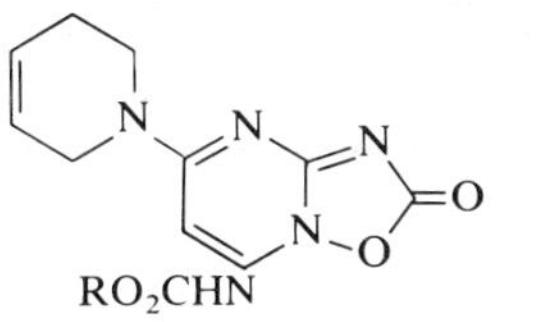 (650) 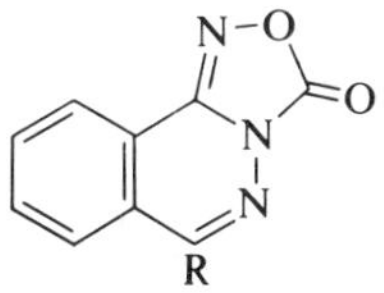 (651)

4.29.10 THIADIAZOLES FUSED WITH AZINES

4.29.10.1 Introduction

1,2,3-, 1,2,4-, 1,3,4- and 1,2,5-thiadiazoles can be fused to a six-membered ring containing the minimum number of nitrogens in 14 different ways; the section on synthesis has been organized accordingly. Further fused systems arise from the diazines and the triazines and their areno and heteroareno homologues.

This is a new and expansive area of heterocyclic chemistry. The thiadiazoloazine ring systems are in general readily accessible and most compounds are chemically stable. Nucleophilic substitution reactions and rearrangements are attracting attention. Known chemistry is almost entirely limited to the fully conjugated ring systems.

The main emphasis has been on synthetic chemistry. This often means preparation of compounds for biological purposes, which is possibly catalyzed by structural similarities to purines; many patents cover biological applications.

4.29.10.2 Structure

4.29.10.2.1 X-Ray diffraction

The X-ray data for 2-ethoxycarbonylimino-2H-[1,2,4]thiadiazolo[2,3-a]pyridine monohydrate show the molecule to be planar except for the ethoxy group which is bent out of plane (Table 27).

In 7-amino[1,2,5]thiadiazolo[3,4-d]pyrimidine the N—S bond (1.622 Å) is substantially shorter than the N—S single bond (1.735 Å) whereas the C—N bonds in the thiadiazole ring are shorter than the corresponding bonds in 9-methyladenine. Comparisons of bond lengths indicate that double bond character of the C—N bond is about 50% in the fused thiadiazole, which indicates an efficient electron delocalization between the ring as opposed to the finding for the oxadiazole analogue when a 75% C—N double bond character is indicated.

The five- and six-membered rings are tilted slightly from coplanarity about their common bond as is often found for purine analogues, probably resulting from the packing forces operating on the molecule in the crystal ⟨71JA7281⟩.

4.29.10.2.2 Molecular spectra

(a) ^1H NMR spectroscopy. Proton chemical shifts and coupling constants for some of the simple systems are listed in Table 28. It is interesting to note the great difference in chemical shift for the two N-methyl groups in the heteroaromatic betaine anhydro-5-hydroxy-3,6-dimethyl-7(6H)-oxo[1,2,3]thiadiazolo[4,5-d]pyrimidinium hydroxide (entry 4), which is largely caused by the charge separation.

(b) UV spectroscopy. UV data for some simple systems are listed in Table 29. It is seen that the charge separation in the heteroaromatic betaine anhydro-5-hydroxy-3,6-dimethyl-7(6H)-oxo[1,2,3]thiadiazolo[4,5-d]pyrimidinium hydroxide (entry 5) results in an absorption maximum of much greater wavelength than in related compounds without charge separation, e.g. entry 6.

Table 27 X-Ray Diffraction Data for Fused Thiadiazoloazines

	Bond lengths (Å) and angles (°)					
	Ring A			Ring B		
2-Ethoxycarbonylimino-2H-[1,2,4]thiadiazolo[2,3-a]pyridine monohydrate ⟨76AX(B)1601⟩	b^1 1.346 b^2 1.320 b^4 1.357 d^1 1.784 i^1 1.737	b^1b^2 111.9 b^1b^4 115.2 b^2d^1 114.1 b^4i^1 112.2 d^1i^1 86.6		a^1 1.341 a^2 1.410 a^3 1.360 a^4 1.393 b^3 1.358 b^4 1.357	a^1a^2 119.4 a^1b^3 119.7 a^2a^3 120.4 a^3a^4 119.3 a^4b^4 118.5 b^3b^4 122.7	
7-Amino[1,2,5]thiadiazolo[3,4-d]pyrimidine[a] ⟨71JA7281⟩	a^2 1.411 (1.426) b^1 1.342 (1.341) b^2 1.341 (1.351) i^1 1.629 (1.615) i^2 1.624 (1.619)	a^2b^1 115.0 (114.8) a^2b^2 113.2 (112.3) b^1i^1 104.9 (105.3) b^2i^2 106.1 (106.2) i^1i^2 100.8 (101.4)		a^1 1.442 (1.451) a^2 1.411 (1.426) b^3 1.383 (1.362) b^4 1.308 (1.322) b^5 1.368 (1.354) b^6 1.342 (1.355)	a^1a^2 118.2 (118.0) a^1b^6 117.6 (116.8) a^2b^3 123.0 (123.5) b^3b^4 113.2 (112.8) b^4b^5 129.5 (130.2) b^5b^6 118.6 (118.7)	

[a] Two unique molecules in the crystal lattice.

(c) *IR spectroscopy.* The use of IR spectroscopy is mainly limited to the analysis for specific substituents in the many different systems possible. It is interesting again to note that heteroaromatic betaines show absorption of their pseudocarbonyl groups at lower frequencies than in closely related compounds where there is no charge separation. The different betaines (**652**) and (**653**) have CO absorption below 1700 cm^{-1} whereas the uracil (**654**) absorbs at 1715 cm^{-1}.

(d) *Mass spectrometry.* The fragmentation patterns may vary greatly; the original literature for the specific system should be consulted.

4.29.10.3 Reactivity

4.29.10.3.1 Survey of reactivity

The thiadiazole rings are π-electron deficient and hence do not react readily with electrophiles at nitrogen or at carbon. The azine ring may be attacked by electrophiles at a nitrogen atom, especially if the ring contains a strongly electron-releasing substituent. Electrophilic attack at carbon is dependent on the presence of electron-releasing substituents.

Substitution chemistry in the fused systems is mainly nucleophilic. The π-electron-deficient thiadiazole promotes substitution in the azine ring; this is very pronounced at C-7 in 3,4-fused 1,2,5-thiadiazoles. Competition with opening of the thiadiazole ring is likely in many cases; especially sensitive are thiadiazolium salts. The fused thiadiazole also facilitates opening of the azine ring. One such reaction is the Dimroth rearrangement which readily occurs in appropriately substituted systems. Proton abstraction in the thiadiazole ring, if possible, may lead to cleavage of a hetero bond.

Reactivity in *C*-attached substituents towards appropriate reagents is increased by the π-electron deficiency of the fused ring system.

4.29.10.3.2 Thermal rearrangements

A reversible Dimroth rearrangement is observed between 1,4-dihydro-7*H*-*v*-triazolo[4,5-*b*]pyridine-7-thione (**655**) and 4-amino[1,2,3]thiadiazolo[4,5-*c*]pyridine (**657**) on heating

Table 28 ^{1}H NMR Data for Fused Thiazoloazines

Compound	Solvent	\multicolumn{7}{c}{^{1}H chemical shifts (p.p.m.)}	Coupling constants (Hz)	Ref.
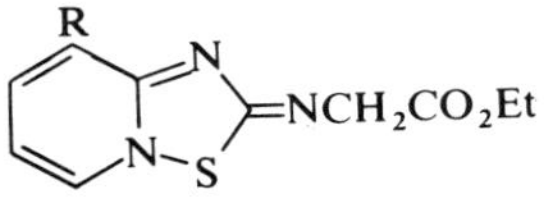		2	3	4	5	6	7	8		
(R-NCH$_2$CO$_2$Et thiazoloazine) R = CH	CDCl$_3$	—	—	—	8.65	7.35	8.05	7.71	4.90 (2-CH$_2$), $J_{5,6}$ 6.0, $J_{6,7}$ 7.0, $J_{7,8}$ 8.0, $J_{5,7}$ 1.6, $J_{6,8}$ 2.0, $J_{5,8}$ 0.8	78H(11)313
R = N	DMSO-d_6	—	—	—	9.19	7.25	8.95	—		77JHC621
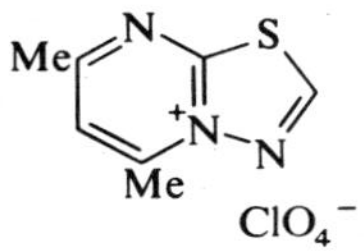 (Me thiazolo ClO$_4^-$)	TFA	9.86	—	—	3.21	7.91	2.94	—		71ZC460
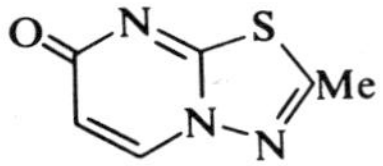 (Me thiazolo)	CDCl$_3$	2.66	—	—	8.10	6.33	—	—	$J_{5,6}$ 7.5	75JHC675
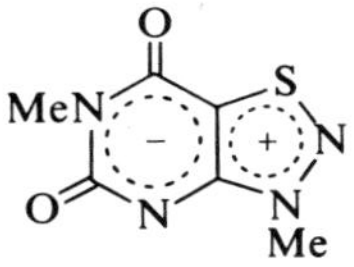 (MeN thiazolo dione)	DMSO-d_6	—	4.27	—	—	3.23	—	—		78JOC1677
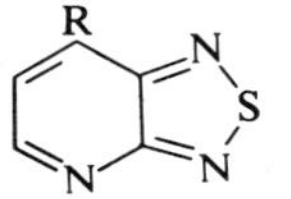 (thiazolopyridine) R = CH	CCl$_4$	—	—	—	9.06	7.55	8.32	—	$J_{5,6}$ 3.9, $J_{5,7}$ 1.9, $J_{6,7}$ 8.9	70RTC5
R = H	CDCl$_3$	—	—	—	9.06	9.06	—	—		76JHC13
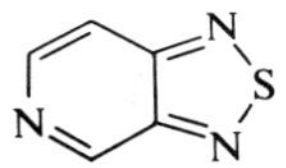 (thiazolopyridine)	CDCl$_3$	—	—	9.39	—	8.43	7.70	—	$J_{4,7}$ 1.2, $J_{6,7}$ 6.3, $J_{4,6}$ 0.1	70RTC5

Table 29 UV Absorption Data for Fused Thiadiazoloazines

Compound	Solvent	λ_{max}[a] (log ε)	Ref.
 [pyridinium fused thiadiazole-NHPh, Br⁻]	CHCl₃	352 (3.98), 292 (4.10)	72AJC993
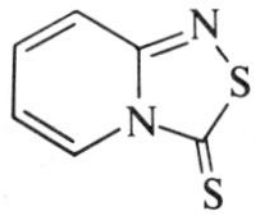 [pyrido-thiadiazole-thione]	MeOH	388 (3.83), 322 (3.72), 368 (3.72), 265 (3.50), 235 (4.07)	71JOC1846
R¹ = Me, R² = H	EtOH	267 (3.36), 213 (4.11)	75JHC675
R¹ = H, R² = OH	Pr^iOH	275[b] (3.75), 239 (4.11), 218 (4.44)	76JHC291
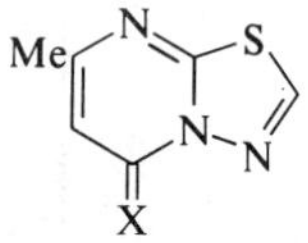 X = O	Pr^iOH	306 (3.90), 255[b] (3.49), 235[b] (3.65), 217 (4.02), 212.5 (4.03)	76JHC291
X = S	H₂O	354.5 (4.14), 297 (3.79), 231.5 (3.95)	78CPB2765
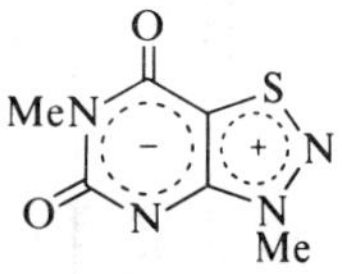 [mesoionic dimethyl fused]	EtOH	405 (3.42), 300 (2.69), 242 (3.81)	78JOC1677
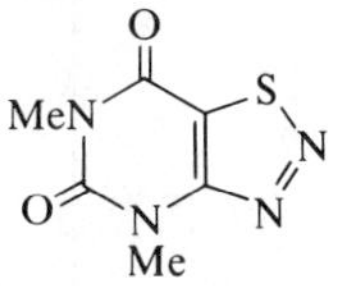 [dimethyl triazolopyrimidinedione]	EtOH	327 (3.79), 242 (3.81)	78JOC1677
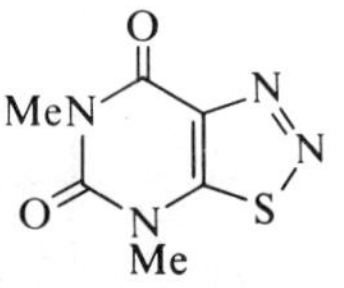 [dimethyl thiadiazolopyrimidinedione]	EtOH	275 (3.92), 230[b] (3.94)	77CPB2790
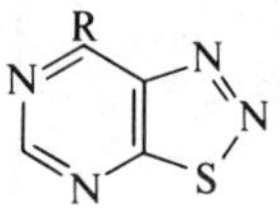 R = OMe	EtOH	280 (3.70), 260 (3.83), 234[b] (3.73)	64JOC2121
R = NH₂	EtOH	313 (3.79), 272[b] (3.51), 263[b] (3.46), 244 (3.69)	64JOC2121
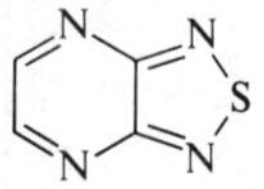 [pyrazino-thiadiazole]	CHCl₃	324 (4.37), 317 (4.31), 311 (4.27), 304 (4.13)	75JHC451
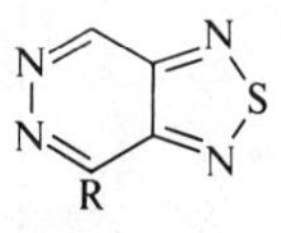 R = OH	EtOH	327 (3.66), 267[b] (3.39), 243 (4.03)	71JHC441
R = SH	EtOH	324 (3.76), 210[b] (4.12), 204 (4.14)	71JHC441
R = SMe	EtOH	301 (3.11), 260 (3.58), 238[b] (3.81), 218 (3.97)	71JHC441

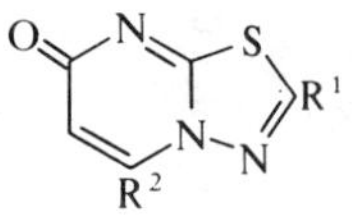

[a] nm. [b] Shoulder.

Table 30 IR Absorption Data for CO Groups in Fused Thiadiazoloazines

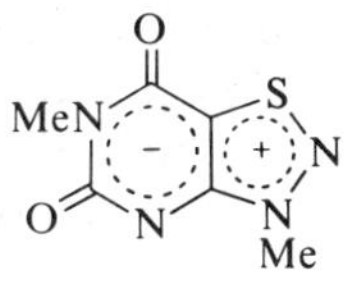 (652)	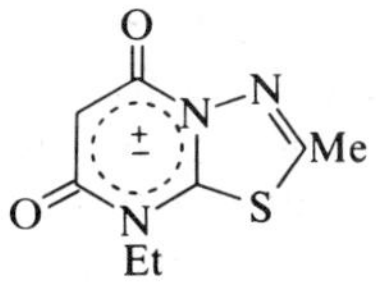 (653)	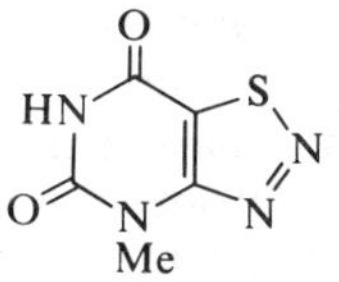 (654)
1690 cm^{-1}	1680, 1630 cm^{-1}	1715 cm^{-1}
Nujol	KBr	Nujol
78JOC1677	73JHC487	78JOC1677

in ethanol. The rearrangement is assumed to involve a diazo-type intermediate (**656**). The reaction is reversed from (**657**) to (**655**) in the presence of base ⟨72JOC3601⟩.

The rearrangement of fused v-triazoles as for the pyridine (**655**) was originally discovered in the pyrimidine series as an easily reversible rearrangement of 9-benzyl-6-mercapto-8-azapurine (**658**) to 7-benzylamino[1,2,3]thiadiazolo[5,4-d]pyrimidine (**659**) ⟨66AG596⟩. Heating drives the reaction over to the thiadiazole (**659**) whereas alkali treatment favors the v-triazole (**658**). The rearrangement may occur very easily as in the case of 8-aza-6-thioguanosine (**660**), which in water solution under N_2 rearranges slowly to N-β-D-ribofuranosyl[1,2,3]thiadiazolo[5,4-d]pyrimidine-5,7-diamine (**661**) without anomerization ⟨76JMC1186⟩. 1-Deaza-8-aza-6-thioguanosine (**662**) also undergoes a similar rearrangement to afford (**663**) on heating in DMF; a minor component was the α-anomer of (**663**) ⟨78JOC4910⟩.

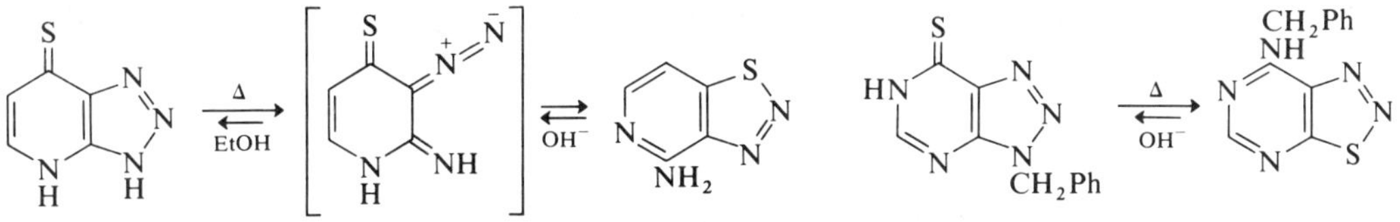

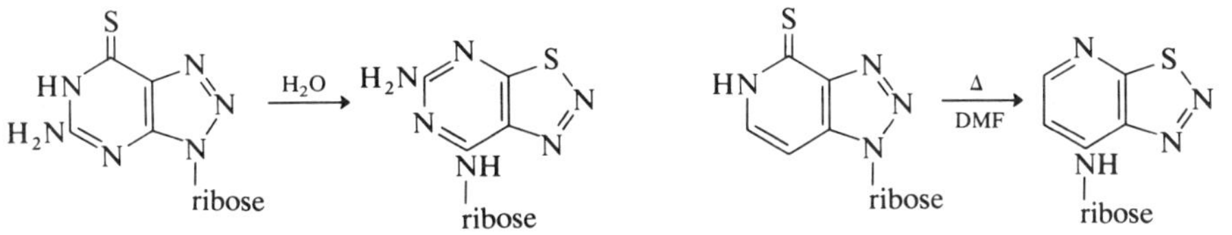

4.29.10.3.3 Electrophilic attack at nitrogen

The nitrogen atoms in the π-deficient azole ring are not very active towards electrophiles. In the azine ring the nitrogens will be attacked by electrophiles especially in the presence of electron-donating substituents. The usual preferential N-alkylation of lactams under alkaline conditions is observed for the uracil (**664**) ⟨65LA(682)90⟩. Acylation is on an amino group as usual in azines, e.g. in the formylation of the 7-amine (**665**) ⟨78JOC960⟩.

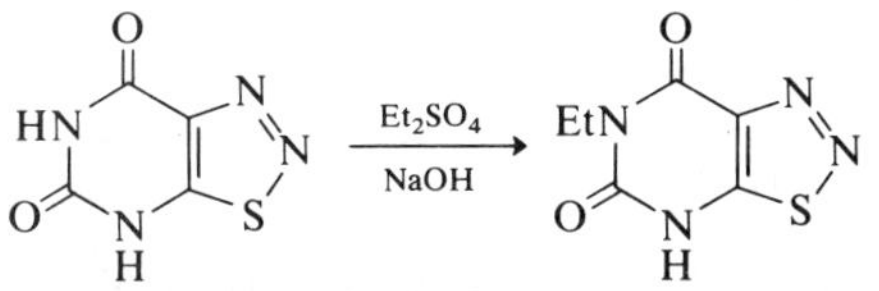

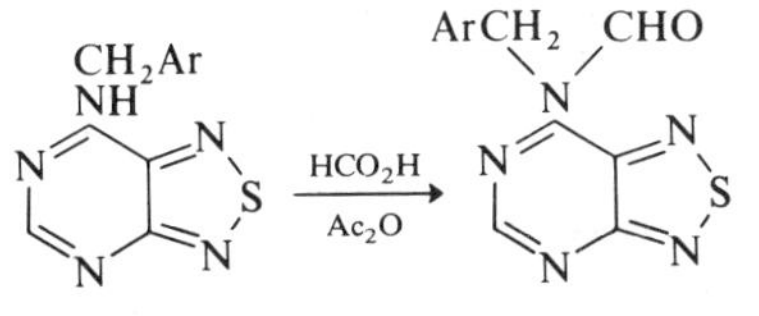

4.29.10.3.4 Electrophilic attack at carbon

Thiadiazoles like oxadiazoles are π-electron deficient systems. When fused to azines, electrophilic substitution on carbon is only to be expected in the presence of strongly electron-donating substituents. Thus the lactams (**666**) and (**667**) are activated for electrophilic substitution ⟨74BCJ2813⟩, whereas in the absence of an activating substituent the electrophile may attack an aromatic side-chain as in the case of (**668**) ⟨70JOC1965⟩.

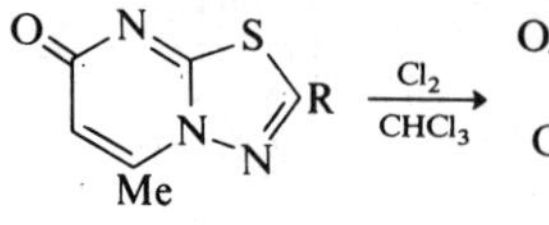

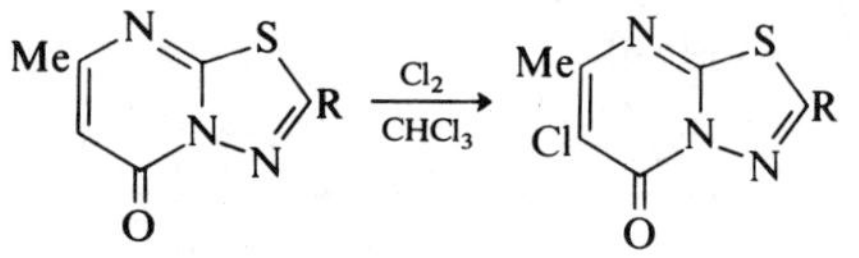

4.29.10.3.5 Nucleophilic attack at carbon

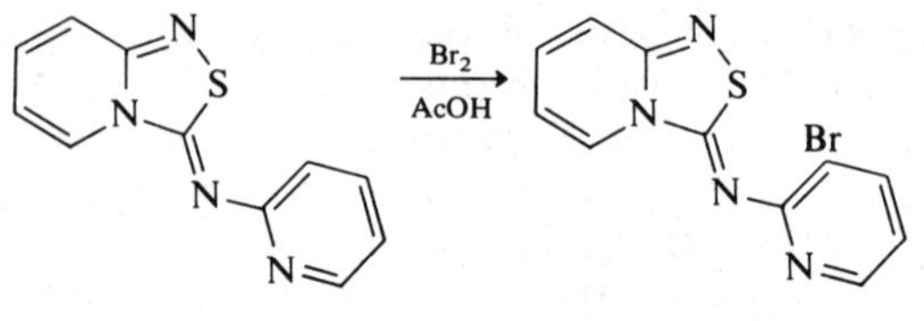

The thiadiazoles are π-electron deficient. This enhances the reactivity of substituents in the azine ring, and this is most pronounced in the 'orthoquinonoid' systems. But the electron deficiencies of these systems make them vulnerable for nucleophilic ring opening reactions. Sometimes the preferred reaction pathway may be difficult to predict. The nucleophilic ring opening reactions are treated in a separate section.

The activation in the orthoquinonoid structure (**669**) from the fusion of 1,2,5-thiadiazole makes the 7-chlorine substituent diplaceable under solvolytic conditions in aqueous acetic acid ⟨70RTC5⟩. The high activation is also manifested in the ease of halogen hydrolysis in the fused pyrazine (**670**) ⟨75JHC451⟩.

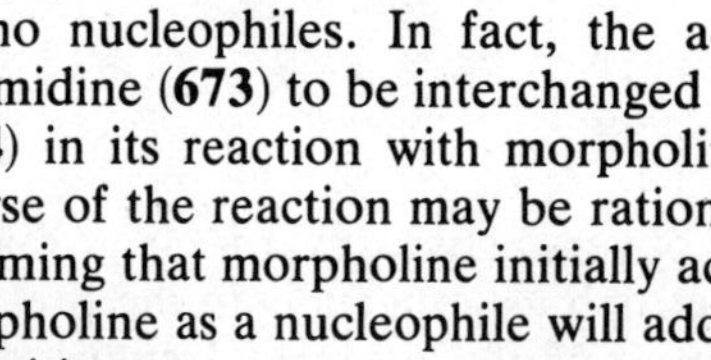

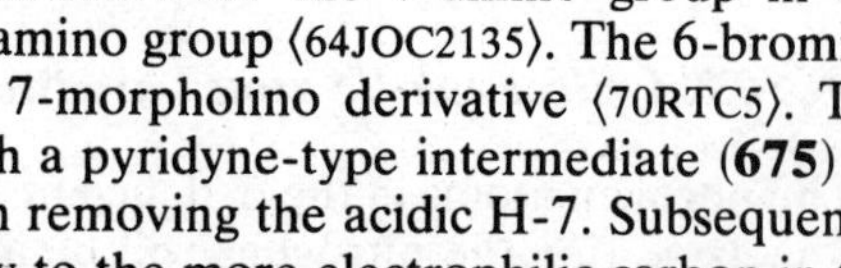

In the most highly activated position, at C-7, the methoxy group in the fused pyrimidine (**671**) ⟨64JOC2121⟩ and even the thioxo group in (**672**) ⟨78JOC1677⟩ can be substituted by amino nucleophiles. In fact, the activation is sufficient for the 7-amino group in the pyrimidine (**673**) to be interchanged with another amino group ⟨64JOC2135⟩. The 6-bromide (**674**) in its reaction with morpholine yields the 7-morpholino derivative ⟨70RTC5⟩. The course of the reaction may be rationalized through a pyridyne-type intermediate (**675**) by assuming that morpholine initially acts as a base in removing the acidic H-7. Subsequently morpholine as a nucleophile will add preferentially to the more electrophilic carbon in the 7-position.

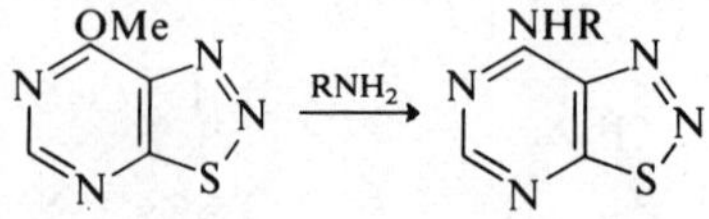

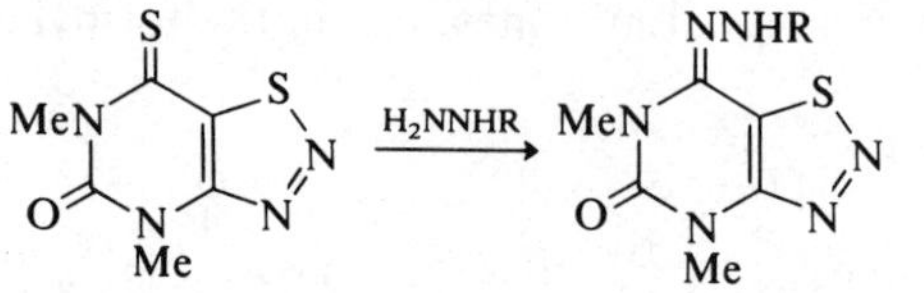

The usual azine reactions with sulfur nucleophiles are observed, *e.g.* chlorine substitution in the pyridazine (**676**). Oxo groups in activated positions are thiated using phosphorus

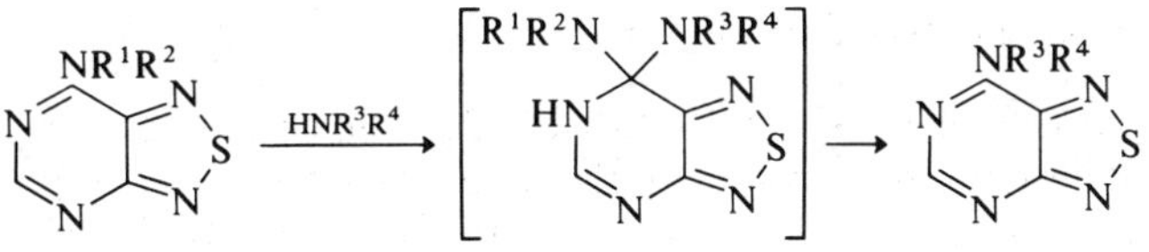

Five-membered Rings Fused with Six-membered Rings 737

(673)

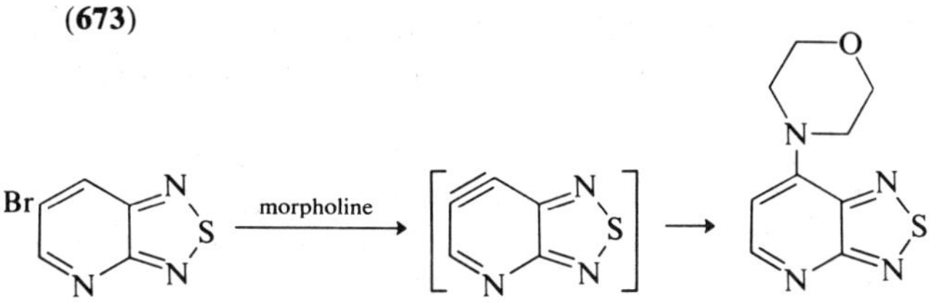

(674) → (675)

pentasulfide, e.g. in pyrimidine (677) ⟨78CPB2765⟩. In the uracil (678; Z = O) the more active 7-oxo group can be selectively thiated ⟨78JOC1677⟩ whereas both oxo groups in the pyridazine (679; Z = O) have been thiated ⟨71MI42900⟩.

Hydroxy groups in activated positions are displaced by chlorine using phosphorus chlorides as for the pyridazine (680) ⟨71MI42900⟩. Halogen interchange reactions are to be expected. Bromination of the chloropyrazine (681) can be effected using bromine in chloroform ⟨75JHC451⟩.

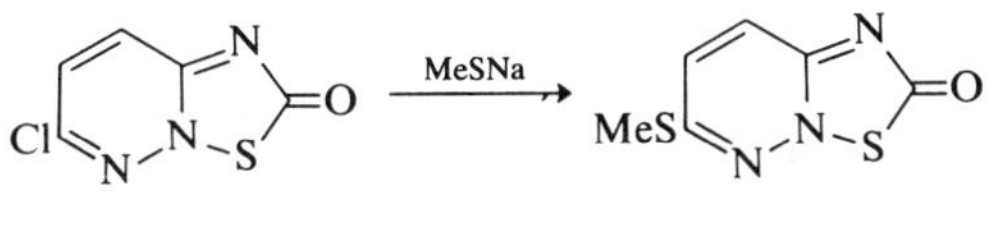

(676)

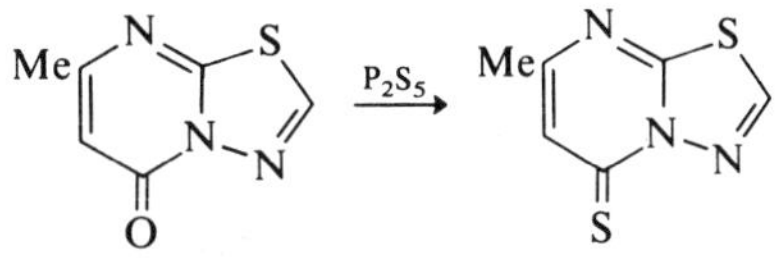

(677)

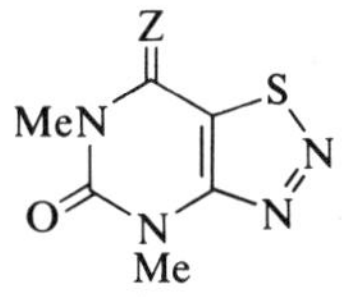

(678) Z = O, S (P_2S_5)

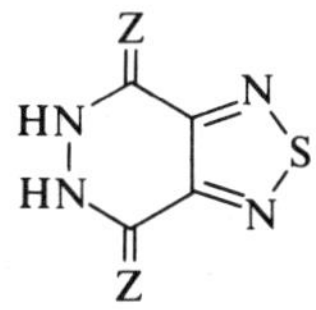

(679) Z = O, S (P_2S_5)

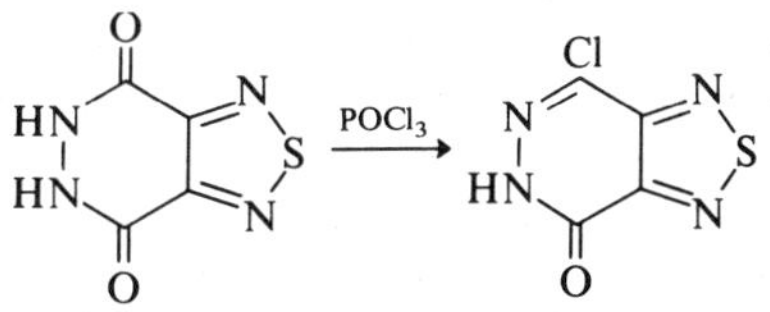

(680)

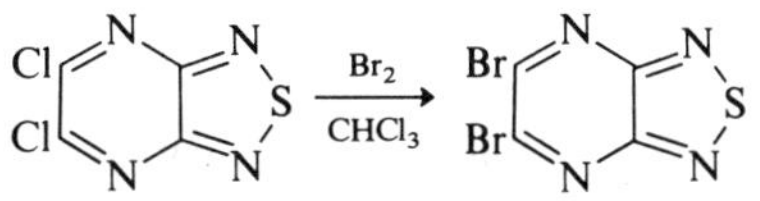

(681)

4.29.10.3.6 Nucleophilic attack with ring opening

Thiadiazolium ions often suffer ring opening after the addition of a nucleophile and especially in derivatives where the sulfur is attached to the azine ring. Thus, in the quinoline (682) addition of an amine at C-3a causes opening of the ring by cleavage of the C—S bond ⟨76IJC(B)176⟩. The mesoionic phenanthridine system (683) is degraded in the same manner ⟨67CC1255⟩. The [1,2,4]thiadiazolo[2,3-a]pyridinium salt (684) undergoes an unusual rearrangement to form 2-substituted benzothiazoles on heating or on treatment with sodium acetate. The reaction is thought to involve nucleophilic attack from the aniline on to the sulfur with subsequent cleavage of the S—N bond ⟨72AJC993⟩.

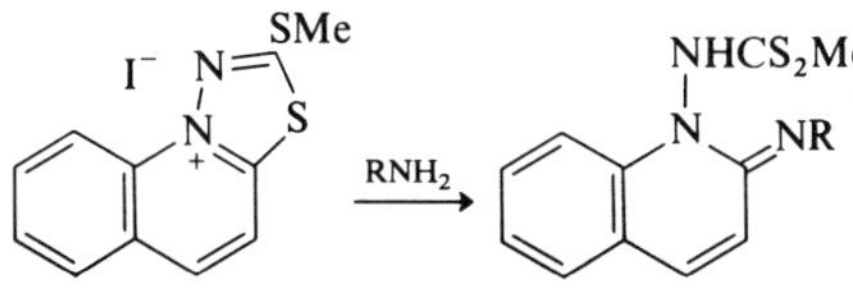

(682)

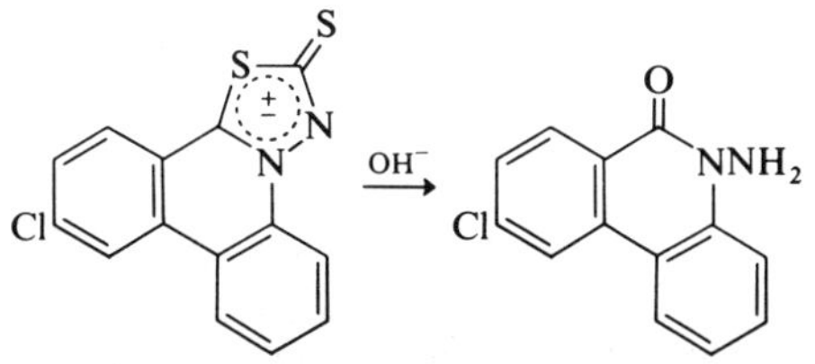

(683)

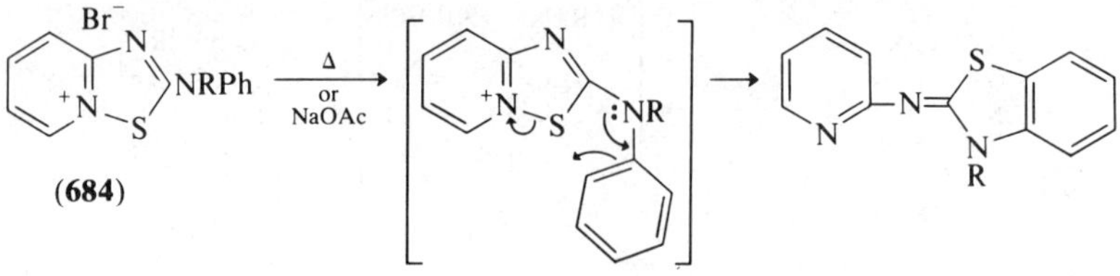

Sometimes a weak acid solution may be sufficient for cleavage of the C—S bond at the point of fusion as in the case of the pyrimidine (**685**) ⟨78CPB2765⟩. The hydrazino analogue (**686**) is cleaved in the same position in weak alkali; the hydrolysis product is recycled *in situ* on to a hydrazine nitrogen. When this ring system carries a 2-alkyl group instead of the hydrazine group as in (**687**) the hydrolysis product corresponds to initial addition of the hydroxide ion at C-2 and subsequent tautomerism ⟨74BCJ2813⟩.

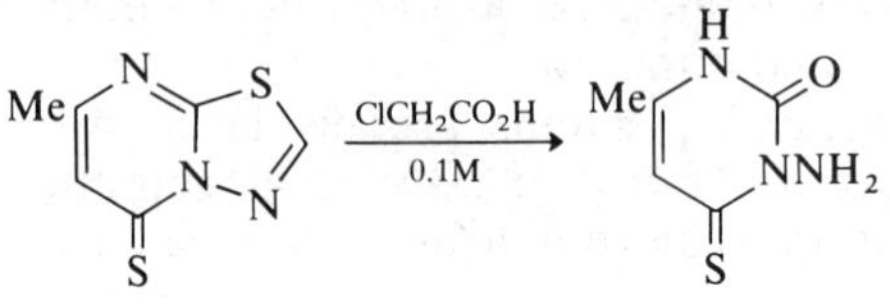

The 3-oxo derivative (**688**) with acid catalysis in ethanol undergoes opening of the ring involving a novel rearrangement and sulfur extrusion. Acid catalysis is required for this reaction to proceed ⟨75JCS(P1)375⟩.

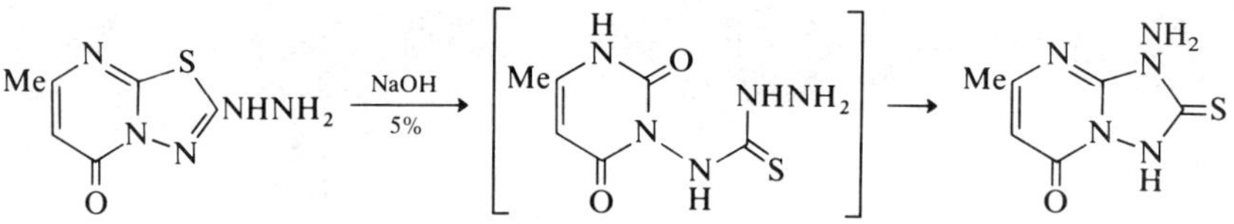

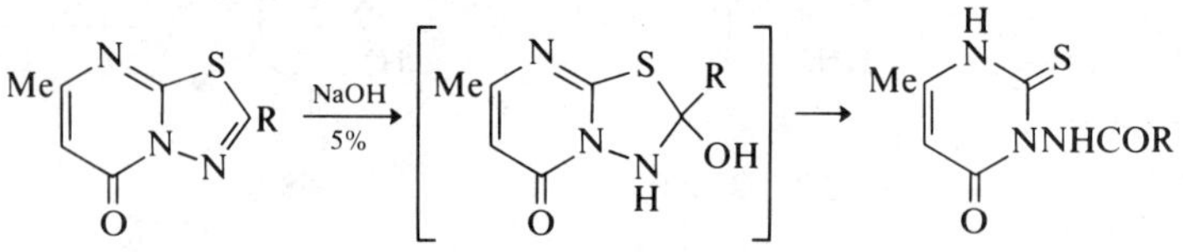

The 4- and 7-positions in 3,4-fused 1,2,5-thiadiazole systems are strongly activated for nucleophilic substitution. Despite this, the 7-chloropyridazine (**689**) reacts with cleavage of the thiadiazole ring between the heteroatoms when heated in alkali ⟨71JHC441⟩. The same type of reaction takes place for the quinoxaline analogue (**690**). The reaction can possibly be rationalized as an addition of water to the sulfur of the electron-deficient thiadiazole ring with subsequent elimination of sulfurous acid. Generation of the latter explains formation of the 4,9-dihydro derivative of (**690**) as a co-product ⟨76JHC13⟩.

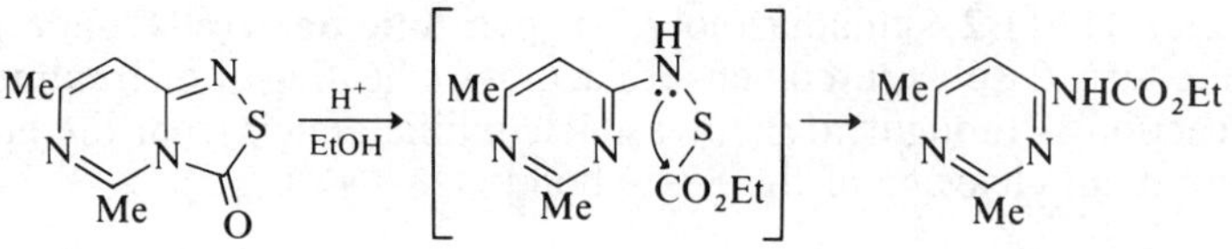

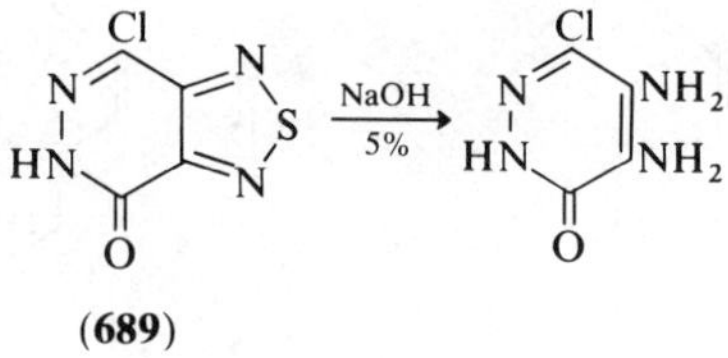

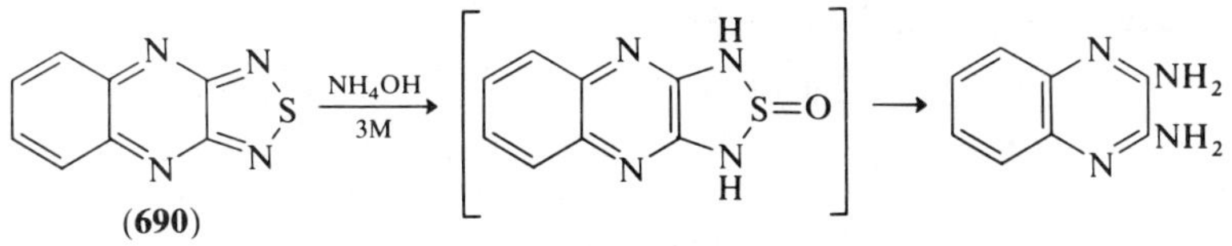

Facile opening of the azine ring also occurs because of the influence from the thiadiazole ring. This may be a competitive reaction during attempts to effect nucleophilic substitution. The triazine betaine (**691**) adds the nucleophile in the azine ring with selective cleavage of the ring at C-5 ⟨73JOC3868⟩.

Under the section dealing with nucleophilic substitution it has been pointed out that the extremely labile 7-methoxy group in the pyrimidine (**671**) is rapidly substituted by amines. In aqueous acid, however, hydrolytic cleavage of the pyrimidine ring occurs ⟨64JOC2121⟩.

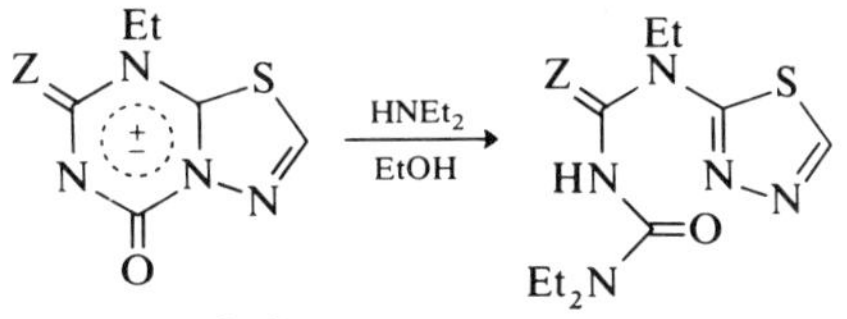

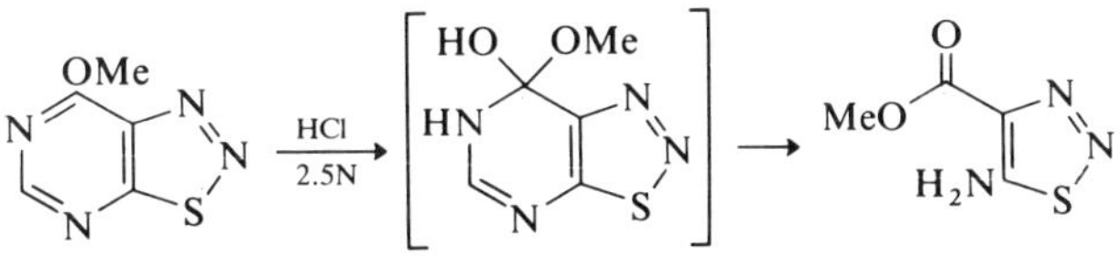

7-Amino derivatives of the fused pyrimidines (**692**) are also cleaved on heating in acid solution with formation of the corresponding thiadiazoloamidines. The ring opening is caused by initial attack of water at C-5, the position between the two azine nitrogens. A minor product from the reaction is the 7-hydroxy derivative which is formed by substitution of the amino group ⟨65JOC2488⟩. On the N,N-dimethyluracil (**693**) the C—N bonds are also cleaved by attack at C-5 whereas in the parent uracil (**694**) the C—N bond cleavage stems from attack at C-7 ⟨64JOC2141⟩.

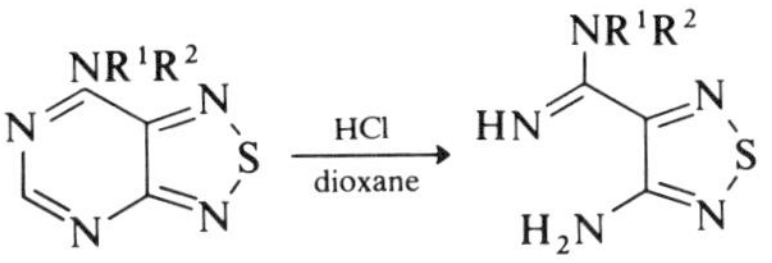

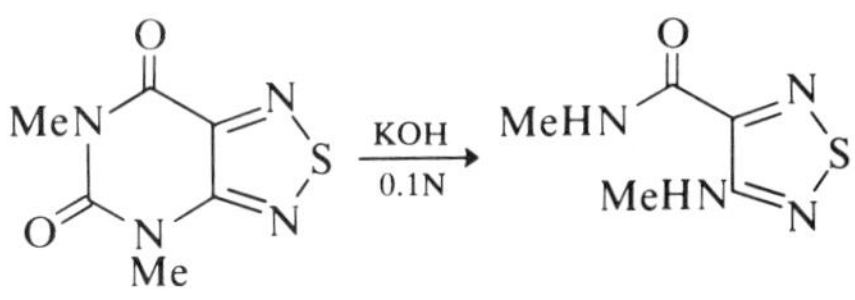

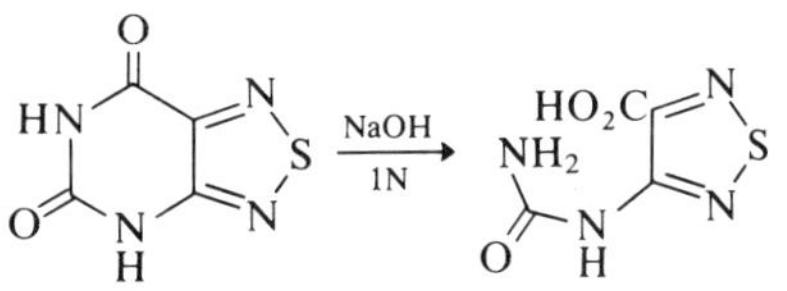

Dimroth-type rearrangements are well documented in fused pyrimidines such as in [1,2,4]triazolo[4,3-a]pyrimidines and [1,2,4]triazolo[4,3-c]pyrimidines under acid or alkaline conditions. The same type of reaction will occur in pyrimidines fused to azoles containing other heteroatoms. Thus, the pyrimidines (**695**) undergo the Dimroth reaction under both acid and alkaline conditions to furnish the rearranged compounds (**696**) ⟨74JOC3783⟩.

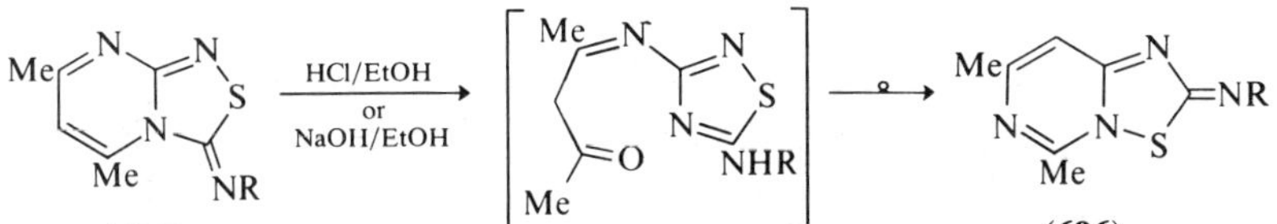

4.29.10.3.7 Nucleophilic attack at hydrogen

A methine proton in the thiadiazole ring is acidic and is readily abstracted by base. In fused 1,3,4-thiadiazoles proton abstraction may lead to cleavage of the N—N bonds as in the pyrimidinium salt (**697**); the thiocyanate (**698**) is formed ⟨74KGS1660⟩. From the isomeric oxo derivatives (**699**) and (**700**) having the same ring system, a common thiocyanate (**701**) is formed ⟨74BCJ2813⟩.

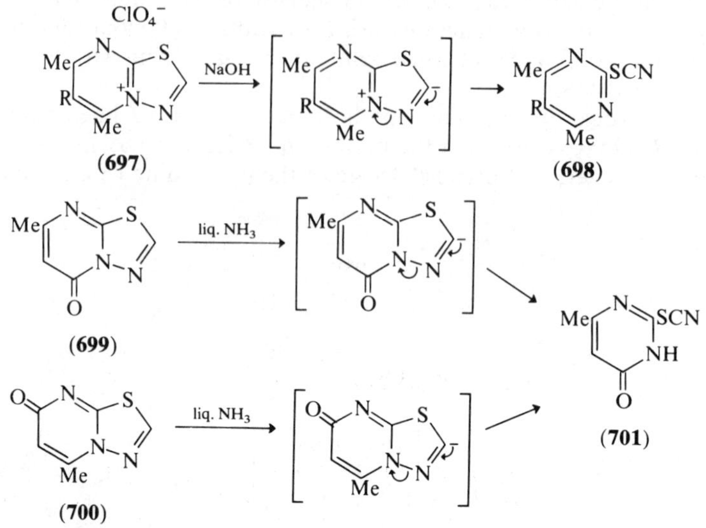

4.29.10.3.8 Other C-linked substituents

Since both rings in the fused system are π-electron deficient, alkyl substituents with at least one α-hydrogen are open for electrophilic attack after initial deprotonation. An example is the aldol-type condensation of (**702**); the 5-methyl group is the more reactive ⟨71ZC460⟩.

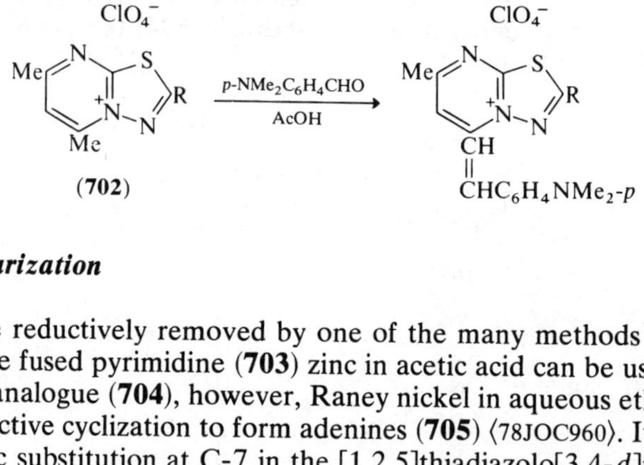

4.29.10.3.9 Desulfurization

The sulfur can be reductively removed by one of the many methods utilized for sulfur heterocycles. For the fused pyrimidine (**703**) zinc in acetic acid can be used ⟨64JOC2135⟩. In the 7-formylamino analogue (**704**), however, Raney nickel in aqueous ethanol was the best reagent for the reductive cyclization to form adenines (**705**) ⟨78JOC960⟩. It has been pointed out that nucleophilic substitution at C-7 in the [1,2,5]thiadiazolo[3,4-d]pyrimidine system easily takes place. This makes the compounds (**704**) readily available as key substances in this convenient approach to the synthesis of 9-substituted adenine derivatives.

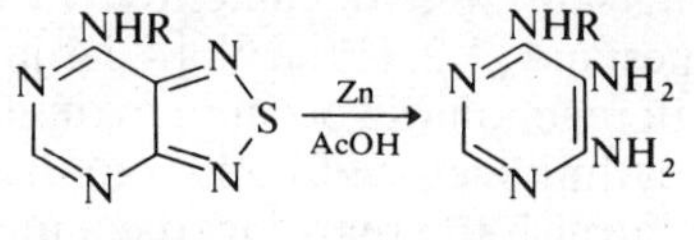

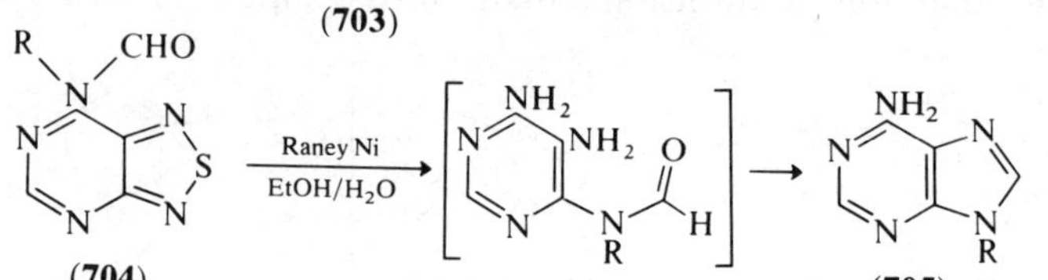

4.29.10.4 Synthesis

4.29.10.4.1 1H-[1,2,3]Thiadiazolo[3,4-a]pyridine

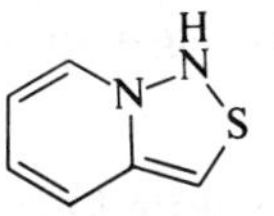

Azine approach. Examples of this heterocyclic ring system have been prepared from 3,4-dihydroisoquinoline azomethine imines (**706**) and sulfenes by 1,3-dipolar cycloaddition reactions. The products are 1,5,6,10b-tetrahydro-3H-[1,2,3]thiadiazolo[4,3-a]isoquinoline 2,2-dioxide derivatives (**707**) ⟨75JOC2260⟩.

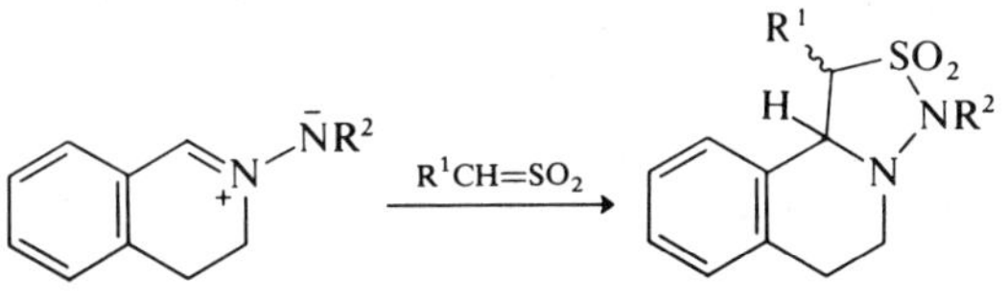

4.29.10.4.2 2H-[1,2,4]Thiadiazolo[2,3-a]pyridine

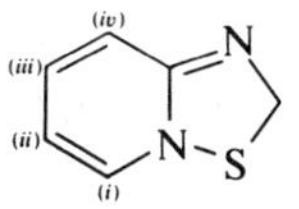

Azine approach. Fused 1,2,4-thiadiazoles are conveniently made by oxidative cyclization to form the N—S bond as the last step in the reaction sequence. Bromine or sulfuryl chloride is commonly used. Cyclization of 2-thioacylaminopyridines (**708**) gives the pyridinium salt ⟨72AJC993⟩ whereas the thioureas (**709**) give 2-imines ⟨78JHC313⟩. Chlorothioformyl chloride converts 2-aminopyridines into the fused 2-oxo derivatives (**710**); acylation of the amino group is thought to be the initial step in the reaction ⟨73JOC1575⟩.

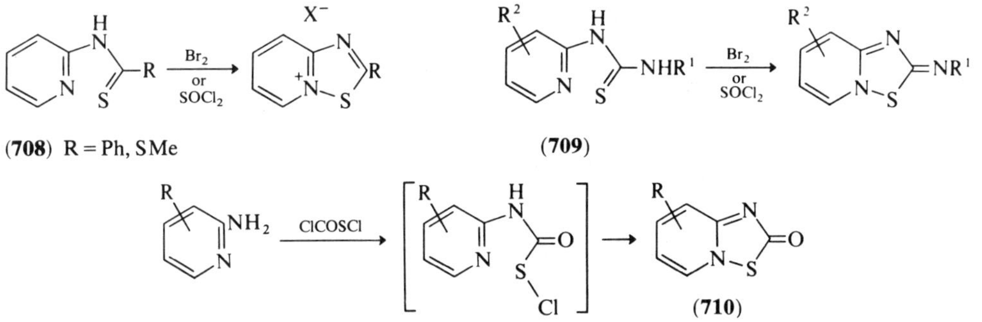

(**708**) R = Ph, SMe

(i) 2H-[1,2,4]Thiadiazolo[2,3-b]pyridazine

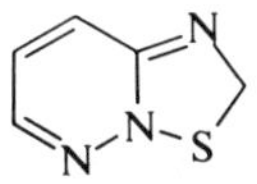

2-Imino derivatives ⟨75OPP55⟩ and 2-oxo derivatives ⟨73JOC1575⟩ can be prepared as for the fused pyridines.

(iii) 2H-[1,2,4]Thiadiazolo[2,3-a]pyrazine

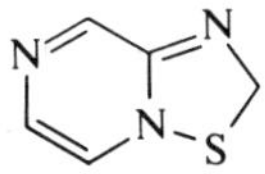

2-Imino derivatives can be prepared as for the fused pyridines ⟨75OPP55⟩.

(iv) 2H-[1,2,4]Thiadiazolo[2,3-a]pyrimidine

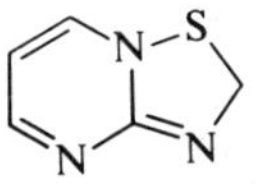

2-Imino derivatives can be prepared as for the fused pyridines ⟨77JHC621⟩.

4.29.10.4.3 3H-[1,2,4]Thiadiazolo[4,3-a]pyridine

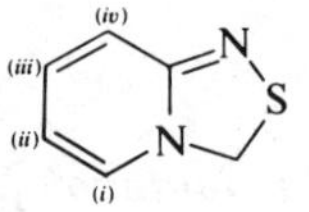

Azine approach. 3-Substituted compounds are readily prepared from α-aminoazines by thiation of the amino group using trichloromethylsulfenyl chloride, the product being subsequently cyclized by divalent nucleophiles. Thus, 2-(trichloromethylthioamino)-pyridines (**711**) are cyclized in the presence of aromatic amines to form 3-imines (**712**; Z = NAr). The products from aliphatic amines are unstable. With sodium hydrogen sulfide, 3-thiones (**712**; Z = S) are formed; enolates furnish carbon analogues (**712**; Z = CR^{1}R^2) ⟨70JOC1965, 71JOC1846⟩.

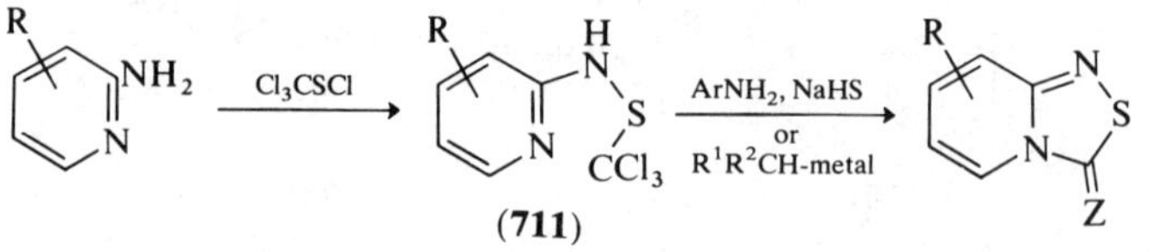

(**711**) → (**712**) Z = NAr, S, CR^{1}R^2

(i) *3H-[1,2,4]Thiadiazolo[4,3-b]pyridazine*

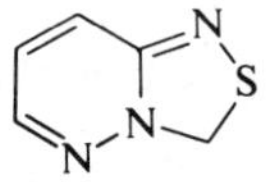

3-Imino derivatives can be prepared as for the fused pyridines ⟨73JOC3087⟩.

(ii) *3H-[1,2,4]Thiadiazolo[4,3-c]pyrimidine*

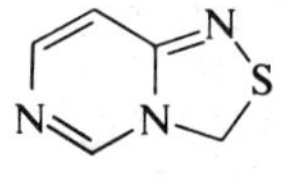

3-Imino derivatives can be prepared as for the fused pyridines ⟨73JOC3087⟩.

(iii) *3H-[1,2,4]Thiadiazolo[4,3-a]pyrazine*

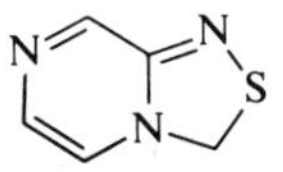

3-Imino derivatives can be prepared as for the fused pyridines ⟨73JOC3087⟩.

(iv) *3H-[1,2,4]Thiadiazolo[4,3-a]pyrimidine*

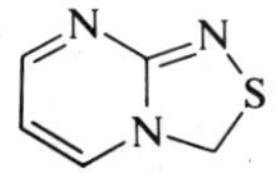

3-Imino derivatives can be prepared as for the fused pyridines ⟨73JOC3087⟩.

4.29.10.4.4 5H-[1,2,4]Thiadiazolo[4,5-a]pyridine

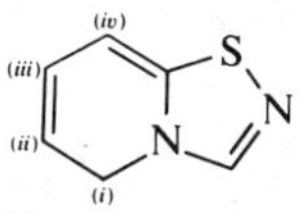

(iv) *5H-[1,2,4]Thiadiazolo[4,5-a]pyrimidine*

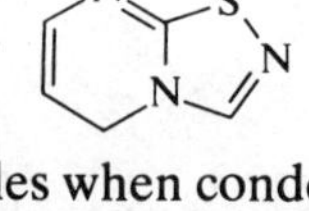

Azole approach. 5-Amino-1,2,4-thiadiazoles when condensed with ethoxymethylenemalonate form fused pyrimidine systems, e.g. (**713**) ⟨59JOC779⟩.

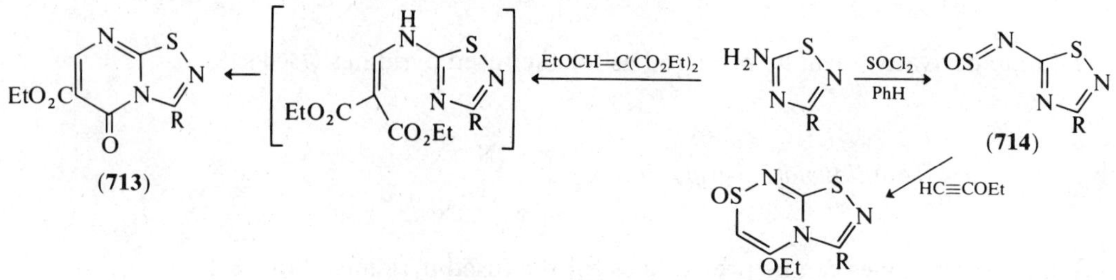

(**713**) ← [intermediate] ← (**714**) → (**715**)

Aromatic N-sulfinylamines form cycloadducts with alkenes and alkynes. This type of reaction can be used to prepare [1,2,4]thiadiazolo[5,4-c][1,2,4]thiadiazine 7-oxides, e.g. (**715**). The 5-sulfinylamino-1,2,4-thiadiazole (**714**) and ethoxyacetylene are reacted together and only the regioisomer (**715**) is formed in the reaction ⟨67CB2159⟩.

4.29.10.4.5 5H-[1,3,4]Thiadiazolo[3,2-a]pyridine

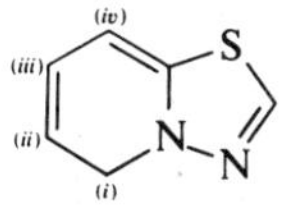

Azine approach. 1-Pyridinimines undergo 1,3-dipolar cycloaddition reactions with thiones. In the reaction between 2-isoquinolinimine and carbon disulfide the mesoionic thiadiazole (**716**) is formed; the formation of (**716**) involves a secondary dehydrogenation of the initial adduct. With diphenyl thionocarbonate, phenoxy group expulsion is succeeded by cyclization leading to the adduct (**717**) ⟨62TL387⟩.

In the reaction of the 2-acylamino isoquinoline (**718**) with phosphorus pentasulfide the straight chain tricyclic compound (**719**) is selectively formed ⟨75BCJ2915⟩.

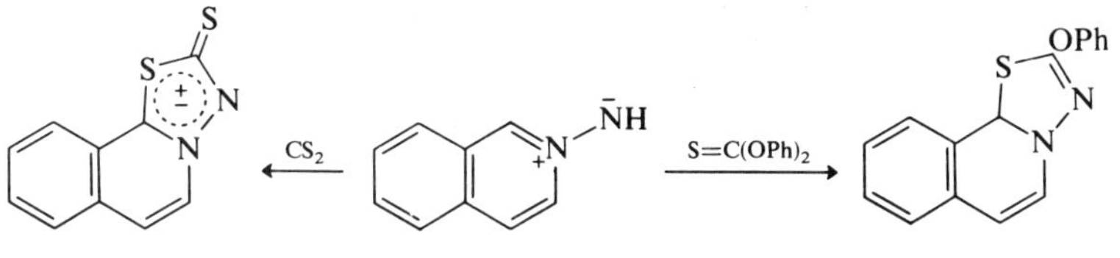

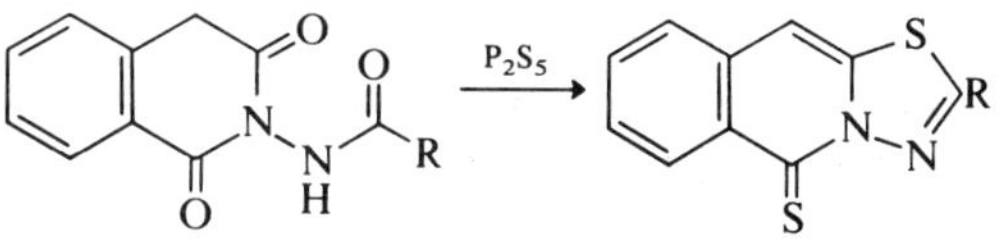

Azole approach. 5-H[1,3,4]Thiadiazolo[3,2-a]pyridin-5-ones (**723**) can be prepared by 1,3-dipolar cycloaddition reactions between electron-deficient alkenic or alkynic dipolarophiles and the thiocarbonyl ylide dipole present in anhydro-5-hydroxy-2-methyl-6-phenylthiazolo[2,3-b][1,3,4]thiadiazolium hydroxide (**720**). Sulfur is extruded from the original acetylene adduct (**722**) whereas H₂S is eliminated from the alkene adduct (**721**) to form the same product (**723**) ⟨79JOC3808⟩.

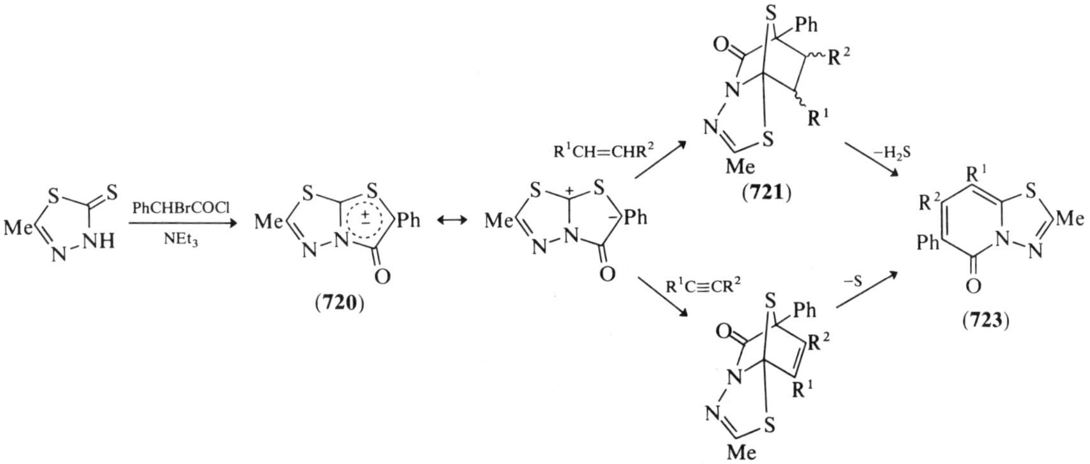

(iv) *5H-[1,3,4]Thiadiazolo[3,2-a]pyrimidine*

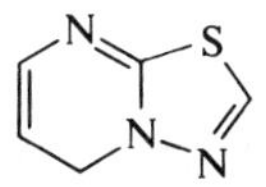

Azine approach. Most widely used are cyclization reactions of N-amino-2-pyrimidinethiones and analogous structures, as in the preparation of the 7H-derivative (**724**) ⟨78ZC66⟩ and the quinazoline (**725**) ⟨70IJC710⟩. Basically, the same reaction conditions are used to prepare the 1,2,4-triazine (**726**) ⟨76JHC117⟩.

Carbon disulfide reacts with the *N*-aminopyrimidine (**727**) to yield the 7-thioxo derivative (**728**), the cyclization reaction being a nucleophilic substitution of the 2-substituent by the initially formed dithiocarbamoyl group.

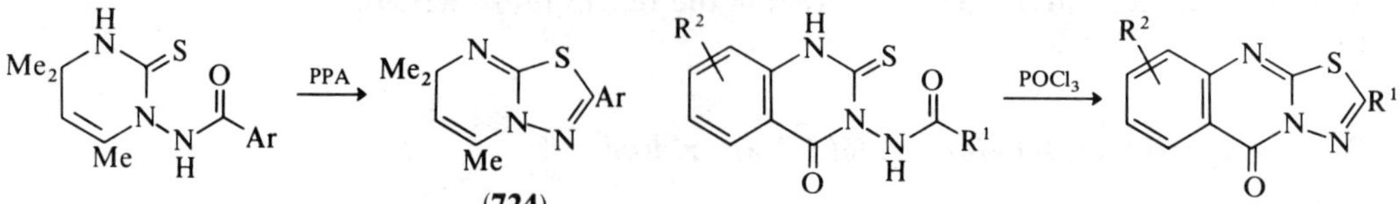

(724) (725) 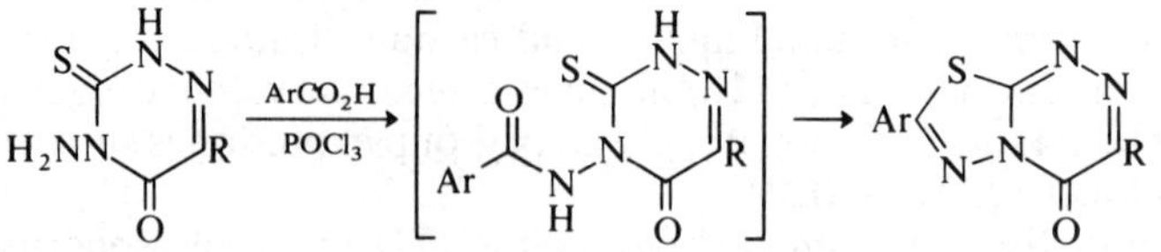 (726)

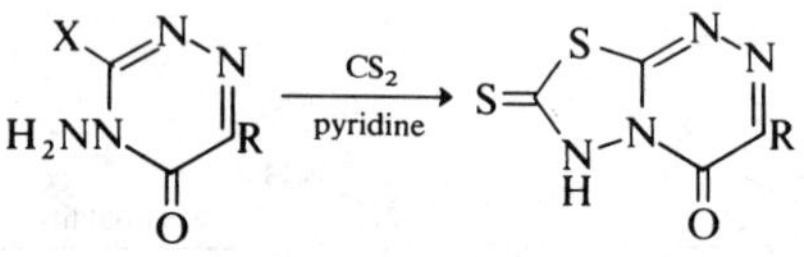

(727) X = Hal   (728)

Azole approach. 2-Amino-1,3,4-thiadiazoles react with appropriately substituted α,β-unsaturated carbonyl compounds to form fused pyrimidines (**729**). The orientation of the substituents in the fusion products suggests that the reaction is initiated by Michael addition of the amino group. 1,3-Dicarbonyl compounds will condense in the same manner to yield the salt (**729**); from a β-keto ester the 5-oxo derivative (**730**) is formed ⟨73ABC1197⟩.

The heteroaromatic pyrimidinedione betaines (**731**) can be prepared simply by heating the 2-aminothiadiazole with an active ester of malonic acid ⟨74JMC1025⟩. When phenoxycarbonyl isocyanate or isothiocyanate is the reactant, the 1,3,5-triazinone mesoions (**732**) are formed ⟨73JOC3868⟩.

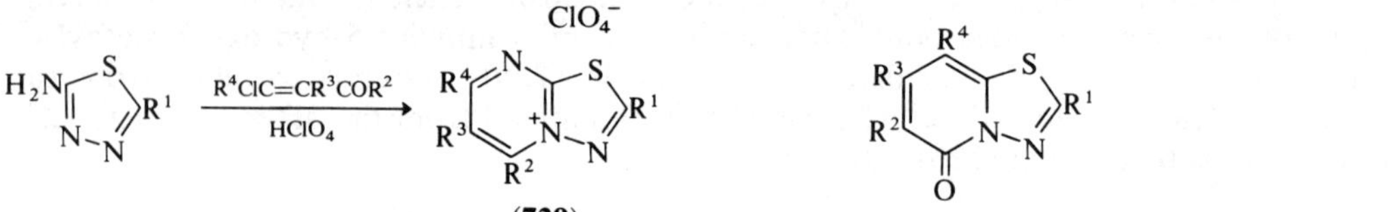

(729) (730)

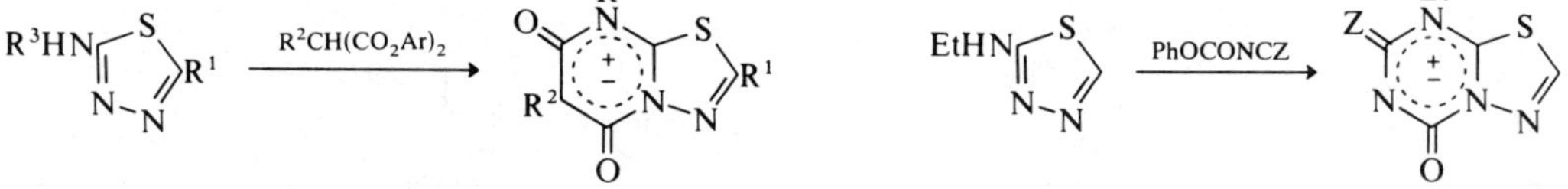

(731) (732) Z = O, S

4.29.10.4.6 1H,3H-[1,3,4]Thiadiazolo[3,4-a]pyridazine

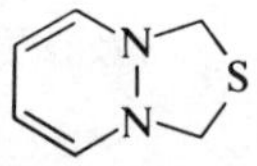

Azole approach. 5,8-Dihydro derivatives of this ring system (**734**) are known. The latter can be prepared by [$_\pi 4 + _\pi 2$] cycloadditions of 1,3,4-thiadiazole-2,5-dione (**723**) and a 1,3-diene. The former is highly reactive and is generated *in situ* by oxidation of 1,3,4-thiadiazolidine-2,5-dione with copper(II) chloride or LTA ⟨74JOC2951⟩.

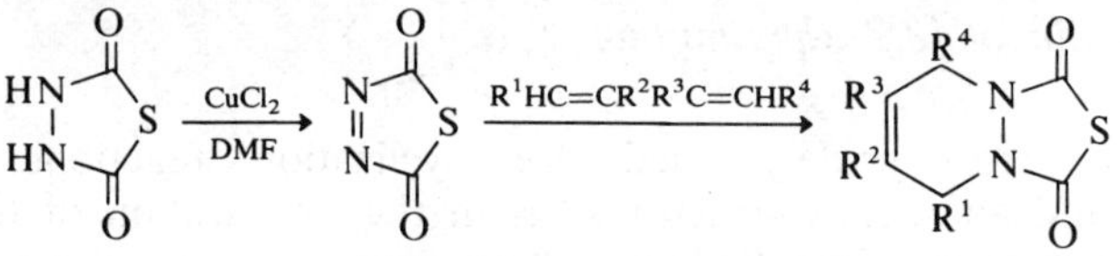

(733) (734)

4.29.10.4.7 2H-[1,2,5]Thiadiazolo[3,4-a]pyridine

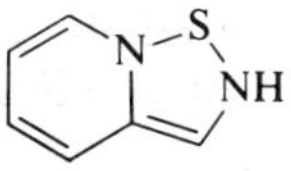

The perhydro 1,1-dioxide (735) has been prepared from 2-aminomethylpiperidine and sulfuryl chloride ⟨62AP621⟩.

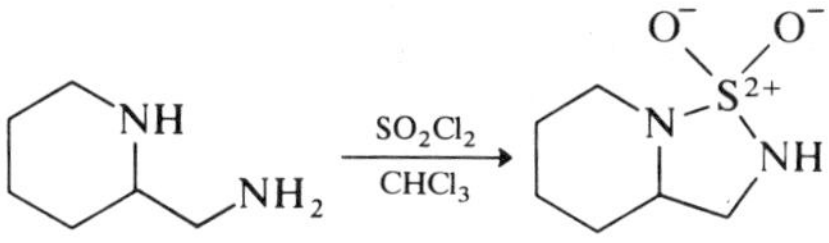

(735)

4.29.10.4.8 [1,2,3]Thiadiazolo[4,5-b]pyridine

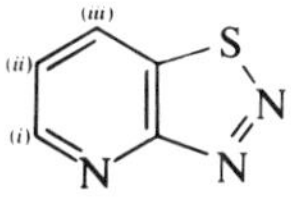

(ii) [1,2,3]Thiadiazolo[4,5-d]pyrimidine

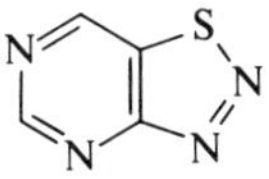

Azine approach. 1,3-Dialkyl-6-hydrazinouracil is susceptible to electrophilic attack at C-5. Treatment of (736; $R^1 = H$) with excess thionyl chloride at room temperature gives the fused uracil (737) in an exothermic reaction. The reaction is thought to involve initial formation of a sulfinyl chloride. Subsequently the electrophilic cyclization occurs at C-5 to form a thiazoline S-oxide which is dehydrated to yield (737). From α,3-dimethyl-6-hydrazinouracil (736; $R^1 = R^3 = Me$, $R^2 = H$) the heteroaromatic betaine (738) is formed ⟨78JOC1677⟩.

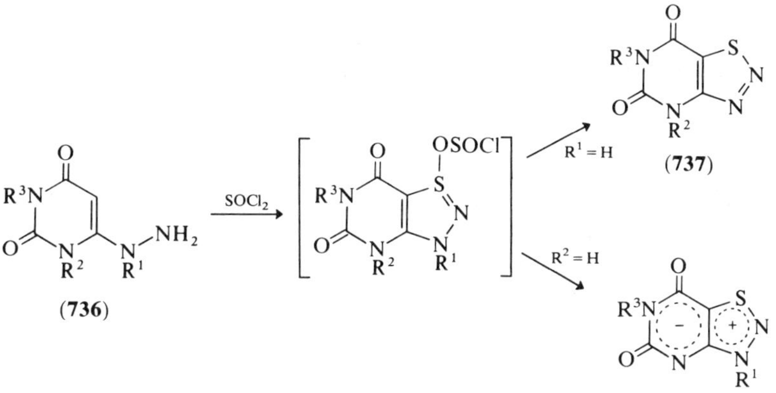

(736) (737) (738)

4.29.10.4.9 [1,2,3]Thiadiazolo[4,5-c]pyridine

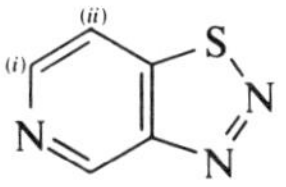

Azine approach. Fused 1,2,3-thiadiazoles are often conveniently prepared from vicinal aminomercaptoarenes by a diazotization reaction; the mercapto group adds to the initially formed diazonium group to form the ring. So far this reaction has not been reported in this series, but a mechanistically similar reaction results when 1,4-dihydro-7H-v-triazolo[4,5-b]pyridine-7-thione (655; R = H) is heated in ethanol to furnish the 1,2,3-thiadiazole (657). Heating the latter under alkaline conditions reverses the reaction with

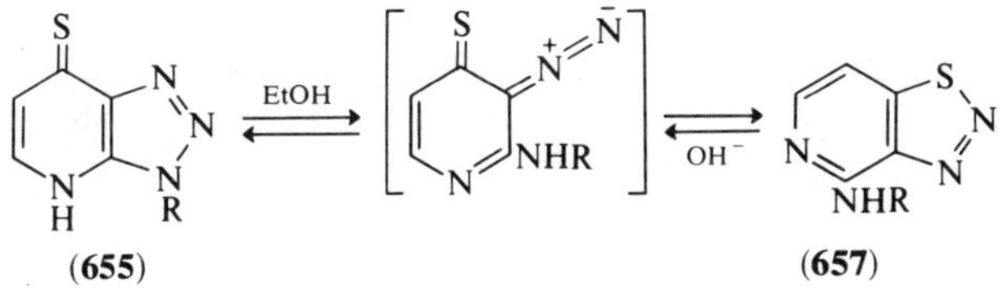

(655) (657)

formation of the *v*-triazole (**655**) ⟨72JOC3601⟩. The same rearrangement takes place from nucleoside analogues ⟨78JOC4910⟩. Further examples are available in Section 4.29.10.3.2.

(i) *[1,2,3]Thiadiazolo[4,5-d]pyridazine*

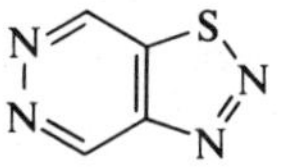

Azole approach. Compounds in this series are available by a phthalazine-type synthesis. A simple example is the reaction of a 4,5-diacylthiadiazole (**739**) with hydrazine ⟨76JHC301⟩.

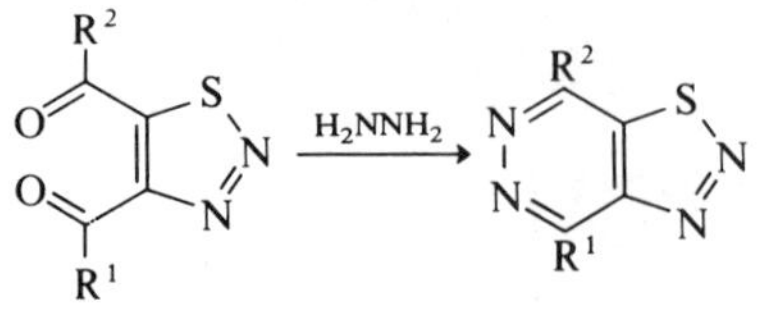

(**739**)

(ii) *[1,2,3]Thiadiazolo[5,4-d]pyrimidine*

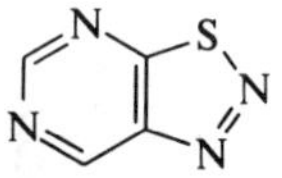

Azine approach. Diazotization of 5-amino-4-mercaptopyrimidines is a simple way of preparing the fused 1,2,3-thiadiazoles (**740**) ⟨64JOC2121⟩.

In Section 4.29.10.3.2 it was shown that *v*-triazoles fused with pyridines and pyrimidines undergo the same rearrangement to form 1,2,3-thiadiazoles. The first reported example was the pyrimidine (**658**) which on heating was transformed into (**659**) ⟨66AG596⟩. The nucleoside (**660**) rearranges to the corresponding derivative of (**659**) in aqueous solution without heating ⟨76JMC1186⟩.

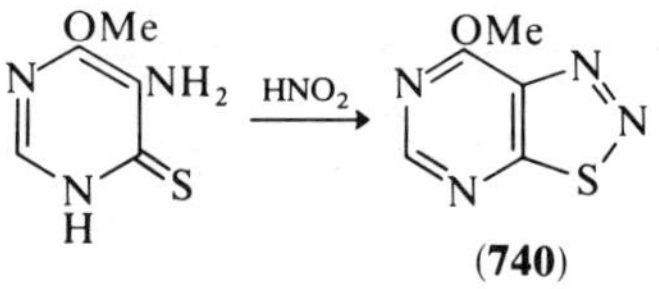

(**740**)

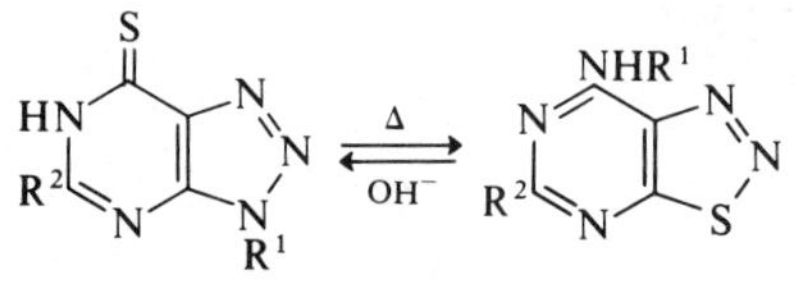

(**658**) $R^1 = CH_2Ph$, $R^2 = H$ (**659**)
(**660**) $R^1 = \beta$-D-ribose, $R^2 = NH_2$

Azole approach. The reaction of (**741**) with formamidine is a typical condensation reaction for pyrimidine fusion ⟨64JOC2121⟩; the cyclization of (**742**) is another example ⟨65LA(682)90⟩.

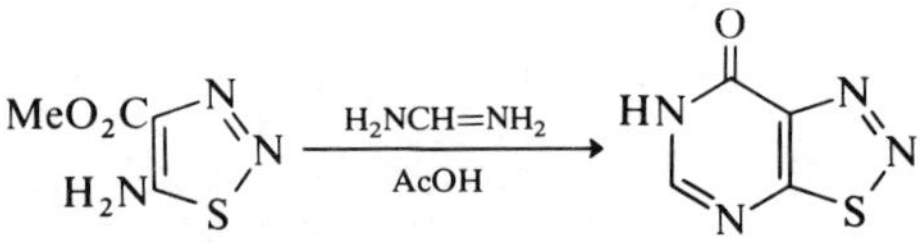

(**741**) 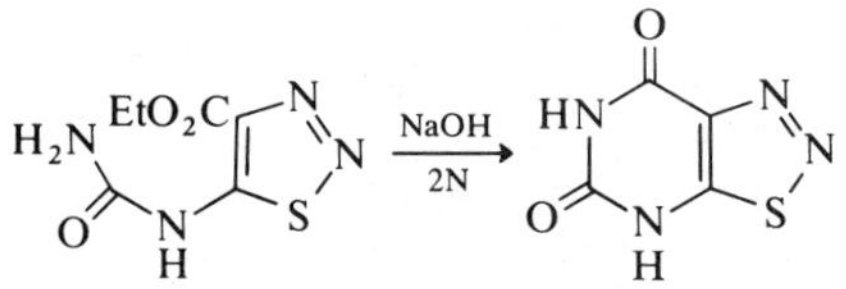   (**742**)

4.29.10.4.10 *[1,2,3]Thiadiazolo[5,4-b]pyridine*

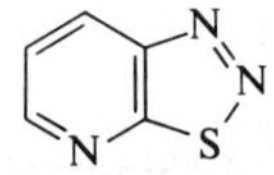

Azine approach. Diazotization of 3-amino-2-mercaptopyridine is a very convenient way for preparing simple members of this ring system (**743**) ⟨74CS(6)222⟩. 7-Amino derivatives (**663**) are available by the isomerism reaction of *v*-triazoles (**662**; R = H) on heating in 1-propanol ⟨72JOC3601⟩. The same transformation occurs in pyridine nucleosides (**662**; R = β-D-ribose) ⟨76JOC1449⟩.

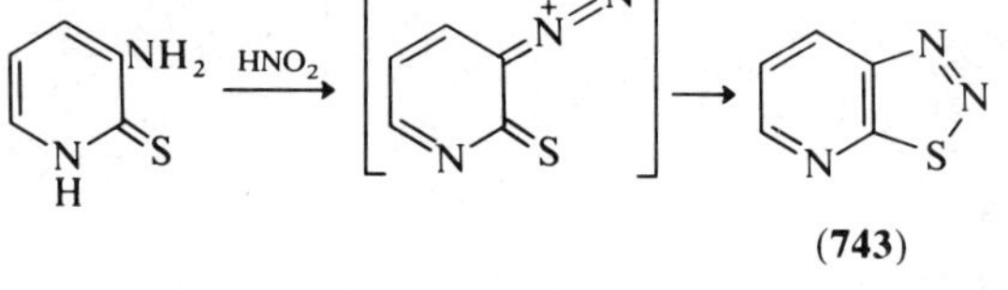

(**743**)

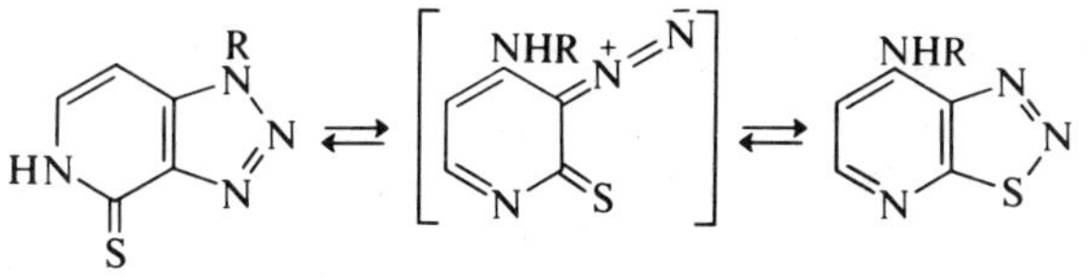

Five-membered Rings Fused with Six-membered Rings 747

(662) R = H, β-D-ribose   (663)

4.29.10.4.11 [1,2,5]Thiadiazolo[3,4-b]pyridine

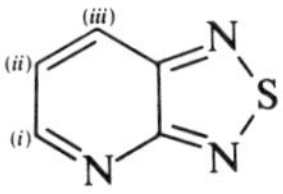

Azine approach. 1,2,5-Thiadiazoles are available from vicinal diamines and thionyl chloride as in the reaction of 2,3-diaminopyridines to form (744) ⟨70RTC5⟩.

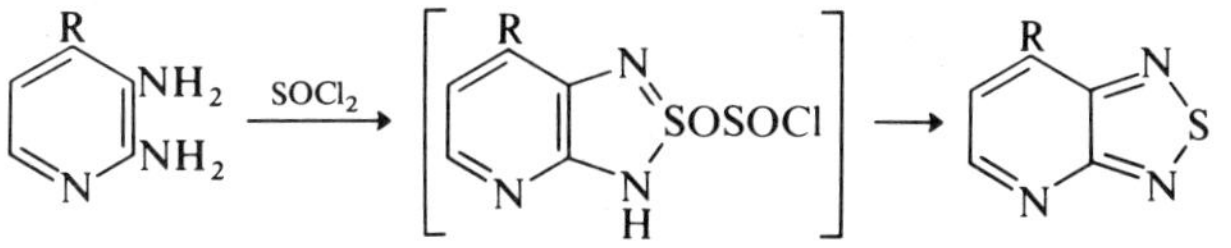

(744)

(ii) *[1,2,5]Thiadiazolo[3,4-d]pyrimidine*

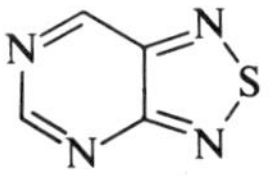

Azine approach. In the cyclization of vicinal diaminopyrimidines either thionyl chloride or sulfinylaniline ⟨64JOC2135⟩ can be used, as for example in the synthesis of (745).

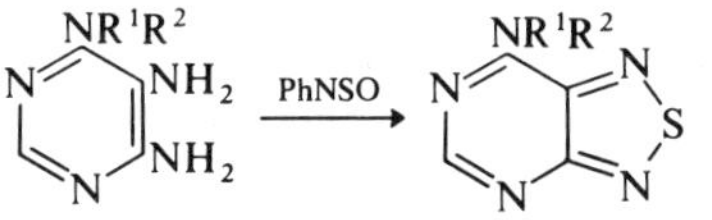

(745)

Azole approach. The cyclocondensation of the thiadiazole (746) with orthoformate ⟨64JOC2135⟩ is a standard pyrimidine fusion reaction.

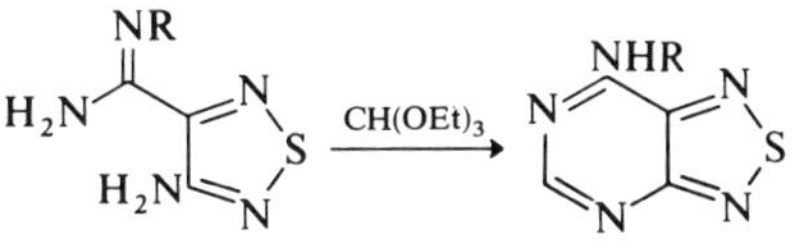

(746)

(iii) *[1,2,5]Thiadiazolo[3,4-b]pyrazine*

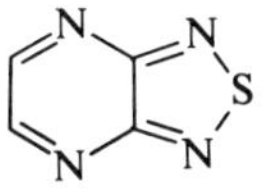

Azole approach. 3,4-Diamino-1,2,5-thiadiazole reacts with 1,2-dicarbonyl compounds to form pyrazines (747) ⟨76JHC13⟩. From the reaction of 1,2,5-thiadiazole 1,1-oxides such as (748) with o-phenylenediamine, the 1,3-dihydro[1,2,5]thiadiazolo[3,4-b]quinoxaline 2,2-dioxide (749) is formed. To understand this reaction it is pointed out that the 1,2,5-thiadiazole 1,1-dioxide ring is to be regarded as alicyclic rather than aromatic and is strongly π-electron deficient. Substituents with leaving properties in the 3,4-positions are therefore readily displaced as in the reaction of (748) ⟨75JOC2743⟩.

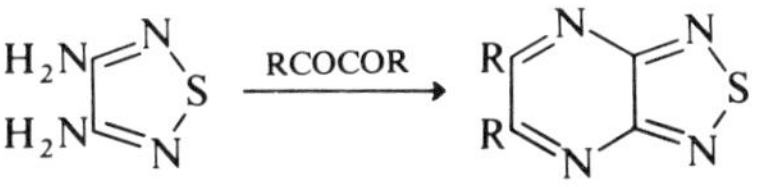

(747)

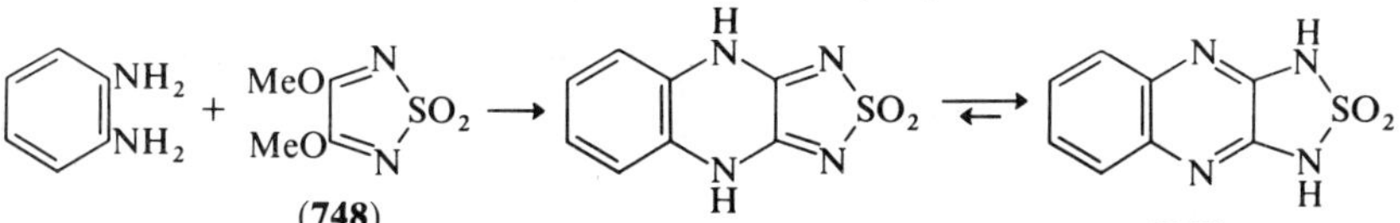

(748)   (749)

4.29.10.4.12 [1,2,5]Thiadiazolo[3,4-c]pyridine

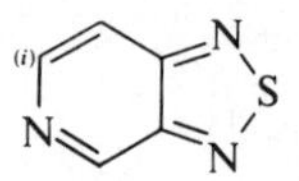

Azole approach. Cyclocondensation in the manner of isoquinoline synthesis using a 3,4-diacyl-1,2,5-thiadiazole and methylamine under the influence of DBU furnishes the pyridines (**750**) ⟨79S687⟩.

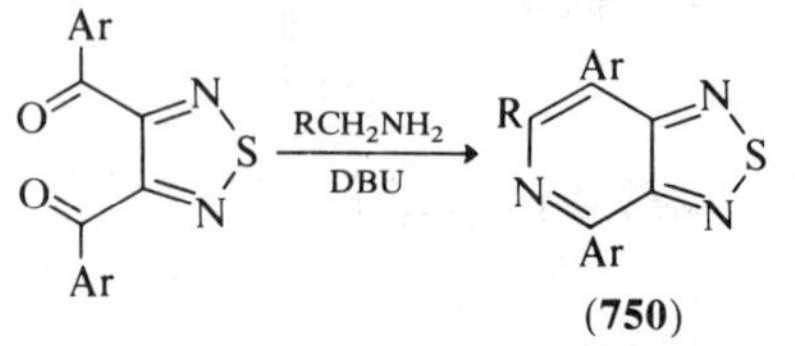

(i) *[1,2,5]Thiadiazolo[3,4-d]pyridazine*

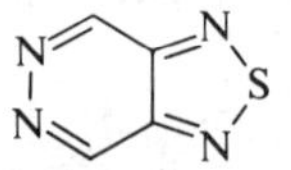

Azole approach. 3,4-Dicyano-1,2,5-thiadiazole with hydrazine yields the 5,7-diamine (**751**) ⟨71JHC441⟩ in a manner analogous to the reaction of *o*-phthalonitrile with hydrazine.

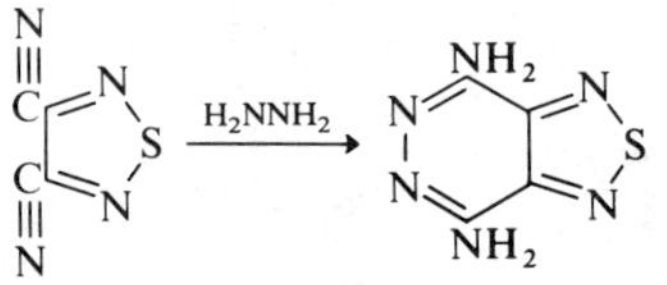

4.29.10.5 Applications

7-Amino derivatives of [1,2,3]thiadiazolo[5,4-*b*]pyridine-6-carboxylic acids and esters (**752**) are claimed to have useful tranquillizing and antiinflammatory properties ⟨76USP3965108⟩.

The heteroaromatic betaines, [1,3,4]thiadiazolo[3,2-*a*]pyrimidine-5,7-diones (**753**), which are isoconjugate with methylated xanthines, are in the same way as the latter found to be inhibitors of cyclic AMP phosphodiesterase ⟨78JPS1762⟩. The *N*-β-D-ribofuranosyl nucleoside (**754**) inhibits the synthesis of RNA and DNA but not of protein. The primary blockade is in the synthesis of purine nucleotides ⟨79MI42902⟩. RNA synthesis is also inhibited by [1,3,4]thiadiazolo[3,2-*a*]pyrimidines of structure (**755**). The activity is attributed to the chemical reactivity at C-2 leading to reactions with an SH or OH group in RNA polymerase ⟨80ABC1923⟩. Compounds with the 7-oxo formula (**756**) are claimed to be useful as immune enhancers ⟨78GEP2712932⟩.

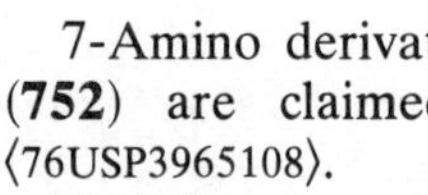

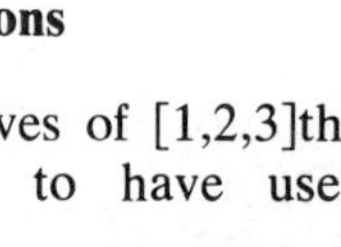

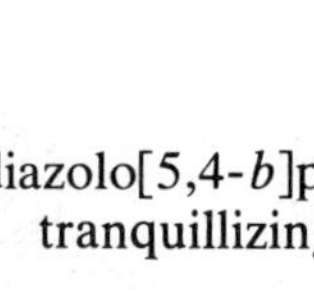

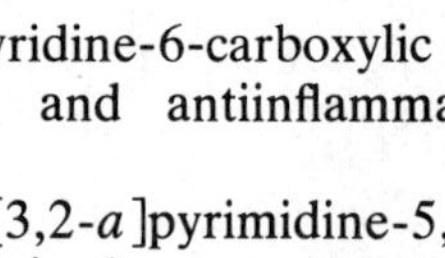

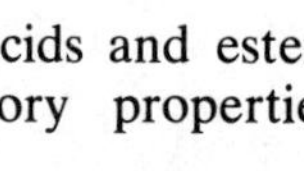

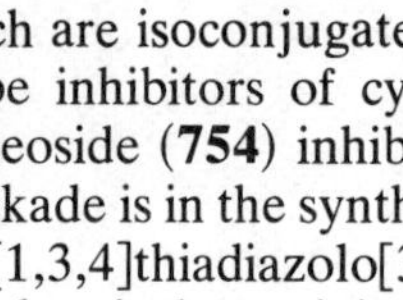

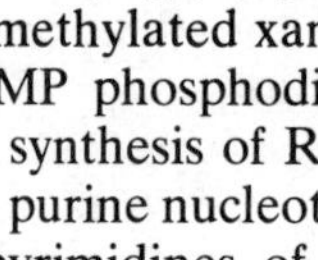

Purinone analogues of formula (**753**) are also active *in vitro* as antimicrobials. Tricyclic oxazines (**757**) are patented as herbicides and fungicides ⟨78USP4080499⟩. 2-Alkylthio[1,2,4]thiadiazolo[2,3-*a*]pyridinium salts (**758**) are anthelmintics for use in warm-blooded animals ⟨73GEP2258279⟩. The triazinediones (**759**) are useful as herbicides ⟨77USP4042372⟩.

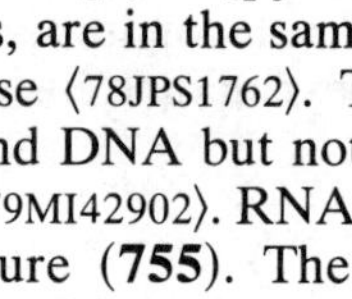

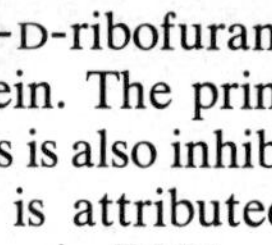

Thiadiazolopyridazines of formula (**760**) when incorporated into varnish give a durable yellow coating on paper ⟨73USP3716534⟩.

4.30

Dioxoles and Oxathioles

A. J. ELLIOTT
Schering-Plough Corporation, Bloomfield

4.30.1 STRUCTURE		750
4.30.1.1 Introduction		750
4.30.1.2 Theoretical Methods		750
4.30.1.3 Molecular Dimensions		750
4.30.1.3.1 X-ray diffraction		750
4.30.1.3.2 Microwave spectroscopy		751
4.30.1.4 Molecular Spectra		752
4.30.1.4.1 ^1H NMR spectra		752
4.30.1.4.2 ^{13}C NMR spectra		754
4.30.1.4.3 UV and IR spectra		756
4.30.1.4.4 Mass spectra		756
4.30.1.4.5 Photoelectron spectra		757
4.30.1.5 Thermodynamic Aspects		757
4.30.1.5.1 Intermolecular forces		757
4.30.1.5.2 Aromaticity		757
4.30.1.5.3 Shape and conformation		758
4.30.1.6 Tautomerism		758
4.30.1.7 Betaine Structures		758
4.30.2 REACTIVITY		759
4.30.2.1 Reactivity at the Ring Atoms		759
4.30.2.1.1 General survey		759
4.30.2.1.2 Thermal and photochemical reactions involving no other species		759
4.30.2.1.3 Reactivity towards electrophiles and oxidants		760
4.30.2.1.4 Reactivity towards nucleophiles and reducing agents		762
4.30.2.1.5 Reactivity towards free radicals, electron-deficient species and at surfaces		764
4.30.2.1.6 Reactions with cyclic transition states		765
4.30.2.2 Reactivity of Substituents		766
4.30.2.2.1 General survey		766
4.30.2.2.2 Fused benzene rings		766
4.30.2.2.3 C-Linked substituents		766
4.30.2.2.4 N-Linked substituents		767
4.30.2.2.5 O-Linked substituents		768
4.30.2.2.6 S-Linked substituents		769
4.30.2.2.7 Halogen substituents		769
4.30.2.2.8 Metal substituents		770
4.30.3 SYNTHESIS		770
4.30.3.1 Ring Synthesis		770
4.30.3.1.1 One-bond formation between heteroatoms		770
4.30.3.1.2 One-bond formation adjacent to a heteroatom		771
4.30.3.1.3 Two-bond formation from [4+1] atom fragments		773
4.30.3.1.4 Two-bond formation from [3+2] atom fragments		774
4.30.3.2 Ring Synthesis by Transformation of Other Heterocycles		775
4.30.3.2.1 Three-membered rings		775
4.30.3.2.2 Four-membered rings		776
4.30.3.2.3 Five-membered rings		777
4.30.3.3 Synthesis of Derivatives		778
4.30.3.3.1 Parent systems		778
4.30.3.3.2 Benzo derivatives		778
4.30.3.3.3 C-Linked substituents		779
4.30.3.3.4 N-Linked substituents		779
4.30.3.3.5 O-Linked substituents		780
4.30.3.3.6 S-Linked substituents		780
4.30.3.3.7 Halogen substituents		780
4.30.4 APPLICATIONS		781

750   *Dioxoles and Oxathioles*

4.30.1 STRUCTURE

4.30.1.1 Introduction

The vast majority of the chemistry known for dioxoles and oxathioles is limited to the 1,3-derivatives. 1,2-Dioxoles are unknown except as the fully saturated systems and 1,2-oxathiolane was only recently reported, although derivatives containing oxidized forms of sulfur have been known for some time. Of the fully aromatic species, only derivatives of the 1,3-oxathiolium ring system and mesoionic 1,3-oxathioliums have been isolated, illustrating the stability imparted by the sulfur atom. In addition, the 1,2-oxathiolium and 1,3-dioxolium cations have been observed in strong acid solution. Of the systems containing one double bond, only 1,3-oxathiole itself is unknown although the isomeric 1,3-oxathiolan-2-ylium cation is well known.

The fully saturated 1,3-dioxolanes and 1,3-oxathiolanes are conformationally labile rings and have been the subject of intense spectral investigation. In particular, 1,3-dioxolanes have a special place in NMR history. It was during a study of 4-substituted 2,2-dimethyl-1,3-dioxolanes that geminal and vicinal couplings in aliphatic systems were shown to have opposite signs, a result which contradicted the theoretical predictions of the time ⟨61JA3901⟩.

4.30.1.2 Theoretical Methods

Relatively little has been reported on theoretical calculations in these systems. Most of the work concerns the shape and conformation of the saturated rings. MO calculations designed to study the degree of ring puckering in 1,3-dioxolanes indicate that a twist conformation with C_2 symmetry is the most stable form. This has a low barrier to pseudorotation into an envelope configuration of C_s symmetry. The calculated energy difference is only 4.35 kJ mol^{-1} and the planar form itself only 4.94 kJ mol^{-1} above the minimum ⟨75JA1358⟩. These conformations are discussed in more detail in Section 4.30.1.5. Such a low energy barrier predicts that pseudorotation should occur freely and this is confirmed by IR and microwave studies ⟨B-77SH(2)68⟩. The X-ray of bi-dioxolane, discussed in Section 4.30.1.3.1, shows a structure mid-way between these conformers. Conformational energy calculations for 1,3-oxathiolane and 2-methyl-1,3-oxathiolane also show that the conformational energy minima are quite shallow, the lowest energy transition states for pseudorotation being of the order of the energy for rotation of ethane (*ca.* 12 kJ mol^{-1}). The energy minima represent very closely the enantiomeric envelope forms in which the sulfur atom is β to the flap atom ⟨74JA2426⟩. This is consistent with the X-ray and NMR studies reported later and is discussed in Section 4.30.1.5.

The INDO optimized geometries and total charge densities for the 1,3-dioxolan-2-ylium cation and its 2-methyl derivative are shown in Figure 1. The calculated C(2)—O bond order of 0.650 ⟨73JOC471⟩ seems quite high but is consistent with IR and NMR studies ⟨65CI(L)691⟩. The calculated charge densities compare favorably with ESCA-derived data for the pentamethyl derivative, also shown in Figure 1 ⟨76CB1837⟩.

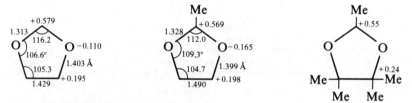

Figure 1 Geometries and charge densities for 1,3-dioxolan-2-ylium cations

4.30.1.3 Molecular Dimensions

4.30.1.3.1 X-ray diffraction

Details of all the X-ray structures of dioxoles and oxathioles studied prior to 1970 have been published ⟨72PMH(5)47⟩. Details of bond lengths and bond angles for the dioxolane ring in (**1**) and the oxathiolane ring in the steroid (**2**) are shown in Table 1. In compound (**1**) the dioxolane ring adopts a solid state conformation mid-way between the theoretically lowest energy half-chair and envelope forms. The oxathiolane ring in (**2**) adopts an envelope conformation with the methylene group adjacent to oxygen and 50 pm outside the plane defined by the other four atoms.

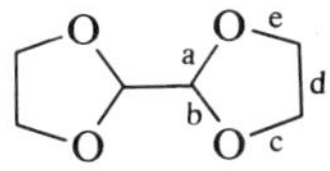

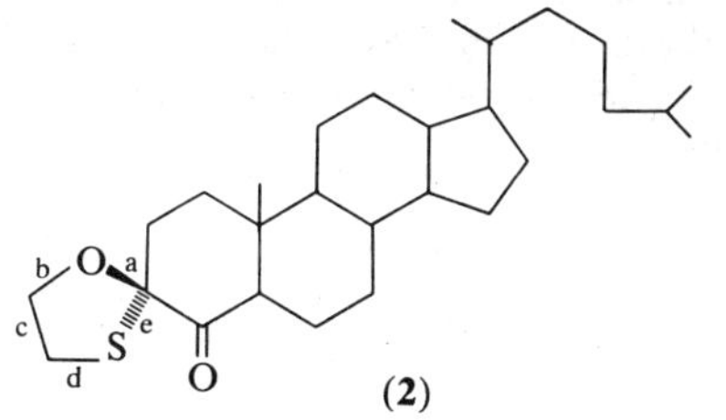

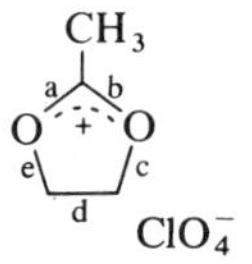

Table 1 Bond Lengths (Å) and Bond Angles (°) of some Dioxoles and Oxathioles

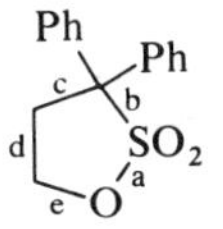

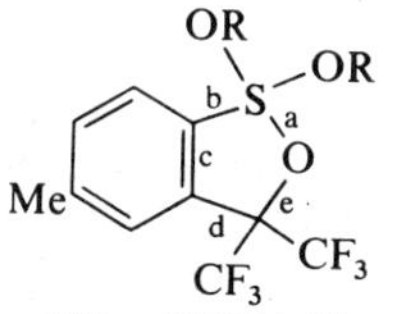

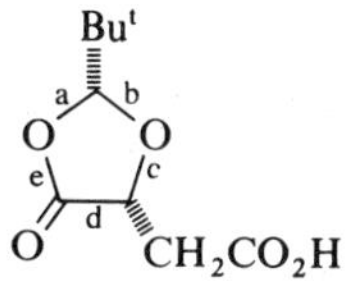

Bond	(**1**)[a]	(**2**)[b]	Compound (**3**)[c]	(**4**)[d]	(**5**)[e]	(**6**)[f]
a	1.402	1.399	1.250	1.533	1.630	1.435
b	1.414	1.852	1.263	1.821	1.777	1.419
c	1.413	1.819	1.470	1.536	1.469	1.442
d	1.534	1.500	1.497	1.503	1.484	1.511
e	1.416	1.425	1.477	1.446	1.430	1.348
ab	105	106.0	116.5	95.2	95.0	105.5
bc	109	91.9	108.9	96.5	106.9	106.2
cd	102	105.2	102.9	108.3	113.6	102.7
de	104	107.8	102.1	109.5	108.0	107.8
ef	111	112.0	109.5	109.5	116.3	109.3

[a] ⟨50ACS1854⟩. [b] ⟨68JOC3535⟩. [c] ⟨76CB1837⟩. [d] ⟨80MI43000⟩.
[e] ⟨79JA1155⟩. [f] ⟨75ACS(A)7⟩.

A number of X-ray structures have been reported since 1970. Representative of these are structures (**3–6**) for which bond lengths and angles are also shown in Table 1. Of particular interest is the finding that 2-methyl-1,3-dioxolan-2-ylium perchlorate (**3**) has a planar ring, indicating the delocalization of charge in the system.

4.30.1.3.2 Microwave spectroscopy

In the microwave spectrum of 1,3-dioxolane (**7**), nine vibrational states have been assigned which indicate that the molecule undergoes pseudorotation ⟨74JSP(49)70⟩. The dipole moment for each vibrational state has been measured and found to be constant, within experimental error, at 1.19 ± 0.03 D. This compares with a value of 1.47 D determined in benzene solution ⟨59JCS3521⟩.

Microwave spectroscopy has also been used to study the shape of ethylene carbonate (**8**) in the gas phase. While X-ray diffraction has shown that the molecule exists in an envelope configuration in the solid state ⟨54AX92⟩, the evidence from studies on its IR ⟨56MI43000⟩ and Raman ⟨62CB2333⟩ spectra suggests that the molecule becomes flat when it is melted, dissolved or vaporized. Data from the microwave spectrum of (**8**), however, can be explained only by the assumption of a non-planar ground state with tunneling through the barrier at the planar configuration ⟨65JA4950⟩. The out-of-plane contribution to the moments of inertia (3.229 u Å^2) is too large to arise from hydrogen atoms only. The introduction of a double bond into (**8**) yields vinylene carbonate (**9**) for which pseudorotation would not be expected, and for which the lowest frequency vibrational mode should be simple ring puckering. A microwave investigation of the ground state has established that (9) is planar and has a dipole moment of 4.51 ± 0.05 D in the vapor phase ⟨54JCP(22)1678⟩. The spectroscopic parameters are consistent with resonating double-bonded structures involving participation of the non-bonded electrons of the oxygen atoms. A later study ⟨66JCP(44)1352⟩ confirms the rather large dipole moment and derives rotational constants and moments of inertia for the vibrational ground state as well as the first three excited states of the ring puckering vibrational mode. The corresponding thione (10) is also planar with a dipole moment of 1.60 ± 0.02 D ⟨80JST(64)137⟩.

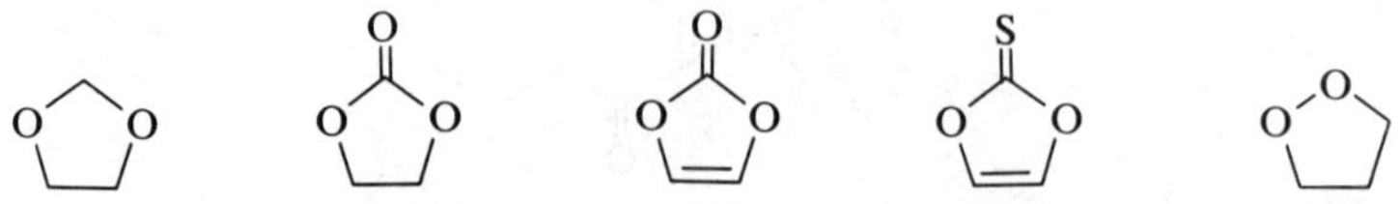

The observed rotational constants in the microwave spectrum of 1,2-dioxolane (11) are fairly well reproduced by the bond lengths of 1.563 (C—C), 1.428 (C—O) and 1.10 Å (C—H) using bond angles of analogous molecules. When the bond angles in the skeletal framework are allowed to adjust slightly, angle values of 105.1 (CCO) and 101.7° (CCC) are obtained. The dihedral angle along the peroxide bond is $50 \pm 2°$ ⟨80TL1649⟩.

4.30.1.4 Molecular Spectra

4.30.1.4.1 ^1H NMR spectra

(i) *Unsaturated rings*

Protonation of vinylene carbonate in a super-acid solvent system gives the 1,3-dioxolium cation (12; equation 1). The ^{1}H NMR spectrum of (12) at −60 °C shows the ring protons at $\delta 8.2$, a deshielding of 1.3 p.p.m. compared with (9) in SO_2 solution. This additional deshielding suggests the presence of a significant ring current in a 6π-electron heteroaromatic system ⟨77JOC2237⟩. The parent 1,3-benzodioxolium ion (14), prepared by acidic treatment of the methoxy compound (13; equation 2), shows two singlets at $\delta 8.1$ and $\delta 8.2$, deshielded approximately 1.0 p.p.m. from the precursor (13), and a sharp singlet for H-2 at $\delta 10.4$.

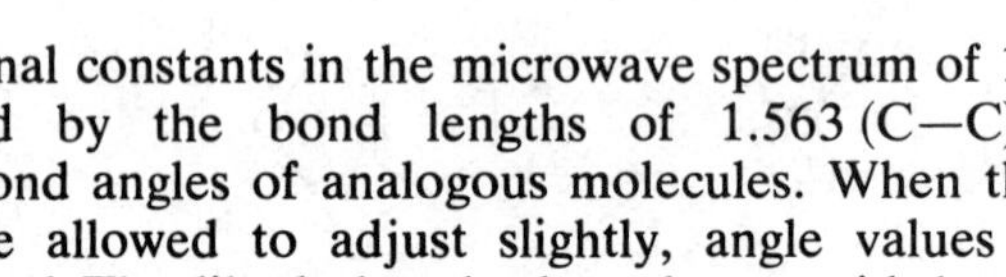

The 1,3-oxathiolium system is more stable and many compounds substituted with stabilizing groups have been reported. Representative among these are the 2,5-diphenyl-1,3-oxathiolium perchlorate (15), which exhibits a resonance at $\delta 8.12$ in deuterated TFA for H-4 ⟨75H(3)217⟩, and the amino compound (16), which is stable to water and in which H-4 is observed at $\delta 7.75$ in D_2O ⟨72CPB304⟩. The 1,2-oxathiolium cation (17) has been observed in TFA–perchloric acid solution ⟨75JCS(P1)2097⟩.

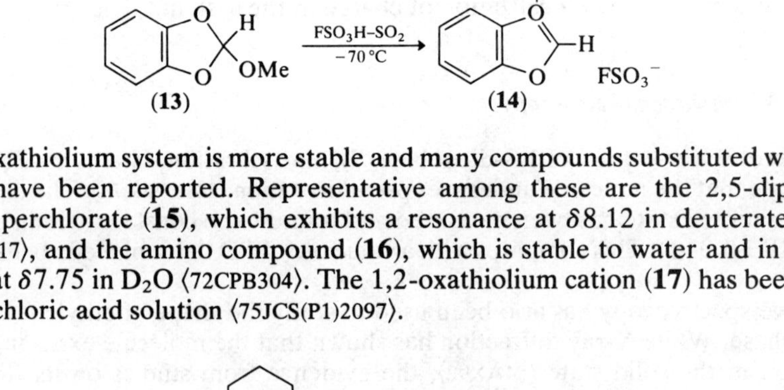

1,3-Dioxole (**18**) has been studied using the ^{13}C satellite peaks to determine the coupling constants ⟨75CJC2734⟩. The results, as determined in benzene solution, are as shown and the long-range coupling is particularly noteworthy. The spectra of 1,3-dioxolan-2-ylium ions have been studied extensively ⟨72CRV357⟩ and detailed analysis of the coupling constants in the unsymmetrical ion (**19**) has revealed typical values, determined in CD$_3$CN, as shown ⟨71CB1264⟩. A quantitative correlation of the effects of *m*- and *p*-aryl substituents of 2-aryl groups on the chemical shift of the 1,3-dioxolan-2-ylium ring protons reveals a linear correlation with Hammett σ values (correlation coefficient 0.966). Using σ$^+$ values a less satisfactory correlation (0.940) is obtained. This suggests that the most important contributor to the resonance hybrid is (**20**) rather than (**21**) ⟨66TL3389⟩. 1,3-Dioxolan-2-ylium ions with hydrogen at the 2-position are rare, although the ^{1}H NMR spectrum of (**22**) in CD$_3$CN has been reported ⟨71CB1264, 71CB1281⟩.

(**18**)
δ 5.38 (2-H)
δ 6.19 (4, 5-H)
$J_{4,5}$ = 1.20 Hz
$J_{2,4}$ = −0.29 Hz

(**19**)
$J_{4,5}$ = 8.0 Hz
$J_{4,5'}$ = 9.0 Hz
$J_{5,5'}$ = 10.0 Hz

(**20**)

(**21**)

(**22**)
δ 9.40 (2-H)
δ 6.22 (4-H)
δ 5.40, 5.60 (5-H)
J_{cis} = 8.0 Hz
J_{trans} = 9.0 Hz
J_{gem} = −10.0 Hz

The C-4 ring protons of the 1,3-oxathiolan-2-ylium cation should be more shielded than in the corresponding oxygen compounds since, apart from the larger size of the sulfur atom, structure (**23**) should have a greater contribution to resonance than (**24**). The chemical shift values obtained for (**25**) and (**26**) support this hypothesis ⟨72CRV357⟩.

(**23**)

(**24**)

(**25**)
δ 5.44 (4-H)
δ 2.29 (5-Me)

(**26**)
δ 4.38 (4-H)
δ 2.37 (5-Me)

(**27**)
alkenic *ca.* δ 7.1
CH$_2$ *ca.* δ 5.3, 5.5 (*J* = 14 Hz)

The ^{1}H NMR spectrum of the cyclic sulfenate ester (**27**) has also been reported ⟨70CJC3704⟩.

(ii) *Saturated rings*

The ^{1}H NMR spectrum of 1,3-dioxolane was first reported in 1959, the coupling constants being derived using the ^{13}C satellite peaks ⟨59MI43000⟩. Other early work on 2-substituted 1,3-dioxolanes led to the conclusion that the ring was puckered and actually existed as a number of rapidly interconverting forms ⟨62CI(L)213⟩. Later studies showed that a progressive rotation (pseudorotation) between the various half-chair and envelope conformations, alternating around the ring, was taking place ⟨65BSB488⟩ and that normal Karplus-type relationships did not apply ⟨65JCS256⟩. Many of the studies of the ^{1}H NMR spectra of 1,3-dioxolanes have centered on the AA'BB' analysis in order to determine the structure of the various substituted derivatives and hence the stereochemistry of the associated diols ⟨B-73NMR382⟩. It is not always easy to decide which conformation is present in solution because of time-averaged coupling constants and chemical shifts obtained through pseudorotation. In general, however, α-substituents cause upfield shifts for synclinal protons and the more remote β-substituents cause a downfield effect on the same protons. The effect of a β-substituent on a *trans* proton is generally an upfield shift of 0.2–0.3 p.p.m. ⟨73OMR(5)299, 75OMR(7)345⟩. These effects have been used to assign relative stereochemistry in systems where both isomers were obtained ⟨77JHC1035⟩. Some ^{1}H NMR data for 1,3-dioxolanes are contained in Table 2. The 1,3-oxathiolane ring is an example of a five-membered ring where certain conformational models illustrate the real situation better than a freely interconverting circuit of pseudorotamers ⟨74JCS(P2)466⟩. The ^{1}H NMR spectral data for substituted 1,3-oxathiolanes may be explained in terms of two preferred envelope conformations ⟨70JA5907⟩. These conformations are discussed in more detail in Section 4.30.1.5. Some ^{1}H NMR data for 1,3-oxathiolanes are collected in Table 3.

Table 2 ^{1}H NMR Data for 1,3-Dioxolanes

Substituent	Solvent	Chemical shifts		Coupling constants (Hz)			Ref.
		$\delta(H\text{-}2)$	$\delta(H\text{-}4,5)$	J_{cis}	J_{trans}	J_{gem}	
None	Neat	—	—	7.3	6.0	—	68T4377
	C_6H_6	4.75	3.48	—	—	—	68T4377
2-Me	CCl_4	—	—	7.20	6.06	7.68	65BSB488
2-Ph	$CDCl_3$	5.65	3.68, 3.82	7.00	6.20	7.75	65BSF3646
2-OMe	C_6H_6	5.66	3.81	6.80	6.28	7.60	68T4377

Table 3 ^{1}H NMR Data for 1,3-Oxathiolanes

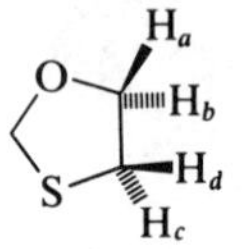

Substituent	Solvent	Chemical shifts				Reference
		$\delta(H_a)$	$\delta(H_b)$	$\delta(H_c)$	$\delta(H_d)$	
None	CCl_4	3.91	3.91	2.90	2.90	67JA4368
2-Et	CCl_4	4.26	3.71	2.94	2.94	70JA5907
2-Ph	C_6H_6	3.96	3.39	2.73	2.56	70JA5907

Substituent	Coupling constants (Hz)						Reference
	J_{ab}	J_{ac}	J_{ad}	J_{bc}	J_{bd}	J_{cd}	
None	—	6.12	5.58	5.58	6.12	—	67JA4368
2-Et	−9.28	5.44	4.01	10.36	4.37	−10.06	70JA5907
2-Ph	−8.59	6.72	2.79	8.39	6.51	−9.31	70JA5907

The ^{1}H NMR spectra of 1,2-dioxolane (**28**) ⟨79JA4290⟩ and 1,2-oxathiolane (**29**) ⟨81CC741⟩ have been reported. No detailed analysis of these spectra has been attempted.

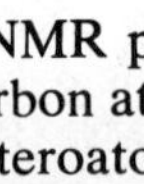

(**28**) δ (CCl_4)
4-H 2.53 ($J = 7$ Hz)
3,5-H 3.92 ($J = 7$ Hz)

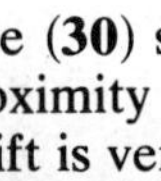

(**29**) δ ($CDCl_3$)
3-H 3.61 ($J = 6.5$ Hz)
4-H 2.20 ($J = 6.5$ Hz)
5-H 3.96 ($J = 6.5$ Hz)

4.30.1.4.2 ^{13}C NMR spectra

(i) *Unsaturated rings*

The ^{13}C NMR parameters for vinylene carbonate (**30**) show significant shielding of the carbonyl carbon atom due most probably to the proximity of the lone pair electrons on the adjacent heteroatoms ⟨77JOC2237⟩. The carbonyl shift is very similar in ethylene carbonate, the saturated analog ⟨B-72MI43000⟩. The alkenic carbons in (**30**) appear at δ 132.2, a typical resonance for sp^2 carbons of unsaturated alkenes and aromatics. On protonation the chemical shifts for the 1,3-dioxolium ion (**31**) are deshielded by 6.1 p.p.m. for the alkenic carbon atoms and 9.5 p.p.m. for the carbonyl carbon atom.

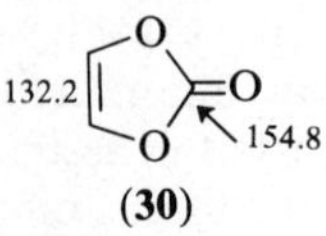

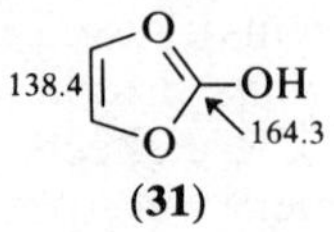

The ^{13}C NMR spectrum of the 1,3-benzodioxolium ion (**33**), prepared in fluorosulfonic acid–sulfur dioxide solution from the methoxy compound (**32**), shows aromatic ring carbons deshielded 6.8 and 10.6 p.p.m. from the corresponding carbon atoms in (**32**). The chemical shift for the carbon between the oxygen atoms is deshielded by 51.9 p.p.m. from the precursor (32) ⟨77JOC2237⟩. These data suggest a significant delocalization of charge over the 6π and 10π systems, although no conclusion can be reached as to the exact nature of the ring currents involved or the relative contribution of resonance forms.

C-1 146.0
C-2 108.0
C-3 121.7
(32)

170.4
C-1 144.4
C-2 114.8
C-3 132.3
(33)

Limited ^{13}C data are available on 1,3-dioxolan-2-ylium cations ⟨76CB1837⟩.

(ii) Saturated rings

Comparison of the ^{13}C NMR resonances for the parent ring systems cyclopentane (34), tetrahydrofuran (35), 1,3-dioxolane (36) and 1,2-oxathiolane (37) shows large downfield shifts of carbons adjacent to oxygen (42.3 p.p.m. for the first oxygen, 27.1 p.p.m. for the second) and an apparent shielding β-effect in (36) of 3.4 p.p.m. if C-4 is considered both α and β to oxygen ⟨79OMR(12)461⟩. The larger sulfur atom in (37) has a relatively small deshielding effect at the α- and β-positions ⟨81CC741⟩.

25.6 (34)   67.9 / 25.8 (35)   95.0 / 64.5 (36)   36.5 / 75.1 / 29.8 (37)

Table 4 shows the ^{13}C chemical shifts of some methyl substituted 1,3-dioxolanes. With simple substitution, such as one or two methyl groups, the 1,3-transannular nonbonded interactions are practically negligible because of the high degree of flexibility of the ring system. The carbon resonances for the methyl groups of *cis*-4,5-dimethyl-1,3-dioxolane are moved upfield by 2.2 p.p.m. compared to the *trans* isomer. This shift is probably due to the dihedral angle of the two vicinal methyl groups of the *cis* compound being smaller than that of the *trans* compound. In tri- and tetra-methyl substituted 1,3-dioxolanes, however, the resonance at ca. δ 27 probably implies the presence of 1,3-transannular nonbonded interactions. Despite the lack of conformational integrity of the ring, a good set of additive parameters has been obtained for 2,4-disubstituted and 2-*cis*-4,5-trisubstituted 1,3-dioxolanes ⟨79OMR(12)461⟩. A comparison of the α-, β- and γ-parameters for methyl substitution and methyl shifts is given in Table 5. As in (34) and (35), the α- and β-shift effects of a methyl substituent are equal, although somewhat smaller in 1,3-dioxolanes. The γ-effect of a methyl substituent is negligible, presumably because of pseudorotation. In this regard it is interesting that of all the 1,3-dioxolanes studied, only the *cis*-4,5-Bu^t-substituted compounds display ^{13}C shift differences between the *cis*- and *trans*-2-alkyl diastereomers of the order of 2 p.p.m. Such systems presumably display some conformational rigidity in the ring framework and can give rise to axial-like and equatorial-like positions of the 2-substituents ⟨79OMR(12)461⟩.

Table 4 ^{13}C NMR Data for 1,3-Dioxolanes

Substituent	C-2	C-4	C-5	2-Me	4-Me	5-Me	Reference
None	95.23	64.81	64.81	—	—	—	79OMR(12)461
2-Oxo	156.07	65.10	65.10	—	—	—	77ACS(B)899
2-Me	101.7	65.0	65.0	19.8	—	—	79OMR(12)461
4-Me	95.0	72.3	71.0	—	18.1	—	79OMR(12)461
cis-2,4-di-Me	101.6	73.0	71.0	20.2	18.9	—	79OMR(12)461
trans-2,4-di-Me	100.7	71.8	72.0	20.2	18.5	—	79OMR(12)461
cis-4,5-di-Me	93.6	74.1	74.1	—	14.7	14.7	79OMR(12)461
trans-4,5-di-Me	94.0	78.8	78.8	—	16.9	16.9	79OMR(12)461
2,2,4-tri-Me	108.7	72.0	70.9	27.2 (cis) 25.9 (trans)	18.6	—	79OMR(12)461
cis-2,2,4,5-tetra-Me	107.2	74.0	74.0	28.7 (cis) 25.8 (trans)	15.6	15.6	79OMR(12)461
trans-2,2,4,5-tetra-Me	107.4	78.3	78.3	27.4	16.9	16.9	79OMR(12)461

Table 5 ^{13}C NMR Additive Parameters (p.p.m.) for Substituted 1,3-Dioxolanes ⟨79OMR(12)461⟩

Parameter	C-2	C-4
α-Me	6.2	7.6
β-Me	—	6.3
γ-Me	0	0.4
Me shift	19.8	18.1

The ^{13}C–^{1}H coupling constants for the ring carbons in 1,3-dioxolanes are 165 Hz (C-2) and 149 Hz (C-4,5) ⟨B-72MI43001⟩.

4.30.1.4.3 UV and IR spectra

The UV spectral data for some 2,5-diaryl-1,3-oxathiolium perchlorates are shown in Table 6. Also shown are the UV and IR data of a series of 2-(disubstituted amino)-5-aryl-1,3-oxathiolium salts. The presence of a strong band at *ca.* 1652 cm^{-1} is explained by a strong contribution from the immonium structure. With one exception, no detailed IR studies have been reported in the series. A complete assignment of the bands in the spectrum of 1,3-dioxolane has been made ⟨59JCS802⟩. The non-planarity of the ring is shown by the broad nature of the bands.

Table 6 IR and UV Data for Oxathiolium Cations

R	R'	X$^-$	λ$_{max}$ (nm) (log ε)	UV solvent	$\nu(C=\overset{+}{O})$	Reference
Ph	H	ClO$_4$	273, 285, 387 (4.16, 4.12, 4.20)	CH$_2$Cl$_2$	—	75H(3)217
p-ClC$_6$H$_4$	H	ClO$_4$	241, 273, 304, 397 (4.13, 4.11, 4.07, 4.33)	CH$_2$Cl$_2$	—	75H(3)217
Piperidino	H	HSO$_4$	213(sh), 220(sh), 278 (4.23, 4.12, 4.30)	H$_2$O	1658, 1652	72CPB304
Piperidino	H	ClO$_4$	212, 220(sh), 278 (4.31, 4.23, 4.29)	EtOH	1651	72CPB304
Me$_2$N—	H	HSO$_4$	211(sh), 218(sh), 274(sh) (4.23, 4.09, 4.26)	H$_2$O	1675	72CPB304
Piperidino	OH	HSO$_4$	216, 283 (4.33, 4.34)	EtOH	1657	72CPB304
Piperidino	NO$_2$	HSO$_4$	232, 325 (4.13, 4.13)	EtOH	1660	72CPB304

4.30.1.4.4 Mass spectra

The mass spectra of 1,3-dioxolanes have been studied extensively ⟨B-60MI43000, B-67MI43000⟩. The most outstanding feature of the spectrum of 1,3-dioxolane is the loss of a hydrogen atom from C-2, confirmed by deuterium labelling, as the main fragmentation process ⟨77ACS(B)227⟩. The relatively low activation energy for this fragmentation is presumably due to the high stability of the delocalized oxonium ion which is formed (equation 3). With 2-substituted 1,3-dioxolanes, an alkyl group is lost in preference to hydrogen and the ease of alkyl group loss increases with chain length and branching. 4-Substituted 1,3-dioxolanes behave differently: 4-methyl-1,3-dioxolane (**38**) shows a substantial [$M-1$] ion which, on first consideration, might be assigned as resulting from the loss of a 2-H atom. Deuterium labelling studies have shown that 5-H atom is lost according to Scheme 1. Since a methyl radical is much more stable than a hydrogen radical, this mechanism explains the observed loss of a methyl radical in the mass spectrum of 4,5-dimethyl-1,3-dioxolane (Scheme 2) ⟨77ACS(B)227⟩.

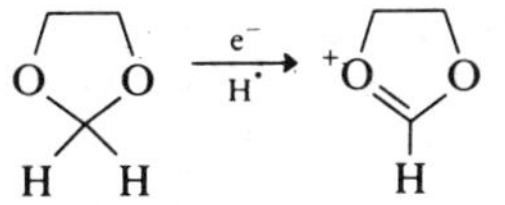

Scheme 1

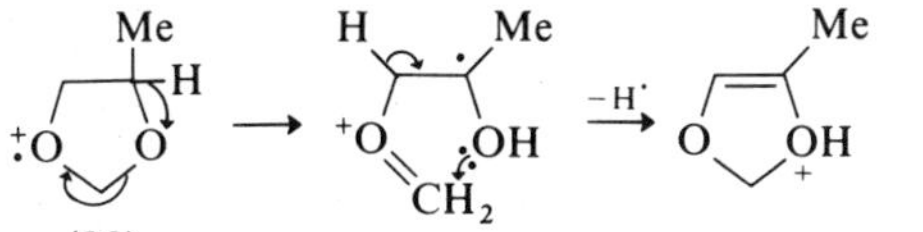

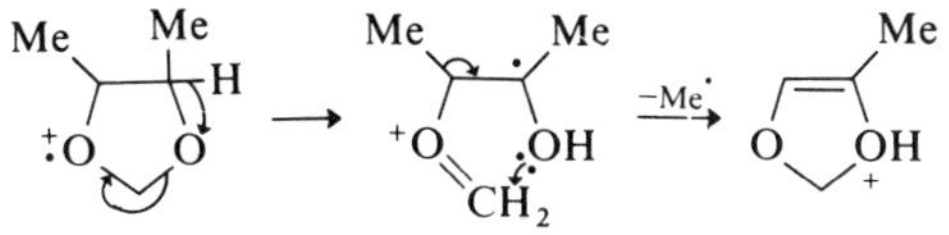

Scheme 2

In contrast, 1,3-oxathiolanes give a substantial molecular ion and a small [M − 1] ion. Fragments tend to retain sulfur and for 1,3-oxathiolane itself the loss of CH_2O is the base peak, the carbon atom being derived from the 2-position ⟨72OMS(6)415⟩.

The main fragmentation pathway for 1,2-dioxolanes has been reported to be the rupture of the peroxide linkage, although no evidence has been presented ⟨77MI43000⟩.

4.30.1.4.5 Photoelectron spectra

Assuming C_2 symmetry, the lone pair orbitals in 1,3-dioxolane transform as a + b. The plus combination ($n_1 + n_2$) has a symmetry. The PE spectra show that a and b are separated by 0.55 eV (10.1, 10.65 eV) for 1,3-dioxolane and 0.49 eV (9.71, 10.20 eV) for 2,2-dimethyl-1,3-dioxolane. Molecular models show that direct overlap of the lone pair orbitals is not likely to be large and hence the splitting is not through-space but rather should be ascribed to a through-bond mechanism. Since methyl substitution at C-2 does not have a pronounced effect on the splitting, the principal interaction is probably via the C—H orbitals of C-4 and C-5. The dominant effect of C-4 and C-5 compared with C-2 probably reflects tighter binding for the latter and a higher electronegativity for C-2 ⟨72JA5599⟩.

The PE spectrum of 3,3,5,5-tetramethyl-1,2-dioxolane has been studied as part of a series of cyclic peroxides. The effect of oxygen lone pair orbital dihedral angle on the ionization potential has been evaluated and the observed splitting of 1.15 eV (9.25, 10.40 eV) is explained by a dihedral angle of 30 ± 5° ⟨75CJC3439⟩.

4.30.1.5 Thermodynamic Aspects

4.30.1.5.1 Intermolecular forces

There has been no effort to study systematically a series of compounds. The lower dioxoles and oxathioles are liquids, insoluble in water but soluble in most organic solvents. 1,3-Dioxole boils at 51 °C at atmospheric pressure, while 2,2-dimethyl substitution raises this to 72–73 °C. No boiling point has been reported for 1,2-dioxolane and the only report for 1,2-oxathiolane indicates that it codistills with chloroform at 20 °C (0.1 mmHg). 1,3-Dioxolane boils at 75 °C (760 mmHg) and the introduction of sulfur raises the boiling point such that 1,3-oxathiolane boils at 132–136 °C (760 mmHg).

4.30.1.5.2 Aromaticity

The available spectroscopic data for dioxolium and oxathiolium cations, as well as mesoionic oxathioliums, indicate that, qualitatively, these compounds should be regarded as aromatic. While these compounds do not possess aromatic stability in the classical sense, the chemical shift data discussed in Section 4.30.1.4.1(i) indicate the presence of a ring current. The 2-dialkylamino-1,3-oxathiolium cations are the most stable derivatives, being inert to aqueous hydrolysis at room temperature. This does not make this series of compounds 'more aromatic' but is best explained by an immonium structure with accompanying delocalization of charge and consequent interruption of the sextet.

4.30.1.5.3 Shape and conformation

From the previous sections on theoretical and spectral aspects of 1,3-dioxolanes, it is clear that the ring is conformationally labile and the data are best explained in terms of a freely pseudorotating system with a small energy barrier between conformers ⟨B-80MI43001⟩. The half-chair (**39**) and envelope (**40**) forms are the two lowest-energy conformers and it is the interconversion of these structures which best describes the pseudorotation ⟨75JA1358⟩. The X-ray structure of (**1**) shows the dioxolane ring has a conformation mid-way between these forms.

Studies on the acid-catalyzed equilibrium of the acetal portion of the ring have shown that the free-energy differences between *cis* and *trans* isomers are small and only very bulky substituents such as *t*-butyl groups show any signs of steric interaction ⟨69TL1775, 70JA5394⟩. Typically these energy differences are 1.2–2.0 kJ mol^{-1} for dialkyl and 1–3 kJ mol^{-1} for trialkyl derivatives ⟨70BSB11, 71BSB215⟩. Such small differences are explained only in terms of a very highly flexible ring. The conformational anchoring effect of two bulky groups is clearly evident, however, in the ^{13}C chemical shift differences of 2-substituents in 1,3-dioxolanes containing *cis*-4,5-di-Bu^t groups ⟨79OMR(12)461⟩ discussed earlier.

The available data on 1,3-oxathiolanes show that the presence of a sulfur atom makes the ring more puckered and that while pseudorotation is occurring, two envelope conformers (**41**) and (**42**) are preferred ⟨70JA5907, 74JCS(P2)466, 77JCS(P2)343⟩. Conformational energy calculations show that an envelope in which the sulfur atom is β to the flap atom (*i.e.* **41** and **42**) represent the minimum energy conformations ⟨74JA2426⟩. The X-ray structure of the spirosteroid (**2**) shows that conformer (**42**) is preferred in this molecule in the solid state.

4.30.1.6 Tautomerism

Little has been reported on tautomerism of dioxoles and oxathioles. The spectral data for a series of 2-dialkylamino-1,3-oxathiolium salts has been interpreted as favoring the immonium structure (**43**) ⟨72CPB304⟩. Similarly, the spectra of several oxathiol-2-ylidene derivatives (**44**) indicate that the dipolar contribution to the resonance hybrid structure is small ⟨73CPB2224⟩.

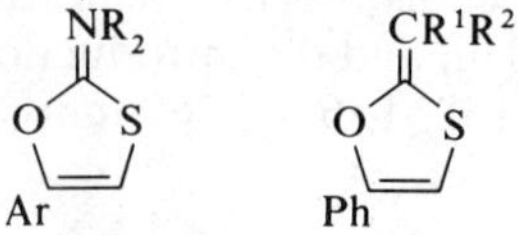

4.30.1.7 Betaine Structures

Two types of mesoionic 1,3-oxathiolium species have been reported. The oxathiolium-5-olate (**45**) is stable and isolable ⟨75CC417⟩, and the presence of the trifluoroacetyl group appears to be necessary for stability, although the 4-phenyl derivative (**46**) may be trapped *in situ* by dipolarophiles ⟨77CPB1471⟩. Compound (**45**), which is unreactive towards cycloaddition reactions, shows two bands in its IR spectrum at 1815 and 1640 cm^{-1} which may be assigned to COCF$_3$ and the ring CO respectively. The oxathiolium-4-olate (**47**) is also too unstable for isolation, but has been trapped by dipolarophiles ⟨75AG(E)422⟩.

In view of the extreme instability of these mesoionic structures it appears unlikely that the corresponding 1,3-dioxolium species will be isolable.

4.30.2 REACTIVITY

4.30.2.1 Reactivity at the Ring Atoms

4.30.2.1.1 General survey

Very little has been reported on aromatic dioxoles and oxathioles and, therefore, they are included together with the saturated analogs rather than treated separately. Their chemistry, when available, is contained at the beginning of each section. The fully aromatic systems are monocations, sensitive to nucleophilic attack at the 2-position. On the other hand, the saturated compounds exhibit chemical reactivity which essentially parallels their acyclic counterparts. Earlier general works are available on 1,3-dioxolanes ⟨B-57MI43000⟩, 1,3-dioxolan-2-ylium cations ⟨72CRV357⟩ and both 1,2- and 1,3-oxathiolanes ⟨66HC(21-1)76⟩. Reactions with stereochemical implications have been covered ⟨B-77SH(2)68, B-77SH(2)314⟩, as have transformations into other heterocyclic systems ⟨B-73MI43000⟩. Also, the use of these rings as protecting groups for carbonyl compounds and their reactivity in the presence of a range of reagents have been tabulated ⟨B-81MI43000⟩.

4.30.2.1.2 Thermal and photochemical reactions involving no other species

Since 1,2-dioxolanes are cyclic peroxides, the thermolysis and photolysis of these compounds is expected to proceed *via* the rupture of the peroxide linkage. The resulting diradical has a short lifetime ($<10^{-7}$ s), and it has been difficult to trap these species and prove their intermediacy. By increasing this lifetime, it is possible to trap the 1,5-diradical before subsequent transformations take place. Thus tetramethyl-1,2-dioxolane (**48**) is destroyed quickly under conditions where the unsaturated dioxolane (**49**) is unchanged. The high thermal stability of the diradical (**50**) may be attributed to the energetically unfavored deketonization step which would result in a vinyl radical center. Thermolysis of (**49**) at 200 °C in the presence of benzhydrol results in the trapping of (**50**), and the quantitative production of the diol (**51**) ⟨75JA926⟩. The photolysis of (**48**) at 350 nm gives (**53**), acetone and 2-butanone. While the intermediacy of (**52**) may account for these, definitive evidence has not been presented. The epoxide (**53**) is the major product, but large amounts of resinous materials are formed ⟨72TL1357⟩.

In contrast, 1,3-dioxolanes are relatively stable to pyrolysis. At temperatures approaching 500 °C, rupture of the ring begins to occur and the products depend to a large part on the substituents present. 1,3-Dioxolane itself yields all possible C_1 and C_2 fragment combinations. A similar myriad of products is seen on the gas-phase photolysis of 1,3-dioxolanes, although liquid-phase photolysis of 1,3-dioxolane at 185 nm proceeds mainly *via* the intermediate diradical (**54**) ⟨78JCS(P2)985⟩. A similar bond cleavage accounts for the decomposition of the carbene (**55**), generated thermally from the corresponding tosylhydrazone. While the observed products of alkene and carbon dioxide could result from a concerted symmetry 'allowed' process, theoretical studies show that a single cleavage is of lower energy. The products, therefore, are derived from diradical (**56**) in a two-step process ⟨81JA2558⟩.

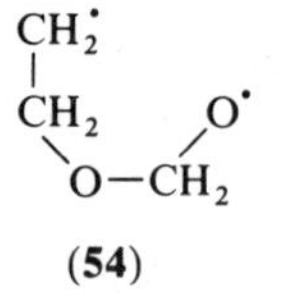

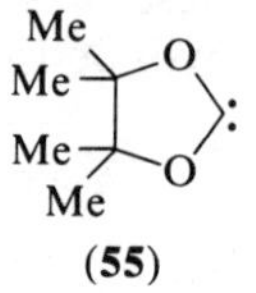

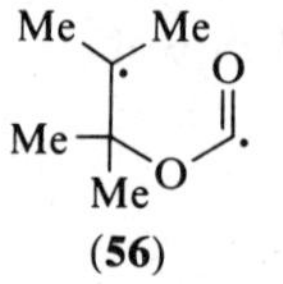

The photochemical and thermolytic breakdown of cyclic sulfones offers a route to cyclopropanes, as shown in equation (4). The utility of the reactions is limited in that an aromatic group (**57**; R = Ar) is required for good yields to be obtained. Flash thermolysis of (**57**; R = Me) at 750 °C gives considerable quantities of acetaldehyde and 2-butene together with methylcyclopropane ⟨79JCS(P1)950⟩.

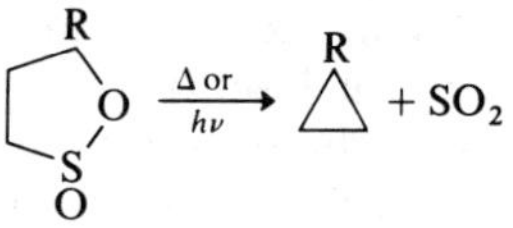

1,3-Oxathioles containing electron-withdrawing groups exhibit interesting thermal reactivity. The oxathiole (**58**) is in thermal equilibrium with its isomer (**60**) at room temperature in organic solvents (Scheme 3). Above 140 °C the intermediate thiocarbonyl ylide (**59**) is itself in equilibrium with the thiirane (**61**) and undergoes desulfurization on treatment with triphenylphosphine to give the alkene ⟨81JA2757⟩.

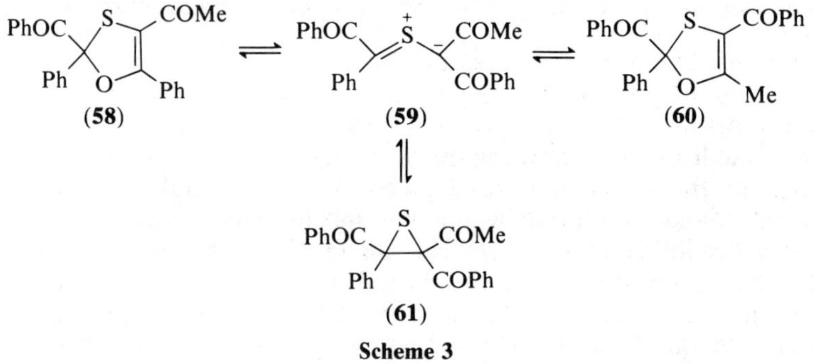

Scheme 3

4.30.2.1.3 Reactivity towards electrophiles and oxidants

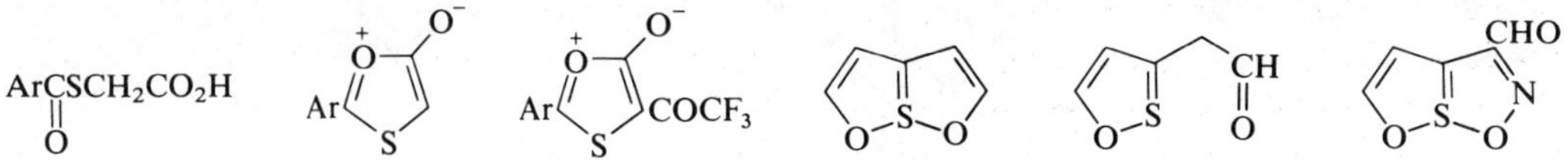

There are only two reports of electrophilic attack on aromatic oxathioles. The cyclization of (**62**) in TFA generates the 1,3-oxathiolium (**63**), which cannot be isolated as such but which suffers acylation to give the stable derivative (**64**). It is unclear whether the nitrosation of thiapentalene (**65**) represents an electrophilic attack on the intact system or whether attack occurs on the ring-opened species (**66**). In the event, nitrosyl hexafluorophosphate treatment of (**65**) gives the aza derivative (**67**), which is formed by ring opening at some stage, followed by rearrangement before subsequent ring closure ⟨74JCS(P1)722⟩.

Since 1,3-dioxolanes are cyclic acetals it follows that they are readily hydrolyzed by acids. The kinetics of this hydrolysis have been extensively reviewed ⟨57MI43000⟩. The acidic hydrolysis of 1,3-oxathiolanes is of theoretical importance in that there are conflicting reports as to whether the C—O or the C—S bond is ruptured first. It is known that protonation of 2,2-dimethyl-1,3-oxathiolane in fluorosulfonic acid leads to the exclusive formation of dication (**68**) ⟨71BSF541⟩. Based on deuterium labelling studies, however, the present data suggest that the C—O bond is broken first in aqueous systems ⟨72TL2569⟩.

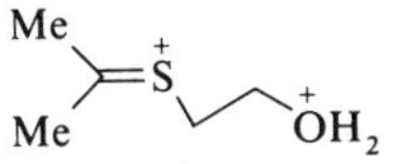

(**68**) 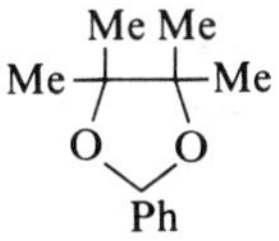 (**69**) 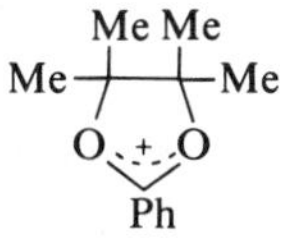 (**70**)

The action of halogens and halogenating agents on 1,3-dioxolanes results in the initial formation of 1,3-dioxolan-2-ylium cations which, in certain cases, may be isolated. Thus halogens and interhalogens oxidize dioxolane (**69**) to the cation (**70**). The presence of the methyl groups is crucial for the isolation of (**70**) and its stability may be attributed to steric hindrance by the methyl groups towards nucleophilic attack at the C-4 or C-5 positions ⟨79CC751⟩. Less hindered dioxolanes yield 2-haloethyl esters, with nucleophilic attack occurring at the least hindered carbon atom, as shown in Scheme 4. This reaction has synthetic importance, especially if trimethyl phosphate is used as a combined solvent and scavenger for the hydrogen halide produced. This eliminates many acid-catalyzed side reactions and leads to high yields of haloesters ⟨79JOC3082⟩. Low-temperature photocatalyzed chlorination of 1,3-dioxolane using an equivalent of chlorine gives 65% of 2-chloroethyl formate together with 7% of the 4-chloro-1,3-dioxolane. When an excess of chlorine is used, penta- and hexa-chloro derivatives are obtained ⟨68JOC2126⟩.

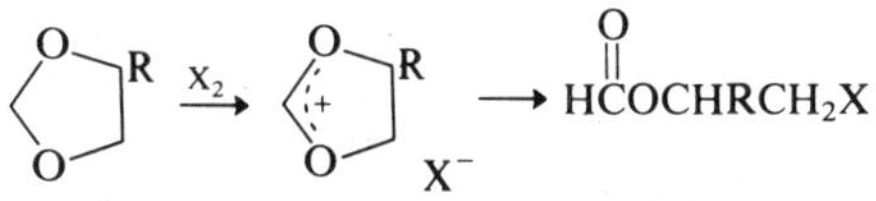

Scheme 4

Halogenation of 1,3-oxathiolanes proceeds *via* initial attack at the sulfur atom to give a halosulfonium salt. The fate of this salt depends on the substituents present on the ring. Halogenation of oxathiolanes derived from ketones and aldehydes having an α-methylene group provides 1,4-oxathienes ⟨65JA3785⟩. A plausible mechanism is shown in Scheme 5. Note that it is the C—S bond which suffers initial cleavage, which is in contrast to the hydrolysis reaction, but which points to the initial involvment of the sulfur atom ⟨65JA3785⟩. In the absence of an α-methylene moiety, the halosulfonium salt suffers attack by halide ion to regenerate the ketone and to yield a β-haloethyl disulfide ⟨76JOC966⟩.

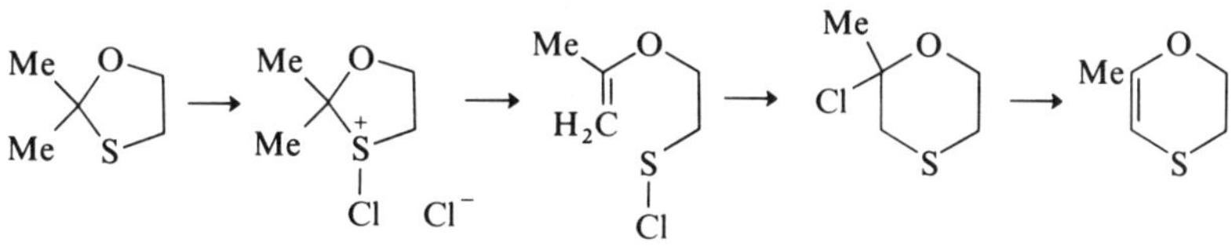

Scheme 5

Lewis acids readily isomerize both 1,3-dioxolanes and 1,3-oxathiolanes in ether solution. The reaction proceeds by coordination with the oxygen atom in the latter case since 1,3-dithiolanes do not isomerize under the same conditions. With trityl carbonium ion, an oxidative cleavage reaction takes place as shown in Scheme 6. Hydride extraction from the 4-position of 2,2-disubstituted 1,3-dioxolanes leads to an α-ketol in a preparatively useful reaction. 1,3-Oxathiolanes are reported to undergo similar cleavage but no mention of products other than regeneration of the ketone has been made ⟨71CC861⟩. Cationic polymerization of 1,3-dioxolane has been initiated by a wide variety of proton acids, Lewis acids and complex catalytic systems. The exact mechanism of the polymerization is still the subject of controversy, as is the structure of the polymer itself. It is unclear if polymerization proceeds to give a linear chain by ring opening or a macrocycle by ring expansion. It is probable that there is no fixed answer since the polymerization is very sensitive to both the catalyst and the impurities present.

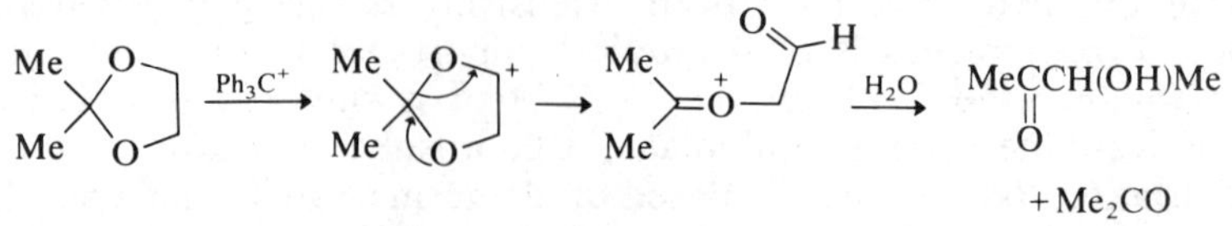

Scheme 6

Both 1,3-dioxolanes and 1,3-oxathiolanes react readily with trimethylsilyl iodide to afford ring-opened products, as shown in Scheme 7. Initial attack at the oxygen atom in 1,3-oxathiolane is expected because of the weak character of the silicon–sulfur bond ⟨78S588⟩. This reaction represents a mode of ring opening quite different from that obtained during halogenation and is probably a reflection of the strength of the silicon–oxygen bond. Both 1,2- and 1,3-oxathiolanes undergo expected oxidation of the sulfur atom. An interesting example of the utility of this reaction is represented by the separation of (**71**) into *cis* and *trans* isomers by the use of the chirality present in sulfoxide (**72**) ⟨80SC725⟩.

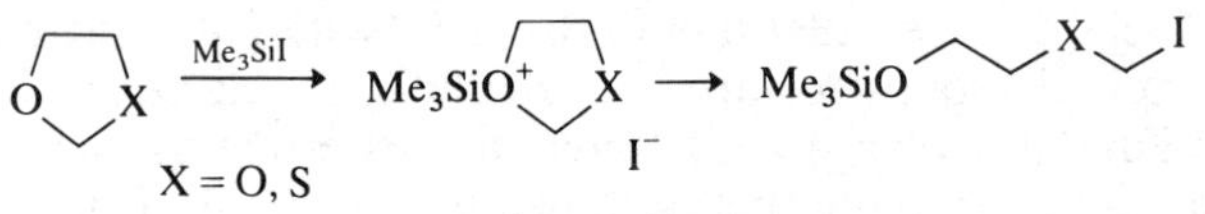

X = O, S

Scheme 7

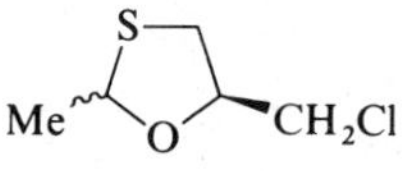

(**71**) 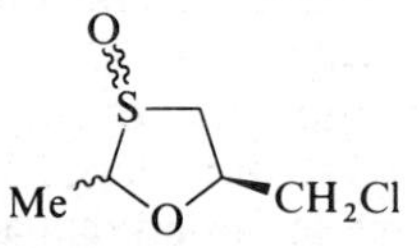   (**72**)

4.30.2.1.4 Reactivity towards nucleophiles and reducing agents

The 1,3-oxathiolium system is attacked by a variety of nucleophiles exclusively at the 2-position, and a number of products is possible depending on the nature of the nucleophile. The cation (**73**) is converted into thiophene derivatives with carbon nucleophiles (Scheme 8) ⟨75H(3)217⟩, while cation (**74**), under similar conditions, gives either thiophenes or oxathiafulvenes (**75**) ⟨71CPB2194, 71TL1137⟩. Amines and hydrazines convert (**74**) into thiazoles and 1,3,4-thiadiazines respectively ⟨78CPB3017⟩. The oxathiolium (**64**) adds water at the 2-position to give (**76**) together with carbon dioxide ⟨75CC417⟩. In the only example studied, the 2-phenyl-1,3-benzodioxolium ion adds water at the 2-position to give 2-hydroxyphenyl benzoate, and it adds phenyl Grignard reagent to give the 1,3-dioxole (**77**) ⟨65AG(E)873⟩.

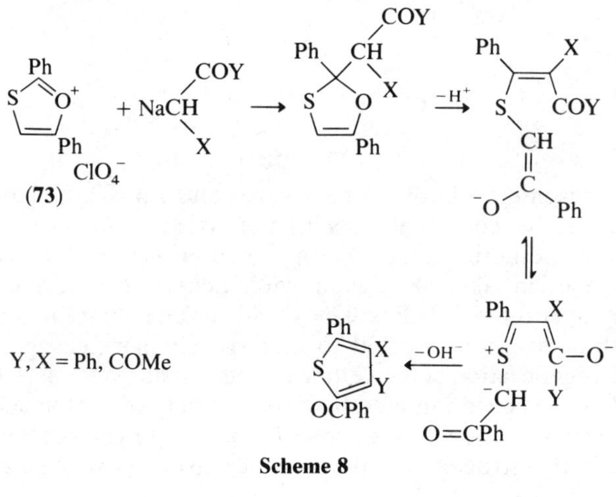

(**73**)

Y, X = Ph, COMe

Scheme 8

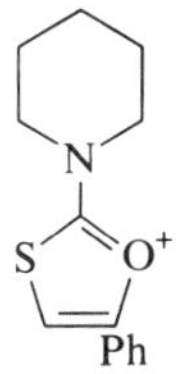

(74) 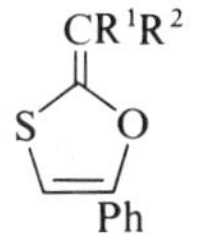 (75)  (76) 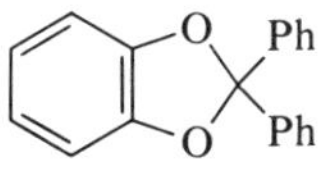 (77)

The sultene (**78**), the first example of a cyclic sulfenate, is stable indefinitely at room temperature but reacts within minutes with water or moist air to give (**79**) ⟨75JA6909⟩. This probably reflects lone pair repulsion between sulfur and oxygen non-bonded electrons in the nearly planar five-membered ring. With this in mind, the hydrolytic stability of (**80**) may reflect some hypervalent bonding interaction between the lone pair of electrons of the hydroxyl oxygen and the sulfur atom, as shown by the dotted line. The non-equivalence of the aromatic protons eliminates a symmetrical structure ⟨77CC521⟩. Strained five-membered cyclic sulfonate esters show a greatly enhanced reactivity at sulfur compared with their acyclic analogs and the sulfonate (**81**) undergoes attack at sulfur even by carboxylate anion in aprotic solvents ⟨81CC632⟩; (**81**) has been proposed as a potential coupling reagent in peptide synthesis.

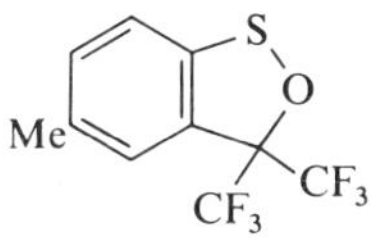

(78) 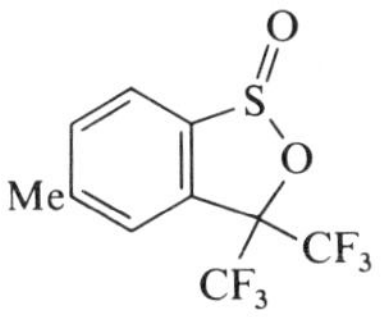 (79) 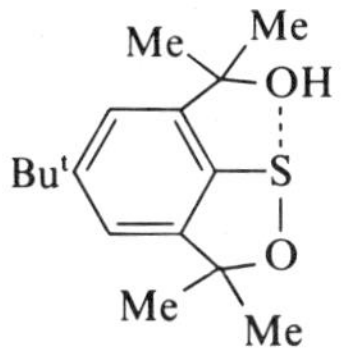 (80) 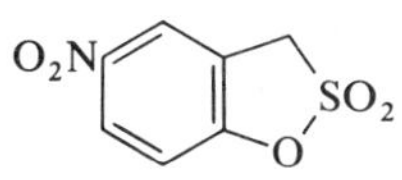 (81)

1,3-Dioxolanes, generally, are stable under Grignard conditions. However, their stability is reduced when the reaction is performed in hot benzene instead of ether, since the reagent is not complexed by the solvent. Attack is usually at C-2 (equation 5), although in some cases attack at a β-hydrogen followed by ring opening may occur ⟨78JOM(144)291⟩ as shown in equation (6). 1,3-Oxathiolanes, although less reactive than dioxolanes, also react with Grignard reagents in benzene solution. The reaction always proceeds with breakage of the C—O bond to form a hydroxysulfide.

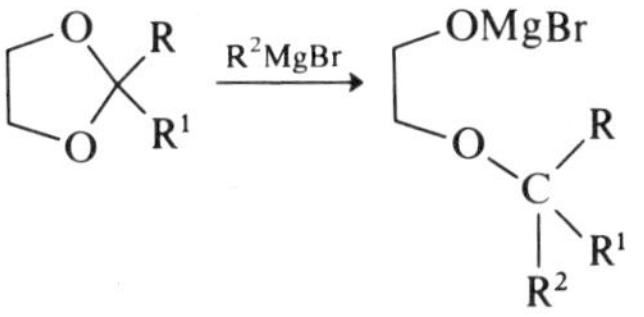

Cyclic sulfinates react with Grignard reagents at sulfur to give hydroxysulfoxides (equation 7). If an excess of Grignard reagent is employed, the product undergoes further reaction to yield mixtures of sulfides. The use of organocopper lithium reagents also effects the conversion of cyclic sulfinates into sulfones. The yields are high, sulfide formation is rare and the reaction proceeds with inversion of stereochemistry at sulfur ⟨76JOC3987⟩.

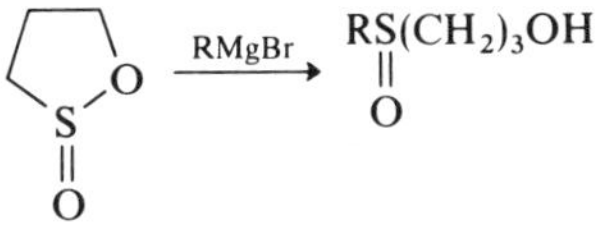

1,3-Dioxolanes are inert to LAH reduction under normal conditions but in the presence of a Lewis acid reduction occurs to yield hydroxy ethers. When aluminum chloride is employed, the reaction requires at least three equivalents for complete reduction. Early reports on the use of boron trifluoride as the Lewis acid stated that it was ineffective. A reevaluation, however, showed that the order of mixing of the reagents was crucial. Best results are obtained by mixing the Lewis acid and the substrate followed by the addition of the LAH. 1,3-Oxathiolanes also react under these conditions with the exclusive cleavage of the C—O bond. This is accounted for by stronger coordination of the oxygen with the Lewis acid ⟨63CJC2671⟩. The LAH reduction of 1,2-oxathiolane 2,2-dioxides proceeds in a manner analogous to their reactions with nucleophiles. When there is no steric hinderance, attack occurs at C-5 with concomitant C—O bond cleavage. In the presence of steric crowding, attack takes place at sulfur and the S—O bond is broken ⟨81JOC101⟩.

1,3-Oxathiolanes and 1,3-dioxolanes containing a 2-phenyl substituent react with alkyl-lithium reagents to produce aryl alkyl ketones and alkenes ⟨65JOC226⟩. The mechanism of this reaction is discussed in Section 4.30.2.2.7.

4.30.2.1.5 Reactivity towards free radicals, electron-deficient species and at surfaces

Radicals abstract a hydrogen atom from the 2-position of 1,3-dioxolanes and 1,3-oxathiolanes. The ESR spectrum of the 2-methyl-1,3-dioxolan-2-yl radical shows temperature-dependent linewidth phenomena indicative of a bent radical center and radical-center inversion ⟨77JCS(P2)1161⟩. Oxygenated radicals are frequently bent and the barrier to inversion of the 2-methoxy-1,3-dioxolan-2-yl radical, which has three oxygen atoms attached to it, is reported to be greater than 40 kJ mol^{-1} ⟨81JA609⟩. The ESR spectra of 2-substituted 1,3-oxathiolan-2-yl radicals, however, show that sulfur exerts a significant delocalizing effect, although the bending effect of the oxygen dominates and the radical is also bent. As expected, the corresponding 1,3-dithiolan-2-yl radicals are planar ⟨79JCS(P2)763⟩.

Little is known of the reactions with free radicals. Singlet oxygen adds to the double bond in 1,3-dioxole with concomitant ring opening to give the diester ⟨71TL1757⟩, and the homolytic addition of 1,3-dioxolane to pyridine in the presence of an iron(III) sulfate–hydrogen peroxide redox system gives a 51% mixture of (**82**) and (**83**) ⟨79KGS990⟩. The photolytic addition of terminal alkenes to the 2-position of 1,3-dioxolanes has also been reported ⟨66CC684⟩.

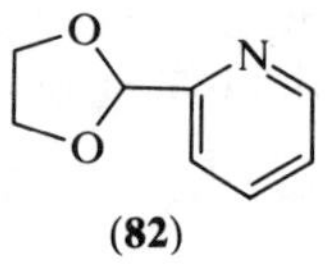

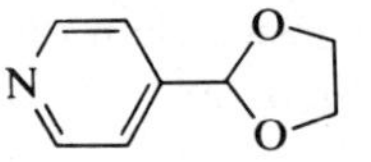

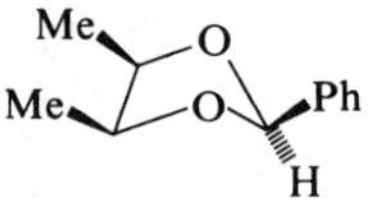

Details of reactions with carbenes are also rare. Carbene, derived photolytically from diazomethane, attacks 1,3-dioxolane at the 2- and the 4-positions. Together with 2- and 4-methyl-1,3-dioxolane, there are obtained lesser amounts of ring-expanded products ⟨69CB1087⟩. Dihalocarbenes insert into the C(2)—H bond of 1,3-dioxolanes in a stereospecific manner and without racemization or inversion. The highest yields are obtained when there is substitution at the 4- and 5-positions and (**84**) is converted into (**85**) in over 80% yield ⟨80JCR(S)95⟩.

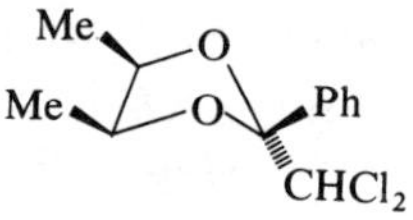

The 1,3-dioxolane ring is usually inert to catalytic hydrogenation unless forcing conditions are employed. Hydrogenolysis over rhodium or palladium catalysts in the presence of an acid catalyst gives mainly hydroxy ethers as shown in equation (8). Raney nickel cleaves the oxathiolane ring (Scheme 9) and converts 1,3-benzoxathioles into phenol derivatives.

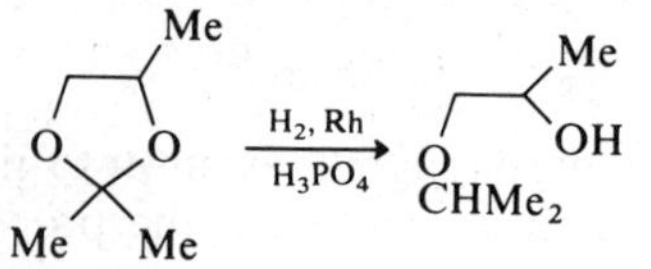

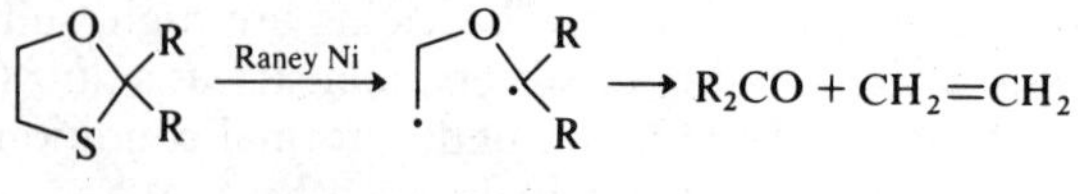

Scheme 9

While dioxolanes are generally inert, sodium in liquid ammonia opens 1,3-oxathiolanes by preferential breakage of the C—S bond, since the developing negative charge is accommodated better by the resulting thiol anion (equation 9) ⟨66JCS(C)415⟩. Little is known about the action of metals on the fully aromatic substrates, although there is a report that zinc dust reduces (86) to the radical (87) with subsequent formation of a dimer (equation 10) ⟨74JHC507⟩.

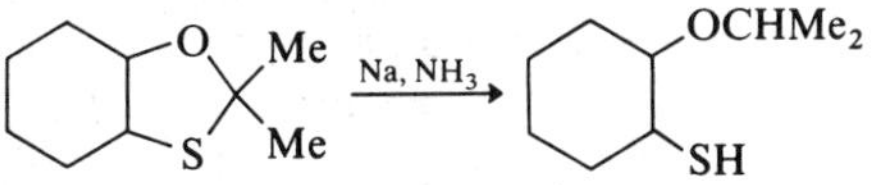

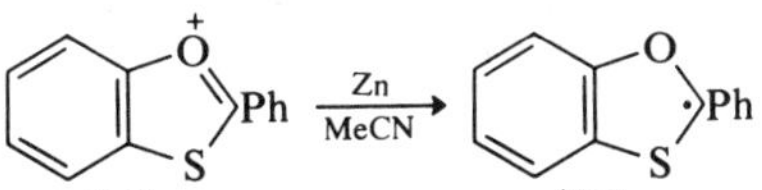

$$\text{(86)} \xrightarrow{\text{Zn, MeCN}} \text{(87)} \tag{10}$$

4.30.2.1.6 Reactions with cyclic transition states

Both of the known oxathiolium systems have been trapped *in situ* by dipolarophiles. Compound (88) reacts with dimethyl acetylenedicarboxylate (DMAD), but not phenyl isocyanate or isothiocyanate, to yield thiophene (89), following loss of carbon dioxide as shown in Scheme 10 ⟨77CPB1471⟩. Similarly, oxathioliums (90) may be trapped by a variety of alkynic dipolarophiles to give furans (91) as shown in Scheme 11. The reaction, which appears to be regiospecific when unsymmetrical alkynes are used, is a useful way of preparing furans containing amine or thioether functionality ⟨75AG(E)422⟩.

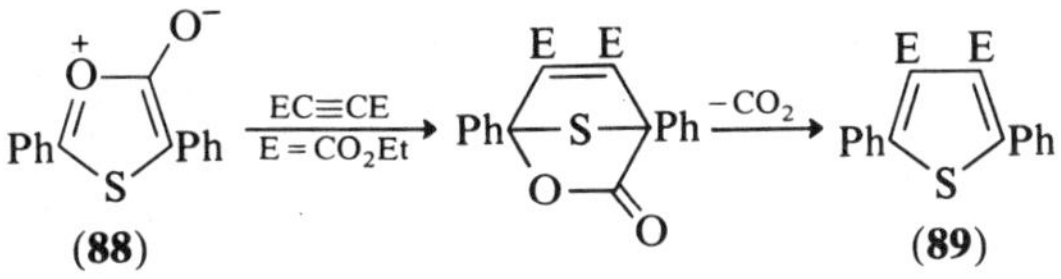

Scheme 10

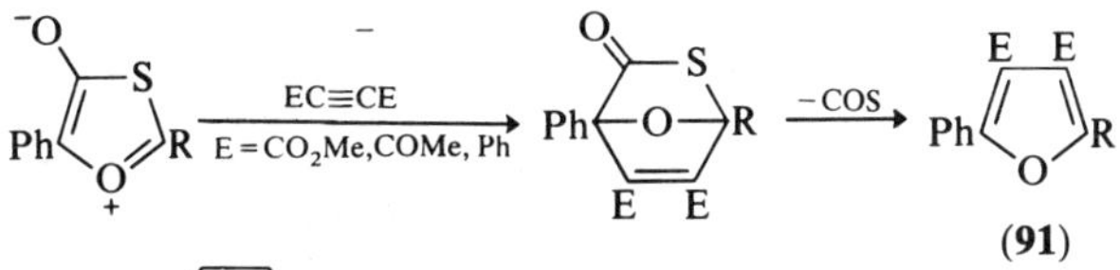

Scheme 11

While not a mesoionic system, (92) undergoes a formal 1,3-dipolar addition to DMAD to yield thiazolone (93) in almost quantitative yield (Scheme 12) ⟨76CC912⟩. It is unlikely that the reaction is concerted. Vinylene carbonate is capable of undergoing the Diels–Alder reaction with dienes, although temperatures in excess of 150 °C are frequently necessary ⟨55JA3789, 62JOC4101⟩. The synthetic potential of vinylene carbonate as a precursor to catechols has yet to be clearly demonstrated, since usually one of the oxygen atoms is eliminated in the aromatization step ⟨79JOC1807⟩. 1,3-Dioxole derivatives should also make interesting dienophiles although little is known of their reactivity in this regard. 1,3-Dioxole and its 2,2-dimethyl derivative are known to add to photoexcited benzene with formation of 1,2-, 1,3- and 1,4-cycloadducts ⟨79CB577⟩.

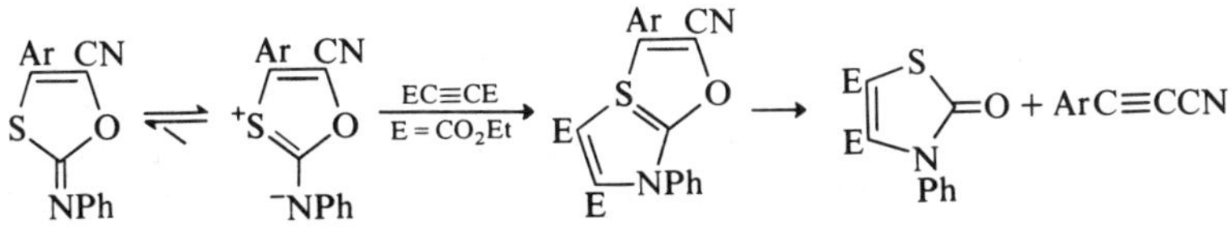

Scheme 12

4.30.2.2 Reactivity of Substituents

4.30.2.2.1 General survey

There have been no reports on the reactions of substituents attached to the fully aromatic systems. There is also a paucity of specific data on the behavior of substituents on the saturated rings, since their reactivity should parallel that of the acyclic analogs. It has not been possible, therefore, to provide a fully systematic account of substituent reactivity. What follows is a review of material published since 1970, together with earlier, important work. Other material, such as it is, may be found in the general references given in Section 4.30.2.1.1.

4.30.2.2.2 Fused benzene rings

In electrophilic substitution reactions of benzo-1,3-dioxoles, only *para* derivatives with respect to the oxygen atoms are formed. This differs from benzo-1,4-dioxanes where some *ortho* isomer is also obtained, and probably reflects the coplanarity of the rings in benzodioxoles resulting in an increased mesomeric effect ⟨81H(15)1395⟩. Benzo-1,3-oxathiole derivatives undergo electrophilic substitution at the 5-position, *i.e. para* to the oxygen atom ⟨53JCS1514⟩, although attack at the sulfur atom may be a complicating factor. 2,2-Disubstituted benzo-1,3-dioxoles are oxidized by lead tetraacetate (LTA) to 5-acetoxy and 5,6-dione derivatives as shown in equation (11) ⟨80AJC527⟩. If the 2-position is not blocked, oxidation occurs there preferentially. The quinones are probably formed *via* a tetraacetoxy intermediate in a manner analogous to the oxidation of naphthyl ethers ⟨79AJC1749⟩. As expected, metallation with *n*-butyllithium takes place *ortho* to the oxygen atom in both systems provided that the 2-position is blocked. When the 2-position is available, there is a competing reaction involving metallation at this position, and with 2-unsubstituted benzo-1,3-oxathioles this is the predominant reaction ⟨77JOM(136)139⟩.

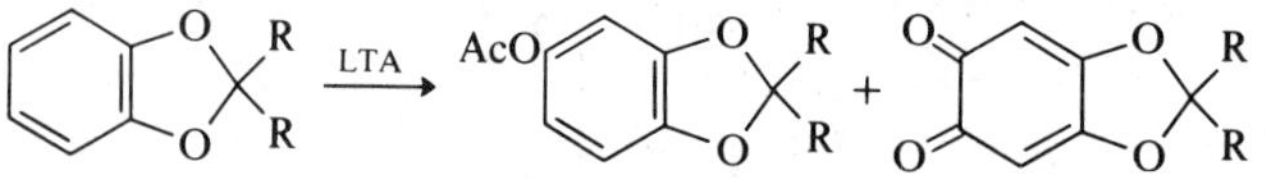

The increased reactivity of strained cyclic sulfonate esters, such as the benzo derivatives, was discussed in Section 4.30.2.1.4. These compounds also show interesting photochemical activity in that they split out sulfur dioxide as shown in equation (12) ⟨71CC383⟩.

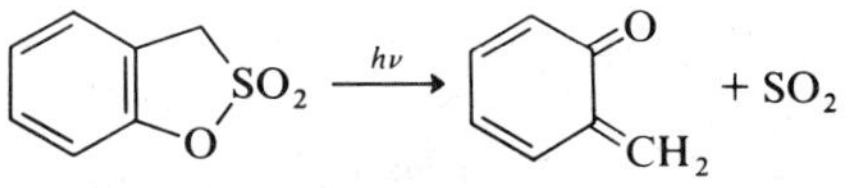

4.30.2.2.3 C-Linked substituents

Alkyl groups attached to the 2-position of the 1,3-dioxolan-2-ylium ring are activated sufficiently to condense with aldehydes (Scheme 13) ⟨80ZOR183⟩.

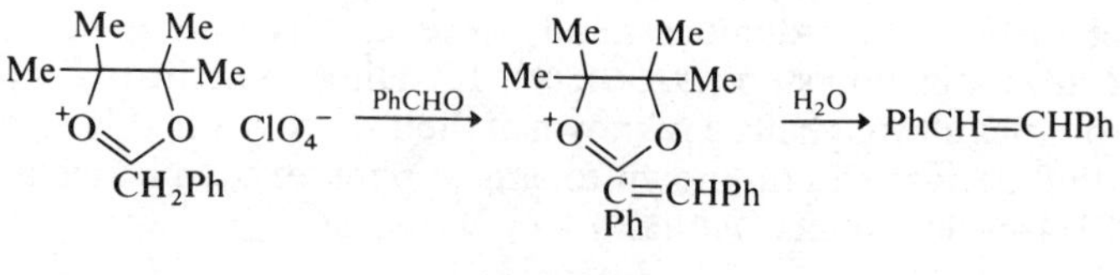

Scheme 13

Phenylselenyl chloride reacts with 2-alkyl-1,3-dioxolanes having at least one α-hydrogen atom to give α-phenylseleno products accompanied by considerable amounts of disubstituted products. It is probable that this reaction proceeds as shown in Scheme 14, with acid catalyzed ring opening to an enol followed by attack of the electrophilic selenium reagent ⟨79S982⟩.

The action of strong base on 1,3-dioxolanes derived from β,γ-unsaturated ketones leads to unstable anions which ring open to enol ethers ⟨80TL1515⟩. This reaction is similar to the

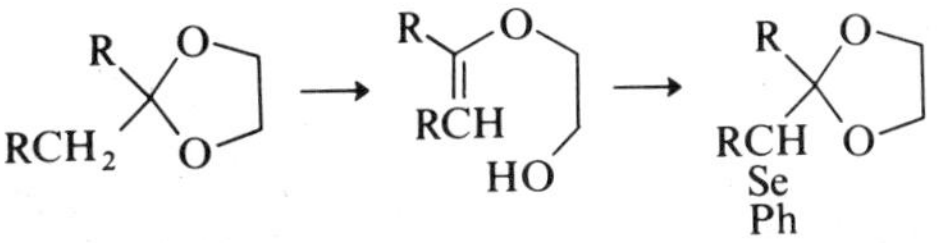

Dioxoles and Oxathioles

ring opening caused by Grignard reagents discussed in Section 4.30.2.1.4. 4-(2-Nitrophenyl)-1,3-dioxolanes undergo photochemical ring opening to regenerate the carbonyl compound as shown in Scheme 15. This method offers the protection of a carbonyl compound which may be regenerated under non-hydrolytic conditions ⟨74CJC187⟩.

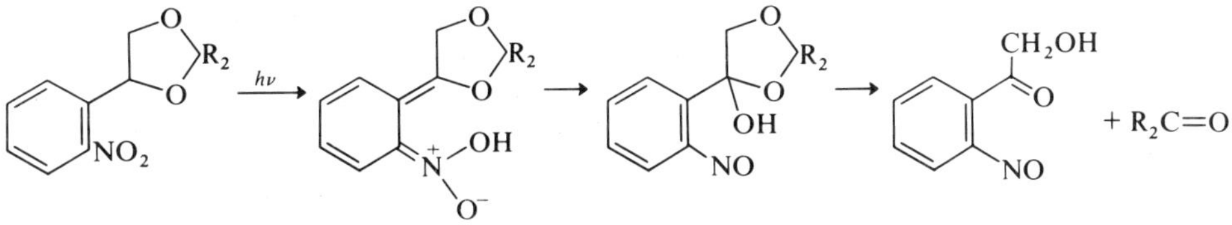

Scheme 15

4.30.2.2.4 N-Linked substituents

As discussed in Section 4.30.2.1.4, the oxathiolium system is attacked by nucleophiles at the 2-position. A series of 2-amino substituted 1,3-oxathiolium cations are known which are probably best described by the imonium structure shown in Scheme 16. These salts react with a variety of nucleophiles at the 2-position and, depending on the nucleophile, may yield products derived from either direct substitution or from ring opening.

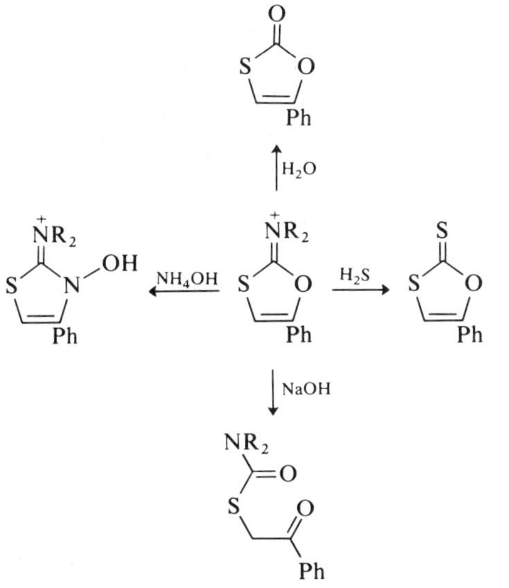

Scheme 16

The *cis*-dioxolane (**94**) on heating to 160 °C gives a *trans*-oxirane as shown in equation (13). The reaction proceeds in high yield for (**94**) while the corresponding *trans* derivative is stable under the same conditions. Both the *cis* and the *trans* compounds are converted stereospecifically into the *cis* and *trans* alkenes, respectively, on treatment with hot acetic anhydride ⟨70TL5223⟩. The reaction for the *trans* compound is shown in equation (14), and the mechanism for the transformation is discussed in Section 4.30.2.2.5.

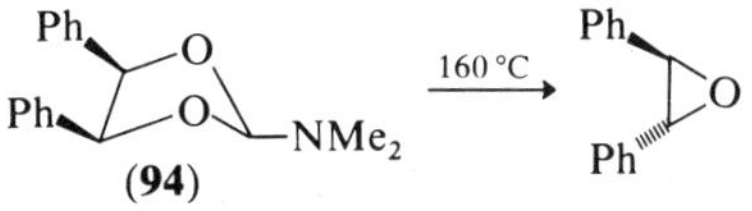

(13)

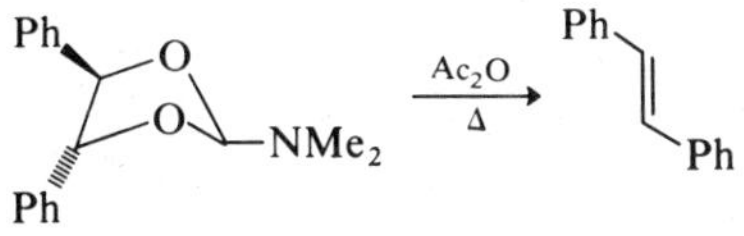

(14)

4.30.2.2.5 O-Linked substituents

Carbonyl derivatives of dioxoles and oxathioles are cyclic carbonates, lactones or sultones with many reactions essentially the same as their acyclic counterparts. The presence of a carbonyl group in the cyclic systems does impart some interesting thermal and photochemical behavior however. 1,2-Dioxolan-3-ones (**95**) photodecarboxylate directly to the 1-oxatrimethylene radical (**96**), which goes on to form the products shown in Scheme 17. The thermal decomposition of (**95**) proceeds *via* initial peroxide bond breakage to give a 1,5-diradical which undergoes stereospecific rearrangement to (**97**) as the principal product ⟨73JOC1434⟩. The dicarbonyl compound (**98**) undergoes a similar photochemical decarboxylation at 77 K to yield the α-lactone (**99**). On warming, (**99**) polymerizes but continued low-temperature photolysis results in decarbonylation and formation of a ketone ⟨72JA1365⟩.

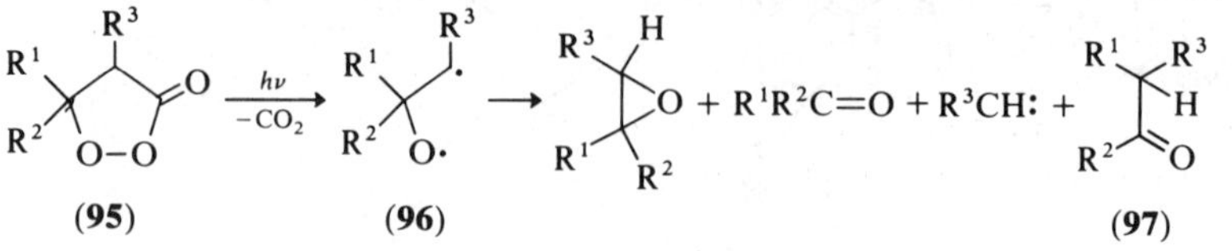

Scheme 17

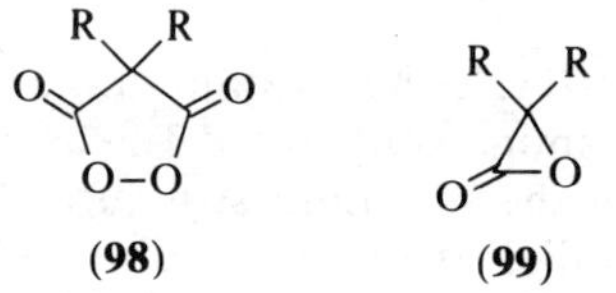

The flash vacuum pyrolysis of 1,3-oxathiolan-5-ones causes a loss of carbon dioxide with concomitant formation of the corresponding thiiranes in nearly quantitative yield (Scheme 18). The reaction is stereospecific and proceeds with clean inversion of configuration, suggesting a concerted mechanism ⟨80JA744⟩. This reaction is probably the simplest way of generating thiocarbonyl ylides. Similarly, the pyrolysis of 1,3-oxathiolan-2-ones also gives thiiranes in high yield, as shown in equation (15). In this case the addition of an alkaline catalyst ensures a smooth reaction which probably proceeds *via* initial attack at the carbonyl group ⟨66HC(21-1)76⟩.

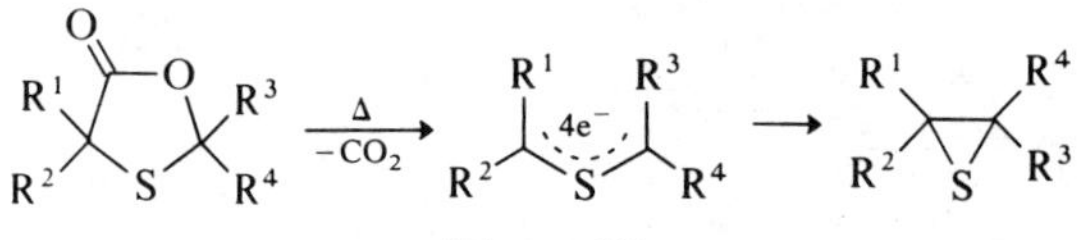

Scheme 18

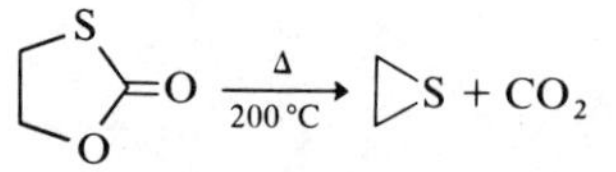

Carbonyl derivatives of 1,3-dioxolanes and 1,3-oxathiolanes react with nucleophiles in a predictable manner. Typical examples are shown in equations (16)–(17). The ring opening of oxathiolan-2-ones and the subsequent generation of ethylene sulfide has been used as a method of mercaptoethylation of amines (equation 18) by employing the amine in excess to trap the sulfide ⟨66HC(21-1)76⟩. The reaction of Grignard reagents with 2-alkoxy-1,3-dioxolanes may result in the displacement of the alkoxide but the reaction is slow and requires high temperatures and long reaction times for significant conversion. If the alkoxide chain contains an ether linkage three or four atoms out on the chain, the reaction is considerably faster and gives high yields. The Grignard reagent presumably becomes part of a chelated complex which activates the molecule ⟨80JCS(P1)756⟩.

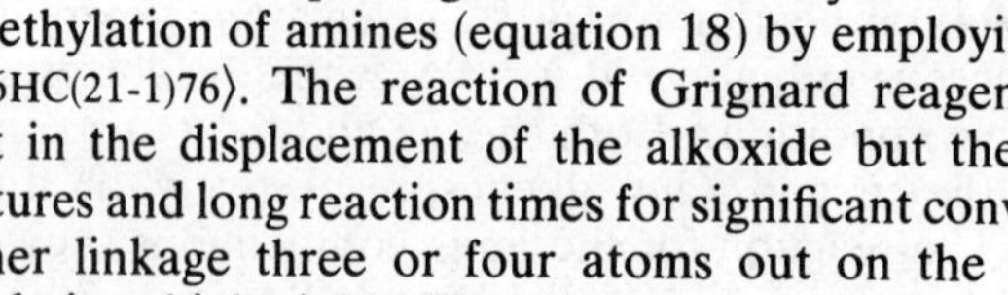

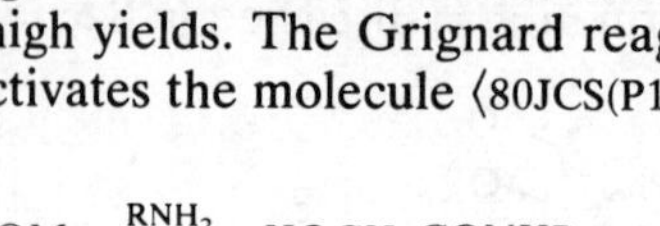

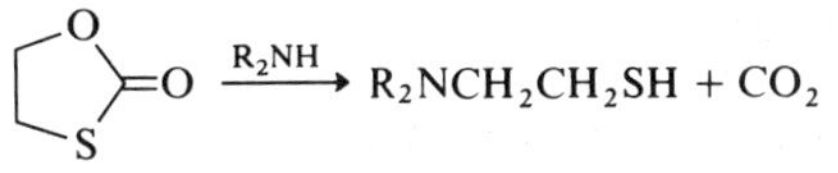

$$\text{[cyclic O-C(=O)-S ethylene]} \xrightarrow{R_2NH} R_2NCH_2CH_2SH + CO_2 \qquad (18)$$

$$\text{[2-OR dioxolane/oxathiolane]} \xrightarrow{HClO_4} \text{[cation]} \; HClO_4^- \qquad (19)$$

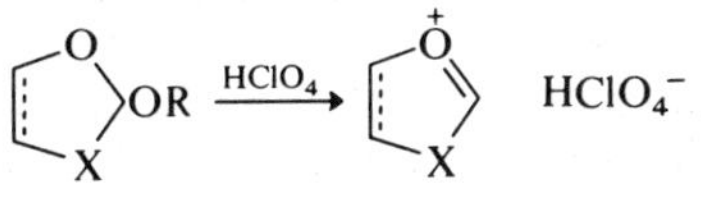

X = O, S

2-Alkoxy derivatives on treatment with strong acids yield the cationic species resulting from the elimination of alkoxide anion as shown in equation (19). The treatment of 2-alkoxy-1,3-dioxolanes with hot carboxylic acids gives alkenes in high yields. The reaction, shown in Scheme 19, is believed to proceed *via* initial formation of the cation followed by hydride loss initiated by the carboxylate anion. The resulting carbene then fragments as discussed in Section 4.30.2.1.2 ⟨68AJC2013⟩. A similar carbene intermediate is thought to be generated in the decomposition of the amine derivatives discussed in Section 4.30.2.2.4 and the sulfur derivatives in Section 4.30.2.2.6.

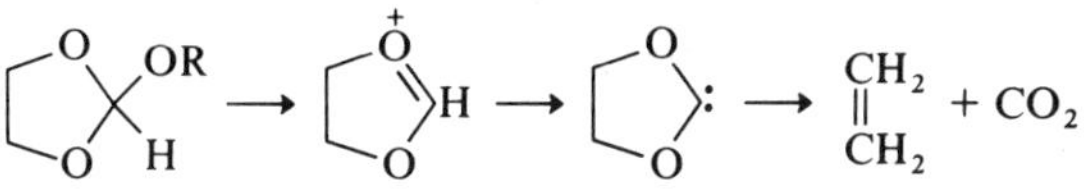

$$\text{[2-OR,2-H dioxolane]} \rightarrow \text{[oxocarbenium]} \rightarrow \text{[carbene]} \rightarrow \begin{array}{c} CH_2 \\ \parallel \\ CH_2 \end{array} + CO_2$$

Scheme 19

4.30.2.2.6 S-Linked substituents

2-Thiocarbonyl derivatives of 1,3-dioxolanes and 1,3-oxathiolanes are readily isomerized to the 2-carbonyl compounds as shown in Scheme 20. Alkylation of the sulfur atom with alkyl halides usually leads to ring-opened products (Scheme 21) ⟨69JOC3011⟩. Most of the other chemistry of the sulfur derivatives has focused on desulfurization and subsequent generation of alkenes. The reaction is shown in equation (20) and proceeds with *cis* elimination *via* carbene intermediate (see Section 4.30.2.2.5) and is usually carried out with a phosphine ⟨73JA7161⟩ or a zero-valent nickel complex ⟨73TL2667⟩.

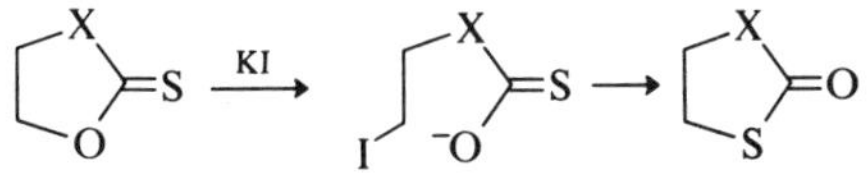

$$\text{[2-thione]} \xrightarrow{KI} \text{[ring-opened iodide]} \rightarrow \text{[2-one with S in ring]}$$

X = O, S

Scheme 20

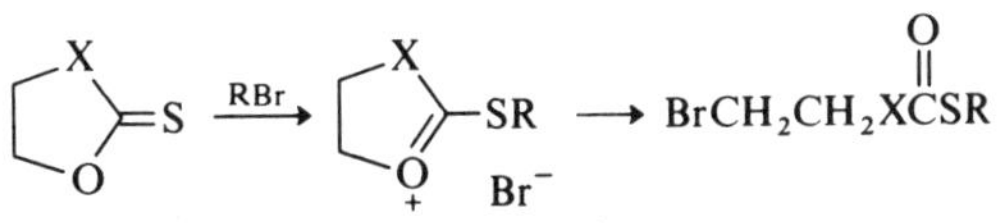

$$\text{[2-thione]} \xrightarrow{RBr} \text{[S-alkyl cation]} \; Br^- \rightarrow BrCH_2CH_2X\overset{O}{\underset{\parallel}{C}}SR$$

Scheme 21

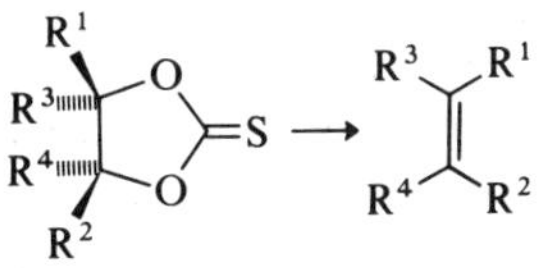

$$\underset{R^2}{\overset{R^1}{\underset{R^{4\cdots}}{R^{3\cdots}}}}\!\!\!\!\!\!\!\!\!\!\!\!\!\!\!\text{[dioxolane-2-thione]} \rightarrow \underset{R^4 \; R^2}{\overset{R^3 \; R^1}{\text{alkene}}} \qquad (20)$$

4.30.2.2.7 Halogen substituents

As discussed in Section 4.30.2.1.3, the chlorination of dioxolanes gives mainly the 2-chloro compound which undergoes ring opening. 4-Chlorodioxolane exhibits the high reactivity expected of a haloether. Photochemical chlorination of ethylene carbonate, in which the 2-position is blocked, gives a high yield of a 4-chloro derivative which is used in the synthesis of vinylene carbonate (Section 4.30.3.3.5).

4.30.2.2.8 Metal substituents

1,2-Oxathiolane 2,2-dioxide undergoes metallation with *n*-butyllithium as expected at the 3-position ⟨81JOC101⟩. The anion may be alkylated with alkyl halides or carbonyl compounds. The isomeric 1,3-oxathiolane 3,3-dioxides also undergo metallation *ortho* to the sulfone and, when the 4-position is blocked, metallation at the 2-position may be used as an efficient conversion of alkyl halides into aldehydes as shown in Scheme 22 ⟨79TL3375⟩.

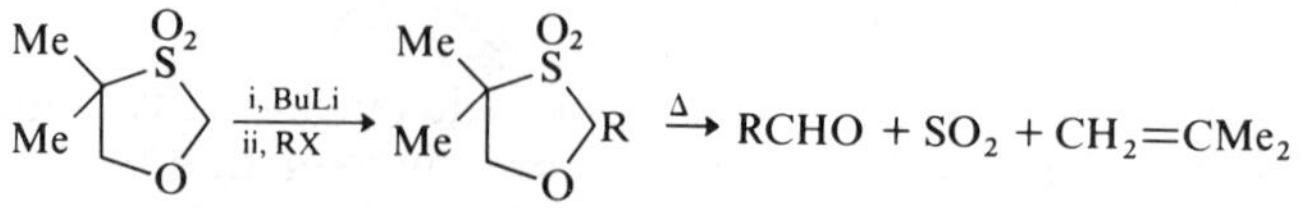

Scheme 22

Metallation of conformationally restricted 1,3-dioxolanes at the 2-position is possible only if the proton can occupy an equatorial-like conformation. The carbanion is apparently destabilized by an antiperiplanar effect of the oxygen lone pair orbitals when the carbanion is forced into an axial position. The same is probably true for 1,3-oxathiolanes and this accounts, in part, for preferential metallation at the 4-position ⟨79TL4159⟩. The metallation of 1,3-dioxolanes at the 2-position produces an anion which is capable of undergoing a symmetry allowed ring opening as shown in equation (21). Treatment of dioxolane (**100**) with *n*-butyllithium gives *trans* alkene (**101**) only ⟨73JCS(P1)2332⟩. Similarly, (**102**) gives only the *cis* alkene. The utility of the reaction is reduced since tri- and tetra-substituted dioxolanes do not fragment to alkene. This is due to conformational restriction in these compounds forcing the 2-phenyl moiety into the equatorial position, and concomitant destabilization of the axial anion. This problem may be circumvented by the use of 2-phenyl-1,3-oxathiolanes in which the 2-proton is more acidic. Oxathiolane (**103**) undergoes metallation and fragmentation on treatment with lithium diethylamide in ether to give tetramethylethylene. The ring opening appears to be concerted since (**104**) gives only *cis*-stilbene ⟨74JCS(P1)433⟩.

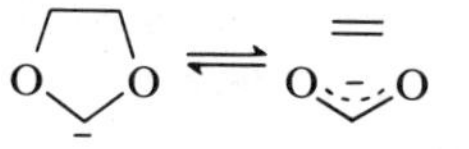

(21)

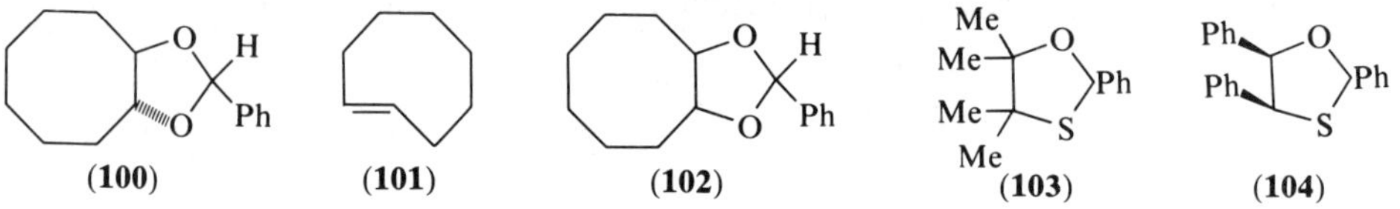

(**100**) (**101**) (**102**) (**103**) (**104**)

4.30.3 SYNTHESIS

4.30.3.1 Ring Synthesis

4.30.3.1.1 One-bond formation between heteroatoms

Only derivatives of the 1,2-oxathiole series have been obtained by direct heteroatom coupling. The successful bond formation involves some type of activation of the sulfur atom in 3-hydroxypropanethiol or its oxidized derivatives. The parent 1,2-oxathiolane system (**106**) was recently reported ⟨81CC741⟩ to be obtained from the vacuum pyrolysis of (**105**), as shown in Scheme 23. A cyclic oligomer is believed to be generated first and (**106**) is formed on standing. Oxidation of (**106**) yields the mono- and di-oxide which have been known for some time. The monoxide may be obtained by the oxidative coupling of 3-hydroxypropane thiol (equation 22). Chlorine is the most frequently used oxidant and

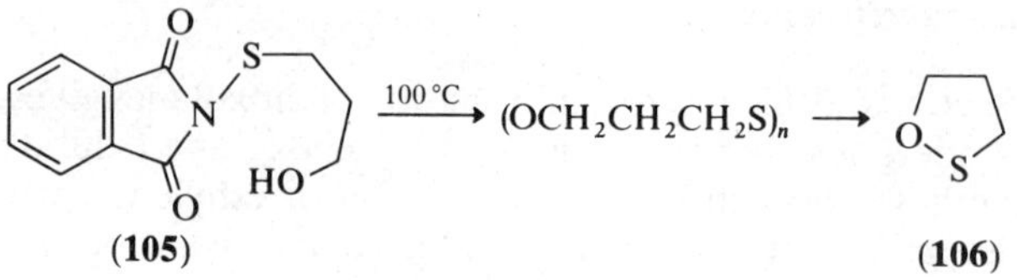

(**105**) $\xrightarrow{100\,°C}$ $(OCH_2CH_2CH_2S)_n$ → (**106**)

Scheme 23

the reaction probably involves the initial formation of a sulfenyl chloride. It is not necessary to use a free thiol since a disulfide linkage is cleaved under the reaction conditions. The amount of chlorine used should be monitored carefully, since the dioxide and ring-opened hydroxysulfonyl chloride have also been observed. In at least one case the use of *N*-chlorosuccinimide is superior to elemental chlorine ⟨81JOC5408⟩.

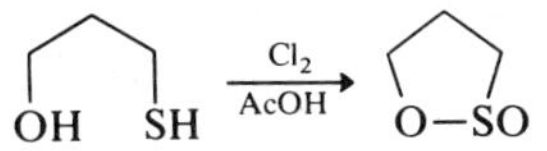

$$\underset{OH \quad SH}{\frown} \xrightarrow[AcOH]{Cl_2} \underset{O-SO}{\frown} \tag{22}$$

The corresponding cyclization of the sulfinyl chloride (**108**) also gives the monoxide, as shown in Scheme 24. In this case the sulfinyl chloride is generated by loss of *t*-butyl chloride from (**107**) in what has been shown to be a general synthesis of cyclic sulfinates ⟨76CJC3012⟩.

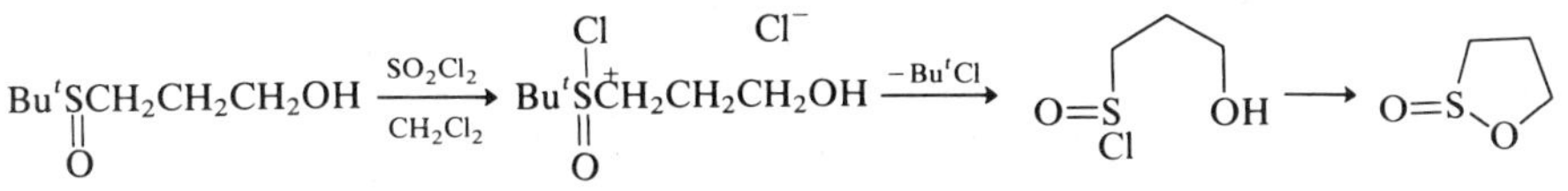

$$Bu^tSCH_2CH_2CH_2OH \xrightarrow[CH_2Cl_2]{SO_2Cl_2} \underset{(107)}{Bu^t\overset{Cl}{\underset{\parallel}{S}}CH_2CH_2CH_2OH} \xrightarrow{-Bu^tCl} \underset{(108)}{\underset{Cl}{\overset{O=S}{\frown}}\text{OH}} \longrightarrow O=S\overset{}{\underset{O}{\frown}}$$

Scheme 24

The dioxides are well known ⟨66HC(21-1)76⟩ and may be obtained from the pyrolysis of 3-hydroxysulfonic acids as shown in equation (23). The unsaturated oxathioles may also be prepared this way (equation 24). The hydroxysulfonic acids are best obtained either by chlorosulfonation of alkanes or by addition of bisulfite to hydroxyalkenes, as shown in equation (25).

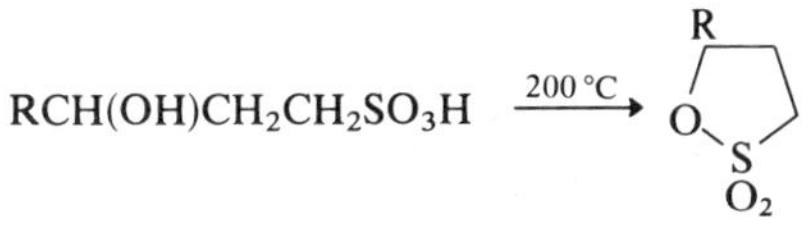

$$RCH(OH)CH_2CH_2SO_3H \xrightarrow{200\,°C} \underset{O_2}{\overset{R}{\underset{S}{\frown}}} \tag{23}$$

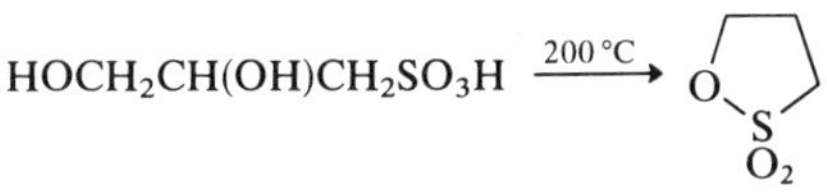

$$HOCH_2CH(OH)CH_2SO_3H \xrightarrow{200\,°C} \underset{O_2}{\overset{}{\underset{S}{\frown}}} \tag{24}$$

$$CH_2=CHCH_2OH \xrightarrow{KHSO_3} \underset{SO_3K}{CH_2CH_2CH_2OH} \tag{25}$$

4.30.3.1.2 One-bond formation adjacent to a heteroatom

Of the possible monocyclic aromatic systems, only derivatives of the 1,3-oxathiolium system have been reported. The 2,5-diaryl derivatives may be obtained by the acid catalyzed cyclization of keto ester (**109**) as shown in Scheme 25. The compounds may be isolated and stored as the perchlorates ⟨75H(3)217⟩. The treatment of 2-alkoxy-1,3-oxathioles with perchloric acid also results in formation of the aromatic systems ⟨74RTC99⟩.

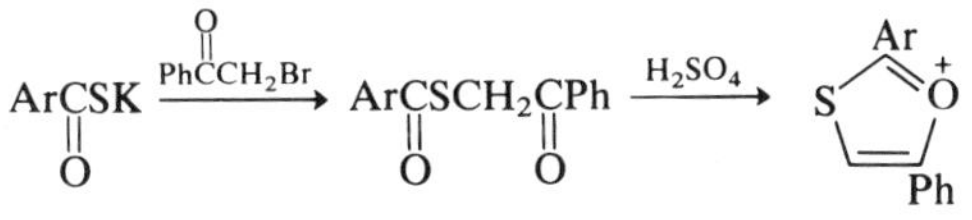

$$ArCSK \xrightarrow{PhCCH_2Br} ArCSCH_2CPh \xrightarrow{H_2SO_4} \underset{Ph}{\overset{Ar}{\underset{S\diagdown O^+}{\frown}}}$$
(109)

Scheme 25

Both mesoionic 1,3-oxathioliums are prepared by similar methods. Cyclization of (**110**) in TFA anhydride (equation 26) gives a 4-acylated product which is isolable and does not undergo cycloaddition reactions ⟨75CC417⟩. On the other hand, if the 4-position is blocked, the product (equation 27) cannot be isolated but may be trapped by dipolarophiles

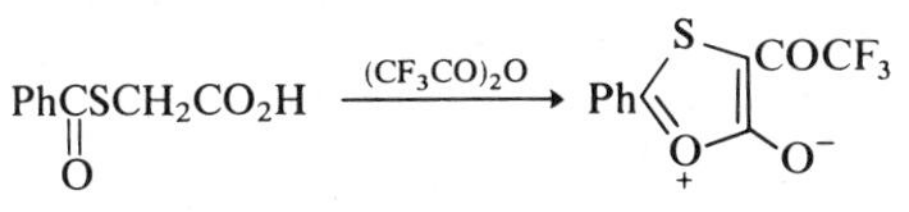

$$PhCSCH_2CO_2H \xrightarrow{(CF_3CO)_2O} Ph\underset{O^+}{\overset{S}{\diagup}}\underset{O^-}{\overset{COCF_3}{\diagdown}} \tag{26}$$
(110)

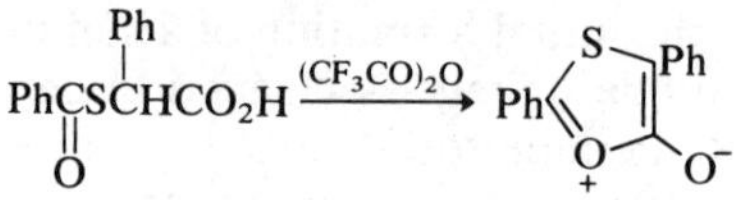

$$PhCSCHCO_2H \xrightarrow{(CF_3CO)_2O} Ph-\underset{O^+}{\overset{S}{\underset{|}{\bigcirc}}}\underset{O^-}{\overset{Ph}{\bigcirc}} \quad (27)$$

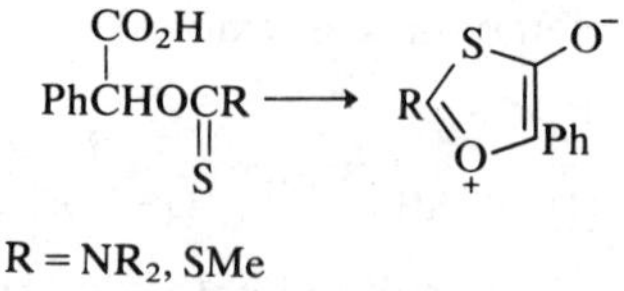

$$PhCHOCR \longrightarrow R-\underset{O^+}{\overset{S}{\underset{|}{\bigcirc}}}\underset{Ph}{\overset{O^-}{\bigcirc}} \quad (28)$$

R = NR₂, SMe

⟨77CPB4171⟩. The isomeric 4-olates (equation 28) are also too unstable for isolation but may be trapped *in situ* ⟨75AG(E)422⟩.

Sulfuric–fluorosulfonic acid mixtures give complete cyclization of allyl and methallyl esters and thiol esters to give the 1,3-dioxolan-2-ylium and 1,3-oxathiolan-2-ylium cations ⟨72CRV357⟩ as shown in equation (29). Such cations are frequently involved in the solvolysis of haloacyl derivatives. This neighboring group effect has important mechanistic and stereochemical consequences, and is frequently encountered when working with carbohydrates.

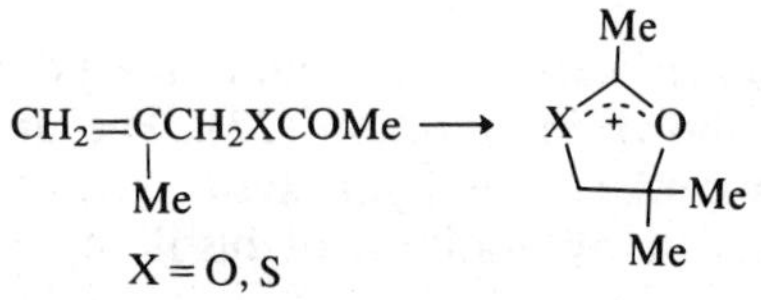

$$CH_2=CCH_2XCOMe \longrightarrow \quad (29)$$

X = O, S

1,2-Dioxolanes may be obtained from the intramolecular cyclization of unsaturated peroxides. For example, the cyclization of compounds typified by (**111**) is achieved by using mercury(II) salts to activate the double bond. As shown in equation (30), the dioxolanes are obtained *via* an exo cyclization as predicted by Baldwin's rules ⟨78JOC4048⟩. Cyclization may be achieved directly on to a saturated carbon atom, either oxidatively (equation 31) if an aromatic group is present ⟨81S237⟩ or by a nucleophilic displacement (equation 32) if a leaving group is present ⟨78JOC1154⟩.

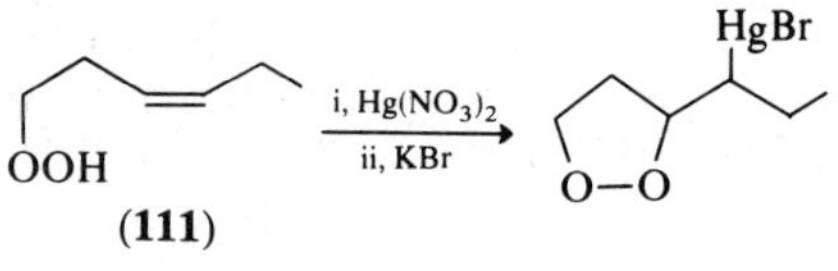

$$(30)$$ 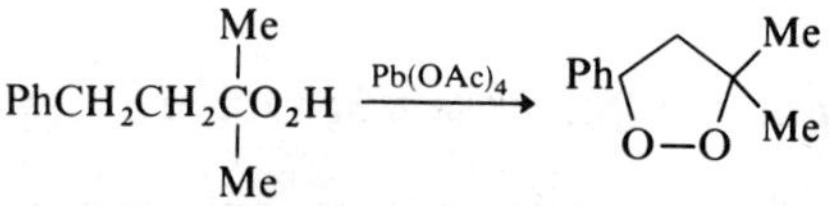
(**111**)

$$PhCH_2CH_2\underset{Me}{\overset{Me}{\underset{|}{C}}}CO_2H \xrightarrow{Pb(OAc)_4} \quad (31)$$

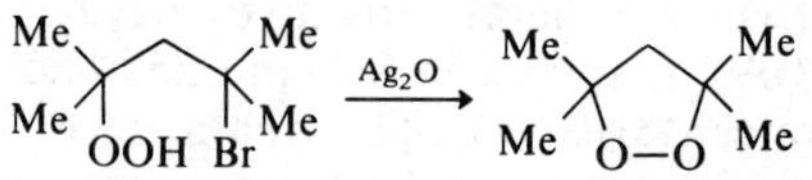

$$(32)$$

1,2-Oxathiolane 2,2-dioxides are obtained from the base catalyzed intramolecular cyclization of mesylates containing a leaving group on the β-carbon atom. The reaction shown in equation (33) is for a dimesylate ⟨70CJC845⟩ although the reaction is quite general ⟨B-77SH(2)314⟩.

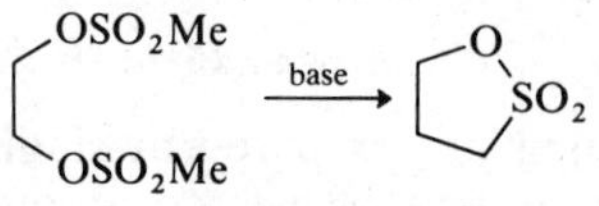

$$\underset{OSO_2Me}{\overset{OSO_2Me}{\bigg\langle}} \xrightarrow{base} \underset{SO_2}{\overset{O}{\bigg\langle}} \quad (33)$$

1,3-Oxathiole 3,3-dioxides may be obtained in a similar manner. When dibromide (**112**) is treated with triethylamine, the expected thiirene 1,1-dioxide is not formed but rather a nucleophilic displacement *via* the enolate oxygen atom occurs, as depicted in Scheme 26. Removal of the bromine and benzoyl moieties yields an efficient synthesis of 5-phenyl-1,3-oxathiole 3,3-dioxide ⟨74JOC2722⟩. This work shows that the previously reported synthesis of 2,3-dibenzoylthiirene dioxide from (**112**) is incorrect.

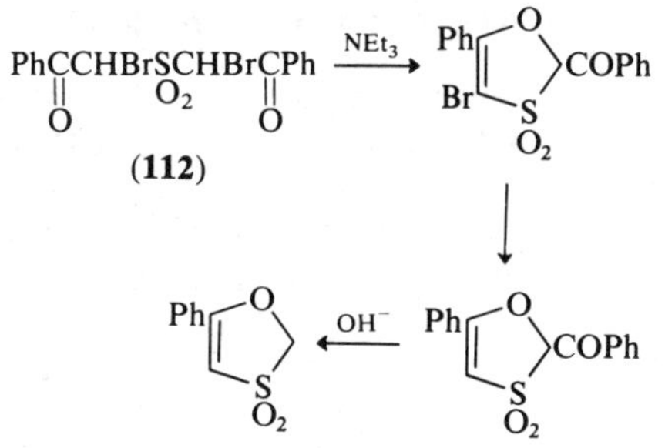

Dioxoles and Oxathioles

$$\text{PhCCHBrSCHBrCPh} \xrightarrow{\text{NEt}_3} \text{[Ph, Br, COPh substituted oxathiole with SO}_2\text{]}$$

(112)

[Scheme shows conversion via OH⁻ to give Ph-substituted oxathiole S,S-dioxide with COPh, then to Ph-substituted oxathiole S,S-dioxide]

Scheme 26

A related reaction, in that the C—O bond is formed in the final step, involves the cyclization of 2-hydroxyphenyl sulfone derivatives as shown in equation (34).

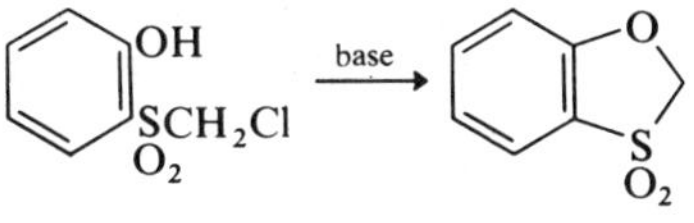

$$\text{2-HOC}_6\text{H}_4\text{SO}_2\text{CH}_2\text{Cl} \xrightarrow{\text{base}} \text{benzo-fused oxathiole S,S-dioxide} \quad (34)$$

4.30.3.1.3 Two-bond formation from [4+1] atom fragments

All reactions of this type involve the addition of two heteroatoms situated on adjacent carbons on to a one-carbon fragment. 1,3-Dioxolanes and 1,3-oxathiolanes are best prepared ⟨81S501⟩ by condensation of a 1,2-glycol or mecaptoalcohol with an aldehyde or a ketone, using an acidic catalyst (equation 35). Protected carbonyl compounds such as enol ethers may also be used. The yields are generally very good and may be quantitative if water is removed by azeotropic distillation. The choice of the acid catalyst depends on the nature of the reactants. Aldehydes are more reactive and the reaction can proceed in the presence of weak acids such as ammonium or calcium chloride. Ketones generally require stronger acids such as sulfuric or *p*-toluenesulfonic acid. Recently, pyridinium tosylate has been shown to be a mild catalyst for the conversion of ketones into 1,3-dioxolanes ⟨79S724⟩. An alternative to an acid catalyzed reaction is the use of a disubstituted methylene derivative under basic conditions, as shown in equation (36). The reaction works best in the preparation of 1,3-oxathiolanes, and a variant of it is the method of choice for the synthesis of 1,3-oxathiolane itself (Section 4.30.3.3.1). 1,2-Glycols also add to alkynes in the presence of mercury(II) salts to give 1,3-dioxolanes ⟨B-57M143000⟩.

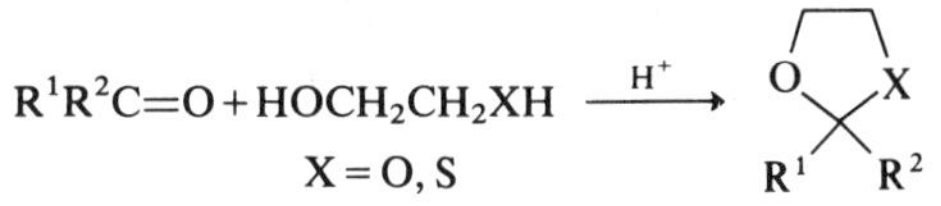

$$R^1R^2C=O + HOCH_2CH_2XH \xrightarrow{H^+} \underset{R^1 \quad R^2}{\overset{O \quad X}{\diagup\!\diagdown}} \quad (35)$$

$$X = O, S$$

$$HOCH_2CH_2XH + CH_2YZ \xrightarrow{\text{base}} \underset{}{\overset{O \quad X}{\diagup\!\diagdown}} \quad (36)$$

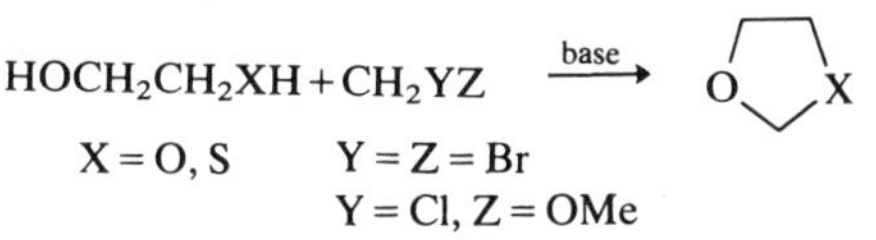

$$X = O, S \quad Y = Z = Br$$
$$Y = Cl, Z = OMe$$

Rings containing unsaturation may be synthesized from fragments in higher oxidation states. 1,3-Dioxolan-2-ylium cations are obtained directly from tertiary 1,2-diols and acyl cations in excellent yields ⟨80ZOR183⟩. The reaction, shown in equation (37), is limited to tertiary alcohols since steric hindrance prevents nucleophilic ring opening by the counterion. Another synthesis of limited scope is the addition of monothiobenzils to diaryldiazomethanes (equation 38). The reaction also works for the dicarbonyl derivatives.

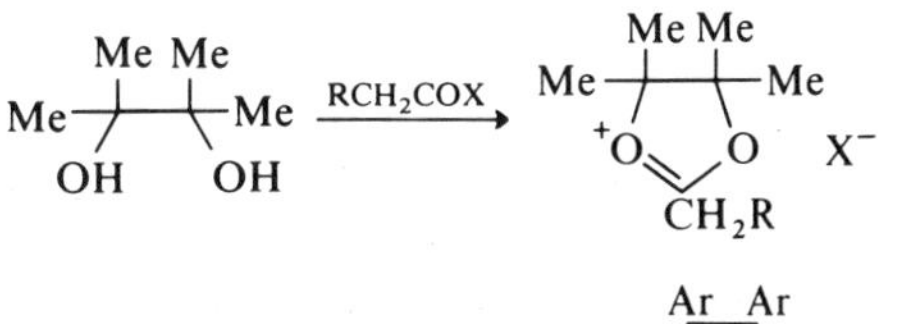

$$\underset{\text{OH} \quad \text{OH}}{\text{Me}_2\text{C}-\text{CMe}_2} \xrightarrow{\text{RCH}_2\text{COX}} \text{[1,3-dioxolan-2-ylium cation with Me}_4\text{, CH}_2\text{R, X}^-\text{]} \quad (37)$$

$$\text{ArC}-\text{CAr} + \text{Ar}_2\text{CN}_2 \rightarrow \text{[oxathiole with Ar}_4\text{]} \quad (38)$$
$$\parallel \quad \parallel$$
$$\text{O} \quad \text{S}$$

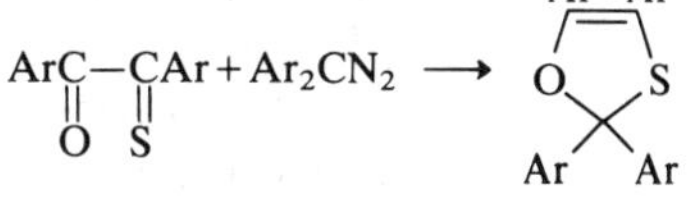

4.30.3.1.4 Two-bond formation from [3+2] atom fragments

In a reaction similar to that shown in equation (38), oxathioles may be obtained when diazobenzil is allowed to react with a thioketone ⟨78JOC3730⟩. A carbene intermediate is presumably not involved since addition of copper(I) sulfate gives vigorous gas evolution and tarry materials. The mechanism probably involves initial nucleophilic attack of the carbonyl oxygen at the thiocarbonyl carbon as shown in Scheme 27.

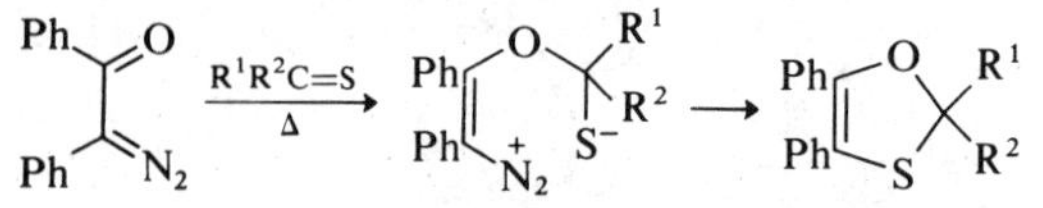

Scheme 27

The reaction of sulfonium ylides with sulfonyl chlorides in the presence of triethylamine (Scheme 28) produces 1,3-dioxoles in fair yields. The mechanism involves an interesting aroyl transfer prior to a final ring-closure reaction ⟨67TL2303⟩.

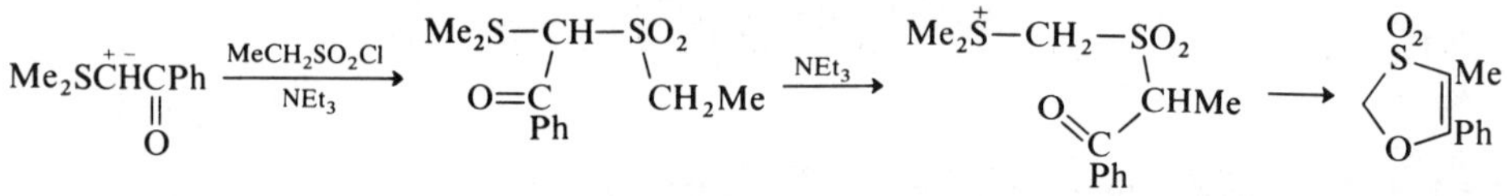

Scheme 28

1,2-Dioxolanes are obtained most conveniently from [3+2] atom couplings. The best general synthesis appears to be from the bis-triflates of 1,3-diols and germanium or tin peroxides in the peroxide transfer reaction shown in equation (39). The yields are usually good and the method is superior to the use of either hydrogen peroxide under basic conditions, or potassium superoxide ⟨79JA4290⟩. The method is applicable to primary and secondary alcohols, but the method of choice for tertiary alcohols is with acidic conditions by an S_N1 process (equation 40). 1,2-Dioxolanes are also obtained from the oxygenation of thiophenol–diene mixtures ⟨79JA7099⟩. The yields of the reactions shown in Scheme 29, are generally 25–30%. The *cis* isomer is obtained as the major product and the reaction thus parallels the biosynthesis of prostaglandin endoperoxides. A more efficient conversion of dienes into 1,2-dioxolanes involves the peroxymercuration reaction shown in equation (41). In this case, equal amounts of *cis* and *trans* compounds are obtained in over 90% yield ⟨76CC94⟩.

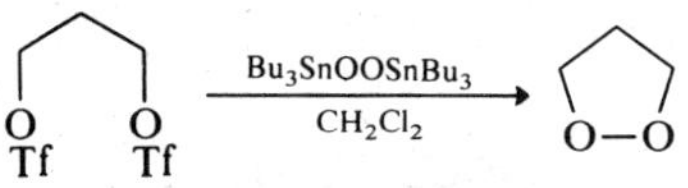

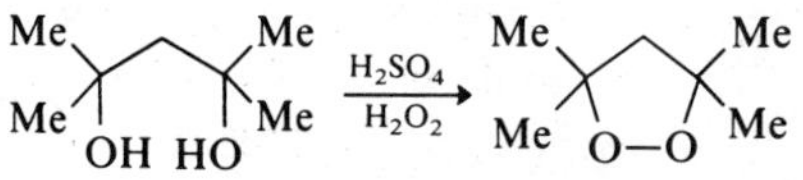

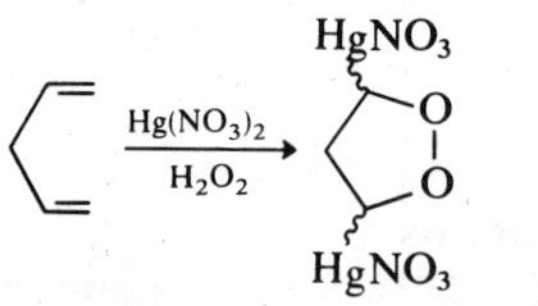

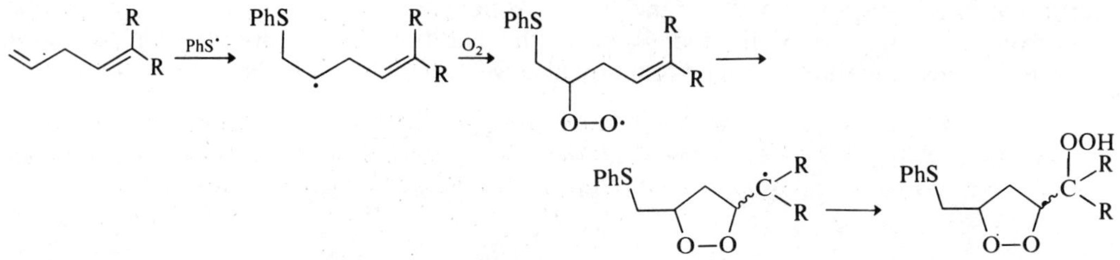

Scheme 29

As discussed in Section 4.30.3.1.3, 1,3-dioxolanes are normally generated from carbonyl compounds and glycols under acidic conditions. Carbonyl derivatives which tend to hydrate easily, for example hexafluoroacetone, need different conditions to form dioxolanes ⟨77MI43001⟩. Typically, anhydrous hexafluoroacetone is allowed to react with 2-chloroethanol (Scheme 30) and the intermediate is cyclized under basic conditions.

$$(CF_3)_2C=O \xrightarrow{ClCH_2CH_2OH} (CF_3)_2C\begin{Bmatrix}OH\\OCH_2CH_2Cl\end{Bmatrix} \xrightarrow{K_2CO_3} (CF_3)_2\begin{pmatrix}O\\O\end{pmatrix}$$

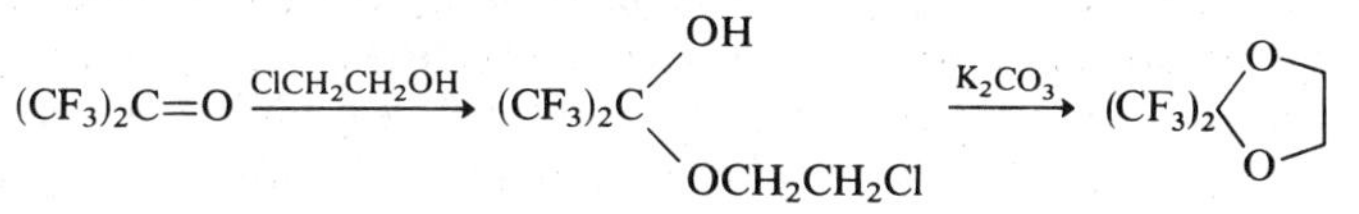

Scheme 30

The reaction of alkenes with sulfur trioxide is an important synthesis of 1,2-oxathiolane 2,2-dioxides. The best conditions involve the use of a dioxane–sulfur trioxide complex at low temperature. The initial addition involves formation of the most stable cation followed by ring closure, as shown in Scheme 31 ⟨74JOC2459⟩.

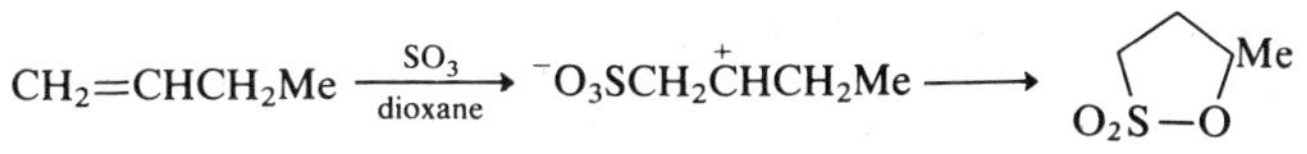

$$CH_2=CHCH_2Me \xrightarrow[\text{dioxane}]{SO_3} {}^-O_3SCH_2\overset{+}{C}HCH_2Me \longrightarrow \underset{O_2S-O}{\overset{}{\bigcirc}}Me$$

Scheme 31

4.30.3.2 Ring Synthesis By Transformation of Other Heterocycles

4.30.3.2.1 Three-membered rings

Oxiranes may be transformed into 1,2-dioxolanes, 1,3-dioxolanes and 1,3-oxathiolanes. In Section 4.30.3.1.2 the intramolecular addition of a hydroperoxide to an activated double bond was discussed. Activation may also be achieved using an oxirane. Thus the hydroperoxide (**113**), obtained from alkene (**111**) by peroxidation, undergoes cyclization under acidic conditions to generate the 1,2-dioxolane as shown in Scheme 32 ⟨78JOC4048⟩. However, the mercury(II) ion catalyzed cyclization of (**111**) discussed in Section 4.30.3.1.2 appears to be a more versatile and efficient process.

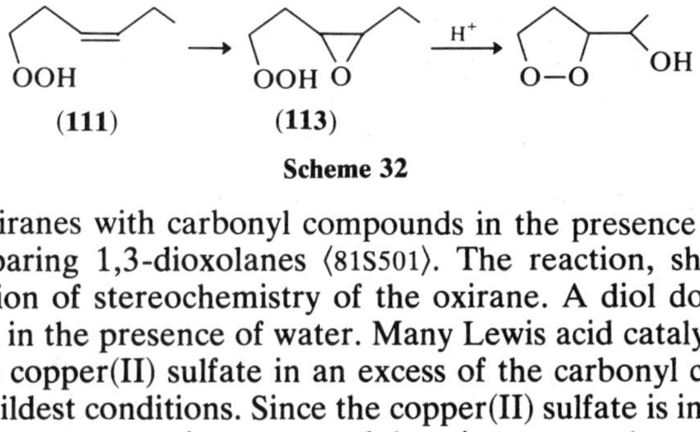

Scheme 32

The reaction of oxiranes with carbonyl compounds in the presence of Lewis acids is an efficient way of preparing 1,3-dioxolanes ⟨81S501⟩. The reaction, shown in Scheme 33, proceeds with inversion of stereochemistry of the oxirane. A diol does not appear to be an intermediate even in the presence of water. Many Lewis acid catalysts are effective, but the use of anhydrous copper(II) sulfate in an excess of the carbonyl compound as solvent probably offers the mildest conditions. Since the copper(II) sulfate is insoluble, the reaction appears to be truly heterogeneous in nature and the mixture must be well stirred ⟨78JOC438⟩.

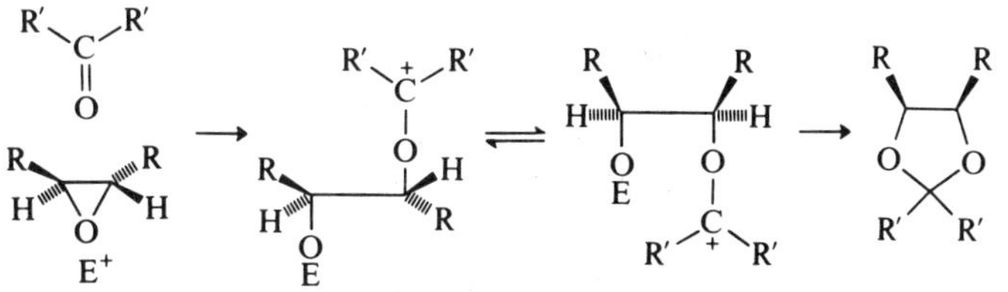

Scheme 33

Under basic conditions, oxiranes react with carbon disulfide exothermically, as shown in Scheme 34, to yield 1,3-oxathiolane-2-thiones. The reaction appears to be quite general although no examples have appeared outside of the patent literature.

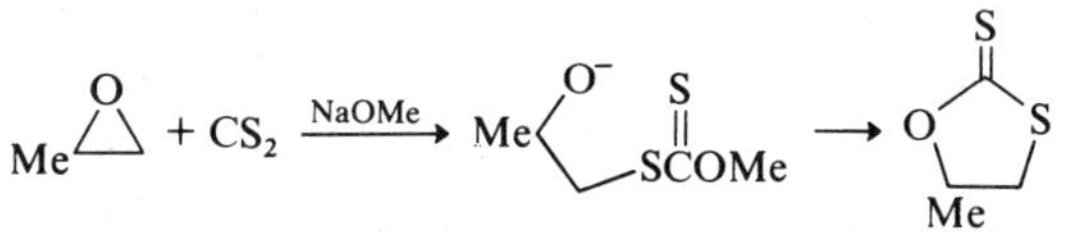

Scheme 34

Carbonyl ylides, generated from the thermolysis of oxiranes, add to carbonyl and thiocarbonyl compounds to produce 1,3-dioxolanes and 1,3-oxathiolanes, respectively. For the reaction to be useful, *i.e.* for the generation of the carbonyl ylide, the oxirane must be substituted by electron-withdrawing substituents, usually cyano groups. Typical examples are shown in Scheme 35. Of special note is the *cis* orientation of the cyano groups in the product, a relationship which follows since the *cis* carbonyl ylide is of lower energy. The reaction appears to have little general applicability since, *a priori*, it is not possible to predict if the cycloadduct will be stable. For example, tetracyanoethylene oxide adds to thiobenzophenone to give an adduct (**114**) which undergoes cycloreversion to yield eventually the alkene (**115**), as shown in Scheme 36 ⟨76H(5)141⟩. This type of reactivity is discussed in Section 4.30.2.1.2. Further, when the oxirane contains an aroyl group, the carbonyl ylide may rearrange to form a dioxole derivative (Scheme 37) ⟨74T2867⟩.

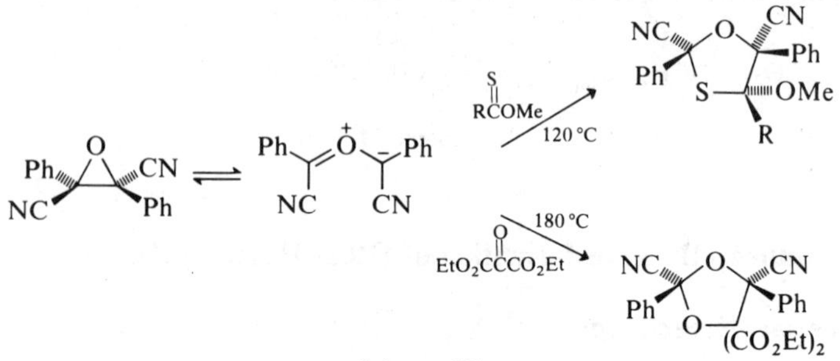

Scheme 35

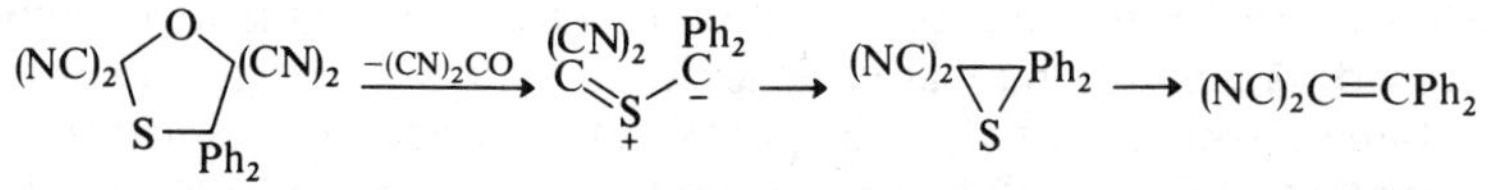

Scheme 36

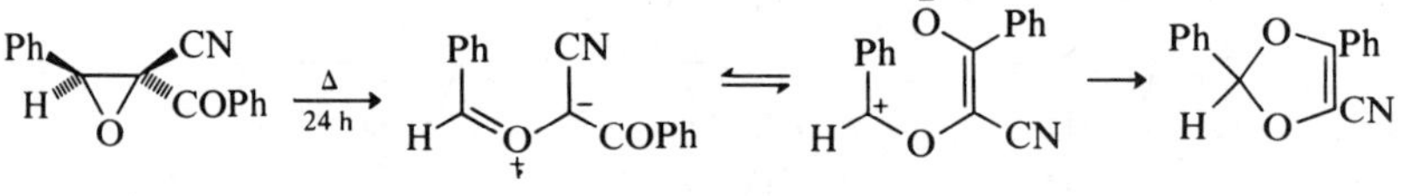

Scheme 37

4.30.3.2.2 Four-membered rings

Thermolysis of thiete 1,1-dioxide (**116**), either in the vapor phase or in benzene solution, gives the oxathiole (**117**) in over 80% yield. The most likely mechanism involves a sulfene intermediate as shown in Scheme 38 ⟨70CJC3704⟩. Not surprisingly, the flash thermolysis of thietane 1,1-dioxide, the saturated analog of (**116**), gives only cyclopropane and propene.

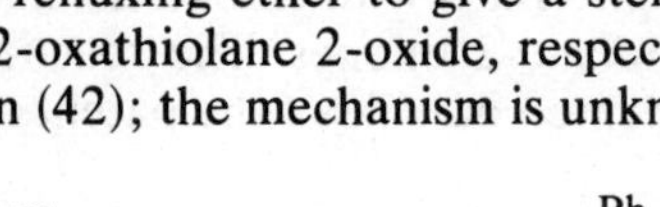

Scheme 38

The transformation of the saturated compounds into 1,2-oxathiolanes may be achieved under basic conditions. For example, *cis*- and *trans*-2,4-diphenylthietane 1,1-dioxides react with magnesium *t*-butoxide in refluxing ether to give a stereospecific ring expansion into *cis*- and *trans*-3,5-diphenyl-1,2-oxathiolane 2-oxide, respectively. The reaction for the *cis* compound is shown in equation (42); the mechanism is unknown ⟨B-73MI43000⟩.

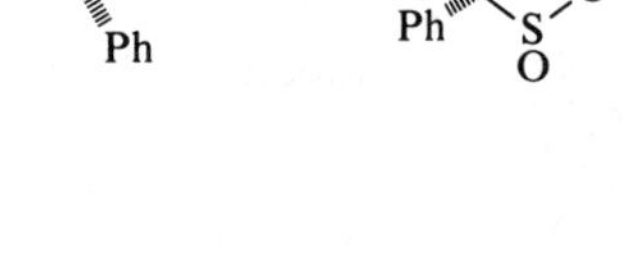

(42)

4.30.3.2.3 Five-membered rings

Thiocarbonyl ylides, generated from the thermolysis of 2,5-disubstituted Δ^3-1,3,4-thiadiazolines, may be trapped by ketenes to give oxathiolane derivatives, as shown in Scheme 39. The reaction appears to be a useful way of obtaining unsaturated derivatives ⟨B-77SH(2)314⟩.

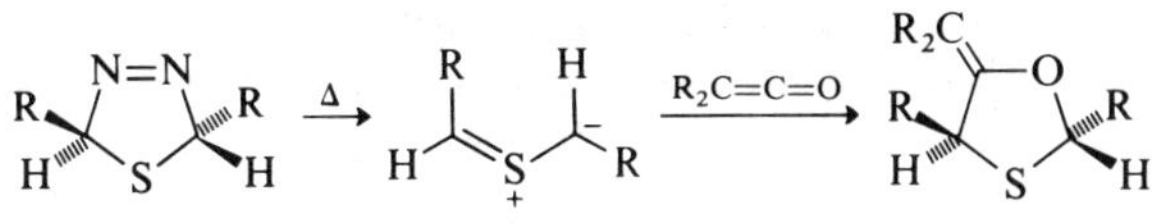

Scheme 39

An interesting conversion of dihydrothiophenes into 1,3-oxathiole derivatives involves conversion of the thiophene into the ylide (**118**), as shown in Scheme 40. On heating in benzene solution overnight, (**118**) begins to decompose into a thiocarbonyl compound which reacts with unchanged (**118**), as shown, to give the oxathiole. The 54% yield is surprisingly high considering the mechanism. ⟨78JSC(P1)1547⟩.

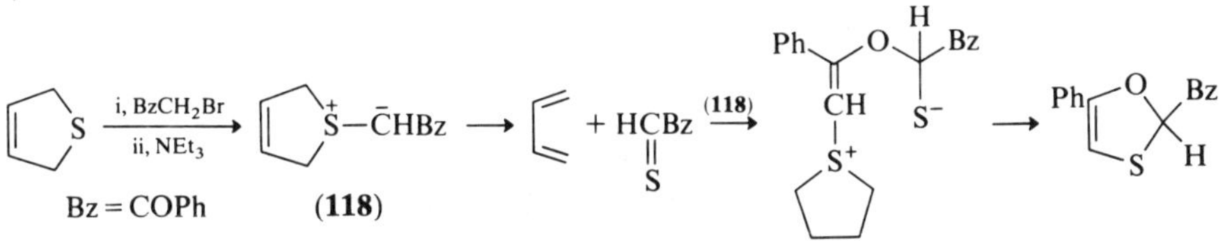

Scheme 40

In what is effectively a desulfurization reaction, 1,2-dithiolane 1,1-dioxide reacts exothermically with aminophosphines such as tris(diethylamino)phosphine in benzene solution to give 1,2-oxathiolane 1-oxide in excellent yield ⟨71JOC322⟩. The reaction, which requires a slight excess of the aminophosphine, is shown in Scheme 41.

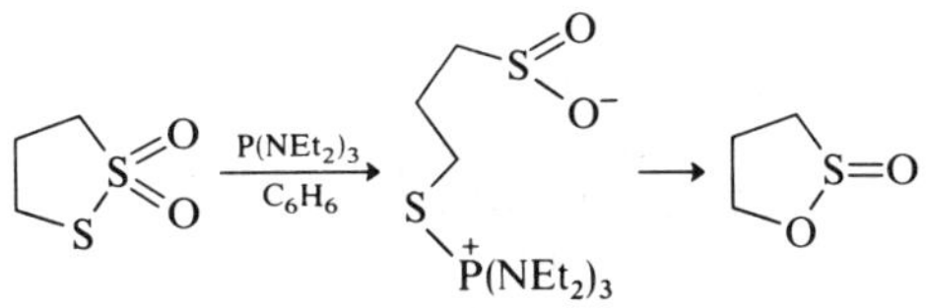

Scheme 41

The ability of furan to undergo both the forward and reverse Diels–Alder reaction has been used to synthesize 1,3-dioxoles, and the reaction appears to have wide applicability. In Scheme 42 the synthesis of 2-phenyl-1,3-dioxole is shown. The key to the process is the dialcohol (**119**), which is obtained from vinylene carbonate as shown. The dialcohol (**119**) has the potential of serving as an intermediate for a variety of 2-substituted 1,3-dioxolanes ⟨73JA7161⟩.

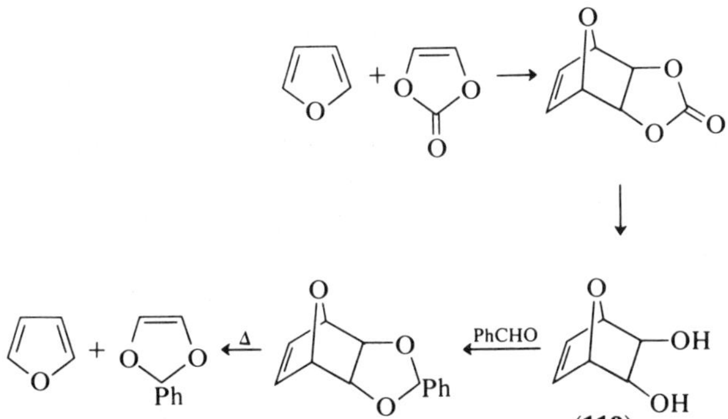

Scheme 42

4.30.3.3 Synthesis of Derivatives

4.30.3.3.1 Parent systems

For many of these systems the parent compounds are unknown or known only either in highly substituted or oxidized forms. In some cases, such as the aromatic 1,3-oxathioles, the mesoionic oxathioles and 1,2-oxathiolane, there is only one synthetic method. The synthesis of these compounds, therefore, will be found in earlier sections. Further, the 1,3-dioxolium and 1,2-oxathiolium cationic species have only been observed in strong acid media and they are discussed in Section 4.30.1.4.1. This present section is directed at those systems where more than one synthetic procedure is available, and indicates the method of choice for synthesis.

1,3-Dioxole is best prepared by the Diels–Alder method outlined in Section 4.30.3.2.3 ⟨61JA3504⟩. 1,3-Dioxolan-2-ylium and 1,3-oxathiolan-2-ylium cations are prepared in a straightforward manner from their 2-alkoxy derivatives as shown in Scheme 43.

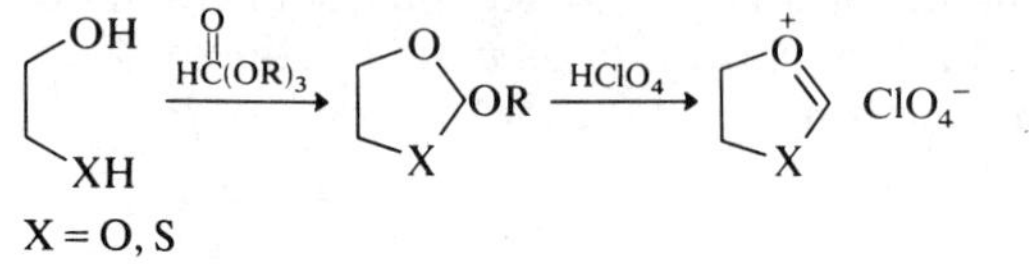

Scheme 43

1,2-Dioxolane is available most conveniently from the peroxide transfer reaction discussed in Section 4.30.3.1.4, while 1,3-dioxolane may be prepared from ethylene glycol and formaldehyde using any acidic catalyst described in Section 4.30.3.1.3. The preparation of 1,3-oxathiolane is best attempted by the method outlined in equation (43) ⟨79TL3263⟩.

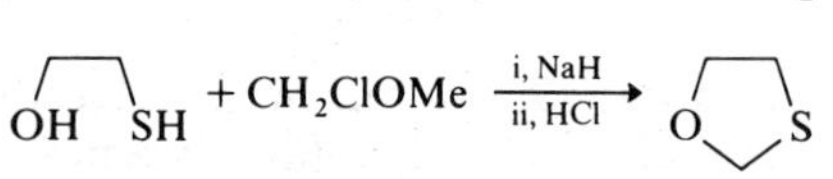

$$\text{(43)}$$

4.30.3.3.2 Benzo derivatives

Only the 1,2-benzodioxole system is unknown. 1,3-Benzodioxole derivatives are found widely in nature, and may be prepared from a catechol and formaldehyde using an acidic catalyst ⟨80AJC675⟩. 2-Alkoxy-1,3-benzodioxoles, dissolved in strong acid, are useful precursors to the 1,3-benzodioxolium cation as indicated in Section 4.30.1.4.1(i). 2-Hydroxythiophenol serves as a starting material for both 1,3-benzoxathiole ⟨76S797⟩ and the 1,3-benzoxathiolium systems ⟨74JHC943⟩, as shown in Scheme 44.

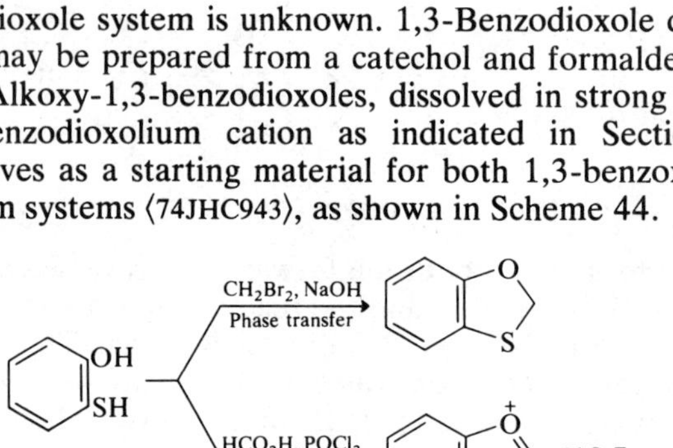

Scheme 44

Only one derivative of the 2,1-benzoxathiole system has been reported in its lowest oxidation state ⟨75JA6909⟩, and involves direct heteroatom coupling as indicated in equation (44). Preparative routes to oxidized forms of 3H-1,2-benzoxathiole ⟨76CJC3012⟩ and 5H-2,1-benzoxathiole ⟨66HC(21-1)76⟩ are shown in equation (45) and Scheme 45, respectively.

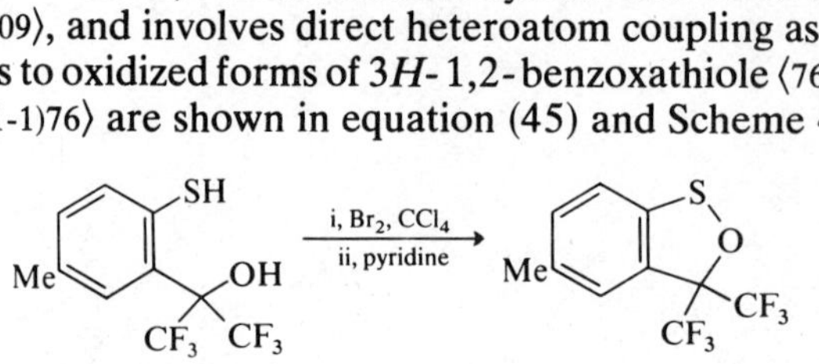

$$\text{(44)}$$

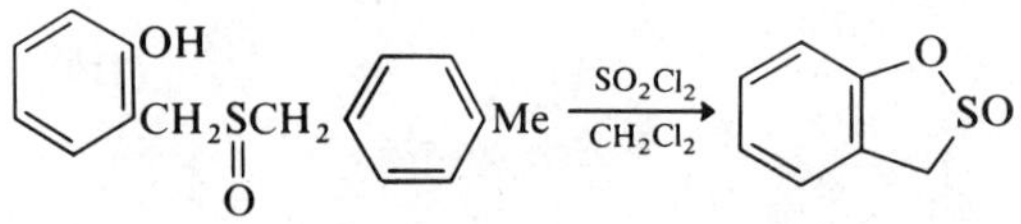

$$\text{(45)}$$

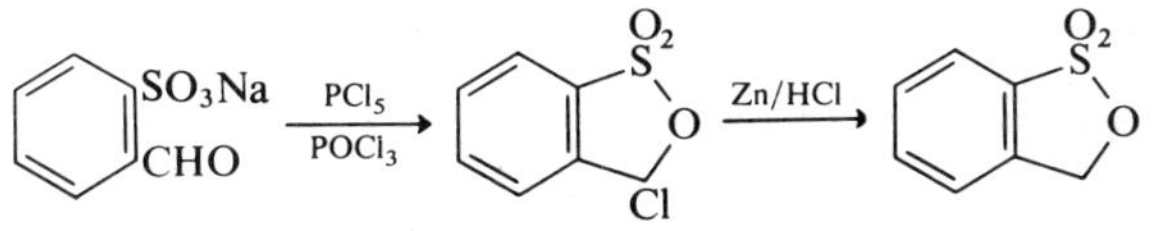

Dioxoles and Oxathioles

Scheme 45

4.30.3.3.3 C-Linked substituents

Most compounds containing *C*-linked substituents are best prepared by total synthesis using a method appropriate for the substitution required. One recent synthesis of 5-methyl-1,3-oxathiole deserves special note, since derivatives of the system are rare ⟨82TL47⟩. The synthesis is shown in Scheme 46.

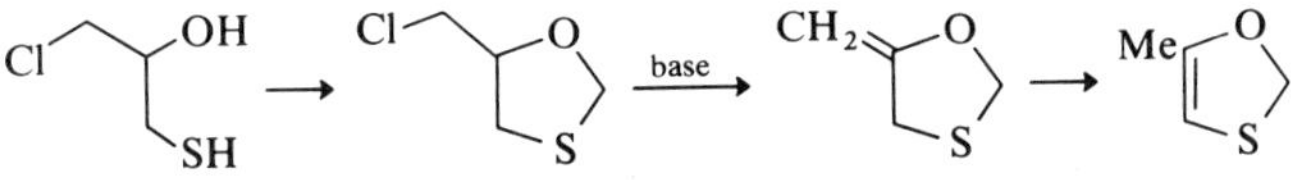

Scheme 46

4.30.3.3.4 N-Linked substituents

Many derivatives of 2-imino-1,3-oxathioles are known, as indicated in Schemes 47 ⟨72CPB304⟩, 48 ⟨67JHC527⟩, 49 ⟨66HC(21-1)76⟩ and 50 ⟨72BCJ1507⟩.

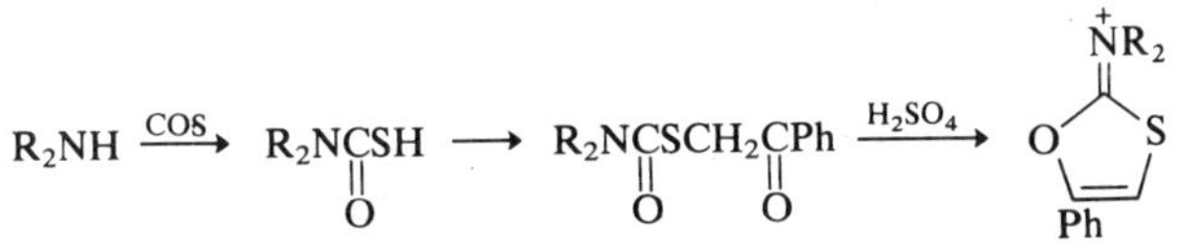

Scheme 47

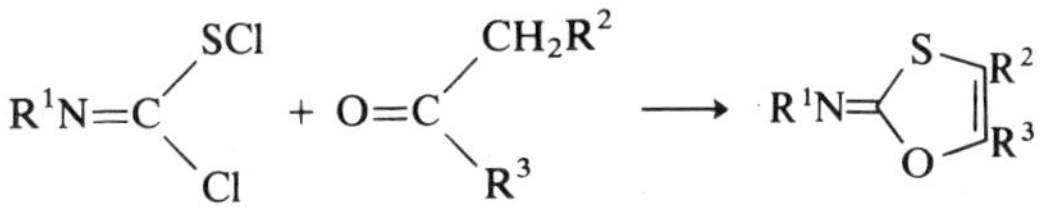

Scheme 48

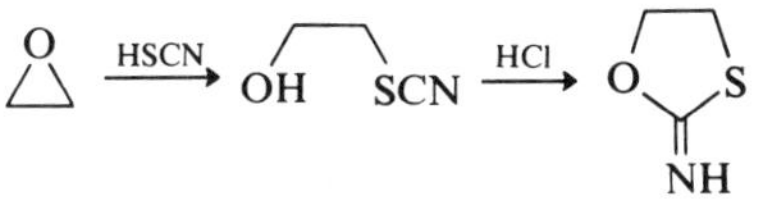

Scheme 49

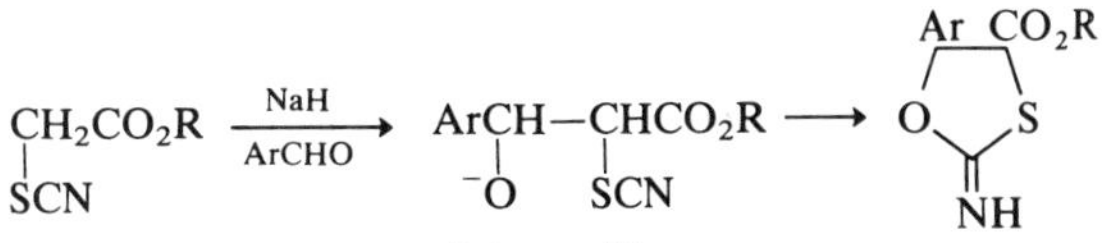

Scheme 50

2-(Amino substituted)-1,3-dioxolanes and -oxathiolanes are readily available as shown in equation (46), while derivatives of the corresponding 2-ylium species are also easily prepared (equation 47) ⟨72CRV357⟩. A 4-amino-1,2-oxathiolane 2-oxide was recently prepared in high yield by total synthesis ⟨81JOC5408⟩.

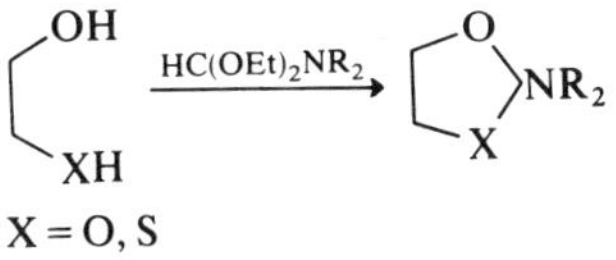

(46)

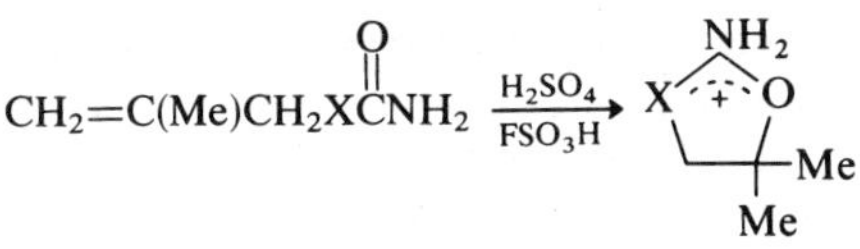

(47)

4.30.3.3.5 O-Linked substituents

Vinylene carbonate is made from ethylene carbonate by chlorination followed by base treatment ⟨55JA3789⟩, as shown in Scheme 51. While ethylene carbonate is available from phosgene and ethylene glycol, the general synthesis of cyclic carbonates shown in equation (48) gives high yields and avoids the use of the toxic gas ⟨79CL1261⟩. 4-Oxo-1,3-dioxolanes are obtained from the condensation of α-hydroxy acids with carbonyl compounds, while 3-oxo-1,2-dioxolanes are available from β-hydroxy acids as shown in Scheme 52 ⟨66JOC2087⟩.

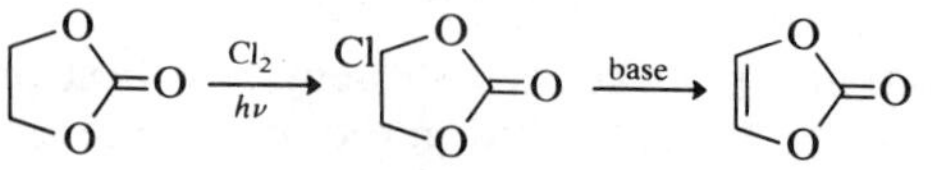

Scheme 51

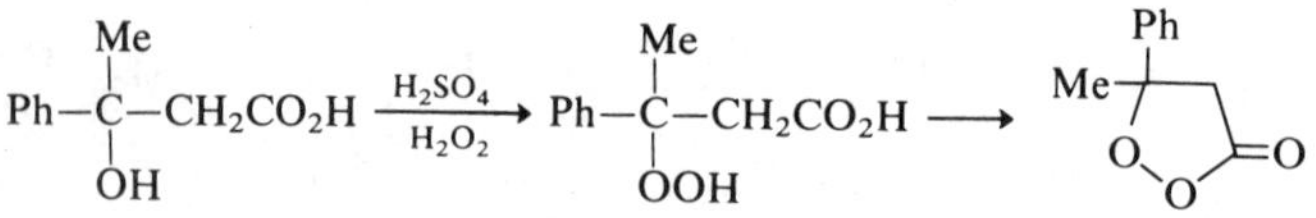

Scheme 52

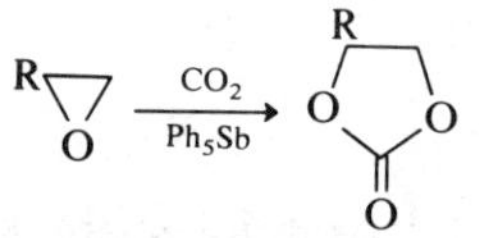

$$(48)$$

2-Alkoxy-1,3-dioxoles are best prepared using the Diels–Alder methods discussed in Section 4.30.3.2.3. The corresponding 1,3-oxathioles are not well known, but have been reported from the reaction of 1-(acylthio)-1-alkynes with alcohols ⟨74RTC99⟩, as shown in equation (49). 1,3-Dioxolanes and 1,3-oxathiolanes containing a 2-alkoxy substituent are easily prepared using an orthoester and a glycol or 2-mercaptoalcohol.

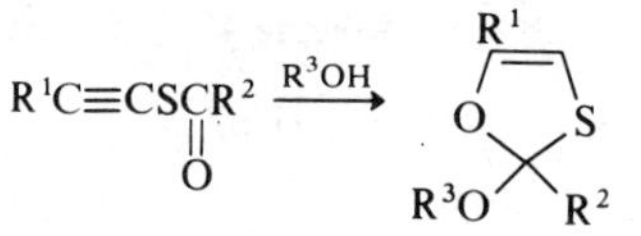

$$R^1C \equiv CSCR^2 \xrightarrow{R^3OH} \quad (49)$$

4.30.3.3.6 S-Linked derivatives

Little is known of the sulfur derivatives of these ring systems. The synthesis of a mesoionic derivative containing a 2-thiomethyl group was discussed in Section 4.30.3.1.2. Other reports of sulfur derivatives have been limited to thiocarbonates, which are available from glycols and mercaptoalcohols on treatment with thiophosgene ⟨69JOC3011⟩. Vinylene thiocarbonate has been prepared using the Diels–Alder method outlined in Section 4.30.3.2.3 by reaction of the dialcohol (**119**) with thionocarbonyl diimidazole and subsequent thermolysis. The direct conversion of vinylene carbonate into the thiocarbonate by treatment with P_4S_{10} gives only low yields ⟨73JA7161⟩.

4.30.3.3.7 Halogen substituents

There is a paucity of data concerning halogen derivatives of these ring systems. The chlorination of 1,3-dioxolane to yield mainly the unstable 2-chloro compound was discussed in Section 4.30.2.1.3. A range of fluorinated 1,3-dioxolanes has been reported from the cobalt trifluoride treatment of 2,2-bis(trifluoromethyl)-1,3-dioxolane ⟨77MI43001⟩. These compounds, on treatment with aluminum chloride, undergo fluorine–chlorine exchange to yield a variety of polyhalogenated compounds. Halogenated derivatives of the 1,2-oxathiolane system are obtained from cyclization of the appropriate sulfonic acid (equation 50). Care must be taken that a second dehydrohalogenation does not take place ⟨66HC(21-1)76⟩. Brominated 1,2-oxathioles are obtained on treatment of allenesulfinates with bromine as shown in Scheme 53 ⟨77TL1753⟩.

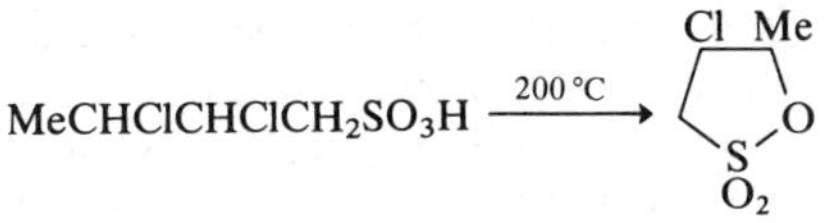

$$\text{MeCHClCHClCH}_2\text{SO}_3\text{H} \xrightarrow{200\,°C} \underset{\underset{O_2}{S}}{\overset{Cl\ Me}{\underset{}{\bigcirc}}}\!\!O \quad (50)$$

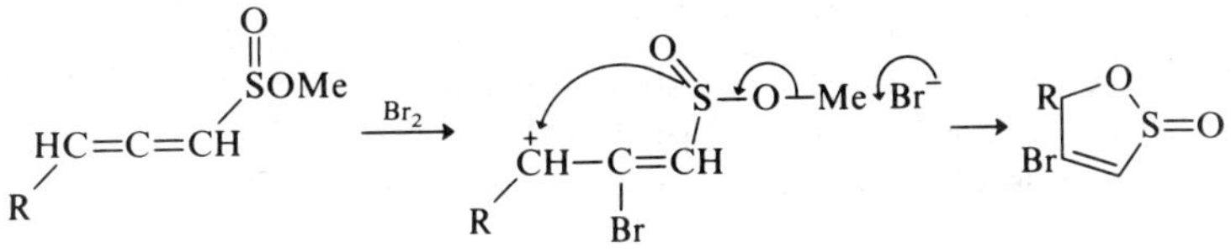

Scheme 53

4.30.4 APPLICATIONS

The benzodioxole ring system is distributed widely in nature and is found in numerous natural products such as safrole and piperonal, as well as a multitude of alkaloids. 1,2-Dioxolanes are intermediates in the arachidonic acid cascade, which is the biochemical pathway from essential fatty acids to prostaglandins and similar hormones. The endoperoxide PGH_2 (121) is believed to be formed on initial oxidation of arachidonic acid (120). PGH_2 has a half-life of 4-5 minutes and is transformed enzymatically into prostaglandins, prostacyclin and the thromboxanes. These compounds are mediators for the control of platelet aggregation, blood vessel dilation and smooth muscle contraction.

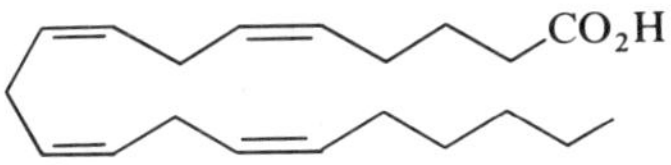

(120) 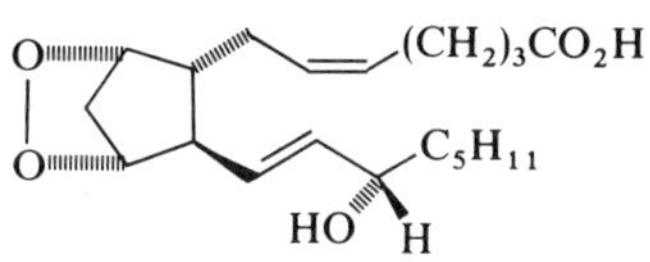   (121) 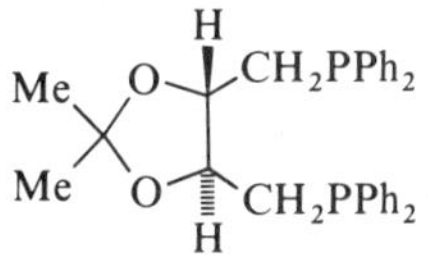   (122)

The main use for 1,3-dioxolanes in synthesis is the protection of functional groups so that transformations in other parts of the molecule may be performed. Depending on the molecule under study, the 1,3-dioxolane ring may be used to protect 1,2-glycol or carbonyl functions. More recently, transition metals having optically active ligands have been shown to be very useful catalysts for homogeneous asymmetric hydrogenation. In particular, rhodium complexes of chiral phosphines are very effective, and one of the most popular chiral phosphines is 2,3-O-isopropylidene-2,3-dihydroxy-1,4-bis(diphenylphosphino)butane, known as DIOP (122). It is commercially available in both the (+) and the (−) form.

There is a vast number of patents covering polymers and copolymers of 1,3-dioxolane. It is unclear if any polymers have found extensive commercial utility. Cation initiated polymers with a degree of polymerization below 10 are mobile or viscous oils, while higher polymers are generally solids. Polymers with molecular weights of 10^4 are tough, cold-drawable solids, similar generally to polyethylene oxide and, since they have solubility in water, they add hydrophilic nature to copolymers. Copolymers with monomers containing active hydrogens such as alcohols, amines and carboxylic acids have been proposed as surface active agents in water. 1,3-Dioxolanes and vinyl compounds such as styrene or isobutene copolymerize to give a low molecular weight liquid with terminal hydroxy groups. These may serve as prepolymers for the preparation of polyurethane elastomers. 2-Vinyl-1,3-dioxolane gives a polymer with pendant dioxolane rings. The ease with which dioxolanes undergo autoxidation to hydroperoxides, and ultimately initiate polymerization of residual double bonds, makes this type of polymer suitable for a wide variety of air-cured coating systems, adhesives and films.

Vinylene carbonate is one of the few 1,2-disubstituted ethylenes that is known to undergo facile radical initiated homopolymerization. Initiation may be by oxygen, peroxides or cobalt-60 γ-radiation. Such polymers are reportedly useful as coatings and films. Vinylene carbonate also copolymerizes with ethylene under high pressure to yield a material with about 10% vinylene carbonate content. This polymer, when blended with polyvinyl chloride, is suitable for injection molding.

The 1,3-dioxolane ring is found in a major antifungal drug, ketoconazole (**123**). Ketoconazole is a broad spectrum, orally active antifungal agent and is used to treat a wide variety of superficial or deep fungal infections. Various workers, especially in Russia, have explored the use of simple oxathiolanes as radioprotectants. For example, the survival rate of mice irradiated with lethal doses of X- or γ-rays was 40% when they were pretreated with 2,2-dimethyl-1,3-oxathiolane. Other oxathiolanes were less active. 1,3-Oxathiolane, administered intravenously to dogs 2 or 3 times daily for 2–4 days before chronic irradiation with polonium-210, increased their life expectancy from 9 months to 3–7 years and alleviated radiation sickness. Some blood indicators were also restored to normal.

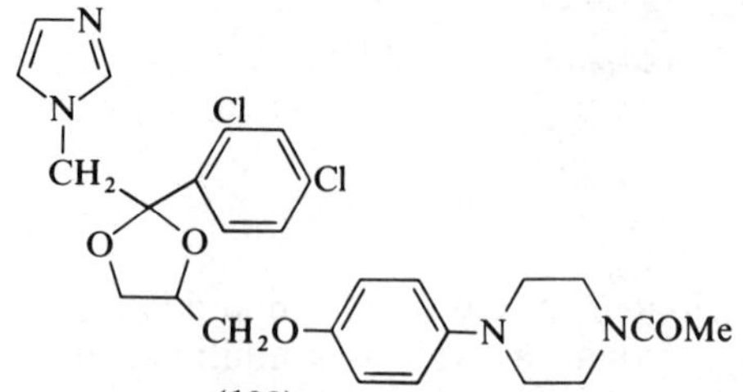

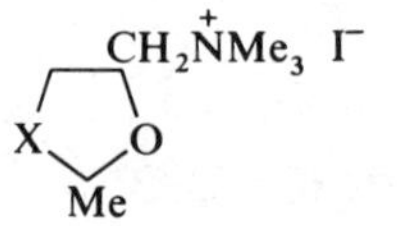

Several 1,3-dioxolanes have been shown to have muscarinic activity. Of these, (**124**; X = O) is a particularly selective agonist at muscarinic receptors, the *cis* being 5–10 times more active than the *trans*. The corresponding oxathiolane (**124**; X = S) also had high activity. A compound of close structural similarity, dioxonium (**125**), is a muscle relaxant in man.

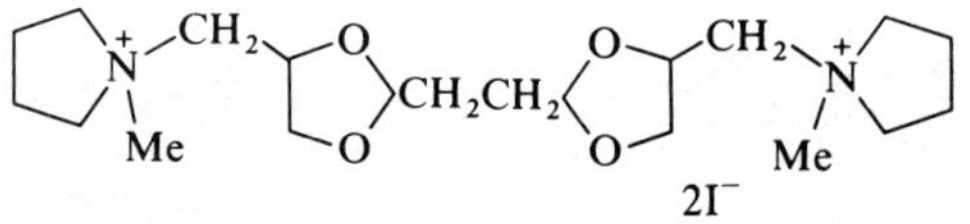

A great number of compounds are known which are sulfonic acid analogs of phenophthaleins. These compounds are usually referred to as phenolsulfonephthaleins although they are better known under their dye or indicator names. Two such examples are phenol red (**126**) and bromophenol blue (**127**).

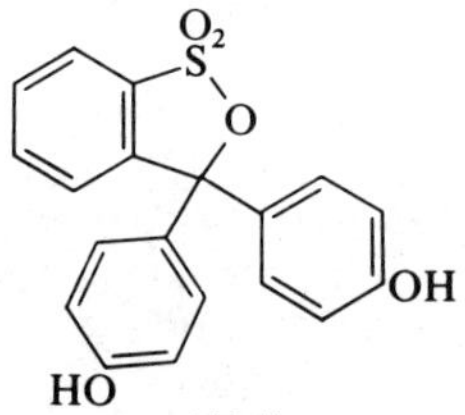

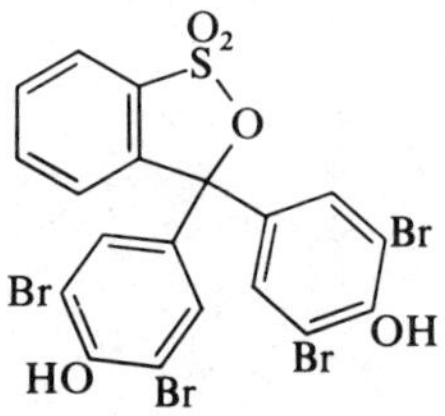

With one important exception, dioxoles and oxathioles do not appear to be a particularly toxic class of compounds. The exception is 1,2-oxathiolane 2,2-dioxide (propane sultone) which produces brain tumors, leukaemias and adenocarcinomas of the small intestine in rats treated with 28 mg kg^{-1} orally twice a week for 60 weeks. If the toxicity is associated with its ability to act as an alkylating agent, then higher homologs should have similar, but reduced, carcinogenic potential.

4.31
1,2-Dithioles

D. M. McKINNON
University of Manitoba

4.31.1 STRUCTURE	783
4.31.1.1 Introduction	*783*
4.31.1.2 Theoretical Studies	*784*
4.31.1.3 Molecular Dimensions	*784*
4.31.1.4 Molecular Spectra	*785*
4.31.1.4.1 NMR spectra	*785*
4.31.1.4.2 UV and visible spectra	*786*
4.31.1.4.3 IR spectra	*786*
4.31.1.4.4 Mass spectrometry	*786*
4.31.1.4.5 Photoelectron spectroscopy	*787*
4.31.1.5 Thermodynamic Aspects	*788*
4.31.1.6 Betaine and Unusual Structures	*788*
4.31.2 REACTIVITY	788
4.31.2.1 Reactivity of 1,2-Dithiolylium Salts	*788*
4.31.2.1.1 Reactivity at the ring	*788*
4.31.2.1.2 Reactivity of substituents	*793*
4.31.2.2 Reactivity of 1,2-Dithioles and 1,2-Dithiolanes	*794*
4.31.2.2.1 Reactivity at the ring	*794*
4.31.2.2.2 Reactivity of substituents	*799*
4.31.3 SYNTHESIS	801
4.31.3.1 Direct Synthesis	*801*
4.31.3.1.1 Direct synthesis by formation of one bond	*801*
4.31.3.1.2 Direct synthesis by formation of two bonds	*803*
4.31.3.1.3 Direct synthesis by formation of three bonds	*804*
4.31.3.2 Ring Synthesis by Transformation of Other Heterocycles	*806*
4.31.3.2.1 From pyran and thiopyrans	*806*
4.31.3.2.2 From thioxins and 1,3-thiazines	*807*
4.31.3.2.3 From 1,3-dithiins	*807*
4.31.3.2.4 From isoxazoles	*807*
4.31.3.2.5 From isothiazoles	*807*
4.31.3.2.6 From 1,3-dithioles	*808*
4.31.3.2.7 From 1,6,6aλ^4-trithiapentalenes	*809*
4.31.3.2.8 From 1,3-dithietes	*809*
4.31.3.3 Best Practical Methods of Synthesis	*809*
4.31.4 USES AND OCCURRENCE	810
4.31.4.1 Natural Products	*810*
4.31.4.2 Compounds of Biological Interest	*810*
4.31.4.3 Compounds of Industrial Interest	*811*
4.31.4.4 Organic Conductors	*811*

4.31.1 STRUCTURE

4.31.1.1 Introduction

The 1,2-dithiole system and related compounds possess two adjacent sulfur atoms in a five-membered ring. The types are known respectively as 1,2-dithiolanes (**1**), 1,2-dithioles (**2**), 1,2-dithiol-3-ones (**3a**), 1,2-dithiole-3-thiones (**3b**), 3-iminodithioles (**3c**), 1,2-dithiol-3-ylidenes (**3d**) and 1,2-dithiolylium (**4**) ions.

(1)   (2)   (3) a; X = O   (4)   (5)
              b; X = S
              c; X = NR
              d; X = CR¹R²

Benzo and other fused derivatives are also known. In addition, the 1,2-dithiole ring is an integral part of the structure of the '1,6,6aλ^4-trithiapentalenes' (**5**) covered in Chapter 4.38. Naming in this chapter will conform mostly to that used in *Chemical Abstracts* except for the dithiolylium cation (**4**) (IUPAC Rule C-83) and also for the trithiapentalenes.

Of the compounds (**1–4**) above, only the ion (**4**) possesses cyclic delocalization and is aromatic. Compound (**1**) reacts mainly as an aliphatic disulfide, while compounds (**2**) and (**3a–d**) may have intermediate properties.

4.31.1.2 Theoretical Studies

These have attempted to determine the nature and extent of the sulfur–sulfur and sulfur–carbon bonding, and the amount of *d*-orbital participation.

For the 1,2-dithiolylium cation (**4**), early studies using a Hückel LCAO MO method gave values of electron densities and free valency. The former were almost unaffected by change in the value of the resonance integral β_{SS}. An SCF LCAO approach indicated a high amount of cyclic conjugation and delocalization of electrons ⟨66AHC(7)39⟩. A later improved Hückel approach on the 3- and 4-phenyl-1,2-dithiolylium cations indicated only a slight perturbation of the π-structure by substitution ⟨80AHC(27)151⟩.

An LCGO approach with only slight *d*-orbital participation gave good agreement with half-wave reduction potentials, but the CNDO studies gave best results when *d*-orbitals were included ⟨80AHC(27)151⟩. Recently, MNDO/3 calculations ⟨78JA7629⟩ gave good agreement of theory and experimental values of molecular dimensions, apart from a small error in the S—S bond. A Perturbation-Graph Theory study gave good agreement with other studies ⟨78T3419⟩.

For 1,2-dithiole-3-thiones (**3b**), LCAO MO methods gave a linear correlation between *n*–π^* transition absorption waves, indicated a negative charge on the exocyclic sulfur atom, and gave information on the effects on phenyl substituents ⟨67KGS758⟩. A modified SCF LCAO method ⟨69T5485⟩, without the explicit inclusion of sulfur *d*-orbitals, gave a good match of calculated and observed spectra using empirical parameters. Semiempirical parameters worked well for the 3-mercapto-1,2-dithiolylium ion (**6**) ⟨68MI43100⟩ and for the mesoionic 1,2-dithiolylium-4-olate (**7**). A CNDO/2 method has been applied, using modified parameters, to 1,2-dithiole-3-thiones (**3b**) and the ion (**6**) ⟨80AHC(27)151⟩.

(6)   (7)

4.31.1.3 Molecular Dimensions

X-Ray structure determinations for 1,2-dithiolylium salts (**4**) ⟨80MI43100⟩ showed S—S bond distances of ~2.02 Å. For 1,2-dithiolylium iodides (**4**; X = I), charge transfer between the cation and iodide was found. 3,5-Diamino-1,2-dithiolylium ions (**8**) and derivatives showed long S—S bonds (~2.07 Å), indicating substantial iminium character, *i.e.* from structures of type (**8a**). The mesoionic 1,2-dithiolylium-4-olates (**7**) have typical dithiolylium bond distances ⟨78AJC297⟩ and are best represented as shown.

For the 5-phenyl-3-(5-phenyl-1,2-dithiol-3-ylidenemethyl)-1,2-dithiolylium ion (**9**), sulfur atoms are almost colinear. A central sulfur–sulfur bond distance of 3.00–3.10 Å was suggested, indicating partial bonding of some type between these atoms ⟨65ACS1253, 68JCS(C)642⟩.

4-Methyl-1,2-dithiole-3-thione (**10**) ⟨66MI43100⟩ had sulfur–sulfur bond distances almost equal to those in acyclic disulfides. Although this may be merely coincidence, because of the different situations of the atoms, observed heats of formation were 71 kJ mol^{-1} higher than calculated values, perhaps due to the disulfide bond torsional stress ⟨72MI43100⟩. A high degree of positive charge delocalization in the ring may be indicated ⟨B-66MI43100⟩. 3-Acylmethylene-1,2-dithioles (**11a**) show sulfur–sulfur and sulfur–oxygen bond distances of 2.106 and 2.382 Å respectively. The former is greater than for simple dithioles, yet less than for 1,6,6aλ^4-trithiapentalenes (**5**) ⟨69ACS1377⟩. Related nitroso compounds (**12**) show decreased sulfur–sulfur bonding ⟨71JCS(B)946, 71JCS(B)952⟩.

For the compound (**13**) the sulfur–sulfur distance was 2.054 Å, in accord with the non-ionic structure (**13a**) ⟨77TL2717⟩. 4-Thioacyl-1,2-dithiole-3-thiones (**14**) show no interaction between the two thione sulfur atoms ⟨77JCS(P2)1854⟩.

4.31.1.4 Molecular Spectra

4.31.1.4.1 NMR spectra

^{1}H NMR data for some 1,2-dithiolylium salts (**4**) have appeared in a number of papers. In the unsubstituted parent 1,2-dithiolylium cation (**4**) the 3(5)-protons absorb at 10.57 p.p.m., and the 4-proton at higher field, at 8.88 p.p.m. ⟨66AHC(7)39⟩. Coupling constants are 5.0 Hz. These values appear to be fairly typical for the ions ⟨73JA4373, 73JCS(P1)2351⟩. Calculated ring charges correlated with observed chemical shifts of ring and substituent protons ⟨71T4705⟩. Protons on 1,2-dithiole-3-thiones (**3b**) and -3-ones (**3a**) absorb at higher field than those on the cations (**4**), although dithiolethiones (**3b**) absorb at lower field than the corresponding dithiolones (**3a**). Protons on the parent dithiolone (**3a**) absorb at 8.30 (H-5) and 6.60 p.p.m. (H-4). On the thione (**3b**) these are at 8.30 and 7.15 p.p.m. with J values of 5.3 and 5.5 Hz respectively ⟨77S802, 73CR(C)(276)1057, 65AJC1211, 68CJC2577, 60CI(L)1568⟩. Protons on imines (**3c**) have similar values ⟨80CPB487⟩. These studies also gave data on the aryl protons of 5-aryl derivatives.

^{13}C Studies on the cation (**4**) and derivatives have been performed ⟨80MI43100, 77OMR(10)43, 78JA7629⟩. In the parent ion, values are 176.72 (C-3,5) and 142.68 (C-4) p.p.m.; in the mesoionic compounds (**7**) the values are similar ⟨80PS(8)79⟩. These shifts have been compared with charge intensities from other methods. Data for some simple dithiol-3-ones (**3a**), -3-thiones (**3b**) and 3-acylmethylene compounds (**11a**) have been listed ⟨75CJC836, 77OMR(9)546⟩. In (**3a**) derivatives the carbonyl carbons absorb at ~193 p.p.m. and in (**3b**) derivatives the thione carbons absorb at 213–217 p.p.m. It is found that the α-effect of a phenyl group at C-5 is twice that at C-4. Results from the compounds (**11**) were not consistent with true bicyclic structures (**11b**). In another study a correlation was drawn between ^{13}C shifts of the carbonyl group and rates of isomerization of the (E) and (Z) isomers of some 3-acylmethylene-1,2-dithioles (**11a**) ⟨77ACS(B)683⟩.

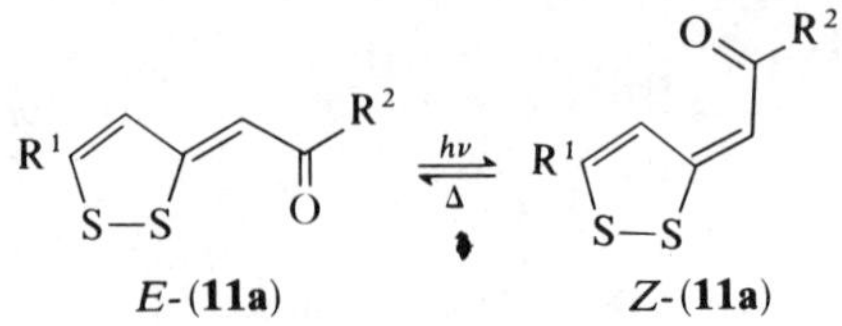

4.31.1.4.2 UV and visible spectra

These data have been tabulated ⟨66AHC(7)39⟩. In the parent ion (**4**) only two absorptions are evident, at 202 and 285 nm. The long wavelength band shows a slight solvatochromic effect. The 3-phenyl and 4-phenyl compounds exhibit two absorptions each, at 287 and 356, and 242 and 345 nm, respectively. Electron releasing substituents in the 3-position provide a greater bathochromic shift than those at the 4-position.

The dithiole monomethinecyanine dyes of type (**9**) absorb strongly around 520–630 nm. In acid these protonate to form dications (**15**) which have similar spectra to 3-methyl-1,2-dithiolylium ions (**16**) ⟨72ZC349⟩. The mesoionic compounds (**7**) exhibit large solvatochromic effects at 460–620 nm, assigned to $\pi^*-\pi$ transitions ⟨80PS(8)79⟩. Some 1,2-dithiolylium salts (**4**) form charge transfer species with the anion of tetracyanopropene which absorb at 500 nm but exhibit large solvatochromic effects ⟨75CR(C)(280)673⟩.

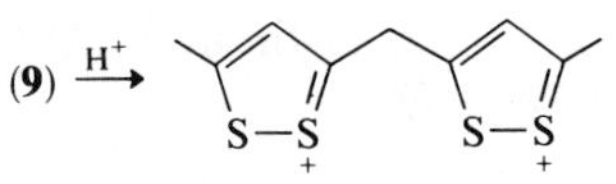

The spectra of a large number of 1,2-dithiol-3-ones (**3a**) and 1,2-dithiole-3-thiones (**3b**) have been tabulated ⟨64MI43100⟩. 1,2-Dithiol-3-one itself shows two maxima at 226 and 310 nm, and 1,2-dithiole-3-thione has four maxima at 229, 251, 336 and 415 nm. Neither of these compounds shows much solvent effect.

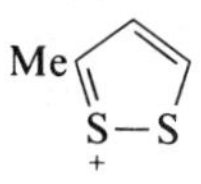

4.31.1.4.3 IR spectra

Studies on 1,2-dithiol-3-ones (**3a**) show carbonyl absorptions around 1640–1670 cm^{-1}; 1,2-dithiole-3-thiones (**3b**) show a thione absorption around 1170 cm^{-1} and imines (**3c**) absorb (C=N) at 1590–1605 cm^{-1} ⟨77S802, 60CI(L)1568, 80CPB487, 74BCJ3084⟩. Analyses of these and other frequencies are also listed ⟨72IJS(A)15, 66CI(L)1962⟩. In 3-acylmethylene-1,2-dithioles (**11**) the frequency of the carbonyl group *cis* to sulfur is considerably displaced from the usual ranges ⟨71AHC(13)161⟩; typical values are 1550 cm^{-1}. On the other hand, *trans* situated groups average 1640 cm^{-1}. This may be attributed to partial sulfur–oxygen bonding. 5-Acylthio-1,2-dithiole-3-thiones (**17**) exhibited no such interaction ⟨76ACS(B)88⟩.

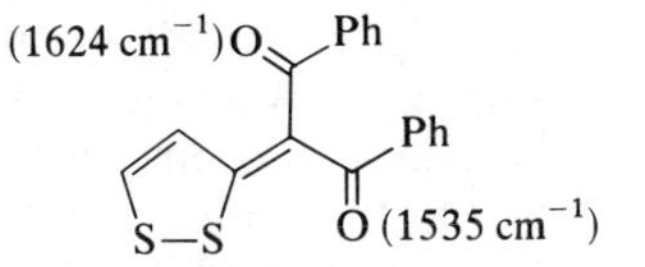

4.31.1.4.4 Mass spectrometry

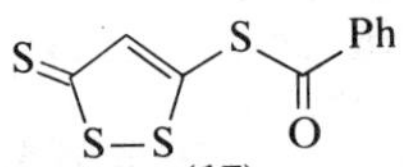

1,2-Dithiolylium bromides (**4**; X = Br) on thermolysis in the ion source of a mass spectrometer give stable 1,2-dithiol-3-yl radicals (**18**). These subsequently form ions corresponding to the molecular ion, which in turn may either lose hydrogen atoms to form 3-alkylidene-1,2-dithioles (**3d**) or lose protons to form stabilized carbenes (**19**) (Scheme 1). Evidence for dimers (**20**) of these is seen in the mass spectrum ⟨74T553⟩, although recent work indicates that the carbenes (**19**) are unstable and thus may not be direct precursors of dimers (**20**) ⟨81JCS(P2)1062⟩. 1,2-Dithiole-3-thiones (**3b**) are also observed from thermal degradation of 3-unsubstituted 1,2-dithiolylium salts (**4**). 3-Alkylthio-1,2-dithiolylium salts

(**21**) may react similarly to the above, but also by loss of alkyl groups from sulfur ⟨74ACS(B)1185⟩.

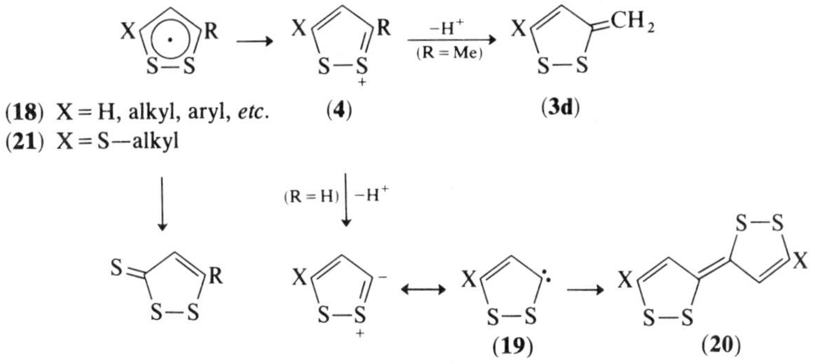

(**18**) X = H, alkyl, aryl, *etc.*
(**21**) X = S—alkyl

Scheme 1

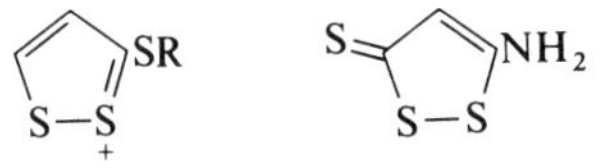

(**21**) (**22**)

1,2-Dithiole-3-ones (**3a**) and -3-thiones (**3b**) provide intense molecular ions. 4-Unsubstituted compounds also show ions of large abundance arising by loss of HS_2 from the parent ion. The 4-substituted compounds show different patterns, and 5-amino-1,2-dithiole-3-thione (**22**) shows initial loss of S_2 from the molecular ion ⟨70BCJ2938⟩. For aryl-1,2-dithiole-3-ones (**23a**) the first fragmentation process involves loss of CO to form 1,2-dithietes (**24**) and thence to thiirenes (**25**). By contrast, methyl-1,2-dithiol-3-ones (**23b**) rearrange to the 1,2-dithiolylium ion (**4**) (Scheme 2) ⟨72ACS250⟩. 3-Acylmethylene-1,2-dithioles (**11a**) show a number of processes, including α-cleavage at the carbonyl group and [*M* − OH] or [*M* − SH] peaks for certain aryl derivatives ⟨80MI43100⟩.

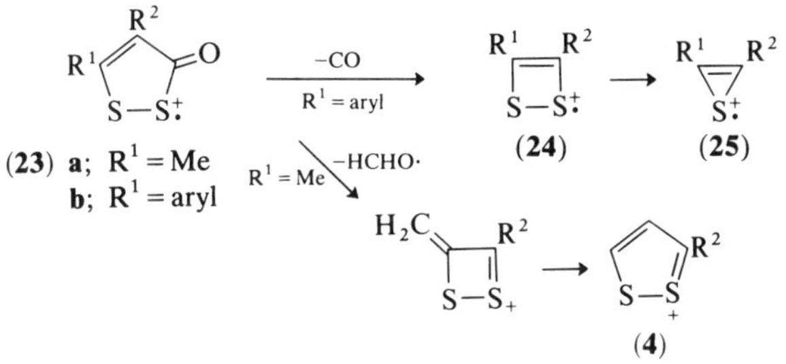

(**23**) a; R^1 = Me
b; R^1 = aryl

Scheme 2

4.31.1.4.5 Photoelectron spectroscopy

One radiation source, the He(Iα) photon energy line, at 21.22 eV, provides vertical and adiabatic ionization potentials. In 1,2-dithiole-3-thione (**3b**), a first ionization band at 8.42 eV has high intensity and is assigned to at least two electronic transitions ⟨75MI43100⟩. Data are also available for other dithiole-3-thiones and two fused dithioles (**26**) and (**27a**) ⟨79MI43100⟩.

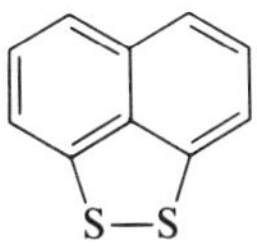

(**26**)
(**27**) a; $R^1 = R^2 = H$ 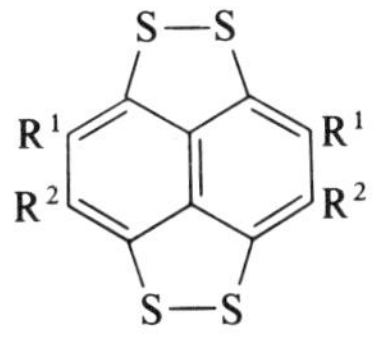
b; $R^1 = R^2 = Cl$
c; $R^1, R^2 = (CH=CH)_2$

Using an X-ray source operating at approximately 10^3 eV, only one S(2*p*) binding energy was found for both sulfur atoms in 1,2-dithiolylium salts (**4**), indicating an almost equal charge distribution. In addition, only one C(1s) line was observed ⟨74JHC105, 71CS(1)183⟩. 1,2-Dithiol-3-ones (**3a**) and -3-thiones (**3b**) give S(2p) binding energies of 164.5 and 164.7 eV respectively. Despite the small spread of values, a correlation has been made between these and attached ^{19}F chemical shifts.

4.31.1.5 Thermodynamic Aspects

1,2-Dithiolanes (**1**), in contrast to other dithioles, are not planar. In the carboxylic acid (**28a**) the CSSC dihedral angle is 26.6° ⟨B-61MI43100⟩, leading to a strain of 67 kJ mol^{-1}. These compounds are unstable and the parent may be obtained only in dilute solution. Lipoic acid (**29**) is more stable, but does polymerize. The instability may account for its particular reactivity ⟨B-61MI43101⟩. 1,2-Dithiol-3-ones (**3a**) and -3-thiones (**3b**) are very stable, as is evident from their modes of formation. Simple 1,2-dithiol-3-imines (**3c**) are also stable, but some readily interconvert to isothiazoline-3-thiones (**30**) (Section 4.31.3.2).

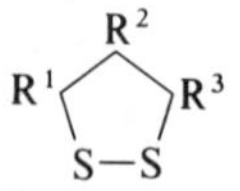

(**28**) a; $R^1 = R^3 = H, R^2 = CO_2H$
b; $R^1 = R^3 = CO_2H, R^2 = H$

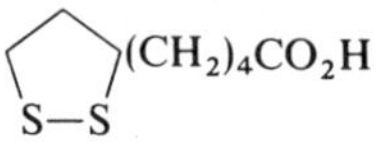

(**29**)

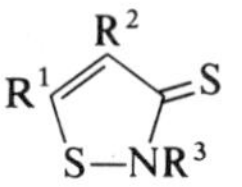

(**30**)

4.31.1.6 Betaine and Unusual Structures

Compounds of interest are the 1,2-dithiolylium-4-olates (**7**), 3-acylmethylene-1,2-dithioles (**11**) and various oxocyclohexadienylidene-1,2-dithioles (**13**) and (**31**). The compounds (**7**) appear by most physical data and chemical properties to be best represented by the structure shown ⟨78AJC297, 80PS(8)79⟩, and for the compounds of type (**11**) data indicate some measure of contributions from structures (**11b**) or (**11c**). However, the exact amount of polar or tetravalent sulfur contributors to each of the structures (**11**), (**13**) and (**31**) is a matter of some discussion ⟨71AHC(13)161, 80MI43100⟩.

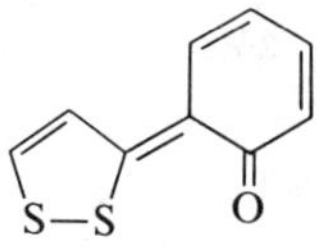

(**31a**) 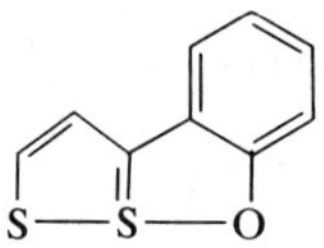   (**31b**) 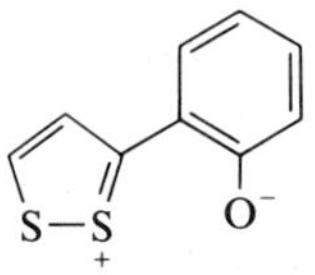   (**31c**)

Absolute values of dipole moments of several dithiol-3-ones and -3-thiones have been calculated using a Guggenheim method ⟨68ZN(B)413⟩. An apparent discrepancy between these and other values ⟨68ZN(B)1547⟩ was due to the latter using another calculation, which gave relative values. When an appropriate correction factor was applied a fairly good match of values was obtained.

Solute–solvent interactions in dithiole-thiones and -ones are attributed to localized electrostatic induction. The strength of the hetero-association complexes is determined by the distribution of the electric field around the carbonyl or thiocarbonyl group ⟨72HCA213⟩.

4.31.2 REACTIVITY

4.31.2.1 Reactivity of 1,2-Dithiolylium Salts

4.31.2.1.1 *Reactivity at the ring*

(i) *General survey*

The 1,2-dithiolylium ions should react readily only with nucleophilic reagents and at ring position 3 (or 5) to form 1,2-dithiole derivatives. These may re-form the 1,2-dithiolylium ring by loss of a leaving group, form an exocyclic double bond, or undergo ring opening (Scheme 3) (see also Section 4.02.1.6) with possible eventual recyclization.

Nucleophilic attack may occur at the least hindered position, not necessarily where there is a good leaving group. In this case, ring opened products are obtained. Other possible reaction modes include loss of a proton (at C-3), nucleophilic attack at ring sulfur, or acceptance of electrons.

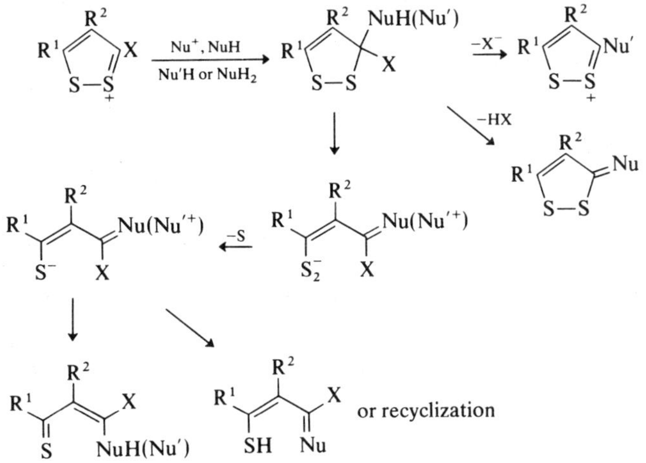

Scheme 3

(ii) Thermal and photochemical reactions

Under the conditions of mass spectrometry, the salts are converted into radicals (Section 4.31.1.4). Radicals (**18**) and anions (**32**) are also obtained by irradiation in a solvent, which provides the reductant ⟨72TL5213⟩. 3,5-Diphenyl-1,2-dithiolylium-4-olate (**33**) produces tetraphenyl-*p*-benzoquinone on thermolysis ⟨78AJC297⟩, presumably by dimerization and loss of sulfur.

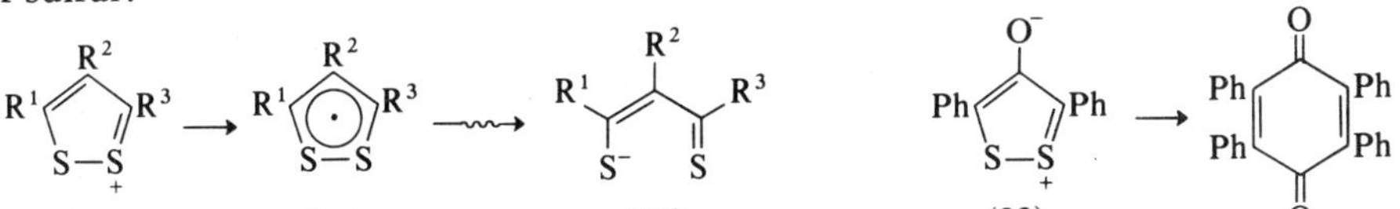

(5) (18) (32) (33)

(iii) Reactions with electrophilic reagents

1,2-Dithiolylium salts (**4**) are resistant to electrophilic attack on the ring ⟨64CR(C)-(258)6446, 74CR(C)(279)237⟩. No work on oxidation products of salts themselves is reported.

(iv) Reactions with nucleophilic reagents

These reactions form one of the most intensively studied areas of dithiole chemistry.

(*a*) *Oxygen nucleophiles.* 1,2-Dithiolylium salts (**4**) decompose rapidly in base solution (Scheme 4). The presence of an intermediate enethiol has been demonstrated ⟨65JCS32⟩. 3-Alkoxy-1,2-dithioles (**34a**) are formed by reaction with alkoxides. 3-Halo-1,2-dithiolylium salts (**35a**) are converted by alcohols or phenols into 3-alkoxy- or 3-aryloxy-1,2-dithiolylium salts (**35b, c**). All of these (**35a, b, c**) react with water or acetic acid to form 1,2-dithiol-3-ones (**3a**) or bis(1,2-dithiol-3-yl) ethers (**36a**) ⟨75ZC395, 72LA(766)1⟩.

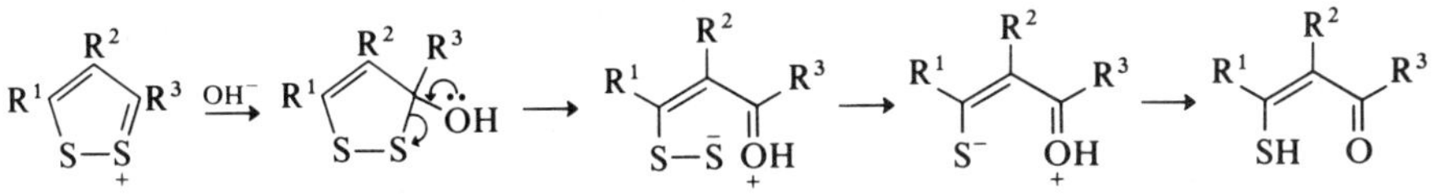

Scheme 4

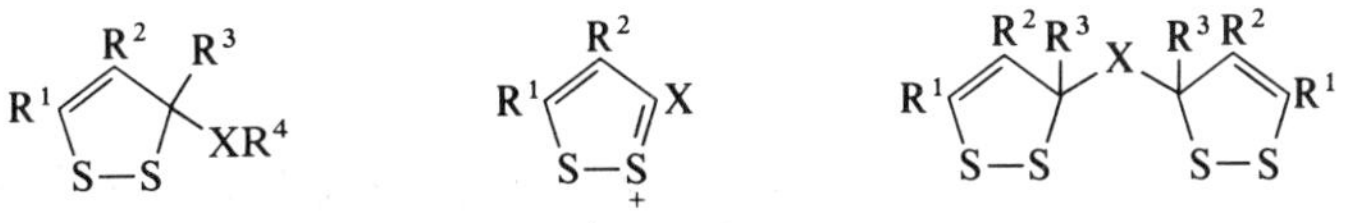

(**34**) a; X = O
b; X = S

(**35**) a; X = Hal
b; X = Oalkyl
c; X = Oaryl

(**36**) a; X = O
b; X = S

(*b*) *Nitrogen nucleophiles.* The products of these reactions depend on the nature of the dithiolylium salt (**4**), the nitrogen nucleophile, and sometimes the reaction medium. While salts bearing no replaceable substituents react with ammonia to form isothiazoles (**37**) ⟨80AHC(27)151, 66AHC(7)39⟩, primary or secondary amines form 3-aminothiones (**38a**) or aminoimines (**38b**) ⟨65JCS32⟩. Tertiary aromatic amines react as carbon nucleophiles (below).

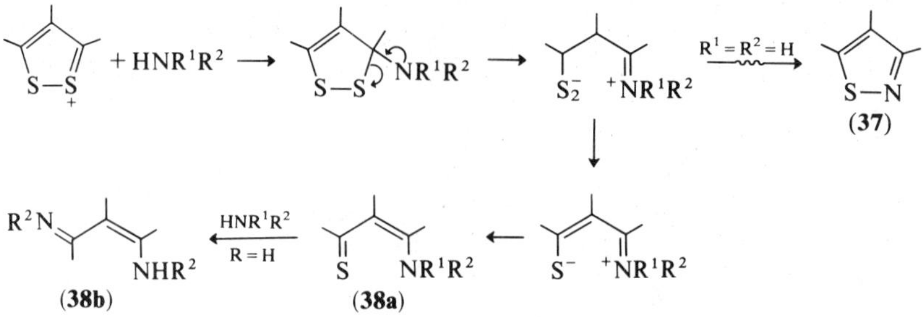

Salts with replaceable substituents at C-3 (or C-5), *e.g.* halogen, *O*-alkyl or *S*-alkyl, react with either retention or opening of the dithiole ring after initial attack at either C-3 or C-5, depending on substitution. Thus 3-alkylthio-5-phenyl-1,2-dithiolylium salts (**39a**) react mainly at C-3, while the 4-phenyl compound (**39b**) by contrast reacts at C-5 ⟨72CJC2568, 75ZC478, 72CR(C)(274)1215⟩. Simple aliphatic amines rarely give 1,2-dithiol-3-imines (**3c**). Instead, isothiazoline-3-thiones (**40**), 1,2-dithiole-3-thiones (**3b**) and aminothioamides (**41**) may result ⟨73CJC3081, 80AHC(27)151⟩. 3-Bromothio-1,2-dithiolylium bromides (**42**), prepared by reaction of 1,2-dithiole-3-thiones (**3b**) with bromine, react with primary amines to give isothiazoline-5-thiones (**43**) or 1,2-dithiole-3-thiones, depending on their substitution ⟨72CJC2568⟩. 1,2-Dithiolylium salts react with hydrazines *via* similar pathways, forming hydrazones and azines when there are replaceable substituents. Otherwise pyrazoles (**44a**) and pyrazolium salts (**44b**) result ⟨66AHC(7)39⟩ by ring opening and recyclization (see also Chapter 4.04).

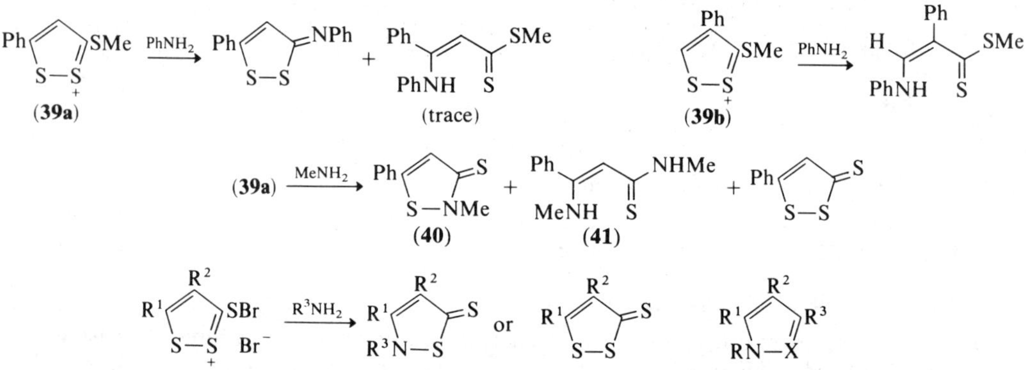

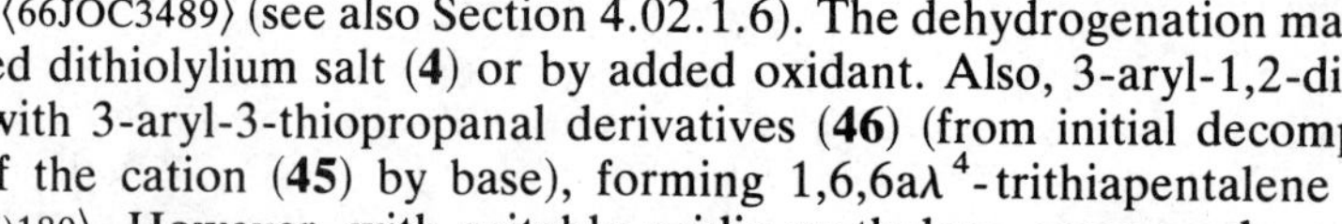

(*c*) *Carbon nucleophiles.* 1,2-Dithiolylium salts (**4**) react at the C-3 position with a large variety of α-methylene ketones or vinyl ethers to form 3-acylmethylene-1,2-dithioles (**11**) (Scheme 5) ⟨66JOC3489⟩ (see also Section 4.02.1.6). The dehydrogenation may be performed by unreacted dithiolylium salt (**4**) or by added oxidant. Also, 3-aryl-1,2-dithiolylium salts (**45**) react with 3-aryl-3-thiopropanal derivatives (**46**) (from initial decomposition of one molecule of the cation (**45**) by base), forming 1,6,6aλ^4-trithiapentalene derivatives (**5**) ⟨68CR(C)(267)180⟩. However, with suitably acidic methylene compounds, another reaction may give thiopyran-2-one (**47a**) (Scheme 6) or -2-thione derivatives (**47b**) ⟨73BSF586, 65JCS32, 80AHC(27)151⟩ and in some cases thiophene derivatives (**48**) ⟨76JPR221⟩,

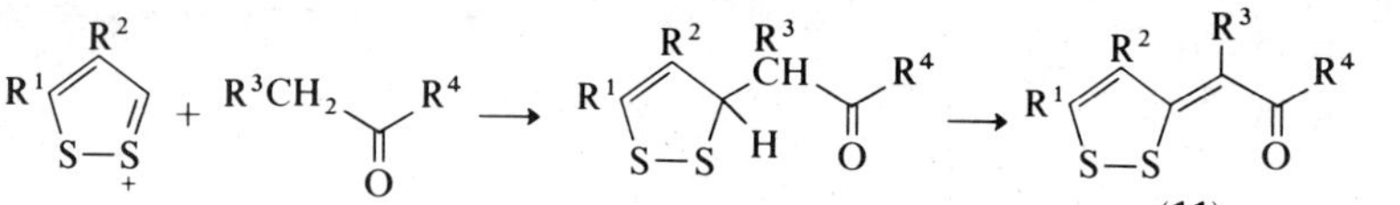

Scheme 5

(45) **(46)**

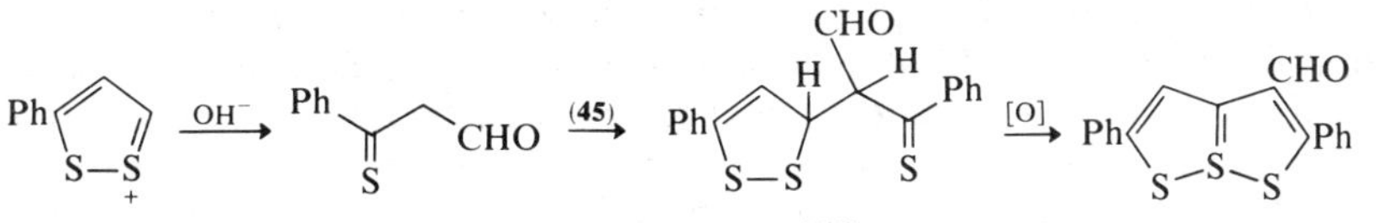

which are probably formed by a mechanism similar to that for the preparation of isothiazoles from 1,2-dithiolylium salts and ammonia.

(47)

Scheme 6

(48)

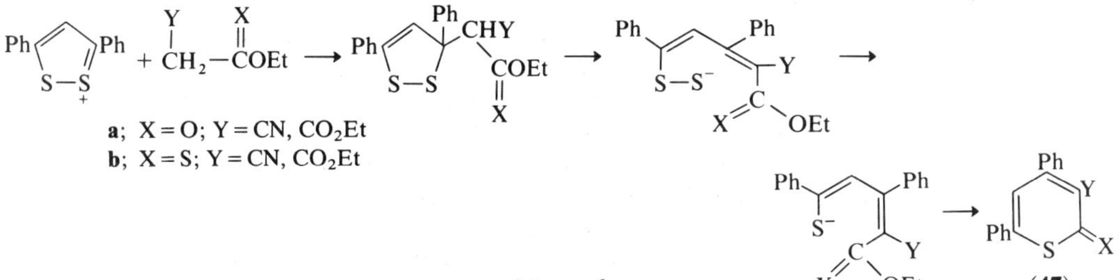

The treatment of 3-halo- (**35a**) or 3-alkylthio-1,2-dithiolylium salts (**21**) with reactive methylene compounds usually gives 3-acylmethylene-1,2-dithioles (**11**) by loss of halide or thiolate. The reaction with sodium benzoylacetate is a convenient source of 3-benzoylmethylene-1,2-dithioles (**49**) ⟨70JCS(C)1202⟩, and malonic acid gives cyanine dyes (type **9**). In some cases, however, alternate reactions give thiopyran (**50**) or thiophene derivatives (**51**) ⟨68JCS(C)642⟩. The product (**51**) is itself a rearrangement product of an initially formed cyanine (**52**). These reactions may be rationalized in terms of initial attack at C-3, followed by ring opening and recyclization (Scheme 7).

Scheme 7

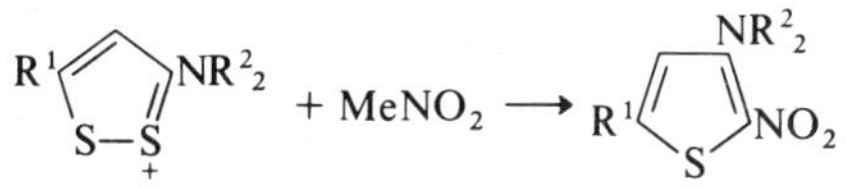

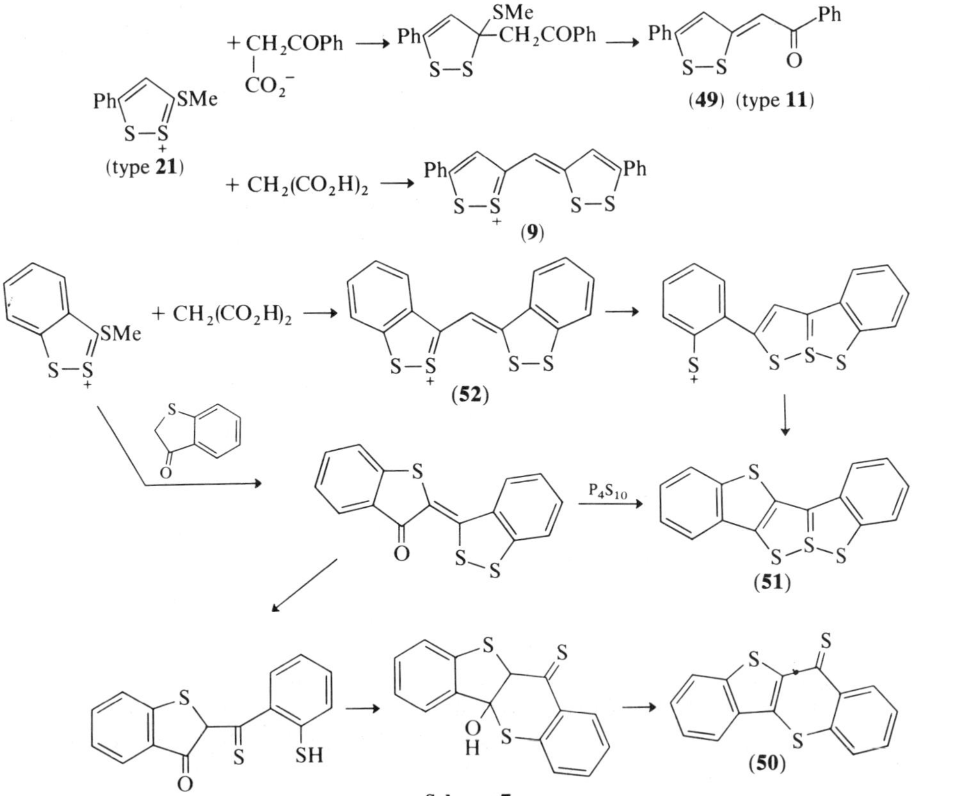

Methylene groups may be conjugatively activated, as for the preparation of the 1,2-dithiol-3-ylidenebutenenitrile (**53**). Other substrates functioning as carbon nucleophiles are phenols, and the anions of cyclopentadienes and imidazoles ⟨66AHC(7)39, 80AHC(27)151⟩. The active methylene group may be part of a dithiolylium side chain, as in the formation of 3-acylmethylene compounds (**11**) from 3-phenacylthio-1,2-dithiolylium salts (**54**) ⟨70JCS(C)1202⟩. Rather than being extruded, the sulfur may in some cases be retained in a disulfide (**55**).

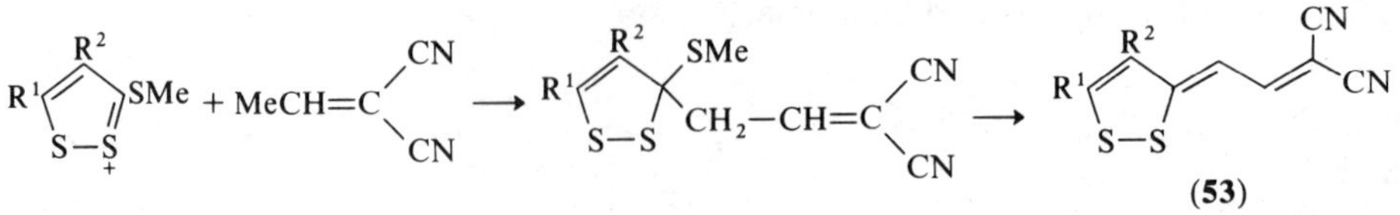

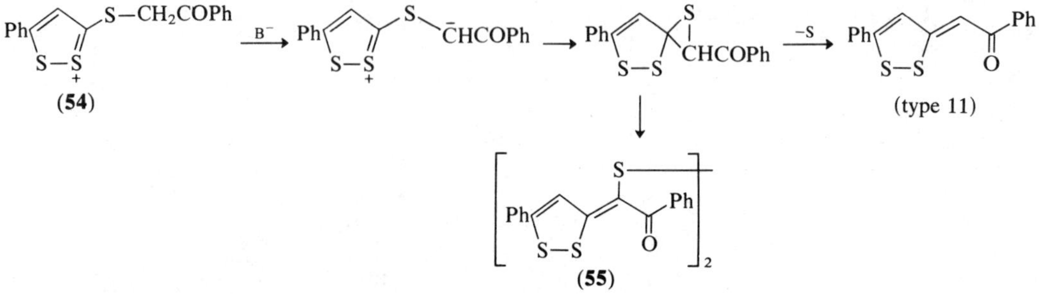

N,N-Dimethylaniline reacts (at its 4-position) on the 3-position of 1,2-dithiolylium salts (**4**), giving either new dithiolylium salts (**56**) or benzothiopyrylium salts (**57**) ⟨66AHC(7)39, 65JCS32⟩, depending on the substitution at C-3 (Scheme 8).

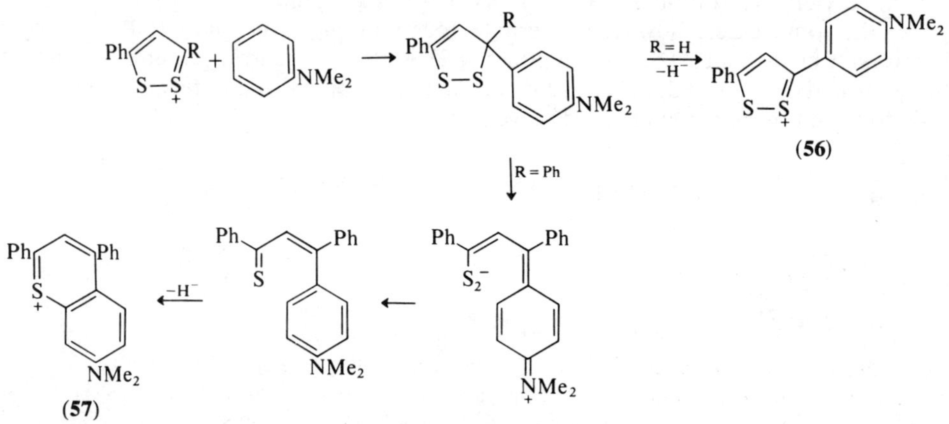

Scheme 8

(*d*) *Sulfur nucleophiles.* 1,2-Dithiolylium salts (**4**) unsubstituted at C-3 react with elemental sulfur in pyridine to give high yields of otherwise inaccessible thiones (**3b**) ⟨63JOC529⟩. With hydrogen sulfide or thiols, bis(1,2-dithiol-3-yl) sulfides (**36b**) or 3-alkylthio-1,2-dithioles (**34b**) are obtained ⟨65JCS32, 67CC353, 72JCS(P1)2305⟩. Thiols may displace halogens, forming 3-alkylthio-1,2-dithiolylium salts (**21**) or 1,2-dithiole-3-thiones (**3b**).

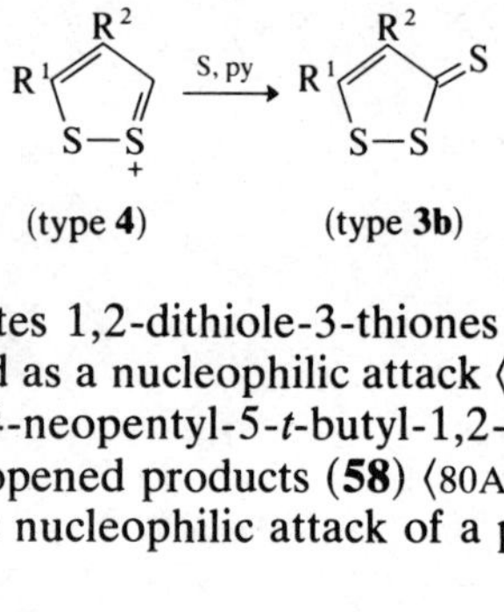

Sodium hydrosulfide regenerates 1,2-dithiole-3-thiones from their mercury(II) chloride complexes. This may be regarded as a nucleophilic attack ⟨66AHC(7)39⟩.

(*e*) *Other nucleophiles.* The 4-neopentyl-5-*t*-butyl-1,2-dithiolylium cation reacts with phosphonate esters to give ring opened products (**58**) ⟨80AHC(27)151⟩. Since this reaction is reversed by acids it is formally a nucleophilic attack of a phosphorus atom on ring sulfur.

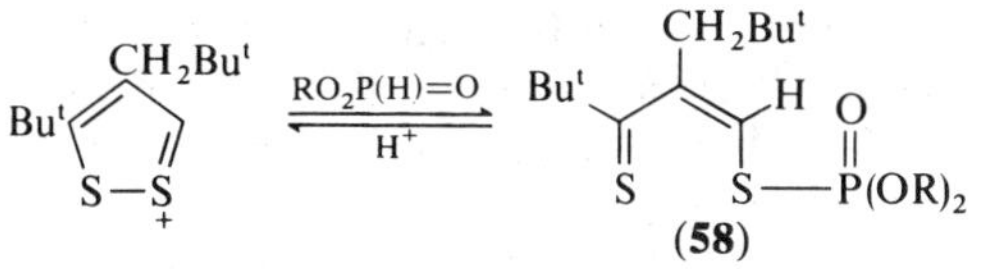

1,2-Dithioles

(58)

(v) Deprotonation reactions

3-Unsubstituted 1,2-dithiolylium salts (4) may undergo proton exchange by deuterium under various conditions ⟨66T2119, 66AHC(7)31⟩, possibly *via* carbene intermediates of type (19).

(vi) Reduction

Electrochemical reduction of the salts (4) provides radicals (18) which dimerize or undergo further reduction to anions (32) or dianions ⟨80MI43100⟩. The reduction potentials are not much affected by substituents. Reduction with zinc in aprotic conditions gives bi(1,2-dithiol-3-yls) (59), and 3-chloro-1,2-dithiolylium salts (35a; X = Cl) are converted into bi(1,2-dithiol-3-ylidenes) (20) ⟨75TL3473⟩. Divalent chromium converts the 3,5-dimethyl-1,2-dithiolylium cation into a dithioacetylacetonate ligand ⟨72AJC2547⟩. The reaction of 3,5-diamino-1,2-dithiolylium salts (8) or alkyl derivatives with thiols provides dithiomalonamides (60) by electron transfer ⟨63ACS163⟩.

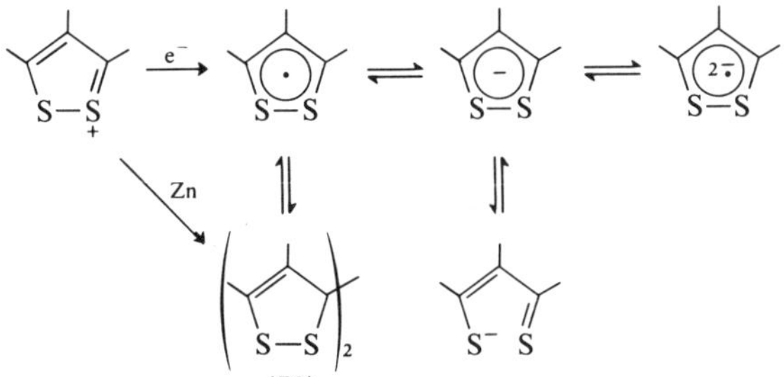

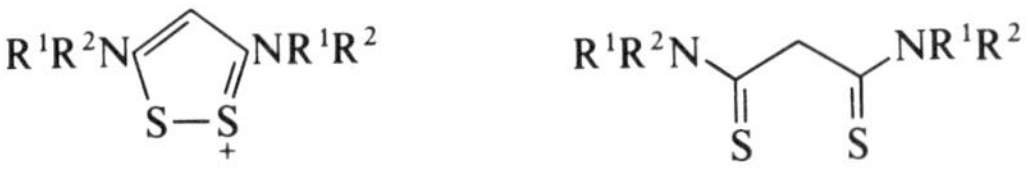

4.31.2.1.2 Reactivity of substituents

(i) General survey

In common with other cationic heterocycles, the charged ring enhances the reactivity of substituents on the ring. Thus these substituents will usually be resistant to electrophilic reagents, and be susceptible to reaction with bases or nucleophilic reagents.

(ii) Reactivity of C-linked substituents

While little work is reported on such reactions of benzo-1,2-dithiolylium salts, aryl-substituted 1,2-dithiolylium salts undergo electrophilic attack in the aromatic ring ⟨74CR(C)-(279)237, 61JA2934⟩. 3-Alkyl-1,2-dithiolylium salts (61) readily deprotonate to 3-alkylidene compounds (3d). In some cases these are stable and isolable, while in others they function as highly reactive nucleophiles because of zwitterionic contributing structures (62). Protons on methyl groups are readily exchanged *via* these forms (see Section 4.02.3.3).

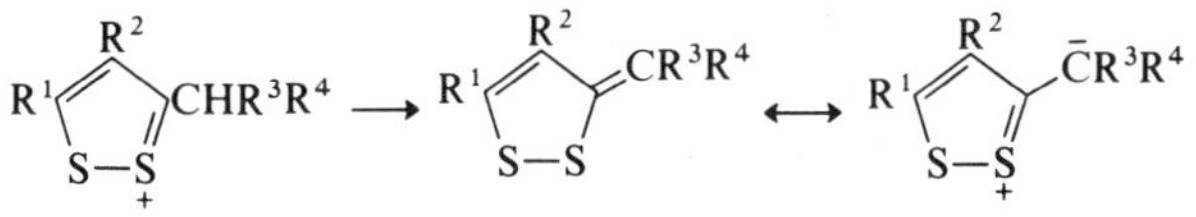

(iii) Reactivity of N-linked substituents

Both 3-amino- and 3,5-diamino-1,2-dithiolylium salts (8) may be readily deprotonated to iminodithioles (3c). Reprotonation may require fairly strong acids, and acylation may be performed with acyl halides (Section 4.02.3.5).

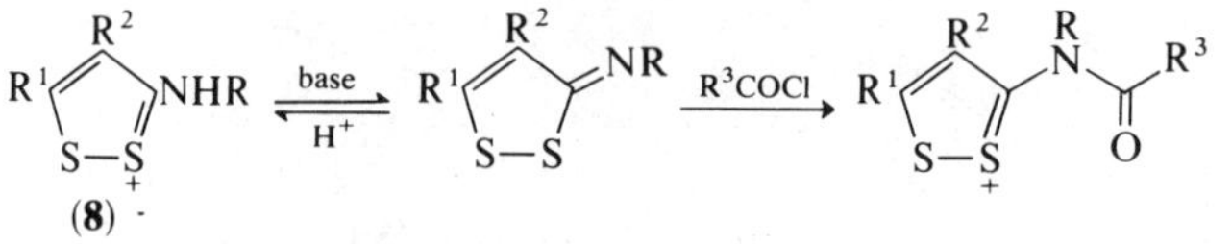

(iv) Reactivity of O-linked substituents

3-Hydroxy-1,2-dithiolylium salts (**63**) are the protonated forms of 1,2-dithiol-3-ones (**3a**), which are readily obtained on deprotonation. Alkylation of the compounds (**3a**) may be effected with strong alkylating agents and nucleophilic attack on the resulting 3-alkoxy-1,2-dithiolylium salts (**35b**) may occur at the alkyl group or on the ring ⟨75ZC478⟩. The mesoionic 1,2-dithiolylium-4-olates (**7**) readily alkylate or protonate on oxygen, forming 4-alkoxy (**64a**) or 4-hydroxy-1,2-dithiolylium salts (**64b**) which are readily isolated and fairly stable ⟨79BSF(2)26⟩.

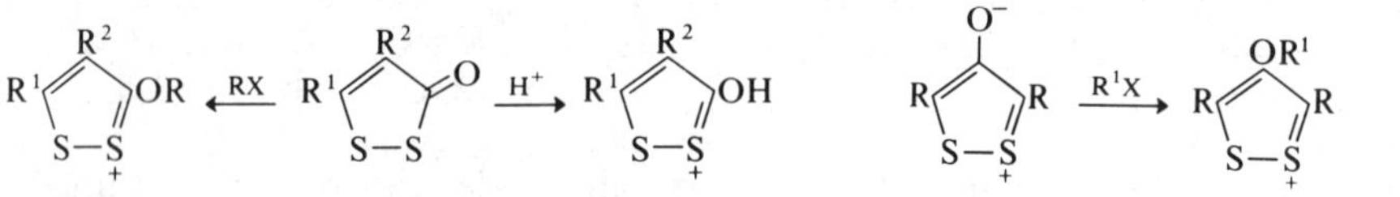

(v) Reactivity of S-linked substituents

3-Alkylthio-1,2-dithiolylium salts (**21**) are made by similar methods to the salts (**35b**) and have similar reactivity, except that, in addition, the exocyclic sulfur may stabilize an adjacent carbanion (see Section 4.02.3.8).

4.31.2.2 Reactivity of 1,2-Dithioles and 1,2-Dithiolanes

4.31.2.2.1 Reactivity at the ring

(i) General survey

By analogy with related heterocycles, these compounds would be expected to exhibit reactions of acyclic disulfides, e.g. reduction and nucleophilic attack. For the dithioles (**3a,b,c**), nucleophilic attack should also occur at C-3, or at C-5, the conjugate position.

(ii) Thermal and photochemical reactions

1,2-Dithiol-3-ones (**3a**) and -3-thiones (**3b**) are often synthesized under thermal conditions and frequently may be distilled without decomposition ⟨65CRV237⟩. Heating some 1,2-dithiol-3-ones (**3a**) with copper powder gives a variety of other heterocycles, but with loss of some sulfur content ⟨74T4113⟩. Monocyclic 1,2-dithiol-3-imines (**3c**) decompose on heating, forming small amounts of 1,2-dithiole-3-thiones ⟨72CJC2568⟩, but benzo derivatives (**65**) are in equilibrium with 2,1-benzisothiazoline-3-thiones (**66**) at elevated temperatures ⟨29JCS1582⟩.

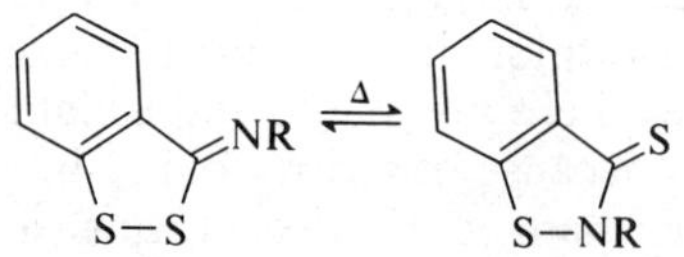

Several thieno[3,2-b]thiophenes (**67**) are produced by pyrolysis of various 1,2-dithioles, including bi(1,2-dithiol-3-ylidenes) (**20**), alkali metal salts of 1,2-dithiole-3-sulfonyl hydrazides (**68**) ⟨75TL3473⟩, and tetrahydrobenzo-1,2-dithiol-3-ones (**69**) ⟨63BSF161⟩.

1,2-Dithioles

1,2-Dithiolane (**1**) itself is unstable, and easily polymerizes ⟨54JA4348⟩, but photolysis of lipoic acid (**29**) in the presence of aldehydes gives a product (**70**) by trapping of radicals ⟨76CC92⟩.

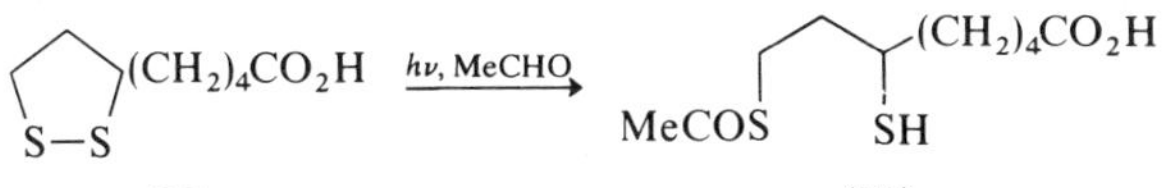

The photolysis of 4-phenyl-1,2-dithiole-3-thione (**71**) gives a dimer (**72**) ⟨73TL1561⟩, but benzo-1,2-dithiolane 2,2-dioxide (**73a**) gives an o-thioquinone methide (**74**) ⟨78JOC3374⟩.

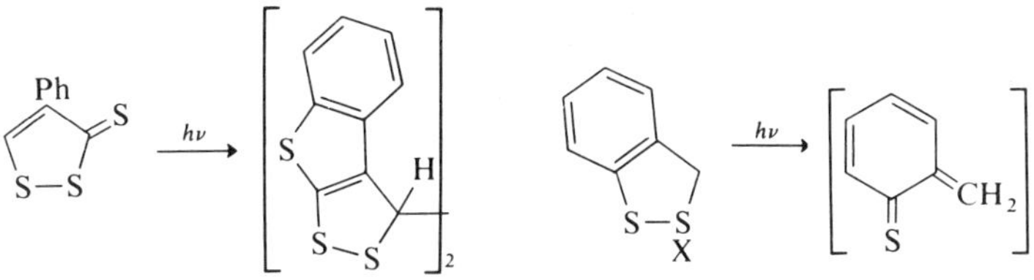

(iii) Reaction with electrophilic reagents

The reaction of 1,2-dithiole-3-thiones (**3b**) with halogens initially gives 1:1 adducts which may then convert into 3-halo-1,2-dithiolylium salts (**35a**) ⟨76PS(1)185⟩. Under certain conditions the ring may also be halogenated in the 4-position ⟨65CRV237, B-67MI43101⟩. 4-Benzoylamino- or 4-thiobenzoylamino-1,2-dithiole-3-thiones (**75a,b**) react with methyl iodide to form oxazoles or thiazoles (**76a,b**). Two possible mechanisms formally involve electrophilic attack on ring sulfur (Scheme 9) ⟨66AJC503⟩. Hydride ion acceptors convert 1,2-dithioles (**2**) and their 4,5-benzo derivatives into the corresponding dithiolylium salts ⟨66AHC(7)39, 62JA2941⟩.

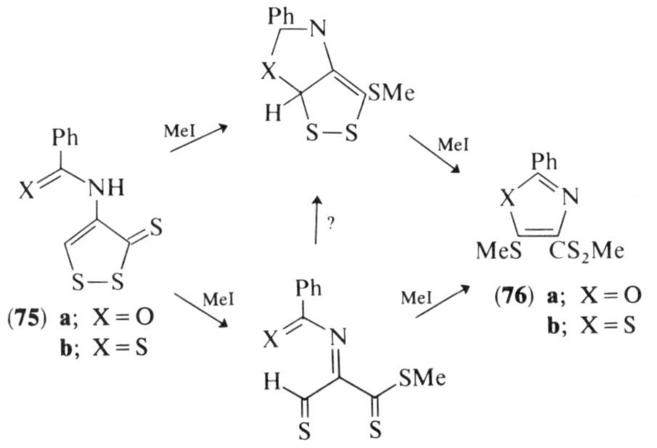

Scheme 9

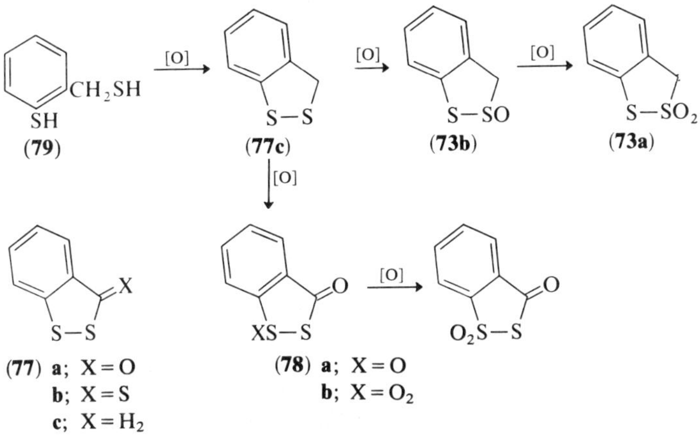

Scheme 10

The oxidation of benzo-1,2-dithiol-3-one (**77a**) with peracids provides a benzo-1,2-dithiolone S-oxide (**78a**) and a dioxide (**78b**). The former is produced by oxidation of 2-mercaptomethylbenzenethiol (**79**), presumably by oxidation of intermediate benzo-1,2-dithiole (**77c**) ⟨78JOC3374⟩. The oxides (**73a,b**) are also formed by oxidation of the dithiole (**77c**) using an appropriate oxidant (Scheme 10). Benzo-1,2-dithiole-3-selenones (**80a**) give rearranged products (**80b**) on oxidation ⟨74H(2)45⟩. Strong oxidation of 1,2-dithioles with nitric or chromic acid causes ring rupture ⟨65CRV237⟩.

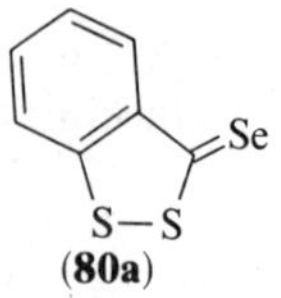

Various 1,2-dithiole derivatives (**3**) undergo electron transfer under chemical or electrochemical conditions, forming various types of 1,2-dithiolylium cation ⟨80MI43100, 80AHC(27)151⟩. These include monocyclic and highly condensed derivatives.

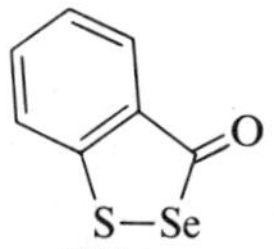

(iv) Reaction with nucleophilic reagents

1,2-Dithiole derivatives (**3a,b,c**) react with hydroxide ion at C-3 to form products derived from ring cleavage. Eventually all sulfur atoms may be lost and β-ketoacid derivatives are obtained (Scheme 11) ⟨65CRV237⟩. However, the intermediate thiolate anion is a good nucleophile and may be intercepted, or, in the case of chlorinated dithioles, react with starting material to give a variety of other heterocycles ⟨65LA(683)132, 75LA1513⟩. 1,2-Dithiolanes undergo ring fission by attack at sulfur ⟨72JOC369⟩.

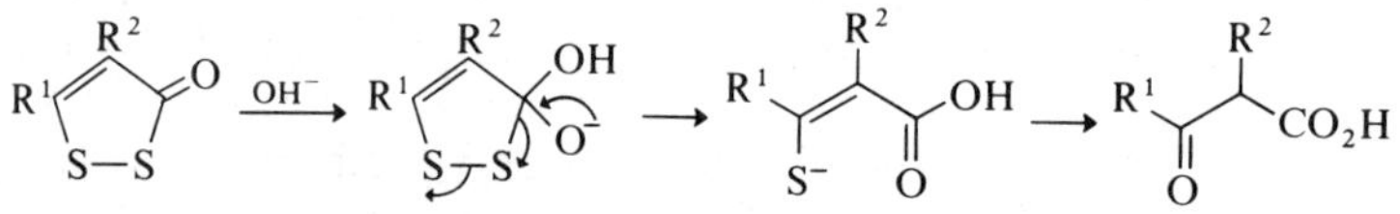

Scheme 11

Various sulfur nucleophiles appear to function as reducing agents, converting the dithioles to thiolate ions which may be intercepted. These reactions may involve an electron transfer or alternatively a nucleophilic mechanism, e.g. Scheme 12 ⟨73BSF1973⟩.

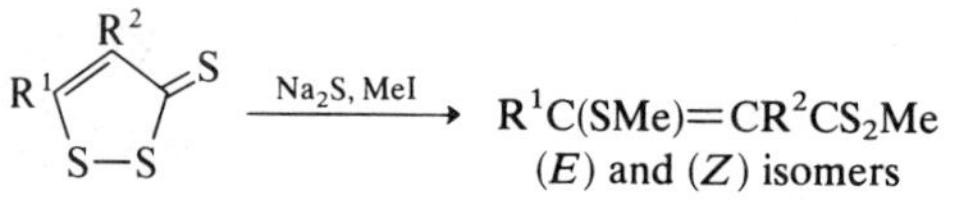

$$R^1C(SMe)=CR^2CS_2Me$$
(E) and (Z) isomers

Scheme 12

Reactions with nitrogen nucleophiles have been intensively studied. Monocyclic 1,2-dithiol-3-ones (**3a**) and -3-thiones (**3b**) react with amines, hydroxylamine, hydrazine, etc. to give imines, oximes, hydrazones etc. ⟨65CRV237⟩. While benzo-1,2-dithiol-3-ones (**77a**) and -3-thiones (**77b**) may react similarly, ring opening reactions are also observed ⟨29JCS1582⟩. Thus 1,2-benzisothiazoline-3-thiones (**66**) are formed by reaction with amines, and diamines give o-thioquinone methides (**81**) or fused isothiazoles (**82**) ⟨79BCJ3640, 78CB2716⟩. When aromatic amines or benzenethiol react with a benzo-1,2-dithiol-3-ylideneselenonium ylide (**83**), complex mixtures of products result from modification of the ring and/or the side chain ⟨74TL1059⟩.

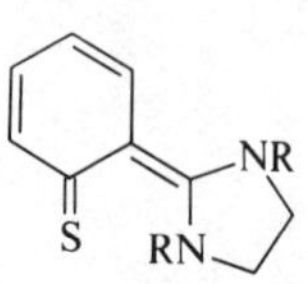

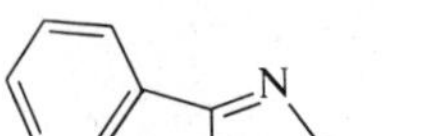

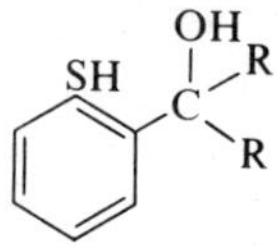

Carbon nucleophiles may react at the sulfur–sulfur bond, or at C-3. Thus Grignard reagents react with monocyclic dithiol-3-ones (**3a**) and -3-imines (**3c**) at S-2 ⟨74LA1261, 73LA247⟩. The product (**84**) of reaction of a Grignard reagent with benzo-1,2-dithiol-3-one (**77a**) ⟨74T4113⟩ could be alternately formed by attack at S-2, at C-3 or at C-5. Phosphonium ylides or other stabilized carbanions react at the 3-position of dithiol-ones

(**3a**) or -thiones (**3b**), forming 3-alkylidene- (**3d**) or 3-acylmethylene-1,2-dithioles (**11**) (Scheme 12a) ⟨78SC315, 71AHC(13)161, 66AHC(7)39⟩.

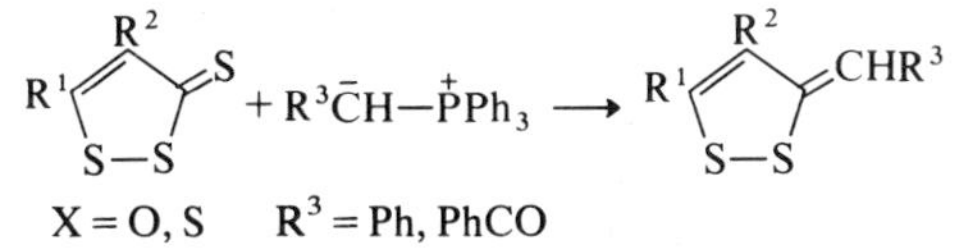

Scheme 12a

Enamines treated with 5-unsubstituted 1,2-dithiole-3-thiones (**3b**) form 2*H*-thiopyran-2-thiones (**85**) ⟨77BSF1142⟩ in a reaction which involves initial attack at C-5 of the dithiole (Scheme 13), and anions of acenaphthenone react at the 5-position of dithiolethiones, eventually forming acenaphthothiopyran-2-thione derivatives ⟨73BSF3334⟩.

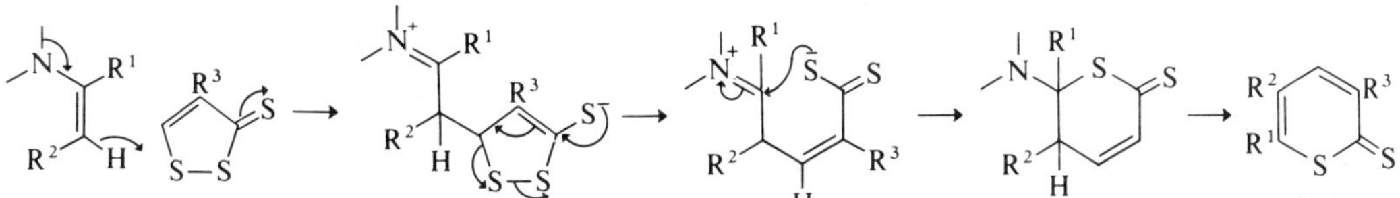

(**85**)

Scheme 13

Trivalent phosphorus compounds cleave dithiolone (**3a**) rings, giving products described as polythioesters ⟨76CA(85)159954⟩, yet β-thioketenes (**86**) result from simple desulfurization reactions ⟨76TL2961⟩ and benzo-1,2-dithiolanes (**77c**) give dimers (**87**) of intermediate *o*-thioquinone methides (**74**) ⟨70JOC3259⟩. Lipoic acid (**29**) derivatives and bi(1,2-dithiol-3-ylidene) (**20**) derivatives give the thietes (**88**) and (**89**), by ring contraction reactions ⟨70JOC3259, 75TL3473⟩.

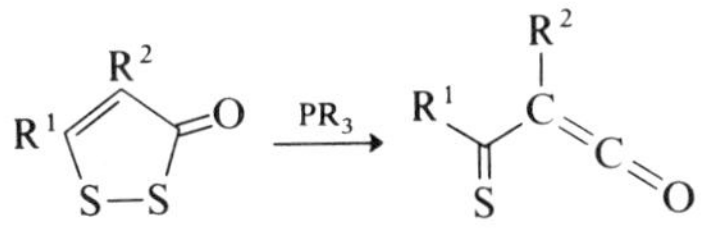

(**86**)

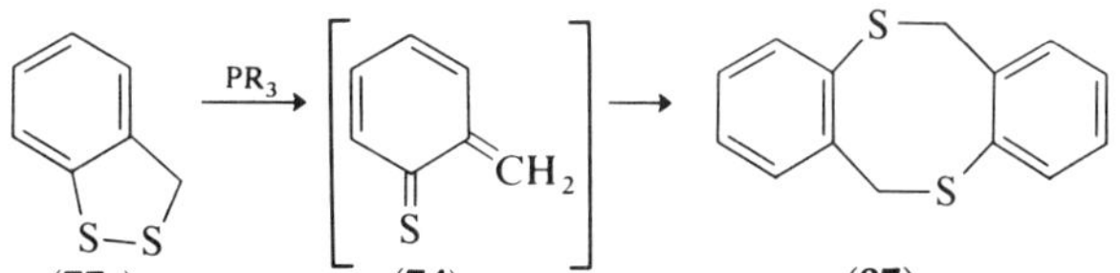

(**77c**) (**74**) (**87**)

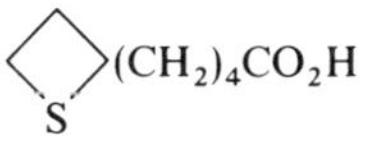

(**88**)

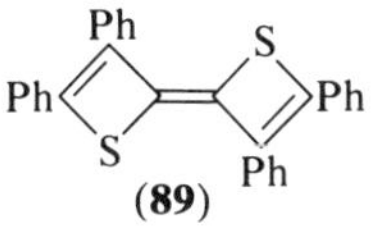

(**89**)

The reduction of benzo-1,2-dithiole-3-thiones (**77b**) to various polythiols ⟨74ACS(B)827, 63LA(661)84⟩ may be accomplished by lithium ⟨78BEP867155⟩, lithium aluminum hydride, or by catalytic hydrogenation, and Raney nickel hydrodesulfurization gives products by complete elimination of sulfur ⟨B-66MI43100⟩.

(*v*) *Reaction with carbenes and nitrenes*

5-Phenyl-1,2-dithiol-3-one (**90a**) reacts with diphenyldiazomethane or ethyl azidoformate to form a thiophene or an isothiazole by insertion and loss of sulfur, but 5-phenyl-1,2-dithiole-3-thione (**90b**) reacts at the thione function to form the dithioles (**92a**) and (**92b**), possibly *via* three-membered rings (**91a, b**) (Scheme 14) ⟨74T4113, 76CJC3879⟩. Similarly, diazoketones react with thiones (**3b**) to form 3-acylmethylene-1,2-dithioles (**11**) ⟨64CI(L)461, 71AHC(13)161⟩. Naphtho[1,8-*cd*][1,2]dithiole (**26**) forms a dithiepin (**93**) with diphenyldiazomethane ⟨72BCJ960⟩, and α-thiocarbene formation is possibly involved in the reaction of the thione (**90b**) with phenylacetylene and sulfur to form the spiran (**94**) ⟨72CC540⟩.

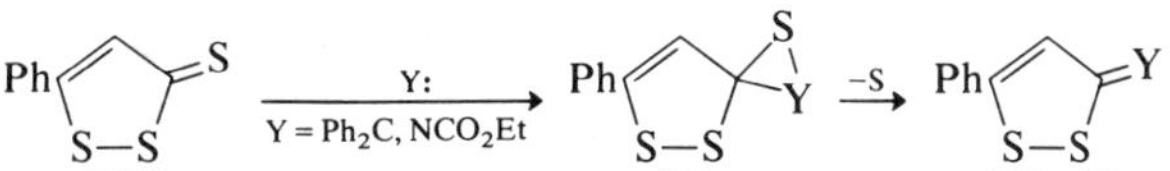

(**90b**) (**91a, b**) (**92a, b**)

Scheme 14

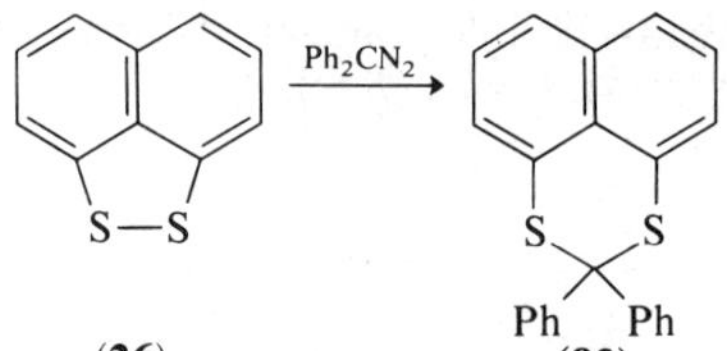

(vi) Reactions involving cyclic transition states

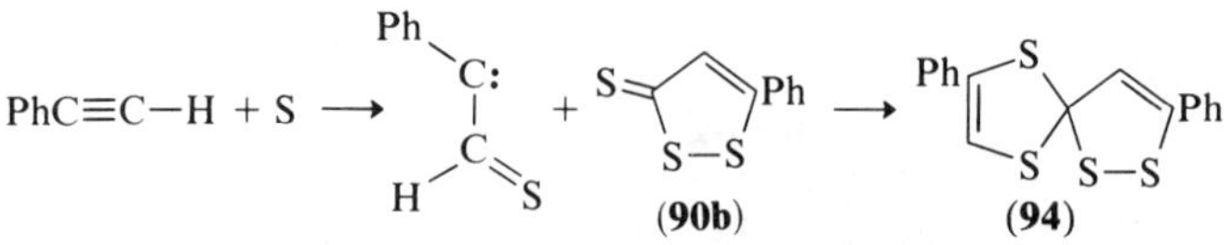

Most of this work has centered on the reaction of 1,2-dithiole-3-thione (**3b**) derivatives with alkenes or alkynes, under photochemical or thermal conditions respectively. However, in some cases dithiol-ones (**3a**) or -imines (**3c**) have been used. The investigations of a large number of workers on dithiolethione/alkyne reactions are summarized in Scheme 15. The thiones first react thermally with the alkynes to form 2-thioacylmethylene-1,3-dithioles (**95**) ⟨69CJC2039, 72JCS(P1)41⟩. These may be isolated, or react further to form diadducts (**96**) when activated alkynes are used. Under appropriate conditions, thials (**95**; $R^2 = H$) may be isolated ⟨75BSF1435⟩; otherwise these also form bis-1,3-dithiol-2-ylidenebutenes (**97**) *via* dimers. Sometimes 1,6,6aλ^4-trithiapentalenes (**5**) are obtained, possibly by rearrangement of the thiones (**95**) in a reaction catalyzed by sulfur, or sulfur-containing reagents ⟨72CC540, 71AHC(13)161, 76BSF120⟩, or by direct reaction of thiones (**3b**) with sulfur and an alkyne.

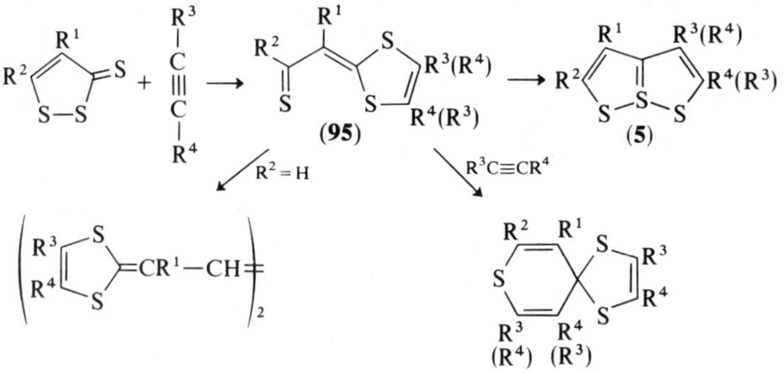

Scheme 15

Benzo-1,2-dithiole-3-thione (**77b**) also reacts, but the monoadducts corresponding to (**95**) have *o*-thioquinone methide structures (*cf.* structure **104**) and form diadducts rapidly. The use of dehydrobenzene (benzyne) gives 2-thioacylmethylenebenzo-1,3-dithioles (**95**; $R^3, R^4 = (CH=CH)_2$).

3-Arylimino-1,2-dithioles (**98**) react in analogous fashion with alkyne esters, forming mono- (**99**) and di-adducts (**100**) ⟨69CJC2039⟩, and the 1,2-dithioles (**3a, b, c**) react with ynamines to give 2-thioacylmethylene-oxathioles, -1,3-dithioles or -thiazoles (**101a, b, c**) ⟨80BSF(2)530⟩.

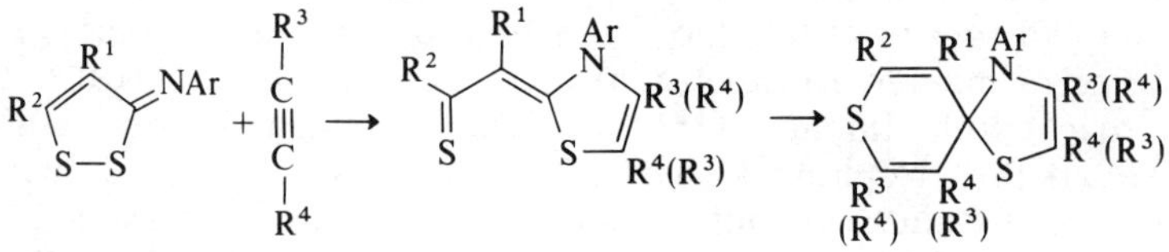

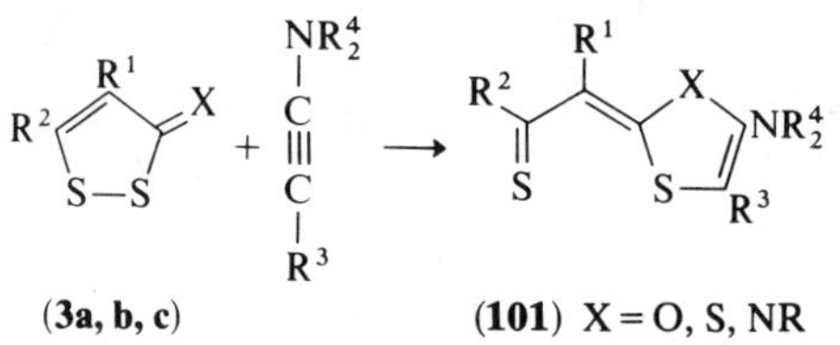

(**101**) X = O, S, NR

The reactions with alkenes appear to be similar. Thus thiones (**3b**) react thermally with alkenes to form 3-thioacylmethylene-1,3-dithiolanes (**102**) ⟨77ZOR2012⟩, but photochemically to form 1,2-dithiolespirothietes (**103**) and side products ⟨78ZOR2459⟩. These spiran types may also be produced in light-catalyzed reactions with alkynes ⟨80ZOR883⟩. Benzo-1,2-dithiole-3-thione (**77b**) and analogous compounds react photochemically with simple alkenes, giving unusual *o*-thiobenzoquinone methides (**104**) which are in equilibrium with dimers ⟨77CJC3763, 75JCS(P1)270⟩.

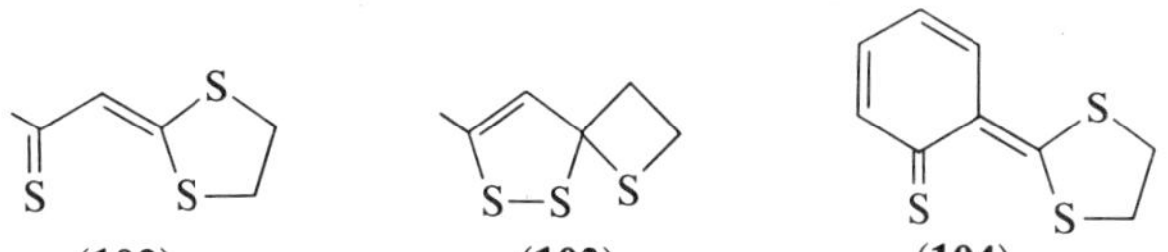

(**102**)   (**103**)   (**104**)

4.31.2.2.2 Reactivity of substituents

(i) General survey

The 1,2-dithioles (**3a, b, c, d**) would be expected to protonate on the exocyclic atom to generate substituted 1,2-dithiolylium salts (**4**) and should thus undergo electrophilic attack at this atom. Also, protons on 5-alkyl derivatives of the dithioles (**3a, b, c**) should be acidic because of conjugative stabilization of an anion and thus undergo typical condensation reactions, *e.g.* with carbonyl compounds.

(ii) C-Linked substituents

As mentioned above (Section 4.31.2.1), 3-alkylidene-1,2-dithioles (**3d**) may possess considerable ionic character consistent with contributions from structures (**62**), and condense with suitable carbonyl or other compounds. Condensation reactions of 3-alkyl-1,2-dithiolylium salts (**105**) ⟨66ZC321, 64CI(L)461, 80AHC(27)151⟩ are readily explained in terms of initial proton transfer from the salt to the substrate, as in the formation of salts of type

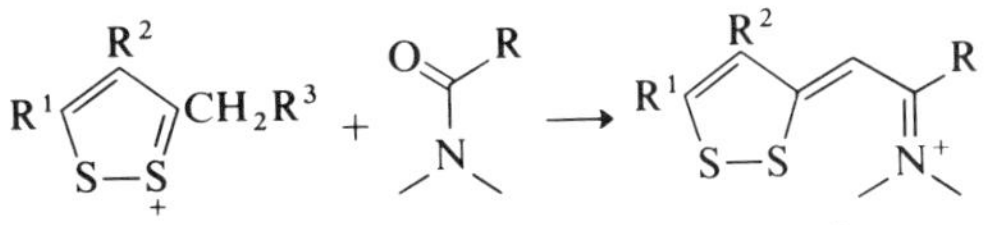

(**105**)   (**106**)

(**106**) from reaction with amides. Interesting products (**9**) and (**107**) are obtained when another 1,2-dithiolylium salt acts as a substrate for the attack at C-3 (Scheme 16) ⟨68JCS(C)642, 77JCS(P1)1511⟩, and ready self-condensation of 5-alkyl-3-alkylthio-1,2-dithiolylium salts is described ⟨64CB1886⟩. The compounds (**3d**) may also react with nitrous acid, or benzenediazonium chloride, forming the dithiole types (**12**) or (**108**) ⟨74JCS(P1)722, 76JCS(P1)228⟩.

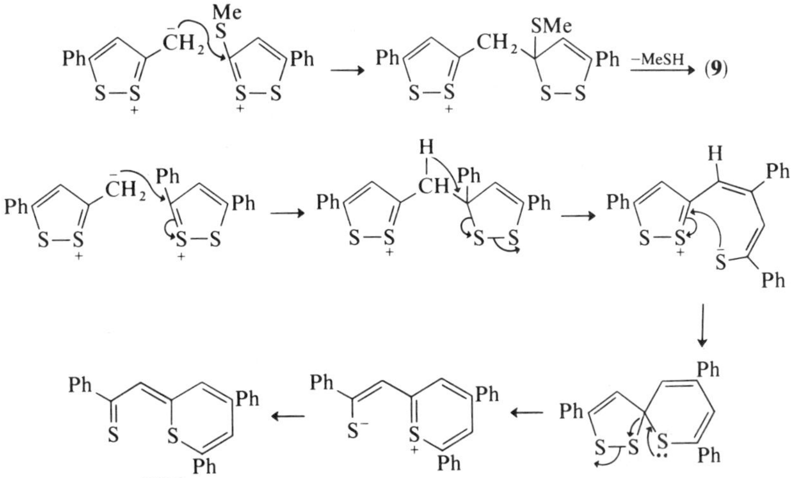

Scheme 16

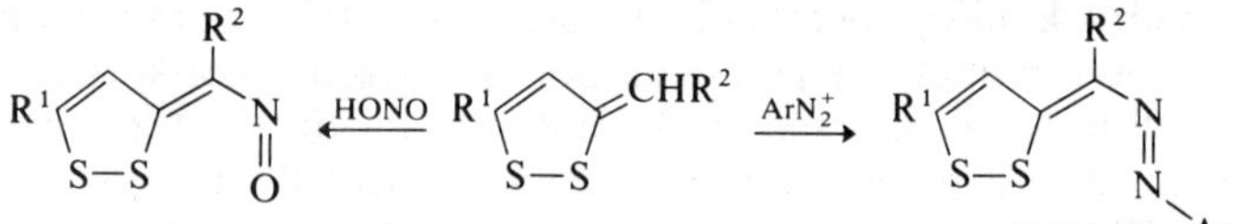

5-Alkyl groups in dithioles (**3a, b**) possess acidic protons and condense with aldehydes to form 5-vinyldithioles (**109**). This type of reactivity has been widely used in the preparation of some 1,6,6aλ^4-trithiapentalenes (**5**) or precursors ⟨71AHC(13)161⟩ by reaction with carbon disulfide (Scheme 17).

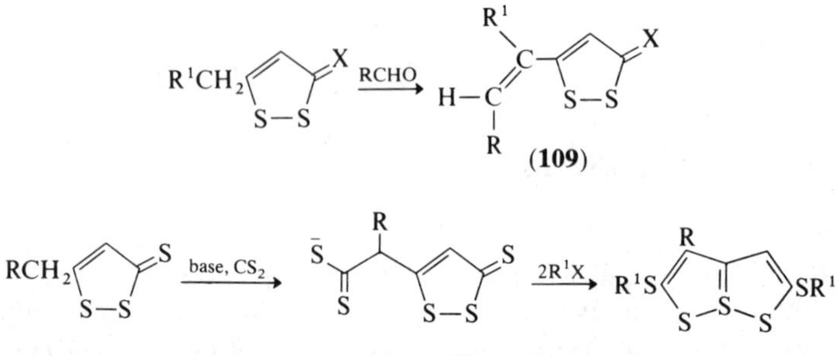

Scheme 17

Tetrahydrobenzo-1,2-dithioles are aromatized on heating with sulfur, and triphenylmethyl fluoroborate converts a tetrahydrobenzodithiole (**110**) (or a trithiapentalene) into a benzodithiolylium salt (**111**) ⟨69CC83⟩. The corresponding ketone reacts similarly.

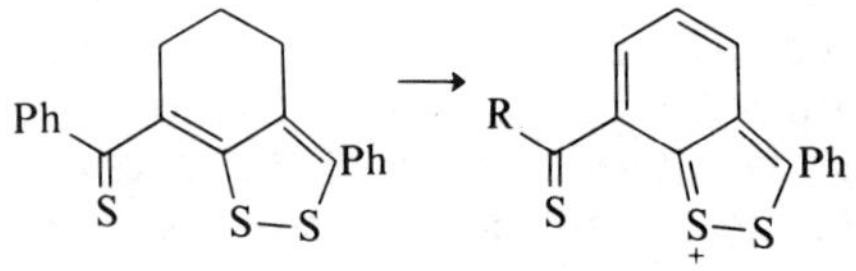

A number of 1,2-dithiolecarboxylic acid derivatives are known. These appear to have typical carboxylic acid reactivity and form acid halides, esters, amides, *etc.* ⟨67ZC275, 72BSF1840⟩, and undergo internal Friedel–Crafts acylation with aromatic substituents ⟨72BSF1840, 73BSF721⟩.

(iii) N-Linked substituents

Aminodithioles, available by a number of methods ⟨66JPR(31)214, 64AJC447⟩, undergo typical reactions including acylation, nitrosation and reaction with isothiocyanates ⟨65LA(681)178, 65AJC61⟩.

(iv) O-Linked substituents

1,2-Dithiole-3-ones (**3a**) readily react with triethyloxonium fluoroborate, forming 3-ethoxydithiolylium salts (**35b**), or with acetic anhydride/perchloric acid to form 3-acyloxy-1,2-dithiolylium salts (**112**) ⟨69ZN(B)577, 65LA(688)150⟩. Thionation with phosphorus pentasulfide is a convenient method for conversion into 1,2-dithiole-3-thiones (**3b**) ⟨65CRV237⟩, although silicon and boron sulfides may also be used.

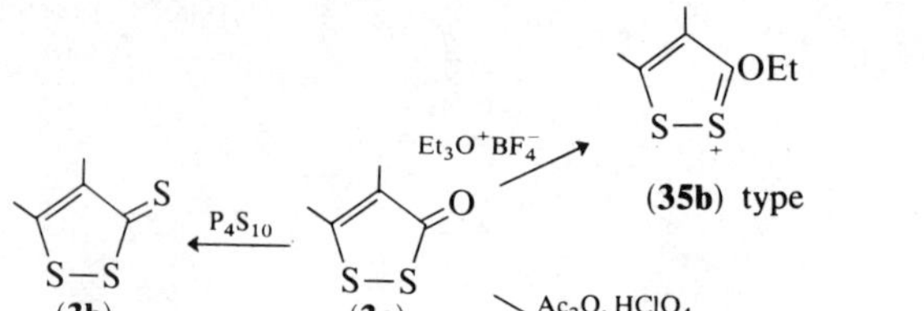

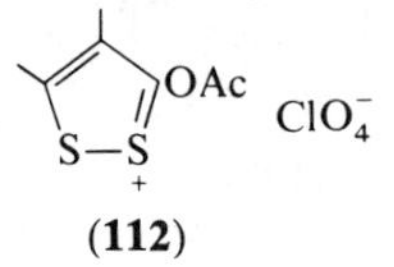

3-Alkoxy-1,2-dithioles (**34a**) and bis(1,2-dithiol-3-yl) ethers (**36a**) are pseudobase forms of 1,2-dithiolylium salts, which they regenerate on acidification ⟨65JCS32⟩. However, the dithiole (**113**) readily loses sulfur to form an acyclic product (**114**). Like the 1,2-dithiole-3-thiones (**3b**) below, 1,2-dithiol-3-ones (**3a**) are readily converted into 3-halo-1,2-dithiolylium salts (**35a**).

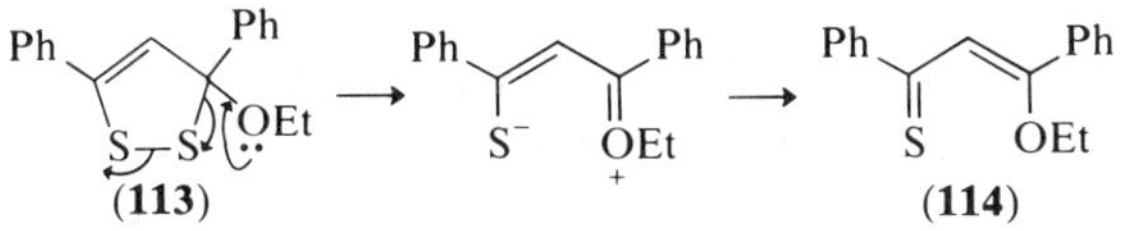

(v) S-Linked substituents

The synthetically important conversion of the readily accessible thiones (**3b**) to 3-unsubstituted dithiolylium salts (**4**) has wide scope ⟨66AHC(7)39, 80AHC(27)151⟩ but fails for benzo-1,2-dithiole-3-thione (**77b**) and for 1,2-dithiole-3-thiones with electron withdrawing groups. These are converted to the corresponding dithiolones (**3a**) instead. Three equivalents of an oxidizing agent, e.g. hydrogen peroxide, are consumed in the reaction, yet the cations (**4**) are at a lower oxidation state than the thiones (**3b**). A plausible reaction sequence is shown (Scheme 18) involving sequential oxidation of thiones to sulfines, sulfenes and conversion to carbenes which are protonated to form the cations (**4**). Certainly sulfines have been identified in photo-oxygenation reactions ⟨80CC598⟩ and in related heterocyclic thione oxidations. The thione sulfur is eventually converted to sulfate. Possibly protonated forms of the above sulfines and sulfenes are also implicated (Section 4.01.3.6). The conversion to dithiolones may also be accomplished using potassium permanganate, mercury(II) acetate or aqueous halogen solutions ⟨76PS(1)185⟩.

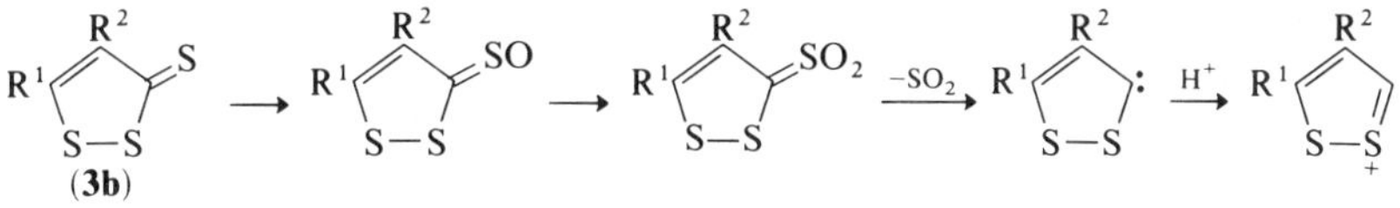

Scheme 18

The thiones (**3b**) are readily alkylated by alkyl halides, forming 3-alkylthio-1,2-dithiolylium salts (**21**; R = alkyl). Aryl halides are usually unreactive but do react under light catalyzed conditions to form 3-arylthio-1,2-dithiolylium salts (**21**; R = aryl) ⟨79ZOR1069⟩.

Molecular halogens react with the thiones to give initially 1:1 adducts, formulated probably as 3-halothio-1,2-dithiolylium halides (**42**). For chlorine, these types convert to 3-chloro-1,2-dithiolylium salts (**35a**; X = Cl) ⟨81AHC(23)151, 76PS(1)185⟩. Other more easily controlled halogen sources may be used, including thionyl chloride, oxalyl chloride, sulfur chloride and phosphorus oxychloride ⟨70LA(742)103⟩. A possible pathway for reaction with oxalyl chloride is shown in Scheme 19.

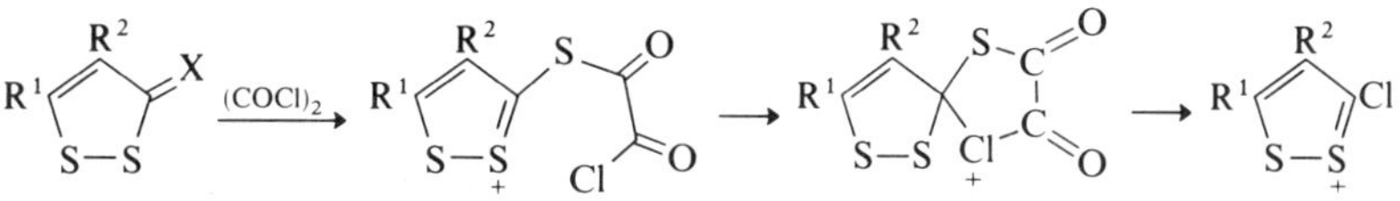

Scheme 19

4.31.3 SYNTHESIS

4.31.3.1 Direct Synthesis

4.31.3.1.1 Direct synthesis by formation of one bond

(i) *Between two sulfur atoms*

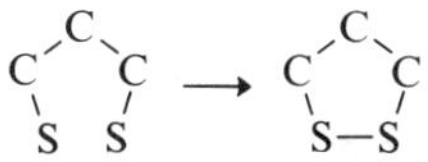

The preparation of 1,2-dithiole derivatives by oxidation of suitable dithiols is quite general, and may be applied to the preparation of 1,2-dithiolane (**1**) ⟨69JOC36⟩, benzo-1,2-dithiole (**77c**) ⟨78JOC3374⟩ and naphtho[1,8-cd][1,2]dithiole (**26**) ⟨65JOC3997⟩ from the corresponding dithiols. Propane-1,3-dithiols form 1,2-dithiole-3-thiones (**3b**) by reaction with sulfur at elevated temperatures ⟨61AG220⟩.

β-Dithio compounds or related compounds form 1,2-dithiolylium salts (**4**) on oxidation in acidic media. Thus dithiomalonamides (**60**) are converted to 3,5-diamino-1,2-dithiolylium salts (**8**) ⟨76JCS(D)455⟩ and 2-thiocyanatoaryl thiones or thioamides (**115**) on acidification form various salts (**116**) possessing the benzo-1,2-dithiolylium nucleus ⟨79TL3339⟩. 1,3-Dithiones and functional derivatives are probably intermediates in some sulfurization reactions (Section 4.31.3.1.3) but are not usually isolated.

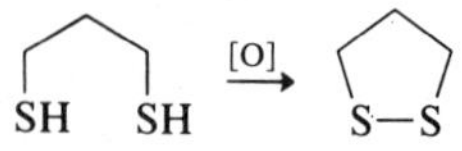

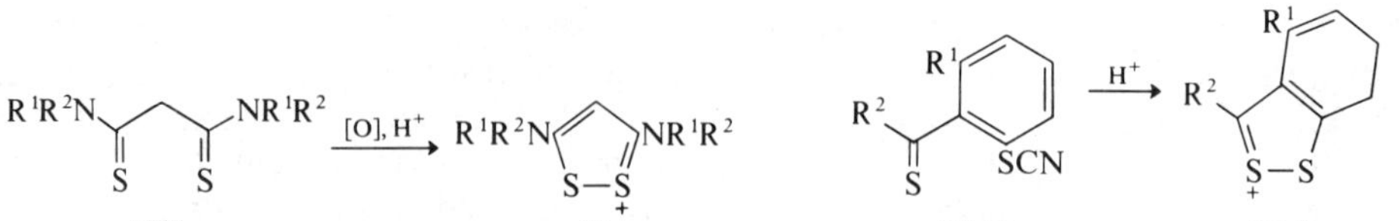

(ii) Between a sulfur and a carbon atom

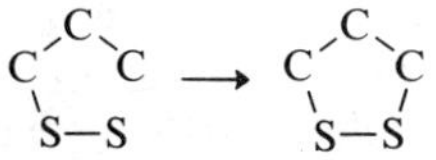

2-Mercaptoacrylic esters (**117a**) and 2-mercaptobenzoate esters (**117b**) react with acetyl chlorosulfide, forming identifiable disulfides (**118**) ⟨59BSF780, 59BSF1670⟩ which undergo acidic cyclization to 1,2-dithiol-3-ones (**3a**) or benzo derivatives (**77a**) in the presence of acid with loss of methyl acetate. This cyclization has also been applied to the preparation of 1,2-dithiolan-3-ones from 3-acyldithiopropanoyl chlorides (**119**) (Scheme 20) ⟨74IZV3836⟩.

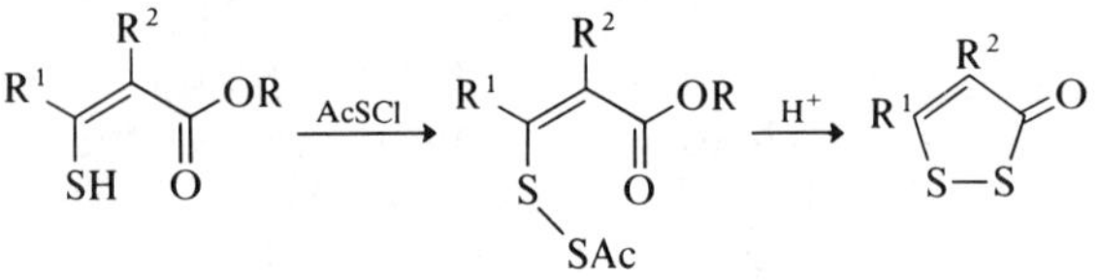

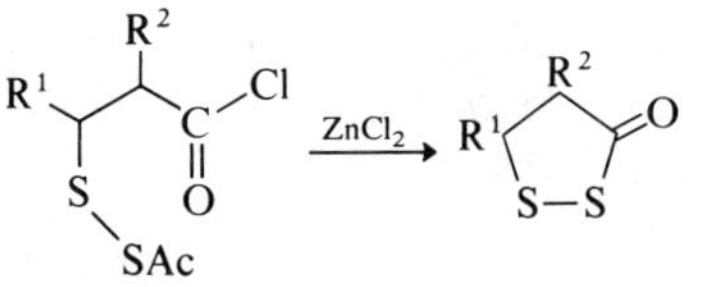

Scheme 20

(iii) Between two carbon atoms

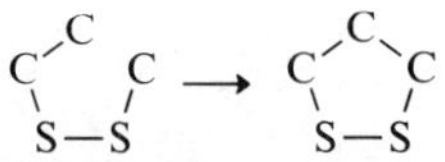

Benzenethiol reacts with trichloromethanesulfenyl chloride to form phenyl trichloromethyl disulfide (**120**), which cyclizes under Friedel–Crafts conditions to 3-chlorobenzo-1,2-dithiolylium chloride (**121**). Hydrolysis of this affords the dithiolone (**77a**) ⟨B-66MI43100⟩.

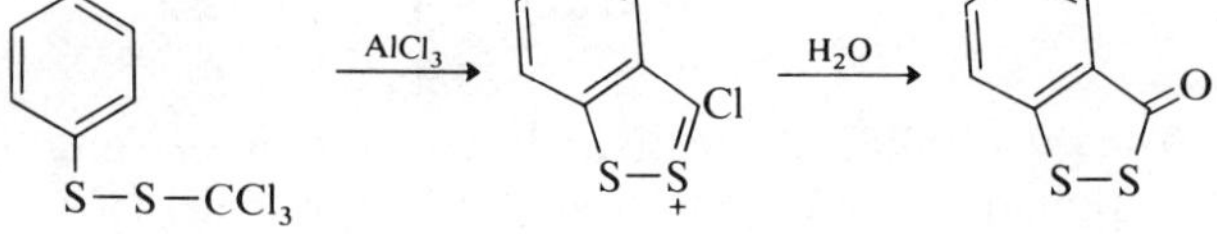

4.31.3.1.2 Direct synthesis by formation of two bonds

(i) From [4+1] units

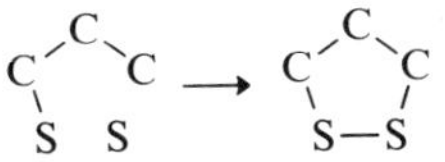

The only synthesis of this type observed is that of a C_3S unit adding to a sulfur atom. Intermediates of this type are probably present in syntheses from three-carbon compounds but in practice yields of dithioles are often higher when some sulfur function is already attached.

The thionation of 1,1-bis(alkylthio)alken-3-ones (**122**) or 3-ketothio acids (**123**) has wide scope in the formation of thiones (**3b**) ⟨62BSF2194⟩. The necessary unit also exists in 2-mercaptobenzoic acid and derivatives, *e.g.* the thionation of 2,2′-dicarboxydiphenyl disulfide (**124**) provides a good preparation of benzo-1,2-dithiole-3-thione (**77b**) ⟨53BSF327⟩. This may be applied to other derivatives ⟨B-66MI43100, 80T3309⟩ and to pyridine analogs ⟨55AG275⟩. While compounds containing thione functions are usually insufficiently stable to be suitable starting materials, β-aminopropenethiones (**125**), which are vinylogous thioamides, are available and give 1,2-dithiolylium salts (**126**) under appropriate conditions ⟨70BSF1918⟩.

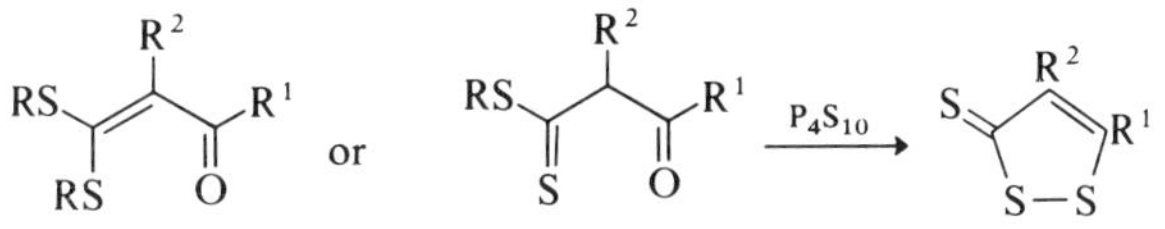

(**122**) (**123**) (**3b**)

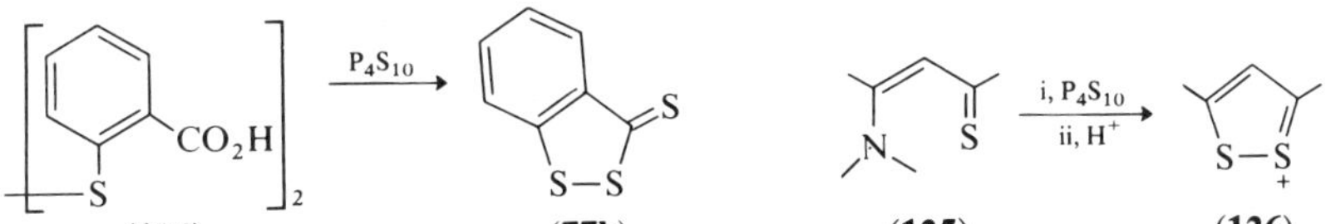

(**124**) (**77b**) (**125**) (**126**)

(ii) From [3+2] units

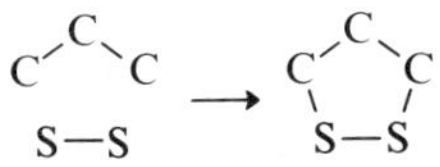

One important method involves the addition of a two-sulfur atom unit to a three-carbon unit. Although hydrogen disulfide may react with phenylpropynoyl chloride to form 5-phenyl-1,2-dithiol-3-one (**90a**) ⟨57AG138⟩, its main use has been in the conversion of aryl 1,3-diketones into 1,2-dithiolylium salts (Scheme 21) ⟨62JCS5104⟩. The reaction has since been extended to aliphatic diketones and 2-chloropropane-1,3-dial ⟨66AHC(7)39, 80AHC(27)151, 68ZN(B)1540⟩. Useful variations of this method substitute hydrogen sulfide under oxidizing conditions, hydrogen polysulfide or diacetyl disulfide for hydrogen disulfide. Synthetic equivalents of β-diketones may also be used.

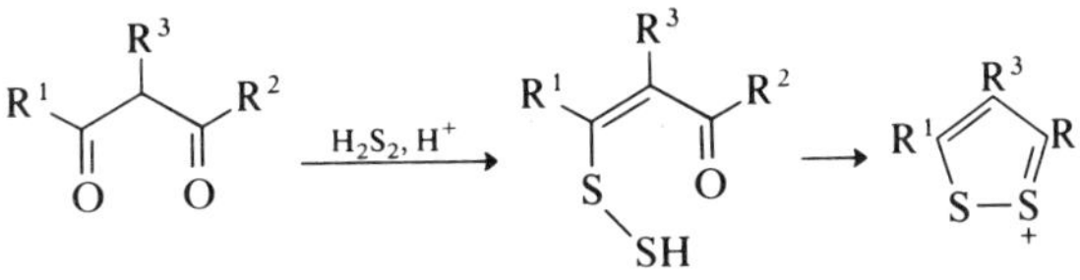

Scheme 21

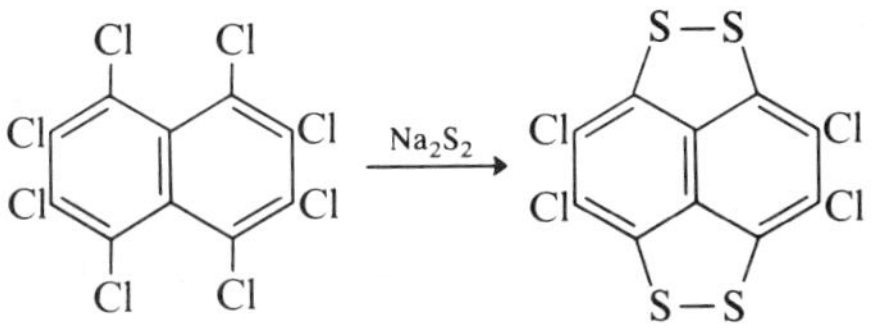

(**27b**)

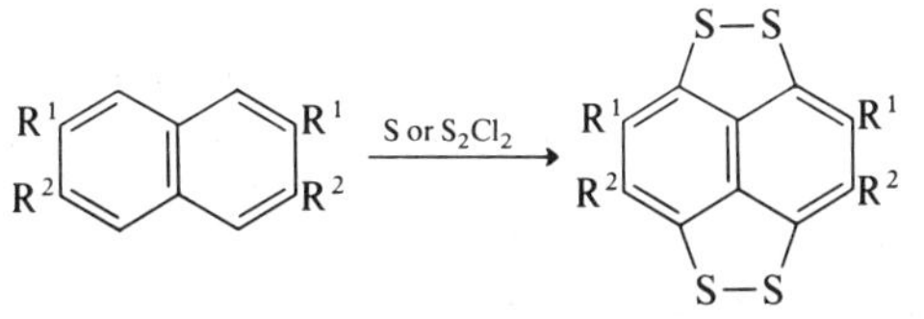

(**27**) **a**; $R^1 = R^2 = H$
c; $R^1, R^2 = (CH{=}CH)_2$

Sodium disulfide reacts with octachloronaphthalene to form tetrachloronaphtho[1,8-cd:4,5-c'd']bis[1,2-dithiole] (**27b**) and naphthalene or tetracene provides related fused 1,2-dithioles (**27a**) and (**27c**) by treatment with sulfur reagents ⟨76JA252⟩.

(iii) From C—C—S and C—S units

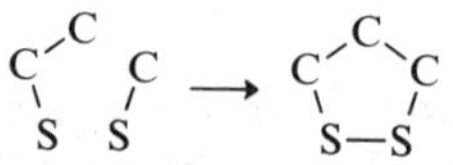

Recently it has been reported that α-thio derivatives of enamines and enol ethers form 5-amino- or 5-alkoxy-1,2-dithiole-3-thiones (**127a, b**) by reaction with carbon disulfide ⟨80ZOR13⟩. Spiran compounds (**128**) are also formed from suitable substrates. The reaction has some similarities to an enamine synthesis below ⟨67AG(E)294⟩.

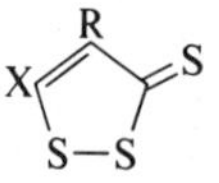

(**127**) a; X = NR¹R²
b; X = OR
c; X = Ph

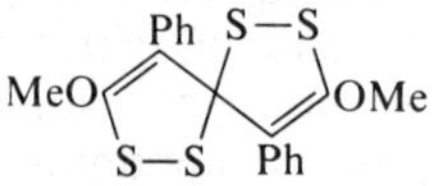

(**128**)

4.31.3.1.3 Direct synthesis by formation of three bonds

(i) From C—C—C and two S units

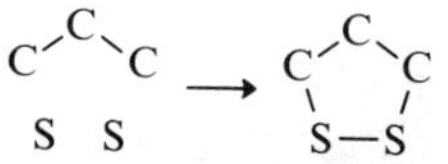

Various 1,2-dithioles are formed from three-carbon compounds by treatment with sulfur, or sulfur-containing reagents. The former reactions need temperature ranges of 180–250 °C. While these may involve multisulfur radicals, here they are treated as involving separate sulfur atoms because of the uncertainty as to the actual nature of the sulfur species. Basic catalysts are used to enhance the production of sulfur radicals, or of ionic species ⟨B-62MI43100⟩.

For alkanes, alkenes and their aryl derivatives, direct sulfurization gives thiones (**3b**). Typical examples are the preparation of 4-methyl- (**10**), 5-(4-methoxyphenyl)- (**129**) and 4-phenyl-1,2-dithiole-3-thiones (**71**) and benzo-1,2-dithiole-3-thione (**77b**) (Scheme 22) ⟨B-66MI43100⟩.

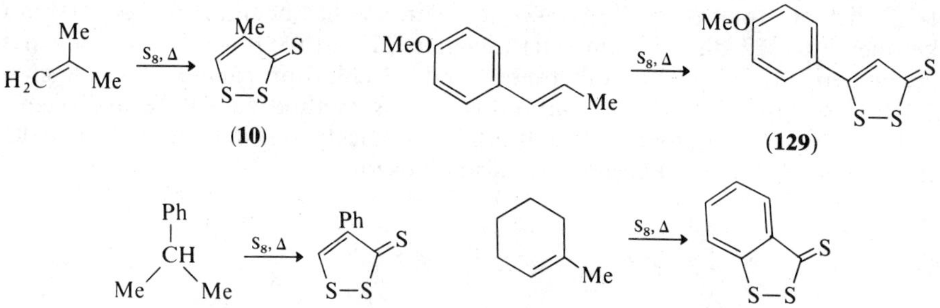

Scheme 22

Sulfurization of α,β-unsaturated esters may give different types of dithioles, depending on the substrate. Thus ethyl cinnamate and fumarate give the dithiolones (**130a, b**), while ethyl 2-methylcinnamate gives the ester (**133**); ethyl cinnamylidenemalonate gives the ester (**132**) ⟨B-67MI43100⟩. Cinnamylideneacetophenone provides a 2-benzoylmethylene-1,2-dithiole derivative (**133**), but cinnamylidenecyclohexanone gives the product (**134**) by further dehydrogenation (Scheme 23).

Phosphorus pentasulfide may also be employed, alone or in combination with sulfur, in which case dithiole-3-thiones are isolated from the reaction. For β-ketoesters, phosphorus

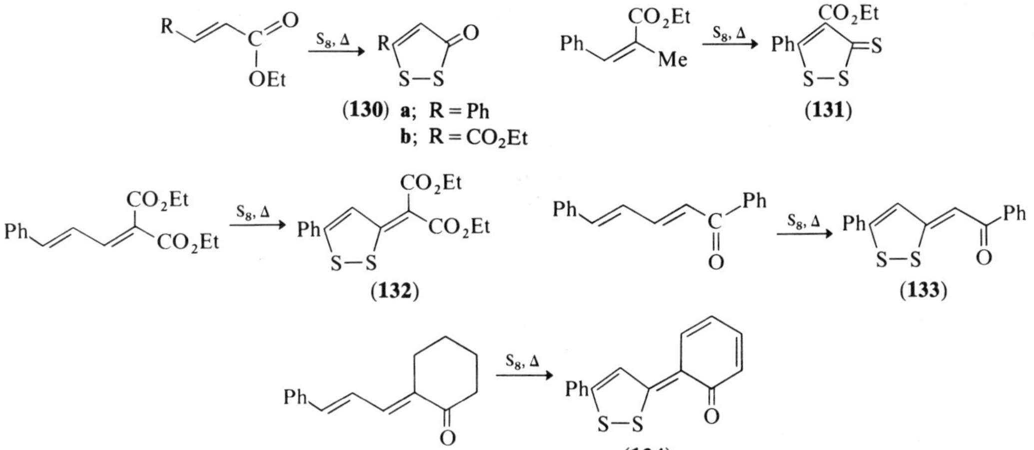

(130) a; R = Ph
b; R = CO_2Et

(131)

(132)

(133)

(134)

Scheme 23

pentasulfide alone is usually employed. Two different reaction intermediates have been suggested ⟨65CRV237, B-66MI43100⟩. Nevertheless, the reaction has wide scope and may be applied to the preparation of aliphatic, aromatic and acid heterocyclic thiones. 2-Acylphenols and their ethers may also be used.

A modification of these reactions is the thionation of *β*-diketones and derivatives. Products are obtained by thionation (*i.e.* treatment with phosphorus pentasulfide) which give 1,2-dithiolylium salts (**4**) by reaction with acid ⟨65LA(682)188⟩, most successfully for aryl derivatives ⟨70BSF1918⟩. An interesting phenalene derivative (**136**) was prepared from 9-ethoxyphenalenone (**135**) ⟨78JA7629⟩.

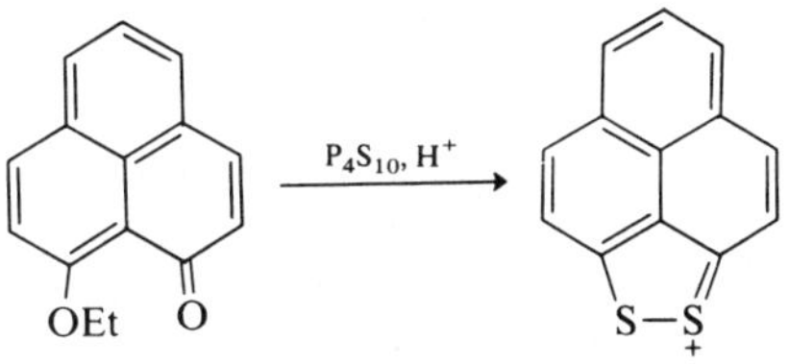

(135) (136)

Hexachloropropene forms 1,2,3-trichloro-1,2-dithiolylium chloride (**137**) by heating with sulfur or other reagents ⟨60AG629⟩. With less halogenated substrates, products are formed that do not contain halogens. Even halogens on aromatic rings may be lost under these conditions. Thus 2-bromotoluene and 2-(2-bromophenyl)propane give the fused dithiole-3-thiones (**77b**) and (**138**) ⟨67KGS633⟩. *β*-Ketonitriles are converted by thionation into imines (**3c**), and elemental sulfur reacts with a dilithioacenaphthylene to form the fused dithiole (**139**) ⟨80TL4565⟩. 1,2,3-Triketones react with hydrogen sulfide, or with phosphorus pentasulfide, forming mesoionic dithioles (**7**) ⟨72BCJ213, 70CB3885⟩. Since 1,1,3,3-tetrabromo(or 1,1,3-tribromo)-2-propanones react with potassium ethyl xanthate to give the same products, 1,3-dithione intermediates appear to be involved (Scheme 23a) ⟨78AJC297⟩. Also, the stepwise base-catalyzed reaction of ethyl carbonate with dimethylthioformamide gives a dithiodiamide (**140**) which may exist as dithiole (**141a**) or (**141b**) ⟨76SC387⟩.

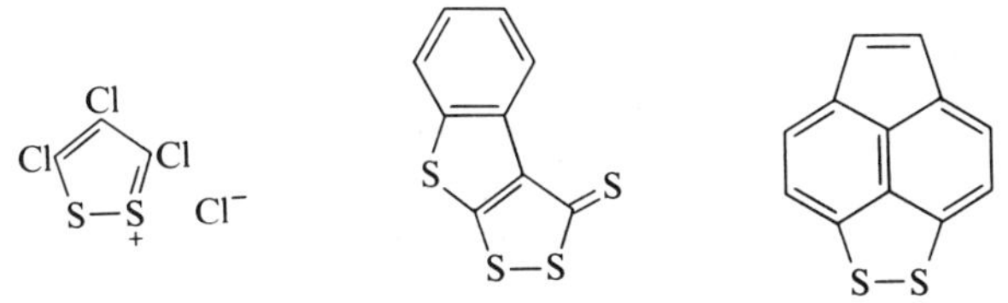

(137) (138) (139)

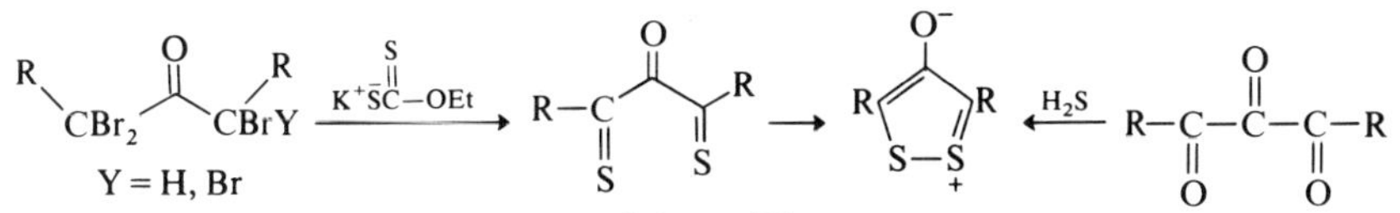

Scheme 23a

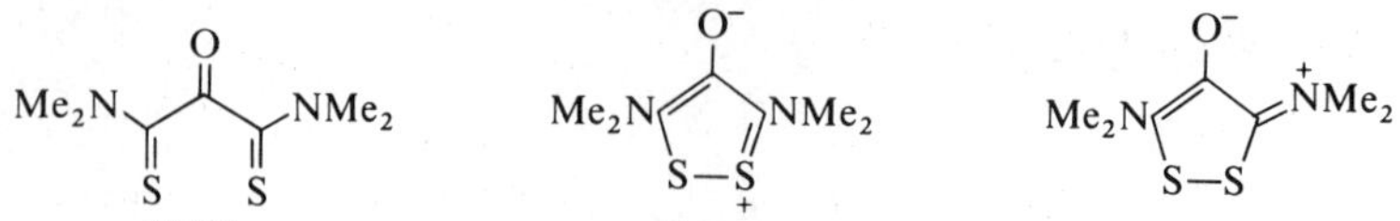

806   *1,2-Dithioles*

(140)   (141a)   (141b)

The readily accessible enamines react with carbon disulfide and sulfur under mild conditions to produce 1,2-dithiole-3-thiones (**3b**) *via* 3-amino dithioacids (Scheme 24). The nucleophilic character of the enamines is necessary for the initial reaction and to activate the sulfur (from S_8) for further insertion ⟨67AG(E)294⟩. Modified procedures produce other heterocycles. The yields of thiones (**3b**) may be low but this versatile reaction produces thiones that may be otherwise difficultly accessible, especially by direct sulfurization procedures, *e.g.* 1-morpholinocyclohexene may be converted into tetrahydrobenzo-1,2-dithiole-3-thione (**142**) in 40% yield. By contrast, direct sulfurization of 1-methylcyclohexene gives benzo-1,2-dithiole-3-thione (**77b**). Even dihydronaphtho-1,2-dithiole-3-thione (**143**) may be made by this procedure, although this compound may be dehydrogenated readily by sulfur at 220 °C.

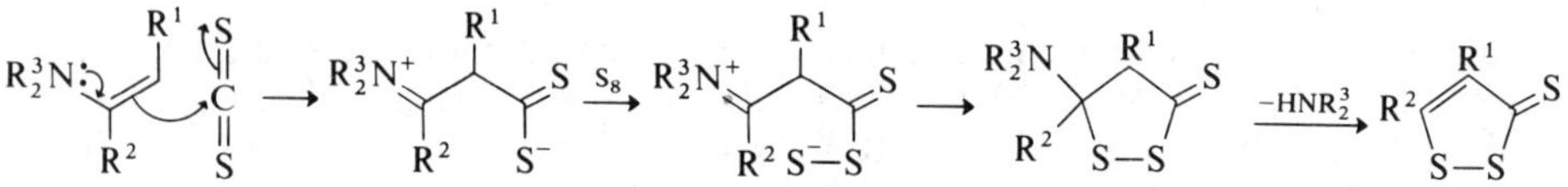

Scheme 24

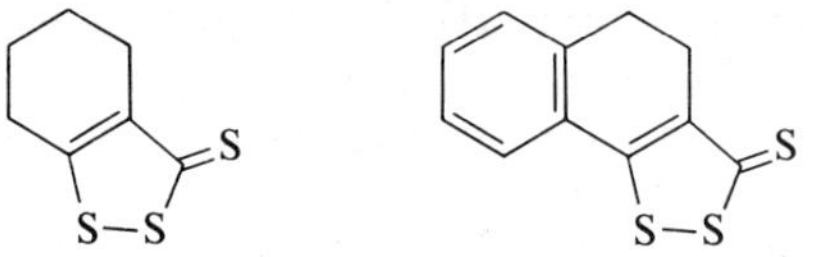

(142)   (143)

Carbon disulfide and sulfur also react with ethyl cyanoacetate or malononitrile, forming 5-amino-1,2-dithiole-3-thiones (**144a, b**) ⟨63ZC26⟩. Carbon disulfide is also used in a number of syntheses of thiones (**3b**) by reaction with dehydrobenzene ⟨70TL629⟩ or with arylsulfonylimines ⟨78JCS(P1)1017⟩. This last reaction has similarities to the enamine method above. Also, dimethyloxosulfonium methylide (**145**) may be thioacylated to the compound (**146**), which reacts with carbon disulfide to give a variety of dithiole-3-thiones (**127a, b, c**) (Scheme 25) ⟨76BCJ3128⟩.

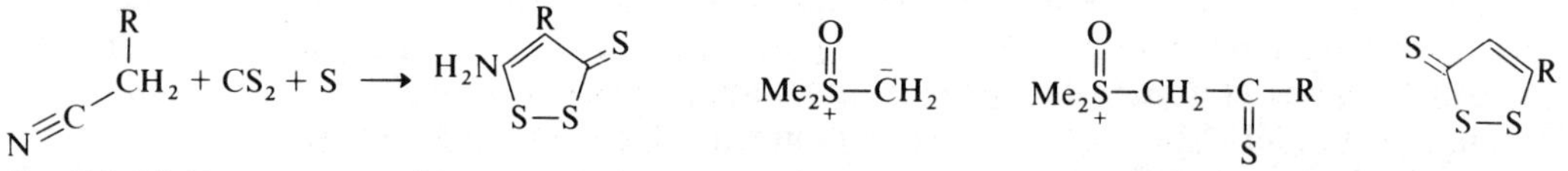

R = CN, CO₂Et   (**144**) a; R = CN   (**145**)   (**146**)   (**127a, b, c**)
                          b; R = CO₂Et

Scheme 25

4.31.3.2 Ring Synthesis by Transformation of Other Heterocycles

4.31.3.2.1 *From pyran and thiopyrans*

4H-Pyran-4-thiones (**147a**) and 4H-thiopyran-4-thiones (**147b**) react with sodium hydrosulfide or sodium hydroxide respectively to give acyclic products which on oxidation form 3-acylmethylene-1,2-dithioles (**11**) (Scheme 26) ⟨70JCS(C)2412⟩.

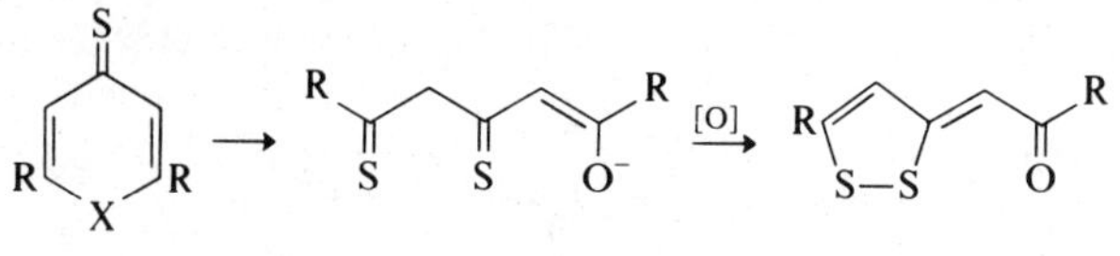

(**147**) X = O, S   (**11**)

Scheme 26

4.31.3.2.2 From 1,3-thioxins and 1,3-thiazines

5,6-Benzo-1,3-thioxin-4-thione (**148**) reacts with phenylhydrazine to form the dithiole hydrazone (**149**), which must arise by initial nucleophilic attack at the thione function ⟨80T3309⟩. The 1,3-thiazine-2,4-dione (**150**) possesses a three-carbon unit with an attached sulfur atom. Thionation with phosphorus pentasulfide gives the unsubstituted 1,2-dithiole-3-thione (**3b**) ⟨70AJC51⟩.

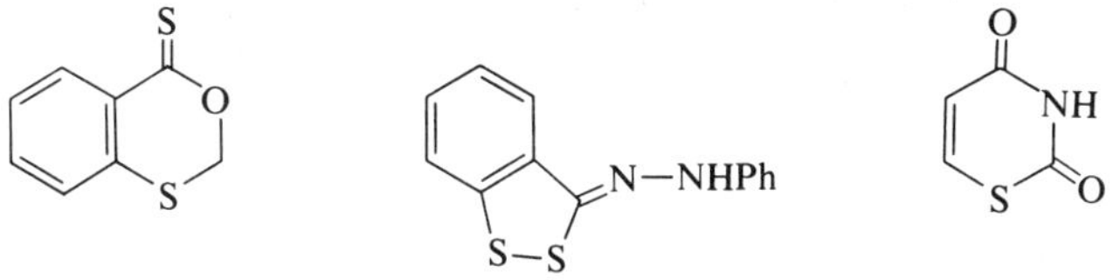

4.31.3.2.3 From 1,3-dithiins

1,3-Dithiins (**151**) are made by reaction of aldehydes with 1,3-dithiols. Treatment of these with suitable electrophilic oxidizing agents, *e.g.* Br$_2$, forms 1,2-dithiolylium ions (**4**) by ring contraction (Scheme 27) ⟨66AHC(7)39⟩. The synthesis may be applied to benzodithiolylium ions as well. A synthesis of the dithiole antibiotic 'holomycin' (Section 4.31.4) from a 1,3-dithiin is related in concept ⟨77JOC2891⟩.

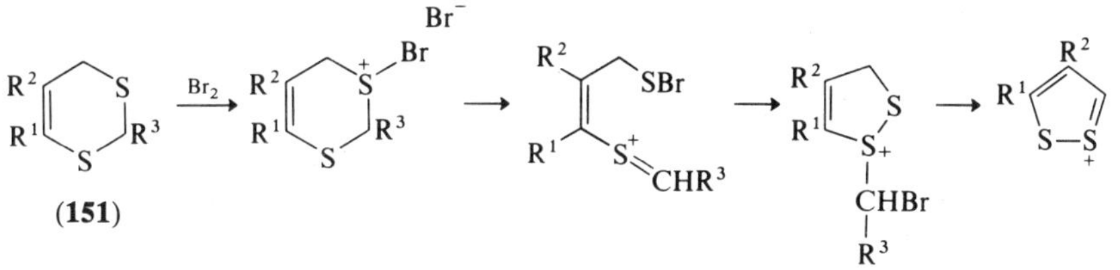

Scheme 27

4.31.3.2.4 From isoxazoles

The oxime of benzoylacetamide forms 5-phenylisoxazole (**152**) ⟨58AC(R)577⟩ on treatment with phosphorus pentasulfide, and on further treatment this is converted into 5-phenyl-1,2-dithiole-3-thione (**90b**). Also, isoxazoline-3-thiones (**153**) react with hydrogen sulfide to form mixtures of dithiol-3-imines (**3c**) and isothiazoline-3-thiones (**40**) ⟨80CPB487⟩.

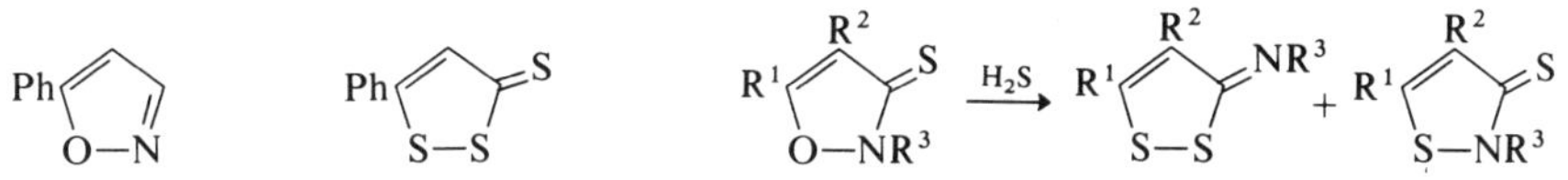

4.31.3.2.5 From isothiazoles

The isothiazole nucleus possesses a three-carbon and sulfur unit, and is a suitable starting material for many dithioles. *N*-Arylisothiazolium salts (**154**) react with sulfur in pyridine, forming 3-arylimino-1,2-dithioles (**156**) possibly *via* β-thioketimine intermediates (**155**), and the reaction of isothiazolium salts with hydrogen sulfide may give bis(1,2-dithiol-3-yl) sulfides (**36b**) ⟨72JCS(P1)2305⟩. A mechanism involving nucleophilic attack on sulfur has been suggested (Scheme 28) ⟨73CJC3081⟩. The reaction of thiolacetic acid with 3-chloro-1,2-benzisothiazole (**157**) gives an acyliminobenzo-1,2-dithiole (**158**) ⟨68CB2472⟩, and 3-chloro-1,2-benzisothiazolium salts (**159**) react to form 1,2-benzisothiazoline-3-thiones (**66**) or benzo-1,2-dithiol-3-imines (**65**), depending on the substitution on nitrogen. These isomers equilibrate at 150 °C ⟨67CB2435⟩. The thiones (**66**) also form benzo-1,2-dithiole-3-thione (**77b**) by reaction with hydrogen sulfide (Scheme 29) ⟨29JCS1582⟩.

Scheme 28

Scheme 29

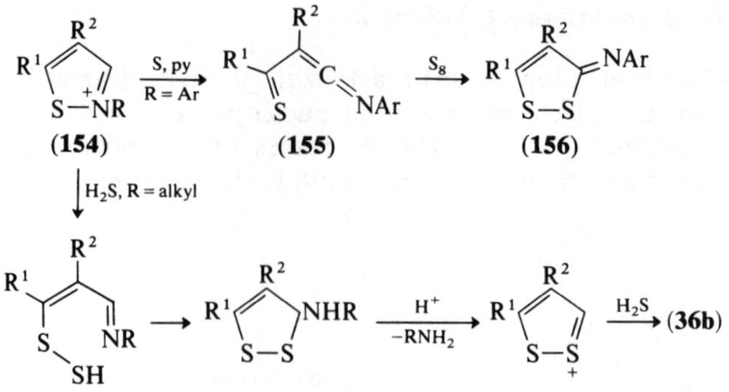

4.31.3.2.6 From 1,3-dithioles

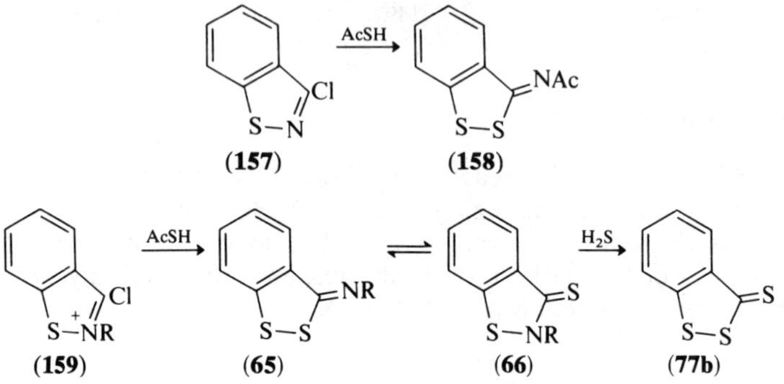

Photochemical rearrangement of some mesoionic 1,3-dithioles (**160**) and (**161**) produces the dithioles (**162**) and (**163**) ⟨80CA(92)197661⟩, and 4,5-dimercapto-1,3-dithiole-3-thione (**164**) salts are reported to rearrange to 4,5-dimercapto-1,2-dithiole-3-thione (**165**) salts ⟨77EGP124044⟩.

Scheme 30

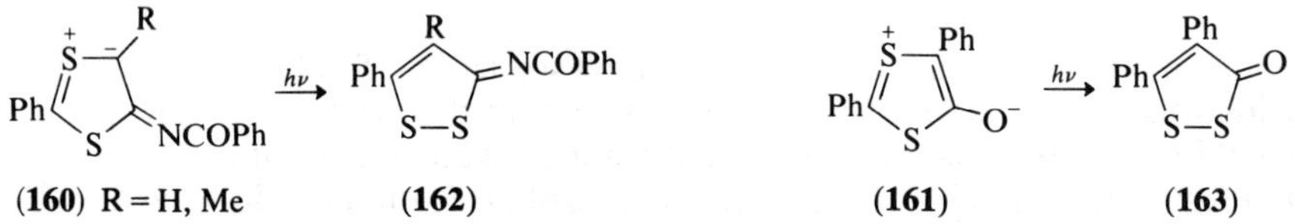

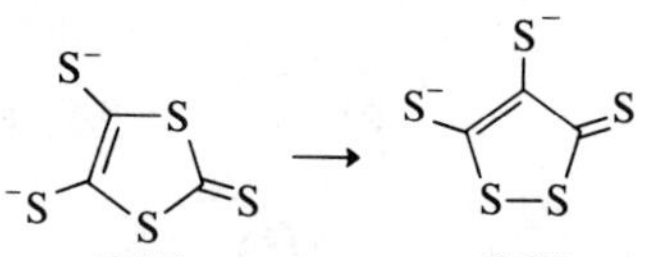

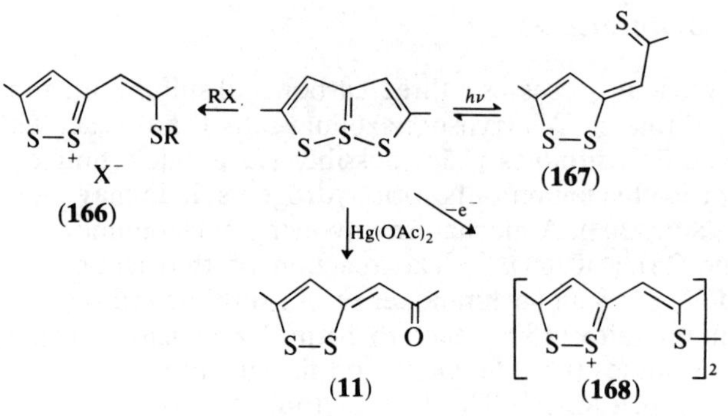

4.31.3.2.7 From 1,6,6aλ⁴-trithiapentalenes

These compounds, discussed separately in Chapter 4.38, may be converted into more recognizable 1,2-dithioles by a number of reagents. Thus mercury(II) acetate treatment produces 3-acylmethylene-1,2-dithioles (**11**), and alkylation gives the 1,2-dithiolylium salts (**166**) ⟨63JA3244, 68JOC2915⟩. This has been recently applied to highly fused trithiapentalenes ⟨80JCR(S)221⟩. Irradiation of trithiapentalenes provides (E)-thioacylmethylene-1,2-dithioles (**167**), which undergo a thermal back reaction ⟨77JCS(P1)994⟩, and electrochemical oxidation provides dimeric disulfides (**168**) (Scheme 30) ⟨72MI43101⟩.

4.31.3.2.8 From 1,3-dithietes

The 2,4-bis(thiobenzoylmethylene)-1,3-dithiete (**169**) has an obvious structural relationship to 4,5-diphenyl-1,2-dithiole-3-thione (**170**), which is formed on thionation ⟨73TL1915⟩.

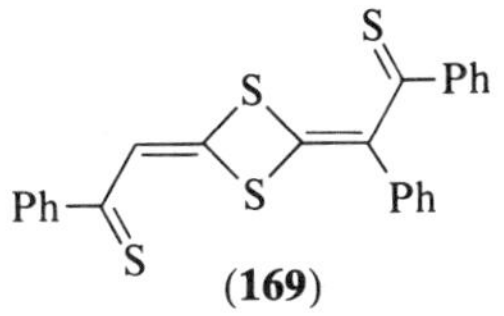

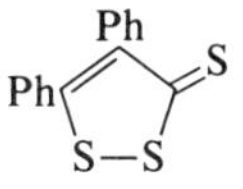

4.31.3.3 Best Practical Methods of Synthesis

The parent 1,2-dithiolylium ion (**4**) is readily prepared by treatment of 1,2-dithiole-3-thione (**3b**; R = R' = H) with hydrogen peroxide in acetic acid ⟨65JCS32⟩. The method may be applied to the alkyl and aryl derivatives with equal success. For cations with 3- and 5-substitution the acid catalyzed reactions of β-dicarbonyl compounds with hydrogen disulfide or equivalent are best ⟨80AHC(27)151⟩, whereas the benzo-1,2-dithiolylium ion (**172**) and related compounds are best prepared by ring contraction of benzo-1,3-dithiins (**171**) ⟨63LA(661)84⟩.

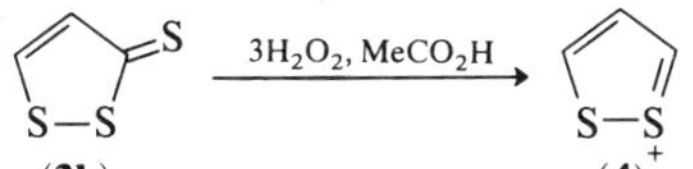

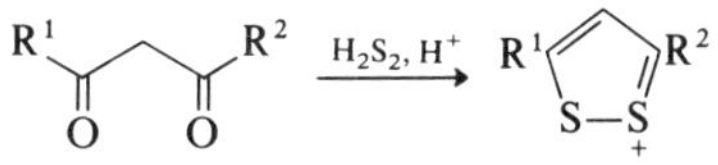

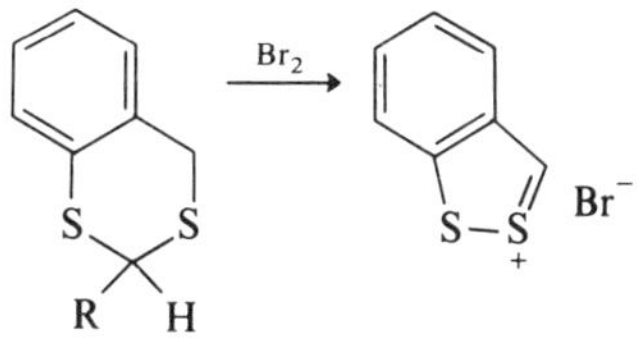

1,2-Dithiolane (**1**) is unstable, but best prepared from the treatment of the lead salt of propane-1,3-dithiol (**173**) with sulfur ⟨73TL655⟩.

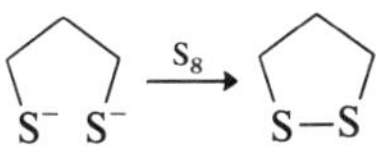

Amino- and imino-dithioles are available in some cases by ring syntheses, but the most general methods involve the reaction of an amine with a 3(or 5)-chloro-1,2-dithiolylium salt or a chloro-1,2-dithiole respectively ⟨80AHC(27)151⟩, and protonation or alkylation of these gives aminodithiolylium salts.

Alkyl- and aryl-1,2-dithiol-3-ones (**3a**) are made by ring synthesis from unsaturated esters ⟨B-66MI43100⟩ and alkylation of these affords 3-alkoxy-1,2-dithiolylium salts (**35b**). The parent (**3a**; $R^1 = R^2 = H$) is best made from the thione (**3b**) by treatment with mercury(II) acetate ⟨63CB2702⟩.

1,2-Dithiole-3-thione (**3b**; $R^1 = R^2 = H$) is made by treatment of 1,1,3,3-tetramethoxypropane with sulfur and phosphorus pentasulfide ⟨77S802⟩. Alkyl- and aryl-1,2-

dithiole-3-thiones are made by direct sulfurization of the corresponding substituted propenes. 1,2-Dithiole-3-thiones are also readily made by thionation of 1,2-dithiol-3-ones. 1,2-Dithiole-3-thiones readily alkylate, forming 3-alkylthio-1,2-dithiolylium salts and treatment with oxalyl chloride yields 3-chloro-1,2-dithiolylium salts (**21**).

3-Chloro-1,2-dithiolylium salts are best made by reaction of 1,2-dithiolyl-3-ones with oxalyl chloride ⟨65LA(688)150⟩. 4-Chlorodithiolylium salts (**35a**; X = Cl) and chlorodithioles are best made by ring synthesis. General interconversions are summarized in Scheme 31.

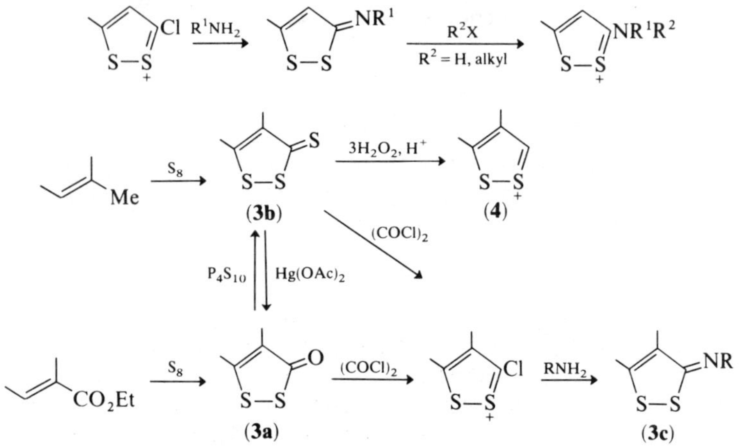

Scheme 31

4.31.4 USES AND OCCURRENCE

4.31.4.1 Natural Products

Some simple dithiolethiones, including the parent (**3b**) and 5-aryl derivatives, have been isolated from *Brassica* species, and some *Streptomyces* strains produce four antibiotics which possess pyrrolo[3,2-c][1,2]dithiole skeletons (**174a–d**). Some of these have been synthesized ⟨B-66MI43100, 77JOC2891⟩. Lipoic acid (**29**) is a growth factor and an essential component of enzyme systems involved in oxidative decarboxylation of pyruvic and related acids ⟨B-61MI43101⟩.

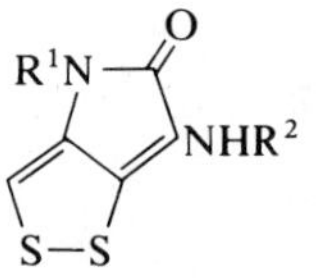

(**174**) a; R^1 = Me, R^2 = COMe
b; R^1 = Me, R^2 = COEt
c; R^1 = Me, R^2 = COCHMe$_2$
d; R^1 = H, R^2 = COMe

4.31.4.2 Compounds of Biological Interest

Apart from the above, several dithioles have biological applications. The 3,5-diamino-1,2-dithiolylium salts (**8**) and derivatives appear to have radio-protective properties ⟨72MI43102⟩. 5-(4-Methoxyphenyl)-1,2-dithiole-3-thione (**129**) appears, among other beneficial properties ⟨65CRV237⟩, to protect against liver damage by carbon tetrachloride ⟨76GEP2625053⟩. Various dithioles, especially halogenated and amino derivatives, have been suggested for use as fungicides and in slime control ⟨B-66MI43100⟩ and as insecticides, nematocides and general pesticides ⟨78GEP2637692, 78GEP2707227, 67FRP1504150⟩. Pyrazinyl-1,2-dithiole-3-thiones have been patented as amebicides and schistosomacides ⟨76GEP2627211⟩.

4.31.4.3 Compounds of Industrial Interest

Many 1,2-dithiole derivatives have been patented or recommended for use as 'pickling' or corrosion inhibitors or as additives to lubricating oils ⟨B-66MI43100⟩. They are efficient in preventing high-temperature fouling of oil feedstock ⟨78USP4116812⟩.

Benzo-1,2-dithiole-3-thione (**77b**) reacts with lithium in THF to produce a complex which is useful in the preparation of lithium hydrides ⟨78BEP867155⟩. Various 3-(4-dimethylaminophenyl-1,2-dithiolylium salts (**175**) and other dithiolylium salts function as dyes for polyacrylonitrile and have been proposed as sensitizers for organic photoconductors ⟨64USP3158621, 66USP3299055, 71USP3530145, 71USP3575968⟩.

4.31.4.4 Organic Conductors

Many 1,2-dithioles and 1,2-dithiolylium ions, especially highly conjugated derivatives, undergo charge transfer with electron acceptors or donors respectively ⟨80AHC(27)151, 80MI43100⟩ to form derivatives with ranges of conductivity. In particular the complex (**176**) shows a high d.c. conductivity ⟨77ACS(B)281⟩, and the complex of the fused dithiole (**27**) with tetracyano-*p*-quinodimethane (TCNQ) shows a conductivity higher than for 'tetrathiafulvalenes' ⟨76JA252⟩ which have recently been the subject of intense research interest.

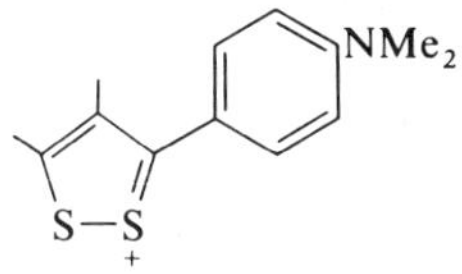

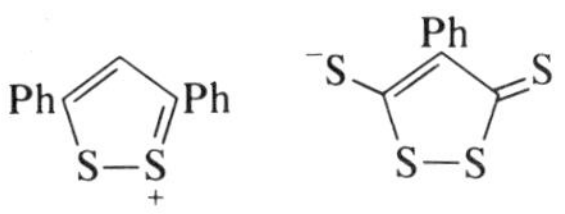

4.32
1,3-Dithioles

H. GOTTHARDT
Universität Wuppertal

4.32.1 INTRODUCTION	813
4.32.2 STRUCTURE	814
4.32.2.1 Theoretical Treatment	814
4.32.2.2 X-Ray Diffraction	815
4.32.2.3 ^1H NMR Spectroscopy	816
4.32.2.4 ^{13}C NMR Spectroscopy	816
4.32.2.5 UV Spectroscopy	817
4.32.2.6 IR Spectroscopy	818
4.32.3 REACTIVITY	819
4.32.3.1 Reactions of 1,3-Dithioles and Mesoionic 1,3-Dithiolones	819
4.32.3.1.1 Thermal and photochemical reactions formally involving no other species	819
4.32.3.1.2 Electrophilic attack at carbon	820
4.32.3.1.3 Oxidation	820
4.32.3.1.4 Nucleophilic attack at carbon	821
4.32.3.1.5 Nucleophilic attack at hydrogen	824
4.32.3.1.6 Reaction with triplet molecular oxygen	824
4.32.3.1.7 Nucleophilic attack at valence tautomeric open-chain ketene	825
4.32.3.1.8 Cycloaddition reactions	825
4.32.3.1.9 Reactions of substituents	831
4.32.3.2 Reactions of 1,3-Dithiolanes	837
4.32.4 SYNTHESES	838
4.32.4.1 Ring Syntheses from Non-Heterocyclic Compounds	838
4.32.4.1.1 Formation of one bond	838
4.32.4.1.2 Formation of two bonds from [4+1] atom fragments	842
4.32.4.1.3 Formation of two bonds from [3+2] atom fragments	844
4.32.4.1.4 Formation of multiple bonds	847
4.32.4.2 Syntheses involving Heterocyclic Compounds	847
4.32.4.2.1 Formation of the 1,3-dithiole ring	847
4.32.4.2.2 Formation of tetrathiafulvalenes	848
4.32.4.3 Syntheses of Parent Compounds	850
4.32.5 APPLICATIONS	850

4.32.1 INTRODUCTION

This chapter deals with 1,3-dithiole compounds such as 1,3-dithiolylium ions (**1**), mesoionic 1,3-dithiol-4-ones (**2**), 1,3-dithioles (**3**), 1,3-dithiolanes (**4**) and the tetrathiafulvalene system (**5**). During the last 15 years the chemistry of 1,3-dithiole compounds has developed considerably. One reason is that tetrathiafulvalene and its derivatives serve as donors in organic charge-transfer salts which exhibit the electrical properties of quasi-one-dimensional metals. For the preparation of such organic metals, 1,3-dithiolylium cations serve as useful synthetic intermediates.

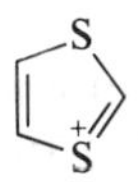

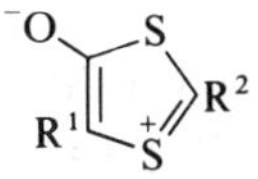

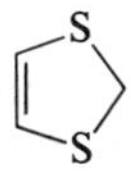

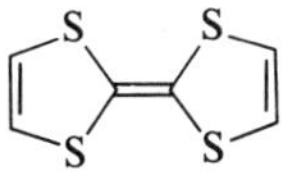

The chemistry of the 1,3-dithiolylium system has been reviewed previously ⟨80AHC(27)151, 66AHC(7)39, 77HC(30)271⟩. The 1,3-dithiolylium ions of the parent compound (**1**), which can be formally derived from their corresponding 1,3-dithioles by abstraction of a hydride ion, possess a 6π-electron system and thus show enhanced stability and exhibit Hückel-type aromatic properties. They can be represented as resonance hybrids with the main resonance contributors (**1a**)–(**1e**). No systematic study of their thermodynamic properties has been made. Data relating to solubilities, melting points, chromatographic behaviour, *etc.* have not been compiled but in many instances information of this type may often be found associated with their synthesis. The ionic system (**1**) has been named according to IUPAC Rule C-83. It is often referred to in the literature and in *Chemical Abstracts* as 1,3-dithiolium.

(**1a**) (**1b**) (**1c**) (**1d**) (**1e**)

The mesoionic 1,3-dithiol-4-ones of type (**2**) are best represented as resonance hybrids of several limiting formulae (**2a**)–(**2e**), of which (**2a**) and (**2b**) are presumably most representative; both formulae are, for example, in good agreement with the direction of the dipole moments. According to the IUPAC nomenclature rules, the mesoionic 1,3-dithiolones (**2**) should be named as 1,3-dithiolylium-4-olates ⟨B-79MI43202⟩.

(**2a**) (**2b**) (**2c**) (**2d**) (**2e**)

The preparation of tetrathiafulvalenes ⟨76S489⟩ and the chemistry of 1,3-dithioles and 1,3-dithiolanes ⟨66HC(21-1)447⟩ have been reviewed.

4.32.2 STRUCTURE

4.32.2.1 Theoretical Treatment

An understanding of certain physical and chemical properties of sulfur-containing heterocycles has been offered by quantum chemical calculations. Some earlier computational results by the Hückel LCAO MO method have been summarized ⟨65AHC(5)1⟩.

Quantum mechanical calculations for the 1,3-dithiolylium ion, benzo-1,3-dithiolylium ion, 1,3-dithiol-2-one and 1,3-dithiole-2-thione have been carried out with the simple LCAO MO method ⟨66AHC(7)39⟩. According to these calculations, the lowest electron density was found at the 2-position of the 1,3-dithiolylium cation, and the C(4)—C(5) bond order corresponded approximately to that of an isolated C=C double bond. A delocalization energy of 105 kJ mol^{-1} has been calculated for this cation. For the benzo-1,3-dithiolylium ion it was found that nucleophilic attack should take place in the 2-position but electrophilic substitution at the 4-position.

A π-electron SCF–MO calculation using the zero differential overlap approximation has been carried out for the 1,3-dithiolylium, thiazolylium and imidazolium ions and their corresponding carbenes at the 2-position ⟨71JHC551⟩. Some computed charges based on orbital densities are given in Table 1. The highest charge densities are at the heteroatoms of the ions. It has been shown that the difference in energy for an ion and its corresponding carbene, which should be a measure of the rate of ionization of the H—C(2) bond, predicts rates in the order 1,3-dithiolylium > thiazolylium > imidazolium. This is in good agreement with experiments, because it is known that thiazolylium ions are deuterated at the 2-position about 3000 times faster than imidazolium ions and 1,3-dithiolylium perchlorate undergoes rapid deuterium exchange at C-2.

On the other hand, a non-empirical calculation of the 1,3-dithiolylium ion using linear combinations of Gaussian orbitals with and without *d*-orbitals showed that almost all of the positive charge is shared by the sulfur and hydrogen atoms, whereas the carbon at the

Table 1 π-Electron SCF–MO Computed Charge Densities ⟨71JHC551⟩

Atom	1,3-Dithiolylium ion	carbene	Thiazolylium ion	carbene	Imidazolium ion	carbene
1	0.44	0.06	0.38	0.04	0.41	0.30
2	0.09	−0.09	0.13	−0.30	0.15	−0.40
3	0.44	0.06	0.45	0.36	0.41	0.30
4	0.02	−0.01	0.09	0.02	0.02	−0.10
5	0.02	−0.01	−0.05	−0.13	0.02	−0.10

2-position possesses a negative charge ⟨72TL4165⟩. Since nucleophilic additions occur exclusively at the 2-position, these results have been reconciled by assuming that the presence of the nucleophilic reagent would be expected to induce an opposite polarization.

Reactivity indexes and π-electron densities of 4-aryl-1,3-dithiolylium ions have also been calculated by the Hückel MO method ⟨70CPB865, 65ZC23⟩.

UV transition energies of 1,3-dithiolylium ions have been calculated by the Pariser–Parr–Pople (PPP) method with neglect of d-orbitals ⟨68MI43200, 68JPC3975⟩. For example, introduction of a mercapto or methylthio group at the 2-position of the 1,3-dithiolylium ion is known to change the UV absorption spectrum drastically: instead of two maxima in the parent ion, there are now three absorption maxima observed. These findings are consistent with the calculated ions ⟨68MI43200⟩. Excellent agreement between observed and calculated UV transition energies of the 1,3-dithiolylium ion was obtained by a semiempirical PPP variable integral method ⟨71MI43200⟩.

Furthermore, the influence of alkyl substituents on the UV transitions of 1,3-dithiolylium ions has been studied by the PPP method and compared with experimental values; the agreement was good ⟨71T4705⟩. These compounds possess two UV absorption maxima of approximately equal intensities. The long wavelength band corresponds to a $\pi \rightarrow \pi^*$ transition with a transition moment in the direction of the two-fold symmetry axis of (**1**), whereas the $\pi \rightarrow \pi^*$ transition of the second band is polarized perpendicular to the first one ⟨71T4705⟩.

The absorption behavior of the fluorescent 2,5-diphenyl-1,3-dithiolylium perchlorate has also been satisfactorily interpreted by PPP calculations ⟨75MI43200⟩.

4.32.2.2 X-Ray Diffraction

Only a few X-ray crystal structure determinations on 1,3-dithioles have been carried out. Some typical bond lengths and bond angles of the tetrathiafulvalene (**5**), 1,3-dithiole-2-thione derivative (**6**) and 1,3-dithiolanes (**7**) and (**8**) are given in Tables 2 and 3, respectively. Also bond lengths and bond angles of the radical cation of tetrathiafulvalene in the tetrathiafulvalene–7,7,8,8-tetracyanoquinodimethane complex have been determined ⟨74AX(B)763⟩.

Table 2 Bond Lengths (Å) of Some 1,3-Dithiole and 1,3-Dithiolane Derivatives

Bond	Compound			
	(**5**)[a]	(**6**)[b]	(**7**)[c]	(**8**)[d]
S(1)—C(2)	1.756	1.733	1.72	1.82
S(1)—C(5)	1.732	1.764	1.83	1.84
S(3)—C(2)	1.758	1.738	1.76	1.86
S(3)—C(4)	1.729	1.757	1.80	1.89
C(4)—C(5)	1.314	1.503	1.54	1.52

[a] ⟨71CC889⟩. [b] ⟨69AX(B)1022⟩. [c] ⟨67TL5311⟩. [d] ⟨54ACS1145⟩.

It is interesting to note that tetrathiafulvalene (**5**) is not planar but rather slightly distorted into a chair conformation ⟨71CC889⟩. Also the five-membered ring of the 1,3-dithiolane

Table 3 Bond Angles (°) of some 1,3-Dithiole and 1,3-Dithiolane Derivatives

Angle	Compound			
	(5)[a]	(6)[b]	(7)[c]	(8)[d]
S(1)—C(2)—S(3)	114.5	116.3	115.5	104
C(2)—S(3)—C(4)	94.3	97.2	96.7	102
S(3)—C(4)—C(5)	118.6	114.8	105.4	109
C(4)—C(5)—S(1)	118.0	113.4	108.3	100
C(5)—S(1)—C(2)	94.5	97.9	94.6	109

[a] ⟨71CC889⟩. [b] ⟨69AX(B)1022⟩. [c] ⟨67TL5311⟩. [d] ⟨54ACS1145⟩.

derivative (8) was found to be slightly puckered, C(4) and C(5) being above and below the plane defined by S(1)—C(2)—S(3) ⟨54ACS1145⟩.

4.32.2.3 ^{1}H NMR Spectroscopy

Interestingly, the ^{1}H NMR spectrum of the 1,3-dithiolylium tetrafluoroborate exhibits the H-2 signals about 2 p.p.m. downfield from the H-4 and H-5 resonances (Table 4). This has been explained in terms of the molecular diagrams of the 1,3-dithiolylium ion which showed an appreciably lower electron density on C-2 than on C-4 and C-5 ⟨66AHC(7)39⟩. Also, correlations have been established between the resonances of methyl protons in methyl-substituted 1,3-dithiolylium ions as well as resonances of ring protons with calculated π-charges of the ring carbons ⟨71T4705⟩.

In 2-p-substituted phenyl derivatives of 4-phenyl-1,3-dithiolylium ions a linear relationship exists between the 5-H resonance signals and the Hammett σ_p values ⟨71T4003⟩. For references to ^{1}H NMR data of variously substituted 1,3-dithiolylium ions such as 2-amino-, 2-alkylthio- and benzo-1,3-dithiolylium ions, see ⟨66AHC(7)39, 80AHC(27)151⟩.

The H-5 resonance of 2-phenyl-1,3-dithiolylium-4-olate (Table 4) appears at significantly higher field than in the 1,3-dithiolylium ion. This is in agreement with the delocalization of the negative charge on O-4 and C-5 which leads to an increased shielding at C-5.

Table 4 ^{1}H NMR Spectral Data of some 1,3-Dithiole Derivatives[a]

Compound	^{1}H Chemical shifts (δ, p.p.m.)		J values (Hz)	Solvent	Ref.
	2	4,5; aromatic	2:4, 2:5		
[1,3-dithiolylium H BF$_4^-$]	11.65	9.67	2.0	CD$_3$CN	74JOC3608
[benzo-1,3-dithiolylium H BF$_4^-$]	11.50	8.06–8.26 8.66–8.86	—	CF$_3$CO$_2$D	76BCJ3567
[2-Ph-1,3-dithiolylium-4-olate]	—	6.06	—	CDCl$_3$	76JOC1724
[1,3-dithiole-2-thione]	—	7.20	—	CS$_2$	66AHC(7)39
[4-Ph-1,3-dithiol-2-one]	—	6.74	—	CS$_2$	66AHC(7)39

[a] Relative to internal TMS.

4.32.2.4 ^{13}C NMR Spectroscopy

Some typical ^{13}C NMR spectral data of 1,3-dithiole derivatives are given in Table 5.

The C-2 resonance signal of the 1,3-dithiolylium ion at δ 179.5 appears at higher field than in the 1,3-dithiolanylium ion (δ 221.2) whereas the opposite holds for the C-4 resonances. These data reflect the greater delocalization of the positive charge in the

Table 5 ^{13}C NMR Spectral Data of some 1,3-Dithiole Derivatives[a]

Compound	C-2 (δ, p.p.m.)	C-4 (δ, p.p.m.)	Solvent	Ref.
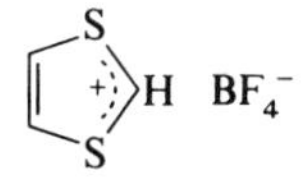 1,3-dithiolylium, H BF$_4^-$	179.5	146.2	CF$_3$CO$_2$D	77CL1133
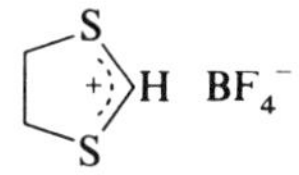 1,3-dithiolanylium, H BF$_4^-$	221.2	46.4	CF$_3$CO$_2$D	77CL1133
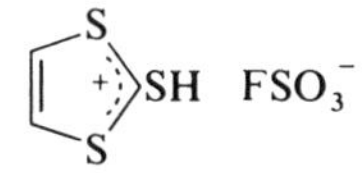 2-SH-1,3-dithiolylium FSO$_3^-$	145.2	140.0	SO$_2$ (−60 °C)	77JOC2237
 2-SMe-1,3-dithiolylium I$^-$	166.4	139.7	CF$_3$CO$_2$D	77CL1133
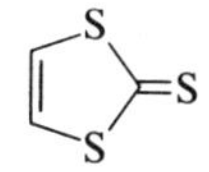 1,3-dithiole-2-thione	140.7	133.7	SO$_2$(−60 °C)	77JOC2237
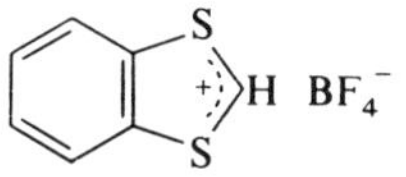 benzo-1,3-dithiolylium H BF$_4^-$	182.4	146.0	CF$_3$CO$_2$D	77CL1133
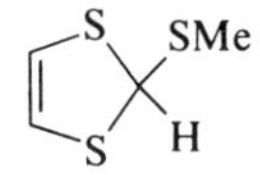 2-SMe-2-H-1,3-dithiole	61.2	115.4	CDCl$_3$	77CL1133

[a] Relative to internal TMS.

1,3-dithiolylium cation. A rough correlation has been obtained between observed ^{13}C chemical shifts of the benzo-1,3-dithiolylium ion and calculated electron densities ⟨77CL1133⟩. On the other hand, the 2-methylthio-1,3-dithiole exhibits C-2 and C-4 resonance signals which are typical of sp^3 and sp^2 carbon atoms, respectively. For further ^{13}C NMR spectral data of 1,3-dithiol-2-ones, 1,3-dithiole-2-thiones and their corresponding saturated compounds, see ⟨81JPR737⟩.

4.32.2.5 UV Spectroscopy

The UV absorption spectrum of 1,3-dithiolylium perchlorate in ethanol/perchloric acid shows two bands, which are shifted to longer wavelength in perchloric acid alone (Table 6). Introduction of alkyl groups, as in trimethyl-1,3-dithiolylium perchlorate, causes a bathochromic shift of the longest wavelength absorption to 305 nm ⟨71T4705⟩.

Table 6 UV Spectral Data of 1,3-Dithiole Compounds

	λ_{max} (nm) (log ε)	Solvent	Ref.
1,3-Dithiolylium perchlorate	212 (3.53) 254 (3.58)	Ethanol/HClO$_4$	71PMH(3)67
	242 (3.80) 264 (3.55)	HClO$_4$	71T4705
Tetrathiafulvalene	303 (4.11) 317 (4.10) 368 (3.28) 450 (2.43)	Cyclohexane	71JA2258
1,3-Dithiolane	207 (3.13) 247 (2.56)	Cyclohexane	71PMH(3)67

An essentially larger effect is caused by aryl groups in the 2-position. For example the 4-methyl-2-phenyl-1,3-dithiolylium ion exhbits the longest wavelength band at 362 nm whereas the 4-phenyl-1,3-dithiolylium ion absorbs at 347 nm ⟨66AHC(7)39⟩. The larger bathochromic shift of 2-aryl groups has been explained by the better delocalization of the positive charge. In the same way, 2-p-substituted phenyl derivatives show an enhanced bathochromic shift of the long wavelength absorptions by electron-donating groups in the order H, Me, OPh, OH, OMe. A good correlation exists between the wavenumber of the longest wavelength maxima of 2-(p-substituted phenyl)-1,3-dithiolylium ions and Hammett σ_p constants ⟨71T4003⟩.

A large number of UV absorption data of 2-substituted 1,3-benzodithiolylium ions is reported in the literature ⟨68TL165, 77S263⟩. For example, 2-phenyl-1,3-benzodithiolylium tetrafluoroborate shows absorption maxima at λ_{max} (log ε) 247sh (3.67), 290 (3.66) and 389 (4.39) nm in acetonitrile/sulfuric acid ⟨77S263⟩. Bathochromic shifts are also observed by introducing electron-donating groups into the p-position of the 2-phenyl group ⟨77S263⟩.

The deeply colored 2,5-diaryl-1,3-dithiolylium-4-olates (2) show intense long wavelength $\pi \to \pi^*$ transitions between 553 and 599 nm (Table 7). According to a PPP calculation, these bands have been assigned to intramolecular charge transfer transitions ⟨73JPR690⟩. The strongest substituent effect is displayed by the dimethylamino function in the p-position of the 2-phenyl group which causes a bathochromic shift of the long wavelength absorption of 46 nm (1388 cm^{-1}). As one can see from the UV data of Table 7, electron-donating p-substituted phenyl groups at the 5-position cause a stronger bathochromic shift than in the 2-position, whereas electron-attracting p-substituted phenyl groups have the larger shift effect at the 2-position. With increasing polarity of the solvent the longest wavelength absorption of the mesoionic dithiolone (2; $R^1 = R^2 = $ Ph) shows a hypsochromic shift which correlates with solvent polarity parameters ⟨76CB740⟩. This negative solvatochromic effect is in agreement with the highly polar mesoionic ground state of (2), which possesses dipole moments between 5.00 and 5.98 D ⟨76CB740⟩.

Table 7 IR and UV Spectral Data of some 1,3-Dithiolylium-4-olates (2) ⟨76CB740, 78CB2021⟩

R^1	R^2	ν(C=O) (cm^{-1})[a]	λ_{max}(log ε) (nm)[b]
Ph	Ph	1580	553 (4.11)
Ph	4-MeOC$_6$H$_4$	1576	563 (4.20)
4-MeOC$_6$H$_4$	Ph	1612, 1595	570 (4.15)
Ph	4-Me$_2$NC$_6$H$_4$	1602, 1592 1568, 1558	599 (4.36)
Ph	4-NO$_2$C$_6$H$_4$	1596, 1578	580 (4.11)
4-NO$_2$C$_6$H$_4$	Ph	1597, 1589	550 (4.25)[c]
Ph	Piperidino	1610, 1584	478 (3.62)
Me	Ph	1576	508 (3.90)
H	Ph	1600	483 (3.78)

[a] KBr. [b] Dioxane. [c] CH$_2$Cl$_2$.

The long wavelength absorption maximum in the UV spectrum of tetrathiafulvalene (5) (Table 6), which exhibits three other absorptions, has low intensity ($\varepsilon = 270$). This band has been assigned to a $\pi \to \sigma^*$ transition ⟨71JA2258⟩.

4.32.2.6 IR Spectroscopy

Only a few IR data of 1,3-dithiolylium cations have been reported ⟨66AHC(7)39, 80AHC(27)151⟩.

The mesoionic 1,3-dithiolones of type (2) show substituent-dependent IR carbonyl stretching vibrations between 1612 and 1558 cm^{-1} (Table 7). These low frequencies are characteristic for this class of compound and are in agreement with the mesoionic structure. Most of the mesoionic 1,3-dithiolones containing p-substituted phenyl groups show split carbonyl absorption bands, presumably as the result of Fermi resonances ⟨76CB740⟩.

Characteristic IR vibrations of 1,3-dithiolan-2-one and 1,3-dithiolane-2-thione are given in Table 8.

1,3-Dithioles 819

Table 8 IR Vibrations of 1,3-Dithiolan-2-one and 1,3-Dithiolane-2-thione ⟨63PMH(2)161⟩

	C=X	scissors	CH$_2$ wag	twist	rock	ν Ring	γ Ring
1,3-Dithiolan-2-one	1638	1434 1422	1275 1254	1158	983	888 826 677 939	—
1,3-Dithiolane-2-thione	1058	1416 1370	1275 1243	1148	983	882 831 670 946	457

4.32.3 REACTIVITY

This section is divided into reactions of fully unsaturated rings like 1,3-dithiolylium ions, mesoionic 1,3-dithiolones, 1,3-dithiol-2-ones and 1,3-dithiole-2-thiones, and reactions of the saturated 1,3-dithiolanes. As a consequence of the positive charge in 1,3-dithiolylium ions, the main reactions of this class of compounds consist of nucleophilic attack at the 2-position, whereas the mesoionic 1,3-dithiolones undergo cycloaddition reactions. Reactivity of the benzo ring in benzo-1,3-dithiole and related systems has not been studied to any extent.

4.32.3.1 Reactions of 1,3-Dithioles and Mesoionic 1,3-Dithiolones

4.32.3.1.1 *Thermal and photochemical reactions formally involving no other species*

(i) *Fragmentation and rearrangement*

In the 1,3-dithiolylium-4-olate series one interesting example of a photochemical reaction is described. Photolysis of (**9**) gave the isomer (**11**) *via* a valence isomerization of (**10**), together with a mixture of (**13**), (**14**), diphenylacetylene and sulfur ⟨79MI43200, 82JCS(P1)1885⟩.

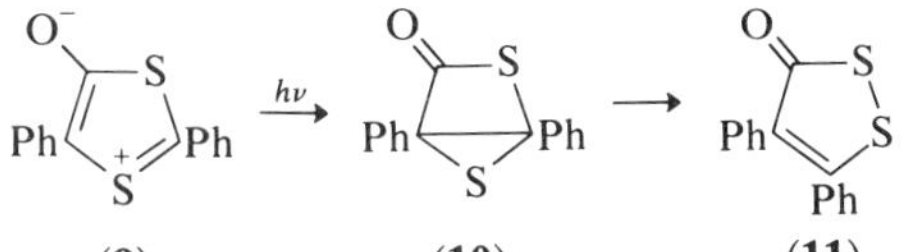

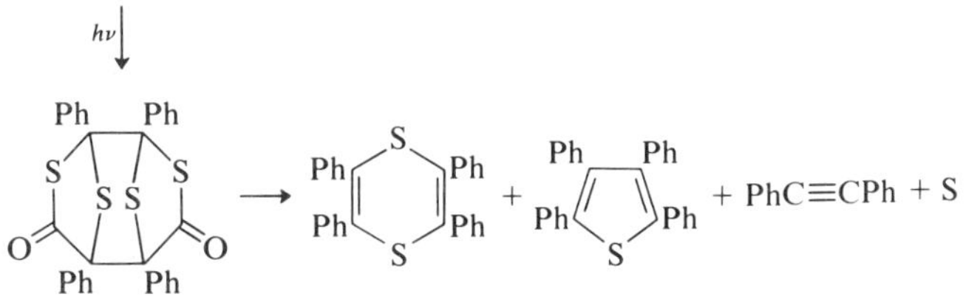

A photochemical decarbonylation reaction has been observed with the 1,3-dithiol-2-one (**15**). Photolysis of (**15**) at room temperature gives the dark-red crystalline dithione derivative (**16**). Further irradiation of (**16**) in dichloromethane at −50 °C yields a yellow product believed to be the dithiete (**17**) ⟨74JA3502⟩.

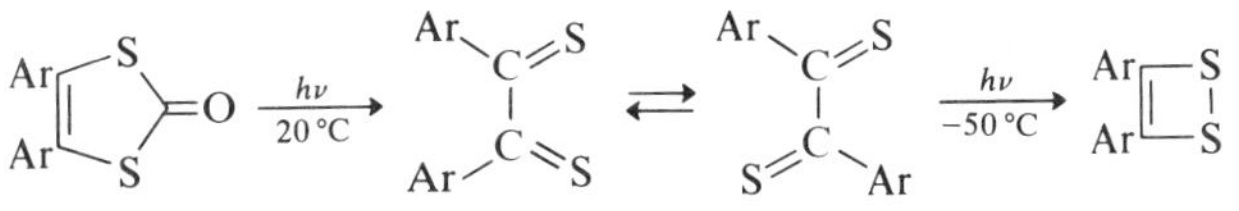

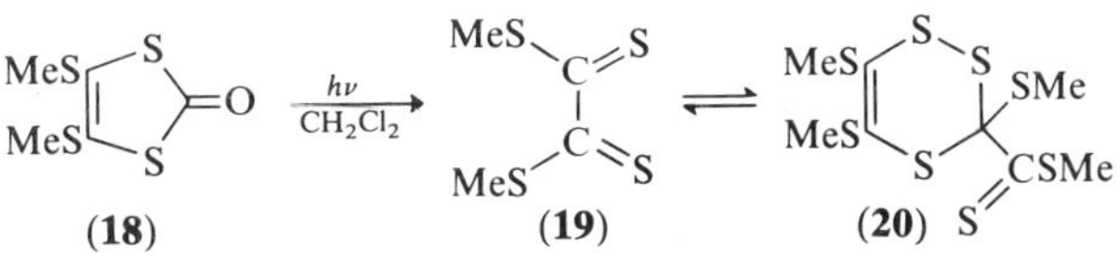

A similar reaction has been observed for the 1,3-dithiolone (**18**) which leads to formation of the tetrathiooxalate (**19**). The latter is in equilibrium with its [4+2] dimer (**20**) in solution ⟨80CB1898⟩.

Interestingly, the antiaromatic thiirene (**22**) has been observed on irradiation of the 1,3-dithiole-2-thione (**21**) in an argon matrix ⟨79NJC149⟩.

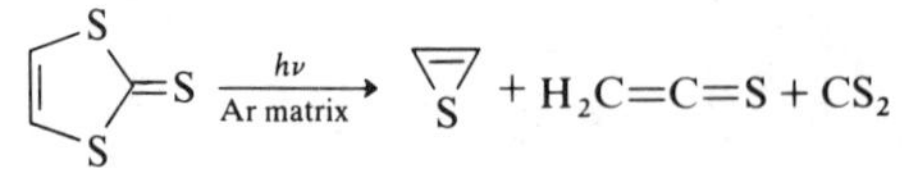

On the other hand, it has been shown ⟨79ZC192⟩ that photolysis of benzo- and aryl-substituted 1,3-dithiolethiones (**23**) produced the tetrathiafulvalene derivatives of type (**24**).

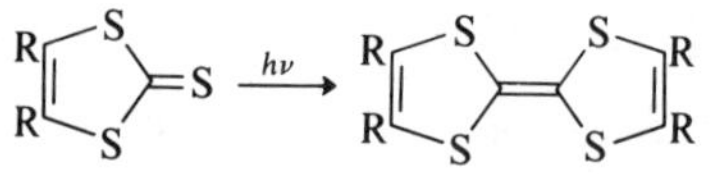

(ii) *Dimerization*

Crystalline dimers of type (**27**) have been synthesized under thermal conditions for the first time from phenylacetic acid derivatives (**25**) ⟨79CB1650⟩. The structure of (**27**; R = Me) was established by X-ray analysis ⟨78TL671⟩. The dimer (**27**; R = Me) is stable in solution, whereas the analogous alkylthio-substituted dimers are in a temperature and solvent dependent equilibrium with the monomeric 1,3-dithiolone (**26**). The monomers (**26**), generated from the dimers (**27**), display the same propensity towards [3+2] cycloaddition as do the mesoionic dithiolones which are generated *in situ* from the precursors (**25**). A photochemical dimerization of (**9**) has also been described ⟨80CL717⟩ (see Section 4.32.3.1.8(iii)).

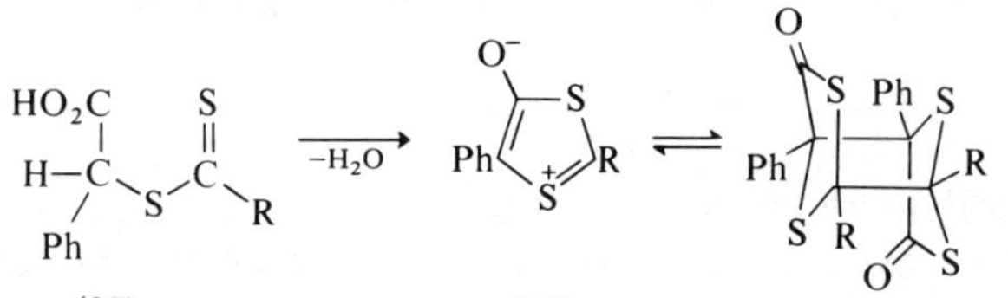

4.32.3.1.2 *Electrophilic attack at carbon*

Since 1,3-dithiolylium ions are positively charged, attack of electrophiles at the ring carbons is very rare. One example, where the mesoionic 1,3-dithiolone is acylated at the 5-position *in situ* during its preparation, occurs in the preparation of (**29**) from its precursor (**28**) ⟨78CB2021⟩.

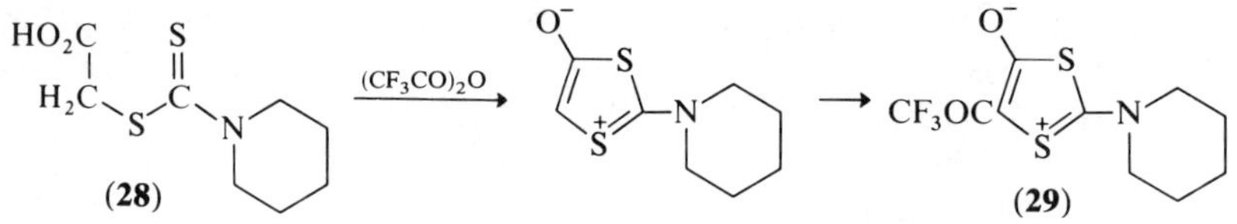

4.32.3.1.3 *Oxidation*

Several 1,3-dithiolylium salts have been prepared by hydride removal from 2*H*-1,3-dithioles. Naturally, this elimination is favored by electron-releasing substituents in the 2-position, whereas 4,5-dicyano-1,3-dithiole remains stable toward oxidizing agents. The reaction of the 1,3-dithioles (**30**) with trityl salts to produce the 1,3-dithiolylium ions (**31**) offers one such example ⟨80AHC(27)151⟩.

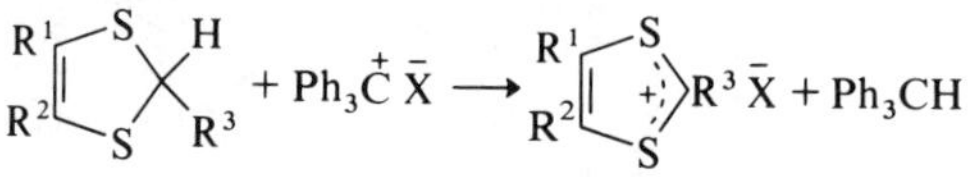

4.32.3.1.4 Nucleophilic attack at carbon

As a 6π-electron system with a positively charged five-membered ring, the 1,3-dithiolylium salts react exclusively at the 2-position with nucleophilic reagents while the 1,3-dithiolylium-4-olates undergo nucleophilic attack at the 4-position. Unlike the 1,2-dithiolylium system which undergoes ring-opening reactions in aqueous hydroxide, the 1,3-dithiolylium ring is stable in strongly basic media.

(i) *Water and other O-nucleophiles*

The benzo-1,3-dithiolylium salts behave towards water like typical Lewis acids. For example, the carbinol (**33**) which is in equilibrium with the 1,3-dithiolylium salt (**32**) reacts with further (**32**) to form the ether (**34**). This reverts to (**32**) on addition of acid ⟨66AHC(7)39⟩.

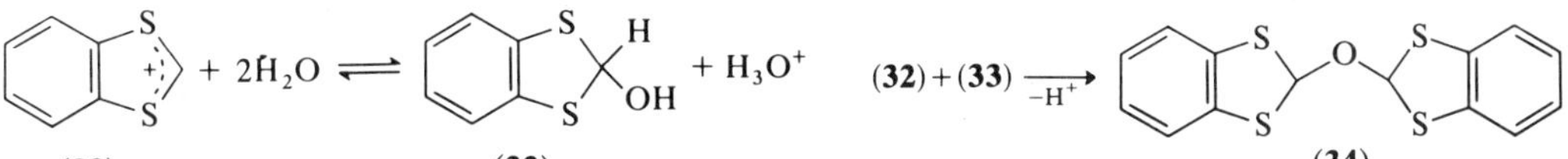

A corresponding ether (**35**) is obtained by treatment of the triphenyl-1,3-dithiolylium salt (**36**) with ethoxide ion. The resulting ether is completely reconverted into (**36**) by treatment with perchloric acid ⟨66AHC(7)39⟩.

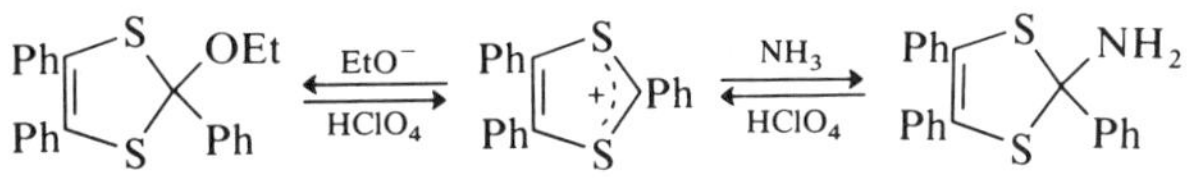

(ii) *N-Nucleophiles*

At room temperature the aminodithiole (**37**) is obtained from the 1,3-dithiolylium salt (**36**) and ammonia in benzene solution, and the reverse reaction is facilitated by treatment with perchloric acid ⟨66AHC(7)39⟩.

Sodium azide can react with 1,3-benzodithiolylium tetrafluoroborate (**32**) in acetonitrile to form the azide (**38**) which on treatment with trityl tetrafluoroborate yields trityl azide and (**32**) ⟨80JOC2024⟩. The last step of this reaction does not lead to the expected 2-azido-1,3-benzodithiolylium salt.

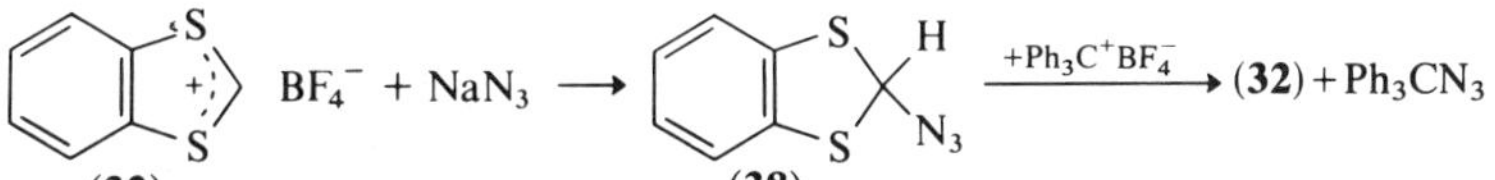

In the case of 1,3-dithiolylium-4-olates (**39**) the reaction with morpholine is initiated by nucleophilic attack at the 4-position of (**39**) with formation of the thiobenzomorpholides (**41**) and mercaptophenylacetomorpholide (**40**); the latter undergoes oxidation to the disulfide (**42**) ⟨78CB3178⟩.

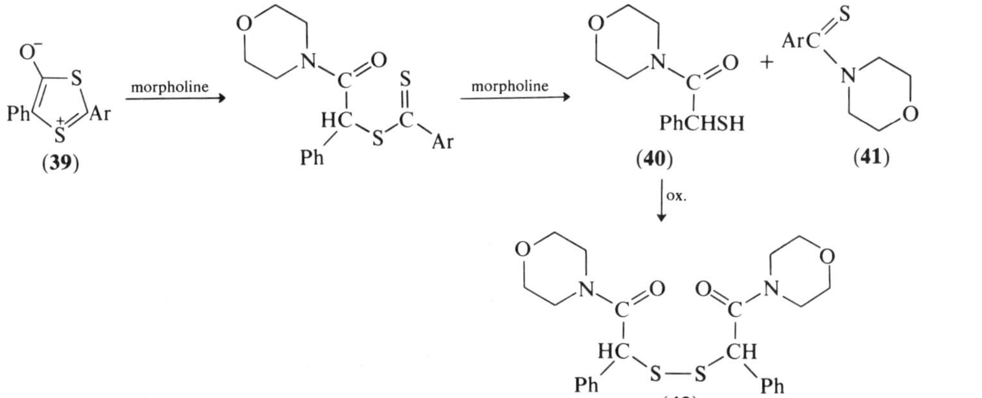

A surprising reaction takes place between the mesoionic dithiolone (**9**) and aniline. After primary attack at the 4-position the ring-opened intermediate (**43**) either recloses with liberation of H₂S in 76% yield to the mesoionic 1,3-thiazolone (**44**), or reacts with a second molecule of aniline to produce the disulfide (**45**; 9%) together with thiobenzanilide (**46**; 5%) ⟨78CB3178⟩.

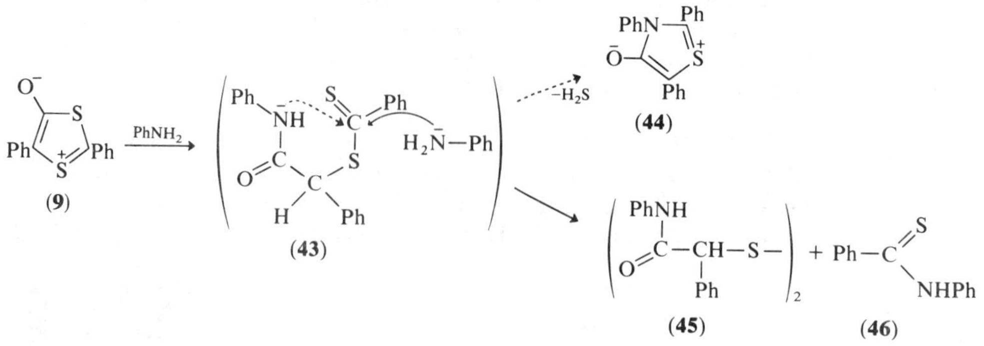

(iii) S-Nucleophiles

Like alkoxide ions, thiolate, dithiocarbamate and *O*-ethyl dithiocarbonate anions easily undergo nucleophilic addition at the 2-position of 1,3-dithiolylium salts (**31**) in boiling ethanol or acetonitrile with formation of derivatives (**47**) ⟨80AHC(27)151⟩.

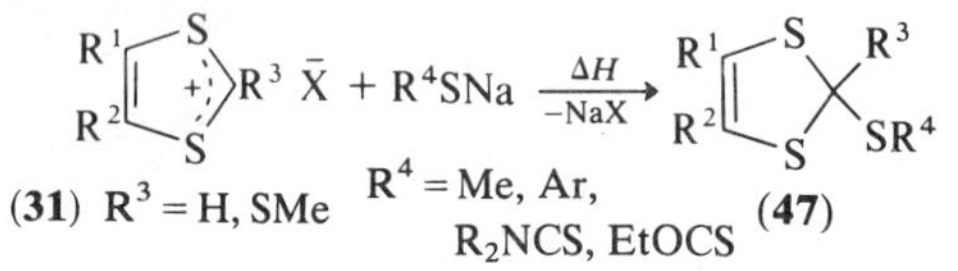

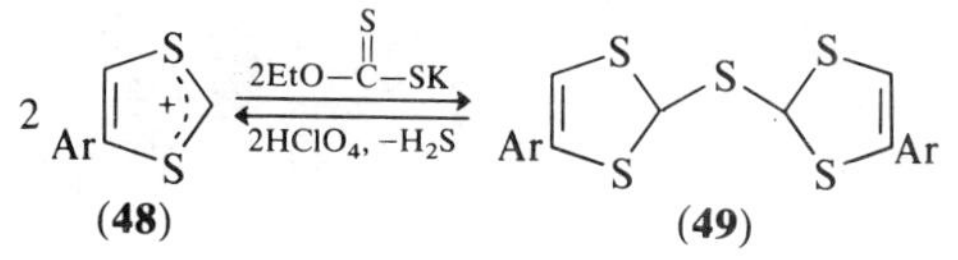

However, reaction of the 4-aryl-1,3-dithiolylium salts (**48**) with 2 moles of potassium *O*-ethyl dithiocarbonate in acetone produces the thioether (**49**) instead of the expected adduct, a result which has been explained in terms of the different solvation effect of the nucleophile in acetone and acetonitrile ⟨77JOC1543⟩. The reaction is reversed by treatment of (**49**) with perchloric acid.

(iv) Carbon nucleophiles

Grignard reagents rapidly add to 1,3-benzodithiolylium perchlorate in dry ether to produce 2-substituted 1,3-benzodithioles which can be hydrolyzed in high yields to aldehydes ⟨76JCS(P1)1886⟩.

Another example results from the synthesis of compound (**53**) which contains the structural elements of a tetrathiafulvalene and of a quinodimethane. Thus treatment of the bis Grignard reagent of (**50**) with the 1,3-benzodithiolylium perchlorate (**51**) leads to (**52**) which is converted into (**53**) by subsequent treatment with trityl fluoroborate and triethylamine ⟨80AG(E)204⟩.

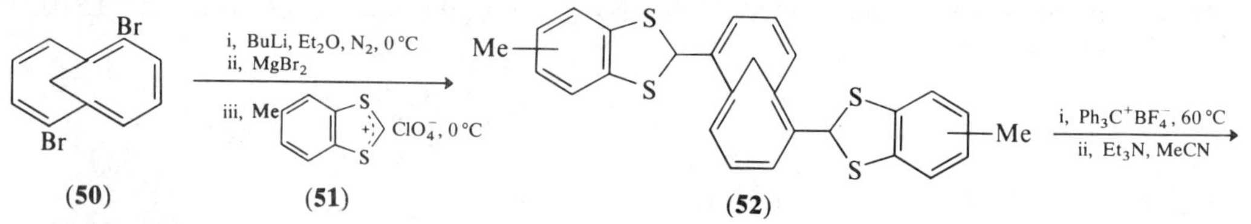

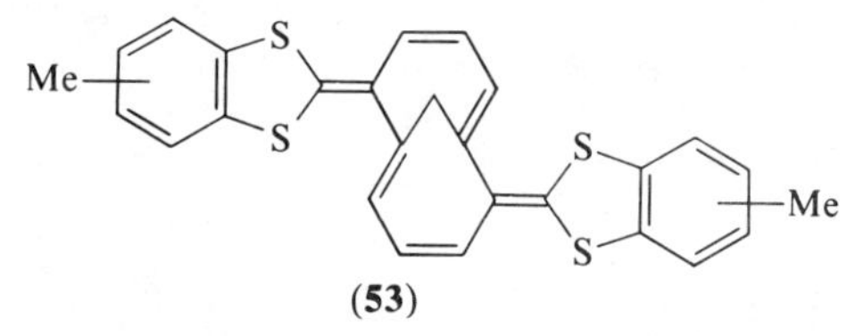

The electron-rich *para*-position of arylamines, phenols, and phenol ethers react at C-2 of 1,3-dithiolylium ions (**54**) leading to compounds of type (**55**). When both *para*-positions are blocked, as in *N,N,N',N'*-tetramethyl-1,4-phenylenediamine, then *ortho* attack is observed ⟨80AHC(27)151⟩.

A colored salt of type (**58**) can be obtained in almost quantitative yield by reaction of the 1,3-benzodithiolylium ion (**51**) with the cinnamic acid derivative (**56**) to yield first the

(54) (55) X = NR_2, OR, OH

compound (**57**) which in turn reacts with a second molecule of (**51**) to form (**58**) and presumably (**59**). Although the reduction product (**59**) has not been isolated, its formation is in agreement with the proposed reaction sequence ⟨66AHC(7)39⟩.

(56) (57) (58) (59)

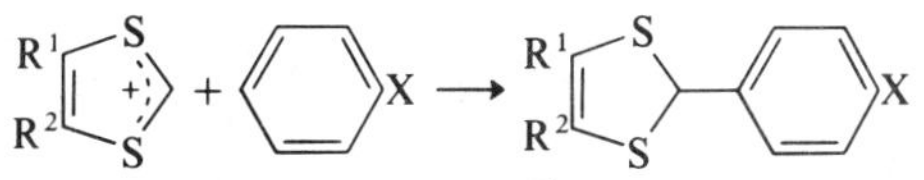

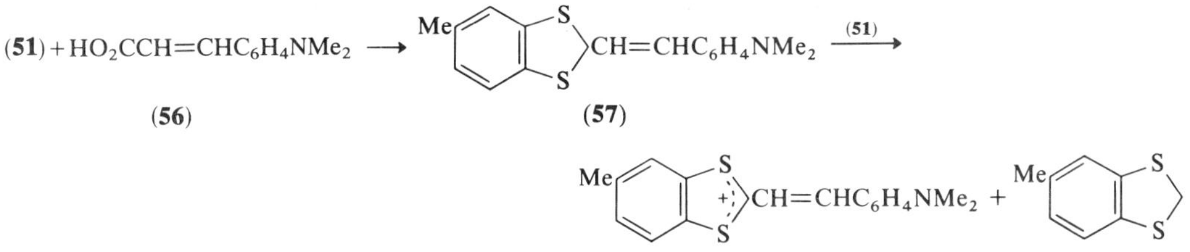

The attack of the cation (**51**) on (**56**) finds parallels in many reactions of 1,3-dithiolylium ions with 1,1-disubstituted alkenes and with cinnamic acid derivatives ⟨59HCA1733, 63HCA2167⟩.

Activated methylene compounds may react with one or two molecules of 1,3-dithiolylium salts to produce (**60**) or (**61**), respectively. For example, malononitrile and 1,3-benzodithiolylium ion (**54**) yields a product of type (**61**), whereas methyl acetate forms the one-to-one product (**60**) ⟨80AHC(27)151⟩.

(54) (60) (61)

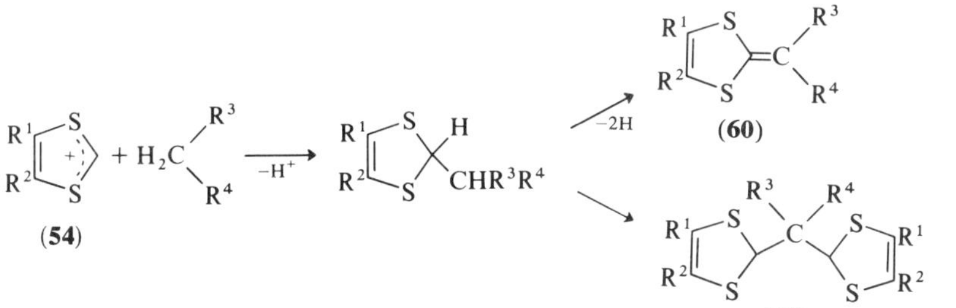

Furthermore, cyclopentenedione (**62**) reacts with 1,3-dithiolylium tetrafluoroborate (**1**) in aqueous ethanol with formation of compound (**63**) which can be easily oxidized using 2,3-dichloro-5,6-dicyanobenzoquinone (DDQ) to produce the dithiafulvalene derivative (**64**) ⟨77CL77⟩.

(1) (62) (63) (64)

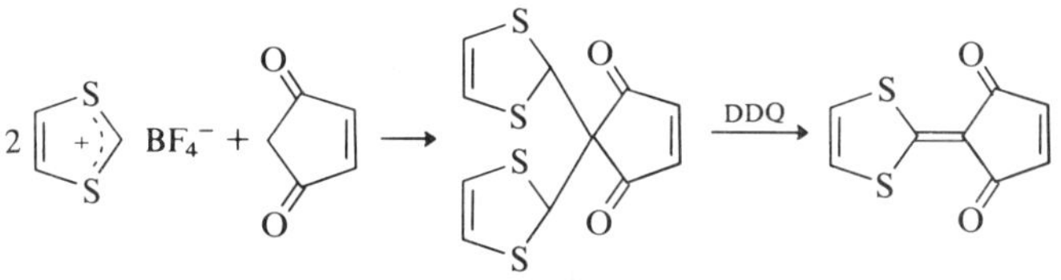

Sulfur ylides can also react with 1,3-dithiolylium salts. One example is given by the treatment of the 1,3-dithiolylium perchlorate (**48**) with the ylide (**65**) in dichloromethane which forms the compound (**66**) in 41% yield ⟨77T1595⟩.

(48) (65) (66)

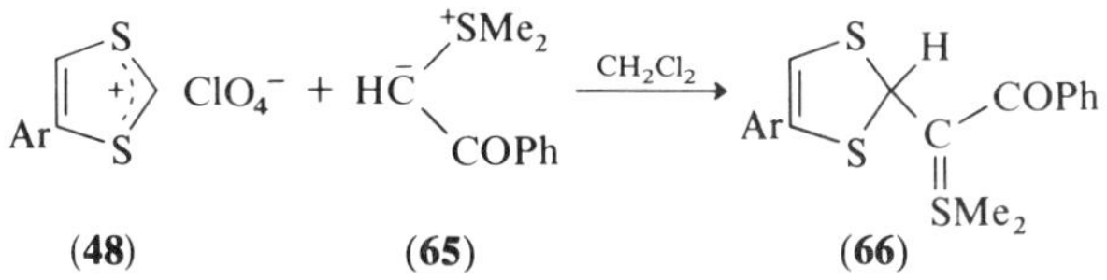

(v) *Reduction*

The 1,3-dithiolylium salts of type (**31**) easily add hydride ions on treatment with sodium borohydride in methanol, ethanol or THF to form the 1,3-dithioles (**67**) ⟨80AHC(27)151⟩.

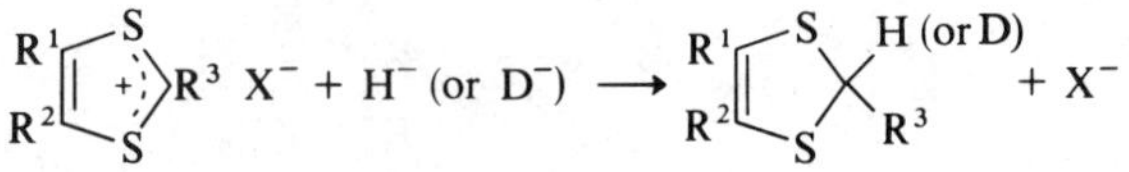

(31) R³ = H, SMe, NR₂, Ar, cyclohexyl   (67)

With sodium in liquid ammonia, 1,3-benzodithiolylium tetrafluoroborate undergoes a reductive cleavage to benzene-1,2-dithiol ⟨76S471⟩.

Coupling reactions sometimes accompany the reduction of 1,3-dithiolylium salts; (1), for example, with zinc dust in an acetic acid/benzene/water mixture leads to formation of (68) ⟨77JOC2778⟩.

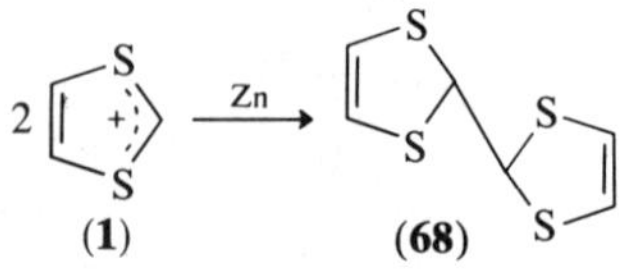

(vi) *Phosphorus nucleophiles*

The reaction of trialkyl- and triaryl-phosphines with 1,3-benzodithiolylium salts leads to formation of phosphonium salts which are deprotonated by treatment with *n*-butyllithium to produce (69). The similar reaction of trialkyl phosphites in the presence of sodium iodide yields dialkyl phosphonates which can be deprotonated to (70). Both (69) and (70) can react further with ketones to give the 2-alkylidene-1,3-dithiole derivatives (71) ⟨80AHC(27)151⟩. The ylide (72) has also been prepared ⟨80H(14)271⟩.

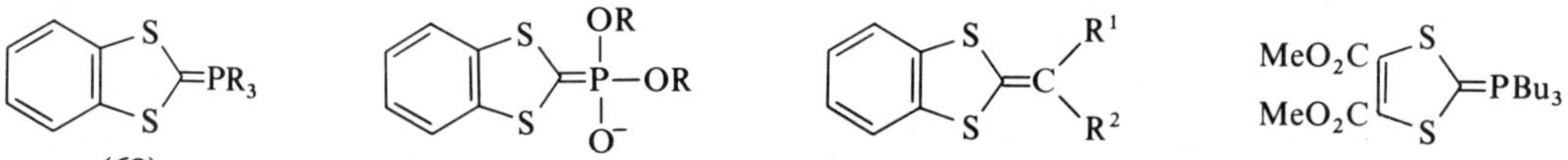

4.32.3.1.5 Nucleophilic attack at hydrogen

As in the 1,3-azolium series, H-2 of 1,3-dithiolylium ions is rather acidic. Thus, 2-unsubstituted 1,3-dithiolylium salts (1) are easily deprotonated by nucleophilic attack at hydrogen in the 2-position. The resulting carbene (73) easily undergoes dimerization with formation of tetrathiafulvalene (5). In CF₃CO₂D/D₂O solution, hydrogen–deuterium exchange occurs with formation of (74) ⟨80AHC(27)151⟩. By coupling reactions of intermediate carbenes of type (73), many tetrathiafulvalenes have been prepared.

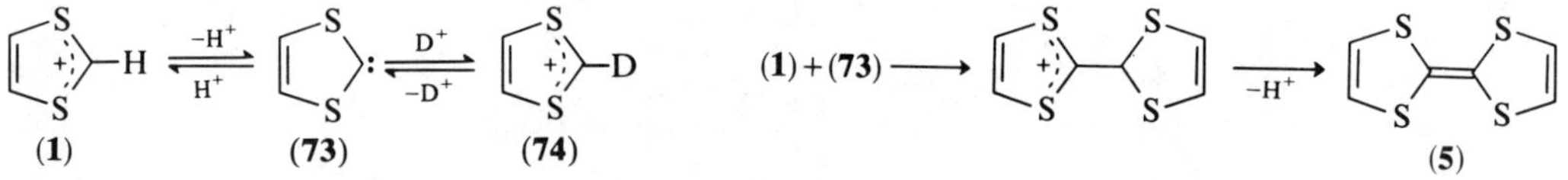

On the other hand, the reactions of 1,3-dithiolylium salts (54) with aromatic aldehydes in the presence of triethylamine proceed with formation of the ketones (75) ⟨80MI43200⟩.

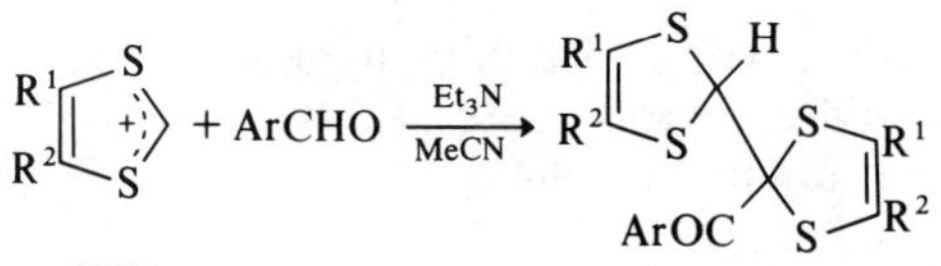

4.32.3.1.6 Reaction with triplet molecular oxygen

Since triplet molecular oxygen behaves like a diradical, an interesting reaction has been observed between triplet oxygen and 5-methyl-2-phenyl-1,3-dithiolylium-4-olate (76). Treatment of (76) in acetonitrile with molecular oxygen at room temperature in the dark results in the formation of (80) in 45% yield.

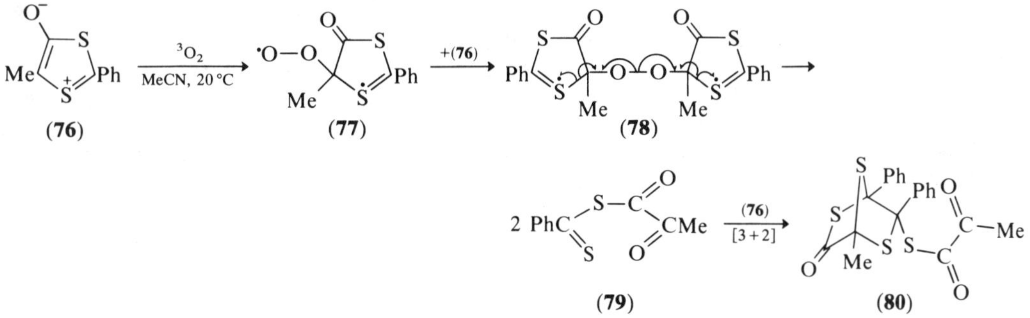

The reaction pathway has been rationalized in terms of an attack of the triplet oxygen at the 5-position of (**76**) to produce the diradical (**77**) which may add to a second molecule of (**76**). The resulting peroxide (**78**) fragments to the dithiobenzoic acid derivative (**79**) which may act as a dipolarophile in the last step of the [3+2] cycloaddition reaction of (**76**). The structure of the final product (**80**) was established by X-ray analysis ⟨81CB285⟩.

4.32.3.1.7 Nucleophilic attack at valence tautomeric open-chain ketene

The mesoionic 1,3-dithiolone (**9**) can also react from the valence tautomeric ketene form of type (**81**) although this equilibrium has not been established by spectral methods. Thus, treatment of the 1,3-dithiolone (**9**) with 3-dimethylamino-2H-azirine (**82**) in boiling acetonitrile produces the oxazoline derivative (**84**; 17–39%) and the 1,2-dithiolone (**11**; 3–22%) ⟨79HCA1236⟩. These surprising results have been rationalized in terms of a nucleophilic attack of the azirine (**82**) on the ketene (**81**) to form the zwitterionic intermediate (**83**) which then collapses to (**84**). But it has not been established whether the isomerization of the 1,3-dithiolone (**9**) to its isomer (**11**) is induced by a thermal or photochemical reaction. In another connection, this isomerization to (**11**) has also been observed in a photochemical reaction ⟨79MI43200⟩.

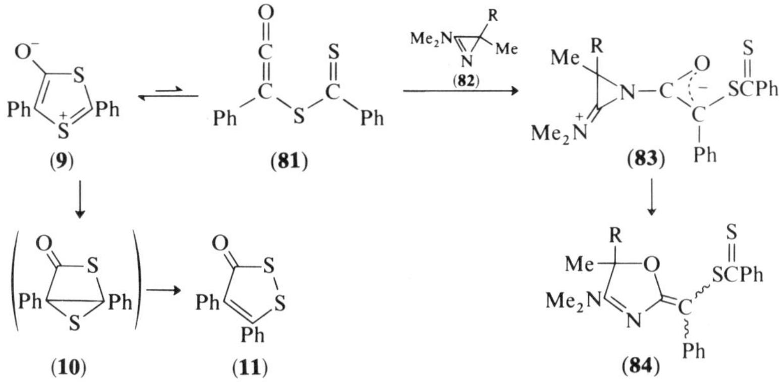

4.32.3.1.8 Cycloaddition reactions

This chapter is subdivided into [3+2], [4+2] and [4+4] cycloaddition reactions. According to the rules of conservation of orbital symmetry, the first two cycloaddition reactions, [3+2] and [4+2], are of the same electronic type, namely [$_\pi 4_s + _\pi 2_s$], and are synchronous, thermally allowed, pericyclic reactions ⟨69AG(E)781⟩.

(i) *1,3-Dipolar cycloaddition reactions of 1,3-dithiolylium-4-olates to alkynes*

The 1,3-dithiolylium-4-olates (**2**) contain the masked 1,3-dipolar system of a cyclic thiocarbonyl ylide, and hence they are preparatively valuable synthons for subsequent heterocyclic syntheses.

For example, the mesoionic 1,3-dithiolones (**2**) combine across the 2,5-position at 90–130 °C with symmetrically substituted alkynes as dipolarophiles with formation of non-isolable primary adducts of type (**85**). The latter fragment in a retro Diels–Alder type reaction with fast evolution of carbon oxysulfide to produce the thiophene derivatives (**86**) of Table 9 ⟨76CB753, 76JOC1724, 78CB2021⟩. The analogous [3+2] cycloaddition reactions of (**9**) with unsymmetrically substituted alkynes such as propiolic ester, phenylpropiolic ester, acetyl phenylacetylene, benzoyl phenylacetylene and phenylacetylene also proceed with formation of thiophene derivatives in high yields ⟨76CB753⟩. Benzyne can be used as a dipolarophile leading to the benzothiophene derivative (**87**) ⟨76JCS(P1)672⟩.

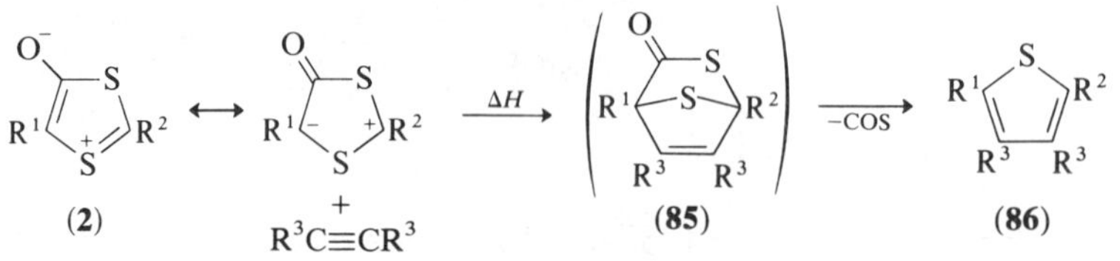

Table 9 Thiophenes (**86**) Synthesized from 1,3-Dithiolylium-4-olates (**2**) and Symmetrically Substituted Alkynes

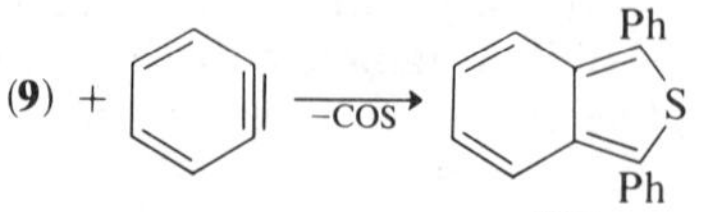

R^1	Thiophene R^2	R^3	Yield (%)	Ref.
H	Ph	CO_2Me	40	76JOC1724
H	4-MeOC$_6$H$_4$	CO_2Me	47	76JOC1724
H	4-ClC$_6$H$_4$	CO_2Me	44	76JOC1724
H	Ph	PhCO	68	76JOC1724
Ph	Ph	CO_2Me	100	76CB753
Ph	4-MeC$_6$H$_4$	CO_2Me	86–89	76CB753
Ph	4-BrC$_6$H$_4$	CO_2Me	75–91	76CB753
Ph	4-MeOC$_6$H$_4$	CO_2Me	90–93	76CB753
Ph	4-Me$_2$NC$_6$H$_4$	CO_2Me	88	76CB753
4-NO$_2$C$_6$H$_4$	Ph	CO_2Me	70	76CB753
Ph	Me	CO_2Me	67	76CB753
Ph	Piperidino	CO_2Me	80	78CB2021
Ph	Morpholino	CO_2Me	87	78CB2021
Ph	Pyrrolidino	CO_2Me	93	78CB2021
Ph	Piperidino	PhCO	83	78CB2021

With unsymmetrically substituted mesoionic 1,3-dithiolones (**2**) and alkynes, the cycloaddition reactions normally yield the two possible isomeric thiophene derivatives. Thus methyl propiolate reacts with (**2a**) to produce an 81:19 ratio of the two possible isomeric thiophenes (**88**) and (**89**) in 90% yield, whereas the same reaction with the isomeric 1,3-dithiolone (**2b**) yields the thiophene derivatives in a 30:70 ratio (99%) ⟨78CB2028⟩. A further example is provided by the reaction of phenylacetylene with (**2a**) which proceeds with formation of the thiophene isomers (**90**) and (**91**) in an 89:11 ratio; however, in the case of (**2b**) and the same substrate, the product isomer ratio is completely reversed (11:89) ⟨78CB2028⟩.

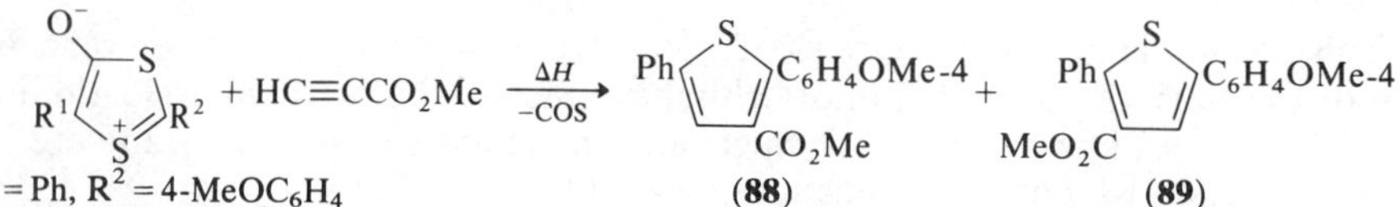

(**2a**) R^1 = Ph, R^2 = 4-MeOC$_6$H$_4$
(**2b**) R^1 = 4-MeOC$_6$H$_4$, R^2 = Ph

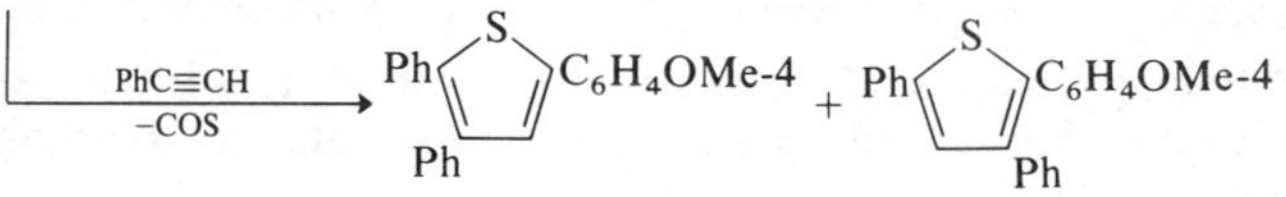

These results clearly indicate that, at least in these cases, the orientation of the dipolarophile during the primary attack at the 2,5-positions of the 1,3-dithiolone is approximately independent of the nature of the substituents at those positions, and that regioselectivity of these cycloaddition reactions is reversed by changing from methyl propiolate to phenylacetylene. The observed regioselectivities during these cycloaddition reactions can be explained in terms of qualitative frontier molecular orbital perturbation theory. It has been established that the 1,3-dipolar cycloaddition reactions of (**2**) and dipolarophiles like propiolate or phenylacetylene are HOMO(dipole)–LUMO(dipolarophile) controlled ⟨78CB2028, 73JA7301⟩. The orientation of unsymmetrically substituted dipolarophiles in the primary cycloadduct is determined by the size of the atomic orbital coefficients at the reaction centers, because the overlap of reaction centers with large atomic orbital coefficients in pairs at one site and small ones in pairs at the other site is energetically more favored than the overlap of large and small ones in pairs on each site. In general, electron-donating substituents ($+M$) at alkynes cause a reduction of the atomic orbital coefficient at the substitution site in the HOMO while the corresponding coefficient in the LUMO is enlarged. On the other hand, electron acceptor groups ($-M$) have the reverse effect on the size of the atomic orbital coefficients, *i.e.* they enlarge the corresponding coefficient in the HOMO and reduce it in the LUMO. Therefore in the HOMO, the 1,3-dithiolone (**2**) possesses a larger atomic orbital coefficient at C-5 than at C-2 as shown in formula (**93**; Scheme 1). Consequently, methyl propiolate combines with (**93**) mainly *via* the primary adduct (**92**) to produce the orientation shown in the thiophene derivative (**95**), whereas phenylacetylene mainly leads to a primary adduct (**94**) from which the thiophene structure (**96**) is derived ⟨78CB2028⟩. These qualitative considerations are in good agreement with the observed isomer ratios.

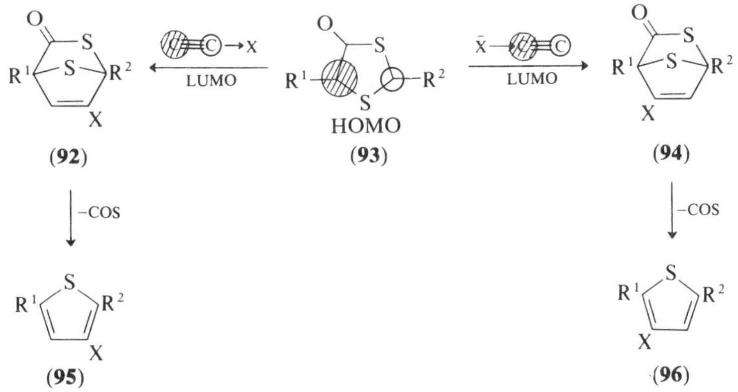

Scheme 1

An interesting example of an intramolecular cycloaddition reaction has been observed in the case of the mesoionic 1,3-dithiolone (**97**) which contains a non-activated alkyne in the same molecule. Thus, on heating (**97**) at 40–45 or 100 °C, the tricyclic thiophene derivative (**99**) is obtained *via* the non-isolable primary adduct (**98**) ⟨81LA347⟩.

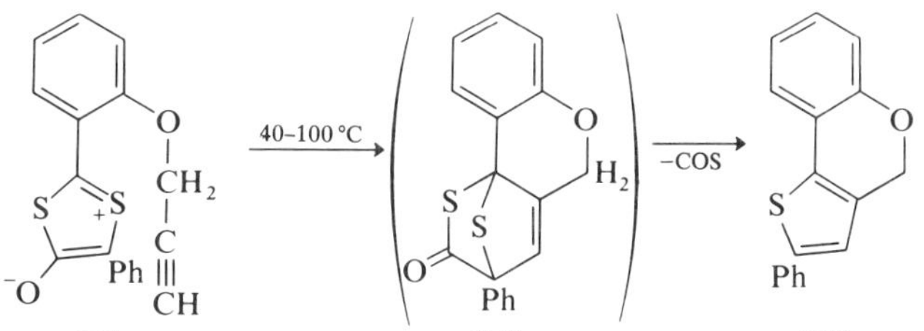

(ii) *1,3-Dipolar cycloaddition reactions of 1,3-dithiolylium-4-olates to alkenes*

The above subsection has shown that the mesoionic 1,3-dithiolones react with alkynes to produce thiophene derivatives exclusively. However, in reactions with alkenes, the primary cycloaddition products do not generally eliminate COS; instead they turn out to be stable under the reaction conditions. Surprisingly, the reaction of (**2**; $R^1 = R^2 = Ph$) with *N*-phenyl maleimide in boiling xylene proceeds with formation of a mixture of the *endo/exo* cycloadducts (**100**) and (**101**; $R^1 = R^2 = Ph$) ⟨78CB3029⟩, whereas at 80 °C (**2**; $R^1 = H$, $R^2 = Ph$, 4-MeOC$_6$H$_4$ or 4-ClC$_6$H$_4$) combines with the same substrate to produce only the *endo* adduct of type (**100**) ⟨76JOC1724⟩. From (**2**; $R^1 = R^2 = Ph$) and maleic anhydride only an *exo*-adduct of 2,7-dithiabicyclo[2.2.1]heptan-3-one structure has been obtained ⟨78CB3029⟩.

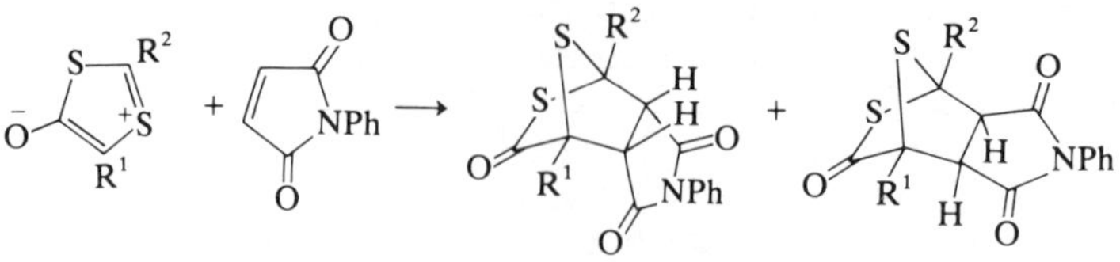

(2)   (100)   (101)

Furthermore, dimethyl maleate combines with (2; R^1 = Ph, Me, R^2 = Ph, 4-MeOC$_6$H$_4$, 4-MeC$_6$H$_4$) to yield the *endo* adducts (102), exclusively. But with dimethyl fumarate and (9), only the *trans* adducts (103) and (104) are obtained in a 69:31 ratio while with *cis*- or *trans*-dibenzoylethylene the *trans* adducts (103) and (104) are formed in a 30:70 ratio ⟨78CB3029⟩.

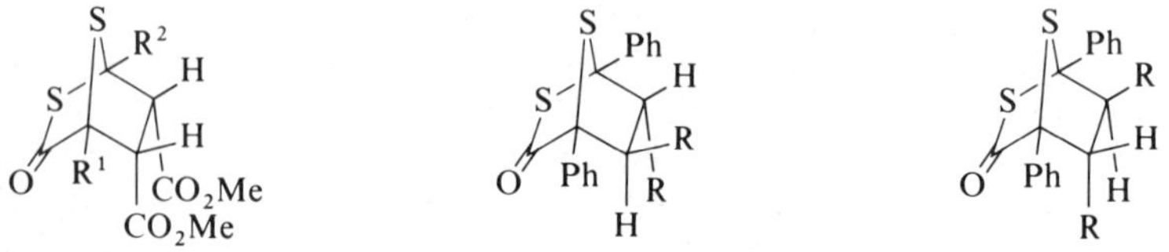

(102)   (103)   (104)

R = CO$_2$Me, COPh

More or less regioselective reactions take place between mesoionic 1,3-dithiolones and unsymmetrically substituted alkenes. For example styrene reacts with (2; R^1 = R^2 = Ph, R^1 = H, R^2 = Ph) to form the bicyclic derivatives (105; R = Ph, H), which contain the phenyl group in the 6-*endo*-position ⟨79LA360⟩. On the other hand, the reaction of methyl acrylate with the diphenyl compound (9) yields the two *endo* adducts (106) and (107) in a 64:36 ratio, and with (2; R^1 = Ph, R^2 = 4-MeC$_6$H$_4$) the isomer ratio was found to be 82:18 ⟨79LA360⟩. Further cycloaddition reactions of mesoionic 1,3-dithiolones were performed with α-methylstyrene, acrylonitrile, ethyl acrylate, methyl cinnamate, tetracyanoethylene and 2,3-dimethylbut-2-ene ⟨79LA360, 76JOC1724⟩.

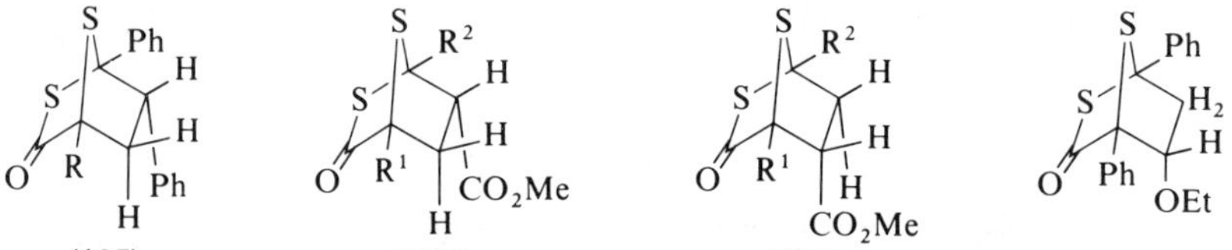

(105)   (106)   (107)   (108)

A reversion of the addition direction has been observed in the case of electron rich alkenes such as enol ethers and enamines. For example, ethyl vinyl ether reacts with the mesoionic compound (9) to produce the cycloadduct (108) in 58% yield. Other examples are the reactions of mesoionic 1,3-dithiolones with cyclohexyl vinyl ether, cyclopenten-1-yl ethyl ether, ethyl isobuten-1-yl ether and *N*-(isobuten-1-yl)morpholine. The observed regioselectivities have been also qualitatively discussed on the basis of MO perturbation theory ⟨79LA360⟩.

Cyclic alkenes also combine with mesoionic 1,3-dithiolones to produce isolable primary cycloadducts. Thus substituted cyclopropenes (109) react with 1,3-dithiolylium-4-olates (2) at 100 °C with formation of the isomeric free *exo* tricyclic compound (110), whereas cyclopropene itself reacts at room temperature to yield the corresponding 1:1 cycloadduct ⟨78CB3037⟩.

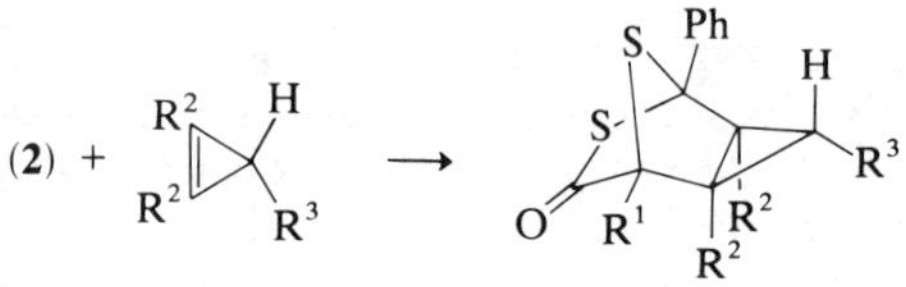

(2) + (109) → (110)

The mesoionic 1,3-dithiolones of type (2) also react with acenaphthylene or with cyclopentene to produce the *endo* adducts (111) or (112), respectively. On the other hand, the reaction of (**2**) with excess *p*-benzoquinone yields in addition to the 1:1 adduct (**113**), the thiophene derivative (**115**), which results from dehydrogenation of (**113**) to (**114**) and subsequent fragmentation ⟨78CB3037⟩.

(**111**) (**112**) (**113**)

(**114**) (**115**)

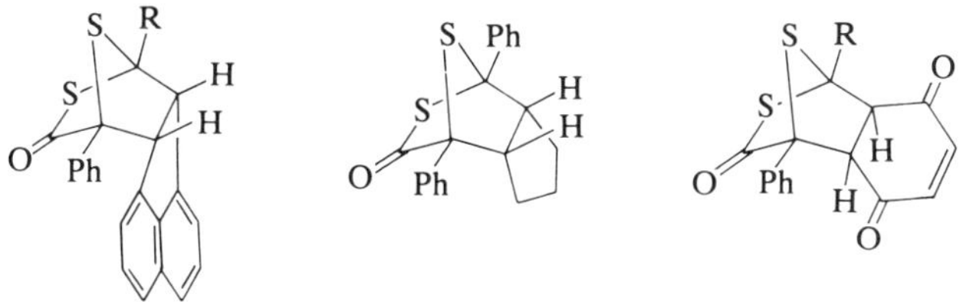

In an analogous reaction of (**2**) with norbornadiene, the tetracyclic adduct (**116**) is formed which on heating undergoes a retro Diels–Alder reaction with formation of the thiophene derivative (**117**). Further cycloaddition reactions of mesoionic 1,3-dithiolones have been carried out with cyclopentadiene, 1,3-cyclohexadiene and 1,5-cyclooctadiene ⟨78CB3037⟩.

(**116**) (**117**)

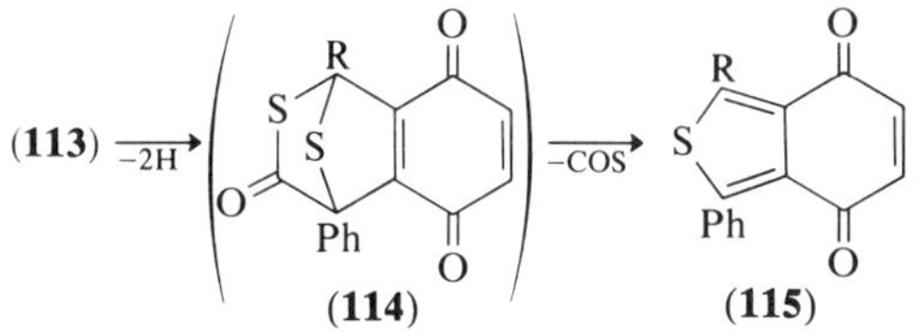

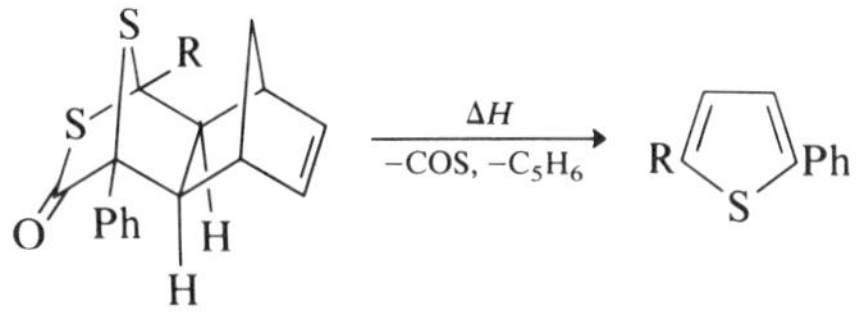

The reaction of (**9**) with dimethyl 7-oxabicyclo[2.2.1]hepta-2,5-diene-2,3-dicarboxylate (**118**) proceeds with formation of the cycloadduct (**119**) in 82% yield. Pyrolysis or photolysis of the adduct (**119**) resulted in a double fragmentation to give the thiophene derivative (**120**) and the furan (**121**) ⟨75CC840⟩. A similar reaction takes place between the mesoionic compound (**9**) and dimethyl 7-*p*-toluenesulfonyl-7-azabicyclo[2.2.1]hepta-2,5-diene-2,3-dicarboxylate which yields a 1:1 adduct of the same type as (**119**) ⟨76BCJ3314⟩.

(**118**) (**119**) (**120**) (**121**)

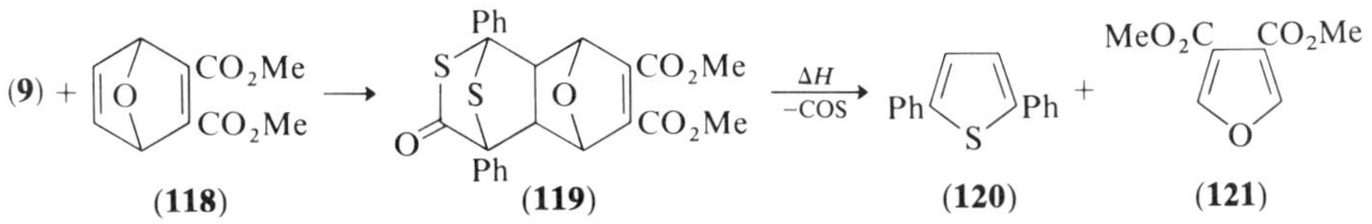

Furthermore, (**9**) combines with the tricyclic triene (**122**) to produce the adduct (**123**) which on heating undergoes a double fragmentation to form the tricyclic derivative (**124**) together with dimethyl phthalate ⟨76JCS(P1)2562⟩.

(**122**) (**123**) (**124**)

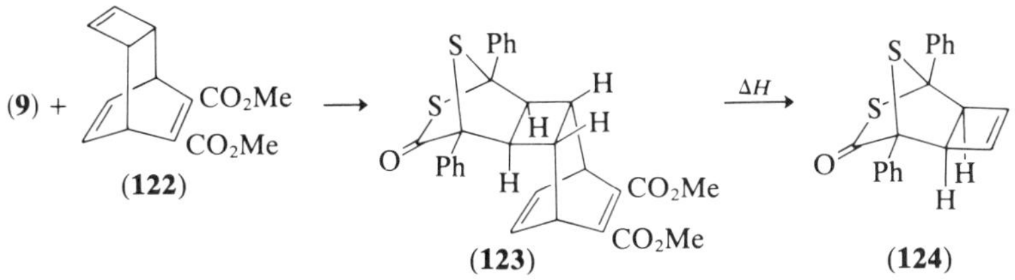

On the other hand, small-ring alkenes such as diphenylcyclopropene derivatives (**125**) may serve as substrates for the preparation of six-membered heterocycles by way of the cycloaddition–extrusion–ring-expansion reaction. Thus, the reaction of (**9**) with (**125a**) leads to formation of (**127a**) *via* the primary adduct (**126**), whereas the reaction of (**9**) with (**125b**) yields (**127b**) and the tricyclic derivative (**128**) ⟨75JCS(P1)632⟩.

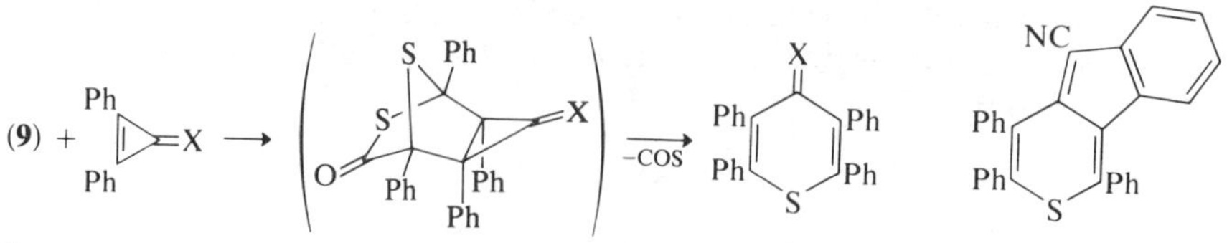

(125a) X = O
(125b) X = C(CN)CO₂Et

An example of an intramolecular cycloaddition reaction was offered by the mesoionic compound (**129**) which contains in the same molecule both a non-activated alkenic function and a cyclic thiocarbonyl ylide system. In an intramolecular [3 + 2] cycloaddition, (**129**) yields at 120 °C the tetracyclic primary adduct (**130**; 90%) ⟨81LA347⟩.

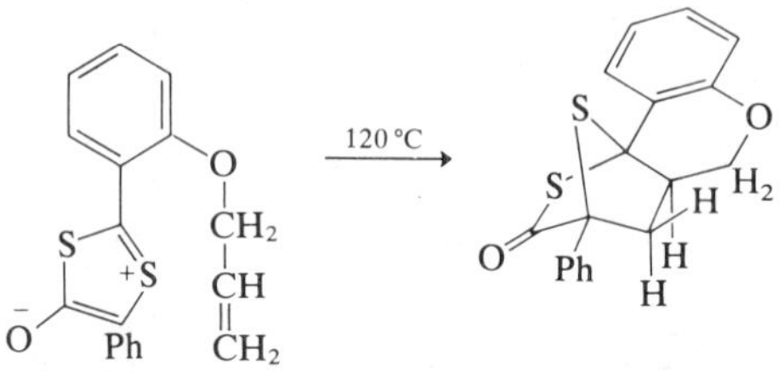

(iii) 1,3-Dipolar cycloaddition reactions of 1,3-dithiolylium-4-olates to hetero double bonds

Some hetero double bond systems have been shown to enter [3 + 2] cycloaddition reactions with the mesoionic 1,3-dithiolones. Thus, the mesoionic 1,3-dithiolones (**2**) react with formaldehyde, prepared *in situ* by depolymerization of paraformaldehyde, with regiospecific formation of the 2-oxa-6,7-dithiabicyclo[2.2.1]heptanone derivatives (**131**). The corresponding reaction of (**2**) with the N=N double bond of dimethyl azodicarboxylate proceeds *via* cycloaddition yielding (**132**), and a similar reaction takes place between (**2**) and 4-phenyl-1,2,4-triazoline-3,5-dione ⟨78CB3171⟩.

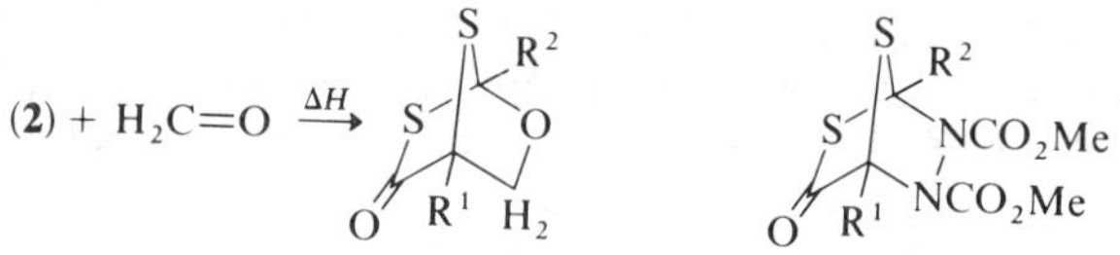

Furthermore, the mesoionic 1,3-dithiolone (**9**) reacts with the C=N double bond of ethyl 3-phenyl-2H-2-azirinecarboxylate (**133**; R = CO₂Et) at 100 °C with regio- and stereospecific formation of the *exo* adduct (**134**; R = CO₂Et; 92%). The second isomer of type (**135**) has not been observed ⟨78CB3171⟩. On the other hand, in boiling xylene the reaction of (**9**) with 2,3-diphenylazirine (**133**; R = Ph) forms the two isomeric adducts (**134**; R = Ph) and (**135**; R = Ph) in 4% and 41% yield, respectively ⟨76JOC1724⟩.

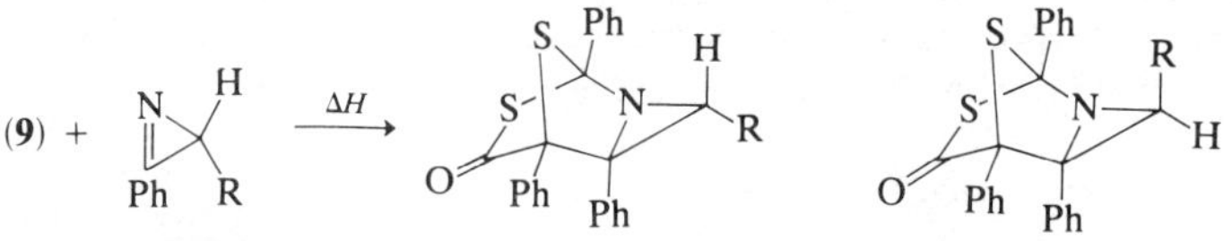

Even photochemically-generated singlet oxygen using Rose Bengal as a triplet sensitizer can serve as a dipolarophile. For example, irradiation with sodium light (λ = 589 nm) of the mesoionic 1,3-dithiolone (**26**; R = Ph) in the presence of oxygen and Rose Bengal in benzene–methanol solution at 10 °C leads to formation of methyl phenylglyoxylate (**138**) and dibenzoyl disulfide (**139**; R = Ph) as a result of a fragmentation of the primarily formed [3 + 2] cycloadduct of type (**136**) *via* (**137**). If Rose Bengal is absent and thus all light is absorbed by the mesoionic compound (**26**), no reaction has been observed. From this result it is obvious that singlet oxygen is responsible for the above reaction ⟨81CB285⟩. A similar reaction takes place with (**26**; R = 4-MeOC₆H₄) and singlet oxygen to produce (**138**) and (**139**; R = 4-MeOC₆H₄) ⟨81CB285⟩.

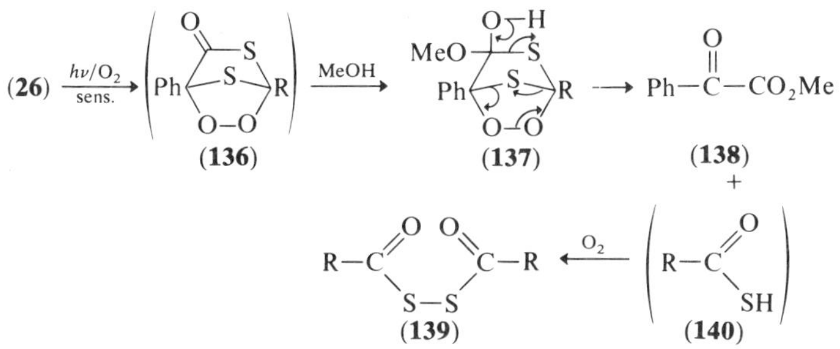

However, irradiation with a tungsten–halogen lamp of the 2,5-diphenyl-1,3-dithiolylium-4-olate in the presence of oxygen in dichloromethane solution allowed isolation of dibenzoyl disulfide (**139**; R = Ph) together with the dimer (**12**; 10%) ⟨80CL717⟩.

(iv) [4 + 2] Cycloaddition reactions of 1,3-dithiolylium-4-olates

Only a few examples of a Diels–Alder type reaction of the valence tautomeric ketene of mesoionic 1,3-dithiolones are known. Thus the reactions of differently substituted mesoionic compounds (**2**) with *o*-chloranil give the unusual adducts (**142**), derived *via* [4 + 2] cycloaddition to the non-detectable valence tautomeric ketene (**141**) which is in equilibrium with (**2**). The reaction rate depends upon the nature of the substituents, for example substituents with $-I$ and/or $-M$ effects lower the reaction rates ⟨81ZN(B)609⟩.

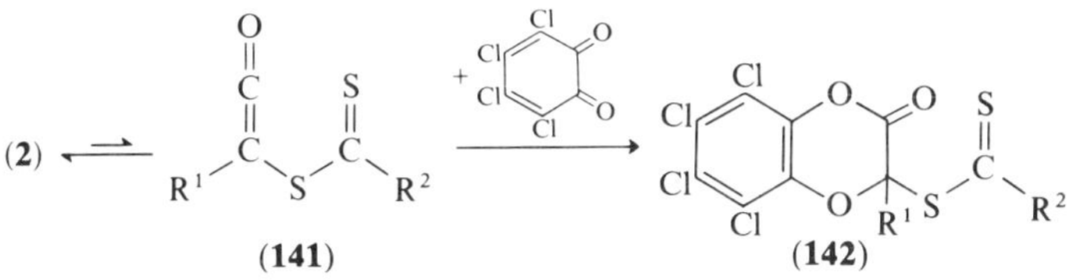

(v) [4 + 4] Cycloaddition reactions of 1,3-dithiolylium-4-olates

Besides the previously mentioned thermal and photochemical dimerizations of mesoionic compounds (see Sections 4.32.3.1.1(ii), 4.32.3.1.8(iii)) only one example of a formal [4 + 4] cycloaddition reaction not involving the same molecule as a substrate is known. Surprisingly, the *in situ* generation in the presence of *o*-chloranil of the mesoionic 1,3-dithiolone, (**144**), from the precursor dithioester (**143**) by treatment with *N,N'*-dicyclohexylcarbodiimide (DCC), proceeds with formation of the cycloadduct (**145**) in 59% yield ⟨81ZN(B)609⟩. In the absence of *o*-chloranil, the mesoionic compound (**144**) normally dimerizes across the 2,5-position to produce the dimer (**27**; R = Me) ⟨79CB1650⟩.

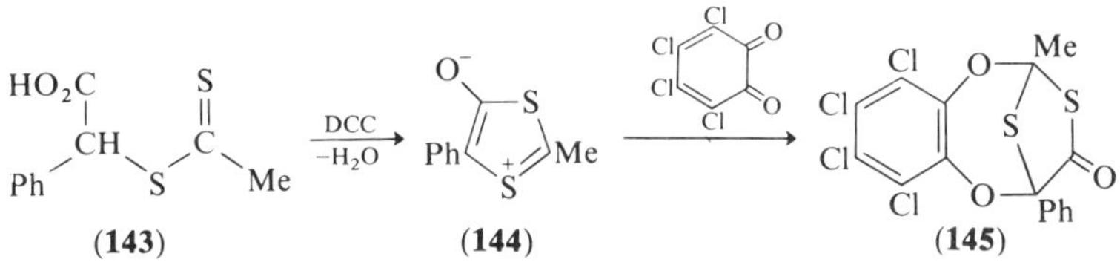

4.32.3.1.9 Reactions of substituents

(i) C-Linked substituents

α-Hydrogens of *C*-linked substituents at the 2-position such as in the 1,3-dithiolylium salts (**146**) exhibit acidic character because of the resonance stabilization of the conjugate base (**147**). Therefore, these salts can react with aldehydes in acetic acid without the aid of a special base to form the condensation products of type (**148**). This reaction has been performed with various aldehydes such as benzaldehydes, cinnamaldehydes, salicylaldehydes and 2-phenyl-4*H*-1-benzopyran-3-aldehyde ⟨80AHC(27)151, 66AHC(7)39⟩. Only when the carbonyl compound is a ketone or pyrone does the addition of strong base become necessary. Methyl groups in the 4- or 5-position are inert to these condensation reactions ⟨66AHC(7)39⟩.

(146) (147)

(148)

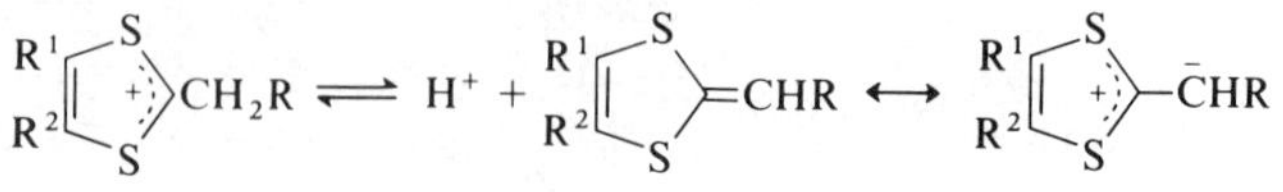

The reaction of 2-methyl-1,3-dithiolylium ion (**151**), in the form of its conjugate base (**147**), with the alkylthio-1,3-dithiolylium salt (**149**) proceeds with formation of the monomethine salt (**150**) ⟨66AHC(7)39⟩. Also trimethine dyes such as (**152**) can be obtained by the reaction of orthoformates with 2-methyl-1,3-dithiolylium ions (**151**) ⟨66AHC(7)39⟩.

(149) (150)

(151) (152)

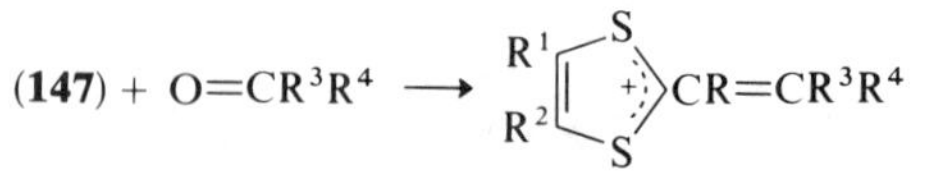

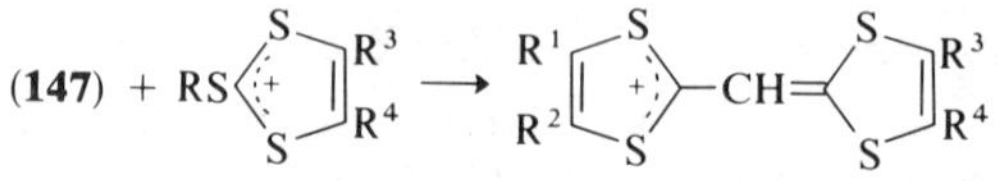

(ii) *N-Linked substituents*

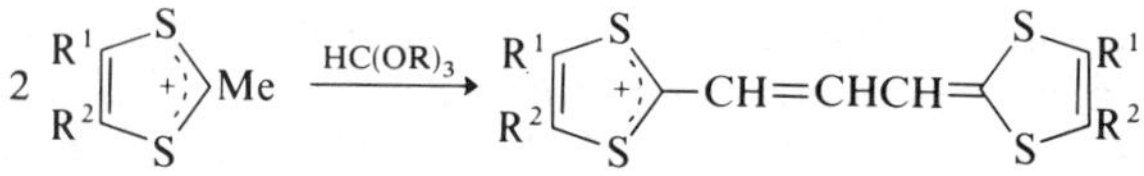

A convenient route to 4-aryl-1,3-dithiolylium salts (**48**) is the treatment of 2-amino-4-aryl-1,3-dithioles (**153**) with acids in ethanol solution. Yields of (**48**) are generally higher than 90% ⟨80AHC(27)151⟩.

(153) (48) $X^- = ClO_4^-, HSO_4^-$ (154) (155)

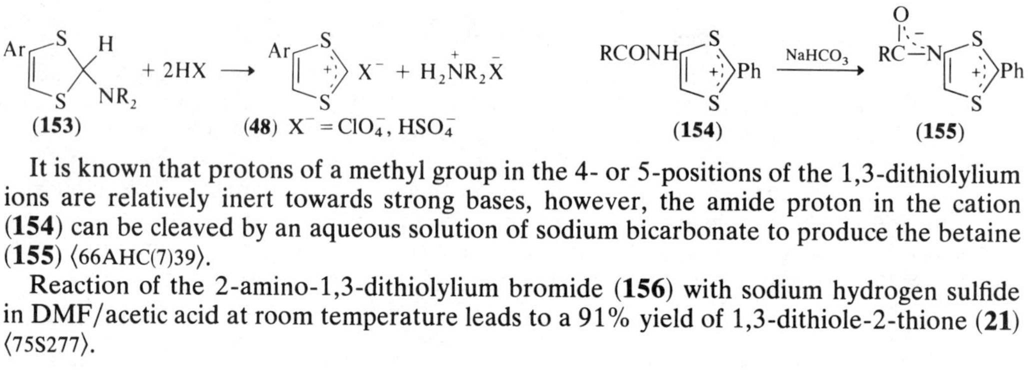

It is known that protons of a methyl group in the 4- or 5-positions of the 1,3-dithiolylium ions are relatively inert towards strong bases, however, the amide proton in the cation (**154**) can be cleaved by an aqueous solution of sodium bicarbonate to produce the betaine (**155**) ⟨66AHC(7)39⟩.

Reaction of the 2-amino-1,3-dithiolylium bromide (**156**) with sodium hydrogen sulfide in DMF/acetic acid at room temperature leads to a 91% yield of 1,3-dithiole-2-thione (**21**) ⟨75S277⟩.

(156) (21)

(48) (157) (158)

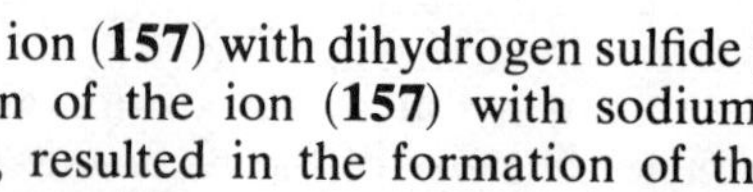

Reaction of the immonium ion (**157**) with dihydrogen sulfide produces the 1,3-dithiole-2-thione (**158**) while reduction of the ion (**157**) with sodium borohydride, followed by treatment with sulfuric acid, resulted in the formation of the 1,3-dithiolylium salt (**48**) ⟨76S489⟩.

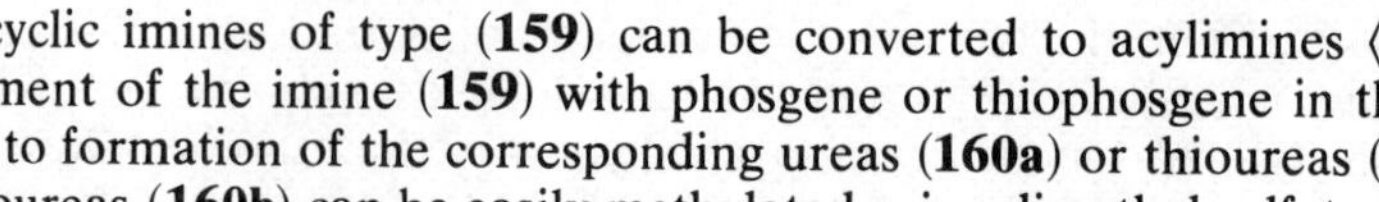

The heterocyclic imines of type (**159**) can be converted to acylimines ⟨76JPR127⟩. For example treatment of the imine (**159**) with phosgene or thiophosgene in the presence of pyridine leads to formation of the corresponding ureas (**160a**) or thioureas (**160b**), respectively. The thioureas (**160b**) can be easily methylated using dimethyl sulfate to produce the 1,3-diazotrimethine cyanines (**161**) and subsequent condensation with the imine (**159**) yields the dyes of type (**162**) ⟨79JPR827⟩.

(159) (160a) X = O
(160b) X = S

(161) (162)

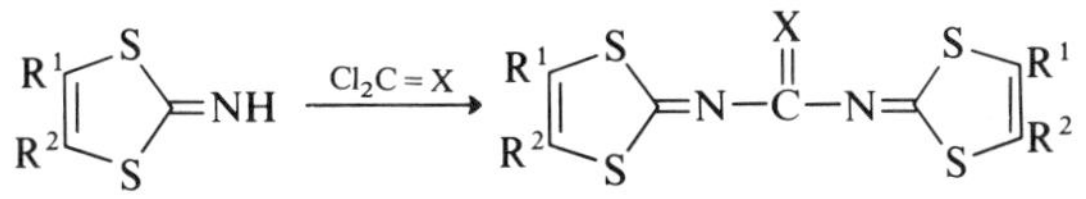

(iii) *O-Linked substituents*

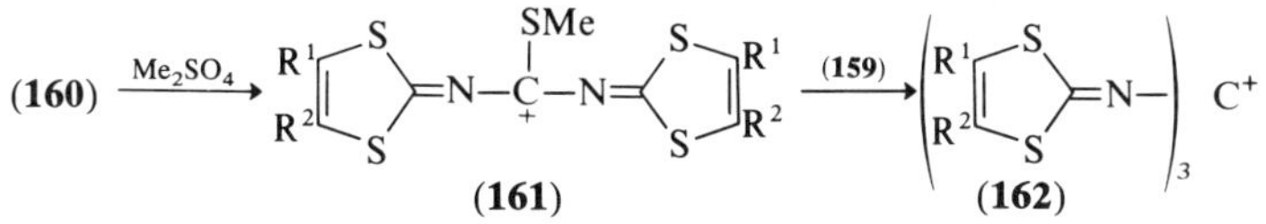

1,3-Dithiol-2-one can be converted into 1,3-dithiole-2-thione in 80% yield by treatment with tetraphosphorus decasulfide ⟨76S489⟩.

2-Alkoxy compounds of type (**163**) can serve as generators of carbenes. Thus, treatment of the methoxy derivative (**163**) with trichloroacetic acid in benzene produces a carbene which in the presence of aromatic aldehydes, forms the dithiole derivatives (**75**) ⟨80MI43200⟩.

Another example is given by the reaction of (**163**) with succinimide in the presence of trichloroacetic acid. In this case, the generated carbene inserts into the N—H bond with formation of the imide (**164**) ⟨80MI43201⟩.

(163) (75)

(164)

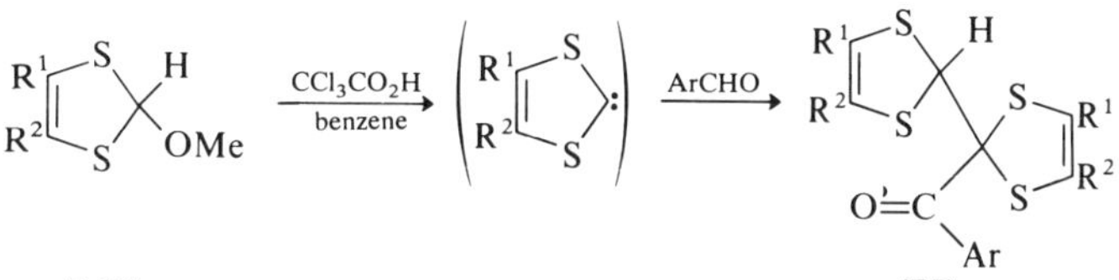

Furthermore, the reactions of the benzodithiole (**165**) with indole derivatives in acetic acid lead to the formation of the substitution products (**166**). These products can be transformed into the dithiolylium ions (**167**) by treatment of (**166**) with trityl tetrafluoroborate in acetonitrile. Finally, the cation (**167**; $R^1 = H$) undergoes deprotonation using triethylamine to produce the dibenzo-1,4-dithia-6-azafulvalene (**168**) ⟨80BCJ1661⟩.

(165) (166)

(167) (168)

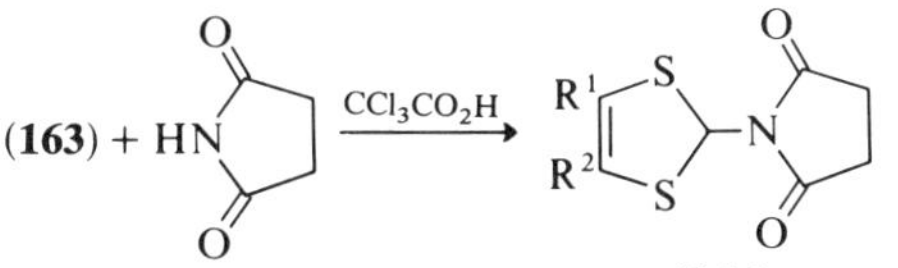

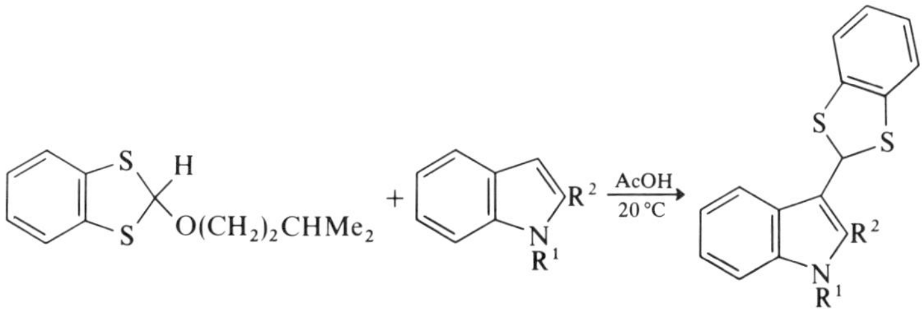

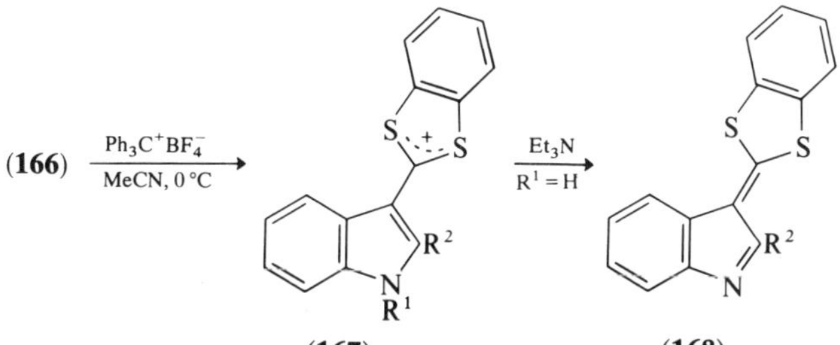

(iv) *S-Linked substituents*

1,3-Dithiole-2-thione derivatives can be readily alkylated at the exocyclic sulfur to form 1,3-dithiolylium salts. For example with the commonly used alkylating reagents methyl iodide, dimethyl sulfate, trimethyloxonium tetrafluoroborate and triethyloxonium tetrafluoroborate, the 1,3-dithiole-2-thiones (**169**) are transformed into the 1,3-dithiolylium salts (**170**) ⟨80AHC(27)151⟩. When R^1 and R^2 are substituents with a strong $-M$ effect, then alkylation is possible only with methyl fluorosulfonate ⟨78JOC678⟩.

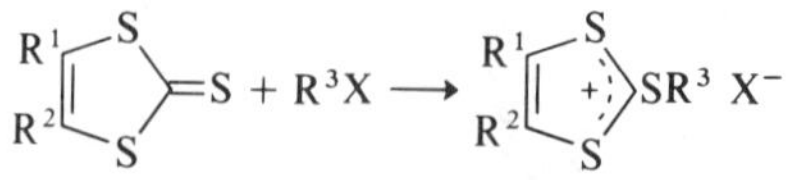

The alkylation of (**21**) with phenacyl or *p*-bromophenacyl bromide to yield the 1,3-dithiolylium salts (**171**) offers a more special case. The methylene group of these salts can be deprotonated to form the unstable thiocarbonyl ylides (**172**) which collapse to the non-isolatable episulfides (**175**). Further reaction of (**175**) by loss of sulfur produces (**174**) whereas episulfide ring-opening and hydrogen shift followed by air oxidation leads to the disulfide (**173**) ⟨80BCJ2281⟩.

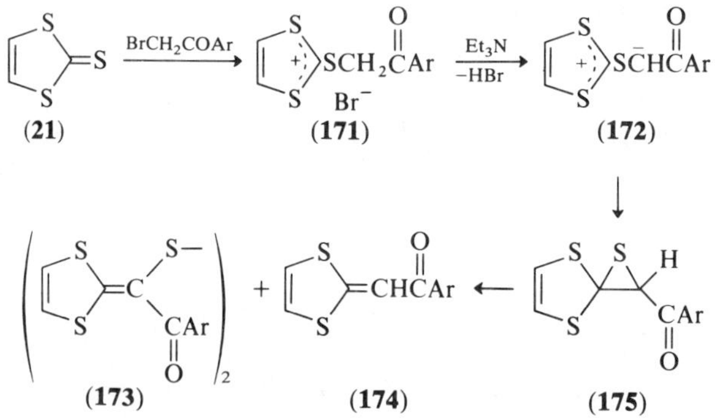

1,3-Dithiole-2-thione (**176**) reacts with *N,N*-dichlorosulfonamides to give the 1,3-dithiol-2-ylidenesulfonamides (**177**) and (**178**) as well as the 1,3-dithiole-2-thione oxide (**179**). It has been established that the primary step in these reactions is nucleophilic attack of the exocyclic sulfur of the 1,3-dithiole-2-thione on the chlorine of the amide to produce an ion pair which can react in three ways to furnish the 1,3-dithiol-2-ylidenesulfonamides ⟨79LA689⟩.

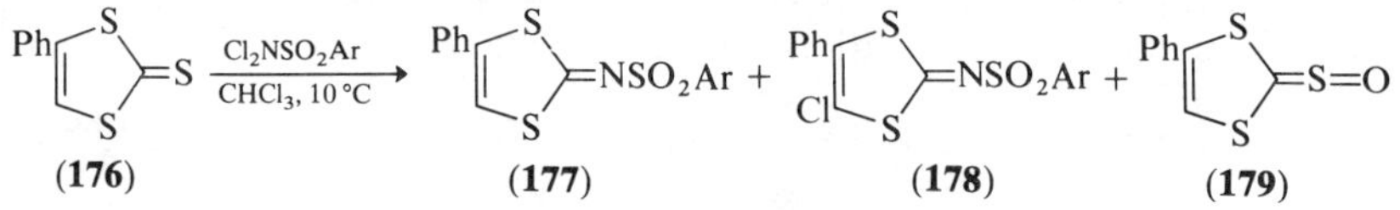

1,3-Dithiole-2-thiones can also undergo coupling reactions with ketones. For example, treatment of the 1,3-dithiole-2-thione (**23**) with the cyclic ketone (**180**) in the presence of trimethyl phosphite in refluxing benzene gives the 1,3-dithiole derivative (**181**) together with the tetrathiafulvalene (TTF) compound (**24**). The 1,3-dithiole (**181**) can be transformed further into the 2-(thiopyran-4-ylidene)-1,3-dithiole (**182**) which is isoelectronic with TTF ⟨77CC687⟩.

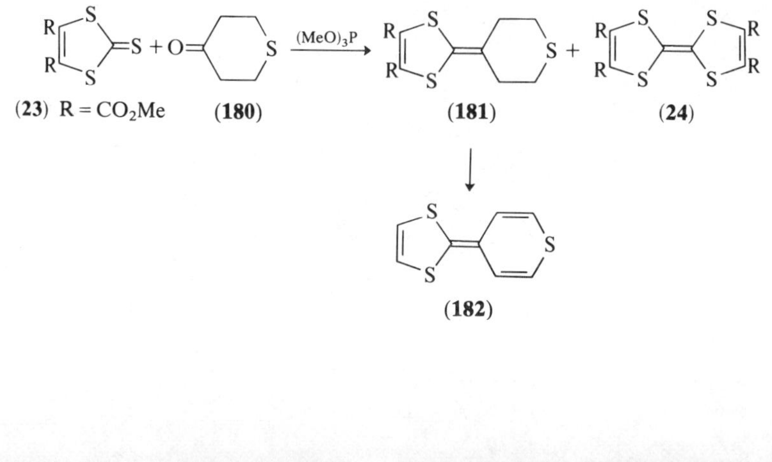

A replacement of sulfur by oxygen takes place on treatment of the 1,3-dithiole-2-thione (**183**) with mercury acetate in acetic acid–chloroform solution to yield the 1,3-dithiol-2-one (**184**) ⟨80CB1898⟩.

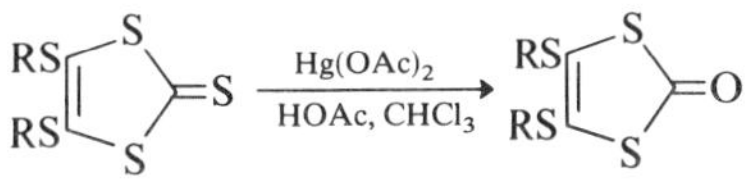

On the other hand, reaction of the 1,3-dithiole-2-thione (**21**) with peracetic acid in acetone at −40 °C produces the parent 1,3-dithiolylium ion (**1**) ⟨74JOC2456⟩.

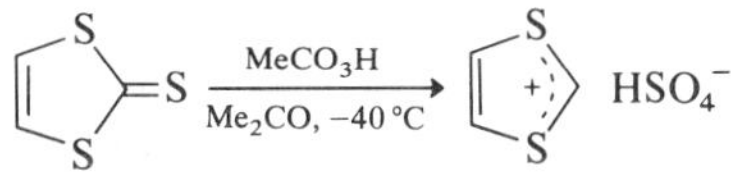

1,3-Dithiole-2-thiones react with photochemically generated singlet oxygen. Thus the main products of the photooxidation of 1,3-dithiole-2-thione (**185**) using methylene blue as a triplet sensitizer are the 1,3-dithiolone (**186**), dibenzoyl sulfide (**187**) and benzil (**188**) besides carbon disulfide, sulfur and sulfur dioxide. It has been shown that the rate of oxidation increases with the capacity of the substituents of (**185**) to increase the electron density in the 1,3-dithiole ⟨77JPR875⟩.

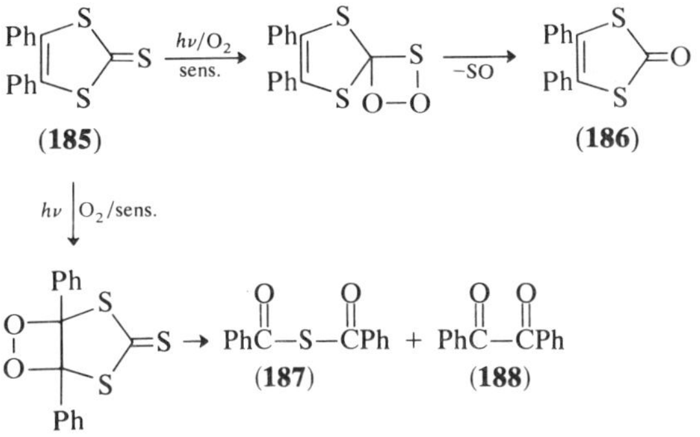

1,3-Dithiole-2-thione can be used for trapping methyl, trifluoromethyl, or tributyltin radicals to form 1,3-dithiolyl radicals of type (**189**) which have been observed by ESR ⟨78JA3868, 80AHC(27)31⟩. Calculation of the spin distribution in the 1,3-dithiolyl radical predicts a very high spin population at the 2-position ⟨80AHC(27)31⟩.

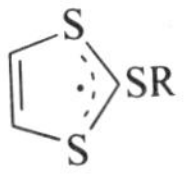

(**189**) R = Me, CF$_3$, SnBu$_3$

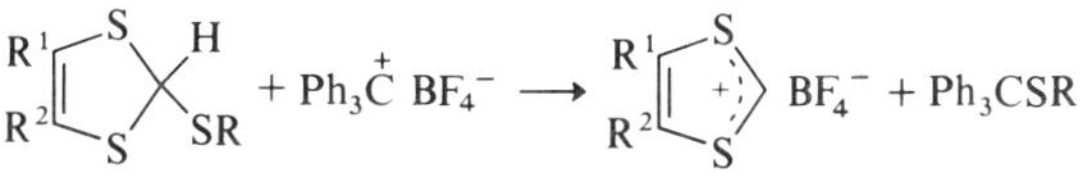

2-Alkylthio-1,3-dithioles (**190**) react with trityl tetrafluoroborate in acetonitrile to produce the 1,3-dithiolylium salts (**54**) in high yields ⟨80AHC(27)151⟩.

A nucleophilic replacement of the methylthio group is observed in the reaction of the 1,3-dithiolylium salt (**192**) with primary aromatic amines which form protonated anils or phenylhydrazones of type (**191**). The similar reaction of (**192**) with secondary aliphatic amines such as piperidine or morpholine leads to formation of (**193**) ⟨66AHC(7)39⟩.

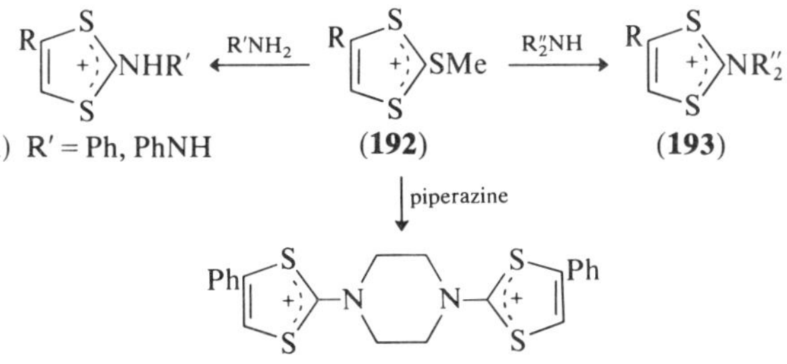

The reaction of 2 moles of 2-methylthio-4-phenyl-1,3-dithiolylium perchlorate (**192**; R = Ph) with piperazine forms the double salt (**194**) ⟨80AHC(27)151⟩.

On the other hand, the addition of pyridine to the 2-methylthio-1,3-dithiolylium ion (**195**) results in transmethylation instead of nucleophilic attack at the 2-position. The reaction products are (**21**) and (**196**) ⟨66AHC(7)39⟩.

(195) (21) (196)

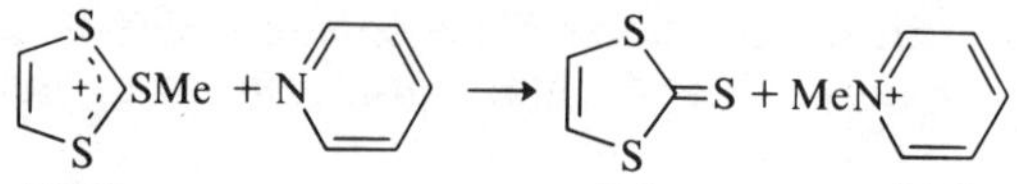

1,4-Dithiafulvalene (**197**) can be obtained by reaction of the 1,3-dithiolylium iodide (**195**) with two moles of the sodium salt of cyclopentadiene in THF ⟨65CB2825⟩.

(195) (197)

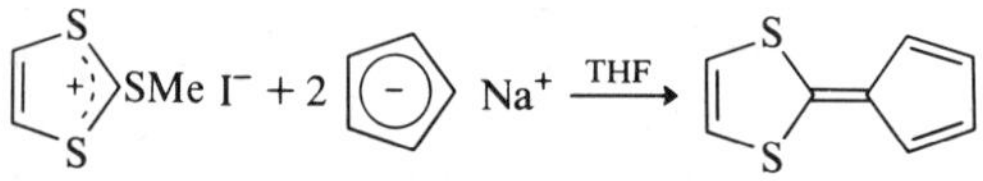

The 1,3-benzodithiolylium salts can react with Grignard reagents in ether to produce 2-substituted 1,3-benzodithioles ⟨76JCS(P1)1886⟩. The similar reaction of 2-methylthio-1,3-benzodithiolylium salts (**198**) with Grignard reagents derived from primary halides yields (**200**), whereas the one-to-one reaction product (**199**) cannot even be isolated when an excess of the 1,3-dithiolylium salt is used ⟨80AHC(27)151⟩.

(198) (199) (200)

(201) (202)

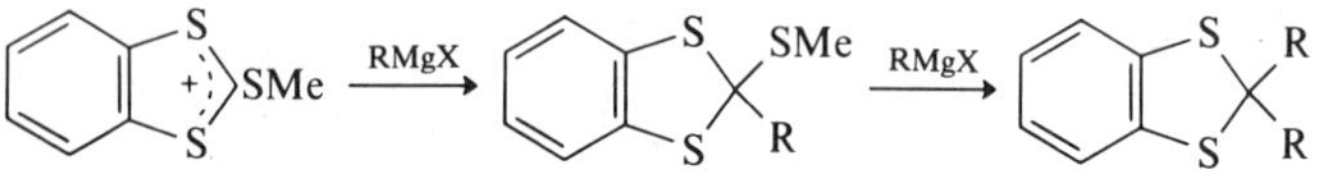

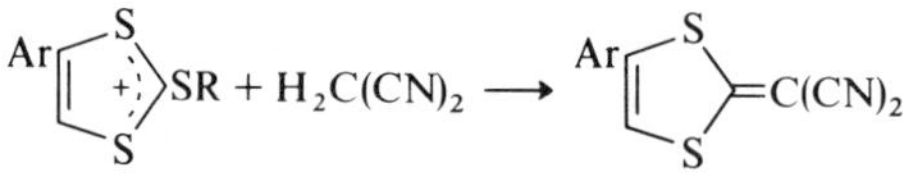

A corresponding reaction takes place between 2-alkylthio-4-aryl-1,3-dithiolylium salts (**201**) and malononitrile which proceeds with formation of the 1,4-dithiafulvene derivatives (**202**) ⟨66AHC(7)39⟩.

Also, interesting 7,10-dithiasesquifulvalene-1′,2-quinone derivatives (**204**) can be obtained by condensation of 2-methylthio-1,3-dithiolylium iodide with α-tropolone derivatives (**203**) ⟨81CL805⟩. With β-tropolone (**205**) and 1,3-dithiolylium iodide (**195**) in the presence of pyridine the reaction proceeds with formation of the 7,10-dithiasesquifulvalene-1,6-quinone (**206**) ⟨77CL287⟩.

(195) (203) (204)

(205) (206)

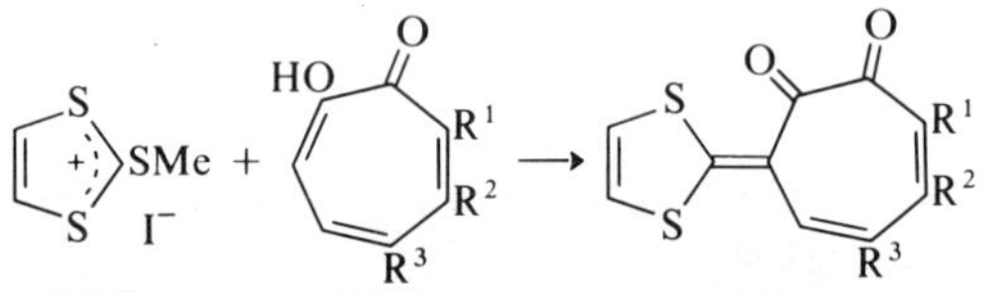

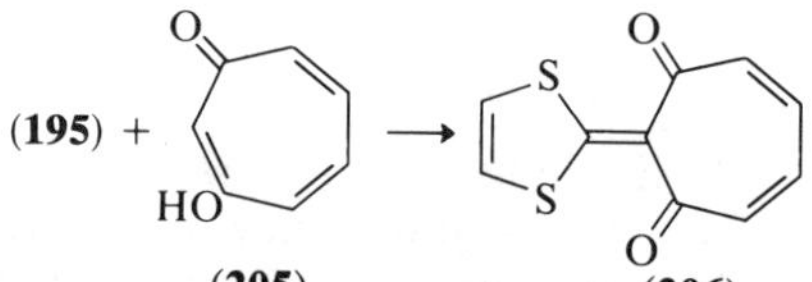

Furthermore, the enamines (**208**) react with 2-methylthio-1,3-dithiolylium salts (**207**) in the presence of triethylamine to give addition products which lead to the ketone (**209a**) upon hydrolysis or the thioketone (**209b**) by hydrosulfolysis ⟨75JPR137⟩.

(207) (208) (**209a**) X = O
(**209b**) X = S

(195) (210) (211)

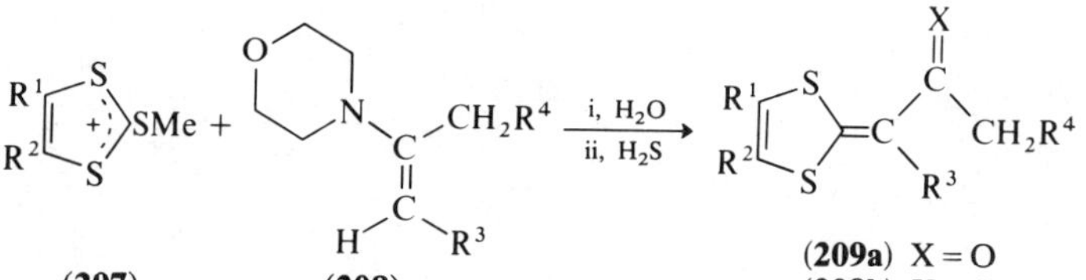

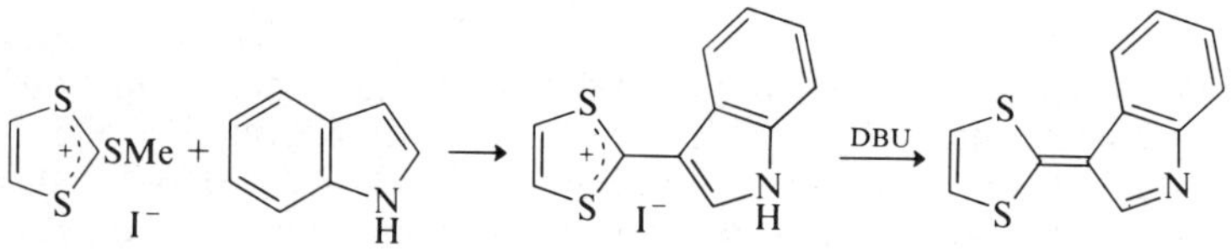

Another nucleophilic displacement of the methylthio group is observed with indole as a reactant. Heating indole with the 2-methylthio-1,3-dithiolylium iodide (**195**) at 50 °C in acetonitrile produces the salt (**210**) in 58% yield. Subsequent treatment of (**210**) with 1,8-diazabicyclo[5.4.0]undec-7-ene (DBU) as a base forms the fulvene derivative (**211**) in quantitative yield ⟨80BCJ1661⟩.

Also *N*-ethyl-1-naphthylamine reacts with 2-methylthio-1,3-dithiolylium iodide (**195**) to form the 1,3-dithiolylium salt (**212**) which can be dehalogenated with base to the quinomethidimine derivative (**213**) ⟨81CC565⟩.

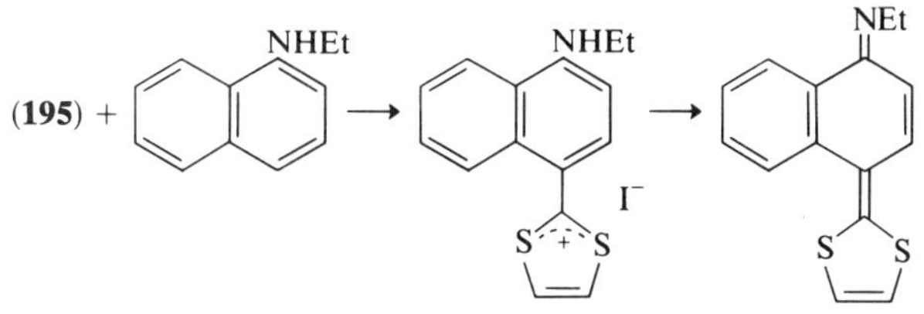

(v) *Metals and metalloid-linked substituents*

The organolithium compound (**214**) is obtained by treatment of the tetrathiafulvalene (**5**) with lithium diisopropylamide (LDA). This lithium compound (**214**) can further react with carbon dioxide, ethyl chloroformate, acetyl chloride, formaldehyde, dimethyl sulfate, and triethyloxonium hexafluorophosphate to produce the correspondingly substituted tetrathiafulvalenes (**215**) ⟨79JOC1476⟩.

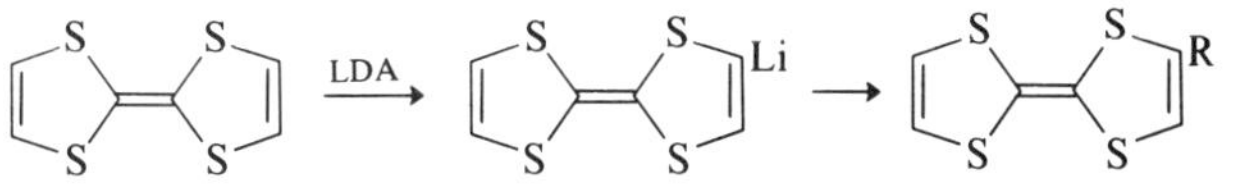

A general synthesis of α-hydroxy esters is based on 2-ethylthio-1,3-benzodithiole which serves as a methoxycarbonyl anion equivalent. Thus reaction of the lithium compound (**216**) with ketones followed by treatment with mercury perchlorate in methanol leads to the formation of α-hydroxy acid derivatives (**217**) ⟨81SC209⟩.

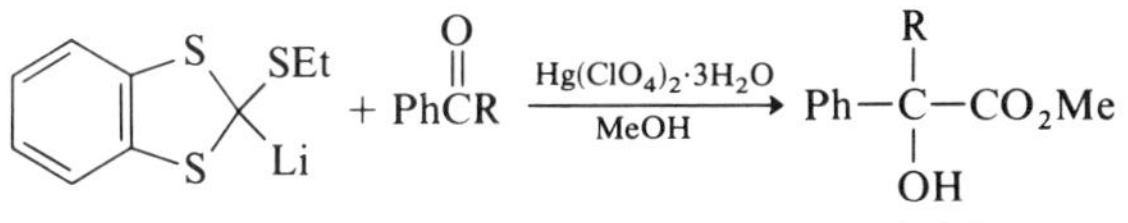

4.32.3.2 Reactions of 1,3-Dithiolanes

1,3-Dithiolanes are quite resistant to both alkaline and acid hydrolysis, however the ring can be cleaved by cadmium carbonate and mercury(II) chloride, a well-known reagent for hydrolyzing thioacetals ⟨53MI43200⟩. Thus the 1,3-dithiolane derivative (**218**) is hydrolyzed by this reagent with formation of the hydroxydithiol (**219**). A similar reaction takes place on reduction of the 1,3-dithiolane derivative (**218**) with sodium and ethanol in liquid ammonia to produce the hydroxydithiol (**219**) ⟨47JCS592⟩.

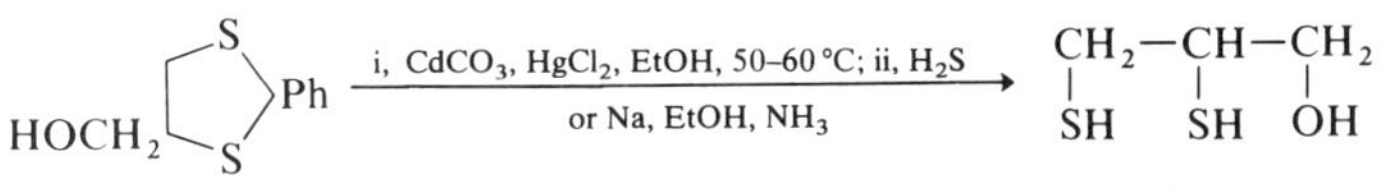

The 1,3-dithiolane ring also can be cleaved by alkylative hydrolysis using different reagents ⟨77S357⟩. For example treatment of the 1,3-dithiolane derivative (**220**) with triethyloxonium tetrafluoroborate followed by hydrolysis with copper(II) sulfate solution and basification gives the ketone (**221**) in 95% yield ⟨77S357⟩.

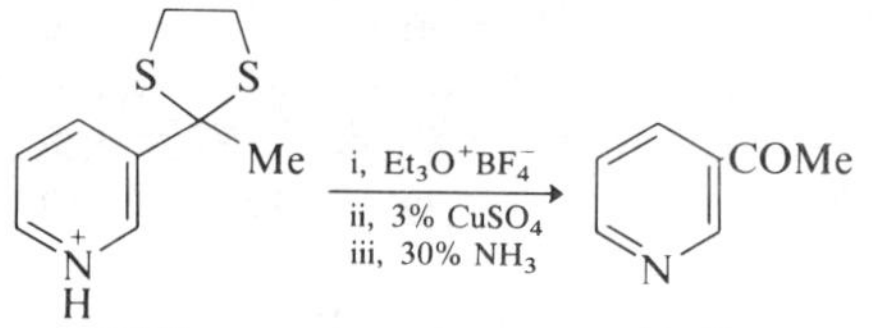

Another alkylative hydrolysis uses 'magic methyl'. Thus, action of the 1,3-dithiolane (**222**) with methyl fluorosulfonate followed by treatment with sodium hydroxide leads to the formation of benzophenone ⟨72S561⟩.

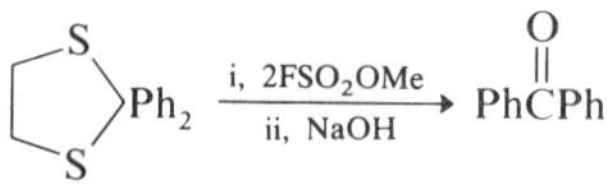

4.32.4 SYNTHESES

Preparative methods are arranged according to the ring syntheses starting from non-heterocyclic compounds by formation of one bond or formation of two bonds from [4+1] or [3+2] atom fragments or formation of multiple bonds. The second subsection deals with the syntheses starting from heterocyclic compounds, and the last subsection reports on the synthesis of some parent compounds.

4.32.4.1 Ring Syntheses from Non-heterocyclic Compounds

4.32.4.1.1 Formation of one bond

(i) *Derivatives containing C-linked substituents*

Several methods are known for the preparation of 1,3-dithiolylium salts from open-chain precursors by formation of one bond. One method is based on acid-catalyzed cyclization in which S-α-oxoalkyl thioesters of type (**223**) are cyclized in a mixture of perchloric acid and glacial acetic acid. Under these reaction conditions the yield of (**31**) is generally poor because some starting material is consumed in converting the monothioester (**223**) into (**224**) and/or (**225**). However, if sufficient sulfur is present in the form of hydrogen sulfide during the cyclization, the yield of (**31**) is raised to 95% ⟨66AHC(7)39⟩.

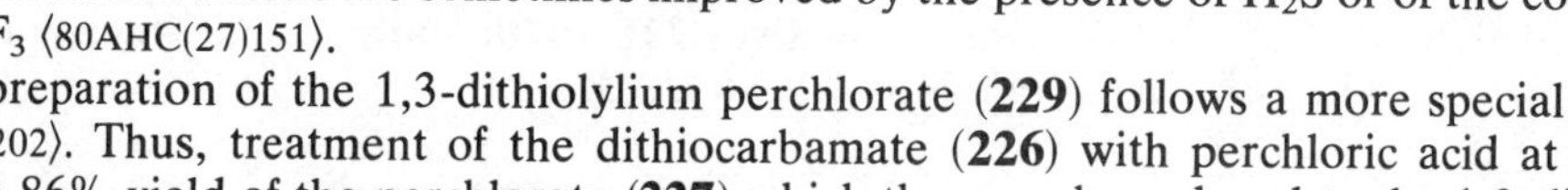

Another method starts with the above-postulated intermediate oxoalkyl dithioesters (**224**). These dithioesters (**224**) can be cyclized to the corresponding 1,3-dithiolylium salts (**31**) in the presence of strong acids by warming either in 70% perchloric acid or concentrated sulfuric acid. The results are sometimes improved by the presence of H_2S or of the complex H_2S/BF_3 ⟨80AHC(27)151⟩.

The preparation of the 1,3-dithiolylium perchlorate (**229**) follows a more special route ⟨80MI43202⟩. Thus, treatment of the dithiocarbamate (**226**) with perchloric acid at 50 °C gives an 86% yield of the perchlorate (**227**) which then can be reduced to the 1,3-dithiole (**228**). Upon further action with perchloric acid in acetone at room temperature, (**228**) is converted into the 1,3-dithiolylium perchlorate (**229**). Also a bis(1,3-dithiolylium) salt (**231**) is obtained by treating the diester (**230**) with cold concentrated sulfuric acid ⟨77TL4607⟩. Furthermore, compounds of structure (**233**) which contain two 1,3-dithiolylium nuclei in the same molecule, can also be prepared by acidic cyclodehydration of the corresponding diester (**232**). These substituted salts (**233**) can be further converted in excellent yields into the salts (**234**), which are unsubstituted at the 2-position. The explosive perchlorate (**235**) can be prepared in a similar manner ⟨80AHC(27)151⟩.

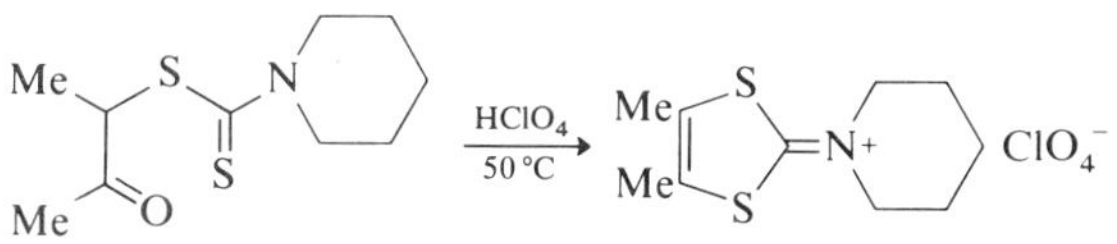

1,3-Dithioles

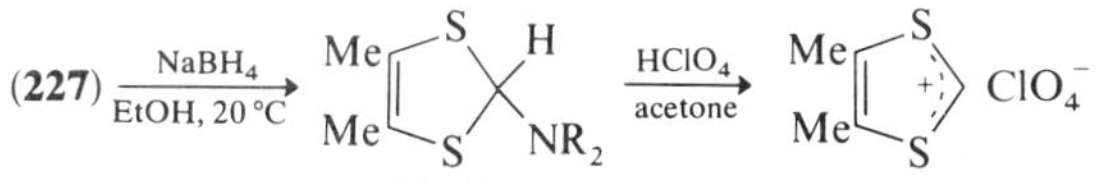

[Structures (226)–(235) with Z = –(CH$_2$)$_3$–, –(CH$_2$)$_4$–, and biphenyl groups]

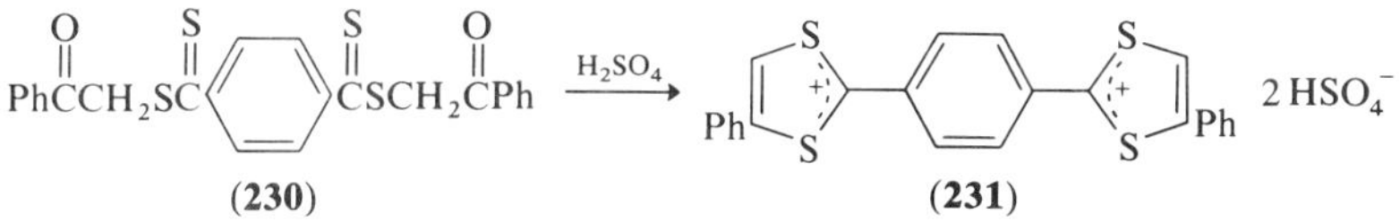

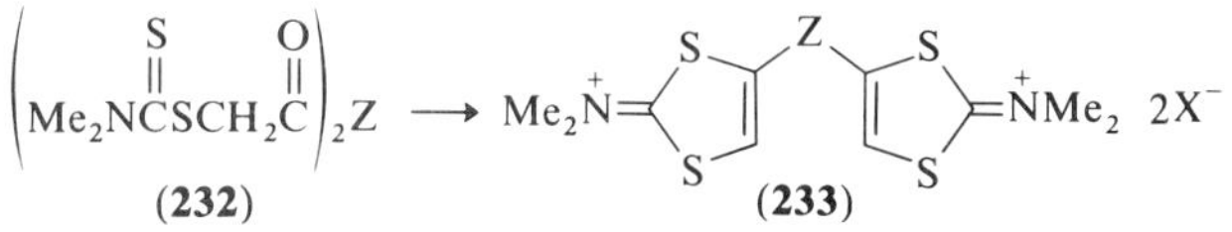

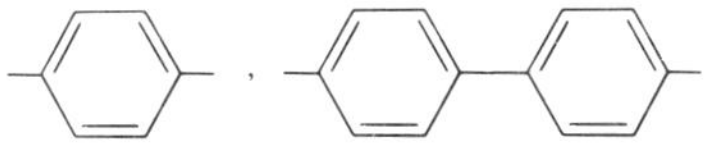

(ii) Derivatives containing N-linked substituents

Cyanoalkyl dithiobenzoates (237) can also undergo cyclization upon treatment with strong acids; this is a route to 4-amino-1,3-dithiolylium ions (236). On the other hand, treatment of (237) with acyl chlorides gives the N-acyl derivatives (238). According to their IR spectra, they exist, at least in part, in the imidol form ⟨65BCJ596, 63BCJ1437⟩.

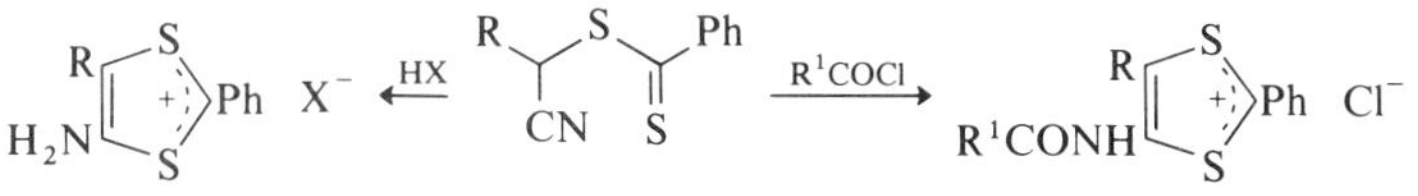

[Structures (236), (237), (238)]

Another entry to 1,3-dithiolylium salts starts from S-vinyl N,N-dialkyl dithiocarbamates (239) which are easily obtained by reaction of sodium dithiocarbamates with 1,2-dibromoethane. Treatment of the dithiocarbamates (239) with bromine gives the 1,3-dithiolylium bromides (242) via the intermediates (240) and (241) ⟨73CL867⟩.

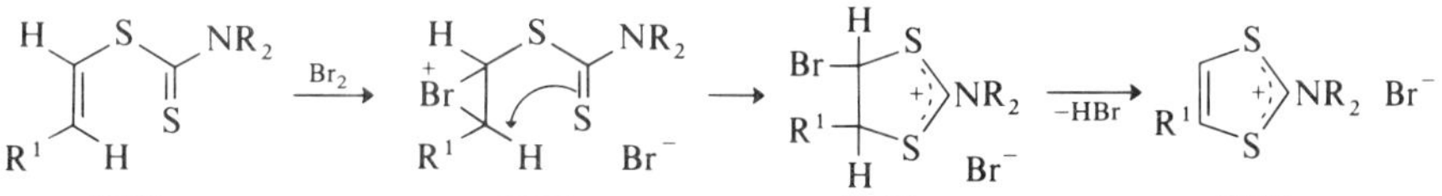

[Structures (239), (240), (241), (242)]

Also, allyl dithiocarbamates (243) react with bromine to form the dithiolanylium salts (244) which on thermolysis eliminate hydrogen bromide to form the 1,3-dithiolylium bromides (245) ⟨80AHC(27)151⟩.

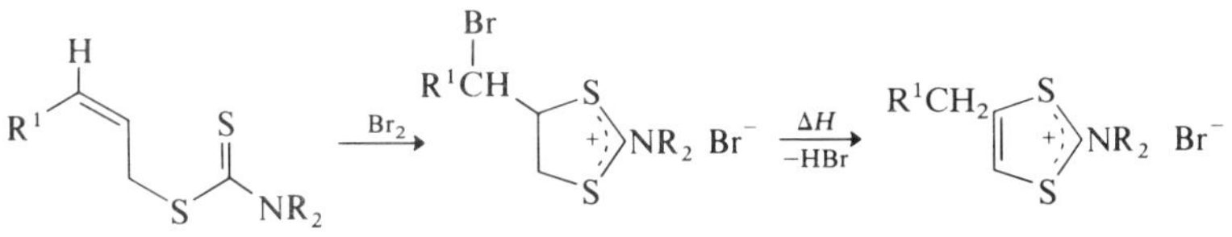

[Structures (243), (244), (245)]

Treatment of the dithiocarbamate (246) first with concentrated sulfuric acid and then with sodium tetraphenylborate produces the salt (156) in 86% yield ⟨80AHC(27)151⟩.

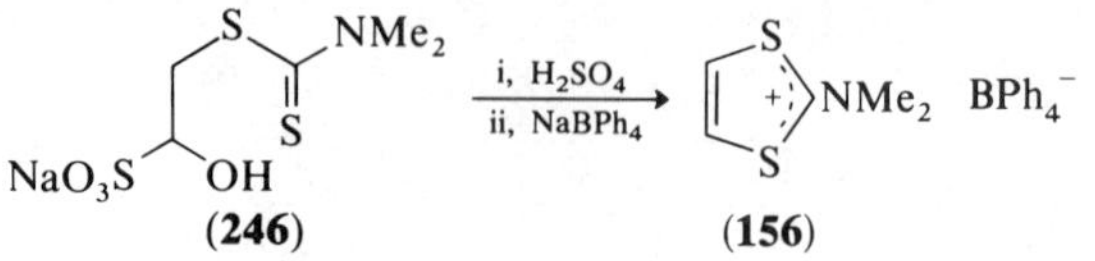

(iii) *Derivatives containing O-linked substituents*

Several 4-hydroxy- and 4-acetoxy-1,3-dithiolylium salts (**248**) can be synthesized by cyclization of α-(thioacylthio)carboxylic acids (**247**) with strong acids like perchloric acid, sulfuric acid or fluorosulfuric acid (Table 10) ⟨64JOC1708, 78CB3178⟩. Using acetic anhydride as the solvent, the 4-acetoxy compounds (**248**) are easily formed.

Table 10 Synthesis of 1,3-Dithiolylium Salts (**248**) from α-(Thioacyl-thio)carboxylic Acids

R^1	R^2	R^3	X^-	Ref.
H	Ph	H	ClO_4^-	64JOC1708
H	NMe_2	H	ClO_4^-	64JOC1708
H	SMe	H	ClO_4^-	64JOC1708
Ph	Ph	H	SO_3F^-	78CB3178
Ph	Ph	Ac	ClO_4^-	78CB3178
H	Ph	Ac	HSO_4^-	78CB3178
Ph	SEt	Ac	ClO_4^-	78CB3178

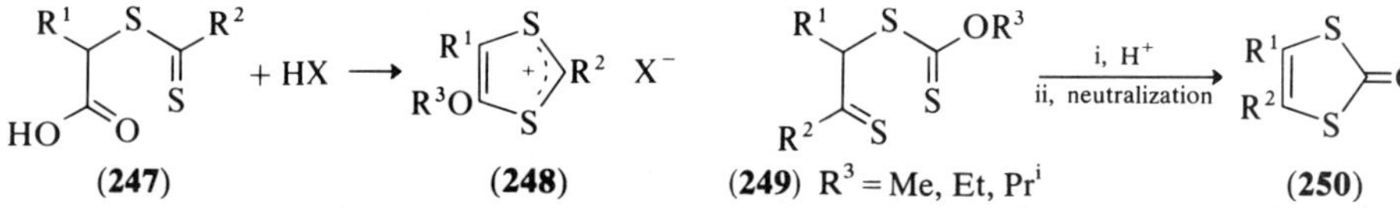

In the presence of acid, β-thioxodithiocarbonates (**249**) cyclize to produce 1,3-dithiol-2-one derivatives (**250**) in fair yields ⟨76S489⟩.

Another preparation of differently substituted 1,3-dithiol-2-ones starts from the unsaturated dithiocarbonates (**251**) which on treatment with iodine lead to the dithiolanone derivatives (**253**) *via* the salts (**252**). Abstraction of hydrogen iodide from (**253**) gives (**254**) and subsequent treatment with trifluoroacetic acid yields the 1,3-dithiol-2-ones (**255**; Table 11). The entire reaction sequence can be easily performed in a few hours ⟨80JOC2959⟩.

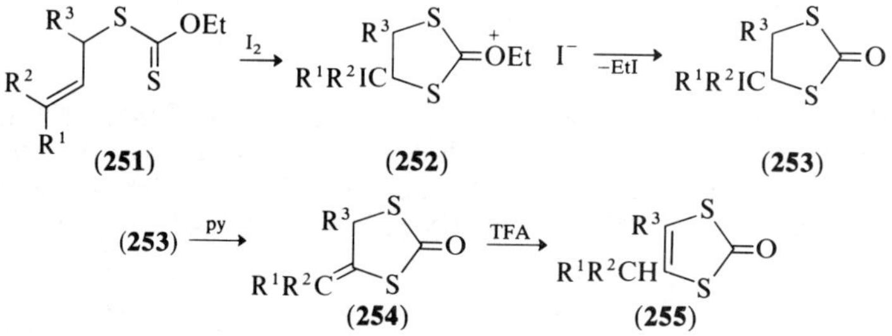

Table 11 Yields of 1,3-Dithiol-2-ones (**255**) Synthesized from Unsaturated Dithiocarbonates (**251**) ⟨80JOC2959⟩

R^1	R^2	R^3	Yield (%)
H	H	H	97
Me	H	H	83
Me	Me	H	94
n-C_5H_{11}	H	H	77
Ph	H	H	86
H	—$(CH_2)_3$—		63

(iv) *Derivatives containing S-linked substituents*

In a very similar way 1,3-dithiole-2-thiones (**260**) can be prepared by action of trithiocarbonates (**256**) with iodine *via* the salts (**257**) which lose hydrogen iodide and isobutene to form the isolated intermediates (**258**). Further treatment of (**258**) with pyridine gives (**259**), and subsequent acid-catalyzed tautomerization produces the thiones (**260**) ⟨80JOC2959⟩.

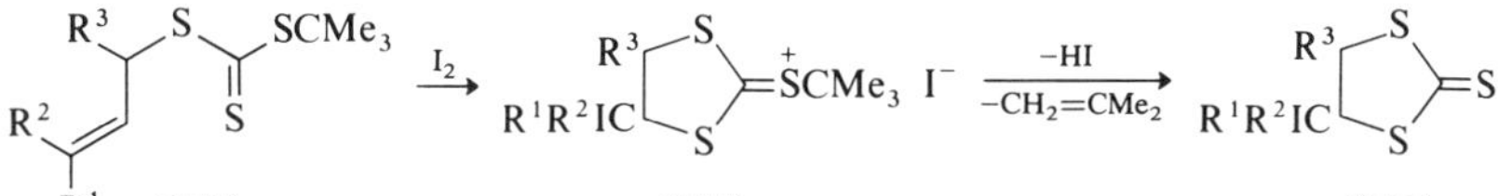

A further efficient and general synthesis of 1,3-dithiole-2-thiones (**169**) with a variety of functional groups is the acid-catalyzed ring closure of β-keto *t*-butyl trithiocarbonates of type (**261**) which are readily available from α-halo ketones and sodium *t*-butyl trithiocarbonate. The intermediates (**262**) fragment into isobutene and 1,3-dithiole-2-thiones (**169**) ⟨80JOC175⟩.

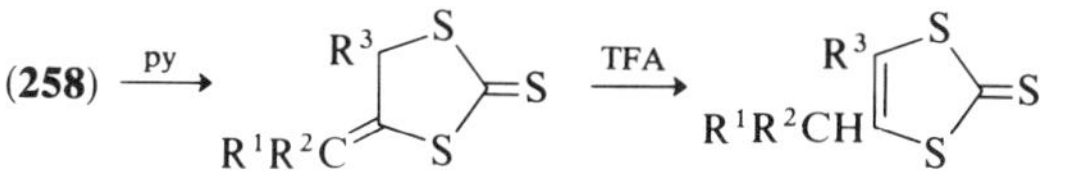

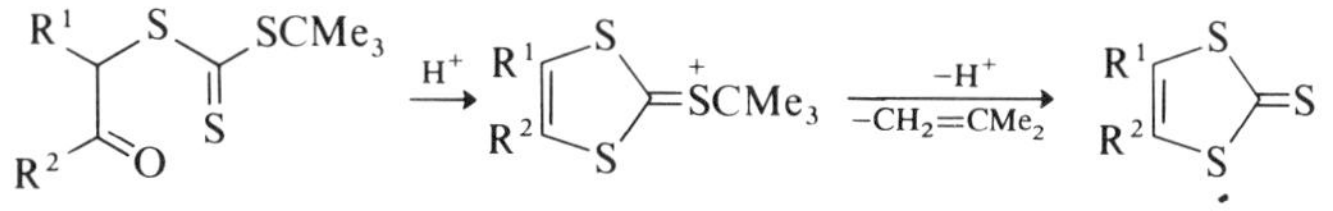

1,3-Dithiole-2-thione derivatives (**169**) can also be prepared by cyclization of *O*-ethyl or *S*-methyl 2-propanoyl dithiocarbonates (**224**) with tetraphosphorus decasulfide in refluxing tetralin or decalin ⟨76S489⟩.

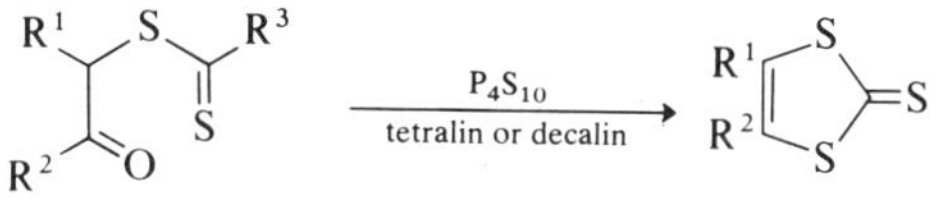

(v) *Mesoionic compounds*

An interesting route to the mesoionic 1,3-dithiolones of type (**2**) is offered by the anhydrocyclization of α-(thioacylthio)- or α-(thiocarbamoylthio)carboxylic acids with trifluoroacetic anhydride or a mixture of acetic anhydride and triethylamine (1:1) at 0–10 °C (Table 12) ⟨81LA347, 78CB2021, 76CB740, 76JOC1724⟩. These easy anhydrocyclizations

Table 12 Synthesis of 1,3-Dithiolylium-4-olates (**2**)

R^1	R^2	Yield (%)	Ref.
H	Ph	33, 77	76JOC1724, 76CB740
H	4-MeOC$_6$H$_4$	80	76JOC1724
H	4-ClC$_6$H$_4$	75	76JOC1724
Me	Ph	34	76CB740
Ph	Ph	90	76CB740
Ph	4-MeC$_6$H$_4$	93	76CB740
4-MeC$_6$H$_4$	Ph	63	76CB740
Ph	4-BrC$_6$H$_4$	35	76CB740
4-BrC$_6$H$_4$	Ph	70	76CB740
Ph	4-MeOC$_6$H$_4$	91, 61	76CB740, 77JOC1633
4-MeOC$_6$H$_4$	Ph	85	76CB740
Ph	4-Me$_2$NC$_6$H$_4$	90	76CB740
Ph	4-NO$_2$C$_6$H$_4$	29	76CB740
4-NO$_2$C$_6$H$_4$	Ph	45	76CB740
Ph	Piperidino	92	78CB2021
Ph	Morpholino	72	78CB2021
Ph	Pyrrolidino	81	78CB2021
CF$_3$CO	Piperidino	70	78CB2021
CO$_2$Et	4-MeOC$_6$H$_4$	45	77JOC1633
Ph	2-CF$_3$CO$_2$C$_6$H$_4$	92	81LA347
Ph	2-HOC$_6$H$_4$	92	81LA347
Ph	2-CH$_2$=CHCH$_2$OC$_6$H$_4$	71	81LA347
Ph	2-CH≡CCH$_2$OC$_6$H$_4$	78	81LA347

presumably proceed *via* the ketene intermediates (**263**) which undergo ring closure to the mesoionic compounds (**2**).

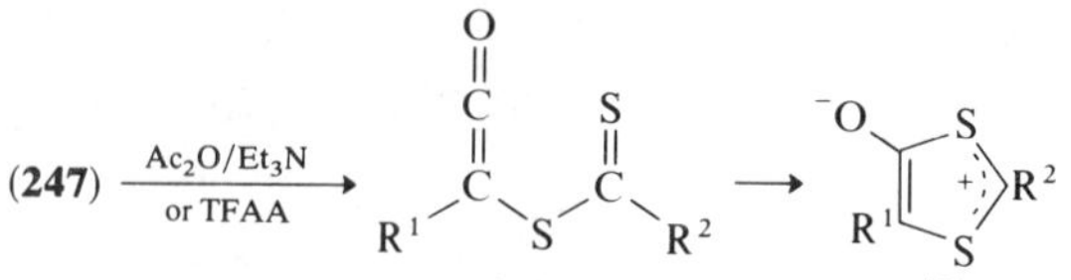

It has been shown ⟨76JOC1724⟩ that a previously described compound ⟨65BCJ596⟩, prepared by treatment of thiobenzoylthioglycollic acid (**264**) with acetic anhydride in the presence of boron trifluoride, is indeed the acylated 1,3-dithiolone (**265**) and not the claimed 2-phenyl-1,3-dithiolylium-4-olate. The latter can be prepared from thiobenzoylthioglycollic acid (**264**) and acetic anhydride–triethylamine ⟨76JOC1724⟩ or acetic anhydride–dicyclohexylethylamine ⟨76CB740⟩.

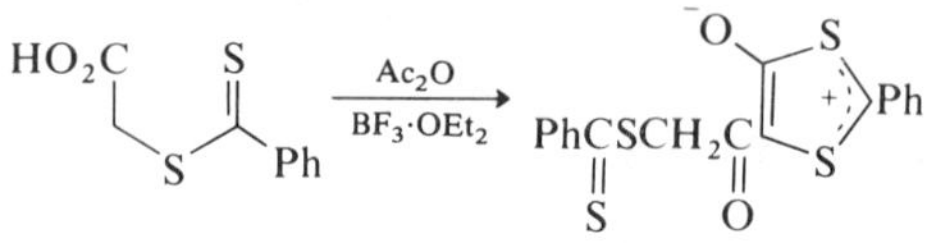

The action of acetyl chloride on the nitrile (**266**) followed by treatment with sodium hydrogen carbonate yields the mesoionic compound (**267**) ⟨78CB2021⟩. Other substituted mesoionic compounds of type (**267**) have been prepared similarly ⟨65BCJ596, 63BCJ1437⟩.

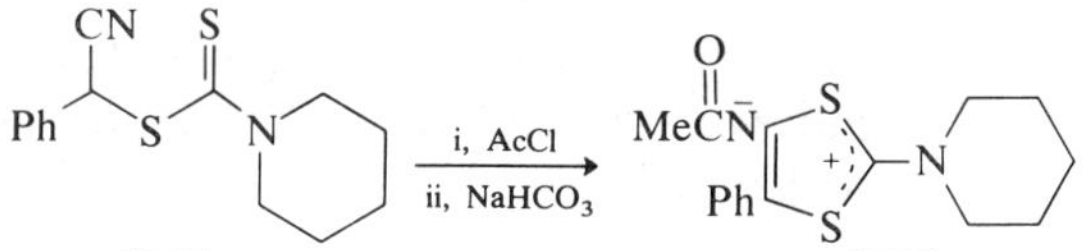

4.32.4.1.2 *Formation of two bonds from [4+1] atom fragments*

(i) *Derivatives containing C-linked substituents*

The 1,3-dithiole ring system can also be prepared from [4+1] atom fragments. For example the ethylene-1,2-dithiol (**268**) reacts with formic acid in the presence of perchloric acid with formation of the 4,5-diphenyl-1,3-dithiolylium perchlorate (**269**) in a 45% yield ⟨69MI43200⟩. A similar route involves benzene-1,2-dithiols (**273**) which condense with carboxylic acids in the presence of phosphorus oxychloride to produce moderate to good yields of 1,3-benzodithiolylium salts (**275**) ⟨80AHC(27)151⟩.

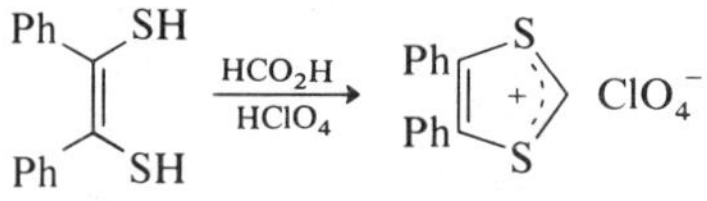

This method also works well with unsaturated dicarboxylic acids containing odd carbon chains such as (**271**) which react with the benzene-1,2-dithiol (**270**) in the presence of phosphorus oxychloride and perchloric acid giving colored methine salts of type (**272**) ⟨66AHC(7)39⟩.

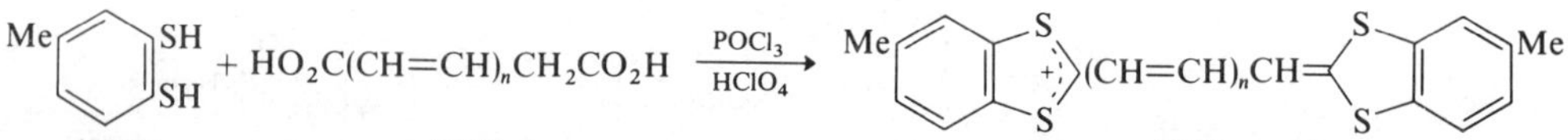

Another entry to 1,3-benzodithiolylium salts (**275**) involves acid-catalyzed condensation of benzene-1,2-dithiols (**273**) with aldehydes followed by oxidation of the intermediates (**274**) using oxidizing reagents. This reaction also works with dialdehydes ⟨80AHC(27)151⟩.

Acyl chlorides may also be condensed with various benzene-1,2-dithiols. For example substituted benzene-1,2-dithiols (**276**) react with acyl chlorides in the presence of perchloric

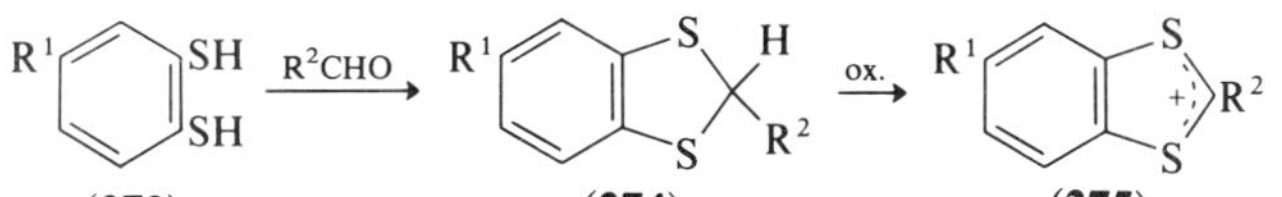

1,3-Dithioles

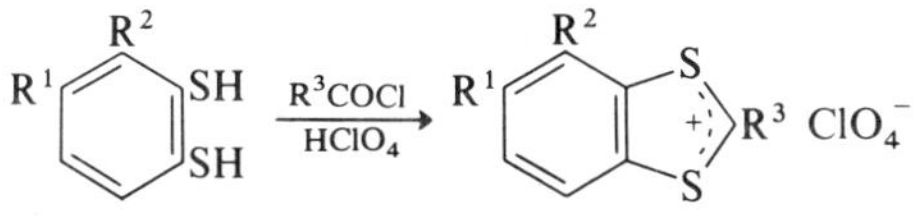

acid giving 1,3-benzodithiolylium perchlorates (**277**) in very good yields ⟨63HCA2167, 59HCA1733, 74BSF510⟩.

Condensation of acetic or benzoic esters with benzene-1,2-dithiol in the presence of tetrafluoroboric acid/ether complex yields 2-methyl- or 2-phenyl-1,3-benzodithiolylium salts, respectively ⟨77S263⟩.

A preparation of 2,2-disubstituted 1,3-benzodithiole derivatives (**278**) involves the reaction of benzene-1,2-dithiols (**273**) with alkynic ketones in the presence of piperidine or potassium carbonate ⟨80S115, 81KGS329⟩.

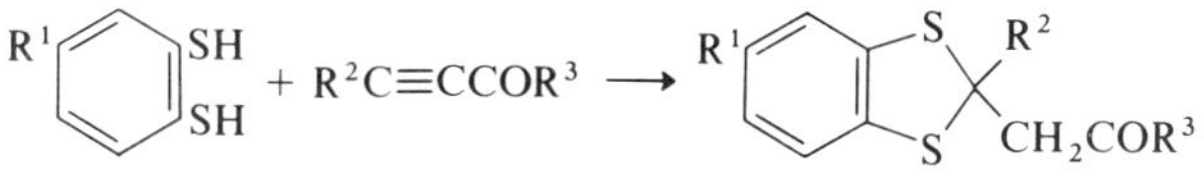

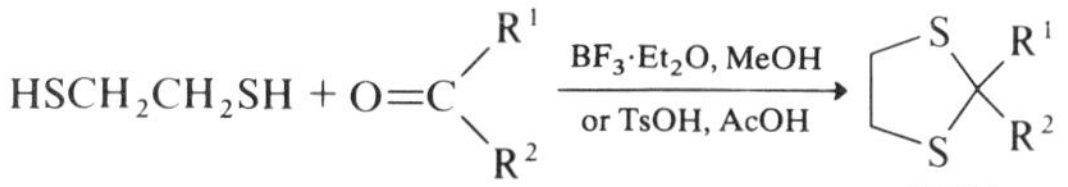

The preparation of the 1,3-dithiolane ring system (**279**) can be carried out by reaction of a carbonyl compound with ethanedithiol in the presence of either boron trifluoride etherate in methanol or *p*-toluenesulfonic acid monohydrate in acetic acid ⟨74S32⟩.

Another method involves the reaction of the bis Bunte salt (**280**) with acetone in the presence of hydrogen chloride at room temperature which proceeds with formation of the 1,3-dithiolane derivative (**281**) ⟨66HC(21-1)447⟩. The syntheses of several substituted 1,3-dithiolanes have been described ⟨66HC(21-1)447⟩.

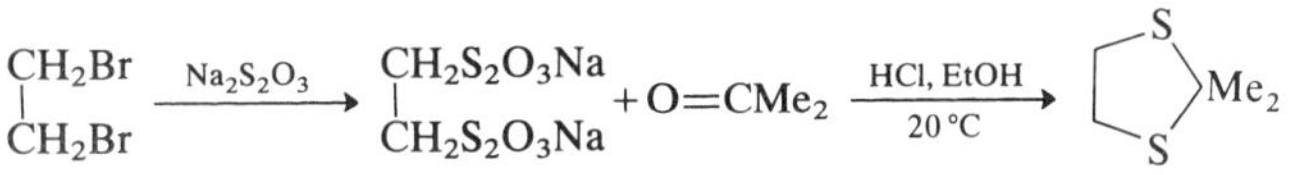

(ii) *Derivatives containing N-linked substituents*

An elegant and versatile route to 2-amino-1,3-dithiolylium salts involves the reaction of cyanogen chloride with benzene-1,2-dithiols (**273**). This reaction needs polar solvents and the catalytic action of acids and ethanol to produce the 2-amino-1,3-benzodithiolylium ions (**282**) ⟨64JOC738, 66AHC(7)39⟩.

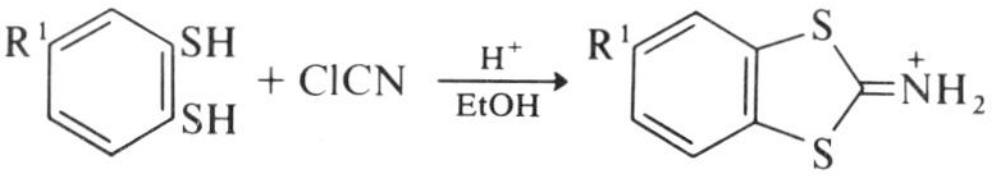

The synthesis of 1,3-dithiolanes (**284**) which contain a 2-dimethylamino group results from the reaction of ethanedithiol with amide acetals of type (**283**) ⟨69BSF332⟩.

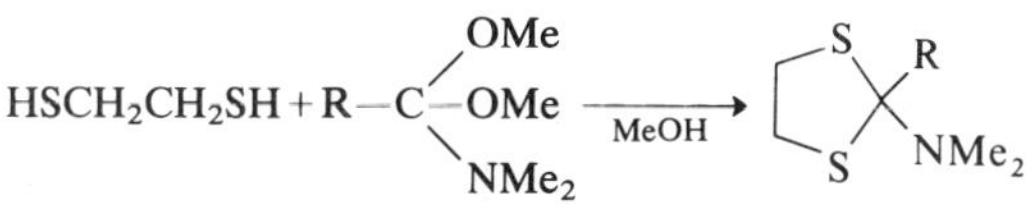

(iii) *Derivatives containing O- or S-linked substituents*

Treatment of 4-methylbenzene-1,2-dithiol (**270**) in methyl chlorodithioformate with 70% perchloric acid forms the 5-methyl-2-methylthio-1,3-benzodithiolylium perchlorate (**285**) in 33% yield ⟨64JOC2877⟩.

(**270**) (**285**)

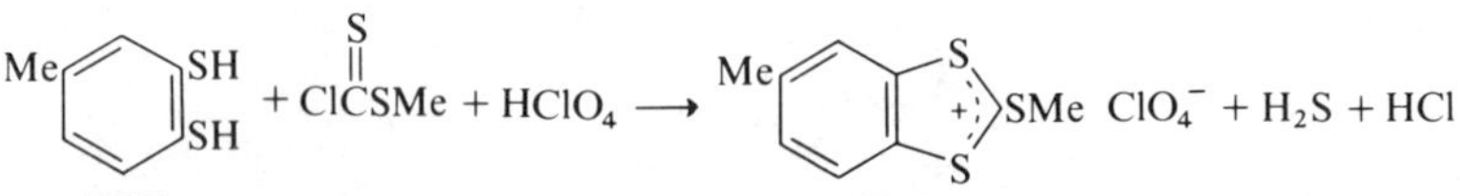

4,5-Dicyano-1,3-dithiol-2-one (**287a**) and the corresponding 2-thione derivative (**287b**) can be obtained by the reaction of the disodium salt (**286**) with phosgene or thiophosgene, respectively ⟨76S489⟩. The preparation of the disodium salt (**286**) utilizes sodium cyanide and carbon disulfide ⟨57CB438⟩.

(**286**) (**287a**) X = O
(**287b**) X = S

(**288**) (**169**)

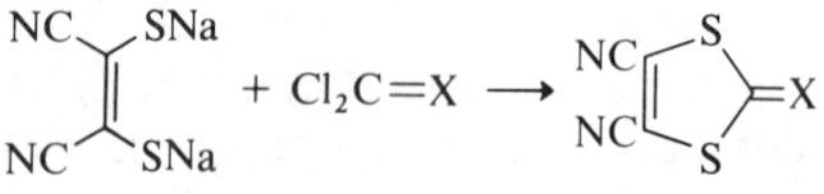

Treatment of the dithiocarbamates (**288**) with dihydrogen sulfide leads to the formation of a variety of substituted 1,3-dithiole-2-thiones (**169**) ⟨75JPR123⟩. The dithiocarbamates (**288**) can be easily prepared from enamines and the corresponding disulfides.

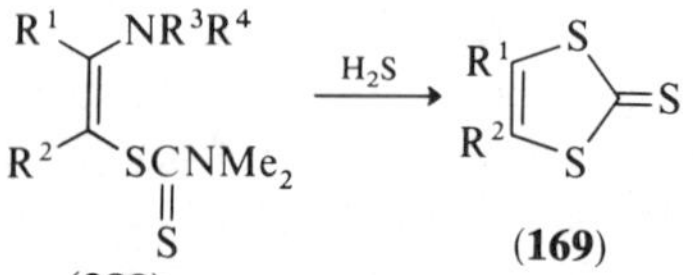

The 1,3-benzodithiole-2-thione derivative (**290**) is obtained by refluxing an aqueous solution of 1,2-benzenedithiol (**289**) and carbon disulfide in the presence of sodium hydroxide ⟨72S29⟩.

(**289**) (**290**)

(**291**)

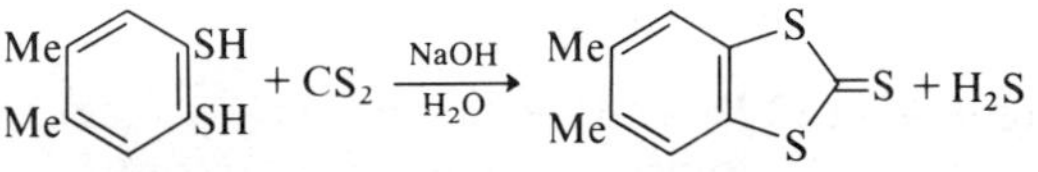

The reactions of mercaptothiolacetic acid with aliphatic or aromatic aldehydes lead to the formation of 1,3-dithiolan-4-one derivatives (**291**) ⟨73JOC3953⟩.

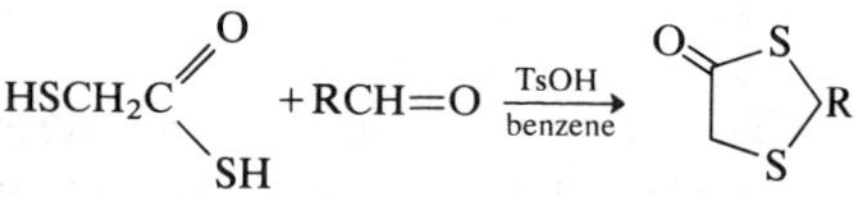

The preparation of 2-methoxy-1,3-dithiolane (**292**) is based on the reaction of dichloromethyl methyl ether with the disodium salt of ethanedithiol in acetonitrile ⟨72HCA75⟩.

Ethanedithiol and phosgene react in ether at room temperature to produce the 1,3-dithiolan-2-one (**293**) in 83% yield ⟨41RTC453⟩. This is probably the best synthesis for (**293**).

(**292**)

(**293**)

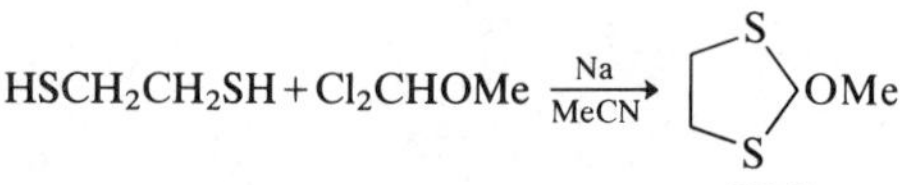

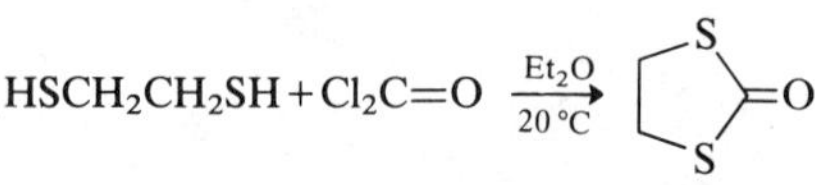

4.32.4.1.3 Formation of two bonds from [3+2] atom fragments

(i) *Derivatives containing C-linked substituents*

Various 1,3-dithiolylium salts (**31**) can be prepared by combining an excess of thioacids or thioesters with α-halo, α-hydroxy or α-mercapto ketones in the presence of hydrogen iodide or perchloric acid ⟨71JPR722⟩.

It has been shown ⟨64JHC163⟩ that α-halo ketones readily form 1,3-dithioles with *gem*-dithiols. A convenient alternative pathway for the synthesis of certain 1,3-dithiol-2-ylidene derivatives (**295**) is the reaction of α-halo ketones with dithioacids of type (**294**) which exist as α,β-unsaturated *gem*-dithiols ⟨65JOC732⟩. For the synthesis of dithiafulvene derivatives *via* (4,5-diethoxycarbonyl-1,3-dithiolyl)tributylphosphonium tetrafluoroborate, see ⟨79JOC930⟩.

(**31**)

(**294**) (**295**)

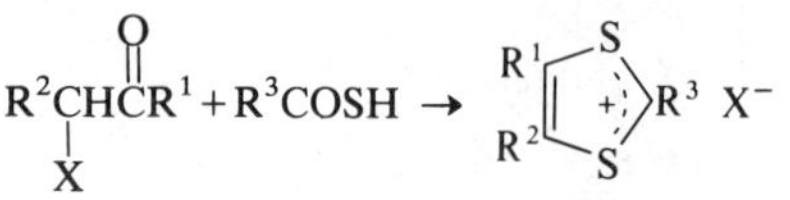

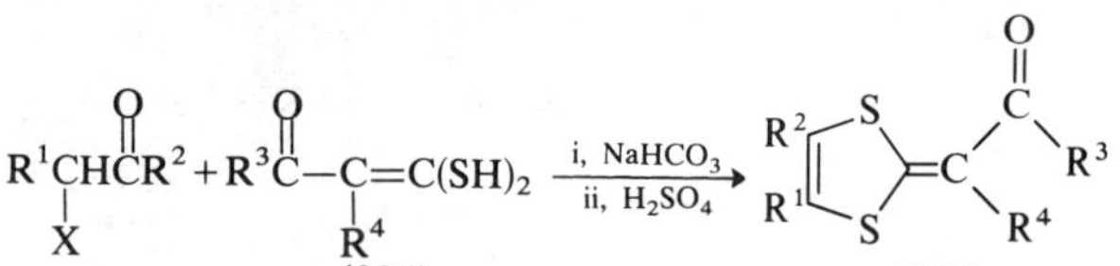

1,3-Dithioles

Alkynes bearing electron-withdrawing substituents react with carbon disulfide to give tetrathiafulvalenes ⟨73JA4379, 70JA1412⟩. For example reaction of perfluorobut-2-yne with carbon disulfide in the presence of trifluoroacetic acid produces the tetrathiafulvalene derivative (**299**). The reaction sequence involves formation of the betaine (**296**) and the carbene (**297**) which is trapped by protonation to give the 1,3-dithiolylium ion (**298**); the latter reacts with the carbene (**297**) to yield (**299**). Since the alkynic compound should bear at least one electron-withdrawing group, this method is limited to preparation of tetrathiafulvalenes bearing electron-withdrawing substituents.

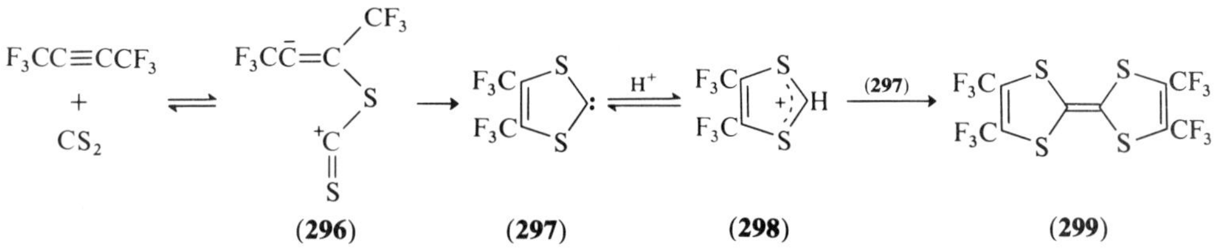

Benzyne also reacts with carbon disulfide to produce the tetrathiafulvalene derivative (**300**) ⟨75S38, 74CC166⟩. Dithioacids react with phenylacetylene to form 1,3-dithioles (**301**) ⟨80TL2247⟩.

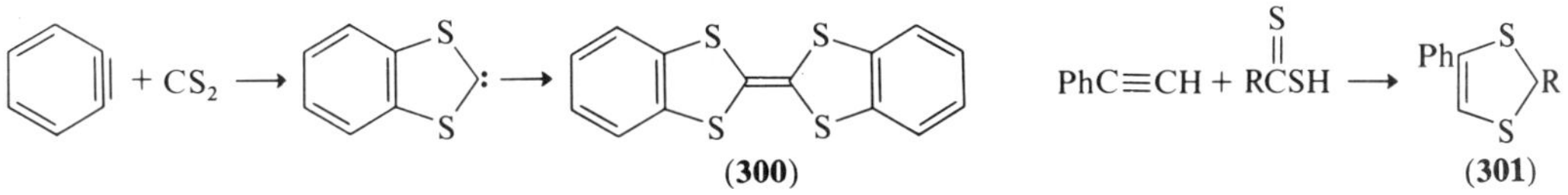

Another method involves anionic cycloaddition of triethylammonium salts of dithioacids or dithiocarbonates to dimethyl acetylenedicarboxylate to produce the 1,3-dithioles (**302**) ⟨81ZOR1788, 80ZOR2616⟩. Also dithiocarboxylic esters react with dimethyl acetylenedicarboxylate giving substituted 1,3-dithioles (**303**) ⟨80ZOR2047, 79ZOR1106⟩.

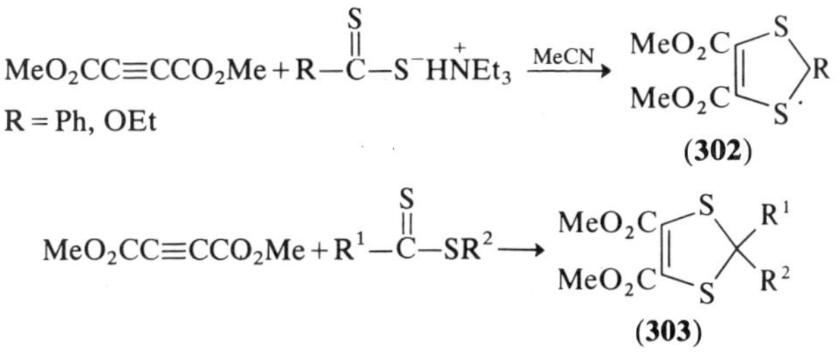

Trialkylphosphines and carbon disulfide give adducts (**304**) which react with alkynoic acids to form the phosphonium betaines (**305**). The latter can be converted into 1,3-dithiole-4-carboxylic acids (**306**) ⟨76BCJ1996⟩. DMAD can also be used in this reaction ⟨79JOC930⟩.

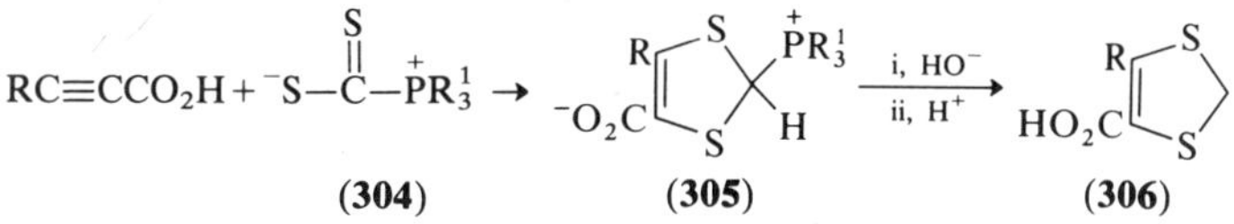

The dianion (**307**), easily prepared from phenols and carbon disulfide in the presence of potassium hydroxide in DMSO, can be alkylated with 1,2-dibromoethane to form the 1,3-dithiole derivatives (**308**) ⟨65LA(684)37⟩.

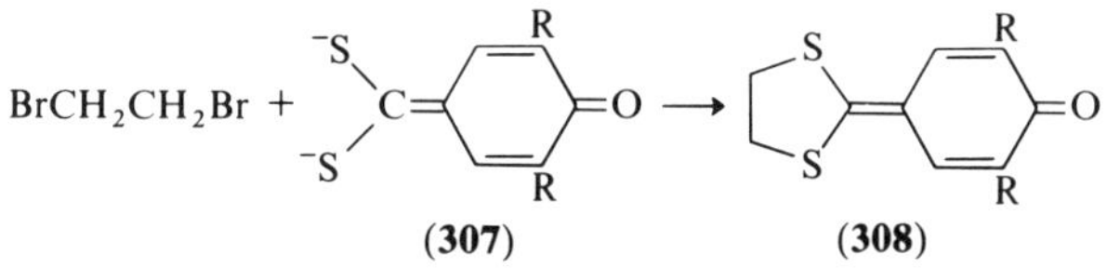

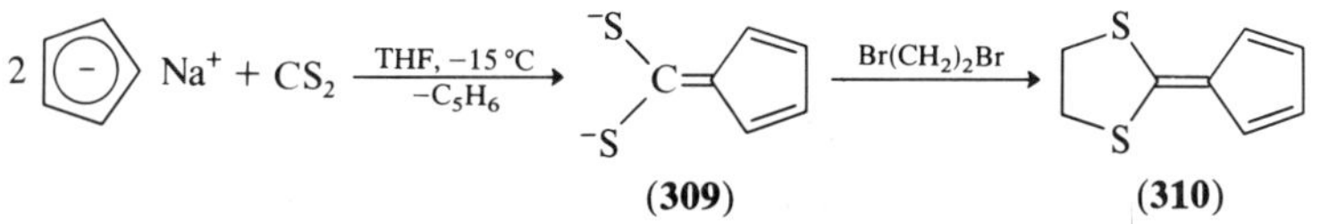

Furthermore, action of two moles of the sodium cyclopentadienide with carbon disulfide gives the dianion (**309**) which can be alkylated with 1,2-dibromoethane to produce the 1,3-dithiole derivative (**310**) ⟨65CB2825⟩.

(ii) Derivatives containing O-linked substituents

A 'one-pot' synthesis of 2-alkoxy-1,3-benzodithioles (**312**) consists of aprotic diazotization of anthranilic acid by alkyl nitrites to give benzenediazonium-2-carboxylate, thermal decomposition to benzyne, reaction of benzyne with carbon disulfide to lead to the carbene (**311**), and addition of alcohols to the carbene ⟨75S38⟩. The alcohols are either generated during the diazotization by alkyl nitrites or are added beforehand to the reaction mixture. Since carbon disulfide is more reactive than alcohols towards benzyne, alkoxybenzenes were not formed. Similarly, substituted 2-alkoxy-1,3-benzodithioles have been prepared ⟨78JCS(P1)468⟩.

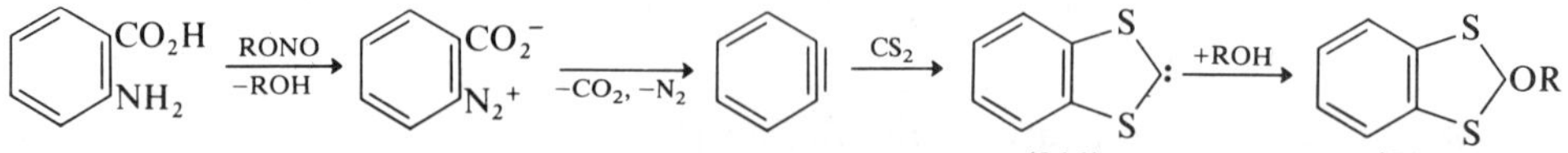

Substituted benzo-1,3-dithiol-2-ones (**315**) are formed, together with products of type (**316**), in the reaction of sodium *N,N*-dimethyldithiocarbamates with 1-chloro-2-nitrobenzenes (**313**) containing additional electron-withdrawing substituents ⟨77JOC1265, 79JOC267⟩. The intermediate dithiocarbamates (**314**) are not isolated.

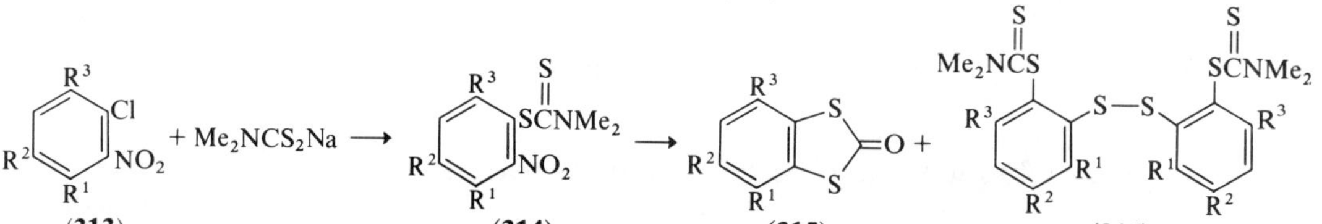

(iii) Derivatives containing S-linked substituents

Introduction of an *S*-linked substituent into the 2-position of 1,3-dithiolylium salts can be carried out by the reaction of 2-chloro-1,2-diphenylethanone with sodium methyl trithiocarbonate in alcohol followed by treatment with perchloric acid to produce 2-methylthio-4,5-diphenyl-1,3-dithiolylium perchlorate (**317**) in 35% yield ⟨65JOC732⟩.

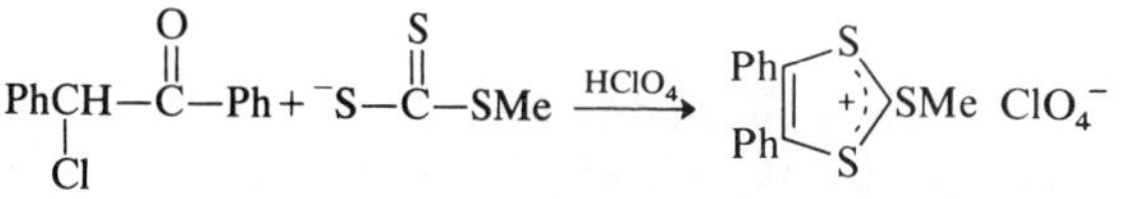

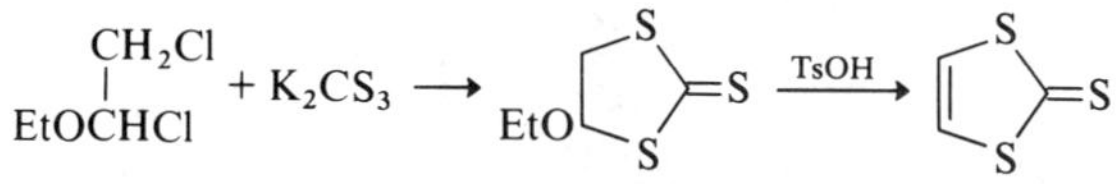

The reaction of 1,2-dichlorodiethyl ether with potassium trithiocarbonate leads to the formation of the cyclic thione (**318**) which on treatment with *p*-toluenesulfonic acid gives 1,3-dithiole-2-thione (**21**) in 61% yield ⟨76CC920⟩. 4-Phenyl-1,3-dithiole-2-thione was prepared by the reaction of 1,2-dibromo-1-phenylethylene with potassium trithiocarbonate ⟨60JPR(11)284⟩.

Substituted 1,3-dithiole-2-thiones (**322**) can be prepared by treatment of propargylic bromides with sodium *t*-butyl trithiocarbonate in acetone at 0 °C giving the trithioesters

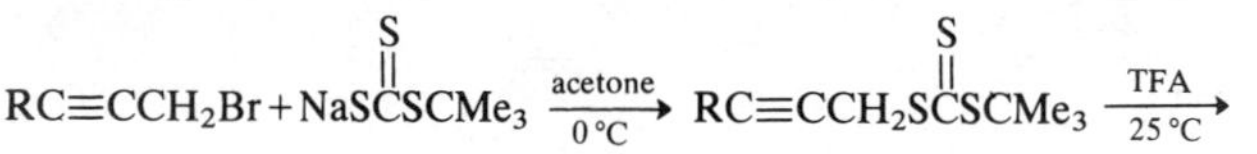

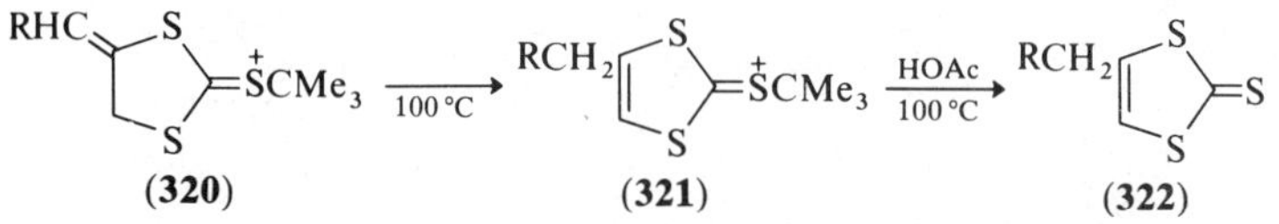

(**319**) in quantitative yields. These trithioesters (**319**) are then heated with one equivalent of trifluoroacetic acid in acetic acid to produce the 1,3-dithiole-2-thiones (**322**) with evolution of isobutene ⟨78TL5161⟩. The intermediates (**320**) and (**321**) can be observed by proceeding stepwise through the reaction sequence. This synthesis allows direct introduction into the 1,3-dithiole-2-thione ring of functional groups that are difficult to introduce by other methods.

An approach to nitro-substituted 1,3-benzodithiole-2-thiones (**325**) is based on the reaction of 1-chloro-2,6-dinitrobenzenes (**323**) with sodium *t*-butyl trithiocarbonate which yields dinitrophenyl *t*-butyl trithiocarbonates (**324**). The latter cyclize on heating in acetic acid with formation of nitro-1,3-benzodithiole-2-thiones (**325**) ⟨80JOC4041⟩.

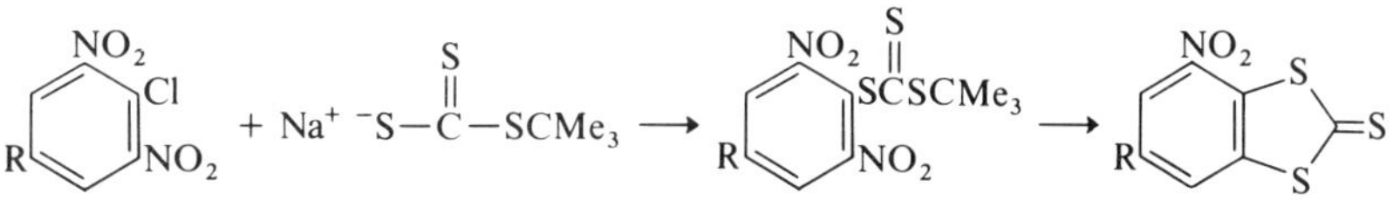

The 1,3-dithiolane-2-thione (**326**) was first prepared by the reaction of ethylene dibromide with sodium trithiocarbonate in ethanol ⟨1862LA(123)83⟩. Presumably the best synthesis of (**326**) is the reaction of ethylene oxide with carbon disulfide and potassium hydroxide in methanol at room temperature ⟨46JCS1050⟩.

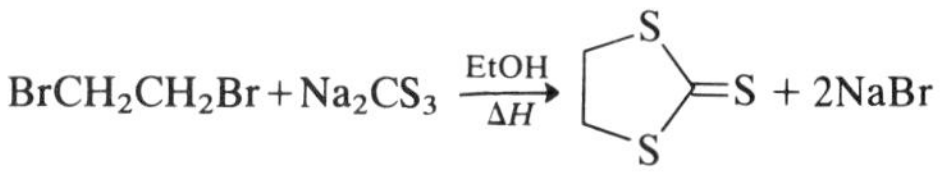

(iv) *Mesoionic 1,3-dithiolones*

Another method of preparing mesoionic 1,3-dithiolones involves the reaction of dithiocarboxylic acids with bielectrophiles. Thus treatment of aromatic dithiocarboxylic acids with α-bromoacyl chlorides in benzene in the presence of triethylamine results in the formation of the mesoionic compounds (**327**) ⟨77JOC1633⟩.

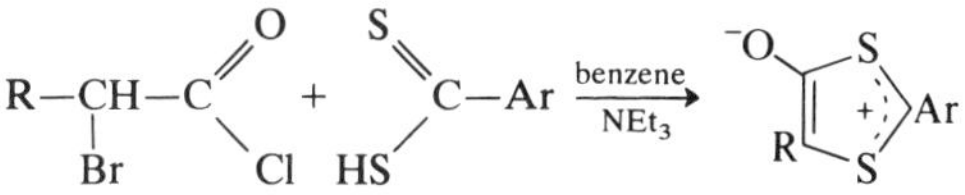

4.32.4.1.4 Formation of multiple bonds

Synthesis of the 1,3-dithiole-2-thione (**21**) also results from the reaction of sodium acetylide with sulfur and carbon disulfide ⟨79MI43201⟩.

The 1,3-dithiole-2-thione derivative (**328**) can be prepared by reduction of carbon disulfide with potassium in DMF under argon and subsequent alkylation with dimethyl sulfate ⟨80CB1898⟩. Other alkylating agents besides dimethyl sulfate can be used.

4.32.4.2 Syntheses involving Heterocyclic Compounds

4.32.4.2.1 Formation of the 1,3-dithiole ring

1,2-Dithiole-3-thione derivatives can be used as starting materials for the synthesis of 1,3-dithioles. For example thermal reactions of 1,2-dithiole-3-thiones (**329**) with alkynic compounds proceed with the formation of 1,3-dithiole derivatives (**330**) ⟨68CB1428, 80ZOR883⟩. It has been shown that this conversion can also be performed photochemically in benzene solution ⟨80ZOR883⟩. Benzyne may also be used as a reactant.

(329) (330)

(331) $R^1 = H$, RCO (332)

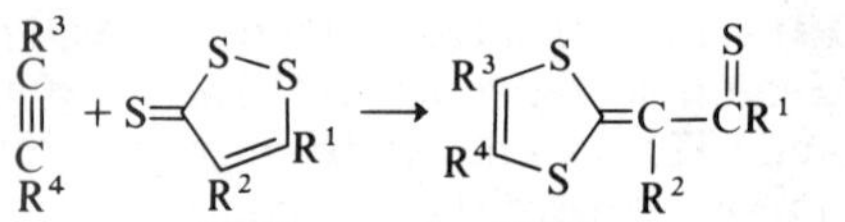

A ring transformation takes place in the reaction of 3-mercapto- or 3-acylthio-3-isothiazoline-5-thiones (**331**) with reactive alkynic compounds in boiling acetonitrile; this reaction produces the 1,3-dithiole derivatives (**332**) in good yields ⟨80JCS(P1)2693, 80H(14)785⟩.

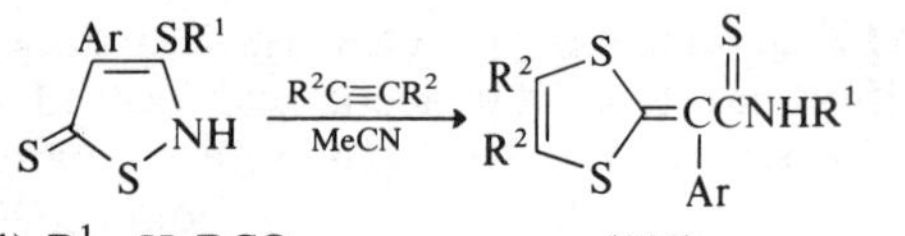

Similarly, reaction of 3-chloroacylthio-3-isothiazoline-5-thiones (**333**) with reactive alkynic compounds proceeds with formation of the 1,3-dithiole derivatives (**334**) which contain thiazolone or 5,6-dihydro-1,3-thiazin-4-one rings ⟨81H(16)595⟩.

(**333**) $n = 1, 2$ (**334**)

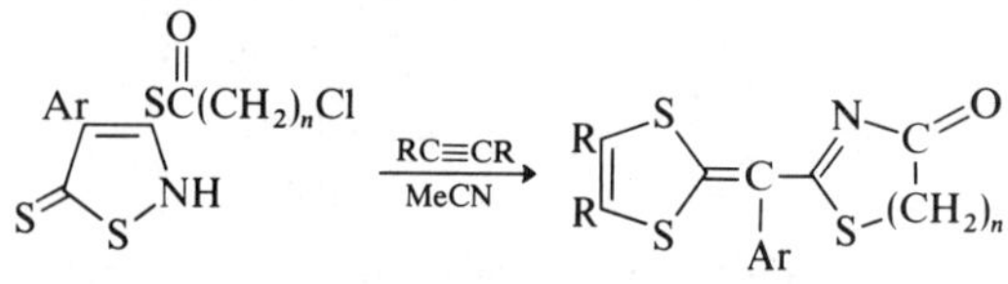

Photocycloaddition–ring-opening reactions have been observed with the heterocyclic systems (**335**). For example photoreactions of 1,2,4-dithiazole-3-thione and 1,2-dithiole-3-thione in the presence of alkenes lead to formation of the 1,3-dithiolanes (**336**) ⟨75JCS(P1)270, 80ZOR443⟩. The reactions are nonstereospecific with respect to the alkene used, both *cis*- and *trans*-but-2-ene affording a mixture of two isomeric 1,3-dithiolanes. It has been suggested that the reaction involves the resonance stabilized biradical (**337**), which is formed from the excited thione (**335**) and the alkene, and its subsequent intramolecular induced cleavage of the S—S bond. In the case of the 1,2-dithiole-3-thione (**335**; X = CMe) this reaction can also be carried out thermally ⟨80ZOR443⟩.

(**335**) (**337**) (**336**) X = N, CH, CMe, CPh

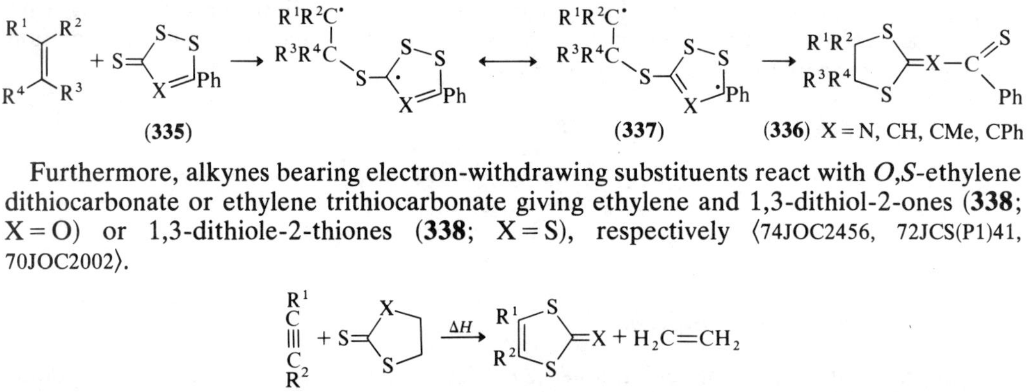

Furthermore, alkynes bearing electron-withdrawing substituents react with *O,S*-ethylene dithiocarbonate or ethylene trithiocarbonate giving ethylene and 1,3-dithiol-2-ones (**338**; X = O) or 1,3-dithiole-2-thiones (**338**; X = S), respectively ⟨74JOC2456, 72JCS(P1)41, 70JOC2002⟩.

X = O, S (**338**)

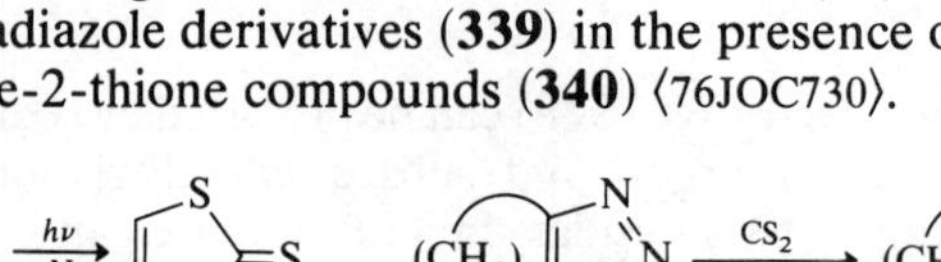

A photochemical path to 1,3-dithiole-2-thione has been reported ⟨61E566⟩. Photolysis of 1,2,3-thiadiazole in carbon disulfide produces an intermediate C_2H_2S fragment which is trapped by carbon disulfide to give the 1,3-dithiole-2-thione (**21**). A similar, albeit thermal, reaction of the 1,2,3-thiadiazole derivatives (**339**) in the presence of carbon disulfide yields at 215 °C the 1,3-dithiole-2-thione compounds (**340**) ⟨76JOC730⟩.

(**21**) (**339**) (**340**) $n = 3$; 41%
$n = 4$; 43%

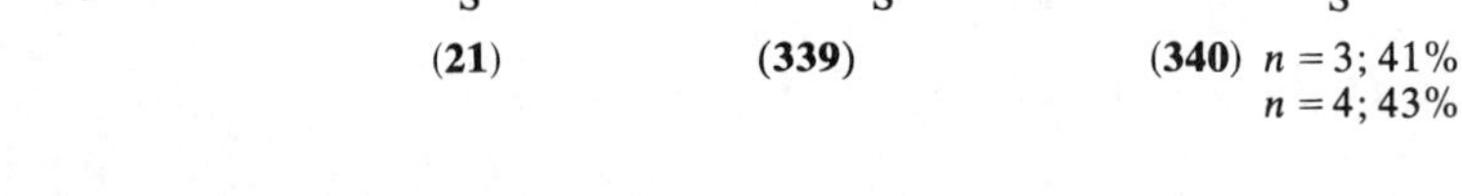

4.32.4.2.2 Formation of tetrathiafulvalenes

The syntheses of tetrathiafulvalenes have been reviewed recently ⟨76S489⟩.

Many tetrathiafulvalenes have been prepared by deprotonation of 1,3-dithiolylium salts with tertiary amines ⟨76S489, 80AHC(27)151⟩. In this reaction, the intermediate resonance stabilized carbene (**73**) reacts with the 1,3-dithiolylium ion (**1**) followed by deprotonation to produce the tetrathiafulvalene (**5**).

(**1**) (**73**) (**5**)

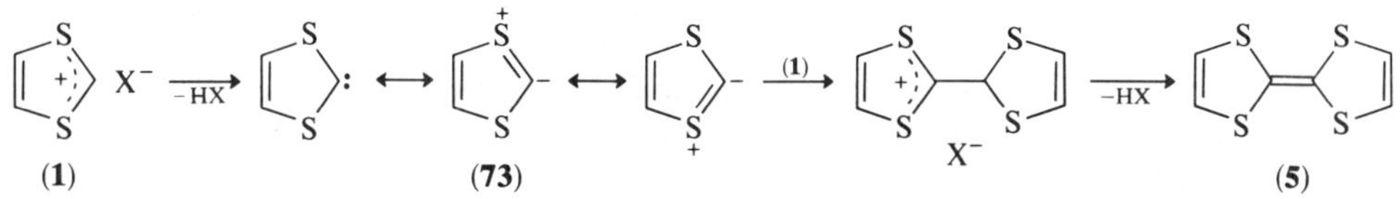

A convenient preparation of tetrathiafulvalenes is achieved by the reduction of 2-methylthio-1,3-dithiolylium salts (**207**) with zinc dust in the presence of bromine ⟨76ZC317⟩. This reaction proceeds with formation of the radical (**341**) which dimerizes to (**342**); the latter produces dimethyl disulfide and the tetrathiafulvalene derivatives (**343**).

(**207**) (**341**) (**342**) (**343**)

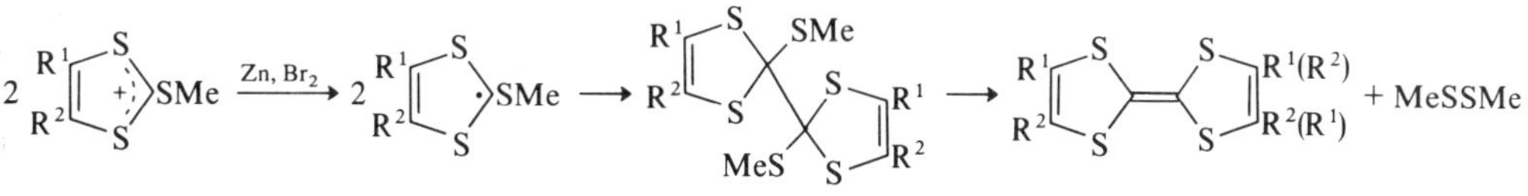

Electrochemical reduction of the 2-ethylthio-1,3-dithiolylium salts (**344**) in acetonitrile gives the corresponding dimers (**345**) which can then pyrolyze to the tetrathiafulvalene derivatives (**24**) in nearly quantitative yields ⟨74JA945⟩. When one considers the total yield, this method is superior to the base-catalyzed dimerization of 1,3-dithiolylium ions. However, for large-scale synthesis, the convenience may be questionable.

(**344**) R = H, Me (**345**) (**24**)

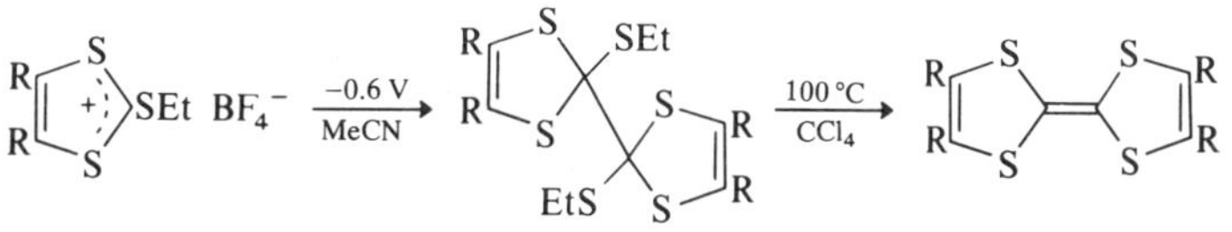

Thermal decomposition of the sodium salt of the tosylhydrazone (**346**) at 180 °C produces the dibenzotetrathiafulvalene (**347**) in 51% yield ⟨74LA403⟩. Also the 2-alkoxy-1,3-benzodithioles (**312**) decompose thermally at 200 °C to give (**347**) in 22–55% yield ⟨75S168⟩.

(**346**) (**347**) (**312**)

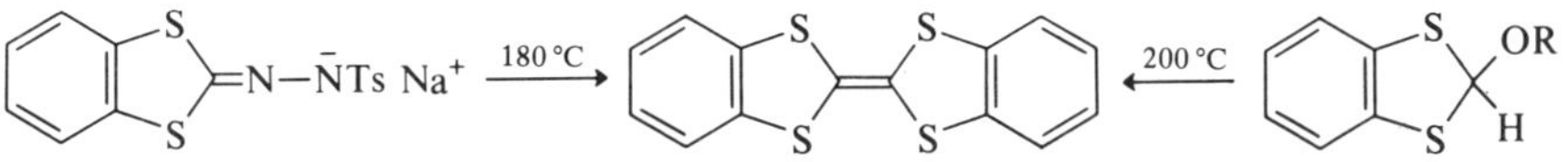

An alternative approach to the synthesis of tetrathiafulvalenes is the desulfurization of 1,3-dithiole-2-thiones with trivalent phosphorus compounds. Thus, triphenylphosphine reacts with the 1,3-dithiole-2-thione (**348**; X = S, R = CF_3) to produce the tetrathiafulvalene (**24**; R = CF_3) in 89% yield ⟨73JA4379, 70JA1412⟩. For (**348**; X = S, R = CN), this reaction works well at 125 °C, but at lower temperatures other products have been observed ⟨75JOC2577⟩. Similarly, trimethyl phosphite reacts with the 1,3-dithiol-2-one (**348**; X = O, R = CN) to form the corresponding tetrathiafulvalene (**24**) in quantitative yield ⟨74CC751⟩.

(**348**) X = O, S (**24**)

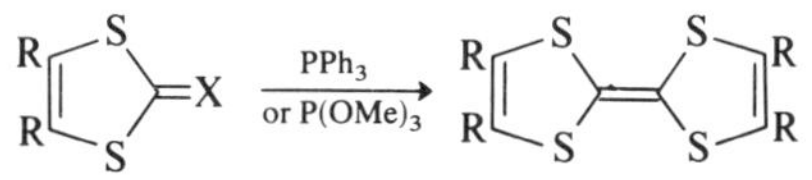

1,3-Dithiolylium salts react with the phosphoranes (**349**) in the presence of excess triethylamine to produce the tetrathiafulvalenes (**351**) ⟨78JOC369⟩. The primary step is formation of the coupling product (**350**); then triethylamine induces elimination of triphenylphosphine. This method allows the preparation of tetrathiafulvalene derivatives from two differently substituted 1,3-dithiolylium cations.

(**349**) (**350**) (**351**)

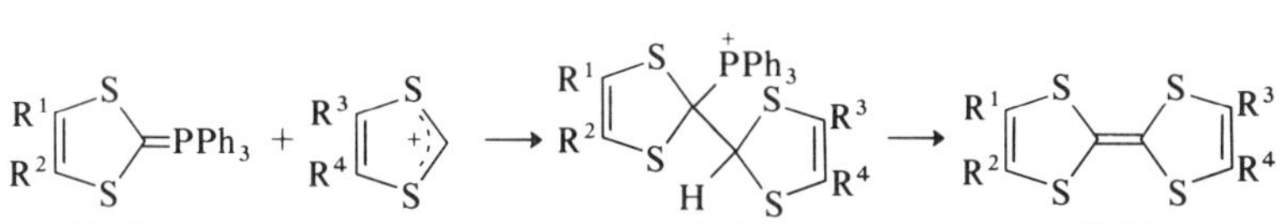

A photochemical one-step synthesis of tetrathiafulvalenes substituted with either electron-donating or -withdrawing groups is the irradiation of 1,3-dithiole-2-thiones (**23**) in the presence of hexabutyldistannane to give the corresponding tetrathiafulvalene derivatives in good yields ⟨76JA7440⟩. It has been shown that a higher yield of (**24**) is obtained when twice the molar amount of hexabutyldistannane is employed. In this reaction the excited thione (**23**) might interact with the tin–tin bond, followed by abstraction of sulfur to form a carbene which then yields (**24**).

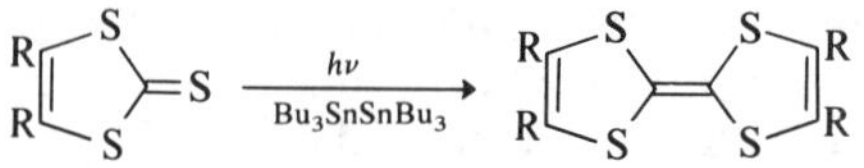

A thermal one-step synthesis affords tetrathiafulvalenes in 20–42% yields when 1,3-dithiole-2-thiones are heated with dicobalt octacarbonyl in boiling benzene or toluene ⟨80CC38⟩. This method has been applied to the synthesis of various substituted tetrathiafulvalenes.

4.32.4.3 Synthesis of Parent Compounds

The unsubstituted 1,3-dithiolylium salt (**1**) can be prepared by reaction of 1,3-dithiole-2-thione (**21**) with peracetic acid in acetone or with hydrogen peroxide in acetic acid ⟨76S489⟩. Another method includes borohydride reduction of 2-methylthio-1,3-dithiolylium salts to the 2-methylthio-1,3-dithiole followed by acidification ⟨76S489⟩.

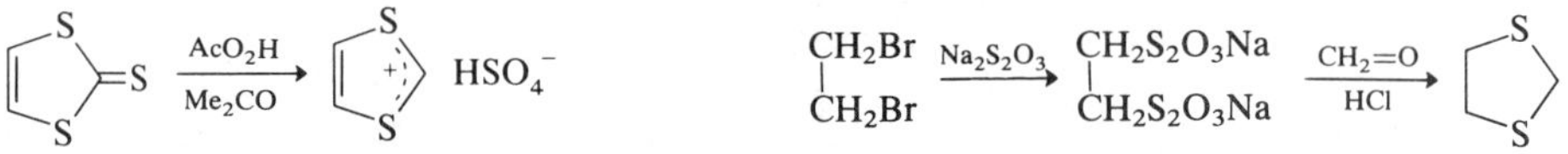

Reaction of the bis Bunte salt (**280**), prepared from ethylene dibromide and sodium thiosulfate, with formaldehyde in the presence of hydrogen chloride produces the 1,3-dithiolane (**4**) ⟨30JCS12⟩.

4.32.5 APPLICATIONS

Since the development of organic metals, tetrathiafulvalene and its derivatives have played an important role as electron donors in charge-transfer salts which show high electrical conductivity. Many of the charge-transfer complexes prepared and their electrical properties have been listed in a review article ⟨76S489⟩. For example the tetrathiafulvalene–7,7,8,8-tetracyanoquinodimethane complex exhibits an electrical conductivity of 652 Ω^{-1} cm^{-1} at room temperature, and a maximum of 1.47×10^4 Ω^{-1} cm^{-1} at 58 K ⟨76S489⟩. Many other acceptors have been used in preparing charge-transfer complexes with tetrathiafulvalene and several of its derivatives. Furthermore, it has been pointed out that the high polarizability and D_{2h} symmetry of tetrathiafulvalene could be of fundamental importance in achieving electrical conductivity in the charge-transfer complexes. For design and synthesis of organic metals see ⟨74ACR232⟩.

4.33

Five-membered Rings containing Three Oxygen or Sulfur Atoms

G. W. FISCHER and T. ZIMMERMANN
Academy of Sciences of the G.D.R., Leipzig

4.33.1 INTRODUCTION	852
4.33.2 STRUCTURE	853
4.33.2.1 Theoretical Investigations	853
4.33.2.2 Structural Methods	854
4.33.2.2.1 Electron and X-ray diffraction	854
4.33.2.2.2 Microwave spectra	855
4.33.2.2.3 Infrared and Raman spectra	858
4.33.2.2.4 Ultraviolet spectra	860
4.33.2.2.5 Photoelectron spectra	861
4.33.2.2.6 Nuclear magnetic resonance spectra	862
4.33.2.2.7 Mass spectra	863
4.33.2.2.8 Other physical methods	864
4.33.3 REACTIVITY	865
4.33.3.1 Thermal and Photochemical Reactions Formally Involving No Other Species	865
4.33.3.1.1 Rearrangement	865
4.33.3.1.2 Thermolysis	866
4.33.3.1.3 Photolysis	869
4.33.3.1.4 Polymerization	869
4.33.3.2 Reactions with Electrophiles	870
4.33.3.2.1 Reactions involving protons or Lewis acids	870
4.33.3.2.2 Electrophilic addition at double bonds	871
4.33.3.2.3 Oxidation without ring cleavage	872
4.33.3.2.4 Oxidative ring cleavage	872
4.33.3.3 Reactions with Nucleophiles	872
4.33.3.3.1 O-Nucleophiles	872
4.33.3.3.2 S-Nucleophiles	876
4.33.3.3.3 N-Nucleophiles	876
4.33.3.3.4 C-Nucleophiles	878
4.33.3.3.5 Halide ions	879
4.33.3.3.6 Reductive ring cleavage	879
4.33.3.4 Reactions with Carbonyl Compounds	880
4.33.3.5 Reactions with Radicals	881
4.33.4 SYNTHESIS	882
4.33.4.1 General Survey of Ring Syntheses	882
4.33.4.1.1 Syntheses from non-heterocyclic compounds	882
4.33.4.1.2 Syntheses by ring transformations	882
4.33.4.2 Syntheses of Various Types of Derivatives	882
4.33.4.2.1 Parent systems (including S-oxides and S,S-dioxides)	882
4.33.4.2.2 Benzo derivatives	887
4.33.4.2.3 C-Linked substituents	888
4.33.4.2.4 N-Linked substituents	891
4.33.4.2.5 O-Linked substituents (including oxo derivatives)	891
4.33.4.2.6 Halogens attached to the ring	892
4.33.5 APPLICATIONS	893
4.33.5.1 Applications in Research and Industry	893
4.33.5.2 Natural Occurrence and Special Modes of Formation	894
4.33.5.3 Biological Activity	895

4.33.1 INTRODUCTION

Five-membered monocyclic rings containing three oxygen or sulfur atoms generally cannot form a conjugated aromatic π-electron system. The 12 conceptually possible saturated parent structures are shown in Scheme 1. According to their composition they may be classified into four groups: C_2O_3 ring systems (trioxolanes), C_2S_3 ring systems (trithiolanes), C_2O_2S ring systems (dioxathiolanes) and C_2OS_2 ring systems (oxadithiolanes). Of the 12 basic structures, monocyclic representatives of the 'mixed' systems (**5**), (**6**) and (**10**) are so far unknown and most of the other systems only occur in the form of derivatives. The existence of (**12**) is in question.

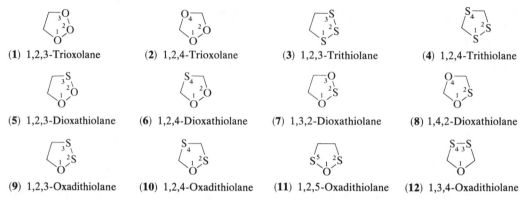

Scheme 1 Saturated five-membered parent ring systems containing three oxygen or sulfur atoms

1,2,3-Trioxolane (**1**) is the parent compound of the so-called primary ozonides, *i.e.* the primary reaction products of ozone with alkenes. They are extremely unstable and rearrange to the more stable ozonides which represent derivatives of 1,2,4-trioxolane (**2**). Also in the C_2S_3 series, derivatives of the 1,2,3-system (**3**) are much less stable and consequentially have been studied less than the corresponding derivatives of the 1,2,4-system (**4**) ('trithioozonides'). Heterocycles of the C_2O_2S and C_2OS_2 type are known principally in the form of their *S*-oxides and *S,S*-dioxides. The largest group of these 'mixed' systems comprises cyclic sulfite and sulfate esters derived from the hypothetical parent 1,3,2-dioxathiolane (**7**). However, few *S*-oxides and *S,S*-dioxides of the hypothetical parent systems (**8**), (**9**) and (**11**) have been reported. The last group represents 1,2-disulfinic and 1,2-disulfonic anhydrides, respectively. Thus, all in all, the most thoroughly investigated five-membered heterocycles containing three O or S atoms are at present the 1,2,4-trioxolanes, 1,2,4-trithiolanes and 1,3,2-dioxathiolanes.

Several saturated parent systems shown in Scheme 1, especially (**2**), (**4**) and (**7**), have been incorporated into spiro compounds or have been fused with a variety of non-aromatic carbocyclic or heterocyclic rings. Unsaturated compounds with an endocyclic carbon–carbon double bond, formally possible in all systems with three adjacent heteroatoms, are only known in very few cases (*e.g.* cyclic sulfites and sulfates of enediols) or in the form of benzo or other condensed systems (mainly cyclic sulfites and sulfates of aromatic 1,2-dihydroxy compounds). For details concerning the various types of derivatives, see Section 4.33.4.2.

Historical landmarks in trioxolane chemistry are: the first ozonolysis of an organic compound, ethylene, by Schönbein (1855); the pioneering work of Harries in isolating defined ozonides and in establishing the ozonolysis reaction as a useful method in organic chemistry (1901–1916); the interpretation of the ozonolysis of alkenes as a multi-step process involving an unstable primary alkene–ozone adduct ('molozonide', formulated as a zwitterionic 1,2-dioxetane 1-oxide) and its isomerization to the isolable 1,2,4-trioxolane-type 'isozonide' by Staudinger (1925); the experimental proof of the 1,2,4-trioxolane structure by an independent synthesis of ozonides from dihydroxydialkyl peroxides (Rieche, 1932); Criegee's three-step mechanism of ozonolysis (1953) and the first preparation of a crystalline primary ozonide (1959) which later actually proved to be a 1,2,3-trioxolane. Stimulated by the possibilities of modern theoretical and experimental methods, the further development of ozonide chemistry is characterized primarily by modifications and refinements of structural and mechanistic concepts ⟨B-78MI43300⟩.

The chemistry of 1,2,4-trithiolanes began, for all practical purposes, in the late 1950s when Asinger *et al.* reported a general synthesis of this ring system (*cf.* Section 4.33.4.2.3).

In recent years, compounds of this type have received considerable interest as flavor components of various foodstuffs and as natural products of certain plant species (*cf.* Section 4.33.5.2). 1,3,2-Dioxathiolane derivatives, in the form of cyclic sulfite and sulfate esters, have been known since the turn of the century; however, details of their structure and chemical behavior have only been studied during the last decades.

As the chemistry of ozonides is the subject of a series of comprehensive review articles ⟨40CRV(27)437, 58CRV925, 67AHC(8)165, 81MI43300⟩, books ⟨B-31MI43300, B-74MI43300, B-78MI43300⟩, more general treatises ⟨52HOU(8)27, 58AG251, B-61MI43300⟩ and summaries covering special aspects ⟨68C392, 75AG765, 80MI43302⟩, the original literature will be quoted in this chapter only in selected cases, emphasis being on more recent work. Similarly, 1,2,4-trithiolanes ⟨66HC(21-1)1⟩, 1,3,2-dioxathiolane 2-oxides ⟨63CRV557, 66HC(21-1)1, 76CRV747⟩ and the 2,2-dioxides ⟨66HC(21-1)1, B-77MI43300⟩ have been reviewed, although mostly in a wider context.

Within each subsection the several structural types will be discussed generally in the order in which they are arranged in Scheme 1.

4.33.2 STRUCTURE

4.33.2.1 Theoretical Investigations

Theoretical studies on the structure of five-membered rings containing three O or S atoms mainly deal with conformational problems. The alicyclic parent system, cyclopentane, is not planar because of the great torsional strain (Pitzer strain) in the planar conformation. The two non-planar conformations with the highest symmetry, the half-chair and the envelope form, as well as intermediate conformations, have nearly the same energy. Cyclopentane is therefore not found in one well-defined conformation, but undergoes a continuous interconversion by pseudorotation. The substitution by oxygen and/or sulfur of three CH_2 moieties of the cyclopentane ring, however, considerably changes the energy situation and increases the probability of preferential conformations being adopted in the resultant ring systems.

In the 1,2,3-trioxolane series the special interest in structural data arises from questions connected with the mechanism and stereochemistry of the ozonolysis reaction (*cf.* Section 4.33.3.1.1). Compounds of this type, owing to their extreme instability, are difficult to investigate by conventional experimental methods, and theoretical approaches utilizing conformational analysis are of particular importance.

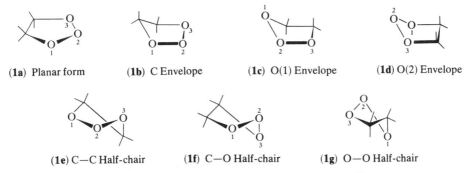

(**1a**) Planar form   (**1b**) C Envelope   (**1c**) O(1) Envelope   (**1d**) O(2) Envelope

(**1e**) C—C Half-chair   (**1f**) C—O Half-chair   (**1g**) O—O Half-chair

Scheme 2 Possible conformations of 1,2,3-trioxolane (**1**)

Scheme 2 shows the possible conformations of 1,2,3-trioxolane (primary ethylene ozonide) (**1**). Various quantum chemical methods have predicted different conformations for (**1**). The extended Hückel LCAO–MO (EHT) calculations ⟨70JA2628⟩ predict that conformation (**1e**) is more stable than (**1b**), whereas by means of the CNDO/2 method ⟨73JA3460⟩ conformations (**1c**) and (**1f**) were found to be lowest in energy. Based on MINDO/3 calculations ⟨77BSB481⟩ a planar ring, slightly distorted from C_{2v} symmetry, was suggested. *Ab initio* studies comparing selected 1,2,3-trioxolane conformations resulted in the following order of stability: (**1g**)>(**1e**)>(**1f**)>(**1a**) ⟨74MI43301⟩, (**1c**)>(**1e**)>(**1a**) ⟨76JA6088⟩, and (**1g**)>(**1f**)>(**1e**)>(**1a**) ⟨80MI43300⟩. However, a thorough *ab initio* study ⟨79JCP(70)1898⟩ using large, augmented basis sets and including all seven possible conformations afforded, after extensive optimization of geometry, the sequence (**1d**)>(**1g**)>(**1c**)>

(**1f**) > (**1b**) > (**1e**) > (**1a**). The symmetrical O(2) envelope form was predicted to represent the most stable conformation. According to these calculations, pseudorotation is hindered by a barrier of 14.7 kJ mol^{-1}, approximately half as high as the barrier to planarity (30.0 kJ mol^{-1}). By means of the same approach which proved to be markedly superior to semiempirical methods (EHT, CNDO/2, MINDO/3), it was also shown ⟨79JCP(70)1911⟩ that (i) primary ozonides of 1-alkenes, *trans*-alkenes and small *cis*-alkenes should preferentially adopt O(2) envelope conformations, while in the case of primary ozonides of *cis*-alkenes with bulky alkyl groups, C—C half-chair conformations seem to be energetically more favored; (ii) only an equatorially placed methyl group leads to a favorable orbital mixing between the π-type MOs of the ring and the substituent.

In the 1,2,4-trioxolane series, owing to the other ring topology (a peroxy bridge is opposite an ether bridge), quantum chemical approaches are more consistent than in the case of five-membered systems containing three adjacent heteroatoms. Thus, calculations performed with the Westheimer–Hendrickson model ⟨69ACS2741⟩, the CNDO/2 method ⟨73JA3460⟩ and the *ab initio* MO theory ⟨79JCP(70)1898⟩ consistently predicted that from the seven possible conformations (**2a**)–(**2g**) shown in Scheme 3, the symmetrical O—O half-chair form (**2g**) corresponds to the conformational minimum. This prediction is in agreement with experimental findings (*cf.* Section 4.33.2.2). The *ab initio* study yielded the following sequence of decreasing stability: (**2g**) > (**2c**) > (**2e**) > (**2b**) > (**2f**) > (**2d**) > (**2a**); for pseudorotation and ring interconversion through the planar form, similar barriers (~13 and 25 kJ mol^{-1}, respectively) were calculated as in the case of 1,2,3-trioxolane (**1**). In agreement with spectroscopic results, 3-methyl-1,2,4-trioxolane as well as *cis*- and *trans*-3,5-dimethyl-1,2,4-trioxolane were also found to prefer likewise the O—O half-chair conformation and to possess at least one methyl group in an equatorial position ⟨79JCP(70)1928⟩.

(a) Planar form (b) C Envelope (c) X(1) Envelope (d) X(4) Envelope

(**2**) X = O
(**4**) X = S

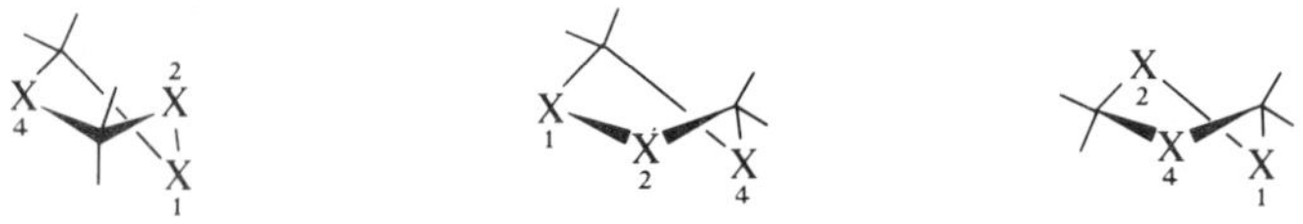

(e) C—X(1) Half-chair (f) C—X(4) Half-chair (g) X—X Half-chair

Scheme 3 Possible conformations of 1,2,4-trioxolane (**2**) and 1,2,4-trithiolane (**4**)

Although quantum chemical calculations have not been carried out on 1,2,3-trithiolane (**3**) or its derivatives, these calculations on 1,2,4-trithiolane (**4**) have shown a gradation of stability (**4g**) > (**4c**) > (**4b**) > (**4d**) > (**4a**) ⟨76JA2741⟩. These calculations show that by analogy to 1,2,4-trioxolane (**2**) the symmetrical S—S half-chair conformation is the most stable, being favored over the envelope form (**4d**) by 13.0 kJ mol^{-1}, which corresponds to the pseudorotational barrier between the symmetrical half-chair forms. The barrier for ring interconversion through the planar form (**4a**) was calculated to be 18.8 kJ mol^{-1}.

Further theoretical calculations, which were performed in association with experimental investigations, will be described in the following subsections.

4.33.2.2 Structural Methods

4.33.2.2.1 *Electron and X-ray diffraction*

Bond lengths and bond angles which were determined for some derivatives of 1,2,4-trioxolane, 1,2,4-trithiolane and 1,3,2-dioxathiolane by electron or X-ray diffraction are summarized in Table 1. In the case of 1,2,4-trioxolane (**2**), these parameters and dihedral angles [O(1)O(2)—C(3)O(4) 42.2°, O(2)C(3)—O(4)C(5) −16.2°, and C(5)O(1)—O(2)C(3) 49.1°] are in satisfactory agreement with an O—O half-chair conformation (**2g**) (*cf.* Scheme 3). However, they cannot entirely rule out an O(4) envelope form like (**2d**). Calculations of conformational energies, using the experimentally obtained molecular parameters, give further support to the conclusion that the 1,2,4-trioxolane ring exists predominantly in the symmetrical O—O half-chair form (**2g**) ⟨69ACS3398⟩. The X-ray crystal structure determination of *trans*-5-anisyl-3-methoxycarbonyl-1,2,4-trioxolane (**13**) unambiguously confirmed the assignment of configuration (being based on ^{1}H NMR and other data), but no definite conclusions could be drawn concerning the conformational problem ⟨70ACS2137⟩.

The X-ray data of 1,2,4-trithiolane-3,5-dione diphenylhydrazone (**14**) ⟨71JCS(B)415⟩ and of the 3,5-bis(*N,N*-diethylimonium) derivative (**15**) show that both compounds are in a puckered conformations, to which in the case of (**15**) an S—S half-chair form was assigned ⟨72MI43300⟩. According to electron diffraction data, gaseous 1,3,2-dioxathiolane 2-oxide (ethylene sulfite) (**16**) seems to be planar, or close to planar, with a pyramidal —SO$_3$— moiety ⟨70DOK(195)1333⟩, whereas alkyl derivatives of (**16**) may adopt markedly puckered conformations. For example, in the 4,5-dimethyl-1,3,2-dioxathiolane 2-oxide (**17**) and its stereoisomer (with both methyl groups in the *trans* position to the S=O group), the dihedral angles O(3)C(4)—C(5)O(1), C(4)C(5)—O(1)S(2) and C(5)O(1)—S(2)O(3) were determined by electron diffraction to be 41.2, 34.6 and 14.0°, respectively, clearly indicating the non-planarity of the ring skeleton. It was also shown that in (**17**) both methyl substituents are pseudoequatorial, whereas in its stereoisomer mentioned above, one methyl is pseudoaxial, the other pseudoequatorial ⟨75BSB775⟩.

1,3,2-Dioxathiolane 2,2-dioxide (ethylene sulfate) (**18**) was found ⟨68JA2970⟩ by X-ray analysis to exist likewise in a puckered conformation, having an angle of 20.6° between the C(4)—C(5) bond and the O(1)S(2)O(3) plane. On the other hand, 1,3,2-dioxathiole 2,2-dioxide (vinylene sulfate) (**19**) is planar within experimental error ⟨68JA2970⟩. However, in the corresponding benzo derivative (catechol sulfate) the sulfur atom lies 0.249 Å above the plane through the six carbon atoms and the two ring oxygen atoms ⟨69JA6604⟩.

4.33.2.2.2 Microwave spectra

Of the five-membered hererocycles containing three O or S ring atoms, mainly the 1,2,4-trioxolanes have been investigated by microwave spectroscopy. Their unambiguous identification by characteristic rotational constants proved to be a valuable analytical aid for elucidating the mechanism of ozonolysis (*cf.* Section 4.33.3.1.1). These constants for a number of 1,2,4-trioxolanes are listed in Table 2, showing also that the various isotopically substituted species of a single compound differ markedly, a fact which is of particular importance not only for mechanistic studies ⟨73JA1348, 80JA4763⟩ but also for detailed structural investigations. Thus, based on the microwave spectra of 1,2,4-trioxolane (ethylene ozonide) and 3-methyl-1,2,4-trioxolane (propylene ozonide), as well as their ^{2}H-, ^{13}C- and ^{18}O-labeled species, the first reliable information on the fine structure of heterocycles of this type was obtained ⟨72JA6337, 74JA348⟩. The structural parameters given in Table 3, especially the dihedral angles, show that both ozonides possess an O—O half-chair conformation (depicted in Scheme 3 and Table 2), confirming suggestions based on electron diffraction (*cf.* Section 4.33.2.2.1). The methyl group in propylene ozonide, which influences the ring geometry only to a very small extent, was found to be in an equatorial position (for consistency with theoretical investigations, see Section 4.33.2.1). In the case of *trans*-3,5-dimethyl-1,2,4-trioxolane (*trans*-2-butene ozonide), where only the normal isotopic species has been studied, the conformational assignment is based on a comparison of observed rotational constants with constants predicted from models possessing the ethylene ozonide ring structure and differing only in the arrangements of the two methyl groups, either axial or equatorial. The agreement of the observed constants with those predicted for diequatorial methyl groups shows unambiguously that *trans*-2-butene ozonide has the same O—O half-chair conformation as ethylene and propylene ozonide ⟨74JA348⟩. The rotational constants of 3-fluoro-1,2,4-trioxolane also correlate with an O—O half-chair form but only if the fluorine atom occupies the axial site ⟨76JA4012⟩. This finding has analogies in other haloheterocyclic systems; indeed, quantum chemical investigations have shown that π-donation from an equatorially oriented fluorine atom destabilizes puckered conformations of both the primary and the final ozonide ⟨81JA3633⟩. 3,3-Difluoro-1,2,4-trioxolane was also found to exist in a single preferred conformation, but in this case the rotational constants do not distinguish unequivocally between a half-chair or envelope conformation

Table 1 Structural Parameters of Derivatives of 1,2,4-Trioxolane, 1,2,4-Trithiolane and 1,3,2-Dioxathiolane Obtained by Electron or X-Ray Diffraction

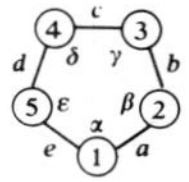

Compound	Bond lengths (Å)					Bond angles (°)					Method[a]	Ref.
	a	b	c	d	e	α	β	γ	δ	ε		
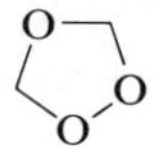 (2)	1.487	1.414	1.414	1.414	1.414	99.2	99.2	105.3	105.9	105.3	E	69ACS3398
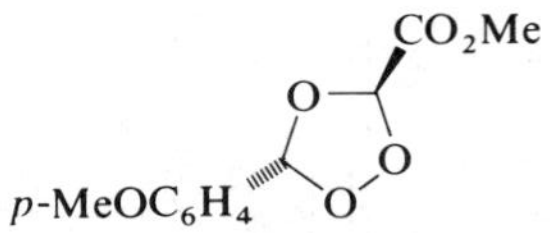 (13)	1.479	1.401	1.468	1.559	1.338	102.2	98.0	107.9	102.1	103.3	X	70ACS2137
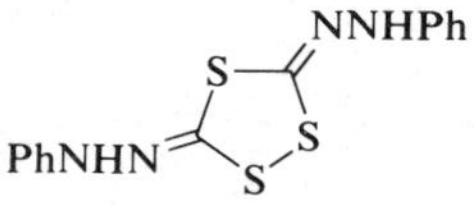 (14)	2.07	1.77	1.77	1.77	1.77	94.8	94.8	115.0	98.5	115.0	X	71JCS(B)415

Structure										Method[a]	Ref.
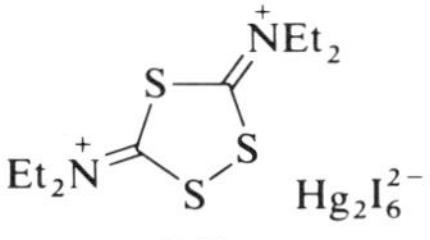 (15)	2.07	1.67	1.77	1.72	1.68	98	99	120	97	123	X	72MI43300
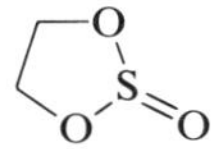 (16)	1.629	1.629	1.438	1.535	1.438	—[b]	102.0	—[b]	—[b]	—[b]	E	70DOK(195)1333
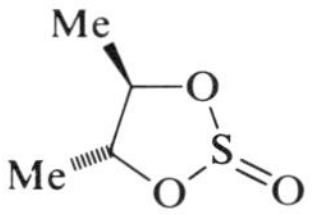 (17)	1.624	1.624	1.434	1.535	1.434	112.3	93.5	112.3	102.4	102.4	E	75BSB775
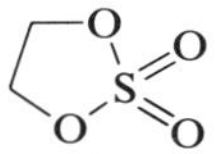 (18)	1.533	1.533	1.459	1.483	1.459	111.5	98.4	111.5	103.7	103.7	X	68JA2970
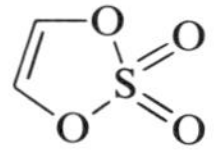 (19)	1.616	1.616	1.341	1.305	1.341	110.2	93.6	110.2	113.0	113.0	X	68JA2970

[a] E = Electron diffraction, X = X-ray diffraction. [b] Not reported.

Table 2 Ground State Rotational Constants of 1,2,4-Trioxolanes and Some of Their Isotopically Substituted Species

Isotopically labeled ring atom	Substituents				Rotational constants (MHz)			Ref.
	R^1	R^2	R^3	R^4	A	B	C	
—	H	H	H	H	8243.84	8093.80	4584.74	72JA6337
^{13}C(3)	H	H	H	H	8233.28	7928.25	4530.18	72JA6337
—	Me	H	H	H	7574.36	3476.33	2644.79	74JA348
—	Me	D	H	H	7326.45	3410.82	2636.66	74JA348
—	Me	H	D	H	7410.52	3372.04	2566.58	74JA348
—	Me	H	H	D	7316.41	3387.18	2609.76	74JA348
^{13}C(3)	Me	H	H	H	7555.92	3460.27	2637.80	74JA348
^{13}C(5)	Me	H	H	H	7515.66	3434.42	2615.48	74JA348
^{18}O(1)	Me	H	H	H	7476.71	3388.34	2583.17	74JA348
^{18}O(2)	Me	H	H	H	7313.40	3475.09	2613.08	74JA348
^{18}O(4)	Me	H	H	H	7292.36	3475.90	2609.88	74JA348
—	Me	H	Me	H	6022.05	2103.08	1695.66	74JA348
—	H	F	H	H	6774.0	3916.3	3122.7	76JA4012
—	H	F	D	H	6582	3791	3040	76JA4012
^{18}O(4)	H	F	H	H	6534.6	3912.9	3071.7	76JA4012
—	F	F	H	H	4966.52	2570.64	2456.79	79JPC1545

Table 3 Structural Parameters of the 1,2,4-Trioxolane ⟨72JA6337⟩ and 3-Methyl-1,2,4-trioxolane ⟨74JA348⟩ Rings Obtained by Microwave Spectroscopy

Parameter			1,2,4-Trioxolane	3-Methyl-1,2,4-trioxolane
Bond lengths[a] (Å)	a		1.470	1.471
	b		1.395	1.399
	c		1.436	1.423
	d		1.436	1.423
	e		1.395	1.411
Bond angles[a] (°)	α		99.23	99.2
	β		99.23	99.7
	γ		106.25	105.6
	δ		102.83	104.6
	ε		106.25	105.7
Dihedral angles[b] (°)		O(1)O(2)—C(3)O(4)	41.27	41.0
		O(2)C(3)—O(4)C(5)	16.60	16.9
		C(3)O(4)—C(5)O(1)	16.60	15.5
		O(4)C(5)—O(1)O(2)	41.27	39.8
		C(5)O(1)—O(2)C(3)	50.24	49.2

[a] For denoting system, see Table 1. [b] Absolute values.

⟨79JPC1545⟩. With all the 1,2,4-trioxolanes studied by microwave spectroscopy, no effects from pseudorotation or internal rotation were observed.

From the Stark shifts of rotational transitions the electric dipole moments of 1,2,4-trioxolane and its 3-methyl, 3,5-dimethyl and 1,1-difluoro derivatives were calculated to be 1.09, 1.29, 1.34 and 2.74 D, respectively.

4.33.2.2.3 Infrared and Raman spectra

The IR spectroscopy of the extremely unstable 1,2,3-trioxolanes (primary ozonides) has been possible only in the recent past when special low-temperature IR cells became available. These cells enabled the preparation of primary ozonides from alkenes and ozone in the spectrometer itself. As shown in Table 4, most of the primary ozonides studied by this technique have characteristic IR bands in three spectral regions, 650–750, 930–990 and 1100–1200 cm^{-1}. Isotopic substitution ($^{16,18}O_3$, $^{18}O_3$, 4,4-D_2, D_4, $^{13}C_2$) of 1,2,3-trioxolane (1) provided a sound basis for the assignment of the following major bands: 409 (OOO) bending, 647 ν_{asym}(OOO), 727 (COO) bending and ν(CO), 846 ν_{sym}(OOO), 927 ν(CO), 983 ν(CO) and (CH$_2$) bending and 1214 cm^{-1} (CH$_2$) bending vibrations ⟨81JA2578⟩. The ν(CO), i.e. C—O stretching vibrations, at 927 and 983 cm^{-1} are reasonable values for single bonds; however, the symmetric and asymmetric O—O—O stretching modes at 647 and 846 cm^{-1}, respectively, are unusually low values for single bonds, consistent with the instability of primary ozonides in which O—O bond rupture occurs readily.

Table 4 IR Absorption Bands of 1,2,3-Trioxolanes (Low-temperature Spectra of Primary Ozonides)

Substituents at positions 4	5	Configuration	IR absorption bands (cm^{-1})	Ref.
H, H	H, H		409, 647, 727, 846, 927, 983, 1214	81JA2578
Me, H	H, H		635, 680, 715, 825, 971, 1110, 1210	72JA4856
Bu, H	H, H		692, 725, 740, 975, 1001, 1230, 1275	79JCS(P2)1644
Me, Me	H, H		670, 698, 790, 980, 1155, 1226	72JA4856
Me, H	Me, H	cis	450, 710, 890, 1019, 1050, 1110, 1180	72JA4856
Me, H	Me, H	trans	690, 705, 900, 995, 1028, 1110, 1185, 1378, 1383	72JA4856
Et, H	Et, H	trans	695, 878, 939, 954, 1010	79JCS(P2)1644
Pr^i, H	Pr^i, H	cis	689, 726, 735, 970, 980, 1020, 1030, 1070	79JCS(P2)1644
Pr^i, H	Pr^i, H	trans	673, 692, 735, 750, 905, 925, 949, 954, 985, 1010	79JCS(P2)1644
Bu^t, H	Bu^t, H	trans	699, 745, 755, 784, 802, 892, 915, 935, 962, 995	79JCS(P2)1644
Me, Me	Me, H		670, 708, 730, 883, 948, 1037, 1168, 1200, 1370	72JA4856
Me, Me	Me, Me		682, 721, 852, 953, 1148, 1197, 1370	72JA4856

In the case of 1,2,4-trioxolane (2), where both the normal species and various isotopic modifications (3,3-D_2, D_4, $^{18}O_3$) were also studied ⟨76JPC1238⟩, all the IR frequencies could be assigned (Table 5) as a result of a normal coordinate analysis. Starting from the molecular geometry determined by microwave spectroscopy (cf. Section 4.33.2.2.2), this analysis led to an excellent agreement between observed and calculated IR frequencies, thus confirming the O—O half-chair conformation of the 1,2,4-trioxolane ring. For a series of substituted 1,2,4-trioxolanes, characteristic IR bands were observed in the regions 890–1000, 1040–1060 and/or 1015–1113 cm^{-1}. The spectra of several pairs of cis and trans isomeric 1,2,4-trioxolanes indicated that the cis isomer absorbs in the range 820–855 cm^{-1} and the trans isomer in the range 1320–1360 cm^{-1} ⟨B-78MI43300⟩.

Table 5 IR Absorption Bands of 1,2,4-Trioxolane (Normal Isotopic Species) in Solid Argon ⟨76JPC1238⟩

Frequency^a (cm^{-1})	Assignment^b	Frequency^a (cm^{-1})	Assignment^b
193m	Ring bending	1143vw	γ_r(CH$_2$)
352vw	Ring pucker	1196m	γ_t(CH$_2$)
689m	ν(CO$_p$), ν(CO$_e$)	1202m	γ_t(CH$_2$)
737vw	ν(CO$_e$), ν(OO), ν(COC)	1346m	γ_w(CH$_2$)
808s	ν(OO), ν(CO$_p$)	1387m	γ_w(CH$_2$)
926w	ν(CO$_p$), δ(OOC)	1483vw	δ(CH$_2$)
952vs	δ(OCO), ν(OO)	2894vs	FR ν(CH)
1029s	δ(OCO), ν(CO$_e$)	2900m	ν(CH)
1078vs	ν(CO$_p$), ν(CO$_e$)	2967s	ν(CH)
1129s	γ_r(CH$_2$)	2973m	ν(CH)

^{a} Some weak or very weak combination vibrations were left out.
^{b} Symbols used: ν stretching vibration, δ scissoring vibration, γ_w wagging vibration, γ_t twisting vibration, γ_r rocking vibration, FR Fermi resonance, O$_p$ peroxidic oxygen, O$_e$ ether oxygen.

1,2,4-Trithiolane (4) and its tetramethyl derivative were investigated by both IR and Raman spectroscopy. For the former, some characteristic IR frequencies were assigned as follows: 505 ν(SS), 668 and 680 ν(CS), 1070, 1085, 1160, 1175 and 1195 γ_r, γ_t, γ_w (CH$_2$), 1392 δ(CH$_2$) (for symbols see Table 5). The analysis of the IR and Raman spectra shows the existence of a single conformer having either the symmetrical S—S half-chair form (**4g**) or the envelope form (**4d**) (*cf.* Scheme 3) ⟨76CJC146⟩; the latter, however, could be excluded by photoelectron spectroscopy (see Section 4.33.2.2.5).

The Raman spectrum of liquid 1,3,2-dioxathiolane 2-oxide (ethylene sulfite) (Table 6) and its solid-state IR spectrum show all 24 vibrations anticipated for a symmetrical envelope conformation ⟨62CB2333⟩. The S=O stretching vibration of substituted derivatives appears in the region 1200–1230 cm^{-1}, depending on the nature of the substituents ⟨56JA454, 56JCS1813⟩, their position with regard to the S=O group and, to a smaller extent, also on the solvent and on the concentration ⟨73JCS(P2)243, 73JCS(P2)1966⟩.

Table 6 Raman Absorption Bands of Liquid 1,3,2-Dioxathiolane 2-Oxide (Ethylene Sulfite) ⟨62CB2333⟩

Frequency (cm^{-1})	Assignment[a]	Frequency (cm^{-1})	Assignment[a]
308	Skeleton	1196	ν(S=O)
463	—	1219	γ_t(CH$_2$)
612	δ(S=O)	1328	γ_w(CH$_2$)
663	δ(CC)	1355	γ_w(CH$_2$)
686	ν_{sym}(S—O)	1464	CH bending
743	ν_{asym}(S—O)	1474	CH bending
852	γ_r(CH$_2$)	2653	ν(CH)[b]
907	ν_{sym}(CCO)	2703	ν(CH)[b]
955	ν_{asym}(CCO)	2909	ν(CH)
1004	ν(CC)	2946	ν(CH)
1118	γ_r(CH$_2$)	2979	ν(CH)
1139	ν(CO)	3024	ν(CH)

[a] For symbols see Table 5.
[b] Uncertain assignment.

1,3,2-Dioxathiolane 2,2-dioxides (cyclic sulfates of 1,2-diols) show, similar to open-chain dialkyl sulfates, the typical IR absorption bands of the —SO$_2$— group in the ranges 1196–1221 and 1365–1401 cm^{-1} ⟨60JCS201⟩. 1,2,3-Oxadithiolan-5-one 2-oxides are characterized by the S=O band at 1170 cm^{-1} and the C=O band at 1800 cm^{-1} ⟨77MI43301⟩.

4.33.2.2.4 Ultraviolet spectra

In the absence of endocyclic or exocyclic double bonds, the UV absorption of five-membered heterocycles containing three O or S atoms resembles, as expected, that of analogous acyclic aliphatic compounds and is mainly due to $n \to \sigma^*$ transitions involving the lone pairs of the heteroatoms. Thus 1,2,4-trioxolane and its alkyl derivatives show in the 200–400 nm region the same absorption curves devoid of maxima as do alkyl peroxides ⟨42LA(553)187, 55CB712⟩; with aryl-substituted 1,2,4-trioxolanes the UV maxima of the aromatic systems predominate ⟨59JCS1071⟩. On the other hand, characteristic absorption bands are observed if the ring possesses endocyclic or exocyclic double bonds or structural elements with possible p_π–d_π interaction like the S → O group (usually formulated as S=O). For example, substituted 3,5-bismethylene-1,2,4-trithiolanes of type (**21**; R^1 = Me, Bu^i, EtO, NH$_2$; R^2 = H, Me, Cl, CN) show, besides the $\pi \to \pi^*$ transition of the unsaturated carbonyl system at 230–250 nm, an intensive UV absorption in the region 320–350 nm which suggests that these heterocycles are planar, or close to planar, and that delocalization may occur through the 3d orbitals of sulfur ⟨66T3001, 70BCJ2938, 71CJC1477⟩. Compared with (**21**), the long-wavelength absorption band of analogously substituted 1,3-dithietanes (**20**) is more intense and has undergone a bathochromic shift, while that of the corresponding

(**20**)   (**21**)   (**22**)

1,2,4,5-tetrathianes (**22**) is less intense and has undergone a hypsochromic shift. This finding indicates that the disulfide linkage is less effective for transmission of delocalization than the sulfide linkage.

Alkyl-substituted 1,3,2-dioxathiolane 2-oxides (cyclic sulfites) show a weak UV absorption at 213–218 nm, attributed to an $n \rightarrow \pi^*$ transition ⟨74JOC2073⟩. In mono-, 4,5-di- and tri-substituted derivatives the sulfur atom is an asymmetric center and the S=O group possesses chiroptical properties which frequently assist in establishing conformational details. Thus the CD spectra of the *cis–trans* isomeric pair (**23**) and (**24**) are in agreement with models having a twist-envelope conformation with axially placed S=O groups and equatorially placed methyl groups ⟨74JOC2073⟩ (*cf.* Section 4.33.2.2.1; for perspective formulas see Section 4.33.2.2.6). From intensity differential curves obtained from UV spectra of the enediol sulfite (**25**; Mes = 2,4,6-trimethylphenyl) and related compounds (*e.g.* hydromesitoin), it was concluded that the enediol sulfite system may be regarded as a chromophore with an absorption maximum at 260 nm ⟨63JOC1075⟩.

(**23**)   (**24**)   (**25**)

4.33.2.2.5 Photoelectron spectra

The vertical ionization potentials of molecules with adjacent heteroatoms (hydrazines, peroxides, disulfides, *etc.*) are known to depend strongly on the torsional angle of the neighboring orbitals. In the case of saturated rings containing such structural elements, PE spectra consequently allow certain conclusions concerning the ring geometry to be made. Thus, as can be seen from Table 7, the experimental ionization potentials obtained for 1,2,4-trioxolane (**2**) and 1,2,4-trithiolane (**4**) correlate with theoretically calculated orbital energies if a heteroatom torsion angle of 50° is taken as a basis. This corresponds to the half-chair conformations (**2g**) and (**4g**) shown in Scheme 3.

Table 7 Vertical Ionization Potentials (IP_v) of 1,2,4-Trioxolane ⟨78JA5584, 79LA1473⟩, 1,2,4-Trithiolane ⟨76CJC146⟩ and 1,3,2-Dioxathiolane 2-Oxide ⟨74CB2299⟩

1,2,4-Trioxolane			1,2,4-Trithiolane			1,3,2-Dioxathiolane 2-oxide
IP_v (eV)	ε^a (eV)	Assignment	IP_v (eV)	ε^b (eV)	Assignment	IP_v (eV)
10.67	10.86	$n_\pi(O)$	8.72	9.25	$n^-(SS)$	10.93
10.96	11.08	$n_\pi(OO)$	9.10	9.80	$n^+(SS) - n(S)$	10.93
	12.00	$n_\sigma(OO)$	10.19	10.99	$n^+(SS) + n(S)$	11.96
12.40						
	12.32	$n_\sigma(O)$	11.06	11.87	$\sigma(CS)$	12.48
	13.16	$\sigma(CH), \sigma(CO)$				13.33
13.38						
	13.19	$\sigma(OO)$				14.6
14.5	14.33	$n_\pi(OO)$				16.1
15	15.26	$\pi(CH_2), \sigma(OO)$				
16.15	16.40	$n_\sigma(OO), \sigma(CO)$				
17	17.50	$\pi(CH_2)$				

^a Orbital energies calculated by MINDO/2 for the O—O half-chair conformation.
^b Orbital energies calculated by CNDO/S for the S—S half-chair conformation.

The PE spectrum of 1,3,2-dioxathiolane 2-oxide (**16**) (ethylene sulfite) shows the first band to have doubled in intensity, indicating that the first and second ionization potentials possess equal energy (*cf.* Table 7). According to CNDO/2 calculations, this finding is consistent with an envelope-like conformation in which the S=O group is in an *endo* (axial) position and the lone pairs of the ring-oxygen atoms are oriented at an angle of about 45° to each other ⟨72AG436⟩.

4.33.2.2.6 Nuclear magnetic resonance spectra

By means of low-temperature ^{1}H NMR studies it could be shown that primary ozonides, corresponding to the suggested 1,2,3-trioxolane structure, actually possess a symmetrical ring system, thus eliminating Staudinger's molozonide formula (*cf.* Section 4.33.1) ⟨B-78MI43300⟩. The ^{1}H NMR spectra of symmetrical 4,5-dialkyl-1,2,3-trioxolanes show only one signal for the two vicinal ring protons, which in the case of the *cis* isomer appears at lower field than that of the corresponding *trans* isomer (see Table 8). In 1,2,4-trioxolane (2) and 1,2,4-trithiolane (4), as well as in their symmetrical 3,5-derivatives, the ring protons are likewise magnetically equivalent: compound (2) shows at −80 °C in acetone-d_6 a singlet at δ 5.20 ⟨79JOC3181⟩ and (4) at −117 °C in CS$_2$ a singlet at δ 4.17 ⟨72T3489⟩. Chemical shifts for the ring protons of some symmetrical 3,5-dialkyl derivatives are summarized in Table 8. As can be seen from these data, the signal for the ring protons is always at slightly lower field in the *cis* isomer. For unsymmetrically substituted derivatives, at least one of these protons will be at lower field in the *cis* isomer. This empirical rule has proved useful in assigning the configuration of *cis* and *trans* isomers, especially in the ozonide series ⟨B-78MI43300⟩, but exceptions are also known (*cf.* Section 4.33.2.2.8). In the case of symmetrical 3,5-dialkyl-1,2,4-trithiolanes, the magnetic equivalence of the ring protons as well as of the alkyl groups was explained in terms of pseudorotation, the barrier of which was estimated to be smaller than 25 kJ mol^{-1} ⟨72T3489⟩.

Table 8 ^{1}H NMR Chemical Shifts (δ)[a] of the Ring Hydrogens of Symmetrical Dialkyl Derivatives of 1,2,3-Trioxolanes ⟨68JOC1629, 66JA4098⟩, 1,2,4-Trioxolanes ⟨67JA2429⟩ and 1,2,4-Trithiolanes ⟨72T3489⟩

Substituent at the ring carbon atoms	1,2,3-Trioxolane		1,2,4-Trioxolane		1,2,4-Trithiolane[b]	
	cis	*trans*	*cis*	*trans*	*cis*	*trans*
Me	4.52 (u)[c]	4.12 (q)[c]	5.30 (q)	5.25 (q)	4.98 (q)	4.80 (q)
Et	4.53 (u)[c]	4.07 (t)[c]	4.98 (t)	4.96 (t)	4.60 (t)	4.56 (t)
Pr^i	—	—	4.77 (d)	4.75 (d)	4.58 (d)	4.50 (d)
Bu^t	—	4.20 (s)[d]	4.70 (s)	4.63 (s)	4.80 (s)	4.60 (s)

[a] Abbreviations used: s singlet, d doublet, t triplet, q quartet, u unresolved.
[b] In CCl$_4$.
[c] In CCl$_2$F$_2$ at −130 °C.
[d] In CFCl$_3$ at −110 °C.

According to more recent ^{1}H and ^{13}C NMR studies ⟨73JCS(P2)243, 75JCS(P2)190, 76CJC1428⟩, 1,3,2-dioxathiolane 2-oxides exist in twist-envelope forms which are interconverted by rapid pseudorotation not involving inversion at sulfur (Scheme 4). For this reason the *cis* vicinal ring protons (H$_A$, H$_C$) as well as the *trans* vicinal ring protons (H$_B$, H$_D$) of ethylene sulfite (16) are magnetically equivalent. In monosubstituted derivatives such as (23) and (24) (*cf.* Scheme 4), conformations having the substituent in an equatorial position are

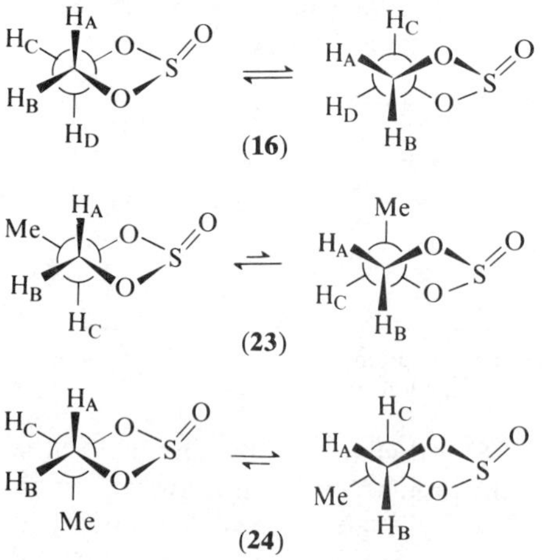

^{1}H NMR parameter (in CCl$_4$)	Compound (16)	(23)	(24)
δ (p.p.m.) H$_A$	4.81	4.22	4.65
H$_B$	4.45	3.81	3.81
H$_C$	4.81	4.57	5.04
H$_D$	4.45	—	—
J (Hz) H$_A$H$_B$	6.58	−8.56	−8.21
H$_A$H$_C$	7.00	9.09	6.16
H$_A$H$_D$	−8.15	—	—
H$_B$H$_C$	−8.15	6.07	6.91
H$_B$H$_D$	7.00	—	—
H$_B$H$_D$	6.58	—	—

Scheme 4 Twist-envelope conformations (Newman projection down C—C bond) and ^{1}H NMR data for the ring hydrogens of 1,3,2-dioxathiolane 2-oxide (16) and its *cis* (23) and *trans* (24) methyl derivatives

found to be predominant (for agreement with CD measurements, see Section 4.33.2.2.4). The assignment of configuration and conformation for two isomeric 4,5-dimethyl-1,3,2-dioxathiolane 2-oxides, being based on ^{1}H NMR studies, was confirmed by an electron diffraction investigation (*cf.* Section 4.33.2.2.1). The interpretation of the ^{1}H NMR spectrum of 1,2,3-oxadithiolane 2-oxide suggests the existence of a single conformation, but it was not possible to arrive at any definite conclusion about its geometry ⟨75ACS(A)414⟩.

Using as a basis a conformational equilibrium analogous to that shown in Scheme 4 for 1,3,2-dioxathiolane 2-oxide (**16**), all the ring protons of the corresponding 2,2-dioxide (**18**) (ethylene sulfate) should be magnetically equivalent. Indeed, the ^{1}H NMR spectrum of (**18**) in CCl_4 consisted only of a single sharp line at δ 4.68 ⟨61JA2105⟩.

4.33.2.2.7 Mass spectra

The fragmentation of the molecular ion of 1,2,4-trioxolanes may occur by two distinct paths (Scheme 5): (i) cleavage of the O—O bond (step a) followed by fission of C—O bonds (steps f and g); (ii) primary cleavage of C—O bonds (steps b and c, respectively) followed by elimination of O_2 (steps d and e, respectively) or repeated C—O fission (steps h and i, respectively). With the exception of the $[M-32]^{\ddagger}$ ion these two major paths give rise to ions having the same m/e ratio. However, the mass spectra of well-defined ^{18}O-labeled ozonides show clearly that these ions are of different origin ⟨69CJC919⟩. Mass spectra therefore play an important role in tracer studies dealing with the mechanism of ozonolysis ⟨74JA1536, 76JA4526, 79JOC3185, 80JA4763⟩. The molecular ions of 1,2,4-trioxolanes are mostly of low intensity or not detectable (*e.g.* in cases where several aryl groups are attached to the ring). *Cis* and *trans* isomers do not differ markedly in their fragmentation patterns.

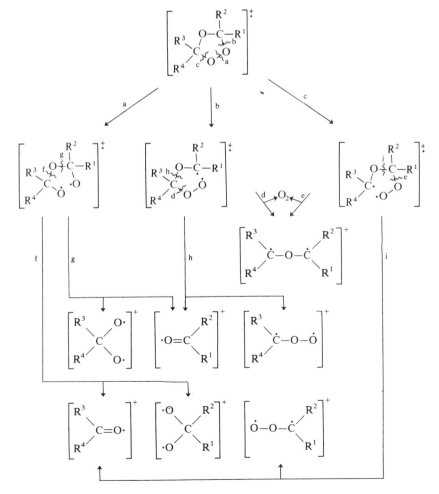

Scheme 5 General fragmentation pattern for 1,2,4-trioxolanes

1,2,4-Trithiolane (4) shows, like other cyclic methylene polysulfides (cf. Chapters 2.25 and 5.18), besides an intensive $M^{\ddot{+}}$ ion the typical ions $[M - CH_2S]^+$, $[\dot{C}H_2S\dot{C}H_2]^+$, $[CH_2S]^+$ and $[CHS]^+$ ⟨67CPB988⟩. 3,5-Dialkyl-1,2,4-trithiolanes which undergo an analogous fragmentation are also characterized by alkenyl ions $[C_nH_{2n-1}]^+$ as well as by elimination of $HS_2\cdot$, alkyl groups and/or alkenes (Scheme 6) ⟨81JPR169⟩.

4.33.2.2.8 Other physical methods

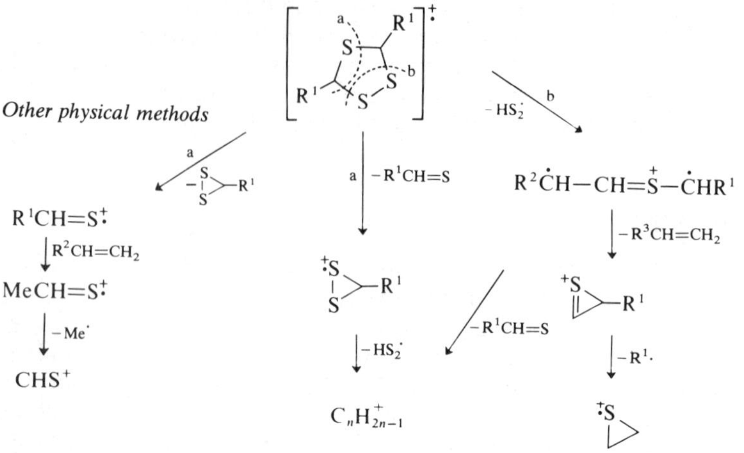

Scheme 6 General fragmentation pattern for symmetrical 3,5-di-n-alkyl-1,2,4-trithiolanes ($R^1 = R^2CH_2$, $R^3CH_2CH_2$, etc.)

In the case of 1,3,2-dioxathiolane 2-oxides (cyclic sulfites of 1,2-diols), four major fragmentation pathways appear to operate: (i) loss of $C_nH_{2n}O$ units; (ii) a pinacol-type rearrangement resulting in expulsion of SO_2 and formation of a ketone or aldehyde rearrangement radical ion; (iii) a second process also affording $[M - SO_2]^{\ddot{+}}$ ions, but in which the daughter ions are related to the corresponding epoxide parent ions; (iv) formation of $[M - HSO_2]^+$ species. Pathway (i) is of significance only in the simple alkyl derivatives, whereas pathway (iii) is typical of aryl-substituted 1,3,2-dioxathiolane 2-oxides. Ions at m/e 48, 50, 64 and 65 in these spectra are due to SO, H_2SO, SO_2 and HSO_2 ions, respectively. In contrast to the pyrolytic fragmentation where *cis* and *trans* isomers sometimes show a marked stereoselectivity (cf. Section 4.33.3.1.2), the mass spectra of such isomer pairs are very similar ⟨68T2949⟩.

Like 1,3,2-dioxathiolane 2-oxides, the corresponding 2,2-dioxides (cyclic sulfates of 1,2-diols) mostly show an intensive $M^{\ddot{+}}$ ion but they fragment quite differently. The mode of fragmentation seems to depend in a non-specific way on the substituents present ⟨77AJC569⟩. 1,2,3-Oxadithiolan-5-one 2-oxide (anhydrosulfite of thioglycolic acid) and its 4,4-dimethyl derivative were reported to show no trace of a molecular ion corresponding to the unfragmented ring ⟨77MI43301⟩.

4.33.2.2.8 Other physical methods

Before the introduction of the more modern physical methods discussed in the preceding subsections, parameters such as molar refraction or the parachor frequently were applied in solving structural problems. Thus, for example, based on refractometric measurements, Rieche in his classic investigations ⟨B-31MI43300, 42LA(553)187⟩ was able to demonstrate that ozonides, according to their suggested 1,2,4-trioxolane structure, contain one ether oxygen and one O—O moiety bonded like that in dialkyl peroxides. Fair agreement between observed and calculated molar refractions of 1,2,4-trioxolanes was obtained if an exaltation of 0.47 for the O—O moiety (average value from numerous measurements of organic peroxides) was taken as a basis (cf. Table 9). Similarly, the parachor of 2-butene ozonide was found to be consistent with a cyclic structure although, owing to the then uncertainty of atomic and ring parachors, no decision between a four- or a five-membered ring could be made ⟨42LA(553)187⟩. From the parachor of the sulfite ester group (121.5) and the usual corrections for the ring and two oxygen atoms, the parachor of 1,3,2-dioxathiolane 2-oxide was calculated to be 194.4, this value being in excellent agreement with the experimental value of 194.3 ⟨50IZV297⟩.

Table 9 Observed and Calculated Molar Refractions of 1,2,4-Trioxolanes

Substituents at position		MR_{obs} (cm^3 mol^{-1})	MR_{calc}^{a} (cm^3 mol^{-1})	Ref.
3	5			
H, H	H, H	14.63	14.89	42LA(553)187
Pr, H	H, H	27.76	28.49	55CB1878
Bu, H	H, H	33.26	33.11	55CB1878
Me, Me	H, H	23.80	23.87	55CB1878
Bu^i, Me	H, H	37.45	37.73	55CB1878
Me, H	Me, H	23.66	23.87	42LA(553)187
Me, H	Pr, H	33.03	33.11	55CB1878
Et, H	Et, H	32.96	33.11	55CB1878
Et, H	Bu, H	42.16	42.35	55CB1878
Pr, H	Pr, H	42.41	42.35	55CB1878

^{a} Calculated on the basis of the following atomic refractions: CH 3.52, CH$_2$ 4.62, CH$_3$ 5.72, O 1.64, and an average exaltation of 0.47 ($\equiv$ 2.11 − 1.64) for the O—O moiety (cf. ⟨55CB1878⟩ and literature quoted therein).

The determination of dipole moments has proved useful in cases where spectroscopic methods did not allow an unequivocal structural assignment to be made. For example, the cis and trans isomers of 3,5-diphenyl-1,2,4-trioxolane (stilbene ozonide), whose ring protons do not follow the general chemical shift rule mentioned in Section 4.33.2.2.6, could be assigned by their dipole moments: $\mu_{cis} = 1.55$ and $\mu_{trans} = 1.44$ D ⟨71CB1807⟩ (for dipole moments obtained from microwave spectra of 1,2,4-trioxolanes, see Section 4.33.2.2.2). The dipole moments of 1,3,2-dioxathiolane 2-oxide ($\mu = 3.65$ ⟨49DOK(69)41⟩, $\mu = 3.74$ D ⟨61JA2105⟩; both in benzene), 1,3,2-dioxathiolane 2,2-dioxide ($\mu = 5.64$ D in dioxane) ⟨68JHC289⟩ and its 4,4,5,5-tetramethyl derivative ($\mu = 6.05$ D in dioxane) ⟨77IZV98⟩ were determined in connection with conformational studies and, like the results of other structural methods, also suggest non-planar ring systems.

4.33.3 REACTIVITY

4.33.3.1 Thermal and Photochemical Reactions Formally Involving No Other Species

4.33.3.1.1 Rearrangement

The most important rearrangement of heterocycles containing three O or S atoms is observed in the ozonolysis of alkenes. According to Criegee ⟨75AG765⟩ this reaction occurs in three discrete steps: (a) formation of a 1,2,3-trioxolane (**26**), the primary ozonide, by cycloaddition of ozone to the alkene double bond; (b) cleavage of (**26**) into a carbonyl compound (**27**) and a zwitterionic carbonyl oxide (**28**) (the 'Criegee zwitterion'); and (c) recombination of (**27**) and (**28**) yielding a 1,2,4-trioxolane (**29**), the final ozonide. All the three steps may be regarded as 1,3-dipolar reactions: steps a and c are [2+3] cycloadditions, and step b is a cycloreversion. The assumption of an intermediate carbonyl oxide (**28**) in addition explains not only various reactions of 1,2,3-trioxolanes in non-inert media or in the presence of other reactants (cf. Section 4.33.3.3), but also accounts for the formation of 1,2,4,5-tetroxanes (**30**). The tetroxanes result mainly from the ozonolysis of tetrasubstituted alkenes which give rise to less reactive ketones as the carbonyl compounds (**27**), thus allowing the intermediate carbonyl oxides (**28**) to dimerize. In practice, the extremely unstable primary ozonides (**26**) are of course not isolated. After treatment of the alkenes in an inert solvent at low temperatures (*e.g.* in CCl_4 at $-78\,°C$) with ozone and warming up the reaction mixture, compounds (**29**) and/or (**30**) are obtained directly. In agreement with the Criegee mechanism, unsymmetrical alkenes, *e.g.* RCH=CHR', besides the 'normally', *i.e.* unsymmetrically, substituted 1,2,4-trioxolanes (possessing the substituents R and R'), result in two kinds of symmetrically substituted ozonides, the so-called cross ozonides (containing the substituents R,R and R',R', respectively). An analogous situation occurs with the 1,2,4,5-tetroxanes (**30**) ⟨72MI43301⟩.

One experimental finding not covered by the original Criegee mechanism is the stereochemistry of the ozonolysis reaction, in particular the fact that the *cis–trans* ratio of ozonides (**29**) formed from 1,2-disubstituted ethylenes or from ones of type $R^1R^2C=CHR^3$ ($R^1 \neq R^2$) depends on the alkene geometry. For this reason, Bailey ⟨B-78MI43300⟩ proposed a stereochemical refinement of the Criegee mechanism which was based on two assumptions. First, the existence of certain stable non-planar conformations of (**26**) and (**29**), and secondly a *syn–anti* isomerism of the carbonyl oxide (**28**). In the case of small alkenes like butene or pentene, however, the observed *cis–trans* ratio cannot be explained by this scheme. Only the most recent quantum chemical investigations, which consider all three steps of the ozonolysis reaction, provide a plausible interpretation of its stereochemical features ⟨81AG934, 81JA3619, 81JA3627, 81JA3633⟩.

In the series of 1,3,2-dioxathiolane 2,2-dioxides, a rearrangement has been observed ⟨79IZV118⟩ which is characterized by insertion of a vinyl group into the ring of the perfluorinated derivative (**31**). This reaction affords the seven-membered cyclic sulfate (**32**).

4.33.3.1.2 Thermolysis

1,2,4-Trioxolanes containing at least one ring hydrogen atom decompose in a quantitative fashion at room temperature or at slightly elevated temperatures in suitable solvents such as alcohols, DMF or DMSO ⟨71CC1094, 72MI43302⟩. Symmetrically disubstituted ozonides (**33**; $R^1 = R^3$, $R^2 = H$) give one mole of acid and one mole of aldehyde; trisubstituted trioxolanes (**33**) yield one mole of acid and one mole of ketone, whereas in the case of monosubstituted or unsymmetrically 3,5-disubstituted systems (**34**; $R^2 = H$ or $R^1 \neq R^2$ respectively), formally two kinds of acids and two kinds of aldehydes are possible. In fact, however, monosubstituted 1,2,4-trioxolanes (**34**; $R^2 = H$) decompose predominantly in one direction (affording R^1CO_2H and formaldehyde). Kinetic investigations ⟨72MI43302⟩ and a radioisotopic dilution study ⟨72CJC4029⟩ suggest a concerted bond cleavage involving solvent molecules in the hydrogen transfer.

The vapor-phase thermolysis of 1,2,4-trioxolanes yields, in principle, the same major products as those obtained on decomposition in solvents. In addition, however, hydrocarbons, CO and CO_2 are formed ⟨68TL5397⟩, probably by subsequent fragmentation of the primary products ⟨62JPR(287)313⟩. Again, from kinetic investigations ⟨72JPC2659⟩, decomposition in monocyclic 1,2,4-trioxolanes of type (**33**) and (**34**) involves concurrent O—O bond cleavage and intramolecular hydrogen transfer. On the other hand, in bicyclic and tetrasubstituted 1,2,4-trioxolanes, synchronous hydrogen transfer is inhibited or is impossible and decomposition takes place *via* a discrete biradical intermediate.

1,2,3-Trithiolanes are not known to undergo thermal rearrangement like 1,2,3-trioxolanes but tend to polymerize (*cf.* Section 4.33.3.1.4). At elevated temperatures they decompose by elimination of sulfur. In this way, for example, from tetrafluoro-1,2,3-trithiolane (35), octafluoro-1,4-dithiane was obtained ⟨62JOC3995⟩. *trans*-4,5-Diphenyl-1,2,3-trithiolane 1,1-dioxide (36) decomposed above its melting point to give a quantitative yield of *trans*-stilbene as well as sulfur and sulfur dioxide ⟨68JCS(C)1612⟩. Pyrolysis of 1,2,4-trithiolane (4) and its 4-oxide (37) likewise proceeds by elimination of sulfur to give thioformaldehyde and thioformaldehyde oxide, respectively ⟨77CC287⟩. The 3,3,5,5-tetraphenyl derivative of (4) decomposes almost quantitatively to thiobenzophenone and sulfur even at 130 °C. In an analogous fashion, 3,5-dispiro derivatives afford cyclic thioketones ⟨66HC(21-1)1⟩.

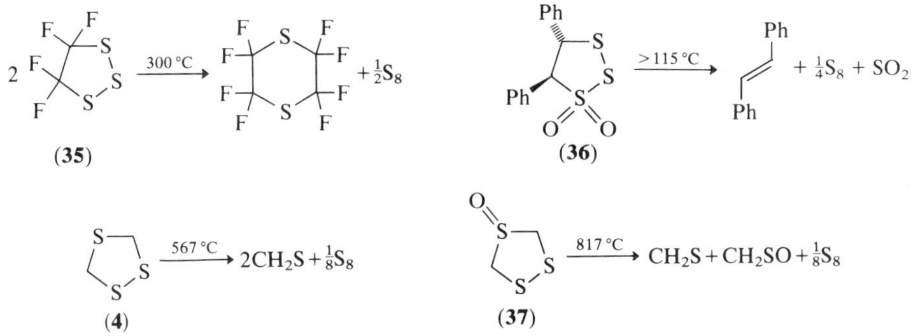

The common feature of the thermolysis of saturated five-membered cyclic sulfites is the elimination of sulfur dioxide (Scheme 7). 4,5-Dimethyl-1,3,2-dioxathiolane 2-oxide (38) decomposes at 275 °C on calcium oxide to a mixture of 2,3-dimethyloxirane (39) and

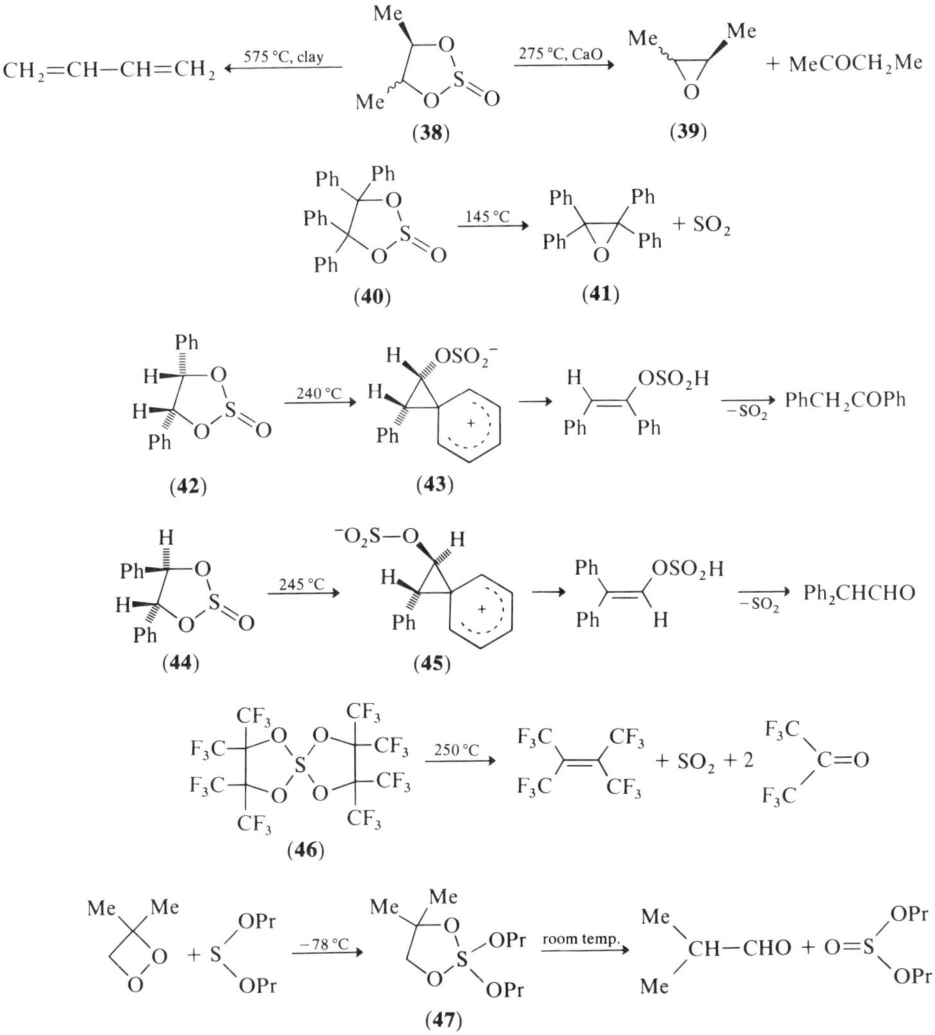

Scheme 7 Thermal decomposition of saturated five-membered cyclic sulfites and orthosulfites.

2-butanone, while passage over clay at 575 °C gives a low yield of butadiene ⟨B-66MI43300⟩. Tetraphenyl-1,3,2-dioxathiolane 2-oxide (**40**) already undergoes fragmentation at 145 °C, yielding tetraphenyloxirane (**41**) almost quantitatively ⟨72JOC2589⟩. The thermolysis of the two isomeric 4,5-diphenyl-1,3,2-dioxathiolane 2-oxides (**42**) and (**44**) shows, unlike their mass spectrometric fragmentation, a marked stereoselectivity. The oxide (**42**) affords in good yield deoxybenzoin, whereas the oxide (**44**) gives almost quantitatively diphenylacetaldehyde ⟨66HC(21-1)1⟩. This difference may be explained by the formation of the bridged phenonium intermediates (**43**) and (**45**). In (**43**) the sulfinate group is *trans* to the phenyl-activated β-hydrogen atom and therefore stabilization is achieved by abstraction of the α-hydrogen atom. In (**45**) the activated β-hydrogen atom and the sulfinate group are *cis* to each other, permitting easy migration of this proton. At temperatures above 200 °C, the sulfurane (**46**), representing a cyclic orthosulfite, undergoes a fragmentation to give perfluorotetramethylethylene, hexafluoroacetone and sulfur dioxide ⟨76JA2895⟩. Another fragmentation pathway is observed in the case of the cyclic orthosulfite (**47**), which is formed at low temperature from 3,3-dimethyl-1,2-dioxetane and dipropyl sulfoxylate. On warming up the reaction mixture to room temperature, decomposition with the production of isobutyraldehyde and dipropyl sulfite occurs ⟨75JA3850⟩.

The thermolysis of cyclic sulfites of enediols and their benzo derivatives is characterized by elimination of sulfur monoxide. Thus 4,5-dimesityl-1,3,2-dioxathiole 2-oxide (**25**) decomposes to mesitil ⟨63JOC1075⟩, and fragmentation of catechol sulfite (**48**) leads *via* *o*-benzoquinone and cyclopentadienone to the dimer (**49**) of the latter ⟨67TL271⟩. In a similar fashion, the thermolysis of 1,2,3-benzoxadithiole 2-oxide at 670 °C resulted in cyclopentadienethione; the intermediate formation of monothiobenzoquinone could be detected by PE spectroscopy ⟨81AG603⟩.

The thermolysis products obtained from 1,3,2-dioxathiolan-4-one 2-oxides (**50**) (cyclic sulfites of α-hydroxy carboxylic acids) depend on the substitution pattern. In the case $R^1 = R^2 = CH_2Cl$ ⟨65JCS791⟩ or $R^1 = Ph$, $R^2 = H$ ⟨70JCS(B)1049⟩, a carbonyl compound, sulfur dioxide and carbon monoxide are formed (path A). If R^1 and R^2 represent *n*-alkyl groups ⟨65JCS791, 66HC(21-1)1, 74MI43302⟩, predominantly polymers of type (**51**) are obtained (path B). The rate-determining process of path B is the primary ring scission with elimination of sulfur dioxide and concurrent ring contraction to form an α-lactone intermediate which then rapidly polymerizes ⟨71JCS(B)1384⟩. Pyrolytic splitting of sulfur trioxide from the 1,4,2-dioxathiolane 2,2-dioxide derivative (**52**) yields the sodium salt (**53**) of (±)-(*cis*-1,2-epoxypropyl)phosphonic acid, whose esters have antibiotic properties ⟨70GEP1924138⟩.

4.33.3.1.3 Photolysis

UV irradiation of simple 3,5-dialkyl-1,2,4-trioxolanes such as (54) leads mainly to the same fission products, though in different amounts, as the thermal decomposition (*cf.* Section 4.33.3.1.2), namely hydrocarbons, aldehydes, acids, carbon monoxide and carbon dioxide. A homolytic cleavage of the O—O bond (step a) followed by double β-scission (step b) has been suggested as a possible mechanism ⟨68TL3291, 68TL5397⟩ and this interpretation is supported by the photochemical formation of cyclopropane from cyclopentene ozonide (55) ⟨68TL3291⟩, and of 1,2,3,4-tetramethylcyclobutene from hexamethylbicyclohexene ozonide (56) ⟨68C392⟩.

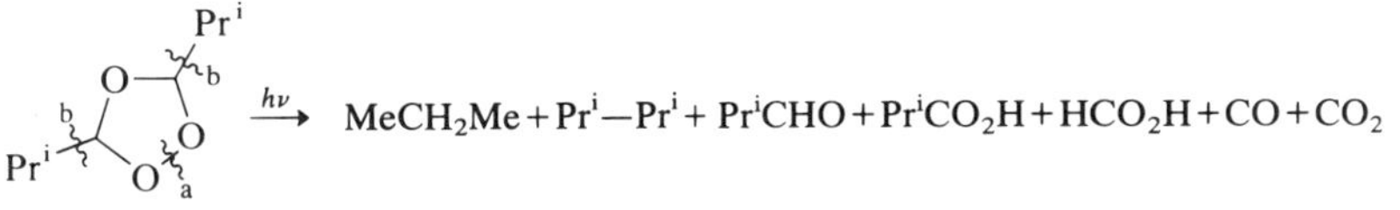

$$(54) \xrightarrow{h\nu} MeCH_2Me + Pr^i\!-\!Pr^i + Pr^iCHO + Pr^iCO_2H + HCO_2H + CO + CO_2$$

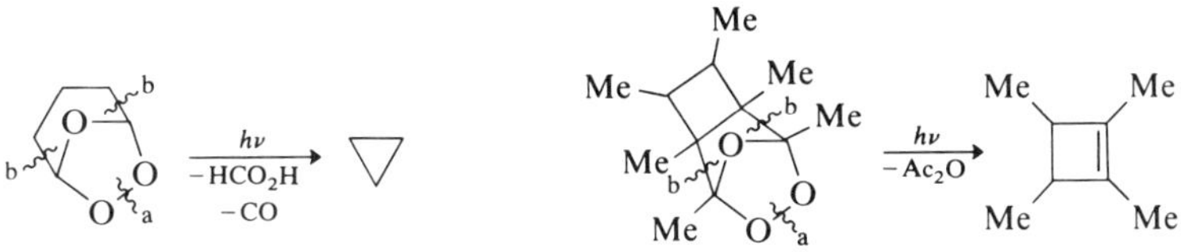

On UV irradiation on silica gel, the 1,2,4-trithiolane derivative (57) undergoes a reversible conversion into the 1,3-dithietane derivative (58) ⟨79MI43300⟩. Aryl-substituted 1,3,2-dioxathiolane 2-oxides, *e.g.* (40), undergo [5 → 1+2+2] photocycloeliminations to give arylcarbenes which can be utilized synthetically ⟨72JOC2589⟩.

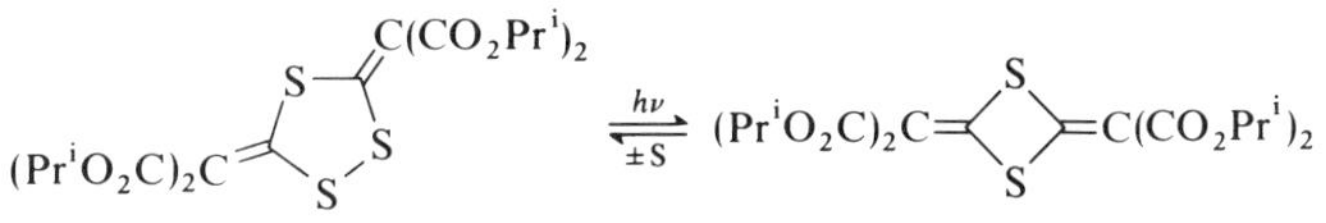

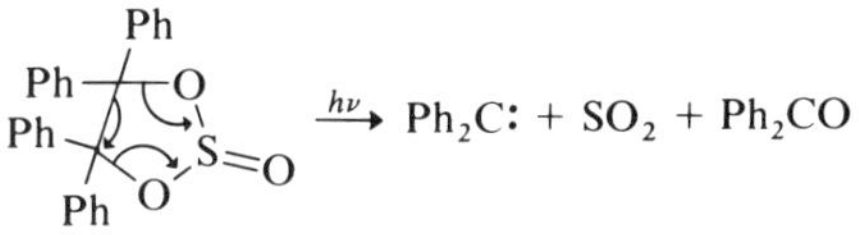

$$(40) \xrightarrow{h\nu} Ph_2C: + SO_2 + Ph_2CO$$

4.33.3.1.4 Polymerization

Besides monomeric ozonides (29) and/or dimeric carbonyl oxides (30) (*cf.* Section 4.33.3.1.1), ozonolysis of alkenes frequently yields, depending on reaction conditions and substitution pattern, more or less large amounts of oligomers or polymers of ozonides, based on their elemental composition ⟨B-31MI43300, 40CRV(27)437, 58CRV925⟩. Since 1,2,4-trioxolanes, once formed, do not tend to polymerize, these polymers most likely arise from the highly reactive primary ozonides and/or their fragmentation products. One possible reaction which could initiate polymerization is the attack of a Criegee zwitterion (28) on the unreacted primary ozonide (26), affording an intermediate of type (59) ⟨67JOC3369⟩. However, to obtain detailed data on structure and mode of formation of polymeric ozonides, further investigations are necessary.

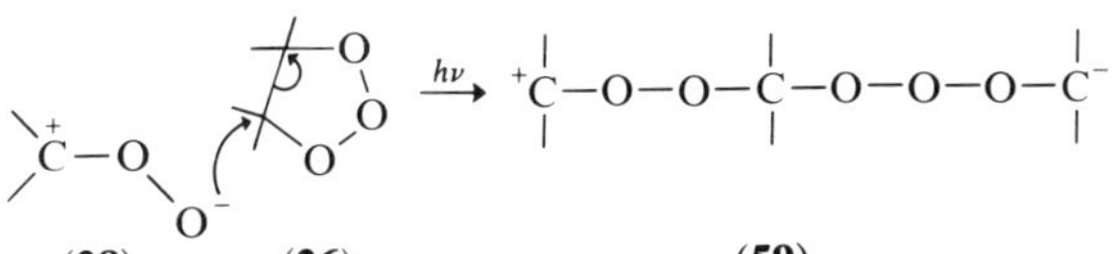

The few known monocyclic 1,2,3-trithiolanes have proved to be rather unstable compounds with a marked tendency towards polymerization. Thus simple alkyl-substituted derivatives such as (60) polymerize at room temperature within some hours to form rubbery, soluble materials having molecular weights up to over 300 000 ⟨77MI43302⟩. Catalyzed by trimethyl phosphite or other mild bases such as acetonitrile, ethanol or acetone, tetrafluoro-1,2,3-trithiolane (35) polymerizes at −80 to −40 °C to a white solid of relatively low molecular weight. Polymerization of (35) also takes place at room temperature without catalysts, but in this case the molecular weight of the polymer is reduced ⟨62JOC3995⟩.

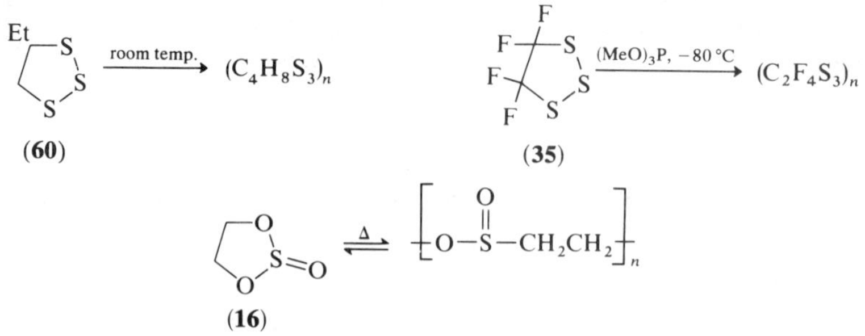

Depending on the method of preparation (cf. Section 4.33.4.2.1), 1,3,2-dioxathiolane 2-oxide (16) or its derivatives are obtained at first in the form of low molecular weight polymers which at elevated temperatures depolymerize to the corresponding monomers ⟨57USP2798877, 61ZOB1332⟩. The polymerization of such cyclic sulfites is catalyzed by bases (e.g. dialkylanilines or pyridines). The formation of polymers on thermolysis of 1,3,2-dioxathiolan-4-one 2-oxides has been described in Section 4.33.3.1.2.

4.33.3.2 Reactions with Electrophiles

4.33.3.2.1 Reactions involving protons or Lewis acids

Similar to the thermal fragmentation in solvents (cf. Section 4.33.3.1.2) and the base-catalyzed cleavage (cf. Section 4.33.3.3.1), the acid-catalyzed decomposition of ozonides (61) usually gives aldehydes and carboxylic acids. The reaction rate depends on the proton-donating strength of the catalyst, e.g. $HClO_4 \gtrsim H_2SO_4 \gg HCl$ ⟨59ACS342⟩ or on its concentration, respectively ⟨75ZOR7⟩. Regarding the point of initial protonation, both the ether oxygen (path A) and a peroxidic oxygen of the 1,2,4-trioxolane ring (path B) have been considered. In the latter case the reaction is believed to proceed via a protonated carbonyl oxide (62), whose rearrangement and deprotonation give rise to the carboxylic acid. As evidence for this mechanism it was cited ⟨75ZOR7⟩ that if acetic acid were used as solvent, acetyloxyalkyl hydroperoxide (63) was obtained as by-product. By analogy to (62) a complex-stabilized zwitterion (65) was suggested to be the crucial intermediate in the reaction of certain 1,2,4-trioxolanes with antimony pentachloride ⟨80JA288⟩; thus from (61; $R = C_5H_{11}$) the corresponding 1,2,4,5-tetroxane (66) was obtained in good yield, probably by reaction of (65) with excess (61) rather than by simple dimerization, cf. ⟨81JA1789⟩.

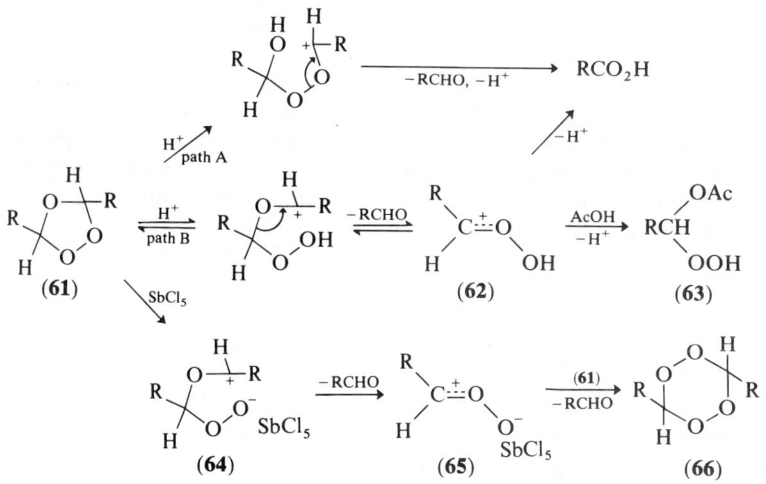

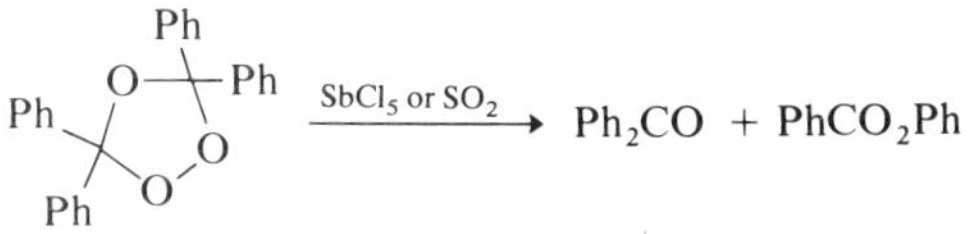

Ph₂CO + PhCO₂Ph ← (67) [SbCl₅ or SO₂]

Depending on the substituents in (**61**) and on the concentration of the catalyst, however, other products may also subsequently be formed. In some cases, SbCl₅-complexed primary intermediates of type (**65**) could even be isolated and characterized by NMR spectroscopy as well as by well-defined chemical reactions ⟨81JA1789⟩. Treatment of tetraphenyl-1,2,4-trioxolane (**67**) with antimony pentachloride or liquid sulfur dioxide, a mild Lewis acid, gave benzophenone and phenyl benzoate in a molar ratio of 1:1 ⟨80JA288⟩. The formation of the ester corresponds to that of the carboxylic acid from (**61**) *via* path A or path B.

On treatment with anhydrous hydrogen chloride in chloroform the cyclic enediol sulfite (**25**) did not yield the substituted desyl chloride (**69**), as one would expect, but was recovered unchanged. This was explained by steric shielding of the enediol double bond towards nucleophilic attack of chloride ion and by additional stabilization due to the formation of the pseudoaromatic system (**68**) ⟨63JOC1075⟩. Photochemical reaction of 1,3,2-dioxathiolane 2-oxide (**16**) (ethylene sulfite) with metal carbonyl derivatives such as MnC₅H₅(CO)₃ and Cr(CO)₆ led to complexes of type (**70**) and (**71**), respectively, in which the sulfur functions as electron donor ⟨65CB2248⟩. Boron trichloride converted 1,3,2-dioxathiolane 2-oxides into the corresponding 3-chloro-1,3,2-dioxaborolane derivatives ⟨62JCS1505⟩. On heating with aqueous acid, tetramethyl-1,3,2-dioxathiolane 2,2-dioxide (**72**) underwent, after protonation and ring opening, a rapid pinacol-type rearrangement to afford pinacolone in good yield ⟨74JOC3415⟩.

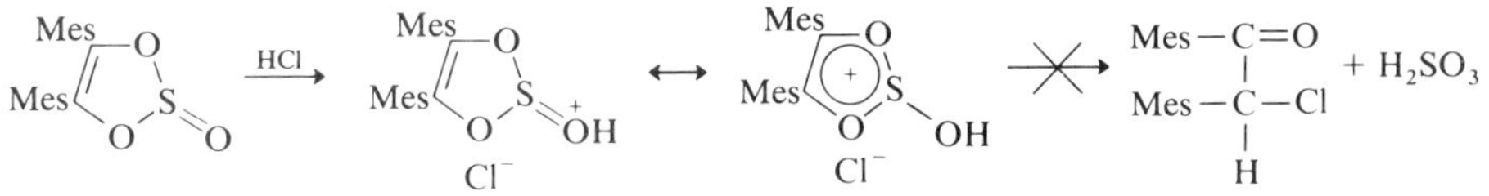

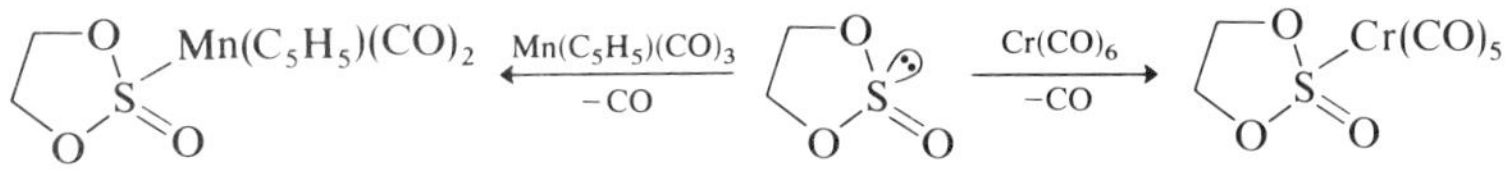

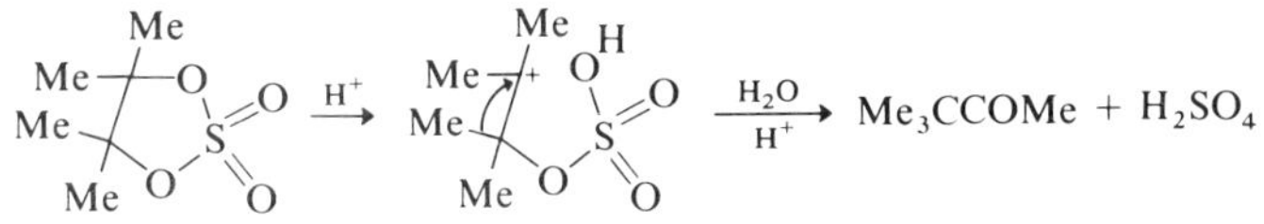

Acid-catalyzed reactions of the systems under discussion with nucleophilic reactants will be dealt with in Section 4.33.3.3.

4.33.3.2.2 Electrophilic addition at double bonds

Examples of this reaction type are the addition of *O,O*-diethyl dithiophosphate at the endocyclic double bond of 1,3,2-dioxathiole 2-oxide (**73**) ⟨62USP3053852⟩ and the addition of bromine or sulfur trioxide at the vinyl group of the cyclic sulfate (**31**) ⟨79IZV118⟩.

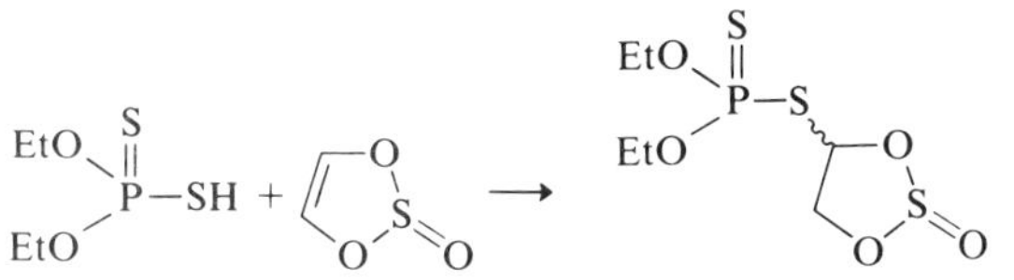

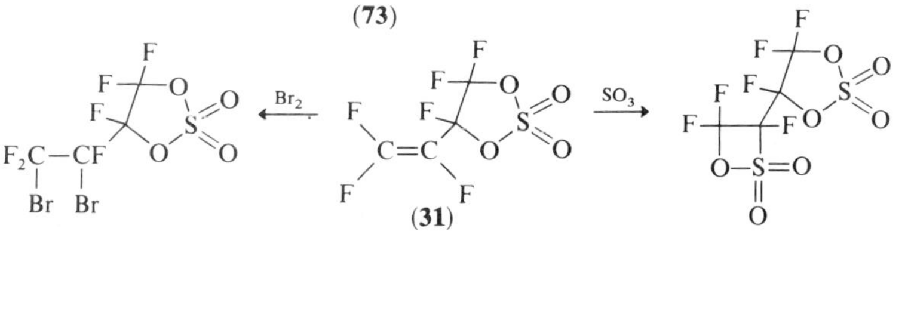

4.33.3.2.3 Oxidation without ring cleavage

Oxidation reactions which take place with retention of the ring skeleton are known only in the 1,2,4-trithiolane and 1,3,2-dioxathiolane 2-oxide series. Oxidation of 1,2,4-trithiolane (**4**) with sodium periodate in aqueous acetone at 0–5 °C gave a mixture of 1-oxide (**74**) and 4-oxide (**37**) in a ratio of 1 : 1 ⟨76JOC2465⟩, while oxidation with hydrogen peroxide in Bu^{t}OH/THF at -30 °C in the presence of a catalytic quantity of V_2O_5 gave (**37**) in pure form ⟨77CC287⟩. The transformation of five-membered cyclic sulfites into the corresponding sulfates has been effected mainly with calcium permanganate as oxidant ⟨63CRV557⟩.

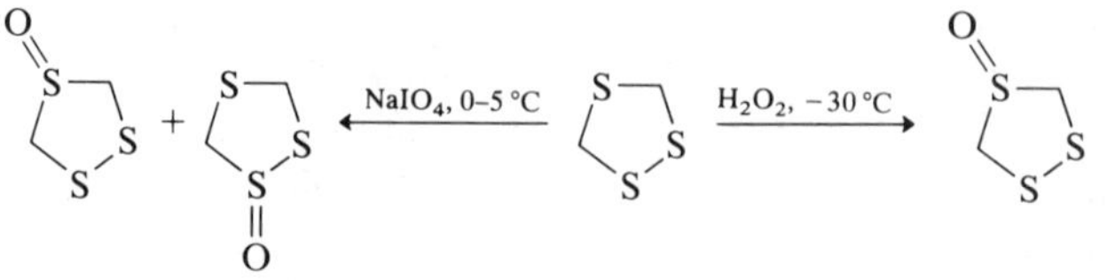

4.33.3.2.4 Oxidative ring cleavage

The oxidative decomposition of ozonides of type (**61**) leads to carboxylic acids and is of considerable synthetic importance (*cf.* Section 4.33.5.1). Amongst the oxidizing agents used are oxygen (usually catalyzed by ozone or metal salts), hydrogen peroxide (especially in alkaline solution), peracids (*e.g.* performic or peracetic acid), chromic acid, potassium permanganate, or silver oxide suspended in sodium hydroxide solution ⟨40CRV(27)437, 58CRV925⟩.

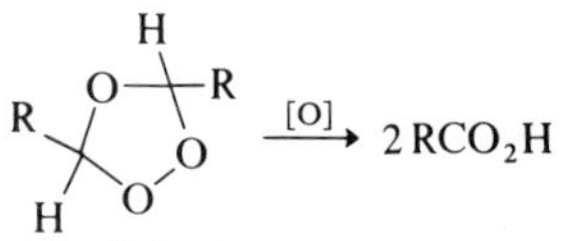

1,2,4-Trithiolanes of type (**75**) proved to be stable against oxidation with hydrogen peroxide in acid medium at room temperature. However, on treatment with chlorine and water in glacial acetic acid, the trithiolane (**75**) was cleaved to 2 moles of benzophenone dichloride and 3 moles of sulfuric acid per mole of cyclic trisulfide ⟨47JOC807⟩. Cyclic sulfites of 1,2-diols likewise undergo an oxidative ring-opening on treating with chlorine, and from 1,3,2-dioxathiolane 2-oxide (**16**) 2-chloroethyl chlorosulfate was obtained ⟨67JCS(C)314⟩. UV irradiation had no effect on either the rate or the nature of the product, indicating that the chlorination does not proceed by a radical mechanism.

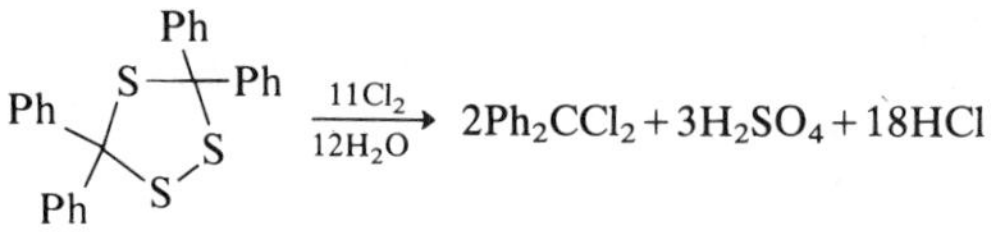

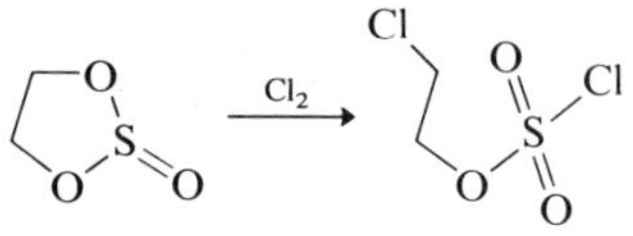

4.33.3.3 Reactions with Nucleophiles

4.33.3.3.1 O-Nucleophiles

If the ozonolysis of an alkene is performed in the presence of an alcohol, then the intermediate Criegee zwitterion (**28**) formed by cycloreversion of the primary ozonide (**26**) (*cf.* Section 4.33.3.1.1) reacts with the alcohol to give an α-alkoxyalkyl hydroperoxide (**76**) ⟨75AG765⟩. This ozonolysis technique is of preparative importance because here the unwelcome formation of dimeric peroxides of type (**30**) as well as of oligomeric ozonides (*cf.* Section 4.33.3.1.4) is avoided. In certain cases the hydroperoxides (**76**) can be dehydrated thermally to carboxylic acid esters ⟨62CJC1189⟩. Addition of acetic acid or hydrogen peroxide to the zwitterion (**28**) has also been observed, giving α,α-bishydroperoxides (**77**) ⟨75AG765⟩. In an analogous fashion, ozonolysis of alkenes in water leads to α-hydroxyalkyl hydroperoxides (**79**) which are, however, unstable and decompose to carboxylic acids and water, or to carbonyl compounds (aldehydes or ketones) and hydrogen peroxide ⟨B-72MI43303⟩. The same end products can be formed by acid hydrolysis of 1,2,4-trioxolanes (**29**) which proceeds likewise *via* the hydroperoxide (**79**). Here the first step consists in the hydrolytic cleavage of an acetal C—O bond, resulting in α,α'-dihydroxydialkyl peroxides (**78**) which could be isolated in some cases ⟨58CRV925⟩.

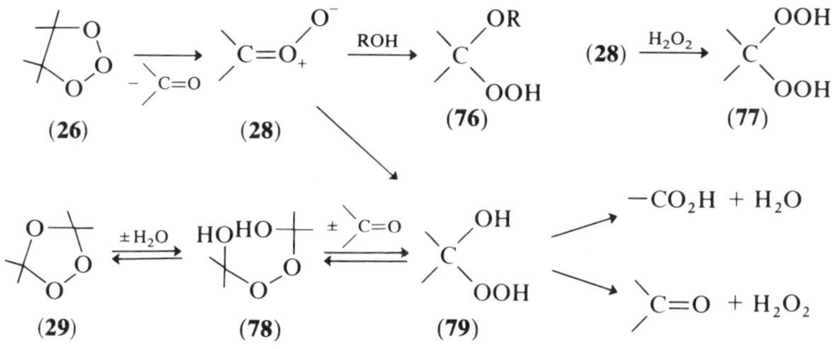

Another role is ascribed to the water in the ozonolysis of α,β-unsaturated carbonyl compounds. The 'abnormal' course of this reaction (cleavage of the alkenic double bond as well as the C—C bond adjacent to the carbonyl group) is rationalized in terms of a fragmentation of the intermediate 1,2,4-trioxolanes (**80**) initiated by hydration of the carbonyl group ⟨67AHC(8)165⟩. The base-catalyzed decomposition of 1,2,4-trioxolanes of type (**33**) apparently involves nucleophilic attack at the ring hydrogen atom to afford a carbonyl compound and the salt of a carboxylic acid ⟨67AHC(8)165⟩, this reaction being closely related to the decomposition of ozonides by solvents (*cf.* Section 4.33.3.1.2). Instead of hydroxyl ions, of course, other basic nucleophiles such as amines may be used ⟨71CC1094⟩.

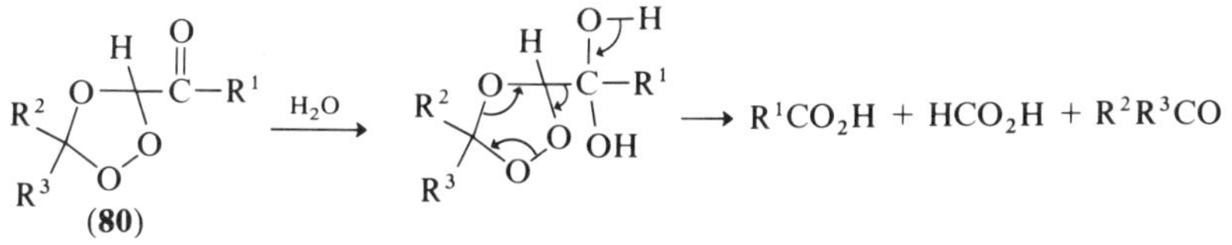

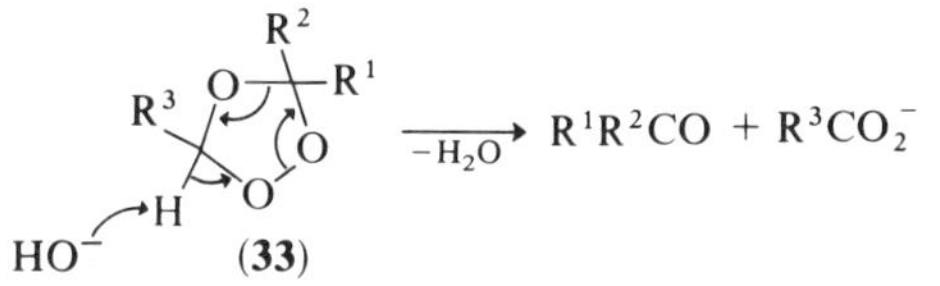

Tetraalkyl-substituted 1,2,4-trithiolanes (**81**) are remarkably stable towards aqueous acids and bases, even at elevated temperatures. On treating with methanolic potassium hydroxide, however, they are cleaved into ketones and potassium sulfide ⟨66HC(21-1)1⟩. Under similar conditions the 1,2,4-trithiolane derivative (**82**) undergoes a ring contraction accompanied by saponification of the ester groups. This reaction probably involves the reductive cleavage of the disulfide linkage by the alcoholic base followed by recyclization with elimination of sulfide ion ⟨71CJC1477⟩. Analogous ring transformations of 3,5-bismethylene-1,2,4-trithiolanes can be effected photochemically (*cf.* Section 4.33.3.1.3) or by desulfurization with triethyl phosphite or triphenylphosphine (*cf.* Section 4.33.3.3.6).

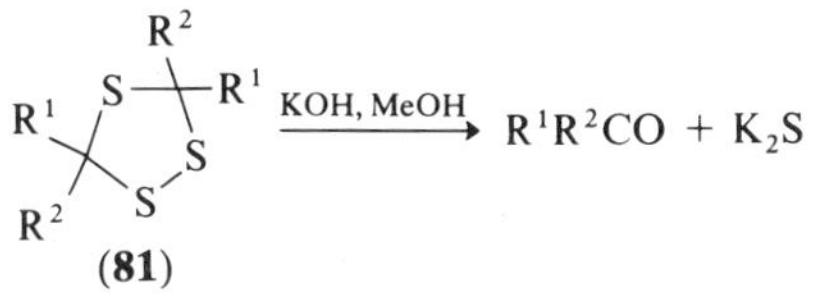

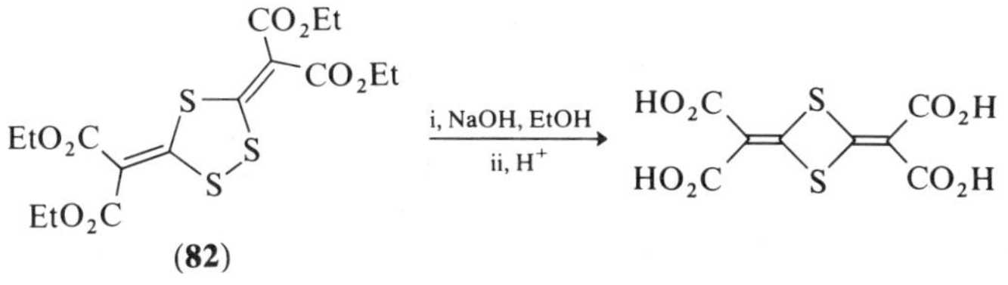

As a rule the acid- and base-catalyzed hydrolysis of 1,3,2-dioxathiolane 2-oxide (**16**) and its monocyclic derivatives ⟨63CRV557, 76CRV747⟩ occurs with S—O bond fission and retention of configuration at carbon. The rate-determining step of the perchloric acid catalyzed hydrolysis is formulated as the addition of water to the conjugate acid (**83**) to afford a tetracoordinate intermediate which, in turn, undergoes a fast ring cleavage. For the hydrolysis catalyzed by mineral acids such as hydrochloric or hydrobromic acid, another mechanism (nucleophilic catalysis, cf. Section 4.33.3.3.5) is tenable. The rate-determining step of the alkaline hydrolysis consists of the attack by hydroxyl ion on sulfur to yield anion (**84**); the subsequent step then proceeds at least 100 times faster. The hydrolysis of catechol sulfite (**48**) was faster than that of (**16**) by a factor of 10^4 in neutral solution and 10^5 in the presence of hydroxide ion. Both (**16**) and (**48**) reacted with alkali 10^3 times faster than their open-chain analogs, dimethyl sulfite and diphenyl sulfite. Because of the close similarity in the heats of hydrolysis of cyclic and open-chain sulfites in both the aliphatic and aromatic series, the kinetic acceleration in the cyclic esters cannot arise from some kind of ring strain but has been attributed to entropy strain ⟨76CRV747⟩.

Reaction of (**16**) with sodium methoxide gave, besides other products, ethylene oxide and sodium methyl sulfite. The surprising absence of 1,2-ethanediol mono- or di-methyl ether among the products suggests that initial alkoxide ion attack occurred exclusively at sulfur, possibly via anion (**85**) or a zwitterion of type $^+CH_2CH_2O^-$. The conjugate acid of (**85**), $HOCH_2CH_2OSO_2Me$, could be detected by ^1H NMR spectroscopy as an intermediate in the transesterification of (**16**) with methanol or of dimethyl sulfite with glycol ⟨76CRV747⟩.

In contrast to the attack of the methoxide ion at sulfur, phenolates react with (**16**) predominantly at carbon to yield aryloxyethyl sulfites (**86**) which are easily converted into the corresponding aryl 2-hydroxyethyl ethers (**87**) ⟨63CRV557⟩. In these reactions, which involve C—O bond fission, the cyclic sulfite acts as an alkylating agent. A similar reaction path underlies the formation of 2-hydroxyethyl acetate from (**16**) and acetic acid ⟨66HC(21-1)1⟩, and of polyesters (**88**) from (**16**) and dicarboxylic acids $X(CO_2H)_2$ in which X is, for example, $(CH_2)_4$ ⟨57BRP781169⟩ or $p-C_6H_4$ ⟨59USP2870127⟩. On reacting 4-chloromethyl-1,3,2-dioxathiolane 2-oxide (**89**) with phenolates, besides the five-membered cyclic sulfites (**90**) the six-membered 5-aryloxy-1,3,2-dioxathiane 2-oxides (**92**) were obtained. This interesting ring expansion probably proceeds via an intermediate of type (**91**) ⟨66HC(21-1)1⟩. 1,3,2-Dioxathiolan-4-one 2-oxides (**50**) decompose rapidly in the presence of water and mineral acids to form α-hydroxy carboxylic acids, the detailed mechanism of hydrolysis being unknown. Nucleophilic attack on (**50**) by alcohols occurs at the acyl carbon atom and, with elimination of sulfur dioxide, gives α-hydroxy carboxylic acid esters ⟨76CRV747⟩. In a similar fashion to (**50**), 1,2,3-oxadithiolan-5-one 2-oxides react rapidly with water to yield the parent α-mercaptocarboxylic acids ⟨77MI43301⟩.

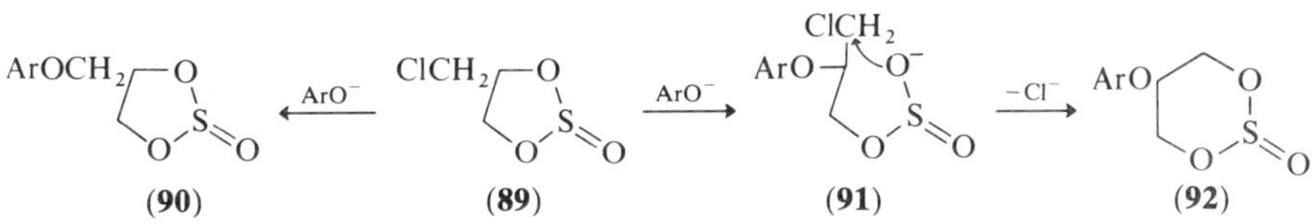

In contrast to 1,3,2-dioxathiolane 2-oxide (**16**), the corresponding 2,2-dioxide (**18**) and their saturated monocyclic derivatives are hydrolyzed, predominantly with C—O bond cleavage, in either acid, neutral or alkaline solution. As shown by experiments with ^{18}O-labeled solvent, it is only under strongly alkaline conditions that (**18**) undergoes S—O bond fission (14%). On the other hand, catechol sulfate is cleaved by alkali exclusively with S—O bond fission. This difference is readily understandable since nucleophilic attack at the aromatic carbon atoms should be extremely improbable. Like five-membered cyclic sulfites, the corresponding cyclic sulfates are in part considerably more reactive than their open-chain analogs, catechol sulfate reacting with alkali 2×10^7 times faster than does diphenyl sulfate ⟨B-77MI43300⟩.

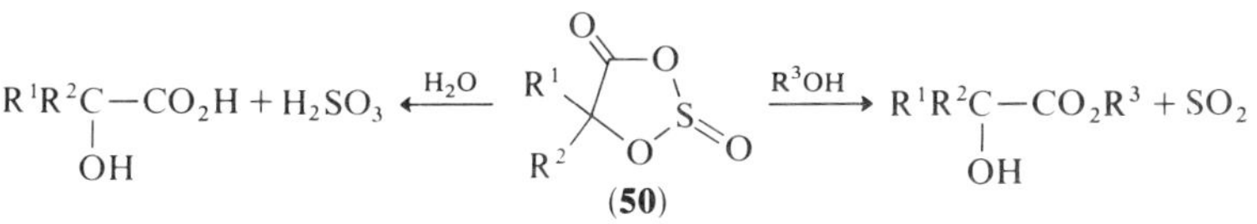

As expected, cyclic sulfates of aliphatic 1,2-diols are powerful alkylating agents towards a series of O-nucleophiles. Thus from (**18**) and phenolates, amine oxides or carboxylates the corresponding alkylation products (**93**), (**94**) and (**95**), respectively, are formed in high yields ⟨72JHC891⟩. Halogen atoms or alkoxy groups attached to the ring carbon atoms may be involved in the nucleophilic displacement. The methanolysis of the perfluorinated cyclic sulfate (**96**) leading to the carboxylic acid ester (**97**) ⟨73IZV2725⟩ and the acid-catalyzed cleavage of the 4,5-dimethoxy derivative (**98**) yielding hexafluorobisacetyl ⟨77JA1214⟩ provide two examples of reactions of this type.

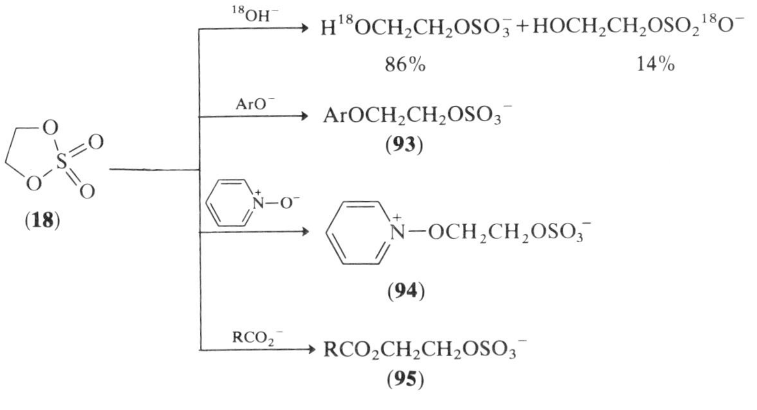

Being the cyclic anhydride of ethane-1,2-disulfinic acid, 1,2,5-oxadithiolane 2,5-dioxide (**99**) hydrolyzed to the parent acid in an almost quantitative fashion on boiling for 1 minute in water ⟨76CRV747⟩. In an analogous manner, 1,2,5-oxadithiolane 2,2,5,5-tetroxide (**100**) and its benzo-, naphtho- and thieno-fused derivatives were hydrolyzed by water or, more rapidly, by hot alkali to the corresponding 1,2-disulfonic acids or their salts, respectively ⟨66HC(21-1)1⟩.

4.33.3.3.2 S-Nucleophiles

Reactions of trioxolanes and trithiolanes with S-nucleophiles have received little attention. With thiophenolates, ethylene sulfite (16) and ethylene sulfate (18) react in the same manner as they do with phenolates (cf. Section 4.33.3.3.1). However, unlike (93), the reaction products of ethylene sulfate with phenolates which are relatively stable in aqueous solution, the corresponding thio derivative (101) readily hydrolyzed to β-thiophenoxyethyl alcohol (103) ⟨72JHC891⟩. This may be rationalized in terms of the well-known neighboring effect in ethyl sulfides having a good leaving group in the β-position. In the present case a cyclic sulfonium ion (102) may be formed which would react rapidly with water to give (103).

4.33.3.3.3 N-Nucleophiles

The ozonolysis of alkenes in liquid ammonia or in aqueous ammonia solutions leads via the primary ozonides (26) and carbonyl oxides (28) to α-aminoalkyl hydroperoxides (104). These products, however, under the reaction conditions readily undergo subsequent reactions, e.g. condensation with the carbonyl product (27) of the ozonolysis to give imines of type (105). In the case of suitable cycloalkenes, especially five-membered alkenes, this condensation step corresponds to a ring expanding recyclization, affording nitrogen heterocycles which in part may be stabilized by aromatization. Thus the 'amozonolysis' of cyclopentadiene leads, via (106) and (107), to pyridine in 18% yield, and indene can be transformed into isoquinoline in 62% yield ⟨64JOC2240⟩. In certain cases α-aminoalkyl hydroperoxides of type (104), or those formed from (28) and primary amines, undergo cyclodehydration to give oxaziridines ⟨66CB3233⟩. N-Nucleophiles such as hydroxylamine, phenylhydrazine or pyridine, on the other hand, may function as reductants if present during ozonolysis (cf. Section 4.33.3.3.6).

Primary, secondary and tertiary amines behave towards 1,2,4-trioxolanes bearing at least one ring hydrogen atom first of all like a proton acceptor, initiating the base-catalyzed fragmentation already discussed in Section 4.33.3.3.1.

Treatment of tetraalkyl-1,2,4-trithiolanes (81) with primary amines at low temperatures (≤0 °C) gave the Schiff's bases (110). This reaction represents the reversal of the Asinger trithiolane synthesis (cf. Section 4.33.4.2.3). At elevated temperatures, besides (110), disulfides of type (111) were formed, possibly by reaction of the primary intermediate (108) with hydrogen sulfide liberated in the step (109) → (110) ⟨59LA(627)195⟩. In the case of unsubstituted 1,2,4-trithiolane (81; $R^1 = R^2 = H$), the primary intermediate (108) decomposed thermally to give thioformamides (112); secondary amines reacted in a similar fashion ⟨67CPB988⟩. The reaction of substituted 3,5-bismethylene-1,2,4-trithiolanes (113) with aliphatic ⟨73GEP2151228⟩ or aromatic primary amines ⟨74EGP104784⟩ gave α-mercaptoenamines (114) or the tautomeric thioamides (115). Bifunctional N-nucleophiles such as arylhydrazines, o-aminophenol or o-phenylenediamine transformed (113) into five- and seven-membered N-heterocycles (116) ⟨73EGP102382⟩ and (117; X = O, NH), respectively ⟨74EGP106041⟩. In these ring interconversions, in which the cyano group is involved, the system (113) functions as a C_3 synthon.

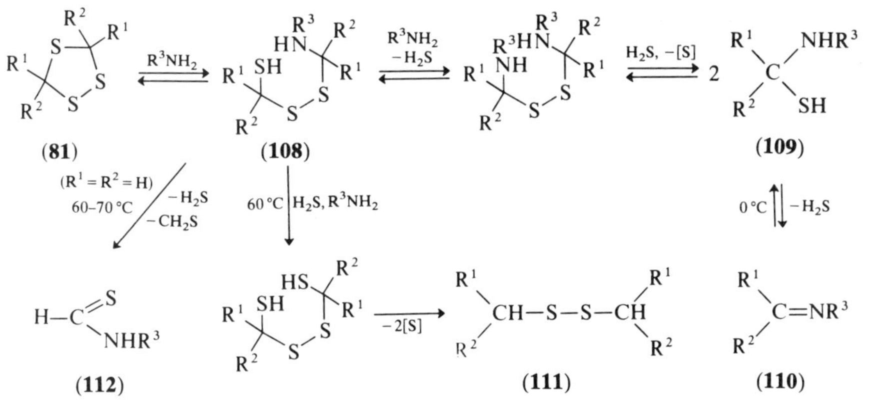

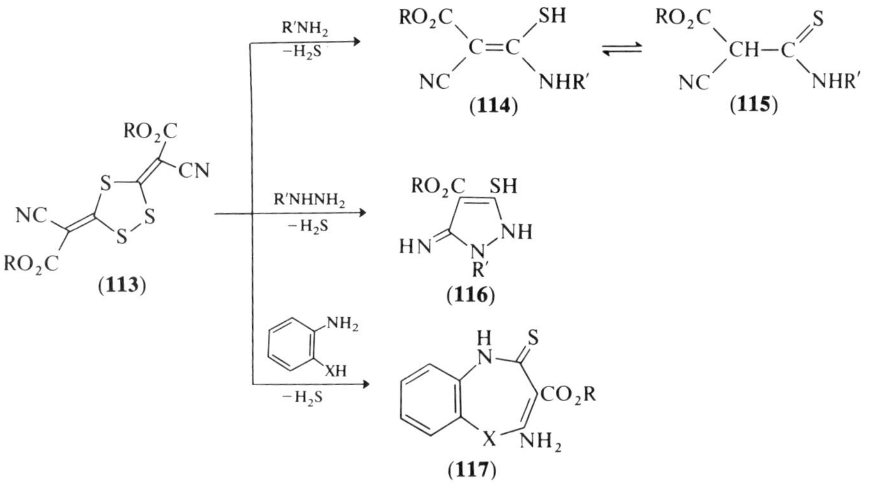

1,3,2-Dioxathiolane 2-oxide (**16**) reacted with primary and secondary amines to give β-aminoethyl sulfites (**118**) or, by subsequent hydrolysis, the corresponding β-aminoethanols (**120**) ⟨66HC(21-1)1⟩. Treatment of (**16**) with tertiary amines resulted in sulfitobetaines (**119**) ⟨71BRP1257851⟩, whose thermal decomposition yielded acetaldehyde among other products ⟨60JOC651⟩.

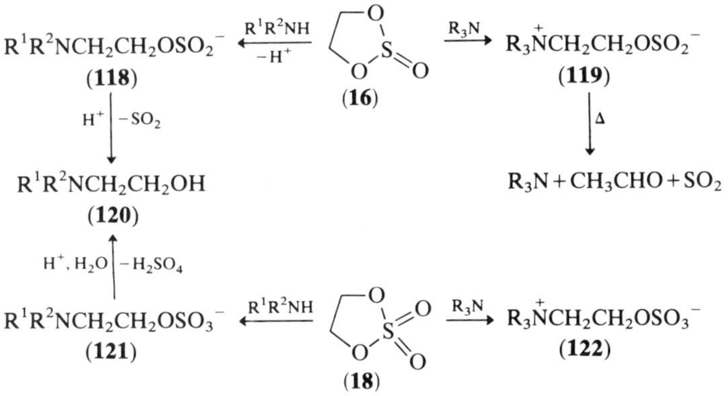

1,3,2-Dioxathiolane 2,2-dioxide (**18**) alkylated ammonia as well as primary and secondary amines to give β-aminoethyl sulfates (**121**) ⟨72JHC891⟩ which can also be hydrolyzed to form (**120**) ⟨73USP3743653⟩. Reaction of (**18**) with tertiary amines afforded sulfatobetaines (**122**) ⟨75JPR943⟩.

1,3,2-Dioxathiolan-4-one 2-oxides (**50**) were converted by aniline into α-hydroxycarboxylic acid anilides (**123**) and sulfur dioxide, while phenylhydrazine liberated the parent acids (**124**) without release of sulfur dioxide ⟨66HC(21-1)1⟩. Analogous to the methanolysis of the perfluorinated cyclic sulfate (**96**) (cf. Section 4.33.3.3.1), the aminolysis of this compound gave the anilide (**125**) ⟨73IZV2725⟩.

(123) (50) (124)

(96) (125)

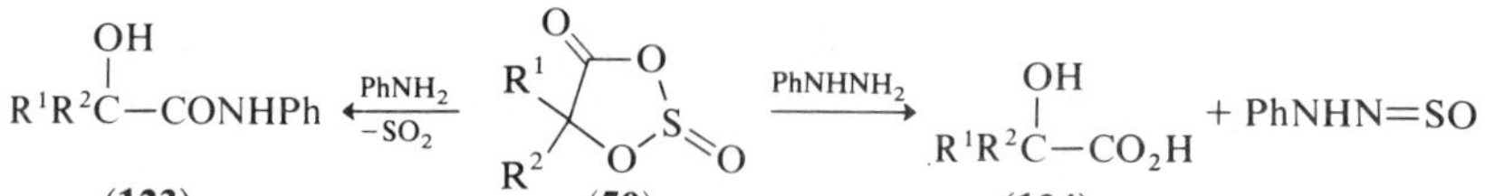

4.33.3.3.4 C-Nucleophiles

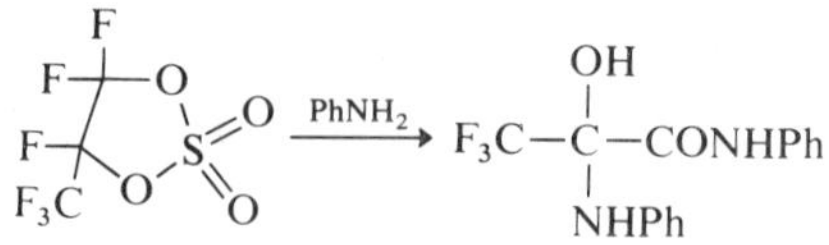

1,2,3-Trioxolanes and 1,2,4-trioxolanes react with C-nucleophiles such as Grignard reagents or organolithium compounds, resulting, as with various other reductants, in reductive ring cleavage (see Section 4.33.3.3.6).

3,5-Dimethyl-1,2,4-trithiolane (**126**) reacted with methyllithium at −60 °C to give intermediate (**127**) which was transformed into 4,6-dimethyl-2,3,5,7-tetrathiaoctane (**128**) in the presence of excess dimethyl disulfide ⟨78HCA2809⟩. Compound (**128**) is a constituent of roasted pork meat (*cf.* Section 4.33.5.2).

(126) (127) (128)

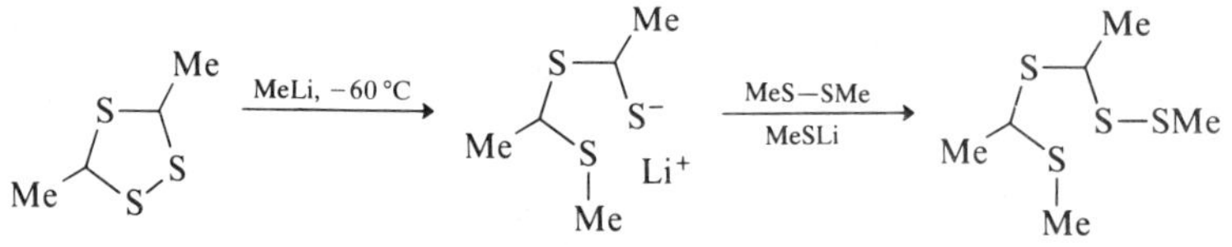

The reaction of phenylmagnesium bromide with 1,3,2-dioxathiolane 2-oxide (**16**) gave 3.4–23% ethylene bromohydrin (**131**) and 42–60% diphenyl sulfoxide (**132**), depending on the conditions of the reaction. The bromohydrin (**131**) arises from a nucleophilic displacement at carbon by bromide ion, presumably *via* (**129**) and (**130**), while the sulfoxide (**132**) is formed by repeated attack of the Grignard reagent at sulfur ⟨56JA454⟩.

(16) (129) (130) (131)

(132)

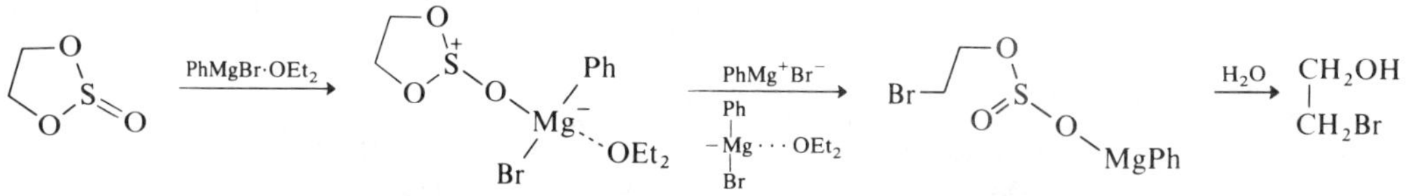

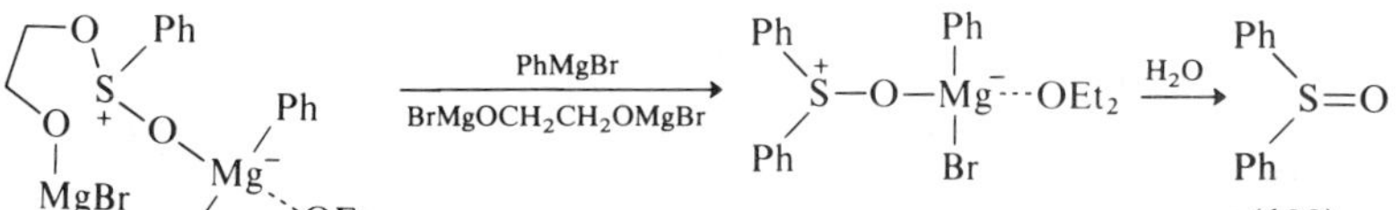

Starting from 1,3,2-dioxathiolane 2,2-dioxide (**18**) and organometallic compounds, β-substituted ethyl sulfates of type (**133**; R=CH(CO_2Me), Ph, PhC≡C) may be synthesized ⟨72JHC891⟩. Reaction of glyoxal sulfate (**134**) with cyclopentadienylsodium followed by treatment with triethylamine led to the fulvene (**135**) ⟨72AG297⟩; the lithium salt of phenalene reacts in a similar fashion ⟨80MI43301⟩.

(18) (133)

(134) (135)

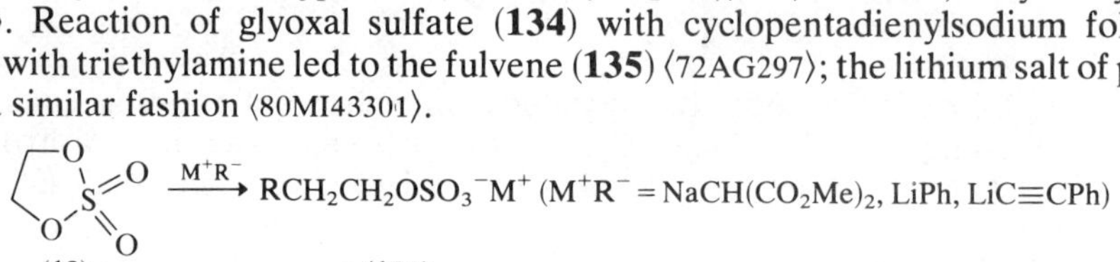

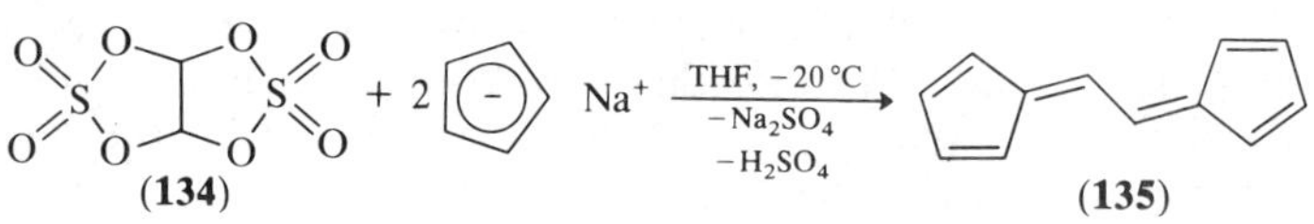

4.33.3.3.5 Halide ions

On reaction with sodium iodide in glacial acetic acid, 1,2,4-trioxolanes undergo a reductive ring cleavage which will be discussed in Section 4.33.3.3.6. In the presence of potassium fluoride, 5,5-difluoro-1,2,4-trioxolan-3-one (**136**) isomerized at low temperatures to give bis(fluoroformyl) peroxide (**137**). Since the reaction is reversible (*cf.* Section 4.33.4.2.5), isomerization occurred with an equilibrium being established ⟨72IC2531⟩.

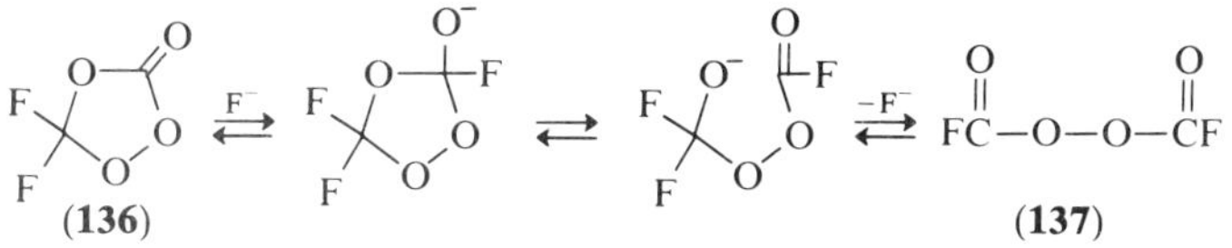

With hydrohalic acids as catalysts, the hydrolysis of 1,3,2-dioxathiolane 2-oxide (**16**) is considered to occur (partly or entirely) *via* a halosulfite intermediate, *e.g.* (**138**; X = Cl). The marked acceleration of the aqueous hydrolysis of catechol cyclic sulfite by small amounts of fluoride ion is likewise attributed to a nucleophilic catalysis ⟨76CRV747⟩. The nucleophilic attack of (**16**) at carbon by halide ion on reaction with phenylmagnesium bromide has already been discussed in Section 4.33.3.3.4.

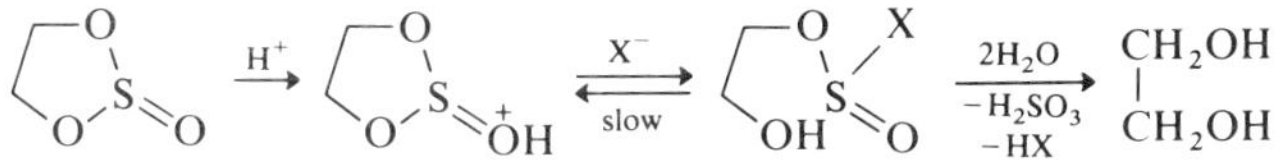

On treating with cesium fluoride in diglyme or acetonitrile, the perfluorinated cyclic sulfate (**96**) underwent an exothermic fragmentation to trifluoro-2-oxopropionyl fluoride (**139**) and sulfuryl fluoride, presumably *via* initial attack at sulfur by fluoride ion ⟨73IZV2725⟩.

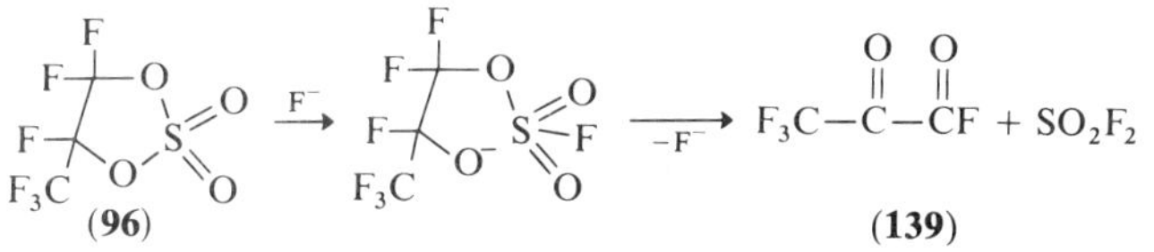

4.33.3.3.6 Reductive ring cleavage

Low-temperature reduction of primary ozonides with isopropylmagnesium halides leads to 1,2-glycols. This reaction provided the chemical evidence that in primary ozonides, based on their 1,2,3-trioxolane structure (**26**), the C—C linkage of the starting alkene was retained. If the ozonolysis is performed in the presence of a reductant which itself is stable to ozone, then one obtains directly carbonyl compounds (aldehydes and/or ketones) without any additional treatment of the ozonolysis mixture. Tetracyanoethylene, for example, may function as such a reducing agent, being oxidized by (**26**) to the epoxide (**140**). Reaction (**26**) → (**140**) represents a convenient detection method for primary ozonides. It also allows for the isolation of the carbonyl compounds in those cases in which rearrangements would otherwise give rise to abnormal ozonolysis products ⟨75AG765⟩.

Similarly, depending on the nature of the reducing agent, 1,2,4-trioxolanes (**29**) are cleaved to yield either alcohols or carbonyl compounds (*cf.* Scheme 8) ⟨40CRV(27)437,

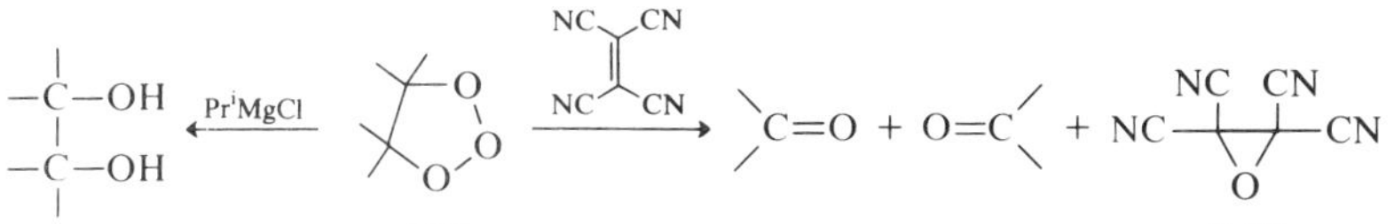

i, LiAlH$_4$, NaBH$_4$, BH$_3$·THF, RMgX or MeLi; ii, Zn, Sn, Mg, Raney Ni, Pd/H$_2$, Fe^{2+}, Sn^{2+}, I$^-$, SO$_2$, HSO$_3^-$, H$_2$PO$_2^-$, HOP(OR)$_2$, HCHO, pyridine, Ph$_3$P, NH$_2$OH, PhNHNH$_2$ or polarographic reduction

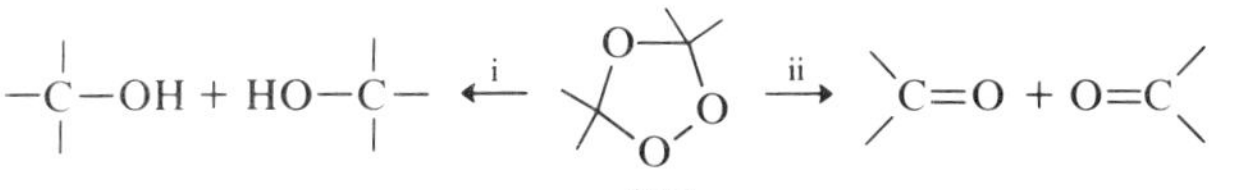

Scheme 8 Reductive cleavage of 1,2,3-trioxolanes and 1,2,4-trioxolanes

58CRV925, B-78MI43300⟩. The well-defined, reductive decomposition of ozonides is of considerable importance not only in its preparative aspects but also for mechanistic studies. Thus triphenylphosphine reduction, which occurs by exclusive attack at the peroxidic oxygen atoms of (29), may be used as a chemical method for determining the location of the ^{18}O label in ozonides ⟨76JOC892, 76JA4526⟩ (physical methods relating to this aspect are discussed in Sections 4.33.2.2.2 and 4.33.2.2.7).

On reduction with lithium aluminum hydride, tetraalkyl-substituted 1,2,4-trithiolanes (81) undergo ring cleavage to give thiols. However, they are stable against trialkyl phosphites or triphenylphosphine. The last reagent, moreover, easily converts 3,5-bismethylene derivatives (141), in which X and/or Y are electron-attracting groups, into the corresponding 1,3-dithietane derivatives (142) ⟨66T3001, 66CC577, 71CJC1477⟩. Ring contractions of this type may also be achieved photochemically (cf. Section 4.33.3.1.3) or by treating with alcoholic alkali (cf. Section 4.33.3.3.1). Reductive cleavage of (141) with zinc and acetic acid leads to, among other products, thioethers (143) and/or sulfur-free compounds (144) ⟨71CJC1477, 71TL1781⟩. Reduction of (81; $R^1 = R^2 = Ph$) with Raney nickel likewise proceeds with desulfurization to give diphenylmethane ⟨47JOC807⟩.

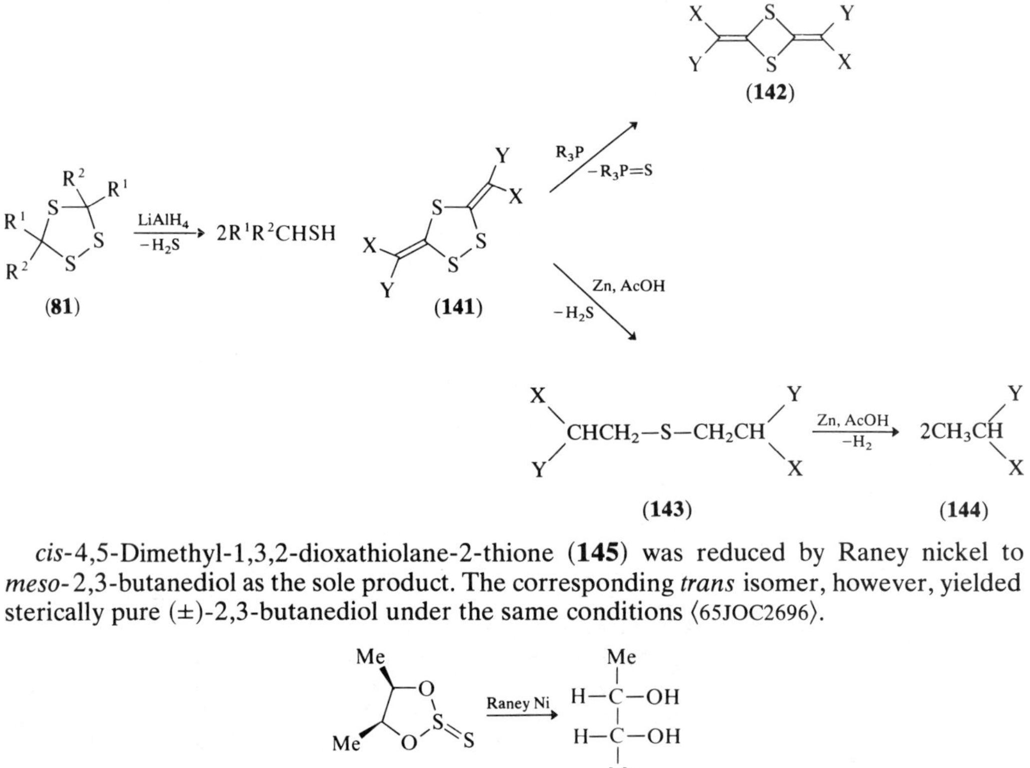

cis-4,5-Dimethyl-1,3,2-dioxathiolane-2-thione (145) was reduced by Raney nickel to meso-2,3-butanediol as the sole product. The corresponding trans isomer, however, yielded sterically pure (±)-2,3-butanediol under the same conditions ⟨65JOC2696⟩.

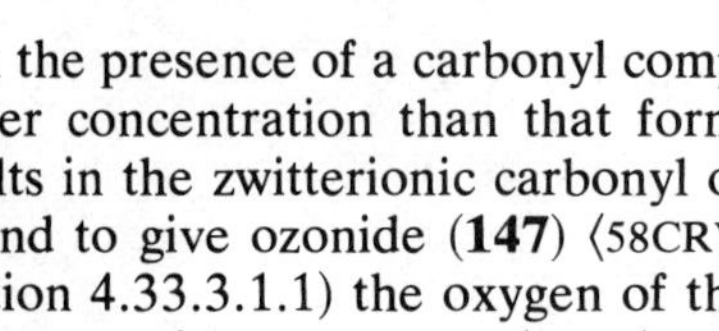

4.33.3.4 Reactions with Carbonyl Compounds

Ozonolysis of an alkene in the presence of a carbonyl compound which is more reactive, or which is present in higher concentration than that formed by cycloreversion of the primary ozonide (146), results in the zwitterionic carbonyl oxide reacting preferably with the added carbonyl compound to give ozonide (147) ⟨58CRV925⟩. In accordance with the Criegee mechanism (cf. Section 4.33.3.1.1) the oxygen of the added carbonyl component is incorporated into (147) as an ether oxygen rather than as a peroxidic oxygen atom

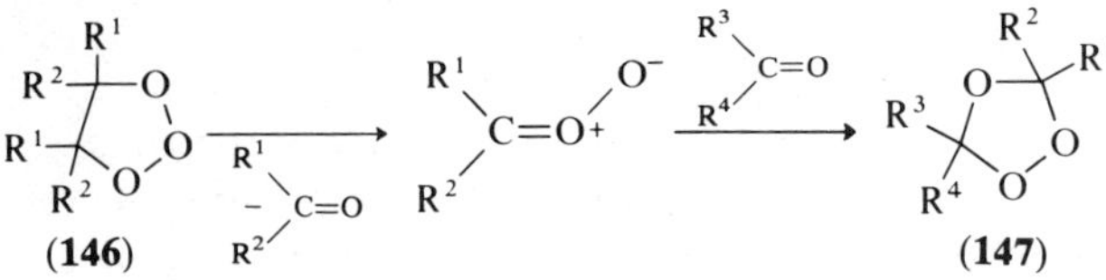

⟨75AG765, B-78MI43300⟩. The outlined method is of preparative importance since it permits the synthesis of 1,2,4-trioxolanes which cannot be obtained by other procedures (*cf.* Section 4.33.4.2.3).

The acid-catalyzed reaction of 1,3,2-dioxathiolane 2-oxide (**16**) with carbonyl compounds leads to 1,3-dioxolanes (**149**), *i.e.* cyclic acetals of glycol ⟨66HC(21-1)1⟩. The ring transformation occurs exclusively with S—O bond fission and a mechanism involving a seven-membered intermediate (**148**) has been postulated. However, in the presence of traces of water the intermediate formation of glycol and its reaction with the carbonyl compound cannot entirely be ruled out ⟨66BCJ1785⟩.

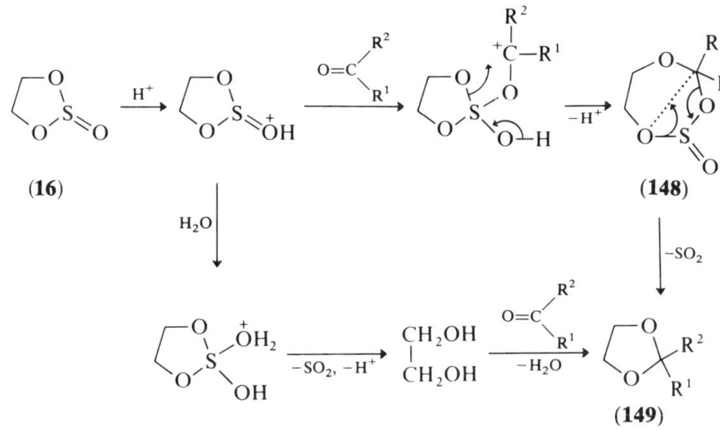

4.33.3.5 Reactions with Radicals

Reactions of trioxolanes and trithiolanes with radical species have received little attention. The action of hydroxyl radicals on 1,3,2-dioxathiolane 2-oxide (**16**) gave rise to the radicals (**150**), (**151**) and (**152**), which were detected by ESR spectroscopy after trapping with the nitromethane *aci* anion ⟨76JCS(P2)1040⟩. The *t*-butoxyl radical likewise attacked (**16**) at sulfur to yield radical (**153**) ⟨77JMR(27)509⟩.

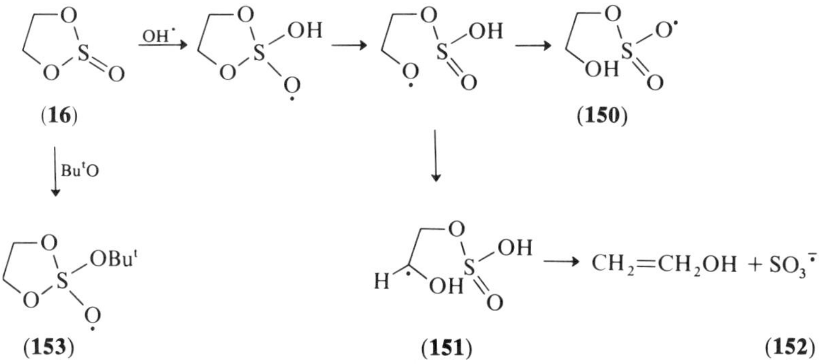

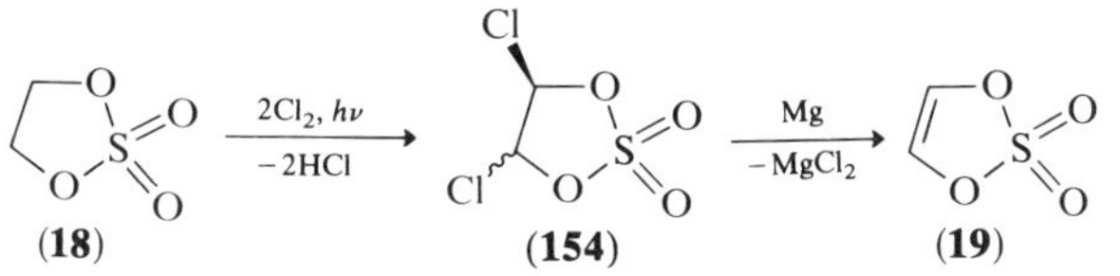

Contrary to the oxidative ring cleavage of (**16**) on treatment with chlorine, a reaction which does not proceed by a radical mechanism (*cf.* Section 4.33.3.2.4), the photochemical chlorination of 1,3,2-dioxathiolane 2,2-dioxide (**18**) to give the 4,5-dichloro derivative (**154**), occurs unambiguously by a radical process. Subsequent dechlorination on (**154**) with magnesium in refluxing THF gave 1,3,2-dioxathiole 2,2-dioxide (**19**) (vinylidene sulfate) ⟨68JA2970⟩.

4.33.4 SYNTHESIS

4.33.4.1 General Survey of Ring Syntheses

4.33.4.1.1 Syntheses from non-heterocyclic compounds

The syntheses of monocyclic five-membered heterocycles containing three oxygen or sulfur atoms can be classified into three groups, according to the number of building blocks required to form the ring in a one-step process:

(i) *One-component syntheses*, in which a suitable starting molecule is cyclized by linkage of two heteroatoms or of a heteroatom and a carbon atom. Rings with three identical adjacent heteroatoms cannot be prepared in this way.

(ii) *Two-component syntheses*, in which the five-membered ring can be built up either by a [2+3] process or by a [4+1] process. The most usual variations of the [2+3] method are C_2+X_3, C_2+X_2Y and $CX+CX_2$, while the [4+1] approach accommodates combinations such as C_2X_2+X, C_2X_2+Y and C_2XY+Y. Ring closure reactions proceeding with C—C bond linkage are unknown in these heterocyclic systems.

(iii) *Multi-component syntheses*, in which the ring is formed via non-isolated intermediates from more than two (identical or different) building blocks, e.g. by the combination $CX+X_2+C$, $2CX+X$ or $2C+3X$. Only systems with a 1,2,4-arrangement of identical heteroatoms have been prepared by these procedures.

The various possibilities for these three types of syntheses are listed schematically in Table 10. Several of these reactions are of special preparative importance and they will be discussed in more detail in the subsection indicated in the last column in Table 10. In all other cases literature references are given.

4.33.4.1.2 Syntheses by ring transformations

Characteristic types of ring transformations leading from three-, four-, five- or six-membered heterocycles to five-membered systems with three oxygen or sulfur atoms are summarized in Table 11. In general, such reactions, because of their limited applicability, are of minor preparative interest when compared with syntheses from non-heterocyclic starting compounds. Exceptions are the ring enlargement of epoxides by sulfur dioxide and the very important rearrangement of primary ozonides (1,2,3-trioxolanes) into ozonides (1,2,4-trioxolanes).

4.33.4.2 Syntheses of Various Types of Derivatives

4.33.4.2.1 Parent systems (including S-oxides and S,S-dioxides)

In this and the following subsections the basic structures will be discussed in the order in which they are arranged in Scheme 1 (cf. Section 4.33.1). In the case of the hypothetical parent systems 1,3,2-dioxathiolane (**7**), 1,4,2-dioxathiolane (**8**), 1,2,3-oxadithiolane (**9**) and 1,2,5-oxadithiolane (**11**), the corresponding S-oxides and S,S-dioxides will be regarded as the individual basic structures.

The sole possibility of preparing the two parent compounds of the trioxolane series consists in the action of ozone on ethylene. When this ozonation was performed under suitable conditions, e.g. in a condensed phase at temperatures below −175 °C, the formation of 1,2,3-trioxolane (**1**) was observed ⟨72JA4856, 81JA2578⟩ (cf. Section 4.33.2.2.4). However, because of its thermal instability, this primary ozonide cannot be isolated but rearranged at temperatures above −100 °C to 1,2,4-trioxolane (**2**), ethylene ozonide, a colorless, explosive liquid extremely sensitive to impact ⟨42LA(583)187⟩. The mechanism of the ring transformation (**1**) → (**2**) has been discussed in Section 4.33.3.1.2.

In the trithiolane series only the parent compound (**4**) with a 1,2,4-arrangement of heteroatoms is known. It was obtained besides some 1,2,4,6-tetrathiepane as a distillable, pale yellow liquid on vigorous stirring of an aqueous solution of $Na_2S_{2.5}$ (prepared from $Na_2S \cdot 9H_2O$ and the corresponding amount of sulfur) with excess methylene chloride according to a multi-component synthesis: $2CH_2Cl_2 + 2Na_2S_x \rightarrow$ (**4**) $+ 4NaCl$ (cf. Table 10)

Table 10 Ring Syntheses of Monocyclic Five-membered Heterocycles Containing Three Oxygen or Sulfur Atoms from Non-heterocyclic Compounds

Type of synthesis	Arrangement of heteroatoms in the starting systems	Starting components utilized	Heterocyclic ring skeleton obtained[a]	Section or reference in which the reaction is cited
One-component syntheses (formation of one bond)	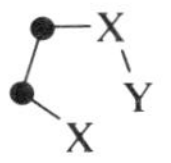	$HO-C-C-O-SO_2-OH$	$\xrightarrow{-H_2O}$ 1,3,2-$[C_2O_2S]$	4.33.4.2.1
		$HO-SO_n-C-C-SO_n-OH$	$\xrightarrow[(n=1\text{ or }2)]{-H_2O}$ 1,2,5-$[C_2OS_2]$	4.33.4.2.1
	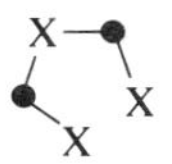	$S=C-S-C=S$	$\xrightarrow{I_2}_{-2I^-}$ 1,2,4-$[C_2S_3]$	4.33.4.2.4
	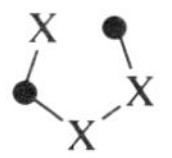	$HO-C-O-O-C-OH$	$\xrightarrow{-H_2O}$ 1,2,4-$[C_2O_3]$	32CB1274
		$HS-C-S-S-C-SH$	$\xrightarrow{-H_2S}$ 1,2,4-$[C_2S_3]$	74AC(R)305
		$S=C-S-S-C=S$	$\xrightarrow[-[S]]{I_2,\ -2I^-}$ 1,2,4-$[C_2S_3]$	4.33.4.2.4
	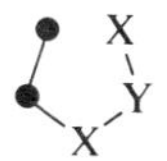	$R_2\overset{+}{S}-C-C-O-SO_2^-$	$\xrightarrow{-R_2S}$ 1,3,2-$[C_2O_2S]$	66GEP1223397
Two-component syntheses (formation of two bonds)	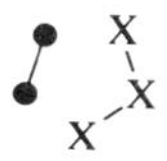	$\text{C}=\text{C} + O_3$	$\longrightarrow$ 1,2,3-$[C_2O_3]$	4.33.4.2.1, 4.33.4.2.3
		$\text{C}=\text{C} + S_x$	$\longrightarrow$ 1,2,3-$[C_2S_3]$	4.33.4.2.6
		$Br-C-C-Br + Na_2S_4$	$\xrightarrow[-[S]]{-2\,NaBr}$ 1,2,3-$[C_2S_3]$	4.33.4.2.3

Table 10 (Continued)

Type of synthesis	Arrangement of heteroatoms in the starting systems	Starting components utilized	Heterocyclic ring skeleton obtained[a]	Section or reference in which the reaction is cited
	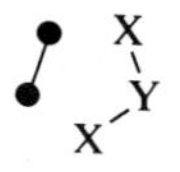 X–Y–X	$\diagup$C=C$\diagdown$ + SO$_n$	$\xrightarrow{(n = 2 \text{ or } 3)}$ 1,3,2-[C$_2$O$_2$S]	4.33.4.2.1 4.33.4.2.3 4.33.4.2.6
		$\diagup$C—CO— + SO$_3$	$\longrightarrow$ 1,3,2-[C$_2$O$_2$S]	74JOC3415
	X–X	AcO—C—C—OAc + (MeO)$_2$SO$_2$	$\xrightarrow{-2\text{ AcOMe}}$ 1,3,2-[C$_2$O$_2$S]	4.33.4.2.1
		Br—C—C—Br + Ag$_2$SO$_4$	$\xrightarrow{-2\text{ AgBr}}$ 1,3,2-[C$_2$O$_2$S]	4.33.4.2.1
	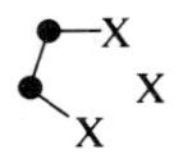	NaS—C=C—SNa + SOCl$_2$	$\xrightarrow{-2\text{ NaCl}}$ 1,2,3-[C$_2$S$_3$]	4.33.4.2.3
		HS—C—C—SH + S(NR$_2$)$_2$	$\xrightarrow{-2\text{ HNR}_2}$ 1,2,3-[C$_2$S$_3$]	4.33.4.2.3
	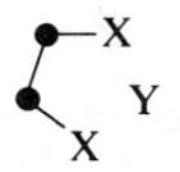 X–Y–X	HO—C=C—OH + SOCl$_2$	$\xrightarrow{-2\text{ HCl}}$ 1,3,2-[C$_2$O$_2$S]	63JOC1075
		HO—C—C—OH + SO$_n$Cl$_2$	$\xrightarrow[(n = 1 \text{ or } 2)]{-2\text{ HCl}}$ 1,3,2-[C$_2$O$_2$S]	4.33.4.2.1 4.33.4.2.3
		HO—C—C—OH + S$_2$Cl$_2$	$\xrightarrow{-2\text{ HCl}}$ 1,3,2-[C$_2$O$_2$S]	4.33.4.2.1 4.33.4.2.3
		HO—C—C—OH + F$_2$S=NCF$_2$CF$_3$	$\xrightarrow{-2\text{ HF}}$ 1,3,2-[C$_2$O$_2$S]	4.33.4.2.4
		HO—C—C—OH + (MeO)$_2$SO	$\xrightarrow{-2\text{ MeOH}}$ 1,3,2-[C$_2$O$_2$S]	4.33.4.2.1
		HO—C—C—OH + SO$_2$	$\xrightarrow[-\text{MeOH}, -\text{N}_2]{\text{CH}_2\text{N}_2}$ 1,3,2-[C$_2$O$_2$S]	60CB1129
	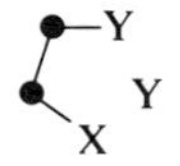 Y–X–X	HO—C—C—SH + SOCl$_2$	$\xrightarrow{-2\text{ HCl}}$ 1,2,3-[C$_2$OS$_2$]	4.33.4.2.1

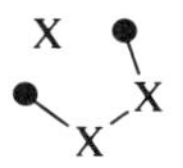

	Cl—C—S—S—C—Cl + Na₂S	$\xrightarrow{-2\text{NaCl}}$ 1,2,4-[C₂S₃]	4.33.4.2.3
	$\begin{matrix}\text{S}^- & \text{S}^- \\ -\text{C} & , = \text{C} \\ \text{SH} & \text{S}^- \end{matrix}$ or $\begin{matrix}\text{SH} \\ \text{C} \\ \text{SH}\end{matrix}$	$\xrightarrow[{-[\text{S}]}]{[\text{O}]}$ 1,2,4-[C₂S₃]	4.33.4.2.3
	CS₂	$[\overset{-}{\text{C}}-\overset{+}{\text{S}}\text{R}_2] \xrightarrow{-[\text{S}]}$ 1,2,4-[C₂S₃]	4.33.4.2.3
	$\overset{-}{>}\text{C}-\overset{+}{\text{N}}=\text{N} + {}^1\text{O}_2 + >\text{C}=\text{O}$	$\xrightarrow{-\text{N}_2}$ 1,2,4-[C₂O₃]	73JA3343
Multi-component syntheses (formation of more than two bonds)			
	2Cl—C—S—Cl + 2Na₂S	$\xrightarrow{-4\text{NaCl}}_{-[\text{S}]}$ 1,2,4-[C₂S₃]	4.33.4.2.3
	3 >C=S	$\xrightarrow[{-\underset{|}{\overset{|}{\text{C}}}=\text{O}}]{[\text{O}]}$ 1,2,4-[C₂S₃]	4.33.4.2.3
	2 >C=O + 2H₂S + S	$\xrightarrow[-2\text{H}_2\text{O}]{[\text{Amine}]}$ 1,2,4-[C₂S₃]	4.33.4.2.3
	2 >CF₂ + 4NaSH	$\xrightarrow[{-[\text{S}]}]{-4\text{NaF}}$ 1,2,4-[C₂S₃]	4.33.4.2.3
	2 >CCl₂ + 2Na₂S$_x$	$\xrightarrow{-4\text{NaCl}}$ 1,2,4-[C₂S₃]	4.33.4.2.1

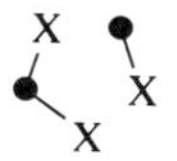

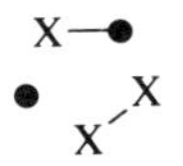

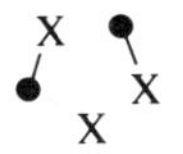

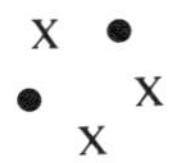

[a] [C₂O₃] trioxolane, [C₂S₃] trithiolane, [C₂O₂S] dioxathiolane, [C₂OS₂] oxadithiolane; *cf.* Scheme 1.

Table 11 Formation of Monocyclic Five-membered Heterocycles containing three Oxygen or Sulfur Atoms by Ring Interconversion

Type of ring transformation	Starting systems	Heterocyclic ring skeleton obtained[a]	Section or reference in which the reaction is cited
Insertion of an —SO$_2$—moiety 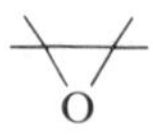		$\xrightarrow{SO_n}$ ($n = 2$ or 3) 1,3,2-[C$_2$O$_2$S]	4.33.4.2.1 4.33.4.2.6
Insertion of a sulfur atom		$\xrightarrow{S(OR)_2}$ 1,3,2-[C$_2$O$_2$S]	4.33.3.1.2
Insertion of sulfur	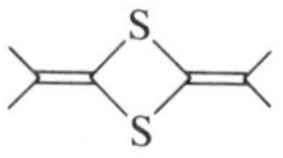	$\xrightarrow{h\nu, S}$ 1,2,4-[C$_2$S$_3$]	80ABC2169
Rearrangement	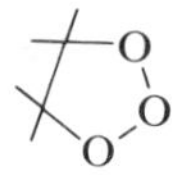	$\xrightarrow{\Delta}$ 1,2,4-[C$_2$O$_3$]	4.33.3.1.1 4.33.4.2.1 4.33.4.2.3
Displacement of a ring heteroatom (Z = GeR$_2$ or AsR)	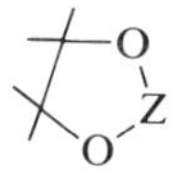	$\xrightarrow[-ZCl_2]{SOCl_2}$ 1,3,2-[C$_2$O$_2$S]	75JOM(101)57
Ring cleavage, partial desulfurization, and recombination of the intermediates	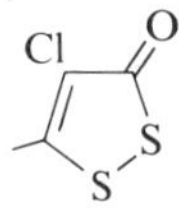	$\xrightarrow{MeONa}$ 1,2,4-[C$_2$S$_3$]	68HCA1421, 75LA1513
Photofragmentation and recyclization of the intermediates	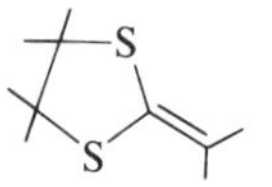	$\xrightarrow{h\nu}$ 1,2,4-[C$_2$S$_3$]	80ABC2169, 75TL555
Desulfurization	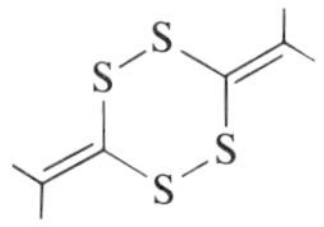	$\xrightarrow{-[S]}$ 1,2,4-[C$_2$S$_3$]	66T3001

[a] See footnote a in Table 10.

⟨67CPB988⟩ (the conversion of (**4**) into its 1-oxide (**74**) and 4-oxide (**37**) has been described in Section 4.33.3.2.3).

The 4,5-unsubstituted 1,3,2-dioxathiolane system (**7**) exists only in the form of its 2-oxide and 2,2-dioxide or their derivatives. 1,3,2-Dioxathiolane 2-oxide (**16**) represents the parent compound of saturated five-membered cyclic sulfites. It is a colorless, distillable liquid which can be prepared according to the methods shown in Scheme 9. The most convenient laboratory technique is the reaction of ethylene glycol with thionyl chloride ⟨66HC(21-1)1⟩. Another possibility is the transesterification of dimethyl sulfite with ethylene glycol ⟨76CRV747⟩. The reaction of ethylene oxide with sulfur dioxide, frequently described in the patent literature ⟨66HC(21-1)1⟩, depends on the reaction conditions. It leads either directly

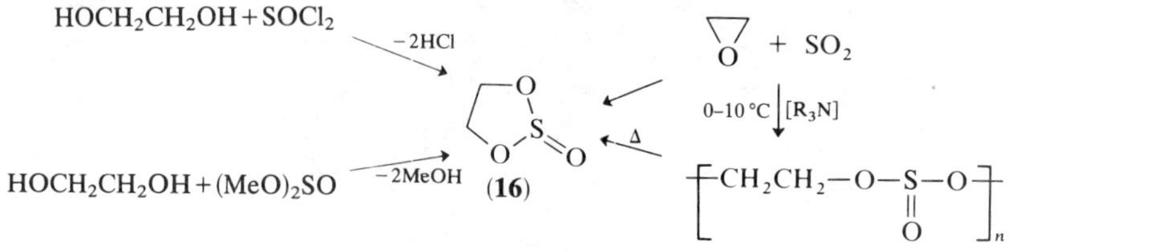

Scheme 9 Methods for preparing 1,3,2-dioxathiolane 2-oxide (ethylene sulfite)

to (**16**) or first to a polymer which depolymerizes on heating to give ethylene sulfite in nearly quantitative yield (*cf.* Section 4.33.3.1.4). 1,3,2-Dioxathiolane-2-thione, the thiono analog of (**16**), was obtained from ethylene glycol and S_2Cl_2 in the presence of triethylamine ⟨65JOC2696⟩.

1,3,2-Dioxathiolane 2,2-dioxide (**18**), the parent compound of saturated five-membered cyclic sulfates, has been prepared from 1,2-ethylene dibromide and silver sulfate in 25% yield as well as from glycol diacetate and dimethyl sulfate in 66% yield ⟨66HC(21-1)1⟩ (*cf.* Scheme 10). Unlike the ready formation of ethylene sulfite (**16**) from ethylene glycol and thionyl chloride or from ethylene oxide and sulfur dioxide, the corresponding reactions with sulfuryl chloride ⟨68JHC289⟩ or sulfur trioxide ⟨66HC(21-1)1⟩ gave only low yields of ethylene sulfate. In contrast, satisfactory results have been obtained by permanganate oxidation of (**16**) in acetone ⟨63JA602⟩ (*cf.* Section 4.33.3.2.3) or by transesterification of (**16**) with sulfuric acid to give 2-hydroxyethylsulfuric acid followed by treatment with thionyl chloride ⟨66HC(21-1)1⟩ (*cf.* Scheme 10). The synthesis of 1,3,2-dioxathiole 2,2-dioxide (**19**), the 4,5-unsaturated derivative of (**18**), from (**18**) *via* chlorination and dechlorination has been described in Section 4.33.3.5.

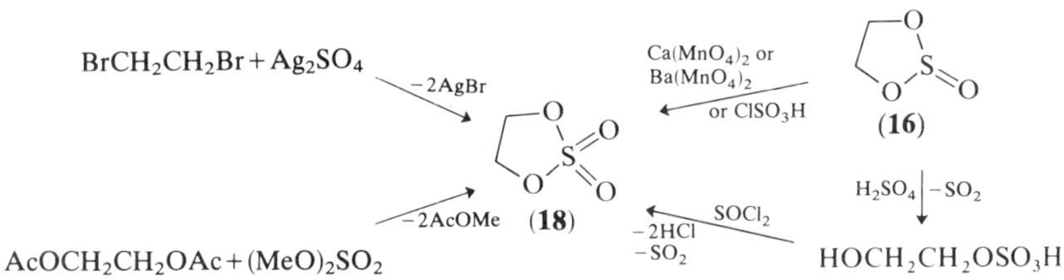

Scheme 10 Methods for preparing 1,3,2-dioxathiolane 2,2-dioxide (ethylene sulfate)

The 2-oxide of 1,2,3-oxadithiolane (**9**) has been prepared *via* a two-component synthesis from 2-hydroxyethanethiol and thionyl chloride in diethyl ether using triethylamine as hydrogen chloride acceptor ⟨75ACS(A)414⟩.

The 2,5-dioxide (**99**) of 1,2,5-oxadithiolane (**11**) represents the cyclic anhydride of 1,2-ethanedisulfinic acid. It was obtained by controlled hydrolysis of 1,2-ethanebis(disulfinyl) chloride ⟨76CRV747⟩. 1,2,5-Oxadithiolane 2,2,5,5-tetroxide (**100**), *i.e.* 1,2-ethanedisulfonic anhydride, has been prepared by heating the acid with thionyl chloride ⟨66HC(21-1)1⟩.

The parent system 1,3,4-oxadithiolane (**12**) was claimed to be formed on reacting α,α'-dichlorodimethyl ether with sodium tetrasulfide in the presence of a base and subsequent steam distillation of the polymer obtained. However, since neither a structural proof nor any properties were described, the existence of this ring system should be considered in doubt ⟨66HC(21-1)1⟩.

4.33.4.2.2 Benzo derivatives

Most compounds of this type are cyclic sulfite and sulfate esters of aromatic 1,2-diols as well as anhydrides of aromatic 1,2-disulfonic acids. The simplest representatives with unsubstituted benzene rings are 1,3,2-benzodioxathiole 2-oxide (**48**) (catechol sulfite), the corresponding 2,2-dioxide (**156**) (catechol sulfate) and 2,1,3-benzoxadithiole 1,1,3,3-tetroxide (**158**) (1,2-benzenedisulfonic anhydride). Compound (**48**) was synthesized by refluxing catechol with thionyl chloride in the presence of pyridine. In a similar fashion, from 2-mercaptophenol 1,2,3-benzoxadithiole 2-oxide was prepared ⟨81AG603⟩. The dioxide (**156**) was obtained in two steps by reaction of catechol monosodium salt with sulfuryl chloride in benzene at 0–10 °C and subsequent reflux of the intermediate (**155**) in the presence of pyridine.

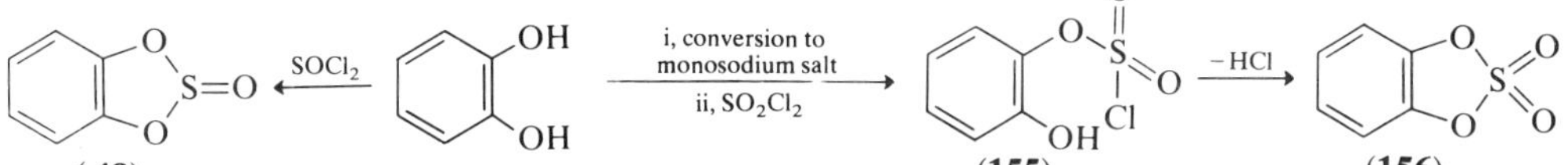

The tetroxide (**158**) was formed on heating the dipotassium salt (**157**) with chlorosulfonic acid. 1,2-Benzenedisulfonic acid, in turn, has been prepared by diazotization of *o*-aminobenzenesulfonic acid, conversion into the mercaptobenzenesulfonic acid, and final oxidation.

Besides the benzo derivatives just described, a series of compounds with a substituted benzene ring as well as those condensed with other aromatic or heteroaromatic systems is known. They all are accessible by the same or similar methods ⟨66HC(21-1)1⟩.

4.33.4.2.3 C-Linked substituents

Like the parent compound (**1**), alkyl- and/or aryl-substituted 1,2,3-trioxolanes can be prepared only through ozonation of the corresponding alkenes. Primary ozonides derived from aliphatic *cis* alkenes are much more unstable and decompose at lower temperature than those formed from the corresponding *trans* alkenes. Under suitable conditions, some of these compounds can be obtained at low temperatures in the form of colorless explosive precipitates ⟨75AG765, B-78MI43300⟩. The most 'stable' monocyclic 4,5-dialkyl derivative and, at the same time, the very first primary ozonide actually isolated in crystalline form, is *trans*-4,5-*t*-butyl-1,2,3-trioxolane. On warming up to −60 °C it rearranged exothermically to give the corresponding 1,2,4-trioxolane derivative ⟨60CB689⟩. The primary ozonide of styrene proved to be likewise relatively 'stable' showing no decay until about −55 °C ⟨79JA2524⟩. A discussion has appeared of more recent quantum chemical investigations on the formation of some alkyl-substituted primary ozonides and their interconversion into final ozonides ⟨81JA3619, 81JA3627⟩.

The synthesis of monocyclic alkyl- and/or aryl-substituted 1,2,4-trioxolanes by ozonolysis of alkenes is only possible when the alkene possesses at least one hydrogen atom at the C—C double bond, otherwise 1,2,4,5-tetroxanes (**30**) (*cf.* Section 4.33.3.1.1), oligomeric ozonides (*cf.* Section 4.33.3.1.4) and/or other subsequent products are formed. Exceptions are tetrasubstituted ethylenes having electron-attracting substituents such as haloalkyl or ester groups at the alkenic double bond. Thus *trans*-1,4-dibromo-2,3-dimethyl-2-butene (**159**) and 2,3-dimethyl crotonic acid methyl ester (**161**) were converted into the corresponding 1,2,4-trioxolanes (**160**) and (**162**), respectively ⟨75AG765, B-78MI43300⟩.

The transformation of tetrasubstituted ethylenes into 1,2,4-trioxolanes may also be achieved if the ozonolysis is carried out in the presence of a 'foreign' carbonyl compound as described in Section 4.33.3.4. With formaldehyde as added carbonyl compound, 3,3-disubstituted derivatives are obtained, whereas in the presence of excess ketone (*e.g.* by using the latter as solvent), the ozonolysis gives rise to tetrasubstituted 1,2,4-trioxolanes which are difficult to prepare by other methods. Reactions (**163**) → (**164**) and (**165**) → (**166**) provide two examples of this versatile 1,2,4-trioxolane synthesis. Unlike the parent system (**2**), alkyl- and/or aryl-substituted 1,2,4-trioxolanes generally are stable, non-explosive compounds. Mixtures of 'crossed' ozonides (*cf.* Section 4.33.3.1.1) or of *cis* and *trans* isomers can be separated by thin layer, column or gas chromatography. The *cis* isomers of symmetrical 3,5-disubstituted 1,2,4-trioxolanes are *meso* forms, whereas the corresponding *trans* isomers represent racemates which in some cases have been resolved into their optical antipodes ⟨75AG765, B-78MI43300⟩.

The principles underlying alternative formation of 1,2,4-trioxolanes by cyclodehydration of α,α'-dihydroxydialkyl peroxides (Rieche's 'ozonide synthesis without ozone'), or by photooxidation of diazo compounds in the presence of aldehydes, are outlined in Table 10.

In the 1,2,3-trithiolane series only a few representatives containing C-linked substituents are known. The 4-ethyl derivative (**60**) was obtained as an unstable, distillable, yellow liquid on treating 1,2-dimercaptobutane (**167**) with diphthalimido sulfide ⟨77UP43300⟩ (for the marked tendency of (**60**) to polymerize, cf. Section 4.33.3.1.4). The reaction of the vicinal dibromides (**168**; R = SO$_3$Na, O(CH$_2$)$_2$SO$_3$Na and S(CH$_2$)$_2$SO$_3$Na) with sodium tetrasulfide resulted in the corresponding 3-alkyl-substituted 1,2,3-trithiolanes (**169**). These are soluble in water and can be characterized by their benzylisothiuronium salts ⟨67UKZ596⟩.

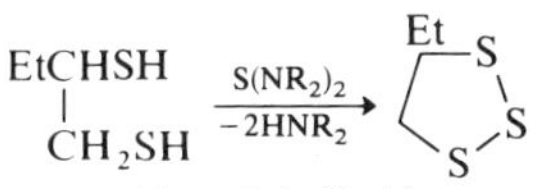

The synthesis of 4,5-dicyano-1,2,3-trithiole 2-oxide (**172**) starts from sodium cyanide and carbon disulfide which via (**170**) gave the disodium salt of 2,3-mercaptomaleonitrile (**171**; M = Na). Treatment of the corresponding silver salt (**171**; M = Ag) with thionyl chloride yielded (**172**) ⟨66HC(21-1)1⟩. Phenylsulfine (**174**), prepared in situ by dehydrohalogenation of phenylmethanesulfinyl chloride (**173**), slowly decomposed in ether solution at room temperature to give cis- and trans-stilbenes, trans-4,5-diphenyl-1,2,3-trithiolane 1,1-dioxide (**36**) and a 5,6-diphenyl-1,2,3,4-tetrathiane dioxide ⟨68JCS(C)1612⟩. The mechanisms of formation of these heterocycles are obscure.

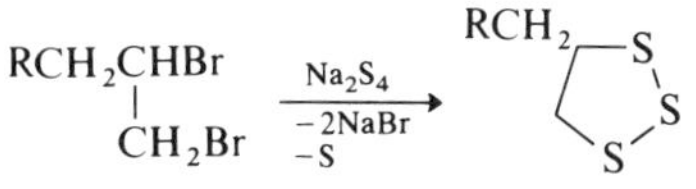

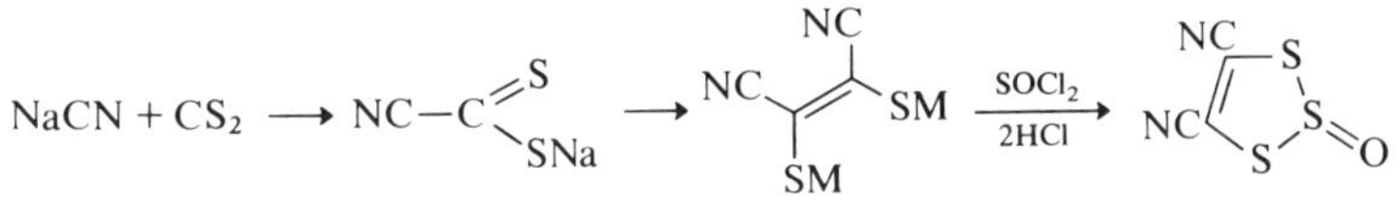

Symmetrical 3,5-dialkyl-1,2,4-trithiolanes (**178**) can be synthesized in reasonable yield by chlorination of dialkyl disulfides (**175**) to α-chloroalkyl sulfenyl chlorides (**176**), which are then reacted with potassium iodide to give di-α-chloroalkyl disulfides (**177**). Subsequent cyclization with sodium sulfide gave (**178**) ⟨72T3489⟩. When (**176**) was treated with one molar equivalent of sodium sulfide, the reductive dimerization and cyclization was effected in one step ⟨78HCA1404⟩. Treatment of perfluoropropene with sodium hydrogen sulfide in THF resulted in the formation of 3,5-bis(2,2,2-trifluoroethyl)-1,2,4-trithiolane (**179**) ⟨72IZV2517⟩.

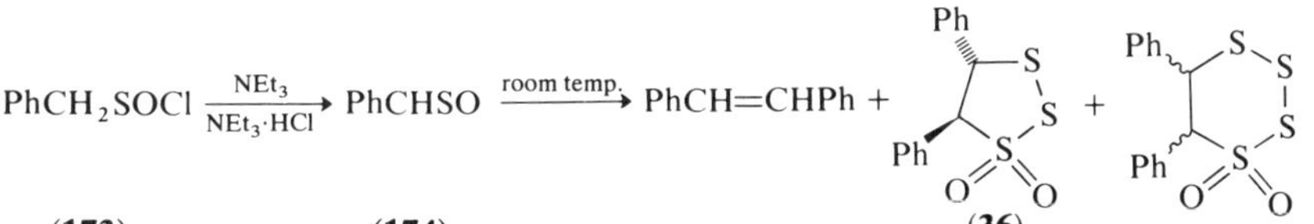

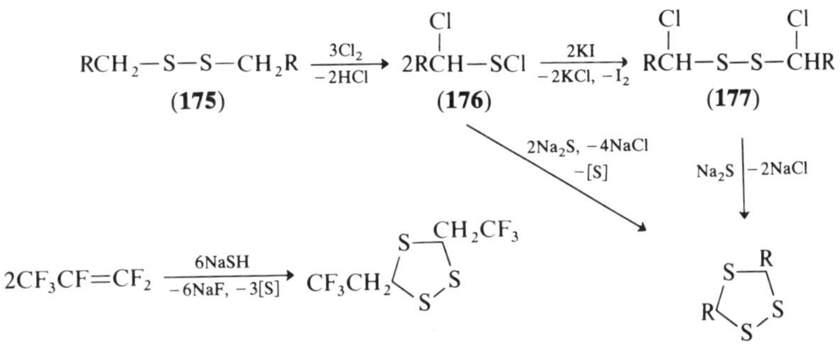

Usual methods for preparing substituted 1,2,4-trithiolanes of type (**81**) are shown in Scheme 11. According to Asinger et al. ⟨59LA(627)195⟩, the common action of hydrogen sulfide and sulfur on ketones at 0 °C in the presence of a primary aliphatic amine leads to

3,3,5,5-tetraalkyl-substituted derivatives. The mechanism of this multi-component synthesis has been discussed in connection with the reverse reaction, the cleavage of (**81**) by primary amines (*cf.* Section 4.33.3.3.3). The reaction also proceeds with some aldehydes, but fails with formaldehyde, acetophenone and benzophenone. Under acid conditions, however, benzophenone readily reacts with hydrogen sulfide in ethanol to give (**81**; $R^1 = R^2 = Ph$) ⟨66HC(21-1)1⟩. Another way to tetrasubstituted derivatives is the action of aerial oxygen ⟨66HC(21-1)1⟩, sulfur ⟨66JPR(306)116⟩, tetrachloro-*o*-quinone ⟨65TL3361⟩ or chloramine-T ⟨71CC179⟩ on thioketones. Geminal dithiols can also be oxidized to 1,2,4-trithiolanes of type (**81**) ⟨62JOC488, 72USP3694498⟩.

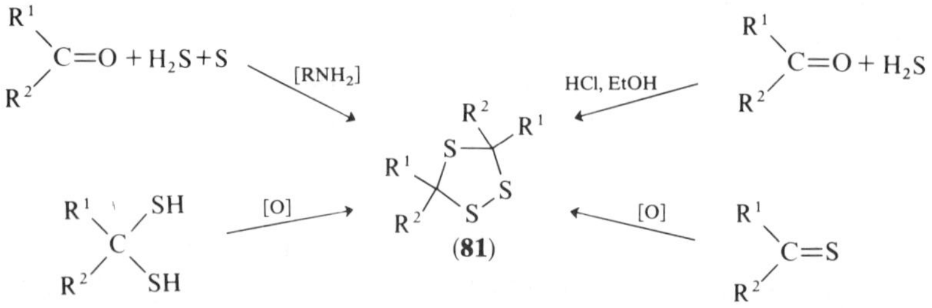

Scheme 11 Methods for preparing 3,3,5,5-tetrasubstituted 1,2,4-trithiolanes

3,5-Bismethylene-1,2,4-trithiolanes (**141**), in which X and/or Y as a rule represent electron-attracting groups such as COR, CO_2R, $CONH_2$, CN, CF_3, *etc.*, are obtained from dithio acids, their anions (**180**) or dianions (**181**) by oxidation with bromine ⟨62CB2861, 70BCJ2938⟩, iodine ⟨62CB2861, 65LA(684)37, 72JOC3226, 76JCS(P1)1706, 77JCS(P1)1273⟩, peroxydisulfate ⟨62CB2861, 71CJC1477⟩ or sulfuric acid ⟨73BSF581⟩. The dithio anions (**180**) and (**181**) can be readily prepared by reaction of the corresponding active methylene compound with carbon disulfide in basic medium. Other methods for synthesizing 1,2,4-trithiolanes of type (**141**) are the treatment of sulfonium ylides (**182**) with carbon disulfide ⟨71TL1781⟩ or the reaction of thioketenes (**183**) with sulfur at elevated temperatures ⟨70JOC3470⟩. The syntheses of 1,2,4-trithiolanes by the ring transformations listed in Table 11 also lead without exception to 3,5-bismethylene derivatives.

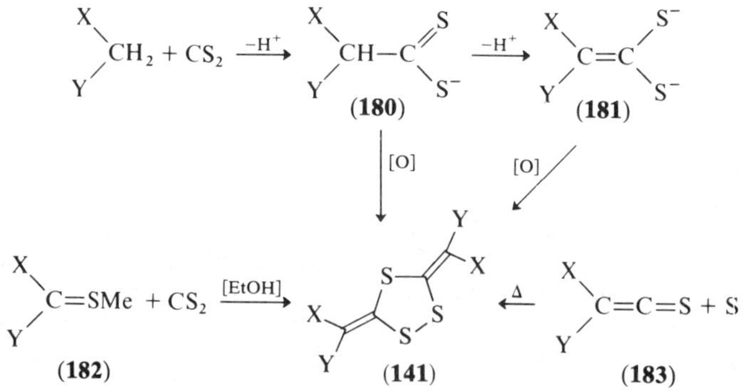

1,3,2-Dioxathiolane 2-oxides containing *C*-linked substituents can be synthesized, in principle, according to the same methods shown in Scheme 9 for preparing the parent compound, ethylene sulfite (**16**). Here also the reaction of the corresponding 1,2-diols (including 1,2-enediols) with thionyl chloride is the most important laboratory method. In the case of polyhydroxy compounds the remaining hydroxy groups frequently are displaced by the chloro substituent ⟨66HC(21-1)1⟩. Treatment of simple aliphatic 1,2-diols such as propylene glycol or *meso-* and (±)-2,3-butanediol with sulfur monochloride (S_2Cl_2) results in the formation of the corresponding 1,3,2-dioxathiolane-2-thiones ⟨65JOC2696⟩. Another route to alkyl-substituted cyclic sulfites consists of the photosulfoxidation of alkenes, *i.e.* their photoreaction with sulfur dioxide and oxygen ⟨74T2053⟩.

Alkyl-substituted 1,3,2-dioxathiolane 2,2-dioxides can be prepared in a similar manner to the parent compound, ethylene sulfate (**18**) ⟨66HC(21-1)1⟩ (*cf.* Scheme 10, Section 4.33.4.2.1). The method of choice is the permanganate oxidation of the corresponding cyclic sulfites (*cf.* Section 4.33.3.2.3) since the direct reaction of 1,2-diols with sulfuryl chloride often proceeds less smoothly than does the reaction with thionyl chloride. 4,5-Diaryl-1,3,2-dioxathiole 2,2-dioxides of type (**186**) are obtained by treatment of 9,10-

Five-membered Rings containing Three Oxygen or Sulfur Atoms 891

dihydroanthracene derivatives (**184**; R = Ph, PhC≡C) with sulfuric acid, half-esters of structure (**185**) being suggested as intermediates in this reaction ⟨66HC(21-1)1⟩.

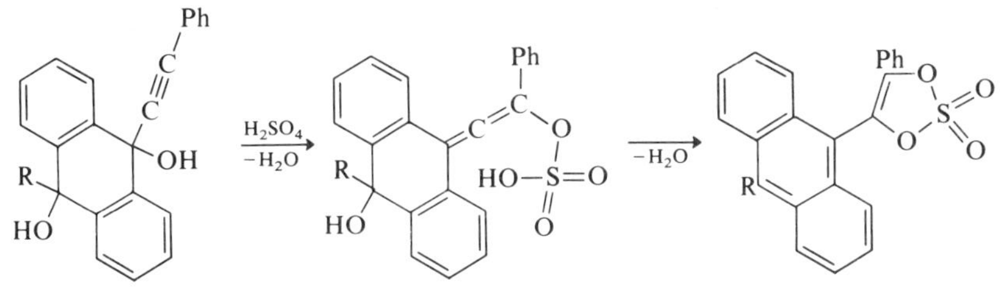

(**184**) (**185**) (**186**)

4.33.4.2.4 N-Linked substituents

Trioxolanes containing *N*-linked substituents appear to be unknown but 1,2,4-trithiolanes of this type are readily accessible. 3,5-Bis(*N,N*-dialkyliminium) salts (**188**) can be prepared according to a one-step synthesis by oxidation either of *N,N,N',N'*-tetraalkylthiuram disulfide complexes such as (**187**) ⟨76MI43300⟩ or of *N,N,N',N'*-tetraalkylthiuram monosulfides (**189**) with halogens ⟨75MI43300⟩. 3,5-Bis(arylhydrazones) (**191**) of 1,2,4-trithiolane-3,5-dione are formed on oxidation of aryldithiocarbazic acids (**190**) ⟨71JCS(B)415⟩ or by desulfurization of 1,2,4,5-tetrathiane-3,6-dione bis(arylhydrazones) (**192**) which can likewise be obtained by oxidation of (**190**) ⟨69G780⟩.

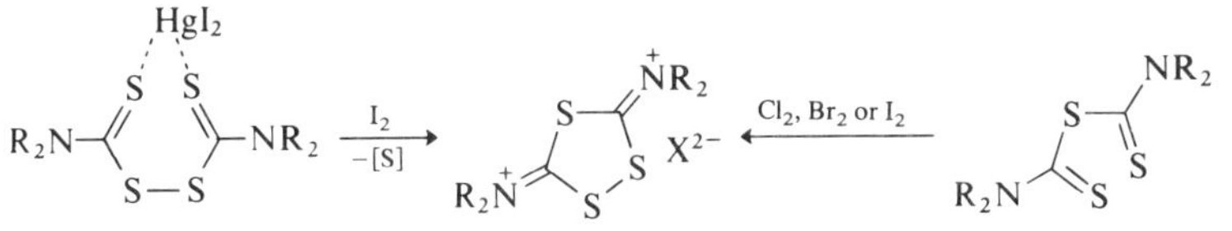

(**187**) (**188**) (**189**)

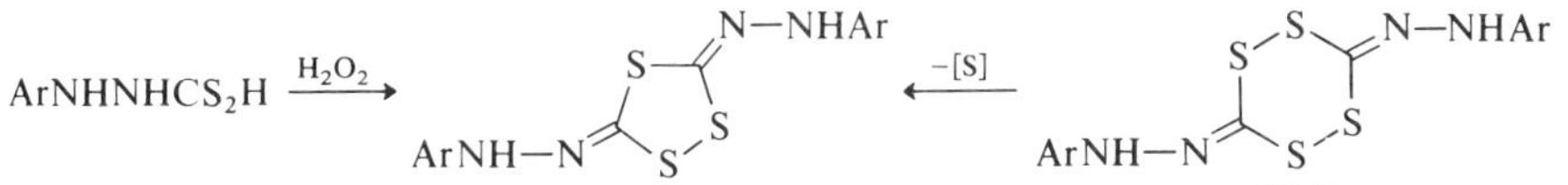

(**190**) (**191**) (**192**)

Pentafluoroethyleneiminosulfur difluoride (**193**) reacted with ethylene glycol in the presence of anhydrous sodium fluoride to give the imine derivative (**194**) of 1,3,2-dioxathiolane ⟨79IC213⟩. α-Hydroxy ketones in fuming sulfuric acid undergo reaction with difluoramine to form substituted 4-difluoroamino-1,3,2-dioxathiolane 2,2-dioxides (**195**) and, under similar conditions, α,β-diketones give substituted 4,5-bis(difluoramino)-1,3,2-dioxathiolane 2,2-dioxides (**196**). Difluoramine is conveniently generated *in situ* by the interaction of *N,N*-difluorourea with sulfuric acid ⟨69JOC917⟩.

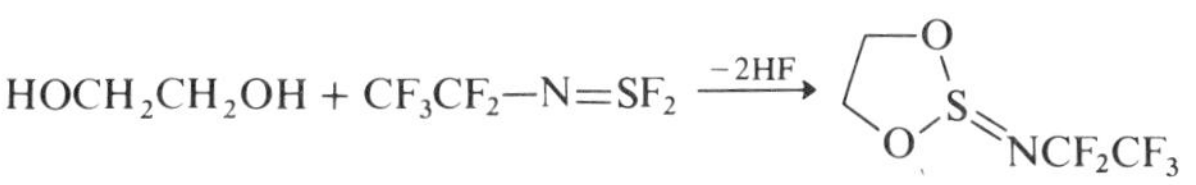

(**193**) (**194**)

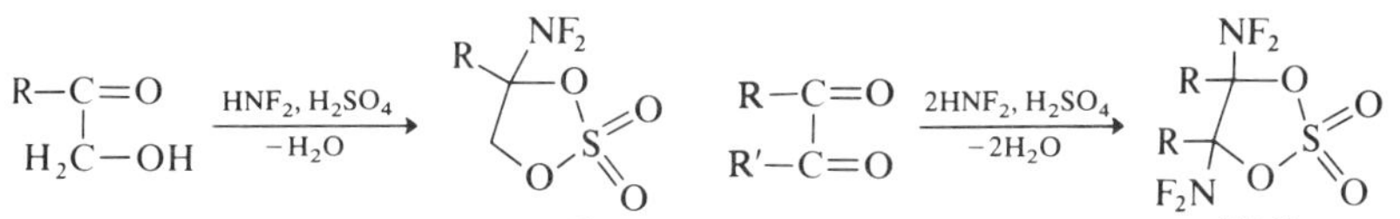

(**195**) (**196**)

4.33.4.2.5 O-Linked substituents (including oxo derivatives)

The sole representative of this type in the trioxolane series is 5,5-difluoro-1,2,4-trioxolan-3-one (**136**), the ozonide of the unknown difluoroketene $F_2C=C=O$. It has been prepared by the reversible isomerization of bis(fluoroformyl) peroxide (**137**) at low temperatures with activated potassium fluoride ⟨72IC2531⟩ as described in Section 4.33.3.3.5.

1,3,2-Dioxathiolan-4-one 2-oxides (**50**) (*cf.* Section 4.33.3.1.2), the so-called anhydrosulfites of α-hydroxy carboxylic acids, are synthesized by reacting the latter with thionyl chloride ⟨66HC(21-1)1⟩. α-Thiocarboxylic acids react analogously, yielding 1,2,3-oxadithiolan-4-one 2-oxides ⟨77MI43301⟩. The bicyclic orthosulfite (**46**) (*cf.* Scheme 7, Section 4.33.3.1.2) was obtained by treatment of perfluoropinacol with excess sulfur dichloride and pyridine in ether ⟨76JA2895⟩. Monocyclic orthosulfites of type (**47**) resulting from 3,3-dialkyl-1,2-dioxetanes and dialkyl sulfoxylates were detected only in solution at low temperatures ⟨75JA3850⟩, since they polymerized at room temperature or decomposed as shown in Scheme 7.

The synthesis of the 4,5-dimethoxy-1,3,2-dioxathiolane 2,2-dioxide (**98**) from perfluoro-2-butene involved reaction with sodium methoxide, resulting in a mixture of *cis* and *trans* isomers of the vinyl diether (**197**). Subsequent action of sulfur trioxide on (**197**) gave, probably *via* the 1,2-oxathietane 2,2-dioxide (**198**), the cyclic sulfate (**98**) ⟨77JA1214⟩, a convenient precursor to hexafluorobiacetyl (*cf.* Section 4.33.3.3.1).

4.33.4.2.6 Halogens attached to the ring

The ozonolysis of chloro- and bromo-substituted alkenes has not resulted in stable, halogenated 1,2,4-trioxolanes but has resulted in a variety of other products derived mostly from subsequent reactions of the acyl halide and halogenated carbonyl oxide expected from a Criegee cleavage of the alkene (*cf.* Section 4.33.3.1.2). Reaction by non-Criegee pathways may also produce products such as epoxides or haloalkanes ⟨72CB3638, 76JA2877⟩. In contrast, ozonolysis of vinyl fluoride ⟨79JOC3181⟩, 1,1-difluoroethylene ⟨79JPC1545, 80JA7572⟩, *cis*- and *trans*-1,2-difluoroethylene ⟨77JA7239⟩ and trifluoroethylene ⟨80JA7572⟩ gave the corresponding fluorinated 1,2,4-trioxolanes (**199**), (**200**), (**201**) (as a mixture of *cis* and *trans* isomers) and (**202**), respectively, among other volatile products (*e.g.* acyl fluorides, carbonyl fluoride and epoxides). The yields decreased in the above order and in the case of perfluoroethylene only traces of tetrafluoro-1,2,4-trioxolane (**203**) were detected ⟨68MI43301⟩. A quantum chemical investigation of the ozonolysis of fluoroalkenes has been published ⟨81JA3633⟩. A fluorinated 1,2,4-trioxolan-3-one (**136**), obtained in a different way, was described in Section 4.33.4.2.5. The direct introduction of halogen atoms such as fluorine, chlorine or bromine into 1,2,4-trioxolanes fails due to the ready cleavage of the ring under these reaction conditions and products such as aldehydes and carboxylic acids are formed ⟨54USP2665280⟩.

The sole representative in the trithiolane series containing halogen substituents in the ring is tetrafluoro-1,2,3-trithiolane (**35**). It is found in 10% yield associated with the major product tetrafluoro-1,2,3,4-tetrathiane (**204**) by reaction of tetrafluoroethylene with the vapors of boiling sulfur at atmospheric pressure ⟨62JOC3995⟩. The compound is rather unstable and rapidly polymerized as described in Section 4.33.3.1.4.

4,4,5-Trifluoro-5-trifluoromethyl-1,3,2-dioxathiolane 2,2-dioxide (**96**) was obtained among other products by reaction of perfluoropropene oxide with sulfur trioxide at 150 °C ⟨73IZV2725⟩. Perfluorobutadiene reacted with sulfur trioxide to give the cyclic sulfate (**31**)

⟨79IZV118⟩ and the formation of (31) is analogous to the reaction (197) → (98) via a four-membered 1,2-oxathietane 2,2-dioxide intermediate. Subsequent products derived from (31) by electrophilic addition reactions at the alkenic double bond have been described in Section 4.33.3.2.2 and the synthesis of 4,5-dichloro-1,3,2-dioxathiolane 2,2-dioxide (154) by chlorination of ethylene sulfate (18) is discussed in Section 4.33.3.5. Cyclic sulfites, on the other hand, cannot be halogenated without ring opening (cf. Section 4.33.3.2.4).

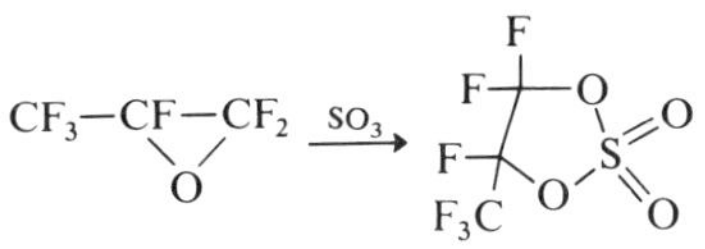

(96)

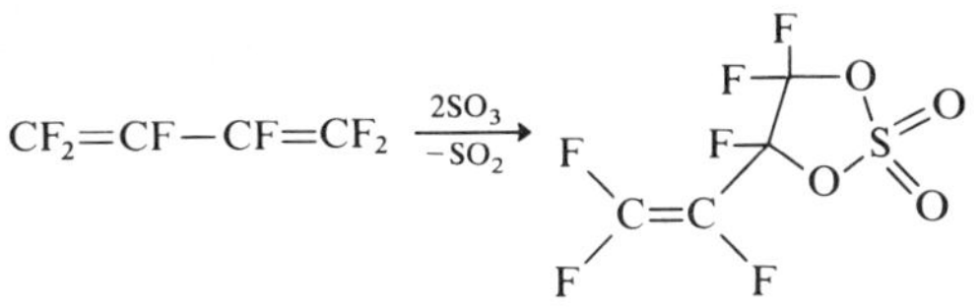

(31)

4.33.5 APPLICATIONS

4.33.5.1 Applications in Research and Industry

The easy formation and cleavage of 1,2,4-trioxolanes in the ozonolysis of alkenes is of both analytical and synthetic significance. No example of a double bond shift during ozonolysis is known and the kinds of ozonolysis product permit conclusions regarding the position of the double bond in the original alkene to be made. This is also true for compounds containing several C—C double bonds. The classical example of such a structural elucidation is that of natural rubber by Harries who obtained, after reductive work up, laevulinic aldehyde as the sole ozonolysis product. Consequently, natural rubber possesses a polyisoprene structure with 1,5-oriented double bonds (cf. Scheme 12). Similarly, the structure of synthetic rubbers and other polymers and copolymers may be investigated and evidence obtained concerning the extent to which a given diene polymerizes by 1,4- or by 1,2-addition, the extent to which head-to-head or head-to-tail polymerization occurs, whether a systematic or completely random distribution of monomers is produced in the polymer during copolymerization, and which of the multiple bonds of a given polyfunctional monomer is involved in the polymerization ⟨58CRV925⟩.

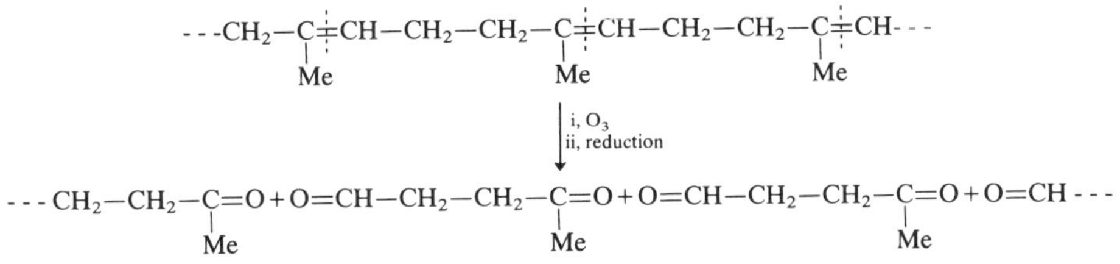

Scheme 12 Structure elucidation of natural rubber by ozonolysis

In laboratory practice, ozonolysis has wide applications as an important synthetic method for the conversion of alkenes into aldehydes or carboxylic acids which are otherwise difficultly accessible ⟨40CRV(27)437, 58CRV925⟩. The ozonolysis of oleic acid followed by oxidative work-up is an industrial preparation of azelaic and pelargonic acids ⟨58CRV925⟩. A more recent patent describes the production of perfluorocarboxylic acids $CF_3(CF_2)_n CO_2H$ (n = 5, 7, 9) by ozonolysis of alkenes of the type $CF_3(CF_2)_n CH=CHR$ (R = H, Me) and subsequent oxidative decomposition of the 1,2,4-trioxolanes formed ⟨77BRP1473807⟩. Ozonolysis also plays an important role in the destruction of unsaturated organic wastes in industrial waters ⟨B-72MI43303, B-78MI43301⟩.

Some derivatives of 1,2,3-trithiolane are of interest as potential starting compounds for sulfur-containing polymers ⟨63USP3088935, 77MI43302⟩ and 1,2,4-trithiolanes have been patented as vulcanization accelerators ⟨60EGP19119⟩, as materials for preparing plastic sulfur compositions ⟨70GEP2004305⟩, and as additives of lubricants for high pressure applications ⟨74GEP2352586⟩.

The possibility of forming 1,3,2-dioxathiolane 2-oxides (cyclic sulfites) from 1,2-diols and thionyl chloride may be used to obtain information on the steric arrangement of the hydroxyl groups ⟨63CRV557⟩. Since the S=O group is a chromophore (cf. Section 4.33.2.2.4), the conversion of 1,2-diols into their cyclic sulfite esters enables a stereochemical investigation of optically active diols by chiroptical methods to be carried out ⟨74JOC2073⟩. Ethylene sulfite has been patented as a hydroxyethylating agent in organic syntheses ⟨48USP2448767⟩, as a spinning solvent for polyacrylonitrile ⟨55USP2706674, 56USP2752318⟩, as starting material for the production of polyesters ⟨57BRP781169, 62BEP610763⟩ or polyurethanes ⟨65BEP659395⟩, and as a component in scrubbing liquids for removing hydrogen sulfide from gaseous mixtures ⟨62BRP902256⟩. Other applications include incorporation in washing liquids for purification of polyalkenes ⟨62BRP903077⟩, in detergents ⟨62GEP1124962⟩, in photographic emulsions ⟨63GEP1159760⟩ and in textile auxiliaries ⟨66GEP1223397⟩. Further patents describe the potential application of ethylene sulfite or derivatives thereof as accelerators for aminoplastic molding compositions ⟨61BRP866440⟩, as vulcanization accelerators ⟨64FRP1379555⟩, as components in hair dying agents ⟨71USP3565571⟩ and as preservatives for black–white developers ⟨73USP3713826, 73GEP2203225⟩. Anthraquinone dyes with a cyclic sulfite ester group may be used for dyeing polyester fibers ⟨72GEP2121525⟩. 1,3,2-Dioxathiolane-2-thiones are reported to increase the lubricating properties of mineral oils ⟨67USP3357993⟩. Finally, 1,3,2-dioxathiolan-4-one 2-oxides (anhydrosulfites of α-hydroxy carboxylic acids) have been patented as monomeric starting materials for the production of homopolymers ⟨57USP2811511⟩ and copolymers ⟨72GEP2162157⟩.

1,3,2-Dioxathiolane 2,2-dioxides (cyclic sulfate esters of 1,2-diols) are used in practice like sultones to introduce acid substituents into nitrogen heterocycles ⟨48BSF1002⟩, especially into cyanine dyes ⟨58GEP1028718⟩. Fluorinated derivatives are useful in the treatment of textiles such as cotton to impart wash and wear characteristics ⟨62USP3055913⟩. The conversion of cyclic sulfates into resinous film-forming polymers has likewise been patented ⟨64USP3154526⟩.

1,2,5-Oxadithiolane 2,2,5,5-tetroxide, *i.e.* 1,2-ethanedisulfonic anhydride, as well as its benzo derivative, may be used to improve the dyeing characteristics of polyesters with basic dyes ⟨71JAP7142631⟩.

4.33.5.2 Natural Occurrence and Special Modes of Formation

Experimental studies have shown that ozone–alkene reactions in the gas phase likewise proceed *via* intermediate formation of 1,2,3-trioxolanes (primary ozonides) whose spontaneous decomposition then as a rule leads to a variety of subsequent products, including 1,2,4-trioxolanes ⟨81JA3807⟩. Because ozone is present in the atmosphere (from 0.02 p.p.m. at sea level up to 0.2 p.p.m. and more in industrial and urbanized areas), reactions of

Table 12 Occurrence of 1,2,4-Trithiolane Derivatives

Substituent at position 3	5	Occurrence	Ref.
H	H[a]	Red alga *Chondria californica*	76JOC2465
Me	Me[b]	Mushroom *Boletus edulis*	73MI43302
		Dry red beans	75MI43301
		Grape leaves	75MI43302
		Potato oil	70MI43300
		Rapeseed protein heated in water	76MI43302
		Roasted filberts	72MI43304
		Pork meat	72TH43300
		Cooked beef	68CI(L)1639
			70MI43301
			72MI43305
		Cooked mutton	79MI43301
		Cooked chicken	76MI43303
Me	Et[b]	Meat extract	B-78MI43303
Me	Pr[i b]	Meat extract	B-78MI43303
Et	Et[b]	Onion *Allium cepa*	79P1397
Ph	Ph[c]	Guinea-hen weed *Petiveria alliacea*	74CC906

[a] Besides 1,2,4-trithiolane, the 1- and 4-oxides were also isolated from *Chondria californica*.
[b] Mixture of *cis* and *trans* isomers.
[c] *cis* Isomer (trithiolaniacin).

this type can proceed wherever alkenes occur or are emitted. For example, the blue haze that forms over some densely vegetated areas is attributed to aerosols formed by the ozone oxidation of the hydrocarbons (mainly terpenes) released by the plants. In industrial and urbanized areas, alkenes arising from automobile exhaust gases and their photochemical reactions are the most important source of unsaturated hydrocarbons. Therefore, ozone–alkene reactions and the products formed play a significant role in environmental chemistry ⟨B-78MI43302⟩ (for biological consequences of these reactions, *cf.* Section 4.33.5.3).

1,2,4-Trithiolanes have been identified in some plant species as well as in the flavor of various food products, especially in those prepared from meat (*cf.* Table 12). These aroma compounds are formed as a rule only during cooking or roasting of meat by reaction of simple degradation products (*e.g.* H_2S and aldehydes) of proteins and carbohydrates. Indeed, a series of model experiments has demonstrated that the thermal decomposition of sulfur-containing amino acids such as cysteine, both in the presence ⟨73MI43300, 73MI43301⟩ and in the absence of carbohydrates ⟨76MI43301⟩ yields, among a variety of other components, typical flavor constituents of heated meat. Thus the most frequently occurring derivative, 3,5-dimethyl-1,2,4-trithiolane (**126**), was shown to result from the oxidation of bis(1-mercaptoethyl) sulfide (**205**) which is a key product formed in the reaction of acetaldehyde with hydrogen sulfide ⟨74MI43303⟩. Reactions of this type play an important role in the manufacture of artificial roasting flavors from cysteine and other amino acids ⟨68USP3365306⟩.

Derivatives of 1,3,2- and 1,4,2-dioxathiolane as well as of 1,2,3- and 1,2,5-oxadithiolane have not been observed in nature.

$$2CH_3CHO + 3H_2S \xrightarrow{-2H_2O} \text{(205)} \xrightarrow[-2[H]]{[O]} \text{(126)}$$

4.33.5.3 Biological Activity

The high toxicity of ozone to man and animals is in part connected with the formation and subsequent reactions of ozonides in the organism. In particular, fatty acid ozonides are suggested to be intermediates in ozone intoxications. Thus methyl oleate ozonide causes *in vitro* the formation of methaemoglobin and loss of cellular thiols. It furthermore catalyzes the formation of disulfide-linked interchain polymers between haemoglobin and ovalbumin and induces, like other fatty acid ozonides, pathological alterations (Heinz bodies) in erythrocytes ⟨75MI43303⟩. This finding may explain the protective effects of antioxidants and thiol-generating systems against ozone-produced toxicants ⟨75MI43304⟩. As possible constituents in air pollution (*cf.* Section 4.33.5.2), ozonides and their decomposition products cannot only injure the health of man and animals but also damage plants ⟨52MI43300, 59MI43300⟩. The powerful germicidal effect of ozonides at very low concentrations may be used for their detection in the air ⟨68MI43300⟩. Ozonized olive oil was found to possess wound-healing ⟨47MI43300⟩ as well as bactericidal and fungicidal properties ⟨47MI43301⟩. Disinfectants containing ozonides have been likewise described ⟨55MI43300, 64GEP1166975⟩.

In the 1,2,4-trithiolane series, relatively little information on biological activity, including toxic effects, is available. The average concentrations of the flavoring substance 3,5-dimethyl-1,2,4-trithiolane (**126**) in food products (maximum up to 0.3 p.p.m.) are generally recognized as safe ⟨78MI43304⟩. 3,5-Bis[bis(isopropoxycarbonyl)methylene]-1,2,4-trithiolane (**57**), which forms as a photochemical degradation product of the fungicide isoprothiolane (diisopropyl 1,3-dithiolan-2-ylidenemalonate), was found to be inactive in the reverse mutation systems of Ames' *Salmonella* strains ⟨79MI43301⟩. Some 1,2,4-trithiolane derivatives have been patented as fungicides ⟨65GEP1192452⟩ and antimalarial agents ⟨71USP3578682, 72USP3694498⟩.

Five-membered cyclic sulfites and sulfates are of toxicological relevance due to their potential bioalkylating properties. Ethylene sulfate (**18**) is more toxic than dimethyl sulfate ⟨75JPR943⟩; like the latter and other alkylating agents, it induces local malignant tumors after subcutaneous injection ⟨74MI43304⟩ and proved to be a weak mutagen both *in vitro* and *in vivo* ⟨77MI43303⟩.

4.34

Dioxazoles, Oxathiazoles and Dithiazoles

M. P. SAMMES
University of Hong Kong

4.34.1 INTRODUCTION	898
4.34.2 STRUCTURE OF THE RING SYSTEMS	898
4.34.2.1 Survey of Possible Structures	898
4.34.2.1.1 Aromatic systems	899
4.34.2.1.2 Non-aromatic systems	899
4.34.2.2 Theoretical Methods	900
4.34.2.3 Structural Methods	900
4.34.2.3.1 X-Ray diffraction	900
4.34.2.3.2 Electron diffraction and microwave spectra	903
4.34.2.3.3 ^1H NMR spectroscopy	904
4.34.2.3.4 ^{13}C NMR spectroscopy	904
4.34.2.3.5 ^{15}N NMR spectroscopy	904
4.34.2.3.6 UV spectroscopy	907
4.34.2.3.7 IR spectroscopy	907
4.34.2.3.8 Mass spectrometry	909
4.34.2.4 Thermodynamic Aspects	909
4.34.2.4.1 Stability and stabilization	909
4.34.2.4.2 Conformation	911
4.34.2.5 Tautomerism	912
4.34.2.6 Other Unusual Structures	912
4.34.3 REACTIVITY OF THE RING SYSTEMS	913
4.34.3.1 General Survey	913
4.34.3.2 Reactions of Aromatic Rings	913
4.34.3.2.1 Thermal and photochemical reactions formally involving no other species	913
4.34.3.2.2 Electrophilic attack at ring nitrogen and carbon	914
4.34.3.2.3 Attack at ring sulfur	915
4.34.3.2.4 Nucleophilic attack at carbon	916
4.34.3.2.5 Reactions involving radicals and at surfaces	918
4.34.3.2.6 Reactions with cyclic transition states	919
4.34.3.3 Reactions of substituents on Aromatic Rings	921
4.34.3.3.1 Fused benzene rings	921
4.34.3.3.2 Alkyl groups	921
4.34.3.3.3 Aryl groups	921
4.34.3.3.4 N-Linked substituents	922
4.34.3.3.5 O-Linked substituents	923
4.34.3.3.6 S-Linked substituents	923
4.34.3.3.7 Halogen atoms	924
4.34.3.3.8 Substituents attached at ring nitrogen	924
4.34.3.4 Reactions at Non-aromatic Rings	924
4.34.3.4.1 Thermal and photochemical reactions formally involving no other species	924
4.34.3.4.2 Electrophilic attack at ring nitrogen and carbon	926
4.34.3.4.3 Attack at ring sulfur	927
4.34.3.4.4 Nucleophilic attack at ring carbon	928
4.34.3.4.5 Attack at ring hydrogen	928
4.34.3.4.6 Reactions at surfaces	929
4.34.3.4.7 Reactions with cyclic transition states	929
4.34.3.5 Reactions of Substituents on Non-aromatic Rings	930
4.34.3.5.1 C-Linked substituents	930
4.34.3.5.2 S-Linked substituents	930
4.34.3.5.3 Halogen atoms	930
4.34.3.5.4 Substituents attached at ring nitrogen	930

4.34.4 SYNTHESES	930
4.34.4.1 Ring Synthesis from Non-heterocyclic Compounds	931
4.34.4.1.1 Formation of one bond between two heteroatoms	931
4.34.4.1.2 Formation of one bond adjacent to a heteroatom	931
4.34.4.1.3 Formation of two bonds: four-atom fragments and carbon	932
4.34.4.1.4 Formation of two bonds: four-atom fragments and nitrogen	932
4.34.4.1.5 Formation of two bonds: four-atom fragments and sulfur	933
4.34.4.1.6 Formation of two bonds: [3+2] atom fragments by cycloaddition	934
4.34.4.1.7 Formation of two bonds: [3+2] atom fragments by other processes	935
4.34.4.1.8 Formation of three bonds	937
4.34.4.1.9 Formation of four bonds	037
4.34.4.2 Ring Synthesis by Transformation of Other Heterocycles	938
4.34.4.2.1 From rings containing less common heteroatoms	938
4.34.4.2.2 From five-membered ring heterocycles with one heteroatom	938
4.34.4.2.3 From five-membered ring heterocycles with two heteroatoms	938
4.34.4.2.4 From three-membered ring heterocycles	939
4:34.4.3 Summary of Best Preparative Methods for the Ring Syntheses	939
4.34.5 APPLICATIONS OF THE RING SYSTEMS	939
4.34.5.1 Agrochemicals	939
4.34.5.1.1 Fungicides	939
4.34.5.1.2 Herbicides and plant growth regulators	944
4.34.5.1.3 Insecticides and insect chemosterilants	944
4.34.5.2 Pharmaceuticals	944
4.34.5.2.1 Antibiotics and antibacterials	944
4.34.5.2.2 Antifertility compounds	944
4.34.5.2.3 CNS activity	945
4.34.5.3 Polymer Chemicals	945
4.34.5.3.1 Masked isocyanates	945
4.34.5.3.2 Others	945
4.34.5.4 Other Applications	946

4.34.1 INTRODUCTION

Although the ring systems described in this chapter have been the subject of more than 700 papers extending over 160 years, they have received little or no attention in previous works on heterocyclic chemistry. The 1,2,4-dithiazole (**1**; $R^1 = R^2 = H$; $X = S$), described in the early literature both as 'xanthane hydride' and as 'isoperthiocyanic acid', must be one of the first heterocycles to have been synthesized; it was reported by Wöhler ⟨1821MI43400⟩ seven years before his famous preparation of urea. It has been the subject of a Russian language review ⟨72MI43400⟩ and the only othe relevant reviews are brief accounts of the related 1,2,4-dithiazoles (**2**; $X = S$) and 1,3,4-dithiazoles (**3**; $X = S$) ⟨70MI43400⟩, and the 1,2,3-benzodithiazolylium (Herz) salts (**4**) ⟨57CRV1011⟩.

Most of the early studies were on 1,2,4-dithiazole derivatives over which there was considerable controversy concerning structures, due in part to facile Dimroth rearrangements with change in pH. Most uncertainties were dispelled by chemical, IR and X-ray studies during the period 1958–1963 ⟨58JA414, 63JCS3165, 61ACS1186⟩; consequently, references in this chapter will be mainly to work post-1960, earlier citations being included where appropriate.

4.34.2 STRUCTURE OF THE RING SYSTEMS

4.34.2.1 Survey of Possible Structures

In the dioxazoles, oxathiazoles and dithiazoles the three heteroatoms may be arranged respectively in four, six and four ways. Examples of all types are known except for the 1,2,3-dioxazoles. Of the seven possible benzo fused systems, only derivatives of 1,2,3-

benzoxathiazole and of 1,2,3- and 1,3,2-benzodithiazoles have been reported, there being no known examples with bridgehead nitrogen.

4.34.2.1.1 Aromatic systems

Aromatic structures can be written for fully unsaturated monocations and for unsaturated rings with exocyclic conjugation; the latter group comprises azolinones and related compounds, and mesoionic compounds (see Section 4.01.1). General structures (**5**)–(**11**) are summarized in Scheme 1. In practice, the majority of systems described fall in the azolinone category, 1,2,3-dithiazolylium, 1,2,4-dithiazolylium and 1,2,3-benzodithiazolylium being the only reported monocationic species, and the 4-aryl-1,3,2-oxathiazolylium-5-olates (**12**) the only class of mesoion. It is possible, however, to draw potentially aromatic structures which are neutral by incorporating S(IV) or S(VI) into the ring. Benzo fused examples having sulfur in either (**13** ⟨77KGS849⟩ and **14** ⟨81TL1175⟩) or both (**15** ⟨63CB1177⟩) oxidation states are known. A number of systems have also been described in which two similar or different rings are fused together to give aza and oxaza analogues of the trithiapentalenes. These are discussed in Chapter 4.38.

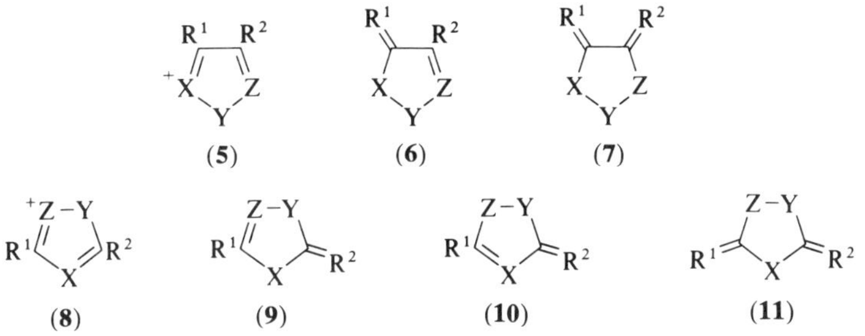

(**5**) (**6**) (**7**) (**8**) (**9**) (**10**) (**11**)

Scheme 1 General structures for azolylium ions, azolinones and azolidinediones

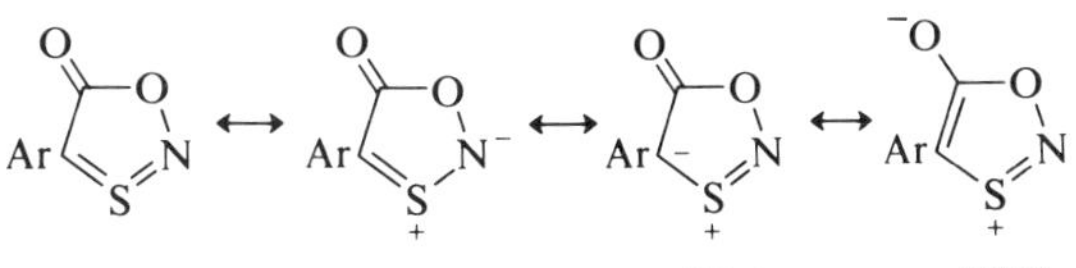

(**12a**) (**12b**) (**12c**) (**12d**)

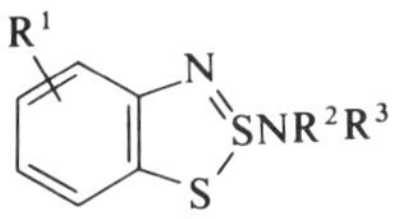

(**13**)

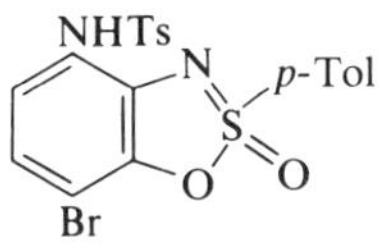

(**14**)

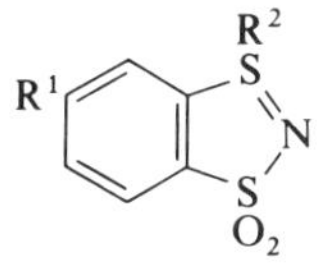

(**15**)

4.34.2.1.2 Non-aromatic systems

Possible dihydro and tetrahydro derivatives for these ring systems are illustrated in Section 4.01.1.3 (Schemes 8 and 9), general structures (**16**)–(**23**) relevant to this discussion being displayed in Scheme 2. Most of these are known though the fully saturated systems are

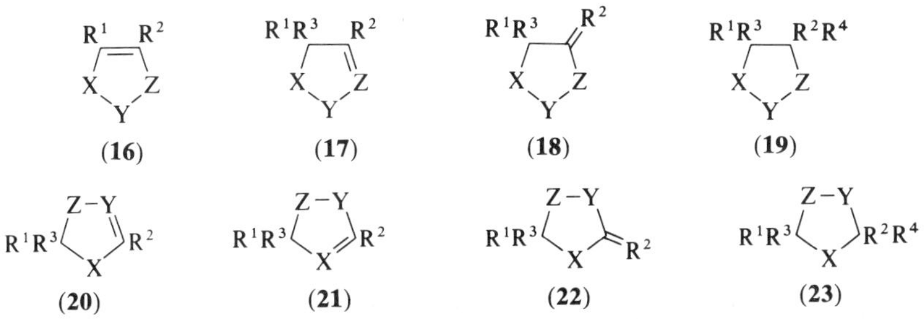

(**16**) (**17**) (**18**) (**19**) (**20**) (**21**) (**22**) (**23**)

Scheme 2 General structures for azolines, azolidinones and azolidines less common. Further structures can be generated by incorporating S(IV) or S(VI) into the ring. Again there are many examples, including structures of type (24) (see *e.g.* ⟨70TL127⟩) in which sulfur is multiply bonded within the ring. Benzo fused systems are all of the 1,2,3-benzazoline type (25).

4.34.2.2 Theoretical Methods

The little work that has been done on theoretical treatments of the structures in this section has been confined to the dithiazoles.

Charge densities and π-bond orders calculated by the SCF–MO method for 1,2,3-benzodithiazolylium (4; R = H) are shown in Figure 1 ⟨78CHE733⟩. The charge densities predict correctly that S(2) and C(6) are the sites for nucleophilic attack on the heterocyclic and the carbocyclic rings, respectively. Protonation does not occur on nitrogen, however, in spite of the indicated slight negative charge. The highest bond orders are predicted for C(4)—C(5) and C(6)—C(7), and this is supported by $^3J_{HH}$ values in the ^{1}H NMR spectrum (see Section 4.34.2.3.3).

Figure 1 The π-electron densities and bond orders in the 1,2,3-benzodithiazolylium cation

For the 3,5-diimino-1,2,4-dithiazole derivatives (26) CNDO/2 calculations have predicted essentially the same molecular geometries as found by X-ray diffraction. Contributions from *d*-orbitals were excluded from the calculations, which predicted a lengthening of the S—S bond relative to normal cyclic values ⟨76ACS(A)397⟩.

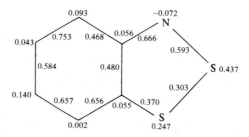

Recent MINDO calculations show that the aminodithiazolethione (30) is thermodynamically more stable than its two tautomers, though only by a small amount. This suggests that the other forms might be involved in its reactions. The calculations also suggest a mechanism for formation of compound (30) from HNCS, different from that of Scheme 48, and proceeding *via* a 1,3-thiazetidine. Experimental results appear to support this mechanism ⟨82JCR(S)65⟩.

4.34.2.3 Structural Methods

4.34.2.3.1 X-ray diffraction

No data appear to have been published for dioxazoles. Oxathiazoles and dithiazoles for which structures have been reported are listed in Table 1.

Table 1 X-Ray Data for Oxathiazole and Dithiazole

Ring	Ring position 1	2	3	4	5	Compounds studied (reference)
C$_2$NOS	O	S	N	—	—	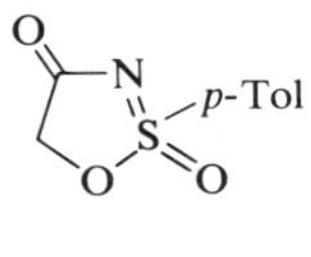 ⟨71TL4243⟩ 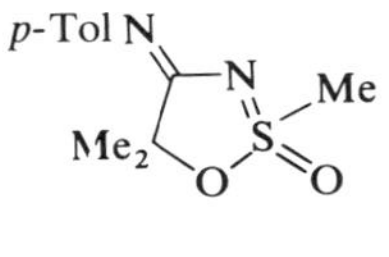 ⟨78AG(E)352⟩ 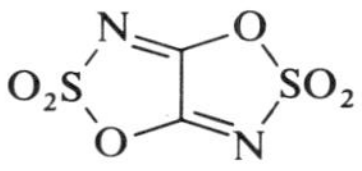 ⟨80MI43400⟩ 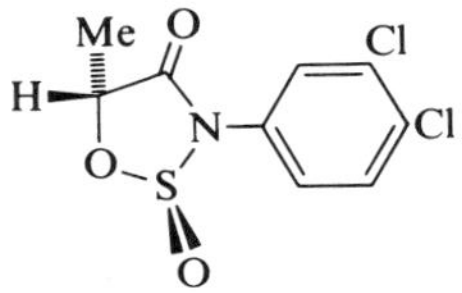 ⟨75JHC393⟩ 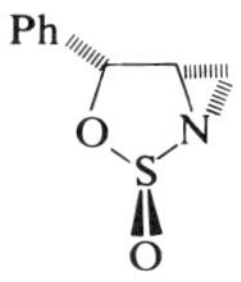 ⟨81CSC1099⟩
C$_2$NOS	O	S	—	N	—	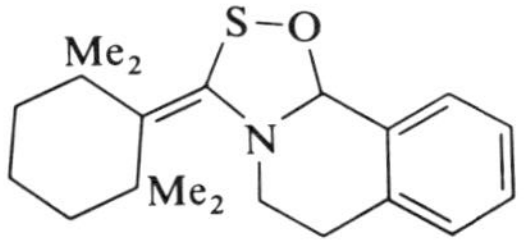 ⟨78AG(E)455⟩
C$_2$NOS	O	N	S	—	—	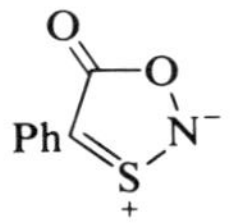 ⟨72G23⟩
C$_2$NOS	O	—	S	N	—	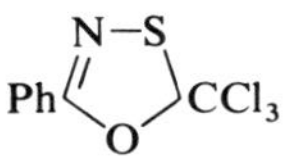 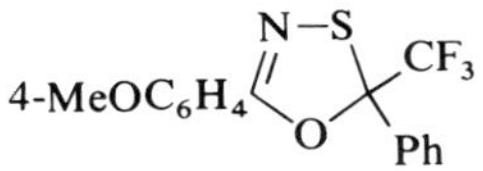 ⟨81JCS(P1)2991⟩
C$_2$NS$_2$	S	S	N	—	—	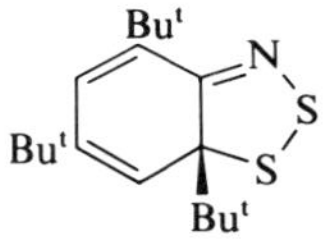 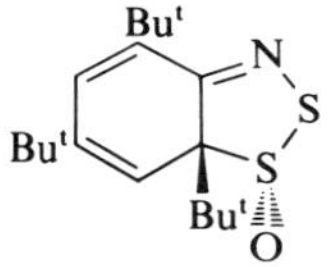 ⟨80AX(B)1466⟩

Table 1 (Continued)

Ring	Ring position 1 2 3 4 5	Compounds studied (reference)
C$_2$NS$_2$	S S — N —	(structures and references as shown)

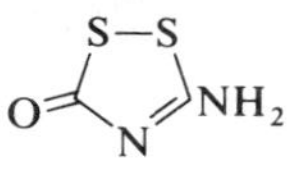

Compounds listed:

- 2-amino-1,3-dithiol-type with C=O: ⟨66ACS754⟩
- 5-amino-1,3-dithiole-2-thione: ⟨63ACS2575, 63AX1157⟩
- N,N-diphenyl-2-amino-1,3-dithiole-2-thione with Ph, NH$_2$: ⟨78AX(B)2570⟩
- Guanidinium dithiolylium salt $H_2N-C(=N)-NH_2$ with S–S ring, X$^-$:
  - X = Cl ⟨66ACS1907⟩
  - X = Br ⟨65ACS1539⟩
  - X = I ⟨58ACS1799, 71AX(B)1687⟩
  - CuCl$_2$ ⟨75CC550⟩
- 2-dimethylamino-3-phenyl-1,3-dithiol derivative: ⟨77PS185⟩
- 2-phenylimino-3-benzyl-1,3-dithiol: ⟨79CSC319⟩
- Bicyclic bis-dithiolo system with R^1, R^2, R^3:

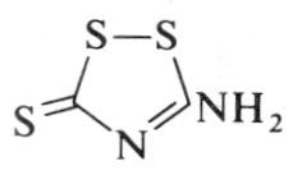

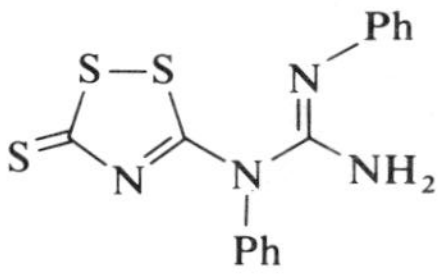

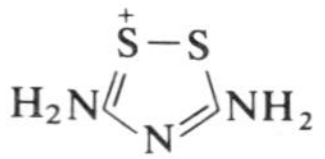

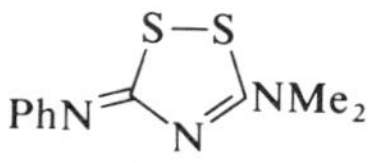

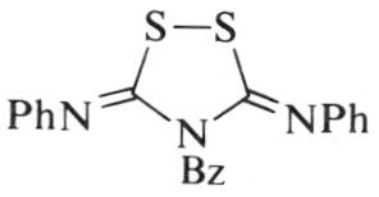

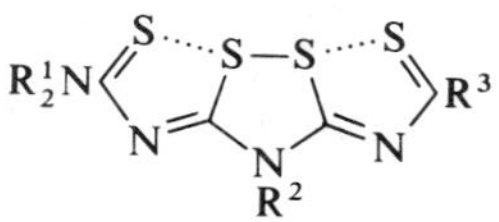

R^1	R^2	R^3	
Me	Me	NMe$_2$	⟨73JA6073⟩
Me	Ph	NMe$_2$	⟨73JA6073⟩
Pr^i	Me	NPr$^i_2$	⟨74ACS(A)989⟩
Pr^i	Me	Ph	⟨77ACS(A)271⟩
Pr^i	Pr^i	NPr$^i_2$	⟨77ACS(A)423⟩

Of particular interest are data for the mesoionic oxathiazole (**12**; Ar = Ph), the exocyclic C—O (1.194 Å) and ring C—S (1.653 Å) bonds being almost pure double bonds, while the ring C—O (1.404 Å) and N—O (1.375 Å) bonds have values similar to isoxazole, and the ring C—C (1.444 Å) bond is unusually long ⟨72G23⟩. This suggests that of the forms (**12a–d**), the first two and the last make, respectively, very large and very small contributions.

In the 1,2,3-dithiazoles (**27**) and (**28**) the ring-function Bu^t group is axial, with the S=O bond in (**28**) *trans* to it ⟨80AX(B)1466⟩. Structures of several 1,2,4-dithiazoles have been reported. The compounds known historically as 'rhodan hydrate' (**29**) ⟨66ACS754⟩, 'xanthane hydride' (**30**) ⟨63ACS2575⟩ and 'thiuret hydroiodide' (**31**; X = I) ⟨58ACS1799⟩ were all shown to have exocyclic amino rather than imino groups, thus removing a long-standing uncertainty. All S—S bond lengths lie between 2.06–2.08 Å, which is significantly longer than the non-cyclic distance (2.044 Å). All C—N bond distances show evidence of conjugation, having values between 1.30–1.35 Å, though the exocyclic C=X bonds in structures (**29**) and (**30**) are almost pure double bonds. In the 3,5-diamino-1,2,4-dithiazolylium salts (**31**) conjugation, and thus delocalization of positive charge, is most pronounced in the C—N part of the molecule, C—S bond distances (1.73–1.77 Å) being substantially longer than in thiophene. The two sulfur atoms are approximately colinear with two halide ions X$^-$, to which they make close approaches ⟨71AX(B)1687⟩. In structures of type (**26**) the four sulfur atoms are nearly colinear, and the skeletal atoms coplanar ⟨73JA6073⟩. The central S—S bond distance (2.16–2.18 Å) is longer than in compounds (**29**)–(**31**), while the outer S—S bond distances (2.62–2.82 Å) are substantially less than the van der Waals contact distance (3.40 Å), indicating a strong interaction. This was confirmed by CNDO/2 calculations (see Section 4.34.2.2), though the S—S interaction in the diazatrithiapentalenes (see Chapter 4.38 for a discussion of this topic) is much stronger. The molecule (**32**) also has a coplanar skeleton and a somewhat long S—S bond length (2.121 Å), though the N · · · S bond distance (2.324 Å) was thought to be too long to indicate any significant interaction ⟨78AX(B)2570⟩.

(**27**) X = :
(**28**) X = O

(**29**) X = O
(**30**) X = S

(**31**)

(**32**)

4.34.2.3.2 Electron diffraction and microwave spectra

The bond distances and angles for 1,3,4-oxathiazolin-2-one (**33**) have been determined from electron diffraction measurements and refined using rotational parameters from microwave spectra ⟨78ACS(A)1005⟩. The molecule is planar, indicating some π-delocalization, though the bond lengths (Table 2; compare Section 4.01.3.2, Table 2) suggest that this is small and that the C=O π-bond does not participate. The C—H and C=O bonds are tilted towards the ring oxygen atom, relative to the ring angle bisectors, by about the same amount as in 1,3,4-oxadiazole.

(**33**)

Table 2 Structure Parameters and Rotational Constants for 1,3,4-Oxathiazolin-2-one (**33**)

Bond lengths (Å)		Bond angles (°)		Rotational parameters (MHz)	
C(2)—O	1.402	C—O—C	110.8	A_0	5587.411
C=O	1.192	O—C=O	122.6	B_0	3645.8735
C—S	1.767	O—C—S	106.3	C_0	2205.4635
S—N	1.690	C—S—N	93.8	Δ^a	0.08236
C=N	1.286	S—N—C	107.9		
C(5)—O	1.356	N—C—O	121.1		
C—H	1.102	H—C—O	114.5		

[a] Inertial defect (μÅ^2).

4.34.2.3.3 ^1H NMR spectroscopy

Representative chemical shifts (δ, relative to internal TMS) for ring protons and certain substituents in structure types (8), (10) and (17)–(24) are given in Table 3. A comparison of ring proton chemical shifts in the cations (34; δ 10.96) and (35; δ 5.90) demonstrates the presence of a ring current in the former ⟨74AP828⟩ which may be compared with the isoelectronic dithiolylium cations (see also Chapters 4.31 and 4.32); the free base (36; Ar = Ph) absorbs at δ 5.70. 'Ethylenethiuram monosulfide', the fungicidal oxidation product of the fungicide 'Nabam', was shown ⟨71TL1317⟩ to have the structure (37) and not the 1,3,6-thiadiazepane-2,7-dithione structure as earlier suggested ⟨60CJC2349⟩, its spectrum showing two coupled triplets at δ 3.93 and 4.52. A comparison between ring δ values for compounds (38; R^1 = Et, R^2 = H; 9.31 p.p.m.) and (39; 10.14 p.p.m.) suggests that there is very little contribution from the oxadiazadithiapentalene structure (38c) to the former ⟨76CB3108⟩. The stereochemistry of fused oxathiazolidines, e.g. (40), and hence that of the aziridine alcohols from which they were derived, has been established by irradiation at the ring methyl frequency and observing the NOE on the proximate aziridine ring substituent ⟨80JHC1009⟩.

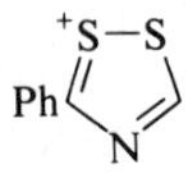

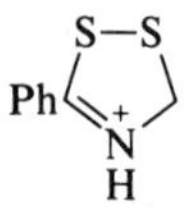

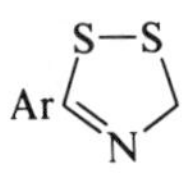

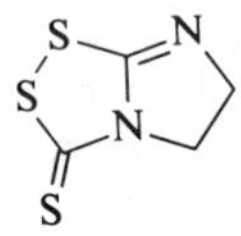

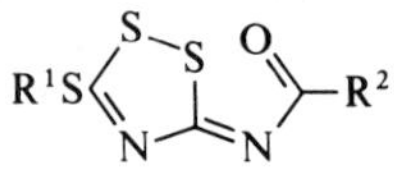

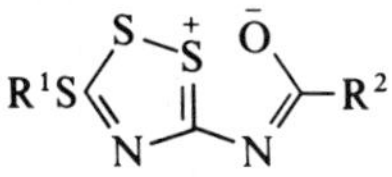

Chemical shifts and coupling constants for the 1,2,3-benzodithiazolylium cation and its selena and diselena analogues (Table 4) indicate some degree of bond fixation and localization of positive charge at C(4) and C(6) relative to C(5), consistent with SCF/MO calculations; the effect increases with the number of selenium atoms.

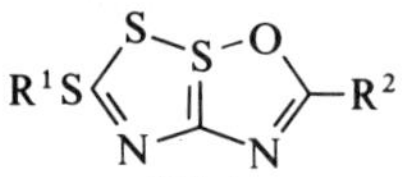

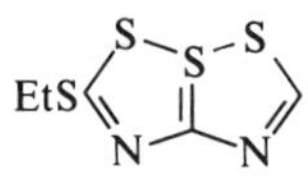

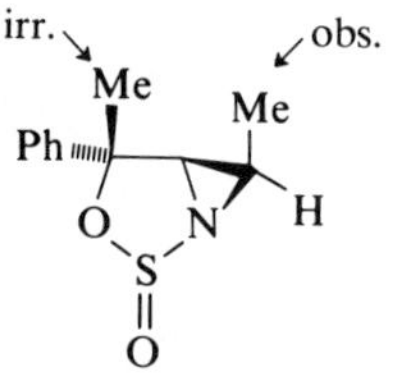

4.34.2.3.4 ^{13}C NMR spectroscopy

Chemical shifts (δ, relative to internal TMS) for ring carbon atoms and certain substituents for a number of dioxazoles, oxathiazoles and dithiazoles are given in Table 5. Different isomers having the general structure (41) are readily distinguished by their C(3) and C(5) chemical shifts ⟨75JOC1728⟩. The 1,2,4-dithiazole (42; R^1 = PhNH, R^2 = Ph) showed only one signal at δ 168.5 for ring carbons, midway between the expected values of 153.5 and 176.3. This demonstrates a rapid tautomeric shift of hydrogen between the exocyclic nitrogen atoms ⟨80JCR(S)266⟩.

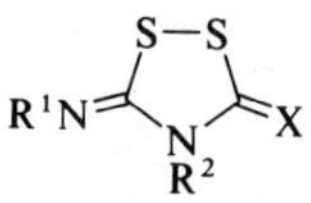

4.34.2.3.5 ^{15}N NMR spectroscopy

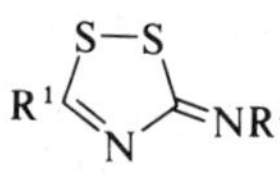

Data have been reported for only two related 1,2,4-dithiazoles (32) ⟨80JCR(S)114⟩ and (30) ⟨78H(11)121⟩. The signal for the ring nitrogen in the former was 285.8 p.p.m. downfield

Table 3 ^{1}H NMR Data for Dioxazole, Oxathiazole and Dithiazole Derivatives

Structure type	Ring atoms X	Y	Z	Substituents R^1	R^2	R^3	R^4	Chemical shifts (δ, p.p.m.) Ring H^a 2	3	4	5	Others		Solvent	Ref.
(8)	N	S	S	Ph	H	—	—	—	—	—	10.96	—	—	TFA	74AP828
(18)	O	S(O)	NPh	H	O	H	—	—	—	—	5.20^b 4.80	—	—	CDCl$_3$	74JHC1
(18)	NPh	S(O)	O	Me	O	H	—	—	—	4.18^c	—	4-Me	1.43^c	(CD$_3$)$_2$SO	77JOM(137)C37
(19)	O	NR	O	H	H	H	H	—	—	3.97^d	3.97^d	—	—	CCl$_4$	80IZV2181
(19)	O	S(O)	NPh	H	H	H	H	—	—	3.88 3.71	4.87^e 4.52	—	—	CDCl$_3$	75ACS(A)414
(19)	NMe	O	S(O)	Ph	Ph	H	PhS	—	—	4.60	—	NMe	2.42	CDCl$_3$	77G283
(19)	S	NMe	S	H	H	H	H	—	—	3.58	3.58	NMe	2.87	CDCl$_3$	69JHC627
(20)	O	N	O	ArCH$_2$	Me	H	—	6.20^f	—	—	—	5-Me	1.97	CDCl$_3$	73CJC1368
(20)	O	N	S	CCl$_3$	Me	H	—	6.30	—	—	—	5-Me	2.12	CDCl$_3$	81JCS(P1)2991
(20)	O	N	SO$_2$	H	R^{1}R^{2}CN	H	—	4.42	—	—	—	—	—	CDCl$_3$	81JHC1309
(21)	S	N	O	MeO$_2$C	Ph	H	—	—	—	—	6.39	—	—	CDCl$_3$	79JHC129
(21)	N	S	S	H	Ph	H	—	—	—	—	5.70	—	—	CCl$_4$	74AP828
(21)	$\overset{+}{\text{NH}}$	S	S	Ph	Ph	H	—	—	—	—	5.90	NH	12.95	(CD$_3$)$_2$SO	74AP828
(22)	S	O	NH	H	(CF$_3$)$_2$C	H	—	—	—	6.05^g	—	NH	6.53^g	CDCl$_3$	70JOC3470
(22)	S	S	NH	H	PhN	H	—	4.75	—	—	—	NH	11.10	(CD$_3$)$_2$SO	77CCC2672
(23)	NH	O	O	Pr^n	Pr^n	H	H	—	—	—	4.63	NH	3.18	CDCl$_3$	69JCS(C)2678
(23)	NH	O	O	Bu^t	Bu^t	H	H	4.96	—	—	4.58	—	—	CDCl$_3$	80JCR(S)122
(23)	NCH$_2$R	S	S	Ph	H	CH$_2$R	H	—	—	—	4.72	—	—	CCl$_4$	74AP828
(24)	O	S(O)Ar	N	H	O	H	Ar	—	—	—	4.80^h 5.00	—	—	CDCl$_3$	70TL127

^{a} Numbering by IUPAC system. ^{b} Diastereotopic, $J = 14$ Hz. ^{c} $J = 7$ Hz. ^{d} AA'BB' system at 100 MHz.
^{e} ABXY system; fully analyzed by LAOCOON 3. ^{f} Triplet, $J = 5$ Hz. ^{g} $J = 12$ Hz. ^{h} AB quartet; no value for J.

Table 4 ^{1}H Chemical Shifts and Coupling Constants for 1,2,3-Benzodithiazolylium Chloride and its Selena and Diselena Analogues ⟨78CHE733⟩

Compound	Chemical shifts^a (δ, p.p.m.)				Coupling constants (Hz)						Ratio
	H-4	H-5	H-6	H-7	$J_{4,5}$	$J_{5,6}$	$J_{6,7}$	$J_{4,6}$	$J_{5,7}$		$J_{5,6}:J_{4,5}$
1,2,3-Benzodithiazolylium	8.98	8.34	8.54	8.93	8.8	7.0	8.2	1.2	1.2		0.80
1,2,3-Benzothiaselenazolylium	8.78	8.09	8.43	8.67	9.0	6.8	8.7	1.2	1.2		0.76
2,1,3-Benzothiaselenazolylium	9.16	8.18	8.42	9.00	8.8	7.0	8.8	1.2	1.4		0.80
1,2,3-Benzodiselenazolylium	9.17	8.06	8.52	8.84	9.2	6.9	8.6	1.2	1.2		0.75

^{a} In TFA.

Table 5 ^{13}C NMR Data for Dioxazole, Oxathiazole and Dithiazole Derivatives

Structure type	Ring atoms X	Ring atoms Y	Ring atoms Z	Substituents R^1	Substituents R^2	Substituents R^3	Substituents R^4	Chemical shifts (δ, p.p.m.) Ring C[a] 2	3	4	5	Others		Solvent	Ref.
(9)	O	S	N	Me	O	—	—	174.2	—	—	158.7	5-Me	16.4	$CDCl_3$	81JCS(P1)2991
(10)	N	S	S	PhNH	S	—	—	—	179.1	—	209.3	—	—	$(CD_3)_2SO$	80JCR(S)266
(11)	NBz	S	S	O	NBz	—	—	—	169.2	—	148.5	—	—	$CDCl_3$	75JOC1728
(11)	NBz	S	S	S	NBz	—	—	—	193.9	—	154.5	—	—	$CDCl_3$	75JOC1728
(11)	NBz	S	S	TsN	NBz	—	—	—	166.1	—	150.5	—	—	$CDCl_3$	75JOC1728
(11)	NBz	S	S	PhN	NPh	—	—	—	153.6	—	153.6	—	—	$CDCl_3$	75JOC1728
(11)	NMe	S	S	PhCON	NPh	—	—	—	173.4	—	155.3	NMe	36.9	$CDCl_3$	78JOC4951
(18)	O	S(O)	NAr	Me	O	H	—	—	—	170.5	76.4	5-Me	17.0	$CDCl_3$	77JOC1015
(19)	O	NR	O	H	H	H	H	—	—	67.6	67.6	—	—	CCl_4	80IZV2181
(19)	O	S(O)	NPh[b]	H	H	H	H	—	—	45.9	70.9	—	—	$CDCl_3$	78BCJ323
(20)	O	N	O	Ph	Ph	Me_2CNH_2	—	119.5	—	—	159.0	—	—	$(CD_3)_2SO$	81KGS124
(20)	O	N	S	CCl_3	Me	H	—	95.9	—	—	157.6	5-Me	15.0	$CDCl_3$	81JCS(P1)2291
(20)	S	N	SO_2	H	R^1R^2CN	H	—	51.6	—	—	161.8	—	—	$CDCl_3$	81JHC1309
(20)	N	S	S	F	Cl	F	—	—	156.19[c]	—	146.36[d]	—	—	$CDCl_3$	71CB2732
(22)	NBz	S	S	Ar	NTs	Ar	—	—	167.9	—	87.0	—	—	$CDCl_3$	76JHC883
(22)	S	NMe	SO_2	H	NPh	H	—	—	145.8	—	46.2	NMe	29.0	$CDCl_3$	78JOC4951
(23)	O	O	NCF_3	CF_3	CF_3	CF_3	CF_3	100.89[e]	—	—	91.26[f]	NCF_3	119.23[g]	—	77AG(E)646
(24)	O	S(O)Ph	N	Me	Bu^tN	Me	Ph	—	—	158.6	96.4	—	—	$CDCl_3$	79TL49

[a] Numbering by IUPAC system. [b] Phenyl substituent parameters for 1,2,3-oxathiazolidine *S*-oxide ring: C-1 +12.2, *ortho* −12.0, *meta* +1.3, *para* −4.8. [c] $^3J_{CF} = 15.9$ Hz. [d] $^1J_{CF} = 275.7$ Hz. [e] $^2J_{CF} = 37.5$ Hz. [f] $^2J_{CF} = 36.1$ Hz. [g] $^1J_{CF} = 270.0$ Hz.

from liquid NH_3 at 25 °C, and while that for the latter was not observed, the substituent nitrogen was shown to be in the amino rather than the imino form from its low field chemical shift (111.3 p.p.m.).

4.34.2.3.6 UV spectroscopy

Values of λ_{max} (nm) and log ε for selected compounds are shown in Table 6. Though only limited data have been published, some conclusions can be drawn. The long wavelength absorption (278 nm) of the cation (31) is substantially shorter than that for non-conjugated cyclic disulfides (330 nm), consistent with participation of the sulfur non-bonding p-electrons in the conjugated system ⟨70JCS(A)1386⟩. All bands in the spectra of the cations (43) are strongly influenced by the nature of the aryl *para*-substituent; a charge transfer band (530–630 nm) is also observed in the presence of $Bu_4N^+I^-$, and undergoes a hypsochromic shift with increasingly electron-donating aryl substituents ⟨77LA1005⟩. The similarity between the spectra of the dithiazoles (1; $R^1 = R^2 =$ Me, X = S) and (30) further confirms that the latter is in the amino rather than the imino form in solution; the spectrum of the ethoxycarbonyl derivative (1; $R^1 = CO_2Et$, $R^2 = H$, $R^3 = S$) is pH dependent, 50% dissociation occurring at pH 7 ⟨73MI43400⟩. The spectrum of the *trans*-dithiazole derivative (44) is very different from that of the *cis*-diazatrithiapentalene form (45), being rather like that of the imine (42; $R^1 =$ PhNH, $R^2 = H$) ⟨64CB2567⟩. A useful comparison of the spectra of the dithiazoles (2 and 3; Ar = Ph, X = O or S) with those for their 1,2- and 1,3-dithiole counterparts, shows that replacement of CH by N brings about only small changes ⟨67TL1013⟩. The mesoions (12) show absorptions near 260 and 410 nm (log ε *ca.* 4.0), both bands being sensitive to the nature of the aryl substituent and the second showing correlation with the empirical polarity parameter E_T for the solvent ⟨72CB188⟩.

(43) (44) (45)

Three regions of absorption are found in the 1,2,3-benzodithiazolylium cations (4), 238–278 nm, 318–375 nm and 422–470 nm, the wavelength of the latter giving a good correlation with the σ^+ substituent constant for the R^1 group ⟨69ZOR153⟩.

4.34.2.3.7 IR spectroscopy

This technique is more useful for identifying functional groups attached to rings than for characterizing the ring systems themselves. However, the dithiazolylium salts (43) all have a characteristic strong absorption near 1400 cm^{-1} ⟨79NKK389⟩, and in the spectrum of compounds (15) ν(S=N) has been assigned to an absorption at 935–980 cm^{-1} ⟨63CB1177⟩. The spectrum of the cation (31) has been fully assigned ⟨70JCS(A)1386⟩ by making use of deuterium exchange, Raman spectra and calculated frequencies based on C_{2v} symmetry and published bond parameters. Similar calculations on the aminothione (30) assuming C_s symmetry also permitted full assignments to be made, but strangely the model used was the imino tautomer ⟨70MI43401⟩.

Data for ν(C=O), ν(C=S) and ν(C=N) for rings containing one or more of these groups are given in Table 7, the effect of substituents on the absorptions being emphasized where possible. Structures of type (9) all show a doublet for ν(C=O) or ν(C=S), the lower frequency band for the oxathiazolone (33) apparently arising from coupling with a combination band ⟨78ACS(A)1005⟩. The spectrum of the mesoionic oxathiazole (12; Ar = Ph) has been interpreted in terms of a highly polarized C=O group (supported by a large dipole moment, 4.5 D ⟨60MI43400⟩) by analogy with the spectra of sydnones ⟨63SA2047⟩, and it has been argued from a comparison of the integral intensity of its C=O band with that for its isomer (46; R^1 = Ph) that the latter is also highly polarized ⟨73ACS2161⟩. This is at variance with the observed short C=O bond lengths for the mesoion (1.194 Å, see Section 4.34.2.3.1) and

Table 6 UV Absorption Maxima for Dioxazole, Oxathiazole and Dithiazole Derivatives

Structure type	Ring atoms			Substituents			λ_{max} (nm) $(\log \varepsilon)$	Solvent	Ref.
	X	Y	Z	R^1	R^2	R^3			
(8)	N	S	S	Ph	Ph	—	241 (4.14), 267 (4.02), 376 (4.30)	CH_2Cl_2	77LA1005
(8)	N	S	S	4-MeC$_6$H$_4$	4-MeC$_6$H$_4$	—	262 (4.10), 285 (4.15), 402 (4.64)	CH_2Cl_2	77LA1005
(8)	N	S	S	4-MeOC$_6$H$_4$	4-MeOC$_6$H$_4$	—	284 (3.93), 298 (4.01), 457 (4.61)	CH_2Cl_2	77LA1005
(8)	N	S	S	NH$_2$	NH$_2$	—	246 (4.22), 278 (3.86)	EtOH	71ZOR2227
(9)	S	S	N	Ph	S	—	257 (4.21), 294 (3.7), 353 (4.12), 3.62 (4.1)	EtOH	67JPR(308)287
(10)	N	O	O	Ph	O	—	255.1 (4.21)	Dioxane	72JPR145
(10)	N	S	S	NH$_2$	S	—	285 (4.35), 345 (3.89)	1.0M HCl	73MI43400
(10)	N	S	S	NMe$_2$	S	—	297 (4.27), 340 (3.90)	1.0M HCl	73MI43400
(10)	N	S	S	NHCO$_2$Et	S	—	253 (3.92), 287 (4.30), 368 (3.95)^a	1.0M HCl	73MI43400
(11)	NCH$_2$R	S	S$^+$	NCH$_2$R	S	—	225 (3.95), 280 (4.26)	EtOH	79JCS(P1)2909
(12)	O	N$^-$	N	O	Ph	COMe	256 (4.06), 410 (3.95)	Dioxane	72CB188
(17)	S	S	N	Me	Me	COMe	236 (3.30), 280 (3.18), 352 (2.53)	Hexane	81BCJ3541
(17)'	S	S	N	Me	Ph	COMe	246 (3.96), 310 (3.52), 348 (3.39), 408 (3.37)	Hexane	81BCJ3541
(20)	O	N	O	Me	Ph	H	214 (3.90), 265 (3.76)	EtOH	67TL331

^{a} At pH 12.0 values become 312 (4.4), 348 (3.8).

for the oxathiazolone (**33**; 1.192 Å, see Section 4.34.2.3.2). Values for exocyclic $\nu(C=O)$ provide evidence for contributions from the diazaoxadithiapentalene form in structures (**38**; 1560–1610 cm^{-1}) ⟨75LA1018⟩, and for a 'no bond' interaction in structure (**47**; $R^1 = Ar = Ph$; 1600 cm^{-1}) though interestingly, the *trans* form (**48**) has $\nu(C=O)$ at the same frequency. The discrepancy between the calculated (6.5 D) and experimental (7.8 D) dipole moments for the former was also taken as evidence for a strong interaction ⟨79CHE1187⟩. Dickoré and coworkers reported that 1,3,4-dithiazolines of types (**49**) and (**50**) can be distinguished as they have $\nu(C=N)$ between 1550–1590 cm^{-1} and near 1630 cm^{-1}, respectively ⟨62AG(E)594⟩. However later workers describe compounds of type (**50**) having $\nu(C=N)$ near 1600 cm^{-1} and these structures seem to be correct from mass spectral fragmentation patterns ⟨77CCC2672⟩. Generally, there is extensive coupling in molecules having the fragments S—C(=X)—Y, where X and Y may be S or N, and bands can rarely be assigned to specific vibrations.

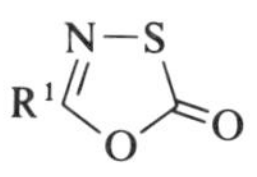

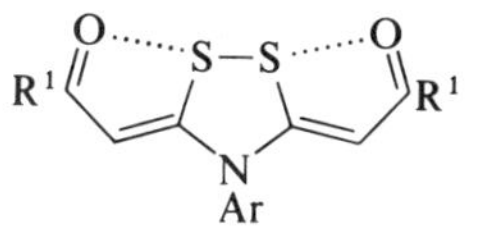

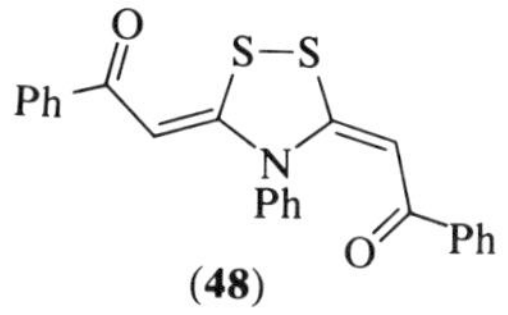

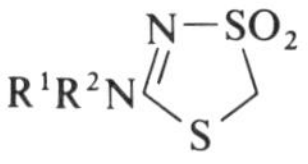

4.34.2.3.8 Mass spectrometry

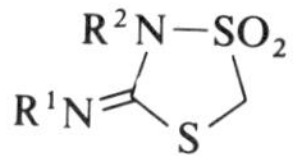

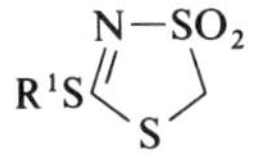

There is considerable variation in fragmentation patterns among the ring systems discussed in this chapter, depending upon the arrangement of heteroatoms, the degree of unsaturation and the nature of the substituents. The oxathiazolone (**46**; $R^1 = Ph$) loses CO followed by NS˙ to give PhCO$^+$ as the base peak, initial loss of CO$_2$ to form a thiazirene as anticipated from its thermal and photochemical behaviour (see Section 4.34.3.2.1) being only a minor pathway. Likewise, the mesoionic compounds (**12**), by analogy with the sydnones (see Section 4.01.3.8), lose first NO˙ and then CO to give ArC≡S$^+$ as the base peak. Structures of type (**24**) fragment both by loss of R^4SO_2 and of R^1R^3CO, while the azolines (**17**), (**20**) and (**21**) fragment by one or more reverse 1,3-dipolar cycloaddition pathways, which may be preceded by loss of group R^1 or R^3 to give an aromatic cation. 1,2,3-Oxathiazoles (**18**) and (**19**) having Y = S(O) extrude SO$_2$, rearrange to a nitrilium ion and then lose RCN. Compounds for which mass spectral data have been interpreted in some detail are given in Table 8.

4.34.2.4 Thermodynamic Aspects

4.34.2.4.1 Stability and stabilization

The ring-fused 1,2,3-dithiazoline (**27**; Scheme 3), exists in equilibrium with the ring-opened form (**52**), the equilibrium moving to the right at lower temperatures and in more polar solvents. In CH$_2$Cl$_2$ $\Delta H = -20.5$ kJ mol^{-1} and $\Delta S = -47.3$ J mol^{-1} K^{-1}; when X = O cyclization does not occur ⟨79BCJ2008⟩. In contrast, there is no evidence for the ring-opened form with 1,2,3-dithiazolines (**54**) ⟨81BCJ3541⟩.

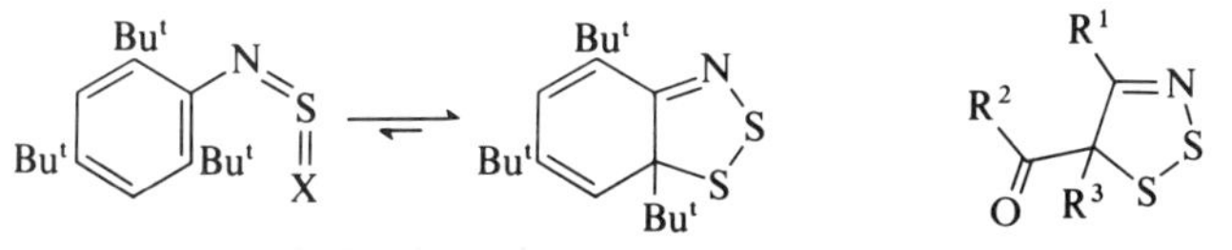

(**52**) X = S
(**53**) X = O

Scheme 3

Table 7 IR Stretching Vibrations (cm^{-1}) for C=O, C=S and Ring C=N in Dioxazole, Oxathiazole and Dithiazole Derivatives

Structure type	Ring atoms X	Y	Z	Substituents R^1	R^2	R^3	ν(C=O)	ν(C=S)	IR Absorptions (cm^{-1}) ν(C=N)	Others	Ref.
(8)	N	S	S	NH$_2$	NH$_2$	—	—	—	1634^a	δ NH$_2$ 1556^a, ρ NH$_2$ 1038	70JCS(A)1386
(9)	O	O	N	Ph	O	—	b	—	c	—	68TL319
(9)	O	O	N	Ph	S	—	—	d	c	—	68TL319
(9)	O	S	N	H	O	—	1809	—	1576	Raman: 1806, 1740, 1573	78ACS(A)1005
(9)	O	S	N	Ph	O	—	1780 1750	—	c	—	67ACS1871
(9)	S	O	N	Mes	O	—	1796 1741	—	c	—	66JOC2417
(10)	N	O	O	Ph	O	—	1815	—	1625	Raman: 1809, 1601	72JPR145
(10)	N	O	O	NArMe	O	—	1802	—	1638	—	74JOC2581
(10)	N	O	S	Me	O	—	1702	—	1540	—	81CB549
(10)	N	S	S	Ph	O	—	1683	—	c	—	66BSF1183
(10)	N	S	S	NH$_2$	O	—	1667	—	1515	δ NH$_2$ 1627, ρ NH$_2$ 1077	63JCS3165
(10)	N	S	S	NMe$_2$	O	—	1685	—	1570	—	81CB549
(10)	N	S	S	Ph	S	—	—	1192	1510	—	67CJC1225
(10)	N	S	S	NH$_2$	S	—	—	1020	1525	H–N–C 1632, 1330	70MI43401
(10)	N	S	S	NHPh	S	—	—	1070	1605	—	80JCR(S)266
(11)	NMe	O	O	O	O	—	1785	—	—	—	81AG(E)784
(11)	NMe	O	S	CHR	O	—	1675	—	—	ν_{NH}(?) 3300, 3170, 3030	77CB285
(11)	NH	S	S	O	O	—	1714^e	—	—	Exocyclic ν(C=N) 1630	71CB2732
(11)	NBz	S	S$^+$	NBz	O	—	1685/1690	—	—	—	79AP1027
(12)	O	N$^-$		Ph	O	—	1738 1704	—	—	—	72CB188
(17)	O	S(O)	N	H	Cl	H	—	—	1623	ν(S=O) 1083	75MI43401
(17)	O	S(O$_2$)	N	H	Cl	H	—	—	1623	ν(SO$_2$) 1157, 1122	75MI43402
(17)	S	S	N	(CH$_2$)$_4$	—	CO$_2$Et	1730	—	1610	ν(C=O) 1730	81BCJ3541
(18)	O	S(O)	NH	Me	O	Me	1700	—	—	ν(NH) 3220, ν(S=O) 1080	75MI43401
(18)	O	S(O)	NAr	Me	O	H	1743	—	—	ν(C=^{18}O) 1710, ν(S=O) 1035	75JHC393
(18)	O	S(O$_2$)	NMe	CF$_3$	O	CF$_3$	1645	—	—	—	74MI43400
(20)	O	N	O	CCl$_3$	Ph	H	—	—	1635	ν(C–O) 1122, ν(N–O) 853	72CB2805
(20)	O	S	S	CCl$_3$	Me	H	—	—	1650	—	81JCS(P1)2991
(20)	O	N	O	CCl$_3$	Ph	H	—	—	1615	—	81JCS(P1)2991
(21)	N	O	O	CO$_2$Me	Ph	H	—	—	1660	ν(C=O) 1750	79JHC129
(21)	N	S	S	H	CF$_3$	CF$_3$	—	—	1626	—	74AP828
(22)	NH	O	O	Ph	O	NEt$_2$	1790	—	1659	ν(NH) 3315	75MI43400
(24)	O	S(O)F	N	CF$_3$	O	CF$_3$	1832 1798	—	—	—	72JPR145
(24)	O	S(O)NEt$_2$	N	CF$_3$	O	CF$_3$	1775	—	—	—	74MI43400

^{a} Assignments from ^{2}H exchange; earlier workers (63JCS3165) reverse assignments. ^{b} Doublet between 1820–1875 cm^{-1}. ^{c} Not recorded. ^{d} Doublet between 1260–1315 cm^{-1}. ^{e} Additional bands at 1775 m and 1660 vs cm^{-1}.

Table 8 Compounds for which Mass Spectra have been Interpreted

Structure type	X	Ring Atoms Y	Z	R^1	R^2	Substituents R^3	R^4	Ref.
(9)	O	S	N	Ph	O	—	—	67ACS1871
(10)	N	S	S	NH_2	S	—	—	69AJC513
(11)	NH	S	S	CR_2	CR_2	—	—	77ZC223
(11)	NR	S	S	NPh	NPh	—	—	75CC24
(11)	S	S	NR	NR	NCSPh	—	—	75JCS(P1)270
(12)	O	N^-	S^+	O	Ar	—	—	81LA1025
(16)	O	S(O)	NTs	Me	Me	—	—	74AP291
(17)	S	S	N	R	R	COR	—	81BCJ3541
(18)	O	S(O)	NAr	Me	O	H	—	74JHC1
(19)	O	S(O)	NAr	Me	H	H	H	72BCJ928
(20)	O	N	O	Ar	Ar	H	—	74OMS(9)1017
(20)	O	N	S	Ar	Ar	H	—	74ZN(B)284
(20)	S	N	O	R	R	R	—	81JCS(P2)2991
(20)	S	N	SO_2	H	NCR_2	H	—	81JHC1309
(21)	N	S	S	H	Ar	H	—	74AP828
(21)	N	S	S	CF_3	R	CF_3	—	75S57
(22)	S	NPh	S	COPh	CR_2	H	—	80JPR407
(22)	S	NR	SO_2	H	NPh	H	—	77CCC2672
(23)	N(NO)	O	O	R	R	R	R	74JOC1791
(24)	O	$S(O)R^4$	N	H	O	H	p-Tol	70TL127
(24)	O	$S(O)R^4$	N	Me	NBu^t	Me	Ph	79TL49

What little evidence there is for aromaticity in structures of types (5), (6) and (8)–(10) has been presented above. A number of stable salts of type (8) are known, in which neither R^1 nor R^2 is a heteroatom, and cation (34) appears to have a significant ring current. The ^{1}H NMR spectrum of the oxathiazolone (33) does not appear to have been reported; its spectrum in a strongly acidic medium would be particularly interesting (compare Section 4.01.4.2.2).

Conductivity measurements on solutions of the salts (4) in anhydrous HCO_2H show them to be fully dissociated ⟨69ZOR153⟩. Ratios of $J_{5,6}:J_{4,5}$ (Table 4) indicate that for 1,2,3-benzodithiazolylium chloride (4; $R^1 = H$), substitution of S(1) by Se has no effect on the aromatic character of the system, whereas similar substitution of S(2) results in greater bond fixation and hence a decrease in aromaticity. This contrasts with data for 1,2,3-benzoselena-, thia- and -oxa-diazoles (see Section 4.01.4.2.2).

4.34.2.4.2 Conformation

Complete analysis of the ^{1}H NMR spectra of the 1,2,3-oxathiazolidine S-oxides (55; Scheme 4) using an LAOCOON3 treatment ⟨75OMR(7)296⟩ has shown that the molecules exist as an equilibrium mixture of two twist envelope forms (56a) and (56b), as do the analogous oxadithiolidine and thiadiazolidine S-oxides ⟨75ACS(A)414⟩. Form (56a) is favoured by aryl meta- and para-substituents, while ortho-substituents favour form (56b) ⟨75BCJ3313⟩. Assignments were assisted by the anisotropic effect of the S=O group, which causes a downfield shift for H_A and H_B relative to H_C and H_D, the direction of this effect being confirmed from an X-ray structure determination on a related oxathiazolidinone ⟨75JHC393⟩. The ring protons in the dioxazolidine (57) show an AA'BB pattern in the ^{1}H NMR spectrum, suggesting that this molecule also exists in a preferred conformation ⟨80IZV2181⟩.

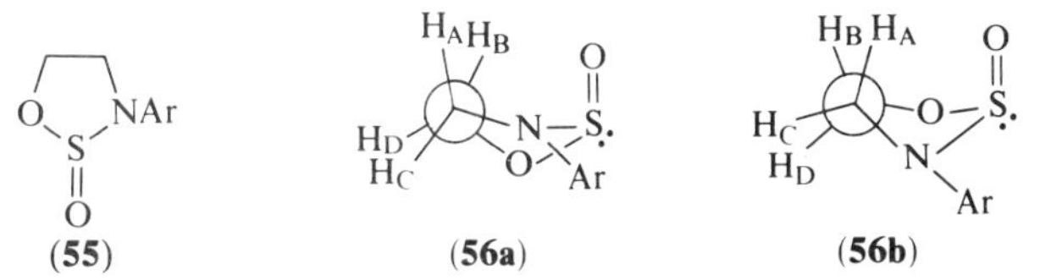

Scheme 4 Conformations of oxathiazolidine S-oxides

(57) (58)

Restricted rotation about the N—N bond is apparent from the ^{1}H NMR spectrum of the dioxazolidines (**58**), the two ring methyl groups for the case having $R^1 = Me$ and $R^2 = Et$ showing separate signals at δ 1.50 and 1.74 ⟨74JOC1791⟩. In the oxathiazoline (**215**; Scheme 46) the electronegativity of the SO_2 group increases the rotational barrier of the Me_2N function relative to the analogous oxazoline, the thermodynamic activation parameters being $\Delta G^{\ddagger}_{326}$ 73.6, $\Delta H^{\ddagger}$ 72.4 and E_a 74.9 kJ mol^{-1}; log A 13.0; and $\Delta S^{\ddagger}$ −4.6 J mol^{-1} K^{-1} ⟨68CC1245⟩.

4.34.2.5 Tautomerism

Until 1960 the structures of certain 1,2,4-dithiazole derivatives were formulated as (**59**) by analogy with cyclic imides, but as a result of the X-ray work of Hordvik and others (Table 1) and the IR studies of Eméleus and coworkers ⟨63JCS3165⟩, molecules having $R^1 = H$ are known to have the structures (**1**; $R^1 = R^2 = H$) in the solid state. Whether an equilibrium exists between the two forms in solution, however, has yet to be determined, and no tautomeric studies appear to have been carried out for examples in which $R^1 \neq H$.

(59)

Some uncertainty remains over the structures of 1,3,4-dithiazoline S,S-dioxides formulated as (**49**) and (**50**). Dickoré and coworkers ⟨62AG(E)594⟩ prepared the same derivative (**49**; $R^1 = R^2 = Me$) both by methylation of the monomethyl compound and by treatment of the thioether (**51**) with Me_2NH, thus showing that alkylation occurs on the exocyclic nitrogen. Later workers, however, claim that alkylation occurs on the ring nitrogen atom, and formulate alkylation products as (**50**), supporting the proposed structures by MS data ⟨77CCC2672⟩. Melting points of compounds (**49**; $R^1 = Ph$, $R^2 = Me$; and $R^1 = Ph$, $R^2 = H$) claimed by the first group differ from those of their counterparts (**50**; $R^1 = Ph$, $R^2 = Me$; and $R^1 = Ph$, $R^2 = H$) claimed by the second group, by 3–5 °C, as do those formulated as (**59**; $R^1 = 4\text{-}ClC_6H_4$, $R^2 = Me$), and prepared unambiguously by Dickoré. Infrared C=N vibrations in these molecules were discussed in Section 4.34.2.3.7.

4.34.2.6 Other Unusual Structures

Remarkably persistent radicals of the general structure (**60**) have been reported recently ⟨80JCS(F1)1490⟩. These structures have been probed by ESR, using both ^{2}H and ^{33}S labelling techniques. The results have been interpreted in terms of a bent envelope conformation (**61**) in which there is poor overlap between the nitrogen orbital carrying the unpaired electron, and the adjacent sulfur orbitals. More recently radicals (**62**), formed from enaminoketones and S_2Cl_2, have been shown from ^{33}S labelling studies to have large spin densities on nitrogen ⟨81ZC324⟩. For the benzo fused analogues (**63**), formed by reduction of the cations (**4**), similar local high spin densities have been indicated, both from MO calculations and from a linear relationship between the hyperfine splitting constant a_N and the Hammett σ_p constants for a range of substituents R^1 ⟨81ZC265⟩. Analogous radicals have been prepared from 1,2-dithiolylium cations (see Chapter 4.31).

(60)   (61)   (62)   (63)

4.34.3 REACTIVITY OF THE RING SYSTEMS

4.34.3.1 General Survey

In this section the reactivity of aromatic and non-aromatic systems are treated separately as in Chapter 4.02. For convenience, the former group is taken to comprise all compounds of structural types (5)–(11), including benzo fused derivatives, even when there is little evidence of aromaticity as is the case with dioxazoles. Likewise, the nonaromatic systems will include dihydro and tetrahydro derivatives of structural types (16)–(24).

The generalizations made about the reactivity of aromatic azolylium ions (see Section 4.02.1.1.2) and aromatic azolones and related compounds (see Section 4.02.1.1.4) apply equally to the ring systems here. In addition, cleavage of the S—S bond on attack by nucleophiles, often with extrusion of sulfur, is an important reaction of 1,2,4-dithiazole derivatives.

4.34.3.2 Reactions of Aromatic Rings

4.34.3.2.1 Thermal and photochemical reactions formally involving no other species

Thermolytic reactions are known for structures of types (9)–(11). Azolone derivatives (64; Scheme 5) fragment by two major pathways, extruding either XCZ when $X = O$ and $Y = S$, or YCZ when $Y = O$ and $X = O$ or S. In the first case, the intermediate nitrile sulfide can be trapped ⟨77JOC1813⟩, but generally breaks down to a nitrile and sulfur, with the exception of the parent (33) which gives HNCS ⟨77JSP(68)169⟩. For the second pathway, YCZ has generally been CO_2, but can be an isocyanate ⟨75T1537⟩; the acyl nitrene fragment then rearranges to isocyanate or isothiocyanate, its intermediacy having been established both by inter- and intra-molecular trapping experiments ⟨68TL319⟩. The thermolysis of azolones (64) having $X = Y = Z = O$ is an important commercial route to isocyanates (see Section 4.34.5.3).

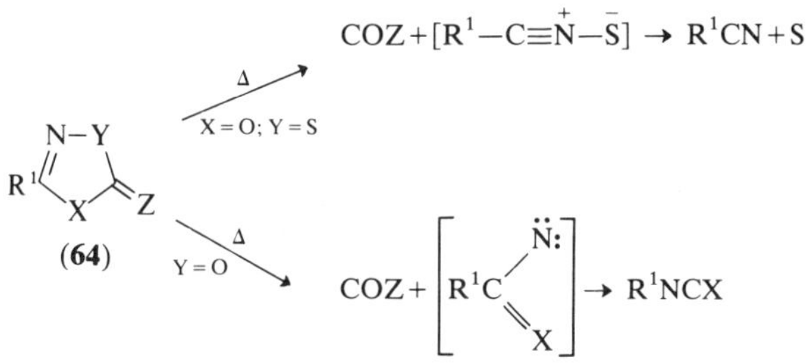

Scheme 5 Thermolysis of azolones I

The related structures (65; Scheme 6) also fragment on heating, the dioxazolone (65; $X = Z = O$) presumably going *via* benzoylnitrene ⟨72JPR145⟩, though the related dithiazolethione (65; $X = Z = S$) ⟨67CJC1225⟩ may well fragment with loss of sulfur by a concerted process.

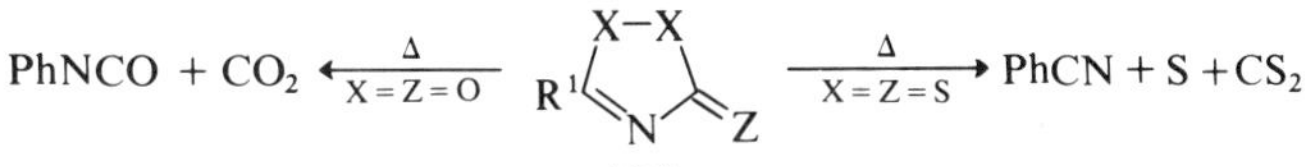

Scheme 6 Thermolysis of azolones II

The photolysis of a range of phenylazoles, including (12), (64; X and/or $Y = S$ and $Z = O$, S or NH), and (65; $X = S$ and $Z = O$ or S) in neat DMAD at the long wavelength band gives high yields of benzonitrile and sulfur, formed *via* 3-phenylthiazirene (see Section 4.02.1.2, Scheme 1) and benzonitrile sulfide. The latter has been trapped with DMAD ⟨78JCS(P1)1445⟩ (see Section 4.02.1.2.1); only minute amounts of PhNCS were detected.

The mesoionic 1,3,2-oxathiazolone (12; Ar = Ph) in fact, gives photo products resulting from two different pathways (Scheme 7). Path B *via* phenylthiazirine (66) predominates with light of wavelength ~400 nm in benzene ⟨81LA1025⟩, whereas in ethanol reversible path A is important, being entirely suppressed on saturation of the solution with NO, and ^{15}N-labelled NO being incorporated into the heterocycle ⟨76T2559⟩. At 10 K in a nitrogen matrix using 367 nm light, isolated molecules ($\nu_{C=O}$ 1750 cm^{-1}) fragment by path A, whereas molecular aggregates ($\nu_{C=O}$ 1710 cm^{-1}) formed on annealing follow path B ⟨76TL873⟩. Intermediate (**67**) has been detected by IR absorption at 2120 cm^{-1}.

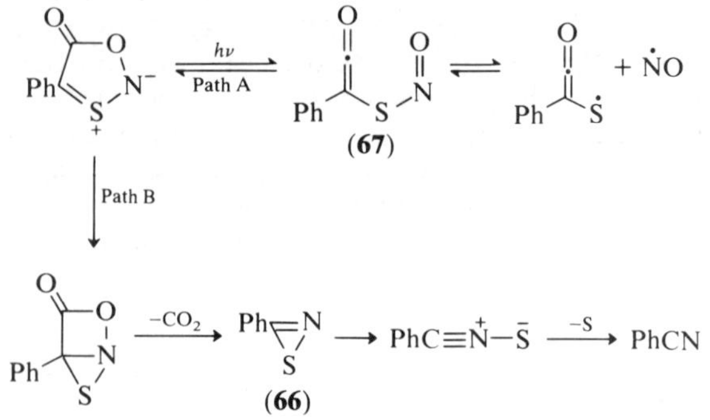

Scheme 7 Photolysis of mesoionic oxathiazole (**12**; R^1 = Ph)

Dioxazolones (**64**; X = Y = Z = O) extrude CO_2 on photolysis, just as they do on thermolysis, and the intermediate acyl nitrene may either be trapped, or may rearrange to an isocyanate. Thus for R^1 = Mes, an intramolecular insertion reaction occurs to give the isoindolinone (**68**), though for the case where R^1 = 2-hydroxyphenyl a Wolff rearrangement precedes cyclization, giving a product different from that from the thermolysis ⟨68TL319⟩. A study into the selectivity of insertion of the nitrene into 2-methylbutane gave the order 3° H > 2° H > 1° H ⟨74TL2565⟩.

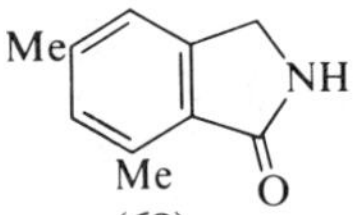

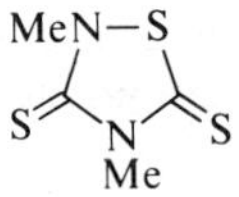

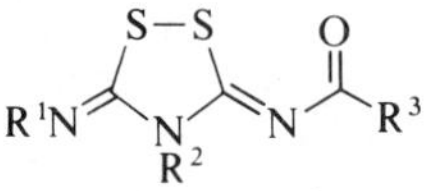

(**70**) R^1 = Me, R^2 = Ph
(**71**) R^1 = Ph, R^2 = Me

A number of thermal rearrangements has been observed with 1,2,4-dithiazole derivatives of types (**10**) and (**11**), and these were the cause of confusion in the early literature as regards structure assignments to products. The iminothione (**41**; R^1 = R^2 = Me, X = S) undergoes a Dimroth rearrangement to the thiadiazolidine (**69**) simply by warming in ethanol ⟨68IJC132⟩, and the diimino analogue (**70**) is readily converted into its isomer (**71**) ⟨78JOC4951⟩ (see also Section 4.34.3.3.4).

The 3-amino-1,2,4-dithiazolimines (**72**) decompose by different pathways depending upon the nature of the group R^3, these being intermolecular for R^3 = Me and intramolecular [leading to the benzothiazole (**73**)] for R^3 = Ph ⟨74JOC2233⟩. The latter rearrangement is also known to occur under acid-catalyzed conditions. Finally, the iminodithiazoline (**44**) isomerizes to the diazatrithiapentalene (**45**) on heating above 120 °C ⟨64CB2567⟩.

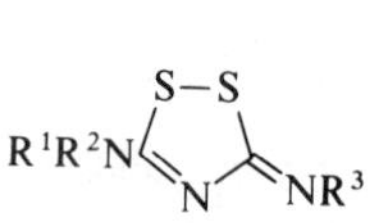

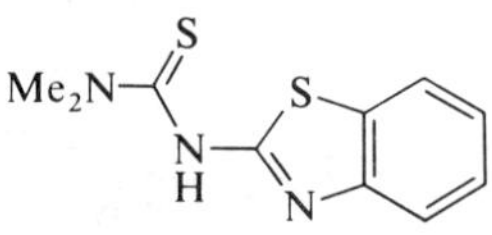

4.34.3.2.2 Electrophilic attack at ring nitrogen and carbon

From a consideration of the charge distribution in structures of types (**5**)–(**11**) this is not expected to occur readily. In practice no examples involving electrophilic attack at carbon have been reported, and although one case of alkylation was claimed to have occurred on ring nitrogen to give the cation (**74**) ⟨72BCJ3572⟩, it was later shown that reaction had in fact occurred on the exocyclic imine nitrogen atom, as is generally the case with alkylations, acylations and protonations of azolones (**6**), (**9**) and (**10**) (see Section 4.34.3.3.4) to give an aromatic cation. The *o*-quinoid 1,2,4-dithiazole (**75**), however, is benzoylated on nitrogen and not on the phenolic oxygen of the tautomer (**76**) to give a benzamide ⟨77ZC223⟩.

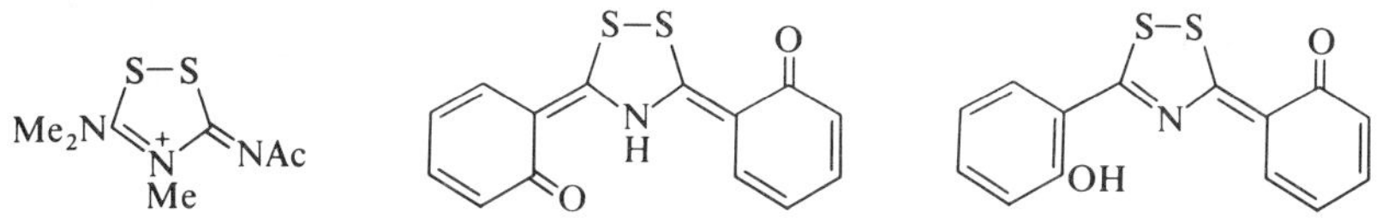

(74) (75) (76)

In the copper(I) chloride complex of the cation (**31**), however, the ring nitrogen atom is coordinated to the copper atom ⟨75CC550⟩, this being in contrast to the related thione (**30**) which coordinates *via* the exocyclic sulfur atom ⟨77MI43400⟩.

4.34.3.2.3 Attack at ring sulfur

Again, very few examples of electrophilic attack at ring sulfur are known. Chlorination of the chlorodithiazolone (**77**) to give CCl_2NCOCl is presumably such a case ⟨71CB2732⟩. 1,2,4-Dithiazoles, having a strong intramolecular S···S 'no bond' interaction (*e.g.* **78**), alkylate at ring sulfur to give the salt (**79**), though for weak interactions, alkylation is at the exocyclic nitrogen atom ⟨74JOC2235⟩ (see Section 4.34.3.3.4).

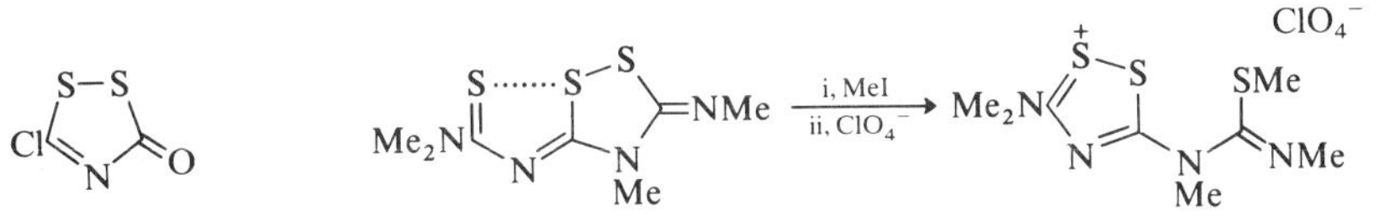

(77) (78) (79)

Nucleophilic attack at sulfur is better documented, especially for electron deficient ring systems, examples of attack by CN^- ⟨79CHE1187⟩, and enolate anion ⟨72BCJ3220⟩ being shown in Scheme 8. Triphenylphosphine attacks the ring sulfur atom adjacent to the thione group in 1,2,4-dithiazole-3-thiones with the formation of thioacylisothiocyanates (Scheme 9) ⟨74CB502⟩. 1,2-Dithiole-3-thiones behave analogously (see Chapter 4.31). Similar reactions also take place both with 1,2,4-dithiazol-3-ones ⟨81CB549⟩ and with 1,2,4-dioxazol-3-ones ⟨72JPR145⟩, leading respectively to thioacyl- and acyl-isocyanates. 5-(2-Pyridyl)-1,2,4-dithiazol-3-one, however, gives a cyclized dipolar ion as the product ⟨81CB808⟩.

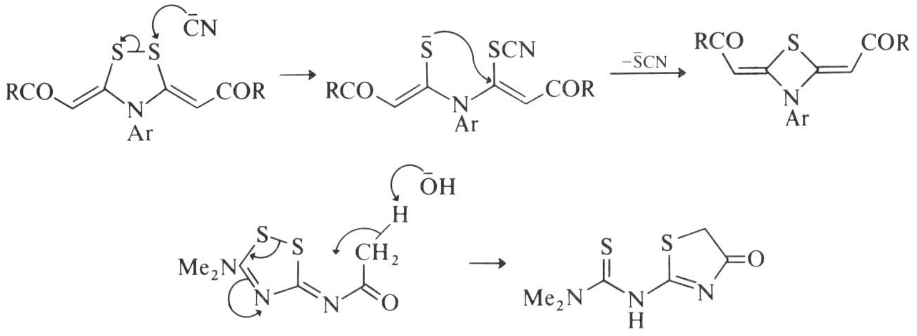

Scheme 8 Nucleophilic attack at sulfur

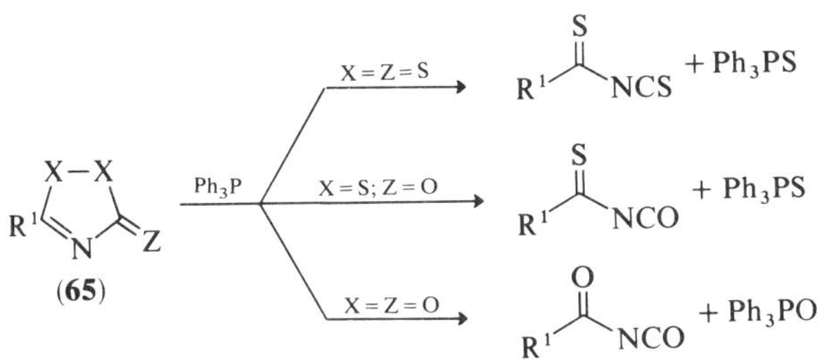

Scheme 9 Action of Ph_3P on azolone derivatives

The dithiazolylium cation (**31**) is reduced reversibly to dithiobiuret (**80**) by a number of thiols, which in turn are oxidized to disulfides. This reaction has been found to be first order in compound (**31**) and second order in the thiol, the rate constant when this is cysteine decreasing from pH 4.60 to 3.50, and the rate being roughly inversely proportional to [H⁺] ⟨64JOC1488⟩. The suggested mechanism is given in Scheme 10, a related reaction having been used to remove a useful protective group from amino acids (Scheme 11) ⟨80JA3084⟩.

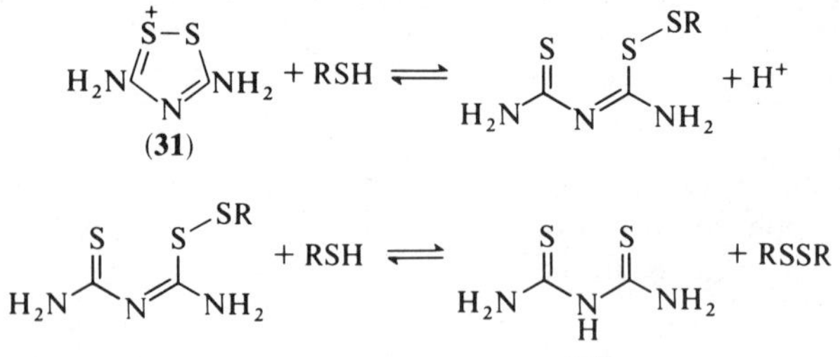

Scheme 10 Reduction of 3,5-diamino-1,2,4-dithiazolylium chloride by thiols

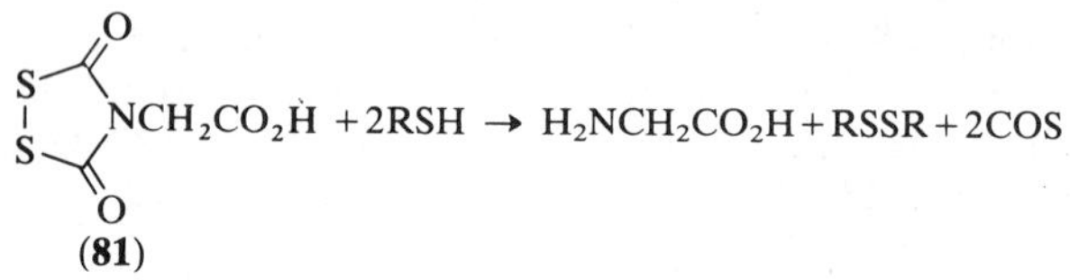

Scheme 11 Deprotection of amino acids

1,2,3-Benzodithiazolylium salts (**4**; Scheme 12) are attacked at sulfur both by secondary amines ⟨77KGS849⟩ and by water, the rate of the latter reaction when R^1 = MeO decreasing with increasing electron donating ability in the order 7-MeO > 4-MeO > 5-MeO > 6-MeO ⟨79KGS1205⟩. The structure of the products was established as (**82**), rather than the S–OH form proposed earlier, by IR absorption near 3450 and 1100 cm⁻¹ ⟨65JOC2763⟩. Two other reactions of salts (**4**) are included in Scheme 12 since the first step may well involve formation of the species (**82**).

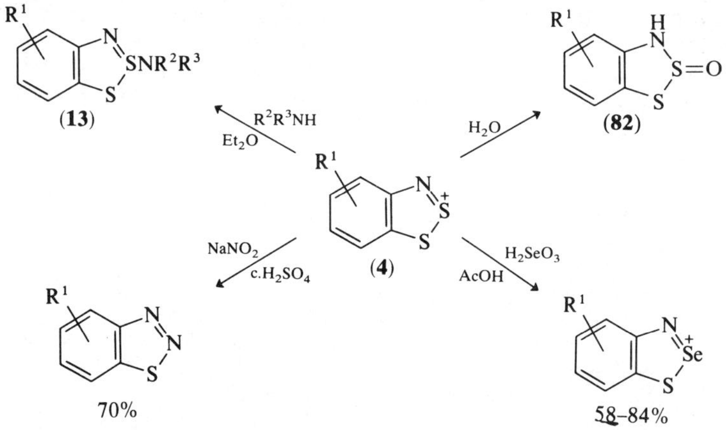

Scheme 12 Attack at sulfur in 1,2,3-benzodithiazolylium salts

Both acids and bases hydrolyze the 1,3,2-benzodithiazole derivatives (**15**) to *o*-sulfinylsulfonamides, attack presumably occurring at the S(IV) atom (compare Scheme 36) ⟨78JOC1218⟩.

4.34.3.2.4 *Nucleophilic attack at carbon*

For unsymmetrical structures (**5**)–(**12**), nucleophilic attack may occur at either ring carbon atom, the preferred site of attack depending on steric and electronic factors, and the fate of the adduct on pathways available for further reaction (see Section 4.02.1.6). For convenience, reactions will be considered by structure type (**8**)–(**12**).

For 1,2,4-dithiazolylium salts containing substituents which are not good leaving groups, nucleophilic addition is followed by ring opening with extrusion of sulfur, the subsequent

Dioxazoles, Oxathiazoles and Dithiazoles 917

fate of the molecule depending upon the nature of Nu⁻ and the reaction conditions. Thus the simple salt (**34**) yields thiobenzoylformamide with water ⟨74AP828⟩, a similar reaction being observed with the diaryl derivatives (**43**) ⟨79EGP135901⟩. These last salts can also give oxadiazoles, triazoles, or thiadiazoles (the latter by oxidation of the intermediate with sulfur) ⟨79NKK389⟩, pyrimidines ⟨78EGP129907⟩ or thioacylamidines (Scheme 13) ⟨77LA1005⟩. The aminodithiazolylium salts (**83**; Scheme 14) react with azide ion at the carbon bearing substituent R^1 for $R^1 = NH_2$, NHMe, NMe_2 and Ph, to give thiadiazoles ⟨71JOC3645⟩. However, with isocyanate ion attack occurs at the other carbon atom when $R^1 = Ph$, and follows a quite different pathway when $R^1 = NMe_2$ ⟨72JOC131⟩. When two different disubstituted amino groups are present, mixtures of products are formed. Mixtures of two products can also be formed with extrusion of sulfur when arylamines react with 3,5-diaminodithiazolylium salts (**84**; Scheme 15). This extensively studied reaction is analogous to that occurring between amines and 1,2-dithiolylium salts (see Chapter 4.31). Prolonged reflux in ethanol when either R^1 or $R^2 = H$ results in oxidation of the ring-opened material by sulfur to a thiadiazole ⟨60JIC151⟩.

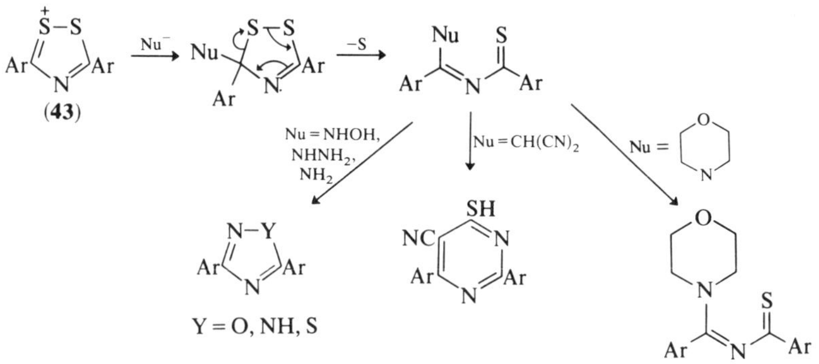

Scheme 13 Nucleophilic attack at carbon in 1,2,4-dithiazolylium salts I

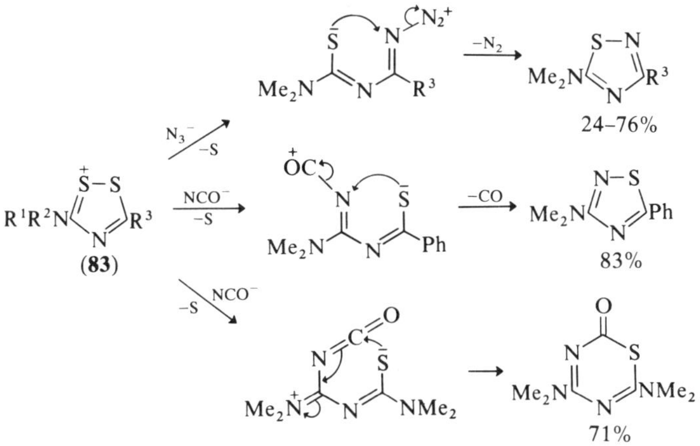

Scheme 14 Nucleophilic attack at carbon in 1,2,4-dithiazolylium salts II

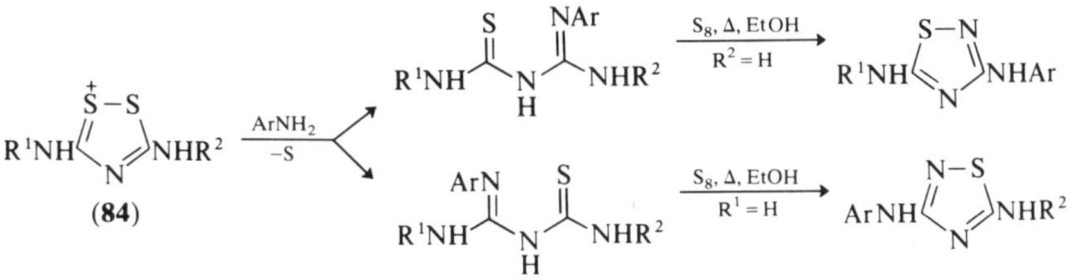

Scheme 15 Action of arylamines on 1,2,4-dithiazolylium salts

Structures of type (**64**; Scheme 16) appear to react exclusively at the exocyclic C=Z bond, giving products from ring opening (**85**) ⟨78ZOB1465⟩ or rearrangement (**86**) ⟨75T1537⟩. Interestingly, the computer program AHMOS correctly predicted the product (**87**) from attack by alkylamines on 5-phenyl-1,3,4-oxathiazol-2-one ⟨77ZC295⟩. The 1,3,4-dioxazolone

(**64**; $R^1 = Ph$, $X = Y = Z = O$) reacts with $MeC \equiv CNEt_2$ to give the isoxazolone (**88**) ⟨73TL233⟩, and 1,3,4-oxathiazolones (**46**) on treatment with $Pd(Ph_3P)_4$ give the heterocycles (**89**), the Pt analogues being formed from $Pt(Ph_3P)_2(C_2H_4)$ ⟨80CB1790⟩. The mechanisms may involve initial attack at the C=O bond.

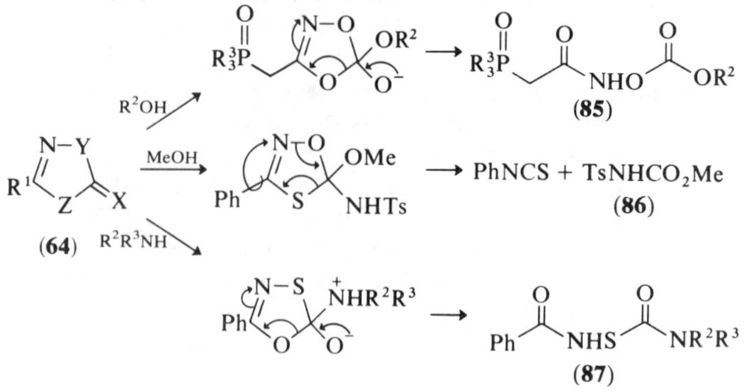

Scheme 16   Nucleophilic attack at carbon in azolones of type (**64**)

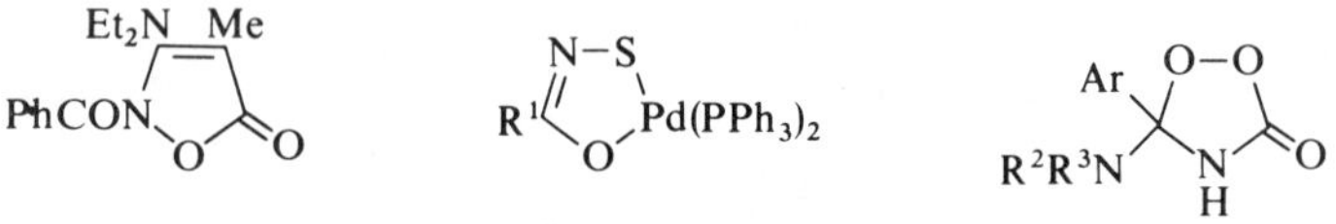

(**88**)   (**89**)   (**90**)

For the 1,2,4-dioxazol-3-one (**65**; $R^1 = Ar$, $X = Z = O$) ⟨72JPR145⟩, nucleophilic addition by secondary amines occurs at the ring C=N to give the adduct (**90**), and while I^- gives a high yield of benzamide, the site of attack in this case is less clear. In contrast, the 1,2,4-dithiazole-3-thione (**30**) is attacked by arylamines at the thione carbon with extrusion of sulfur, in a reaction analogous to that in Scheme 15. A similar reaction occurs with the isomeric 1,2,4-dithiazolidine derivatives (**41**; X = O, S) ⟨72IJC1138⟩.

The mesoionic 1,3,2-oxathiazolone (**12**) behaves in its reaction with sulfide ion (Scheme 17) ⟨76MI43400⟩ more as structure (**12b**) than the generally accepted 'azolylium-5-olate' form (**12d**), consistent with X-ray bond parameters (see Section 4.34.2.3.1).

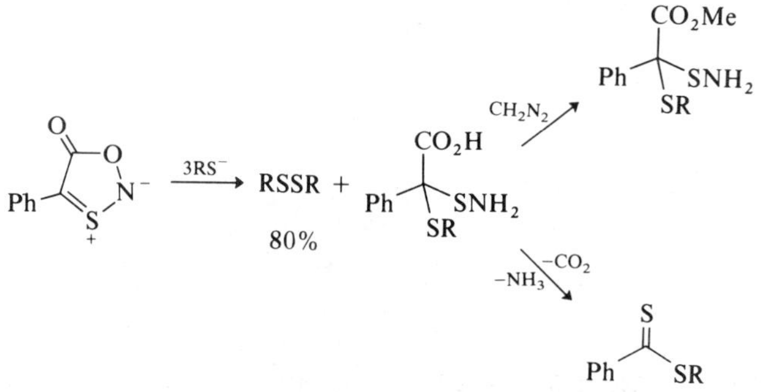

Scheme 17   Nucleophilic attack on mesoionic oxathiazole (**12**; $R^1 = Ph$)

4.34.3.2.5 Reactions involving radicals and at surfaces

Hydrogenation of the 1,3,4-dioxazol-2-ones (**64**; $X = Y = Z = O$) over Raney nickel using 1 mol H_2, results in cleavage of the N—O bond, the products being $RCONH_2$ and CO_2 ⟨51CB688⟩.

Dithiazolylium salts and related compounds with S—S bonds are readily reduced both to thiols and to radicals. Thus, the thione (**30**) is reduced quantitatively to thiourea and CS_2 by Sn/HCl ⟨1897JCS607⟩, and for the electrochemical reduction of the salt (**31**) to dithiobiuret (**80**) at a Pt electrode, the half-cell potential $E'_0 - 0.251$ V at pH 0 decreases with increasing pH, indicating that the equilibrium favours the cation (**31**) ⟨47JA2632⟩. The same cation, when refluxed in EtOH with dithiomalonamide, is converted quantitatively to dithiobiuret (**80**) with the coproduction of the 3,5-diamino-1,2-dithiolylium cation, showing that (**31**) is more electrophilic than the latter ⟨71ZOR2227⟩. Reduction of the dithiazolidine (**37**) at a Hg electrode gave a value of −0.45 V relative to saturated calomel ⟨73TL939⟩.

Reduction of the 1,2,3-benzodithiazolylium cations (**4**) to the radicals (**63**) was referred to in Section 4.34.3.2; the reaction can be reversed with Cl_2.

4.34.3.2.6 Reactions with cyclic transition states

The thermolysis of the mesoionic 1,3,2-oxathiazol-5-ones (**12**) in DMAD gives high yields of the isothiazoles (**92**) *via* the intermediate adducts (**91**; Scheme 18), whereas in contrast, photolysis gives low yields of the isomeric compounds (**93**) ⟨81LA1025⟩ *via* the nitrile sulfide (Section 4.34.3.2.1). Unsymmetrical alkynes give a mixture of two products, *e.g.* (**94**) and (**95**), one being favoured presumably on electronic rather than steric grounds. Interestingly, alkenes also give isothiazoles *via* an oxidation step; a mechanism has been suggested ⟨72CB196⟩ (see also Chapter 4.17).

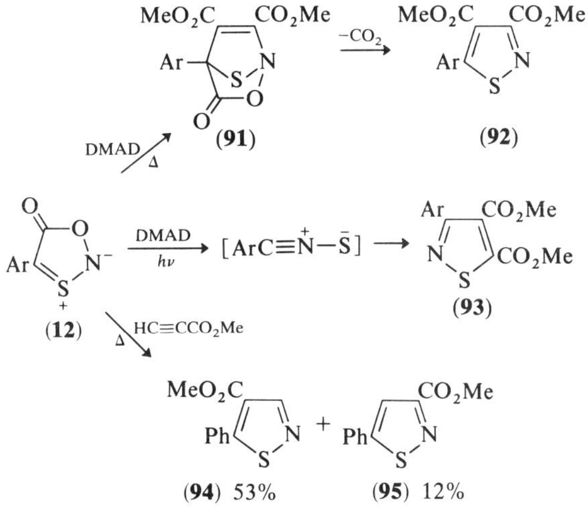

Scheme 18 Cycloaddition reactions of mesoionic oxathiazoles (**12**)

The isomeric 1,3,4-oxathiazol-2-ones (**46**) also give adducts with DMAD and with other alkynes above 130 °C (Scheme 19). However, kinetic measurements indicate that CO_2 is lost to give a nitrile sulfide prior to cycloaddition of the alkyne, no intermediate analogous to (**91**) being formed. Ethyl propynoate gives equal amounts of both isomeric product isothiazoles, whereas ethyl phenylpropynoate shows some selectivity ⟨78JOC3742⟩. The observed order of reactivity, DMAD > HC≡CCO$_2$Et > PhC≡CCO$_2$Et > MeC≡CCO$_2$Et is in the order of increasing LUMO energies, with a superimposed rate-retarding steric

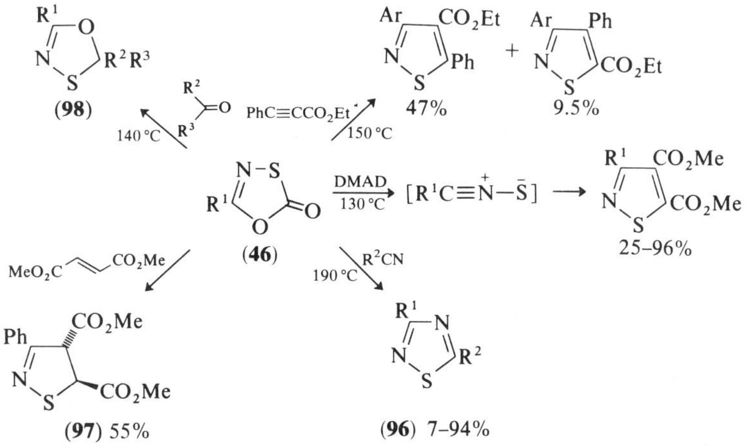

Scheme 19 Cycloaddition reactions of 1,3,4-oxathiazol-2-ones

effect with PhC≡CCO$_2$Et. Cycloadditions have also been achieved with nitriles to give thiadiazoles (**96**) ⟨81JOC771⟩ (highest yields being found for R^1 = Ar, and electrophilic R^2), with a wide range of electrophilic alkenes with retention of configuration to give, for example, compound (**97**) ⟨78JOC3736⟩, and with electrophilic aldehydes and ketones to give oxathiazolines (**98**). In the latter case, competing fragmentation of the nitrile sulfide to R^{1}CN and S is serious with electron withdrawing R^1 groups ⟨81JCS(P1)2991⟩.

Cycloadditions involving a ring sulfur atom and an exocyclic C=X bond are also known with 1,2,4-dithiazole derivatives, resulting in the formation of a new five-membered ring and cleavage of the ring S—S bond (Scheme 20) ⟨74JOC2228⟩. The scope of the reactions is shown in Schemes 21 and 22. With the thiones (**2**; X = S) unsymmetrical alkynes add regiospecifically to give the dithioles (**99**), both for R^1 = CO$_2$Et and for R^1 = Ar^2 and R^2 = H, the latter example giving also the azatrithiapentalene (**100**) formed as a result of further rearrangement ⟨67BSF2865⟩. Benzyne gives a benzodithiole and varying (small) amounts of the benzo fused analogue of the azatrithiapentalene (**100**), depending upon the method of benzyne generation ⟨76BSF120⟩. Reaction with excess DMAD leads to the adduct (**101**), formed by a further Diels–Alder reaction on the intermediate (**99**; R^1 = R^2 = CO$_2$Me) which has been isolated ⟨67CJC1225⟩. These reactions are analogous to those shown by 1,2-dithiole-3-thiones (see Chapter 4.31) ⟨69CJC2039⟩.

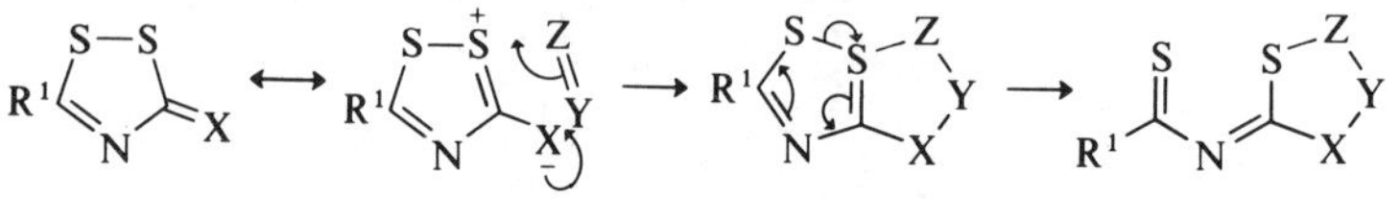

Scheme 20 Cycloaddition reactions of 1,2,4-dithiazolones

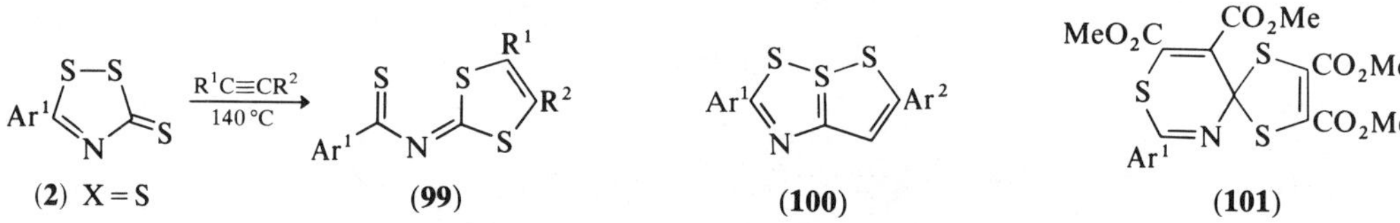

Scheme 21 Reactions of 1,2,4-dithiazole-3-thiones with alkynes

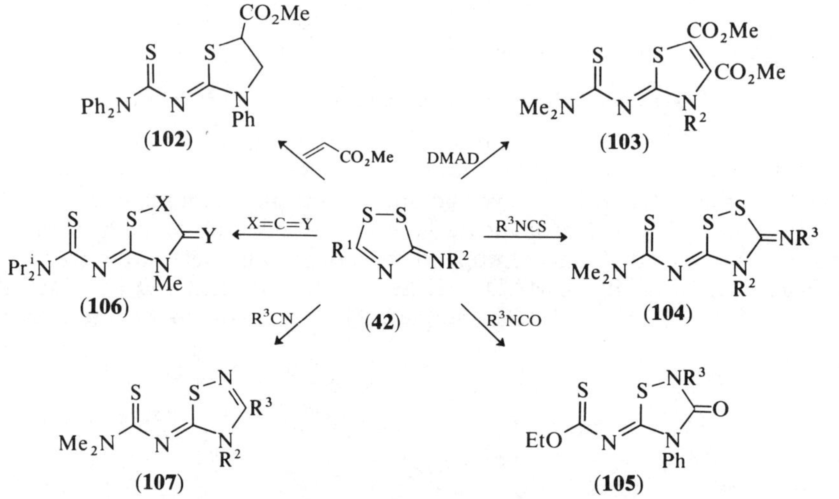

Scheme 22 Cycloaddition reactions of 1,2,4-dithiazol-3-imines

Likewise, the imines (**42**; Scheme 22) give a range of products, though R^1 has generally been an electron-releasing substituent. Unsymmetrical alkenes add regiospecifically to give high yields of adducts, *e.g.* **102**, in a reaction which also proceeds for R^1 = Mes. However, diethyl *trans*-but-2-enedioate gives a mixture of stereoisomeric products ⟨75TL3387⟩. Addition of DMAD or isothiocyanates gives adducts (**103**) and (**104**) respectively, thioacylisothiocyanates leading to the interesting compounds (**26**) having S ··· S 'no bond' interactions ⟨74JOC2228⟩. Isocyanates, however, add across the C=N bond forming thiadiazolones (**105**) ⟨76CB848⟩. Interestingly, CS$_2$, COS and CO$_2$ all add in high yields to give structures (**106**) where X = S for COS ⟨77CB285⟩, while nitriles form the thiadiazoles (**107**) ⟨74JOC2225⟩.

Finally, the dithiazolidine (**37**) adds CS_2 to form a symmetrical diazatrithiapentalene ⟨79JCS(P1)2909⟩.

Photolysis of the thione (**2**; Ar = Ph, X = S) in the presence of cycloalkenes or 2,3-dimethylbut-2-ene produces the adducts (**108**) formed non-stereospecifically by the mechanism suggested in Scheme 23. 1,2-Dithiole-3-thiones behave similarly. In benzene in the absence of alkene the products are (**109**) and (**110**), though the mechanism for their formation is not clear ⟨75JCS(P1)270⟩.

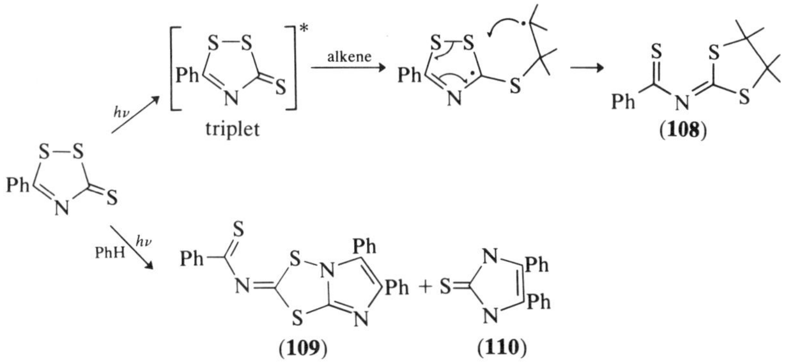

Scheme 23 Photolysis of 3-phenyl-1,2,4-dithiazoline-5-thione

4.34.3.3 Reactions of Substituents on Aromatic Rings

The general principles governing the reactivity of substituents attached to aromatic azole rings are given in Section 4.02.3.1. In addition, for 1,2,4-dithiazole derivatives, suitably placed substituents at the 3- or 5-positions can lead to 'no-bond' interactions, formation of azatrithiapentalene derivatives, or even cleavage of the S—S bond.

4.34.3.3.1 Fused benzene rings

Reactions at fused benzene rings have been reported only for the 1,2,3-benzodithiazolylium salts (**4**). The parent (**4**; R^1 = H) is chlorinated quantitatively at the 6-position by $S_2Cl_2/SOCl_2$, this being the most electron deficient site from SCF/MO calculations (see Section 4.34.3.2), and the chlorine atom is readily displaced by benzenamine. For 4-, 5-, or 6-chloro or -methoxy substituents, only those in the 6-position are readily displaced by benzenamine, the chloro compound reacting eight times faster than the methoxy compound ⟨76KGS183⟩. A nitro group in the 4-position can be replaced by a Cl group in refluxing S_2Cl_2 when a 6-MeO, but not a 5-MeO, group is present. The latter is too deactivating for the reaction to occur ⟨70CHE327⟩.

4.34.3.3.2 Alkyl groups

Little is known about the reactivity of alkyl substituents on the heterocycles of this chapter. The 1,3,4-dioxazole derivative (**111**) is cleaved by strong acid, the reaction possibly proceeding *via* protonation of the side chain ⟨70IZV2140⟩.

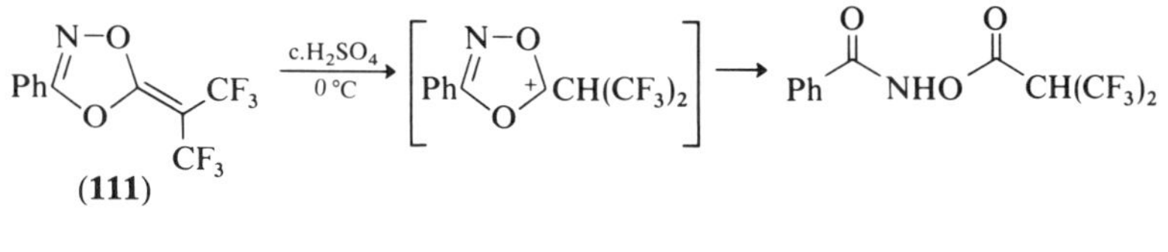

4.34.3.3.3 Aryl groups

The 5-aryl-1,3,4-oxathiazol-2-ones (**46**; R^1 = Ar) undergo a number of reactions at the aryl substituent without destruction of the heterocyclic ring, an example being hydrolysis of a p-CCl$_3$ group to CO$_2$H. Nitration of the parent (**46**; R^1 = Ph) at 25 °C using AcONO$_2$/Ac$_2$O gives all three mononitro derivatives, partial rate factors relative to benzene being o_f 0.005, m_f 0.003 and p_f 0.018. When the benzene ring bears methyl or fluoro substituents, the nitro group enters mostly *ortho* to the substituent, but for nitration of the 4-CH$_2$Cl derivative, a mixture of 45% 2-NO$_2$ and 55% 3-NO$_2$ derivatives was obtained. Substituent constants for the benzene ring bearing the oxathiazolone group are σ_m 0.49, σ_m^+ 0.42, σ_p 0.64, σ_p^+ 0.29, σ_M^o 0.14 and σ_I 0.44 ⟨73ACS2161⟩.

4.34.3.3.4 N-Linked substituents

3-Amino- or -imino-1,2,4-dithiazole derivatives undergo a Dimroth rearrangement to 1,2,4-thiadiazoles on heating with ammonia in ethanol (see also Section 4.34.3.2.1), the reaction being reversed by acids. Thus, structures (**41**; R^1 = R^2 = Me, X = S) and (**69**) are readily interconverted ⟨1895LA(285)175⟩.

As pointed out in Section 4.34.3.2.2, alkylation, acylation and protonation take place in 1,2,4-dithiazole derivatives at the exocyclic rather than at the ring nitrogen atom giving products analogous to those from imino-1,2-dithioles (see Chapter 4.31). However, in rings containing a thione group, alkylation occurs at sulfur rather than at nitrogen (see Section 4.34.3.3.6).

Alkylation of the aminoimines (**72**) provides a convenient route to the salts (**83**; R^1 = R^2 = Me, R^3 = dialkylamino). However, although 3-bromopropene gives initially an isolable bromide (**112**; Scheme 24), this readily rearranges to the thiazolidine (**113**) ⟨74JOC2225⟩.

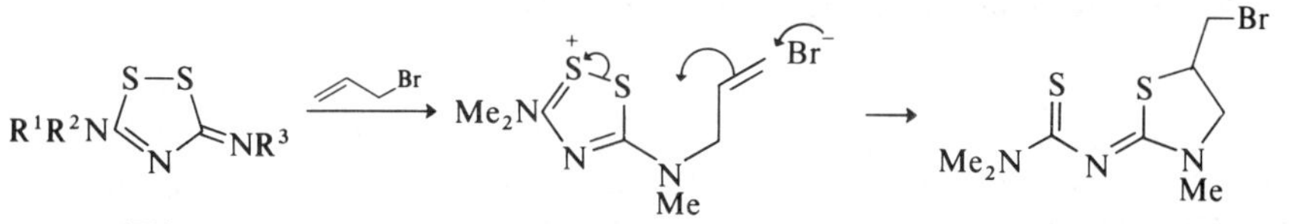

Scheme 24 Alkylation of 3-amino-5-imino-1,2,4-dithiazoles with 3-bromopropene

Acylation of the 1,2,4-dithiazole (**30**) leads to simple *N*-acyl derivatives, which show amide ν(C=O) at 1680–1695 cm^{-1} in their IR spectra ⟨75LA1018⟩. Structures of type (**42**) give more complex reactions, however. Thus the dithiazole (**42**; R^1 = PhNH, R^2 = H) is acylated to yield the thiourea derivative (**44**) using phenylisothiocyanate and pyridine at room temperature, but the diazatrithiapentalene (**45**) is formed in boiling xylene. Isocyanates lead to the derivatives (**114**) (compare product **105**, Scheme 22, when R^2 ≠ H), which like their thio analogues (**44**) can be reconverted to the parent on heating with acid (Scheme 25) ⟨64CB2567⟩. Alkylsulfonyl chlorides initially form sulfonamides (**115**) but under the reaction conditions these rearrange to the 1,3,4-dithiazoline derivatives (**116**) by S_N attack at ring sulfur in a reaction similar to the second example in Scheme 8. Use of TsCl leads to isolation of an intermediate salt analogous to (**115**) ⟨77TL1729⟩.

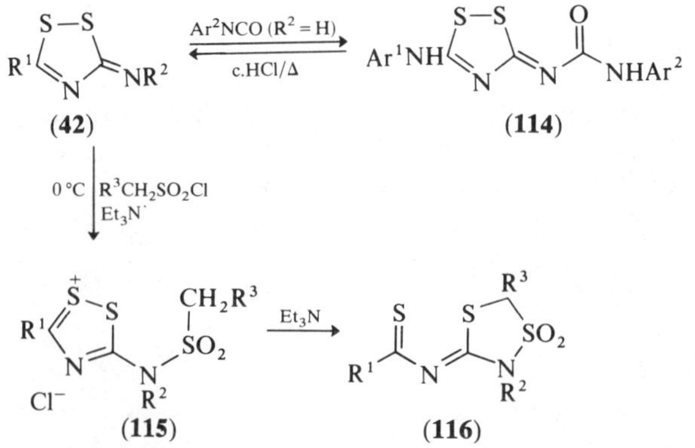

Scheme 25 Acylation of exocyclic imino groups

1,2,4-Dithiazolylium salts (**83**), having R^3 = NH$_2$, react with ketones ⟨67IJC216⟩, presumably *via* an intermediate enamine, *e.g.* (**117**), to give 1,3-thiazoles (**118**; Scheme 26). Mild base treatment yields the free imine (**72**; $R^3 = H$), but this is unstable, and readily decomposes with loss of sulfur to give derivatives of *N*-cyanothiourea ⟨62JIC263⟩. The thione (**30**) and its *N*-acetyl derivative behave similarly on treatment with potassium hydroxide ⟨04LA(331)265⟩, as does the oxo analogue (**29**) ⟨19LA(419)217⟩.

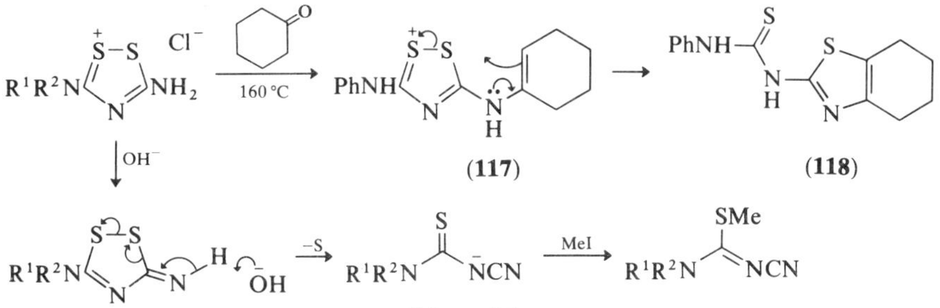

Scheme 26

Amino derivatives of 1,2,4-dithiazoles lacking a free NH group can be hydrolyzed without ring rupture. Thus the imine (**119**) is converted into the oxo derivative (**120**) with H_2SO_4 ⟨76JHC883⟩, and the guanidine (**32**) into the benzenamine derivative (**1**; $R^1 = Ph$, $R^2 = H$, $X = S$) with strong base ⟨80JCR(S)266⟩.

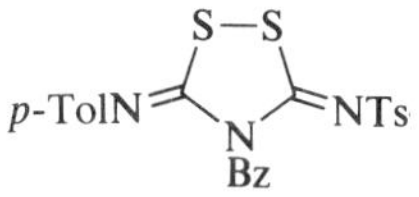

4.34.3.3.5 *O-Linked substituents*

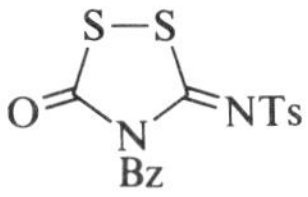

Few reactions have been studied involving *O*-linked substituents. Three reactions in which nucleophilic addition at an exocyclic carbonyl group is followed by ring opening are given in Scheme 16, and structures (**65**; R = Ar, X = S, Z = O) can be converted into the analogous thiones with P_4S_{10}, the reaction being reversed with mercury(II) acetate ⟨66BSF1183⟩. In a preparation of the interesting dithiazoledione (**122**), the ethoxy derivative (**121**) was treated with strong acid ⟨70AG(E)54⟩. The product has a remarkably complex IR spectrum, both in the carbonyl region, and above 3000 cm^{-1} (Table 7).

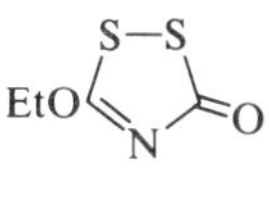

4.34.3.3.6 *S-Linked substituents*

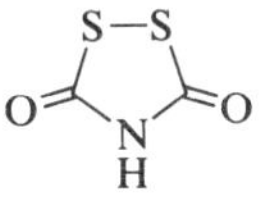

1,2,4-Dithiazole-3-thiones of types (**1**) and (**2**; X = S) are alkylated at sulfur to give thioalkoxythiazolylium salts (**123**), suitable reagents being alkyl iodides, dimethyl sulfate, and triethyloxonium tetrafluoroborate. The SR group is readily displaced: by secondary amines, providing an alternative route to cations (**83**; R^3 = dialkylamino) ⟨72JHC447⟩; by primary aryl amines to give imines (**2**; X = NAr^2), which are reported to react further producing structures (**124**; Scheme 27) ⟨70CJC2142⟩ [compare structure (**90**)]; with pyrroles at either the α or β positions with the formation of, for example, structure (**125**) ⟨68AG(E)296⟩; and with active methylene compounds to yield, for example, the enone (**126**), which can be transformed into the azatrithiapentalene (**127**) ⟨67CB4027⟩.

N-Acyl derivatives of the thione (**30**) are also alkylated at the exocyclic sulfur atom, the product acylimines (**38a**) showing evidence of contributions from structures (**38b**) and (**38c**), since their IR carbonyl frequencies are found near 1570 cm^{-1} ⟨75LA1018⟩. Structures of this type are readily converted into diazatrithiapentales with P_4S_{10}, *e.g.* (**38**) → (**39**) ⟨76CB3108⟩.

Cycloadditions with nitrile oxides occur across the C=S bond both with 1,3,5-oxathiazole-2-thiones ⟨66JOC2417⟩ and 1,3,4-dithiazole-2-thiones (Scheme 28) ⟨67BSF2239⟩, the reaction proceeding *via* an interesting spiro compound which has been isolated in some cases.

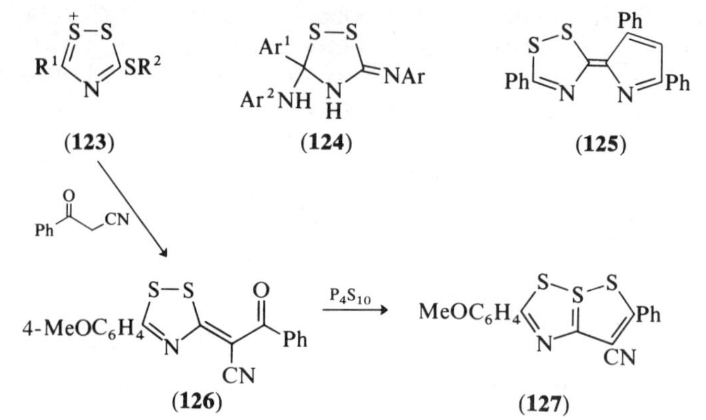

Scheme 27 Reactions of thialkoxy 1,2,4-dithiazolylium salts

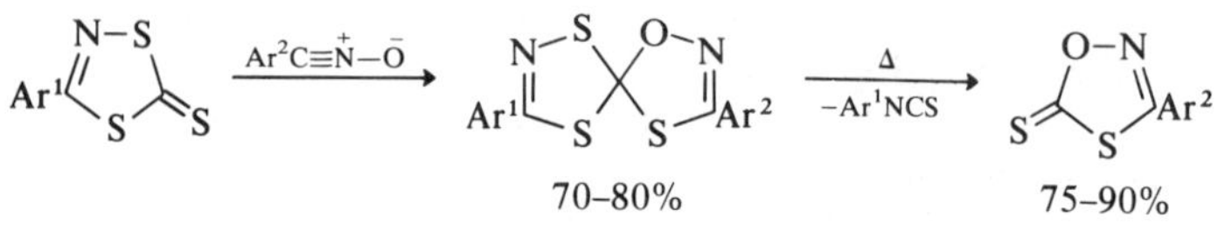

Scheme 28 Cycloaddition with nitrile oxides

The aryl 1,2,4-dithiazole-3-thiones (**2**; X = S) are oxidized with peracetic acid or with chlorine to the corresponding dithiazolones ⟨70CJC2142⟩ in marked contrast to 1,2-dithiole-3-thiones which give 1,2-dithiolylium salts (see Chapter 4.31). The conversion also occurs with mercury(II) acetate ⟨69LA(727)22⟩. Additionally, the thione group has been displaced with *N,N*-dimethylhydrazine ⟨75ZOB485⟩ and with an NCO_2Et group from N_3CO_2Et ⟨76CJC3879⟩.

4.34.3.3.7 Halogen atoms

4,5-Dichloro-1,2,3-dithiazolylium chloride (**128**) is attacked at the most electrophilic carbon atom by *O*-, *S*- and *N*-nucleophiles to yield the products (**129**) ⟨80GEP2848221⟩. The chlorodithiazolone (**77**) can be hydrolyzed to the dione (**122**) using aqueous propanone ⟨71CB2732⟩.

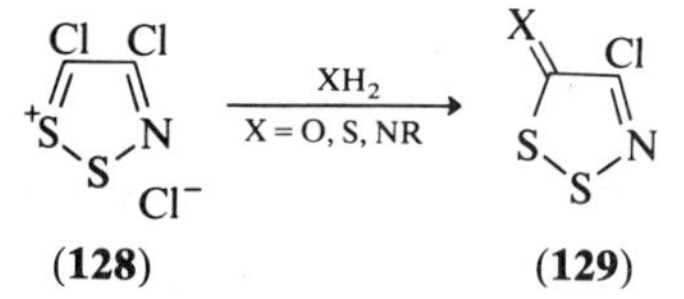

4.34.3.3.8 Substituents attached at ring nitrogen

Only structures of type (**11**) can bear a substituent at nitrogen, and few reactions have been reported. However, the protected amino acids (**81**) may be generated from their methyl, *t*-butyl, or trimethylsilyl esters by treatment with strong acid or water ⟨77JA7363⟩.

4.34.3.4 Reactions at Non-aromatic Rings

This section comprises reactions of ring systems (**16**)–(**24**) and benzo fused derivatives, dihydro compounds preceding tetrahydro derivatives.

4.34.3.4.1 Thermal and photochemical reactions formally involving no other species

Azolines of general structure (**130**; Scheme 29) fragment of thermolysis in a manner analogous to that for the azolones (**64**; Scheme 5). When X = O and Y = S, high yields of ketone and nitrile are formed, the latter via an intermediate nitrile sulfide, which has been trapped with alkynes and nitriles ⟨80CC714⟩ (see Section 4.34.3.4.7). For the isomeric structures having X = S and Y = O, fragmentation gives a carbonyl compound together with a thioacyl nitrene, which rearranges to an isothiocyanate, both products being formed in high yields. Substituents R^1 may be alkyl or aryl and R^2 and R^3 may be alkyl, aryl, SR or OR, making this a very general reaction ⟨72CB2815⟩. Further, since the azolines are prepared from a nitrile oxide and a C=S-containing compound, the sequence of, for example, Scheme 30 provides an attractive route from C=S to C=O in compounds sensitive to the usual reagents ⟨79CB1873⟩. For conjugated C=S compounds, the stereochemistry at the C=C bond, however, may be lost ⟨73BSF1973⟩. Dioxazolines (**130**; X = Y = O) fragment analogously giving ketones and isocyanates. For reactions studied in $PhNO_2$ and Bu_2^nNH, the process was first order, having $E_a \approx 125$ kJ mol^{-1}, and the ease of formation of ketones decreasing in the order $Ph_2CO > PhCOMe > Me_2CO >$ cyclohexanone. When R^1 was substituted aryl, the nature of the substituent had little effect ⟨67BCJ664⟩.

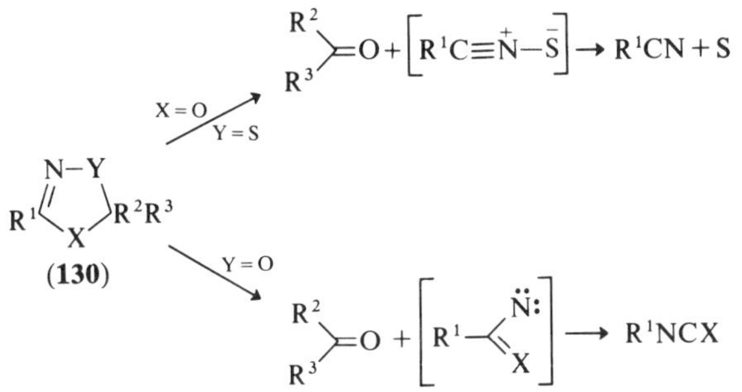

Scheme 29 Thermolysis of Δ^4-1,3,4-azolines

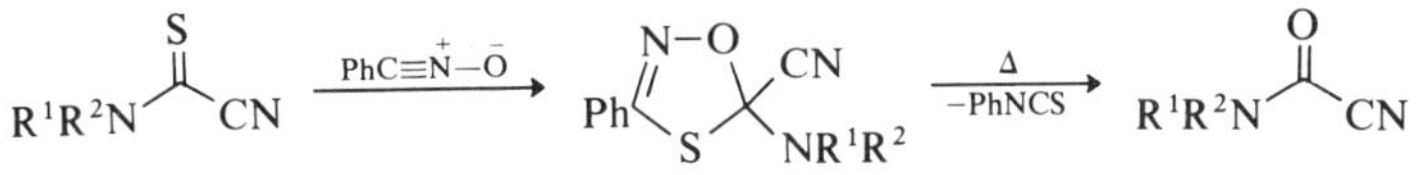

Scheme 30 Conversion of C=S to C=O

On heating in benzene, the dioxazoline (**131**; $R^2 =$ Me; Scheme 31) is converted into the triacylamine (**132**), whereas the protio analogue (**131**; $R^2 =$ H) rearranges at room temperature to the imido ester (**133**) ⟨79JHC129⟩. The products could result from O—O bond fission and a concomitant acyl or hydride shift.

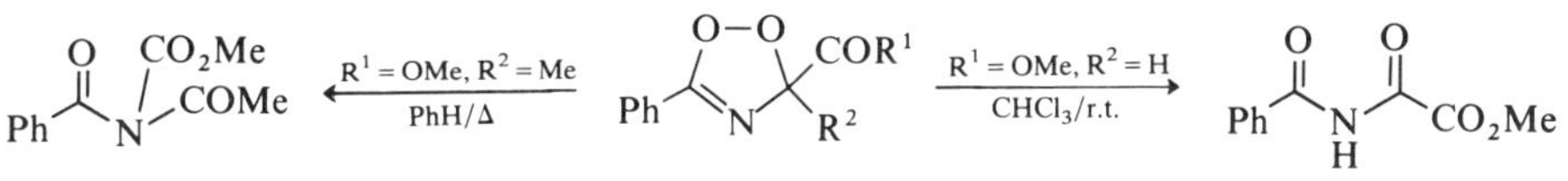

Scheme 31 Thermolysis of dioxazolines

The quantitative thermal rearrangement of the iminooxathiazoline (**134**) to the sulfonylamidine (**135**) may occur by a concerted 1,7 hydride shift, or via a dipolar intermediate from ring fission. A kinetics study in several solvents was not able to distinguish between the two possibilities, but in benzene-d_6 activation parameters were $E_a = 68.2$ kJ mol^{-1} and $\Delta S^\ddagger = -134$ J mol^{-1} K^{-1} ⟨79TL49⟩.

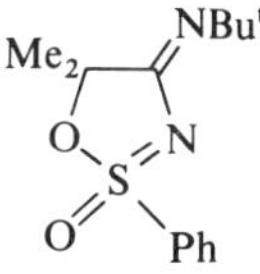

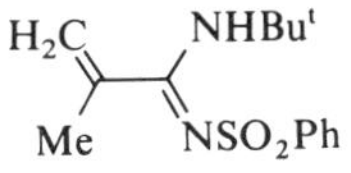

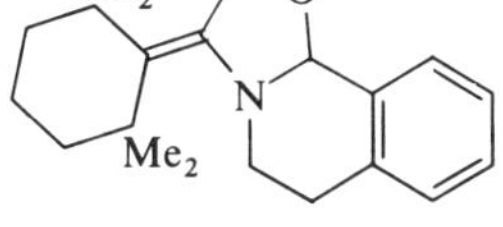

A rare example (**136**) of the 1,2,4-oxathiazolidine ring system rearranges to the oxazolidine (**137**) with a half life of 14 h in CDCl$_3$ at 52 °C, cleavage of the S—O bond leading either to an intermediate diradical or to a dipolar ion ⟨78AG(E)455⟩.

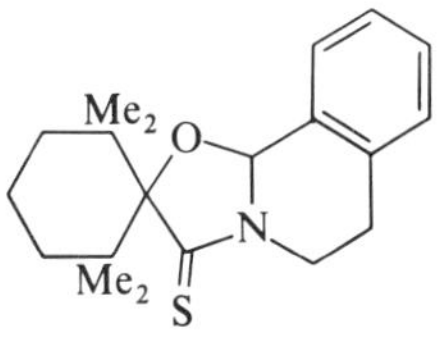

The thermal decompositions of a number of azolidines have been studied (Scheme 32). Thus 1,2,4-dioxazolidines, e.g. (**138**), rearrange to cyano acids ⟨69JCS(C)2671⟩ while their

N-nitroso derivatives (**58**) fragment both thermally and photochemically to give high yields of ketones and N_2O ⟨74JOC1791⟩. 1,3,2-Dioxazolidines, *e.g.* (**139**), fragment into a nitrene and a carbonyl compound ⟨71JA2463⟩; the 1,3,4-dioxazolidine (**140**) gives a nitrone ⟨80JCR(S)122⟩; and 1,2,3-oxathiazolidine *S*-oxides (**55**) extrude SO_2 and form piperazines in low yield ⟨68BCJ1925⟩.

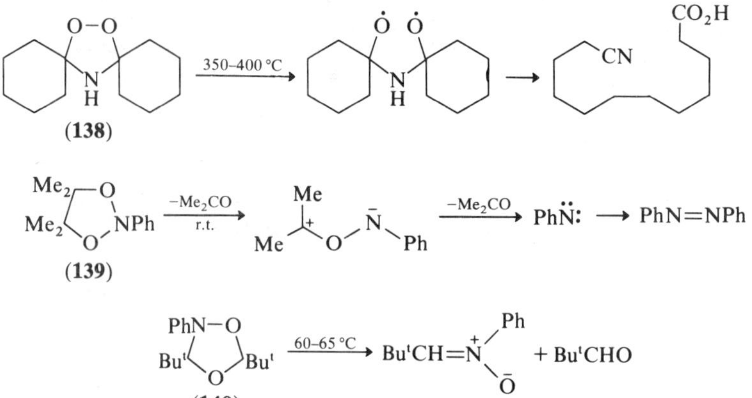

Scheme 32 Fragmentations of dioxazolidines

A remarkably stable benzothiazet-2-yl radical (**141**) is formed from the 1,2,3-benzodithiazoline *S*-oxide (**82**; R^1 = H) at 90–100 °C in biphenyl ⟨78ZC323⟩. The 4,6-dichloro analogue, however, gives the violet pigment (**142**), thus making it useful in thermal copiers, and the related *S*-oxide (**28**) gives a mixture of at least three products in which the aromaticity of the benzene ring has been regenerated ⟨79BCJ3615⟩.

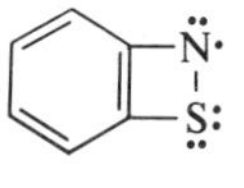

4.34.3.4.2 Electrophilic attack at ring nitrogen and carbon

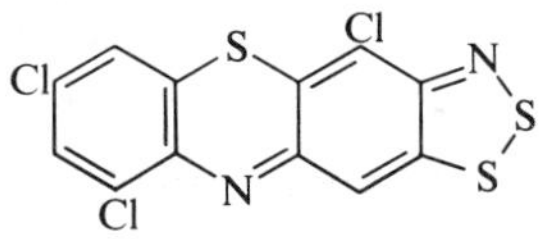

Of the structures (**16**)–(**24**), only those of type (**16**) are expected to react readily with electrophiles at carbon. However, little is known of such reactions, though the oxathiazolines (**143**) are reported to give dibromo derivatives ⟨74AP291⟩.

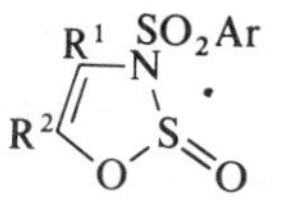

Electrophilic attack at nitrogen is better documented, stereochemical inversion at sulfur of an ephedrine derivative from its *cis* to its *trans* form (**144**) with HCl probably occurring *via* initial *N*-protonation, followed by ring opening involving S_N attack with Cl^- ⟨73JA6349⟩. The rate constant for the acid-catalyzed hydrolysis of the oxathiazolidine (**145**) was maximum in 3.0M $HClO_4$, and the mechanism, different from that for ethylene sulfite, was deduced to be as shown in Scheme 33, with $\Delta S^{\ddagger} = -80.0 \pm 6.0$ J mol^{-1} K^{-1} ⟨75JOC949⟩.

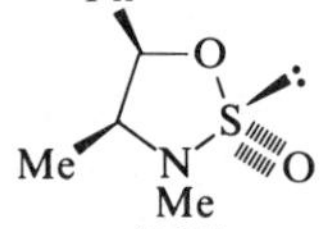

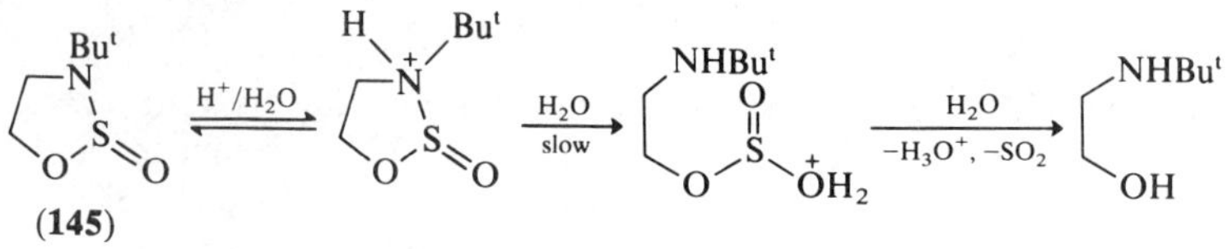

Scheme 33 Acid-catalyzed hydrolysis of 1,2,3-oxathiazolidine *S*-oxide

1,3,4-Diathiazolidines (**50**) having $R^2 = H$ are claimed to be alkylated at ring nitrogen either directly using diazomethane or methyl iodide, or by alkylation of the sodium salt ⟨77CCC2672⟩ (see Section 4.34.2.5). The 1,2,3-oxathiazolidinone (**146**) is sufficiently acidic to form a silver salt, which also gives an *N*-methyl derivative with methyl iodide ⟨74MI43400⟩. The disulfonimide (**147**; X = H) is an acid of strength comparable to HCl in water, and forms a number of metal salts, of which the sodium salt is readily converted into *N*-alkyl derivatives ⟨57MI43400⟩, and the silver salt chlorinated (or brominated) to *N*-halo compounds ⟨69JOC3434⟩. 1,2,4-Dioxazoles of general structure (**148**) can be acylated to urea derivatives with PhNCO ⟨69JCS(C)2678⟩, and also nitrosated to give *N*-nitroso compounds (**58**) ⟨74JOC1791⟩.

(**146**)   (**147**)   (**148**)

4.34.3.4.3 Attack at ring sulfur

Electrophilic attack at ring sulfur in oxathiazoles and dithiazoles by certain oxidizing agents leads to sulfoxy and sulfonyl compounds. For example, the *S*-oxide (**149**) may be prepared from the 1,3,4-oxathiazoline with acidic $KMnO_4$ ⟨72CB2815⟩ and the 1,2,3-oxathiazolidine *S,S*-dioxide (**151**) from the *S*-oxide (**150**) with peracetic acid ⟨76MI43401⟩. Oxidation of the 1,2,3-dithiazoline (**27**) with MCPBA gives the *S*-oxide (**28**) together with the benzodithiazoline *S*-oxide (**152**) and the sulfinylamine (**53**), the latter two compounds also forming on thermolysis of the *S*-oxide (**28**) ⟨79BCJ3615⟩. Thermolysis of the dithiazoline (**27**) in air also gives the *S*-oxide (**152**) together with 2,4,6-tri-*t*-butylbenzenamine, the mechanism being ionic when catalyzed by TsOH, but radical in its absence, as indicated by ESR signals ⟨79BCJ2002⟩. Perhaps the latter proceeds *via* species analogous to the radical (**63**). Compounds of general structure (**82**) may be converted back to 1,2,3-benzodithiazolylium cations (**4**) by TFA, boron trifluoride etherate, or perchloric acid in acetic acid ⟨66JHC518⟩. The latter two methods give stable Herz salts.

(**149**)   (**150**)   (**151**)   (**152**)

Nucleophilic attack at ring sulfur may displace an exocyclic substituent, or lead to S—O or S—S bond cleavage and ring opening. An example of the former is conversion of the 1,2,3-oxathiazolinone (**153**; X = F) into the *N*-amino derivative (**153**; X = NEt₂) with Et₂NH, or into the structure (**146**) with water ⟨74MI43400⟩. Grignard reagents attack the ephedrine derivative (**144**) to give low yields of, for example, the sulfinamide (**154**) with total retention of configuration at sulfur. The coproduction of substantial amounts of (*p*-Tol)₂SO is virtually eliminated, and the yield of sulfinamide considerably increased by use of methyl- or phenyl-lithium, but the reaction is less stereoselective ⟨73JA6349⟩. Ring opening of the 1,2,4-dithiazoline (**36**) has been observed with PhMgBr to give the thioamide (**155**) ⟨74AP828⟩, and the action of Ph₃P on the 1,2,3-dithiazoline (**27**) leads to bis(tri-*t*-butylphenyl) sulfodiimide ⟨80BCJ205⟩. In a reaction analogous to that of Scheme 9, the 1,2,4-dioxazoline (**131**; R^1 = OMe, R^2 = Me) is converted by Ph₃P into the acylimido ester (**156**) ⟨77TL447⟩. Finally, mild hydrolysis of the 1,2,3-benzodithiazolines (**82**), formed in

(**153**)   (**154**)   (**155**)   (**156**)

turn from Herz salts (**4**; Scheme 12), provides one of the most general routes to *o*-aminobenzenethiols ⟨57CRV1011⟩.

4.34.3.4.4 Nucleophilic attack at ring carbon

1,3,4-Dioxazolines of general structure (**157**) are remarkably stable towards hydrolytic ring fission. Thus, the triphenyl derivative resists 40% NaOH solution and boiling with 10% HCl ⟨56MI43400⟩, though it is cleaved to benzoic acid and diphenylmethanone on boiling with 60% HI. However, for $R^1 = R^2 =$ alkyl or $R^1 = H$ ⟨72CB2805⟩, the ring is cleaved by acetic acid to the arylhydroxamic acid and a carbonyl compound. For the latter class ($R^1 = H$), the reaction follows a different course with strong base (see Section 4.34.3.4.5).

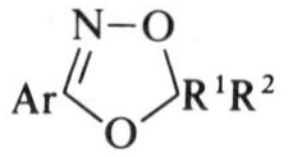

(**157**)

The 1,2,4-dioxazolidine (**138**) is converted with NaOMe/MeOH into caprolactam and cyclohexanone *via* a sequence involving enlargement of one ring ⟨69JCS(C)2663⟩, though 3,5-diheptyl-1,2,4-dioxazolidine is reported only to give heptanal on reaction with aqueous NaOH ⟨80ZOR782⟩.

The oxathiazolinone (**158**; Scheme 34) is cleaved by aqueous base to the hydroxyamide (**159**), and by ·pyridine to the dipolar ion (**160**) ⟨70TL127⟩, and while the oxathiazoline dioxides (**161**) react analogously with pyridine or Ph$_3$P to give dipolar ions, *e.g.* (**162**), convertible into salts in high yields with HClO$_4$, the intermediate ions formed with secondary amines fragment to give salts such as (**163**) ⟨64LA(671)135⟩.

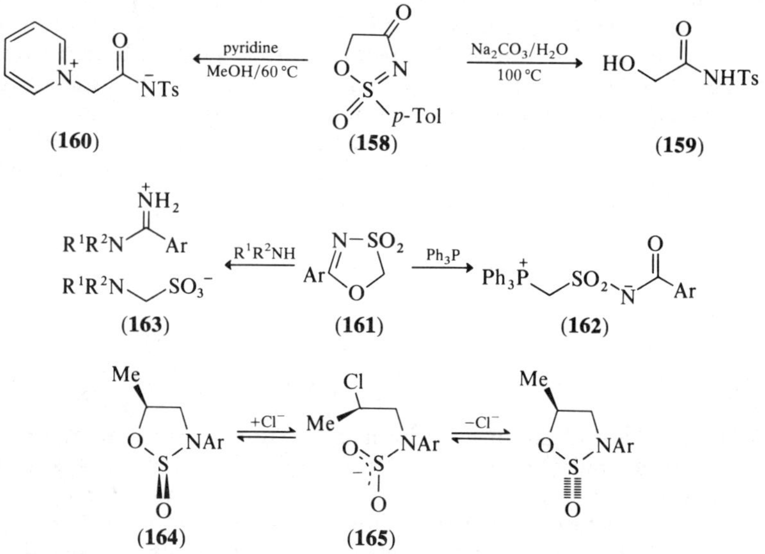

Scheme 34 Nucleophilic attack at ring carbon

Treatment of the oxathiazolidine (**55**) at 140 °C with PhNH$_2$ or PhCO$_2$H leads to S_N attack at the oxy carbon, with loss of SO$_2$ and formation of 2-substituted aminoethanes ⟨68BCJ1925⟩. Likewise, *C*-methyl derivatives, *e.g.* (**164**), are catalytically isomerized at sulfur by halide ions to an equilibrium mixture of *cis* and *trans* isomers *via* the intermediates (**165**) ⟨75BCJ929⟩.

Few reactions with complex metal hydrides have been reported, but the dioxazoline (**157**; Ar = $R^1 = R^2 =$ Ph) is reduced by LAH in THF to Ph$_2$CHOH and BzNH$_2$ ⟨56MI43400⟩.

4.34.3.4.5 Attack at ring hydrogen

Oxidation of the 1,2,4-dithiazoline (**36**) to the cation (**34**) with trityl tetrafluoroborate ⟨74AP828⟩ is an example of electrophilic attack at ring hydrogen. Nucleophilic attack with

Bu^{t}O$^-$ on the 1,3,4-dioxazoline (**157**; Ar = Ph, R^1 = CO$_2$Et, R^2 = H) results in fragmentation to oxalate and benzonitrile ⟨72CB2805⟩, and while formation of butanal and butanamide from 3,5-dipropyl-1,2,4-dioxazolidine on heating with Et$_3$N ⟨69JCS(C)2678⟩ may be a related reaction, the mechanism may also be analogous to that for the formation of caprolactam from the dioxazolidine (**138**) (see previous section).

For the 1,2,3-benzodithiazolines (**82**), $-pK_a$ for the NH proton shows a linear correlation with the Hammett σ constants for a range of 6-substituents, with $\rho = 6.35$ ⟨79KGS1205⟩.

4.34.3.4.6 Reactions at surfaces

The stability of the 1,3,4-dioxazoline ring system (**157**) is further illustrated by its resistance to reduction by Zn/AcOH, Na/EtOH and Al/Hg/Et$_2$O ⟨56MI43400⟩, and by Pt/H$_2$ (see Section 4.34.3.5.1). The rings of other dioxazole derivatives are cleaved by catalytic hydrogenation, the 1,2,4-dioxazole (**131**; R^1 = OMe, R^2 = Me) to the amido ester (**166**) ⟨77TL447⟩, the 1,3,2-dioxazoles (**167**) to cis-cyclohexane-1,2-diol and an arylamine ⟨71JA2463⟩, and the 1,3,4-dioxazole (**140**) to Bu^{t}CHO and Bu^{t}CH$_2$NHPh ⟨80JCR(S)122⟩, all products being formed in high yields. There appear to be no data on the reduction of oxathiazole and dithiazole derivatives having structure types (**16**)–(**24**).

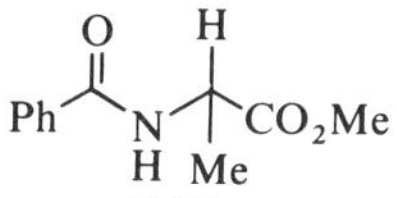

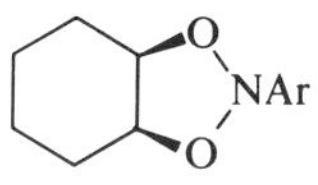

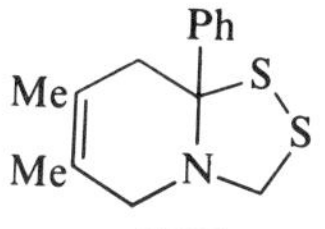

4.34.3.4.7 Reactions with cyclic transition states

A Diels–Alder reaction has been observed between the dithiazolinium cation (**35**) and 2,3-dimethylbutadiene to yield a dithiazolidinium salt, which in turn is readily deprotonated with Et$_3$N to the free base (**168**) ⟨74AP828⟩.

Thermolysis of the 1,3,4-oxathiazolines (**130**; R^1 = Ar, X = O, Y = S) gives a ketone and a nitrile sulfide (see Section 4.34.3.4.1; Scheme 29). The latter can be trapped with DMAD to give isothiazoles (**93**) or with EtO$_2$CCN to yield 1,2,4-thiadiazoles (**96**; R^2 = CO$_2$Et). Trapping with ethyl propynoate leads to a mixture of the two possible isomeric isothiazoles, their ratio (1.31–1.34) being apparently independent of R^2, R^3 and of the aryl substituent ⟨80CC714⟩ (compare with Section 4.34.3.2.6).

Cycloadditions involving the ring sulfur and the exocyclic imino group of the 1,3,4-dithiazolidines (**169**) are formally analogous to the generalized reaction of Scheme 20, except that a sulfene is extruded in the last step. Thus, PhNCO and PhNCS give products (**170**) and (**171**) resulting from different modes of cycloaddition, as in Scheme 22, and t-butylcyanoketene leads to the thiazolidine (**172**) ⟨78JOC4951⟩. The iminooxathiazoline (**173**) reacts with isocyanates forming oxazolidines (**175**), together with increasing amounts of the isomeric imidazolidines (**176**; X = O) as the solvent polarity increases (Scheme 35). Isothiocyanates and ketenes give analogues of structure (**176**); the intermediate is thought to be a dipolar ion (**174**) ⟨79JOC3991⟩.

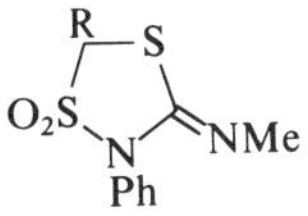

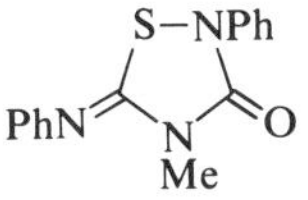

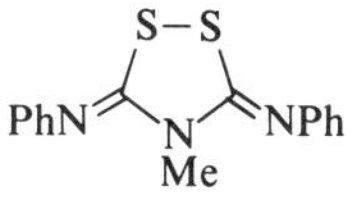

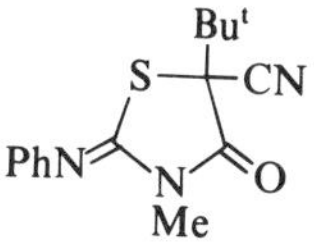

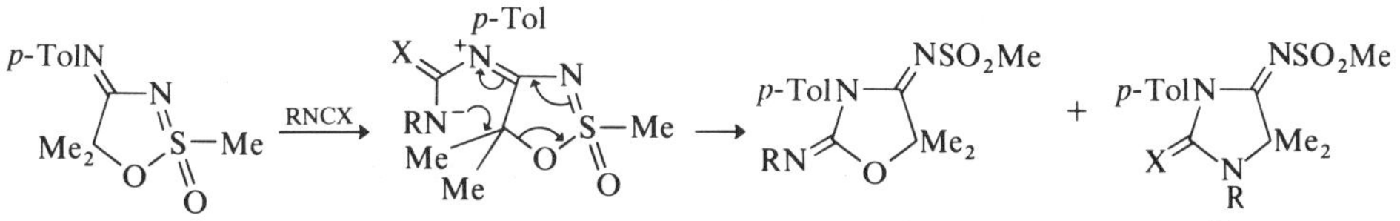

Scheme 35

4.34.3.5 Reactions of Substituents on Non-aromatic Rings

4.34.3.5.1 C-Linked substituents

The high degree of stability of the 1,3,4-dioxazoline ring system (**157**) has permitted numerous transformations to be carried out on substituents at the 2-position. The *cis*-enone derivative (**177**) can be isomerized to the *trans* form with I_2 or tertiary amines, and secondary amines add by a Michael-type reaction to the C=C bond. Selective hydrogenation of this bond is also possible over Pt, though ring cleavage occurs with excess hydrogen ⟨80JCS(P1)2408⟩. A haloform reaction on the methyl ketone (**178**; R = Me) yields the corresponding carboxylic acid, which with CH_2N_2 gives the methyl ester, convertible into the amide with NH_3 ⟨72CB2805⟩. Hydrazine transforms the ethyl ester (**178**; R = OEt) into the hydrazide without ring opening, and this with nitrous acid yields the azide. The ethyl ester can also be hydrolyzed with 40% KOH in EtOH to the carboxylic acid, and converted *via* the acid chloride into a highly effective penicillin derivative ⟨65USP3222360⟩ (see Section 4.34.5.2).

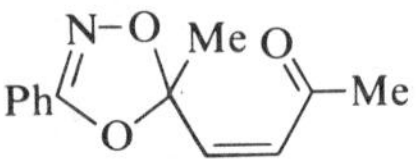

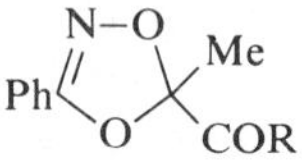

4.34.3.5.2 S-Linked substituents

The methyl sulfide (**51**) and related alkyl and acyl derivatives, which are readily prepared from the sodium sulfide, show reactions typical of isodithiocarbamates. Secondary amines, for example, give the dialkylamino derivatives (**49**) ⟨62AG(E)594⟩.

4.34.3.5.3 Halogen atoms

Imido chlorides such as the oxathiazolines (**179**) behave typically, reacting with water to form cyclic imides, and with amines and sulfur nucleophiles to yield the expected derivatives ⟨75MI43401⟩. The chlorodithiazolone (**77**) is itself prepared by hydrolysis of the chlorodifluoro compound (**180**) ⟨71CB2732⟩.

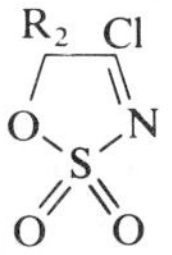

4.34.3.5.4 Substituents attached at ring nitrogen

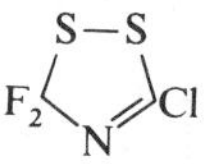

Designed as a potential halogenating agent, the *N*-chlorodisulfonimide (**147**; X = Cl) introduces chlorine into the *ortho*- and *para*-positions of PhMe, but not into the side chain. Reaction with alkenes gives addition products such as the *N*-(2-chlorocyclohexyl) derivative from cyclohexene. The *N*-hydroxy analogue (**147**; X = OH) can be *O*-methylated or -acetylated with, respectively, CH_2N_2 and AcCl ⟨69JOC3434⟩, and while aqueous SO_2 reduces it to the parent imide, treatment with 3M NH_3 followed by acidification gave the first reported example of an aryl 1,2-disulfinic anhydride ⟨81JOC2691⟩.

4.34.4 SYNTHESES

Since thirteen of the possible fourteen ring systems, and three of the possible seven benzo fused systems are known, many in several oxidation states and with a wide range of substituents, it is only possible in this section to discuss the more general approaches to synthesis of the rings. Assembly of the rings will be described from non-cyclic fragments in Section 4.34.4.1 and by transformations of other rings in Section 4.34.4.2, mechanisms being discussed where appropriate. When references are not given in the text, these will be found in Section 4.34.4.3 where there will be tabulated methods for many individual systems, together with literature references.

4.34.4.1 Ring Synthesis from Non-heterocyclic Compounds

4.34.4.1.1 Formation of one bond between two heteroatoms

1,3,2-Benzodithiazole derivatives (**15**) are prepared in good yields by oxidative bromination of the sulfonamides (**181**; Scheme 36), or by dehydration of the sulfoxides (**182**) with $SOCl_2$ ⟨63CB1177⟩. Lower yields, together with the sulfoxides (**182**), are obtained on thermolysis of 2-phenylmercaptobenzenesulfonyl azide, though the 2-phenoxy analogue gives a dibenzooxathiazepine ⟨78JOC1218⟩.

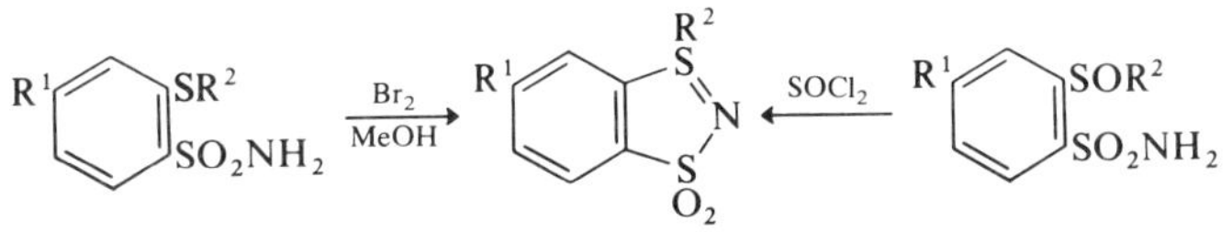

Scheme 36 Synthesis of 1,3,2-benzodithiazole derivatives

The formation of the S—S bond in 1,2,4-dithiazole derivatives (**183**) by oxidative cyclization of the dithioimides (**184**; Scheme 37; Route 1) has been known for 90 years ⟨1893LA(275)20, p. 43⟩ and is still an important method; 1,2,4-dithiazolylium salts can also be prepared this way. The preferred oxidant has been I_2 ⟨69JCS(C)194⟩, though Br_2, Cl_2, H_2SO_4, H_2O_2, $KMnO_4$ and $FeCl_3$ have all been used. An alternative approach developed by Verma and coworkers ⟨63IJC300⟩ involves oxidation of the thioethers (**185**; Route 2) with Br_2 or I_2, and generally having R^3 = Bz, though substituted Bz, CH_2=CHCH$_2$ and PhCOCH$_2$ also give good results. Apparently Route 1 has been used to prepare only one 1,2,4-oxathiazole (**186**), the oxidizing agent being aqueous H_2O_2/NaOH ⟨73MI43401⟩.

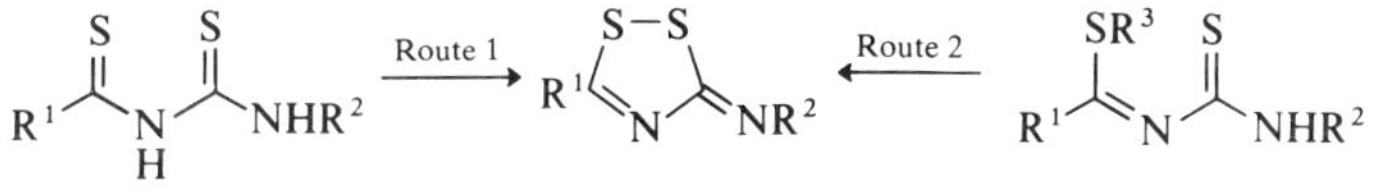

Scheme 37 Synthesis of 1,2,4-dithiazole derivatives by oxidative cyclization

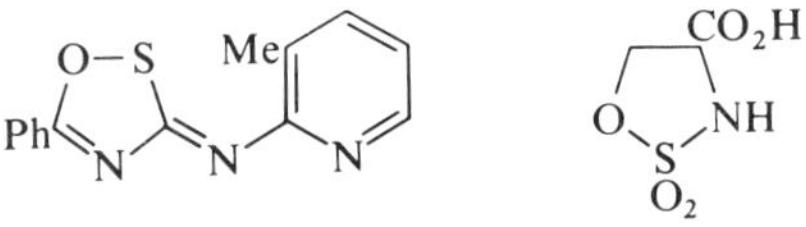

L-Serine *O*-sulfonate has been cyclized to the 1,2,3-oxathiazolidine (**187**) using DCC in DMF ⟨67BCJ1554⟩.

4.34.4.1.2 Formation of one bond adjacent to a heteroatom

The unusual 1,2,3-benzoxathiazole (**14**) was prepared by Tl(III) oxidation of the bis(tosylamide) of 4-bromobenzene-1,2-diamine; treatment of the sulfonamidines (**188**) with $KOBu^t$ to give the 1,2,3-oxathiazolines (**189**) is a related reaction.

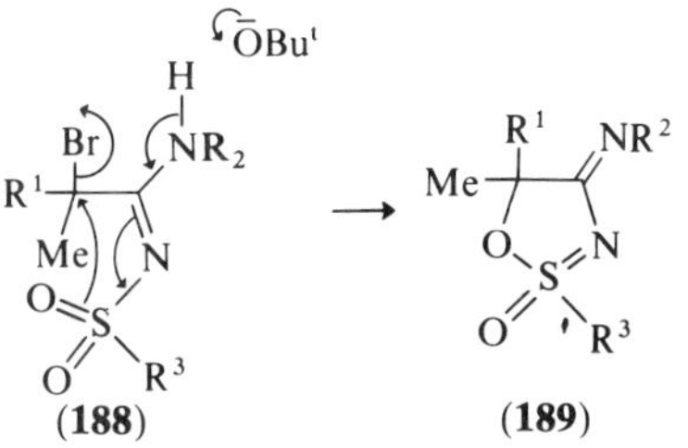

Other intramolecular cyclizations accompanied by elimination are the preparation of the 1,2,4-dioxazolin-3-ones (**190**) by HCO$_2$H dehydrative cyclization of a precursor percarbamic acid ⟨72JPR145⟩; formation of the 1,3,4-dioxazol-5-ones (**191**) from the hydroxamic ester (**192**) ⟨78JAP78127479⟩; and preparation of the 1,3,2-dioxazolidine (**57**) by base treatment of the precursor *N*-methoxy-*N*-hydroxyethoxy compound ⟨80IZV2181⟩.

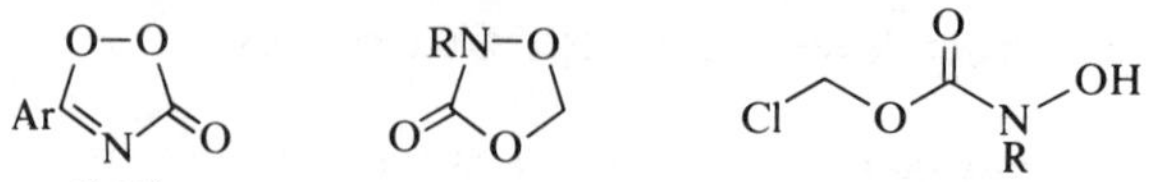

(**190**) (**191**) (**192**)

An electrocyclic ring closure occurs in one preparative method for the 1,3,4-oxathiazolines (**98**; R^1 = Ar). Base treatment of the chloroacylsulfenamine (**193**; Scheme 38) at low temperature gives the unstable intermediate (**194**), which has been isolated in one instance, but which normally cyclizes spontaneously ⟨74JOC2885⟩.

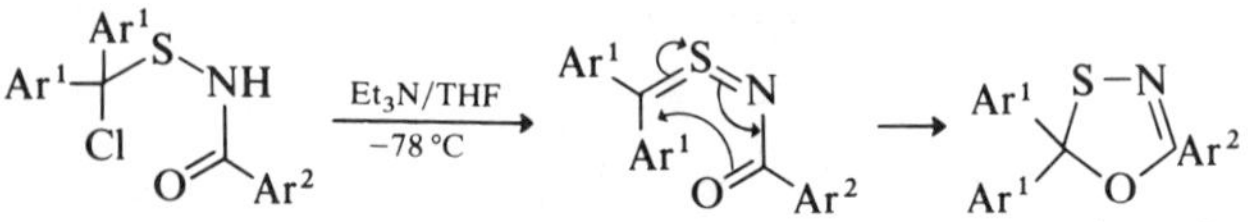

(**193**)   (**194**)   (**98**) R^1 = Ar^2

Scheme 38

4.34.4.1.3 Formation of two bonds: 4-atom fragments and carbon

Hydroxamic acids (**195**) having X = O or S cyclize with reagents capable of undergoing nucleophilic attack twice at the same carbon atom (Scheme 39) giving 1,3,4-dioxazole or -oxathiazole derivatives. The dichloride reagents of Route 3 may have Y = O, S, or NSO$_2$R, and in the products, X may be O or S ⟨72GEP2059990⟩ while by Route 4, Y may be Cl or OEt, and the R groups alkyl or aryl ⟨67BCJ664⟩. By Route 5 ⟨80ZOB1014⟩, chloroenones lead to 1,3,4-dioxazoline ketones in which R^2 may be alkyl or aryl, and alkynes activated by aryl and/or ester groups give appropriately substituted dioxazolines and oxathiazolines ⟨75H(3)563⟩ by Route 6. Reactions, especially by Routes 3 and 5, are often carried out in the presence of base, but some proceed spontaneously at room temperature, and others on thermolysis.

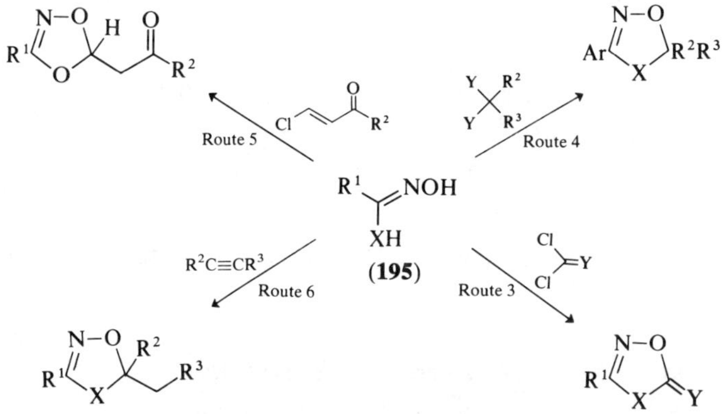

Scheme 39 Cyclization of hydroxamic acids

A carbon atom may also be inserted into a ring as a carbene or carbenoid fragment, and the oxathiazolinones (**24**; R^2 = O) have generally been prepared from a diazoalkane R^{1}R^{3}CN$_2$ and a sulfonylisocyanate R^{4}SO$_2$NCO ⟨74MI43400⟩.

4.34.4.1.4 Formation of two bonds: 4-atom fragments and nitrogen

Benzenedisulfonimide derivatives (**147**; X = H or Ar) are accessible *via* the bis sulfonyl chloride and NH$_3$ ⟨69JOC3434⟩ or arylamines. The *N*-hydroxy compound (**147**; X = OH) is prepared by the action of HNO$_2$ on the disulfinic acid ⟨81JOC2691⟩.

Reactions at high dilution between the appropriate bis sulfenyl chlorides and primary amines have led in low yields to the 1,3,2-dithiazole derivatives (**196**) and (**197**), and nitrosation of substituted 2-mercapto-2-phenylacetic acids in Ac$_2$O gives mesoionic compounds (**12**) ⟨81LA1025⟩.

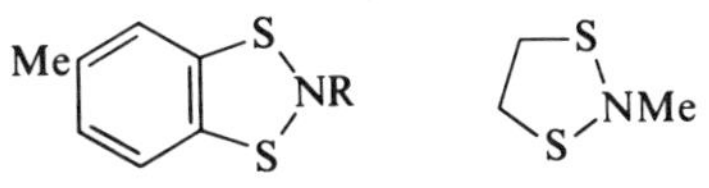

(**196**)   (**197**)

4.34.4.1.5 Formation of two bonds: 4-atom fragments and sulfur

The action of SOCl$_2$ on *o*-aminothiophenols leads to 1,2,3-benzodithiazolylium salts (**4**), presumably *via* the sulfinyl compounds (**82**) ⟨65JOC2763⟩. In an analogous reaction, the *p*-tosylamide of *o*-aminophenol is converted into the 1,2,3-benzoxathiazoline (**198**).

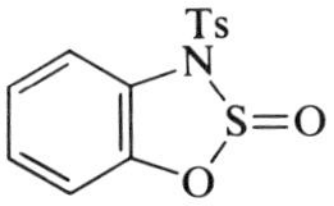

(**198**)

A number of other 1,2,3-oxathiazole *S*-oxide derivatives is also available (Scheme 40) on treatment of various substituted ethanes having a vicinal oxygen and nitrogen atom with SOCl$_2$. The analogous *S,S*-dioxides are formed from cyanohydrins with SO$_2$Cl$_2$. The preparation of fungicidal 1,2,3-dithiazolines (**129**; X = NAr) from ArNHC(S)CN and SCl$_2$ is analogous to the reaction between SOCl$_2$ and cyanohydrins in Scheme 40 ⟨77USP4059590⟩. 1,2,4-Dithiazole derivatives are also available by the action of P$_4$S$_{10}$ on acyl- or thioacyl-imines or acylisothiocyanates, and S$_8$ on thioacylisocyanates (Scheme 41).

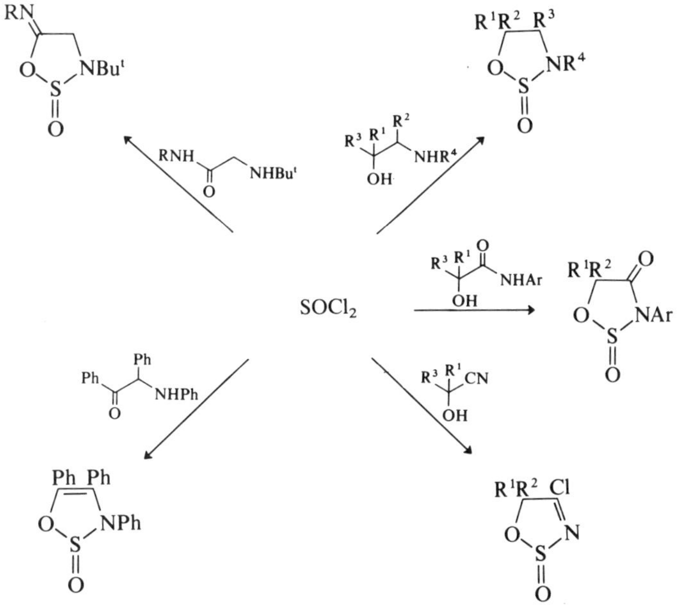

Scheme 40 Cyclization of thionyl chloride to 1,2,3-oxathiazole *S*-oxides

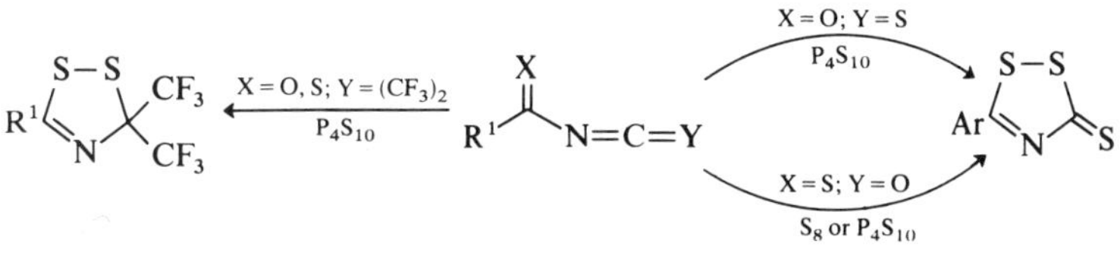

Scheme 41 Formation of 1,2,4-dithiazole derivatives

4.34.4.1.6 Formation of two bonds: [3+2] atom fragments by cycloaddition

1,3-Dipolar cycloadditions have been used successfully in assembling a number of these ring systems. Nitrile oxides, generated from α-chloro oximes and base, add to a wide range of aliphatic and aromatic aldehydes and ketones, but not to esters (Scheme 42), yielding 1,3,4-dioxazolines, and to thiocarbonyl compounds yielding 1,3,5-oxathiazolines, in which R^2 and R^3 may be alkyl, aryl, OR, SR, Ts or $SiMe_3$. Thioamides, however, give products which fragment spontaneously to amides and isocyanates ⟨72CB2815⟩, an exception being that shown in Scheme 30. Kinetic studies for thioketones having R^2 and R^3 as p-substituted aryl groups show the reaction to be first order in each substrate (second order overall), and a plot of $\log k$ against σ for six different p-substituents when R^1 = Mes gave $\rho = 1.02$, showing a small substituent effect. For $R^2 = R^3 = Ph$, $E_a = 33.0$ kJ mol^{-1} and $\Delta S^{\ddagger} = -116$ J mol^{-1} K^{-1}, while for the p-methoxy compounds, the corresponding values were 38.5 kJ mol^{-1} and -110 J mol^{-1} K^{-1}. The data were interpreted in terms of a concerted 1,3-dipolar cycloaddition, with a highly ordered but charge-separated transition state, in which the C—O bond is forming slightly ahead of the C—S bond ⟨71JCS(B)2096⟩. Nitrile oxides also add across the C=S bond of TsNCS and CS_2 (Scheme 42), but in the latter case a second mole of ArCNO reacts across the product exocyclic C=S bond (compare Scheme 28) to give a final product having X = O ⟨66JOC2417⟩. Ketenes and thioketenes yield 1,3,4-dioxazoles and 1,3,5-oxathiazoles, respectively; and sulfines are converted stereospecifically into the 1,3,5-oxathiazoline S-oxides (**199**; $\nu_{S=O}$ 1080 cm^{-1}), except for the sulfine from fluorene, which gives mostly the isomeric 1,2,5-oxathiazoline S-oxide (**200**; $\nu_{S=O}$ 1155 cm^{-1}) ⟨77JCS(P1)1468⟩.

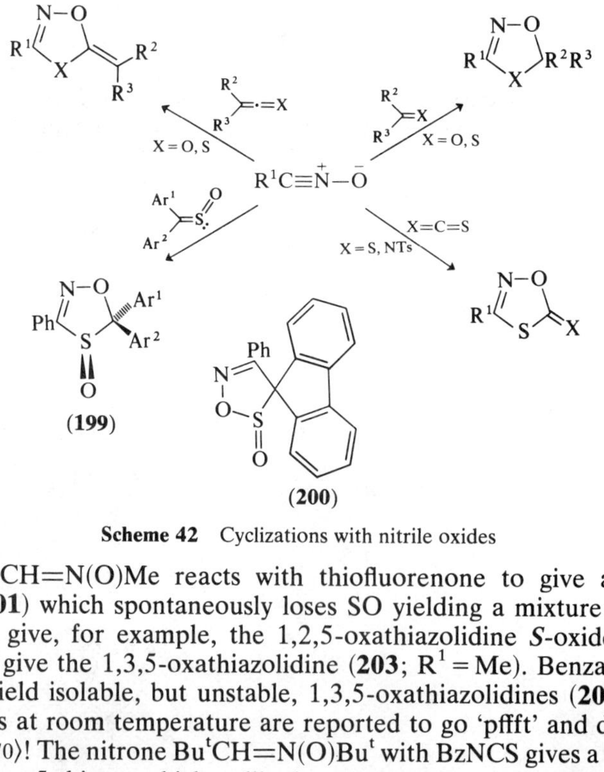

Scheme 42 Cyclizations with nitrile oxides

The nitrone PhCH=N(O)Me reacts with thiofluorenone to give a transient 1,2,5-oxathiazolidine (**201**) which spontaneously loses SO yielding a mixture of products; with certain sulfines to give, for example, the 1,2,5-oxathiazolidine S-oxide (**202**), and with $(CF_3)_2C=C=S$ to give the 1,3,5-oxathiazolidine (**203**; R^1 = Me). Benzaldoximes with the same thioketene yield isolable, but unstable, 1,3,5-oxathiazolidines (**203**; R^1 = H), which after a few minutes at room temperature are reported to go 'pffft' and disappear in a puff of smoke ⟨70JOC3470⟩! The nitrone $Bu^tCH=N(O)Bu^t$ with BzNCS gives a transient 3-imino-1,2,4-oxathiazolidine-5-thione, which unlike its more stable 1,2,4-dithiazolidine analogues (Section 4.34.3.2.1) undergoes a Dimroth rearrangement spontaneously to an isomeric 1,2,4-thiadiazolidine ⟨79AP1027⟩.

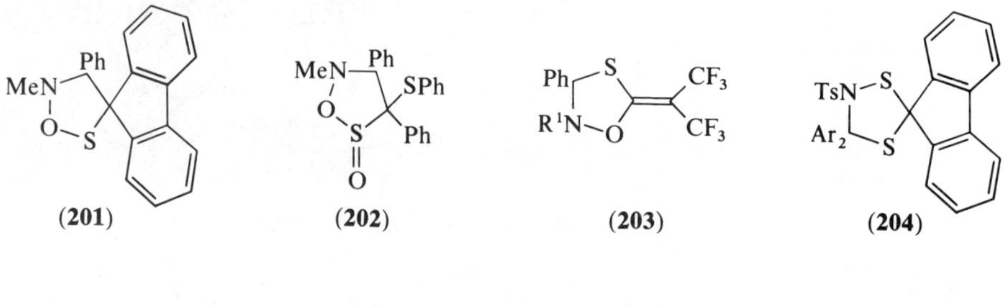

Other thermal 1,3-dipolar cycloadditions are the reactions between nitrile sulfides and electron-deficient aldehydes and ketones forming 1,3,4-oxathiazolines (**98**; Scheme 19); between aryl thioketones and thiofluorenone S-tosylimide to yield the 1,3,4-dithiazolidines (**204**); and between thioketene S-oxides and imines, yielding in one case the 1,2,4-oxathiazolidine (**136**).

Addition of thermally-generated ethoxycarbonylnitrene to 1,4-naphthoquinone in the presence of O_2 gives a small amount of a 1,3,4-dioxazolidine derivative, thought to arise from an initially-formed 1,2,3-dioxazolidine by a process analogous to the molozonide–ozonide rearrangement ⟨77CJC2363⟩.

The reaction between alkyl azides and isothiocyanates produces an isolable thiatriazoline (**205**; Scheme 43), which was thought to lose N_2 on thermolysis to give an unstable intermediate (**206**), since this has apparently been trapped by a number of C=S containing compounds to give 1,2,4-dithiazolidine derivatives ⟨76JHC883⟩. Further reaction with isothiocyanate, however, occurred in three different ways, and depended on the nature of R^3NCS, and on the reaction conditions. For ArNCS, the relative amounts of products (**207**), (**208**) and (**209**) depended upon the aryl *para*-substituent, and the heating time. The thiadiazolidine (**207**) was also convertible into the other products *via* a Dimroth rearrangement ⟨77JOC1159⟩. More recent work, however, suggests a mechanism for addition to the thiatriazoline analogous to that in Scheme 20 with loss of N_2 in the last step, this being supported by a small solvent effect and a moderately negative $\Delta S^{\neq}$. Products (**208**) and (**209**) having R^3 = acyl were interconvertible, and the action of sulfenes on the thiatriazoline (**207**) gave the 1,3,4-dithiazolidines (**169**) ⟨78JOC4951⟩.

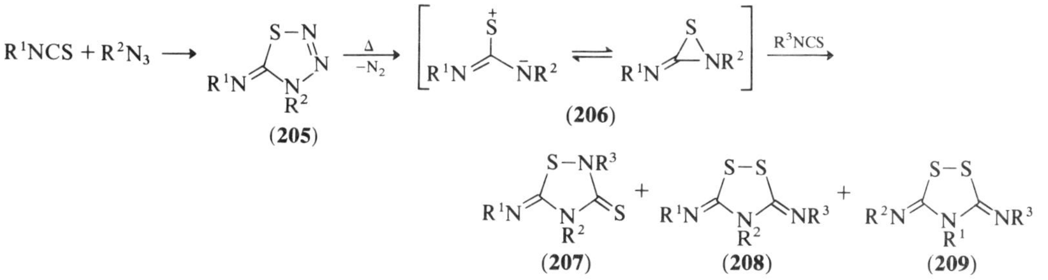

Scheme 43

Benzoyl azide cyclizes photochemically ($\lambda > 300$ nm) with ketones to give intermediate oxatriazolines, which lose N_2 and cyclize to 1,3,4-dioxazolines (Scheme 44). The coproduct oxaziridine (**210**) which has been prepared by other methods ⟨67CB2593⟩ thermally rearranges to the dioxazoline. Light of wavelength 254 nm, however, yields mostly PhNCO, with a small amount of the dioxazoline formed *via* benzoylnitrene ⟨76TL4325⟩.

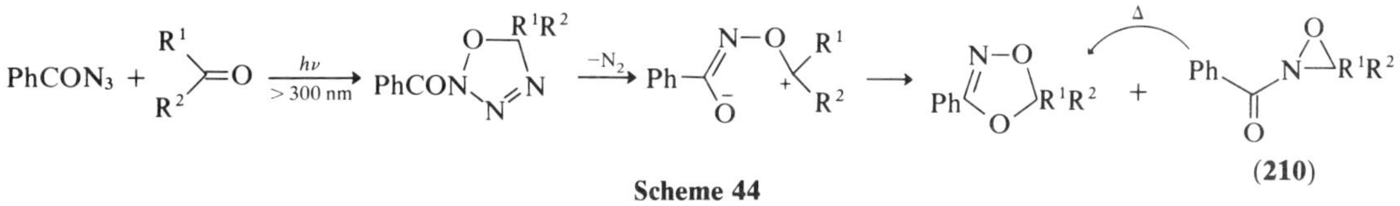

Scheme 44

Nitrobenzene adds photochemically to alkenes at low temperatures to yield 1,3,2-dioxazolidines, *e.g.* (**167**), the mechanism apparently involving addition of the nitro n, π^* triplet in a two-stage electrophilic process ⟨71JA2463⟩. A *thermal* cycloaddition occurs with the strained *cis,trans*-1,5-cyclooctadiene, and for substituted nitrobenzenes rate constants at 25 °C in MeCN gave ρ values of +2.2 and +2.4, respectively, when correlated with σ_p and σ_m substituent constants. The addition transition state appears to resemble starting materials, the rate increasing with increasing electron deficiency at nitrogen ⟨76AG(E)372⟩.

4.34.4.1.7 *Formation of two bonds: [3+2] atom fragments by other processes*

The most extensively studied benzo fused ring system within the scope of this chapter is the 1,2,3-benzodithiazolylium cation of general structure (**4**). Although it was first reported 60 years ago ⟨22GEP360690⟩, and the field (Herz salts) later reviewed ⟨57CRV1011⟩, its controversial mechanism of formation is still not settled. Reaction between primary amines and S₂Cl₂ in various solvents at 70–80 °C leads directly to salts (**4**) as chlorides. Amines having a displaceable *para*-substituent, or PhNH₂ itself, give salts (**4**) substituted by Cl *para* to the nitrogen atom. The reaction also works with ArNHOH, ArNHNH₂, ArNO and ArNSO suggesting a common intermediate ⟨66MI43400⟩, this apparently being an *N*-sulfinylaniline from the observed formation of an equilibrium mixture of structures (**52**) and (**27**; Scheme 3) from 2,4,6-tri-*t*-butylbenzenamine ⟨79BCJ2008⟩. Additionally, the parent (**4**; $R^1 = H$), itself formed by brief treatment of *o*-aminothiophenol with SOCl₂, can be chlorinated at the 6-position with S₂Cl₂ which suggested that chlorination is a nucleophilic rather than an electrophilic process, and that in the Herz reaction it almost certainly occurs after ring formation ⟨69CHE421⟩. Recently, 1,2,3-dithiazolines (**54**) were prepared analogously from enaminoketones, in good yields for $R^3 = Me$, but not for $R^3 = H$ ⟨81BCJ3541⟩. It now appears that in this latter case radicals (**62**) are formed, and analogous structures (**63**) have been prepared by reduction of the cations (**4**) ⟨81ZC324⟩, suggesting their possible intermediacy in the Herz reaction. It is highly probable that the Herz reaction proceeds *via* structure (**211**), but whether conversion into the cation (**4**) is *via* a radical (**212**) or an ionic (**213**) ⟨78ZC323⟩ intermediate remains to be determined. The former intermediate (**212**) is intuitively the more attractive, and may also be involved in the displacement of the 6-substituent by Cl. Interestingly, 4-amino-1,2-naphthoquinone gives the isolable dihydro compound (**214**) with S₂Cl₂ in AcOH, and this can be oxidized to a radical of type (**62**).

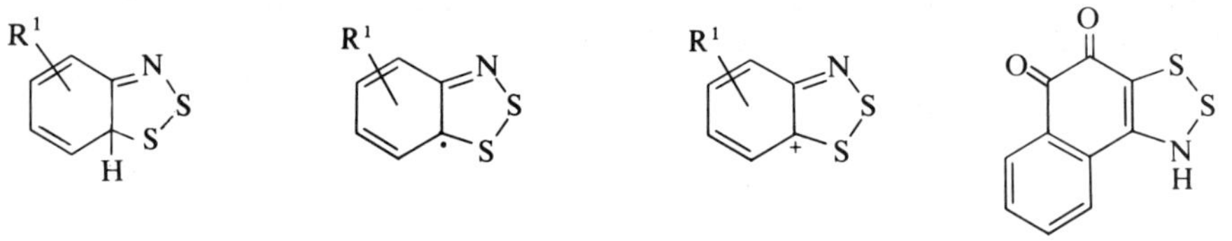

(**211**)   (**212**)   (**213**)   (**214**)

A very general method for preparing 1,3,4-oxathiazole and 1,2,4- and 1,3,4-dithiazole derivatives from sulfenyl chlorides and amides is shown in Scheme 45, and some of this work has been reviewed ⟨70AG(E)54⟩. As is apparent, the nature of the sulfenyl chloride, and particularly of the amide, strongly influences the course of the reaction.

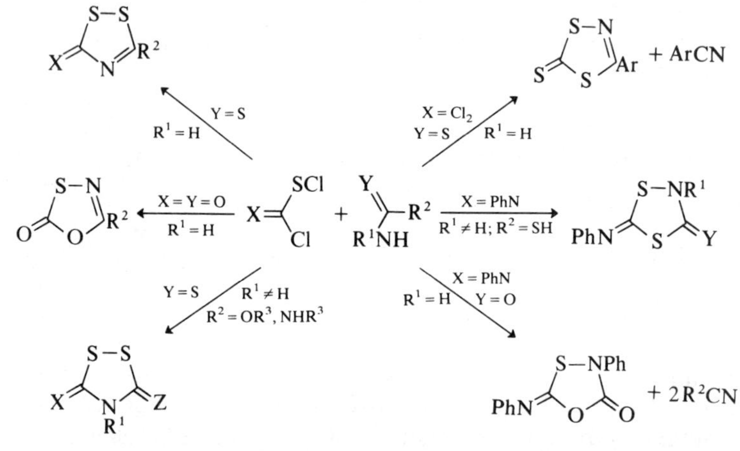

Scheme 45 Cyclizations of chlorosulfenyl chlorides

Structures (**41**; $R^1 = R^2$, $X = O$) known as 'isothiocyanate oxides' are readily prepared by Cl₂ or Br₂ oxidation of isothiocyanates, followed by hydrolysis ⟨79AP1027⟩, the intermediate possibly being a halodithiazolylium salt ⟨67AG(E)649⟩. Chlorination of FCSNCS leads by a related reaction to the difluorochloro compound (**180**) and its trichloro analogue. The corresponding thiones (**41**; $X = S$) have been made by oxidative addition of CS₂ to isothiuronium salts, the mechanism being related to that of Scheme 37, Route 2 ⟨68IJC132⟩.

Oxidation of thioamides with I₂, Br₂ or H₂O₂ to diimino disulfides, followed by acid-catalyzed cyclization with elimination of amine, is an important route to symmetrically substituted 1,2,4-dithiazolylium salts (**43**) ⟨79NKK389⟩, and structures (**47**) ⟨79CHE1187⟩. The preparation of the 3-phenyl-1,2,4-thiazolines (**35**), and thus (**36**), is a related reaction.

1,3,4-Oxathiazoline and 1,3,4-dithiazoline *S,S*-dioxides are readily accessible, having a range of 2-substituents from $ClCH_2SO_2NH_2$ (Scheme 46); *N*-sulfinylanilines provide a two-atom fragment for 1,2,3-oxathiazolidines (Scheme 47), substituted oxiranes giving a mixture of two geometric isomers ⟨71BCJ3073⟩.

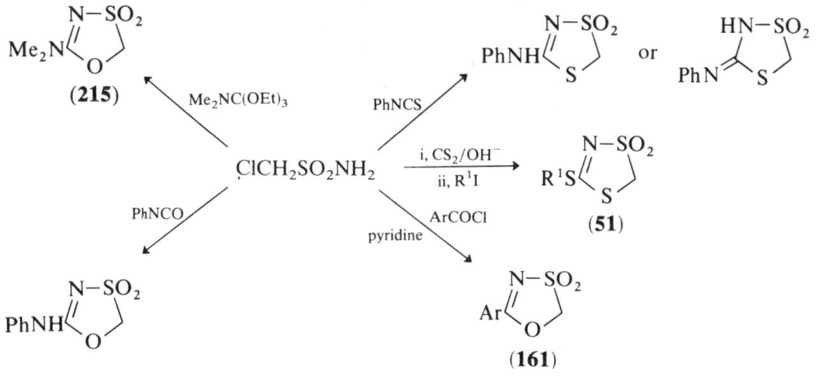

Scheme 46 Cyclizations of chloromethanesulfonamide

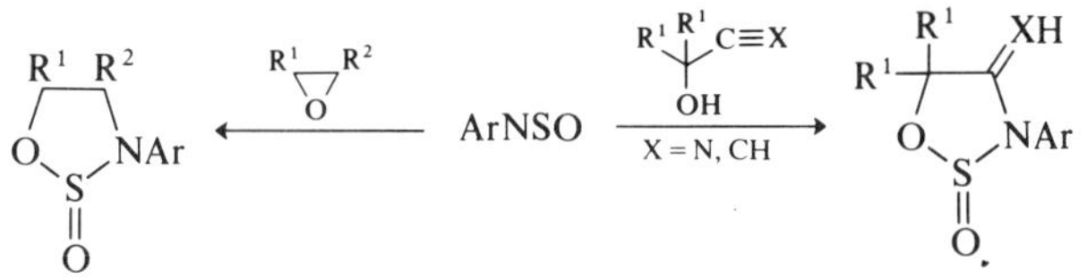

Scheme 47 Cyclizations with *N*-sulfinylamines

Unsymmetrical 1,2,4-dioxazolidines have been prepared by photooxygenation of imines ⟨71JCS(C)160⟩, and both 1,2,4-dioxazolidine-3,5-diones and 1,2,4-dithiazolines are available by base-catalyzed addition of H_2X_2 to 1,3-dihalo compounds.

4.34.4.1.8 Formation of three bonds

Some methods for the synthesis of 1,2,3- and 1,2,4-dithiazole derivatives are considered in this category. The cation (**128**) is formed by heating $ClCH_2CN$ with S_2Cl_2, while those of type (**83**; R^3 = dialkylamino) have been assembled from *N,N*-dimethylthionocarbamic acid derivatives with BrCN or KSCN and a secondary amine ⟨66NEP6500213⟩. Analogues of the dithiazolidinedione (**81**) having a range of *N*-substituents are accessible from ClCOSCl and amino acid esters ⟨77JA7363⟩ or substituted formamides.

Wöhler's original method ⟨1821MI43400⟩ for preparing the aminodithiazolethione (**30**) by action of acid on SCN^- with the coproduction of HCN was shown ⟨69AJC513⟩ to have a rate proportional to $[SCN]^3$ and a linear dependence on the Hammett acidity function h_-, a mechanism being suggested. An alternative mechanism (Scheme 48) was later proposed by another group, who also reviewed the early work on the preparation and reactions of this compound ⟨73MI43400⟩.

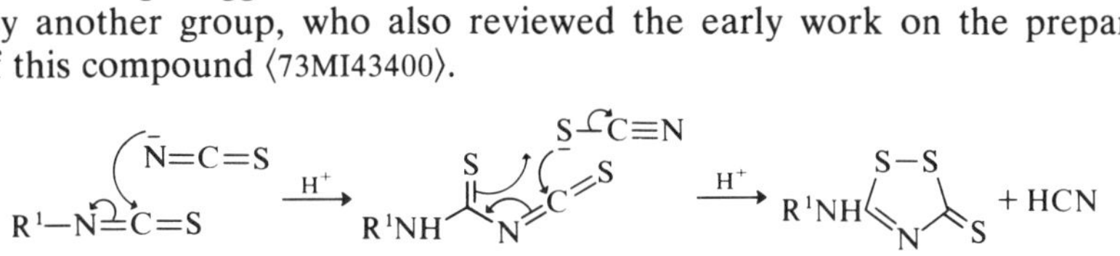

Scheme 48 Acid-catalyzed cyclization of thiocyanate

4.34.4.1.9 Formation of four bonds

The action of NH_3 and 30% H_2O_2 on saturated aldehydes and ketones, catalyzed by sodium EDTA, gives good yields of the symmetrical 1,2,4-dioxazolidines (**148**) though the reaction fails with large ring cyclic ketones. These compounds have also been prepared by ozonolysis of alkenes in the presence of NH_3 ⟨80ZOR782⟩. The 1,3,4-dioxazolidine (**140**) was prepared from PhNHOH and two mol of Bu^{t}CHO, while its analogues (**216**) were formed from a nitroso and a diazo compound, the intermediate oxaziridine reacting further with hexafluoropropanone ⟨79IZV1788⟩.

4.34.4.2 Ring Synthesis by Transformation of Other Heterocycles

4.34.4.2.1 From rings containing less common heteroatoms

2,2,2-Trimethoxy-1,3,2-dioxaphospholenes react with arylsulfonylsulfinylamides, with extrusion of $(MeO)_3PO$ to give 1,2,3-oxathiazoline *S*-oxides (**143**), a mixture of two isomers being obtained when $R^1 \neq R^2$. The action of $SOCl_2$ on the oxazagermolidine (**217**) leads to the *N*-methyl analogue of the oxathiazolidine (**55**) ⟨75JOM(88)C35⟩, while the oxo derivative (**218**) with $SOCl_2$ provides the only known route to the 1,2,3-oxathiazolidin-5-one *S*-oxide (**219**). A number of phosphorus and arsenic heterocycles can also be prepared from these and related germolidines (see also Chapters 1.17 and 1.18).

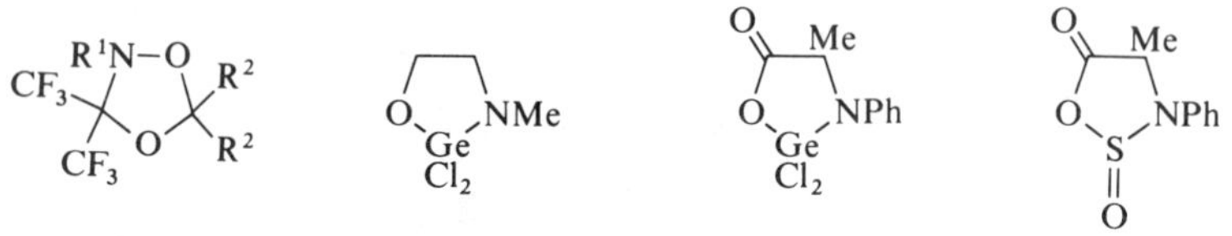

(216) (217) (218) (219)

4.34.4.2.2 From five-membered ring heterocycles with one heteroatom

Oxidation of aryl hydroxamic acids in the presence of excess 2,5-dimethylfuran leads to transient acyl nitroso compounds, which undergo cycloaddition and subsequent rearrangement to 1,3,4-dioxazolines (**177**) having a side chain with *cis* configuration (Scheme 49) ⟨80JCS(P1)2408⟩.

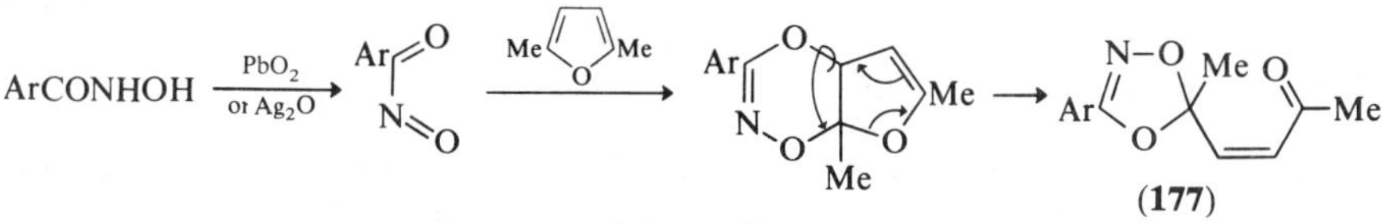

Scheme 49

4.34.4.2.3 From five-membered ring heterocycles with two heteroatoms

The methylene blue-sensitized photochemical addition of singlet O_2 to 5-alkoxyoxazoles in the presence of DABCO gives high yields of 1,2,4-dioxazolines (**131**) *via* a 1,2-addition of oxygen (Scheme 50). The DABCO is believed to inhibit the more commonly observed 1,4-cycloaddition pathway ⟨77S572⟩.

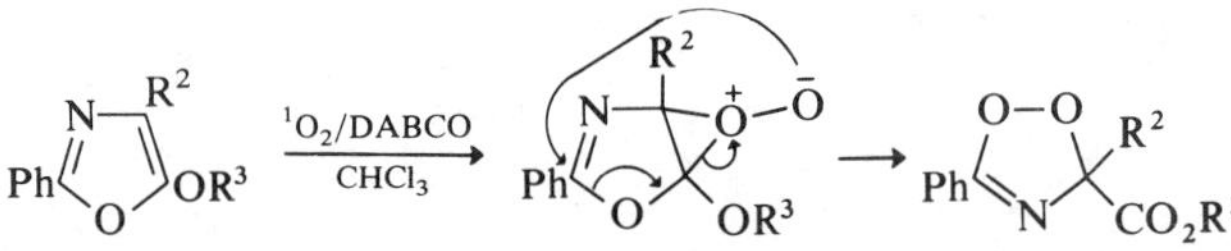

Scheme 50

One of the earliest preparations of the 1,2,4-dithiazole (**30**) was by acid treatment of a concentrated aqueous solution of 2-thioxo-1,3-thiazolidin-4-one ('rhodanic acid') ⟨1842LA(43)76⟩ or of one of its salts; Ac_2O with the ammonium salt gave the *N*-acetyl derivative ⟨1873CB902⟩. Thermolysis of 2-phenyl-1,3-thiazoline-4,5-dione leads *via* loss of CO to thiobenzoylisocyanate, which with sulfur or P_4S_{10} yields a dithiazolinethione (Scheme 41).

The formation of a number of oxathiazole and dithiazole derivatives by ring transformations of 1,2,4-dithiazolinimines (**42**) are illustrated in Schemes 22 and 25. Analogous reactions (Scheme 51) with thiadiazolimines (**220**) lead to the 1,2,4-dithiazoline-5-thione

(**32**) ⟨80JCR(S)114⟩, the 1,2,4-dithiazolidines (**221**) ⟨79CB517⟩ and the 1,3,4-dithiazoline (**222**) ⟨81JHC1309⟩.

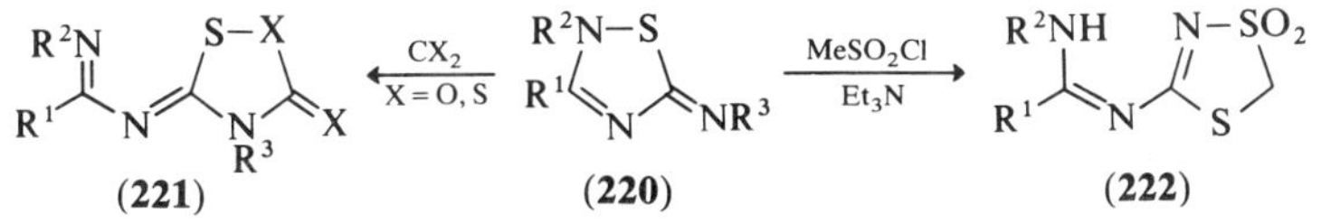

Scheme 51

4.34.4.2.4 From three-membered ring heterocycles

Finally, arylhydroxamic acids react with the azirine (**223**) forming 1,3,4-dioxazolines (**224**) in high yields ⟨81KGS124⟩, and the configuration of chiral aziridinyl alcohols, *e.g.* (**225**), has been established by reaction with $SOCl_2$ to yield oxathiazolidines, *e.g.* (**226**). These configurations were established by ^1H NMR spectroscopy ⟨77MI43401⟩. The transformation of an acyloxaziridine to a 1,3,4-dioxazoline is shown in Scheme 44.

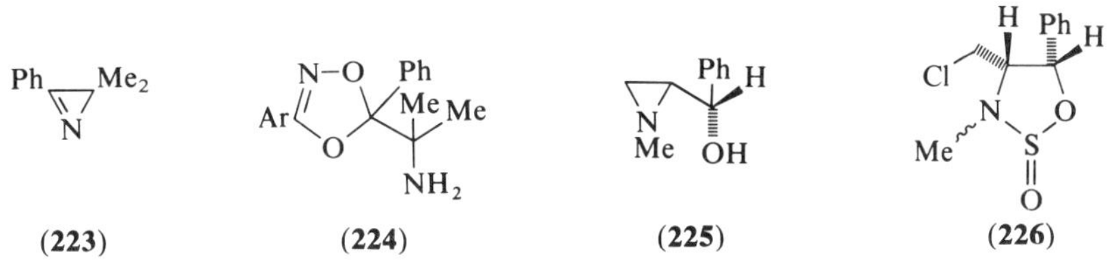

4.34.4.3 Summary of Best Preparative Methods for Ring Syntheses

In this section are tabulated the best synthetic routes to the ring systems of this chapter, taking into account simplicity, yield and scope. None of the parent ring systems appear to have been reported.

The data are arranged for convenience in the following order: benzo fused systems (Table 9), dioxazoles (Table 10), oxathiazoles (Table 11) and dithiazoles (Table 12). In all tables, the 'Method' column refers to the section in the text where the method is described, and often illustrated in a scheme. Yields have been quoted where available, though in some cases, especially where the preparation is from a patent or a less common journal, data are not given. In Tables 10–12 information is arranged in order of increasing separation of heteroatoms, and within each of these subsections by decreasing oxidation level. Substituents follow the order used throughout these volumes.

4.34.5 APPLICATIONS OF THE RING SYSTEMS

A number of dioxazoles, oxathiazoles and dithiazoles have been found to have useful biological activity, suggesting applications as agrochemicals and pharmaceuticals. Others have found use in industry as corrosion inhibitors, rubber chemicals, and polymer intermediates, a particularly important class being the 1,3,4-dioxazol-2-ones on account of their facile conversion into isocyanates. The most versatile compound must surely be 3-amino-1,2,4-dithiazole-5-thione (**30**), which has been used as a bactericide, fungicide, slime inhibitor, corrosion inhibitor, rubber accelerator ⟨74GEP2420441⟩ and in a colorimetric determination of vitamin K_1 ⟨58ACS347⟩. Both it and its *N*-acetyl derivative have shown antitumour activity. A useful commercial preparation has been claimed ⟨71USP3584003⟩.

4.34.5.1 Agrochemicals

4.34.5.1.1 Fungicides

The 1,3,4-oxathiazolones (**46**) have been patented as fungicides ⟨67BRP1079348⟩, and the isomeric 4-phenyl-1,3,5-oxathiazolone (**227**; Ar = Ph) is an effective fungal growth inhibitor

Table 9 Best Synthetic Methods for Benzoxathiazole and Benzodithiazole Derivatives

Aromatic systems						Non-aromatic systems (25)						
Structure type	R^1	Yield (%)	Method	Ref.		Structure type	X	Y	Z	R^1	Yield (%)	Method	Ref.
(14)	—	70	4.34.4.1.2	81TL1175		O	S(O)	NTs	H	81	4.34.4.1.5	79CB1012
(4)	H	56	4.34.4.1.7	69CHE421		S	S(O)	NH	various	—	4.34.3.2.3	79KGS1205
(4)	6-Ph	~100	4.34.4.1.7	31JA1891		S	NR	S	5-Me	30-50	4.34.4.1.4	79JHC182
(4)	6-NR^2R^3	65-96	4.34.3.3.1	77KGS1499		SO_2	NH	SO_2	H	79	4.34.4.1.4	69JOC3434
(4)	6-Cl	87-98	4.34.4.1.7	70CHE327		SO_2	NOH	SO_2	H	92	4.34.4.1.4	81JOC2691
(13)	Various	—	4.34.3.2.3	77KGS849		SO_2	NR	SO_2	H	26-57	4.34.3.4.2	57MI43400
(15)	6-Me, 6-NO_2	58-92	4.34.4.1.4	63CB1177								

Table 10 Best Synthetic Methods for Dioxazole Derivatives[a]

Structure type	Ring atoms			Substituents[a]				Yield (%)	Method	Ref.
	X	Y	Z	R^1	R^2	R^3	R^4			
Heteroatoms 1,2,4										
(10)	N	O	O	Ar	O	—	—	10-76	4.34.4.1.2	72JPR145
(11)	NMe	O	O	O	O	—	—	—	4.34.4.1.7	81AG(E)784
(21)	N	O	O	Ph or Alk	Ph or Alk	CO_2Alk	—	73-98	4.34.4.2.3	77S572
(22)	NH	O	O	Ar	O	NR_2	—	83-97	4.34.3.2.4	72JPR145
(23)	NH	O	O	Alk^1	Alk^1	Alk^2	Alk^2	33-93[b]	4.34.4.1.9	69JCS(C)2678
(23)	N(NO)	O	O	Alk^1	Alk^1	Alk^2	Alk^2	50-81	4.34.3.4.2	74JOC1791
Heteroatoms 1,3,2										
(19)	O	NR	O	H	H	H	H	29-55	4.34.4.1.2	80IZV2181
(19)	O	NAr	O	H	$(CH_2)_2CH=CH(CH_2)_2$		H	30-40	4.34.4.1.6	76AG(E)372
Heteroatoms 1,3,4										
(9)	O	O	N	Ph	$(CF_3)_2C$	—	—	68	4.34.4.1.6	70IZV2140
(9)	O	O	N	Ar or Alk	O	—	—	51-90	4.34.4.1.3	68TL319
(9)	O	O	N	Ar	S	—	—	74-93	4.34.4.1.3	68TL319
(20)	O	O	O	CH_2R	Me, Ph	H, CO_2Me	—	15-50	4.34.4.1.3	73CJC1368
(20)	O	O	O	Various	Ar	Various	—	41-91	4.34.4.1.6	72CB2805
(22)	O	NAr	O	H	O	H	—	46[c]	4.34.4.1.2	78IAP78127479
(23)	O	O	NBu^t	Bu^t	Bu^t	H	H	86	4.34.4.1.9	80ICR(S)122
(23)	O	O	NCF_3	CF_3	CF_3	CF_3	CF_3	25	4.34.4.1.9	77AG(E)646

[a] Alk = alkyl substituent; R = various substituents. [b] Crude yields. For $Alk^1 Alk^2 = (CH_2)_5$, yield pure product = 78%. [c] Ar = Ph.

Table 11 Best Synthetic Methods for Oxathiazole Derivatives

Structure type	X	Ring atoms Y	Z	R¹	R²	Substituents[a] R³	R⁴	Yield (%)	Method	Ref.
Heteroatoms 1,2,3										
(16)	O	S(O)	NSO₂R	Me or Ph	Me or Ph	—	—	52–59	4.34.4.2.1	74AP291
(17)	O	S(O)	N	H, Me	Cl	H, Me	—	51, 70	4.34.4.1.5	75MI43401
(17)	O	S(O₂)	N	H, Me	Cl	H, Me	—	68, 85	4.34.4.1.5	75MI43402
(18)	O	S(O)	NAr	H, Me	CH₂	H, Me	—	50–63	4.34.4.1.7	76MI43401
(18)	O	S(O₂)	NAr	Me	CH₂	Me	—	67.6	4.34.3.4.3	76MI43401
(18)	O	S(O)	NBuᵗ	H	NAr	H	—	35–80	4.34.4.1.5	77IJC(B)133
(18)	O	S(O)	NAr	H, Me	NH	H, Me	—	43–68	4.34.4.1.7	76MI43402
(18)	O	S(O)	NAr	H, Alk	O	H, Alk	—	45–73	4.34.4.1.5	74JHC1
(18)	NPh	S(O)	O	Me	O	H	—	35	4.34.4.2.1	77JOM(137)C37
(19)	O	S(O)	NR	H, Me, Ph	H	Various	H	33–99	4.34.4.1.5	69JOC175
(19)	O	S(O₂)	NH	H	CO₂H	H	H	16	4.34.4.1.1	67BCJ1554
(24)	O	S(O)R⁴	N	Me, H	NR	Me	Me, Ph	40–83	4.34.4.1.2	79TL49
(24)	O	S(O)R⁴	N	H	O	H	Ar, OAr	52–79	4.34.4.1.3	70TL127
(24)	O	S(O)R⁴	N	CF₃	O	CF₃	Various	67–72	4.34.4.1.3	74MI43400
Heteroatoms 1,2,4										
(10)	N	S	O	Ph	NAr	—	—	—	4.34.4.1.1	73MI43401
(11)	NR	S	O	O	CHR	—	—	67–71	4.34.4.2.3	77CB285
(11)	NMe	S	O	O	NCSNPrⁱ₂	—	—	100	4.34.3.2.6	77CB285
(22)	NR	S	O	Ar	CR₂	H	—	70	4.34.4.1.6	78AG(E)455
Heteroatoms 1,2,5										
(17)	S(O)	O	N	Arᵇ	Ph	Arᵇ	—	69	4.34.4.1.6	77JCS(P1)1468
(19)	S(O)	O	NMe	Ph	Ph	H	PhS	10	4.34.4.1.6	77G283
Heteroatoms 1,3,2										
(6)	O	N⁻	S⁺	O	Ar	—	—	67–90	4.34.4.1.4	72CB188
Heteroatoms 1,3,4										
(9)	O	S	N	H	O	—	—	58	4.34.4.1.7	80CB1790
(9)	O	S	NPh	Alk, Ar	O	—	—	76–98	4.34.4.1.7	69LA(726)110
(11)	O	S	S	O	NPh	—	—	—	4.34.4.1.7	70AG(E)54
(20)	O	N	S(O₂)	CX₃ᶜ	Alk, Ar	H, CX₃ᶜ, Ph	—	18–76	4.34.3.2.6	81JCS(P1)2991
(20)	O	N	S(O₂)	H	Ar	H	—	49–93	4.34.4.1.7	64LA(671)135
(20)	O	N	S(O₂)	H	NMe₂	H	—	—	4.34.4.1.7	68CC1245
(20)	O	N	S(O₂)	H	NHPh	H	—	—	4.34.4.1.7	62AG(E)594
Heteroatoms 1,3,5										
(9)	S	O	N	Ar	C(CN)R	—	—	56–87	4.34.4.1.6	66AG(E)970
(9)	S	O	N	Various	NTs	—	—	34–98	4.34.4.1.6	75T1537
(9)	S	O	O	Ar	O	—	—	—	4.34.4.1.3	72GEP2059990
(9)	S	O	O	Ar	S	—	—	—	4.34.4.1.3	72GEP2059990
(20)	S	N	O	α-C₁₀H₇	Me, Et	MeS	—	49, 31	4.34.4.1.6	72CB2815
(20)	S	N	O	Various	Ar	Various	—	46–94	4.34.4.1.6	72CB2815
(20)	S	N	O	CN	Ar	RCO(Ar)N	—	62–77	4.34.4.1.6	79CB1873
(20)	S	N	O	Ph	Ph	Me₃Si	—	60	4.34.4.1.6	81CC822
(20)	S(O)	N	O	Ph	Ph	Ph	—	—	4.34.3.4.3	72CB2815
(22)	S	O	NR	Ar	(CF₃)₂C	H	—	31–90	4.34.4.1.6	70JOC3470

[a] Alk = alkyl substituent; R = various substituents. [b] R¹R³ = 2,2'-biphenylene, compound (200). [c] X = Cl, F.

Table 12 Best Synthetic Methods for Dithiazole Derivatives

Structure type	\multicolumn{3}{c}{Ring atoms}			\multicolumn{4}{c}{Substituents[a]}				Yield (%)	Method	Ref.
	X	Y	Z	R^1	R^2	R^3	R^4			

Heteroatoms 1,2,3

Structure type	X	Y	Z	R^1	R^2	R^3	R^4	Yield (%)	Method	Ref.
(5)	S	S	N	Cl	Cl	—	—	—	4.34.4.1.8	80GEP2848221
(6)	S	S	N	NR	Cl	—	—	—	4.34.3.3.7	80GEP2848221
(6)	S	S	N	O	Cl	—	—	—	4.34.3.3.7	80GEP2848221
(6)	S	S	N	S	Cl	—	—	—	4.34.3.3.7	80GEP2848221
(16)	S	S	NH	COCOAr	Ar	—	—	39	4.34.3.1.5	64LA(675)151
(17)	S	S	N	Alk	Alk	COR	—	73–83	4.34.4.1.7	81BCJ3541
(17)	S	S	N	Alk	OMe	COR	—	35–53	4.34.4.1.7	81BCJ3541

Heteroatoms 1,2,4

Structure type	X	Y	Z	R^1	R^2	R^3	R^4	Yield (%)	Method	Ref.
(8)	N	S	S	Ph	H	—	—	100	4.34.3.4.5	74AP828
(8)	N	S	S	Ar	Ar	—	—	46–98	4.34.4.1.7	79NKK389
(8)	N	S	S	Ar	NR_2	—	—	30–40	4.34.4.1.1	77EGP126401
(8)	N	S	S	Ar	SMe	—	—	61–91	4.34.3.3.6	70CJC2142
(8)	N	S	S	NH_2	NH_2	—	—	87	4.34.4.1.1	63ACS163
(8)	N	S	S	NHAr	NH_2	—	—	—	4.34.4.1.1	69JCS(C)194
(8)	N	S	S	NMe_2	NR_2	—	—	90	4.34.4.1.1	73JAP7326763
(8)	N	S	S	NR_2	SR	—	—	41–91	4.34.3.3.4	74JOC2225
(8)	N	S	S	NH_2	SR	—	—	27–100	4.34.4.1.1	54JA1158
(8)	N	S	S	NR_2	CR_2[b]	—	—	46–84	4.34.3.3.6	72JHC447
(10)	N	S	S	Ph	NR	—	—	16–51	4.34.4.1.1	68AG(E)296
(10)	N	S	S	Me	NR	—	—	63–68	4.34.4.1.1	80JIC1166
(10)	N	S	S	Ph	NR	—	—	58–73	4.34.4.1.1	79IJC(B)284
(10)	N	S	S	Alk, Ar[c]	O	—	—	51–78	4.34.4.1.7	81CB549
(10)	N	S	S	Ar	S	—	—	23–70	4.34.4.1.5	66BSF1183
(10)	N	S	S	NR_2	NR	—	—	80–96	4.34.4.1.1	72CB1568
(10)	N	S	S	NRMe	NCOR	—	—	84	4.34.3.3.4	73JAP7328470
(10)	N	S	S	NH_2	O	—	—	—	4.34.4.1.7	19LA(419)217
(10)	N	S	S	NHPh	O	—	—	27	4.34.4.1.1	76CB848
(10)	N	S	S	NH_2	S	—	—	54	4.34.4.1.8	70JOC1665
(10)	N	S	S	NR_2	S	—	—	42–71	4.34.4.1.8	72JHC447
(10)	N	S	S	NHCOR	S	—	—	50–90	4.34.4.1.8	73MI43400
(10)	N	S	S	OEt	NPh	—	—	93	4.34.4.1.1	76CB848

(10)	N	S	S	OEt	O	—	—	—	4.34.4.1.7	70AG(E)54
(10)	N	S	S	SBz	NAr	—	—	—	4.34.4.1.7	75JIC237
(10)	N	S	S	SR	NCOR	—	—	45–90	4.34.3.3.6	75LA1018
(10)	N	S	S	Cl	O	—	—	72	4.34.3.5.3	71CB2732
(11)	NAr	S	S	CHCOR	CHCOR	—	—	26–80[d]	4.34.4.1.7	79CHE1187
(11)	NMe	S	S	CR_2[e]	NMe	—	—	—	4.34.4.1.7	81T2451
(11)	NR	S	S	NPh	NPh	—	—	29–65	4.34.4.1.6	75CC24
(11)	NBz	S	S	NR	NTs	—	—	50–93	4.34.4.1.6	75JOC1728
(11)	NR	S	S	NR	NCSR	—	—	12–89	4.34.4.1.6	76CB848
(11)	NR	S	S	NR	O	—	—	—	4.34.4.1.7	40JCS191
(11)	NPh	S	S	NSO_2R	O	—	—	—	4.34.4.1.7	67AG(E)649
(11)	NR	S	S	NR	S	—	—	—	4.34.4.1.7	72IJC1138
(11)	NR	S	S	$NCSNMe_2$	S	—	—	68–94	4.34.3.2.6	74JOC2228
(11)	NH	S	S	O	O	—	—	—	4.34.3.3.5	70AG(E)54
(11)	NR	S	S	O	O	—	—	35–75	4.34.4.1.7	68BRP1136737
(21)	N	S	S	H	Ar	H	—	38–85	4.34.4.1.7	74AP828
(21)	N	S	S	CF_3	R	CF_3	—	42–85	4.34.4.1.5	77CB2114
(21)	N	S	S	Cl, F	Cl	Cl, F	—	89, 94	4.34.4.1.7	71CB2732
(22)	NBz	S	S	4-MeOC_6H_4	NTs	4-MeOC_6H_4	—	56	4.34.4.1.6	76JHC883
(23)	NR[f]	S	S	Ph	H	R[f]	H	83	4.34.3.4.7	74AP828
Heteroatoms 1,3,2										
(19)	S	NMe	S	H	H	H	H	5	4.34.4.1.4	69JHC627
Heteroatoms 1,3,4										
(9)	S	S	N	Ar	O	—	—	75–90	4.34.3.3.6	67BSF2239
(9)	S	S	N	Ar	S	—	—	15–25	4.34.4.1.7	67BSF2239
(11)	S	S	NAr	NBz	NPh	—	—	—	4.34.4.1.7	80IJC(B)970
(11)	S	S	NAr	O	NPh	—	—	60–80	4.34.4.1.7	80IJC(B)970
(11)	S	S	NAr	S	NPh	—	—	—	4.34.4.1.7	80IJC(B)970
(20)	S	N	$S(O_2)$	H	NR_2	H	—	60–75	4.34.4.1.7	62AG(E)594
(20)	S	N	$S(O_2)$	H	SR	H	—	—	4.34.3.5.2	62AG(E)594
(22)	S	NAr	S	COAr	C(CN)COAr	H	—	39–57	g	80JPR407
(22)	S	NR	$S(O_2)$	H, Me, Ph	NCSR	H	—	—	4.34.3.3.4	77TL1729
(23)	S	S	NTs	Ar	Ar[h]	Ar	Ar[h]	13–88	4.34.4.1.6	80BCJ1023

[a] Alk = alkyl substituents; R = various substituents. [b] CR_2 = disubstituted 2*H*- or 3*H*-pyrrylidene. [c] Includes heteroaryl substituents, also CO_2Et. [d] Mixture of *cis* and *trans* isomers. [e] CR_2 = 2-oxo-3-bornylidene (*cis* and *trans* mixture). [f] RR = $CH_2(Me)C{=}C(Me)CH_2$, compound (**168**). [g] Unique method—see literature. [h] R^2R^4 = 2,2′-biphenylene, compound (**204**).

⟨72GEP2059990⟩. Most antifungal activity, however, has been observed among dithiazole derivatives, the 1,2,3-dithiazolimines (**129**; X = NAr) being effective against *Trichophyton mertagrophytes* ⟨80GEP2848221⟩, and the 1,2,4-dithiazolines (**36**) against *T. dermatophytes* ⟨74AP828⟩. The oxidative degradation product (**37**) of 'Nabam' is also a fungicide in its own right.

4.34.5.1.2 Herbicides and plant growth regulators

High pre-emergent selectivity is found with the oxathiazolone (**227**; Ar = 2,6-$Cl_2C_6H_3$), radish, oats and mustard being controlled without damage to cotton, wheat, and millet ⟨63BEP632072⟩. The dithiazolidine *S,S*-dioxides (**50**) show low pre- and post-emergent activity ⟨77CCC2680⟩, while certain benzodithiazolines (**82**) have good post-emergent properties ⟨74USP3808222⟩, and the dithiazolylium salts (**228**) are useful defoliants for cotton ⟨66NEP6500213⟩.

The oxathiazolimine (**186**) is a root growth inhibitor which has been evaluated against linseed ⟨73MI43401⟩, and the dithiazolidine (**47**; R^1 = Me, Ar = 4-ClC_6H_4) has been claimed as a plant growth stimulant ⟨80MIP43400⟩.

4.34.5.1.3 Insecticides and insect chemosterilants

Broad spectrum insecticidal properties are claimed for the 1,3,4-dioxazolidinones (**191**) ⟨78JAP78127479⟩, while aminodithiazolines (**49**) having R^1 = substituted aryl and R^2 = H are useful against plant eating insects ⟨76GEP2608488⟩. Certain diaminodithiazolylium salts (**228**) ⟨76USP3944670⟩ and the aminodithiazolimines (**72**; R^3 = Ac or $CONHBu^t$) are effective acaricides, the latter also having ovicidal properties ⟨76USP3956303⟩.

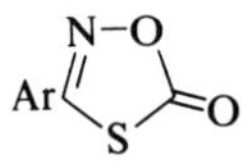

(**227**) 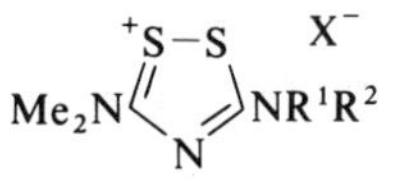   (**228**)

An important class of chemosterilants are the dithiazolylium salts (**228**), many of which show high activity against *Musca domestica*, the example having NR^1R^2 = 4-methylpiperazino being selectively active against males ⟨73MI43402⟩ (see also Chapter 1.07).

4.34.5.2 Pharmaceuticals

4.34.5.2.1 Antibiotics and antibacterials

The synthetic penicillins (**229**) are highly effective antibiotics ⟨65USP3222360⟩. Antibacterial activity has been found with 1,2,3-oxathiazolidine *S*-oxides bearing C_{10} or C_{12} alkyl side chains, *e.g.* (**230**) ⟨76GEP2456874⟩, and with aryl dithiazolethiones (**2**; X = S) having a *meta*-halo substituent ⟨80JAP80118478⟩, while the oxathiazolimine (**231**) is useful as an agricultural antibacterial ⟨77JAP77148076⟩.

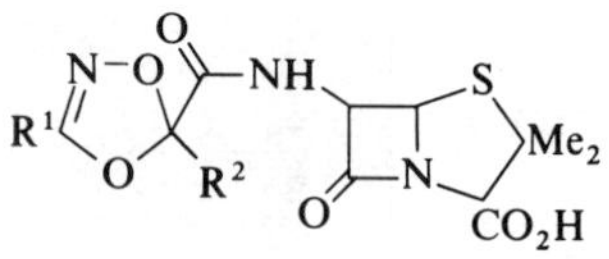

(**229**) 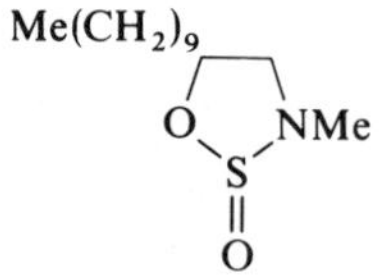   (**230**) 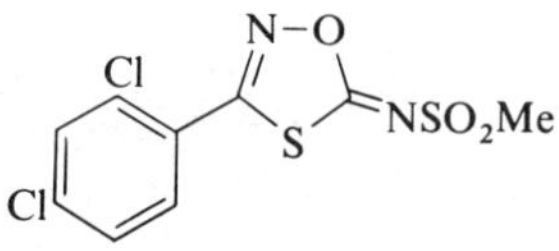   (**231**)

4.34.5.2.2 Antifertility compounds

The dithiazolylium salts (**228**), besides being insect chemosterilants, cause atrophy of the prostate gland and seminal vesicles in male rats, as well as inhibiting ovulation and interrupting pregnancy in females. The results of a detailed study have recently been published ⟨80MI43401⟩. Antiandrogenic activity against rats has been observed in the oxathiazolidinone (**232**) ⟨77FRP2315922⟩.

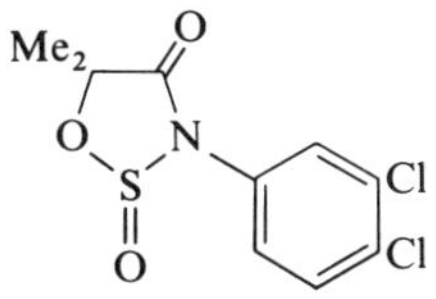

(**232**)

4.34.5.2.3 CNS activity

Certain oxathiazolines are useful ataraxics with low toxicity. An example is the chlorobenzylamine derivative (**233**) ⟨68SAP6804299⟩. Modest CNS depressant activity has also been found in certain 1,3,4-dioxazolines (**234**) ⟨77JPS772⟩ and in 1,2,3-oxathiazolidin-5-imine S-oxides (**235**) ⟨77IJC(B)133⟩ (see also Chapter 1.06).

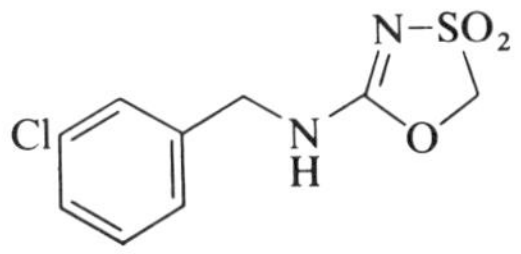

(**233**) 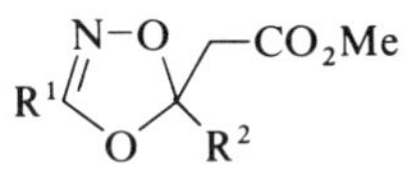   (**234**) 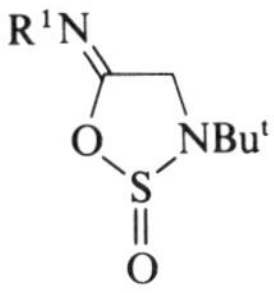   (**235**)

4.34.5.3 Polymer Chemicals

4.34.5.3.1 Masked isocyanates

1,3,4-Dioxazole derivatives (**64**; X = Y = O, Z = O, S) lose COZ on pyrolysis to give isocyanates (Scheme 5), and being themselves readily prepared from hydroxamic acids, they are important industrial precursors to simple carbamates, polyurethanes, and copolymers. Additionally, the CO_2 released may function as a blowing agent. Representative examples from a very considerable patent literature are bis(dioxazoles) for polyurethanes ⟨76USP3979401⟩, 2-propenoate esters for polymerization ⟨76GEP2447370⟩, ethenyl esters for copolymerization ⟨79USP4133813⟩, blowing agents ⟨80EUP7499⟩, and an important synthesis of the 5-ethenyl compound (**236**; Scheme 52) ⟨73USP3737435⟩, which can readily be converted into N-ethenyl O-methyl carbamate (**237**) ⟨78MIP43400⟩. Methods for polymerizations and copolymerizations of a number of masked isocyanates have been described ⟨75MI43403⟩.

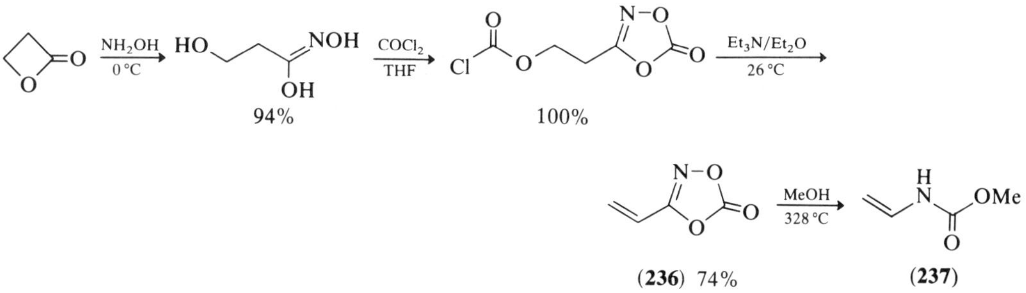

(**236**) 74%   (**237**)

Scheme 52

4.34.5.3.2 Others

The sulfonimide (**238**) has been copolymerized with $CH_2=CHCN$ and $CH_2=C(Me)CO_2Me$ to produce a useful fibre ⟨75GEP2346566⟩, and the bis(oxathiazolone) (**239**), as a nitrile sulfide precursor (Scheme 5), has been used in the vulcanization of SBR ⟨77GEP2638029⟩.

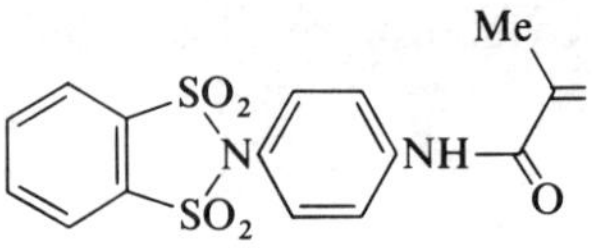

(238) 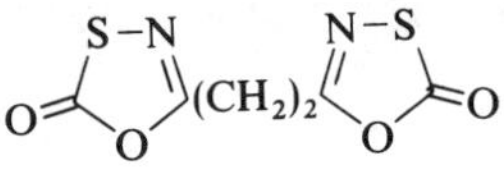   (239)

4.34.5.4 Other Applications

Monosubstituted aminodithiazolethiones (**1**; $R^2 = H$, $X = S$) are corrosion inhibitors for ferrous metals in pickling baths, the best compounds having $R^1 = R^3CH(OH)$ ⟨77GEP2606788⟩. Related dithiazoles, having $R^1 =$ long-chain fatty acyl, are effective oxidation inhibitors in high pressure lubricating oils ⟨73USP3753908⟩.

Salts of types (**84**) and (**228**) have been used as photographic development accelerators ⟨78MI43400⟩. While the 1,2,4-dithiazolidine (**41**; $R^1 = R^2 = Ph$, $X = S$) is a useful silver halide emulsion sensitizer, which causes no increase in fogging ⟨79JAP7970822⟩.

The disulfonimidecoumarin derivative (**240**) is a bleach-resistant fluorescent whitener.

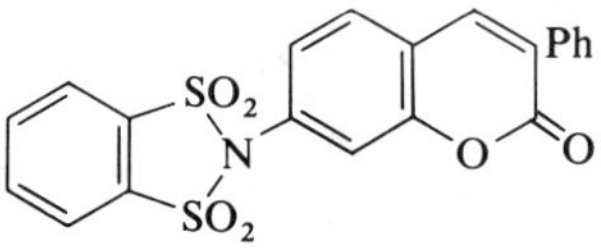

(240)

4.35

Five-membered Rings containing One Selenium or Tellurium Atom and One Other Group VI Atom and their Benzo Derivatives

M. R. DETTY
Eastman Kodak Company, Rochester

4.35.1 INTRODUCTION	948
4.35.2 MOLECULAR STRUCTURE	950
4.35.2.1 Theoretical Investigations	950
4.35.2.2 X-ray Diffraction	950
4.35.3 MOLECULAR SPECTROSCOPY	951
4.35.3.1 Ultraviolet–Visible Absorption Spectroscopy	951
4.35.3.2 Photoelectron Spectroscopy	952
4.35.3.3 Infrared and Raman Spectroscopy	954
4.35.3.4 Mass Spectrometry	955
4.35.3.5 Nuclear Magnetic Resonance Spectroscopy	956
4.35.3.5.1 ^1H NMR spectroscopy	957
4.35.3.5.2 ^{13}C NMR spectroscopy	959
4.35.3.5.3 ^{125}Te spectroscopy	960
4.35.3.6 Electronic Paramagnetic Resonance Spectroscopy	961
4.35.4 THERMODYNAMIC ASPECTS	961
4.35.4.1 Intermolecular Forces	961
4.35.4.1.1 Melting points, boiling points and solubility	961
4.35.4.2 Stability and Stabilization	961
4.35.4.2.1 Thermal chemistry	961
4.35.4.2.2 Cyclic voltammetry	961
4.35.4.2.3 Aromaticity	962
4.35.5 REACTIVITY	962
4.35.5.1 Thermal and Photochemical Reactions Formally Involving No Other Species	962
4.35.5.1.1 Rearrangement	962
4.35.5.1.2 Thermolysis	963
4.35.5.1.3 Photochemical reactions	963
4.35.5.1.4 Polymerization	963
4.35.5.2 Reactions with Electrophiles	964
4.35.5.2.1 Reactions involving protons or Lewis acids	964
4.35.5.2.2 Reactions involving alkylating agents	964
4.35.5.3 Reactions with Nucleophiles	964
4.35.5.3.1 Reactions with bases	964
4.35.5.3.2 Reactions with sodium borohydride	965
4.35.5.3.3 Reactions with sulfur and selenium nucleophiles	965
4.35.5.3.4 Reactions with phosphorus nucleophiles	965
4.35.5.3.5 Reactions with halide and pseudohalide nucleophiles	966
4.35.5.3.6 Cycloaddition reactions	966
4.35.6 SYNTHESIS	967
4.35.6.1 General Survey of Ring Syntheses	967
4.35.6.2 One-component Syntheses	967
4.35.6.3 [1+4] Two-component Syntheses	969
4.35.6.4 [2+3] Two-component Syntheses	970
4.35.6.5 Multicomponent Syntheses	971
4.35.7 APPLICATIONS	971

4.35.1 INTRODUCTION

Five-membered monocyclic rings containing one selenium or one tellurium atom and one other chalcogen (Group VIB) atom (oxygen, sulfur, selenium, tellurium) form 47 different ring systems, considering the fully saturated systems, those systems with one carbon–carbon double bond, and the aromatic dichalcogenolylium cations. These systems are summarized in Scheme 1. According to their composition, these systems may be classified into seven groups: C_3OSe systems (oxaselenolane, oxaselenole, oxaselenolylium), C_3OTe systems (oxatellurolane, oxatellurole, oxatellurolylium), C_3SSe systems (thiaselenolane, thiaselenole, thiaselenolylium), C_3STe systems (thiatellurolane, thiatellurole, thiatellurolylium), C_3Se_2 systems (diselenolane, diselenole, diselenolylium), C_3SeTe systems (selenatellurolane, selenatellurole, selenatellurolylium) and C_3Te_2 systems (ditellurolane, ditellurole, ditellurolylium) with 1,2- and 1,3-isomers possible in each system as well as double bond isomers in the 1,2-dichalcogenoles. Many of these systems remain unknown, particularly the chalcogenatellurolane, oxatellurole, 1,2-thiatellurole, 1,2-selenatellurole, 1,3-oxatellurolylium, 1,2-thiatellurolylium, and 1,2- and 1,3-selenatellurolylium ring systems. For the known systems, many of the parents as well as derivatives including benzo fused analogs have been prepared and studied.

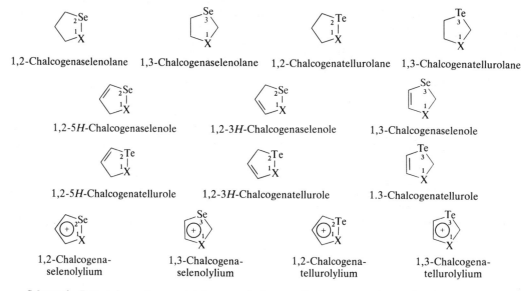

Scheme 1 Parent ring systems containing one selenium or tellurium atom and one other chalcogen atom

The selenium containing systems have been represented in the literature for a considerable number of years. 1,2-Diselenolane (**1**) was first reported by Hagelberg in 1890 ⟨1890CB1090⟩. The 1,2-diselenolanes and 1,2-thiaselenolanes have been of interest as monomers for forming polydiselenides and mixed polydisulfides and polydiselenides ⟨30JCS1497, 57ACS911, 61AK(18)143, 62AK(19)195, 72ANY(192)25⟩.

The first tellurium system to be represented was the 1,2-ditellurole (**2**), reported in 1977 by Meinwald et al. ⟨77JA255⟩. Thus, the tellurium systems represent an emerging area of heterocyclic chemistry.

The 1,2- and 1,3-dichalcogenolylium ions have been of interest as 6π aromatic systems isoelectronic with the 1,2- and 1,3-dithiolylium ions (see Chapter 4.32) ⟨67ACS1991, 75TL1259, 82TL1531⟩. However, the primary interest in 1,3-dichalcogenolylium ions as well as in the 1,3-dichalcogenolane and 1,3-dichalcogenole systems has been to use these compounds to synthesize 1,4,5,8-tetrahydro-1,4,5,8-tetrachalcogenafulvalenes which have been useful as donors to form organic conductors and semiconductors (see Chapter 1.13).

The structure, reactivity and synthesis of the 1,4,5,8-tetrahydro-1,4,5,8-tetrachalcogenafulvalenes containing selenium and tellurium are also considered in this chapter. These structures are summarized in Scheme 2 and correspond to $C_6S_2Se_2$ systems (dithiadiselenafulvalenes), C_6SSe_3 systems (thiatriselenafulvalenes), C_6Se_4 systems (tetraselenafulvalenes) and C_6Te_4 systems (tetratellurafulvalenes).

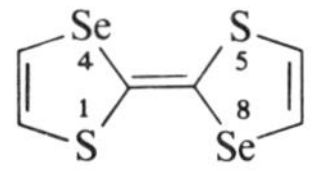

trans-Dithiadiselenafulvalene 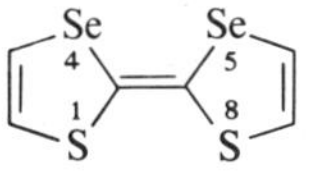   *cis*-Dithiadiselenafulvalene

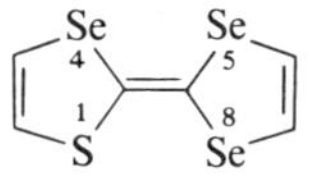

Thiatriselenafulvalene 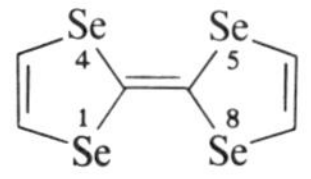   Tetraselenafulvalene 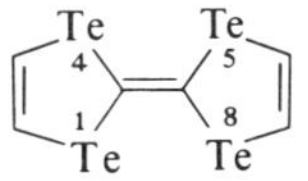   Tetratellurafulvalene

Scheme 2 Known tetrachalcogenafulvalene systems containing selenium or tellurium

The syntheses and properties of organic conductors have been extensively reviewed ⟨79ACR79, B-78MI43500, 78ANY(313)1⟩ and will not be discussed here.

Selenium and tellurium containing heterocycles are fundamentally different from their oxygen and sulfur containing counterparts. Selenium and tellurium have filled $3d$ and $4d$ orbitals, respectively, making them much larger than oxygen, sulfur and carbon (Table 1). The larger size of these atoms results in poorer overlap with an adjoining carbon π-framework. Both selenium and tellurium are more electropositive than carbon (Table 1), which changes the polarity of the carbon–chalcogen bond as one goes down the chalcogen elements in the periodic table. Tellurium is a boundary atom between the metals and the non-metals and is more metallic in its behavior than selenium, sulfur and oxygen. The metallic behavior of tellurium has been documented for a thermal four-electron electrocyclic rearrangement in the conversion of 1,3-ditellurolylium to 1,2-ditellurolylium ions ⟨82TL1531⟩.

Table 1 Covalent Radii and Electronegativities for Carbon and the Chalcogen Atoms

Element	Covalent radius (Å)	Electronegativity (Pauling)
C	0.77	2.5
O	0.73	3.5
S	1.02	2.5
Se	1.16	2.4
Te	1.36	2.1

The inorganic chemistry of selenium and tellurium is important to the study of the heterocycles in this chapter. Both selenium and tellurium coordinate highly electronegative atoms, which leads to hypervalent bonds and unusual structures unprecedented in sulfur heterocycles ⟨78ANY(313)1⟩. While dioxaselenapentalene (**3**) is protonated with perchloric acid to give the ionic oxaselenolylium perchlorate (**4**) ⟨75JCS(P1)2097⟩, oxaselenolylium chloride (**5**) has a covalent Se—Cl bond ⟨83JA883⟩. Oxatellurolylium chloride (**6**) has a similar structure. Upon treatment with silver fluoroborate, compound (**6**) gives a species which abstracts fluoride from the fluoroborate anion to give the covalent oxatellurolylium fluoride (**7**) ⟨83JA875⟩.

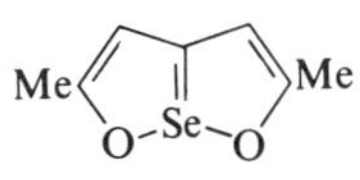

(3) 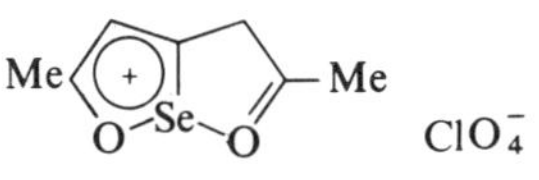   (4) 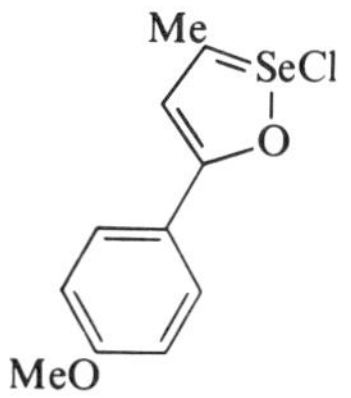   (5) 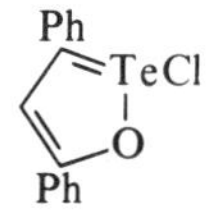   (6) 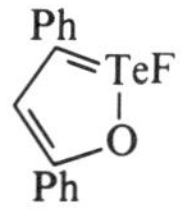   (7)

4.35.2 MOLECULAR STRUCTURE

4.35.2.1 Theoretical Investigations

Very few structural investigations have been made for five-membered heterocycles containing selenium or tellurium and one other chalcogen atom by either theoretical methods or X-ray diffraction. The lone theoretical treatment ⟨81JST(71)1⟩ was a MINDO/3 optimization of molecular geometry for 1,3-thiaselenole-2-thione and -2-selenone and for 1,3-diselenole-2-selenone (8) and -2-thione. These molecules were predicted to be planar. The calculated bond lengths and angles for (8) are shown in Scheme 3.

MINDO/3

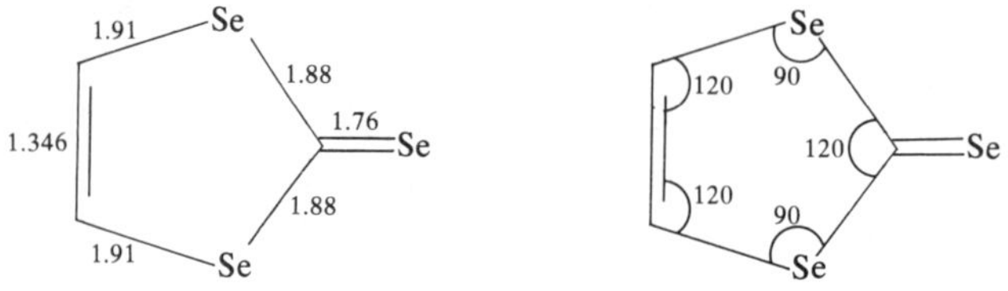

X-ray diffraction

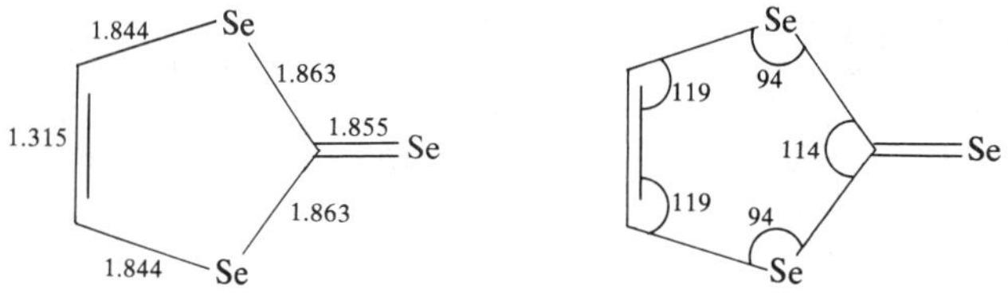

X-ray diffraction

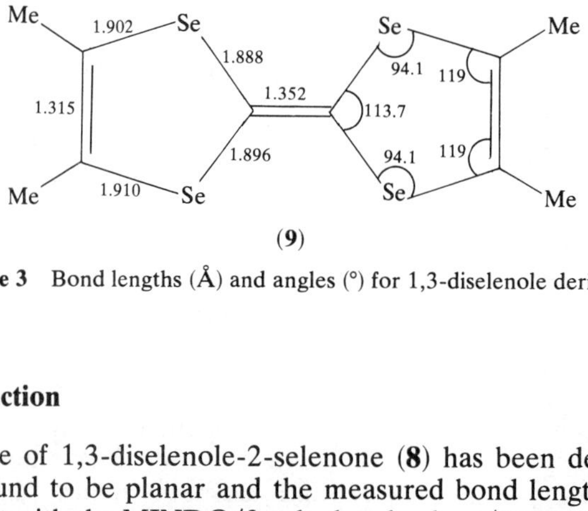

Scheme 3 Bond lengths (Å) and angles (°) for 1,3-diselenole derivatives

4.35.2.2 X-ray Diffraction

The X-ray structure of 1,3-diselenole-2-selenone (8) has been determined ⟨76IZV179⟩. The molecule was found to be planar and the measured bond lengths and angles for (8) are in good agreement with the MINDO/3 calculated values ⟨81JST(71)1⟩.

The X-ray structure of a related compound, tetramethyltetraselenafulvalene (9), has also been determined ⟨79AX(B)772⟩ and the appropriate bond lengths and angles for (9) are also shown in Scheme 3. Interestingly, the bond lengths and angles for (9) are in better agreement with the MINDO/3 calculated values for (8) than are the observed values for (8). In both of these compounds the Se—C(sp^2) bond length is between 1.84 and 1.91 Å and C—Se—C angles are approximately 90°.

An X-ray structure has been reported for dithiadiselenafulvalene ⟨78MI43501⟩. Unfortunately, this planar compound possessed a disordered array, presumably due to a mixture of cis and trans isomers (Scheme 2) or to two different orientations of one isomer.

The structure of hexamethylenetetratellurafulvalene (10) has been determined by X-ray diffraction ⟨82CC1316⟩. This planar compound was found to have C—Te—C bond angles of 90.2° and Te—C—Te bond angles of 115.2°. The Te—C bond length (2.098 Å) was appropriately longer than the corresponding Se—C (1.89 Å) and S—S bond lengths (1.75 Å) in (9) and 2,3,6,7-tetramethyl-1,4,5,8-tetrathiafulvalene, respectively ⟨79AX(B)772⟩.

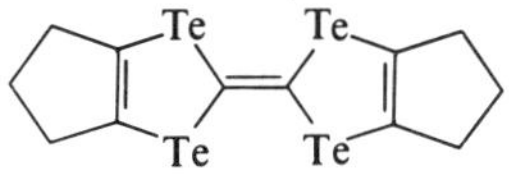

The mode of packing of (**10**) was quite unexpected when compared with those of other tetrachalcogenafulvalenes. The packing of (**10**) was in a zigzag chain arrangement of adjacent molecules with neighboring chains approximately perpendicular; the other tetrachalcogenafulvalenes have parallel chains.

The 1,2-dichalcogenoles have been examined by X-ray diffraction only for fused aromatic systems. The structure of the 1,2-thiaselenole ring system has been determined for naptho[1,8-cd][1,2]thiaselenole (**11**) ⟨77JA7743⟩. The compound is planar with a S—Se bond length of 2.246, C—S 1.80, and C—Se 1.88 Å, and C—Se—S and C—S—Se bond angles of approximately 90°. Similarly, 5,6,11,12-tetratellurotetracene (**12**) was shown to be planar with C—Te bond lengths of 2.113, Te—Te of 2.673 Å, and C—Te—Te angles of 87.6° ⟨82MI43500⟩.

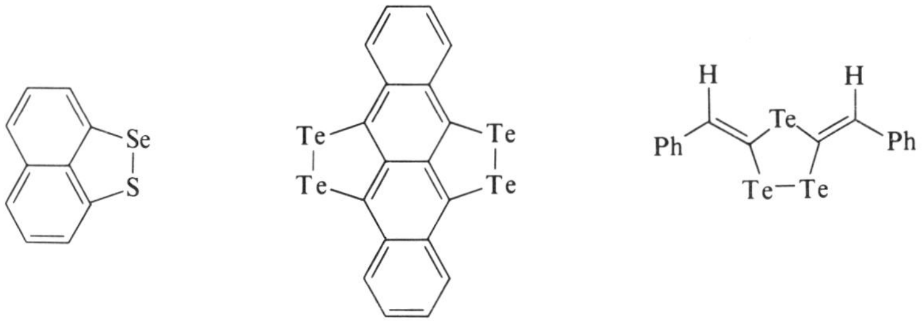

The X-ray structure of a tritellurole, cis-3,5-dibenzylidene-1,2,4-tritellurole (**13**), has been reported ⟨81TL4199⟩. The tritellurole was shown to have a non-planar heterocyclic system in which the ditelluride bond forms a dihedral angle of 35.9° with the other three ring atoms. The Te—C bond lengths averaged 2.125 and the Te—Te bond length was shown to be 2.710 Å.

The structure of the 1,3-diselenolylium ion has been investigated for radical-cation salts of tetraselenafulvalenes, materials useful as organic conductors and semiconductors. These compounds are discussed in Chapter 1.13.

The 1,2-oxatellurolylium system and benzo fused derivatives are the only other dichalcogenolylium systems to be examined by X-ray techniques. 3-Phenyl-5-(p-methoxyphenyl)-1,2-oxatellurolylium chloride (**14**) was found to be planar with a nearly linear (170°) O—Te—Cl bond ⟨83JA883⟩. The Te—Cl bond was found to be 2.48 in length and the Te—O bond 2.19 Å, indicating approximately 50% covalent character to these bonds. (In solution, conductance studies have shown (**14**) to be less than 0.5% ionized.) The bonding to tellurium in (**14**) was found to be trigonal bipyramidal with oxygen and chlorine occupying axial positions. Several benzo fused derivatives (**15**) have been examined and found to have similar bonding, although aromaticity in the benzene ring is maintained and the Te—O bond is somewhat longer ⟨74AX(B)139, 79AX(B)849⟩.

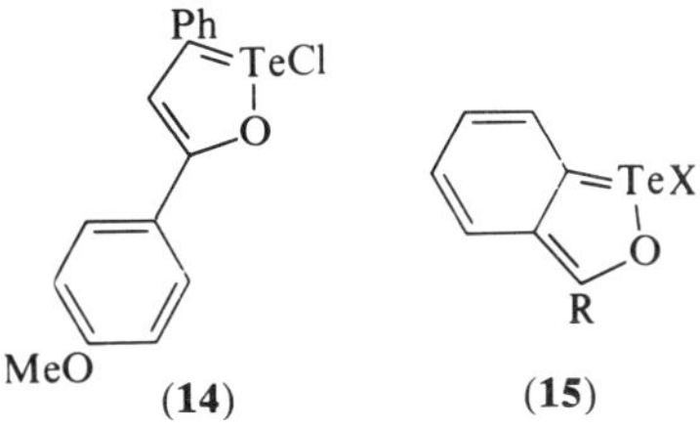

4.35.3 MOLECULAR SPECTROSCOPY

4.35.3.1 Ultraviolet–Visible Absorption Spectroscopy

The dominant feature of the UV–visible absorption spectra of five-membered rings containing one selenium or one tellurium and one other chalcogen atom is the sequential bathochromic shift in all systems as the chalcogen atoms become heavier. In the dichalcogenolanes the $n \to \sigma^*$ transition sequentially shifts from 330 nm for dithiolane (**16a**) to 386 nm for thiaselenolane (**16b**) to 440 nm for diselenolane (**16c**) ⟨61AK(18)143⟩. These absorptions are at lower energies than for acyclic disulfides and diselenides. The 1,3-dichalcogenole-2-thiones (**17a–c**) and -2-selenones (**18a–c**) show similar trends. The planar nature of these compounds allows overlap of adjacent π-frameworks with the heteroatoms and lower energy $n \to \pi^*$ and $\pi \to \pi^*$ transitions. Thus, (**17a**) absorbs at 430 ($\varepsilon = 160$) and 359 nm ($\varepsilon = 11\,000$), (**17b**) absorbs at 440 ($\varepsilon = 100$) and 368 nm ($\varepsilon = 15\,500$) and (**17c**) absorbs at 455 ($\varepsilon = 100$) and 377 nm ($\varepsilon = 12\,100$) ⟨75JOC387⟩. The corresponding selenones are at longer wavelength with (**18b**) absorbing at 530 ($\varepsilon = 300$) and 402 nm ($\varepsilon = 11\,800$) and (**18c**) absorbing at 555 ($\varepsilon = 400$) and 413 nm ($\varepsilon = 14\,100$) ⟨75JOC387⟩. The 1,3-dichalcogenolylium ions show similar behavior with (**19a**) absorbing at 252 ($\varepsilon = 4050$), (**19b**) at 258 ($\varepsilon = 3700$) and (**19c**) at 262 nm ($\varepsilon = 3400$) ⟨75TL1259⟩. Alkyl and aryl substitution have small effects on the energy of absorption ⟨75JOC746⟩.

(a) X = Y = S; (b) X = S, Y = Se; (c) X = Y = Se

A detailed theoretical study of the $n \to \pi^*$ and $\pi \to \pi^*$ transitions of (**17a–c**) has been made using the SCF MO–LCAO method and the Pariser, Parr and Pople (PPP) modification. The calculated and experimental values are in excellent agreement ⟨79MI43500⟩. The calculated and experimental values of (**17c**), (**18a**) and (**18c**) have also been determined in polyethylene ⟨77JA2855⟩.

The tetrachalcogenafulvalenes show similar trends, with tetramethyltetrathiafulvalene (**20**) showing absorption bands at 315 ($\varepsilon = 14\,000$), 327 ($\varepsilon = 13\,800$) and 473 nm ($\varepsilon = 248$). Tetramethyltetraselenafulvalene (**9**) gave similar bands at 299 ($\varepsilon = 12\,600$), 508 ($\varepsilon = 200$) and 534 nm ($\varepsilon = 176$). Hexamethylenetetratellurafulvalene (**10**) absorbed at 320 ($\varepsilon = 15\,560$), 335 ($\varepsilon = 19\,350$), 410 ($\varepsilon = 9115$), 610 ($\varepsilon = 189$) and 653 nm ($\varepsilon = 121$) ⟨82JA1154⟩. The stepwise introduction of selenium to give dithiadiselenafulvalene resulted in intermediate absorption frequencies between tetrathia- and tetraselena-fulvalene ⟨80JOC2632⟩.

The 1,2-dichalcogenoles as represented by naphtho[1,8-*cd*]-[1,2]diselenole (**21**) and -[1,2]-ditellurole (**2**) displayed a 30 nm difference between (**21**) (380 nm, $\varepsilon = 16\,000$) and (**2**) (410 nm, $\varepsilon = 10\,000$) ⟨77JA255⟩. Similarly, 1,2-oxaselenolylium chloride (**22**) and 1,2-oxatellurolylium chloride (**14**) showed a 57 nm difference between (**22**) (378 nm, $\varepsilon = 28\,000$) and (**14**) (435 nm, $\varepsilon = 30\,200$) ⟨83JA875, 83JA883⟩.

4.35.3.2 Photoelectron Spectroscopy

The photoelectron spectra of various 1,2- and 1,3-dichalcogenoles and tetrachalcogenafulvalenes have been examined. The most detailed study involved the naphtho[1,8-*cd*][1,2]dichalcogenoles (**23**), $C_{10}H_6XY$ compounds with XY=SS, SeS, SeSe, TeS, TeSe and TeTe ⟨81CB2622⟩. These compounds were all found to exhibit similar low first vertical ionization potentials at about 7.1 eV. The π-ionizations were discernible from the σ-ionizations and were measured for all systems (Table 2). The observed values were found to be in good agreement with values reproduced by a molecular state parametrized HMO model. The expected inductive shift ($-\alpha_X$) to lower π-ionization energies in selenium and tellurium containing derivatives was found to be counteracted by a reduced second order perturbation ($-\beta_{XY}$).

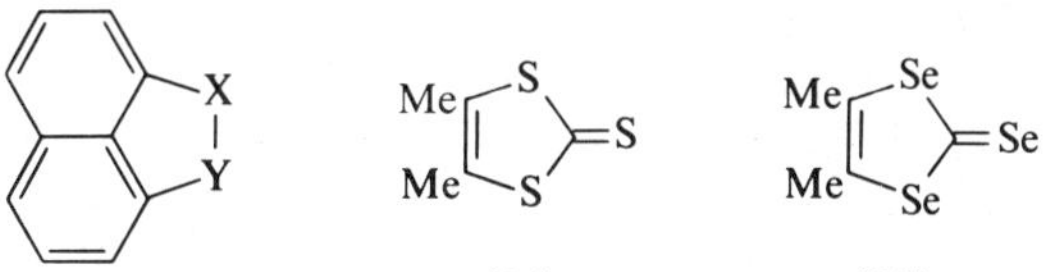

(23) (24) (25)

Table 2 Vertical Ionization Potentials (IP$_v$) and Calculated Orbital Energies (ε) for Naphtho[1,3-cd]-[1,2]dichalcogenoles (23) ⟨81CB2622⟩

X—Y	IP$_v$ or ε (eV)	Band 1	2	3	4	5	6
S—S	IP$_v$	7.14	8.91	9.07	9.35	11.15	12.0
	ε	7.09	9.10	9.31	9.40	11.28	—
Se—S	IP$_v$	7.14	8.91	9.10	9.30	11.00	11.7
Se—Se	IP$_v$	7.06	8.95	9.09	9.20	10.95	11.6
	ε	7.20	9.01	9.10	9.18	10.88	—
Te—S	IP$_v$	7.03	8.78	8.91	9.29	10.60	10.9
	ε	7.20	8.55	8.90	9.11	10.44	—
Te—Se	IP$_v$	7.05	8.70	8.90	9.00	10.60	10.9
Te—Te	IP$_v$	7.10	8.80	9.00	9.10	10.60	11.0

The photoelectron spectra of the 1,3-dichalcogenole-2-thiones (**17a–c**) and 1,3-dichalcogenole-2-selenones (**18a–c**) as well as the dimethyl derivatives (**24**) and (**25**) have been measured and are compiled in Table 3 ⟨77JA2855⟩. Very slight differences in the first six ionization bands were observed for these compounds. Those bands showing the greatest effect upon changing heteroatoms are those which are largely localized on the heteroatoms. The PE spectra of the closely related tetrathiafulvalene (**26**), dithiadiselenafulvalene (**27**) and tetraselenafulvalene (**28**) gave similar results (Table 4) ⟨77MI43500⟩.

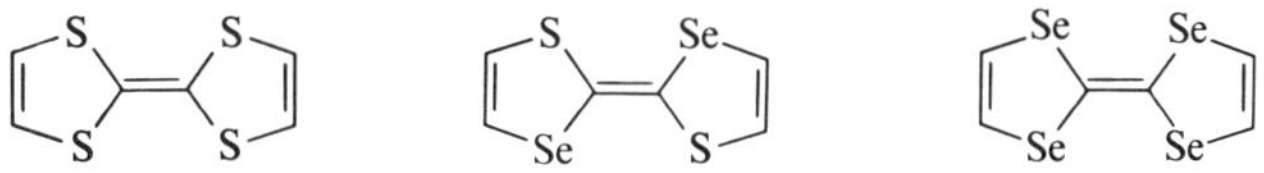

(26) (27) (28)

Table 3 Vertical Ionization Potentials (IP$_v$) for 1,3-Dichalcogenole-2-thiones and 1,3-Dichalcogenole-2-selenones ⟨77JA2855⟩

Compound	IP$_v$ (eV) [assignment]					
	1 [b$_2(\sigma)$]	2 [B$_1(\pi)$]	3 [a$_2(\pi)$]	4 [b$_1(\pi)$]	5 [b$_2(\sigma)$]	6 [a$_1(\sigma)$]
(**17a**)	8.33	8.56	10.60	10.90	12.00	12.37
(**17b**)	8.28	8.52	10.17	10.75	11.72	12.12
(**17c**)	8.24	8.47	9.80	10.60	11.37	11.89
(**18a**)	7.81	8.06	10.54	10.54	11.66	11.98
(**18b**)	7.83	8.08	10.01	10.40	11.51	11.77
(**18c**)	7.85	8.08	9.71	10.18	11.26	11.53
(**24**)	7.96	8.29	10.15	10.33	11.50	11.50
(**25**)	7.68	7.90	9.46	9.73	10.90	11.17

Table 4 Vertical Ionization Potentials (IP$_v$) for Tetrathiafulvalene, Dithiadiselenafulvalene and Tetraselenafulvalene ⟨77MI43500⟩

Compound	IP$_v$ (eV) [assignment]					
	1 [B$_{1u}$]	2 [B$_{3g}$]	3 [A$_u$]	4 [B$_{2g}$]	5 [B$_{1u}$]	6 [B$_{1g}(\sigma)$]
(**26**)	6.70	8.47	9.63	10.03	10.49	11.01
(**27**)	6.75	8.42	9.29	9.65	10.38	10.74
(**28**)	6.90	8.32	9.08	9.41	10.40	10.40

4.35.3.3 Infrared and Raman Spectroscopy

The IR spectra of the compounds of this chapter have been determined and studied in some detail. The 1,2-diselenolanes and 1,2-thiaselenolanes show C—Se stretching bands around 500 cm^{-1} ⟨62AK(19)195⟩. Similarly, 1,3-diselenolane-2-selenone (**29**) ⟨75SA(A)191⟩ and 1,3-diselenolane-2-ylidenemalononitrile (**30**) ⟨75SA(A)1371⟩ show C—Se stretching bands between 550 and 560 cm^{-1}, while Se—C—Se stretching bands appear between 850 and 930 cm^{-1}. These assignments and others for these compounds were made by comparing IR and Raman spectra of both (**29**) and (**30**) and their d_4 derivatives.

Complete assignment of the IR bands of the 1,3-dichalcogenole-2-thiones (**17a–c**) and 1,3-dichalcogenole-2-selenones (**18a–c**) has been made ⟨81JST(71)1⟩. These data are compiled in Table 5. Alkyl and aryl substituents at C-4 and C-5 change these values slightly ⟨75JOC746⟩.

Table 5 Correlation of the Fundamental Frequencies (cm^{-1}) of 1,3-Dichalcogenole-2-thiones and 1,3-Dichalcogenole-2-selenones in Potassium Bromide Pellets ⟨81JST(71)1⟩

	(17a)	(17b)	(18a)	(18b)	(18c)	Main assignment[a]
In-plane	3090	3080	3080	3075	3060	ν(CH)
	3075	3040	3064	3040	3040	ν(CH)
	1396	1398	1382	1396	1395	ν(C=C)
	1250	1245	1244	1243	1218	δ(CH)
		1228		1222	1221	
	1070	1092	1044	1070	1060	δ(CH)
		1076				
	1050	1030	940	928	890	ν(C=X)
	1015					
	890	828	906	840	770	ν(C(2)—Y)
	820	782	818	788	730	ν(C(4)—Y)
	655	660	670	664	494	ν(C(4)—Y)
	508	436	492	405	386	ν(C(5)—Z)
	460	420	430	382	265	δ(C(5)—Z)
	376	332	265	265	232	ring def.
	242	211	174	160	142	δ(C=X)
Out-of-plane	730	680	745	681	661	π(CH)
	665	660	670	664	652	π(CH)
	410	420	395	420	326	τ(C=C)
	395	350	295	280	290	π(C=X)
				290		
	77	94	80	86	80	τ(ring)

[a] X, Y, Z = S/Se; ν, stretching; δ, in-plane bending; π, out-of-plane bending; τ, torsion.

Raman scattering spectra have been determined for tetrathiafulvalene (**26**), tetraselenafulvalene (**28**) and tetramethyltetraselenafulvalene (**9**) ⟨82MI43501⟩. The frequencies and assignment for these compounds are compiled in Table 6. The Raman spectra of various radical cation salts were determined as well. Both the IR and Raman spectra of these compounds were simple, reflecting the high degree of symmetry in these molecules.

In the IR spectrum of tetratellurafulvalene (**10**) the inner C=C stretching frequency is observed at 1560 cm^{-1}, significantly different from the corresponding stretching frequency of 1618 cm^{-1} observed for hexamethylenetetraselenafulvalene (**31**) ⟨82JA1154⟩. This weakening of the endocyclic C=C bond was attributed to transannular nonbonded Te–Te interactions. Other strong bands for (**10**) were observed at 1475 and 1080 cm^{-1}.

The 1,3-dichalcogenolylium fluoroborates (**19a–c**) showed several characteristic bands in their IR spectra. As KBr pellets, strong bands were observed between 1450 and 1455, between 1000 and 1160 (fluoroborate anion), between 910 and 935, between 760 and 812 and between 700 and 740 cm^{-1} ⟨75TL1259⟩.

The 1,2-diselenolylium chloride (**32a**) showed characteristic IR bands at 1630, 1604 and 1330 cm^{-1} ⟨67ACS1991⟩. As in the case of 1,2-dithiolylium salts, the corresponding iodide

Table 6 Raman lines (cm^{-1}) of Tetrathiafulvalene (**26**), Tetraselenafulvalene (**28**) and Tetramethyltetraselenafulvalene (**9**)[a] ⟨82MI43501⟩

(**26**)	(**28**)	(**9**)	Normal coordinate[a]
		173	
2563	231w	180w	ring def.
	260m	263s	
316vs	272s	276s	
420m	379m	328m	Me in (**9**)
469m	451m	453vs	ν(C—S), inner in (**26**)
480w	480w	472m	Me in (**9**)
612w	599m	603w	
740m	718m	682m	
800w		916m	
1010w	1010w	1018w	
1086vs	1075w		ν(CH) in (**26**)
		1167w	
1246		1225w	combination tones
		1348w	and Me in (**9**)
	1430w	1420w	
	1480w	1444w	
1520vs	1526vs	1503vs	ν(C=C) center
		1530w	
1556vs	1549vs	1539vs	ν(C=C) ring
1569s		1589w	

[a] ν, stretching.

showed a shift of these bands to lower frequency at 1610, 1590 and 1330 cm^{-1}. The 1,2-oxaselenolylium chlorides (**5**) and (**33**) containing a covalent Cl—Se bond show a characteristic, strong C—O stretch between 1530 and 1545 cm^{-1} ⟨83JA883⟩. Similarly, the 1,2-oxatellurolylium halides and pseudohalides (*e.g.* **6** and **7**) display strong C—O stretching bands between 1510 and 1530 cm^{-1} ⟨83JA875⟩.

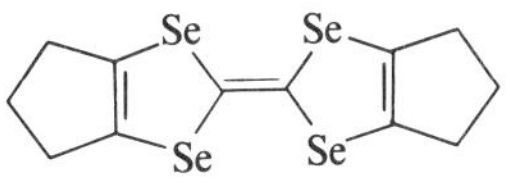

(**31**) 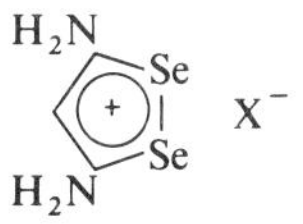   (**32**) **a**; X = Cl   (**33**)
            b; X = I

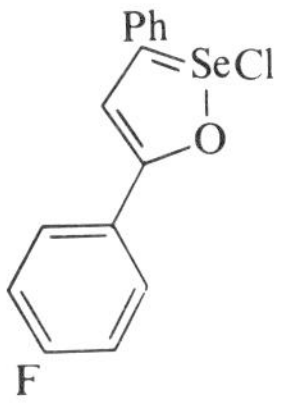

4.35.3.4 Mass Spectrometry

The mass spectra of selenium and tellurium containing molecules are particularly diagnostic due to the abundance of isotopes for each element (Table 7). Thus, selenium has isotopes between nominal masses 74 and 82 while tellurium has isotopes between nominal masses 120 and 130 ⟨B-61MI43500⟩. For molecules containing more than one selenium or tellurium, the observed molecular ion clusters upon comparison with calculated spectra are essentially unequivocal for the numbers of selenium or tellurium atoms involved (Table 8) ⟨77JA255, 77JA7743, 80TL4565, 82CC336⟩.

The mass spectra of 13 tetraselenafulvalenes, dithiadiselenafulvalenes and tetrathiafulvalenes have been examined in detail ⟨78OMS121⟩. The tetraselenafulvalenes were found to lose selenium much more readily than the tetrathiafulvalenes were found to lose sulfur. The tetraselenafulvalenes readily lost an alkyne fragment while the tetrathiafulvalenes lost an SCR radical. The diathiadiselenafulvalenes were found to lose an SeC$_2$R$_2$ fragment exclusively. A rearrangement to a 1,3-diselenole-2-selenone radical-cation was found to occur by selenium migration (Scheme 4). In the mass spectrum of 4,5-dimethyl-1,3-diselenole-2-selenone (**25**), the molecular ion is the base peak representing 36% of the total ionization ⟨78OMS121⟩.

Table 7 Isotopic Distribution for Selenium and Tellurium
⟨B-61MI43500⟩

	Selenium		Tellurium	
Nominal mass	Relative abundance		Nominal mass	Relative abundance
74	1.75		120	0.26
75	0.00		121	0.00
76	18.11		122	7.13
77	15.21		123	2.52
78	47.21		124	13.37
79	0.00		125	20.27
80	100.00		126	54.26
81	0.00		127	0.00
82	18.45		128	92.20
			129	0.00
			130	100.00

Table 8 Molecular Ion Clusters for Five-membered Heterocycles Containing More Than One Selenium or Tellurium Atom

(21) ⟨77JA255⟩			(35) ⟨77JA7743⟩			(2) ⟨77JA255⟩			(35) ⟨82CC336⟩		
m/e	Obs.	(Calc.)[a]	m/e	Obs.	(Calc.)[a]	m/e	Obs.	(Calc.)[a]	m/e	Obs.	(Calc.)[a]
278	4	(4)	324	1	(1)	373	2	(2)	674	1.6	(1.5)
279	5	(5)	325	1	(1)	374	7	(6)	675	2.3	(2.1)
280	18	(20)	326	5	(5)	375	5	(5)	676	4.6	(4.3)
281	10	(14)	327	6	(5)	376	17	(17)	677	5.4	(5.5)
282	47	(51)	328	20	(19)	377	16	(15)	678	11.2	(10.4)
283	32	(31)	329	16	(15)	378	36	(36)	679	12.8	(12.0)
284	88	(88)	330	40	(40)	379	26	(25)	680	21.7	(21.8)
285	23	(15)	331	27	(27)	380	67	(66)	681	22.3	(22.2)
286	100	(100)	332	83	(81)	381	30	(28)	682	41.1	(39.8)
287	19	(11)	333	23	(21)	382	100	(100)	683	34.5	(35.0)
288	39	(32)	334	100	(100)	383	11	(11)	684	64.3	(62.4)
289	5	(4)	335	12	(11)	384	92	(93)	686	87.2	(85.2)
			336	76	(77)	385	9	(10)	687	47.6	(47.9)
			337	8	(9)	386	50	(51)	688	100.0	(100.0)
			338	12	(13)	387	5	(6)	689	38.1	(37.6)
			339	1	(1)				690	97.1	(97.1)
									691	23.1	(22.7)
									692	70.2	(71.8)
									693	10.5	(11.1)
									694	35.6	(36.0)
									695	5.6	(5.6)
									696	9.7	(10.0)
									697	1.4	(1.5)

[a] Relative intensities.

(34)   (35)

4.35.3.5 Nuclear Magnetic Resonance Spectroscopy

The NMR spectra of selenium and tellurium containing heterocycles are influenced by the NMR active isotopes ^{77}Se and ^{125}Te. Both isotopes occur in about 7% natural abundance and enriched isotope mixtures of each metal are available. ^{77}Se–^{1}H, ^{77}Se–^{13}C, ^{125}Te–^{1}H and ^{125}Te–^{13}C satellites may be observed upon careful scrutiny of spectra. It is possible to obtain both ^{77}Se and ^{125}Te NMR spectra of compounds containing these elements.

The chemical shifts reported in this section are in parts per million (p.p.m.) from tetramethylsilane (TMS), where the chemical shift of TMS protons and carbons are δ 0.00

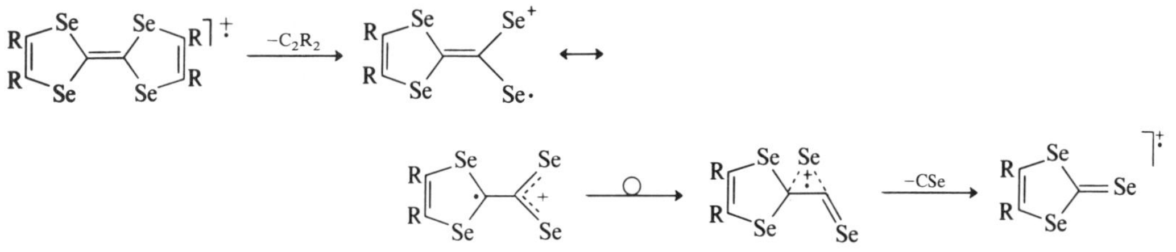

Scheme 4 Rearrangement pathway from tetraselenafulvalene radical cation to 1,3-diselenole-2-selenone radical cation

for ^{1}H NMR and ^{13}C NMR, respectively. Positive δ values are shifts to lower field while negative δ values are shifts to higher field.

4.35.3.5.1 ^{1}H NMR spectroscopy

The dichalcogenolanes have been little examined by ^{1}H NMR spectroscopy. The ^{1}H NMR spectrum of 1,3-diselenolane has been reported as a four-proton singlet at δ 3.4 and a 2-proton singlet at δ 3.85 ⟨79JOC4689⟩. The ^{1}H NMR spectrum of 1,3-diselenolane-2-selenone has been reported as well ⟨75SA(A)191⟩. The ^{1}H NMR spectrum of the cyclic seleninic ester (**36**), a 1,2-oxaselenolane Se-oxide, has also been described ⟨80CS(16)24⟩. In fluorosulfonic acid, 1,3-diselenolane was observed to give a complex spectrum consistent with monoprotonation ⟨79JOC4689⟩.

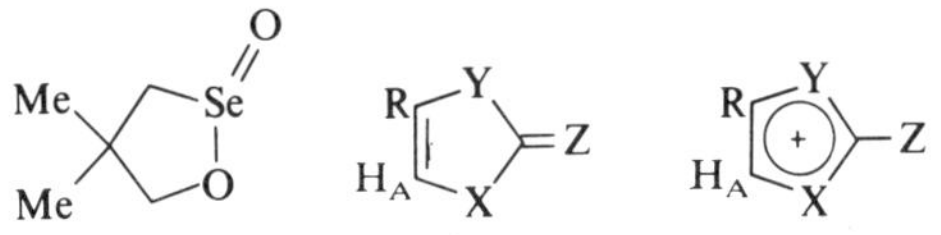

The 1,3-dichalcogenole compounds have been studied by ^{1}H NMR in considerably more detail. Representative chemical shifts for these compounds as described by (**37**) are compiled in Table 9. Several trends are apparent from the data. The chemical shift of protons attached to C-2 is little influenced by the nature of the two chalcogen atoms attached to C-2, showing less than 0.3 p.p.m. change from 1,3-dithioles to 1,3-ditelluroles. The chemical shifts of protons attached to C-4 and C-5 are much more sensitive to the chalcogen atoms in both positions, moving to lower field with increasing size of the chalcogen atoms. Thus, the alkenic protons of 1,3-ditellurole are at 2.75 p.p.m. lower field than the protons of 1,3-dithiole. A linear dependence of the chemical shift of the alkenic protons with the size of the chalcogen atom in 1,3-dichalcogenol-2-ones, -2-thiones and -2-selenones has been recognized ⟨81JPR737⟩. Protons attached to C-4 are influenced by substituents attached to C-5.

The ^{1}H NMR spectra of various 1,3-dichalcogenolylium ions of structure (**38**) are compiled in Table 10. The chemical shifts of all ring protons are quite sensitive to the adjoining chalcogen atoms. The chemical shift of the proton attached to C-2 is δ 11.43 in 1,3-dithiolylium fluoroborate ⟨75TL1259⟩ and is δ 15.00 in 1,3-ditellurolylium fluoroborate ⟨82TL1531⟩, while the protons attached to C-4 and C-5 resonate at δ 9.45 in 1,3-dithiolylium fluoroborate and at δ 11.81 in 1,3-ditellurolylium fluoroborate.

The coupling constants are somewhat different between protons in the 1,3-dichalcogenoles and 1,3-dichalcogenolylium ions. While no apparent coupling is observed between the protons attached to C-2 and C-4 or C-2 and C-5 in the dichalcogenoles, these same protons show 1-2 Hz coupling in the dichalcogenolylium ions ⟨75TL1259, 82TL1531⟩. Coupling between protons attached to C-4 and C-5 is of the order of 7.5 Hz in the 1,3-dichalcogenoles and 6.0-6.5 Hz in the 1,3-dichalcogenolylium ions. Allylic coupling of 1.0-2.0 Hz is observed between protons on C-4 or C-5 and alkyl substituents on C-5 or C-4.

The ^{1}H NMR chemical shifts of various 1,2-dichalcogenolylium systems are compiled in Table 11. These compounds are of two types: ionic, as in structure (**39**), and covalent, as in structure (**40**). The chemical shifts of ring protons and substituents appear to be more sensitive to the ring heteroatoms than to the ionic or covalent nature of the compound. Coupling between protons attached to C-3 and C-4 or C-4 and C-5 appears to be of the

Table 9 ^{1}H NMR Spectral Data for 1,3-Dichalcogenoles of Structure (37)[a]

R	X	Y	Z	$\delta(H_A)$	$\delta(H_B)$	$\delta(R)$	Solvent	Ref.
H	S	S	H_B, H_B	6.08	4.50	—	$CDCl_3$	83TL237
H	Te	Te	H_B, H_B	8.83	4.70	—	$CDCl_3$	83TL237
Me	S	S	H_B, H_B	5.53	4.43	1.86	CCl_4	81RTC10
Me	S	Se	H_B, H_B	5.80	4.43	2.03	CCl_4	81RTC10
Me	S	Te	H_B, H_B	6.16	4.73	2.10	CCl_4	81RTC10
Me	Te	Te	H_B, H_B	8.07	4.73	2.23	$CDCl_3$	83TL237
Ph	S	S	H_B, H_B	6.33	4.57	7.40	$CDCl_3$	83TL237
Ph	S	Se	H_B, H_B	6.40	4.40	7.20	CCl_4	81RTC10
Ph	S	Te	H_B, H_B	6.70	4.66	7.13	CCl_4	81RTC10
Ph	Te	Te	H_B, H_B	8.57	4.53	7.25	$CDCl_3$	82TL1531
H	S	S	H_B, SMe	5.97	6.07	—	$CDCl_3$	75TL1259
H	S	Se	H_B, SMe	6.40	6.27	6.68	$CDCl_3$	75TL1259
H	Se	Se	H_B, SeMe	7.12	6.37	—	$CDCl_3$	75TL1259
Me	Se	S	H_B, SMe	6.22	6.13	2.07	$CDCl_3$	75TL1259
Ph	Te	Te	H_B, Me	8.75	5.12	7.30	$CDCl_3$	83TL237
Ph	S	S	O	7.00	—	—	$CDCl_3$	81JPR737
Ph	Se	S	O	7.62	—	—	$CDCl_3$	81JPR737
Ph	Se	Se	O	7.63	—	—	$CDCl_3$	81JPR737
H	S	Se	S	7.26	—	7.80	$CDCl_3$	75JOC746
H	Se	Se	S	7.92	—	—	$CDCl_3$	75JOC746
CO_2Et	S	Se	S	7.93	—	—	$CDCl_3$	80JHC549
CO_2Et	Se	S	S	8.68	—	—	$CDCl_3$	80JHC549
Me	Se	Se	S	7.23	—	2.42	$CDCl_3$	75JOC746
H	S	Se	Se	7.42	—	8.01	$CDCl_3$	75JOC746
H	Se	Se	Se	7.92	—	—	$CDCl_3$	75JOC746
Me	Se	Se	Se	7.42	—	2.42	$CDCl_3$	75JOC746
Ph	Se	Se	Se	7.87	—	7.36	$CDCl_3$	75JOC746
Ph	Se	S	NCO_2Et	7.70	—	7.48	$CDCl_3$	80JHC117
Me	Se	S	NPh	6.53	—	2.17	$CDCl_3$	80JHC117
Ph	Te	Te	CH_2Ph	8.85	7.86	cis	$(CD_3)_2SO$	81CC828
				8.77	7.91	trans	$(CD_3)_2SO$	81CC828

[a] In p.p.m. relative to TMS.

Table 10 ^{1}H NMR Spectral Data for 1,3-Dichalcogenolylium Ions of Structure (38)

R	X	Y	Z	$\delta(H_A)$	$\delta(H_B)$	$\delta(R)$	Solvent	Ref.
H	S	S	H_B	9.45	11.43	—	CF_3CO_2D	75TL1259
H	S	Se	H_B	9.48	12.43	10.18	CF_3CO_2D	75TL1259
H	Se	Se	H_B	10.22	13.41	—	CF_3CO_2D	75TL1259
H	Te	Te	H_B	11.81	15.00	—	CD_3CN	82TL1531
Me	Se	S	H_B	9.63	12.27	3.07	CF_3CO_2D	75TL1259
H	S	S	SMe	8.82	—	—	DMSO-d_6	75TL1259
H	S	Se	SMe	8.82	—	9.23	DMSO-d_6	75TL1259
H	Se	Se	SeMe	9.45	—	—	DMSO-d_6	75TL1259
Me	Se	S	SMe	8.82	—	2.67	DMSO-d_6	75TL1259
Me	Se	Se	$N(CH_2)_5$	7.83	—	2.56	$(CD_3)_2CO$	75JOC746
Ph	Se	Se	$N(CH_2)_5$	7.89	—	7.59	CF_3CO_2H	75JOC746

order of 6.6–7.5 Hz. The chemical shifts of protons attached to C-3 and C-5 are at significantly lower field than protons attached to C-4, indicative of positive charge delocalization at C-3 and C-5. This is similar to the expected charge delocalization in the allyl cation ⟨82TL1531, 83JA875, 83JA883⟩.

(39)   (40)

The ^{1}H NMR spectrum of tetraselenafulvalene (28) has been reported as a singlet at δ 7.25 ⟨74JA7376⟩. Both cis- and trans-dithiadiselenafulvalenes (27) showed pairs of doublets

Table 11 ^{1}H NMR Spectral Data for 1,2-Dichalcogenolylium Ions of Structure (39) or (40)

R^1	R^2	X	Y	Z	$\delta(H_A)$	Chemical shifts $\delta(R^1)$	$\delta(R^2)$	Solvent	Ref.
CH$_2$COMe	Me	O	Se	ClO$_4$	7.81	5.10 (2H) 2.77 (3H)	2.81	CF$_3$CO$_2$H	75JCS(P1)2097
Ph	p-MeOC$_6$H$_4$	O	Se	Cl	7.52	7.32	7.90 (2H) 6.88 (2H)	CDCl$_3$	83JA883
Me	p-MeOC$_6$H$_4$	O	Se	Cl	7.60	2.88	7.95 (2H) 6.95 (2H)	CDCl$_3$	83JA883
Me	p-FC$_6$H$_4$	O	Se	Cl	7.35	2.53	7.85 (2H) 7.05 (2H)	CDCl$_3$	83JA883
H	Ph	O	Te	Cl	8.67	11.15	8.05 (2H) 7.55 (3H)	CDCl$_3$	83JA875
H	Ph	O	Te	I	8.47	11.30	8.10 (2H) 7.60 (3H)	CDCl$_3$	83JA875
Me	Ph	O	Te	Cl	8.40	2.97	8.15 (2H) 7.55 (3H)	CDCl$_3$	83JA875
Me	Ph	O	Te	Br	8.35	3.00	8.04 (2H) 7.50 (3H)	CDCL$_3$	83JA875
Me	Ph	O	Te	I	8.30	3.05	8.10 (2H) 7.60 (3H)	CDCl$_3$	83JA875
Ph	Ph	O	Te	F	8.53	7.50	8.10 (2H) 7.57 (3H)	CDCl$_3$	83JA875
Ph	Ph	O	Te	Cl	8.42	7.45	8.08 (2H) 7.50 (3H)	CDCl$_3$	83JA875
Ph	Ph	O	Te	I	8.23	7.45	8.13 (2H) 7.50 (3H)	CDCl$_3$	83JA875
Ph	Ph	O	Te	CF$_3$CO$_2$	8.50	7.40	8.07 (2H) 7.53 (3H)	CDCl$_3$	83JA875
H$_B$	H$_A$	Te	Te	BF$_4$	10.31	13.83	—	CD$_3$CN	82TL1531
H	Me	Te	Te	BF$_4$	10.20	13.40	—	CD$_3$CN	82TL1531
H	Bu	Te	Te	BF$_4$	10.15	13.25	—	CD$_3$CN	82TL1531
H	Ph	Te	Te	BF$_4$	10.61	13.30	—	CD$_3$CN	82TL1531

in their ^{1}H NMR spectra ⟨80JOC2632⟩. The chemical shifts of one isomer are at δ 6.82 and 6.60 ($J = 6.5$ Hz), while the chemical shifts of the other isomer are at δ 6.71 and 6.50 ($J = 6.5$ Hz).

^{1}H NMR spectra have been useful for physical organic chemistry aspects of the compounds discussed in this chapter. The kinetics of the thermal 1,3-ditellurolylium to 1,2-ditellurolylium ion rearrangement were monitored by ^{1}H NMR spectroscopy at 283 K and 298 K in CD$_3$CN ⟨82TL1531⟩.

Two-dimensional NMR spectroscopy is a new tool for the spectroscopist in studying oriented molecules ⟨B-79MI43501⟩. Two-dimensional ^{1}H NMR spectroscopy has been applied to an ordered array of 1-thia-3-selenole-2-thione in N-(p-methoxybenzylidene)-p-(n-butyl)aniline ⟨80JMR(37)349⟩. For oriented AB spectra, one-dimensional and two-dimensional spectra are not identical.

4.35.3.5.2 ^{13}C NMR spectroscopy

Very little information is available for the ^{13}C NMR spectra of dichalcogenolanes. The ^{13}C NMR spectra of 1,3-diselenolane-2-thione and 1,3-diselenolane-2-selenone (29) have been reported with methylene carbon chemical shifts in CDCl$_3$ at δ 42.9 and 45.5, respectively, a thione carbon chemical shift of δ 228.7 and a selenone carbon chemical shift of δ 222.5 ⟨81JPR737⟩.

The 1,3-dichalcogenoles have been studied in more detail and ^{13}C NMR data for the ring carbons of structure (41) are compiled in Table 12. Several trends are apparent. For sp^3 hybridized C-2, the ^{13}C NMR chemical shift of C-2 is moved to higher field as the two adjoining heteroatoms increase in size. The shift is quite dramatic, as is seen in going from 4-methyl-1,3-dithiole (C-2, δ 37.0) to 4-methyl-1,3-thiatellurole (C-2, −40.6) ⟨81RTC10, 82TL1531⟩. The chemical shifts of C-4 and C-5 show two opposing trends as the size of the chalcogen atoms increases. As the heteroatom attached to C-4 or C-5 becomes larger, a shift to higher field is observed. However, as the chalcogen atom two bonds removed becomes larger, a lower field shift is observed. This is most likely due to changes in the electron density distribution. For 1,3-dichalcogenol-2-ones, -2-thiones and -2-selenones, the ^{13}C NMR chemical shifts move to lower field in the series CO, CSe, CS ⟨81JPR737⟩. This observation is best explained by changes in the electron density distribution, not by variations in the ΔE term of the ^{13}C chemical shift. The chemical shifts of C-4 and C-5 with sp^2 hybridized C-2 move to lower field as the size of either chalcogen atom increases.

The protonation of diselenafulvenes (42) and (43) with trifluoroacetic acid to give 1,3-diselenolylium ions has been investigated by ^{13}C NMR ⟨79JHC1303⟩. Upon protonation the chemical shifts of the ring carbons of (42) move as follows: C-2 from δ 105.9 to 232.8; C-4 and C-5 from δ 128.0 to 159.2. For (43) the chemical shifts of the ring carbons move as follows: C-2 from δ 125.2 to δ 231.2; C-4 from δ 135.8 to 169.8; C-5 from δ 114.2 to 141.3. These shifts are indicative that the positive charge in 1,3-diselenolylium ions is more localized on C-2 than on C-4 and C-5 ⟨79JHC1303⟩.

(41) (42) (43)

Table 12 ^{13}C NMR Spectral Data for 1,3-Dichalcogenoles of Structure (41) in CDCl$_3$

R^1	R^2	X	Y	Z	$\delta(C$-2$)$	$\delta(C$-4$)$	$\delta(C$-5$)$	Ref.
H	H	Te	Te	H, H	−39.0	120.0	120.0	82TL1531
Me	H	S	S	H, H	37.0	130.7	111.4	81RTC10
Me	H	S	Se	H, H	27.1	129.2	115.5	81RTC10
Me	H	S	Te	H, H	3.9	117.8	125.4	81RTC10
Me	H	Te	Te	H, H	−40.6	136.1	112.4	82TL1531
Ph	H	S	S	H, H	36.1	134.8	113.0	81RTC10
Ph	H	S	Se	H, H	25.9	134.5	116.6	81RTC10
Ph	H	S	Te	H, H	2.8	128.3	122.9	81RTC10
Ph	H	Te	Te	H, H	−20.7	143.1	141.6	82TL1531
Ph	p-MeOC$_6$H$_4$	O	Se	Ph, ArCO	99.5	107.9	143.4	78CZ361
Ph	p-MeOC$_6$H$_4$	O	Se	O	168.5	115.6	141.2	78CZ361
H	H	S	S	O	194.1	118.5	118.5	81JPR737
H	H	Se	Se	O	192.2	124.3	124.3	81JPR737
H	Ph	S	S	O	192.4	11.7	134.8	81JPR737
H	Ph	S	Se	O	191.1	115.4	136.1	81JPR737
H	Ph	Se	Se	O	188.7	117.5	141.3	81JPR737
H	H	S	S	S	213.3	129.4	129.4	81JPR737
H	H	Se	Se	S	218.2	134.4	134.4	81JPR737
H	Ph	S	S	S	212.4	122.0	146.4	81JPR737
H	Ph	S	Se	S	214.6	125.6	147.5	81JPR737
H	Ph	Se	Se	S	216.8	127.5	151.9	81JPR737
H	H	S	S	Se	203.7	134.4	134.4	81JPR737
H	H	Se	Se	Se	207.7	139.5	139.5	81JPR737
H	Ph	S	Se	Se	202.9	126.5	151.6	81JPR737
H	Ph	S	Se	Se	204.9	130.2	152.9	81JPR737
H	Ph	Se	Se	Se	206.7	132.1	157.5	81JPR737
Me	Me	Se	Se	O	188.1	128.5	128.5	81JPR737
Me	Me	Se	Se	S	216.5	139.8	139.8	81JPR737
Me	Me	Se	Se	Se	206.4	145.0	145.0	81JPR737
—(CH$_2$)$_4$—		S	S	C(CH$_2$)$_4$	118.0	124.5	124.5	79JHC1303
—(CH$_2$)$_4$—		Se	Se	C(CH$_2$)$_4$	105.9	128.0	128.0	79JHC1303
Ph	H	S	S	CHPh	132.6	134.3	111.7	79JHC1303
Ph	H	Se	Se	CHPh	125.2	135.8	114.2	79JHC1303

The ^{13}C NMR spectra of 1,2-oxatellurolylium halides of structure (39) indicate shift ranges of δ 178.5–189.1 for C-3, δ 124.3–125.3 for C-4 and δ 189.8–191.4 for C-5 ⟨83JA875⟩. This is consistent with positive charge delocalization to C-3 and C-5 in such systems.

4.35.3.5.3 ^{125}Te spectroscopy

The ^{125}Te chemical shifts of tellurium in 1,2-oxatellurolylium halides (44), (45) and (46) have been reported as δ 911.0, 858.0 and 677.0 p.p.m., respectively, from bis(diethyldithiocarbamate)tellurium(II) ⟨83JA875⟩. The ^{125}Te chemical shift is a sensitive probe for the oxidation state of tellurium ⟨80JOM(192)183⟩. The ^{125}Te chemical shifts of all three compounds fall in the range expected for Te(IV) compounds.

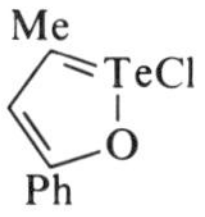

(44) 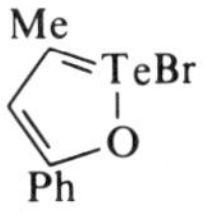   (45) 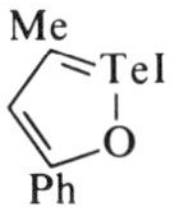   (46)

4.35.3.6 Electronic Paramagnetic Resonance Spectroscopy

The naphtho[1,8-*cd*][1,2]dichalcogenoles (23), with XY = SS, SeS, SeSe, TeS, TeSe, TeTe, were easily oxidized to radical cations with aluminum chloride in dichloromethane ⟨81CB2622⟩. The radical cations generated (XY = SS to TeSe) were stable at room temperature. The EPR spectra of these compounds indicate that the chalcogen bridges dominate the radical cation species regarding spin population as well as charge distribution. The EPR spectra reflect the effect of increasing spin–orbit interaction in the order sulfur, selenium, tellurium, with increasing line widths and linearly increasing g values from 2.0086 for the dithiole to 2.0409 for the selenatellurole ⟨81CB2622⟩.

The EPR spectrum of the tetracyanoquinodimethane salt of (10) has been measured ⟨82JA1154⟩. This material gave a g value of 2.0039 and a line width of 4 G.

4.35.4 THERMODYNAMIC ASPECTS

4.35.4.1 Intermolecular Forces

4.35.4.1.1 *Melting points, boiling points and solubility*

X-ray studies have shown significant intermolecular Te–Te interactions in the crystalline state for both non-cyclic and heterocyclic systems ⟨83JA875, 82CC1316, 82MI43500, 81TL4199⟩. However, this observation has not been correlated to changes in the melting points, boiling points or solubilities of tellurium containing heterocycles. In fact, little information is available correlating the melting points, boiling points and solubilities of any of the classes of compounds in this chapter.

4.35.4.2 Stability and Stabilization

4.35.4.2.1 *Thermal chemistry*

Poly(propyl diselenide) upon heating gives 1,2-diselenolane, indicating a thermodynamic preference for the small ring at higher temperature. Upon cooling, the melt reforms the polydiselenide ⟨62AK(19)195⟩.

The 1,3-ditellurolylium ions undergo a thermal rearrangement to give 1,2-ditellurolylium ions ⟨82TL1531⟩. This is indicative that the 1,2-ditellurolylium ion is thermodynamically preferred.

4.35.4.2.2 *Cyclic voltammetry*

The electrochemical oxidation potentials of various tetrachalcogenafulvalenes are compiled in Table 13. As the size of the heteroatom increases, the gap between $E^{1/2}$ for the first oxidation potential and $E^{1/2}$ for the second oxidation potential narrows. This is presumably due to more localization of the radical cation on the heteroatom lone pairs. As the size of the heteroatoms increases, overlap with the adjoining π-framework becomes less efficient.

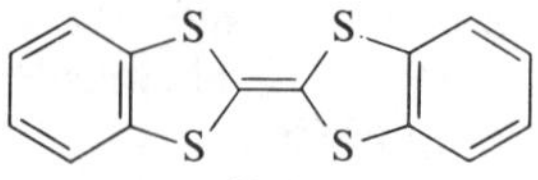

(47) 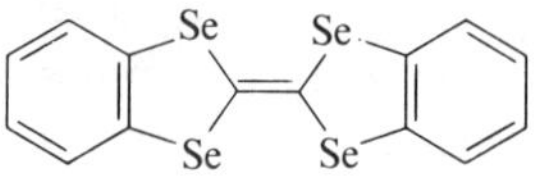   (48)

Table 13 Electrochemical Oxidation Potentials of Tetrachalcogenafulvalenes at a Platinum Electrode Surface *vs.* Standard Calomel Electrode

Compound	Solvent	Electrolyte	$E_1^{1/2}$(V)	$E_2^{1/2}$ (V)	ΔE (V)	Ref.
(26)	MeCN	Et$_4$NClO$_4$	0.33	0.70	0.37	75JA2921
(27)	MeCN	Et$_4$NClO$_4$	0.40	0.72	0.32	75JA2921
(28)	MeCN	Et$_4$NClO$_4$	0.48	0.76	0.28	75JA2921
(20)	PhCN	Bu$_4$NAsF$_4$	0.24	0.73	0.49	82JA1154
(9)	PhCN	Bu$_4$NAsF$_6$	0.42	0.81	0.39	82JA1154
(10)	PhCN	Bu$_4$NAsF$_6$	0.40	0.69	0.29	82JA1154
(47)	CH$_2$Cl$_2$	Bu$_4$NBF$_4$	0.71	1.14	0.43	82CC336
(48)	CH$_2$Cl$_2$	Bu$_4$NBF$_4$	0.78	1.17	0.39	82CC336
(35)	CH$_2$Cl$_2$	Bu$_4$NBF$_4$	0.71	1.05	0.34	82CC336

4.35.4.2.3 Aromaticity

The 1,2- and 1,3-dichalcogenolylium ions exhibit ^{1}H NMR chemical shifts that are consistent with a diamagnetic ring current in these molecules (Section 4.35.3.5.1). These compounds are formally 6π heteroaromatics and would be expected to display some aromatic character.

4.35.5 REACTIVITY

4.35.5.1 Thermal and Photochemical Reactions Formally Involving No Other Species

4.35.5.1.1 Rearrangement

The photochemical rearrangement of 1,3-dithiolylium ions to 1,2-dithiolylium ions was described in Chapter 4.32. In the 1,3-ditellurolylium ions (49), a rapid, thermal rearrangement to the 1,2-ditellurolylium ions (50) occurs, presumably by the sequence shown in Scheme 5 ⟨82TL1531⟩. The salient features of this mechanism are the proposed thermal four-electron electrocyclic reaction as well as the positioning of substituents in the rearranged product. The kinetics of the rearrangement of the parent cation (R = H) were determined at 283 and 298 K by ^{1}H NMR. The activation parameters were determined to be: ΔH = 33.4 kJ mol^{-1}, ΔS = 196 J K^{-1} mol^{-1} and ΔG = 93.2 kJ mol^{-1} ⟨82TL1531⟩. These values are consistent with a facile process involving significant ordering in the transition state. For other cations (R = Me, Bu, Ph), only the 1,2-ditellurolylium ions were observed, indicating an accelerated rate of rearrangement.

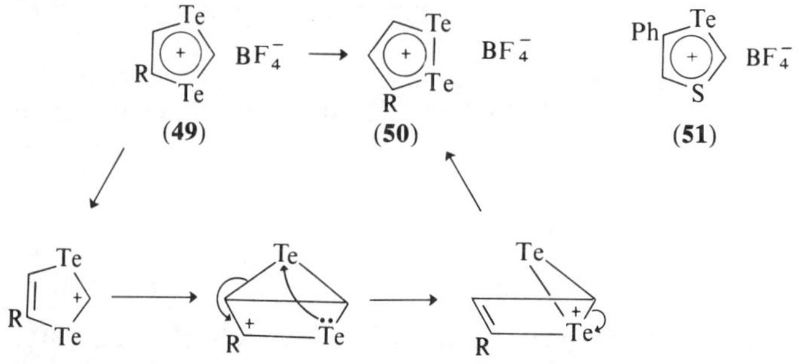

(49)   (50)   (51)

Scheme 5 1,3-Ditellurolylium cation to 1,2-ditellurolylium cation rearrangement

The unusual thermal behavior of the 1,3-ditellurolylium ions can be attributed to the more metallic characteristics of tellurium relative to sulfur and to tellurium–tellurium interactions. Interestingly, the 4-phenyl-1,3-thiatellurolylium ion (51) does not undergo a thermal rearrangement to the 1,2-thiatellurolylium ion.

4.35.5.1.2 Thermolysis

Thermolyses of 1,3-dichalcogenole-2-thiones and -2-selenones lead formally to carbenes by loss of elemental sulfur and selenium, respectively. These species dimerize to produce 1,4,5,8-tetrachalcogenafulvalenes (Scheme 6). In practice this transformation is carried out in the presence of a sulfur scavenger such as a phosphite or phosphine. Examples of this transformation are compiled in Table 14.

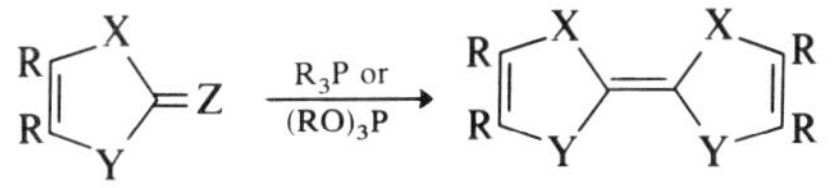

Scheme 6 Thermolyses of 1,3-dichalcogenole-2-thiones and -2-selenones to give tetrachalcogenafulvalenes

Table 14 Thermolyses of 1,3-Dichalcogenole-2-thiones and -2-selenones to give Tetrachalcogenafulvalenes according to Scheme 6

X	Y	Z	R	Scavenger	Yield (%)	Ref.
Se	Se	S	Me	$(EtO)_3P$	30	74CC937
S	Se	Se	H	$(MeO)_3P$	80	80JOC2632
S	Se	Se	CO_2Me	Ph_3P	50	80JOC2632
Se	Se	Se	H	Ph_3P	70–80	74JA7376, 77JA5909
Se	Se	Se	H	$(MeO)_3P$	70–80	74JA7376, 77JA5909
Se	Se	Se	SeMe	$(MeO)_3P$	10	76CC148
Se	Se	Se	Ph	$(MeO)_3P$	30–60	75JOC746
Se	Se	Se	Me	$(MeO)_3P$	83	80CC866
Se	Se	Se	$(CH_2)_3$	$(EtO)_3P$	—	80CC867

An interesting side reaction was observed when certain 1,3-diselenole-2-thiones were heated with trimethyl phosphite ⟨77CC835⟩. Instead of formation of tetraselenafulvalenes, thiatriselenafulvalenes were formed as shown in Scheme 7. The phosphite most likely initiates a rearrangement of the 1,3-diselenole-2-thiones to 1,3-thiaselenole-2-selenones.

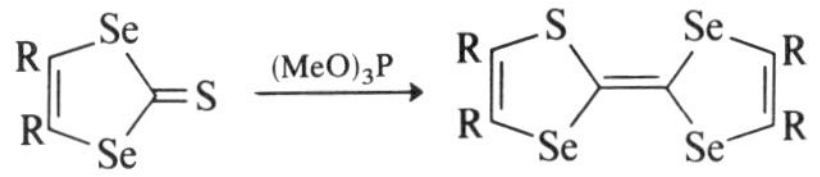

$R = Me, CF_3, CO_2Me$

Scheme 7 Thiatriselenafulvalenes from 1,3-diselenole-2-thiones and trimethyl phosphite

4.35.5.1.3 Photochemical reactions

The photochemical couplings of 4,5-diphenyl- and 4-phenyl-1,3-diselenole-2-thione to give tetraphenyl- and diphenyl-tetraselenafulvalene, respectively, and elemental sulfur have been reported ⟨79ZC192⟩. In this same report the irradiation of 4,5-diphenyl-1,3-thiaselenol-2-one (52) in the presence of nickel tetracarbonyl gave the nickel thiaselenane (53).

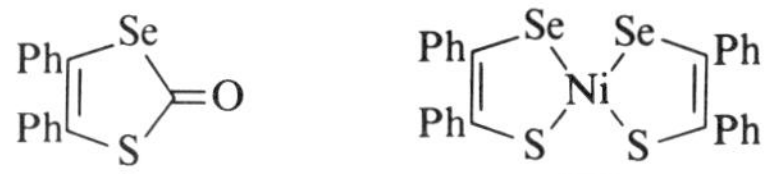

(52)   (53)

4.35.5.1.4 Polymerization

The 1,2-diselenolanes exist in equilibrium with their polydiselenide polymers ⟨30JCS1497, 57ACS911, 61AK(18)143, 62AK(19)195⟩. Interestingly, the monocyclic diselenolanes are favored in the melt while the polymers are favored at lower temperatures.

4.35.5.2 Reactions with Electrophiles

4.35.5.2.1 Reactions involving protons or Lewis acids

A variety of 1,3-diselenolylium and 1,3-thiaselenolylium ions has been generated by the reaction of protons with 1,3-diselenoles and 1,3-thiaselenoles, respectively. Protonation of an exocyclic double bond at the 2-position with trifluoroacetic acid has generated 1,3-diselenolylium cations (**54**) and (**55**) ⟨79JHC1303⟩. The driving force for such reactions is presumably the heteroatom stabilization of the positive charge or formation of the aromatic sextet.

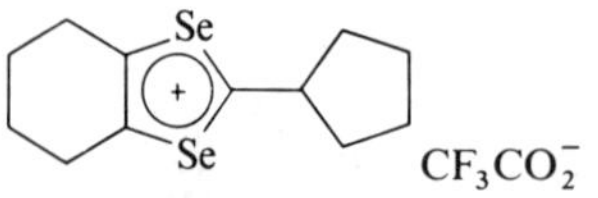

Similarly, proton induced elimination of neutral molecules has also been utilized to generate 1,3-diselenolylium and 1,3-thiaselenolylium ions. As shown in Scheme 8, protonation of 2-methylthio- or 2-methylseleno-1,3-dichalcogenoles with fluoroboric acid results in loss of methanethiol or methaneselenol and in formation of the corresponding 1,3-dichalcogenolylium ions ⟨80H(14)271, 75TL1259⟩.

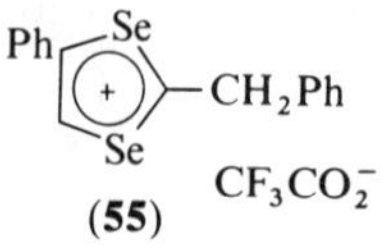

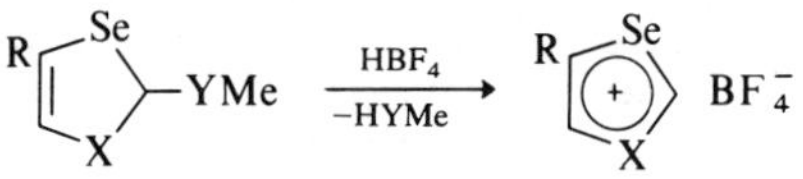

X = Y = S; R = Ph, Me, H
X = Y = Se; R = H

Scheme 8

The Lewis acid triphenylmethyl fluoroborate (trityl fluoroborate) has been used to effect net hydride abstraction from 1,3-ditelluroles to give 1,3-ditellurolylium ions (Scheme 9) ⟨82TL1531⟩. There is some question as to whether the loss of hydride occurs by a single two-electron process or by two separate one-electron oxidations coupled with a proton loss.

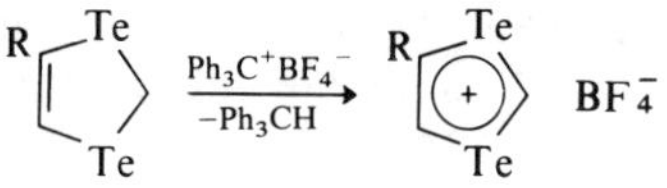

R = H, Me, Bu, Ph

Scheme 9

4.35.5.2.2 Reactions involving alkylating agents

1,3-Diselenole-2-selenones and 1,3-thiaselenole-2-thiones have been alkylated with methyl iodide and methyl fluorosulfonate ⟨80H(14)271, 75TL1259⟩. The resulting 2-methylseleno-1,3-diselenolylium (**56**) and 2-methylthio-1,3-thiaselenolylium (**57**) salts can be isolated in good yield.

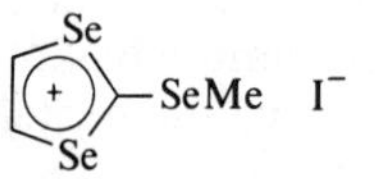

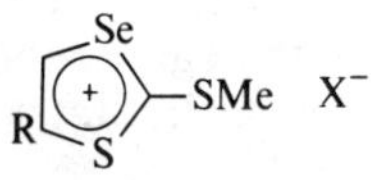

(57) **a**, R = H, X = I
b; R = Me, X = I
c; R = Ph, X = FSO$_3$

4.35.5.3 Reactions with Nucleophiles

4.35.5.3.1 Reactions with bases

The reactions of 1,3-dithioles with strong bases have been reported for 1,3-benzodithioles ⟨78TL2345⟩ and 2-(O,O-dimethylphosphonyl) or 2-tributylphosphonium substituted 1,3-

dithioles ⟨71JA4961, 76TL3695, 77S861, 78BCJ2674, 79JOC930⟩. Surprisingly, little information is available for the reactions of selenium and tellurium containing dichalcogenoles with bases. Phosphonium salt (**58**) upon treatment with benzaldehyde and potassium hydroxide in methanol gave thiaselenafulvene (**59**) as a mixture of E and Z isomers ⟨80H(14)271⟩. Other thiaselenafulvenes were prepared in 80–93% yield by this method as well.

(**58**)   (**59**)   (**60**) E = PhCHOH, O, Me   (**61**)

The direct metalation of 1,3-ditellurole (**60**; E = H) gave quite different results than the direct metalation of 4-phenyl-1,3-ditellurole (**61**; E = H) with lithium diisopropylamide (LDA) ⟨83TL237⟩. The parent system was lithiated at a vinylic position to give vinyl substituted products of structure (**60**) following electrophilic capture with benzaldehyde, methanol-*O*-d and methyl iodide. Identical results were obtained with 1,3-dithiole and 4-phenyl-1,3-dithiole with LDA.

These results suggest that the phenyl group in the 4-position of 1,3-dichalcogenoles sterically shields the 5-position. With the strong and hindered base LDA, protons in the 2-position are easily removed to give the products of kinetic control. In the absence of this substituent the thermodynamically preferrred vinyl anion is formed. With weaker bases such as sodium methoxide or potassium *t*-butoxide in protic solvents, preferential formation of the vinyl anion is observed even with 4-phenyl substituents ⟨83TL237⟩.

4.35.5.3.2 Reactions with sodium borohydride

1,2-Thiaselenolane-4-carboxylic acid (**16b**) was reduced by sodium borohydride to give the sulfide-selenide (**62**) ⟨67JOC3931⟩. Interestingly, 1,2-diselenolane (**16c**) was not reduced under identical reaction conditions.

(**62**)   (**63**)   (**64**)   (**65**)   (**66**)

Sodium borohydride acts as a nucleophilic reducing agent with 1,3-dichalcogenolylium ions to give reduction products by hydride addition to the 2-position ⟨75TL1259, 80H(14)271⟩. The dichalcogenoles (**63**)–(**66**) were obtained by reduction of the corresponding 1,3-dichalcogenolylium fluoroborates or fluorosulfonates.

4.35.5.3.3. Reactions with sulfur and selenium nucleophiles

2-(*N,N*-Dialkylamino)-1,3-diselenolylium salts react with hydrogen selenide and hydrogen sulfide to give 1,3-diselenole-2-selenones and -2-thiones, respectively ⟨74CC937, 80CC866, 80CC867⟩. Examples of these transformations include the conversion of (**67**) and (**68**) into selenone (**25**) and thione (**69**).

(**67**)   (**68**)   (**69**)

4.35.5.3.4 Reactions with phosphorus nucleophiles

Very few reactions of the compounds in this chapter with phosphorus nucleophiles have been reported. The coupling reactions of 1,3-dichalcogenole-2-selenones and -2-thiones with triaryl- and trialkyl-phosphines and with trialkyl phosphites were considered in Section 4.35.5.1.2. 1,3-Thiaselenolylium ions (**19b**) and (**70**) react with triphenylphosphine to give phosphonium salts (**71**) and (**58**), respectively ⟨80H(14)271⟩.

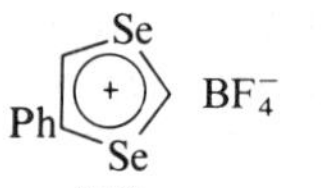

4.35.5.3.5 Reactions with halide and pseudohalide nucleophiles

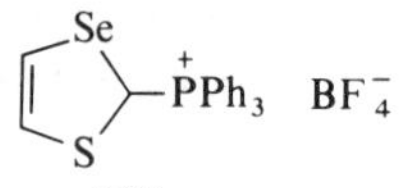

1,2-Oxatellurolylium chlorides undergo a variety of nucleophilic substitution reactions with halides and pseudohalides, as shown in Scheme 10 ⟨83JA875⟩. The most interesting feature of these reactions is the removal of fluoride from the fluoroborate anion by (**6**) to give (**7**) upon treatment with silver fluoroborate. The driving force for this reaction is the placement of highly electronegative atoms in the axial positions of the trigonal bipyramid surrounding tellurium.

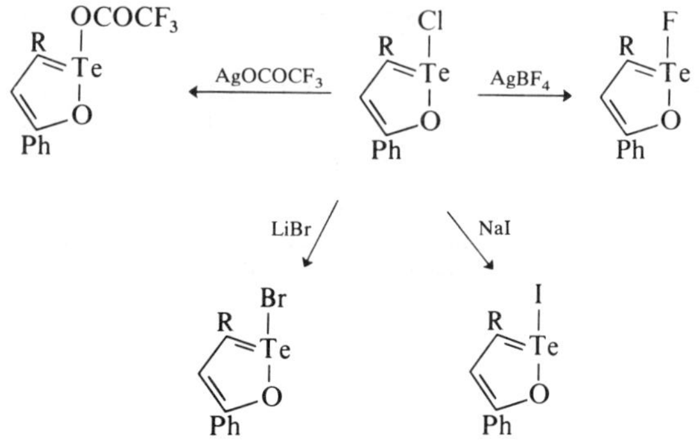

Scheme 10 Reactions of 1,2-oxatellurolylium chloride with halides and pseudohalides

4.35.5.3.6 Cycloaddition reactions

1,3-Diselenolane-2-selenone (**29**) and 1,3-diselenolane-2-thione (**72**) undergo what are formally [3+2] dipolar cycloaddition reactions with diethyl acetylenedicarboxylate with consequent loss of ethylene to give 1,3-diselenole-2-selenone (**73**) and 1,3-thiaselenole-2-selenone (**74**) ⟨76IZV179, 80JOC2632⟩.

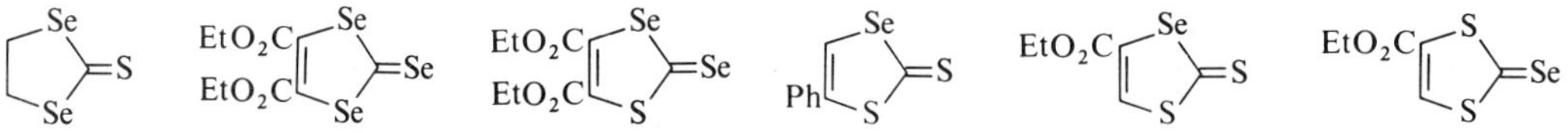

1,3-Thiaselenole-2-thiones undergo similar reactions. The cycloaddition of 5-phenyl-1,3-thiaselenole-2-thione (**75**) with ethyl propiolate gave two [3+2] cycloaddition products: thione (**76**) and selenone (**77**) ⟨80JHC549⟩. Phenylacetylene was the neutral molecule lost in these reactions.

The thione carbon–sulfur bond in 1,3-thiaselenole-2-thiones has been used as a dipolarophile for ethyl diazoacetate, phenyl azide and ethyl azidoformate ⟨80JHC117⟩. The proposed course of these reactions is illustrated for 5-phenyl-1,3-thiaselenole-2-thione (**75**)

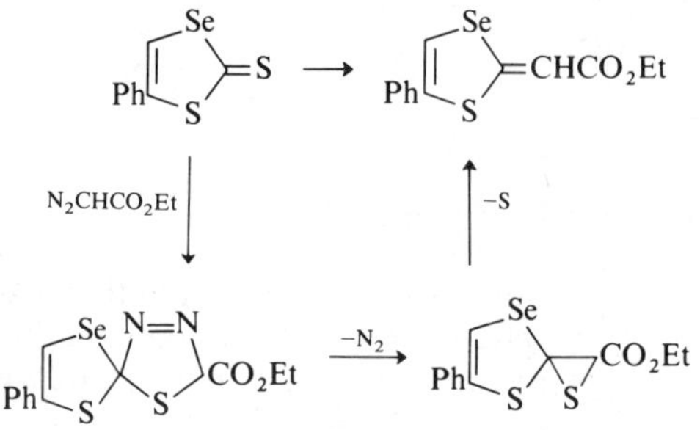

Scheme 11 Cycloaddition of 5-phenyl-1,3-thiaselenole-2-thione with ethyl diazoacetate

and ethyl diazoacetate in Scheme 11. The product of this reaction is a thiaselenafulvene. 5-Alkyl- and 5-aryl-1,3-thiaselenole-2-thiones (**78**) and phenyl azide give 2-imino-1,3-thiaselenoles (**79**). With (**78**) and ethyl azidoformate, 2-ethoxycarbonylimino-1,3-thiaselenoles (**80**) are formed.

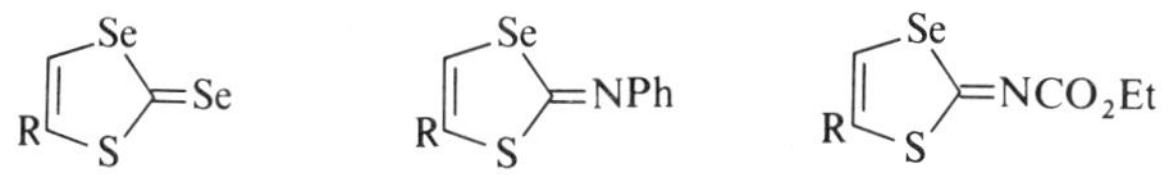

(**78**)   (**79**)   (**80**)

4.35.6 SYNTHESIS

4.35.6.1 General Survey of Ring Syntheses

The syntheses of monocyclic five-membered heterocycles containing one selenium or tellurium atom and one other chalcogen atom can be classified into four groups, according to the number and types of building blocks required to form the ring in a one-step process:

(i) *One-component syntheses*, in which a suitable starting molecule is cyclized by joining of two heteroatoms or by joining of one heteroatom and one carbon atom.

(ii) *[1+4] Two-component syntheses*, in which a heteroatom is joined with a four-atom unit in which the termini are two carbon atoms or one carbon atom and one heteroatom or in which a carbon atom is joined to a four-atom unit with two heteroatoms as termini.

(iii)) *[2+3] Two component syntheses*, in which a two-atom unit consisting of two heteroatoms, one heteroatom and one carbon atom or two carbon atoms is joined with a three-atom unit consisting of three carbon atoms, two carbon atoms and one heteroatom or one carbon atom and two heteroatoms.

(iv) *Multicomponent syntheses*, in which the ring is formed from non-isolated intermediates from more than two building blocks.

The various possibilities for these reactions are compiled in Table 15. The specifics of these reactions are discussed in the following sections. Relatively few syntheses have been reported, reflecting the rather recent interest in most of the compounds of this chapter.

The syntheses discussed in this section are all ring-forming reactions. Transformations at the carbons or heteroatoms of the five-membered ring are discussed in Section 4.35.5, the section on reactivity. Similarly, many of the coupling reactions used to prepare tetrachalcogenafulvalenes are also discussed in Section 4.35.5.

4.35.6.2 One-component Syntheses

The oxidative coupling of propane-1,3-diselenols or their dianions have been quite useful for the preparation of 1,2-diselenolanes. Similar reactions have been useful for the preparation of 1,2-thiaselenolanes. Acidic hydrolyses of 1,3-diselenocyanates generate propane-1,3-diselenols which oxidize to 1,2-diselenolanes ⟨37RTC691, 62AK(19)195⟩. Reductive removal of benzyl groups from 1,3-bis(benzylseleno)propanes with sodium in liquid ammonia gives 1,2-diselenolanes by oxidative coupling of the resulting dianions ⟨62AK(19)195⟩. Hydrogen peroxide induced oxidative coupling of 1,3-bis(selenosulfonates) has also been used to prepare 1,2-diselenolanes ⟨71BSB639⟩. These reactions are summarized in Scheme 12.

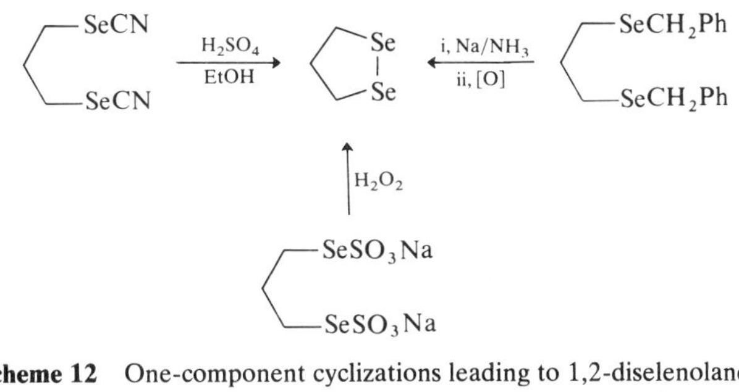

Scheme 12   One-component cyclizations leading to 1,2-diselenolanes

Table 15 Ring Syntheses of Five-membered Heterocycles containing One Selenium or Tellurium and One Other Chalcogen Atom

Type of synthesis	Starting array of atoms	Product array of atoms	Ring systems made in this manner
One-component	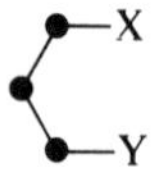	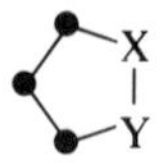	1,2-Thiaselenolane 1,2-Diselenolane 1,2-Oxaselenolylium 1,2-Oxatellurolylium 1,2-Diselenolylium
	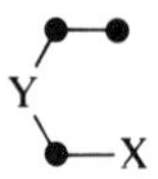	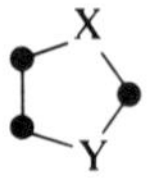	1,3-Diselenolylium
[1+4] Two-component	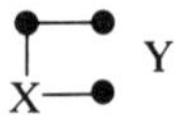	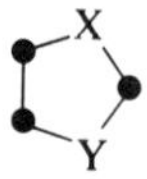	1,3-Thiaselenole 1,3-Thiatellurole 1,3-Ditellurole
	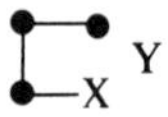	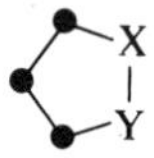	1,2-Thiaselenole 1,2-Thiatellurole
	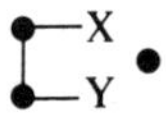	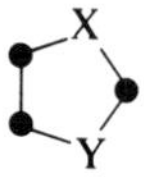	1,3-Diselenolane
[2+3] Two-component	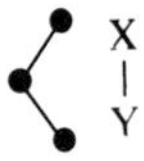	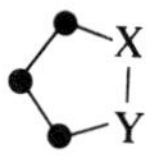	1,2-Diselenolane 1,2-Diselenole 1,2-Ditellurole
	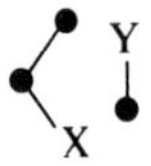	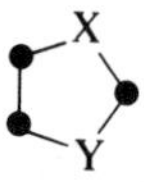	1,3-Oxaselenole 1,3-Thiaselenole 1,3-Diselenole
	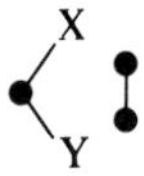	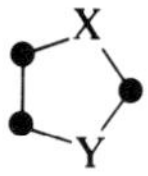	1,3-Thiaselenolane 1,3-Diselenolane 1,4,5,8-Tetraselenafulvalene
Multicomponent	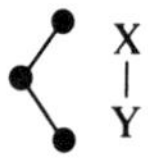	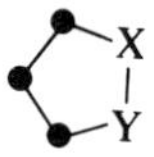	1,2-Thiaselenole 1,2-Thiatellurole 1,2-Selenatellurole 1,2-Diselenole 1,2-Ditellurole
	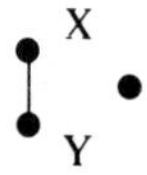		1,4,5,8-Tetratellurafulvalene
	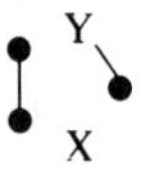	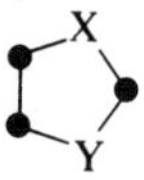	1,3-Diselenole
	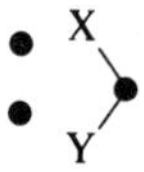		1,4,5,8-Tetraselenafulvalene

1,2-Diselenolylium ions have been prepared by halogen induced oxidation of propane-1,3-diselenones ⟨67ACS1991⟩. 3,5-Diamino-1,2-diselenolylium chloride (**32a**) and iodide (**32b**) were prepared in this manner.

1,2-Oxaselenolylium chlorides and 1,2-oxatellurolylium chlorides can be prepared by cyclization of 3-arylseleno- or 3-aryltelluro-propenoyl chlorides ⟨83JA875, 83JA883⟩. These thermal or Lewis acid induced cyclizations involve attack of the acylium cation on the *ipso* carbon of the aromatic ring followed by nucleophilic attack of chloride on selenium or tellurium to complete the aryl migration with subsequent bond reorganization to the heterocycle. This sequence is shown in Scheme 13.

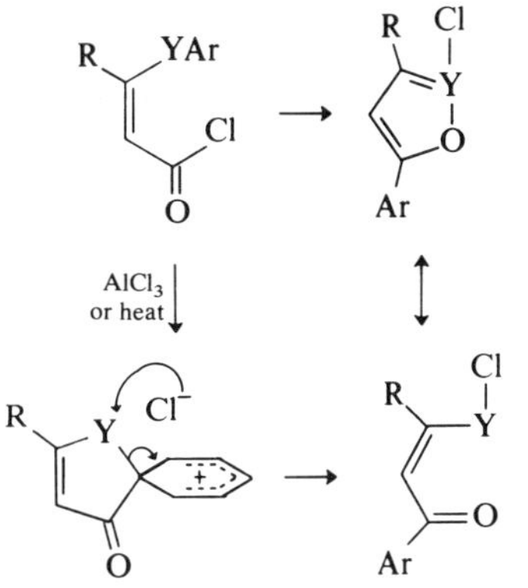

Scheme 13 1,2-Oxaselenolylium and 1,2-oxatellurolylium chlorides from propenoyl chlorides

1,3-Diselenolylium ions have been prepared by acid induced cyclization and dehydration of oxo esters of *N,N*-dialkyldiselenocarbamates ⟨75JOC746, 77CC505, 80CC866, 80CC867⟩. As shown in Scheme 14, protonation of the carbonyl oxygen results in carbon–selenium bond formation by electrophilic attack on selenium. Acid induced dehydration generates the 1,3-diselenolylium ion in approximately 90% yield.

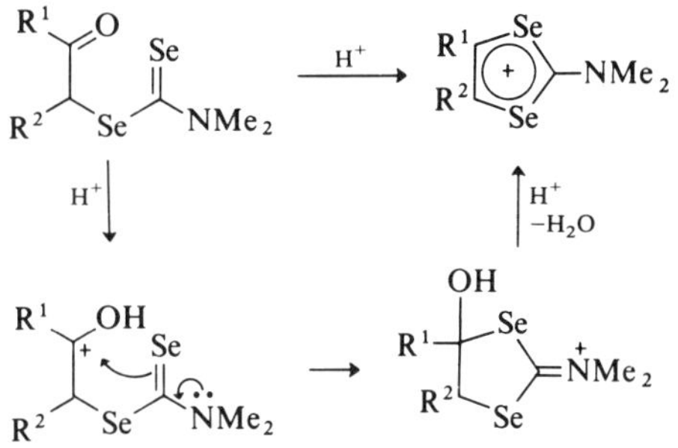

Scheme 14 1,3-Diselenolylium cations from diselenocarbamate cyclization

4.35.6.3 [1+4] Two-component Syntheses

1,3-Diselenolane (**81**) has been prepared by the condensation of paraformaldehyde with ethane-1,2-diselenol ⟨79JOC4689⟩. The condensation of ethane-1,2-diselenol with (thiocarbonyl)diimidazole (**82**) gave 1,3-diselenolane-2-thione (**72**) ⟨80JOC2632⟩.

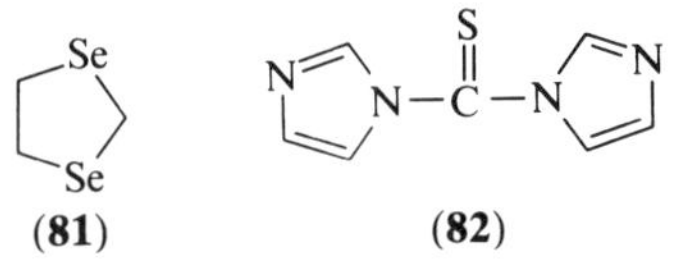

The additions of disodium telluride, dilithium telluride and disodium selenide to chloromethyl alkynyl sulfides and tellurides give a variety of 1,3-dichalcogenoles as shown in Scheme 15 ⟨81RTC10, 82TL1531⟩. The preparation of the 1,3-ditelluroles required the use of dilithium telluride under carefully controlled conditions.

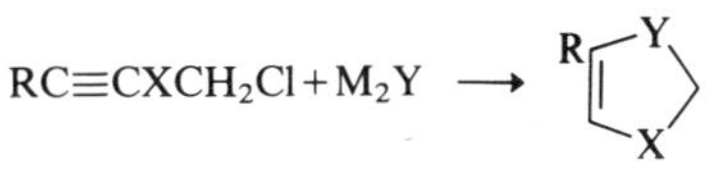

$$RC{\equiv}CXCH_2Cl + M_2Y \rightarrow$$

X = S; Y = Se, Te; R = Me, Bu, Bu^t, Ph
X = Y = Te; R = H, Me, Bu, Ph

Scheme 15 [1+4] Two-component synthesis of 1,3-dichalcogenoles

The naphtho[1,8-*cd*]-[1,2]thiaselenole (**11**) and -[1,2]thiatellurole (**83**) can be prepared from 8-chloronaphthalene-1-thiol (**84**) as shown in Scheme 16 ⟨77JA7743⟩. Treatment of (**84**) with two equivalents of *n*-butyllithium gives dianion (**85**). The addition of selenium or tellurium metal to (**85**) gives (**11**) or (**83**), respectively, in 65 and 60% yields.

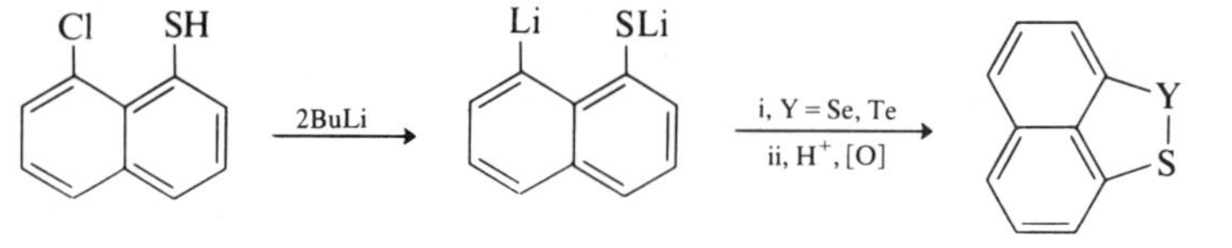

970   Rings containing One Se or Te Atom and One Other Group VI Atom

Scheme 16 Preparation of naphtho[1,8-cd]-[1,2]thiaselenole and -[1,2]thiatellurole

4.35.6.4 [2+3] Two-component Syntheses

The disodium diselenide and disodium ditelluride units have been used with three-carbon units bearing halogens on the 1,3-positions to give five-membered rings. Thus, disodium diselenide displaces the two bromine atoms in 1,3-dibromo-2-methyl-2-phenylpropane (**86**) to give 4-methyl-4-phenyl-1,2-diselenolane (**87**) ⟨62AK(19)195⟩. The reaction of disodium ditelluride with tetrachlorotetracene (**88**) gave a mixture of bis(ditellura)tetracene (**12**) and various ditellurotetracenes (**89**) ⟨82MI43500⟩.

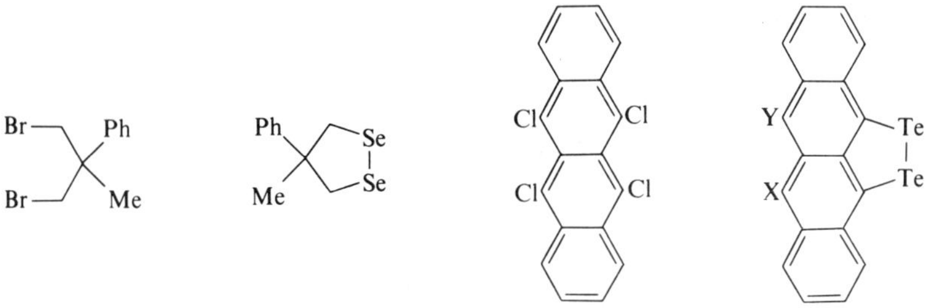

The joining of a two-atom unit consisting of a heteroatom and a carbon atom with a three-atom unit consisting of a terminal heteroatom and two carbon atoms has been useful for the preparation of a variety of 1,3-dichalcogenoles. The addition of carbon disulfide to aryl and alkyl alkynyl selenide anions gives 5-aryl- or 5-alkyl-1,3-thiaselenole-2-thiones (**90**) ⟨74JA7376, 76IZV179, 77S764⟩. The addition of phenyl isoselenocyanate to the sodium salts of phenyl alkynyl sulfide and phenyl alkynyl selenide gave 2-phenylimino-1,3-thiaselenole (**91**) and 2-phenylimino-1,3-diselenole (**92**) ⟨71JPR804⟩.

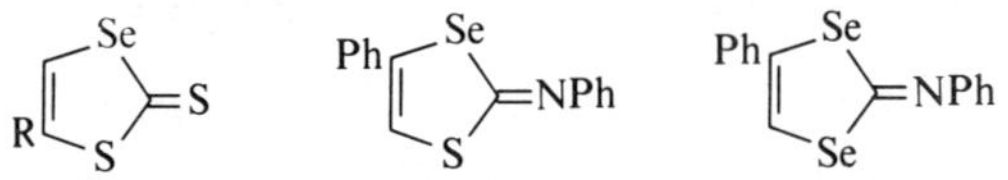

An unusual [2+3] two-component synthesis is observed in the addition of potassium O-ethyl diselenoxanthate to an α,α-dibromoketone as shown in Scheme 17. The α,α-dibromoketone functions as the three-atom unit to give 1,3-oxaselenol-2-one (**93**) in 61% yield and 1,3-oxaselenole (**94**) in 30% yield ⟨78CZ361⟩.

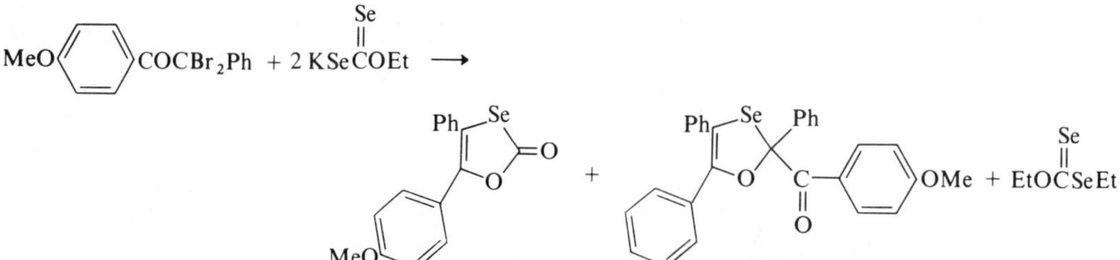

Scheme 17 Addition of potassium O-ethyl diselenoxanthate to α,α-dibromoketones

[2+3] Cyclizations involving 1,2-dibromoethane as the two-atom unit have been successful for the preparation of 1,3-thiaselenolane and 1,3-diselenolane derivatives. The base induced additions of active methylene compounds to carbon diselenide generate 1,3-diselenide dianions which add to 1,2-dibromoethane to give structures such as (**30**), (**95**) and (**96**) ⟨70ACS3213⟩. Similarly, *N*,*N*-dialkyl-diselenocarbamates (**97**) and -thiaselenocarbamates (**98**) add to 1,2-dibromoethane to give 2-(*N*,*N*-dialkyliminium)-1,3-thiaselenolanes (**99**) and 2-(*N*,*N*-dialkyliminium)-1,3-diselenolanes (**100**) ⟨72BCJ489⟩.

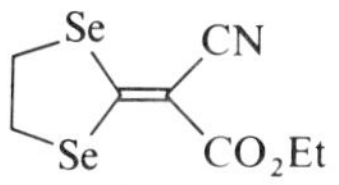

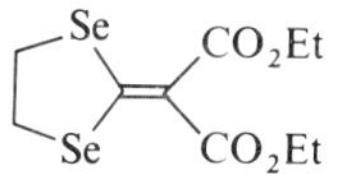

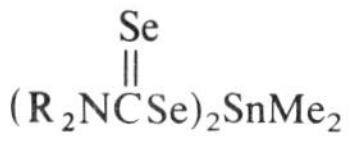

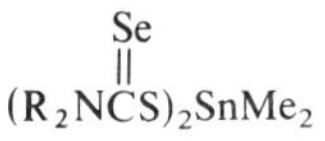

Methyl propiolate and DMAD add to carbon diselenide under high pressure to give 1,4,5,8-tetraselenafulvalenes (**101**) and (**102**). The dimerization of a 1,3-diselenole-2-carbene is implicated ⟨81CC669⟩. The alkynes function as the two-atom unit while carbon diselenide is the three-atom unit.

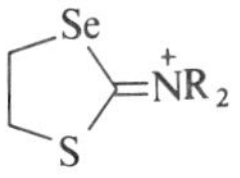

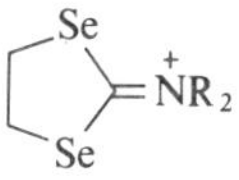

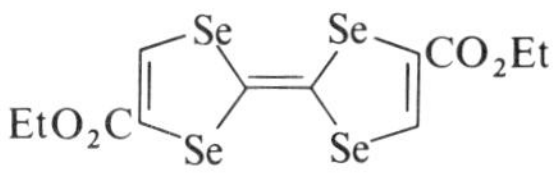

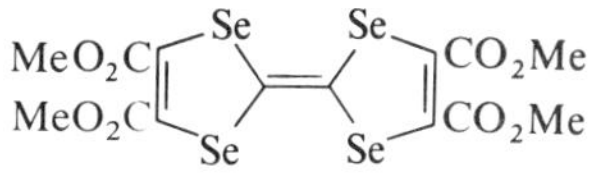

4.35.6.5 Multicomponent Syntheses

The majority of multicomponent syntheses have involved the generation of an organometallic species which then reacts with sulfur, selenium or tellurium. Repetition of this cycle or addition of another component then leads to the heterocyclic system. Dibenzotetratellurafulvalene (**35**) and hexamethylenetetratellurafulvalene (**10**) have been generated by the addition of tetrachloroethylene to the corresponding dilithium ditellurides (**103**) and (**104**). The dilithium ditellurides were prepared by the addition of tellurium metal to *o*-dilithiobenzene ⟨82CC336⟩ and 1,2-dilithiocyclopentene ⟨82JA1154⟩.

The stepwise lithiation of 1,8-dibromonaphthalene and related derivatives with subsequent capture of sulfur, selenium or tellurium after each lithiation has generated naphtho[1,8-*cd*][1,2]dichalcogenoles (**2**), (**11**), (**21**), (**34**) and (**83**) ⟨77JA7743⟩. The closely related compounds (**105**) and (**106**) have been generated in a similar fashion ⟨80TL4565⟩.

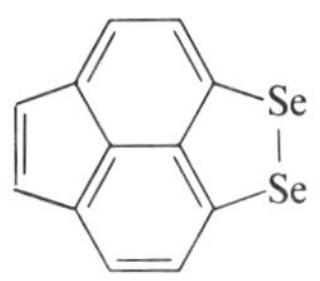

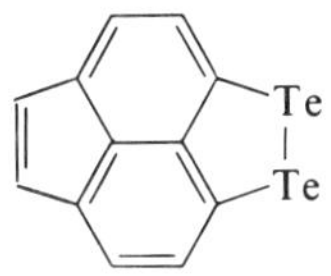

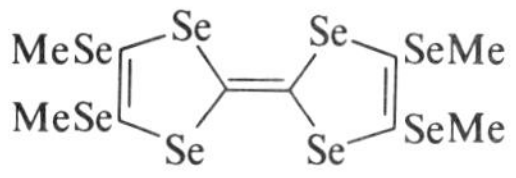

The electrochemical reduction of carbon diselenide in the presence of methyl iodide generated tetraselenafulvalene (**107**) ⟨76CC148⟩. Six carbon diselenide molecules were necessary for each molecule of product generated.

4.35.7 APPLICATIONS

The compounds of this chapter have been utilized in the study of organic conductors and semiconductors. The materials have served as electron donors or as intermediates to form electron donors. Chapter 1.13 considers the organic conductors and semiconductors in more detail.

4.36

Two Fused Five-membered Heterocyclic Rings: (i) Classical Systems

K. H. PILGRAM
Shell Development Co., Modesto

4.36.1 INTRODUCTION	974
4.36.2 MOLECULAR DIMENSIONS	974
4.36.2.1 *X-ray Diffraction*	974
4.36.2.2 *Molecular Spectra*	976
4.36.2.2.1 ^1H *NMR spectroscopy*	976
4.36.2.2.2 ^{13}C *NMR spectroscopy*	977
4.36.2.2.3 ^{11}B *NMR spectroscopy*	977
4.36.2.2.4 *Ultraviolet spectroscopy*	977
4.36.2.2.5 *Infrared spectroscopy*	978
4.36.2.2.6 *Mass spectrometry*	979
4.36.2.3 *Theoretical Methods*	979
4.36.3 GENERAL SURVEY OF STRUCTURE AND REACTIVITY	980
4.36.3.1 *Tautomerism*	980
4.36.3.1.1 *Azido–tetrazole*	980
4.36.3.1.2 *Lactam–lactim*	980
4.36.3.1.3 *Amine–imine*	980
4.36.3.1.4 *Thiol–thione*	980
4.36.3.1.5 *Methylene–amino*	980
4.36.3.1.6 *Ring–chain*	980
4.36.3.2 *Reactions Involving Electrophilic Attack*	981
4.36.3.3 *Reactions Involving Nucleophilic Attack*	981
4.36.3.4 *Ring Openings and Transformations*	981
4.36.3.4.1 *Cleavage of N—O bonds*	982
4.36.3.4.2 *Cleavage of N—N bonds*	982
4.36.3.4.3 *Cleavage of C—N bonds*	982
4.36.3.4.4 *Cleavage of C—S bonds*	984
4.36.3.4.5 *Cleavage of P—P bonds*	984
4.36.3.4.6 *Cleavage of C—O bonds*	984
4.36.4 SYNTHESIS OF (5,5)-FUSED RING SYSTEMS	985
4.36.4.1 *Nucleophilic Type Displacements*	985
4.36.4.1.1 *Reactions with 1,1-bielectrophiles*	985
4.36.4.1.2 *Reactions with 1,2-bielectrophiles*	986
4.36.4.1.3 *Reactions with 1,3-bielectrophiles*	987
4.36.4.1.4 *Reactions with 1,4-bielectrophiles*	987
4.36.4.2 *Aldol-type Condensations*	988
4.36.4.2.1 *Formation of C—C bonds*	988
4.36.4.2.2 *Formation of C—O bonds*	989
4.36.4.2.3 *Formation of C—N bonds*	989
4.36.4.2.4 *Formation of C—S bonds*	994
4.36.4.2.5 *Formation of N—N bonds*	996
4.36.4.3 *Bimolecular Dipolar Cycloadditions*	996
4.36.4.3.1 *1,3-Dipoles of the propargyl–allenyl anion type*	996
4.36.4.3.2 *1,3-Dipoles of the allyl anion type*	1000
4.36.4.3.3 *1,3-Dipoles without a double bond*	1005
4.36.4.3.4 *Cycloadditions to electron-deficient alkenes and alkynes*	1005
4.36.4.4 *Intramolecular Cycloadditions*	1007
4.36.4.4.1 *1,3-Dipolar cycloadditions*	1007
4.36.4.4.2 *1.5-Dipolar and higher order cycloadditions*	1008
4.36.4.5 *Conjugate Cycloadditions*	1010
4.36.4.5.1 *Reactions involving N-nucleophiles*	1010
4.36.4.5.2 *Reactions involving S-nucleophiles*	1011
4.36.4.5.3 *Reactions involving O-nucleophiles*	1011

4.36.4.6 Intramolecular Alkylations	1012
4.36.4.6.1 C—N bond formation	1012
4.36.4.6.2 C—S bond formation	1013
4.36.4.6.3 C—O bond formation	1013
4.36.4.7 Oxidative Ring Closures	1014
4.36.4.7.1 C—N bond formation	1014
4.36.4.7.2 C—S bond formation	1014
4.36.4.7.3 C—O bond formation	1015
4.36.4.7.4 N—N bond formation	1015
4.36.4.7.5 N—S bond formation	1016
4.36.4.7.6 S—S bond formation	1016
4.36.4.8 Electrophilic Ring Closures	1016
4.36.4.9 Ring Interconversions	1017
4.36.4.9.1 Formation by ring expansion (3→5, 4→5)	1017
4.36.4.9.2 Formation by ring contraction (7→5, 6→5)	1018
4.36.4.9.3 Formation by rearrangement (5⇌5)	1020
4.36.4.10 Simultaneous Formation of Two Rings	1021
4.36.5 (5,5)-FUSED HETEROCYCLES OF BIOLOGICAL INTEREST	1023

4.36.1 INTRODUCTION

Interest in (5,5)-fused ring systems developed in separate, albeit closely related, directions. In the first, chemists continue their quest for new compounds having unique biological, chemical and physical properties. In the second, emphasis is on the classification of aromaticity based upon molecular conformity with Hückel's rule. An increasing understanding of the fundamental chemistry of heterocyclic ring systems has led to considerable advances in this field, as evidenced by the growing list of applications as biocides and in industrial processes.

No general review of (5,5)-fused heterocycles exists, though much of the work before 1973 is on (5,5)-fused ring systems with a bridgehead nitrogen atom ⟨61HC(15-1)63, 77HC(30)1⟩. More recent reviews deal with heteropentalenes ⟨77HC(30)317⟩, particularly azapentalenes ⟨78AHC(22)183⟩.

The discussion in this Chapter is limited to (5,5)-fused ring systems containing at least one heteroatom per ring. No distinction is made between 'aromatic' and saturated systems. In the section dealing with their synthesis, an attempt has been made to classify syntheses as a means of understanding some of the principles involved ⟨B-80MI43600⟩ and of handling the vast number of ring systems with large variation in the number and type of peripheral heteroatoms. Additionally, syntheses are listed under headings of specific bond formation, thereby focusing attention on the similarity of many synthetic procedures.

While the aim of this chapter is to survey and correlate published work, it is hoped that general and specific points will stimulate further work in this fertile field.

4.36.2 MOLECULAR DIMENSIONS

4.36.2.1 X-ray Diffraction

X-ray analyses of (5,5)-fused heterocycles are being reported in increasing numbers. The structure of Clerodin, the bitter principle of *Clerodendron infortunatum* which contains a furo[2,3-*b*]furan ring moiety (**1**), was elucidated by an X-ray study ⟨61MI43600⟩.

The structure of the reaction product derived from the interaction of 2-amino-Δ^2-thiazoline with chlorothioformyl chloride (chloroformylsulfur chloride) in chloroform solution was shown to be that of 5,6-dihydrothiazolo[2,3-*c*][1,2,4]thiadiazol-3-one (**2**). The dihydrothiazole ring of (**2**) adopts an irregular half-chair conformation, whereas the thiadiazole ring deviates little from planarity, there being a very slight tendency towards a shallow envelope conformation in which the bridgehead nitrogen is the unique atom ⟨81JCS(P2)789⟩. Similarly, the structure of the minor reaction product (**3**) of 5,5-diphenyl-2-thiohydantoin with ethylene dibromide has been verified by an X-ray study ⟨81JCS(P2)789⟩.

The structure of 2,2,5,5-tetramethyl-2,3,5,6-tetrahydrothieno[3,2-*b*]thiophene-3,6-dione (see equations 64, 73), the first compound containing the basic chromophoric system of the thioindigo dyes, was also determined with the aid of X-ray analysis ⟨77CB1421⟩.

Conclusive evidence for the thiazole ring opening of (±)-2,3,5,6-tetrahydro-6-phenyl-imidazo[2,1-b]thiazole (**78**) (tetramisole) by lithium diethylamide was obtained by a single-crystal X-ray diffraction study ⟨78JHC307⟩.

In order to verify the empirical rule that the chemical shifts of N-methyl protons in 1-methylpyrazolo[1,5-d]tetrazoles should appear at lower frequencies than those in 3-methylpyrazolo[1,5-d]pyrazoles ⟨73CJC2315⟩, the corresponding 1,6-dimethyl-7-ethoxycarbonyl derivative (**4**), a 3a-azapentalene having only nitrogen heteroatoms, was examined by X-ray crystallography. Considering both calculated and experimental bond lengths, the pyrazole ring portion appears to be more aromatic than the tetrazolic counterpart ⟨78JHC395⟩.

3a,6a-Dihydroisoxazolo[5,4-d]isoxazoles (see equation 65) are among the products generated from photolysis of pyridazine N,N'-dioxides, as evidenced by an X-ray study ⟨79T1267⟩.

The formal criss-cross cycloaddition product of cyanogen with sulfur trioxide (molar ratio 1:2) has been shown to be 1,2,3-oxathiazolo[5,4-d][1,2,3]oxathiazole 2,2,5,5-tetroxide (**5**). The S—O bond in the ring is extremely long at 1.71 Å, while the exocyclic S=O bonds are shortened to 1.39 Å, suggesting a description of the molecule in terms of bond–nonbond resonance, the resonance formulae (**5a**) and (**5b**) also providing some rationalization for the short C—O bond (1.29 Å) ⟨79AG(E)223⟩.

The cycloaddition reactions of nitrile oxides with several substituted cyclopentadienones led to the formation of only one regioisomer; the cyclopenta[2,3-d]isoxazol-4-one (**6**) structure for five 1:1 adducts was fully supported by X-ray analysis ⟨79JHC731⟩.

Bicyclic phosphoranes of type (**7**) are structurally adapted to be in tautomeric equilibrium with open forms such as (**8**), in which the phosphorus atom is in valence state 3. The nitrogen atom is planar in (**7**) but becomes pyramidal in (**8**) and would be expected to recover its donor properties. Tautomer (**8**) has never been detected spectroscopically ⟨72TL2969⟩. Through the action of various atoms or cations, the bicyclic phosphoranes can be converted into coordination adducts of tautomer (**8**). The bidentate character of (**8**) has been demonstrated by X-ray diffraction of the complex (**9**) obtained from (**7**) and Rh(CO)$_2$Cl$_2$ in toluene ⟨79JA2234⟩.

Unequivocal structural proof of 1,1,4,4-tetramethyl-1H,4H-thieno[3,4-c]thiophene (see Scheme 19) was obtained by X-ray crystallographic analysis ⟨80TL3617⟩. Methyl N-vinyl carbamate adds benzonitrile oxide to yield a mono- or bis-adduct, depending upon the experimental conditions. The *syn* stereochemistry of the bis-adduct (**10**) has been elucidated by an X-ray structure determination which showed that the carbamate moiety orients its NH above the isoxazolidine ring and toward the oxygen of the *syn*-oxadiazolinic ring (NH ···· O is 3.08 Å) ⟨80JCR(S)4341⟩.

Oxidation of the fused heterocycle (**82**) with potassium permanganate is accompanied by cleavage of the P—P bridge to give the 2,4,6,8-tetraaza-1,5-diphosphabicyclo[3.3.1]nonane 1,5-dioxide (**83**), as evidenced by X-ray crystallography ⟨81CB2132⟩. However, oxidation of (**82**) with tetrachloro-*o*-benzoquinone leads to the formation of the corresponding diphosphorane (**11**) without breaking the P—P bond. This is the first compound known to contain an axial $\lambda^5 P-\lambda^5 P$ bond ⟨82JA2919⟩.

Crystal structures for a number of 'bimanes' (see Scheme 6) have been reported ⟨81JOC1666⟩. The formation from 3-(1-pyrrolidinyl)thiophenes and dimethyl acetylenedicarboxylate (DMAD) of 6,7,7a,8-tetrahydro-5*H*-thieno[2,3-*b*]pyrrolizines (**226**) always occurs stereospecifically, as shown by the X-ray structure of (**226**) ⟨81JOC424⟩. It is also shown that in 1,3a,6,6a-tetrahydropyrrolo[3,4-*c*]pyrazoles (see equation 70), both of the *cis*-fused five-membered rings are in the envelope form with six peripheral side groups directly bonded to them ⟨81BCJ41⟩.

An examination of the X-ray structures of two element homologous bicyclic compounds, $Bu^t_6 P_8$ and $Bu^t_6 As_8$, provides the first example showing that homologous phosphanes and arsanes may also have entirely different structures. Strong transannular interactions between the Bu^t groups prevent the 2,2′,3,3′,4,4′-Bu^t_6-1,1′-bicyclotetraphosphane having the 2,3,4,6,7,8-Bu^t_6-bicyclo[3.3.0]octane (**12**) structure like the octaarsane. In (**12**) both five-membered rings have twisted envelope conformations. The spectroscopic findings admit two configurations with the Bu^t groups within each five-membered ring next to the zero bridge in the *cis* position. X-ray structure analysis shows the presence of the sterically preferred isomer with the Bu^t groups in the 3,7-positions arranged *trans* with respect to the free electron pairs of the bridgehead arsenic atoms ⟨81AG(E)406⟩.

To determine the three-dimensional structure of *N*-phenylthiolano[3,4-*d*]thiazolidine-2-thione 5,5-dioxide (**13**), an X-ray diffraction study was carried out. The thiazolidine ring is planar, while the thiolane ring has an envelope configuration, presumably the result of the effect of the thiazolidine ring portion ⟨82CHE668⟩.

4.36.2.2 Molecular Spectra

In many instances the spectral data of (5,5)-fused ring systems, although useful for structural verification, show no unifying features. Reference to the ^1H NMR, ^{13}C NMR, UV, IR and mass spectra of most (5,5)-fused heterocycles that have been examined is beyond the scope of this Chapter, but several illustrative references to spectral data that are of special value in particular applications, such as structure elucidation, are included in this section. Original publications should be consulted for data related to a particular structure.

4.36.2.2.1 ^1H NMR spectroscopy

The coupling constant of the pyrazole protons (2.3 Hz) of pyrazolo[5,1-*c*]-*s*-triazole (**14**) approximates that of a 4,5-coupling (rather than 3,4), indicating that the tautomer (**14b**) predominates ⟨70CB3284⟩. The NMR characteristics of the isomeric thiazolo[2,3-*c*]-*s*-triazole (**15**) and thiazolo[3,2-*b*]-*s*-triazole (**16**) ring systems are particularly useful for structural determination ⟨71JOC10, 74JHC459, 76JHC1225⟩.

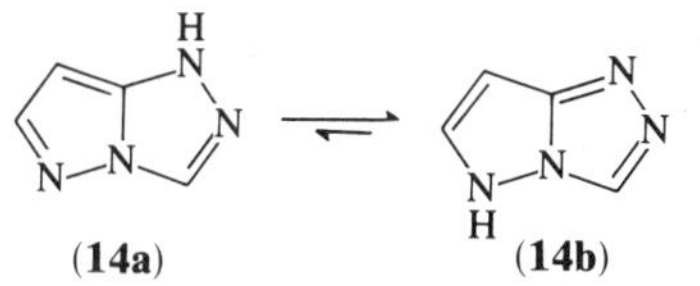

(14a) ⇌ (14b) 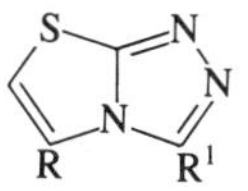   (15) 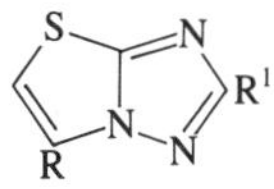   (16)

Methylation of the bicyclic pyrazolo[1,5-d]tetrazole anion gave 1-methyl- and 3-methyl-pyrazolo[1,5-d]tetrazole. The structures were assigned on the basis of ^{1}H NMR spectroscopy using the relationship indicated by Butler ⟨73CJC2315⟩ that the chemical shift of N-methyl groups increases in the following order for the structural units: =C—N(Me)—N= < =N—N(Me)—N= ⟨76JHC379⟩.

The chemical shift of the Δ^3-pyrroline methyl group in the cyclazine (17) was diagnostic, occurring at δ 1.67 instead of in the δ 2.2–2.6 range observed for pyrrole methyl groups ⟨80JOC5396⟩.

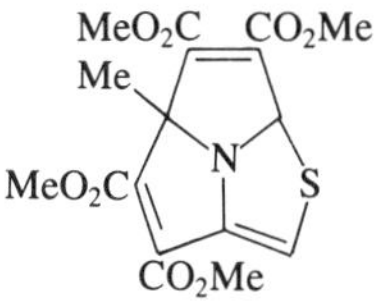

(17) 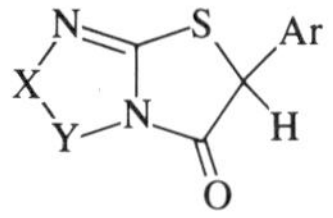   (18) 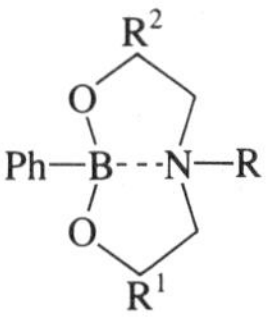   (19)

In condensed mesoionic thiazolones carrying an acidic hydrogen atom on the ring-junction carbon atom, this hydrogen migrates on to the carbon bearing the aryl group to give the tautomeric imidazothiazole (18), as shown by ^{1}H NMR spectroscopy (δ 5.5–5.8(s)) ⟨81S981⟩.

4.36.2.2.2 ^{13}C NMR spectroscopy

The ^{13}C NMR spectrum of the condensation product of thiazolium N-ylide (194) with DMAD provided definitive evidence in support of the pyrrolo[2,1-c][1,4]thiazine (196) structure. Particularly important is the absence of an absorption that could be assigned to the tertiary C-5 in (195), this absorption being anticipated at δ 65.5 ⟨76JOC187⟩.

A method has been developed for distinguishing between the possible structural types of adducts for guanidine and amidine derivatives with DMAD using ^{13}C NMR spectroscopy. The 5,6-dihydroimidazo[2,1-b]thiazol-3(2H)-one (237) is of interest since it contains an sp^3 carbon atom attached to two nitrogen, one oxygen, and one sulfur atom and this leads to very low-field ^{13}C resonances (δ 114.6) ⟨81JCS(P1)415⟩.

4.36.2.2.3 ^{11}B NMR spectroscopy

Boron-11 NMR spectroscopy can be used for structural elucidation of complex boron containing heterocycles. Hydrolytic stabilities of dioxazabora heterocycles (19) and their corresponding N Mannich bases with different carboxamides have been determined by ^{11}B NMR studies ⟨82M1025⟩.

4.36.2.2.4 Ultraviolet spectroscopy

The similarity of the UV spectra of s-triazolo[3,2-c]-s-triazoles (see equation 44) and 3-aryl-s-triazoles indicates a lack of electronic interaction between the two fused s-triazole rings. In 1H-2,6-diphenyl-s-triazolo[3,2-c]-s-triazole, the UV spectrum exhibits the resultant effects of the two triazole rings each substituted with an aryl group ⟨66JHC119⟩.

The most conclusive evidence that isomeric ring systems were involved in the dehydrative ring closure of 3-benzylidenehydrazino-s-triazoles (see equation 57) came from a study of their UV spectra. The [5,1-c] (20) isomers absorbed at shorter wavelengths, but with relatively higher intensities, than their [3,4-c] isomers (21) ⟨68JOC143⟩.

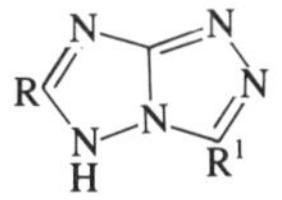

(20) 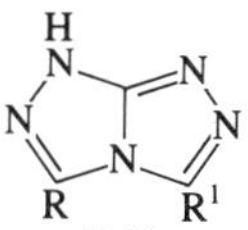   (21) 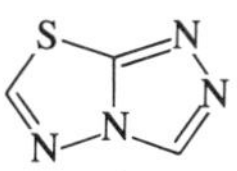   (22) 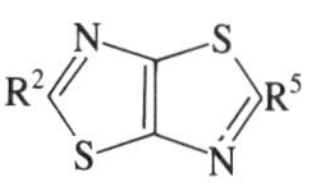   (23)

The UV absorption of the s-triazolo[3,4-b][1,3,4]thiadiazole nucleus (**22**), as represented by the 3,6-dimethyl derivative, occurs at 251 nm (log ε 3.45). The introduction of a 6-amino substituent results in a 13 nm hypsochromic shift of the absorption maximum, whereas a 6-thiol group causes a shift to 285 nm ⟨66JOC3528⟩.

2,5-Disubstituted thiazolo[5,4-d]thiazoles (**23**) containing aromatic and heterocyclic functions are bright yellow or orange solids which are fluorescent in dilute solution, and which absorb strongly in the UV with principal emission maxima in the range 440–470 nm. The UV absorption spectra exhibit three peaks in the 350–400 nm region, although the exact positions of the maxima are dependent on the nature of the substituents at the C-2 and C-5 positions. The introduction of heterocyclic groups at C-2 and C-5 causes bathochromic shifts which are greater than any observed with carboxylic substituents as the conjugation is extended. The observed shifts are in the order: 2-thienyl<2-benzoxazolyl< 2-benzimidazolyl ⟨70JHC457⟩.

In contrast to the IR spectra, the UV spectra of (**56**; A, B = O) in neutral, acidic and alkaline ethanol solution enable one to distinguish tetrasubstituted from trisubstituted and symmetrical and unsymmetrical disubstituted representatives (**56**) ⟨71CHE1028⟩.

Most syn-bimanes (see Scheme 6) which have been synthesized are very strongly fluorescent (emission maxima between 338 and 520 nm, quantum yield 0.7–0.9) ⟨78JA6516⟩. The absorption maxima show that the S_0–S_1 electronic transition is moderately sensitive to the nature of the substituents ⟨75CHE666⟩.

4.36.2.2.5 Infrared spectroscopy

Numerous applications of IR spectral data have been reported for derivatives of (5,5)-fused ring systems, especially in connection with the problem of tautomeric equilibria. For example, the IR spectra of 5,6,7,8-tetrahydro-1H,3H-pyrrolo[1,2-c][1,3]oxazole-1,3-diones (**24**; X=O) and corresponding thiazole-1,3-diones (**24**; X = S) show no carbonyl absorption, but broad bands at 3300 cm^{-1} due to the associated OH group, indicating that the fused structures (**24**) exist in the enolic forms rather than their diketonic tautomers ⟨81BJC1844⟩.

An inspection of the IR spectra of the condensation products derived from 4-phenyl-2-ethoxy-Δ^1-pyrroline with diethyl aminomalonate indicates that the hydroxyimidazole (**25a**) predominates ⟨82JHC193⟩.

The IR absorption of the carbonyl group in the fused γ-lactone (**26**) is at 1780 cm^{-1}, in accord with a peculiar characteristic of γ-lactones ⟨80JHC609⟩.

The absence of the characteristic N_3 band of azides at 2100–2200 cm^{-1} (CH_2Cl_2) and the presence of an absorption at 1080 cm^{-1} are in accord with the fused tetrazole structure of (**27**), the reaction product of 4-phenyl-2-ethoxy-Δ^1-pyrroline with sodium azide in acetic acid ⟨82JHC193⟩. Tetrazoles fused with five-membered heterocycles are generally in the azido form. For example, in thiazolo[2,3-e]tetrazole the azido form (**28**) is the predominant structure, whereas in the 3-methyl analogue the tetrazole (**29**) form predominates ⟨75BSB1189⟩. The predominance of the azido (**30**) form in 6-methylisoxazolo[3,2-e]tetrazole has been reported ⟨77CJC1728⟩. 6-Phenyl-1,3,4-thiadiazolo[3,2-d]tetrazole was shown to have the tetrazole structure (**31**) in the solid state (KBr disk) and the azide structure (**32**) in solution ($CHCl_3$) ⟨58CPB382⟩.

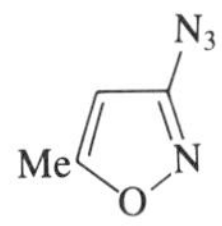

(30) 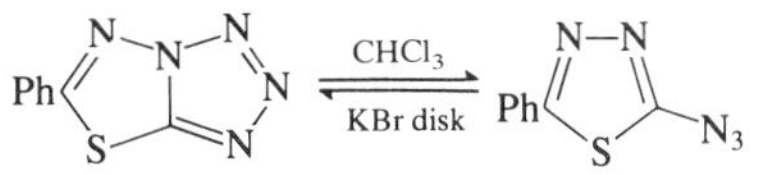 (31) ⇌ CHCl₃ / KBr disk (32)

4.36.2.2.6 Mass spectrometry

The [PhCNNAr]⁺ ions in the mass spectrum of the 3,4-diaryl-*s*-triazolo[3,4-*c*]-*s*-triazole (**33**) confirm the integrity of the original 1,2,4-triazole ring of (**33**) ⟨76ACS(B)463⟩.

The structure of the bicyclic oxidation product of dehydroascorbic acid bis(phenylhydrazone) has been confirmed by detailed high-resolution mass spectroscopy as (**34**) ⟨80JHC1181⟩.

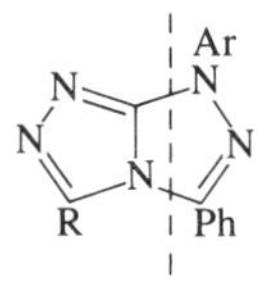

(33) 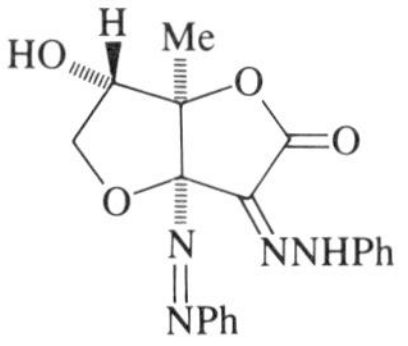 (34) 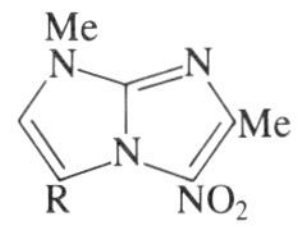 (35)

4.36.2.3 Theoretical Methods

Very little has been reported on the theoretical aspects of most (5,5)-fused heterocycles, especially classical systems. Hückel molecular orbital (HMO) calculations for 1,6-dimethyl- and 1,6-dimethyl-5-nitroimidazo[1,2-*a*]imidazole have been used to assign the site (R) of nitration of the 1,6-dimethyl-5-nitro derivative (**35**) ⟨73JOC1955⟩.

HMO calculations for furo[3,2-*b*]pyrrole (**36**) indicate a compound without a well pronounced aromatic character (order of double bonds: 0.98–0.73; order of single bonds: 0.44–0.48). The introduction into the 5-position of an electron-accepting substituent (CO_2H, CO_2R) increased the aromatic character of the skeleton. Based upon the calculated indices of chemical reactivity, electrophilic attack is anticipated at carbons 5 and 2, and nucleophilic attack at the heteroatoms of the skeleton: O > N ⟨81CCC2949⟩.

A thione-thiol structure in the crystalline state and in aprotic solvents was established for the [1,2,4]triazolo[3,4-*b*][1,3,4]thiadiazole (**37**), in accordance with calculations of the π-bonding and solvation energies by the Pariser–Parr–Pople (PPP) method and also on the basis of IR and UV spectral information ⟨75CHE304⟩. The electron-density distribution in four tautomeric forms, calculated by the PPP method, also indicates that the ring-junction carbon atom is the most reactive one with respect to nucleophilic attack ⟨75CHE500⟩.

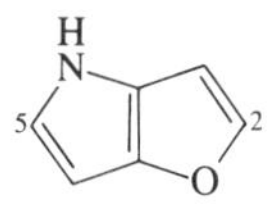

(36) 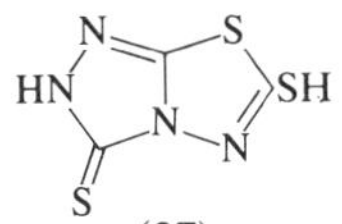 (37) 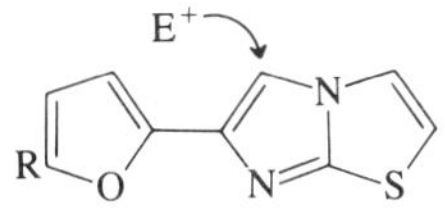 (38)

According to quantum-chemical calculations within the PPP approximation, the 5-position of the imidazo[2,1-*b*]thiazole (**38**) system has the highest capacity for electrophilic substitution reactions ⟨75CHE45⟩. This was confirmed by experiments ⟨72CHE1223⟩.

The π-electron structures and energies of the singlet π–π^* transitions for a number of 1*H*-pyrrolo[1,2-*a*]imidazoles (**39**), 1*H*-pyrrolo[1,2-*b*]-*s*-triazoles (**40**) and 1*H*-pyrrolo[2,1-*c*]-*s*-triazoles (**41**) were calculated by the MO LCAO method within the semiempirical self-consistent field (SCF) approximation. A comparison of the data shows that the maximum

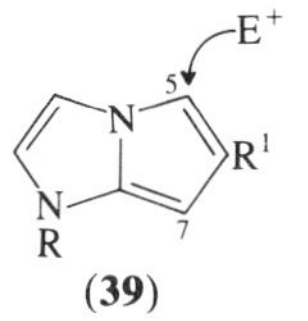

(39) 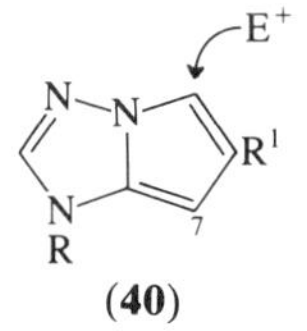 (40) 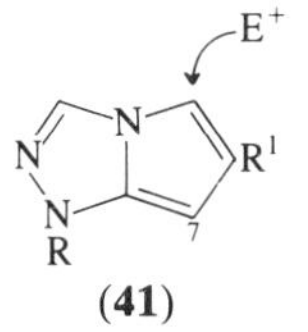 (41)

π-electron densities are localized on C-5 and C-7. Reactivity indices indicate that C-5 will primarily undergo electrophilic attack, followed only by C-7 ⟨74CHE230⟩.

Calculations for the cyclization of 2-azidoimidazole (**42**) and its anion (**44**), using the MINDO semiempirical SCF–MO method, predict the bicyclic anion (**45**) to be 70.6 kJ mol^{-1} relatively more stable than the neutral bicyclic molecule (**43**). Although the possible equilibrium between (**42**) and (**43**) is shifted to the azido form, there is sufficient evidence to support the role of the more efficient delocalization of the negative charge on the nitrogen atoms of the tetrazole ring as the principal cause for the experimentally observed shifting of equilibrium (**45**) ⇌ (**44**) to the tetrazole form (**45**) ⟨79JHC685⟩.

4.36.3 GENERAL SURVEY OF STRUCTURE AND REACTIVITY

4.36.3.1 Tautomerism

4.36.3.1.1 Azido–tetrazole

In five-membered α-azido N-heterocycles, the effect of the azole ring is completely in favor of the azide form. All azidoazoles which are known have spectroscopic properties typical of azido derivatives. However, an azidomethine unit as part of the thiazole and 1,3,4-thiadiazole rings may exist in either the cyclized or open-chain form under normal conditions; both forms are usually detected ⟨77AHC(21)323, 75BSB1189⟩.

In general, shifting the azido–tetrazole equilibrium of azapentalenic systems to the tetrazole form does not depend on the whole electron-attracting effect of a group, but it is mainly governed by its resonance electron-withdrawing ability ⟨76JHC33⟩.

4.36.3.1.2 Lactam–lactim

Where lactam–lactim tautomerism exists, the lactam form is usually considered to predominate as a result of work on other systems ⟨63AHC(2)1, 50HCA273⟩.

4.36.3.1.3 Amine–imine

In compounds with amine–imine potential, the amine form is usually considered to predominate as a result of work on other systems ⟨63AHC(2)1⟩.

4.36.3.1.4 Thiol–thione

Spectral data for the (5,5)-systems studied in this respect show the thione form to prevail ⟨77HC(30)1⟩.

4.36.3.1.5 Methylene–amino

The location of the maverick proton on the pyrrolo[1,2-*a*]imidazole ring system is on carbon. NMR spectral evidence favors the methylene and not the NH form ⟨77HC(30)1⟩.

4.36.3.1.6 Ring–chain

The tautomeric structure (**47**), 3,3-dimethyl-5-ethyloxalyl-2*H*-thiazolo[2,3-*c*]-[1,2,4]triazole oxime, was assigned on the basis of a singlet methyl resonance, different from the doublet seen for (**46**) ⟨78JHC401⟩.

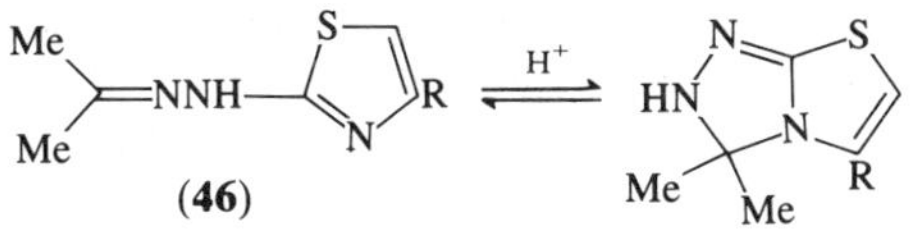

The ring–chain tautomerism of 5-acylalkyl substituted imidazoles has been confirmed by IR techniques. The weak absorption bands of CO and NH groups that are characteristic for open-chain form (**48**) are observed along with the absorption bands of the OH groups of cyclic form (**49**). According to ^{1}H NMR spectral measurements, both (**49a**) and (**49b**) exist in solution ⟨72CHE1017⟩.

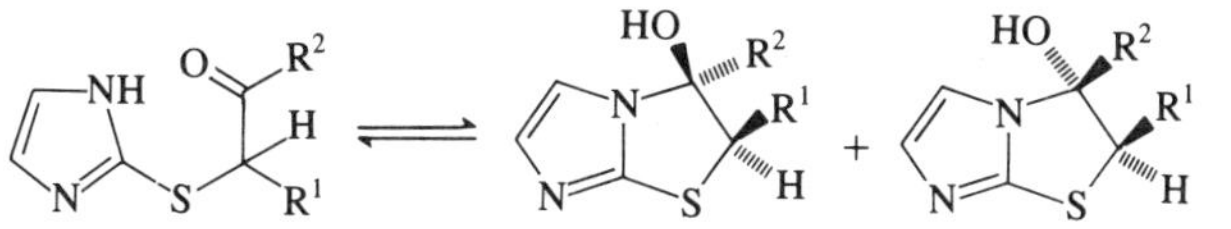

4.36.3.2 Reactions Involving Electrophilic Attack

While there are several reports concerning electrophilic substitution on to (5,5)-fused heterocycles, very few of these involve a study with the parent system. The π-excessive systems (**50**), (**51**), (**52**) and (**53**) were found to be susceptible to attack by electrophilic reagents at the positions indicated, leading to alkylation, formylation (Vilsmeier–Haack reaction), acylation, tritylation, metalation, tricyanoethylation, halogenation, thiocyanation, nitrosation, nitration and diazo coupling ⟨77HC(30)1⟩.

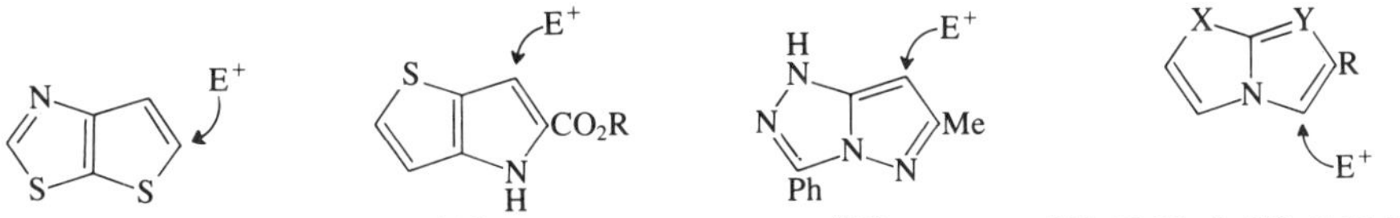

There have been some studies of the diazo coupling and carbonyl reactions with imidazo[2,1-*b*]thiazol-3(2*H*)-ones ⟨67CHE706⟩, 2,3-dihydrothiazolo[4,3-*c*]-*s*-triazol-5-ones ⟨76ZN(B)853⟩ and 2,3-dihydroimidazo[2,1-*b*]thiazol-3(2*H*)-ones ⟨81ZN(B)501⟩. Oxidation of 2,3-dihydrothieno[3,2-*b*]pyrrole with chloranil to thieno[3,2-*b*]pyrrole has been reported ⟨72CHE1428⟩.

4.36.3.3 Reactions Involving Nucleophilic Attack

Many of the nucleophilic substitution reactions described are halogen displacements by cyanide, nitrite, piperidine and morpholine which take place most readily when the halogen is activated by an electron-withdrawing group in the molecule. In the desulfurization of 3-mercaptothiazolo[3,4-*b*]-*s*-triazole with Raney nickel or with peroxide, thiazolo[3,4-*b*]-*s*-triazoles were obtained. Reduction of 2-chloroimidazo[2,1-*b*]thiazole with phosphorus and hydriodic acid gave the parent heterocycle. Several *C*-acylated (5,5)-fused ring systems were deacylated in aqueous hydrochloric acid. *N*-Debenzylation of pyrrolo[1,2-*a*]imidazoles and imidazo[1,2-*a*]imidazoles with sodium in anhydrous ammonia gave the expected compounds, as well as dihydro derivatives ⟨77HC(30)1⟩. Dihydroisoxazolo[5,4-*d*]isoxazoles are reduced by metal hydrides to the tetrahydro derivatives (equation 1) ⟨79JOC3524⟩.

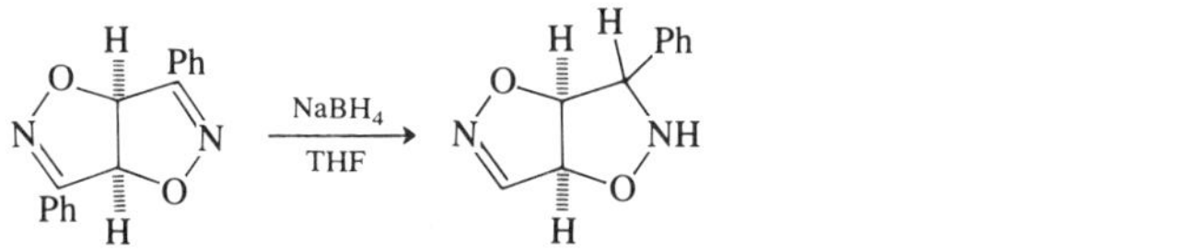

4.36.3.4 Ring Openings and Transformations

Ring opening reactions are of considerable interest, particularly in view of the trend of using heterocycles as sources for other organic fragments. Cleavage of (5,5)-fused ring systems is usually accomplished using similar degradative techniques as in the monocyclic aromatic and reduced series. The properties of fully reduced (5,5)-fused ring systems are usually similar to those of the fully reduced analogous monocyclic ring systems and their acyclic analogues.

4.36.3.4.1 Cleavage of N—O bonds

Bicyclic isoxazolines obtained by 1,3-dipolar cycloaddition of 1-pyrroline *N*-oxides with DMAD undergo ring fission at room temperature to give pyrroline derivatives (equation 2) ⟨66CC607⟩. This ring fission is similar to that of isoxazolium salts but requires no external nucleophile ⟨61JA1007⟩.

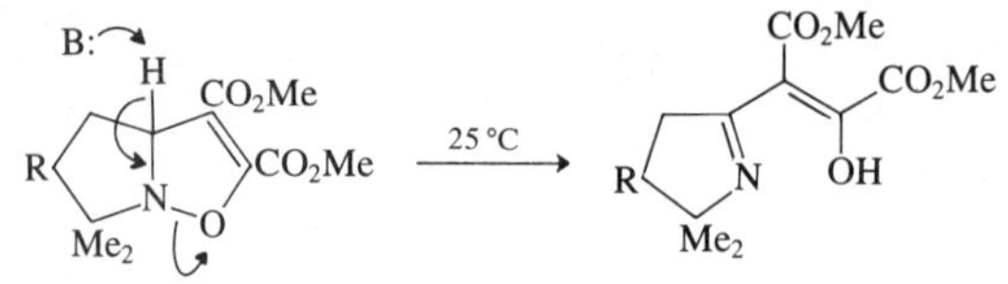

(2)

Isoxazolo[1,2-*b*]pyrazolones are easily cleaved by hydrogenolysis and hydrolysis to give 4-pyrazolones. Catalytic hydrogenation of (**54**; R=Me) yielded the α-hydroxy ester (**55**; R′ = CH(OH)CO_2Me), and similar treatment of (**54**; R = Ph) gave (**55**; R′ = CHO). The base-catalyzed ring opening of (**54**) gave (**55**; R′ = CO_2Me) ⟨71JOC19⟩.

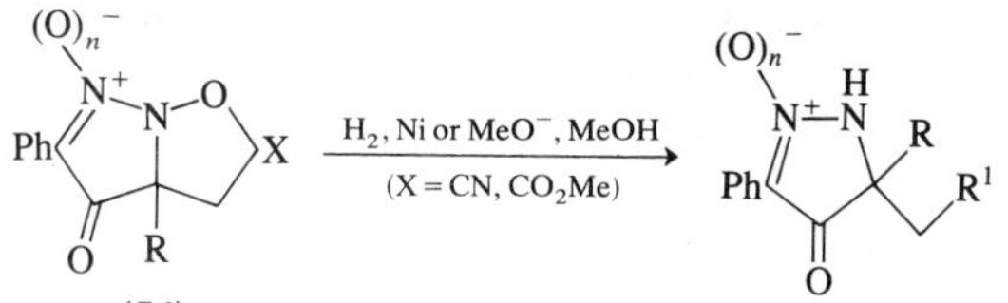

(**54**) (**55**)

4.36.3.4.2 Cleavage of N—N bonds

Fission of the N—N bond of tetrahydropyrazolo[4,5-*b*][1,2,4]oxadiazoles forming 5-substituted 1,2,4-oxadiazoles (equation 3) ⟨79JHC311⟩ occurs with 12N hydrochloric acid solution.

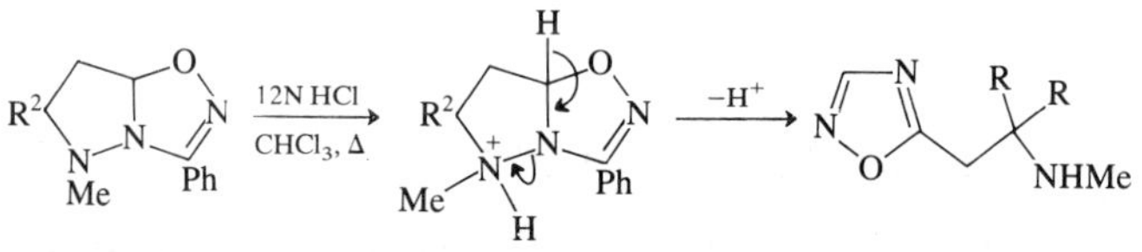

(3)

Metal hydride reduction of the dione (**56**; A = O, B = H_2) leads to the tetrahydro derivative (**56**; A = B = H_2). The N—N bond of (**56**) is cleaved when ethanolic solutions are refluxed in the presence of Raney nickel, to give octahydro-1,5-diazocine derivatives (**57**). The N—N bond of (**56**; A = B = O) is more resistant to hydrogenolysis than is the similar bond in (**56**; A = B = H_2) ⟨76S349, 65CB3228⟩.

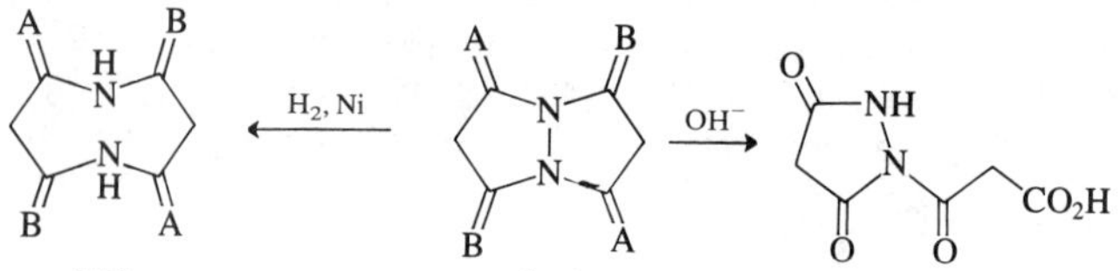

(**57**) (**56**) (**58**)

4.36.3.4.3 Cleavage of C—N bonds

The C—N bond of (**56**; A = B = O) is easily cleaved in alkaline media to form pyrazolidine-3,5-diones (**58**) ⟨71CHE1028⟩. The product of the reaction of hydroxide ion with a typical *syn*-bimane was shown to be the 2-pyrazolinonylacrylic acid ⟨82JOC4222⟩.

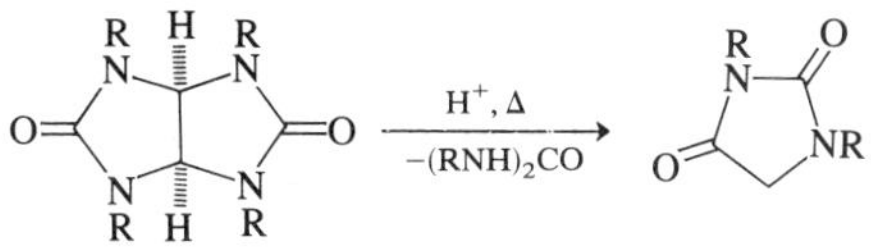

In the presence of strong acids and elevated temperature, glycolurils may be converted into hydantoins (equation 4) ⟨68JPR(309)78⟩.

The thiazolothiadiazolium perchlorate (**59**) is converted by base into the dithiadiazocine (**60**) as shown ⟨67AG(E)629⟩.

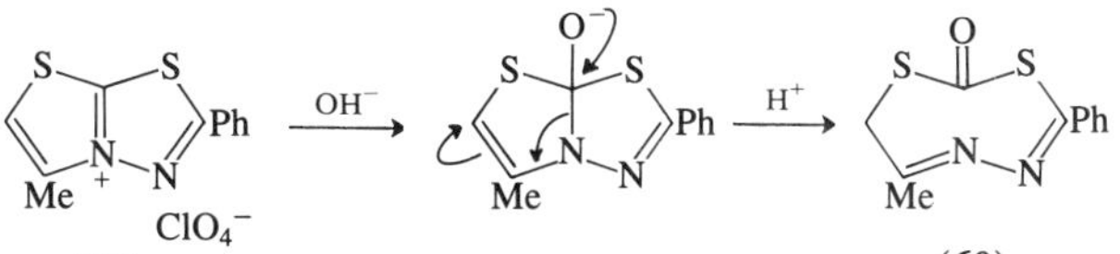

Base catalyzed Hofmann elimination carried out on the quaternary ammonium salts (**61**) gives (**62**). Good nucleophiles, but relatively weak bases, add at room temperature to the activated double bond of (**62**), affording *trans*-4,5-dihydro-1,2,3-triazoles (**63**) ⟨80JHC267⟩.

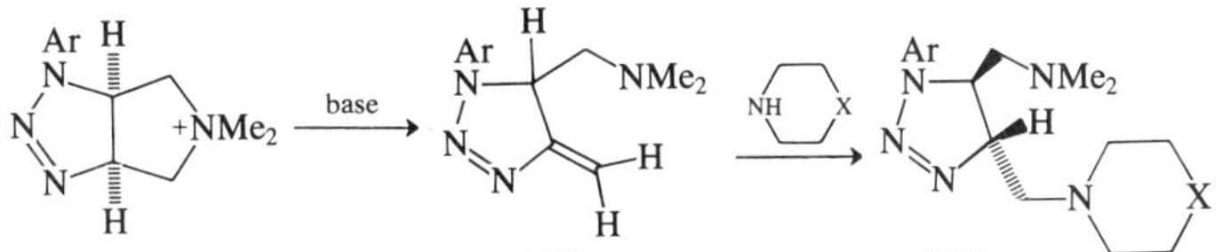

The pyrazolo[1,2-*a*]pyrazole ring system is susceptible to thermolysis, photolysis, reduction and hydrogenolysis. Two successive electrocyclic ring opening processes are responsible for the formation of the heteropolyene (**66**) when (**64**) is heated; the *C*-vinylazomethine imine (**65**) is proposed as the intermediate ⟨75TL1125⟩.

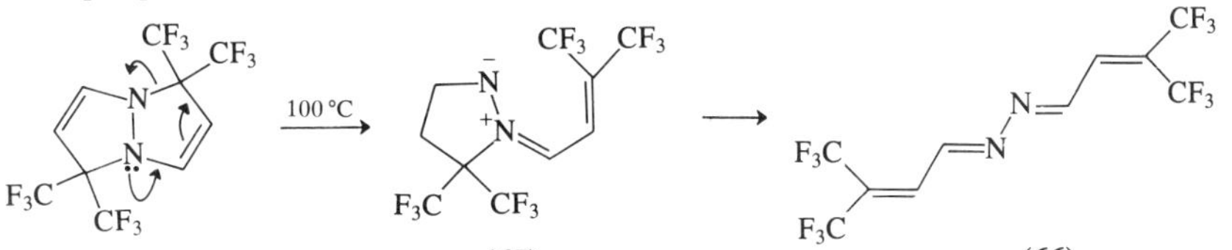

In addition to its stereoselectivity, the synthesis of chiral α-hydroxy aldehydes (**69**) from the aminal (**68**) has the advantage that the chiral auxiliary reagent (**67**) is easily prepared from (*S*)-proline, and that it may be recovered unchanged after use ⟨79CL705⟩.

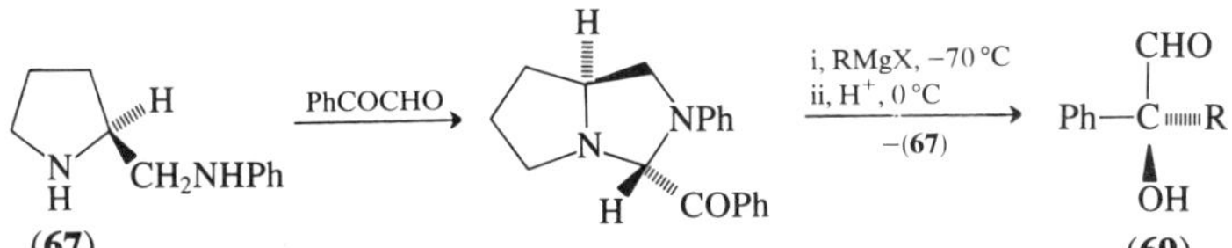

Pyrrolo[3,4-*d*]-*s*-triazoles (**70**) ⟨56JA145⟩, phospholo[2,3-*c*]pyrazoles (**72**) ⟨65JCS2184⟩ and 3*H*-[1,2,3]diazaphospholo[4,3-*c*][1,2,4]diazaphosphole (**74**) ⟨B-81MI43600, 77IZV1453⟩, all of which are prepared by bimolecular 1,3-dipolar cycloaddition reactions, lose nitrogen under mild conditions to give aziridines (**71**), cyclopropanes (**73**) and phosphiranes (**75**), respectively.

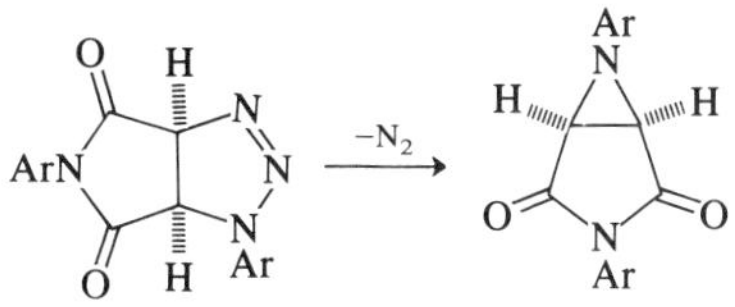

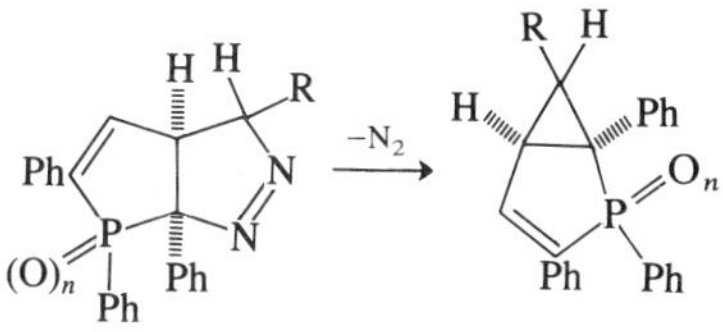

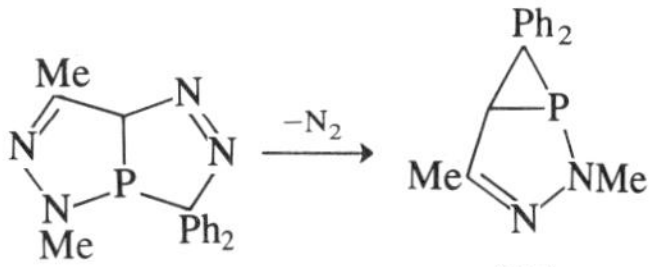

4.36.3.4.4 Cleavage of C—S bonds

Desulfurization of 2,3-dihydrothieno[3,2-b]pyrrole (**76**) with Raney nickel forms the 2-ethylpyrrole (**77**) ⟨72CHE1428⟩.

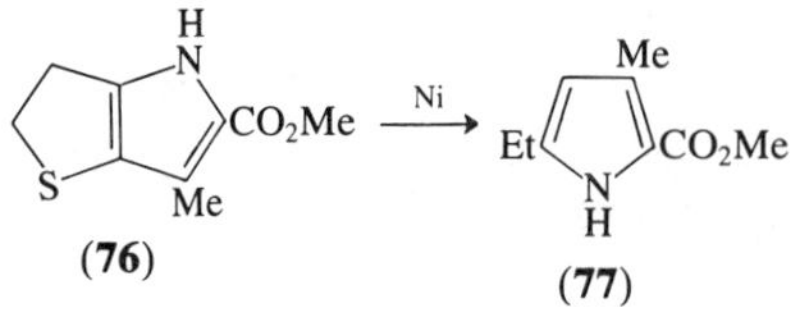

In alkaline solution, tetramisole (**78**) forms (**79**) ⟨66JMC545⟩; sulfur and nitrogen nucleophiles react analogously ⟨77JHC607⟩. Lithium diethylamide also ruptures the thiazole ring ⟨78JHC307⟩.

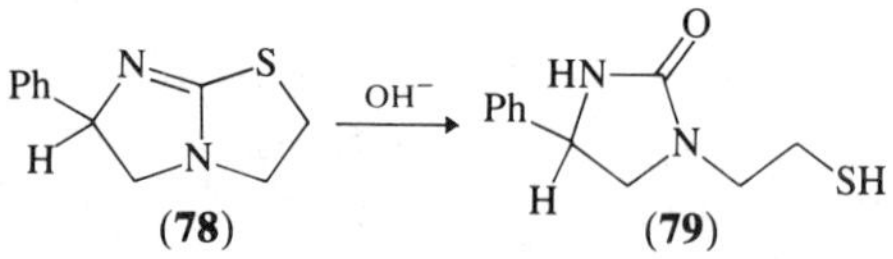

Imidazo[2,1-b][1,3,4]thiadiazoles undergo hydrazinolysis to give 1-amino-2-mercaptoimidazoles ⟨67ZC341⟩. Destructive hydrazinolysis of s-triazolo[3,4-b][1,3,4]thiadiazoles (**37**) leads to opening of the thiadiazole ring during attack by the nucleophile on the ring-junction carbon atom. The ease of secondary attack leading to (**80**) and (**81**) is determined by the charge on these ring carbon atoms ⟨75CHE500⟩.

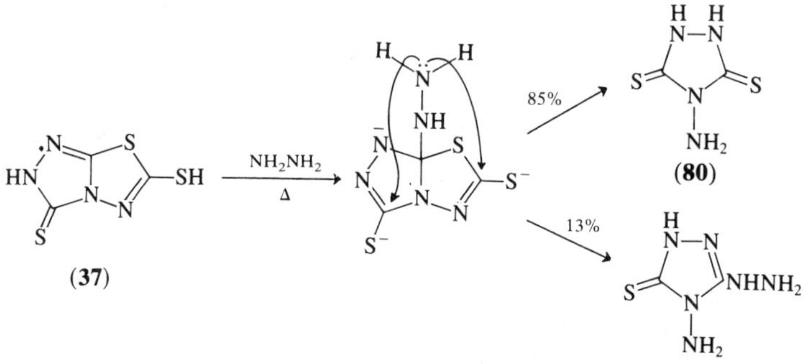

4.36.3.4.5 Cleavage of P—P bonds

Oxidation of the trivalent phosphorus atoms in the tetrasubstituted [1,4,2,3]diazadiphospholo[3,2-b]diazadiphosphole (**82**) is accompanied by rupture of the P—P bond to give the 9-oxa-2,4,6,8-tetraaza-1,5-diphosphabicyclo[3.3.1]nonane-3,7-dione 1,5-dioxide (**83**), whereas reaction with sulfur at 180–200 °C gives the corresponding 2,3-disulfide without cleaving the P—P bond ⟨81CB2132⟩.

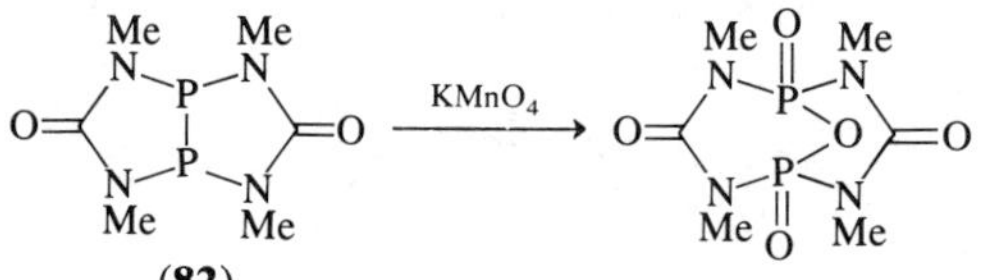

4.36.3.4.6 Cleavage of C—O bonds

Bicyclic amide acetals have synthetic utility, as illustrated by the catalytic reduction of (**84**; $R^1 = H$, $R^2 = Pr^n$) leading to amino alcohols (**85**); reaction of (**84**; $R^1 = OH$, $R^2 = H$) with formic acid involves double ring fission to give (**86**) ⟨71S16, 78MI43600⟩.

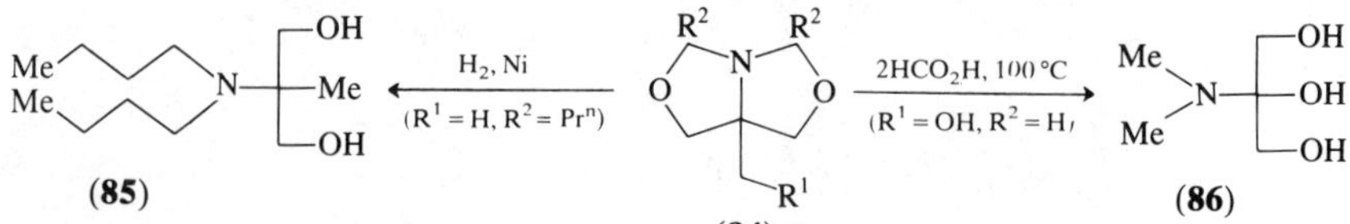

Heating of N-acetyl-S-(α-ketoalkyl)cysteines with acetic anhydride gives 3-acetamido-2,5-dihydrothiophenes (**88**); transfer of a proton from CO_2H to N of intermediate (**87**) would lead to a structure which is arranged to undergo a fragmentation reaction with loss of CO_2 to give (**88**) ⟨79JOC825⟩.

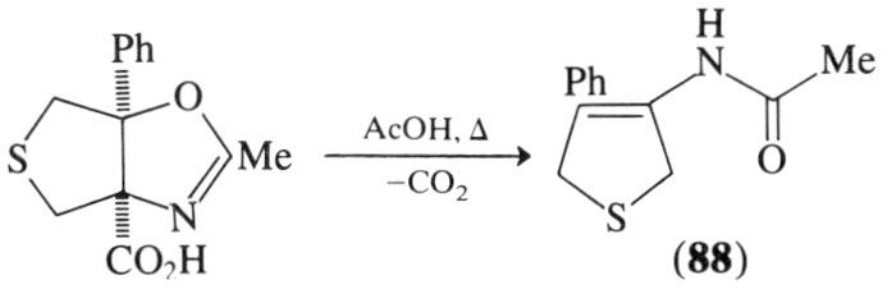

4,6-Diphenylthieno[3,4-*d*]dioxol-2-one 5,5-dioxide (**89**) is used to prepare active esters of BOC amino acids (**91**) via intermediates (**90**) ⟨79AG(E)307⟩.

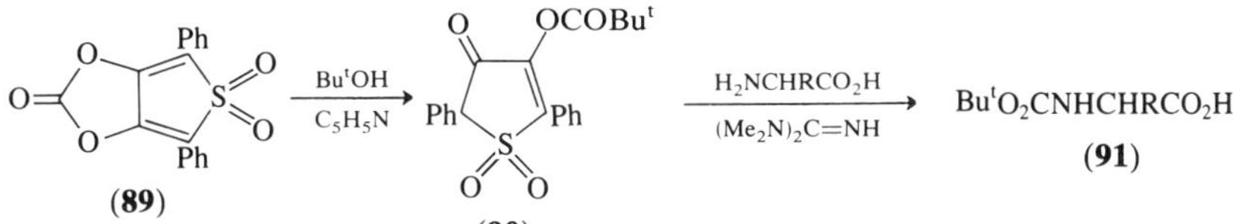

The glyoxaline (**92**) readily loses benzaldehyde when treated with cold HCl to give (**93**), which was converted into β-phenylserine ester (**94**) ⟨51JCS3479⟩. Prolonged boiling with H$_2$O converts the mercaptoglyoxaline (**95**) into (**96**) ⟨54JCS3283⟩.

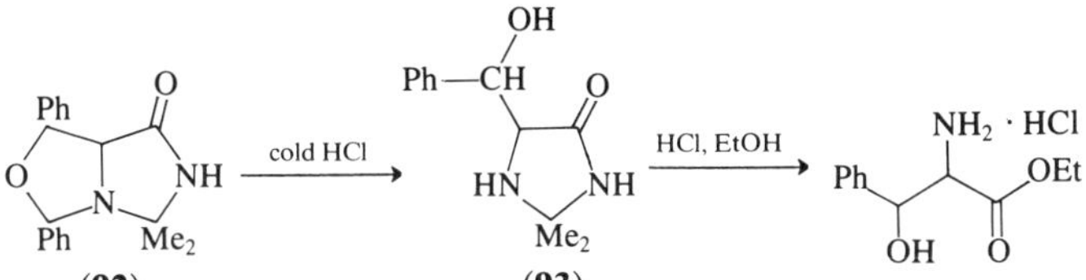

The tetrahydrofuran ring of (**97**) is cleaved with refluxing 50% H$_2$SO$_4$ to give the carbinol (**98**) ⟨68TL743⟩.

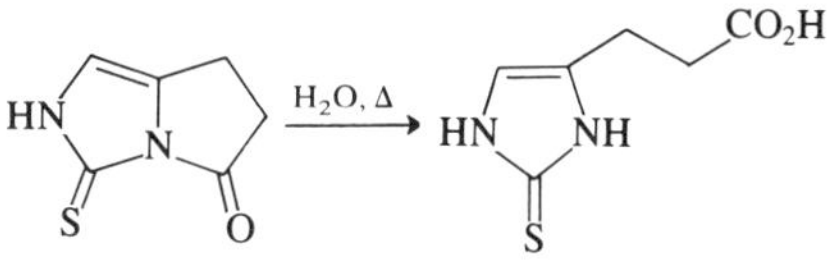

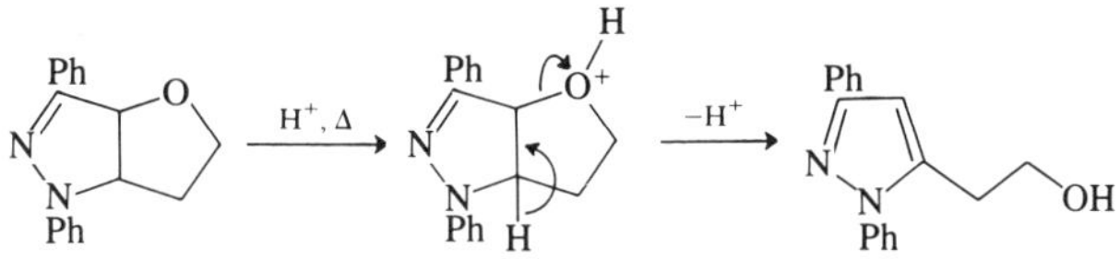

4.36.4 SYNTHESIS OF (5,5)-FUSED RING SYSTEMS

4.36.4.1 Nucleophilic Type Displacements

Bifunctional molecules which favor bimolecular combinations have been widely used, with many variations in experimental details, for the preparation of fused ring systems. The electrophile can be either a reagent or preformed ring component.

4.36.4.1.1 Reactions with 1,1-bielectrophiles

The last step in a stereospecific total synthesis of (±)-biotin consists of the facile reaction of the precursor diamine with phosgene (equation 5) ⟨77JA6754⟩. The reaction of a cyclic amidrazone precursor with phosgeniminium chloride gives thiazolo[3,2-*b*]-*s*-triazoles (equation 6) ⟨73AG(E)405⟩. Treatment of 4,5-diaminopyrazoles with thionyl chloride forms pyrazolo[3,4-*c*][1,2,5]thiadiazoles (equation 7) ⟨68JMC1164⟩.

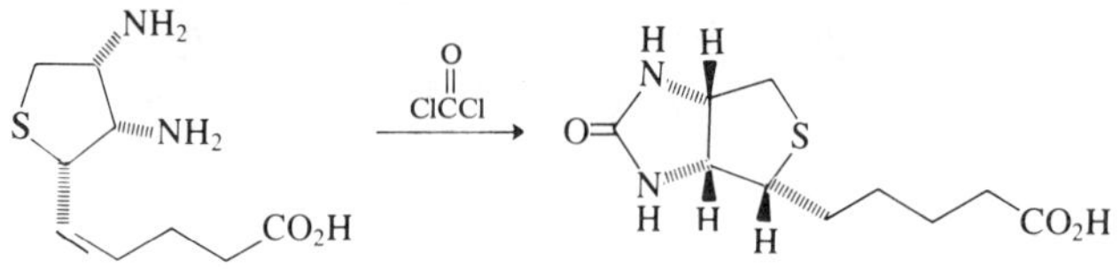

Thieno-1,3-dithiolane-2-thione ⟨62ACS105⟩ was prepared by a convenient one-flask reaction in which 3,4-dibromothiophene was twice treated with BuLi followed by elemental sulfur at −78 °C to provide the dilithio intermediate. This sequence was followed by the addition of an excess of CS_2 and 2N NaOH (equation 8). Alternatively, the dilithio intermediate could be protonated, and the resulting dithiole allowed to react with thiocarbonyldiimidazole ⟨81CC920⟩.

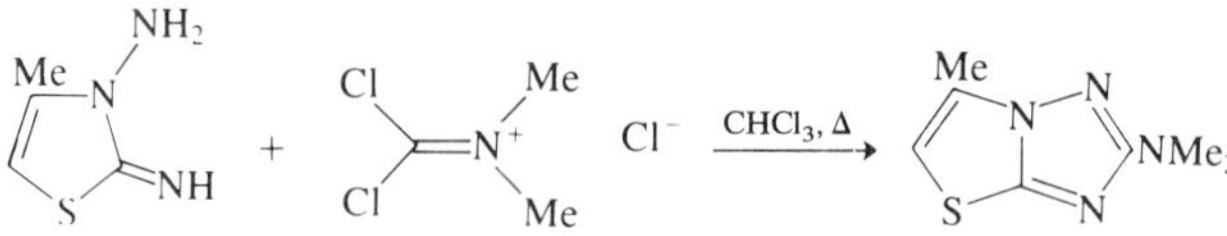

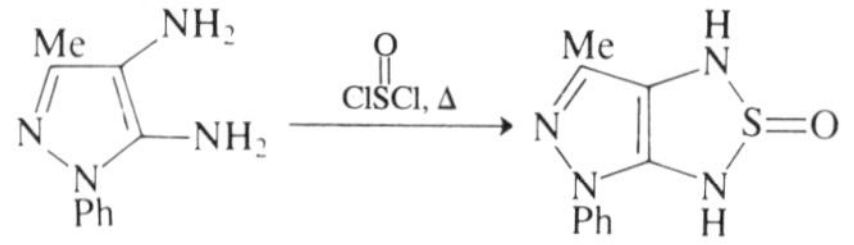

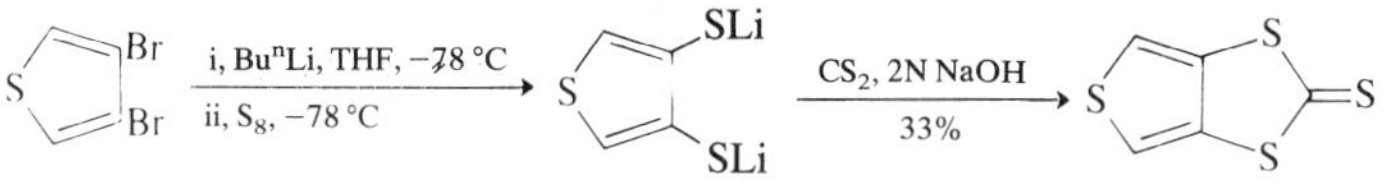

4.36.4.1.2 Reactions with 1,2-bielectrophiles

Amino derivatives of isoxazole, 1,3,4-oxadiazole, thiazole, 1,2,4-thiadiazole and 1,3,4-thiadiazole containing the amino group as part of a partial amidine structure react with trichloromethanesulfenyl chloride to give isolable trichloromethanesulfenamides which react with substituted anilines to yield fused thiadiazoles of general structure shown in equation (9). The method has been used to prepare 3H-isoxazolo[3,2-c]-, 3H-thiadiazolo[2,3-c]-, 3H-1,3,4-thiadiazolo[2,3-c]- and 3H-[1,2,4]thiadiazolo[4,3-d]-[1,2,4]thiadiazoles, as well as 3H-[1,2,4]thiadiazolo[3,4-b][1,3,4]oxadiazoles ⟨75JOC2600⟩.

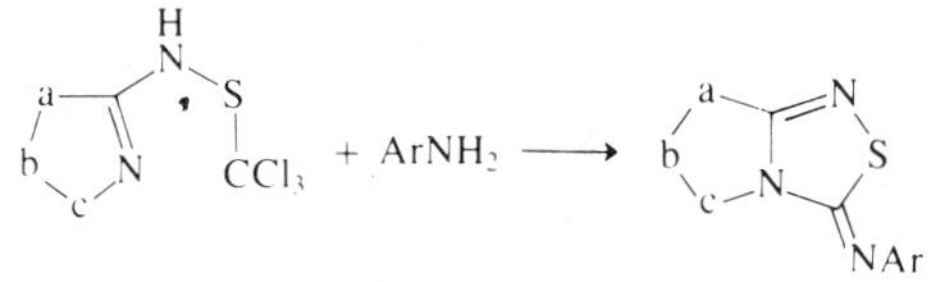

Imidazoline-2-thiones and imidazolidine-2-thiones react with 1,2-dibromoethane in the presence of aqueous sodium hydroxide and a charge transfer catalyst (CTC), or with α-cyanobenzyl benzenesulfonate, to give the corresponding imidazo[3,1-b]thiazoles (equation 10) ⟨80JHC393, 61JOC2715⟩. In an analogous reaction, the condensation of 5,5-diphenyl-2-thiohydantoin with ethylene dibromide yielded two isomeric products, (**99**; major) and (**100**; minor) ⟨81JCS(P2)789⟩.

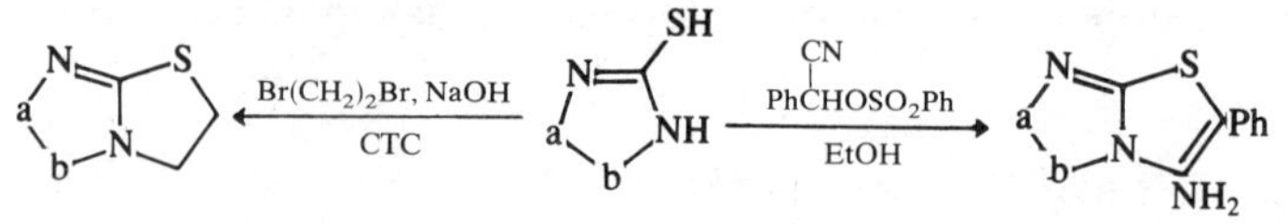

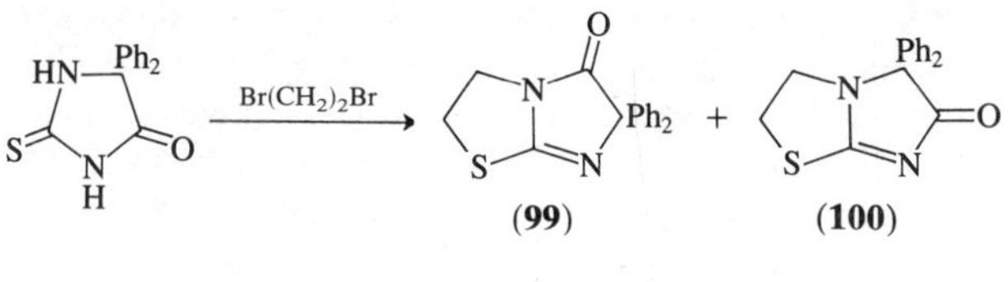

Condensations of α-amino N-heterocycles with chlorothioformyl chloride can give rise to either (2,3)-fused 1,2,4-thiadiazolones (**101**) ⟨73JOC1575⟩ or (3,4)-fused 1,2,4-thiadiazolones (**102**) ⟨75JCS(P1)375⟩. The fused heterocycle (**101**) is formed when the reaction is carried out in a proton-acceptor solvent (THF), which is capable of solvating the exocyclic amine group. In contrast, (**102**) is formed in a proton-donor solvent (CHCl$_3$), which would tend to solvate the tertiary amino group, leaving the exocyclic amino group free to react.

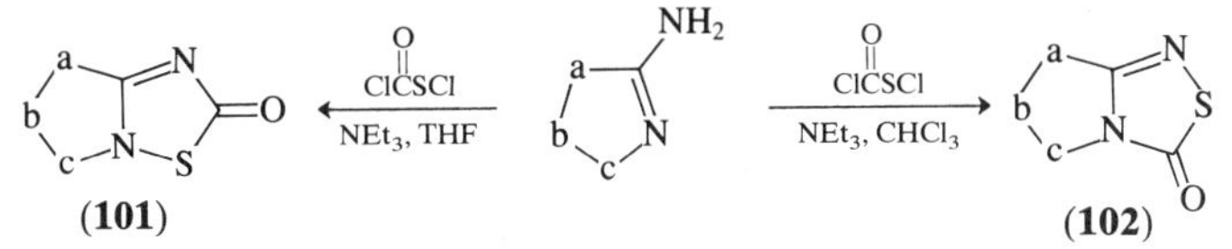

Owing to their high reactivity, acylated 2-aminothiophenes react readily with oxalyl chloride in the absence of catalysts to give thieno[2,3-b]pyrroles (equation 11). Thieno[3,2-b]pyrroles were prepared in a similar fashion from 3-acylaminothiophenes ⟨75CHE666⟩.

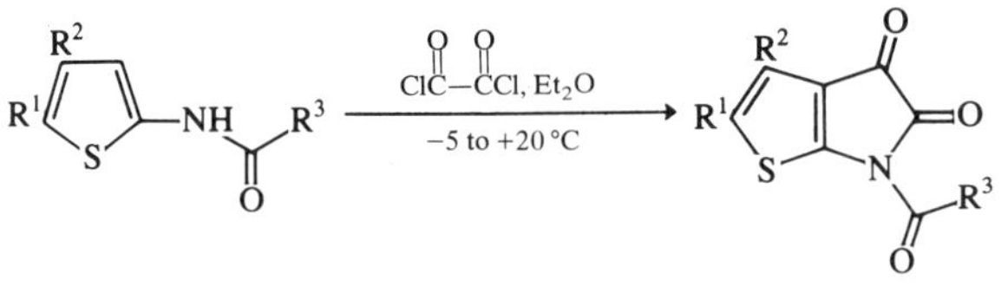

1,2-Diimonium salts have proven to be a useful tool for the preparation of amino-functionalized 2-imidazoline derivatives (equation 12) ⟨80JHC1041⟩.

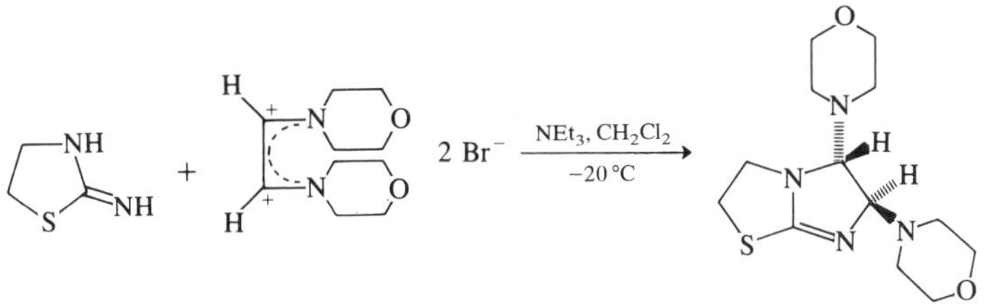

Cyclizations involving reactions of α-amino N-heterocycles with α-halocarbonyl compounds leading to similar ring systems by forming two new C—N bonds will be discussed in more detail in Section 4.36.4.2.3.

4.36.4.1.3 Reactions with 1,3-bielectrophiles

An acyl group in the 4-position of a pyrazole activates a chlorine atom in the 5-position, which reacts under mild conditions with hydrazine to form pyrazolo[4,5-b]pyrazoles (equation 13) ⟨65CHE271⟩.

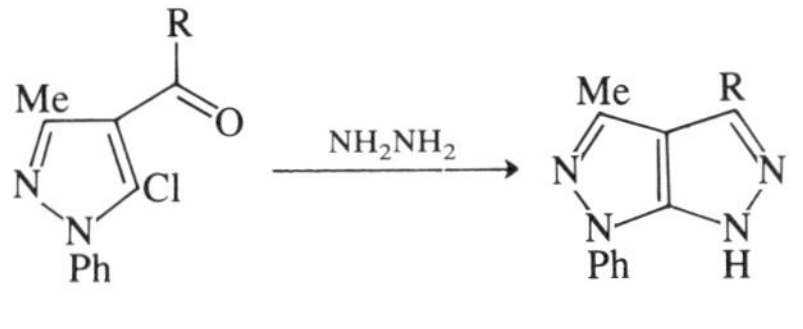

4.36.4.1.4 Reactions with 1,4-bielectrophiles

The reaction of 3,4-bis(halomethyl)thiophenes with sodium sulfide and primary amines gives 1H,3H-thieno[3,4-c]thiophenes (**103**) and 5,6-dihydro-4H-thieno[3,4-c]pyrroles (**104**), respectively. Some of the cyclizations are adversely affected by *ortho* substituents (X) ⟨69JOC333⟩. The method has been used to prepare 1,3,4,6-tetrahydrothieno[3,4-c]pyrrole 2,2-dioxides ⟨73JHC785⟩.

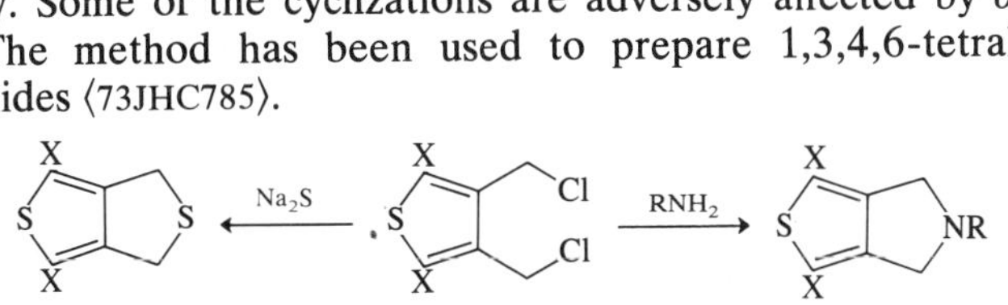

2-Bromo-N-(2-chloroethyl)pyrazoles react with ammonia and primary amines to give 2,3-dihydroimidazo[1,2-a]imidazoles (equation 14) ⟨70CHE1177⟩.

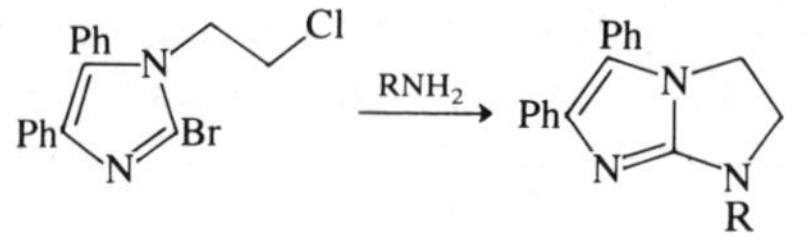

$$\text{(14)}$$

4.36.4.2 Aldol-type Condensations

An important and common method for lengthening and shortening the carbon chains of organic molecules, both chemically and biochemically, is the aldol-type reaction. Fischer and Tafel (in 1887) first described an alkali-catalyzed condensation of two trioses to form a hexose.

Condensation reactions of the aldol type play an important part in heterocyclic chemistry. There are a large number of condensation reactions that are closely related to the aldol condensation. Each of these reactions has its own name: Claisen, Dieckmann, Doebner, Knoevenagel, Perkin, to mention a few, but the chemistry is essentially the same as that of the aldol condensation.

4.36.4.2.1 Formation of C—C bonds

The pyrrolylthioacetate (**105**) undergoes Dieckmann ring closure when treated with sodium hydride to give the thieno[2,3-b]pyrrole-2,4-dicarboxylate (**106**) ⟨65JOC184⟩.

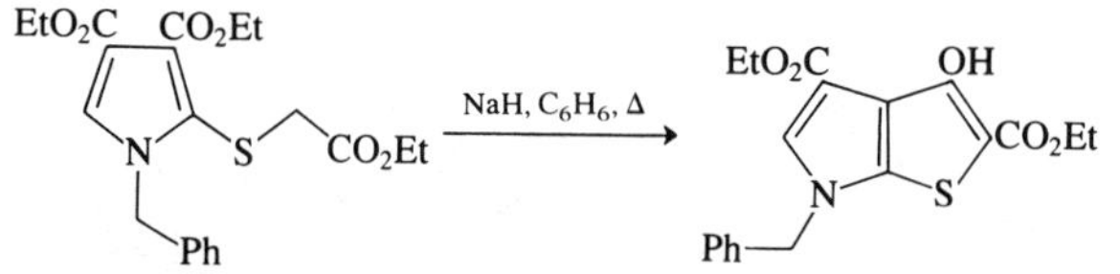

(**105**)   (**106**) 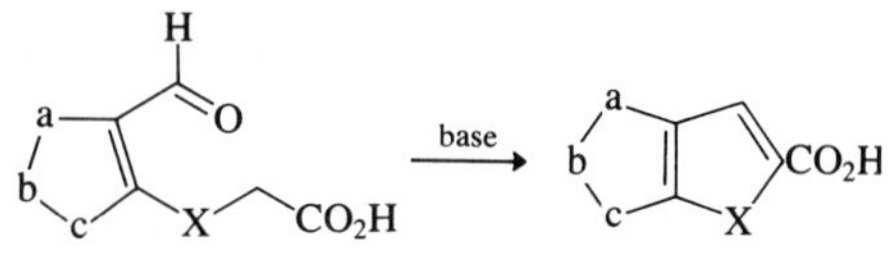   (**107**)   (**108**)

Thioglycolic acid derivatives (**107**) which contain an 'active methylene' group and a formyl group undergo cyclodehydration when treated with base (Knoevenagel reaction), leading to fused thiophenecarboxylic acids (**108**) which decarboxylate when heated above their melting point. The method has been used to prepare thieno[3,2-b]thiophenes ⟨72CHE392⟩, thieno[2,3-b]pyrroles ⟨73CHE521⟩ and thieno[2,3-d]thiazoles ⟨69CHE567⟩ and the respective selenium heterocycles (X=Se).

Thieno[2,3-b]pyrroles have been prepared from 2-thienylhydrazines and carbonyl compounds under the conditions of 'Fischer indolization' ⟨77S487⟩. The Bu^{t}OCO group is eliminated, and the intermediate hydrazone can undergo ring closure (equation 15). Thieno[3,2-b]pyrroles were prepared similarly from acylated 3-hydrazinothiophenes ⟨75CHE1133⟩.

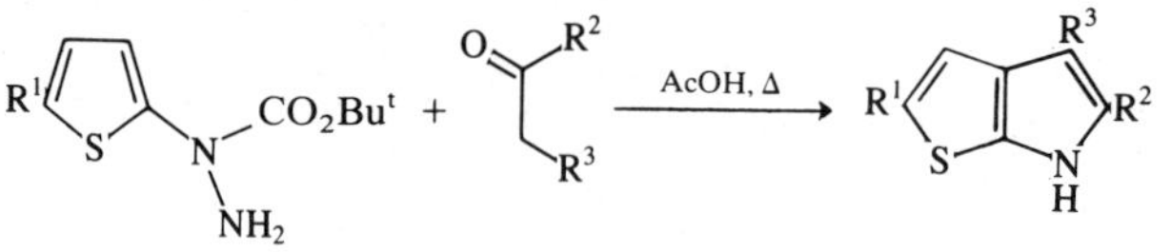

$$\text{(15)}$$

The pyrolysis of lead salts of dicarboxylic acids gives ketones, as exemplified by the formation of 2-methyl-2-cyclopentano[4,3-d]thiazol-5-one (equation 16) ⟨63JPR(292)285⟩.

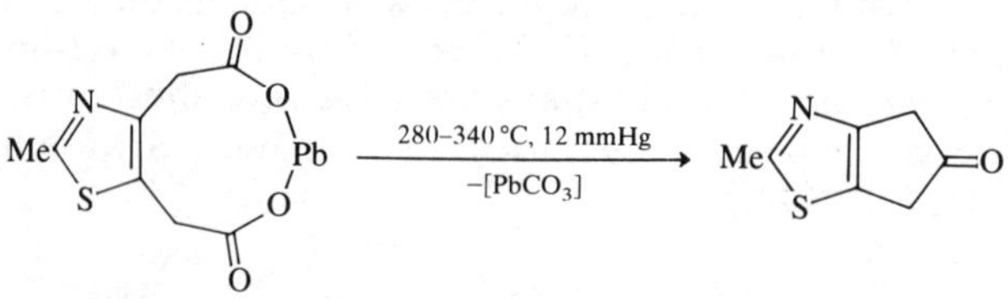

$$\text{(16)}$$

The zinc chloride assisted cyclo-condensation of ethyl 3-acetyl-1,3-oxazolidine-4-carboxylates proceeds at 180 °C to give derivatives of pyrrolo[1,2-c]oxazole (Scheme 1) (X = O); pyrrolo[1,2-c]thiazoles (X=S) were prepared similarly ⟨81BCJ1844⟩.

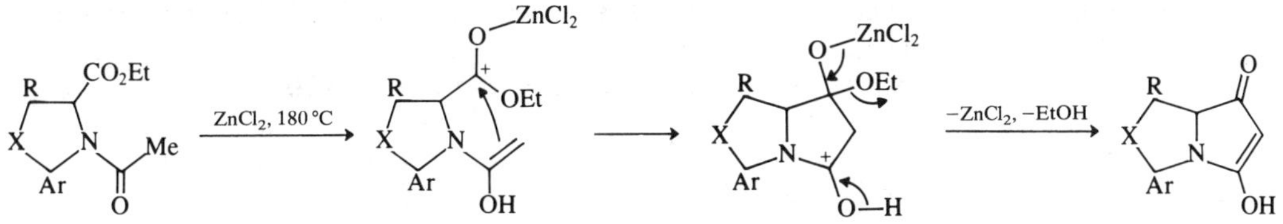

4.36.4.2.2 Formation of C—O bonds

β-Ketoximes are readily dehydrated by acetic anhydride to give isoxazolo[4,5-d]isoxazoles (equation 17) ⟨59G571⟩.

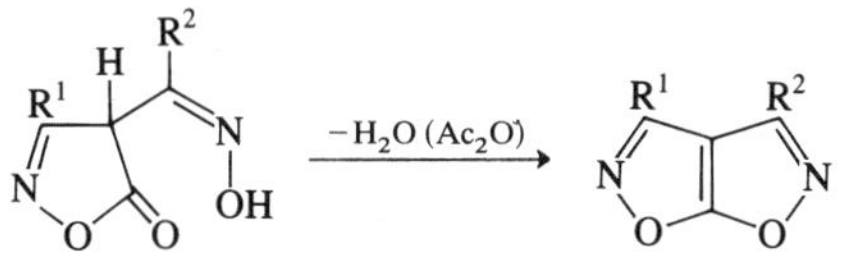

$$\text{(17)}$$

Reactions of nitriles with nucleophilic reagents normally require catalysis by acids or bases. Heating iminodiethanols with aliphatic nitriles in the presence of catalytic amounts of sodium alkoxide leads to oxazolo[2,3-b]oxazoles ('bicyclic amide acetals'), in a single process via oxazolidine intermediates (equation 18) ⟨73AG(E)1055⟩.

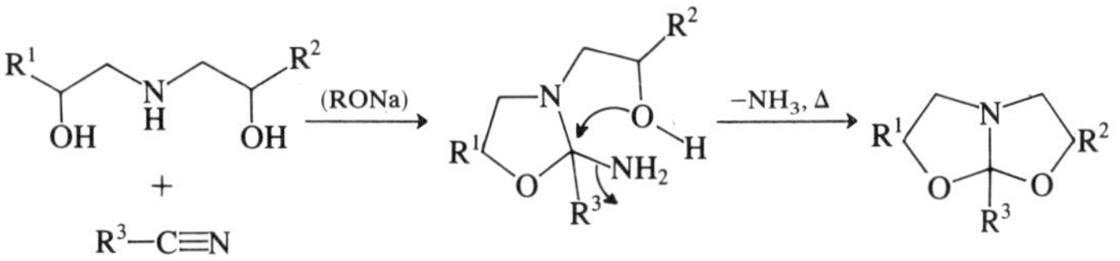

$$\text{(18)}$$

4.36.4.2.3 Formation of C—N bonds

As reported by Mannich ⟨29AP(267)699⟩, reaction of ethyl cyclopentanone-2-carboxylate with phenylhydrazine and subsequent cyclization of the phenylhydrazone at elevated temperature under strongly basic conditions gives tetrahydro-2-phenylcyclopentapyrazol-3-one (equation 19) (a = b = CH$_2$). The method has been used to prepare thieno[3,4-c]pyrazol-3-ones, thieno[3,2-c]pyrazol-3-ones and pyrrolo[3,4-c]pyrazolones ⟨71JMC454, 71JMC1129⟩.

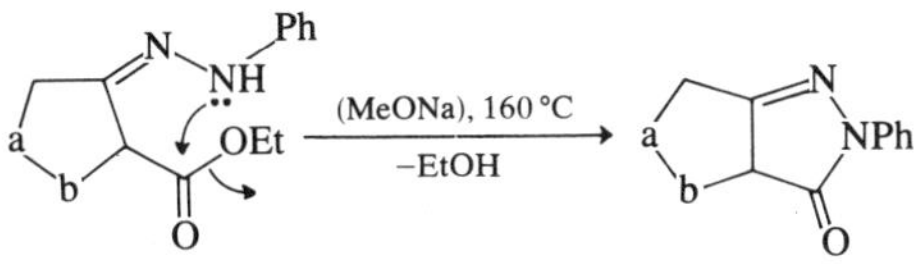

$$\text{(19)}$$

Condensation of acetyl-tetronic or -tetramic acids with methyl- or phenyl-hydrazine leads to 3-(1-hydrazinoethyl)-tetronic and -tetramic acid, respectively, which upon dehydration form 1-substituted 4-oxo-1,4-dihydro-6H-furo[3,4-c]pyrazoles (X=O) and -pyrrolo[3,4-c]pyrazoles (X=NH) (equation 20) ⟨82SC431⟩.

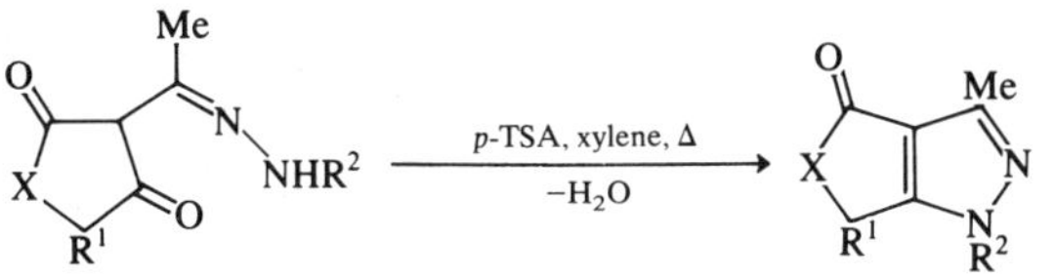

$$\text{(20)}$$

2,3-Dihydro-3-oxothieno[3,4-d]isothiazole 1,1-dioxide was obtained from methyl 4-sulfamoylthiophene-3-carboxylate in the presence of sodium methoxide (equation 21) ⟨80JOC617⟩.

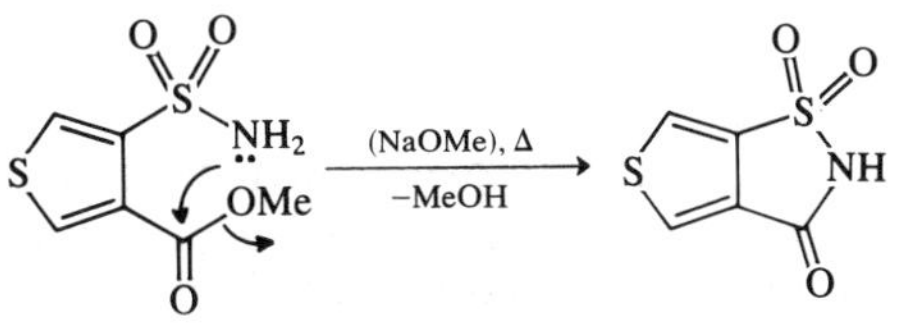

$$\text{(21)}$$

5-Phenylthiophene-2,3-quinone-2-oxime (**109**) reacts with aromatic aldehydes under the conditions of the Diels oxazole synthesis ⟨15CB897⟩ to give the thieno[2,3-d]oxazoles (**111**). It has been represented that (**109**) is first reduced to the imine (**110**), which then reacts with a second equivalent of aldehyde to give (**111**) ⟨55JA5370⟩. Similar reactions of 1,3-diphenyl-5-iminoimidazolidine-2,4-dithione and the corresponding 2-one-4-thione with benzaldehyde leading to 4,6-dihydro-2,4,6-triphenyl-5H-imidazo[4,5-d]thiazoles are BF$_3$ catalyzed ⟨80JOC3748⟩.

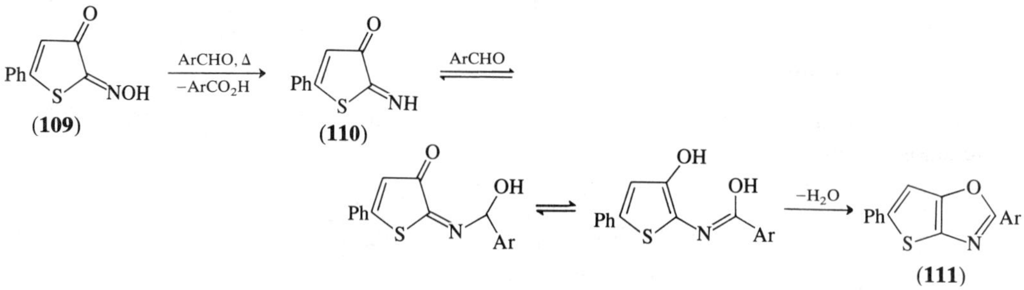

4-Phenylhydrazono-2-thiazolidinones are cyclized upon treatment with aliphatic and aromatic aldehydes to give thiazolo[4,3-c][1,2,4]triazol-5-ones (equation 22) ⟨76ZN(B)380, 76ZN(B)853⟩. Similar reactions of 2-(and 4-)phenylhydrazono-1-phenylhydantoins with aliphatic and aromatic aldehydes gave imidazo[2,3-c][1,2,4]triazoles and imidazo[4,3-c]-[1,2,4]triazoles ⟨82H(19)1375⟩.

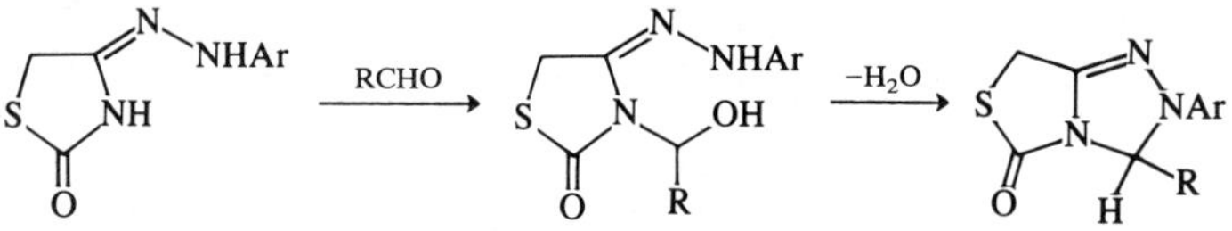

(22)

3-Amino-2-imino-4-methylthiazoline hydrochloride (**112**) and aminium salts (**114**) react with carboxylic acid anhydrides to give thiazolo[3,2-b]-s-triazoles (**113**) ⟨76JHC1225, 74JHC459⟩.

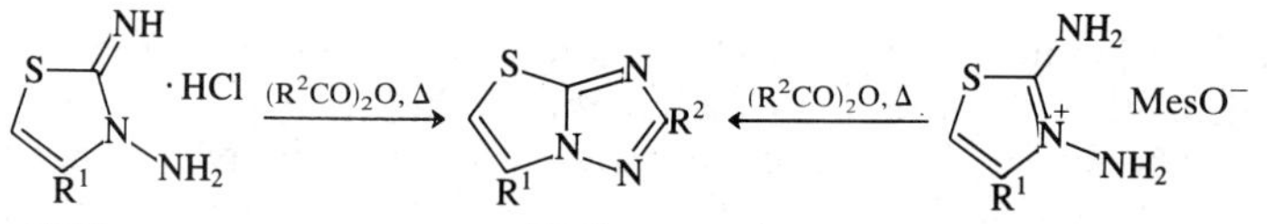

Dehydration of 1-(2-hydroxyethyl)-5-aminopyrazoles gives imidazo[1,2-b]pyrazole derivatives (equation 23) ⟨61USP2989537, 80JHC73⟩.

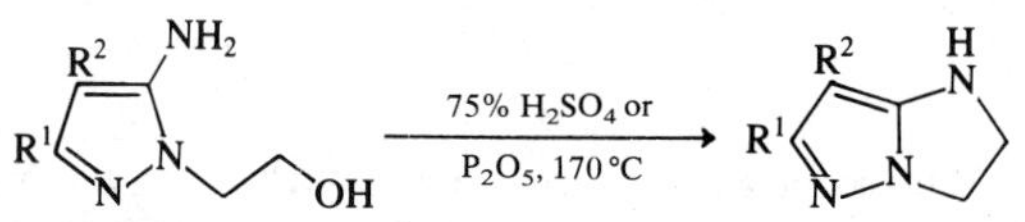

(23)

The reaction of diacetonitrile with hydrazinoacetaldehyde diethyl acetal followed by acid-catalyzed cyclo-condensation of intermediate (**115**) leads to the formation of imidazo[1,2-b]pyrazoles (**116**) ⟨73JHC411⟩. Similarly, α-amino aldehydes are sufficiently nucleophilic to undergo uncatalyzed addition to the carbon–nitrogen triple bond of cyanamide to give intermediates analogous to (**115**) which undergo cyclodehydration when treated with conc. HCl to give imidazo[1,2-a]imidazoles ⟨56JCS307⟩.

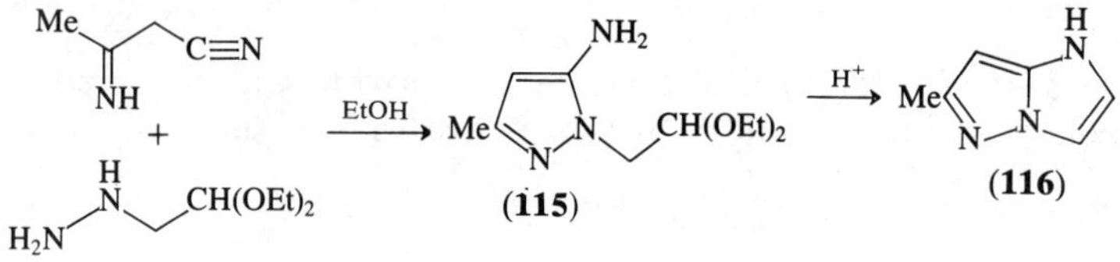

α-Aminoacetals react readily with 2-alkoxy-Δ^1-pyrrolines to give cyclic amidines which cyclize under thermal conditions to 6,7-dihydro-5H-pyrrolo[1,2-a]imidazoles (**117**). α-Aminoalkyl aryl ketones and ethyl carbazate react similarly with 2-alkyl-Δ^1-pyrrolines to give cyclic amidines which undergo thermal ring closure leading to 6,7-dihydro-5H-pyrrolo[1,2-a]imidazoles and 1H-pyrrolo[2,1-c]-s-triazol-3-ones ⟨82JHC193⟩. Hydrazinolysis of the N-(benzylideneamino)imidazolidine derivative (118) with 2,4-dinitrophenylhydrazine (DNPH) in the presence of 5–10% H_2SO_4 in ethanol gave the 2,3-dihydro-1H-imidazo[1,2-b]pyrazole derivatives (120) and (121) via intermediate (119) ⟨80JHC1413⟩.

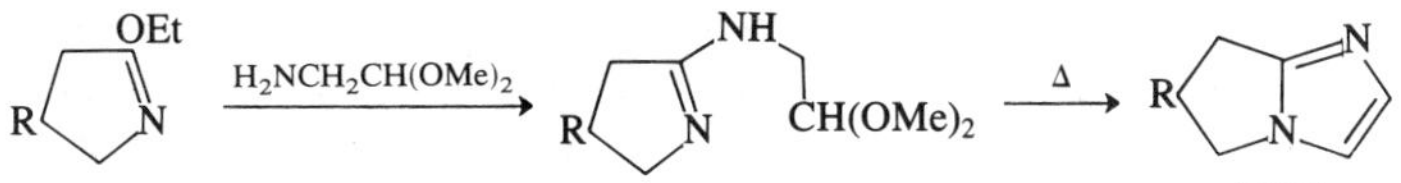

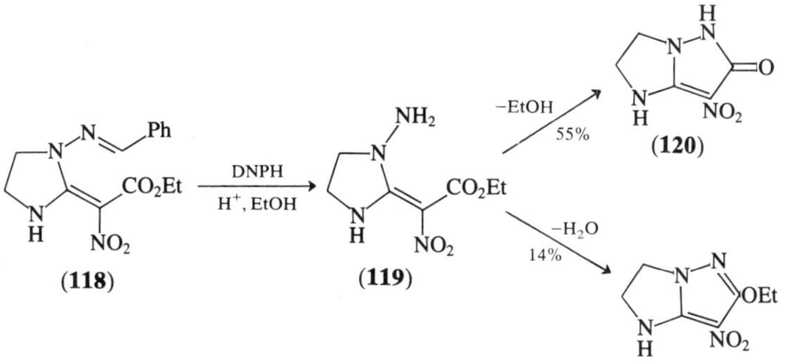

4,5-Disubstituted imidazoles have been N- and C-acylated by aryl isocyanates, leading to imidazo[1,2-c]hydantoins (equation 24) ⟨59CB550, 77JOC3925⟩.

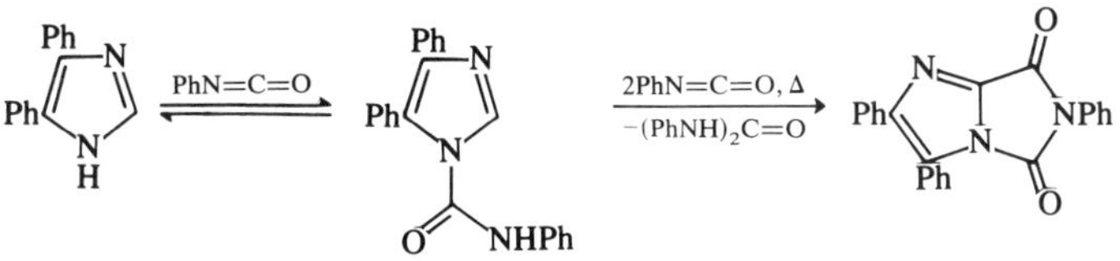

A key step in the synthesis of pyrazolo[2,3-d]imidazoles (124) involves diazotization of hydrazide (122), followed by Curtius rearrangement to give the intermediate isocyanate (123) which cyclizes to (124) ⟨75JHC595⟩.

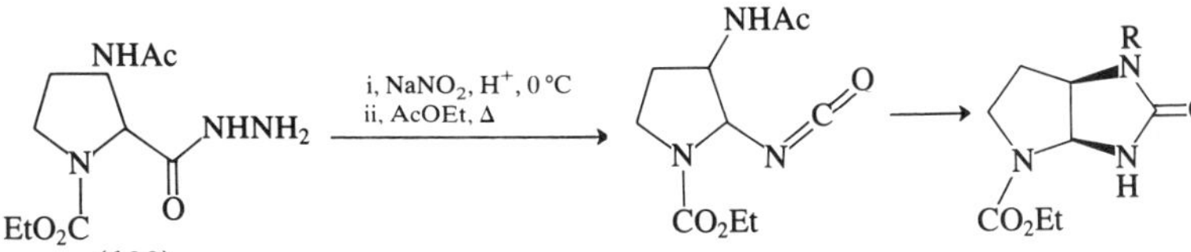

Carbon disulfide, cyanogen bromide and orthoesters are effective cyclization agents, although orthoesters usually require longer reaction periods. The reaction of 4-amino-s-triazole-3-thiols with carbon disulfide or cyanogen bromide is a particularly effective way of obtaining s-triazolo[3,4-b][1,3,4]thiadiazole-6-thiols (equation 25) ⟨66JOC3528⟩. These reagents have been used to prepare members of the thiazolo[2,3-c]-s-triazole ⟨71JOC10⟩, thiazolo[3,2-b]-s-triazole ⟨76JHC1225⟩ and s-triazolo[5,1-c]-s-triazole families ⟨68JOC143⟩.

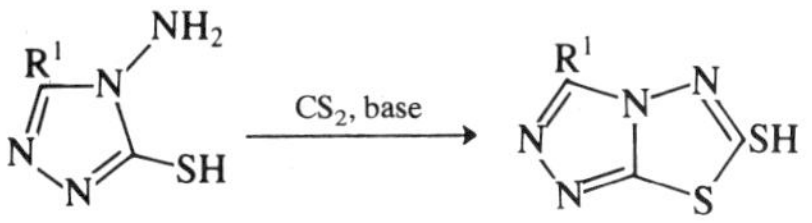

The ability of nucleophiles to add in 1,4-fashion to α,β-enones (Michael reaction) is of great importance for heterocyclizations. Esters of α,β-unsaturated acids generally undergo reaction with hydrazine to give pyrazolines. Treatment of the reagents in the molar ratio 2:1, at 140 °C, results in the formation of pyrazolo[1,2-a]pyrazole-1,5-diones (equation 26) ⟨65CB3228⟩. Members of the same ring system have been prepared from 4-methylpyrazolidine with ethyl crotonate or 3-methylpyrazolin-5-one with ethyl acetoacetate and PCl_3 or PBr_3 as the condensing agents ⟨57T201, 80JA4983⟩.

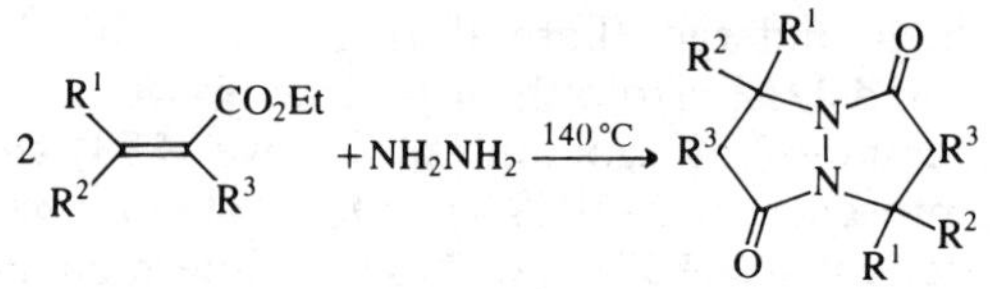

2-Thiohydantoins undergo ready condensation with α,β-unsaturated nitriles to give pyrrolo[1,2-c]imidazoles (equation 27). The same products were obtained when the corresponding 5-methylene-2-thiohydantoins were treated with malononitrile ⟨82S502⟩. A likely pathway has been suggested ⟨81S531⟩.

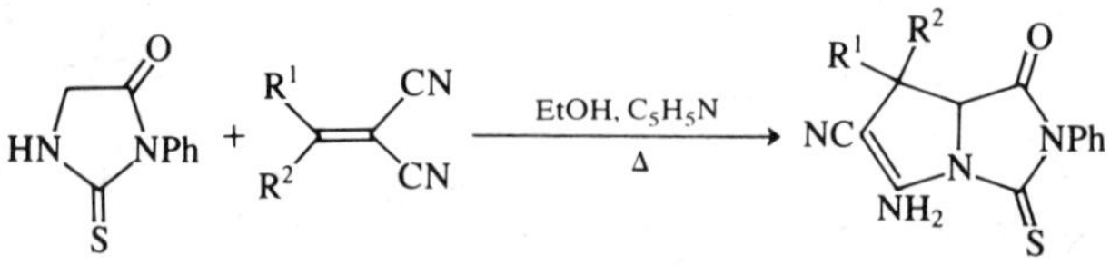

Treatment of the α-aminoaldehyde (**125**) with potassium thiocyanate gave 3H-imidazole-2-thione (**126**) (Marckwald synthesis), which produced the S-acetyl derivative of (**127**); hydrolysis gave (**127**) itself ⟨54JCS3283⟩.

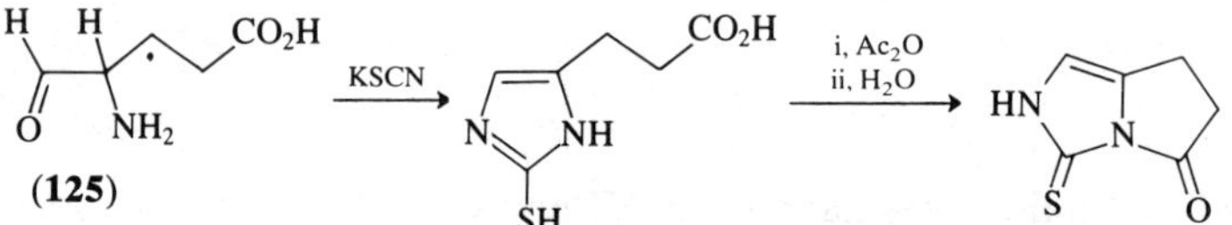

Five-membered N-heterocycles of general structure (**128**), carrying an α-XH grouping, react with α-halocarbonyl compounds to give quaternary salts (**129**) and (**131**) which readily undergo cyclodehydration to give (5,5)-fused pyrroles (**130**) and imidazoles (**132**). With (**128**; X = S) and α-halocarbonyl compounds, cationic-fused thiazoles (**134**) are obtained by treatment of the intermediates (**133**) with a strong acid (HA).

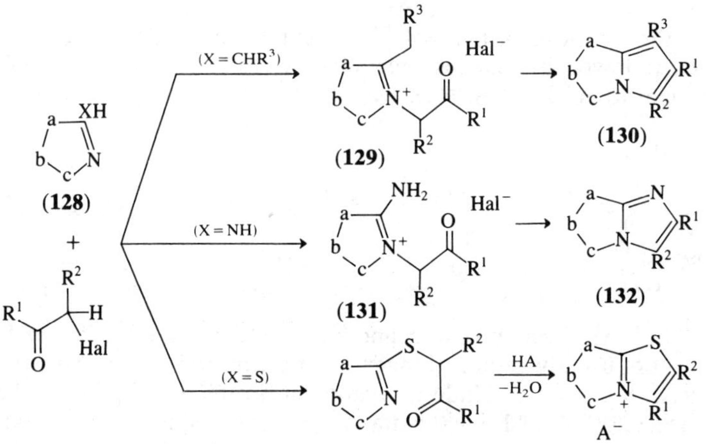

The method has been used to prepare pyrrolo[2,1-b]thiazoles, pyrrolo[1,2-a]imidazoles, pyrrolo[1,2-b]triazoles, 1H-pyrazolo[1,5-b]imidazoles, imidazo[2,1-b]thiazoles, imidazo[2,1-b][1,3,4]oxadiazoles, imidazo[2,1-b][1,3,4]thiadiazoles, imidazo[2,1-b]-[1,3,4]selenadiazoles, imidazo[1,2-d][1,2,4]thiadiazoles, imidazo[1,2-a]imidazoles, 1H-imidazo[1,2-d]tetrazoles and 1H-imidazo[1,2-b]-s-triazoles. The cationic-fused thiazoles (**134**) include 6,7-dihydro-5H-pyrrolo[2,1-b]thiazolium, thiazolo[2,3-b]thiazolium, 2,3-dihydrothiazolo[2,3-b]thiazolium, thiazolo[2,3-b]oxazolium, 2,3-dihydrothiazolo[2,3-b]oxazolium, 6,7-dihydro-5H-imidazo[2,1-b]thiazolium and thiazolo[3,2-d]tetrazolium. In general, oxazolium salts are less resistant toward alkaline hydrolysis than the corresponding thiazolium salts ⟨77HC(30)1⟩.

Reductive ring opening of isoxazoles (**135**) followed by cyclodehydration of the intermediates (**136**) offers an alternative route to imidazo[2,1-b]thiazoles (**137**) ⟨74JHC91⟩.

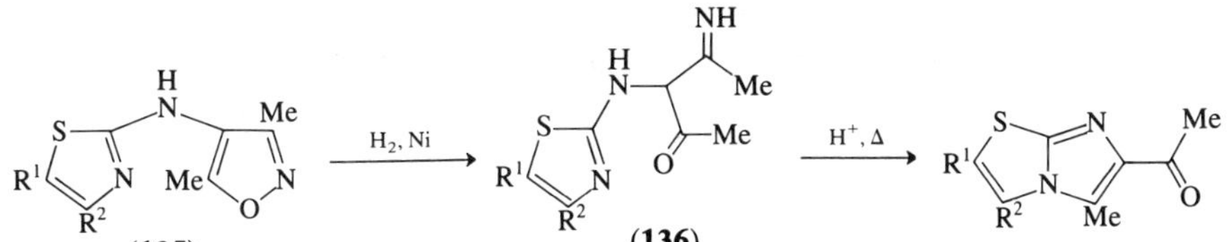

(135) (136) (137)

5,6-Dihydro- and 2,3,5,6-tetrahydro-imidazo[2,1-*b*]thiazoles have been prepared in large numbers by condensation of phenacyl bromides with thiazolines followed by acetylation, sodium borohydride reduction, and ring closure with thionyl chloride and acetic anhydride (Scheme 2) ⟨77JMC563, 66JMC545⟩.

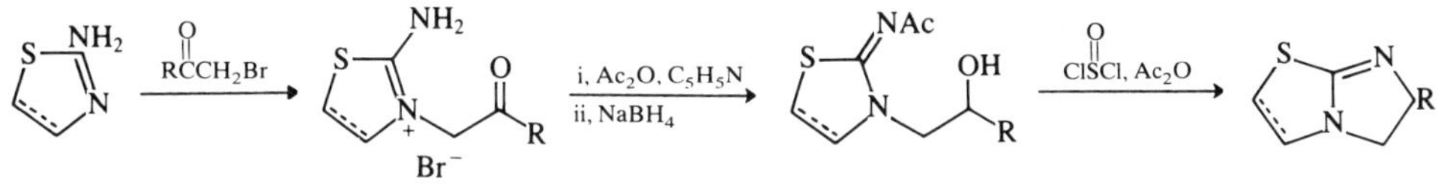

Scheme 2

The most commonly used method for the preparation of fused thiazoles involves the reaction of α-mercapto N-heterocyclic compounds of type (**138**) with an α-halocarbonyl compound or ester to give *S*-alkylated intermediates (**139**) which can be dehydrated to (**140**). When R^2 is alkoxy, thiazolones (**141**) are formed (Hantzsch synthesis). Strong dehydrating agents are necessary to cyclize aldehydes and ketones (**139**) to fused thiazoles. The method has been used to prepare (dihydro) imidazo[2,1-*b*]thiazoles and thiazolo[3,2-*b*]-*s*-triazoles ⟨80JHC1321, 78JHC401, 82IJC(B)243⟩.

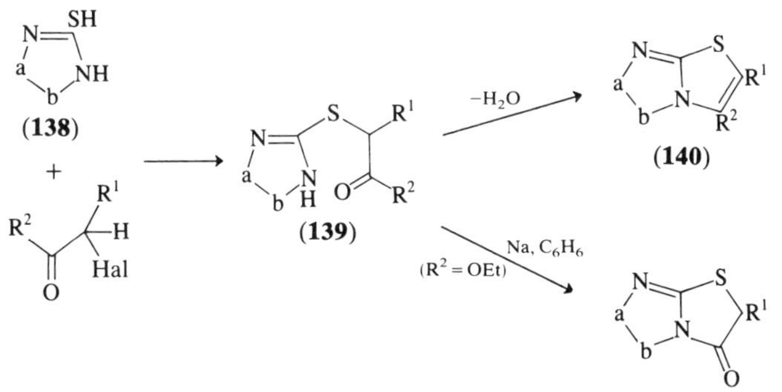

(138) (139) (140) (141)

The cyclo-condensation of an α-dicarbonyl compound with the NH group of a urea, the SH group of a thiourea or the NH group of a guanidine, in accordance with the principle of α-ureidoalkylation, provides a route to tetrahydroimidazo[4,5-*d*]imidazoles ⟨59CRV667, 73S243⟩. For example, the 4,5-dihydroxyimidazoline (**142**), most likely formed initially in the condensation of benzil with guanidine, reacts with a second equivalent of guanidine to give (**145**). Both (**143**) ⟨79JOC818⟩ and the dianion (**144**) ⟨63CB168⟩ have been suggested as intermediates.

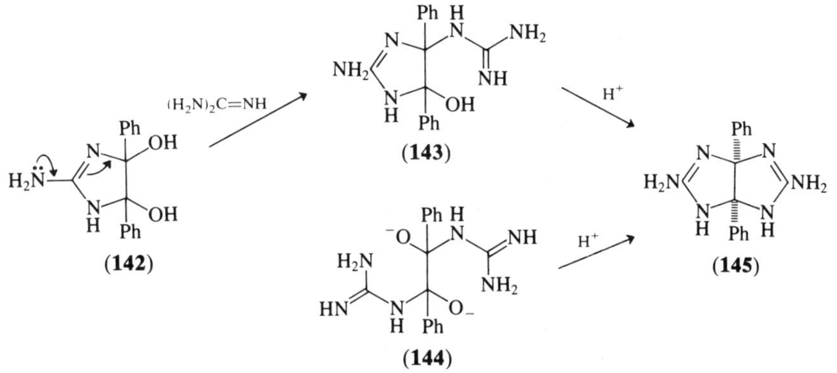

(142) (143) (144) (145)

Amidinium salts (**146**) when heated were initially thought to give imidazolines (**147**) which then underwent dehydrative cyclization to imidazo[4,5-*d*]imidazoles (**148**) ⟨72JHC1429⟩. This reaction sequence has been reinterpreted in terms of an initial triazine (**147a**) which with POCl₃ underwent dehydration to 2-cyano-4,6-diphenyl-1,3,5-triazine (**148a**) ⟨78JHC1055⟩. The triazine structure for this thermolysis product was initially ruled out because of a lack of infrared ν_{CN} absorption but it could be detected at higher concentrations. The ^{13}C NMR spectrum of (**148a**) showed δ_{CN} at 114.9 p.p.m., in confirmation of the revised structure.

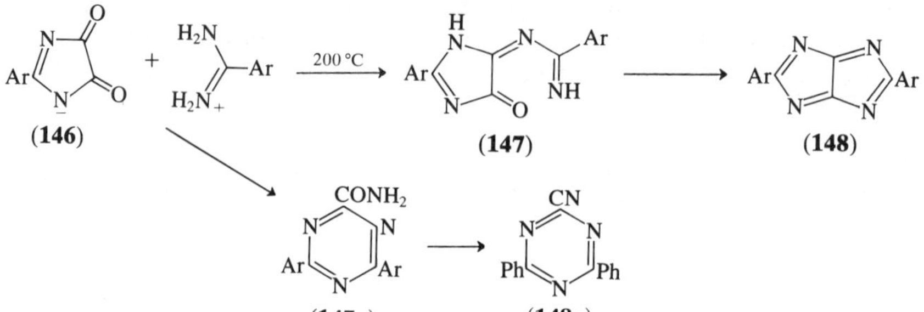

With methylurea (and urea) the main products of reaction with biacetyl are the bicyclic compounds (**152**) and (**153**). In the proposed mechanism the crucial intermediate is (**150**), formed by elimination of water in acid conditions from the diol (**149**). If (**150**) is protonated on the hydroxy group, a good leaving group (H₂O) is formed and displacement by a second methylurea molecule forms (**151**). Ring closure occurs by addition across the carbon–nitrogen double bond, probably by *N*-protonation and generation of a carbonium ion which reacts with the nucleophilic NH₂ group. Attack of protonated (**150**) by the other end of the methylurea molecule gives (**153**) ⟨81JCS(P2)310⟩.

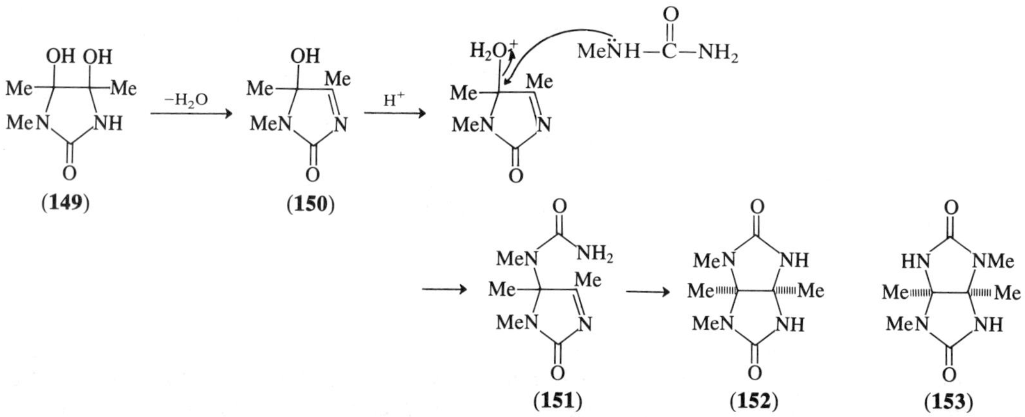

4.36.4.2.4 Formation of C—S bonds

Phosphorus pentasulfide (P₄S₁₀) is widely used as a thiating reagent for specific functional groups. The Gabriel reaction ⟨10CB1283⟩ of α-acylaminocarbonyl compounds with P₄S₁₀ has been used to prepare thieno[2,3-*d*]thiazoles ⟨70CHE1515⟩, pyrazolo[5,4-*d*]thiazoles ⟨74CHE813⟩ and pyrazolo[4,5-*d*]thiazoles ⟨65CHE165⟩ (equation 28).

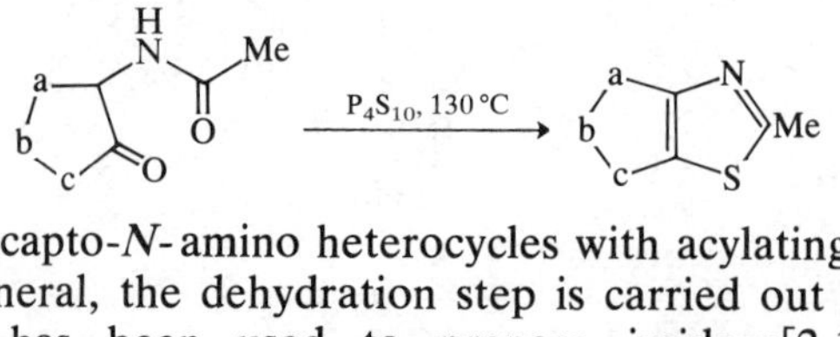

The reaction of α-mercapto-*N*-amino heterocycles with acylating agents leads to fused 1,3,4-thiadiazoles. In general, the dehydration step is carried out with phosphorus oxychloride. The method has been used to prepare imidazo[2,1-*b*][1,3,4]thiadiazoles ⟨63LA(663)113⟩ and s-triazolo[3,4-*b*][1,3,4]thiadiazoles ⟨64CI(L)1919⟩ (equation 29).

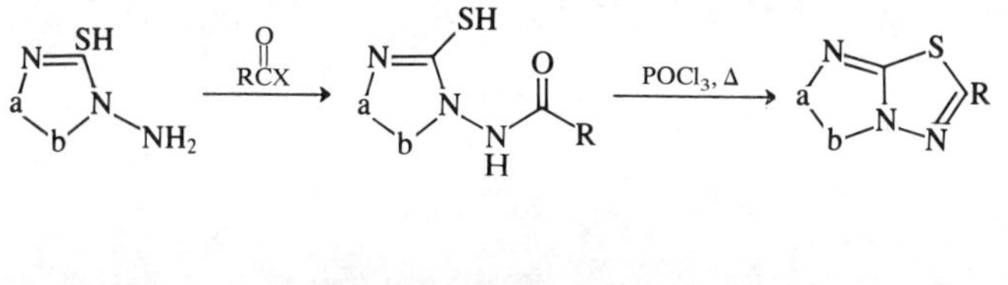

Methyl thiacyclopentylidene-3-cyanoacetate (**154**) reacts with sulfur and catalytic amounts of diethylamine to give either (**155**) or the thieno [2,3-*b*]thiophene derivative (**156**), depending upon the reaction temperature. Hydrazine hydrate reduction of (**155**) gives (**156**) ⟨73JPR39⟩. The reaction is a special case of the Gewald reaction ⟨66CB94⟩, which consists of the simultaneous interaction of a carbonyl compound, a methylene-activated nitrile and sulfur in the presence of a secondary or tertiary amine, leading to α-aminothiophenes (Scheme 3, path a). Alkylidenemalononitriles prepared from the corresponding carbonyl compounds and nitriles (Knoevenagel–Cope reaction) ⟨41JA3452⟩ react similarly with sulfur in the presence of amine (path b). The simultaneous interaction of carbonyl compound, nitrile and sulfur gives generally higher yields than the reaction of α-mercaptocarbonyl compounds with nitriles (Asinger reaction, path c) ⟨64AG(E)19⟩, supporting the view that in the Gewald reaction it is the alkylidenemalononitrile which is formed first, followed by the reversible thiolation and irreversible cyclization steps.

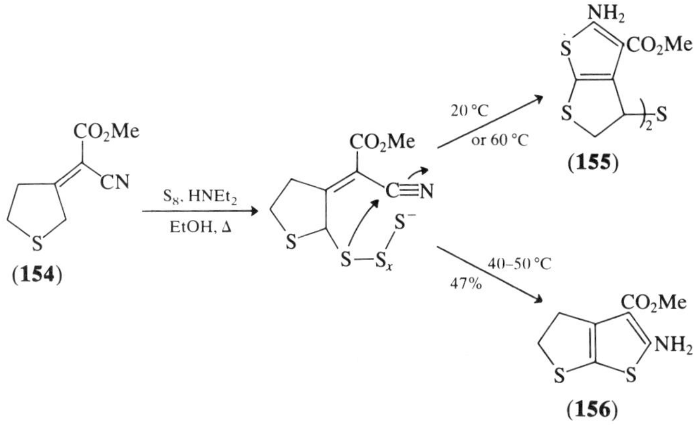

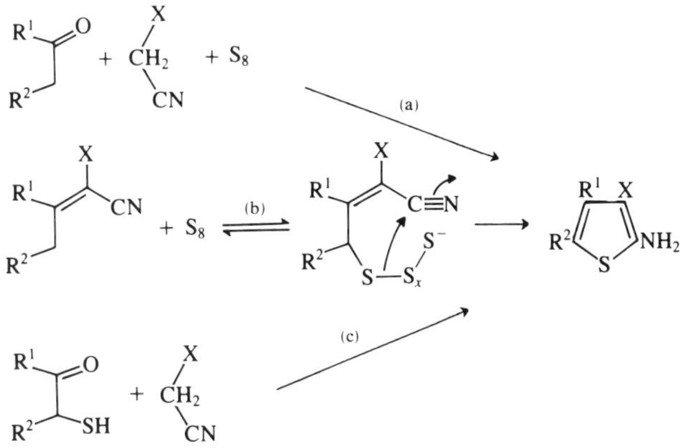

Scheme 3

Thiophene-2-acrylic acids (**157**) react with thionyl chloride in the presence of pyridine to form 3-chlorothieno[3,2-*b*]thiophene-2-carbonyl chlorides (**159**) ⟨72JHC879⟩. The initial step in this reaction appears to be the addition of $SOCl_2$ to the electron deficient alkene moiety of (**157**) to give (**158**). Pummerer-type rearrangement of (**158**) containing the π-electron accepting COCl group at the α-carbon atom may be the most probable reaction pathway leading to (**159**) ⟨75JOC3037⟩.

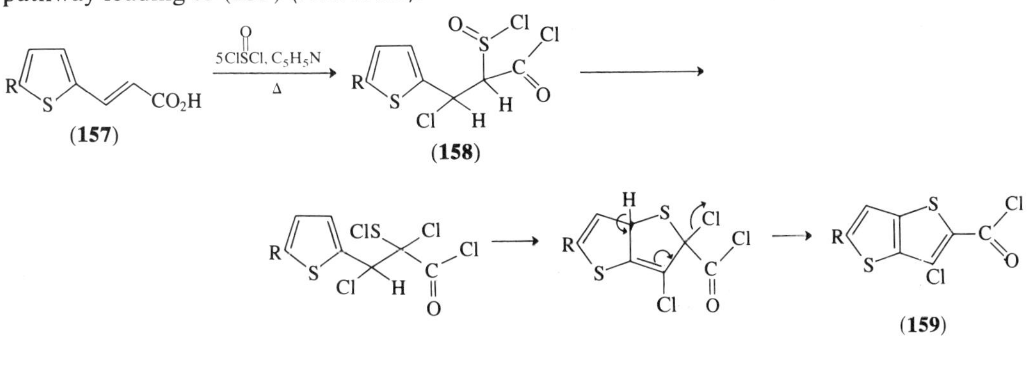

4.36.4.2.5 Formation of N—N bonds

There are several syntheses of fused *vic*-triazoles which depend on the treatment of 1,2-diamino-substituted heterocycles with sodium nitrite. The method has been used to prepare isoxazolo[3,4-*d*][1,2,3]triazoles, isothiazolo[5,4-*b*][1,2,3]triazoles ⟨72AHC(14)36⟩ and pyrazolo[3,4-*d*][1,2,3]triazoles ⟨10LA(375)297, 75JHC279⟩ (equation 30).

$$\text{(30)}$$

Although the elaboration of the 2-phenyl-2*H*-triazole ring by the oxidative elimination of aniline from osazones in carbohydrates is well documented ⟨44JOC470⟩, the first example of a cyclodehydration of readily available tetronic acid osazones was only recently reported with the conversion of dehydroascorbic acid into (**160**; $R^1 = H$, $R^2 = CH(OH)CH_2OH$) and its subsequent cyclization to the corresponding 2,6-dihydrofuro[3,4-*d*][1,2,3]triazole ⟨71MI43600⟩. Generalization of the above procedure to 3-unsubstituted tetronic acids by sequential coupling with benzenediazonium sulfate, oximation and cyclodehydration of (**160**) in acetic anhydride afforded furo[3,4-*d*][1,2,3]triazoles (**161**) ⟨79S977⟩.

The reaction of 4-phenyl-2-ethoxy-Δ^1-pyrroline (**162**) with sodium azide in glacial acetic acid at 60 °C gave 6-phenyl-6,7-dihydro-5*H*-pyrrolo[1,2-*a*]tetrazole (**163**) ⟨82JHC193⟩.

4.36.4.3 Bimolecular Dipolar Cycloadditions

1,3-Dipolar addition reactions of suitably substituted 1,3-dipoles to dipolarophiles offer a wide range of utility in the synthesis of fused heterocyclic ring systems ⟨63AG(E)565⟩. The 1,3-dipole combines in a cycloaddition with a multiple bond system (the dipolarophile) to form an uncharged five-membered ring. In most instances of 1,3-dipolar cycloaddition reactions when two isomers are possible as a result of the use of unsymmetrical reagents, one isomer usually predominates, often to the exclusion of the other. The stereospecificity and regioselectivity point to a highly ordered transition state. It is only within the last 10 years that a better understanding of the phenomena reactivity, stereoselectivity and regioselectivity has begun to emerge. The preferential formation of certain regioisomers may appear obvious in many cases by considering electronic and steric effects, but a consistent explanation of the orientation has been made possible by applying MO perturbation theory which predicts bonding between the atoms with the larger and smaller coefficients, respectively ⟨B-76MI43600, B-77MI43600⟩. Valence-bond formulae can be very helpful for understanding reactivity. Indeed, the weights of octet zwitterionic structures are directly connected to the ambident nucleophilicity of some 1,3-dipoles ⟨76JOC403⟩. Qualitative interpretation of minimal and extended basis set valence-bond functions support the importance of the diradical structures in alkyl-like dipoles. Propargyl–allenyl-like dipoles are best described as zwitterions ⟨82JA66⟩.

4.36.4.3.1 1,3-Dipoles of the propargyl–allenyl anion type

(i) *Nitrile oxides*

1,3-Dipolar cycloaddition of nitrile oxides to unsaturated five-membered carbocyclic and heterocyclic systems gives fused 1,2-oxazolines. Steric effects are apparently of far greater importance than electronic ones, especially with regard to the orientation of the dipolarophile towards the nitrile oxide in the cycloaddition reaction. Originally, only one orientation (164) was believed to play a role, but recent investigations have demonstrated that in several cases the 'inverse' addition product (165) may also be observed ⟨77CHE524⟩.

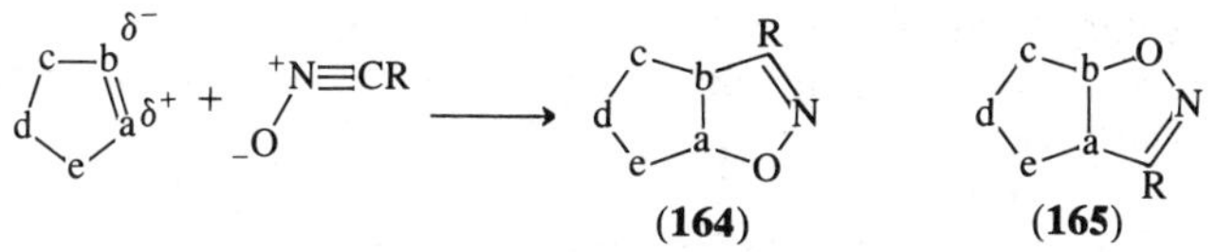

Dipolarophiles containing heteroatoms are often less reactive than C=C unsaturated analogues ⟨B-71MI43602⟩. The orientation in this case is unambiguous and obeys the principle of maximum gain in σ-bond energy, viz. the reactants join in that direction that better allows compensation of the π-bond energy lost by the energy of the two new σ-bonds formed. Thus the electronegative end of the heterodipolarophile becomes joined to the carbon atom of the nitrile oxide.

Beginning in 1946 when it was shown that treatment of acid chlorides of hydroxamic acids with alkali hydroxide gave nitrile oxides ⟨46G148⟩, numerous isoxazole derivatives have been prepared. This method lead to isoxazolo[4,5-d]isoxazoles (equation 31) ⟨59G571⟩.

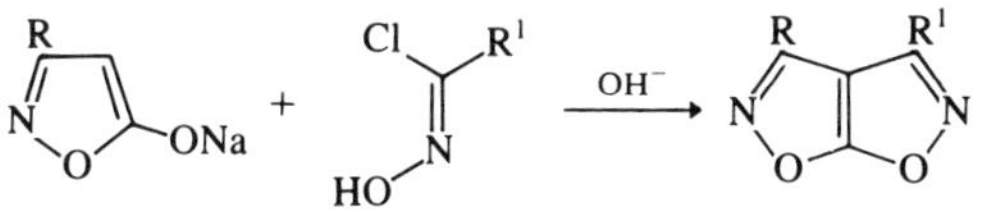

Exposure of ethylene to an ethereal solution of formonitrile oxide gave the unstable 3-anti-aldoximino-2-oxazoline which was converted into the syn-oxime by ethanolic hydrogen chloride. The syn-oxime undergoes base-catalyzed cyclization to give Δ^3-isoxazolidino[2,3-b][1,2,5]oxadiazoline (equation 32) ⟨74JHC63⟩.

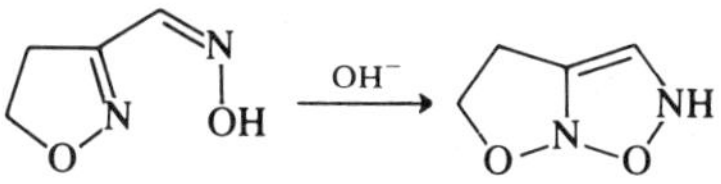

With 2,3-(and 2,5-)dihydrofuran and nitrile oxides, the cycloadducts formed are furo[3,2-d]isoxazoles and furo[3,4-d]isoxazoles, respectively. Whereas the cycloaddition of nitrile oxides with open-chain α,β-unsaturated esters often yielded a mixture of structural isomeric 2,5-dihydro-1,2-oxazoles, the reaction with butenolides always led to the 4-oxo derivative ⟨70S365⟩. The double bonds of furan, pyrrole and thiophene do not react with benzonitrile oxide (BNO) under standard conditions. However, cycloaddition can be assisted by treating the heterocyclic compound with a nitrile oxide generated in situ. Furan, for example, yields as main product either (166), arising from attack of the electrophilic carbon of BNO on the α-position of furan, or the bis-adduct in which a second equivalent of BNO has added to the vinyl ether bond of (166) ⟨66TL2911⟩. 2-Methyl-1,3-oxazoline reacts with BNO under mild conditions to give the oxazolo[3,2-d][1,2,4]oxadiazole (167) ⟨71AG(E)810⟩. The reaction of BNO with 1-phenyl-1-oxo-3-phospholene gave phospholo[3,4-d]isoxazole (168) ⟨67DOK1075⟩. Cycloaddition of p-chlorobenzonitrile oxide (ClBNO) takes place only at the C=P bond of 2-acetyl-5-methyl-1,2,3-diazaphosphole to yield the regioisomer (169).

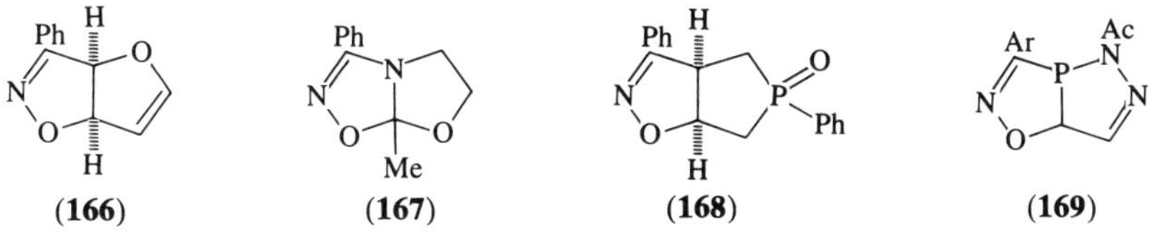

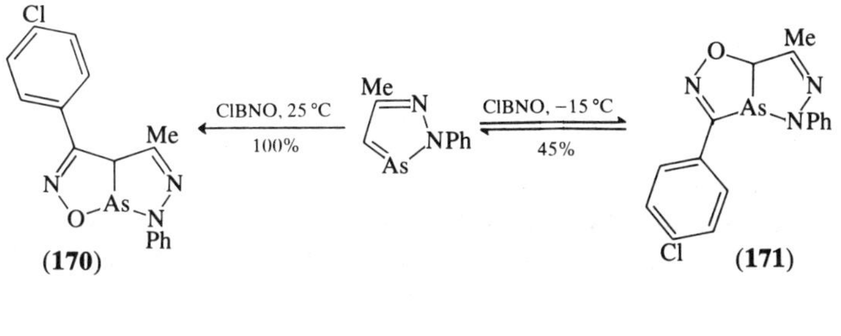

Treatment of the same nitrile oxide with 5-methyl-2-phenyl-1,2,3-diazaarsole in toluene at room temperature afforded quantitatively the stable regioisomer (170). However, the regioisomer (171) was produced by carrying out the same reaction below −15 °C ⟨81CC1131⟩. In toluene solution at room temperature the adduct (171) (kinetic product) is converted quantitatively into the regioisomer (170) (thermodynamic product). The occurrence of reversibility in 1,3-dipolar cycloaddition reactions indicates that care must be taken in the interpretation of experimental results by perturbation theory in order to make sure that one is dealing with the kinetic rather than the thermodynamic product.

Pyrazolidino[3,4-d]isoxazoles, separable by silica chromatography, are formed stereoselectively and stereospecifically from 3-pyrazolines and BNO ⟨77JHC253⟩. 2-Pyrazolines reacted similarly with BNO to give stereoisomeric tetrahydropyrazolo[4,5-b]-[1,2,4]oxadiazoles ⟨79JHC311⟩.

From 3,4-dibromotetrahydrothiophene 1,1-dioxide in the presence of pyridine and mesitonitrile oxide is obtained 4,4-dioxo-3-(2,4,6-trimethylphenyl)-3a,6a-dihydrothieno[2,3-d]isoxazoline (172). A second equivalent of nitrile oxide leads to the formation of the 2:1 adduct (173). NMR analysis and crystallographic studies show the formation of the adduct where the regioselectivity corresponds to the oxygen atom of the dipoles bonded to the carbon atom β to the sulfone group, the 'endo' nature of the addition, and the anti situation of the two rings in the diadduct (173) ⟨81JOC3502⟩.

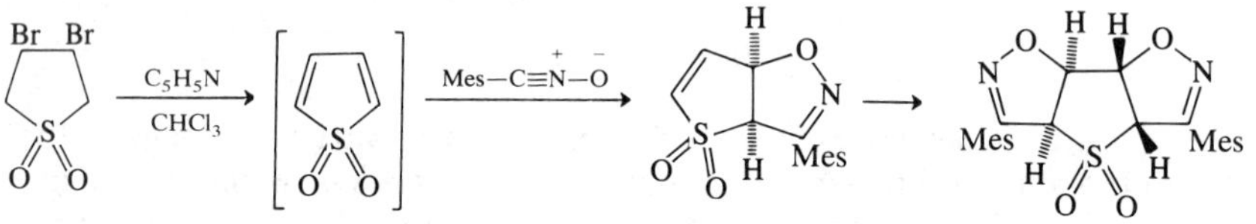

(ii) *Nitrile imines*

The 1,3-dipolar cycloaddition of nitrile imines to unsaturated five-membered ring systems is a versatile method for the stereoselective and regioselective synthesis of fused 2-pyrazolines and pyrazoles ⟨62T3, 81JCR(S)135⟩. Both furan and 2,5-dihydrofurans enter into cycloaddition reactions with nitrile imines generated from hydrazidoyl halides and triethylamine ⟨80JHC833⟩ to form derivatives of furo[2,3-d]pyrazoles ⟨68TL743⟩ and furo[3,4-d]pyrazoles ⟨71CHE139⟩, respectively. The electrophilic carbon atom of the nitrile imine attacks the α-position of the furan ring, thus controlling the orientation of the cycloaddition.

Cycloaddition to the diazaphosphole (174) involves the P=C bond, as evidenced by nitrile imine addition resulting in the formation of 3H-[1,2,3]diazaphospholo[4,3-c][1,2,4]diazaphospholes (175) and (176) ⟨B-81MI43600⟩.

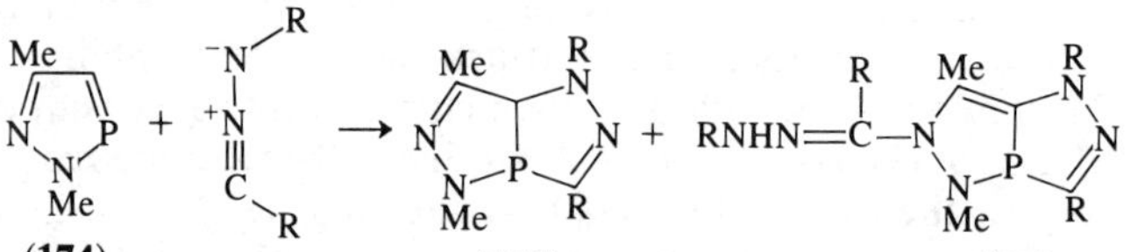

Addition reactions of 2-substituted N-methylpyrroles with C-acetyl-N-phenylnitrilimine gave double cycloaddition products, spiro-cycloadducts and noncyclic bis-adducts ⟨78JHC293⟩.

C,N-Diarylnitrile imines react with carbonyl-stabilized sulfonium ylides to give furo[2,3-d]pyrazoles (Scheme 4) ⟨70TL605⟩.

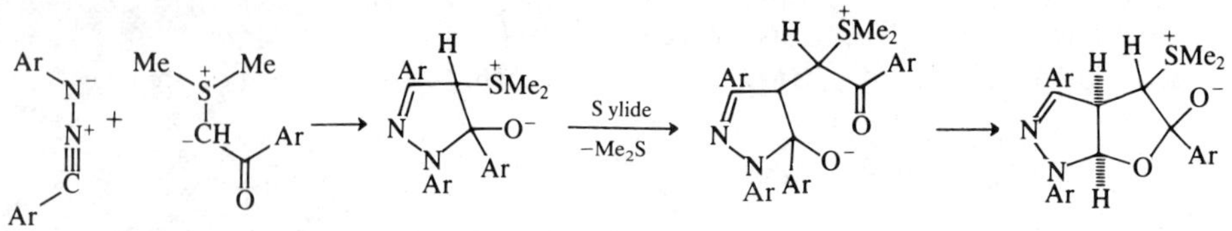

Scheme 4

(iii) *Nitrile sulfides*

Whereas nitrile ylides, nitrile imines and nitrile oxides have all been utilized frequently in 1,3-dipolar cycloaddition reactions to form heterocycles ⟨63AG(E)565⟩, nitrile sulfides were missing from this series of 1,3-dipoles until 1970 ⟨70TL1381⟩. In addition to thermolysis ⟨78JOC3736⟩ and photolysis ⟨75JA6197⟩ of 1,3,4-oxathiazol-2-one (**177**), nitrile sulfides have been generated by photolysis of 4-phenyl-1,3,2-oxathiazolylium-5-olate (**178**) ⟨72CB188⟩, thermolysis of (*N*-benzylimino)sulfur difluoride (**179**) in the presence of sodium fluoride ⟨75TL1739⟩, decomposition of the 1,3,4-oxathiazoline (**180**) at room temperature ⟨73JA279⟩ and thermolysis of *N*-thiocarbonylsulfimides (**181**) at 50 °C ⟨76BCJ3124⟩ (see also Chapters 4.17 and 4.34).

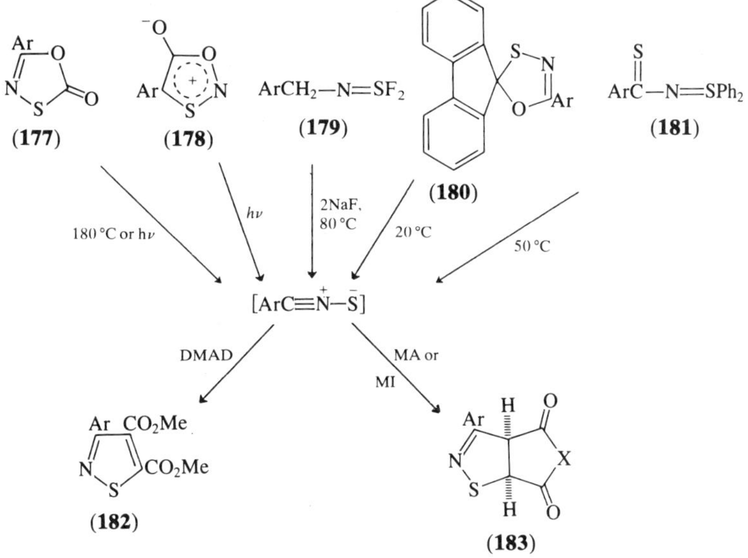

For example, thermolysis of (**177**) in the presence of DMAD resulted in formation of the dimethyl isothiazoledicarboxylate (**182**) (>90% yield). Benzonitrile sulfides, generated by the rare 1,3-elimination of two equivalents of hydrogen fluoride from (**179**), reacted with maleic anhydride (MA) to give dihydrofuro[3,4-*d*]isothiazole-4,6-diones (**183**; X = O).

(iv) *Diazoalkanes*

The adducts obtained from diazoalkanes with dipolarophiles are generally unstable and lose nitrogen at low temperatures, leading to three-membered ring compounds (see Section 4.36.3.4.3). The mechanism of this synthetic route to three-membered rings *via* pyrazoline intermediates must be carefully distinguished from that of the addition of carbenes to unsaturated compounds ⟨63AG(E)565⟩. 2-Methoxycarbonyl-1-methyl-2-phospholene oxide is also reactive to 1,3-dipoles, as evidenced by its condensation with ethyl diazoacetate leading to phospholo[2,3-*c*]pyrazole derivatives (equation 33) ⟨72TL299⟩.

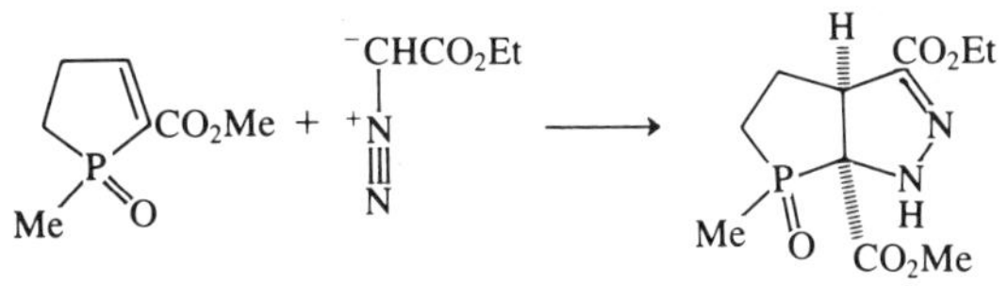

2-(Dichloroacetamido)-1-methyl-5-nitroimidazole undergoes a 1,3-dipolar cycloaddition reaction with diazomethane to give (dichloroacetimino)tetrahydroimidazo[4,5-*c*]pyrazoles in about 10% each (Scheme 5). The failure to observe cycloaddition with 1,2-dimethyl-5-nitroimidazole underlines the role of the *N*-methylated imidazole intermediate (**184**), a typical non-aromatic α, β-unsaturated nitro compound acting as a dipolarophile ⟨80TL4757⟩.

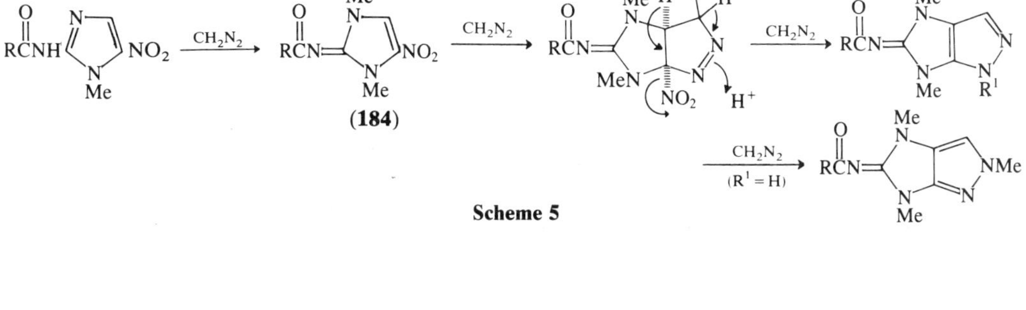

Scheme 5

Reaction of 4-chloro-2-pyrazolin-5-ones with potassium carbonate proceeds along two pathways, leading to diketo derivatives of pyrazolo[1,2-a]pyrazole (*syn-* and *anti-* 'dioxabimanes'). A mechanism involving the intermediate formation of diazacyclopentadienone (**185**) which reacts with its potential extended diazoalkane functionality in dipolar cycloaddition fashion to give *syn*-dioxabimanes (major product) and *anti*-dioxabimanes (minor product) has been proposed (Scheme 6) ⟨80JA4983⟩.

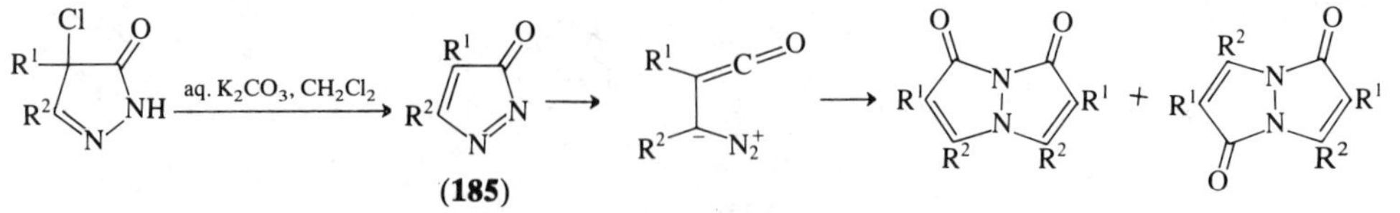

Scheme 6

(v) *Azides*

Most of the important general methods for forming 1*H*-1,2,3-triazole derivatives involve the thermal 1,3-dipolar cycloaddition of hydrazoic acid or organic azides to suitable dipolarophiles ⟨69CRV345, 74AHC(16)33⟩. For example, the cycloaddition of aryl azides to 1-alkyl-2,5-dihydropyrroles gives pyrrolo[3,4-*d*][1,2,3]triazoles (equation 34) ⟨80JHC267⟩.

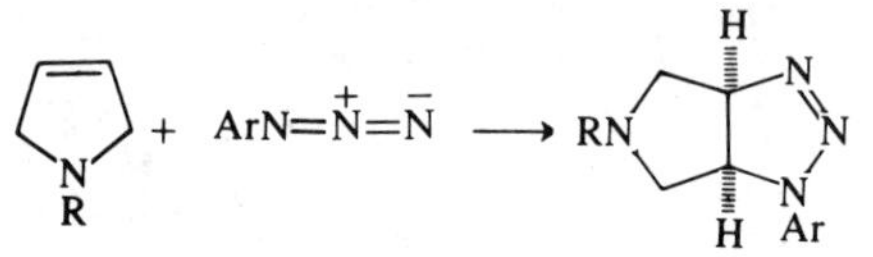

(34)

4.36.4.3.2 1,3-Dipoles of the allyl anion type

(i) *Azomethine oxides (nitrones)*

Although nitrones provide only one member of an extensive series of 1,3-dipoles, they can be considered of special interest because they are simple derivatives of carbonyl compounds. There are two main aspects of the synthetic utility of nitrone cycloaddition reactions. The first is the construction of a variety of five-membered heterocyclic ring systems. The second is the extension of carbonyl chemistry, and depends on the decomposition of a cycloadduct so as to perform a useful transformation of a carbonyl group. In general, cycloaddition reactions only occur readily with aldo-nitrones, but some of the more powerful dipolarophiles also add to keto-nitrones. Acyclic nitrones are generally less reactive than aliphatic cyclic analogues in cycloaddition reactions ⟨63AG(E)565, 65QR329, 64CRV473, 75S205, 78AHC(21)207⟩.

The most interesting aspect of these cycloadditions is their high regioselectivity. However, the direction of cycloaddition may well be subject to kinetic and thermodynamic control. For example, when 2,2-dimethylpyrrolidine *N*-oxide and ethyl acrylate were mixed at room temperature (thermodynamic control), the pyrrolidinoisoxazolidine (**187**) was formed quantitatively. When the mixture of the two reagents was heated at 100 °C, the kinetically controlled 1:1 cycloadduct (**186**) was obtained in 98% yield ⟨63JCS4693⟩.

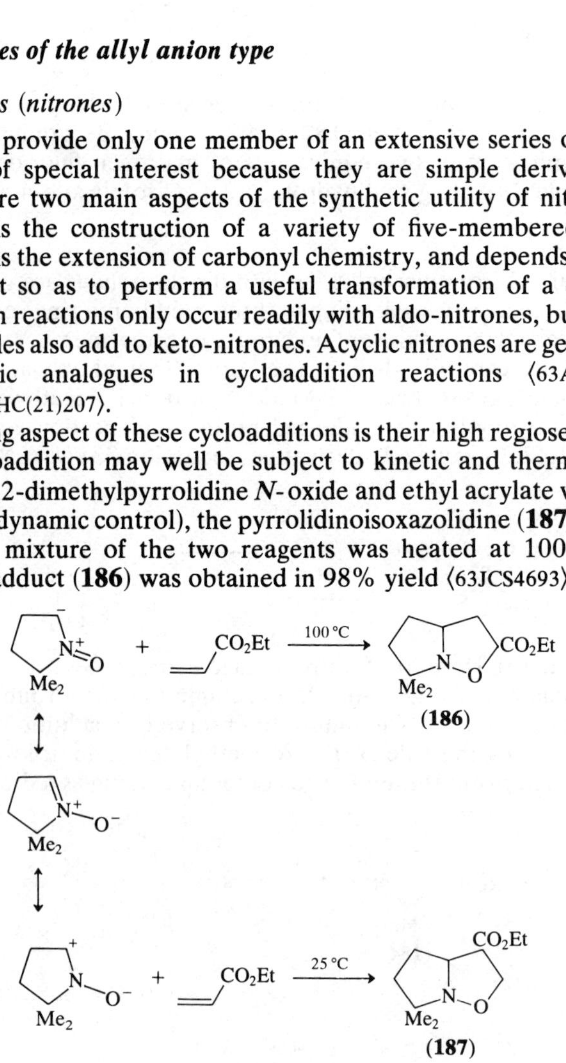

Similar reactions have been carried out with variously substituted pyrroline 1-oxides, imidazole 1-oxides, isoxazoline N-oxides (nitronic esters) and 3,4-diazacyclopentadienone N-oxides in combination with a large variety of alkenic and alkynic dipolarophiles, aryl isocyanates, aryl isothiocyanates and N-sulfinylamines, leading to pyrrolidinoisoxazoles, pyrrolo[1,2-b][1,2,4]oxadiazoles, pyrrolo[2,1-d][1,2,3,5]oxathiadiazoles, isoxazolo[2,3-b]isoxazoles and isoxazolo[1,2-b]pyrazoles.

The reaction of 5,5-dimethyl-1-pyrroline 1-oxide with dimethylketene N-phenylimine leading to pyrrolo[1,2-a]imidazol-2(3H)-ones (**190**) proceeds *via* initial formation of zwitterion (**188**). Subsequent sigmatropic rearrangement of (**188**) gives zwitterion (**189**), capable of undergoing ring closure to (**190**) ⟨79JOC4543⟩.

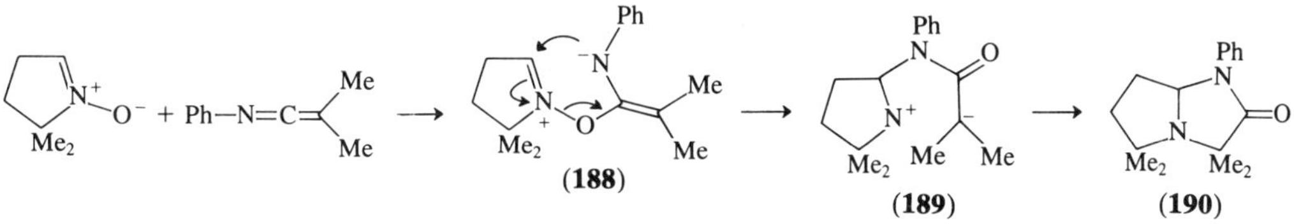

Whereas diarylnitrones, N-alkoxynitrones and C-aroyl-N-arylnitrones react exothermically with most alkenes and alkynes, the cycloadditions of diphenylnitrone with 2,5-diphenyl-1,2,3-diazaarsole and 1,1-diphenylsilacyclopent-3-ene, leading to the 1,2,3-diazaarsolo[3,4-d][1,2,4]oxazaarsole (equation 35) ⟨79DOK(246)1130⟩ and silacyclopenta[3,4-d]isoxazole derivatives (equation 36) ⟨80JGU1638⟩, respectively, proceeded very sluggishly (several months in THF).

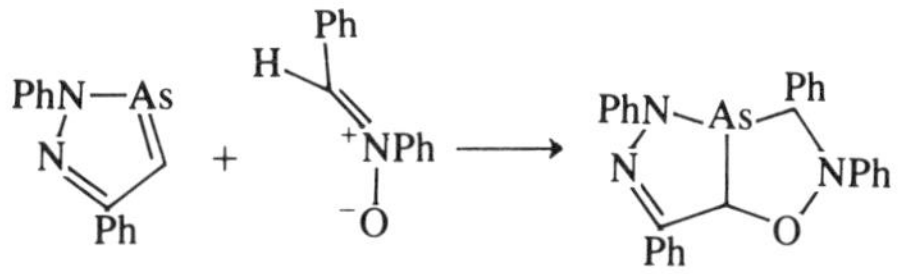

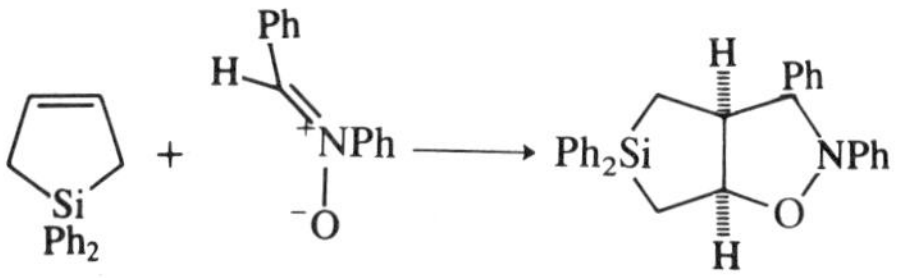

(ii) *Azomethine imines*

The 1,3-dipolar cycloaddition reaction of azomethine imines and sydnones leading to pyrazolidines, pyrazolines and pyrazoles is applicable to a wide variety of C—C unsaturated compounds and imines ⟨73S469⟩. For example, azomethine imines, readily prepared by heating diaziridines substituted in the 1- and 2-positions with electron-withdrawing groups or from 3-pyrazolidinones with carbonyl compounds, undergo reaction with phenyl isocyanate and isothiocyanate to give pyrazolo[1,2-a][1,2,4]triazoles. With azomethine imines and DMAD, pyrazolo[1,2-a]pyrazoles are formed (Scheme 7) ⟨68CB3287, 70JPR161⟩.

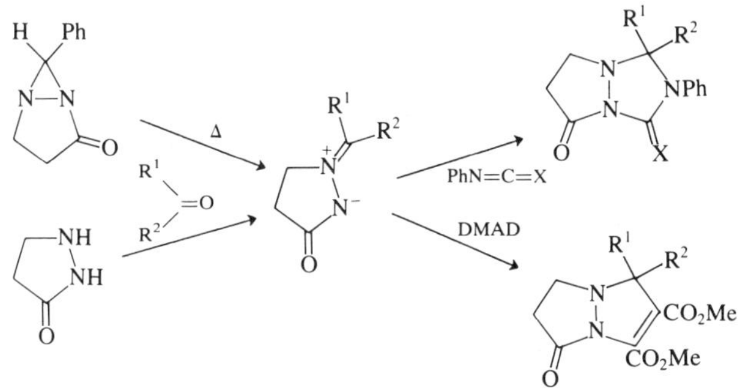

Scheme 7

1-(Diphenylmethylene)-3-oxo-1,2-diazetidinium inner salts react with DMAD to give fused aza-β-lactams. Extrusion of CO leads to new azomethine imines which can be trapped by a second equivalent of DMAD to give pyrazolo[1,2-a]pyrazoles in quantitative yield (Scheme 8) ⟨81JA7743⟩.

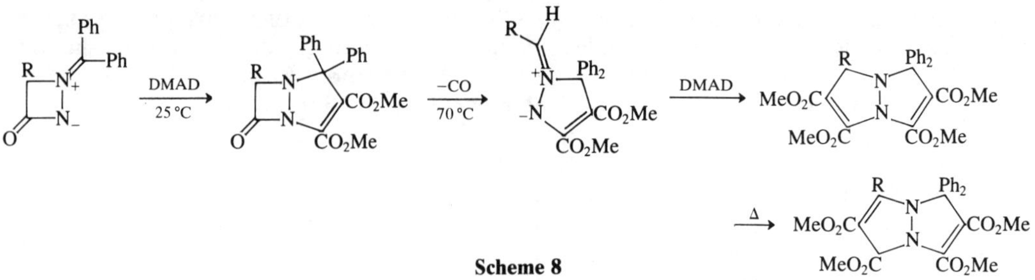

Scheme 8

The so-called criss-cross additions of azines involve a sequence of two 1,3-dipolar cycloaddition reactions, as exemplified by the reactions of hexafluoroacetone azine and benzaldazine with DMAD and phenyl isocyanate (Scheme 9) ⟨76S349⟩.

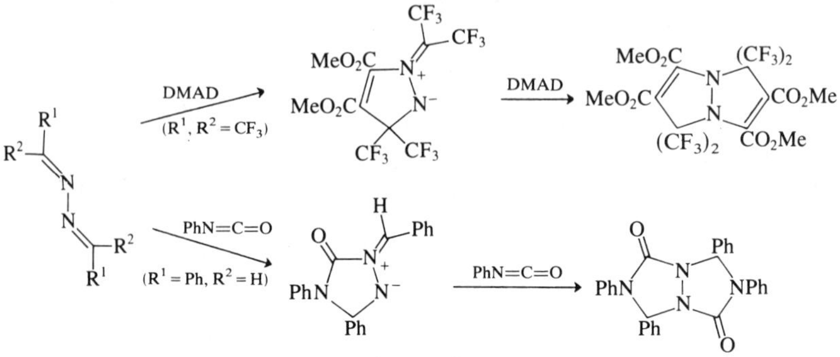

Scheme 9

The addition product of N-nitrosopyrrolidine with phenyllithium gives a compound which reacts with DMAD to give a pyrrolo[1,2-b]pyrazole, possibly via an intermediate azomethine imine (equation 37) ⟨73JOC4259⟩.

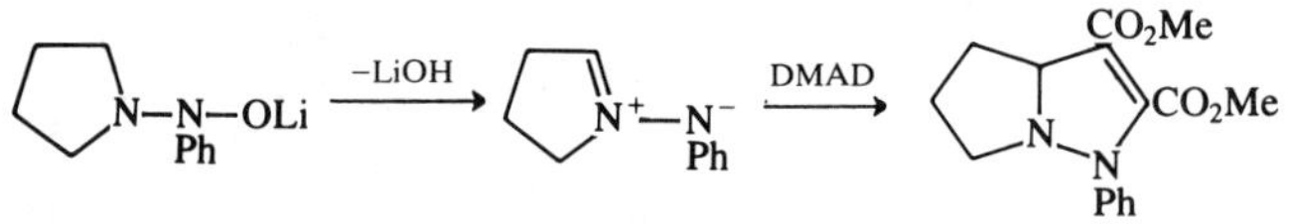

$$ \tag{37} $$

However, the reaction of the thiazolium N-iminoylide (**191**), generated *in situ* from the corresponding 3-aminothiazolium mesitylsulfonate and triethylamine, with DMAD gave the pyrazole (**192**) rather than the reported 7,9a-dihydrothiazolo[3,2-b][1,2]diazepine (**193**) ⟨74CPB482⟩. A 1:2 adduct was formed, but rearrangement had occurred during the reaction ⟨77JOC1648⟩.

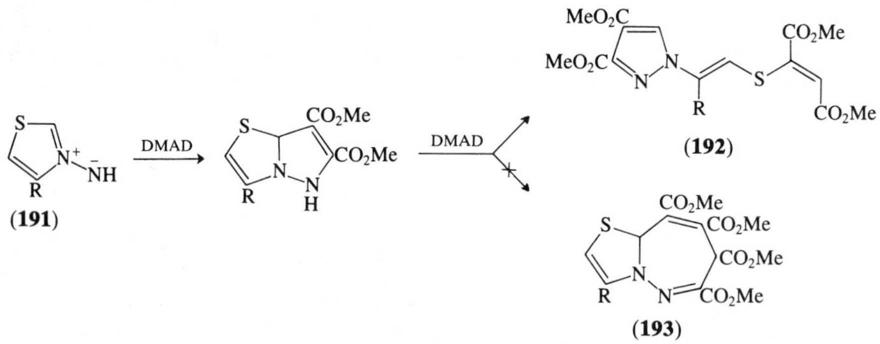

Lactones of azocarboxylic acids are remarkably reactive. In the presence of phenyl isocyanate, the imino isocyanate formed from reactive 2-hydrazono-Δ^3-1,3,4-oxadiazolines via a 1,3-dipolar cycloreversion is intercepted to give [1,2,4]triazolo[1,2-a]-[1,2,4]triazole-1,3,5-triones by means of two subsequent [2+3] cycloadditions via azomethine imine intermediates (Scheme 10) ⟨76T2685⟩.

Azomethine imines have been postulated as intermediates in the reaction of aryl-substituted diazo compounds with N-phenyltriazolinedione (PTAD) and DMAD, leading to pyrazolo[1,2-a][1,2,4]triazoles (Scheme 11) ⟨81TL2535⟩.

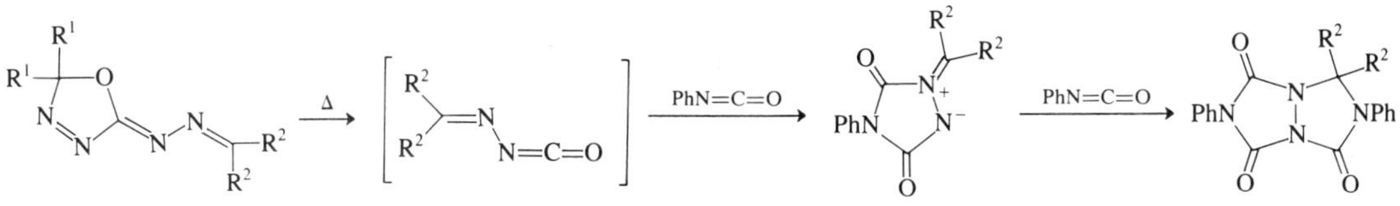

Scheme 10

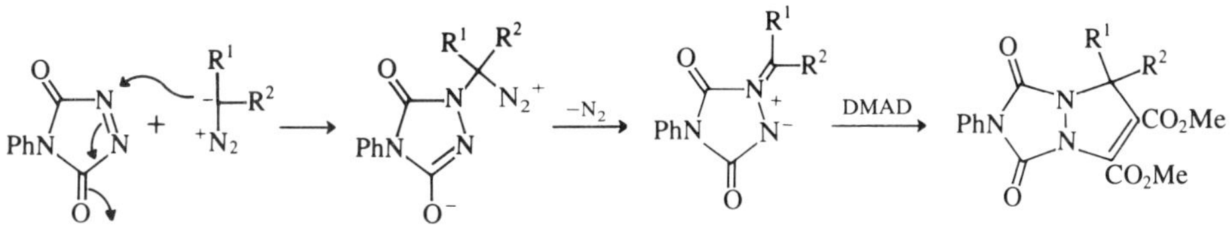

Scheme 11

(iii) Azomethine ylides

Compounds having a carbanion attached to a nitrogen atom bearing a positive charge undergo 1,3-cycloaddition reactions to give rise to a variety of heterocyclic compounds ⟨73S469⟩. Imidazoles quaternized with phenacyl bromide generated the corresponding imidazolium N-ylide when treated with potassium carbonate. The intermediate dihydropyrrolo[1,2-a]imidazole formed by reaction with ethyl propiolate (EP) is dehydrogenated by excess EP to give a pyrrolo[1,2-a]imidazole (equation 38) ⟨68JA3830⟩.

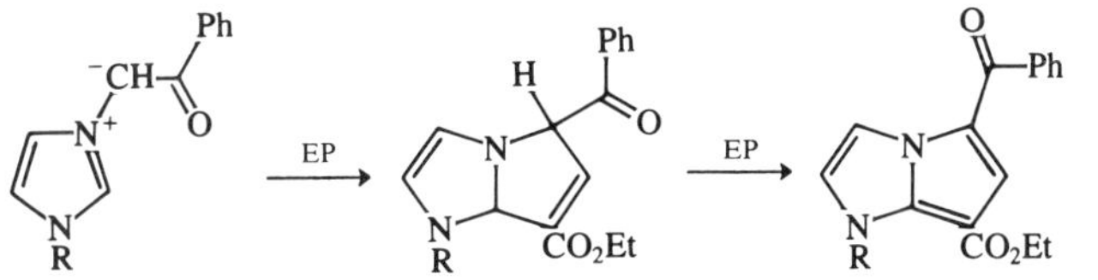

(38)

In contrast, condensation of thiazolium N-ylides (**194**) with dipolarophiles such as DMAD gives 1-hydroxy-4-methyl-1H-pyrrolo[2,1-c][1,4]thiazine-7,8-dicarboxylates (**196**). The intermediate vinyl sulfide formed from (**195**) by rotation and condensation at the carbonyl group gives rise to the hemithioacetal (**196**), the formation of the aromatic pyrrole ring providing the driving force for the rearrangement ⟨76JOC187⟩. With thiazolium ylides (**194**; R = CH_2CH_2OH) and α,β-unsaturated esters and nitriles, fused thiazolines (**197**) are formed which undergo conjugate cycloaddition to (**198**) when purified by silica chromatography ⟨81TL2727⟩.

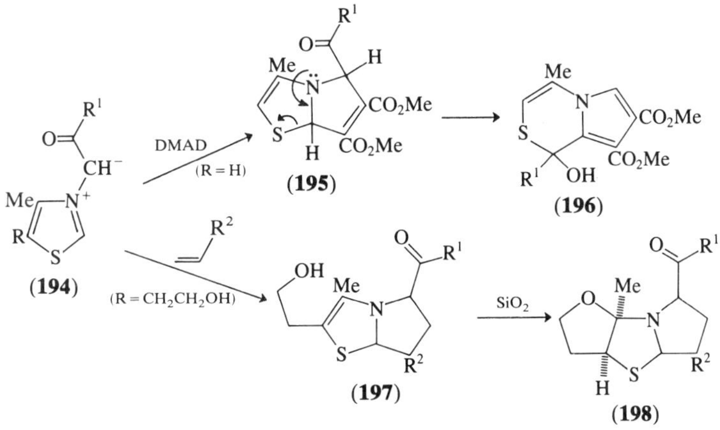

Thermolysis and photolysis of solutions of *endo*-2,4,6-triphenyl-1,3-diazabicyclo[3.1.0]hex-3-ene in benzene containing DMAD, diethyl fumarate or maleate gives pyrrolo[1,2-c]imidazole derivatives *via* azomethine ylide intermediates. Those from reaction with dimethyl fumarate are outlined in Scheme 12 ⟨73JOC284⟩.

(iv) Azimines

The cyclic azimine obtained by oxidation of benzil bis(phenylhydrazone) reacted with DMAD and EP to give 1H-pyrazolo[1,5-c][1,2,3]triazoles (equation 39) ⟨72T3987⟩.

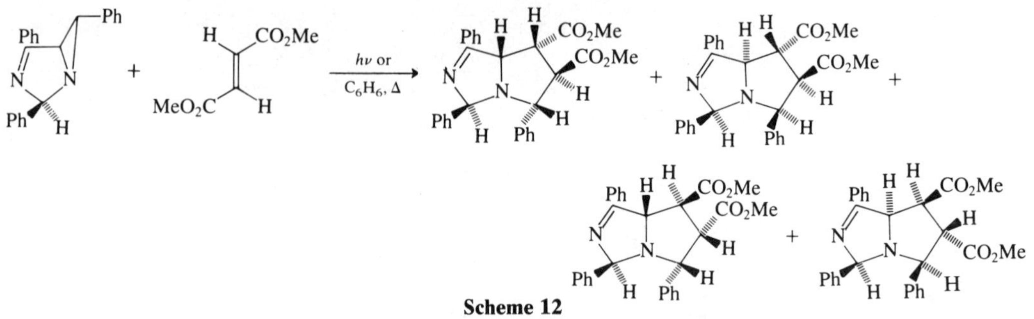

[Scheme 12 – structures]

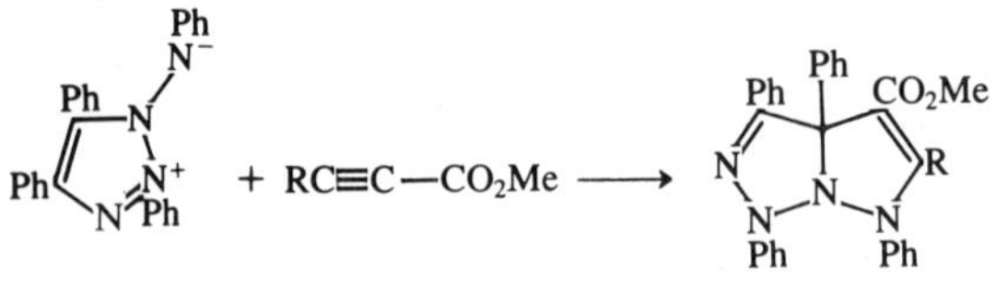

[Equation (39) – structures]

(v) Carbonyl ylides

Carbonyl ylides are not sufficiently stable to permit their isolation. An oxiranimine (iminolactone) (**200**) and a carbonyl ylide (**201**) have been postulated as reactive intermediates in the thermal reaction of the heterocyclic dione (**199**) with phenyl isocyanide, leading to 4,6a-diphenyl-6-(phenylimino)-6,6a-dihydro-1H-furo[3,4-b]pyrrole-2,3-dione (**202**) ⟨76LA2071, 80AG(E)276⟩.

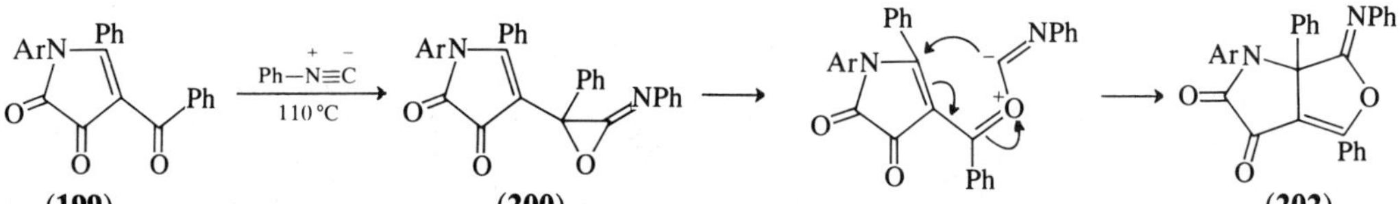

(vi) Thiocarbonyl ylides

Numerous structures containing the thiocarbonyl ylide dipole are conceivable. Incorporation of the thiocarbonyl ylide dipole into a bicyclic heterocyclic system is possible by the conversion of the cyclic thione (**203**) into the ring-fused mesoionic system (**204**). The thiocarbonyl ylide dipole (**205**) undergoes cycloaddition with both alkenic and alkynic electron-poor dipolarophiles in refluxing benzene or xylene so that, after extrusion of hydrogen sulfide or sulfur, respectively, from the initial 1:1 cycloadducts (**206**) and (**207**), a ring-fused pyridinone is formed. The method has been used for the annelation of pyridinones to the imidazole, 1,2,4-triazole, thiazole and 1,3,4-thiadiazole systems

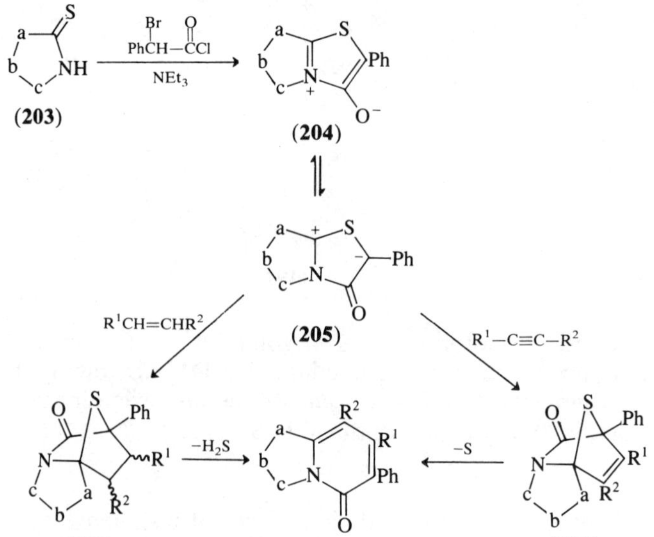

⟨79JOC3803⟩. Good yields of these condensed mesoionic thiazolones (**204**) are obtained by simply reacting stoichiometric amounts of 2,2-dicyanooxiranes with the cyclic thioamides (**203**) ⟨81S981⟩.

4.36.4.3.3 *1,3-Dipoles without a double bond*

Cyclic imidates (**208**; X = O) and thioimidates (**208**; X = S) react with ketene in sulfur dioxide to give oxazolo[2,3-*b*]thiazol-3(2*H*)-one 1,1-dioxide (**210**; X = O) and the corresponding thiazole analog. It appears possible that the reaction produces as the initial product the open-chain 1,3-dipole (**209**), which is not octet-stabilized ⟨69JHC729⟩.

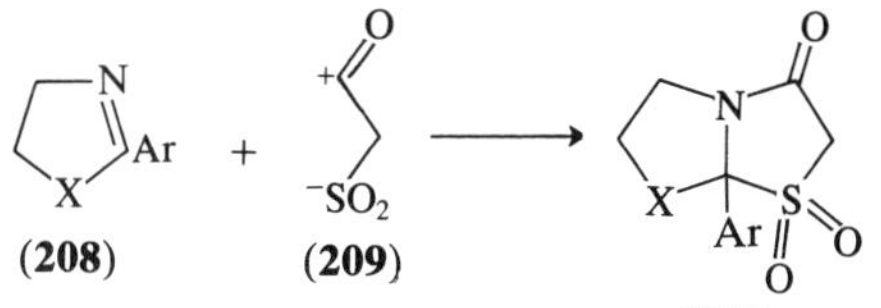

4.36.4.3.4 *Cycloadditions to electron-deficient alkenes and alkynes*

Alkynic esters participate as dipolarophiles in 1,3-dipolar cycloadditions and as dienophiles in Diels–Alder additions. They participate in [2+2] cycloadditions with alkenes, and they also undergo facile addition reactions with several nucleophiles to give (5,5)-fused heterocyclic ring systems ⟨76AHC(19)279⟩.

With dehydrodithizone and several types of alkenes and alkynes, the original assignments ⟨71CC490⟩ of the adducts were later revised and in many cases confirmed by X-ray crystallography ⟨76JCS(P1)1673, 77JCS(P1)1612⟩. In general, electron-rich multiple bond compounds give products derived from the acyclic valence-bond isomer (**211**) of dehydrodithizone, whereas electron-poor alkenes and alkynes yield initial 1,3-dipolar cycloadducts (**213**) and (**214**), respectively, of the mesoionic form (**212**). Both types of products contain the unknown tetrazoline structure, which is presumably destabilized by the presence of eight π-electrons. They undergo spontaneous ring-opening to give the 1,5-dipolar azimines (**215**) and (**216**). When the reaction of (**212**) with benzyne is carried out under mild conditions (70 °C), the azimine (**217**) is obtained. At 150 °C, 2-phenylazobenzothiazole (**218**) is formed as the result of the formal loss of phenylnitrene from (**217**) ⟨81JCS(P1)2692⟩. The initial cycloadduct (**219**) of *N*-ethylmaleimide and (**212**) apparently underwent oxidation and subsequent loss of phenylnitrene under the reaction conditions to give 5-ethyl-2-phenylazo-5*H*-pyrrolo[3,4-*d*]thiazole-4,6-dione (**220**) ⟨81JCS(P1)2692⟩.

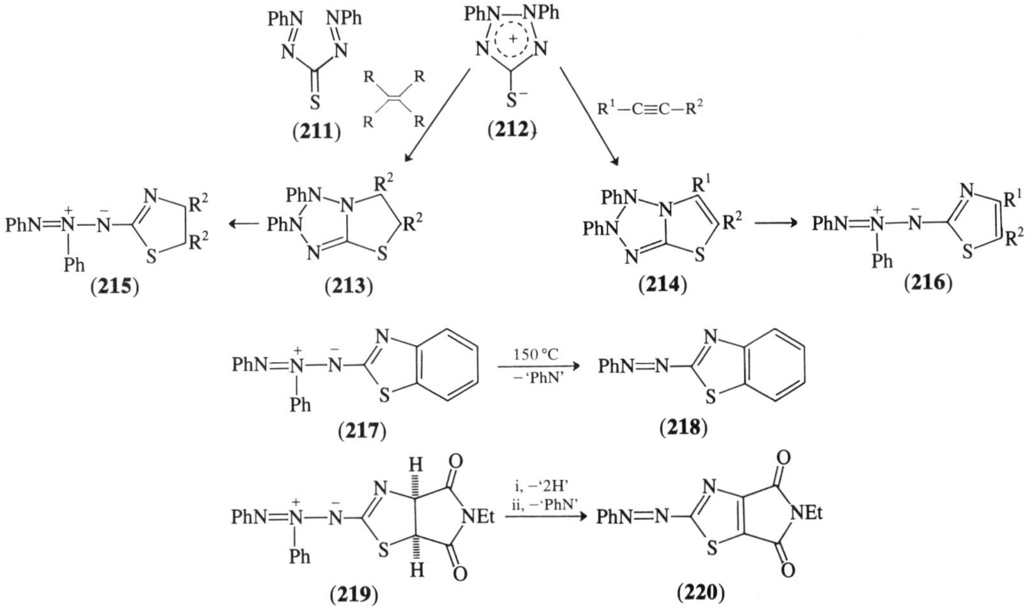

Thermal [2+2] cycloadditions can take place by either the concerted formation of the two σ-bonds or by a nonconcerted reaction *via* a diradical or 1,4-dipolar intermediate. Theoretical predictions about the mechanism are not unequivocal ⟨69AG(E)781, 74AG(E)751⟩. In protic solvents with DMAD and 3-(1-pyrrolidinyl)thiophenes (**221**) acting as pseudo enaminas, 6,7,7a,8-tetrahydro-5*H*-thieno[2,3-*b*]pyrrolizines (**226**) are formed by intramolecular abstraction of hydrogen in the initially formed 1,4-dipolar intermediate (**223**). The initial step is the nucleophilic addition of (**221**) to the electron-poor triple bond of DMAD to give an intermediate (**222**) that is the tied ion-pair form of a 1,4-dipolar intermediate. This step will be rate determining since it involves the loss of the thiophene aromaticity. The second step is solvent dependent. In polar solvents, rotation around the newly formed σ-bond takes place to give the solvent-stabilized charge-separated ion pair (**223**). A developing positive charge on nitrogen facilitates hydrogen abstraction *via* an acid–base reaction to give (**224**). The azomethine ylide (**224**) then undergoes a hydrogen shift, probably *via* the solvent, to give the conjugated 1,5-dipolar tautomer (**225**), which finally undergoes a symmetry-allowed disrotatory electrocyclization to give (**226**). The formation of (**226**) always occurs stereospecifically. The CH_2 ester group and the pyrrolidinyl group have the expected *cis* configuration ⟨81JOC424⟩.

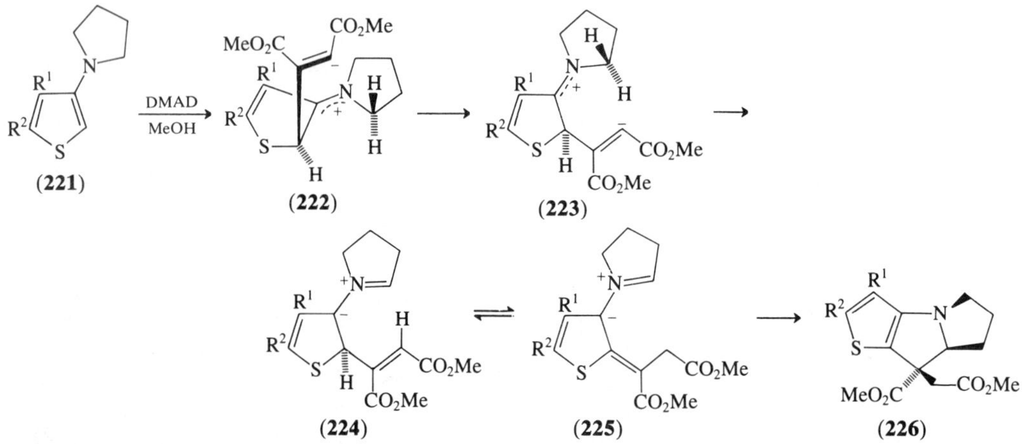

Reaction of thiazolidine-4-carboxylic acid and refluxing acetic anhydride ⟨70CB2611⟩ in the presence of DMAD gave 1*H*,3*H*-pyrrolo[1,2-*c*]thiazole (equation 40) ⟨80JOC5396⟩.

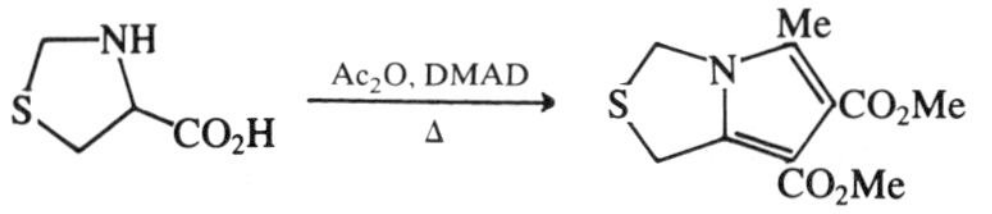

N-Imidyliminophosphoranes (**227**) with DMAD gave phosphoranes (**228**) which spontaneously underwent ring closure with the least sterically hindered imidyl carbonyl group to give pyrrolo[1,2-*b*]pyrazole derivatives (**229**) ⟨81T2595⟩.

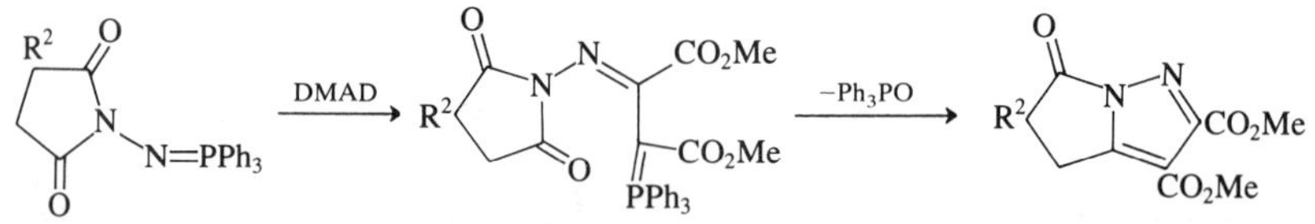

Allyltrimethylsilane reacts rapidly with PTAD to generate a 1,4-dipolar species (**230**) leading to three products, including (**232**), different from those expected from the usual ene reaction ⟨81JOC614⟩. In the formation of the pyrazolo[1,2-*a*]-*s*-triazole (**232**), a three-membered ring intermediate (**231**), a co-contributor to the resonance hybrid along with (**230**), may be involved. A direct transformation of (**230**) to (**232**) appears to be rather difficult, since it includes transformation of a secondary carbonium ion to a primary one.

Ethylenethiourea reacted with DMAD to give 5,6-dihydroimidazo[2,1-*b*]thiazol-3(2*H*)-one (**233**), as shown by unambiguous synthesis from imidazothiazolone (**234**) ⟨71JCS(C)3062⟩. On the other hand, *N*-methylimidazolidine-2-thione and DMAD in methanol gave (**237**), presumably *via* intermediate (**236**), formed by addition of methanol to (**235**) ⟨81JCS(P1)415⟩.

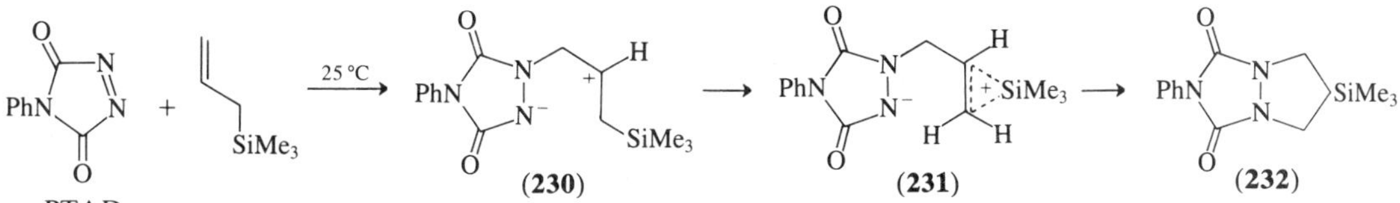

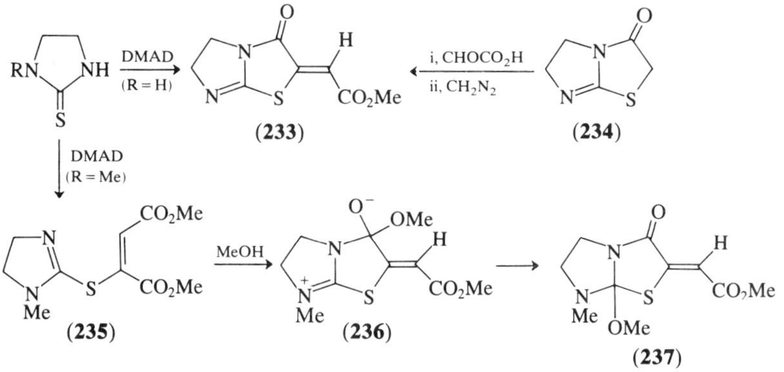

4.36.4.4 Intramolecular Cycloadditions

4.36.4.4.1 1,3-Dipolar cycloadditions

1,3-Dipoles bearing a functional group able to behave as a dipolarophile represent a general scheme for the synthesis of fused ring heterocycles (equation 41) ⟨79MI43600, 77AG(E)10⟩. This useful synthetic idea was first investigated by LeBel et al. ⟨59JA6334⟩ using molecules in which the nitrone moiety was separated by a propene chain from the alkene. The stereospecific ring closure reactions were suggested to be initiated by electrophilic attack of the nitrone on the alkene, the nitrone carbon thus resembling a carbenium ion center. No distinction was made between a one-step process or a two-stage mechanism involving a rapid second step.

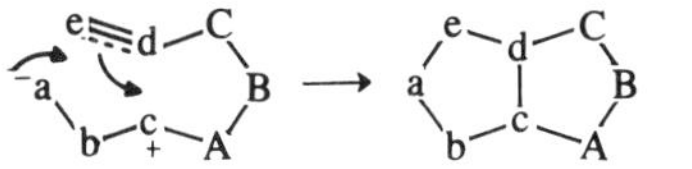

$$\tag{41}$$

The concept of intramolecular 1,3-dipolar cycloaddition reaction has been extended to include azides, azomethine imines, nitrile oxides, nitrile imines and azomethine ylides. Such reactions are summarized in Table 1.

Table 1 Intramolecular 1,3-Dipolar Cycloaddition Reactions

1,3-Dipole	Starting material Chain	Dipolarophile	Product	Ref.
$\bar{N}=\overset{+}{N}=N-$	$-\underset{R^2}{\overset{R^1}{C}}-CH_2CH_2-$	$-CH=CH_2$	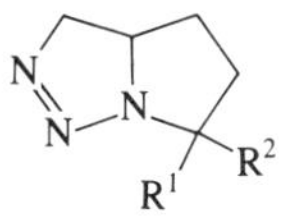	65JA749
$PhCH=\overset{+}{N}-\bar{N}-$ $\quad\quad\quad\; CH_2Ph$	$-\overset{O}{\overset{\|}{C}}-CH_2CH_2-$	$-CH=CH_2$	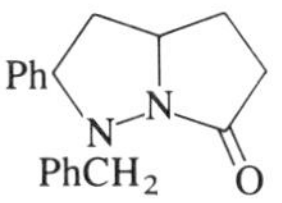	72TL1707, 81JA239
$\bar{O}-\overset{+}{N}\equiv C-$	$-\overset{O}{\overset{\|}{C}}-O-CH_2-$	$-CH=C\overset{R^1}{\underset{R^2}{\diagdown}}$	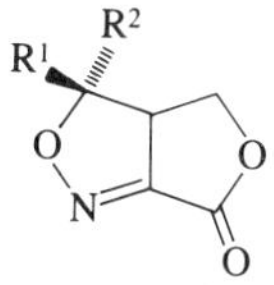	80JHC609
$\bar{O}-\overset{+}{N}\equiv C-$	$-\overset{O}{\overset{\|}{C}}-O-\underset{R^3}{\overset{R^2}{C}}-$	$-C\equiv CR^1$		75S666

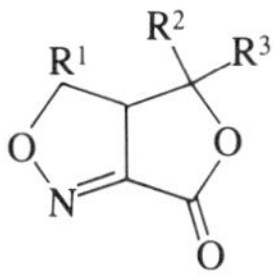

Table 1 (continued)

1,3-Dipole	Starting material Chain	Dipolarophile	Product	Ref.
$R^1\bar{N}-\overset{+}{N}\equiv C$	$-\underset{R^4}{\overset{R^3}{\underset{\|}{\overset{\|}{C}}}}-O-\overset{O}{\overset{\|}{C}}-$	$-C\equiv CR^2$	[pyrazolo-furanone with R^1N, N, R^2, R^3, R^4, O]	75S666, 76SC269
$R^1\bar{N}-\overset{+}{N}\equiv C$	$-\overset{O}{\overset{\|}{C}}-NHCH_2-$	$-C\equiv CH$	[pyrazolo-pyrrolinone with R^1N, NH]	76SC269
$\underset{Me\ Ph}{\bar{C}H_2-\overset{+}{N}=C}$	$\underset{Ph}{-CH-N}\underset{Me\ Me}{-\overset{Me}{\overset{\|}{C}}}-$	$-CH=CH_2$	[bicyclic pyrrolidine with H, Me, Me, NMe, N, Me, Ph]	77TL3437
$Ar C\equiv \overset{+}{N}-\bar{N}$	$-CH_2CH_2CH_2-$	$-CH=CHX$	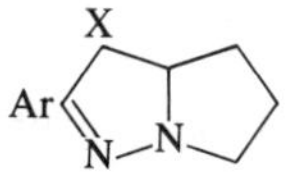 [bicyclic with Ar, X, N, N]	82JOC4256

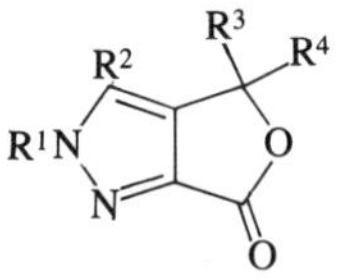

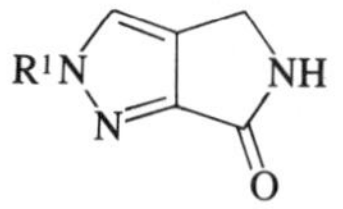

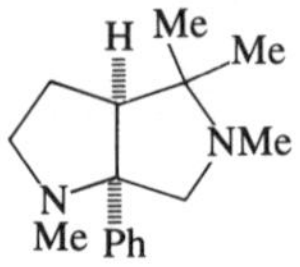

The effect of an internal allylic asymmetric center on the course of the intramolecular cycloaddition reaction has been studied ⟨82TL2081⟩. Subjection of the (Z)-nitroalkene (**238**) to phenyl isocyanate–triethylamine led to the desired cycloaddition product (**240**) in high yield (>98% purity), whereas the corresponding (E)-nitroalkene gave a 3:1 mixture of two cycloadducts. These results are understandable in terms of $A^{1,3}$ strain present in the transition state for cycloaddition ⟨68CRV375⟩. For the cis-alkene (**238**), one mode of cycloaddition (**241**) would involve a severe Me–Me interaction whereas the other (**239**) contains a less serious Me–H interaction ⟨82TL2081⟩.

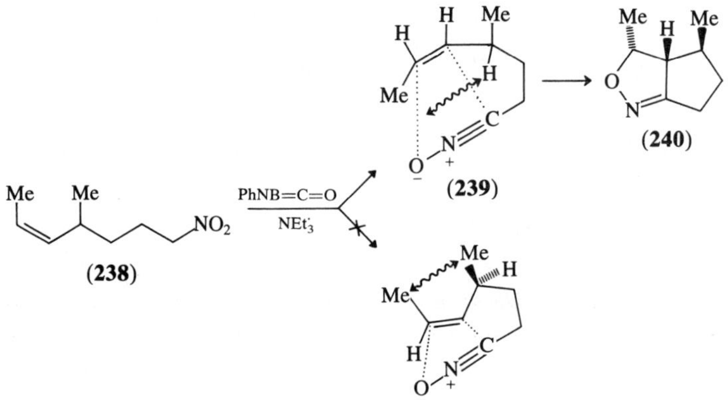

4.36.4.4.2 1,5-Dipolar and higher order cycloadditions

The concept of 1,5-dipolar cyclization gives rise to a general method for the preparation of many (5,5)-fused heterocycles. 1,5-Dipoles are derived from 1,3-dipoles by conjugation with different double bond systems (equation 42) ⟨79CRV181⟩.

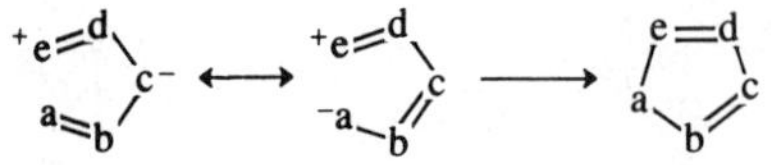

$$\underset{a=b}{\overset{+e=d}{\diagdown}}c^- \longleftrightarrow \underset{-a\diagup b}{\overset{+e=d}{\diagdown}}c \longrightarrow \underset{a\diagdown b}{\overset{e=d}{\diagdown}}c \qquad (42)$$

Heterocyclic hydrazonyl halides which possess an amino-hydrogen undergo ring closure via 1,3-dipolar nitrile imines which can also be written as 1,5-dipolar ions. 1,5-Electrocyclization then gives the fused product (Scheme 13). The method has been used for the preparation of 6,7-dihydro-3-aryl-5H-imidazo[2,1-c]-s-triazoles ⟨72JCS(P2)1887⟩,

3-aryl-6-phenyl-5H-s-triazolo[4,3-b]-s-triazoles ⟨74JCS997⟩, s-triazolo[4,3-d]tetrazoles ⟨68JCS(C)1711⟩, s-triazolo[3,4-b][1,3,4]oxadiazoles ⟨71TL1729⟩, 3,5-diaryl-s-triazolo[3,4-c]-s-triazoles ⟨76ACS(B)463⟩ and pyrazolo[1,5-c][1,2,4]triazoles ⟨77JHC227⟩.

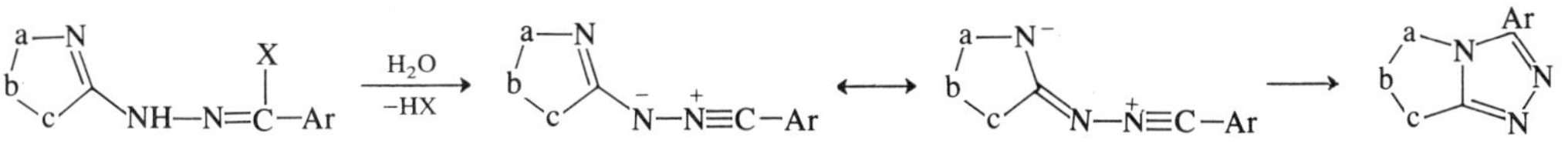

Scheme 13

Thermal decompositions of aryl azides containing α,β-unsaturated substituents result in a cyclization ⟨64CRV149, 79MI43601⟩. The nitrene formed by rapid loss of nitrogen can also be written as a 1,5-dipolar ion. 1,5-Dipolar cycloaddition then leads to the fused pyrrole derivative. A second versatile method for the generation of reactive azene intermediates consists of the reductive deoxygenation of aromatic and heteroaromatic nitro compounds with a trialkyl phosphite at elevated temperature ⟨62MI43600⟩. For example, thermolysis of substituted α-azidoacrylates gives fused pyrroles in high yields (equation 43). The procedure has been used to prepare furo[3,2-b]pyrroles ⟨81CCC2564⟩, thieno[3,2-b]pyrroles, thieno[2,3-b]pyrroles, pyrrolo[3,2-b]pyrroles ⟨72M194, 76CJC1074⟩, pyrrolo[3,2-d]thiazoles and pyrrolo[3,2-d]selenazoles ⟨79JHC1563⟩.

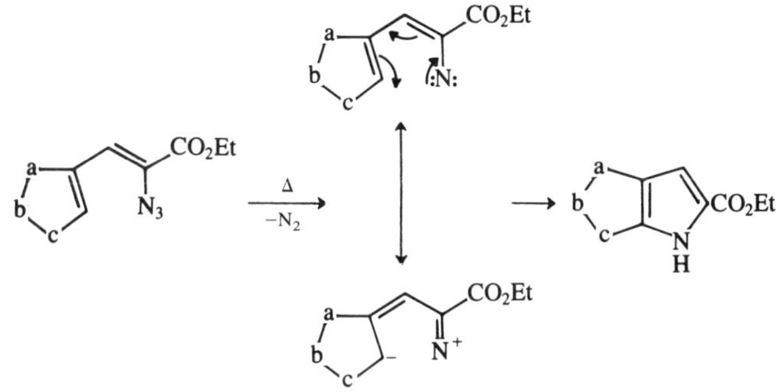

(43)

The major thermolysis product of the azidothiophene (**242**) is diethyl thieno[3,2-b]pyrrole-5,6-dicarboxylate (**244**; 76%), probably formed by a [1,5]-shift of the ester group in (**243**) ⟨81CC550⟩.

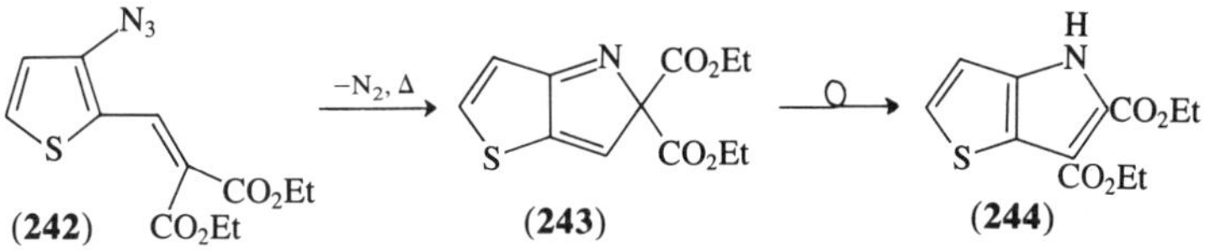

(**242**)   (**243**)   (**244**)

The azene intermediate generated at 115 °C from 3-azido-4-benzylideneamino-5-methyl-s-triazole underwent an intramolecular cyclization at the azomethine linkage to form 1H-6-methyl-2-phenyl-s-triazolo[3,2-c]-s-triazole (equation 44) ⟨66JHC119⟩.

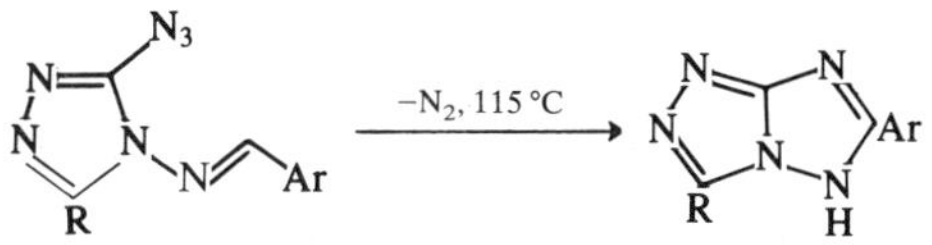

(44)

Reductive deoxygenation of nitrostyrylthiophenes using triethyl phosphite at 180 °C represents a facile and general synthesis of 5-arylthieno[2,3-b]pyrroles and thieno[3,2-b]pyrroles via reactive azene intermediates (equation 45) ⟨73S313⟩.

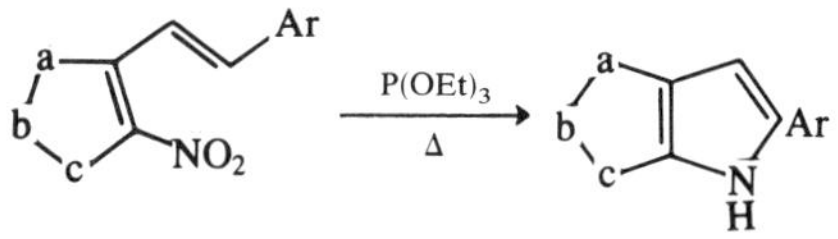

(45)

In the [7+1]-cycloreaction of the diazopyrazole (**245**) with diazoalkanes, the diazoalkane reacts ylide-like as a 1-nucleophile-1-electrophile: the carbon atom first behaves as a nucleophile and then as an electrophile (equation 46). Other 1-nucleophile-1-electrophiles such as ylides of phosphorus, nitrogen and sulfur react analogously with (**245**) ⟨79TL1567⟩.

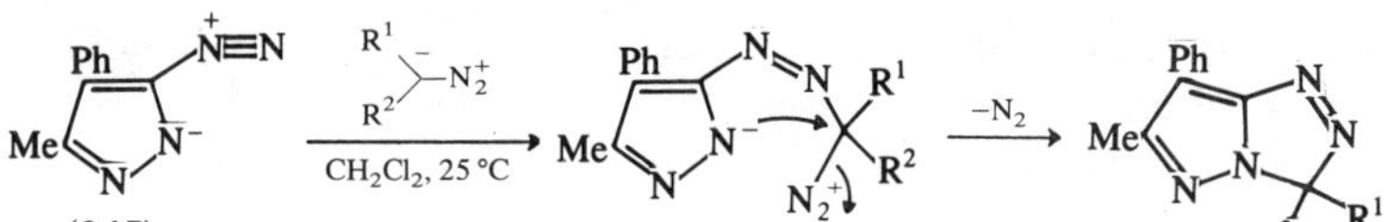

1010   Classical Heteropentalenes

(46)

4.36.4.5 Conjugate Cycloadditions

4.36.4.5.1 Reactions involving N-nucleophiles

Upon heating in alkaline or acidic medium, 3-aryl-5-(2-propargylthio)-s-triazoles (246) cyclize to give mixtures of thiazolo[3,2-b]-s-triazoles (247) and thiazolo[2,3-b]-s-triazoles (248) ⟨79S52⟩. 3-Allenylthio-5-phenyl-s-triazole reacted analogously in the presence of sodium methoxide to give (247; Ar = Ph) ⟨81IJC(B)161⟩.

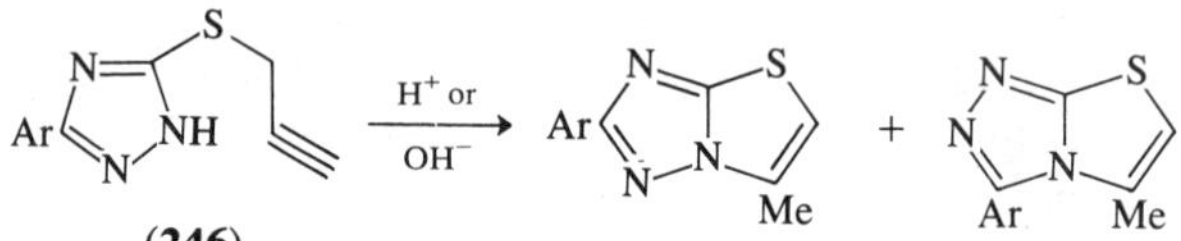

Ketone semicarbazones undergo acid catalyzed reversible cyclization to 3,3-disubstituted 1,2,4-triazolidinones ⟨76JHC1257⟩. The only example available in the literature where a fused triazoline was obtained involves the nitrosation of ethyl 2-isopropylidenehydrazonothiazole-4-acetate, resulting in the formation of a 3,3-dimethyl-2H-s-triazolo[3,4-b]thiazole (Scheme 14) ⟨78JHC401⟩. The reaction may be classified as a 1,5-electrocyclization.

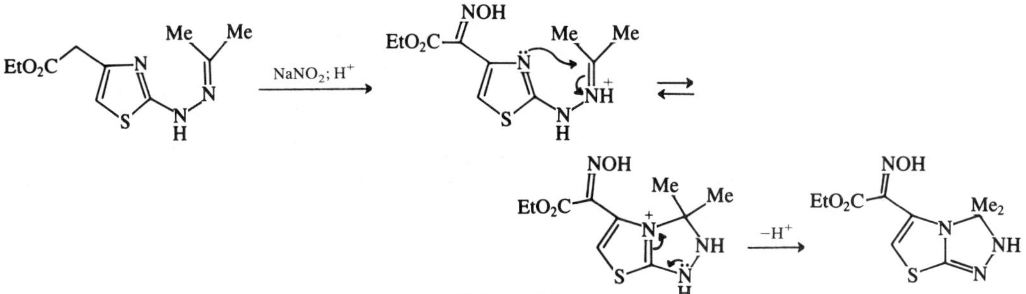

Scheme 14

Treatment of the thiazolium salt (249) with base gives the imidazo[2,1-b]thiazole (250) ⟨64CPB813⟩. Bromination of the hydrochloride of (251) in the presence of KOAc proceeds with simultaneous cyclization to give 3-oxo-6-bromomethyl-2,3,5,6-tetrahydroimidazo[2,1-b]thiazoles (252) ⟨73CHE395⟩. Imidazole-4-carbaldehyde reacts with vinyltriphenylphosphonium bromide in the presence of NaH to give 5H-pyrrolo[1,2-c]imidazole, presumably via an allenic intermediate (equation 47). Imidazole-2-carbaldehyde reacts analogously ⟨76JHC111⟩.

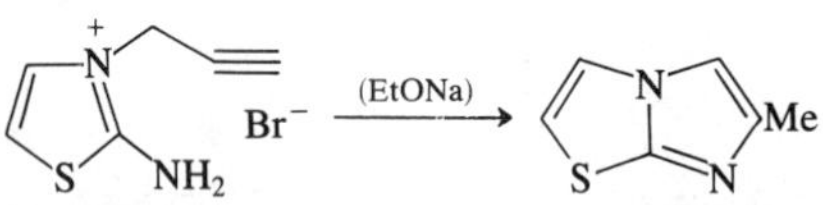

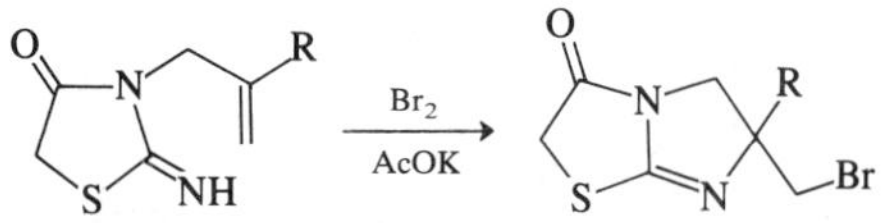

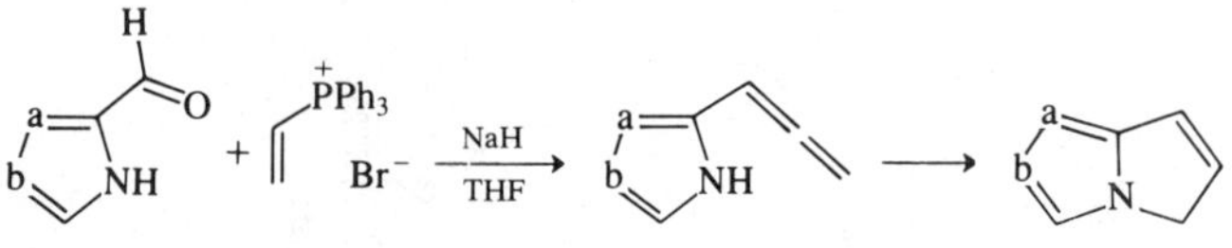

(47)

Thiocyanation of 2-(acyl)aminothiophenes gives 3-thiocyano derivatives which cyclize to 2-(acyl)aminothieno[2,3-d]thiazoles ⟨73JPR539⟩. 5-Amino-3-phenyl-4-thiocyanatopyrazole, under a variety of conditions, afforded 2-amino-6-phenylthiazolo[4,5-c]pyrazole (equation 48) ⟨79JHC61⟩.

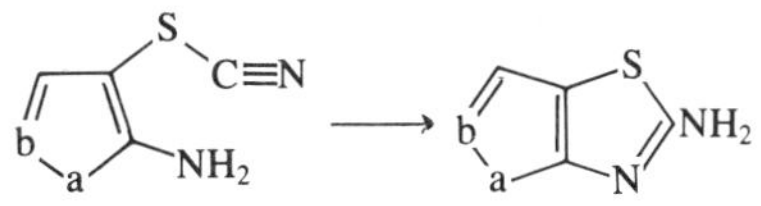

$$\text{(48)}$$

Treatment of the substituted aminosulfolene (**253**) with phenyl isothiocyanate in refluxing benzene–pyridine (2:1) gave the perhydrothieno[3,4-*d*]imidazole-2-thione 5,5-dioxide (**254**) ⟨82CHE328⟩. Reaction of 4-halo-2-thiolene (or -thiolane) 1,1-dioxides with dithiocarbamates proceeded similarly to give perhydrothieno[3,4-*d*]thiazolidine-2-thione 5,5-dioxides ⟨82CHE668⟩.

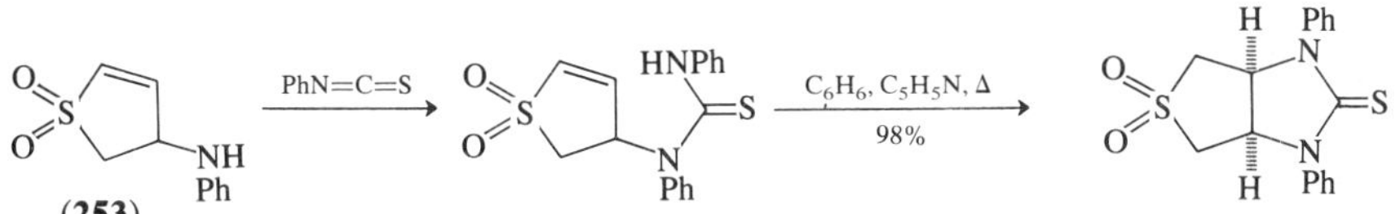

Arylidenehippuric acid azlactones condense with phenylhydrazine to give 3-aryl-5-phenyl-3,3a-dihydropyrazolo[4,3-*d*]oxazoles (see equation 51; X = NPh) ⟨82JIC(B)867⟩.

4.36.4.5.2 Reactions involving S-nucleophiles

Treatment of 4-allyl-1,2,4-triazole-3-thiones with bromine in chloroform is accompanied by cyclization to give the hydrobromide salts of thiazolo[2,3-*c*]-*s*-triazole derivatives (equation 49). 3-Allyl-2-thiohydantoins react analogously ⟨75CHE740⟩.

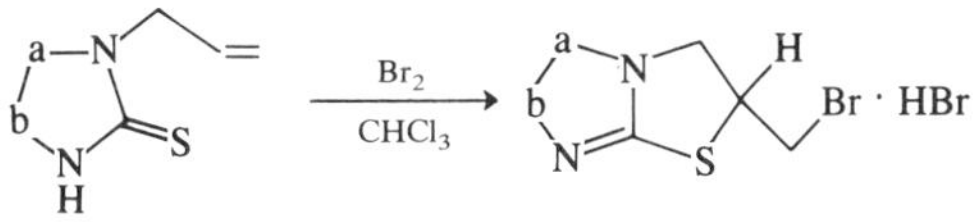

$$\text{(49)}$$

2-(and 3-)(Propargylthio)thiophenes undergo cyclization when heated (160–170 °C) in polar solvents such as DMSO, DMF, HMPT, quinoline or *N,N*-diethylaniline. The highest ratios of thienothiophenes and thienothiopyrans (80:20) were obtained with diisopropylamine as catalyst and DMSO as solvent at a temperature of about 145 °C. With the same amine, DMF gave only the thienopyrans ⟨71IJS(B)85⟩. The formation of thienothiopyran possibly proceeds *via* an enolization of the primary Claisen product, a subsequent [1,5]-H shift and an electrocyclization. The role of the polar solvent may consist of promoting the enolization of the allenic dithiolactone. The thienothiophene is more likely produced by a competing nucleophilic attack of S⁻ at the *sp*-atom of the allenic system (Scheme 15). The general features of these mechanisms are in accordance with literature data concerning the formation of benzopyrans and benzofurans from phenyl propargylic ethers and *o*-allenylphenols ⟨68HCA1510⟩.

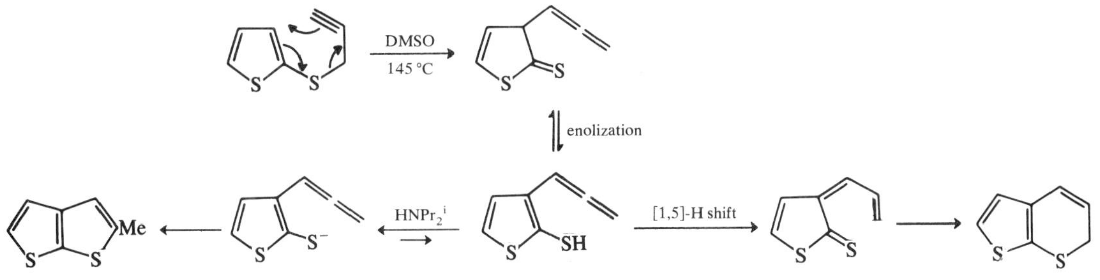

Scheme 15

4.36.4.5.3 Reactions involving O-nucleophiles

The condensation of ethyl 3-cyano-4-hydroxy-3-pyrroline-1-carboxylate with hydroxylamine followed by base treatment of the resulting oxime gives ethyl 3-amino-4*H*-pyrrolo[3,4-*c*]isoxazole-5(6*H*)-carboxylate (equation 50) ⟨68JMC453⟩. Arylidenehippuric acid azlactones condense with hydroxylamine to give 3-aryl-5-phenyl-3,3a-dihydrooxazolo[5,4-c]isoxazoles (equation 51; X = O) ⟨82JIC(B)867⟩.

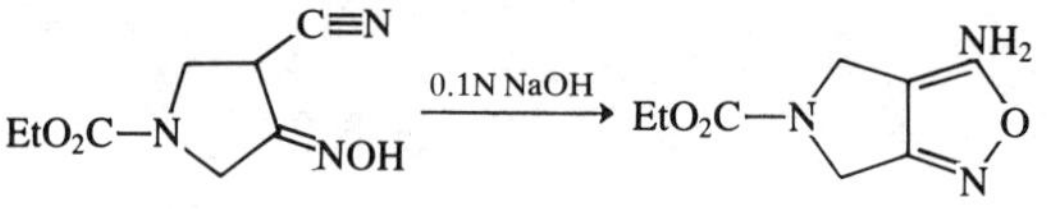

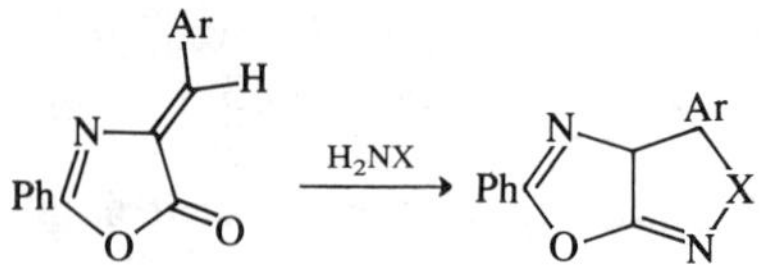

4.36.4.6 Intramolecular Alkylations

4.36.4.6.1 C—N bond formation

Treatment of the dihydropyrazole (**255**) with thionyl chloride gives 2,3,6,7-tetrahydro-1H-pyrazolo[1,2-a]pyrazol-4-ium chlorides (**256**). The imonium group is readily reduced with LiAlH$_4$ to give (**257**), whereas base-catalyzed air oxidation at C-3 in (**256**) followed by loss of H$_2$O or H$_2$O$_2$ gives (**258**) ⟨68JOC3941⟩.

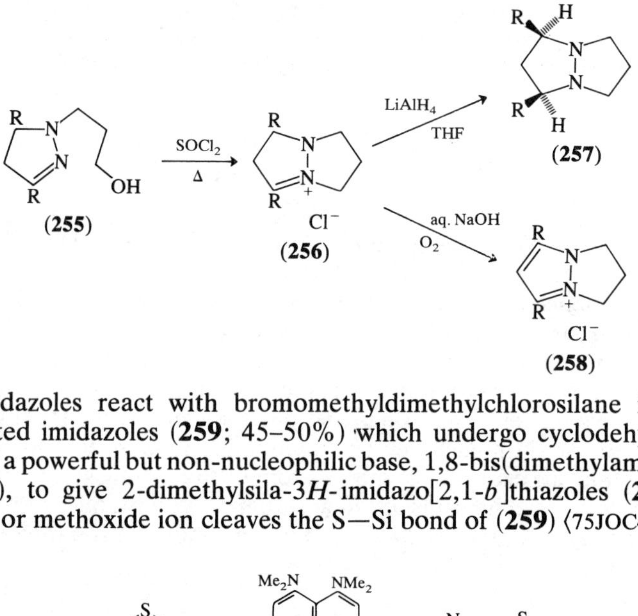

2-Mercaptoimidazoles react with bromomethyldimethylchlorosilane in THF to give bromomethylsilated imidazoles (**259**; 45–50%) which undergo cyclodehydrobromination in the presence of a powerful but non-nucleophilic base, 1,8-bis(dimethylamino)naphthalene ('proton sponge'), to give 2-dimethylsila-3H-imidazo[2,1-b]thiazoles (**260**) in 80–90% yield. Hydroxide or methoxide ion cleaves the S—Si bond of (**259**) ⟨75JOC437⟩.

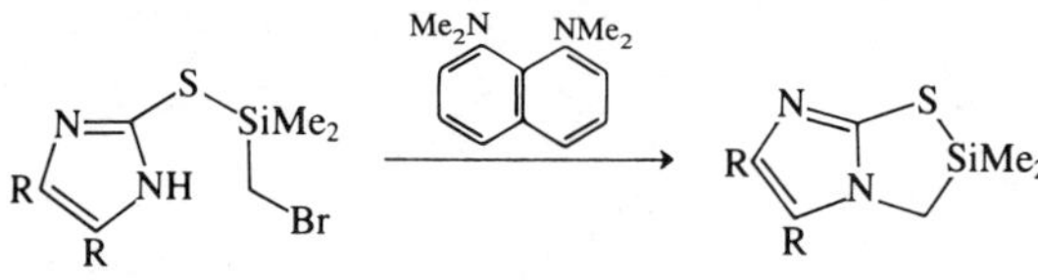

The 2-iminothiazolidine obtained from 2-aminothiazoline and 2-phenyl-1-(p-toluenesulfonyl)aziridine (**261**) readily underwent ring closure in the presence of concentrated H$_2$SO$_4$ to give 6-phenyl-2,3,5,6-tetrahydroimidazo[2,1-b]thiazole (**262**) ⟨69JHC751⟩. Similar C–N fissions of sulfonamides rather than the normal N—S fissions in strong acid have been observed when it is possible to generate either a tertiary or secondary aryl carbonium ion by such a fission ⟨59CRV1077⟩.

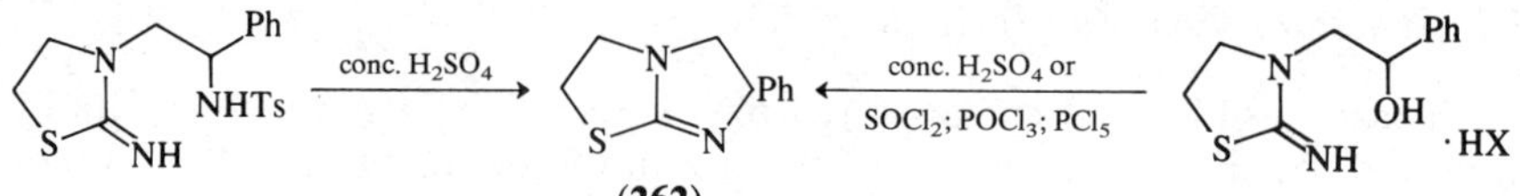

Conditions for the ring closure of readily available 3-(2-hydroxy-2-phenylethyl)-2-iminotetrahydro-1,3-thiazole (**263**) ⟨82S511⟩ to (**262**), the racemate of the anthelmintic tetramisole, involve (a) dehydration with concentrated sulfuric acid or (b) reaction with thionyl chloride, phosphorus oxychloride or phosphorus pentachloride in the presence of base followed by dehydrochlorination ⟨66JMC545, 68JOC1350⟩.

4.36.4.6.2 C—S bond formation

The reaction of 2-(chloromethyl)pyrrolidine with CS_2 proceeded at room temperature in DMF containing K_2CO_3 to give hexahydropyrrolo[1,2-c]thiazole-3-thione (equation 52) ⟨63JOC981⟩.

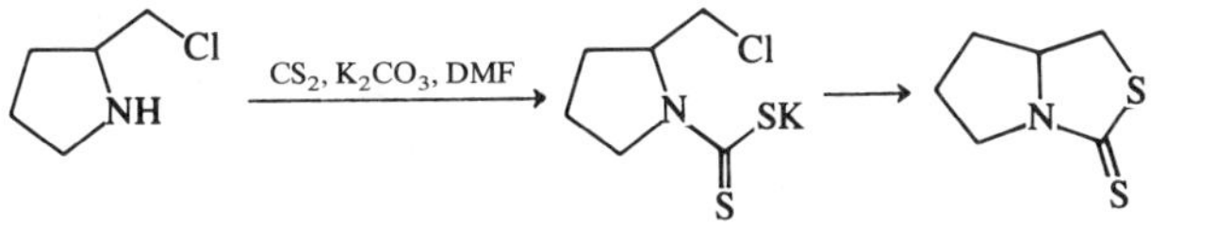

The reaction of 2-chloroethyl isothiocyanate with ethylene sulfide in the presence of $Et_4N^+Br^-$ (20 d, 20 °C) gave an 83% yield of 2,3,5,6-tetrahydrothiazolo[2,3-b]thiazolium chloride (**265**) by way of intermediate (**264**) ⟨71CHE1534⟩. The identical ring system is formed on treatment of N,N-bis(2-chloroethyl)amine and CS_2 in aqueous NaOH ⟨63JPR(291)101⟩. With 2-chloroethyl isothiocyanate and N-phenylethyleneimine, in the absence of a catalyst, imidazo[2,1-b]thiazolium chloride (**266**) is obtained in 60% yield.

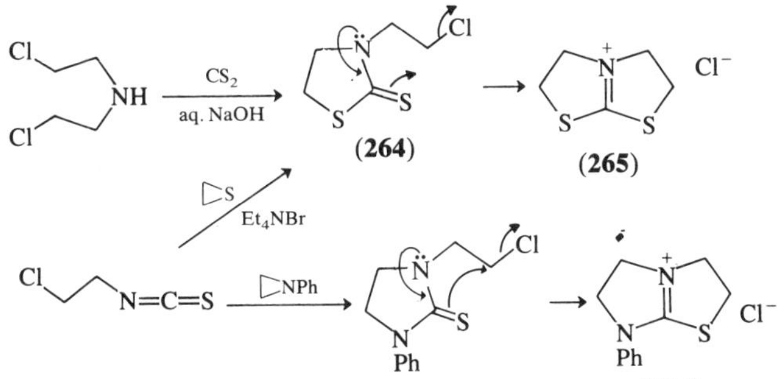

Displacement of nitro groups from aromatic nuclei by nucleophiles is a common occurrence ⟨78T2057⟩. Thermal decomposition of 2,4-dinitro-3-thienyl N,N-dimethyldithiocarbamate (**267**) in refluxing glacial acetic acid gave 6-nitrothieno[2,3-d][1,3]dithiol-2-one (**268**) in 6% yield and not the isomeric 1-nitrothieno[3,4-d][1,3]dithiol-2-one structure as shown by NMR spectroscopy ⟨81JHC1581⟩.

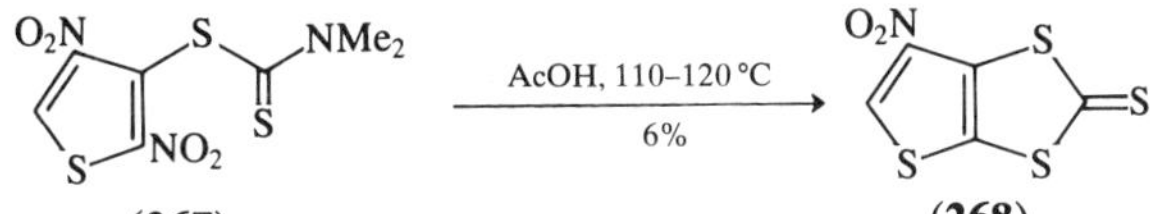

4.36.4.6.3 C—O bond formation

2,3,5,6-Tetrahydroimidazo[2,1-b]oxazole is the product resulting from the intramolecular O-alkylation of 3-(2-chloroethyl)imidazolidin-2-one ⟨57JA5276⟩. Bicyclic hydantoins and dihydroimidazo[2,1-b]oxazoles were prepared analogously from the corresponding mesylates (X = OSO_2Me) in the presence of NaH (equation 53) ⟨77JHC511, 79JMC1030⟩. Treatment of 2,5-dichloro-3-chloromethyl-4-hydroxymethylthiophene with $KOBu^t$ furnished 4,6-dichloro-1H,3H-thieno[3,4-c]furan (equation 54) ⟨69JOC340⟩.

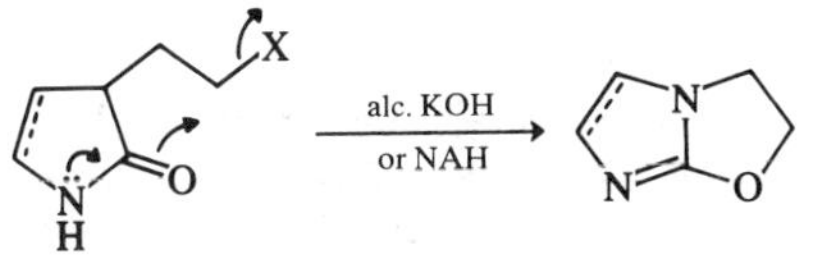

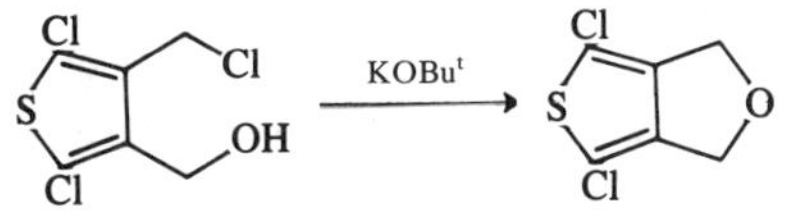

Alkali treatment of the reaction product formed from benzhydrazide benzenesulfonate and DMF on long heating did not result in ring closure to 7,8-dihydro-2,5-diphenyl-s-triazolo[3,4-b][1,3,4]oxadiazolium benzenesulfonate but gave instead the ring-opened product, 1-benzoyl-3-formamidobenzhydrazidine (equation 55) ⟨61CRV87, 61CI(L)2049⟩.

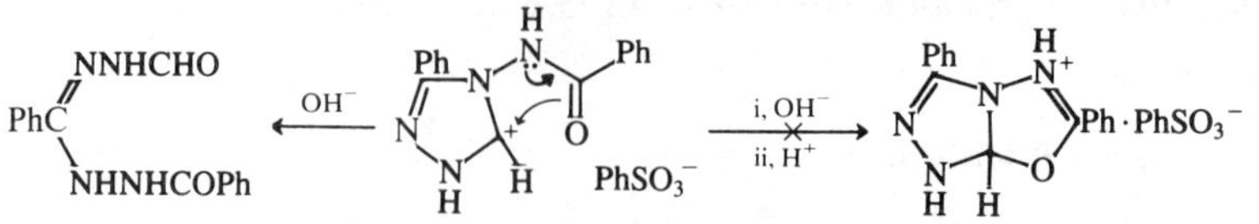

$$(55)$$

The reaction of 5-bromo-3-nitro-1,2,4-triazole and 3,5-dinitro-1,2,4-triazole with a variety of oxiranes yielded the expected 1-substituted imidazoles, and also resulted in the formation of 5,6-dihydrooxazolo[3,2-b]-s-triazoles upon treatment with base ⟨75CHE612⟩. The proposed pathway involves proton abstraction from the imidazole and subsequent attack of the oxirane on the N-anion followed by cyclization in a concerted fashion (equation 56). 2,4(5)-Dinitroimidazole reacts analogously with oxiranes to give isomeric nitro-imidazo[2,1-b]oxazoles in good overall yield ⟨81JMC601⟩.

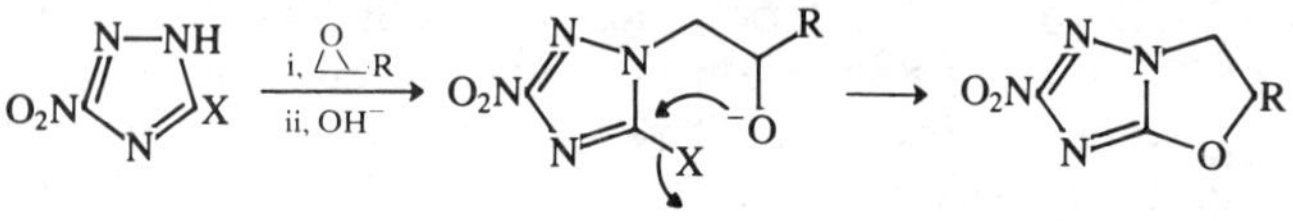

$$(56)$$

4.36.4.7 Oxidative Ring Closures

Uniquely substituted preformed five-membered ring components can undergo intramolecular cyclization with the aid of oxidizing reagents to form a new C—N, C—S, C—O, N—N or S—S bond. Oxidizing reagents commonly used include halogens (Cl_2, Br_2, I_2), sulfur chlorides (S_2Cl_2, $SOCl_2$, SO_2Cl_2), NBS, O_2, peroxides (H_2O_2, $(PhCO)_2O_2$), lead tetraacetate, mercury(II), copper(II) and iron(III) salts.

4.36.4.7.1 C—N bond formation

The oxidation of hydrazones of general structure (**269**) has been frequently used to prepare derivatives of s-triazolo[3,2-c]-s-triazole ⟨68JOC143⟩, oxazolo[2,3-c]-s-triazole and thiazolo[2,3-c]-s-triazole ⟨65JPR(30)280⟩, 1H-s-triazolo[4,3-d]tetrazole and 3H-s-triazolo[4,3-d]tetrazole ⟨68JCS(C)1711⟩ (equation 57).

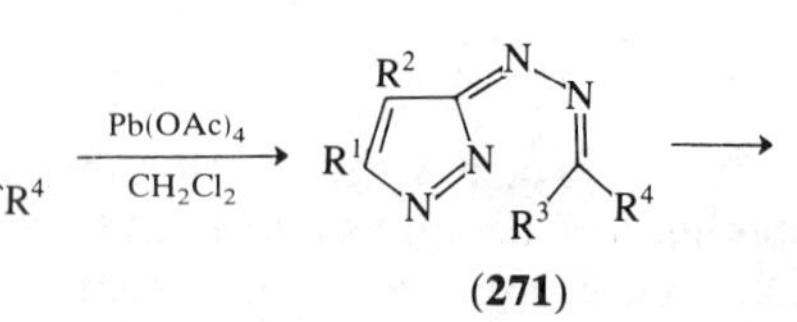

$$(57)$$

(**269**)

3H-Pyrazolo[5,1-c][1,2,4]triazoles (**272**) have been prepared by lead tetraacetate oxidation of arylhydrazones (**270**) via azine intermediates (**271**) ⟨79TL1567⟩.

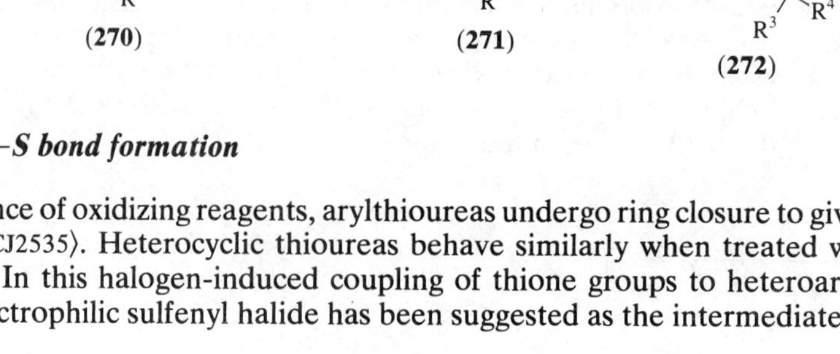

(**270**)   (**271**)   (**272**)

4.36.4.7.2 C—S bond formation

In the presence of oxidizing reagents, arylthioureas undergo ring closure to give condensed thiazoles ⟨70BCJ2535⟩. Heterocyclic thioureas behave similarly when treated with bromine in acetic acid. In this halogen-induced coupling of thione groups to heteroaromatic compounds, an electrophilic sulfenyl halide has been suggested as the intermediate ⟨71AJC1229⟩.

The sulfenyl bromide would add to the 2,3-bond followed by elimination of HBr. The method has been used for the preparation of thieno[3,2-d]thiazoles ⟨78JHC81⟩ and pyrazolo[3,4-d]pyrazoles ⟨76GEP2429195⟩ (equation 58).

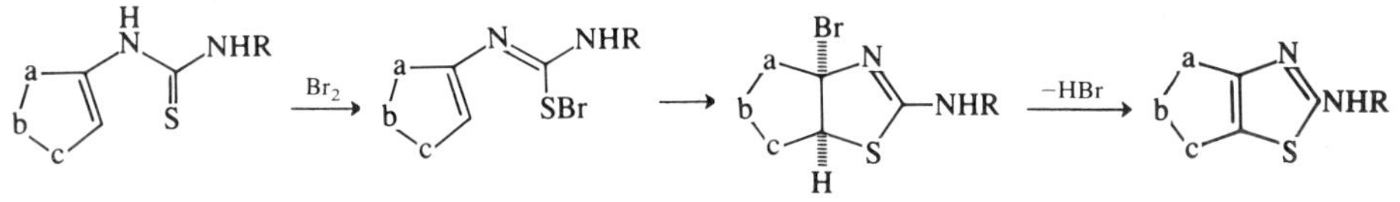

$$\qquad (58)$$

4-Formylpyrazoline-5-thiones undergo cyclo-condensation with compounds containing an 'active' methyl or methylene group to give thieno[2,3-c]pyrazoles ⟨74TL4069⟩. The addition of 0.5 molar equivalent of dibenzoyl peroxide increases the yield from 29 to 77% (equation 59).

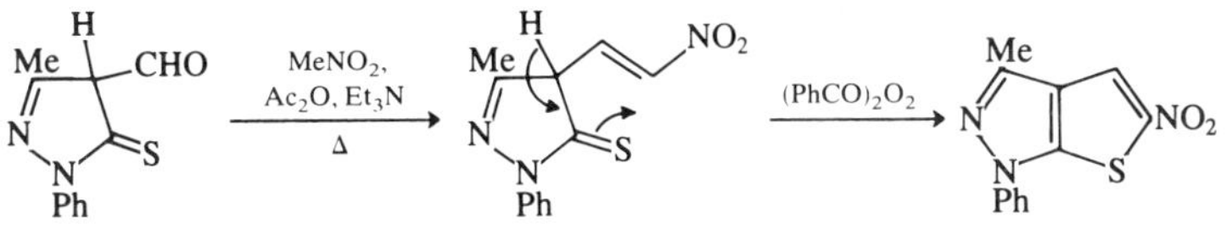

$$\qquad (59)$$

4.36.4.7.3 C—O bond formation

The mercury(II) acetate oxidation of pyrrolidino alcohols can be controlled to give fair yields of bicyclic oxazolidines in which the masked carbonyl carbon atom is at the ring junction ⟨60JA5148⟩. The course of the reaction is regarded as an internal nucleophilic attack by the alcohol group on the ternary iminium salt formed in the initial oxidation step (equation 60).

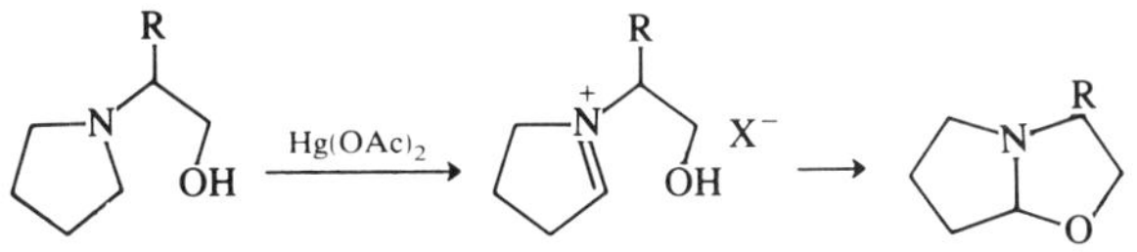

$$\qquad (60)$$

4.36.4.7.4 N—N bond formation

Treatment of hydrazones of general structure (273) with NBS leads to v-triazolo[5,1-b]thiazolium and imidazolium salts. The imidazolium salts (X = NH) lose HBr in aqueous pyridine to give v-triazolo[1,5-a]imidazoles (equation 61) ⟨67AG(E)261⟩. The benzoxazole analogous to (273) failed to cyclize.

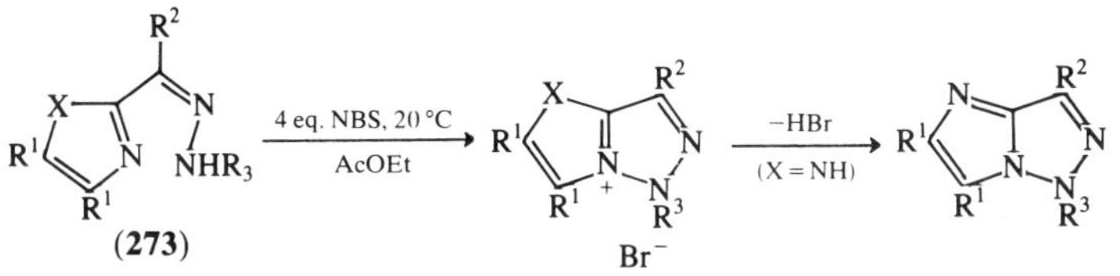

$$\qquad (61)$$

v-Triazoles are formed by cyclization of aromatic o-aminoazo derivatives under oxidative conditions ⟨21CB2191⟩. Template cyclization of 5-amino-4-arylazopyrazoles in refluxing DMF and in the presence of K_2CO_3 and $NiNO_3$ gave 1H-pyrazolo[4,5-d]-s-triazoles in low (20%) yield, in addition to the formation of macrocyclic chelates (equation 62) ⟨80CHE1160⟩.

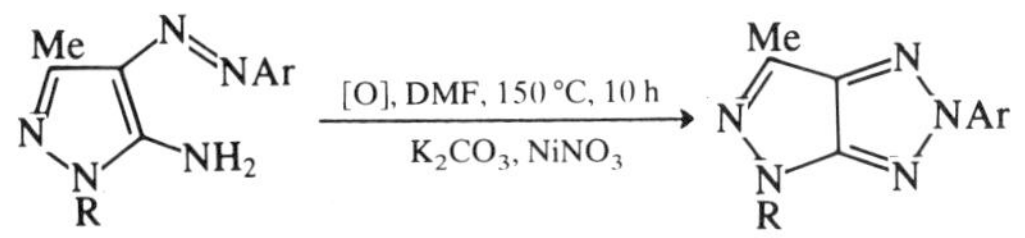

$$\qquad (62)$$

One of four products resulting from the lead tetraacetate oxidation of the dihydrazone (274) in the presence of phenyl azide is tetramethylthieno[3,4-d][1,2,3]triazole (276); the bis(diazo) compound (275) is the presumed intermediate ⟨81JA2868⟩.

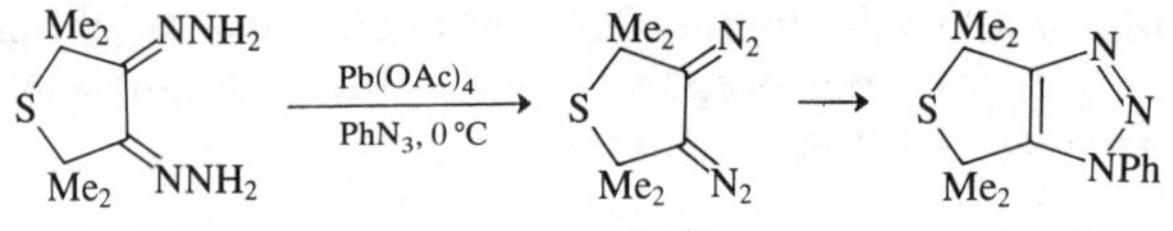

(274) **(275)** **(276)**

4.36.4.7.5 *N—S bond formation*

The formation of (5,5)-fused ring systems involving the formation of a new N—S bond depends on the oxidation of precursors which may be represented by 'imino-enethiols' as one of their tautomeric forms. For example, bromine oxidation of thiazol-2-ylthiourea (**277**) gives 2-aminothiazolo[3,2-*b*][1,2,4]thiadiazolium bromide (**278**) ⟨71JPR1148⟩. Similarly, the oxidation of 2-amino-3-aminothiocarbonylthiophenes (**279**) with H_2O_2 in ethanol–pyridine gives 3-aminothieno[2,3-*d*][2,1]thiazoles (**280**) ⟨73JPR539⟩.

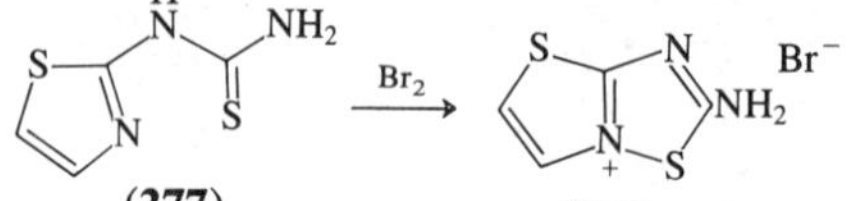

(277) **(278)**

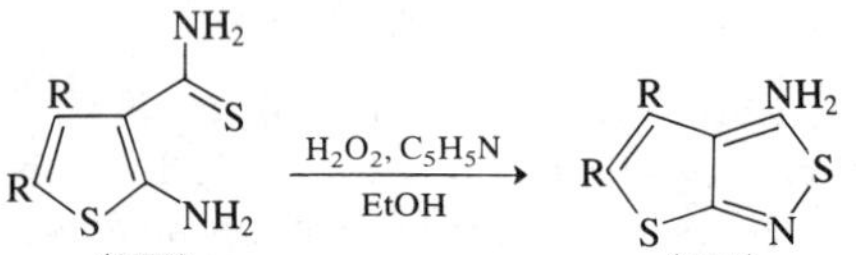

(279) **(280)**

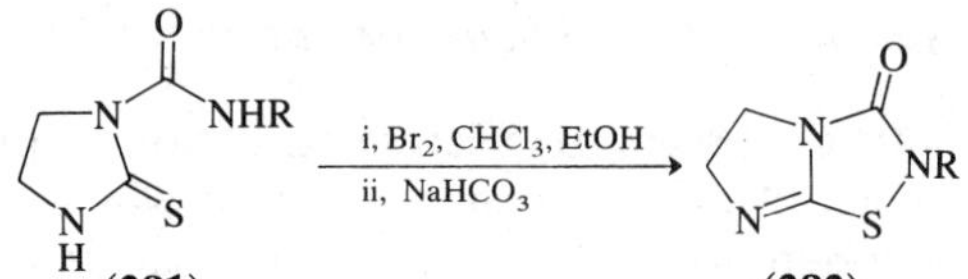

(281) **(282)**

The oxidation of *N*-carbamoylimidazolidine-2-thiones (**281**) has been used to prepare 2,3,5,6-tetrahydroimidazo[1,2-*d*][1,2,4]thiadiazol-3-ones (**282**) which undergo cycloaddition reactions with heterocumulenes to give a variety of heteropentalenes ⟨81T2485⟩.

4.36.4.7.6 *S—S bond formation*

Mild oxidation (aeration) of an aqueous solution of disodium ethylenebisdithiocarbamate ('Nabam') containing manganese sulfate gives 5,6-dihydroimidazo[2,1-*c*][1,2,4]dithiazole-3-thione (**283**) ⟨71TL1317⟩.

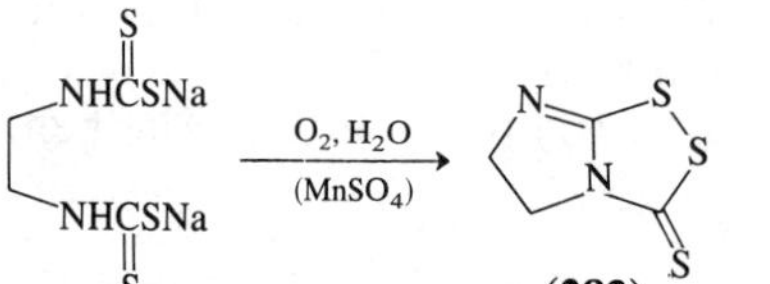

(283)

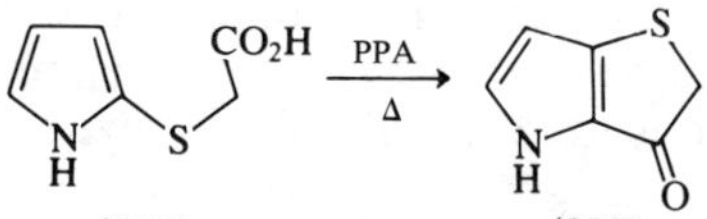

(284) **(285)**

4.36.4.8 Electrophilic Ring Closures

The cyclization of (2-pyrrolylthio)acetic acid (**284**) with polyphosphoric acid (PPA) is accompanied by a rearrangement to give (**285**) ⟨57JOC1500, 61JOC2615⟩. This rearrangement is of a more general nature, as evidenced by the cyclization of (2-thienylthio)acetic acid with concentrated sulfuric acid, leading to 2*H*-thieno[3,2-*b*]thiophen-3-one, identical with the product obtained upon cyclization of (3-thienylthio)acetic acid (Scheme 16) ⟨62ACS155⟩.

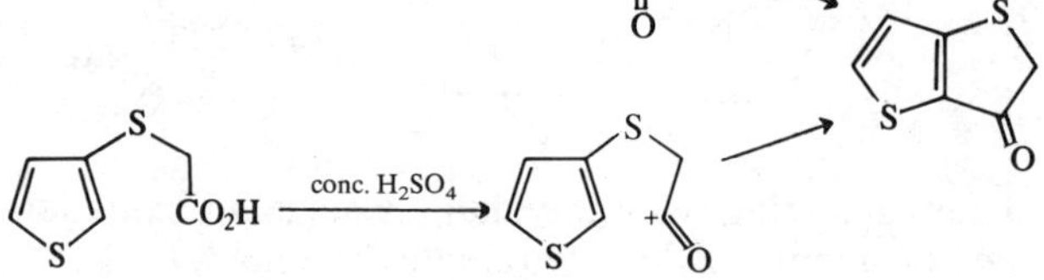

Scheme 16

The rearrangement involves electrophilic substitution by the cationoid carbonyl carbon at the ring α-carbon carrying the sulfur atom with simultaneous rearrangement of the sulfur to the 3-position. The much greater reactivity of the α-position over the β-position in electrophilic substitution is responsible for the observed rearrangement.

The cyclization of (2-acetamido-5-thiazolylthio)acetic acid with refluxing phosphorus oxychloride is not accompanied by rearrangement (equation 63) ⟨56AC(R)275⟩.

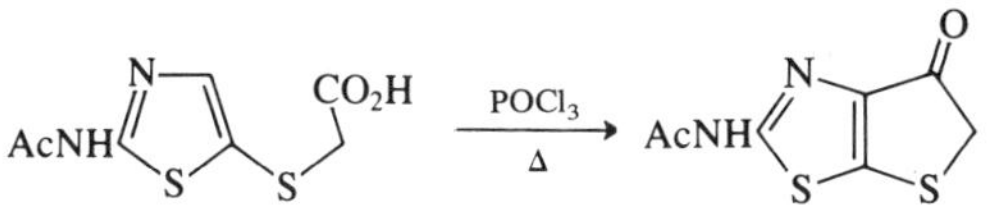

4.36.4.9 Ring Interconversions

A variety of mono- and bi-cyclic heterocycles undergo ring interconversions leading to (5,5)-fused heterocycles. Both thermal and photolytic, and acid- and base-catalyzed, rearrangements, ring enlargements and ring contractions are known.

4.36.4.9.1 Formation by ring expansion (3→5, 4→5)

The pyrolytic rearrangement of acylaziridines follows one of two courses ⟨62AG(E)528⟩. At 235–240 °C, 6-benzoyl-3-oxa-6-azabicyclo[3.1.0]hexane (**286**) rearranges to give 2-phenyl-*cis*-tetrahydrofuro[3,4-*d*]oxazoline (**290**). The mechanism of this reaction may either be concerted, involving a rather strained, four-center transition state (**287**), or it may possibly involve an intermediate tight ion pair (**288**). The identical oxazoline (**290**) was obtained from (**286**) and NaI in acetonitrile at room temperature. Evidence has been presented that this type of rearrangement occurs *via* an iodoamide or iodoamide ion (**289**) ⟨66JOC59⟩.

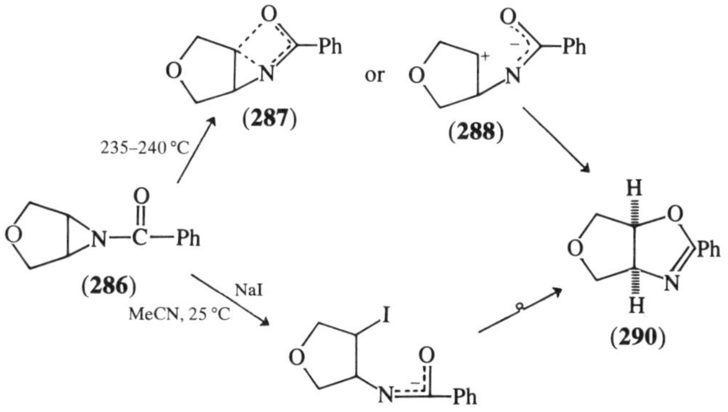

The thermally stable 1,2-diethylene-3-(3-nitrophenyl)guanidine (**291**) rearranges to 1-(3-nitrophenyl)-2,3,5,6-tetrahydro-1*H*-imidazo[1,2-*a*]imidazole (**292**) when treated with NaI in refluxing acetone. Attempted reduction (LiAlH$_4$) or oxidation (Hg(OAc)$_2$) of the guanidine moiety was unsuccessful ⟨70ZN(B)1191⟩.

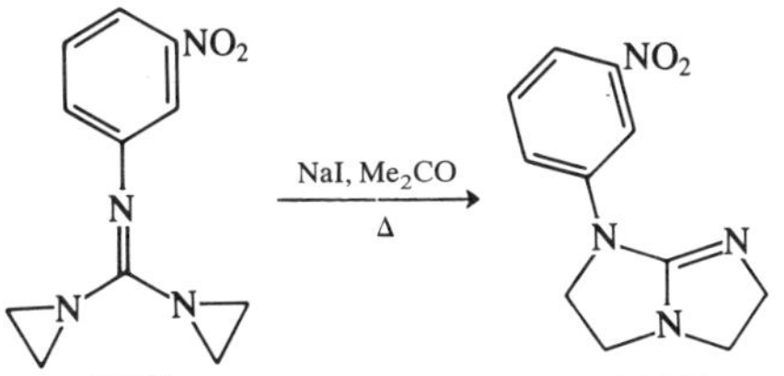

Before interest developed in the structure of penicillin, the imidazo[5,1-*b*]thiazole ring system was unknown, and even today, almost all of the examples which have been described are those prepared in connection with penicillin studies. Very mild acidic hydrolysis converts benzylpenicillin methyl ester (**293**) into benzylpenicillic acid (**294**) by a mechanism suggested by Woodward. When the methyl ester is heated in hydrocarbon solvents for short periods, an imidazo[2,1-b]thiazole ('methyl benzyl penillonate') (295) is produced ⟨B-49MI43600⟩.

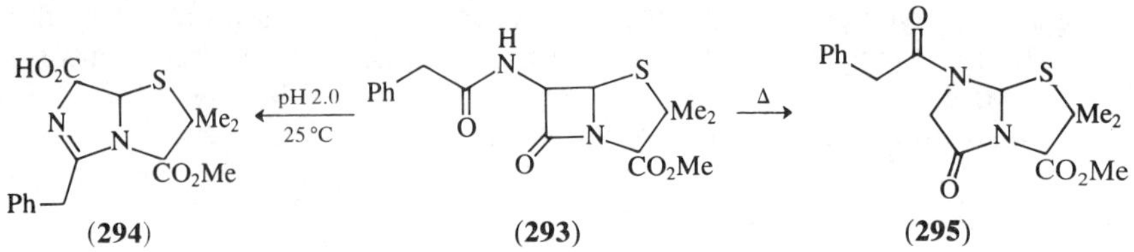

The attempted oxidative dimerization of 2,2-dimethyl-3-thietanone with alkaline ferricyanide gave 2,2,5,5-tetramethyl-2,3,5,6-tetrahydrothieno[3,2-b]thiophene-3,6-dione (equation 64) ⟨77CB1421⟩. The mechanism of this oxidative rearrangement is largely unclear. However, similar rearrangements resulting in ring enlarged structures involving carbenoid and radical intermediates are known ⟨66TL4671⟩.

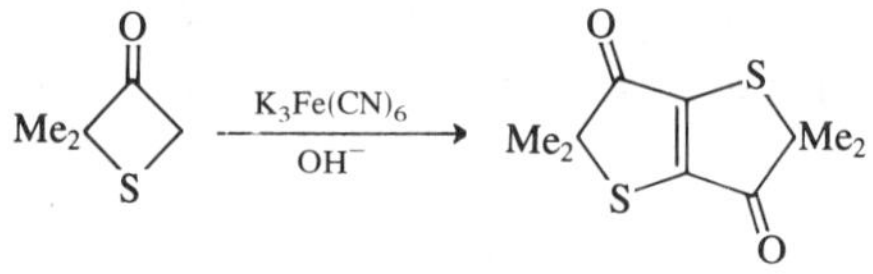

4.36.4.9.2 Formation by ring contraction (7 → 5, 6 → 5)

Pyridazine 1,2-dioxides undergo photoisomerization to 3a,6a-dihydroisoxazolo[5,4-d]isoxazoles. The proposed mechanism involves cleavage of the N—N bond to give bis(iminoxyl) radicals which rearrange and recyclize (equation 65) ⟨79T1267⟩.

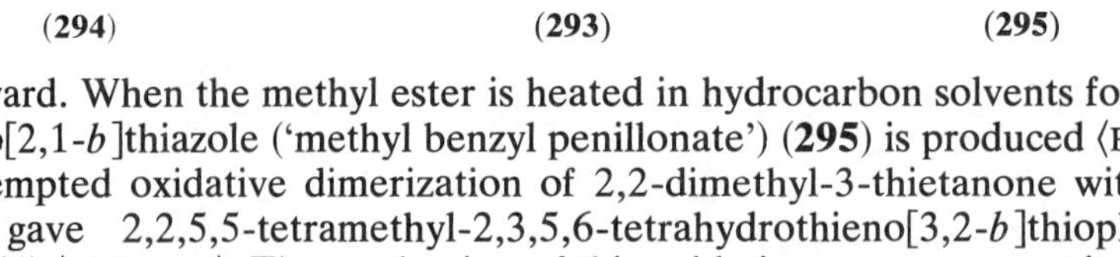

Photolysis of impact- and shock-sensitive o-quinone diazides proceeds in dilute HCl to give pyrazolo[3,2-c]-s-triazole-7-carboxylic acids (Süs rearrangement ⟨44LA(556)65⟩) which are easily decarboxylated ⟨72JPR55⟩. The sequence (Scheme 17) involves the intermediate formation of a ketocarbene which rearranges to a ketene (Wolff rearrangement) followed by reaction with water. Irradiation at 254 nm in the presence of aliphatic alcohols gives the corresponding esters ⟨79JHC195⟩. The diazepinones prepared from 3,4-diamino-1,2,4-triazoles and ethyl acetoacetate react with acetic anhydride to give N-acetylated pyrazolo[3,2-c]-s-triazoles which, in base, are readily deacetylated ⟨74JHC751⟩.

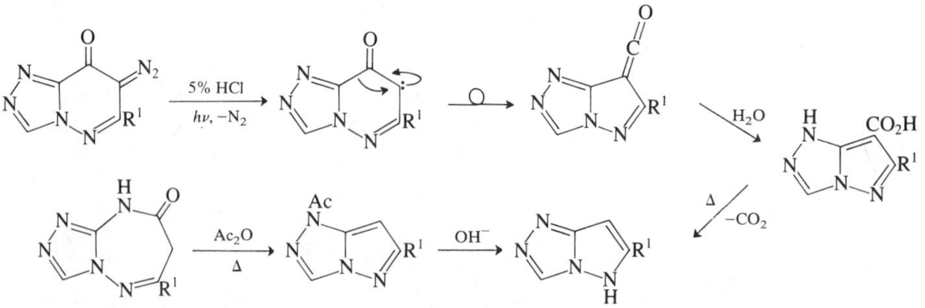

Scheme 17

Photolysis of ylidic thieno[3,2-c][1,2λ^4]thiazines, which are stable sulfinimides, gives thieno[3,2-b]pyrroles, suggesting that the methyl(and phenyl)thio group undergoes very rapid [1,5]-sigmatropic shifts (equation 66) ⟨81CC927⟩.

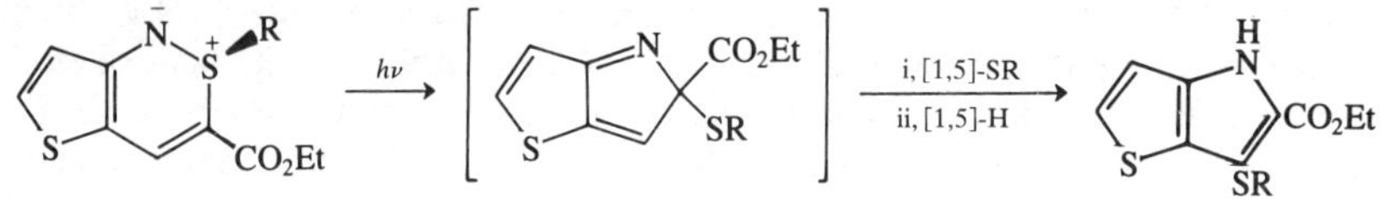

The betaine anhydro-2-hydroxy-8-methyl-4-oxo-3-phenyl-4H-imidazo[2,1-b][1,3]thiazinium hydroxide undergoes thermal rearrangement to 2,3-dihydro-7-hydroxy-2-methyl-5-oxo-6-phenyl-3(2H),5H-pyrrolo[1,2-c]imidazolethione. A process involving an initial fission of the 1,2-bond and electrophilic ring closure of the resultant ketene intermediate into the 4-position of the imidazole nucleus is most likely involved in the rearrangement (equation 67) ⟨80JOC2474⟩.

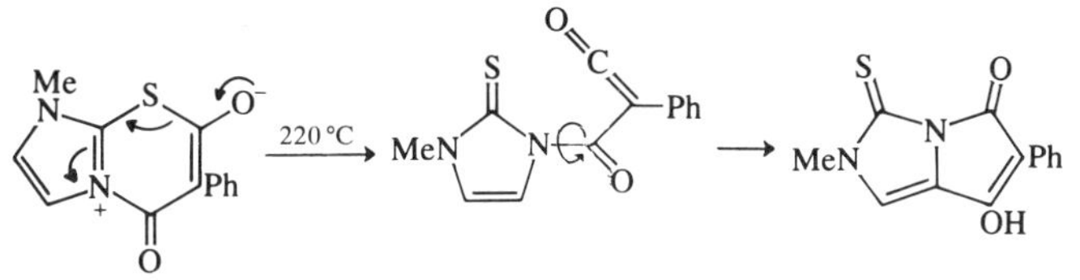

(67)

6,7-Dihydro-2H-thiazolo[2,3-c][1,2,4]triazine-3,4-dione rearranges in dilute base to give an unstable acid which decarboxylates on acidification of the sodium salt to give 5,6-dihydrothiazolo[2,3-c]-s-triazole (equation 68) ⟨81CB1200⟩. Kinetic evidence has been put forward in favor of covalently hydrated intermediates in the acid-catalyzed rearrangement of triazolo[4,3-a]pyrazines to 1H-imidazo[2,1-c]-s-triazoles. The intermediate triazole has been isolated and characterized (equation 69) ⟨72JCS(P2)4⟩.

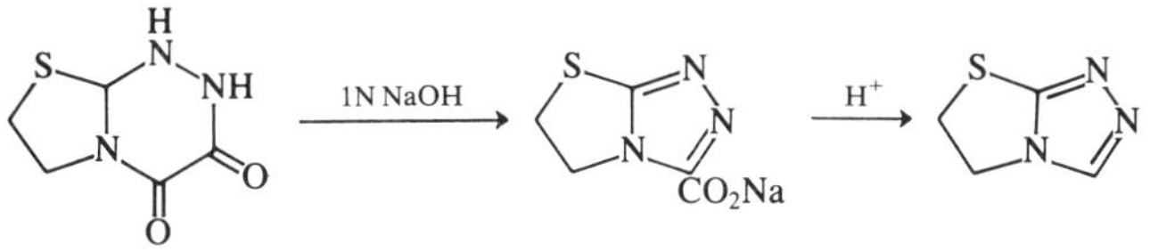

(68)

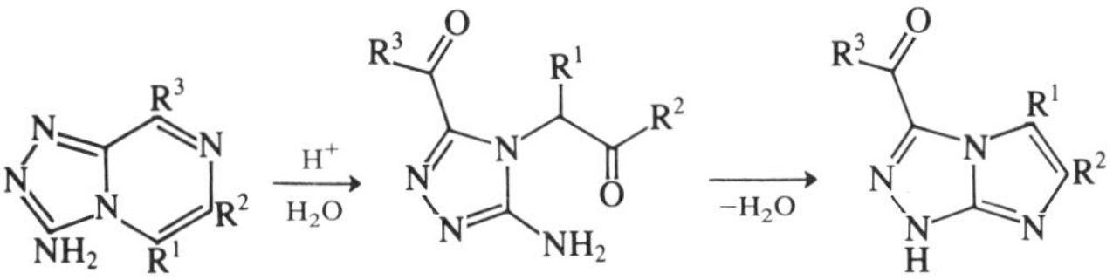

(69)

N^2-Methylated-2,5-dihydro-1,2,4-triazines (X = O, S) react readily with DMAD in refluxing toluene to give 1,3a,6,6a-tetrahydropyrrolo[3,4-c]pyrazoles (equation 70), as evidenced by X-ray crystallography. The five-membered rings are in the envelope form. The groups which protrude from the fused ring to the same side take conformations so as to avoid close interactions among the six peripheral groups ⟨80JOC4587, 81BCJ41⟩.

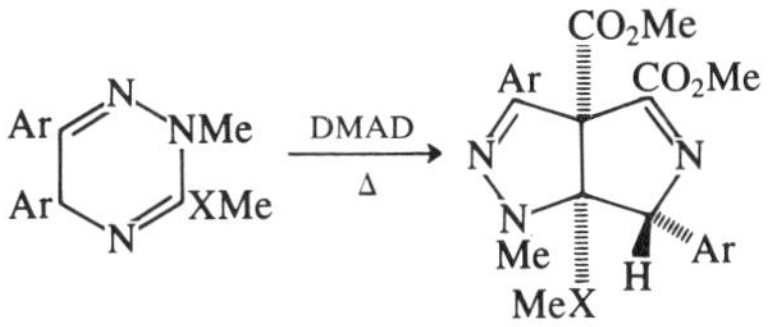

(70)

The structure of the antibiotic holomycin bears a formal resemblance to that of the penicillins. Both substances are bicyclic lactams incorporating a sulfur heteroatom and have an acylamine side chain. One synthetic scheme was designed around one key reaction, contraction of the dithioacetal (**296**) to disulfide (**297**) following the general method developed by Kishi and coworkers ⟨73JA6490⟩. Reaction of (**296**) with m-chloroperbenzoic acid followed by perchloric acid gave (**297**). Treatment of (**298**) with anhydrous HF and anisole gave (**299**; R = CO_2Me). Holomycin (**299**; R = H) was obtained in a single step from the methyl ester by cleavage and concomitant decarboxylation with LiI in pyridine, a reaction presumably requiring isomerization of the double bond α,β to the carboxyl group to the β,γ position ⟨77JOC2891⟩.

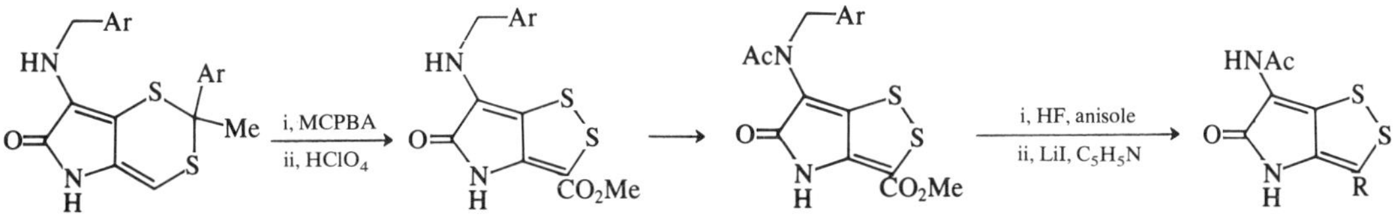

(**296**)   (**297**)   (**298**)   (**299**)

Disrotatory photolysis of the bicyclic aziridine (**300**) occurred readily to give the azomethine imine (**301**); reaction in the presence of DMAD in dioxane gave (**302**). Photolysis of (**301**) in the presence of DMAD gave the isomeric photoproduct (**303**) along with 4% of (**302**) ⟨69AG(E)604⟩.

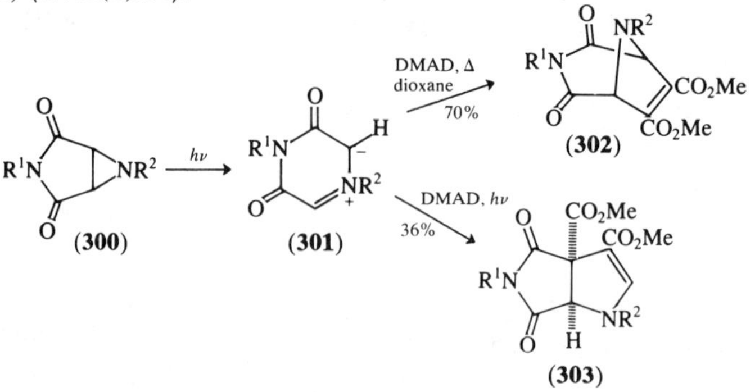

4.36.4.9.3 Formation by rearrangement (5→5)

The photolysis of enaminonitriles provides a convenient and direct one-step synthesis of imidazoles. 1,6-Dihydroimidazo[4,5-*d*]imidazole was prepared by the photochemical rearrangement of 3-aminopyrazole-4-carbonitrile ⟨76JOC19⟩ and by photolysis of 4-aminoimidazole-5-carbonitrile ⟨74JA2014⟩ (equation 71). The precise nature of the reaction intermediates (if any) formed on irradiation of enaminonitriles remains to be determined. However, the direct photochemical inversion of the nitrile to isocyanide is a possibility.

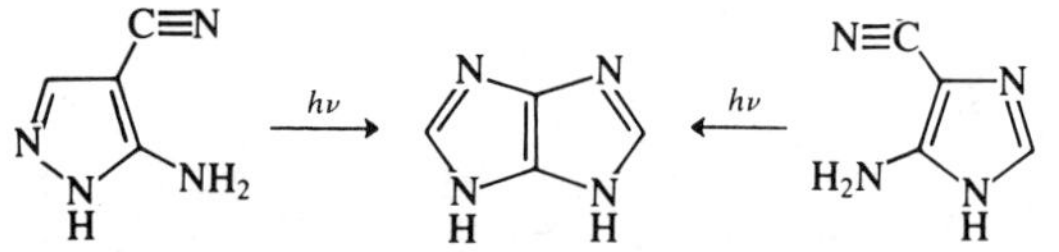

$$(71)$$

The α-diazoketone of 2,2,5,5-tetramethylthiolane-3,4-dione (**304**) undergoes both thermally and photochemically induced rearrangement. The major product on thermally induced decomposition is the thietanone (**305**; 99.3%). In sharp contrast, on irradiation in benzene only the fused ring compound (**307**) is formed ⟨80JOC4804⟩. The thermal reaction of (**304**) clearly proceeds through a variant of the Wolff rearrangement, involving a selective shift of the sulfur bridge to provide (**305**). The enol acetate (**307**) could arise if (**304**) is converted into (**305**), which undergoes rapid secondary photolysis to give the cumulene (**306**) which cycloadds with (**304**) with subsequent loss of nitrogen to provide (**307**).

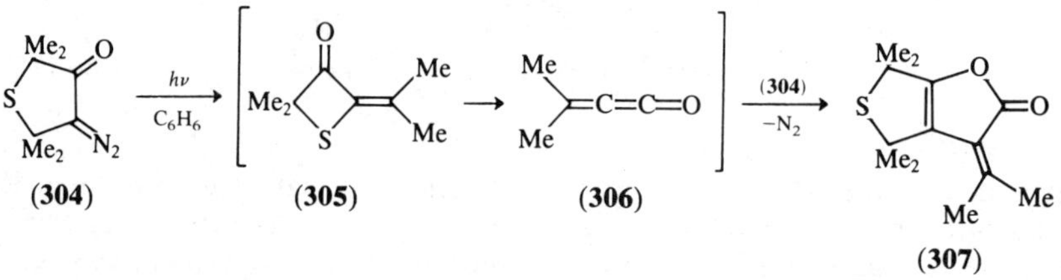

Heating *N*-acetyl-*S*-(α-ketoalkyl)cysteine (**308**) for a brief period with acetic anhydride gives dihydro-2-methylthieno[3,4-*d*]oxazole-3a(4*H*)-carboxylic acids (**310**) ⟨79JOC825⟩. While the mechanism has not been investigated in detail, the isolation of several compounds suggests the existence of the intermediate spiro compound (**309**). Rearrangement to (**310**) is feasible since the anhydride character of (**309**) is changed to the ester character of (**310**).

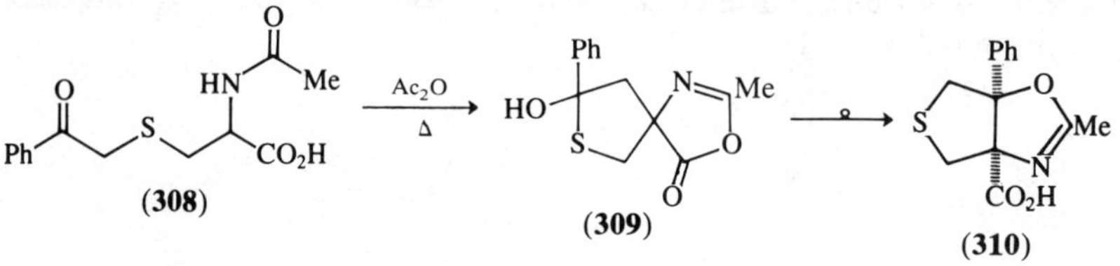

The major products resulting from an interaction of trithiones with 2,3-diphenylcyclopropenone (and -thione) are thieno[3,2-*b*]furans and thieno[3,2-*b*]thiophenes ⟨82LA315⟩.

The reaction involves nucleophilic attack of thione-sulfur on to the C-2 of the cyclopropenone to give a 1,3-dipolar intermediate which reacts as indicated to give isolable spiro compounds ⟨82TL3059⟩ which undergo thermal rearrangement (equation 72).

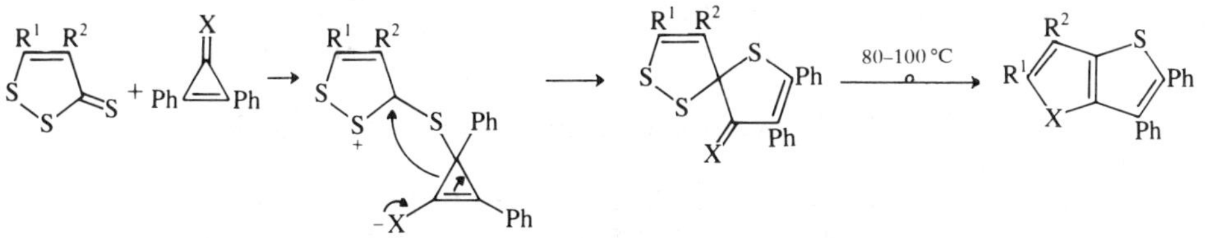

When the 2-(5-oxo-2-oxazolin-4-yl)thiazolidine-4-carboxylate (**311**) is heated in benzene, the penicillic acid derivative (**312**) is formed. This is a result of attack of the thiazolidine nitrogen at the imino-ether rather than at the carbonyl group of the oxazolone ring to form the five-membered ring in preference to a four-membered ring ⟨74JHC823⟩.

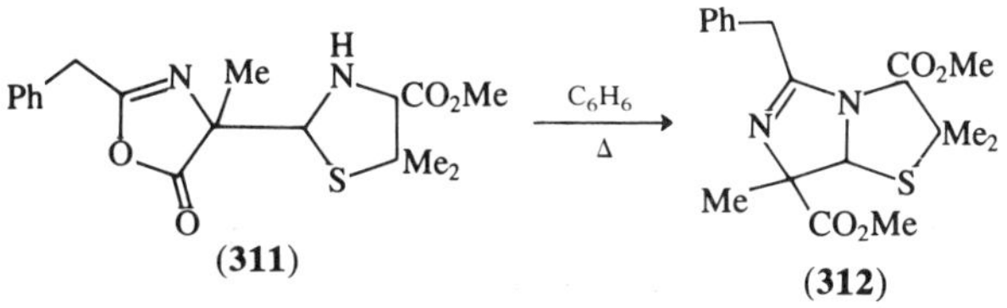

The condensation of 2-(3-oxobutylthio)-5-ethyl-1,3,4-oxadiazole with ammonia leading to a thiazolo[2,3-c]-s-triazole involves an intramolecular ring transformation (Scheme 18) ⟨82JOC2757⟩.

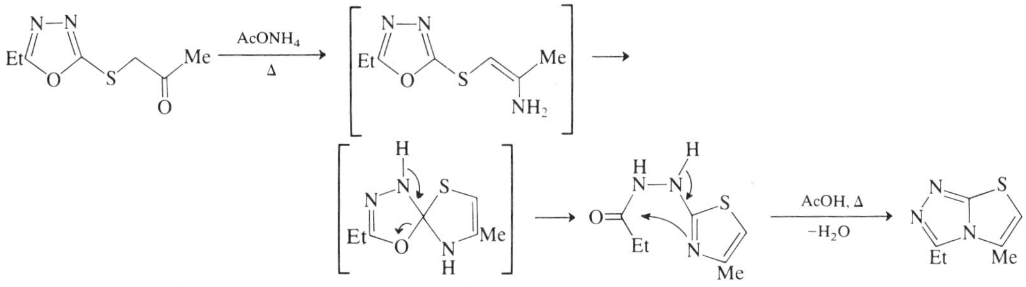

Scheme 18

4.36.4.10 Simultaneous Formation of Two Rings

The formation of fused five-membered ring systems most often involves the building of one ring on to an existing five-membered ring. However, there are instances in which both rings are formed simultaneously during the course of a reaction. Although none could ever be isolated, a one-ring intermediate may most plausibly be written in certain cases, and indirect evidence obtained as to its probable nature. Both saturated and fully conjugated (5,5)-fused ring systems may be formed simultaneously, involving a cross section of concepts including nucleophilic-type displacements, aldol-type condensations and oxidative ring closures.

2,4,5,7-Tetrabromooctane-3,6-dione acting as a bis(1,4-bielectrophile) reacts with sodium sulfide to give 2,2,5,5-tetramethyl-2,3,5,6-tetrahydrothieno[3,2-b]thiophene-3,6-dione, the first compound containing the basic chromophoric system of the thioindigo dyes with *trans-s-trans-s-trans* configuration of the carbonyl groups. The leuko form was not isolated as it is readily oxidized by air (equation 73) ⟨77CB1421⟩.

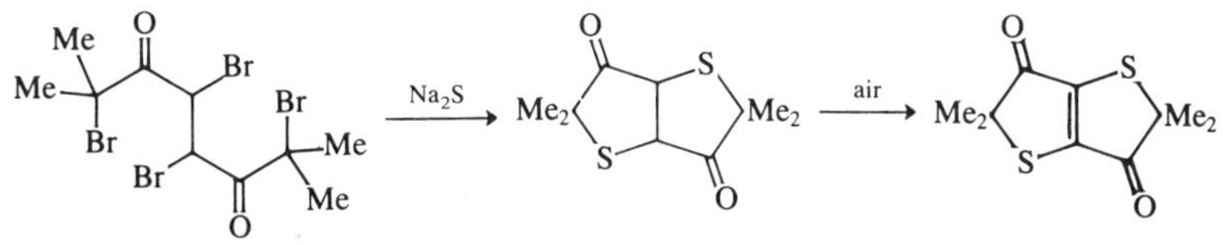

Diphosphane reacts with 1,3-disilylated ureas acting as 1,3-bielectrophiles (molar ratio 1:2) to give [1,4,2,3]diazadiphospholo[2,3-b][1,4,2,3]diazadiphospholes (equation 74).

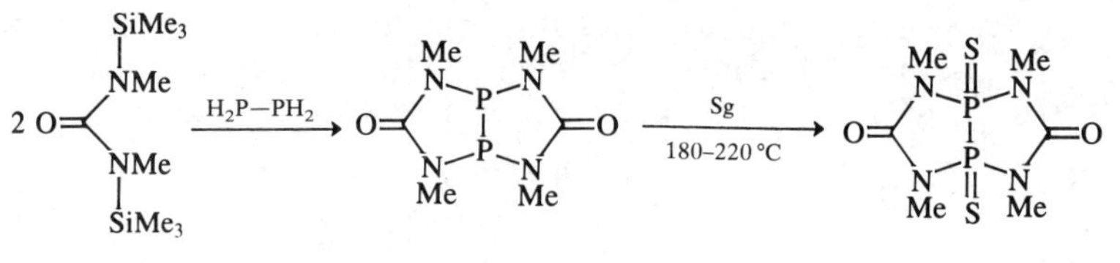

Reaction with sulfur at 180–220 °C gives the corresponding 2,3-disulfide without cleaving the P—P bond ⟨81CB2132⟩.

The condensation product of dithiooxamide with benzaldehyde has been shown to be 2,5-diphenylthiazolo[5,4-d]thiazole (equation 75) ⟨60JA2719⟩. Its formation involves dehydration of the reactants and further dehydrogenation. The reaction is comparable in scope with condensations such as the Perkin reaction, and many 2,5-diarylthiazolo[5,4-d]thiazoles have been prepared from a variety of aryl aldehydes. Condensation products were not obtained from simple aliphatic aldehydes ⟨71JMC743⟩.

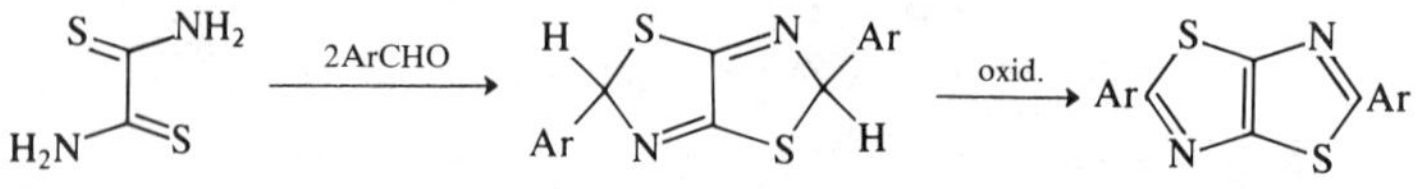

Bis(hydroxymethyl)phosphines react with hydrazine hydrate Mannich-like to give [1,2,4]diazaphospholo[1,2-a][1,2,4]diazaphospholes (**313**) ⟨81TL229⟩.

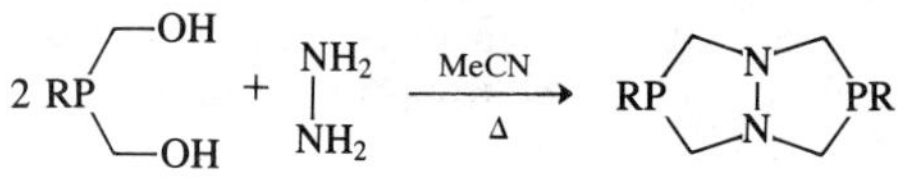

Bicyclic phosphoranes containing the P(H)R group are formed in the reaction of diethanolamine with RP(NEt$_2$)$_2$ ⟨72TL2969⟩. Bicyclic phosphoranes containing the PH$_2$ group have been prepared from diethanolamine with dineopentyl phosphinite ⟨81JCR(S)4036⟩. Conformational studies show that in bicyclic phosphoranes the nitrogen atom has a planar geometry; the five-membered rings are blocked in an envelope conformation; the equatorial plane containing CPH(N) is a symmetry plane of the molecule, and the folds of the two envelopes are oriented in the same way. The direction of the folds of the two envelopes is dependent upon substitution on carbon ⟨81JCS(P2)19⟩.

The reaction of bis(β-mercaptoethyl)phosphane with tris(dimethylamino)-phosphane or -arsane leads to stable fused heterocycles containing phosphorus–phosphorus and arsene–phosphorus linkages (equation 76) ⟨82TL1345⟩.

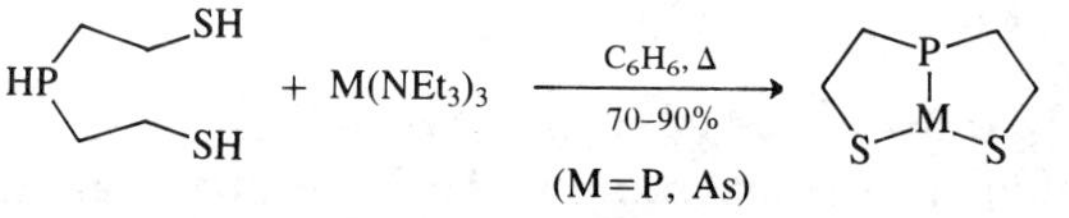

In the reaction of N-organyldipropargylamines with halogenoboranes both C≡C triple bonds insert into B—X bonds to form the [1,2]azaborolo[1,2-a][1,2]azaborole system by intramolecular coordination (equation 77) ⟨81CB2519⟩.

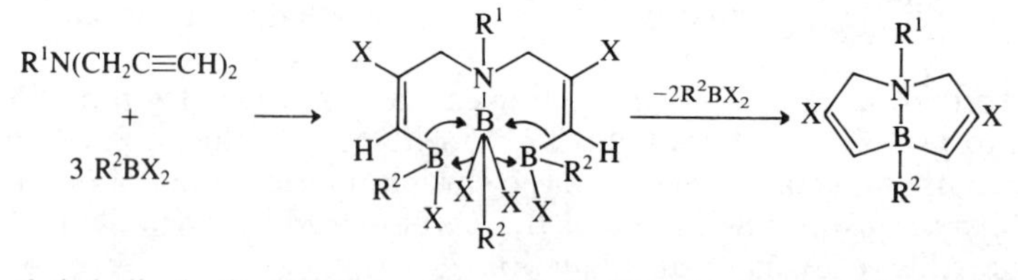

Oxidation of 1,4-disubstituted 2-butene-1,4-dione dioximes with lead tetraacetate or phenyliodoso bis(trifluoroacetate) leads to mixtures containing 3,6-disubstituted dihydroisoxazolo[5,4-d]isoxazoles and 3,6-disubstituted pyridazine 1,2-dioxides (equation 78)

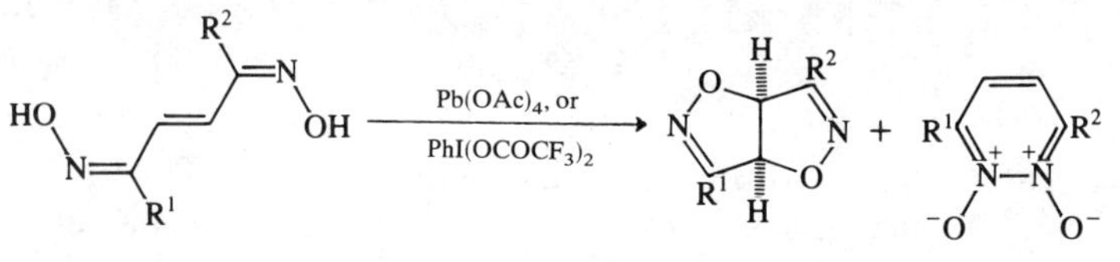

⟨79JOC3524⟩. Bis(iminoxyl) radicals are postulated as precursors since it is known that oximes generate iminoxyl radicals under oxidative conditions (Section 4.36.4.9.2) ⟨73CRV93⟩.

1,1,4,4-Tetramethyl-1H,4H-thieno[3,4-c]thiophene has been synthesized by the action of lithium methoxide on γ,γ-dimethylallenyl thiocyanate (**314**). Although undetected in the course of the reaction, the formation of bis-γ,γ-dimethylallenyl disulfide as key intermediate is suggested. The rearrangement to a conjugated dienic dithial intermediate by either a concerted [3,3]-sigmatropic rearrangement, or by a diradical mechanism appears plausible. Finally, rotation around the central C—C σ-bond by 180° brings the molecule into the requisite conformation for the operation of a double intramolecular Michael-type addition to give the product (Scheme 19) ⟨80TL3617⟩.

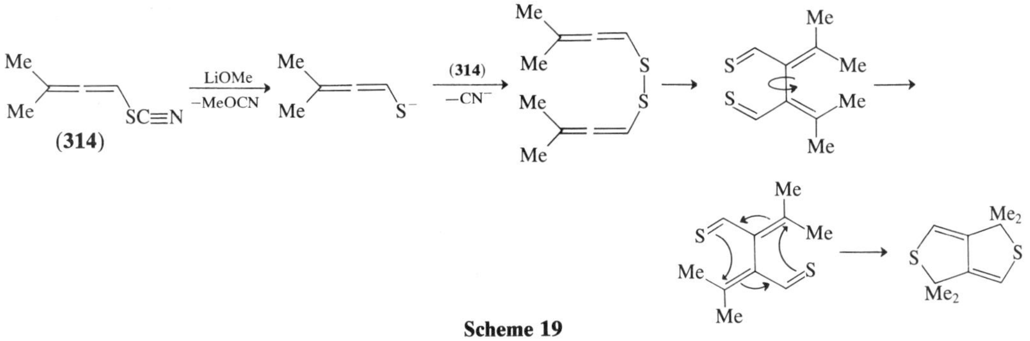

Scheme 19

3-Benzoylimino-6-mercapto[1,2]dithiolo[4,3-c]isothiazole is formed by double ring closure from sodium cyanodithioformate and benzoyl chloride (equation 79) ⟨81EGP148341⟩.

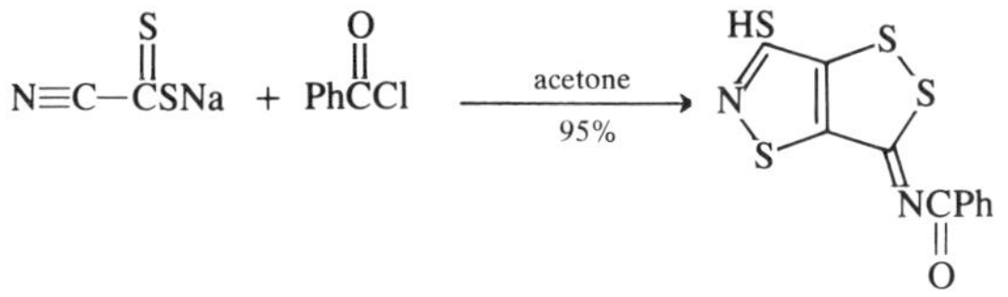

$$N \equiv C - \overset{S}{\underset{\parallel}{C}} SNa + Ph\overset{O}{\underset{\parallel}{C}}Cl \xrightarrow[95\%]{\text{acetone}} \text{product} \quad (79)$$

Thieno[2,3-c]isothiazole-3-carboxylic acid was the compound obtained by the reaction of ethyl 3-cyano-5,5-diethoxy-2-oxopentanoate with phosphorus pentachloride in refluxing toluene, previously assumed ⟨55JA4069⟩ to yield thieno[3,4-d]isothiazole-4-carboxylic acid. The proposed mechanism involves the cyclization of the intermediate 3-cyano-2,5-dithioxopentanoate as shown in Scheme 20. The parent heterocycle, which is a weak base and fails to give quaternary salts with alkyl halides, was obtained by decarboxylation of the acid and also by independent synthesis ⟨82AJC385⟩.

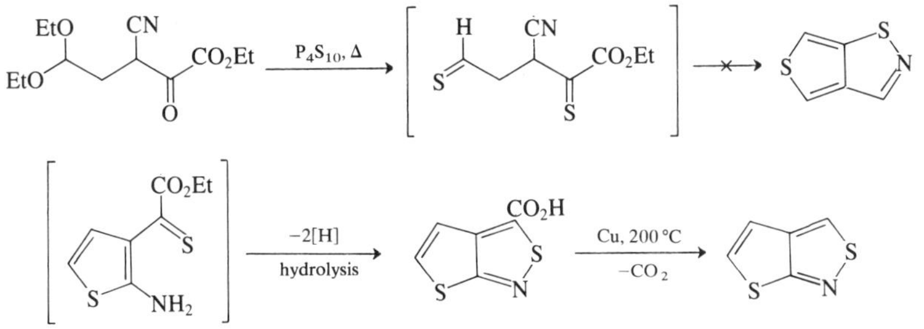

Scheme 20

4.36.5 (5,5)-FUSED HETEROCYCLES OF BIOLOGICAL INTEREST

An increasing understanding of the fundamental chemistry of heterocyclic ring systems has enabled chemists to use heterocycles (1) as sources for smaller fragments of synthetic value (Section 4.36.3.4) and (2) incorporate heterocyclic rings into a wide variety of compounds with biological activity. Investigation of the biological properties of (5,5)-fused heterocycles has proceeded mainly in two directions: (1) natural products, their synthetic analogues and reaction products and (2) the pharmacological actions of compounds containing heterocyclic ring systems. Naturally occurring compounds and those produced by microorganisms have attracted special attention.

For example, 1,2-dithiolo[4,3-*b*]pyrrole derivatives such as holomycin, thiolutin, aureothricin and isobutyropyrrothine are antibiotic substances produced by certain *Streptomyces* species. They are the first recognized examples of microbiologically active unsaturated lactams. In the series of aflatoxins, the presence of the furo[2,3-*b*]furan ring portion is necessary for the toxicities to be effective.

The enzymatic role of the thieno[3,4-*d*]imidazole derivative biotin as the coenzyme for the transfer of carbon dioxide in carboxylation reactions is well established. Strong acidic hydrolysis of biocytin, a naturally occurring complex of biotin, yields biotin and L-lysine.

(±)-2,3,5,6-Tetrahydro-6-phenylimidazo[2,1-*b*]thiazole hydrochloride (tetramisole) is a broad-spectrum anthelmintic; the D-form is an antidepressant; the L-form (levamisole) is known to possess immunoregulatory or immunostimulatory activity. Some phenacylthioimidazolines which are tautomeric to cyclic carbinolamines are very potent antagonists of reserpine-induced hypothermia in mice. The corresponding 3-aryl-5,6-dihydroimidazo[2,1-*b*]thiazoles are also potent antidepressants. 5,6-Bis(4-methylsulfinylphenyl)-2,3-dihydroimidazo[2,1-*b*]thiazole is used especially in the treatment of rheumatic arthritis.

6,7-Dihydro-3-(5-nitro-2-furanyl)-5*H*-imidazo[2,1-*b*]thiazolium chloride (furazolium chloride) is a topical antibacterial agent. Substituted imidazo[1,2-*c*]thiazole derivatives are antihypertensives, CNS stimulants, and appetite suppressants. 2*H*-1,3-Dithiolo[4,5-*c*]pyrazole-5-thiones and imidazo[5,1-*b*]thiazoles are fungicidal with low toxicity to plants, animals and humans.

Pyrazolo[2,3-*a*]imidazole possesses sedative and antipyretic action. Imidazo[2,1-*b*]-[1,3,4]thiadiazoles are antithrombotic and thrombolytic agents. Furo[3,2-*c*]isoxazol-5-ones and their enantiomeric or racemic mixtures are used in the production of antitumor agents obtained from methyl-L-acosaminide and -daunosaminide.

Isosorbide is a diuretic; the dinitrate is a vasodilator. Hexahydrofuro[3,2-*b*]pyrrole derivatives are useful as bronchodilators and thrombocyte aggregation inhibitors. 3-Amino-4*H*-pyrrolo[3,4-*c*]isoxazoles produce hypotension. 3,4,5,6-Tetrahydro-2*H*-cyclopentathiazol-2-one is an effective analgesic.

2,3-Dihydro-3-oxothieno[3,4-*d*]isothiazole 1,1-dioxide, its [2,3-*d*] and [3,2-*d*] isomers are artificial sweeteners (thiophenesaccharins). The [3,4-*d*] isomer is 1000 times sweeter than sucrose and does not have the bitter metallic after-taste of saccharin.

4,6-Dihydro-2*H*-thieno[3,4-*c*]pyrazol-3-ylcarbamates show insecticidal, acaricidal and nematocidal activity.

5-Phenylpyrrolidino[2,1-*b*][1,3,4]oxadiazolines are useful pharmaceuticals.

In addition to the demonstrated biological usefulness of the selected number of fused heterocycles listed in Table 2, increasing numbers of (5,5)-fused heterocycles containing a variety of heteroatoms are continuously evaluated and patented for potential use in various sectors of industry, including agricultural chemicals, dyes, pharmaceuticals and polymers.

Table 2 Biologically Active (5,5)-Fused Heterocycles

Structure	Name or ring system	Activity	Ref.
	Holomycin (R^1 = Me, R^2 = H)	Antimicrobial	59HCA563
	Thiolultin ($R^1 = R^2$ = Me)	Antibiotic	B-76MI43600
	Aureothricin (R^1 = Et, R^2 = Me)	Antibiotic	B-76MI43600
	Isobutyropyrrothine ($R^1 = Pr^i$, R^2 = Me)	Antibiotic	B-76MI43600
	Aflatoxins (R = H, OH)	Carcinogen, antitumor agent	B-76MI43600
	Biotin (R = OH)	Coenzyme	71MI43601
	Biocytin (R = $NH(CH_2)_4CH(NH_2)CO_2H$)		B-76MI43600

Table 2 (*continued*)

Structure	Name or ring system	Activity	Ref.
	Tetramisole	Anthelmintic	B-76MI43600
	Dexamisole, D-(+)-form	Antidepressant	B-76MI43600
	Levamisole, L-(−)-form	Immunoregulator	B-76MI43600
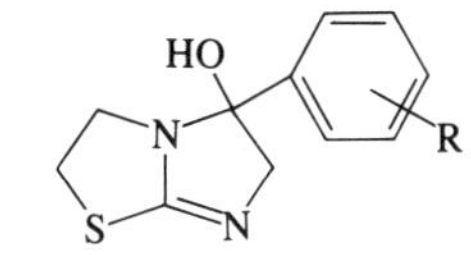	Imidazo[2,1-b]thiazole	Antidepressant	71JMC977
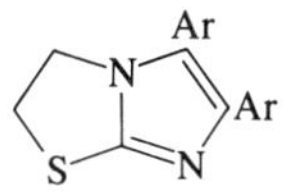	Imidazo[2,1-b]thiazole	Antirheumatic	81USP4263311
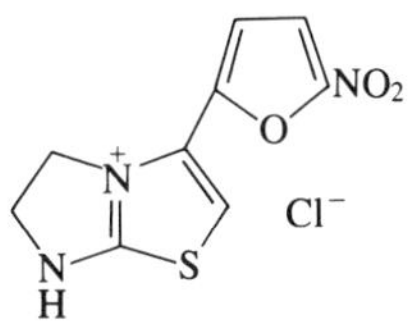	Furazolium chloride	Antibacterial	B-76MI43600
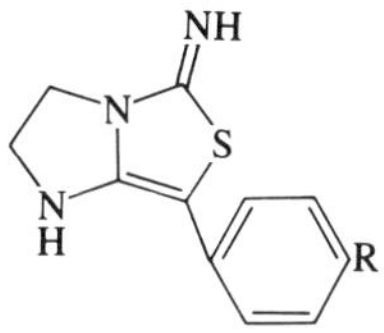	Imidazo[1,2-c]thiazole	R=Hal: CNS stimulant; R=H: antihypotensive	71USP3555039
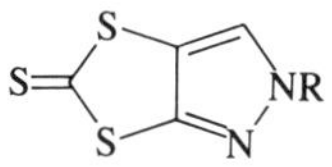	1,3-Dithiolo[4,5-d]pyrazole	Fungicide	81USP4275073
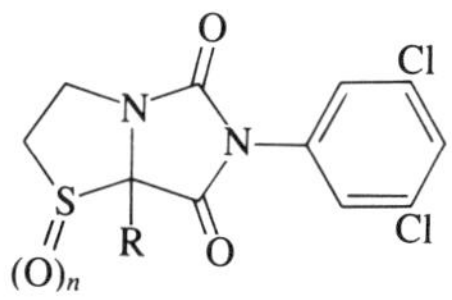	Imidazo[5,1-b]thiazole	Fungicide	80USP4229461
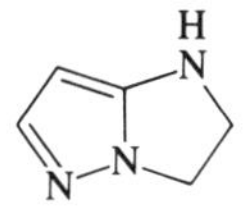	Pyrazolo[2,3-a]imidazole	Sedative, antipyretic	61USP2989537
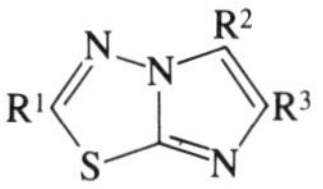	Imidazo[2,1-b]-[1,3,4]thiadiazole	Antithrombic	81USP4265898
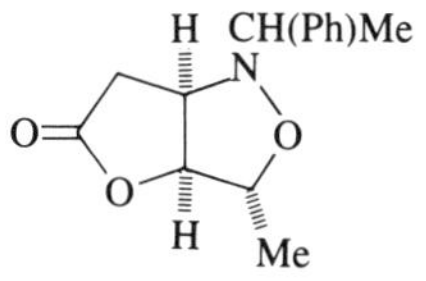	Furo[3,2-c]isoxazole	Antitumor agent intermediate	81USP4252964
	Isosorbide (R=H)	Diuretic	B-76MI43600
	Isosorbide dinitrate (R=NO$_2$)	Coronary vasodilator	B-76MI43600
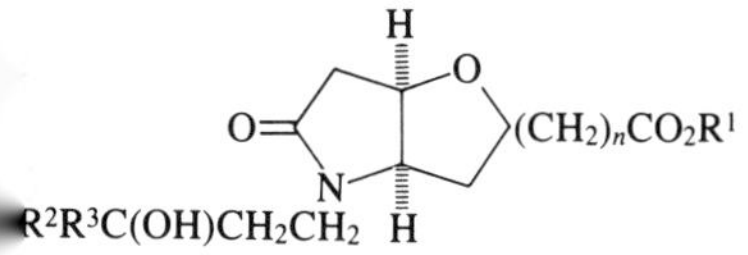	Furo[3,2-b]pyrrole	Bronchodilator	81USP4262008

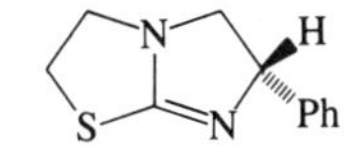

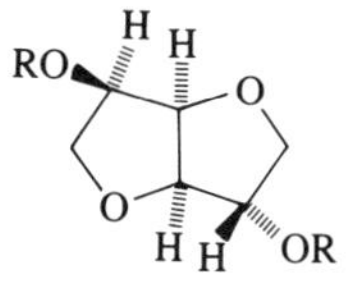

Table 2 ((continued))

Structure	Name or ring system	Activity	Ref.
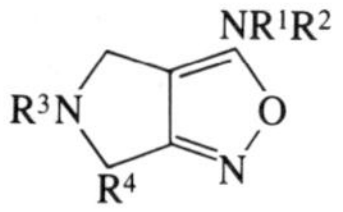	Pyrrolo[3,4-c]isoxazole	Produces hypotension	68JMC453
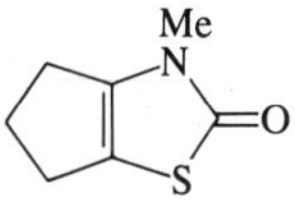	Cyclopentathiazol-2-one	Analgesic	57JA1516
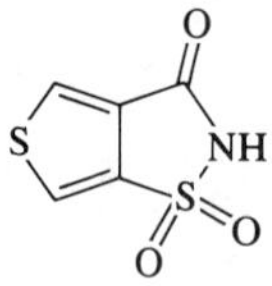	Thiophenesaccharin	Sweetener	76GEP2534689
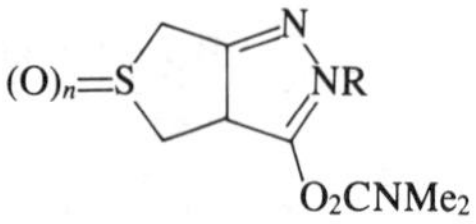	Thieno[3,4-c]pyrazole	Insecticide, acaricide, nematocide	81GEP3003019
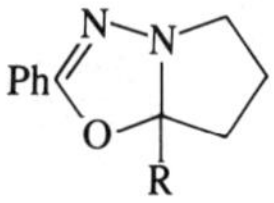	Pyrrolidino[1,2-b]-[1,3,4]oxadiazole	Drug, agricultural chemical	81JAP8125390

4.37

Two Fused Five-membered Heterocyclic Rings: (ii) Non-classical Systems

C. A. RAMSDEN
May & Baker Ltd., Dagenham

4.37.1	HETEROPENTALENE MESOMERIC BETAINES OF TYPES A, B, C AND D: CLASSIFICATION AND NOMENCLATURE	1028
4.37.2	STRUCTURE OF HETEROPENTALENE MESOMERIC BETAINES	1029
	4.37.2.1 Electronic Structure	1029
	4.37.2.2 Spectroscopic Properties	1031
	4.37.2.3 X-Ray Diffraction	1031
4.37.3	REACTIVITY: TYPE A SYSTEMS	1033
	4.37.3.1 General Features	1033
	4.37.3.2 Pyrrolo[3,4-c]pyrazoles	1034
	4.37.3.3 Thieno[3,4-c]pyrazoles	1034
	4.37.3.4 Thieno[3,4-c]isothiazoles	1034
	4.37.3.5 Pyrazolo[4,3-c]pyrazoles	1034
	4.37.3.6 Thieno[3,4-c][1,2,5]oxadiazoles	1035
	4.37.3.7 Thieno[3,4-d][1,2,3]triazoles	1036
	4.37.3.8 Thieno[3,4-c][1,2,5]thiadiazoles	1036
	4.37.3.9 [1,2,3]Triazolo[4,5-d][1,2,3]triazoles	1036
	4.37.3.10 [1,2,3]Triazolo[4,5-c][1,2,5]oxadiazoles	1037
	4.37.3.11 [1,2,3]Triazolo[4,5-c][1,2,5]thiadiazoles	1037
	4.37.3.12 [1,2,3]Triazolo[4,5-c][1,2,5]selenadiazoles	1037
	4.37.3.13 [1,2,5]Thiadiazolo[3,4-c][1,2,5]thiadiazole	1037
	4.37.3.14 [1,2,5]Selenadiazolo[3,4-c][1,2,5]thiadiazole	1037
4.37.4	REACTIVITY: TYPE B SYSTEMS	1037
	4.37.4.1 General Features	1037
	4.37.4.2 Pyrazolo[1,2-a]pyrazoles	1038
	4.37.4.3 Pyrazolo[1,2-a][1,2,3]triazoles	1038
	4.37.4.4 [1,2,3]Triazolo[1,2-b][1,2,3]triazoles	1039
	4.37.4.5 [1,2,3]Triazolo[1,2-a][1,2,3]triazoles	1040
4.37.5	REACTIVITY: TYPE C SYSTEMS	1040
	4.37.5.1 General Features	1040
	4.37.5.2 Pyrrolo[1,2-c]imidazoles	1040
	4.37.5.3 Pyrrolo[1,2-c]thiazoles	1040
	4.37.5.4 Pyrazolo[2,3-c]imidazoles	1041
	4.37.5.5 Pyrazolo[2,3-c]thiazoles	1041
	4.37.5.6 Imidazolo[1,2-c]thiazoles	1041
	4.37.5.7 Imidazolo[1,2-c][1,2,3]triazoles	1042
	4.37.5.8 Imidazolo[1,2-d][1,2,4]triazoles	1042
4.37.6	REACTIVITY: TYPE D SYSTEMS	1042
	4.37.6.1 General Features	1042
4.37.7	SYNTHESIS: TYPE A SYSTEMS	1042
	4.37.7.1 General Methods	1042
	4.37.7.2 Pyrrolo[3,4-c]pyrazoles	1043
	4.37.7.3 Thieno[3,4-c]pyrazoles	1043
	4.37.7.4 Thieno[3,4-c]isothiazoles	1043
	4.37.7.5 Pyrazolo[4,3-c]pyrazoles	1044
	4.37.7.6 Thieno[3,4-c][1,2,5]oxadiazoles	1044
	4.37.7.7 Thieno[3,4-d][1,2,3]triazoles	1044
	4.37.7.8 Thieno[3,4-c][1,2,5]thiadiazoles	1044
	4.37.7.9 [1,2,3]Triazolo[4,5-d][1,2,3]triazoles	1044

4.37.7.10 [1,2,3]Triazolo[4,5-c][1,2,5]oxadiazoles — 1045
4.37.7.11 [1,2,3]Triazolo[4,5-c][1,2,5]thiadiazoles — 1045
4.37.7.12 [1,2,3]Triazolo[4,5-c][1,2,5]selenadiazoles — 1045
4.37.7.13 [1,2,5]Thiadiazolo[3,4-c][1,2,5]thiadiazole — 1045
4.37.7.14 [1,2,5]Selenadiazolo[3,4-c][1,2,5]thiadiazole — 1045

4.37.8 SYNTHESIS: TYPE B SYSTEMS — 1045
    4.37.8.1 General Methods — 1045
    4.37.8.2 Pyrazolo[1,2-a]pyrazoles — 1046
    4.37.8.3 Pyrazolo[1,2-a][1,2,3]triazoles — 1046
    4.37.8.4 [1,2,3]Triazolo[1,2-b][1,2,3]triazoles — 1046
    4.37.8.5 [1,2,3]Triazolo[1,2-a][1,2,3]triazoles — 1047

4.37.9 SYNTHESIS: TYPE C SYSTEMS — 1047
    4.37.9.1 General Methods — 1047
    4.37.9.2 Pyrrolo[1,2-c]imidazoles — 1047
    4.37.9.3 Pyrrolo[1,2-c]thiazoles — 1047
    4.37.9.4 Pyrazolo[2,3-c]imidazoles — 1047
    4.37.9.5 Pyrazolo[2,3-c]thiazoles — 1047
    4.37.9.6 Imidazolo[1,2-c]thiazoles — 1048
    4.37.9.7 Imidazolo[1,2-c][1,2,3]triazoles — 1048
    4.37.9.8 Imidazolo[1,2-d][1,2,4]triazoles — 1048

4.37.10 SYNTHESIS: TYPE D SYSTEMS — 1048
    4.37.10.1 Anhydrocyclopenta[d]thiazolium hydroxides — 1048

4.37.1 HETEROPENTALENE MESOMERIC BETAINES OF TYPES A, B, C AND D: CLASSIFICATION AND NOMENCLATURE

This chapter describes heterocyclic systems which are isoconjugate with the pentalenyl dianion (**13**) and which cannot be represented by classical Kekulé structures. Molecules of this type (*e.g.* **9–12**) can only be represented by dipolar structures, or possibly by structures involving tetracovalent sulfur atoms, and are described as conjugated mesomeric betaines or non-classical heteropentalenes. The structure and chemistry of this type of heterocycle have been the subject of two earlier reviews ⟨77T3203, 77HC(30)317⟩ and their classification as conjugated mesomeric betaines and relationship to other dipolar heterocycles has also been discussed ⟨83UP43700⟩.

It has been demonstrated that there are 10 possible general types of neutral heteropentalene which are isoconjugate with the pentalenyl dianion (**13**). Four of these general types (**1–4**) are mesomeric betaines and they are conveniently described as heteropentalene mesomeric betaines of type A (**1**), type B (**2**), type C (**3**) and type D (**4**) ⟨77T3203⟩. The difference between these four classes can be appreciated by considering the structures (**5**)–(**8**) in which a–h represent suitably substituted carbon or heteroatoms and the superscripts indicate the origin of the 10 π-electrons. Structures (**9**)–(**12**) are known examples of these classes of molecule.

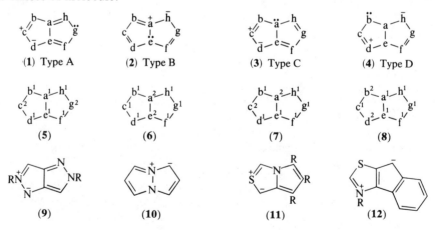

Most of the known heteropentalene mesomeric betaines are of type A and type B but interest has recently begun to focus on type C systems. At present only a single example of a type D heteropentalene has been reported. In their structure and chemistry, the systems of types A, B and C have many common features which arise from similarities in their chemical bonding (see Section 4.37.2.1). In contrast, systems of type D can be expected to have different chemical properties.

The heterocycles in this chapter are named according to the IUPAC method. For example the systems (9) are named as pyrazolo[4,3-c]pyrazoles and compound (10) is pyrazolo[1,2-a]pyrazole.

4.37.2 STRUCTURE OF HETEROPENTALENE MESOMERIC BETAINES

4.37.2.1 Electronic Structure

It is instructive to consider the heteropentalene mesomeric betaines as perturbations of the pentalenyl dianion (13) ⟨77T3203⟩. The HOMO of the dianion (13) is a non-bonding molecular orbital (NBMO) (14) and inspection of the structures of the elementary type A, B and C heteropentalenes (15)–(17) demonstrates that the heteroatoms (X and Y) are exclusively associated with positions at which the NBMO of the pentadienyl anion vanishes. To a first approximation, the monocentric perturbations introduced by the heteroatoms will have no effect on the NBMO energy and type A, B and C systems can be expected to be associated with HOMOs which are high in energy and which have the topology of a NBMO (i.e. 18–20). Aza substitution at positions where the NBMO does not vanish will, of course, lower the energy of the HOMO. The orbital characteristics of type D mesomeric betaines (21) are quite different to those of types A–C. In this case the heteroatoms (X and Y) introduce a large perturbation of the pentalenyl NBMO and the nature of the HOMO (22) will be significantly different.

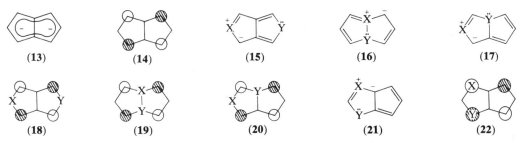

Figure 1 shows the relationship between the π-MOs of the pentalenyl dianion and heteropentalene mesomeric betaines, as calculated by the Hückel (HMO) method ⟨74BCJ1490, 82UP43701⟩. The calculated orbital energies and symmetries are in good agreement with qualitative conclusions. A feature worthy of note is the reordering of the energies of the unoccupied orbitals: the LUMOs of type A and type B systems are calculated to have different symmetries. The high energy HOMO of the type A–C systems is clearly discernible and type A systems are calculated to have a particularly small HOMO–LUMO gap. HMO calculations on polyaza type A and B systems indicate that, as expected, aza substitution at positions 1, 3, 4 and 6 lowers the HOMO energy ⟨74BCJ1490⟩. The calculated effect of aza substitution on type A systems is shown in Figure 2. Note that the HOMO–LUMO gap is calculated to increase with increasing aza substitution.

The HMO calculations described above provide some useful insight into the electronic structure of the heteropentalene mesomeric betaines. However, the HMO method must be applied with some caution to non-alternant systems or heterosystems. It is reassuring to note, therefore, that the Pariser–Parr–Pople (PPP) and CNDO/2 ⟨77T3203⟩ methods have also been employed to study this class of mesomeric betaine and the results are in good agreement with the conclusions of the HMO method. In recent studies the electronic and PE spectra of type A systems have been rationalized using PPP and CNDO/2 models ⟨76JA7187, 78JOC3893, 79CB260⟩ (see also Section 4.37.2.2 and Chapter 3.18).

The question of the extent of sulfur 3d-orbital participation in the bonding of some type A and C heteropentalenes is an interesting one. In valence bond terms d-orbital participation is introduced by assuming that canonical forms involving tetracovalent sulfur (e.g. 23 ↔ 24) make a significant contribution to the structure. In terms of MO theory, d-orbitals are introduced by mixing the HOMO (i.e. pseudo-NBMO) with a sulfur d_{yz}-orbital (25). The

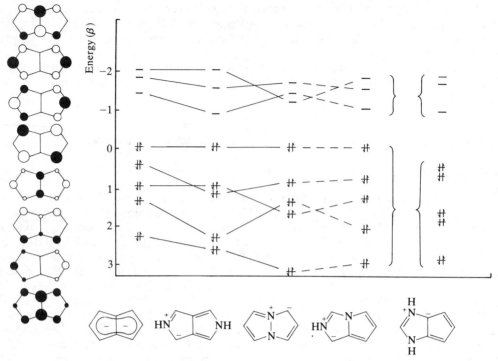

Figure 1 π-Molecular orbitals of representative heteropentalene mesomeric betaines calculated using the Hückel method

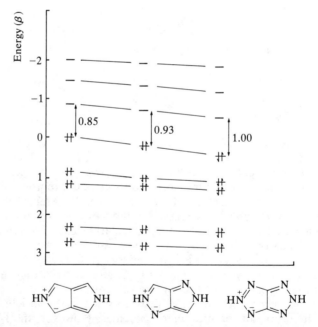

Figure 2 The effect of aza substitution on the π-molecular orbitals of type A heteropentalenes ⟨74BCJ1490⟩

factors which may influence the participation of d-orbitals have been discussed in a review ⟨77T3203⟩. Recent studies of the electronic structure of thieno[3,4-c]thiophenes (**23**; X=S) suggest that d-orbitals do not make an important contribution to the ground state ⟨76JA7187, 78JOC3893⟩.

4.37.2.2 Spectroscopic Properties

The PE spectra of tetraphenylthieno[3,4-c]thiophene (**26**) and triphenylthieno[3,4-c]isothiazole (**29**) have been recorded ⟨76JA7187, 78JOC3893, 79CB260⟩ and the low first ionization energies (**26**; 6.2 eV) and (**29**; 6.9 eV) are consistent with the high HOMO energy expected from theory (see Section 4.37.2.1). Aza substitution (**29**) lowers the HOMO energy (ca. 0.7 eV).

Absorption spectra of a number of type A systems have been measured and these are summarized in Table 1. The parent systems (**26**)–(**28**) are highly coloured suggesting a small HOMO–LUMO gap. Aza substitution results in a hypsochromic shift of the first absorption band presumably due to an increase in frontier orbital separation resulting from differential perturbation of HOMO and LUMO. These aspects of the spectra are consistent with the HMO model (Section 4.37.2.1). Absorption spectra of type B systems are not well documented. Pyrazolo[1,2-a]pyrazole (**10**) is a colourless compound and shows an absorption band at 284 nm in ethanol solution ⟨66JA1992⟩.

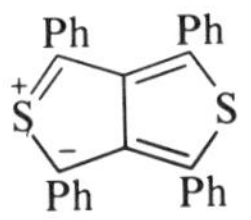

(**26**)

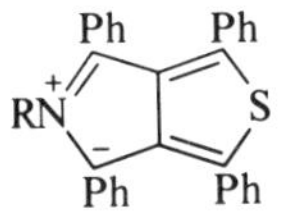

(**27**) R = Ph
(**28**) R = Me

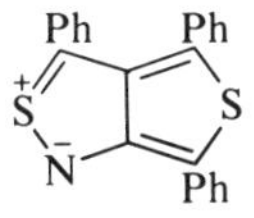

(**29**)

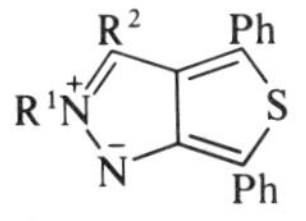

(**30**) R^1 = Me, R^2 = H
(**31**) R^1 = Ph, R^2 = Me
(**32**) R^1 = Ph, R^2 = H

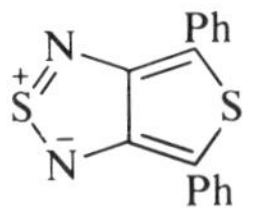

(**33**)

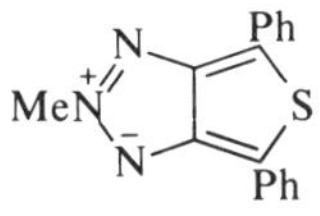

(**34**)

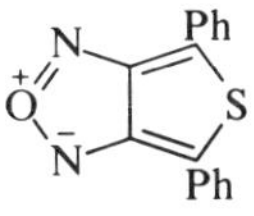

(**35**)

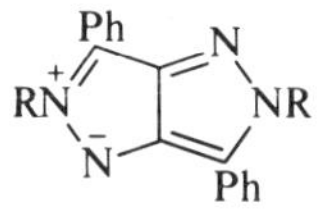

(**36**) R = p-ClC$_6$H$_4$

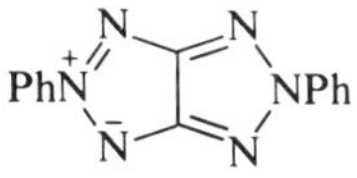

(**37**)

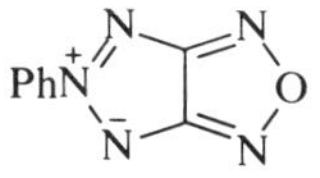

(**38**)

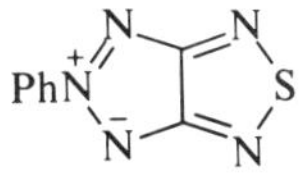

(**39**)

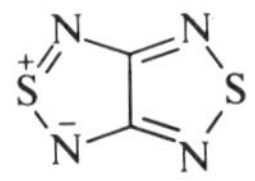

(**40**)

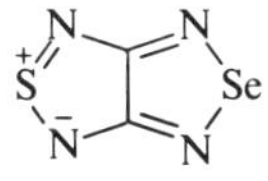

(**41**)

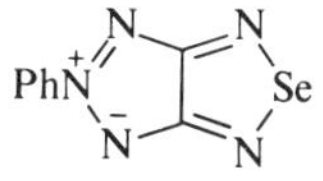

(**42**)

The IR and NMR spectra of these molecules have no unifying features. MS has been shown to be a useful technique for structural verification ⟨77HC(30)317⟩.

4.37.2.3 X-Ray Diffraction

The crystal structure of tetraphenylthieno[3,4-c]thiophene (**26**) has been determined (see Chapter 3.18).

The planar structure of [1,2,5]thiadiazolo[3,4-c][1,2,5]thiadiazole (**43**) has been confirmed and the geometry is in accord with its symmetrical mesomeric betaine structure (C—C, 1.44; N—C, 1.35; N—S, 1.62 Å; ∠N—C—C, 114.0°; ∠C—N—S, 104.4°; ∠N—S—N, 103.2°) ⟨75JOC2749⟩.

Studies of the crystal structures of four [1,2,3]triazolo[1,2-b][1,2,3]triazoles (**44**) have been reported. The dibromo derivative (**44**; R^1 = Me, R^2 = Br) has a geometry consistent with its representation as a mesomeric betaine [N(1)—N(2), 1.36; N(2)—N(3), 1.40; N(2)—C, 1.40; N(3)—C, 1.37; C—C, 1.33 Å] and similar results have been obtained for the rubidium salts (**44**; R^1 = Me, R^2 = CO$_2^-$Rb$^+$) and (**44**; R^1 = CO$_2^-$Rb$^+$, R^2 = H) ⟨72PMH(5)283⟩. The dibenzo derivative (**108**; R = H) has been shown to be planar ⟨63AX64⟩.

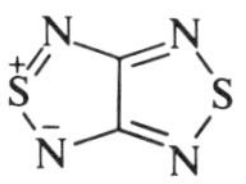

(**43**)

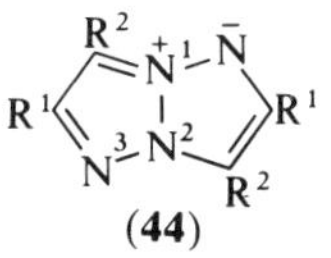

(**44**)

Table 1 Physical Characteristics and Absorption Spectra of Type A Systems

Compound	M.p. (°C)	Appearance	Solvent	λ_{max}^{a} (log ε)	Ref.
(26)	—	Purple needles	CHCl$_3$	258 (4.30), 262sh (4.27), 292 (4.15), 551 (3.92)	74JA4268
(27)	257–258	Reddish purple needles	C$_2$H$_4$Cl$_2$	265 (4.44), 295 (4.42), 553 (4.04)	73JA2558
(28)	212–214	Red needles	C$_2$H$_4$Cl$_2$	247 (4.34), 260 (4.35), 345sh (3.82), 526 (3.86)	74JA1817
(29)	210–212	Red needles	CHCl$_3$	256 (4.41), 533 (3.15)	74JA4268
	—		MeOH	237sh (4.15), 256sh (4.20), 275 (4.25), 297sh (3.97), 329sh (3.73), 526 (3.85)	77HC(30)317
	—	Purple needles	CH$_2$Cl$_2$	243 (4.12), 280 (4.41), 300 (4.17), 346 (3.91), 528 (4.12)	79CB260
(30)	133–135	Orange prisms	MeOH	264 (4.40), 313sh (3.70), 455sh (4.41), 465 (4.44)	74JA4276
(31)	189–191	Brown plates	CHCl$_3$	289 (4.33), 322sh (3.79), 478 (4.33)	74JA4276
(32)	200–202	Brick red needles	MeOH	250 (4.12), 276 (4.21), 298 (4.18), 497 (4.06)	74JA4276
(33)	146	Purple needles	CH$_2$Cl$_2$	275 (4.22), 312 (4.33), 330 (4.30), 558 (3.94)	69JA6891
(34)	158–160	Orange prisms	CHCl$_3$	265 (4.29), 292sh (4.06), 313 (4.00), 322sh (3.98), 466 (4.24)	77HC(30)317
(35)	161–165	Blue crystals	Insoluble	—	77H(6)1173
(36)	328	Pale yellow crystals	CHCl$_3$	292 (4.37), 398 (4.33)	74BCJ946
(37)	249	Colourless leaflets	EtOH	231 (4.11), 342 (4.62)	74BCJ1493
(38)	165.5–166.5	Yellow crystals	EtOH	234 (3.59), 319 (4.27)	74BCJ1493
(39)	181–182.5	Yellow crystals	EtOH	235 (3.61), 337 (4.47)	74BCJ1493
(40)	115.7–116	Colourless prisms	MeOH	215 (3.17), 317 (4.32)	75JOC2749
(41)	—	Orange solid		—	76JHC13
(42)	203.8–204.5	Yellow crystals	EtOH	238 (3.47), 365 (4.52)	74BCJ1493

^{a} nm.

4.37.3 REACTIVITY: TYPE A SYSTEMS

4.37.3.1 General Features

It is useful to consider the cycloaddition reactions of these systems in general terms. Other aspects of their chemistry have not been studied in sufficient detail to justify a general treatment.

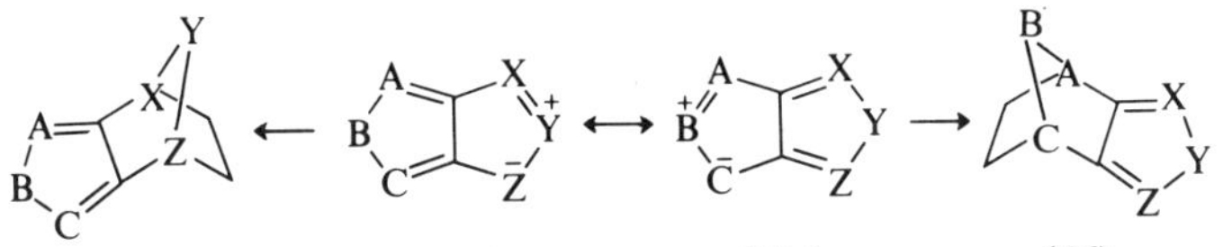

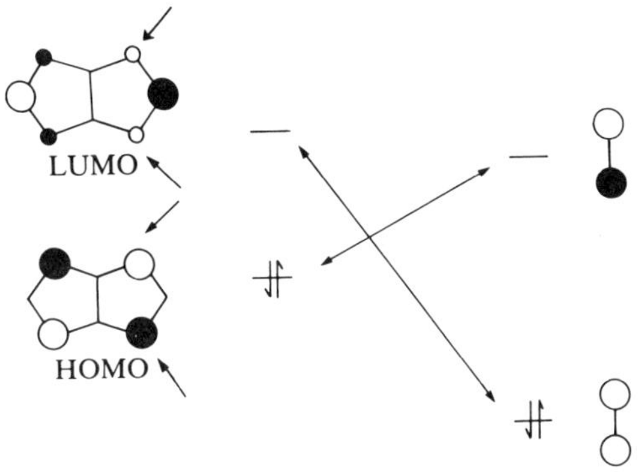

Figure 3 Frontier orbital interactions in the thermal 1,3-dipolar cycloadditions of type A mesomeric betaines

The type A systems can be regarded as 'masked' 1,3-dipoles and an important feature of their chemistry is participation in 1,3-dipolar cycloaddition reactions (**45a → 46**). In accord with frontier orbital theory these heteropentalenes (high energy HOMO) are particularly reactive towards electron-deficient 1,3-dipolarophiles (low energy LUMO). The second order HOMO-LUMO interactions for cycloaddition are shown in Figure 3. A large bonding interaction between betaine HOMO and dipolarophile LUMO leads to transition state stabilization. Aza substitution lowers the energy of the betaine HOMO and decreases the reactivity by reducing the frontier orbital interaction. Accordingly, the parent systems (*e.g.* **26–28**) are reactive 1,3-dipoles whereas the pyrazolo[4,3-*c*]pyrazoles (*e.g.* **36**) and the triazolo[4,5-*c*]triazoles (*e.g.* **37**) do not appear to show any 1,3-dipolar reactivity.

An interesting aspect of the type A heteropentalenes is the fact that each molecule is associated with two 1,3-dipolar fragments (**45a ↔ 45b**) and, in principle, unsymmetrical systems can form two types of cycloadduct (**46** or **47**). In some cases the kinetically controlled product (**46**) is obtained at low temperature and the thermodynamically controlled product (**47**) is obtained at higher temperatures (see thieno[3,4-*c*]pyrroles, Chapter 3.18). For a given set of reaction conditions cycloaddition is usually site specific. For example, the non-classical thiophene derivatives of general structure (**48**) usually add across the thiocarbonyl ylide fragment. This site selectivity is probably determined by the relative size of the HOMO coefficients at the alternative sites of addition.

Addition of alkynic 1,3-dipolarophiles to the thiocarbonyl ylide systems (**48**) gives the adducts (**49**) which readily lose sulfur giving the benzoheterocycles (**50**) and this is a useful synthetic route to these systems. The heterocycles (**50**) have also been made by elimination of hydrogen sulfide or methylamine from analogous alkenic adducts.

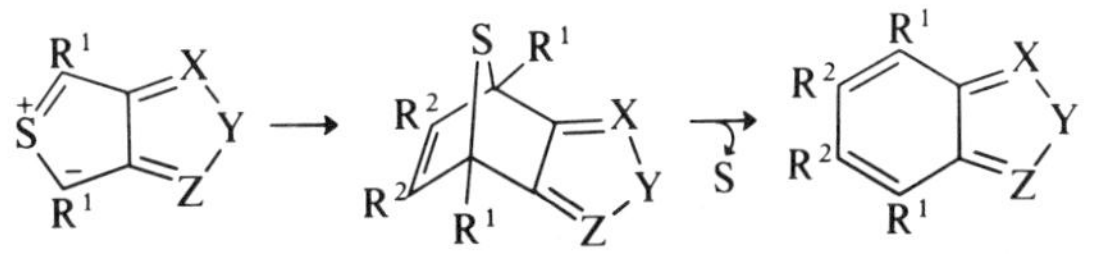

4.37.3.2 Pyrrolo[3,4-c]pyrazoles

The pyrrolo[3,4-c]benzopyrazole (**51**; R^1=Ph, R^2 = Me, R^3 = CO$_2$Et) reacts with DMAD giving the cycloadduct (**52**; R^1 = Ph, R^2 = Me, R^3 = CO$_2$Et). Reaction with *N*-phenylmaleimide in boiling xylene slowly gives compound (**53**), presumably formed by elimination of methylamine from the primary adduct ⟨79JOC622⟩.

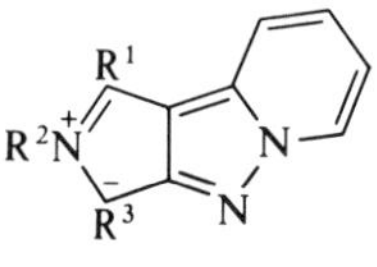

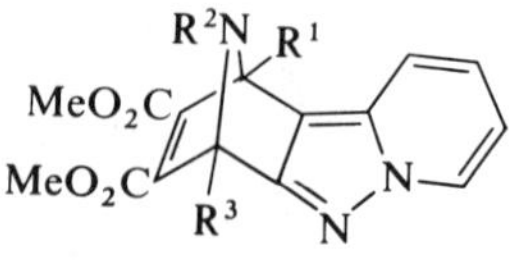

4.37.3.3 Thieno[3,4-c]pyrazoles

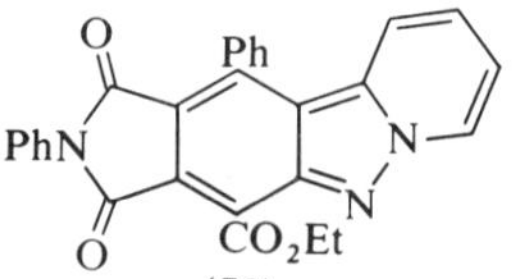

In their cycloaddition reactions the thieno[3,4-c]pyrazoles (**54**) react exclusively as thiocarbonyl ylides. *N*-Phenylmaleimide gives the *endo* cycloadducts (**55**); a low yield of the *exo* cycloadduct (**56**) was also encountered in one case. The triphenyl derivative (**54**; $R^1 = R^2 = R^3$ = Ph, R^4 = H) and fumaronitrile give a primary adduct which readily eliminates hydrogen sulfide to give the 2*H*-indazole (**57**; $R^1 = R^2 = R^3$ = Ph, R^4 = H, R = CN). The primary adducts (**58**) formed by addition of alkynic dipolarophiles to the betaines (**54**) also give 2*H*-indazole derivatives (**57**) by spontaneous loss of elemental sulfur ⟨74JA4276⟩. The ring-fused derivatives (**59**) show similar chemical behavior ⟨80JOC90⟩.

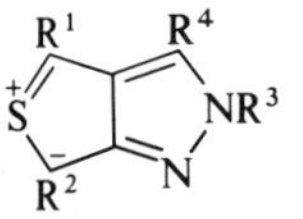

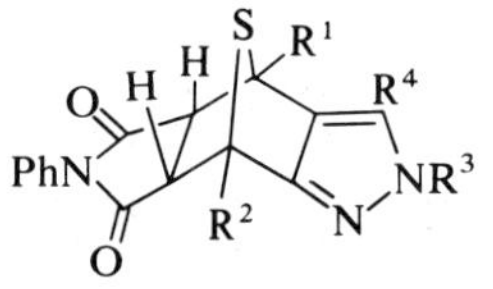

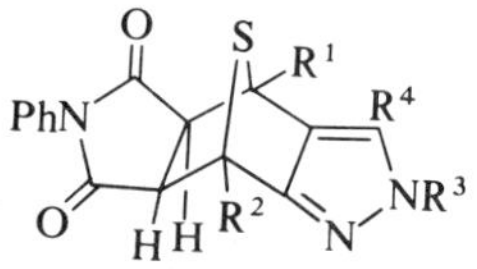

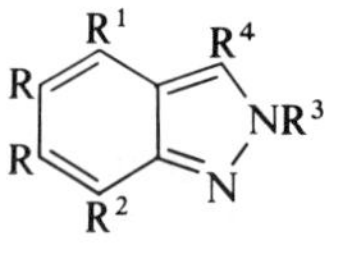

4.37.3.4 Thieno[3,4-c]isothiazoles

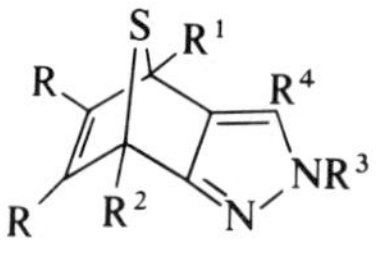

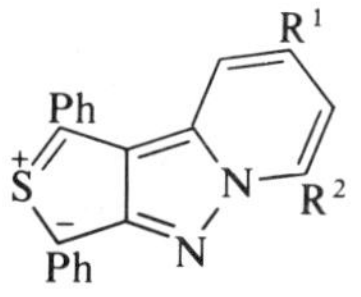

Like the thieno[3,4-c]pyrazoles (**54**), triphenylthieno[3,4-c]isothiazole (**60**; $R^1=R^2=R^3$ = Ph) reacts exclusively as a thiocarbonyl ylide with 1,3-dipolarophiles. Addition of DMAD gives the benzisothiazole (**61**; $R^4 = R^5$ = CO$_2$Me; 86%) by elimination of sulfur from the non-isolable primary adduct. Similarly, methyl propiolate gives a mixture of the regioisomers (**61**; R^4 = H, R^5 = CO$_2$Me) and (**61**; R^4 = CO$_2$Me, R^5 = H). Alkenic dipolarophiles give the adducts (**62**) in good yield. Treatment of these adducts (**62**) with base results in elimination of hydrogen sulfide giving the benzisothiazoles (**61**) ⟨79CB266⟩.

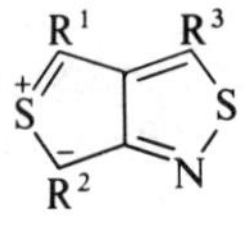

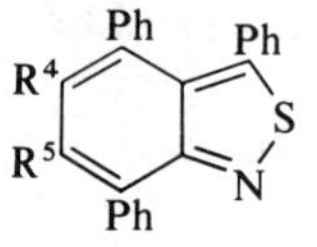

4.37.3.5 Pyrazolo[4,3-c]pyrazoles

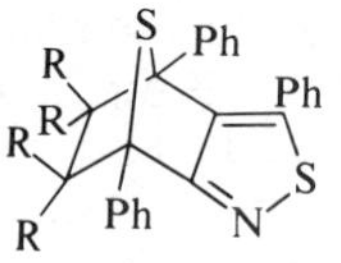

Cycloaddition reactions of pyrazolo[4,3-c]pyrazoles (**63**) have not been reported.
Oxidation of the tetraaryl derivative (**63**; R^1=*p*-ClC$_6$H$_4$, R^2 = Ph) with potassium permanganate or peroxyacetic acid gives the 3-aroyl-4-arylazopyrazole (**64**; R^1 = *p*-ClC$_6$H$_4$,

R² = Ph) ⟨74BCJ946, 73CL455⟩. The phenylazo compound (**63**; R¹ = Ph, R² = N=NPh) is oxidized to the diazoxy derivative (**63**; R¹ = Ph, R² = N=NOPh) by hydrogen peroxide and to the dinitro derivative (**63**; R¹ = Ph, R² = NO₂) by nitric acid ⟨71LA(744)88⟩.

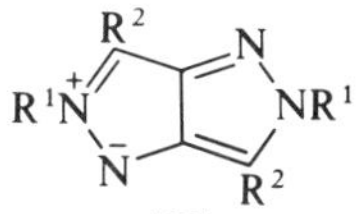

(63) 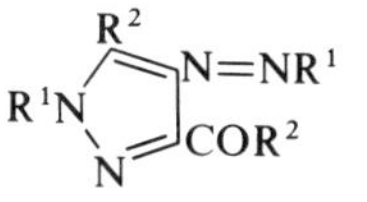   (64) 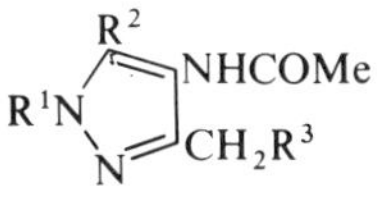   (65)

Catalytic reduction of the derivative (**63**; R¹ = p-ClC₆H₄, R² = Ph) in acetic acid solution gives the pyrazoles (**65**; R¹ = p-ClC₆H₄, R² = Ph, R³ = C₆H₁₁; 25%) and (**65**; R¹ = p-ClC₆H₄, R² = R³ = C₆H₁₁; 10%) ⟨74BCJ946⟩.

Electrophilic substitutions of the phenyl substituents of compound (**63**; R¹ = p-ClC₆H₄, R² = Ph) have been described: nitration gives compound (**63**; R¹ = p-ClC₆H₄, R² = p-NO₂C₆H₄) and similarly bromination gives bromo derivatives. Treatment of compound (**63**; R¹ = p-ClC₆H₄, R² = Ph) with methyl iodide slowly gives the methiodide (**66**; R¹ = p-ClC₆H₄, R² = Ph) ⟨74BCJ946⟩.

Vacuum pyrolysis (500 °C) of compound (**63**; R¹ = p-ClC₆H₄, R² = Ph) gives α-(p-chlorophenylimino)phenylacetonitrile (**67**; R¹ = p-ClC₆H₄, R² = Ph; 30%) ⟨74BCJ946⟩.

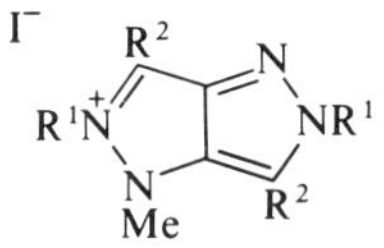

(66) 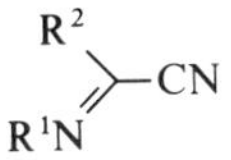   (67)

4.37.3.6 Thieno[3,4-c][1,2,5]oxadiazoles

Cycloaddition reactions of the diphenyl derivative (**68**; R¹ = R² = Ph) have received detailed attention. Addition occurs exclusively across the thiocarbonyl ylide fragment. N-Phenylmaleimide gives a mixture of the *endo* (**69**; 34%) and *exo* (**70**; 40%) adducts. Similar adducts are formed by other maleimides, maleic anhydride, dimethyl fumarate and dimethyl maleate. Base-catalyzed elimination of hydrogen sulfide from the adducts (**69**) or (**70**) gives the benzo[c]oxadiazole derivative (**71**; R¹ = R² = R³ = Ph). Thermolysis of the *endo* adduct (**69**) results in cleavage of the oxadiazole ring giving the nitrile oxide (**72**) which is trapped by DMAD forming the isoxazole (**73**; 74%). Similar behaviour of the *exo* adduct (**70**) is also observed ⟨77H(6)1173, 79CL1029, 80H(14)423, 81H(16)789⟩.

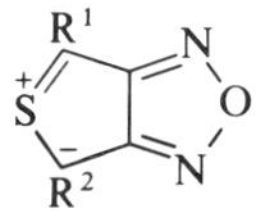

(68) 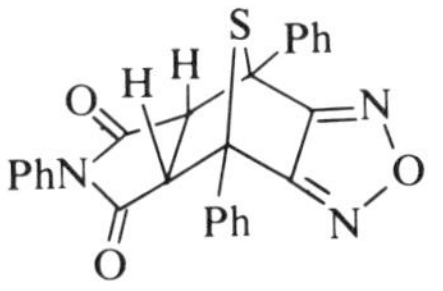   (69) 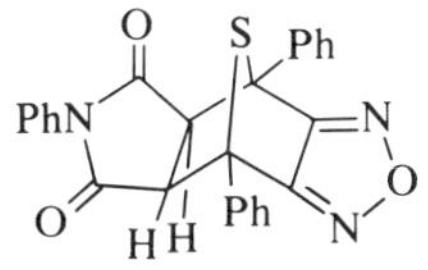   (70)

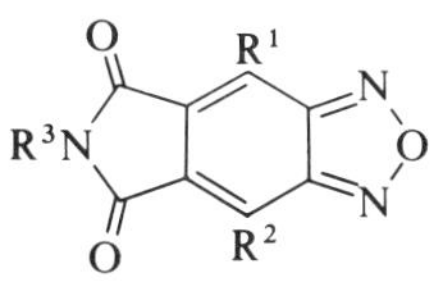

(71) 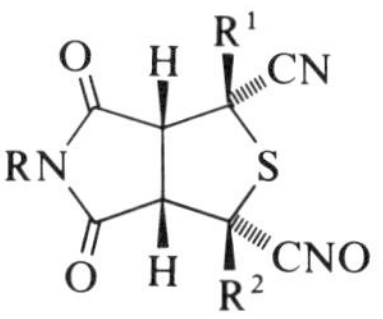   (72) 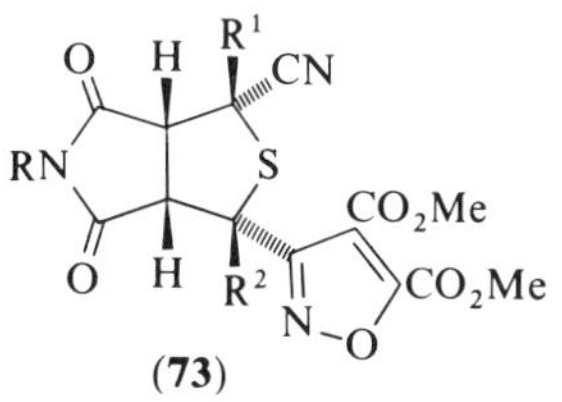   (73)

Reaction with alkynic 1,3-dipolarophiles gives the primary adducts (**74**) which are not isolated. Thermal cleavage of the oxadiazole ring gives the nitrile oxide valence tautomer

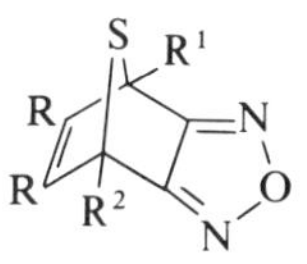

(74) 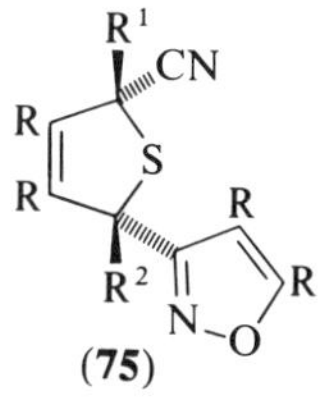   (75) 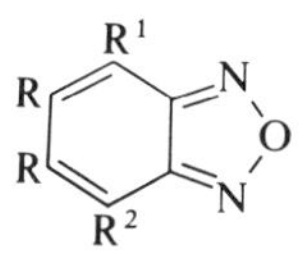   (76)

which is trapped by a second alkyne molecule giving the major product (75). Alternatively, desulfurization of the adduct (74) gives benzoxadiazoles (76) as a minor product ⟨80JOC2956⟩.

Cycloaddition reactions with 6,6-diphenylfulvene and tropone have also been reported ⟨80CL1031⟩.

4.37.3.7 Thieno[3,4-d][1,2,3]triazoles

4,6-Diphenyl-2-methylthieno[3,4-d][1,2,3]triazole (77; $R^1=R^2=Ph$, $R^3=Me$) apparently does not undergo cycloaddition with alkenic or alkynic dipolarophiles ⟨77HC(30)317⟩.

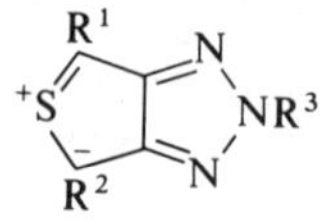

(77)

4.37.3.8 Thieno[3,4-c][1,2,5]thiadiazoles

In the absence of 1,3-dipolarophiles the parent molecule (78; $R^1=R^2=H$) gives the dimer (79; $R^1=R^2=H$) ⟨79TL4493⟩. A similar dimer (79; $R^1=R^2=Ph$) is obtained by irradiation of the diphenyl derivative (78; $R^1=R^2=Ph$) ⟨69JA6891⟩.

Addition of N-phenylmaleimide to compound (78; $R^1=R^2=Ph$) gives a mixture of the exo and endo adducts (80; $R^1=R^2=Ph$). When pure samples of these are heated in boiling xylene, thermal interconversion of the adducts is observed indicating that under these conditions retro-cycloaddition occurs ⟨69JA6891, 79CL1029⟩. The unsubstituted derivative (78; $R^1=R^2=H$) and N-phenylmaleimide yielded only the endo adduct (80; $R^1=R^2=H$) ⟨79TL4493⟩.

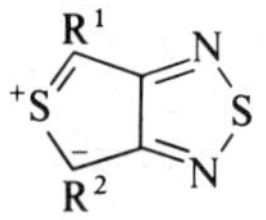

(78) 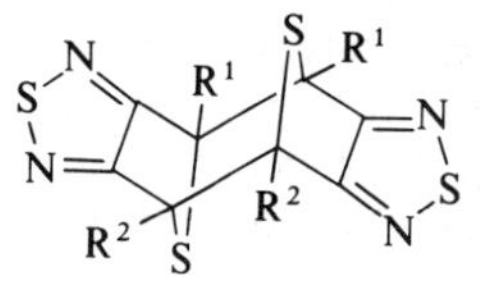   (79) 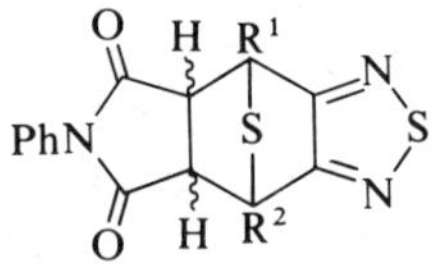   (80)

In the presence of alkynes the thieno[3,4-c][1,2,5]thiadiazoles (78) give benzo-thiadiazoles (81) in good yield. These products are formed by desulfurization of the primary adducts (82) ⟨79TL4493, 80JOC2956⟩. Cycloaddition with 6,6-diphenylfulvene has also been reported ⟨80CL1031⟩.

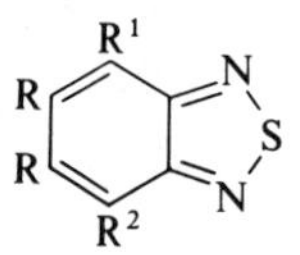

(81) 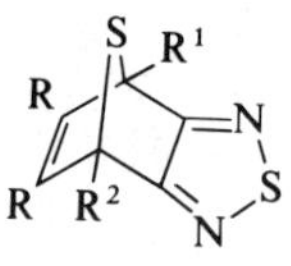   (82)

4.37.3.9 [1,2,3]Triazolo[4,5-d][1,2,3]triazoles

The chemistry of this system (83) remains unexplored. Catalytic reduction of the diphenyl derivative (83; $R^1=R^2=Ph$) gave mono- and di-cyclohexyl derivatives (83; $R^1=C_6H_{11}$, $R^2=Ph$ or C_6H_{11}) ⟨70BCJ3587⟩.

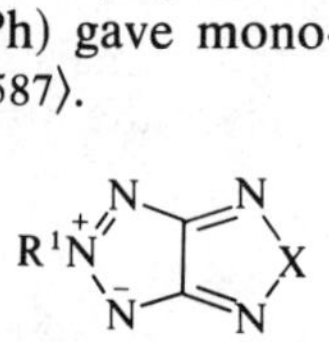

(83) X = NR^2
(84) X = O
(85) X = S
(86) X = Se

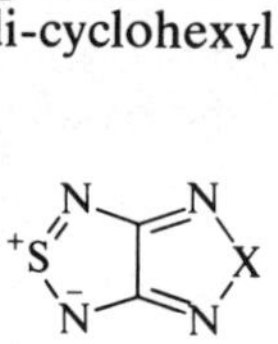

(87) X = S
(88) X = Se

4.37.3.10 [1,2,3]Triazolo[4,5-c][1,2,5]oxadiazoles

Chemical reactions of this system (**84**) have not been reported.

4.37.3.11 [1,2,3]Triazolo[4,5-c][1,2,5]thiadiazoles

The phenyl derivative (**85**; R^1=Ph) is reduced by LAH giving 2-phenyl-4,5-diamino-1,2,3-triazole ⟨74BCJ1493⟩.

4.37.3.12 [1,2,3]Triazolo[4,5-c][1,2,5]selenadiazoles

No chemical reactions of this system (**86**) have been reported.

4.37.3.13 [1,2,5]Thiadiazolo[3,4-c][1,2,5]thiadiazole

Hydrolysis of compound (**87**) gives 3,4-diamino-1,2,5-thiadiazole, oxamide and sulfur. In dilute ammonium hydroxide the yield of 3,4-diamino-1,2,5-thiadiazole is almost quantitative ⟨75JOC2749⟩.

4.37.3.14 [1,2,5]Selenadiazolo[3,4-c][1,2,5]thiadiazole

The chemistry of this system (**88**) has not been investigated.

4.37.4 REACTIVITY: TYPE B SYSTEMS

4.37.4.1 General Features

The general chemistry of the type B heteropentalenes (**90**) has not been widely explored. Like type A systems (Section 4.37.3.1) they are electron rich (high energy HOMO) and can be regarded as masked 1,3-dipolarophiles. They are reactive towards electron-deficient species and the main features of their known chemistry are (i) electrophilic substitution (*e.g.* **90** → **89**) and (ii) cycloaddition reaction with electron-deficient 1,3-dipolarophiles (*e.g.* **90** → **91**). Reactivity is reduced by aza substitution.

The frontier orbital interaction diagram for 1,3-dipolar cycloaddition is shown in Figure 4.

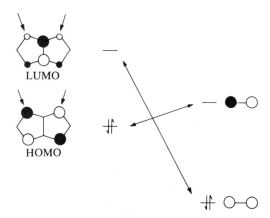

Figure 4 Frontier orbital interactions in the thermal 1,3-dipolar cycloadditions of type B heteropentalenes

4.37.4.2 Pyrazolo[1,2-a]pyrazoles

The pyrazolopyrazoles (92) are unstable in air giving coloured oxidation products. Stable derivatives are obtained by electrophilic substitution at the 1- and 3-positions. Typically, acetic anhydride gives the 1,3-diacetyl derivatives (93) and similar substitutions occur using benzoyl chloride or cyanogen chloride ⟨66JA5588⟩.

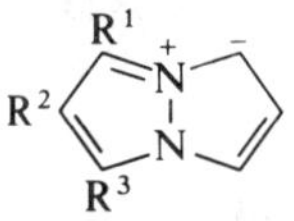

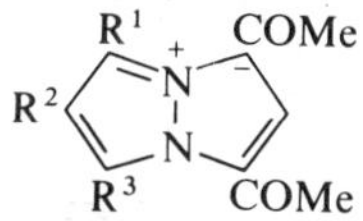

1,3-Dipolar cycloaddition of DMAD to the heteropentalenes (95) gives an interesting variety of products (Scheme 1). In the presence of palladium–charcoal, the 1-benzoyl-2-phenyl derivative (95; R^1 = COPh, R^2 = Ph) gives the 8-azacycl[2.2.2]azine (97; R^1 = COPh, R^2 = Ph) via the primary adduct (94; R^1 = COPh, R^2 = Ph) ⟨66PNA(55)1385⟩. In the absence of the dehydrogenation catalyst and using two moles of DMAD, the derivative (95; R^1 = COPh, R^2 = Ph) gives the pyrazole (99; R^1 = COPh, R^2 = Ph; 65%) by rearrangement of the 1:2 cycloadduct (96; R^1 = COPh, R^2 = Ph) ⟨78CL1093⟩. Reaction of the parent system (95; R^1 = R^2 = H) with excess DMAD gives the product (98; 30%) presumably formed by rearrangement of the adduct (94; R^1 = C(CO$_2$Me)=CHCO$_2$Me) ⟨78CL1093⟩.

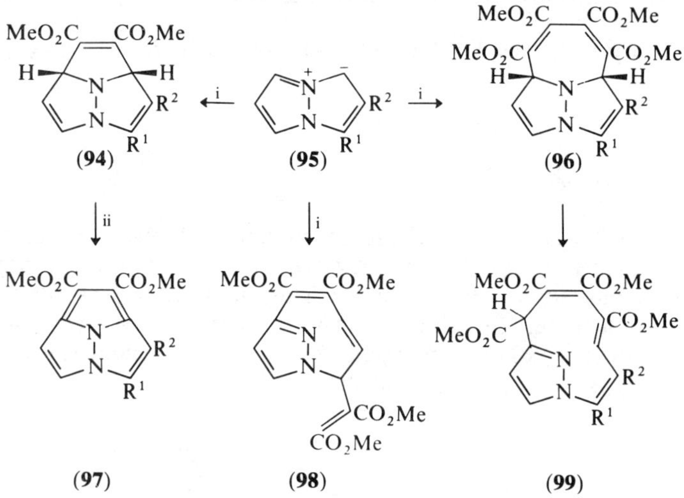

i, DMAD; ii, Pd/C

Scheme 1

4.37.4.3 Pyrazolo[1,2-a][1,2,3]triazoles

The derivatives (100; R^1 = R^2 = R^3 = R^4 = H, R^5 = Me, Ph) undergo electrophilic substitution at position 3. Acetic anhydride gives the 3-acetyl compounds (100; R^1 = COMe, R^2 = R^3 = R^4 = H, R^5 = Me, Ph) in quantitative yield and similarly, using nitrous acid, the 3-nitroso derivatives are obtained. Treatment of compounds (100; R^1 = R^2 = R^3 = R^4 = H, R^5 = Me, Ph) with equimolar amounts of DMAD results in addition across the azomethine fragment giving the 1:1 cycloadducts (101; R = Me, Ph) in high yield. Oxidation of these adducts (101) with 2,3-dichloro-5,6-dicyano-1,4-benzoquinone (DDQ) gives the diazacycl[2.2.2]azines (102; R = Me, Ph) ⟨78TL1291⟩.

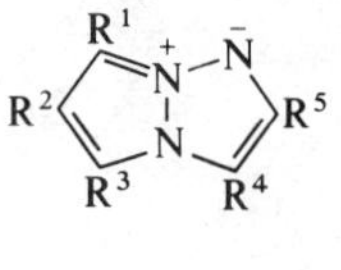

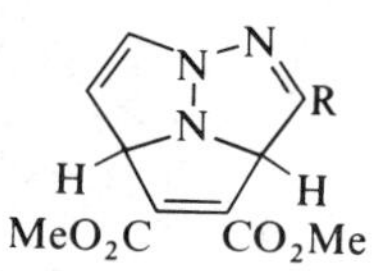

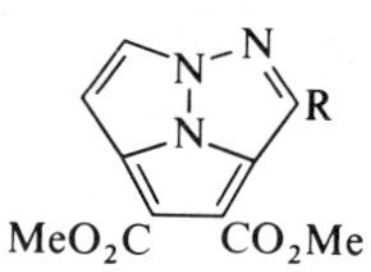

The dibenzo derivatives (**103**; $R^2 = H$) similarly undergo electrophilic substitution and cycloaddition reactions. Typically, under Vilsmeier–Haack conditions the 7-formyl (**103**; $R^2 = CHO$) and the 7-acetyl (**103**; $R^2 = COMe$) derivatives are formed from the unsubstituted compounds (**103**; $R^2 = H$). With strong acids the dibenzo compounds (**103**) form salts ⟨73CL175⟩.

With alkynic 1,3-dipolarophiles the dibenzo derivatives (**103**) react as azomethine imines giving the adducts (**104**). Regiospecific addition is observed when monosubstituted alkynes are used. In some reactions the cycloadduct (**104**) is accompanied by a Michael adduct (*e.g.* **103**; $R^2 = C(CO_2Me)=CHCO_2Me$). Reaction of compound (**103**; $R^1 = Me$, $R^2 = H$) with two equivalents of DMAD gives a product which appears to be the 2:1 cycloadduct (**105**; $R^1 = Me$) ⟨73TL597, 74H(2)27⟩.

(**103**)   (**104**)   (**105**)

Fused heteroaromatic derivatives of the pyrazolotriazoles (**100**) have also been reported ⟨80JCS(P1)2904, 81JCS(P1)4, 1821⟩ and give interesting addition products with DMAD ⟨80JCR(M)4801⟩.

4.37.4.4 [1,2,3]Triazolo[1,2-*b*][1,2,3]triazoles

Electrophilic substitution of this ring system readily occurs. The dimethyl derivative (**106**; $R^1 = H$, $R^2 = Me$) upon treatment with bromine in acetic acid solution gives the dibromo compound (**106**; $R^1 = Br$, $R^2 = Me$) and Vilsmeier formylation gives compound (**106**; $R^1 = CHO$, $R^2 = Me$). The derivative (**106**; $R^1 = H$, $R^2 = Me$) also forms a picrate and yields a 1:1 complex with silver nitrate ⟨63CB1827⟩. The ester functions of compounds (**106**; $R^1 = CO_2Et$, $R^2 = Me$ and $R^1 = CO_2Me$, $R^2 = Ph$) have been transformed into a wide variety of substituents ⟨57CB2411, 63CB1827, 67GEP1245386, 71GEP1620103⟩. The acid (**106**; $R^1 = CO_2H$, $R^2 = Me$), whose structure has been confirmed by X-ray analysis of its rubidium salt (see Section 4.37.2.3), is oxidized to 4-methyl-1,2,3-triazole-5-carboxylic acid by potassium permanganate ⟨63CB1827⟩.

(**106**)   (**107**)

The monobenzo compound (**107**; $R^1 = R^2 = H$) forms a stable salt with methyl iodide. Reaction with TCNE results in elimination of hydrogen cyanide and formation of the tricyanovinyl derivative (**107**; $R^1 = C(CN)=C(CN)_2$, $R^2 = H$) ⟨67JA2626⟩.

The dibenzo derivative (**108**; $R = H$) has been studied in some detail. Electrophilic substitution (*e.g.* halogenation, nitration) gives predominantly the disubstituted derivatives (*e.g.* **108**; $R = NO_2$). Methylation gives an *N*-methyl salt which is thermally unstable. Oxidation by peracetic acid yields the nitroso derivative (**109**; $R = NO$) and reduction using LAH or copper(I) cyanide gives *o*-aminophenyl-2*H*-benzotriazole (**109**; $R = NH_2$) ⟨67JA2618, 67JA2626⟩. These molecules do not equilibrate with the valence tautomeric dibenzo-1,2,5,6-tetraazacyclooctatetraenes (**110**) ⟨67JA2618⟩.

(**108**)   (**109**)   (**110**)

4.37.4.5 [1,2,3]Triazolo[1,2-a][1,2,3]triazoles

The bicyclic species (**111**) are unknown but some reactions of the monobenzo (**112**) and dibenzo (**113**) species have been described and they closely resemble those of the isomeric betaines (**107**) and (**108**) (Section 4.37.4.4). Electrophilic substitution occurs on the benzenoid rings of compound (**113**) and treatment with methyl iodide slowly gives an adduct. The monobenzo derivative is more reactive, readily giving a methiodide and rapidly undergoing electrophilic substitution reactions. Reaction of compound (**112**; R = H) with TCNE results in elimination of hydrogen cyanide and formation of the tricyanovinyl derivative (**112**; R = C(CN)=C(CN)$_2$) ⟨67JA2633⟩.

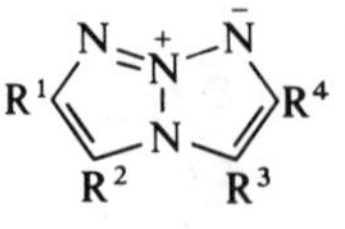

(**111**) 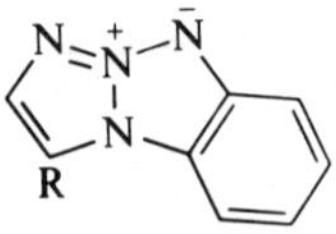   (**112**) 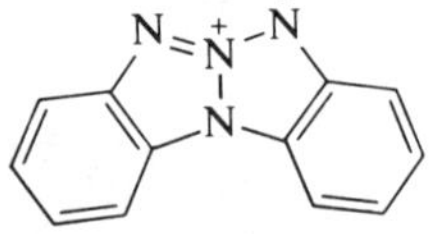   (**113**)

4.37.5 REACTIVITY: TYPE C SYSTEMS

4.37.5.1 General Features

Too little is known about the chemistry of type C heteropentalene mesomeric betaines to justify a discussion of their general features. In their chemical reactions, type C systems are expected to show a mixture of type A and type B character ⟨77T3203⟩ and the limited information which is available supports this view.

4.37.5.2 Pyrrolo[1,2-c]imidazoles

Base-catalyzed deuteration of the 5- and 7-positions of *N*-methyl-5*H*-pyrrolo[1,2-c]imidazolium iodide (**115**) occurs *via* the mesomeric betaine (**114**; R = Me). This species (**114**; R = Me) has been trapped by DMAD to give the 2:1 adduct (**116**) — a reaction which is characteristic of type B heteropentalene mesomeric betaines ⟨82UP43700⟩.

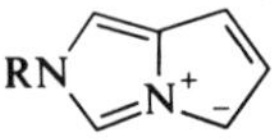

(**114**) 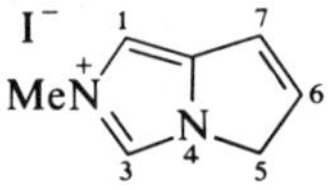   (**115**) 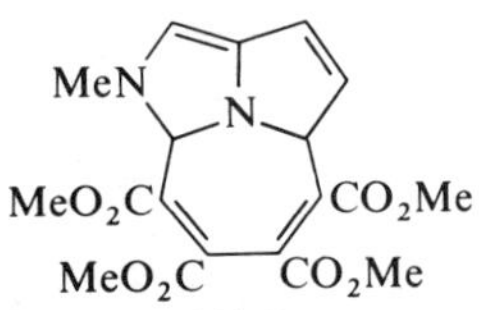   (**116**)

4.37.5.3 Pyrrolo[1,2-c]thiazoles

The pyrrolothiazole (**117**; R^1 = R^2 = CO$_2$Me, R^3 = Me) reacts either as a thiocarbonyl ylide (**117a**) or an azomethine ylide (**117b**) depending upon the nature of the dipolarophile. Reaction with *N*-phenylmaleimide gives a mixture of the *exo* and *endo* adducts (**118**) which upon treatment with sodium methoxide yield the indolizine derivative (**119**). In contrast, reaction of the species (**117**; R^1 = R^2 = CO$_2$Me, R^3 = Me) with DMAD gives the alternative type of 1,3-dipolar cycloadduct (**120**) ⟨80JOC5396⟩.

The benzo derivative (**121**) is reported to undergo spontaneous intramolecular cyclization to compound (**122**) ⟨77CL1237⟩.

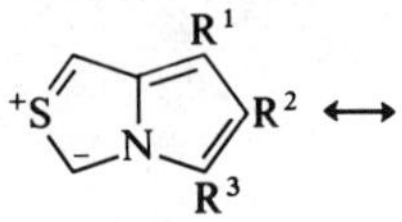

(**117a**) 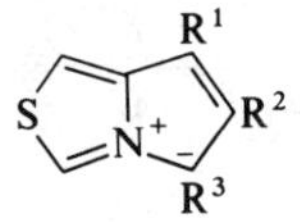   (**117b**) 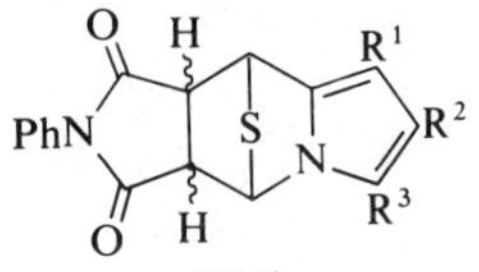   (**118**)

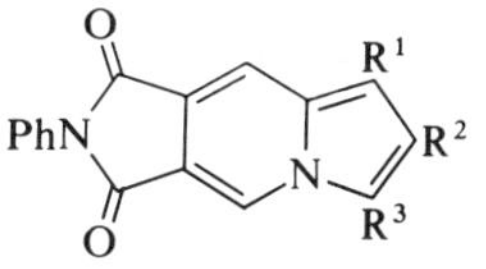

(119) 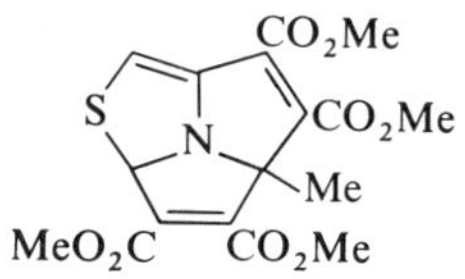 (120) 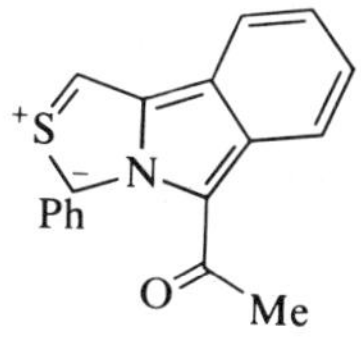 (121) 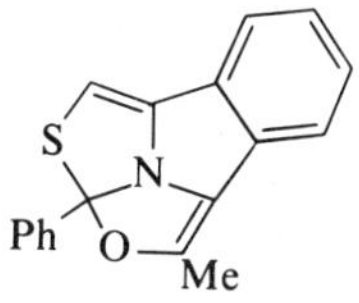 (122)

4.37.5.4 Pyrazolo[2,3-c]imidazoles

The derivative (123) reacts with DMAD to give a 2:1 adduct which has been assigned the structure (124) ⟨82UP43 00⟩.

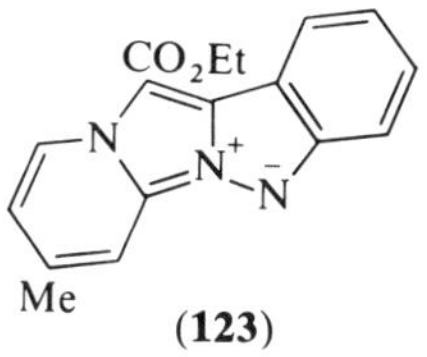

(123) 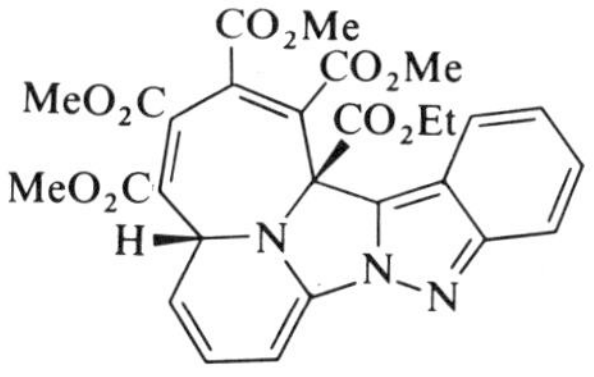 (124)

4.37.5.5 Pyrazolo[2,3-c]thiazoles

With N-phenylmaleimide, the derivative (125) reacts as a thiocarbonyl ylide giving the cycloadduct (126; 65%). When DMAD is used as 1,3-dipolarophile, the 2:1 adduct (127; 82%) is formed. This type C system, therefore, is clearly exhibiting both type A and type B character ⟨76JOC129⟩.

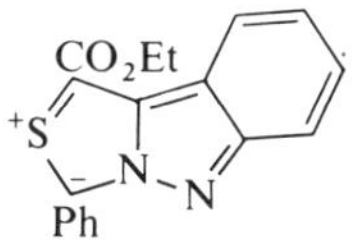

(125) 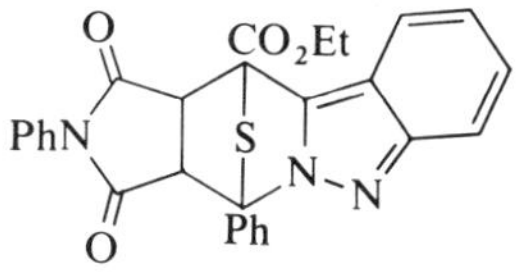 (126) 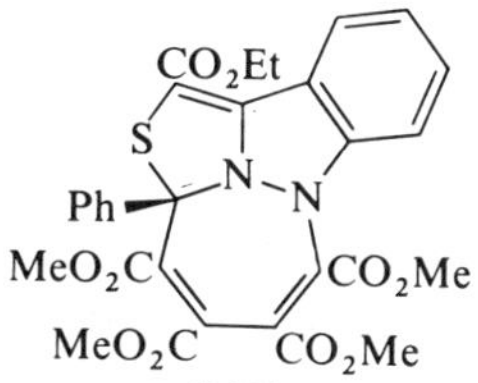 (127)

4.37.5.6 Imidazolo[1,2-c]thiazoles

The benzo derivative (128) reacts as a thiocarbonyl ylide. Addition of N-(p-tolyl)maleimide gives a mixture of the *exo* (71%) and *endo* (16%) adducts (129; Ar = p-tolyl), which in hot acetic acid eliminate hydrogen sulfide giving the pyrido[1,2-a]benzimidazole (130; Ar = p-tolyl). Analogous 1:1 cycloadducts (131) are formed with dimethyl maleate, dimethyl fumarate, methyl crotonate and methyl acrylate. In contrast to the transformation (129) → (130), treatment of the adducts (131; R = H, Me) with hot acetic acid gives the tetracyclic compounds (133) *via* the benzimidazole derivatives (132; R = H, Me). Reaction with alkynic 1,3-dipolarophiles gives pyrido[1,2-a]benzimidazole (134) by desulfurization of the primary adducts ⟨80CL1369⟩.

Cycloadditions of compound (128) with 6,6-diphenylfulvene, 8,8-dicyanoheptafulvene and tropone have also been reported ⟨81CL213⟩.

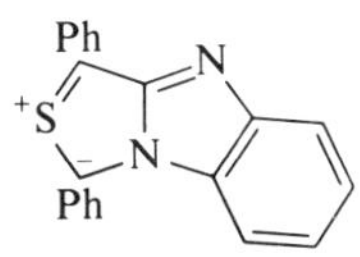

(128) 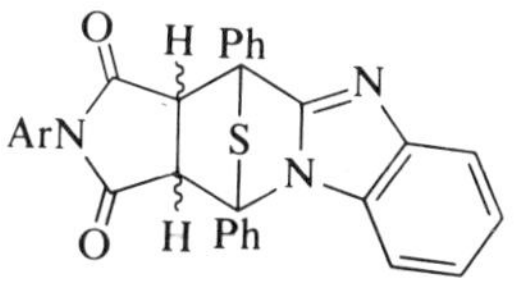 (129) 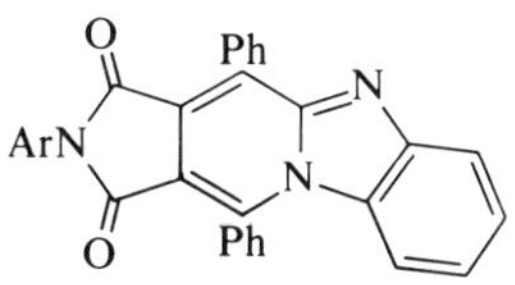 (130)

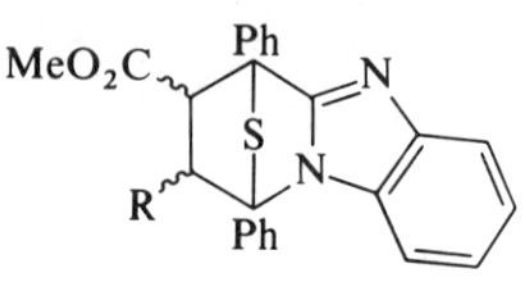

4.37.5.7 Imidazolo[1,2-c][1,2,3]triazoles

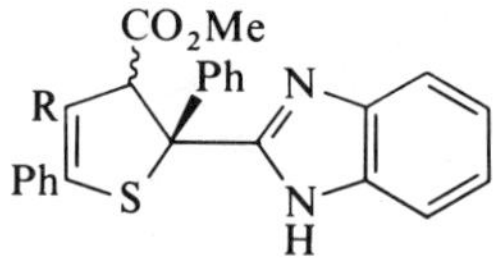

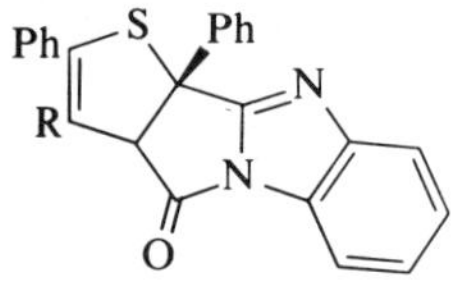

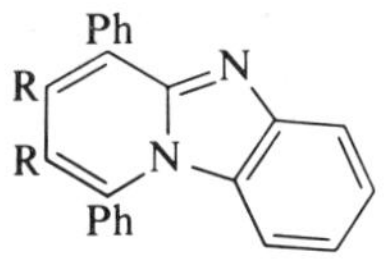

No chemical reactions of this system have been reported.

4.37.5.8 Imidazolo[1,2-d][1,2,4]triazoles

No chemical reactions of this system have been reported.

4.37.6 REACTIVITY: TYPE D SYSTEMS

4.37.6.1 General Features

Chemical reactions of type D heteropentalene mesomeric betaines have not been reported.

4.37.7 SYNTHESIS: TYPE A SYSTEMS

4.37.7.1 General Methods

Four methods of preparing type A systems merit general consideration.

(i) Treatment of the heterocyclic diketones (**136**) with phosphorus pentasulfide in a boiling solvent (pyridine, xylene or dioxane) gives mesomeric betaines associated with a thiophene fragment (**137**), usually in good yield. Systems prepared in this way include thieno[3,4-c]pyrroles and thieno[3,4-c]thiophenes (Chapter 3.18), thieno[3,4-c]pyrazoles (Section 4.37.7.3), thieno[3,4-c][1,2,5]oxadiazoles (Section 4.37.7.6), thieno[3,4-d]-[1,2,3]triazoles (Section 4.37.7.7) and thieno[3,4-c][1,2,5]thiadiazoles (Section 4.37.7.8).

The heterocyclic diketones (**136**) can be prepared by addition of diketoacetylenes (RCOC≡CCOR) to the appropriate mesoionic heterocycle (**135**) ⟨B-79MI43700⟩. Dibenzoylacetylene is commonly used for this purpose and leads to the diphenyl heteropentalenes (**137**; R = Ph).

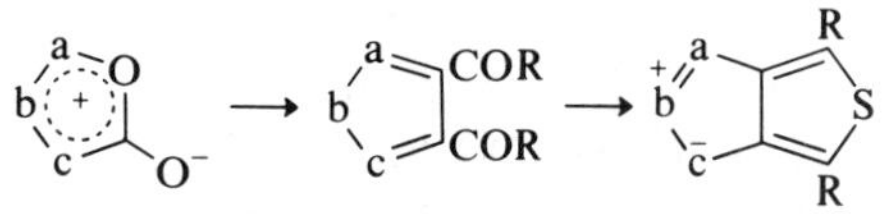

(ii) An alternative route to some condensed thiophenes (**140**) involves Pummerer dehydration (hot Ac$_2$O) of the sulfoxides (**139**) which are obtained by oxidation of the sulfides (**138**). This route (**139** → **140**) uses milder conditions than the alternative method (**136** → **137**) and if a 1,3-dipolarophile is added to the reaction mixture, sensitive systems can be trapped *in situ* as cycloadducts.

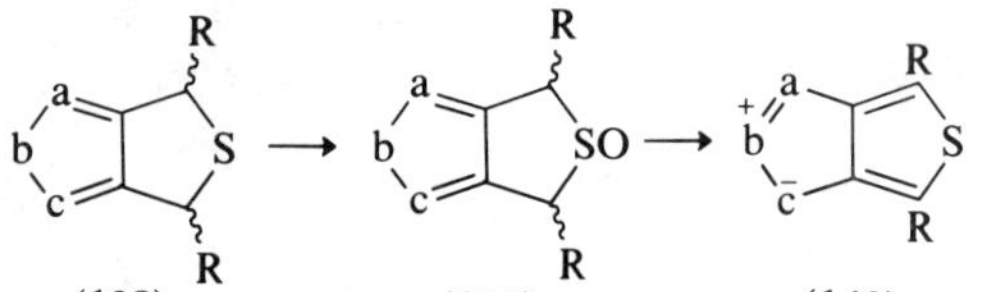

(iii) A useful route to the triazolo heteropentalenes (**143**) involves generating the nitrene intermediates (**142**) which cyclize to the desired betaines (**143**). Suitable nitrenes (**142**) have been formed either by lead tetraacetate oxidation of the amines (**141**; R = NH_2) or by thermal decomposition of the azides (**141**; R = N_3).

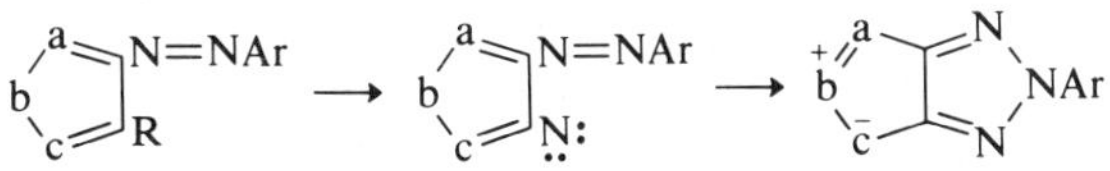

(**141**) (**142**) (**143**)

Similar reactions involving nitrene or carbene insertions have been used to prepare other type A systems.

(iv) Reaction of heterocyclic diamines (**144**) with sulfur monochloride, selenous acid or selenium oxychloride has been used to prepare the sulfur and selenium systems (**145**; X = S or Se).

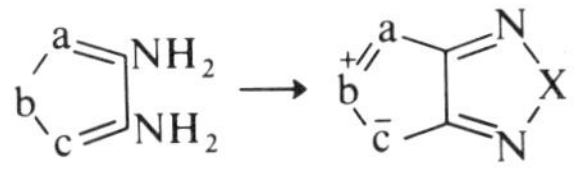

(**144**) (**145**)

4.37.7.2 Pyrrolo[3,4-*c*]pyrazoles

The derivative (**146**; R^1=Ph, R^2 = Me, R^3 = CO_2Et) has been obtained as golden yellow crystals by heating the nitropyrrole (**147**) with triethyl phosphite in xylene solution ⟨77HC(30)317⟩.

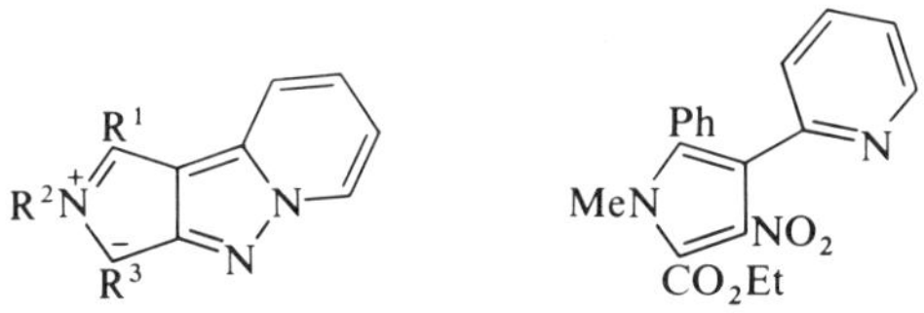

(**146**) (**147**)

4.37.7.3 Thieno[3,4-*c*]pyrazoles

The diphenyl derivatives (**150**; X=NR) have been obtained in good yield by treatment of dibenzoylpyrazoles (**149**; X = NR) with phosphorus pentasulfide in hot pyridine. The pyrazole precursors (**149**; X = NR) are conveniently obtained by reaction of the appropriate sydnone (**148**; X = NR) with dibenzoylacetylene ⟨74JA4276⟩.

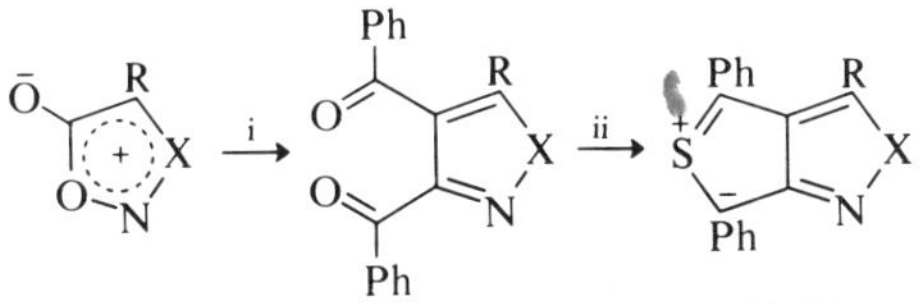

(**148**) (**149**) (**150**)

i, PhCOC≡CCOPh; ii, P_4S_{10}–pyridine

Scheme 2

4.37.7.4 Thieno[3,4-*c*]isothiazoles

The triphenyl compound (**150**; X=S, R = Ph) has been prepared in high yield by phosphorus pentasulfide–pyridine treatment of the dibenzoyl isothiazole (**149**; X = S, R = Ph) ⟨77HC(30)317, 79CB260⟩. The isothiazole precursor has been obtained from the mesoionic 1,3,2-oxathiazol-5-olate (**148**; X = S, R = Ph) by treatment with either dibenzoylacetylene ⟨79CB260⟩ or dibenzoylethylene ⟨77HC(30)317⟩.

4.37.7.5 Pyrazolo[4,3-c]pyrazoles

Tetraaryl derivatives (**151**; R^1 and R^2=Ar) are obtained in good yield by thermal dimerization of arylazoethynylarenes ($Ar^1N=NC\equiv CAr^2$) ⟨74BCJ946⟩. In a similar reaction, bis-arylazo derivatives (**151**; R^1 = Ar, R^2 = N=NR^1) are formed by dimerization of bis-arylazoacetylenes (ArCN=NC≡CN=NAr) generated *in situ* by dehydrohalogenation of 1,2-dichloroglyoxal-bis(arylhydrazones) (ArNHN=CClCCl=NNHAr) ⟨71LA(744)88⟩.

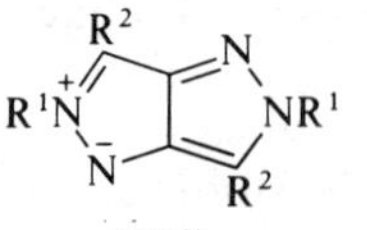

(**151**) 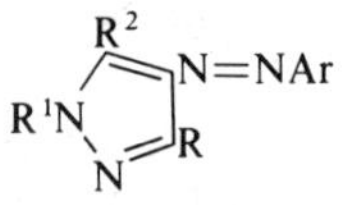 (**152**)

In an alternative approach, good yields of tetraaryltetraazapentalenes (**151**) have been obtained by prolonged treatment of 3-benzoyl-4-arylazopyrazoles (**152**; R = COPh) with hot triethyl phosphite. This reaction may well occur *via* a carbene intermediate (**152**; R = CPh). In a similar cyclization selenium dioxide oxidation of the 3-benzylpyrazoles (**152**; R = CH₂Ph) gives low yields of the heteropentalenes (**151**) ⟨73CL455, 74CL951⟩.

4.37.7.6 Thieno[3,4-c][1,2,5]oxadiazoles

Dehydration of the sulfoxide (**155**; R=Ph, X = O) in acetic anhydride gives a 40% yield of the diphenyl derivative (**153**; R = Ph, X = O) which has also been obtained by treatment of 3,4-dibenzoyl-1,2,5-oxadiazole (**154**; R = Ph, X = O) with phosphorus pentasulfide in pyridine ⟨77H(6)1173⟩. Other authors have claimed that phosphorus pentasulfide treatment of compound (**154**; R = Ph, X = O) gives only the thieno[3,4-c][1,2,5]thiadiazole (**153**; R=Ph, X = S) (Section 4.37.7.8) ⟨77HC(30)317⟩.

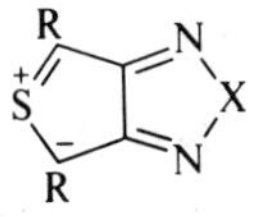

(**153**) 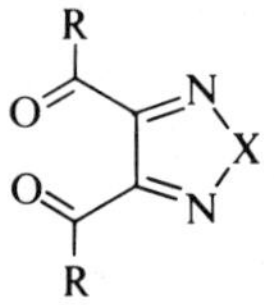 (**154**) 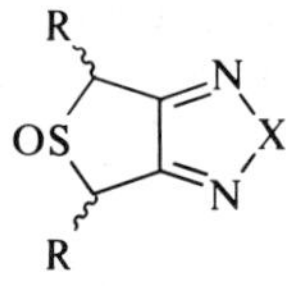 (**155**)

4.37.7.7 Thieno[3,4-d][1,2,3]triazoles

Phosphorus pentasulfide–pyridine treatment of 4,5-dibenzoyl-2-methyl-1,2,3-triazole (**154**; R=Ph, X = NMe) is reported to give the derivative (**153**; R = Ph, X = NMe) as orange prisms ⟨77HC(30)317⟩.

4.37.7.8 Thieno[3,4-c][1,2,5]thiadiazoles

A 78% yield of the diphenyl derivative (**153**; R=Ph, X = S) has been obtained by treatment of 3,4-dibenzoyl-1,2,5-thiadiazole (**154**; R = Ph, X = S) with phosphorus pentasulfide ⟨69JA6891⟩. The unsubstituted derivative (**153**; R = H, X = S) has been generated *in situ* by dehydration of the sulfoxide (**155**; R = H, X = S) using acetic anhydride ⟨79TL4493⟩.

4.37.7.9 [1,2,3]Triazolo[4,5-d][1,2,3]triazoles

Gentle heating of the azides (**157**; R=N₃), obtained from the amines (**157**; R = NH₂) by diazotization followed by treatment with sodium azide, gives high yields of the diaryl

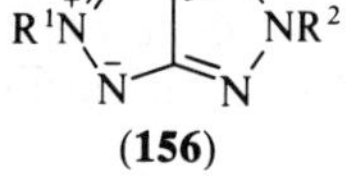

(**156**) 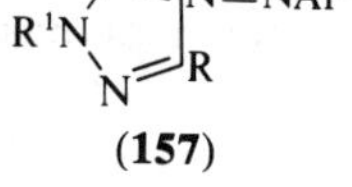 (**157**) 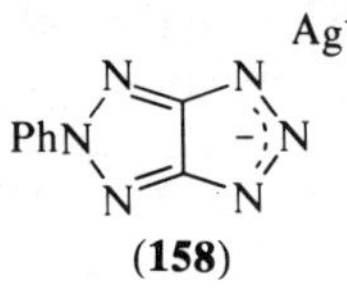 (**158**)

derivatives (**156**; R¹ and R² = Ar). The *N*-methyl derivative (**156**; R¹=Ph, R²=Me) has been obtained by methylation (MeI) of the silver salt (**158**) ⟨70BCJ3587⟩.

4.37.7.10 [1,2,3]Triazolo[4,5-*c*][1,2,5]oxadiazoles

Treatment of 3,4-diamino-1,2,5-oxadiazole with nitrosobenzene gives 3-amino-4-phenylazo-1,2,5-oxadiazole (**160**) which upon oxidation with lead tetraacetate gives the phenyl derivative (**159**; R=Ph) ⟨74BCJ1493⟩.

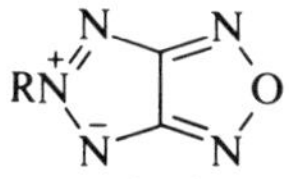

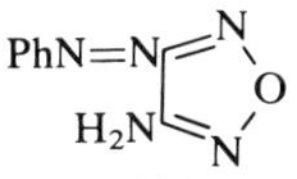

4.37.7.11 [1,2,3]Triazolo[4,5-*c*][1,2,5]thiadiazoles

The phenyl derivative (**161**; R=Ph, X=S) has been prepared in 38% yield by reaction of 4,5-diamino-2-phenyl-1,2,3-triazole (**162**) with sulfur monochloride ⟨74BCJ1493⟩.

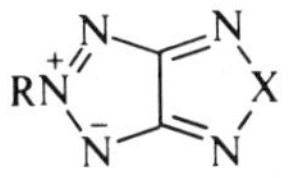

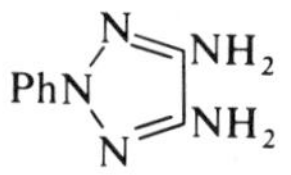

4.37.7.12 [1,2,3]Triazolo[4,5-*c*][1,2,5]selenadiazoles

Reaction of compound (**162**) with selenous acid (H₂SeO₃) gives the phenyl derivative (**161**; R=Ph, X=Se) in low yield ⟨74BCJ1493⟩.

4.37.7.13 [1,2,5]Thiadiazolo[3,4-*c*][1,2,5]thiadiazole

Treatment of the diamine (**164**) with SCl–DMF, SCl₂–DMF or SOCl₂–pyridine gives the thiadiazolothiadiazole (**163**; X=S). Alternatively, the same product can be obtained by treatment of the dioxime (**165**) with SCl₂–DMF ⟨75JOC2749⟩.

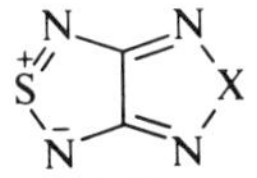

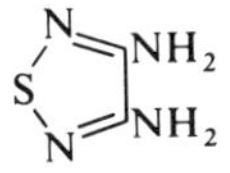

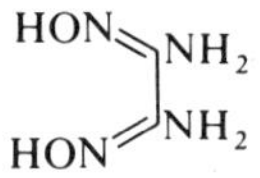

4.37.7.14 [1,2,5]Selenadiazolo[3,4-*c*][1,2,5]thiadiazole

The selenadiazolothiadiazole (**163**; X=Se) is reported to be formed from compound (**164**) and selenium oxychloride ⟨76JHC13⟩.

4.37.8 SYNTHESIS: TYPE B SYSTEMS

4.37.8.1 General Methods

A useful route to benzo derivatives (**167**) involves intramolecular cyclization of the nitrenes (**166**). Other methods which have been devised for the synthesis of type B systems are not versatile enough to justify generalization.

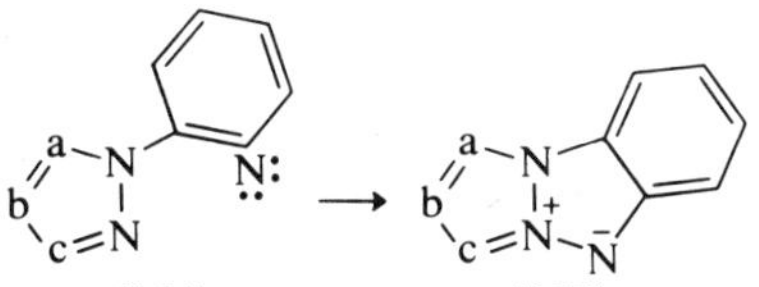

4.37.8.2 Pyrazolo[1,2-a]pyrazoles

Bromination of the *N*-allylpyrazoles (**168**) results in thermal cyclization to the salts (**169**), which with aqueous alkali give the mesomeric betaines (**170**). These derivatives (**170**; R and R^1 = H, Me, Br) are stable in solution but extremely air sensitive ⟨66JA5588, 66JA1992⟩. In an alternative approach, cyclodehydration of the salts (**171**) gives high yields of the yellow crystalline derivatives (**172**) ⟨65JA528⟩.

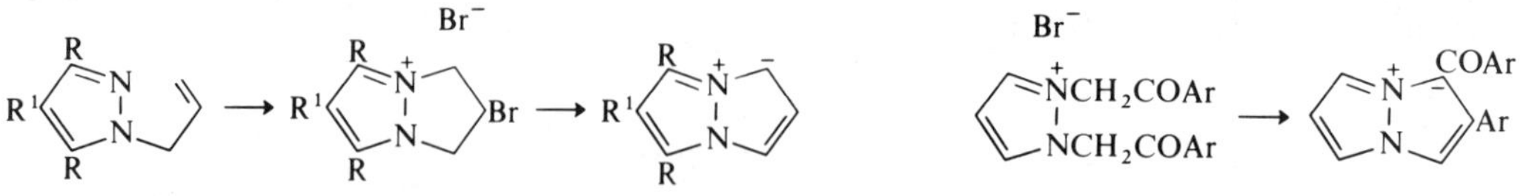

4.37.8.3 Pyrazolo[1,2-a][1,2,3]triazoles

The phenyl derivative (**173**; R^1 = Ph, R^2 = H) has been obtained (66%) by *N*-amination of 1-phenacylpyrazole followed by base-catalyzed cyclodehydration. Using a different approach the 1-acyl-2-methyl compound (**173**; R^1 = Me, R^2 = COMe) is formed (58%) by condensation of 1-aminopyrazole with 3-chloropentane-2,4-dione [CHCl(COMe)$_2$]. Treatment of the derivative (**173**; R^1 = Me, R^2 = COMe) with hot concentrated hydrochloric acid gives the 2-methyl derivative (**173**; R^1 = Me, R^2 = H) ⟨78TL1291⟩.

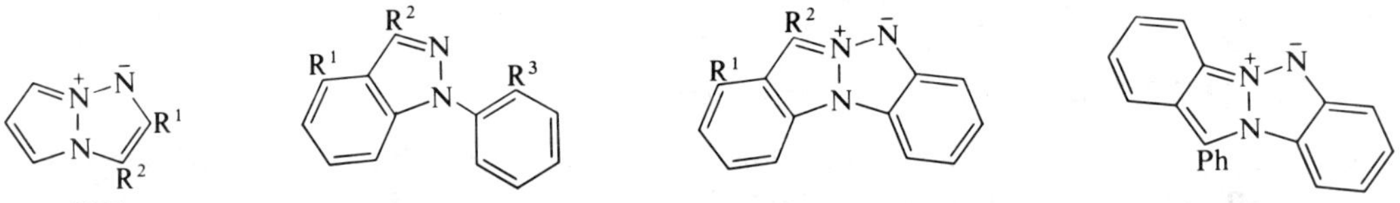

The dibenzo derivatives (**175**) are formed in low yield by treatment of 1-(*o*-nitroaryl)indazoles (**174**; R^3 = NO$_2$) with triethyl phosphite ⟨71JHC707⟩. A similar route to monobenzo derivatives has been described ⟨65JHC218, 76TL925, 81CL331⟩. An alternative route to the dibenzo derivative (**175**; R^1 = R^2 = H) involves photolysis of 1-(*o*-azidophenyl)indazole (**174**; R^1 = R^2 = H, R^3 = N$_3$) ⟨71JHC707⟩.

Reductive cyclization of 2-(*o*-benzoylphenyl)benzotriazole with triethyl phosphite gives the phenyl derivative of the isomeric dibenzo system (**176**) ⟨74CL951⟩.

4.37.8.4 [1,2,3]Triazolo[1,2-b][1,2,3]triazoles

A large number of derivatives of the triazolo[1,2-b]triazoles (**178**) have been prepared either by self-condensation of aroyl- or acyl-hydrazones (**177** → **178**) or by condensation of dihydrazones with dicarbonyl derivatives (**179** → **178**) ⟨57CB2411, 63CB1827, 67GEP1245386, 71GEP1620103⟩.

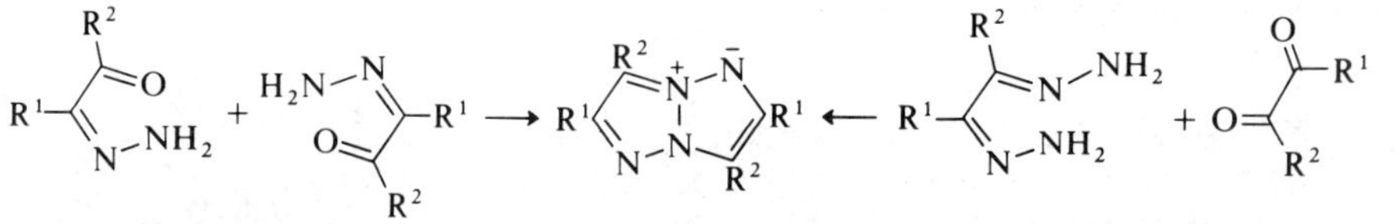

The dibenzo derivative (**181**) has been obtained by cyclization of *o*-nitrophenyl-2*H*-benzotriazole (**180**; R = NO$_2$) using triethyl phosphite ⟨67JA2633⟩ or by thermolysis of the corresponding azide (**180**; R = N$_3$) ⟨67JA2618⟩.

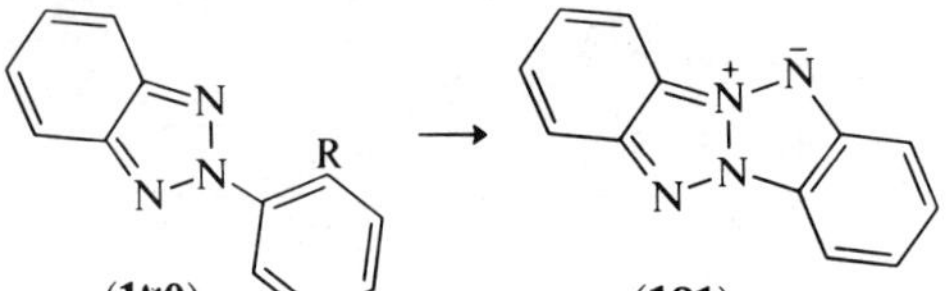

4.37.8.5 [1,2,3]Triazolo[1,2-a][1,2,3]triazoles

The dibenzo heterocycle (183) has been prepared by methods analogous to those used for the isomeric system (181). Cyclodeoxygenation of o-nitrophenyl-1H-benzotriazole (182; R = NO₂) using triethyl phosphite gives the colourless crystalline heterocycle (183) which is also formed by thermolysis (180 °C) of the azide (182; R = N₃) ⟨67JA2618, 67JA2633⟩. Monobenzo derivatives have been similarly prepared ⟨67JA2633⟩.

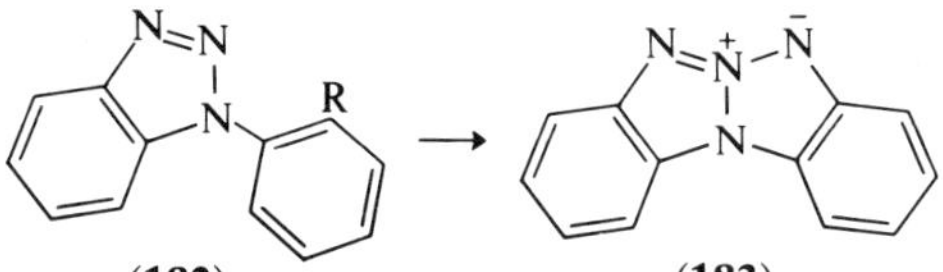

4.37.9 SYNTHESIS: TYPE C SYSTEMS

4.37.9.1 General Methods

Too few examples of type C heteropentalenes are known to justify a general discussion.

4.37.9.2 Pyrrolo[1,2-c]imidazoles

The N-methyl derivative (114; R = Me) (Section 4.37.5.2) has been generated *in situ* by base treatment of N-methyl-5H-pyrrolo[1,2-c]imidazolium iodide (115) ⟨82UP43700⟩.

4.37.9.3 Pyrrolo[1,2-c]thiazoles

The derivative (184; $R^1 = R^2 = CO_2Me$, $R^3 = Me$) has been generated *in situ* by dehydration of the sulfoxide (185) using acetic anhydride ⟨80JOC5396⟩. *In situ* generation of benzo derivatives has also been reported ⟨77CL1237⟩.

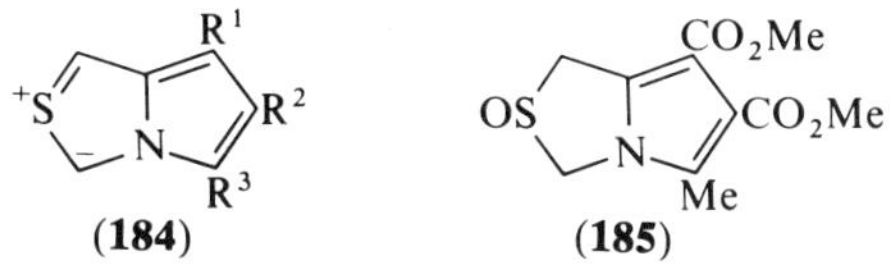

4.37.9.4 Pyrazolo[2,3-c]imidazoles

The derivative (123) (Section 4.37.5.4) has been obtained as orange needles in 31% yield by cyclodeoxygenation [P(OEt)₃] of ethyl 7-methyl-2-(2'-nitrophenyl)imidazo[1,2-a]pyridine-3-carboxylate ⟨82UP43700⟩.

4.37.9.5 Pyrazolo[2,3-c]thiazoles

Deoxygenation of ethyl 4-(2-nitrophenyl)-2-phenylthiazole-5-carboxylate (187) using triethyl phosphite gives a 24% yield of the derivative (186; $R^1 = CO_2Et$, $R^2 = Ph$) which is obtained as maroon needles ⟨76JOC129⟩.

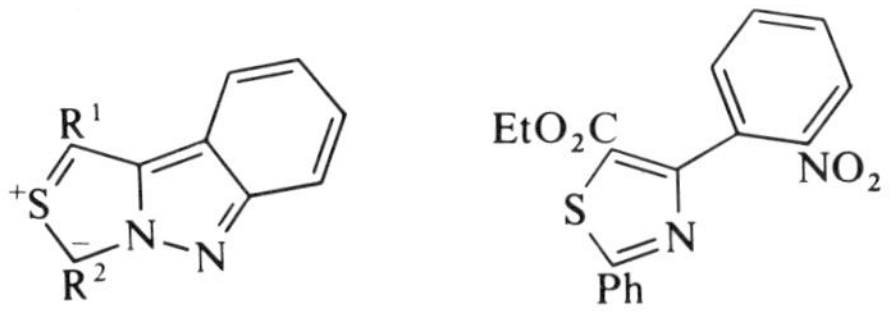

4.37.9.6 Imidazolo[1,2-c]thiazoles

The benzo derivative (**188**) has been obtained as reddish violet needles (49%) by treatment of 1,2-dibenzoylbenzimidazole (**189**) with phosphorus pentasulfide in hot toluene ⟨80CL1369⟩. Treatment of 4-amino-2,5-diphenylthiazole with phenacyl bromide is reported to give the triphenyl derivative (**190**) ⟨80KGS1695⟩.

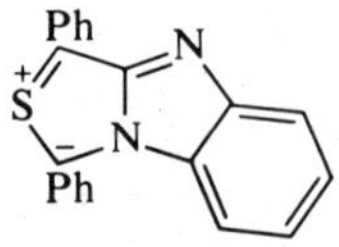

(**188**) 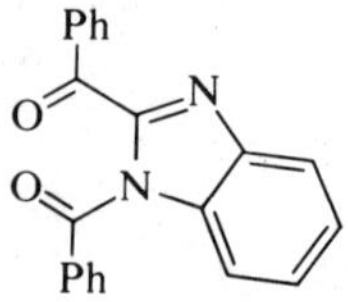   (**189**) 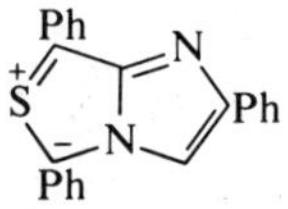   (**190**)

4.37.9.7 Imidazolo[1,2-c][1,2,3]triazoles

Treatment of 4-amino-1,5-diphenyl-1,2,3-triazole with phenacyl bromide is reported to give the derivative (**191**) ⟨80KGS1695⟩.

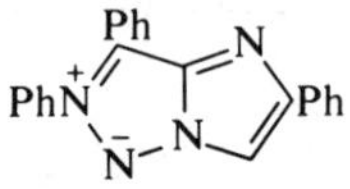

(**191**)

4.37.9.8 Imidazolo[1,2-d][1,2,4]triazoles

Derivatives of the type (**192**) have been obtained from the appropriately fused 3-amino-1,2,4-triazole and phenacyl bromide ⟨80KGS1695⟩.

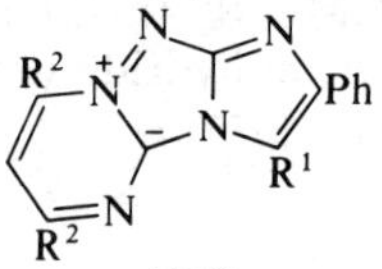

(**192**)

4.37.10 SYNTHESIS: TYPE D SYSTEMS

4.37.10.1 Anhydrocyclopenta[d]thiazolium hydroxides

The purple derivative (**193**) has been obtained by methylation of 2-phenylindeno[1,2-d]thiazole and subsequent treatment with base ⟨65TL1421⟩.

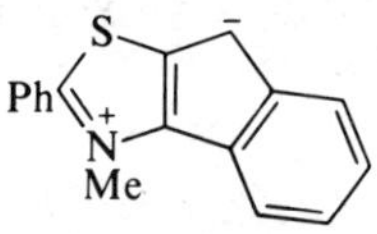

(**193**)

4.38

Two Fused Five-membered Heterocyclic Rings: (iii) 1,6,6aλ^4-Trithiapentalenes and Related Systems

N. LOZAC'H
Université de Caen

4.38.1 INTRODUCTION		1050
4.38.1.1 Scope		1050
4.38.1.2 Nomenclature		1051
4.38.2 STRUCTURE		1052
4.38.2.1 Theoretical Methods		1052
4.38.2.2 Molecular Dimensions		1052
4.38.2.3 NMR Spectroscopy		1056
4.38.2.4 Visible and UV Spectroscopy		1056
4.38.2.5 IR Spectroscopy		1056
4.38.2.6 Dipole Moments		1057
4.38.2.7 Photoelectron Spectroscopy		1057
4.38.3 REACTIVITY AT RING ATOMS		1057
4.38.3.1 Reactivity at Atoms 1,6,6a		1057
4.38.3.1.1 Isomerization		1057
4.38.3.1.2 Protonation		1058
4.38.3.1.3 Metalation		1058
4.38.3.1.4 Alkylation		1058
4.38.3.1.5 S-Oxidation or reduction		1059
4.38.3.2 Electrophilic Attack at Atoms Other than 1,6,6a		1059
4.38.3.2.1 Halogenation		1059
4.38.3.2.2 Nitration		1059
4.38.3.2.3 Nitrosation		1059
4.38.3.2.4 Diazo coupling		1060
4.38.3.2.5 Formylation		1060
4.38.3.2.6 Miscellaneous		1060
4.38.3.3 Nucleophilic Attack at Atoms Other than 1,6,6a		1060
4.38.3.3.1 Replacement of a substituent		1060
4.38.3.3.2 Modification of the sequence 1,6,6a		1060
4.38.3.3.3 Rearrangement to γ-pyrones		1061
4.38.4 REACTIVITY AT SUBSTITUENTS		1061
4.38.4.1 Hydrolysis of Ester or Cyano Groups		1061
4.38.4.2 Oxidation or Reduction		1061
4.38.4.3 Carbanion or Related Reactions in Side Chains		1061
4.38.4.4 Reactivity of Amino Groups		1062
4.38.5 SYNTHESIS BY RING FORMATION		1062
4.38.5.1 Formation of Two Rings		1062
4.38.5.1.1 From one fragment		1062
4.38.5.1.2 From [7+1] atom fragments		1062
4.38.5.1.3 From [6+2] atom fragments		1063
4.38.5.1.4 From [6+1+1] atom fragments		1063
4.38.5.1.5 From [3+2+2] atom fragments		1064
4.38.5.1.6 From [5+1+1+1] atom fragments		1064
4.38.5.1.7 From [4+2+1+1] atom fragments		1064
4.38.5.1.8 From [3+2+2+1] atom fragments		1064
4.38.5.2 Formation of One Ring		1065
4.38.5.2.1 From one fragment		1065
4.38.5.2.2 From [4+1] atom fragments		1065

4.38.5.2.3 From [3+2] atom fragments   1066
4.38.5.2.4 From [2+3] atom fragments   1067
4.38.5.3 Formation of Extended Structures   1067
4.38.5.3.1 Structures containing two or three trithiapentalene systems   1067
4.38.5.3.2 Structures with more than three sulfur atoms in line   1068

4.38.6 SYNTHESIS BY RING TRANSFORMATION   1069
4.38.6.1 From γ-Pyrones or their Chalcogen Derivatives   1069
4.38.6.2 From 1,3aλ⁴,4-Trithiapentalenes   1069
4.38.6.3 From Thiirane Intermediates   1069
4.38.6.4 From 1,2,3,4-Thiatriazoles   1070
4.38.6.5 From 3-(1,2-Dithiol-3-ylidene)methyl-1,2-dithiolylium Cations   1070

4.38.1 INTRODUCTION

4.38.1.1 Scope

Systematic study of 1,6,6aλ⁴-trithiapentalenes and of related structures was initiated in 1958 by two independent observations, an X-ray structure determination of 2,5-dimethyl-1,6,6aλ⁴-trithiapentalene and an IR absorption study of 2,5-dimethyl-1-oxa-6,6aλ⁴-dithiapentalene ⟨71AHC(13)161, p.162⟩. The name 1,6,6aλ⁴-trithiapentalene designates structure (**1**) which accounts reasonably well for the properties of these compounds although it should be noted that the length of the S—S bonds is distinctly longer than that of a single covalent S—S bond and may be considered as corresponding to a bond order of about 0.5. Formula (**2**), indicating partial bonding with dotted lines, describes perhaps more adequately the real situation but, in the following text, the more classical formula (**1**) is preferred for the sake of simplicity.

Although the first correct structure for (**1**) was suggested in 1958, compounds belonging to this class had been obtained much earlier. For instance, the 2,5-dimethyl derivative was described, but with a wrong formula, in 1925 and some anthracene dyes containing this bicyclic system had been prepared before 1958, but their structures were not established.

Since 1958, the structure of 1,6,6aλ⁴-trithiapentalenes and of related compounds has been the subject of much study and discussion. The essential question was whether such a compound is a mixture of valence isomers in rapid equilibrium (equation 1) or is best represented by a symmetrical formula (**1**) or by resonance forms (equation 2). In fact equation (2) or formula (**1**) are equivalent and imply a C_{2v} symmetry, while equation (1) implies a C_s type of structure.

These various structures raise the question of the aromaticity of the 1,6,6aλ⁴-trithiapentalene system. From a qualitative viewpoint, thermal stability and electrophilic attack on atoms 3 and 4 may be considered as consistent with an aromatic character. Furthermore, NMR deshielding of ring protons ⟨71AHC(13)161, p. 230⟩ suggests a strong ring current similar to those of aromatic compounds.

More precise numerical data, such as heats of combustion, are not available but theoretical methods (see Section 4.38.2.1) suggest that 1,6,6aλ⁴-trithiapentalenes exhibit a naphthalene-like π-electron delocalization. It should be noted, however, that the resonance shown in equation (2) implies a π- and σ-bonding delocalization, while the aromaticity concept is usually related to a delocalization affecting π-bonds only.

Interest in the bonding pattern of 1,6,6aλ^4-trithiapentalenes has led to the study of various related structures. The simplest consists of the presence of more than one trithiapentalene system in the same compound. Some examples of this type are known.

Another type of structure is obtained by the replacement of sulfur by other elements in the characteristic 1,6,6a triatomic sequence. Very early it had been shown that α-(1,2-dithiol-3-ylidene) ketones exhibit very peculiar properties somewhat similar to those of 1,6,6aλ^4-trithiapentalenes, the preferred configuration having the sulfur and oxygen atoms almost in line, a fact suggesting that some sort of bonding exists between the 'carbonyl' oxygen and a sulfur atom. Many more such structures have been obtained, containing not only carbon, hydrogen and sulfur but also, combinations of oxygen, nitrogen, selenium and/or tellurium, as indicated in formula (3). In this formula the atoms or groups Y and Y' can be identical or different, and the same applies to Z^1, Z^2, Z^3 and Z^4.

(3) X = S, Se or Te
Y or Y' = O, S, Se or NR
Z^1, Z^2, Z^3, Z^4 = N or CR'

(4)

For the sake of consistency, in this chapter formulae of type (3) are used thoughout although in some cases, such as α-(1,2-dithiol-3-ylidene) ketones, more classical formulae such as (4) are often preferred on the grounds that the O–S interaction is relatively weak.

A more challenging aspect of the study of the trithiapentalene structure has been the synthesis of polycyclic systems containing more than three sulfur atoms in line. For these compounds, partial bonds of different strengths are involved and the use of formulae with quadricovalent sulfur does not help very much. In such cases, formulae are drawn with dicovalent sulfur atoms, with single bonds between those sulfur atoms whose distance is relatively short, with the optional addition of dotted lines to indicate the weakest bonds, as shown in formulae (5), (6) and (7). In this chapter we shall omit these dotted lines, the partial bonding between sulfur atoms being implied by the fact that the four or five sulfur atoms are aligned in the formula.

(5)   (6)   (7)

4.38.1.2 Nomenclature

Nomenclature follows the recommendations of the International Union of Pure and Applied Chemistry (IUPAC) ⟨B-79MI43800⟩. For the description of fused heterocycles, two methods may be used, the fusion method (IUPAC Rule B-3) and the replacement nomenclature (IUPAC Rule B-4). In this chapter, the latter is used because it shows more clearly the relationship between the various structures under consideration.

In replacement nomenclature, the name of a heterocyclic compound is formed by prefixing 'a' terms, such as 'oxa', 'thia', 'aza', *etc.* to the name of the corresponding cyclic hydrocarbon. As IUPAC Rule B-4 states that the replacement method should be applied to a homocyclic hydrocarbon, we use pentalene for this purpose. The replacement procedure forbids the use of heterocyclic parent structures and for this reason the name 'thiathiophthene', which is found in the literature, is not in compliance with IUPAC recommendations. Another name which is sometimes found is 'thiothiophthene', which is even worse because of an incorrect use of the prefix 'thio'.

In Tables 1 and 2 the necessity to derive all structures from a pentalene skeleton leads to the fact that, for some compounds, the replacements or substitutions indicated, though unambiguous, do not correspond to an IUPAC recommended name. These deviations are due to one of two different reasons. First, the parent polycyclic hydrocarbon may be larger than the pentalene system and, in the tables, supplementary rings have been denoted by bridges, an extension of the normal scope of IUPAC Rule A-34.1. Second, the presence of heteroatoms outside the characteristic triatomic sequence may modify the numbering. When such deviations exist for a given compound, the name following IUPAC recommendations is given in a footnote.

It should be noted that the ending 'ylium' has been used, according to IUPAC recommendations, for denoting the formation of a cation by loss of a hydride ion instead of the ending 'ium' which is increasingly being reserved for the addition of a proton. For this reason, the term 1,2-dithiolylium cation has been used for the structure often named 1,2-dithiolium cation ⟨80AHC(27)151⟩.

4.38.2. STRUCTURE

4.38.2.1 Theoretical Methods

A large quantity of work has been devoted in recent years to the nature of the bonding in $1,6,6a\lambda^4$-trithiapentalenes and related compounds. The electron distribution in these molecules has been a subject of controversy but, at least for $1,6,6a\lambda^4$-trithiapentalenes themselves, it is now generally assumed that such a distribution involves a delocalized 10π-electron system similar to that of naphthalene.

It appears that the conclusions drawn from theoretical calculations depend significantly upon the type of approximation chosen and should always be checked with experimental spectroscopic evidence.

CNDO/2 semiempirical calculations have been applied to 2-oxa-$6,6a\lambda^4$-dithiapentalene and to 8-oxa-$1,8a\lambda^4$-dithiacyclopent[a]indene. In accordance with IR data, these calculations confirm the existence of a partial bond between the oxygen atom and the sulfur atom 8a ⟨72CR(C)(275)909⟩.

CNDO/2 calculations have been carried out on $1,6,6a\lambda^4$-trithiapentalene and on its 2- and 3-methyl and 2- and 3-phenyl derivatives which are consistent with experimental results. A 2-methyl group and a twisted 2-phenyl group cause a lengthening of the bond between the sulfur atoms 1 and 6a. A 4-phenyl group affects this S—S bond to a smaller degree ⟨72CC222⟩.

Using a linear combination of gaussian orbitals (LCGO) approach to the Hartree–Fock method, calculations on the ground-state wave-functions have been made for the $1,6,6a\lambda^4$-trithia-, the 1-oxa-$6,6a\lambda^4$-dithia-, the 1,6-dioxa-$6a\lambda^4$-thia- and the $1,6a\lambda^4$-dithia-6-azapentalenes ⟨74JCS(P2)1885⟩. According to these calculations, two S—S bonds exist in the first compound and a weak S—N bond in the last one. The calculation does not show the existence of an S—O bond in the oxadithiapentalene. It appears also that d-orbitals are not appreciably involved in the bonding of this series of molecules and, according to bond energy analysis, these molecules have little resonance energy.

Ground-state potential curves related to the central (6a) atom have been established from CNDO/2 calculations for $1,6,6a\lambda^4$-trithiapentalene, 1,6-dithia-$6a\lambda^4$-selenapentalene and $1,6,6a\lambda^4$-triselenapentalene ⟨81IC399⟩. The curve for trithiapentalene has a very flat minimum corresponding to a large vibrational amplitude of the sulfur atom 6a, within a relatively rigid framework; in contrast, for the two selenium compounds, the ground-state potential curves are distinctly narrower. This conclusion is supported by the fact that, in crystals, intermolecular interactions affect the position of a 6a selenium atom less than the position of a 6a sulfur atom ⟨72ACS2139⟩.

4.38.2.2 Molecular Dimensions

Interatomic distance is one of the most important features of a chemical bond and the particular problem raised by $1,6,6a\lambda^4$-trithiapentalenes prompted numerous structural studies on these compounds and their analogues.

As indicated in the introduction, one particularly important question is whether the $1,6,6a\lambda^4$-trithiapentalene system is symmetrical (C_{2v} symmetry) or not (C_s symmetry). As will be seen from the available data, the S(1)—S(6a) and S(6)—S(6a) distances are not always equal in crystals of symmetrically substituted structures. As these discrepancies may be attributed to intermolecular forces in the crystal lattice, measurements in the gaseous state are of particular significance. Unfortunately, the very low volatility of these compounds renders such experiments difficult and very few data on trithiapentalenes in the gaseous state are available. However it has been possible to study electron diffraction of $1,6,6a\lambda^4$-trithiapentalene in the gaseous state and the findings are in good agreement with a C_{2v} symmetry ⟨74JA289⟩.

The first X-ray structure determination of a $1,6,6a\lambda^4$-trithiapentalene dates back to 1958 and since that time the number of such studies shows the interest raised by this type of compound. Some data have been collected before ⟨71AHC(13)161, p. 212⟩ and Tables 1 and 2 contain more recent results together with the most significant previous ones.

Table 1 Partial Bonding in $1,6,6a\lambda^4$-Trithiapentalenes

Substituents and 'aza' replacements[a]	Bond lengths (Å) 1–6a	6a–6	Ref.
None	2.363	2.363	73ACS411
2-Methyl	2.424	2.301	75ACS(A)136
2,5-Dimethyl	2.358	2.358	69CC137
2,4-Diphenyl	2.499	2.218	69ACS1852
2,5-Diphenyl	2.304	2.362	71ACS1583
3,4-Diphenyl	2.232	2.434	69CC1014
2,3,4-Triphenyl	2.270	2.375	71ACS1822
2,3,4,5-Tetraphenyl	2.312	2.312	73ACS2666
2-p-Dimethylaminophenyl-4-phenyl	2.348	2.350	72ACS3114
2-Methyl-4-phenyl	2.481	2.242	71ACS1835
3-Methyl-2,5-diphenyl	2.255	2.398	72ACS1297
3,4-Dimethyl-2,5-diphenyl	2.303	2.303	73ACS379
3,4-Ethano-2,5-diphenyl[b]	2.351	2.351	73ACS382
3,4-Propano-2,5-diphenyl[c]	2.288	2.329	72ACS2140
2,5-Diphenyl-3,4-diaza[d]	2.319	2.328	73ACS510
2,5-Dianilino-3,4-diaza[e]	2.225	2.475	72CC543
3-Amino-2-methylthio-5-phenyl	2.375	2.266	72CC836

[a] Unless otherwise indicated, the names of these compounds, based on IUPAC Rule B-4, are obtained by prefixing the names of the substituents, if any, to '$1,6,6a\lambda^4$-trithiapentalene'. [b] 2,5-Diphenyl-3,4-dihydro-$1,6,6a\lambda^4$-trithiacyclopenta[cd]pentalene. [c] 1,4-Diphenyl-6,7-dihydro-$5H$-$2,2a\lambda^4$,3-trithiacylopent[cd]indene. [d] 2,5-Diphenyl-$3,3a\lambda^4$,4-trithia-1,6-diazapentalene. [e] 2,5-Dianilino-$3,3a\lambda^4$,4-trithia-1,6-diazapentalene.

When speaking of the nature of a bond we should first compare the observed interatomic distance with the expected covalent single bond length and with the van der Waals contact distance, some chemical bonding being assumed if the interatomic distance is smaller than this van der Waals contact distance. Table 3 compares observed interatomic distances, taken from Tables 1 and 2, with the expected single bond distances computed according to Pauling ⟨B-60MI43800⟩ and the contact distances computed according to Bondi ⟨64JPC441⟩.

These expected covalent bond distances and, to an even greater extent, the van der Waals contact distances are not defined with great precision. Nevertheless if the distance between two atoms is distinctly larger than the expected single bond distance and clearly smaller than the computed contact distance, some interaction should exist between these two atoms and one way of describing this interaction is to consider it as a partial covalent single bond.

In the case of $1,6,6a\lambda^4$-trithiapentalenes, where a large number of measurements can be compared (Table 1), the mean length of S—S bonds is 2.338 Å with a mean absolute deviation of 0.055 Å (2.35%), whereas the mean absolute deviation for the sum of the two S—S bonds is 0.036 Å, only 0.77% of the mean sum. It is clear that when comparing two compounds much smaller variations can be expected for the distance S(1)—S(6) than for the position of S(6a) between S(1) and S(6).

In the crystals, intermolecular forces have a significant effect on the twist angle of phenyl substituents as well as on the position of the S(6a) atom between the S(1) and S(6) atoms.

Table 2 Partial Bonding in Compounds Related to 1,6,6aλ^4-Trithiapentalenes

$$Y-X-Y$$

1	Triatomic sequence 6a	6	Substituents and 'aza' replacements[a]	Bond lengths (Å) 1–6a	6a–6	Ref.
S	Se	S	None	2.446	2.446	71ACS1895
S	Se	S	3,4-Dimethyl	2.414	2.414	73ACS360
S	Se	S	2,5-Diphenyl	2.419	2.433	72ACS2139
S	S	Se	2,4-Diphenyl	2.492	2.333	66JA5045
Se	Se	Se	None	2.579	2.586	71ACS2507
Se	Se	Se	3,4-Propano[b]	2.554	2.568	73ACS485
O	S	S	2,5-Dimethyl	2.41	2.12	61MI43800
O	S	S	3,5-Diphenyl	2.382	2.106	69ACS1377
O	S	S	2-p-Dimethylaminophenyl-4-phenyl	2.440 / 2.287	2.103 / 2.107	75AX(B)30
O	S	S	2,3-(1,3-Butadieno)-5-phenyl[c]	2.184	2.137	72TL4687
O	S	S	3-Benzoyl-5-phenyl-2-aza	2.034	2.178	69JA781
O	S	O	3,4-(2,2-Dimethylpropano)-2,5-diaza[d]	1.875	1.878	73TL1565
O	S	O	2,3:5,4-Bis(2-methoxy-1,3-butadieno)[e]	1.878	1.879	71AX(B)1073
O	Se	O	3,4-(2,2-Dimethylpropano)-2,5-diaza[f]	2.017	2.030	71CC594
S	S	N	3,4,6-Triphenyl	2.396	1.871	72CJC324
S	S	N	2,4-Diphenyl-6-(3-quinolinyl)	2.364	1.887	71CJC167
S	S	N	2-t-Butyl-6-phenyl-5-aza[g]	2.426	1.841	77ACS(A)292
S	S	N	3,4-Dimethyl-6-phenyl-5-aza[h]	2.493	1.779	77ACS(A)412
N	S	N	1,6-Dimethyl-3,4-propano[i]	1.901	1.948	72ACS343

[a] Unless otherwise stated, the names of these compounds, based on IUPAC Rule B-4, are obtained by prefixing the names of the substituents and the replacement terms, if any, to the appropriate polyheterapentalene name.
[b] 6,7-Dihydro-5H-2,2aλ^4,3-triselenacyclopent[cd]indene.
[c] 2-Phenyl-8-oxa-1,8aλ^4-dithiacyclopent[a]indene.
[d] 6,6-Dimethyl-6,7-dihydro-5H-2,3-dioxa-2aλ^4-thia-1,4-diazacyclopent[cd]indene.
[e] 2,7-Dimethoxy-9,10-dioxa-9aλ^4-thiaindeno[1,2-a]indene.
[f] 6,6-Dimethyl-6,7-dihydro-5H-2,3-dioxa-2aλ^4-selena-1,4-diazacyclopent[cd]indene.
[g] 5-t-Butyl-1-phenyl-6,6aλ^4-dithia-1,2-diazapentalene.
[h] 3,4-Dimethyl-1-phenyl-6,6aλ^4-dithia-1,2-diazapentalene.
[i] 2,3-Dimethyl-6,7-dihydro-5H-2aλ^4-thia-2,3-diazacyclopent[cd]indene.

Table 3 Partial Bonding in Triatomic Sequences Compared with Single Covalent Bond Lengths and van der Waals Contact Distances

Triatomic sequence	Atoms considered	Observed distances (Å)	Computed single bond lengths[a] (Å)	Computed van der Waals contact distances[b] (Å)
S—S—S	S—S	2.2–2.5	2.08	3.60
S—S—Se	S—S	2.4–2.5	2.08	3.60
S—S—O	S—S	2.1–2.2	2.08	3.60
S—S—N	S—S	2.3–2.4	2.08	3.60
Se—Se—Se	Se—Se	2.5–2.6	2.34	3.80
S—S—Se	S—Se	2.3–2.4	2.21	3.70
S—Se—S	S—Se	2.4–2.5	2.21	3.70
S—S—O	S—O	2.0–2.5	1.70	3.22
O—S—O	S—O	1.8–1.9	1.70	3.22
S—S—N	S—N	1.8–1.9	1.74	3.35
N—S—N	S—N	1.9–2.0	1.74	3.35

[a] B-60MI43800. [b] 64JPC441.

Different substituent groups perturb the bonding of the three sulfur atoms to different degrees ⟨72MI43800⟩. The importance of intermolecular forces in defining the molecular dimensions is clearly shown by the fact that, in some crystals, the same molecule can exist in two dimensionally different forms, as in the case of 2-p-dimethylamino-4-phenyl-1-oxa-6,6aλ^4-dithiapentalene (Table 2) ⟨75AX(B)30⟩.

Table 4 Bond Lengths in some Multisulfur Compounds (8), (9), (10) and (11)

Compound	Substituents R	R'	Bond lengths (Å) a–b	b–c	c–d	d–e	Ref.
(8a)	Ph	Bu^t	2.14	2.62	2.55	2.16	70ACS1464
(8b)	Bu^t	Bu^t	2.183	2.580	2.583	2.172	73ACS2517
(9)	Ph	Bu^t	2.327	2.110	2.856	2.064	73ACS3881
(10)	Bu^t	—	2.063	2.863	2.062	—	72ACS873
(11a)	Bu^t	—	2.482	2.209	2.965	—	71ACS3577
(11b)	p-MeOC$_6$H$_4$	—	{2.429, 2.563}	2.225, 2.165	2.920, 2.974	—	74ACS(A)499

^{a} For this compound two different molecules exist in the crystal.

The results given in Table 2 and the comparisons with single covalent bonds and van der Waals contact distances show that other elements can lead to the particular type of interaction observed between sulfur atoms in 1,6,6aλ^4-trithiapentalenes. From these results and other studies it appears that in these series selenium behaves very much like sulfur and, generally speaking, S—N bonds are more similar to S—S bonds than are S—O bonds. Such a general statement, however, should be treated with some reservations as the structure of the rest of the molecule may influence significantly the bonding pattern of the triatomic sequence.

X-Ray diffraction determination has shown the existence of various structures in which four or even five sulfur atoms are practically in a straight line and in which the various distances between neighbouring sulfur atoms are rather different, but significantly smaller than the van der Waals contact distance. Earlier results have been previously reviewed ⟨71AHC(13)161, p. 217⟩ and more recent studies confirming these findings are given in Table 4.

For all the compounds in Table 4, it appears that the longest S—S bond lengths found (2.5–3.0 Å) are significantly less than the estimated van der Waals contact distance (3.60 Å) while the shortest S—S bond lengths, while sometimes near to the single bond distance (2.08 Å) may be larger than it (2.2 Å). This clearly indicates bond orders between zero and unity.

More precise comparisons may be made on symmetrical structures for which a theoretical bond order may be calculated by averaging the possible canonical forms with, for S$_5$ compounds, the additional assumption that all canonical forms written with divalent sulfur atoms are equivalent.

In order to avoid conjugation and twist effects, for this comparison two compounds having aliphatic substituents only have been selected, namely 2,5-dimethyl-1,6,6aλ^4-trithiapentalene and compound (**8b**) of Table 4. For the first of these compounds, the order of S—S bonds is 0.5 and the bond length (Table 1) is 2.358 Å. For (**8b**), the order of bonds a–b and d–e is 0.67 and the bond lengths are 2.183 and 2.172 Å, respectively; the order of the bonds b–c and c–d is 0.33 and the bond lengths are 2.580 and 2.583 Å, respectively. It appears that there is a rather good correlation between this assumed bond order and the observed bond length:

Bond order:	0.33	0.50	0.67	1.00
Bond length (Å):	2.58	2.36	2.18	2.08

4.38.2.3 NMR Spectroscopy

Proton magnetic resonance has been widely used for identification and structure determination of 1,6,6aλ^4-trithiapentalenes and similar compounds such as 1-oxa-6,6aλ^4-dithiapentalenes and 6aλ^4-thia-1,6-diazapentalenes. Chemical shifts and coupling constants have been collected ⟨71AHC(13)161, p. 229⟩. Experiments have also been carried out using ^{13}C NMR ⟨74TL2783⟩.

All results show that for symmetrically substituted compounds, even at low temperatures, the molecules have a total magnetic symmetry on the NMR time-scale, for ^{1}H as well as for ^{13}C NMR. These results are consistent with either a very rapid equilibrium between valence isomers or with a C_{2v} symmetry, which is now generally accepted.

In 1,6,6aλ^4-trithia- and in 1-oxa-6,6aλ^4-dithia-pentalenes, ring protons are much more deshielded than thiophene or furan ring protons at similar positions relative to sulfur or oxygen. For instance, in 1,6,6aλ^4-trithiapentalene in CDCl$_3$, the chemical shifts are 9.18 p.p.m. for protons 2 and 5, and 7.96 p.p.m. for protons 3 and 4. For thiophene, the corresponding chemical shifts are 7.19 p.p.m. for protons 2 and 5, and 7.04 for protons 3 and 4. This strong deshielding has been attributed, at least in part, to an aromatic ring current.

A correlation has been established between S 2p-electron energies measured by ESCA and ^{19}F NMR chemical shifts in some 5-(p-fluorophenyl)-1-oxa-6,6aλ^4-dithiapentalenes. It appears that, in these cases, the fluorine chemical shift is a convenient evaluation of the electrical charge of the heterocyclic ring to which the p-fluorophenyl substituent is attached ⟨74T2537⟩.

4.38.2.4 Visible and UV Spectroscopy

1,6,6aλ^4-Trithiapentalenes have a strong absorption band near 500 nm which is responsible for the orange to red color of these compounds and is assigned to a $\pi \to \pi^*$ transition. In the UV region there is another band near 260 nm ⟨71AHC(13)161, p. 224⟩.

The bands of the absorption spectra are only slightly affected by the substituents of the trithiapentalene system and therefore constitute a rather good and practical structural proof. This applies also to extended structures where UV–visible spectra show structural relationships very clearly ⟨71BSF4426, 71BSF4429⟩.

Compared with 1,6,6aλ^4-trithiapentalenes, the visible absorption band of 1-oxa-6,6aλ^4-dithiapentalenes is stronger but appears at a lower wavelength, near 450 nm, and therefore confers a lighter color to these compounds. With 1,6aλ^4-dithia-6-selenapentalenes, the opposite is observed, the visible absorption band is weaker and occurs at longer wavelengths than for the trithia analogue ⟨71AHC(13)161, p. 224⟩.

4.38.2.5 IR Spectroscopy

IR spectra have played an important rôle in the structural study of the class of compounds described in this chapter as 1-oxa-6,6aλ^4-dithiapentalenes and often named in the literature as α-(1,2-dithiol-3-ylidene) ketones. The prominent feature of the IR spectra of these compounds is the absence of the usual carbonyl band in the 1620–1720 cm^{-1} range and the presence of two or more strong bands in the 1500–1610 cm^{-1} range, a fact considered as the result of a short distance O–S interaction.

Owing to the fact that the 'carbonyl' frequency falls in the region of C=C absorption, the identification of the C—O stretching vibration has sometimes been established by isotopic substitution with ^{18}O ⟨71AHC(13)161, p. 220, 72BSF1385, 73BSF3339⟩.

As the experimental S—O bond distances given in Section 4.38.2.2 are relatively long, these compounds have been often named as α-(1,2-dithiol-3-ylidene) ketones in the Z configuration. In this chapter we have preferred the oxadithiapentalene names because of the similarities over the whole family of polyheterapentalenes. Supporting this point of view is the fact that if X-ray structure determinations are not available, the IR spectra are a good proof for some O–S interaction or bonding ⟨71AHC(13)161, p. 218⟩. In fact, for O–S interactions, a correlation has been found between IR and X-ray crystallographic data ⟨71CC1352⟩.

4.38.2.6 Dipole Moments

In conjunction with IR spectra, dipole moments have been used for testing the bonding in 1-oxa-6,6aλ^4-dithiapentalenes for which three major canonical forms (**12**), (**13**) and (**14**) can be considered ⟨71AHC(13)161, p. 222⟩.

Experimental dipole moments are generally between 3.7 and 4.5 D, while for (**14**) the expected dipole moment would be about 13 D. This fact shows the importance of the forms (**12**) and/or (**13**) and the low carbonyl character of the C—O bond, shown by IR spectra, offers further support for formula (**12**).

(**12**)   (**13**)   (**14**)

On the other hand, dipole moments of 1-oxa-6,6aλ^4-dithiapentalenes are significantly larger than for the corresponding 1,6,6aλ^4-trithiapentalene, a fact indicating that the S–O interaction is weaker than the corresponding S–S interaction.

4.38.2.7 Photoelectron Spectroscopy

Photoelectron spectroscopy using electrons expelled by He (584 Å) or Mg–K$_\alpha$ X-rays (9.869 Å) has been applied to 1,6,6aλ^4-trithiapentalenes and to 1-oxa-6,6aλ^4-dithia, 1,6-dithia-6aλ^4-selena and 1,6,6aλ^4-triselena analogues.

For the study of the trithiapentalene structure, X-ray irradiation (ESCA) is particularly interesting because of the very short time scale of the phenomenon. Thus it could be hoped, in the case of a rapid equilibrium, to prove by ESCA the existence of isomers which cannot be detected by other methods, such as NMR, whose time scale is considerably larger.

After experiments on crystallized samples ⟨71CS(1)183⟩, it appeared that the spectra could be interpreted as comprising either three signals of equal intensity and equal half-width, or two signals with the intensity ratio 2:1, the half-width of the stronger signal being much larger than that of the other. The first interpretation favours a rapid equilibrium between two valence tautomers. The second interpretation corresponds to a vibrational broadening of the signals from atoms 1 and 6, due essentially to the vibration of atom 6a between atoms 1 and 6.

Further experiments on gaseous trithiapentalene supported the second interpretation ⟨77MI43800⟩. This has been corroborated by a comparative study with 1,6-dithia-6aλ^4-selenapentalene and with 1,6,6aλ^4-triselenapentalene which shows that the vibrational broadening of the peaks corresponding to the terminal atoms (1 and 6) of the sequence is greatly diminished when the central (6a) atom is selenium ⟨81IC399⟩.

This is also consistent with the fact that intermolecular forces in crystals (and twist of phenyl groups) affect the position of the central atom in an S—Se—S sequence to a lesser extent than in an S—S—S sequence; compare for instance the S—S—S sequence in Table 1 ⟨71ACS1583⟩ and the S—Se—S sequence in Table 2 ⟨72ACS2139⟩.

The preceding results with compounds containing the S—S—S, S—Se—S and Se—Se—Se sequences have been confirmed by gas phase ESCA studies of 1,6-dioxa-6aλ^4-thia-2,5-diazapentalene and of its 6aλ^4-selena and 6aλ^4-tellura analogues ⟨80JA1783⟩.

4.38.3 REACTIVITY AT RING ATOMS

4.38.3.1 Reactivity at Atoms 1,6,6a

4.38.3.1.1 Isomerization

Photoisomerization of a 1-oxa-6,6aλ^4-dithiapentalene leads to the *trans* structure which reverts rapidly to the original *cis* compound by a thermal process (equation 3). The rate of the thermal reversal in a polymethacrylate matrix is much smaller than in cyclohexane or ethanol ⟨73JCS(P1)2837, 74AG(E)349, 77ACS(B)683⟩. This fact rules out the interpretation of the photoisomerization as a valence isomerism as described in equation (4). Effectively, the reaction rate for a valence isomerization, which does not imply an important movement of atoms, would not be considerably affected by a change in the viscosity of the solvent.

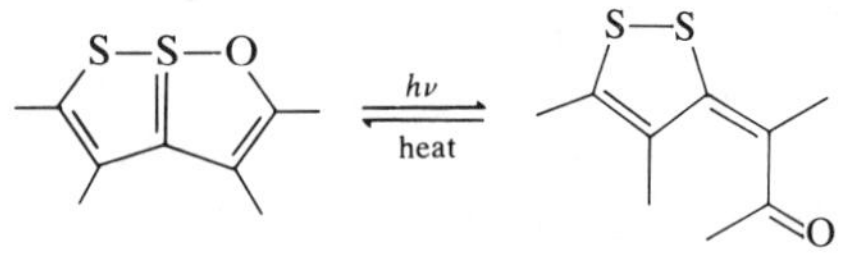

(3)

(4)

4.38.3.1.2 Protonation

1,6,6aλ^4-Trithiapentalenes are protonated by strong acids and in the presence of water the resulting dithiolylium cation is hydrolyzed to a 1-oxa-6,6aλ^4-dithiapentalene ⟨71AHC(13)161, p. 179⟩.

Depending on their structure, 1-oxa-6,6aλ^4-dithiapentalenes are more or less easily protonated by perchloric acid and crystalline perchlorates may be obtained ⟨71AHC(13)161, p. 198⟩. Protonation of 1,6-dioxa-6aλ^4-thiapentalenes or of their selenium analogues may occur on oxygen or on carbon atoms 3 or 4, as shown by ^{1}H NMR and deuterium exchange ⟨75JCS(P1) 2097⟩.

4.38.3.1.3 Metalation

Coordination compounds of nickel ⟨79AG(E)683⟩ and cobalt ⟨70BCJ3604⟩ with trithiapentalenes have been described. 2,5-Diphenyl-1,6,6aλ^4-trithiapentalene reacts with palladium dichloride, in the presence of triphenylphosphine, giving a coordination compound (15) in which a benzene ring is directly linked to palladium ⟨79AG(E)684⟩.

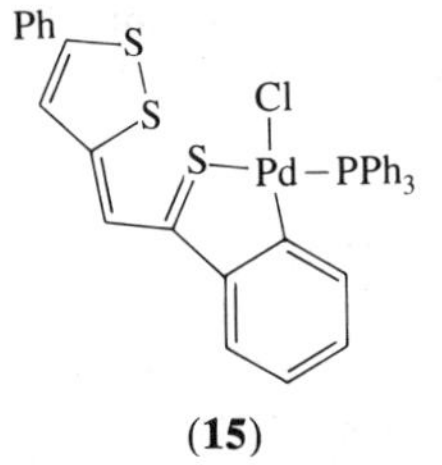

(15)

Mercury (II) acetate reacts with 1,6,6aλ^4-trithiapentalenes or their 3-aza derivatives to give the corresponding 1-oxa-6,6aλ^4-dithiapentalenes or 1-oxa-6,6aλ^4-dithia-3-azapentalenes ⟨71AHC(13)161 p. 179, 199⟩. This reaction proceeds first by acetoxymercuration of a sulfur atom, followed by hydrolysis of the intermediate dithiolylium cation as indicated in Scheme 1.

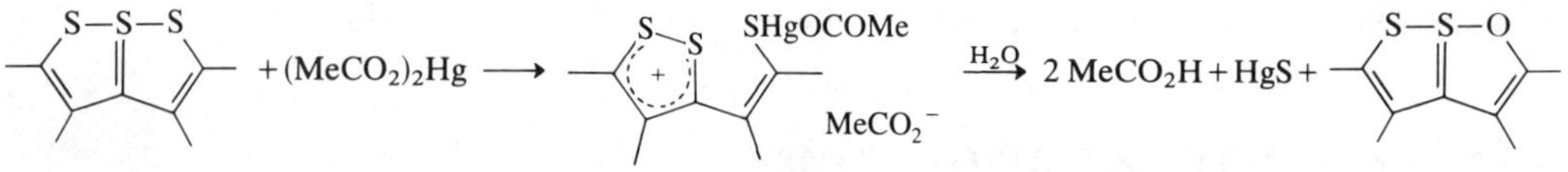

Scheme 1

S—S—N sequences are similarly attacked, giving O—S—N sequences ⟨78JCS(P1)195⟩. Compounds containing an O—S—O sequence behave differently, acetoxymercuration occurring at positions 3 or 4 as indicated in Section 4.38.3.2.6.

4.38.3.1.4 Alkylation

In 1,6,6aλ^4-trithiapentalenes, the sulfur atoms numbered 1 or 6 are much less reactive than the thione sulfur of 1,2-dithiole-3-thiones. Nevertheless in some cases they may be alkylated leading to a compound which may be considered as a dithiolylium cation. Methyl iodide and triethyloxonium tetrafluoroborate have been used as methylating agents ⟨71AHC(13)161, p. 201⟩.

1,6aλ^4-Dithia-6-azapentalenes may be alkylated on sulfur atom 1 either by methyl iodide ⟨74JCS(P1)242⟩ or by methyl fluorosulfonate ⟨79JCS(P1)926⟩.

4.38.3.1.5 S-Oxidation or reduction

Peracetic acid oxidizes 1,6,6aλ^4-trithiapentalenes to 1-oxa-6,6aλ^4-dithiapentalenes, most probably with intermediate formation of *S*-oxides ⟨71AHC(13)161, p. 180⟩.

A trithiapentalene derived from anthraquinone has been reduced by alkaline sodium dithionite and the monothiodiphenol has been methylated to (16) ⟨78IJC(B)673⟩.

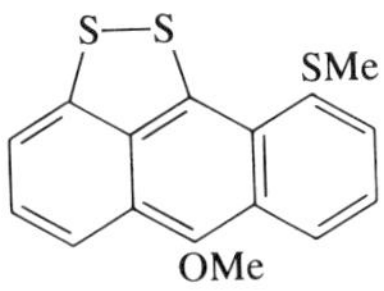

(16)

4.38.3.2 Electrophilic Attack at Atoms Other than 1,6,6a

Electrophilic reagents generally attack carbon atoms 3 or 4 which are particularly reactive in 1,6-dioxa-6aλ^4-thiapentalenes.

4.38.3.2.1 Halogenation

Chlorination by sulfuryl chloride occurs at position 3 for 1-oxa-6,6aλ^4-dithiapentalenes ⟨71AHC(13)161, p. 181⟩. Bromination of 1,6-dioxa-6aλ^4-thiapentalene occurs readily and leads to the 3,4-dibromo derivative. With the same compound, iodine and silver acetate give the 3-iodo and the 3,4-diiodo derivatives ⟨72CC1283⟩.

1,6,6aλ^4-Trithiapentalenes ⟨71JCS(C)963⟩ and 6,6aλ^4-dithia-1,2-diazapentalenes ⟨77JCS(P1)848⟩ are brominated in a similar way.

4.38.3.2.2 Nitration

Few examples are known and sometimes nitrosation occurs in place of an expected nitration of a 1,6,6aλ^4-trithiapentalene ⟨71AHC(13)161, p. 194⟩. When a 6,6aλ^4-dithia-1,2-diazapentalene has a methyl group in position 3, nitration destroys the thiadiazole ring and a nitrodithiole is obtained ⟨77JCS(P1)848⟩.

4.38.3.2.3 Nitrosation

A detailed account of nitrosation reactions is available ⟨74JCS(P1)722⟩. Generally, nitrosation which occurs at position 3 or 4 is accompanied by a rearrangement of the triatomic sequence 1,6,6a, with formation of a 1-oxa-6,6aλ^4-dithia-2-azapentalene, as indicated in Scheme 2. 1,6-Dioxa-6aλ^4-thiapentalene reacts similarly with nitrosyl hexafluorophosphate.

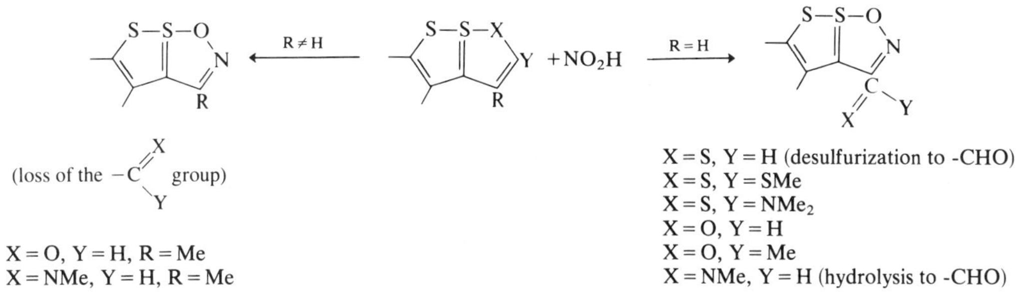

(loss of the $-C\begin{smallmatrix}X\\\\Y\end{smallmatrix}$ group)

X = O, Y = H, R = Me
X = NMe, Y = H, R = Me

X = S, Y = H (desulfurization to -CHO)
X = S, Y = SMe
X = S, Y = NMe$_2$
X = O, Y = H
X = O, Y = Me
X = NMe, Y = H (hydrolysis to -CHO)

Scheme 2

4.38.3.2.4 Diazo coupling

This reaction has some analogy with nitrosation in so far as it is generally accompanied by a rearrangement leading to a 1,2-diaza derivative as shown in equation (5) ⟨76JCS(P1)880⟩. When the starting product is substituted at position 3 or 4 where the electrophilic attack occurs, the rearrangement takes place with elimination of atoms 1 and 2 or 5 and 6, respectively, as indicated in Scheme 3 ⟨79JCS(P1)926⟩.

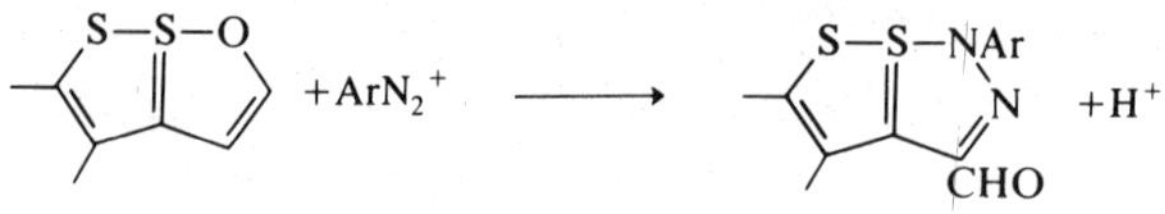

$$\text{S—S—O structure} + ArN_2^+ \longrightarrow \text{S—S—NAr structure with CHO} + H^+ \quad (5)$$

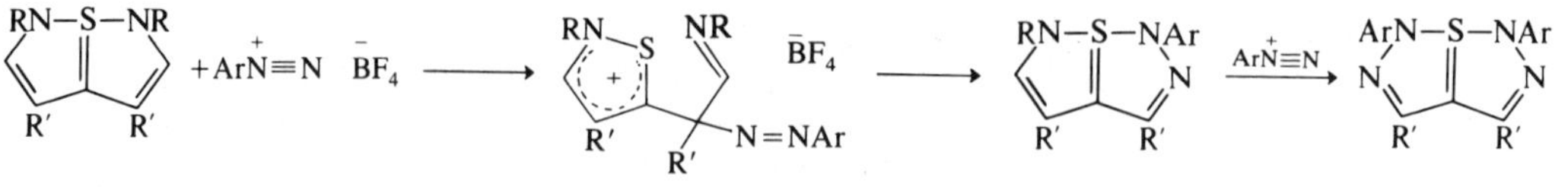

Scheme 3

4.38.3.2.5 Formylation

The Vilsmeier–Haack formylation, using DMF and phosphorus oxychloride has been applied to: $1,6,6a\lambda^4$-trithiapentalenes ⟨71AHC(13)161, p. 194, 72BSF4181⟩, 1-oxa-$6,6a\lambda^4$-dithiapentalenes ⟨71AHC(13)161, p. 181⟩ and $6,6a\lambda^4$-dithia-1,2-diazapentalenes ⟨77JCS(P1)848⟩.

In 3-iodo-1,6-dioxa-$6a\lambda^4$-thiapentalene, butyllithium replaces iodine by lithium and the resulting organometallic species reacts with DMF leading to 1,6-dioxa-$6a\lambda^4$-thiapentalene-3-carbaldehyde ⟨77JCS(P1)854⟩.

4.38.3.2.6 Miscellaneous

1,6-Dioxa-$6a\lambda^4$-thiapentalene is tritylated in position 3 by trityl perchlorate and is acetoxymercurated in positions 3 and 4 simultaneously by mercury (II) acetate ⟨77JCS(P1)854⟩.

4.38.3.3 Nucleophilic Attack on Atoms Other than 1,6,6a

Nucleophilic reagents generally attack at positions 2 or 5. Their action may be limited to the replacement of a substituent (generally SMe) or may lead to a modification of the triatomic sequence (1,6,6a) or may even cause a rearrangement to a γ-pyrone.

4.38.3.3.1 Replacement of a substituent

Methylthio groups are replaced by ethoxy groups on reaction with sodium ethoxide. In the same way, aliphatic amines may replace methylthio groups by alkylamino substituents ⟨71AHC(13)161, p. 194, 71JCS(C)963, 74JCS(P1)722⟩.

4.38.3.3.2 Modification of the sequence 1,6,6a

Phosphorus pentasulfide has often been used for converting an O—S—S sequence into an S—S—S one ⟨71AHC(13)161, p. 192, 77AG(E)403, 80JOC3909⟩. Similarly, an O—S—O sequence can be transformed into an O—S—S one and an O—Se—Se sequence into an S—Se—Se one. All these reactions involve an attack at position 2 by a nucleophilic sulfur atom, while the oxygen atom in position 1 is removed by fixation on phosphorus. In the same way, phosphorus pentaselenide converts O—S—S sequences into Se—S—S ones.

Methylamine attacks position 2 in 1,6,6aλ^4-trithiapentalenes, giving 1,6aλ^4-dithia-6-azapentalenes ⟨73JCS(P1)2351⟩, and in 1,6-dioxa-6aλ^4-thiapentalenes, giving 1-oxa-6aλ^4-thia-6-azapentalenes ⟨77JCS(P1)854⟩.

4.38.3.3.3 Rearrangement to γ-pyrones

In DMF, 1,6,6aλ^4-trithiapentalenes having a free 2- or 5-position are attacked by hydrosulfide or hydroxide ions to give a γ-dithiopyrone ⟨71AHC(13)161, p. 196⟩.

4.38.4 REACTIVITY AT SUBSTITUENTS

4.38.4.1 Hydrolysis of Ester or Cyano Groups

These reactions have been observed with substituents in position 3 or 4. 3-Alkoxycarbonyl- or 3-cyano-1,6,6aλ^4-trithiapentalenes have been hydrolyzed by heating with hydrogen chloride or bromide in acetic acid and decarboxylation follows immediately ⟨71AHC(13)161, p. 194⟩.

In basic medium a 3-alkoxycarbonyl-1-oxa-6,6aλ^4-trithiapentalene is also hydrolyzed and the resulting salt, acidified with caution, yields the corresponding carboxylic acid which easily loses carbon dioxide on heating ⟨71AHC(13)161, p. 182⟩.

4.38.4.2 Oxidation or Reduction

Few examples have been described. A thioketo group in a condensed ring has been oxidized to a keto group using potassium permanganate; it did not attack the fused trithiapentalene system ⟨71AHC(13)161, p. 195⟩.

In 1,6,6aλ^4-trithiapentalenes or in 1-oxa-6,6aλ^4-dithiapentalenes having a 3,4-propano bridge, triphenylmethyl tetrafluoroborate oxidizes the cyclohexane ring to a benzene ring with formation of a cation ⟨71AHC(13)161, p. 198⟩.

Nitro groups in the 3-position are reduced to amino groups by tin(II) chloride and hydrogen chloride ⟨72CC836⟩.

4.38.4.3 Carbanion or Related Reactions in Side Chains

In 1,6,6aλ^4-trithiapentalenes and related compounds, carbon atoms in the 2- or 5-position are electrophilic and consequently a methylene group substituted in one of these positions has distinctly acidic properties.

In acidic medium, the reaction intermediate is probably the enethiol (**17**). An example of such a reaction is the condensation of the thio-DMF, in the presence of phosphorus oxychloride, with a methyl group at the 2-position. An enamine is obtained, in poor yields, with elimination of hydrogen sulfide ⟨71JCS(C)2829⟩.

(**17**)

In mildly basic conditions (pyridine), a methylene group in the 2-position may react with aromatic aldehydes to give a 2-styryl-1,6,6aλ^4-trithiapentalene ⟨71AHC(13)161, p. 194⟩. As the trithiapentalene system is not affected in this case, it is likely that a free carbanion is not the intermediate and that a push–pull mechanism is involved.

In contrast strong bases such as sodium *t*-butoxide lead to a free anion which can undergo a rearrangement giving a thiophene derivative ⟨71AHC(13)161, p. 195⟩. This reaction implies a rotation around the C(2)—C(3) bond in the trithiapentalene, as shown in Scheme 4. If this rotation is forbidden, for instance by a 3,4-propano bridge, the carbanion does not rearrange and can react with carbon disulfide, a key step in the first synthesis of extended structures containing five sulfur atoms in a row ⟨71AHC(13)161, p. 205⟩.

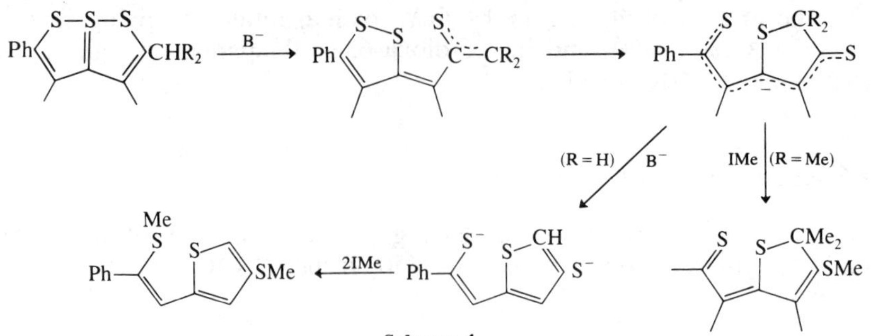

[Scheme 4]

4.38.4.4 Reactivity of Amino Groups

In 1,6,6aλ^4-trithiapentalenes, an amino group in the 2- or 5-position is rather acidic, which is consistent with the properties of methylene groups at the same places, but amino groups in the 3- or 4-position are distinctly basic and can be acetylated ⟨72CC836⟩.

4.38.5 SYNTHESIS BY RING FORMATION

Preparative methods are classified according to the number of fragments necessary for the construction of the bicyclic system. When there is doubt concerning the mechanism and the number of independent fragments used in building the cyclic system, the mechanism involving the smallest possible number of fragments has been chosen arbitrarily.

4.38.5.1 Formation of Two Rings

4.38.5.1.1 From one fragment

Bromine oxidation of N,N'-diacylthioureas leads to the formation of 3,4-dioxa-3aλ^4-thia-1,6-diazapentalenes (equation 6) ⟨80JOC3909⟩.

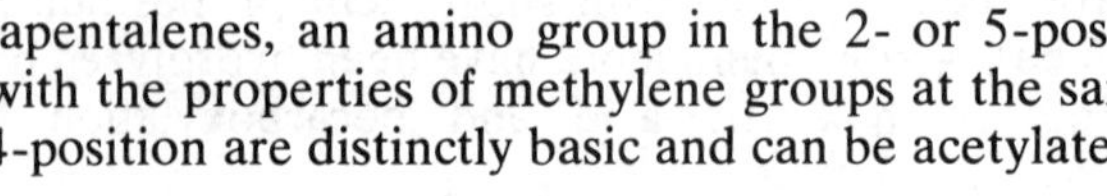

$$\text{[equation 6]}$$

4.38.5.1.2 From [7+1] atom fragments

Cyclization of α,α'-bis(dimethylthiomethylene) ketones by phosphorus pentasulfide leads to 2,5-bis(methylthio)-1,6,6aλ^4-trithiapentalenes (equation 7) ⟨71AHC(13)161, p. 185⟩. Condensation of carbon disulfide with pyrrolidin-2-one in basic medium, followed by methylation of the reaction product, leads to methyl 3-(dimethylthiomethylene)-2-oxopyrrolidine-1-dithiocarboxylate. This compound, when treated with phosphorus pentasulfide and perchloric acid, yields the 2,5-bis(methylthio)-1,6,6aλ^4-trithia-2aλ^5-azacyclopenta[cd]pentalen-2a-ylium cation. With ε-caprolactam similar results are obtained ⟨76BSF1200⟩.

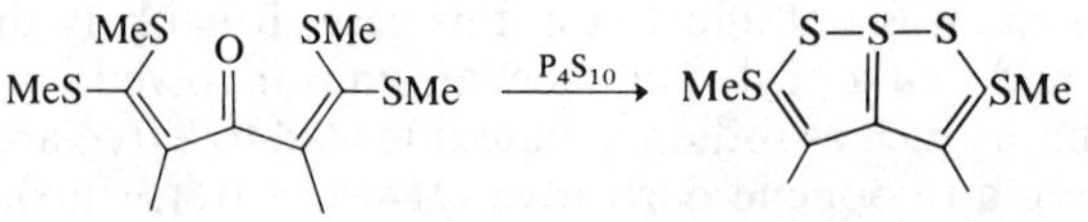

$$\text{[equation 7]}$$

Reaction of 1,8-dihydroxy-4,5-dinitroanthraquinone with sodium sulfide in boiling DMF gives a blue compound (**18**) to which a thiadiazapentalene structure has been assigned ⟨78IJC(B)673⟩.

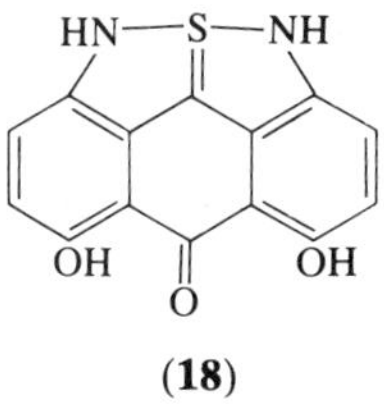

(18)

Dioximes or bis(arylhydrazones) of β-diketones, react with chlorides or oxides of sulfur, selenium or tellurium, to produce various polyheterapentalenes, as indicated in Scheme 5.

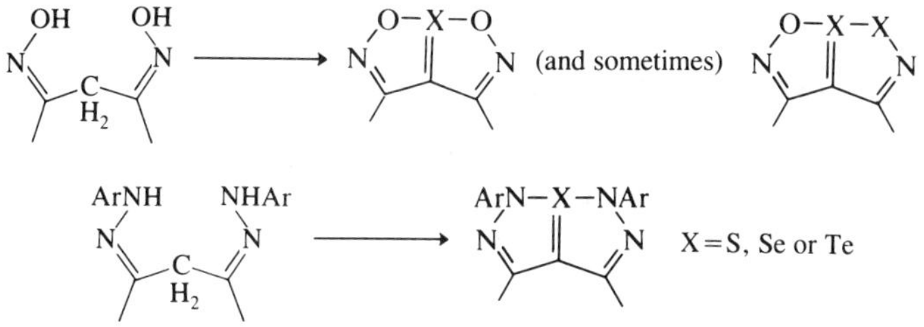

Reagents	Sequences formed	Ref.	Sequences formed	Ref.
SCl_2 or S_2Cl_2	O—S—O and O—S—S	72TL1835, 79BSF(2)199	N—S—N	79BSF(2)205
SeO_2 or Se_2Cl_2	O—Se—O	79BSF(2)199, 49JCS274, 70BSF4517	N—Se—N	79BSF(2)205
TeO_2 or $TeCl_4$	O—Te—O	79BSF(2)199	N—Te—N	79BSF(2)205

Scheme 5

4.38.5.1.3 From [6+2] atom fragments

Good yields of 1-oxa-6,6aλ^4-dithiapentalenes are obtained by reacting sodium disulfide at pH 9 with α,γ-dialkynic carbonyl compounds (equation 8) ⟨71AHC(13)161, p. 172⟩.

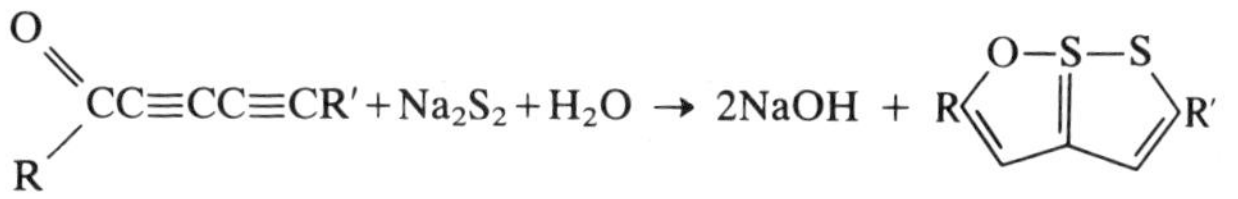

$$\text{RCOC≡CC≡CR'} + Na_2S_2 + H_2O \rightarrow 2NaOH + \text{[O—S—S ring]R'} \tag{8}$$

4.38.5.1.4 From [6+1+1] atom fragments

Although reaction of phosphorus pentasulfide with β,δ-triketones generally leads to 1,6,6aλ^4-trithiapentalenes, in some cases the reaction can be limited to the formation of a 1-oxa-6,6aλ^4-dithiapentalene. When the same reaction is applied to a β,δ-diketophenol, the phenolic oxygen is not touched and only the keto oxygens are replaced by sulfur as shown in equation (9) ⟨71AHC(13)161, p. 168⟩. These reactions are not a simple replacement of oxygen by sulfur; dehydrogenation is also implied.

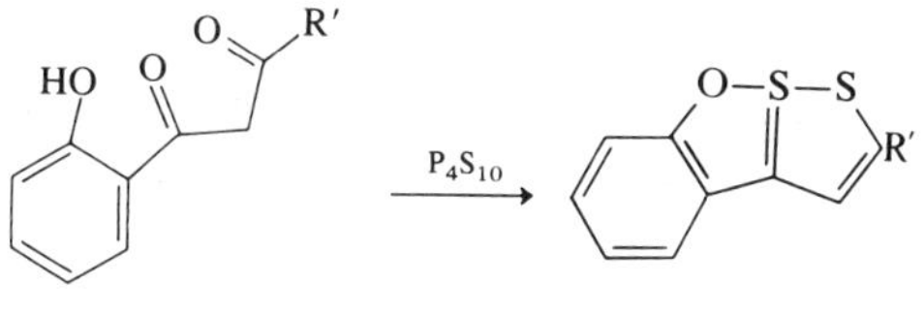

$$\text{(β,δ-diketophenol)} \xrightarrow{P_4S_{10}} \text{(1-oxa-dithiapentalene product)} \tag{9}$$

Reaction of phosphorus pentasulfide on a β,δ-diketophenol methyl ether or thioether yields respectively a 1-oxa-6,6aλ^4-dithiapentalene or a 1,6,6aλ^4-trithiapentalene as shown in equation (10) ⟨73BSF277⟩.

α,γ-Diethylenic ketones or esters react with elemental sulfur at approximately 200 °C to give a 1-oxa-6,6aλ^4-dithiapentalene, as shown in equation (11), with dehydrogenation of cyclohexane rings if present ⟨71AHC(13)161, p. 170⟩.

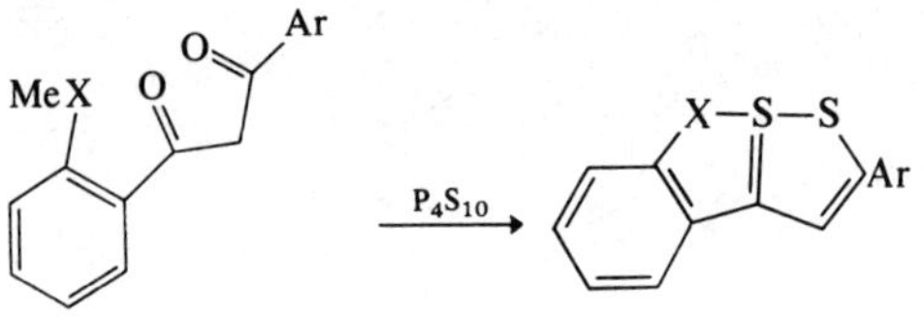

4.38.5.1.5 From [3+2+2] atom fragments

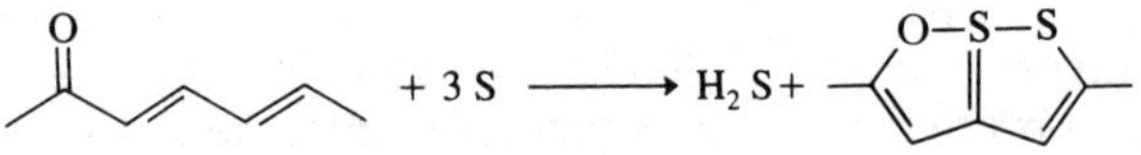

Captan [N-(trichloromethylthio)-1,2,3,6-tetrahydrophthalimide] reacts with resorcinol with the formation of a condensed ring, 1,6-dioxa-6aλ^4-dithiapentalene as indicated in equation (12). Similar results are obtained when captan is replaced by thiophosgene ⟨72T2183⟩.

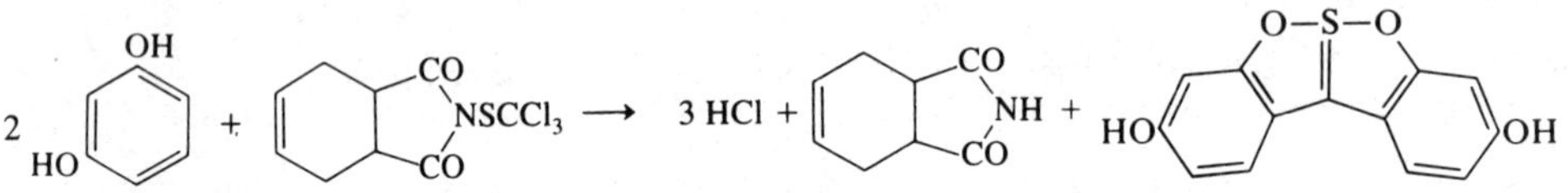

4.38.5.1.6 From [5+1+1+1] atom fragments

Several 1,6,6aλ^4-trithiapentalenes have been obtained by reacting β,δ-triketones with phosphorus pentasulfide as indicated in equation (13) ⟨71AHC(13)161, p. 182, 71BSF4418⟩.

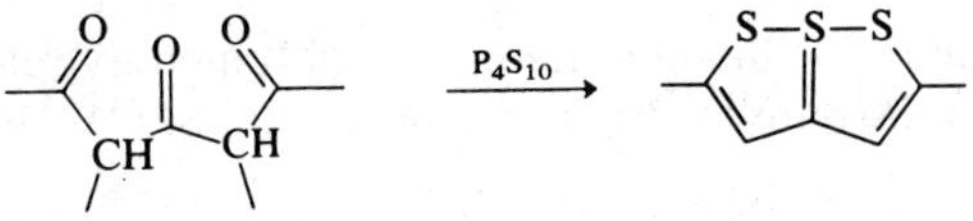

Reaction of sodium sulfide in DMF with 1,8-dichloroanthraquinone gives the green compound (19) ⟨78IJC(B)673⟩ and the selenium analogue is obtained using cadmium selenide. Under similar conditions, 2,2',6,6'-tetrachlorobenzophenone gives the purple compound (20) in low yields ⟨80JCR(S)221⟩.

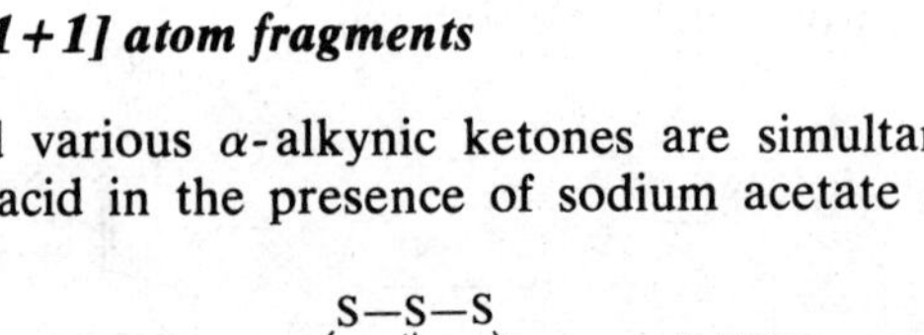

4.38.5.1.7 From [4+2+1+1] atom fragments

Pentane-2,4-dione and various α-alkynic ketones are simultaneously sulfurated and acetylated by thioacetic acid in the presence of sodium acetate as shown in Scheme 6 ⟨71AHC(13)161, p. 183⟩.

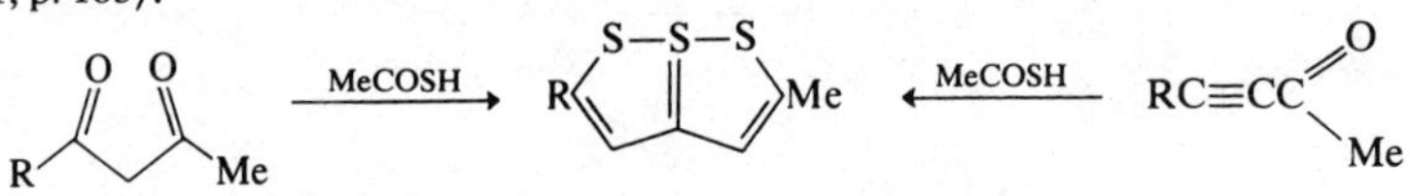

Scheme 6

4.38.5.1.8 From [3+2+2+1] atom fragments

Acyl chlorides react with the cyanodithioformate anion to give, through relatively complex reactions, compounds with the structures (21) and (22) ⟨77ZC221⟩.

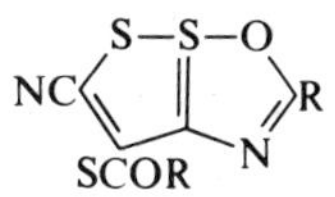

1,6,6aλ⁴-Trithiapentalenes and Related Systems   1065

(21)

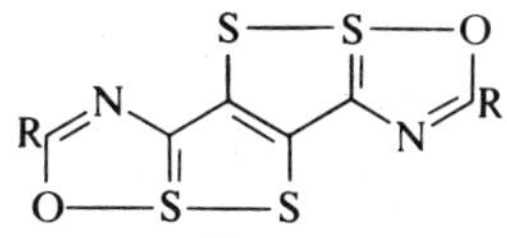

(22)

4.38.5.2 Formation of One Ring

In these syntheses, the starting material already contains one of the pentagonal heterocycles constituting the resulting polyheterapentalene. When denoting the fragments whose assembly will create the second heterocycle, only the atoms of the first cycle that will be common with the second are counted. The fragment containing these atoms is listed first when describing the ring formation.

4.38.5.2.1 From one fragment

Hydrazones or the oxime of isothiazole-5-carbaldehyde, reacting with a methylating agent, lead to various polyheterapentalenes as indicated in Scheme 7 ⟨79JCS(P1)2340⟩. In this reaction, the acid YH is neutralized by sodium carbonate.

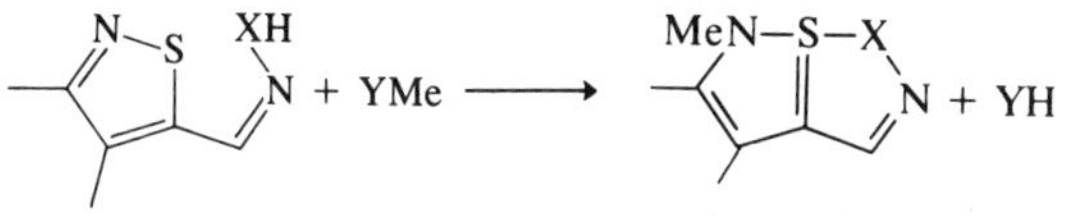

X = O, NMe, NPh, NC₆H₄Me-p; Y = FSO₃, p-MeC₆H₄SO₃

Scheme 7

4.38.5.2.2 From [4+1] atom fragments

3-(2-Dialkylaminovinyl)-1,2-dithiolylium cations and similar structures such as 3-(2-alkylthiovinyl)-1,2-dithiolylium cations or 5-(2-alkoxyvinyl)isothiazolium cations react with various nucleophiles to give 1,6,6aλ⁴-trithiapentalenes or analogous compounds, as indicated in Table 5.

Table 5 Nucleophilic Attack on 3-(2-Dimethylaminovinyl)-1,2-dithiolylium Cations and Related Structures

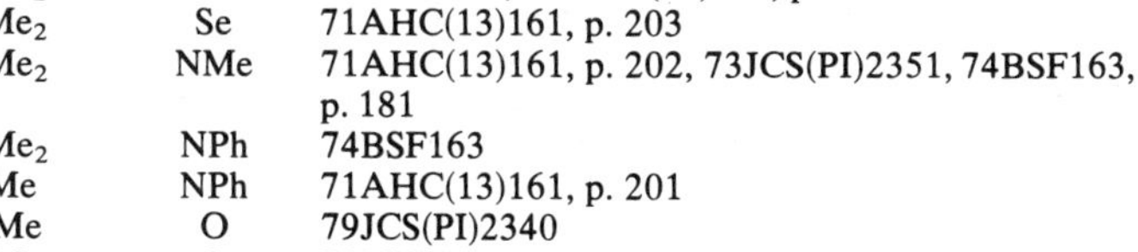

X	Y	Z	Ref.
S	NMe₂	O	71AHC(13)161, p. 181, 72BSF4576, 80AHC(27)151, p. 172
S	NMe₂	S	72BSF4576, 71AHC(13)161, p. 193
S	NMe₂	Se	71AHC(13)161, p. 203
S	NMe₂	NMe	71AHC(13)161, p. 202, 73JCS(PI)2351, 74BSF163, 80AHC(27)151, p. 181
S	NMe₂	NPh	74BSF163
S	SMe	NPh	71AHC(13)161, p. 201
NMe	OMe	O	79JCS(PI)2340
NMe	OMe	S	79JCS(P1)2340
NMe	OMe	NMe	79JCS(P1)2340
NMe	SMe	NMe	71AHC(13)161, p. 202

Similar reactions have been observed with 5-(2-methylthiovinyl)-1,2,3-thiadiazolium cations as shown in equation (14) ⟨79JCS(P1) 926⟩.

1-Oxa-6,6aλ⁴-dithiapentalenes and 1,6,6aλ⁴-trithiapentalenes have been obtained respectively by hydrolysis or sulfhydrolysis of (1,2-dithiol-3-ylidene)malononitriles ⟨71AHC(13)161, p. 181, 193⟩.

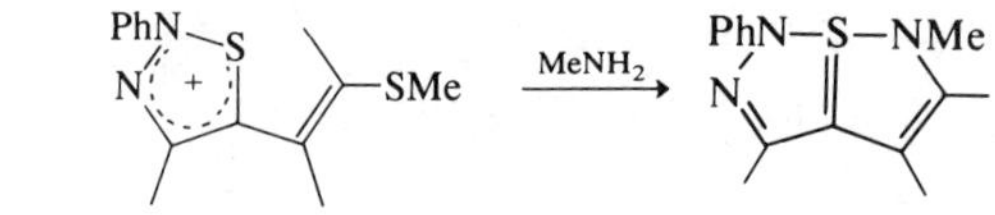

4.38.5.2.3 From [3+2] atom fragments

Carbanions derived from 5-alkyl-1,2-dithiole-3-thiones react with carbon oxysulfide or disulfide, or with phenyl isothiocyanate, to give dianions which can be methylated as shown in Scheme 8.

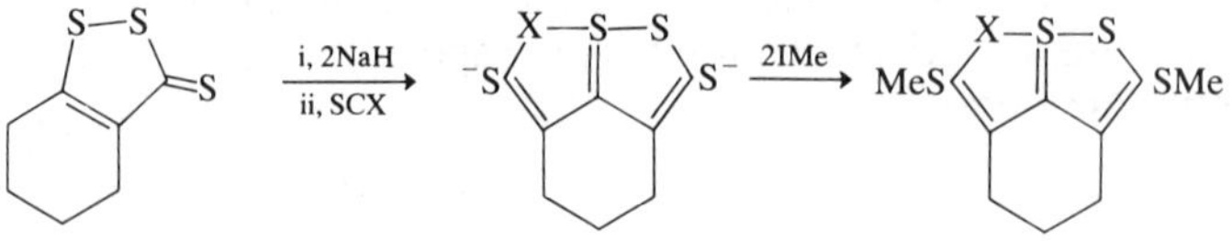

X = O ⟨71AHC(13)161, p. 172, 74BSF140⟩; S ⟨71AHC(13)161, p. 186, 74BSF137⟩; NPh ⟨76BSF1499⟩

Scheme 8

3-Alkyl-1,2-dithiolylium cations, or rather their conjugate bases, react with electrophiles such as dithioesters ⟨71AHC(13)161, p. 187⟩, diazonium cations ⟨80AHC(27)151, p. 191, 78JCS(P1)195⟩ or nitrous acid ⟨80AHC(27)151, p. 191, 72JCS(P1)1360, 74JCS(P1)722⟩ giving various types of polyheterapentalenes. 3-Alkyl-1,2-diselenolylium cations react in the same way with nitrous acid ⟨72JCS(P1)1360⟩. Similarly, 1,2-dithiol-3-imines react with acyl chlorides ⟨71AHC(13)161, p. 199, 71BSF4002⟩ or aryl isothiocyanates ⟨67CB4027⟩ and 1,2,4-dithiazol-3-imines react with aryl isothiocyanates ⟨71AHC(13)161, p. 200⟩. These various reactions are summarized in Scheme 9.

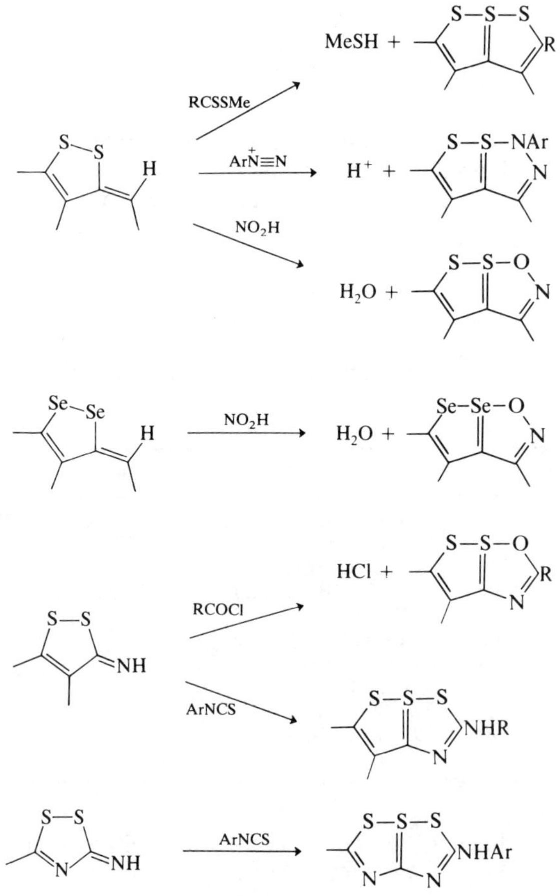

Scheme 9

4.38.5.2.4 *From [2+3] atom fragments*

1,2-Dithiolylium cations (see Chapter 4.31) unsubstituted on the 3- or 5-position react with active methylene compounds, in most cases with simultaneous oxidation, as shown in equation (15). This reaction has been performed with a large variety of ketones, acylic or cyclic, β-diketones, β-keto dithioesters and α-cyano ketones ⟨71AHC(13)161, pp. 174, 188, 80AHC(27)151, p. 183⟩.

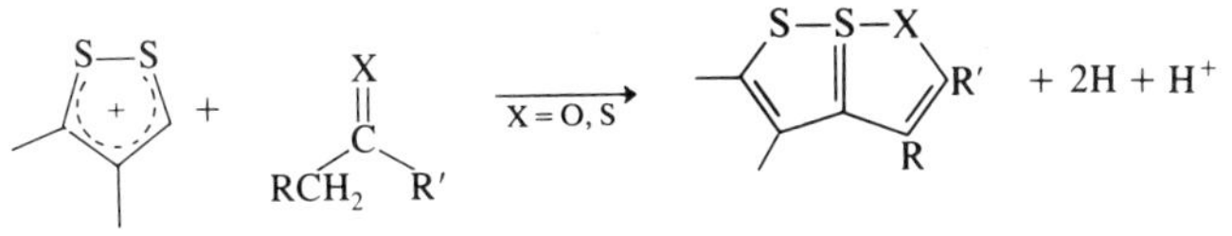

$$\text{(15)}$$

1,2-Dithiolylium cations having a negative leaving group in the 3- or 5-position react also with compounds having an active methylene group, such as ketones, β-diketones, β-keto esters, β-keto nitriles, α-nitro ketones, β-keto dithioesters, and α-nitro dithioesters. *ortho*-Positions in phenols can also react in this way ⟨71AHC(13)161, pp. 175–177, 189, 190, 72CC836, 80AHC(27)151, pp. 186–188⟩. These reactions are summarized in equation (16). Similar reactions have been performed with 3-methylthio-1,2,4-dithiazolylium cations ⟨67CB4027⟩.

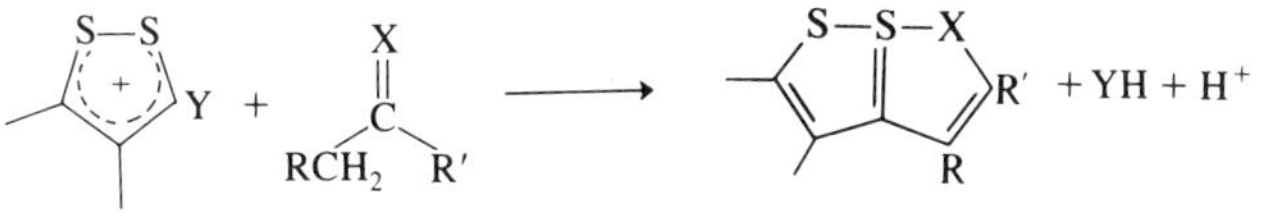

$$\text{(16)}$$

X = O or S; Y = Cl or SMe

3-Oxa-3aλ^4,4-dithia-1-azapentalenes are obtained, as shown in equation (17), by reacting 1,2-dithiole-3-thiones with acyl isothiocyanates ⟨71BSF4002⟩.

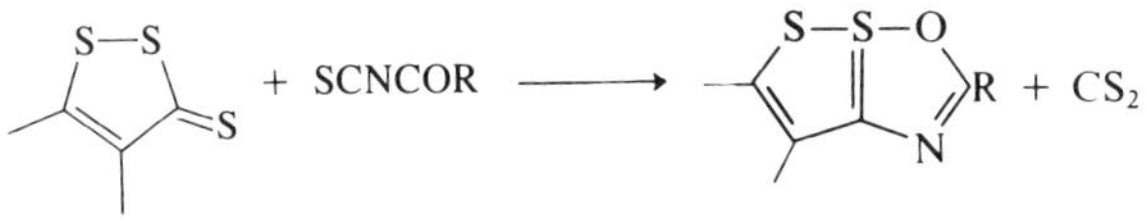

$$\text{(17)}$$

4.38.5.3 Formation of Extended Structures

Using some of the methods already indicated, extended structures of two types have been obtained. Structures containing two or three independent trithiapentalene systems; and structures containing more than three sulfur atoms in line, namely, four or five atoms.

4.38.5.3.1 *Structures containing two or three trithiapentalene systems*

Reaction of sodium sulfide with 1,4,5,8-tetrachloroanthraquinone in boiling DMF gives a green dye apparently identical with the commercial vat dye Olive Heliane F-JR. Formula (**23**) has been assigned to this compound. The commercial dye Indanthrene Olive GG, obtained by reaction of sulfur monochloride with anthracene is probably the tetrachloro derivative of (**23**) ⟨78IJC(B)673⟩. Reaction of sulfur with pentacene or hexacene gives compounds to which structures containing two trithiapentalene systems have been assigned ⟨72JCS(P1)1310⟩.

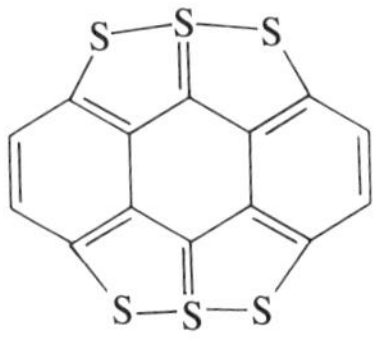

(**23**) 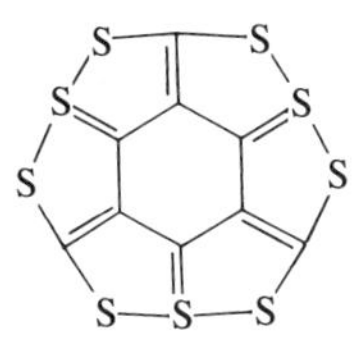   (**24**)

1,3,5-Trichloro-2,4,6-tris(dimethylaminomethyl)benzene, reacting with sulfur in boiling DMF, gives in low yield a compound C_9S_9 ⟨74JCS(P1)866⟩ whose structure (**24**) has been established by X-ray diffraction ⟨74CC800⟩.

The synthesis of the compound (**22**) containing two fused 3-oxa-3aλ^4,4-dithia-1-azapentalene systems has been described in Section 4.38.5.1.8.

4.38.5.3.2 Structures with more than three sulfur atoms in line

Cations containing four sulfur atoms in line are obtained by reacting a 3-methylthio-1,2-dithiolylium cation with the conjugate base of a 3-alkyl-1,2-dithiolylium cation as shown in equation (18) ⟨71AHC(13)161, p. 205, 73BSF1659, 80AHC(27)151, p. 191⟩.

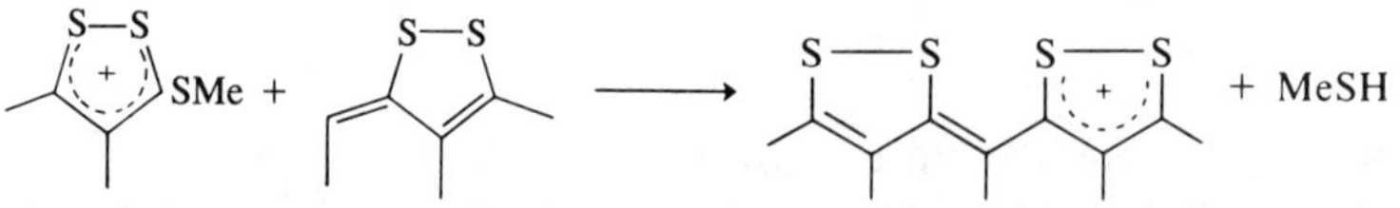

Compounds having five sulfur atoms in line have been prepared in various ways. In one method, the starting material is a trithiapentalene and two supplementary sulfur atoms are added successively as indicated in Scheme 10 ⟨71AHC(13)161, p. 205⟩.

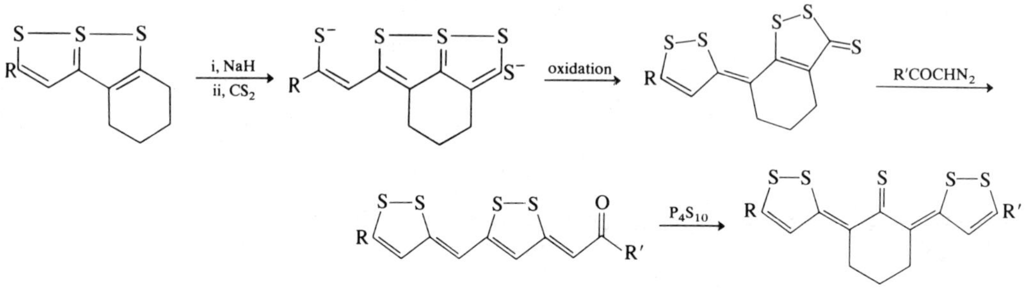

Scheme 10

Another method consists of the double condensation of 3-chloro- or 3-methylthio-1,2-dithiolylium cations with a carbonyl compound. In the structure thus obtained, the carbonyl group is afterwards transformed into a thiocarbonyl group by reaction with phosphorus pentasulfide.

This method has been applied to urea, as indicated in Scheme 11 ⟨71AHC(13)161, p. 207⟩ and to 4-hydroxy-3*H*-pyran-2,6-dione, as shown in Scheme 12 ⟨77CC851⟩.

Scheme 11

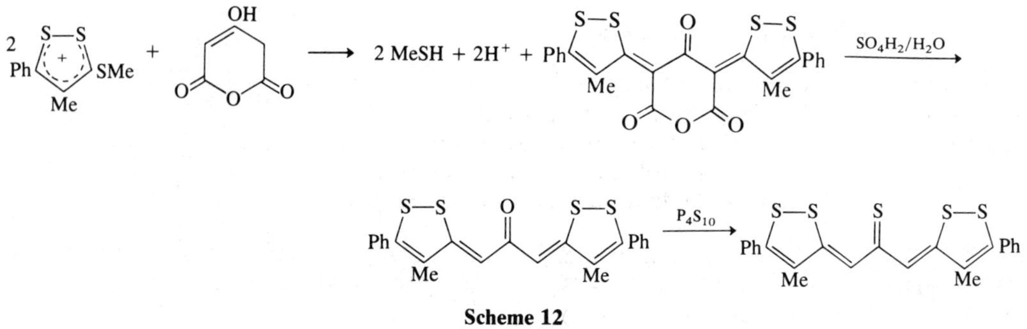

Scheme 12

4.38.6 SYNTHESIS BY RING TRANSFORMATION

4.38.6.1 From γ-Pyrones or their Chalcogen Derivatives

This reaction is closely related to the formation of two rings from β,δ-triketones. The mechanism involves first the opening of the pyrone ring by a strong nucleophile (OH^-, SH^-, SeH^-), followed by oxidation of the open intermediate into a $1,6,6a\lambda^4$-triheterapentalene as shown in equation (19). The oxidation may result from exposure to air but is often realized with a mild oxidizing agent such as potassium ferricyanide ⟨71AHC(13)161, p. 167, 183, 202⟩.

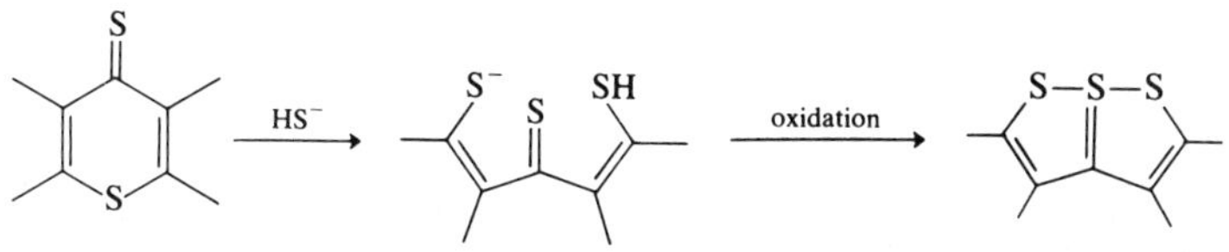

$$\text{(19)}$$

Oxidation of pyran-4-thiones by thallium (III) trifluoracetate leads, as shown in equation (20), to $1,6$-dioxa-$6a\lambda^4$-thiapentalenes ⟨72CC1283⟩, a type of structure which does not seem obtainable by the method described in the preceding paragraph.

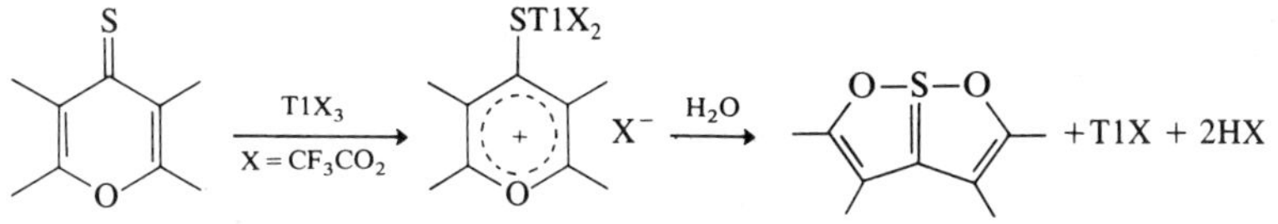

$$\text{(20)}$$

Pyranopyrans derived from 4,9-dioxaphenanthrene-1,10-dione, treated with boron or silicon sulfides have given various condensed ring compounds containing the $1,6,6a\lambda^4$-trithia-, the 1-oxa-$6,6a\lambda^4$-dithia- or the 1,6-dioxa-$6a\lambda^4$-thiapentalene ring systems ⟨73JCS(P1)1022⟩.

4.38.6.2 From $1,3a\lambda^4,4$-Trithiapentalenes

The starting material, which can also be considered as an α-(1,3-dithiol-3-ylidene) thioketone, is obtained by reacting alkynes with 1,2-dithiole-3-thiones. These compounds when heated with elemental sulfur or sulfur-containing catalysts rearrange to $1,6,6a\lambda^4$-trithiapentalenes as shown in equation (21). A mechanism has been suggested for this reaction ⟨72CC540⟩ but it should be noted that in the reaction of alkynes on 1,2-dithiole-3-thiones or on 1,2,4-dithiazole-3-thiones the final $1,6,6a\lambda^4$-trithiapentalene is often obtained directly ⟨71AHC(13)161, p. 191, 199⟩.

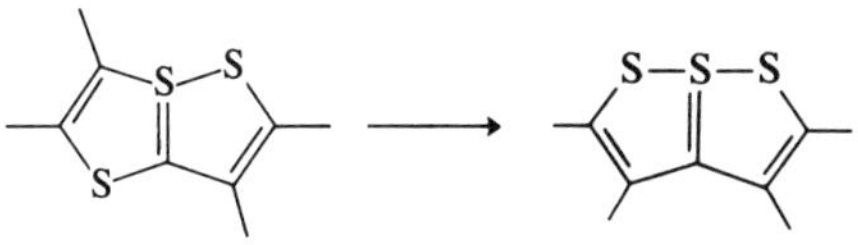

$$\text{(21)}$$

Similar reactions have been observed when alkynes are replaced by benzyne ⟨74CR(C)(279)259⟩.

4.38.6.3 From Thiirane Intermediates

Thiiranes are assumed to be intermediates in the formation of 1-oxa-$6,6a\lambda^4$-dithiapentalenes from 1,2-dithiole-3-thiones reacting either with diazo ketones or with α-halogeno ketones with subsequent desulfuration in basic medium ⟨71AHC(13)161, pp. 177–179⟩.

In the first case, there is probably an initial reaction of the dithiolethione with the acylcarbene resulting from the thermal decomposition of a diazo ketone (equation 22).

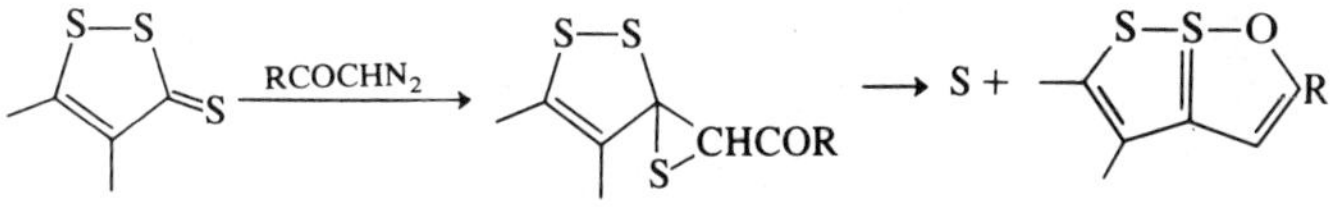

$$\text{(22)}$$

In the second case, the 3-(acylmethylthio)-1,2-dithiolylium cation is first deprotonated to a thiirane which finally yields an oxadithiapentalene according to equation (23). In this case, sulfur extrusion is not always complete and a disulfide (25) may also be formed.

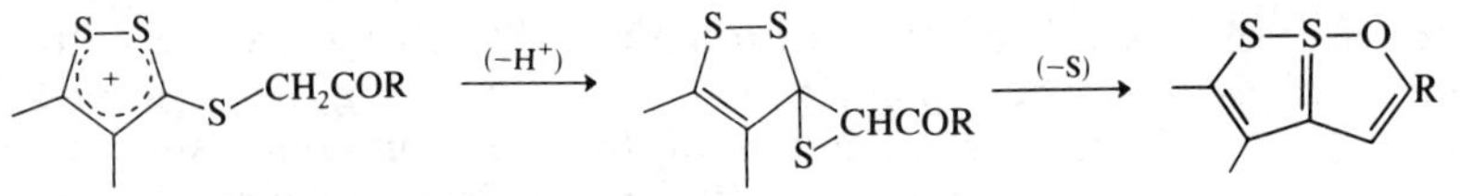

$$\text{dithiolylium-S-CH}_2\text{COR} \xrightarrow{(-H^+)} \text{thiirane intermediate} \xrightarrow{(-S)} \text{oxadithiapentalene-R} \qquad (23)$$

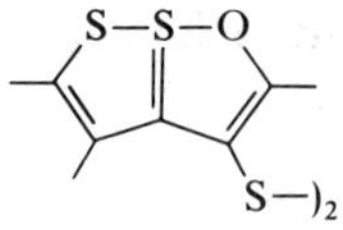

(25)

4.38.6.4 From 1,2,3,4-Thiatriazoles

Two molecules of arenecarbonyl chloride or of arenecarboximidoyl chloride can react, in one or two steps, giving, according to the nature of the acyl chloride(s), the following bicyclic systems, as shown in Scheme 13 ⟨77AG(E)403, 77CC143⟩: 3,4-dioxa-3aλ^4-thia-1,6-diazapentalene, 3-oxa-3aλ^4-thia-1,4,6-triazapentalene and 3aλ^4-thia-1,3,4,6-tetraazapentalene.

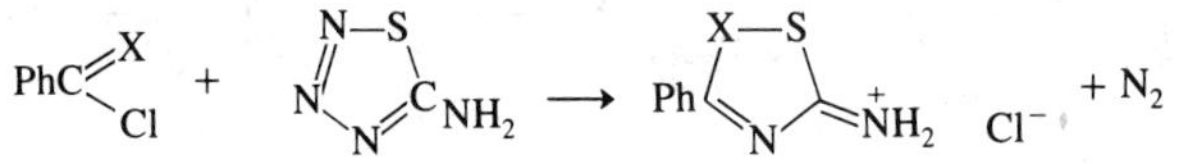

$$\text{PhC(X)Cl} + \text{thiatriazole-NH}_2 \longrightarrow \text{Ph-C=N-NH}_2^+ \text{Cl}^- + N_2$$

(26)

$$(26) + \text{PhC(Y)Cl} \longrightarrow 2\text{HCl} + \text{Ph-(X-S-Y)-Ph}$$

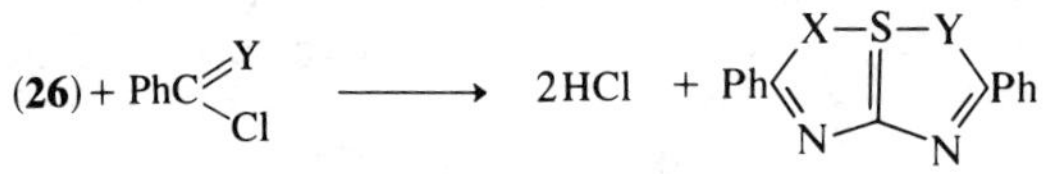

X = Y = O; X = O, Y = NPh; X = Y = NPh

Scheme 13

4.38.6.5 From 3-(1,2-Dithiol-3-ylidene)methyl-1,2-dithiolylium Cations

Pyridine induced deprotonation (Scheme 14) of these cations generally leads to a red condensed-ring trithiapentalene (27) sometimes accompanied by a blue isomer to which formula (28) has been assigned ⟨80AHC(27)151, p. 177⟩.

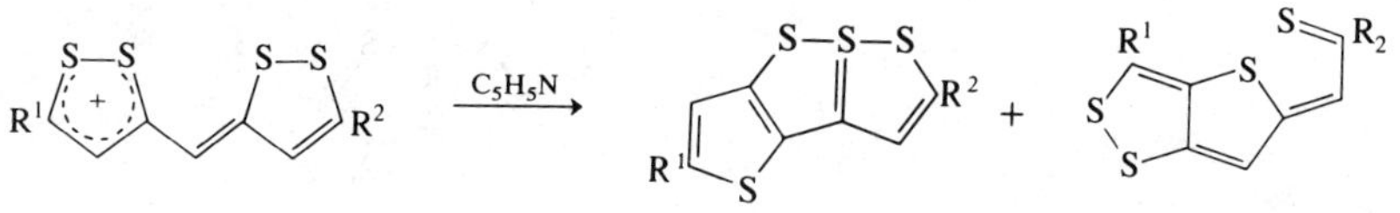

$$\text{cation with } R^1, R^2 \xrightarrow{C_5H_5N} \text{(27)} + \text{(28)}$$

Scheme 14

References

EXPLANATION OF THE REFERENCE SYSTEM

Throughout this work, references are designated by a number–letter coding of which the first two numbers denote tens and units of the year of publication, the next one to three letters denote the journal, and the final numbers denote the page. This code appears in the text each time a reference is quoted; the advantages of this system are outlined in the Introduction (Chapter 1.01). The system is based on that previously used in the following two monographs: (a) A. R. Katritzky and J. M. Lagowski, 'Chemistry of the Heterocyclic N-Oxides', Academic Press, New York, 1971; (b) J. Elguero, C. Marzin, A. R. Katritzky and P. Linda, 'The Tautomerism of Heterocycles', in 'Advances in Heterocyclic Chemistry', Supplement 1, Academic Press, New York, 1976.

The following additional notes apply:

1. A list of journals which have been assigned codes is given (in alphabetical order) together with their codes immediately following these notes. Journal names are abbreviated throughout by the CASSI (Chemical Abstracts Service Source Index) system.

2. A list of journal codes in alphabetical order, together with the journals to which they refer, is given on the end papers of each volume.

3. Each volume contains all the references cited *in that volume*; no separate lists are given for individual chapters.

4. The list of references is arranged in order of (a) year, (b) journal in alphabetical order of journal code, (c) part letter or number if relevant, (d) volume number if relevant, (e) page number.

5. In the reference list the code is followed by (a) the complete literature citation in the conventional manner and (b) the number(s) of the page(s) on which the reference appears, whether in the text or in tables, schemes, *etc.*

6. For non-twentieth century references the year is given in full in the code.

7. For journals which are published in separate parts, the part letter or number is given (when necessary) in parentheses immediately after the journal code letters.

8. Journal volume numbers are *not* included in the code numbers unless more than one volume was published in the year in question, in which case the volume number is included in parentheses immediately after the journal code letters.

9. Patents are assigned appropriate three letter codes.

10. Frequently cited books are assigned codes, but the whole code is now prefixed by the letter 'B-'.

11. Less common journals and books are given the code 'MI' for miscellaneous.

12. Where journals have changed names, the same code is used throughout, *e.g.* CB refers both to *Chem. Ber.* and to *Ber. Dtsch. Chem. Ges.*

Journals

Acc. Chem. Res.	ACR
Acta Chem. Scand., Ser. B	ACS(B)
Acta Chim. Acad. Sci. Hung.	ACH
Acta Crystallogr., Part B	AX(B)
Adv. Phys. Org. Chem.	APO
Agric. Biol. Chem.	ABC

Angew. Chem.	AG
Angew. Chem., Int. Ed. Engl.	AG(E)
Ann. Chim. (Rome)	AC(R)
Ann. N.Y. Acad. Sci.	ANY
Arch. Pharm. (Weinheim, Ger.)	AP
Ark. Kemi	AK
Arzneim.-Forsch.	AF
Aust. J. Chem.	AJC
Biochem. Biophys. Res. Commun.	BBR
Biochemistry	B
Biochem. J.	BJ
Biochim. Biophys. Acta	BBA
Br. J. Pharmacol.	BJP
Bull. Acad. Pol. Sci., Ser. Sci. Chim.	BAP
Bull. Acad. Sci. USSR, Div. Chem. Sci.	BAU
Bull. Chem. Soc. Jpn.	BCJ
Bull. Soc. Chim. Belg.	BSB
Bull. Soc. Chim. Fr., Part 2	BSF(2)
Can. J. Chem.	CJC
Chem. Abstr.	CA
Chem. Ber.	CB
Chem. Heterocycl. Compd. (Engl. Transl.)	CHE
Chem. Ind. (London)	CI(L)
Chem. Lett.	CL
Chem. Pharm. Bull.	CPB
Chem. Rev.	CRV
Chem. Scr.	CS
Chem. Soc. Rev.	CSR
Chem.-Ztg.	CZ
Chimia	C
Collect. Czech. Chem. Commun.	CCC
Coord. Chem. Rev.	CCR
C.R. Hebd. Seances Acad. Sci., Ser. C	CR(C)
Cryst. Struct. Commun.	CSC
Diss. Abstr. Int. B	DIS(B)
Dokl. Akad. Nauk SSSR	DOK
Experientia	E
Farmaco Ed. Sci.	FES
Fortschr. Chem. Org. Naturst.	FOR
Gazz. Chim. Ital.	G
Helv. Chim. Acta	HCA
Heterocycles	H
Hoppe-Seyler's Z. Physiol. Chem.	ZPC
Indian J. Chem., Sect. B	IJC(B)
Inorg. Chem.	IC
Int. J. Sulfur Chem., Part B	IJS(B)
Izv. Akad. Nauk SSSR, Ser. Khim.	IZV
J. Am. Chem. Soc.	JA
J. Biol. Chem.	JBC
J. Chem. Phys.	JCP
J. Chem. Res. (S)	JCR(S)
J. Chem. Soc. (C)	JCS(C)
J. Chem. Soc., Chem. Commun.	CC
J. Chem. Soc., Dalton Trans.	JCS(D)
J. Chem. Soc., Faraday Trans. 1	JCS(F1)
J. Chem. Soc., Perkin Trans. 1	JCS(P1)
J. Gen. Chem. USSR (Engl. Transl.)	JGU
J. Heterocycl. Chem.	JHC
J. Indian Chem. Soc.	JIC
J. Magn. Reson.	JMR

J. Med. Chem.	JMC
J. Mol. Spectrosc.	JSP
J. Mol. Struct.	JST
J. Organomet. Chem.	JOM
J. Org. Chem.	JOC
J. Org. Chem. USSR (Engl. Transl.)	JOU
J. Pharm. Sci.	JPS
J. Phys. Chem.	JPC
J. Prakt. Chem.	JPR
Khim. Geterotsikl. Soedin.	KGS
Kristallografiya	K
Liebigs Ann. Chem.	LA
Monatsh. Chem.	M
Naturwissenschaften	N
Nippon Kagaku Kaishi	NKK
Nouv. J. Chim.	NJC
Org. Magn. Reson.	OMR
Org. Mass Spectrom.	OMS
Org. Prep. Proced. Int.	OPP
Org. React.	OR
Org. Synth.	OS
Org. Synth., Coll. Vol.	OSC
Phosphorus Sulfur	PS
Phytochemistry	P
Proc. Indian Acad. Sci., Sect. A	PIA(A)
Proc. Natl. Acad. Sci. USA	PNA
Pure Appl. Chem.	PAC
Q. Rev., Chem. Soc	QR
Recl. Trav. Chim. Pays-Bas	RTC
Rev. Roum. Chim.	RRC
Russ. Chem. Rev. (Engl. Transl.)	RCR
Spectrochim. Acta, Part A	SA(A)
Synth. Commun.	SC
Synthesis	S
Tetrahedron	T
Tetrahedron Lett.	TL
Ukr. Khim. Zh. (Russ. Ed.)	UKZ
Yakugaku Zasshi	YZ
Z. Chem.	ZC
Zh. Obshch. Khim.	ZOB
Zh. Org. Khim.	ZOR
Z. Naturforsch., Teil B	ZN(B)

Book Series

'Advances in Heterocyclic Chemistry'	AHC
'Chemistry of Heterocyclic Compounds' [Weissberger–Taylor series]	HC
'Methoden der Organischen Chemie (Houben-Weyl)'	HOU
'Organic Compounds of Sulphur, Selenium, and Tellurium' [R. Soc. Chem. series]	SST
'Physical Methods in Heterocyclic Chemistry'	PMH

Specific Books

Q. N. Porter and J. Baldas, 'Mass Spectromety of Heterocyclic Compounds', Wiley, New York, 1971	MS
T. J. Batterham, 'NMR Spectra of Simple Heterocycles', Wiley, New York, 1973	NMR

'Photochemistry of Heterocyclic Compounds', ed. O. Buchardt, Wiley, New York, 1976   PH

W. L. F. Armarego, 'Stereochemistry of Heterocyclic Compounds', Wiley, New York, 1977, parts 1 and 2   SH

Patents

Belg. Pat.	BEP
Br. Pat.	BRP
Eur. Pat.	EUP
Fr. Pat.	FRP
Ger. (East) Pat.	EGP
Ger. Pat.	GEP
Neth. Pat.	NEP
Jpn. Pat.	JAP
Jpn. Kokai	JAP(K)
S. Afr. Pat.	SAP
U.S. Pat.	USP

Other Publications

All Other Books and Journals ('Miscellaneous')	MI
All Other Patents	MIP
Personal Communications	PC
Theses	TH
Unpublished Results	UP

VOLUME 6 REFERENCES

1821MI43400	F. Wöhler; *Ann. Phys.*, 1821, **69**, 273.	898, 937
1842LA(43)76	Völckel; *Liebigs Ann. Chem.*, 1842, **43**, 76.	938
1862LA(123)83	A. Husemann; *Liebigs Ann. Chem.*, 1862, **123**, 83.	847
1873CB902	M. Nencki and W. Leppert; *Ber.*, 1873, **6**, 902.	938
1876CB1524	A. Ladenburg; *Ber.*, 1876, **9**, 1524.	178
1881JPR353	O. Prinz; *J. Prakt. Chem.*, 1881, **24**, 353.	120
1882CB2105	P. Friedländer and R. Henriques; *Ber.*, 1882, **15**, 2105.	3
1884CB1685	F. Tiemann and P. Krüger; *Ber.*, 1884, **17**, 1685.	382
1887CB3118	A. Hantzsch and J. H. Weber; *Ber.*, 1887, **20**, 3118.	178, 294
1888CB1149	L. Claisen and O. Lowmann; *Ber.*, 1888, **21**, 1149.	3, 61
1888LA(249)27	L. Arapides; *Liebigs Ann. Chem.*, 1888, **249**, 27.	299
1888LA(249)31	V. Traumann; *Liebigs Ann. Chem.*, 1888, **249**, 31.	296
1889LA(250)294	G. Hofmann; *Liebigs Ann. Chem.*, 1889, **250**, 294.	342
1890CB157	S. Gabriel and P. Heymann; *Ber.*, 1890, **23**, 157.	306
1890CB1090	L. Hagelberg; *Ber.*, 1890, **23**, 1090.	948
1890LA(260)79	L. Wolff; *Liebigs Ann. Chem.*, 1890, **260**, 79.	394
1891CB130	L. Claisen and R. Stock; *Ber.*, 1891, **24**, 130.	83
1891JCS410	W. R. Dunstan and T. S. Dymond; *J. Chem. Soc.*, 1891, 410.	76
1892CB1498	W. R. Cathcart and V. Meyer; *Ber.*, 1892, **25**, 1498.	3, 114

1892CB2607	J. Tcherniac; *Ber.*, 1892, **25**, 2607.	299
1892CB3048	W. Michels; *Ber.*, 1892, **25**, 3048.	345
1892CB3291	W. R. Cathcart and V. Meyer, *Ber.*, 1892, **25**, 3291.	3
1893CB2897	St. von Kostanecki; *Ber.*, 1893, **26**, 2897.	411
1893G(23)437	A. Miolati; *Gazz. Chim. Ital.*, 1893, **23**, 437.	298
1893G(24)65	G. Marchesini; *Gazz. Chim. Ital.*, 1893, **24**, 65.	298
1893LA(275)20	E. Fromm; *Liebigs Ann. Chem.*, 1893, **275**, 20.	931
1893LA(277)162	L. Claisen; *Liebigs Ann. Chem.*, 1893, **277**, 162.	85
1895CB957	H. Rupe and F. Schneider; *Ber*, 1895, **28**, 957.	3
1895CB1925	O. Widman; *Ber.*, 1895, **28**, 1925.	604
1895LA(285)166	M. Freund and E. Asbrand; *Liebigs Ann. Chem.*, 1895, **285**, 166.	922
1896CB1686	M. Busch and J. Becker; *Ber.*, 1896, **29**, 1686.	601, 604
1896CB2161	H. von Pechmann; *Ber.*, 1896, **29**, 2161.	604
1896CB2491	M. Freund and H. P. Schwarz; *Ber.*, 1896, **29**, 2491.	581
1896CB2500	M. Freund and A. Schander; *Ber.*, 1896, **29**, 2500.	581
1897JCS607	F. D. Chattaway and H. P. Stevens; *J. Chem. Soc.*, 1897, **71**, 607.	918
1899JPR(39)107	T. F. Curtius; *J. Prakt. Chem.*, 1899, **39**, 107.	366
01CB3973	T. Posner; *Ber.*, 1901, **34**, 3973.	84
03CB3664	L. Claisen; *Ber.*, 1903, **36**, 3664.	82
03LA(329)225	H. Wieland; *Liebigs Ann. Chem.*, 1903, **329**, 225.	395
04CB966	E. Bamberger; *Ber.*, 1904, **37**, 966.	120
04CB2550	W. Becker and J. Meyer; *Ber.*, 1904, **37**, 2550.	355
04JPR(69)509	R. Stolle and L. Gutmann; *J. Prakt. Chem.*, 1904, **69**, 509.	357
04LA(331)265	A. Hantzsch and M. Wolvekamp; *Liebigs Ann. Chem.*, 1904, **331**, 265.	923
B-04MI42200	A. Werner; 'Lehrbuch der Stereochemie', Fischer, Jena, 1904, p. 260.	395
07CB479	F. Schlotterbeck; *Ber.*, 1907, **40**, 479.	367
09CB59	L. Claisen; *Ber.*, 1909, **42**, 59.	85
09CB1886	J. Schmidt and K. Th. Widmann; *Ber.*, 1909, **42**, 1886.	85
10CB134	S. Gabriel; *Ber.*, 1910, **43**, 134.	302
10CB1283	S. Gabriel; *Ber.*, 1910, **43**, 1283.	302, 994
10LA(375)297	H. Wieland and H. Gmelin; *Liebigs Ann. Chem.*, 1910, **375**, 297.	996
11CB2016	F. Heim; *Ber.*, 1911, **44**, 2016.	76
11CB2409	J. Scheiber; *Ber.*, 1911, **44**, 2409.	120
13CB92	H. Bauer, *Ber.*, 1913, **46**, 92.	344
14CB3163	M. Bachstez; *Ber.*, 1914, **47**, 3163.	302
15CB897	O. Diels and D. Riley; *Ber.*, 1915, **48**, 897.	223, 990
15G12	G. Ponzio; *Gazz. Chim. Ital.*, 1915, **45**(2), 12.	604
16CB1110	S. Gabriel; *Ber.*, 1916, **49**, 1110.	308
16G56	G. Ponzio; *Gazz. Chim. Ital.*, 1916, **46**(2), 56.	604
16JPR(93)183	B. Rassow, W. Döhle and E. Reim; *J. Prakt. Chem.*, 1916, **93**, 183.	274
17CB804	S. Gabriel and H. Ohle; *Ber.*, 1917, **50**, 804.	307
18CB192	S. Bodforss; *Ber.*, 1918, **51**, 192.	3
19LA(419)217	E. Söderbäck; *Liebigs Ann. Chem.*, 1919, **419**, 217.	923, 942
20JA1055	D. E. Worrall; *J. Am. Chem. Soc.*, 1920, **42**, 1055.	86
21CB2191	M. P. Schmidt and A. Hagenböcker; *Ber.*, 1921, **54**, 2191.	1015
21HCA239	H. Staudinger; *Helv. Chim. Acta*, 1921, **4**, 239	366
22GEP360690	R. Herz; *Ger. Pat.* 360 690 (1922) (*Chem. Zentr.*, 1922, **4**, 948).	935
23HCA102	H. Rupe and G. Grünholz; *Helv. Chim. Acta*, 1923, **6**, 102.	41
24JA1733	E. P. Kohler; *J. Am. Chem. Soc.*, 1924, **46**, 1733.	86

24JA2105	E. P. Kohler and G. R. Barrett; *J. Am. Chem. Soc.*, 1924, **46**, 2105.	76, 84
26LA(437)162	K. Von Auwers, H. Bundesmann and F. Weiners; *Liebigs Ann. Chem.*, 1926, **437**, 162.	120
26LA(449)63	H. Lindemann and H. Thiele; *Liebigs Ann. Chem.*, 1926, **449**, 63.	23
27CB1736	J. Meisenheimer, O. Senn and P. Zimmermann; *Ber.*, 1927, **60**, 1736.	115
28CB1118	F. Arndt and B. Eisert; *Ber.*, 1928, **61**, 1118.	367
29AP(267)699	C. Mannich; *Arch. Pharm. (Weinheim, Ger.)*, 1929, **267**, 699.	989
29CB1449	M. Busch and W. Schmidt; *Ber.*, 1929, **62**, 1449.	601, 602, 603, 604
29G930	A. Quilico and M. Freri; *Gazz. Chim. Ital.*, 1929, **59**, 930.	85
29JCS1582	E. W. McClelland, L. A. Warren and J. H. Jackson; *J. Chem. Soc.*, 1929, 1582.	794, 796, 807
30JCS12	D. T. Gibson; *J. Chem. Soc.*, 1930, 12.	850
30JCS1497	G. T. Morgan and F. H. Burstall; *J. Chem. Soc.*, 1930, 1497.	948, 963
31G759	A. Quilico; *Gazz. Chim. Ital.*, 1931, **61**, 759.	86
31JA1891	J. M. F. Leaper; *J. Am. Chem. Soc.*, 1931, **53**, 1891.	940
B-31MI43300	A. Rieche; 'Alkylperoxyde und Ozonide', Steinkopff, Dresden, 1931.	853, 864, 869
32CB1274	A. Rieche and R. Meister; *Ber.*, 1932, **65**, 1274.	883
32G436	A. Quilico and M. Freri; *Gazz. Chim. Ital.*, 1932, **62**, 436.	30
32G503	A. Quilico; *Gazz. Chim. Ital.*, 1932, **62**, 503.	604
32G912	A. Quilico; *Gazz. Chim. Ital.*, 1932, **62**, 912.	602, 604
33G269	A. Quilico; *Gazz. Chim. Ital.*, 1933, **63**, 269.	604
33G471	G. Ponzio; *Gazz. Chim. Ital.*, 1933, **63**, 471.	604
33G862	A. Quilico and R. Justoni; *Gazz. Chim. Ital.*, 1933, **63**, 862.	604
34JA2190	A. H. Blatt and W. L. Hawkins; *J. Am. Chem. Soc.*, 1934, **56**, 2190.	87
35G201	A. Quilico and R. Justoni; *Gazz. Chim. Ital.*, 1935, **65**, 201.	604
35JA1079	H. Wenker; *J. Am. Chem. Soc.*, 1935, **57**, 1079.	307
35JCS899	J. C. Earl and A. W. Mackney; *J. Chem. Soc.*, 1935, 899.	367
35JCS1762	C. Hasan and R. F. Hunter; *J. Chem. Soc.*, 1935, 1762.	344
B-35MI41600	Ricther-Anschutz; 'Chemie der Kohlenstoffverbindungen II,' 12th edn., part 2, Akad. Verlags Gesselschaft, Leipzig, 1935, p. 308.	120
35YZ233	S. Keimatsu and I. Satoda; *Yakugaku Zasshi*, 1935, **55**, 233 (*Chem. Abstr.*, 1937, **31**, 6661).	354
36G819	G. Ponzio; *Gazz. Chim. Ital.*, 1936, **66**, 819.	409
37G589	A. Quilico and R. Fusco; *Gazz. Chim. Ital.*, 1937, **67**, 589.	71, 85
37G779	T. Ajello; *Gazz. Chim. Ital.*, 1937, **67**, 779.	60
37RTC691	H. J. Backer and H. J. Winter; *Recl. Trav. Chim. Pays-Bas*, 1937, **56**, 691.	967
38G109	A. Quilico and L. Panizzi; *Gazz. Chim. Ital.*, 1938, **68**, 109.	52
38G566	S. Cusmano and G. Massara; *Gazz. Chim. Ital.*, 1938, **68**, 566.	85
38G665	R. Rusco and C. Musante; *Gazz. Chim. Ital.*, 1938, **68**, 665.	357
38G792	T. Ajello and S. Cusmano; *Gazz. Chim. Ital.*, 1938, **68**, 792.	60
38JA1198	D. E. Worrall; *J. Am. Chem. Soc.*, 1938, **60**, 1198.	86
39G322	L. Panizzi; *Gazz. Chim.. Ital.*, 1939, **69**, 332.	71, 85
39G391	T. Ajello and S. Cusmano; *Gazz. Chim. Ital.*, 1939, **69**, 391.	60
39LA(540)83	W. Borsche and W. Scriba; *Liebigs Ann. Chem.*, 1939, **540**, 83.	114
40CRV(27)437	L. Long; *Chem. Rev.*, 1940, **27**, 437.	853, 869, 872, 879, 893
40G89	L. Panizzi; *Gazz. Chim. Ital.*, 1940, **70**, 89.	71, 85
40G119	L. Panizzi; *Gazz. Chim. Ital.*, 1940, **70**, 119.	85
40G770	T. Ajello and S. Cusmano; *Gazz. Chim. Ital.*, 1940, **70**, 770.	60
40JA2604	S. B. Lippincott; *J. Am. Chem. Soc.*, 1940, **62**, 2604.	76, 83
40JCS191	G. M. Dyson and T. Harrington; *J. Chem. Soc.*, 1940, 191.	943
41CB(B)1407	E. Ochiai and T. Nishizawa; *Chem. Ber.*, 1941, **74B**, 1407 (*Chem. Abstr.*, 1942, **36**, 5475).	274
41G553	C. Musante; *Gazz. Chim. Ital.*, 1941, **71**, 553.	85
41JA3452	A. C. Cope, C. M. Hofmann, C. Wyckoff and E. Hardenbergh; *J. Am. Chem. Soc.*, 1941, **63**, 3452.	995
41LA(546)273	W. Borsche and M. Wagner-Roemmich; *Liebigs Ann. Chem.*, 1941, **546**, 273.	114
41RTC453	H. J. Backer and G. L. Wiggerink; *Recl. Trav. Chim. Pays-Bas*, 1941, **60**, 453.	844

42G458	A. Quilico and L. Panizzi; *Gazz. Chim. Ital.*, 1942, **72**, 458.	85
42LA(553)187	A. Rieche, R. Meister and H. Sauthoff; *Liebigs Ann. Chem.*, 1942, **553**, 187.	860, 864, 865, 882
43G99	L. Panizzi; *Gazz. Chim. Ital.*, 1943, **73**, 99.	85
43JCS654	A. Schonberg and A. Mostafa; *J. Chem. Soc.*, 1943, 654.	35
44JOC470	C. S. Hudson; *J. Org. Chem.*, 1944, **9**, 470.	996
44LA(556)65	O. Süs; *Liebigs Ann. Chem.*, 1944, **556**, 65.	1018
44MI41600	K. A. Jensen and A. Friediger; *Kgl. Danske Videnskab. Selskab. Math. Fys. Medd.*, 1944, **1**, 416 (*Chem. Abstr.*, 1945, **39**, 2068).	120
45CRV(37)401	R. H. Wiley; *Chem. Rev.*, 1945, **37**, 401.	178
45JA134	R. P. Barnes and A. S. Spriggs; *J. Am. Chem. Soc.*, 1945, **67**, 134.	63
45JA1745	W. S. Johnson and W. E. Shelberg; *J. Am. Chem. Soc.*, 1945, **67**, 1745.	62
45JA2242	R. M. Dodson and L. C. King; *J. Am. Chem. Soc.*, 1945, **67**, 2242.	296
46G1	A. Quilico and M. Freri; *Gazz. Chim. Ital.*, 1946, **76**, 1.	60
46G87	A. Quilico, R. Fusco and V. Rosnati; *Gazz. Chim. Ital.*, 1946, **76**, 87.	85
46G148	A. Quilico and G. Speroni; *Gazz. Chim. Ital.*, 1946, **76**, 148.	71, 84, 85, 997
46G259	A. Quilico and M. Simonetta; *Gazz. Chim. Ital.*, 1946, **76**, 259.	604
46JCS953	K. Bowden and E. R. H. Jones; *J. Chem. Soc.*, 1946, 953.	83
46JCS1050	C. C. J. Culvenor, W. Davies and K. H. Pausacker; *J. Chem. Soc.*, 1946, 1050.	847
46OR(3)198	H. E. Carter; *Org. React.*, 1946, **3**, 198.	199, 226
47G206	L. Panizzi; *Gazz. Chim. Ital.*, 1947, **77**, 206.	30, 85
47G332	T. Ajello and B. Tornetta; *Gazz. Chim. Ital.*, 1947, **77**, 332.	60
47G556	L. Panizzi and E. Monti; *Gazz. Chim. Ital.*, 1947, **77**, 556.	83
47G586	A. Quilico and M. Simonetta; *Gazz. Chim. Ital.*, 1947, **77**, 586.	85
47JA2632	P. W. Preisler and M. M. Bateman; *J. Am. Chem. Soc.*, 1947, **69**, 2632.	918
47JCS96	J. W. Cornforth and R. H. Cornforth; *J. Chem. Soc.*, 1947, 96.	178
47JCS592	L. A. Stocken; *J. Chem. Soc.*, 1947, 592.	837
47JCS1594	A. Cook, I. Heilbron and A. L. Levy; *J. Chem. Soc.*, 1947, 1594.	290, 300
47JOC807	E. Campaigne and W. B. Reid, Jr.; *J. Org. Chem.*, 1947, **12**, 807.	872, 880
47MI43300	G. Cronheim; *J. Am. Pharm. Assoc., Sci. Ed.*, 1947, **36**, 274.	895
47MI43301	G. Cronheim; *J. Am. Pharm. Assoc., Sci. Ed.*, 1947, **36**, 278.	895
48BSF1002	J. Lichtenberger and R. Lichtenberger; *Bull. Soc. Chim. Fr.*, 1948, 1002.	894
48BSF1021	A. Pullman and J. Metzger; *Bull. Soc. Chim. Fr.*, 1948, 1021 (*Chem. Abstr.*, 1949, **43**, 2511).	255
48G630	S. Cusmano and S. Giambrone; *Gazz. Chim. Ital.*, 1948, **78**, 630.	85
48G764	S. Cusmano; *Gazz. Chim. Ital.*, 1948, **73**, 764.	85
48JCS2269	J. C. Earl, E. M. W. Leake and R. J. W. LeFevre; *J. Chem. Soc.*, 1948, 2269.	368
48MI42400	H. H. Hodgson and D. P. Dodgson; *J. Soc. Dyers Colour.*, 1948, **64**, 65.	448, 458
48USP2448767	W. W. Carlson; *U.S. Pat.* 2 448 767 (1948) (*Chem. Abstr.*, 1949, **43**, 673).	894
48YZ191	J. Haginiwa; *Yakugaku Zasshi*, 1948, **68**, 191 (*Chem. Abstr.*, 1953, **47**, 8074).	342
49CB257	A. Dornow and K. Peterlein; *Chem. Ber.*, 1949, **82**, 257.	83
49DOK(69)41	B. A. Arbuzov and T. G. Shavsha; *Dokl. Akad. Nauk SSSR*, 1949, **69**, 41.	865
49G683	C. Musante and R. Berretti; *Gazz. Chim. Ital.*, 1949, **79**, 683.	85
49G703	A. Quilico and G. S. D'Alcontres; *Gazz. Chim. Ital.*, 1949, **79**, 703.	83, 85
49JA4059	J. C. Sheehan and P. T. Izzo; *J. Am. Chem. Soc.*, 1949, **71**, 4059.	225
49JCS274	F. E. King and D. G. I. Felton; *J. Chem. Soc.*, 1949, 274.	360, 361, 1063
49JCS746	R. A. W. Hill and L. E. Sutton; *J. Chem. Soc.*, 1949, 746.	368
B-49MI41800	J. W. Cornforth; in 'The Chemistry of Penicillin', Princeton University Press, Princeton, 1949.	178, 207
B-49MI41900	A. H. Cook and I. Heilbron; in 'The Chemistry of Penicillin', Princeton University Press, 1949, chap. 25, pp. 921–972.	320
B-49MI43600	'The Chemistry of Penicillin', ed. H. T. Clarke, J. R. Johnson and R. Robinson; Princeton University Press, 1949.	1018
49YZ566	J. Haginiwa; *Yakugaku Zasshi*, 1949, **69**, 566 (*Chem. Abstr.*, 1950, **44**, 4465).	341
50ACS1584	S. Furberg and O. Hassel; *Acta Chem. Scand.*, 1950, **4**, 1584.	751
50G140	A. Quilico and G. S. D'Alconties; *Gazz. Chim. Ital.*, 1950, **80**, 140.	119
50HCA273	C. A. Grob and P. Ankli; *Helv. Chim. Acta*, 1950, **33**, 273.	980
50IZV297	B. A. Arbuzov and V. S. Vinogradova; *Izv. Akad. Nauk SSSR, Ser. Khim.*, 1950, 297.	864
51AX193	V. Lazzati; *Acta Crystallogr.*, 1951, **4**, 193.	524
51CB688	G. Beck; *Chem. Ber.*, 1951, **84**, 688.	918
51JCS2999	F. S. Bridson-Jones, G. D. Buckley, L. H. Cross and A. P. Driver; *J. Chem. Soc.*, 1951, 2999.	367
51JCS3016	G. D. Buckley and W. J. Levy; *J. Chem. Soc.*, 1951, 3016.	366

51JCS3479	A. C. Davis and A. L. Levy; *J. Chem. Soc.*, 1951, 3479.	985
51MI42100	L. E. Orgel, T. L. Cottrell, W. Dick and L. E. Sutton; *Trans. Faraday Soc.*, 1951, 47, 113.	368
51OS(31)14	P. A. S. Smith and J. H. Boyer; *Org. Synth.*, 1951, 31, 14.	424
52CB1122	H. Beyer, H. Höhn and W. Lässig; *Chem. Ber.*, 1952, 85, 1122. (*Chem. Abstr.*, 1953, 47, 11 183).	297
52HC(4)3	L. L. Bambas; *Chem. Heterocycl. Compd.*, 1952, 4, 3.	448, 458
52HC(4)81	L. L. Bambas; *Chem. Heterocycl. Compd.*, 1952, 4, 81.	546, 555, 568
52HC(4)297	L. L. Bambas; *Chem. Heterocycl. Compd.*, 1952, 4, 297.	132, 150, 152, 159, 171
52HOU(8)27	R. Criegee; *Methoden Org. Chem.* (*Houben-Weyl*), 1952, 8, 27.	853
52JA4906	R. Adams and I. J. Pachter; *J. Am. Chem. Soc.*, 1952, 74, 4906.	656, 663
52JCS3197	F. M. Hamer; *J. Chem. Soc.*, 1952, 3197.	341
52MI43300	A. J. Haagen-Smit, E. F. Darley, M. Zaitlein, H. Hull and W. Noble; *Plant Physiol.*, 1952, 27, 18.	898
53BSF327	L. Legrand, Y. Mollier and N. Lozac'h; *Bull. Soc. Chim. Fr.*, 1953, 327,	803
53CB888	M. Seyhan; *Chem. Ber.*, 1953, 86, 888.	341
53CR(237)906	J. Metzger and P. Baily; *C.R. Hebd. Seances Acad. Sci.*, 1953, 237, 906.	342
53CRV(53)309	E. D. Bergmann; *Chem. Rev.*, 1953, 53, 309.	213, 229
53JCP(21)767	R. Pariser and R. G. Parr; *J. Chem. Phys.*, 1953, 21, 767.	516
53JCS1514	D. Greenwood and H. A. Stevenson; *J. Chem. Soc.*, 1953, 1514.	766
53LA28	S. Hünig and O. Boes; *Liebigs Ann. Chem.*, 1953, 579, 28.	604
53MI42600	J. A. Pople; *Trans. Faraday Soc.*, 1953, 49, 1375.	516
53MI43200	V. Hach; *Chem. Listy*, 1953, 47, 227 (*Chem. Abstr.*, 1955, 49, 172).	837
54ACS1145	L. B. Brahde; *Acta Chem. Scand.*, 1954, 8, 1145.	815, 816
54AX92	C. J. Brown; *Acta Crystallogr.*, 1954, 7, 92.	751
54CB68	J. Goerdeler, A. Huppertz and K. Wember; *Chem. Ber.*, 1954, 87, 68.	482
54CJC288	A. Beelik and W. H. Brown; *Can. J. Chem.*, 1954, 32, 288.	74
54IZV47	N. K. Kochetkov; *Izv. Akad. Nauk SSSR, Otd. Khim. Nauk*, 1954, 47.	63, 83
54JA1158	R. E. Allen, R. S. Shelton and M. G. Van Campen, Jr.; *J. Am. Chem. Soc.*, 1954, 76, 1158.	942
54JA4348	J. A. Bartrop, P. M. Hayes and M. Calvin; *J. Am. Chem. Soc.*, 1954, 76, 4348.	795
54JCP(22)1678	G. R. Slayton, J. W. Simmons and J. H. Goldstein; *J. Chem. Phys.*, 1954, 22, 1678.	752
54JCS665	G. Shaw and G. Sugowdz; *J. Chem. Soc.*, 1954, 665.	34
54JCS3283	R. A. F. Bullerwell, A. Lawson and H. V. Morley; *J. Chem. Soc.*, 1954, 3283.	985, 992
54USP2665280	J. O. Knobloch and J. W. Sparks; *U.S. Pat.* 2 665 280 (1954) (*Chem. Abstr.*, 1955, 49, 378).	892
55AG275	A. Lüttringhaus, R. Cordes and U. Schmidt; *Angew. Chem.*, 1955, 67, 275.	803
55AG395	W. Franke and G. Kraft; *Angew. Chem.*, 1955, 67, 395.	63
55CB712	R. Criegee and G. Paulig; *Chem. Ber.*, 1955, 88, 712.	880
55CB1878	R. Criegee, A. Kerckow and H. Zinke; *Chem. Ber.*, 1955, 88, 1878.	865
55CI(L)119	J. M. Tien and I. M. Hunsberger; *Chem. Ind.* (*London*), 1955, 119.	375
55JA1280	J. H. Boyer and F. C. Canter; *J. Am. Chem. Soc.*, 1955, 77, 1280.	603
55JA1843	J. Fugger, J. M. Tien and I. M. Hunsberger; *J. Am. Chem. Soc.*, 1955, 77, 1843.	371
55JA3789	M. S. Newman and R. W. Addor; *J. Am. Chem. Soc.*, 1955, 77, 3789.	765, 780
55JA4069	R. G. Jones; *J. Am. Chem. Soc.*, 1955, 77, 4069.	1023
55JA5359	C. D. Hurd and R. I. Mori; *J. Am. Chem. Soc.*, 1955, 77, 5359.	460
55JA5370	C. M. Selwitz and A. I. Kosak; *J. Am. Chem. Soc.*, 1955, 77, 5370.	990
55JA6604	J. M. Tien and I. M. Hunsberger; *J. Am. Chem. Soc.*, 1955, 77, 6604.	395
55JCS1468	E. S. Lane and C. Williams; *J. Chem. Soc.*, 1955, 1468.	526
55JOC1342	R. G. Jones and C. W. Whitehead; *J. Org. Chem.*, 1955, 20, 1342.	63, 85
55MI43300	H. J. Ferlin and J. V. Karabinos; *Trans. Illinois State Acad. Sci.*, 1955, 47, 86 (*Chem. Abstr.*, 1955, 49, 16 070).	895
55QR150	A. Baltazzi; *Q. Rev., Chem. Soc.*, 1955, 9, 150.	199, 226
55USP2706674	G. M. Rothrock; *U.S. Pat.* 2 706 674 (1955) (*Chem. Abstr.*, 1955, 49, 10 636).	894
55YZ677	T. Iwatsu; *Yakugaku Zasshi*, 1955, 75, 677 (*Chem. Abstr.*, 1956, 50, 3459).	315
56AC(R)275	N. Cagnoli and A. Ricci; *Ann. Chim.* (*Rome*), 1956, 46, 275.	1017
56CB1534	J. Goerdeler, J. Ohm and O. Tegtmeyer; *Chem. Ber.*, 1956, 89, 1534.	489, 492, 506
56CB2825	R. Gompper and H. Herlinger; *Chem. Ber.*, 1956, 89, 2825.	185
56CI(L)846	L. L. Ingraham and F. H. Westheimer; *Chem. Ind.* (*London*), 1956, 846.	262
56CI(L)R28	R. Breslow; *Chem. Ind.* (*London*), 1956, R28.	262
56JA145	A. Mustafa, S. M. A. D. Zayed and S. Khattab; *J. Am. Chem. Soc.*, 1956, 78, 145.	983
56JA454	H. H. Szmant and W. Emerson; *J. Am. Chem. Soc.*, 1956, 78, 454.	860, 878
56JA5124	J. H. Boyer and J. A. Hernandez; *J. Am. Chem. Soc.*, 1956, 78, 5124.	602, 603, 604
56JCS307	A. Lawson; *J. Chem. Soc.*, 1956, 307.	990
56JCS1813	P. B. D. De la Mare, D. J. Millen, J. G. Pritchard and D. Watson; *J. Chem. Soc.*, 1956, 1813.	860
56JCS2063	A. R. Katritzky; *J. Chem. Soc.*, 1956, 2063.	717, 723

References

56MI41900	J. Metzger and H. Plank; *Chim. Ind. (Milan)*, 1956, **75**, 929 (*Chem. Abstr.*, 1957, **51**, 4358).	321, 323
56MI41901	J. Metzger and H. Plank; *Chim. Ind. (Milan)*, 1956, **75**, 1290.	321, 323
56MI43000	C. L. Angell; *Trans. Faraday Soc.*, 1956, **52**, 1178.	751
56MI43400	O. Exner; *Chem. Listy*, 1956, **50**, 779 (*Chem. Abstr.*, 1956, **50**, 15 477).	928, 929
56USP2752318	H. D. De Witt; *U.S. Pat.* 2 752 318 (1956) (*Chem. Abstr.*, 1956, **50**, 13 509).	894
57ACS911	G. Bergson and G. Claeson; *Acta Chem. Scand.*, 1957, **11**, 911.	948, 963
57AG138	A. Lüttringhaus, U. Schmidt and H. Alpes; *Angew. Chem.*, 1957, **69**, 138.	803
57BRP781169	Farbwerke Hoechst A. G., *Br. Pat.* 781 169 (1957) (*Chem. Abstr.*, 1958, **52**, 1682).	874, 894
57CB438	G. Bähr and G. Schleitzer; *Chem. Ber.*, 1957, **90**, 438.	844
57CB2411	R. Pfleger and H.-G. Hahn; *Chem. Ber.*, 1957, **90**, 2411.	1039, 1046
57CI(L)893	R. Breslow; *Chem. Ind. (London)*, 1957, 893.	262
57CI(L)1650	R. Fusco and S. Rossi; *Chem. Ind. (London)*, 1957, 1650.	86
57CRV1011	W. K. Warburton; *Chem. Rev.*, 1957, **57**, 1011.	898, 928, 935
57IZV949	N. K. Kochetkov, E. E. NiFantjev and A. N. Nesmeyanov; *Izv. Akad. Nauk SSSR, Otd. Khim. Nauk*, 1957, 949.	63
57JA1516	G. DeStevens, H. A. Luts and J. A. Schneider; *J. Am. Chem. Soc.*, 1957, **79**, 1516.	1026
57JA5276	A. F. McKay, G. Y. Paris and M. E. Kreling; *J. Am. Chem. Soc.*, 1957, **79**, 5276.	1013
57JA5667	D. G. Doherty, R. Shapira and W. T. Burnett, Jr.; *J. Am. Chem. Soc.*, 1957, **79**, 5667.	312
57JCS1556	A. Lawson and C. E. Searle; *J. Chem. Soc.*, 1957, 1556.	306
57JCS1652	G. M. Badger and N. Kowanko; *J. Chem. Soc.*, 1957, 1652.	257
57JCS4625	J. Fraser and E. Tittenson; *J. Chem. Soc.*, 1957, 4625.	659
57JOC1500	D. S. Matteson and H. R. Snyder; *J. Org. Chem.*, 1957, **22**, 1500.	1016
57LA(609)143	S. Hünig and K. H. Fritsch; *Liebigs Ann. Chem.*, 1957, **609**, 143 (*Chem. Abstr.*, 1958, **52**, 7296).	328
57MI41600	R. A. Abramovitch; *Proc. Chem. Soc.*, 1957, 8.	120
57MI41800	G. Ya. Kondrat'eva; *Khim. Nauka Prom.*, 1957, **2**, 666 (*Chem. Abstr.*, 1958, **52**, 6345).	178, 186, 195
B-57MI41801	J. W. Cornforth; in 'Heterocyclic Compounds', ed. R. C. Elderfield; Wiley, New York, 1957, vol. 5, p. 298.	178, 186, 199, 213, 215, 226
B-57MI41802	J. W. Cornforth; in 'Heterocyclic Compounds', ed. R. C. Elderfield; Wiley, New York, 1957, vol. 5, p. 418.	178, 186, 199, 215
B-57MI41900	J. M. Sprague and A. H. Land; in 'Heterocyclic Compounds', ed. R. C. Elderfield; Wiley, New York, 1957, vol. 5, p. 484.	293
B-57MI43000	R. C. Elderfield and F. W. Short; in 'Heterocyclic Compounds', ed. R. C. Elderfield; Wiley, New York, 1957, vol. 5, p. 1.	759, 761, 773
57MI43400	J. Y. Masuda and G. H. Hamar; *J. Am. Pharm. Assoc.*, 1957, **46**, 61 (*Chem. Abstr.*, 1957, **51**, 5298).	927, 940
57OS(37)1	F. B. Mallory; *Org. Synth.*, 1957, **37**, 1.	425
57QR15	W. Baker and W. D. Ollis; *Q. Rev., Chem. Soc.*, 1957, **11**, 15.	371
57T201	S. Veibel and H. Lillelund; *Tetrahedron*, 1957, **1**, 201.	991
57USP2798877	M. J. Viard; *U.S. Pat.* 2 798 877 (1957) (*Chem. Abstr.*, 1958, **52**, 1249).	870
57USP2811511	T. Alderson; *U.S. Pat.* 2 811 511 (1957) (*Chem. Abstr.*, 1958, **52**, 2452).	894
58AC(R)577	B. Tornetta; *Ann. Chim. (Rome)*, 1958, **48**, 577.	807
58ACS347	K. Schilling and H. Dam; *Acta Chem. Scand.*, 1958, **12**, 347.	939
58ACS1799	O. Foss and O. Tjomsland; *Acta Chem. Scand.*, 1958, **12**, 1799.	902, 903
58AG251	A. Rieche; *Angew. Chem.*, 1958, **70**, 251.	853
58AG667	F. Asinger and M. Thiel; *Angew. Chem.*, 1958, **70**, 667 (*Chem. Abstr.*, 1959, **53**, 10 249).	312, 315
58CI(L)461	A. Lawson and D. H. Miles; *Chem. Ind. (London)*, 1958, 461.	207
58CPB382	M. Kanaoka; *Chem. Pharm. Bull.*, 1958, **6**, 382.	978
58CRV925	P. S. Bailey; *Chem. Rev.*, 1958, **58**, 925.	853, 869, 872, 873, 880, 893
58G149	P. Grünanger and S. Mangiapan; *Gazz. Chim. Ital.*, 1958, **88**, 149.	85
58G879	C. Musante and S. Fatutta; *Gazz. Chim. Ital.*, 1958, **88**, 879.	31
58GEP1028718	J. Brunken and J. Müller; *Ger. Pat.* 1 028 718 (1958) (*Chem. Abstr.*, 1960, **54**, 19 240).	894
58HCA548	K. Hoegerle; *Helv. Chim. Acta*, 1958, **41**, 548.	717, 720
58JA414	C. K. Bradsher, F. C. Brown, E. F. Sinclair and S. T. Webster; *J. Am. Chem. Soc.*, 1958, **80**, 414.	898
58JA3719	R. Breslow; *J. Am. Chem. Soc.*, 1958, **80**, 3719.	326
58JOC2002	W. Asker, A. Mustafa, M. K. Hilmy and M. A. Allam; *J. Org. Chem.*, 1958, **23**, 2002.	151
58MI41600	P. Grünanger; *Atti Accad. Naz. Lincei, Cl. Sci. Fis. Mat. Nat. Rend.*, 1958, **24**, 163 (*Chem. Abstr.*, 1958, **52**, 18 376).	104
58MI41601	W. Lampe and J. Smolinska; *Bull. Acad. Pol. Sci.*, 1958, **6**, 481.	50
58MI41602	C. D. Nenitzescu, E. Cioranescu and L. Birladeanu; *Commun. Acad. Rep. Populare Romine*, 1958, **8**, 775 (*Chem. Abstr.*, 1959, **53**, 18 003).	35
59AC(R)2083	C. Caradonna, M. L. Stein and M. Ikram; *Ann. Chim. (Rome)*, 1959, **49**, 2083 (*Chem. Abstr.*, 1960, **54**, 19 646).	40

59ACS342	E. Bernatek and F. Thoresen; *Acta Chem. Scand.*, 1959, **13**, 342.	870
59BSF780	P. Raoul and J. Vialle; *Bull. Soc. Chim. Fr.*, 1959, 780.	802
59BSF1398	A. Thuiller and J. Vialle; *Bull. Soc. Chim. Fr.*, 1959, 1398.	88
59BSF1670	P. Raoul and J. Vialle; *Bull. Soc. Chim. Fr.*, 1959, 1670.	802
59CB550	R. Gompper, E. Hoyer and H. Herlinger; *Chem. Ber.*, 1959, **92**, 550.	990
59CPB725	A. Koshiro; *Chem. Pharm. Bull.*, 1959, **7**, 725.	662
59CRV429	J. V. R. Kaufman and J. P. Picard; *Chem. Rev.*, 1959, **59**, 429.	394
59CRV667	C. Lempert; *Chem. Rev.*, 1959, **59**, 667.	993
59CRV1077	S. Searles and S. Nukina; *Chem. Rev.*, 1959, **59**, 1077.	1012
59DIS(19)3136	L. M. Weinstock; *Diss. Abstr.* 1959, **19**, 3136.	520, 537, 539, 541
59DIS(20)1593	D. Shew; *Diss. Abstr.*, 1959, **20**, 1593.	538
59G571	A. Quilico, G. Gaudiano and L. Merlini; *Gazz. Chim. Ital.*, 1959, **89**, 571.	989, 997
59G1511	R. Scarpati and G. Speroni; *Gazz. Chim. Ital.*, 1959, **89**, 1511.	104
59G1525	R. Scarpati and P. Sorrentino; *Gazz. Chim. Ital.*, 1959, **89**, 1525.	91
59G2466	G. Gaudiano, A. Ricca and L. Merlini; *Gazz. Chim. Ital.*, 1959, **89**, 2466.	88
59HCA563	L. Ettlinger, E. Gäumann, R. Hütter, W. Keller-Schierlein, F. Kradolfer, L. Neipp, V. Prelog and H. Zähner; *Helv. Chim. Acta*, 1959, **42**, 563.	1024
59HCA1733	L. Soder and R. Wizinger; *Helv. Chim. Acta*, 1959, **42**, 1733.	823, 843
59JA6334	N. A. LeBel and J. J. Whang; *J. Am. Chem. Soc.*, 1959, **81**, 6334.	1007
59JCS55	W. Baker; *J. Chem. Soc.*, 1959, 55.	367
59JCS257	J. M. Tedder and G. Theaker; *J. Chem. Soc.*, 1959, 257.	367
59JCS802	S. A. Barker, E. J. Bourne, R. M. Pinkard and D. H. Whiffen; *J. Chem. Soc.*, 1959, 802.	756
59JCS1071	F. M. Dean, D. R. Randell and G. Winfield; *J. Chem. Soc.*, 1959, 1071.	860
59JCS3061	A. Adams and R. Slack; *J. Chem. Soc.*, 1959, 3061.	144, 641
59JCS3521	C. W. N. Cumper and A. I. Vogel; *J. Chem. Soc.*, 1959, 3521.	751
59JOC779	C. F. H. Allen, H. R. Beifuss, D. M. Burness, G. A. Reynolds, J. F. Tinker and J. A. Van Allen; *J. Org. Chem.*, 1959, **24**, 779.	742
59LA(627)195	F. Asinger, M. Thiel and G. Lipfert; *Liebigs Ann. Chem.*, 1959, **627**, 195.	876, 889
59MI41600	P. Grünanger, C. Gandini and A. Quilico; *Rend. Ist. Lomb. Sci., I, Cl. Sci. Mat. Nat.*, 1959, **93A**, 467 (*Chem. Abstr.*, 1961, **55**, 10 417).	89
59MI41601	P. Grünanger and P. V. Finzi; *Atti Accad. Naz. Lincei, Cl. Sci. Fis. Mat. Nat. Rend.*, 1959, **26**, 386 (*Chem. Abstr.*, 1960, **54**, 3379).	89
B-59MI42500	A. Albert; 'Heterocyclic Chemistry', Athlone Press, London, 1959.	464
59MI43000	N. Sheppard and J. J. Turner; *Proc. R. Soc., Ser. A*, 1959, **252**, 506.	753
59MI43300	G. W. Todd; *Plant Physiol.*, 1959, **33**, 416.	895
59USP2870127	G. E. Ham; *U.S. Pat.* 2 870 127 (1959) (*Chem. Abstr.*, 1959, **53**, 6643).	874
59USP2908688	T. S. Gardner, J. Lee and E. Wenis; *U.S. Pat.* 2 908 688 (1959).	85
59YZ836	H. Yasuda; *Yakugaku Zasshi*, 1959, **79**, 836.	63, 85
60AG629	F. Boberg; *Angew. Chem.*, 1960, **72**, 629.	805
60CB689	R. Criegee and G. Schröder; *Chem. Ber.*, 1960, **93**, 689.	888
60CB963	J. Goerdeler, K. Deselaers and A. Ginsberg; *Chem. Ber.*, 1960, **93**, 963.	483
60CB1129	G. Hesse and S. Majmudar; *Chem. Ber.*, 1960, **93**, 1129.	884
60CB1208	H. Bredereck, H. Herlinger and E. H. Schweizer; *Chem. Ber.*, 1960, **93**, 1208.	83
60CB1389	H. Bredereck, R. Gompper and F. Reich; *Chem. Ber.*, 1960, **93**, 1389.	182
60CB2353	W. Kirmse; *Chem. Ber.*, 1960, **93**, 2353.	582, 593
60CI(L)1568	E. Klingsberg; *Chem. Ind. (London)*, 1960, 1568.	785, 786
60CJC2349	G. D. Thorn; *Can. J. Chem.*, 1960, **38**, 2349.	904
60DOK598	N. K. Kochetkov, S. D. Sokolov, N. M. Vagurtova and E. E. Nitantjev; *Dokl. Akad. Nauk SSSR*, 1960, **133**, 598.	25
60EGP19119	F. Asinger, M. Thiel and G. Lipfert; *Ger. (East) Pat.* 19 119 (1960) (*Chem. Abstr.*, 1962, **56**, 1459).	893
60G347	G. S. D'Alcontres and L. LoVecchio; *Gazz. Chim. Ital.*, 1960, **90**, 347.	37
60JA2719	J. R. Johnson and R. Ketcham; *J. Am. Chem. Soc.*, 1960, **82**, 2719.	1022
60JA5148	N. J. Leonard and W. K. Musker; *J. Am. Chem. Soc.*, 1960, **82**, 5148.	1015
60JA5339	T. Mukaiyama and T. Hoshino; *J. Am. Chem. Soc.*, 1960, **82**, 5339.	423
60JCS201	J. S. Brimacombe, A. B. Foster, E. B. Hancock, W. G. Overend and M. Stacey; *J. Chem. Soc.*, 1960, 201.	860
60JCS3234	F. Kurzer and S. A. Taylor; *J. Chem. Soc.*, 1960, 3234.	508
60JIC151	S. N. Dixit; *J. Indian Chem. Soc.*, 1960, **37**, 151.	917
60JOC651	R. G. Gillis; *J. Org. Chem.*, 1960, **25**, 651.	877
60JOC1160	W. R. Vaughan and J. L. Spencer; *J. Org. Chem.*, 1960, **25**, 1160.	91, 92
60JPR(11)284	F. Runge, Z. El-Hewehi, H. J. Renner and E. Taeger; *J. Prakt. Chem.*, 1960, **11**, 284.	846
60LA(633)49	W. Walter; *Liebigs Ann. Chem.*, 1960, **633**, 49.	492
B-60MI43000	H. J. Beynon; 'Mass Spectrometry and its Applications to Organic Chemistry,' Elsevier, Amsterdam, 1960, p. 368.	756
60MI43400	T. Bacchetti and A. Alemagna; *Atti Acad. Nazl. Lincei, Rend. Classe Sci. Fis., Mat. e Nat.*, 1960, **28**, 646 (*Chem. Abstr.*, 1961, **55**, 9382).	907
B-60MI43800	L. Pauling; 'The Nature of the Chemical Bond', Cornell University Press, Ithaca, 3rd edn., 1960, p. 224.	1053, 1054

60ZOB600	S. V. Sokolov and I. Ya. Postovski; *Zh. Obshch. Khim.*, 1960, **30**, 600 (*Chem. Abstr.*, 1960, **54**, 24 658).	40
60ZOB698	V. F. Vasil'eva, V. G. Yashunskii and M. N. Shchukina; *Zh. Obschch. Khim.*, 1960, **30**, 698.	376
60ZOB954	N. K. Kochetkov and E. D. Khomutova; *Zh. Obshch. Khim.*, 1960, **30**, 954.	63, 83
61ACS1104	K. A. Jensen and C. Pedersen; *Acta Chem. Scand.*, 1961, **15**, 1104.	593
61ACS1186	A. Hordvik; *Acta Chem. Scand.*, 1961, **15**, 1186.	898
61AG220	R. Mayer and U. Kubasch; *Angew. Chem.*, 1961, **73**, 220.	802
61AK(18)143	G. Bergson and A. Biezais; *Ark. Kemi*, 1961, **18**, 143.	948, 952, 963
61AP769	H. J. Roth and M. Schwarz; *Arch. Pharm.* (*Weinheim, Ger.*), 1961, **294**, 769.	63, 80, 84
61BBA(46)576	B. Pullman and C. Spanjaard; *Biochim. Biophys. Acta*, 1961, **46**, 576 (*Chem. Abstr.*, 1961, **55**, 20 044).	238, 240
61BRP866440	C. P. Vale, S. Gutter and W. Wilson; *Br. Pat.* 866 440 (1961) (*Chem. Abstr.*, 1961, **55**, 24 105).	894
61CB1682	J. Goerdeler and M. Budnowski; *Chem. Ber.*, 1961, **94**, 1682.	471
61CB1956	F. Korte and K. Storiko; *Chem. Ber.*, 1961, **94**, 1956.	42
61CCC156	R. Zahradník and J. Koutecký; *Collect. Czech. Chem. Commun.*, 1961, **26**, 156.	236, 255, 264, 449, 464, 516
61CI(L)2049	D. R. Liljegren and K. T. Potts; *Chem. Ind.* (*London*), 1961, 2049.	1014
61CRV87	K. T. Potts; *Chem. Rev.*, 1961, **61**, 87.	1014
61E566	R. Huisgen and V. Weberndörfer; *Experientia*, 1961, **17**, 566.	848
61G47	P. Bravo, G. Gaudiano, A. Quilico and A. Ricca; *Gazz. Chim. Ital.*, 1961, **91**, 47.	58, 67, 86, 87
61HC(15-1)63	W. L. Mosby; *Chem. Heterocycl. Compd.*, 1961, **15-1**, 63.	974
61HCA865	H. A. Christ, P. Diehl, H. R. Schneider and H. Dahn; *Helv. Chim. Acta*, 1961, **44**, 865.	398
61JA1007	R. B. Woodward and R. A. Olofson; *J. Am. Chem. Soc.*, 1961, **83**, 1007.	982
61JA2105	J. G. Pritchard and P. C. Lauterbur; *J. Am. Chem. Soc.*, 1961, **83**, 2105.	863, 865
61JA2934	E. Klingsberg; *J. Am. Chem. Soc.*, 1961, **83**, 2934.	793
61JA3504	N. D. Field; *J. Am. Chem. Soc.*, 1961, **83**, 3504.	778
61JA3805	J. F. Bunnett and S. Y. Yih; *J. Am. Chem. Soc.*, 1961, **83**, 3805.	115
61JA3901	R. R. Frazer, R. U. Lemieux and J. D. Stevens; *J. Am. Chem. Soc.*, 1961, **83**, 3901.	750
61JA4475	R. A. Bonham and F. A. Momany; *J. Am. Chem. Soc.*, 1961, **83**, 4475.	514, 516, 517, 581
61JCS2825	E. R. Ward and W. H. Poesche; *J. Chem. Soc.*, 1961, 2825.	274
61JOC432	G. N. Walker and M. A. Moore; *J. Org. Chem.*, 1961, **26**, 432.	186
61JOC1644	E. Lieber, C. B. Lawyer and J. P. Trivedi; *J. Org. Chem.*, 1961, **26**, 1644.	594
61JOC2615	S. Gronowitz, A. B. Hörnfeldt, B. Gestblom and R. A. Hoffman; *J. Org. Chem.*, 1961, **26**, 2615.	1016
61JOC2715	E. C. Taylor, G. A. Berchtold, N. A. Goeckner and F. G. Stroehmann; *J. Org. Chem.*, 1961, **26**, 2715.	986
61JOC2976	R. Cramer and W. R. McClellan; *J. Org. Chem.*, 1961, **26**, 2976.	84
61JOC3802	V. Boekelheide and R. Scharrer; *J. Org. Chem.*, 1961, **26**, 3802.	630
61JOC4684	J. H. Boyer and S. E. Ellzey, Jr.; *J. Org. Chem.*, 1961, **26**, 4684.	730
61JOC5221	P. A. S. Smith and D. H. Kenny; *J. Org. Chem.*, 1961, **26**, 5221.	593
61JPC1279	R. A. Robinson and V. E. Bower; *J. Phys. Chem.*, 1961, **65**, 1279.	62
61JSP(7)58	B. Bak, D. Christensen, L. Hansen-Naygaard and J. Rastrup-Andersen; *J. Mol. Spectrosc.*, 1961, **7**, 58.	514
61MI41600	A. R. Katritzky and S. Øksne; *Proc. Chem. Soc.*, 1961, 387.	87
B-61MI42200	J. H. Boyer; 'Heterocyclic Compounds', ed. R. C. Elderfield; Wiley, New York, 1961, vol. 7, p. 463.	394, 400, 415, 418
B-61MI42300	'Heterocyclic Compounds', ed. R. C. Elderfield; Wiley, New York, 1961, vol. 7, p. 525.	428, 436, 437, 439
B-61MI42400	W. R. Sherman; in 'Heterocyclic Compounds', ed. R. C. Elderfield; Wiley, New York, 1961, vol. 7, p. 541.	448, 454, 456, 458, 459, 460, 461
B-61MI42600	W. R. Sherman; in 'Heterocyclic Compounds', ed. R. C. Elderfield; Wiley, New York, 1961, vol. 7, Chap. 7, p. 579.	514, 537
B-61MI42700	W. R. Sherman; in 'Heterocyclic Compounds', ed. R. C. Elderfield; Wiley, New York, 1961, vol. 7, p. 587.	546, 568
B-61MI43100	O. Foss; in 'Organic Sulfur Compounds', ed. N. Kharasch; Pergamon, Oxford, 1961, vol. 1, p. 75.	788
B-61MI43101	L. J. Reed; in 'Organic Sulfur Compounds', ed. N. Kharasch; Pergamon, Oxford, 1961, vol. 1, p. 443.	788, 810
B-61MI43300	E. G. E. Hawkins; 'Organic Peroxides: Their Formation and Reactions', Spon, London, 1961, p. 229.	853
B-61MI43500	'Table of Atomic Masses,' ed. J. W. Guthrie; monograph SCR-245A from US Atomic Energy Commission 1961.	955, 956
61MI43600	G. A. Sim, T. A. Hamor, I. C. Paul and J. M. Robertson; *Proc. Chem. Soc.*, 1961, 75.	974
61MI43800	M. Mammi, R. Bardi, G. Traverso and S. Bezzi; *Nature* (*London*), 1961, **192**, 1282.	1054
61USP2989537	J. Druey, P. Schmidt and K. Eichenberger; *U.S. Pat.* 2 989 537 (1961) (*Chem. Abstr.*, 1961, **55**, 25 990).	990, 1025

61ZOB1332	G. A. Razuvaev, V. S. Etlis and L. N. Grobov; *Zh. Obshch. Khim.*, 1961, **31**, 1332.	870
61ZOB1962	T. I. Abramovich, I. P. Gragerov and V. V. Perekalin; *Zh. Obshch. Khim.*, 1961, **31**, 1962 (*Chem. Abstr.*, 1961, **55**, 27 373).	276
62ACS105	S. Gronowitz and P. Moses; *Acta Chem. Scand.*, 1962, **16**, 105.	986
62ACS155	S. Gronowitz and P. Moses; *Acta Chem. Scand.*, 1962, **16**, 155.	1016
62AG135	G. Kresze, A. Maschke, R. Albrecht, K. Bederke, H. P. Patzschke, H. Smalla and A. Trede; *Angew. Chem.*, 1962, **74**, 135.	492, 506
62AG753	F. Wille, A. Ascherl, G. Kaupp and C. Capeller; *Angew. Chem.*, 1962, **74**, 753.	337
62AG(E)48	R. Huisgen, R. Grashey, H. Gotthardt and R. Schmidt; *Angew. Chem., Int. Ed. Engl.*, 1962, **1**, 48.	376
62AG(E)49	R. Huisgen, H. Gotthardt and R. Grashey; *Angew. Chem., Int. Ed. Engl.*, 1962, **1**, 49.	376
62AG(E)528	H. W. Heine; *Angew. Chem., Int. Ed. Engl.*, 1962, **1**, 528.	1017
62AG(E)594	K. Dickoré, R. Wegler and K. Sasse; *Angew. Chem., Int. Ed. Engl.*, 1962, **1**, 594.	909, 912, 930, 941, 943
62AG(E)662	H. Bredereck and R. Bangert; *Angew. Chem., Int. Ed. Engl.*, 1962, **1**, 662.	178
62AK(19)195	G. Bergson; *Ark. Kemi*, 1962, **19**, 195.	948, 954, 961, 962, 967, 970
62AP621	K. Winterfeld and W. Haering; *Arch. Pharm. (Weinheim, Ger.)*, 1962, **295**, 621.	745
62BCJ766	M. Hashimoto and M. Ohta; *Bull. Chem. Soc. Jpn.*, 1962, **35**, 766.	604
62BEP610763	Fabriek Van Chemische Producten N.V., *Belg. Pat.* 610 763 (1962) (*Chem. Abstr.*, 1962, **57**, 13 992).	894
62BRP902256	R. Jowitt; *Br. Pat.* 902 256 (1962) (*Chem. Abstr.*, 1962, **57**, 12 808).	894
62BRP903077	'Montecatini' Societa Generale per l'Industria Mineraria e Chimica, *Br. Pat.* 903 077 (1962) (*Chem. Abstr.*, 1962, **57**, 11 396).	894
62BSF2194	A. Thuillier and J. Vialle; *Bull. Soc. Chim. Fr.*, 1962, 2194.	803
62BSF2215	R. Paul and S. Tchelitcheff; *Bull. Soc. Chim. Fr.*, 1962, 2215.	70, 83, 91
62CB2333	A. Simon and G. Heintz; *Chem. Ber.*, 1962, **95**, 2333.	751, 860
62CB2861	R. Gompper and W. Töpfl; *Chem. Ber.*, 1962, **95**, 2861.	890
62CI(L)213	R. J. Abraham, K. A. McLauchlin, L. D. Hall and L. Hough; *Chem. Ind. (London)*, 1962, 213.	753
62CJC882	J. F. King and T. Durst; *Can. J. Chem.*, 1962, **40**, 882.	51
62CJC1189	D. G. M. Diaper and D. L. Mitchell; *Can. J. Chem.*, 1962, **40**, 1189.	872
62G501	S. Cabiddu, G. Gaudiano and A. Quilico; *Gazz. Chim. Ital.*, 1962, **92**, 501.	112
62GEP1124962	W. Rosenthal; *Ger. Pat.* 1 124 962 (1962) (*Chem. Abstr.*, 1962, **57**, 9978).	894
62HC(17)1	A. Quilico; *Chem. Heterocycl. Compd.*, 1962, **17**, 1.	2, 3, 4, 5, 8, 9, 11, 7, 36, 39, 40, 54, 55, 61, 62, 64, 66, 71, 82, 83, 84, 85, 86, 87, 88, 89, 91, 93, 96, 102, 103, 104, 105, 106, 107, 108, 112, 113, 120
62HC(17)263	L. C. Behr; *Chem. Heterocycl. Compd.*, 1962, **17**, 263.	429, 430, 445
62HC(17)283	L. C. Behr; *Chem. Heterocycl. Compd.*, 1962, **17**, 283.	394, 415, 418
62HCA504	P. Diehl, H. A. Christ and F. B. Mallory; *Helv. Chim. Acta*, 1962, **45**, 504.	398
62HCA2441	H. U. Daeniker and J. Druey; *Helv. Chim. Acta*, 1962, **45**, 2441.	377
62JA2941	E. Klingsberg and A. M. Schreiber; *J. Am. Chem. Soc.*, 1962, **84**, 2941.	795
62JCS1505	J. Charalambous, H. J. Davis, M. J. Frazer and W. Gerrard; *J. Chem. Soc.*, 1962, 1505.	871
62JCS4234	G. A. Mina, L. Rateb and G. Soliman; *J. Chem. Soc.*, 1962, 4234.	85
62JCS5104	D. Leaver, W. A. H. Robertson and D. M. McKinnon; *J. Chem. Soc.*, 1962, 5104.	803
62JIC263	S. N. Dixit; *J. Indian Chem. Soc.*, 1962, **39**, 263.	923
62JOC488	E. Campaigne and B. E. Edwards; *J. Org. Chem.*, 1962, **27**, 488.	890
62JOC869	W. G. Toland; *J. Org. Chem.*, 1962, **27**, 869.	501
62JOC2160	H. Ulrich, J. N. Tilley and A. A. Sayigh; *J. Org. Chem.*, 1962, **27**, 2160.	39, 41, 104
62JOC2899	S. H. Chu and H. G. Mautner; *J. Org. Chem.*, 1962, **27**, 2899.	345
62JOC3683	H. G. Garg; *J. Org. Chem.*, 1962, **27**, 3683.	120
62JOC3995	C. G. Krespan and W. R. Brasen; *J. Org. Chem.*, 1962, **27**, 3995.	867, 870, 892
62JOC4101	P. Yates and J. E. Hyre; *J. Org. Chem.*, 1962, **27**, 4101.	765
62JPR(287)313	A. Greiner and V. Müller; *J. Prakt. Chem.*, 1962, **287**, 313.	866
62LA(658)128	F. Weygand, W. Steglich and H. Tanner; *Liebigs. Ann. Chem.*, 1962, **658**, 128.	186
62LA(658)169	R. Huisgen, R. Grashey, M. Seidel, H. Knupfer and R. Schmidt; *Liebigs Ann. Chem.*, 1962, **658**, 169.	609
62M1441	E. Ziegler and R. Wolf; *Monatsh. Chem.*, 1962, **93**, 1441.	727
62MI41600	A. F. Aboulezz and R. Quelet; *J. Chem. U.A.R.*, 1962, **5**, 137 (*Chem. Abstr.*, 1965, **63**, 11 411).	120
62MI42900	B. Glowiak and R. Kulik; *Rocz. Chem.*, 1962, **36**, 959 (*Chem. Abstr.*, 1963, **58**, 5631).	728
B-62MI43100	W. A. Pryor; 'Mechanisms of Sulfur Reactions', McGraw-Hill, New York, 1962, p. 7.	804
62MI43600	J. I. G. Cadogan and M. Cameron-Wood; *Proc. Chem. Soc.*, 1962, 361.	1009
62T3	R. Huisgen, M. Seidel, G. Wallbillich and H. Knupfer; *Tetrahedron*, 1962, **17**, 3.	998
62TL387	R. Huisgen, R. Grashey and R. Krischke; *Tetrahedron Lett.*, 1962, 387.	743
62USP3053852	H. W. Coover, Jr. and R. L. McConnel; *U.S. Pat.* 3 053 852 (1962) (*Chem. Abstr.*, 1963, **58**, 8911).	871
62USP3055907	H. D. Brown and L. H. Sarett; *U.S. Pat.* 3 055 907 (1962) (*Chem. Abstr.*, 1963, **58**, 2456).	543

62USP3055913	L. O. Moore and J. W. Clark; *U.S. Pat.* 3 055 913 (1962) (*Chem. Abstr.*, 1963, **58**, 1347).	894
62USP3056781	*U.S. Pat.* 3 056 781 (1962) (*Chem. Abstr.*, 1963, **58**, 5702).	628
62USP3066147	M. Carmack and L. M. Weinstock; *U.S. Pat.* 3 066 147 (1962) (*Chem. Abstr.*, 1963, **58**, 7949).	543
62ZC69	G. Blankenstein and K. Möckel; *Z. Chem.* 1962, **2**, 69.	428, 439
62ZOB1878	G. I. Derkach, G. F. Dregval and A. V. Kirsanov; *Zh. Obshch. Khim.*, 1962, **32**, 1878.	159
62ZOB2961	V. T. Klimko, T. V. Protopopova, N. V. Smirnova and A. P. Skoldinov; *Zh. Obshch. Khim.*, 1962, **32**, 2961.	83
63AC(R)1697	D. Dal Monte and E. Sandri; *Ann. Chim.* (*Rome*), 1963, **53**, 1697.	532
63ACS163	K. A. Jensen, H. R. Baccaro and O. Buchardt; *Acta Chem. Scand.*, 1963, **17**, 163.	793, 942
63ACS2575	A. Hordvik; *Acta Chem. Scand.*, 1963, **17**, 2575.	902, 903
63AG742	R. Huisgen; *Angew. Chem.*, 1963, **75**, 742.	66
63AG(E)565	R. Huisgen; *Angew. Chem., Int. Ed. Engl.*, 1963, **2**, 565.	996, 998, 999, 1000
63AHC(2)1	A. R. Katritzky and J. M. Lagowski; *Adv. Heterocycl. Chem.*, 1963, **2**, 1.	980
63AHC(2)365	N. K. Kochetkov and S. D. Sokolov; *Adv. Heterocycl. Chem.*, 1963, **2**, 365.	2, 3, 4, 5, 22, 24, 26, 27, 29, 30, 31, 33, 50, 52, 53, 55, 57, 58, 59, 61, 62, 83, 84, 86, 87
63AP389	W. Schneider and K. Schilken; *Arch. Pharm.* (*Weinheim, Ger.*), 1963, **296**, 389.	666
63AX64	M. E. Burke, R. A. Sparks and K. N. Trueblood; *Acta Crystallogr.*, 1963, **16**, 64.	1031
63AX471	H. Barnighausen, F. Jellinek, J. Munnick and A. Vos; *Acta Crystallogr.*, 1963, **16**, 471.	369
63AX1157	R. H. Stanford, Jr.; *Acta Crystallogr.*, 1963, **16**, 1157.	902
63BCJ1150	S. Umezawa and S. Zen; *Bull. Chem. Soc. Jpn.*, 1963, **36**, 1150.	77, 85
63BCJ1437	M. Ohta and M. Sugiyama; *Bull. Chem. Soc. Jpn.*, 1963, **36**, 1437.	839, 842
63BEP632072	K. Sasse, R. Wegler and L. Eue; *Belg. Pat.* 632 072 (1963) (*Chem. Abstr.*, 1964, **61**, 3119).	944
63BSF161	M. Ebel, L. Legrand and N. Lozac'h; *Bull. Soc. Chim. Fr.*, 1963, 161.	794
63CB168	M. Lempert-Sreter, V. Solt and K. Lempert; *Chem. Ber.*, 1963, **96**, 168.	993
63CB534	J. Goerdeler and W. Roth; *Chem. Ber.*, 1963, **96**, 534.	480
63CB1088	H.-D. Stachel; *Chem. Ber.*, 1963, **96**, 1088.	86, 104
63CB1177	A. W. Wagner and R. Banholzer; *Chem. Ber.*, 1963, **96**, 1177.	899, 907, 931, 940
63CB1289	S. Goerdeler, D. Gross and H. Klinke; *Chem. Ber.*, 1963, **96**, 1289.	355
63CB1827	R. Pfleger, E. Garthe and K. Rauer; *Chem. Ber.*, 1963, **96**, 1827.	1039, 1046
63CB2029	G. Habermehl; *Chem. Ber.*, 1963, **96**, 2029.	185
63CB2519	T. Kauffmann and H. Marhan; *Chem. Ber.*, 1963, **96**, 2519.	606, 607, 610
63CB2702	R. Mayer and J. Faust; *Chem. Ber.*, 1963, **96**, 2702.	809
63CI(L)1926	F. H. C. Stewart and N. Danieli; *Chem. Ind.* (*London*), 1963, 1926.	370
63CJC2671	B. E. Leggetter and R. K. Brown; *Can. J. Chem.*, 1963, **41**, 2671.	764
63CRV557	H. F. van Woerden; *Chem. Rev.*, 1963, **63**, 557.	853, 872, 874, 893
63GEP1146494	M. Matter, C. Vogel and R. Bosshard; *Ger. Pat.* 1 146 494 (1963) (*Chem. Abstr.*, 1963, **59**, 10 058).	106
63GEP1159760	H. Gernert, *Ger. Pat.* 1 159 760 (1963) (*Chem. Abstr.*, 1964, **60**, 6385).	894
63HCA805	H. U. Daeniker and J. Druey; *Helv. Chim. Acta*, 1963, **46**, 805.	367
63HCA2167	R. Wizinger and D. Dürr; *Helv. Chim. Acta*, 1963, **46**, 2167.	823, 843
63IJC300	V. K. Verma; *Indian J. Chem.*, 1963, **1**, 300.	931
63JA602	E. T. Kaiser, M. Panar and F. H. Westheimer; *J. Am. Chem. Soc.*, 1963, **85**, 602.	887
63JA3244	E. Klingsberg; *J. Am. Chem. Soc.*, 1963, **85**, 3244.	809
63JA3553	V. Dobyns and L. Pierce; *J. Am. Chem. Soc.*, 1963, **85**, 3553.	514, 515, 516, 581
63JCS197	R. K. Harris, A. R. Katritzky, S. Øksne, A. S. Bailey and W. G. Paterson; *J. Chem. Soc.*, 1963, 197.	397
63JCS2032	D. Buttimore, D. H. Jones, R. Slack and K. R. H. Wooldridge; *J. Chem. Soc.*, 1963, 2032.	140
63JCS3165	H. J. Emeléus, A. Haas and N. Sheppard; *J. Chem. Soc.*, 1963, 3165.	898, 910, 912
63JCS3168	H. J. Emeléus, A. Haas and N. Sheppard; *J. Chem. Soc.*, 1963, 3168.	468
63JCS4566	L. E. A. Godfrey and F. Kurzer; *J. Chem. Soc.*, 1963, 4566.	484
63JCS4693	G. R. Delpierre and M. Lamchen; *J. Chem. Soc.*, 1963, 4693.	1000
63JGU1714	V. G. Pesin, A. M. Khaletskii and V. A. Sergeev; *J. Gen. Chem. USSR* (*Engl. Transl.*), 1963, **33**, 1714.	531
63JOC529	E. Klingsberg; *J. Org. Chem.*, 1963, **28**, 529.	792
63JOC981	J. R. Piper and T. P. Johnston; *J. Org. Chem.*, 1963, **28**, 981.	1013
63JOC1075	Y. Okumura; *J. Org. Chem.*, 1963, **28**, 1075.	861, 868, 871, 884
63JOC2581	A. Richardson, Jr.; *J. Org. Chem.*, 1963, **28**, 2581.	708
63JPR(291)101	W. Schulze, G. Letsch and H. Willitzer; *J. Prakt. Chem.*, 1963, **291**, 101.	1013
63JPR(292)285	M. Muehlstaedt and E. Bordes; *J. Prakt. Chem.*, 1963, **292**, 285.	988
63LA(661)84	A. Lüttringhaus, M. Mohr and N. Engelhard; *Liebigs Ann. Chem.*, 1963, **661**, 84.	797, 809
63LA(663)113	T. Pyl, F. Waschk and H. Beyer; *Liebigs Ann. Chem.*, 1963, **663**, 113.	994
63MI41600	P. Bravo; *Chim. Ind.* (*Milan*), 1963, **45**, 1239.	55, 57, 58, 67, 86, 87
63MI41601	N. Campbell and H. F. Andrew; *Proc. R. Soc. Edinburgh, Sect. A*, 1963–64, **66**, 252 (*Chem. Abstr.*, 1965, **62**, 16 157).	24
63PMH(1)1	A. Albert; *Phys. Methods Heterocycl. Chem.*, 1963, **1**, 1.	4

63PMH(2)61	S. F. Mason; *Phys. Methods Heterocycl. Chem.*, 1963, **2**, 61.	519
63PMH(2)161	A. R. Katritzky and A. P. Ambler; *Phys. Methods Heterocycl. Chem.*, 1963, **2**, 161.	4, 5, 819
63PMH(2)229	A. R. Katritzky and P. Ambler; *Phys. Methods Heterocycl. Chem.*, 1963, **2**, 229.	370, 379
63SA1145	G. Adembri, G. Speroni and S. Califano; *Spectrochim. Acta*, 1963, **19**, 1145.	5
63SA2047	A. Alemagna; *Spectrochim. Acta*, 1963, **19**, 2047.	907
63T169	A. T. Balaban, I. Bally, P. T. Frangopol, M. Bacescu, E. Cioranescu and L. Birladeanu; *Tetrahedron*, 1963, **19**, 169.	182
63UP41700	D. F. Muggleton; Authors' Laboratories, 1963, unpublished results.	144
63USP3088935	C. G. Krespan; *U.S. Pat.* 3 088 935 (1963) (*Chem. Abstr.*, 1963, **59**, 10 096).	893
63ZC26	R. Mayer and K. Gewald; *Z. Chem.*, 1963, **3**, 26.	806
63ZC388	E. Bulka and K-D. Ahlers; *Z. Chem.*, 1963, **3**, 388.	341
63ZOB3667	M. O. Kolosova; *Zh. Obshch. Khim.*, 1963, **33**, 3667 (*Chem. Abstr.*, 1964, **60**, 8022).	290
63ZOB3699	V. G. Yashunskii, L. E. Kholodov and E. M. Peresleni; *Zh. Obshch. Khim.*, 1963, **33**, 3699.	371
64ACS825	K. A. Jensen, A. Holm and B. Thorkilsen; *Acta Chem. Scand.*, 1964, **18**, 825.	595
64ACS871	J. Sandstrom; *Acta Chem. Scand.*, 1964, **18**, 871.	552
64AG303	D. Martin; *Angew. Chem.*, 1964, **76**, 303.	595
64AG(E)19	F. Asinger, W. Schäfer, K. Halcour, A. Saus and H. Triem; *Angew. Chem., Int. Ed. Engl.*, 1964, **3**, 19.	995
64AHC(3)263	K. A. Jensen and C. Pedersen; *Adv. Heterocycl. Chem.*, 1964, **3**, 263.	580, 581, 583, 584, 588, 591, 593, 594, 595, 598
64AJC119	W. D. Crow and J. H. Hodgkin; *Aust. J. Chem.*, 1964, **17**, 119.	232
64AJC447	R. F. C. Brown and I. D. Rae; *Aust. J. Chem.*, 1964, **17**, 447.	800
64BSF2888	J. Metzger, H. Larivé, R. Dennilauler, R. Baralle and C. Gaurat; *Bull. Soc. Chim. Fr.*, 1964, 2888 (*Chem. Abstr.*, 1965, **62**, 9267).	277
64CB159	G. Drefahl and H. H. Hoerhold; *Chem. Ber.*, 1964, **97**, 159.	36, 91
64CB1886	R. Mayer and H. Hartmann; *Chem. Ber.*, 1964, **97**, 1886.	799
64CB2023	F. Weygand, W. Steglich, D. Mayer and W. von Philipsborn; *Chem. Ber.*, 1964, **97**, 2023.	186
64CB2567	H. Behringer and D. Weber; *Chem. Ber.*, 1964, **97**, 2567.	907, 914, 922
64CB2689	D. Martin; *Chem. Ber.*, 1964, **97**, 2689.	595
64CI(L)461	D. Leaver and D. M. McKinnon; *Chem. Ind. (London)*, 1964, 461.	797, 799
64CI(L)1919	K. T. Potts and R. M. Huseby; *Chem. Ind. (London)*, 1964, 1919.	994
64CJC43	C. N. R. Rao and R. Venkataraghavan; *Can. J. Chem.*, 1964, **42**, 43.	452, 465
64CJC2375	H. J. Anderson, D. J. Barnes and Z. M. Khan; *Can. J. Chem.*, 1964, **42**, 2375.	254
64CPB813	I. Iwai and T. Hiraoka; *Chem. Pharm. Bull.*, 1964, **12**, 813.	1010
64CR(258)4579	H. Najer, J. Menin and J.-F. Giudicelli; *C. R. Hebd. Seances Acad. Sci.*, 1964, **258**, 4579.	430, 431
64CR(258)6446	J. Teste and B. Antoine; *C. R. Hebd. Seances Acad. Sci.*, 1964, **258**, 6446.	789
64CRV129	F. H. C. Stewart; *Chem. Rev.*, 1964, **64**, 129.	367, 370, 371, 372, 377
64CRV149	R. A. Abramovitch and B. A. Davis; *Chem. Rev.*, 1964, **64**, 149.	1009
64CRV473	J. Hamer and A. Macaluso; *Chem. Rev.*, 1964, **64**, 473.	1000
64FRP1379555	Chemische Werke Huels A.-G., *Fr. Pat.* 1 379 555 (1964) (*Chem. Abstr.*, 1965, **63**, 3147).	894
64GEP1166975	K. Gaebelein; *Ger. Pat.* 1 166 975 (1964) (*Chem. Abstr.* 1964, **60**, 15 686).	895
64JA107	J. B. Hendrickson, R. Rees and J. F. Templeton; *J. Am. Chem. Soc.*, 1964, **86**, 107.	475
64JA162	F. A. Momany and R. A. Bonham; *J. Am. Chem. Soc.*, 1964, **86**, 162.	514, 515, 516
64JA2861	J. M. Ross and W. C. Smith; *J. Am. Chem. Soc.*, 1964, **86**, 2861.	538, 541
64JCP(43)166	E. Saegebarth and A. P. Cox; *J. Chem. Phys.*, 1965, **43**, 166.	514, 515
64JCS446	M. P. L. Caton, D. H. Jones, R. Slack and K. R. H. Wooldridge; *J. Chem. Soc.*, 1964, 446.	140, 144
64JCS906	W. V. Farrar; *J. Chem. Soc.*, 1964, 906.	600, 604
64JCS3114	D. H. Jones, R. Slack and K. R. H. Wooldridge; *J. Chem. Soc.*, 1964, 3114.	140
64JHC163	E. Campaigne and F. Haaf; *J. Heterocycl. Chem.*, 1964, **1**, 163.	844
64JOC738	R. W. Addor; *J. Org. Chem.*, 1964, **29**, 738.	843
64JOC1488	J. F. Roesler, J. Leslie and G. Gorin; *J. Org. Chem.*, 1964, **29**, 1488.	916
64JOC1582	M. E. Kuehne, S. J. Weaver and P. Franz; *J. Org. Chem.*, 1964, **29**, 1582.	633
64JOC1708	E. Campaigne, R. D. Hamilton and N. W. Jacobsen; *J. Org. Chem.*, 1964, **29**, 1708.	840
64JOC2116	E. C. Taylor and E. E. Garcia; *J. Org. Chem.*, 1964, **29**, 2116.	620, 627
64JOC2121	E. C. Taylor and E. E. Garcia; *J. Org. Chem.*, 1964, **29**, 2121.	734, 736, 739, 746
64JOC2135	Y. F. Shealy and C. A. O'Dell; *J. Org. Chem.*, 1964, **29**, 2135.	736, 740, 747
64JOC2141	Y. F. Shealy and J. D. Clayton; *J. Org. Chem.*, 1964, **29**, 2141.	739
64JOC2240	M. I. Fremery and E. K. Fields; *J. Org. Chem.*, 1964, **29**, 2240.	876
64JOC2877	E. Campaigne and R. D. Hamilton; *J. Org. Chem.*, 1964, **29**, 2877.	844
64JPC441	A. Bondi; *J. Phys. Chem.*, 1964, **68**, 441.	1053, 1054
64LA(671)135	K. Dickoré; *Liebigs Ann. Chem.*, 1964, **671**, 135.	928, 941
64LA(675)151	R. Gompper, H. Euchner and H. Kast; *Liebigs Ann. Chem.*, 1964, **675**, 151.	942
64MI41600	Y. Ogawa; *Shinogi Kenyusko Nempo*, 1964, **14**, 50 (*Chem. Abstr.*, 1965, **62**, 16 786).	129
64MI43100	R. Mayer, P. Rosmus and J. Fabian; *J. Chromatogr.*, 1964, **15**, 153.	786
64NEP6407011	F. Hoffmann–La Roche & Co. A.-G., *Neth. Appl.* 6 407 011 (1964) (*Chem. Abstr.*, 1965, **63**, 583).	123
64RCR508	E. P. Nesynov and A. P. Grekov; *Russ. Chem. Rev. (Engl. Transl.)*, 1964, **33**, 508.	428, 444

64RTC877	B. Zwanenburg, W. E. Weening and J. Strating; *Recl. Trav. Chim. Pays-Bas*, 1964, **83**, 877.	444
64T159	H. Kano and E. Yamazaki; *Tetrahedron*, 1964, **20**, 159.	60
64T461	H. Kano and E. Yamazaki; *Tetrahedron*, 1964, **20**, 461.	55, 60
64TL887	P. Rajagopalan; *Tetrahedron Lett.*, 1964, 887.	607, 609
64TL1477	W. D. Crow and N. J. Leonard; *Tetrahedron Lett.*, 1964, 1477.	142
64TL2829	D. Martin; *Tetrahedron Lett.*, 1964, 2829.	595
64TL3815	M. P. Cava and R. H. Schlessinger; *Tetrahedron Lett.*, 1964, 3815.	529
64USP3154526	D. L. Klass and J. E. King; *U.S. Pat.* 3 154 526 (1964) (*Chem. Abstr.*, 1965, **62**, 1820).	894
64USP3156704	R. B. Davis; *U.S. Pat.* 3 156 704 (1964) (*Chem. Abstr.*, 1965, **62**, 2743).	123
64USP3158621	E. Klingsberg; (Am. Cyanamid Co.), *U.S. Pat.* 3 158 621 (1964) (*Chem. Abstr.*, 1965, **62**, 10 575).	811
64USP3261870	A. I. Rachlin; *U.S. Pat.* 3 261 870 (1964) (*Chem. Abstr.*, 1966, **65**, 15 277).	122
65ACS438	K. A. Jensen, M. Due and A. Holm; *Acta Chem. Scand.*, 1965, **19**, 438.	595
65ACS1215	S. Gronowitz, B. Mathiasson, R. Dahlbom, B. Holmberg and K. A. Jensen; *Acta Chem. Scand.*, 1965, **19**, 1215.	248, 287
65ACS1253	A. Hordvik; *Acta Chem. Scand.*, 1965, **19**, 1253.	784
65ACS1539	A. Hordvik and S. Joys; *Acta Chem. Scand.*, 1965, **19**, 1539.	902
65AG348	K. Pilgram and F. Korte; *Angew. Chem.*, 1965, **77**, 348.	595
65AG(E)239	V. Bertini and P. Pino; *Angew. Chem., Int. Ed. Engl.*, 1965, **4**, 239.	540, 541
65AG(E)873	K. Dimroth, P. Heinrich and K. Schromm; *Angew. Chem., Int. Ed. Engl.*, 1965, **4**, 873.	762
65AHC(4)75	R. Filler; *Adv. Heterocycl. Chem.*, 1965, **4**, 75.	186, 199
65AHC(4)107	R. Slack and K. R. H. Wooldridge; *Adv. Heterocycl. Chem.*, 1965, **4**, 107.	132, 133, 142, 144, 147, 148, 149, 151, 155, 156, 157, 158, 162, 163, 164, 166, 168, 169, 175
65AHC(5)1	R. Zahradnik; *Adv. Heterocycl. Chem.*, 1965, **5**, 1.	814
65AHC(5)119	F. Kurzer; *Adv. Heterocycl. Chem.*, 1965, **5**, 119.	464, 465, 467, 468, 469, 470, 471, 474, 477, 478, 480, 481, 482, 483, 484, 485, 487, 488, 492, 493, 494, 495, 496, 502, 503, 505, 506, 507
65AJC61	R. F. C. Brown, I. D. Rae and S. Sternhell; *Aust. J. Chem.*, 1965, **18**, 61.	800
65AJC1211	R. F. C. Brown, I. D. Rae and S. Sternhell; *Aust. J. Chem.*, 1965, **18**, 1211.	785
65AX942	I. Ambats and R. E. Marsh; *Acta Crystallogr.*, 1965, **19**, 942.	180
65BCJ596	M. Ohta and M. Sugiyama; *Bull. Chem. Soc. Jpn.*, 1965, **38**, 596.	839, 842
65BEP659236	*Belg. Pat.* 659 236 (1965) (*Chem. Abstr.*, 1966, **64**, 3749).	710
65BEP659395	F. Blomeyer, W. Zecher and H. Holtschmidt; *Belg. Pat.* 659 395 (1965) (*Chem. Abstr.*, 1965, **63**, 18 470).	894
65BSB488	F. Alderweireldt and M. Anteunis; *Bull. Soc. Chim. Belg.*, 1965, **74**, 488.	753, 754
65BSF3646	D. Gagnaire and J. B. Robert; *Bull. Soc. Chim. Fr.*, 1965, 3646.	754
65CB1111	H. A. Staab and A. Mannschreck; *Chem. Ber.*, 1965, **98**, 1111.	136
65CB1562	H. Musso and H. Schröder; *Chem. Ber.*, 1965, **98**, 1562.	54, 125
65CB1831	D. Klamann, W. Koser, P. Weyerstahl and M. Fligge; *Chem. Ber.*, 1965, **98**, 1831.	421
65CB2059	D. Martin and W. Mucke; *Chem. Ber.*, 1965, **98**, 2059.	595
65CB2063	P. Reich and D. Martin; *Chem. Ber.*, 1965, **98**, 2063.	595
65CB2248	W. Strohmeier, J. F. Guttenberger and G. Popp; *Chem. Ber.*, 1965, **98**, 2248.	871
65CB2825	R. Gompper and E. Kutter; *Chem. Ber.*, 1965, **98**, 2825.	836, 846
65CB3020	U. Türck and H. Behringer; *Chem. Ber.*, 1965, **98**, 3020.	73, 83, 84
65CB3228	H. Stetter and K. Findeisen; *Chem. Ber.*, 1965, **98**, 3228.	982, 991
65CC408	K.-H. Wunsch, H. Linke, A. J. Boulton and M. Altaf-ur-Rahman; *Chem. Commun.*, 1965, 408.	24
65CC624	N. P. Buu-Hoï, P. Jacquignon and M. Mangane; *Chem. Commun.*, 1965, 624.	522
65CHE165	Z. I. Miroshnichenko and M. A. Alperovich; *Chem. Heterocycl. Compd. (Engl. Transl.)*, 1965, **1**, 165.	994
65CHE271	I. I. Grandberg, S. V. Tabak, N. I. Bobrova, A. N. Kost and L. G. Vasina; *Chem. Heterocycl. Compd. (Engl. Transl.)*, 1965, **1**, 271.	987
65CI(L)36	Y. S. Rao, T. M. Muzyczko, P. Owen and R. Filler; *Chem. Ind. (London)*, 1965, 36.	105
65CI(L)691	J. A. Magnuson, C. A. Hirt and P. J. Lauer; *Chem. Ind. (London)*, 1965, 691.	750
65CPB248	M. Fujimoto and M. Sakai; *Chem. Pharm. Bull.*, 1965, **13**, 248.	80
65CR(260)2839	M. Hedayatullah and L. Denivelle; *C.R. Hebd. Seances Acad. Sci.*, 1965, **260**, 2839.	595
65CRV237	P. S. Landis; *Chem. Rev.*, 1965, **65**, 237.	794, 795, 796, 800 805, 810
65EGP37461	H. H. Hoerhold; *Ger. (East) Pat.* 37 461 (1965) (*Chem. Abstr.*, 1965, **63**, 10 062).	91
65FES686	G. Pagliarini, G. Cignarella and E. Testa; *Farmaco Ed. Sci.*, 1965, **20**, 686 (*Chem. Abstr.*, 1966, **64**, 17 569).	129
65FRP84686	Rhone-Poulenc; *Fr. Addn.* 84 686 (1965) (*Chem. Abstr.*, 1965, **63**, 4332).	95
65GI478	G. Renzi and V. D. Piaz; *Gazz. Chim. Ital.*, 1965, **95**, 1478.	628
65GEP1192452	R. Gompper, H. Herlinger and F. Grewe; *Ger. Pat.* 1 192 452 (1965) (*Chem. Abstr.*, 1965, **63**, 3565).	895
65HCA1973	K. Michel, H. Gerlach-Gerber, Ch. Vogel and M. Matter; *Helv. Chim. Acta*, 1965, **48**, 1973.	113
65JA528	T. W. G. Solomons, F. W. Fowler and J. Calderazzo; *J. Am. Chem. Soc.*, 1965, **87**, 528.	1046

65JA749	A. L. Logothetis; *J. Am. Chem. Soc.*, 1965, **87**, 749.	1007
65JA2743	G. L. Schmir; *J. Am. Chem. Soc.*, 1965, **87**, 2743.	271
65JA3785	G. E. Wilson, Jr.; *J. Am. Chem. Soc.*, 1965, **87**, 3785.	761
65JA4950	I. Wang, C. O. Britt and J. E. Boggs; *J. Am. Chem. Soc.*, 1965, **89**, 4950.	751
65JA5800	C. Ainsworth; *J. Am. Chem. Soc.*, 1965, **87**, 5800.	427, 429
65JAP6520705	H. Kugota and M. Tanaka; *Jpn. Pat.* 65 20 705 (1965) (*Chem. Abstr.*, 1966, **64**, 20 916).	129
65JCP(43)166	E. Saegebarth and A. P. Cox; *J. Chem. Phys.*, 1965, **43**, 166.	396
65JCS32	D. Leaver, D. M. McKinnon and W. A. H. Robertson; *J. Chem. Soc.*, 1965, 32.	789, 790
65JCS256	R. J. Abraham; *J. Chem. Soc.*, 1965, 256.	753
65JCS791	J. B. Rose and C. K. Warren; *J. Chem. Soc.*, 1965, 791.	868
65JCS2184	I. G. M. Campbell, R. C. Cookson, M. B. Hocking and A. N. Hughes; *J. Chem. Soc.*, 1965, 2184.	983
65JCS2248	E. R. Ward and C. H. Williams; *J. Chem. Soc.*, 1965, 2248.	274
65JCS5166	D. L. Pain and R. Slack; *J. Chem. Soc.*, 1965, 5166.	456, 462
65JCS6004	C. W. Haigh, M. H. Palmer and B. Semple; *J. Chem. Soc.*, 1965, 6004.	518
65JCS6769	A. J. Nunn and J. T. Ralph; *J. Chem. Soc.*, 1965, 6769.	526
65JCS7277	A. Holland, R. Slack, T. F. Warren and D. Buttimore; *J. Chem. Soc.*, 1965, 7277.	637, 640, 643
65JHC97	R. W. Balsiger, J. A. Montgomery and T. P. Johnston; *J. Heterocycl. Chem.*, 1965, **2**, 97.	702, 704
65JHC218	B. M. Lynch and Y.-Y. Hung; *J. Heterocycl. Chem.*, 1965, **2**, 218.	1046
65JHC253	T. Ichikawa, T. Kato and T. Takenishi; *J. Heterocycl. Chem.*, 1965, **2**, 253.	729
65JIC733	K. C. Joshi and A. K. Jaukar; *J. Indian Chem. Soc.*, 1965, **42**, 733.	93
65JMC515	R. F. Meyer, B. L. Cummings, P. Bass and H. O. J. Collier; *J. Med. Chem.*, 1965, **8**, 515.	144
65JMC550	K. Kishor, R. C. Arora and S. S. Parmar; *J. Med. Chem.*, 1965, **8**, 550.	124
65JOC184	R. K. Olsen and H. R. Snyder; *J. Org. Chem.*, 1965, **30**, 184.	988
65JOC226	K. D. Berlin, B. S. Rathore and M. Peterson; *J. Org. Chem.*, 1965, **30**, 226.	764
65JOC491	C. S. Dewey and R. A. Bafford; *J. Org. Chem.*, 1965, **30**, 491.	318
65JOC495	C. S. Dewey and R. A. Bafford; *J. Org. Chem.*, 1965, **30**, 495.	318
65JOC567	W. G. Finnegan and R. A. Henry; *J. Org. Chem.*, 1965, **30**, 567.	603, 604
65JOC732	E. Campaigne and F. Haaf; *J. Org. Chem.*, 1965, **30**, 732.	844, 846
65JOC1104	J. L. Pinkus, G. G. Woodyard and T. Cohen; *J. Org. Chem.*, 1965, **30**, 1104.	123
65JOC1506	R. G. Hiskey and S. J. Dominianni; *J. Org. Chem.*, 1965, **30**, 1506.	706
65JOC1854	R. A. Olofson and J. S. Michelman; *J. Org. Chem.*, 1965, **30**, 1854.	397, 402, 415, 518, 526
65JOC1916	S. C. J. Fu, E. Chinoporos and H. Terzian; *J. Org. Chem.*, 1965, **30**, 1916.	700
65JOC2454	D. L. Klayman; *J. Org. Chem.*, 1965, **30**, 2454.	345
65JOC2488	Y. F. Shealy and C. A. O'Dell; *J. Org. Chem.*, 1965, **30**, 2488.	739
65JOC2660	W. D. Crow and N. J. Leonard; *J. Org. Chem.*, 1965, **30**, 2660.	144
65JOC2696	Q. E. Thompson, M. M. Crutchfield and M. W. Dietrich; *J. Org. Chem.*, 1965, **30**, 2696.	880, 887, 890
65JOC2763	L. D. Huestis, M. L. Walsh and N. Hahn; *J. Org. Chem.*, 1965, **30**, 2763.	916
65JOC2809	C. Grundmann and J. M. Dean; *J. Org. Chem.*, 1965, **30**, 2809.	421
65JOC3229	D. Taub, R. D. Hoffsommer, C. H. Kuo and N. L. Wendler; *J. Org. Chem.*, 1965, **30**, 3229.	48
65JOC3997	A. Zweig and A. K. Hoffmann; *J. Org. Chem.*, 1965, **30**, 3997.	801
65JPR(30)280	H. Beyer, E. Bulker and K. Dittrich; *J. Prakt. Chem.*, 1965, **30**, 280.	1014
65KGS328	L. E. Kholodov and V. G. Yashunskii; *Khim. Geterotsikl. Soedin.*, 1965, 328.	371
65LA(681)178	F. Boberg; *Liebigs Ann. Chem.*, 1965, **681**, 178.	800
65LA(682)90	D. Martin and W. Mucke; *Liebigs Ann. Chem.*, 1965, **682**, 90.	735, 746
65LA(682)188	H. Behringer and A. Grimm; *Liebigs Ann. Chem.*, 1965, **682**, 188.	805
65LA(683)132	F. Boberg; *Liebigs Ann. Chem.*, 1965, **683**, 132.	796
65LA(684)37	R. Gompper, R. R. Schmidt and E. Kutter; *Liebigs Ann. Chem.*, 1965, **684**, 37.	845, 890
65LA(685)176	H. Gehlen and K. Möckel; *Liebigs Ann. Chem.*, 1965, **685**, 176.	434
65LA(687)191	C. Gründmann, V. Mini, J. M. Dean and H.-D. Frommeld, *Liebigs Ann. Chem.*, 1965, **687**, 191.	119
65LA(688)150	J. Faust and R. Mayer; *Liebigs Ann. Chem.*, 1965, **688**, 150.	800, 810
B-65MI41600	'Heilbrons Dictionary of Organic Compounds', 4th edn., Oxford Univ. Press, New York, 1965, p. 253.	120
65MI41800	E. Tubaro; *Boll. Chim. Farm.*, 1965, **104**, 602.	233
65MI42100	F. Eloy; *Fortschr. Chem. Forsch.*, 1965, **4**, 807.	382
65NKK526	M. Kawana, M. Yoshioka, S. Miyaji, H. Katoaka, Y. Omote and N. Sugiyaina; *Nippon Kagaku Zasshi*, 1965, **86**, 526 (*Chem. Abstr.*, 1965, **63**, 11 479).	120
65QR329	G. R. Delpierre and M. Lamchen; *Q. Rev., Chem. Soc.*, 1965, **19**, 329.	1000
65RRC1035	A. Silberg and Z. Frenkel; *Rev. Roum. Chim.*, 1965, **10**, 1035 (*Chem. Abstr.*, 1966, **64**, 12 641).	120, 121
65T817	G. Bianchi and P. Grünanger; *Tetrahedron*, 1965, **21**, 817.	78
65T3019	D. S. Kemp and R. B. Woodward; *Tetrahedron*, 1965, **21**, 3019.	117
65TL1421	G. V. Boyd; *Tetrahedron Lett.*, 1965, 1421.	1048
65TL3361	A. Schoenberg and B. Koenig; *Tetrahedron Lett.*, 1965, 3361.	890
65USP3222360	J. Fried; *U.S. Pat.* 3 222 360 (1965) (*Chem. Abstr.*, 1966, **64**, 8197).	930, 944
65ZC23	A. Mehlhorn, J. Fabian and R. Mayer; *Z. Chem.*, 1965, **5**, 23.	815

66ACS57	J. Sandström and I. Wennerbeck; *Acta Chem. Scand.*, 1966, **20**, 57.	433, 552
66ACS754	A. Hordvik; *Acta Chem. Scand.*, 1966, **20**, 754.	902, 903
66ACS1907	A. Hordvik and J. Sletten; *Acta Chem. Scand.*, 1966, **20**, 1907.	902
66ACS2107	K. A. Jensen, A. Holm, C. Wentrup and J. Møller; *Acta Chem. Scand.*, 1966, **20**, 2107.	595
66AF1034	A. E. Wilder Smith; *Arzneim.-Forsch.*, 1966, **16**, 1034.	445
66AG596	A. Albert and K. Tratt; *Angew. Chem.*, 1966, **78**, 596.	735, 746
66AG(E)875	W. Seeliger, E. Aufderhaar, W. Diepers, R. Feinauer, R. Nehring, W. Thier and H. Hellmann; *Angew. Chem., Int. Ed. Engl.*, 1966, **5**, 875.	665
66AG(E)970	K. Dickoré and R. Wegler; *Angew. Chem., Int. Ed. Engl.*, 1966, **5**, 970.	571, 941
66AG(E)1042	G. Kresze and C. Seyfried; *Angew. Chem., Int. Ed. Engl.*, 1966, **5**, 1042.	541
66AHC(7)39	H. Prinzbach and E. Futterer; *Adv. Heterocycl. Chem.*, 1966, **7**, 39.	784, 785, 786, 790, 792, 793, 795, 797, 801, 803, 807, 814, 816, 818, 821, 823, 831, 832, 835, 836, 838, 842, 843
66AHC(7)183	A. Hetzheim and K. Möckel; *Adv. Heterocycl. Chem.*, 1966, **7**, 183.	428, 429, 433, 434, 435, 436, 438, 445
66AJC503	R. F. C. Brown, I. D. Rae, J. S. Shannon, S. Sternhell and J. M. Swan; *Aust. J. Chem.*, 1966, **19**, 503.	795
66AP817	W. Schneider and E. Kaemmerer; *Arch. Pharm. (Weinheim, Ger.)*, 1966, **299**, 817.	663
66BCJ1125	M. Ohno and N. Naruse; *Bull. Chem. Soc. Jpn.*, 1966, **39**, 1125.	125
66BCJ1248	H. Kato, K. Tanaka and M. Ohta; *Bull. Chem. Soc. Jpn.*, 1966, **39**, 1248.	693
66BCJ1785	M. Kobayashi, A. Yabe and R. Kiritani; *Bull. Chem. Soc. Jpn.*, 1966, **39**, 1785.	881
66BSB243	C. Draguet and M. Renson; *Bull. Soc. Chim. Belg.*, 1966, **75**, 243.	346
66BSF1183	J.-L. Derocque and J. Vialle; *Bull. Soc. Chim. Fr.*, 1966, 1183.	910, 923, 942
66BSF2395	H. J. M. Dou and J. Metzger; *Bull. Soc. Chim. Fr.*, 1966, 2395 (*Chem. Abstr.*, 1966, **65**, 18 467).	255
66BSF2857	R. Arnaud, M. Gelus, J. C. Malet and J. B. Monnier; *Bull. Soc. Chim. Fr.*, 1966, 2857 (*Chem. Abstr.*, 1967, **66**, 37 815).	246
66BSF3537	E. J. Vincent, R. Phan-Tan-Luu and J. Metzger; *Bull. Soc. Chim. Fr.*, 1966, 3537 (*Chem. Abstr.*, 1967, **66**, 104 623).	246
66CB94	K. Gewald, E. Schinke and H. Böttcher; *Chem. Ber.*, 1966, **99**, 94.	995
66CB3233	M. Schulz, A. Rieche and D. Becker; *Chem. Ber.*, 1966, **99**, 3233.	876
66CC491	J. I. G. Cadogan, R. K. Mackie and M. J. Todd; *Chem. Commun.*, 1966, 491.	120
66CC577	M. S. Raasch; *Chem. Commun.*, 1966, 577.	880
66CC607	R. Grigg; *Chem. Commun.*, 1966, 607.	982
66CC684	D. Elad and I. Rosenthal; *Chem. Commun.*, 1966, 684.	764
66CC689	D. W. Kurtz and H. Schechter; *Chem. Commun.*, 1966, 689.	97
66CI(L)1962	J. Fabian and R. Mayer; *Chem. Ind. (London)*, 1966, 1962.	786
66CJC2465	C. E. Hall and A. Taurins; *Can. J. Chem.*, 1966, **44**, 2465.	696
66CPB89	K. Sirakawa, O. Aki, S. Tsushima and K. Konishi; *Chem. Pharm. Bull.*, 1966, **14**, 89.	58, 87
66CPB1277	I. Iwai and N. Nakamura; *Chem. Pharm. Bull.*, 1966, **14**, 1277.	64, 86, 87, 88
66CR(C)(262)1017	C. Broquet and A. Tchoukarine; *C.R. Hebd. Seances Acad. Sci., Ser. C*, 1966, **262**, 1017 (*Chem. Abstr.*, 1966, **64**, 19 590).	237
66CR(C)(263)1333	R. Meyer and J. Metzger; *C.R. Hebd. Seances Acad. Sci., Ser. C*, 1966, **263**, 1333 (*Chem. Abstr.*, 1967, **66**, 54 922).	245
66DIS(B)102	D. R. Eckroth; Ph.D. Thesis, Princeton Univ., 1966 (*Diss. Abstr. Int. B*, 1966, **27**, 102).	26, 35, 120, 124
66EGP52668	H. Gehlen and M. Just; *Ger. (East) Pat.* 52 668 (1966) (*Chem. Abstr.*, 1968, **68**, 68 996).	442
66GI1046	M. C. Aversa, G. Cum and M. Crisafull; *Gazz. Chim. Ital.*, 1966, **96**, 1046.	6
66GEP1223397	H. Distler; *Ger. Pat.* 1 223 397 (1966) (*Chem. Abstr.*, 1966, **65**, 20 008).	883, 894
66HC(21-1)1	D. S. Breslow and H. Skolnik; *Chem. Heterocycl. Compd.*, 1966, **21-1**, 1.	853, 867, 868, 873, 874, 875, 877, 881, 886, 887, 888, 889, 890, 891, 892
66HC(21-1)76	D. S. Breslow and H. Skolnik; *Chem. Heterocycl. Compd.*, 1966, **21-1**, 76.	759, 768
66HC(21-1)447	D. S. Breslow and H. Skolnik; *Chem. Heterocycl. Compd.*, 1966, **21-1**, 447.	814, 824
66HCA2426	H. Dahn and H. Moll; *Helv. Chim. Acta*, 1966, **49**, 2426.	626
66JA1992	T. W. G. Solomons and C. F. Voigt; *J. Am. Chem. Soc.*, 1966, **88**, 1992.	1031, 1046
66JA4098	P. S. Bailey, J. A. Thompson and B. A. Shoulders; *J. Am. Chem. Soc.*, 1966, **88**, 4098.	862
66JA4263	R. A. Olofson and J. M. Landesberg; *J. Am. Chem. Soc.*, 1966, **88**, 4263.	137
66JA4265	R. A. Olofson, J. M. Landesberg, K. N. Houk and J. S. Michelman; *J. Am. Chem. Soc.*, 1966, **88**, 4265.	137
66JA5045	J. H. Van Den Hende and E. Klingsberg; *J. Am. Chem. Soc.*, 1966, **88**, 5045.	1054
66JA5588	S. Trofimenko; *J. Am. Chem. Soc.*, 1966, **88**, 5588.	1038, 1046
66JAP6616384	S. Noguchi and K. Morita; *Jpn. Pat.* 66 16 384 (1966) (*Chem. Abstr.*, 1966, **66**, 46 525).	96
66JCP(44)1352	K. L. Dorris, C. O. Britt and J. E. Boggs; *J. Chem. Phys.*, 1966, **44**, 1352.	752
66JCS(B)127	R. J. Abraham and W. A. Thomas; *J. Chem. Soc. (B)*, 1966, 127.	246
66JCS(B)339	G. M. Clarke, R. Grigg and D. H. Williams; *J. Chem. Soc. (B)*, 1966, 339.	244
66JCS(C)415	E. D. Brown, S. M. Iqbal and L. N. Owen; *J. Chem. Soc. (C)*, 1966, 415.	765
66JCS(C)1361	M. N. G. James and K. J. Watson; *J. Chem. Soc. (C)*, 1966, 1361.	327

66JHC27	C. K. Bradsher and D. F. Lohr, Jr.; *J. Heterocycl. Chem.*, 1966, **3**, 27.	679, 692
66JHC119	H. H. Takimoto, G. C. Denault and S. Hotta; *J. Heterocycl. Chem.*, 1966, **3**, 119.	977, 1009
66JHC155	J. K. Stille, F. W. Harris and M. A. Bedford; *J. Heterocycl. Chem.*, 1966, **3**, 155.	376
66JHC418	T. A. Crabb and R. F. Newton; *J. Heterocycl. Chem.*, 1966, **3**, 418.	666
66JHC518	L. D. Huestis, I. Emery and E. Steffensen; *J. Heterocycl. Chem.*, 1966, **3**, 518.	927
66JIC585	H. Singh, A. Kaur and K. S. Narang; *J. Indian Chem. Soc.*, 1966, **43**, 585.	709
66JMC478	H. L. Yale and K. Losee; *J. Med. Chem.*, 1966, **9**, 478.	445
66JMC545	A. H. M. Raeymaekers, F. T. N. Allewijn, J. Vandenberk, P. J. A. Demoen, T. T. T. Van Offenwert and P. A. J. Janssen; *J. Med. Chem.*, 1966, **9**, 545.	984, 993, 1013
66JOC59	P. E. Fanta and E. N. Walsh; *J. Org. Chem.*, 1966, **31**, 59.	1017
66JOC2039	R. B. Woodward and D. J. Woodman; *J. Org. Chem.*, 1966, **31**, 2039.	21, 74
66JOC2087	F. D. Greene, W. Adam and G. A. Knudsen, Jr.; *J. Org. Chem.*, 1966, **31**, 2087.	780
66JOC2417	W. O. Foye and J. M. Kauffman; *J. Org. Chem.*, 1966, **31**, 2417.	910, 923, 934
66JOC3193	N. J. Doorenbos and L. Milewich; *J. Org. Chem.*, 1966, **31**, 3193.	62, 83
66JOC3442	C. Ainsworth and R. E. Hackler; *J. Org. Chem.*, 1966, **31**, 3442.	428
66JOC3489	E. Klingsberg; *J. Org. Chem.*, 1966, **31**, 3489.	790
66JOC3528	K. T. Potts and R. M. Huseby; *J. Org. Chem.*, 1966, **31**, 3528.	564, 978, 991
66JOC4235	C. Grundmann and H.-D. Frommeld; *J. Org. Chem.*, 1966, **31**, 4235.	67
66JPR(31)214	K. Gewald; *J. Prakt. Chem.*, 1966, **31**, 214.	800
66JPR(304)26	K. Gewald; *J. Prakt. Chem.*, 1966, **304**, 26.	686, 697
66JPR(306)116	J. Morgenstern and R. Mayer; *J. Prakt. Chem.*, 1966, **306**, 116.	890
66JPS807	L. B. Kier and E. B. Roche; *J. Pharm. Sci.*, 1966, **55**, 807.	368
66JSP(19)283	B. Bak, L. Nygaard, E. J. Pedersen and J. Rastrup-Andersen; *J. Mol. Spectrosc.*, 1966, **19**, 283.	516, 556, 581
66LA(698)149	M. Wilk, H. Schwab and J. Rochlitz; *Liebigs Ann. Chem.*, 1966, **698**, 149.	36
66MI41600	M. Arbasino and P. V. Finzi; *Ric. Sci.*, 1966, **36**, 1339 (*Chem. Abstr.*, 1967, **67**, 100 041).	89
66MI41601	F. Monforte and G. LoVecchio; *Atti Accad. Peloritana Pericolanti Cl. Sci. Fis. Mat. Nat.*, 1966, **49**, 169 (*Chem. Abstr.*, 1968, **69**, 43 835).	89
66MI41602	G. Cum, G. LoVecchio and M. C. Aversa; *Atti Accad. Peloritana Pericolanti Cl. Sci. Fis. Mat. Nat.*, 1966, **49**, 151 (*Chem. Abstr.*, 1968, **69**, 43 836).	40
66MI42500	J. Mostecky; *Sb. Vys. Sk. Chem. Technol. Praze, Technol. Paliv.*, 1966, **9**, 31 (*Chem. Abstr.*, 1967, **66**, 94 961).	511
66MI42600	N. K. Ray and P. T. Narasimhan; *Theor. Chim. Acta*, 1966, **5**, 401.	516
B-66MI43100	N. Lozac'h and J. Vialle; in 'Organic Sulfur Compounds', ed. N. Kharasch and C. Y. Meyers; Pergamon, Oxford, 1966, vol. 2, p. 257.	785, 797, 802, 803, 805, 809, 810, 811
66MI43400	N. Arsenescu and O. Maior; *Rev. Chim.*, 1966, **17**, 172 (*Chem. Abstr.*, 1966, **65**, 8895).	936
66NEP6500213	Hercules Powder Co., *Neth. Pat.* 65 00 213 (1966) (*Chem. Abstr.*, 1966, **65**, 15 389).	937, 944
66PNA(55)1385	V. Boekelheide and N. A. Fedoruk; *Proc. Natl. Acad. Sci. USA*, 1966, **55**, 1385.	1038
66RCR388	L. A. Kazitsyna, B. S. Kikot and A. V. Upadysheva; *Russ. Chem. Rev. (Engl. Transl.)*, 1966, **35**, 388.	366, 718, 728
66SA1417	B. Šoptrajanov and G. E. Ewing; *Spectrochim. Acta*, 1966, **22**, 1417.	521
66T2119	R. A. Olofson, J. M. Landesberg, R. O. Berry, D. Leaver, W. A. H. Robertson and D. M. McKinnon; *Tetrahedron*, 1966, **22**, 2119.	144, 793
66T3001	A. J. Kirby; *Tetrahedron*, 1966, **22**, 3001.	860, 880, 886
66T(S7)49	M. Altaf-ur-Rahman and A. J. Boulton; *Tetrahedron*, 1966, suppl. 7, 49.	26
66T(S7)415	R. B. Woodward and R. A. Olofson; *Tetrahedron*, 1966, suppl. 7, 415.	31
66TL405	R. Huisgen, H. Blaschke and E. Brunn; *Tetrahedron Lett.*, 1966, 405.	607
66TL2887	A. J. Boulton, P. B. Ghosh and A. R. Katritzky; *Tetrahedron Lett.*, 1966, 2887.	419
66TL2911	A. C. Coda, P. Gruenanger and G. Veronesi; *Tetrahedron Lett.*, 1966, 2911.	997
66TL3389	D. A. Tomalia and H. Hart; *Tetrahedron Lett.*, 1966, 3389.	753
66TL4043	C. H. Krauch, J. Kuhls and H.-J. Piek; *Tetrahedron Lett.*, 1966, 4043.	373
66TL4671	T. Kubota, N. Ichikawa, K. Matsuo and S. Shibata; *Tetrahedron Lett.*, 1966, 4671.	1018
66TL5451	A. R. Gagneux and R. Goeschke; *Tetrahedron Lett.*, 1966, 5451.	43
66TL6009	M. F. Saettone and A. Marsili; *Tetrahedron Lett.*, 1966, 6009.	203
66USP3299055	E. Klingsberg; (Am. Cyanamid Co.), *U.S. Pat.* 3 299 055 (1966) (*Chem. Abstr.*, 1967, **66**, 105 899).	811
66USP3852999	W. M. Hutchinson; *U.S. Pat.* 3 852 999 (1966) (*Chem. Abstr.*, 1965, **63**, 10 062).	92
66ZC321	R. Mayer and H. Hartmann; *Z. Chem.*, 1966, **6**, 312.	799
66ZOR1766	I. K. Vagina and V. N. Cristokletov; *Zh. Org. Khim.*, 1966, **2**, 1766 (*Chem. Abstr.*, 1967, **66**, 46 353).	91
66ZOR2225	V. A. Tartakovskii, A. A. Onishchenko, V. A. Smirnyagin and S. S. Novikov; *Zh. Org. Khim.*, 1966, **2**, 2225 (*Chem. Abstr.*, 1967, **66**, 75 940).	59
67ACS1871	A. Senning and P. Kelly; *Acta Chem. Scand.*, 1967, **21**, 1871.	910, 911
67ACS1991	K. A. Jensen and U. Henriksen; *Acta Chem. Scand.*, 1967, **21**, 1991.	948, 954, 968
67AG(E)261	A. Messmer and A. Gelléri; *Angew. Chem., Int. Ed. Engl.*, 1967, **6**, 261.	1015
67AG(E)294	R. Mayer and K. Gewald; *Angew. Chem., Int. Ed. Engl.*, 1967 **6**, 294.	804, 806
67AG(E)361	B. Föhlisch, R. Braun and K. W. Schultze; *Angew. Chem., Int. Ed. Engl.*, 1967, **6**, 361.	570

67AG(E)364	L. M. Weinstock, P. Davis, D. M. Mulvey and J. C. Schaeffer; *Angew. Chem., Int. Ed. Engl.*, 1967, **6**, 364.	518, 539
67AG(E)456	R. Huisgen and M. Christl; *Angew. Chem., Int. Ed. Engl.*, 1967, **6**, 456.	83
67AG(E)629	G. Ege; *Angew. Chem., Int. Ed. Engl.*, 1967, **6**, 629.	983
67AG(E)649	E. Kuhle, B. Anders and G. Zumach; *Angew. Chem., Int. Ed. Engl.*, 1967, **6**, 649.	936, 943
67AG(E)709	E. Winterfeldt and W. Krohn; *Angew. Chem., Int. Ed. Engl.*, 1967, **6**, 709.	99
67AG(E)1084	W. Mack; *Angew. Chem., Int. Ed. Engl.*, 1967, **6**, 1084.	501
67AHC(8)165	M. Schulz and K. Kirschke; *Adv. Heterocycl. Chem.*, 1967, **8**, 165.	853, 873
67AHC(8)277	K.-H. Wünsch and A. J. Boulton; *Adv. Heterocycl. Chem.*, 1967, **8**, 277.	3, 4, 16, 21, 23, 24, 25, 26, 28, 48, 51, 52, 53, 54, 56, 59, 114, 115, 116, 117, 118, 119, 120, 121, 122, 123, 124, 125, 126, 127
67BCJ664	H. Nohira, K. Inoue, H. Hattori, T. Okawa and T. Mukaiyama; *Bull. Chem. Soc. Jpn.*, 1967, **40**, 664.	925, 932
67BCJ1554	Y. Noda; *Bull. Chem. Soc. Jpn.*, 1967, **40**, 1554.	931, 941
67BCJ2608	T. Sasaki and T. Yoshioka; *Bull. Chem. Soc. Jpn.*, 1967, **40**, 2608.	89
67BRP1079348	Farbenfabriken Bayer A.-G., *Br. Pat.* 1 079 348 (1967) (*Chem. Abstr.*, 1968, **68**, 69 000).	939
67BRP1083607	Bayer, *Br. Pat.* 1 083 607 (1967) (*Chem. Abstr.*, 1968, **68**, 95 805).	510
67BSF422	M. Hedayatullah; *Bull. Soc. Chim. Fr.*, 1967, 422.	595
67BSF571	P. A. Laurent; *Bull. Soc. Chim. Fr.*, 1967, 571.	229
67BSF846	G. Vernin and J. Metzger; *Bull. Soc. Chim. Fr.*, 1967, 846 (*Chem. Abstr.*, 1967, **67**, 7782).	246
67BSF1948	P. Bastianelli, M. Chanon and J. Metzger; *Bull. Soc. Chim. Fr.*, 1967, 1948 (*Chem. Abstr.*, 1967, **67**, 73 544).	299
67BSF2040	H. Najer, R. Giudicelli and J. Menin; *Bull. Soc. Chim. Fr.*, 1967, 2040.	185
67BSF2239	D. Noël and J. Vialle; *Bull. Soc. Chim. Fr.*, 1967, 2239.	923, 943
67BSF2865	G. Lang and J. Vialle; *Bull. Soc. Chim. Fr.*, 1967, 2865.	920
67BSF3283	R. Phan-Tan-Luu, L. Bouscasse, E. J. Vincent and J. Metzger; *Bull. Soc. Chim., Fr.*, 1967, 3283 (*Chem. Abstr.*, 1968, **68**, 44 384).	239, 240
67BSF4134	J. Crousier and J. Metzger; *Bull. Soc. Chim. Fr.*, 1967, 4134 (*Chem. Abstr.*, 1968, **69**, 10 391).	275
67BSF4465	R. Meyer and J. Metzger; *Bull. Soc. Chim. Fr.*, 1967, 4465 (*Chem. Abstr.*, 1968, **68**, 108 142).	246
67BSF4583	R. Meyer and J. Metzger; *Bull. Soc. Chim. Fr.*, 1967, 4583 (*Chem. Abstr.*, 1968, **68**, 90 410).	246
67CB1802	R. Sustmann, R. Huisgen and H. Huber; *Chem. Ber.*, 1967, **100**, 1802.	6
67CB2159	H. Beecken; *Chem. Ber.*, 1967, **100**, 2159.	732
67CB2170	H. Beecken; *Chem. Ber.*, 1967, **100**, 2170.	537
67CB2435	H. Boshagen, H. Feltkamp and W. Geiger; *Chem. Ber.*, 1967, **100**, 2435.	807
67CB2585	A. Dornow and H. Pietsch; *Chem. Ber.*, 1967, **100**, 2585.	726
67CB2593	E. Schmitz and S. Schramm; *Chem. Ber.*, 1967, **100**, 2593.	935
67CB4027	H. Behringer and D. Bender; *Chem. Ber.*, 1967, **100**, 4027	923, 1066, 1067
67CC353	H. Newman and R. B. Angier; *Chem. Commun.*, 1967, 353.	792
67CC1255	R. M. Moriarty, J. M. Kliegman and R. B. Desai; *Chem. Commun.*, 1967, 1255.	737
67CC1272	A. W. Murray and K. Vaughan; *Chem. Commun.*, 1967, 1272.	658
67CHE662	V. G. Pesin and V. A. Sergeev; *Chem. Heterocycl. Compd.* (*Engl. Transl.*), 1967, **3**, 662.	532
67CHE706	I. A. Mazur and P. M. Kochergin; *Chem. Heterocycl. Compd.* (*Engl. Transl.*), 1967, **3**, 706.	981
67CJC1225	J. W. MacDonald and D. M. McKinnon; *Can. J. Chem.*, 1967, **45**, 1225.	910, 913, 920
67CPB366	S. Minami and J. Matsumoto; *Chem. Pharm. Bull.*, 1967, **15**, 366.	91
67CPB988	K. Morita and S. Kobayashi, *Chem. Pharm. Bull.*, 1967, **15**, 988.	864, 876, 886
67CPB1025	Y. Kishida, T. Hiraoka, J. Ide, A. Terada and N. Nakamura; *Chem. Pharm. Bull.*, 1967, **15**, 1025.	87
67CRV197	M. E. Dyen and D. Swern; *Chem. Rev.*, 1967, **67**, 197.	213, 229
67DIS(B)(27)4330	I. W. Stapleton; *Diss. Abstr. Int. B*, 1967, **27**, 4330.	518, 520, 523, 525
67DOK(172)1075	B. A. Arbuzov, A. O. Vizel, A. P. Petrov and Y. Y. Samitov; *Dokl. Akad. Nauk SSSR*, 1967, **172**, 1075.	997
67DOK(173)1321	B. A. Arbuzov, E. N. Dianova, V. S. Vinogradova and Yu. Yu. Samitov; *Dokl. Akad. Nauk SSSR*, 1967, **173**, 1321 (*Chem. Abstr.*, 1968, **68**, 12 911).	40
67FRP1504150	Shell Internationale Research Maatschappij N.V., *Fr. Pat.* 1 504 150 (1967) (*Chem. Abstr.*, 1969, **70**, 46 426).	810
67G173	V. Bertini, A. DeMunno and P. Pino; *Gazz. Chim. Ital.*, 1967, **97**, 173.	29
67G185	V. Bertini, A. DeMunno and P. Pino; *Gazz. Chim. Ital.*, 1967, **97**, 185.	29
67G379	E. Benedetti, G. Sbrana and V. Bertini; *Gazz. Chim. Ital.*, 1967, **97**, 379.	521
67G1614	V. Bertini and A. De Munno; *Gazz. Chim. Ital.*, 1967, **97**, 1614.	540, 541
67G1870	V. Bertini; *Gazz. Chim. Ital.*, 1967, **97**, 1870.	524, 539
67GEP1245386	R. Pfleger, E. Garthe and K. Rauer; (Chemische Fabrik GmbH), *Ger. Pat.* 1 245 386 (1967) (*Chem. Abstr.* 1968, **68**, 69 005).	1039, 1046
67IJC216	A. K. Bhattacharya; *Indian J. Chem.*, 1967, **5**, 216.	922
67JA2077	F. W. Fowler, A. Hassner and L. A. Levy; *J. Am. Chem. Soc.*, 1967, **89**, 2077.	75, 84

67JA2429	R. W. Murray, R. D. Youssefyeh and P. R. Story; *J. Am. Chem. Soc.*, 1967, **89**, 2429.	862
67JA2618	R. A. Carboni, J. C. Kauer, J. E. Castle and H. E. Simmons; *J. Am. Chem. Soc.*, 1967, **89**, 2618.	1039, 1046, 1047
67JA2626	R. A. Carboni, J. C. Kauer, W. R. Hatchard and R. J. Harder; *J. Am. Chem. Soc.*, 1967, **89**, 2626.	1039
67JA2633	J. C. Kauer and R. A. Carboni; *J. Am. Chem. Soc.*, 1967, **89**, 2633.	1040, 1046, 1047
67JA2743	D. S. Kemp and S. W. Chien; *J. Am. Chem. Soc.*, 1967, **89**, 2743.	32
67JA4368	D. J. Pasto, F. M. Klein and T. W. Doyle; *J. Am. Chem. Soc.*, 1967, **89**, 4368.	754
67JA5977	W. E. Thiessen and H. Hope; *J. Am. Chem. Soc.*, 1967, **89**, 5977.	367, 369
67JAP6708028	Toyrana Chem., *Jpn. Pat.* 67 08 028 (1967) (*Chem. Abstr.*, 1967, **67**, 54 137).	510
67JCS(A)771	N. M. Atherton, J. N. Ockwell and R. Dietz; *J. Chem. Soc. (A)*, 1967, 771.	522
67JCS(B)1117	M. G. Newton, M. C. McDaniel, J. E. Baldwin and I. C. Paul; *J. Chem. Soc. (B)*, 1967, 1117.	669
67JCS(C)314	P. A. Bristow, R. G. Jones and J. G. Tillett; *J. Chem. Soc. (C)*, 1967, 314.	872
67JCS(C)515	G. Jones and D. G. Jones; *J. Chem. Soc. (C)*, 1967, 515.	673
67JCS(C)2005	A. J. Boulton, A. R. Katritzky and A. M. Hamid; *J. Chem. Soc. (C)*, 1967, 2005.	60
67JCS(C)2364	M. S. El Shanta, R. M. Scrowston and M. V. Twigg; *J. Chem. Soc. (C)*, 1967, 2364.	141
67JHC54	G. Adembri and R. Nesi; *J. Heterocycl. Chem.*, 1967, **4**, 54.	88
67JHC66	C. K. Bradsher and M. F. Zinn; *J. Heterocycl. Chem.*, 1967, **4**, 66.	652, 657
67JHC71	C. K. Bradsher and D. F. Lohr, Jr.; *J. Heterocycl. Chem.*, 1967, **4**, 71.	679, 689
67JHC139	E. Barni, G. Di Modica and A. Gasco; *J. Heterocycl. Chem.*, 1967, **4**, 139.	340
67JHC445	D. M. Mulvey and L. M. Weinstock; *J. Heterocycl. Chem.*, 1967, **4**, 445.	536
67JHC527	G. Ottmann, G. D. Vickers and H. Hooks, Jr.; *J. Heterocycl. Chem.*, 1967, **4**, 527.	779
67JHC533	F. DeSarlo and G. Dini; *J. Heterocycl. Chem.*, 1967, **4**, 533.	42
67JHC577	H. F. Andrew and C. K. Bradsher; *J. Heterocycl. Chem.*, 1967, **4**, 577.	695
67JOC1387	G. W. Moersch, E. L. Wittle and W. A. Neuklis; *J. Org. Chem.*, 1967, **32**, 1387.	39
67JOC1566	R. J. Timmons and L. S. Wittenbrook; *J. Org. Chem.*, 1967, **32**, 1566.	498
67JOC1899	E. C. Taylor, D. R. Eckroth and J. Bartulin; *J. Org. Chem.*, 1967, **32**, 1899.	35
67JOC2823	L. M. Weinstock, P. Davis, B. Handelsman and R. Tull; *J. Org. Chem.*, 1967, **32**, 2823.	537, 538
67JOC3318	B. T. Gillis and M. P. LaMontagne; *J. Org. Chem.*, 1967, **32**, 3318.	436, 443
67JOC3369	F. L. Greenwood and H. Rubinstein; *J. Org. Chem.*, 1967, **32**, 3369.	869
67JOC3931	W. H. H. Günther; *J. Org. Chem.*, 1967, **32**, 3931.	965
67JOM(9)19	A. Cogoli and P. Gruenanger; *J. Organomet. Chem.*, 1967, **9**, 19.	106
67JPR(308)287	J. Fabian and E. Fanghanel; *J. Prakt. Chem.*, 1967, **308**, 287.	908
67JPR(309)97	K. Gewald, P. Blauschmit and R. Mayer; *J. Prakt. Chem.*, 1967, **309**, 97 (*Chem. Abstr.*, 1967, **66**, 85 721).	304
67JPS149	L. B. Kier and E. B. Roche; *J. Pharm. Sci.*, 1967, **56**, 149.	371, 378
67JPS608	E. C. Foernzler and A. N. Martin; *J. Pharm. Sci.*, 1967, **56**, 608.	547
67KGS297	E. I. Fedin, Z. V. Todres and L. S. Efros; *Khim. Geterotsikl. Soedin.*, 1967, 297.	518
67KGS633	M. G. Voronkov, T. V. Lapina and E. P. Popova; *Khim. Geterotsikl. Soedin.*, 1967, 633.	805
67KGS758	M. G. Voronkov, V. I. Minkin, O. A. Osipov, M. G. Kogan and T. Lapina; *Khim. Geterotsikl. Soedin.*, 1967, 758 (*Chem. Abstr.*, 1968, **68**, 86 745).	784
67KGS811	S. P. Solodovnikov and Z. V. Todres; *Khim. Geterotsikl. Soedin.*, 1967, 811.	522
67MI41600	G. S. D'Alcontres, G. Cum and M. Gattuso; *Ric. Sci.*, 1967, **37**, 750 (*Chem. Abstr.*, 1968, **68**, 95 733).	108
67MI41601	A. Dondoni and F. Taddu; *Bull. Sci. Fac. Chim. Ind. Bologna*, 1967, **25**, 145.	6
67MI41602	G. Gaudiano, C. Ticozzi, A. Umani-Ronchi and A. Selva; *Chim. Ind. (Milan)*, 1967, **49**, 1343 (*Chem. Abstr.*, 1968, **69**, 2941).	94
67MI41700	J. H. Griffiths, A. Wardley, V. E. Williams, N. L. Owen and J. Sheridan; *Nature (London)*, 1967, **216**, 1301.	136
B-67MI41800	R. Huisgen; in 'Aromaticity', Chemical Society Special Publication, Chemical Society, London, 1967, no. 21, p. 51.	178, 206
67MI41900	M. Gelus, P. M. Vay and G. Berthier; *Theor. Chim. Acta*, 1967, **9**, 182 (*Chem. Abstr.*, 1968, **68**, 33 422).	237
67MI42200	D. Klamann and W. Siemens; *Chem.-Ing.-Tech.*, 1967, **39**, 511.	425
67MI42300	R. Milcent; *Ann. Chim. (Paris)*, 1967, **2**, 169 (*Chem. Abstr.*, 1968, **68**, 68 935).	438
67MI42600	W. D. Moseley, Jr., J. Ladik and O. Martensson; *Theor. Chim. Acta*, 1967, **8**, 18.	516
67MI42700	L. Giammanco; *Atti Accad. Sci., Lett. Arti. Palermo Part I*, 1965–66, **20**, 313 (publ. 1967) (*Chem. Abstr.*, 1968, **68**, 114 509).	575
67MI42800	W. Wojciak, R. Solecki and Z. Kurzawa; *Chem. Anal. (Warsaw)*, 1967, **12**, 849.	598
B-67MI43000	H. Budzikiewicz, C. Djerassi and D. H. Williams; 'Mass Spectrometry of Organic Compounds', Holden-Day, San Francisco, 1967, pp. 257, 265.	756
B-67MI43100	N. Lozac'h; in 'Organosulfur Chemistry', ed. M. J. Janssen; Interscience, New York, 1967, p. 179.	804
B-67MI43101	E. Klingsberg; in 'Organosulfur Chemistry', ed. M. J. Janssen; Interscience, New York, 1967, p. 171.	795
67NEP6606579	Roussel-UCLAF, *Neth. Appl.* 6 606 579 (1967) (*Chem. Abstr.*, 1967, **66**, 115 858).	36
67NEP6610627	Olin Mathieson, *Neth. Pat.* 66 10 627 (1967) (*Chem. Abstr.*, 1968, **68**, 29 699).	497, 507
67RTC670	H. E. Wijers, L. Brandsma and J. F. Arens; *Recl. Trav. Chim. Pays-Bas*, 1967, **86**, 670.	593

67T675	G. Desimoni, P. Grünanger and P. Vita Finzi; *Tetrahedron*, 1967, **23**, 675.	620, 622, 623, 626
67T681	G. Desimoni and P. Vita Finzi; *Tetrahedron*, 1967, **23**, 681.	620, 623
67T687	G. Desimoni, P. Grünanger and P. Vita Finzi; *Tetrahedron*, 1967, **23**, 687.	627
67T831	F. DeSarlo; *Tetrahedron*, 1967, **23**, 831.	104
67T1379	E. Fahr, K. Döppert and K. Königsdorfer; *Tetrahedron*, 1967, **23**, 1379.	443
67T2001	D. S. Kemp; *Tetrahedron*, 1967, **23**, 2001.	32
67T4697	G. Adembri, P. Sarti-Fantoni, F. DeSio and P. F. Franchini; *Tetrahedron*, **23**, 4697.	84
67TL271	D. C. DeJongh, R. Y. Van Fossen and C. F. Bourgeois; *Tetrahedron Lett.*, 1967, 271.	868
67TL331	S. Morrocchi, R. Ricca and L. Velo; *Tetrahedron Lett.*, 1967, 331.	908
67TL1013	H. Behringer and D. Diechmann; *Tetrahedron Lett.*, 1967, 1013.	907
67TL2303	H. Nozaki, M. Takaku and Y. Hayashi; *Tetrahedron Lett.*, 1967, 2303.	773
67TL3501	R. W. Hoffmann and H. J. Luthardt; *Tetrahedron Lett.*, 1967, 3501.	442
67TL3653	H. Ahlbrecht and F. Kroehnke; *Tetrahedron Lett.*, 1967, 3653.	663
67TL4313	W. Schäfer and H. Schlude; *Tetrahedron Lett.*, 1967, 4313.	57
67TL5311	W. H. Schmidt and A. Tulinsky; *Tetrahedron Lett.*, 1967, 5311.	815, 816
67UKZ596	V. N. Fedoseeva and V. E. Petrun'kin; *Ukr. Khim. Zh. (Russ. Ed.)*, 1967, **33**, 596.	889
67USP3322749	L. B. Crast, Jr., *U.S. Pat.* 3 322 749 (1967) (*Chem. Abstr.*, 1968, **69**, 10 452).	542
67USP3324141	Olin Mathieson, *U.S. Pat.* 3 324 141 (1967) (*Chem. Abstr.*, 1967, **67**, 73 611).	509
67USP3357993	E. T. Quentin; *U.S. Pat.* 3 357 993 (1967) (*Chem. Abstr.*, 1968, **68**, 39 097).	894
67ZC275	J. Faust, H. Spies and R. Mayer; *Z. Chem.*, 1967, **7**, 275.	800
67ZC341	A. Sitte, H. Paul and G. Hilgetag; *Z. Chem.*, 1967, **7**, 341.	984
67ZOR821	G. N. Bondorev, U. S. Ryzhov, V. N. Cristokletov and A. A. Petrov; *Zh. Org. Chem.*, 1967, **3**, 821.	89, 90
67ZOR942	S. A. Zotova and V. G. Yashunskii; *Zh. Org. Khim.*, 1967, **3**, 942.	371
67ZOR980	V. A. Tartakovskii, O. A. Luk'yanov, N. I. Shlykova and S. S. Novikov; *Zh. Org. Khim.*, 1967, **3**, 980 (*Chem. Abstr.*, 1967, **67**, 100 039).	95
67ZOR1532	S. D. Sokolov; *Zh. Org. Khim.* 1967, **3**, 1532.	39
68AC(R)121	V. Sprio and R. Pirisi; *Ann. Chim. (Rome)*, 1968, **58**, 121.	624
68AC(R)189	M. Grifantini, F. Gaultieri and M. L. Stein; *Ann. Chim. (Rome)*, 1968, **58**, 189 (*Chem. Abstr.*, 1968, **69**, 27 352).	40, 42
68ACS2051	P. Markov and P. N. Skancke; *Acta Chem. Scand.*, 1968, **22**, 2051.	516
68ACS2719	P. L. Kumler, C. L. Pedersen and O. Buchardt; *Acta Chem. Scand.*, 1968, **22**, 2719.	98
68AG(E)296	R. Gompper and R. Weiss; *Angew. Chem., Int. Ed. Engl.*, 1968, **7**, 296.	923, 942
68AG(E)811	L. Karle; *Angew. Chem., Int. Ed. Engl.*, 1968, **7**, 811.	339
68AHC(9)107	L. M. Weinstock and P. J. Pollack; *Adv. Heterocycl. Chem.*, 1968, **9**, 107. 517, 519, 520, 521, 523, 525, 526, 528, 529, 533, 535, 536, 537, 538, 540, 542	514, 515,
68AHC(9)165	J. Sandstrom; *Adv. Heterocycl. Chem.*, 1968, **9**, 165.	546, 548, 549, 552, 557, 561, 562
68AJC1665	J. H. Bowie, R. A. Eade and J. C. Earl; *Aust. J. Chem.*, 1968, 1665.	371
68AJC2013	J. S. Josan and F. W. Eastwood; *Aust. J. Chem.*, 1968, **21**, 2013.	769
68AP611	K. Hartke and L. Peshkar; *Arch. Pharm. (Weinheim, Ger.)*, 1968, **301**, 611.	639
68BCJ1925	H. Takei, H. Shimizu, M. Higo and T. Mukaiyama; *Bull. Chem. Soc. Jpn.*, 1968, **41**, 1925.	926, 928
68BCJ2212	T. Sasaki and T. Yoshioka; *Bull. Chem. Soc. Jpn.*, 1968, **41**, 2212.	68
68BRP1116198	Badische Anilin- und Soda-Fabrik A.-G., *Br. Pat.* 1 116 198 (1968) (*Chem. Abstr.*, 1968, **69**, 86 995).	495, 508
68BRP1136737	G. Zunach, W. Weiss and E. Kühle; *Br. Pat.* 1 136 737 (1968) (*Chem. Abstr.*, 1969, **70**, 77 951).	943
68BSF4636	G. Kille and J. P. Fleury; *Bull. Soc. Chim. Fr.*, 1968, 4636.	186
68C392	R. Criegee; *Chimia*, 1968, **22**, 392.	853, 869
68CB371	W. Wucherpfennig; *Chem. Ber.*, 1968, **101**, 371.	537
68CB536	R. Huisgen, H. Gotthardt and R. Grashey; *Chem. Ber.*, 1968, **101**, 536.	376, 601
68CB552	H. Gotthardt and R. Huisgen; *Chem. Ber.*, 1968, **101**, 552.	376
68CB829	R. Huisgen, R. Grashey and H. Gotthardt; *Chem. Ber.*, 1968, **101**, 829.	376
68CB839	R. Huisgen and H. Gotthardt; *Chem. Ber.*, 1968, **101**, 839.	376
68CB1056	H. Gotthardt, R. Huisgen and R. Knorr; *Chem. Ber.*, 1968, **101**, 1056.	376
68CB1059	R. Huisgen and H. Gotthardt; *Chem. Ber.*, 1968, **101**, 1059.	376
68CB1428	H. Behringer, D. Bender, J. Falkenberg and R. Wiedenmann; *Chem. Ber.*, 1968, **101**, 1428.	847
68CB2472	H. Boshagen and W. Geiger; *Chem. Ber.*, 1968, **101**, 2472.	807
68CB3287	H. Dorn and A. Otto; *Chem. Ber.*, 1968, **101**, 3287.	1001
68CC499	A. R. McCarthy, W. D. Ollis and C. A. Ramsden; *Chem. Commun.*, 1968, 499.	575
68CC672	K. T. Potts and C. Sapino, Jr.; *Chem. Commun.*, 1968, 672.	573
68CC977	T. S. Cantrell and W. S. Haller; *Chem. Commun.*, 1968, 977.	91, 525
68CC1245	H. J. Jackobsen and A. Senning; *Chem. Commun.*, 1968, 1245.	912, 941
68CHE186	V. G. Pesin, V. A. Sergeev and M. G. Nikulina; *Chem. Heterocycl. Compd. (Engl. Transl.)*, 1968, **4**, 186.	533
68CI(L)1639	S. S. Chang, C. Hirai, B. R. Reddy, K. O. Herz, A. Kato and G. Sima; *Chem. Ind. (London)*, 1968, 1639.	894
68CJC691	C. E. Hall and A. Taurins; *Can. J. Chem.*, 1968, **46**, 691.	684, 691

68CJC1057	R. Raap and R. G. Micetich; *Can. J. Chem.*, 1968, **46**, 1057.	457, 460
68CJC2251	R. Raap; *Can. J. Chem.*, 1968, **46**, 2251.	457
68CJC2255	R. Raap; *Can. J. Chem.*, 1968, **46**, 2255.	442
68CJC2577	R. F. C. Brown, L. Radom, S. Sternhell and I. D. Rae; *Can. J. Chem.*, 1968, **46**, 2577.	785
68CPB117	I. Adachi and H. Kano; *Chem. Pharm. Bull.*, 1968, **16**, 117.	633
68CPB544	T. Naito, J. Okumura and K. Kasai; *Chem. Pharm. Bull.*, 1968, **16**, 544.	538
68CPB741	S. Sugiura, E. Suzuki, T. Naito and S. Inoue; *Chem. Pharm. Bull.*, 1968, **16**, 741.	700
68CPB750	E. Suzuki, S. Sugiura, T. Naito and S. Inoue; *Chem. Pharm. Bull.*, 1968, **16**, 750.	686, 687
68CPB1700	T. Okamoto and H. Takahashi; *Chem. Pharm. Bull.*, 1968, **16**, 1700.	629
68CR(C)(266)714	H. Dou, A. Friedmann, G. Vernin and J. Metzger; *C.R. Hebd. Seances Acad. Sci., Ser. C*, 1968, **266**, 714 (*Chem. Abstr.*, 1968, **69**, 43 182).	255
68CR(C)(267)114	R. Meyer, G. Bourrely and J. Metzger; *C.R. Hebd. Seances Acad. Sci., Ser. C*, 1968, **267**, 114 (*Chem. Abstr.*, 1968, **69**, 70 354).	245
68CR(C)(267)180	J. Bignebat and H. Quiniou; *C. R. Hebd. Seances Acad. Sci., Ser. C*, 1968, **267**, 180.	790
68CRV375	F. Johnson; *Chem. Rev.*, 1968, **68**, 375.	1008
68G42	M. E. Aversa, G. Cum and M. Crisafulli; *Gazz. Chim. Ital.*, 1968, **98**, 42.	6
68G48	G. Gaudino and P. P. Ponto; *Gazz. Chim. Ital.*, 1968, **98**, 48.	94
68G74	G. Bianchi, A. Cogoli and G. R. Augusto; *Gazz. Chim. Ital.*, 1968, **98**, 74.	36
68G331	G. Bianchi, A. Galli and R. Gandolfi; *Gazz. Chim. Ital.*, 1968, **98**, 331.	37
68HCA241	E. Cherbuliez, O. Espejo, B. Willhalm and J. Rabinowitz; *Helv. Chim. Acta*, 1968, **51**, 241.	708
68HCA1421	J. Bader; *Helv. Chim. Acta*, 1968, **51**, 1421.	886
68HCA1510	J. Zsindely and H. Schmid; *Helv. Chim. Acta*, 1968, **51**, 1510.	1011
68IJC132	M. G. Paranjpe; *Indian J. Chem.*, 1968, **6**, 132.	914, 936
68IZV2666	S. B. Savvin and Y. G. Rosovskii; *Izv. Akad. Nauk SSSR, Ser. Khim*; 1968, 2666 (*Chem. Abstr.*, 1969, **70**, 68 239).	328
68JA2970	F. P. Boer, J. J. Flynn, E. T. Kaiser, O. R. Zaborsky, D. A. Tomalia, A. E. Young and Y. C. Tong; *J. Am. Chem. Soc.*, 1968, **90**, 2970.	855, 857, 881
68JA3830	V. Boekelheide and N. A. Fedoruk; *J. Am. Chem. Soc.*, 1968, **90**, 3830.	266, 1003
68JA5325	J. E. Baldwin, R. G. Pudussery, A. K. Qureshi and B. Sklarz; *J. Am. Chem. Soc.*, 1968, **90**, 5325.	44
68JCS(B)1516	Altaf-ur-Rahman, A. J. Boulton, D. P. Clifford and G. J. T. Tiddy; *J. Chem. Soc. (B)*, 1968, 1516.	718, 730
68JCS(C)642	D. B. J. Easton, D. Leaver and D. M. McKinnon; *J. Chem. Soc. (C)*, 1968, 642.	784, 791, 799
68JCS(C)1088	T. L. Hough and G. Jones; *J. Chem. Soc. (C)*, 1968, 1088.	730
68JCS(C)1612	A. M. Hamid and S. Trippett; *J. Chem. Soc. (C)*, 1968, 1612.	867, 889
68JCS(C)1711	R. N. Butler and F. L. Scott; *J. Chem. Soc. (C)*, 1968, 1711.	1009, 1014
68JCS(C)1774	K. M. Johnston and R. G. Shotter; *J. Chem. Soc. (C)*, 1968, 1774.	63, 74, 84
68JCS(C)2871	R. G. Cooks and P. Sykes; *J. Chem. Soc. (C)*, 1968, 2871.	316
68JHC49	G. Bianchi, R. Gandolfi and P. Pruenanger; *J. Heterocycl. Chem.*, 1968, **5**, 49.	38
68JHC277	L. B. Kier and M. K. Scott; *J. Heterocycl. Chem.*, 1968, **5**, 277.	573
68JHC289	I. J. Tyminski and K. K. Andersen; *J. Heterocycl. Chem.*, 1968, **5**, 289.	865, 887
68JHC331	L. Bauer and C. S. Mahajanshetti; *J. Heterocycl. Chem.*, 1968, **5**, 331.	690, 698
68JHC881	M. W. Barker and J. H. Gardner; *J. Heterocycl. Chem.*, 1968, **5**, 881.	113
68JMC70	R. Raap and R. G. Micetich; *J. Med. Chem.*, 1968, **11**, 70.	144
68JMC453	S. M. Gadekar, S. Nibi, B. D. Johnson, E. Cohen and J. R. Cummings; *J. Med. Chem.*, 1968, **11**, 453.	1011, 1026
68JMC1164	W. B. Wright, Jr. and H. J. Brabander; *J. Med. Chem.*, 1968, **11**, 1164.	985
68JOC143	K. T. Potts and C. Hirsch; *J. Org. Chem.*, 1968, **33**, 143.	977, 991, 1014
68JOC1350	L. D. Spicer, M. W. Bullock, M. Garber, W. Groth, J. J. Hand, D. W. Long, J. L. Sawyer and R. S. Wayne; *J. Org. Chem.*, 1968, **33**, 1350.	1013
68JOC1629	L. J. Durham and F. L. Greenwood; *J. Org. Chem.*, 1968, **33**, 1629.	862
68JOC2086	C. Temple, Jr., C. L. Kussner and J. A. Montgomery; *J. Org. Chem.*, 1968, **33**, 2086.	717, 729
68JOC2126	J. Jonas, T. P. Forrest, M. qKratochvíl and H. Gross; *J. Org. Chem.*, 1968, **33**, 2126.	761
68JOC2397	D. J. Woodman; *J. Org. Chem.*, 1968, **33**, 2397.	21
68JOC2544	G. Asato; *J. Org. Chem.*, 1968, **33**, 2544.	255
68JOC2880	R. Kwok and P. Pranc; *J. Org. Chem.*, 1968, **33**, 2880.	19
68JOC2915	E. Klingsberg; *J. Org. Chem.*, 1968, **33**, 2915.	809
68JOC3535	A. Cooper and D. A. Norton; *J. Org. Chem.*, 1968, **33**, 3535.	751
68JOC3941	W. J. Houlihan and W. J. Theuer; *J. Org. Chem.*, 1968, **33**, 3941.	1012
68JPC3975	J. Fabian, A. Mehlhorn and R. Zahradnik; *J. Phys. Chem.*, 1968, **72**, 3975.	815
68JPR(309)78	W. Dietz and R. Mayer; *J. Prakt. Chem.*, 1968, **309**, 78.	983
68KGS360	S. P. Solodovnikov and Z. V. Todres; *Khim. Geterotsikl. Soedin.*, 1968, 360.	400
68LA(711)174	G. Tuchtenhagen and K. Ruehlmann; *Leibigs Ann. Chem.*, 1968, **711**, 174.	538, 539
68M2534	P. Margaretha and O. E. Polansky; *Monatsh. Chem.*, 1968, **99**, 2534.	114
68MI41600	G. Aum, M. C. Aversa, N. Vicella, G. Nicola and M. Gattuso; *Atti Soc. Peloritana Sci. Fis. Mat. Nat.*, 1968, **14**, 413.	6
68MI41601	G. Lo Vecchio and M. Gattuso; *Atti Soc. Peloritana Sci. Fis. Mat. Nat.*, 1968, **14**, 439 (*Chem. Abstr.*, 1971, **74**, 3538).	104

68MI41602	A. L. Fridman and F. A. Gabitov; *Otkrytiya Izobret., Prom. Obraztsy, Tovarnye Znaki*, 1968, **45**, 29.	103
68MI41603	B. Unterhalt; *Pharm. Zentralb.*, 1968, **107**, 356.	93
68MI41604	A. Battaglia, D. Arturo and A. Dondoni; *Ric. Sci.*, 1968, **38**, 201 (*Chem. Abstr.*, 1968, **69**, 67 277).	89
68MI41700	H. H. Otto; *Pharm. Zentralb.*, 1968, **107**, 444 (*Chem. Abstr.*, 1968, **69**, 106 601).	149
68MI41701	J. P. Kintzinger and J. M. Lehn; *Mol. Phys.*, 1968, **14**, 133.	136
68MI41900	R. Guglielmetti, M. Mossé, J. C. Metras and J. Metzger; *J. Chim. Phys. Phys. Chim. Biol.*, 1968, **65**, 454 (*Chem. Abstr.*, 1969, **70**, 28 197).	245, 250
68MI41901	P. Goursot and E. F. Westrum, Jr.; *J. Chem. Eng. Data*, 1968, **13**, 468 (*Chem. Abstr.*, 1968, **69**, 110 623).	245
68MI41902	P. Goursot and E. F. Westrum, Jr.; *J. Chem. Eng. Data*, 1968, **13**, 471 (*Chem. Abstr.*, 1968, **69**, 110 624).	245
68MI41903	R. Pohloudek-Fabini and E. Schröpl; *Pharmazie*, 1968, **23**, 561 (*Chem. Abstr.*, 1969, **70**, 47 346).	310
B-68MI42600	A. D. Josey; 'Abstract, 155th Meeting of the American Chemical Society, Division of Organic Chemistry, San Francisco', 1968, p. 14.	541
68MI42700	I. Simiti, L. Proinov, I. Schwartz and L. Gilau; *Rev. Chim.* (*Bucharest*), 1968, **19**, 413 (*Chem. Abstr.*, 1969, **70**, 37 723).	555
68MI42800	K. Sundaram and W. P. Purcell; *Int. J. Quant. Chem.*, 1968, **2**, 145.	599
68MI43100	J. Fabian, K. Fabian and H. Hartmann; *Theor. Chim. Acta*, 1968, **12**, 319.	784
68MI43200	J. Fabian, K. Fabian and H. Hartmann; *Theor. Chim. Acta*, 1968, **12**, 319.	815
68MI43300	H. E. Druett and L. P. Packman; *Nature* (*London*), 1968, **218**, 699.	895
68MI43301	F. Gozzo and G. Camaggi; *Chim. Ind.* (*Milan*), 1968, **50**, 197.	892
68NKK951	H. Wada and G. Nakagawa; *Nippon Kagaku Zasshi*, 1968, **89**, 951 (*Chem. Abstr.*, 1969, **70**, 16 854).	328
68RRC909	E. Popper, F. Eugenia and P. Marcu; *Rev. Roum. Chim.*, 1968, **13**, 909 (*Chem. Abstr.* 1968, **69**, 113 156).	555
68SA(A)361	T. R. Manley and D. A. Williams; *Spectrochim. Acta, Part A*, 1968, **24**, 361.	5, 184, 556
68SA(A)1869	N. M. D. Brown and P. Bladon; *Spectrochim. Acta, Part A*, 1968, **24**, 1869.	524
68SAP6804299	K. Dickoré and F. Hoffmeister; *S. Afr. Pat.* 68 04 299 (1968) (*Chem. Abstr.*, 1969, **71**, 81 372).	945
68T2949	P. Brown and C. Djerassi; *Tetrahedron*, 1968, **24**, 2949.	864
68T3209	A. Alemagna, T. Bacchetti and P. Beltrame; *Tetrahedron*, 1968, **24**, 3209.	564
68T4377	C. Altona and A. P. M. Van Der Veek; *Tetrahedron*, 1968, **24**, 4377.	754
68T4907	G. Desimoni and G. Minoli; *Tetrahedron*, 1968, **24**, 4907.	620, 626
68T5059	A. T. Balaban; *Tetrahedron*, 1968, **24**, 5059.	79
68T6577	N. M. D. Brown and P. Bladon; *Tetrahedron*, 1968, **24**, 6577.	518
68TL165	R. Gompper and H.-U. Wagner; *Tetrahedron Lett.*, 1968, 165.	818
68TL319	J. Sauer and K. K. Mayer; *Tetrahedron Lett.*, 1968, 319.	910, 913, 914, 940
68TL325	J. Sauer and K. K. Mayer; *Tetrahedron Lett.*, 1968, 325.	431
68TL743	P. Caramella; *Tetrahedron Lett.*, 1968, 743.	985, 998
68TL2281	C. J. Michejda; *Tetrahedron Lett.*, 1968, 2281.	437
68TL3291	P. R. Story, W. H. Morrison, III, T. K. Hall, J.-C. Farine and C. E. Bishop; *Tetrahedron Lett.*, 1968, 3291.	869
68TL3375	A. T. Nielsen and T. G. Archibald; *Tetrahedron Lett.*, 1968, 3375.	102
68TL3557	J. V. Paukstelis and R. M. Hammaker; *Tetrahedron Lett.*, 1968, 3557.	187
68TL5209	M. Christl and R. Huisgen; *Tetrahedron Lett.*, 1968, 5209.	89
68TL5397	P. R. Story, T. K. Hall, W. H. Morrison, III and J.-C. Farine; *Tetrahedron Lett.*, 1968, 5397.	866, 869
68TL5759	K. Kotera, Y. Takano, A. Matsuura and K. Kitahonoki; *Tetrahedron Lett.*, 1968, 5759.	36
68TL5789	K. Kotera, Y. Takano, M. Yoshihero, A. Matsuura and K. Kitahontri; *Tetrahedron Lett.*, 1968, 5789.	36
68TL5881	R. Grashey, M. Baumann and W.-D. Lubos; *Tetrahedron Lett.*, 1968, 5881.	573
68USP3365306	M. A. Perret; *U.S. Pat.* 3 365 306 (1968) (*Chem. Abstr.*, 1968, **68**, 86 295).	895
68YZ1437	T. Noguchi; *Yakugaku Zasshi*, 1968, **88**, 1437 (*Chem. Abstr.*, 1969, **70**, 77 873).	510
68ZC170	J. Faust; *Z. Chem.*, 1968, **8**, 170.	43
68ZN(B)413	H. F. Eicke, F. Boberg and J. Knoop; *Z. Naturforsch., Teil B*, 1968, **23**, 413.	788
68ZN(B)1540	M. Schmidt and H. Schulz; *Z. Naturforsch., Teil B*, 1968, **23**, 1540.	803
68ZN(B)1547	E. Kuss; *Z. Naturforsch., Teil B*, 1968, **23**, 1547,	788
68ZOB1248	I. G. Kolokol'tseva, V. N. Chistokletov, B. I. Ionin and A. A. Petrov; *Zh. Obshch. Khim.*, 1968, **38**, 1248 (*Chem. Abstr.*, 1968, **69**, 96 834).	89
68ZOB1820	I. G. Kolokol'tseva, V. N. Chistokletov, M. D. Stadnichuk and A. A. Petrov; *Zh. Obshch. Khim.*, 1968, **38**, 1820 (*Chem. Abstr.*, 1969, **70**, 78 067).	89, 90
68ZOR236	V. A. Tartakovskii, Z. Ya. Lapshina, I. A. Sovost'yanova and S. S. Novikov; *Zh. Org. Khim.*, 1968, **4**, 236 (*Chem. Abstr.*, 1968, **68**, 95 734).	95
68ZOR2259	A. L. Fridman and F. A. Gabitov; *Zh. Org. Khim.*, 1968, **4**, 2259 (*Chem. Abstr.*, 1969, **70**, 68 226).	103
69ACH(62)179	R. Bognar, I. Farkas, L. Szilagyi, M. Menyhart, E. N. Nemes and I. F. Szabo; *Acta Chim. Acad. Sci. Hung.*, 1969, **62**, 179 (*Chem. Abstr.*, 1970, **72**, 90 801).	295

69ACR17	O. H. Griffith and A. S. Waggoner; *Acc. Chem. Res.*, 1969, **2**, 17.	213
69ACS371	K. Undheim and V. Nordal; *Acta Chem. Scand.*, 1969, **23**, 371.	680
69ACS1377	A. Hordvik, E. Sletten and J. Sletten; *Acta Chem. Scand.*, 1969, **23**, 1377.	785, 1054
69ACS1567	K. A. Jensen, A. Holm and J. Wolff-Jensen; *Acta Chem. Scand.*, 1969, **23**, 1567.	595
69ACS1704	K. Undheim, V. Nordal and K. Tjønneland; *Acta Chem. Scand.*, 1969, **23**, 1704.	675
69ACS1852	A. Hordvik, E. Sletten and J. Sletten; *Acta Chem. Scand.*, 1969, **23**, 1852.	1053
69ACS1966	K. Undheim and V. Nordal; *Acta Chem. Scand.*, 1969, **23**, 1966.	675
69ACS2437	K. Undheim and J. Røe; *Acta Chem. Scand.*, 1969, **23**, 2437.	676, 680
69ACS2741	H. M. Seip; *Acta Chem. Scand.*, 1969, **23**, 2741.	854
69ACS2879	G. Kjellin and J. Sandström; *Acta Chem. Scand.*, 1969, **23**, 2879.	247
69ACS2888	G. Kjellin and J. Sandstrom; *Acta Chem. Scand.*, 1969, **23**, 2888.	185
69ACS3139	B. N. Cyvin and S. J. Cyvin; *Acta Chem. Scand.*, 1969, **23**, 3139.	516
69ACS3398	A. Almenningen, P. Kolsaker, H. M. Seip and T. Willadsen; *Acta Chem. Scand.*, 1969, **23**, 3398.	855, 856
69ACS3407	S. J. Cyvin, B. N. Cyvin, G. Hagen and P. Markov; *Acta Chem. Scand.*, 1969, **23**, 3407.	516
69AG543	C. A. Grob; *Angew. Chem.*, 1969, **81**, 543.	585
69AG(E)74	O. Westphal and A. Joos; *Angew. Chem., Int. Ed. Engl.*, 1969, **8**, 74.	693
69AG(E)604	R. Huisgen, H. Mäder and E. Brunn; *Angew. Chem., Int. Ed. Engl.*, 1969, **8**, 604.	1020
69AG(E)781	R. B. Woodward and R. Hoffmann; *Angew. Chem., Int. Ed. Engl.*, 1969, **8**, 781.	825, 1006
69AHC(10)1	A. J. Boulton and P. B. Ghosh; *Adv. Heterocycl. Chem.*, 1969, **10**, 1.	394, 398, 424
69AJC513	W. H. Hall and I. R. Wilson; *Aust. J. Chem.*, 1969, **22**, 513.	911, 937
69AJC2497	A. W. K. Chan and W. D. Crow; *Aust. J. Chem.*, 1969, **22**, 2497.	640
69AX(B)1022	B. Krebs and D. F. Koenig; *Acta Crystallogr., Part B*, 1969, **25**, 1022.	815, 816
69AX(B)2257	Y. Okaya; *Acta Crystallogr., Part B*, 1969, **25**, 2257.	135, 143
69AX(B)2349	L. Cavalca, G. F. Gasparri, A. Mangia and G. Pelizzi; *Acta Crystallogr., Part B*, 1969, **25**, 2349.	135
69BCJ1152	S. Watanabe; *Bull. Chem. Soc. Jpn.*, 1969, **42**, 1152.	150
69BCJ2310	A. Chinone, S. Sato, T. Mase and M. Ohta; *Bull. Chem. Soc. Jpn.*, 1969, **42**, 2310.	186
69BCJ2973	T. Matsuura and I. Saito; *Bull. Chem. Soc. Jpn.*, 1969, **42**, 2973 (*Chem. Abstr.*, 1970, **72**, 39 061).	257, 269
69BRP1147546	Ciba Geigy, *Br. Pat.* 1 147 546 (1969) (*Chem. Abstr.*, 1969, **71**, 71 915).	510
69BSB299	J. Jadot, J. Casimir and R. Warin; *Bull. Soc. Chim. Belg.*, 1969, **78**, 299 (*Chem. Abstr.*, 1969, **71**, 102 205).	327
69BSF332	C. Feugeas and D. Olschwang; *Bull. Soc. Chim. Fr.*, 1969, 332.	843
69BSF870	J. F. Giudicelli, J. Menin and H. Najer; *Bull. Soc. Chim. Fr.*, 1969, 870.	430
69BSF874	J. F. Giudicelli, J. Menin and H. Najer; *Bull. Soc. Chim. Fr.*, 1969, 874.	430
69BSF3970	E. Miler Srenger; *Bull. Soc. Chim. Fr.*, 1969, 3970 (*Chem. Abstr.*, 1970, **72**, 60 181).	238
69CB351	H. Kunzek and G. Barnikow; *Chem. Ber.*, 1969, **102**, 351 (*Chem. Abstr.*, 1969, **70**, 67 777).	323
69CB568	H. Quast and E. Schmitt; *Chem. Ber.*, 1969, **102**, 568 (*Chem. Abstr.*, 1969, **70**, 87 665).	322
69CB736	R. Huisgen, R. Hauck, R. Grashey and H. Seidl; *Chem. Ber.*, 1969, **102**, 736.	631
69CB904	H. Seidl, R. Huisgen and R. Knorr; *Chem. Ber.*, 1969, **102**, 904.	631
69CB1087	W. Kirmse and M. Buschhoff; *Chem. Ber.*, 1969, **102**, 1087.	764
69CB1468	W. Saenger and H. Hettler; *Chem. Ber.*, 1969, **102**, 1468.	135
69CB1961	W. Geiger, H. Boeshagen and H. Medenwald; *Chem. Ber.*, 1969, **102**, 1961.	140, 146
69CB2346	E. Winterfeldt, W. Krohn and H. U. Stracke; *Chem. Ber.*, 1969, **102**, 2346.	99
69CB3775	H. Böshagen and W. Geiger; *Chem. Ber.*, 1969, **102**, 3775.	56
69CC83	E. I. G. Brown, D. Leaver and T. J. Rawlings; *Chem. Commun.*, 1969, 83.	800
69CC137	F. Leung and S. C. Nyburg; *Chem. Commun.*, 1969, 137.	1053
69CC1014	P. L. Johnson and I. C. Paul; *Chem. Commun.*, 1969, 1014.	1053
69CC1062	J. Castells and A. Colombo; *Chem. Commun.*, 1969, 1062.	100, 101
69CC1129	K. T. Potts, E. Houghton and U. P. Singh; *Chem. Commun.*, 1969, 1129.	269
69CHE53	V. G. Pesin and V. A. Sergeev; *Chem. Heterocycl. Compd.* (*Engl. Transl.*), 1969, **5**, 53.	536
69CHE180	V. G. Pesin; *Chem. Heterocycl. Compd.* (*Engl. Transl.*), 1969, **5**, 180.	521
69CHE421	B. K. Strelets and L. S. Efros; *Chem. Heterocycl. Compd.* (*Engl. Transl.*), 1969, **5**, 421.	936, 940
69CHE567	I. Ya. Kvitko; *Chem. Heterocycl. Compd.* (*Engl. Transl.*), 1969, **5**, 567.	988
69CJC919	J. Castonguay, M. Bertrand, J. Carles, S. Fliszár and Y. Rousseau; *Can. J. Chem.*, 1969, **47**, 919.	863
69CJC2039	J. M. Buchshriber, D. M. McKinnon and M. Ahmed; *Can. J. Chem.*, 1969, **47**, 2039.	798, 920
69CJC3557	J. W. Lown, G. Dallas and T. W. Maloney; *Can. J. Chem.*, 1969, **47**, 3557.	316
69CPB1598	S. Takahashi and H. Kano; *Chem. Pharm. Bull.*, 1969, **17**, 1598 (*Chem. Abstr.*, 1969, **71**, 913 627).	254
69CPB1815	M. Kamiya; *Chem. Pharm. Bull.*, 1969, **17**, 1815.	182
69CPB1821	M. Kamiya, S. Katayama and Y. Akahori; *Chem. Pharm. Bull.*, 1969, **17**, 1821.	518
69CPB2201	I. Adachi and H. Kano; *Chem. Pharm. Bull.*, 1969, **17**, 2201.	28, 29, 50
69CR(C)(268)870	J. Roggero and C. Divorne; *C.R. Hebd. Seance. Acad. Sci., Ser. C*, 1969, **268**, 870 (*Chem. Abstr.*, 1969, **70**, 106 423).	271

69CR(C)(269)1560	A. Friedmann; *C.R. Hebd. Seances Acad. Sci., Ser. C*, 1969, **269**, 1560 (*Chem. Abstr.*, 1970, **72**, 78 934).	255
69CRV345	G. L'abbe; *Chem. Rev.*, 1969, **69**, 345.	1000
69FRP1555414	Rohm and Haas Co., *Fr. Pat.* 1 555 414 (1969) (*Chem. Abstr.*, 1970, **72**, 43 651).	147, 148, 160, 161
69FRP1581417	Badische Anilin- und Soda-Fabrik A.-G., *Fr. Pat.* 1 581 417 (1969) (*Chem. Abstr.*, 1970, **73**, 57 150).	510
69G780	L. Cambi, G. Bargigia, L. Colombo and E. Dubini Paglia; *Gazz. Chim. Ital.*, 1969, **99**, 780.	891
69G1107	S. Cabiddu and V. Solinas; *Gazz. Chim. Ital.*, 1969, **99**, 1107.	67, 68, 86
69GEP1814116	K. Morita, N. Hashimoto and K. Matsumura; *Ger. Offen.* 1 814 116 (1969) (*Chem. Abstr.*, 1969, **71**, 124 415).	76, 87
69GEP1915644	T. Vitali, R. Ponci and F. Berteccini; *Ger. Offen.* 1 915 644 (1969) (*Chem. Abstr.*, 1970, **72**, 31 781).	129
69IJC97	N. K. Ray and P. T. Narasimhan; *Indian J. Chem.*, 1969, **7**, 97.	522
69IJC760	R. Madhavan and V. R. Srinivasan; *Indian J. Chem.*, 1969, **7**, 760.	432
69IJC767	H. S. Chaudhary and H. K. Pujari; *Indian J. Chem.*, 1969, **7**, 767.	695
69JA781	P. L. Johnson and I. C. Paul; *J. Am. Chem. Soc.*, 1969, **91**, 781.	1054
69JA5835	T. Nishiguchi, H. Tochio, A. Nabeya and Y. Iwakura; *J. Am. Chem. Soc.*, 1969, **91**, 5835.	309
69JA5841	T. Nishiguchi, H. Tochio, A. Nabeya and Y. Iwakura; *J. Am. Chem. Soc.*, 1969, **91**, 5841.	309
69JA6604	F. P. Boer and J. J. Flynn; *J. Am. Chem. Soc.*, 1969, **91**, 6604.	855
69JA6891	J. D. Bowler and R. H. Schlessinger; *J. Am. Chem. Soc.*, 1969, **91**, 6891.	534, 535, 538, 1032, 1036, 1044
69JCS(B)270	D. J. Brown and P. B. Ghosh; *J. Chem. Soc. (B)*, 1969, 270.	181, 184
69JCS(B)681	A. S. Bailey, C. J. W. Gutch, J. M. Peach and W. A. Waters; *J. Chem. Soc. (B)*, 1969, 681.	400
69JCS(B)1185	A. R. McCarthy, W. D. Ollis, A. N. M. Barnes, L. E. Sutton and C. Ainsworth; *J. Chem. Soc. (B)*, 1969, 1185.	429, 441
69JCS(B)1194	C. W. Atkin, A. N. M. Barnes, P. G. Edgerley and L. E. Sutton; *J. Chem. Soc. (B)*, 1969, 1194.	549
69JCS(C)194	J. S. Davidson; *J. Chem. Soc. (C)*, 1969, 194.	931, 942
69JCS(C)385	T. Kametani, T. Yamanaka and K. Ogasawara; *J. Chem. Soc. (C)*, 1969, 385.	662
69JCS(C)707	D. G. Jones and G. Jones; *J. Chem. Soc. (C)*, 1969, 707.	640, 673, 676, 679
69JCS(C)772	R. S. Atkinson and C. W. Rees; *J. Chem. Soc. (C)*, 1969, 772.	201
69JCS(C)1117	G. C. Barrett and A. R. Khokhar; *J. Chem. Soc. (C)*, 1969, 1117.	310
69JCS(C)2189	M. Davis and A. W. White; *J. Chem. Soc. (C)*, 1969, 2189.	137
69JCS(C)2270	J. Adamson, E. C. Campbell and E. E. Glover; *J. Chem. Soc. (C)*, 1969, 2270.	695
69JCS(C)2663	E. G. E. Hawkins; *J. Chem. Soc. (C)*, 1969, 2663.	928
69JCS(C)2671	E. G. E. Hawkins; *J. Chem. Soc. (C)*, 1969, 2671.	928
69JCS(C)2678	E. G. E. Hawkins; *J. Chem. Soc. (C)*, 1969, 2678.	905, 927, 929, 940
69JHC123	T. H. Kinstle and L. J. Darlage; *J. Heterocycl. Chem.*, 1969, **6**, 123.	117
69JHC199	J. A. White and R. C. Anderson; *J. Heterocycl. Chem.*, 1969, **6**, 199.	137
69JHC279	G. Casini, F. Gaultieri and M. L. Stein; *J. Heterocycl. Chem.*, 1969, **6**, 279.	118
69JHC317	R. C. Bertelson, K. D. Glanz and D. B. McQuain; *J. Heterocycl. Chem.*, 1969, **6**, 317.	409
69JHC575	H. J. M. Dou, G. Vernin and J. Metzger; *J. Heterocycl. Chem.*, 1969, **6**, 575.	255
69JHC627	W. H. Mueller and M. Dines; *J. Heterocycl. Chem.*, 1969, **6**, 627.	905, 943
69JHC635	M. Wilhelm and P. Schmidt; *J. Heterocycl. Chem.*, 1969, **6**, 635.	325
69JHC729	A. de Souza Gomez and M. M. Joullié; *J. Heterocycl. Chem.*, 1969, **6**, 729.	319, 1005
69JHC745	N. C. Rose; *J. Heterocycl. Chem.*, 1969, **6**, 745.	154
69JHC751	T. Bailey, T. P. Seden and R. W. Turner; *J. Heterocycl. Chem.*, 1969, **6**, 751.	1012
69JHC835	W. A. Remers, G. J. Gibs and M. J. Weiss; *J. Heterocycl. Chem.*, 1969, **6**, 835.	575
69JIC779	C. B. Chowdhury and R. Basu; *J. Indian Chem. Soc.*, 1969, **46**, 779 (*Chem. Abstr.*, 1969, **71**, 128 907).	516, 547
69JOC36	L. Field and R. B. Barbee; *J. Org. Chem.*, 1969, **34**, 36.	801
69JOC175	J. A. Deyrup and C. L. Moyer; *J. Org. Chem.*, 1969, **34**, 175.	941
69JOC333	D. J. Zwanenburg and H. Wynberg; *J. Org. Chem.*, 1969, **34**, 333.	987
69JOC340	D. J. Zwanenburg and H. Wynberg; *J. Org. Chem.*, 1969, **34**, 340.	1013
69JOC917	K. F. Mueller and M. J. Cziesla; *J. Org. Chem.*, 1969, **34**, 917.	891
69JOC919	H. Watanabe, C. L. Mao, I. T. Barnish and C. R. Hauser; *J. Org. Chem.*, 1969, **34**, 919.	146, 166
69JOC984	A. T. Nielsen and T. G. Archibald; *J. Org. Chem.*, 1969, **34**, 984.	59, 97, 103
69JOC999	W. W. Paudler and A. G. Zeiler; *J. Org. Chem.*, 1969, **34**, 999.	65
69JOC1618	S. Kaufmann, L. Tokes, J. W. Murphy and P. Crabbé; *J. Org. Chem.*, 1969, **34**, 1618.	75
69JOC2053	S. C. Mutha and R. Ketcham; *J. Org. Chem.*, 1969, **34**, 2053.	308
69JOC2129	C. K. Bradsher, R. D. Brandau, J. E. Boliek and T. L. Hough; *J. Org. Chem.*, 1969, **34**, 2129.	657
69JOC2562	R. Seltzer; *J. Org. Chem.* 1969, **34**, 2562.	503
69JOC2981	D. J. Woodman, C. H. Borman, N. Tontapanish and P. M. Stonebraker; *J. Org. Chem.*, 1969, **34**, 2981.	41

69JOC2985	M. Davis and A. W. White; *J. Org. Chem.*, 1969, **34**, 2985.	141
69JOC3011	F. N. Jones and S. Andreades; *J. Org. Chem.*, 1969, **34**, 3011.	769, 780
69JOC3233	P. R. West and J. Warkentin; *J. Org. Chem.*, 1969, **34**, 3233.	437
69JOC3248	D. T. Manning and H. A. Coleman; *J. Org. Chem.*, 1969, **34**, 3248.	100, 101
69JOC3285	I. M. Goldman; *J. Org. Chem.*, 1969, **34**, 3285.	697
69JOC3434	J. B. Hendrickson, S. Okano and R. K. Bloom; *J. Org. Chem.*, 1969, **34**, 3434.	927, 930, 932, 940
69JPR118	G. Mueller, H. Frischleder and M. Muehlstaedt; *J. Prakt. Chem.*, 1969, **311**, 118.	108
69JPR408	J. Beger, D. Schöde and J. Vogel; *J. Prakt. Chem.*, 1969, **311**, 408.	308
69JPR646	H. G. O. Becker, J. Witthauer, N. Sauder and G. West; *J. Prakt. Chem.*, 1969, **311**, 646.	429, 444
69JSP(30)459	D. M. Levine, W. D. Krugh and L. P. Gold; *J. Mol. Spectrosc.*, 1969, **30**, 450.	356
69KGS62	L. Braier, P. V. Petrovskii, Z. V. Todres and E. I. Fedin; *Khim. Geterotsikl. Soedin.*, 1969, 62.	518
69KGS235	V. G. Pesin; *Khim. Geterotsikl. Soedin.*, 1969, 235.	521
69LA(726)110	F. Becke and J. Gnad; *Liebigs Ann. Chem.*, 1969, **726**, 110.	941
69LA(727)22	W. Walter and P. M. Hell; *Liebigs Ann. Chem.*, 1969, **727**, 22.	924
69M602	H. Wittmann, E. Ziegler, F. Eichenseer and G. Dworak; *Monatsh. Chem.*, 1969, **100**, 602.	120
69M959	H. Wittmann, E. Ziegler, H. Sterk and G. Dworak; *Monatsh. Chem.*, 1969, **100**, 959.	540
69MI41800	W. Steglich; *Fortschr. Chem. Forsch.*, 1969, **12**, 77.	186, 199
69MI41900	M. A. Soulié, P. Goursot, A. Penelouz and J. Metzger; *J. Chim. Phys. Phys. Chim. Biol.*, 1969, **66**, 607 (*Chem. Abstr.*, 1969, **71**, 42 995).	244
69MI41901	P. Goursot and E. F. Westrum, Jr.; *J. Chem. Eng. Data*, 1969, **14**, 1.	245
69MI41902	D. K. Lewis; *J. Sci. Food Agric.*, 1969, **20**, 185.	328
69MI41903	D. M. Foulkes; *Nature (London)*, 1969, **221**, 582.	328
B-69MI42100	M. Ohta and H. Kato; in 'Nonbenzenoid Aromatic Compounds', ed. J. P. Snyder; Academic Press, New York, 1969, pp. 117–248.	373, 376
69MI42600	D. Dal Monte, E. Sandri, L. Di Nunno, S. Florio and P. E. Todesco; *Chim. Ind. (Milan)*, 1969, **51**, 987.	533
69MI43200	K. M. Pazdro; *Rocz. Chem.*, 1969, **43**, 1089.	842
69OMR(1)311	A. J. Boulton, P. J. Halls and A. R. Katritzky; *Org. Magn. Reson.*, 1969, **1**, 311.	524
69OMS(2)1117	H. Hettler, H. M. Schiebel and H. Budzikiewicz; *Org. Mass Spectrom.*, 1969, **2**, 1117.	143
69OPP255	M. Carmack, I. W. Stapleton and R. Y. Wen; *Org. Prep. Proced. Int.*, 1969, **1**, 255.	523, 525, 531
69RCR540	M. Ya. Karspeiskii and V. L. Florentev; *Russ. Chem. Rev. (Engl. Transl.)*, 1969, **38**, 540.	195
69RTC204	H. W. van Meeteren and H. C. van der Plas; *Recl. Trav. Chim. Pays-Bas*, 1969, **88**, 204.	391
69SA(A)1027	F. L. Tobiason and J. H. Goldstein; *Spectrochim. Acta, Part A*, 1969, **25**, 1027.	518
69T191	D. L. Klayman and G. W. A. Milne; *Tetrahedron*, 1969, **25**, 191.	347
69T389	D. N. McGregor, U. Corbin, J. E. Swigor and L. C. Cheney; *Tetrahedron*, 1969, **25**, 389.	33
69T4277	H. H. Hoerhold and H. Eibisch; *Tetrahedron*, 1969, **25**, 4277.	531, 539
69T5485	R. A. Johnstone and S. D. Ward; *Tetrahedron*, 1969, **25**, 5485.	784
69TL447	T. Minami, O. Aoki, H. Miki, Y. Ohshiro and T. Agawa; *Tetrahedron Lett.*, 1969, 447.	539
69TL543	Y. Makisumi and T. Sasatani; *Tetrahedron Lett.*, 1969, 543.	39, 59
69TL1011	M. Benedek-Vamos and R. Promel; *Tetrahedron Lett.*, 1969, 1011.	674
69TL1775	E. L. Eliel and W. E. Willy; *Tetrahedron Lett.*, 1969, 1775.	758
69TL2709	J. R. Owen; *Tetrahedron Lett.*, 1969, 2709.	277
69TL4117	G. Kresze and H. Grill; *Tetrahedron Lett.*, 1969, 4117.	539
69TL4627	M. Moriarty and R. Mukherjee; *Tetrahedron Lett.*, 1969, 4627.	559
69TL4817	M. Dines and M. L. Scheinbaum; *Tetrahedron Lett.*, 1969, 4817.	73, 87
69TL4875	I. Adachi, K. Harada and H. Kano; *Tetrahedron Lett.*, 1969, 4875.	28, 44, 99
69TL5105	I. Lalezari, A. Shafiee and M. Yalpani; *Tetrahedron Lett.*, 1969, 5105.	352
69YZ699	Y. Usui; *Yakugaku Zasshi*, 1969, **89**, 699 (*Chem. Abstr.*, 1969, **71**, 69 601).	328
69ZC121	K. A. Jensen; *Z. Chem.*, 1969, **9**, 121.	580
69ZC241	M. Wahren, *Z. Chem.*, 1969, **9**, 241.	589
69ZN(B)577	J. Faust and J. Fabian; *Z. Naturforsch., Teil B*, 1969, **24**, 577.	800
69ZOB54	Z. N. Timofeeva, M. V. Petrova, M. Z. Girshovich and A. V. El'tsov; *Zh. Obshch. Khim.*, 1969, **39**, 54 (*Chem. Abstr.*, 1969, **70**, 83 603).	265
69ZOR153	B. K. Strelets and L. S. Efros; *Zh. Org. Khim.*, 1969, **5**, 153 (*Chem. Abstr.*, 1969, **70**, 86 889).	907, 911
69ZOR1179	N. A. Pogorzhelskaya, I. A. Maretina and A. A. Petrov; *Zh. Org. Khim.*, 1969, **5**, 1179.	65, 83
70ABC780	Y. Yamada, N. Seki, T. Kitahari, M. Takahashi and M. Matsui; *Agric. Biol. Chem.*, 1970, **35**, 780 (*Chem. Abstr.*, 1970, **73**, 35 265).	327
70AC(R)801	D. Dal Monte, E. Sandri and W. Cere; *Ann. Chim. (Rome)*, 1970, **60**, 801.	523
70ACS179	U. Anthoni, C. Larsen and P. H. Nielsen; *Acta Chem. Scand.*, 1970, **24**, 179.	554, 573

70ACS1423	E. Falch and T. Natvig; *Acta Chem. Scand.*, 1970, **24**, 1423.	702
70ACS1464	J. Sletten; *Acta Chem. Scand.*, 1970, **24**, 1464.	1055
70ACS1512	C. Christophersen and A. Holm; *Acta Chem. Scand.*, 1970, **24**, 1512.	595
70ACS2137	P. Groth; *Acta Chem. Scand.*, 1970, **24**, 2137.	855, 856
70ACS2525	P. Markov and R. Stølevik; *Acta Chem. Scand.*, 1970, **24**, 2525.	547, 581
70ACS2956	K. Undheim and K. R. Reistad; *Acta Chem. Scand.*, 1970, **24**, 2956.	673, 679
70ACS3129	K. Undheim and A. Eidem; *Acta Chem. Scand.*, 1970, **24**, 3129 (*Chem. Abstr.*, 1971, **74**, 141 616).	329
70ACS3213	K. A. Jensen and L. Henriksen; *Acta Chem. Scand.*, 1970, **24**, 3213.	971
70ACS3435	C. L. Pedersen, N. Harrit and O. Buchardt; *Acta Chem. Scand.*, 1970, **24**, 3435.	79
70ACS3729	P. Stenson; *Acta Chem. Scand.*, 1970, **24**, 3729.	238
70AG(E)54	G. Zumach and E. Kühle; *Angew. Chem., Int. Ed. Engl.*, 1970, **9**, 54.	496, 508, 923, 936, 941, 943
70AG(E)62	G. Zumach and E. Kuhle; *Angew. Chem., Int. Ed. Engl.*, 1970, **9**, 62.	493, 494
70AG(E)464	I. Lalezari, A. Shafiee and M. Yalpani; *Angew. Chem., Int. Ed. Engl.*, 1970, **9**, 464.	349, 353
70AHC(12)213	H. Lund; *Adv. Heterocycl. Chem.*, 1970, **12**, 213.	265
70AJC51	E. N. Cain and R. N. Warrener; *Aust. J. Chem.*, 1970, **23**, 51.	807
70AP501	H. Gehlen and B. Simon; *Arch. Pharm. (Weinheim, Ger.)*, 1970, **303**, 501.	727
70AP625	K. Hartke and B. Seib; *Arch. Pharm. (Weinheim, Ger.)*, 1970, **303**, 625 (*Chem. Abstr.*, 1970, **73**, 98 859).	303
70BCJ268	M. Kamiya and Y. Akahori; *Bull. Chem. Soc. Jpn.*, 1970, **43**, 268.	522
70BCJ1905	N. Yasuoka, N. Kasai, T. Minami, Y. Ohshiro, T. Agawa and M. Kakudo; *Bull. Chem. Soc. Jpn.*, 1970, **43**, 1905.	515, 539
70BCJ2281	Y. Ohtsuka and K. Sugimoto; *Bull. Chem. Soc. Jpn.*, 1970, **43**, 2281.	655
70BCJ2535	Y. Iwakura and K. Kurita; *Bull. Chem. Soc. Jpn.*, 1970, **43**, 2535.	1014
70BCJ2938	M. Yokoyama; *Bull. Chem. Soc. Jpn.*, 1970, **43**, 2938.	787, 860, 890
70BCJ3305	Y. Ohtsuka and K. Sugimoto; *Bull. Chem. Soc. Jpn.*, 1970, **43**, 3305.	692, 655, 661
70BCJ3587	M. Yoshida, A. Matsumoto and O. Simamura; *Bull. Chem. Soc. Jpn.*, 1970, **43**, 3587.	1036, 1045
70BCJ3604	A. Furuhashi; *Bull. Chem. Soc. Jpn.*, 1970, **43**, 3604.	1058
70BCJ3909	Y. Ohtsuka; *Bull. Chem. Soc., Jpn.*, 1970, **43**, 3909.	650, 652, 655, 661
70BRP1190359	A. G. Oesterreichische Stickstoffwerke, *Br. Pat.* 1 190 359 (1970) (*Chem. Abstr.*, **73**, 45 521).	535
70BSB11	Y. Rommelaere and M. Anteunis; *Bull. Soc. Chim. Belg.*, 1970, **79**, 11.	758
70BSF1918	G. Duguay and H. Quiniou; *Bull. Soc. Chim. Fr.*, 1970, 1918.	803, 805
70BSF1978	R. Jacquier, C. Petrus, F. Petrus and J. Verducci; *Bull. Soc. Chim. Fr.*, 1970, 1978.	64, 74, 106
70BSF2685	R. Jacquier, C. Petrus, F. Petrus and J. Verducci; *Bull. Soc. Chim. Fr.*, 1970, 2685.	104
70BSF2705	G. Vernin, H. J. M. Dou, G. Loridan and J. Metzger; *Bull. Soc. Chim. Fr.*, 1970, 2705 (*Chem. Abstr.*, 1970, **73**, 119 890).	274
70BSF4517	D. Paquer, M. Perrier and J. Vialle; *Bull. Soc. Chim. Fr.*, 1970, 4517.	360, 1063
70C134	A. R. Katritzky; *Chimia*, 1970, **24**, 134.	11
70CB112	J. Goerdeler and M. Roegler; *Chem. Ber.*, 1970, **103**, 112.	144
70CB123	H. Böshagen and W. Geiger; *Chem. Ber.*, 1970, **103**, 123.	56
70CB1805	J. Goerdeler and P. Mertens; *Chem. Ber.*, 1970, **103**, 1805.	480, 481, 496
70CB1918	H. Reimlinger, J. J. M. Vandewalle, G. S. D. King, W. R. F. Lingier and R. Merenyi; *Chem. Ber.*, 1970, **103**, 1918.	606, 607, 610
70CB1934	H. Reimlinger, J. J. M. Vandewalle and W. R. F. Lingier; *Chem. Ber.*, 1970, **103**, 1934.	608, 609
70CB2581	H. O. Bayer, R. Huisgen, R. Knorr and F. C. Schaefer; *Chem. Ber.*, 1970, **103**, 2581.	182
70CB2611	R. Huisgen, H. Gotthardt, H. O. Bayer and F. C. Schaefer; *Chem. Ber.*, 1970, **103**, 2611.	1006
70CB3058	H. Böhme and G. Dähler; *Chem. Ber.*, 1970, **103**, 3058 (*Chem. Abstr.*, 1970, **73**, 130 925).	320
70CB3166	H. Böschagen, W. Geiger and H. Medenwald; *Chem. Ber.*, 1970, **103**, 3166.	142
70CB3196	G. Schmidt, H. U. Stracke and E. Winterfeldt; *Chem. Ber.*, 1970, **103**, 3196.	44, 99
70CB3284	H. Reimlinger and R. Merenyi; *Chem. Ber.*, 1970, **101**, 3284.	976
70CB3370	G. Häfelinger; *Chem. Ber.*, 1970, **103**, 3370.	395
70CB3885	A. Schonberg and E. Frese; *Chem. Ber.*, 1970, **103**, 3885.	805
70CC386	M. Kojima and M. Maeda; *Chem. Commun.*, 1970, 386.	250
70CC1591	H. Kato, T. Shiba, H. Yoshida and S. Fujimori; *Chem. Commun.*, 1970, 1591.	602
70CHE141	V. F. Vasil'eva, R. A. Zagrutdinova, E. M. Peresleni and V. G. Yashunskii; *Chem. Heterocycl. Compd. (Engl. Transl.)*, 1970, **6**, 141.	431, 439
70CHE327	B. K. Strelets, L. S. Efros and Z. V. Todres; *Chem. Heterocycl. Compd. (Engl. Transl.)*, 1970, **6**, 327.	921, 940
70CHE1177	E. E. Mikhlina, V. Ya. Vorob'eva and L. N. Yakhontov; *Chem. Heterocycl. Compd.*, 1970, **6**, 1177.	988
70CHE1515	P. I. Abramenko and V. G. Zhiryakov; *Chem. Heterocycl. Compd. (Engl. Transl.)*, 1970, **6**, 1515.	994
70CI(L)624	L. Jurd; *Chem. Ind. (London)*, 1970, 624.	44, 99
70CJC467	R. G. Micetich; *Can. J. Chem.*, 1970, **48**, 467.	89, 100, 101
70CJC845	T. Durst and K.-C. Tin; *Can. J. Chem.*, 1970, **48**, 845.	772
70CJC2006	R. G. Micetich; *Can. J. Chem.*, 1970, **48**, 2006.	49, 478, 527, 535

70CJC2142	M. Ahmed and D. M. McKinnon; *Can. J. Chem.*, 1970, **48**, 2142.	923, 924, 942
70CJC3059	J. M. Prokipcak and P. A. Forte; *Can. J. Chem.*, 1970, **48**, 3059.	419
70CJC3554	A. L. Lee, D. Mackay and E. L. Manery; *Can. J. Chem.*, 1970, **48**, 3554.	266
70CJC3704	J. F. King, P. de Mayo, C. L. McIntosh, K. Piers and D. J. H. Smith; *Can. J. Chem.*, 1970, **48**, 3704.	753, 776
70CJC3753	R. G. Micetich; *Can. J. Chem.*, 1970, **48**, 3753.	89, 90
70CPB128	K. Masuda, Y. Imashiro and T. Kaneko; *Chem. Pharm. Bull.*, 1970, **18**, 128.	377
70CPB865	A. Takamizawa and K. Hirai; *Chem. Pharm. Bull.*, 1970, **18**, 865.	815
70CPB1176	S. Takahashi, S. Hashimoto and H. Kano; *Chem. Pharm. Bull.*, 1970, **18**, 1176 (*Chem. Abstr.*, 1970, **73**, 45 390).	267
70CPB1233	M. Sekiya, J. Suzuki and Y. Kakiya; *Chem. Pharm. Bull.*, 1970, **18**, 1233.	648, 649, 652, 653
70CPB1696	S. Kubota, Y. Koida, T. Kosaka and O. Kirino; *Chem. Pharm. Bull.*, 1970, **18**, 1696.	570
70CPB2242	M. Sekiya and J. Suzuki; *Chem. Pharm. Bull.*, 1970, **18**, 2242.	648, 649, 652, 653, 661
70CR(C)(270)1677	J. C. Poite, A. Perichaut and J. Roggero; *C. R. Hebd. Seances Acad. Sci., Ser. C*, 1970, **270**, 1677.	141
70CR(C)(271)17	A. Friedmann, A. Cormons and J. Metzger; *C.R. Hebd. Seances Acad. Sci., Ser. C*, 1970, **271**, 17 (*Chem. Abstr.*, 1970, **73**, 103 824).	240
70CR(C)(271)1468	P. Battioni, L. Vo-Quang, J. C. Raymond and Y. Vo-Quang; *C. R. Hebd. Seances Acad. Sci., Ser. C*, 1970, **271**, 1468 (*Chem. Abstr.*, 1971, **74**, 53 611).	108
70CRV593	A. Lawson and R. B. Tinkler; *Chem. Rev.*, 1970, **70**, 593.	448, 529, 540
70DOK(195)1333	B. A. Arbuzov, V. A. Naumov, N. M. Zaripov and L. D. Pronicheva; *Dokl. Akad. Nauk SSSR*, 1970, **195**, 1333.	857
70E1169	A. Quilico; *Experientia*, 1970, **26**, 1169.	66, 67, 68
70FRP94493	Sociéte des Usines Chemiques Rhone-Poulenc; *Fr. Addn.* 94 493 (1970) (*Chem. Abstr.*, 1970, **72**, 111 455).	107
70G461	L. Minale, E. Fattorusso, S. De Stefano and R. Nicolaus; *Gazz. Chim. Ital.*, 1970, **100**, 461 (*Chem. Abstr.*, 1970, **73**, 116 507).	327
70G629	L. Cavalca, A. Gaetani, A. Mangia and G. Pelizzi; *Gazz. Chim. Ital.*, 1970, **100**, 629.	135, 143
70G1144	G. F. Bettinetti and A. Gamba; *Gazz. Chim. Ital.*, 1970, **100**, 1144.	89
70GEP1907407	Badische Anilin- und Soda-Fabrick A.-G., *Ger. Pat.* 1 907 407 (1970) (*Chem. Abstr.*, 1971, **74**, 4680).	510
70GEP1924138	R. A. Firestone; *Ger. Offen.* 1 924 138 (1970) (*Chem. Abstr.*, 1970, **72**, 111 613).	868
70GEP1950990	*Ger. Pat.* 1 950 990 (1970) (*Chem. Abstr.*, 1970, **73**, 56 118).	637
70GEP2003575	*Ger. Pat.* 2 003 575 (1970) (*Chem. Abstr.*, 1970, **73**, 110 912).	668
70GEP2004305	C. Esclamadon, J. B. Signouret and Y. Labat; *Ger. Offen.* 2 004 305 (1970) (*Chem. Abstr.*, 1970, **73**, 111 043).	893
70GEP2006131	Ciba Geigy, *Ger. Pat.* 2 006 131 (1970) (*Chem. Abstr.*, 1971, **74**, 4640).	510
70HCA1883	A. R. Gagneux and R. Meier; *Helv. Chim. Acta*, 1970, **53**, 1883.	116
70IJC710	S. K. Modi, V. Kumar and K. S. Narang; *Indian J. Chem.*, 1970, **8**, 710.	743
70IJC796	K. T. Borkhade, M. G. Marath and G. Madhov; *Indian J. Chem.*, 1970, **8**, 796.	93
70IZV2140	D. P. Del'tsova, S. O. Koshtoyan and Y. V. Zeifman; *Izv. Akad. Nauk SSSR, Ser. Khim.*, 1970, 2140 (*Chem. Abstr.*, 1971, **74**, 76 378).	921, 940
70JA1412	H. D. Hartzler; *J. Am. Chem. Soc.*, 1970, **92**, 1412.	845, 849
70JA2628	J. Renard and S. Fliszár; *J. Am. Chem. Soc.*, 1970, **92**, 2628.	853
70JA3133	G. A. Olah, D. P. Kelly and N. Suciu; *J. Am. Chem. Soc.*, 1970, **92**, 3133.	371
70JA4340	H. Gotthardt, R. Huisgen and H. O. Bayer; *J. Am. Chem. Soc.*, 1970, **92**, 4340.	186
70JA5394	W. E. Willy, G. Binsch and E. L. Eliel; *J. Am. Chem. Soc.*, 1970, **92**, 5394.	758
70JA5707	J. Crosby and G. E. Lienhard; *J. Am. Chem. Soc.*, 1970, **92**, 5707.	326
70JA5907	G. E. Wilson, Jr., M. G. Huang and F. A. Bovey; *J. Am. Chem. Soc.*, 1970, **92**, 5907.	754, 758
70JCP(52)2787	L. Krause and M. A. Whitehead; *J. Chem. Phys.*, 1970, **52**, 2787.	515
70JCS(A)1386	R. T. Bailey; *J. Chem. Soc. (A)*, 1970, 1386.	907, 910
70JCS(B)636	A. J. Boulton, P. J. Halls and A. R. Katritzky; *J. Chem. Soc. (B)*, 1970, 636.	716, 718, 729
70JCS(B)1049	B. W. Evans, D. J. Fenn and B. J. Tighe; *J. Chem. Soc. (B)*, 1970, 1049.	868
70JCS(C)1165	B. J. Wakefield and D. J. Wright; *J. Chem. Soc. (C)*, 1970, 1165.	89
70JCS(C)1202	E. I. G. Brown, D. Leaver and D. M. McKinnon; *J. Chem. Soc. (C)*, 1970, 1202.	791, 792
70JCS(C)1397	G. V. Boyd and S. R. Dando; *J. Chem. Soc. (C)*, 1970, 1397.	431, 440, 726
70JCS(C)1429	A. Lawson and R. B. Tinkler; *J. Chem. Soc. (C)*, 1970, 1429.	500
70JCS(C)1485	G. V. Boyd and P. H. Wright; *J. Chem. Soc. (C)*, 1970, 1485.	648, 649, 652, 653, 654, 655, 657
70JCS(C)1825	T. Nishiwaki, A. Nakano and H. Matsuoka; *J. Chem. Soc. (C)*, 1970, 1825.	12
70JCS(C)1938	R. H. Good and G. Jones; *J. Chem. Soc. (C)*, 1970, 1938.	649, 658
70JCS(C)2250	P. Kirby, S. B. Soloway, J. H. Davies and S. B. Webb; *J. Chem. Soc. (C)*, 1970, 2250.	460
70JCS(C)2412	J. G. Dingwall, D. H. Reid and J. D. Symon; *J. Chem. Soc. (C)*, 1970, 2412.	806
70JCS(C)2478	J. A. Baker and P. V. Chatfield; *J. Chem. Soc. (C)*, 1970, 2478.	687, 689, 697
70JCS(C)2660	D. R. Eckroth and T. G. Cochran; *J. Chem. Soc. (C)*, 1970, 2660.	121
70JCS(C)2993	F. L. Scott; *J. Chem. Soc. (C)*, 1970, 2993.	94
70JHC123	M. Gotz and K. Grozinger; *J. Heterocycl. Chem.*, 1970, **7**, 123.	377
70JHC457	D. A. Thomas; *J. Heterocycl. Chem.*, 1970, **7**, 457.	978
70JHC629	K. Pilgram, M. Zupan and R. Skiles; *J. Heterocycl. Chem.*, 1970, **7**, 629.	532

70JMC196	T. L. Thomas, M. Fedorchuk, B. V. Shetty and F. E. Anderson; *J. Med. Chem.*, 1970, **13**, 196.	603
70JMC1250	H. G. Garg and P. P. Singh; *J. Med. Chem.*, 1970, **13**, 1250.	105
70JOC806	R. V. Kendall and R. A. Olofson; *J. Org. Chem.*, 1970, **35**, 806.	356, 357
70JOC1165	K. Pilgram; *J. Org. Chem.*, 1970, **35**, 1165.	538
70JOC1662	A. J. Boulton and R. C. Brown; *J. Org. Chem.*, 1970, **35**, 1662.	48
70JOC1665	R. Seltzer and W. J. Considine; *J. Org. Chem.*, 1970, **35**, 1665.	942
70JOC1806	C. F. Beam, M. C. D. Dyer, R. A. Schwarz and C. R. Hauser; *J. Org. Chem.*, 1970, **35**, 1806.	71, 72
70JOC1965	K. T. Potts and R. Armbruster; *J. Org. Chem.*, 1970, **35**, 1965.	490, 499, 736, 742
70JOC2002	B. R. O'Connor and F. N. Jones; *J. Org. Chem.*, 1970, **35**, 2002.	848
70JOC2065	K. P. Park, C.-Y. Shiue and L. B. Clapp; *J. Org. Chem.*, 1970, **35**, 2065.	95, 96
70JOC2440	R. V. Coombs and G. E. Hardtmann; *J. Org. Chem.*, 1970, **35**, 2440.	21, 126
70JOC3130	L. D. Quin and D. O. Pinion; *J. Org. Chem.*, 1970, **35**, 3130.	41, 105
70JOC3259	D. N. Harpp and J. G. Gleason; *J. Org. Chem.*, 1970, **35**, 3259.	797
70JOC3470	M. S. Raasch; *J. Org. Chem.*, 1970, **35**, 3470	584, 593, 890, 905, 934, 941
70JOC3768	S. P. McManus, J. T. Carroll and C. U. Pittman, Jr.; *J. Org. Chem.*, 1970, **35**, 3768.	311
70JPR161	M. Schulz and G. West; *J. Prakt. Chem.*, 1970, **312**, 161.	1001
70JPR776	K. Gewald, H. Spies and R. Mayer; *J. Prakt. Chem.*, 1970, **312**, 776 (*Chem. Abstr.*, 1971, **74**, 141 622).	303
70JPS835	G. J. Krol, J. F. Carney and B. T. Kho; *J. Pharm. Sci.*, 1970, **59**, 835.	556
70JST(5)236	G. H. Schmid; *J. Mol. Struct.*, 1970, **5**, 236.	368, 369
70KGS1505	L. Kh. Vinograd and N. N. Suvorov; *Khim. Geterotsikl. Soedin.*, 1970, 1505 (*Chem. Abstr.*, 1971, **74**, 53 610).	103
70KGS(S2)303	V. S. Garkusha-Bozhko and O. P. Shvaika; *Khim. Geterotsikl. Soedin., Sb. 2*, 1970, 303 (*Chem. Abstr.*, 1972, **76**, 140 656).	436
70LA(731)142	K. H. Mayer and D. Lauerer; *Liebigs Ann. Chem.*, 1970, **731**, 142.	557
70LA(732)195	H. G. Aurich; *Liebigs Ann. Chem.*, 1970, **732**, 195.	43
70LA(738)60	G. Griss and H. Machleidt; *Liebigs Ann. Chem.*, 1970, **738**, 60.	634
70LA(742)103	R. Wiedermann, W. von Gentzkow and F. Boberg; *Liebigs Ann. Chem.*, 1970, **742**, 103.	801
70LA(742)128	K.-H. Magosch and R. Feinauer; *Liebigs Ann. Chem.*, 1970, **742**, 128.	665
70M383	H. Wittmann, A. Wohlkoenig, H. Sterk and E. Ziegler; *Monatsh. Chem.*, 1970, **101**, 383.	664
70M704	A. T. Balaban, I. Zugravescu, S. Avramovici and W. Silhan, *Monatsh. Chem.*, 1970, **101**, 704.	93
70M1109	W. Kloetzer, H. Bretschneider, E. Fritz, R. Reiner and G. Bader; *Monatsh. Chem.*, 1970, **101**, 1109.	65, 86
70MI41600	H. Schulz and S. J. Wakil; *Anal. Biochem.*, 1970, **37**, 457.	104
70MI41601	N. Yoshida, K. Tomita and K. Wachi; *Sankyo Kenkyusko Nempo*, 1970, **22**, 80 (*Chem. Abstr.*, 1971, **75**, 63 325).	101
B-70MI41800	'The Current Status of Liquid Scintillation Counting', ed. E. D. Bransome; Grune and Stratton, New York, 1970.	233
70MI41900	R. Evers; *J. Polym. Sci., Part A-1, Polym. Chem.*, 1970, **8**, 563.	330
70MI42300	H. Kovacs, A. D. Delman and B. B. Simms; *J. Polym. Sci., Part A1*, 1970, **8**, 869 (*Chem. Abstr.*, 1970, **72**, 112 182).	446
70MI42600	H. K. Y. Lau and P. F. Lott; *Talanta*, 1970, **17**, 717.	526
70MI42700	H. Kubo, R. Sato, I. Hamura and T. Ohi; *J. Agr. Food Chem.*, 1970, **18**, 60 (*Chem. Abstr.*, 1970, **72**, 53 955).	576
70MI42701	K. Kariyone, H. Harada, M. Kurita and T. Takano; *J. Antibiotics*, 1970, **23**, 131.	576
70MI43300	R. G. Buttery, R. M. Seifert and L. C. Ling; *J. Agr. Food Chem.*, 1970, **18**, 538.	894
70MI43301	J. Pokorny; *Prum. Potravin.*, 1970, **21**, 262 (*Chem. Abstr.*, 1970, **73**, 119 365).	894
70MI43400	J. Vialle; *Quart. Rep. Sulfur Chem.*, 1970, **5**, 151 (*Chem. Abstr.*, 1970, **73**, 109 715).	898
70MI43401	R. S. Lebedev, A. V. Korshunov, R. P. Chumakova and V. E. Volkov; *Izv. Vysshikh Uchebrykh. Zavedenii Fizika*, 1970, **13**, 50 (*Chem. Abstr.*, 1971, **74**, 81 294).	907, 910
70N395	J. J. van Daalen and J. Daams; *Naturwissenschaften*, 1970, **57**, 395.	543
70OMS(3)1341	H. Ogura, S. Sugimoto and T. Itoh; *Org. Mass Spectrom.*, 1970, **3**, 1341.	183
70RCR923	V. G. Pesin; *Russ. Chem. Rev. (Engl. Transl.)*, 1970, **39**, 923.	514, 525, 526, 528
70RTC5	G. H. Harts, K. B. De Roos and C. A. Salemink; *Recl. Trav. Chim. Pays-Bas*, 1970, **89**, 5.	533, 538, 733, 736, 747
70S344	C. Grundmann; *Synthesis*, 1970, 344.	3, 66, 67, 69, 70, 83, 88, 89
70S365	R. Metelli and G. F. Bettinetti; *Synthesis*, 1970, 365.	997
70SA(A)2057	D. H. Christensen and T. Stroyer-Hansen; *Spectrochim. Acta, Part A*, 1970, **26**, 2057 (*Chem. Abstr.*, 1970, **73**, 135 723).	552
70SST(1)378	F. Kurzer; *Org. Compd. Sulphur, Selenium, Tellurium*, 1970, **1**, 378–409.	328
70SST(1)410	F. Kurzer; *Org. Compd. Sulphur, Selenium, Tellurium*, 1970, **1**, 410–423.	328, 330
70SST(1)444	F. Kurzer; *Org. Compd. Sulphur, Selenium, Tellurium*, 1970, **1**, 444.	462
70SST(1)449	F. Kurzer; *Org. Compd. Sulphur, Selenium, Tellurium*, 1970, **1**, 449.	546, 562
70T539	K. Kotera, Y. Takano, A. Matsuura and K. Kitahonoki; *Tetrahedron*, 1970, **26**, 539.	36
70T685	R. A. Coburn, J. M. Landesberg, D. S. Kemp and R. A. Olofson; *Tetrahedron*, 1970, **26**, 685.	525

70T2497	A. W. K. Chan, W. D. Crow and I. Gosney; *Tetrahedron*, 1970, **26**, 2497.	145
70T2759	K. Torssell; *Tetrahedron*, 1970, **26**, 2759.	244
70T4641	P. D. Klemmensen, J. Z. Mortensen and S. O. Lawesson; *Tetrahedron*, 1970, **26**, 4641.	319
70T5113	G. Bianchi, R. Gandolfi and P. Gruenanger; *Tetrahedron*, 1970, **26**, 5113.	89, 90
70TL127	G. Lohaus; *Tetrahedron Lett.*, 1970, 127.	900, 905, 911, 928, 941
70TL169	W. Steglich, G. Höfle, L. Wilschowitz and G. C. Barrett; *Tetrahedron Lett.*, 1970, 169.	249
70TL605	Y. Hayashi, T. Watanabe and R. Oda; *Tetrahedron Lett.*, 1970, 605.	998
70TL629	E. K. Fields and S. Meyerson; *Tetrahedron Lett.*, 1970, 629.	806
70TL1117	W. Oppolzer and K. Keller; *Tetrahedron Lett.*, 1970, 1117.	633
70TL1381	J. E. Franz and L. L. Black; *Tetrahedron Lett.*, 1970, 1381.	500, 506, 999
70TL1427	H. Grill and G. Kresze; *Tetrahedron Lett.*, 1970, 1427.	539
70TL2993	R. J. MacConaill and F. L. Scott; *Tetrahedron Lett.*, 1970, 2993.	94
70TL3223	P. Bravo, G. Gaudiano and C. Ticozzi; *Tetrahedron Lett.*, 1970, 3223.	101
70TL4473	D. J. Woodman and P. M. Stonebraker; *Tetrahedron Lett.*, 1970, 4473.	42
70TL5083	R. Grashey, M. Baumann and R. Hamprecht; *Tetrahedron Lett.*, 1970, 5083.	573
70TL5223	F. W. Eastwood, K. J. Harrington, J. S. Josan and J. L. Pura; *Tetrahedron Lett.*, 1970, 5223.	767
70TL5261	J. L. Neumeyer and F. E. Granchelli; *Tetrahedron Lett.*, 1970, 5261.	659
70UKZ483	S. I. Shul'ga and V. A. Chuiguk; *Ukr. Khim. Zh.* (*Russ. Ed.*), 1970, **36**, 483 (*Chem. Abstr.*, 1970, **73**, 77 196).	696
70USP3493556	Eastman Kodak Co., *U.S. Pat.* 3 493 556 (1970) (*Chem. Abstr.*, 1970, **73**, 36 555).	577
70USP3524859	*U.S. Pat.* 3 524 859 (1970) (*Chem. Abstr.*, 1970, **73**, 131 006).	722
70ZC406	J. Sandstrom; *Z. Chem.*, 1970, **10**, 406 (*Chem. Abstr.*, 1971, **74**, 13 070).	547
70ZC432	C. Roussel, A. Babadjamian, M. Chanon and J. Metzger; *Z. Chem.*, 1970, **10**, 432 (*Chem. Abstr.*, 1971, **74**, 42 305).	314
70ZN(B)1191	E. V. Dehmlow and H. J. Westendorf; *Z. Naturforsch., Teil B*, 1970, **25**, 1191.	1017
70ZOB2612	I. G. Kolokol'tseva, V. N. Cristokletov and A. A. Petrov; *Zh. Obshch. Khim.*, 1970, **40**, 2612 (*Chem. Abstr.*, 1971, **75**, 20 477).	89
70ZOR2273	E. S. Levchenko and I. N. Berzina; *Zh. Org. Khim.*, 1970, **6**, 2273.	168
71AC(R)587	G. Werber, F. Buccheri and M. L. Marino; *Ann. Chim.* (*Rome*), 1971, **61**, 587.	441
71ACS1	A. Eidem, K. Undheim and K. R. Reistad; *Acta Chem. Scand.*, 1971, **25**, 1.	309
71ACS5	K. Undheim and P. O. Tveita; *Acta Chem. Scand.*, 1971, **25**, 5.	673, 679, 681
71ACS118	P. Groth; *Acta Chem. Scand.*, 1971, **25**, 118.	670, 672
71ACS625	C. Christophersen and S. Treppendahl; *Acta Chem. Scand.*, 1971, **25**, 625.	601, 602, 604
71ACS1160	C. Christophersen; *Acta Chem. Scand.*, 1971, **25**, 1160.	595
71ACS1162	C. Christophersen; *Acta Chem. Scand.*, 1971, **25**, 1162.	595
71ACS1353	N. Thorup; *Acta Chem. Scand.*, 1971, **25**, 1353.	670
71ACS1583	A. Hordvik; *Acta Chem. Scand.*, 1971, **25**, 1583.	1053, 1057
71ACS1822	A. Hordvik; *Acta Chem. Scand.*, 1971, **25**, 1822.	1053
71ACS1835	A. Hordvik and K. Julshamn; *Acta Chem. Scand.*, 1971, **25**, 1835.	1053
71ACS1895	A. Hordvik and K. Julshamn; *Acta Chem. Scand.*, 1971, **25**, 1895.	1054
71ACS2015	C. Christophersen and A. Holm; *Acta Chem. Scand.*, 1971, **25**, 2015.	595
71ACS2507	A. Hordvik and K. Julshamn; *Acta Chem. Scand.*, 1971, **25**, 2507.	1054
71ACS2739	I. N. Bodjesen, J. H. Høg, J. T. Nielsen, I. B. Petersen and K. Schaumburg; *Acta Chem. Scand.*, 1971, **25**, 2739.	243, 244
71ACS3577	J. Sletten; *Acta Chem. Scand.*, 1971, **25**, 3577.	1055
71AG(E)76	P. Stoss and G. Satzinger; *Angew. Chem., Int. Ed. Engl.*, 1971, **10**, 76.	168
71AG(E)810	K. H. Magosch and R. Feinauer; *Angew. Chem., Int. Ed. Engl.*, 1971, **10**, 810.	997
71AHC(13)161	N. Lozac'h; *Adv. Heterocycl. Chem.*, 1971, **13**, 161–234.	786, 788, 797, 798, 800, 1050, 1053, 1055, 1056, 1057, 1058, 1059, 1060, 1061, 1062, 1063, 1064, 1065, 1066, 1067, 1068, 1069
71AJC1229	S. Beveridge and R. L. N. Harris; *Aust. J. Chem.*, 1971, **24**, 1229.	1014
71AJC2405	R. K. Buckley, M. Davis and K. S. L. Srivastava; *Aust. J. Chem.*, 1971, **24**, 2405.	141
71AP687	R. Neidlein and J. Tauber; *Arch. Pharm.* (*Weinheim, Ger.*), 1971, **304**, 687.	587, 596
71AX(B)1073	R. D. Gilardi and I. L. Karle; *Acta Crystallogr., Part B*, 1971, **27**, 1073.	1054
71AX(B)1388	M. Mathew and G. J. Palenik; *Acta Crystallogr., Part B*, 1971, **27**, 1388.	396
71AX(B)1687	P. F. Rodesiler and E. L. Amma; *Acta Crystallogr., Part B*, 1971, **27**, 1687.	902, 903
71AX(B)1775	A. Corradi-Bonamartini, M. Nardelli, C. Palmieri and C. Pelizzi; *Acta Crystallogr., Part B*, 1971, **27**, 1775.	136
71BCJ803	T. Sasaki and T. Yoshioka; *Bull. Chem. Soc. Jpn.*, 1971, **44**, 803.	68
71BCJ1667	Y. Huseya, A. Chinone and M. Ohta; *Bull. Chem. Soc. Jpn.*, 1971, **44**, 1667.	374
71BCJ3073	T. Nishiyama and F. Yamada; *Bull. Chem. Soc. Jpn.*, 1971, **44**, 3073.	937
71BRP1257851	Badische Anilin- und Soda-Fabrik A.-G., *Br. Pat.* 1 257 851 (1971) (*Chem. Abstr.*, 1972, **76**, 58 380).	877
71BSB215	G. Lemiere and M. Anteunis; *Bull. Soc. Chim. Belg.*, 1971, **80**, 215.	758
71BSB639	A. Geens and M. Anteunis; *Bull. Soc. Chim. Belg.*, 1971, **80**, 639.	967
71BSF541	F. Guinot, G. Lamaty and H. Münsch; *Bull. Soc. Chim. Fr.*, 1971, 541.	761
71BSF1103	G. Vernin, J. C. Poite, J. Metzger, J. P. Aune and H. J. M. Dou; *Bull. Soc. Chim. Fr.*, 1971, 1103.	147
71BSF1491	M. Robba and Y. Le Guen; *Bull. Soc. Chim. Fr.*, 1971, 1491.	674, 679, 699

71BSF1902	C. Roussel, R. Gallo, M. Chanon and J. Metzger; *Bull. Soc. Chim. Fr.*, 1971, 1902 (*Chem. Abstr.*, 1971, **75**, 48 966).	314
71BSF3537	M. Robba and Y. Le Guen; *Bull. Soc. Chim. Fr.*, 1971, 3537.	686, 687
71BSF3664	R. Jacquier, F. Petrus, J. Verducci and Y. Vidal; *Bull. Soc. Chim. Fr.*, 1971, 3664.	106
71BSF4002	A. Grandin and J. Vialle; *Bull. Soc. Chim. Fr.*, 1971, 4002.	1066, 1067
71BSF4021	J. Roggero and M. Audibert; *Bull. Soc. Chim. Fr.*, 1971, 4021 (*Chem. Abstr.*, 1972, **76**, 113 121).	307
71BSF4310	M. Baule, R. Vivaldi, J. C. Poite, H. J. M. Dou, G. Vernin and J. Metzger; *Bull. Soc. Chim. Fr.*, 1971, 4310.	133
71BSF4418	M. Stavaux; *Bull. Soc. Chim. Fr.*, 1971, 4418.	1064
71BSF4426	M. Stavaux; *Bull. Soc. Chim. Fr.*, 1971, 4426.	1056
71BSF4429	M. Stavaux; *Bull. Soc. Chim. Fr.*, 1971, 4429.	1056
71BSF4591	M. Perrier and J. Vialle; *Bull. Soc. Chim. Fr.*, 1971, 4591.	360, 361
71CB639	H. Reimlinger, W. R. F. Lingier and J. J. M. Vandewalle; *Chem. Ber.*, 1971, **104**, 639.	608, 609
71CB1264	H. Paulsen and H. Behre; *Chem. Ber.*, 1971, **104**, 1264.	753
71CB1281	H. Paulsen and H. Behre; *Chem. Ber.*, 1971, **104**, 1281.	753
71CB1807	R. Criegee and H. Korber; *Chem. Ber.*, 1971, **104**, 1807.	865
71CB2134	H. Hofmann, R. Wagner and J. Uhl; *Chem. Ber.*, 1971, **104**, 2134.	185
71CB2732	G. Dahms, A. Haas and W. Klug; *Chem. Ber.*, 1971, **104**, 2732.	906, 910, 915, 924, 930, 943
71CB3039	L. Capuano and H. J. Schrepfer; *Chem. Ber.*, 1971, **104**, 3039.	664
71CC179	M. M. Campbell and D. M. Evgenios; *Chem. Commun.*, 1971, 179.	890
71CC314	N. H. Nilsson, C. Jacobsen and A. Senning; *Chem. Commun.*, 1971, 314.	595
71CC329	F. A. Vingiello, M. P. Rorer and M. A. Ogliaruso; *Chem. Commun.*, 1971, 329.	310
71CC383	O. L. Chapman and C. L. McIntosh; *Chem. Commun.*, 1971, 383.	766
71CC490	P. Rajagopalan and P. Penev; *Chem. Commun.*, 1971, 490.	1005
71CC594	R. J. S. Beer, J. R. Hatton, E. C. Llaguno and I. C. Paul; *Chem. Commun.*, 1971, 594.	360, 361, 1054
71CC837	W. L. Mosby and M. L. Vega; *Chem. Commun.*, 1971, 837.	565
71CC861	D. H. R. Barton, P. D. Magnus, G. Smith and D. Zurr; *Chem. Commun.*, 1971, 861.	761
71CC889	W. F. Cooper, N. C. Kenney, J. W. Edmonds, A. Nagel, F. Wudl and P. Coppens; *Chem. Commun.*, 1971, 889.	815, 816
71CC1094	R. M. Ellam and J. M. Padbury; *Chem. Commun.*, 1971, 1094.	866, 873
71CC1222	W. D. Ollis and C. A. Ramsden; *Chem. Commun.*, 1971, 1222.	573
71CC1223	W. D. Ollis and C. A. Ramsden; *Chem. Commun.*, 1971, 1223.	441
71CC1318	K. Hirai and T. Ishiba; *Chem. Commun.*, 1971, 1318.	304
71CC1352	R. Pinel, Y. Mollier, E. C. Llaguno and I. C. Paul; *Chem. Commun.*, 1971, 1352.	1056
71CC1368	E. Firkins and A. W. Richardson; *Chem. Commun.*, 1971, 1368.	520
71CC1436	R. M. Moriarty, R. Mukherjee, J. L. Flippen and J. Karle; *Chem. Commun.*, 1971, 1436.	573
71CC1623	J. P. Schaefer and S. K. Arora; *Chem. Commun.*, 1971, 1623.	515
71CCC2640	M. Procházka, O. Ryba and D. Lím; *Collect. Czech. Chem. Commun.*, 1971, **36**, 2640.	377
71CHE139	K. Yu. Novitskii, N. K. Sadovaya and L. M. Trutneva; *Chem. Heterocycl. Compd. (Engl. Transl.)*, 1971, **7**, 139.	998
71CHE561	V. A. Sergeev and V. G. Pesin; *Chem. Heterocycl. Compd. (Engl. Transl.)*, 1971, **7**, 561.	532
71CHE1028	B. L. Moldaver and M. P. Papirnick; *Chem. Heterocycl. Compd. (Engl. Transl.)*, 1971, **7**, 1028.	978, 982
71CHE1534	A. P. Sineokov and V. S. Kutyreva; *Chem. Heterocycl. Compd. (Engl. Transl.)*, 1971, **7**, 1534.	1013
71CJC167	F. Leung and S. C. Nyburg; *Can. J. Chem.*, 1971, **49**, 167.	1054
71CJC956	J. C. Panizzi, G. Davidovics, R. Guglielmetti, G. Mille, J. Metzger and J. Chouteau; *Can. J. Chem.*, 1971, **49**, 956.	241
71CJC1477	P. Yates and T. R. Lynch; *Can. J. Chem.*, 1971, **49**, 1477.	860, 873, 880, 890
71CJC2254	J. L. Meyer, G. Davidovics, J. Chouteau, J. C. Poite and J. Roggero; *Can. J. Chem.*, 1971, **49**, 2254.	141
71CPB559	K. Masuda, T. Kamiya and K. Kashiwa; *Chem. Pharm. Bull.*, 1971, **19**, 559.	603, 604
71CPB1389	N. Nakamura, Y. Kishida and N. Ishida; *Chem. Pharm. Bull.*, 1971, **19**, 1389.	106
71CPB2194	K. Hirai and T. Ishiba; *Chem. Pharm. Bull.*, 1971, **19**, 2194.	762
71CR(C)(272)854	G. Vernin, H. Dou and J. Metzger; *C.R. Hebd. Seances Acad. Sci., Ser. C*, 1971, **272**, 854 (*Chem. Abstr.*, 1971, **74**, 125 546).	278
71CRV483	J. A. Frump; *Chem. Rev.*, 1971, **71**, 483.	211, 227
71CS(1)183	B. J. Lindberg, S. Högberg, G. Malmsten, J. E. Bergmark, Ö. Nilsson, S. E. Karlsson, A. Fahlman, U. Gelius, R. Pinel, M. Stavaux, Y. Mollier and N. Lozac'h; *Chem. Scr.*, 1971, **1**, 183.	788, 1057
71DIS(B)4483	L. J. Darlage; Ph.D., Thesis, Iowa State Univ., 1971 (*Diss. Abstr. Int. B*, 1972, **32**, 4483).	5, 7, 17, 18, 19, 117
71FRP2043473	Agence Nationale de Valorisation de la Recherche and Institut National de la Recherche Agronomique; *Fr. Pat.* 2 043 473 (1971) (*Chem. Abstr.*, 1972, **76**, 14 341).	157, 170

71G259	V. Bertini and A. De Munno; *Gazz. Chim. Ital.*, 1971, **101**, 259.	541
71GEP1620103	R. Pfleger and E. Garthe; (Chemische Fabrik GmbH), *Ger. Pat.* 1 620 103 (1971) (*Chem. Abstr.* 1971, **75**, 49 090).	1039, 1046
71GEP1920245	H. J. Sturm and H. Armbrust; *Ger. Offen.* 1 920 245 (1971) (*Chem. Abstr.*, 1971, **74**, 22 816).	97
71GEP2024393	P. F. H. Freeman, J. M. Pedlar and I. T. Kay; *Ger. Offen.* 2 024 393 (1971) (*Chem. Abstr.*, 1971, **74**, 53 830).	105
71GEP2062928	A. Lang; *Ger. Pat.* 2 062 928 (1971) (*Chem. Abstr.*, 1971, **75**, 110 690).	421
71HCA1275	M. Märky, H.-J. Hansen and H. Schmid; *Helv. Chim. Acta*, 1971, **54**, 1275.	373
71HCA2111	E. Giovannini, J. Rosales and B. DeSouza; *Helv. Chim. Acta*, 1971, **54**, 2111.	18
71HCA2916	M. Georgarakis, T. Doppler, M. Maerky, H. J. Hansen and H. Schmid; *Helv. Chim. Acta*, 1971, **54**, 2916.	17, 18
71IJC1311	S. K. Shah, M. P. Patel and B. N. Mankad; *Indian J. Chem.*, 1971, **9**, 1311.	115
71IJC1355	J. M. Bachhawat and N. K. Mathur; *Indian J. Chem.*, 1971, **9**, 1335.	174
71IJS(B)85	L. Brandsma, P. J. W. Schuijl, D. Schuijl-Laros, J. Meijer and H. E. Wijers; *Int. J. Sulfur Chem., Part B*, 1971, 85.	1011
71IZV362	D. P. Deltsova, E. S. Ananyan and N. P. Gambaryan; *Izv. Akad. Nauk SSSR, Ser. Khim.*, 1971, 362.	67, 68
71JA1543	R. A. Olofson, R. K. Vander Meer and S. Stournas; *J. Am. Chem. Soc.*, 1971, **93**, 1543.	29, 33
71JA2258	D. L. Coffen, J. Q. Chambers, D. R. Williams, P. E. Garrett and N. D. Canfield; *J. Am. Chem. Soc.*, 1971, **93**, 2258.	817, 818
71JA2463	J. L. Charlton, C. C. Liao and P. De Mayo; *J. Am. Chem. Soc.*, 1971, **93**, 2463.	926, 929, 935
71JA4961	H. D. Hartzler; *J. Am. Chem. Soc.*, 1971, **93**, 4961.	965
71JA7045	P. Haake, L. P. Bausher and J. P. McNeal; *J. Am. Chem. Soc.*, 1971, **93**, 7045.	279
71JA7281	E. Shefter, B. E. Evans and E. C. Taylor; *J. Am. Chem. Soc.*, 1971, **93**, 7281.	515, 713, 715, 719, 731, 732
71JAP7100026	T. Motomiya and Y. Iwakura; *Jpn. Pat.* 71 00 026 (1971) (*Chem. Abstr.*, 1971, **74**, 141 743).	91
71JAP7142631	K. Matsukura and K. Takagi; *Jpn. Pat.* 71 42 631 (1971) (*Chem. Abstr.*, 1972, **77**, 103 168).	894
71JCS(B)415	G. Casalone and A. Mugnoli; *J. Chem. Soc. (B)*, 1971, 415.	855, 856, 891
71JCS(B)946	P. L. Johnson, K. I. G. Reid and I. C. Paul; *J. Chem. Soc. (B)*, 1971, 946.	785
71JCS(B)952	K. I. G. Reid and I. C. Paul; *J. Chem. Soc. (B)*, 1971, 952.	785
71JCS(B)1384	G. P. Blackbourn and B. J. Tighe; *J. Chem. Soc. (B)*, 1971, 1384.	868
71JCS(B)2096	A. Battaglia, A. Dondoni, C. Maccagnani and G. Mazzanti; *J. Chem. Soc. (B)*, 1971, 2096.	934
71JCS(B)2209	D. Dal Monte, E. Sandri, L. Di Nunno, S. Florio and P. E. Todesco; *J. Chem. Soc. (B)*, 1971, 2209.	532, 533
71JCS(B)2365	A. G. Burton, P. P. Forsythe, C. D. Johnson and A. R. Katritzky; *J. Chem. Soc. (B)*, 1971, 2365.	21, 22, 148
71JCS(C)86	F. DeSarlo, G. Dini and P. Lacrimini; *J. Chem. Soc. (C)*, 1971, 86.	104
71JCS(C)160	E. G. E. Hawkins; *J. Chem. Soc. (C)*, 1971, 160.	937
71JCS(C)225	G. V. Boyd and S. R. Dando; *J. Chem. Soc. (C)*, 1971, 225.	434
71JCS(C)409	G. V. Boyd and A. J. H. Summers; *J. Chem. Soc. (C)*, 1971, 409.	434
71JCS(C)584	R. J. MacConaill and F. L. Scott; *J. Chem. Soc. (C)*, 1971, 584.	94
71JCS(C)598	A. M. Knowles, A. Lawson, G. V. Boyd and R. A. Newberry; *J. Chem. Soc. (C)*, 1971, 598.	202
71JCS(C)963	R. J. S. Beer, D. Cartwright, R. J. Gait and H. Harris; *J. Chem. Soc. (C)*, 1971, 963.	1059, 1060
71JCS(C)974	J. S. Griffiths, C. F. Beam and C. R. Hauser; *J. Chem. Soc. (C)*, 1971, 974.	71, 72, 88
71JCS(C)993	D. L. Forster, T. L. Gilchrist and C. W. Rees; *J. Chem. Soc. (C)*, 1971, 993.	539
71JCS(C)1196	G. Jones and R. H. Good; *J. Chem. Soc. (C)*, 1971, 1196.	619, 620, 625
71JCS(C)1211	A. S. Bailey, M. W. Heaton and J. I. Murphy; *J. Chem. Soc. (C)*, 1971, 1211.	716, 730
71JCS(C)1527	G. R. Brown and W. R. Dyson; *J. Chem. Soc. (C)*, 1971, 1527.	680, 681, 683, 685, 686, 687, 707
71JCS(C)1615	D. W. Dunwell and D. Evans; *J. Chem. Soc. (C)*, 1971, 1615.	655
71JCS(C)1747	T. Caronna, R. Galli, V. Malatesta and F. Minisci; *J. Chem. Soc. (C)*, 1971, 1747.	265
71JCS(C)2094	D. Evans and D. W. Dunwell; *J. Chem. Soc. (C)*, 1971, 2094.	673, 679, 681, 689
71JCS(C)2314	G. V. Boyd and S. R. Dando; *J. Chem. Soc. (C)*, 1971, 2314.	439
71JCS(C)2644	T. Nishiwaki, T. Saito, S. Onomura and K. Kondo; *J. Chem. Soc. (C)*, 1971, 2644.	12
71JCS(C)2829	G. Duguay, D. H. Reid, K. O. Wade and R. G. Webster; *J. Chem. Soc. (C)*, 1971, 2829.	1061
71JCS(C)3062	R. B. Blackshires and C. J. Sharpe; *J. Chem. Soc. (C)*, 1971, 3062.	1006
71JCS(C)3873	G. V. Boyd and S. R. Dando; *J. Chem. Soc. (C)*, 1971, 3873.	720
71JCS(C)3994	E. Haddock, P. Kirby and A. W. Johnson; *J. Chem. Soc. (C)*, 1971, 3994.	154
71JHC89	D. J. McCaustland, W. H. Burton and C. C. Cheng; *J. Heterocycl. Chem.*, 1971, **8**, 89.	603
71JHC441	D. Pichler and R. N. Castle; *J. Heterocycl. Chem.*, 1971, **8**, 441.	536, 538, 734, 738, 748
71JHC503	V. D. Patil and L. B. Townsend; *J. Heterocycl. Chem.*, 1971, **8**, 503.	654, 660
71JHC551	H. C. Sorensen and L. L. Ingraham; *J. Heterocycl. Chem.*, 1971, **8**, 551.	814, 815

71JHC571	S. N. Lewis, G. A. Miller, M. Hausman and E. C. Szamborski; *J. Heterocycl. Chem.*, 1971, **8**, 571.	136, 137
71JHC591	S. N. Lewis, G. A. Miller, M. Hausman and E. C. Szamborski; *J. Heterocycl. Chem.*, 1971, **8**, 591.	142
71JHC707	O. Tsuge and H. Samura; *J. Heterocycl. Chem.*, 1971, **8**, 707.	1046
71JHC835	I. Lalezari and A. Shafiee; *J. Heterocycl. Chem.*, 1971, **8**, 835.	356, 358, 359
71JHC899	K. Pilgrim and F. Görgen; *J. Heterocycl. Chem.*, 1971, **8**, 899.	595
71JHC1011	A. Shafiee and I. Lalezari; *J. Heterocycl. Chem.*, 1971, **8**, 1011.	347
71JIC843	M. S. Solanki and J. P. Trivedi; *J. Indian Chem. Soc.*, 1971, **48**, 843.	584, 594, 595
71JMC10	L. M. Werbel and J. R. Battaglia; *J. Med. Chem.*, 1971, **14**, 10.	328
71JMC454	V. J. Bauer, R. P. Williams and S. R. Safir; *J. Med. Chem.*, 1971, **14**, 454.	989
71JMC458	G. M. Sieger, W. C. Barringer and J. E. Krueger; *J. Med. Chem.*, 1971, **14**, 458.	576
71JMC743	R. Ketcham and S. Mah; *J. Med. Chem.*, 1971, **14**, 743.	1022
71JMC977	C. J. Sharpe, R. S. Shadbolt, A. Ashford and J. W. Ross; *J. Med. Chem.*, 1971, **14**, 977.	1025
71JMC1075	G. Crank and M. Foulis; *J. Med. Chem.*, 1971, **14**, 1075.	658
71JMC1129	V. J. Bauer and S. R. Safir; *J. Med. Chem.*, 1971, **14**, 1129.	989
71JOC5	J. V. Burakevich, A. M. Lore and G. P. Volpp; *J. Org. Chem.*, 1971, **36**, 5.	397
71JOC10	K. T. Potts and S. Husain; *J. Org. Chem.*, 1971, **36**, 10.	976, 991
71JOC14	W. A. Thaler and J. R. McDivitt; *J. Org. Chem.*, 1971, **36**, 14.	497, 503, 507
71JOC19	J. P. Freeman and M. J. Hoare; *J. Org. Chem.*, 1971, **36**, 19.	982
71JOC207	K. Pilgram and M. Zupan; *J. Org. Chem.*, 1971, **36**, 207.	532
71JOC222	R. W. Franck and S. J. M. Gilligan; *J. Org. Chem.*, 1971, **36**, 222.	657
71JOC322	D. N. Harpp, J. G. Gleason and D. K. Ash; *J. Org. Chem.*, 1971, **36**, 322.	777
71JOC1068	T. A. Foglia, L. M. Gregory, G. Maerker and S. F. Osman; *J. Org. Chem.*, 1971, **36**, 1068.	318
71JOC1088	L. J. Darlage, T. H. Kinstle and C. L. McIntosh; *J. Org. Chem.*, 1971, **36**, 1088.	17
71JOC1543	R. A. Olofson, R. K. VanderMeer and S. Stournas; *J. Org. Chem.*, 1971, **93**, 1543.	4, 5, 6, 21
71JOC1589	C. S. Angadiyavar and M. V. George; *J. Org. Chem.*, 1971, **36**, 1589.	373
71JOC1846	K. T. Potts and R. Armbruster; *J. Org. Chem.*, 1971, **36**, 1846.	491, 734, 742
71JOC2155	Y. H. Chiang; *J. Org. Chem.*, 1971, **36**, 2155.	422
71JOC2784	G. Stork, M. Ohashi, H. Kamachi and H. Kakisawa; *J. Org. Chem.*, 1971, **36**, 2784.	34
71JOC2836	I. Lalezari, A. Shafiee and M. Yalpani; *J. Org. Chem.*, 1971, **36**, 2836.	348, 349, 352, 353
71JOC3211	E. C. Taylor, G. P. Beardsley and Y. Maki; *J. Org. Chem.*, 1971, **36**, 3211.	412, 418 713, 717, 718, 719, 729
71JOC3465	J. E. Oliver; *J. Org. Chem.*, 1971, **36**, 3465.	502, 917
71JOC3470	A. Ius, C. Parini, G. Sportoletti, G. Vecchio and G. Ferrara; *J. Org. Chem.*, 1971, **36**, 3470.	90
71JOC3885	A. P. Krapcho, D. R. Rao, M. P. Silvon and B. Abegaz; *J. Org. Chem.*, 1971, **36**, 3885.	556, 572
71JPR722	K. Fabian and H. Hartmann; *J. Prakt. Chem.*, 1971, **313**, 722.	844
71JPR745	M. Muehlstaedt and B. Schulze; *J. Prakt. Chem.*, 1971, **313**, 745.	89
71JPR804	V. H. Spies, K. Gewald and R. Mayer; *J. Prakt. Chem.*, 1971, **313**, 804.	970
71JPR1148	G. Barnikow and J. Bodeker; *J. Prakt. Chem.*, 1971, **313**, 1148.	497, 1016
71JST(9)163	L. Nygaard, R. Lykke Hansen and G. O. Soerensen; *J. Mol. Struct.*, 1971, **9**, 163 (*Chem. Abstr.*, 1972, **76**, 19 755).	548, 556
71JST(9)222	L. Nygaard, E. Asmussen, J. H. Høg, R. C. Maheshwari, C. H. Nielsen, I. B. Petersen, J. Rastrup-Andersen and C. O. Soerensen; *J. Mol. Struct.*, 1971, **9**, 222.	239
71JST(9)321	O. Martensson; *J. Mol. Struct.*, 1971, **9**, 321.	368, 369
71KGS905	V. N. Artemov and O. P. Shvaika; *Khim. Geterotsikl. Soedin.*, 1971, 905 (*Chem. Abstr.*, 1972, **76**, 140 741).	433
71LA(744)88	C. Grundmann, S. K. Datta and R. F. Sprecher; *Liebigs Ann. Chem.*, 1971, **744**, 88.	1035, 1044
71LA(748)143	H. Stamm and J. Hoenicke; *Liebigs Ann. Chem.*, 1971, **748**, 143.	631
71M321	F. Asinger, A. Saus, H. Offermans, D. Neuray and K. H. Lim; *Monatsh. Chem.*, 1971, **102**, 321 (*Chem. Abstr.*, 1971, **74**, 111 983).	318
71M550	H. Paul and A. Sitte; *Monatsh. Chem.*, 1971, **102**, 550 (*Chem. Abstr.*, 1971, **74**, 141 901).	567
71MI41600	G. S. Shchegoleva and V. A. Barkhash; *Izv. Sib. Otd. Akad. Nauk SSSR, Ser. Khim. Nauk*, 1971, **6**, 123 (*Chem. Abstr.*, 1972, **77**, 48 316).	115
71MI41601	K. Tada, Y. Numata and T. Katsumura; *J. Appl. Polym. Sci.*, 1971, **15**, 117.	107
71MI41602	N. Sampei and K. Tomita; *Sankyo Kenkyusho Nempo*, 1971, **23**, 245 (*Chem. Abstr.*, 1972, **76**, 153 655).	106
71MI41603	A. Sammour, Y. Akhnookh and H. Jahine; *U.A.R.J. Chem.*, 1971, **14**, 213 (*Chem. Abstr.*, 1972, **77**, 152 060).	98
71MI41604	A. Branski; *Neft. Gaz. Ikh. Prod.*, 1971, 172 (*Chem. Abstr.*, 1972, **77**, 164 572).	89
B-71MI41605	J. R. Hoggett, R. B. Moodi, J. R. Penton and K. Schofield; 'Nitration and Aromatic Reactivity,' Cambridge University Press, Cambridge, 1971, p. 206.	11, 23
71MI41606	T. Nguyen, N. C. Nguyen and J. Weimann; *Ann. Chim. (Paris)*, 1971, **6**, 235 (*Chem. Abstr.*, 1972, **76**, 3771).	94
71MI41700	R. Raap; *J. Antibiotics*, 1971, **24**, 695 (*Chem. Abstr.*, 1972, **76**, 14 409).	155

71MI41701	T. Onaka and T. Oikawa; *ITSUU Kenkyusho Nempo*, 1971, 53 (*Chem. Abstr.*, 1972, **77**, 48 320).	162, 163
71MI41900	E. M. White, E. Rapaport, H. M. Seliger and T. A. Hopkins; *Bioorg. Chem.*, 1971, **1**, 92.	327
B-71MI42100	C. Grundmann and P. Grunanger; 'The Nitrile Oxides', Springer-Verlag, Berlin, 1971, pp. 44ff.	389
B-71MI42200	C. Grundmann and P. Grünanger; 'The Nitrile Oxides', Springer-Verlag, Berlin, 1971, ch. 3.	422
71MI42500	H. N. Miller and R. T. DeNeve; *Plant Dis. Rep.*, 1971, **55**, 587 (*Chem. Abstr.*, 1972, **76**, 772).	509
71MI42600	R. H. Hanson and C. E. Meloan; *Inorg. Nucl. Chem. Lett.*, 1971, **7**, 461.	526
71MI42601	R. H. Hanson and C. E. Meloan; *Inorg. Nucl. Chem. Lett.*, 1971, **7**, 467.	526
71MI42602	J. Marn, B. Stanovnik and M. Tisler; *Croat. Chem. Acta*, 1971, **43**, 101 (*Chem. Abstr.*, 1971, **75**, 98 523).	538
71MI42603	M. Carmack, L. M. Weinstock, D. Shew, F. H. Marquardt, R. Y. Wen, I. W. Stapleton and R. W. Street; *Proc. Indiana Acad. Sci.*, 1971, **80**, 164.	533
71MI42900	J. Marn, B. Stanovnik and M. Tisler; *Croat. Chem. Acta*, 1971, **43**, 101 (*Chem. Abstr.*, 1971, **75**, 98 523).	737
71MI43200	Z. Yoshida and T. Kobayashi; *Theor. Chim. Acta*, 1971, **20**, 216.	815
71MI43600	M. El Sekily; *Carbohydr. Res.*, 1971, **49**, 141.	996
71MI43601	J. Moss and M. D. Lane; *Adv. Enzymol.*, 1971, **35**, 321.	1024
B-71MI43602	Ch. Grundmann and P. Gruenanger; 'The Nitride Oxides', Springer, Berlin, 1971.	997
B-71MS	Q. N. Porter and J. Baldas; 'Mass Spectrometry of Heterocyclic Compounds', Wiley, New York, 1971.	399, 429
71NKK867	S. Kambe, T. Hayashi, H. Yasuda and A. Sakura; *Nippon Kagaku Zasshi*, 1971, **92**, 867 (*Chem. Abstr.*, 1972, **77**, 5397).	320
71OPP163	R. G. Micetich; *Org. Prep. Proced. Int.*, 1971, **3**, 163.	457
71OPP167	R. G. Micetich and R. Raap; *Org. Prep. Proced. Int.*, 1971, **3**, 167.	164
71PMH(3)1	A. Albert; *Phys. Methods Heterocycl. Chem.*, 1971, **3**, 1.	10
71PMH(3)67	W. L. F. Armarego; *Phys. Methods Heterocycl. Chem.*, 1971, **3**, 67.	817
71PMH(3)180	W. L. F. Armarego; *Phys. Methods Heterocycl. Chem.*, 1971, **3**, 180.	452, 552
71PMH(3)223	G. Spiteller; *Phys. Methods Heterocycl. Chem.*, 1971, **3**, 223.	6
71PMH(4)21	E. A. C. Lucken; *Phys. Methods Heterocycl. Chem.*, 1971, **4**, 21.	6
71PMH(4)237	J. Kraft and S. Walker; *Phys. Methods Heterocycl. Chem.*, 1971, **4**, 237.	4
71PMH(4)239	J. Kraft and S. Walker; *Phys. Methods Heterocycl. Chem.*, 1971, **4**, 239.	368
71PMH(4)265	A. R. Katritzky and P. J. Taylor; *Phys. Methods Heterocycl. Chem.*, 1971, **4**, 265.	4, 5, 398
71S16	R. Feinauer; *Synthesis*, 1971, **16**.	984
71S433	P. Caramella and E. Cereda; *Synthesis*, 1971, 433.	69, 70, 84
71T19	L. V. Grobrovsky and G. L. Schmir; *Tetrahedron*, 1971, **27**, 19.	270
71T187	H. Meier and E. Voigt; *Tetrahedron*, 1972, **28**, 187.	353
71T711	P. Crabbé, L. A. Maldonado and I. Sanchez; *Tetrahedron*, 1971, **27**, 711.	78, 116
71T3447	U. Olthoff, R. Huettenrauch and K. Matthey; *Tetrahedron*, 1971, **27**, 3447.	701, 706
71T4003	K. Hirai; *Tetrahedron*, 1971, **27**, 4003.	816, 818
71T4117	R. Neidlein and H. Reuter; *Tetrahedron*, 1971, **27**, 4117.	494
71T4407	M. E. Rennekamp, J. V. Paukstelis and R. G. Cooks; *Tetrahedron*, 1971, **27**, 4407.	187
71T4449	M. Gotz and K. Grozinger; *Tetrahedron*, 1971, **27**, 4449.	377
71T4705	K. Fabian, H. Hartmann, J. Fabian and R. Mayer; *Tetrahedron*, 1971, **27**, 4705.	785, 815, 816, 817
71TL477	R. Huisgen, V. Martin-Ramos and W. Scheer; *Tetrahedron Lett.*, 1971, 477.	318
71TL1075	D. E. L. Carrington, K. Clarke and R. M. Scrowston; *Tetrahedron Lett.*, 1971, 1075.	162
71TL1137	K. Hirai; *Tetrahedron Lett.*, 1971, 1137.	762
71TL1281	H. Gotthardt; *Tetrahedron Lett.*, 1971, 1281.	156
71TL1315	J. Ashby and H. Suschitzky; *Tetrahedron Lett.*, 1971, 1315.	170
71TL1317	C. W. Pluijgers, J. W. Vonk and G. D. Thorn; *Tetrahedron Lett.*, 1971, 1317.	904, 1016
71TL1729	F. L. Scott, T. M. Lambe and R. N. Butler; *Tetrahedron Lett.*, 1971, 1729.	1009
71TL1757	A. P. Schaap; *Tetrahedron Lett.*, 1971, 1757.	764
71TL1781	Y. Hayashi, T. Akazawa, K. Yamamoto and R. Oda; *Tetrahedron Lett.*, 1971, 1781.	880, 890
71TL2143	C. W. Bird and C. K. Wong; *Tetrahedron Lett.*, 1971, 2143.	536
71TL2749	H. Gotthardt and F. Reiter; *Tetrahedron Lett.*, 1971, 2749.	374
71TL4343	D. Kobelt, E. F. Paulus and G. Lohaus; *Tetrahedron Lett.*, 1971, 4343.	901
71USP3530145	E. Klingsberg; (Am. Cyanamid Co.), *U.S. Pat.* 3 530 145 (1970) (*Chem. Abstr.*, 1971, **74**, 4696).	811
71USP3555039	R. E. Manning; *U.S. Pat.* 3 555 039 (1971).	1025
71USP3565571	G. Reese, P. Berth and K. J. Boosen; *U.S. Pat.* 3 565 571 (1971) (*Chem. Abstr.*, 1972, **77**, 105 518).	894
71USP3575968	E. Klingsberg; (Am. Cyanamid Co.), *U.S. Pat.* 3 575 968 (1971) (*Chem. Abstr.*, 1971, **75**, 37 915).	811
71USP3578682	A. B. Ash, C. L. Stevens and A. Markovac; *U.S. Pat.* 3 578 682 (1971) (*Chem. Abstr.*, 1971, **75**, 36 044).	895
71USP3579506	R and L Molecular Research Ltd.; *U.S. Pat.* 3 579 506 (1971) (*Chem. Abstr.*, 1971, **75**, 36 084).	156

71USP3584003	R. Seltzer and W. J. Considine; *U.S. Pat.* 3 584 003 (1971) (*Chem. Abstr.*, 1971, **75**, 63 796).	939
71USP3615633	D. A. Brooks; *U.S. Pat.* 3 615 633 (1971) (*Chem. Abstr.*, 1973, **78**, 36 228).	446
71USP3630745	*U.S. Pat.* 3 630 745 (1971) (*Chem. Abstr.*, 1972, **76**, 106 437).	710
71ZC460	H. Hartmann; *Z. Chem.*, 1971, **11**, 460.	733, 740
71ZOR1309	A. L. Fridman, F. A. Gabitov and A. D. Nikolaeva; *Zh. Org. Khim.*, 1971, **7**, 1309 (*Chem. Abstr.*, 1971, **75**, 98 477).	103
71ZOR2227	M. B. Kolesova, L. I. Maksimova and A. V. El'tsov; *Zh. Org. Khim.*, 1971, **7**, 2227 (*Chem. Abstr.*, 1972, **76**, 25 183).	908, 919
72ACS250	C. Th. Pedersen and J. Moller; *Acta Chem. Scand.*, 1972, **26**, 250.	787
72ACS343	A. Hordvik and K. Julshamn; *Acta Chem. Scand.*, 1972, **26**, 343.	1054
72ACS459	U. Svanholm; *Acta Chem. Scand.*, 1972, **26**, 459.	549
72ACS858	C. Christophersen and S. Treppendahl; *Acta. Chem. Scand.*, 1972, **26**, 858.	603, 604
72ACS873	J. Sletten; *Acta Chem. Scand.*, 1972, **26**, 873.	1055
72ACS947	K. Berg-Nielsen, T. Stensrud and E. Bernatek; *Acta Chem. Scand.*, 1972, **26**, 947.	686, 707
72ACS1297	A. Hordvik, O. Sjølset and L. J. Saethre; *Acta Chem. Scand.*, 1972, **26**, 1297.	1053
72ACS1847	T. Groenneberg and K. Undheim; *Acta Chem. Scand.*, 1972, **26**, 1847.	682
72ACS2139	A. Hordvik, T. S. Rimala and L. J. Saethre; *Acta Chem. Scand.*, 1972, **26**, 2139.	1052, 1054, 1057
72ACS2140	B. Birknes, A. Hordvik and L. J. Saethre; *Acta Chem. Scand.*, 1972, **26**, 2140.	1053
72ACS3114	A. Hordvik and L. J. Saethre; *Acta Chem. Scand.*, 1972, **26**, 3114.	1053
72ACS3131	P. Groth; *Acta Chem. Scand.*, 1972, **26**, 3131.	670
72AG297	H. Sauter and H. Prinzbach; *Angew. Chem.*, 1972, **84**, 297 (*Angew. Chem., Int. Ed. Engl.*, 1972, **11**, 296).	878
72AG436	H. Bock and B. Soloukis; *Angew. Chem.*, 1972, **84**, 436 (*Angew. Chem., Int. Ed. Engl.*, 1972, **11**, 436).	861
72AHC(14)1	K. R. H. Wooldridge; *Adv. Heterocycl. Chem.*, 1972, **14**, 1.	132, 142, 147, 148, 150, 151, 152, 155, 156, 157, 158, 159, 160, 161, 162, 163, 164, 166, 167, 168, 169, 170, 171, 172, 174, 175
72AHC(14)36	K. R. H. Wooldridge; *Adv. Heterocycl. Chem.*, 1972, **14**, 36 (footnote 147).	996
72AHC(14)43	M. Davis; *Adv. Heterocycl. Chem.*, 1972, **14**, 43.	132, 137, 143, 147, 148, 149, 150, 151, 154, 155, 157, 158, 159, 160, 161, 162, 163, 166, 167, 168, 169, 170, 171, 174, 175
72AJC993	R. L. N. Harris; *Aust. J. Chem.*, 1972, **25**, 993.	734, 737, 741
72AJC2547	G. A. Heath, R. L. Martin and A. F. Masters; *Aust. J. Chem.*, 1972, **25**, 2547.	793
72ANY(192)25	W. H. H. Günther and M. N. Salzman; *Ann. N.Y. Acad. Sci.*, 1972, **192**, 25.	948
72AP359	H. Stamm and J. Hoenicke; *Arch. Pharm.* (*Weinheim, Ger.*), 1972, **305**, 359 (*Chem. Abstr.*, 1972, **77**, 75 167).	105
72AP833	T. Denzel and H. Hoehn; *Arch. Pharm.* (*Weinheim, Ger.*), 1972, **305**, 833.	622, 628
72AP902	F. Meissner and K. Hartke; *Arch. Pharm.* (*Weinheim, Ger.*), 1972, **305**, 902.	166
72AX(B)1116	D. Britton and W. E. Noland; *Acta Crystallogr., Part B*, 1972, **28**, 1116.	396
72AX(B)1207	A. Corradi-Bonamartini, M. Nardelli and C. Palmieri; *Acta Crystallogr., Part B*, 1972, **28**, 1207.	136
72AX(B)1584	T. C. Downie, W. Harrison, E. S. Raper and M. A. Hepworth; *Acta Crystallogr., Sect. B*, 1972, **28**, 1584.	548
72BCJ213	A. Chinone, K. Inouye and M. Ohta; *Bull. Chem. Soc. Jpn.*, 1972, **45**, 213.	805
72BCJ489	K. Tanaka and T. Tanaka; *Bull. Chem. Soc. Jpn.*, 1972, **45**, 489.	971
72BCJ928	T. Nishiyama, Y. Fujimoto and F. Yamada; *Bull. Chem. Soc. Jpn.*, 1972, **45**, 928.	911
72BCJ960	S. Tamagaki and S. Oae; *Bull. Chem. Soc. Jpn.*, 1972, **45**, 960.	797
72BCJ1507	T. Hayashi; *Bull. Chem. Soc., Jpn.*, 1972, **45**, 1507.	779
72BCJ1846	H. Fujita, R. Endo, A. Aoyama and T. Ichii; *Bull. Chem. Soc. Jpn.*, 1972, **45**, 1846.	65
72BCJ3202	Y. Huseya, A. Chinone and M. Ohta; *Bull. Chem. Soc. Jpn.*, 1972, **45**, 3202.	374
72BCJ3220	I. Iwataki and A. Ueda; *Bull. Chem. Soc. Jpn.*, 1972, **45**, 3220.	915
72BCJ3572	I. Iwataki; *Bull. Chem. Soc. Jpn.*, 1972, **45**, 3572.	914
72BRP1266542	Hokko Chem. Ind. Co. Ltd., *Br. Pat.* 1 266 542 (1972) (*Chem. Abstr.* 1972, **77**, 5474).	577
72BRP1285333	*Br. Pat.* 1 285 333 (1972) (*Chem. Abstr.*, 1972, **77**, 126 671).	730
72BSB279	C. Draguet and M. Renson; *Bull. Soc. Chim. Belg.*, 1972, **81**, 279.	346
72BSB289	C. Draguet and M. Renson; *Bull. Soc. Chim. Belg.*, 1972, **81**, 289.	346
72BSB295	C. Draguet and M. Renson; *Bull. Soc. Chim. Belg.*, 1972, **81**, 295.	346
72BSB303	C. Draguet and M. Renson; *Bull. Soc. Chim. Belg.*, 1972, **81**, 303.	346
72BSF162	J. C. Poite, J. Roggero, H. J. M. Dou, G. Vernin and J. Metzger; *Bull. Soc. Chim. Fr.*, 1972, 162.	133, 148
72BSF330	A. Belly, R. Jacquier, F. Petrus and J. Verducci; *Bull. Soc. Chim. Fr.*, 1972, 330.	93
72BSF1040	R. Garnier, R. Faure, A. Babadjamian and E. J. Vincent; *Bull. Soc. Chim. Fr.*, 1972, 1040 (*Chem. Abstr.*, 1972, **77**, 11 964).	237
72BSF1055	L. Bouscasse, M. Chanon, R. Phan-Tan-Luu, E. J. Vincent and J. Metzger; *Bull. Soc. Chim. Fr.*, 1972, 1055 (*Chem. Abstr.*, 1972, **77**, 87 606).	239
72BSF1173	G. Vernin, H. J. M. Dou and J. Metzger; *Bull. Soc. Chim. Fr.*, 1972, 1173.	153
72BSF1385	R. Pinel and Y. Mollier; *Bull. Soc. Chim. Fr.*, 1972, 1385.	1056
72BSF1840	C. Trebaul; *Bull. Soc. Chim. Fr.*, 1972, 1840.	800
72BSF2296	J. C. Poite, J. Julien, E. J. Vincent and J. Roggero; *Bull. Soc. Chim. Fr.*, 1972, 2296.	160

72BSF2365	H. Jubault and D. Peltier; *Bull. Soc. Chim. Fr.*, 1972, 2365.	125
72BSF2679	J. P. Aune, R. Phan-Tan-Luu, E. J. Vincent and J. Metzger; *Bull. Soc. Chim. Fr.*, 1972, 2679 (*Chem. Abstr.*, 1973, **78**, 3389).	237
72BSF3862	Y. Ferré, E. J. Vincent, H. Larivé and J. Metzger; *Bull. Soc. Chim. Fr.*, 1972, 3862 (*Chem. Abstr.*, 1973, **78**, 57 518).	237, 240
72BSF4181	J. Bignebat and H. Quiniou; *Bull. Soc. Chim. Fr.*, 1972, 4181.	1060
72BSF4576	C. Metayer, G. Duguay and H. Quiniou; *Bull. Soc. Chim. Fr.*, 1972, 4576.	1065
72C501	H. G. Buehrer; *Chimia*, 1972, **26**, 501.	214
72CB188	H. Gotthardt; *Chem. Ber.*, 1972, **105**, 188.	907, 908, 910, 941, 999
72CB196	H. Gotthardt; *Chem. Ber.*, 1972, **105**, 196.	919
72CB1307	K. Bunge, R. Huisgen, R. Raab and H. J. Sturm; *Chem. Ber.*, 1972, **105**, 1307 (*Chem. Abstr.*, 1972, **77**, 19 577).	305
72CB1568	J. Goerdeler and J. Ulmen; *Chem. Ber.*, 1972, **105**, 1568.	942
72CB2462	S. L. Spassov, J. N. Stefanovsky, B. J. Kurtev and G. Fodor; *Chem. Ber.*, 1972, **105**, 2462.	181
72CB2805	R. Huisgen and W. Mack; *Chem. Ber.*, 1972, **105**, 2805.	910, 928, 929, 930, 940
72CB2815	R. Huisgen and W. Mack; *Chem. Ber.*, 1972, **105**, 2815.	925, 927, 934, 941
72CB2841	H. Blaschke, E. Brunn, R. Huisgen and W. Mack; *Chem. Ber.*, 1972, **105**, 2841.	606, 607, 609
72CB3638	K. Griesbaum and J. Brüggemann; *Chem. Ber.*, 1972, **105**, 3638.	892
72CC222	L. K. Hansen, A. Hordvik and L. J. Saethre; *J. Chem. Soc., Chem. Commun.*, 1972, 222.	1052
72CC418	M. Kawanisi and K. Matsunaga; *J. Chem. Soc., Chem. Commun.*, 1972, 418.	349
72CC498	H. Kato, T. Shiba and Y. Miki; *J. Chem. Soc., Chem. Commun.*, 1972, 498.	374
72CC540	S. Davidson and D. Leaver; *J. Chem. Soc., Chem. Commun.*, 1972, 540.	1069
72CC543	A. Hordvik and P. Oftedal; *J. Chem. Soc., Chem. Commun.*, 1972, 543.	1053
72CC836	A. J. Barnett, R. J. S. Beer, B. V. Karaoghlanian, E. C. Llaguno and I. C. Paul; *J. Chem. Soc., Chem. Commun.*, 1972, 836.	1053, 1061, 1062, 1067
72CC896	T. Matsuura and Y. Ito; *J. Chem. Soc., Chem. Commun.*, 1972, 896.	271
72CC1117	C. S. Wu, W. A. Szarek and J. K. N. Jones; *J. Chem. Soc., Chem. Commun.*, 1972, 1117.	421
72CC1283	D. H. Reid and R. G. Webster; *J. Chem. Soc., Chem. Commun.*, 1972, 1283.	1059, 1069
72CC1300	R. M. Moriarty and A. Chin; *J. Chem. Soc., Chem. Commun.*, 1972, 1300.	565
72CCC2273	J. Ciernik; *Collect. Czech. Chem. Commun.*, 1972, **37**, 2273 (*Chem. Abstr.*, 1973, **78**, 5378).	276
72CHE297	V. A. Sergeev, V. G. Pesin and M. M. Kotikova; *Chem. Heterocycl. Compd. (Engl. Transl.)*, 1972, **8**, 297.	532
72CHE392	V. I. Shvedov, V. K. Vasil'eva and A. N. Grinev; *Chem. Heterocycl. Compd. (Engl. Transl.)*, 1972, **8**, 392.	988
72CHE943	V. A. Kosobutskii, G. I. Kagan, V. K. Belyakov and O. G. Tarakanov; *Chem. Heterocycl. Compd. (Engl. Transl.)*, 1972, **8**, 943.	428
72CHE1017	L. M. Alekseeva, E. M. Peresleni, Y. N. Sheinker, P. M. Kochergin, A. N. Krasovskii and B. V. Kurmaz; *Chem. Heterocycl. Compd. (Engl. Transl.)*, 1972, **8**, 1017.	981
72CHE1223	N. O. Saldabol, L. L. Zeligman, S. A. Giller, Y. Y. Popelis, A. E. Abele and L. N. Alekseeva; *Chem. Heterocycl. Compd. (Engl. Transl.)*, 1972, **8**, 1223.	979
72CHE1428	V. I. Shvedov, L. B. Altukhova, Yu. I. Trofimkin and A. N. Grinev; *Chem. Heterocycl. Compd. (Engl. Transl.)*, 1972, **8**, 1428.	981, 984
72CJC324	F. Leung and S. C. Nyburg; *Can. J. Chem.*, 1972, **50**, 324.	1054
72CJC627	A. W. Richardson; *Can. J. Chem.*, 1972, **50**, 627.	521
72CJC2088	B. J. Forrest and A. W. Richardson; *Can. J. Chem.*, 1972, **50**, 2088.	398
72CJC2326	S. L. Lee, A. M. Cameron and J. Warkentin; *Can. J. Chem.*, 1972, **50**, 2326.	437
72CJC2568	G. E. Bachers, D. M. McKinnon and J. M. Buchshriber; *Can. J. Chem.*, 1972, **50**, 2568.	790, 794
72CJC3079	D. E. Horning and J. M. Muchowski; *Can. J. Chem.*, 1972, **50**, 3079.	428, 430, 442
72CJC3248	P. Knittel, S. L. Lee and J. Warkentin; *Can. J. Chem.*, 1972, **50**, 3248.	442
72CJC4029	D. R. Kerur and D. G. M. Diaper; *Can. J. Chem.*, 1972, **50**, 4029.	866
72CPB304	K. Hirai and T. Ishiba; *Chem. Pharm. Bull.*, 1972, **20**, 304.	752, 756, 758, 779
72CPB1663	H. Saikachi, N. Shimojo and Y. Uehara; *Chem. Pharm. Bull.*, 1972, **20**, 1663.	442
72CPB2209	Y. Nakagawa, O. Aki and K. Sirakawa; *Chem. Pharm. Bull.*, 1972, **20**, 2209.	21, 29, 126
72CPB2372	O. Aki, Y. Nakagawa and K. Sirakawa; *Chem. Pharm. Bull.*, 1972, **20**, 2372.	149
72CR(B)(274)532	G. Mille, G. Davidovics and J. Chouteau; *C. R. Hebd. Seances Acad. Sci., Ser. B*, 1972, **274**, 532.	182
72CR(C)(274)1215	G. Le Coustumer and Y. Mollier; *C. R. Hebd. Seances Acad. Sci., Ser. C*, 1972, **274**, 1215.	790
72CR(C)(274)1871	J. Julien, J. C. Poite, G. Salmona and E. J. Vincent; *C.R. Hebd. Seances Acad. Sci., Ser. C*, 1972, **274**, 1871.	142
72CR(C)(275)909	R. Pinel, Y. Mollier, J. P. Barbeyrac and G. Pfister-Guillouzo; *C. R. Hebd. Seances Acad. Sci., Ser. C*, 1972, **275**, 909.	1052
72CRV357	C. U. Pittman, Jr., S. P. McManus and J. W. Larsen; *Chem. Rev.*, 1972, **72**, 357.	753, 759, 772, 779
72G23	G. D. Andreetti, G. Bocelli, L. Cavalca and P. Sgarabotto; *Gazz. Chim. Ital.*, 1972, **102**, 23.	901, 903
72G223	A. Selva and V. Vettori; *Gazz. Chim. Ital.*, 1973, **103**, 223.	6

72G395	P. Bravo, G. Gaudiano and C. Ticozzi; *Gazz. Chim. Ital.*, 1972, **102**, 395.	101
72GEP2049691	W. Lorenz, H. Boeshagen, I. Hammann and W. Stendel; *Ger. Offen.* 2 049 691 (1972) (*Chem. Abstr.*, 1972, **77**, 34 483).	130
72GEP2059990	K. Sasse and P. E. Frohberger; *Ger. Pat.* 2 059 990 (1972) (*Chem. Abstr.*, 1972, **77**, 110 567).	932, 941, 944
72GEP2121525	K. Maier; *Ger. Offen.* 2 121 525 (1972) (*Chem. Abstr.*, 1973, **78**, 45 048).	894
72GEP2162157	M. Goodman and G. S. Kirshenbaum; *Ger. Offen.* 2 162 157 (1972) (*Chem. Abstr.*, 1973, **78**, 125 762).	894
72GEP2223648	Hoffmann-La Roche and Co. A.G.; *Ger. Pat.* 2 223 648 (1972) (*Chem. Abstr.*, 1973, **78**, 43 533).	155
72HCA75	P. Stütz and P. A. Stadler; *Helv. Chim. Acta*, 1972, **55**, 75.	844
72HCA213	H. F. Eicke and H. Christen; *Helv. Chim. Acta*, 1972, **55**, 213.	788
72HCA1730	T. Doppler, H.-J. Hansen and H. Schmid; *Helv. Chim. Acta*, 1972, **55**, 1730.	18
72IC2531	D. Pilipovich, C. J. Schack and R. D. Wilson; *Inorg. Chem.*, 1972, **11**, 2531.	879, 891
72IJC318	M. P. Mahajan, S. K. Vasudeva and N. K. Ralhan; *Indian J. Chem.*, 1972, **10**, 318 (*Chem. Abstr.*, 1972, **77**, 101 444).	316
72IJC881	M. N. Gudi and M. V. George; *Indian J. Chem.*, 1972, **10**, 881.	152
72IJC1138	M. G. Paranjpe and A. S. Mahajan; *Indian J. Chem.*, 1972, **10**, 1138.	918, 943
72IJC1194	N. S. Ramegowda, C. K. Narang, J. M. Bachhawat and N. K. Mathur; *Indian J. Chem.*, 1972, **10**, 1194.	174
72IJS(A)15	D. Gentric and P. Saumagne; *Int. J. Sulfur Chem.*, Part A, 1972, **2**, 15.	786
72IJS(A)25	A. De Munno, V. Bertini and G. Denti; *Int. J. Sulfur Chem.*, Part A, 1972, **2**, 25.	525, 535
72IZV2517	S. R. Sterlin, L. G. Zhuravkova, B. L. Dyatkin and I. L. Knunyants; *Izv. Akad. Nauk SSSR, Ser. Khim.*, 1972, 2517.	889
72JA1365	O. L. Chapman, P. W. Wojtkowski, W. Adam, O. Rodriquez and R. Rucktäschel; *J. Am. Chem. Soc.*, 1972, **94**, 1365.	768
72JA4856	L. A. Hull, I. C. Hisatsune and J. Heicklen; *J. Am. Chem. Soc.*, 1972, **94**, 4856.	859, 882
72JA4952	J. H. Hall, F. E. Behr and R. L. Reed; *J. Am. Chem. Soc.*, 1972, **94**, 4952.	122
72JA5599	D. A. Sweigart and D. W. Turner; *J. Am. Chem. Soc.*, 1972, **94**, 5599.	757
72JA6337	C. W. Gillies and R. L. Kuczkowski; *J. Am. Chem. Soc.*, 1972, **94**, 6337.	855, 858
72JA7108	S. F. Nelsen and P. J. Hintz; *J. Am. Chem. Soc.*, 1972, **94**, 7108.	725, 727
72JA7180	H. H. Wasserman, F. J. Vinick and Y. C. Chang; *J. Am. Chem. Soc.*, 1972, **94**, 7180.	198
72JA9128	G. Büchi and J. C. Vederas; *J. Am. Chem. Soc.*, 1972, **94**, 9128.	33
72JAP7207549	Kumiai Chem. Ind. Co. Ltd. *Jpn. Pat.* 72 07 549 (1972) (*Chem. Abstr.*, 1972, **77**, 5491).	577
72JAP7212350	*Jpn. Pat.* 72 12 350 (1972) (*Chem. Abstr.*, 1972, **77**, 34 595).	709
72JAP7230833	Mitsui Toatsu Chem. Co. Ltd., *Jpn. Pat.* 72 30 833 (1972) (*Chem. Abstr.*, 1973, **78**, 155 430).	510
72JAP(K)7217781	Sankyo Co. Ltd.; *Jpn. Kokai*, 72 17 781 (1972) (*Chem. Abstr.*, 1972, **77**, 140 107).	152
72JAP(K)7242659	H. Kano, M. Ogata and Y. Watanabe; *Jpn. Kokai* 72 42 659 (1972) (*Chem. Abstr.*, 1973, **78**, 97 623).	119, 129
72JCS(P1)41	D. B. J. Easton, D. Leaver and T. J. Rawlings; *J. Chem. Soc., Perkin Trans. 1*, 1972, 41.	798, 848
72JCS(P1)90	T. Nishiwaki and K. Kondo; *J. Chem. Soc., Perkin Trans. 1*, 1972, 90.	40
72JCS(P1)437	G. S. King, P. D. Magnus and H. S. Rzepa; *J. Chem. Soc., Perkin Trans. 1*, 1972, 437.	36, 37
72JCS(P1)914	G. V. Boyd and P. H. Wright; *J. Chem. Soc., Perkin Trans. 1*, 1972, 914.	183
72JCS(P1)1140	G. V. Boyd and P. H. Wright; *J. Chem. Soc., Perkin Trans. 1*, 1972, 1140.	204
72JCS(P1)1310	E. P. Goodings, D. A. Mitchard and G. Owen; *J. Chem. Soc., Perkin Trans. 1*, 1972, 1310.	1067
72JCS(P1)1360	J. G. Dingwall, A. R. Dunn, D. H. Reid and K. O. Wade; *J. Chem., Soc., Perkin Trans. 1*, 1972, 1360.	362, 1066
72JCS(P1)1432	M. D. Scott; *J. Chem. Soc., Perkin Trans. 1*, 1972, 1432.	159, 160
72JCS(P1)1587	J. Ackrell, M. Altaf-ur-Rahman, A. J. Boulton and R. C. Brown; *J. Chem. Soc., Perkin Trans. 1*, 1972, 1587.	81
72JCS(P1)2165	T. L. Gilchrist, P. G. Mente and C. W. Rees; *J. Chem. Soc., Perkin Trans. 1*, 1972, 2165.	349
72JCS(P1)2305	P. Sykes and H. Ullah; *J. Chem. Soc., Perkin. Trans. 1*, 1972, 2305.	792, 807
72JCS(P1)2441	R. H. Good, G. Jones and J. R. Phipps; *J. Chem. Soc., Perkin Trans. 1*, 1972, 2441.	630
72JCS(P1)3006	D. E. L. Carrington, K. Clarke, C. G. Hughes and R. M. Scrowston; *J. Chem. Soc., Perkin Trans. 1*, 1972, 3006.	149, 152, 155
72JCS(P2)4	S. Nicholson, G. J. Stacey and P. J. Taylor; *J. Chem. Soc., Perkin Trans. 2*, 1972, 4.	1019
72JCS(P2)565	M. Davis, M. F. Mackay and W. A. Denne; *J. Chem. Soc., Perkin Trans. 2*, 1972, 565.	133, 135, 137
72JCS(P2)1682	A. R. Katritzky and Y. Takeuchi; *J. Chem. Soc., Perkin Trans. 2*, 1972, 1682.	518
72JCS(P2)1887	A. F. Hegarty, J. O'Driscoll, J. K. O'Halloran and F. L. Scott; *J. Chem. Soc., Perkin Trans. 2*, 1972, 1887.	1008
72JCS(P2)1914	P. Beltrame, P. L. Beltrame, A. Filippi and G. Zecchi; *J. Chem. Soc., Perkin Trans. 2*, 1972, 1914.	89, 90
72JCS(P2)2001	E. C. Llaguno and I. C. Paul; *J. Chem. Soc., Perkin Trans. 2*, 1972, 2001.	360
72JCS(P2)2125	E. Gaetani, T. Vitali, A. Mangia, M. Nardelli and G. Pelizzi; *J. Chem. Soc., Perkin Trans. 2*, 1972, 2125.	134

72JHC107	G. Werber, F. Buccheri and F. Maggio; *J. Heterocycl. Chem.*, 1972, **9**, 107.	432
72JHC183	C. F. Beam, R. S. Foote and C. R. Hauser; *J. Heterocycl. Chem.*, 1972, **9**, 183.	71, 72, 84
72JHC427	R. E. Rondeau, M. A. Berwick and H. M. Rosenberg; *J. Heterocycl. Chem.*, 1972, **9**, 427.	6
72JHC447	J. E. Oliver, R. T. Brown and N. L. Redfearn; *J. Heterocycl. Chem.*, 1972, **9**, 447.	923, 942
72JHC651	K. T. Potts and R. Armbruster; *J. Heterocycl. Chem.*, 1972, **9**, 651.	466
72JHC879	W. B. Wright, Jr.; *J. Heterocycl. Chem.*, 1972, **9**, 879.	995
72JHC891	D. A. Tomalia and J. C. Falk; *J. Heterocycl. Chem.*, 1972, **9**, 891.	875, 876, 877, 878
72JHC1109	E. F. Elslager and N. F. Haley; *J. Heterocycl. Chem.*, 1972, **9**, 1109.	538
72JHC1189	P. Crabbé, J. Haro, C. Rius and E. Santos; *J. Heterocycl. Chem.*, 1972, **9**, 1189.	98
72JHC1411	I. Lalezari, A. Shafiee and M. Yalpani; *J. Heterocycl. Chem.*, 1972, **9**, 1411.	353
72JHC1429	H. R. Kwasnik, J. E. Oliver and R. T. Brown; *J. Heterocycl. Chem.*, 1972, **9**, 1429.	994
72JMC315	J. E. Oliver, S. C. Chang, R. T. Brown, J. B. Stokes and A. B. Bořkovec; *J. Med. Chem.*, 1972, **15**, 315.	537
72JMC651	B. K. Wasson, W. K. Gibson, R. S. Stuart, H. W. R. Williams and C. H. Yates; *J. Med. Chem.*, 1972, **15**, 651.	536, 542
72JMC1203	A. Richardson, Jr. and F. J. McCarty; *J. Med. Chem.*, 1972, **15**, 1203.	696
72JOC131	J. E. Oliver, B. A. Bierl and J. M. Ruth; *J. Org. Chem.*, 1972, **37**, 131.	917
72JOC318	G. Just and P. Rossy; *J. Org. Chem.*, 1972, **37**, 318.	320
72JOC369	J. P. Danehy and V. J. Elia; *J. Org. Chem.*, 1972, **37**, 369.	796
72JOC1707	R. A. Coburn and J. P. O'Donnell; *J. Org. Chem.*, 1972, **37**, 1707.	713, 717, 718, 723
72JOC1842	S. M. Katzman and J. Moffat; *J. Org. Chem.*, 1972, **37**, 1842.	401
72JOC2259	T. J. Curphey and K. S. Prasad; *J. Org. Chem.*, 1972, **37**, 2259.	469
72JOC2498	D. D. Chapman; *J. Org. Chem.*, 1972, **37**, 2498.	624, 626
72JOC2583	H. Dounchis; *J. Org. Chem.*, 1972, **37**, 2583.	661
72JOC2587	V. Bertini, A. De Munno and A. Marraccini; *J. Org. Chem.*, 1972, **37**, 2587.	541
72JOC2589	G. W. Griffin and A. Manmade; *J. Org. Chem.*, 1972, **37**, 2589.	868, 869
72JOC2686	S. D. Nelson, Jr., D. J. Kasparian and W. F. Trager; *J. Org. Chem.*, 1972, **37**, 2686.	67, 70
72JOC2983	W. J. Tuman and L. Bauer; *J. Org. Chem.*, 1972, **37**, 2983.	619
72JOC3226	E. Klingsberg; *J. Org. Chem.*, 1972, **37**, 3226.	890
72JOC3601	C. Temple, Jr., B. H. Smith and J. A. Montgomery; *J. Org. Chem.*, 1972, **37**, 3601.	735, 746
72JOC3810	T. Minami, K. Yamataka, Y. Ohshiro, T. Agawa, N. Yasuoka and N. Kasai; *J. Org. Chem.*, 1972, **37**, 3810.	539
72JOC4045	J. Buter, S. Wassenaar and R. M. Kellogg; *J. Org. Chem.*, 1972, **37**, 4045.	572
72JOC4136	R. W. Begland and D. R. Hartter; *J. Org. Chem.*, 1972, **37**, 4136.	538
72JOC4401	R. A. Wohl and D. F. Headley; *J. Org. Chem.*, 1972, **37**, 4401.	310
72JOM(44)325	J. C. Weis and W. Beck; *J. Organomet. Chem.*, 1972, **44**, 325.	139, 153
72JPC2659	L. A. Hull, I. C. Hisatsune and J. Heicklen; *J. Phys. Chem.*, 1972, **76**, 2659.	866
72JPR55	H. G. O. Becker and H. Boettcher; *J. Prakt. Chem.*, 1972, **314**, 55.	1018
72JPR145	E. Hoeft and S. Ganschow; *J. Prakt. Chem.*, 1972, **314**, 145.	908, 910, 913, 915, 918, 932, 940
72JPR815	M. H. Elnagdi, N. A. L. Kassab, M. E. E. Sobhy, M. R. Hanza and M. U. Wahby; *J. Prakt. Chem.*, 1972, **314**, 815.	106
72JST(12)197	M. H. Palmer, A. J. Gaskell and M. S. Barber; *J. Mol. Struct.*, 1972, **12**, 197.	369
72LA(759)107	M. Haake, H. Fode and B. Eichenauer; *Liebigs Ann. Chem.*, 1972, **759**, 107.	540
72LA(762)154	H. G. Aurich, G. Blinne and W. Dersch; *Liebigs Ann. Chem.*, 1972, **762**, 154.	105
72LA(764)94	P. Thieme, M. Patsch and H. Koenig; *Liebigs Ann. Chem.*, 1972, **764**, 94.	553
72LA(766)1	F. Boberg and W. von Gentzkow; *Liebigs Ann. Chem.*, 1972, **766**, 1.	789
72M194	H. Hemetsberger and D. Knittel; *Monatsh. Chem.*, 1972, **103**, 194.	1009
72MI41600	A. Jaszkowska and B. Serafinowa; *Rocz. Chem.*, 1972, **46**, 2051.	123
72MI41601	M. Makosza and A. Zielinska; *Rocz. Chem.*, 1972, **46**, 955.	123
72MI41602	G. A. Shvekhgeimer, E. V. Arslanov and A. Baranski; *Rocz. Chem.*, 1972, **46**, 2381.	89
72MI41603	A. Jaszkowska and B. Serafinowa; *Rocz. Chem.*, 1972, **46**, 2051.	123
72MI41604	G. A. Shvekhgeimer, N. I. Sobtsova and A. Baranski; *Rocz. Chem.*, 1972, **46**, 1735 (*Chem. Abstr.*, 1973, **78**, 72 285).	111
72MI41605	G. A. Shvekhgeimer, N. I. Sobtsova and A. Baranski; *Rocz. Chem.*, 1972, **46**, 1543.	39, 95
72MI41606	K. Eiter and N. Joop; *Naturwissenschaften*, 1972, **59**, 468.	87
72MI41607	N. S. Dokunikhin and S. A. Sokolov; *Zh. Vses. Khim. Obshch.*, 1972, **17**, 695 (*Chem. Abstr.*, 1973, **78**, 58 284).	122
72MI41608	G. LoVecchio; *Atti Accad. Peloritana Pericolanti Cl. Sci. Fis. Mat. Nat.*, 1972, **52**, 217 (*Chem. Abstr.*, 1974, **80**, 13 319).	108
72MI41609	G. LoVecchio; *Atti Accad. Peloritana Pericolanti Cl. Sci. Fis. Mat. Nat.*, 1972 **52**, 207 (*Chem. Abstr.*, 1974, **80**, 13 317).	108
72MI41610	A. L. Fridman and F. A. Gabitov; *Otkrytiya Izobret., Prom. Obraztsy, Tovarnye Znaki*, 1972, **49**, 92 (*Chem. Abstr.*, 1973, **78**, 29 752).	103
72MI41611	K. C. Jacob, G. U. Jadhav and M. N. Vakharia; *Pesticides*, 1972, **6**, 94 (*Chem. Abstr.*, 1973, **78**, 147 867).	93
72MI41612	Y. Iwakura, F. Toda, H. Suzuki, N. Kusakawa and K. Yagi; *J. Polym. Sci., Part A-1*, 1972, **10**, 1133.	54
72MI41613	A. H. Harash, M. H. Elnagdi and N. S. T. Hussein; *Egypt. J. Chem.*, 1972, **15**, 201.	40
72MI41614	Yu. Devi, K. S. R. Krishna Mohan Rao and N. V. S. Rao; *Curr. Sci.*, 1972, **41**, 773 (*Chem. Abstr.*, 1973, **78**, 43 330).	40

72MI41615	Y. Iwakura, K. Uno, S. Shiraishi, M. Yuyama and Y. Kihara; *Yuki Gosei Kagaku Kyokai Shi*, 1972, **30**, 894 (*Chem. Abstr.*, 1973, **79**, 65 548).	39
72MI41616	I. Yoshio, S. Shiraishi and M. Yuyama; *Yuki Gosei Kagaku Kyokai Shi*, 1972, **30**, 889 (*Chem. Abstr.*, 1973, **79**, 5289).	39
72MI41700	G. Salmona, Y. Ferré, E. J. Vincent and D. Megy; *J. Chim. Phys., Phys. Chim. Biol.*, 1972, **69**, 1292 (*Chem. Abstr.*, 1972, **77**, 163 866).	133
72MI42300	Y. Iwakura, F. Toda, H. Suzuki, N. Kusakawa and K. Yagi; *J. Polym. Sci., Part A1*, 1972, **10**, 1133 (*Chem. Abstr.*, 1972, **77**, 62 340).	438
72MI42500	G. J. Muller, M. B. Linn and J. B. Sinclair; *Plant Dis. Rep.*, 1972, **56**, 1054 (*Chem. Abstr.*, 1973, **78**, 80 711).	509
72MI42600	C. L. Khetrapal and A. V. Patankar; *Mol. Cryst. Liq. Cryst.*, 1972, **15**, 367.	517
72MI42800	G. S. Johar; *Talanta*, 1972, **19**, 1461.	598
72MI42900	B. Bovio and S. Locchi; *J. Cryst. Mol. Struct.*, 1972, **2**, 89.	618
72MI42901	B. Bovio and S. Locchi; *J. Cryst. Mol. Struct.*, 1972, **2**, 251.	618
B-72MI43000	G. C. Levy and G. L. Nelson; 'Carbon-13 Nuclear Magnetic Resonance for Organic Chemists', Wiley, New York, 1972, p. 59.	754
B-72MI43001	J. B. Strothers; 'Carbon-13 NMR Spectroscopy', Academic, New York, 1972, p. 342.	756
72MI43100	G. Geiseler and H. J. Rauh; *Z. Phys. Chem.* 1972, **249**, 376.	785
72MI43101	C. Th. Pedersen, O. Hammerich and V. D. Parker; *J. Electroanal. Chem. Interfacial Electrochem.*, 1972, **38**, 479.	809
72MI43102	G. N. Alikseva, A. V. Eltsov, M. B. Dolesova, L. I. Maksimova and A. M. Rusanov; *Khim. Farm. Zh.*, 1972, **6**, 23.	810
72MI43300	P. T. Beurskens, W. P. J. H. Bosman and J. A. Cras; *J. Cryst. Mol. Struct.*, 1972, **2**, 183.	855, 857
72MI43301	R. W. Murray, J. W.-P. Lin and D. A. Grumke; *Adv. Chem. Ser.*, 1972, **112**, 9.	866
72MI43302	R. Criegee and H. Korber; *Adv. Chem. Ser.*, 1972, **112**, 22.	866
B-72MI43303	P. S. Bailey; in 'Ozone in Water and Wastewater Treatment', ed. F. L. Evans; Ann Arbor Science, Ann Arbor, 1972, pp. 29–59.	873, 893
72MI43304	T. E. Kinlin, R. Muralidhara, A. O. Pittet, A. Sanderson and J. P. Walradt; *J. Agric. Food Chem.*, 1972, **20**, 1021.	894
72MI43305	H. W. Brinkman, H. Copier, J. J. M. De Leuw and S. B. Tjan; *J. Agric. Food Chem.*, 1972, **20**, 177.	894
72MI43400	N. V. Zhorkin, V. A. Ignatov and G. A. Blokh; *Vop. Khim., Khim. Tekhnol.*, 1972, 176 (*Chem. Abstr.*, 1973, **78**, 124 470).	898
72MI43800	A. Hordvik and L. J. Saethre; *Isr. J. Chem.*, 1972, **10**, 239.	1054
72MIP41600	R. G. Micetich and R. U. Lemieux; *Can. Pat.* 905 409 (1972) (*Chem. Abstr.*, 1974, **81**, 3913).	128
72NKK1452	Y. Iwakura, K. Uno, Y. Kihara, M. Setsu and M. Ginnai; *Nippon Kagaku Zasshi*, 1972, 1452 (*Chem. Abstr.*, 1972, **77**, 139 863).	100, 101
72OMS(6)415	G. Conde-Caprace and J. E. Collin; *Org. Mass Spectrom.*, 1972, **6**, 415.	757
72OMS(6)1321	A. Croisy, P. Jacquinon, R. Weber and M. Renson; *Org. Mass Spectrom.*, 1972, **6**, 1321.	143
72PMH(5)1	P. J. Wheatley; *Phys. Methods Heterocycl. Chem.*, 1972, **5**, 1.	4, 238
72PMH(5)18	P. J. Wheatley; *Phys. Methods Heterocycl. Chem.*, 1972, **5**, 18.	369
72PMH(5)47	P. J. Wheatley; *Phys. Methods Heterocycl. Chem.*, 1972, **5**, 47.	750
72PMH(5)283	P. J. Wheatley; *Phys. Methods Heterocycl. Chem.* 1972, **5**, 283.	1031
72RTC552	J. Bus; *Recl. Trav. Chim. Pays-Bas*, 1972, **91**, 552.	400
72RTC711	T. Doornbos and H. G. Peer; *Recl. Trav. Chim. Pays-Bas*, 1972, **91**, 711. (*Chem. Abstr.*, 1972, **77**, 61 869).	308
72S29	E. Klingsberg; *Synthesis*, 1972, 29.	844
72S561	T.-L. Ho and C. M. Wong; *Synthesis*, 1972, 561.	838
72T303	A. S. Angeloni, V. Ceré, V. Dal Monte, E. Sandri and G. Scapini; *Tetrahedron*, 1972, **28**, 303.	518
72T637	M. Witanowski, L. Stefaniak, H. Januszewski, Z. Grabowski and G. A. Webb; *Tetrahedron*, 1972, **28**, 637.	133, 139, 182
72T1223	T. Greibrokk and K. Undheim; *Tetrahedron*, 1972, **28**, 1223.	680
72T1353	K. P. Zeller, H. Meier and E. Muller; *Tetrahedron*, 1972, **5**, 1353.	451
72T2183	H. Pomerantz, L. J. Miller, R. Barron, E. Hansen, D. Mastbrook and I. Egry; *Tetrahedron*, 1972, **28**, 2183.	1064
72T2799	V. Galasso and N. Trinajstić; *Tetrahedron*, 1972, **28**, 2799.	237
72T3141	A. Lablache-Combier and A. Pollet; *Tetrahedron*, 1972, **28**, 3141.	147
72T3271	V. Ceré, V. Dal Monte and E. Sandri; *Tetrahedron*, 1972, **28**, 3271.	402
72T3295	F. Eiden and W. Loewe; *Tetrahedron*, 1972, **28**, 3295.	116
72T3489	S. B. Tjan, J. C. Haakman, C. J. Teunis and H. G. Peer; *Tetrahedron*, 1972, **28**, 3489.	862, 889
72T3845	P. Bravo, G. Gaudiano, P. P. Ponti and C. Ticozzi; *Tetrahedron*, 1972, **28**, 3845.	95
72T3987	K. B. Sukumaran, C. S. Angadiyavar and M. V. George; *Tetrahedron*, 1972, **28**, 3987.	1003
72T4737	O. Tsuge and S. Kanemasa; *Tetrahedron*, 1972, **28**, 4737.	676
72TH43300	J. W. Swain; Thesis, Univ. Missouri, 1972 (*Diss. Abstr. Int. B*, 1973, **34**, 1143).	894
72TL299	L. D. Quin and S. G. Borleske; *Tetrahedron Lett.*, 1972, 299.	999
72TL445	H. Meier and I. Menzel; *Tetrahedron Lett.*, 1972, 445.	351

72TL1357	W. Adam and N. Duran; *Tetrahedron Lett.*, 1972, 1357.	759
72TL1429	R. M. Kellog; *Tetrahedron Lett.*, 1972, 1429.	250
72TL1707	W. Oppolzer; *Tetrahedron Lett.*, 1972, 1707.	1007
72TL1835	R. J. S. Beer and A. J. Poole; *Tetrahedron Lett.*, 1972, 1835.	1063
72TL2569	F. Guinot and G. Lamaty; *Tetrahedron Lett.*, 1972, 2569.	761
72TL2969	D. Houalla, J. F. Brazier and R. Wolf; *Tetrahedron Lett.*, 1972, 2969.	975, 1022
72TL3169	G. Scherowsky and B. Kundu; *Tetrahedron Lett.*, 1972, 3169.	436, 443
72TL3469	M. Altaf-ur-Rahman, A. J. Boulton and D. Middleton; *Tetrahedron Lett.*, 1972, 3469.	98
72TL4087	A. Padwa, D. Dean and J. Smolanoff; *Tetrahedron Lett.*, 1972, 4087.	310
72TL4165	M. H. Palmer and R. H. Findlay; *Tetrahedron Lett.*, 1972, 4165.	815
72TL4687	A. L. Llaguno, I. C. Paul, R. Pinel and Y. Mollier; *Tetrahedron Lett.*, 1972, 4687.	1054
72TL5213	C. Th. Pedersen and C. Lohse; *Tetrahedron Lett.*, 1972, 5213.	789
72USP3629245	D. Bertin and L. Nedelec; *U.S. Pat.* 3 629 245 (1972) (*Chem. Abstr.*, 1972, **76**, 86 006).	101
72USP3634442	M. and T. Chemicals Inc., *U.S. Pat.* 3 634 442 (1972) (*Chem. Abstr.*, 1972, **76**, 99 833).	487
72USP3642897	G. E. Hardtmann; *U.S. Pat.* 3 642 897 (1972) (*Chem. Abstr.*, 1972, **76**, 153 341).	129
72USP3691183	Esso Research and Engineering Co., *U.S. Pat.* 3 691 183 (1972) (*Chem. Abstr.*, 1972, **77**, 152 191).	509
72USP3692794	Ciba-Geigy Corp., *U.S. Pat.* 3 692 794 (1972) (*Chem. Abstr.*, 1972, **77**, 152 239).	509
72USP3694498	A. B. Ash and C. L. Stevens; *U.S. Pat.* 3 694 498 (1972) (*Chem. Abstr.*, 1972, **77**, 164 228).	890, 895
72ZC349	J. Fabian and H. Hartmann; *Z. Chem.*, 1972, **12**, 349.	786
72ZOB750	B. A. Arbuzov, E. N. Dianova and V. S. Vinogradova; *Zh. Obshch. Khim.*, 1972, **42**, 750 (*Chem. Abstr.*, 1972, **77**, 126 777).	40
72ZOB2049	V. Sh. Tsveniashvili, V. N. Gaprindashvili and N. S. Khavtasi; *Zh. Obshch. Khim.*, 1972, **42**, 2049.	402
72ZOR1419	L. M. Andreeva, K. V. Altukhov and V. V. Perekalin; *Zh. Org. Khim.*, 1972, **8**, 1419 (*Chem. Abstr.*, 1972, **77**, 126 479).	111
73ABC1197	T. Okabe, M. Maekawa and E. Taniguchi; *Agric. Biol. Chem.*, 1973, **37**, 1197.	744
73AC(R)613	S. Mangiavacchi and M. Scotton; *Ann. Chim.* (*Rome*), 1973, **63**, 613.	57
73ACS360	A. Hordvik, T. S. Rimala and L. J. Saethre; *Acta Chem. Scand.*, 1973, **27**, 360.	1054
73ACS379	A. Hordvik, O. Sjølset and L. J. Saethre; *Acta Chem. Scand.*, 1973, **27**, 379.	1053
73ACS382	B. Birknes, A. Hordvik and L. J. Saethre; *Acta Chem. Scand.*, 1973, **27**, 382.	1053
73ACS391	H. Lund; *Acta Chem. Scand.*, 1973, **27**, 391.	549, 560
73ACS411	L. K. Hansen and A. Hordvik; *Acta Chem. Scand.*, 1973, **27**, 411.	1053
73ACS485	A. Hordvik and J. A. Porten; *Acta Chem. Scand.*, 1973, **27**, 485.	1054
73ACS510	A. Hordvik and L. M. Milje; *Acta Chem. Scand.*, 1973, **27**, 510.	1053
73ACS1749	K. Undheim and R. Lie; *Acta Chem. Scand.*, 1973, **27**, 1749.	675, 680
73ACS2161	A. Senning and J. S. Rasmussen; *Acta Chem. Scand.*, 1973, **27**, 2161.	907, 922
73ACS2517	R. Kristensen and J. Sletten; *Acta Chem. Scand.*, 1973, **27**, 2517.	1055
73ACS2666	O. Hjellum and A. Hordvik; *Acta Chem. Scand.*, 1973, **27**, 2666.	1053
73ACS2802	P. Krogsggaard-Larsen, S. B. Christensen and H. Hjeds; *Acta Chem. Scand.*, 1973, **27**, 2802.	64, 88, 104
73ACS3881	J. Sletten and M. Velsvik; *Acta Chem. Scand.*, 1973, **27**, 3881.	1055
73AG(E)405	F. Hervens and H. G. Viehe; *Angew. Chem., Int. Ed. Engl.*, 1973, **12**, 405.	985
73AG(E)1055	K. Burzin and R. Feinauer; *Angew. Chem., Int. Ed. Engl.*, 1973, **12**, 1055.	989
73AHC(15)233	H. Hettler; *Adv. Heterocycl. Chem.*, 1973, **15**, 233. 132, 144, 148, 149, 150, 151, 152, 154, 155, 159, 160, 161, 162, 167, 168, 173, 174	
73AJC1683	K. J. Bird, I. D. Rae and A. M. White; *Aust. J. Chem.*, 1973, **26**, 1683.	419
73AJC1763	G. D. Beresford, R. C. Cambie and K. P. Mathai; *Aust. J. Chem.*, 1973, **26**, 1763.	78
73AJC1949	L. W. Deady; *Aust. J. Chem.*, 1973, **26**, 1949.	21, 148
73AJC2705	I. F. Eckhard, K. Lehtonen, T. Staub and L. A. Summers; *Aust. J. Chem.*, 1973, **26**, 2705.	128
73AP134	G. Zinner and W. Kilwing; *Arch. Pharm.* (*Weinheim, Ger.*), 1973, **306**, 134.	438, 444
73AP284	H. Wollweber and R. Hiltmann; *Arch. Pharm.* (*Weinheim, Ger.*), 1973, **306**, 284.	666
73AP746	T. Denzel and H. Hoehn; *Arch. Pharm.* (*Weinheim, Ger.*), 1973, **306**, 746.	624
73AX(B)43	A. Braibanti, M. A. Pellinghelli, A. Tiripicchio and C. M. Tiripicchio; *Acta Crystallogr., Part B*, 1973, **29**, 43.	135
73BCJ310	C. Kashima, S. Tobe, N. Sugiyama and M. Yamamoto; *Bull. Chem. Soc. Jpn.*, 1973, **46**, 310.	49
73BCJ506	Y. Ohtsuka; *Bull. Chem. Soc. Jpn.*, 1973, **46**, 506.	655, 661
73BCJ3304	H.-J. Tien, K. Kanda, A. Chinone and M. Ohta; *Bull. Soc. Chem. Jpn.*, 1973, **46**, 3304.	376
73BCJ3533	C. Kashima and Y. Tsuda; *Bull. Chem. Soc. Jpn.*, 1973, **46**, 3533.	49
73BSF254	M. Golfier and R. Milcent; *Bull. Soc. Chim. Fr.*, 1973, 254.	431
73BSF277	D. Barillier, C. Gy, P. Rioult and J. Vialle; *Bull. Soc. Chim. Fr.*, 1973, 277.	1063
73BSF581	F. Clesse and H. Quiniou; *Bull. Soc. Chim. Fr.*, 1973, 581.	890
73BSF586	F. Clesse, J. P. Pradere and H. Quiniou; *Bull. Soc. Chim. Fr.*, 1973, 586.	790

Ref	Citation	Pages
73BSF677	P. Jacquignon, G. Maréchal, M. Renson, A. Ruwet and D. P. Hien; *Bull. Soc. Chim. Fr.*, 1973, 677.	518
73BSF721	C. Trebaul; *Bull. Soc. Chim. Fr.*, 1973, 721.	800
73BSF1138	J. L. Olive, C. Petrus and F. Petrus; *Bull. Soc. Chim. Fr.*, 1973, 1138.	98
73BSF1390	A. Belly, C. Petrus and F. Petrus; *Bull. Soc. Chim. Fr.*, 1973, 1390.	39
73BSF1659	C. Lemarié-Retour, M. Stavaux and N. Lozac'h; *Bull. Soc. Chim. Fr.*, 1973, 1659.	1068
73BSF1743	G. Vernin, C. Riou, H. J. M. Dou, L. Bouscasse, J. Metzger and G. Loridan; *Bull. Soc. Chim. Fr.*, 1973, 1743.	133, 144, 147, 250
73BSF1822	G. Vernin, J. C. Poite, C. Riou, H. J. M. Dou, J. Metzger and G. Vernin; *Bull. Soc. Chim. Fr.*, 1973, 1822.	163
73BSF1973	J. Maignan and J. Vialle; *Bull. Soc. Chim. Fr.*, 1973, 1973.	796, 925
73BSF3079	F. Petrus, J. Verducci and Y. Vidal; *Bull. Soc. Chim. Fr.*, 1973, 3079.	40, 41
73BSF3334	Nguyen Kim Son, R. Pinel and Y. Mollier; *Bull. Soc. Chim Fr.*, 1973, 3334.	797
73BSF3339	D. Festal, J. Tison, Nguyen Kim Son, R. Pinel and Y. Mollier; *Bull. Soc. Chim. Fr.*, 1973, 3339.	1056
73CB220	T. Kauffmann and H. Fischer; *Chem. Ber.*, 1973, **106**, 220.	660
73CB332	W. Müller, U. Kraatz and F. Korte; *Chem. Ber.*, 1973, **106**, 332.	43
73CB1496	J. Goerdeler and R. Schimpf; *Chem. Ber.*, 1973, **106**, 1496.	309, 499
73CB3291	R. Huisgen and M. Christi; *Chem. Ber.*, 1973, **106**, 3291.	91, 92
73CB3391	A. Haas and V. Plass; *Chem. Ber.*, 1973, **106**, 3391.	496
73CC150	G. G. Abott and D. Leaver; *J. Chem. Soc., Chem. Commun.*, 1973, 150.	635, 636, 638, 639
73CC226	R. Thomas and D. J. Williams; *J. Chem. Soc., Chem. Commun.*, 1973, 226.	667
73CC349	R. Grigg, R. Hayes, J. L. Jackson and T. J. King; *J. Chem. Soc., Chem. Commun.*, 1973, 349.	35
73CC524	R. K. Howe and J. E. Franz; *J. Chem. Soc., Chem. Commun.*, 1973, 524.	156
73CHE395	T. A. Kranitskaya, I. V. Smolanka and A. L. Vais; *Chem. Heterocycl. Compd. (Engl. Transl.)*, 1973, **9**, 395.	1010
73CHE521	I. Ya. Kvitko and N. B. Sokolova; *Chem. Heterocycl. Compd. (Engl. Transl.)*, 1973, **9**, 521.	988
73CHE1216	A. Y. Lazaris, S. M. Shmuilovich and A. N. Egorochkin; *Chem. Heterocycl. Compd. (Engl. Transl.)*, 1973, **9**, 1216.	435, 441
73CI(L)1162	H. Ullah and P. Sykes; *Chem. Ind. (London)*, 1973, 1162.	152
73CJC680	A. W. Richardson; *Can. J. Chem.*, 1973, **51**, 680.	521
73CJC1368	F. M. F. Chen and T. P. Forrest; *Can. J. Chem.*, 1973, **51**, 1368.	905, 940
73CJC1741	A. Taurins and V. T. Khouw; *Can. J. Chem.*, 1973, **51**, 1741.	635, 636, 637, 641, 642, 644
73CJC2315	R. N. Butler; *Can. J. Chem.*, 1973, **51**, 2315.	975, 977
73CJC2349	D. E. Horning and J. M. Muchowski; *Can. J. Chem.*, 1973, **51**, 2349.	536
73CJC2353	G. Lacasse and J. M. Muchowski; *Can. J. Chem.*, 1973, **51**, 2353.	465, 466, 468, 486
73CJC3081	D. M. McKinnon and M. E. Hassan; *Can. J. Chem.*, 1973, **51**, 3081.	790, 807
73CL175	O. Tsuge and H. Samura; *Chem. Lett.*, 1973, 175.	1039
73CL185	K. Maekawa and E. Kuwano; *Chem. Lett.*, 1973, 185.	56
73CL455	A. Matsumoto, J. H. Lee, M. Yoshida and O. Simamura; *Chem. Lett.*, 1973, 455.	1035, 1044
73CL867	K. Hiratani, H. Shiono and M. Okawara; *Chem. Lett.*, 1973, 867.	839
73CL917	T. Fuchigami and K. Odo; *Chem. Lett.*, 1973, 917.	473, 498
73CPB1327	K. Ikeda, T. Sumi, K. Yokoi and Y. Mizuno; *Chem. Pharm. Bull.*, 1973, **21**, 1327.	660
73CPB2224	K. Hirai and H. Sugimoto; *Chem. Pharm. Bull.*, 1973, **21**, 2224.	758
73CPB2396	M. Nagano, M. Oshige, T. Matsui, J. Tobitsuka and K. Oyamada; *Chem. Pharm. Bull.* 1973, **21**, 2396.	465
73CR(B)(276)31	G. Mille, G. Davidovics, J. C. Poite and M. Guiliano; *C.R. Hebd. Seances Acad. Sci., Ser. B*, 1973, **276**, 31.	141
73CR(C)(276)1057	A. Dorange, F. Tonnard and F. Venien; *C. R. Hebd. Seances Acad. Sci., Ser C*, 1973, **276**, 1057.	785
73CRV93	R. N. Butler, F. L. Scott and T. A. F. O'Mahony; *Chem. Rev.*, 1973, **73**, 93.	1023
73DIS(B)1434	P. M. Stonebraker; *Diss. Abstr. Int. B*, 1973, **34**, 1434 (*Chem. Abstr.*, 1973, **79**, 145 582).	4, 43
73EGP102382	K. Peseke; *Ger. (East) Pat.* 102 382 (1973) (*Chem. Abstr.*, 1974, **81**, 25 658).	876
73G219	G. Renzi, V. D. Piaz and C. Musante; *Gazz. Chim. Ital.*, 1973, **103**, 219.	627
73GEP2151228	A. Grafe and H. Liebig; *Ger. Offen.* 2 151 228 (1973) (*Chem. Abstr.*, 1973, **79**, 66 050).	876
73GEP2155753	*Ger. Pat.* 2 155 753 (1973) (*Chem. Abstr.*, 1973, **79**, 42 516).	725
73GEP2155963	*Ger. Pat.* 2 155 963 (1973) (*Chem. Abstr.*, 1973, **79**, 32 085).	710
73GEP2203225	D. J. Sykes, H. Kroll and T. R. Finch; *Ger. Offen.* 2 203 225 (1973) (*Chem. Abstr.*, 1973, **79**, 151 610).	894
73GEP2205065	*Ger. Pat.* 2 205 065 (1973) (*Chem. Abstr.*, 1973, **79**, 146 512).	710
73GEP2210633	*Ger. Pat.* 2 210 633 (1973) (*Chem. Abstr.*, 1973, **79**, 146 519).	709
73GEP2223421	*Ger. Pat.* 2 223 421 (1973) (*Chem. Abstr.*, 1974, **80**, 37 148).	710
73GEP2249099	*Ger. Pat.* 2 249 099 (1973) (*Chem. Abstr.*, 1973, **79**, 32 094).	644
73GEP2249163	*Ger. Pat.* 2 249 163 (1973) (*Chem. Abstr.*, 1973, **79**, 32 095).	629, 634
73GEP2258279	*Ger. Pat.* 2 258 279 (1973) (*Chem. Abstr.*, 1973, **79**, 92 236).	748
73GEP2263774	L. Credali, G. Baruzzi and M. Russo; *Ger. Pat.* 2 263 774 (1973) (*Chem. Abstr.*, 1973, **79**, 116 199).	426, 543

Ref	Citation	Pages
73GEP2313256	H. Nishimura, M. Shimizu, H. Uno, T. Hirooka, Y. Masuda and M. Kurokawa; *Ger. Offen.* 2 313 256 (1973) (*Chem. Abstr.*, 1973, **79**, 137 159).	129
73GEP2325043	Roussel-UCLAF, *Ger. Pat.* 2 325 043 (1973) (*Chem. Abstr.*, 1974, **80**, 59 932).	159
73HCA2588	H. Giezendanner, H. J. Rosenkranz, H. J. Hansen and H. Schmid; *Helv. Chim. Acta*, 1973, **56**, 2588.	37
73IJC1	A. H. Harhash, M. H. Elnagdi, V. S. Hussein and S. M. Fahmy; *Indian J. Chem.*, 1973, **11**, 1.	40
73IJC128	A. H. Harhash, M. H. Elnagdi and S. O. Abdallah; *Indian J. Chem.*, 1973, **11**, 128 (*Chem. Abstr.*, 1973, **79**, 31 974).	319
73IJC541	S. S. Kumari, K. S. R. Krishna Mohan Rao and N. V. Subba Rao; *Indian J. Chem.*, 1973, **11**, 541.	48, 116
73IJC609	J. M. Bachhawat, A. K. Koul, B. Prashad, N. S. Ramegowda, C. K. Narang and N. K. Mathur; *Indian J. Chem.*, 1973, **11**, 609.	174
73IJC732	T. RadhaVakula and V. R. Srinivasan; *Indian J. Chem.*, 1973, **11**, 732.	438
73IJC1017	S. Bhattacharya and A. Bhaumik; *Indian J. Chem.*, 1973, **11**, 1017.	428
73IZV203	S. L. Ioffe, L. M. Makarenkova, L. M. Shitkin, M. V. Kashutina and V. A. Tartakovskii; *Izv. Akad. Nauk SSSR, Ser. Khim.*, 1973, 203 (*Chem. Abstr.*, 1973, **78**, 136 367).	111
73IZV2725	I. L. Knunyants, V. V. Shokina and E. I. Mysov; *Izv. Akad. Nauk SSSR, Ser. Khim.*, 1973, 2725.	875, 877, 879, 892
73JA279	E. M. Burgess and H. R. Penton, Jr.; *J. Am. Chem. Soc.*, 1973, **95**, 279.	999
73JA919	J. P. Ferris, F. R. Antonucci and R. W. Trimmer; *J. Am. Chem. Soc.*, 1973, **95**, 919.	16, 17
73JA1348	R. P. Lattimer, C. W. Gillies and R. L. Kuczkowski; *J. Am. Chem. Soc.*, 1973, **95**, 1348.	855
73JA2390	W. M. Horspool, J. R. Kershaw, A. W. Murray and G. M. Stevenson; *J. Am. Chem. Soc.*, 1973, **95**, 2390.	124
73JA2392	W. E. McEwen, M. A. Calabro, I. C. Mineo and I. C. Wang; *J. Am. Chem. Soc.*, 1973, **95**, 2392.	659
73JA2558	M. P. Cava, N. M. Pollack and G. A. Dieterle; *J. Am. Chem. Soc.*, 1973, **95**, 2558.	1032
73JA3343	R. W. Murray and A. Suzui; *J. Am. Chem. Soc.*, 1973, **95**, 3343.	885
73JA3408	A. I. Meyers and G. N. Knaus; *J. Am. Chem. Soc.*, 1973, **95**, 3408.	275
73JA3460	R. A. Rouse; *J. Am. Chem. Soc.*, 1973, **95**, 3460.	853
73JA3807	A. Babadjamian, M. Chanon, R. Gallo and J. Metzger; *J. Am. Chem. Soc.*, 1973, **95**, 3807.	254
73JA4373	K. Bechgaard, V. D. Parker and C. Th. Pedersen; *J. Am. Chem. Soc.*, 1973, **95**, 4373.	785
73JA4379	H. D. Hartzler; *J. Am. Chem. Soc.*, 1973, **95**, 4379.	845, 849
73JA5003	R. M. Forbis and K. L. Rinehart, Jr.; *J. Am. Chem. Soc.*, 1973, **95**, 5003.	650, 656, 666, 667
73JA6073	J. L. Flippen; *J. Am. Chem. Soc.*, 1973, **95**, 6073.	902, 903
73JA6349	F. Wudl and T. B. K. Lee; *J. Am. Chem. Soc.*, 1973, **95**, 6349.	926, 927
73JA6490	Y. Kishi, T. Fukuyama and S. Nakatsuka; *J. Am. Chem. Soc.*, 1973, **95**, 6490.	1019
73JA7161	W. K. Anderson and R. H. Dewey; *J. Am. Chem. Soc.*, 1973, **95**, 7161.	769, 777, 780
73JA7301	K. N. Houk, J. Sims, C. R. Watts and L. J. Luskus; *J. Am. Chem. Soc.*, 1973, **95**, 7301.	827
73JA7692	C. R. Johnson and E. R. Janiga; *J. Am. Chem. Soc.*, 1973, **95**, 7692.	165
73JAP7326763	S. Kano and I. Iwataki; *Jpn. Pat.* 73 26 763 (1973) (*Chem. Abstr.*, 1973, **78**, 159 623).	942
73JAP7328470	I. Iwataki, S. Kano, M. Kaeriyama and S. Kosaka; *Jpn. Pat.* 73 28 470 (1973) (*Chem. Abstr.*, 1973, **78**, 159 610).	942
73JAP7329329	*Jpn. Pat.* 73 29 329 (1973) (*Chem. Abstr.*, 1974, **80**, 27 261).	723, 730
73JAP(K)7391092	*Jpn. Kokai* 73 91 092 (1973) (*Chem. Abstr.*, 1974, **81**, 120 669).	694, 709
73JCS(F2)1155	R. A. Johnstone and F. A. Mellon; *J. Chem. Soc., Faraday Trans. 2*, 1973, **69**, 1155.	517
73JCS(P1)356	K. Clarke, C. G. Hughes and R. M. Scrowston; *J. Chem. Soc., Perkin Trans. 1*, 1973, 356.	167
73JCS(P1)465	D. C. Cook and A. Lawson; *J. Chem. Soc., Perkin Trans. 1*, 1973, 465.	40
73JCS(P1)1022	F. M. Dean, J. Goodchild and A. W. Hill; *J. Chem. Soc., Perkin Trans. 1*, 1973, 1022.	1069
73JCS(P1)1148	G. Bianchi, C. DeMicheli, R. Gandolfi, P. Gruenanger, P. V. Fiuzi and O. V. de Pava; *J. Chem. Soc., Perkin Trans. 1*, 1973, 1148.	89
73JCS(P1)1357	R. N. Butler, T. M. Lambe, J. C. Tobin and F. L. Scott; *J. Chem. Soc., Perkin Trans. 1*, 1973, 1357.	439
73JCS(P1)1863	M. Davis, E. Homfeld and K. S. L. Srivastava; *J. Chem. Soc., Perkin Trans. 1*, 1973, 1863.	148, 174
73JCS(P1)1954	L. Di Nunno, S. Florio and P. E. Todesco; *J. Chem. Soc., Perkin Trans. 1*, 1973, 1954.	419
73JCS(P1)2220	P. Crabbé, A. Villarino and J. M. Muchowski; *J. Chem. Soc., Perkin Trans. 1*, 1973, 2220.	116
73JCS(P1)2332	J. N. Hines, M. J. Peagram, E. J. Thomas and G. H. Whitham; *J. Chem. Soc., Perkin Trans. 1*, 1973, 2332.	770
73JCS(P1)2351	J. G. Dingwall, A. S. Ingram, D. H. Reid and J. D. Symon; *J. Chem. Soc., Perkin Trans. 1*, 1973, 2351.	785, 1061, 1065
73JCS(P1)2503	L. VanRompuy, N. Schamp, N. De Kempe and R. VanParijs; *J. Chem. Soc., Perkin Trans. 1*, 1973, 2503.	39

73JCS(P1)2580	Y. Tamura, Y. Miki, Y. Sumida and M. Ikeda; *J. Chem. Soc., Perkin Trans. 1,* 1973, 2580.	75
73JCS(P1)2837	C. Th. Pedersen and C. Lohse; *J. Chem. Soc., Perkin Trans. 1,* 1973, 2837.	1057
73JCS(P1)4372	G. L. Buchanan, R. A. Raphael and I. W. J. Stiel; *J. Chem. Soc., Perkin Trans. 1,* 1973, 4372.	126
73JCS(P2)243	C. H. Green and D. G. Hellier; *J. Chem. Soc., Perkin Trans. 2,* 1973, 243.	860, 862
73JCS(P2)1093	G. Vernin, H. J. M. Dou and J. Metzger; *J. Chem. Soc., Perkin Trans. 2,* 1973, 1093.	266
73JCS(P2)1542	J. R. Grunwell and H. S. Baker; *J. Chem. Soc., Perkin Trans. 2,* 1973, 1542.	517
73JCS(P2)1732	M. R. Arshadi and M. Shabrang; *J. Chem. Soc., Perkin Trans. 2,* 1973, 1732.	349
73JCS(P2)1966	C. H. Green and D. G. Hellier; *J. Chem. Soc., Perkin Trans. 2,* 1973, 1966.	860
73JHC99	E. O. Snoke and F. D. Popp; *J. Heterocycl. Chem.,* 1973, **10**, 99.	659
73JHC181	E. Abushanab, D. Y. Lee and L. Goodman; *J. Heterocycl. Chem.,* 1973, **10**, 181.	620, 622, 626
73JHC249	J. A. Skorcz, J. T. Suh and R. L. Germershausen; *J. Heterocycl. Chem.,* 1973, **10**, 249.	154, 166
73JHC267	R. Weber and M. Renson; *J. Heterocycl. Chem.,* 1973, **10**, 267.	335, 337
73JHC411	J. Elguero, R. Jacquier and S. Mignonac-Mondon; *J. Heterocycl. Chem.,* 1973, **10**, 411.	990
73JHC413	A. H. Albert, R. K. Robins and D. E. O'Brien; *J. Heterocycl. Chem.,* 1973, **10**, 413.	160
73JHC487	R. A. Coburn and R. A. Glennon; *J. Heterocycl. Chem.,* 1973, **10**, 487.	672, 680, 681, 684, 688, 691, 735
73JHC587	A. Gasco, V. Mortarini, G. Rua and A. Serafino; *J. Heterocycl. Chem.,* 1973, **10**, 587.	406
73JHC611	E. F. Elslager, J. Johnson and L. M. Werbel; *J. Heterocycl. Chem.,* 1973, **10**, 611.	510
73JHC655	I. Lalezari, A. Shafiee and H. Golgolab; *J. Heterocycl. Chem.,* 1973, **10**, 655.	349
73JHC669	P. Bravo and P. P. Ponti; *J. Heterocycl. Chem.,* 1973, **10**, 669.	71, 88
73JHC773	F. L. Tobiason, L. Huestis, C. Chandler, S. E. Pedersen and P. Peters; *J. Heterocycl. Chem.,* 1973, **10**, 773.	400, 517
73JHC785	R. M. Ottenbrite and P. V. Alston; *J. Heterocycl. Chem.,* 1973, **10**, 785.	987
73JHC953	I. Lalezari, A. Shafiee, F. Rabet and M. Yalpani; *J. Heterocycl. Chem.,* 1973, **10**, 953.	349
73JMC512	J. W. Scott and A. Boris; *J. Med. Chem.,* 1973, **16**, 512.	53, 156
73JMC978	M. Hatanaka and T. Ishimaru; *J. Med. Chem.,* 1973, **16**, 978.	155, 162, 164
73JMC1170	O. B. T. Nielsen, C. K. Nielsen and P. W. Feit; *J. Med. Chem.,* 1973, **16**, 1170.	152
73JOC20	T. R. Williams and D. J. Cram; *J. Org. Chem.,* 1973, **38**, 20.	166
73JOC284	A. Padwa and E. Glazer; *J. Org. Chem.,* 1973, **38**, 284.	1003
73JOC465	L. S. Wittenbrook, G. L. Smith and R. J. Timmons; *J. Org. Chem.,* 1973, **38**, 465.	465, 486
73JOC471	C. U. Pittman, Jr., T. B. Patterson, Jr. and L. D. Kispert; *J. Org. Chem.,* 1973, **38**, 471.	750
73JOC1054	A. J. Boulton and S. S. Mathur; *J. Org. Chem.,* 1973, **38**, 1054.	81, 400, 418
73JOC1356	D. C. De Jongh and M. L. Thomson; *J. Org. Chem.,* 1973, **38**, 1356.	183
73JOC1434	W. Adam and N. Duran; *J. Org. Chem.,* 1973, **38**, 1434.	768
73JOC1575	K. Pilgram and R. D. Skiles; *J. Org. Chem.,* 1973, **38**, 1575.	491, 495, 741, 987
73JOC1955	L. F. Miller and R. E. Bambury; *J. Org. Chem.,* 1973, **38**, 1955.	979
73JOC2238	E. C. Taylor, S. F. Martin, Y. Maki and G. P. Beardsley; *J. Org. Chem.,* 1973, **38**, 2238.	722
73JOC2407	H. H. Wasserman and F. J. Vinick; *J. Org. Chem.,* 1973, **38**, 2407.	217
73JOC2916	E. Van Loock, J.-M. Vandensavel, G. L'abbé and G. Smets; *J. Org. Chem.,* 1973, **38**, 2916.	583, 586, 596
73JOC3087	K. T. Potts and J. Kane; *J. Org. Chem.,* 1973, **38**, 3087.	499, 742
73JOC3316	D. S. Noyce and S. A. Fike; *J. Org. Chem.,* 1973, **38**, 3316.	280
73JOC3868	R. A. Coburn and B. Bhooshan; *J. Org. Chem.,* 1973, **38**, 3868.	679, 681, 688, 694, 739, 744
73JOC3947	R. A. Coburn, B. Bhooshan and R. A. Glennon; *J. Org. Chem.,* 1973, **38**, 3947.	568
73JOC3953	S. Satsumabayashi, H. Takahashi, T. Tanaka and S. Motoki; *J. Org. Chem.,* 1973, **38**, 3953.	844
73JOC4259	P. R. Farina and H. Tieckelmann; *J. Org. Chem.,* 1973, **38**, 4259.	1002
73JOC4324	Y. Tamura, K. Sumoto, H. Matsushima, H. Taniguchi and M. Ikeda; *J. Org. Chem.,* 1973, **38**, 4324.	75, 83, 84, 85
73JOC4383	C. O. Okafor; *J. Org. Chem.,* 1973, **38**, 4383.	699
73JPR39	K. Gewald and J. Schael; *J. Prakt. Chem.,* 1973, **315**, 39.	995
73JPR155	E. Bulka and D. Ehlers; *J. Prakt. Chem.,* 1973, **315**, 155.	358
73JPR164	E. Bulka, D. Ehlers and H. Storm; *J. Prakt. Chem.,* 1973, **315**, 164.	358
73JPR185	T. Radha Vakula, V. Ranga Rao and V. R. Srinivasan; *J. Prakt. Chem.,* 1973, **315**, 185.	440
73JPR497	H. Hartmann, H. Schäfer and K. Gewald; *J. Prakt. Chem.,* 1973, **315**, 497.	305
73JPR510	E. Bulka and D. Ehlers; *J. Prakt. Chem.,* 1973, **315**, 510.	358
73JPR539	K. Gewald, M. Hentschel and R. Heickel; *J. Prakt. Chem.,* 1973, **315**, 539.	1010, 1016
73JPR587	E. Kleinpeter, R. Borsdorf, G. Bach and J. von Grossmann; *J. Prakt. Chem.,* 1973, **315**, 587.	339
73JPR690	J. Fabian; *J. Prakt. Chem.,* 1973, **315**, 690.	818
73JPR765	E. Kleinpeter and R. Borsdorf; *J. Prakt. Chem.,* 1973, **315**, 765.	237
73JPR791	G. Westphal and R. Schmidt; *J. Prakt. Chem.,* 1973, **315**, 791.	413, 417

73JPR915	G. Wagner and B. Dietzsch; *J. Prakt. Chem.*, 1973, **315**, 915.	560
73JPS839	A. Shafiee, I. Lalezari, S. Yazdani and R. Pournorouz; *J. Pharm. Sci.*, 1973, **62**, 839.	358
73KGS930	O. P. Shvaika, V. N. Artemov, V. E. Kononenko and S. N. Baranov; *Khim. Geterotsikl. Soedin.*, 1973, 930 (*Chem. Abstr.*, 1973, **79**, 115 502).	346
73KGS1016	G. N. Dorofeenko, A. V. Koblik, B. A. Tertov and T. I. Polyakova; *Khim. Geterotsikl. Soedin.*, 1973, 1016 (*Chem. Abstr.*, 1973, **79**, 137 021).	79
73KGS1334	Yu. S. Shabarov and S. S. Mochalov; *Khim. Geterotsikl. Soedin.*, 1973, 1334 (*Chem. Abstr.*, 1974, **80**, 27 148).	122
73LA190	K. Harsanyi, C. Gonczi and D. Korbonits; *Liebigs Ann. Chem.*, 1973, 190.	725
73LA247	F. Boberg and W. von Gentzkow; *Liebigs Ann. Chem.*, 1973, 247.	796
73LA256	F. Boberg and W. von Gentzkow; *Liebigs Ann. Chem.*, 1973, 256.	65, 86
73LA578	U. Heep; *Liebigs Ann. Chem.*, 1973, 578.	71
73LA1018	K. Grohe and H. Heitzer; *Liebigs Ann. Chem.*, 1973, 1018.	697
73LA1644	K. Hartke and G. Goelz; *Liebigs Ann. Chem.*, 1973, 1644.	641
73MI41600	Z. Jermanowska and W. Basenski; *Rocz. Chem.*, 1973, **47**, 1785.	93, 117
73MI41601	V. I. Kelarev, A. Baranski and G. A. Shvekhgeimer; *Rocz. Chem.*, 1973, **47**, 1669.	91
73MI41602	S. J. Hong; *Kwahak Konghak*, 1973, **11**, 404 (*Chem. Abstr.*, 1974, **81**, 7293).	130
73MI41603	K. Takasaki, M. Kaneko, T. Fujii and H. Kobayashi; *Nippon Yakurigaku Zasshi*, 1973, **69**, 977 (*Chem. Abstr.*, 1975, **83**, 528).	128
73MI41604	A. Ya. Strakov, M. B. Andaburskaya, A. M. Moiseenkov and A. A. Akhrem; *Latv. PSR Zinat. Akad. Vestis. Kim. Ser.*, 1973, 330 (*Chem. Abstr.*, 1973, **79**, 92 147).	125
73MI41605	T. N. Grigorova; *Sovrem. Probl. Khim.*, 1973, 51 (*Chem. Abstr.*, 1974, **81**, 136 029).	98, 103
73MI41606	W. S. Hamilton and D. A. Ayers; *J. Chem. Eng. Data*, 1973, **18**, 366; W. S. Hamilton and G. M. Mitchell; *ibid.*, 1973, **18**, 36.	10
B-73MI41607	'Aromatic and Heteroaromatic Chemistry', The Chemical Society, London, 1973–1979, vols. 1–7.	3
73MI41700	Z. Machón, A. D. Inglot and E. Wolna; *Arch. Immunol. Ther. Exp.*, 1973, **21**, 883 (*Chem. Abstr.*, 1974, **80**, 128 065).	156
73MI41900	R. Arnaud, D. Faramond-Baud and M. Gelus; *Theor. Chim. Acta*, 1973, **31**, 335.	237
73MI41901	D. Bares, M. A. Soulié and J. Metzger; *J. Chim. Phys. Phys. Chim. Biol.*, 1973, 1531 (*Chem. Abstr.*, 1974, **80**, 52 983).	246
73MI42100	P. E. Keeley, C. H. Carter and J. H. Miller; *Weed Sci.*, 1973, **21**, 327.	391
73MI42200	V. N. Sololenko and S. P. Suchilina; *Vop. Khim. Khim. Tekhnol.*, 1972, no. 27, 107 (*Chem. Abstr.*, 1973, **78**, 136 183).	412
73MI42300	V. J. Ram and H. N. Pandey; *Agric. Biol. Chem.*, 1973, **37**, 2191 (*Chem. Abstr.*, 1974, **80**, 14 880).	445
73MI42301	V. N. Sokolenko, S. P. Suchilina and V. T. Dorofeev; *Izv. Vyssh. Ucheb. Zaved., Khim. Khim. Tekhnol*, 1973, **16**, 1462 (*Chem. Abstr.*, 1974, **80**, 37 491).	446
73MI42400	S. I. Ramsby, S. O. öGren, S. S. Ross and N. E. Stjernstrom; *Acta Pharm. Suec.* 1973, **10**, 285.	462
73MI42600	R. N. MacDonald and W. H. Sharkey; *J. Polym. Sci., Polym. Chem. Ed.*, 1973, **11**, 2519.	543
73MI42601	L. Krause and M. A. Whitehead; *Mol. Phys.*, 1973, **25**, 99.	515
73MI42602	N. L. Klimasenko, L. A. Chetkina and G. A. Gol'der; *Zh. Strukt. Khim.*, 1973, **14**, 515.	516
73MI42700	H. Thielemann and M. Paepke; *Fresenius' Z. Anal. Chem.*, 1973, **266**, 128 (*Chem. Abstr.*, 1974, **80**, 7014).	555
B-73MI43000	H. C. Van Der Plas; 'Ring Transformations of Heterocycles,' Academic, New York, 1973, vol. 1, pp. 246, 334.	759, 776
73MI43300	F. Ledl and T. Severin; *Chem. Mikrobiol. Technol. Lebensm.*, 1973, **2**, 155.	895
73MI43301	E. J. Mulders; *Z. Lebensm. Unters.-Forschung*, 1973, **152**, 193.	895
73MI43302	A. F. Thomas; *J. Agric. Food Chem.*, 1973, **21**, 955.	894
73MI43400	A. J. Lawson and C. R. Meloy; *Michigan Academician*, 1973, **5**, 465 (*Chem. Abstr.*, 1973, **79**, 92 116).	907, 908, 937, 942
73MI43401	B. T. Brown and R. L. N. Harris; *Pestic. Sci.*, 1973, **4**, 215 (*Chem. Abstr.*, 1973, **78**, 144 228).	931, 941, 944
73MI43402	J. E. Oliver, R. T. Brown, R. L. Fye and A. B. Bořkovec; *J. Agric. Food Chem.*, 1973, **21**, 753.	944
73MIP41600	Yu. Shabaraov and S. S. Mochalov; *USSR Pat.* 367 099 (1973) (*Chem. Abstr.*, 1973, **79**, 5324).	122
B-73NMR382	T. J. Batterham; 'NMR Spectra of Simple Heterocycles,' Wiley, New York, 1973, p. 382.	753
73OMR(5)295	R. Cahill and T. A. Crabb; *Org. Magn. Reson.*, 1973, **5**, 295.	666
73OMR(5)299	F. Borremans, M. Anteunis and F. Anteunis-De Ketelaere; *Org. Magn. Reson.*, 1973, **5**, 299.	753
73OMS(7)327	A. Selva and E. Gaetani; *Org. Mass Spectrom.*, 1973, **7**, 327.	143
73OMS(7)463	J. Julien, E. J. Vincent, J. C. Poite and J. Roggero; *Org. Mass Spectrom.*, 1973, **7**, 463.	142, 143
73OMS(7)555	A. Shoeb, S. P. Popli and R. Gopalchari; *Org. Mass Spectrom.*, 1973, **7**, 555.	467
73OPP197	M. Davis, E. Homfeld and T. Paproth; *Org. Prep. Proced. Int.*, 1973, **5**, 197.	168
73OS(53)70	J. E. McMurry; *Org. Synth.*, 1973, **53**, 70.	52

73OSC(5)962	C. J. Thoman and D. J. Voaden; *Org. Synth. Coll. Vol.*, 1973, **5**, 962.	377
73S156	C. Grundmann and G. F. Kite; *Synthesis*, 1973, 156.	67, 391
73S243	H. Petersen; *Synthesis*, 1973, 243.	993
73S313	K. Srinivasan, K. K. Balasubramanian and S. Swaminathan; *Synthesis*, 1973, 313.	1009
73S469	C. G. Stuckwisch; *Synthesis*, 1973, 469.	1001, 1003
73SA(A)1393	D. H. Christensen, P. W. Jensen, J. T. Nielsen and O. F. Nielsen; *Spectrochim. Acta, Part A*, 1973, **29**, 1393.	398
73SST(2)556	F. Kurzer; *Org. Compd. Sulphur, Selenium, Tellurium*, 1973, **2**, 556.	149, 150, 151, 152, 161, 162, 166, 168, 169, 170, 171
73SST(2)587	F. Kurzer; *Org. Compd. Sulphur, Selenium, Tellurium*, 1973, **2**, 587–652.	328, 330
73SST(2)653	F. Kurzer; *Org. Compd. Sulphur, Selenium, Tellurium*, 1973, **2**, 653–682.	328, 330
73SST(2)717	F. Kurzer; *Org. Compd. Sulphur, Selenium, Tellurium*, 1973, **2**, 717.	451, 454, 459
73SST(2)725	F. Kurzer; *Org. Compd. Sulphur, Selenium, Tellurium*, 1973, **2**, 725.	546, 562, 576
73T2683	M. Lj. Mihailovic, Lj. Lorenc, Z. Maksimovic and J. Kalvoda; *Tetrahedron*, 1973, **29**, 2683.	111
73T2937	P. Henklein, G. Westphal and R. Kraft; *Tetrahedron*, 1973, **29**, 2937.	439, 727
73T3085	P. A. Clark, R. Gleiter and E. Heilbronner; *Tetrahedron*, 1973, **29**, 3085.	395, 517
73T3861	H.-R. Schulten, H. D. Beckey, G. Eckhardt and S. H. Doss; *Tetrahedron*, 1973, **29**, 3861.	143
73T3985	N. S. Ramegowda, M. N. Modi, A. K. Koul, J. M. Bora, C. K. Narang and N. K. Mathur; *Tetrahedron*, 1973, **29**, 3985.	148
73T4291	W. Mueller, V. Kraatz and F. Korte; *Tetrahedron*, 1973, **29**, 4291.	40
73TL233	R. Gompper and J. Stetter; *Tetrahedron Lett.*, 1973, 233.	81, 918
73TL453	R. Gree, F. Tonnard and R. Carrie; *Tetrahedron Lett.*, 1973, 453.	110
73TL485	S. Rossi and E. Duranti; *Tetrahedron Lett.*, 1973, 485.	66, 87
73TL597	O. Tsuge and H. Samura; *Tetrahedron Lett.*, 1973, 597.	1039
73TL655	R. H. Cragg and A. F. Weston; *Tetrahedron Lett.*, 1973, 655.	809
73TL939	M. Alvarez, R. W. Creekmore and R. T. Rosen; *Tetrahedron Lett.*, 1973, 939.	919
73TL1561	P. deMayo and H. Y. Ng; *Tetrahedron Lett.*, 1973, 1561.	795
73TL1565	E. C. Llaguno and I. C. Paul; *Tetrahedron Lett.*, 1973, 1565.	360, 1054
73TL1915	H. Behringer and E. Meinetsberger; *Tetrahedron Lett.*, 1973, 1915.	809
73TL2159	R. B. Morin, E. M. Gordon, T. McGrath and R. Shuman; *Tetrahedron Lett.*, 1973, 2159.	167
73TL2195	C. L. Pedersen, O. Buchardt, S. Larsen and K. J. Watson; *Tetrahedron Lett.*, 1973, 2195.	79, 85
73TL2283	T. Matsuura and Y. Ito; *Tetrahedron Lett.*, 1973, 2283.	37
73TL2667	M. F. Semmelhack and R. D. Stauffer; *Tetrahedron Lett.*, 1973, 2667.	769
73TL3589	L. Thijs, A. Wagenaar, E. M. M. Van Rens and B. Zwanenburg; *Tetrahedron Lett.*, 1973, 3589.	556, 572
73TL5009	B. Zwanenburg and A. Wagenaar; *Tetrahedron Lett.*, 1973, 5009.	571
73TL5075	K. Maeda, T. Hosokawa, S. Murahasi and I. Moritani; *Tetrahedron Lett.*, 1973, 5075.	75
73TL5213	R. B. Morin, E. M. Gordon and J. R. Lake; *Tetrahedron Lett.*, 1973, 5213.	165, 167
73USP3713826	D. J. Sykes, H. Kroll and T. R. Finch; *U.S. Pat.* 3 713 826 (1973) (*Chem. Abstr.*, 1973, **78**, 117 585).	894
73USP3716534	*U.S. Pat.* 3 716 534 (1973) (*Chem. Abstr.*, 1973, **78**, 137 957).	748
73USP3720684	Velsicol Chemical Corp., *U.S. Pat.* 3 720 684 (1973) (*Chem. Abstr.*, 1973, **79**, 5343).	510
73USP3736328	S. Wittenbrook and R. J. Timmons; *U.S. Pat.* 3 736 328 (1973) (*Chem. Abstr.*, 1973, **79**, 42 517).	509
73USP3737434	*U.S. Pat.* 3 737 434 (1973) (*Chem. Abstr.*, 1973, **79**, 42 574).	610
73USP3737435	E. H. Burk, Jr. and D. D. Carlos; *U.S. Pat.* 3 737 435 (1973) (*Chem. Abstr.*, 1973, **79**, 79 470).	945
73USP3743653	S. Marburg; *U.S. Pat.* 3 743 653 (1973) (*Chem. Abstr.*, 1973, **79**, 92 226).	877
73USP3753908	L. De Vries and B. R. Kennedy; *U.S. Pat.* 3 753 908 (1973) (*Chem. Abstr.*, 1974, **80**, 50 199).	946
73USP3755571	Stauffer Chemical Co., *U.S. Pat.* 3 755 571 (1973) (*Chem. Abstr.*, 1974, **80**, 34 393).	510
73USP3772290	*U.S. Pat.* 3 772 290 (1973) (*Chem. Abstr.*, 1974, **80**, 48 027).	710
73ZC134	E. Lippmann, D. Reifegerste and E. Kleinpeter; *Z. Chem.*, 1973, **13**, 134.	590
73ZOB1179	Yu. N. Kukushkin, S. A. D'yachenko, R. A. Vlasova and N. P. Glazyuk; *Zh. Obshch. Khim.*, 1973, **43**, 1179 (*Chem. Abstr.*, 1973, **79**, 61 010).	526
73ZOB1715	V. M. Kashutina, S. L. Ioffe, V. M. Shoitkin, M. O. Cherskaya, V. A. Korenevskii and V. A. Tartakovskii; *Zh. Obshch. Khim.*, 1973, **43**, 1715 (*Chem. Abstr.*, 1973, **79**, 126 558).	95
73ZOR269	K. V. Altukhov, E. V. Ratsino and V. V. Perekalin; *Zh. Org. Khim.*, 1973, **9**, 269.	111
73ZOR2502	L. N. Markovskii, G. S. Fedyuk, E. S. Levchenko and A. V. Kirsanov; *Zh. Org. Khim.*, 1973, **9**, 2502.	541
74ABC2205	M. Nakagawa, T. Nakamura and K. Tomita; *Agr. Biol. Chem.*, 1974, **38**, 2205.	13
74AC(R)131	S. Facchetti and G. F. Bettinetti; *Ann. Chim. (Rome)*, 1974, **64**, 131 (*Chem. Abstr.*, 1975, **83**, 192 033).	6
74AC(R)305	A. Tajana, D. Nardi and R. Cappelletti; *Ann. Chim. (Rome)*, 1974, **64**, 305.	883
74ACR232	A. F. Garito and A. J. Heeger; *Acc. Chem. Res.*, 1974, **7**, 232.	850
74ACS(A)499	J. Sletten; *Acta Chem. Scand., Ser. A*, 1974, **28**, 499.	1055

74ACS(A)989	J. Sletten; *Acta Chem. Scand., Ser. A*, 1974, **28**, 989.	902
74ACS(B)308	L. Brehm, P. Krogsggaard-Larsen and H. Hjeds; *Acta Chem. Scand., Ser. B*, 1974, **28**, 308.	4
74ACS(B)533	P. Krogsggaard-Larsen and H. Hjeds; *Acta Chem. Scand., Ser. B*, 1974, **28**, 533.	64
74ACS(B)625	L. Brehm, P. Krogsggaard-Larsen and H. Hjeds; *Acta Chem. Scand., Ser. B*, 1974, **28**, 625.	4
74ACS(B)636	P. Krogsggaard-Larsen and S. B. Christensen; *Acta Chem. Scand., Ser. B*, 1974, **28**, 636.	52, 53
74ACS(B)827	P. E. Iversen and H. Lund; *Acta Chem. Scand., Ser. B*, 1974, **28**, 827.	797
74ACS(B)1185	C. Th. Pedersen, H. L. Huaman and J. Moller; *Acta Chem. Scand., Ser. B*, 1974, **28**, 1185.	787
74AG(E)349	C. Th. Pedersen; *Angew. Chem., Int. Ed. Engl.*, 1974, **13**, 349.	1057
74AG(E)751	N. D. Epiotis; *Angew. Chem., Int. Ed. Engl.*, 1974, **13**, 751.	1006
74AHC(16)1	J. A. Elvidge, J. R. Jones, C. O'Brien, E. A. Evans and H. C. Sheppard; *Adv. Heterocycl. Chem.*, 1974, **16**, 1.	457
74AHC(16)10	J. A. Elvidge, J. R. Jones, C. O'Brien, E. A. Evans and H. C. Sheppard; *Adv. Heterocycl. Chem.*, 1974, **16**, 10.	261
74AHC(16)14	J. A. Elvidge, J. R. Jones and C. O'Brien; *Adv. Heterocycl. Chem.*, 1974, **16**, 14.	562
74AHC(16)33	T. L. Gilchrist and G. E. Gymer; *Adv. Heterocycl. Chem.*, 1974, **16**, 33.	1000
74AHC(17)99	R. Lakhan and B. Ternai; *Adv. Heterocycl. Chem.*, 1974, **17**, 99.	178, 215
74AJC1221	M. Davis, L. W. Deady and E. Homfeld; *Aust. J. Chem.*, 1974, **27**, 1221.	21, 54, 145
74AJC1917	M. Davis, L. W. Deady and E. Homfeld; *Aust. J. Chem.*, 1974, **27**, 1917.	401, 526
74AP291	R. Neidlein and R. Mosebach; *Arch. Pharm. (Weinheim, Ger.)*, 1974, **307**, 291.	911, 926, 941
74AP828	H. Böhme and K. H. Ahrens; *Arch. Pharm. (Weinheim, Ger.)*, 1974, **307**, 828.	904, 905
74AX(B)139	M. Baiwir, G. Llabrès, O. Dideberg, L. Dupont and J. L. Piette; *Acta Crystallogr., Part B*, 1974, **30**, 139.	951
74AX(B)763	T. J. Kistenmacher, T. E. Phillips and D. O. Cowan; *Acta Crystallogr., Part B*, 1974, **30**, 763.	815
74AX(B)1123	J. L. Flippen; *Acta Crystallogr., Part B*, 1974, **30**, 1123.	671
74AX(B)1642	T. LaCour; *Acta Crystallogr., Part B*, 1974, **30**, 1642.	547
74B5358	J. M. Duclos and P. Haake; *Biochemistry*, 1974, **13**, 5358.	327
74BCJ785	T. Kitazume and N. Ishikawa; *Bull. Chem. Soc. Jpn.*, 1974, **47**, 785.	186, 228
74BCJ946	A. Matsumoto, J. H. Lee, M. Yoshida and O. Simamura; *Bull. Chem. Soc. Jpn.*, 1974, **47**, 946.	1032, 1035, 1044
74BCJ1490	A. Matsumoto, J. H. Lee, M. Yoshida and O. Simamura; *Bull. Chem. Soc. Jpn.*, 1974, **47**, 1490.	1029, 1030
74BCJ1493	A. Matsumoto, M. Yoshida and O. Simamura; *Bull. Chem. Soc. Jpn.*, 1974, **47**, 1493.	414, 518, 538, 1032, 1037, 1045
74BCJ2813	T. Okabe, E. Taniguchi and K. Maekawa; *Bull. Chem. Soc. Jpn.*, 1974, **47**, 2813.	736, 738, 740
74BCJ3084	S. Tamagaki, K. Sakaki and S. Oae; *Bull. Chem. Soc. Jpn.*, 1974, **47**, 3084.	786
74BRP1334882	Sankyo Co. Ltd., *Br. Pat.* 1 334 882 (1974) (*Chem. Abstr.*, 1974, **80**, 47 971).	43
74BRP1474691	J. Crosby, R. M. Paton, J. Tanner and R. A. C. Rennie; *Br. Pat.* 1 474 691 (1974).	426
74BRP1521690	J. Crosby, R. A. C. Rennie and R. M. Paton; *Br. Pat.* 1 521 690 (1974) (*Chem. Abstr.*, 1977, **85**, 124 937).	425
74BSF137	J. L. Burgot and J. Vialle; *Bull. Soc. Chim. Fr.*, 1974, 137.	1066
74BSF140	J. L. Burgot; *Bull. Soc. Chim. Fr.*, 1974, 140.	1066
74BSF163	C. Metayer, G. Duguay and H. Quiniou; *Bull. Soc. Chim. Fr.*, 1974, 163.	1065
74BSF510	P. Appriou and R. Guglielmetti; *Bull. Soc. Chim. Fr.*, 1974, 510.	843
74BSF725	R. Escale, F. Petrus and J. Verducci; *Bull. Soc. Chim. Fr.*, 1974, 725.	100
74BSF1025	A. Belly, C. Petrus and F. Petrus; *Bull. Soc. Chim. Fr.*, 1974, 1025.	98
74BSF1479	C. Aspisi, C. Petrus and F. Petrus; *Bull. Soc. Chim. Fr.*, 1974, 1479.	89
74BSF1580	A. Étienne, A. Le Bere, G. Lonchambon, G. Lochey and B. Cucumel; *Bull. Soc. Chim. Fr.*, 1974, 1580.	477, 500
74BSF1651	R. Jacquier, J. L. Olive, C. Petrus and F. Petrus; *Bull. Soc. Chim. Fr.*, 1974, 1651.	112
74BSF2507	P. Duguay, C. Metayer, H. Quiniou and J. Bourrigaud; *Bull. Soc. Chim. Fr.*, 1974, 2507.	166
74CB13	H. G. Aurich and G. Blinne; *Chem. Ber.*, 1974, **107**, 13.	32, 43, 105
74CB502	J. Goerdeler, J. Haag, C. Lindner and R. Losch; *Chem. Ber.*, 1974, **107**, 502.	152, 166, 915
74CB2299	H. Bock and B. Solouki; *Chem. Ber.*, 1974, **107**, 2299.	861
74CB3036	H. Dickopp; *Chem. Ber.*, 1974, **107**, 3036.	376
74CB3574	E. Schaumann, E. Kausch and W. Walter; *Chem. Ber.*, 1974, **107**, 3574.	313
74CB3717	L. Birkofer and R. Stilke; *Chem. Ber.*, 1974, **107**, 3717.	68
74CC166	J. Nakayama; *J. Chem. Soc., Chem. Commun.*, 1974, 166.	845
74CC358	M. Ruccia, N. Vivona and G. Cusmano; *J. Chem. Soc., Chem. Commun.*, 1974, 358.	504
74CC585	R. F. Cookson and A. C. Richardson; *J. Chem. Soc., Chem. Commun.*, 1974, 585.	525
74CC621	D. Mackay and L. L. Wong; *J. Chem. Soc., Chem. Commun.*, 1974, 621.	726
74CC751	M. G. Miles, J. D. Wilson, D. J. Dahm and J. H. Wagenknecht; *J. Chem. Soc., Chem. Commun.*, 1974, 751.	849
74CC800	L. K. Hansen and A. Hordvik; *J. Chem. Soc., Chem. Commun.*, 1974, 800.	1067
74CC906	E. K. Adesogan; *J. Chem. Soc., Chem. Commun.*, 1974, 906.	894
74CC937	K. Bechgaard, D. O. Cowan and A. N. Bloch; *J. Chem. Soc., Chem. Commun.*, 1974, 937.	963, 965

74CHE230	L. I. Savranskii, V. A. Kovtunenko and F. S. Babichev; *Chem. Heterocycl. Compd. (Engl. Transl.)*, 1974, **10**, 230.	980
74CHE813	Z. I. Moskalenko and G. P. Shumelyak; *Chem. Heterocycl. Compd. (Engl. Transl.)*, 1974, **10**, 813.	994
74CJC100	A. W. Richardson; *Can. J. Chem.*, 1974, **52**, 100.	520
74CJC187	J. Hébert and D. Gravel; *Can. J. Chem.*, 1974, **52**, 187.	767
74CJC833	R. E. Wasylishen, J. B. Rowbotham and T. Schaefer; *Can. J. Chem.*, 1974, **52**, 833.	5, 133, 137
74CJC1738	M. S. Chauhan, M. E. Hassan and D. M. McKinnon; *Can. J. Chem.*, 1974, **52**, 1738.	149, 161
74CL951	J. H. Lee, A. Matsumoto, M. Yoshida and O. Simamura; *Chem. Lett.*, 1974, 951.	1044, 1046
74CPB54	I. Adachi, R. Miyazaki and H. Kano; *Chem. Pharm. Bull.*, 1974, **22**, 54.	99
74CPB61	I. Adachi, K. Harada, R. Miyazaki and H. Kano; *Chem. Pharm. Bull.*, 1974, **22**, 61.	28, 29, 44, 51, 99
74CPB70	I. Adachi, R. Miyazaki and H. Kano; *Chem. Pharm. Bull.*, 1974, **22**, 70.	29, 44, 45, 98
74CPB342	R. Marumoto, Y. Yoshioka and M. Honjo; *Chem. Pharm. Bull.*, 1974, **22**, 342.	688
74CPB477	S. Zen and E. Kaji; *Chem. Pharm. Bull.*, 1974, **22**, 477.	74, 102
74CPB482	H. Koga, M. Hirobe and T. Okamoto; *Chem. Pharm. Bull.*, 1974, **22**, 482.	1002
74CPB1990	M. Furukawa, T. Yuki and S. Hayashi; *Chem. Pharm. Bull.*, 1974, **22**, 1990.	93
74CR(C)(279)237	A. Dorange and F. Venien; *C. R. Hebd. Seances Acad. Sci., Ser. C*, 1974, **279**, 237.	789, 793
74CR(C)(279)259	J. M. Decrouen, D. Paquer and R. Pou; *C. R. Hebd. Seances, Acad. Sci., Ser. C*, 1974, **279**, 259.	1069
74CS(6)222	O. Thorstad and K. Undheim; *Chem. Scr.*, 1974, **6**, 222.	366, 718, 728, 746
74CSC397	V. Debugeard, J. L. Boudour and J. C. Messaga; *Cryst. Struct. Commun.*, 1974, **3**, 397.	4
74CSC535	M. A. Pellinghelli, A. Tiripicchio and C. M. Tiripicchio; *Cryst. Struct. Commun.*, 1974, **3**, 535.	134
74DIS(B)147	R. E. Sheads; Ph.D. Thesis, Univ. of Maryland, 1973 (*Diss. Abstr. Int. B*, 1974, **35**, 147).	5, 6, 21
74DOK109	M. V. Kashutina, S. L. Ioffe and V. A. Tartakovskii; *Dokl. Akad. Nauk SSSR*, 1974, **218**, 109 (*Chem. Abstr.*, 1975, **82**, 43 227).	47, 95, 110
74E405	M. Shimizu, K. Yoshida, T. Karasawa, M. Masuda, M. Oka, T. Ito, C. Kamei, M. Hori, Y. Sohji and K. Furukawa; *Experientia*, 1974, **30**, 405.	129
74EGP104784	K. Peseke; *Ger. (East) Pat.* 104 784 (1974) (*Chem. Abstr.*, 1974, **81**, 151 812).	876
74EGP106041	K. Peseke; *Ger. (East) Pat.* 106 041 (1974) (*Chem. Abstr.*, 1974, **81**, 169 569).	876
74FRP2197513	*Fr. Pat.* 2 197 513 (1974) (*Chem. Abstr.*, 1975, **82**, 112 101).	709
74FRP2223371	*Fr. Pat.* 2 223 371 (1974) (*Chem. Abstr.*, 1975, **82**, 171 031).	709
74G715	V. Parreni, R. Pepino and E. Belgodere; *Gazz. Chim. Ital.*, 1974, **104**, 715.	41, 104
74GEP2231098	Bayer A.G.; *Ger. Pat.* 2 231 098 (1974) (*Chem. Abstr.*, 1974, **80**, 82 951).	168
74GEP2232468	*Ger. Pat.* 2 232 468 (1974) (*Chem. Abstr.*, 1974, **80**, 83 067).	724
74GEP2252122	*Ger. Pat.* 2 252 122 (1974) (*Chem. Abstr.*, 1974, **81**, 13 543).	668
74GEP2253555	*Ger. Pat.* 2 253 555 (1974) (*Chem. Abstr.*, 1974, **81**, 25 692).	668
74GEP2330109	*Ger. Pat.* 2 330 109 (1974) (*Chem. Abstr.*, 1974, **80**, 95 916).	661
74GEP2342213	K. Streiff; *Ger. Offen.* 2 342 213 (1974) (*Chem. Abstr.*, 1975, **82**, 110 544).	128
74GEP2352586	Y. Labat, Y. Mordelet and J. B. Signouret; *Ger. Offen.* 2 352 586 (1974) (*Chem. Abstr.*, 1975, **83**, 134 819).	893
74GEP2403357	J. Thomas; *Ger. Offen.* 2 403 357 (1974) (*Chem. Abstr.*, 1974, **81**, 136 153).	445
74GEP2420441	K. F. W. Lautenschlaeger; *Ger. Pat.* 2 420 441 (1974) (*Chem. Abstr.*, 1975, **82**, 141 393).	939
74H(2)27	O. Tsuge and H. Samura; *Heterocycles*, 1974, **2**, 27.	1039
74H(2)45	S. Tamagaki, K. Sakaki and S. Oae; *Heterocycles*, 1974, **2**, 45.	796
74HCA376	W. Heinzelmann and M. Markey; *Helv. Chim. Acta*, 1974, **57**, 376.	16, 29, 59
74IJC134	G. Bhaskaraiah; *Indian J. Chem.*, 1974, **12**, 134.	502
74IZV845	Zh. A. Krasnaya, T. S. Stytsenko, E. P. Prokof'ev, I. P. Yakovlev and V. F. Kucherov; *Izv. Akad. Nauk SSSR, Ser. Khim.*, 1974, 845 (*Chem. Abstr.*, 1974, **81**, 49 608).	102, 107
74IZV908	N. Singh and K. Krishan; *Izv. Akad. Nauk SSSR, Ser. Khim.*, 1974, 908 (*Chem. Abstr.*, 1974, **81**, 37 496).	108
74IZV2651	I. E. Chlenov, I. M. Petrova and V. A. Tartakovskii; *Izv. Akad. Nauk SSSR*, 1974, **23**, 2651.	25
74IZV3836	T. P. Valilev, M. G. Lin'kova, O. V. Kil'disheva and I. L. Knunyants; *Izv. Akad. Nauk SSSR, Ser. Khim.*, 1974, **700**, 3836.	802
74JA289	Q. Shen and K. Hedberg; *J. Am. Chem. Soc.*, 1974, **96**, 289.	1053
74JA348	R. P. Lattimer, R. L. Kuczkowski and C. W. Gillies; *J. Am. Chem. Soc.*, 1974, **96**, 348.	855, 858
74JA945	P. R. Moses and J. Q. Chambers; *J. Am. Chem. Soc.*, 1974, **96**, 945.	849
74JA1536	C. W. Gillies, R. P. Lattimer and R. L. Kuczkowski; *J. Am. Chem. Soc.*, 1974, **96**, 1536.	863
74JA1817	M. P. Cava, M. A. Sprecker and W. R. Hall; *J. Am. Chem. Soc.*, 1974, **96**, 1817.	1032
74JA2014	J. P. Ferris and F. R. Antonucci; *J. Am. Chem. Soc.*, 1974, **96**, 2014.	13, 1020
74JA2426	G. E. Wilson, Jr.; *J. Am. Chem. Soc.*, 1974, **96**, 2426.	750, 758
74JA2916	S. F. Nelsen, G. R. Weisman, P. J. Hintz, D. Olp and M. R. Fahey; *J. Am. Chem. Soc.*, 1974, **96**, 2916.	429
74JA3502	W. Kusters and P. deMayo; *J. Am. Chem. Soc.*, 1974, **96**, 3502.	819

74JA3973	G. L'abbé, E. Van Loock, R. Albert, S. Toppet, G. Verhelst and G. Smets; *J. Am. Chem. Soc.*, 1974, **96**, 3973.	583, 596
74JA4268	K. T. Potts and D. McKeough; *J. Am. Chem. Soc.*, 1974, **96**, 4268.	1032
74JA4276	K. T. Potts and D. McKeough; *J. Am. Chem. Soc.*, 1974, **96**, 4276.	1032, 1033, 1043
74JA6217	M. Arbelot, J. Metzger, M. Chanon, C. Guimon and G. Pfister-Guillouzo; *J. Am. Chem. Soc.*, 1974, **96**, 6217.	241
74JA6768	A. Krantz and J. Laureni; *J. Am. Chem. Soc.*, 1974, **96**, 6768.	455
74JA7376	E. M. Engler and V. V. Patel; *J. Am. Chem. Soc.*, 1974, **96**, 7376.	958, 963, 970
74JA7394	J. R. Davidson, A. K. Burnham, B. M. Siegel, P. Beak and W. H. Flygare; *J. Am. Chem. Soc.*, 1974, **96**, 7394.	4, 8
74JAP7418899	Sankyo Co. Ltd., *Jpn. Pat.* 74 18 899 (1974) (*Chem. Abstr.*, 1974, **80**, 133 446).	510
74JAP7441196	H. Tawada, K. Meguro, Y. Kuwatq, Y. Hara and Y. Usui; *Jpn. Pat.* 74 41 196 (1974) (*Chem. Abstr.*, 1975, **82**, 170 384).	122
74JAP74117630	Sankyo Co. Ltd., *Jpn. Pat.* 74 117 630 (1974) (*Chem. Abstr.*, 1975, **83**, 2261).	509
74JAP(K)7472252	K. Tomita, T. Murakami, T. Honma, Y. Yamazaki, Y. Morisawa and H. Takagi; *Jpn. Kokai* 74 72 252 (1974) (*Chem. Abstr.*, 1975, **83**, 114 367).	43
74JAP(K)7480062	I. Iijima, M. Miyazaki and T. Tanaka; *Jpn. Kokai* 74 80 062 (1974) (*Chem. Abstr.*, 1975, **82**, 156 261).	128
74JAP(K)74107845	Y. Yamazaki and K. Tomita; *Jpn. Kokai* 74 107 845 (1974) (*Chem. Abstr.*, 1975, **82**, 120 096).	129
74JAP(K)74116054	H. Nishimura, M. Shimizu, H. Uno, H. Tetsuo, Y. Masuda and M. Kurokawa; *Jpn. Kokai* 74 116 054 (1974) (*Chem. Abstr.*, 1975, **83**, 28 211).	129
74JAP(K)74117462	H. Nishimura, H. Kinugasa, J. Kashima, M. Shimizu, Y. Takase and A. Minami; *Jpn. Kokai* 74 117 462 (1974) (*Chem. Abstr.*, 1975, **83**, 10 040).	98
74JAP(K)74134672	I. Iijima, M. Miyazaki and T. Tanaka; *Jpn. Kokai* 74 134 672 (1974) (*Chem. Abstr.*, 1975, **83**, 10 043).	128
74JCP(61)253	J. R. Durig, S. Riethmiller and Y. S. Li; *J. Chem. Phys.*, 1974, **61**, 253.	181
74JCP(61)1494	L. Guibé, M. Gourdji and A. Péneau; *J. Chem. Phys.*, 1974, **61**, 1494.	429
74JCS(P1)30	A. Caplin; *J. Chem. Soc., Perkin Trans. 1*, 1974, 30.	348, 353
74JCS(P1)158	G. Jones and J. R. Phipps; *J. Chem. Soc., Perkin Trans. 1*, 1974, 158.	630
74JCS(P1)242	A. S. Ingram, D. H. Reid and J. D. Symon; *J. Chem. Soc., Perkin Trans. 1*, 1974, 242.	1059
74JCS(P1)433	M. Jones, P. Temple, E. J. Thomas and G. H. Whitham; *J. Chem. Soc., Perkin Trans. 1*, 1974, 433.	770
74JCS(P1)627	A. R. McCarthy, W. D. Ollis and C. A. Ramsden; *J. Chem. Soc., Perkin Trans. 1*, 1974, 627.	442
74JCS(P1)645	W. D. Ollis and C. A. Ramsden; *J. Chem. Soc., Perkin Trans. 1*, 1974, 645.	429, 600
74JCS(P1)722	R. M. Christie, A. S. Ingram, D. H. Reid and R. G. Webster; *J. Chem. Soc., Perkin Trans. 1*, 1974, 722.	761, 799, 1059, 1060, 1066
74JCS(P1)866	J. P. Brown and T. B. Gay; *J. Chem. Soc., Perkin Trans. 1*, 1974, 866.	1067
74JCS(P1)1572	R. J. Stoodley and R. B. Wilkins; *J. Chem. Soc., Perkin Trans. 1*, 1974, 1572.	667
74JCS(P1)1694	J. I. G. Cadogan, R. J. Scott, R. D. Gee and I. Gosney; *J. Chem. Soc., Perkin Trans. 1*, 1974, 1694.	419
74JCS(P1)1867	T. Nishiwaki, K. Azechi and F. Fujiyama; *J. Chem. Soc., Perkin Trans. 1*, 1974, 1867.	15
74JCS(P1)2589	R. A. Abramovitch, E. M. Smith, M. Humber, B. Purtschert, P. C. Srinivasan and G. M. Singer; *J. Chem. Soc., Perkin Trans. 1*, 1974, 2589.	151
74JCS(P2)389	A. G. Burton, A. R. Katritzky, M. Konya and H. O. Tarhan; *J. Chem. Soc., Perkin Trans. 2*, 1974, 389.	23
74JCS(P2)399	S. Clementi, P. P. Forsythe, C. D. Johnson, A. R. Katritzky and B. Terem; *J. Chem. Soc., Perkin Trans. 2*, 1974, 399.	21, 22, 133, 136, 145
74JCS(P2)420	M. H. Palmer, R. H. Findley and A. J. Gaskell; *J. Chem. Soc., Perkin Trans. 2*, 1974, 420.	517
74JCS(P2)466	R. Keskinen, A. Nikkilä, K. Pihlaja and F. G. Riddell; *J. Chem. Soc., Perkin Trans. 2*, 1974, 466.	753, 758
74JCS(P2)532	M. Mathew and G. J. Palenik; *J. Chem. Soc., Perkin Trans. 2*, 1974, 532.	548
74JCS(P2)875	M. T. W. Hearn and K. T. Potts; *J. Chem. Soc., Perkin Trans. 2*, 1974, 875.	370
74JCS(P2)997	A. F. Hegarty, P. Quain, T. A. F. O'Mahoy and F. L. Scott; *J. Chem. Soc., Perkin Trans. 2*, 1974, 997.	1009
74JCS(P2)1096	C. L. Raston, A. H. White, A. C. Willis and J. N. Varghese; *J. Chem. Soc., Perkin Trans. 2*, 1974, 1096.	465
74JCS(P2)1132	J. J. Guy and T. A. Hamor; *J. Chem. Soc., Perkin Trans. 2*, 1974, 1132.	647
74JCS(P2)1171	L. Di Nunno, S. Florio and P. E. Todesco; *J. Chem. Soc., Perkin Trans. 1*, 1974, 1171.	410
74JCS(P2)1301	P. Beltrame, P. L. Beltrame, M. G. Cattania and G. Zecchi; *J. Chem. Soc., Perkin Trans. 2*, 1974, 1301.	89, 90
74JCS(P2)1409	K. Simon, K. Sasvari, P. Dvortsak, K. Horvath and K. Harsanyi; *J. Chem. Soc., Perkin Trans. 2*, 1974, 1409.	4
74JCS(P2)1419	T. A. Crabb and M. J. Hall; *J. Chem. Soc., Perkin Trans. 2*, 1974, 1419.	663
74JCS(P2)1591	A. Dondoni and G. Barbaro; *J. Chem. Soc., Perkin Trans. 2*, 1974, 1591.	68, 83
74JCS(P2)1885	M. H. Palmer and R. H. Findlay; *J. Chem. Soc., Perkin Trans. 2*, 1974, 1885.	1052
74JHC1	J. P. Chupp; *J. Heterocycl. Chem.*, 1974, **11**, 1.	905, 911, 941
74JHC63	J. Fajkos and J. A. Edwards; *J. Heterocycl. Chem.*, 1974, **11**, 63.	91, 92, 997

74JHC91	V. Spiro, O. Migliara and E. Ajello; *J. Heterocycl. Chem.*, 1974, **11**, 91.	992
74JHC105	S. W. Schneller and W. E. Swartz; *J. Heterocycl. Chem.*, 1974, **11**, 105.	788
74JHC125	R. Y. Ning, W. Y. Chen and L. H. Sternbach; *J. Heterocycl. Chem.*, 1974, **11**, 125.	122
74JHC343	B. Modarai, M. H. Ghanderhari, H. Massoumi, A. Shafiee, I. Lalezari and A. Badali; *J. Heterocycl. Chem.*, 1974, **11**, 343.	457
74JHC395	J. A. Van Allen and G. A. Reynolds; *J. Heterocycl. Chem.*, 1974, **11**, 395.	124, 125
74JHC455	A. Martani, A. Fravolini and G. Grandolini; *J. Heterocycl. Chem.*, 1974, **11**, 455.	624
74JHC459	Y. Tamura, H. Hayashi, E. Saeki, J. H. Kim and M. Ikeda; *J. Heterocycl. Chem.*, 1974, **11**, 459.	976, 990
74JHC507	I. Degani, R. Fochi and P. Tundo; *J. Heterocycl. Chem.*, 1974, **11**, 507.	765
74JHC751	R. M. Claramunt, J. M. Fabregà and J. Elguero; *J. Heterocycl. Chem.*, 1974, **11**, 751.	1018
74JHC763	E. O. Sherman, Jr., S. M. Lambert and K. Pilgram; *J. Heterocycl. Chem.*, 1974, **11**, 763.	402
74JHC777	K. Pilgram and R. D. Skiles; *J. Heterocycl. Chem.*, 1974, **11**, 777.	533
74JHC813	K. Pilgram and M. Zupan; *J. Heterocycl. Chem.*, 1974, **11**, 813.	410
74JHC823	M. R. Bell, S. D. Clemens, R. Oesterlin and J. A. Carlson; *J. Heterocycl. Chem.*, 1974, **11**, 823.	1021
74JHC835	K. Pilgram; *J. Heterocycl. Chem.*, 1974, **11**, 835.	536
74JHC885	A. Walser, T. Flynn and R. I. Fryer; *J. Heterocycl. Chem.*, 1974, **11**, 895.	60, 117
74JHC943	L. Costa, I. Degani, R. Fochi and P. Tundo; *J. Heterocycl. Chem.*, 1974, **11**, 943.	778
74JHC1011	M. Davis, L. W. Deady and E. Homfeld; *J. Heterocycl. Chem.* 1974, **11**, 1011.	339, 518
74JIC453	S. K. P. Sinha and M. P. Thakur; *J. Indian Chem. Soc.*, 1974, **51**, 453.	664, 665
74JMC34	J. R. E. Hoover, G. L. Dunn, J. R. Jakas, L. L. Lam, J. J. Taggart, J. R. Guarini and L. Phillips; *J. Med. Chem.*, 1974, **17**, 34.	156
74JMC1025	R. A. Coburn, R. A. Glennon and Z. F. Chmielewicz; *J. Med. Chem.*, 1974, **17**, 1025.	744
74JMC1282	V. D. Patil, D. S. Wise, L. B. Townsend and A. Bloch; *J. Med. Chem.*, 1974, **17**, 1282.	653, 667
74JOC957	M. Komatsu, Y. Ohshiro, K. Yasuda, S. Ichijima and T. Agawa; *J. Org. Chem.*, 1974, **39**, 957.	505
74JOC962	R. K. Howe and J. E. Franz; *J. Org. Chem.*, 1974, **39**, 962.	465, 466, 468, 479, 500
74JOC1189	G. Knaus and A. I. Meyers; *J. Org. Chem.*, 1974, **39**, 1189.	438, 564
74JOC1192	G. Knaus and A. I. Meyers; *J. Org. Chem.*, 1974, **39**, 1192.	276
74JOC1210	T. Minami and T. Agawa; *J. Org. Chem.*, 1974, **39**, 1210.	165
74JOC1221	G. L'abbe and G. Mathys; *J. Org. Chem.*, 1974, **39**, 1221.	69, 70, 84
74JOC1235	R. W. Begland, D. R. Hartter, D. S. Donald, A. Cairncross and W. A. Sheppard; *J. Org. Chem.*, 1974, **39**, 1235.	538
74JOC1791	N. Durán; *J. Org. Chem.*, 1974, **39**, 1791.	911, 912, 926, 927, 940
74JOC1795	R. A. Abramovitch and G. M. Singer; *J. Org. Chem.*, 1974, **39**, 1795.	148
74JOC1819	D. L. Klayman and T. S. Woods; *J. Org. Chem.*, 1974, **39**, 1819.	347, 680
74JOC1976	A. Padwa and E. Chen; *J. Org. Chem.*, 1974, **39**, 1976.	15
74JOC2073	V. Usieli, A. Pilersdorf, S. Shor, J. Katzhendler and S. Sarel; *J. Org. Chem.*, 1974, **39**, 2073.	861, 894
74JOC2225	J. E. Oliver and A. B. DeMilo; *J. Org. Chem.*, 1974, **39**, 2225.	920, 922, 942
74JOC2228	J. E. Oliver and R. T. Brown; *J. Org. Chem.*, 1974, **39**, 2228.	920, 943
74JOC2233	J. E. Oliver and J. L. Flippen; *J. Org. Chem.*, 1974, **39**, 2233.	914
74JOC2235	J. E. Oliver; *J. Org. Chem.*, 1974, **39**, 2235.	502, 915
74JOC2294	V. Bertini, A. De Munno, A. Menconi and A. Fissi; *J. Org. Chem.*, 1974, **39**, 2294.	527
74JOC2336	D. Seyferth and H.-M. Shih; *J. Org. Chem.*, 1974, **39**, 2336.	442, 443
74JOC2456	L. R. Melby, H. D. Hartzler and W. A. Sheppard; *J. Org. Chem.*, 1974, **39**, 2456.	835, 848
74JOC2459	M. D. Robbins and C. D. Broaddus; *J. Org. Chem.*, 1974, **39**, 2459.	775
74JOC2581	K. Ishizumi, S. Inaba and H. Yamamoto; *J. Org. Chem.*, 1974, **39**, 2581.	910
74JOC2722	U. Jacobsson, T. Kempe and T. Norin; *J. Org. Chem.*, 1974, **39**, 2722.	772
74JOC2885	E. M. Burgess and H. R. Penton, Jr.; *J. Org. Chem.*, 1974, **39**, 2885.	932
74JOC2951	S. W. Mojé and P. Beak; *J. Org. Chem.*, 1974, **39**, 2951.	555, 563, 744
74JOC3415	J. C. Sheehan and U. Zoller; *J. Org. Chem.*, 1974, **39**, 3415.	871, 884
74JOC3608	F. Wudl and M. L. Kaplan; *J. Org. Chem.*, 1974, **39**, 3608.	816
74JOC3770	G. L'abbé, S. Toppet, G. Verhelst and C. Martens; *J. Org. Chem.*, 1974, **39**, 3770.	583
74JOC3783	K. T. Potts and J. Kane; *J. Org. Chem.*, 1974, **39**, 3783.	490, 739
74JOC3906	M. H. Ghandehari, D. Davalian, M. Yalpani and M. H. Partovi; *J. Org. Chem.*, 1974, **39**, 3906.	350
74JPR684	H. Schaefer and K. Gewald; *J. Prakt. Chem.*, 1974, **316**, 684.	693
74JPR851	A. Tzschach and K. Kellner; *J. Prakt. Chem.*, 1974, **316**, 851.	164
74JPS628	I. Lalezari, A. Shafiee and S. Yazdany; *J. Pharm. Sci.*, 1974, **63**, 628.	462
74JSP(49)70	P. A. Baron and D. O. Harris; *J. Mol. Spectrosc.*, 1974, **49**, 70.	751
74JST(22)389	D. Bouin, C. Roussel, M. Chanon and J. Metzger; *J. Mol. Struct.*, 1974, **22**, 389.	239
74KGS453	K. K. Zhigulev, R. A. Khmel'nitskii, S. D. Sokolov and N. M. Przheval'skii; *Khim. Geterotsikl. Soedin.*, 1974, 453 (*Chem. Abstr.*, 1974, **81**, 48 959).	6
74KGS457	K. K. Zhigulev, R. A. Khmel'nitskii and M. A. Panina; *Khim. Geterotsikl. Soedin.*, 1974, 457 (*Chem. Abstr.*, 1974, **81**, 36 887).	6
74KGS571	A. L. Fridman, F. A. Gabitov, V. D. Surkov and V. S. Zalesov; *Khim. Geterotsikl. Soedin.*, 1974, 571 (*Chem. Abstr.*, 1974, **81**, 37 497).	102

74KGS597	S. D. Sokolov, T. N. Egorova and I. M. Yudintseva; *Khim. Geterotsikl. Soedin.*, 1974, 597.	22, 52
74KGS602	S. D. Sokolov, T. N. Egorova and P. V. Petrovskii; *Khim. Geterotsikl. Soedin.*, 1974, 602.	51
74KGS755	K. K. Zhiguev, S. D. Sokolov and R. A. Khmel'nitskii; *Khim. Geterotsikl. Soedin.*, 1974, 755 (*Chem. Abstr.*, 1975, **82**, 3553).	6
74KGS901	A. A. Akhrem, V. A. Khripach and F. A. Lakhvich; *Khim. Geterotsikl. Soedin.*, 1974, 901 (*Chem. Abstr.*, 1974, **81**, 105 372).	118
74KGS1435	V. A. Chuiguk, D. I. Sheiko and V. G. Glushakov; *Khim. Geterotsikl. Soedin.*, 1974, 1435 (*Chem. Abstr.*, 1975, **82**, 43 317).	359
74KGS1660	V. A. Chuiguk and Y. M. Volouenko; *Khim. Geterotsikl. Soedin.*, 1974, 1660 (*Chem. Abstr.*, 1975, **82**, 112 026).	740
74KGS1697	S. D. Sokolov and T. N. Egorova; *Khim. Geterotsikl. Soedin.*, 1974, 1697.	52
74LA403	G. Scherowsky and J. Weiland; *Liebigs Ann. Chem.*, 1974, 403.	849
74LA593	K.-D. Kampe; *Liebigs Ann. Chem.*, 1974, 593.	665
74LA996	H. Witte and W. Seeliger; *Liebigs Ann. Chem.*, 1974, 996.	228
74LA1261	F. Boberg, M. Ghoudikian and M. H. Khorgami; *Liebigs Ann. Chem.*, 1974, 1261.	796
74M882	F. Sauter, W. Dienhammer and K. Danksegmueller; *Monatsh. Chem.*, 1974, **105**, 882.	703
74MI41600	A. Cristini, G. Ponticelli and C. Preti; *J. Inorg. Nucl. Chem.*, 1974, **36**, 2473.	20
74MI41601	G. O. Potts, A. L. Beyler and H. P. Schane; *Fertil. Steril.*, 1974, **25**, 367.	129
74MI41602	H. P. Schane and G. O. Potts; *Fertil. Steril.*, 1974, **25**, 363.	129
74MI41603	D. G. Martin, L. J. Hanka and G. L. Neil; *Cancer Chemother. Rep., Part 1*, 1974, **58**, 935 (*Chem. Abstr.*, 1975, **82**, 132 911).	129
74MI41604	B. H. Shapiro and A. S. Goldman; *Proc. Soc. Exp. Biol. Med.*, 1975, **149**, 896 (*Chem. Abstr.*, 1975, **83**, 158 288).	128
74MI41605	Zh. A. Krasnaya, T. S. Stytsenko, E. P. Prokof'ev, I. P. Yakovlev and V. F. Kucherov; *Otkrytiya Izobret., Prom. Obraztsy, Tovarynye Znaki*, 1974, **51**, 58 (*Chem. Abstr.*, 1974, **81**, 77 898).	102
74MI41606	V. I. Esafov, V. I. Azarova and I. N. Tugarinova; *Izv. Vyssh. Ucheb. Zaved., Khim. Khim. Tekhnol.*, 1974, **17**, 65 (*Chem. Abstr.*, 1974, **80**, 82 766).	99
B-74MI41607	G. A. Shvekhgeimer and A. Baranski; in 'Tezisy Vses. Soveshch. Khim. Nitrosoedin., 5th', 1974, p. 62, 65 (*Chem. Abstr.*, 1977, **87**, 53 126, 53 125).	89
B-74MI41608	G. A. Shvekhgeimer and A. Baranski; in 'Tezisy Vses. Soveshch. Khim. Nitrosoedin., 5th,' 1974, p. 8 (*Chem. Abstr.*, 1977, **87**, 5033).	89
74MI41609	N. De Kimpe, N. Schamp, R. Van Parijs and L. Van Rompuy; *Meded. Fac. Landbouwwet. Rjkksuniv. Gent.*, 1974, **39**, 1478 (*Chem. Abstr.*, 1975, **83**, 97 100).	39
74MI41610	R. N. Manzo, P. Catania and M. Martinez de Bertorello; *Anal. Chem.*, 1974, **46**, 1141.	11
B-74MI41611	P. H. VanderVoort and W. L. Driessen; in 'Proc. Int. Conf. Coord. Chem., 16th', 1974, **R43**, 2 (*Chem. Abstr.*, 1976, **85**, 55 940).	7
74MI41612	C. Pouchan, S. Senez, J. Raymond and H. Sauvaitre; *J. Chim. Phys. Phys. Chim. Biol.*, 1974, **71**, 525 (*Chem. Abstr.*, 1974, **81**, 31 448).	5
74MI41700	J. M. Essery, U. Corbin, V. Sprancmanis, L. B. Crast, Jr., R. G. Graham, P. F. Misco, Jr., D. Willner, D. N. McGregor and L. C. Cheney; *J. Antibiotics*, 1974, **27**, 573 (*Chem. Abstr.*, 1975, **82**, 11 679).	157
74MI41800	K. Pihlaja; *Kem.-Kemi*, 1974, **1**, 492.	185
74MI41900	N. Nakamura, Y. Itaka, H. Sakakibara and H. Umezawa; *J. Antibiotics*, 1974, **27**, 894.	238
74MI42300	D. Ghiran, I. Schwartz and I. Simiti; *Farmacia (Bucharest)*, 1974, **22**, 141 (*Chem. Abstr.*, 1975, **82**, 43 274).	445
74MI42600	R. N. MacDonald, A. Cairncross, J. B. Sieja and W. H. Sharkey; *J. Polym. Sci., Polym. Chem. Ed.*, 1974, **12**, 663.	543
74MI42800	E. A. Neves and D. W. Franco; *J. Inorg. Nucl. Chem.*, 1974, **36**, 3851.	595
B-74MI43300	S. D. Razumovskii and G. E. Zaikov; 'Ozone and its Reactions with Organic Compounds' (in Russian), Nauka, Moscow, 1974.	853
74MI43301	R. A. Rouse; *Int. J. Quantum Chem.*, 1974, **8S**, 201.	833
74MI43302	A. J. Crone and B. J. Tighe; *Br. Polym. J.*, 1974, **6**, 79.	868
74MI43303	M. Boelens, L. M. van der Linde, P. J. de Valois, H. M. van Dort and H. C. Takken; *J. Agric. Food Chem.*, 1974, **22**, 1071.	895
74MI43304	B. L. Van Duuren, B. M. Goldschmidt, C. Katz, I. Seidman and J. S. Paul; *J. Natl. Cancer Inst.*, 1974, **53**, 695.	895
74MI43400	H. Steinbeisser, R. Mews and O. Glemser; *Z. Anorg. Allg. Chem.*, 1974, **406**, 299.	910, 927, 932, 941
74MIP41601	M. V. Kashutina, S. L. Ioffe and V. A. Tartakovskii; *USSR Pat.* 430 105 (*Chem. Abstr.*, 1974, **81**, 63 762).	46, 110
74NKK1539	N. Matsumura, Y. Otsuji and E. Imoto; *Nippon Kagaku Kaishi*, 1974, 1539 (*Chem. Abstr.*, 1975, **82**, 57 595).	164
74OMR(6)430	G. W. H. Cheeseman and C. J. Turner; *Org. Magn. Reson.*, 1974, **6**, 430.	518, 524
74OMS(9)149	A. Maquestiau, Y. Van Haverbeke, C. De Meyer and R. Flammang; *Org. Mass Spectrom.*, 1974, **9**, 149.	183
74OMS(9)181	P. Wolkoff and S. Hammerum; *Org. Mass Spectrom.*, 1974, **9**, 181 (*Chem. Abstr.*, 1974, **81**, 62 710).	552, 553

74OMS(9)1017	A. Selva, A. Citterio, E. Pella and R. Tonani; *Org. Mass Spectrom.*, 1974, **9**, 1017.	911
74OMS(9)1161	A. Selva, U. Vettori and E. Gaetani; *Org. Mass Spectrom.*, 1974, **9**, 1161.	142
74PMH(6)53	J. Sheridan, *Phys. Methods Heterocycl. Chem.*, 1974, **6**, 53.	429
74RTC58	R. Peereboom, H. C. van der Plas and A. Koudijs; *Recl. Trav. Chim. Pays-Bas*, 1974, **93**, 58.	80
74RTC99	H. Wevers and W. Drenth; *Rec. Trav. Chim. Pays-Bas*, 1974, **93**, 99.	771, 780
74RTC139	H. duCrocq, R. J. J. C. Lousberg and C. A. Salemink; *Recl. Trav. Chim. Pays-Bas*, 1974, **93**, 139.	51, 120, 129
74RTC225	H. C. van der Plas, M. C. Vollering, H. Jongejan and B. Zuurdeeg; *Recl. Trav. Chim. Pays-Bas*, 1974, **93**, 225.	80
74S22	G. Ege and E. Beisiegel; *Synthesis*, 1974, 22.	529
74S30	W. Schaefer, H. W. Moore and A. Aguado; *Synthesis*, 1974, 30.	75, 126
74S32	J. R. Williams and G. M. Sarkisian; *Synthesis*, 1974, 32.	843
74S120	B. Stanovnik and M. Tisler; *Synthesis*, 1974, 120.	699
74S294	P. Dubs and M. Pesaro; *Synthesis*, 1974, 294.	312
74S346	T. George and R. Tahilramani; *Synthesis*, 1974, 346.	679, 681, 694
74T63	S. Ranganathan and H. H. Raman; *Tetrahedron*, 1974, **30**, 63.	98
74T221	P. Henklein, R. Kraft and G. Westphal; *Tetrahedron*, 1974, **30**, 221.	439, 727
74T351	K. Brocklehurst and K. Williamson; *Tetrahedron*, 1974, **30**, 351.	185
74T409	K. Masuda and T. Okutani; *Tetrahedron*, 1974, **30**, 409.	373
74T553	C. Th. Pedersen and J. Moller; *Tetrahedron*, 1974, **30**, 553.	786
74T863	L. Di Nunno, S. Florio and P. E. Todesco; *Tetrahedron*, 1974, **30**, 863.	410
74T879	C. Riou, J. C. Poite, G. Vernin and J. Metzger; *Tetrahedron*, 1974, **30**, 879.	171, 250
74T1121	M. Joucla, J. Hamelin and R. Carrie; *Tetrahedron*, 1974, **30**, 1121.	47
74T1365	A. McKillop and R. J. Kobylecki; *Tetrahedron*, 1974, **30**, 1365.	67
74T2053	P. W. Jones and A. H. Adelman; *Tetrahedron*, 1974, **30**, 2053.	890
74T2537	B. J. Lindberg, R. Pinel and Y. Mollier; *Tetrahedron*, 1974, **30**, 2537.	1056
74T2765	M. Bianchi, A. Butti and S. Rossi; *Tetrahedron*, 1974, **30**, 2765.	80, 105
74T2867	B. Moisan, A. Robert and A. Foucaud; *Tetrahedron*, 1974, **30**, 2867.	776
74T3677	D. S. Kemp, S.-W. Wang, R.-C. Mollan, S.-L. Hsia and P. N. Confalone; *Tetrahedron*, 1974, **30**, 3677.	32
74T3723	M. Makosza, M. Jagusztyn-Grochowska, M. Ludwikow and M. Jawdosiuk; *Tetrahedron*, 1974, **30**, 3723.	390
74T3765	G. DeMicheli, R. Gandolfi and P. Gruenager; *Tetrahedron*, 1974, **30**, 3765.	89
74T3839	A. S. Angeloni, D. Dal Monte, S. Pollicino, E. Sandri and G. Scapini; *Tetrahedron*, 1974, **30**, 3839.	518, 531
74T3955	D. S. Kemp, S.-W. Wang, J. Rebek, Jr., R. C. Mollan, C. Banquer and G. Subramanyam; *Tetrahedron*, 1974, **30**, 3955.	32
74T4113	M. A.-F. Elkaschef, F. M. E. Abdel-Megeid and A. A. El-Barbary; *Tetrahedron*, 1974, **30**, 4113.	794, 796, 797
74T4171	G. Vernin, M. A. Lebreton, H. J. M. Dou, J. Metzger and G. Vernin; *Tetrahedron*, 1974, **30**, 4171.	266
74TL253	M. H. Palmer and R. H. Findlay; *Tetrahedron Lett.*, 1974, 253.	246
74TL375	K. H. Grellman and E. Tauer; *Tetrahedron Lett.*, 1974, 375.	16
74TL387	R. Jacquier, F. Petrus, J. Verducci and Y. Vidal; *Tetrahedron Lett.*, 1974, 387.	101
74TL1059	S. Tamagaki, K. Sakaki and S. Oae; *Tetrahedron Lett.*, 1974, 1059.	796
74TL1863	N. Suzuki and Y. Izawa; *Tetrahedron Lett.*, 1974, 1863.	307
74TL2565	E. Eibler and J. Sauer; *Tetrahedron Lett.*, 1974, 2565.	914
74TL2643	K. K. Balasubramanian and B. Venugopalan; *Tetrahedron Lett.*, 1974, 2643.	124
74TL2645	K. K. Balasubramanian and B. Venugopalan; *Tetrahedron Lett.*, 1974, 2645.	124
74TL2783	R. D. Lapper and A. J. Poole; *Tetrahedron Lett.*, 1974, 2783.	1056
74TL2793	S. Auricchio, S. Morrocchi and A. Ricca; *Tetrahedron Lett.*, 1974, 2793.	33
74TL2899	P. Koch and E. Perrotti; *Tetrahedron Lett.*, 1974, 2899.	320
74TL3319	P. K. Klaus, P. Hofbauer and W. Rieder; *Tetrahedron Lett.*, 1974, 3319.	165
74TL3871	M. Golfier and R. Milcent; *Tetrahedron Lett.*, 1974, 3871.	432
74TL3875	M. Golfier, M.-G. Guillerez and R. Milcent; *Tetrahedron Lett.*, 1974, 3875.	431, 432
74TL4069	K. J. Brown and O. Meth-Cohn; *Tetrahedron Lett.*, 1974, 4069.	1015
74TL4359	C. F. Joshua and P. K. Ramdas; *Tetrahedron Lett.*, 1974, 4359.	121
74UKZ633	V. A. Chuiguk and E. A. Leshchenko; *Ukr. Khim. Zh.* (Russ. Ed.), 1974, **40**, 633 (*Chem. Abstr.*, 1974, **81**, 105 438).	658
74USP3786028	R. N. MacDonald; *U.S. Pat.* 3 786 028 (1974) (*Chem. Abstr.*, 1975, **83**, 28 814).	543
74USP3793296	W. J. I. Bracke; *U.S. Pat.* 3 793 296 (1974) (*Chem. Abstr.*, 1974, **81**, 26 219).	130
74USP3808222	W. C. Doyle, Jr.; *U.S. Pat.* 3 808 222 (1974) (*Chem. Abstr.*, 1974, **81**, 13 518).	944
74USP3819354	S. T. D. Gough; *U.S. Pat.* 3 819 354 (1974) (*Chem. Abstr.*, 1975, **82**, 12 277).	543
74USP3850933	*U.S. Pat.* 3 850 933 (1974) (*Chem. Abstr.*, 1975, **82**, 140 163).	706
74USP3854000	G. K. Kohn and M. S. Singer; *U.S. Pat.* 3 854 000 (1974) (*Chem. Abstr.*, 1975, **82**, 140 144).	543
74ZC305	S. Leistner and G. Wagner; *Z. Chem.*, 1974, **14**, 305.	575
74ZC470	J. Liebscher and H. Hartmann; *Z. Chem.*, 1974, **14**, 470.	304
74ZN(B)284	M. Haake, B. Eichenauer and K. H. Ahrens; *Z. Naturforsch., Teil B*, 1974, **29**, 284.	911
74ZOB2553	V. Ignato, N. V. Zharkin, G. A. Blokh, R. A. Akchuring and L. M. Agafonova; *Zh. Obshch. Khim.*, 1974, **44**, 2553.	487

75ACR361	K. N. Houk; *Acc. Chem. Res.*, 1975, **8**, 361.	368
75ACS(A)7	M. V. Gabrielson; *Acta Chem. Scand., Ser. A*, 1975, **29**, 7.	751
75ACS(A)45	T. Ottersen, C. Christophersen and S. Treppendahl; *Acta Chem. Scand., Ser. A*, 1975, **29**, 45.	599
75ACS(A)136	L. J. Saethre and A. Hordvik; *Acta Chem. Scand., Ser. A*, 1975, **29**, 136.	1053
75ACS(A)414	P. Albriktsen and M. Bjorøy; *Acta Chem. Scand., Ser. A*, 1975, **29**, 414.	863, 887, 905, 911
75ACS(A)799	T. Ottersen; *Acta. Chem. Scand., Ser. A*, 1975, **29**, 799.	599
75ACS(B)483	C. L. Pedersen and J. Møller; *Acta Chem. Scand., Ser. B*, 1975, **29**, 483.	400, 522
75ACS(B)622	F. C. V. Larsson, S. O. Lawesson, I. Jardine and R. I. Reed; *Acta Chem. Scand., Ser. B*, 1975, **29**, 622.	522
75ACS(B)1092	K. Berg-Nielsen; *Acta Chem. Scand., Ser. B*, 1975, **29**, 1092.	695
75AG765	R. Criegee; *Angew. Chem.*, 1975, **87**, 765 (*Angew. Chem., Int. Ed. Engl.*, 1975, **14**, 745).	853, 865, 872, 879, 881, 888
75AG(E)248	G. Seybold and C. Heibl; *Angew. Chem., Int. Ed. Engl.*, 1975, **14**, 248.	454
75AG(E)422	H. Gotthardt, M. C. Weisshuhn and K. Dörhöfer; *Angew. Chem., Int. Ed. Engl.*, 1975, **14**, 422.	759, 765, 772
75AG(E)762	R. Neidlein and P. Leinberger; *Angew. Chem., Int. Ed. Engl.*, 1975, **14**, 762.	540
75AJC129	M. Davis, L. W. Deady, E. Homfeld and S. Pogany; *Aust. J. Chem.*, 1975, **28**, 129.	145, 159
75AJC151	G. C. Brophy, D. J. Collins, J. J. Hobbs and S. Sternhell; *Aust. J. Chem.*, 1975, **28**, 151.	557, 572
75AJC207	R. Kazlauskas and J. T. Pinhey; *Aust. J. Chem.*, 1975, **28**, 207.	95
75AP75	W. Schunack; *Arch. Pharm.* (*Weinheim, Ger.*), 1975, **308**, 75.	128
75AX(B)30	L. J. Saethre and A. Hordvik; *Acta Crystallogr., Part B*, 1975, 30.	1054
75AX(B)2310	C. Foces-Foces, F. H. Cano, S. Garcia-Blanco and J. G. Rodriguez; *Acta Crystallogr., Part B*, 1975, **31**, 2310.	713, 715
75BCJ310	T. Fuchigami and K. Odo; *Bull. Chem. Soc. Jpn.*, 1975, **48**, 310.	473, 498
75BCJ929	T. Nishiyama and F. Yamada; *Bull. Chem. Soc. Jpn.*, 1975, **48**, 929.	928
75BCJ1675	S. Torii, M. Ukida and H. Tanaka; *Bull. Chem. Soc. Jpn.*, 1975, **48**, 1675 (*Chem. Abstr.*, 1975, **83**, 164 045).	108
75BCJ2915	M. Takahashi, T. Shinoda, H. Osada and T. Nakajima; *Bull. Chem. Soc. Jpn.*, 1975, **48**, 2915.	743
75BCJ3313	F. Yamada, T. Nishiyama and H. Samukawa; *Bull. Chem. Soc. Jpn.*, 1975, **48**, 3313.	911
75BJ(151)417	T. Stuchbury, M. Shipton, R. Norris, J. P. G. Malthouse, K. Brocklehurst, J. A. L. Herbert, and H. Suschitzky; *Biochem. J.*, 1975, **151**, 417.	426
75BRP1395929	G. J. Durant, J. C. Emmett and P. R. Ganellin; *Br. Pat.* 1 395 929 (1975) (*Chem. Abstr.*, 1975, **83**, 114 385).	128
75BRP1414503	D. B. Baird, J. L. Leng and D. F. Newton; *Br. Pat.* 1 414 503 (1975) (*Chem. Abstr.*, 1976, **84**, 123 387).	130
75BSB207	A. Maquestiau, Y. VanHaverbeke, R. Flammang and J. Pierard; *Bull. Soc. Chim. Belg.*, 1975, **84**, 207.	7
75BSB775	H. J. Geise and E. Van Laere; *Bull. Soc. Chim. Belg.*, 1975, **84**, 775.	855, 857
75BSB1189	J. Elguero, R. Faure, J. P. Galy and E. J. Vincent; *Bull. Soc. Chim. Belg.*, 1975, **84**, 1189.	978, 980
75BSF1127	G. L'abbé; *Bull. Soc. Chim. Fr.*, 1975, 1127.	586
75BSF1319	R. Gree and R. Carrie; *Bull. Soc. Chim. Fr.*, 1975, 1319.	110
75BSF1435	H. Davy and J. Vialle; *Bull. Soc. Chim. Fr.*, 1975, 1635.	798
75C350	R. Verbruggen and H. G. Viehe; *Chimia*, 1975, **29**, 350.	69, 87
75CC24	D. M. Revitt; *J. Chem. Soc., Chem. Commun.*, 1975, 24.	589, 911, 943
75CC42	F. McCapra and Z. Razavi; *J. Chem. Soc., Chem. Commun.*, 1975, 42.	324
75CC417	K. T. Potts, J. Kane, E. Carnahan and U. P. Singh; *J. Chem. Soc., Chem. Commun.*, 1975, 417.	758, 762, 771
75CC550	D. L. Dunning, B. L. Kindberg, C. F. Plese, E. H. Griffith and E. L. Amma; *J. Chem. Soc., Chem. Commun.*, 1975, 550.	902, 915
75CC617	M. P. Cava and L. E. Saris; *J. Chem. Soc., Chem. Commun.*, 1975, 617.	343
75CC621	V. Calò, L. Lopez, L. Marchese and G. Pesce; *J. Chem. Soc., Chem. Commun.*, 1975, 621.	329
75CC840	H. Matsukubo and H. Kato; *J. Chem. Soc., Chem. Commun.*, 1975, 840.	829
75CHE45	N. O. Saldabol, L. L. Zeligman, Y. Y. Popelis and S. A. Giller; *Chem. Heterocycl. Compd.* (*Engl. Transl.*), 1975, **11**, 45.	979
75CHE304	E. G. Kovalev, A. D. Sinegibskaya, I. Y. Postovskii and S. L. Mertsalov; *Chem. Heterocycl. Compd.* (*Engl. Transl.*), 1975, **11**, 304.	979
75CHE500	I. Ya. Postovskii, A. A. Sinegibskaya and E. G. Kovalev; *Chem. Heterocycl. Compd.* (*Engl. Transl.*), 1975, **11**, 500.	979, 984
75CHE612	T. P. Kofman, V. I. Manuilova, M. S. Pevzner and T. N. Timofeeva; *Chem. Heterocycl. Compd.* (*Engl. Transl.*), 1975, **11**, 612.	1014
75CHE666	V. I. Shvedov, V. K. Vasil'eva, I. A. Kharizomenova and A. N. Grinev; *Chem. Heterocycl. Compd.* (*Engl. Transl.*), 1975, **11**, 666.	978, 987
75CHE740	S. M. Khripak, M. M. Tsitsika and I. V. Smolanka; *Chem. Heterocycl. Compd.* (*Engl. Transl.*), 1975, **11**, 740.	1011
75CHE1133	V. I. Shvedov, Yu. I. Trofimkin, V. K. Vasil'eva and A. N. Grinev; *Chem. Heterocycl. Compd.* (*Engl. Transl.*), 1975, **11**, 1133.	988
75CI(L)1018	P. Bravo and C. Ticozzi; *Chem. Ind.* (*London*), 1975, 1018.	53

References

75CJC596	R. E. Wasylishen, T. R. Clem and E. D. Becker; *Can. J. Chem.*, 1975, **53**, 596.	133, 136, 138
75CJC836	N. Plavac, I. W. J. Still, M. S. Chauhan and D. M. McKinnon; *Can. J. Chem.*, 1975, **53**, 836.	138, 785
75CJC913	B. V. Badami and G. S. Puranik; *Can. J. Chem.*, 1975, **53**, 913.	376
75CJC1336	M. S. Chauhan and D. M. McKinnon; *Can. J. Chem.*, 1975, **53**, 1336.	120, 124
75CJC1642	G. Mille, J. C. Poite, J. Chouteau and J. Metzger; *Can. J. Chem.*, 1975, **53**, 1642.	136, 140, 141
75CJC1677	R. Faure, J. R. Llinas, E. J. Vincent and M. Rajzmann; *Can. J. Chem.*, 1975, **53**, 1677.	133, 138, 139
75CJC2734	T. Schaefer, K. Chum, D. McKinnon and M. S. Chauhan; *Can. J. Chem.*, 1975, **53**, 2734.	753
75CJC3439	R. S. Brown; *Can. J. Chem.*, 1975, **53**, 3439.	757
75CL965	H. Sato, T. Kusumi, K. Imaye and H. Kakisawa; *Chem. Lett.*, 1975, 965.	46
75CPB2104	I. Ito, N. Oda, T. Kato and K. Ota; *Chem. Pharm. Bull.*, 1975, **23**, 2104.	653
75CPB2654	T. Taguchi, S. Morita and Y. Kawazoe; *Chem. Pharm. Bull.*, 1975, **23**, 2654.	541
75CR(C)(280)673	J. M. Catel and Y. Mollier; *C. R. Hebd. Seances Acad. Sci., Ser. C*, 1975, **280**, 673.	786
75CR(C)(281)1	D. Bouin-Roubaud and J. Metzger; *C.R. Hebd. Seances Acad. Sci., Ser. C*, 1975, **281**, 1 (*Chem. Abstr.*, 1975, **83**, 209 936).	239
75CRV389	I. J. Turchi and M. J. S. Dewar; *Chem. Rev.*, 1975, **75**, 389.	178, 179, 215, 232
75CS(8A)39	E. Bulka; *Chem. Scr.*, 1975, **8A**, 39.	341, 342, 345
75CSC561	G. Chiari, D. Viterbo, A. G. Manfredotti and G. Guastini; *Cryst. Struct. Commun.*, 1975, **4**, 561.	396
75FES128	G. Cantarelli, M. Carissimi and F. Ravenna; *Farmaco Ed. Sci.*, 1975, **30**, 128 (*Chem. Abstr.*, 1975, **83**, 9880).	128
75FES992	E. Abignente, P. DeCaprariis and M. L. Stein; *Farmaco Ed. Sci.*, 1975, **30**, 992.	624
75FRP2244462	C. Fauran, C. Douzon, G. Raynaud and B. Pourrias; *Fr. Pat.* 2 244 462 (1975) (*Chem. Abstr.*, 1975, **83**, 179 072).	433
75G91	P. Bravo and C. Ticozzi; *Gazz. Chim. Ital.*, 1975, **105**, 91.	73
75G723	V. Ceri, D. Dal Monte, S. Pollicino and E. Sandri; *Gazz. Chim. Ital.*, 1975, **105**, 723.	412
75GEP2336978	*Ger. Pat.* 2 336 978 (1975) (*Chem. Abstr.*, 1975, **83**, 12 166).	644
75GEP2346566	F. Bentz and G. Nischk; *Ger. Pat.* 2 346 566 (1975) (*Chem. Abstr.*, 1975, **83**, 165 739).	945
75GEP2365391	Teikoku Hormone Mfg. Co. Ltd., *Ger. Offen.* 2 365 391 (1975) (*Chem. Abstr.*, 1975, **82**, 4236).	111
75GEP2424691	A. Cambon, F. Jeanneaux and C. Massyn; *Ger. Offen.* 2 424 691 (1975) (*Chem. Abstr.*, 1975, **82**, 72 972).	100, 101
75GEP2445884	H. Nakao, H. Yanagisawa, M. Nagano, B. Shimizu, S. Sugawara and M. Kaneko; *Ger. Offen.* 2 445 884 (1975) (*Chem. Abstr.*, 1975, **83**, 97 330).	128
75GEP2450053	J. C. Saunders and W. R. N. Williamson; *Ger. Offen.* 2 450 053 (1975) (*Chem. Abstr.*, 1975, **83**, 97 263).	116, 129
75GEP2500485	Ciba Geigy A.-G., *Ger. Pat.* 2 500 485 (1975) (*Chem. Abstr.*, 1976, **84**, 440 685).	510
75GEP2514984	D. G. Martin and L. J. Hanka; *Ger. Offen.* 2 514 984 (1975) (*Chem. Abstr.*, 1976, **84**, 41 994).	129
75H(3)217	K. Hirai and T. Ishiba; *Heterocycles*, 1975, **3**, 217.	752, 756, 762, 771
75H(3)563	P. Rajagopalan and C. N. Talaty; *Heterocycles*, 1975, **3**, 563.	932
75H(3)651	K. L. Stuart; *Heterocycles*, 1975, **3**, 651.	394, 400, 415, 418, 426
75IJC241	C. P. Joshua and P. N. K. Nambisan; *Indian J. Chem.*, 1975, **13**, 241.	493
75IJC532	M. B. Devani, C. J. Shishoo, S. D. Patel, B. Mukherji and A. C. Padhya; *Indian J. Chem.*, 1975, **13**, 532 (*Chem. Abstr.*, 1975, **83**, 147 416).	315
75IZV1451	S. I. Al'ber, G. V. Lagodzinskaya, G. B. Manelis and E. B. Fel'dman; *Izv. Akad. Nauk SSSR, Ser. Khim.*, 1975, 1451 (*Chem. Abstr.*, 1975, **83**, 95 875).	110
75IZV1870	V. I. Pepekin, Yu. N. Matyushin, A. D. Nikolaeva, A. P. Kirsanov, L. V. Platonova and Yu A. Lebedev; *Izv. Akad. Nauk SSSR, Ser. Khim.*, 1975, 1870.	414
75IZV2348	V. M. Shitkin, S. L. Ioffe, Yu. D. Kuznetsov and V. A. Tartakovskii; *Izv. Akad. Nauk SSSR, Ser. Khim.*, 1975, 2348.	110
75JA245	J. G. Burr, W. A. Summers and Y. J. Lee; *J. Am. Chem. Soc.*, 1975, **97**, 245.	410
75JA926	W. Adam and J. Arce; *J. Am. Chem. Soc.*, 1975, **97**, 926.	759
75JA1358	D. Cremer and J. A. Pople; *J. Am. Chem. Soc.*, 1975, **97**, 1358.	750, 758
75JA2921	E. M. Engler, F. B. Kaufman, D. C. Green, C. E. Klots and R. N. Compton; *J. Am. Chem. Soc.*, 1975, **97**, 2921.	962
75JA3850	B. S. Campbell, D. B. Denney, D. Z. Denney and L. S. Shih; *J. Am. Chem. Soc.*, 1975, **97**, 3850.	868, 892
75JA5008	S. Nakatsuka, H. Tanino and Y. Kishi; *J. Am. Chem. Soc.*, 1975, **97**, 5008.	311
75JA6197	A. Holm, N. Harrit and N. Toubro; *J. Am. Chem. Soc.*, 1975, **97**, 6197.	171, 999
75JA6484	A. Padwa, E. Chen and A. Ku; *J. Am. Chem. Soc.*, 1975, **97**, 6484.	13
75JA6909	G. W. Astrologes and J. C. Martin; *J. Am. Chem. Soc.*, 1975, **97**, 6909.	763, 778
75JAP7589535	Kumiai Chem. Ind. Co. Ltd., *Jpn. Pat.* 75 89 535 (1975) (*Chem. Abstr.*, 1975, **83**, 173 914).	509
75JAP(K)7504068	Daiichi Seiyaku Co. Ltd.; *Jpn. Kokai* 75 04 068 (1975) (*Chem. Abstr.*, 1975, **83**, 114 373).	162
75JAP(K)7512080	Daiichi Seiyaku Co. Ltd.; *Jpn. Kokai* 75 12 080 (1975) (*Chem. Abstr.*, 1975, **83**, 10 060).	156

75JAP(K)7515840	H. Ono, H. Katsuyama and O. Watarai; *Jpn. Kokai* 75 15 840 (1975) (*Chem. Abstr.*, 1976, **84**, 44 997).	107
75JAP(K)7535165	A. Kotone; *Jpn. Kokai* 75 35 165 (1965) (*Chem. Abstr.*, 1975, **83**, 179 068).	433
75JAP(K)7547975	M. Nagano, K. Tomita and H. Takagi; *Jpn. Kokai* 75 47 975 (1975) (*Chem. Abstr.*, 1975, **83**, 131 575).	128
75JAP(K)7595272	N. Ito and S. Saijo; *Jpn. Kokai* 75 95 272 (1975) (*Chem. Abstr.*, 1976, **84**, 105 567).	128
75JCP(63)2560	O. L. Stiefvater; *J. Chem. Phys.*, 1975, **63**, 2560.	6, 8
75JCS(P1)270	R. Okazaki, F. Ishii, K. Okawa, K. Ozawa and N. Inamoto; *J. Chem. Soc., Perkin Trans. 1*, 1975, 270.	799, 848, 911, 921
75JCS(P1)375	D. Baldwin and P. van den Broek; *J. Chem. Soc., Perkin Trans. 1*, 1975, 375.	489, 495, 738, 987
75JCS(P1)632	H. Matsukubo and H. Kato; *J. Chem. Soc., Perkin Trans. 1*, 1975, 632.	829
75JCS(P1)693	M. A. Khan and F. K. Rafla; *J. Chem. Soc., Perkin Trans. 1*, 1975, 693.	624, 628, 630
75JCS(P1)1342	P. Crabbé, J. Haro, C. Rius and E. Santos; *J. Chem. Soc., Perkin Trans. 1*, 1975, 1342.	98
75JCS(P1)1959	V. Skaric, B. Djuras and V. Turjak-Zebic; *J. Chem. Soc., Perkin Trans. 1*, 1975, 1959.	125
75JCS(P1)2097	D. H. Reid and R. G. Webster; *J. Chem. Soc., Perkin Trans. 1*, 1975, 2097.	752, 949, 959, 1058
75JCS(P1)2115	G. M. Buchan and A. B. Turner; *J. Chem. Soc., Perkin Trans. 1*, 1975, 2115.	6
75JCS(P1)2190	G. Adembri, A. Camparini, F. Ponticelli and P. Tedeschi; *J. Chem. Soc., Perkin Trans. 1*, 1975, 2190.	619, 620, 621, 623, 626
75JCS(P1)2271	R. Hull, P. J. Van den Broek and M. L. Swain; *J. Chem. Soc., Perkin Trans. 1*, 1975, 2271.	638
75JCS(P1)2340	A. R. Forrester, R. H. Thomson and S.-O. Woo; *J. Chem. Soc., Perkin Trans. 1*, 1975, 2340.	108
75JCS(P1)2348	A. R. Forrester, R. H. Thomson and S.-O. Woo; *J. Chem. Soc., Perkin Trans. 1*, 1975, 2348	108
75JCS(P2)190	C. H. Green and D. G. Hellier; *J. Chem. Soc., Perkin Trans. 2*, 1975, 190.	862
75JCS(P2)974	M. H. Palmer and R. H. Findlay; *J. Chem. Soc., Perkin Trans. 2*, 1975, 974.	517
75JCS(P2)1191	V. J. Baker, A. R. Katritzky and J.-P. Majoral; *J. Chem. Soc., Perkin Trans. 2*, 1975, 1191.	430
75JCS(P2)1223	M. H. Palmer and R. H. Findlay; *J. Chem. Soc., Perkin Trans. 2*, 1975, 1223.	517
75JCS(P2)1280	G. V. Boyd, C. G. Davies, J. D. Donaldson, J. Silver and P. H. Wright; *J. Chem. Soc., Perkin Trans. 2*, 1975, 1280.	180, 184
75JCS(P2)1469	L. Di Nunno, S. Florio and P. E. Todesco; *J. Chem. Soc., Perkin Trans. 2*, 1975, 1469.	411
75JCS(P2)1614	A. R. Katritzky, C. Ogretir, H. O. Tarhan, H. J. M. Dou and J. V. Metzger; *J. Chem. Soc., Perkin Trans. 2*, 1975, 1614.	255
75JCS(P2)1620	A. R. Katritzky, H. O. Tarhan and B. Terem; *J. Chem. Soc., Perkin Trans. 2*, 1975, 1620.	133, 148
75JCS(P2)1627	A. R. Katritzky, M. Konya, H. O. Tarhan and A. G. Burton; *J. Chem. Soc., Perkin Trans. 2*, 1975, 1627.	22, 23, 52, 87
75JHC37	L. S. Wittenbrook; *J. Heterocycl. Chem.*, 1975, **12**, 37.	388
75JHC49	R. Kalish, E. Broger, G. F. Field, T. Anton, T. V. Steppe and L. H. Sternbach; *J. Heterocycl. Chem.*, 1975, **12**, 49.	164
75JHC175	M. W. Barker and C. J. Wierengo; *J. Heterocycl. Chem.*, 1975, **12**, 175.	113
75JHC279	E. Gonzalez, R. Sarlin and J. Elguero; *J. Heterocycl. Chem.*, 1975, **12**, 279.	996
75JHC337	L. Kramberger, P. Lorencak, S. Polanc, B. Verček, B. Stanovnik, M. Tišler and F. Povazanec; *J. Heterocycl. Chem.*, 1975, **12**, 337.	697
75JHC393	J. P. Chupp and D. J. Dahm; *J. Heterocycl. Chem.*, 1975, **12**, 393.	901, 910, 911
75JHC451	Y. C. Tong; *J. Heterocycl. Chem.*, 1975, **12**, 451.	538, 734, 736, 737
75JHC595	H. C. Wormser and W.-H. Chiu; *J. Heterocycl. Chem.*, 1975, **12**, 595.	991
75JHC605	H. Meier and H. Buhl; *J. Heterocycl. Chem.*, 1975, **12**, 605.	454
75JHC607	G. L'abbé, E. Van Loock, G. Verhelst and S. Toppet; *J. Heterocycl. Chem.*, 1975, **12**, 607.	589
75JHC639	M. Perrier, R. Pinel and S. Vialle; *J. Heterocycl. Chem.*, 1975, **12**, 639.	361
75JHC675	A. Shafiee and I. Lalezari; *J. Heterocycl. Chem.*, 1975, **12**, 675.	342, 345, 359, 733, 734
75JHC775	K. O. Alt, E. Christen and C. D. Weis; *J. Heterocycl. Chem.*, 1975, **12**, 775.	659
75JHC801	H. Golgolab and I. Lalezari; *J. Heterocycl. Chem.*, 1975, **12**, 801.	349
75JHC829	A. P. Komin and M. Carmack; *J. Heterocycl. Chem.*, 1975, **12**, 829.	536, 538
75JHC841	G. Weber, F. Buccheri and N. Vivona; *J. Heterocycl. Chem.*, 1975, **12**, 841.	560
75JHC877	G. M. Singerman; *J. Heterocycl. Chem.*, 1975, **12**, 877.	151, 164
75JHC883	R. C. Anderson and Y. Y. Hsiao; *J. Heterocycl. Chem.*, 1975, **12**, 883.	635, 641
75JHC985	N. Vivona, G. Cusmano, M. Ruccia and D. Spinelli; *J. Heterocycl. Chem.*, 1975, **12**, 985.	385
75JHC1005	J. A. Van Allan, J. D. Mee, C. A. Maggiulli and R. S. Henion; *J. Heterocycl. Chem.*, 1975, **12**, 1005.	259
75JHC1091	R. Weber and M. Renson; *J. Heterocycl. Chem.*, 1975, **12**, 1091.	339
75JHC1191	N. P. Peet and S. Sunder; *J. Heterocycl. Chem.*, 1975, **12**, 1191.	459, 460
75JIC231	U. J. Ram and H. N. Pandey; *J. Indian Chem. Soc.*, 1975, **52**, 231.	40

75JIC237	A. S. Mahajan and M. G. Paranjpe; *J. Indian Chem. Soc.*, 1975, **52**, 237.	943
75JOC387	E. M. Engler and V. V. Patel; *J. Org. Chem.*, 1975, **40**, 387.	952
75JOC431	A. Holm, K. Schaumburg, N. Dahlberg, C. Christophersen and J. P. Snyder; *J. Org. Chem.*, 1975, **40**, 431.	581, 587, 593
75JOC437	H. Alper and M. S. Wolin; *J. Org. Chem.*, 1975, **40**, 437.	1012
75JOC526	C. Kashima; *J. Org. Chem.*, 1975, **40**, 526.	34
75JOC746	K. Bechgaard, D. O. Cowan, A. N. Bloch and L. Henriksen; *J. Org. Chem.*, 1975, **40**, 746.	952, 954, 958, 963, 969
75JOC949	S. Cox, O. M. H. El Dusouqui, W. McCormack and J. G. Tillett; *J. Org. Chem.*, 1975, **40**, 949.	926
75JOC955	M. Winn; *J. Org. Chem.*, 1975, **40**, 955.	149
75JOC1525	W. B. Renfrow and M. Devadoss; *J. Org. Chem.*, 1975, **40**, 1525.	167
75JOC1559	W. A. Summers, J. Y. Lee and J. G. Burr; *J. Org. Chem.*, 1975, **40**, 1559.	410
75JOC1713	D. Lipkin and E. G. Lovett; *J. Org. Chem.*, 1975, **40**, 1713.	648, 649, 652, 656, 664
75JOC1728	G. L'abbé, G. Verhelst, C. C. Yu and S. Toppet; *J. Org. Chem.*, 1975, **40**, 1728.	505, 904, 906, 943
75JOC1734	F. J. Villani, J. Hannon, E. A. Wefer, T. A. Mann and J. B. Morton; *J. Org. Chem.*, 1975, **40**, 1734.	80
75JOC2000	D. L. Klayman and T. S. Woods; *J. Org. Chem.*, 1975, **40**, 2000.	706
75JOC2021	A. I. Meyers and J. Durandetta; *J. Org. Chem.*, 1975, **40**, 2021.	329
75JOC2025	A. I. Meyers, J. Durandetta and R. Munavu; *J. Org. Chem.*, 1975, **40**, 2025.	271
75JOC2143	H. P. Albrecht, D. B. Repke and J. G. Moffatt; *J. Org. Chem.*, 1975, **40**, 2143.	68
75JOC2260	W. E. Truce and J. R. Allison; *J. Org. Chem.*, 1975, **40**, 2260.	741
75JOC2476	C. L. Schmidt and L. B. Townsend; *J. Org. Chem.*, 1975, **40**, 2476.	679, 683, 689
75JOC2577	M. G. Miles, J. S. Wager, J. D. Wilson and A. R. Siedle; *J. Org. Chem.*, 1975, **40**, 2577.	849
75JOC2600	K. T. Potts and J. Kane; *J. Org. Chem.*, 1975, **40**, 2600.	491, 499, 567, 986
75JOC2604	M. H. Elnagdi, M. R. H. Elmoghayar, E. A. A. Hafez and H. H. Alnima; *J. Org. Chem.*, 1975, **40**, 2604.	64
75JOC2743	R. Y. Wen, A. P. Komin, R. W. Street and M. Carmack; *J. Org. Chem.*, 1975, **40**, 2743.	523, 529, 531, 533, 540, 747
75JOC2749	A. P. Komin, R. W. Street and M. Carmack; *J. Org. Chem.*, 1975, **40**, 2749.	516, 519, 528, 535, 538, 1031, 1032, 1037, 1045
75JOC2880	I. Yavari, S. Esfandiari, A. J. Mostashari and P. W. W. Hunter; *J. Org. Chem.*, 1975, **40**, 2880.	81, 401
75JOC3037	T. Higa and A. J. Krubsack; *J. Org. Chem.*, 1975, **40**, 3037.	995
75JOC3112	C. J. Wilkerson and F. D. Greene; *J. Org. Chem.*, 1975, **40**, 3112.	434, 444
75JOC3141	C. Temple, Jr., B. H. Smith and J. A. Montgomery; *J. Org. Chem.*, 1975, **40**, 3141.	660
75JOC3381	D. S. Noyce and B. B. Sandel; *J. Org. Chem.*, 1975, **40**, 3381.	155, 156, 164
75JOM(88)C35	D. Dousse, H. Lavayssiere and J. Satge; *J. Organomet. Chem.*, 1975, **88**, C35.	938
75JOM(101)57	F. Kober and W. J. Rühl; *J. Organomet. Chem.*, 1975, **101**, 57.	886
75JOU1737	D. Ya. Movshovich, V. N. Sheinker, A. D. Garnovskii and O. A. Osipov; *J. Org. Chem. USSR (Engl. Transl.)*, 1975, **11**, 1737.	182
75JPR123	E. Fanghänel; *J. Prakt. Chem.*, 1975, **317**, 123.	844
75JPR137	E. Fanghänel; *J. Prakt. Chem.*, 1975, **317**, 137.	836
75JPR771	H. Schäfer and K. Gewald; *J. Prakt. Chem.*, 1975, **317**, 771.	166
75JPR943	G. W. Fischer, R. Jentzsch, V. Kasanzewa and F. Riemer; *J. Prakt. Chem.*, 1975, **317**, 943.	877, 895
75JPR959	K. Gewald, U. Schlegel and H. Schaefer; *J. Prakt. Chem.*, 1975, **317**, 959.	642
75JPS1371	W. O. Foye, J. M. Kauffman, J. J. Lanzillo and E. F. LaSala; *J. Pharm. Sci.*, 1975, **64**, 1371.	699
75JST(25)329	M. Guiliano, G. Davidovics, J. Chouteau, J. L. Larice and J. P. Roggero; *J. Mol. Struct.*, 1975, **25**, 329.	242
75JST(25)343	M. Guiliano, G. Davidovics, J. Chouteau, J. L. Larice and J. P. Roggero; *J. Mol. Struct.*, 1975, **25**, 343.	242
75KGS162	T. Lesiak and S. Nielek; *Khim. Geterotsikl. Soedin.*, 1975, 162 (*Chem. Abstr.*, 1975, **82**, 170 772).	93
75KGS180	G. A. Shvekhgeimer and O. A. Vol'skaya; *Khim. Geterotsikl. Soedin.*, 1975, 180 (*Chem. Abstr.*, 1975, **82**, 156 161).	89
75KGS275	Yu. I. Akulin, B. K. Strelets and L. S. Efros; *Khim. Geterotsikl. Soedin.*, 1975, 275 (*Chem. Abstr.*, 1975, **82**, 156 191).	360
75KGS1055	A. M. Gyulmaliev, I. V. Stankevich and Z. V. Todres; *Khim. Geterotsikl. Soedin.*, 1975, 1055.	400, 522
75KGS1195	Yu. S. Shabaraov, S. S. Mochalov, A. N. Fedatov and V. V. Kalashnikov; *Khim. Geterotsikl. Soedin.*, 1975, 1195 (*Chem. Abstr.*, 1976, **84**, 30 946).	122
75KGS1292	M. I. Komendantov and R. R. Bekmukhametov; *Khim. Geterotsikl. Soedin.*, 1975, 1292 (*Chem. Abstr.*, 1976, **84**, 4823).	15
75KGS1493	J. P. Bachkovskii and V. A. Chuiguk; *Khim. Geterotsikl. Soedin.*, 1975, 1493 (*Chem. Abstr.*, 1976, **84**, 74 214).	727
75LA1018	D. Wobig; *Liebigs Ann. Chem.*, 1975, 1018.	909, 922, 923, 943
75LA1029	C. Grundmann, G. W. Nickel and R. K. Bansal; *Liebigs Ann. Chem.*, 1975, 1029.	413
75LA1257	H. P. Braun, K.-P. Zeller and H. Meier; *Liebigs Ann. Chem.*, 1975, 1257.	451

75LA1513	F. Boberg and M. Ghoudikian; *Liebigs Ann. Chem.*, 1975, 1513.	796, 886
75LA1961	B. Junge; *Liebigs Ann. Chem.*, 1975, 1961.	481, 496
75LA1994	H. Hagen and H. Fleig; *Liebigs Ann. Chem.*, 1975, 1994.	154, 161
75M1291	A. Sitte, R. Wessel and H. Paul; *Monatsh. Chem.*, 1975, **106**, 1291 (*Chem. Abstr.*, 1976, **84**, 74 189).	559
75MI41600	M. Massacesi, G. Ponticelli and C. Preti; *J. Inorg. Nucl. Chem.*, 1975, **37**, 1641.	20
75MI41601	R. Pinna, G. Ponticelli and C. Preti; *J. Inorg. Nucl. Chem.*, 1975, **37**, 1681.	20
75MI41602	M. Ogawa and Y. Ohta; *Kagaku To Seibutsu*, 1975, **13**, 404 (*Chem. Abstr.*, 1975, **83**, 189 124).	129
75MI41603	M. Pigini, M. Gianella, F. Gualieri, C. Melechcorre, P. Balla and L. Anglucci; *Eur. J. Med. Chem. – Chim. Ther.*, 1975, **10**, 29, 33.	129
75MI41604	R. Krogsggaard-Larsen; *J. Neurochem.*, 1975, **25**, 797, 803.	128
75MI41605	F. Freeman, P. L. Wilson and B. H. Kazan; *Exp. Parasitol.*, 1975, **38**, 181.	128
75MI41606	K. Imai, M. Matsuura, H. Furukawa and Y. Hayashi; *Nippon Yakurigaku Zasshi*, 1975, **71**, 691 (*Chem. Abstr.*, 1976, **85**, 13 791).	128
75MI41607	R. Fusco, L. Garanti and G. Zeechi; *Chem. Ind. (Milan)*, 1975, 16.	127
75MI41608	Yu. V. Svetkin, N. A. Ahmanova, F. L. Mal'tseva and G. P. Muratova; *Tr. Ural'sk. Un-tov. Org. Khim.*, 1975, 124 (*Chem. Abstr.*, 1977, **86**, 155 556).	108
75MI41609	C. Caristi, G. Cimino, A. Ferlazzo and M. Gattuso; *Atti Soc. Perloritana Sci. Fis. Mat. Nat.*, 1975, **21**, 71 (*Chem. Abstr.*, 1977, **87**, 167 932).	108
75MI41610	N. Bregant, A. Nuri and M. Stromar; *Croat. Chem. Acta*, 1975, **47**, 595 (*Chem. Abstr.*, 1976, **85**, 5543).	104
75MI41611	C. Massyn and A. Cambon; *J. Fluorine Chem.*, 1975, **5**, 67.	100
75MI41612	N. Anjum, A. Rahman and L. B. Clapp; *Pak. J. Sci. Res.*, 1975, **27**, 174 (*Chem. Abstr.*, 1978, **89**, 109 195).	95
75MI41613	I. M. Roushdi, A. Ibrahim and N. S. Habib; *Egypt. J. Pharm. Sci.*, 1975, **16**, 415 (*Chem. Abstr.*, 1978, **89**, 163 473).	40
75MI41614	O. Grushina; *V. sb., Vopr. Teor. Mat. Metody Resheniya Zadach*, 1975, 86 (*Chem. Abstr.*, 1978, **89**, 128 856).	7
75MI41615	O. L. Stiefvater, P. Noesberger and J. Sheridan; *Chem. Phys.*, 1975, **9**, 435.	6, 8
75MI41616	V. I. Esafov and V. I. Azarova; *Izv. Vyssh. Uchebn. Zaved., Khim. Khim. Tekhnol.*, 1975, **18**, 148 (*Chem. Abstr.*, 1975, **83**, 9871).	93, 99
75MI41617	A. H. Harhash, M. H. Elnagdi and N. A. Kassab; *J. Chem. Eng. Data*, 1975, **20**, 120.	40
75MI41700	J. Heindl, E. Schröder and H. W. Kelm; *Eur. J. Med. Chem., Chim. Ther.*, 1975, **10**, 591 (*Chem. Abstr.*, 1975, **84**, 164 664).	157, 158
75MI41701	M. Nakamura, K. Watabe, T. Kirigaya, Y. Yazawa, A. Watabe, Y. Suzuki and T. Kawamura; *Shokuhin Eiseigaku Zasshi*, 1975, **16**, 264 (*Chem. Abstr.*, 1976, **84**, 88 058).	152, 160
75MI41702	K. H. Pannell, C. C. Y. Lee, C. Párkányi and R. Redfearn; *Inorg. Chim. Acta*, 1975, **12**, 127 (*Chem. Abstr.*, 1975, **82**, 105 729).	153
B-75MI41703	G. Salmona; in 'Proceedings of the 18th International Spectroscopy Colloquium', 1975, vol. 2, p. 382 (*Chem. Abstr.*, 1977, **87**, 183 798).	133, 134
75MI41704	G. Mille, J. Chouteau and J. Metzger; *J. Chim. Phys., Phys. Chim. Biol.*, 1975, **72**, 31 (*Chem. Abstr.*, 1975, **82**, 154 682).	140, 141
B-75MI41800	'Lasers in Physical Chemistry and Biophysics', ed. J. Joussot-Dubien; Elsevier, Amsterdam, 1975.	233
75MI41900	R. Meyer, M. Meyer, D. Bares and E. J. Vincent; *Thermochim. Acta*, 1975, **11**, 211 (*Chem. Abstr.*, 1975, **82**, 145 949).	327
75MI42100	P. Dubus, B. Decroix, J. Morel and P. Pastour; *Ann. Chim. (Paris)*, 1975, **10**, 331.	382
75MI42200	C. Merritt, Jr., C. DiPietro, C. W. Hand, J. H. Cornell and D. E. Remy; *J. Chromatogr.*, 1975, **112**, 301.	419
75MI42201	M. Russo; *Kunststoffe*, 1975, **65**, 346 (*Chem. Abstr.*, 1975, **83**, 179 654).	426
75MI42600	V. V. Dovlatyan and R. S. Mirzoyan; *Arm. Khim. Zh.*, 1975, **28**, 412 (*Chem. Abstr.*, 1975, **83**, 206 174).	539
75MI42601	R. H. Schieferstein and K. Pilgram; *J. Agric. Food. Chem.*, 1975, **23**, 393.	543
75MI42800	E. A. Neves and D. W. Franco; *J. Inorg. Nucl. Chem.*, 1975, **37**, 277.	592
B-75MI42900	P. A. Stadler and P. Stütz; in 'The Alkaloids', ed. R. H. F. Manske; Academic Press, New York, 1975, vol. 15, p. 1.	664, 667
75MI43100	D. Gonbeau, C. Guimon, J. Deschamps and G. Pfister-Guillouzo; *J. Electron Spectrosc.*, 1975, **6**, 99.	787
75MI43200	R. Mayer, J. Sühnel, H. Hartmann and J. Fabian; *Z. Phys. Chem. (Leipzig)*, 1975, **256**, 792.	815
75MI43300	H. H. Carbacho and L. L. Victoriano; *J. Inorg. Nucl. Chem.*, 1975, **37**, 1327.	891
75MI43301	R. G. Buttery, R. M. Seifert and L. C. Ling; *J. Agric. Food Chem.*, 1975, **23**, 516.	894
75MI43302	H. L. Wildenradt, E. N. Christensen, B. Stackler, A. Caputi, K. Slinkard and K. Scutt; *Am. J. Enol. Vitic.*, 1975, **26**, 148.	894
75MI43303	D. B. Menzel, R. J. Slaughter, A. M. Bryant and H. O. Jansequi; *Arch. Environ. Health*, 1975, **30**, 296.	895
75MI43304	D. B. Menzel, R. J. Slaughter, A. M. Bryant and H. O. Jansequi; *Arch. Environ. Health*, 1975, **30**, 234.	895
75MI43400	K. E. Peterman and J. M. Shreeve; *J. Fluorine Chem.*, 1975, **6**, 83.	910

75MI43401	V. V. Dovlatyan and R. S. Mirzoyan; *Arm. Khim. Zhur.*, 1975, **28**, 233 (*Chem. Abstr.*, 1975, **83**, 58 725).	910, 930, 941
75MI43402	V. V. Dovlatyan and R. S. Mirzoyan; *Arm. Khim. Zhur.*, 1975, **28**, 311 (*Chem. Abstr.*, 1975, **83**, 178 938).	910, 941
75MI43403	H. Naarmann and H. Pohlemann; *Makromol. Chem., Suppl. 1*, 1975, 71 (*Chem. Abstr.*, 1975, **83**, 131 962).	945
75MIP41600	K. Samula, E. Jurkowska-Kowalczyk, B. Cichy, L. Krzywosinski and B. Borkowska; *Pol. Pat.* 78 637 (1975) (*Chem. Abstr.*, 1976, **85**, 94 347).	97, 116
75MIP41601	A. L. Fridman, F. A. Gabitov, V. D. Surkov and V. S. Zalesov; *USSR Pat.* 472 939 (1975) (*Chem. Abstr.*, 1975, **83**, 164 161).	102
75OMR(7)296	J. Devillers, M. Cornus, J. Navech and L. Cazaux; *Org. Magn. Reson.*, 1975, **7**, 296.	911
75OMR(7)345	M. Anteunis and D. Danneels; *Org. Magn. Reson.*, 1975, **7**, 345.	753
75OPP55	B. Koren, B. Stanovnik and M. Tišler; *Org. Prep. Proced. Int.*, 1975, **7**, 55.	489, 497, 741
75OPP179	E. V. P. Tao and C. F. Christie, Jr.; *Org. Prep. Proced. Int.*, 1975, **7**, 179.	568
75S20	T. Nishiwaki; *Synthesis*, 1975, 20.	55
75S38	J. Nakayama; *Synthesis*, 1975, 38.	845, 846
75S52	R. Neidlein and K. Salzmann; *Synthesis*, 1975, 52.	505
75S57	K. Burger, J. Albanbauer and W. Strych; *Synthesis*, 1975, 57.	911
75S165	M. Furukawa, T. Suda, A. Tsukamoto and S. Hayashi; *Synthesis*, 1975, 165.	149, 174
75S168	J. Nakayama; *Synthesis*, 1975, 168.	849
75S205	D. St. C. Black, R. F. Crozier and V. C. Davis; *Synthesis*, 1975, 205.	1000
75S277	Y. Ueno, A. Nakayama and M. Okawara; *Synthesis*, 1975, 277.	832
75S415	K. Ley and F. Seng; *Synthesis*, 1975, 415.	394, 407, 721
75S664	V. D. Piaz, S. Pinzauti and P. Lacrimini; *Synthesis*, 1975, 664.	71, 87
75S666	L. Garanti, A. Sala and G. Zecchi; *Synthesis*, 1975, 666.	1007, 1008
75S749	Y. S. Rao and R. Filler; *Synthesis*, 1975, 749.	185
75SA(A)191	L. Henriksen, P. H. Nielsen, G. Borch and P. Klaboe; *Spectrochim. Acta, Part A*, 1975, **31**, 191.	954, 957
75SA(A)1115	G. Mille, J. Metzger, C. Pouchan and M. Chaillet; *Spectrochim. Acta, Part A*, 1975, **31**, 1115.	141
75SA(A)1371	L. Henriksen, P. H. Nielsen, G. Borch and P. Klaboe; *Spectrochim. Acta, Part A*, 1975, **31**, 1371.	954
75SC143	R. E. Hackler and T. W. Balko; *Synth. Commun.*, 1975, **5**, 143.	311
75SST(3)541	F. Kurzer; *Org. Compd. Sulphur, Selenium, Tellurium*, 1975, **3**, 541.	132, 148, 150, 151, 152, 153, 161, 163, 164, 165, 166, 167, 169, 170, 171, 172, 175
75SST(3)566	F. Kurzer; *Org. Compd. Sulphur, Selenium, Tellurium*, 1975, **3**, 566–616.	328
75SST(3)617	F. Kurzer; *Org. Compd. Sulphur, Selenium, Tellurium*, 1975, **3**, 617–641.	328
75SST(3)670	F. Kurzer; *Org. Compd. Sulphur, Selenium, Tellurium*, 1975, **3**, 670.	454, 456, 458, 459
75SST(3)687	F. Kurzer; *Org. Compd. Sulphur, Selenium, Tellurium*, 1975, **3**, 687.	546, 562
75T831	L. H. Zalkow and R. H. Hill; *Tetrahedron*, 1975, **31**, 831.	100
75T1373	Y. Ito and T. Matsuura; *Tetrahedron*, 1975, **31**, 1373.	37
75T1537	J. M. Borsus, G. L'abbé and G. Smets; *Tetrahedron*, 1975, **31**, 1537.	913, 917, 941
75T1783	A. Holm, L. Carlsen, S.-O. Lawesson and H. Kolind-Andersen; *Tetrahedron*, 1975, **31**, 1783.	582, 587, 597
75T1861	G. Zvilichovsky; *Tetrahedron*, 1975, **31**, 1861.	61
75T2884	L. Jurd; *Tetrahedron*, 1975, **31**, 2884.	79
75T3069	R. G. Shotter, D. Sesardic and P. H. Wright; *Tetrahedron*, 1975, **31**, 3069.	94
75TL163	A. S. Shawali and A. O. Abdelhamid; *Tetrahedron Lett.*, 1975, 163.	570
75TL455	K. Akiba, T. Tsuchiya, M. Ochiumi and N. Inamoto; *Tetrahedron Lett.*, 1975, 455.	475
75TL459	K. Akiba, M. Ochiumi, T. Tsuchiya and N. Inamoto; *Tetrahedron Lett.*, 1975, 459.	475
75TL555	H.-J. Kyi and K. Praefke; *Tetrahedron Lett.*, 1975, 555.	886
75TL1125	K. Burger, H. Schickaneder and W. Therm; *Tetrahedron Lett.*, 1975, 1125.	983
75TL1259	E. M. Engler and V. V. Patel; *Tetrahedron Lett.*, 1975, 1259.	948, 952, 954, 958, 964, 965
75TL1739	J. R. Grunwell and S. L. Dye; *Tetrahedron Lett.*, 1975, 1739.	999
75TL1969	N. Sonoda, G. Yamamoto, K. Natsukawa, K. Kondo and S. Murai; *Tetrahedron Lett.*, 1975, 1969.	321
75TL1985	R. Gallo, A. Lidén, C. Roussel, J. Sandström and J. Metzger; *Tetrahedron Lett.*, 1975, 1985.	243
75TL2131	N. Alexandrou, E. Coutouli and A. Varvoglis; *Tetrahedron Lett.*, 1975, 2131.	95, 119
75TL2337	R. Jacquier, J. L. Olive, C. Petrus and F. Petrus; *Tetrahedron Lett.*, 1975, 2337.	97, 111
75TL3283	J. B. Hill; *Tetrahedron Lett.*, 1975, 3283.	104
75TL3387	J. Goerdeler and H. W. Linden; *Tetrahedron Lett.*, 1975, 3387.	920
75TL3473	H. Behringer and E. Meinetsberger; *Tetrahedron Lett.*, 1975, 3473.	793, 794, 797
75TL3519	V. Dryanska and C. Ivanov; *Tetrahedron Lett.*, 1975, 3519.	276
75TL3577	A. B. Bulacinski, E. F. V. Scriven and H. Suschitzky; *Tetrahedron Lett.*, 1975, 3577.	403
75TL4123	M. Fukumura, N. Hamma and T. Nakagome; *Tetrahedron Lett.*, 1975, 4123.	149
75USP3852299	W. M. Hutchinson; *U.S. Pat.* 3 852 299 (1975) (*Chem. Abstr.*, 1975, **83**, 58 788).	91
75USP3859296	Monsanto Co., *U.S. Pat.* 3 859 296 (1975) (*Chem. Abstr.*, 1975, **82**, 98 007).	509

75USP3869467	R. W. Guthrie, R. W. Kierstead and R. A. LeMahieu; *U.S. Pat.* 3 869 467 (1975) (*Chem. Abstr.*, 1975, **83**, 43 609).	128
75USP3873531	*U.S. Pat.* 3 873 531 (1975) (*Chem. Abstr.*, 1975, **83**, 12 226).	668
75USP3874873	FMC Corp., *U.S. Pat.* 3 874 873 (1975) (*Chem. Abstr.*, 1973, **78**, 97 663).	462
75USP3879533	J. B. Carr and D. K. Hass; *U.S. Pat.* 3 879 533, 3 879 532 (1975) (*Chem. Abstr.*, 1975, **83**, 108 627, 72 183).	128
75USP3896140	R. A. Plepys and Z. Jezic; *U.S. Pat.* 3 896 140 (1975) (*Chem. Abstr.*, 1976, **84**, 4936).	128
75USP3899502	Monsanto Co., *U.S. Pat.* 3 899 502 (1975) (*Chem. Abstr.*, 1975, **83**, 194 869).	511
75USP3904537	Lubrizol Corp., *U.S. Pat.* 3 904 537 (1975) (*Chem. Abstr.*, 1975, **84**, 7432).	511
75USP3904619	Monsanto Co., *U.S. Pat.* 3 904 619 (1975) (*Chem. Abstr.*, 1976, **84**, 6300).	511
75USP3911132	R. V. Coombs and G. E. Hardtmann; *U.S. Pat.* 3 911 132 (1975) (*Chem. Abstr.*, 1976, **84**, 31 040).	21, 126, 129
75YZ87	M. Hamana and S. Kumadaki; *Yakugaku Zasshi*, 1975, **95**, 87 (*Chem. Abstr.*, 1975, **83**, 9734).	723
75YZ326	T. Takemoto, T. Nakajima, S. Arihara and K. Koike; *Yakugaki Zasshi*, 1975, **95**, 326 (*Chem. Abstr.*, 1975, **83**, 43 706).	390
75ZC18	K. Gewald, P. Bellmann and H. J. Jansch; *Z. Chem.*, 1975, **15**, 18.	158
75ZC57	J. Dost and R. Leisner; *Z. Chem.*, 1975, **15**, 57.	388
75ZC395	J. Faust, B. Bartho and R. Mayer; *Z. Chem.*, 1975, **15**, 395.	789
75ZC478	J. Faust, *Z. Chem.*, 1975, **15**, 478.	167, 790, 794
75ZN(A)1279	P. J. Mjoberg, W. M. Ralowski and S. O. Ljunggren; *Z. Naturforsch., Teil A*, 1975, **30**, 1279.	8
75ZN(C)183	G. Schaefer, A. Trebst and K. H. Buechel; *Z. Naturforsch., Teil C.*, 1975, **30**, 183 (*Chem. Abstr.*, 1975, **83**, 91 476).	577
75ZOB485	O. M. Astakhova, G. S. Supin, O. S. Martynova, N. P. Lyalyakina and N. I. Shvetsov-Shilovskii; *Zh. Obshch. Khim.*, 1975, **45**, 485 (*Chem. Abstr.*, 1975, **83**, 9914).	924
75ZOB2090	L. N. Volovel'skii, N. V. Popova, M. Ya. Yakovleva and V. G. Khukhryanskii; *Zh. Obshch. Khim.*, 1975, **45**, 2090 (*Chem. Abstr.*, 1976, **84**, 74 496).	97, 128
75ZOB2746	A. P. Rakov, V. M. Filippov and G. F. Andreev; *Zh. Obshch. Khim.*, 1975, **45**, 2746 (*Chem. Abstr.*, 1976, **84**, 59 682).	89
75ZOR7	Yu. N. Yur'ev, S. D. Razumovskii, L. B. Berezova and E. S. Zelikman; *Zh. Org. Khim.*, 1975, **11**, 7.	876
76ABC759	H. Singh and L. S. D. Yadav; *Agric. Biol. Chem.*, 1976, **40**, 759.	598
76AC(R)57	A. Selva and P. Traldi; *Ann. Chim. (Rome)*, 1976, **66**, 57.	379
76ACR287	R. B. Moodie and K. Schofield; *Acc. Chem. Res.*, 1976, **9**, 287.	532
76ACR371	A. Padwa; *Acc. Chem. Res.*, 1976, **9**, 371.	223
76ACS(A)351	T. Ottersen; *Acta Chem. Scand., Ser. A*, 1976, **30**, 351.	581
76ACS(A)397	J. Sletten; *Acta Chem. Scand., Ser. A*, 1976, **30**, 397.	900
76ACS(A)997	L. Carlsen and A. Holm; *Acta Chem. Scand., Ser. A*, 1976, **30**, 997.	596
76ACS(B)88	N. Loyaza and C. Th. Pedersen; *Acta Chem. Scand, Ser. B*, 1976, **30**, 88.	786
76ACS(B)463	P. Wolkoff; *Acta Chem. Scand., Ser. B*, 1976, **30**, 463.	979, 1009
76ACS(B)600	J. Møller, R. M. Cristie, C. T. Pedersen and D. H. Reid; *Acta. Chem. Scand., Ser. B*, 1976, **30**, 600.	362
76ACS(B)675	C. L. Pedersen; *Acta Chem. Scand., Ser. B*, 1976, **30**, 675.	522, 525, 539
76ACS(B)781	J. Lykkeberg and P. Krogsgaard-Larsen; *Acta Chem. Scand., Ser. B.*, 1976, **30**, 781.	148, 157, 158, 160
76ACS(B)837	P. Wolkoff and S. Hammerum; *Acta Chem. Scand., Ser. B*, 1976, **30**, 837.	571
76AG510	G. L'abbé and G. Verhelst; *Angew. Chem.*, 1976, **88**, 510.	590
76AG(E)372	J. Leitich; *Angew. Chem., Int. Ed. Engl.*, 1976, **15**, 372.	935, 940
76AG(E)489	G. L'abbé and G. Verhelst; *Angew. Chem., Int. Ed. Engl.* 1976, **15**, 489.	501
76AHC(19)1	W. D. Ollis and C. A. Ramsden; *Adv. Heterocycl. Chem.*, 1976, **19**, 1.	184, 206, 448, 598, 599, 600, 601, 605
76AHC(19)279	M. V. George, S. K. Khetan and R. K. Gupta; *Adv. Heterocycl. Chem.*, 1976, **19**, 279.	1005
76AHC(20)65	L. B. Clapp; *Adv. Heterocycl. Chem.*, 1976, **20**, 65.	379, 380, 382, 384, 385, 386, 390, 391
76AHC(20)145	A. Holm; *Adv. Heterocycl. Chem.*, 1976, **20**, 145.	580, 581, 582, 583, 584, 585, 586, 588, 591, 592, 595, 597
76AHC(S1)1	J. Elguero, C. Marzin, A. R. Katritzky and P. Linda; *Adv. Heterocycl. Chem.*, 1976, Suppl. 1, 1.	11, 145, 160
76AHC(S1)266	J. Elguero, C. Marzin, A. R. Katritzky and P. Linda; *Adv. Heterocycl. Chem.*, 1976, Suppl. 1, 266.	467, 468
76AHC(S1)380	J. Elguero, C. Marzin, A. R. Katritzky and P. Linda; *Adv. Heterocycl. Chem.*, 1976, Suppl. 1, 380.	557
76AJC1745	L. W. Deady and D. C. Stillman; *Aust. J. Chem.*, 1976, **29**, 1745.	144, 148
76AP928	K. C. Liu, J. Y. Tuan and B. J. Shik; *Arch. Pharm. (Weinheim, Ger.)*, 1976, **309**, 928.	695
76AP935	H. Stamm and H. Steudle; *Arch. Pharm. (Weinheim, Ger.)*, 1976, **309**, 935.	113
76AP1014	H. Stamm and H. Steudle; *Arch. Pharm. (Weinheim, Ger.)*, 1976, **309**, 1014.	99

76AX(B)1074	M. Mellini and S. Merlino; *Acta Crystallogr., Part B*, 1976, **32**, 1074.	515
76AX(B)1079	M. Mellini and S. Merlino; *Acta Crystallogr., Part B*, 1976, **32**, 1079.	395
76AX(B)1317	G. Pèpe and M. Pierrot; *Acta Crystallogr., Part B*, 1976, **32**, 1317 (*Chem. Abstr.*, 1976, **85**, 39 514).	238
76AX(B)1321	G. Pèpe and M. Pierrot, *Acta Crystallogr., Part B*, 1976, **32**, 1321 (*Chem. Abstr.*, 1976, **85**, 27 578).	238
76AX(B)1601	I. Leban; *Acta Crystallogr., Part B*, 1976, **32**, 1601.	732
76BCJ954	H. Kanazawa, Y. Matsuura, N. Tanaka, M. Kakudo, T. Komoto and T. Kawai; *Bull. Chem. Soc. Jpn.*, 1976, **49**, 954.	185
76BCJ1138	O. Tsuge and A. Torii; *Bull. Chem. Soc. Jpn.*, 1976, **49**, 1138.	108
76BCJ1996	C. U. Pittman, Jr. and M. Narita; *Bull. Chem. Soc. Jpn.*, 1976, **49**, 1996.	845
76BCJ2254	C. Kashima, M. Uemori, Y. Tsuda and Y. Omote; *Bull. Chem. Soc. Jpn.*, 1976, **49**, 2254.	49
76BCJ2815	H. Sato, T. Kusumi, K. Imaye and H. Kakisawa; *Bull. Chem. Soc. Jpn.*, 1976, **49**, 2815.	46, 110
76BCJ3124	H. Yoshida, H. Taketani, T. Ogata and S. Inokawa; *Bull. Chem. Soc. Jpn.*, 1976, **49**, 3124.	169, 999
76BCJ3128	H. Yoshida, T. Yao, T. Ogata and S. Inokawa; *Bull. Chem. Soc. Jpn.*, 1976, **49**, 3128.	806
76BCJ3173	K. Ueda, M. Igaki and F. Toda; *Bull. Chem. Soc. Jpn.*, 1976, **49**, 3173.	606, 609
76BCJ3314	H. Matsukubo and H. Kato; *Bull. Chem. Soc. Jpn.*, 1976, **49**, 3314.	829
76BCJ3567	J. Nakayama, K. Fujiwara and M. Hoshino; *Bull. Chem. Soc. Jpn.*, 1976, **49**, 3567.	816
76BSB35	Y. Van Haverbeke, A. Maquestiau, R. N. Muller and M. L. Stamanne, *Bull. Soc. Chim. Belg.*, 1976, **85**, 35.	380
76BSF120	D. Paquer and R. Pou; *Bull. Soc. Chim. Fr.*, 1976, 120.	798, 920
76BSF1043	L. Carbonnel and J. C. Rosso; *Bull. Soc. Chim. Fr.*, 1976, 1043.	144
76BSF1124	R. Weber and M. Renson; *Bull. Soc. Chim. Fr.*, 1976, 1124.	337
76BSF1200	J. C. Meslin and G. Duguay; *Bull. Soc. Chim. Fr.*, 1976, 1200.	1062
76BSF1499	J. L. Burgot and J. Vialle; *Bull. Soc. Chim. Fr.*, 1976, 1499.	1066
76BSF1589	J. L. Olive, C. Petrus and F. Petrus; *Bull. Soc. Chim. Fr.*, 1976, 1589.	112
76CA(85)159954	T. P. Vasil'eva, M. G. Lin'kova, O. V. Kil'disheva and I. L. Knunyants; *Chem. Abstr.*, 1976, **85**, 159 954.	797
76CB139	J. C. Jochims and A. Abu-Taba; *Chem. Ber.*, 1976, **109**, 139 (*Chem. Abstr.*, 1976, **84**, 90 052).	318
76CB740	H. Gotthardt, M. C. Weisshuhn and B. Christl; *Chem. Ber.*, 1976, **109**, 740.	818, 841, 842
76CB753	H. Gotthardt, M. C. Weisshuhn and B. Christl; *Chem. Ber.*, 1976, **109**, 753.	826
76CB848	J. Goerdeler, W. Kunnes and F. M. Panshiri; *Chem. Ber.*, 1976, **109**, 848.	920, 942, 943
76CB1837	H. Paulsen and R. Dammeyer; *Chem. Ber.*, 1976, **109**, 1837.	750, 751, 755
76CB2442	R. Appel, J. R. Lundehn and E. Lassmann; *Chem. Ber.*, 1976, **109**, 2442.	538
76CB2648	G. Höfle, W. Steglich and H. Daniel; *Chem. Ber.*, 1976, **109**, 2648.	186
76CB3108	J. Goerdeler, H. Hohage and I. Zeid; *Chem. Ber.*, 1976, **109**, 3108.	904, 923
76CB3326	H. Böshagen; *Chem. Ber.*, 1976, **100**, 3326.	58
76CB3646	H. Pauls and F. Kröhnke; *Chem. Ber.*, 1976, **109**, 3646.	657
76CB3653	H. Pauls and F. Kröhnke; *Chem. Ber.*, 1976, **109**, 3653.	673
76CB3668	T. Kappe and W. Golser; *Chem. Ber.*, 1976, **109**, 3668.	691
76CC92	M. M. Takagi, S. Goto and T. Matsuda; *J. Chem. Soc., Chem. Commun.*, 1976, 92.	795
76CC94	A. J. Bloodworth and M. E. Loveitt; *J. Chem. Soc., Chem. Commun.*, 1976, 94.	774
76CC148	E. M. Engler, D. C. Green and J. Q. Chambers; *J. Chem. Soc., Chem. Commun.*, 1976, 148.	963, 971
76CC209	J. Kalvoda and H. Kaufmann; *J. Chem. Soc., Chem. Commun.*, 1976, 209.	108
76CC210	H. Kaufmann and J. Kalvoda; *J. Chem. Soc., Chem. Commun.*, 1976, 210.	108
76CC240	J. A. Chapman, J. Crosby, C. A. Cummings and R. A. C. Rennie; *J. Chem. Soc., Chem. Commun.*, 1976, 240.	81
76CC306	R. N. Hanley, W. D. Ollis and C. A. Ramsden; *J. Chem. Soc., Chem. Commun.*, 1976, 306.	601, 603, 604
76CC307	R. N. Hanley, W. D. Ollis and C. A. Ramsden; *J. Chem. Soc., Chem. Commun.*, 1976, 307.	601
76CC771	R. A. Abramovitch, K. M. More, I. Shinkai and P. C. Srinivasan; *J. Chem. Soc., Chem. Commun.*, 1976, 771.	149, 155
76CC795	J. E. Baldwin, C. Hoskins and L. Kruse; *J. Chem. Soc., Chem. Commun.*, 1976, 795.	91, 92
76CC912	M. Baudy and A. Robert; *J. Chem. Soc., Chem. Commun.*, 1976, 912.	765
76CC920	C. H. Chen; *J. Chem. Soc., Chem. Commun.*, 1976, 920.	846
76CHE61	F. S. Mikhailitsyn and A. F. Bekhli; *Chem. Heterocycl. Compd. (Engl. Transl.)*, 1976, **12**, 61.	538
76CHE289	G. I. Eremeva, B. Kh. Strelets and L. S. Efros; *Chem. Heterocycl. Compd. (Engl. Transl.)*, 1976, **12**, 289.	526
76CHE711	O. P. Shvaika, N. I. Korotkikh, G. F. Tereshchenko and N. A. Kovach; *Chem. Heterocycl. Compd. (Engl. Transl.)*, 1976, **12**, 711.	433
76CJC146	M. F. Guimon, C. Guimon, F. Metras and G. Pfister-Guillouzo; *Can. J. Chem.*, 1976, **54**, 146.	860, 861
76CJC1074	M. Farnier, S. Soth and P. Fournari; *Can. J. Chem.*, 1976, **54**, 1074.	1009
76CJC1428	G. W. Buchanan and D. G. Hellier; *Can. J. Chem.*, 1976, **54**, 1428.	862
76CJC2804	J. P. Devlin; *Can. J. Chem.*, 1976, **54**, 2804.	717, 721

76CJC3012	N. K. Sharma, F. De Reinach-Hirtzbach and T. Durst; *Can. J. Chem.*, 1976, **54**, 3012.	771, 778
76CJC3850	G. Mille, J. Chouteau, T. Avignon and L. Bouscasse; *Can. J. Chem.*, 1976, **54**, 3850.	141
76CJC3879	M. S. Chauhan and D. M. McKinnon; *Can. J. Chem.*, 1976, **54**, 3879.	161, 797, 924
76CL723	K.-Y. Akiba, T. Tsuchiya, N. Inamoto and K.-I. Onuma; *Chem. Lett.*, 1976, 723.	475
76CPB235	Y. Maki; *Chem. Pharm. Bull.*, 1976, **24**, 235.	717, 719, 720
76CPB632	H. Uno, M. Kurokawa, K. Natsuka, Y. Yamato and H. Nishimura; *Chem. Pharm. Bull.*, 1976, **24**, 632.	129
76CPB979	Y. Furukawa and S. Shima; *Chem. Pharm. Bull.*, 1976, **24**, 979.	636, 638
76CPB1106	T. Kurihara, T. Sakaguchi and H. Hirano; *Chem. Pharm. Bull.*, 1976, **24**, 1106.	124
76CPB1197	N. Nagahara, K. Takagi and T. Ueda; *Chem. Pharm. Bull.*, 1976, **24**, 1197.	384
76CPB1299	K. Mizuyama, Y. Matsuo, Y. Tominaga, Y. Matsuda and G. Kobayashi; *Chem. Pharm. Bull.*, 1976, **24**, 1299.	688
76CPB1757	S. Hashimoto, M. Shizu and S. Takahashi; *Chem. Pharm. Bull.*, 1976, **24**, 1757.	128
76CPB2532	M. Sakamoto, K. Miyazawa and Y. Tomimatsu; *Chem. Pharm. Bull.*, 1976, **24**, 2532.	359
76CPB2673	T. Karasawa, K. Furukawa, K. Yoshida and M. Shimizu; *Chem. Pharm. Bull.*, 1976, **24**, 2673.	129
76CPB2889	M. Sakamoto, M. Shibano, K. Miyazawa, M. Suzuki and T. Tomimatso; *Chem. Pharm. Bull.*, 1976, **24**, 2889.	197
76CPB2918	T. Ito, K. Yoshida and M. Shimizu; *Chem. Pharm. Bull.*, 1976, **24**, 2918.	129
76CPB3001	S. Kishimoto, S. Noguchi and K. Masuda; *Chem. Pharm. Bull.*, 1976, **24**, 3001.	376
76CR(C)(283)401	A. Carpy, J.-C. Colleter, M. Gadret, M. Goursolle and J. M. Leger; *C. R. Hebd. Seances Acad. Sci., Ser. C*, 1976, **283**, 401.	516
76CRV747	J. G. Tillett; *Chem. Rev.*, 1976, **76**, 747.	853, 874, 875, 879, 886, 887
76CSC67	M. Cannas, G. Marongiu and R. Destro; *Cryst. Struct. Commun.*, 1976, **5**, 67 (*Chem. Abstr.*, 1976, **84**, 114 628).	91
76CSC71	M. Cannas, G. Marongiu and R. Destro; *Cryst. Struct. Commun.*, 1976, **5**, 71 (*Chem. Abstr.*, 1976, **84**, 114 629).	91
76CSC75	M. Cannas, G. Marongiu and R. Destro; *Cryst. Struct. Commun.*, 1976, **5**, 75 (*Chem. Abstr.*, 1976, **84**, 114 630).	89
76CSC113	M. Calleri, G. Chiasi, A. Chiesa Villa and C. Guastini; *Cryst. Struct. Commun.*, 1976, **5**, 113.	396
76CSC329	A. G. Manfredotti, G. Guastini, M. Calleri and D. Viterbo; *Cryst. Struct. Commun.*, 1976, **5**, 329.	396
76EGP119791	*Ger. (East) Pat.* 119 791 (1976) (*Chem. Abstr.*, 1977, **86**, 106 608).	498, 507
76EGP122249	M. Mühlstädt, R. Brämer and B. Schulze; *Ger. (East) Pat.* 122 249 (1976) (*Chem. Abstr.*, 1977, **87**, 39 468).	169
76FES393	B. Cavalleri, G. Volpe, B. Rosselli del Turco and A. Diena, *Farmaco Ed. Sci.*, 1976, **31**, 393.	384
76FRP2027211	M. Barreau, C. Cotrel and C. Jeanmart; (Rhone Poulenc S.A.), *Ger. Pat.* 2 627 211 (1976) (*Chem. Abstr.*, 1977, **86**, 121 373).	810
76G769	M. T. Lugari-Mangia and G. Pelizzi; *Gazz. Chim. Ital.*, 1976, **106**, 769.	134, 159, 160
76G823	S. Auricchio, R. Colle, S. Morocchi and A. Ricca; *Gazz. Chim. Ital.*, 1976, **106**, 823.	49
76GEP2264979	*Ger. Pat.* 2 264 979 (1976) (*Chem. Abstr.*, 1976, **84**, 150 656).	709
76GEP2429195	H. Eilingsfeld and R. Niess; *Ger. Pat.* 2 429 195 (1976) (*Chem. Abstr.*, 1976, **84**, 135 649).	1015
76GEP2447370	H. Eilingsfeld and H. Naarmann; *Ger. Pat.* 2 447 370 (1976) (*Chem. Abstr.*, 1976, **85**, 63 633).	945
76GEP2456874	H. U. Stracke and G. Koppensteiner; *Ger. Pat.* 2 456 874 (1976) (*Chem. Abstr.*, 1976, **85**, 112 656).	944
76GEP2525023	G. Heubach, G. Hoerlein and B. Sachse; *Ger. Offen.* 2 525 023 (1976) (*Chem. Abstr.*, 1977, **86**, 106 569).	128
76GEP2529292	J. Katsuhe, T. Kobayashi, K. Tomoto, Y. Takebayashi, K. Sasajima, S. Inaba and H. Yamamoto; *Ger. Offen.* 2 529 292 (1976) (*Chem. Abstr.*, 1976, **84**, 135 627).	120, 129
76GEP2534689	O. Hromatka and D. Binder; *Ger. Pat.* 2 534 689 (1976) (*Chem. Abstr.*, 1976, **85**, 5612).	1026
76GEP2541115	Eli Lilly and Co., *Ger. Pat.* 2 541 115 (1976) (*Chem. Abstr.* 1976, **85**, 33 024).	576
76GEP2608488	V. Dittrich, W. Toepfl and O. Kristiansen; *Ger. Pat.* 2 608 488 (1976) (*Chem. Abstr.*, 1977, **86**, 29 512).	944
76GEP2625053	K. Taninaka, H. Kurono and T. Sakai; (Nihon Nohyaku Co. Ltd.), *Ger. Pat.* 2 625 053 (1976) (*Chem. Abstr.*, 1977, **86**, 96 001).	810
76H(4)767	M. J. Haddadin and C. H. Issidorides; *Heterocycles*, 1976, **4**, 767.	394, 407, 721
76H(5)95	R. A. Abramovitch, K. M. More, I. Shinkai and P. C. Srinivasan; *Heterocycles*, 1976, **5**, 95.	152
76H(5)109	J. Palmer, J. L. Roberts, P. S. Rutledge and P. D. Woodgate; *Heterocycles*, 1976, **5**, 109.	108, 109
76H(5)141	R. Huisgen and S. Sustmann; *Heterocycles*, 1976, **5**, 141.	776
76HCA1593	K. Rüefenacht, H. Kristinsson and G. Mattern; *Helv. Chim. Acta*, 1976, **59**, 1593.	648, 653, 654, 657, 659, 668
76HCA2074	K. Dietliker, P. Gilgen, H. Heimgartner and H. Schmid; *Helv. Chim. Acta*, 1976, **59**, 2074.	13

76HCA2727	W. Heinzelmann and P. Gilgen; *Helv. Chim. Acta*, 1976, **59**, 2727.	401
76IJC(B)176	P. B. Talukdar, S. K. Sengupta and A. K. Datta; *Indian J. Chem., Sect. B*, 1976, **14**, 176.	737
76IJC(B)391	S. Rajappa, G. B. Advani and R. Screenivasan; *Indian J. Chem., Sect. B*, 1976, **14**, 391.	641
76IJC(B)394	S. Rajappa and R. Sreenivasan; *Indian J. Chem., Sect. B*, 1976, **14**, 394.	156, 157
76IJC(B)1001	K. S. Sharma, R. Prasad and V. Singh; *Indian J. Chem., Sect. B*, 1976, **14**, 1001.	533
76IZV179	R. N. Lyubovskaya, Ya. D. Lipshan, O. N. Krasochka and L. O. Atovmyan; *Izv. Akad. Nauk SSSR, Ser. Khim.*, 1976, 179.	950, 966, 970
76IZV1903	R. G. Kostyanovskii and V. F. Rudchenko; *Izv. Akad. Nauk SSSR, Ser. Khim.*, 1976, 1903 (*Chem. Abstr.*, 1977, **86**, 29 694).	110
76IZV2779	A. B. Arbuzov, Yu. Yu. Samitov, E. N. Dianova and A. F. Lisin; *Izv. Akad. Nauk SSSR, Ser. Khim.*, 1976, 2779 (*Chem. Abstr.*, 1977, **86**, 155 745).	108
76JA252	F. Wudl, D. E. Schafer and B. Miller; *J. Am. Chem. Soc.*, 1976, **98**, 252.	804, 811
76JA299	J. M. Liesch, J. A. McMillan, R. C. Pandey, I. C. Paul, K. L. Rinehart, Jr. and F. Reusser; *J. Am. Chem. Soc.*, 1976, **98**, 299.	669
76JA965	R. S. Glass and J. R. Duchek; *J. Am. Chem. Soc.*, 1976, **98**, 965.	136
76JA1260	U. Berg, R. Gallo, J. Metzger and M. Chanon; *J. Am. Chem. Soc.*, 1976, **98**, 1260.	254
76JA2741	N. L. Allinger, M. J. Hickey and J. Kao; *J. Am. Chem. Soc.*, 1976, **98**, 2741.	854
76JA2847	C. Roussel, A. Lidén, M. Chanon, J. Metzger and J. Sandström; *J. Am. Chem. Soc.*, 1976, **98**, 2847.	243
76JA2877	K. Griesbaum and P. Hofmann; *J. Am. Chem. Soc.*, 1976, **98**, 2877.	892
76JA2895	G. W. Astrologes and J. C. Martin; *J. Am. Chem. Soc.*, 1976, **98**, 2895.	868, 892
76JA3916	R. C. Wheland and J. L. Gillson; *J. Am. Chem. Soc.*, 1976, **98**, 3916.	351, 354
76JA4012	R. P. Lattimer, U. Mazur and R. L. Kuczkowski; *J. Am. Chem. Soc.*, 1976, **98**, 4012.	855, 858
76JA4526	D. P. Higley and R. W. Murray; *J. Am. Chem. Soc.*, 1976, **98**, 4526.	863, 880
76JA6036	D. J. Woodman, P. M. Stonebraker and L. Weiler; *J. Am. Chem. Soc.*, 1976, **98**, 6036.	42
76JA6088	P. C. Hiberty; *J. Am. Chem. Soc.*, 1976, **98**, 6088.	853
76JA7187	C. Müller, A. Schweig, M. P. Cava and M. V. Lakshmikantham; *J. Am. Chem. Soc.* 1976, **98**, 7187.	1029, 1030, 1031
76JA7440	Y. Ueno, A. Nakayama and M. Okawara; *J. Am. Chem. Soc.*, 1976, **98**, 7440.	850
76JAP(K)7644637	N. Tsuji, K. Tomita, S. Yamanoto, T. Honma and Y. Takahi; *Jpn. Kokai* 76 44 637 (1976) (*Chem. Abstr.*, 1977, **86**, 5440).	43
76JAP(K)7659858	S. Sumimoto, H. Yukinaga, I. Ishizuka and J. Sugita; *Jpn. Kokai* 76 59 858 (1976) (*Chem. Abstr.*, 1976, **85**, 192 703).	130
76JAP(K)7663192	*Jpn. Kokai* 76 63 192 (1976) (*Chem. Abstr.*, 1977, **86**, 5442).	634
76JAP(K)76110032	M. Umino and S. Yamashita; *Jpn. Kokai* 76 110 032 (1976) (*Chem. Abstr.*, 1977, **86**, 51 569).	129
76JAP(K)76136666	H. Uno, K. Natsuki and M. Kurokawa; *Jpn. Kokai* 76 136 666 (1976) (*Chem. Abstr.*, 1977, **87**, 39 459).	129
76JCP(64)4475	D. R. Jones, C. H. Wang, D. H. Christensen and O. F. Nielson; *J. Chem. Phys.*, 1976, **64**, 4475.	521
76JCS(D)455	L. Menabue and G. C. Pellacani; *J. Chem. Soc., Dalton Trans.*, 1976, 455.	802
76JCS(P1)228	R. M. Christie and D. H. Reid; *J. Chem. Soc., Perkin Trans. 1*, 1976, 228.	362, 363, 799
76JCS(P1)570	V. Bertini, A. DeMunno and M. Pocci; *J. Chem. Soc., Perkin Trans. 1*, 1976, 570.	27, 54, 130
76JCS(P1)584	A. G. W. Baxter and R. J. Stoodley; *J. Chem. Soc., Perkin Trans. 1*, 1976, 584.	667
76JCS(P1)619	P. D. Croce and D. Pocar; *J. Chem. Soc., Perkin Trans. 1*, 1976, 619.	25, 70, 85
76JCS(P1)672	H. Kato, S. Nakazawa, T. Kiyosawa and K. Hirakawa; *J. Chem. Soc., Perkin Trans. 1*, 1976, 672.	826
76JCS(P1)783	T. Sato, K. Yamamoto, K. Fukui, K. Saito, K. Hayakawa and S. Yoshiie; *J. Chem. Soc., Perkin Trans. 1*, 1976, 783.	13
76JCS(P1)880	R. M. Christie and D. H. Reid; *J. Chem. Soc., Perkin Trans. 1*, 1976, 880.	1060
76JCS(P1)1327	R. Nutiu and A. J. Boulton; *J. Chem. Soc., Perkin Trans. 1*, 1976, 1327.	717, 718
76JCS(P1)1518	G. Bianchi, C. De Micheli and R. Gandolfi; *J. Chem. Soc., Perkin Trans. 1*, 1976, 1518.	89
76JCS(P1)1673	G. V. Boyd, T. Norris and P. F. Lindley; *J. Chem. Soc., Perkin Trans. 1*, 1976, 1673.	1005
76JCS(P1)1694	G. S. D'Alcontres, C. Caristi, A. Ferlazzo and M. Gattuso; *J. Chem. Soc., Perkin Trans. 1*, 1976, 1694.	25
76JCS(P1)1706	T. Takeshima, N. Fukuda, T. Ishii and M. Muraoka; *J. Chem. Soc., Perkin Trans. 1*, 1976, 1706.	890
76JCS(P1)1886	I. Degani and R. Fochi; *J. Chem. Soc., Perkin Trans. 1*, 1976, 1886.	822, 836
76JCS(P1)2079	T. G. Back, D. H. R. Barton, M. R. Britten-Kelly and F. S. Guziec, Jr.; *J. Chem. Soc., Perkin Trans. 1*, 1976, 2079.	357, 358
76JCS(P1)2562	H. Matsukubo and H. Kato; *J. Chem. Soc., Perkin Trans. 1*, 1976, 2562.	829
76JCS(P2)203	T. A. Crabb and M. J. Hall; *J. Chem. Soc., Perkin Trans. 2*, 1976, 203.	663
76JCS(P2)548	M. J. S. Dewar and I. J. Turchi; *J. Chem. Soc., Perkin Trans. 2*, 1976, 548.	184
76JCS(P2)1040	B. C. Gilbert, H. A. H. Laue, R. O. C. Norman and R. C. Sealy; *J. Chem. Soc., Perkin Trans. 2*, 1976, 1040.	881
76JHC13	A. P. Komin and M. Carmack; *J. Heterocycl. Chem.*, 1976, **13**, 13.	536, 537, 538, 733, 738, 747, 1032, 1045
76JHC33	M. Rull and J. Vilarrasa; *J. Heterocycl. Chem.*, 1977, **14**, 33.	980

76JHC111	I. Antonini, P. Franchetti, M. Grifantini and S. Martelli; *J. Heterocycl. Chem.*, 1976, **13**, 111.	1010
76JHC117	A. Shafiee, I. Lalezari and M. Mirrashed; *J. Heterocycl. Chem.*, 1976, **13**, 117.	568, 743
76JHC169	G. R. Revankar and R. K. Robins; *J. Heterocycl. Chem.*, 1976, **13**, 169.	485, 494
76JHC171	R. A. Abramovitch, I. Shinkai and R. Van Dahm; *J. Heterocycl. Chem.*, 1976, **13**, 171.	723
76JHC291	R. F. Lauer and G. Zenchoff; *J. Heterocycl. Chem.*, 1976, **13**, 291.	566, 570, 734
76JHC301	A. Shafiee; *J. Heterocycl. Chem.*, 1976, **13**, 301.	746
76JHC379	E. Alcalde and R. M. Claramunt; *J. Heterocycl. Chem.*, 1976, **13**, 379.	977
76JHC409	V. Dal Piaz, S. Pinzauti and P. Lacrimini; *J. Heterocycl. Chem.*, 1976, **13**, 409.	625
76JHC449	C. A. Park, C. F. Beam, E. M. Kaiser, R. J. Kaufman, F. E. Henoch and C. R. Hauser; *J. Heterocycl. Chem.*, 1976, **13**, 449.	71, 72, 94
76JHC491	J. Bourdais, D. Abenhaim, B. Sabourault and A. Lorre; *J. Heterocycl. Chem.*, 1976, **13**, 491.	674
76JHC607	R. M. Sandifer, L. M. Shaffer, W. M. Hollinger, D. C. Reames and C. F. Beam; *J. Heterocycl. Chem.*, 1976, **13**, 607.	71, 72, 85
76JHC661	T. Kurihara, T. Sakaguchi and H. Hirano; *J. Heterocycl. Chem.*, 1976, **13**, 661.	124
76JHC825	D. Clerin, J.-P. Fleury and H. Fritz; *J. Heterocycl. Chem.*, 1976, **13**, 825.	80
76JHC883	G. L'abbé and C. C. Yu; *J. Heterocycl. Chem.*, 1976, **13**, 883.	906, 923, 935, 943
76JHC1097	E. D. Weiler and G. A. Miller; *J. Heterocycl. Chem.*, 1976, **13**, 1097.	164
76JHC1109	P. Dalla Croce; *J. Heterocycl. Chem.*, 1976, **13**, 1109.	108
76JHC1131	W. A. Szarek, C. Depew and J. K. N. Jones; *J. Heterocycl. Chem.*, 1976, **13**, 1131.	159
76JHC1225	K. Pilgram and G. E. Pollard; *J. Heterocycl. Chem.*, 1976, **13**, 1225.	976, 990, 991
76JHC1257	K. H. Pilgram, R. D. Skiles and G. E. Pollard; *J. Heterocycl. Chem.*, 1976, **13**, 1257.	1010
76JHC1321	E. D. Weiler, G. A. Miller and M. Hausman; *J. Heterocycl. Chem.*, 1976, **13**, 1321.	148, 167
76JIC181	C. Mohan, G. S. Saharia and H. R. Sharma; *J. Indian Chem. Soc.*, 1976, **53**, 181.	128
76JIC439	N. N. Ghosh and P. C. Chaudhury; *J. Indian Chem. Soc.*, 1976, **53**, 439.	20
76JIC779	N. N. Ghosh and P. C. Chaudhury; *J. Indian Chem. Soc.*, 1976, **53**, 779.	20
76JMC524	S. C. Bell and P. H. L. Wei; *J. Med. Chem.*, 1976, **19**, 524.	709
76JMC1186	R. D. Elliott and J. A. Montgomery; *J. Med. Chem.*, 1976, **19**, 1186.	735, 746
76JMC1221	G. W. Adelstein, C. H. Yen, E. Z. Dajani and R. G. Bianchi; *J. Med. Chem.*, 1976, **19**, 1221.	445
76JOC13	J. P. Ferris and R. W. Trimmer; *J. Org. Chem.*, 1973, **41**, 13.	13, 118
76JOC19	J. P. Ferris and R. W. Trimmer; *J. Org. Chem.*, 1976, **41**, 19.	1020
76JOC122	A. Rahman and L. B. Clapp; *J. Org. Chem.*, 1976, **41**, 122.	95
76JOC129	K. T. Potts and J. L. Marshall; *J. Org. Chem.*, 1976, **41**, 129.	1041, 1047
76JOC187	K. T. Potts, D. R. Choudhury and T. R. Westby; *J. Org. Chem.*, 1976, **41**, 187.	267, 977, 1003
76JOC403	R. Huisgen; *J. Org. Chem.*, 1976, **41**, 403.	67, 996
76JOC620	J. E. Franz, R. K. Howe and H. K. Pearl; *J. Org. Chem.*, 1976, **41**, 620.	478, 500
76JOC729	F. Malek-Yazdi and M. Yalpani; *J. Org. Chem.*, 1976, **41**, 729.	351
76JOC730	H. K. Spencer and M. P. Cava; *J. Org. Chem.*, 1976, **41**, 730.	848
76JOC892	K. L. Gallaher and R. L. Kuczkowski; *J. Org. Chem.*, 1976, **41**, 892.	880
76JOC966	G. E. Wilson, Jr. and M.-G. Huang; *J. Org. Chem.*, 1976, **41**, 966.	761
76JOC1296	J. E. Franz and H. K. Pearl; *J. Org. Chem.*, 1976, **41**, 1296.	67, 81
76JOC1325	J. C. Grivas; *J. Org. Chem.*, 1976, **41**, 1325.	150
76JOC1449	J. A. May, Jr. and L. B. Townsend; *J. Org. Chem.*, 1976, **41**, 1449.	746
76JOC1724	K. T. Potts, D. R. Choudhury, A. J. Elliot and U. P. Singh; *J. Org. Chem.*, 1976, **41**, 1724.	816, 826, 827, 828, 830, 841, 842
76JOC2465	S. J. Wratten and D. J. Faulkner; *J. Org. Chem.*, 1976, **41**, 2465.	872, 894
76JOC3040	E. A. Isukul, R. Ranson and J. G. Tillett; *J. Org. Chem.*, 1976, **41**, 3040.	602
76JOC3121	L. M. Weinstock, D. M. Mulvey and R. Tull; *J. Org. Chem.*, 1976, **41**, 3121.	537, 542
76JOC3763	W. H. Pirkle and P. L. Gravel; *J. Org. Chem.*, 1976, **41**, 3763.	441
76JOC3987	D. N. Harpp, S. M. Vines, J. P. Montillier and T. H. Chan; *J. Org. Chem.*, 1976, **41**, 3987.	763
76JOC4033	M. Clagett, A. Gooch, P. Graham, N. Holy, B. Mains and J. Strunk; *J. Org. Chem.*, 1976, **41**, 4033.	103
76JOU1102	V. N. Yandovskii; *J. Org. Chem. USSR (Engl. Transl.)*, 1976, **12**, 1102.	442
76JPC1238	H. Kühne and H. H. Günthard; *J. Phys. Chem.*, 1976, **80**, 1238.	859
76JPC1786	C. L. Kwan, M. Carmack and J. K. Kochi; *J. Phys. Chem.*, 1976, **80**, 1786.	523
76JPR127	E. Fanghänel; *J. Prakt. Chem.*, 1976, **318**, 127.	161, 832
76JPR161	J. Faust and R. Mayer; *J. Prakt. Chem.*, 1976, **318**, 161.	167
76JPR221	B. Bartho, J. Faust, R. Pohl and R. Mayer; *J. Prakt. Chem.*, 1976, **318**, 221.	790
76JPR658	M. Z. El-Sabee, M. H. El-Nagdy and G. M. Habashy; *J. Prakt. Chem.*, 1976, **318**, 658.	40
76JPR779	K. Gewald, H. Schaefer and U. Schlegel; *J. Prakt. Chem.*, 1976, **318**, 779.	636, 638, 639, 642
76JPR823	B.-M. Neumann, H.-G. Henning, D. Gloyna and M. Bandlow; *J. Prakt. Chem.*, 1976, **318**, 823.	376
76JPS304	A. Shafiee, I. Lalezari, S. Yazdani, F. M. Shahbazian and T. Partovi; *J. Pharm. Sci.*, 1976, **65**, 304.	352
76JPS1408	C. F. Beam, K. D. Shealy, C. E. Harris, N. L. Shealy, L. W. Dasher, W. M. Hollinger, R. M. Sandifer and D. C. Reames; *J. Pharm. Sci.*, 1976, **65**, 1408.	125

76KGS183	B. K. Strelets, L. S. Efros and Y. I. Akulin; *Khim. Geterotsikl. Soedin.*, 1976, 183 (*Chem. Abstr.*, 1976, **85**, 32 925).	921
76KGS625	A. A. Akhrem, F. A. Lakhvich, V. A. Khirpach, I. B. Klebanovich and A. G. Pozdeev; *Khim. Geterotsikl. Soedin.*, 1976, 625 (*Chem. Abstr.*, 1976, **85**, 46 478).	89
76KGS886	S. S. Mochalov, T. P. Surikova and Yu. S. Shabaravo; *Khim. Geterotsikl. Soedin.*, 1976, 886 (*Chem. Abstr.*, 1976, **85** 192 606).	122
76KGS891	A. A. Akrem, F. A. Lakhvich, V. A. Khripach and I. I. Petrusevich; *Khim. Geterotsikl. Soedin.*, 1976, 891 (*Chem. Abstr.*, 1976, **85**, 192 607).	36
76KGS1029	S. D. Sokolov and G. B. Tikhomirova; *Khim. Geterotsikl. Soedin.*, 1976, 1029 (*Chem. Abstr.*, 1977, **86**, 42 976).	11
76KGS1361	L. S. Efros, B. K. Strelets and Yu. I. Akulin; *Khim. Geterotsikl. Soedin.*, 1976, 1361 (*Chem. Abstr.*, 1977, **86**, 72 527).	360
76LA13	D. Rosenberg and P. Strehlke; *Liebigs Ann. Chem.*, 1976, 13.	128
76LA1997	M. Kloft and D. Hoppe; *Liebigs Ann. Chem.*, 1976, 1997 (*Chem. Abstr.*, 1977, **86**, 72 506).	311
76LA2071	E. Ziegler, G. Kollenz and W. Ott; *Liebigs Ann. Chem.*, 1976, 2071.	1004
76LA2122	U. Schöllkopf, P. H. Porsch and E. Blume; *Liebigs Ann. Chem.*, 1976, 2122 (*Chem. Abstr.*, 1977, **86**, 55 330).	305
76M763	P. Stuetz and P. A. Stadler; *Monatsh. Chem.*, 1976, **107**, 763.	664
76MI41600	J. Suwinski; *Rocz. Chem.*, 1976, **50**, 2005.	117
76MI41601	W. Basinski and Z. Jerzmanowska; *Rocz. Chem.*, 1976, **50**, 1067	79
76MI41602	M. Biddau, G. Devoto, M. Massacesi and G. Ponticelli; *Transition Met. Chem.*, 1976, **1**, 295.	20
76MI41603	R. Pinna, G. Ponticelli, C. Preti and G. Tosi; *Transition Met. Chem.*, 1976, **1**, 173.	20
76MI41604	M. Biddau, G. Devoto, M. Massacesi and G. Ponticelli; *Transition Met. Chem.*, 1976, **1**, 99.	20
76MI41605	G. Devoto, G. Ponticelli, C. Preti and G. Tosi; *J. Inorg. Nucl. Chem.*, 1976, **38**, 1744.	20
76MI41606	M. Massacesi, G. Ponticelli and C. Preti; *J. Inorg. Nucl. Chem.*, 1976, **38**, 1556.	20
76MI41607	J. M. J. Trouchet and J. Poncet; *Carbohydr. Res.*, 1976, **46**, 119.	129
76MI41608	A. Ya. Strakov, M. T. Opmane and E. Gudriniece; *Latv. PSR Zinat. Akad. Vestis Khim. Ser.*, 1976, 234 (*Chem. Abstr.*, 1976, **85**, 46 561).	115
76MI41609	P. Van Brandt and R. Wille; *Bull. Cl. Sci. Acad., R. Belg.*, 1976, **62**, 430 (*Chem. Abstr.*, 1977, **86**, 170 343).	104
76MI41610	G. Renzi, G. Bramanti, L. Bausi and M. Maoggi; *Boll. Chim. Farm.*, 1976, **115**, 575 (*Chem. Abstr.*, 1977, **86**, 105 777).	40
76MI41611	K. S. R. Krishna Mohan Rao and Y. U. Devi; *Proc. Indian Acad. Sci., Sect. A*, 1976, **84**, 79.	40
76MI41612	D. A. Murature, J. D. Perez, M. M. De Bertorello and H. E. Bertorello; *An. Asoc. Quim. Argent.*, 1976, **64**, 337 (*Chem. Abstr.*, 1978, **89**, 128 847).	15
76MI41613	G. LoVecchio, F. Foti, F. Risitano and G. Grassi; *Chem. Ind. (Milan)*, 1976, **58**, 450 (*Chem. Abstr.*, 1977, **86**, 139 961).	89
76MI41700	Z. Machoń; *Arch. Immunol. Ther. Exp.*, 1976, **24**, 863 (*Chem. Abstr.*, 1977, **86**, 121 233).	158
76MI41701	A. Avalos, R. M. Claramunt and R. Granados; *An. Quim.*, 1976, **72**, 922 (*Chem. Abstr.*, 1977, **87**, 183 791).	136, 146, 158
76MI41702	L. K. Gibbons, E. F. Koldenhoven, A. A. Nethery, R. E. Montgomery and W. P. Purcell; *J. Agric. Food Chem.*, 1976, **24**, 203 (*Chem. Abstr.*, 1976, **84**, 131 319).	158
76MI41703	P. P. Singh, L. P. Pathak and S. A. Khan; *J. Inorg. Nucl. Chem.*, 1976, **38**, 475 (*Chem. Abstr.*, 1976, **84**, 144 068).	132
76MI41704	C. L. Greenstock, G. W. Ruddock and P. Neta; *Radiat. Res.*, 1976, **66**, 472 (*Chem. Abstr.*, 1976, **85**, 43 084).	134
B-76MI41800	L. J. Berliner; 'Spin Labelling Theory and Applications', Academic, New York, 1976.	213
76MI42200	T. J. Hopen, R. C. Briner, H. G. Sadler and R. L. Smith; *J. Forensic Sci.*, 1976, **21**, 842.	426
76MI42300	P. H. Singh and L. D. S. Yadav; *Agric. Biol. Chem.*, 1976, **40**, 759 (*Chem. Abstr.*, 1976, **85**, 5568).	445
76MI42301	S. Giri, H. Singh and L. D. S. Yadav; *Agric. Biol. Chem.*, 1976, **40**, 17 (*Chem. Abstr.*, 1976, **84**, 121 736).	445
76MI42400	O. L. Stiefvater; *Chem. Phys.*, 1976, **13**, 73.	450
76MI42401	D. L. Gil and C. F. Wilkinson; *Pestic. Biochem. Physiol.*, 1976, **6**, 338.	461
76MI42500	Y. Akazawa; *Shin Noyaku*, 1976, **30**, 13 (*Chem. Abstr.*, 1978, **88**, 131 826).	509
76MI42501	V. A. Ignatov, G. A. Blokh, Yu. S. Rudei, N. V. Zherkin, R. A. Akchurina and L. H. Agafanova; *Izv. Vyssh. Uch. Zaved. Khim. Khim. Tekhnol.*, 1976, **19**, 290 (*Chem. Abstr.*, 1976, **85**, 7000).	511
76MI42600	W. von Niessen, W. P. Kreamer and L. S. Cederbaum; *J. Electron Spectrosc. Relat. Phenom.*, 1976, **8**, 179.	516
76MI42601	P. M. Solozhenkin, V. Sh. Tsveniashvili, E. V. Semenov and N. S. Khavtasi; *Dokl. Akad. Nauk Tadzh. SSR*, 1976, **19**, 38.	523
76MI42700	J. A. Nelson, L. M. Rose and L. L. Bennett, Jr.; *Cancer Res.*, 1976, **36**, 1375 (*Chem. Abstr.*, 1976, **84**, 173 639).	576

76MI42701	F. Champagnol; *J. Chromatogr.*, 1976, **120**, 489.	555
76MI42800	*76 U.S. Agric. Res. Serv. South Reg.* ARS-S-131 (*Chem. Abstr.*, 1977, **86**, 151 443).	598
76MI42900	G. Garcia-Munoz, R. Madronero, C. Ochoa, M. Stud and W. Pfleiderer; *An. R. Acad. Farm.*, 1976, **42**, 327 (*Chem. Abstr.*, 1977, **86**, 155 616).	729
76MI43300	J. Willemse, J. A. Cras and P. J. H. A. M. van de Leemput; *Inorg. Nucl. Chem. Lett.*, 1976, **12**, 258.	891
76MI43301	F. Ledl; *Z. Lebensm. Unters.-Forsch.*, 1976, **161**, 125.	895
76MI43302	I. H. Qvist and E. C. F. von Sydow; *J. Agric. Food Chem.*, 1976, **24**, 437.	894
76MI43303	R. P. Horvat; *J. Agric. Food Chem.*, 1976, **24**, 953.	894
76MI43400	A. Alemagna and T. Bacchetti; *Chim. Ind. (Milan)*, 1976, **58**, 616 (*Chem. Abstr.*, 1977, **86**, 72 113).	918
76MI43401	V. V. Dovlatyan and R. S. Mirzoyan; *Arm. Khim. Zh.*, 1976, **29**, 959 (*Chem. Abstr.*, 1977, **86**, 171 346).	927, 941
76MI43402	V. V. Dovlatyan and R. S. Mirzoyan; *Arm. Khim. Zh.*, 1976, **29**, 764 (*Chem. Abstr.*, 1977, **86**, 106 492).	941
B-76MI43600	I. Fleming; 'Frontier Orbitals and Organic Chemical Reactions', Wiley, London, 1976.	996, 1024, 1025, 1026
B-76MI43601	'The Merck Index', 9th edn., ed. M. Windholz; Merck, Rahway, N.J., 1976.	000
76MIP42300	N. Yandovskii, P. M. Adrov, and I. A. Zamorina; *U.S.S.R. Pat.* 523 092 (1976) (*Chem. Abstr.*, 1977, **86**, 29 825).	445
76MIP42301	W. G. Brouwer, E. J. MacPherson, R. B. Ames and R. W. Niedermyer; *Can. Pat.* 1 002 336 (1976) (*Chem. Abstr.*, 1977, **86**, 171 458).	445
76MIP42302	H. R. Meyer; *Swiss Pat.* 577 536 (1976) (*Chem. Abstr.*, 1976, **85**, 125 807).	446
76NKK315	H. Fukuda, T. Endo and M. Okawara; *Nippon Kagaku Kaishi*, 1976, 315 (*Chem. Abstr.*, 1976, **85**, 78 054).	432
76OMR(8)56	F. Terrier, F. Millot, A. P. Chatrousse, M. J. Pouet and M. P. Simonnin; *Org. Magn. Reson.*, 1976, **8**, 56.	411
76OMR(8)158	F. A. L. Anet and I. Yauari; *Org. Magn. Reson.*, 1976, **8**, 158.	6
76OMR(8)226	J. Gainer, G. A. Howarth, W. Hoyle and S. M. Roberts; *Org. Magn. Reson.*, 1976, **8**, 226.	6
76OMS304	R. Kraft, P. Henklein, G. Etzold, H.-J. Zöpfl, G. Westphal and R. Schmidt; *Org. Mass Spectrom.*, 1976, **11**, 304.	429
76OMS1047	H.-E. Audier, M. Fetizon, Y. Henry and T. Prange; *Org. Mass Spectrom.*, 1976, **11**, 1047.	183
76OPP87	C. Kashima, Y. Omote, K. Kawada and Y. Tsuda; *Org. Prep. Proced. Int.*, 1976, 87.	49
76PMH(6)1	E. Heilbronner, J. P. Maier and E. Haselbach; *Phys. Methods Heterocycl. Chem.*, 1976, **6**, 1.	5
76PMH(6)53	J. Sheridan; *Phys. Methods Heterocycl. Chem.*, 1976, **6**, 53.	8
76PS(1)185	S. M. Loosmore and D. M. McKinnon; *Phosphorus Sulfur*, 1976, **1**, 185.	795, 801
76RTC67	D. N. Reinhoudt and C. G. Kouwenhoven; *Recl. Trav. Chim. Pays-Bas*, 1976, **95**, 67.	157
76S270	G. Bartoli and G. Rosini; *Synthesis*, 1976, 270.	274
76S273	J. Liebscher and H. Hartmann; *Synthesis*, 1976, 273.	339
76S281	J. M. Patterson; *Synthesis*, 1976, 281.	44
76S349	T. Wagner-Jauregg; *Synthesis*, 1976, 349.	982, 1001
76S403	J. Liebscher and H. Hartmann; *Synthesis*, 1976, 403.	304
76S469	K.-D. Kampe; *Synthesis*, 1976, 469.	652, 665
76S471	I. Degani and R. Fochi; *Synthesis*, 1976, 471.	824
76S489	M. Narita and C. U. Pittman, Jr.; *Synthesis*, 1976, 489.	814, 832, 833, 840, 841, 844, 848, 850
76S591	A. F. Cockerill, A. Deacon, R. G. Harrison, D. J. Osborne, D. M. Prime, W. J. Ross, A. Todd and J. P. Verge; *Synthesis*, 1976, 591.	222
76S612	G. A. Shvekhgeimer, A. Baranski and M. Grzegozek; *Synthesis*, 1976, 612.	69, 70, 73, 83
76S681	G. D. Hartmann and L. M. Weinstock; *Synthesis*, 1976, 681.	305
76S730	M. J. Spitulnik; *Synthesis*, 1976, 730.	323
76S736	E. I. Sanchez and M. J. Fumarola; *Synthesis*, 1976, 736.	148, 174
76S797	S. Cabiddu, A. Maccioni and M. Secci; *Synthesis*, 1976, 797.	778
76SA(A)971	J. AW. Anthonsen, D. H. Christensen, J. T. Nielsen and O. F. Nielsen; *Spectrochim. Acta, Part A*, 1976, **32**, 971.	429
76SA(A)1779	C. Preti, G. Tosi, M. Massacesi and G. Ponticelli; *Spectrochim. Acta, Part A*, 1976, **32**, 1779.	20
76SC269	L. Garanti, A. Sala and G. Zecchi; *Synth. Commun.*, 1976, **6**, 269.	1008
76SC387	E. Campaigne, G. Skowronski, R. A. Forsch and J. C. Beckmann; *Synth. Commun.*, 1976, **6**, 387.	805
76T399	O. Attanasi, G. Bartoli and P. E. Todesco; *Tetrahedron*, 1976, **32**, 399.	674, 689
76T579	G. C. Barrett and R. Walker; *Tetrahedron*, 1976, **32**, 579.	249
76T583	G. C. Barrett and R. Walker; *Tetrahedron*, 1976, **32**, 583.	305
76T675	R. Gree, F. Tonnard and R. Carrie; *Tetrahedron*, 1976, **32**, 675.	108
76T683	R. Gree and R. Carrie; *Tetrahedron*, 1976, **32**, 683.	110
76T745	C. Christophersen and P. Carlsen; *Tetrahedron*, 1976, **32**, 745.	584, 595
76T1277	V. Ceré, S. Pollicino, E. Sandri and G. Scapini; *Tetrahedron*, 1976, **32**, 1277.	403, 412, 518

76T1369	R. Escale, R. Jacquier, B. Ly, F. Petrus and J. Verducci; *Tetrahedron*, 1976, **32**, 1369.	100
76T2559	A. Holm, N. Harrit and N. H. Toubro; *Tetrahedron*, 1976, **32**, 2559.	914
76T2685	K. Ramakrishnan, J. B. Fulton and J. Warkentin; *Tetrahedron*, 1976, **32**, 2685.	1002
76TL873	R. Dunkin, M. Poliakoff, J. J. Turner, N. Harrit and A. Holm; *Tetrahedron Lett.*, 1976, 873.	914
76TL925	I. M. McRobbie, O. Meth-Cohn and H. Suschitzky; *Tetrahedron Lett.*, 1976, 925.	1046
76TL1825	T. Kurihara and M. Mori; *Tetrahedron Lett.*, 1976, 1825.	97, 111
76TL2163	H. Gotthardt and F. Reiter; *Tetrahedron Lett.*, 1976, 2163.	152
76TL2961	J. Goerdeler and K.-H. Kohler; *Tetrahedron Lett.*, 1976, 2961.	797
76TL3463	M. Maeda, S. Ito and M. Kojima; *Tetrahedron Lett.*, 1976, 3463.	693
76TL3615	J. Bergman, J. O. Lindstroem, J. Abrahamsson and E. Hadler; *Tetrahedron Lett.*, 1976, 3615.	714, 724
76TL3695	K. Ishikawa, K. Akiba and N. Inamoto; *Tetrahedron Lett.*, 1976, 3695.	965
76TL3893	A. A. Akhrem, F. A. Lakhvich, V. A. Khripach and I. B. Klebanovich; *Tetrahedron Lett.*, 1976, 3893.	33
76TL3931	C. W. Rees, R. C. Storr and P. J. Whittle; *Tetrahedron Lett.*, 1976, 3931.	97
76TL4325	E. Eibler, J. Skura and J. Sauer; *Tetrahedron Lett.*, 1976, 4325.	935
76UKZ261	V. A. Chuiguk and N. A. Parkhomenko; *Ukr. Khim. Zh.* (*Russ. Ed.*), 1976, **42**, 261 (*Chem. Abstr.*, 1976, **85**, 21 276).	625
76USP3933823	*U.S. Pat.* 3 933 823 (1976) (*Chem. Abstr.*, 1976, **84**, 135 628).	634
76USP3944670	R. F. Bellina; *U.S. Pat.* 3 944 670 (1976) (*Chem. Abstr.*, 1976, **85**, 15 371).	944
76USP3956303	G. A. Bullock and S. S. Sharp; *U.S. Pat.* 3 956 303 (1976) (*Chem. Abstr.*, 1976, **85**, 78 137).	944
76USP3956343	L. S. Crawley and S. R. Safir; *U.S. Pat.* 3 956 343 (*Chem. Abstr.*, 1976, **85**, 63 052).	91
76USP3957805	W. J. Fanshawe, G. E. Wiegand, L. S. Crawley and S. R. Safir; *U.S. Pat.* 3 957 805 (1976) (*Chem. Abstr.*, 1976, **85**, 143 084).	128
76USP3957808	Rohm and Haas Co.; *U.S. Pat.* 3 957 808 (1976) (*Chem. Abstr.*, 1976, **85**, 160 071).	148
76USP3959301	Gulf Res. and Dev. Co., *U.S. Pat.* 3 959 301 (1976) (*Chem. Abstr.*, 1976, **85**, 46 695).	577
76USP3965108	*U.S. Pat.* 3 965 108 (1976) (*Chem. Abstr.*, 1977, **86**, 5467).	748
76USP3966748	C. M. Hofmann and S. R. Safir; *U.S. Pat.* 3 966 748 (1976) (*Chem. Abstr.*, 1977, **86**, 29 820).	128
76USP3979401	E. H. Burk, Jr. and D. D. Carlos; *U.S. Pat.* 3 979 401 (1976) (*Chem. Abstr.*, 1977, **86**, 29 827).	945
76ZC270	J. Faust; *Z. Chem.*, 1976, **16**, 270.	39, 105
76ZC317	E. Fanghänel, L. van Hinh and G. Schukat; *Z. Chem.*, 1976, **16**, 317.	849
76ZN(A)1681	O. L. Stiefvater; *Z. Naturforsch., Teil A*, 1976, **31**, 1681.	465, 472
76ZN(B)380	N. A. E. Kassab, S. O. A. Alla and H. A. R. Ead; *Z. Naturforsch., Teil B*, 1976, **31**, 380.	990
76ZN(B)853	N. A. E. Kassab, S. O. A. Alla and H. A. R. Ead; *Z. Naturforsch., Teil B*, 1976, **31**, 853.	981, 990
76ZOR2028	T. D. Zheved and K. V. Altukhov; *Zh. Org. Khim.*, 1976, **12**, 2028 (*Chem. Abstr.*, 1977, **86**, 16 585).	47
76ZOR2095	R. I. Bodina, E. S. Lipina and V. V. Perekalin; *Zh. Org. Khim.*, 1976, **12**, 2095 (*Chem. Abstr.*, 1977, **86**, 71 800).	110
77AC(R)371	A. Selva, P. Traldi, L. F. Zerilli and G. G. Gallo; *Ann. Chim.* (*Rome*), 1977, **67**, 371.	379
77AC(R)621	A. Selva and P. Traldi; *Ann. Chim.* (*Rome*), 1977, **67**, 621.	379
77ACH(94)403	P. Lugosi, G. Doleschall, L. Parkanyi and A. Kalman; *Acta Chim. Acad. Sci. Hung.*, 1977, **94**, 403 (*Chem. Abstr.*, 1978, **88**, 152 474).	105
77ACS(A)271	J. Sletten and O. C. Loekke; *Acta Chem. Scand., Ser. A*, 1977, **31**, 271.	902
77ACS(A)292	L. K. Hansen and K. Tomren; *Acta Chem. Scand., Ser. A*, 1977, **31**, 292.	1054
77ACS(A)340	P. Groth; *Acta Chem. Scand., Ser. A*, 1977, **31**, 340.	671
77ACS(A)412	L. P. Darmo and L. K. Hansen; *Acta Chem. Scand., Ser. A*, 1977, **31**, 412.	1054
77ACS(A)423	J. Sletten and O. C. Loekke; *Acta Chem. Scand., Ser. A*, 1977, **31**, 423.	902
77ACS(B)167	E. Falch; *Acta Chem. Scand., Ser. B*, 1977, **31**, 167.	685, 702, 707
77ACS(B)184	J. Becher and J. P. Jacobsen; *Acta Chem. Scand., Ser. B*, 1977, **31**, 184.	93
77ACS(B)227	J. U. R. Nielsen, S. E. Jørgensen, N. Frederiksen, R. B. Jensen, G. Schroll and D. H. Williams; *Acta Chem. Scand., Ser. B*, 1977, **31**, 227.	756
77ACS(B)264	S. Treppendahl and P. Jakobsen; *Acta Chem. Scand., Ser. B*, 1977, **31**, 264.	569
77ACS(B)281	O. Simonsen, N. Loyaza and C. T. Pedersen; *Acta Chem. Scand., Ser. B*, 1977, **31**, 281.	811
77ACS(B)460	R. Perregaard, B. S. Pedersen and S. O. Lawesson; *Acta Chem. Scand., Ser. B.*, 1977, **31**, 460.	160
77ACS(B)683	C. Lohse and C. Th. Pedersen; *Acta Chem. Scand., Ser. B*, 1977, **31**, 683.	785, 1057
77ACS(B)687	A. Holm, C. Christophersen, T. Ottersen, H. Hope and A. Christensen; *Acta Chem. Scand., Ser. B*, 1977, **31**, 687.	584
77ACS(B)848	C. L. Pedersen, N. Harrit, M. Poliakoff and I. Dunkin; *Acta Chem. Scand., Ser. B*, 1977, **31**, 848.	355, 525
77ACS(B)899	K. Pihlaja and K. Rossi; *Acta Chem. Scand., Ser. B*, 1977, **31**, 899.	755

77ACS(B)919	G. A. Ulsaker, F. G. Evans and K. Undheim; *Acta Chem. Scand., Ser. B*, 1977, **31**, 919.	675
77AF760	D. B. Reisner, B. J. Ludwig, F. J. Stiefel, S. Gister, M. Meyer, L. S. Powell and R. D. Sofia; *Arzneim.-Forsch.*, 1977, **27**, 760.	631, 634
77AF766	D. B. Reisner, B. J. Ludwig, E. Simon, T. Dejneka and R. D. Sofia; *Arzneim.-Forsch.*, 1977, **27**, 766.	632, 634
77AG420	G. L'abbé, G. Verhelst and G. Vermeulen; *Angew. Chem.*, 1977, **89**, 420.	591
77AG(E)10	W. Oppolzer; *Angew. Chem., Int. Ed. Engl.*, 1977, **16**, 10.	1007
77AG(E)403	G. L'abbé, G. Verhelst and G. Vermeulen; *Angew. Chem., Int. Ed. Engl.*, 1977, **16**, 403.	1060, 1070
77AG(E)646	J. Varwig and R. Mews; *Angew. Chem., Int. Ed. Engl.*, 1977, **16**, 646.	906, 940
77AHC(21)175	R. Filler and Y. S. Rao; *Adv. Heterocycl. Chem.*, 1977, **21**, 175.	186, 199
77AHC(21)207	Y. Takeuchi and F. Furusaki; *Adv. Heterocycl. Chem.*, 1977, **21**, 207.	2, 9, 45, 46, 47, 108, 110, 111
77AHC(21)323	R. N. Butler; *Adv. Heterocycl. Chem.*, 1977, **21**, 323.	980
77AJC563	C. P. Joshua and K. N. Rajasekharan; *Aust. J. Chem*, 1977, **30**, 563.	467
77AJC569	E. J. Lloyd and Q. N. Porter; *Aust. J. Chem.*, 1977, **30**, 569.	864
77AJC1625	M. Davis, J. L. McVicars and T. G. Paproth; *Aust. J. Chem.*, 1977, **30**, 1625.	161
77AJC1815	M. Davis, M. C. Dereani, J. L. McVicars and I. J. Morris; *Aust. J. Chem.*, 1977, **30**, 1815.	160
77AJC1847	U. R. Kalkote and D. D. Goswami; *Aust. J. Chem.*, 1977, **30**, 1847.	117
77AJC1855	T. G. Burrowes, W. R. Jackson, S. Faulks and I. Sharp; *Aust. J. Chem.*, 1977, **30**, 1855.	38
77AJC2225	R. L. N. Harris and J. L. Huppatz; *Aust. J. Chem.*, 1977, **30**, 2225.	129, 445
77AP242	H. Böhme, K. H. Ahrens and E. Tippmann; *Arch. Pharm. (Weinheim, Ger.)*, 1977, **310**, 242.	231
77AP873	H. Stamm and H. Steudle; *Arch. Pharm. (Weinhein, Ger.)*, 1977, **310**, 873.	113
77AX(B)99	G. Argay, A. Kálmán, D. Lazar, R. Ribár and G. Tóth; *Acta Crystallogr., Part B*, 1977, **33**, 99 (*Chem. Abstr.*, 1977, **86**, 82 064).	238
77AX(B)3685	M. Calleri, L. Bonaccorti and D. Viterbo; *Acta Crystallogr., Part B*, 1977, **33**, 3685.	515
77BCJ3268	S. Inoue, N. Asai, G. Yasuda and T. Hori; *Bull. Chem. Soc. Jpn.*, 1977, **50**, 3268.	375
77BCJ3281	K. Oe, M. Tashiro and O. Tsuge; *Bull. Chem. Soc. Jpn.*, 1977, **50**, 3281.	435
77BEP853648	E.I. du Pont Nemours and Co.; *Belg. Pat.* 853 648 (1977) (*Chem. Abstr.*, 1978, **89**, 18 537).	161
77BRP1473807	H. Ukihashi, T. Hayashi and Y. Tahasaki; *Br. Pat.* 1 473 807 (1977) (*Chem. Abstr.*, 1977, **87**, 133 898).	893
77BSB95	J. Elguero, R. Faure, R. Lazaro and E. J. Vincent; *Bull. Soc. Chim. Belg.*, 1977, **86**, 95 (*Chem. Abstr.*, 1977, **86**, 170 261).	243
77BSB399	V. J. Ram and H. N. Pandey; *Bull. Soc. Chim. Belg.*, 1977, **86**, 399 (*Chem. Abstr.*, 1977, **87**, 135 208).	569
77BSB481	G. Klopman and P. Andreozzi; *Bull. Soc. Chim. Belg.*, 1977, **86**, 481.	853
77BSF1142	F. Ishii, M. Stavaux and N. Lozac'h; *Bull. Soc. Chim. Fr.*, 1977, 1142.	797
77CB285	J. Goerdeler, R. Buechler and S. Solyom; *Chem. Ber.*, 1977, **110**, 285.	910, 920, 941
77CB1225	G. Seybold and C. Heibl; *Chem. Ber.*, 1977, **110**, 1225.	454
77CB1421	U. Luhmann, F. G. Wentz and W. Lüttke; *Chem. Ber.*, 1977, **110**, 1421.	974, 1018, 1021
77CB2114	K. Burger, R. Ottlinger and J. Albanbauer; *Chem. Ber.*, 1977, **110**, 2114.	943
77CB3149	R. Neidlein, P. Leinberger, A. Gieren and B. Dederer; *Chem. Ber.*, 1977, **110**, 3149.	515, 540
77CB3205	R. Appel and J.-R. Lundehn; *Chem. Ber.*, 1977, **110**, 3205.	539
77CC7	T. Sakakibara and R. Sudoh; *J. Chem. Soc., Chem. Commun.*, 1977, 7.	103
77CC143	R. J. S. Beer and I. Hart; *J. Chem. Soc., Chem. Commun.*, 1977, 143.	591, 1070
77CC287	H. Bock, B. Solouki, S. Mohmand, E. Block and L. K. Revelle; *J. Chem. Soc., Chem. Commun.*, 1977, 287.	867, 872
77CC303	C. Belzecki and I. Panfil; *J. Chem. Soc., Chem. Commun.*, 1977, 303.	108, 109
77CC316	M. Fiorentino, L. Testaferri, M. Tiecco and L. Troisi; *J. Chem. Soc., Chem. Commun.*, 1977, 316.	265
77CC505	P. Shu, A. N. Bloch, T. F. Carruthers and D. O. Cowan; *J. Chem. Soc., Chem. Commun.*, 1977, 505.	969
77CC521	P. H. W. Lau and J. C. Martin; *J. Chem. Soc., Chem. Commun.*, 1977, 521.	763
77CC687	D. J. Sandman, A. P. Fisher, III, T. J. Holmes and A. J. Epstein; *J. Chem. Soc., Chem. Commun.*, 1977, 687.	834
77CC749	L. Lorenc, I. Juranic and M. Lj. Mihailovic; *J. Chem. Soc., Chem. Commun.*, 1977, 749.	46
77CC835	E. M. Engler, V. V. Patel and R. R. Schumaker; *J. Chem. Soc., Chem. Commun.*, 1977, 835.	963
77CC851	E. G. Frandsen, *J. Chem. Soc., Chem. Commun.*, 1977, 851.	1068
77CC856	H. Arai, A. Ohsawa, K. Saiki, H. Igeta, A. Tsuji and T. Akimoto; *J. Chem. Soc., Chem. Commun.*, 1977, 856.	82
77CCC811	S. Nešpurek and M. Šorm; *Collect. Czech. Chem. Commun.*, 1977, **42**, 811.	375
77CCC1557	A. Martvoň, M. Uher, Š. Stankovský and J. Surá; *Collect. Czech. Chem. Commun.*, 1977, **42**, 1557.	594
77CCC2060	P. Zahradnik and J. Leska; *Collect. Czech. Chem. Commun.*, 1977, **42**, 2060.	516
77CCC2672	M. Zbirovský and R. Seifert; *Collect. Czech. Chem. Commun.*, 1977, **42**, 2672.	905, 909, 911, 912, 927

Ref	Citation	Pages
77CCC2680	M. Zbirovský, R. Seifert and Š. Truchlik; *Collect. Czech. Chem. Commun.*, 1977, **42**, 2680.	944
77CCC2945	L. Floch, A. Martvoň, M. Uher, J. Leško and W. Weis; *Collect. Czech. Chem. Commun.*, 1977, **42**, 2945.	594
77CHE524	A. A. Akhrem, F. A. Lakhvich, V. A. Khripack, I. B. Klebaovich and A. G. Pozdeev; *Chem. Heterocycl. Compd. (Engl. Transl.)*, 1977, **12**, 524.	997
77CHE954	Y. B. Vysotskii, N. A. Kovach and O. P. Shvaika; *Chem. Heterocycl. Compd. (Engl. Transl.)*, 1977, **13**, 954.	428
77CJC619	R. E. Wasylishen and H. M. Hutton; *Can. J. Chem.*, 1977, **55**, 619.	133, 139
77CJC1123	D. M. McKinnon, M. E. R. Hassan and M. Chauhan; *Can. J. Chem.*, 1977, **55**, 1123.	151, 152
77CJC1728	R. Faure, J. P. Galy, E. J. Vincent, J. P. Fayet, P. Mauret, M. C. Vertut and J. Elguero; *Can. J. Chem.*, 1977, **55**, 1728.	249, 978
77CJC2302	G. Mille, M. Guiliano and J. Chouteau; *Can. J. Chem.*, 1977, **55**, 2302.	141
77CJC2363	M. S. Chauhan, D. M. McKinnon and R. G. Cooke; *Can. J. Chem.*, 1977, **55**, 2363.	935
77CJC3736	D. Rousseau and A. Taurins; *Can. J. Chem.*, 1977, **55**, 3736.	713, 716, 717, 723
77CJC3763	P. deMayo and H. Y. Ng; *Can. J. Chem.*, 1977, **55**, 3763.	799
77CL77	J. Nakayama, M. Ishihara and M. Hoshino; *Chem. Lett.*, 1977, 77.	823
77CL245	T. Takeda, M. Ueda and T. Mukaiyama; *Chem. Lett.*, 1977, 245.	436, 443
77CL287	J. Nakayama, M. Ishihara and M. Hoshino; *Chem. Lett.*, 1977, 287.	836
77CL1133	K. Sakamoto, N. Nakamura, M. Oki, J. Nakayama and M. Hoshino; *Chem. Lett.*, 1977, 1133.	817
77CL1195	O. Seshimoto, T. Kumagai, K. Shimizu and T. Mukai; *Chem. Lett.*, 1977, 1195.	37
77CL1207	O. Tsuge, K. Oe, and M. Tashiro; *Chem. Lett.*, 1977, 1207.	433
77CL1237	H. Shimoharada, S. Ikeda, S. Kajigaeshi and S. Kanemasa; *Chem. Lett.*, 1977, 1237.	1040, 1047
77CPB299	K. Arakawa, T. Miyasaka and K. Satoh; *Chem. Pharm. Bull.*, 1977, **25**, 299.	673, 694
77CPB491	T. Kato, N. Oda and I. Ito; *Chem. Pharm. Bull.*, 1977, **25**, 491.	660
77CPB1471	K. Masuda, J. Adachi and K. Nomura; *Chem. Pharm. Bull.*, 1977, **25**, 1471.	759, 765, 772
77CPB2790	S. Nishigaki, K. Shimizu and K. Senga; *Chem. Pharm. Bull.*, 1977, **25**, 2790.	734
77CPB2974	R. Marumoto and Y. Furukawa; *Chem. Pharm. Bull.*, 1977, **25**, 2974.	621, 625, 629
77CR(C)(284)73	M. Hamdi and V. Herault; *C.R. Hebd. Seances Acad. Sci., Ser. C*, 1977, **284**, 73 (*Chem. Abstr.*, 1977, **86**, 189 661).	660
77CR(C)(285)257	C. Rocheville-Divorne and J. P. Roggero; *C.R. Hebd. Seances Acad. Sci., Ser. C*, 1977, **285**, 257 (*Chem. Abstr.*, 1978, **88**, 22 733).	308
77CSC553	G. Malmros and A. Wagner; *Cryst. Struct. Commun.*, 1977, **6**, 553.	646
77CZ35	R. Neidlein and P. Leinberger; *Chem.-Ztg.*, 1977, **101**, 35.	540
77CZ154	G. Zinner and U. Krueger; *Chem.-Ztg.*, 1977, **101**, 154.	390
77CZ302	G. Zinner and T. Krause; *Chem.-Ztg.*, 1977, **101**, 302.	444
77DIS(B)(38)1721	A. B. Nelson; *Diss. Abstr. Int. B*, 1977, **38**, 1721.	377
77EGP124044	E. Hoyer, G. Steimecke, C. Schroeter, G. Fischer and W. Walther; *Ger. (East) Pat.* 124 044 (*Chem. Abstr.*, 1978, **88**, 50 836).	808
77EGP126401	J. Liebscher and H. Hartmann; *Ger. (East) Pat.* 126 401 (1977) (*Chem. Abstr.*, 1978, **88**, 62 394).	942
77FRP2315922	Roussel-UCLAF, *Fr. Pat.* 2 315 922 (1977) (*Chem. Abstr.*, 1977, **87**, 184 512).	945
77G1	L. Pentimalli, G. Milani, F. Biavati; *Gazz. Chim. Ital.*, 1977, **107**, 1.	480
77G283	B. F. Bonini, G. Maccagnani, G. Mazzanti, P. Pedrini and B. Zwanenburg; *Gazz. Chim. Ital.*, 1977, **107**, 283.	905, 941
77GEP2533605	Bayer A.-G., *Ger. Pat.* 2 533 605 (1977) (*Chem. Abstr.*, 1977, **86**, 171 462).	577
77GEP2606788	V. E. Privalov, E. I. Vail, A. M. Chanin, E. I. Gromov, V. M. Petropol'skaya and L. M. Agarkova; *Ger. Pat.* 2 606 788 (1977) (*Chem. Abstr.*, 1978, **88**, 125 205).	946
77GEP2625026	W. Sauerteig, W. Himmelmann, R. Meyer, E. Ranz and W. Pelz; *Ger. Offen.* 2 625 026 (1977) (*Chem. Abstr.*, 1978, **88**, 129 015).	130
77GEP2638029	J. Crosby; *Ger. Pat.* 2 638 029 (1977) (*Chem. Abstr.*, 1977, **86**, 156 824).	945
77GEP2639189	C. B. C. Boyce and S. B. Webb; *Ger. Offen.* 2 639 189 (1977) (*Chem. Abstr.*, 1977, **87**, 23 258).	129
77GEP2640652	I. Adacli, M. Ueda and S. Kimoto; *Ger. Offen.* 2 640 652 (1977) (*Chem. Abstr.*, 1977, **87**, 39 460); *Swiss Pat.* 605 856 (1978) (*Chem. Abstr.*, 1979, **90**, 87 436).	129
77GEP2653800	M. Sugiyama and A. Ogawa; *Ger. Offen.* 2 653 800 (1977) (*Chem. Abstr.*, 1977, **87**, 93 521).	130
77GEP2701853	*Ger. Pat.* 2 701 853 (1977) (*Chem. Abstr.*, 1977, **87**, 135 310).	709
77GEP2718707	*Ger. Pat.* 2 718 707 (1977) (*Chem. Abstr.*, 1978, **88**, 50 843).	644
77GEP2723688	J. D. Davenport, B. A. Driekorn and A. F. Elsasser; *Ger. Offen.* 2 723 688 (1977) (*Chem. Abstr.*, 1978, **88**, 132 015).	128
77H(6)107	N. Vivona and G. Cusmano; *Heterocycles*, 1977, **6**, 107.	384, 720, 724
77H(6)143	P. Gilgen, H. Heimgartner and H. Schmid; *Heterocycles*, 1977, **6**, 143.	223
77H(6)933	M. Tashiro, S. Mataka and K. Takahashi; *Heterocycles*, 1977, **6**, 933.	541
77H(6)1173	O. Tsuge, T. Takata and M. Noguchi; *Heterocycles*, 1977, **6**, 1173.	415, 417, 1032, 1035, 1044
77H(6)1599	T. Mukai, H. Saiki, T. Miyashi and Y. Ikegami; *Heterocycles*, 1977, **6**, 1599.	107
77H(6)1919	K. Senga, J. Sato, K. Shimizu and S. Nishigaki; *Heterocycles*, 1977, **6**, 1919.	660

Ref	Citation	Pages
77H(6)1925	K. Senga, Y. Kanamori and S. Nishigaki; *Heterocycles*, 1977, **6**, 1925.	660
77H(6)1985	T. Honma and Y. Tada; *Heterocycles*, 1977, **6**, 1985.	157, 536
77H(7)51	H. Yamanaka, T. Sakamoto and A. Shiozawa; *Heterocycles*, 1977, **7**, 51.	26, 54, 619, 627
77H(7)73	K. L. Turner and S. Turner; *Heterocycles*, 1977, **7**, 73.	439
77H(7)201	J. Palmer, P. S. Rutledge and P. D. Woodgate; *Heterocycles*, 1977, **7**, 201.	6
77H(7)241	C. Kashima, N. Mukai, Y. Yamamoto, Y. Tsuda and Y. Omote; *Heterocycles*, 1977, **7**, 241.	50, 51
77H(7)247	D. J. Woodman, W. H. Campbell and E. F. DeRose; *Heterocycles*, 1977, **7**, 247.	42
77H(8)387	Y. Kobayashi, I. Kumadaki and T. Yoshida; *Heterocycles*, 1977, **8**, 387.	44, 99
77H(8)461	H. Abe, T. Takaishi, Y. Ito and T. Okuda; *Heterocycles*, 1977, **8**, 461.	708
77HC(30)1	J. A. Paolini; *Chem. Heterocycl. Compd.*, 1977, **30**, 1.	974, 980, 981, 992
77HC(30)271	R. D. Hamilton and E. Campaigne; *Chem. Heterocycl. Compd.*, 1977, **30**, 271.	814
77HC(30)317	K. T. Potts; *Chem. Heterocycl. Compd.*, 1977, **30**, 317.	944, 1028, 1031, 1032, 1036, 1043, 1044
77HCA215	E. Davin-Pretelli, M. Guiliano, G. Mille, J. Chouteau, R. Guglielmetti and C. Gelebart; *Helv. Chim. Acta*, 1977, **60**, 215 (*Chem. Abstr.*, 1977, **86**, 154 763).	242
77HCA426	A. Vasella; *Helv. Chim. Acta*, 1977, **60**, 426.	108
77HCA1087	H. Meier, H. Heimgartner and H. Schmid; *Helv. Chim. Acta*, 1977, **60**, 1087.	376
77IJC(B)133	V. P. Arya, R. S. Grewal, C. L. Kaul, J. David and V. Honkan; *Indian J. Chem., Sect. B*, 1977, **15**, 133 (*Chem. Abstr.*, 1977, **87**, 201 415).	314, 941, 945
77IJC(B)473	V. P. Arya, J. David, R. S. Grewal, C. L. Kaul, R. H. Mizzoni, S. Rajappa and S. J. Shenoy; *Indian J. Chem., Sect. B*, 1977, **15**, 473.	156
77IJC(B)490	L. Azhakumoni, C. P. Joshua and K. N. Rajasekharan; *Indian J. Chem., Sect. B*, 1977, **15**, 490.	465
77IJC(B)499	R. Mukherjee and M. M. Moriarty; *Indian J. Chem., Sect. B*, 1977, **15**, 499 (*Chem. Abstr.*, 1978, **88**, 37 707).	573
77IJC(B)848	M. Rai, K. Krishnan and A. Singh; *Indian J. Chem., Sect. B*, 1977, **15**, 848.	389
77IJC(B)886	S. Rajappa, G. B. Advani and R. Sreenivasan; *Indian J. Chem., Sect. B*, 1977, **15**, 886.	149, 639
77IJC(B)1051	K. A. Thakar and A. B. Dumir; *Indian J. Chem., Sect. B*, 1977, **15**, 1051.	129
77IJC(B)1056	K. A. Thakar and B. M. Bhawal; *Indian J. Chem., Sect. B*, 1977, **15**, 1056.	48
77IJC(B)1058	K. A. Thakar, D. D. Goswami and B. M. Bhawal; *Indian J. Chem., Sect. B*, 1977, **15**, 1058.	48, 116
77IJC(B)1061	K. A. Thakar and B. M. Bhawal; *Indian J. Chem., Sect. B*, 1977, **15**, 1061.	23, 48
77IZV98	E. N. Klimovitskii, L. K. Yuldasheva, A. N. Vereshchagin, G. N. Sergeeva, S. G. Vul'fson and B. A. Arbuzov; *Izv. Akad. Nauk SSSR, Ser. Khim.*, 1977, 98.	865
77IZV211	V. M. Shitkin, I. E. Chlenov and V. A. Tartakovskii; *Izv. Akad. Nauk SSSR, Ser. Khim.*, 1977, 211 (*Chem. Abstr.*, 1977, **86**, 188 997).	633
77IZV716	R. G. Kostyanovskii and V. F. Rudchenko; *Izv. Akad. Nauk SSSR, Ser. Khim.*, 1977, 716 (*Chem. Abstr.*, 1977, **87**, 39 347).	110
77IZV1453	B. A. Arbuzov and E. N. Dianova; *Izv. Akad. Nauk SSSR, Ser. Khim.*, 1977, 1453.	983
77IZV1687	R. G. Kostyanovskii, V. F. Rudchenko and G. V. Shustov; *Izv. Akad. Nauk SSSR, Ser. Khim.*, 1977, 1687 (*Chem. Abstr.*, 1977, **87**, 167 427).	10
77IZV2266	V. M. Shitkin, S. L. Ioffe, M. V. Kashutina and V. A. Tartakovskii; *Izv. Akad. Nauk SSSR, Ser. Khim.*, 1977, 2266 (*Chem. Abstr.*, 1978, **88**, 50 151).	10
77JA255	J. Meinwald, D. Dauplaise, F. Wudl and J. J. Hauser; *J. Am. Chem. Soc.*, 1977, **99**, 255.	948, 952, 955, 956
77JA1214	C. G. Krespan, B. E. Smart and E. G. Howard; *J. Am. Chem. Soc.*, 1977, **99**, 1214.	875, 892
77JA1645	J. M. Liesch and K. L. Rinehart, Jr.; *J. Am. Chem. Soc.*, 1977, **99**, 1645.	709
77JA1663	H. Bock, B. Solouki, G. Bert and P. Rosmus; *J. Am. Chem. Soc.*, 1977, **99**, 1663.	454
77JA2855	J. Spanget-Larsen, R. Gleiter, M. Kobayashi, E. M. Engler, P. Shu and D. O. Cowan; *J. Am. Chem. Soc.*, 1977, **99**, 2855.	952, 953
77JA4647	A. M. Nadzan and K. L. Rinehart, Jr.; *J. Am. Chem. Soc.*, 1977, **99**, 4647.	651
77JA4842	A. Krantz and J. Laureni; *J. Am. Chem. Soc.*, 1977, **99**, 4842.	350, 455
77JA5909	E. M. Engler, B. A. Scott, S. Etemad, T. Penney and V. V. Patel; *J. Am. Chem. Soc.*, 1977, **99**, 5909.	963
77JA6667	R. Gree and R. Carrie; *J. Am. Chem. Soc.*, 1977, **99**, 6667.	100
77JA6754	M. Marx, F. Marti, J. Reisdorff, R. Sandmeier and S. Clark; *J. Am. Chem. Soc.*, 1977, **99**, 6754.	423, 985
77JA7239	C. W. Gillies; *J. Am. Chem. Soc.*, 1977, **99**, 7329.	892
77JA7363	G. Barany and R. B. Merrifield; *J. Am. Chem. Soc.*, 1977, **99**, 7363.	924, 937
77JA7743	J. Meinwald, D. Dauplaise and J. Clardy; *J. Am. Chem. Soc.*, 1977, **99**, 7743.	951, 955, 956, 969, 971
77JAP77148076	K. Agihara, M. Mizuno and A. Nakata; *Jpn. Pat.* 77 148 076 (1977) (*Chem. Abstr.*, 1978, **88**, 152 625).	944
77JAP(K)7725028	Inst. Phys. Chem. Res., *Jpn. Kokai* 77 25 028 (1977) (*Chem. Abstr.*, 1977, **87**, 147 054).	577
77JAP(K)7731070	I. Adachi, M. Ueda and S. Kimoto; *Jpn. Kokai* 77 31 070 (1977) (*Chem. Abstr.*, 1978, **88**, 62 379).	129
77JAP(K)7736663	Nippon Soda Co. Ltd.; *Jpn. Kokai* 77 36 663 (*Chem. Abstr.*, 1977, **87**, 152 177).	160
77JCR(M)2813	D. Bartholomew and I. T. Kay; *J. Chem. Res. (M)*, 1977, 2813.	609, 612
77JCR(M)2826	D. Bartholomew and I. T. Kay; *J. Chem. Res. (M)*, 1977, 2826.	608
77JCR(S)238	D. Bartholomew and I. T. Kay; *J. Chem. Res. (S)*, 1977, 238.	609, 612

77JCR(S)239	D. Bartholomew and I. T. Kay; *J. Chem. Res. (S)*, 1977, 239.	608
77JCS(P1)239	M. Maeda and M. Kojima; *J. Chem. Soc., Perkin Trans. 1*, 1977, 239.	189
77JCS(P1)848	R. M. Christie and D. H. Reid; *J. Chem. Soc., Perkin Trans. 1*, 1977, 848.	1059, 1060
77JCS(P1)854	D. H. Reid and R. G. Webster; *J. Chem. Soc., Perkin Trans. 1*, 1977, 854.	1060, 1061
77JCS(P1)916	D. H. R. Barton and W. A. Bubb; *J. Chem. Soc., Perkin Trans. 1*, 1977, 916.	169, 541
77JCS(P1)971	G. Adembri, A. Camparini, F. Ponticelli and P. Tedeschi; *J. Chem. Soc., Perkin Trans. 1*, 1977, 971.	13
77JCS(P1)994	C. Th. Pedersen and C. Lohse; *J. Chem. Soc., Perkin Trans. 1*, 1977, 994.	809
77JCS(P1)1196	R. Faragher and T. L. Gilchrist; *J. Chem. Soc., Perkin Trans. 1*, 1977, 1196.	73, 77
77JCS(P1)1273	M. Muraoko, T. Yamamoto, S. Yamaguchi, F. Tonosaki, T. Takeshima and N. Fukada; *J. Chem. Soc., Perkin Trans. 1*, 1977, 1273.	890
77JCS(P1)1468	B. F. Bonini, G. Maccagnani, G. Mazzanti, H. P. M. M. Ambrosius and B. Zwanenburg; *J. Chem. Soc., Perkin Trans. 1*, 1977, 1468.	934, 941
77JCS(P1)1511	E. I. G. Brown, D. Leaver and D. M. McKinnon; *J. Chem. Soc., Perkin Trans. 1*, 1977, 1511.	799
77JCS(P1)1612	G. V. Boyd, T. Norris, P. F. Lindley and M. M. Mahmoud; *J. Chem. Soc., Perkin Trans. 1*, 1977, 1612.	1005
77JCS(P1)1616	N. Vivona, G. Cusmano and G. Macaluso; *J. Chem. Soc., Perkin Trans. 1*, 1977, 1616.	54, 60, 403, 413, 417, 504
77JCS(P1)1791	S. Crook and P. Sykes; *J. Chem. Soc., Perkin Trans. 1*, 1977, 1791.	470, 471, 484
77JCS(P1)2154	A. Corsaro, U. Chiacchio and G. Purrello; *J. Chem. Soc., Perkin Trans. 1*, 1977, 2154.	70
77JCS(P1)2222	G. Bianchi, C. De Micheli, A. Gamba, R. Gandolfi and B. Rezzani; *J. Chem. Soc., Perkin Trans. 1*, 1977, 2222.	89, 90
77JCS(P2)47	G. Bianchi, L. Casotti, D. Passadore and N. Stabile; *J. Chem. Soc., Perkin Trans. 2*, 1977, 47.	11, 23, 48
77JCS(P2)343	R. Keskinen, A. Nikkilä and K. Pihlaja; *J. Chem. Soc., Perkin Trans. 2*, 1977, 343.	758
77JCS(P2)561	C. A. Veracini, A. De Munno, V. Bertini, M. Longeri and G. Chidichimo; *J. Chem. Soc., Perkin Trans. 2*, 1977, 561.	517
77JCS(P2)724	M. J. S. Dewar and I. J. Turchi; *J. Chem. Soc., Perkin Trans. 2*, 1977, 724.	188
77JCS(P2)1015	E. Gentric, J. Lauransan, C. Roussel and J. Metzger; *J. Chem. Soc., Perkin Trans. 2*, 1977, 1015.	248
77JCS(P2)1114	T. Vitali, E. Gaetani, F. Ronchini, M. Nardelli and G. Pelizzi; *J. Chem. Soc., Perkin Trans. 2*, 1977, 1114.	155, 167, 170
77JCS(P2)1121	A. Demunno, V. Bertini and F. Lucchesini; *J. Chem. Soc., Perkin Trans. 2*, 1977, 1121.	29
77JCS(P2)1161	C. Gaze and B. C. Gilbert; *J. Chem. Soc., Perkin Trans. 2*, 1977, 1161.	764
77JCS(P2)1332	J. L. McVicars, M. F. Mackay and M. Davis; *J. Chem. Soc., Perkin Trans. 2*, 1977, 1332.	135
77JCS(P2)1854	P.-T. Cheng and S. C. Nyburg; *J. Chem. Soc., Perkin Trans. 2*, 1977, 1854.	785
77JHC37	P. Bravo and G. Gaviraghi; *J. Heterocycl. Chem.*, 1977, **14**, 37.	49
77JHC129	H. W. Altland and G. A. Molander; *J. Heterocycl. Chem.*, 1977, **14**, 129.	698
77JHC181	F. DeSarlo, R. Cencioni, G. Renzi and L. Bausi; *J. Heterocycl. Chem.*, 1977, **14**, 181.	104
77JHC227	M. H. Elnagdi, M. R. H. Emoghayar, E. M. Kandeel and M. K. A. Ibrahim; *J. Heterocycl. Chem.*, 1977, **14**, 227.	1009
77JHC253	J. P. Gilbert, C. Petrus and F. Petrus; *J. Heterocycl. Chem.*, 1977, **14**, 253.	998
77JHC317	M. Davis, R. Lakhan and B. Ternai; *J. Heterocycl. Chem.*, 1977, **14**, 317.	219
77JHC345	Y.-i Lin and S. A. Lang, Jr.; *J. Heterocycl. Chem.*, 1977, **14**, 345.	63, 83
77JHC401	A. M. Kirkien-Rzeszotarski and W. J. Rzeszotarski; *J. Heterocycl. Chem.*, 1977, **14**, 401.	552
77JHC435	A. Camparini, F. Ponticelli and P. Tedeschi; *J. Heterocycl. Chem.*, 1977, **14**, 4359, 620, 627
77JHC511	K. Okada, J. A. Kelley and J. S. Driscoll; *J. Heterocycl. Chem.*, 1977, **14**, 511.	1013
77JHC515	G. L'abbe, G. Verhelst, L. Huybrechts and S. Toppet; *J. Heterocycl. Chem.*, 1977, **14**, 515.	550, 553
77JHC523	T. Kurihara, M. Mori and Y. Sakamoto; *J. Heterocycl. Chem.*, 1977, **14**, 523.	39, 93
77JHC531	L. S. Crawley and W. J. Fanshawe; *J. Heterocycl. Chem.*, 1977, **14**, 531.	74
77JHC567	A. Shafiee, I. Lalezari, M. Mirrashed and D. Nercesian; *J. Heterocycl. Chem.*, 1977, **14**, 567.	349, 352
77JHC607	D. C. H. Bigg and S. R. Purvis; *J. Heterocycl. Chem.*, 1977, **14**, 607.	984
77JHC621	B. Koren, B. Stanovnik and M. Tišler; *J. Heterocycl. Chem.*, 1977, **14**, 621.	733, 741
77JHC627	E. D. Weiler, R. B. Petigara, M. H. Wolfersberger and G. A. Miller; *J. Heterocycl. Chem.*, 1977, **14**, 627.	147
77JHC725	E. D. Weiler, M. Hausman and G. A. Miller; *J. Heterocycl. Chem.*, 1977, **14**, 725.	144, 156, 157, 158, 159, 162
77JHC745	S. I. Pennanen; *J. Heterocycl. Chem.*, 1977, **14**, 745.	349
77JHC823	G. Werber, F. Buccheri and M. Gentile; *J. Heterocycl. Chem.*, 1977, **14**, 823.	561
77JHC853	G. Werber, F. Buccheri, M. Gentile and L. Librici; *J. Heterocycl. Chem.*, 1977, **14**, 853.	569
77JHC951	D. Donati, M. Fiorenza, E. Moschi and P. Sarti-Fantoni; *J. Heterocycl. Chem.*, 1977, **14**, 951.	141
77JHC963	S. Mataka, K. Takahashi and M. Tashiro; *J. Heterocycl. Chem.*, 1977, **14**, 963.	541
77JHC1035	K. Pilgrim and M. Zupan; *J. Heterocycl. Chem.*, 1977, **14**, 1035.	753

77JHC1045	A. Petrič, B. Stanovnik and M. Tišler; *J. Heterocycl. Chem.*, 1977, **14**, 1045.	686, 690, 698
77JHC1063	H. Zinnes, R. A. Comes and J. Shavel, Jr., *J. Heterocycl. Chem.*, 1977, **14**, 1063.	170
77JHC1263	G. Werber, F. Buccheri and M. Gentile; *J. Heterocycl. Chem.*, 1977, **14**, 1263.	565
77JHC1289	J. F. Hansen and S. A. Strong; *J. Heterocycl. Chem.*, 1977, **14**, 1289.	75, 94
77JHC1385	G. Werber, F. Buccheri, R. Noto and M. Gentile; *J. Heterocycl. Chem.*, 1977, **14**, 1385.	439, 441
77JHC1417	G. L'abbé, S. Toppet, A. Willocx and G. Mathys; *J. Heterocycl. Chem.*, 1977, **14**, 1417.	583, 596
77JIC536	N. N. Ghosh, G. N. Mukhopadhyay and P. C. Chaudhury; *J. Indian Chem. Soc.*, 1977, **54**, 536.	20
77JIC875	K. A. Thakar and B. M. Bhawal; *J. Indian Chem. Soc.*, 1977, **54**, 875.	48, 116, 128
77JIC1143	H. Singh and L. D. S. Yadav; *J. Indian Chem. Soc.*, 1977, **54**, 1143.	445
77JMC563	K. K. Bhargava, M. H. Lee, Y.-M. Huang, L. S. Cunningham, K. C. Agrawal and A. C. Sartorelli; *J. Med. Chem.*, 1977, **20**, 563.	993
77JMC934	J. B. Carr, H. G. Durham and D. K. Hass; *J. Med. Chem.*, 1977, **20**, 934.	25, 55, 58, 74, 86, 87, 88, 128
77JMC965	J. W. Hines, Jr. and C. H. Stammer; *J. Med. Chem.*, 1977, **20**, 965.	74
77JMC1572	B. Blank, N. W. Ditullio, L. Deviney, J. T. Roberts, A. Magnani, M. Billig and H. L. Saunders; *J. Med. Chem.*, 1977, **20**, 1572.	699
77JMR(27)509	W. B. Gara, B. P. Roberts, C. M. Kirk, B. C. Gilbert and R. O. C. Norman; *J. Magn. Reson.*, 1977, **27**, 509.	881
77JOC338	I. Lalezari, A. Shafiee and M. Yalpani; *J. Org. Chem.*, 1977, **38**, 338.	350
77JOC575	L. Benati, P. C. Montevecchi and G. Zanardi; *J. Org. Chem.*, 1977, **42**, 575.	453, 454
77JOC897	S. N. Balasubrahmanyam, A. S. Radhakrishna, A. J. Boulton and T. Kan-Woon; *J. Org. Chem.*, 1977, **42**, 897.	48, 407
77JOC1015	J. A. Deyrup and H. L. Gingrich; *J. Org. Chem.*, 1977, **42**, 1015.	539, 906
77JOC1035	J. D. Mee; *J. Org. Chem.*, 1977, **42**, 1035.	281
77JOC1159	G. L'abbé, G. Verhelst and S. Toppet; *J. Org. Chem.*, 1977, **42**, 1159.	588, 589, 935
77JOC1265	K. Rasheed and J. D. Warkentin; *J. Org. Chem.*, 1977, **42**, 1265.	846
77JOC1356	R. Bengelmans and C. Morin; *J. Org. Chem.*, 1977, **42**, 1356.	79
77JOC1364	D. T. Connor, P. A. Young and M. von Strandtmann; *J. Org. Chem.*, 1977, **42**, 1364.	619, 620, 625
77JOC1543	K. Hirai, H. Sugimoto and T. Ishiba; *J. Org. Chem.*, 1977, **42**, 1543.	822
77JOC1555	R. M. Srivastava and I. M. Brinn; *J. Org. Chem.*, 1977, **42**, 1555.	378, 379, 389, 391
77JOC1633	K. T. Potts, S. J. Chen, J. Kane and J. L. Marshall; *J. Org. Chem.*, 1977, **42**, 1633.	841, 847
77JOC1644	K. T. Potts, F. Huang and R. K. Khattak; *J. Org. Chem.*, 1977, **42**, 1644.	344
77JOC1648	K. T. Potts and D. R. Choudhury; *J. Org. Chem.*, 1977, **42**, 1648.	1002
77JOC1791	R. Y. Ning, J. F. Blaunt, P. B. Madan and R. I. Fryer; *J. Org. Chem.*, 1977, **42**, 1791.	120
77JOC1813	R. K. Howe, T. A. Gruner and J. E. Franz; *J. Org. Chem.*, 1977, **42**, 1813.	500, 506, 913
77JOC2237	G. A. Olah and J. L. Grant; *J. Org. Chem.*, 1977, **42**, 2237.	752, 754, 755, 817
77JOC2525	K. T. Potts, S. J. Chen and J. Szmuszkovicz; *J. Org. Chem.*, 1977, **42**, 2525.	693
77JOC2778	A. Kruger and F. Wudl; *J. Org. Chem.*, 1977, **42**, 2778.	824
77JOC2891	J. E. Ellis, J. H. Fried, I. T. Harrison, E. Rapp and C. H. Ross; *J. Org. Chem.*, 1977, **42**, 2891.	807, 810, 1019
77JOC3725	J. R. Bartels-Keith, M. T. Burgess and J. M. Stevenson; *J. Org. Chem.*, 1977, **42**, 3725.	340, 356, 358, 549, 550, 582, 583
77JOC3925	E. P. Papadopoulos; *J. Org. Chem.*, 1977, **42**, 3925.	991
77JOC3929	N. F. Haley; *J. Org. Chem.*, 1977, **42**, 3929.	20, 21, 51
77JOM(136)139	S. Cabiddu, A. Maccioni, P. P. Piras and M. Secci; *J. Organomet. Chem.*, 1977, **136**, 139.	766
77JOM(137)C37	H. Lavayssiere, G. Dousse and J. Satge; *J. Organomet. Chem.*, 1977, **137**, C37.	905, 941
77JPR875	E. Fanghänel and G. Lutze; *J. Prakt. Chem.*, 1977, **319**, 875.	835
77JPS772	N. D. Heindel, W. P. Fives and R. A. Carrano; *J. Pharm. Sci.*, 1977, **66**, 772 (*Chem. Abstr.*, 1977, **87**, 135 207).	945
77JSP(68)169	B. Bak, O. J. Nielsen and H. Svanholt; *J. Mol. Spectrosc.*, 1977, **68**, 169.	913
77JST(39)189	M. H. Palmer, R. H. Findlay, J. N. A. Ridyard, A. Barrie and P. Swift; *J. Mol. Struct.*, 1977, **39**, 189.	449, 467, 517
77JST(40)191	M. H. Palmer, R. H. Findlay and R. G. Egdell; *J. Mol. Struct.*, 1977, **40**, 191.	5, 183, 184, 399, 517
77KGS30	I. V. Vigalok, A. V. Ostrovskaya, N. V. Svetlakov, T. V. Zykova and N. A. Zhikhareva; *Khim. Geterotsikl. Soedin.*, 1977, 30.	412
77KGS849	Y. I. Akulin, B. K. Strelets and L. S. Efros; *Khim. Geterotsikl. Soedin.*, 1977, 849 (*Chem. Abstr.*, 1977, **87**, 184 436).	899, 916, 940
77KGS1110	N. N. Zatsepina, I. F. Tupitsyn, A. I. Belyashova, A. A. Kane, N. S. Kolodina and G. N. Sudakova; *Khim. Geterotsikl. Soedin.*, 1977, 1110.	395
77KGS1499	Yu. I. Akulin, B. K. Strelets and L. S. Efros; *Khim. Geterotsikl. Soedin.*, 1977, 1499 (*Chem. Abstr.*, 1978, **88**, 89 587).	359, 940
77LA159	E. Regel; *Liebigs Ann. Chem.*, 1977, 159.	432
77LA1005	J. Liebscher and H. Hartmann; *Liebigs Ann. Chem.*, 1977, 1005.	907, 908, 917
77LA1347	G. Trickes, H. P. Braun and H. Meir; *Liebigs Ann. Chem.*, 1977, 1347.	455, 457
77LA1888	H. Meyer, F. Bossert and H. Horstmann; *Liebigs Ann. Chem.*, 1977, 1888.	706
77M665	H. Paul, A. Sitte and R. Wessel; *Monatsh. Chem*; 1977, **108**, 665.	560

77MI41600	H. Oka and K. Tomita; *Sankyo Kenkyusko Nempo*, 1977, **29**, 99 (*Chem. Abstr.*, 1978, **88**, 170 017).	56
77MI41601	H. Masago, M. Yoshikawa, M. Fukada and N. Nakanishi; *Phytopathology*, 1977, **67**, 425.	129
77MI41602	G. Devoto, M. Massacesi, R. Pinna and G. Ponticelli; *Transition Met. Chem.*, 1977, **2**, 236.	20
77MI41603	M. Biddau, M. Massacesi, R. Pinna and G. Ponticelli; *Transition Met. Chem.*, 1977, **2**, 5.	20
77MI41604	G. Devoto, M. Massacesi, G. Ponticelli and R. Ruggeri; *J. Inorg. Nucl. Chem.*, 1977, **39**, 355.	20
77MI41605	G. Devoto, M. Massacesi, G. Ponticelli and C. Preti; *J. Inorg. Nucl. Chem.*, 1977, **39**, 271.	20
77MI41606	G. Ponticelli; *J. Inorg. Nucl. Chem.*, 1977, **39**, 45.	20
77MI41607	W. L. Driessen and P. H. van der Voort; *Inorg. Chim. Acta*, 1977, **21**, 217.	20
77MI41608	N. H. Lauersen and K. H. Wilson; *Obstet. Gynecol.*, 1977, **50**, 91.	129
77MI41609	R. Kadlubowski, A. Ochecka and W. Basinski; *Wiad. Parazytol.*, 1977, **23**, 637 (*Chem. Abstr.*, 1978, **88**, 183 767).	20, 128
77MI41610	J. Suwinski; *Zesz. Nauk. Politech. Slask. Chem.*, 1977, **82**, 3 (*Chem. Abstr.*, 1972, **90**, 72 132).	93
77MI41611	K. H. Grellmann and E. Tauer; *J. Photochem.*, 1977, **6**, 365.	17, 18
B-77MI41612	R. R. Bekmukhametov; in 'Tenzisy Dokl.-Resp. Konf. Molodykh Uch.-Khim., 2nd', 1977, vol. 1, p. 49 (*Chem. Abstr.*, 1978, **89**, 162 830).	15
77MI41613	S. L. Srivastava; *Indian J. Phys., Sect. B*, 1977, **51**, 1.	8
77MI41614	R. Vilceu, R. Lazar, F. Irinei and Z. Costea; *Bull. Inst. Politeh. "Gheorghe Gheorghilu-Dej" Bucuresti, Ser. Chim.-Met.*, 1977, **39**, 13 (*Chem. Abstr.*, 1978, **88**, 55 675).	7
77MI41615	O. G. Grushina, I. I. Furlei and V. I. Khvostenko; *Teor. Eksp. Khim.*, 1977, **13**, 534 (*Chem. Abstr.*, 1977, **87**, 183 637).	7
77MI41616	K. J. Oliver, T. N. Waters, D. F. Cook and C. E. F. Rickard; *Inorg. Chim. Acta*, 1977, **24**, 85.	4, 20
77MI41700	L. Amoretti and L. Zappia; *Ateneo Parmense, Acta Nat.*, 1977, **13**, 3 (*Chem. Abstr.*, 1977, **87**, 135 177).	162
77MI41701	N. N. Voznesenskaya, L. A. Oksent'evich, E. N. Teleshov and A. N. Pravednikov; *Vysokomol. Soedin., Ser. B*, 1977, **19**, 285 (*Chem. Abstr.*, 1977, **87**, 23 904).	167
77MI41702	S. Senez, G. Mille and J. Chouteau; *J. Chim. Phys., Phys. Chim. Biol.*, 1977, **74**, 207 (*Chem. Abstr.*, 1977, **87**, 21 761).	141
77MI41800	T. Saegusa, Y. Kimura and S. Kobayashi; *Macromolecules*, 1977, **10**, 236.	211
77MI41801	K. Burger, A. Meffert and S. Bauer; *J. Fluorine Chem.*, 1977, **10**, 57.	213, 230
77MI41900	C. Guimon, G. Pfister-Guillouzo and J. L. Larice; *J. Chim. Phys. Phys. Chim. Biol.*, 1977, **74**, 1097 (*Chem. Abstr.*, 1978, **88**, 136 022).	242
77MI42000	C. DeMarco, R. Coccia, A. Rinaldi and D. Cavallini; *Ital. J. Biochem.*, 1977, **26**, 51.	346
77MI42100	R. J. Lemire and P. G. Sears; *J. Chem. Eng. Data*, 1977, **22**, 376.	368, 371
77MI42200	H. Tondys and J. Lange; *Rocz. Chem.*, 1977, **51**, 1531 (*Chem. Abstr.*, 1978, **88**, 50 735).	412, 413
B-77MI42201	G. Bianchi, C. De Micheli and R. Gandolfi; 'The Chemistry of Double-bonded Functional Groups', ed. S. Patai; Wiley, London, 1977, p. 369.	421
77MI42300	T. Kakitani and H. Kakitani; *Theor. Chim. Acta*, 1977, **46**, 259 (*Chem. Abstr.*, 1978, **88**, 104 488).	428
77MI42301	F. M. E. Abdel-Megeid, M. A. F. Elkaschef and A. A. G. Ghattas; *Egypt. J. Chem.*, 1977, **20**, 279 (*Chem. Abstr.*, 1980, **93**, 26 347).	440
77MI42302	Y. V. Karabanov, A. K. Konstantinova, S. N. Kukota, M. O. Lozinskii and V. P. Forsyuk; *Fiziol. Akt. Veshchestva*, 1977, **9**, 40 (*Chem. Abstr.*, 1978, **88**, 46 215).	445
77MI42400	D. L. Gil and C. F. Wilkinson; *Pestic. Biochem. Physiol.*, 1977, **7**, 183.	462
77MI42600	T.-S. Lin and J. R. Braun; *Chem. Phys.*, 1977, **26**, 403.	517
77MI42700	S. A. A. Zaidi, A. S. Farooqi, D. K. Varshney, V. Islam and K. S. Siddiqi; *J. Inorg. Nucl. Chem.*, 1977, **39**, 581.	565
77MI42800	E. A. Neves, D. W. Franco and P. F. Romanelli; *Anal. Chim. Acta*, 1977, **92**, 393.	598
77MI42801	D. W. Franco, E. A. Neves and J. F. de Andrade; *Anal. Lett.*, 1977, **10**, 243.	598
77MI42802	L. Floch, A. Martvoň and M. Košik; *J. Therm. Anal.*, 1977, **12**, 407.	584
B-77MI42803	K. A. Jensen and A. Holm; in 'The Chemistry of Cyanates and their Thioderivatives', ed. S. Patai; Wiley, 1977, part 1, p. 569.	584
77MI42901	M. Akashi, H. Futagawa, Y. Inaki, K. Kondo and K. Takemoto; *Nucleic Acids Res., Spec. Publ.*, 1977, **3**, 7.	664
77MI43000	N. Duran and J. Sanabria; *Rev. Latinoam. Quim.*, 1977, **8**, 74.	757
77MI43001	P. D. Bagnall, W. Bella and K. Pearson; *J. Fluorine Chem.*, 1977, **9**, 359.	775, 780
B-77MI43300	E. T. Kaiser; in 'Organic Chemistry of Sulfur', ed. S. Oae; Plenum, New York, 1977, p. 649.	853, 875
77MI43301	M. Ali, S. Roy and B. J. Tighe; *J. Appl. Chem. Biotechnol.*, 1977, **27**, 696.	860, 864, 874, 892
77MI43302	J. A. Moore, J. E. Kelly, D. N. Harpp and T. G. Back; *Macromolecules*, 1977, **10**, 718.	870, 893
77MI43303	R. Braun, G. W. Fischer and J. Schoeneich; *Chem.-Biol. Interactions*, 1977, **19**, 241.	895

77MI43400	D. A. Edwards, R. Richards, R. E. Myers and R. A. Walton; *Inorg. Chim. Acta*, 1977, **23**, 215.	915
77MI43401	P. Baret, A. Boucherle, H. Handel, A. Million, J. L. Pierre, A. Stenger, M. Charveron and H. Lauressergues; *Eur. J. Med. Chem.*, 1977, **12**, 149 (*Chem. Abstr.*, 1977, **87**, 38 711).	939
77MI43500	A. Schweig, N. Thon and E. M. Engler; *J. Electron Spectrosc. Relat. Phenom.*, 1977, 335.	953
B-77MI43600	K. N. Houk; in 'Pericyclic Reactions', Academic Press, New York, 1977, **2**, 181.	996
77MI43800	L. J. Saethre, S. Svensson, N. Mårtensson, U. Gelius, P. Å. Malmquist, E. Basilier and K. Siegbahn; *Chem. Phys.*, 1977, **20**, 431.	1057
77MIP41600	F. J. Ugalde and J. R. Blanco; *Span. Pat.* 452 681 (1977) (*Chem. Abstr.*, 1978, **89**, 109 446).	129
77MIP41601	P. Borowicz; *Pol. Pat.* 92 735, 94 382 (1977) (*Chem. Abstr.*, 1979, **90**, 103 946, 103 945).	128
77MIP41602	S. D. Sokolov and V. V. Paramonova; *U.S.S.R. Pat.* 551 330 (1977) (*Chem. Abstr.*, 1977, **87**, 68 337).	99
77OMR(9)546	C. T. Pedersen, E. K. Frandsen and K. Schaumberg; *Org. Magn. Reson.*, 1977, **9**, 546.	785
77OMR(10)43	E. K. Frandsen and J. P. Jacobson; *Org. Magn. Reson.*, 1977, **10**, 43.	785
77OMS65	J.-L. Aubagnac and D. Bourgeon; *Org. Mass Spectrom.*, 1977, **12**, 65.	11
77OMS628	R. Neidlein, P. Leinberger and A. Hotzel; *Org. Mass. Spectrom.*, 1977, **12**, 628.	522
77PIA(A)(86)265	S. A. Kudchadker and C. N. R. Rao; *Proc. Indian Acad. Sci., Sect. A*, 1977, **86**, 265.	582
77PS(3)185	J. L. Flippen, *Phosphorus Sulfur*, 1977, **3**, 185.	902
77RRC1413	P. A. Laurent; *Rev. Roum. Chim.*, 1977, **22**, 1413.	213
77S63	R. Neidlein and P. Leinberger; *Synthesis*, 1977, 63.	540
77S263	I. Degani and R. Fochi; *Synthesis*, 1977, 263.	818, 843
77S357	B.-T. Gröbel and D. Seebach; *Synthesis*, 1977, 357.	837
77S407	M. S. Manhas, H. P. S. Chawla, S. G. Amin and A. K. Bose; *Synthesis*, 1977, 407.	174
77S487	D. Binder, G. Habison and C. R. Noe; *Synthesis*, 1977, 487.	988
77S572	M. L. Graziano, M. R. Iesce, A. Carotenuto and R. Scarpati; *Synthesis*, 1977, 572.	938, 940
77S764	A. Shafiee, I. Lalezari and F. Savabi; *Synthesis*, 1977, 764.	351, 970
77S765	A. Shafiee, I. Lalezari and F. Savabi; *Synthesis*, 1977, 765.	351
77S802	E. Meinetsberger, A. Schöffer and H. Behringer; *Synthesis*, 1977, 802.	785, 786, 809
77S837	A. Barco, S. Benetti, G. P. Pollini and P. G. Baraldi; *Synthesis*, 1977, 837.	78
77S861	K. Akiba, K. Ishikawa and N. Inamoto; *Synthesis*, 1977, 861.	965
B-77SH(2)68	W. L. F. Armarego; 'Stereochemistry of Heterocyclic Compounds,' Wiley, New York, 1977, part 2, p. 68.	750, 759
B-77SH(2)314	W. L. F. Armarego; 'Stereochemistry of Heterocyclic Compounds,' Wiley, New York, 1977, part 2, p. 314.	759, 772, 777
77SST(4)339	F. Kurzer; *Org. Compd. Sulphur, Selenium, Tellurium*, 1977, **4**, 339.	132, 147, 148, 150, 151, 152, 155, 157, 158, 159, 161, 162, 164, 165, 167, 168, 169, 170, 171, 172, 174, 175
77SST(4)354	B. Iddon and P. Alowe; *Org. Compd. Sulphur, Selenium, Tellurium*, 1977, **4**, 354–385.	328
77SST(4)386	B. Iddon and P. Alowe; *Org. Compd. Sulphur, Selenium, Tellurium*, 1977, **4**, 386–397.	328
77SST(4)417	F. Kurzer; *Org. Compd. Sulphur, Selenium, Tellurium*, 1977, **4**, 417.	448, 454
77SST(4)422	F. Kurzer; *Org. Compd. Sulphur, Selenium, Tellurium*, 1977, **4**, 422.	464
77SST(4)431	F. Kurzer; *Org. Compd. Sulphur, Selenium, Tellurium*, 1977, **4**, 431.	546; 556, 560, 562, 576
77T449	H. Buehl, B. Seitz and H. Meier; *Tetrahedron*, 1977, **33**, 449.	453, 454
77T855	L. Di Nunno and S. Florio; *Tetrahedron*, 1977, **33**, 855.	533
77T1057	S. Rajappa, B. G. Advani and R. Sreenivasan; *Tetrahedron*, 1977, **33**, 1057.	149
77T1595	K. Hirai, H. Sugimoto and T. Ishiba; *Tetrahedron*, 1977, **33**, 1595.	823
77T2231	L. Carlsen, A. Holm, J. P. Snyder, E. Koch and B. Stilkerieg; *Tetrahedron*, 1977, **33**, 2231.	581
77T2571	L. Stefaniak; *Tetrahedron*, 1977, **33**, 2571.	370, 377
77T3009	R. A. Firestone; *Tetrahedron*, 1977, **33**, 3009.	67
77T3203	C. A. Ramsden; *Tetrahedron*, 1977, **33**, 3203.	1028, 1029, 1030, 1040
77TH41600	R. K. Vander Meer; Ph.D. Thesis, Penn. State Univ., 1977.	21
77TL251	E. Gentric, J. Lauransan, C. Roussel and J. Metzger; *Tetrahedron Lett.*, 1977, 251.	248
77TL447	M. L. Graziano, A. Carotenuto, M. R. Iesce and R. Scarpati; *Tetrahedron Lett.*, 1977, 447.	927, 929
77TL735	H. Abe, M. Ikeda, T. Takaishi, Y. Ito and T. Okuda; *Tetrahedron Lett.*, 1977, 735.	687, 709
77TL1351	E. Schaumann, S. Grabley, K. D. Seidel and E. Kausch; *Tetrahedron Lett.*, 1977, 1351.	313
77TL1729	H. W. Linden and J. Goerdeler; *Tetrahedron Lett.*, 1977, 1729.	922, 943
77TL1753	S. Braverman and D. Reisman; *Tetrahedron Lett.*, 1977, 1753.	780
77TL2095	G. Scherowsky, K. Dunnbier and G. Hoefle; *Tetrahedron Lett.*, 1977, 2095.	562
77TL2643	T. Wooldridge and T. D. Roberts; *Tetrahedron Lett.*, 1977, 2643.	454
77TL2717	K.-T. H. Wei, I. C. Paul, G. Le Coustumer, R. Pinel and Y. Mollier; *Tetrahedron Lett.*, 1977, 2717.	785

77TL3437	C. L. Deyrup, J. A. Deyrup and M. Hamilton; *Tetrahedron Lett.*, 1977, 3437.	1008
77TL3759	R. Baker and M. S. Nobbs; *Tetrahedron Lett.*, 1977, 3759.	108
77TL3981	C. L. Pedersen and N. Hacker; *Tetrahedron Lett.*, 1977, 3981.	355
77TL4587	P. Espinasse, G. Killé, A. Kalt and G. Nansé; *Tetrahedron Lett.*, 1977, 4587.	186
77TL4607	Y. Ueno, M. Bahry and M. Okawara; *Tetrahedron Lett.*, 1977, 4607.	838
77TL4619	H. Saiki, T. Miyashi, T. Mukai and Y. Ikegami; *Tetrahedron Lett.*, 1977, 4619.	14, 107
77UP43300	D. N. Harpp and T. G. Back; unpublished results, cited in ref. 77MI43302.	889
77USP4006007	Monsanto Co.; *U.S. Pat.* 4 006 007 (1977) (*Chem. Abstr.*, 1977, **87**, 5948).	171
77USP4010176	P. Kulsa and C. S. Rooney; *U.S. Pat.* 4 010 176 (1977) (*Chem. Abstr.*, 1977, **87**, 23 279).	108, 128
77USP4017738	M. Hyman, Jr.; *U.S. Pat.* 4 017 738 (1977) (*Chem. Abstr.*, 1977, **87**, 30 914).	446
77USP4018774	R. K. Varma and C. M. Cimarusti; *U.S. Pat.* 4 018 774 (*Chem. Abstr.*, 1977, **87**, 118 017).	108
77USP4018781	*U.S. Pat.* 4 018 781 (1977) (*Chem. Abstr.*, 1977, **87**, 39 463).	634
77USP4031227	Sandoz-Wander Inc.; *U.S. Pat.* 4 031 227 (1977) (*Chem. Abstr.*, 1977, **87**, 102 318).	165
77USP4032322	FMC Corp.; *U.S. Pat.* 4 032 322 (1977) (*Chem. Abstr.*, 1977, **87**, 117 849).	158
77USP4032644	J. Nadelson; *U.S. Pat.* 4 032 644 (1977) (*Chem. Abstr.*, 1977, **87**, 102 314).	128
77USP4038396	*U.S. Pat.* 4 038 396 (1977) (*Chem. Abstr.*, 1979, **90**, 137 799).	696
77USP4042372	*U.S. Pat.* 4 042 372 (1977) (*Chem. Abstr.*, 1977, **87**, 184 555).	748
77USP4044018	K. Tomita, T. Murakami, Y. Yamazaki and T. Honna; *U.S. Pat.* 4 044 018 (1977) (*Chem. Abstr.*, 1978, **88**, 22 879).	129
77USP4049813	J. Nadelson; *U.S. Pat.* 4 049 813 (1977) (*Chem. Abstr.*, 1978, **88**, 6862).	128
77USP4057639	Olin Corp., *U.S. Pat.* 4 057 639 (1977) (*Chem. Abstr.*, 1978, **88**, 59 437).	510
77USP4059590	J. E. Moore; *U.S. Pat.* 4 059 590 (1977) (*Chem. Abstr.*, 1978, **88**, 50 874).	933
77USP4225721	Sogo Pharm. KK, *U.S. Pat.* 4 225 721 (1977).	74
77YZ422	K. Satoh, T. Miyasaka and K. Arakawa; *Yakugaku Zasshi*, 1977, **97**, 422 (*Chem. Abstr.*, 1977, **87**, 102 255).	655, 658
77ZC15	I. M. Ismail, R. Jacobi and W. Sauer; *Z. Chem.*, 1977, **17**, 15.	698
77ZC221	A. Reiter, P. Hansen and H. U. Kibbel; *Z. Chem.*, 1977, **17**, 221.	1064
77ZC223	S. Leistner, G. Wagner and M. Ackermann; *Z. Chem.*, 1977, **17**, 223 (*Chem. Abstr.*, 1977, **87**, 117 830).	911, 915
77ZC295	G. Westphal, A. Klebsch, A. Weise, U. Sternberg and A. Otto; *Z. Chem.*, 1977, **17**, 295 (*Chem. Abstr.*, 1977, **87**, 184 437).	917
77ZN(B)443	H. H. Zoorob, H. A. Hammouda and E. Ismail; *Z. Naturforsch., Teil B*, 1977, **32**, 443.	93
77ZOB1888	Yu. N. Kukushkin, S. A. Simanova, V. K. Krylov and V. V. Strukov; *Zh. Obshch. Khim.*, 1977, **47**, 1888.	526
77ZOR462	E. N. Glibin, B. V. Tsukerman, S. S. Tsymbalova and O. F. Ginzburg; *Zh. Org. Khim.*, 1977, **13**, 462 (*Chem. Abstr.*, 1977, **87**, 23 123).	103
77ZOR2012	V. N. Drozd, G. S. Bogomolova; *Zh. Org. Khim.*, 1977, **13**, 2012.	799
77ZOR2495	E. V. Ratsino, K. V. Altukhov, V. V. Perekalin and O. N. Fedorishcheva; *Zh. Org. Khim.*, 1977, **13**, 2495 (*Chem. Abstr.*, 1978, **88**, 89 224).	111
78ACH(97)69	V. Szabo, J. Borda and L. Losonczi; *Acta Chim. Acad. Sci. Hung.*, 1978, **97**, 69.	79
78ACR375	A. I. Meyers; *Acc. Chem. Res.*, 1978, **11**, 375.	212, 228
78ACS(A)1005	B. Bak, O. Nielsen, H. Svanholt, A. Almenningen, O. Bastiansen, L. Fernholt, G. Gundersen, C. J. Nielsen, B. N. Cyvin and S. T. Cyvin; *Acta Chem. Scand., Ser. A*, 1978, **32**, 1005.	903, 907, 910
78ACS(B)70	P.-O. Ranger, G. A. Ulsaker and K. Undheim; *Acta Chem. Scand., Ser. B*, 1978, **32**, 70.	675
78ACS(B)118	K. Torssell and O. Zeuthen; *Acta Chem. Scand., Ser. B*, 1978, **32**, 118.	47, 95, 110
78ACS(B)625	C. L. Pedersen, C. Lohse and M. Poliakoff; *Acta Chem. Scand., Ser. B*, 1978, **32**, 625.	525
78ACS(B)651	T. Laerum, G. A. Ulsaker and K. Undheim; *Acta Chem. Scand., Ser. B*, 1978, **32**, 651.	673, 677
78AG(E)352	G. L'abbé, C. C. Yu, J. P. Declercq, G. Germain and M. Van Meerssche; *Angew. Chem., Int. Ed. Engl.*, 1978, **17**, 352.	901
78AG(E)450	R. Lohmar and W. Steglich; *Angew. Chem., Int. Ed. Engl.*, 1978, **17**, 450.	202
78AG(E)455	E. Schaumann, J. Ehlers and U. Behrens; *Angew. Chem., Int. Ed. Engl.*, 1978, **17**, 455.	901, 925, 941
78AHC(21)207	Y. Takeuchi and F. Furusaki; *Adv. Heterocycl. Chem.*, 1977, **21**, 207.	1000
78AHC(22)183	J. Elguero, R. M. Claramunt and A. J. H. Summers; *Adv. Heterocycl. Chem.*, 1978, **22**, 183.	974
78AHC(23)171	I. J. Fletcher and A. E. Siegrist; *Adv. Heterocycl. Chem.*, 1978, **23**, 171.	52, 215
78AHC(23)265	R. M. Acheson and N. F. Elmore; *Adv. Heterocycl. Chem.*, 1978, **23**, 265.	190
78AJC113	J. T. Pinhey, E. Rizzardo and G. C. Smith; *Aust. J. Chem.*, 1978, **31**, 113.	104
78AJC297	J. R. Cannon, K. T. Potts, C. L. Raston, A. F. Sierakowski and A. H. White; *Aust. J. Chem.*, 1978, **31**, 297.	784, 788, 789, 805
78AJC2239	D. St. C. Black, R. F. Crozier and I. D. Rae; *Aust. J. Chem.*, 1978, **31**, 2239.	108, 109
78ANY(313)1	J. S. Miller and A. J. Epstein; *Ann. N.Y. Acad. Sci.*, 1978, **313**, 1.	949
78AP817	E. Roeder and J. Pigulla; *Arch. Pharm. (Weinheim, Ger.)*, 1978, **311**, 817.	93
78AX(B)2570	A. R. Butler, C. Glidewell and D. C. Liles; *Acta Crystallogr., Part B*, 1978, **34**, 2570.	902, 903
78AX(B)2953	M. Calleri, G. Chiari and D. Viterbo; *Acta Crystallogr., Part B*, 1978, **34**, 2953.	396

78AX(B)3803	S. Larsen; *Acta Crystallogr., Sect. B*, 1978, **34**, 3803.	548
78BAP291	L. Stefaniak; *Bull. Acad. Pol. Sci., Ser. Sci. Chim.*, 1978, **26**, 291.	519
78BCJ323	T. Nishiyama, T. Mizuno and F. Yamada; *Bull. Chem. Soc. Jpn.*, 1978, **51**, 323.	906
78BCJ1261	T. Kusumi, H. Kakisawa, S. Suzuki, K. Harada and C. Kashima; *Bull. Chem. Soc. Jpn.*, 1978, **51**, 1261.	91
78BCJ1484	M. Kurabayashi and C. Grundmann; *Bull. Soc. Chem. Jpn.*, 1978, **51**, 1484.	389
78BCJ2674	K. Akiba, K. Ishikawa and N. Inamoto; *Bull. Chem. Soc. Jpn.*, 1978, **51**, 2674.	965
78BEP866987	*Belg. Pat.* 866 987 (1978) (*Chem. Abstr.*, 1979, **90**, 52 211).	710
78BEP867155	Studiengesellschaft Kohle mbH, *Belg. Pat.* 867 155 (1978) (*Chem. Abstr.*, 1979, **90**, 121 789).	797, 811
78BSB391	J. Gorissen and H. G. Viehe; *Bull. Soc. Chim. Belg.*, 1978, **87**, 391.	86
78CB1915	M. Preiss; *Chem. Ber.*, 1978, **111**, 1915.	539
78CB2021	H. Gotthardt and C. M. Weisshuhn; *Chem. Ber.*, 1978, **111**, 2021.	818, 820, 826, 841, 842
78CB2028	H. Gotthardt, and C. M. Weisshuhn; *Chem. Ber.*, 1978, **111**, 2028.	826, 827
78CB2716	R. W. Hoffmann and S. Goldmann; *Chem. Ber.*, 1978, **111**, 2716.	796
78CB3029	H. Gotthardt and B. Christl; *Chem. Ber.*, 1978, **111**, 3029.	827, 828
78CB3037	H. Gotthardt, C. M. Weisshuhn and B. Christl; *Chem. Ber.*, 1978, **111**, 3037.	828, 829
78CB3171	H. Gotthardt and C. M. Weisshuhn; *Chem. Ber.*, 1978, **111**, 3171.	830
78CB3178	H. Gotthardt and C. M. Weisshuhn; *Chem. Ber.*, 1978, **111**, 3178.	821, 822, 840
78CB3423	H. Peterson and H. Meier; *Chem. Ber.*, 1978, **111**, 3423.	354
78CC113	J. F. Barnes, R. M. Paton, P. L. Ashcroft, R. Bradbury, J. Crosby, C. J. Joyce, D. R. Holmes and J. A. Milner; *J. Chem. Soc., Chem. Commun.*, 1978, 113.	405, 421
78CC652	A. R. Butler, C. Glidewell and D. C. Liles; *J. Chem. Soc., Chem. Commun.*, 1978, 652.	465, 474
78CC971	K. K. Knapp, P. C. Keller and J. V. Rund; *J. Chem. Soc., Chem. Commun.*, 1978, 971.	254
78CCC2298	P. Kristian and L. Kniezo; *Collect. Czech. Chem. Commun.*, 1978, **43**, 2298.	346
78CHE733	Y. I. Akulin, M. M. Gel'mont, B. K. Strelets and L. S. Efros; *Chem. Heterocycl. Compd.* (*Engl. Transl.*), 1978, **14**, 733.	900, 905
78CI(L)92	J. S. Davidson and S. S. Dhami; *Chem. Ind. (London)*, 1978, 92.	720
78CJC308	G. A. MacAlpine and J. Warkentin; *Can. J. Chem.*, 1978, **56**, 308.	437
78CJC722	P. C. Belanger; *Can. J. Chem.*, 1978, **56**, 722.	536
78CJC1319	A. J. Paine and N. H. Werstiuk; *Can. J. Chem.*, 1978, **56**, 1319.	428
78CL1093	K. Matsumoto and T. Uchida; *Chem. Lett.*, 1978, 1093.	1038
78CPB549	H. Uno and M. Kurokawa; *Chem. Pharm. Bull.*, 1978, **26**, 549.	56
78CPB765	K. Senga, J. Sato and S. Nishigaki; *Chem. Pharm. Bull.*, 1978, **26**, 765.	655, 686
78CPB2122	T. Ueda, K. Miura and T. Kasai; *Chem. Pharm. Bull.*, 1978, **26**, 2122.	724
78CPB2497	S. Nishigaki, Y. Kanamori and K. Senga; *Chem. Pharm. Bull.*, 1978, **26**, 2497.	621, 623, 624, 629, 660
78CPB2765	T. Tsuji and Y. Otsuka; *Chem. Pharm. Bull.*, 1978, **26**, 2765.	734, 737, 738
78CPB3017	K. Hirai and T. Ishiba; *Chem. Pharm. Bull.*, 1978, **26**, 3017.	762
78CPB3254	E. Kaji, K. Harada and S. Zen; *Chem. Pharm. Bull.*, 1978, **26**, 3254.	85
78CPB3498	H. Uno and M. Kurokawa; *Chem. Pharm. Bull.*, 1978, **26**, 3498.	24
78CPB3888	H. Uno and M. Kurokawa; *Chem. Pharm. Bull.*, 1978, **26**, 3888.	155
78CPB3896	T. Uno, K. Takagi and M. Tomoeda; *Chem. Pharm. Bull.*, 1978, **26**, 3896.	402, 528, 538
78CR(C)(286)613	G. Mille, S. Senez and J. Chouteau; *C.R. Hebd. Seances Acad. Sci., Ser. C*, 1978, **286**, 613 (*Chem. Abstr.*, 1979, **90**, 5709).	9
78CZ264	C. Skoetsch and E. Breitmaier; *Chem.-Ztg.*, 1978, **102**, 264.	43
78CZ361	C. Bak, K. Praefcke and L. Henriksen; *Chem. Ztg.*, 1978, **102**, 361.	960, 970
78EGP129907	J. Liebscher; *Ger. (East) Pat.* 129 907 (1978) (*Chem. Abstr.*, 1978, **89**, 109 569).	917
78FRP2392981	F. Guigues, J. Mourier and D. Demozay; *Fr. Demande* 2 392 981 (1978) (*Chem. Abstr.*, 1979, **91**, 193 293).	129
78GEP2637692	E. Enders, I. Hammann, W. Brandes, P. Kraus and W. Stendel; (Bayer A.-G.), *Ger. Pat.* 2 637 692 (1978) (*Chem. Abstr.* 1978, **88**, 190 800).	810
78GEP2703492	T. Takayanagi; *Ger. Offen.* 2 703 492 (1978) (*Chem. Abstr.*, 1978, **89**, 163 561).	129
78GEP2706398	*Ger. Pat.* 2 706 398 (1978) (*Chem. Abstr.*, 1979, **90**, 6391).	709
78GEP2707227	E. Enders, I. Hammann, W. Brandes, P. Kraus and W. Stendel; (Bayer A.-G.), *Ger. Pat.* 2 707 227 (1978) (*Chem. Abstr.*, 1978, **89**, 197 524).	810
78GEP2711382	F. F. Frickel, D. Lenke and J. Gries; *Ger. Offen.* 2 711 382 (1978) (*Chem. Abstr.*, 1978, **89**, 215 399).	129
78GEP2712932	*Ger. Pat.* 2 712 932 (1978) (*Chem. Abstr.*, 1979, **90**, 38 957).	748
78GEP2725379	E. Bayer and K. Geckeler; *Ger. Pat.* 2 725 379 (1978) (*Chem. Abstr.*, 1979, **90**, 120 480).	148
78GEP2727146	Reckitt and Colman Products Ltd., *Ger. Pat.* 2 727 146 (1978) (*Chem. Abstr.*, 1978, **88**, 105 357).	577
78GEP2745246	Fujisawa Pharm. Co. Ltd., *Ger. Pat.* 2 745 246 (1978) (*Chem. Abstr.*, 1978, **89**, 43 461).	155
78GEP2747122	M. R. G. Leeming and J. K. Stubbs; *Ger. Offen.* 2 747 122 (1978) (*Chem. Abstr.*, 1978, **89**, 43 424).	128
78GEP2804518	*Ger. Pat.* 2 804 518 (1978) (*Chem. Abstr.*, 1978, **89**, 197 595).	730
78GEP2804519	*Ger. Pat.* 2 804 519 (1978) (*Chem. Abstr.*, 1979, **91**, 211 434).	724
78GEP2808842	R. Boesch; *Ger. Offen.* 2 808 842 (1978) (*Chem. Abstr.*, 1979, **90**, 23 060).	439, 440

78GEP2812367	Ger. Pat. 2 812 367 (1978) (*Chem. Abstr.*, 1979, **90**, 82 129).	634
78GEP2825194	Y. Makisumi, A. Murabayashi and T. Sasatani; Ger. Pat. 2 825 194 (1978) (*Chem. Abstr.*, 1979, **90**, 103 939).	65, 86
78H(9)185	Y. Yamamoto and Y. Azuma; *Heterocycles*, 1978, **9**, 185.	80
78H(9)457	M. A. Abow-Gharbia, M. M. Joullie and I. Muira; *Heterocycles*, 1978, **9**, 457.	113
78H(9)1223	K. Hirai and T. Ishiba; *Heterocycles*, 1978, **9**, 1223.	304
78H(10)57	M. Yoshifuji, R. Nagase, T. Kawashima and N. Inamoto; *Heterocycles*, 1978, **10**, 57.	218
78H(10)257	T. Kusumi, S. Takahashi, Y. Sato and H. Kakisawa; *Heterocycles*, 1978, **10**, 257.	109
78H(11)121	K. L. Williamson and J. D. Roberts; *Heterocycles*, 1978, **11**, 121.	550, 551, 557, 904
78H(11)187	K. Kikuchi, Y. Maki, M. Hayashi and N. Murakoshi; *Heterocycles*, 1978, **11**, 187.	74, 101
78H(11)313	B. Verček, B. Stanovnik and M. Tišler; *Heterocycles*, 1978, **11**, 313.	733
78HCA108	H. Balli and L. Felder; *Helv. Chim. Acta*, 1978, **61**, 108.	158
78HCA1072	M. Guiliano, E. Davin-Pretelli, G. Mille, J. Chouteau and R. Guglielmetti; *Helv. Chim. Acta*, 1978, **61**, 1072 (*Chem. Abstr.*, 1978, **89**, 41 745).	242
78HCA1404	P. Dubs and M. Joho; *Helv. Chim. Acta*, 1978, **61**, 1404.	889
78HCA1477	M. Märky, H. Meier, A. Wunderli, H. Heimgartner, H. Schmid and H.-J. Hansen; *Helv. Chim. Acta*, 1978, **61**, 1477.	374
78HCA2419	H. Link; *Helv. Chim. Acta*, 1978, **61**, 2419.	445
78HCA2809	P. Dubs and M. Joho; *Helv. Chim. Acta*, 1978, **61**, 2809.	878
78IJC(B)57	M. A. El-Maghraby and M. A. Abbady; *Indian J. Chem., Sect. B*, 1978, **16**, 57.	93
78IJC(B)146	R. B. Mitra, G. H. Kulkarni and G. S. Shirwaiker; *Indian J. Chem., Sect. B*, 1978, **16**, 146.	431, 435
78IJC(B)673	N. R. Ayyangar, S. R. Purao and B. D. Tilak; *Indian J. Chem., Sect. B*, 1978, **16**, 673.	1059, 1062, 1064, 1067
78IZV313	V. V. Zverev, I. Sh. Saifullin and G. P. Sharnin; *Izv. Akad. Nauk SSSR, Ser. Khim.*, 1978, 313.	400
78IZV850	V. F. Rudchenko, A. O. D'iyachenko, I. I. Chervin, A. B. Zolotoi and L. D. Atovmyan; *Izv. Akad. Nauk SSSR, Ser. Khim.*, 1978, 850 (*Chem. Abstr.*, 1978, **89**, 107 785).	4, 6
78IZV1149	I. E. Chlenov, Yu. B. Salamanov, B. N. Khasapov, V. M. Shitkin, N. F. Karpenko, O. S. Chizhov and V. A. Tartakovskii; *Izv. Akad. Nauk SSSR, Ser. Khim.*, 1978, 1149 (*Chem. Abstr.*, 1978, **89**, 109 305).	98
78IZV1881	D. P. Del'tsova, Z. V. Safronova, N. P. Gambaryan, M. Yu. Antipin and Yu. T. Struchkov; *Izv. Akad. Nauk SSSR, Ser. Khim.*, 1978, 1881 (*Chem. Abstr.*, 1978, **89**, 215 280).	108
78IZV2551	I. E. Chlenov, I. M. Petrova, B. N. Khasapov, N. F. Karpenko, A. U. Stepanyants, O. S. Chizhov and V. A. Tartakovskii; *Izv. Akad. Nauk SSSR, Ser. Khim.*, 1978, 2551 (*Chem. Abstr.*, 1979, **90**, 103 909).	108
78IZV2588	B. A. Arbuzov, A. F. Lisin, E. N. Dianova and Yu. Yu. Samitov; *Izv. Akad. Nauk SSSR, Ser. Khim.*, 1978, 2588 (*Chem. Abstr.*, 1979, **90**, 87 580).	108, 109
78JA3638	J. J. Tufariello and G. B. Mullen; *J. Am. Chem. Soc.*, 1978, **100**, 3638.	109
78JA3868	D. Forrest and K. U. Ingold; *J. Am. Chem. Soc.*, 1978, **100**, 3868.	835
78JA4208	M. Braun, G. Buchi and D. F. Bushey; *J. Am. Chem. Soc.*, 1978, **100**, 4208.	386
78JA4260	N. Shimizu and P. D. Bartlett; *J. Am. Chem. Soc.*, 1978, **100**, 4260.	437, 443
78JA5584	R. S. Brown and R. W. Marcinko; *J. Am. Chem. Soc.*, 1978, **100**, 5584.	861
78JA6516	E. M. Kosower, B. Pazhenchevsky and E. Hershkowitz; *J. Am. Chem. Soc.*, 1978, **100**, 6516.	978
78JA7629	R. C. Haddon, F. Wudl, M. L. Kaplan, J. H. Marshall, R. E. Cais and F. B. Bramwell; *J. Am. Chem. Soc.*, 1978, **100**, 7629.	784, 785, 805
78JA7927	R. V. Hoffman, G. Orphanides and H. Schechter; *J. Am. Chem. Soc.*, 1978, **100**, 7927.	15
78JAP78127479	H. Sugano, K. Ikeda, M. Yasui and T. Harada; *Jpn. Pat.* 78 127 479 (1978) (*Chem. Abstr.*, 1979, **90**, 137 829).	932, 940, 944
78JAP(K)7825566	K. Tomita, H. Oka and Y. Kondo; *Jpn. Kokai* 78 25 566 (1978) (*Chem. Abstr.*, 1978, **89**, 109 445).	130
78JAP(K)7844592	*Jpn. Kokai* 78 44 592 (1978) (*Chem. Abstr.*, 1978, **89**, 109 566).	710
78JAP(K)7863376	K. Tomita and S. Sugai; *Jpn. Kokai* 78 63 376 (1978) (*Chem. Abstr.*, 1979, **90**, 6387).	43
78JAP(K)7879862	H. Uno, M. Kurokawa and Y. Takase; *Jpn. Kokai* 78 79 862 (1978) (*Chem. Abstr.*, 1979, **90**, 23 023).	129
78JAP(K)7884992	*Jpn. Kokai* 78 84 992 (1978) (*Chem. Abstr.*, 1979, **90**, 23 121).	709
78JAP(K)7891134	*Jpn. Kokai* 78 91 134 (1978) (*Chem. Abstr.*, 1979, **90**, 49 637).	709
78JAP(K)7892768	Sumitomo Chemical Co. Ltd.; *Jpn. Kokai* 78 92 768 (1978) (*Chem. Abstr.*, 1979, **90**, 72 180).	175
78JAP(K)78135971	K. Tomita and T. Murakami; *Jpn. Kokai* 78 135 971 (*Chem. Abstr.*, 1979, **90**, 152 161).	43
78JCR(M)0855	A. R. Butler; *J. Chem. Res. (M)*, 1978, 0855.	474
78JCR(M)2038	B. Iddon, H. Suschitzky, A. W. Thompson, B. J. Wakefield and D. J. Wright; *J. Chem. Res. (M)*, 1978, 2038.	68
78JCR(S)164	J. P. Gilbert, C. Petrus and F. Petrus; *J. Chem. Res. (S)*, 1978, 164.	89
78JCR(S)192	J. Galluci, M. LeBlanc and J. G. Riess; *J. Chem. Res. (S)*, 1978, 192.	89
78JCR(S)240	F. Tonnard, D. Gree and J. Hamelin; *J. Chem. Res. (S)*, 1978, 240.	108, 109

78JCR(S)407	H. Singh and C. S. Gandhi; *J. Chem. Res. (S)*, 1978, 407.	693
78JCS(P1)45	D. J. Humphreys, C. E. Newall, G. H. Phillips and G. A. Smith; *J. Chem. Soc., Perkin Trans. 1*, 1978, 45.	557, 574
78JCS(P1)195	R. M. Christie, D. H. Reid, R. Walker and R. G. Webster; *J. Chem. Soc., Perkin Trans. 1*, 1978, 195.	1058, 1066
78JCS(P1)378	M. G. Barlow, R. N. Haszeldine and J. A. Pickett; *J. Chem. Soc., Perkin Trans. 2*, 1978, 378.	433
78JCS(P1)468	J. Nakayama, E. Seki and M. Hoshino; *J. Chem. Soc., Perkin Trans., 1*, 1978, 468.	846
78JCS(P1)600	R. N. Hanley, W. D. Ollis, C. A. Ramsden, G. Rowlands and L. E. Sutton; *J. Chem. Soc., Perkin Trans. 1*, 1978, 600.	599
78JCS(P1)746	A. Holm, N. Harrit and I. Trabjerg; *J. Chem. Soc., Perkin Trans. 1*, 1978, 746.	585
78JCS(P1)817	M. M. Campbell, R. G. Harcus and K. H. Nelson; *J. Chem. Soc., Perkin Trans. 1*, 1978, 817.	706
78JCS(P1)1006	C. D. Campbell, C. W. Rees, M. R. Bryce, M. D. Cooke, P. Hanson and J. M. Vernon; *J. Chem. Soc., Perkin Trans. 1*, 1978, 1006.	529
78JCS(P1)1017	M. Muraoka, T. Yamamoto, T. Ebisawa, W. Koyabashi and T. Takeshima; *J. Chem. Soc., Perkin Trans. 1*, 1978, 1017.	169, 806
78JCS(P1)1440	N. H. Toubro and A. Holm; *J. Chem. Soc., Perkin Trans. 1*, 1978, 1440.	586, 588, 596
78JCS(P1)1445	A. Holm and N. H. Toubro; *J. Chem. Soc., Perkin Trans. 1*, 1978, 1445.	585, 913
78JCS(P1)1547	U. J. Kempe, T. Kempe and T. Norin; *J. Chem. Soc., Perkin Trans. 1*, 1978, 1547.	777
78JCS(P2)613	A. R. Katritzky, S. Clementi, G. Milletti and G. V. Sebastiani; *J. Chem. Soc., Perkin Trans. 2*, 1978, 613.	148
78JCS(P2)985	E. Cetinkaya, H.-P. Schuchmann and C. Von Sonntag; *J. Chem. Soc., Perkin Trans. 2*, 1978, 985.	760
78JHC81	L. Grehn; *J. Heterocycl. Chem.*, 1978, **15**, 81.	1015
78JHC293	M. Ruccia, N. Vivona and G. Cusmano; *J. Heterocycl. Chem.*, 1978, **15**, 293.	998
78JHC307	R. C. Haltiwanger and D. S. Watt; *J. Heterocycl. Chem.*, 1978, **15**, 307.	974, 984
78JHC313	B. Verček and B. Stanovnik; *J. Heterocycl. Chem.*, 1978, **11**, 313.	741
78JHC395	E. Alcalde, R. M. Claramunt, J. Elguero and C. P. Saunderson-Huber; *J. Heterocycl. Chem.*, 1978, **15**, 395.	975
78JHC401	E. Campaigne and T. P. Selby; *J. Heterocycl. Chem.*, 1978, **15**, 401.	980, 993, 1010
78JHC501	I. Lalezari and S. Sadeghi-Milani; *J. Heterocycl. Chem.*, 1978, **15**, 501.	348, 353
78JHC529	A. H. Albert, D. E. O'Brien and R. K. Robins; *J. Heterocycl. Chem.*, 1978, **15**, 529.	154, 170
78JHC695	J. Rokach and P. Hamel; *J. Heterocycl. Chem.*, 1978, **14**, 695.	159, 160
78JHC849	H. N. Al-Jallo and M. A. Muniem; *J. Heterocycl. Chem.*, 1978, **15**, 849.	687
78JHC865	R. Weber, J-L. Piette and M. Renson; *J. Heterocycl. Chem.*, 1978, **15**, 865.	335, 336, 337, 328, 344
78JHC1055	S. Braun and K. Hafner; *J. Heterocycl. Chem.*, 1978, **15**, 1055.	994
78JHC1145	N. Bregant and I. Perina; *J. Heterocycl. Chem.*, 1978, **15**, 1145.	65, 85
78JHC1373	H. L. Yale and E. R. Spitzmiller; *J. Heterocycl. Chem.*, 1978, **15**, 1373.	384
78JHC1515	A. Alemagna and T. Bacchetti; *J. Heterocycl. Chem.*, 1978, **15**, 1515.	575
78JHC1519	R. B. Silverman; *J. Heterocycl. Chem.*, 1978, **15**, 1519.	66, 85, 87
78JHC1527	I. Iijima and K. C. Rice; *J. Heterocycl. Chem.*, 1978, **15**, 1527.	638, 642
78JHC1838	J. H. Looker, N. A. Khatri, R. B. Patterson and C. A. Kingsbury; *J. Heterocycl. Chem.*, 1978, **15**, 1383.	451
78JIC108	A. K. Sengupta and O. P. Bajaj; *J. Indian Chem. Soc.*, 1978, **55**, 108.	445
78JMC489	B. Blank, N. W. DiTullio, A. J. Krogh and H. L. Saunders; *J. Med. Chem.*, 1978, **21**, 489.	692, 709
78JMC496	R. N. Hanson, R. W. Giese, M. A. Davis and S. M. Costello; *J. Med. Chem.*, 1978, **21**, 496.	347
78JMC1100	H. Jones, M. W. Fordice, R. B. Greenwald, J. Hannah, A. Jacobs, W. V. Ruyle, G. L. Walford and T. Y. Shen; *J. Med. Chem.*, 1978, **21**, 1100.	156
78JMC1158	R. L. Clark, A. A. Pessolano, B. Witzel, T. Lanza, T. Y. Shen, C. G. Van Arman and E. A. Risley; *J. Med. Chem.*, 1978, **21**, 1158.	659, 662, 668
78JOC79	W. H. Koster, J. E. Dolfini, B. Toeplitz and J. Z. Gougoutas; *J. Org. Chem.*, 1978, **43**, 79.	135
78JOC341	N. Sato and J. Adachi; *J. Org. Chem.*, 1978, **43**, 341.	720, 722, 729
78JOC369	N. C. Gonnella and M. P. Cava; *J. Org. Chem.*, 1978, **43**, 369.	849
78JOC419	H. Aoyama, T. Hasegawa, M. Watabe, H. Shiraishi and Y. Omote; *J. Org. Chem.*, 1978, **43**, 419.	230
78JOC438	R. P. Hanzlik and M. Leinwetter; *J. Org. Chem.*, 1978, **43**, 438.	775
78JOC672	C. N. Filer, F. E. Granchelli, A. H. Soloway and J. L. Neumeyer; *J. Org. Chem.*, 1978, **43**, 672.	651, 652
78JOC678	N. F. Haley; *J. Org. Chem.*, 1978, **43**, 678.	834
78JOC960	G. H. Hartman, S. E. Biffar, L. M. Weinstock and R. Tull; *J. Org. Chem.*, 1978, **43**, 960.	528, 533, 538, 735, 740
78JOC1154	W. Adam, A. Birke, C. Cádiz, S. Díaz and A. Rodríguez; *J. Org. Chem.*, 1978, **43**, 1154.	772
78JOC1218	R. A. Abramovitch, C. I. Azogu, I. T. McMaster and D. P. Vanderpool; *J. Org. Chem.*, 1978, **43**, 1218.	916, 931
78JOC1233	N. F. Haley; *J. Org. Chem.*, 1978, **43**, 1233.	20, 21, 29, 51, 74
78JOC1604	J. R. Beck and J. A. Yahner; *J. Org. Chem.*, 1978, **43**, 1604.	166

78JOC1677	K. Senga, M. Ichiba and S. Nishigaki; *J. Org. Chem.*, 1978, **43**, 1677.	608, 611, 733, 734, 735, 736, 737, 745
78JOC2020	P. A. Wade; *J. Org. Chem.*, 1978, **43**, 2020.	39, 96
78JOC2037	C. Wentrup, A. Damerius and W. Reichen; *J. Org. Chem.*, 1978, **43**, 2037.	431
78JOC2487	J. Font, M. Torres, H. E. Gunning and O. P. Strausz; *J. Org. Chem.*, 1978, **43**, 2487.	451, 452, 455
78JOC2490	M. Torres, A. Clement, J. E. Bertie, H. E. Gunning and O. P. Strausz; *J. Org. Chem.*, 1978, **43**, 2490.	455
78JOC2500	M. S. Raasch; *J. Org. Chem.*, 1978, **43**, 2500.	171, 587, 597
78JOC2542	I. Yavari, R. E. Botto and J. D. Roberts; *J. Org. Chem.*, 1978, **43**, 2542.	398, 519
78JOC2700	K. T. Potts and D. R. Choudhury; *J. Org. Chem.*, 1978, **43**, 2700.	691
78JOC3015	G. N. Barber and R. A. Olofson; *J. Org. Chem.*, 1978, **43**, 3015.	71, 72, 84
78JOC3374	A. G. Hortmann, A. J. Aron and A. K. Bhattacharya; *J. Org. Chem.*, 1978, **43**, 3374.	795, 796, 801
78JOC3730	S. Mataka, S. Ishi-i and M. Tashiro; *J. Org. Chem.*, 1978, **43**, 3730.	774
78JOC3736	R. K. Howe, T. A. Gruner, L. G. Carter, L. L. Black and J. E. Franz; *J. Org. Chem.*, 1978, **43**, 3736.	156, 500, 920, 999
78JOC3742	R. K. Howe and J. E. Franz; *J. Org. Chem.*, 1978, **43**, 3742.	163, 500, 919
78JOC3893	R. Gleiter, R. Bartetzko, G. Brähler and H. Bock; *J. Org. Chem.*, 1978, **43**, 3893.	1029, 1030, 1031
78JOC4042	D. Spinelli, R. Noto, G. Consiglio, G. Werber and F. Buccheri; *J. Org. Chem.*, 1978, **43**, 4042.	564
78JOC4048	J. R. Nixon, M. A. Cudd and N. A. Porter; *J. Org. Chem.*, 1978, **43**, 4048.	772, 775
78JOC4154	E. C. Taylor and E. Wachsen; *J. Org. Chem.*, 1978, **43**, 4154.	148, 640
78JOC4693	I. I. Schuster, S. H. Doss and J. D. Roberts; *J. Org. Chem.*, 1978, **43**, 4693.	133, 140
78JOC4816	A. Holm, L. Carlsen and E. Larsen; *J. Org. Chem.*, 1978, **43**, 4816.	584
78JOC4910	B. L. Cline, R. P. Panzica and L. B. Townsend; *J. Org. Chem.*, 1978, **43**, 4910.	735, 746
78JOC4951	G. L'abbé, A. Timmerman, C. Martens and S. Toppet; *J. Org. Chem.*, 1978, **43**, 4951.	590, 906, 914, 929, 935
78JOM(144)291	G. Westera, C. Blomberg and F. Bickelhaupt; *J. Organomet. Chem.*, 1978, **144**, 291.	763
78JPC463	J. Mason, W. Van Bronswijk and O. Glemser; *J. Phys. Chem.*, 1978, **82**, 463.	515, 519
78JPR206	E. Banschke and G. Tomaschewski; *J. Prakt. Chem.*, 1978, **320**, 206.	376
78JPR585	W. D. Rudorf and M. Augustin; *J. Prakt. Chem.*, 1978, **320**, 585.	86, 88
78JPS1336	I. Lalezari, A. Shafiee, J. Khorrami and A. Soltani; *J. Pharm. Sci.*, 1978, **67**, 1336.	348, 349
78JPS1507	S. K. Chaudhary, M. Chaudhary, A. Chaudhari and S. S. Parmar; *J. Pharm. Sci.*, 1978, **67**, 1507.	445
78JPS1762	R. A. Glennon, M. E. Rogers, R. G. Bass and S. B. Ryan; *J. Pharm. Sci.*, 1978, **67**, 1762.	709, 748
78JST(43)33	M. H. Palmer and S. M. F. Kennedy; *J. Mol. Struct.*, 1978, **43**, 33.	134, 395, 399, 517
78JST(43)203	M. H. Palmer and S. M. F. Kennedy; *J. Mol. Struct.*, 1978, **43**, 203.	134, 183, 184
78JST(48)205	G. Salmona, R. Faure, E. J. Vincent, C. Guimon and G. Pfister-Guillouzo; *J. Mol. Struct.*, 1978, **48**, 205.	133, 134
78JST(48)227	M. M. Borel, A. Leclaire, G. le Coustumer and Y. Mollier; *J. Mol. Struct.*, 1978, **48**, 227.	140
78JST(50)233	M. Guiliano, G. Mille and J. Chouteau; *J. Mol. Struct.*, 1978, **50**, 233.	242
78JST(50)247	G. Mille, J. L. Meyer and J. Chouteau; *J. Mol. Struct.*, 1978, **50**, 247.	242
78KGS324	F. A. Gabitov, O. B. Kremleva and A. L. Fridman; *Khim. Geterotsikl. Soedin.*, 1978, 324 (*Chem. Abstr.*, 1978, **89**, 43 207).	103
78KGS327	S. D. Sokolov and G. B. Tikhomirova; *Khim. Geterotsikl. Soedin.*, 1978, 327 (*Chem. Abstr.*, 1978, **89**, 23 565).	4, 10
78KGS917	I. N. Azerbaev, L. A. Tsoi, S. T. Chlopankulova, A. B. Asmanova, and V. I. Artyukhin; *Khim. Geterotsikl. Soedin.*, 1978, 917 (*Chem. Abstr.*, 1978, **89**, 197 415).	345
78KGS969	V. M. Neplyuev, T. A. Sinenko and P. S. Pel'kis; *Khim. Geterotsikl. Soedin.*, 1978, 969.	65, 88
78KGS1053	M. I. Komendantov, R. R. Bekmukhametov and R. R. Kostikov; *Khim. Geterotsikl Soedin.*, 1978, 1053.	15
78KGS1196	A. V. Eremeev, V. G. Andrianov and I. P. Piskunova; *Khim. Geterotsikl. Soedin.*, 1978, 1196.	414, 417
78KGS1632	D. E. Balode, R. E. Valters and S. P. Valtere; *Khim. Geterotsikl. Soedin.*, 1978, 1632.	146
78MI41600	J. Kulig and B. Lenarcik; *Pol. J. Chem.*, 1978, **52**, 477.	20, 71
78MI41602	D. N. Nicolaides and A. G. Catsaounis; *Chem. Chron., New Ser.*, 1978, **7**, 189.	70
78MI41603	H. P. Schane, A. J. Anzalone and G. O. Potts; *Fertil. Steril.*, 1978, **29**, 692.	129
78MI41604	H. P. Schane, J. E. Creange, A. J. Anzalone and G. O. Potts; *Fertil. Steril.*, 1978, **30**, 343.	129
78MI41605	J. E. Creange, H. P. Schane, A. J. Anzalone and G. O. Potts; *Fertil. Steril.*, 1978, **30**, 86.	129
B-78MI41606	I. S. Levina, A. V. Kamernitskii, E. I. Mortikova, T. N. Galakhova, V. M. Shitkin and B. S. El'yanov; in '5th Tezisy Dokl.-Sov.-Indiiskii Simp. Khim. Prir. Soedin.', 1978, p. 46 (*Chem. Abstr.*, 1980, **93**, 186 648).	110
78MI41607	K. Tada and F. Toda; *Yuki Gosei Kagaku Kyokaishi*, 1978, **36**, 620 (*Chem. Abstr.*, 1978, **89**, 196 549).	108

78MI41608	S. Watarai, H. Katsuyama, A. Umehara and H. Sato; *J. Polym. Sci., Polym. Chem. Ed.*, 1978, **16**, 2039.	107
78MI41609	A. Lllamas, A. Gonzalez and E. Martinez; *Carbohydr. Res.*, 1978, **67**, 515.	101
78MI41610	A. Rahman, N. Razzaq and A. Jabbar; *Pak. J. Sci. Res.*, 1978, **30**, 91 (*Chem. Abstr.*, 1979, **90**, 203 967).	95
78MI41611	E. Domagalina and T. Slawik; *Pol. J. Pharmacol. Pharm.*, 1978, **30**, 717 (*Chem. Abstr.*, 1979, **91**, 74 510).	56, 129
B-78MI41612	I. Juranie, L. Lorenc and M. L. Mihailovic; in 'Proc. 7th IUPAC Symp. Photochem.', 1978, p. 193 (*Chem. Abstr.*, 1979, **90**, 151 267).	46
78MI41613	K. Matsumoto and N. Tsuji; *Sankyo Kenkyusho Nempo*, 1978, **30**, 207 (*Chem. Abstr.*, 1979, **90**, 146 988).	43
78MI41614	S. E. Lowe and J. Sheridan; *Chem. Phys. Lett.*, 1978, **58**, 79.	6
78MI41615	D. G. McCormick and W. S. Hamilson; *J. Chem. Thermodyn.*, 1978, **10**, 275.	10
78MI41700	C. Rufer, F. Bahlmann and J. F. Kapp; *Eur. J. Med. Chem., Chim. Ther.*, 1978, **13**, 193 (*Chem. Abstr.*, 1978, **89**, 108 923).	149
78MI41701	J. Kulig and B. Lenarcik; *Pol. J. Chem.*, 1978, **52**, 477 (*Chem. Abstr.*, 1978, **89**, 31 719).	132
78MI41702	L. Stefaniak; *Bull. Acad. Pol. Sci., Ser. Sci. Chim.*, 1978, **26**, 291 (*Chem. Abstr.*, 1978, **89**, 128 706).	139
78MI41800	R. D. Gordon; *Spectrosc. Lett.*, 1978, **11**, 607.	182
78MI41801	T. Saegusa, A. Yamada, H. Taoda and S. Kobayashi; *Macromolecules*, 1978, **11**, 435.	211
78MI41900	M. Guiliano, G. Mille, T. Avignon and J. Chouteau; *J. Raman Spectrosc.*, 1978, **7**, 214 (*Chem. Abstr.*, 1979, **90**, 5439).	242
78MI42000	B. Bak, O. J. Nielsen, H. S. Svanhold and A. Holm; *Chem. Phys. Lett.*, 1978, **55**, 36.	350
78MI42200	A. De Munno, V. Bertini, P. Rasero, N. Picci and L. Bonfanti; *Atti Accad. Naz. Lincei, Cl. Sci. Fis. Mat. Nat. Rend.*, 1978, **64**, 385 (*Chem. Abstr.*, 1979, **91**, 139 958).	401
78MI42300	N. Jaiswal, B. R. Pandey, K. Raman, J. P. Barthwal, K. Kishor and K. P. Bhargava; *Indian J. Pharm. Sci.*, 1978, **40**, 202 (*Chem. Abstr.*, 1979, **90**, 132 589).	445
78MI42600	T.-S. Lin and J. R. Braun; *Chem. Phys.*, 1978, **28**, 379.	517
78MI42601	A. De Munno, V. Bertini, P. Rasero, N. Picci and L. Bonfanti; *Atti Accad. Naz. Lincei, Cl. Sci. Fis. Mat. Nat. Rend.*, 1978, **64**, 385 (*Chem. Abstr.*, 1979, **91**, 139 958).	535
78MI42700	M. El Dareer, K. F. Tillery and D. L. Hill; *Cancer Treat. Rep.*, 1978, **62**, 75 (*Chem. Abstr.*, 1978, **88**, 163 795).	576
78MI42701	L. Stefaniak; *Bull. Acad. Pol. Sci.*, 1978, **26**, 291.	551
78MI42702	N. B. Singh and J. Singh; *J. Therm. Anal.* 1978, **14**, 229 (*Chem. Abstr.*, 1979, **90**, 15 677).	556
B-78MI42800	M. Sassi, N. T. Stradiotto, D. W. Franco and E. A. Neves; in 'Symp. Bras. Eletroquim. Eletroanal.,' 1978, p. 35.	592
B-78MI42801	E. H. Nitto, N. R. Stradiotto and D. W. Franco; in 'Symp. Bras. Eletroquim. Eletroanal.,' 1978, p. 105.	592
B-78MI42802	W. L. Polito, E. A. Neves, D. W. Franco and T. Tamura; in 'Symp. Bras. Eletroquim. Eletroanal.,' 1978, p. 127.	592
78MI42900	P. Gayral, J. Bourdais, A. Lorre, D. Abenheim, F. Dusset, M. Pommies and G. Fouret; *Eur. J. Med. Chem. – Chim. Ther.*, 1978, **13**, 171.	686
78MI42901	H. Wollweber, U. Pohl and K. Stoepel; *Eur. J. Med. Chem. – Chim. Ther.*, 1978, **13**, 141.	708
B-78MI43300	P. S. Bailey; 'Ozonation in Organic Chemistry, Vol. 1, Olefinic Compounds', Academic, New York, 1978. 852, 853, 859, 862, 866, 880, 881, 888, 895	
B-78MI43301	'Verfahrensberichte zur physikalisch-chemischen Behandlung von Abwässern. 8. Bericht: Abwasserreinigung durch Ozonisierung,' ed. Verband der Chemischen Industrie, Köln, 1978.	893
B-78MI43302	R. A. Bailey, H. M. Clark, J. P. Ferris, S. Krause and R. L. Strong; 'Chemistry of the Environment', Academic, New York, 1978.	895
B-78MI43303	I. Flament, B. Willhalm and G. Ohloff; in 'Flavor of Foods and Beverages', ed. G. Charalambous and G. E. Inglett; Academic, New York, 1978, p. 15.	894
78MI43304	B. L. Oser and R. A. Ford; *Food Technol.*, 1978, 60.	895
78MI43400	H. W. Atland and I. A. Olivares; *Photogr. Sci. Eng.*, 1978, **22**, 263 (*Chem. Abstr.*, 1978, **89**, 223 928).	946
78MI43500	Z. G. Soos; *J. Chem. Ed.*, 1978, **55**, 546.	949
78MI43501	V. K. Bel'skii and D. Voet; *J. Cryst. Mol. Struct.*, 1978, **8**, 9.	950
78MI43600	S. Wawzonek; *OPPI Briefs*, 1978, **16B**, 737.	984
78MIP41600	A. Svab and J. Leiner; *Czech. Pat.* 175 335 (1978) (*Chem. Abstr.*, 1979, **90**, 168 579).	64, 85
78MIP43400	K. E. Fuger; *Can. Pat.* 1 029 027 (1978) (*Chem. Abstr.*, 1978, **89**, 108 219).	945
78NKK1256	T. Kinoshita, S. Sato, Y. Furukawa and C. Tamura; *Nippon Kagaku Kaishi*, 1978, 1256.	466
78OMR(11)385	L. Stefaniak; *Org. Magn. Reson.*, 1978, **11**, 385.	139, 182, 515, 519
78OMS14	R. M. Srivastava; *Org. Mass Spectrom.*, 1978, **13**, 14.	380
78OMS119	G. Salmona and E. J. Vincent; *Org. Mass Spectrom.*, 1978, **13**, 119.	142

78OMS121	J. R. Andersen, H. Egsgaard, E. Larsen, K. Bechgaard and E. M. Engler; *Org. Mass Spectrom.*, 1978, **13**, 121.	955
78OMS379	M. R. Arshadi; *Org. Mass Spectrom.*, 1978, **13**, 379.	399, 522
78OPP59	B. Stanovnik, M. Tišler and B. Valenčič; *Org. Prep. Proced. Int.*, 1978, **10**, 59.	594
78OS(58)106	N. A. LeBel and D. Hwang; *Org. Synth.*, 1978, **58**, 106.	108
78RRC1541	A. Abdel Maksoud, G. Hosni, O. Hassan and S. Shafik; *Rev. Roum. Chim.*, 1978, **23**, 1541 (*Chem. Abstr.*, 1979, **91**, 39 067).	93
78S43	A. A. Akhrem, F. A. Lakhvich, V. A. Khripach and A. G. Pozdeyev; *Synthesis*, 1978, 43.	118
78S55	K. Fukunaga; *Synthesis*, 1978, 55.	77
78S535	D. Daniil, U. Merkle and H. Meier; *Synthesis*, 1978, 535.	437
78S588	M. E. Jung, M. A. Mazurek and R. M. Lim; *Synthesis*, 1978, 588.	762
78S768	V. I. Cohen; *Synthesis*, 1978, 768.	355
78S829	B. Singh and G. Y. Lesher; *Synthesis*, 1978, 829.	65, 85
78SA(A)877	M. Witanowski, L. Stefaniak, A. Grabowska and G. A. Webb; *Spectrochim. Acta, Part A*, 1978, **34**, 877.	519
78SC315	R. S. Tewari and K. C. Gupta; *Synth. Commun.*, 1978, **8**, 315.	797
78SST(5)431	M. Davis; *Org. Compd. Sulphur, Selenium, Tellurium*, 1978, **5**, 431.	460
78T213	I. D. Entwistle, T. Gilkerson, R. A. W. Johnstone and R. P. Telford; *Tetrahedron*, 1978, **34**, 213.	536
78T453	A. Martvoň, L. Floch and S. Sekretár; *Tetrahedron*, 1978, **34**, 453.	594
78T481	A. Melhorn, B. Schwenzer, H. J. Brückner and K. Schwetlick; *Tetrahedron*, 1978, **34**, 481.	237
78T989	J. Becher, C. Dreier, E. G. Frandsen and A. S. Wengel; *Tetrahedron*, 1978, **34**, 989.	635, 636, 642
78T1571	A. V. Chapman, M. J. Cook, A. R. Katritzky, M. H. Abraham, A. F. Daniel deNamor; *Tetrahedron*, 1978, **34**, 1571.	4
78T2057	J. R. Beck; *Tetrahedron*, 1978, **34**, 2057.	1013
78T2459	G. Leroy, M. T. Nguyen and M. Sana; *Tetrahedron*, 1978, **34**, 2459.	108, 109
78T3419	W. C. Herndon and C. Parkanyi; *Tetrahedron*, 1978, **34**, 3419.	516, 784
78TL13	E. J. Corey and D. L. Boger; *Tetrahedron Lett.*, 1978, 13.	329
78TL671	H. Gotthardt, C. M. Weisshuhn, O. M. Huss and D. J. Brauer; *Tetrahedron Lett.*, 1978, 671.	820
78TL841	P. Rademacher and B. Freckmann; *Tetrahedron Lett.*, 1978, 841.	5, 10
78TL1291	H. Koga, M. Hirobe and T. Okamoto; *Tetrahedron Lett.*, 1978, 1291.	1038, 1046
78TL1887	E. Tighineanu, F. Chiraleu and D. Răileanu; *Tetrahedron Lett.*, 1978, 1887.	654
78TL2309	R. K. Smalley, R. H. Smith and H. Suschitzky; *Tetrahedron Lett.*, 1978, 2309.	52, 122, 123
78TL2325	T. Sheradsky and S. Avramovici-Grisaru; *Tetrahedron Lett.*, 1978, 2325.	223
78TL2345	S. Ncube, A. Pelter, K. Smith, P. Blatcher and S. Warren; *Tetrahedron Lett.*, 1978, 2345.	964
78TL2753	J. J. Tufariello and R. C. Gatrone; *Tetrahedron Lett.*, 1978, 2753.	630, 631
78TL3129	V. Jaeger and W. Schwab; *Tetrahedron Lett.*, 1978, 3129.	38
78TL3133	V. Jaeger, V. Buss and W. Schwab; *Tetrahedron Lett.*, 1978, 3133.	36
78TL3985	K. R. Fountan and G. Gerhardt; *Tetrahedron Lett.*, 1978, 3985.	113
78TL4647	J. J. Tufariello and S. A. Ali; *Tetrahedron Lett.*, 1978, 4647.	631
78TL5161	N. F. Haley; *Tetrahedron Lett.*, 1978, 5161.	847
78USP4053481	J. B. Hill; *U.S. Pat.* 4 053 481 (*Chem. Abstr.*, 1978, **88**, 37 783).	104
78USP4066769	J. B. Wright; *U.S. Pat.* 4 066 769 (1978) (*Chem. Abstr.*, 1978, **88**, 105 313).	128
78USP4069226	M. S. Kablaoui, J. M. Larkin and R. E. Reid; *U.S. Pat* 4 069 226 (*Chem. Abstr.*, 1978, **88**, 121 157).	96
78USP4075205	C. A. Wilson and C. E. Mixan; *U.S. Pat.* 4 075 205 (1978) (*Chem. Abstr.*, 1978, **88**, 170 195).	543
78USP4075207	*U.S. Pat.* 4 075 207 (1978) (*Chem. Abstr.*, 1978, **88**, 190 900).	710
78USP4080499	*U.S. Pat.* 4 080 499 (1978) (*Chem. Abstr.*, 1978, **89**, 43 506).	748
78USP4092327	R. G. Duranleau; *U.S. Pat.* 4 092 327 (1978) (*Chem. Abstr.*, 1978, **89**, 146 892).	95, 96, 130
78USP4094986	J. Rokach and G. W. Reader; *U.S. Pat.* 4 094 986 (1978) (*Chem. Abstr.*, 1978, **89**, 163 579).	543
78USP4104375	L. W. Fancher; *U.S. Pat.* 4 104 375 (1978) (*Chem. Abstr.*, 1979, **90**, 55 103).	130
78USP4107059	Pennwalt Corp., *U.S. Pat.* 4 107 059 (1978) (*Chem. Abstr.*, 1978, **90**, 154 577).	511
78USP4107377	Olin Corp., *U.S. Pat.* 4 107 377 (1978) (*Chem. Abstr.*, 1979, **90**, 87 447).	494, 507
78USP4115095	Monsanto, *U.S. Pat.* 4 115 095 (1978) (*Chem. Abstr.*, 1979, **90**, 72 202).	509
78USP4116812	R. L. Godar and P. R. Corneli; (Petrolite Corp.), *U.S. Pat.* 4 116 812 (1978) (*Chem. Abstr.*, 1979, **90**, 57 733).	811
78USP4127584	J. Rokach, E. J. Cragoe, Jr., C. S. Rooney and G. W. Reader; *U.S. Pat.* 4 127 584 (1978) (*Chem. Abstr.*, 1979, **90**, 87 474).	543
78ZC66	D. Heydenhauss, G. Jaenecke, H. Voigt and F. Hofmann; *Z. Chem.*, 1978, **18**, 66.	743
78ZC262	H.-G. Henning, B.-M. Neumann and L. Alder; *Z. Chem.*, 1978, **18**, 262.	376
78ZC323	R. Mayer, S. Bleisch and G. Domschke; *Z. Chem.*, 1978, **18**, 323 (*Chem. Abstr.*, 1979, **90**, 56 311).	926, 936
78ZC336	H. Buchwald and K. Rühlmann; *Z. Chem.*, 1978, **18**, 336.	611

78ZN(A)145	A. Kumar, J. Sheridan and O. L. Stiefvater; *Z. Naturforsch., Teil A*, 1978, **33**, 145.	181
78ZN(A)549	A. Kumar, J. Sheridan and O. L. Stiefvater; *Z. Naturforsch., Teil A*, 1978, **33**, 549.	181
78ZN(A)1511	O. L. Stiefvater; *Z. Naturforsch., Teil A*, 1978, **33**, 1511.	514
78ZN(A)1518	O. L. Stiefvater; *Z. Naturforsch., Teil A*, 1978, **33**, 1518.	514
78ZN(B)316	W. Winter, U. Plücken and H. Meier; *Z. Naturforsch., Teil B*, 1978, **33**, 316.	450, 455, 456
78ZN(B)1056	L. A. Summers; *Z. Naturforsch., Teil. B*, 1978, **33**, 1056.	57
78ZN(B)1120	W. L. Driessen and P. L. A. Everstijn; *Z. Naturforsch., Teil B*, 1978, **33**, 1120.	403
78ZOB418	A. D. Garnovskii, B. G. Gribov and S. D. Sokolov; *Zh. Obshch. Khim.*, 1978, **48**, 418.	20
78ZOB1465	O. A. Mukhacheva, M. A. Shchelkunova and A. I. Razumov; *Zh. Obshch. Khim.*, 1978, **48**, 1465 (*Chem. Abstr.*, 1978, **89**, 180 100).	917
78ZOR216	V. P. Martynova, E. R. Zakhs and A. V. El'tsov; *Zh. Org. Khim.*, 1978, **14**, 216 (*Chem. Abstr.*, 1978, **88**, 170 022).	684
78ZOR862	E. S. Levchenko and T. N. Dubinina; *Zh. Org. Khim.*, 1978, **14**, 862.	148, 164
78ZOR1255	I. V. Vigalok, A. V. Ostrovskaya, G. G. Petrova and Ya. A. Levin; *Zh. Org. Khim.*, 1978, **14**, 1255.	412
78ZOR1693	Ya. D. Samuilov, S. E. Salov'eva, T. F. Girutskaya and A. I. Konovalov; *Zh. Org. Khim.*, 1978, **14**, 1693.	108, 109
78ZOR2000	L. N. Volovel'skii, S. A. Korotkov, N. V. Popova and Z. P. Shchechenko; *Zh. Org. Khim.*, 1978, **14**, 2000 (*Chem. Abstr.*, 1979, **90**, 22 865).	98
78ZOR2003	M. I. Shevchuk, A. F. Tolochko, M. G. Bal'on and M. V. Khalaturnik; *Zh. Org. Khim.*, 1978, **14**, 2003.	70, 71
78ZOR2459	V. N. Drozd, G. S. Bogomolova and Y. M. Udachin; *Zh. Org. Khim.*, 1978, **14**, 2459.	799
79AC(R)81	A. Selva, R. Stoyanova, U. Vettori and S. Auricchio; *Ann. Chim. (Rome)*, 1979, **69**, 81.	6
79ACR79	J. B. Torrance; *Acc. Chem. Res.*, 1979, **12**, 79.	949
79ACR396	J. J. Tufariello; *Acc. Chem. Res.*, 1979, **12**, 396.	3
79AF362	R. Gericke and W. Rogalski; *Arzneim.-Forsch.*, 1979, **29**, 362.	598
79AF728	A. Wahab; *Arzneim.-Forsch.*, 1979, **29**, 728.	598
79AG91	V. Jaeger, H. Grund and W. Schwab; *Angew. Chem.*, 1979, **91**, 91.	36
79AG503	C. Wentrup, G. Gerecht and H. Briehl; *Angew. Chem.*, 1979, **91**, 503.	42
79AG(E)167	J. Fischer and W. Steglich; *Angew. Chem., Int. Ed. Engl.*, 1979, **18**, 167.	200
79AG(E)223	H. W. Roesky, N. Amin, G. Reumers, A. Gieren, U. Riemann and B. Dederer; *Angew. Chem., Int. Ed. Engl.*, 1979, **18**, 223.	975
79AG(E)307	G. Schnorrenberg and W. Steglich; *Angew. Chem., Int. Ed. Engl.*, 1979, **18**, 307.	985
79AG(E)683	B. Bogdanovic, C. Kruger and O. Kuzmin; *Angew. Chem., Int. Ed. Engl.*, 1979, **18**, 683.	1058
79AG(E)684	B. Bogdanovic, C. Kruger and P. Locatelli; *Angew. Chem., Int. Ed. Engl.*, 1979, **18**, 684.	1058
79AG(E)707	T. Mukaiyama; *Angew. Chem., Int. Ed. Engl.*, 1979, **18**, 707.	193
79AG(E)941	A. Senning; *Angew. Chem., Int. Ed. Engl.*, 1979, **18**, 941.	503
79AHC(25)46	J. W. Bunting; *Adv. Heterocycl. Chem.*, 1979, **25**, 46.	258
79AHC(25)147	B. J. Wakefield and D. J. Wright; *Adv. Heterocycl. Chem.* 1979, **25**, 147.	2, 3, 4, 5, 11, 12, 13, 15, 20, 21, 27, 29, 31, 33, 34, 54, 56, 57, 59, 60, 62, 66, 67, 68, 78
79AJC1749	E. R. Cole, G. Crank and B. J. Stapleton; *Aust. J. Chem.*, 1979, **32**, 1749.	766
79AP977	B. G. Ugarkar, B. V. Badami and G. S. Puranik; *Arch. Pharm. (Weinheim, Ger.)*, 1979, **312**, 977.	373
79AP1027	E. Eghtessad and G. Zinner; *Arch. Pharm. (Weinheim, Ger.)*, 1979, **312**, 1027.	910, 934, 936
79AX(B)437	J. Hašek, J. Obrda, K. Huml, S. Nešpurek and M. Šorm; *Acta Crystallogr., Part B*, 1979, **35**, 437.	375
79AX(B)468	M. A. Hamid, A. Hempel and S. E. Hull; *Acta Crystallogr., Part B*, 1979, **35**, 468.	618
79AX(B)470	M. A. Hamid and A. Hempel; *Acta Crystallogr., Part B*, 1979, **35**, 470.	618
79AX(B)471	A. Hempel and M. A. Hamid; *Acta Crystallogr., Part B*, 1979, **35**, 471.	4
79AX(B)772	T. J. Kistenmacher, T. J. Emge, P. Shu and D. O. Cowan; *Acta Crystallogr., Part B*, 1979, **35**, 772.	950
79AX(B)849	P. L. Dupont, O. Dideberg, J. Lamotte and J. L. Piette; *Acta Crystallogr., Part B*, 1979, **35**, 849.	951
79AX(B)2256	L. Golič, I. Leban, B. Stanovik and M. Tišler; *Acta Crystallogr., Part B*, 1979, **35**, 2256.	375
79AX(B)2449	J. Hašek, J. Obrda, K. Huml, S. Nešpurek and M. Šorm; *Acta Crystallogr., Part B*, 1979, **35**, 2449.	369
79AX(B)3076	D. Britton and J. M. Olson; *Acta Crystallogr., Part B*, 1979, **35**, 3076.	396
79AX(B)3114	L. Golič, I. Leban, B. Stanovik and M. Tisler; *Acta Crystallogr., Part B*, 1979, **35**, 3114.	449
79BCJ2002	Y. Inagaki, R. Okazaki and N. Inamoto; *Bull. Chem. Soc. Jpn.*, 1979, **52**, 2002.	927
79BCJ2008	Y. Inagaki, R. Okazaki, N. Inamoto and K. Yamada; *Bull. Chem. Soc. Jpn.*, 1979, **52**, 2008.	909, 936

79BCJ3615	Y. Inagaki, R. Okazaki and N. Inamoto; *Bull. Chem. Soc. Jpn.*, 1979, **52**, 3615.	926, 927
79BCJ3640	K. T. Kang, R. Okazaki, and N. Inamoto; *Bull. Chem. Soc. Jpn.*, 1979, **52**, 3640.	796
79BCJ3763	Y. Inouye, Y. Watanabe, S. Takahashi and H. Kakisawa; *Bull. Chem. Soc. Jpn.*, 1979, **52**, 3763.	108, 109
79BRP1548397	M. C. Neville, W. J. Ross and A. Todd; *Br. Pat.* 1 548 397 (1979) (*Chem. Abstr.*, 1980, **92**, 58 761).	128
79BRP1550440	S. Valenti; *Br. Pat.* 1 550 440 (1979) (*Chem. Abstr.*, 1980, **92**, 199 758).	438, 446
79BRP1550828	*Br. Pat.* 1 550 828 (1979) (*Chem. Abstr.*, 1980, **92**, 112 246).	639
79BRP2015878	*Br. Pat.* 2 015 878 (1979) (*Chem. Abstr.*, 1980, **93**, 26 443).	605
79BRP2018247	M. Yoshimoto, H. Miyazawa, T. Nishimura, A. Ando, N. Nakamura and H. Nakao; *Br. Pat. Appl.* 2 018 247 (1979) (*Chem. Abstr.*, 1980, **93**, 26 449).	128
79BSB107	G. L'abbé, A. Willocx, J.-P. Declercq, G. Germain and M. Van Meerssche; *Bull. Soc. Chim. Belg.*, 1979, **88**, 107.	580
79BSB245	G. L'abbé, M. Komatsu, C. Martens, S. Toppet, J.-P. Declercq, G. Germain and M. Van Meerssche; *Bull. Soc. Chim. Belg.*, 1979, **88**, 245.	589
79BSF(2)26	D. Barillier; *Bull. Soc. Chim. Fr.*, Part 2, 1979, 26.	138, 160, 794
79BSF(2)199	M. Perrier and J. Vialle; *Bull. Soc. Chim. Fr.*, Part 2, 1979, 199.	361, 1063
79BSF(2)205	M. Perrier and J. Vialle; *Bull. Soc. Chim. Fr.*, Part 2, 1979, 205.	1063
79CB260	H. Gotthardt, F. Reiter, R. Gleiter and R. Bartetzko; *Chem. Ber.*, 1979, **112**, 260.	1029, 1031, 1032, 1043
79CB266	H. Gotthardt and F. Reiter; *Chem. Ber.*, 1979, **112**, 266.	1034
79CB517	J. Goerdler and W. Löbach; *Chem. Ber.*, 1979, **112**, 517.	939
79CB577	J. Mattay, H. Leismann and H.-D. Scharf; *Chem. Ber.*, 1979, **112**, 577.	765
79CB1012	L. Capuano, G. Urhahn and A. Willmes; *Chem. Ber.*, 1979, **112**, 1012.	539, 940
79CB1102	A. Q. Hussein, S. Herzberger and J. C. Jochims; *Chem. Ber.*, 1979, **112**, 1102.	594
79CB1193	H. Gotthardt and F. Reiter; *Chem. Ber.*, 1979, **112**, 1193.	376
79CB1288	J. Goerdeler, J. Haag and W. Löbach; *Chem. Ber.*, 1979, **112**, 1288.	148, 149
79CB1635	H. Gotthardt and F. Reiter; *Chem. Ber.*, 1979, **112**, 1635.	374
79CB1650	H. Gotthardt, O. M. Huss and C. M. Weisshuhn; *Chem. Ber.*, 1979, **112**, 1650.	820, 831
79CB1769	E. Schaumann, J. Ehlers, W.-R. Förster and G. Adiwidjaja; *Chem. Ber.*, 1979, **112**, 1769.	454
79CB1873	K. Friedrich and M. Zamkanei; *Chem. Ber.*, 1979, **112**, 1873.	925, 941
79CB1956	A. Q. Hussein and J. C. Jochims; *Chem. Ber.*, 1979, **112**, 1956.	594
79CB2120	W. Kirmse, O. Schnurr and H. Jendralla; *Chem. Ber.*, 1979, **112**, 2120.	213
79CB3282	C. Skoetsch and E. Breitmaier; *Chem. Ber.*, 1979, **112**, 3282.	621, 649, 651, 652, 662
79CB3286	H. Boeshagen and W. Geiger; *Chem. Ber.*, 1979, **112**, 3286.	162
79CB3623	H. Franke, H. Grasshoff and G. Scherowsky; *Chem. Ber.*, 1979, **112**, 3623.	436
79CB3728	H. Buehl, U. Timm and H. Meier; *Chem. Ber.*, 1979, **112**, 3728.	455
79CC99	A. Holm, C. Berg, C. Bjerre, B. Bak and H. Svanholt; *J. Chem. Soc., Chem. Commun.*, 1979, 99.	350
79CC751	A. Goosen and C. W. McCleland; *J. Chem. Soc., Chem. Commun.*, 1979, 751.	761
79CC786	J. Rokach and P. Hamel; *J. Chem. Soc., Chem. Commun.*, 1979, 786.	147
79CC1135	H. C. Hansen and A. Senning; *J. Chem. Soc., Chem. Commun.*, 1979, 1135.	595
79CHE1187	A. N. Borisevich and P. S. Pel'kis; *Chem. Heterocycl. Compd. (Engl. Trans.)*, 1979, **15**, 1187.	909, 915, 936, 943
79CI(L)319	J. B. Bremner and K. N. Winzenberg; *Chem. Ind. (London)*, 1979, 319.	654
79CJC207	D. M. McKinnon, M. E. Hassan and M. S. Chauhan; *Can. J. Chem.*, 1979, **57**, 207.	160
79CJC3168	H. Hiemstra, H. A. Houwing, O. Possel and A. M. van Leusen; *Can. J. Chem.*, 1979, **57**, 3168.	182
79CL9	N. Kamigata, T. Saegusa, S. Fujie and M. Kobayashi; *Chem. Lett.*, 1979, 9.	147
79CL705	T. Mukaiyama, Y. Sakito and M. Asami; *Chem. Lett.*, 1979, 705.	983
79CL1029	O. Tsuge, T. Takata and I. Ueda; *Chem. Lett.*, 1979, 1029.	534, 1035, 1036
79CL1117	T. Mukaiyama, K. Kawata, A. Sasaki and M. Asami; *Chem. Lett.*, 1979, 1117.	193
79CL1261	H. Matsuda, A. Ninagawa and R. Nomura; *Chem. Lett.*, 1979, 1261.	780
79CL1265	K. Inomata, H. Kinoshita, H. Fukuda, O. Miyano, Y. Yamashiro and H. Kotake; *Chem. Lett.*, 1979, 1265.	174
79CPB410	N. Suzuki and R. Dohmori; *Chem. Pharm. Bull.*, 1979, **27**, 410.	683, 687, 689, 699, 710
79CPB880	A. Kosasayama, K. Higashi and F. Ishikawa; *Chem. Pharm. Bull.*, 1979, **27**, 880.	665
79CPB1147	H. Takahashi, N. Nimura and H. Ogura; *Chem. Pharm. Bull.*, 1979, **27**, 1147.	643
79CPB1207	K. Kubo, N. Ito, Y. Isomura, I. Sozu, H. Homma and M. Murakami; *Chem. Pharm. Bull.*, 1979, **27**, 1207.	701
79CPB2261	T. Hisano, M. Ichikawa, T. Matsuoka, H. Hagiwara, K. Muraoka, T. Komori, K. Harano, Y. Ida and A. T. Christensen; *Chem. Pharm. Bull.*, 1979, **27**, 2261.	643, 667
79CPB2398	K. Tomita, S. Sugai, T. Kobayashi and T. Murakami; *Chem. Pharm. Bull.*, 1979, **27**, 2398.	43, 57, 86
79CPB2415	K. Tomita, S. Sugai and M. Saito; *Chem. Pharm. Bull.*, 1979, **27**, 2415.	58, 88
79CRV181	E. C. Taylor and I. J. Turchi; *Chem. Rev.*, 1979, **79**, 181.	218, 1008
79CSC319	M. J. Begley, T. J. King and D. B. Sowerby; *Cryst. Struct. Commun.*, 1979, **8**, 319.	902
79CSC415	G. Pelizzi, P. Tarasconi and G. Ponticelli; *Cryst. Struct. Commun.*, 1979, **8**, 415.	4
79DOK(244)615	A. A. Akhrem, F. A. Lakhvich, V. A. Kirpack and I. B. Klebanovich; *Dokl. Akad. Nauk SSSR*, 1979, **244**, 615.	36
79DOK(246)1130	B. A. Arbuzov, E. N. Dianova and N. A. Chadeva; *Dokl. Akad. Nauk SSSR*, 1979, **246**, 1130.	1001

79EGP135901	J. Liebscher and H. Hartmann; *Ger. (East) Pat.* 135 901 (1979) (*Chem. Abstr.*, 1979, **91**, 193 019).	917
79EGP136963	Akad. Wissen. DDR., *Ger. (East) Pat.* 136 963 (1979) (*Chem. Abstr.*, 1980, **92**, 41 963).	577
79EUP2666	G. M. Shutske, L. L. Setescak and R. C. Allen; *Eur. Pat. Appl.* 2666 (1979) (*Chem. Abstr.*, 1980, **92**, 94 378).	114, 115, 129
79FRP2422667	Sankyo Co. Ltd., *Fr. Pat.* 2 422 667 (1979) (*Chem. Abstr.*, 1980, **93**, 8193).	128
79GEP2728523	Schering A.-G., *Ger. Pat.* 2 728 523 (*Chem. Abstr.* 1979, **90**, 137 833).	462
79GEP2734866	BASF A.G.; *Ger. Pat.* 2 734 866 (1979) (*Chem. Abstr.*, 1979, **90**, 204 087).	169
79GEP2818998	P. C. Thieme, H. Theobald, A. Franke and R. Huber; *Ger. Pat.* 2 818 998, 2 818 999 (1979) (*Chem. Abstr.*, 1979, **92**, 94 381, 94 382).	128
79GEP2851471	M. Iwasaki; *Ger. Offen.* 2 851 471 (1979) (*Chem. Abstr.*, 1979, **91**, 202 241).	446
79GEP2900708	T. Nakagawa, Y. Watanabe and K. Ohmori; *Ger. Offen.* 2 900 708 (1979) (*Chem. Abstr.*, 1979, **91**, 152 744).	129
79GEP2909025	T. Honna, M. Tanaka, S. Yamada and H. Miyake; *Ger. Offen.* 2 909 025 (1979) (*Chem. Abstr.*, 1979, **92**, 41 927).	128
79H(12)239	K. Nagahara and A. Takada; *Heterocycles*, 1979, **12**, 239.	384
79H(12)485	H. Okuda, Y. Tominaga, Y. Matsuda and G. Kobayashi; *Heterocycles*, 1979, **12**, 485.	636, 637, 643
79H(12)1343	C. Kashima; *Heterocycles*, 1979, **12**, 1343.	49, 50, 83
79H(13)297	Y. Yamamoto and K. Akiba; *Heterocycles*, 1979, **13**, 297.	592
79HC(34-1)1	J. V. Metzger; *Chem. Heterocycl. Compd.*, 1979, **34-1**, 1.	236, 240, 241, 245, 252, 262, 267, 292
79HC(34-2)1	J. V. Metzger; *Chem. Heterocycl. Compd.*, 1979, **34-2**, 1.	247, 248, 252, 291, 328
79HC(34-3)1	J. V. Metzger; *Chem. Heterocycl. Compd.*, 1979, **34-3**, 1.	277, 330
79HCA185	E. Giovannini, B. F. S. E. de Souza and F. S. E. Bernardo; *Helv. Chim. Acta*, 1979, **62**, 185.	18
79HCA198	E. Giovannini, B. F. S. E. de Souza and F. S. E. Bernardo; *Helv. Chim. Acta*, 1979, **62**, 198.	18
79HCA271	T. Doppler, H. Schmid and H. J. Hansen; *Helv. Chim. Acta*, 1979, **62**, 271.	18
79HCA304	T. Doppler, H. Schmid and H. J. Hansen; *Helv. Chim. Acta*, 1979, **62**, 304.	18, 19
79HCA314	T. Doppler, H. Schmid and H. J. Hansen; *Helv. Chim. Acta*, 1979, **62**, 314.	11, 17
79HCA391	B. Jackson, H. Schmid and H. J. Hansen; *Helv. Chim. Acta*, 1979, **62**, 391.	147
79HCA1236	J. Lukáč and H. Heimgartner; *Helv. Chim. Acta*, 1979, **62**, 1236.	825
79IC213	O. Glemser and J. M. Shreeve; *Inorg. Chem.*, 1979, **18**, 213.	539, 891
79IJC(A)191	N. B. Singh and J. Singh; *Indian J. Chem., Sect. A*, 1979, **17**, 191 (*Chem. Abstr.*, 1979, **91**, 167 635).	565
79IJC(B)4	H. Singh, K. B. Lal and S. Malhotra; *Indian J. Chem., Sect. B*, 1979, **17**, 4.	687
79IJC(B)13	K. S. Sharma, R. P. Singh and V. Singh; *Indian J. Chem., Sect. B*, 1979, **17**, 13.	534
79IJC(B)284	R. Rai and V. K. Verma; *Indian J. Chem., Sect. B*, 1979, **18**, 284.	942
79IJC(B)371	K. A. Thakar, B. M. Bhawal and A. B. Dumir; *Indian J. Chem., Sect. B*, 1979, **18**, 371.	25, 48
79IJC(B)486	J. N. Shah and B. D. Tilak; *Indian J. Chem., Sect. B*, 1979, **18**, 486.	696
79IZV118	N. B. Kazmina, I. L. Knunyants, G. M. Kuz'yants, E. I. Mysov and E. P. Lur'e; *Izv. Akad. Nauk SSSR, Ser. Khim.*, 1979, 118.	866, 871, 893
79IZV131	A. I. Mishchenko, A. V.]Prosyanik, A. P. Pleshkova, M. D. Isobaev, V. I. Markov and R. G. Kostyanovskii; *Izv. Akad. Nauk SSR, Ser. Khim.*, 1979, 131 (*Chem. Abstr.*, 1979, **90**, 186 844).	108
79IZV1059	N. A. Akmanova, Kh. F. Sagitdinova and T. F. Petrushina; *Izv. Vyssh. Uchebn. Zaved., Khim. Khim. Tekhnol.*, 1979, **22**, 1059 (*Chem. Abstr.*, 1980, **92**, 76 371).	111
79IZV1788	D. P. Del'tsova, N. P. Gambaryan and E. P. Lur'e; *Izv. Akad. Nauk SSSR, Ser. Khim.*, 1979, 1788 (*Chem. Abstr.*, 1980, **92**, 6475).	938
79JA1054	R. C. Kelly, I. Schletter, S. J. Stein and W. Wierenga; *J. Am. Chem. Soc.*, 1979, **101**, 1054.	98
79JA1155	E. F. Perozzi and J. C. Martin; *J. Am. Chem. Soc.*, 1979, **101**, 1155.	751
79JA1319	P. A. Wade and H. R. Hinney; *J. Am. Chem. Soc.*, 1979, **101**, 1319.	36, 39, 91, 92
79JA2234	C. Pradat and I. G. Riess; *J. Am. Chem. Soc.*, 1979, **101**, 2234.	975
79JA2524	I. C. Hisatune, K. Shinoda and J. Heicklen; *J. Am. Chem. Soc.*, 1979, **101**, 2524.	888
79JA3607	P. S. Mariano and A. A. Leone; *J. Am. Chem. Soc.*, 1979, **101**, 3607.	648, 649, 651, 652, 656, 663
79JA3976	H. Murai, M. Torres and O. P. Strausz; *J. Am. Chem. Soc.*, 1979, **101**, 3976.	455
79JA4290	M. F. Salomon and R. G. Salomon; *J. Am. Chem. Soc.*, 1979, **101**, 4290.	754, 774
79JA6981	T. Graafland, A. Wagenaar, A. J. Kirby and J. B. F. N. Engberts; *J. Am. Chem. Soc.*, 1979, **101**, 6981.	151, 163, 165
79JA7099	A. L. J. Beckwith and R. D. Wagner; *J. Am. Chem. Soc.*, 1979, **101**, 7099.	774
79JAP7970822	K. Hokujo and S. Enado; *Jpn. Pat.* 79 70 822 (1979) (*Chem. Abstr.*, 1979, **91**, 202 161).	946
79JAP(K)7901314	H. Kano, H. Ogata and J. Yukinaga; *Jpn. Kokai* 79 01 314 (1979) (*Chem. Abstr.*, 1979, **91**, 74 591).	129
79JAP(K)7909278	R. Honna, K. Ogawa, O. Miyoshi, S. Hashimoto and T. Suzue; *Jpn. Kokai* 79 09 278 (1979) (*Chem. Abstr.*, 1979, **91**, 20 483).	128
79JAP(K)7912371	R. Motokuni, S. Yamada, K. Toratani, S. Hashimoto and T. Suzue; *Jpn. Kokai* 79 12 371 (1979) (*Chem. Abstr.*, 1979, **90**, 186 927).	128

79JAP(K)7914968	R. Honna, K. Ogawa, M. Tanaka, S. Yamada, K. Toratani, S. Hashimoto and T. Suzue; *Jpn. Kokai* 79 14 968 (1979) (*Chem. Abstr.*, 1980, **92**, 41 920).	128
79JAP(K)7930171	R. Honna, K. Ogawa, M. Tanaka, S. Yamada, K. Toratani, S. Hashimoto and T. Suzue; *Jpn. Kokai* 79 30 171 (1979) (*Chem. Abstr.*, **91**, 1979, 74 595).	128
79JAP(K)7934728	H. Kano and S. Takahashi; *Jpn. Kokai* 79 34 728 (1979) (*Chem. Abstr.*, 1980, **93**, 150 236).	128
79JAP(K)7944665	R. Honna, K. Ogawa, M. Tanaka, S. Yamada, K. Toratani, S. Hashimoto and T. Suzue; *Jpn. Kokai* 79 44 665 (1979) (*Chem. Abstr.*, 1979, **92**, 6518).	128
79JAP(K)7944686	R. Honna, K. Ogawa, K. Toratani, S. Hashimoto and T. Suzue; *Jpn. Kokai* 79 44 686 (1979) (*Chem. Abstr.*, 1979, **91**, 175 401).	128
79JAP(K)7944696	*Jpn. Kokai* 79 44 696 (1979) (*Chem. Abstr.*, 1979, **91**, 175 325).	634
79JAP(K)7944723	J. Yukinaga, S. Sumimoto, I. Ishizuka and M. Sugita; *Jpn. Kokai* 79 44 723 (1979) (*Chem. Abstr.*, 1980, **93**, 63 609).	129
79JAP(K)7946779	S. Sumimoto, I. Ishizuka, J. Sugita and H. Yukinaga; *Jpn. Kokai* 79 46 779 (1979) (*Chem. Abstr.*, 1979, **91**, 140 834).	129
79JAP(K)7952074	K. Tomita and M. Mizugai; *Jpn. Kokai* 79 52 074 (1979) (*Chem. Abstr.*, 1980, **92**, 41 921).	129
79JAP(K)7966685	K. Honna, S. Hashimoto and T. Suzue; *Jpn. Kokai* 79 66 685 (1979) (*Chem. Abstr.*, 1979, **91**, 193 292).	128
79JAP(K)7973773	R. Honna, M. Tanaka, S. Yamada, K. Ogawa and S. Hashimoto; *Jpn. Kokai* 79 73 773 (1979) (*Chem. Abstr.*, 1979, **91**, 193 294).	128
79JAP(K)7973774	M. Nagano, J. Sakai, M. Mizukai, N. Nakamura, E. Misaka, S. Kobayashi and K. Tomita; *Jpn. Kokai* 79 73 774 (1979) (*Chem. Abstr.*, 1980, **92**, 41 922).	128
79JAP(K)7984592	C. Eguchi, N. Yasuda, H. Iwagami, E. Takigawa, M. Okutsu, T. Onuki and T. Nakamiya; *Jpn. Kokai* 79 84 592 (1979) (*Chem. Abstr.*, 1979, **91**, 157 733).	128
79JAP(K)79106466	Nippon Soda Co. Ltd., *Jpn. Kokai* 79 106 466 (1979) (*Chem. Abstr.*, 1980, **92**, 58 755).	64, 88
79JAP(K)79119490	*Jpn. Kokai* 79 119 490 (1979) (*Chem. Abstr.*, 1980, **92**, 76 535).	710
79JCP(70)1898	D. Cremer; *J. Chem. Phys.*, 1979, **70**, 1898.	853, 854
79JCP(70)1911	D. Cremer; *J. Chem. Phys.*, 1979, **70**, 1911.	854
79JCP(70)1928	D. Cremer; *J. Chem. Phys.*, 1979, **70**, 1928.	854
79JCR(S)54	H. Grund and V. Jaeger; *J. Chem. Res. (S)*, 1979, 54.	38
79JCR(S)311	G. Bianchi and M. DeAmici; *J. Chem. Res. (S)*, 1979, 311.	37, 78
79JCR(S)314	J. F. Barnes, M. J. Barrow, M. M. Harding, R. M. Paton, P. L. Ashcroft, J. Crosby and C. J. Joyce; *J. Chem. Res. (S)*, 1979, 314.	420
79JCR(S)395	K. Clarke, B. Gleadhill and R. M. Scrowston; *J. Chem. Res. (S)*, 1979, 395.	155, 159, 162, 172
79JCS(D)371	W. Rigby, P. M. Bailey, J. A. McCleverty and P. M. Maitlis; *J. Chem. Soc., Dalton Trans.*, 1979, 371.	595
79JCS(P1)539	N. Ardabilchi, A. O. Fitton, J. R. Frost, F. K. Oppong-Boachie, A. H. b. A. Hadi and A. b. M. Sharif; *J. Chem. Soc., Perkin Trans. 1*, 1979, 539.	228
79JCS(P1)652	A. G. M. Barrett, D. H. R. Barton, J. R. Falck, D. Papaioannou and D. W. Widdowson; *J. Chem. Soc., Perkin Trans. 1*, 1979, 652.	229
79JCS(P1)732	R. N. Hanley, W. D. Ollis and C. A. Ramsden; *J. Chem. Soc., Perkin Trans. 1*, 1979, 732.	582, 597, 600, 601,
79JCS(P1)736	R. N. Hanley, W. D. Ollis and C. A. Ramsden; *J. Chem. Soc., Perkin Trans. 1*, 1979, 736.	599, 600, 601, 603, 604
79JCS(P1)741	R. N. Hanley, W. D. Ollis and C. A. Ramsden; *J. Chem. Soc., Perkin Trans. 1*, 1979, 741.	600, 601, 603, 605
79JCS(P1)744	R. N. Hanley, W. D. Ollis, C. A. Ramsden and I. S. Smith; *J. Chem. Soc., Perkin Trans. 1*, 1979, 744.	600, 605
79JCS(P1)747	R. N. Hanley, W. D. Ollis and C. A. Ramsden; *J. Chem. Soc., Perkin Trans. 1*, 1979, 747.	600
79JCS(P1)926	R. M. Christie, D. H. Reid and R. Wolfe-Murray; *J. Chem. Soc., Perkin Trans. 1*, 1979, 926.	1059, 1060, 1065
79JCS(P1)950	T. Durst, J. D. Finlay and D. J. H. Smith; *J. Chem. Soc., Perkin Trans. 1*, 1979, 950.	760
79JCS(P1)960	A. Holm, J. J. Christiansen and C. Lohse; *J. Chem. Soc., Perkin Trans. 1*, 1979, 960.	586
79JCS(P1)1013	S. Cerrini, W. Fedeli, F. Mazza, G. Lucente, M. P. Paradisi and A. Romeo; *J. Chem. Soc., Perkin Trans. 1*, 1979, 1013.	646
79JCS(P1)1150	T. Laerum and K. Undheim; *J. Chem. Soc., Perkin Trans. 1*, 1979, 1150.	673, 677, 678, 679, 680
79JCS(P1)1430	M. Cinquini, F. Cozzi and M. Pelosi; *J. Chem. Soc., Perkin Trans. 1*, 1979, 1430.	229
79JCS(P1)2340	A. G. Briggs, J. Cryzewski and D. H. Reid; *J. Chem. Soc., Perkin Trans. 1*, 1979, 2340.	1065
79JCS(P1)2905	S. Mataka, K. Takahashi, S. Ishii and M. Tashiro; *J. Chem. Soc., Perkin Trans. 1*, 1979, 2905.	541
79JCS(P1)2909	R. J. S. Beer, N. H. Holmes and A. Naylor; *J. Chem. Soc., Perkin Trans. 1*, 1979, 2909.	908, 921
79JCS(P2)533	Z. Said and J. G. Tillett; *J. Chem. Soc., Perkin Trans. 2*, 1979, 533.	600, 602
79JCS(P2)572	C. A. Veracini, A. DeMunno, G. Chidichimo, M. Longeri and V. Bertini; *J. Chem. Soc., Perkin Trans. 2*, 1979, 572.	6

79JCS(P2)763	C. Gaze and B. C. Gilbert; *J. Chem. Soc., Perkin Trans. 2*, 1979, 763.	764
79JCS(P2)1644	B. Mile, G. W. Morris and W. G. Alcock; *J. Chem. Soc., Perkin Trans. 2*, 1979, 1644.	859
79JCS(P2)1665	F. Mossini, M. R. Mingiardi, E. Gaetani, M. Nardelli and G. Pelizzi; *J. Chem. Soc., Perkin Trans. 2*, 1979, 1665.	162
79JCS(P2)1751	T. J. King, P. N. Preston, J. S. Suffolk and K. Turnbull; *J. Chem. Soc., Perkin Trans. 2*, 1979, 1751.	369, 373, 712, 714
79JHC49	G. Adembei, A. Camparini, F. Ponticelli and P. Tedeschi; *J. Heterocycl. Chem.*, 1979, **16**, 49.	623
79JHC61	M. H. Elnagdi, S. M. Fahmy, M. R. H. Elmoghayar and E. M. Kandeel; *J. Heterocycl. Chem.*, 1979, **16**, 61.	1010
79JHC129	M. L. Graziano, M. R. Iesce and R. Scarpati; *J. Heterocycl. Chem.*, 1979, **16**, 129.	905, 910, 925
79JHC145	G. Werber, F. Buccheri, N. Vivona and R. Bianchini; *J. Heterocycl. Chem.*, 1979, **16**, 145.	445
79JHC183	C. H. Chen and B. A. Donatelli; *J. Heterocycl. Chem.*, 1979, **16**, 183.	940
79JHC195	B. Stanovnik, M. Tišler, B. Kirn and I. Kovač; *J. Heterocycl. Chem.*, 1979, **16**, 195.	1018
79JHC311	J. P. Gilbert, C. Petrus and F. Petrus; *J. Heterocycl. Chem.*, 1979, **16**, 311.	981, 998
79JHC365	V. I. Cohen; *J. Heterocycl. Chem.*, 1979, **16**, 365.	357
79JHC685	S. Olivella and J. Vilarrasa; *J. Heterocycl. Chem.*, 1979, **16**, 685.	980
79JHC689	M. Iwao and T. Kuraishi; *J. Heterocycl. Chem.*, 1979, **16**, 689.	414, 417
79JHC731	N. G. Argyropoulos and N. E. Alexandron; *J. Heterocycl. Chem.*, 1979, **16**, 731.	975
79JHC877	N. P. Peet and P. B. Anzeveno; *J. Heterocycl. Chem.*, 1979, **16**, 877.	665
79JHC895	S. D. Ziman; *J. Heterocycl. Chem.*, 1979, **16**, 895.	611
79JHC903	R. A. Glennon, R. G. Bass and E. Schubert; *J. Heterocycl. Chem.*, 1979, **16**, 903.	702
79JHC961	L. Taliani and J. Perronnet; *J. Heterocycl. Chem.*, 1979, **16**, 961.	465, 485, 498, 499, 507
79JHC1009	S. Mataka, K. Takahashi, Y. Yamada and M. Tashiro; *J. Heterocycl. Chem.*, 1979, **16**, 1009.	541
79JHC1059	C. V. Greco and J. R. Mehta; *J. Heterocycl. Chem.*, 1979, **16**, 1059.	375
79JHC1097	M. El-Deek, K. El-Badry and S. M. Abdel-Wahhab; *J. Heterocycl. Chem.*, 1979, **16**, 1097.	439
79JHC1109	M. H. Elnagdi, S. M. Fahmy, E. A. A. Hafez, M. R. H. Elmoghayar and S. A. R. Amer; *J. Heterocycl. Chem.*, 1979, **16**, 1109.	627
79JHC1197	A. A. Santilli and R. L. Morris; *J. Heterocycl. Chem.*, 1979, **16**, 1197.	387
79JHC1249	T. J. McCord, D. R. Smith, J. K. Swan, A. M. Goebel, D. E. Thornton, C. C. Yaksheand and A. L. Davis; *J. Heterocycl. Chem.*, 1979, **16**, 1249.	120
79JHC1277	R. Friary and B. R. Sunday; *J. Heterocycl. Chem.*, 1979, **16**, 1277.	116
79JHC1295	U. Timm and H. Meier; *J. Heterocycl. Chem.*, 1979, **16**, 1295.	454
79JHC1303	U. Timm, U. Plücken, H. Petersen and H. Meier; *J. Heterocycl. Chem.*, 1979, **16**, 1303.	960, 964
79JHC1405	I. Lalezari, A. Shafiee and S. Sadeghi-Milani; *J. Heterocycl. Chem.*, 1979, **16**, 1405.	349, 353
79JHC1469	G. Palazzo, L. Baiocchi and G. Picconi; *J. Heterocycl. Chem.*, 1979, **16**, 1469.	381
79JHC1477	L. Baiocchi, G. Picconi and G. Palazzo; *J. Heterocycl. Chem.*, 1979, **16**, 1477.	381
79JHC1555	R. C. Boruah, P. Devi and J. S. Sandhu; *J. Heterocycl. Chem.*, 1979, **16**, 1555.	36
79JHC1563	A. Shafiee, A. Mazloumi and V. I. Cohen; *J. Heterocycl. Chem.*, 1979, **16**, 1563.	344, 1009
79JHC1589	G. P. Zecchini and M. P. Paradisi; *J. Heterocycl. Chem.*, 1979, **16**, 1589.	648, 650, 653, 666
79JHC1611	C. Bellec, D. Bertin, R. Colau, S. Deswarte, P. Maitte and C. Viel; *J. Heterocycl. Chem.*, 1979, **16**, 1611.	75
79JHC1617	J. D. Warren, V. J. Lee and R. B. Angier; *J. Heterocycl. Chem.*, 1979, **16**, 1617.	538
79JHC1657	C. Bellec, D. Bertin, R. Colau, S. Deswarte, P. Maitte and C. Viel; *J. Heterocycl. Chem.*, 1979, **16**, 1657.	75
79JIC1230	A. K. Sengupta, M. Garg and U. Chandra; *J. Indian Chem. Soc.*, 1979, **56**, 1230.	445
79JIC1251	D. V. Rao and P. Lingaiah; *J. Indian Chem. Soc.*, 1979, **56**, 1251.	10
79JMC237	J. Rokach, P. Hamel, N. R. Hunter, G. Reader, C. S. Rooney, P. S. Anderson, E. J. Cragoe, Jr., and L. R. Mandel; *J. Med. Chem.*, 1979, **22**, 237.	148, 157
79JMC944	R. B. Meyer, Jr. and E. B. Skibo; *J. Med. Chem.*, 1979, **22**, 944.	538
79JMC1030	C. Li, M. H. Lee and A. C. Sartorelli; *J. Med. Chem.*, 1979, **22**, 1030.	1013
79JMC1214	G. S. Lewis and P. H. Nelson; *J. Med. Chem.*, 1979, **22**, 1214.	453, 458, 462
79JMC1554	J. C. Saunders and W. R. N. Williamson; *J. Med. Chem.*, 1979, **22**, 1554.	116, 129
79JMR(36)227	M. Witanowski, L. Stefaniak and G. A. Webb; *J. Magn. Reson.*, 1979, **36**, 227.	370, 377
79JOC267	K. Rasheed and J. D. Warkentin; *J. Org. Chem.*, 1979, **44**, 267.	846
79JOC285	C. N. Filer, F. E. Granchelli, P. Perri and J. L. Neumeyer; *J. Org. Chem.*, 1979, **44**, 285.	663
79JOC302	E. C. Taylor and A. J. Cocuzza; *J. Org. Chem.*, 1979, **44**, 302.	719
79JOC510	M. J. Sanders, S. L. Dye, A. G. Miller and J. R. Grunwell; *J. Org. Chem.*, 1979, **44**, 510.	138, 156, 169
79JOC622	K. T. Potts, S. K. Datta and J. L. Marshall; *J. Org. Chem.*, 1979, **44**, 622.	1033
79JOC626	K. T. Potts and J. L. Marshall; *J. Org. Chem.*, 1979, **44**, 626.	186, 208
79JOC818	T. Nishimura and K. Kitajima; *J. Org. Chem.*, 1979, **44**, 818.	993
79JOC825	G. F. Field; *J. Org. Chem.*, 1979, **44**, 825.	984, 1020
79JOC835	E. J. Fornefeld and A. J. Pike; *J. Org. Chem.*, 1979, **44**, 835.	108
79JOC873	A. Silveira and S. K. Satra; *J. Org. Chem.*, 1979, **44**, 873.	41
79JOC930	M. Sato, N. C. Gonnella and M. P. Cava; *J. Org. Chem.*, 1979, **44**, 930.	844, 845, 965

79JOC1118	J. Rokach, P. Hamel, Y. Girard and G. Reader; *J. Org. Chem.*, 1979, **44**, 1118.	158, 159, 160, 162, 527, 538
79JOC1212	C. Belzecki and I. Panfil; *J. Org. Chem.*, 1979, **44**, 1212.	108, 109
79JOC1476	D. C. Green; *J. Org. Chem.*, 1979, **44**, 1476.	837
79JOC1807	E. A. Harrison, Jr.; *J. Org. Chem.*, 1979, **44**, 1807.	765
79JOC1819	N. A. LeBel, M. E. Post and D. Hwang; *J. Org. Chem.*, 1979, **44**, 1819.	46
79JOC2395	H. C. Berk and J. E. Franz; *J. Org. Chem.*, 1979, **44**, 2395.	376
79JOC2632	D. O. Lambeth and D. W. Swank; *J. Org. Chem.*, 1979, **44**, 2632.	139
79JOC2796	G. Zecchi; *J. Org. Chem.*, 1979, **44**, 2796.	89, 90, 108
79JOC2957	V. Bhat, V. M. Dixit, B. G. Ugarker, A. M. Trozzolo and M. V. George; *J. Org. Chem.*, 1979, **44**, 2957.	375
79JOC2961	M. A. Abou-Gharbia and M. M. Joullié; *J. Org. Chem.*, 1979, **44**, 2961.	113
79JOC3082	S. D. Venkataramu, J. H. Cleveland and D. E. Pearson; *J. Org. Chem.*, 1979, **44**, 3082.	761
79JOC3181	U. Mazur, R. P. Lattimer, A. Lopata and R. L. Kuczkowski; *J. Org. Chem.*, 1979, **44**, 3181.	862, 892
79JOC3185	U. Mazur and R. L. Kuczkowski; *J. Org. Chem.*, 1979, **44**, 3185.	863
79JOC3524	A. Ohsawa, H. Arai, H. Igeta, T. Akimoto and A. Tsuji; *J. Org. Chem.*, 1979, **44**, 3524.	981, 1023
79JOC3803	K. T. Potts and S. Kanemasa; *J. Org. Chem.*, 1979, **44**, 3803.	1005
79JOC3808	K. T. Potts and S. Kanemasa; *J. Org. Chem.*, 1979, **44**, 3808.	743
79JOC3840	G. Kaugars and V. L. Rizzo; *J. Org. Chem.*, 1979, **44**, 3840.	592
79JOC3991	G. L'abbé, C. C. Yu and S. Toppet; *J. Org. Chem.*, 1979, **44**, 3991.	929
79JOC4160	Y. Lin, S. A. Lang, Jr., M. F. Lovell and N. A. Perkinson; *J. Org. Chem.*, 1979, **44**, 4160.	387
79JOC4213	J. J. Tufariello, S. A. Ali and H. O. Klingele; *J. Org. Chem.*, 1979, **44**, 4213.	109
79JOC4543	O. Tsuge, H. Watanabe, K. Masuda and M. M. Yousif; *J. Org. Chem.*, 1979, **44**, 4543.	1001
79JOC4689	J. B. Lambert, M. W. Majchrzak and D. Stec, III; *J. Org. Chem.*, 1979, **44**, 4689.	957, 969
79JOM(166)25	H. Buchwald and K. Ruehlmann; *J. Organomet. Chem.*, 1979, **166**, 25.	538
79JOM(166)265	F. Babudsi, L. Di Nunno, S. Florio, G. Marchese and F. Naso; *J. Organomet. Chem.*, 1979, **166**, 265.	411
79JPC1545	C. W. Gillies, S. P. Sponseller and R. L. Kuczkowski; *J. Phys. Chem.*, 1979, **83**, 1545.	858, 892
79JPR71	K. Gewald and J. Oelsner; *J. Prakt. Chem.*, 1979, **321**, 71.	643
79JPR260	W. Walek and K. Götzschel; *J. Prakt. Chem.*, 1979, **321**, 260.	687
79JPR320	G. Ciurdaru and M. Ciuciu; *J. Prakt. Chem.*, 1979, **321**, 320.	341
79JPR827	E. Fanghänel and A. M. Richter; *J. Prakt. Chem.*, 1979, **321**, 827.	832
79JPR1021	W. D. Rudorf, A. Schierhorn and M. Augustin; *J. Prakt. Chem.*, 1979, **321**, 1021.	149, 166
79JPS1156	W. J. A. Vandenheuvel, B. H. Arison, T. W. Miller, P. Kulsa, P. Eskola, H. Mrozik, A. K. Miller, H. Skeggs, S. B. Zimmerman and B. M. Miller; *J. Pharm. Sci.*, 1979, **68**, 1156.	129
79JST(51)87	T.-K. Ha; *J. Mol. Struct.*, 1979, **51**, 87.	4, 395, 428
79JST(51)133	M. Hotoka and P. Pyykkö; *J. Mol. Struct.*, 1979, **51**, 133.	185
79JST(55)143	B. A. Sastry, S. M. Asadulla, K. V. G. Reddy, G. Ponticelli and M. Massacesi; *J. Mol. Struct.*, 1979, **55**, 143.	7
79KGS38	I. G. Markova, M. K. Polievktov and S. D. Sokolov; *Khim. Geterotsikl. Soedin.*, 1979, 38.	402
79KGS599	A. V. Prosyanik, A. I. Mishchenko, N. L. Zaichenko, Ya. Z. Zorin, R. G. Kostyanovskii and V. I. Markov; *Khim Geterotsikl. Soedin.*, 1979, 599 (*Chem. Abstr.*, 1979, **91**, 175 249).	108
79KGS746	A. F. Rukasov, V. P. Tashchi, Yu. A. Kondrat'ev, Yu. A. Baskakov and Yu. G. Putsykin; *Khim. Geterotsikl. Soedin.*, 1979, 746 (*Chem. Abstr.*, 1979, **91**, 140 755).	25, 46
79KGS990	I. A. Borisova, V. V. Zorin, S. S. Zlotskii and D. L. Rakhmankulov; *Khim. Geterotsikl. Soedin.*, 1979, 990 (*Chem. Abstr.*, 1979, **91**, 157 640).	764
79KGS1189	V. N. Sheinker, T. V. Lifintseva, S. B. Bulgarevich, S. M. Vinogradova, S. D. Sokolov, A. D. Garnovskii and O. A. Osipov; *Khim Geterotsikl. Soedin.*, 1979, 1189 (*Chem. Abstr.*, 1980, **92**, 128 276).	4
79KGS1205	B. K. Strelets, M. M. Gel'mont, Y. I. Akulin and L. S. Efros; *Khim. Geterotsikl. Soedin.*, 1979, 1205 (*Chem. Abstr.*, 1980, **92**, 128 024).	916, 929, 940
79LA63	H. Gotthardt and F. Reiter; *Liebigs Ann. Chem.*, 1979, 63.	369, 370
79LA200	D. Braun and J. Weinert; *Liebigs Ann. Chem.*, 1979, 200.	229
79LA219	I. Hoppe and V. Schollkopf; *Liebigs Ann. Chem.*, 1979, 219.	30
79LA360	H. Gotthardt, C. M. Weisshuhn and B. Christl; *Liebigs Ann. Chem.*, 1979, 360.	828
79LA689	F. Boberg, U. Puttins and G.-J. Wentrup; *Liebigs Ann. Chem.*, 1979, 689.	834
79LA1473	P. Rademacher and W. Elling; *Liebigs Ann. Chem.*, 1979, 1473.	861
79LA1534	K. Gewald and P. Bellmann; *Liebigs Ann. Chem.*, 1979, 1534.	156, 158
79LA1547	K. Burger and M. Eggersdorfer; *LiebigsAnn. Chem.*, 1979, 1547.	227
79LA1734	E. Schaumann, J. Ehlers and H. Mrotzek; *Liebigs Ann. Chem.*, 1979, 1734.	454
79MI41600	W. Basinski and Z. Jerzmanowska; *Pol. J. Chem.*, 1979, **53**, 229.	78, 79
79MI41601	K. C. Joshi, V. N. Pathak and V. Grover; *Pharmazie*, 1979, **34(2)**, 68.	63, 64, 84
79MI41603	D. N. Nicolaides, M. A. Kanetakis and K. E. Litinas; *Chem. Chron., New Ser.*, 1979, **8**, 187.	70

79MI41604	J. S. Holcenberg; *Cancer Treat. Rep.*, 1979, **63**, 1109.	129
79MI41605	N. C. Tye, L. Horseman, F. C. Wright and I. A. Pullar; *Eur. J. Pharmacol.*, 1979, **55**, 103.	129
79MI41606	G. L. Neil, A. E. Berger, R. P. McPartland, G. B. Grindley and A. Bloch; *Cancer Res.*, 1979, **39**, 852.	129
79MI41607	A. I. Csapo and B. Resch; *J. Steroid Biochem.*, 1979, **11**, 963 (*Chem. Abstr.*, 1979, **91**, 168 782).	128
79MI41608	J. P. Triveda, T. P. Gandhi, G. F. Shah and V. C. Patel; *Indian J. Pharm. Sci.*, 1979, **41**, 28.	128
B-79MI41609	T. Yamamori and I. Adachi; in '12th Fukusokan Kagaku Toronkai Koen Yoshishu', 1979, p. 121 (*Chem. Abstr.*, 1981, **94**, 65 513).	117
79MI41610	S. Blechert; *Nachr. Chem., Tech. Lab.*, 1979, **27**, 173 (*Chem. Abstr.*, 1979, **91**, 5127).	108
B-79MI41611	T. Kato, T. Chiba, M. Sato, N. Katagiri, Y. Suzuki, R. Sato and T. Yashima; in '12th Fukusokan Kagaku Toronkai Koen Yoshishu', 1979, p. 36 (*Chem. Abstr.*, 1980, **93**, 26 176).	104
79MI41612	L. N. Volovel'skii, I. B. Skachek and I. I. Kuz'menko; *Khim. Prir. Soedin.*, 1979, 168 (*Chem. Abstr.*, 1980, **92**, 111 219).	97
B-79MI41613	K. Harada, E. Kaji and S. Zen; in '12th Fukusokan Kagaku Toronkai Koen Yoshishu', 1979, p. 271 (*Chem. Abstr.*, 1980, **93**, 26 351).	91, 92
79MI41614	E. Domagalina and T. Slawik; *J. Therm. Anal.*, 1979, **15**, 257.	56
B-79MI41615	P. Zalupsky, T. Kumagai and T. Mukai; in 'Kokagaku Toronkai Koen Yoshishu', 1979, p. 158 (*Chem. Abstr.*, 1980, **92**, 214 554).	38
B-79MI41616	T. Kumagai, Y. Kawamura, K. Shimizu and T. Mukai; in '12th Koen Yoshishu-Hibenzenkei Hokozoku Kagaku Toronkai Kozo Yuki Kagaku Toronkai', 1979, p. 317 (*Chem. Abstr.*, 1980, **92**, 197 557).	38
79MI41617	G. Pelizzi; *Transition Met. Chem.*, 1979, **4**, 199.	4
79MI41619	V. K. Strivastava, B. R. Pandey, R. C. Gupta and K. Kishor; *Pharmazie*, 1979, **34**, 638.	40
79MI41620	V. N. Pathak and V. Grover; *Pharmazie*, 1979, **34**, 568.	128
79MI41700	P. P. Singh, A. K. Srivastava and L. P. Pathak; *J. Coord. Chem.*, 1979, **9**, 65 (*Chem. Abstr.*, 1979, **91**, 101 251).	132
B-79MI41701	G. C. Levy and R. L. Lichter; 'Nitrogen-15 Nuclear Magnetic Resonance Spectroscopy', Wiley, New York, 1979, p. 28.	140
B-79MI41800	M. M. Campbell; in 'Comprehensive Organic Chemistry', ed. D. H. R. Barton and W. D. Ollis; Pergamon, Oxford, 1979, vol. 4, p. 961.	178
79MI41801	M. Baudet and M. Gelbcke; *Anal. Lett.*, 1979, **12(B6)**, 641.	181, 185
79MI41802	T. Jaworski, T. Mizerski and A. Krokilowska; *Pol. J. Chem.*, 1979, **53**, 1799.	196
79MI42000	C. DeMarco, C. Cini, R. Coccia and C. Blarzino; *Ital. J. Biochem.*, 1979, **28**, 104.	346
79MI42001	R. Hanson and M. Davis; *J. Labelled Compd. Radiopharm.*, 1979, **16**, 31.	353, 354
79MI42100	M. Polgar, L. Vereczkey, L. Szporny, G. Czira, J. Tamas, E. Gacs-Baitz and S. Holly; *Xenobiotica*, 1979, **9**, 511.	378
B-79MI42101	'New Aspects in the Therapy of Ischemic Heart Disease', ed. W. Lochner and F. Bender; Urban and Schwarzenbach, Munich, 1979 (*Chem. Abstr.*, 1980, **91**, 151 536).	378
79MI42200	R. Bartnik and S. Lesniak; *Pol. J. Chem.*, 1979, **53**, 1781 (*Chem. Abstr.*, 1980, **92**, 180 552).	401, 416
79MI42201	P. E. Cassidy and N. C. Fawcett; *J. Macromol. Sci., Rev. Macromol. Chem.*, 1979, **C17**, 209.	426
79MI42300	S. P. Suman and S. C. Bahel; *Agric. Biol. Chem.*, 1979, **43**, 1339 (*Chem. Abstr.*, 1979, **91**, 118 443).	440, 445
B-79MI42400	M. M. Campbell; in 'Comprehensive Organic Chemistry', D. H. R. Barton and W. D. Ollis; ed. Pergamon, Oxford, 1979, vol. 4, p. 1033.	448, 453, 454, 460, 461
B-79MI42500	'Pesticide Manual', ed. C. R. Worthing; British Crop Protection Council, 6th edn., 1979, p. 254.	509
B-79MI42501	'Comprehensive Organic Chemistry', ed. D. H. R. Barton and W. D. Ollis; Pergamon, Oxford, 1979, vol. 4, pp. 1033–1047.	464
B-79MI42502	'Aromatic and Heteroaromatic Chemistry', Chemical Society, London, vol. 7 and preceding volumes.	464
79MI42600	B. Kasterka; *Chem. Anal. (Warsaw)*, 1979, **24**, 329 (*Chem. Abstr.*, 1979, **91**, 150 619).	537
79MI42601	P. E. Cassidy and N. C. Fawcett; *J. Macromol. Sci., Rev. Macromol. Chem.*, 1979, **C17**, 209.	543
B-79MI42700	M. M. Campbell; in 'Comprehensive Organic Chemistry', ed. D. H. R. Barton and W. D. Ollis; Pergamon, Oxford, 1979, vol. 4, p. 961.	546, 563, 576
B-79MI42701	C. A. Ramsden; in 'Comprehensive Organic Chemistry', ed. D. H. R. Barton and W. D. Ollis; Pergamon, Oxford, 1979, vol. 4, p. 1207.	546, 573
79MI42800	E. F. de A. Neves, W. L. Polito, T. Tamura and D. W. Franco; *Anal. Lett.*, 1979, **12**, 741.	598
79MI42801	M. J. Zaworotko, J. L. Atwood and L. Floch; *J. Cryst. Mol. Struct.*, 1979, **9**, 173.	582
79MI42802	E. Koch and B. Stilkerieg; *J. Therm. Anal.*, 1979, **17**, 395.	585
79MI42803	E. A. Neves and D. W. Franco; *Talanta*, 1979, **26**, 81.	598
79MI42900	P. Krogsgaard-Larsen and T. Roldskov-Christiansen; *Eur. J. Med. Chem. – Chim. Ther.*, 1979, **14**, 157.	633, 634

79MI42901	S. A. Ibrahim, S. Dine, F. S. Soliman, S. G. Farid and I. M. Labouta; *Pharmazie*, 1979, **34**, 392.	634, 694, 695
79MI42902	L. L. Bennett, Jr., L. M. Rose, P. W. Allan, D. Smithers, D. J. Adamson, R. D. Elliott and J. A. Montgomery; *Mol. Pharmacol.*, 1979, **16**, 981.	748
79MI43100	R. Gleiter and J. Spanget-Larsen; *Top. Curr. Chem.*, 1979, **86**, 139.	787
79MI43200	H. Tezuka, T. Shiba, N. Aoki, K. Iijima and H. Kato; *Kokagaku Toronkai Koen Yoshishu*, 1979, **8** (Japan), Chem. Soc. Jpn., Tokyo (*Chem. Abstr.*, 1980, **92**, 197 661).	819, 825
79MI43201	H.-F. Li, S.-Z. Li, E.-S. Yao and C. Ye; *Tzu Jan Tsa Chih*, 1979, **2**, 136 (*Chem. Abstr.*, 1979, **91**, 39 393).	847
B-79MI43202	J. Rigaudy and S. P. Klesney; 'Nomenclature of Organic Chemistry', Pergamon, Oxford, 1979, p. 142.	814
79MI43300	M. Eto, S.-S. Chou and E. Taniguchi; *J. Pesticide Sci.*, 1979, **4**, 379.	869
79MI43301	L. N. Nixon, E. Wong, C. B. Johnson and E. J. Birch; *J. Agric. Food Chem.*, 1979, **27**, 355.	895
79MI43500	O. M. Tsygleva, I. M. Gella and I. V. Krivoshei; *Teor. Eksp. Khim.*, 1979, **15**, 441.	952
B-79MI43501	C. L. Khetrapal, A. Kumar, A. C. Kunwar, P. C. Mathias and K. V. Ramanathan; in 'Proceedings of the International Conference on Liquid Crystals', ed. S. Chandrasekhar; Heyden, London, 1979, p. 469.	959
79MI43600	A. Padwa; *New Synth. Methods*, 1979, **5**, 25.	1007
79MI43601	G. L'abbe; *New Synth. Methods*, 1979, **5**, 1.	1009
B-79MI43700	C. A. Ramsden; in 'Comprehensive Organic Chemistry', ed. D. H. R. Barton and W. D. Ollis; Pergamon Press, Oxford, 1979, vol. 4, p. 1171.	1042
B-79MI43800	'Nomenclature of Organic Chemistry', ed. J. Rigaudy and S. P. Klesney; Pergamon, Oxford, 1979.	1051
79MIP41600	J. L. Huppatz; *Aust. Pat.* 497 898 (1979) (*Chem. Abstr.*, 1979, **90**, 181 597).	129
79MIP41601	T. A. Yagudeev, R. B. Baisakalova, A. N. Nurgalieva and S. Zh. Zhumagaliev; *USSR Pat.* 703 531 (1979) (*Chem. Abstr.*, 1980, **93**, 46 644).	98
79MIP42300	H. R. Meyer; *Swiss Pat.* 610 478 (1979) (*Chem. Abstr.*, 1980, **92**, 182 529).	446
79NJC149	M. Torres, A. Clement, H. E. Gunning and O. P. Strausz; *Nouv. J. Chim.*, 1979, **3**, 149.	820
79NKK389	I. Shibuya; *Nippon Kagaku Kaishi*, 1979, 389.	907, 917, 936, 942
79OMR(12)178	S. Icli; *Org. Magn. Reson.*, 1979, **12**, 178.	181
79OMR(12)461	E. L. Eliel, V. S. Rao and K. M. Pietrusiewicz; *Org. Magn. Reson.*, 1979, **12**, 461.	755, 756, 758
79P1397	H. Kameoka and Y. Demizu; *Phytochemistry*, 1979, **18**, 1397.	894
79PAC261	A. M. Trozzolo, T. M. Leslie, A. S. Sarpotdar, R. D. Small, G. J. Ferrandi, T. DoMinh and R. L. Hartless; *Pure Appl. Chem.*, 1979, **51**, 261.	375
79PAC1347	U. Schöllkopf; *Pure Appl. Chem.*, 1979, **51**, 1347.	220
79RCR289	S. D. Sokolov; *Russ. Chem. Rev. (Engl. Trans.)*, 1979, **48**, 289.	132, 133
79RRC111	N. Barbulescu; *Rev. Roum. Chim.*, 1979, **30**, 111 (*Chem. Abstr.*, 1979, **91**, 19 292).	108
79S35	H. Vorbrüggen and K. Krolikiewicz; *Synthesis*, 1979, 35.	594
79S52	A. Mignot, H. Moskowitz and M. Miocque; *Synthesis*, 1979, 52.	1010
79S66	V. I. Cohen; *Synthesis*, 1979, 66.	342
79S102	H. Kristinsson; *Synthesis*, 1979, 102.	438
79S370	C. Skoetsch and E. Breitmaier; *Synthesis*, 1979, 370.	642
79S449	C. Skoetsch, I. Kohlmeyer and E. Breitmaier; *Synthesis*, 1979, 449.	54, 620
79S470	K. Masuda, Y. Arai and M. Itoh; *Synthesis*, 1979, 470.	456
79S524	S. Mataka, A. Hosoki, K. Takahashi and M. Tashiro; *Synthesis*, 1979, 524.	541
79S687	S. Mataka, K. Takahashi and M. Tashiro; *Synthesis*, 1979, 687.	413, 538, 730, 748
79S724	R. Sterzycki; *Synthesis*, 1979, 724.	773
79S977	P. Pollet and S. Gelin; *Synthesis*, 1979, 977.	413, 416, 996
79S979	V. Bertini, F. Lucchesini and A. De Munno; *Synthesis*, 1979, 979.	527, 539
79S982	K. C. Nicolaou, R. L. Magolda and W. J. Sipio; *Synthesis*, 1979, 982.	766
79SST(5)345	M. Davis; *Org. Compd. Sulphur, Selenium, Tellurium*, 1979, **5**, 345.	132, 147, 148, 149, 152, 155, 158, 159, 160, 161, 165, 166, 167, 168, 169, 170, 171, 175
79SST(5)358	B. Iddon and P. A. Lowe; *Org. Compd. Sulphur, Selenium, Tellurium*, 1979, **5**, 358–392.	267, 328
79SST(5)393	B. Iddon and P. A. Lowe; *Org. Compd. Sulphur, Selenium, Tellurium*, 1979, **5**, 393–407.	328
79SST(5)440	M. Davis; *Org. Compd. Sulphur, Selenium, Tellurium*, 1979, **5**, 440.	546
79T213	R. G. Kostyanovsky, V. F. Rudchenko, O. A. D'yachenko, I. I. Chervin, A. B. Zolotoi and L. O. Atovmyan; *Tetrahedron*, 1979, **35**, 213.	47, 59
79T241	J. W. Barton, M. C. Goodland, K. J. Gould, J. F. W. McOmie, W. R. Mound and S. A. Saleh; *Tetrahedron*, 1979, **35**, 241.	412
79T409	C. Bjerre, C. Christophersen, B. Hansen, N. Harrit, F. M. Nicolaisen and A. Holm; *Tetrahedron*, 1979, **35**, 409.	602, 604
79T647	H. Stamm and H. Steudle; *Tetrahedron*, 1979, **35**, 647.	113
79T1267	A. Ohsawa, H. Arai, H. Igeta, T. Akimoto, A. Tsuji and Y. Iitaka; *Tetrahedron*, 1979, **35**, 1267.	98, 975, 1018
79T1771	P. Battistoni, P. Bruni and G. Fava; *Tetrahedron*, 1979, **35**, 1771.	632
79T2009	W. Klyne, P. M. Scopes, N. Berova, J. N. Stefanovsky and B. J. Kurtev; *Tetrahedron*, 1979, **35**, 2009.	185

79TL49	G. L'abbé and A. Verbruggen; *Tetrahedron Lett.*, 1979, 49.	906, 911, 925, 941
79TL745	C. L. Pedersen; *Tetrahedron Lett.*, 1979, 745.	419
79TL1217	W. Rasshofer and F. Voegtle; *Tetrahedron Lett.*, 1979, 1217.	148
79TL1281	J. Rokach, P. Hamel, Y. Girard and G. Reader; *Tetrahedron Lett.*, 1979, 1281.	527
79TL1567	G. Ege and K. Gilbert; *Tetrahedron Lett.*, 1979, 1567.	1009
79TL2443	W. R. Mitchell and R. M. Paton; *Tetrahedron Lett.*, 1979, 2443.	91, 404
79TL2461	S. Mohr; *Tetrahedron Lett.*, 1979, 2461.	201
79TL2785	R. C. White, J. Scoby and T. D. Roberts; *Tetrahedron Lett.*, 1979, 2785.	454
79TL2937	D. C. Billington and B. T. Golding; *Tetrahedron Lett.*, 1979, 2937.	165
79TL2961	S.-I. Hayashi, M. Nair, D. J. Houser and H. Schechter; *Tetrahedron Lett.*, 1979, 2961.	15
79TL3139	S. Mohr; *Tetrahedron Lett.*, 1979, 3139.	201
79TL3263	G. E. Keyser, J. D. Bryant and J. R. Barrio; *Tetrahedron Lett.*, 1979, 6263.	778
79TL3339	R. C. Davis, T. J. Grinter, D. Leaver and R. M. O'Neil; *Tetrahedron Lett.*, 1979, 3339.	802
79TL3375	G. W. Gokel, H. M. Gerdes, D. E. Miles, J. M. Hufnall and G. A. Zerby; *Tetrahedron Lett.*, 1979, 3375.	770
79TL4159	A. I. Meyers, A. L. Campbell, A. G. Abatjoglou and E. L. Eliel; *Tetrahedron Lett.*, 1979, 4159.	770
79TL4167	S. A. Ali, P. A. Senaratne, C. R. Illig, H. Meckler and J. J. Tufariello; *Tetrahedron Lett.*, 1979, 4167.	631
79TL4493	M. Behforouz and R. Benrashid; *Tetrahedron Lett.*, 1979, 4493.	534, 1036, 1044
79TL4685	G. Kumar, K. Rajogopalan, S. Swaminathan and K. K. Balasubramanian; *Tetrahedron Lett.*, 1979, 4685.	15
79TL4687	R. K. Smalley, R. H. Smith and H. Suschitzky; *Tetrahedron Lett.*, 1979, 4687.	51
79TL4903	S. A. Tischler and L. Weiler; *Tetrahedron Lett.*, 1979, 4903.	40
79USP4133813	K. E. Fuger and M. N. Sheng; *U.S. Pat.* 4 133 813 (1979) (*Chem. Abstr.*, 1979, **90**, 169 552).	945
79USP4139366	R. K. Howe; *U.S. Pat.* 4 139 366 (1979) (*Chem. Abstr.*, 1979, **90**, 168 580).	129
79USP4140693	E.R. Squibb & Sons Inc.; *U.S. Pat.* 4 140 693 (1979) (*Chem. Abstr.*, 1979, **90**, 204 108).	160
79USP4143044	Olin Corp., *U.S. Pat.* 4 143 044 (1979) (*Chem. Abstr.*, 1979, **90**, 186 962).	509
79USP4144047	J. E. Franz and R. K. Howe; *U.S. Pat.* 4 144 047 (1979) (*Chem. Abstr.*, 1979, **91**, 15 174).	129
79USP4160027	R. G. Christiansen; *U.S. Pat.* 4 160 027 (1979) (*Chem. Abstr.*, 1979, **91**, 211 680).	128
79USP4163057	J. Nadelson; *U.S. Pat.* 4 163 057 (1979) (*Chem. Abstr.*, 1980, **92**, 41 959).	128
79USP4172079	R. F. Love and R. G. Duranleau; *U.S. Pat.* 4 172 079 (1979) (*Chem. Abstr.*, 1980, **92**, 76 487).	130
79USP4178447	Polaroid Corp.; *U.S. Pat.* 4 178 447 (1979) (*Chem. Abstr.*, 1980, **93**, 8159).	151
79ZC192	E. Fanghänel and H. Poleschner; *Z. Chem.*, 1979, **19**, 192.	820, 963
79ZC452	J. Wrubel; *Z. Chem.*, 1979, **19**, 452.	23, 58
79ZN(A)220	M. Redshaw, M. H. Palmer and R. H. Findlay; *Z. Naturforsch., Teil A*, 1979, **34**, 220.	136, 395, 400, 515
79ZN(B)235	B. Wrackmeyer; *Z. Naturforsch., Teil B*, 1979, **34**, 235.	6
79ZOB1322	M. K. Polievktov, I. G. Morkova and S. D. Sokolov; *Zh. Obshch. Khim.*, 1979, **49**, 1322 (*Chem. Abstr.*, 1979, **91**, 123 160).	11
79ZOR735	L. A. Demina, G. K. Khisamutdinov, S. V. Tkachev and A. A. Fainzil'berg; *Zh. Org. Khim.*, 1979, **15**, 735.	71, 72, 96
79ZOR1069	V. N. Drozd, G. S. Bogomolova and M. Y. Udachin; *Zh. Org. Khim.*, 1979, **15**, 1069.	801
79ZOR1106	V. N. Drozd; *Zh. Org. Khim.*, 1979, **15**, 1106 (*Chem. Abstr.*, 1979, **91**, 193 229).	845
79ZOR2408	L. N. Baeva, L. A. Demina, T. V. Trusova, G. G. Furin and G. K. Khisamutdinov; *Zh. Org. Khim.*, 1979, **15**, 2408.	71, 72, 96
79ZOR2436	G. Kh. Khisamutdinov, L. A. Demina, G. E. Cherkosava and V. G. Klimenko; *Zh. Org. Khim.*, 1979, **15**, 2436 (*Chem. Abstr.*, 1980, **92**, 110 903).	37, 106
80ABC1923	M. Suiko, E. Taniguchi and K. Maekawa; *Agric. Biol. Chem.*, 1980, **44**, 1923.	748
80ABC2169	S.-S. Chou, E. Taniguchi and M. Eto; *Agric. Biol. Chem.*, 1980, **44**, 2169.	886
80AF477	Y. Masuda, T. Karasawa, Y. Shiraishi, M. Hori, K. Yoshida and M. Shimizu; *Arzneim.-Forsch.*, 1980, **30**, 477.	129
80AF603	T. Ito, M. Hori, Y. Masuda, K. Yoshida and M. Shimizu; *Arzneim.-Forsch.*, 1980, **30**, 603.	129
80AG277	G. L'abbé; *Angew. Chem.*, 1980, **92**, 277.	586
80AG743	H. W. Winter and C. Wentrup; *Angew. Chem.*, 1980, **92**, 743.	42
80AG(E)204	R. Neidlein and H. Zeiner; *Angew. Chem., Int. Ed. Engl.*, 1980, **19**, 204.	822
80AG(E)276	G. L'abbe; *Angew. Chem., Int. Ed. Engl.*, 1980, **19**, 276.	1004
80AG(E)564	H. M. Berstermann, R. Harder, H. W. Winter and C. Wentrup; *Angew. Chem., Int. Ed. Engl.*, 1980, **19**, 564.	199
80AHC(27)31	P. Hanson; *Adv. Heterocycl. Chem.*, 1980, **27**, 31.	429, 835
80AHC(27)151	N. Lozac'h and M. Stavaux; *Adv. Heterocycl. Chem.*, 1980, **27**, 151.	171, 784, 790, 792, 796, 799, 801, 803, 809, 811, 814, 816, 818, 820, 822, 823, 824, 831, 832, 834, 835, 836, 838, 839, 842, 848, 1052, 1065, 1066, 1067, 1068, 1070

80AJC527	E. R. Cole, G. Crank and H. T. H. Minh; *Aust. J. Chem.*, 1980, **33**, 527.	766
80AJC675	E. R. Cole, G. Crank and H. T. H. Minh; *Aust. J. Chem.*, 1980, **33**, 675.	778
80AP39	G. Zinner; *Arch. Pharm.* (*Weinheim, Ger.*), 1980, **313**, 39.	113
80AX(B)971	U. Rychlewska; *Acta Crystallogr., Part B*, 1980, **36**, 971.	671
80AX(B)1229	K. Sasvari, L. Párkányi, G. Hajos, H. Hess and W. Schwarz; *Acta Crystallogr., Part B*, 1980, **36**, 1229.	669
80AX(B)1466	F. Iwasaki; *Acta Crystallogr., Part B*, 1980, **36**, 1466.	901, 903
80AX(B)1626	P. A. McCallum, H. M. N. H. Irving, A. T. Hutton and L. R. Nassimbeni; *Acta Crystallogr., Sect. B*, 1980, **36**, 1626.	548
80BCJ205	Y. Inagaki, T. Hosogai, R. Okazaki and N. Inamoto; *Bull. Chem. Soc. Jpn.*, 1980, **53**, 205.	927
80BCJ1023	T. Saito, I. Oikawa and S. Motoki; *Bull. Chem. Soc. Jpn.*, 1980, **53**, 1023.	943
80BCJ1185	M. Takahashi and M. Kurosawa; *Bull. Chem. Soc. Jpn.*, 1980, **53**, 1185.	358
80BCJ1661	J. Nakayama, M. Imura and M. Hoshino; *Bull. Chem. Soc. Jpn.*, 1980, **53**, 1661.	833, 837
80BCJ2281	J. Nakayama, T. Takemasa and M. Hoshino; *Bull. Chem. Soc. Jpn.*, 1980, **53**, 2281.	834
80BRP2024218	P. T. Haken and S. B. Webb; *Br. Pat. Appl.* 2 024 218 (1980) (*Chem. Abstr.*, 1980, **93**, 132 471).	108
80BSB247	K. A. Jørgensen, R. Shabana, S. Scheibye and S.-O. Lawesson; *Bull. Soc. Chim. Belg.*, 1980, **89**, 247.	405
80BSB773	N. V. Onyamboko, R. Weber, N. Dereu, M. Renson and C. Paulmier; *Bull. Soc. Chim. Belg.*, 1980, **89**, 773.	335, 337
80BSF(1)451	M. M. Osman, M. A. Makhyoun and A. B. Tadros; *Bull. Soc. Chim. Fr., Part 1*, 1980, 451.	552, 565
80BSF(2)151	C. Paulmier; *Bull. Soc. Chim. Fr., Part 2*, 1980, 151.	344
80BSF(2)163	R. Colau and C. Viel; *Bull. Soc. Chim. Fr., Part 2*, 1980, 163.	75, 87
80BSF(2)530	A. Dibo, M. Stavaux and N. Lozac'h; *Bull. Soc. Chim. Fr., Part 2*, 1980, 530.	798
80CA(92)197661	H. Tezuka, T. Shiba, N. Aoki, K. Iijima and H. Kato; *Chem. Abstr.*, 1980, **92**, 197 661.	808
80CB221	M. Foerterer and P. Rademacher; *Chem. Ber.*, 1980, **113**, 221.	430
80CB1195	H. Junek, B. Thierrichter and G. Lukas; *Chem. Ber.*, 1980, **113**, 1195.	71, 628
80CB1507	H. Gnichtel and E. Boehringer; *Chem. Ber.*, 1980, **113**, 1507.	101
80CB1790	W. Beck, E. Leidl, M. Keubler and U. Nagel; *Chem. Ber.*, 1980, **113**, 1790.	918, 941
80CB1830	H. Dickopp; *Chem. Ber.*, 1980, **113**, 1830.	373
80CB1898	K. Hartke, T. Kissel, J. Quante and R. Matusch; *Chem. Ber.*, 1980, **113**, 1898.	820, 835, 847
80CB2490	H. Boeshagen and W. Geiger; *Chem. Ber.*, 1980, **113**, 2490.	162
80CB2852	E. Öhler and E. Zbiral; *Chem. Ber.*, 1980, **113**, 2852.	63, 83, 84
80CB3187	H. Bock, S. Aygen, P. Rosmus and B. Solouki; *Chem. Ber.*, 1980, **113**, 3187.	348, 350
80CC38	G. Le Coustumer and Y. Mollier; *J. Chem. Soc., Chem. Commun.*, 1980, 38.	850
80CC421	A. J. Boulton and T. G. Tsoungeas; *J. Chem. Soc., Chem. Commun.*, 1980, 421.	6, 22, 23, 59, 118
80CC471	M. Lancaster and D. J. H. Smith; *J. Chem. Soc., Chem. Commun.*, 1980, 471.	164, 165
80CC503	H. M. Berstermann, K. P. Netsch and C. Wentrup; *J. Chem. Soc., Chem. Commun.*, 1980, 503.	200
80CC598	S. Tamagaki and K. Hotta; *J. Chem. Soc., Chem. Commun.*, 1980, 598.	801
80CC714	R. M. Paton, F. M. Robertson and J. F. Ross; *J. Chem. Soc., Chem. Commun.*, 1980, 714.	925, 929
80CC826	C. Kashima, S. Shirai, N. Yoshiwara and Y. Omote; *J. Chem. Soc., Chem. Commun.*, 1980, 826.	63
80CC866	F. Wudl and D. Nalewajek; *J. Chem. Soc., Chem. Commun.*, 1980, 866.	963, 965, 969
80CC867	L.-Y. Chiang, T. O. Poehler, A. N. Bloch and D. O. Cowan; *J. Chem. Soc., Chem. Commun.*, 1980, 866.	963, 965, 969
80CC1054	T. Sasaki, K. Hayakawa and S. Nishida; *J. Chem. Soc., Chem. Commun.*, 1980, 1054.	15
80CCC2329	M. Marchalín and A. Martvoň; *Collect. Czech. Chem. Commun.*, 1980, **45**, 2329.	582, 594
80CHE1160	V. M. Dziomko, B. K. Berestevich, A. V. Kessenikh, R. S. Kuzanyan and L. V. Shmelev; *Chem. Heterocycl. Compd.* (*Engl. Transl.*), 1980, **16**, 1160.	1015
80CI(L)665	W. R. Mitchell and R. M. Paton; *Chem. Ind.* (*London*), 1980, 665.	405
80CL401	T. Nishiwaki, E. Kawamura, N. Abe and M. Iori; *Chem. Lett.*, 1980, 401.	161
80CL717	H. Kato, K. Tani, H. Kurumisawa and Y. Tamura; *Chem. Lett.*, 1980, 717.	820, 831
80CL1031	O. Tsuge, T. Takata and M. Noguchi; *Chem. Lett.*, 1980, 1031.	534, 1036
80CL1369	O. Tsuge, H. Shiraishi and T. Takata; *Chem. Lett.*, 1980, 1369.	1041, 1048
80CPB103	S. Sugai and K. Tomita; *Chem. Pharm. Bull.*, 1980, **28**, 103.	44
80CPB479	E. Kaji and S. Zen; *Chem. Pharm. Bull.*, 1980, **28**, 479.	77, 96
80CPB487	S. Sugai and K. Tomita; *Chem. Pharm. Bull.*, 1980, **28**, 487.	44, 137, 785, 786, 787
80CPB552	S. Sugai and K. Tomita; *Chem. Pharm. Bull.*, 1980, **28**, 552.	57, 88
80CPB567	Y. Okamoto, K. Takagi and T. Ueda; *Chem. Pharm. Bull.*, 1980, **28**, 567.	80
80CPB761	N. Suzuki; *Chem. Pharm. Bull.*, 1980, **28**, 761.	662
80CPB1832	T. Sakamoto, H. Yamanaka, A. Shiozawa, W. Tanaka and H. Miyazaki; *Chem. Pharm. Bull.*, 1980, **28**, 1832.	54
80CPB1909	T. Uno, K. Takagi and M. Tomoeda; *Chem. Pharm. Bull.*, 1980, **28**, 1909.	533, 538
80CPB2083	S. Tanaka, K. Wachi and A. Terada; *Chem. Pharm. Bull.*, 1980, **28**, 2083.	82

80CPB2629	M. Iwanami, T. Maeda, M. Fujimoto, Y. Nagano, N. Nagano, A. Yamazaki, T. Shibanuma, K. Tamazawa and K. Yano; *Chem. Pharm. Bull.*, 1980, **28**, 2629.	149, 157, 161, 172
80CPB3296	K. Harada, E. Kaji and S. Zen; *Chem. Pharm. Bull.*, 1980. **28**, 3296.	67, 389
80CS(16)24	B. Lindgren; *Chem. Scr.*, 1980, **16**, 24.	957
80CZ111	R. L. Neidlein, P. Leinberger and W. Lehr; *Chem.-Ztg.*, 1980, **104**, 111.	525
80DOK(255)917	V. G. Prokudin, G. M. Nazin and G. B. Manelis; *Dokl. Akad. Nauk SSSR*, 1980, **255**, 917.	400, 405
80EUP7499	R. Kirchmayr, W. Fussenegger and H. Illy; *Eur. Pat.* 7 499 (1980) (*Chem. Abstr.*, 1980, **93**, 47 881).	945
80EUP7643	*Eur. Pat. Appl.* 7643 (1980) (*Chem. Abstr.*, 1980, **93**, 168 252).	724
80G233	B. Tornetta, G. Ronsisvalle, E. Bousquet, F. Guerrera and M. A. Siracusa; *Gazz. Chim. Ital.*, 1980, **110**, 233.	160
80GEP1574430	Fisons Ltd., *Ger. Pat.* 1 574 430 (1980) (*Chem. Abstr.*, 1981, **95**, 43 124).	509
80GEP2815956	H. Knupfer and C. W. Schellhammer; *Ger. Offen.* 2 815 956 (1980) (*Chem. Abstr.*, 1980, **92**, 94 406).	38
80GEP2848221	R. Appel, H. Janssen, I. Haller and M. Plempel; *Ger. Pat.* 2 848 221 (1980) (*Chem. Abstr.*, 1980, **93**, 186 358).	924, 942, 944
80GEP2852869	H. J. Knops, W. Brandes and V. Paul; *Ger. Pat.* 2 852 869 (1980) (*Chem. Abstr.*, 1980, **93**, 220 755).	543
80GEP2853196	Schering A.-G., *Ger. Pat.* 2 853 196 (1980) (*Chem. Abstr.*, 1980, **93**, 220 754).	577
80GEP2854438	F. J. Kaemmerer and R. Schleyerbach; *Ger. Pat.* 2 854 438 and 2 854 439 (1980) (*Chem. Abstr.*, 1980, **93**, 239 392 and 239 393).	65, 128
80GEP2919293	J. Stetter, K. Ditgens, R. Thomas, L. Eue and R. R. Schmidt; *Ger. Pat.* 2 919 293 (1980) (*Chem. Abstr.*, 1981, **94**, 156 934).	412
80H(14)15	N. Kawahara, M. Katsuyama, T. Itoh and H. Ogura; *Heterocycles*, 1980, **14**, 15.	190
80H(14)19	T. Hisano, T. Matsuoka, M. Ichikawa and M. Hamana; *Heterocycles*, 1980, **14**, 19.	723
80H(14)185	R. A. Jones and M. T. P. Marriott; *Heterocycles*, 1980, **14**, 185.	91
80H(14)271	M. V. Lakshmikantham and M. P. Cava; *Heterocycles*, 1980, **14**, 271.	824, 964, 965
80H(14)423	O. Tsuge and T. Takata; *Heterocycles*, 1980, **14**, 423.	400, 415, 1035
80H(14)785	T. Nishiwaki, E. Kawamura, N. Abe and M. Iori; *Heterocycles*, 1980, **14**, 785.	152, 848
80H(14)1279	S. K. Talapatra, P. Chaudhuri and B. Talapatra; *Heterocycles*, 1980, **14**, 1279.	426
80H(14)1319	Z. Witczak; *Heterocycles*, 1980, **14**, 1319.	98
80HCA588	K. Murato, T. Yatsunami and S. Iwasaki; *Helv. Chim. Acta*, 1980, **63**, 588.	429, 444
80HCA653	K.-H. Pfoertner and J. Foricher; *Helv. Chim. Acta*, 1980, **63**, 653.	374
80HCA832	J. W. Tilley, H. Ramuz, P. Levitan and J. F. Blount; *Helv. Chim. Acta*, 1980, **63**, 832.	388
80HCA841	J. W. Tilley, H. Ramuz, P. Levitan and J. F. Blount; *Helv. Chim. Acta*, 1980, **63**, 841.	388
80HCA1706	W. Oppolzer and J. I. Grayson; *Helv. Chim. Acta*, 1980, **63**, 1706.	108
80IJC(B)(19)667	P. V. Indukumari and C. P. Joshua; *Ind. J. Chem., Sect. B*, 1980, **19**, 667.	493
80IJC(B)406	E. A. Soliman and M. A. I. Salem; *Indian J. Chem., Sect. B*, 1980, **19**, 406.	119
80IJC(B)571	M. A. Elkasaby and M. A. I. Salem; *Indian J. Chem., Sect. B*, 1980, **19**, 571.	37, 116, 117
80IJC(B)970	R. T. Jadhav, N. M. Nimdeokar and M. G. Paranjpe; *Indian J. Chem., Sect. B*, 1980, **19**, 970.	943
80IZV207	B. A. Arbuzov, A. F. Lisin and E. V. Dianova; *Izv. Akad. Nauk SSSR, Ser. Khim.*, 1980, 207 (*Chem. Abstr.*, 1980, **92**, 146 854).	108
80IZV715	B. A. Arbuzov, A. F. Lisin and E. N. Dianova; *Izv. Akad. Nauk SSSR, Ser. Khim.*, 1980, 715 (*Chem. Abstr.*, 1980, **93**, 186 452).	108, 109
80IZV1694	V. F. Rudchenko, V. G. Shtamburg and R. G. Kostyanowskii; *Izv. Akad. Nauk SSSR, Ser. Khim.*, 1980, 1694 (*Chem. Abstr.*, 1980, **93**, 186 226).	112
80IZV1893	A. V. Kamernitskii, I. S. Levina, V. M. Shitkin and B. S. El'yanov; *Izv. Akad. Nauk SSSR, Ser. Khim.*, 1980, 1893 (*Chem. Abstr.*, 1981, **94**, 47 596).	110
80IZV2181	V. F. Rudchenko, V. G. Shtamburg, S. S. Nasibov, I. I. Chervin and R. G. Kostyanovskii; *Izv. Akad. Nauk SSSR, Ser. Khim.*, 1980, 2181 (*Chem. Abstr.*, 1981, **94**, 30 669).	905, 906, 911, 932, 940
80JA288	M. Miura and M. Nojima; *J. Am. Chem. Soc.*, 1980, **102**, 288.	870, 871
80JA744	T. B. Cameron and H. W. Pinnick; *J. Am. Chem. Soc.*, 1980, **102**, 744.	768
80JA1153	B. Nader, R. W. Franck and S. M. Weinreb; *J. Am. Chem. Soc.*, 1980, **102**, 1153.	666
80JA1372	G. Chidichimo, G. Cum, F. Lelj, G. Sindona and N. Uccella; *J. Am. Chem. Soc.*, 1980, **102**, 1372.	42
80JA1649	J. Hine and R. A. Evangelista; *J. Am. Chem. Soc.*, 1980, **102**, 1649.	47
80JA1763	A. Komornicki, J. D. Goddard and H. F. Schaefer, III; *J. Am. Chem. Soc.*, 1980, **102**, 1763.	67
80JA1783	L. J. Saethre, N. Mårtensson, S. Svensson, P. Å. Malmquist, U. Gelius and K. Siegbahn; *J. Am. Chem. Soc.*, 1980, **102**, 1783.	1057
80JA3084	G. Barany and R. B. Merrifield; *J. Am. Chem. Soc.*, 1980, **102**, 3084.	916
80JA3588	J. B. Lambert and M. W. Majchrzak; *J. Am. Chem. Soc.*, 1980, **102**, 3588.	187
80JA4265	A. P. Kozikowski and H. Ishida; *J. Am. Chem. Soc.*, 1980, **102**, 4265.	45, 112, 423
80JA4763	G. D. Fong and R. L. Kuczkowski; *J. Am. Chem. Soc.*, 1980, **102**, 4763.	855, 863
80JA4983	E. M. Kosower and B. Pazhenchevsky; *J. Am. Chem. Soc.*, 1980, **102**, 4983.	991, 1000
80JA7572	J. W. Agopovich and C. W. Gillies; *J. Am. Chem. Soc.*, 1980, **102**, 7572.	892

80JAP8027042	Hodogaya Chemical Co. Ltd., *Jpn. Pat.* 80 27 042 (1980) (*Chem. Abstr.*, 1980, **93**, 232 719).	445
80JAP80118478	Kuraray Co. Ltd., *Jpn. Pat.* 80 118 478 (1980) (*Chem. Abstr.*, 1981, **94**, 192 344).	944
80JAP(K)8017386	*Jpn. Kokai* 80 17 386 (1980) (*Chem. Abstr.*, 1980, **93**, 26 458).	724
80JAP(K)8019209	T. Sasaki; *Jpn. Kokai* 80 19 209 (1980) (*Chem. Abstr.*, 1980, **93**, 167 744).	108
80JAP(K)8028946	Kowa Co. Ltd., *Jpn. Kokai* 80 28 946 (1980) (*Chem. Abstr.*, 1980, **93**, 114 536).	577
80JAP(K)8045607	T. Takayanagi; *Jpn. Kokai* 80 45 607, 45 608 (1980) (*Chem. Abstr.*, 1980, **93**, 133 819, 204 626).	129
80JAP(K)8083766	Sankyo Co. Ltd., *Jpn. Kokai* 80 83 766 (1980) (*Chem. Abstr.*, 1981, **94**, 3074).	43
80JAP(K)8085590	*Jpn. Kokai* 80 85 590 (1980) (*Chem. Abstr.*, 1981, **94**, 65 692).	730
80JBC(25)6734	J. Y. Tso, S. G. Bower and H. Zalkin; *J. Biol. Chem.*, 1980, **25**, 6734.	129
80JCP(72)4242	M. E. Gress, S. H. Linn, Y. Ono, H. F. Prest and C. Y. Ng; *J. Chem. Phys.* 1980, **72**, 4242.	503
80JCR(M)4801	A. Albini, G. F. Bettinetti, G. Minoli and R. Oberti; *J. Chem. Res. (M)*, 1980, 4801.	1039
80JCR(S)95	K. Steinbeck; *J. Chem. Res. (S)*, 1980, 95.	764
80JCR(S)114	A. R. Butler, C. Glidewell, I. Hussain and P. R. Maw; *J. Chem. Res. (S)*, 1980, 114.	904, 939
80JCR(S)122	F. De Sarlo and A. Brandi; *J. Chem. Res. (S)*, 1980, 122.	905, 926, 929, 940
80JCR(S)197	K. Clarke, B. Gleadhill and R. M. Scrowston; *J. Chem. Res. (S)*, 1980, 197.	154
80JCR(S)221	S. Davidson, T. J. Grinter, D. Leaver and J. H. Steven; *J. Chem. Res. (S)*, 1980, 221.	809, 1064
80JCR(S)266	A. R. Butler and I. Hussain; *J. Chem. Res. (S)*, 1980, 266.	904, 906, 910, 923
80JCR(S)348	F. M. Albine, D. Vitali, R. Oberti and P. Caramella; *J. Chem. Res. (S)*, 1980, 348 (*Chem. Abstr.*, 1981, **94**, 15 625).	89
80JCR(S)407	A. R. Butler and I. Hussain; *J. Chem. Res. (S)*, 1980, 407.	493
80JCR(S)4341	F. M. Albini, D. Vitali, R. Oberti and P. Caramella; *J. Chem. Res. (S)*, 1980, 4341.	975
80JCS(F1)1490	S. A. Fairhurst and L. H. Sutcliffe; *J. Chem. Soc., Faraday Trans. 1*, 1980, 1490.	912
80JCS(P1)20	C. V. Greco and J. R. Mehta; *J. Chem. Soc., Perkin Trans. 1*, 1980, 20.	373
80JCS(P1)385	F. DeSarlo; *J. Chem. Soc., Perkin Trans. 1*, 1980, 385.	228
80JCS(P1)574	L. Motoyoshiya, M. Nishijima, I. Yamamoto, H. Gotoh, Y. Katsube, Y. Ohshiro and T. Agawa; *J. Chem. Soc., Perkin Trans. 1*, 1980, 574.	549, 553, 573
80JCS(P1)756	R. P. Houghton and A. D. Morgan; *J. Chem. Soc., Perkin Trans. 1*, 1980, 756.	768
80JCS(P1)1029	K. Clarke, W. R. Fox and R. M. Scrowston; *J. Chem. Soc., Perkin Trans. 1*, 1980, 1029.	160, 161
80JCS(P1)1635	A. Corsaro, U. Chiacchio, A. Campagnini and G. Purrello; *J. Chem. Soc., Perkin Trans. 1*, 1980, 1635.	389
80JCS(P1)1667	R. Nesi, S. Chimichi, M. Scotton, A. Degl'Innocenti and G. Adembri; *J. Chem. Soc., Perkin Trans. 1*, 1980, 1667.	157
80JCS(P1)2249	G. Zanotti, F. Filira, A. Del Pra, G. Cavicchioni, A. C. Veronesene and F. D'Angeli; *J. Chem. Soc., Perkin Trans. 1*, 1980, 2249.	230
80JCS(P1)2408	D. Mackay, Le H. Dao and J. M. Dust; *J. Chem. Soc., Perkin Trans. 1*, 1980, 2408.	930, 938
80JCS(P1)2693	T. Nishiwaki, E. Kawamura, N. Abe and M. Iori; *J. Chem. Soc., Perkin Trans. 1*, 1980, 2693.	138, 149, 161, 166, 848
80JCS(P1)2904	A. Albini, G. F. Bettinetti, G. Minoli and S. Pietra; *J. Chem. Soc., Perkin Trans. 1*, 1980, 2904.	1039
80JCS(P2)421	M. L. Schenetti, F. Taddei, L. Greci, L. Marchetti, G. Milani, G. D. Andreetti, G. Bolelli and P. Sgarabotto; *J. Chem. Soc., Perkin Trans. 2*, 1980, 421.	567
80JCS(P2)1096	D. Viterbo, R. Calvino and A. Serafino; *J. Chem. Soc., Perkin Trans. 2*, 1980, 1096.	403
80JCS(P2)1437	Z. Said and J. G. Tillett; *J. Chem. Soc., Perkin Trans. 2*, 1980, 1437.	600, 602
80JCS(P2)1792	N. S. Ooi and D. A. Wilson; *J. Chem. Soc., Perkin Trans. 2*, 1980, 1792.	386
80JGU117	Y. D. Samuilov, S. E. Solov'eva and A. I. Konovalov; *J. Gen. Chem. USSR (Engl. Transl.)*, 1980, **50**, 117.	390
80JGU1638	N. A. Akmanova, F. A. Akbutina, R. F. Talipov, Kh. F. Sagitdinova and V. P. Yur'ev; *J. Gen. Chem. USSR (Engl. Transl.)*, 1980, **50**, 1638.	1001
80JHC39	S. Jerumanis and A. Lemieux; *J. Heterocycl. Chem.*, 1980, **17**, 39.	707
80JHC65	J. Gray and D. R. Waring; *J. Heterocycl. Chem.*, 1980, **17**, 65.	167
80JHC73	M. H. Elnagdi, E. A. A. Hafez, H. A. El Fahham and E. M. Kandeel; *J. Heterocycl. Chem.*, 1980, **17**, 73.	990
80JHC117	A. Shafiee, M. Vosooghi and R. Asgharian; *J. Heterocycl. Chem.*, 1980, **17**, 117.	958, 966
80JHC213	M. M. El-Abadelah, A. A. Anani, Z. H. Khan and A. M. Hassan; *J. Heterocycl. Chem.*, 1980, **17**, 213.	411
80JHC267	D. Pocar, L. M. Rossi and P. Trimarco; *J. Heterocycl. Chem.*, 1980, **17**, 267.	983, 1000
80JHC299	G. Zvilichovsky, M. David and E. Nemes; *J. Heterocycl. Chem.*, 1980, **17**, 299.	113, 114
80JHC385	A. H. Albert, D. E. O'Brien and P. K. Robins; *J. Heterocycl. Chem.*, 1980, **17**, 385.	148, 161
80JHC393	H. J. M. Dou, M. Ludwikow, P. Hassanaly, J. Kister and J. Metzger; *J. Heterocycl. Chem.*, 1980, **17**, 393.	986
80JHC475	J. F. Hansen, Y. I. Kim, S. E. McCrotty, S. A. Strong and D. E. Zimmer; *J. Heterocycl. Chem.*, 1980, **17**, 475.	94
80JHC533	B. Danylec and M. Davis; *J. Heterocycl. Chem.*, 1980, **17**, 533.	154
80JHC537	B. Danylec and M. Davis; *J. Heterocycl. Chem.*, 1980, **17**, 537.	154, 538
80JHC545	A. Shafiee, M. Vosooghi and I. Lalezari; *J. Heterocycl. Chem.*, 1980, **17**, 545.	351

80JHC549	A. Shafiee and F. Assadi; *J. Heterocycl. Chem.*, 1980, **17**, 549.	958, 966
80JHC585	R. Martinez and E. Cortes; *J. Heterocycl. Chem.*, 1980, **17**, 585.	7
80JHC607	T. J. Kress and S. M. Costantino; *J. Heterocycl. Chem.*, 1980, **17**, 607.	568
80JHC609	L. Garanti and G. Zecchi; *J. Heterocycl. Chem.*, 1980, **17**, 609.	68, 978, 1007
80JHC621	P. Sarti-Fantoni, D. Donati, M. Fiorenza, E. Moschi and V. Dal Piaz; *J. Heterocycl. Chem.*, 1980, **17**, 621.	50
80JHC727	J. Perronnet, P. Girault and J. P. Demoute; *J. Heterocycl. Chem.*, 1980, **17**, 727.	112
80JHC763	E. M. Beccali, A. Marchesini and B. Gioia; *J. Heterocycl. Chem.*, 1980, **17**, 763.	107
80JHC833	A. S. Shawali and C. Párkányi; *J. Heterocycl. Chem.*, 1980, **17**, 833.	000
80JHC1009	A. Laurent, A. Marsura and J. L. Pierre; *J. Heterocycl. Chem.*, 1980, **17**, 1009.	904
80JHC1041	S. Meola, E. Rivera, R. Stradi and B. Gioia; *J. Heterocycl. Chem.*, 1980, **17**, 1041.	987
80JHC1181	H. S. El Khadem, E. S. H. El Ashry, D. L. Jaeger, G. P. Kreishman and R. L. Foltz; *J. Heterocycl. Chem.*, 1980, **17**, 1181.	979
80JHC1185	H. M. Hassaneen, A. Shetta and A. S. Shawali; *J. Heterocycl. Chem.*, 1980, **17**, 1185.	357
80JHC1217	S. Auricchio, S. Bruckner, L. M. Giunchi, V. A. Kozinsky and O. V. Zelenskaja; *J. Heterocycl. Chem.*, 1980, **17**, 1217.	451, 456
80JHC1245	R. N. Hanson and M. A. Davis; *J. Heterocycl. Chem.*, 1980, **17**, 1245.	353
80JHC1281	C. B. Schapira, I. A. Perillo and S. Lamdan; *J. Heterocycl. Chem.*, 1980, **17**, 1281.	149
80JHC1321	H. Moskowitz, A. Mignot and M. Miocque; *J. Heterocycl. Chem.*, 1980, **17**, 1321.	993
80JHC1413	K. Pilgram; *J. Heterocycl. Chem.*, 1980, **17**, 1413.	991
80JHC1469	D. Leppard and H. Sauter; *J. Heterocycl. Chem.*, 1980, **17**, 1469.	551, 553
80JHC1629	E. Belgodere, R. Bossio, V. Parrini and R. Pepino; *J. Heterocycl. Chem.*, 1980, **17**, 1629.	186
80JHC1639	H. Meier and O. Zimmer; *J. Heterocycl. Chem.*, 1980, **17**, 1639.	459
80JHC1645	A. Tsolomitis and C. Sandris; *J. Heterocycl. Chem.*, 1980, **17**, 1645.	149, 152, 168
80JHC1681	S. Mataka, A. Hosoki, K. Takahashi and M. Tashiro; *J. Heterocycl. Chem.*, 1980, **17**, 1681.	541
80JHC1777	T. Nakano, W. Rodríguez, S. Z. de Roche, J. M. Larrauri, C. Rivas and C. Pérez; *J. Heterocycl. Chem.*, 1980, **17**, 1777.	155
80JIC1166	R. Rai and V. K. Verma; *J. Indian. Chem. Soc.*, 1980, **57**, 1166.	942
80JIC1243	M. El-Deek, E. El-Sawi and K. El-Badry; *J. Indian Chem. Soc.*, 1980, **57**, 1243.	439
80JMC65	J. J. Baldwin, E. L. Engelhardt, R. Hirschmann, G. S. Ponticello, J. G. Atkinson, B. K. Wasson, C. S. Sweet and A. Scriabine; *J. Med. Chem.*, 1980, **23**, 65.	175
80JMC690	J. W. H. Watthey, M. Desai, R. Rutledge and R. Dotson; *J. Med. Chem.*, 1980, **23**, 690.	391
80JMC1245	R. B. Brundrett; *J. Med. Chem.*, 1980, **23**, 1245.	367
80JMR(37)349	C. L. Khetrapal, A. Kumar, A. C. Kunwar, P. C. Mathias and K. V. Ramanathan; *J. Magn. Reson.*, 1980, **37**, 349.	959
80JOC90	K. T. Potts, H. P. Youzwak and S. J. Zurawel, Jr.; *J. Org. Chem.*, 1980, **45**, 90.	1034
80JOC175	N. F. Haley and M. W. Fichtner; *J. Org. Chem.*, 1980, **45**, 175.	841
80JOC479	T. Sasaki, T. Manabe and S. Nishida; *J. Org. Chem.*, 1980, **45**, 479.	376
80JOC482	R. J. Baker, S.-K. Chiu, C. Klein, J. W. Timberlake, L. M. Trefonas and R. Majeste; *J. Org. Chem.*, 1980, **45**, 482.	541
80JOC529	T. M. Balthazor and R. A. Flores; *J. Org. Chem.*, 1980, **45**, 529.	68, 69, 70
80JOC617	P. A. Rossy, W. Hoffmann and N. Müller; *J. Org. Chem.*, 1980, **45**, 617.	989
80JOC1473	S. Kubota, Y. Ueda, K. Fujikane, K. Toyooka and M. Shibuya; *J. Org. Chem.*, 1980, **45**, 1473.	569
80JOC2024	J. Nakayama, K. Fujiwara and M. Hoshino; *J. Org. Chem.*, 1980, **45**, 2024.	821
80JOC2474	K. T. Potts, R. Ehlinger and S. Kanemasa; *J. Org. Chem.*, 1980, **45**, 2474.	1019
80JOC2632	M. V. Lakshmikantham and M. P. Cava; *J. Org. Chem.*, 1980, **45**, 2632.	348, 351, 353, 952, 959, 963, 966, 969
80JOC2785	A. I. Meyers and J. Slade; *J. Org. Chem.*, 1980, **45**, 2785.	212
80JOC2912	A. I. Meyers and J. Slade; *J. Org. Chem.*, 1980, **45**, 2912.	212
80JOC2956	O. Tsuge and T. Takata; *J. Org. Chem.*, 1980, **45**, 2956.	534, 1036
80JOC2959	N. F. Haley and M. W. Fichtner; *J. Org. Chem.*, 1980, **45**, 2959.	840, 841
80JOC3748	R. Ketcham and E. Schaumann; *J. Org. Chem.*, 1980, **45**, 3748.	990
80JOC3750	Y. Lin, S. A. Lang, Jr. and S. R. Petty; *J. Org. Chem.*, 1980, **45**, 3750.	465, 466, 499, 506
80JOC3753	M. J. Sanders and J. R. Grunwell; *J. Org. Chem.*, 1980, **45**, 3753.	169
80JOC3827	L. E. Crane, G. P. Beardsley and Y. Maki; *J. Org. Chem.*, 1980, **45**, 3827.	412, 719, 722
80JOC3909	J. O. Gardiner; *J. Org. Chem.*, 1980, **45**, 3909.	1060, 1062
80JOC3916	K. C. Liu, B. R. Shelton and R. K. Howe; *J. Org. Chem.*, 1980, **45**, 3916.	67
80JOC4041	K. Rasheed and J. D. Warkentin; *J. Org. Chem.*, 1980, **45**, 4041.	847
80JOC4065	A. Padwa, T. Caruso and S. Nahm; *J. Org. Chem.*, 1980, **45**, 4065.	431
80JOC4158	E. Coutouli-Argyropoulou and N. E. Alexandrou; *J. Org. Chem.*, 1980, **45**, 4158.	102
80JOC4587	T. Sasaki, K. Minamoto and K. Harada; *J. Org. Chem.*, 1980, **45**, 4587.	1019
80JOC4804	J. Bolster and R. M. Kellogg; *J. Org. Chem.*, 1980, **45**, 4804.	1020
80JOC4857	Y.-i Lin and S. A. Lang, Jr.; *J. Org. Chem.*, 1980, **45**, 4857.	73, 83
80JOC4860	G. Kornis, P. J. Marks and C. G. Chidester; *J. Org. Chem.*, 1980, **45**, 4860.	548
80JOC5095	T. Kiguchi, J. L. Schuppiser, J. C. Schwaller and J. Streiht; *J. Org. Chem.*, 1980, **45**, 5095.	716, 720, 726
80JOC5130	O. Tsuge, S. Urano and K. Oe; *J. Org. Chem.*, 1980, **45**, 5130.	594
80JOC5396	J. M. Kane; *J. Org. Chem.*, 1980, **45**, 5396.	977, 1006, 1040, 1047

80JOM(192)183	N. Zumbulyadis and H. J. Gysling; *J. Organomet. Chem.*, 1980, **192**, 183.	961
80JOM(195)275	R. Nesi, A. Ricci, M. Taddei and P. Tedeschi; *J. Organomet. Chem.*, 1980, **195**, 275.	6, 59
80JPR273	R. Beckert and R. Mayer; *J. Prakt. Chem.*, 1980, **322**, 273.	531, 539, 541
80JPR407	H. Dehne and P. Krey; *J. Prakt. Chem.*, 1980, **322**, 407.	911, 943
80JPR933	K. Dimitrowa, J. Hauschild, H. Zaschke and H. Schubert; *J. Prakt. Chem.*, 1980, **322**, 933.	555
80JPR1021	K. Gewald, W. Radke and U. Hain; *J. Prakt. Chem.*, 1980, **322**, 1021.	160
80JST(64)137	K. Iqball, N. L. Owen and J. Sheridan; *J. Mol. Struct.*, 1980, **64**, 137.	752
80JST(69)151	M. V. Andreocci, F. A. Devillanova, C. Furlani, G. Mattogno, G. Verani and R. Zanoni; *J. Mol. Struct.*, 1980, **69**, 151.	183
80KGS621	L. T. Gorb, A. D. Kachkovskii, N. N. Romanov, I. S. Shpileva and A. I. Tolmachev; *Khim. Geterotsikl. Soedin.*, 1980, 621 (*Chem. Abstr.*, 1980, **93**, 134 416).	684
80KGS853	V. L. Savel'ev, O. S. Artamonova, T. G. Afanas'eva and V. A. Zagorevskii; *Khim. Geterotsikl. Soedin.*, 1980, 853 (*Chem. Abstr.*, 1980, **93**, 186 256).	538
80KGS1255	P. B. Terentev, N. P. Lomakina, M. I. Rakhimi, K. D. Riad, Ya. B. Zelikhover and A. N. Kost; *Khim. Geterotsikl. Soedin.*, 1980, 1255.	195
80KGS1695	V. A. Chiuguk and A. G. Maidannik; *Khim. Geterotsikl. Soedin.*, 1980, 1695.	1048
80LA80	H. Grund and V. Jaeger; *Liebigs Ann. Chem.*, 1980, 80.	38
80LA101	V. Jaeger and V. Buss; *Liebigs Ann. Chem.*, 1980, 101.	36, 112
80LA122	V. Jaeger, V. Buss and W. Schwab; *Liebigs Ann. Chem.*, 1980, 122.	36
80LA542	S. Linke, J. Kurz, D. Lipinski and W. Gau; *Liebigs Ann. Chem.*, 1980, 542.	625
80LA1216	M. Hagen, R. D. Kohler and H. Fleig; *Liebigs Ann. Chem.*, 1980, 1216 (*Chem. Abstr.*, 1980, **93**, 220 674).	574
80LA1376	G. Heubach; *Liebigs Ann. Chem.*, 1980, 1376.	608, 609
80LA1557	U. Plücken, W. Winter and H. Meier; *Liebigs Ann. Chem.*, 1980, 1557.	396, 397, 421
80LA1573	H. Griengl, G. Prischl and A. Bleikolm; *Liebigs Ann. Chem.*, 1980, 1573.	213
80LA1623	K. Gewald, P. Bellmann and H. J. Jaensch; *Liebigs Ann. Chem.*, 1980, 1623.	76, 85, 86
80M407	D. Binder, C. R. Noe, J. Nussbaumer and B. C. Prager; *Monatsh. Chem.*, 1980, **111**, 407.	721, 729
80MI41600	A. Baranski, G. A. Shvelihgeimer and N. I. Kirillova; *Pol. J. Chem.*, 1980, **54**, 23 (*Chem. Abstr.*, 1980, **93**, 95 165).	95
80MI41601	A. Baranski; *Pol. J. Chem.*, 1980, **54**, 103.	89, 90
80MI41602	A. V. Kamernitskii, I. S. Levina, A. I. Terekhina and G. L. Gritsina; *Khim.-Farm. Zh.*, 1980, **14**, 37 (*Chem. Abstr.*, 1980, **93**, 72 078).	108, 110
80MI41603	B. A. Sastry, S. M. Asadulla, G. Ponticelli and M. Massacesi; *J. Inorg. Nucl. Chem.*, 1980, **42**, 833.	7
80MI41604	G. Miele, M. Guilano and J. M. Angelli; *J. Raman Spectrosc.*, 1980, **9**, 339.	5
80MI41606	M. Lozanovic, M. Kupinic, N. Blazevic and D. Kolbah; *Acta Pharm. Jugoslav.*, 1980, **30**, 189.	73
B-80MI41700	P. A. Lowe; 'Heterocyclic Chemistry', Royal Society of Chemistry, London, 1980, vol. 1, p. 109.	132, 144, 147, 148, 149, 151, 152, 154, 158, 162, 164, 165, 166, 167, 168, 169, 171, 173, 175
80MI41701	P. P. Singh, L. P. Pathak and S. K. Srivastava; *J. Inorg. Nucl. Chem.*, 1980, **42**, 533 (*Chem. Abstr.*, 1980, **93**, 142 136).	132
80MI41702	J. Gieldanowski, S. H. Kowalczyk-Bronisz, Z. Machoń, A. Szary and B. Blaszczyk; *Arch. Immunol. Ther. Exp.*, 1980, **28**, 393 (*Chem. Abstr.*, 1980, **93**, 230 880).	175
B-80MI41800	A. Padwa; 'Rearrangements of Ground and Excited States', Academic, New York, 1980, vol. 3, p. 501.	189
80MI41801	A. M. van Leusen; *Lect. Heterocycl. Chem.*, 1980, **5**, 111 (Supplementary issue of *J. Heterocycl. Chem.*, 1980, **17**).	220
80MI41802	C. D. Withrow; *Adv. Neurol.*, 1980, **27**, 577.	233
B-80MI41900	P. A. Lowe; in 'Heterocyclic Chemistry', ed. H. Suschitzky and O. Meth-Cohn; Chemical Society, London, 1980, vol. 1, pp. 119–139.	328
80MI42000	B. Marcewicz-Rojewska and S. Bilinski; *Acta Pol. Pharm.*, 1980, **37**, 159 (*Chem. Abstr.*, 1981, **95**, 7160).	345
80MI42100	G. Eber, S. Schneider and F. Dörr; *Ber. Bunsenges. Phys. Chem.*, 1980, **84**, 281.	374
80MI42101	G. J. Gross and M. Ghoraibeh; *Eur. J. Pharmacol.*, 1980, **67**, 111.	378
80MI42102	A. I. Machula, N. K. Barkov and V. P. Fisenko; *Farmakol. Toksikol.*, 1980, **43**, 16 (*Chem. Abstr.*, 1980, **92**, 140 649).	378
80MI42103	R. A. Altshuler and V. A. Parshin; *Farmakol. Toksikol.*, 1980, **43**, 153.	378
80MI42104	R. A. Altshuler and M. D. Mashkovsky; *Farmakol. Toksikol.*, 1980, **43**, 345.	378
80MI42105	I. G. Veksler, V. N. Ryabukta, K. P. Balitsky, R. A. Altshuler and M. D. Mashkovsky; *Farmakol. Toksikol.*, 1980, **43**, 349.	378
80MI42106	I. S. Mozorov; *Farmakol. Toksikol.*, 1980, **43**, 540.	378
80MI42107	M. D. Mashkovsky, S. B. Seredenin, R. A. Altshuler, A. A. Vedernikov, B. A. Badyshtov and N. I. Andreeva; *Khim.-Farm. Zh.*, 1980, **14**, 7.	378
80MI42108	P. A. Majid, P. J. F. DeFeyter, E. E. Van der Wall, R. Wardeh and J. P. Roos; *New Engl. J. Med.*, 1980, **302**, 1.	378
80MI42109	B. J. Johnson; *Weed Sci.*, 1980, **28**, 378.	391
80MI42110	A. Selvi and S. Facchetti; *Adv. Mass Spectrom., Part A*, 1980, **8**, 723.	371

80MI42111	R. J. Lemaire and P. G. Sears; *J. Solution Chem.*, 1980, **9**, 553.	371
80MI42200	M. R. Arshadi; *J. Chem. Thermodyn.*, 1980, **12**, 903.	400
80MI42201	A. Fundaro and M. C. Cassone; *Boll. Soc. Ital. Biol. Sper.*, 1980, **56**, 2364 (*Chem. Abstr.*, 1981, **94**, 168 824).	425
80MI42202	O. Eidelman and Z. I. Cabantchik; *Anal. Biochem.*, 1980, **106**, 335.	426
80MI42203	Ya. L. Kostyukovskii, F. A. Medvedev and D. B. Melamed; *Zh. Anal. Khim.*, 1980, **35**, 551.	426
80MI42204	A. V. Ostrovskaya, I. Vigalok and N. V. Svetlakov; *Otkrytiya, Izobret., Prom. Obraztsy, Tovarnye Znaki*, 1980, 142 (*Chem. Abstr.*, 1981, **94**, 85 092).	426
80MI42400	D. L. Gil, J. Ferreira and B. Reynafarie; *Xenobiotica*, 1980, **10**, 7.	462
80MI42600	D. J. Tocco, A. E. W. Duncan, F. A. De Luna, J. L. Smith, R. W. Walker and W. J. A. Vandenheuvel; *Drug Metab. Disposit.*, 1980, **8**, 236.	542
80MI42700	K.Lu and T. Loo; *Cancer Chemother. Pharmacol.*, 1980, **4**, 275 (*Chem. Abstr.*, 1981, **94**, 95 653).	576
80MI42800	D. W. Franco, L. H. Mazo, E. A. Neves and W. L. Polito; *Ann. Acad. Bras. Cienc.*, 1980, **52**, 261.	592
80MI42900	K. Undheim; *Adv. Mass Spectrom.*, 1980, **8**, 776.	682
80MI42901	E. L. Crane, Y. Maki and P. G. Beardsley; *J. Carbohydr., Nucleosides, Nucleotides*, 1980, **7**, 281.	719
80MI42902	H. Wollweber, R. Hiltmann, K. Stoepel and H. G. Kroneberg; *Eur. J. Med. Chem. – Chim. Ther.*, 1980, **15**, 111.	668
80MI43000	G. H. Pettit, A. T. H. Lenstra, H. J. Geise, D. G. Hellier and A. M. Phillips; *Israel J. Chem.*, 1980, **20**, 133.	751
B-80MI43001	F. G. Riddel; 'The Conformational Analysis of Heterocyclic Compounds,' Academic, New York, 1980, p. 56.	758
80MI43100	C. Th. Pedersen; *Sulfur Reports*, 1980, **1**, 1. 784, 785, 787, 788, 793, 796, 811
80MI43200	D. Buza and W. Gradowska; *Pol. J. Chem.*, 1980, **54**, 717.	824, 833
80MI43201	D. Buza and W. Gradowska; *Pol. J. Chem.*, 1980, **54**, 145.	833
80MI43202	D. Buza and W. Gradowska; *Pol. J. Chem.*, 1980, **54**, 2379.	838
80MI43300	P. Ruoff, J. Almlöf and S. Saebboe; *Chem. Phys. Lett.*, 1980, **72**, 489.	853
80MI43301	I. Murata and K. Nakasuji; *Is. J. Chem.*, 1980, **20**, 244.	878
80MI43302	S. D. Razumovskii and G. E. Zaikov; *Usp. Khim.*, 1980, **49**, 2345.	853
80MI43400	A. von Gieren, U. Riemann and B. Dederer; *Z. Anorg. Allg. Chem.*, 1980, **468**, 15.	901
80MI43401	D. W. Hahn, R. M. McConnell and J. L. McGuire; *Contraception*, 1980, **21**, 529 (*Chem. Abstr.*, 1980, **93**, 126 123).	945
B-80MI43600	K. T. Potts; 'Modern Methods of Heterocyclic Chemistry', Rensselaer Polytechnic Institute, Troy, New York, 1980.	974
80MIP41600	R. Kulboszek, K. Slon and R. Przybylik; *Pol. Pat.* 106 899 (1980) (*Chem. Abstr.*, 1981, **94**, 47 312).	122, 130
80MIP43400	A. N. Borisevich, Y. V. Karabanov, P. S. Pel'kis, V. P. Borisenko, A. S. Bragina, A. M. Vnukovskii and N. V. Rogovoi; *USSR Pat.* 753 412 (1980) (*Chem. Abstr.*, 1980, **93**, 232 716).	944
80NJC527	E. Gentric, J. Lauransan, C. Roussel and J. Metzger; *Nouv. J. Chim.*, 1980, **4**, 527.	380
80OMR(13)159	T. A. Crabb; *Org. Magn. Reson.*, 1980, **13**, 159.	700
80OMR(13)274	L. Stefaniak, M. Witanowski, B. Kamienski and G. A. Webb; *Org. Magn. Res.*, 1980, **13**, 274.	371
80OMR(14)356	M. Witanowski, L. Stefaniak, S. Biernat and G. A. Webb; *Org. Magn. Reson.*, 1980, **14**, 356.	398
80OMR(14)515	D. M. Rackham, S. E. Morgan and W. R. N. Williamson; *Org. Magn. Reson.*, 1980, **14**, 515.	551, 553
80OMS573	M. L. Deem; *Org. Mass. Spectrom.*, 1980, **15**, 573.	379
80PS(8)79	D. Barillier; *Phosphorus Sulfur*, 1980, **8**, 79.	785, 786, 788
80RCR28	V. G. Yashunskii and L. E. Kholodov; *Russ. Chem. Rev.* (*Engl. Transl.*), 1980, **49**, 28 (*Uspekhi Khim.*, 1980, **49**, 54).	378
80RTC278	C. Schenk, M. L. Beekes, J. A. M. van der Drift and T. J. de Boer; *Recl. Trav. Chim. Pays-Bas*, 1980, **99**, 278.	390
80S55	M. Pinza, G. Pifferi and F. Nasi; *Synthesis*, 1980, 55.	227
80S115	M. N. Basyouni, M. T. Omar and E. A. Ghali; *Synthesis*, 1980, 115.	843
80S387	D. Barbry, D. Couturier and G. Ricart; *Synthesis*, 1980, 387.	706
80S842	S. Mataka, T. Takahashi, M. Tashiro and Y. Tsuda; *Synthesis*, 1980, 842.	538, 830
80SA(A)143	M. Massacessi, G. Paschina, G. Ponticelli, M. N. Chary and B. A. Sastry; *Spectrochim. Acta, Part A*, 1980, **36**, 143.	20
80SA(A)199	F. A. Devillanova, G. Verani, K. R. Gayathri Devi and D. N. Sathyanarayana; *Spectrochim. Acta, Part A*, 1980, **36**, 199.	183
80SC725	M. Pigini, M. Giannella and F. Gualtieri; *Synth. Commun.*, 1980, **10**, 725.	762
80T1245	V. Bertini, F. Lucchesini and A. De Munno; *Tetrahedron*, 1980, **36**, 1245.	527, 538
80T3309	A. A. El-Barbary, K. Clausen and S. O. Lawesson; *Tetrahedron*, 1980, **36**, 3309.	803, 807
80TL229	A. A. Hagedorn, III, B. J. Miller and J. O. Nagy; *Tetrahedron Lett.*, 1980, **21**, 229. 67, 91, 92
80TL619	S. Oida, A. Yoshida, T. Hayashi, E. Nakayama, S. Sato and E. Ohki; *Tetrahedron Lett.*, 1980, 619.	136
80TL1457	S. Heitz, M. Durgeat and M. Guyot; *Tetrahedron Lett.*, 1980, **21**, 1457.	464
80TL1515	K. Steinbeck and B. Osterwinter; *Tetrahedron Lett.*, **21**, 1980, 1515.	766

80TL1649	T. Kondo, M. Tanimoto, M. Matsumoto, K. Notomato, Y. Achiba and K. Kimura; *Tetrahedron Lett.*, 1980, **21**, 1649.	752
80TL1719	C. Jouitteau, P. LePerchec, A. Forestiere and B. Sillion; *Tetrahedron Lett.*, 1980, **21**, 1719.	213
80TL2101	S. Brueckner, G. Fronza, L. M. Giunchi, V. A. Kozinski and O. V. Zelenskaja; *Tetrahedron Lett.*, 1980, 2101.	450, 456
80TL2247	G. Levesque and A. Mahjoub; *Tetrahedron Lett.*, 1980, 2247.	845
80TL2429	H-J. Federsel and J. Bergman; *Tetrahedron Lett.*, 1980, 2429.	341
80TL2995	H. P. Figeys and R. Jammar; *Tetrahedron Lett.*, 1980, 2995.	158
80TL3419	A. Padwa, H. L. Gingrich and R. Lim; *Tetrahedron Lett.*, 1980, **21**, 3419.	209
80TL3447	R. A. Reamer, M. Sletzinger and I. Shinkai; *Tetrahedron Lett.*, 1980, **21**, 3447.	110
80TL3617	S. Braverman, M. Freund and I. Goldberg; *Tetrahedron Lett.*, 1980, **21**, 3617.	975, 1023
80TL3627	G. S. Reddy and M. V. Bhatt; *Tetrahedron Lett.*, 1980, **21**, 3627.	196
80TL3755	F. Pochat; *Tetrahedron Lett.*, 1980, **21**, 3755.	65, 87, 88
80TL4025	M. Alajarin and P. Molina; *Tetrahedron Lett.*, 1980, **21**, 4025.	713, 717, 726
80TL4565	L.-Y. Chiang and J. Meinwald; *Tetrahedron Lett.*, 1980, 4565.	805, 955, 971
80TL4757	V. Sudarsanam, K. Nagarajan, K. R. Rao and S. J. Shenoy; *Tetrahedron Lett.*, 1980, 4757.	999
80UKZ637	I. P. Bachkovskii, A. P. Mikhailovskii and V. A. Chuiguk; *Ukr. Khim. Zh. (Russ. Ed.)*, 1980, **46**, 637 (*Chem. Abstr.*, 1980, **93**, 168 222).	401, 727
80USP4187099	J. E. Franz and R. K. Howe; *U.S. Pat.* 4 187 099 (1980) (*Chem. Abstr.*, 1980, **92**, 175 775).	129
80USP4207089	Olin Corp., *U.S. Pat.* 4 207 089 (1980) (*Chem. Abstr.*, 1980, **93**, 204 666).	509
80USP4207090	Olin Corp., *U.S. Pat.* 4 207 090 (1980) (*Chem. Abstr.*, 1980, **93**, 239 931).	509
80USP4212861	H. Adolphi, K. Kiehs and H. Theobald; *U.S. Pat.* 4 212 861 (1980) (*Chem. Abstr.*, 1980, **93**, 220 727).	130
80USP4229461	T. Shigematsu, K. Yoshida, M. Nakazawa, H. Kasugai and M. Tsuda; *U.S. Pat.* 4 229 461 (1980).	1025
80ZC18	J. Wrubel; *Z. Chem.*, 1980, **20**, 18.	23, 58
80ZC19	H. Voigt; *Z. Chem.*, 1980, **20**, 19.	93, 94
80ZC413	R. Evers, E. Fischer and M. Pulkenat; *Z. Chem.*, 1980, **20**, 413 (*Chem. Abstr.*, 1981, **94**, 121 422).	570
80ZN(A)712	J. Wiese and D. H. Sutter; *Z. Naturforsch., Teil A*, 1980, **35**, 712.	145
80ZN(B)568	P. Paetzold and G. Schimmel; *Z. Naturforsch., Teil. B*, 1980, **35**, 568.	376
80ZOB1014	O. A. Mukhacheva, M. A. Shchelkunova and A. I. Razumov; *Zh. Obshch. Khim.*, 1980, **50**, 1014 (*Chem. Abstr.*, 1980, **93**, 168 344).	932
80ZOR13	N. A. Benina, M. L. Petrow and A. A. Petrow; *Zh. Org. Khim.*, 1980, **16**, 13.	804
80ZOR183	L. V. Mezheritskaya, E. S. Matskovskaya and G. N. Dorofeenko; *Zh. Org. Khim.*, 1980, **16**, 183 (*Chem. Abstr.*, 1980, **93**, 95 160).	766, 773
80ZOR443	V. N. Drozd, Yu. M. Udachin, V. V. Sergeichuk and G. S. Bogomolova; *Zh. Org. Khim.*, 1980, **16**, 443.	848
80ZOR782	G. G. Filina, T. A. Bortyan, A. T. Menyailo and M. V. Pospelov; *Zh. Org. Khim.*, 1980, **16**, 782 (*Chem. Abstr.*, 1980, **93**, 94 519).	928, 937
80ZOR883	V. N. Drozd, Yu. M. Udachin, G. S. Bogomolova and V. V. Sergeichuk; *Zh. Org. Khim.*, 1980, **16**, 883.	799, 847
80ZOR1328	T. D. Mechkov, I. G. Sulimov, N. Y. Usik, I. Mladenov and V. V. Perekalin; *Zh. Org. Khim.*, 1980, **16**, 1328.	66, 87
80ZOR2047	V. N. Drozd and O. A. Popova; *Zh. Org. Khim.*, 1980, **16**, 2047 (*Chem. Abstr.*, 1981, **94**, 139 664).	845
80ZOR2185	Yu. A. Sharanin; *Zh. Org. Khim.*, 1980, **16**, 2185.	221
80ZOR2616	V. N. Drozd and O. A. Popova; *Zh. Org. Khim.*, 1980, **16**, 2616 (*Chem. Abstr.*, 1981, **94**, 156 794).	845
81AG603	R. Schulz and A. Schweig; *Angew. Chem.*, 1981, **93**, 603 (*Angew. Chem., Int. Ed. Engl.*, 1981, **20**, 570).	868, 887
81AG934	D. Cremer; *Angew. Chem.*, 1981, **93**, 934 (*Angew. Chem., Int. Ed. Engl.*, 1981, **20**, 888).	866
81AG(E)406	M. Baudler, J. Hellmann, P. Bachmann, K. F. Tebble, R. Fröhlich and M. Feher; *Angew. Chem., Int. Ed. Engl.*, 1981, **20**, 406.	976
81AG(E)784	H. Hageman; *Angew. Chem., Int. Ed. Engl.*, 1981, **20**, 784.	910, 940
81AHC(28)183	M. R. Bryce, and J. M. Vernon; *Adv. Heterocycl. Chem.*, 1981, **28**, 183.	171
81AHC(28)231	C. Wentrup; *Adv. Heterocycl. Chem.*, 1981, **28**, 231.	455
81AHC(29)1	R. K. Smalley; *Adv. Heterocycl. Chem.*, 1981, **29**, 1.	3, 20, 28, 31, 35, 37, 48
81AHC(29)141	M. Ruccia, N. Vivona and D. Spinelli; *Adv. Heterocycl. Chem.*, 1981, **29**, 141	60, 385, 401, 403
81AHC(29)251	A. Gasco and A. J. Boulton; *Adv. Heterocycl. Chem.*, 1981, **29**, 251.	394, 395, 396, 397, 398, 400, 404, 407, 408, 409, 420, 424, 426
81AP10	M. Neitzel and G. Zinner; *Arch. Pharm. (Weinheim, Ger.)*, 1981, **314**, 10.	390
81AP193	G. Zinner and M. Heitmann; *Arch. Pharm. (Weinheim, Ger.)*, 1981, **314**, 193.	441
81AP294	M. Neitzel and G. Zinner; *Arch. Pharm. (Weinheim, Ger.)*, 1981, **314**, 294.	390
81AP470	K. G. Upadhya, B. V. Badami and G. S. Puranik; *Arch. Pharm. (Wienheim, Ger.)*, 1981, **314**, 470.	373

81AP503	S. B. Havanur, B. V. Badami and G. S. Puranik; *Arch. Pharm. (Weinheim, Ger.)*, 1981, **314**, 503.	373
81BCJ41	H. Ayato, I. Tanaka, T. Yamane, T. Ashida, T. Sasaki, K. Minamoto and K. Harada; *Bull. Chem. Soc. Jpn.*, 1981, **54**, 41.	976, 1019
81BCJ1844	M. Z. A. Badr, M. M. Aly, A. M. Fahmy and M. E. Y. Mansour; *Bull. Chem. Soc. Jpn.*, 1981, **54**, 1844.	978, 988
81BCJ3541	R. Okazaki, K. Inoue and N. Inamoto; *Bull. Chem. Soc. Jpn.*, 1981, **54**, 3541.	908, 909, 910, 911, 936, 942
81BSB63	G. L'abbé, D. Sorgeloos, S. Toppet, G. S. D. King and L. Van Meervelt; *Bull. Soc. Chim. Belg.*, 1981, **90**, 63.	594
81BSB89	G. L'abbé and G. Vermeulen; *Bull. Soc. Chim. Belg.*, 1981, **90**, 89.	591, 592
81BSF(2)449	M. Kaafarani, M. P. Crozet and J. M. Surzur; *Bull. Soc. Chim. Fr.*, Part 2, 1981, 449.	321
81CB80	R. Neidlein and W. Lehr; *Chem. Ber.*, 1981, **114**, 80.	540
81CB285	H. Gotthardt, O. M. Huss and S. Schoy-Tribbensee; *Chem. Ber.*, 1981, **114**, 285.	825, 830
81CB536	J. Goerdeler, C. Lindner and F. Zander; *Chem. Ber.*, 1981, **114**, 536.	152
81CB549	J. Goerdeler and K. Nandi; *Chem. Ber.*, 1981, **114**, 549.	910, 915, 942
81CB787	H. Quast and F. Kees; *Chem. Ber.*, 1981, **114**, 787.	572
81CB802	H. Quast and F. Kees; *Chem. Ber.*, 1981, **114**, 802.	556, 558, 574
81CB808	J. Goerdeler and K. Nandi; *Chem. Ber.*, 1981, **114**, 808.	915
81CB1200	K.-H. Ongania; *Chem. Ber.*, 1981, **114**, 1200.	1019
81CB1737	H. Gotthardt and F. Reiter; *Chem. Ber.*, 1981, **114**, 1737.	374
81CB2132	W. S. Sheldrick, S. Pohl, H. Zamankhan, M. Banek, D. Amirzadeh-Asl and H. W. Roesky; *Chem. Ber.*, 1981, **114**, 2132.	975, 984, 1022
81CB2382	M. Lorch and H. Meier; *Chem. Ber.*, 1981, **114**, 2382.	353
81CB2450	H. Gotthardt and F. Reiter; *Chem. Ber.*, 1981, **114**, 2450.	374
81CB2519	A. Meller, F. J. Hirninger, M. Noltemeyer and W. Mariggele; *Chem., Ber.*, 1981, **114**, 2519.	1022
81CB2580	H. Wegmann and W. Steglich; *Chem. Ber.*, 1981, **114**, 2580.	202
81CB2622	H. Bock, G. Brähler, D. Dauplaise and J. Meinwald; *Chem. Ber.*, 1981, **114**, 2622.	952, 953, 961
81CB2938	O. Zimmer and H. Meier; *Chem. Ber.*, 1981, **114**, 2938.	353, 460
81CC241	O. Meth-Cohn and S. Rhouati; *J. Chem. Soc., Chem. Commun.*, 1981, 241.	223
81CC365	J. H. Boyer and C. Huang; *J. Chem. Soc., Chem. Commun.*, 1981, 365.	405
81CC510	Y. Uchida and S. Kozuka; *J. Chem. Soc., Chem. Commun.*, 1981, 510.	167
81CC550	C. J. Moody, C. W. Rees and S. C. Tsoi; *J. Chem. Soc., Chem. Commun.*, 1981, 550.	170, 1009
81CC565	J. Nakayama, N. Matsumaru and M. Hoshino; *J. Chem. Soc., Chem. Commun.*, 1981, 565.	837
81CC632	M. Wakselman and F. Acher; *J. Chem. Soc., Chem. Commun.*, 1981, 632.	763
81CC669	Y. Okamato and P. S. Wojciechowski; *J. Chem. Soc., Chem. Commun.*, 1981, 669.	971
81CC741	A. P. Davis and G. H. Whitham; *J. Chem. Soc., Chem. Commun.*, 1981, 741.	754, 755, 770
81CC822	B. F. Bonini, G. Mazzanti, S. Sarti, P. Zanirato and G. Maccagnani; *J. Chem. Soc., Chem. Commun.*, 1981, 822.	941
81CC828	M. V. Lakshmikantham, M. P. Cava, M. Albeck, L. Engman, F. Wadl and E. Aharon-Shalom; *J. Chem. Soc., Chem. Commun.*, 1981, 828.	958
81CC920	P. Shu, L. Chiang, T. Emge, D. Holt, T. Kistenmacher, M. Lee, J. Stokes, T. Poehler, A. Bloch and D. Cowan; *J. Chem. Soc., Chem. Commun.*, 1981 920.	986
81CC927	C. J. Moody, C. W. Rees, S. C. Tsoi and D. J. Williams; *J. Chem. Soc., Chem. Commun.*, 1981, 927.	1018
81CC1131	R. Carrie, Y. Y. C. Yeung Lam Ko, F. de Sarlo and A. Brandi; *J. Chem. Soc., Chem. Commun.*, 1981, 1131.	998
81CCC2564	A. Krutosikova, J. Kovac, M. Dandarova, J. Lesko and S. Ferik; *Collect. Czech. Chem. Commun.*, 1981, **46**, 2564.	1009
81CCC2949	A. Kurtosikova, J. Kovac, M. Chudoleova and D. Ilavsky; *Collect. Czech. Chem. Commun.*, 1981, **45**, 2949.	979
81CL213	O. Tsuge, H. Shiraishi and M. Noguchi; *Chem. Lett.*, 1981, 213.	1041
81CL331	A. Albini, G. F. Bettinetti and G. Minoli; *Chem. Lett.*, 1981, 331.	1046
81CL805	K. Takahashi, K. Takase and Y. Noda; *Chem. Lett.*, 1981, 805.	836
81CL1457	K. Inomata, H. Yamada and H. Kotake; *Chem. Lett.*, 1981, 1457.	174
81CPB1743	K. Masuda, J. Adachi, H. Moria and K. Nomura; *Chem. Pharm. Bull.*, 1981, **29**, 1743.	456
81CPB3543	H. Yamanaka, M. Shiraiwa, E. Yamamoto and T. Sakamoto; *Chem. Pharm. Bull.*, 1981, **29**, 3543.	162
81CSC1099	R. Faure, H. Loiseleur, R. Bartnik, S. Lesniak and A. Laurent; *Cryst. Struct. Commun.*, 1981, **10**, 1099.	901
81EGP148341	H. U. Kibbel, U. Ohnmacht and J. Teller; *Ger. (East) Pat.* 148 341 (1981) (*Chem. Abstr.*, 1982, **96**, 20 092).	1023
81FRP2457289	Ciba Geigy, *Fr. Pat.* 2 457 289 (1981) (*Chem. Abstr.*, 1981, **94**, 17 5123).	509
81G71	A. Braibanti and M. T. Lugari-Mangia; *Gazz. Chim. Ital.*, 1981, **111**, 71.	145, 146
81G167	A. J. Boulton, D. E. Coe and P. G. Tsoungas; *Gazz. Chim. Ital.*, 1981, **111**, 167.	408
81G289	F. Bellesia, R. Grandi, U. M. Pagnoni and R. Trave; *Gazz. Chim. Ital.*, 1981, **111**, 289.	460

Ref	Citation	Pages
81GEP3003019	F. Maurer, I. Hamann and B. Homeyer; *Ger. Pat.* 3 003 019 (1981) (*Chem. Abstr.*, 1981, **95**, 150 656).	1026
81GEP3033169	R. R. Crenshaw and A. A. Algieri; *Ger. Pat.* 3 033 169 (1981) (*Chem. Abstr.*, 1981, **95**, 62 220).	542
81H(15)341	F. Yoneda, T. Tachibana, J. Tanoue, T. Yano and Y. Sakuma; *Heterocycles*, 1981, **15**, 341.	721
81H(15)1349	K. Undheim; *Heterocycles*, 1981, **15**, 1349.	682, 683, 684, 685, 687, 688, 689, 690, 692, 693, 701, 702, 703, 704, 705, 706, 709
81H(15)1395	V. K. Daukas, G. V. Purvaneckas, E. B. Udrenaite, V. L. Gineityte and A. V. Barauskaite; *Heterocycles*, 1981, **15**, 1395.	766
81H(16)35	J.-P. Maffrand; *Heterocycles*, 1981, **16**, 35.	376, 722
81H(16)145	C. Kashima, N. Yoshihara and S. Shirai; *Heterocycles*, 1981, **16**, 145.	63, 84
81H(16)156	T. Nishiwaki, E. Kawamura, N. Abe and M. Iori; *Heterocycles*, 1981, **16**, 156.	152
81H(16)209	O. Tsuge, M. Noguchi and H. Moriyama; *Heterocycles*, 1981, **16**, 209.	206
81H(16)595	T. Nishiwaki, E. Kawamura, N. Abe, H. Kochi, Y. Sasaoka and K. Soneda; *Heterocycles*, 1981, **16**, 595.	152, 848
81H(16)789	O. Tsuge, T. Takata and M. Noguchi; *Heterocycles*, 1981, **16**, 789.	1035
81H(16)1561	S. Karady, J. S. Amato, D. Dortmund and L. M. Weinstock; *Heterocycles*, 1981, **16**, 1561.	525, 529, 530, 531, 539
81H(16)1565	S. Karady, J. S. Amato, D. Dortmund, R. A. Reamer and L. M. Weinstock; *Heterocycles*, 1981, **16**, 1565.	525, 529, 531
81H(16)1855	A. Bhattacharjya; *Heterocycles*, 1981, **16**, 1855.	113
81IC399	L. J. Saethre, P. Å. Malmquist, N. Mårtensson, S. Svensson, U. Gelius and K. Siegbahn; *Inorg. Chem.*, 1981, **20**, 399.	1052, 1057
81IJC(B)111	P. S. Rao and V. Veeranagaiah; *Indian J. Chem., Sect. B*, 1981, **20**, 111.	538
81IJC(B)161	V. P. Upadhyaya and V. R. Srinivasan; *Indian J. Chem., Sect. B*, 1981, **20**, 161.	1010
81IJC(B)322	G. S. Reddy and M. V. Bhatt; *Indian J. Chem., Sect. B*, 1981, **20**, 322.	222
81JA239	P. A. Jacobi, A. Brownstein, M. Martinelli and K. Grozinger; *J. Am. Chem. Soc.*, 1981, **103**, 239.	1007
81JA609	V. Malatesta and K. U. Ingold; *J. Am. Chem. Soc.*, 1981, **103**, 609.	764
81JA1540	I. Ernest, W. Holick, G. Rihs, D. Schomburg, G. Shoham, D. Wenkert and R. B. Woodward; *J. Am. Chem. Soc.*, 1981, **103**, 1540.	535
81JA1789	M. Miura, M. Nojima, S. Kusabayashi and S. Nagase; *J. Am. Chem. Soc.*, 1981, **103**, 1789.	870, 871
81JA2558	D. Feller, E. R. Davidson and W. T. Borden; *J. Am. Chem. Soc.*, 1981, **103**, 2558.	760
81JA2578	C. K. Kohlmiller and L. Andrews; *J. Am. Chem. Soc.*, 1981, **103**, 2578.	859, 881
81JA2757	K. Oka, A. Dobashi and S. Hara; *J. Am. Chem. Soc.*, 1981, **103**, 2757.	760
81JA2868	J. M. Bolster and R. M. Kellogg; *J. Am. Chem. Soc.*, 1981, **103**, 2868.	1015
81JA3619	D. Cremer; *J. Am. Chem. Soc.*, 1981, **103**, 3619.	866, 888
81JA3627	D. Cremer; *J. Am. Chem. Soc.*, 1981, **103**, 3627.	866, 888
81JA3633	D. Cremer; *J. Am. Chem. Soc.*, 1981, **103**, 3633.	855, 866, 892
81JA3807	R. I. Martinez, J. T. Herron and R. E. Huie; *J. Am. Chem. Soc.*, 1981, **103**, 3807.	894
81JA4278	A. I. Meyers and Y. Yamamoto; *J. Am. Chem. Soc.*, 1981, **103**, 4278.	212
81JA7032	I. Kalwinsch, L. Xingya, J. Gottstein and R. Huisgen; *J. Am. Chem. Soc.*, 1981, **103**, 7032.	572
81JA7743	E. C. Taylor, N. F. Haley and R. J. Clemens; *J. Am. Chem. Soc.*, 1981, **103**, 7743.	1001
81JAP81125390	Y. Kaneoka; *Jap. Pat.* 81 125 390 (1981) (*Chem. Abstr.*, 1982, **96**, 85 558).	1026
81JCR(S)135	G. Bianchi, R. Gandolfi and C. Demicheli; *J. Chem. Res. (S)*, 1981, 135.	998
81JCR(S)4036	K. Bruzik, W. Stec, D. Houalla and R. Wolf; *J. Chem. Res. (S)*, 1981, 4036.	1022
81JCS(P1)4	A. Albini, G. F. Bettinetti and G. Minoli; *J. Chem. Soc., Perkin Trans. 1*, 1981, 4.	1039
81JCS(P1)360	S. H. Askari, S. F. Moss and D. R. Taylor; *J. Chem. Soc., Perkin Trans. 1*, 1981, 360.	553, 570
81JCS(P1)415	R. M. Acheson and J. D. Wallis; *J. Chem. Soc., Perkin Trans. 1*, 1981, 415.	977, 1006
81JCS(P1)607	M. R. Bryce, C. D. Reynolds, P. Henson and J. M. Vernon; *J. Chem. Soc., Perkin Trans. 1*, 1981, 607.	334, 529
81JCS(P1)1033	K. Masuda, J. Adachi and K. Nomura; *J. Chem. Soc., Perkin Trans. 1*, 1981, 1033.	531, 536, 538
81JCS(P1)1037	Y. Tamura, Y. Takebe, S. M. M. Bayomi, C. Mukai, M. Ikeda, M. Murase and M. Kise; *J. Chem. Soc., Perkin Trans. 1*, 1981, 1037.	170
81JCS(P1)1401	T. Nishiwaki, E. Kawamura, N. Abe and M. Iori; *J. Chem. Soc., Perkin Trans. 1*, 1981, 1401.	161
81JCS(P1)1591	K. Masuda, J. Adachi, H. Nate, H. Takahata and K. Nomura; *J. Chem. Soc., Perkin Trans. 1*, 1981, 1591.	456
81JCS(P1)1703	G. Adembri, A. Camparini, F. Ponticelli and P. Tedeschi; *J. Chem. Soc., Perkin Trans. 1*, 1981, 1703.	386
81JCS(P1)1821	A. Albini, G. F. Bettinetti and G. Minoli; *J. Chem. Soc., Perkin Trans. 1*, 1981, 1821.	1039
81JCS(P1)2692	K. T. Potts, A. J. Elliott, G. R. Titus, D. Al-Hilal, P. F. Lindley, G. V. Boyd and T. Norris; *J. Chem. Soc., Perkin Trans. 1*, 1981, 2692.	1005
81JCS(P1)2952	H. Mastalerz and M. S. Gibson; *J. Chem. Soc., Perkin Trans. 1*, 1981, 2952.	554
81JCS(P1)2991	A. M. Damas, R. O. Gould, M. M. Harding, R. M. Paton, J. F. Ross and J. Crosby; *J. Chem. Soc., Perkin Trans. 1*, 1981, 2991.	901, 905, 906, 910, 911, 920, 941
81JCS(P2)19	C. Bomaingue, D. Houalla, M. Sanchez and R. Wolf; *J. Chem. Soc., Perkin. Trans. 2*, 1981, 19.	1022
81JCS(P2)310	A. R. Butler and I. Hussain; *J. Chem. Soc., Perkin Trans. 2*, 1981, 310.	994

81JCS(P2)789	A. F. Cameron, I. R. Cameron and F. D. Duncanson; *J. Chem. Soc., Perkin Trans. 2*, 1981, 789.	974, 986
81JCS(P2)1062	C. T. Pedersen, J. Oddershede and J. R. Sabin; *J. Chem. Soc., Perkin Trans. 2*, 1981, 1062.	786
81JCS(P2)1240	R. Calvino, A. Gasco, A. Serafino and D. Viterbo; *J. Chem. Soc., Perkin Trans. 2*, 1981, 1240.	396, 406, 414
81JHC37	M. J. Dimsdale; *J. Heterocycl. Chem.*, 1981, **18**, 37.	386
81JHC205	R. N. Hanson and M. A. Davis; *J. Heterocycl. Chem.*, 1981, **18**, 205.	343
81JHC409	D. A. Kennedy and L. A. Summers; *J. Heterocycl. Chem.*, 1981, **18**, 409.	564
81JHC437	P. D. Clark and D. M. McKinnon; *J. Heterocycl. Chem.*, 1981, **18**, 437.	161, 171
81JHC723	V. Frenna, N. Vivona, D. Spinelli and G. Consiglio; *J. Heterocycl. Chem.*, 1981, **18**, 723.	385
81JHC789	A. Shafiee, G. Kiaeay and M. Vosooghi; *J. Heterocycl. Chem.*, 1981, **18**, 789.	345
81JHC885	W. J. Hammar and M. R. Rustad; *J. Heterocycl. Chem.*, 1981, **18**, 885.	190
81JHC1197	L. L. Whitfield, Jr. and E. P. Papadopoulos; *J. Heterocycl. Chem.*, 1981, **18**, 1197.	387
81JHC1247	D. L. Boger and C. E. Brotherton; *J. Heterocycl. Chem.*, 1981, **18**, 1247.	412, 535, 536
81JHC1309	G. L'abbé, G. Vermeulen, S. Toppet, G. S. D. King, J. Aerts and L. Sengier; *J. Heterocycl. Chem.*, 1981, **18**, 1309.	905, 906, 911, 939
81JHC1335	G. N. Jham, R. N. Hanson, R. W. Giese and P. Vouros; *J. Heterocycl. Chem.*, 1981, **18**, 1335.	340
81JHC1485	P. Catsoulacos and C. Camoutsis; *J. Heterocycl. Chem.*, 1981, **18**, 1485.	167
81JHC1581	K. Rasheed and J. D. Warkentin; *J. Heterocycl. Chem.*, 1981, **18**, 1581.	1013
81JMC601	R. K. Sehgal, M. W. Webb and K. C. Agrawal; *J. Med. Chem.*, 1981, **24**, 601.	1014
81JMR(42)337	H. J. Jakobsen and S. Deshmukh; *J. Magn. Reson.*, 1981, **42**, 337.	139, 140
81JOC101	M. B. Smith and J. Wolinsky; *J. Org. Chem.*, 1981, **46**, 101.	764, 770
81JOC312	D. R. Brittelli and G. A. Boswell, Jr.; *J. Org. Chem.*, 1981, **46**, 312.	421
81JOC316	D. R. Brittelli and G. A. Boswell, Jr.; *J. Org. Chem.*, 1981, **46**, 316.	67, 81, 87, 405
81JOC424	D. N. Reinhoudt, J. Geevers, W. P. Trompenaars, S. Harkema and G. J. van Hummel; *J. Org. Chem.*, 1981, **46**, 424.	976, 1006
81JOC614	S. Ohashi, W. E. Ruch and G. B. Butler; *J. Org. Chem.*, 1981, **46**, 614.	1006
81JOC771	R. K. Howe and B. R. Shelton; *J. Org. Chem.*, 1981, **46**, 771.	478, 500, 920
81JOC1410	B. H. Lipshutz and R. W. Hungate; *J. Org. Chem.*, 1981, **46**, 1410.	193
81JOC1666	E. M. Kosower, B. Pazhenchevsky, H. Dodink, H. Kanety and D. Faust; *J. Org. Chem.*, 1981, **46**, 1666.	976
81JOC2069	A. M. van Leusen, H. J. Jeuring, J. Wildeman and S. P. J. M. van Nispen; *J. Org. Chem.*, 1981, **46**, 2069.	220
81JOC2691	J. L. Kice and S. T. Liao; *J. Org. Chem.*, 1981, **46**, 2691.	930, 932, 940
81JOC3256	M. I. El-Sheikh, A. Marks and E. R. Biehl; *J. Org. Chem.*, 1981, **46**, 3256.	217
81JOC3502	A. Bened, R. Durand, D. Pioch, P. Geneste, J. P. Declereq, G. Germain, J. Rambaud and R. Rogues; *J. Org. Chem.*, 1981, **46**, 3502.	998
81JOC3575	B. D. Dean and W. E. Truce; *J. Org. Chem.*, 1981, **46**, 3575.	165
81JOC3742	J. R. Falck, S. Manna and O. C. Mioskowski; *J. Org. Chem.*, 1981, **46**, 3742.	211
81JOC4065	K. T. Potts, R. D. Cody and R. J. Dennis; *J. Org. Chem.*, 1981, **46**, 4065.	538, 606, 609, 612
81JOC4567	J. F. Bunnett and P. Singh; *J. Org. Chem.*, 1981, 4567.	608
81JOC4998	P. C. Montevecchi and A. Tundo; *J. Org. Chem.*, 1981, **46**, 4998.	458
81JOC5408	R. M. J. Liskamp, H. J. M. Zeegers and H. C. J. Ottenheijm; *J. Org. Chem.*, 1981, **46**, 5408.	771, 779
81JPR169	H.-P. Kruse and H. Anger; *J. Prakt. Chem.*, 1981, **323**, 169.	864
81JPR279	A. Preiss, W. Walek and S. Dietzel; *J. Prakt. Chem.*, 1981, **323**, 279.	466
81JPR737	H. Poleschner, R. Radeglia and E. Fanghänel; *J. Prakt. Chem.*, 1981, **323**, 737.	817, 957, 958, 959, 960
81JST(71)1	K. R. Gayathri Devi, D. N. Sathyanarayana and E. M. Engler; *J. Mol. Struct.*, 1981, **71**, 1.	950, 954
81JST(74)343	C. L. Khetrapal and A. C. Kunwar; *J. Mol. Struct.*, 1981, **74**, 343.	518
81KGS35	I. V. Tselinskii, S. F. Mel'nikova, S. N. Vergizov and G. M. Frolova; *Khim. Geterotsikl. Soedin.*, 1981, 35.	401
81KGS124	A. V. Eremeev, R. S. El'kinson, E. Liepins and V. Imuns; *Khim. Geterotsikl. Soedin.*, 1981, 124 (*Chem. Abstr.*, 1981, **94**, 192 237).	906, 939
81KGS329	V. N. Elokhina, A. S. Nakhmanovich, R. V. Karnaukhova, I. D. Kalikhman and M. G. Voronkov; *Khim. Geterotsikl. Soedin.*, 1981, 329.	843
81KGS1011	L. S. Peshakova, V. B. Kalcheva and D. A. Simov; *Khim. Geterotsikl. Soedin.*, 1981, 1011.	185
81KGS1209	R. E. Valters, D. E. Balode, R. B. Kampare and S. P. Valtere; *Khim. Geterotsikl. Soedin.*, 1981, 1209.	146
81LA191	R. Becker and W. Rohr; *Liebigs Ann. Chem.*, 1981, 191.	390
81LA347	H. Gotthardt and O. M. Huss; *Liebigs Ann. Chem.*, 1981, 347.	827, 830, 841
81LA1025	H. Gotthardt, F. Reiter and C. Kromer; *Liebigs Ann. Chem.*, 1981, 1025.	911, 913, 919, 933
81LA1361	L. Capuano, F. Braun, J. Lorenz, R. Zander and J. Bender; *Liebigs Ann. Chem.*, 1981, 1361.	611
81MI41700	K. J. Ahmed, A. Habib, S. Z. Haider, K. M. A. Malik and M. B. Hursthouse; *Inorg. Chim. Acta*, 1981, **56**, L37 (*Chem. Abstr.*, 1981, **95**, 196 560).	136

81MI41701	G. Mille, J. C. Poite, M. Guiliano and J. Chouteau; *Spectroscopy Lett.*, 1981, **14**, 271.	141
81MI41702	A. De; *Prog. Med. Chem.*, 1981, **18**, 117 (*Chem. Abstr.* 1982, **96**, 154 817).	175
81MI41800	I. J. Turchi; *Product Research and Development*, 1981, **20**, 32.	178, 206, 223, 232
81MI41900	M. Petitjean, G. Vernin and J. Metzger; *Industries Alimentaires et Agricoles*, 1981, **98**, 741.	327
81MI42100	J. C. Watkins, J. Davies, R. H. Evans, A. A. Francis and A. W. Jones, *Adv. Biochem. Psychopharmacol.*, 1981, **27**, 263.	390
81MI42101	J. Davis and J. C. Watkins; *Adv. Biochem. Psychopharmacol.*, 1981, **27**, 275.	390
81MI42102	H. Shinozaki and M. Ishida; *Adv. Biochem. Psychopharmacol.*, 1981, **27**, 327.	390
81MI42103	N. A. Anis, R. B. Clark, K. A. F. Gration and P. N. R. Usherwood; *J. Physiol.*, 1981, **345**, 312.	390
81MI42104	R. M. Menges and S. Tamez; *Weed Sci.*, 1981, **29**, 74.	391
81MI42200	P. Ghosh, B. Ternai and M. Whitehouse; *Med. Chem. Rev.*, 1981, **1**, 159.	394, 399, 410, 411, 425, 426
81MI42201	V. G. Prokudin, G. M. Nazin and V. V. Dubikhin; *Kinet. Katal.*, 1981, **22**, 871 (*Chem. Abstr.*, 1981, **95**, 186 411).	400, 404
81MI42202	L. M. Smith, H. M. McConnell, B. A. Smith and P. J. Wallace; *Biophys. J.*, 1981, **33**, 139.	426
81MI42600	A. Montiel; *Analusis*, 1981, **9**, 102 (*Chem. Abstr.*, 1981, **95**, 67 686).	537
81MI42700	M. R. Mahmoud, R. Abdel Hamide and F. Abdel Goad; *Z. Phys. Chem. (Leipzig)*, 1981, **262**, 551 (*Chem. Abstr.*, 1981, **95**, 79 479).	552
B-81MI42701	M. Witanowski, L. Stefaniak and G. A. Webb; in 'Annual Reports on NMR Spectroscopy', ed. G. A. Webb; Academic, London, 1981, vol. 11B, p. 319.	551
81MI42900	M. Gerold, R. Eigenmann, F. Hefti, A. Daum and G. Haeusler; *J. Pharmacol. Exp. Ther.*, 1981, **21**, 624.	730
B-81MI43000	T. W. Green; 'Protective Groups in Organic Synthesis,' Wiley, New York, 1981, p. 114.	759
81MI43300	V. N. Odinokov and G. A. Tolstikov; *Usp. Khim.*, 1981, **50**, 1207.	853
B-81MI43600	A. Schmidpeter and J. Högel; in '8th Int. Congr. Heterocycl. Chem., Graz', 1981, p. 181.	983, 998
81MIP859361	V. Z. Laishev, M. L. Petrov and A. A. Petrov; *USSR Pat.* 859 361 (*Chem. Abstr.*, 1982, **96**, P34 860).	351
81OMS29	G. Bouchoux, Y. Hoppilliard, M. Golfier and M. G. Guillerez; *Org. Mass Spectrom.*, 1981, **16**, 29.	553
81RTC10	R. S. Sukhai, W. Verboom, J. Meijer, M. J. M. Schoufs and L. Brandsma; *Recl. Trav. Chim. Pays-Bas*, 1981, **100**, 10.	958, 959, 960, 969
81S237	H. Kroff and H. Von Wallis; *Synthesis*, 1981, 237.	772
81S316	K. S. Sharma, S. Kumari and R. P. Singh; *Synthesis*, 1981, 316.	538
81S501	F. A. J. Meskens; *Synthesis*, 1981, 501.	773, 775
81S531	S. Kambe and K. Saito; *Synthesis*, 1981, 531.	992
81S981	M. Bandy-Floch and A. Robert; *Synthesis*, 1981, 981.	977, 1005
81S991	M. Ueda, K. Seki and Y. Imai; *Synthesis*, 1981, 991.	201
81SC209	K. Fuji, M. Ueda, K. Sumi and E. Fujita; *Synth. Commun.*, 1981, **11**, 209.	837
81T1415	G. Ronsisvalle, F. Guerrera and M. A. Siracusa; *Tetrahedron*, 1981, **37**, 1415.	65, 86, 386
81T2181	J. E. Baldwin, A. L. J. Beckwith, A. P. Davis, G. Procter and K. A. Singleton; *Tetrahedron*, 1981, **37**, 2181.	165, 167, 172
81T2451	A. M. Lamazouere and J. Sotiropoulos; *Tetrahedron*, 1981, **37**, 2451.	943
81T2485	R. J. S. Beer, H. Singh, D. Wright and L. K. Hansen; *Tetrahedron*, 1981, **37**, 2485.	1016
81T2595	P. Merot, C. Gadreau and A. Foucoud; *Tetrahedron*, 1981, **37**, 2595.	1006
81T3377	M. Sindler-Kulyk, D. C. Neckers and J. R. Blount; *Tetrahedron*, 1981, **37**, 3377.	153
81T3627	A. Mehlhorn, F. Fratev and V. Monev; *Tetrahedron*, 1981, **37**, 3627.	133, 144
81T3867	R. J. S. Beer and D. Wright; *Tetrahedron*, 1981, **37**, 3867.	168
81TL229	G. Märkl and G. Yu. Jin; *Tetrahedron Lett.*, 1981, **22**, 229.	1022
81TL525	M. Sindler-Kulyk and D. C. Neckers; *Tetrahedron Lett.*, 1981, **22**, 525.	153
81TL529	M. Sindler-Kulyk and D. C. Neckers; *Tetrahedron Lett.*, 1981, **22**, 529.	153
81TL1175	K. S. Y. Lau and D. I. Basiulis; *Tetrahedron Lett.*, 1981, 1175.	899, 940
81TL2435	A. Padwa, M. Akiba, L. A. Cohen and J. G. MacDonald; *Tetrahedron Lett.*, 1981, **22**, 2435.	199, 200
81TL2535	W. Bethäuser, M. Regitz and W. Theis; *Tetrahedron Lett.*, 1981, **22**, 2535.	1002
81TL2727	G. A. Kraus and J. O. Nagy; *Tetrahedron Lett.*, 1981, 2727.	1003
81TL3163	A. I. Meyers and J. P. Lawson; *Tetrahedron Lett.*, 1981, **22**, 3163.	193
81TL3371	R. A. Whitney and E. S. Nicholas; *Tetrahedron Lett.*, 1981, 3371.	405
81TL3699	D. J. Brunelle; *Tetrahedron Lett.*, 1981, **22**, 3699.	49, 50, 83
81TL4199	M. V. Lakshmikantham, M. P. Cava, M. Albeck, L. Engman, P. Carroll, J. Bergman and F. Wudl; *Tetrahedron Lett.*, 1981, 4199.	951, 961
81UP41700	A. H. Sykes; 1981, unpublished results.	175
81USP4246126	Lubrizol Corp., *U.S. Pat.* 4 246 126 (1981) (*Chem. Abstr.* 1981, **94**, 142 505).	577
81USP4252964	M. R. Uskokovic and P. M. Wovkulich; *U.S. Pat.* 4 252 964 (1981).	1025
81USP4254265	Olin Corp., *U.S. Pat.* 4 254 265 (1981) (*Chem. Abstr.*, 1981, **95**, 25 073).	509
81USP4262008	Beecham Group Ltd. *U.S. Pat.* 4 262 008 (1980) (*Chem. Abstr.*, 1980, **93**, 239 387).	1025
81USP4263311	P. E. Bender; *U.S. Pat.* 4 263 311 (1981) (*Chem. Abstr.*, 1981, **95**, 97 797).	1025

81USP4265898	H. Hortsmann, K. Meng, F. Seuter and E. Moeller; *U.S. Pat.* 4 265 898 (1979) (*Chem. Abstr.*, 1980, **92**, 215 440).	1025
81USP4275073	W. K. Moberg; *U.S. Pat.* 4 275 073 (1981) (*Chem. Abstr.*, 1981, **95**, 150 654).	1025
81USP4276298	Merck and Co. Inc.; *U.S. Pat.* 4 276 298 (1981) (*Chem. Abstr.*, 1981, **95**, 203 929).	149
81USP4283543	Morton-Norwich Inc., *U.S. Pat.* 4 283 543 (1981) (*Chem. Abstr.*, 1981, **95**, 187 269).	576
81USP4288576	Hercules Inc., *U.S. Pat.* 4 288 576 (1981) (*Chem. Abstr.*, 1981, **95**, 188 481).	577
81ZC102	K. Peseke; *Z. Chem.*, 1981, **21**, 102.	596
81ZC182	R. Socher, C. Csongár, I. Müller and G. Tomaschewski; *Z. Chem.*, 1981, **21**, 182.	443
81ZC185	H. Paul and G. Huschert; *Z. Chem.*, 1981, **21**, 185 (*Chem. Abstr.*, 1981, **95**, 132 764).	567
81ZC265	E. Gey, J. Fabian, S. Bleisch, G. Domschke and R. Mayer; *Z. Chem.*, 1981, **21**, 265 (*Chem. Abstr.*, 1981, **95**, 167 956).	912
81ZC324	R. Mayer, G. Domschke, S. Bleisch and A. Bartl; *Z. Chem.*, 1981, **21**, 324 (*Chem. Abstr.*, 1982, **96**, 35 164).	912, 936
81ZC326	M. Mühlstädt, B. Schulze and I. Schubert; *Z. Chem.*, 1981, **21**, 326.	170
81ZN(B)501	A. F. A. Shalaby, M. A. A. Aziz and S. S. M. Boghdadi; *Z. Naturforsch., Teil B*, 1981, **36**, 501.	981
81ZN(B)609	W. Friedrichsen, W.-D. Schröer, I. Schwarz and A. Böttcher; *Z. Naturforsch., Teil B*, 1981, **36**, 609.	831
81ZN(B)1017	H. Meier, J. Zountsas and O. Zimmer; *Z. Naturforsch, Teil B*, 1981, **36**, 1017.	348
81ZN(B)1640	K. G. Jensen and E. B. Pedersen; *Z. Naturforsch., Teil B*, 1981, **36**, 1640.	150, 160
81ZOB1192	S. A. Dyachenko, M. I. Burenova, M. P. Papirnik, V. G. Pesin, N. V. Ostashkova and A. I. Stetsenko; *Zh. Obshch. Khim.*, 1981, **51**, 1912.	403
81ZOR667	V. Z. Laishev, V. Z. Laishev and A. A. Petrov; *Zh. Org. Khim.*, 1981, **17**, 667 (*Chem. Abstr.*, 1981, **95**, 80 827).	351
81ZOR861	G. D. Solodyuk, M. D. Boldyrev, B. V. Gidaspov and V. D. Nikolaev; *Zh. Org. Khim.*, 1981, **17**, 861.	414
81ZOR1047	L. I. Vereshchagin, L. P. Kirikkova, N. S. Bukina and G. A. Gareev; *Zh. Org. Khim.*, 1981, **17**, 1047.	414
81ZOR1123	I. V. Tselinskii, S. F. Melnikova and S. N. Vergizov; *Zh. Org. Khim.*, 1981, **17**, 1123.	414
81ZOR1435	L. D. Sychkova, O. L. Kalinkina and Y. S. Shabarov; *Zh. Org. Khim.*, 1981, **17**, 1435.	82
81ZOR1788	M. L. Petrov, V. A. Bobylev and A. A. Petrov; *Zh. Org. Khim.*, 1981, **17**, 1788.	845
82AJC385	F. C. James and H. D. Krebs; *Aust. J. Chem.*, 1982, **35**, 385.	1023
82CB2135	M. Geisel and R. Mews; *Chem. Ber.*, 1982, **115**, 2135.	537
82CC60	G. Tennant and C. W. Yacomeni; *J. Chem. Soc., Chem. Commun.*, 1982, 60.	406
82CC188	D. M. Evans and D. R. Taylor; *J. Chem. Soc., Chem. Commun.*, 1982, 188.	554, 557
82CC267	G. Tennant and G. M. Wallace; *J. Chem. Soc., Chem. Commun.*, 1982, 267.	406
82CC299	M. Bryce, P. Hanson and J. M. Vernon; *J. Chem. Soc., Chem. Commun.*, 1982, 299.	529
82CC336	K. Lerstrup, D. Talham, A. Bloch, T. Poehler and D. Cowan; *J. Chem. Soc., Chem. Commun.*, 1982, 336.	955, 956, 962, 971
82CC380	J. Roussilhe, B. Despax, A. Lopex and N. Paillous; *J. Chem. Soc., Chem. Commun.*, 1982, 380.	189
82CC1316	P. J. Carroll, M. V. Lakshmikantham, M. P. Cava, F. Wudl, E. Aharon-Shalom and S. D. Cox; *J. Chem. Soc., Chem. Commun.*, 1982, 1316.	950, 961
82CHE328	G. I. Khaskin, T. E. Bezmenova and P. G. Dulnev; *Chem. Heterocycl. Compd. (Engl. Transl.)*, 1982, **17**, 328.	1011
82CHE668	T. E. Bezmenova, R. I. Khaskin, V. I. Slutskii, P. G. Dulnev, L. N. Zakharov, V. I. Kulishov and Y. T. Struchkov; *Chem. Heterocycl. Compd. (Engl. Transl.)*, 1982, **17**, 668.	976, 1011
82H(19)57	A. S. Shawali, H. M. Hassaneen, A. Shetta, A. Osman and F. Abdel-Galil; *Heterocycles*, 1982, **19**, 57.	547
82H(19)1063	J. H. Boyer and T. P. Pillai; *Heterocycles*, 1982, **19**, 1063.	413
82H(19)1375	H. A. Daboun and M. A. Abdel-Aziz; *Heterocycles*, 1982, **19**, 1375.	990
82IJC(B)243	J. Mohan; *Indian J. Chem., Sect. B*, 1982, **21**, 243.	993
82JA66	P. C. Hiberty and G. Ohanessian; *J. Am. Chem. Soc.*, 1982, **104**, 66.	996
82JA1154	F. Wudl and E. Aharon-Shalom; *J. Am. Chem. Soc.*, 1982, **104**, 1154.	952, 954, 961, 962, 971
82JA1375	J. S. Amato, S. Karady, R. A. Reamer, H. B. Schlegel, J. P. Springer and L. M. Weinstock; *J. Am. Chem. Soc.*, 1982, **104**, 1375.	515, 530
82JA2919	H. W. Roesky, D. Amirzadeh-Asl and W. S. Sheldrick; *J. Am. Chem. Soc.*, 1982, **104**, 2919.	976
82JCR(S)65	A. R. Butler and C. Glidewell; *J. Chem. Res. (S)*, 1982, 65.	900
82JCS(P1)351	P. Molina, M. Alajarin, A. Arques and R. Benzal; *J. Chem. Soc., Perkin Trans. 1*, 1982, 351.	573
82JCS(P1)1885	H. Kato, T. Shiba, N. Aoki, H. Iijima and H. Tezuka; *J. Chem. Soc., Perkin Trans. 1*, 1982, 1885.	819
82JHC193	M. Langlois, C. Guillonneau, T. VoVan, J. P. Meingau and J. Maillard; *J. Heterocycl. Chem.*, 1982, **19**, 193.	978, 991, 996
82JHC427	R. Calvino, R. Fruttero, A. Gasco, V. Mortarini and S. Aime; *J. Heterocycl. Chem.*, 1982, **19**, 427.	398

82JIC(B)867	M. A. Elmagraby and A. A. E. Hassan; *J. Indian Chem. Soc., Sect. B*, 1982, **59**, 867.	1011, 1012
82JMC207	W. Lumma, P. S. Anderson, J. J. Baldwin, W. A. Bolhoffer, C. N. Habecker, J. M. Hirshfield, A. M. Pietruszkiewicz, W. C. Randall, M. L. Torchiana, S. F. Britcher, B. V. Clineschmidt, G. H. Denny, R. Hirschmann, J. M. Hoffman, B. T. Phillips and K. B. Streeter; *J. Med. Chem.*, 1982, **25**, 207.	542
82JMC210	A. A. Algieri, G. M. Luke, R. T. -Standridge, M. Brown, R. A. Partyka and R. R. Crenshaw; *J. Med. Chem.*, 1982, **25**, 210.	542
82JOC723	M. J. Haddadin, A. M. Kattan and J. P. Freeman; *J. Org. Chem.*, 1982, **47**, 723.	225
82JOC2527	T. Sasaki, E. Ito and I. Shimizu; *J. Org. Chem.*, 1982, **47**, 2757.	1021
82JOC4222	H. Kanety and E. M. Kosower; *J. Org. Chem.*, 1982, **47**, 4222.	982
82JOC4256	E. Sato, Y. Kanaoka and A. Padwa; *J. Org. Chem.*, 1982, **47**, 4256.	1008
82LA14	H. Böshagen and W. Geiger; *Liebigs Ann. Chem.*, 1982, 14.	151, 173
82LA315	H. Behringer and E. Meinetsberger; *Liebigs Ann. Chem.*, 1982, 315.	1020
82M1025	R. Csuk, H. Honig and C. Romanin; *Monatsh. Chem.*, 1982, **113**, 1025.	977
B-82MI41600	'USAN and the USP Dictionary of Drug Names', ed. M. C. Griffiths; U. S. Pharmacopoeia Convention, Rockville, Md., 1982.	127
82MI42700	K. N. Johri and B. S. Arora; *Thermochim. Acta* 1982, **54**, 237 (*Chem. Abstr.*, 1982, **96**, 228 047).	556
82MI43500	D. J. Sandman, J. C. Stark and B. M. Foxman; *Organometallics*, 1982, **1**, 739.	951, 961, 970
82MI43501	K. Iwahana, H. Kuzmany, F. Wudl and E. Aharon-Shalom; *Mol. Cryst. Liq. Cryst.*, 1982, **79**, 39.	954, 955
82OMR(18)159	G. Aranda, M. Dessolin, M. Golfier and M. G. Guillerez; *Org. Magn. Reson.*, 1982, **18**, 159.	554
82PC42600	J. Rokach and Y. Girard; personal communication, 1982.	527
82S502	H. A. F. Daboun, S. E. Abdou, M. M. Hussein and M. H. Elnagdi; *Synthesis*, 1982, 502.	992
82S511	J. J. C. Dockx, A. G. M. DeKamp and R. W. H. Albert; *Synthesis*, 1982, 511.	1013
82S677	R. C. Boruah and J. S. Sandhu; *Synthesis*, 1982, 677.	36
82SC431	S. Gelin, B. Chantegrel and M. Chabannet; *Synth. Commun.*, 1982, **12**, 431.	989
82TL47	J. Mattay and H.-D. Scharf; *Tetrahedron Lett.*, 1982, **23**, 47.	779
82TL1345	K. Jurkschat, C. Mügge, A. Tzschach, W. Uhlig and A. Zschunke; *Tetrahedron Lett.*, 1982, **23**, 1345.	1022
82TL1531	S. L. Bender, M. R. Detty and N. F. Haley; *Tetrahedron Lett.*, 1982, 1531.	948, 949, 957, 958, 959, 960, 961, 962, 964, 969
82TL2081	A. P. Kozikowski and Y. Y. Chen; *Tetrahedron Lett.*, 1982, **23**, 2081.	1008
82TL3059	C. Jenny, D. Obrecht and H. Heimgartner; *Tetrahedron Lett.*, 1982, **23**, 3059.	1021
82UP41700	B. J. Peart; 1982, unpublished results.	138, 140, 141
82UP42000	I. Lalezari; unpublished results, 1982.	351
82UP42200	R. M. Paton; unpublished observations.	398
82UP42600	S. Karady, J. S. Amato, R. A. Reamer and L. M. Weinstock; unpublished results, 1982.	530
82UP42800	M. Jacobsen, L. Henriksen and A. Holm; unpublished results.	581, 587, 588, 598
82UP42900	H. Breivik and K. Undheim; unpublished results, 1982.	678
82UP43700	W. D. Ollis, S. P. Stanforth and C. A. Ramsden; unpublished results, 1982.	1040, 1041, 1047
82UP43701	C. A. Ramsden; unpublished calculations, 1982.	1029
83JA875	M. R. Detty, B. J. Murray, D. L. Smith and N. Zumbulyadis; *J. Am. Chem. Soc.*, 1983, **105**, 875.	949, 952, 955, 958, 959, 960, 961, 966, 968
83JA883	M. R. Detty and B. J. Murray; *J. Am. Chem. Soc.*, 1983, **105**, 883.	949, 951, 952, 955, 958, 959, 968
83TL237	S. L. Bender, M. R. Detty, M. W. Fichtner and N. F. Haley; *Tetrahedron Lett.*, 1983, 237.	958, 965
83UP43700	W. D. Ollis, S. P. Stanforth and C. A. Ramsden; unpublished results, 1983	1028

JOURNAL CODES FOR REFERENCES
For explanation of the reference system, see p. 1071.

ABC	Agric. Biol. Chem.	CS	Chem. Scr.
ACH	Acta Chim. Acad. Sci. Hung.	CSC	Cryst. Struct. Commun.
ACR	Acc. Chem. Res.	CSR	Chem. Soc. Rev.
AC(R)	Ann. Chim. (Rome)	CZ	Chem.-Ztg.
ACS	Acta Chem. Scand.	DIS	Diss. Abstr.
ACS(B)	Acta Chem. Scand., Ser. B	DIS(B)	Diss. Abstr. Int. B
AF	Arzneim.-Forsch.	DOK	Dokl. Akad. Nauk SSSR
AG	Angew. Chem.	E	Experientia
AG(E)	Angew. Chem., Int. Ed. Engl.	EGP	Ger. (East) Pat.
AHC	Adv. Heterocycl. Chem.	EUP	Eur. Pat.
AJC	Aust. J. Chem.	FES	Farmaco Ed. Sci.
AK	Ark. Kemi	FOR	Fortschr. Chem. Org. Naturst.
ANY	Ann. N.Y. Acad. Sci.	FRP	Fr. Pat.
AP	Arch. Pharm. (Weinheim, Ger.)	G	Gazz. Chim. Ital.
APO	Adv. Phys. Org. Chem.	GEP	Ger. Pat.
AX	Acta Crystallogr.	H	Heterocycles
AX(B)	Acta Crystallogr., Part B	HC	Chem. Heterocycl. Compd. [Weissberger–Taylor series]
B	Biochemistry		
BAP	Bull. Acad. Pol. Sci., Ser. Sci. Chim.	HCA	Helv. Chim. Acta
		HOU	Methoden Org. Chem. (Houben-Weyl)
BAU	Bull. Acad. Sci. USSR, Div. Chem. Sci.	IC	Inorg. Chem.
BBA	Biochim. Biophys. Acta	IJC	Indian J. Chem.
BBR	Biochem. Biophys. Res. Commun.	IJC(B)	Indian J. Chem., Sect. B
BCJ	Bull. Chem. Soc. Jpn.	IJS	Int. J. Sulfur Chem.
BEP	Belg. Pat.	IJS(B)	Int. J. Sulfur Chem., Part B
BJ	Biochem. J.	IZV	Izv. Akad. Nauk SSSR, Ser. Khim.
BJP	Br. J. Pharmacol.	JA	J. Am. Chem. Soc.
BRP	Br. Pat.	JAP	Jpn. Pat.
BSB	Bull. Soc. Chim. Belg.	JAP(K)	Jpn. Kokai
BSF	Bull. Soc. Chim. Fr.	JBC	J. Biol. Chem.
BSF(2)	Bull. Soc. Chim. Fr., Part 2	JCP	J. Chem. Phys.
C	Chimia	JCR(S)	J. Chem. Res. (S)
CA	Chem. Abstr.	JCS	J. Chem. Soc.
CB	Chem. Ber.	JCS(C)	J. Chem. Soc. (C)
CC	J. Chem. Soc., Chem. Commun.	JCS(D)	J. Chem. Soc., Dalton Trans.
CCC	Collect. Czech. Chem. Commun.	JCS(F1)	J. Chem. Soc., Faraday Trans. 1
CCR	Coord. Chem. Rev.	JCS(P1)	J. Chem. Soc., Perkin Trans. 1
CHE	Chem. Heterocycl. Compd. (Engl. Transl.)	JGU	J. Gen. Chem. USSR (Engl. Transl.)
CI(L)	Chem. Ind. (London)	JHC	J. Heterocycl. Chem.
CJC	Can. J. Chem.	JIC	J. Indian Chem. Soc.
CL	Chem. Lett.	JMC	J. Med. Chem.
CPB	Chem. Pharm. Bull.	JMR	J. Magn. Reson.
CR	C.R. Hebd. Seances Acad. Sci.	JOC	J. Org. Chem.
CR(C)	C.R. Hebd. Seances Acad. Sci., Ser. C	JOM	J. Organomet. Chem.
		JOU	J. Org. Chem. USSR (Engl. Transl.)
CRV	Chem. Rev.		